Algebra and Trigonometry 2e

SENIOR CONTRIBUTING AUTHOR
JAY ABRAMSON, ARIZONA STATE UNIVERSITY

openstax™

ISBN: 978-1-71149-404-3

OpenStax
Rice University
6100 Main Street MS-375
Houston, Texas 77005

To learn more about OpenStax, visit https://openstax.org.
Individual print copies and bulk orders can be purchased through our website.

HARDCOVER BOOK ISBN-13	**978-1-711494-04-3**
B&W PAPERBACK BOOK ISBN-13	**978-1-711494-03-6**
DIGITAL VERSION ISBN-13	**978-1-951693-40-4**
ORIGINAL PUBLICATION YEAR	**2021**

1 2 3 4 5 6 7 8 9 10 RS 21

Printed by
XanEdu
4750 Venture Drive, Suite 400
Ann Arbor, MI 48108
800-562-2147
www.xanedu.com

OpenStax

OpenStax provides free, peer-reviewed, openly licensed textbooks for introductory college and Advanced Placement® courses and low-cost, personalized courseware that helps students learn. A nonprofit ed tech initiative based at Rice University, we're committed to helping students access the tools they need to complete their courses and meet their educational goals.

Rice University

OpenStax, OpenStax CNX, and OpenStax Tutor are initiatives of Rice University. As a leading research university with a distinctive commitment to undergraduate education, Rice University aspires to path-breaking research, unsurpassed teaching, and contributions to the betterment of our world. It seeks to fulfill this mission by cultivating a diverse community of learning and discovery that produces leaders across the spectrum of human endeavor.

RICE

Philanthropic Support

OpenStax is grateful for the generous philanthropic partners who advance our mission to improve educational access and learning for everyone. To see the impact of our supporter community and our most updated list of partners, please visit openstax.org/impact.

Arnold Ventures
Chan Zuckerberg Initiative
Chegg, Inc.
Arthur and Carlyse Ciocca Charitable Foundation
Digital Promise
Ann and John Doerr
Bill & Melinda Gates Foundation
Girard Foundation
Google Inc.
The William and Flora Hewlett Foundation
The Hewlett-Packard Company
Intel Inc.
Rusty and John Jaggers
The Calvin K. Kazanjian Economics Foundation
Charles Koch Foundation
Leon Lowenstein Foundation, Inc.
The Maxfield Foundation
Burt and Deedee McMurtry
Michelson 20MM Foundation
National Science Foundation
The Open Society Foundations
Jumee Yhu and David E. Park III
Brian D. Patterson USA-International Foundation
The Bill and Stephanie Sick Fund
Steven L. Smith & Diana T. Go
Stand Together
Robin and Sandy Stuart Foundation
The Stuart Family Foundation
Tammy and Guillermo Treviño
Valhalla Charitable Foundation
White Star Education Foundation
Schmidt Futures
William Marsh Rice University

Contents

Preface

About OpenStax

OpenStax is part of Rice University, which is a 501(c)(3) nonprofit charitable corporation. As an educational initiative, it's our mission to transform learning so that education works for every student. Through our partnerships with philanthropic organizations and our alliance with other educational resource companies, we're breaking down the most common barriers to learning. Because we believe that everyone should and can have access to knowledge.

About OpenStax Resource

Customization

Algebra and Trigonometry 2e is licensed under a Creative Commons Attribution 4.0 International (CC BY) license, which means that you can distribute, remix, and build upon the content, as long as you provide attribution to OpenStax and its content contributors.

Because our books are openly licensed, you are free to use the entire book or pick and choose the sections that are most relevant to the needs of your course. Feel free to remix the content by assigning your students certain chapters and sections in your syllabus, in the order that you prefer. You can even provide a direct link in your syllabus to the sections in the web view of your book.

Instructors also have the option of creating a customized version of their OpenStax book. The custom version can be made available to students in low-cost print or digital form through their campus bookstore. Visit your book page on openstax.org for more information.

Art attribution

In *Algebra and Trigonometry 2e*, most photos and third-party illustrations contain attribution to their creator, rights holder, host platform, and/or license within the caption. Because the art is openly licensed, anyone may reuse the art as long as they provide the same attribution to its original source. To maximize readability and content flow, mathematical expressions that are rendered as art do not include attribution in the text. This art can be assumed to be developed by OpenStax and can be reused under the CC-BY license with attribution.

Errata

All OpenStax textbooks undergo a rigorous review process. However, like any professional-grade textbook, errors sometimes occur. Since our books are web based, we can make updates periodically when deemed pedagogically necessary. If you have a correction to suggest, submit it through the link on your book page on openstax.org. Subject matter experts review all errata suggestions. OpenStax is committed to remaining transparent about all updates, so you will also find a list of past errata changes on your book page on openstax.org.

Format

You can access this textbook for free in web view or PDF through openstax.org, and for a low cost in print.

About *Algebra and Trigonometry 2e*

Algebra and Trigonometry 2e provides a comprehensive exploration of algebraic principles and meets scope and sequence requirements for a typical introductory algebra and trigonometry course. The modular approach and the richness of content ensures that the book meets the needs of a variety of courses. *Algebra and Trigonometry 2e* offers a wealth of examples with detailed, conceptual explanations, building a strong foundation in the material before asking students to apply what they've learned.

Coverage and Scope

In determining the concepts, skills, and topics to cover, we engaged dozens of highly experienced instructors with a range of student audiences. The resulting scope and sequence proceeds logically while allowing for a significant amount of flexibility in instruction.

Chapters 1 and 2 provide both a review and foundation for study of functions that begins in Chapter 3. The authors recognize that while some institutions may find this material a prerequisite, other institutions have told us that they have a cohort that needs the prerequisite skills built into the course.

- Chapter 1: Prerequisites
- Chapter 2: Equations and Inequalities

Chapters 3-6: The Algebraic Functions

- Chapter 3: Functions

- Chapter 4: Linear Functions
- Chapter 5: Polynomial and Rational Functions
- Chapter 6: Exponential and Logarithm Functions

Chapters 7-10: A Study of Trigonometry

- Chapter 7: The Unit Circle: Sine and Cosine Functions
- Chapter 8: Periodic Functions
- Chapter 9: Trigonometric Identities and Equations
- Chapter 10: Further Applications of Trigonometry

Chapters 11-13: Further Study in Algebra and Trigonometry

- Chapter 11: Systems of Equations and Inequalities
- Chapter 12: Analytic Geometry
- Chapter 13: Sequences, Probability, and Counting Theory

Development Overview

OpenStax *Algebra and Trigonometry 2e* is the product of a collaborative effort by a group of dedicated authors, editors, and instructors whose collective passion for this project has resulted in a text that is remarkably unified in purpose and voice. Special thanks is due to our Lead Author, Jay Abramson of Arizona State University, who provided the overall vision for the book and oversaw the development of each and every chapter, drawing up the initial blueprint, reading numerous drafts, and assimilating field reviews into actionable revision plans for our authors and editors.

The collective experience of our author team allowed us to pinpoint the subtopics, exceptions, and individual connections that give students the most trouble. The textbook is therefore replete with well-designed features and highlights which help students overcome these barriers. As the students read and practice, they are coached in methods of thinking through problems and internalizing mathematical processes.

Accuracy of the Content

We understand that precision and accuracy are imperatives in mathematics, and undertook a dedicated accuracy program led by experienced faculty.Examples, art, problems, and solutions were reviewed by dedicated faculty, with a separate team evaluating the answer key and solutions.

The text also benefits from years of usage by thousands of faculty and students. A core aspect of the second edition revision process included consolidating and ensuring consistency with regard to any errata and corrections that have been during in the series' extensive usage and incorporation into homework systems.

Changes to the Second Edition

The *Algebra and Trigonometry 2e* revision focused on mathematical clarity and accuracy as well as inclusivity. Examples, Exercises, and Solutions were reviewed by multiple faculty experts. All improvement suggestions and errata updates, driven by faculty and students from several thousand colleges, were considered and unified across the different formats of the text.

OpenStax and our authors are aware of the difficulties posed by shifting problem and exercise numbers when textbooks are revised. In an effort to make the transition to the 2nd edition as seamless as possible, we have minimized any shifting of exercise numbers.

The revision also focused on supporting inclusive and welcoming learning experiences. The introductory narratives, example and problem contexts, and even many of the names used for fictional people in the text were all reviewed using a diversity, equity, and inclusion framework. Several hundred resulting revisions improve the balance and relevance to the students using the text, while maintaining a variety of applications to diverse careers and academic fields. In particular, explanations of scientific and historical aspects of mathematics have been expanded to include more contributors. For example, the authors added additional historical and multicultural context regarding what is widely known as Pascal's Triangle, and similarly added details regarding the international process of decoding the Enigma machine (including the role of Polish college students). Several chapter opening narratives and in-chapter references are completely new, and contexts across all chapters were specifically reviewed for equity in gender representation and connotation.

Finally, prior to the release of this edition, OpenStax published a series of Corequisite Skillsheets to support different models and approaches to instruction. These remain available, and are described in more detail below.

Pedagogical Foundations and Features

Learning Objectives

Each chapter is divided into multiple sections (or modules), each of which is organized around a set of learning objectives. The learning objectives are listed explicitly at the beginning of each section and are the focal point of every instructional element

Narrative text

Narrative text is used to introduce key concepts, terms, and definitions, to provide real-world context, and to provide transitions between topics and examples. Throughout this book, we rely on a few basic conventions to highlight the most important ideas:

- Key terms are boldfaced, typically when first introduced and/or when formally defined.
- Key concepts and definitions are called out in a blue box for easy reference.

Examples

Each learning objective is supported by one or more worked examples, that demonstrate the problem-solving approaches that students must master. The multiple Examples model different approaches to the same type of problem, or introduce similar problems of increasing complexity.

All Examples follow a simple two- or three-part format. The question clearly lays out a mathematical problem to solve. The Solution walks through the steps, usually providing context for the approach—in other words, why the instructor is solving the problem in a specific manner. Finally, the Analysis (for select Examples) reflects on the broader implications of the Solution just shown. Examples are followed by a Try It question, as explained below.

Figures

Algebra and Trigonometry 2e contains more than 2000 figures and illustrations, the vast majority of which are graphs and diagrams. Art throughout the text adheres to a clear, understated style, drawing the eye to the most important information in each figure while minimizing visual distractions. Color contrast is employed with discretion to distinguish between the different functions or features of a graph.

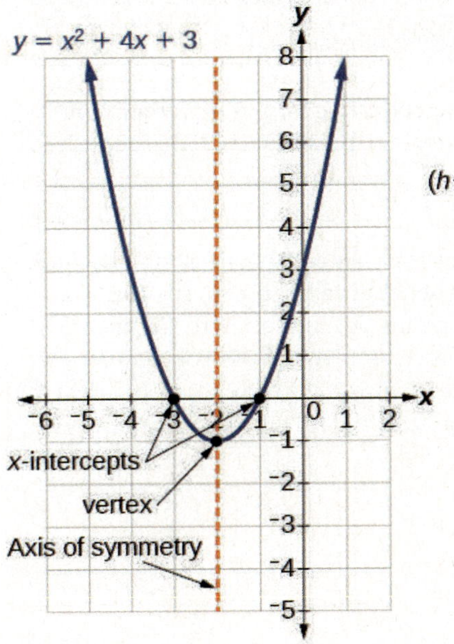

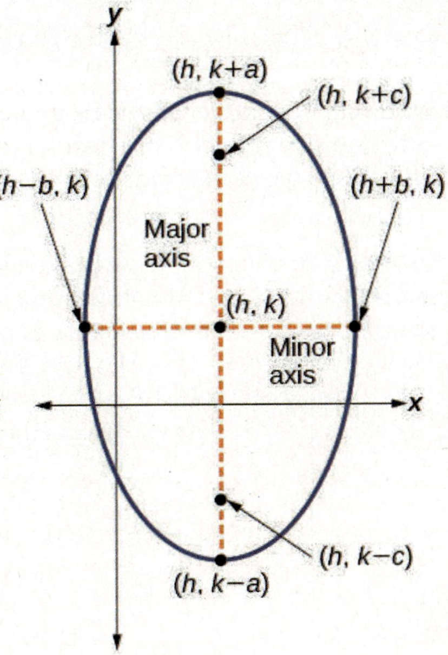

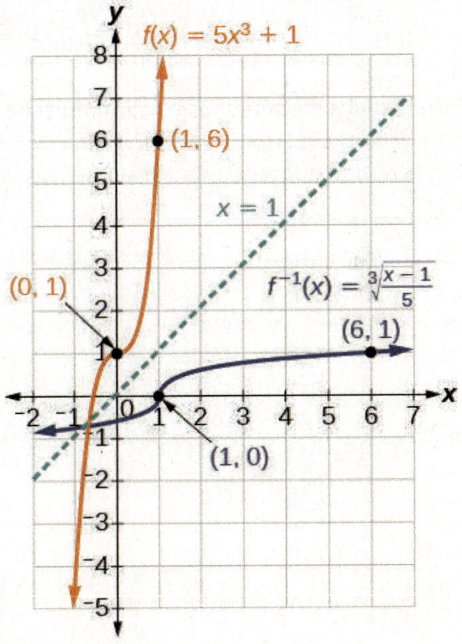

Supporting Features

Several elements contribute to and check understanding.

- A **"How To"** is a list of steps necessary to solve a certain type of problem. A How To typically precedes an Example that proceeds to demonstrate the steps in action.
- A **"Try It"** exercise immediately follows an Example or a set of related Examples, providing the student with an immediate opportunity to solve a similar problem. In the PDF and the Web View version of the text, answers to the Try It exercises are located in the Answer Key.
- A **Q&A** may appear at any point in the narrative, but most often follows an Example. This feature pre-empts misconceptions by posing a commonly asked yes/no question, followed by a detailed answer and explanation.
- The **"Media"** icon appears at the conclusion of each section, just prior to the Section Exercises. This icon marks a list

of links to online video tutorials that reinforce the concepts and skills introduced in the section.

While we have selected tutorials that closely align to our learning objectives, we did not produce these tutorials, nor were they specifically produced or tailored to accompany *Algebra and Trigonometry 2e*.

Section Exercises

Each section of every chapter concludes with a well-rounded set of exercises that can be assigned as homework or used selectively for guided practice. With over 6300 exercises across the 13 chapters, instructors should have plenty from which to choose.

Section Exercises are organized by question type, and generally appear in the following order:

- **Verbal** questions assess conceptual understanding of key terms and concepts.
- **Algebraic** problems require students to apply algebraic manipulations demonstrated in the section.
- **Graphical** problems assess students' ability to interpret or produce a graph.
- **Numeric** problems require the student to perform calculations or computations.
- **Technology** problems encourage exploration through use of a graphing utility, either to visualize or verify algebraic results or to solve problems via an alternative to the methods demonstrated in the section.
- **Extensions** pose problems more challenging than the Examples demonstrated in the section. They require students to synthesize multiple learning objectives or apply critical thinking to solve complex problems.
- **Real-World Applications** present realistic problem scenarios from fields such as physics, geology, biology, finance, and the social sciences.

Chapter Review Features

Each chapter concludes with a review of the most important takeaways, as well as additional practice problems that students can use to prepare for exams.

- **Key Terms** provides a formal definition for each bold-faced term in the chapter.
- **Key Equations** presents a compilation of formulas, theorems, and standard-form equations.
- **Key Concepts** summarizes the most important ideas introduced in each section, linking back to the relevant Example(s) in case students need to review.
- **Chapter Review Exercises** include 40-80 practice problems that recall the most important concepts from each section.
- **Practice Test** includes 25-50 problems assessing the most important learning objectives from the chapter. Note that the practice test is not organized by section, and may be more heavily weighted toward cumulative objectives as opposed to the foundational objectives covered in the opening sections.

Corequisite Support

Each *Algebra and Trigonometry 2e* section is paired with a thoughtfully developed, topically aligned skills module that prepares students for the course material. Sharon North (St. Louis Community College) developed a coordinated set of support resources, which provide review, instruction, and practice for algebra students. The author team identified foundational skills and concepts, then mapped them to each module. The corequisite sections include conceptual overviews, worked examples, and guided practice; they incorporate relevant material from OpenStax's Developmental Math series. They are available as separate, openly accessible downloads from the student and instructor resources pages accompanying the text.

Additional Resources

Student and Instructor Resources

We've compiled additional resources for both students and instructors, including Getting Started Guides, instructor solution manual, Corequisite skillsheets, and PowerPoint slides. Instructor resources require a verified instructor account, which can be requested on your openstax.org log-in. Take advantage of these resources to supplement your OpenStax book.

Community Hubs

OpenStax partners with the Institute for the Study of Knowledge Management in Education (ISKME) to offer Community Hubs on OER Commons—a platform for instructors to share community-created resources that support OpenStax books, free of charge. Through our Community Hubs, instructors can upload their own materials or download resources to use in their own courses, including additional ancillaries, teaching material, multimedia, and relevant course content. We encourage instructors to join the hubs for the subjects most relevant to your teaching and research as an opportunity both to enrich your courses and to engage with other faculty. To reach the Community Hubs, visit www.oercommons.org/hubs/openstax.

Technology partners

As allies in making high-quality learning materials accessible, our technology partners offer optional low-cost tools that are integrated with OpenStax books. To access the technology options for your text, visit your book page on openstax.org.

About the Authors

Senior Contributing Author

Jay Abramson, Arizona State University

Jay Abramson has been teaching Precalculus for over 35 years, the last 20 at Arizona State University, where he is a principal lecturer in the School of Mathematics and Statistics. His accomplishments at ASU include co-developing the university's first hybrid and online math courses as well as an extensive library of video lectures and tutorials. In addition, he has served as a contributing author for two of Pearson Education's math programs, NovaNet Precalculus and Trigonometry. Prior to coming to ASU, Jay taught at Texas State Technical College and Amarillo College. He received Teacher of the Year awards at both institutions.

Contributing Authors

Valeree Falduto, Palm Beach State College
Rachael Gross, Towson University
David Lippman, Pierce College
Melonie Rasmussen, Pierce College
Rick Norwood, East Tennessee State University
Nicholas Belloit, Florida State College Jacksonville
Jean-Marie Magnier, Springfield Technical Community College
Harold Whipple
Christina Fernandez

Reviewers

Phil Clark, Scottsdale Community College
Michael Cohen, Hofstra University
Charles Conrad, Volunteer State Community College
David French, Tidewater Community College
Matthew Goodell, SUNY Ulster
Lance Hemlow, Raritan Valley Community College
Dongrin Kim, Arizona State University
Cynthia Landrigan, Eerie Community College
Wendy Lightheart, Lane Community College
Chinenye Ofodile, Albany State University
Carl Penziul, Tompkins-Cortland Community College
Sandra Nite, Texas A&M University
Eugenia Peterson, Richard J. Daley College
Rhonda Porter, Albany State University
Michael Price, University of Oregon
Steven Purtee, Valencia College
William Radulovich, Florida State College Jacksonville
Camelia Salajean, City Colleges of Chicago
Katy Shields, Oakland Community College
Nathan Schrenk, ECPI University
Pablo Suarez, Delaware State University
Allen Wolmer, Atlanta Jewish Academy

The following faculty contributed to the development of OpenStax *Precalculus*, the text from which this product was updated and derived.

Precalculus Reviewers

Nina Alketa, Cecil College
Kiran Bhutani, Catholic University of America
Brandie Biddy, Cecil College
Lisa Blank, Lyme Central School

Bryan Blount, Kentucky Wesleyan College
Jessica Bolz, The Bryn Mawr School
Sheri Boyd, Rollins College
Sarah Brewer, Alabama School of Math and Science
Charles Buckley, St. Gregory's University
Michael Cohen, Hofstra University
Kenneth Crane, Texarkana College
Rachel Cywinski, Alamo Colleges
Nathan Czuba
Srabasti Dutta, Ashford University
Kristy Erickson, Cecil College
Nicole Fernandez, Georgetown University / Kent State University
David French, Tidewater Community College
Douglas Furman, SUNY Ulster
Lance Hemlow, Raritan Valley Community College
Erinn Izzo, Nicaragua Christian Academy
John Jaffe
Jerry Jared, Blue Ridge School
Stan Kopec, Mount Wachusett Community College
Kathy Kovacs
Cynthia Landrigan, Erie Community College
Sara Lenhart, Christopher Newport University
Wendy Lightheart, Lane Community College
Joanne Manville, Bunker Hill Community College
Karla McCavit, Albion College
Cynthia McGinnis, Northwest Florida State College
Lana Neal, University of Texas at Austin
Rhonda Porter, Albany State University
Steven Purtee, Valencia College
William Radulovich, Florida State College Jacksonville
Alice Ramos, Bethel College
Nick Reynolds, Montgomery Community College
Amanda Ross, A. A. Ross Consulting and Research, LLC
Erica Rutter, Arizona State University
Sutandra Sarkar, Georgia State University
Willy Schild, Wentworth Institute of Technology
Todd Stephen, Cleveland State University
Scott Sykes, University of West Georgia
Linda Tansil, Southeast Missouri State University
John Thomas, College of Lake County
Diane Valade, Piedmont Virginia Community College
Allen Wolmer, Atlanta Jewish Academy

1 PREREQUISITES

Credit: Andreas Kambanls

Chapter Outline

Introduction to Prerequisites

It's a cold day in Antarctica. In fact, it's always a cold day in Antarctica. Earth's southernmost continent, Antarctica experiences the coldest, driest, and windiest conditions known. The coldest temperature ever recorded, over one hundred degrees below zero on the Celsius scale, was recorded by remote satellite. It is no surprise then, that no native human population can survive the harsh conditions. Only explorers and scientists brave the environment for any length of time.

Measuring and recording the characteristics of weather conditions in Antarctica requires a use of different kinds of numbers. For tens of thousands of years, humans have undertaken methods to tally, track, and record numerical information. While we don't know much about their usage, the Lebombo Bone (dated to about 35,000 BCE) and the Ishango Bone (dated to about 20,000 BCE) are among the earliest mathematical artifacts. Found in Africa, their clearly deliberate groupings of notches may have been used to track time, moon cycles, or other information. Performing calculations with them and using the results to make predictions requires an understanding of relationships among numbers. In this chapter, we will review sets of numbers and properties of operations used to manipulate numbers. This understanding will serve as prerequisite knowledge throughout our study of algebra and trigonometry.

1.1 Real Numbers: Algebra Essentials

Learning Objectives

In this section, you will:

- Classify a real number as a natural, whole, integer, rational, or irrational number.
- Perform calculations using order of operations.
- Use the following properties of real numbers: commutative, associative, distributive, inverse, and identity.
- Evaluate algebraic expressions.
- Simplify algebraic expressions.

It is often said that mathematics is the language of science. If this is true, then an essential part of the language of mathematics is numbers. The earliest use of numbers occurred 100 centuries ago in the Middle East to count, or enumerate items. Farmers, cattle herders, and traders used tokens, stones, or markers to signify a single quantity—a sheaf of grain, a head of livestock, or a fixed length of cloth, for example. Doing so made commerce possible, leading to improved communications and the spread of civilization.

Three to four thousand years ago, Egyptians introduced fractions. They first used them to show reciprocals. Later, they used them to represent the amount when a quantity was divided into equal parts.

But what if there were no cattle to trade or an entire crop of grain was lost in a flood? How could someone indicate the existence of nothing? From earliest times, people had thought of a "base state" while counting and used various symbols to represent this null condition. However, it was not until about the fifth century CE in India that zero was added to the number system and used as a numeral in calculations.

Clearly, there was also a need for numbers to represent loss or debt. In India, in the seventh century CE, negative numbers were used as solutions to mathematical equations and commercial debts. The opposites of the counting numbers expanded the number system even further.

Because of the evolution of the number system, we can now perform complex calculations using these and other categories of real numbers. In this section, we will explore sets of numbers, calculations with different kinds of numbers, and the use of numbers in expressions.

Classifying a Real Number

The numbers we use for counting, or enumerating items, are the **natural numbers**: 1, 2, 3, 4, 5, and so on. We describe them in set notation as $\{1, 2, 3, ...\}$ where the ellipsis (...) indicates that the numbers continue to infinity. The natural numbers are, of course, also called the *counting numbers*. Any time we enumerate the members of a team, count the coins in a collection, or tally the trees in a grove, we are using the set of natural numbers. The set of **whole numbers** is the set of natural numbers plus zero: $\{0, 1, 2, 3, ...\}$.

The set of **integers** adds the opposites of the natural numbers to the set of whole numbers: $\{..., -3, -2, -1, 0, 1, 2, 3, ...\}$. It is useful to note that the set of integers is made up of three distinct subsets: negative integers, zero, and positive integers. In this sense, the positive integers are just the natural numbers. Another way to think about it is that the natural numbers are a subset of the integers.

$$\underbrace{\ldots, -3, -2, -1,}_{\text{negative integers}} \quad \underbrace{0,}_{\text{zero}} \quad \underbrace{1, 2, 3, \cdots}_{\text{positive integers}}$$

The set of **rational numbers** is written as $\left\{\frac{m}{n} \,\middle|\, m \text{ and } n \text{ are integers and } n \neq 0\right\}$. Notice from the definition that rational numbers are fractions (or quotients) containing integers in both the numerator and the denominator, and the denominator is never 0. We can also see that every natural number, whole number, and integer is a rational number with a denominator of 1.

Because they are fractions, any rational number can also be expressed in decimal form. Any rational number can be represented as either:

ⓐ a terminating decimal: $\frac{15}{8} = 1.875$, or ⓑ a repeating decimal: $\frac{4}{11} = 0.36363636\ldots = 0.\overline{36}$

We use a line drawn over the repeating block of numbers instead of writing the group multiple times.

EXAMPLE 1

Writing Integers as Rational Numbers

Write each of the following as a rational number.

ⓐ 7 ⓑ 0 ⓒ -8

Solution

Write a fraction with the integer in the numerator and 1 in the denominator.

ⓐ $7 = \frac{7}{1}$ ⓑ $0 = \frac{0}{1}$ ⓒ $-8 = -\frac{8}{1}$

TRY IT #1 Write each of the following as a rational number.

ⓐ 11 ⓑ 3 ⓒ -4

EXAMPLE 2

Identifying Rational Numbers

Write each of the following rational numbers as either a terminating or repeating decimal.

ⓐ $-\frac{5}{7}$ ⓑ $\frac{15}{5}$ ⓒ $\frac{13}{25}$

Solution

Write each fraction as a decimal by dividing the numerator by the denominator.

ⓐ $-\frac{5}{7} = -0.\overline{714285}$, a repeating decimal ⓑ $\frac{15}{5} = 3$ (or 3.0), a terminating decimal

ⓒ $\frac{13}{25} = 0.52$, a terminating decimal

TRY IT #2 Write each of the following rational numbers as either a terminating or repeating decimal.

ⓐ $\frac{68}{17}$ ⓑ $\frac{8}{13}$ ⓒ $-\frac{17}{20}$

Irrational Numbers

At some point in the ancient past, someone discovered that not all numbers are rational numbers. A builder, for instance, may have found that the diagonal of a square with unit sides was not 2 or even $\frac{3}{2}$, but was something else. Or a garment maker might have observed that the ratio of the circumference to the diameter of a roll of cloth was a little bit more than 3, but still not a rational number. Such numbers are said to be *irrational* because they cannot be written as fractions. These numbers make up the set of **irrational numbers**. Irrational numbers cannot be expressed as a fraction of two integers. It is impossible to describe this set of numbers by a single rule except to say that a number is irrational if it is not rational. So we write this as shown.

$$\{h \mid h \text{ is not a rational number}\}$$

EXAMPLE 3

Differentiating Rational and Irrational Numbers

Determine whether each of the following numbers is rational or irrational. If it is rational, determine whether it is a terminating or repeating decimal.

ⓐ $\sqrt{25}$ ⓑ $\frac{33}{9}$ ⓒ $\sqrt{11}$ ⓓ $\frac{17}{34}$ ⓔ $0.3033033303333\ldots$

Solution

ⓐ $\sqrt{25}$: This can be simplified as $\sqrt{25} = 5$. Therefore, $\sqrt{25}$ is rational.

ⓑ $\frac{33}{9}$: Because it is a fraction of integers, $\frac{33}{9}$ is a rational number. Next, simplify and divide.

$$\frac{33}{9} = \frac{\overset{11}{\cancel{33}}}{\underset{3}{\cancel{9}}} = \frac{11}{3} = 3.\overline{6}$$

So, $\frac{33}{9}$ is rational and a repeating decimal.

ⓒ $\sqrt{11}$: This cannot be simplified any further. Therefore, $\sqrt{11}$ is an irrational number.

ⓓ $\frac{17}{34}$: Because it is a fraction of integers, $\frac{17}{34}$ is a rational number. Simplify and divide.

$$\frac{17}{34} = \frac{\overset{1}{\cancel{17}}}{\underset{2}{\cancel{34}}} = \frac{1}{2} = 0.5$$

So, $\frac{17}{34}$ is rational and a terminating decimal.

ⓔ $0.3033033303333\ldots$ is not a terminating decimal. Also note that there is no repeating pattern because the group of 3s increases each time. Therefore it is neither a terminating nor a repeating decimal and, hence, not a rational number. It is an irrational number.

TRY IT #3 Determine whether each of the following numbers is rational or irrational. If it is rational, determine whether it is a terminating or repeating decimal.

ⓐ $\frac{7}{77}$ ⓑ $\sqrt{81}$ ⓒ $4.27027002700027\ldots$ ⓓ $\frac{91}{13}$ ⓔ $\sqrt{39}$

Real Numbers

Given any number *n*, we know that *n* is either rational or irrational. It cannot be both. The sets of rational and irrational numbers together make up the set of **real numbers**. As we saw with integers, the real numbers can be divided into three subsets: negative real numbers, zero, and positive real numbers. Each subset includes fractions, decimals, and irrational numbers according to their algebraic sign (+ or –). Zero is considered neither positive nor negative.

The real numbers can be visualized on a horizontal number line with an arbitrary point chosen as 0, with negative numbers to the left of 0 and positive numbers to the right of 0. A fixed unit distance is then used to mark off each integer (or other basic value) on either side of 0. Any real number corresponds to a unique position on the number line.The converse is also true: Each location on the number line corresponds to exactly one real number. This is known as a one-to-one correspondence. We refer to this as the **real number line** as shown in Figure 1.

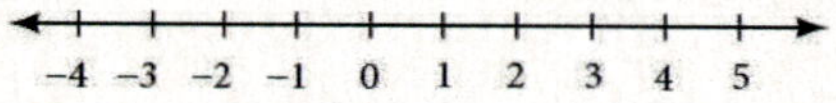

Figure 1 The real number line

EXAMPLE 4

Classifying Real Numbers

Classify each number as either positive or negative and as either rational or irrational. Does the number lie to the left or the right of 0 on the number line?

ⓐ $-\frac{10}{3}$ ⓑ $\sqrt{5}$ ⓒ $-\sqrt{289}$ ⓓ -6π ⓔ $0.615384615384\ldots$

Solution

ⓐ $-\frac{10}{3}$ is negative and rational. It lies to the left of 0 on the number line.

ⓑ $\sqrt{5}$ is positive and irrational. It lies to the right of 0.

ⓒ $-\sqrt{289} = -\sqrt{17^2} = -17$ is negative and rational. It lies to the left of 0.

ⓓ -6π is negative and irrational. It lies to the left of 0.

ⓔ $0.615384615384\ldots$ is a repeating decimal so it is rational and positive. It lies to the right of 0.

TRY IT #4 Classify each number as either positive or negative and as either rational or irrational. Does the number lie to the left or the right of 0 on the number line?

ⓐ $\sqrt{73}$ ⓑ $-11.411411411\ldots$ ⓒ $\frac{47}{19}$ ⓓ $-\frac{\sqrt{5}}{2}$ ⓔ 6.210735

Sets of Numbers as Subsets

Beginning with the natural numbers, we have expanded each set to form a larger set, meaning that there is a subset relationship between the sets of numbers we have encountered so far. These relationships become more obvious when seen as a diagram, such as Figure 2.

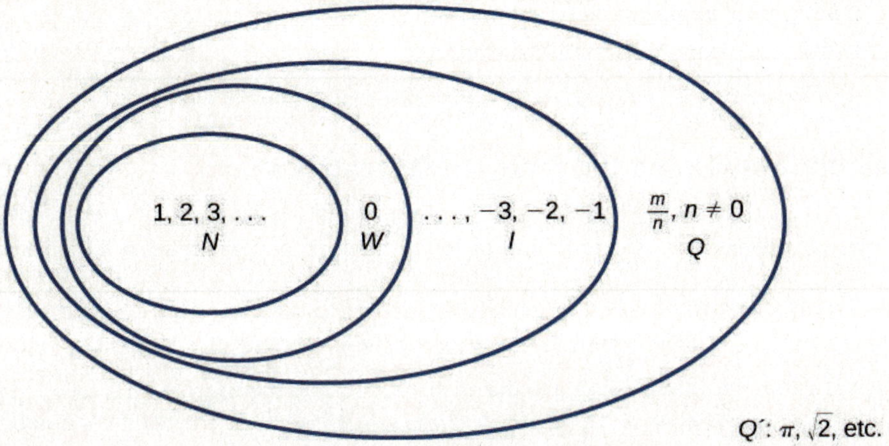

Figure 2 Sets of numbers
N: the set of natural numbers
W: the set of whole numbers
I: the set of integers
Q: the set of rational numbers
Q´: the set of irrational numbers

Sets of Numbers

The set of **natural numbers** includes the numbers used for counting: $\{1, 2, 3, ...\}$.

The set of **whole numbers** is the set of natural numbers plus zero: $\{0, 1, 2, 3, ...\}$.

The set of **integers** adds the negative natural numbers to the set of whole numbers: $\{..., -3, -2, -1, 0, 1, 2, 3, ...\}$.

The set of **rational numbers** includes fractions written as $\left\{\frac{m}{n} \mid m \text{ and } n \text{ are integers and } n \neq 0\right\}$.

The set of **irrational numbers** is the set of numbers that are not rational, are nonrepeating, and are nonterminating: $\{h \mid h \text{ is not a rational number}\}$.

EXAMPLE 5

Differentiating the Sets of Numbers
Classify each number as being a natural number (*N*), whole number (*W*), integer (*I*), rational number (*Q*), and/or irrational number (*Q'*).

ⓐ $\sqrt{36}$ ⓑ $\frac{8}{3}$ ⓒ $\sqrt{73}$ ⓓ -6 ⓔ $3.2121121112\ldots$

Solution

	N	*W*	*I*	*Q*	*Q'*
a. $\sqrt{36} = 6$	X	X	X	X	
b. $\frac{8}{3} = 2.\overline{6}$				X	
c. $\sqrt{73}$					X
d. −6			X	X	
e. 3.2121121112...					X

> **TRY IT** #5 Classify each number as being a natural number (N), whole number (W), integer (I), rational number (Q), and/or irrational number (Q').
>
> ⓐ $-\frac{35}{7}$ ⓑ 0 ⓒ $\sqrt{169}$ ⓓ $\sqrt{24}$ ⓔ $4.763763763\ldots$

Performing Calculations Using the Order of Operations

When we multiply a number by itself, we square it or raise it to a power of 2. For example, $4^2 = 4 \cdot 4 = 16$. We can raise any number to any power. In general, the **exponential notation** a^n means that the number or variable a is used as a factor n times.

$$a^n = \overbrace{a \cdot a \cdot a \cdot \ldots \cdot a}^{n \text{ factors}}$$

In this notation, a^n is read as the nth power of a,or a to the n where a is called the **base** and n is called the **exponent.** A term in exponential notation may be part of a mathematical expression, which is a combination of numbers and operations. For example, $24 + 6 \cdot \frac{2}{3} - 4^2$ is a mathematical expression.

To evaluate a mathematical expression, we perform the various operations. However, we do not perform them in any random order. We use the **order of operations**. This is a sequence of rules for evaluating such expressions.

Recall that in mathematics we use parentheses ( ), brackets [ ], and braces { } to group numbers and expressions so that anything appearing within the symbols is treated as a unit. Additionally, fraction bars, radicals, and absolute value bars are treated as grouping symbols. When evaluating a mathematical expression, begin by simplifying expressions within grouping symbols.

The next step is to address any exponents or radicals. Afterward, perform multiplication and division from left to right and finally addition and subtraction from left to right.

Let's take a look at the expression provided.

$$24 + 6 \cdot \frac{2}{3} - 4^2$$

There are no grouping symbols, so we move on to exponents or radicals. The number 4 is raised to a power of 2, so simplify 4^2 as 16.

$$24 + 6 \cdot \tfrac{2}{3} - 4^2$$
$$24 + 6 \cdot \tfrac{2}{3} - 16$$

Next, perform multiplication or division, left to right.

$$24 + 6 \cdot \tfrac{2}{3} - 16$$
$$24 + 4 - 16$$

Lastly, perform addition or subtraction, left to right.

$$24 + 4 - 16$$
$$28 - 16$$
$$12$$

Therefore, $24 + 6 \cdot \frac{2}{3} - 4^2 = 12$.

For some complicated expressions, several passes through the order of operations will be needed. For instance, there may be a radical expression inside parentheses that must be simplified before the parentheses are evaluated. Following the order of operations ensures that anyone simplifying the same mathematical expression will get the same result.

Order of Operations

Operations in mathematical expressions must be evaluated in a systematic order, which can be simplified using the acronym **PEMDAS**:

P(arentheses)
E(xponents)
M(ultiplication) and **D**(ivision)
A(ddition) and **S**(ubtraction)

HOW TO

Given a mathematical expression, simplify it using the order of operations.

Step 1. Simplify any expressions within grouping symbols.
Step 2. Simplify any expressions containing exponents or radicals.
Step 3. Perform any multiplication and division in order, from left to right.
Step 4. Perform any addition and subtraction in order, from left to right.

EXAMPLE 6

Using the Order of Operations

Use the order of operations to evaluate each of the following expressions.

ⓐ $(3 \cdot 2)^2 - 4(6 + 2)$ ⓑ $\frac{5^2-4}{7} - \sqrt{11 - 2}$ ⓒ $6 - |5 - 8| + 3(4 - 1)$ ⓓ $\frac{14-3\cdot2}{2\cdot5-3^2}$

ⓔ $7(5 \cdot 3) - 2\left[(6 - 3) - 4^2\right] + 1$

Solution

ⓐ

$$\begin{aligned} (3 \cdot 2)^2 - 4(6 + 2) &= (6)^2 - 4(8) && \text{Simplify parentheses.} \\ &= 36 - 4(8) && \text{Simplify exponent.} \\ &= 36 - 32 && \text{Simplify multiplication.} \\ &= 4 && \text{Simplify subtraction.} \end{aligned}$$

ⓑ

$$\begin{aligned} \tfrac{5^2-4}{7} - \sqrt{11 - 2} &= \tfrac{5^2-4}{7} - \sqrt{9} && \text{Simplify grouping symbols (radical).} \\ &= \tfrac{5^2-4}{7} - 3 && \text{Simplify radical.} \\ &= \tfrac{25-4}{7} - 3 && \text{Simplify exponent.} \\ &= \tfrac{21}{7} - 3 && \text{Simplify subtraction in numerator.} \\ &= 3 - 3 && \text{Simplify division.} \\ &= 0 && \text{Simplify subtraction.} \end{aligned}$$

Note that in the first step, the radical is treated as a grouping symbol, like parentheses. Also, in the third step, the fraction bar is considered a grouping symbol so the numerator is considered to be grouped.

ⓒ

$$\begin{aligned} 6 - |5 - 8| + 3(4 - 1) &= 6 - |-3| + 3(3) && \text{Simplify inside grouping symbols.} \\ &= 6 - 3 + 3(3) && \text{Simplify absolute value.} \\ &= 6 - 3 + 9 && \text{Simplify multiplication.} \\ &= 3 + 9 && \text{Simplify subtraction.} \\ &= 12 && \text{Simplify addition.} \end{aligned}$$

ⓓ

$$\begin{aligned}\frac{14-3\cdot2}{2\cdot5-3^2} &= \frac{14-3\cdot2}{2\cdot5-9} && \text{Simplify exponent.}\\ &= \frac{14-6}{10-9} && \text{Simplify products.}\\ &= \frac{8}{1} && \text{Simplify differences.}\\ &= 8 && \text{Simplify quotient.}\end{aligned}$$

In this example, the fraction bar separates the numerator and denominator, which we simplify separately until the last step.

ⓔ

$$\begin{aligned}7(5\cdot3)-2\left[(6-3)-4^2\right]+1 &= 7(15)-2\left[(3)-4^2\right]+1 && \text{Simplify inside parentheses.}\\ &= 7(15)-2(3-16)+1 && \text{Simplify exponent.}\\ &= 7(15)-2(-13)+1 && \text{Subtract.}\\ &= 105+26+1 && \text{Multiply.}\\ &= 132 && \text{Add.}\end{aligned}$$

TRY IT #6 Use the order of operations to evaluate each of the following expressions.

ⓐ $\sqrt{5^2-4^2}+7(5-4)^2$ ⓑ $1+\frac{7\cdot5-8\cdot4}{9-6}$ ⓒ $\left|1.8-4.3\right|+0.4\sqrt{15+10}$
ⓓ $\frac{1}{2}\left[5\cdot3^2-7^2\right]+\frac{1}{3}\cdot9^2$ ⓔ $\left[(3-8)^2-4\right]-(3-8)$

Using Properties of Real Numbers

For some activities we perform, the order of certain operations does not matter, but the order of other operations does. For example, it does not make a difference if we put on the right shoe before the left or vice-versa. However, it does matter whether we put on shoes or socks first. The same thing is true for operations in mathematics.

Commutative Properties

The **commutative property of addition** states that numbers may be added in any order without affecting the sum.

$$a+b=b+a$$

We can better see this relationship when using real numbers.

$$(-2)+7=5 \quad \text{and} \quad 7+(-2)=5$$

Similarly, the **commutative property of multiplication** states that numbers may be multiplied in any order without affecting the product.

$$a\cdot b=b\cdot a$$

Again, consider an example with real numbers.

$$(-11)\cdot(-4)=44 \quad \text{and} \quad (-4)\cdot(-11)=44$$

It is important to note that neither subtraction nor division is commutative. For example, $17-5$ is not the same as $5-17$. Similarly, $20\div5\neq5\div20$.

Associative Properties

The **associative property of multiplication** tells us that it does not matter how we group numbers when multiplying. We can move the grouping symbols to make the calculation easier, and the product remains the same.

$$a(bc)=(ab)c$$

Consider this example.

$$(3\cdot4)\cdot5=60 \quad \text{and} \quad 3\cdot(4\cdot5)=60$$

The **associative property of addition** tells us that numbers may be grouped differently without affecting the sum.

$$a+(b+c)=(a+b)+c$$

This property can be especially helpful when dealing with negative integers. Consider this example.

$$[15+(-9)]+23=29 \quad \text{and} \quad 15+[(-9)+23]=29$$

Are subtraction and division associative? Review these examples.

$$\begin{aligned} 8-(3-15) &\stackrel{?}{=} (8-3)-15 \\ 8-(-12) &= 5-15 \\ 20 &\neq -10 \end{aligned} \qquad \begin{aligned} 64 \div (8 \div 4) &\stackrel{?}{=} (64 \div 8) \div 4 \\ 64 \div 2 &\stackrel{?}{=} 8 \div 4 \\ 32 &\neq 2 \end{aligned}$$

As we can see, neither subtraction nor division is associative.

Distributive Property

The **distributive property** states that the product of a factor times a sum is the sum of the factor times each term in the sum.

$$a \cdot (b+c) = a \cdot b + a \cdot c$$

This property combines both addition and multiplication (and is the only property to do so). Let us consider an example.

$$\begin{aligned} 4 \cdot [12 + (-7)] &= 4 \cdot 12 + 4 \cdot (-7) \\ &= 48 + (-28) \\ &= 20 \end{aligned}$$

Note that 4 is outside the grouping symbols, so we distribute the 4 by multiplying it by 12, multiplying it by –7, and adding the products.

To be more precise when describing this property, we say that multiplication distributes over addition. The reverse is not true, as we can see in this example.

$$\begin{aligned} 6+(3 \cdot 5) &\stackrel{?}{=} (6+3) \cdot (6+5) \\ 6+(15) &\stackrel{?}{=} (9) \cdot (11) \\ 21 &\neq 99 \end{aligned}$$

A special case of the distributive property occurs when a sum of terms is subtracted.

$$a-b = a+(-b)$$

For example, consider the difference $12-(5+3)$. We can rewrite the difference of the two terms 12 and $(5+3)$ by turning the subtraction expression into addition of the opposite. So instead of subtracting $(5+3)$, we add the opposite.

$$12+(-1) \cdot (5+3)$$

Now, distribute -1 and simplify the result.

$$\begin{aligned} 12-(5+3) &= 12+(-1) \cdot (5+3) \\ &= 12+[(-1) \cdot 5+(-1) \cdot 3] \\ &= 12+(-8) \\ &= 4 \end{aligned}$$

This seems like a lot of trouble for a simple sum, but it illustrates a powerful result that will be useful once we introduce algebraic terms. To subtract a sum of terms, change the sign of each term and add the results. With this in mind, we can rewrite the last example.

$$\begin{aligned} 12-(5+3) &= 12+(-5-3) \\ &= 12+(-8) \\ &= 4 \end{aligned}$$

Identity Properties

The **identity property of addition** states that there is a unique number, called the additive identity (0) that, when added to a number, results in the original number.

$$a+0=a$$

The **identity property of multiplication** states that there is a unique number, called the multiplicative identity (1) that, when multiplied by a number, results in the original number.

$$a \cdot 1 = a$$

For example, we have $(-6)+0=-6$ and $23 \cdot 1 = 23$. There are no exceptions for these properties; they work for every real number, including 0 and 1.

Inverse Properties

The **inverse property of addition** states that, for every real number *a*, there is a unique number, called the additive inverse (or opposite), denoted by (−*a*), that, when added to the original number, results in the additive identity, 0.

$$a + (-a) = 0$$

For example, if $a = -8$, the additive inverse is 8, since $(-8) + 8 = 0$.

The **inverse property of multiplication** holds for all real numbers except 0 because the reciprocal of 0 is not defined. The property states that, for every real number *a*, there is a unique number, called the multiplicative inverse (or reciprocal), denoted $\frac{1}{a}$, that, when multiplied by the original number, results in the multiplicative identity, 1.

$$a \cdot \frac{1}{a} = 1$$

For example, if $a = -\frac{2}{3}$, the reciprocal, denoted $\frac{1}{a}$, is $-\frac{3}{2}$ because

$$a \cdot \frac{1}{a} = \left(-\frac{2}{3}\right) \cdot \left(-\frac{3}{2}\right) = 1$$

Properties of Real Numbers

The following properties hold for real numbers *a*, *b*, and *c*.

	Addition	Multiplication
Commutative Property	$a + b = b + a$	$a \cdot b = b \cdot a$
Associative Property	$a + (b + c) = (a + b) + c$	$a\,(bc) = (ab)\,c$
Distributive Property	$a \cdot (b + c) = a \cdot b + a \cdot c$	
Identity Property	There exists a unique real number called the additive identity, 0, such that, for any real number *a* $a + 0 = a$	There exists a unique real number called the multiplicative identity, 1, such that, for any real number *a* $a \cdot 1 = a$
Inverse Property	Every real number a has an additive inverse, or opposite, denoted −*a*, such that $a + (-a) = 0$	Every nonzero real number *a* has a multiplicative inverse, or reciprocal, denoted $\frac{1}{a}$, such that $a \cdot \left(\frac{1}{a}\right) = 1$

EXAMPLE 7

Using Properties of Real Numbers

Use the properties of real numbers to rewrite and simplify each expression. State which properties apply.

ⓐ $3 \cdot 6 + 3 \cdot 4$ ⓑ $(5 + 8) + (-8)$ ⓒ $6 - (15 + 9)$ ⓓ $\frac{4}{7} \cdot \left(\frac{2}{3} \cdot \frac{7}{4}\right)$ ⓔ $100 \cdot [0.75 + (-2.38)]$

Solution

ⓐ

$$\begin{aligned} 3\cdot 6+3\cdot 4 &= 3\cdot(6+4) && \text{Distributive property.}\\ &= 3\cdot 10 && \text{Simplify.}\\ &= 30 && \text{Simplify.}\end{aligned}$$

ⓑ

$$\begin{aligned} (5+8)+(-8) &= 5+[8+(-8)] && \text{Associative property of addition.}\\ &= 5+0 && \text{Inverse property of addition.}\\ &= 5 && \text{Identity property of addition.}\end{aligned}$$

ⓒ

$$\begin{aligned} 6-(15+9) &= 6+[(-15)+(-9)] && \text{Distributive property.}\\ &= 6+(-24) && \text{Simplify.}\\ &= -18 && \text{Simplify.}\end{aligned}$$

ⓓ

$$\begin{aligned} \tfrac{4}{7}\cdot\left(\tfrac{2}{3}\cdot\tfrac{7}{4}\right) &= \tfrac{4}{7}\cdot\left(\tfrac{7}{4}\cdot\tfrac{2}{3}\right) && \text{Commutative property of multiplication.}\\ &= \left(\tfrac{4}{7}\cdot\tfrac{7}{4}\right)\cdot\tfrac{2}{3} && \text{Associative property of multiplication.}\\ &= 1\cdot\tfrac{2}{3} && \text{Inverse property of multiplication.}\\ &= \tfrac{2}{3} && \text{Identity property of multiplication.}\end{aligned}$$

ⓔ

$$\begin{aligned} 100\cdot[0.75+(-2.38)] &= 100\cdot 0.75+100\cdot(-2.38) && \text{Distributive property.}\\ &= 75+(-238) && \text{Simplify.}\\ &= -163 && \text{Simplify.}\end{aligned}$$

TRY IT #7 Use the properties of real numbers to rewrite and simplify each expression. State which properties apply.

ⓐ $\left(-\frac{23}{5}\right)\cdot\left[11\cdot\left(-\frac{5}{23}\right)\right]$ ⓑ $5\cdot(6.2+0.4)$ ⓒ $18-(7-15)$
ⓓ $\frac{17}{18}+\left[\frac{4}{9}+\left(-\frac{17}{18}\right)\right]$ ⓔ $6\cdot(-3)+6\cdot 3$

Evaluating Algebraic Expressions

So far, the mathematical expressions we have seen have involved real numbers only. In mathematics, we may see expressions such as $x+5$, $\frac{4}{3}\pi r^3$, or $\sqrt{2m^3n^2}$. In the expression $x+5$, 5 is called a **constant** because it does not vary and x is called a **variable** because it does. (In naming the variable, ignore any exponents or radicals containing the variable.) An **algebraic expression** is a collection of constants and variables joined together by the algebraic operations of addition, subtraction, multiplication, and division.

We have already seen some real number examples of exponential notation, a shorthand method of writing products of the same factor. When variables are used, the constants and variables are treated the same way.

$$\begin{aligned} (-3)^5 &= (-3)\cdot(-3)\cdot(-3)\cdot(-3)\cdot(-3) & x^5 &= x\cdot x\cdot x\cdot x\cdot x\\ (2\cdot 7)^3 &= (2\cdot 7)\cdot(2\cdot 7)\cdot(2\cdot 7) & (yz)^3 &= (yz)\cdot(yz)\cdot(yz)\end{aligned}$$

In each case, the exponent tells us how many factors of the base to use, whether the base consists of constants or variables.

Any variable in an algebraic expression may take on or be assigned different values. When that happens, the value of the algebraic expression changes. To evaluate an algebraic expression means to determine the value of the expression for a given value of each variable in the expression. Replace each variable in the expression with the given value, then simplify the resulting expression using the order of operations. If the algebraic expression contains more than one variable, replace each variable with its assigned value and simplify the expression as before.

EXAMPLE 8

Describing Algebraic Expressions

List the constants and variables for each algebraic expression.

ⓐ $x + 5$ ⓑ $\frac{4}{3}\pi r^3$ ⓒ $\sqrt{2m^3n^2}$

Solution

	Constants	Variables
a. $x + 5$	5	x
b. $\frac{4}{3}\pi r^3$	$\frac{4}{3}, \pi$	r
c. $\sqrt{2m^3n^2}$	2	m, n

TRY IT #8 List the constants and variables for each algebraic expression.

ⓐ $2\pi r(r + h)$ ⓑ $2(L + W)$ ⓒ $4y^3 + y$

EXAMPLE 9

Evaluating an Algebraic Expression at Different Values

Evaluate the expression $2x - 7$ for each value for x.

ⓐ $x = 0$ ⓑ $x = 1$ ⓒ $x = \frac{1}{2}$ ⓓ $x = -4$

Solution

ⓐ Substitute 0 for x.

$$\begin{aligned} 2x - 7 &= 2(0) - 7 \\ &= 0 - 7 \\ &= -7 \end{aligned}$$

ⓑ Substitute 1 for x.

$$\begin{aligned} 2x - 7 &= 2(1) - 7 \\ &= 2 - 7 \\ &= -5 \end{aligned}$$

ⓒ Substitute $\frac{1}{2}$ for x.

$$\begin{aligned} 2x - 7 &= 2\left(\frac{1}{2}\right) - 7 \\ &= 1 - 7 \\ &= -6 \end{aligned}$$

ⓓ Substitute -4 for x.

$$\begin{aligned} 2x - 7 &= 2(-4) - 7 \\ &= -8 - 7 \\ &= -15 \end{aligned}$$

TRY IT #9 Evaluate the expression $11 - 3y$ for each value for y.

ⓐ $y = 2$ ⓑ $y = 0$ ⓒ $y = \frac{2}{3}$ ⓓ $y = -5$

EXAMPLE 10

Evaluating Algebraic Expressions

Evaluate each expression for the given values.

ⓐ $x + 5$ for $x = -5$ ⓑ $\frac{t}{2t-1}$ for $t = 10$ ⓒ $\frac{4}{3}\pi r^3$ for $r = 5$ ⓓ $a + ab + b$ for $a = 11, b = -8$

ⓔ $\sqrt{2m^3n^2}$ for $m = 2, n = 3$

Solution

ⓐ Substitute -5 for x.

$$\begin{aligned} x + 5 &= (-5) + 5 \\ &= 0 \end{aligned}$$

ⓑ Substitute 10 for t.

$$\begin{aligned} \frac{t}{2t-1} &= \frac{(10)}{2(10)-1} \\ &= \frac{10}{20-1} \\ &= \frac{10}{19} \end{aligned}$$

ⓒ Substitute 5 for r.

$$\begin{aligned} \frac{4}{3}\pi r^3 &= \frac{4}{3}\pi(5)^3 \\ &= \frac{4}{3}\pi(125) \\ &= \frac{500}{3}\pi \end{aligned}$$

(d) Substitute 11 for a and −8 for b.

$$\begin{aligned} a + ab + b &= (11) + (11)(-8) + (-8) \\ &= 11 - 88 - 8 \\ &= -85 \end{aligned}$$

(e) Substitute 2 for m and 3 for n.

$$\begin{aligned} \sqrt{2m^3n^2} &= \sqrt{2(2)^3(3)^2} \\ &= \sqrt{2(8)(9)} \\ &= \sqrt{144} \\ &= 12 \end{aligned}$$

> TRY IT #10 Evaluate each expression for the given values.

(a) $\frac{y+3}{y-3}$ for $y = 5$ (b) $7 - 2t$ for $t = -2$ (c) $\frac{1}{3}\pi r^2$ for $r = 11$
(d) $\left(p^2q\right)^3$ for $p = -2, q = 3$ (e) $4(m - n) - 5(n - m)$ for $m = \frac{2}{3}, n = \frac{1}{3}$

Formulas

An **equation** is a mathematical statement indicating that two expressions are equal. The expressions can be numerical or algebraic. The equation is not inherently true or false, but only a proposition. The values that make the equation true, the solutions, are found using the properties of real numbers and other results. For example, the equation $2x + 1 = 7$ has the solution of 3 because when we substitute 3 for x in the equation, we obtain the true statement $2(3) + 1 = 7$.

A **formula** is an equation expressing a relationship between constant and variable quantities. Very often, the equation is a means of finding the value of one quantity (often a single variable) in terms of another or other quantities. One of the most common examples is the formula for finding the area A of a circle in terms of the radius r of the circle: $A = \pi r^2$. For any value of r, the area A can be found by evaluating the expression πr^2.

EXAMPLE 11

Using a Formula

A right circular cylinder with radius r and height h has the surface area S (in square units) given by the formula $S = 2\pi r(r + h)$. See Figure 3. Find the surface area of a cylinder with radius 6 in. and height 9 in. Leave the answer in terms of π.

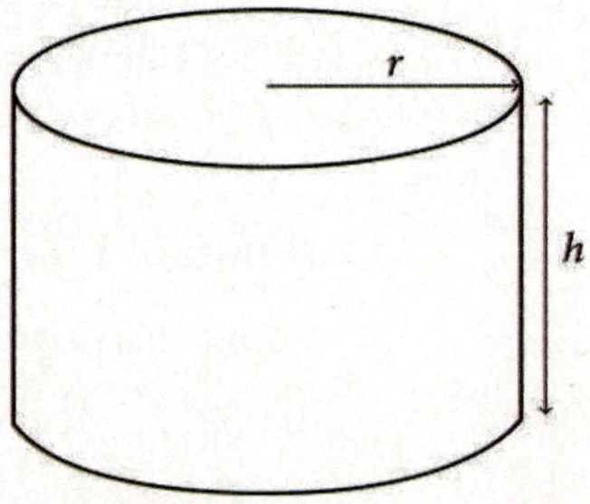

Figure 3 Right circular cylinder

Solution

Evaluate the expression $2\pi r(r + h)$ for $r = 6$ and $h = 9$.

$$\begin{aligned} S &= 2\pi r(r + h) \\ &= 2\pi(6)[(6) + (9)] \\ &= 2\pi(6)(15) \\ &= 180\pi \end{aligned}$$

The surface area is 180π square inches.

> TRY IT #11 A photograph with length L and width W is placed in a mat of width 8 centimeters (cm). The area of the mat (in square centimeters, or cm^2) is found to be $A = (L + 16)(W + 16) - L \cdot W$. See Figure 4. Find the area of a mat for a photograph with length 32 cm and width 24 cm.

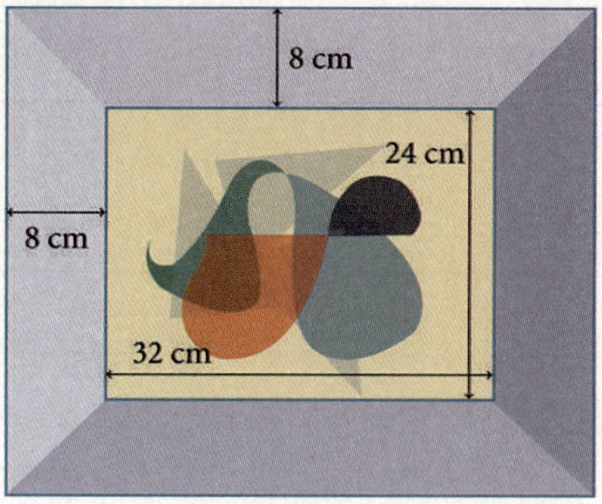

Figure 4

Simplifying Algebraic Expressions

Sometimes we can simplify an algebraic expression to make it easier to evaluate or to use in some other way. To do so, we use the properties of real numbers. We can use the same properties in formulas because they contain algebraic expressions.

EXAMPLE 12

Simplifying Algebraic Expressions

Simplify each algebraic expression.

ⓐ $3x - 2y + x - 3y - 7$ ⓑ $2r - 5(3 - r) + 4$ ⓒ $\left(4t - \frac{5}{4}s\right) - \left(\frac{2}{3}t + 2s\right)$ ⓓ $2mn - 5m + 3mn + n$

Solution

ⓐ

$$\begin{aligned} 3x - 2y + x - 3y - 7 &= 3x + x - 2y - 3y - 7 && \text{Commutative property of addition.} \\ &= 4x - 5y - 7 && \text{Simplify.} \end{aligned}$$

ⓑ

$$\begin{aligned} 2r - 5(3 - r) + 4 &= 2r - 15 + 5r + 4 && \text{Distributive property.} \\ &= 2r + 5r - 15 + 4 && \text{Commutative property of addition.} \\ &= 7r - 11 && \text{Simplify.} \end{aligned}$$

ⓒ

$$\begin{aligned} \left(4t - \tfrac{5}{4}s\right) - \left(\tfrac{2}{3}t + 2s\right) &= 4t - \tfrac{5}{4}s - \tfrac{2}{3}t - 2s && \text{Distributive property.} \\ &= 4t - \tfrac{2}{3}t - \tfrac{5}{4}s - 2s && \text{Commutative property of addition.} \\ &= \tfrac{10}{3}t - \tfrac{13}{4}s && \text{Simplify.} \end{aligned}$$

ⓓ

$$\begin{aligned} 2mn - 5m + 3mn + n &= 2mn + 3mn - 5m + n && \text{Commutative property of addition.} \\ &= 5mn - 5m + n && \text{Simplify.} \end{aligned}$$

> **TRY IT** #12 Simplify each algebraic expression.
>
> ⓐ $\frac{2}{3}y - 2\left(\frac{4}{3}y + z\right)$ ⓑ $\frac{5}{t} - 2 - \frac{3}{t} + 1$ ⓒ $4p(q - 1) + q(1 - p)$
> ⓓ $9r - (s + 2r) + (6 - s)$

EXAMPLE 13

Simplifying a Formula

A rectangle with length L and width W has a perimeter P given by $P = L + W + L + W$. Simplify this expression.

Solution

$$\begin{aligned} P &= L + W + L + W & \\ P &= L + L + W + W & \text{Commutative property of addition} \\ P &= 2L + 2W & \text{Simplify} \\ P &= 2(L + W) & \text{Distributive property} \end{aligned}$$

TRY IT #13 If the amount P is deposited into an account paying simple interest r for time t, the total value of the deposit A is given by $A = P + Prt$. Simplify the expression. (This formula will be explored in more detail later in the course.)

MEDIA

Access these online resources for additional instruction and practice with real numbers.

Simplify an Expression. (http://openstax.org/l/simexpress)
Evaluate an Expression 1. (http://openstax.org/l/ordofoper1)
Evaluate an Expression 2. (http://openstax.org/l/ordofoper2)

1.1 SECTION EXERCISES

Verbal

1. Is $\sqrt{2}$ an example of a rational terminating, rational repeating, or irrational number? Tell why it fits that category.

2. What is the order of operations? What acronym is used to describe the order of operations, and what does it stand for?

3. What do the Associative Properties allow us to do when following the order of operations? Explain your answer.

Numeric

For the following exercises, simplify the given expression.

4. $10 + 2 \times (5 - 3)$

5. $6 \div 2 - \left(81 \div 3^2\right)$

6. $18 + (6 - 8)^3$

7. $-2 \times \left[16 \div (8 - 4)^2\right]^2$

8. $4 - 6 + 2 \times 7$

9. $3(5 - 8)$

10. $4 + 6 - 10 \div 2$

11. $12 \div (36 \div 9) + 6$

12. $(4 + 5)^2 \div 3$

13. $3 - 12 \times 2 + 19$

14. $2 + 8 \times 7 \div 4$

15. $5 + (6 + 4) - 11$

16. $9 - 18 \div 3^2$

17. $14 \times 3 \div 7 - 6$

18. $9 - (3 + 11) \times 2$

19. $6 + 2 \times 2 - 1$

20. $64 \div (8 + 4 \times 2)$

21. $9 + 4\left(2^2\right)$

22. $(12 \div 3 \times 3)^2$

23. $25 \div 5^2 - 7$

24. $(15 - 7) \times (3 - 7)$

25. $2 \times 4 - 9(-1)$

26. $4^2 - 25 \times \frac{1}{5}$

27. $12(3 - 1) \div 6$

Algebraic

For the following exercises, evaluate the expressions using the given variable.

28. $8(x+3)-64$ for $x=2$

29. $4y+8-2y$ for $y=3$

30. $(11a+3)-18a+4$ for $a=-2$

31. $4z-2z(1+4)-36$ for $z=5$

32. $4y(7-2)^2+200$ for $y=-2$

33. $-(2x)^2+1+3$ for $x=2$

34. For the $8(2+4)-15b+b$ for $b=-3$

35. $2(11c-4)-36$ for $c=0$

36. $4(3-1)x-4$ for $x=10$

37. $\frac{1}{4}\left(8w-4^2\right)$ for $w=1$

For the following exercises, simplify the expression.

38. $4x+x(13-7)$

39. $2y-(4)^2y-11$

40. $\frac{a}{2^3}(64)-12a\div 6$

41. $8b-4b(3)+1$

42. $5l\div 3l\ \times\ (9-6)$

43. $7z-3+z\ \times\ 6^2$

44. $4\ \times\ 3+18x\div 9-12$

45. $9(y+8)-27$

46. $\left(\frac{9}{6}t-4\right)2$

47. $6+12b-3\ \times\ 6b$

48. $18y-2(1+7y)$

49. $\left(\frac{4}{9}\right)^2\ \times\ 27x$

50. $8(3-m)+1(-8)$

51. $9x+4x(2+3)-4(2x+3x)$

52. $5^2-4(3x)$

Real-World Applications

For the following exercises, consider this scenario: Fred earns \$40 at the community garden. He spends \$10 on a streaming subscription, puts half of what is left in a savings account, and gets another \$5 for walking his neighbor's dog.

53. Write the expression that represents the number of dollars Fred keeps (and does not put in his savings account). Remember the order of operations.

54. How much money does Fred keep?

For the following exercises, solve the given problem.

55. According to the U.S. Mint, the diameter of a quarter is 0.955 inches. The circumference of the quarter would be the diameter multiplied by π. Is the circumference of a quarter a whole number, a rational number, or an irrational number?

56. Jessica and her roommate, Adriana, have decided to share a change jar for joint expenses. Jessica put her loose change in the jar first, and then Adriana put her change in the jar. We know that it does not matter in which order the change was added to the jar. What property of addition describes this fact?

For the following exercises, consider this scenario: There is a mound of g pounds of gravel in a quarry. Throughout the day, 400 pounds of gravel is added to the mound. Two orders of 600 pounds are sold and the gravel is removed from the mound. At the end of the day, the mound has 1,200 pounds of gravel.

57. Write the equation that describes the situation.

58. Solve for *g*.

For the following exercise, solve the given problem.

59. Ramon runs the marketing department at their company. Their department gets a budget every year, and every year, they must spend the entire budget without going over. If they spend less than the budget, then the department gets a smaller budget the following year. At the beginning of this year, Ramon got $2.5 million for the annual marketing budget. They must spend the budget such that $2{,}500{,}000 - x = 0$. What property of addition tells us what the value of *x* must be?

Technology

For the following exercises, use a graphing calculator to solve for x. Round the answers to the nearest hundredth.

60. $0.5(12.3)^2 - 48x = \frac{3}{5}$

61. $(0.25 - 0.75)^2 x - 7.2 = 9.9$

Extensions

62. If a whole number is not a natural number, what must the number be?

63. Determine whether the statement is true or false: The multiplicative inverse of a rational number is also rational.

64. Determine whether the statement is true or false: The product of a rational and irrational number is always irrational.

65. Determine whether the simplified expression is rational or irrational: $\sqrt{-18 - 4(5)(-1)}$.

66. Determine whether the simplified expression is rational or irrational: $\sqrt{-16 + 4(5) + 5}$.

67. The division of two natural numbers will always result in what type of number?

68. What property of real numbers would simplify the following expression: $4 + 7(x - 1)$?

1.2 Exponents and Scientific Notation

Learning Objectives

In this section, you will:

- Use the product rule of exponents.
- Use the quotient rule of exponents.
- Use the power rule of exponents.
- Use the zero exponent rule of exponents.
- Use the negative rule of exponents.
- Find the power of a product and a quotient.
- Simplify exponential expressions.
- Use scientific notation.

Mathematicians, scientists, and economists commonly encounter very large and very small numbers. But it may not be obvious how common such figures are in everyday life. For instance, a pixel is the smallest unit of light that can be perceived and recorded by a digital camera. A particular camera might record an image that is 2,048 pixels by 1,536 pixels, which is a very high resolution picture. It can also perceive a color depth (gradations in colors) of up to 48 bits per frame, and can shoot the equivalent of 24 frames per second. The maximum possible number of bits of information used to film a one-hour (3,600-second) digital film is then an extremely large number.

Using a calculator, we enter $2{,}048 \times 1{,}536 \times 48 \times 24 \times 3{,}600$ and press ENTER. The calculator displays 1.304596316E13. What does this mean? The "E13" portion of the result represents the exponent 13 of ten, so there are a maximum of approximately 1.3×10^{13} bits of data in that one-hour film. In this section, we review rules of exponents first and then apply them to calculations involving very large or small numbers.

Using the Product Rule of Exponents

Consider the product $x^3 \cdot x^4$. Both terms have the same base, x, but they are raised to different exponents. Expand each expression, and then rewrite the resulting expression.

$$\begin{aligned} x^3 \cdot x^4 &= \overbrace{x \cdot x \cdot x}^{3 \text{ factors}} \cdot \overbrace{x \cdot x \cdot x \cdot x}^{4 \text{ factors}} \\ &= \overbrace{x \cdot x \cdot x \cdot x \cdot x \cdot x \cdot x}^{7 \text{ factors}} \\ &= x^7 \end{aligned}$$

The result is that $x^3 \cdot x^4 = x^{3+4} = x^7$.

Notice that the exponent of the product is the sum of the exponents of the terms. In other words, when multiplying exponential expressions with the same base, we write the result with the common base and add the exponents. This is the *product rule of exponents.*

$$a^m \cdot a^n = a^{m+n}$$

Now consider an example with real numbers.

$$2^3 \cdot 2^4 = 2^{3+4} = 2^7$$

We can always check that this is true by simplifying each exponential expression. We find that 2^3 is 8, 2^4 is 16, and 2^7 is 128. The product $8 \cdot 16$ equals 128, so the relationship is true. We can use the product rule of exponents to simplify expressions that are a product of two numbers or expressions with the same base but different exponents.

The Product Rule of Exponents

For any real number a and natural numbers m and n, the product rule of exponents states that

$$a^m \cdot a^n = a^{m+n}$$

EXAMPLE 1

Using the Product Rule

Write each of the following products with a single base. Do not simplify further.

ⓐ $t^5 \cdot t^3$ ⓑ $(-3)^5 \cdot (-3)$ ⓒ $x^2 \cdot x^5 \cdot x^3$

Solution

Use the product rule to simplify each expression.

ⓐ $t^5 \cdot t^3 = t^{5+3} = t^8$ ⓑ $(-3)^5 \cdot (-3) = (-3)^5 \cdot (-3)^1 = (-3)^{5+1} = (-3)^6$ ⓒ $x^2 \cdot x^5 \cdot x^3$

At first, it may appear that we cannot simplify a product of three factors. However, using the associative property of multiplication, begin by simplifying the first two.

$$x^2 \cdot x^5 \cdot x^3 = \left(x^2 \cdot x^5\right) \cdot x^3 = \left(x^{2+5}\right) \cdot x^3 = x^7 \cdot x^3 = x^{7+3} = x^{10}$$

Notice we get the same result by adding the three exponents in one step.

$$x^2 \cdot x^5 \cdot x^3 = x^{2+5+3} = x^{10}$$

TRY IT #1 Write each of the following products with a single base. Do not simplify further.

ⓐ $k^6 \cdot k^9$ ⓑ $\left(\frac{2}{y}\right)^4 \cdot \left(\frac{2}{y}\right)$ ⓒ $t^3 \cdot t^6 \cdot t^5$

Using the Quotient Rule of Exponents

The *quotient rule of exponents* allows us to simplify an expression that divides two numbers with the same base but different exponents. In a similar way to the product rule, we can simplify an expression such as $\frac{y^m}{y^n}$, where $m > n$. Consider the example $\frac{y^9}{y^5}$. Perform the division by canceling common factors.

$$\begin{aligned} \frac{y^9}{y^5} &= \frac{y \cdot y \cdot y \cdot y \cdot y \cdot y \cdot y \cdot y \cdot y}{y \cdot y \cdot y \cdot y \cdot y} \\ &= \frac{\cancel{y} \cdot \cancel{y} \cdot \cancel{y} \cdot \cancel{y} \cdot \cancel{y} \cdot y \cdot y \cdot y \cdot y}{\cancel{y} \cdot \cancel{y} \cdot \cancel{y} \cdot \cancel{y} \cdot \cancel{y}} \\ &= \frac{y \cdot y \cdot y \cdot y}{1} \\ &= y^4 \end{aligned}$$

Notice that the exponent of the quotient is the difference between the exponents of the divisor and dividend.

$$\frac{a^m}{a^n} = a^{m-n}$$

In other words, when dividing exponential expressions with the same base, we write the result with the common base and subtract the exponents.

$$\frac{y^9}{y^5} = y^{9-5} = y^4$$

For the time being, we must be aware of the condition $m > n$. Otherwise, the difference $m - n$ could be zero or negative. Those possibilities will be explored shortly. Also, instead of qualifying variables as nonzero each time, we will simplify matters and assume from here on that all variables represent nonzero real numbers.

The Quotient Rule of Exponents

For any real number a and natural numbers m and n, such that $m > n$, the quotient rule of exponents states that

$$\frac{a^m}{a^n} = a^{m-n}$$

EXAMPLE 2

Using the Quotient Rule

Write each of the following products with a single base. Do not simplify further.

ⓐ $\frac{(-2)^{14}}{(-2)^9}$ ⓑ $\frac{t^{23}}{t^{15}}$ ⓒ $\frac{\left(z\sqrt{2}\right)^5}{z\sqrt{2}}$

Solution

Use the quotient rule to simplify each expression.

ⓐ $\frac{(-2)^{14}}{(-2)^9} = (-2)^{14-9} = (-2)^5$ ⓑ $\frac{t^{23}}{t^{15}} = t^{23-15} = t^8$ ⓒ $\frac{\left(z\sqrt{2}\right)^5}{z\sqrt{2}} = \left(z\sqrt{2}\right)^{5-1} = \left(z\sqrt{2}\right)^4$

TRY IT #2 Write each of the following products with a single base. Do not simplify further.

ⓐ $\frac{s^{75}}{s^{68}}$ ⓑ $\frac{(-3)^6}{-3}$ ⓒ $\frac{\left(ef^2\right)^5}{\left(ef^2\right)^3}$

Using the Power Rule of Exponents

Suppose an exponential expression is raised to some power. Can we simplify the result? Yes. To do this, we use the *power rule of exponents.* Consider the expression $\left(x^2\right)^3$. The expression inside the parentheses is multiplied twice because it has an exponent of 2. Then the result is multiplied three times because the entire expression has an exponent of 3.

$$\begin{aligned}
\left(x^2\right)^3 &= \overbrace{\left(x^2\right)\cdot\left(x^2\right)\cdot\left(x^2\right)}^{3\text{ factors}} \\
&= \overbrace{\left(\overbrace{x\cdot x}^{2\text{ factors}}\right)\cdot\left(\overbrace{x\cdot x}^{2\text{ factors}}\right)\cdot\left(\overbrace{x\cdot x}^{2\text{ factors}}\right)}^{3\text{ factors}} \\
&= x\cdot x\cdot x\cdot x\cdot x\cdot x \\
&= x^6
\end{aligned}$$

The exponent of the answer is the product of the exponents: $\left(x^2\right)^3 = x^{2\cdot 3} = x^6$. In other words, when raising an exponential expression to a power, we write the result with the common base and the product of the exponents.

$$\left(a^m\right)^n = a^{m\cdot n}$$

Be careful to distinguish between uses of the product rule and the power rule. When using the product rule, different terms with the same bases are raised to exponents. In this case, you add the exponents. When using the power rule, a term in exponential notation is raised to a power. In this case, you multiply the exponents.

Product Rule				Power Rule		
$5^3\cdot 5^4$	$= 5^{3+4}$	$= 5^7$	but	$\left(5^3\right)^4$	$= 5^{3\cdot 4}$	$= 5^{12}$
$x^5\cdot x^2$	$= x^{5+2}$	$= x^7$	but	$\left(x^5\right)^2$	$= x^{5\cdot 2}$	$= x^{10}$
$(3a)^7\cdot(3a)^{10}$	$= (3a)^{7+10}$	$= (3a)^{17}$	but	$\left((3a)^7\right)^{10}$	$= (3a)^{7\cdot 10}$	$= (3a)^{70}$

The Power Rule of Exponents

For any real number a and positive integers m and n, the power rule of exponents states that

$$\left(a^m\right)^n = a^{m\cdot n}$$

EXAMPLE 3

Using the Power Rule

Write each of the following products with a single base. Do not simplify further.

ⓐ $\left(x^2\right)^7$ ⓑ $\left((2t)^5\right)^3$ ⓒ $\left((-3)^5\right)^{11}$

Solution

Use the power rule to simplify each expression.

ⓐ $(x^2)^7 = x^{2\cdot 7} = x^{14}$ ⓑ $((2t)^5)^3 = (2t)^{5\cdot 3} = (2t)^{15}$ ⓒ $((-3)^5)^{11} = (-3)^{5\cdot 11} = (-3)^{55}$

TRY IT #3 Write each of the following products with a single base. Do not simplify further.

ⓐ $((3y)^8)^3$ ⓑ $(t^5)^7$ ⓒ $((-g)^4)^4$

Using the Zero Exponent Rule of Exponents

Return to the quotient rule. We made the condition that $m > n$ so that the difference $m - n$ would never be zero or negative. What would happen if $m = n$? In this case, we would use the *zero exponent rule of exponents* to simplify the expression to 1. To see how this is done, let us begin with an example.

$$\frac{t^8}{t^8} = \frac{\cancel{t^8}}{\cancel{t^8}} = 1$$

If we were to simplify the original expression using the quotient rule, we would have

$$\frac{t^8}{t^8} = t^{8-8} = t^0$$

If we equate the two answers, the result is $t^0 = 1$. This is true for any nonzero real number, or any variable representing a real number.

$$a^0 = 1$$

The sole exception is the expression 0^0. This appears later in more advanced courses, but for now, we will consider the value to be undefined.

The Zero Exponent Rule of Exponents

For any nonzero real number a, the zero exponent rule of exponents states that

$$a^0 = 1$$

EXAMPLE 4

Using the Zero Exponent Rule

Simplify each expression using the zero exponent rule of exponents.

ⓐ $\frac{c^3}{c^3}$ ⓑ $\frac{-3x^5}{x^5}$ ⓒ $\frac{\left(j^2k\right)^4}{\left(j^2k\right)\cdot\left(j^2k\right)^3}$ ⓓ $\frac{5\left(rs^2\right)^2}{\left(rs^2\right)^2}$

Solution

Use the zero exponent and other rules to simplify each expression.

ⓐ

$$\begin{aligned}\frac{c^3}{c^3} &= c^{3-3}\\ &= c^0\\ &= 1\end{aligned}$$

ⓑ

$$\begin{aligned}\frac{-3x^5}{x^5} &= -3\cdot\frac{x^5}{x^5}\\ &= -3\cdot x^{5-5}\\ &= -3\cdot x^0\\ &= -3\cdot 1\\ &= -3\end{aligned}$$

ⓒ

$$\begin{aligned}\frac{(j^2k)^4}{(j^2k)\cdot(j^2k)^3} &= \frac{(j^2k)^4}{(j^2k)^{1+3}} && \text{Use the product rule in the denominator.}\\ &= \frac{(j^2k)^4}{(j^2k)^4} && \text{Simplify.}\\ &= (j^2k)^{4-4} && \text{Use the quotient rule.}\\ &= (j^2k)^0 && \text{Simplify.}\\ &= 1\end{aligned}$$

ⓓ

$$\begin{aligned}\frac{5(rs^2)^2}{(rs^2)^2} &= 5(rs^2)^{2-2} && \text{Use the quotient rule.}\\ &= 5(rs^2)^0 && \text{Simplify.}\\ &= 5\cdot 1 && \text{Use the zero exponent rule.}\\ &= 5 && \text{Simplify.}\end{aligned}$$

> **TRY IT** #4 Simplify each expression using the zero exponent rule of exponents.
>
> ⓐ $\frac{t^7}{t^7}$ ⓑ $\frac{(de^2)^{11}}{2(de^2)^{11}}$ ⓒ $\frac{w^4\cdot w^2}{w^6}$ ⓓ $\frac{t^3\cdot t^4}{t^2\cdot t^5}$

Using the Negative Rule of Exponents

Another useful result occurs if we relax the condition that $m > n$ in the quotient rule even further. For example, can we simplify $\frac{h^3}{h^5}$? When $m < n$ —that is, where the difference $m - n$ is negative—we can use the *negative rule of exponents* to simplify the expression to its reciprocal.

Divide one exponential expression by another with a larger exponent. Use our example, $\frac{h^3}{h^5}$.

$$\begin{aligned}\frac{h^3}{h^5} &= \frac{h\cdot h\cdot h}{h\cdot h\cdot h\cdot h\cdot h}\\ &= \frac{\cancel{h}\cdot\cancel{h}\cdot\cancel{h}}{\cancel{h}\cdot\cancel{h}\cdot\cancel{h}\cdot h\cdot h}\\ &= \frac{1}{h\cdot h}\\ &= \frac{1}{h^2}\end{aligned}$$

If we were to simplify the original expression using the quotient rule, we would have

$$\begin{aligned}\frac{h^3}{h^5} &= h^{3-5}\\ &= h^{-2}\end{aligned}$$

Putting the answers together, we have $h^{-2} = \frac{1}{h^2}$. This is true for any nonzero real number, or any variable representing a nonzero real number.

A factor with a negative exponent becomes the same factor with a positive exponent if it is moved across the fraction bar—from numerator to denominator or vice versa.

$$a^{-n} = \frac{1}{a^n} \quad \text{and} \quad a^n = \frac{1}{a^{-n}}$$

We have shown that the exponential expression a^n is defined when n is a natural number, 0, or the negative of a natural number. That means that a^n is defined for any integer n. Also, the product and quotient rules and all of the rules we will look at soon hold for any integer n.

The Negative Rule of Exponents

For any nonzero real number a and natural number n, the negative rule of exponents states that

$$a^{-n} = \frac{1}{a^n}$$

EXAMPLE 5

Using the Negative Exponent Rule

Write each of the following quotients with a single base. Do not simplify further. Write answers with positive exponents.

ⓐ $\frac{\theta^3}{\theta^{10}}$ ⓑ $\frac{z^2 \cdot z}{z^4}$ ⓒ $\frac{\left(-5t^3\right)^4}{\left(-5t^3\right)^8}$

Solution

ⓐ $\frac{\theta^3}{\theta^{10}} = \theta^{3-10} = \theta^{-7} = \frac{1}{\theta^7}$ ⓑ $\frac{z^2 \cdot z}{z^4} = \frac{z^{2+1}}{z^4} = \frac{z^3}{z^4} = z^{3-4} = z^{-1} = \frac{1}{z}$

ⓒ $\frac{\left(-5t^3\right)^4}{\left(-5t^3\right)^8} = \left(-5t^3\right)^{4-8} = \left(-5t^3\right)^{-4} = \frac{1}{\left(-5t^3\right)^4}$

TRY IT #5 Write each of the following quotients with a single base. Do not simplify further. Write answers with positive exponents.

ⓐ $\frac{(-3t)^2}{(-3t)^8}$ ⓑ $\frac{f^{47}}{f^{49} \cdot f}$ ⓒ $\frac{2k^4}{5k^7}$

EXAMPLE 6

Using the Product and Quotient Rules

Write each of the following products with a single base. Do not simplify further. Write answers with positive exponents.

ⓐ $b^2 \cdot b^{-8}$ ⓑ $(-x)^5 \cdot (-x)^{-5}$ ⓒ $\frac{-7z}{(-7z)^5}$

Solution

ⓐ $b^2 \cdot b^{-8} = b^{2-8} = b^{-6} = \frac{1}{b^6}$ ⓑ $(-x)^5 \cdot (-x)^{-5} = (-x)^{5-5} = (-x)^0 = 1$

ⓒ $\frac{-7z}{(-7z)^5} = \frac{(-7z)^1}{(-7z)^5} = (-7z)^{1-5} = (-7z)^{-4} = \frac{1}{(-7z)^4}$

TRY IT #6 Write each of the following products with a single base. Do not simplify further. Write answers with positive exponents.

ⓐ $t^{-11} \cdot t^6$ ⓑ $\frac{25^{12}}{25^{13}}$

Finding the Power of a Product

To simplify the power of a product of two exponential expressions, we can use the *power of a product rule of exponents,* which breaks up the power of a product of factors into the product of the powers of the factors. For instance, consider $(pq)^3$. We begin by using the associative and commutative properties of multiplication to regroup the factors.

$$\begin{aligned}(pq)^3 &= \overbrace{(pq)\cdot(pq)\cdot(pq)}^{3 \text{ factors}} \\ &= p\cdot q\cdot p\cdot q\cdot p\cdot q \\ &= \overbrace{p\cdot p\cdot p}^{3 \text{ factors}}\cdot\overbrace{q\cdot q\cdot q}^{3 \text{ factors}} \\ &= p^3\cdot q^3\end{aligned}$$

In other words, $(pq)^3 = p^3 \cdot q^3$.

The Power of a Product Rule of Exponents

For any real numbers a and b and any integer n, the power of a product rule of exponents states that

$$(ab)^n = a^n b^n$$

EXAMPLE 7

Using the Power of a Product Rule

Simplify each of the following products as much as possible using the power of a product rule. Write answers with positive exponents.

ⓐ $(ab^2)^3$ ⓑ $(2t)^{15}$ ⓒ $(-2w^3)^3$ ⓓ $\frac{1}{(-7z)^4}$ ⓔ $(e^{-2}f^2)^7$

Solution

Use the product and quotient rules and the new definitions to simplify each expression.

ⓐ $(ab^2)^3 = (a)^3 \cdot (b^2)^3 = a^{1\cdot 3} \cdot b^{2\cdot 3} = a^3 b^6$ ⓑ $(2t)^{15} = (2)^{15} \cdot (t)^{15} = 2^{15}t^{15} = 32,768t^{15}$

ⓒ $(-2w^3)^3 = (-2)^3 \cdot (w^3)^3 = -8 \cdot w^{3\cdot 3} = -8w^9$ ⓓ $\frac{1}{(-7z)^4} = \frac{1}{(-7)^4\cdot(z)^4} = \frac{1}{2{,}401z^4}$

ⓔ $(e^{-2}f^2)^7 = (e^{-2})^7 \cdot (f^2)^7 = e^{-2\cdot 7} \cdot f^{2\cdot 7} = e^{-14}f^{14} = \frac{f^{14}}{e^{14}}$

TRY IT #7 Simplify each of the following products as much as possible using the power of a product rule. Write answers with positive exponents.

ⓐ $(g^2h^3)^5$ ⓑ $(5t)^3$ ⓒ $(-3y^5)^3$ ⓓ $\frac{1}{(a^6b^7)^3}$ ⓔ $(r^3s^{-2})^4$

Finding the Power of a Quotient

To simplify the power of a quotient of two expressions, we can use the *power of a quotient rule,* which states that the power of a quotient of factors is the quotient of the powers of the factors. For example, let's look at the following example.

$$(e^{-2}f^2)^7 = \frac{f^{14}}{e^{14}}$$

Let's rewrite the original problem differently and look at the result.

$$\begin{aligned}(e^{-2}f^2)^7 &= \left(\frac{f^2}{e^2}\right)^7 \\ &= \frac{f^{14}}{e^{14}}\end{aligned}$$

It appears from the last two steps that we can use the power of a product rule as a power of a quotient rule.

$$\begin{aligned}\left(e^{-2}f^2\right)^7 &= \left(\frac{f^2}{e^2}\right)^7 \\ &= \frac{\left(f^2\right)^7}{\left(e^2\right)^7} \\ &= \frac{f^{2\cdot 7}}{e^{2\cdot 7}} \\ &= \frac{f^{14}}{e^{14}}\end{aligned}$$

The Power of a Quotient Rule of Exponents

For any real numbers a and b and any integer n, the power of a quotient rule of exponents states that

$$\left(\frac{a}{b}\right)^n = \frac{a^n}{b^n}$$

EXAMPLE 8

Using the Power of a Quotient Rule

Simplify each of the following quotients as much as possible using the power of a quotient rule. Write answers with positive exponents.

ⓐ $\left(\frac{4}{z^{11}}\right)^3$ ⓑ $\left(\frac{p}{q^3}\right)^6$ ⓒ $\left(\frac{-1}{t^2}\right)^{27}$ ⓓ $\left(j^3k^{-2}\right)^4$ ⓔ $\left(m^{-2}n^{-2}\right)^3$

Solution

ⓐ $\left(\frac{4}{z^{11}}\right)^3 = \frac{(4)^3}{\left(z^{11}\right)^3} = \frac{64}{z^{11\cdot 3}} = \frac{64}{z^{33}}$ ⓑ $\left(\frac{p}{q^3}\right)^6 = \frac{(p)^6}{\left(q^3\right)^6} = \frac{p^{1\cdot 6}}{q^{3\cdot 6}} = \frac{p^6}{q^{18}}$

ⓒ $\left(\frac{-1}{t^2}\right)^{27} = \frac{(-1)^{27}}{\left(t^2\right)^{27}} = \frac{-1}{t^{2\cdot 27}} = \frac{-1}{t^{54}} = -\frac{1}{t^{54}}$ ⓓ $\left(j^3k^{-2}\right)^4 = \left(\frac{j^3}{k^2}\right)^4 = \frac{\left(j^3\right)^4}{\left(k^2\right)^4} = \frac{j^{3\cdot 4}}{k^{2\cdot 4}} = \frac{j^{12}}{k^8}$

ⓔ $\left(m^{-2}n^{-2}\right)^3 = \left(\frac{1}{m^2n^2}\right)^3 = \frac{(1)^3}{\left(m^2n^2\right)^3} = \frac{1}{\left(m^2\right)^3\left(n^2\right)^3} = \frac{1}{m^{2\cdot 3}\cdot n^{2\cdot 3}} = \frac{1}{m^6n^6}$

> **TRY IT** #8 Simplify each of the following quotients as much as possible using the power of a quotient rule. Write answers with positive exponents.

ⓐ $\left(\frac{b^5}{c}\right)^3$ ⓑ $\left(\frac{5}{u^8}\right)^4$ ⓒ $\left(\frac{-1}{w^3}\right)^{35}$ ⓓ $\left(p^{-4}q^3\right)^8$ ⓔ $\left(c^{-5}d^{-3}\right)^4$

Simplifying Exponential Expressions

Recall that to simplify an expression means to rewrite it by combing terms or exponents; in other words, to write the expression more simply with fewer terms. The rules for exponents may be combined to simplify expressions.

EXAMPLE 9

Simplifying Exponential Expressions

Simplify each expression and write the answer with positive exponents only.

ⓐ $\left(6m^2n^{-1}\right)^3$ ⓑ $17^5 \cdot 17^{-4} \cdot 17^{-3}$ ⓒ $\left(\frac{u^{-1}v}{v^{-1}}\right)^2$ ⓓ $\left(-2a^3b^{-1}\right)\left(5a^{-2}b^2\right)$

ⓔ $\left(x^2\sqrt{2}\right)^4\left(x^2\sqrt{2}\right)^{-4}$ ⓕ $\frac{\left(3w^2\right)^5}{\left(6w^{-2}\right)^2}$

Solution

ⓐ

$$\begin{aligned}(6m^2n^{-1})^3 &= (6)^3(m^2)^3(n^{-1})^3 && \text{The power of a product rule}\\ &= 6^3m^{2\cdot 3}n^{-1\cdot 3} && \text{The power rule}\\ &= 216m^6n^{-3} && \text{Simplify.}\\ &= \frac{216m^6}{n^3} && \text{The negative exponent rule}\end{aligned}$$

ⓑ

$$\begin{aligned}17^5\cdot 17^{-4}\cdot 17^{-3} &= 17^{5-4-3} && \text{The product rule}\\ &= 17^{-2} && \text{Simplify.}\\ &= \frac{1}{17^2}\text{ or }\frac{1}{289} && \text{The negative exponent rule}\end{aligned}$$

ⓒ

$$\begin{aligned}\left(\frac{u^{-1}v}{v^{-1}}\right)^2 &= \frac{(u^{-1}v)^2}{(v^{-1})^2} && \text{The power of a quotient rule}\\ &= \frac{u^{-2}v^2}{v^{-2}} && \text{The power of a product rule}\\ &= u^{-2}v^{2-(-2)} && \text{The quotient rule}\\ &= u^{-2}v^4 && \text{Simplify.}\\ &= \frac{v^4}{u^2} && \text{The negative exponent rule}\end{aligned}$$

ⓓ

$$\begin{aligned}(-2a^3b^{-1})(5a^{-2}b^2) &= -2\cdot 5\cdot a^3\cdot a^{-2}\cdot b^{-1}\cdot b^2 && \text{Commutative and associative laws of multiplication}\\ &= -10\cdot a^{3-2}\cdot b^{-1+2} && \text{The product rule}\\ &= -10ab && \text{Simplify.}\end{aligned}$$

ⓔ

$$\begin{aligned}\left(x^2\sqrt{2}\right)^4\left(x^2\sqrt{2}\right)^{-4} &= \left(x^2\sqrt{2}\right)^{4-4} && \text{The product rule}\\ &= \left(x^2\sqrt{2}\right)^0 && \text{Simplify.}\\ &= 1 && \text{The zero exponent rule}\end{aligned}$$

ⓕ

$$\begin{aligned}\frac{(3w^2)^5}{(6w^{-2})^2} &= \frac{(3)^5\cdot(w^2)^5}{(6)^2\cdot(w^{-2})^2} && \text{The power of a product rule}\\ &= \frac{3^5w^{2\cdot 5}}{6^2w^{-2\cdot 2}} && \text{The power rule}\\ &= \frac{243w^{10}}{36w^{-4}} && \text{Simplify.}\\ &= \frac{27w^{10-(-4)}}{4} && \text{The quotient rule and reduce fraction}\\ &= \frac{27w^{14}}{4} && \text{Simplify.}\end{aligned}$$

TRY IT #9 Simplify each expression and write the answer with positive exponents only.

ⓐ $(2uv^{-2})^{-3}$ ⓑ $x^8\cdot x^{-12}\cdot x$ ⓒ $\left(\frac{e^2f^{-3}}{f^{-1}}\right)^2$ ⓓ $(9r^{-5}s^3)(3r^6s^{-4})$

ⓔ $\left(\frac{4}{9}tw^{-2}\right)^{-3}\left(\frac{4}{9}tw^{-2}\right)^3$ ⓕ $\frac{(2h^2k)^4}{(7h^{-1}k^2)^2}$

Using Scientific Notation

Recall at the beginning of the section that we found the number 1.3×10^{13} when describing bits of information in digital images. Other extreme numbers include the width of a human hair, which is about 0.00005 m, and the radius of an electron, which is about 0.0000000000047 m. How can we effectively work read, compare, and calculate with numbers such as these?

A shorthand method of writing very small and very large numbers is called **scientific notation**, in which we express numbers in terms of exponents of 10. To write a number in scientific notation, move the decimal point to the right of the first digit in the number. Write the digits as a decimal number between 1 and 10. Count the number of places *n* that you moved the decimal point. Multiply the decimal number by 10 raised to a power of *n*. If you moved the decimal left as in a very large number, *n* is positive. If you moved the decimal right as in a small large number, *n* is negative.

For example, consider the number 2,780,418. Move the decimal left until it is to the right of the first nonzero digit, which is 2.

$$2{,}780{,}418 \longrightarrow \overbrace{2.780418}^{\text{6 places left}}$$

We obtain 2.780418 by moving the decimal point 6 places to the left. Therefore, the exponent of 10 is 6, and it is positive because we moved the decimal point to the left. This is what we should expect for a large number.

$$2.780418 \times 10^6$$

Working with small numbers is similar. Take, for example, the radius of an electron, 0.0000000000047 m. Perform the same series of steps as above, except move the decimal point to the right.

$$0.00000000000047 \longrightarrow \overbrace{00000000000004.7}^{\text{13 places right}}$$

Be careful not to include the leading 0 in your count. We move the decimal point 13 places to the right, so the exponent of 10 is 13. The exponent is negative because we moved the decimal point to the right. This is what we should expect for a small number.

$$4.7 \times 10^{-13}$$

Scientific Notation

A number is written in **scientific notation** if it is written in the form $a \times 10^n$, where $1 \leq |a| < 10$ and n is an integer.

EXAMPLE 10

Converting Standard Notation to Scientific Notation

Write each number in scientific notation.

ⓐ Distance to Andromeda Galaxy from Earth: 24,000,000,000,000,000,000,000 m

ⓑ Diameter of Andromeda Galaxy: 1,300,000,000,000,000,000,000 m

ⓒ Number of stars in Andromeda Galaxy: 1,000,000,000,000

ⓓ Diameter of electron: 0.0000000000094 m

ⓔ Probability of being struck by lightning in any single year: 0.00000143

Solution

ⓐ

$$24{,}000{,}000{,}000{,}000{,}000{,}000{,}000 \text{ m}$$
$$24{,}000{,}000{,}000{,}000{,}000{,}000{,}000 \text{ m}$$
$$\leftarrow 22 \text{ places}$$
$$2.4 \times 10^{22} \text{ m}$$

ⓑ

$1{,}300{,}000{,}000{,}000{,}000{,}000{,}000\ \text{m}$

$1{,}300{,}000{,}000{,}000{,}000{,}000{,}000\ \text{m}$

$\leftarrow 21\ \text{places}$

$1.3 \times 10^{21}\ \text{m}$

ⓒ

$1{,}000{,}000{,}000{,}000$

$1{,}000{,}000{,}000{,}000$

$\leftarrow 12\ \text{places}$

1×10^{12}

ⓓ

$0.00000000000094\ \text{m}$

$0.00000000000094\ \text{m}$

$\rightarrow 13\ \text{places}$

$9.4 \times 10^{-13}\ \text{m}$

ⓔ

0.00000143

0.00000143

$\rightarrow 6\ \text{places}$

1.43×10^{-6}

Analysis

Observe that, if the given number is greater than 1, as in examples a–c, the exponent of 10 is positive; and if the number is less than 1, as in examples d–e, the exponent is negative.

TRY IT #10 Write each number in scientific notation.

ⓐ U.S. national debt per taxpayer (April 2014): $152,000

ⓑ World population (April 2014): 7,158,000,000

ⓒ World gross national income (April 2014): $85,500,000,000,000

ⓓ Time for light to travel 1 m: 0.00000000334 s

ⓔ Probability of winning lottery (match 6 of 49 possible numbers): 0.0000000715

Converting from Scientific to Standard Notation

To convert a number in **scientific notation** to standard notation, simply reverse the process. Move the decimal n places to the right if n is positive or n places to the left if n is negative and add zeros as needed. Remember, if n is positive, the value of the number is greater than 1, and if n is negative, the value of the number is less than one.

EXAMPLE 11

Converting Scientific Notation to Standard Notation

Convert each number in scientific notation to standard notation.

ⓐ 3.547×10^{14} ⓑ -2×10^{6} ⓒ 7.91×10^{-7} ⓓ -8.05×10^{-12}

Solution

ⓐ	ⓑ	ⓒ	ⓓ
3.547×10^{14}	-2×10^{6}	7.91×10^{-7}	-8.05×10^{-12}
3.54700000000000	-2.000000	0000007.91	-000000000008.05
$\rightarrow 14\ \text{places}$	$\rightarrow 6\ \text{places}$	$\rightarrow 7\ \text{places}$	$\rightarrow 12\ \text{places}$
$354{,}700{,}000{,}000{,}000$	$-2{,}000{,}000$	0.000000791	-0.00000000000805

> **TRY IT** #11 Convert each number in scientific notation to standard notation.

ⓐ 7.03×10^5 ⓑ -8.16×10^{11} ⓒ -3.9×10^{-13} ⓓ 8×10^{-6}

Using Scientific Notation in Applications

Scientific notation, used with the rules of exponents, makes calculating with large or small numbers much easier than doing so using standard notation. For example, suppose we are asked to calculate the number of atoms in 1 L of water. Each water molecule contains 3 atoms (2 hydrogen and 1 oxygen). The average drop of water contains around 1.32×10^{21} molecules of water and 1 L of water holds about 1.22×10^4 average drops. Therefore, there are approximately $3 \cdot (1.32 \times 10^{21}) \cdot (1.22 \times 10^4) \approx 4.83 \times 10^{25}$ atoms in 1 L of water. We simply multiply the decimal terms and add the exponents. Imagine having to perform the calculation without using scientific notation!

When performing calculations with scientific notation, be sure to write the answer in proper scientific notation. For example, consider the product $(7 \times 10^4) \cdot (5 \times 10^6) = 35 \times 10^{10}$. The answer is not in proper scientific notation because 35 is greater than 10. Consider 35 as 3.5 × 10. That adds a ten to the exponent of the answer.

$$(35) \times 10^{10} = (3.5 \times 10) \times 10^{10} = 3.5 \times \left(10 \times 10^{10}\right) = 3.5 \times 10^{11}$$

EXAMPLE 12

Using Scientific Notation

Perform the operations and write the answer in scientific notation.

ⓐ $(8.14 \times 10^{-7})(6.5 \times 10^{10})$ ⓑ $(4 \times 10^5) \div (-1.52 \times 10^9)$ ⓒ $(2.7 \times 10^5)(6.04 \times 10^{13})$
ⓓ $(1.2 \times 10^8) \div (9.6 \times 10^5)$ ⓔ $(3.33 \times 10^4)(-1.05 \times 10^7)(5.62 \times 10^5)$

Solution

ⓐ

$(8.14 \times 10^{-7})(6.5 \times 10^{10})$	$=$	$(8.14 \times 6.5)(10^{-7} \times 10^{10})$	Commutative and associative properties of multiplication
	$=$	$(52.91)(10^3)$	Product rule of exponents
	$=$	5.291×10^4	Scientific notation

ⓑ

$(4 \times 10^5) \div (-1.52 \times 10^9)$	$=$	$\left(\frac{4}{-1.52}\right)\left(\frac{10^5}{10^9}\right)$	Commutative and associative properties of multiplication
	$\approx$	$(-2.63)(10^{-4})$	Quotient rule of exponents
	$=$	-2.63×10^{-4}	Scientific notation

ⓒ

$(2.7 \times 10^5)(6.04 \times 10^{13})$	$=$	$(2.7 \times 6.04)(10^5 \times 10^{13})$	Commutative and associative properties of multiplication
	$=$	$(16.308)(10^{18})$	Product rule of exponents
	$=$	1.6308×10^{19}	Scientific notation

ⓓ

$(1.2 \times 10^8) \div (9.6 \times 10^5)$	$=$	$\left(\frac{1.2}{9.6}\right)\left(\frac{10^8}{10^5}\right)$	Commutative and associative properties of multiplication
	$=$	$(0.125)(10^3)$	Quotient rule of exponents
	$=$	1.25×10^2	Scientific notation

ⓔ

$$\begin{aligned}(3.33 \times 10^4)(-1.05 \times 10^7)(5.62 \times 10^5) &= [3.33 \times (-1.05) \times 5.62](10^4 \times 10^7 \times 10^5) \\ &\approx (-19.65)(10^{16}) \\ &= -1.965 \times 10^{17}\end{aligned}$$

> **TRY IT** #12 Perform the operations and write the answer in scientific notation.

ⓐ $(-7.5 \times 10^8)(1.13 \times 10^{-2})$ ⓑ $(1.24 \times 10^{11}) \div (1.55 \times 10^{18})$
ⓒ $(3.72 \times 10^9)(8 \times 10^3)$ ⓓ $(9.933 \times 10^{23}) \div (-2.31 \times 10^{17})$
ⓔ $(-6.04 \times 10^9)(7.3 \times 10^2)(-2.81 \times 10^2)$

EXAMPLE 13

Applying Scientific Notation to Solve Problems

In April 2014, the population of the United States was about 308,000,000 people. The national debt was about $17,547,000,000,000. Write each number in scientific notation, rounding figures to two decimal places, and find the amount of the debt per U.S. citizen. Write the answer in both scientific and standard notations.

Solution

The population was $308{,}000{,}000 = 3.08 \times 10^8$.

The national debt was $\$17{,}547{,}000{,}000{,}000 \approx \$1.75 \times 10^{13}$.

To find the amount of debt per citizen, divide the national debt by the number of citizens.

$$\begin{aligned}(1.75 \times 10^{13}) \div (3.08 \times 10^8) &= \left(\frac{1.75}{3.08}\right) \cdot \left(\frac{10^{13}}{10^8}\right) \\ &\approx 0.57 \times 10^5 \\ &= 5.7 \times 10^4\end{aligned}$$

The debt per citizen at the time was about $\$5.7 \times 10^4$, or $57,000.

TRY IT #13 An average human body contains around 30,000,000,000,000 red blood cells. Each cell measures approximately 0.000008 m long. Write each number in scientific notation and find the total length if the cells were laid end-to-end. Write the answer in both scientific and standard notations.

MEDIA

Access these online resources for additional instruction and practice with exponents and scientific notation.

Exponential Notation (http://openstax.org/l/exponnot)
Properties of Exponents (http://openstax.org/l/exponprops)
Zero Exponent (http://openstax.org/l/zeroexponent)
Simplify Exponent Expressions (http://openstax.org/l/exponexpres)
Quotient Rule for Exponents (http://openstax.org/l/quotofexpon)
Scientific Notation (http://openstax.org/l/scientificnota)
Converting to Decimal Notation (http://openstax.org/l/decimalnota)

1.2 SECTION EXERCISES

Verbal

1. Is 2^3 the same as 3^2? Explain.
2. When can you add two exponents?
3. What is the purpose of scientific notation?
4. Explain what a negative exponent does.

Numeric

For the following exercises, simplify the given expression. Write answers with positive exponents.

5. 9^2

6. 15^{-2}

7. $3^2 \times 3^3$

8. $4^4 \div 4$

9. $\left(2^2\right)^{-2}$

10. $(5-8)^0$

11. $11^3 \div 11^4$

12. $6^5 \times 6^{-7}$

13. $\left(8^0\right)^2$

14. $5^{-2} \div 5^2$

For the following exercises, write each expression with a single base. Do not simplify further. Write answers with positive exponents.

15. $4^2 \times 4^3 \div 4^{-4}$

16. $\frac{6^{12}}{6^9}$

17. $\left(12^3 \times 12\right)^{10}$

18. $10^6 \div \left(10^{10}\right)^{-2}$

19. $7^{-6} \times 7^{-3}$

20. $\left(3^3 \div 3^4\right)^5$

For the following exercises, express the decimal in scientific notation.

21. 0.0000314

22. 148,000,000

For the following exercises, convert each number in scientific notation to standard notation.

23. 1.6×10^{10}

24. 9.8×10^{-9}

Algebraic

For the following exercises, simplify the given expression. Write answers with positive exponents.

25. $\frac{a^3 a^2}{a}$

26. $\frac{mn^2}{m^{-2}}$

27. $\left(b^3 c^4\right)^2$

28. $\left(\frac{x^{-3}}{y^2}\right)^{-5}$

29. $ab^2 \div d^{-3}$

30. $\left(w^0 x^5\right)^{-1}$

31. $\frac{m^4}{n^0}$

32. $y^{-4}\left(y^2\right)^2$

33. $\frac{p^{-4} q^2}{p^2 q^{-3}}$

34. $(l \times w)^2$

35. $\left(y^7\right)^3 \div x^{14}$

36. $\left(\frac{a}{2^3}\right)^2$

37. $(25m) \div ({}^{5}_{0}m)$

38. $\frac{(16\sqrt{x})^2}{y^{-1}}$

39. $\frac{2^3}{(3a)^{-2}}$

40. $\left(ma^6\right)^2 \frac{1}{m^3 a^2}$

41. $\left(b^{-3} c\right)^3$

42. $\left(x^2 y^{13} \div y^0\right)^2$

43. $(9z^3)^{-2}y$

Real-World Applications

44. To reach escape velocity, a rocket must travel at the rate of 2.2×10^6 ft/min. Rewrite the rate in standard notation.

45. A dime is the thinnest coin in U.S. currency. A dime's thickness measures 1.35×10^{-3} m. Rewrite the number in standard notation.

46. The average distance between Earth and the Sun is 92,960,000 mi. Rewrite the distance using scientific notation.

47. A terabyte is made of approximately 1,099,500,000,000 bytes. Rewrite in scientific notation.

48. The Gross Domestic Product (GDP) for the United States in the first quarter of 2014 was $\$1.71496 \times 10^{13}$. Rewrite the GDP in standard notation.

49. One picometer is approximately 3.397×10^{-11} in. Rewrite this length using standard notation.

50. The value of the services sector of the U.S. economy in the first quarter of 2012 was $10,633.6 billion. Rewrite this amount in scientific notation.

Technology

For the following exercises, use a graphing calculator to simplify. Round the answers to the nearest hundredth.

51. $\left(\frac{12^3 m^{33}}{4^{-3}}\right)^2$

52. $17^3 \div 15^2 x^3$

Extensions

For the following exercises, simplify the given expression. Write answers with positive exponents.

53. $\left(\frac{3^2}{a^3}\right)^{-2}\left(\frac{a^4}{2^2}\right)^2$

54. $(6^2 - 24)^2 \div \left(\frac{x}{y}\right)^{-5}$

55. $\frac{m^2 n^3}{a^2 c^{-3}} \cdot \frac{a^{-7} n^{-2}}{m^2 c^4}$

56. $\left(\frac{x^6 y^3}{x^3 y^{-3}} \cdot \frac{y^{-7}}{x^{-3}}\right)^{10}$

57. $\left(\frac{(ab^2 c)^{-3}}{b^{-3}}\right)^2$

58. Avogadro's constant is used to calculate the number of particles in a mole. A mole is a basic unit in chemistry to measure the amount of a substance. The constant is 6.0221413×10^{23}. Write Avogadro's constant in standard notation.

59. Planck's constant is an important unit of measure in quantum physics. It describes the relationship between energy and frequency. The constant is written as $6.62606957 \times 10^{-34}$. Write Planck's constant in standard notation.

1.3 Radicals and Rational Exponents

Learning Objectives

In this section, you will:

- Evaluate square roots.
- Use the product rule to simplify square roots.
- Use the quotient rule to simplify square roots.
- Add and subtract square roots.
- Rationalize denominators.
- Use rational roots.

A hardware store sells 16-ft ladders and 24-ft ladders. A window is located 12 feet above the ground. A ladder needs to be purchased that will reach the window from a point on the ground 5 feet from the building. To find out the length of ladder needed, we can draw a right triangle as shown in Figure 1, and use the Pythagorean Theorem.

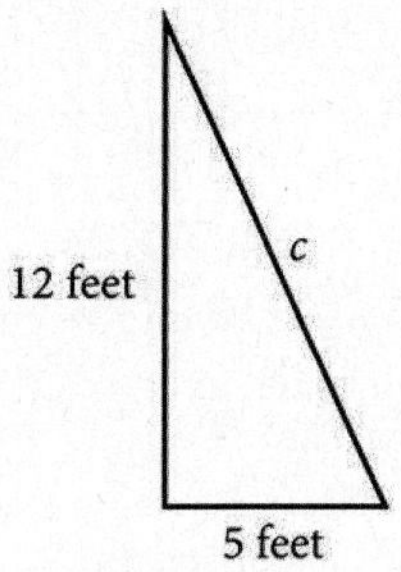

Figure 1

$$\begin{aligned} a^2 + b^2 &= c^2 \\ 5^2 + 12^2 &= c^2 \\ 169 &= c^2 \end{aligned}$$

Now, we need to find out the length that, when squared, is 169, to determine which ladder to choose. In other words, we need to find a square root. In this section, we will investigate methods of finding solutions to problems such as this one.

Evaluating Square Roots

When the square root of a number is squared, the result is the original number. Since $4^2 = 16$, the square root of 16 is 4. The square root function is the inverse of the squaring function just as subtraction is the inverse of addition. To undo squaring, we take the square root.

In general terms, if a is a positive real number, then the square root of a is a number that, when multiplied by itself, gives a. The square root could be positive or negative because multiplying two negative numbers gives a positive number. The **principal square root** is the nonnegative number that when multiplied by itself equals a. The square root obtained using a calculator is the principal square root.

The principal square root of a is written as $\sqrt{a}$. The symbol is called a **radical**, the term under the symbol is called the **radicand**, and the entire expression is called a **radical expression**.

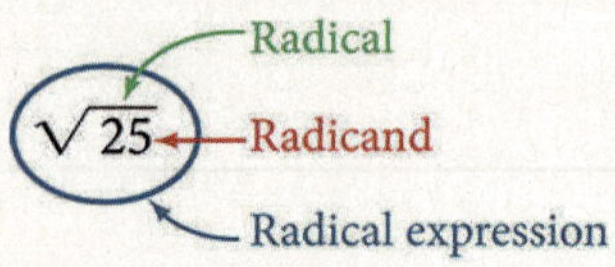

Principal Square Root

The **principal square root** of a is the nonnegative number that, when multiplied by itself, equals a. It is written as a **radical expression,** with a symbol called a **radical** over the term called the **radicand**: $\sqrt{a}$.

Q&A **Does $\sqrt{25} = \pm 5$?**

No. Although both 5^2 and $(-5)^2$ are 25, the radical symbol implies only a nonnegative root, the principal square root. The principal square root of 25 is $\sqrt{25} = 5$.

EXAMPLE 1

Evaluating Square Roots
Evaluate each expression.

ⓐ $\sqrt{100}$ ⓑ $\sqrt{\sqrt{16}}$ ⓒ $\sqrt{25+144}$ ⓓ $\sqrt{49} - \sqrt{81}$

Solution

ⓐ $\sqrt{100} = 10$ because $10^2 = 100$ ⓑ $\sqrt{\sqrt{16}} = \sqrt{4} = 2$ because $4^2 = 16$ and $2^2 = 4$

ⓒ $\sqrt{25+144} = \sqrt{169} = 13$ because $13^2 = 169$ ⓓ $\sqrt{49} - \sqrt{81} = 7 - 9 = -2$ because $7^2 = 49$ and $9^2 = 81$

Q&A **For $\sqrt{25+144}$, can we find the square roots before adding?**

No. $\sqrt{25} + \sqrt{144} = 5 + 12 = 17$. This is not equivalent to $\sqrt{25+144} = 13$. The order of operations requires us to add the terms in the radicand before finding the square root.

TRY IT #1 Evaluate each expression.

ⓐ $\sqrt{225}$ ⓑ $\sqrt{\sqrt{81}}$ ⓒ $\sqrt{25-9}$ ⓓ $\sqrt{36} + \sqrt{121}$

Using the Product Rule to Simplify Square Roots

To simplify a square root, we rewrite it such that there are no perfect squares in the radicand. There are several properties of square roots that allow us to simplify complicated radical expressions. The first rule we will look at is the *product rule for simplifying square roots,* which allows us to separate the square root of a product of two numbers into the product of two separate rational expressions. For instance, we can rewrite $\sqrt{15}$ as $\sqrt{3} \cdot \sqrt{5}$. We can also use the product rule to express the product of multiple radical expressions as a single radical expression.

The Product Rule for Simplifying Square Roots

If a and b are nonnegative, the square root of the product ab is equal to the product of the square roots of a and b.

$$\sqrt{ab} = \sqrt{a} \cdot \sqrt{b}$$

HOW TO

Given a square root radical expression, use the product rule to simplify it.

1. Factor any perfect squares from the radicand.
2. Write the radical expression as a product of radical expressions.
3. Simplify.

EXAMPLE 2

Using the Product Rule to Simplify Square Roots

Simplify the radical expression.

ⓐ $\sqrt{300}$ ⓑ $\sqrt{162a^5b^4}$

Solution

ⓐ

$\sqrt{100 \cdot 3}$	Factor perfect square from radicand.
$\sqrt{100} \cdot \sqrt{3}$	Write radical expression as product of radical expressions.
$10\sqrt{3}$	Simplify.

ⓑ

$\sqrt{81a^4b^4 \cdot 2a}$	Factor perfect square from radicand.
$\sqrt{81a^4b^4} \cdot \sqrt{2a}$	Write radical expression as product of radical expressions.
$9a^2b^2\sqrt{2a}$	Simplify.

TRY IT #2 Simplify $\sqrt{50x^2y^3z}$.

HOW TO

Given the product of multiple radical expressions, use the product rule to combine them into one radical expression.

1. Express the product of multiple radical expressions as a single radical expression.
2. Simplify.

EXAMPLE 3

Using the Product Rule to Simplify the Product of Multiple Square Roots

Simplify the radical expression.

$\sqrt{12} \cdot \sqrt{3}$

Solution

$\sqrt{12 \cdot 3}$	Express the product as a single radical expression.
$\sqrt{36}$	Simplify.
6	

TRY IT #3 Simplify $\sqrt{50x} \cdot \sqrt{2x}$ assuming $x > 0$.

Using the Quotient Rule to Simplify Square Roots

Just as we can rewrite the square root of a product as a product of square roots, so too can we rewrite the square root of a quotient as a quotient of square roots, using the *quotient rule for simplifying square roots.* It can be helpful to separate the numerator and denominator of a fraction under a radical so that we can take their square roots separately.

We can rewrite $\sqrt{\frac{5}{2}}$ as $\frac{\sqrt{5}}{\sqrt{2}}$.

The Quotient Rule for Simplifying Square Roots

The square root of the quotient $\frac{a}{b}$ is equal to the quotient of the square roots of a and b, where $b \neq 0$.

$$\sqrt{\frac{a}{b}} = \frac{\sqrt{a}}{\sqrt{b}}$$

HOW TO

Given a radical expression, use the quotient rule to simplify it.

1. Write the radical expression as the quotient of two radical expressions.
2. Simplify the numerator and denominator.

EXAMPLE 4

Using the Quotient Rule to Simplify Square Roots
Simplify the radical expression.

$$\sqrt{\frac{5}{36}}$$

Solution

$\frac{\sqrt{5}}{\sqrt{36}}$	Write as quotient of two radical expressions.
$\frac{\sqrt{5}}{6}$	Simplify denominator.

TRY IT #4 Simplify $\sqrt{\frac{2x^2}{9y^4}}$.

EXAMPLE 5

Using the Quotient Rule to Simplify an Expression with Two Square Roots
Simplify the radical expression.

$$\frac{\sqrt{234x^{11}y}}{\sqrt{26x^7y}}$$

Solution

$\sqrt{\frac{234x^{11}y}{26x^7y}}$	Combine numerator and denominator into one radical expression.
$\sqrt{9x^4}$	Simplify fraction.
$3x^2$	Simplify square root.

TRY IT #5 Simplify $\frac{\sqrt{9a^5b^{14}}}{\sqrt{3a^4b^5}}$.

Adding and Subtracting Square Roots

We can add or subtract radical expressions only when they have the same radicand and when they have the same radical type such as square roots. For example, the sum of $\sqrt{2}$ and $3\sqrt{2}$ is $4\sqrt{2}$. However, it is often possible to simplify radical expressions, and that may change the radicand. The radical expression $\sqrt{18}$ can be written with a 2 in the radicand, as $3\sqrt{2}$, so $\sqrt{2}+\sqrt{18}=\sqrt{2}+3\sqrt{2}=4\sqrt{2}$.

HOW TO

Given a radical expression requiring addition or subtraction of square roots, simplify.

1. Simplify each radical expression.
2. Add or subtract expressions with equal radicands.

EXAMPLE 6

Adding Square Roots

Add $5\sqrt{12}+2\sqrt{3}$.

Solution

We can rewrite $5\sqrt{12}$ as $5\sqrt{4\cdot 3}$. According the product rule, this becomes $5\sqrt{4}\sqrt{3}$. The square root of $\sqrt{4}$ is 2, so the expression becomes $5\,(2)\,\sqrt{3}$, which is $10\sqrt{3}$. Now the terms have the same radicand so we can add.

$$10\sqrt{3}+2\sqrt{3}=12\sqrt{3}$$

TRY IT #6 Add $\sqrt{5}+6\sqrt{20}$.

EXAMPLE 7

Subtracting Square Roots

Subtract $20\sqrt{72a^3b^4c} - 14\sqrt{8a^3b^4c}$.

Solution

Rewrite each term so they have equal radicands.

$$\begin{aligned}20\sqrt{72a^3b^4c} &= 20\sqrt{9}\sqrt{4}\sqrt{2}\sqrt{a}\sqrt{a^2}\sqrt{(b^2)^2}\sqrt{c}\\ &= 20(3)(2)|a|b^2\sqrt{2ac}\\ &= 120|a|b^2\sqrt{2ac}\end{aligned}$$

$$\begin{aligned}14\sqrt{8a^3b^4c} &= 14\sqrt{2}\sqrt{4}\sqrt{a}\sqrt{a^2}\sqrt{(b^2)^2}\sqrt{c}\\ &= 14(2)|a|b^2\sqrt{2ac}\\ &= 28|a|b^2\sqrt{2ac}\end{aligned}$$

Now the terms have the same radicand so we can subtract.

$$120|a|b^2\sqrt{2ac}-28|a|b^2\sqrt{2ac}=92|a|b^2\sqrt{2ac}$$

TRY IT #7 Subtract $3\sqrt{80x} - 4\sqrt{45x}$.

Rationalizing Denominators

When an expression involving square root radicals is written in simplest form, it will not contain a radical in the denominator. We can remove radicals from the denominators of fractions using a process called *rationalizing the denominator.*

We know that multiplying by 1 does not change the value of an expression. We use this property of multiplication to change expressions that contain radicals in the denominator. To remove radicals from the denominators of fractions, multiply by the form of 1 that will eliminate the radical.

For a denominator containing a single term, multiply by the radical in the denominator over itself. In other words, if the denominator is $b\sqrt{c}$, multiply by $\frac{\sqrt{c}}{\sqrt{c}}$.

For a denominator containing the sum or difference of a rational and an irrational term, multiply the numerator and denominator by the conjugate of the denominator, which is found by changing the sign of the radical portion of the denominator. If the denominator is $a + b\sqrt{c}$, then the conjugate is $a - b\sqrt{c}$.

HOW TO

Given an expression with a single square root radical term in the denominator, rationalize the denominator.

a. Multiply the numerator and denominator by the radical in the denominator.
b. Simplify.

EXAMPLE 8

Rationalizing a Denominator Containing a Single Term

Write $\frac{2\sqrt{3}}{3\sqrt{10}}$ in simplest form.

Solution

The radical in the denominator is $\sqrt{10}$. So multiply the fraction by $\frac{\sqrt{10}}{\sqrt{10}}$. Then simplify.

$$\frac{2\sqrt{3}}{3\sqrt{10}} \cdot \frac{\sqrt{10}}{\sqrt{10}}$$
$$\frac{2\sqrt{30}}{30}$$
$$\frac{\sqrt{30}}{15}$$

TRY IT #8 Write $\frac{12\sqrt{3}}{\sqrt{2}}$ in simplest form.

HOW TO

Given an expression with a radical term and a constant in the denominator, rationalize the denominator.

1. Find the conjugate of the denominator.
2. Multiply the numerator and denominator by the conjugate.
3. Use the distributive property.
4. Simplify.

EXAMPLE 9

Rationalizing a Denominator Containing Two Terms

Write $\frac{4}{1+\sqrt{5}}$ in simplest form.

Solution

Begin by finding the conjugate of the denominator by writing the denominator and changing the sign. So the conjugate of $1+\sqrt{5}$ is $1-\sqrt{5}$. Then multiply the fraction by $\frac{1-\sqrt{5}}{1-\sqrt{5}}$.

$$\frac{4}{1+\sqrt{5}} \cdot \frac{1-\sqrt{5}}{1-\sqrt{5}}$$

$$\frac{4-4\sqrt{5}}{-4} \quad \text{Use the distributive property.}$$

$$\sqrt{5}-1 \quad \text{Simplify.}$$

TRY IT #9 Write $\frac{7}{2+\sqrt{3}}$ in simplest form.

Using Rational Roots

Although square roots are the most common rational roots, we can also find cube roots, 4th roots, 5th roots, and more. Just as the square root function is the inverse of the squaring function, these roots are the inverse of their respective power functions. These functions can be useful when we need to determine the number that, when raised to a certain power, gives a certain number.

Understanding *n*th Roots

Suppose we know that $a^3=8$. We want to find what number raised to the 3rd power is equal to 8. Since $2^3=8$, we say that 2 is the cube root of 8.

The *n*th root of a is a number that, when raised to the *n*th power, gives a. For example, -3 is the 5th root of -243 because $(-3)^5=-243$. If a is a real number with at least one *n*th root, then the **principal *n*th root** of a is the number with the same sign as a that, when raised to the *n*th power, equals a.

The principal *n*th root of a is written as $\sqrt[n]{a}$, where n is a positive integer greater than or equal to 2. In the radical expression, n is called the **index** of the radical.

Principal n th Root

If a is a real number with at least one *n*th root, then the **principal *n*th root** of a, written as $\sqrt[n]{a}$, is the number with the same sign as a that, when raised to the *n*th power, equals a. The **index** of the radical is n.

EXAMPLE 10

Simplifying *n*th Roots

Simplify each of the following:

ⓐ $\sqrt[5]{-32}$ ⓑ $\sqrt[4]{4}\cdot\sqrt[4]{1,024}$ ⓒ $-\sqrt[3]{\frac{8x^6}{125}}$ ⓓ $8\sqrt[4]{3}-\sqrt[4]{48}$

Solution

ⓐ $\sqrt[5]{-32}=-2$ because $(-2)^5=-32$

ⓑ First, express the product as a single radical expression. $\sqrt[4]{4,096}=8$ because $8^4=4,096$

ⓒ

$$\frac{-\sqrt[3]{8x^6}}{\sqrt[3]{125}} \quad \text{Write as quotient of two radical expressions.}$$

$$\frac{-2x^2}{5} \quad \text{Simplify.}$$

ⓓ

$8\sqrt[4]{3} - 2\sqrt[4]{3}$	Simplify to get equal radicands.
$6\sqrt[4]{3}$	Add.

TRY IT #10 Simplify.

ⓐ $\sqrt[3]{-216}$ ⓑ $\dfrac{3\sqrt[4]{80}}{\sqrt[4]{5}}$ ⓒ $6\sqrt[3]{9,000} + 7\sqrt[3]{576}$

Using Rational Exponents

Radical expressions can also be written without using the radical symbol. We can use rational (fractional) exponents. The index must be a positive integer. If the index n is even, then a cannot be negative.

$$a^{\frac{1}{n}} = \sqrt[n]{a}$$

We can also have rational exponents with numerators other than 1. In these cases, the exponent must be a fraction in lowest terms. We raise the base to a power and take an nth root. The numerator tells us the power and the denominator tells us the root.

$$a^{\frac{m}{n}} = \left(\sqrt[n]{a}\right)^m = \sqrt[n]{a^m}$$

All of the properties of exponents that we learned for integer exponents also hold for rational exponents.

Rational Exponents

Rational exponents are another way to express principal nth roots. The general form for converting between a radical expression with a radical symbol and one with a rational exponent is

$$a^{\frac{m}{n}} = \left(\sqrt[n]{a}\right)^m = \sqrt[n]{a^m}$$

HOW TO

Given an expression with a rational exponent, write the expression as a radical.

1. Determine the power by looking at the numerator of the exponent.
2. Determine the root by looking at the denominator of the exponent.
3. Using the base as the radicand, raise the radicand to the power and use the root as the index.

EXAMPLE 11

Writing Rational Exponents as Radicals

Write $343^{\frac{2}{3}}$ as a radical. Simplify.

Solution

The 2 tells us the power and the 3 tells us the root.

$$343^{\frac{2}{3}} = \left(\sqrt[3]{343}\right)^2 = \sqrt[3]{343^2}$$

We know that $\sqrt[3]{343} = 7$ because $7^3 = 343$. Because the cube root is easy to find, it is easiest to find the cube root before squaring for this problem. In general, it is easier to find the root first and then raise it to a power.

$$343^{\frac{2}{3}} = \left(\sqrt[3]{343}\right)^2 = 7^2 = 49$$

TRY IT #11 Write $9^{\frac{5}{2}}$ as a radical. Simplify.

EXAMPLE 12

Writing Radicals as Rational Exponents

Write $\frac{4}{\sqrt[7]{a^2}}$ using a rational exponent.

Solution

The power is 2 and the root is 7, so the rational exponent will be $\frac{2}{7}$. We get $\frac{4}{a^{\frac{2}{7}}}$. Using properties of exponents, we get

$$\frac{4}{\sqrt[7]{a^2}} = 4a^{\frac{-2}{7}}.$$

TRY IT #12 Write $x\sqrt{(5y)^9}$ using a rational exponent.

EXAMPLE 13

Simplifying Rational Exponents

Simplify:

ⓐ $5\left(2x^{\frac{3}{4}}\right)\left(3x^{\frac{1}{5}}\right)$ ⓑ $\left(\frac{16}{9}\right)^{-\frac{1}{2}}$

Solution

ⓐ

$30x^{\frac{3}{4}}x^{\frac{1}{5}}$	Multiply the coefficients.
$30x^{\frac{3}{4}+\frac{1}{5}}$	Use properties of exponents.
$30x^{\frac{19}{20}}$	Simplify.

ⓑ

$\left(\frac{9}{16}\right)^{\frac{1}{2}}$	Use definition of negative exponents.
$\sqrt{\frac{9}{16}}$	Rewrite as a radical.
$\frac{\sqrt{9}}{\sqrt{16}}$	Use the quotient rule.
$\frac{3}{4}$	Simplify.

TRY IT #13 Simplify $(8x)^{\frac{1}{3}}\left(14x^{\frac{6}{5}}\right)$.

MEDIA

Access these online resources for additional instruction and practice with radicals and rational exponents.

Radicals (http://openstax.org/l/introradical)
Rational Exponents (http://openstax.org/l/rationexpon)
Simplify Radicals (http://openstax.org/l/simpradical)
Rationalize Denominator (http://openstax.org/l/rationdenom)

1.3 SECTION EXERCISES

Verbal

1. What does it mean when a radical does not have an index? Is the expression equal to the radicand? Explain.
2. Where would radicals come in the order of operations? Explain why.
3. Every number will have two square roots. What is the principal square root?
4. Can a radical with a negative radicand have a real square root? Why or why not?

Numeric

For the following exercises, simplify each expression.

5. $\sqrt{256}$
6. $\sqrt{\sqrt{256}}$
7. $\sqrt{4(9+16)}$
8. $\sqrt{289}-\sqrt{121}$
9. $\sqrt{196}$
10. $\sqrt{1}$
11. $\sqrt{98}$
12. $\sqrt{\frac{27}{64}}$
13. $\sqrt{\frac{81}{5}}$
14. $\sqrt{800}$
15. $\sqrt{169}+\sqrt{144}$
16. $\sqrt{\frac{8}{50}}$
17. $\frac{18}{\sqrt{162}}$
18. $\sqrt{192}$
19. $14\sqrt{6}-6\sqrt{24}$
20. $15\sqrt{5}+7\sqrt{45}$
21. $\sqrt{150}$
22. $\sqrt{\frac{96}{100}}$
23. $\left(\sqrt{42}\right)\left(\sqrt{30}\right)$
24. $12\sqrt{3}-4\sqrt{75}$
25. $\sqrt{\frac{4}{225}}$
26. $\sqrt{\frac{405}{324}}$
27. $\sqrt{\frac{360}{361}}$
28. $\frac{5}{1+\sqrt{3}}$
29. $\frac{8}{1-\sqrt{17}}$
30. $\sqrt[4]{16}$
31. $\sqrt[3]{128}+3\sqrt[3]{2}$
32. $\sqrt[5]{\frac{-32}{243}}$
33. $\frac{15\sqrt[4]{125}}{\sqrt[4]{5}}$
34. $3\sqrt[3]{-432}+\sqrt[3]{16}$

Algebraic

For the following exercises, simplify each expression.

35. $\sqrt{400x^4}$

36. $\sqrt{4y^2}$

37. $\sqrt{49p}$

38. $(144p^2q^6)^{\frac{1}{2}}$

39. $m^{\frac{5}{2}}\sqrt{289}$

40. $9\sqrt{3m^2} + \sqrt{27}$

41. $3\sqrt{ab^2} - b\sqrt{a}$

42. $\frac{4\sqrt{2n}}{\sqrt{16n^4}}$

43. $\sqrt{\frac{225x^3}{49x}}$

44. $3\sqrt{44z} + \sqrt{99z}$

45. $\sqrt{50y^8}$

46. $\sqrt{490bc^2}$

47. $\sqrt{\frac{32}{14d}}$

48. $q^{\frac{3}{2}}\sqrt{63p}$

49. $\frac{\sqrt{8}}{1-\sqrt{3x}}$

50. $\sqrt{\frac{20}{121d^4}}$

51. $w^{\frac{3}{2}}\sqrt{32} - w^{\frac{3}{2}}\sqrt{50}$

52. $\sqrt{108x^4} + \sqrt{27x^4}$

53. $\frac{\sqrt{12x}}{2+2\sqrt{3}}$

54. $\sqrt{147k^3}$

55. $\sqrt{125n^{10}}$

56. $\sqrt{\frac{42q}{36q^3}}$

57. $\sqrt{\frac{81m}{361m^2}}$

58. $\sqrt{72c} - 2\sqrt{2c}$

59. $\sqrt{\frac{144}{324d^2}}$

60. $\sqrt[3]{24x^6} + \sqrt[3]{81x^6}$

61. $\sqrt[4]{\frac{162x^6}{16x^4}}$

62. $\sqrt[3]{64y}$

63. $\sqrt[3]{128z^3} - \sqrt[3]{-16z^3}$

64. $\sqrt[5]{1{,}024c^{10}}$

Real-World Applications

65. A guy wire for a suspension bridge runs from the ground diagonally to the top of the closest pylon to make a triangle. We can use the Pythagorean Theorem to find the length of guy wire needed. The square of the distance between the wire on the ground and the pylon on the ground is 90,000 feet. The square of the height of the pylon is 160,000 feet. So the length of the guy wire can be found by evaluating $\sqrt{90{,}000 + 160{,}000}$. What is the length of the guy wire?

66. A car accelerates at a rate of $6 - \frac{\sqrt{4}}{\sqrt{t}}$ m/s^2 where t is the time in seconds after the car moves from rest. Simplify the expression.

Extensions

For the following exercises, simplify each expression.

67. $\frac{\sqrt{8}-\sqrt{16}}{4-\sqrt{2}} - 2^{\frac{1}{2}}$

68. $\frac{4^{\frac{3}{2}} - 16^{\frac{3}{2}}}{8^{\frac{1}{3}}}$

69. $\frac{\sqrt{mn^3}}{a^2\sqrt{c^{-3}}} \cdot \frac{a^{-7}n^{-2}}{\sqrt{m^2c^4}}$

70. $\frac{a}{a-\sqrt{c}}$

71. $\frac{x\sqrt{64y}+4\sqrt{y}}{\sqrt{128y}}$

72. $\left(\frac{\sqrt{250x^2}}{\sqrt{100b^3}}\right)\left(\frac{7\sqrt{b}}{\sqrt{125x}}\right)$

73. $\sqrt{\frac{\sqrt[3]{64}+\sqrt[4]{256}}{\sqrt{64}+\sqrt{256}}}$

1.4 Polynomials

Learning Objectives

In this section, you will:

- Identify the degree and leading coefficient of polynomials.
- Add and subtract polynomials.
- Multiply polynomials.
- Use FOIL to multiply binomials.
- Perform operations with polynomials of several variables.

Maahi is building a little free library (a small house-shaped book repository), whose front is in the shape of a square topped with a triangle. There will be a rectangular door through which people can take and donate books. Maahi wants to find the area of the front of the library so that they can purchase the correct amount of paint. Using the measurements of the front of the house, shown in Figure 1, we can create an expression that combines several variable terms, allowing us to solve this problem and others like it.

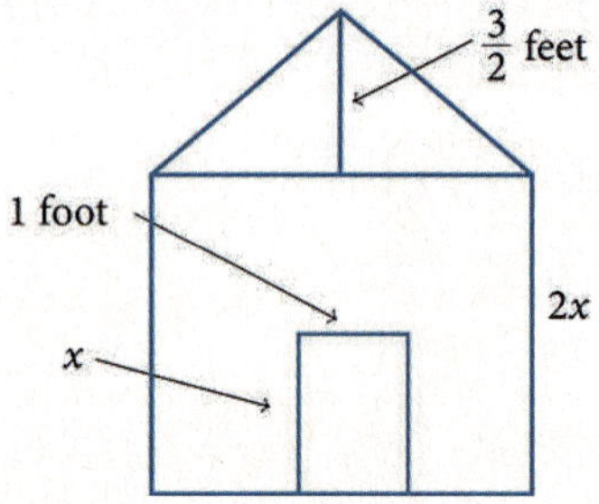

Figure 1

First find the area of the square in square feet.

$$\begin{aligned} A &= s^2 \\ &= (2x)^2 \\ &= 4x^2 \end{aligned}$$

Then find the area of the triangle in square feet.

$$\begin{aligned} A &= \tfrac{1}{2}bh \\ &= \tfrac{1}{2}(2x)\left(\tfrac{3}{2}\right) \\ &= \tfrac{3}{2}x \end{aligned}$$

Next find the area of the rectangular door in square feet.

$$\begin{aligned} A &= lw \\ &= x \cdot 1 \\ &= x \end{aligned}$$

The area of the front of the library can be found by adding the areas of the square and the triangle, and then subtracting the area of the rectangle. When we do this, we get $4x^2 + \frac{3}{2}x - x$ ft^2, or $4x^2 + \frac{1}{2}x$ ft^2.

In this section, we will examine expressions such as this one, which combine several variable terms.

Identifying the Degree and Leading Coefficient of Polynomials

The formula just found is an example of a **polynomial**, which is a sum of or difference of terms, each consisting of a variable raised to a nonnegative integer power. A number multiplied by a variable raised to an exponent, such as 384π, is known as a **coefficient**. Coefficients can be positive, negative, or zero, and can be whole numbers, decimals, or fractions. Each product $a_i x^i$, such as $384\pi w$, is a **term of a polynomial**. If a term does not contain a variable, it is called a *constant*.

A polynomial containing only one term, such as $5x^4$, is called a **monomial**. A polynomial containing two terms, such as $2x - 9$, is called a **binomial**. A polynomial containing three terms, such as $-3x^2 + 8x - 7$, is called a **trinomial**.

We can find the **degree** of a polynomial by identifying the highest power of the variable that occurs in the polynomial. The term with the highest degree is called the **leading term** because it is usually written first. The coefficient of the leading term is called the **leading coefficient**. When a polynomial is written so that the powers are descending, we say that it is in standard form.

Leading coefficient Degree

$$a_n x^n + \dots + a_2 x^2 + a_1 x + a_0$$

Leading term

Polynomials

A **polynomial** is an expression that can be written in the form

$$a_n x^n + \dots + a_2 x^2 + a_1 x + a_0$$

Each real number a_i is called a **coefficient**. The number a_0 that is not multiplied by a variable is called a *constant*. Each product $a_i x^i$ is a **term of a polynomial**. The highest power of the variable that occurs in the polynomial is called the **degree** of a polynomial. The **leading term** is the term with the highest power, and its coefficient is called the **leading coefficient**.

HOW TO

Given a polynomial expression, identify the degree and leading coefficient.

1. Find the highest power of x to determine the degree.
2. Identify the term containing the highest power of x to find the leading term.
3. Identify the coefficient of the leading term.

EXAMPLE 1

Identifying the Degree and Leading Coefficient of a Polynomial

For the following polynomials, identify the degree, the leading term, and the leading coefficient.

ⓐ $3 + 2x^2 - 4x^3$ ⓑ $5t^5 - 2t^3 + 7t$ ⓒ $6p - p^3 - 2$

Solution

ⓐ The highest power of x is 3, so the degree is 3. The leading term is the term containing that degree, $-4x^3$. The leading coefficient is the coefficient of that term, -4.

ⓑ The highest power of t is 5, so the degree is 5. The leading term is the term containing that degree, $5t^5$. The leading coefficient is the coefficient of that term, 5.

Ⓒ The highest power of p is 3, so the degree is 3. The leading term is the term containing that degree, $-p^3$, The leading coefficient is the coefficient of that term, -1.

TRY IT #1 Identify the degree, leading term, and leading coefficient of the polynomial $4x^2 - x^6 + 2x - 6$.

Adding and Subtracting Polynomials

We can add and subtract polynomials by combining like terms, which are terms that contain the same variables raised to the same exponents. For example, $5x^2$ and $-2x^2$ are like terms, and can be added to get $3x^2$, but $3x$ and $3x^2$ are not like terms, and therefore cannot be added.

HOW TO

Given multiple polynomials, add or subtract them to simplify the expressions.

1. Combine like terms.
2. Simplify and write in standard form.

EXAMPLE 2

Adding Polynomials

Find the sum.

$$\left(12x^2 + 9x - 21\right) + \left(4x^3 + 8x^2 - 5x + 20\right)$$

Solution

$4x^3 + \left(12x^2 + 8x^2\right) + (9x - 5x) + (-21 + 20)$	Combine like terms.
$4x^3 + 20x^2 + 4x - 1$	Simplify.

Analysis

We can check our answers to these types of problems using a graphing calculator. To check, graph the problem as given along with the simplified answer. The two graphs should be equivalent. Be sure to use the same window to compare the graphs. Using different windows can make the expressions seem equivalent when they are not.

TRY IT #2 Find the sum.

$$\left(2x^3 + 5x^2 - x + 1\right) + \left(2x^2 - 3x - 4\right)$$

EXAMPLE 3

Subtracting Polynomials

Find the difference.

$$\left(7x^4 - x^2 + 6x + 1\right) - \left(5x^3 - 2x^2 + 3x + 2\right)$$

Solution

$7x^4 - x^2 + 6x + 1 - 5x^3 + 2x^3 - 2$	Distribute negative sign.
$7x^4 - 5x^3 + x^2 + 6x - 3x + 1 - 2$	Group like terms.
$7x^4 - 5x^3 + x^2 + 3x - 1$	Combine/simplify.

Analysis

Note that finding the difference between two polynomials is the same as adding the opposite of the second polynomial to the first.

TRY IT #3 Find the difference.

$$\left(-7x^3 - 7x^2 + 6x - 2\right) - \left(4x^3 - 6x^2 - x + 7\right)$$

Multiplying Polynomials

Multiplying polynomials is a bit more challenging than adding and subtracting polynomials. We must use the distributive property to multiply each term in the first polynomial by each term in the second polynomial. We then combine like terms. We can also use a shortcut called the **FOIL** method when multiplying binomials. Certain special products follow patterns that we can memorize and use instead of multiplying the polynomials by hand each time. We will look at a variety of ways to multiply polynomials.

Multiplying Polynomials Using the Distributive Property

To multiply a number by a polynomial, we use the distributive property. The number must be distributed to each term of the polynomial. We can distribute the 2 in $2(x + 7)$ to obtain the equivalent expression $2x + 14$. When multiplying polynomials, the distributive property allows us to multiply each term of the first polynomial by each term of the second. We then add the products together and combine like terms to simplify.

HOW TO

Given the multiplication of two polynomials, use the distributive property to simplify the expression.

1. Multiply each term of the first polynomial by each term of the second.
2. Combine like terms.
3. Simplify.

EXAMPLE 4

Multiplying Polynomials Using the Distributive Property

Find the product.

$$(2x + 1)\left(3x^2 - x + 4\right)$$

Solution

$2x\left(3x^2 - x + 4\right) + 1\left(3x^2 - x + 4\right)$	Use the distributive property.
$\left(6x^3 - 2x^2 + 8x\right) + \left(3x^2 - x + 4\right)$	Multiply.
$6x^3 + \left(-2x^2 + 3x^2\right) + (8x - x) + 4$	Combine like terms.
$6x^3 + x^2 + 7x + 4$	Simplify.

Analysis

We can use a table to keep track of our work, as shown in Table 1. Write one polynomial across the top and the other down the side. For each box in the table, multiply the term for that row by the term for that column. Then add all of the terms together, combine like terms, and simplify.

	$3x^2$	$-x$	$+4$
$2x$	$6x^3$	$-2x^2$	$8x$

Table 1

$+1$	$3x^2$	$-x$	4

Table 1

TRY IT #4 Find the product.

$$(3x + 2)\left(x^3 - 4x^2 + 7\right)$$

Using FOIL to Multiply Binomials

A shortcut called FOIL is sometimes used to find the product of two binomials. It is called FOIL because we multiply the **f**irst terms, the **o**uter terms, the **i**nner terms, and then the **l**ast terms of each binomial.

First terms Last terms

$$(ax + b)(cx + d) = acx^2 + adx + bcx + bd$$

Inner terms

Outer terms

The FOIL method arises out of the distributive property. We are simply multiplying each term of the first binomial by each term of the second binomial, and then combining like terms.

HOW TO

Given two binomials, use FOIL to simplify the expression.

1. Multiply the first terms of each binomial.
2. Multiply the outer terms of the binomials.
3. Multiply the inner terms of the binomials.
4. Multiply the last terms of each binomial.
5. Add the products.
6. Combine like terms and simplify.

EXAMPLE 5

Using FOIL to Multiply Binomials

Use FOIL to find the product.

$$(2x - 18)(3x + 3)$$

Solution

Find the product of the first terms.

$2x - 18 \quad 3x + 3 \qquad 2x \cdot 3x = 6x^2$

Find the product of the outer terms.

$2x - 18 \quad 3x + 3 \qquad 2x \cdot 3 = 6x$

Find the product of the inner terms.

$2x - 18 \quad 3x + 3 \qquad -18 \cdot 3x = -54x$

Find the product of the last terms.

$2x - 18 \quad 3x + 3$	$-18 \cdot 3 = -54$
$6x^2 + 6x - 54x - 54$	Add the products.
$6x^2 + (6x - 54x) - 54$	Combine like terms.
$6x^2 - 48x - 54$	Simplify.

TRY IT #5 Use FOIL to find the product.

$(x + 7)(3x - 5)$

Perfect Square Trinomials

Certain binomial products have special forms. When a binomial is squared, the result is called a **perfect square trinomial**. We can find the square by multiplying the binomial by itself. However, there is a special form that each of these perfect square trinomials takes, and memorizing the form makes squaring binomials much easier and faster. Let's look at a few perfect square trinomials to familiarize ourselves with the form.

$$\begin{aligned} (x+5)^2 &= x^2 + 10x + 25 \\ (x-3)^2 &= x^2 - 6x + 9 \\ (4x-1)^2 &= 16x^2 - 8x + 1 \end{aligned}$$

Notice that the first term of each trinomial is the square of the first term of the binomial and, similarly, the last term of each trinomial is the square of the last term of the binomial. The middle term is double the product of the two terms. Lastly, we see that the first sign of the trinomial is the same as the sign of the binomial.

Perfect Square Trinomials

When a binomial is squared, the result is the first term squared added to double the product of both terms and the last term squared.

$$(x + a)^2 = (x + a)(x + a) = x^2 + 2ax + a^2$$

HOW TO

Given a binomial, square it using the formula for perfect square trinomials.

1. Square the first term of the binomial.
2. Square the last term of the binomial.
3. For the middle term of the trinomial, double the product of the two terms.
4. Add and simplify.

EXAMPLE 6

Expanding Perfect Squares

Expand $(3x - 8)^2$.

Solution

Begin by squaring the first term and the last term. For the middle term of the trinomial, double the product of the two terms.

$$(3x)^2 - 2(3x)(8) + (-8)^2$$

Simplify.

$$9x^2 - 48x + 64.$$

> TRY IT #6 Expand $(4x - 1)^2$.

Difference of Squares

Another special product is called the **difference of squares**, which occurs when we multiply a binomial by another binomial with the same terms but the opposite sign. Let's see what happens when we multiply $(x + 1)(x - 1)$ using the FOIL method.

$$\begin{aligned}(x + 1)(x - 1) &= x^2 - x + x - 1 \\ &= x^2 - 1\end{aligned}$$

The middle term drops out, resulting in a difference of squares. Just as we did with the perfect squares, let's look at a few examples.

$$\begin{aligned}(x + 5)(x - 5) &= x^2 - 25 \\ (x + 11)(x - 11) &= x^2 - 121 \\ (2x + 3)(2x - 3) &= 4x^2 - 9\end{aligned}$$

Because the sign changes in the second binomial, the outer and inner terms cancel each other out, and we are left only with the square of the first term minus the square of the last term.

Q&A **Is there a special form for the sum of squares?**

No. The difference of squares occurs because the opposite signs of the binomials cause the middle terms to disappear. There are no two binomials that multiply to equal a sum of squares.

Difference of Squares

When a binomial is multiplied by a binomial with the same terms separated by the opposite sign, the result is the square of the first term minus the square of the last term.

$$(a + b)(a - b) = a^2 - b^2$$

HOW TO

Given a binomial multiplied by a binomial with the same terms but the opposite sign, find the difference of squares.

1. Square the first term of the binomials.
2. Square the last term of the binomials.
3. Subtract the square of the last term from the square of the first term.

EXAMPLE 7

Multiplying Binomials Resulting in a Difference of Squares

Multiply $(9x + 4)(9x - 4)$.

Solution

Square the first term to get $(9x)^2 = 81x^2$. Square the last term to get $4^2 = 16$. Subtract the square of the last term from the square of the first term to find the product of $81x^2 - 16$.

TRY IT #7 Multiply $(2x + 7)(2x - 7)$.

Performing Operations with Polynomials of Several Variables

We have looked at polynomials containing only one variable. However, a polynomial can contain several variables. All of the same rules apply when working with polynomials containing several variables. Consider an example:

$(a + 2b)(4a - b - c)$	
$a(4a - b - c) + 2b(4a - b - c)$	Use the distributive property.
$4a^2 - ab - ac + 8ab - 2b^2 - 2bc$	Multiply.
$4a^2 + (-ab + 8ab) - ac - 2b^2 - 2bc$	Combine like terms.
$4a^2 + 7ab - ac - 2bc - 2b^2$	Simplify.

EXAMPLE 8

Multiplying Polynomials Containing Several Variables

Multiply $(x + 4)(3x - 2y + 5)$.

Solution

Follow the same steps that we used to multiply polynomials containing only one variable.

$x(3x - 2y + 5) + 4(3x - 2y + 5)$	Use the distributive property.
$3x^2 - 2xy + 5x + 12x - 8y + 20$	Multiply.
$3x^2 - 2xy + (5x + 12x) - 8y + 20$	Combine like terms.
$3x^2 - 2xy + 17x - 8y + 20$	Simplify.

TRY IT #8 Multiply $(3x - 1)(2x + 7y - 9)$.

MEDIA

Access these online resources for additional instruction and practice with polynomials.

Adding and Subtracting Polynomials (http://openstax.org/l/addsubpoly)
Multiplying Polynomials (http://openstax.org/l/multiplpoly)
Special Products of Polynomials (http://openstax.org/l/specialpolyprod)

1.4 SECTION EXERCISES

Verbal

1. Evaluate the following statement: The degree of a polynomial in standard form is the exponent of the leading term. Explain why the statement is true or false.

2. Many times, multiplying two binomials with two variables results in a trinomial. This is not the case when there is a difference of two squares. Explain why the product in this case is also a binomial.

3. You can multiply polynomials with any number of terms and any number of variables using four basic steps over and over until you reach the expanded polynomial. What are the four steps?

4. State whether the following statement is true and explain why or why not: A trinomial is always a higher degree than a monomial.

Algebraic

For the following exercises, identify the degree of the polynomial.

5. $7x - 2x^2 + 13$

6. $14m^3 + m^2 - 16m + 8$

7. $-625a^8 + 16b^4$

8. $200p - 30p^2 m + 40m^3$

9. $x^2 + 4x + 4$

10. $6y^4 - y^5 + 3y - 4$

For the following exercises, find the sum or difference.

11. $\left(12x^2 + 3x\right) - \left(8x^2 - 19\right)$

12. $\left(4z^3 + 8z^2 - z\right) + \left(-2z^2 + z + 6\right)$

13. $\left(6w^2 + 24w + 24\right) - \left(3w^2 - 6w + 3\right)$

14. $\left(7a^3 + 6a^2 - 4a - 13\right) + \left(-3a^3 - 4a^2 + 6a + 17\right)$

15. $\left(11b^4 - 6b^3 + 18b^2 - 4b + 8\right) - \left(3b^3 + 6b^2 + 3b\right)$

16. $\left(49p^2 - 25\right) + \left(16p^4 - 32p^2 + 16\right)$

For the following exercises, find the product.

17. $(4x + 2)(6x - 4)$

18. $\left(14c^2 + 4c\right)\left(2c^2 - 3c\right)$

19. $\left(6b^2 - 6\right)\left(4b^2 - 4\right)$

20. $(3d - 5)(2d + 9)$

21. $(9v - 11)(11v - 9)$

22. $\left(4t^2 + 7t\right)\left(-3t^2 + 4\right)$

23. $(8n - 4)\left(n^2 + 9\right)$

For the following exercises, expand the binomial.

24. $(4x + 5)^2$

25. $(3y - 7)^2$

26. $(12 - 4x)^2$

27. $(4p + 9)^2$

28. $(2m - 3)^2$

29. $(3y - 6)^2$

30. $(9b + 1)^2$

For the following exercises, multiply the binomials.

31. $(4c + 1)(4c - 1)$

32. $(9a - 4)(9a + 4)$

33. $(15n - 6)(15n + 6)$

34. $(25b + 2)(25b - 2)$

35. $(4 + 4m)(4 - 4m)$

36. $(14p + 7)(14p - 7)$

37. $(11q - 10)(11q + 10)$

For the following exercises, multiply the polynomials.

38. $\left(2x^2 + 2x + 1\right)(4x - 1)$

39. $\left(4t^2 + t - 7\right)\left(4t^2 - 1\right)$

40. $(x - 1)\left(x^2 - 2x + 1\right)$

41. $(y - 2)\left(y^2 - 4y - 9\right)$

42. $(6k - 5)\left(6k^2 + 5k - 1\right)$

43. $\left(3p^2 + 2p - 10\right)(p - 1)$

44. $(4m - 13)\left(2m^2 - 7m + 9\right)$

45. $(a + b)(a - b)$

46. $(4x - 6y)(6x - 4y)$

47. $(4t - 5u)^2$

48. $(9m + 4n - 1)(2m + 8)$

49. $(4t - x)(t - x + 1)$

50. $\left(b^2 - 1\right)\left(a^2 + 2ab + b^2\right)$

51. $(4r - d)(6r + 7d)$

52. $(x + y)\left(x^2 - xy + y^2\right)$

Real-World Applications

53. A developer wants to purchase a plot of land to build a house. The area of the plot can be described by the following expression: $(4x + 1)(8x - 3)$ where x is measured in meters. Multiply the binomials to find the area of the plot in standard form.

54. A prospective buyer wants to know how much grain a specific silo can hold. The area of the floor of the silo is $(2x + 9)^2$. The height of the silo is $10x + 10$, where x is measured in feet. Expand the square and multiply by the height to find the expression that shows how much grain the silo can hold.

Extensions

For the following exercises, perform the given operations.

55. $(4t - 7)^2(2t + 1) - \left(4t^2 + 2t + 11\right)$

56. $(3b + 6)(3b - 6)\left(9b^2 - 36\right)$

57. $\left(a^2 + 4ac + 4c^2\right)\left(a^2 - 4c^2\right)$

1.5 Factoring Polynomials

Learning Objectives

In this section, you will:

- Factor the greatest common factor of a polynomial.
- Factor a trinomial.
- Factor by grouping.
- Factor a perfect square trinomial.
- Factor a difference of squares.
- Factor the sum and difference of cubes.
- Factor expressions using fractional or negative exponents.

Imagine that we are trying to find the area of a lawn so that we can determine how much grass seed to purchase. The lawn is the green portion in Figure 1.

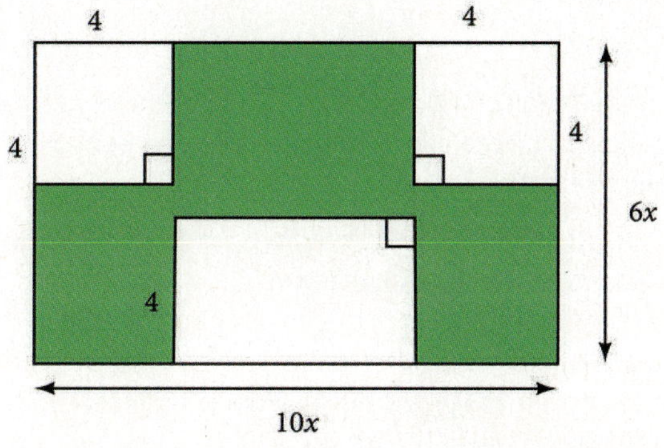

Figure 1

The area of the entire region can be found using the formula for the area of a rectangle.

$$\begin{aligned} A &= lw \\ &= 10x \cdot 6x \\ &= 60x^2 \text{ units}^2 \end{aligned}$$

The areas of the portions that do not require grass seed need to be subtracted from the area of the entire region. The two square regions each have an area of $A = s^2 = 4^2 = 16$ units². The other rectangular region has one side of length $10x - 8$ and one side of length 4, giving an area of $A = lw = 4(10x - 8) = 40x - 32$ units². So the region that must be subtracted has an area of $2(16) + 40x - 32 = 40x$ units².

The area of the region that requires grass seed is found by subtracting $60x^2 - 40x$ units². This area can also be expressed in factored form as $20x(3x - 2)$ units². We can confirm that this is an equivalent expression by multiplying.

Many polynomial expressions can be written in simpler forms by factoring. In this section, we will look at a variety of methods that can be used to factor polynomial expressions.

Factoring the Greatest Common Factor of a Polynomial

When we study fractions, we learn that the **greatest common factor** (GCF) of two numbers is the largest number that divides evenly into both numbers. For instance, 4 is the GCF of 16 and 20 because it is the largest number that divides evenly into both 16 and 20 The GCF of polynomials works the same way: $4x$ is the GCF of $16x$ and $20x^2$ because it is the largest polynomial that divides evenly into both $16x$ and $20x^2$.

When factoring a polynomial expression, our first step should be to check for a GCF. Look for the GCF of the coefficients, and then look for the GCF of the variables.

Greatest Common Factor

The **greatest common factor** (GCF) of polynomials is the largest polynomial that divides evenly into the polynomials.

HOW TO

Given a polynomial expression, factor out the greatest common factor.

1. Identify the GCF of the coefficients.
2. Identify the GCF of the variables.
3. Combine to find the GCF of the expression.
4. Determine what the GCF needs to be multiplied by to obtain each term in the expression.
5. Write the factored expression as the product of the GCF and the sum of the terms we need to multiply by.

EXAMPLE 1

Factoring the Greatest Common Factor

Factor $6x^3y^3 + 45x^2y^2 + 21xy$.

Solution

First, find the GCF of the expression. The GCF of 6, 45, and 21 is 3. The GCF of x^3, x^2, and x is x. (Note that the GCF of a set of expressions in the form x^n will always be the exponent of lowest degree.) And the GCF of y^3, y^2, and y is y. Combine these to find the GCF of the polynomial, $3xy$.

Next, determine what the GCF needs to be multiplied by to obtain each term of the polynomial. We find that $3xy(2x^2y^2) = 6x^3y^3$, $3xy(15xy) = 45x^2y^2$, and $3xy(7) = 21xy$.

Finally, write the factored expression as the product of the GCF and the sum of the terms we needed to multiply by.

$$(3xy)\left(2x^2y^2 + 15xy + 7\right)$$

Analysis

After factoring, we can check our work by multiplying. Use the distributive property to confirm that $(3xy)\left(2x^2y^2 + 15xy + 7\right) = 6x^3y^3 + 45x^2y^2 + 21xy$.

TRY IT #1 Factor $x\left(b^2 - a\right) + 6\left(b^2 - a\right)$ by pulling out the GCF.

Factoring a Trinomial with Leading Coefficient 1

Although we should always begin by looking for a GCF, pulling out the GCF is not the only way that polynomial expressions can be factored. The polynomial $x^2 + 5x + 6$ has a GCF of 1, but it can be written as the product of the factors $(x + 2)$ and $(x + 3)$.

Trinomials of the form $x^2 + bx + c$ can be factored by finding two numbers with a product of c and a sum of b. The trinomial $x^2 + 10x + 16$, for example, can be factored using the numbers 2 and 8 because the product of those numbers is 16 and their sum is 10. The trinomial can be rewritten as the product of $(x + 2)$ and $(x + 8)$.

Factoring a Trinomial with Leading Coefficient 1

A trinomial of the form $x^2 + bx + c$ can be written in factored form as $(x + p)(x + q)$ where $pq = c$ and $p + q = b$.

Q&A **Can every trinomial be factored as a product of binomials?**

No. Some polynomials cannot be factored. These polynomials are said to be prime.

HOW TO

Given a trinomial in the form $x^2 + bx + c$, factor it.

1. List factors of c.
2. Find p and q, a pair of factors of c with a sum of b.
3. Write the factored expression $(x + p)(x + q)$.

EXAMPLE 2

Factoring a Trinomial with Leading Coefficient 1
Factor $x^2 + 2x - 15$.

Solution
We have a trinomial with leading coefficient 1, $b = 2$, and $c = -15$. We need to find two numbers with a product of -15 and a sum of 2. In the table below, we list factors until we find a pair with the desired sum.

Factors of -15	Sum of Factors
$1, -15$	-14
$-1, 15$	14
$3, -5$	-2
$-3, 5$	2

Now that we have identified p and q as -3 and 5, write the factored form as $(x - 3)(x + 5)$.

Analysis
We can check our work by multiplying. Use FOIL to confirm that $(x - 3)(x + 5) = x^2 + 2x - 15$.

Q&A **Does the order of the factors matter?**

No. Multiplication is commutative, so the order of the factors does not matter.

> **TRY IT** #2 Factor $x^2 - 7x + 6$.

Factoring by Grouping

Trinomials with leading coefficients other than 1 are slightly more complicated to factor. For these trinomials, we can **factor by grouping** by dividing the *x* term into the sum of two terms, factoring each portion of the expression separately, and then factoring out the GCF of the entire expression. The trinomial $2x^2 + 5x + 3$ can be rewritten as $(2x + 3)(x + 1)$ using this process. We begin by rewriting the original expression as $2x^2 + 2x + 3x + 3$ and then factor each portion of the expression to obtain $2x(x + 1) + 3(x + 1)$. We then pull out the GCF of $(x + 1)$ to find the factored expression.

Factor by Grouping

To factor a trinomial in the form $ax^2 + bx + c$ by grouping, we find two numbers with a product of ac and a sum of b. We use these numbers to divide the x term into the sum of two terms and factor each portion of the expression separately, then factor out the GCF of the entire expression.

HOW TO

Given a trinomial in the form $ax^2 + bx + c$, factor by grouping.

1. List factors of ac.
2. Find p and q, a pair of factors of ac with a sum of b.
3. Rewrite the original expression as $ax^2 + px + qx + c$.
4. Pull out the GCF of $ax^2 + px$.
5. Pull out the GCF of $qx + c$.
6. Factor out the GCF of the expression.

EXAMPLE 3

Factoring a Trinomial by Grouping
Factor $5x^2 + 7x - 6$ by grouping.

Solution
We have a trinomial with $a = 5, b = 7$, and $c = -6$. First, determine $ac = -30$. We need to find two numbers with a product of -30 and a sum of 7. In the table below, we list factors until we find a pair with the desired sum.

Factors of -30	Sum of Factors
$1, -30$	-29
$-1, 30$	29
$2, -15$	-13
$-2, 15$	13
$3, -10$	-7
$-3, 10$	7

So $p = -3$ and $q = 10$.

$5x^2 - 3x + 10x - 6$	Rewrite the original expression as $ax^2 + px + qx + c$.
$x(5x - 3) + 2(5x - 3)$	Factor out the GCF of each part.
$(5x - 3)(x + 2)$	Factor out the GCF of the expression.

Analysis
We can check our work by multiplying. Use FOIL to confirm that $(5x - 3)(x + 2) = 5x^2 + 7x - 6$.

TRY IT #3 Factor

ⓐ $2x^2 + 9x + 9$ ⓑ $6x^2 + x - 1$

Factoring a Perfect Square Trinomial

A perfect square trinomial is a trinomial that can be written as the square of a binomial. Recall that when a binomial is squared, the result is the square of the first term added to twice the product of the two terms and the square of the last term.

$$a^2 + 2ab + b^2 = (a + b)^2$$

and

$$a^2 - 2ab + b^2 = (a - b)^2$$

We can use this equation to factor any perfect square trinomial.

Perfect Square Trinomials

A perfect square trinomial can be written as the square of a binomial:

$$a^2 + 2ab + b^2 = (a + b)^2$$

HOW TO

Given a perfect square trinomial, factor it into the square of a binomial.

1. Confirm that the first and last term are perfect squares.
2. Confirm that the middle term is twice the product of ab.
3. Write the factored form as $(a + b)^2$.

EXAMPLE 4

Factoring a Perfect Square Trinomial
Factor $25x^2 + 20x + 4$.

Solution
Notice that $25x^2$ and 4 are perfect squares because $25x^2 = (5x)^2$ and $4 = 2^2$. Then check to see if the middle term is twice the product of $5x$ and 2. The middle term is, indeed, twice the product: $2(5x)(2) = 20x$. Therefore, the trinomial is a perfect square trinomial and can be written as $(5x + 2)^2$.

TRY IT #4 Factor $49x^2 - 14x + 1$.

Factoring a Difference of Squares

A difference of squares is a perfect square subtracted from a perfect square. Recall that a difference of squares can be rewritten as factors containing the same terms but opposite signs because the middle terms cancel each other out when the two factors are multiplied.

$$a^2 - b^2 = (a + b)(a - b)$$

We can use this equation to factor any differences of squares.

Differences of Squares

A difference of squares can be rewritten as two factors containing the same terms but opposite signs.

$$a^2 - b^2 = (a + b)(a - b)$$

HOW TO

Given a difference of squares, factor it into binomials.

1. Confirm that the first and last term are perfect squares.
2. Write the factored form as $(a + b)(a - b)$.

EXAMPLE 5

Factoring a Difference of Squares
Factor $9x^2 - 25$.

Solution
Notice that $9x^2$ and 25 are perfect squares because $9x^2 = (3x)^2$ and $25 = 5^2$. The polynomial represents a difference of squares and can be rewritten as $(3x + 5)(3x - 5)$.

TRY IT #5 Factor $81y^2 - 100$.

Q&A **Is there a formula to factor the sum of squares?**

No. A sum of squares cannot be factored.

Factoring the Sum and Difference of Cubes

Now, we will look at two new special products: the sum and difference of cubes. Although the sum of squares cannot be factored, the sum of cubes can be factored into a binomial and a trinomial.

$$a^3 + b^3 = (a + b)\left(a^2 - ab + b^2\right)$$

Similarly, the difference of cubes can be factored into a binomial and a trinomial, but with different signs.

$$a^3 - b^3 = (a - b)\left(a^2 + ab + b^2\right)$$

We can use the acronym SOAP to remember the signs when factoring the sum or difference of cubes. The first letter of each word relates to the signs: **S**ame **O**pposite **A**lways **P**ositive. For example, consider the following example.

$$x^3 - 2^3 = (x - 2)\left(x^2 + 2x + 4\right)$$

The sign of the first 2 is the *same* as the sign between $x^3 - 2^3$. The sign of the $2x$ term is *opposite* the sign between $x^3 - 2^3$. And the sign of the last term, 4, is *always positive*.

Sum and Difference of Cubes

We can factor the sum of two cubes as

$$a^3 + b^3 = (a + b)\left(a^2 - ab + b^2\right)$$

We can factor the difference of two cubes as

$$a^3 - b^3 = (a - b)\left(a^2 + ab + b^2\right)$$

HOW TO

Given a sum of cubes or difference of cubes, factor it.

1. Confirm that the first and last term are cubes, $a^3 + b^3$ or $a^3 - b^3$.
2. For a sum of cubes, write the factored form as $(a + b)\left(a^2 - ab + b^2\right)$. For a difference of cubes, write the factored form as $(a - b)\left(a^2 + ab + b^2\right)$.

EXAMPLE 6

Factoring a Sum of Cubes

Factor $x^3 + 512$.

Solution

Notice that x^3 and 512 are cubes because $8^3 = 512$. Rewrite the sum of cubes as $(x + 8)\left(x^2 - 8x + 64\right)$.

Analysis

After writing the sum of cubes this way, we might think we should check to see if the trinomial portion can be factored further. However, the trinomial portion cannot be factored, so we do not need to check.

TRY IT #6 Factor the sum of cubes: $216a^3 + b^3$.

EXAMPLE 7

Factoring a Difference of Cubes

Factor $8x^3 - 125$.

Solution

Notice that $8x^3$ and 125 are cubes because $8x^3 = (2x)^3$ and $125 = 5^3$. Write the difference of cubes as $(2x - 5)\left(4x^2 + 10x + 25\right)$.

Analysis

Just as with the sum of cubes, we will not be able to further factor the trinomial portion.

TRY IT #7 Factor the difference of cubes: $1{,}000x^3 - 1$.

Factoring Expressions with Fractional or Negative Exponents

Expressions with fractional or negative exponents can be factored by pulling out a GCF. Look for the variable or exponent that is common to each term of the expression and pull out that variable or exponent raised to the lowest power. These expressions follow the same factoring rules as those with integer exponents. For instance, $2x^{\frac{1}{4}} + 5x^{\frac{3}{4}}$ can be factored by pulling out $x^{\frac{1}{4}}$ and being rewritten as $x^{\frac{1}{4}}\left(2 + 5x^{\frac{1}{2}}\right)$.

EXAMPLE 8

Factoring an Expression with Fractional or Negative Exponents

Factor $3x(x+2)^{\frac{-1}{3}} + 4(x+2)^{\frac{2}{3}}$.

Solution

Factor out the term with the lowest value of the exponent. In this case, that would be $(x+2)^{-\frac{1}{3}}$.

$$(x+2)^{-\frac{1}{3}}(3x+4(x+2)) \quad \text{Factor out the GCF.}$$
$$(x+2)^{-\frac{1}{3}}(3x+4x+8) \quad \text{Simplify.}$$
$$(x+2)^{-\frac{1}{3}}(7x+8)$$

TRY IT #8 Factor $2(5a-1)^{\frac{3}{4}} + 7a(5a-1)^{-\frac{1}{4}}$.

MEDIA

Access these online resources for additional instruction and practice with factoring polynomials.

Identify GCF (http://openstax.org/l/findgcftofact)
Factor Trinomials when a Equals 1 (http://openstax.org/l/facttrinom1)
Factor Trinomials when a is not equal to 1 (http://openstax.org/l/facttrinom2)
Factor Sum or Difference of Cubes (http://openstax.org/l/sumdifcube)

1.5 SECTION EXERCISES

Verbal

1. If the terms of a polynomial do not have a GCF, does that mean it is not factorable? Explain.

2. A polynomial is factorable, but it is not a perfect square trinomial or a difference of two squares. Can you factor the polynomial without finding the GCF?

3. How do you factor by grouping?

Algebraic

For the following exercises, find the greatest common factor.

4. $14x + 4xy - 18xy^2$

5. $49mb^2 - 35m^2ba + 77ma^2$

6. $30x^3y - 45x^2y^2 + 135xy^3$

7. $200p^3m^3 - 30p^2m^3 + 40m^3$

8. $36j^4k^2 - 18j^3k^3 + 54j^2k^4$

9. $6y^4 - 2y^3 + 3y^2 - y$

For the following exercises, factor by grouping.

10. $6x^2 + 5x - 4$

11. $2a^2 + 9a - 18$

12. $6c^2 + 41c + 63$

13. $6n^2 - 19n - 11$

14. $20w^2 - 47w + 24$

15. $2p^2 - 5p - 7$

For the following exercises, factor the polynomial.

16. $7x^2 + 48x - 7$

17. $10h^2 - 9h - 9$

18. $2b^2 - 25b - 247$

19. $9d^2 - 73d + 8$

20. $90v^2 - 181v + 90$

21. $12t^2 + t - 13$

22. $2n^2 - n - 15$

23. $16x^2 - 100$

24. $25y^2 - 196$

25. $121p^2 - 169$

26. $4m^2 - 9$

27. $361d^2 - 81$

28. $324x^2 - 121$

29. $144b^2 - 25c^2$

30. $16a^2 - 8a + 1$

31. $49n^2 + 168n + 144$

32. $121x^2 - 88x + 16$

33. $225y^2 + 120y + 16$

34. $m^2 - 20m + 100$

35. $25p^2 - 120p + 144$

36. $36q^2 + 60q + 25$

For the following exercises, factor the polynomials.

37. $x^3 + 216$

38. $27y^3 - 8$

39. $125a^3 + 343$

40. $b^3 - 8d^3$

41. $64x^3 - 125$

42. $729q^3 + 1331$

43. $125r^3 + 1{,}728s^3$

44. $4x(x-1)^{-\frac{2}{3}} + 3(x-1)^{\frac{1}{3}}$

45. $3c(2c+3)^{-\frac{1}{4}} - 5(2c+3)^{\frac{3}{4}}$

46. $3t(10t+3)^{\frac{1}{3}} + 7(10t+3)^{\frac{4}{3}}$

47. $14x(x+2)^{-\frac{2}{5}} + 5(x+2)^{\frac{3}{5}}$

48. $9y(3y-13)^{\frac{1}{5}} - 2(3y-13)^{\frac{6}{5}}$

49. $5z(2z-9)^{-\frac{3}{2}} + 11(2z-9)^{-\frac{1}{2}}$

50. $6d(2d+3)^{-\frac{1}{6}} + 5(2d+3)^{\frac{5}{6}}$

Real-World Applications

For the following exercises, consider this scenario:

Charlotte has appointed a chairperson to lead a city beautification project. The first act is to install statues and fountains in one of the city's parks. The park is a rectangle with an area of $98x^2 + 105x - 27$ m^2*, as shown in the figure below. The length and width of the park are perfect factors of the area.*

$l \times w = 98x^2 + 105x - 27$

51. Factor by grouping to find the length and width of the park.

52. A statue is to be placed in the center of the park. The area of the base of the statue is $4x^2 + 12x + 9\text{m}^2$. Factor the area to find the lengths of the sides of the statue.

53. At the northwest corner of the park, the city is going to install a fountain. The area of the base of the fountain is $9x^2 - 25\text{m}^2$. Factor the area to find the lengths of the sides of the fountain.

For the following exercise, consider the following scenario:

A school is installing a flagpole in the central plaza. The plaza is a square with side length 100 yd. as shown in the figure below. The flagpole will take up a square plot with area $x^2 - 6x + 9$ *yd²*.

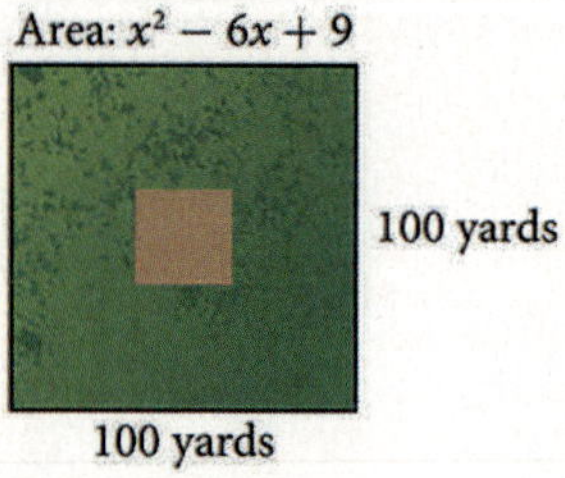

54. Find the length of the base of the flagpole by factoring.

Extensions

For the following exercises, factor the polynomials completely.

55. $16x^4 - 200x^2 + 625$

56. $81y^4 - 256$

57. $16z^4 - 2{,}401a^4$

58. $5x(3x+2)^{-\frac{2}{4}} + (12x+8)^{\frac{3}{2}}$

59. $\left(32x^3 + 48x^2 - 162x - 243\right)^{-1}$

1.6 Rational Expressions

Learning Objectives

In this section, you will:

- Simplify rational expressions.
- Multiply rational expressions.
- Divide rational expressions.
- Add and subtract rational expressions.
- Simplify complex rational expressions.

A pastry shop has fixed costs of $\$280$ per week and variable costs of $\$9$ per box of pastries. The shop's costs per week in terms of x, the number of boxes made, is $280 + 9x$. We can divide the costs per week by the number of boxes made to determine the cost per box of pastries.

$$\frac{280 + 9x}{x}$$

Notice that the result is a polynomial expression divided by a second polynomial expression. In this section, we will explore quotients of polynomial expressions.

Simplifying Rational Expressions

The quotient of two polynomial expressions is called a **rational expression**. We can apply the properties of fractions to rational expressions, such as simplifying the expressions by canceling common factors from the numerator and the denominator. To do this, we first need to factor both the numerator and denominator. Let's start with the rational expression shown.

$$\frac{x^2 + 8x + 16}{x^2 + 11x + 28}$$

We can factor the numerator and denominator to rewrite the expression.

$$\frac{(x+4)^2}{(x+4)(x+7)}$$

Then we can simplify that expression by canceling the common factor $(x + 4)$.

$$\frac{x+4}{x+7}$$

HOW TO

Given a rational expression, simplify it.

1. Factor the numerator and denominator.
2. Cancel any common factors.

EXAMPLE 1

Simplifying Rational Expressions

Simplify $\frac{x^2-9}{x^2+4x+3}$.

Solution

$\frac{(x+3)(x-3)}{(x+3)(x+1)}$	Factor the numerator and the denominator.
$\frac{x-3}{x+1}$	Cancel common factor $(x + 3)$.

Analysis

We can cancel the common factor because any expression divided by itself is equal to 1.

Q&A **Can the x^2 term be cancelled in Example 1?**

No. A factor is an expression that is multiplied by another expression. The x^2 term is not a factor of the numerator or the denominator.

TRY IT #1 Simplify $\frac{x-6}{x^2-36}$.

Multiplying Rational Expressions

Multiplication of rational expressions works the same way as multiplication of any other fractions. We multiply the numerators to find the numerator of the product, and then multiply the denominators to find the denominator of the product. Before multiplying, it is helpful to factor the numerators and denominators just as we did when simplifying rational expressions. We are often able to simplify the product of rational expressions.

HOW TO

Given two rational expressions, multiply them.

1. Factor the numerator and denominator.

2. Multiply the numerators.
3. Multiply the denominators.
4. Simplify.

EXAMPLE 2

Multiplying Rational Expressions

Multiply the rational expressions and show the product in simplest form:

$$\frac{x^2+4x-5}{3x+18}\cdot\frac{2x-1}{x+5}$$

Solution

$\frac{(x+5)(x-1)}{3(x+6)}\cdot\frac{(2x-1)}{(x+5)}$	Factor the numerator and denominator.
$\frac{(x+5)(x-1)(2x-1)}{3(x+6)(x+5)}$	Multiply numerators and denominators.
$\frac{\cancel{(x+5)}(x-1)(2x-1)}{3(x+6)\cancel{(x+5)}}$	Cancel common factors to simplify.
$\frac{(x-1)(2x-1)}{3(x+6)}$	

TRY IT #2 Multiply the rational expressions and show the product in simplest form:

$$\frac{x^2+11x+30}{x^2+5x+6}\cdot\frac{x^2+7x+12}{x^2+8x+16}$$

Dividing Rational Expressions

Division of rational expressions works the same way as division of other fractions. To divide a rational expression by another rational expression, multiply the first expression by the reciprocal of the second. Using this approach, we would rewrite $\frac{1}{x}\div\frac{x^2}{3}$ as the product $\frac{1}{x}\cdot\frac{3}{x^2}$. Once the division expression has been rewritten as a multiplication expression, we can multiply as we did before.

$$\frac{1}{x}\cdot\frac{3}{x^2}=\frac{3}{x^3}$$

HOW TO

Given two rational expressions, divide them.

1. Rewrite as the first rational expression multiplied by the reciprocal of the second.
2. Factor the numerators and denominators.
3. Multiply the numerators.
4. Multiply the denominators.
5. Simplify.

EXAMPLE 3

Dividing Rational Expressions

Divide the rational expressions and express the quotient in simplest form:

$$\frac{2x^2+x-6}{x^2-1}\div\frac{x^2-4}{x^2+2x+1}$$

Solution

$\frac{2x^2+x-6}{x^2-1} \cdot \frac{x^2+2x+1}{x^2-4}$	Rewrite as multiplication.
$\frac{(2x-3)(x+2)}{(x+1)(x-1)} \cdot \frac{(x+1)^2}{(x+2)(x-2)}$	Factor.
$\frac{(2x-3)(x+2)(x+1)^2}{(x+1)(x-1)(x+2)(x-2)}$	Multiply.
$\frac{(2x-3)(x+1)}{(x-1)(x-2)}$	Cancel common factors to simplify.

TRY IT #3 Divide the rational expressions and express the quotient in simplest form:

$$\frac{9x^2-16}{3x^2+17x-28} \div \frac{3x^2-2x-8}{x^2+5x-14}$$

Adding and Subtracting Rational Expressions

Adding and subtracting rational expressions works just like adding and subtracting numerical fractions. To add fractions, we need to find a common denominator. Let's look at an example of fraction addition.

$$\begin{aligned} \frac{5}{24}+\frac{1}{40} &= \frac{25}{120}+\frac{3}{120} \\ &= \frac{28}{120} \\ &= \frac{7}{30} \end{aligned}$$

We have to rewrite the fractions so they share a common denominator before we are able to add. We must do the same thing when adding or subtracting rational expressions.

The easiest common denominator to use will be the **least common denominator**, or LCD. The LCD is the smallest multiple that the denominators have in common. To find the LCD of two rational expressions, we factor the expressions and multiply all of the distinct factors. For instance, if the factored denominators were $(x+3)(x+4)$ and $(x+4)(x+5)$, then the LCD would be $(x+3)(x+4)(x+5)$.

Once we find the LCD, we need to multiply each expression by the form of 1 that will change the denominator to the LCD. We would need to multiply the expression with a denominator of $(x+3)(x+4)$ by $\frac{x+5}{x+5}$ and the expression with a denominator of $(x+4)(x+5)$ by $\frac{x+3}{x+3}$.

HOW TO

Given two rational expressions, add or subtract them.

1. Factor the numerator and denominator.
2. Find the LCD of the expressions.
3. Multiply the expressions by a form of 1 that changes the denominators to the LCD.
4. Add or subtract the numerators.
5. Simplify.

EXAMPLE 4

Adding Rational Expressions

Add the rational expressions:

$$\frac{5}{x}+\frac{6}{y}$$

Solution

First, we have to find the LCD. In this case, the LCD will be xy. We then multiply each expression by the appropriate form of 1 to obtain xy as the denominator for each fraction.

$$\frac{5}{x}\cdot\frac{y}{y}+\frac{6}{y}\cdot\frac{x}{x}$$
$$\frac{5y}{xy}+\frac{6x}{xy}$$

Now that the expressions have the same denominator, we simply add the numerators to find the sum.

$$\frac{6x+5y}{xy}$$

Analysis

Multiplying by $\frac{y}{y}$ or $\frac{x}{x}$ does not change the value of the original expression because any number divided by itself is 1, and multiplying an expression by 1 gives the original expression.

EXAMPLE 5

Subtracting Rational Expressions

Subtract the rational expressions:

$$\frac{6}{x^2+4x+4}-\frac{2}{x^2-4}$$

Solution

$\frac{6}{(x+2)^2}-\frac{2}{(x+2)(x-2)}$	Factor.
$\frac{6}{(x+2)^2}\cdot\frac{x-2}{x-2}-\frac{2}{(x+2)(x-2)}\cdot\frac{x+2}{x+2}$	Multiply each fraction to get LCD as denominator.
$\frac{6(x-2)}{(x+2)^2(x-2)}-\frac{2(x+2)}{(x+2)^2(x-2)}$	Multiply.
$\frac{6x-12-(2x+4)}{(x+2)^2(x-2)}$	Apply distributive property.
$\frac{4x-16}{(x+2)^2(x-2)}$	Subtract.
$\frac{4(x-4)}{(x+2)^2(x-2)}$	Simplify.

Q&A **Do we have to use the LCD to add or subtract rational expressions?**

No. Any common denominator will work, but it is easiest to use the LCD.

TRY IT #4 Subtract the rational expressions: $\frac{3}{x+5}-\frac{1}{x-3}$.

Simplifying Complex Rational Expressions

A complex rational expression is a rational expression that contains additional rational expressions in the numerator, the denominator, or both. We can simplify complex rational expressions by rewriting the numerator and denominator as single rational expressions and dividing. The complex rational expression $\frac{a}{\frac{1}{b}+c}$ can be simplified by rewriting the numerator as the fraction $\frac{a}{1}$ and combining the expressions in the denominator as $\frac{1+bc}{b}$. We can then rewrite the expression as a multiplication problem using the reciprocal of the denominator. We get $\frac{a}{1}\cdot\frac{b}{1+bc}$, which is equal to $\frac{ab}{1+bc}$.

HOW TO

Given a complex rational expression, simplify it.

1. Combine the expressions in the numerator into a single rational expression by adding or subtracting.
2. Combine the expressions in the denominator into a single rational expression by adding or subtracting.
3. Rewrite as the numerator divided by the denominator.
4. Rewrite as multiplication.
5. Multiply.
6. Simplify.

EXAMPLE 6

Simplifying Complex Rational Expressions

Simplify: $\frac{y+\frac{1}{x}}{\frac{x}{y}}$.

Solution

Begin by combining the expressions in the numerator into one expression.

$y \cdot \frac{x}{x} + \frac{1}{x}$	Multiply by $\frac{x}{x}$ to get LCD as denominator.
$\frac{xy}{x} + \frac{1}{x}$	
$\frac{xy+1}{x}$	Add numerators.

Now the numerator is a single rational expression and the denominator is a single rational expression.

$$\frac{\frac{xy+1}{x}}{\frac{x}{y}}$$

We can rewrite this as division, and then multiplication.

$\frac{xy+1}{x} \div \frac{x}{y}$	
$\frac{xy+1}{x} \cdot \frac{y}{x}$	Rewrite as multiplication.
$\frac{y(xy+1)}{x^2}$	Multiply.

TRY IT #5 Simplify: $\frac{\frac{x}{y}-\frac{y}{x}}{y}$

Q&A **Can a complex rational expression always be simplified?**

Yes. We can always rewrite a complex rational expression as a simplified rational expression.

MEDIA

Access these online resources for additional instruction and practice with rational expressions.

Simplify Rational Expressions (http://openstax.org/l/simpratexpress)
Multiply and Divide Rational Expressions (http://openstax.org/l/multdivratex)
Add and Subtract Rational Expressions (http://openstax.org/l/addsubratex)
Simplify a Complex Fraction (http://openstax.org/l/complexfract)

1.6 SECTION EXERCISES

Verbal

1. How can you use factoring to simplify rational expressions?

2. How do you use the LCD to combine two rational expressions?

3. Tell whether the following statement is true or false and explain why: You only need to find the LCD when adding or subtracting rational expressions.

Algebraic

For the following exercises, simplify the rational expressions.

4. $\frac{x^2-16}{x^2-5x+4}$

5. $\frac{y^2+10y+25}{y^2+11y+30}$

6. $\frac{6a^2-24a+24}{6a^2-24}$

7. $\frac{9b^2+18b+9}{3b+3}$

8. $\frac{m-12}{m^2-144}$

9. $\frac{2x^2+7x-4}{4x^2+2x-2}$

10. $\frac{6x^2+5x-4}{3x^2+19x+20}$

11. $\frac{a^2+9a+18}{a^2+3a-18}$

12. $\frac{3c^2+25c-18}{3c^2-23c+14}$

13. $\frac{12n^2-29n-8}{28n^2-5n-3}$

For the following exercises, multiply the rational expressions and express the product in simplest form.

14. $\frac{x^2-x-6}{2x^2+x-6}\cdot\frac{2x^2+7x-15}{x^2-9}$

15. $\frac{c^2+2c-24}{c^2+12c+36}\cdot\frac{c^2-10c+24}{c^2-8c+16}$

16. $\frac{2d^2+9d-35}{d^2+10d+21}\cdot\frac{3d^2+2d-21}{3d^2+14d-49}$

17. $\frac{10h^2-9h-9}{2h^2-19h+24}\cdot\frac{h^2-16h+64}{5h^2-37h-24}$

18. $\frac{6b^2+13b+6}{4b^2-9}\cdot\frac{6b^2+31b-30}{18b^2-3b-10}$

19. $\frac{2d^2+15d+25}{4d^2-25}\cdot\frac{2d^2-15d+25}{25d^2-1}$

20. $\frac{6x^2-5x-50}{15x^2-44x-20}\cdot\frac{20x^2-7x-6}{2x^2+9x+10}$

21. $\frac{t^2-1}{t^2+4t+3}\cdot\frac{t^2+2t-15}{t^2-4t+3}$

22. $\frac{2n^2-n-15}{6n^2+13n-5}\cdot\frac{12n^2-13n+3}{4n^2-15n+9}$

23. $\frac{36x^2-25}{6x^2+65x+50}\cdot\frac{3x^2+32x+20}{18x^2+27x+10}$

For the following exercises, divide the rational expressions.

24. $\frac{3y^2-7y-6}{2y^2-3y-9}\div\frac{y^2+y-2}{2y^2+y-3}$

25. $\frac{6p^2+p-12}{8p^2+18p+9}\div\frac{6p^2-11p+4}{2p^2+11p-6}$

26. $\frac{q^2-9}{q^2+6q+9}\div\frac{q^2-2q-3}{q^2+2q-3}$

27. $\frac{18d^2+77d-18}{27d^2-15d+2}\div\frac{3d^2+29d-44}{9d^2-15d+4}$

28. $\frac{16x^2+18x-55}{32x^2-36x-11}\div\frac{2x^2+17x+30}{4x^2+25x+6}$

29. $\frac{144b^2-25}{72b^2-6b-10}\div\frac{18b^2-21b+5}{36b^2-18b-10}$

30. $\frac{16a^2-24a+9}{4a^2+17a-15}\div\frac{16a^2-9}{4a^2+11a+6}$

31. $\frac{22y^2+59y+10}{12y^2+28y-5}\div\frac{11y^2+46y+8}{24y^2-10y+1}$

32. $\frac{9x^2+3x-20}{3x^2-7x+4}\div\frac{6x^2+4x-10}{x^2-2x+1}$

For the following exercises, add and subtract the rational expressions, and then simplify.

33. $\frac{4}{x} + \frac{10}{y}$

34. $\frac{12}{2q} - \frac{6}{3p}$

35. $\frac{4}{a+1} + \frac{5}{a-3}$

36. $\frac{c+2}{3} - \frac{c-4}{4}$

37. $\frac{y+3}{y-2} + \frac{y-3}{y+1}$

38. $\frac{x-1}{x+1} - \frac{2x+3}{2x+1}$

39. $\frac{3z}{z+1} + \frac{2z+5}{z-2}$

40. $\frac{4p}{p+1} - \frac{p+1}{4p}$

41. $\frac{x}{x+1} + \frac{y}{y+1}$

For the following exercises, simplify the rational expression.

42. $\frac{\frac{6}{y} - \frac{4}{x}}{y}$

43. $\frac{\frac{2}{a} + \frac{7}{b}}{b}$

44. $\frac{\frac{x}{4} - \frac{p}{8}}{p}$

45. $\frac{\frac{3}{a} + \frac{b}{6}}{\frac{2b}{3a}}$

46. $\frac{\frac{3}{x+1} + \frac{2}{x-1}}{\frac{x-1}{x+1}}$

47. $\frac{\frac{a}{b} - \frac{b}{a}}{\frac{a+b}{ab}}$

48. $\frac{\frac{2x}{3} + \frac{4x}{7}}{\frac{x}{2}}$

49. $\frac{\frac{2c}{c+2} + \frac{c-1}{c+1}}{\frac{2c+1}{c+1}}$

50. $\frac{\frac{x}{y} - \frac{y}{x}}{\frac{x}{y} + \frac{y}{x}}$

Real-World Applications

51. Brenda is placing tile on her bathroom floor. The area of the floor is $15x^2 - 8x - 7$ ft^2. The area of one tile is $x^2 - 2x + 1$ft^2. To find the number of tiles needed, simplify the rational expression: $\frac{15x^2-8x-7}{x^2-2x+1}$.

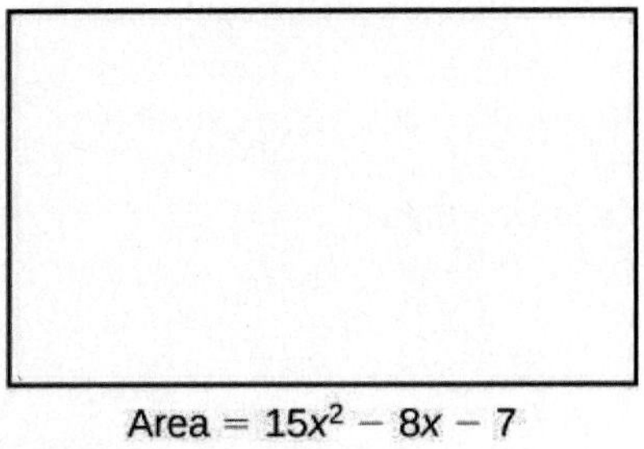

52. The area of Lijuan's yard is $25x^2 - 625$ ft^2. A patch of sod has an area of $x^2 - 10x + 25$ ft^2. Divide the two areas and simplify to find how many pieces of sod Lijuan needs to cover her yard.

53. Elroi wants to mulch his garden. His garden is $x^2 + 18x + 81$ ft^2. One bag of mulch covers $x^2 - 81$ ft^2. Divide the expressions and simplify to find how many bags of mulch Elroi needs to mulch his garden.

Extensions

For the following exercises, perform the given operations and simplify.

54. $\frac{x^2+x-6}{x^2-2x-3} \cdot \frac{2x^2-3x-9}{x^2-x-2} \div \frac{10x^2+27x+18}{x^2+2x+1}$

55. $\frac{\frac{3y^2-10y+3}{3y^2+5y-2} \cdot \frac{2y^2-3y-20}{2y^2-y-15}}{y-4}$

56. $\frac{\frac{4a+1}{2a-3} + \frac{2a-3}{2a+3}}{\frac{4a^2+9}{a}}$

57. $\frac{x^2+7x+12}{x^2+x-6} \div \frac{3x^2+19x+28}{8x^2-4x-24} \div \frac{2x^2+x-3}{3x^2+4x-7}$

Chapter Review

Key Terms

algebraic expression constants and variables combined using addition, subtraction, multiplication, and division

associative property of addition the sum of three numbers may be grouped differently without affecting the result; in symbols, $a+(b+c)=(a+b)+c$

associative property of multiplication the product of three numbers may be grouped differently without affecting the result; in symbols, $a\cdot(b\cdot c)=(a\cdot b)\cdot c$

base in exponential notation, the expression that is being multiplied

binomial a polynomial containing two terms

coefficient any real number a_i in a polynomial in the form $a_n x^n + ... + a_2 x^2 + a_1 x + a_0$

commutative property of addition two numbers may be added in either order without affecting the result; in symbols, $a+b=b+a$

commutative property of multiplication two numbers may be multiplied in any order without affecting the result; in symbols, $a\cdot b=b\cdot a$

constant a quantity that does not change value

degree the highest power of the variable that occurs in a polynomial

difference of squares the binomial that results when a binomial is multiplied by a binomial with the same terms, but the opposite sign

distributive property the product of a factor times a sum is the sum of the factor times each term in the sum; in symbols, $a\cdot(b+c)=a\cdot b+a\cdot c$

equation a mathematical statement indicating that two expressions are equal

exponent in exponential notation, the raised number or variable that indicates how many times the base is being multiplied

exponential notation a shorthand method of writing products of the same factor

factor by grouping a method for factoring a trinomial in the form ax^2+bx+c by dividing the x term into the sum of two terms, factoring each portion of the expression separately, and then factoring out the GCF of the entire expression

formula an equation expressing a relationship between constant and variable quantities

greatest common factor the largest polynomial that divides evenly into each polynomial

identity property of addition there is a unique number, called the additive identity, 0, which, when added to a number, results in the original number; in symbols, $a+0=a$

identity property of multiplication there is a unique number, called the multiplicative identity, 1, which, when multiplied by a number, results in the original number; in symbols, $a\cdot 1=a$

index the number above the radical sign indicating the nth root

integers the set consisting of the natural numbers, their opposites, and 0: $\{\ldots,-3,-2,-1,0,1,2,3,\ldots\}$

inverse property of addition for every real number a, there is a unique number, called the additive inverse (or opposite), denoted $-a$, which, when added to the original number, results in the additive identity, 0; in symbols, $a+(-a)=0$

inverse property of multiplication for every non-zero real number a, there is a unique number, called the multiplicative inverse (or reciprocal), denoted $\frac{1}{a}$, which, when multiplied by the original number, results in the multiplicative identity, 1; in symbols, $a\cdot\frac{1}{a}=1$

irrational numbers the set of all numbers that are not rational; they cannot be written as either a terminating or repeating decimal; they cannot be expressed as a fraction of two integers

leading coefficient the coefficient of the leading term

leading term the term containing the highest degree

least common denominator the smallest multiple that two denominators have in common

monomial a polynomial containing one term

natural numbers the set of counting numbers: $\{1,2,3,\ldots\}$

order of operations a set of rules governing how mathematical expressions are to be evaluated, assigning priorities to operations

perfect square trinomial the trinomial that results when a binomial is squared

polynomial a sum of terms each consisting of a variable raised to a nonnegative integer power

principal *n*th root the number with the same sign as a that when raised to the nth power equals a

principal square root the nonnegative square root of a number a that, when multiplied by itself, equals a

radical the symbol used to indicate a root

radical expression an expression containing a radical symbol

radicand the number under the radical symbol

rational expression the quotient of two polynomial expressions
rational numbers the set of all numbers of the form $\frac{m}{n}$, where m and n are integers and $n \neq 0$. Any rational number may be written as a fraction or a terminating or repeating decimal.
real number line a horizontal line used to represent the real numbers. An arbitrary fixed point is chosen to represent 0; positive numbers lie to the right of 0 and negative numbers to the left.
real numbers the sets of rational numbers and irrational numbers taken together
scientific notation a shorthand notation for writing very large or very small numbers in the form $a \times 10^n$ where $1 \leq |a| < 10$ and n is an integer
term of a polynomial any $a_i x^i$ of a polynomial in the form $a_n x^n + ... + a_2 x^2 + a_1 x + a_0$
trinomial a polynomial containing three terms
variable a quantity that may change value
whole numbers the set consisting of 0 plus the natural numbers: $\{0, 1, 2, 3, ...\}$

Key Equations

Rules of Exponents
For nonzero real numbers a and b and integers m and n

Product rule	$a^m \cdot a^n = a^{m+n}$
Quotient rule	$\frac{a^m}{a^n} = a^{m-n}$
Power rule	$(a^m)^n = a^{m \cdot n}$
Zero exponent rule	$a^0 = 1$
Negative rule	$a^{-n} = \frac{1}{a^n}$
Power of a product rule	$(a \cdot b)^n = a^n \cdot b^n$
Power of a quotient rule	$\left(\frac{a}{b}\right)^n = \frac{a^n}{b^n}$

perfect square trinomial	$(x + a)^2 = (x + a)(x + a) = x^2 + 2ax + a^2$
difference of squares	$(a + b)(a - b) = a^2 - b^2$

difference of squares	$a^2 - b^2 = (a + b)(a - b)$
perfect square trinomial	$a^2 + 2ab + b^2 = (a + b)^2$
sum of cubes	$a^3 + b^3 = (a + b)(a^2 - ab + b^2)$
difference of cubes	$a^3 - b^3 = (a - b)(a^2 + ab + b^2)$

Key Concepts

1.1 Real Numbers: Algebra Essentials

- Rational numbers may be written as fractions or terminating or repeating decimals. See Example 1 and Example 2.
- Determine whether a number is rational or irrational by writing it as a decimal. See Example 3.
- The rational numbers and irrational numbers make up the set of real numbers. See Example 4. A number can be classified as natural, whole, integer, rational, or irrational. See Example 5.

- The order of operations is used to evaluate expressions. See Example 6.
- The real numbers under the operations of addition and multiplication obey basic rules, known as the properties of real numbers. These are the commutative properties, the associative properties, the distributive property, the identity properties, and the inverse properties. See Example 7.
- Algebraic expressions are composed of constants and variables that are combined using addition, subtraction, multiplication, and division. See Example 8. They take on a numerical value when evaluated by replacing variables with constants. See Example 9, Example 10, and Example 12
- Formulas are equations in which one quantity is represented in terms of other quantities. They may be simplified or evaluated as any mathematical expression. See Example 11 and Example 13.

1.2 Exponents and Scientific Notation

- Products of exponential expressions with the same base can be simplified by adding exponents. See Example 1.
- Quotients of exponential expressions with the same base can be simplified by subtracting exponents. See Example 2.
- Powers of exponential expressions with the same base can be simplified by multiplying exponents. See Example 3.
- An expression with exponent zero is defined as 1. See Example 4.
- An expression with a negative exponent is defined as a reciprocal. See Example 5 and Example 6.
- The power of a product of factors is the same as the product of the powers of the same factors. See Example 7.
- The power of a quotient of factors is the same as the quotient of the powers of the same factors. See Example 8.
- The rules for exponential expressions can be combined to simplify more complicated expressions. See Example 9.
- Scientific notation uses powers of 10 to simplify very large or very small numbers. See Example 10 and Example 11.
- Scientific notation may be used to simplify calculations with very large or very small numbers. See Example 12 and Example 13.

1.3 Radicals and Rational Exponents

- The principal square root of a number a is the nonnegative number that when multiplied by itself equals a. See Example 1.
- If a and b are nonnegative, the square root of the product ab is equal to the product of the square roots of a and b See Example 2 and Example 3.
- If a and b are nonnegative, the square root of the quotient $\frac{a}{b}$ is equal to the quotient of the square roots of a and b See Example 4 and Example 5.
- We can add and subtract radical expressions if they have the same radicand and the same index. See Example 6 and Example 7.
- Radical expressions written in simplest form do not contain a radical in the denominator. To eliminate the square root radical from the denominator, multiply both the numerator and the denominator by the conjugate of the denominator. See Example 8 and Example 9.
- The principal nth root of a is the number with the same sign as a that when raised to the nth power equals a. These roots have the same properties as square roots. See Example 10.
- Radicals can be rewritten as rational exponents and rational exponents can be rewritten as radicals. See Example 11 and Example 12.
- The properties of exponents apply to rational exponents. See Example 13.

1.4 Polynomials

- A polynomial is a sum of terms each consisting of a variable raised to a non-negative integer power. The degree is the highest power of the variable that occurs in the polynomial. The leading term is the term containing the highest degree, and the leading coefficient is the coefficient of that term. See Example 1.
- We can add and subtract polynomials by combining like terms. See Example 2 and Example 3.
- To multiply polynomials, use the distributive property to multiply each term in the first polynomial by each term in the second. Then add the products. See Example 4.
- FOIL (First, Outer, Inner, Last) is a shortcut that can be used to multiply binomials. See Example 5.
- Perfect square trinomials and difference of squares are special products. See Example 6 and Example 7.
- Follow the same rules to work with polynomials containing several variables. See Example 8.

1.5 Factoring Polynomials

- The greatest common factor, or GCF, can be factored out of a polynomial. Checking for a GCF should be the first step in any factoring problem. See Example 1.
- Trinomials with leading coefficient 1 can be factored by finding numbers that have a product of the third term and a sum of the second term. See Example 2.

- Trinomials can be factored using a process called factoring by grouping. See Example 3.
- Perfect square trinomials and the difference of squares are special products and can be factored using equations. See Example 4 and Example 5.
- The sum of cubes and the difference of cubes can be factored using equations. See Example 6 and Example 7.
- Polynomials containing fractional and negative exponents can be factored by pulling out a GCF. See Example 8.

1.6 Rational Expressions

- Rational expressions can be simplified by cancelling common factors in the numerator and denominator. See Example 1.
- We can multiply rational expressions by multiplying the numerators and multiplying the denominators. See Example 2.
- To divide rational expressions, multiply by the reciprocal of the second expression. See Example 3.
- Adding or subtracting rational expressions requires finding a common denominator. See Example 4 and Example 5.
- Complex rational expressions have fractions in the numerator or the denominator. These expressions can be simplified. See Example 6.

Exercises

Review Exercises

Real Numbers: Algebra Essentials

For the following exercises, perform the given operations.

1. $(5-3\cdot 2)^2-6$

2. $64\div(2\cdot 8)+14\div 7$

3. $2\cdot 5^2+6\div 2$

For the following exercises, solve the equation.

4. $5x+9=-11$

5. $2y+4^2=64$

For the following exercises, simplify the expression.

6. $9(y+2)\div 3\cdot 2+1$

7. $3m(4+7)-m$

For the following exercises, identify the number as rational, irrational, whole, or natural. Choose the most descriptive answer.

8. 11

9. 0

10. $\frac{5}{6}$

11. $\sqrt{11}$

Exponents and Scientific Notation

For the following exercises, simplify the expression.

12. $2^2\cdot 2^4$

13. $\frac{4^5}{4^3}$

14. $\left(\frac{a^2}{b^3}\right)^4$

15. $\frac{6a^2\cdot a^0}{2a^{-4}}$

16. $\frac{(xy)^4}{y^3}\cdot\frac{2}{x^5}$

17. $\frac{4^{-2}x^3y^{-3}}{2x^0}$

18. $\left(\frac{2x^2}{y}\right)^{-2}$

19. $\left(\frac{16a^3}{b^2}\right)\left(4ab^{-1}\right)^{-2}$

20. Write the number in standard notation: 2.1314×10^{-6}

21. Write the number in scientific notation: 16,340,000

Radicals and Rational Expressions

For the following exercises, find the principal square root.

22. $\sqrt{121}$

23. $\sqrt{196}$

24. $\sqrt{361}$

25. $\sqrt{75}$

26. $\sqrt{162}$

27. $\sqrt{\frac{32}{25}}$

28. $\sqrt{\frac{80}{81}}$

29. $\sqrt{\frac{49}{1250}}$

30. $\frac{2}{4+\sqrt{2}}$

31. $4\sqrt{3}+6\sqrt{3}$

32. $12\sqrt{5}-13\sqrt{5}$

33. $\sqrt[5]{-243}$

34. $\frac{\sqrt[3]{250}}{\sqrt[3]{-8}}$

Polynomials

For the following exercises, perform the given operations and simplify.

35. $\left(3x^3+2x-1\right)+\left(4x^2-2x+7\right)$

36. $(2y+1)-\left(2y^2-2y-5\right)$

37. $\left(2x^2+3x-6\right)+\left(3x^2-4x+9\right)$

38. $\left(6a^2+3a+10\right)-\left(6a^2-3a+5\right)$

39. $(k+3)(k-6)$

40. $(2h+1)(3h-2)$

41. $(x+1)\left(x^2+1\right)$

42. $(m-2)\left(m^2+2m-3\right)$

43. $(a+2b)(3a-b)$

44. $(x+y)(x-y)$

Factoring Polynomials

For the following exercises, find the greatest common factor.

45. $81p+9pq-27p^2q^2$

46. $12x^2y+4xy^2-18xy$

47. $88a^3b+4a^2b-144a^2$

For the following exercises, factor the polynomial.

48. $2x^2-9x-18$

49. $8a^2+30a-27$

50. $d^2-5d-66$

51. $x^2+10x+25$

52. y^2-6y+9

53. $4h^2-12hk+9k^2$

54. $361x^2-121$

55. p^3+216

56. $8x^3-125$

57. $64q^3-27p^3$

58. $4x(x-1)^{-\frac{1}{4}}+3(x-1)^{\frac{3}{4}}$

59. $3p(p+3)^{\frac{1}{3}}-8(p+3)^{\frac{4}{3}}$

60. $4r(2r-1)^{-\frac{2}{3}} - 5(2r-1)^{\frac{1}{3}}$

Rational Expressions

For the following exercises, simplify the expression.

61. $\frac{x^2-x-12}{x^2-8x+16}$

62. $\frac{4y^2-25}{4y^2-20y+25}$

63. $\frac{2a^2-a-3}{2a^2-6a-8} \cdot \frac{5a^2-19a-4}{10a^2-13a-3}$

64. $\frac{d-4}{d^2-9} \cdot \frac{d-3}{d^2-16}$

65. $\frac{m^2+5m+6}{2m^2-5m-3} \div \frac{2m^2+3m-9}{4m^2-4m-3}$

66. $\frac{4d^2-7d-2}{6d^2-17d+10} \div \frac{8d^2+6d+1}{6d^2+7d-10}$

67. $\frac{10}{x} + \frac{6}{y}$

68. $\frac{12}{a^2+2a+1} - \frac{3}{a^2-1}$

69. $\frac{\frac{1}{d}+\frac{2}{c}}{\frac{6c+12d}{dc}}$

70. $\frac{\frac{3}{x}-\frac{7}{y}}{\frac{2}{x}}$

Practice Test

For the following exercises, identify the number as rational, irrational, whole, or natural. Choose the most descriptive answer.

1. -13

2. $\sqrt{2}$

For the following exercises, evaluate the expression.

3. $2(x+3)-12; x=2$

4. $y(3+3)^2-26; y=1$

5. Write the number in standard notation: 3.1415×10^6

6. Write the number in scientific notation: 0.0000000212.

For the following exercises, simplify the expression.

7. $-2 \cdot (2+3 \cdot 2)^2 + 144$

8. $4(x+3)-(6x+2)$

9. $3^5 \cdot 3^{-3}$

10. $\left(\frac{2}{3}\right)^3$

11. $\frac{8x^3}{(2x)^2}$

12. $\left(16y^0\right)2y^{-2}$

13. $\sqrt{441}$

14. $\sqrt{490}$

15. $\sqrt{\frac{9x}{16}}$

16. $\frac{\sqrt{121b^2}}{1+\sqrt{b}}$

17. $6\sqrt{24}+7\sqrt{54}-12\sqrt{6}$

18. $\frac{\sqrt[3]{-8}}{\sqrt[4]{625}}$

19. $\left(13q^3+2q^2-3\right)-\left(6q^2+5q-3\right)$

20. $\left(6p^2+2p+1\right)+\left(9p^2-1\right)$

21. $(n-2)\left(n^2-4n+4\right)$

22. $(a-2b)(2a+b)$

For the following exercises, factor the polynomial.

23. $16x^2-81$

24. $y^2+12y+36$

25. $27c^3-1331$

26. $3x(x-6)^{-\frac{1}{4}}+2(x-6)^{\frac{3}{4}}$

For the following exercises, simplify the expression.

27. $\frac{2z^2+7z+3}{z^2-9}\cdot\frac{4z^2-15z+9}{4z^2-1}$

28. $\frac{x}{y}+\frac{2}{x}$

29. $\frac{\frac{a}{2b}-\frac{2b}{9a}}{\frac{3a-2b}{6a}}$

2 EQUATIONS AND INEQUALITIES

From the air, a landscape of circular crop fields may seem random, but they are laid out and irrigated very precisely. Farmers and irrigation providers combining agricultural science, engineering, and mathematics to achieve the most productive and efficient array. (Credit: Modification of "Aerial Phot of Center Pivot Irrigations Systems (1)" by Soil Science/flickr)

Chapter Outline

Introduction to Equations and Inequalities

Irrigation is a critical aspect of agriculture, which can expand the yield of farms and enable farming in areas not naturally viable for crops. But the materials, equipment, and the water itself are expensive and complex. To be efficient and productive, farm owners and irrigation specialists must carefully lay out the network of pipes, pumps, and related equipment. The available land can be divided into regular portions (similar to a grid), and the different sizes of irrigation systems and conduits can be installed within the plotted area.

2.1 The Rectangular Coordinate Systems and Graphs

Learning Objectives

In this section, you will:

- Plot ordered pairs in a Cartesian coordinate system.
- Graph equations by plotting points.
- Graph equations with a graphing utility.
- Find x-intercepts and y-intercepts.
- Use the distance formula.
- Use the midpoint formula.

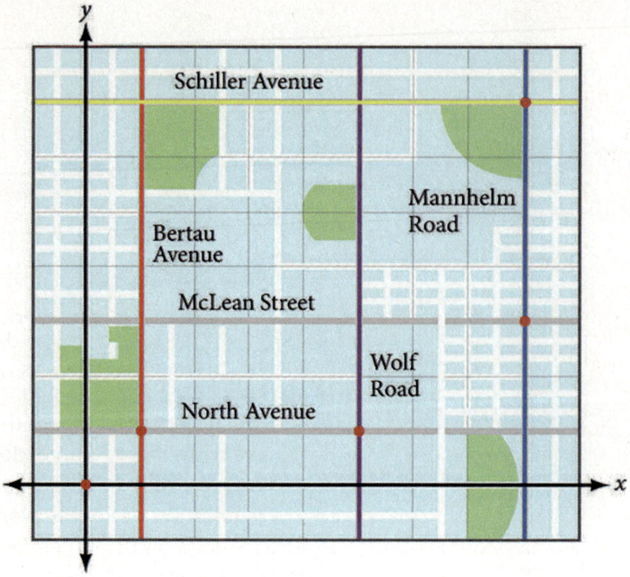

Figure 1

Tracie set out from Elmhurst, IL, to go to Franklin Park. On the way, she made a few stops to do errands. Each stop is indicated by a red dot in Figure 1. Laying a rectangular coordinate grid over the map, we can see that each stop aligns with an intersection of grid lines. In this section, we will learn how to use grid lines to describe locations and changes in locations.

Plotting Ordered Pairs in the Cartesian Coordinate System

An old story describes how seventeenth-century philosopher/mathematician René Descartes, while sick in bed, invented the system that has become the foundation of algebra. According to the story, Descartes was staring at a fly crawling on the ceiling when he realized that he could describe the fly's location in relation to the perpendicular lines formed by the adjacent walls of his room. He viewed the perpendicular lines as horizontal and vertical axes. Further, by dividing each axis into equal unit lengths, Descartes saw that it was possible to locate any object in a two-dimensional plane using just two numbers—the displacement from the horizontal axis and the displacement from the vertical axis.

While there is evidence that ideas similar to Descartes' grid system existed centuries earlier, it was Descartes who introduced the components that comprise the **Cartesian coordinate system**, a grid system having perpendicular axes. Descartes named the horizontal axis the ***x*-axis** and the vertical axis the ***y*-axis**.

The Cartesian coordinate system, also called the rectangular coordinate system, is based on a two-dimensional plane consisting of the *x*-axis and the *y*-axis. Perpendicular to each other, the axes divide the plane into four sections. Each section is called a **quadrant**; the quadrants are numbered counterclockwise as shown in Figure 2

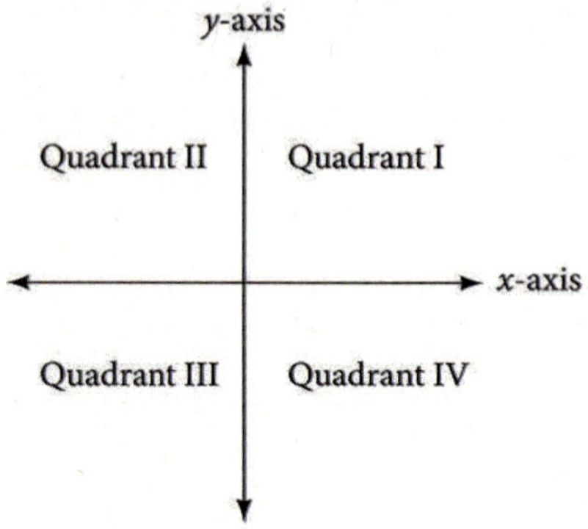

Figure 2

The center of the plane is the point at which the two axes cross. It is known as the **origin**, or point $(0, 0)$. From the origin, each axis is further divided into equal units: increasing, positive numbers to the right on the *x*-axis and up the *y*-axis; decreasing, negative numbers to the left on the *x*-axis and down the *y*-axis. The axes extend to positive and negative infinity as shown by the arrowheads in Figure 3.

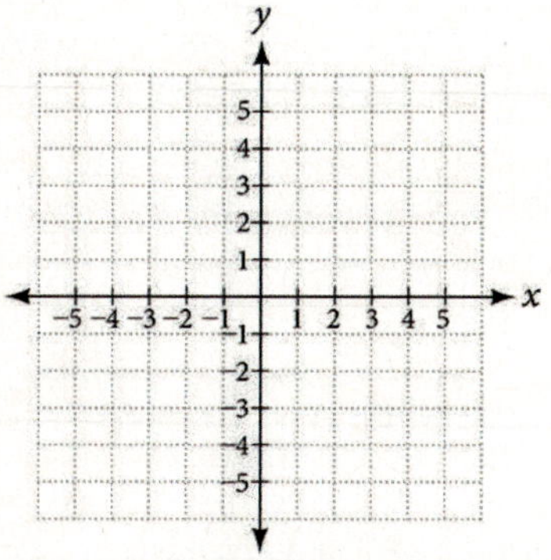

Figure 3

Each point in the plane is identified by its ***x*-coordinate**, or horizontal displacement from the origin, and its ***y*-coordinate**, or vertical displacement from the origin. Together, we write them as an **ordered pair** indicating the combined distance from the origin in the form (x, y). An ordered pair is also known as a coordinate pair because it consists of *x*- and *y*-coordinates. For example, we can represent the point $(3, -1)$ in the plane by moving three units to the right of the origin in the horizontal direction, and one unit down in the vertical direction. See Figure 4.

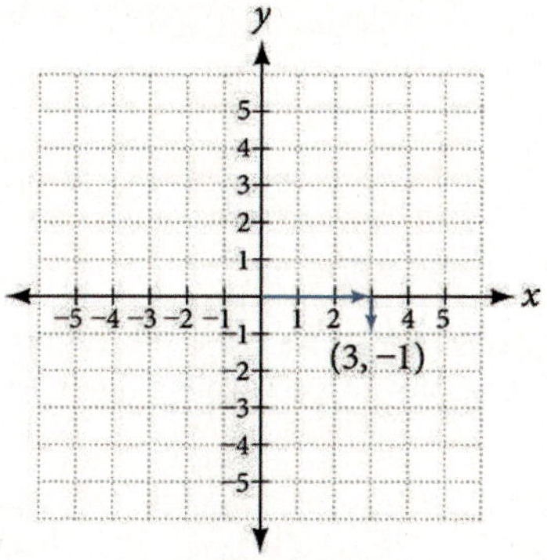

Figure 4

When dividing the axes into equally spaced increments, note that the *x*-axis may be considered separately from the *y*-axis. In other words, while the *x*-axis may be divided and labeled according to consecutive integers, the *y*-axis may be divided and labeled by increments of 2, or 10, or 100. In fact, the axes may represent other units, such as years against the balance in a savings account, or quantity against cost, and so on. Consider the rectangular coordinate system primarily as a method for showing the relationship between two quantities.

Cartesian Coordinate System

A two-dimensional plane where the

- *x*-axis is the horizontal axis
- *y*-axis is the vertical axis

A point in the plane is defined as an ordered pair, (x, y), such that *x* is determined by its horizontal distance from the origin and *y* is determined by its vertical distance from the origin.

EXAMPLE 1

Plotting Points in a Rectangular Coordinate System
Plot the points $(-2, 4)$, $(3, 3)$, and $(0, -3)$ in the plane.

Solution
To plot the point $(-2, 4)$, begin at the origin. The *x*-coordinate is –2, so move two units to the left. The *y*-coordinate is 4, so then move four units up in the positive *y* direction.

To plot the point $(3, 3)$, begin again at the origin. The *x*-coordinate is 3, so move three units to the right. The *y*-coordinate is also 3, so move three units up in the positive *y* direction.

To plot the point $(0, -3)$, begin again at the origin. The *x*-coordinate is 0. This tells us not to move in either direction along the *x*-axis. The *y*-coordinate is –3, so move three units down in the negative *y* direction. See the graph in Figure 5.

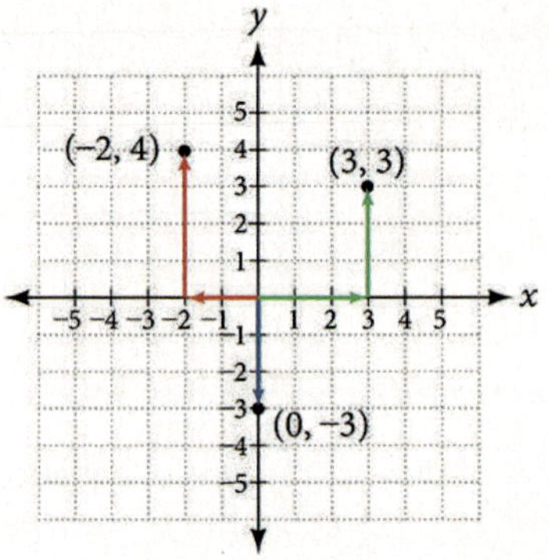

Figure 5

Analysis

Note that when either coordinate is zero, the point must be on an axis. If the *x*-coordinate is zero, the point is on the *y*-axis. If the *y*-coordinate is zero, the point is on the *x*-axis.

Graphing Equations by Plotting Points

We can plot a set of points to represent an equation. When such an equation contains both an *x* variable and a *y* variable, it is called an **equation in two variables**. Its graph is called a **graph in two variables**. Any graph on a two-dimensional plane is a graph in two variables.

Suppose we want to graph the equation $y = 2x - 1$. We can begin by substituting a value for *x* into the equation and determining the resulting value of *y*. Each pair of *x*- and *y*-values is an ordered pair that can be plotted. Table 1 lists values of *x* from –3 to 3 and the resulting values for *y*.

x	$y = 2x - 1$	(x, y)
-3	$y = 2(-3) - 1 = -7$	$(-3, -7)$
-2	$y = 2(-2) - 1 = -5$	$(-2, -5)$
-1	$y = 2(-1) - 1 = -3$	$(-1, -3)$
0	$y = 2(0) - 1 = -1$	$(0, -1)$
1	$y = 2(1) - 1 = 1$	$(1, 1)$
2	$y = 2(2) - 1 = 3$	$(2, 3)$
3	$y = 2(3) - 1 = 5$	$(3, 5)$

Table 1

We can plot the points in the table. The points for this particular equation form a line, so we can connect them. See Figure 6. This is not true for all equations.

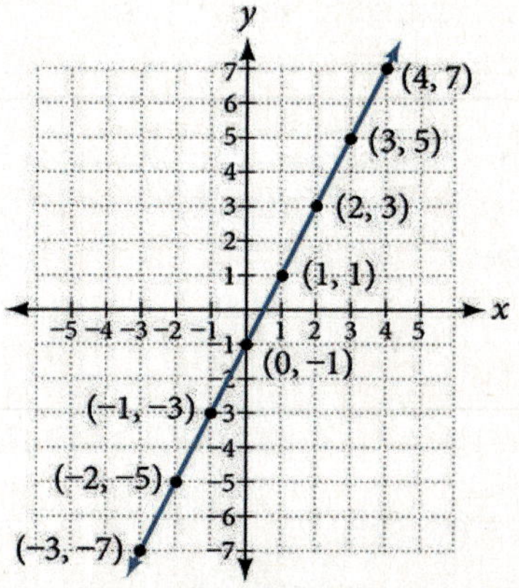

Figure 6

Note that the *x*-values chosen are arbitrary, regardless of the type of equation we are graphing. Of course, some situations may require particular values of *x* to be plotted in order to see a particular result. Otherwise, it is logical to choose values that can be calculated easily, and it is always a good idea to choose values that are both negative and positive. There is no rule dictating how many points to plot, although we need at least two to graph a line. Keep in mind, however, that the more points we plot, the more accurately we can sketch the graph.

HOW TO

Given an equation, graph by plotting points.

1. Make a table with one column labeled *x*, a second column labeled with the equation, and a third column listing the resulting ordered pairs.
2. Enter *x*-values down the first column using positive and negative values. Selecting the *x*-values in numerical order will make the graphing simpler.
3. Select *x*-values that will yield *y*-values with little effort, preferably ones that can be calculated mentally.
4. Plot the ordered pairs.
5. Connect the points if they form a line.

EXAMPLE 2

Graphing an Equation in Two Variables by Plotting Points

Graph the equation $y = -x + 2$ by plotting points.

Solution

First, we construct a table similar to Table 2. Choose *x* values and calculate *y*.

x	$y = -x + 2$	(x, y)
-5	$y = -(-5) + 2 = 7$	$(-5, 7)$
-3	$y = -(-3) + 2 = 5$	$(-3, 5)$
-1	$y = -(-1) + 2 = 3$	$(-1, 3)$
0	$y = -(0) + 2 = 2$	$(0, 2)$
1	$y = -(1) + 2 = 1$	$(1, 1)$

Table 2

3	$y = -(3) + 2 = -1$	$(3, -1)$
5	$y = -(5) + 2 = -3$	$(5, -3)$

Table 2

Now, plot the points. Connect them if they form a line. See Figure 7

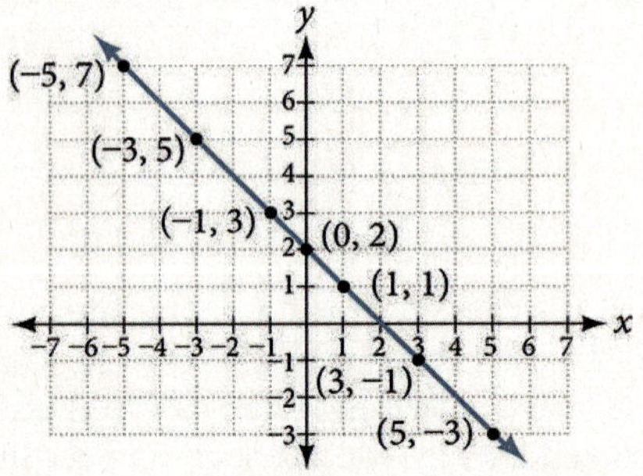

Figure 7

> TRY IT #1 Construct a table and graph the equation by plotting points: $y = \frac{1}{2}x + 2$.

Graphing Equations with a Graphing Utility

Most graphing calculators require similar techniques to graph an equation. The equations sometimes have to be manipulated so they are written in the style $y =$______. The TI-84 Plus, and many other calculator makes and models, have a mode function, which allows the window (the screen for viewing the graph) to be altered so the pertinent parts of a graph can be seen.

For example, the equation $y = 2x - 20$ has been entered in the TI-84 Plus shown in Figure 8**a.** In Figure 8**b,** the resulting graph is shown. Notice that we cannot see on the screen where the graph crosses the axes. The standard window screen on the TI-84 Plus shows $-10 \leq x \leq 10$, and $-10 \leq y \leq 10$. See Figure 8**c.**

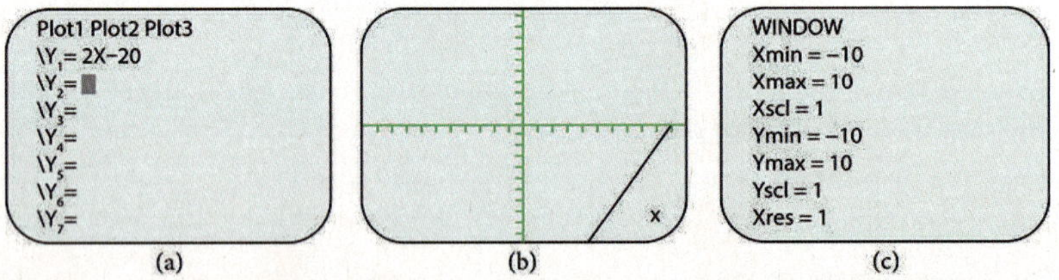

Figure 8 a. Enter the equation. b. This is the graph in the original window. c. These are the original settings.

By changing the window to show more of the positive *x*-axis and more of the negative *y*-axis, we have a much better view of the graph and the *x*- and *y*-intercepts. See Figure 9**a** and Figure 9**b.**

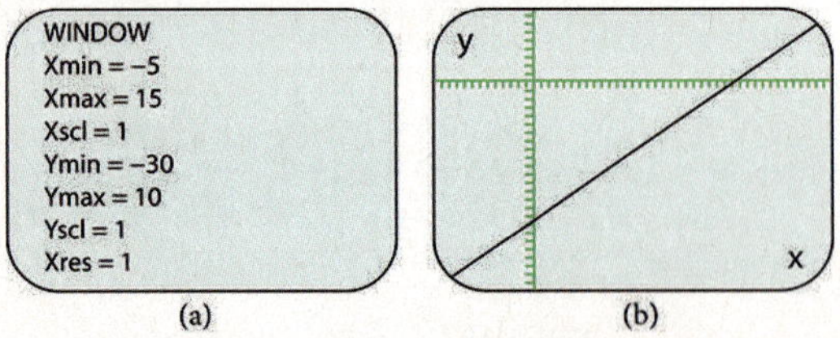

Figure 9 a. This screen shows the new window settings. b. We can clearly view the intercepts in the new window.

EXAMPLE 3

Using a Graphing Utility to Graph an Equation

Use a graphing utility to graph the equation: $y = -\frac{2}{3}x + \frac{4}{3}$.

Solution

Enter the equation in the *y*= function of the calculator. Set the window settings so that both the *x*- and *y*- intercepts are showing in the window. See Figure 10.

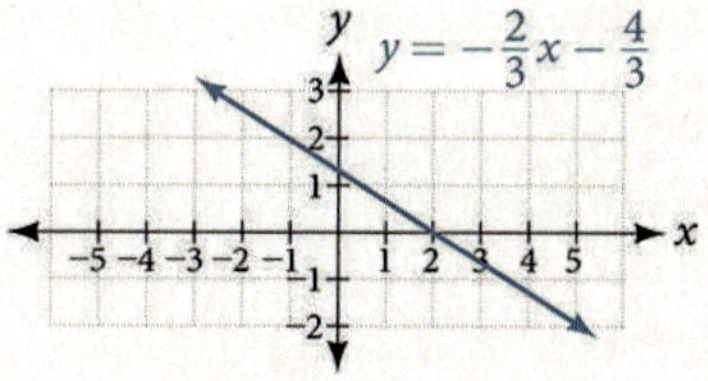

Figure 10

Finding *x*-intercepts and *y*-intercepts

The **intercepts** of a graph are points at which the graph crosses the axes. The **x-intercept** is the point at which the graph crosses the *x*-axis. At this point, the *y*-coordinate is zero. The **y-intercept** is the point at which the graph crosses the *y*-axis. At this point, the *x*-coordinate is zero.

To determine the *x*-intercept, we set *y* equal to zero and solve for *x*. Similarly, to determine the *y*-intercept, we set *x* equal to zero and solve for *y*. For example, lets find the intercepts of the equation $y = 3x - 1$.

To find the *x*-intercept, set $y = 0$.

$$\begin{aligned} y &= 3x - 1 \\ 0 &= 3x - 1 \\ 1 &= 3x \\ \frac{1}{3} &= x \\ &\left(\frac{1}{3}, 0\right) \qquad x\text{–intercept} \end{aligned}$$

To find the *y*-intercept, set $x = 0$.

$$\begin{aligned} y &= 3x - 1 \\ y &= 3(0) - 1 \\ y &= -1 \\ &(0, -1) \qquad y\text{–intercept} \end{aligned}$$

We can confirm that our results make sense by observing a graph of the equation as in Figure 11. Notice that the graph crosses the axes where we predicted it would.

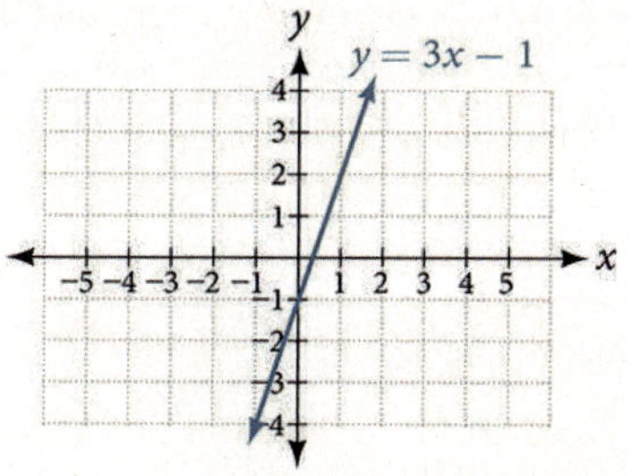

Figure 11

Given an equation, find the intercepts.

- Find the *x*-intercept by setting $y = 0$ and solving for x.

- Find the y-intercept by setting $x = 0$ and solving for y.

EXAMPLE 4

Finding the Intercepts of the Given Equation

Find the intercepts of the equation $y = -3x - 4$. Then sketch the graph using only the intercepts.

Solution

Set $y = 0$ to find the x-intercept.

$$\begin{aligned} y &= -3x - 4 \\ 0 &= -3x - 4 \\ 4 &= -3x \\ -\tfrac{4}{3} &= x \\ \left(-\tfrac{4}{3}, 0\right) & \qquad x\text{–intercept} \end{aligned}$$

Set $x = 0$ to find the y-intercept.

$$\begin{aligned} y &= -3x - 4 \\ y &= -3(0) - 4 \\ y &= -4 \\ (0, -4) & \qquad y\text{–intercept} \end{aligned}$$

Plot both points, and draw a line passing through them as in Figure 12.

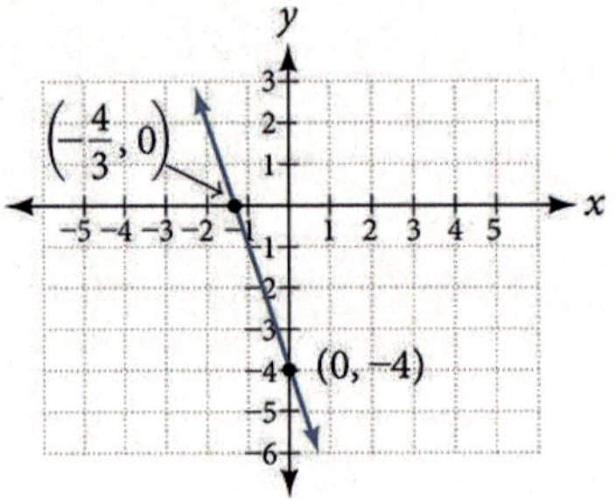

Figure 12

TRY IT #2 Find the intercepts of the equation and sketch the graph: $y = -\frac{3}{4}x + 3$.

Using the Distance Formula

Derived from the Pythagorean Theorem, the **distance formula** is used to find the distance between two points in the plane. The Pythagorean Theorem, $a^2 + b^2 = c^2$, is based on a right triangle where a and b are the lengths of the legs adjacent to the right angle, and c is the length of the hypotenuse. See Figure 13.

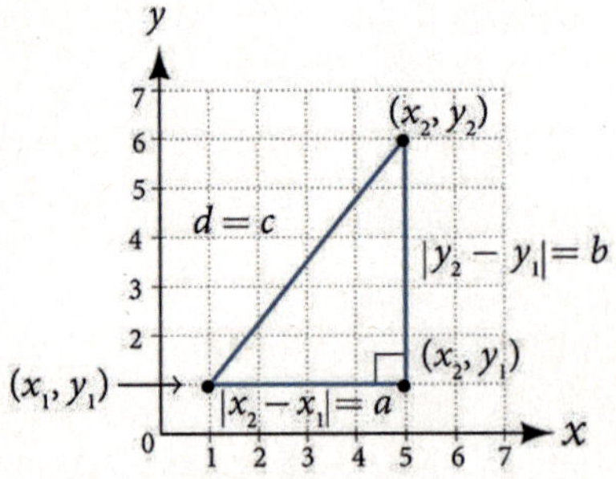

Figure 13

The relationship of sides $|x_2 - x_1|$ and $|y_2 - y_1|$ to side *d* is the same as that of sides *a* and *b* to side *c*. We use the absolute value symbol to indicate that the length is a positive number because the absolute value of any number is positive. (For example, $|-3| = 3$.) The symbols $|x_2 - x_1|$ and $|y_2 - y_1|$ indicate that the lengths of the sides of the triangle are positive. To find the length *c*, take the square root of both sides of the Pythagorean Theorem.

$$c^2 = a^2 + b^2 \to c = \sqrt{a^2 + b^2}$$

It follows that the distance formula is given as

$$d^2 = (x_2 - x_1)^2 + (y_2 - y_1)^2 \to d = \sqrt{(x_2 - x_1)^2 + (y_2 - y_1)^2}$$

We do not have to use the absolute value symbols in this definition because any number squared is positive.

The Distance Formula

Given endpoints (x_1, y_1) and (x_2, y_2), the distance between two points is given by

$$d = \sqrt{(x_2 - x_1)^2 + (y_2 - y_1)^2}$$

EXAMPLE 5

Finding the Distance between Two Points

Find the distance between the points $(-3, -1)$ and $(2, 3)$.

Solution

Let us first look at the graph of the two points. Connect the points to form a right triangle as in Figure 14.

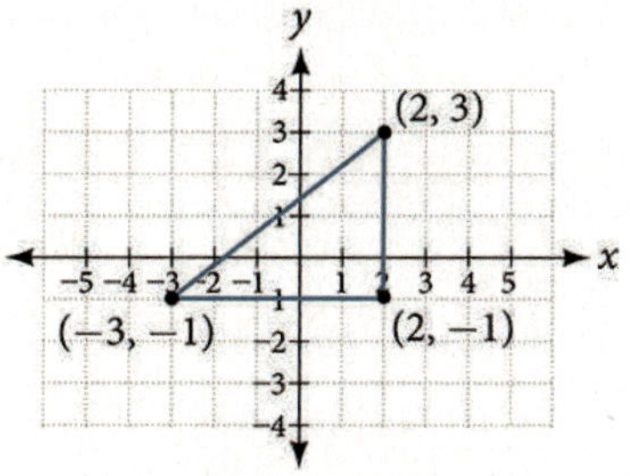

Figure 14

Then, calculate the length of *d* using the distance formula.

$$\begin{aligned} d &= \sqrt{(x_2 - x_1)^2 + (y_2 - y_1)^2} \\ d &= \sqrt{(2 - (-3))^2 + (3 - (-1))^2} \\ &= \sqrt{(5)^2 + (4)^2} \\ &= \sqrt{25 + 16} \\ &= \sqrt{41} \end{aligned}$$

TRY IT #3 Find the distance between two points: $(1, 4)$ and $(11, 9)$.

EXAMPLE 6

Finding the Distance between Two Locations

Let’s return to the situation introduced at the beginning of this section.

Tracie set out from Elmhurst, IL, to go to Franklin Park. On the way, she made a few stops to do errands. Each stop is indicated by a red dot in Figure 1. Find the total distance that Tracie traveled. Compare this with the distance between her starting and final positions.

Solution

The first thing we should do is identify ordered pairs to describe each position. If we set the starting position at the origin, we can identify each of the other points by counting units east (right) and north (up) on the grid. For example, the first stop is 1 block east and 1 block north, so it is at $(1, 1)$. The next stop is 5 blocks to the east, so it is at $(5, 1)$. After that, she traveled 3 blocks east and 2 blocks north to $(8, 3)$. Lastly, she traveled 4 blocks north to $(8, 7)$. We can label these points on the grid as in Figure 15.

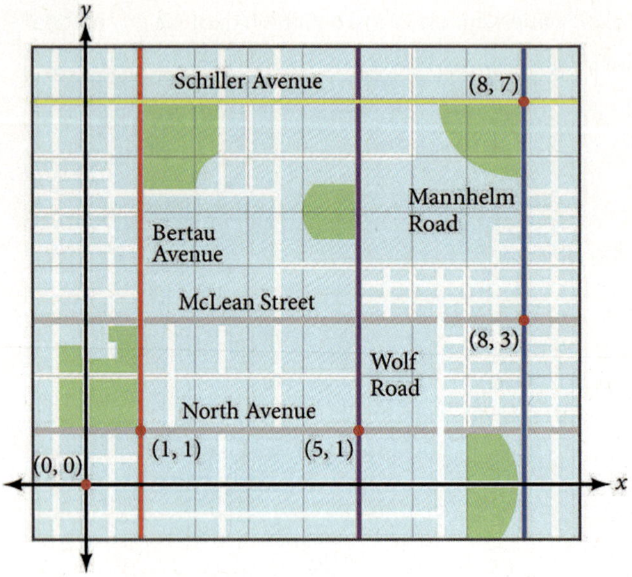

Figure 15

Next, we can calculate the distance. Note that each grid unit represents 1,000 feet.

- From her starting location to her first stop at $(1, 1)$, Tracie might have driven north 1,000 feet and then east 1,000 feet, or vice versa. Either way, she drove 2,000 feet to her first stop.
- Her second stop is at $(5, 1)$. So from $(1, 1)$ to $(5, 1)$, Tracie drove east 4,000 feet.
- Her third stop is at $(8, 3)$. There are a number of routes from $(5, 1)$ to $(8, 3)$. Whatever route Tracie decided to use, the distance is the same, as there are no angular streets between the two points. Let's say she drove east 3,000 feet and then north 2,000 feet for a total of 5,000 feet.
- Tracie's final stop is at $(8, 7)$. This is a straight drive north from $(8, 3)$ for a total of 4,000 feet.

Next, we will add the distances listed in Table 3.

From/To	Number of Feet Driven
$(0, 0)$ to $(1, 1)$	2,000
$(1, 1)$ to $(5, 1)$	4,000
$(5, 1)$ to $(8, 3)$	5,000
$(8, 3)$ to $(8, 7)$	4,000
Total	15,000

Table 3

The total distance Tracie drove is 15,000 feet, or 2.84 miles. This is not, however, the actual distance between her starting and ending positions. To find this distance, we can use the distance formula between the points $(0, 0)$ and $(8, 7)$.

$$\begin{aligned} d &= \sqrt{(8-0)^2 + (7-0)^2} \\ &= \sqrt{64+49} \\ &= \sqrt{113} \\ &\approx 10.63 \text{ units} \end{aligned}$$

At 1,000 feet per grid unit, the distance between Elmhurst, IL, to Franklin Park is 10,630.14 feet, or 2.01 miles. The distance formula results in a shorter calculation because it is based on the hypotenuse of a right triangle, a straight diagonal from the origin to the point $(8, 7)$. Perhaps you have heard the saying "as the crow flies," which means the shortest distance between two points because a crow can fly in a straight line even though a person on the ground has to travel a longer distance on existing roadways.

Using the Midpoint Formula

When the endpoints of a line segment are known, we can find the point midway between them. This point is known as the midpoint and the formula is known as the **midpoint formula**. Given the endpoints of a line segment, (x_1, y_1) and (x_2, y_2), the midpoint formula states how to find the coordinates of the midpoint M.

$$M = \left(\frac{x_1 + x_2}{2}, \frac{y_1 + y_2}{2}\right)$$

A graphical view of a midpoint is shown in Figure 16. Notice that the line segments on either side of the midpoint are congruent.

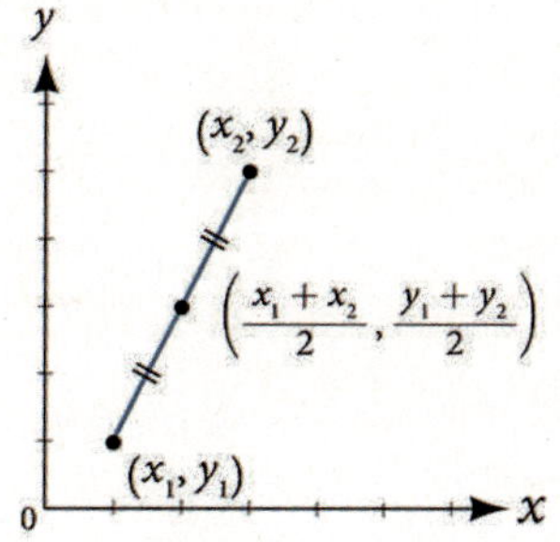

Figure 16

EXAMPLE 7

Finding the Midpoint of the Line Segment

Find the midpoint of the line segment with the endpoints $(7, -2)$ and $(9, 5)$.

Solution

Use the formula to find the midpoint of the line segment.

$$\begin{aligned} \left(\frac{x_1+x_2}{2}, \frac{y_1+y_2}{2}\right) &= \left(\frac{7+9}{2}, \frac{-2+5}{2}\right) \\ &= \left(8, \frac{3}{2}\right) \end{aligned}$$

TRY IT #4 Find the midpoint of the line segment with endpoints $(-2, -1)$ and $(-8, 6)$.

EXAMPLE 8

Finding the Center of a Circle

The diameter of a circle has endpoints $(-1, -4)$ and $(5, -4)$. Find the center of the circle.

Solution

The center of a circle is the center, or midpoint, of its diameter. Thus, the midpoint formula will yield the center point.

$$\left(\frac{x_1+x_2}{2}, \frac{y_1+y_2}{2}\right)$$

$$\left(\frac{-1+5}{2}, \frac{-4-4}{2}\right) = \left(\frac{4}{2}, -\frac{8}{2}\right) = (2, -4)$$

MEDIA

Access these online resources for additional instruction and practice with the Cartesian coordinate system.

Plotting points on the coordinate plane (http://openstax.org/l/coordplotpnts)
Find x and y intercepts based on the graph of a line (http://openstax.org/l/xyintsgraph)

2.1 SECTION EXERCISES

Verbal

1. Is it possible for a point plotted in the Cartesian coordinate system to not lie in one of the four quadrants? Explain.

2. Describe the process for finding the *x*-intercept and the *y*-intercept of a graph algebraically.

3. Describe in your own words what the *y*-intercept of a graph is.

4. When using the distance formula $d = \sqrt{(x_2 - x_1)^2 + (y_2 - y_1)^2}$, explain the correct order of operations that are to be performed to obtain the correct answer.

Algebraic

For each of the following exercises, find the x-intercept and the y-intercept without graphing. Write the coordinates of each intercept.

5. $y = -3x + 6$

6. $4y = 2x - 1$

7. $3x - 2y = 6$

8. $4x - 3 = 2y$

9. $3x + 8y = 9$

10. $2x - \frac{2}{3} = \frac{3}{4}y + 3$

For each of the following exercises, solve the equation for y in terms of x.

11. $4x + 2y = 8$

12. $3x - 2y = 6$

13. $2x = 5 - 3y$

14. $x - 2y = 7$

15. $5y + 4 = 10x$

16. $5x + 2y = 0$

For each of the following exercises, find the distance between the two points. Simplify your answers, and write the exact answer in simplest radical form for irrational answers.

17. $(-4, 1)$ and $(3, -4)$

18. $(2, -5)$ and $(7, 4)$

19. $(5, 0)$ and $(5, 6)$

20. $(-4, 3)$ and $(10, 3)$

21. Find the distance between the two points given using your calculator, and round your answer to the nearest hundredth.

$(19, 12)$ and $(41, 71)$

For each of the following exercises, find the coordinates of the midpoint of the line segment that joins the two given points.

22. $(-5, -6)$ and $(4, 2)$

23. $(-1, 1)$ and $(7, -4)$

24. $(-5, -3)$ and $(-2, -8)$

25. $(0, 7)$ and $(4, -9)$

26. $(-43, 17)$ and $(23, -34)$

Graphical

For each of the following exercises, identify the information requested.

27. What are the coordinates of the origin?

28. If a point is located on the *y*-axis, what is the *x*-coordinate?

29. If a point is located on the *x*-axis, what is the *y*-coordinate?

For each of the following exercises, plot the three points on the given coordinate plane. State whether the three points you plotted appear to be collinear (on the same line).

30. $(4, 1)\,(-2, -3)\,(5, 0)$

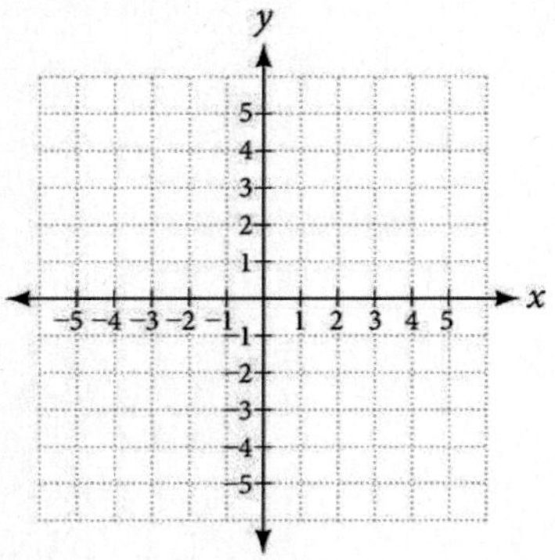

31. $(-1, 2)(0, 4)(2, 1)$

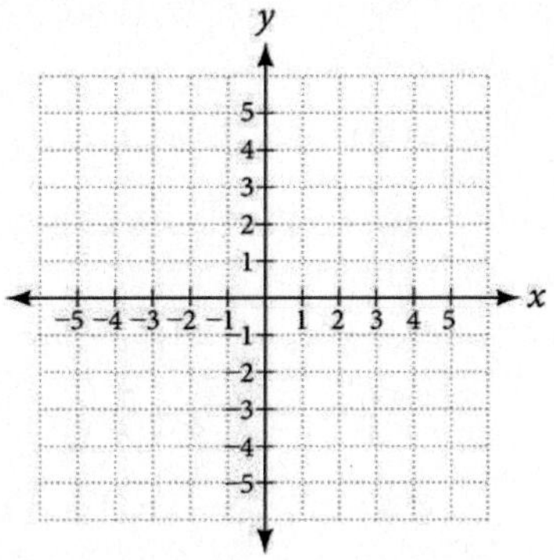

32. $(-3, 0)\,(-3, 4)\,(-3, -3)$

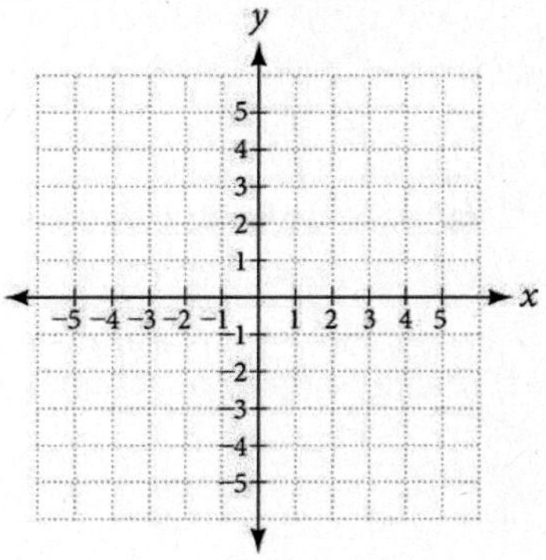

33. Name the coordinates of the points graphed.

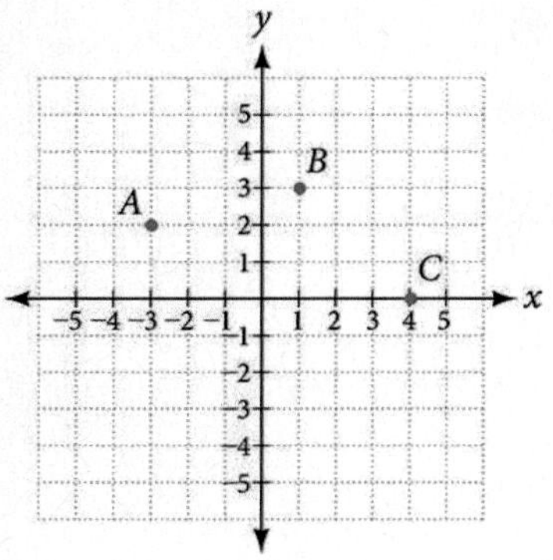

34. Name the quadrant in which the following points would be located. If the point is on an axis, name the axis.

ⓐ $(-3, -4)$ ⓑ $(-5, 0)$
ⓒ $(1, -4)$ ⓓ $(-2, 7)$
ⓔ $(0, -3)$

For each of the following exercises, construct a table and graph the equation by plotting at least three points.

35. $y = \frac{1}{3}x + 2$

36. $y = -3x + 1$

37. $2y = x + 3$

Numeric

For each of the following exercises, find and plot the x- and y-intercepts, and graph the straight line based on those two points.

38. $4x - 3y = 12$

39. $x - 2y = 8$

40. $y - 5 = 5x$

41. $3y = -2x + 6$

42. $y = \frac{x-3}{2}$

For each of the following exercises, use the graph in the figure below.

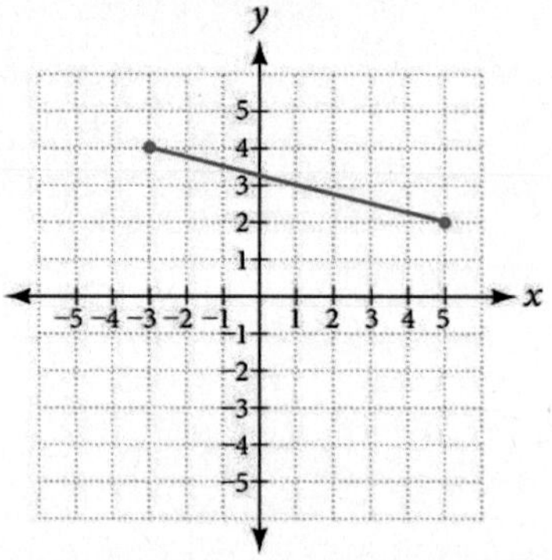

43. Find the distance between the two endpoints using the distance formula. Round to three decimal places.

44. Find the coordinates of the midpoint of the line segment connecting the two points.

45. Find the distance that $(-3, 4)$ is from the origin.

46. Find the distance that $(5, 2)$ is from the origin. Round to three decimal places.

47. Which point is closer to the origin?

Technology

For the following exercises, use your graphing calculator to input the linear graphs in the Y= graph menu.

After graphing it, use the 2nd CALC button and 1:value button, hit enter. At the lower part of the screen you will see "x=" and a blinking cursor. You may enter any number for x and it will display the y value for any x value you input. Use this and plug in x = 0, thus finding the y-intercept, for each of the following graphs.

48. $Y_1 = -2x + 5$

49. $Y_1 = \frac{3x-8}{4}$

50. $Y_1 = \frac{x+5}{2}$

For the following exercises, use your graphing calculator to input the linear graphs in the Y= graph menu.

*After graphing it, use the 2nd **CALC** button and 2:zero button, hit **ENTER**. At the lower part of the screen you will see "left bound?" and a blinking cursor on the graph of the line. Move this cursor to the left of the x-intercept, hit **ENTER**. Now it says "right bound?" Move the cursor to the right of the x-intercept, hit **ENTER**. Now it says "guess?" Move your cursor to the left somewhere in between the left and right bound near the x-intercept. Hit **ENTER**. At the bottom of your screen it will display the coordinates of the x-intercept or the "zero" to the y-value. Use this to find the x-intercept.*

Note: With linear/straight line functions the zero is not really a "guess," but it is necessary to enter a "guess" so it will search and find the exact x-intercept between your right and left boundaries. With other types of functions (more than one x-intercept), they may be irrational numbers so "guess" is more appropriate to give it the correct limits to find a very close approximation between the left and right boundaries.

51. $Y_1 = -8x + 6$

52. $Y_1 = 4x - 7$

53. $Y_1 = \frac{3x+5}{4}$ Round your answer to the nearest thousandth.

Extensions

54. Someone drove 10 mi directly east from their home, made a left turn at an intersection, and then traveled 5 mi north to their place of work. If a road was made directly from the home to the place of work, what would its distance be to the nearest tenth of a mile?

55. If the road was made in the previous exercise, how much shorter would the person's one-way trip be every day?

56. Given these four points: $A(1, 3)$, $B(-3, 5)$, $C(4, 7)$, and $D(5, -4)$ find the coordinates of the midpoint of line segments $\overline{AB}$ and $\overline{CD}$.

57. After finding the two midpoints in the previous exercise, find the distance between the two midpoints to the nearest thousandth.

58. Given the graph of the rectangle shown and the coordinates of its vertices, prove that the diagonals of the rectangle are of equal length.

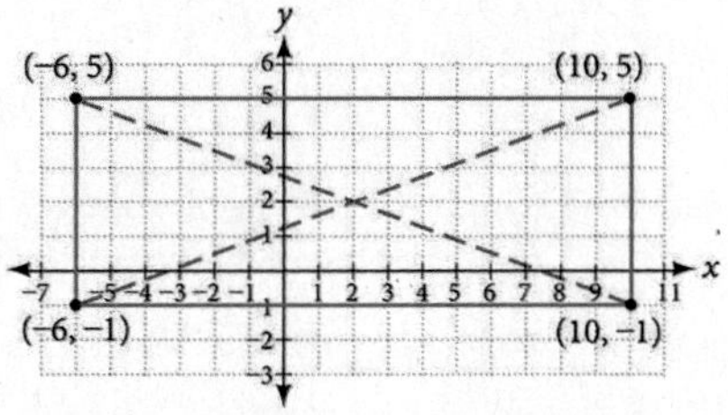

59. In the previous exercise, find the coordinates of the midpoint for each diagonal.

Real-World Applications

60. The coordinates on a map for San Francisco are $(53, 17)$ and those for Sacramento are $(128, 78)$. Note that coordinates represent miles. Find the distance between the cities to the nearest mile.

61. If San Jose's coordinates are $(76, -12)$, where the coordinates represent miles, find the distance between San Jose and San Francisco to the nearest mile.

62. A small craft in Lake Ontario sends out a distress signal. The coordinates of the boat in trouble were $(49, 64)$ One rescue boat is at the coordinates $(60, 82)$ and a second Coast Guard craft is at coordinates $(58, 47)$. Assuming both rescue craft travel at the same rate, which one would get to the distressed boat the fastest?

63. A person on the top of a building wants to have a guy wire extend to a point on the ground 20 ft from the building. To the nearest foot, how long will the wire have to be if the building is 50 ft tall?

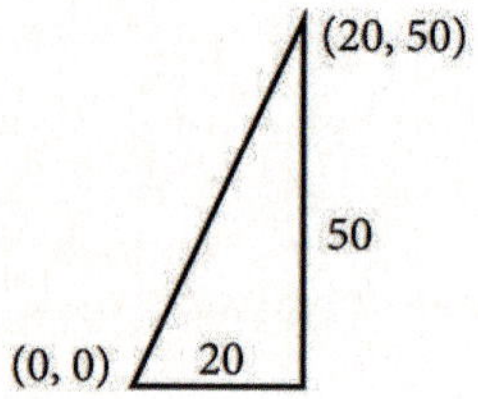

64. If we rent a truck and pay a \$75/day fee plus \$.20 for every mile we travel, write a linear equation that would express the total cost per day y, using x to represent the number of miles we travel. Graph this function on your graphing calculator and find the total cost for one day if we travel 70 mi.

2.2 Linear Equations in One Variable

Learning Objectives

In this section, you will:

- Solve equations in one variable algebraically.
- Solve a rational equation.
- Find a linear equation.
- Given the equations of two lines, determine whether their graphs are parallel or perpendicular.
- Write the equation of a line parallel or perpendicular to a given line.

Caroline is a full-time college student planning a spring break vacation. To earn enough money for the trip, she has taken a part-time job at the local bank that pays \$15.00/hr, and she opened a savings account with an initial deposit of \$400 on January 15. She arranged for direct deposit of her payroll checks. If spring break begins March 20 and the trip will cost approximately \$2,500, how many hours will she have to work to earn enough to pay for her vacation? If she can only work 4 hours per day, how many days per week will she have to work? How many weeks will it take? In this section, we will investigate problems like this and others, which generate graphs like the line in Figure 1.

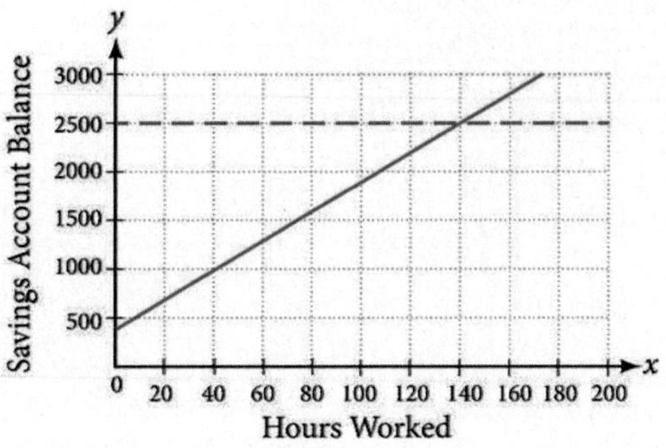

Figure 1

Solving Linear Equations in One Variable

A **linear equation** is an equation of a straight line, written in one variable. The only power of the variable is 1. Linear equations in one variable may take the form $ax + b = 0$ and are solved using basic algebraic operations.

We begin by classifying linear equations in one variable as one of three types: identity, conditional, or inconsistent. An **identity equation** is true for all values of the variable. Here is an example of an identity equation.

$$3x = 2x + x$$

The **solution set** consists of all values that make the equation true. For this equation, the solution set is all real numbers because any real number substituted for x will make the equation true.

A **conditional equation** is true for only some values of the variable. For example, if we are to solve the equation $5x + 2 = 3x - 6$, we have the following:

$$\begin{aligned} 5x + 2 &= 3x - 6 \\ 2x &= -8 \\ x &= -4 \end{aligned}$$

The solution set consists of one number: $\{-4\}$. It is the only solution and, therefore, we have solved a conditional equation.

An **inconsistent equation** results in a false statement. For example, if we are to solve $5x - 15 = 5(x - 4)$, we have the following:

$$\begin{aligned} 5x - 15 &= 5x - 20 & & \\ 5x - 15 - 5x &= 5x - 20 - 5x & & \text{Subtract } 5x \text{ from both sides.} \\ -15 &\neq -20 & & \text{False statement} \end{aligned}$$

Indeed, $-15 \neq -20$. There is no solution because this is an inconsistent equation.

Solving linear equations in one variable involves the fundamental properties of equality and basic algebraic operations. A brief review of those operations follows.

Linear Equation in One Variable

A linear equation in one variable can be written in the form

$$ax + b = 0$$

where a and b are real numbers, $a \neq 0$.

HOW TO

Given a linear equation in one variable, use algebra to solve it.

The following steps are used to manipulate an equation and isolate the unknown variable, so that the last line reads $x =$ __________, if x is the unknown. There is no set order, as the steps used depend on what is given:

1. We may add, subtract, multiply, or divide an equation by a number or an expression as long as we do the same thing to both sides of the equal sign. Note that we cannot divide by zero.
2. Apply the distributive property as needed: $a(b+c) = ab + ac$.
3. Isolate the variable on one side of the equation.
4. When the variable is multiplied by a coefficient in the final stage, multiply both sides of the equation by the reciprocal of the coefficient.

EXAMPLE 1

Solving an Equation in One Variable

Solve the following equation: $2x + 7 = 19$.

Solution

This equation can be written in the form $ax + b = 0$ by subtracting 19 from both sides. However, we may proceed to solve the equation in its original form by performing algebraic operations.

$$\begin{aligned} 2x+7 &= 19 \\ 2x &= 12 && \text{Subtract 7 from both sides.} \\ x &= 6 && \text{Multiply both sides by } \tfrac{1}{2} \text{ or divide by 2.} \end{aligned}$$

The solution is 6.

TRY IT #1 Solve the linear equation in one variable: $2x + 1 = -9$.

EXAMPLE 2

Solving an Equation Algebraically When the Variable Appears on Both Sides

Solve the following equation: $4(x-3) + 12 = 15 - 5(x+6)$.

Solution

Apply standard algebraic properties.

$$\begin{aligned} 4(x-3)+12 &= 15-5(x+6) \\ 4x-12+12 &= 15-5x-30 && \text{Apply the distributive property.} \\ 4x &= -15-5x && \text{Combine like terms.} \\ 9x &= -15 && \text{Place } x\text{-terms on one side and simplify.} \\ x &= -\tfrac{15}{9} && \text{Multiply both sides by } \tfrac{1}{9}\text{, the reciprocal of 9.} \\ x &= -\tfrac{5}{3} \end{aligned}$$

Analysis

This problem requires the distributive property to be applied twice, and then the properties of algebra are used to reach the final line, $x = -\frac{5}{3}$.

TRY IT #2 Solve the equation in one variable: $-2(3x-1) + x = 14 - x$.

Solving a Rational Equation

In this section, we look at rational equations that, after some manipulation, result in a linear equation. If an equation contains at least one rational expression, it is a considered a **rational equation**.

Recall that a rational number is the ratio of two numbers, such as $\frac{2}{3}$ or $\frac{7}{2}$. A rational expression is the ratio, or quotient, of two polynomials. Here are three examples.

$$\frac{x+1}{x^2-4},\ \frac{1}{x-3},\ \text{or}\ \frac{4}{x^2+x-2}$$

Rational equations have a variable in the denominator in at least one of the terms. Our goal is to perform algebraic operations so that the variables appear in the numerator. In fact, we will eliminate all denominators by multiplying both sides of the equation by the least common denominator (LCD).

Finding the LCD is identifying an expression that contains the highest power of all of the factors in all of the denominators. We do this because when the equation is multiplied by the LCD, the common factors in the LCD and in each denominator will equal one and will cancel out.

EXAMPLE 3

Solving a Rational Equation

Solve the rational equation: $\frac{7}{2x} - \frac{5}{3x} = \frac{22}{3}$.

Solution

We have three denominators; $2x$, $3x$, and 3. The LCD must contain $2x$, $3x$, and 3. An LCD of $6x$ contains all three denominators. In other words, each denominator can be divided evenly into the LCD. Next, multiply both sides of the equation by the LCD $6x$.

$$\begin{aligned}
(6x)\left(\frac{7}{2x} - \frac{5}{3x}\right) &= \left(\frac{22}{3}\right)(6x) \\
(6x)\left(\frac{7}{2x}\right) - (6x)\left(\frac{5}{3x}\right) &= \left(\frac{22}{3}\right)(6x) && \text{Use the distributive property.} \\
(\cancel{6x})\left(\frac{7}{\cancel{2x}}\right) - (\cancel{6x})\left(\frac{5}{\cancel{3x}}\right) &= \left(\frac{22}{\cancel{3}}\right)(\cancel{6}x) && \text{Cancel out the common factors.} \\
3(7) - 2(5) &= 22(2x) && \text{Multiply remaining factors by each numerator.} \\
21 - 10 &= 44x \\
11 &= 44x \\
\frac{11}{44} &= x \\
\frac{1}{4} &= x
\end{aligned}$$

A common mistake made when solving rational equations involves finding the LCD when one of the denominators is a binomial—two terms added or subtracted—such as $(x+1)$. Always consider a binomial as an individual factor—the terms cannot be separated. For example, suppose a problem has three terms and the denominators are x, $x-1$, and $3x-3$. First, factor all denominators. We then have x, $(x-1)$, and $3(x-1)$ as the denominators. (Note the parentheses placed around the second denominator.) Only the last two denominators have a common factor of $(x-1)$. The x in the first denominator is separate from the x in the $(x-1)$ denominators. An effective way to remember this is to write factored and binomial denominators in parentheses, and consider each parentheses as a separate unit or a separate factor. The LCD in this instance is found by multiplying together the x, one factor of $(x-1)$, and the 3. Thus, the LCD is the following:

$$x(x-1)3 = 3x(x-1)$$

So, both sides of the equation would be multiplied by $3x(x-1)$. Leave the LCD in factored form, as this makes it easier to see how each denominator in the problem cancels out.

Another example is a problem with two denominators, such as x and x^2+2x. Once the second denominator is factored as $x^2+2x = x(x+2)$, there is a common factor of x in both denominators and the LCD is $x(x+2)$.

Sometimes we have a rational equation in the form of a proportion; that is, when one fraction equals another fraction and there are no other terms in the equation.

$$\frac{a}{b} = \frac{c}{d}$$

We can use another method of solving the equation without finding the LCD: cross-multiplication. We multiply terms by crossing over the equal sign.

If $\frac{a}{b} = \frac{c}{d}$, then $\frac{a}{b} \times \frac{c}{d}$.

Multiply $a\,(d)$ and $b\,(c)$, which results in $ad = bc$.

Any solution that makes a denominator in the original expression equal zero must be excluded from the possibilities.

Rational Equations

A **rational equation** contains at least one rational expression where the variable appears in at least one of the denominators.

HOW TO

Given a rational equation, solve it.

1. Factor all denominators in the equation.
2. Find and exclude values that set each denominator equal to zero.
3. Find the LCD.
4. Multiply the whole equation by the LCD. If the LCD is correct, there will be no denominators left.
5. Solve the remaining equation.
6. Make sure to check solutions back in the original equations to avoid a solution producing zero in a denominator.

EXAMPLE 4

Solving a Rational Equation without Factoring

Solve the following rational equation:

$$\frac{2}{x} - \frac{3}{2} = \frac{7}{2x}$$

Solution

We have three denominators: x, 2, and $2x$. No factoring is required. The product of the first two denominators is equal to the third denominator, so, the LCD is $2x$. Only one value is excluded from a solution set, 0. Next, multiply the whole equation (both sides of the equal sign) by $2x$.

$$\begin{aligned}
2x\left(\tfrac{2}{x} - \tfrac{3}{2}\right) &= \left(\tfrac{7}{2x}\right)2x \\
2\cancel{x}\left(\frac{2}{\cancel{x}}\right) - \cancel{2}x\left(\frac{3}{\cancel{2}}\right) &= \left(\frac{7}{\cancel{2x}}\right)\cancel{2x} && \text{Distribute } 2x. \\
2(2) - 3x &= 7 && \text{Denominators cancel out.} \\
4 - 3x &= 7 \\
-3x &= 3 \\
x &= -1 \\
& \quad \text{or } \{-1\}
\end{aligned}$$

The proposed solution is −1, which is not an excluded value, so the solution set contains one number, -1, or $\{-1\}$ written in set notation.

TRY IT #3 Solve the rational equation: $\frac{2}{3x} = \frac{1}{4} - \frac{1}{6x}$.

EXAMPLE 5

Solving a Rational Equation by Factoring the Denominator

Solve the following rational equation: $\frac{1}{x} = \frac{1}{10} - \frac{3}{4x}$.

Solution

First find the common denominator. The three denominators in factored form are x, $10 = 2 \cdot 5$, and $4x = 2 \cdot 2 \cdot x$. The smallest expression that is divisible by each one of the denominators is $20x$. Only $x = 0$ is an excluded value. Multiply the whole equation by $20x$.

$$\begin{aligned} 20x\left(\tfrac{1}{x}\right) &= \left(\tfrac{1}{10} - \tfrac{3}{4x}\right)20x \\ 20 &= 2x - 15 \\ 35 &= 2x \\ \tfrac{35}{2} &= x \end{aligned}$$

The solution is $\frac{35}{2}$.

TRY IT #4 Solve the rational equation: $-\frac{5}{2x} + \frac{3}{4x} = -\frac{7}{4}$.

EXAMPLE 6

Solving Rational Equations with a Binomial in the Denominator

Solve the following rational equations and state the excluded values:

ⓐ $\frac{3}{x-6} = \frac{5}{x}$ ⓑ $\frac{x}{x-3} = \frac{5}{x-3} - \frac{1}{2}$ ⓒ $\frac{x}{x-2} = \frac{5}{x-2} - \frac{1}{2}$

Solution

ⓐ

The denominators x and $x - 6$ have nothing in common. Therefore, the LCD is the product $x(x - 6)$. However, for this problem, we can cross-multiply.

$$\begin{aligned} \tfrac{3}{x-6} &= \tfrac{5}{x} \\ 3x &= 5(x-6) && \text{Distribute.} \\ 3x &= 5x - 30 \\ -2x &= -30 \\ x &= 15 \end{aligned}$$

The solution is 15. The excluded values are 6 and 0.

ⓑ

The LCD is $2(x - 3)$. Multiply both sides of the equation by $2(x - 3)$.

$$\begin{aligned} 2(x-3)\left(\tfrac{x}{x-3}\right) &= \left(\tfrac{5}{x-3} - \tfrac{1}{2}\right)2(x-3) \\ \frac{2\cancel{(x-3)}x}{\cancel{x-3}} &= \frac{2\cancel{(x-3)}5}{\cancel{x-3}} - \frac{\cancel{2}(x-3)}{\cancel{2}} \\ 2x &= 10 - (x-3) \\ 2x &= 10 - x + 3 \\ 2x &= 13 - x \\ 3x &= 13 \\ x &= \tfrac{13}{3} \end{aligned}$$

The solution is $\frac{13}{3}$. The excluded value is 3.

Ⓒ

The least common denominator is $2(x-2)$. Multiply both sides of the equation by $x(x-2)$.

$$\begin{aligned} 2(x-2)\left(\frac{x}{x-2}\right) &= \left(\frac{5}{x-2}-\frac{1}{2}\right)2(x-2) \\ 2x &= 10-(x-2) \\ 2x &= 12-x \\ 3x &= 12 \\ x &= 4 \end{aligned}$$

The solution is 4. The excluded value is 2.

> **TRY IT** #5 Solve $\frac{-3}{2x+1}=\frac{4}{3x+1}$. State the excluded values.

EXAMPLE 7

Solving a Rational Equation with Factored Denominators and Stating Excluded Values

Solve the rational equation after factoring the denominators: $\frac{2}{x+1}-\frac{1}{x-1}=\frac{2x}{x^2-1}$. State the excluded values.

✓ **Solution**

We must factor the denominator x^2-1. We recognize this as the difference of squares, and factor it as $(x-1)(x+1)$. Thus, the LCD that contains each denominator is $(x-1)(x+1)$. Multiply the whole equation by the LCD, cancel out the denominators, and solve the remaining equation.

$$\begin{aligned} (x-1)(x+1)\left(\frac{2}{x+1}-\frac{1}{x-1}\right) &= \left(\frac{2x}{(x-1)(x+1)}\right)(x-1)(x+1) & \\ 2(x-1)-1(x+1) &= 2x & \\ 2x-2-x-1 &= 2x & \text{Distribute the negative sign.} \\ -3-x &= 0 & \\ -3 &= x & \end{aligned}$$

The solution is -3. The excluded values are 1 and -1.

> **TRY IT** #6 Solve the rational equation: $\frac{2}{x-2}+\frac{1}{x+1}=\frac{1}{x^2-x-2}$.

Finding a Linear Equation

Perhaps the most familiar form of a linear equation is the slope-intercept form, written as $y=mx+b$, where $m=\text{slope}$ and $b=y\text{-intercept}$. Let us begin with the slope.

The Slope of a Line

The **slope** of a line refers to the ratio of the vertical change in *y* over the horizontal change in *x* between any two points on a line. It indicates the direction in which a line slants as well as its steepness. Slope is sometimes described as rise over run.

$$m=\frac{y_2-y_1}{x_2-x_1}$$

If the slope is positive, the line slants to the right. If the slope is negative, the line slants to the left. As the slope increases, the line becomes steeper. Some examples are shown in Figure 2. The lines indicate the following slopes: $m=-3$, $m=2$, and $m=\frac{1}{3}$.

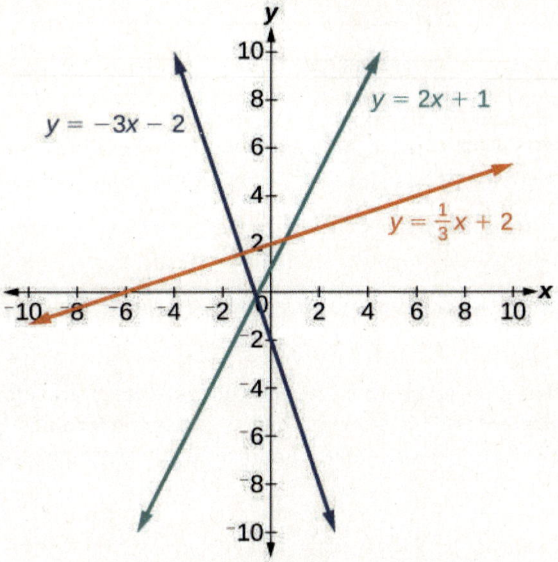

Figure 2

The Slope of a Line

The slope of a line, *m*, represents the change in *y* over the change in *x*. Given two points, (x_1, y_1) and (x_2, y_2), the following formula determines the slope of a line containing these points:

$$m = \frac{y_2 - y_1}{x_2 - x_1}$$

EXAMPLE 8

Finding the Slope of a Line Given Two Points

Find the slope of a line that passes through the points $(2, -1)$ and $(-5, 3)$.

Solution

We substitute the *y*-values and the *x*-values into the formula.

$$\begin{aligned} m &= \frac{3-(-1)}{-5-2} \\ &= \frac{4}{-7} \\ &= -\frac{4}{7} \end{aligned}$$

The slope is $-\frac{4}{7}$.

Analysis

It does not matter which point is called (x_1, y_1) or (x_2, y_2). As long as we are consistent with the order of the *y* terms and the order of the *x* terms in the numerator and denominator, the calculation will yield the same result.

TRY IT #7 Find the slope of the line that passes through the points $(-2, 6)$ and $(1, 4)$.

EXAMPLE 9

Identifying the Slope and *y*-intercept of a Line Given an Equation

Identify the slope and *y*-intercept, given the equation $y = -\frac{3}{4}x - 4$.

Solution

As the line is in $y = mx + b$ form, the given line has a slope of $m = -\frac{3}{4}$. The y-intercept is $b = -4$.

Analysis

The y-intercept is the point at which the line crosses the y-axis. On the y-axis, $x = 0$. We can always identify the y-intercept when the line is in slope-intercept form, as it will always equal b. Or, just substitute $x = 0$ and solve for y.

The Point-Slope Formula

Given the slope and one point on a line, we can find the equation of the line using the point-slope formula.

$$y - y_1 = m(x - x_1)$$

This is an important formula, as it will be used in other areas of college algebra and often in calculus to find the equation of a tangent line. We need only one point and the slope of the line to use the formula. After substituting the slope and the coordinates of one point into the formula, we simplify it and write it in slope-intercept form.

The Point-Slope Formula

Given one point and the slope, the point-slope formula will lead to the equation of a line:

$$y - y_1 = m(x - x_1)$$

EXAMPLE 10

Finding the Equation of a Line Given the Slope and One Point

Write the equation of the line with slope $m = -3$ and passing through the point $(4, 8)$. Write the final equation in slope-intercept form.

Solution

Using the point-slope formula, substitute -3 for m and the point $(4, 8)$ for (x_1, y_1).

$$\begin{aligned} y - y_1 &= m(x - x_1) \\ y - 8 &= -3(x - 4) \\ y - 8 &= -3x + 12 \\ y &= -3x + 20 \end{aligned}$$

Analysis

Note that any point on the line can be used to find the equation. If done correctly, the same final equation will be obtained.

TRY IT #8 Given $m = 4$, find the equation of the line in slope-intercept form passing through the point $(2, 5)$.

EXAMPLE 11

Finding the Equation of a Line Passing Through Two Given Points

Find the equation of the line passing through the points $(3, 4)$ and $(0, -3)$. Write the final equation in slope-intercept form.

Solution

First, we calculate the slope using the slope formula and two points.

$$\begin{aligned} m &= \frac{-3-4}{0-3} \\ &= \frac{-7}{-3} \\ &= \frac{7}{3} \end{aligned}$$

Next, we use the point-slope formula with the slope of $\frac{7}{3}$, and either point. Let's pick the point $(3, 4)$ for (x_1, y_1).

$$\begin{aligned} y - 4 &= \tfrac{7}{3}(x - 3) \\ y - 4 &= \tfrac{7}{3}x - 7 \qquad \text{Distribute the } \tfrac{7}{3}. \\ y &= \tfrac{7}{3}x - 3 \end{aligned}$$

In slope-intercept form, the equation is written as $y = \frac{7}{3}x - 3$.

Analysis

To prove that either point can be used, let us use the second point $(0, -3)$ and see if we get the same equation.

$$\begin{aligned} y - (-3) &= \tfrac{7}{3}(x - 0) \\ y + 3 &= \tfrac{7}{3}x \\ y &= \tfrac{7}{3}x - 3 \end{aligned}$$

We see that the same line will be obtained using either point. This makes sense because we used both points to calculate the slope.

Standard Form of a Line

Another way that we can represent the equation of a line is in standard form. Standard form is given as

$$Ax + By = C$$

where A, B, and C are integers. The *x*- and *y*-terms are on one side of the equal sign and the constant term is on the other side.

EXAMPLE 12

Finding the Equation of a Line and Writing It in Standard Form

Find the equation of the line with $m = -6$ and passing through the point $\left(\frac{1}{4}, -2\right)$. Write the equation in standard form.

Solution

We begin using the point-slope formula.

$$\begin{aligned} y - (-2) &= -6\left(x - \tfrac{1}{4}\right) \\ y + 2 &= -6x + \tfrac{3}{2} \end{aligned}$$

From here, we multiply through by 2, as no fractions are permitted in standard form, and then move both variables to the left aside of the equal sign and move the constants to the right.

$$\begin{aligned} 2(y + 2) &= \left(-6x + \tfrac{3}{2}\right)2 \\ 2y + 4 &= -12x + 3 \\ 12x + 2y &= -1 \end{aligned}$$

This equation is now written in standard form.

TRY IT #9 Find the equation of the line in standard form with slope $m = -\frac{1}{3}$ and passing through the point $\left(1, \frac{1}{3}\right)$.

Vertical and Horizontal Lines

The equations of vertical and horizontal lines do not require any of the preceding formulas, although we can use the formulas to prove that the equations are correct. The equation of a vertical line is given as

$$x = c$$

where *c* is a constant. The slope of a vertical line is undefined, and regardless of the *y*-value of any point on the line, the *x*-coordinate of the point will be *c*.

Suppose that we want to find the equation of a line containing the following points: $(-3, -5)$, $(-3, 1)$, $(-3, 3)$, and $(-3, 5)$. First, we will find the slope.

$$m = \frac{5-3}{-3-(-3)} = \frac{2}{0}$$

Zero in the denominator means that the slope is undefined and, therefore, we cannot use the point-slope formula. However, we can plot the points. Notice that all of the *x*-coordinates are the same and we find a vertical line through $x = -3$. See Figure 3.

The equation of a horizontal line is given as

$$y = c$$

where *c* is a constant. The slope of a horizontal line is zero, and for any *x*-value of a point on the line, the *y*-coordinate will be *c*.

Suppose we want to find the equation of a line that contains the following set of points: $(-2, -2)$, $(0, -2)$, $(3, -2)$, and $(5, -2)$. We can use the point-slope formula. First, we find the slope using any two points on the line.

$$\begin{aligned} m &= \frac{-2-(-2)}{0-(-2)} \\ &= \frac{0}{2} \\ &= 0 \end{aligned}$$

Use any point for (x_1, y_1) in the formula, or use the *y*-intercept.

$$\begin{aligned} y-(-2) &= 0(x-3) \\ y+2 &= 0 \\ y &= -2 \end{aligned}$$

The graph is a horizontal line through $y = -2$. Notice that all of the *y*-coordinates are the same. See Figure 3.

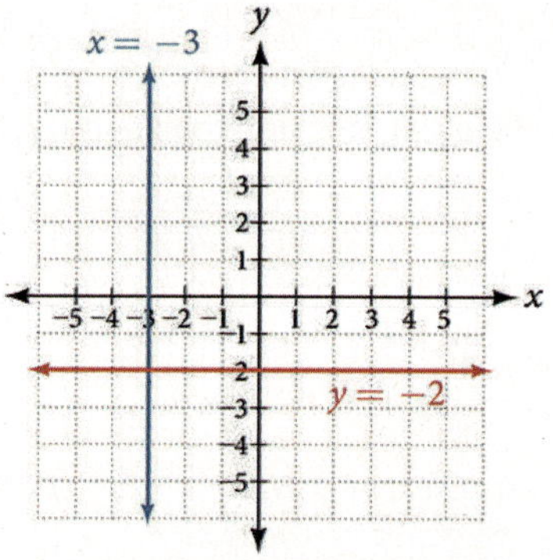

Figure 3 The line $x = -3$ is a vertical line. The line $y = -2$ is a horizontal line.

EXAMPLE 13

Finding the Equation of a Line Passing Through the Given Points

Find the equation of the line passing through the given points: $(1, -3)$ and $(1, 4)$.

Solution

The *x*-coordinate of both points is 1. Therefore, we have a vertical line, $x = 1$.

TRY IT #10 Find the equation of the line passing through $(-5, 2)$ and $(2, 2)$.

Determining Whether Graphs of Lines are Parallel or Perpendicular

Parallel lines have the same slope and different *y*-intercepts. Lines that are parallel to each other will never intersect. For example, Figure 4 shows the graphs of various lines with the same slope, $m = 2$.

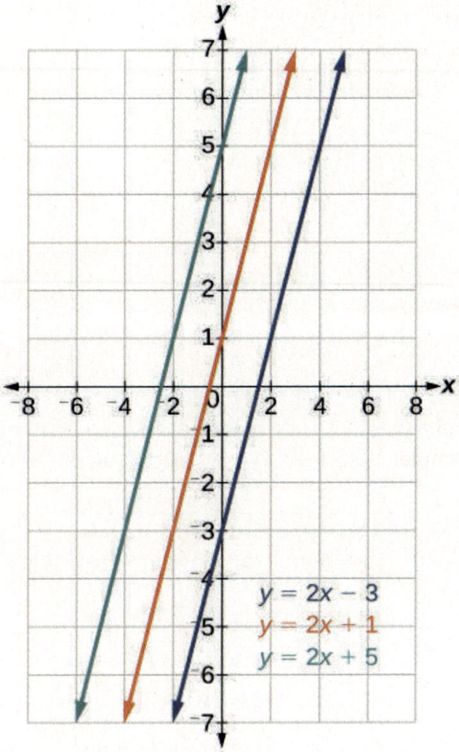

Figure 4 Parallel lines

All of the lines shown in the graph are parallel because they have the same slope and different *y*-intercepts.

Lines that are perpendicular intersect to form a $90°$ -angle. The slope of one line is the negative reciprocal of the other. We can show that two lines are perpendicular if the product of the two slopes is -1 : $m_1 \cdot m_2 = -1$. For example, Figure 5 shows the graph of two perpendicular lines. One line has a slope of 3; the other line has a slope of $-\frac{1}{3}$.

$$\begin{aligned} m_1 \cdot m_2 &= -1 \\ 3 \cdot \left(-\tfrac{1}{3}\right) &= -1 \end{aligned}$$

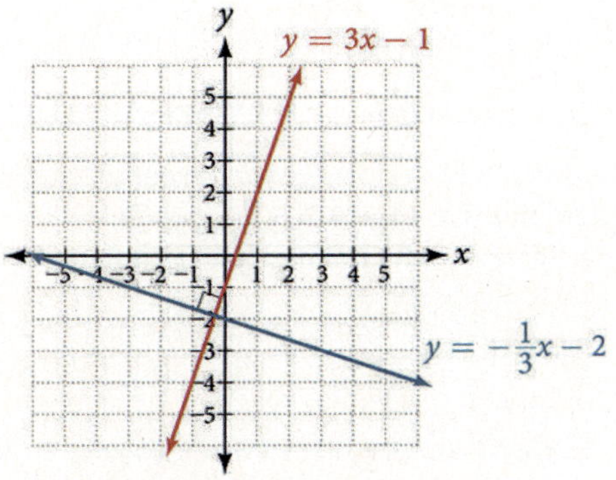

Figure 5 Perpendicular lines

EXAMPLE 14

Graphing Two Equations, and Determining Whether the Lines are Parallel, Perpendicular, or Neither

Graph the equations of the given lines, and state whether they are parallel, perpendicular, or neither: $3y = -4x + 3$ and $3x - 4y = 8$.

Solution

The first thing we want to do is rewrite the equations so that both equations are in slope-intercept form.

First equation:

$$\begin{aligned} 3y &= -4x + 3 \\ y &= -\frac{4}{3}x + 1 \end{aligned}$$

Second equation:

$$\begin{aligned} 3x - 4y &= 8 \\ -4y &= -3x + 8 \\ y &= \frac{3}{4}x - 2 \end{aligned}$$

See the graph of both lines in Figure 6

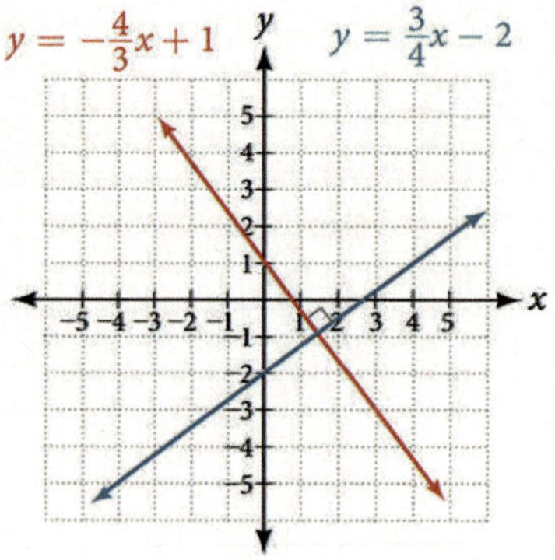

Figure 6

From the graph, we can see that the lines appear perpendicular, but we must compare the slopes.

$$\begin{aligned} m_1 &= -\frac{4}{3} \\ m_2 &= \frac{3}{4} \\ m_1 \cdot m_2 &= \left(-\frac{4}{3}\right)\left(\frac{3}{4}\right) = -1 \end{aligned}$$

The slopes are negative reciprocals of each other, confirming that the lines are perpendicular.

> **TRY IT** #11 Graph the two lines and determine whether they are parallel, perpendicular, or neither: $2y - x = 10$ and $2y = x + 4$.

Writing the Equations of Lines Parallel or Perpendicular to a Given Line

As we have learned, determining whether two lines are parallel or perpendicular is a matter of finding the slopes. To write the equation of a line parallel or perpendicular to another line, we follow the same principles as we do for finding the equation of any line. After finding the slope, use the point-slope formula to write the equation of the new line.

HOW TO

Given an equation for a line, write the equation of a line parallel or perpendicular to it.

1. Find the slope of the given line. The easiest way to do this is to write the equation in slope-intercept form.
2. Use the slope and the given point with the point-slope formula.
3. Simplify the line to slope-intercept form and compare the equation to the given line.

EXAMPLE 15

Writing the Equation of a Line Parallel to a Given Line Passing Through a Given Point

Write the equation of line parallel to a $5x + 3y = 1$ and passing through the point $(3, 5)$.

Solution

First, we will write the equation in slope-intercept form to find the slope.

$$\begin{aligned} 5x + 3y &= 1 \\ 3y &= -5x + 1 \\ y &= -\tfrac{5}{3}x + \tfrac{1}{3} \end{aligned}$$

The slope is $m = -\frac{5}{3}$. The *y*-intercept is $\frac{1}{3}$, but that really does not enter into our problem, as the only thing we need for two lines to be parallel is the same slope. The one exception is that if the *y*-intercepts are the same, then the two lines are the same line. The next step is to use this slope and the given point with the point-slope formula.

$$\begin{aligned} y - 5 &= -\tfrac{5}{3}(x - 3) \\ y - 5 &= -\tfrac{5}{3}x + 5 \\ y &= -\tfrac{5}{3}x + 10 \end{aligned}$$

The equation of the line is $y = -\frac{5}{3}x + 10$. See Figure 7.

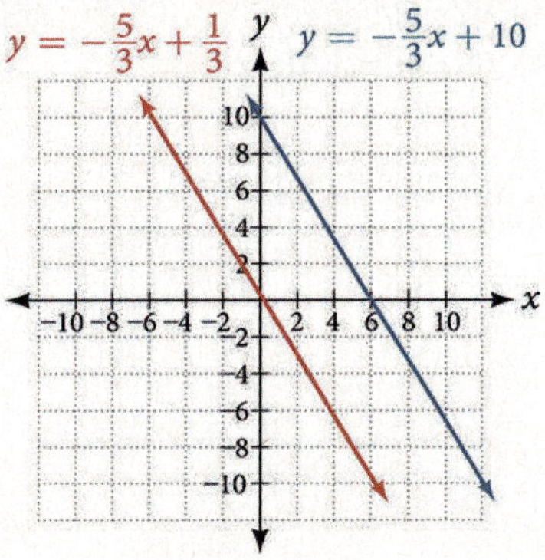

Figure 7

TRY IT #12 Find the equation of the line parallel to $5x = 7 + y$ and passing through the point $(-1, -2)$.

EXAMPLE 16

Finding the Equation of a Line Perpendicular to a Given Line Passing Through a Given Point

Find the equation of the line perpendicular to $5x - 3y + 4 = 0$ and passing through the point $(-4, 1)$.

Solution

The first step is to write the equation in slope-intercept form.

$$\begin{aligned} 5x - 3y + 4 &= 0 \\ -3y &= -5x - 4 \\ y &= \tfrac{5}{3}x + \tfrac{4}{3} \end{aligned}$$

We see that the slope is $m = \frac{5}{3}$. This means that the slope of the line perpendicular to the given line is the negative reciprocal, or $-\frac{3}{5}$. Next, we use the point-slope formula with this new slope and the given point.

$$\begin{aligned} y - 1 &= -\tfrac{3}{5}(x - (-4)) \\ y - 1 &= -\tfrac{3}{5}x - \tfrac{12}{5} \\ y &= -\tfrac{3}{5}x - \tfrac{12}{5} + \tfrac{5}{5} \\ y &= -\tfrac{3}{5}x - \tfrac{7}{5} \end{aligned}$$

MEDIA

Access these online resources for additional instruction and practice with linear equations.

Solving rational equations (http://openstax.org/l/rationaleqs)
Equation of a line given two points (http://openstax.org/l/twopointsline)
Finding the equation of a line perpendicular to another line through a given point (http://openstax.org/l/findperpline)
Finding the equation of a line parallel to another line through a given point (http://openstax.org/l/findparaline)

2.2 SECTION EXERCISES

Verbal

1. What does it mean when we say that two lines are parallel?

2. What is the relationship between the slopes of perpendicular lines (assuming neither is horizontal nor vertical)?

3. How do we recognize when an equation, for example $y = 4x + 3$, will be a straight line (linear) when graphed?

4. What does it mean when we say that a linear equation is inconsistent?

5. When solving the following equation:

$\frac{2}{x-5} = \frac{4}{x+1}$

explain why we must exclude $x = 5$ and $x = -1$ as possible solutions from the solution set.

Algebraic

For the following exercises, solve the equation for x.

6. $7x + 2 = 3x - 9$

7. $4x - 3 = 5$

8. $3(x + 2) - 12 = 5(x + 1)$

9. $12 - 5(x + 3) = 2x - 5$

10. $\frac{1}{2} - \frac{1}{3}x = \frac{4}{3}$

11. $\frac{x}{3} - \frac{3}{4} = \frac{2x+3}{12}$

12. $\frac{2}{3}x + \frac{1}{2} = \frac{31}{6}$

13. $3(2x - 1) + x = 5x + 3$

14. $\frac{2x}{3} - \frac{3}{4} = \frac{x}{6} + \frac{21}{4}$

15. $\frac{x+2}{4} - \frac{x-1}{3} = 2$

For the following exercises, solve each rational equation for x. State all x-values that are excluded from the solution set.

16. $\frac{3}{x} - \frac{1}{3} = \frac{1}{6}$

17. $2 - \frac{3}{x+4} = \frac{x+2}{x+4}$

18. $\frac{3}{x-2} = \frac{1}{x-1} + \frac{7}{(x-1)(x-2)}$

19. $\frac{3x}{x-1} + 2 = \frac{3}{x-1}$

20. $\frac{5}{x+1} + \frac{1}{x-3} = \frac{-6}{x^2-2x-3}$

21. $\frac{1}{x} = \frac{1}{5} + \frac{3}{2x}$

For the following exercises, find the equation of the line using the point-slope formula. Write all the final equations using the slope-intercept form.

22. $(0, 3)$ with a slope of $\frac{2}{3}$

23. $(1, 2)$ with a slope of $-\frac{4}{5}$

24. x-intercept is 1, and $(-2, 6)$

25. y-intercept is 2, and $(4, -1)$

26. $(-3, 10)$ and $(5, -6)$

27. $(1, 3)$ and $(5, 5)$

28. parallel to $y = 2x + 5$ and passes through the point $(4, 3)$

29. perpendicular to $3y = x - 4$ and passes through the point $(-2, 1)$.

For the following exercises, find the equation of the line using the given information.

30. $(-2, 0)$ and $(-2, 5)$

31. $(1, 7)$ and $(3, 7)$

32. The slope is undefined and it passes through the point $(2, 3)$.

33. The slope equals zero and it passes through the point $(1, -4)$.

34. The slope is $\frac{3}{4}$ and it passes through the point $(1, 4)$.

35. $(-1, 3)$ and $(4, -5)$

Graphical

For the following exercises, graph the pair of equations on the same axes, and state whether they are parallel, perpendicular, or neither.

36. $y = 2x + 7$
$y = -\frac{1}{2}x - 4$

37. $3x - 2y = 5$
$6y - 9x = 6$

38. $y = \frac{3x+1}{4}$
$y = 3x + 2$

39. $x = 4$
$y = -3$

Numeric

For the following exercises, find the slope of the line that passes through the given points.

40. $(5, 4)$ and $(7, 9)$

41. $(-3, 2)$ and $(4, -7)$

42. $(-5, 4)$ and $(2, 4)$

43. $(-1, -2)$ and $(3, 4)$

44. $(3, -2)$ and $(3, -2)$

For the following exercises, find the slope of the lines that pass through each pair of points and determine whether the lines are parallel or perpendicular.

45. $(-1, 3)$ and $(5, 1)$
$(-2, 3)$ and $(0, 9)$

46. $(2, 5)$ and $(5, 9)$
$(-1, -1)$ and $(2, 3)$

Technology

For the following exercises, express the equations in slope intercept form (rounding each number to the thousandths place). Enter this into a graphing calculator as Y1, then adjust the ymin and ymax values for your window to include where the y-intercept occurs. State your ymin and ymax values.

47. $0.537x - 2.19y = 100$

48. $4{,}500x - 200y = 9{,}528$

49. $\frac{200-30y}{x} = 70$

Extensions

50. Starting with the point-slope formula $y - y_1 = m(x - x_1)$, solve this expression for x in terms of x_1, y, y_1, and m.

51. Starting with the standard form of an equation $Ax + By = C$ solve this expression for y in terms of A, B, C and x. Then put the expression in slope-intercept form.

52. Use the above derived formula to put the following standard equation in slope intercept form: $7x - 5y = 25$.

53. Given that the following coordinates are the vertices of a rectangle, prove that this truly is a rectangle by showing the slopes of the sides that meet are perpendicular.

$(-1, 1), (2, 0), (3, 3)$ and $(0, 4)$

54. Find the slopes of the diagonals in the previous exercise. Are they perpendicular?

Real-World Applications

55. The slope for a wheelchair ramp for a home has to be $\frac{1}{12}$. If the vertical distance from the ground to the door bottom is 2.5 ft, find the distance the ramp has to extend from the home in order to comply with the needed slope.

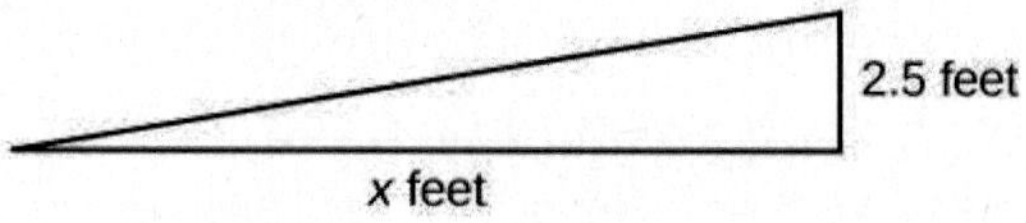

56. If the profit equation for a small business selling x number of item one and y number of item two is $p = 3x + 4y$, find the y value when $p = \$453$ and $x = 75$.

For the following exercises, use this scenario: The cost of renting a car is \$45/wk plus \$0.25/mi traveled during that week. An equation to represent the cost would be $y = 45 + .25x$, where x is the number of miles traveled.

57. What is your cost if you travel 50 mi?

58. If your cost were $\$63.75$, how many miles were you charged for traveling?

59. Suppose you have a maximum of \$100 to spend for the car rental. What would be the maximum number of miles you could travel?

2.3 Models and Applications

Learning Objectives

In this section, you will:

- Set up a linear equation to solve a real-world application.
- Use a formula to solve a real-world application.

Figure 1 Credit: Kevin Dooley

Neka is hoping to get an A in his college algebra class. He has scores of 75, 82, 95, 91, and 94 on his first five tests. Only the final exam remains, and the maximum of points that can be earned is 100. Is it possible for Neka to end the course with an A? A simple linear equation will give Neka his answer.

Many real-world applications can be modeled by linear equations. For example, a cell phone package may include a monthly service fee plus a charge per minute of talk-time; it costs a widget manufacturer a certain amount to produce x widgets per month plus monthly operating charges; a car rental company charges a daily fee plus an amount per mile driven. These are examples of applications we come across every day that are modeled by linear equations. In this section, we will set up and use linear equations to solve such problems.

Setting up a Linear Equation to Solve a Real-World Application

To set up or model a linear equation to fit a real-world application, we must first determine the known quantities and define the unknown quantity as a variable. Then, we begin to interpret the words as mathematical expressions using mathematical symbols. Let us use the car rental example above. In this case, a known cost, such as \$0.10/mi, is multiplied by an unknown quantity, the number of miles driven. Therefore, we can write $0.10x$. This expression represents a variable cost because it changes according to the number of miles driven.

If a quantity is independent of a variable, we usually just add or subtract it, according to the problem. As these amounts do not change, we call them fixed costs. Consider a car rental agency that charges \$0.10/mi plus a daily fee of \$50. We can use these quantities to model an equation that can be used to find the daily car rental cost C.

$$C = 0.10x + 50$$

When dealing with real-world applications, there are certain expressions that we can translate directly into math. Table 1 lists some common verbal expressions and their equivalent mathematical expressions.

Verbal	Translation to Math Operations
One number exceeds another by *a*	$x, \ x+a$
Twice a number	$2x$
One number is *a* more than another number	$x, \ x+a$
One number is *a* less than twice another number	$x, \ 2x-a$
The product of a number and *a*, decreased by *b*	$ax-b$
The quotient of a number and the number plus *a* is three times the number	$\frac{x}{x+a} = 3x$

Table 1

Verbal	Translation to Math Operations
The product of three times a number and the number decreased by b is c	$3x(x-b)=c$

Table 1

HOW TO

Given a real-world problem, model a linear equation to fit it.

1. Identify known quantities.
2. Assign a variable to represent the unknown quantity.
3. If there is more than one unknown quantity, find a way to write the second unknown in terms of the first.
4. Write an equation interpreting the words as mathematical operations.
5. Solve the equation. Be sure the solution can be explained in words, including the units of measure.

EXAMPLE 1

Modeling a Linear Equation to Solve an Unknown Number Problem

Find a linear equation to solve for the following unknown quantities: One number exceeds another number by 17 and their sum is 31. Find the two numbers.

Solution

Let x equal the first number. Then, as the second number exceeds the first by 17, we can write the second number as $x + 17$. The sum of the two numbers is 31. We usually interpret the word *is* as an equal sign.

$$\begin{aligned} x + (x + 17) &= 31 \\ 2x + 17 &= 31 \qquad \text{Simplify and solve.} \\ 2x &= 14 \\ x &= 7 \\ \\ x + 17 &= 7 + 17 \\ &= 24 \end{aligned}$$

The two numbers are 7 and 24.

TRY IT #1 Find a linear equation to solve for the following unknown quantities: One number is three more than twice another number. If the sum of the two numbers is 36, find the numbers.

EXAMPLE 2

Setting Up a Linear Equation to Solve a Real-World Application

There are two cell phone companies that offer different packages. Company A charges a monthly service fee of $34 plus $.05/min talk-time. Company B charges a monthly service fee of $40 plus $.04/min talk-time.

ⓐ Write a linear equation that models the packages offered by both companies.
ⓑ If the average number of minutes used each month is 1,160, which company offers the better plan?
ⓒ If the average number of minutes used each month is 420, which company offers the better plan?
ⓓ How many minutes of talk-time would yield equal monthly statements from both companies?

Solution

ⓐ The model for Company *A* can be written as $A = 0.05x + 34$. This includes the variable cost of $0.05x$ plus the monthly service charge of $34. Company *B*'s package charges a higher monthly fee of $40, but a lower variable cost of $0.04x$. Company *B*'s model can be written as $B = 0.04x + \$40$.

ⓑ

If the average number of minutes used each month is 1,160, we have the following:

$$\begin{aligned}\text{Company } A &= 0.05(1,160) + 34\\ &= 58 + 34\\ &= 92\end{aligned}$$

$$\begin{aligned}\text{Company } B &= 0.04(1,160) + 40\\ &= 46.4 + 40\\ &= 86.4\end{aligned}$$

So, Company *B* offers the lower monthly cost of $86.40 as compared with the $92 monthly cost offered by Company *A* when the average number of minutes used each month is 1,160.

ⓒ

If the average number of minutes used each month is 420, we have the following:

$$\begin{aligned}\text{Company } A &= 0.05(420) + 34\\ &= 21 + 34\\ &= 55\end{aligned}$$

$$\begin{aligned}\text{Company } B &= 0.04(420) + 40\\ &= 16.8 + 40\\ &= 56.8\end{aligned}$$

If the average number of minutes used each month is 420, then Company *A* offers a lower monthly cost of $55 compared to Company *B*'s monthly cost of $56.80.

ⓓ

To answer the question of how many talk-time minutes would yield the same bill from both companies, we should think about the problem in terms of (x, y) coordinates: At what point are both the *x*-value and the *y*-value equal? We can find this point by setting the equations equal to each other and solving for *x*.

$$\begin{aligned}0.05x + 34 &= 0.04x + 40\\ 0.01x &= 6\\ x &= 600\end{aligned}$$

Check the *x*-value in each equation.

$$\begin{aligned}0.05(600) + 34 &= 64\\ 0.04(600) + 40 &= 64\end{aligned}$$

Therefore, a monthly average of 600 talk-time minutes renders the plans equal. See Figure 2

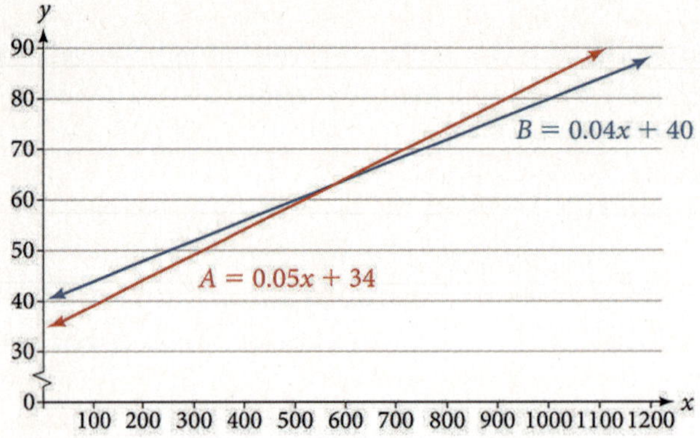

Figure 2

TRY IT #2 Find a linear equation to model this real-world application: It costs ABC electronics company \$2.50 per unit to produce a part used in a popular brand of desktop computers. The company has monthly operating expenses of \$350 for utilities and \$3,300 for salaries. What are the company's monthly expenses?

Using a Formula to Solve a Real-World Application

Many applications are solved using known formulas. The problem is stated, a formula is identified, the known quantities are substituted into the formula, the equation is solved for the unknown, and the problem's question is answered. Typically, these problems involve two equations representing two trips, two investments, two areas, and so on. Examples of formulas include the **area** of a rectangular region, $A = LW$; the **perimeter** of a rectangle, $P = 2L + 2W$; and the **volume** of a rectangular solid, $V = LWH$. When there are two unknowns, we find a way to write one in terms of the other because we can solve for only one variable at a time.

EXAMPLE 3

Solving an Application Using a Formula

It takes Andrew 30 min to drive to work in the morning. He drives home using the same route, but it takes 10 min longer, and he averages 10 mi/h less than in the morning. How far does Andrew drive to work?

Solution

This is a distance problem, so we can use the formula $d = rt$, where distance equals rate multiplied by time. Note that when rate is given in mi/h, time must be expressed in hours. Consistent units of measurement are key to obtaining a correct solution.

First, we identify the known and unknown quantities. Andrew's morning drive to work takes 30 min, or $\frac{1}{2}$ h at rate r. His drive home takes 40 min, or $\frac{2}{3}$ h, and his speed averages 10 mi/h less than the morning drive. Both trips cover distance d. A table, such as Table 2, is often helpful for keeping track of information in these types of problems.

	d	r	t
To Work	d	r	$\frac{1}{2}$
To Home	d	$r - 10$	$\frac{2}{3}$

Table 2

Write two equations, one for each trip.

$$\begin{aligned} d &= r\left(\tfrac{1}{2}\right) && \text{To work} \\ d &= (r-10)\left(\tfrac{2}{3}\right) && \text{To home} \end{aligned}$$

As both equations equal the same distance, we set them equal to each other and solve for r.

$$\begin{aligned} r\left(\tfrac{1}{2}\right) &= (r-10)\left(\tfrac{2}{3}\right) \\ \tfrac{1}{2}r &= \tfrac{2}{3}r - \tfrac{20}{3} \\ \tfrac{1}{2}r - \tfrac{2}{3}r &= -\tfrac{20}{3} \\ -\tfrac{1}{6}r &= -\tfrac{20}{3} \\ r &= -\tfrac{20}{3}(-6) \\ r &= 40 \end{aligned}$$

We have solved for the rate of speed to work, 40 mph. Substituting 40 into the rate on the return trip yields 30 mi/h. Now we can answer the question. Substitute the rate back into either equation and solve for d.

$$\begin{aligned} d &= 40\left(\tfrac{1}{2}\right) \\ &= 20 \end{aligned}$$

The distance between home and work is 20 mi.

Analysis

Note that we could have cleared the fractions in the equation by multiplying both sides of the equation by the LCD to solve for r.

$$\begin{aligned} r\left(\tfrac{1}{2}\right) &= (r-10)\left(\tfrac{2}{3}\right) \\ 6 \times r\left(\tfrac{1}{2}\right) &= 6 \times (r-10)\left(\tfrac{2}{3}\right) \\ 3r &= 4(r-10) \\ 3r &= 4r - 40 \\ -r &= -40 \\ r &= 40 \end{aligned}$$

TRY IT #3 On Saturday morning, it took Jennifer 3.6 h to drive to her mother's house for the weekend. On Sunday evening, due to heavy traffic, it took Jennifer 4 h to return home. Her speed was 5 mi/h slower on Sunday than on Saturday. What was her speed on Sunday?

EXAMPLE 4

Solving a Perimeter Problem

The perimeter of a rectangular outdoor patio is 54 ft. The length is 3 ft greater than the width. What are the dimensions of the patio?

Solution

The perimeter formula is standard: $P = 2L + 2W$. We have two unknown quantities, length and width. However, we can write the length in terms of the width as $L = W + 3$. Substitute the perimeter value and the expression for length into the formula. It is often helpful to make a sketch and label the sides as in Figure 3.

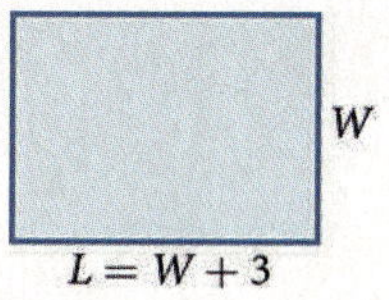

Figure 3

Now we can solve for the width and then calculate the length.

$$\begin{aligned} P &= 2L + 2W \\ 54 &= 2(W+3) + 2W \\ 54 &= 2W + 6 + 2W \\ 54 &= 4W + 6 \\ 48 &= 4W \\ 12 &= W \\ (12+3) &= L \\ 15 &= L \end{aligned}$$

The dimensions are $L = 15$ ft and $W = 12$ ft.

TRY IT #4 Find the dimensions of a rectangle given that the perimeter is 110 cm and the length is 1 cm more than twice the width.

EXAMPLE 5

Solving an Area Problem

The perimeter of a tablet of graph paper is 48 in. The length is 6 in. more than the width. Find the area of the graph paper.

Solution

The standard formula for area is $A = LW$; however, we will solve the problem using the perimeter formula. The reason we use the perimeter formula is because we know enough information about the perimeter that the formula will allow us to solve for one of the unknowns. As both perimeter and area use length and width as dimensions, they are often used together to solve a problem such as this one.

We know that the length is 6 in. more than the width, so we can write length as $L = W + 6$. Substitute the value of the perimeter and the expression for length into the perimeter formula and find the length.

$$\begin{aligned} P &= 2L + 2W \\ 48 &= 2(W+6) + 2W \\ 48 &= 2W + 12 + 2W \\ 48 &= 4W + 12 \\ 36 &= 4W \\ 9 &= W \\ (9+6) &= L \\ 15 &= L \end{aligned}$$

Now, we find the area given the dimensions of $L = 15$ in. and $W = 9$ in.

$$\begin{aligned} A &= LW \\ A &= 15(9) \\ &= 135 \text{ in.}^2 \end{aligned}$$

The area is 135 in.^{2}.

TRY IT #5 A game room has a perimeter of 70 ft. The length is five more than twice the width. How many ft^2 of new carpeting should be ordered?

EXAMPLE 6

Solving a Volume Problem

Find the dimensions of a shipping box given that the length is twice the width, the height is 8 inches, and the volume is

1,600 in.^{3}.

Solution

The formula for the volume of a box is given as $V = LWH$, the product of length, width, and height. We are given that $L = 2W$, and $H = 8$. The volume is 1,600 cubic inches.

$$\begin{aligned} V &= LWH \\ 1{,}600 &= (2W)W(8) \\ 1{,}600 &= 16W^2 \\ 100 &= W^2 \\ 10 &= W \end{aligned}$$

The dimensions are $L = 20$ in., $W = 10$ in., and $H = 8$ in.

Analysis

Note that the square root of W^2 would result in a positive and a negative value. However, because we are describing width, we can use only the positive result.

MEDIA

Access these online resources for additional instruction and practice with models and applications of linear equations.

Problem solving using linear equations (http://openstax.org/l/lineqprobsolve)
Problem solving using equations (http://openstax.org/l/equationprsolve)
Finding the dimensions of area given the perimeter (http://openstax.org/l/permareasolve)
Find the distance between the cities using the distance = rate * time formula (http://openstax.org/l/ratetimesolve)
Linear equation application (Write a cost equation) (http://openstax.org/l/lineqappl)

2.3 SECTION EXERCISES

Verbal

1. To set up a model linear equation to fit real-world applications, what should always be the first step?

2. Use your own words to describe this equation where n is a number: $5(n + 3) = 2n$

3. If the total amount of money you had to invest was $2,000 and you deposit x amount in one investment, how can you represent the remaining amount?

4. If a carpenter sawed a 10-ft board into two sections and one section was n ft long, how long would the other section be in terms of n ?

5. If Bill was traveling v mi/h, how would you represent Daemon's speed if he was traveling 10 mi/h faster?

Real-World Applications

For the following exercises, use the information to find a linear algebraic equation model to use to answer the question being asked.

6. Mark and Don are planning to sell each of their marble collections at a garage sale. If Don has 1 more than 3 times the number of marbles Mark has, how many does each boy have to sell if the total number of marbles is 113?

7. Beth and Ann are joking that their combined ages equal Sam's age. If Beth is twice Ann's age and Sam is 69 yr old, what are Beth and Ann's ages?

8. Ruden originally filled out 8 more applications than Hanh. Then each boy filled out 3 additional applications, bringing the total to 28. How many applications did each boy originally fill out?

For the following exercises, use this scenario: Two different telephone carriers offer the following plans that a person is considering. Company A has a monthly fee of \$20 and charges of \$.05/min for calls. Company B has a monthly fee of \$5 and charges \$.10/min for calls.

9. Find the model of the total cost of Company A's plan, using m for the minutes.

10. Find the model of the total cost of Company B's plan, using m for the minutes.

11. Find out how many minutes of calling would make the two plans equal.

12. If the person makes a monthly average of 200 min of calls, which plan should for the person choose?

For the following exercises, use this scenario: A wireless carrier offers the following plans that a person is considering. The Family Plan: \$90 monthly fee, unlimited talk and text on up to 8 lines, and data charges of \$40 for each device for up to 2 GB of data per device. The Mobile Share Plan: \$120 monthly fee for up to 10 devices, unlimited talk and text for all the lines, and data charges of \$35 for each device up to a shared total of 10 GB of data. Use P for the number of devices that need data plans as part of their cost.

13. Find the model of the total cost of the Family Plan.

14. Find the model of the total cost of the Mobile Share Plan.

15. Assuming they stay under their data limit, find the number of devices that would make the two plans equal in cost.

16. If a family has 3 smart phones, which plan should they choose?

For exercises 17 and 18, use this scenario: A retired woman has $50,000 to invest but needs to make $6,000 a year from the interest to meet certain living expenses. One bond investment pays 15% annual interest. The rest of it she wants to put in a CD that pays 7%.

17. If we let x be the amount the woman invests in the 15% bond, how much will she be able to invest in the CD?

18. Set up and solve the equation for how much the woman should invest in each option to sustain a $6,000 annual return.

19. Two planes fly in opposite directions. One travels 450 mi/h and the other 550 mi/h. How long will it take before they are 4,000 mi apart?

20. Ben starts walking along a path at 4 mi/h. One and a half hours after Ben leaves, his sister Amanda begins jogging along the same path at 6 mi/h. How long will it be before Amanda catches up to Ben?

21. Fiora starts riding her bike at 20 mi/h. After a while, she slows down to 12 mi/h, and maintains that speed for the rest of the trip. The whole trip of 70 mi takes her 4.5 h. For what distance did she travel at 20 mi/h?

22. A chemistry teacher needs to mix a 30% salt solution with a 70% salt solution to make 20 qt of a 40% salt solution. How many quarts of each solution should the teacher mix to get the desired result?

23. Raúl has $20,000 to invest. His intent is to earn 11% interest on his investment. He can invest part of his money at 8% interest and part at 12% interest. How much does Raúl need to invest in each option to make get a total 11% return on his $20,000?

For the following exercises, use this scenario: A truck rental agency offers two kinds of plans. Plan A charges $75/wk plus $.10/mi driven. Plan B charges $100/wk plus $.05/mi driven.

24. Write the model equation for the cost of renting a truck with plan A.

25. Write the model equation for the cost of renting a truck with plan B.

26. Find the number of miles that would generate the same cost for both plans.

27. If Tim knows he has to travel 300 mi, which plan should he choose?

For the following exercises, use the formula given to solve for the required value.

28. $A = P(1 + rt)$ is used to find the principal amount *P* deposited, earning *r*% interest, for *t* years. Use this to find what principal amount *P* David invested at a 3% rate for 20 yr if $A = \$8{,}000$.

29. The formula $F = \frac{mv^2}{R}$ relates force (F), velocity (v), mass , and resistance (m). Find R when $m = 45$, $v = 7$, and $F = 245$.

30. $F = ma$ indicates that force (*F*) equals mass (*m*) times acceleration (*a*). Find the acceleration of a mass of 50 kg if a force of 12 N is exerted on it.

31. $Sum = \frac{1}{1-r}$ is the formula for an infinite series sum. If the sum is 5, find r.

For the following exercises, solve for the given variable in the formula. After obtaining a new version of the formula, you will use it to solve a question.

32. Solve for W: $P = 2L + 2W$

33. Use the formula from the previous question to find the width, W, of a rectangle whose length is 15 and whose perimeter is 58.

34. Solve for f : $\frac{1}{p} + \frac{1}{q} = \frac{1}{f}$

35. Use the formula from the previous question to find f when $p = 8 \text{and } q = 13.$

36. Solve for m in the slope-intercept formula: $y = mx + b$

37. Use the formula from the previous question to find m when the coordinates of the point are $(4, 7)$ and $b = 12.$

38. The area of a trapezoid is given by $A = \frac{1}{2}h\left(b_1 + b_2\right).$ Use the formula to find the area of a trapezoid with $h = 6, \ \ b_1 = 14, \ \text{ and } b_2 = 8.$

39. Solve for h: $A = \frac{1}{2}h\left(b_1 + b_2\right)$

40. Use the formula from the previous question to find the height of a trapezoid with $A = 150, b_1 = 19,$ and $b_2 = 11.$

41. Find the dimensions of an American football field. The length is 200 ft more than the width, and the perimeter is 1,040 ft. Find the length and width. Use the perimeter formula $P = 2L + 2W.$

42. Distance equals rate times time, $d = rt$. Find the distance Tom travels if he is moving at a rate of 55 mi/h for 3.5 h.

43. Using the formula in the previous exercise, find the distance that Susan travels if she is moving at a rate of 60 mi/h for 6.75 h.

44. What is the total distance that two people travel in 3 h if one of them is riding a bike at 15 mi/h and the other is walking at 3 mi/h?

45. If the area model for a triangle is $A = \frac{1}{2}bh$, find the area of a triangle with a height of 16 in. and a base of 11 in.

46. Solve for h: $A = \frac{1}{2}bh$

47. Use the formula from the previous question to find the height to the nearest tenth of a triangle with a base of 15 and an area of 215.

48. The volume formula for a cylinder is $V = \pi r^2 h$. Using the symbol π in your answer, find the volume of a cylinder with a radius, r, of 4 cm and a height of 14 cm.

49. Solve for h: $V = \pi r^2 h$

50. Use the formula from the previous question to find the height of a cylinder with a radius of 8 and a volume of 16π

51. Solve for r: $V = \pi r^2 h$

52. Use the formula from the previous question to find the radius of a cylinder with a height of 36 and a volume of 324π.

53. The formula for the circumference of a circle is $C = 2\pi r$. Find the circumference of a circle with a diameter of 12 in. (diameter = $2r$). Use the symbol π in your final answer.

54. Solve the formula from the previous question for π. Notice why π is sometimes defined as the ratio of the circumference to its diameter.

2.4 Complex Numbers

Learning Objectives

In this section, you will:

- Add and subtract complex numbers.
- Multiply and divide complex numbers.
- Simplify powers of i

Figure 1

Discovered by Benoit Mandelbrot around 1980, the Mandelbrot Set is one of the most recognizable fractal images. The image is built on the theory of self-similarity and the operation of iteration. Zooming in on a fractal image brings many surprises, particularly in the high level of repetition of detail that appears as magnification increases. The equation that generates this image turns out to be rather simple.

In order to better understand it, we need to become familiar with a new set of numbers. Keep in mind that the study of mathematics continuously builds upon itself. Negative integers, for example, fill a void left by the set of positive integers. The set of rational numbers, in turn, fills a void left by the set of integers. The set of real numbers fills a void left by the set of rational numbers. Not surprisingly, the set of real numbers has voids as well. In this section, we will explore a set of numbers that fills voids in the set of real numbers and find out how to work within it.

Expressing Square Roots of Negative Numbers as Multiples of i

We know how to find the square root of any positive real number. In a similar way, we can find the square root of any negative number. The difference is that the root is not real. If the value in the radicand is negative, the root is said to be an imaginary number. The imaginary number i is defined as the square root of -1.

$$\sqrt{-1} = i$$

So, using properties of radicals,

$$i^2 = \left(\sqrt{-1}\right)^2 = -1$$

We can write the square root of any negative number as a multiple of i. Consider the square root of -49.

$$\begin{aligned}\sqrt{-49} &= \sqrt{49\cdot(-1)}\\ &= \sqrt{49}\sqrt{-1}\\ &= 7i\end{aligned}$$

We use $7i$ and not $-7i$ because the principal root of 49 is the positive root.

A complex number is the sum of a real number and an imaginary number. A complex number is expressed in standard form when written $a + bi$ where a is the real part and b is the imaginary part. For example, $5 + 2i$ is a complex number. So, too, is $3 + 4i\sqrt{3}$.

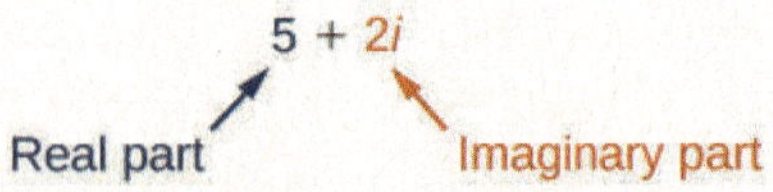

Imaginary numbers differ from real numbers in that a squared imaginary number produces a negative real number. Recall that when a positive real number is squared, the result is a positive real number and when a negative real number is squared, the result is also a positive real number. Complex numbers consist of real and imaginary numbers.

Imaginary and Complex Numbers

A **complex number** is a number of the form $a + bi$ where

- a is the real part of the complex number.
- b is the imaginary part of the complex number.

If $b = 0$, then $a + bi$ is a real number. If $a = 0$ and b is not equal to 0, the complex number is called a pure imaginary number. An **imaginary number** is an even root of a negative number.

HOW TO

Given an imaginary number, express it in the standard form of a complex number.

1. Write $\sqrt{-a}$ as $\sqrt{a}\sqrt{-1}$.
2. Express $\sqrt{-1}$ as i.
3. Write $\sqrt{a}\cdot i$ in simplest form.

EXAMPLE 1

Expressing an Imaginary Number in Standard Form

Express $\sqrt{-9}$ in standard form.

Solution

$$\begin{aligned}\sqrt{-9} &= \sqrt{9}\sqrt{-1}\\ &= 3i\end{aligned}$$

In standard form, this is $0 + 3i$.

TRY IT #1 Express $\sqrt{-24}$ in standard form.

Plotting a Complex Number on the Complex Plane

We cannot plot complex numbers on a number line as we might real numbers. However, we can still represent them graphically. To represent a complex number, we need to address the two components of the number. We use the **complex plane**, which is a coordinate system in which the horizontal axis represents the real component and the vertical axis represents the imaginary component. Complex numbers are the points on the plane, expressed as ordered pairs

(a, b), where a represents the coordinate for the horizontal axis and b represents the coordinate for the vertical axis.

Let's consider the number $-2 + 3i$. The real part of the complex number is -2 and the imaginary part is 3. We plot the ordered pair $(-2, 3)$ to represent the complex number $-2 + 3i$, as shown in Figure 2.

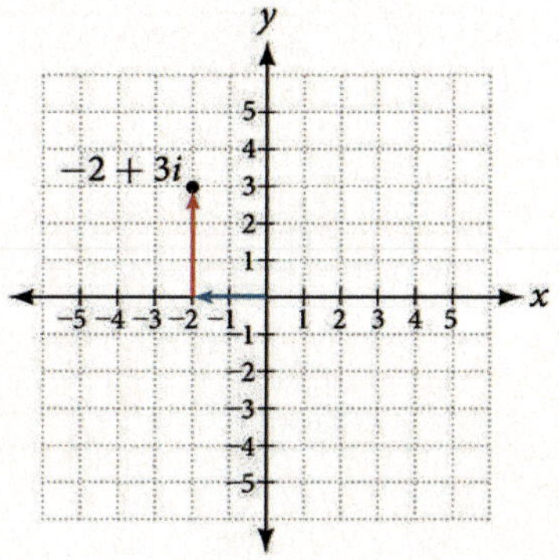

Figure 2

Complex Plane

In the complex plane, the horizontal axis is the real axis, and the vertical axis is the imaginary axis, as shown in Figure 3.

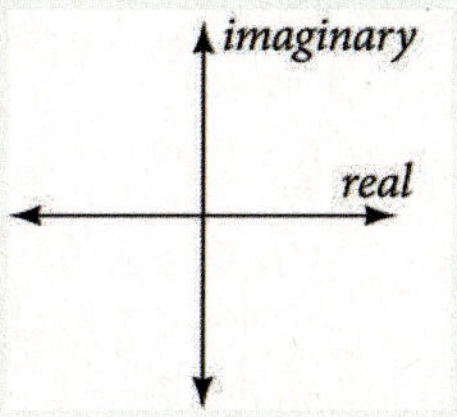

Figure 3

HOW TO

Given a complex number, represent its components on the complex plane.

1. Determine the real part and the imaginary part of the complex number.
2. Move along the horizontal axis to show the real part of the number.
3. Move parallel to the vertical axis to show the imaginary part of the number.
4. Plot the point.

EXAMPLE 2

Plotting a Complex Number on the Complex Plane

Plot the complex number $3 - 4i$ on the complex plane.

Solution

The real part of the complex number is 3, and the imaginary part is –4. We plot the ordered pair $(3, -4)$ as shown in Figure 4.

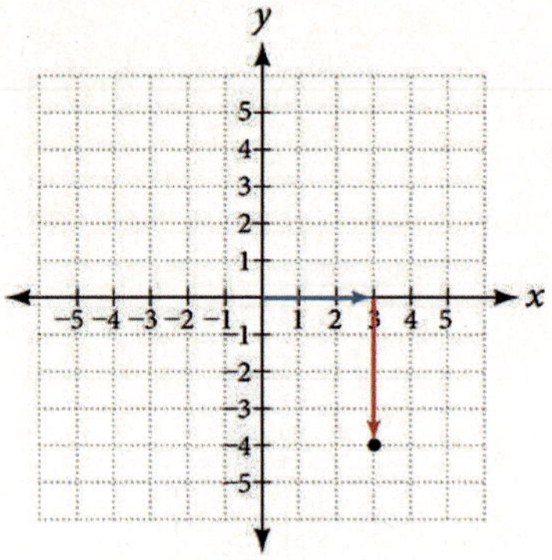

Figure 4

> **TRY IT** #2 Plot the complex number $-4 - i$ on the complex plane.

Adding and Subtracting Complex Numbers

Just as with real numbers, we can perform arithmetic operations on complex numbers. To add or subtract complex numbers, we combine the real parts and then combine the imaginary parts.

Complex Numbers: Addition and Subtraction

Adding complex numbers:

$$(a + bi) + (c + di) = (a + c) + (b + d)\,i$$

Subtracting complex numbers:

$$(a + bi) - (c + di) = (a - c) + (b - d)\,i$$

HOW TO

Given two complex numbers, find the sum or difference.

1. Identify the real and imaginary parts of each number.
2. Add or subtract the real parts.
3. Add or subtract the imaginary parts.

EXAMPLE 3

Adding and Subtracting Complex Numbers

Add or subtract as indicated.

ⓐ $(3 - 4i) + (2 + 5i)$ ⓑ $(-5 + 7i) - (-11 + 2i)$

Solution

We add the real parts and add the imaginary parts.

ⓐ

$$\begin{aligned}(3 - 4i) + (2 + 5i) &= 3 - 4i + 2 + 5i \\ &= 3 + 2 + (-4i) + 5i \\ &= (3 + 2) + (-4 + 5)i \\ &= 5 + i\end{aligned}$$

ⓑ

$$\begin{aligned}(-5 + 7i) - (-11 + 2i) &= -5 + 7i + 11 - 2i \\ &= -5 + 11 + 7i - 2i \\ &= (-5 + 11) + (7 - 2)i \\ &= 6 + 5i\end{aligned}$$

> **TRY IT** #3 Subtract $2 + 5i$ from $3 - 4i$.

Multiplying Complex Numbers

Multiplying complex numbers is much like multiplying binomials. The major difference is that we work with the real and imaginary parts separately.

Multiplying a Complex Number by a Real Number

Lets begin by multiplying a complex number by a real number. We distribute the real number just as we would with a binomial. Consider, for example, $3(6 + 2i)$:

$$\begin{aligned} 3(6 + 2i) &= (3 \cdot 6) + (3 \cdot 2i) && \text{Distribute.} \\ &= 18 + 6i && \text{Simplify.} \end{aligned}$$

HOW TO

Given a complex number and a real number, multiply to find the product.

1. Use the distributive property.
2. Simplify.

EXAMPLE 4

Multiplying a Complex Number by a Real Number

Find the product $4(2 + 5i)$.

Solution

Distribute the 4.

$$\begin{aligned} 4(2 + 5i) &= (4 \cdot 2) + (4 \cdot 5i) \\ &= 8 + 20i \end{aligned}$$

> **TRY IT** #4 Find the product: $\frac{1}{2}(5 - 2i)$.

Multiplying Complex Numbers Together

Now, let's multiply two complex numbers. We can use either the distributive property or more specifically the FOIL method because we are dealing with binomials. Recall that FOIL is an acronym for multiplying First, Inner, Outer, and Last terms together. The difference with complex numbers is that when we get a squared term, i^2, it equals -1.

$$\begin{aligned} (a + bi)(c + di) &= ac + adi + bci + bdi^2 \\ &= ac + adi + bci - bd && i^2 = -1 \\ &= (ac - bd) + (ad + bc)i && \text{Group real terms and imaginary terms.} \end{aligned}$$

HOW TO

Given two complex numbers, multiply to find the product.

1. Use the distributive property or the FOIL method.
2. Remember that $i^2 = -1$.
3. Group together the real terms and the imaginary terms

EXAMPLE 5

Multiplying a Complex Number by a Complex Number

Multiply: $(4 + 3i)(2 - 5i)$.

Solution

$$\begin{aligned}(4+3i)(2-5i) &= 4(2) - 4(5i) + 3i(2) - (3i)(5i) \\ &= 8 - 20i + 6i - 15\left(i^2\right) \\ &= (8+15) + (-20+6)i \\ &= 23 - 14i\end{aligned}$$

TRY IT #5 Multiply: $(3 - 4i)(2 + 3i)$.

Dividing Complex Numbers

Dividing two complex numbers is more complicated than adding, subtracting, or multiplying because we cannot divide by an imaginary number, meaning that any fraction must have a real-number denominator to write the answer in standard form $a + bi$. We need to find a term by which we can multiply the numerator and the denominator that will eliminate the imaginary portion of the denominator so that we end up with a real number as the denominator. This term is called the complex conjugate of the denominator, which is found by changing the sign of the imaginary part of the complex number. In other words, the complex conjugate of $a + bi$ is $a - bi$. For example, the product of $a + bi$ and $a - bi$ is

$$\begin{aligned}(a+bi)(a-bi) &= a^2 - abi + abi - b^2i^2 \\ &= a^2 + b^2\end{aligned}$$

The result is a real number.

Note that complex conjugates have an opposite relationship: The complex conjugate of $a + bi$ is $a - bi$, and the complex conjugate of $a - bi$ is $a + bi$. Further, when a quadratic equation with real coefficients has complex solutions, the solutions are always complex conjugates of one another.

Suppose we want to divide $c + di$ by $a + bi$, where neither a nor b equals zero. We first write the division as a fraction, then find the complex conjugate of the denominator, and multiply.

$$\frac{c+di}{a+bi} \text{ where } a \neq 0 \text{ and } b \neq 0$$

Multiply the numerator and denominator by the complex conjugate of the denominator.

$$\frac{(c+di)}{(a+bi)} \cdot \frac{(a-bi)}{(a-bi)} = \frac{(c+di)(a-bi)}{(a+bi)(a-bi)}$$

Apply the distributive property.

$$= \frac{ca - cbi + adi - bdi^2}{a^2 - abi + abi - b^2i^2}$$

Simplify, remembering that $i^2 = -1$.

$$\begin{aligned}&= \frac{ca - cbi + adi - bd(-1)}{a^2 - abi + abi - b^2(-1)} \\ &= \frac{(ca+bd) + (ad-cb)i}{a^2+b^2}\end{aligned}$$

The Complex Conjugate

The **complex conjugate** of a complex number $a + bi$ is $a - bi$. It is found by changing the sign of the imaginary part of the complex number. The real part of the number is left unchanged.

- When a complex number is multiplied by its complex conjugate, the result is a real number.

- When a complex number is added to its complex conjugate, the result is a real number.

EXAMPLE 6

Finding Complex Conjugates

Find the complex conjugate of each number.

ⓐ $2 + i\sqrt{5}$ ⓑ $-\frac{1}{2}i$

Solution

ⓐ The number is already in the form $a + bi$. The complex conjugate is $a - bi$, or $2 - i\sqrt{5}$.

ⓑ We can rewrite this number in the form $a + bi$ as $0 - \frac{1}{2}i$. The complex conjugate is $a - bi$, or $0 + \frac{1}{2}i$. This can be written simply as $\frac{1}{2}i$.

Analysis

Although we have seen that we can find the complex conjugate of an imaginary number, in practice we generally find the complex conjugates of only complex numbers with both a real and an imaginary component. To obtain a real number from an imaginary number, we can simply multiply by i.

TRY IT #6 Find the complex conjugate of $-3 + 4i$.

HOW TO

Given two complex numbers, divide one by the other.

1. Write the division problem as a fraction.
2. Determine the complex conjugate of the denominator.
3. Multiply the numerator and denominator of the fraction by the complex conjugate of the denominator.
4. Simplify.

EXAMPLE 7

Dividing Complex Numbers

Divide: $(2 + 5i)$ by $(4 - i)$.

Solution

We begin by writing the problem as a fraction.

$$\frac{(2 + 5i)}{(4 - i)}$$

Then we multiply the numerator and denominator by the complex conjugate of the denominator.

$$\frac{(2 + 5i)}{(4 - i)} \cdot \frac{(4 + i)}{(4 + i)}$$

To multiply two complex numbers, we expand the product as we would with polynomials (using FOIL).

$$\begin{aligned}
\frac{(2+5i)}{(4-i)} \cdot \frac{(4+i)}{(4+i)} &= \frac{8+2i+20i+5i^2}{16+4i-4i-i^2} \\
&= \frac{8+2i+20i+5(-1)}{16+4i-4i-(-1)} && \text{Because } i^2 = -1. \\
&= \frac{3+22i}{17} \\
&= \frac{3}{17} + \frac{22}{17}i && \text{Separate real and imaginary parts.}
\end{aligned}$$

Note that this expresses the quotient in standard form.

Simplifying Powers of i

The powers of i are cyclic. Let's look at what happens when we raise i to increasing powers.

$$\begin{aligned} i^1 &= i \\ i^2 &= -1 \\ i^3 &= i^2 \cdot i = -1 \cdot i = -i \\ i^4 &= i^3 \cdot i = -i \cdot i = -i^2 = -(-1) = 1 \\ i^5 &= i^4 \cdot i = 1 \cdot i = i \end{aligned}$$

We can see that when we get to the fifth power of i, it is equal to the first power. As we continue to multiply i by increasing powers, we will see a cycle of four. Let's examine the next four powers of i.

$$\begin{aligned} i^6 &= i^5 \cdot i = i \cdot i = i^2 = -1 \\ i^7 &= i^6 \cdot i = i^2 \cdot i = i^3 = -i \\ i^8 &= i^7 \cdot i = i^3 \cdot i = i^4 = 1 \\ i^9 &= i^8 \cdot i = i^4 \cdot i = i^5 = i \end{aligned}$$

The cycle is repeated continuously: $i, -1, -i, 1,$ every four powers.

EXAMPLE 8

Simplifying Powers of i

Evaluate: i^{35}.

Solution

Since $i^4 = 1$, we can simplify the problem by factoring out as many factors of i^4 as possible. To do so, first determine how many times 4 goes into 35: $35 = 4 \cdot 8 + 3$.

$$i^{35} = i^{4 \cdot 8+3} = i^{4 \cdot 8} \cdot i^3 = \left(i^4\right)^8 \cdot i^3 = 1^8 \cdot i^3 = i^3 = -i$$

TRY IT #7 Evaluate: i^{18}

Q&A **Can we write i^{35} in other helpful ways?**

As we saw in Example 8, we reduced i^{35} to i^3 by dividing the exponent by 4 and using the remainder to find the simplified form. But perhaps another factorization of i^{35} may be more useful. Table 1 shows some other possible factorizations.

Factorization of i^{35}	$i^{34} \cdot i$	$i^{33} \cdot i^2$	$i^{31} \cdot i^4$	$i^{19} \cdot i^{16}$
Reduced form	$\left(i^2\right)^{17} \cdot i$	$i^{33} \cdot (-1)$	$i^{31} \cdot 1$	$i^{19} \cdot \left(i^4\right)^4$
Simplified form	$(-1)^{17} \cdot i$	$-i^{33}$	i^{31}	i^{19}

Table 1

Each of these will eventually result in the answer we obtained above but may require several more steps than our earlier method.

MEDIA

Access these online resources for additional instruction and practice with complex numbers.

Adding and Subtracting Complex Numbers (http://openstax.org/l/addsubcomplex)
Multiply Complex Numbers (http://openstax.org/l/multiplycomplex)
Multiplying Complex Conjugates (http://openstax.org/l/multcompconj)
Raising *i* to Powers (http://openstax.org/l/raisingi)

2.4 SECTION EXERCISES

Verbal

1. Explain how to add complex numbers.

2. What is the basic principle in multiplication of complex numbers?

3. Give an example to show that the product of two imaginary numbers is not always imaginary.

4. What is a characteristic of the plot of a real number in the complex plane?

Algebraic

For the following exercises, evaluate the algebraic expressions.

5. If $y = x^2 + x - 4$, evaluate y given $x = 2i$.

6. If $y = x^3 - 2$, evaluate y given $x = i$.

7. If $y = x^2 + 3x + 5$, evaluate y given $x = 2 + i$.

8. If $y = 2x^2 + x - 3$, evaluate y given $x = 2 - 3i$.

9. If $y = \frac{x+1}{2-x}$, evaluate y given $x = 5i$.

10. If $y = \frac{1+2x}{x+3}$, evaluate y given $x = 4i$.

Graphical

For the following exercises, plot the complex numbers on the complex plane.

11. $1 - 2i$

12. $-2 + 3i$

13. i

14. $-3 - 4i$

Numeric

For the following exercises, perform the indicated operation and express the result as a simplified complex number.

15. $(3 + 2i) + (5 - 3i)$

16. $(-2 - 4i) + (1 + 6i)$

17. $(-5 + 3i) - (6 - i)$

18. $(2 - 3i) - (3 + 2i)$

19. $(-4 + 4i) - (-6 + 9i)$

20. $(2 + 3i)(4i)$

21. $(5 - 2i)(3i)$

22. $(6 - 2i)(5)$

23. $(-2 + 4i)(8)$

24. $(2 + 3i)(4 - i)$

25. $(-1 + 2i)(-2 + 3i)$

26. $(4 - 2i)(4 + 2i)$

27. $(3+4i)(3-4i)$

28. $\frac{3+4i}{2}$

29. $\frac{6-2i}{3}$

30. $\frac{-5+3i}{2i}$

31. $\frac{6+4i}{i}$

32. $\frac{2-3i}{4+3i}$

33. $\frac{3+4i}{2-i}$

34. $\frac{2+3i}{2-3i}$

35. $\sqrt{-9}+3\sqrt{-16}$

36. $-\sqrt{-4}-4\sqrt{-25}$

37. $\frac{2+\sqrt{-12}}{2}$

38. $\frac{4+\sqrt{-20}}{2}$

39. i^8

40. i^{15}

41. i^{22}

Technology

For the following exercises, use a calculator to help answer the questions.

42. Evaluate $(1+i)^k$ for $k = 4, 8,$ and 12. Predict the value if $k = 16$.

43. Evaluate $(1-i)^k$ for $k = 2, 6,$ and 10. Predict the value if $k = 14$.

44. Evaluate $(1+i)^k - (1-i)^k$ for $k = 4, 8,$ and 12. Predict the value for $k = 16$.

45. Show that a solution of $x^6 + 1 = 0$ is $\frac{\sqrt{3}}{2} + \frac{1}{2}i$.

46. Show that a solution of $x^8 - 1 = 0$ is $\frac{\sqrt{2}}{2} + \frac{\sqrt{2}}{2}i$.

Extensions

For the following exercises, evaluate the expressions, writing the result as a simplified complex number.

47. $\frac{1}{i} + \frac{4}{i^3}$

48. $\frac{1}{i^{11}} - \frac{1}{i^{21}}$

49. $i^7\left(1+i^2\right)$

50. $i^{-3} + 5i^7$

51. $\frac{(2+i)(4-2i)}{(1+i)}$

52. $\frac{(1+3i)(2-4i)}{(1+2i)}$

53. $\frac{(3+i)^2}{(1+2i)^2}$

54. $\frac{3+2i}{2+i} + (4+3i)$

55. $\frac{4+i}{i} + \frac{3-4i}{1-i}$

56. $\frac{3+2i}{1+2i} - \frac{2-3i}{3+i}$

2.5 Quadratic Equations

Learning Objectives

In this section, you will:

- Solve quadratic equations by factoring.
- Solve quadratic equations by the square root property.
- Solve quadratic equations by completing the square.
- Solve quadratic equations by using the quadratic formula.

Figure 1

The computer monitor on the left in Figure 1 is a 23.6-inch model and the one on the right is a 27-inch model. Proportionally, the monitors appear very similar. If there is a limited amount of space and we desire the largest monitor possible, how do we decide which one to choose? In this section, we will learn how to solve problems such as this using four different methods.

Solving Quadratic Equations by Factoring

An equation containing a second-degree polynomial is called a quadratic equation. For example, equations such as $2x^2 + 3x - 1 = 0$ and $x^2 - 4 = 0$ are quadratic equations. They are used in countless ways in the fields of engineering, architecture, finance, biological science, and, of course, mathematics.

Often the easiest method of solving a quadratic equation is factoring. Factoring means finding expressions that can be multiplied together to give the expression on one side of the equation.

If a quadratic equation can be factored, it is written as a product of linear terms. Solving by factoring depends on the zero-product property, which states that if $a \cdot b = 0$, then $a = 0$ or $b = 0$, where *a* and *b* are real numbers or algebraic expressions. In other words, if the product of two numbers or two expressions equals zero, then one of the numbers or one of the expressions must equal zero because zero multiplied by anything equals zero.

Multiplying the factors expands the equation to a string of terms separated by plus or minus signs. So, in that sense, the operation of multiplication undoes the operation of factoring. For example, expand the factored expression $(x - 2)(x + 3)$ by multiplying the two factors together.

$$\begin{aligned}(x-2)(x+3) &= x^2 + 3x - 2x - 6 \\ &= x^2 + x - 6\end{aligned}$$

The product is a quadratic expression. Set equal to zero, $x^2 + x - 6 = 0$ is a quadratic equation. If we were to factor the equation, we would get back the factors we multiplied.

The process of factoring a quadratic equation depends on the leading coefficient, whether it is 1 or another integer. We will look at both situations; but first, we want to confirm that the equation is written in standard form, $ax^2 + bx + c = 0$, where *a*, *b*, and *c* are real numbers, and $a \neq 0$. The equation $x^2 + x - 6 = 0$ is in standard form.

We can use the zero-product property to solve quadratic equations in which we first have to factor out the greatest common factor (GCF), and for equations that have special factoring formulas as well, such as the difference of squares, both of which we will see later in this section.

The Zero-Product Property and Quadratic Equations

The **zero-product property** states

$$\text{If } a \cdot b = 0, \quad \text{then } a = 0 \ \text{ or } b = 0,$$

where *a* and *b* are real numbers or algebraic expressions.

A **quadratic equation** is an equation containing a second-degree polynomial; for example

$$ax^2 + bx + c = 0$$

where *a*, *b*, and *c* are real numbers, and if $a \neq 0$, it is in standard form.

Solving Quadratics with a Leading Coefficient of 1

In the quadratic equation $x^2 + x - 6 = 0$, the leading coefficient, or the coefficient of x^2, is 1. We have one method of factoring quadratic equations in this form.

HOW TO

Given a quadratic equation with the leading coefficient of 1, factor it.

1. Find two numbers whose product equals *c* and whose sum equals *b*.
2. Use those numbers to write two factors of the form $(x+k)$ or $(x-k)$, where *k* is one of the numbers found in step 1. Use the numbers exactly as they are. In other words, if the two numbers are 1 and -2, the factors are $(x+1)(x-2)$.
3. Solve using the zero-product property by setting each factor equal to zero and solving for the variable.

EXAMPLE 1

Factoring and Solving a Quadratic with Leading Coefficient of 1

Factor and solve the equation: $x^2+x-6=0$.

Solution

To factor $x^2+x-6=0$, we look for two numbers whose product equals -6 and whose sum equals 1. Begin by looking at the possible factors of -6.

$$1\cdot(-6)$$
$$(-6)\cdot 1$$
$$2\cdot(-3)$$
$$3\cdot(-2)$$

The last pair, $3\cdot(-2)$ sums to 1, so these are the numbers. Note that only one pair of numbers will work. Then, write the factors.

$$(x-2)(x+3)=0$$

To solve this equation, we use the zero-product property. Set each factor equal to zero and solve.

$$\begin{aligned}(x-2)(x+3)&=0\\(x-2)&=0\\x&=2\\(x+3)&=0\\x&=-3\end{aligned}$$

The two solutions are 2 and -3. We can see how the solutions relate to the graph in Figure 2. The solutions are the *x*-intercepts of $y=x^2+x-6=0$.

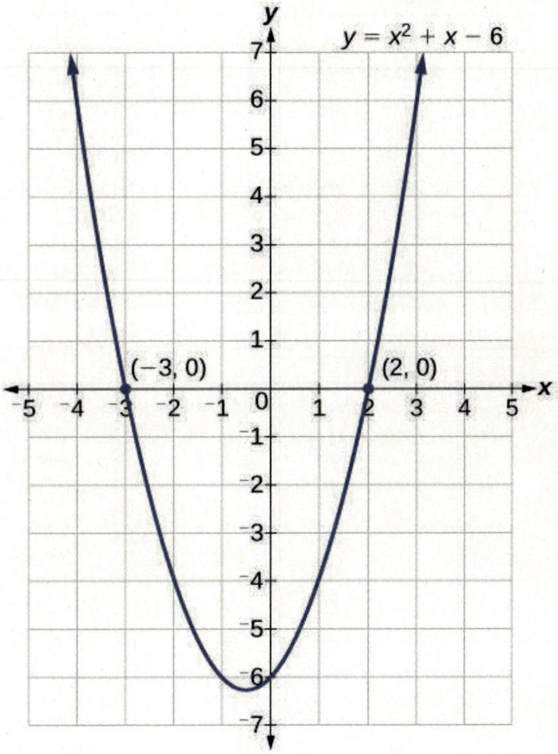

Figure 2

> **TRY IT** #1 Factor and solve the quadratic equation: $x^2 - 5x - 6 = 0$.

EXAMPLE 2

Solve the Quadratic Equation by Factoring

Solve the quadratic equation by factoring: $x^2 + 8x + 15 = 0$.

Solution

Find two numbers whose product equals 15 and whose sum equals 8. List the factors of 15.

$$\begin{array}{l} 1 \cdot 15 \\ 3 \cdot 5 \\ (-1) \cdot (-15) \\ (-3) \cdot (-5) \end{array}$$

The numbers that add to 8 are 3 and 5. Then, write the factors, set each factor equal to zero, and solve.

$$\begin{aligned} (x+3)(x+5) &= 0 \\ (x+3) &= 0 \\ x &= -3 \\ (x+5) &= 0 \\ x &= -5 \end{aligned}$$

The solutions are -3 and -5.

> **TRY IT** #2 Solve the quadratic equation by factoring: $x^2 - 4x - 21 = 0$.

EXAMPLE 3

Using the Zero-Product Property to Solve a Quadratic Equation Written as the Difference of Squares
Solve the difference of squares equation using the zero-product property: $x^2 - 9 = 0$.

Solution

Recognizing that the equation represents the difference of squares, we can write the two factors by taking the square root of each term, using a minus sign as the operator in one factor and a plus sign as the operator in the other. Solve using the zero-factor property.

$$\begin{aligned} x^2 - 9 &= 0 \\ (x-3)(x+3) &= 0 \\ \\ (x-3) &= 0 \\ x &= 3 \\ \\ (x+3) &= 0 \\ x &= -3 \end{aligned}$$

The solutions are 3 and -3.

TRY IT #3 Solve by factoring: $x^2 - 25 = 0$.

Solving a Quadratic Equation by Factoring when the Leading Coefficient is not 1

When the leading coefficient is not 1, we factor a quadratic equation using the method called grouping, which requires four terms. With the equation in standard form, let's review the grouping procedures:

1. With the quadratic in standard form, $ax^2 + bx + c = 0$, multiply $a \cdot c$.
2. Find two numbers whose product equals ac and whose sum equals b.
3. Rewrite the equation replacing the bx term with two terms using the numbers found in step 1 as coefficients of x.
4. Factor the first two terms and then factor the last two terms. The expressions in parentheses must be exactly the same to use grouping.
5. Factor out the expression in parentheses.
6. Set the expressions equal to zero and solve for the variable.

EXAMPLE 4

Solving a Quadratic Equation Using Grouping
Use grouping to factor and solve the quadratic equation: $4x^2 + 15x + 9 = 0$.

Solution

First, multiply $ac : 4(9) = 36$. Then list the factors of 36.

$$\begin{gathered} 1 \cdot 36 \\ 2 \cdot 18 \\ 3 \cdot 12 \\ 4 \cdot 9 \\ 6 \cdot 6 \end{gathered}$$

The only pair of factors that sums to 15 is $3 + 12$. Rewrite the equation replacing the b term, $15x$, with two terms using 3 and 12 as coefficients of x. Factor the first two terms, and then factor the last two terms.

$$\begin{aligned} 4x^2 + 3x + 12x + 9 &= 0 \\ x(4x+3) + 3(4x+3) &= 0 \\ (4x+3)(x+3) &= 0 \end{aligned}$$

Solve using the zero-product property.

$$\begin{aligned}(4x+3)(x+3) &= 0\\ \\ (4x+3) &= 0\\ x &= -\tfrac{3}{4}\\ \\ (x+3) &= 0\\ x &= -3\end{aligned}$$

The solutions are $-\frac{3}{4}$, and -3. See Figure 3.

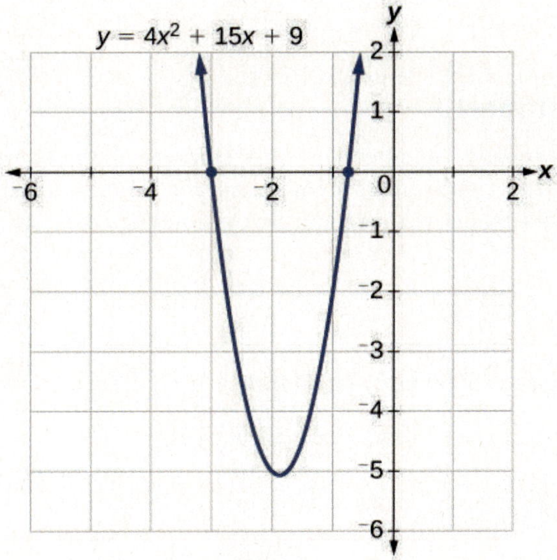

Figure 3

> TRY IT #4 Solve using factoring by grouping: $12x^2 + 11x + 2 = 0$.

EXAMPLE 5

Solving a Polynomial of Higher Degree by Factoring

Solve the equation by factoring: $-3x^3 - 5x^2 - 2x = 0$.

Solution

This equation does not look like a quadratic, as the highest power is 3, not 2. Recall that the first thing we want to do when solving any equation is to factor out the GCF, if one exists. And it does here. We can factor out $-x$ from all of the terms and then proceed with grouping.

$$\begin{aligned}-3x^3 - 5x^2 - 2x &= 0\\ -x\left(3x^2 + 5x + 2\right) &= 0\end{aligned}$$

Use grouping on the expression in parentheses.

$$\begin{aligned}-x\left(3x^2 + 3x + 2x + 2\right) &= 0\\ -x[3x(x+1) + 2(x+1)] &= 0\\ -x(3x+2)(x+1) &= 0\end{aligned}$$

Now, we use the zero-product property. Notice that we have three factors.

$$\begin{aligned} -x &= 0 \\ x &= 0 \\ 3x+2 &= 0 \\ x &= -\tfrac{2}{3} \\ x+1 &= 0 \\ x &= -1 \end{aligned}$$

The solutions are $0, -\frac{2}{3}$, and -1.

TRY IT #5 Solve by factoring: $x^3 + 11x^2 + 10x = 0$.

Using the Square Root Property

When there is no linear term in the equation, another method of solving a quadratic equation is by using the **square root property**, in which we isolate the x^2 term and take the square root of the number on the other side of the equals sign. Keep in mind that sometimes we may have to manipulate the equation to isolate the x^2 term so that the square root property can be used.

The Square Root Property

With the x^2 term isolated, the square root property states that:

$$\text{if } x^2 = k, \text{then } x = \pm\sqrt{k}$$

where k is a nonzero real number.

HOW TO

Given a quadratic equation with an x^2 term but no x term, use the square root property to solve it.

1. Isolate the x^2 term on one side of the equal sign.
2. Take the square root of both sides of the equation, putting a $\pm$ sign before the expression on the side opposite the squared term.
3. Simplify the numbers on the side with the $\pm$ sign.

EXAMPLE 6

Solving a Simple Quadratic Equation Using the Square Root Property
Solve the quadratic using the square root property: $x^2 = 8$.

Solution
Take the square root of both sides, and then simplify the radical. Remember to use a $\pm$ sign before the radical symbol.

$$\begin{aligned} x^2 &= 8 \\ x &= \pm\sqrt{8} \\ &= \pm 2\sqrt{2} \end{aligned}$$

The solutions are $2\sqrt{2}, -2\sqrt{2}$.

EXAMPLE 7

Solving a Quadratic Equation Using the Square Root Property
Solve the quadratic equation: $4x^2 + 1 = 7$.

Solution
First, isolate the x^2 term. Then take the square root of both sides.

$$\begin{aligned} 4x^2 + 1 &= 7 \\ 4x^2 &= 6 \\ x^2 &= \tfrac{6}{4} \\ x &= \pm\frac{\sqrt{6}}{2} \end{aligned}$$

The solutions are $\frac{\sqrt{6}}{2}$, and $-\frac{\sqrt{6}}{2}$.

TRY IT #6 Solve the quadratic equation using the square root property: $3(x-4)^2 = 15$.

Completing the Square

Not all quadratic equations can be factored or can be solved in their original form using the square root property. In these cases, we may use a method for solving a quadratic equation known as **completing the square**. Using this method, we add or subtract terms to both sides of the equation until we have a perfect square trinomial on one side of the equal sign. We then apply the square root property. To complete the square, the leading coefficient, *a*, must equal 1. If it does not, then divide the entire equation by *a*. Then, we can use the following procedures to solve a quadratic equation by completing the square.

We will use the example $x^2 + 4x + 1 = 0$ to illustrate each step.

1. Given a quadratic equation that cannot be factored, and with $a = 1$, first add or subtract the constant term to the right side of the equal sign.

$$x^2 + 4x = -1$$

2. Multiply the *b* term by $\frac{1}{2}$ and square it.

$$\begin{aligned} \tfrac{1}{2}(4) &= 2 \\ 2^2 &= 4 \end{aligned}$$

3. Add $\left(\frac{1}{2}b\right)^2$ to both sides of the equal sign and simplify the right side. We have

$$\begin{aligned} x^2 + 4x + 4 &= -1 + 4 \\ x^2 + 4x + 4 &= 3 \end{aligned}$$

4. The left side of the equation can now be factored as a perfect square.

$$\begin{aligned} x^2 + 4x + 4 &= 3 \\ (x+2)^2 &= 3 \end{aligned}$$

5. Use the square root property and solve.

$$\begin{aligned} \sqrt{(x+2)^2} &= \pm\sqrt{3} \\ x + 2 &= \pm\sqrt{3} \\ x &= -2 \pm \sqrt{3} \end{aligned}$$

6. The solutions are $-2 + \sqrt{3}$, and $-2 - \sqrt{3}$.

EXAMPLE 8

Solving a Quadratic by Completing the Square

Solve the quadratic equation by completing the square: $x^2 - 3x - 5 = 0$.

Solution

First, move the constant term to the right side of the equal sign.

$$x^2 - 3x = 5$$

Then, take $\frac{1}{2}$ of the *b* term and square it.

$$\begin{aligned}\tfrac{1}{2}(-3) &= -\tfrac{3}{2} \\ \left(-\tfrac{3}{2}\right)^2 &= \tfrac{9}{4}\end{aligned}$$

Add the result to both sides of the equal sign.

$$\begin{aligned}x^2 - 3x + \left(-\tfrac{3}{2}\right)^2 &= 5 + \left(-\tfrac{3}{2}\right)^2 \\ x^2 - 3x + \tfrac{9}{4} &= 5 + \tfrac{9}{4}\end{aligned}$$

Factor the left side as a perfect square and simplify the right side.

$$\left(x - \frac{3}{2}\right)^2 = \frac{29}{4}$$

Use the square root property and solve.

$$\begin{aligned}\sqrt{\left(x - \tfrac{3}{2}\right)^2} &= \pm\sqrt{\tfrac{29}{4}} \\ \left(x - \tfrac{3}{2}\right) &= \pm\tfrac{\sqrt{29}}{2} \\ x &= \tfrac{3}{2} \pm \tfrac{\sqrt{29}}{2}\end{aligned}$$

The solutions are $\frac{3+\sqrt{29}}{2}$ and $\frac{3-\sqrt{29}}{2}$.

> **TRY IT** #7 Solve by completing the square: $x^2 - 6x = 13$.

Using the Quadratic Formula

The fourth method of solving a quadratic equation is by using the quadratic formula, a formula that will solve all quadratic equations. Although the quadratic formula works on any quadratic equation in standard form, it is easy to make errors in substituting the values into the formula. Pay close attention when substituting, and use parentheses when inserting a negative number.

We can derive the quadratic formula by completing the square. We will assume that the leading coefficient is positive; if it is negative, we can multiply the equation by -1 and obtain a positive *a*. Given $ax^2 + bx + c = 0, a \neq 0$, we will complete the square as follows:

1. First, move the constant term to the right side of the equal sign:

$$ax^2 + bx = -c$$

2. As we want the leading coefficient to equal 1, divide through by *a*:

$$x^2 + \frac{b}{a}x = -\frac{c}{a}$$

3. Then, find $\frac{1}{2}$ of the middle term, and add $\left(\frac{1}{2}\frac{b}{a}\right)^2 = \frac{b^2}{4a^2}$ to both sides of the equal sign:

$$x^2 + \frac{b}{a}x + \frac{b^2}{4a^2} = \frac{b^2}{4a^2} - \frac{c}{a}$$

4. Next, write the left side as a perfect square. Find the common denominator of the right side and write it as a single fraction:

$$\left(x+\frac{b}{2a}\right)^2=\frac{b^2-4ac}{4a^2}$$

5. Now, use the square root property, which gives

$$\begin{aligned} x+\frac{b}{2a} &= \pm\sqrt{\frac{b^2-4ac}{4a^2}} \\ x+\frac{b}{2a} &= \frac{\pm\sqrt{b^2-4ac}}{2a} \end{aligned}$$

6. Finally, add $-\frac{b}{2a}$ to both sides of the equation and combine the terms on the right side. Thus,

$$x=\frac{-b\pm\sqrt{b^2-4ac}}{2a}$$

The Quadratic Formula

Written in standard form, $ax^2+bx+c=0$, any quadratic equation can be solved using the **quadratic formula**:

$$x=\frac{-b\pm\sqrt{b^2-4ac}}{2a}$$

where *a*, *b*, and *c* are real numbers and $a\neq 0$.

HOW TO

Given a quadratic equation, solve it using the quadratic formula

1. Make sure the equation is in standard form: $ax^2+bx+c=0$.
2. Make note of the values of the coefficients and constant term, a, b, and c.
3. Carefully substitute the values noted in step 2 into the equation. To avoid needless errors, use parentheses around each number input into the formula.
4. Calculate and solve.

EXAMPLE 9

Solve the Quadratic Equation Using the Quadratic Formula

Solve the quadratic equation: $x^2+5x+1=0$.

Solution

Identify the coefficients: $a=1, b=5, c=1$. Then use the quadratic formula.

$$\begin{aligned} x &= \frac{-(5)\pm\sqrt{(5)^2-4(1)(1)}}{2(1)} \\ &= \frac{-5\pm\sqrt{25-4}}{2} \\ &= \frac{-5\pm\sqrt{21}}{2} \end{aligned}$$

EXAMPLE 10

Solving a Quadratic Equation with the Quadratic Formula

Use the quadratic formula to solve $x^2+x+2=0$.

Solution

First, we identify the coefficients: $a = 1, b = 1$, and $c = 2$.

Substitute these values into the quadratic formula.

$$\begin{aligned} x &= \frac{-b \pm \sqrt{b^2 - 4ac}}{2a} \\ &= \frac{-(1) \pm \sqrt{(1)^2 - (4) \cdot (1) \cdot (2)}}{2 \cdot 1} \\ &= \frac{-1 \pm \sqrt{1 - 8}}{2} \\ &= \frac{-1 \pm \sqrt{-7}}{2} \\ &= \frac{-1 \pm i\sqrt{7}}{2} \end{aligned}$$

The solutions to the equation are $\frac{-1+i\sqrt{7}}{2}$ and $\frac{-1-i\sqrt{7}}{2}$

TRY IT #8 Solve the quadratic equation using the quadratic formula: $9x^2 + 3x - 2 = 0$.

The Discriminant

The quadratic formula not only generates the solutions to a quadratic equation, it tells us about the nature of the solutions when we consider the discriminant, or the expression under the radical, $b^2 - 4ac$. The discriminant tells us whether the solutions are real numbers or complex numbers, and how many solutions of each type to expect. Table 1 relates the value of the discriminant to the solutions of a quadratic equation.

Value of Discriminant	Results
$b^2 - 4ac = 0$	One rational solution (double solution)
$b^2 - 4ac > 0$, perfect square	Two rational solutions
$b^2 - 4ac > 0$, not a perfect square	Two irrational solutions
$b^2 - 4ac < 0$	Two complex solutions

Table 1

The Discriminant

For $ax^2 + bx + c = 0$, where a, b, and c are real numbers, the **discriminant** is the expression under the radical in the quadratic formula: $b^2 - 4ac$. It tells us whether the solutions are real numbers or complex numbers and how many solutions of each type to expect.

EXAMPLE 11

Using the Discriminant to Find the Nature of the Solutions to a Quadratic Equation

Use the discriminant to find the nature of the solutions to the following quadratic equations:

ⓐ $x^2 + 4x + 4 = 0$ ⓑ $8x^2 + 14x + 3 = 0$ ⓒ $3x^2 - 5x - 2 = 0$ ⓓ $3x^2 - 10x + 15 = 0$

Solution

Calculate the discriminant $b^2 - 4ac$ for each equation and state the expected type of solutions.

ⓐ
$x^2 + 4x + 4 = 0$

$b^2 - 4ac = (4)^2 - 4(1)(4) = 0$. There will be one rational double solution.

ⓑ
$8x^2 + 14x + 3 = 0$

$b^2 - 4ac = (14)^2 - 4(8)(3) = 100$. As 100 is a perfect square, there will be two rational solutions.

ⓒ
$3x^2 - 5x - 2 = 0$

$b^2 - 4ac = (-5)^2 - 4(3)(-2) = 49$. As 49 is a perfect square, there will be two rational solutions.

ⓓ
$3x^2 - 10x + 15 = 0$

$b^2 - 4ac = (-10)^2 - 4(3)(15) = -80$. There will be two complex solutions.

Using the Pythagorean Theorem

One of the most famous formulas in mathematics is the **Pythagorean Theorem**. It is based on a right triangle, and states the relationship among the lengths of the sides as $a^2 + b^2 = c^2$, where a and b refer to the legs of a right triangle adjacent to the $90°$ angle, and c refers to the hypotenuse. It has immeasurable uses in architecture, engineering, the sciences, geometry, trigonometry, and algebra, and in everyday applications.

We use the Pythagorean Theorem to solve for the length of one side of a triangle when we have the lengths of the other two. Because each of the terms is squared in the theorem, when we are solving for a side of a triangle, we have a quadratic equation. We can use the methods for solving quadratic equations that we learned in this section to solve for the missing side.

The Pythagorean Theorem is given as

$$a^2 + b^2 = c^2$$

where a and b refer to the legs of a right triangle adjacent to the $90°$ angle, and c refers to the hypotenuse, as shown in Figure 4.

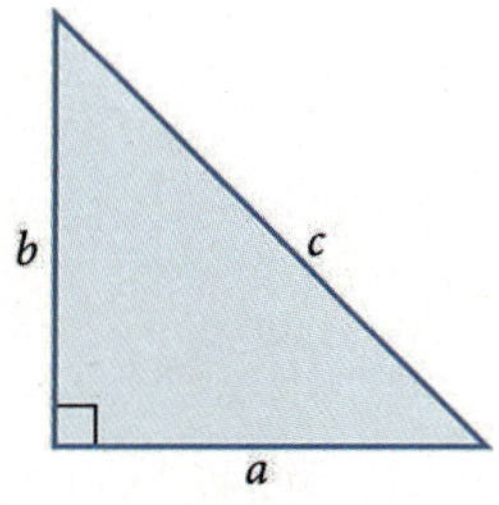

Figure 4

EXAMPLE 12

Finding the Length of the Missing Side of a Right Triangle
Find the length of the missing side of the right triangle in Figure 5.

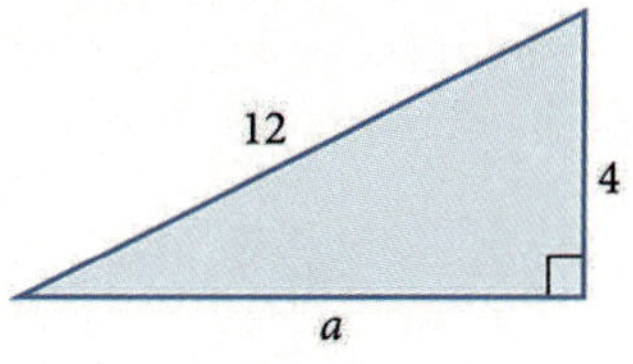

Figure 5

Solution

As we have measurements for side *b* and the hypotenuse, the missing side is *a*.

$$\begin{aligned} a^2 + b^2 &= c^2 \\ a^2 + (4)^2 &= (12)^2 \\ a^2 + 16 &= 144 \\ a^2 &= 128 \\ a &= \sqrt{128} \\ &= 8\sqrt{2} \end{aligned}$$

TRY IT #9 Use the Pythagorean Theorem to solve the right triangle problem: Leg *a* measures 4 units, leg *b* measures 3 units. Find the length of the hypotenuse.

MEDIA

Access these online resources for additional instruction and practice with quadratic equations.

Solving Quadratic Equations by Factoring (http://openstax.org/l/quadreqfactor)
The Zero-Product Property (http://openstax.org/l/zeroprodprop)
Completing the Square (http://openstax.org/l/complthesqr)
Quadratic Formula with Two Rational Solutions (http://openstax.org/l/quadrformrat)
Length of a leg of a right triangle (http://openstax.org/l/leglengthtri)

2.5 SECTION EXERCISES

Verbal

1. How do we recognize when an equation is quadratic?

2. When we solve a quadratic equation, how many solutions should we always start out seeking? Explain why when solving a quadratic equation in the form $ax^2 + bx + c = 0$ we may graph the equation $y = ax^2 + bx + c$ and have no zeroes (x-intercepts).

3. When we solve a quadratic equation by factoring, why do we move all terms to one side, having zero on the other side?

4. In the quadratic formula, what is the name of the expression under the radical sign $b^2 - 4ac$, and how does it determine the number of and nature of our solutions?

5. Describe two scenarios where using the square root property to solve a quadratic equation would be the most efficient method.

Algebraic

For the following exercises, solve the quadratic equation by factoring.

6. $x^2 + 4x - 21 = 0$

7. $x^2 - 9x + 18 = 0$

8. $2x^2 + 9x - 5 = 0$

9. $6x^2 + 17x + 5 = 0$

10. $4x^2 - 12x + 8 = 0$

11. $3x^2 - 75 = 0$

12. $8x^2 + 6x - 9 = 0$

13. $4x^2 = 9$

14. $2x^2 + 14x = 36$

15. $5x^2 = 5x + 30$

16. $4x^2 = 5x$

17. $7x^2 + 3x = 0$

18. $\frac{x}{3} - \frac{9}{x} = 2$

For the following exercises, solve the quadratic equation by using the square root property.

19. $x^2 = 36$

20. $x^2 = 49$

21. $(x - 1)^2 = 25$

22. $(x - 3)^2 = 7$

23. $(2x + 1)^2 = 9$

24. $(x - 5)^2 = 4$

For the following exercises, solve the quadratic equation by completing the square. Show each step.

25. $x^2 - 9x - 22 = 0$

26. $2x^2 - 8x - 5 = 0$

27. $x^2 - 6x = 13$

28. $x^2 + \frac{2}{3}x - \frac{1}{3} = 0$

29. $2 + z = 6z^2$

30. $6p^2 + 7p - 20 = 0$

31. $2x^2 - 3x - 1 = 0$

For the following exercises, determine the discriminant, and then state how many solutions there are and the nature of the solutions. Do not solve.

32. $2x^2 - 6x + 7 = 0$

33. $x^2 + 4x + 7 = 0$

34. $3x^2 + 5x - 8 = 0$

35. $9x^2 - 30x + 25 = 0$

36. $2x^2 - 3x - 7 = 0$

37. $6x^2 - x - 2 = 0$

For the following exercises, solve the quadratic equation by using the quadratic formula. If the solutions are not real, state No Real Solution.

38. $2x^2 + 5x + 3 = 0$

39. $x^2 + x = 4$

40. $2x^2 - 8x - 5 = 0$

41. $3x^2 - 5x + 1 = 0$

42. $x^2 + 4x + 2 = 0$

43. $4 + \frac{1}{x} - \frac{1}{x^2} = 0$

Technology

*For the following exercises, enter the expressions into your graphing utility and find the zeroes to the equation (the x-intercepts) by using **2nd CALC 2:zero**. Recall finding zeroes will ask left bound (move your cursor to the left of the zero,enter), then right bound (move your cursor to the right of the zero,enter), then guess (move your cursor between the bounds near the zero, enter). Round your answers to the nearest thousandth.*

44. $Y_1 = 4x^2 + 3x - 2$

45. $Y_1 = -3x^2 + 8x - 1$

46. $Y_1 = 0.5x^2 + x - 7$

47. To solve the quadratic equation $x^2 + 5x - 7 = 4$, we can graph these two equations

$$Y_1 = x^2 + 5x - 7$$
$$Y_2 = 4$$

and find the points of intersection. Recall 2nd CALC 5:intersection. Do this and find the solutions to the nearest tenth.

48. To solve the quadratic equation $0.3x^2 + 2x - 4 = 2$, we can graph these two equations

$$Y_1 = 0.3x^2 + 2x - 4$$
$$Y_2 = 2$$

and find the points of intersection. Recall 2nd CALC 5:intersection. Do this and find the solutions to the nearest tenth.

Extensions

49. Beginning with the general form of a quadratic equation, $ax^2 + bx + c = 0$, solve for x by using the completing the square method, thus deriving the quadratic formula.

50. Show that the sum of the two solutions to the quadratic equation is $-\frac{b}{a}$.

51. A person has a garden that has a length 10 feet longer than the width. Set up a quadratic equation to find the dimensions of the garden if its area is 119 ft.^{2}. Solve the quadratic equation to find the length and width.

52. Abercrombie and Fitch stock had a price given as $P = 0.2t^2 - 5.6t + 50.2$, where t is the time in months from 1999 to 2001. ($t = 1$ is January 1999). Find the two months in which the price of the stock was \$30.

53. Suppose that an equation is given $p = -2x^2 + 280x - 1000$, where x represents the number of items sold at an auction and p is the profit made by the business that ran the auction. How many items sold would make this profit a maximum? Solve this by graphing the expression in your graphing utility and finding the maximum using 2nd CALC maximum. To obtain a good window for the curve, set x [0,200] and y [0,10000].

Real-World Applications

54. A formula for the normal systolic blood pressure for a man age A, measured in mmHg, is given as $P = 0.006A^2 - 0.02A + 120$. Find the age to the nearest year of a man whose normal blood pressure measures 125 mmHg.

55. The cost function for a certain company is $C = 60x + 300$ and the revenue is given by $R = 100x - 0.5x^2$. Recall that profit is revenue minus cost. Set up a quadratic equation and find two values of x (production level) that will create a profit of \$300.

56. A falling object travels a distance given by the formula $d = 5t + 16t^2$ ft, where t is measured in seconds. How long will it take for the object to travel 74 ft?

57. A vacant lot is being converted into a community garden. The garden and the walkway around its perimeter have an area of 378 ft^2. Find the width of the walkway if the garden is 12 ft. wide by 15 ft. long.

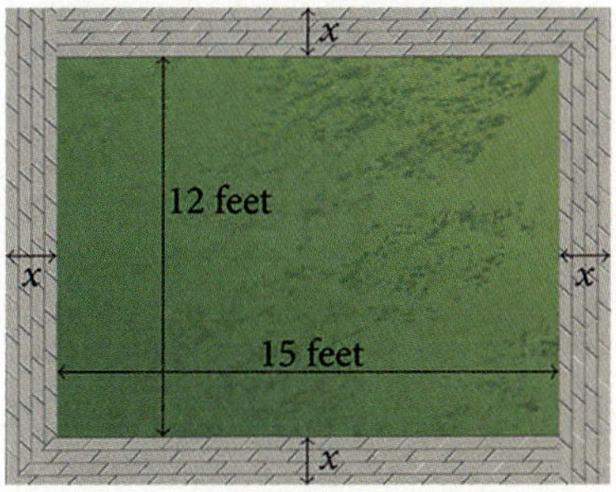

58. An epidemiological study of the spread of a certain influenza strain that hit a small school population found that the total number of students, P, who contracted the flu t days after it broke out is given by the model $P = -t^2 + 13t + 130$, where $1 \leq t \leq 6$. Find the day that 160 students had the flu. Recall that the restriction on t is at most 6.

2.6 Other Types of Equations

Learning Objectives

In this section, you will:

- Solve equations involving rational exponents.
- Solve equations using factoring.
- Solve radical equations.
- Solve absolute value equations.
- Solve other types of equations.

We have solved linear equations, rational equations, and quadratic equations using several methods. However, there are many other types of equations, and we will investigate a few more types in this section. We will look at equations involving rational exponents, polynomial equations, radical equations, absolute value equations, equations in quadratic form, and some rational equations that can be transformed into quadratics. Solving any equation, however, employs the same basic algebraic rules. We will learn some new techniques as they apply to certain equations, but the algebra never changes.

Solving Equations Involving Rational Exponents

Rational exponents are exponents that are fractions, where the numerator is a power and the denominator is a root. For example, $16^{\frac{1}{2}}$ is another way of writing $\sqrt{16}$; $8^{\frac{1}{3}}$ is another way of writing $\sqrt[3]{8}$. The ability to work with rational exponents is a useful skill, as it is highly applicable in calculus.

We can solve equations in which a variable is raised to a rational exponent by raising both sides of the equation to the reciprocal of the exponent. The reason we raise the equation to the reciprocal of the exponent is because we want to eliminate the exponent on the variable term, and a number multiplied by its reciprocal equals 1. For example, $\frac{2}{3}\left(\frac{3}{2}\right) = 1$,

$3\left(\frac{1}{3}\right) = 1$, and so on.

Rational Exponents

A rational exponent indicates a power in the numerator and a root in the denominator. There are multiple ways of writing an expression, a variable, or a number with a rational exponent:

$$a^{\frac{m}{n}} = \left(a^{\frac{1}{n}}\right)^m = \left(a^m\right)^{\frac{1}{n}} = \sqrt[n]{a^m} = \left(\sqrt[n]{a}\right)^m$$

EXAMPLE 1

Evaluating a Number Raised to a Rational Exponent

Evaluate $8^{\frac{2}{3}}$.

Solution

Whether we take the root first or the power first depends on the number. It is easy to find the cube root of 8, so rewrite $8^{\frac{2}{3}}$ as $\left(8^{\frac{1}{3}}\right)^2$.

$$\begin{aligned} \left(8^{\frac{1}{3}}\right)^2 &= (2)^2 \\ &= 4 \end{aligned}$$

TRY IT #1 Evaluate $64^{-\frac{1}{3}}$.

EXAMPLE 2

Solve the Equation Including a Variable Raised to a Rational Exponent

Solve the equation in which a variable is raised to a rational exponent: $x^{\frac{5}{4}} = 32$.

Solution

The way to remove the exponent on x is by raising both sides of the equation to a power that is the reciprocal of $\frac{5}{4}$, which is $\frac{4}{5}$.

$$\begin{aligned} x^{\frac{5}{4}} &= 32 \\ \left(x^{\frac{5}{4}}\right)^{\frac{4}{5}} &= (32)^{\frac{4}{5}} \\ x &= (2)^4 && \text{The fifth root of 32 is 2.} \\ &= 16 \end{aligned}$$

TRY IT #2 Solve the equation $x^{\frac{3}{2}} = 125$.

EXAMPLE 3

Solving an Equation Involving Rational Exponents and Factoring

Solve $3x^{\frac{3}{4}} = x^{\frac{1}{2}}$.

Solution

This equation involves rational exponents as well as factoring rational exponents. Let us take this one step at a time. First, put the variable terms on one side of the equal sign and set the equation equal to zero.

$$\begin{aligned} 3x^{\frac{3}{4}} - \left(x^{\frac{1}{2}}\right) &= x^{\frac{1}{2}} - \left(x^{\frac{1}{2}}\right) \\ 3x^{\frac{3}{4}} - x^{\frac{1}{2}} &= 0 \end{aligned}$$

Now, it looks like we should factor the left side, but what do we factor out? We can always factor the term with the lowest exponent. Rewrite $x^{\frac{1}{2}}$ as $x^{\frac{2}{4}}$. Then, factor out $x^{\frac{2}{4}}$ from both terms on the left.

$$\begin{aligned} 3x^{\frac{3}{4}} - x^{\frac{2}{4}} &= 0 \\ x^{\frac{2}{4}}\left(3x^{\frac{1}{4}} - 1\right) &= 0 \end{aligned}$$

Where did $x^{\frac{1}{4}}$ come from? Remember, when we multiply two numbers with the same base, we add the exponents. Therefore, if we multiply $x^{\frac{2}{4}}$ back in using the distributive property, we get the expression we had before the factoring, which is what should happen. We need an exponent such that when added to $\frac{2}{4}$ equals $\frac{3}{4}$. Thus, the exponent on x in the parentheses is $\frac{1}{4}$.

Let us continue. Now we have two factors and can use the zero factor theorem.

$$\begin{aligned} x^{\frac{2}{4}}\left(3x^{\frac{1}{4}} - 1\right) &= 0 \\ x^{\frac{2}{4}} &= 0 \\ x &= 0 \\ 3x^{\frac{1}{4}} - 1 &= 0 \\ 3x^{\frac{1}{4}} &= 1 \\ x^{\frac{1}{4}} &= \frac{1}{3} && \text{Divide both sides by 3.} \\ \left(x^{\frac{1}{4}}\right)^4 &= \left(\frac{1}{3}\right)^4 && \text{Raise both sides to the reciprocal of } \frac{1}{4}. \\ x &= \frac{1}{81} \end{aligned}$$

The two solutions are 0 and $\frac{1}{81}$.

TRY IT #3 Solve: $(x+5)^{\frac{3}{2}} = 8$.

Solving Equations Using Factoring

We have used factoring to solve quadratic equations, but it is a technique that we can use with many types of polynomial equations, which are equations that contain a string of terms including numerical coefficients and variables. When we are faced with an equation containing polynomials of degree higher than 2, we can often solve them by factoring.

Polynomial Equations

A polynomial of degree n is an expression of the type

$$a_n x^n + a_{n-1}x^{n-1} + \quad \cdot \quad \cdot \quad \cdot \quad + a_2x^2 + a_1x + a_0$$

where n is a positive integer and $a_n, \ldots, a_0$ are real numbers and $a_n \neq 0$.

Setting the polynomial equal to zero gives a **polynomial equation**. The total number of solutions (real and complex) to a polynomial equation is equal to the highest exponent *n*.

EXAMPLE 4

Solving a Polynomial by Factoring

Solve the polynomial by factoring: $5x^4 = 80x^2$.

Solution

First, set the equation equal to zero. Then factor out what is common to both terms, the GCF.

$$\begin{aligned} 5x^4 - 80x^2 &= 0 \\ 5x^2\left(x^2 - 16\right) &= 0 \end{aligned}$$

Notice that we have the difference of squares in the factor $x^2 - 16$, which we will continue to factor and obtain two solutions. The first term, $5x^2$, generates, technically, two solutions as the exponent is 2, but they are the same solution.

$$\begin{aligned} 5x^2 &= 0 \\ x &= 0 \\ x^2 - 16 &= 0 \\ (x-4)(x+4) &= 0 \\ x &= 4 \\ x &= -4 \end{aligned}$$

The solutions are 0 (double solution), 4, and -4.

Analysis

We can see the solutions on the graph in Figure 1. The *x*-coordinates of the points where the graph crosses the *x*-axis are the solutions—the *x*-intercepts. Notice on the graph that at the solution 0, the graph touches the *x*-axis and bounces back. It does not cross the *x*-axis. This is typical of double solutions.

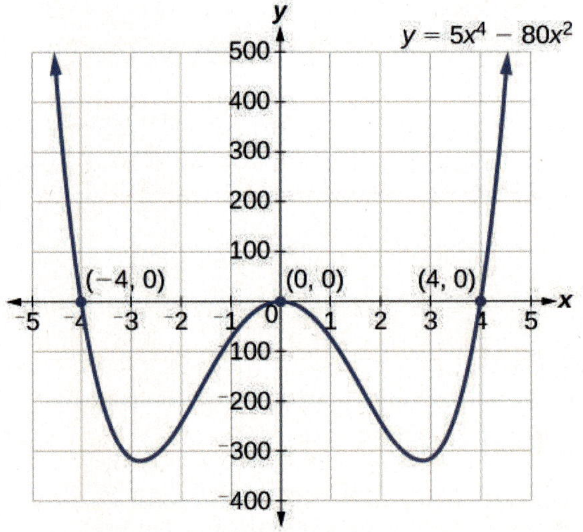

Figure 1

TRY IT #4 Solve by factoring: $12x^4 = 3x^2$.

EXAMPLE 5

Solve a Polynomial by Grouping

Solve a polynomial by grouping: $x^3 + x^2 - 9x - 9 = 0$.

Solution

This polynomial consists of 4 terms, which we can solve by grouping. Grouping procedures require factoring the first two terms and then factoring the last two terms. If the factors in the parentheses are identical, we can continue the process and solve, unless more factoring is suggested.

$$\begin{aligned} x^3 + x^2 - 9x - 9 &= 0 \\ x^2(x+1) - 9(x+1) &= 0 \\ \left(x^2 - 9\right)(x+1) &= 0 \end{aligned}$$

The grouping process ends here, as we can factor $x^2 - 9$ using the difference of squares formula.

$$\begin{aligned} \left(x^2 - 9\right)(x+1) &= 0 \\ (x-3)(x+3)(x+1) &= 0 \\ x &= 3 \\ x &= -3 \\ x &= -1 \end{aligned}$$

The solutions are $3, -3$, and -1. Note that the highest exponent is 3 and we obtained 3 solutions. We can see the solutions, the x-intercepts, on the graph in Figure 2.

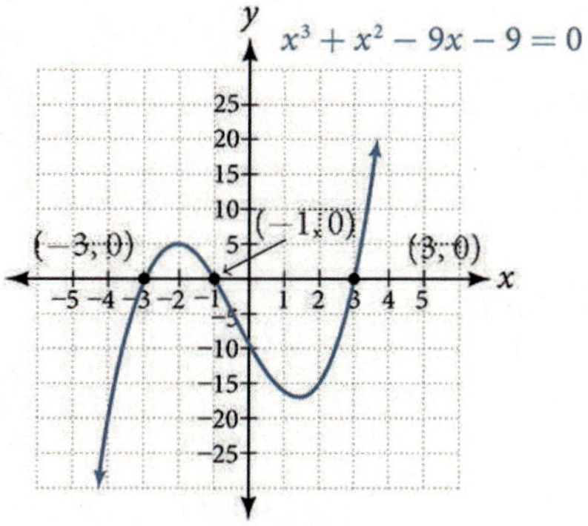

Figure 2

Analysis

We looked at solving quadratic equations by factoring when the leading coefficient is 1. When the leading coefficient is not 1, we solved by grouping. Grouping requires four terms, which we obtained by splitting the linear term of quadratic equations. We can also use grouping for some polynomials of degree higher than 2, as we saw here, since there were already four terms.

Solving Radical Equations

Radical equations are equations that contain variables in the radicand (the expression under a radical symbol), such as

$$\begin{aligned} \sqrt{3x+18} &= x \\ \sqrt{x+3} &= x - 3 \\ \sqrt{x+5} - \sqrt{x-3} &= 2 \end{aligned}$$

Radical equations may have one or more radical terms, and are solved by eliminating each radical, one at a time. We have to be careful when solving radical equations, as it is not unusual to find **extraneous solutions**, roots that are not, in fact, solutions to the equation. These solutions are not due to a mistake in the solving method, but result from the process of raising both sides of an equation to a power. However, checking each answer in the original equation will confirm the true solutions.

Radical Equations

An equation containing terms with a variable in the radicand is called a **radical equation**.

HOW TO

Given a radical equation, solve it.

1. Isolate the radical expression on one side of the equal sign. Put all remaining terms on the other side.
2. If the radical is a square root, then square both sides of the equation. If it is a cube root, then raise both sides of the equation to the third power. In other words, for an *n*th root radical, raise both sides to the *n*th power. Doing so eliminates the radical symbol.
3. Solve the remaining equation.
4. If a radical term still remains, repeat steps 1–2.
5. Confirm solutions by substituting them into the original equation.

EXAMPLE 6

Solving an Equation with One Radical

Solve $\sqrt{15-2x}=x$.

Solution

The radical is already isolated on the left side of the equal side, so proceed to square both sides.

$$\begin{aligned}\sqrt{15-2x} &= x\\ \left(\sqrt{15-2x}\right)^2 &= (x)^2\\ 15-2x &= x^2\end{aligned}$$

We see that the remaining equation is a quadratic. Set it equal to zero and solve.

$$\begin{aligned}0 &= x^2+2x-15\\ &= (x+5)(x-3)\\ x &= -5\\ x &= 3\end{aligned}$$

The proposed solutions are -5 and 3. Let us check each solution back in the original equation. First, check $x=-5$.

$$\begin{aligned}\sqrt{15-2x} &= x\\ \sqrt{15-2(-5)} &= -5\\ \sqrt{25} &= -5\\ 5 &\neq -5\end{aligned}$$

This is an extraneous solution. While no mistake was made solving the equation, we found a solution that does not satisfy the original equation.

Check $x=3$.

$$\begin{aligned}\sqrt{15-2x} &= x\\ \sqrt{15-2(3)} &= 3\\ \sqrt{9} &= 3\\ 3 &= 3\end{aligned}$$

The solution is 3.

> **TRY IT** #5 Solve the radical equation: $\sqrt{x+3} = 3x - 1$

EXAMPLE 7

Solving a Radical Equation Containing Two Radicals

Solve $\sqrt{2x+3} + \sqrt{x-2} = 4$.

Solution

As this equation contains two radicals, we isolate one radical, eliminate it, and then isolate the second radical.

$$\begin{aligned} \sqrt{2x+3} + \sqrt{x-2} &= 4 \\ \sqrt{2x+3} &= 4 - \sqrt{x-2} && \text{Subtract } \sqrt{x-2} \text{ from both sides.} \\ \left(\sqrt{2x+3}\right)^2 &= \left(4 - \sqrt{x-2}\right)^2 && \text{Square both sides.} \end{aligned}$$

Use the perfect square formula to expand the right side: $(a-b)^2 = a^2 - 2ab + b^2$.

$$\begin{aligned} 2x+3 &= (4)^2 - 2(4)\sqrt{x-2} + \left(\sqrt{x-2}\right)^2 \\ 2x+3 &= 16 - 8\sqrt{x-2} + (x-2) \\ 2x+3 &= 14 + x - 8\sqrt{x-2} && \text{Combine like terms.} \\ x - 11 &= -8\sqrt{x-2} && \text{Isolate the second radical.} \\ (x-11)^2 &= \left(-8\sqrt{x-2}\right)^2 && \text{Square both sides.} \\ x^2 - 22x + 121 &= 64(x-2) \end{aligned}$$

Now that both radicals have been eliminated, set the quadratic equal to zero and solve.

$$\begin{aligned} x^2 - 22x + 121 &= 64x - 128 \\ x^2 - 86x + 249 &= 0 \\ (x-3)(x-83) &= 0 && \text{Factor and solve.} \\ x &= 3 \\ x &= 83 \end{aligned}$$

The proposed solutions are 3 and 83. Check each solution in the original equation.

$$\begin{aligned} \sqrt{2x+3} + \sqrt{x-2} &= 4 \\ \sqrt{2x+3} &= 4 - \sqrt{x-2} \\ \sqrt{2(3)+3} &= 4 - \sqrt{(3)-2} \\ \sqrt{9} &= 4 - \sqrt{1} \\ 3 &= 3 \end{aligned}$$

One solution is 3.

Check $x = 83$.

$$\begin{aligned} \sqrt{2x+3} + \sqrt{x-2} &= 4 \\ \sqrt{2x+3} &= 4 - \sqrt{x-2} \\ \sqrt{2(83)+3} &= 4 - \sqrt{(83-2)} \\ \sqrt{169} &= 4 - \sqrt{81} \\ 13 &\neq -5 \end{aligned}$$

The only solution is 3. We see that 83 is an extraneous solution.

> **TRY IT** #6 Solve the equation with two radicals: $\sqrt{3x+7} + \sqrt{x+2} = 1$.

Solving an Absolute Value Equation

Next, we will learn how to solve an absolute value equation. To solve an equation such as $|2x - 6| = 8$, we notice that the absolute value will be equal to 8 if the quantity inside the absolute value bars is 8 or -8. This leads to two different equations we can solve independently.

$$\begin{aligned} 2x - 6 &= 8 \\ 2x &= 14 \\ x &= 7 \end{aligned} \qquad \text{or} \qquad \begin{aligned} 2x - 6 &= -8 \\ 2x &= -2 \\ x &= -1 \end{aligned}$$

Knowing how to solve problems involving absolute value functions is useful. For example, we may need to identify numbers or points on a line that are at a specified distance from a given reference point.

Absolute Value Equations

The absolute value of x is written as $|x|$. It has the following properties:

$$\begin{aligned} &\text{If } x \geq 0, \quad \text{then } |x| = x. \\ &\text{If } x < 0, \quad \text{then } |x| = -x. \end{aligned}$$

For real numbers A and B, an equation of the form $|A| = B$, with $B \geq 0$, will have solutions when $A = B$ or $A = -B$. If $B < 0$, the equation $|A| = B$ has no solution.

An **absolute value equation** in the form $|ax + b| = c$ has the following properties:

$$\begin{aligned} &\text{If } c < 0, |ax + b| = c \text{ has no solution.} \\ &\text{If } c = 0, |ax + b| = c \text{ has one solution.} \\ &\text{If } c > 0, |ax + b| = c \text{ has two solutions.} \end{aligned}$$

HOW TO

Given an absolute value equation, solve it.

1. Isolate the absolute value expression on one side of the equal sign.
2. If $c > 0$, write and solve two equations: $ax + b = c$ and $ax + b = -c$.

EXAMPLE 8

Solving Absolute Value Equations

Solve the following absolute value equations:

ⓐ $|6x + 4| = 8$ ⓑ $|3x + 4| = -9$ ⓒ $|3x - 5| - 4 = 6$ ⓓ $|-5x + 10| = 0$

Solution

ⓐ

$|6x + 4| = 8$

Write two equations and solve each:

$$\begin{aligned} 6x + 4 &= 8 \\ 6x &= 4 \\ x &= \frac{2}{3} \end{aligned} \qquad \begin{aligned} 6x + 4 &= -8 \\ 6x &= -12 \\ x &= -2 \end{aligned}$$

The two solutions are $\frac{2}{3}$ and -2.

ⓑ

$|3x + 4| = -9$

There is no solution as an absolute value cannot be negative.

ⓒ

$|3x - 5| - 4 = 6$

Isolate the absolute value expression and then write two equations.

$$\begin{aligned} |3x - 5| - 4 &= 6 \\ |3x - 5| &= 10 \end{aligned}$$

$$\begin{aligned} 3x - 5 &= 10 \\ 3x &= 15 \\ x &= 5 \end{aligned} \qquad \begin{aligned} 3x - 5 &= -10 \\ 3x &= -5 \\ x &= -\tfrac{5}{3} \end{aligned}$$

There are two solutions: 5, and $-\frac{5}{3}$.

ⓓ

(d) $|-5x + 10| = 0$

The equation is set equal to zero, so we have to write only one equation.

$$\begin{aligned} -5x + 10 &= 0 \\ -5x &= -10 \\ x &= 2 \end{aligned}$$

There is one solution: 2.

> **TRY IT** #7 Solve the absolute value equation: $|1 - 4x| + 8 = 13$.

Solving Other Types of Equations

There are many other types of equations in addition to the ones we have discussed so far. We will see more of them throughout the text. Here, we will discuss equations that are in quadratic form, and rational equations that result in a quadratic.

Solving Equations in Quadratic Form

Equations in quadratic form are equations with three terms. The first term has a power other than 2. The middle term has an exponent that is one-half the exponent of the leading term. The third term is a constant. We can solve equations in this form as if they were quadratic. A few examples of these equations include $x^4 - 5x^2 + 4 = 0$, $x^6 + 7x^3 - 8 = 0$, and $x^{\frac{2}{3}} + 4x^{\frac{1}{3}} + 2 = 0$. In each one, doubling the exponent of the middle term equals the exponent on the leading term. We can solve these equations by substituting a variable for the middle term.

Quadratic Form

If the exponent on the middle term is one-half of the exponent on the leading term, we have an **equation in quadratic form**, which we can solve as if it were a quadratic. We substitute a variable for the middle term to solve equations in quadratic form.

HOW TO

Given an equation quadratic in form, solve it.

1. Identify the exponent on the leading term and determine whether it is double the exponent on the middle term.
2. If it is, substitute a variable, such as *u*, for the variable portion of the middle term.
3. Rewrite the equation so that it takes on the standard form of a quadratic.
4. Solve using one of the usual methods for solving a quadratic.
5. Replace the substitution variable with the original term.

6. Solve the remaining equation.

EXAMPLE 9

Solving a Fourth-degree Equation in Quadratic Form

Solve this fourth-degree equation: $3x^4 - 2x^2 - 1 = 0$.

Solution

This equation fits the main criteria, that the power on the leading term is double the power on the middle term. Next, we will make a substitution for the variable term in the middle. Let $u = x^2$. Rewrite the equation in *u*.

$$3u^2 - 2u - 1 = 0$$

Now solve the quadratic.

$$\begin{aligned} 3u^2 - 2u - 1 &= 0 \\ (3u+1)(u-1) &= 0 \end{aligned}$$

Solve each factor and replace the original term for *u*.

$$\begin{aligned} 3u + 1 &= 0 \\ 3u &= -1 \\ u &= -\tfrac{1}{3} \\ x^2 &= -\tfrac{1}{3} \\ x &= \pm i\sqrt{\tfrac{1}{3}} \end{aligned}$$

$$\begin{aligned} u - 1 &= 0 \\ u &= 1 \\ x^2 &= 1 \\ x &= \pm 1 \end{aligned}$$

The solutions are $\pm i\sqrt{\frac{1}{3}}$ and ± 1.

TRY IT #8 Solve using substitution: $x^4 - 8x^2 - 9 = 0$.

EXAMPLE 10

Solving an Equation in Quadratic Form Containing a Binomial

Solve the equation in quadratic form: $(x+2)^2 + 11(x+2) - 12 = 0$.

Solution

This equation contains a binomial in place of the single variable. The tendency is to expand what is presented. However, recognizing that it fits the criteria for being in quadratic form makes all the difference in the solving process. First, make a substitution, letting $u = x + 2$. Then rewrite the equation in *u*.

$$\begin{aligned} u^2 + 11u - 12 &= 0 \\ (u+12)(u-1) &= 0 \end{aligned}$$

Solve using the zero-factor property and then replace *u* with the original expression.

$$\begin{aligned} u + 12 &= 0 \\ u &= -12 \\ x + 2 &= -12 \\ x &= -14 \end{aligned}$$

The second factor results in

$$\begin{aligned} u-1 &= 0 \\ u &= 1 \\ x+2 &= 1 \\ x &= -1 \end{aligned}$$

We have two solutions: -14, and -1.

TRY IT #9 Solve: $(x-5)^2 - 4(x-5) - 21 = 0$.

Solving Rational Equations Resulting in a Quadratic

Earlier, we solved rational equations. Sometimes, solving a rational equation results in a quadratic. When this happens, we continue the solution by simplifying the quadratic equation by one of the methods we have seen. It may turn out that there is no solution.

EXAMPLE 11

Solving a Rational Equation Leading to a Quadratic

Solve the following rational equation: $\frac{-4x}{x-1} + \frac{4}{x+1} = \frac{-8}{x^2-1}$.

Solution

We want all denominators in factored form to find the LCD. Two of the denominators cannot be factored further. However, $x^2-1 = (x+1)(x-1)$. Then, the LCD is $(x+1)(x-1)$. Next, we multiply the whole equation by the LCD.

$$\begin{aligned} (x+1)(x-1)\left[\frac{-4x}{x-1} + \frac{4}{x+1}\right] &= \left[\frac{-8}{(x+1)(x-1)}\right](x+1)(x-1) \\ -4x(x+1) + 4(x-1) &= -8 \\ -4x^2 - 4x + 4x - 4 &= -8 \\ -4x^2 + 4 &= 0 \\ -4\left(x^2-1\right) &= 0 \\ -4(x+1)(x-1) &= 0 \\ x &= -1 \\ x &= 1 \end{aligned}$$

In this case, either solution produces a zero in the denominator in the original equation. Thus, there is no solution.

TRY IT #10 Solve $\frac{3x+2}{x-2} + \frac{1}{x} = \frac{-2}{x^2-2x}$.

MEDIA

Access these online resources for additional instruction and practice with different types of equations.

Rational Equation with no Solution (http://openstax.org/l/rateqnosoln)
Solving equations with rational exponents using reciprocal powers (http://openstax.org/l/ratexprecpexp)
Solving radical equations part 1 of 2 (http://openstax.org/l/radeqsolvepart1)
Solving radical equations part 2 of 2 (http://openstax.org/l/radeqsolvepart2)

2.6 SECTION EXERCISES

Verbal

1. In a radical equation, what does it mean if a number is an extraneous solution?

2. Explain why possible solutions *must* be checked in radical equations.

3. Your friend tries to calculate the value $-9^{\frac{3}{2}}$ and keeps getting an ERROR message. What mistake are they probably making?

4. Explain why $|2x+5|=-7$ has no solutions.

5. Explain how to change a rational exponent into the correct radical expression.

Algebraic

For the following exercises, solve the rational exponent equation. Use factoring where necessary.

6. $x^{\frac{2}{3}}=16$

7. $x^{\frac{3}{4}}=27$

8. $2x^{\frac{1}{2}}-x^{\frac{1}{4}}=0$

9. $(x-1)^{\frac{3}{4}}=8$

10. $(x+1)^{\frac{2}{3}}=4$

11. $x^{\frac{2}{3}}-5x^{\frac{1}{3}}+6=0$

12. $x^{\frac{7}{3}}-3x^{\frac{4}{3}}-4x^{\frac{1}{3}}=0$

For the following exercises, solve the following polynomial equations by grouping and factoring.

13. $x^3+2x^2-x-2=0$

14. $3x^3-6x^2-27x+54=0$

15. $4y^3-9y=0$

16. $x^3+3x^2-25x-75=0$

17. $m^3+m^2-m-1=0$

18. $2x^5-14x^3=0$

19. $5x^3+45x=2x^2+18$

For the following exercises, solve the radical equation. Be sure to check all solutions to eliminate extraneous solutions.

20. $\sqrt{3x-1}-2=0$

21. $\sqrt{x-7}=5$

22. $\sqrt{x-1}=x-7$

23. $\sqrt{3t+5}=7$

24. $\sqrt{t+1}+9=7$

25. $\sqrt{12-x}=x$

26. $\sqrt{2x+3}-\sqrt{x+2}=2$

27. $\sqrt{3x+7}+\sqrt{x+2}=1$

28. $\sqrt{2x+3}-\sqrt{x+1}=1$

For the following exercises, solve the equation involving absolute value.

29. $|3x-4|=8$

30. $|2x-3|=-2$

31. $|1-4x|-1=5$

32. $|4x+1|-3=6$

33. $|2x-1|-7=-2$

34. $|2x+1|-2=-3$

35. $|x+5|=0$

36. $-|2x+1|=-3$

For the following exercises, solve the equation by identifying the quadratic form. Use a substitute variable and find all real solutions by factoring.

37. $x^4-10x^2+9=0$

38. $4(t-1)^2-9(t-1)=-2$

39. $\left(x^2-1\right)^2+\left(x^2-1\right)-12=0$

40. $(x+1)^2-8(x+1)-9=0$

41. $(x-3)^2-4=0$

Extensions

For the following exercises, solve for the unknown variable.

42. $x^{-2}-x^{-1}-12=0$

43. $\sqrt{|x|^2}=x$

44. $t^{10}-2t^5+1=0$

45. $\left|x^2+2x-36\right|=12$

Real-World Applications

For the following exercises, use the model for the period of a pendulum, T, such that $T=2\pi\sqrt{\frac{L}{g}}$, where the length of the pendulum is L and the acceleration due to gravity is g.

46. If the acceleration due to gravity is 9.8 m/s^2 and the period equals 1 s, find the length to the nearest cm (100 cm = 1 m).

47. If the gravity is 32 ft/s^2 and the period equals 1 s, find the length to the nearest in. (12 in. = 1 ft). Round your answer to the nearest in.

For the following exercises, use a model for body surface area, BSA, such that $BSA=\sqrt{\frac{wh}{3600}}$, where w = weight in kg and h = height in cm.

48. Find the height of a 72-kg female to the nearest cm whose $BSA=1.8$.

49. Find the weight of a 177-cm male to the nearest kg whose $BSA=2.1$.

2.7 Linear Inequalities and Absolute Value Inequalities

Learning Objectives

In this section, you will:

- Use interval notation
- Use properties of inequalities.
- Solve inequalities in one variable algebraically.
- Solve absolute value inequalities.

Figure 1

It is not easy to make the honor roll at most top universities. Suppose students were required to carry a course load of at least 12 credit hours and maintain a grade point average of 3.5 or above. How could these honor roll requirements be expressed mathematically? In this section, we will explore various ways to express different sets of numbers, inequalities, and absolute value inequalities.

Using Interval Notation

Indicating the solution to an inequality such as $x \geq 4$ can be achieved in several ways.

We can use a number line as shown in Figure 2. The blue ray begins at $x = 4$ and, as indicated by the arrowhead, continues to infinity, which illustrates that the solution set includes all real numbers greater than or equal to 4.

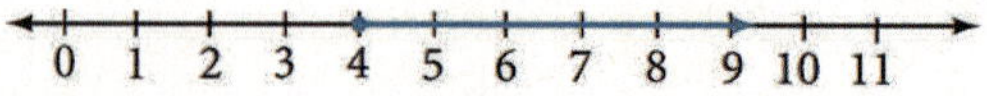

Figure 2

We can use set-builder notation: $\{x|x \geq 4\}$, which translates to "all real numbers *x* such that *x* is greater than or equal to 4." Notice that braces are used to indicate a set.

The third method is **interval notation**, in which solution sets are indicated with parentheses or brackets. The solutions to $x \geq 4$ are represented as $\left[4, \infty\right)$. This is perhaps the most useful method, as it applies to concepts studied later in this course and to other higher-level math courses.

The main concept to remember is that parentheses represent solutions greater or less than the number, and brackets represent solutions that are greater than or equal to or less than or equal to the number. Use parentheses to represent infinity or negative infinity, since positive and negative infinity are not numbers in the usual sense of the word and, therefore, cannot be "equaled." A few examples of an **interval**, or a set of numbers in which a solution falls, are $[-2, 6)$, or all numbers between -2 and 6, including -2, but not including 6; $(-1, 0)$, all real numbers between, but not including -1 and 0; and $\left(-\infty, 1\right]$, all real numbers less than and including 1. Table 1 outlines the possibilities.

Set Indicated	Set-Builder Notation	Interval Notation
All real numbers between *a* and *b*, but not including *a* or *b*	$\{x\|a < x < b\}$	(a, b)
All real numbers greater than *a*, but not including *a*	$\{x\|x > a\}$	$\left(a, \infty\right)$
All real numbers less than *b*, but not including *b*	$\{x\|x < b\}$	$\left(-\infty, b\right)$
All real numbers greater than *a*, including *a*	$\{x\|x \geq a\}$	$\left[a, \infty\right)$
All real numbers less than *b*, including *b*	$\{x\|x \leq b\}$	$\left(-\infty, b\right]$
All real numbers between *a* and *b*, including *a*	$\{x\|a \leq x < b\}$	$[a, b)$
All real numbers between *a* and *b*, including *b*	$\{x\|a < x \leq b\}$	$(a, b]$

Table 1

Set Indicated	Set-Builder Notation	Interval Notation
All real numbers between *a* and *b*, including *a* and *b*	$\{x \mid a \le x \le b\}$	$[a, b]$
All real numbers less than *a* or greater than *b*	$\{x \mid x < a \text{ or } x > b\}$	$(-\infty, a) \cup (b, \infty)$
All real numbers	$\{x \mid x \text{ is all real numbers}\}$	$(-\infty, \infty)$

Table 1

EXAMPLE 1

Using Interval Notation to Express All Real Numbers Greater Than or Equal to *a*

Use interval notation to indicate all real numbers greater than or equal to -2.

Solution

Use a bracket on the left of -2 and parentheses after infinity: $[-2, \infty)$. The bracket indicates that -2 is included in the set with all real numbers greater than -2 to infinity.

TRY IT #1 Use interval notation to indicate all real numbers between and including -3 and 5.

EXAMPLE 2

Using Interval Notation to Express All Real Numbers Less Than or Equal to *a* or Greater Than or Equal to *b*

Write the interval expressing all real numbers less than or equal to -1 or greater than or equal to 1.

Solution

We have to write two intervals for this example. The first interval must indicate all real numbers less than or equal to 1. So, this interval begins at $-\infty$ and ends at -1, which is written as $(-\infty, -1]$.

The second interval must show all real numbers greater than or equal to 1, which is written as $[1, \infty)$. However, we want to combine these two sets. We accomplish this by inserting the union symbol, ∪, between the two intervals.

$$(-\infty, -1] \cup [1, \infty)$$

TRY IT #2 Express all real numbers less than -2 or greater than or equal to 3 in interval notation.

Using the Properties of Inequalities

When we work with inequalities, we can usually treat them similarly to but not exactly as we treat equalities. We can use the addition property and the multiplication property to help us solve them. The one exception is when we multiply or divide by a negative number; doing so reverses the inequality symbol.

Properties of Inequalities

Addition Property If $a < b$, then $a + c < b + c$.

Multiplication Property If $a < b$ and $c > 0$, then $ac < bc$.
If $a < b$ and $c < 0$, then $ac > bc$.

These properties also apply to $a \le b$, $a > b$, and $a \ge b$.

EXAMPLE 3

Demonstrating the Addition Property

Illustrate the addition property for inequalities by solving each of the following:

1. ⓐ $x - 15 < 4$
2. ⓑ $6 \geq x - 1$
3. ⓒ $x + 7 > 9$

Solution

The addition property for inequalities states that if an inequality exists, adding or subtracting the same number on both sides does not change the inequality.

ⓐ

$$\begin{aligned} x - 15 &< 4 \\ x - 15 + 15 &< 4 + 15 \quad \text{Add 15 to both sides.} \\ x &< 19 \end{aligned}$$

ⓑ

$$\begin{aligned} 6 &\geq x - 1 \\ 6 + 1 &\geq x - 1 + 1 \quad \text{Add 1 to both sides.} \\ 7 &\geq x \end{aligned}$$

ⓒ

$$\begin{aligned} x + 7 &> 9 \\ x + 7 - 7 &> 9 - 7 \quad \text{Subtract 7 from both sides.} \\ x &> 2 \end{aligned}$$

> **TRY IT** #3 Solve: $3x{-}2 < 1$.

EXAMPLE 4

Demonstrating the Multiplication Property

Illustrate the multiplication property for inequalities by solving each of the following:

1. ⓐ $3x < 6$
2. ⓑ $-2x - 1 \geq 5$
3. ⓒ $5 - x > 10$

Solution

ⓐ

$$\begin{aligned} 3x &< 6 \\ \tfrac{1}{3}(3x) &< (6)\tfrac{1}{3} \\ x &< 2 \end{aligned}$$

ⓑ

$$\begin{aligned} -2x - 1 &\geq 5 \\ -2x &\geq 6 \\ \left(-\tfrac{1}{2}\right)(-2x) &\geq (6)\left(-\tfrac{1}{2}\right) \quad \text{Multiply by } -\tfrac{1}{2}. \\ x &\leq -3 \quad \text{Reverse the inequality.} \end{aligned}$$

ⓒ

$$\begin{aligned} 5 - x &> 10 \\ -x &> 5 \\ (-1)(-x) &> (5)(-1) \quad \text{Multiply by } -1. \\ x &< -5 \quad \text{Reverse the inequality.} \end{aligned}$$

> **TRY IT** #4 Solve: $4x + 7 \geq 2x - 3$.

Solving Inequalities in One Variable Algebraically

As the examples have shown, we can perform the same operations on both sides of an inequality, just as we do with equations; we combine like terms and perform operations. To solve, we isolate the variable.

EXAMPLE 5

Solving an Inequality Algebraically

Solve the inequality: $13 - 7x \geq 10x - 4$.

Solution

Solving this inequality is similar to solving an equation up until the last step.

$13 - 7x \geq 10x - 4$	
$13 - 17x \geq -4$	Move variable terms to one side of the inequality.
$-17x \geq -17$	Isolate the variable term.
$x \leq 1$	Dividing both sides by -17 reverses the inequality.

The solution set is given by the interval $\left(-\infty, 1\right]$, or all real numbers less than and including 1.

TRY IT #5 Solve the inequality and write the answer using interval notation: $-x + 4 < \frac{1}{2}x + 1$.

EXAMPLE 6

Solving an Inequality with Fractions

Solve the following inequality and write the answer in interval notation: $-\frac{3}{4}x \geq -\frac{5}{8} + \frac{2}{3}x$.

Solution

We begin solving in the same way we do when solving an equation.

$-\frac{3}{4}x \geq -\frac{5}{8} + \frac{2}{3}x$	
$-\frac{3}{4}x - \frac{2}{3}x \geq -\frac{5}{8}$	Put variable terms on one side.
$-\frac{9}{12}x - \frac{8}{12}x \geq -\frac{5}{8}$	Write fractions with common denominator.
$-\frac{17}{12}x \geq -\frac{5}{8}$	
$x \leq -\frac{5}{8}\left(-\frac{12}{17}\right)$	Multiplying by a negative number reverses the inequality.
$x \leq \frac{15}{34}$	

The solution set is the interval $\left(-\infty, \frac{15}{34}\right]$.

TRY IT #6 Solve the inequality and write the answer in interval notation: $-\frac{5}{6}x \leq \frac{3}{4} + \frac{8}{3}x$.

Understanding Compound Inequalities

A **compound inequality** includes two inequalities in one statement. A statement such as $4 < x \leq 6$ means $4 < x$ and $x \leq 6$. There are two ways to solve compound inequalities: separating them into two separate inequalities or leaving the compound inequality intact and performing operations on all three parts at the same time. We will illustrate both methods.

EXAMPLE 7

Solving a Compound Inequality

Solve the compound inequality: $3 \leq 2x + 2 < 6$.

Solution

The first method is to write two separate inequalities: $3 \leq 2x + 2$ and $2x + 2 < 6$. We solve them independently.

$$\begin{array}{lcr} 3 \le 2x+2 & \text{and} & 2x+2<6 \\ 1 \le 2x & & 2x<4 \\ \frac{1}{2} \le x & & x<2 \end{array}$$

Then, we can rewrite the solution as a compound inequality, the same way the problem began.

$$\frac{1}{2} \le x < 2$$

In interval notation, the solution is written as $\left[\frac{1}{2}, 2\right)$.

The second method is to leave the compound inequality intact, and perform solving procedures on the three parts at the same time.

$$\begin{array}{ll} 3 \le 2x+2<6 & \\ 1 \le 2x<4 & \text{Isolate the variable term, and subtract 2 from all three parts.} \\ \frac{1}{2} \le x<2 & \text{Divide through all three parts by 2.} \end{array}$$

We get the same solution: $\left[\frac{1}{2}, 2\right)$.

> TRY IT #7 Solve the compound inequality: $4 < 2x - 8 \le 10$.

EXAMPLE 8

Solving a Compound Inequality with the Variable in All Three Parts

Solve the compound inequality with variables in all three parts: $3 + x > 7x - 2 > 5x - 10$.

Solution

Let's try the first method. Write two inequalities:

$$\begin{array}{rcr} 3+x>7x-2 & \text{and} & 7x-2>5x-10 \\ 3>6x-2 & & 2x-2>-10 \\ 5>6x & & 2x>-8 \\ \frac{5}{6}>x & & x>-4 \\ x<\frac{5}{6} & & -4<x \end{array}$$

The solution set is $-4 < x < \frac{5}{6}$ or in interval notation $\left(-4, \frac{5}{6}\right)$. Notice that when we write the solution in interval notation, the smaller number comes first. We read intervals from left to right, as they appear on a number line. See Figure 3.

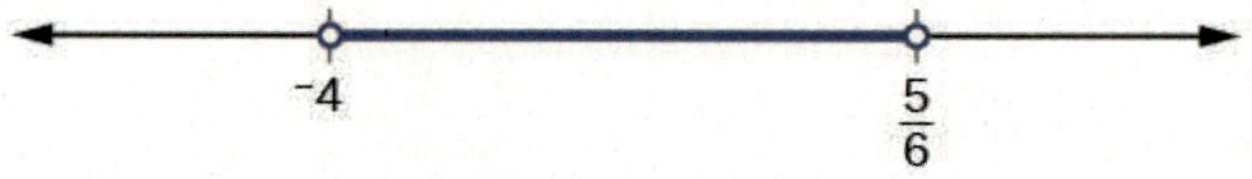

Figure 3

> TRY IT #8 Solve the compound inequality: $3y < 4 - 5y < 5 + 3y$.

Solving Absolute Value Inequalities

As we know, the absolute value of a quantity is a positive number or zero. From the origin, a point located at $(-x, 0)$ has an absolute value of x, as it is x units away. Consider absolute value as the distance from one point to another point. Regardless of direction, positive or negative, the distance between the two points is represented as a positive number or zero.

An absolute value inequality is an equation of the form

$$|A| < B, \quad |A| \le B, \quad |A| > B, \quad \text{or} \quad |A| \ge B,$$

Where *A*, and sometimes *B*, represents an algebraic expression dependent on a variable *x*. Solving the inequality means finding the set of all x -values that satisfy the problem. Usually this set will be an interval or the union of two intervals and will include a range of values.

There are two basic approaches to solving absolute value inequalities: graphical and algebraic. The advantage of the graphical approach is we can read the solution by interpreting the graphs of two equations. The advantage of the algebraic approach is that solutions are exact, as precise solutions are sometimes difficult to read from a graph.

Suppose we want to know all possible returns on an investment if we could earn some amount of money within $200 of $600. We can solve algebraically for the set of *x*-values such that the distance between x and 600 is less than or equal to 200. We represent the distance between x and 600 as $|x-600|$, and therefore, $|x-600| \leq 200$ or

$$\begin{gathered} -200 \leq x-600 \leq 200 \\ -200+600 \leq x-600+600 \leq 200+600 \\ 400 \leq x \leq 800 \end{gathered}$$

This means our returns would be between $400 and $800.

To solve absolute value inequalities, just as with absolute value equations, we write two inequalities and then solve them independently.

Absolute Value Inequalities

For an algebraic expression *X*, and $k>0$, an **absolute value inequality** is an inequality of the form

$$\begin{aligned} |X|<k \quad & \text{is equivalent to} -k<X<k \\ |X|>k \quad & \text{is equivalent to} X<-k \quad \text{or} X>k \end{aligned}$$

These statements also apply to $|X| \leq k$ and $|X| \geq k$.

EXAMPLE 9

Determining a Number within a Prescribed Distance

Describe all values x within a distance of 4 from the number 5.

Solution

We want the distance between x and 5 to be less than or equal to 4. We can draw a number line, such as in Figure 4, to represent the condition to be satisfied.

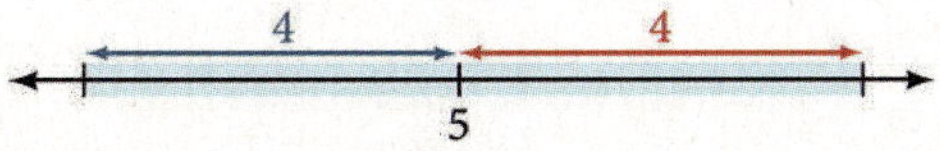

Figure 4

The distance from x to 5 can be represented using an absolute value symbol, $|x-5|$. Write the values of x that satisfy the condition as an absolute value inequality.

$$|x-5| \leq 4$$

We need to write two inequalities as there are always two solutions to an absolute value equation.

$$\begin{aligned} x-5 &\leq 4 \quad \text{and} \quad & x-5 &\geq -4 \\ x &\leq 9 & x &\geq 1 \end{aligned}$$

If the solution set is $x \leq 9$ and $x \geq 1$, then the solution set is an interval including all real numbers between and including 1 and 9.

So $|x-5| \leq 4$ is equivalent to $[1,9]$ in interval notation.

TRY IT #9 Describe all *x*-values within a distance of 3 from the number 2.

EXAMPLE 10

Solving an Absolute Value Inequality

Solve $|x-1| \leq 3$.

Solution

$$|x-1| \leq 3$$

$$-3 \leq x-1 \leq 3$$

$$-2 \leq x \leq 4$$

$$[-2, 4]$$

EXAMPLE 11

Using a Graphical Approach to Solve Absolute Value Inequalities

Given the equation $y = -\frac{1}{2}|4x-5| + 3$, determine the x-values for which the y-values are negative.

Solution

We are trying to determine where $y < 0$, which is when $-\frac{1}{2}|4x-5| + 3 < 0$. We begin by isolating the absolute value.

$$-\frac{1}{2}|4x-5| < -3 \qquad \text{Multiply both sides by } -2\text{, and reverse the inequality.}$$

$$|4x-5| > 6$$

Next, we solve for the equality $|4x-5| = 6$.

$$4x-5=6 \qquad\qquad 4x-5=-6$$

$$4x=11 \qquad \text{or} \qquad 4x=-1$$

$$x=\frac{11}{4} \qquad\qquad x=-\frac{1}{4}$$

Now, we can examine the graph to observe where the y-values are negative. We observe where the branches are below the x-axis. Notice that it is not important exactly what the graph looks like, as long as we know that it crosses the horizontal axis at $x = -\frac{1}{4}$ and $x = \frac{11}{4}$, and that the graph opens downward. See Figure 5.

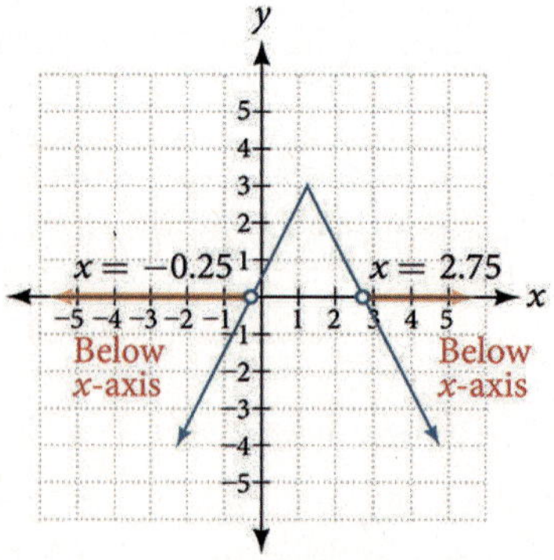

Figure 5

> **TRY IT** #10 Solve $-2|k-4| \leq -6$.

MEDIA

Access these online resources for additional instruction and practice with linear inequalities and absolute value inequalities.

Interval notation (http://openstax.org/l/intervalnotn)
How to solve linear inequalities (http://openstax.org/l/solvelinineq)
How to solve an inequality (http://openstax.org/l/solveineq)
Absolute value equations (http://openstax.org/l/absvaleq)
Compound inequalities (http://openstax.org/l/compndineqs)
Absolute value inequalities (http://openstax.org/l/absvalineqs)

2.7 SECTION EXERCISES

Verbal

1. When solving an inequality, explain what happened from Step 1 to Step 2:

Step 1 $\quad -2x > 6$
Step 2 $\quad x < -3$

2. When solving an inequality, we arrive at:

$$x + 2 < x + 3$$
$$2 < 3$$

Explain what our solution set is.

3. When writing our solution in interval notation, how do we represent all the real numbers?

4. When solving an inequality, we arrive at:

$$x + 2 > x + 3$$
$$2 > 3$$

Explain what our solution set is.

5. Describe how to graph $y = |x - 3|$

Algebraic

For the following exercises, solve the inequality. Write your final answer in interval notation.

6. $4x - 7 \leq 9$

7. $3x + 2 \geq 7x - 1$

8. $-2x + 3 > x - 5$

9. $4(x + 3) \geq 2x - 1$

10. $-\frac{1}{2}x \leq -\frac{5}{4} + \frac{2}{5}x$

11. $-5(x - 1) + 3 > 3x - 4 - 4x$

12. $-3(2x + 1) > -2(x + 4)$

13. $\frac{x+3}{8} - \frac{x+5}{5} \geq \frac{3}{10}$

14. $\frac{x-1}{3} + \frac{x+2}{5} \leq \frac{3}{5}$

For the following exercises, solve the inequality involving absolute value. Write your final answer in interval notation.

15. $|x + 9| \geq -6$

16. $|2x + 3| < 7$

17. $|3x - 1| > 11$

18. $|2x + 1| + 1 \leq 6$

19. $|x - 2| + 4 \geq 10$

20. $|-2x + 7| \leq 13$

21. $|x - 7| < -4$

22. $|x - 20| > -1$

23. $\left|\frac{x-3}{4}\right| < 2$

For the following exercises, describe all the x-values within or including a distance of the given values.

24. Distance of 5 units from the number 7

25. Distance of 3 units from the number 9

26. Distance of 10 units from the number 4

27. Distance of 11 units from the number 1

For the following exercises, solve the compound inequality. Express your answer using inequality signs, and then write your answer using interval notation.

28. $-4 < 3x + 2 \leq 18$

29. $3x + 1 > 2x - 5 > x - 7$

30. $3y < 5 - 2y < 7 + y$

31. $2x - 5 < -11 \quad \text{or} \quad 5x + 1 \geq 6$

32. $x + 7 < x + 2$

Graphical

For the following exercises, graph the function. Observe the points of intersection and shade the x-axis representing the solution set to the inequality. Show your graph and write your final answer in interval notation.

33. $|x - 1| > 2$

34. $|x + 3| \geq 5$

35. $|x + 7| \leq 4$

36. $|x - 2| < 7$

37. $|x - 2| < 0$

For the following exercises, graph both straight lines (left-hand side being y1 and right-hand side being y2) on the same axes. Find the point of intersection and solve the inequality by observing where it is true comparing the y-values of the lines.

38. $x + 3 < 3x - 4$

39. $x - 2 > 2x + 1$

40. $x + 1 > x + 4$

41. $\frac{1}{2}x + 1 > \frac{1}{2}x - 5$

42. $4x + 1 < \frac{1}{2}x + 3$

Numeric

For the following exercises, write the set in interval notation.

43. $\{x|-1 < x < 3\}$

44. $\{x|x \geq 7\}$

45. $\{x|x < 4\}$

46. $\{\ x|\ x\ \text{ is all real numbers}\}$

For the following exercises, write the interval in set-builder notation.

47. $\left(-\infty, 6\right)$

48. $\left(4, \infty\right)$

49. $[-3, 5)$

50. $[-4, 1] \cup [9, \infty)$

For the following exercises, write the set of numbers represented on the number line in interval notation.

51.

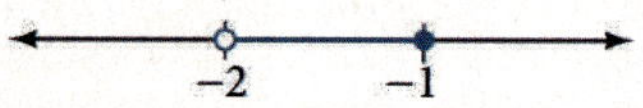

52.

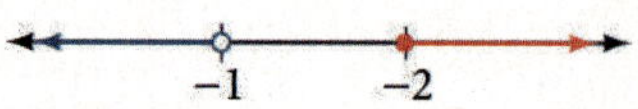

53.

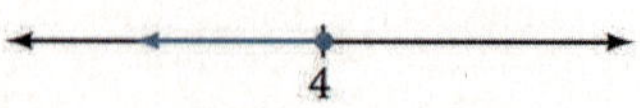

Technology

For the following exercises, input the left-hand side of the inequality as a Y1 graph in your graphing utility. Enter y2 = the right-hand side. Entering the absolute value of an expression is found in the MATH menu, Num, 1:abs(. Find the points of intersection, recall (2nd CALC 5:intersection, 1st curve, enter, 2nd curve, enter, guess, enter). Copy a sketch of the graph and shade the x-axis for your solution set to the inequality. Write final answers in interval notation.

54. $|x+2|-5<2$

55. $\frac{-1}{2}|x+2|<4$

56. $|4x+1|-3>2$

57. $|x-4|<3$

58. $|x+2|\geq 5$

Extensions

59. Solve $|3x+1|=|2x+3|$

60. Solve $x^2-x>12$

61. $\frac{x-5}{x+7}\leq 0, x\neq -7$

62. $p=-x^2+130x-3000$ is a profit formula for a small business. Find the set of x-values that will keep this profit positive.

Real-World Applications

63. In chemistry the volume for a certain gas is given by $V=20T$, where V is measured in cc and T is temperature in °C. If the temperature varies between 80°C and 120°C, find the set of volume values.

64. A basic cellular package costs $20/mo. for 60 min of calling, with an additional charge of $.30/min beyond that time.. The cost formula would be $C=20+.30(x-60)$. If you have to keep your bill no greater than $50, what is the maximum calling minutes you can use?

Chapter Review

Key Terms

absolute value equation an equation in which the variable appears in absolute value bars, typically with two solutions, one accounting for the positive expression and one for the negative expression

area in square units, the area formula used in this section is used to find the area of any two-dimensional rectangular region: $A = LW$

Cartesian coordinate system a grid system designed with perpendicular axes invented by René Descartes

completing the square a process for solving quadratic equations in which terms are added to or subtracted from both sides of the equation in order to make one side a perfect square

complex conjugate a complex number containing the same terms as another complex number, but with the opposite operator. Multiplying a complex number by its conjugate yields a real number.

complex number the sum of a real number and an imaginary number; the standard form is $a + bi$, where a is the real part and b is the complex part.

complex plane the coordinate plane in which the horizontal axis represents the real component of a complex number, and the vertical axis represents the imaginary component, labeled *i*.

compound inequality a problem or a statement that includes two inequalities

conditional equation an equation that is true for some values of the variable

discriminant the expression under the radical in the quadratic formula that indicates the nature of the solutions, real or complex, rational or irrational, single or double roots.

distance formula a formula that can be used to find the length of a line segment if the endpoints are known

equation in two variables a mathematical statement, typically written in *x* and *y*, in which two expressions are equal

equations in quadratic form equations with a power other than 2 but with a middle term with an exponent that is one-half the exponent of the leading term

extraneous solutions any solutions obtained that are not valid in the original equation

graph in two variables the graph of an equation in two variables, which is always shown in two variables in the two-dimensional plane

identity equation an equation that is true for all values of the variable

imaginary number the square root of $-1 : i = \sqrt{-1}$.

inconsistent equation an equation producing a false result

intercepts the points at which the graph of an equation crosses the *x*-axis and the *y*-axis

interval an interval describes a set of numbers within which a solution falls

interval notation a mathematical statement that describes a solution set and uses parentheses or brackets to indicate where an interval begins and ends

linear equation an algebraic equation in which each term is either a constant or the product of a constant and the first power of a variable

linear inequality similar to a linear equation except that the solutions will include sets of numbers

midpoint formula a formula to find the point that divides a line segment into two parts of equal length

ordered pair a pair of numbers indicating horizontal displacement and vertical displacement from the origin; also known as a coordinate pair, (x, y)

origin the point where the two axes cross in the center of the plane, described by the ordered pair $(0, 0)$

perimeter in linear units, the perimeter formula is used to find the linear measurement, or outside length and width, around a two-dimensional regular object; for a rectangle: $P = 2L + 2W$

polynomial equation an equation containing a string of terms including numerical coefficients and variables raised to whole-number exponents

Pythagorean Theorem a theorem that states the relationship among the lengths of the sides of a right triangle, used to solve right triangle problems

quadrant one quarter of the coordinate plane, created when the axes divide the plane into four sections

quadratic equation an equation containing a second-degree polynomial; can be solved using multiple methods

quadratic formula a formula that will solve all quadratic equations

radical equation an equation containing at least one radical term where the variable is part of the radicand

rational equation an equation consisting of a fraction of polynomials

slope the change in *y*-values over the change in *x*-values

solution set the set of all solutions to an equation

square root property one of the methods used to solve a quadratic equation, in which the x^2 term is isolated so that the square root of both sides of the equation can be taken to solve for *x*

volume in cubic units, the volume measurement includes length, width, and depth: $V = LWH$

***x*-axis** the common name of the horizontal axis on a coordinate plane; a number line increasing from left to right

x-coordinate the first coordinate of an ordered pair, representing the horizontal displacement and direction from the origin
x-intercept the point where a graph intersects the *x*-axis; an ordered pair with a *y*-coordinate of zero
y-axis the common name of the vertical axis on a coordinate plane; a number line increasing from bottom to top
y-coordinate the second coordinate of an ordered pair, representing the vertical displacement and direction from the origin
y-intercept a point where a graph intercepts the *y*-axis; an ordered pair with an *x*-coordinate of zero
zero-product property the property that formally states that multiplication by zero is zero, so that each factor of a quadratic equation can be set equal to zero to solve equations

Key Equations

quadratic formula	$x = \frac{-b \pm \sqrt{b^2 - 4ac}}{2a}$

Key Concepts

2.1 The Rectangular Coordinate Systems and Graphs

- We can locate, or plot, points in the Cartesian coordinate system using ordered pairs, which are defined as displacement from the *x*-axis and displacement from the *y*-axis. See Example 1.
- An equation can be graphed in the plane by creating a table of values and plotting points. See Example 2.
- Using a graphing calculator or a computer program makes graphing equations faster and more accurate. Equations usually have to be entered in the form *y*=_____. See Example 3.
- Finding the *x*- and *y*-intercepts can define the graph of a line. These are the points where the graph crosses the axes. See Example 4.
- The distance formula is derived from the Pythagorean Theorem and is used to find the length of a line segment. See Example 5 and Example 6.
- The midpoint formula provides a method of finding the coordinates of the midpoint dividing the sum of the *x*-coordinates and the sum of the *y*-coordinates of the endpoints by 2. See Example 7 and Example 8.

2.2 Linear Equations in One Variable

- We can solve linear equations in one variable in the form $ax + b = 0$ using standard algebraic properties. See Example 1 and Example 2.
- A rational expression is a quotient of two polynomials. We use the LCD to clear the fractions from an equation. See Example 3 and Example 4.
- All solutions to a rational equation should be verified within the original equation to avoid an undefined term, or zero in the denominator. See Example 5 and Example 6 and Example 7.
- Given two points, we can find the slope of a line using the slope formula. See Example 8.
- We can identify the slope and *y*-intercept of an equation in slope-intercept form. See Example 9.
- We can find the equation of a line given the slope and a point. See Example 10.
- We can also find the equation of a line given two points. Find the slope and use the point-slope formula. See Example 11.
- The standard form of a line has no fractions. See Example 12.
- Horizontal lines have a slope of zero and are defined as $y = c$, where *c* is a constant.
- Vertical lines have an undefined slope (zero in the denominator), and are defined as $x = c$, where *c* is a constant. See Example 13.
- Parallel lines have the same slope and different *y*-intercepts. See Example 14 and Example 15.
- Perpendicular lines have slopes that are negative reciprocals of each other unless one is horizontal and the other is vertical. See Example 16.

2.3 Models and Applications

- A linear equation can be used to solve for an unknown in a number problem. See Example 1.
- Applications can be written as mathematical problems by identifying known quantities and assigning a variable to unknown quantities. See Example 2.
- There are many known formulas that can be used to solve applications. Distance problems, for example, are solved using the $d = rt$ formula. See Example 3.
- Many geometry problems are solved using the perimeter formula $P = 2L + 2W$, the area formula $A = LW$, or the

volume formula $V = LWH$. See Example 4, Example 5, and Example 6.

2.4 Complex Numbers

- The square root of any negative number can be written as a multiple of i. See Example 1.
- To plot a complex number, we use two number lines, crossed to form the complex plane. The horizontal axis is the real axis, and the vertical axis is the imaginary axis. See Example 2.
- Complex numbers can be added and subtracted by combining the real parts and combining the imaginary parts. See Example 3.
- Complex numbers can be multiplied and divided.
  - To multiply complex numbers, distribute just as with polynomials. See Example 4 and Example 5.
  - To divide complex numbers, multiply both numerator and denominator by the complex conjugate of the denominator to eliminate the complex number from the denominator. See Example 6 and Example 7.
- The powers of i are cyclic, repeating every fourth one. See Example 8.

2.5 Quadratic Equations

- Many quadratic equations can be solved by factoring when the equation has a leading coefficient of 1 or if the equation is a difference of squares. The zero-product property is then used to find solutions. See Example 1, Example 2, and Example 3.
- Many quadratic equations with a leading coefficient other than 1 can be solved by factoring using the grouping method. See Example 4 and Example 5.
- Another method for solving quadratics is the square root property. The variable is squared. We isolate the squared term and take the square root of both sides of the equation. The solution will yield a positive and negative solution. See Example 6 and Example 7.
- Completing the square is a method of solving quadratic equations when the equation cannot be factored. See Example 8.
- A highly dependable method for solving quadratic equations is the quadratic formula, based on the coefficients and the constant term in the equation. See Example 9 and Example 10.
- The discriminant is used to indicate the nature of the roots that the quadratic equation will yield: real or complex, rational or irrational, and how many of each. See Example 11.
- The Pythagorean Theorem, among the most famous theorems in history, is used to solve right-triangle problems and has applications in numerous fields. Solving for the length of one side of a right triangle requires solving a quadratic equation. See Example 12.

2.6 Other Types of Equations

- Rational exponents can be rewritten several ways depending on what is most convenient for the problem. To solve, both sides of the equation are raised to a power that will render the exponent on the variable equal to 1. See Example 1, Example 2, and Example 3.
- Factoring extends to higher-order polynomials when it involves factoring out the GCF or factoring by grouping. See Example 4 and Example 5.
- We can solve radical equations by isolating the radical and raising both sides of the equation to a power that matches the index. See Example 6 and Example 7.
- To solve absolute value equations, we need to write two equations, one for the positive value and one for the negative value. See Example 8.
- Equations in quadratic form are easy to spot, as the exponent on the first term is double the exponent on the second term and the third term is a constant. We may also see a binomial in place of the single variable. We use substitution to solve. See Example 9 and Example 10.
- Solving a rational equation may also lead to a quadratic equation or an equation in quadratic form. See Example 11.

2.7 Linear Inequalities and Absolute Value Inequalities

- Interval notation is a method to indicate the solution set to an inequality. Highly applicable in calculus, it is a system of parentheses and brackets that indicate what numbers are included in a set and whether the endpoints are included as well. See Table 1 and Example 2.
- Solving inequalities is similar to solving equations. The same algebraic rules apply, except for one: multiplying or dividing by a negative number reverses the inequality. See Example 3, Example 4, Example 5, and Example 6.
- Compound inequalities often have three parts and can be rewritten as two independent inequalities. Solutions are given by boundary values, which are indicated as a beginning boundary or an ending boundary in the solutions to the two inequalities. See Example 7 and Example 8.

- Absolute value inequalities will produce two solution sets due to the nature of absolute value. We solve by writing two equations: one equal to a positive value and one equal to a negative value. See Example 9 and Example 10.
- Absolute value inequalities can also be solved by graphing. At least we can check the algebraic solutions by graphing, as we cannot depend on a visual for a precise solution. See Example 11.

Exercises

Review Exercises

The Rectangular Coordinate Systems and Graphs

For the following exercises, find the x-intercept and the y-intercept without graphing.

1. $4x - 3y = 12$

2. $2y - 4 = 3x$

For the following exercises, solve for y in terms of x, putting the equation in slope–intercept form.

3. $5x = 3y - 12$

4. $2x - 5y = 7$

For the following exercises, find the distance between the two points.

5. $(-2, 5)\,(4, -1)$

6. $(-12, -3)\,(-1, 5)$

7. Find the distance between the two points $(-71{,}432)$ and $(511{,}218)$ using your calculator, and round your answer to the nearest thousandth.

For the following exercises, find the coordinates of the midpoint of the line segment that joins the two given points.

8. $(-1, 5)$ and $(4, 6)$

9. $(-13, 5)$ and $(17, 18)$

For the following exercises, construct a table and graph the equation by plotting at least three points.

10. $y = \frac{1}{2}x + 4$

11. $4x - 3y = 6$

Linear Equations in One Variable

For the following exercises, solve for x.

12. $5x + 2 = 7x - 8$

13. $3(x + 2) - 10 = x + 4$

14. $7x - 3 = 5$

15. $12 - 5(x + 1) = 2x - 5$

16. $\frac{2x}{3} - \frac{3}{4} = \frac{x}{6} + \frac{21}{4}$

For the following exercises, solve for x. State all x-values that are excluded from the solution set.

17. $\frac{x}{x^2-9} + \frac{4}{x+3} = \frac{3}{x^2-9}$
$x \neq 3, -3$

18. $\frac{1}{2} + \frac{2}{x} = \frac{3}{4}$

For the following exercises, find the equation of the line using the point-slope formula.

19. Passes through these two points: $(-2, 1), (4, 2)$.

20. Passes through the point $(-3, 4)$ and has a slope of $-\frac{1}{3}$.

21. Passes through the point $(-3, 4)$ and is parallel to the graph $y = \frac{2}{3}x + 5$.

22. Passes through these two points: $(5, 1), (5, 7)$.

Models and Applications

For the following exercises, write and solve an equation to answer each question.

23. The number of male fish in the tank is five more than three times the number of females. If the total number of fish is 73, how many of each sex are in the tank?

24. A landscaper has 72 ft. of fencing to put around a rectangular garden. If the length is 3 times the width, find the dimensions of the garden.

25. A truck rental is $25 plus $.30/mi. Find out how many miles Ken traveled if his bill was $50.20.

Complex Numbers

For the following exercises, use the quadratic equation to solve.

26. $x^2 - 5x + 9 = 0$

27. $2x^2 + 3x + 7 = 0$

For the following exercises, name the horizontal component and the vertical component.

28. $4 - 3i$

29. $-2 - i$

For the following exercises, perform the operations indicated.

30. $(9 - i) - (4 - 7i)$

31. $(2 + 3i) - (-5 - 8i)$

32. $2\sqrt{-75} + 3\sqrt{25}$

33. $\sqrt{-16} + 4\sqrt{-9}$

34. $-6i(i - 5)$

35. $(3 - 5i)^2$

36. $\sqrt{-4} \cdot \sqrt{-12}$

37. $\sqrt{-2}\left(\sqrt{-8} - \sqrt{5}\right)$

38. $\frac{2}{5-3i}$

39. $\frac{3+7i}{i}$

Quadratic Equations

For the following exercises, solve the quadratic equation by factoring.

40. $2x^2 - 7x - 4 = 0$

41. $3x^2 + 18x + 15 = 0$

42. $25x^2 - 9 = 0$

43. $7x^2 - 9x = 0$

For the following exercises, solve the quadratic equation by using the square-root property.

44. $x^2 = 49$

45. $(x - 4)^2 = 36$

For the following exercises, solve the quadratic equation by completing the square.

46. $x^2 + 8x - 5 = 0$

47. $4x^2 + 2x - 1 = 0$

For the following exercises, solve the quadratic equation by using the quadratic formula. If the solutions are not real, state No real solution.

48. $2x^2 - 5x + 1 = 0$

49. $15x^2 - x - 2 = 0$

For the following exercises, solve the quadratic equation by the method of your choice.

50. $(x - 2)^2 = 16$

51. $x^2 = 10x + 3$

Other Types of Equations

For the following exercises, solve the equations.

52. $x^{\frac{3}{2}} = 27$

53. $x^{\frac{1}{2}} - 4x^{\frac{1}{4}} = 0$

54. $4x^3 + 8x^2 - 9x - 18 = 0$

55. $3x^5 - 6x^3 = 0$

56. $\sqrt{x + 9} = x - 3$

57. $\sqrt{3x + 7} + \sqrt{x + 2} = 1$

58. $|3x - 7| = 5$

59. $|2x + 3| - 5 = 9$

Linear Inequalities and Absolute Value Inequalities

For the following exercises, solve the inequality. Write your final answer in interval notation.

60. $5x - 8 \leq 12$

61. $-2x + 5 > x - 7$

62. $\frac{x-1}{3} + \frac{x+2}{5} \leq \frac{3}{5}$

63. $|3x + 2| + 1 \leq 9$

64. $|5x - 1| > 14$

65. $|x - 3| < -4$

For the following exercises, solve the compound inequality. Write your answer in interval notation.

66. $-4 < 3x + 2 \leq 18$

67. $3y < 1 - 2y < 5 + y$

For the following exercises, graph as described.

68. Graph the absolute value function and graph the constant function. Observe the points of intersection and shade the *x*-axis representing the solution set to the inequality. Show your graph and write your final answer in interval notation.

$|x+3| \geq 5$

69. Graph both straight lines (left-hand side being y1 and right-hand side being y2) on the same axes. Find the point of intersection and solve the inequality by observing where it is true comparing the *y*-values of the lines. See the interval where the inequality is true.

$x+3 < 3x-4$

Practice Test

1. Graph the following: $2y = 3x + 4$.

2. Find the *x*- and *y*-intercepts for the following: $2x - 5y = 6$

3. Find the *x*- and *y*-intercepts of this equation, and sketch the graph of the line using just the intercepts plotted.

$3x - 4y = 12$

4. Find the exact distance between $(5, -3)$ and $(-2, 8)$. Find the coordinates of the midpoint of the line segment joining the two points.

5. Write the interval notation for the set of numbers represented by $\{x | x \leq 9\}$.

6. Solve for *x*: $5x + 8 = 3x - 10$.

7. Solve for *x*: $3(2x-5) - 3(x-7) = 2x - 9$.

8. Solve for *x*: $\frac{x}{2} + 1 = \frac{4}{x}$

9. Solve for *x*: $\frac{5}{x+4} = 4 + \frac{3}{x-2}$.

10. The perimeter of a triangle is 30 in. The longest side is 2 less than 3 times the shortest side and the other side is 2 more than twice the shortest side. Find the length of each side.

11. Solve for *x*. Write the answer in simplest radical form.

$\frac{x^2}{3} - x = -\frac{1}{2}$

12. Solve: $3x - 8 \leq 4$.

13. Solve: $|2x+3| < 5$.

14. Solve: $|3x-2| \geq 4$.

For the following exercises, find the equation of the line with the given information.

15. Passes through the points $(-4, 2)$ and $(5, -3)$.

16. Has an undefined slope and passes through the point $(4, 3)$.

17. Passes through the point $(2, 1)$ and is perpendicular to $y = -\frac{2}{5}x + 3$.

18. Add these complex numbers: $(3 - 2i) + (4 - i)$.

19. Simplify: $\sqrt{-4} + 3\sqrt{-16}$.

20. Multiply: $5i\,(5 - 3i)$.

21. Divide: $\frac{4-i}{2+3i}$.

22. Solve this quadratic equation and write the two complex roots in $a + bi$ form: $x^2 - 4x + 7 = 0$.

23. Solve: $(3x - 1)^2 - 1 = 24$.

24. Solve: $x^2 - 6x = 13$.

25. Solve: $4x^2 - 4x - 1 = 0$

26. Solve: $\sqrt{x - 7} = x - 7$

27. Solve: $2 + \sqrt{12 - 2x} = x$

28. Solve: $(x - 1)^{\frac{2}{3}} = 9$

For the following exercises, find the real solutions of each equation by factoring.

29. $2x^3 - x^2 - 8x + 4 = 0$

30. $(x + 5)^2 - 3(x + 5) - 4 = 0$

3 FUNCTIONS

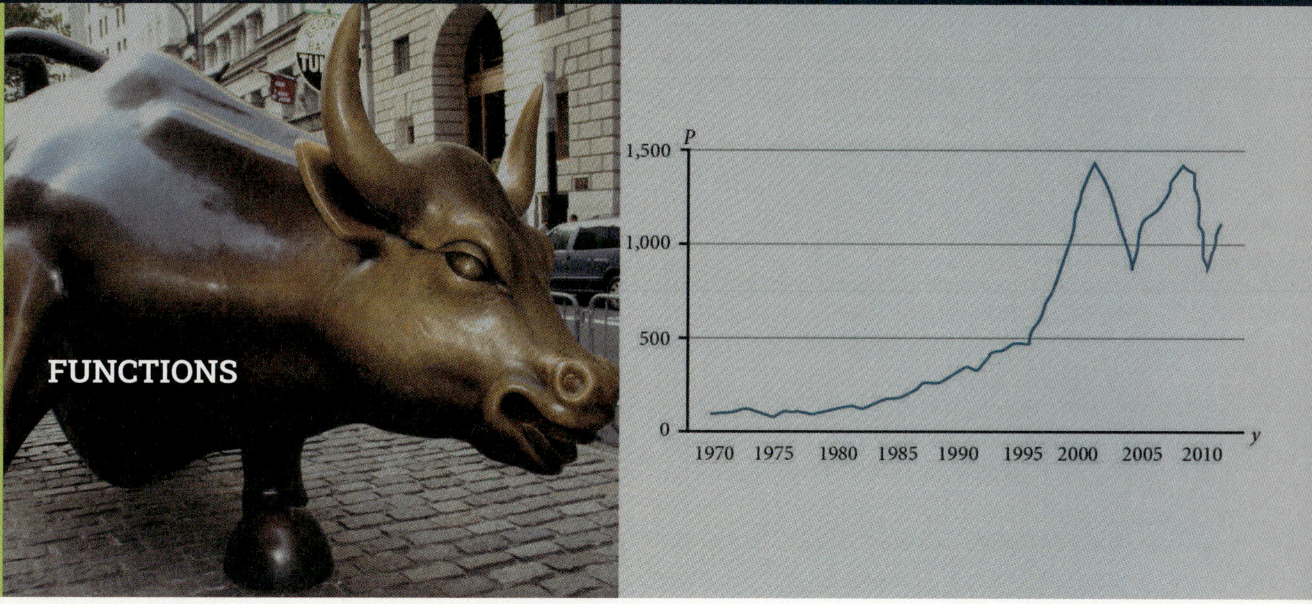

Standard and Poor's Index with dividends reinvested (credit "bull": modification of work by Prayitno Hadinata; credit "graph": modification of work by MeasuringWorth)

Chapter Outline

Introduction to Functions

Toward the end of the twentieth century, the values of stocks of Internet and technology companies rose dramatically. As a result, the Standard and Poor's stock market average rose as well. The graph above tracks the value of that initial investment of just under $100 over the 40 years. It shows that an investment that was worth less than $500 until about 1995 skyrocketed up to about $1100 by the beginning of 2000. That five-year period became known as the "dot-com bubble" because so many Internet startups were formed. As bubbles tend to do, though, the dot-com bubble eventually burst. Many companies grew too fast and then suddenly went out of business. The result caused the sharp decline represented on the graph beginning at the end of 2000.

Notice, as we consider this example, that there is a definite relationship between the year and stock market average. For any year we choose, we can determine the corresponding value of the stock market average. In this chapter, we will explore these kinds of relationships and their properties.

3.1 Functions and Function Notation

Learning Objectives

In this section, you will:

- Determine whether a relation represents a function.
- Find the value of a function.
- Determine whether a function is one-to-one.
- Use the vertical line test to identify functions.
- Graph the functions listed in the library of functions.

A jetliner changes altitude as its distance from the starting point of a flight increases. The weight of a growing child increases with time. In each case, one quantity depends on another. There is a relationship between the two quantities that we can describe, analyze, and use to make predictions. In this section, we will analyze such relationships.

Determining Whether a Relation Represents a Function

A **relation** is a set of ordered pairs. The set of the first components of each ordered pair is called the **domain** and the set of the second components of each ordered pair is called the **range**. Consider the following set of ordered pairs. The first numbers in each pair are the first five natural numbers. The second number in each pair is twice that of the first.

$$\{(1,\ 2),\ (2,\ 4),\ (3,\ 6),\ (4,\ 8),\ (5,\ 10)\}$$

The domain is $\{1,\ 2,\ 3,\ 4,\ 5\}$. The range is $\{2,\ 4,\ 6,\ 8,\ 10\}$.

Note that each value in the domain is also known as an **input** value, or **independent variable**, and is often labeled with the lowercase letter x. Each value in the range is also known as an **output** value, or **dependent variable**, and is often labeled lowercase letter y.

A function f is a relation that assigns a single value in the range to each value in the domain. In other words, no x-values are repeated. For our example that relates the first five natural numbers to numbers double their values, this relation is a function because each element in the domain, $\{1,\ 2,\ 3,\ 4,\ 5\}$, is paired with exactly one element in the range, $\{2,\ 4,\ 6,\ 8,\ 10\}$.

Now let's consider the set of ordered pairs that relates the terms "even" and "odd" to the first five natural numbers. It would appear as

$$\{(\text{odd},\ 1),\ (\text{even},\ 2),\ (\text{odd},\ 3),\ (\text{even},\ 4),\ (\text{odd},\ 5)\}$$

Notice that each element in the domain, $\{\text{even},\ \text{odd}\}$ is *not* paired with exactly one element in the range, $\{1,\ 2,\ 3,\ 4,\ 5\}$. For example, the term "odd" corresponds to three values from the range, $\{1,\ 3,\ 5\}$ and the term "even" corresponds to two values from the range, $\{2,\ 4\}$. This violates the definition of a function, so this relation is not a function.

Figure 1 compares relations that are functions and not functions.

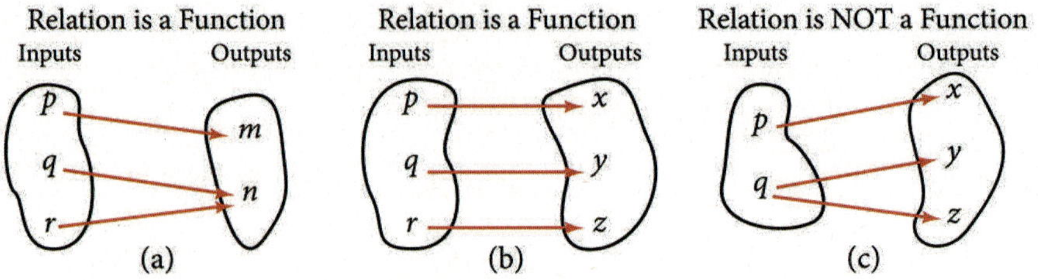

Figure 1 (a) This relationship is a function because each input is associated with a single output. Note that input q and r both give output n. (b) This relationship is also a function. In this case, each input is associated with a single output. (c) This relationship is not a function because input q is associated with two different outputs.

Function

A **function** is a relation in which each possible input value leads to exactly one output value. We say "the output is a function of the input."

The **input** values make up the **domain**, and the **output** values make up the **range**.

HOW TO

Given a relationship between two quantities, determine whether the relationship is a function.

1. Identify the input values.
2. Identify the output values.
3. If each input value leads to only one output value, classify the relationship as a function. If any input value leads

to two or more outputs, do not classify the relationship as a function.

EXAMPLE 1

Determining If Menu Price Lists Are Functions

The coffee shop menu, shown below, consists of items and their prices.

ⓐ Is price a function of the item?

ⓑ Is the item a function of the price?

Solution

ⓐ Let's begin by considering the input as the items on the menu. The output values are then the prices.

Each item on the menu has only one price, so the price is a function of the item.

ⓑ Two items on the menu have the same price. If we consider the prices to be the input values and the items to be the output, then the same input value could have more than one output associated with it. See the image below.

Therefore, the item is a not a function of price.

EXAMPLE 2

Determining If Class Grade Rules Are Functions

In a particular math class, the overall percent grade corresponds to a grade point average. Is grade point average a function of the percent grade? Is the percent grade a function of the grade point average? Table 1 shows a possible rule for assigning grade points.

Percent grade	0–56	57–61	62–66	67–71	72–77	78–86	87–91	92–100
Grade point average	0.0	1.0	1.5	2.0	2.5	3.0	3.5	4.0

Table 1

Solution

For any percent grade earned, there is an associated grade point average, so the grade point average is a function of the percent grade. In other words, if we input the percent grade, the output is a specific grade point average.

In the grading system given, there is a range of percent grades that correspond to the same grade point average. For example, students who receive a grade point average of 3.0 could have a variety of percent grades ranging from 78 all the way to 86. Thus, percent grade is not a function of grade point average.

TRY IT #1 Table 2[1] lists the five greatest baseball players of all time in order of rank.

Player	Rank
Babe Ruth	1
Willie Mays	2
Ty Cobb	3
Walter Johnson	4
Hank Aaron	5

Table 2

ⓐ Is the rank a function of the player name? ⓑ Is the player name a function of the rank?

Using Function Notation

Once we determine that a relationship is a function, we need to display and define the functional relationships so that we can understand and use them, and sometimes also so that we can program them into computers. There are various ways of representing functions. A standard function notation is one representation that facilitates working with functions.

To represent "height is a function of age," we start by identifying the descriptive variables h for height and a for age. The letters $f,\ g,$ and h are often used to represent functions just as we use $x,\ y,$ and z to represent numbers and $A,\ B,$ and C to represent sets.

$$\begin{array}{ll} h \text{ is } f \text{ of } a & \text{We name the function } f;\ \text{height is a function of age.} \\ h = f(a) & \text{We use parentheses to indicate the function input.} \\ f(a) & \text{We name the function } f;\ \text{the expression is read as "} f \text{ of } a\text{."} \end{array}$$

Remember, we can use any letter to name the function; the notation $h(a)$ shows us that h depends on a. The value a must be put into the function h to get a result. The parentheses indicate that age is input into the function; they do not indicate multiplication.

We can also give an algebraic expression as the input to a function. For example $f(a+b)$ means "first add *a* and *b*, and the result is the input for the function *f*." The operations must be performed in this order to obtain the correct result.

---

1 http://www.baseball-almanac.com/legendary/lisn100.shtml. Accessed 3/24/2014.

Function Notation

The notation $y = f(x)$ defines a function named f. This is read as "y is a function of x." The letter x represents the input value, or independent variable. The letter y, or $f(x)$, represents the output value, or dependent variable.

EXAMPLE 3

Using Function Notation for Days in a Month

Use function notation to represent a function whose input is the name of a month and output is the number of days in that month. Assume that the domain does not include leap years.

Solution

The number of days in a month is a function of the name of the month, so if we name the function f, we write $\text{days} = f(\text{month})$ or $d = f(m)$. The name of the month is the input to a "rule" that associates a specific number (the output) with each input.

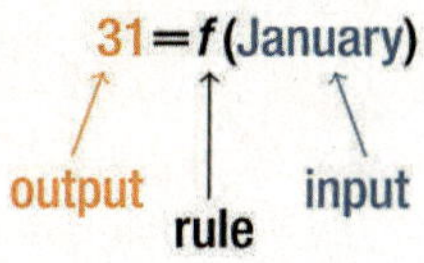

Figure 2

For example, $f(\text{March}) = 31$, because March has 31 days. The notation $d = f(m)$ reminds us that the number of days, d (the output), is dependent on the name of the month, m (the input).

Analysis

Note that the inputs to a function do not have to be numbers; function inputs can be names of people, labels of geometric objects, or any other element that determines some kind of output. However, most of the functions we will work with in this book will have numbers as inputs and outputs.

EXAMPLE 4

Interpreting Function Notation

A function $N = f(y)$ gives the number of police officers, N, in a town in year y. What does $f(2005) = 300$ represent?

Solution

When we read $f(2005) = 300$, we see that the input year is 2005. The value for the output, the number of police officers (N), is 300. Remember, $N = f(y)$. The statement $f(2005) = 300$ tells us that in the year 2005 there were 300 police officers in the town.

TRY IT #2 Use function notation to express the weight of a pig in pounds as a function of its age in days d.

Q&A **Instead of a notation such as $y = f(x)$, could we use the same symbol for the output as for the function, such as $y = y(x)$, meaning "y is a function of x?"**

Yes, this is often done, especially in applied subjects that use higher math, such as physics and engineering. However, in exploring math itself we like to maintain a distinction between a function such as f, which is a rule or procedure, and the output y we get by applying f to a particular input x. This is why we usually use notation such as $y = f(x)$, $P = W(d)$, and so on.

Representing Functions Using Tables

A common method of representing functions is in the form of a table. The table rows or columns display the corresponding input and output values. In some cases, these values represent all we know about the relationship; other times, the table provides a few select examples from a more complete relationship.

Table 3 lists the input number of each month (January = 1, February = 2, and so on) and the output value of the number of days in that month. This information represents all we know about the months and days for a given year (that is not a leap year). Note that, in this table, we define a days-in-a-month function f where $D = f(m)$ identifies months by an integer rather than by name.

Month number, m (input)	1	2	3	4	5	6	7	8	9	10	11	12
Days in month, D (output)	31	28	31	30	31	30	31	31	30	31	30	31

Table 3

Table 4 defines a function $Q = g(n)$. Remember, this notation tells us that g is the name of the function that takes the input n and gives the output Q.

n	1	2	3	4	5
Q	8	6	7	6	8

Table 4

Table 5 displays the age of children in years and their corresponding heights. This table displays just some of the data available for the heights and ages of children. We can see right away that this table does not represent a function because the same input value, 5 years, has two different output values, 40 in. and 42 in.

Age in years, a (input)	5	5	6	7	8	9	10
Height in inches, h (output)	40	42	44	47	50	52	54

Table 5

HOW TO

Given a table of input and output values, determine whether the table represents a function.

1. Identify the input and output values.
2. Check to see if each input value is paired with only one output value. If so, the table represents a function.

EXAMPLE 5

Identifying Tables that Represent Functions

Which table, Table 6, Table 7, or Table 8, represents a function (if any)?

Input	Output
2	1
5	3
8	6

Table 6

Input	Output
–3	5
0	1
4	5

Table 7

Input	Output
1	0
5	2
5	4

Table 8

Solution

Table 6 and Table 7 define functions. In both, each input value corresponds to exactly one output value. Table 8 does not define a function because the input value of 5 corresponds to two different output values.

When a table represents a function, corresponding input and output values can also be specified using function notation.

The function represented by Table 6 can be represented by writing

$$f(2) = 1, f(5) = 3, \text{and } f(8) = 6$$

Similarly, the statements

$$g(-3) = 5, \quad g(0) = 1, \text{and } g(4) = 5$$

represent the function in Table 7.

Table 8 cannot be expressed in a similar way because it does not represent a function.

TRY IT #3 Does Table 9 represent a function?

Input	Output
1	10
2	100
3	1000

Table 9

Finding Input and Output Values of a Function

When we know an input value and want to determine the corresponding output value for a function, we *evaluate* the function. Evaluating will always produce one result because each input value of a function corresponds to exactly one output value.

When we know an output value and want to determine the input values that would produce that output value, we set the output equal to the function's formula and *solve* for the input. Solving can produce more than one solution because different input values can produce the same output value.

Evaluation of Functions in Algebraic Forms

When we have a function in formula form, it is usually a simple matter to evaluate the function. For example, the function $f(x) = 5 - 3x^2$ can be evaluated by squaring the input value, multiplying by 3, and then subtracting the product from 5.

HOW TO

Given the formula for a function, evaluate.

1. Substitute the input variable in the formula with the value provided.
2. Calculate the result.

EXAMPLE 6

Evaluating Functions at Specific Values

Evaluate $f(x) = x^2 + 3x - 4$ at:

ⓐ 2 ⓑ a ⓒ $a + h$ ⓓ Now evaluate $\frac{f(a+h)-f(a)}{h}$

Solution

Replace the x in the function with each specified value.

ⓐ Because the input value is a number, 2, we can use simple algebra to simplify.

$$\begin{aligned} f(2) &= 2^2 + 3(2) - 4 \\ &= 4 + 6 - 4 \\ &= 6 \end{aligned}$$

ⓑ In this case, the input value is a letter so we cannot simplify the answer any further.

$$f(a) = a^2 + 3a - 4$$

With an input value of $a + h$, we must use the distributive property.

$$\begin{aligned} f(a+h) &= (a+h)^2 + 3(a+h) - 4 \\ &= a^2 + 2ah + h^2 + 3a + 3h - 4 \end{aligned}$$

ⓒ In this case, we apply the input values to the function more than once, and then perform algebraic operations on the result. We already found that

$$f(a+h) = a^2 + 2ah + h^2 + 3a + 3h - 4$$

and we know that

$$f(a) = a^2 + 3a - 4$$

Now we combine the results and simplify.

$$\begin{aligned} \frac{f(a+h)-f(a)}{h} &= \frac{(a^2+2ah+h^2+3a+3h-4)-(a^2+3a-4)}{h} \\ &= \frac{2ah+h^2+3h}{h} \\ &= \frac{h(2a+h+3)}{h} && \text{Factor out } h. \\ &= 2a + h + 3 && \text{Simplify.} \end{aligned}$$

EXAMPLE 7

Evaluating Functions

Given the function $h(p) = p^2 + 2p$, evaluate $h(4)$.

Solution

To evaluate $h(4)$, we substitute the value 4 for the input variable p in the given function.

$$\begin{aligned} h(p) &= p^2 + 2p \\ h(4) &= (4)^2 + 2(4) \\ &= 16 + 8 \\ &= 24 \end{aligned}$$

Therefore, for an input of 4, we have an output of 24.

TRY IT #4 Given the function $g(m) = \sqrt{m - 4}$, evaluate $g(5)$.

EXAMPLE 8

Solving Functions

Given the function $h(p) = p^2 + 2p$, solve for $h(p) = 3$.

Solution

$$\begin{aligned} h(p) &= 3 \\ p^2 + 2p &= 3 && \text{Substitute the original function } h(p) = p^2 + 2p. \\ p^2 + 2p - 3 &= 0 && \text{Subtract 3 from each side.} \\ (p + 3)(p - 1) &= 0 && \text{Factor.} \end{aligned}$$

If $(p + 3)(p - 1) = 0$, either $(p + 3) = 0$ or $(p - 1) = 0$ (or both of them equal 0). We will set each factor equal to 0 and solve for p in each case.

$$\begin{aligned} (p + 3) = 0, &\quad p = -3 \\ (p - 1) = 0, &\quad p = 1 \end{aligned}$$

This gives us two solutions. The output $h(p) = 3$ when the input is either $p = 1$ or $p = -3$. We can also verify by graphing as in Figure 3. The graph verifies that $h(1) = h(-3) = 3$ and $h(4) = 24$.

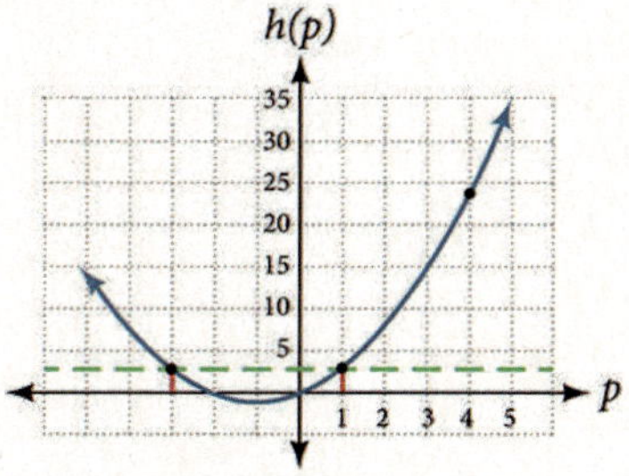

p	−3	−2	0	1	4
$h(p)$	3	0	0	3	24

Figure 3

TRY IT #5 Given the function $g(m) = \sqrt{m - 4}$, solve $g(m) = 2$.

Evaluating Functions Expressed in Formulas

Some functions are defined by mathematical rules or procedures expressed in equation form. If it is possible to express the function output with a formula involving the input quantity, then we can define a function in algebraic form. For example, the equation $2n + 6p = 12$ expresses a functional relationship between n and p. We can rewrite it to decide if p

is a function of n.

HOW TO

Given a function in equation form, write its algebraic formula.

1. Solve the equation to isolate the output variable on one side of the equal sign, with the other side as an expression that involves *only* the input variable.
2. Use all the usual algebraic methods for solving equations, such as adding or subtracting the same quantity to or from both sides, or multiplying or dividing both sides of the equation by the same quantity.

EXAMPLE 9

Finding an Equation of a Function

Express the relationship $2n + 6p = 12$ as a function $p = f(n)$, if possible.

Solution

To express the relationship in this form, we need to be able to write the relationship where p is a function of n, which means writing it as $p = [\text{expression involving } n]$.

$$\begin{aligned} 2n + 6p &= 12 \\ 6p &= 12 - 2n && \text{Subtract } 2n \text{ from both sides.} \\ p &= \tfrac{12-2n}{6} && \text{Divide both sides by 6 and simplify.} \\ p &= \tfrac{12}{6} - \tfrac{2n}{6} \\ p &= 2 - \tfrac{1}{3}n \end{aligned}$$

Therefore, p as a function of n is written as

$$p = f(n) = 2 - \frac{1}{3}n$$

Analysis

It is important to note that not every relationship expressed by an equation can also be expressed as a function with a formula.

EXAMPLE 10

Expressing the Equation of a Circle as a Function

Does the equation $x^2 + y^2 = 1$ represent a function with x as input and y as output? If so, express the relationship as a function $y = f(x)$.

Solution

First we subtract x^2 from both sides.

$$y^2 = 1 - x^2$$

We now try to solve for y in this equation.

$$\begin{aligned} y &= \pm\sqrt{1 - x^2} \\ &= +\sqrt{1 - x^2} \text{ and } -\sqrt{1 - x^2} \end{aligned}$$

We get two outputs corresponding to the same input, so this relationship cannot be represented as a single function $y = f(x)$.

TRY IT #6 If $x - 8y^3 = 0$, express y as a function of x.

Q&A **Are there relationships expressed by an equation that do represent a function but which still cannot be represented by an algebraic formula?**

Yes, this can happen. For example, given the equation $x = y + 2^y$, *if we want to express* y *as a function of* x, *there is no simple algebraic formula involving only* x *that equals* y. *However, each* x *does determine a unique value for* y, *and there are mathematical procedures by which* y *can be found to any desired accuracy. In this case, we say that the equation gives an implicit (implied) rule for* y *as a function of* x, *even though the formula cannot be written explicitly.*

Evaluating a Function Given in Tabular Form

As we saw above, we can represent functions in tables. Conversely, we can use information in tables to write functions, and we can evaluate functions using the tables. For example, how well do our pets recall the fond memories we share with them? There is an urban legend that a goldfish has a memory of 3 seconds, but this is just a myth. Goldfish can remember up to 3 months, while the beta fish has a memory of up to 5 months. And while a puppy's memory span is no longer than 30 seconds, the adult dog can remember for 5 minutes. This is meager compared to a cat, whose memory span lasts for 16 hours.

The function that relates the type of pet to the duration of its memory span is more easily visualized with the use of a table. See Table 10.[2]

Pet	Memory span in hours
Puppy	0.008
Adult dog	0.083
Cat	16
Goldfish	2160
Beta fish	3600

Table 10

At times, evaluating a function in table form may be more useful than using equations.Here let us call the function P. The domain of the function is the type of pet and the range is a real number representing the number of hours the pet's memory span lasts. We can evaluate the function P at the input value of "goldfish." We would write $P(\text{goldfish}) = 2160$. Notice that, to evaluate the function in table form, we identify the input value and the corresponding output value from the pertinent row of the table. The tabular form for function P seems ideally suited to this function, more so than writing it in paragraph or function form.

HOW TO

Given a function represented by a table, identify specific output and input values.

1. Find the given input in the row (or column) of input values.
2. Identify the corresponding output value paired with that input value.
3. Find the given output values in the row (or column) of output values, noting every time that output value appears.
4. Identify the input value(s) corresponding to the given output value.

2 http://www.kgbanswers.com/how-long-is-a-dogs-memory-span/4221590. Accessed 3/24/2014.

EXAMPLE 11

Evaluating and Solving a Tabular Function

Using Table 11,

ⓐ Evaluate $g(3)$.

ⓑ Solve $g(n) = 6$.

n	1	2	3	4	5
$g(n)$	8	6	7	6	8

Table 11

Solution

ⓐ Evaluating $g(3)$ means determining the output value of the function g for the input value of $n = 3$. The table output value corresponding to $n = 3$ is 7, so $g(3) = 7$.

ⓑ Solving $g(n) = 6$ means identifying the input values, n, that produce an output value of 6. The table below shows two solutions: 2 and 4.

n	1	2	3	4	5
$g(n)$	8	6	7	6	8

When we input 2 into the function g, our output is 6. When we input 4 into the function g, our output is also 6.

TRY IT #7 Using the table from **Evaluating and Solving a Tabular Function** above, evaluate $g(1)$.

Finding Function Values from a Graph

Evaluating a function using a graph also requires finding the corresponding output value for a given input value, only in this case, we find the output value by looking at the graph. Solving a function equation using a graph requires finding all instances of the given output value on the graph and observing the corresponding input value(s).

EXAMPLE 12

Reading Function Values from a Graph

Given the graph in Figure 4,

ⓐ Evaluate $f(2)$.

ⓑ Solve $f(x) = 4$.

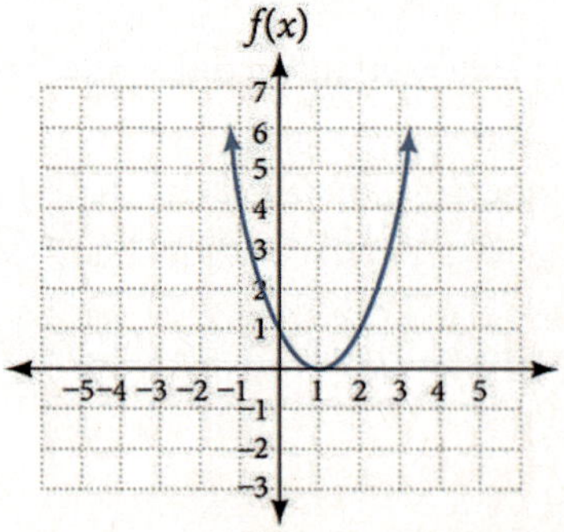

Figure 4

Solution

ⓐ To evaluate $f(2)$, locate the point on the curve where $x = 2$, then read the *y*-coordinate of that point. The point has coordinates $(2, 1)$, so $f(2) = 1$. See Figure 5.

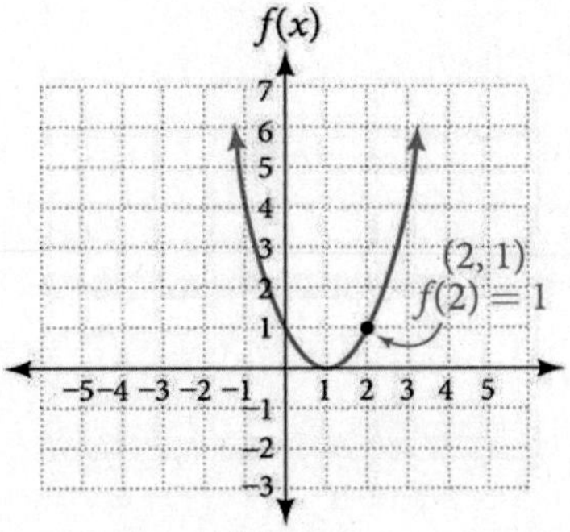

Figure 5

ⓑ To solve $f(x) = 4$, we find the output value 4 on the vertical axis. Moving horizontally along the line $y = 4$, we locate two points of the curve with output value 4: $(-1, 4)$ and $(3, 4)$. These points represent the two solutions to $f(x) = 4$: -1 or 3. This means $f(-1) = 4$ and $f(3) = 4$, or when the input is -1 or 3, the output is 4. See Figure 6.

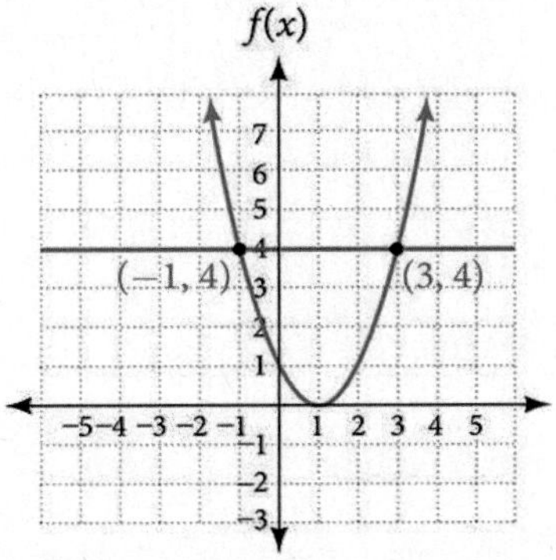

Figure 6

TRY IT #8 Using Figure 4, solve $f(x) = 1$.

Determining Whether a Function is One-to-One

Some functions have a given output value that corresponds to two or more input values. For example, in the stock chart shown in the figure at the beginning of this chapter, the stock price was \$1000 on five different dates, meaning that there were five different input values that all resulted in the same output value of \$1000.

However, some functions have only one input value for each output value, as well as having only one output for each input. We call these functions one-to-one functions. As an example, consider a school that uses only letter grades and decimal equivalents, as listed in Table 12.

Letter grade	Grade point average
A	4.0
B	3.0
C	2.0

Table 12

Letter grade	Grade point average
D	1.0

Table 12

This grading system represents a one-to-one function, because each letter input yields one particular grade point average output and each grade point average corresponds to one input letter.

To visualize this concept, let's look again at the two simple functions sketched in Figure 1**(a)** and Figure 1**(b)**. The function in part (a) shows a relationship that is not a one-to-one function because inputs q and r both give output n. The function in part (b) shows a relationship that is a one-to-one function because each input is associated with a single output.

One-to-One Function

A **one-to-one function** is a function in which each output value corresponds to exactly one input value.

EXAMPLE 13

Determining Whether a Relationship Is a One-to-One Function
Is the area of a circle a function of its radius? If yes, is the function one-to-one?

Solution
A circle of radius r has a unique area measure given by $A = \pi r^2$, so for any input, r, there is only one output, A. The area is a function of radius r.

If the function is one-to-one, the output value, the area, must correspond to a unique input value, the radius. Any area measure A is given by the formula $A = \pi r^2$. Because areas and radii are positive numbers, there is exactly one solution: $\sqrt{\frac{A}{\pi}}$. So the area of a circle is a one-to-one function of the circle's radius.

TRY IT #9

ⓐ Is a balance a function of the bank account number?
ⓑ Is a bank account number a function of the balance?
ⓒ Is a balance a one-to-one function of the bank account number?

TRY IT #10

Evaluate the following:
ⓐ If each percent grade earned in a course translates to one letter grade, is the letter grade a function of the percent grade?
ⓑ If so, is the function one-to-one?

Using the Vertical Line Test

As we have seen in some examples above, we can represent a function using a graph. Graphs display a great many input-output pairs in a small space. The visual information they provide often makes relationships easier to understand. By convention, graphs are typically constructed with the input values along the horizontal axis and the output values along the vertical axis.

The most common graphs name the input value x and the output value y, and we say y is a function of x, or $y = f(x)$ when the function is named f. The graph of the function is the set of all points (x, y) in the plane that satisfies the equation $y = f(x)$. If the function is defined for only a few input values, then the graph of the function is only a few points, where the x-coordinate of each point is an input value and the y-coordinate of each point is the corresponding output value. For example, the black dots on the graph in Figure 7 tell us that $f(0) = 2$ and $f(6) = 1$. However, the set of all points (x, y) satisfying $y = f(x)$ is a curve. The curve shown includes $(0, 2)$ and $(6, 1)$ because the curve passes through those points.

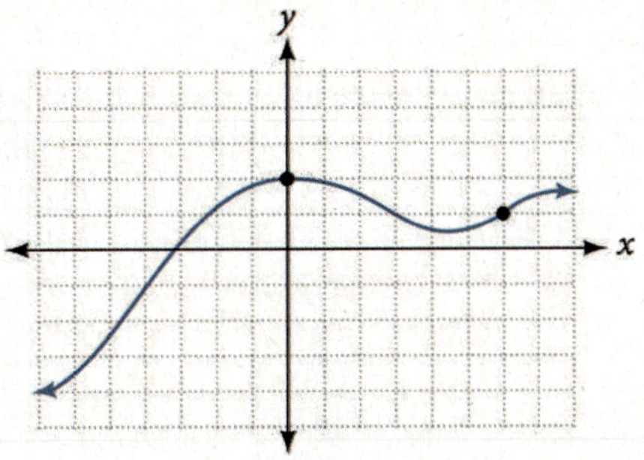

Figure 7

The **vertical line test** can be used to determine whether a graph represents a function. If we can draw any vertical line that intersects a graph more than once, then the graph does *not* define a function because a function has only one output value for each input value. See Figure 8.

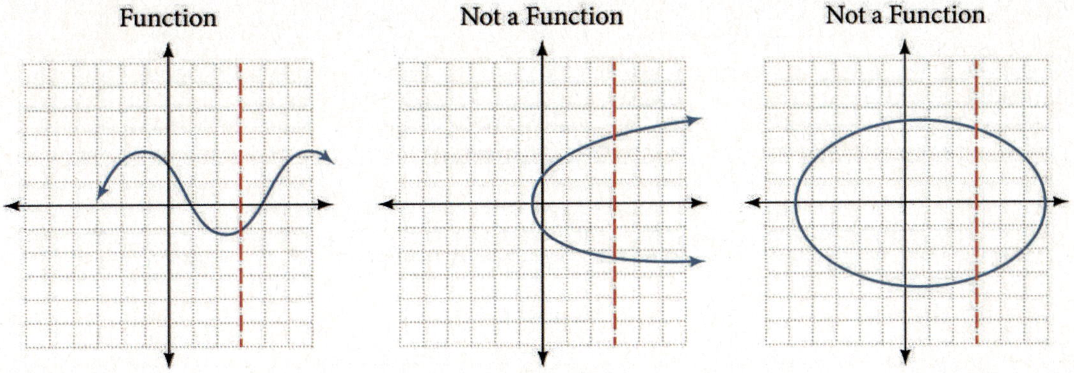

Figure 8

HOW TO

Given a graph, use the vertical line test to determine if the graph represents a function.

1. Inspect the graph to see if any vertical line drawn would intersect the curve more than once.
2. If there is any such line, determine that the graph does not represent a function.

EXAMPLE 14

Applying the Vertical Line Test

Which of the graphs in Figure 9 represent(s) a function $y = f(x)$?

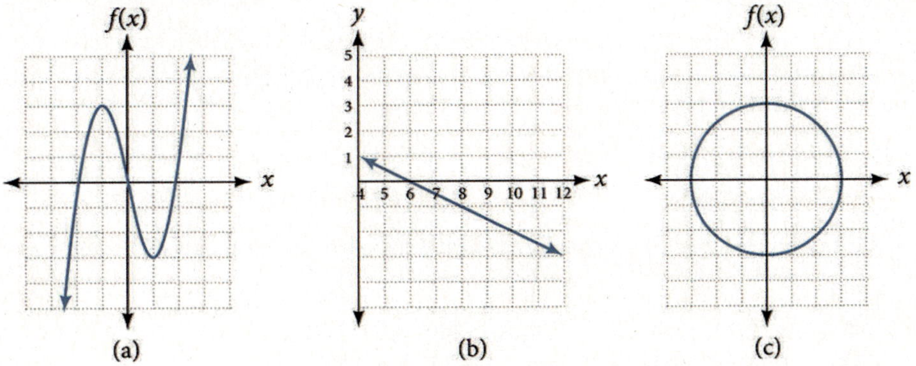

Figure 9

Solution

If any vertical line intersects a graph more than once, the relation represented by the graph is not a function. Notice that any vertical line would pass through only one point of the two graphs shown in parts (a) and (b) of Figure 9. From this we can conclude that these two graphs represent functions. The third graph does not represent a function because, at most x-values, a vertical line would intersect the graph at more than one point, as shown in Figure 10.

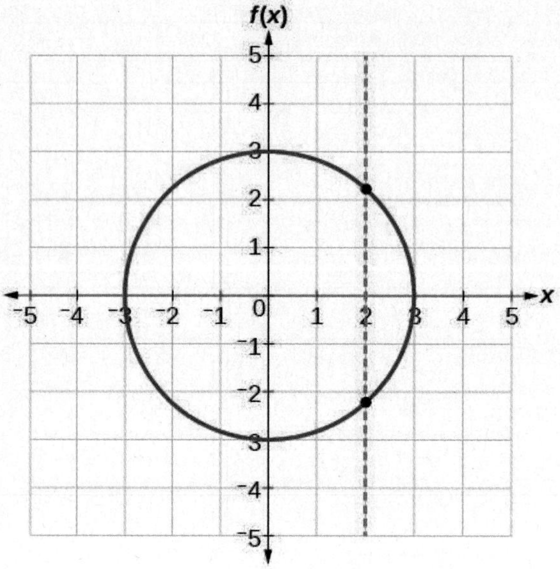

Figure 10

TRY IT #11 Does the graph in Figure 11 represent a function?

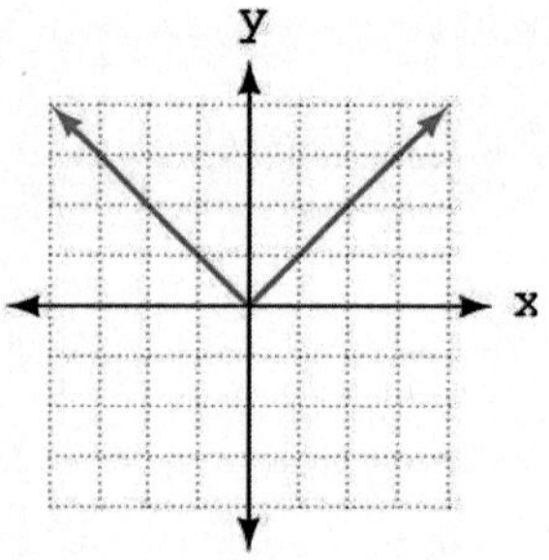

Figure 11

Using the Horizontal Line Test

Once we have determined that a graph defines a function, an easy way to determine if it is a one-to-one function is to use the **horizontal line test**. Draw horizontal lines through the graph. If any horizontal line intersects the graph more than once, then the graph does not represent a one-to-one function.

HOW TO

Given a graph of a function, use the horizontal line test to determine if the graph represents a one-to-one function.

1. Inspect the graph to see if any horizontal line drawn would intersect the curve more than once.
2. If there is any such line, determine that the function is not one-to-one.

EXAMPLE 15

Applying the Horizontal Line Test

Consider the functions shown in Figure 9**(a)** and Figure 9**(b)**. Are either of the functions one-to-one?

Solution

The function in Figure 9**(a)** is not one-to-one. The horizontal line shown in Figure 12 intersects the graph of the function at two points (and we can even find horizontal lines that intersect it at three points.)

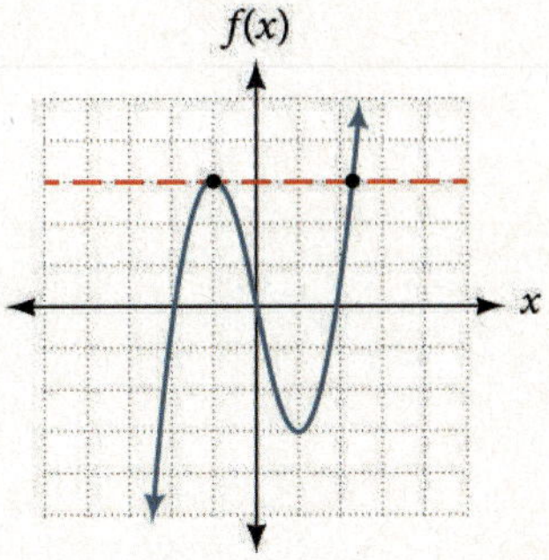

Figure 12

The function in Figure 9**(b)** is one-to-one. Any horizontal line will intersect a diagonal line at most once.

TRY IT #12 Is the graph shown in Figure 9 one-to-one?

Identifying Basic Toolkit Functions

In this text, we will be exploring functions—the shapes of their graphs, their unique characteristics, their algebraic formulas, and how to solve problems with them. When learning to read, we start with the alphabet. When learning to do arithmetic, we start with numbers. When working with functions, it is similarly helpful to have a base set of building-block elements. We call these our "toolkit functions," which form a set of basic named functions for which we know the graph, formula, and special properties. Some of these functions are programmed to individual buttons on many calculators. For these definitions we will use x as the input variable and $y = f(x)$ as the output variable.

We will see these toolkit functions, combinations of toolkit functions, their graphs, and their transformations frequently throughout this book. It will be very helpful if we can recognize these toolkit functions and their features quickly by name, formula, graph, and basic table properties. The graphs and sample table values are included with each function shown in Table 13.

Toolkit Functions		
Name	**Function**	**Graph**
Constant	$f(x) = c$, where c is a constant	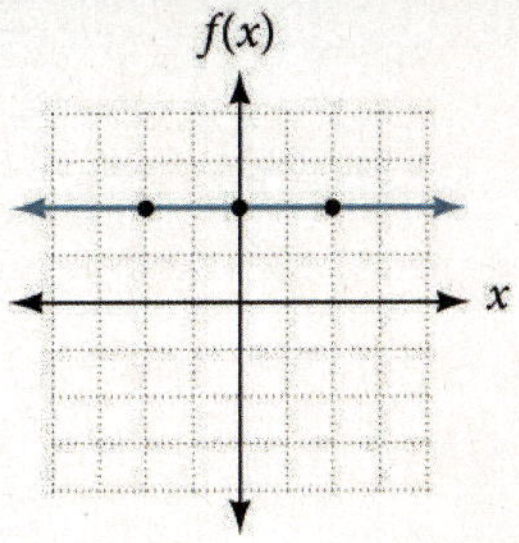

x	$f(x)$
−2	2
0	2
2	2

Table 13

<table>
<tr><th colspan="3">Toolkit Functions</th></tr>
<tr><th>Name</th><th>Function</th><th>Graph</th></tr>
<tr><td>Identity</td><td>$f(x) = x$</td><td>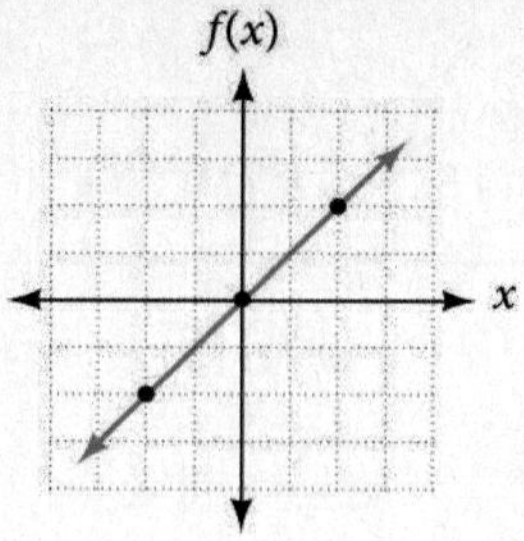

<table><tr><th>x</th><th>f(x)</th></tr><tr><td>–2</td><td>–2</td></tr><tr><td>0</td><td>0</td></tr><tr><td>2</td><td>2</td></tr></table></td></tr>
<tr><td>Absolute value</td><td>$f(x) = |x|$</td><td>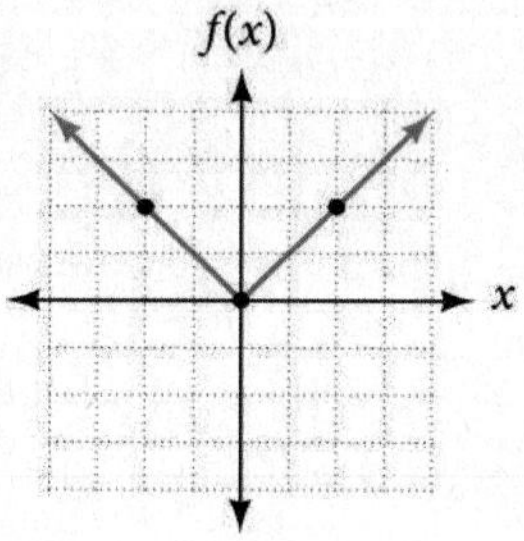

<table><tr><th>x</th><th>f(x)</th></tr><tr><td>–2</td><td>2</td></tr><tr><td>0</td><td>0</td></tr><tr><td>2</td><td>2</td></tr></table></td></tr>
<tr><td>Quadratic</td><td>$f(x) = x^2$</td><td>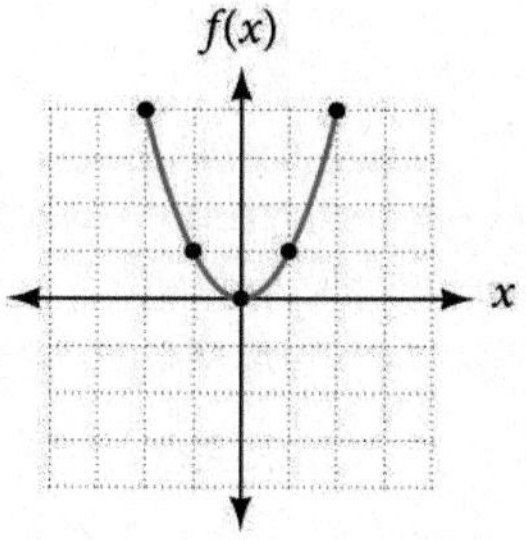

<table><tr><th>x</th><th>f(x)</th></tr><tr><td>–2</td><td>4</td></tr><tr><td>–1</td><td>1</td></tr><tr><td>0</td><td>0</td></tr><tr><td>1</td><td>1</td></tr><tr><td>2</td><td>4</td></tr></table></td></tr>
<tr><td>Cubic</td><td>$f(x) = x^3$</td><td>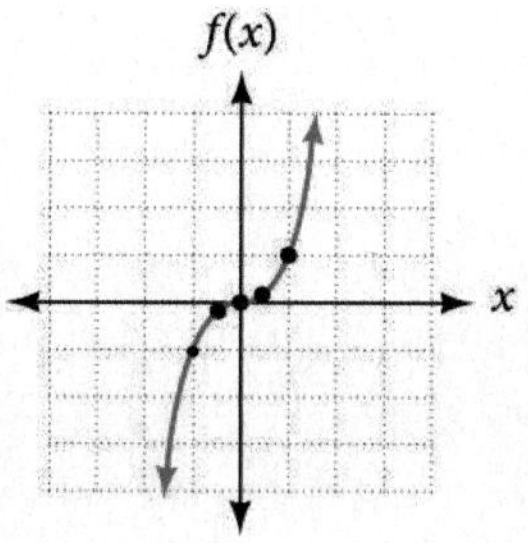

<table><tr><th>x</th><th>f(x)</th></tr><tr><td>–1</td><td>–1</td></tr><tr><td>–0.5</td><td>–0.125</td></tr><tr><td>0</td><td>0</td></tr><tr><td>0.5</td><td>0.125</td></tr><tr><td>1</td><td>1</td></tr></table></td></tr>
<tr><td>Reciprocal</td><td>$f(x) = \frac{1}{x}$</td><td>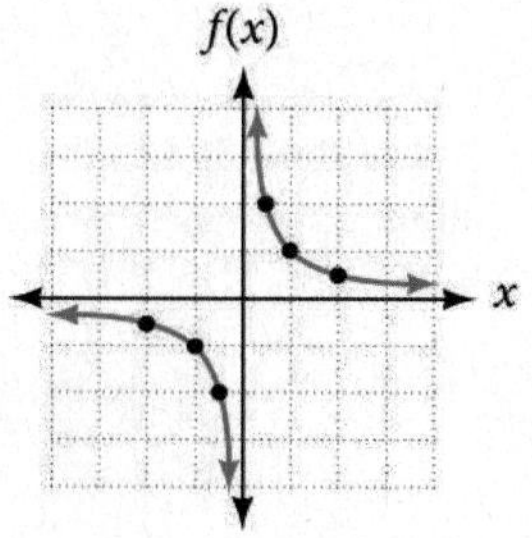

<table><tr><th>x</th><th>f(x)</th></tr><tr><td>–2</td><td>–0.5</td></tr><tr><td>–1</td><td>–1</td></tr><tr><td>–0.5</td><td>–2</td></tr><tr><td>0.5</td><td>2</td></tr><tr><td>1</td><td>1</td></tr><tr><td>2</td><td>0.5</td></tr></table></td></tr>
</table>

Table 13

<table>
<tr><th colspan="3">Toolkit Functions</th></tr>
<tr><th>Name</th><th>Function</th><th>Graph</th></tr>
<tr><td>Reciprocal squared</td><td>$f(x) = \frac{1}{x^2}$</td><td>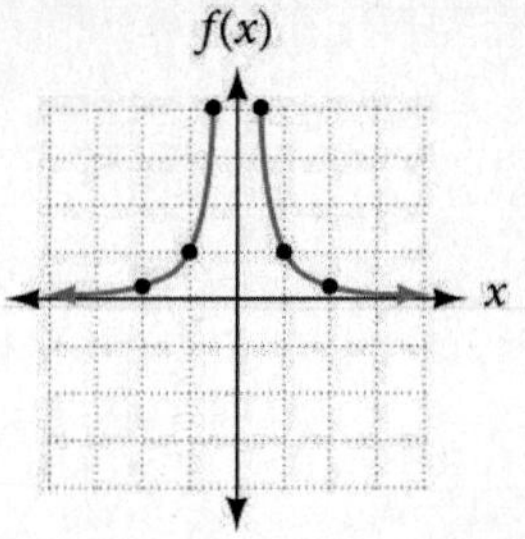

<table>
<tr><th>x</th><th>f(x)</th></tr>
<tr><td>–2</td><td>0.25</td></tr>
<tr><td>–1</td><td>1</td></tr>
<tr><td>–0.5</td><td>4</td></tr>
<tr><td>0.5</td><td>4</td></tr>
<tr><td>1</td><td>1</td></tr>
<tr><td>2</td><td>0.25</td></tr>
</table></td></tr>
<tr><td>Square root</td><td>$f(x) = \sqrt{x}$</td><td>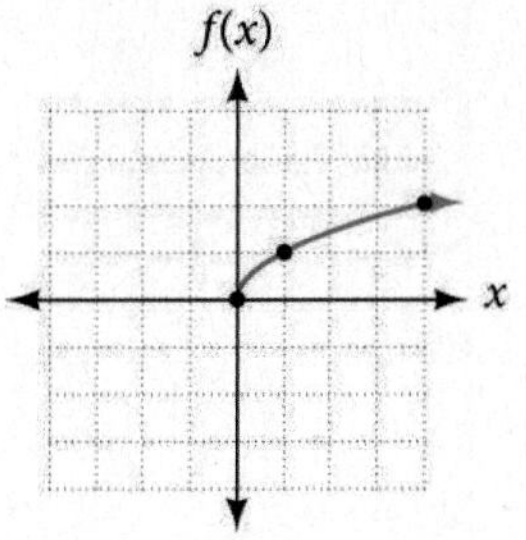

<table>
<tr><th>x</th><th>f(x)</th></tr>
<tr><td>0</td><td>0</td></tr>
<tr><td>1</td><td>1</td></tr>
<tr><td>4</td><td>2</td></tr>
</table></td></tr>
<tr><td>Cube root</td><td>$f(x) = \sqrt[3]{x}$</td><td>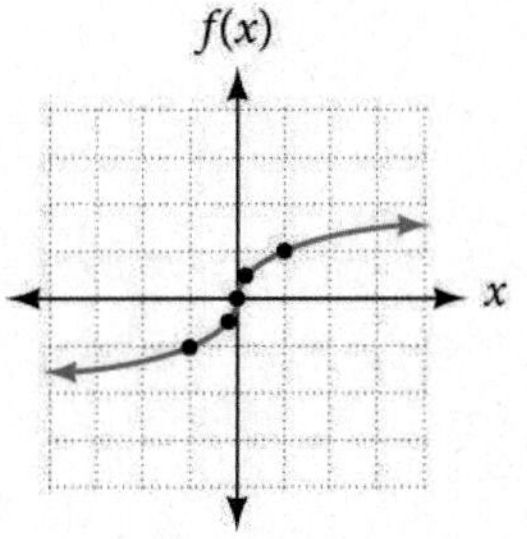

<table>
<tr><th>x</th><th>f(x)</th></tr>
<tr><td>–1</td><td>–1</td></tr>
<tr><td>–0.125</td><td>–0.5</td></tr>
<tr><td>0</td><td>0</td></tr>
<tr><td>0.125</td><td>0.5</td></tr>
<tr><td>1</td><td>1</td></tr>
</table></td></tr>
</table>

Table 13

MEDIA

Access the following online resources for additional instruction and practice with functions.

[Determine if a Relation is a Function (http://openstax.org/l/relationfunction)](http://openstax.org/l/relationfunction)
[Vertical Line Test (http://openstax.org/l/vertlinetest)](http://openstax.org/l/vertlinetest)
[Introduction to Functions (http://openstax.org/l/introtofunction)](http://openstax.org/l/introtofunction)
[Vertical Line Test on Graph (http://openstax.org/l/vertlinegraph)](http://openstax.org/l/vertlinegraph)
[One-to-one Functions (http://openstax.org/l/onetoone)](http://openstax.org/l/onetoone)
[Graphs as One-to-one Functions (http://openstax.org/l/graphonetoone)](http://openstax.org/l/graphonetoone)

3.1 SECTION EXERCISES

Verbal

1. What is the difference between a relation and a function?

2. What is the difference between the input and the output of a function?

3. Why does the vertical line test tell us whether the graph of a relation represents a function?

4. How can you determine if a relation is a one-to-one function?

5. Why does the horizontal line test tell us whether the graph of a function is one-to-one?

Algebraic

For the following exercises, determine whether the relation represents a function.

6. $\{(a, b), \ (c, d), \ (a, c)\}$

7. $\{(a, b), (b, c), (c, c)\}$

For the following exercises, determine whether the relation represents y as a function of x.

8. $5x + 2y = 10$

9. $y = x^2$

10. $x = y^2$

11. $3x^2 + y = 14$

12. $2x + y^2 = 6$

13. $y = -2x^2 + 40x$

14. $y = \frac{1}{x}$

15. $x = \frac{3y+5}{7y-1}$

16. $x = \sqrt{1 - y^2}$

17. $y = \frac{3x+5}{7x-1}$

18. $x^2 + y^2 = 9$

19. $2xy = 1$

20. $x = y^3$

21. $y = x^3$

22. $y = \sqrt{1 - x^2}$

23. $x = \pm\sqrt{1 - y}$

24. $y = \pm\sqrt{1 - x}$

25. $y^2 = x^2$

26. $y^3 = x^2$

For the following exercises, evaluate $f(-3), f(2), f(-a), -f(a), f(a + h)$.

27. $f(x) = 2x - 5$

28. $f(x) = -5x^2 + 2x - 1$

29. $f(x) = \sqrt{2 - x} + 5$

30. $f(x) = \frac{6x-1}{5x+2}$

31. $f(x) = |x - 1| - |x + 1|$

32. Given the function $g(x) = 5 - x^2$, evaluate $\frac{g(x+h)-g(x)}{h}, h \neq 0$.

33. Given the function $g(x) = x^2 + 2x$, evaluate $\frac{g(x)-g(a)}{x-a}, x \neq a$.

34. Given the function $k(t) = 2t - 1$:
    ⓐ Evaluate $k(2)$.
    ⓑ Solve $k(t) = 7$.

35. Given the function $f(x) = 8 - 3x$:
    ⓐ Evaluate $f(-2)$.
    ⓑ Solve $f(x) = -1$.

36. Given the function $p(c) = c^2 + c$:
    ⓐ Evaluate $p(-3)$.
    ⓑ Solve $p(c) = 2$.

37. Given the function $f(x) = x^2 - 3x$:
    ⓐ Evaluate $f(5)$.
    ⓑ Solve $f(x) = 4$.

38. Given the function $f(x) = \sqrt{x + 2}$:
    ⓐ Evaluate $f(7)$.
    ⓑ Solve $f(x) = 4$.

39. Consider the relationship $3r + 2t = 18$.

ⓐ Write the relationship as a function $r = f(t)$.

ⓑ Evaluate $f(-3)$.

ⓒ Solve $f(t) = 2$.

Graphical

For the following exercises, use the vertical line test to determine which graphs show relations that are functions.

40.

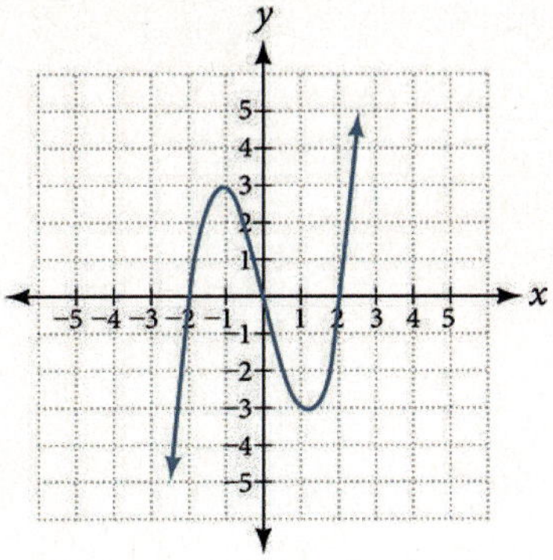

41.

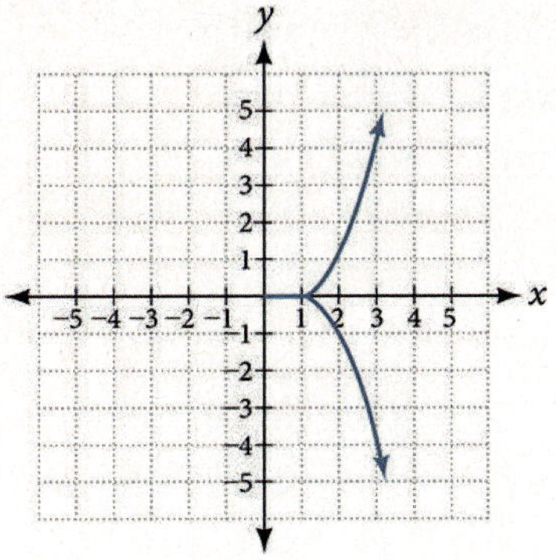

42.

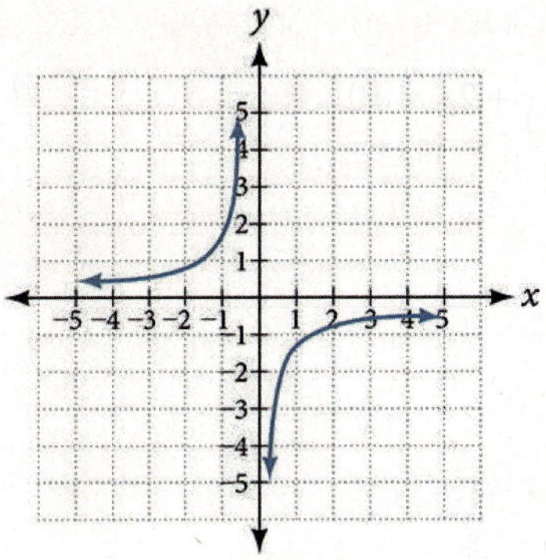

43.

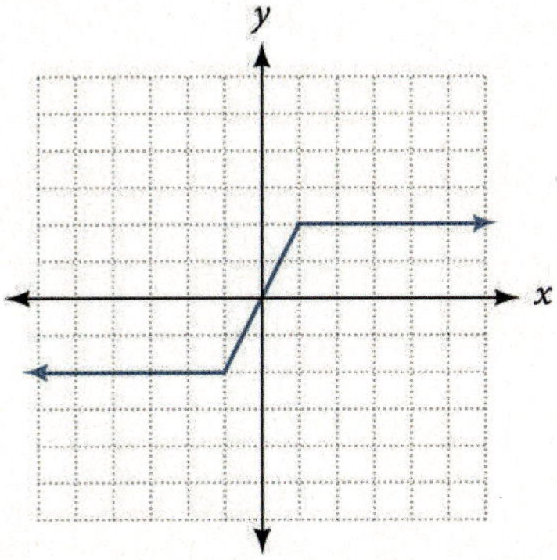

44.

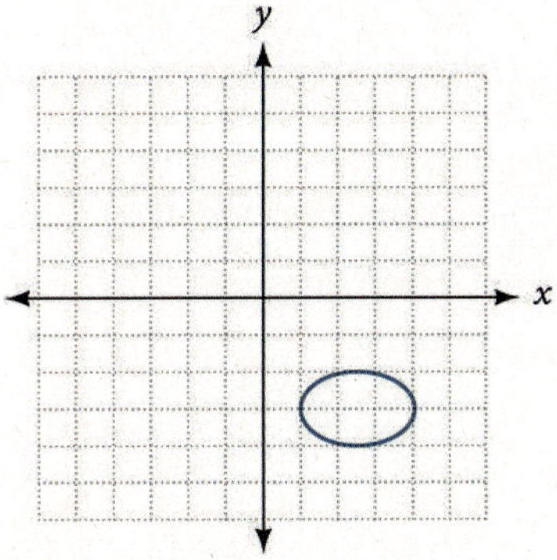

45.

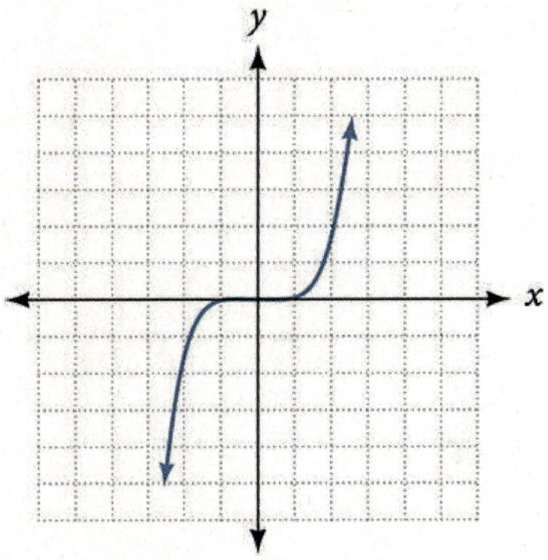

46.

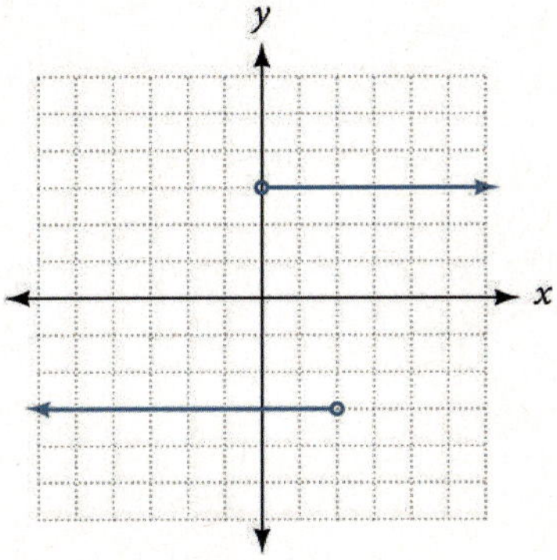

47.

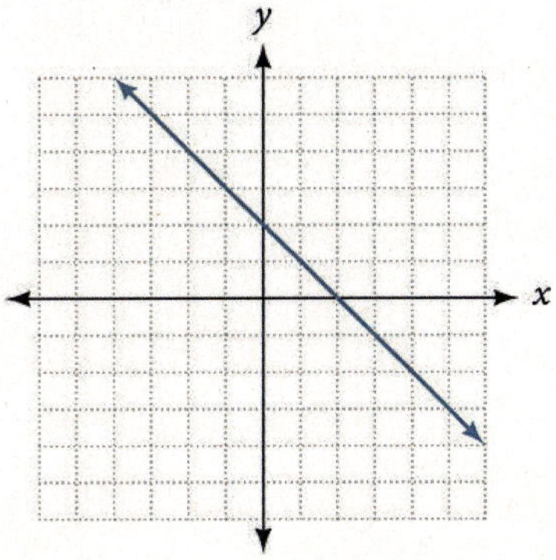

48.

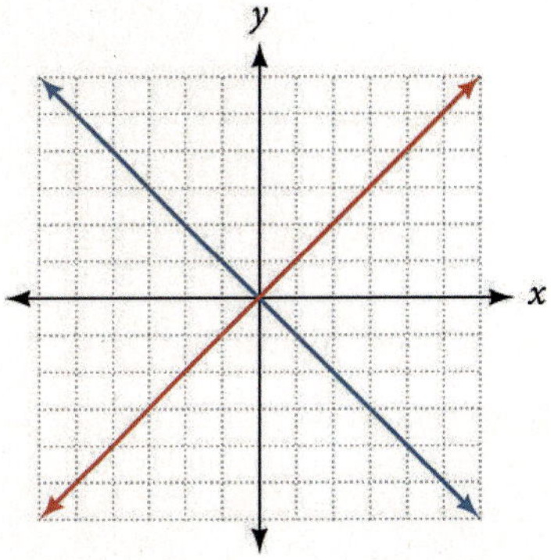

49.

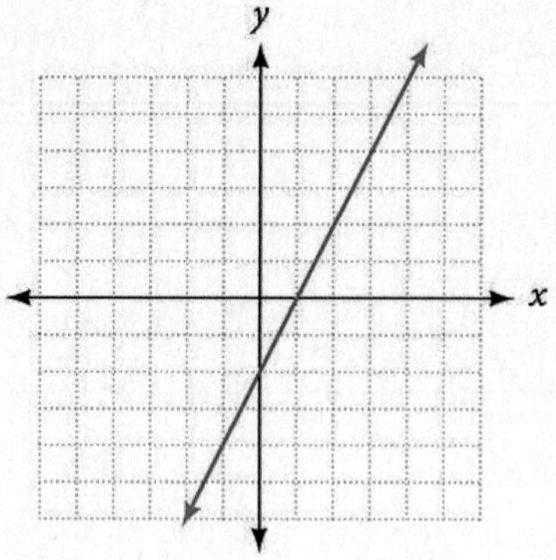

50.

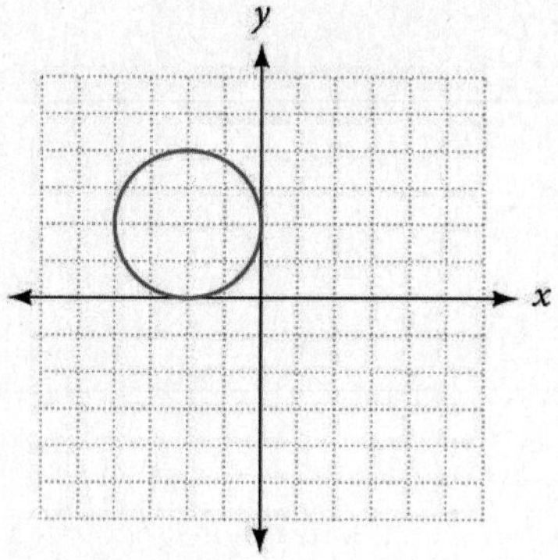

51.

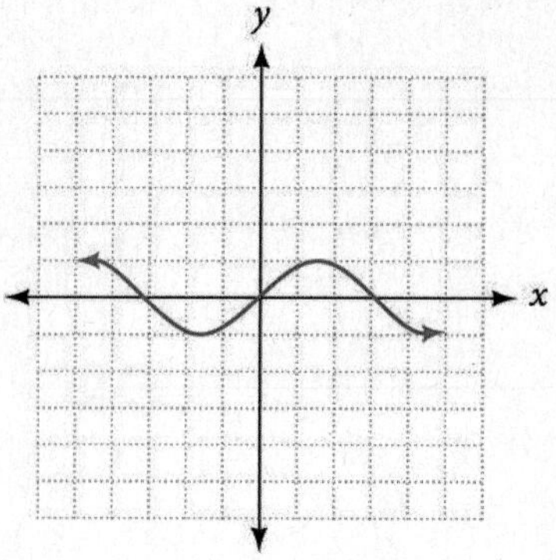

52. Given the following graph,

ⓐ Evaluate $f(-1)$.

ⓑ Solve for $f(x) = 3$.

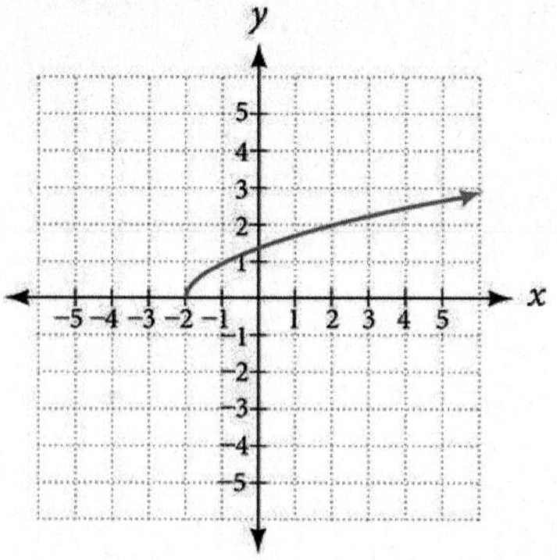

53. Given the following graph,

ⓐ Evaluate $f(0)$.

ⓑ Solve for $f(x) = -3$.

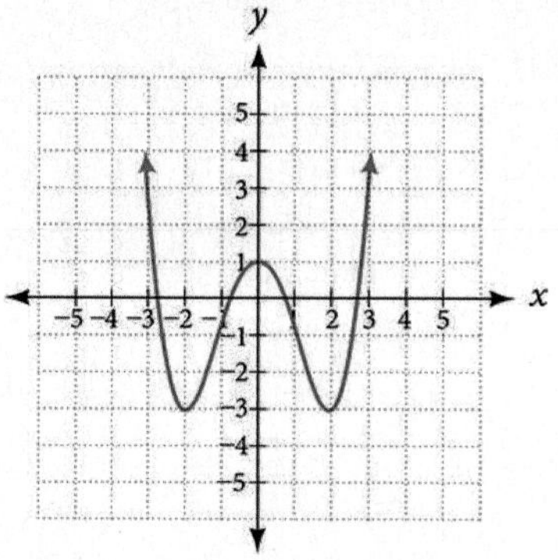

54. Given the following graph,

ⓐ Evaluate $f(4)$.

ⓑ Solve for $f(x) = 1$.

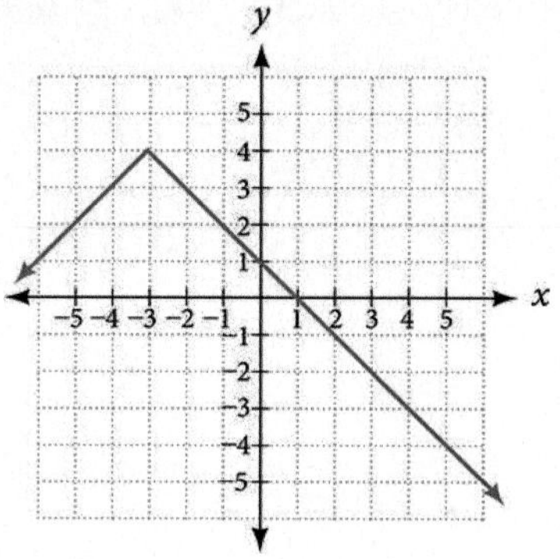

For the following exercises, determine if the given graph is a one-to-one function.

55.

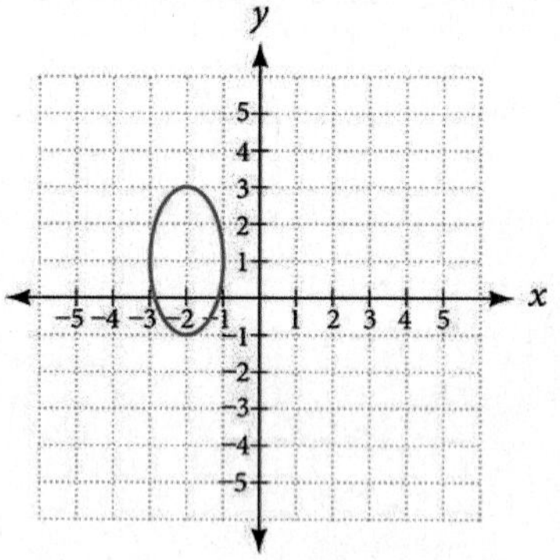

56.

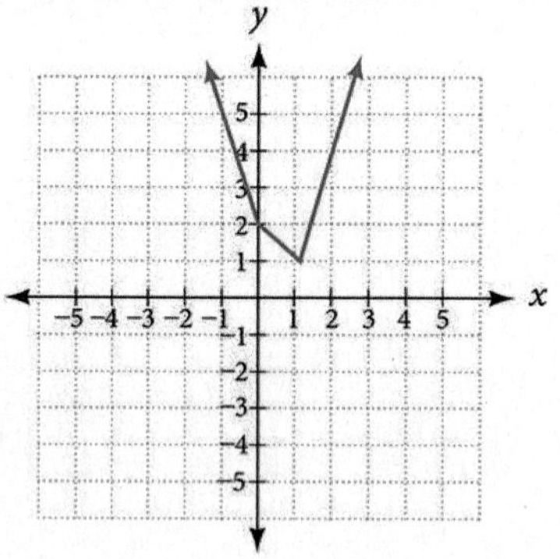

57.

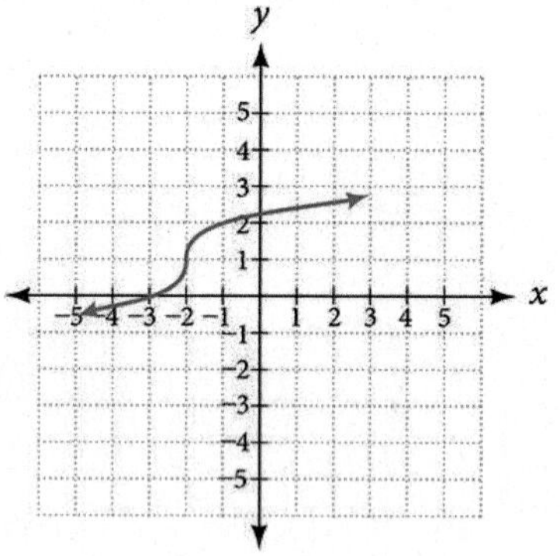

58.

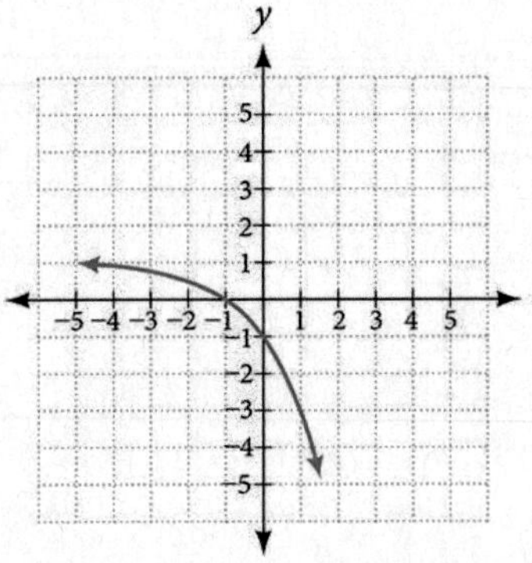

59.

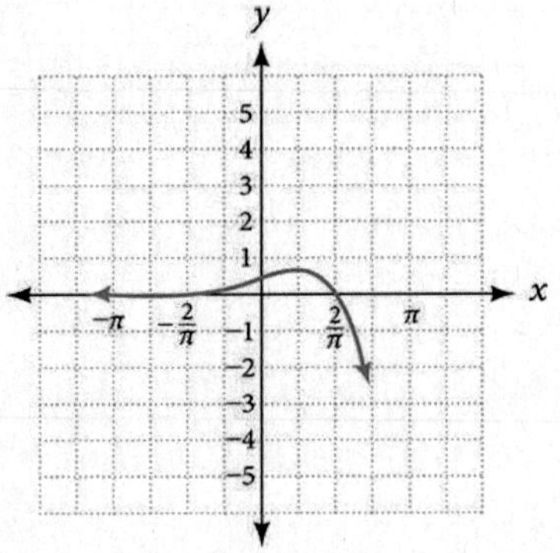

Numeric

For the following exercises, determine whether the relation represents a function.

60. $\{(-1,-1),(-2,-2),(-3,-3)\}$

61. $\{(3,4),(4,5),(5,6)\}$

62. $\{(2,5),(7,11),(15,8),(7,9)\}$

For the following exercises, determine if the relation represented in table form represents y as a function of x.

63.

x	5	10	15
y	3	8	14

64.

x	5	10	15
y	3	8	8

65.

x	5	10	10
y	3	8	14

For the following exercises, use the function f represented in the table below.

x	0	1	2	3	4	5	6	7	8	9
$f(x)$	74	28	1	53	56	3	36	45	14	47

Table 14

66. Evaluate $f(3)$.

67. Solve $f(x) = 1$.

For the following exercises, evaluate the function f at the values $f(-2)$, $f(-1)$, $f(0)$, $f(1)$, and $f(2)$.

68. $f(x) = 4 - 2x$

69. $f(x) = 8 - 3x$

70. $f(x) = 8x^2 - 7x + 3$

71. $f(x) = 3 + \sqrt{x+3}$

72. $f(x) = \frac{x-2}{x+3}$

73. $f(x) = 3^x$

For the following exercises, evaluate the expressions, given functions $f, g,$ and h :

$f(x) = 3x - 2$
$g(x) = 5 - x^2$
$h(x) = -2x^2 + 3x - 1$

74. $3f(1) - 4g(-2)$

75. $f\left(\frac{7}{3}\right) - h(-2)$

Technology

For the following exercises, graph $y = x^2$ on the given domain. Determine the corresponding range. Show each graph.

76. $[-0.1,\ 0.1]$

77. $[-10,\ 10]$

78. $[-100, 100]$

For the following exercises, graph $y = x^3$ on the given domain. Determine the corresponding range. Show each graph.

79. $[-0.1,\ 0.1]$

80. $[-10,\ 10]$

81. $[-100,\ 100]$

For the following exercises, graph $y = \sqrt{x}$ on the given domain. Determine the corresponding range. Show each graph.

82. $[0,\ 0.01]$

83. $[0,\ 100]$

84. $[0,\ 10{,}000]$

For the following exercises, graph $y = \sqrt[3]{x}$ on the given domain. Determine the corresponding range. Show each graph.

85. $[-0.001, 0.001]$

86. $[-1000, 1000]$

87. $[-1{,}000{,}000, 1{,}000{,}000]$

Real-World Applications

88. The amount of garbage, G, produced by a city with population p is given by $G = f(p)$. G is measured in tons per week, and p is measured in thousands of people.

ⓐ The town of Tola has a population of 40,000 and produces 13 tons of garbage each week. Express this information in terms of the function f.
ⓑ Explain the meaning of the statement $f(5) = 2$.

89. The number of cubic yards of dirt, D, needed to cover a garden with area a square feet is given by $D = g(a)$.

ⓐ A garden with area 5000 ft^2 requires 50 yd^3 of dirt. Express this information in terms of the function g.
ⓑ Explain the meaning of the statement $g(100) = 1$.

90. Let $f(t)$ be the number of ducks in a lake t years after 1990. Explain the meaning of each statement:

ⓐ $f(5) = 30$
ⓑ $f(10) = 40$

91. Let $h(t)$ be the height above ground, in feet, of a rocket t seconds after launching. Explain the meaning of each statement:

ⓐ $h(1) = 200$
ⓑ $h(2) = 350$

92. Show that the function $f(x) = 3(x - 5)^2 + 7$ is not one-to-one.

3.2 Domain and Range

Learning Objectives

In this section, you will:

- Find the domain of a function defined by an equation.
- Graph piecewise-defined functions.

Horror and thriller movies are both popular and, very often, extremely profitable. When big-budget actors, shooting locations, and special effects are included, however, studios count on even more viewership to be successful. Consider five major thriller/horror entries from the early 2000s—*I am Legend*, *Hannibal*, *The Ring*, *The Grudge*, and *The Conjuring*. Figure 1 shows the amount, in dollars, each of those movies grossed when they were released as well as the ticket sales for horror movies in general by year. Notice that we can use the data to create a function of the amount each movie earned or the total ticket sales for all horror movies by year. In creating various functions using the data, we can identify different independent and dependent variables, and we can analyze the data and the functions to determine the domain and range. In this section, we will investigate methods for determining the domain and range of functions such as these.

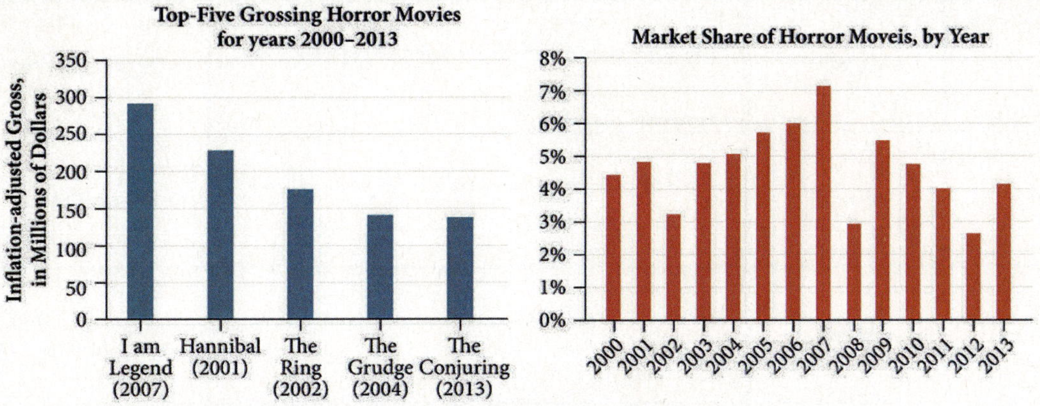

Figure 1 Based on data compiled by www.the-numbers.com.[3]

Finding the Domain of a Function Defined by an Equation

In Functions and Function Notation, we were introduced to the concepts of domain and range. In this section, we will practice determining domains and ranges for specific functions. Keep in mind that, in determining domains and ranges, we need to consider what is physically possible or meaningful in real-world examples, such as tickets sales and year in the horror movie example above. We also need to consider what is mathematically permitted. For example, we cannot include any input value that leads us to take an even root of a negative number if the domain and range consist of real numbers. Or in a function expressed as a formula, we cannot include any input value in the domain that would lead us to divide by 0.

We can visualize the domain as a "holding area" that contains "raw materials" for a "function machine" and the range as another "holding area" for the machine's products. See Figure 2.

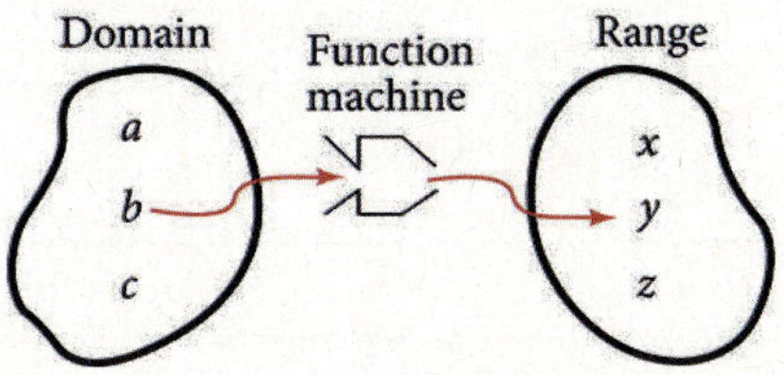

Figure 2

We can write the domain and range in **interval notation**, which uses values within brackets to describe a set of numbers. In interval notation, we use a square bracket [when the set includes the endpoint and a parenthesis (to indicate that the endpoint is either not included or the interval is unbounded. For example, if a person has $100 to spend, they would need to express the interval that is more than 0 and less than or equal to 100 and write $(0,\ \ 100]$. We will discuss interval notation in greater detail later.

Let's turn our attention to finding the domain of a function whose equation is provided. Oftentimes, finding the domain

3 The Numbers: Where Data and the Movie Business Meet. "Box Office History for Horror Movies." http://www.the-numbers.com/market/genre/Horror. Accessed 3/24/2014

of such functions involves remembering three different forms. First, if the function has no denominator or an odd root, consider whether the domain could be all real numbers. Second, if there is a denominator in the function's equation, exclude values in the domain that force the denominator to be zero. Third, if there is an even root, consider excluding values that would make the radicand negative.

Before we begin, let us review the conventions of interval notation:

- The smallest number from the interval is written first.
- The largest number in the interval is written second, following a comma.
- Parentheses, (or), are used to signify that an endpoint value is not included, called exclusive.
- Brackets, [or], are used to indicate that an endpoint value is included, called inclusive.

See Figure 3 for a summary of interval notation.

Inequality	Interval Notation	Graph on Number Line	Description
$x > a$	(a, ∞)		x is greater than a
$x < a$	$(-\infty, a)$		x is less than a
$x \geq a$	$[a, \infty)$		x is greater than or equal to a
$x \leq a$	$(-\infty, a]$		x is less than or equal to a
$a < x < b$	(a, b)		x is strictly between a and b
$a \leq x < b$	$[a, b)$		x is between a and b, to include a
$a < x \leq b$	$(a, b]$		x is between a and b, to include b
$a \leq x \leq b$	$[a, b]$	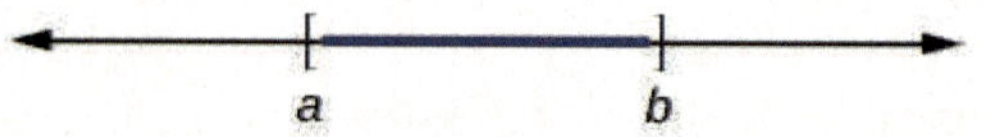	x is between a and b, to include a and b

Figure 3

EXAMPLE 1

Finding the Domain of a Function as a Set of Ordered Pairs

Find the domain of the following function: $\{(2,\ 10), (3,\ 10), (4,\ 20), (5,\ 30), (6,\ 40)\}$.

Solution

First identify the input values. The input value is the first coordinate in an ordered pair. There are no restrictions, as the ordered pairs are simply listed. The domain is the set of the first coordinates of the ordered pairs.

$$\{2, 3, 4, 5, 6\}$$

TRY IT #1 Find the domain of the function:

$$\{(-5, 4), (0, 0), (5, -4), (10, -8), (15, -12)\}$$

HOW TO

Given a function written in equation form, find the domain.

1. Identify the input values.
2. Identify any restrictions on the input and exclude those values from the domain.
3. Write the domain in interval form, if possible.

EXAMPLE 2

Finding the Domain of a Function

Find the domain of the function $f(x) = x^2 - 1$.

Solution

The input value, shown by the variable x in the equation, is squared and then the result is lowered by one. Any real number may be squared and then be lowered by one, so there are no restrictions on the domain of this function. The domain is the set of real numbers.

In interval form, the domain of f is $\left(-\infty, \infty\right)$.

TRY IT #2 Find the domain of the function: $f(x) = 5 - x + x^3$.

HOW TO

Given a function written in an equation form that includes a fraction, find the domain.

1. Identify the input values.
2. Identify any restrictions on the input. If there is a denominator in the function's formula, set the denominator equal to zero and solve for x . If the function's formula contains an even root, set the radicand greater than or equal to 0, and then solve.
3. Write the domain in interval form, making sure to exclude any restricted values from the domain.

EXAMPLE 3

Finding the Domain of a Function Involving a Denominator

Find the domain of the function $f(x) = \frac{x+1}{2-x}$.

Solution

When there is a denominator, we want to include only values of the input that do not force the denominator to be zero. So, we will set the denominator equal to 0 and solve for x.

$$\begin{aligned} 2 - x &= 0 \\ -x &= -2 \\ x &= 2 \end{aligned}$$

Now, we will exclude 2 from the domain. The answers are all real numbers where $x < 2$ or $x > 2$ as shown in Figure 4. We can use a symbol known as the union, ∪, to combine the two sets. In interval notation, we write the solution: $(-\infty, 2) \cup \left(2, \infty\right)$.

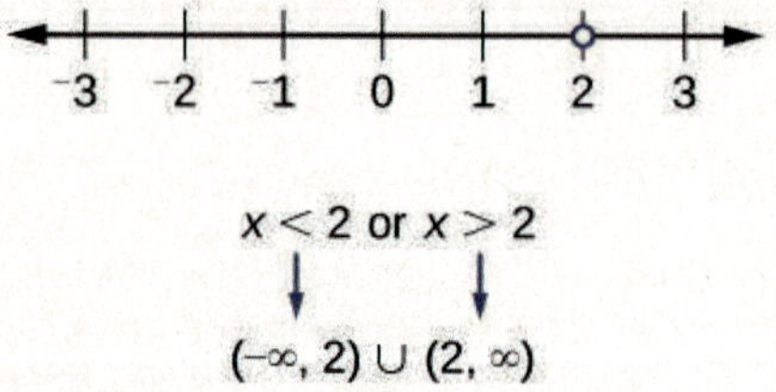

Figure 4

TRY IT #3 Find the domain of the function: $f(x) = \frac{1+4x}{2x-1}$.

HOW TO

Given a function written in equation form including an even root, find the domain.

1. Identify the input values.
2. Since there is an even root, exclude any real numbers that result in a negative number in the radicand. Set the radicand greater than or equal to zero and solve for x.
3. The solution(s) are the domain of the function. If possible, write the answer in interval form.

EXAMPLE 4

Finding the Domain of a Function with an Even Root

Find the domain of the function $f(x) = \sqrt{7 - x}$.

Solution

When there is an even root in the formula, we exclude any real numbers that result in a negative number in the radicand.

Set the radicand greater than or equal to zero and solve for x.

$$\begin{aligned} 7 - x &\geq 0 \\ -x &\geq -7 \\ x &\leq 7 \end{aligned}$$

Now, we will exclude any number greater than 7 from the domain. The answers are all real numbers less than or equal to 7, or $(-\infty, 7]$.

TRY IT #4 Find the domain of the function $f(x) = \sqrt{5 + 2x}$.

Q&A **Can there be functions in which the domain and range do not intersect at all?**

Yes. For example, the function $f(x) = -\frac{1}{\sqrt{x}}$ *has the set of all positive real numbers as its domain but the set of all negative real numbers as its range. As a more extreme example, a function's inputs and outputs can be completely different categories (for example, names of weekdays as inputs and numbers as outputs, as on an attendance chart), in such cases the domain and range have no elements in common.*

Using Notations to Specify Domain and Range

In the previous examples, we used inequalities and lists to describe the domain of functions. We can also use inequalities, or other statements that might define sets of values or data, to describe the behavior of the variable in **set-builder notation**. For example, $\{x|10 \le x < 30\}$ describes the behavior of x in set-builder notation. The braces { } are read as "the set of," and the vertical bar | is read as "such that," so we would read $\{x|10 \le x < 30\}$ as "the set of x-values such that 10 is less than or equal to x, and x is less than 30."

Figure 5 compares inequality notation, set-builder notation, and interval notation.

	Inequality Notation	Set-builder Notation	Interval Notation
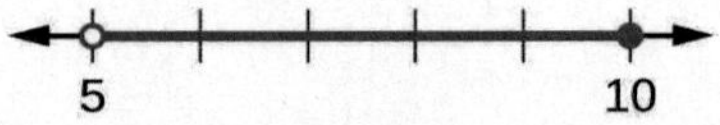	$5 < h \le 10$	$\{h \mid 5 < h \le 10\}$	$(5, 10]$
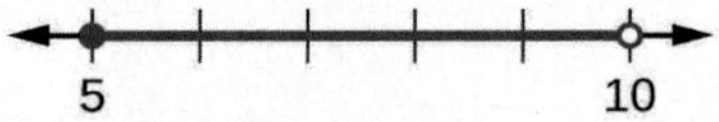	$5 \le h < 10$	$\{h \mid 5 \le h < 10\}$	$[5, 10)$
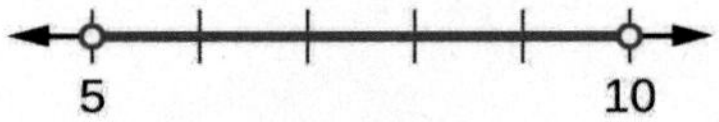	$5 < h < 10$	$\{h \mid 5 < h < 10\}$	$(5, 10)$
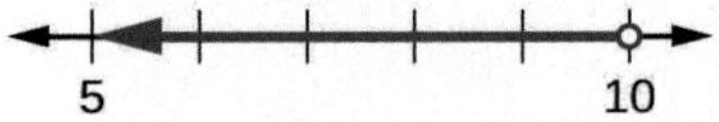	$h < 10$	$\{h \mid h < 10\}$	$(-\infty, 10)$
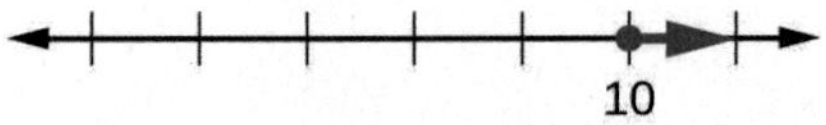	$h \ge 10$	$\{h \mid h \ge 10\}$	$[10, \infty)$
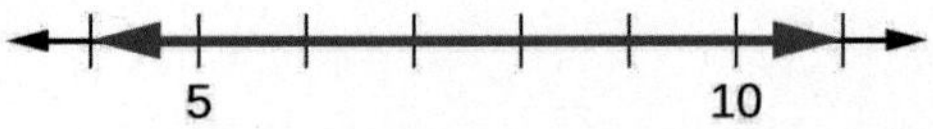	All real numbers	$\mathbb{R}$	$(-\infty, \infty)$

Figure 5

To combine two intervals using inequality notation or set-builder notation, we use the word "or." As we saw in earlier examples, we use the union symbol, ∪, to combine two unconnected intervals. For example, the union of the sets $\{2, 3, 5\}$ and $\{4, 6\}$ is the set $\{2, 3, 4, 5, 6\}$. It is the set of all elements that belong to one *or* the other (or both) of the original two sets. For sets with a finite number of elements like these, the elements do not have to be listed in ascending order of numerical value. If the original two sets have some elements in common, those elements should be listed only once in the union set. For sets of real numbers on intervals, another example of a union is

$$\{x| \quad |x| \ge 3\} = \left(-\infty, -3\right] \cup \left[3, \infty\right)$$

Set-Builder Notation and Interval Notation

Set-builder notation is a method of specifying a set of elements that satisfy a certain condition. It takes the form $\{x| \text{ statement about } x\}$ which is read as, "the set of all x such that the statement about x is true." For example,

$$\{x|4 < x \le 12\}$$

Interval notation is a way of describing sets that include all real numbers between a lower limit that may or may not be included and an upper limit that may or may not be included. The endpoint values are listed between brackets or parentheses. A square bracket indicates inclusion in the set, and a parenthesis indicates exclusion from the set. For example,

$$(4, 12]$$

HOW TO

Given a line graph, describe the set of values using interval notation.

1. Identify the intervals to be included in the set by determining where the heavy line overlays the real line.
2. At the left end of each interval, use [with each end value to be included in the set (solid dot) or (for each excluded end value (open dot).
3. At the right end of each interval, use] with each end value to be included in the set (filled dot) or) for each excluded end value (open dot).
4. Use the union symbol ∪ to combine all intervals into one set.

EXAMPLE 5

Describing Sets on the Real-Number Line

Describe the intervals of values shown in Figure 6 using inequality notation, set-builder notation, and interval notation.

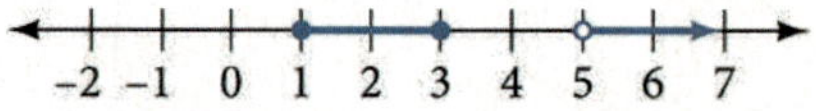

Figure 6

Solution

To describe the values, x, included in the intervals shown, we would say, " x is a real number greater than or equal to 1 and less than or equal to 3, or a real number greater than 5."

Inequality	$1 \le x \le 3$ or $x > 5$
Set-builder notation	$\{x \mid 1 \le x \le 3 \text{ or } x > 5\}$
Interval notation	$[1, 3] \cup (5, \infty)$

Remember that, when writing or reading interval notation, using a square bracket means the boundary is included in the set. Using a parenthesis means the boundary is not included in the set.

TRY IT #5 Given Figure 7, specify the graphed set in

ⓐ words ⓑ set-builder notation ⓒ interval notation

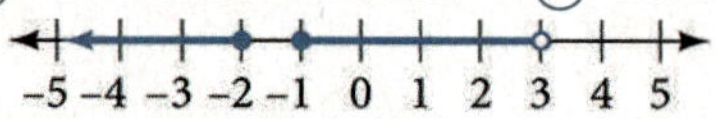

Figure 7

Finding Domain and Range from Graphs

Another way to identify the domain and range of functions is by using graphs. Because the domain refers to the set of possible input values, the domain of a graph consists of all the input values shown on the *x*-axis. The range is the set of possible output values, which are shown on the *y*-axis. Keep in mind that if the graph continues beyond the portion of the graph we can see, the domain and range may be greater than the visible values. See Figure 8.

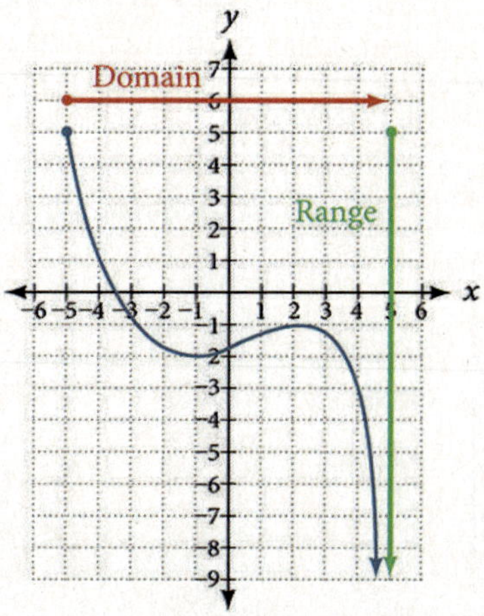

Figure 8

We can observe that the graph extends horizontally from -5 to the right without bound, so the domain is $\left[-5, \infty\right)$. The vertical extent of the graph is all range values 5 and below, so the range is $(-\infty, 5]$. Note that the domain and range are always written from smaller to larger values, or from left to right for domain, and from the bottom of the graph to the top of the graph for range.

EXAMPLE 6

Finding Domain and Range from a Graph

Find the domain and range of the function f whose graph is shown in Figure 9.

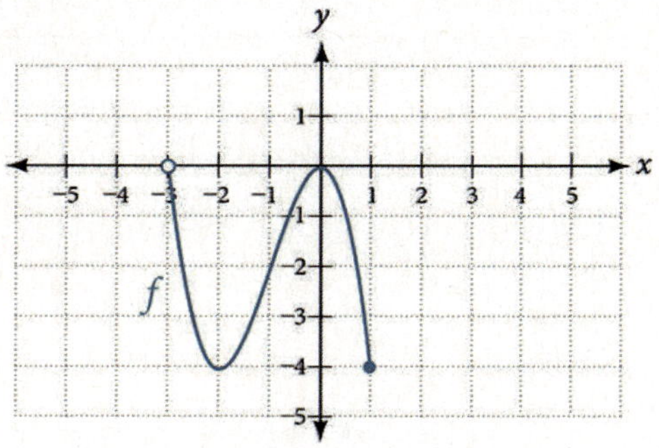

Figure 9

Solution

We can observe that the horizontal extent of the graph is –3 to 1, so the domain of f is $(-3, 1]$.

The vertical extent of the graph is 0 to –4, so the range is $[-4, 0)$. See Figure 10.

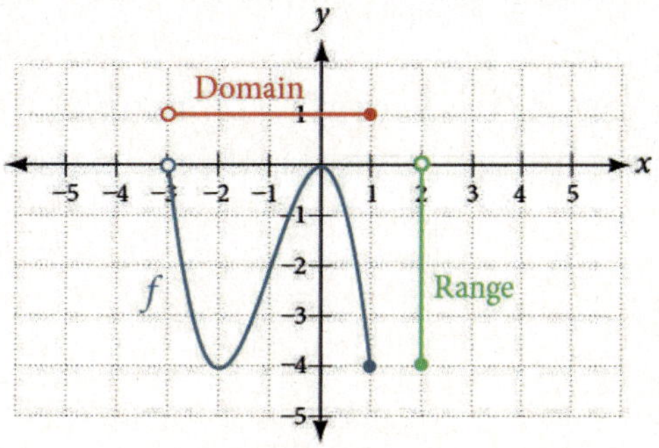

Figure 10

EXAMPLE 7

Finding Domain and Range from a Graph of Oil Production

Find the domain and range of the function f whose graph is shown in Figure 11.

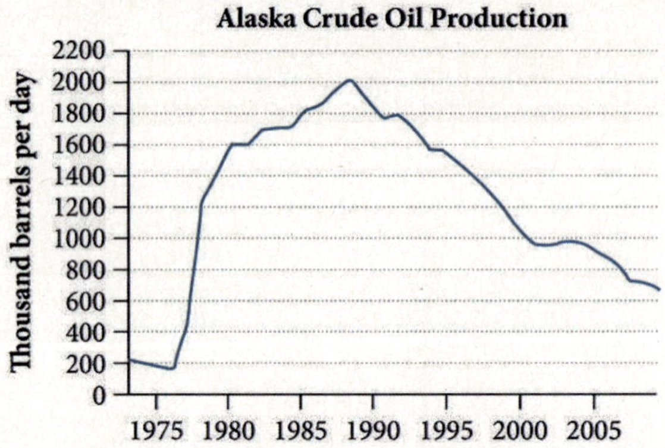

Figure 11 (credit: modification of work by the U.S. Energy Information Administration)[4]

Solution

The input quantity along the horizontal axis is "years," which we represent with the variable t for time. The output quantity is "thousands of barrels of oil per day," which we represent with the variable b for barrels. The graph may continue to the left and right beyond what is viewed, but based on the portion of the graph that is visible, we can determine the domain as $1973 \leq t \leq 2008$ and the range as approximately $180 \leq b \leq 2010$.

In interval notation, the domain is [1973, 2008], and the range is about [180, 2010]. For the domain and the range, we approximate the smallest and largest values since they do not fall exactly on the grid lines.

TRY IT #6 Given Figure 12, identify the domain and range using interval notation.

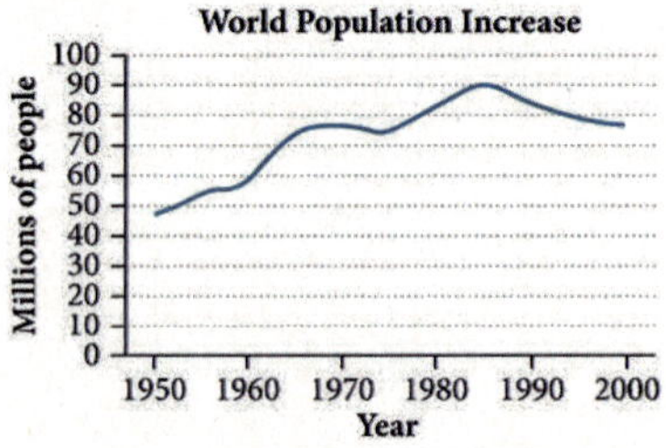

Figure 12

Q&A **Can a function's domain and range be the same?**

Yes. For example, the domain and range of the cube root function are both the set of all real numbers.

Finding Domains and Ranges of the Toolkit Functions

We will now return to our set of toolkit functions to determine the domain and range of each.

4 http://www.eia.gov/dnav/pet/hist/LeafHandler.ashx?n=PET&s=MCRFPAK2&f=A.

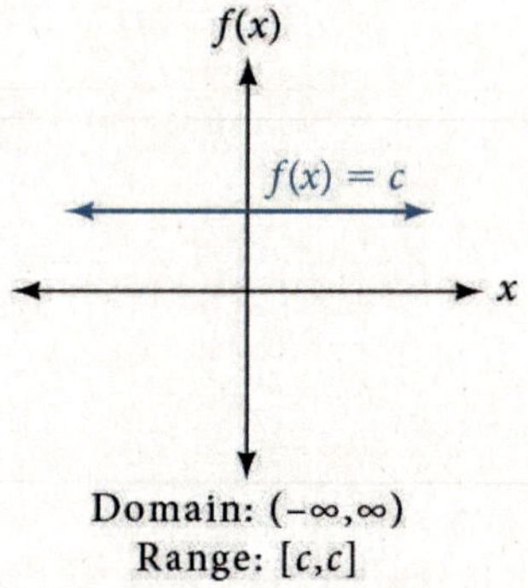

Figure 13 For the **constant function** $f(x) = c$, the domain consists of all real numbers; there are no restrictions on the input. The only output value is the constant c, so the range is the set $\{c\}$ that contains this single element. In interval notation, this is written as $[c, c]$, the interval that both begins and ends with c.

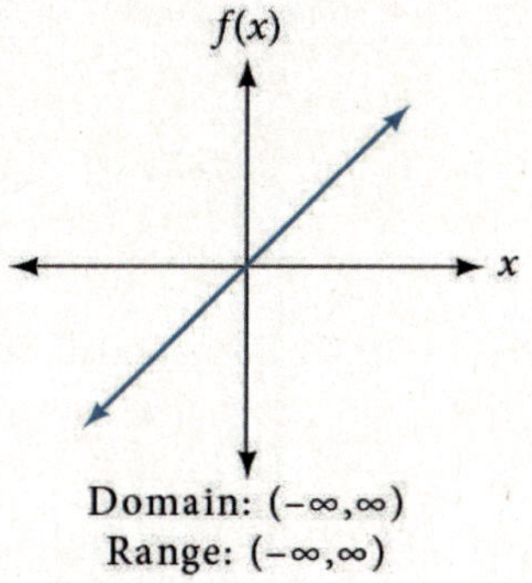

Figure 14 For the **identity function** $f(x) = x$, there is no restriction on x. Both the domain and range are the set of all real numbers.

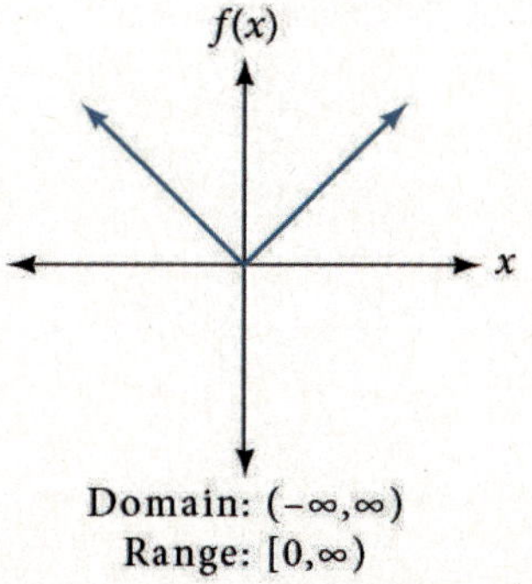

Figure 15 For the **absolute value function** $f(x) = |x|$, there is no restriction on x. However, because absolute value is defined as a distance from 0, the output can only be greater than or equal to 0.

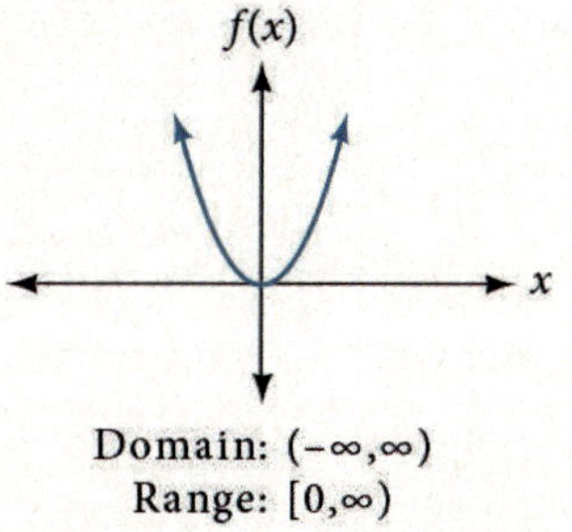

Figure 16 For the **quadratic function** $f(x) = x^2$, the domain is all real numbers since the horizontal extent of the graph is the whole real number line. Because the graph does not include any negative values for the range, the range is only nonnegative real numbers.

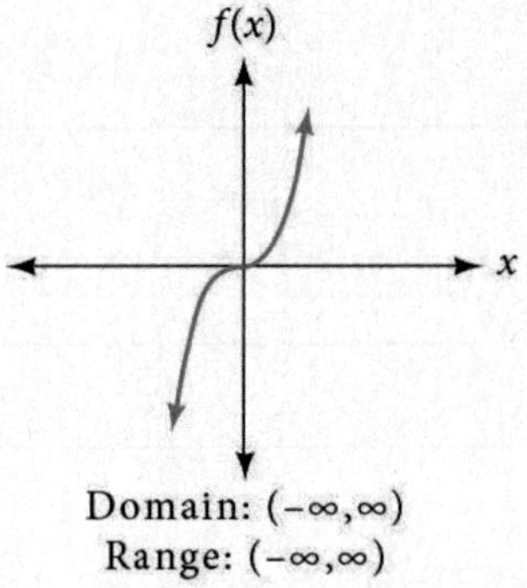

Figure 17 For the **cubic function** $f(x) = x^3$, the domain is all real numbers because the horizontal extent of the graph is the whole real number line. The same applies to the vertical extent of the graph, so the domain and range include all real numbers.

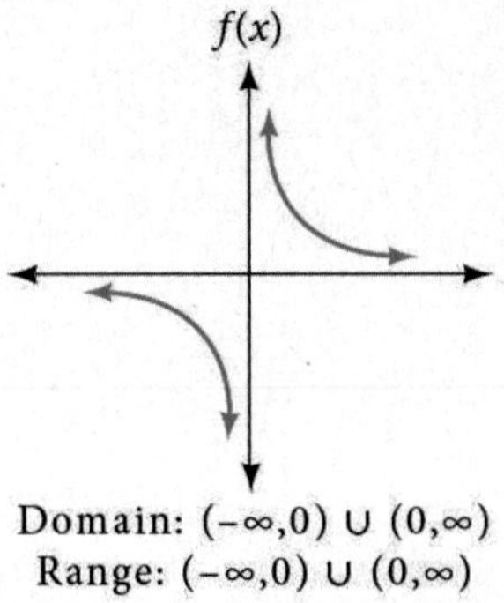

Figure 18 For the **reciprocal function** $f(x) = \frac{1}{x}$, we cannot divide by 0, so we must exclude 0 from the domain. Further, 1 divided by any value can never be 0, so the range also will not include 0. In set-builder notation, we could also write $\{x|\ x \neq 0\}$, the set of all real numbers that are not zero.

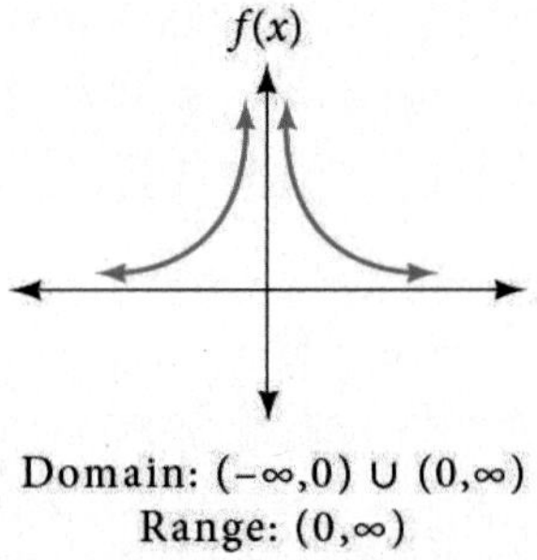

Figure 19 For the **reciprocal squared function** $f(x) = \frac{1}{x^2}$, we cannot divide by 0, so we must exclude 0 from the domain. There is also no x that can give an output of 0, so 0 is excluded from the range as well. Note that the output of this function is always positive due to the square in the denominator, so the range includes only positive numbers.

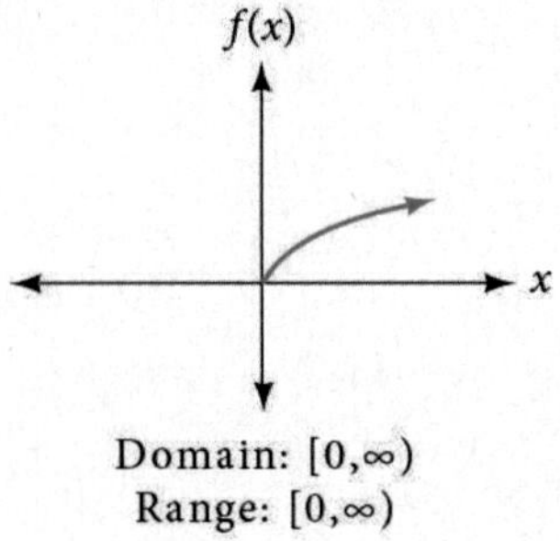

Figure 20 For the **square root function** $f(x) = \sqrt{x}$, we cannot take the square root of a negative real number, so the domain must be 0 or greater. The range also excludes negative numbers because the square root of a positive number x is defined to be positive, even though the square of the negative number $-\sqrt{x}$ also gives us x.

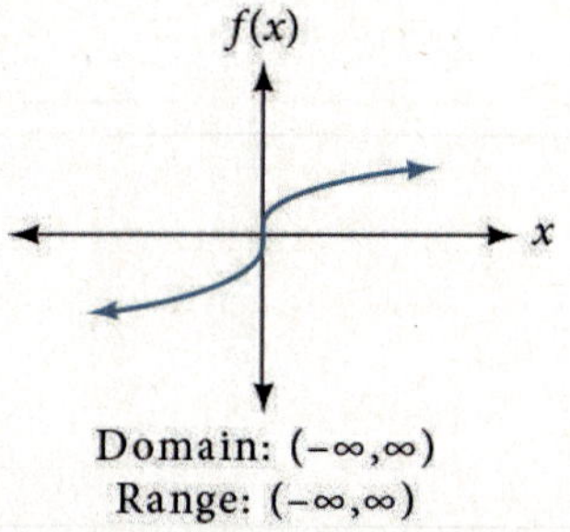

Figure 21 For the **cube root function** $f(x) = \sqrt[3]{x}$, the domain and range include all real numbers. Note that there is no problem taking a cube root, or any odd-integer root, of a negative number, and the resulting output is negative (it is an odd function).

HOW TO

Given the formula for a function, determine the domain and range.

1. Exclude from the domain any input values that result in division by zero.
2. Exclude from the domain any input values that have nonreal (or undefined) number outputs.
3. Use the valid input values to determine the range of the output values.
4. Look at the function graph and table values to confirm the actual function behavior.

EXAMPLE 8

Finding the Domain and Range Using Toolkit Functions

Find the domain and range of $f(x) = 2x^3 - x$.

Solution

There are no restrictions on the domain, as any real number may be cubed and then subtracted from the result.

The domain is $\left(-\infty, \infty\right)$ and the range is also $\left(-\infty, \infty\right)$.

EXAMPLE 9

Finding the Domain and Range

Find the domain and range of $f(x) = \frac{2}{x+1}$.

Solution

We cannot evaluate the function at -1 because division by zero is undefined. The domain is $\left(-\infty, -1\right) \cup \left(-1, \infty\right)$.

Because the function is never zero, we exclude 0 from the range. The range is $\left(-\infty, 0\right) \cup \left(0, \infty\right)$.

EXAMPLE 10

Finding the Domain and Range

Find the domain and range of $f(x) = 2\sqrt{x+4}$.

Solution

We cannot take the square root of a negative number, so the value inside the radical must be nonnegative.

$$x + 4 \geq 0 \text{ when } x \geq -4$$

The domain of $f(x)$ is $[-4, \infty)$.

We then find the range. We know that $f(-4) = 0$, and the function value increases as x increases without any upper limit. We conclude that the range of f is $\left[0, \infty\right)$.

Analysis

Figure 22 represents the function f.

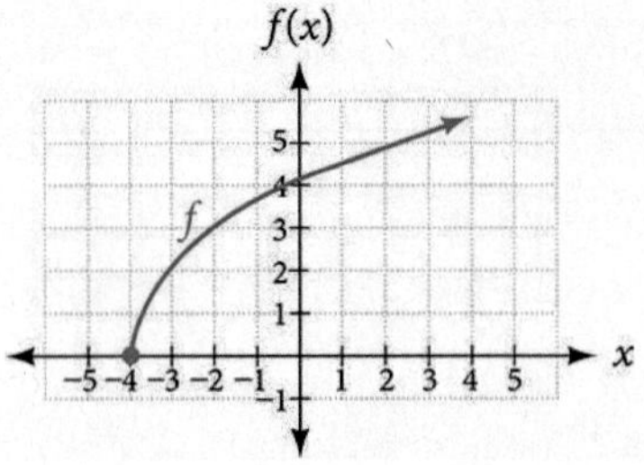

Figure 22

TRY IT #7 Find the domain and range of $f(x) = -\sqrt{2 - x}$.

Graphing Piecewise-Defined Functions

Sometimes, we come across a function that requires more than one formula in order to obtain the given output. For example, in the toolkit functions, we introduced the absolute value function $f(x) = |x|$. With a domain of all real numbers and a range of values greater than or equal to 0, absolute value can be defined as the magnitude, or modulus, of a real number value regardless of sign. It is the distance from 0 on the number line. All of these definitions require the output to be greater than or equal to 0.

If we input 0, or a positive value, the output is the same as the input.

$$f(x) = x \text{ if } x \geq 0$$

If we input a negative value, the output is the opposite of the input.

$$f(x) = -x \text{ if } x < 0$$

Because this requires two different processes or pieces, the absolute value function is an example of a piecewise function. A **piecewise function** is a function in which more than one formula is used to define the output over different pieces of the domain.

We use piecewise functions to describe situations in which a rule or relationship changes as the input value crosses certain "boundaries." For example, we often encounter situations in business for which the cost per piece of a certain item is discounted once the number ordered exceeds a certain value. Tax brackets are another real-world example of piecewise functions. For example, consider a simple tax system in which incomes up to \$10,000 are taxed at 10%, and any additional income is taxed at 20%. The tax on a total income S would be $0.1S$ if $S \leq \$10{,}000$ and $\$1000 + 0.2(S - \$10{,}000)$ if $S > \$10{,}000$.

Piecewise Function

A **piecewise function** is a function in which more than one formula is used to define the output. Each formula has its own domain, and the domain of the function is the union of all these smaller domains. We notate this idea like this:

$$f(x) = \begin{cases} \text{formula 1} & \text{if } x \text{ is in domain 1} \\ \text{formula 2} & \text{if } x \text{ is in domain 2} \\ \text{formula 3} & \text{if } x \text{ is in domain 3} \end{cases}$$

In piecewise notation, the absolute value function is

$$|x| = \begin{cases} x & \text{if } x \geq 0 \\ -x & \text{if } x < 0 \end{cases}$$

HOW TO

Given a piecewise function, write the formula and identify the domain for each interval.

1. Identify the intervals for which different rules apply.
2. Determine formulas that describe how to calculate an output from an input in each interval.
3. Use braces and if-statements to write the function.

EXAMPLE 11

Writing a Piecewise Function

A museum charges \$5 per person for a guided tour with a group of 1 to 9 people or a fixed \$50 fee for a group of 10 or more people. Write a function relating the number of people, n, to the cost, C.

Solution

Two different formulas will be needed. For n-values under 10, $C = 5n$. For values of n that are 10 or greater, $C = 50$.

$$C(n) = \begin{cases} 5n & \text{if} \quad 0 < n < 10 \\ 50 & \text{if} \quad n \geq 10 \end{cases}$$

Analysis

The function is represented in Figure 23. The graph is a diagonal line from $n = 0$ to $n = 10$ and a constant after that. In this example, the two formulas agree at the meeting point where $n = 10$, but not all piecewise functions have this property.

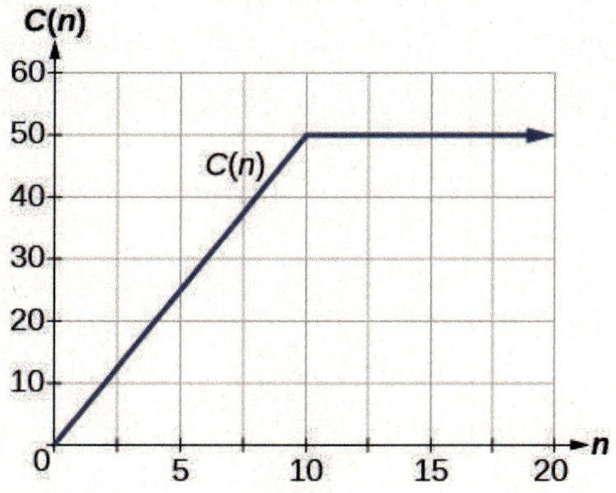

Figure 23

EXAMPLE 12

Working with a Piecewise Function

A cell phone company uses the function below to determine the cost, C, in dollars for g gigabytes of data transfer.

$$C(g) = \begin{cases} 25 & \text{if} \quad 0 < g < 2 \\ 25 + 10(g - 2) & \text{if} \quad g \geq 2 \end{cases}$$

Find the cost of using 1.5 gigabytes of data and the cost of using 4 gigabytes of data.

Solution

To find the cost of using 1.5 gigabytes of data, $C(1.5)$, we first look to see which part of the domain our input falls in. Because 1.5 is less than 2, we use the first formula.

$$C(1.5) = \$25$$

To find the cost of using 4 gigabytes of data, $C(4)$, we see that our input of 4 is greater than 2, so we use the second formula.

$$C(4) = 25 + 10(4 - 2) = \$45$$

Analysis

The function is represented in Figure 24. We can see where the function changes from a constant to a shifted and stretched identity at $g = 2$. We plot the graphs for the different formulas on a common set of axes, making sure each formula is applied on its proper domain.

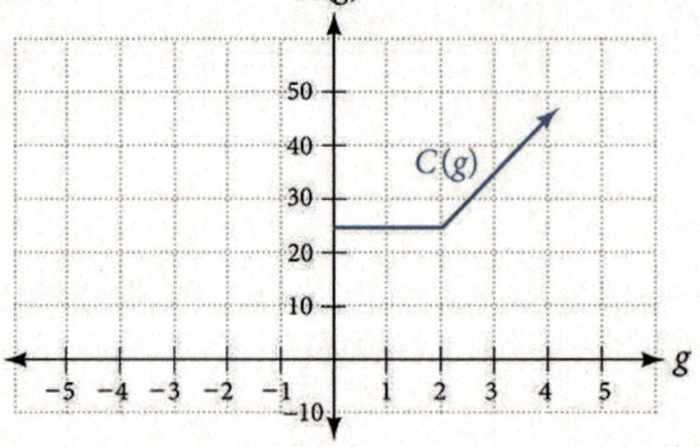

Figure 24

HOW TO

Given a piecewise function, sketch a graph.

1. Indicate on the x-axis the boundaries defined by the intervals on each piece of the domain.
2. For each piece of the domain, graph on that interval using the corresponding equation pertaining to that piece. Do not graph two functions over one interval because it would violate the criteria of a function.

EXAMPLE 13

Graphing a Piecewise Function

Sketch a graph of the function.

$$f(x) = \begin{cases} x^2 & \text{if} \quad x \leq 1 \\ 3 & \text{if} \quad 1 < x \leq 2 \\ x & \text{if} \quad x > 2 \end{cases}$$

Solution

Each of the component functions is from our library of toolkit functions, so we know their shapes. We can imagine graphing each function and then limiting the graph to the indicated domain. At the endpoints of the domain, we draw open circles to indicate where the endpoint is not included because of a less-than or greater-than inequality; we draw a closed circle where the endpoint is included because of a less-than-or-equal-to or greater-than-or-equal-to inequality.

Figure 25 shows the three components of the piecewise function graphed on separate coordinate systems.

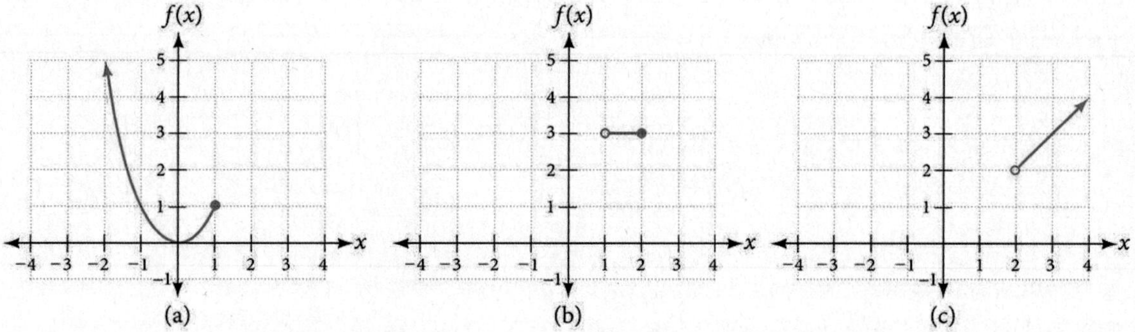

Figure 25 (a) $f(x) = x^2$ if $x \leq 1$; (b) $f(x) = 3$ if $1 < x \leq 2$; (c) $f(x) = x$ if $x > 2$

Now that we have sketched each piece individually, we combine them in the same coordinate plane. See Figure 26.

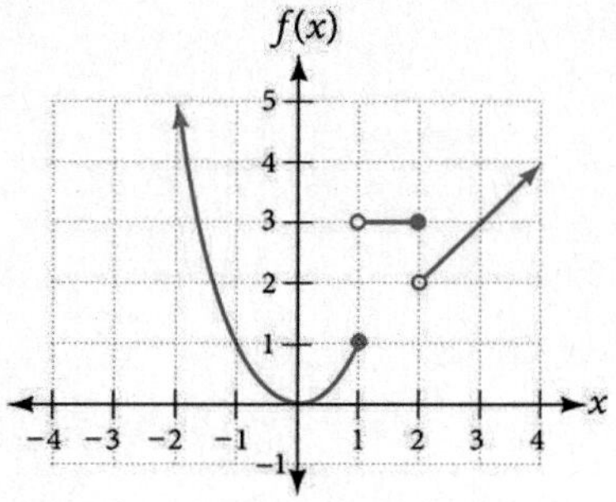

Figure 26

Analysis

Note that the graph does pass the vertical line test even at $x = 1$ and $x = 2$ because the points $(1, 3)$ and $(2, 2)$ are not part of the graph of the function, though $(1, 1)$ and $(2, 3)$ are.

TRY IT #8 Graph the following piecewise function.

$$f(x) = \begin{cases} x^3 & \text{if} \quad x < -1 \\ -2 & \text{if} \quad -1 < x < 4 \\ \sqrt{x} & \text{if} \quad x > 4 \end{cases}$$

Q&A **Can more than one formula from a piecewise function be applied to a value in the domain?**

No. Each value corresponds to one equation in a piecewise formula.

MEDIA

Access these online resources for additional instruction and practice with domain and range.

Domain and Range of Square Root Functions (http://openstax.org/l/domainsqroot)
Determining Domain and Range (http://openstax.org/l/determinedomain)
Find Domain and Range Given the Graph (http://openstax.org/l/drgraph)
Find Domain and Range Given a Table (http://openstax.org/l/drtable)
Find Domain and Range Given Points on a Coordinate Plane (http://openstax.org/l/drcoordinate)

3.2 SECTION EXERCISES

Verbal

1. Why does the domain differ for different functions?

2. How do we determine the domain of a function defined by an equation?

3. Explain why the domain of $f(x) = \sqrt[3]{x}$ is different from the domain of $f(x) = \sqrt{x}$.

4. When describing sets of numbers using interval notation, when do you use a parenthesis and when do you use a bracket?

5. How do you graph a piecewise function?

Algebraic

For the following exercises, find the domain of each function using interval notation.

6. $f(x) = -2x(x-1)(x-2)$

7. $f(x) = 5 - 2x^2$

8. $f(x) = 3\sqrt{x-2}$

9. $f(x) = 3 - \sqrt{6-2x}$

10. $f(x) = \sqrt{4-3x}$

11. $f(x) = \sqrt{x^2+4}$

12. $f(x) = \sqrt[3]{1-2x}$

13. $f(x) = \sqrt[3]{x-1}$

14. $f(x) = \frac{9}{x-6}$

15. $f(x) = \frac{3x+1}{4x+2}$

16. $f(x) = \frac{\sqrt{x+4}}{x-4}$

17. $f(x) = \frac{x-3}{x^2+9x-22}$

18. $f(x) = \frac{1}{x^2-x-6}$

19. $f(x) = \frac{2x^3-250}{x^2-2x-15}$

20. $\frac{5}{\sqrt{x-3}}$

21. $\frac{2x+1}{\sqrt{5-x}}$

22. $f(x) = \frac{\sqrt{x-4}}{\sqrt{x-6}}$

23. $f(x) = \frac{\sqrt{x-6}}{\sqrt{x-4}}$

24. $f(x) = \frac{x}{x}$

25. $f(x) = \frac{x^2-9x}{x^2-81}$

26. Find the domain of the function $f(x) = \sqrt{2x^3 - 50x}$ by:

ⓐ using algebra.

ⓑ graphing the function in the radicand and determining intervals on the *x*-axis for which the radicand is nonnegative.

Graphical

For the following exercises, write the domain and range of each function using interval notation.

27.

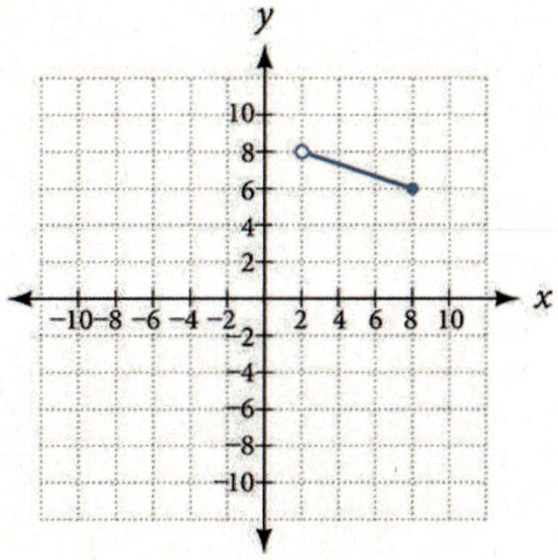

28.

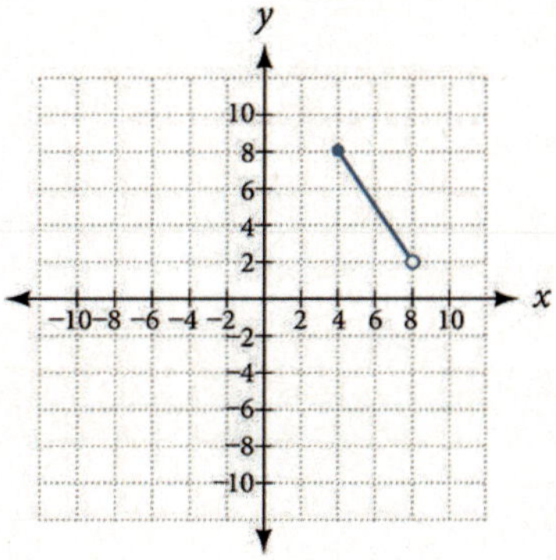

29.

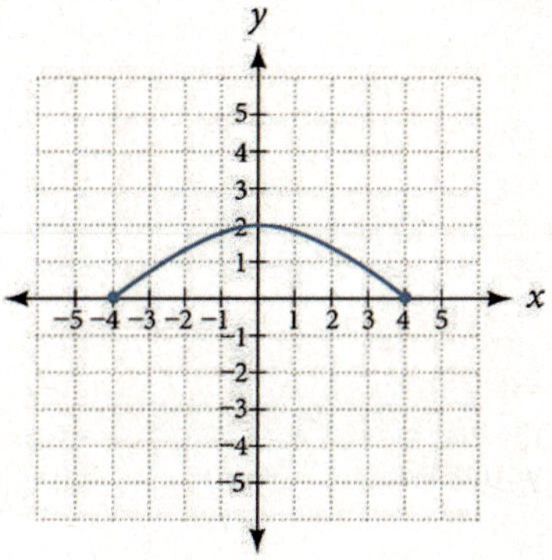

30.

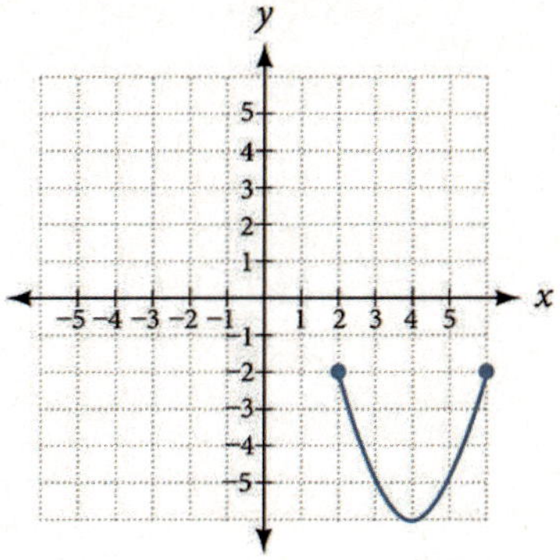

31.

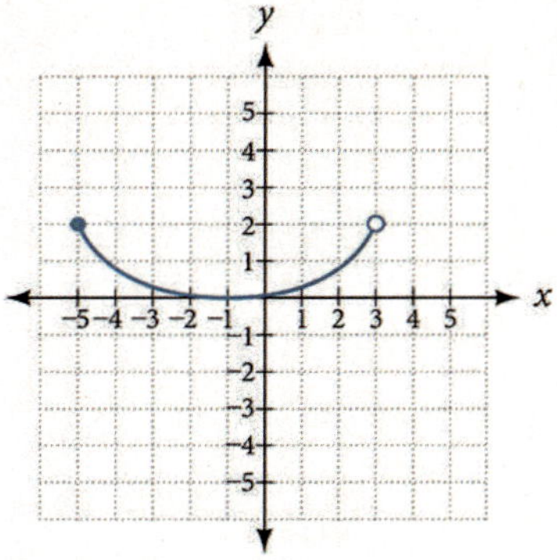

32.

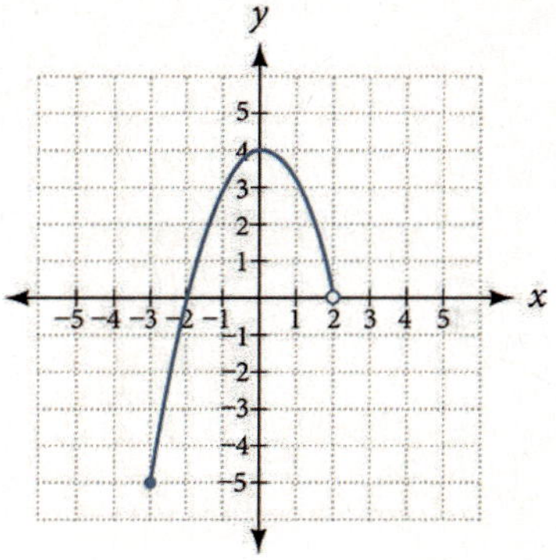

33.

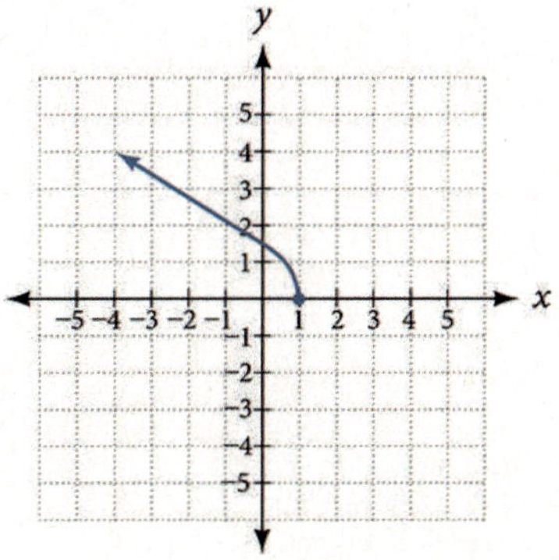

34.

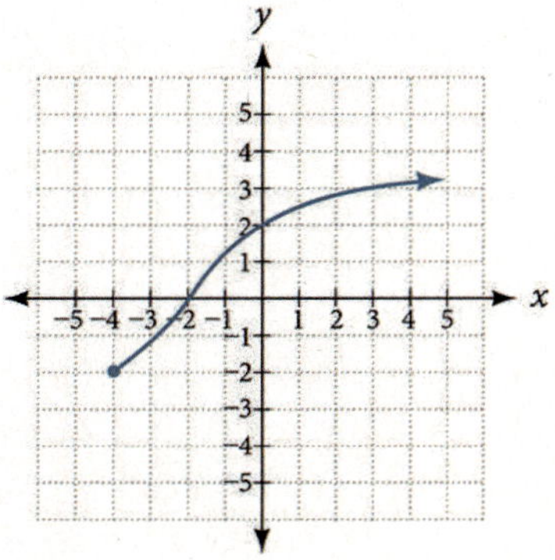

35.

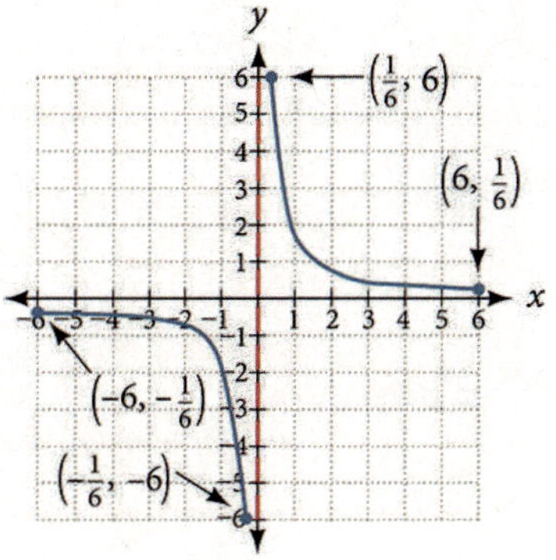

36.

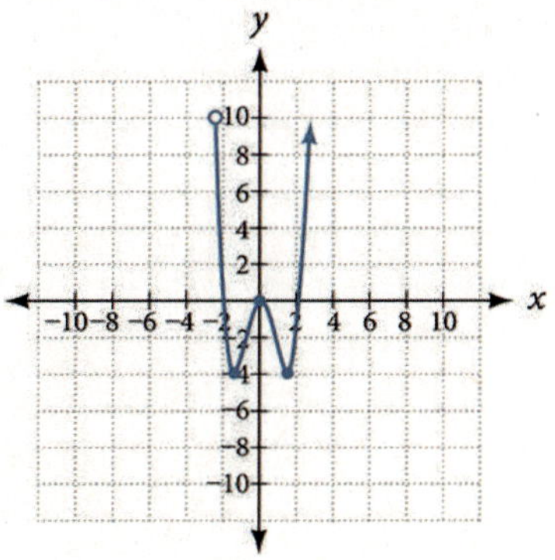

37.

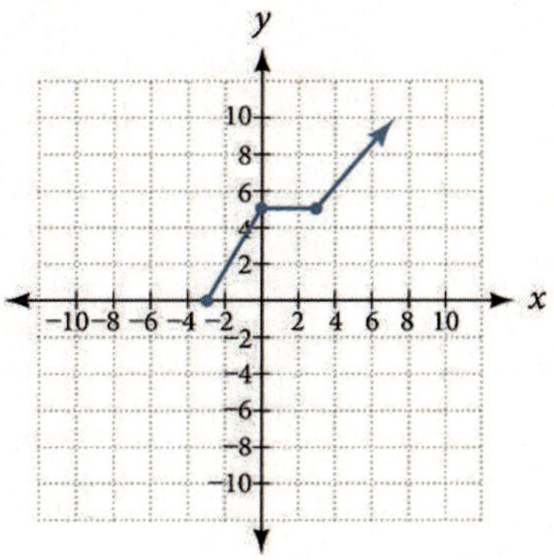

For the following exercises, sketch a graph of the piecewise function. Write the domain in interval notation.

38. $f(x) = \begin{cases} x+1 & \text{if } x < -2 \\ -2x-3 & \text{if } x \geq -2 \end{cases}$

39. $f(x) = \begin{cases} 2x-1 & \text{if } x < 1 \\ 1+x & \text{if } x \geq 1 \end{cases}$

40. $f(x) = \begin{cases} x+1 & \text{if } x < 0 \\ x-1 & \text{if } x > 0 \end{cases}$

41. $f(x) = \begin{cases} 3 & \text{if } x < 0 \\ \sqrt{x} & \text{if } x \geq 0 \end{cases}$

42. $f(x) = \begin{cases} x^2 & \text{if } x < 0 \\ 1-x & \text{if } x > 0 \end{cases}$

43. $f(x) = \begin{cases} x^2 & \text{if } x < 0 \\ x+2 & \text{if } x \geq 0 \end{cases}$

44. $f(x) = \begin{cases} x+1 & \text{if } x < 1 \\ x^3 & \text{if } x \geq 1 \end{cases}$

45. $f(x) = \begin{cases} |x| & \text{if } x < 2 \\ 1 & \text{if } x \geq 2 \end{cases}$

Numeric

For the following exercises, given each function f, evaluate $f(-3)$, $f(-2)$, $f(-1)$, and $f(0)$.

46. $f(x) = \begin{cases} x+1 & \text{if } x < -2 \\ -2x-3 & \text{if } x \geq -2 \end{cases}$

47. $f(x) = \begin{cases} 1 & \text{if } x \leq -3 \\ 0 & \text{if } x > -3 \end{cases}$

48. $f(x) = \begin{cases} -2x^2+3 & \text{if } x \leq -1 \\ 5x-7 & \text{if } x > -1 \end{cases}$

For the following exercises, given each function f, evaluate $f(-1)$, $f(0)$, $f(2)$, and $f(4)$.

49. $f(x) = \begin{cases} 7x+3 & \text{if } x < 0 \\ 7x+6 & \text{if } x \geq 0 \end{cases}$

50. $f(x) = \begin{cases} x^2-2 & \text{if } x < 2 \\ 4+|x-5| & \text{if } x \geq 2 \end{cases}$

51. $f(x) = \begin{cases} 5x & \text{if } x < 0 \\ 3 & \text{if } 0 \leq x \leq 3 \\ x^2 & \text{if } x > 3 \end{cases}$

For the following exercises, write the domain for the piecewise function in interval notation.

52. $f(x) = \begin{cases} x+1 & \text{if } x < -2 \\ -2x-3 & \text{if } x \geq -2 \end{cases}$

53. $f(x) = \begin{cases} x^2-2 & \text{if } x < 1 \\ -x^2+2 & \text{if } x > 1 \end{cases}$

54. $f(x) = \begin{cases} 2x-3 & \text{if } x < 0 \\ -3x^2 & \text{if } x \geq 2 \end{cases}$

Technology

55. Graph $y = \frac{1}{x^2}$ on the viewing window $[-0.5, -0.1]$ and $[0.1, 0.5]$. Determine the corresponding range for the viewing window. Show the graphs.

56. Graph $y = \frac{1}{x}$ on the viewing window $[-0.5, -0.1]$ and $[0.1,\ 0.5]$. Determine the corresponding range for the viewing window. Show the graphs.

Extension

57. Suppose the range of a function f is $[-5,\ 8]$. What is the range of $|f(x)|$?

58. Create a function in which the range is all nonnegative real numbers.

59. Create a function in which the domain is $x > 2$.

Real-World Applications

60. The height h of a projectile is a function of the time t it is in the air. The height in feet for t seconds is given by the function $h(t) = -16t^2 + 96t$. What is the domain of the function? What does the domain mean in the context of the problem?

61. The cost in dollars of making x items is given by the function $C(x) = 10x + 500$.

ⓐ The fixed cost is determined when zero items are produced. Find the fixed cost for this item.
ⓑ What is the cost of making 25 items?
ⓒ Suppose the maximum cost allowed is \$1500. What are the domain and range of the cost function, $C(x)$?

3.3 Rates of Change and Behavior of Graphs

Learning Objectives

In this section, you will:

- Find the average rate of change of a function.
- Use a graph to determine where a function is increasing, decreasing, or constant.
- Use a graph to locate local maxima and local minima.
- Use a graph to locate the absolute maximum and absolute minimum.

Gasoline costs have experienced some wild fluctuations over the last several decades. Table 1[5] lists the average cost, in dollars, of a gallon of gasoline for the years 2005–2012. The cost of gasoline can be considered as a function of year.

y	2005	2006	2007	2008	2009	2010	2011	2012
$C(y)$	2.31	2.62	2.84	3.30	2.41	2.84	3.58	3.68

Table 1

If we were interested only in how the gasoline prices changed between 2005 and 2012, we could compute that the cost per gallon had increased from \$2.31 to \$3.68, an increase of \$1.37. While this is interesting, it might be more useful to look at how much the price changed *per year*. In this section, we will investigate changes such as these.

Finding the Average Rate of Change of a Function

The price change per year is a **rate of change** because it describes how an output quantity changes relative to the change in the input quantity. We can see that the price of gasoline in Table 1 did not change by the same amount each year, so the rate of change was not constant. If we use only the beginning and ending data, we would be finding the **average rate of change** over the specified period of time. To find the average rate of change, we divide the change in the output value by the change in the input value.

$$\begin{aligned}\text{Average rate of change} &= \frac{\text{Change in output}}{\text{Change in input}}\\ &= \frac{\Delta y}{\Delta x}\\ &= \frac{y_2 - y_1}{x_2 - x_1}\\ &= \frac{f(x_2) - f(x_1)}{x_2 - x_1}\end{aligned}$$

The Greek letter Δ (delta) signifies the change in a quantity; we read the ratio as "delta-y over delta-x" or "the change in y divided by the change in x." Occasionally we write Δf instead of Δy, which still represents the change in the function's

5 http://www.eia.gov/totalenergy/data/annual/showtext.cfm?t=ptb0524. Accessed 3/5/2014.

output value resulting from a change to its input value. It does not mean we are changing the function into some other function.

In our example, the gasoline price increased by \$1.37 from 2005 to 2012. Over 7 years, the average rate of change was

$$\frac{\Delta y}{\Delta x} = \frac{\$1.37}{7 \text{ years}} \approx 0.196 \text{ dollars per year}$$

On average, the price of gas increased by about 19.6¢ each year.

Other examples of rates of change include:

- A population of rats increasing by 40 rats per week
- A car traveling 68 miles per hour (distance traveled changes by 68 miles each hour as time passes)
- A car driving 27 miles per gallon (distance traveled changes by 27 miles for each gallon)
- The current through an electrical circuit increasing by 0.125 amperes for every volt of increased voltage
- The amount of money in a college account decreasing by \$4,000 per quarter

Rate of Change

A rate of change describes how an output quantity changes relative to the change in the input quantity. The units on a rate of change are "output units per input units."

The average rate of change between two input values is the total change of the function values (output values) divided by the change in the input values.

$$\frac{\Delta y}{\Delta x} = \frac{f(x_2) - f(x_1)}{x_2 - x_1}$$

HOW TO

Given the value of a function at different points, calculate the average rate of change of a function for the interval between two values x_1 and x_2.

1. Calculate the difference $y_2 - y_1 = \Delta y$.
2. Calculate the difference $x_2 - x_1 = \Delta x$.
3. Find the ratio $\frac{\Delta y}{\Delta x}$.

EXAMPLE 1

Computing an Average Rate of Change

Using the data in Table 1, find the average rate of change of the price of gasoline between 2007 and 2009.

Solution

In 2007, the price of gasoline was \$2.84. In 2009, the cost was \$2.41. The average rate of change is

$$\begin{aligned} \frac{\Delta y}{\Delta x} &= \frac{y_2 - y_1}{x_2 - x_1} \\ &= \frac{\$2.41 - \$2.84}{2009 - 2007} \\ &= \frac{-\$0.43}{2 \text{ years}} \\ &= -\$0.22 \text{ per year} \end{aligned}$$

Analysis

Note that a decrease is expressed by a negative change or "negative increase." A rate of change is negative when the output decreases as the input increases or when the output increases as the input decreases.

> **TRY IT** #1 Using the data in Table 1, find the average rate of change between 2005 and 2010.

EXAMPLE 2

Computing Average Rate of Change from a Graph

Given the function $g(t)$ shown in Figure 1, find the average rate of change on the interval $[-1, 2]$.

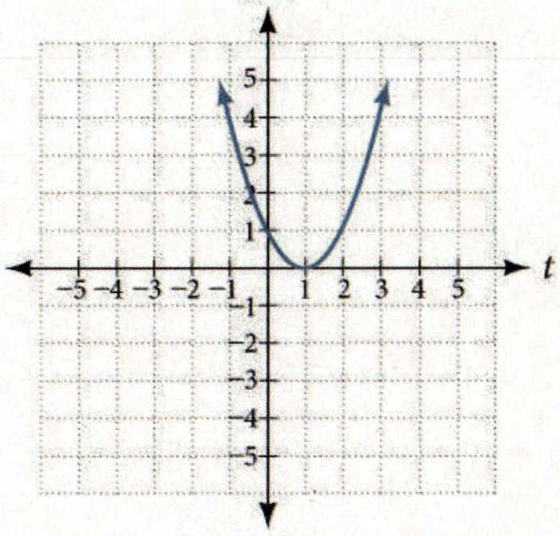

Figure 1

Solution

At $t = -1$, Figure 2 shows $g(-1) = 4$. At $t = 2$, the graph shows $g(2) = 1$.

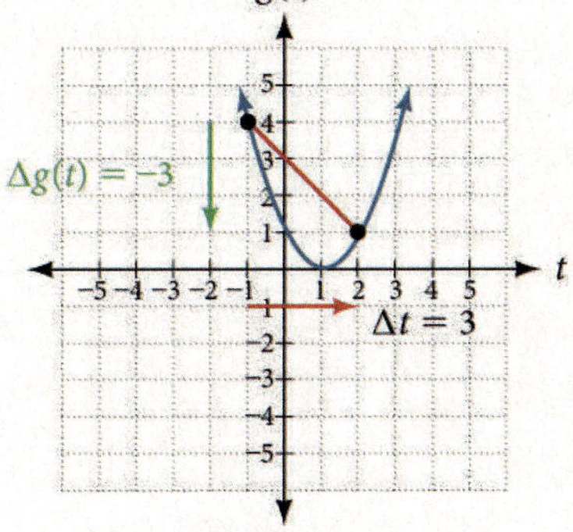

Figure 2

The horizontal change $\Delta t = 3$ is shown by the red arrow, and the vertical change $\Delta g(t) = -3$ is shown by the turquoise arrow. The average rate of change is shown by the slope of the orange line segment. The output changes by –3 while the input changes by 3, giving an average rate of change of

$$\frac{1 - 4}{2 - (-1)} = \frac{-3}{3} = -1$$

Analysis

Note that the order we choose is very important. If, for example, we use $\frac{y_2 - y_1}{x_1 - x_2}$, we will not get the correct answer. Decide which point will be 1 and which point will be 2, and keep the coordinates fixed as (x_1, y_1) and (x_2, y_2).

EXAMPLE 3

Computing Average Rate of Change from a Table

After picking up a friend who lives 10 miles away and leaving on a trip, Anna records her distance from home over time. The values are shown in Table 2. Find her average speed over the first 6 hours.

t (hours)	0	1	2	3	4	5	6	7
$D(t)$ (miles)	10	55	90	153	214	240	292	300

Table 2

Solution

Here, the average speed is the average rate of change. She traveled 282 miles in 6 hours.

$$\begin{aligned}\frac{292-10}{6-0} &= \frac{282}{6}\\ &= 47\end{aligned}$$

The average speed is 47 miles per hour.

Analysis

Because the speed is not constant, the average speed depends on the interval chosen. For the interval [2,3], the average speed is 63 miles per hour.

EXAMPLE 4

Computing Average Rate of Change for a Function Expressed as a Formula

Compute the average rate of change of $f(x) = x^2 - \frac{1}{x}$ on the interval $[2,\ 4]$.

Solution

We can start by computing the function values at each endpoint of the interval.

$$\begin{aligned}f(2) &= 2^2 - \frac{1}{2} & f(4) &= 4^2 - \frac{1}{4}\\ &= 4 - \frac{1}{2} & &= 16 - \frac{1}{4}\\ &= \frac{7}{2} & &= \frac{63}{4}\end{aligned}$$

Now we compute the average rate of change.

$$\begin{aligned}\text{Average rate of change} &= \frac{f(4)-f(2)}{4-2}\\ &= \frac{\frac{63}{4}-\frac{7}{2}}{4-2}\\ &= \frac{\frac{49}{4}}{2}\\ &= \frac{49}{8}\end{aligned}$$

TRY IT #2 Find the average rate of change of $f(x) = x - 2\sqrt{x}$ on the interval $[1,\ 9]$.

EXAMPLE 5

Finding the Average Rate of Change of a Force

The electrostatic force F, measured in newtons, between two charged particles can be related to the distance between the particles d, in centimeters, by the formula $F(d) = \frac{2}{d^2}$. Find the average rate of change of force if the distance between the particles is increased from 2 cm to 6 cm.

Solution

We are computing the average rate of change of $F(d) = \frac{2}{d^2}$ on the interval $[2, 6]$.

$$\begin{aligned}
\text{Average rate of change} &= \frac{F(6)-F(2)}{6-2} \\
&= \frac{\frac{2}{6^2}-\frac{2}{2^2}}{6-2} && \text{Simplify.} \\
&= \frac{\frac{2}{36}-\frac{2}{4}}{4} \\
&= \frac{-\frac{16}{36}}{4} && \text{Combine numerator terms.} \\
&= -\frac{1}{9} && \text{Simplify}
\end{aligned}$$

The average rate of change is $-\frac{1}{9}$ newton per centimeter.

EXAMPLE 6

Finding an Average Rate of Change as an Expression

Find the average rate of change of $g(t) = t^2 + 3t + 1$ on the interval $[0,\ a]$. The answer will be an expression involving a in simplest form.

Solution

We use the average rate of change formula.

$$\begin{aligned}
\text{Average rate of change} &= \frac{g(a)-g(0)}{a-0} && \text{Evaluate.} \\
&= \frac{\left(a^2+3a+1\right)-\left(0^2+3\left(0\right)+1\right)}{a-0} && \text{Simplify.} \\
&= \frac{a^2+3a+1-1}{a} && \text{Simplify and factor.} \\
&= \frac{a(a+3)}{a} && \text{Divide by the common factor } a. \\
&= a+3
\end{aligned}$$

This result tells us the average rate of change in terms of a between $t = 0$ and any other point $t = a$. For example, on the interval $[0, 5]$, the average rate of change would be $5 + 3 = 8$.

TRY IT #3 Find the average rate of change of $f(x) = x^2 + 2x - 8$ on the interval $[5, a]$ in simplest forms in terms of a.

Using a Graph to Determine Where a Function is Increasing, Decreasing, or Constant

As part of exploring how functions change, we can identify intervals over which the function is changing in specific ways. We say that a function is increasing on an interval if the function values increase as the input values increase within that interval. Similarly, a function is decreasing on an interval if the function values decrease as the input values increase over that interval. The average rate of change of an increasing function is positive, and the average rate of change of a decreasing function is negative. Figure 3 shows examples of increasing and decreasing intervals on a function.

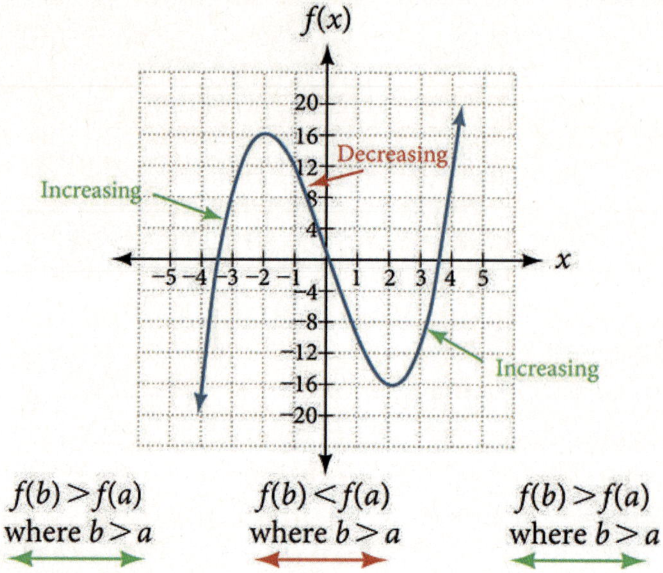

Figure 3 The function $f(x) = x^3 - 12x$ is increasing on $\left(-\infty, \ -2\right) \cup \ \left(2, \ \infty\right)$ and is decreasing on $(-2, \ 2)$.

While some functions are increasing (or decreasing) over their entire domain, many others are not. A value of the input where a function changes from increasing to decreasing (as we go from left to right, that is, as the input variable increases) is the location of a **local maximum**. The function value at that point is the local maximum. If a function has more than one, we say it has local maxima. Similarly, a value of the input where a function changes from decreasing to increasing as the input variable increases is the location of a **local minimum**. The function value at that point is the local minimum. The plural form is "local minima." Together, local maxima and minima are called **local extrema**, or local extreme values, of the function. (The singular form is "extremum.") Often, the term *local* is replaced by the term *relative*. In this text, we will use the term *local*.

Clearly, a function is neither increasing nor decreasing on an interval where it is constant. A function is also neither increasing nor decreasing at extrema. Note that we have to speak of *local* extrema, because any given local extremum as defined here is not necessarily the highest maximum or lowest minimum in the function's entire domain.

For the function whose graph is shown in Figure 4, the local maximum is 16, and it occurs at $x = -2$. The local minimum is -16 and it occurs at $x = 2$.

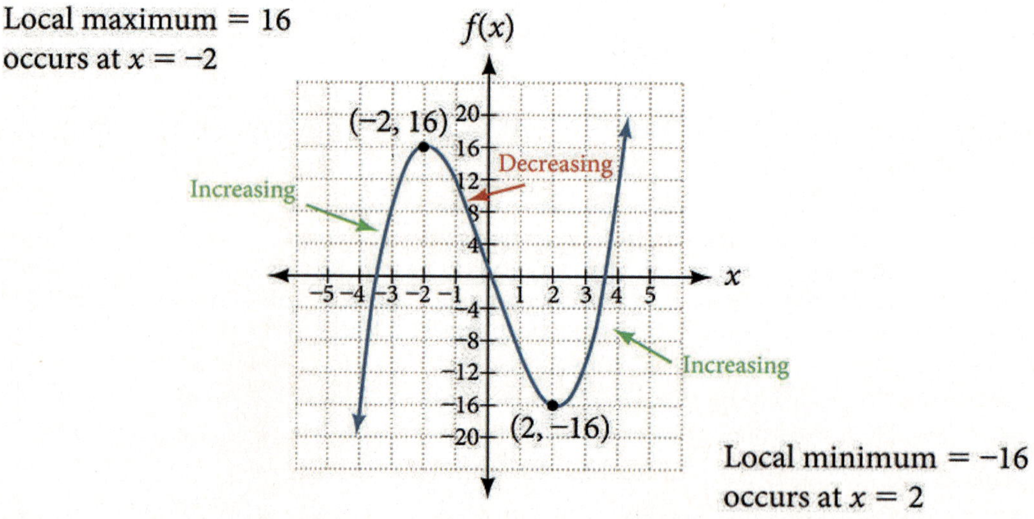

Figure 4

To locate the local maxima and minima from a graph, we need to observe the graph to determine where the graph attains its highest and lowest points, respectively, within an open interval. Like the summit of a roller coaster, the graph of a function is higher at a local maximum than at nearby points on both sides. The graph will also be lower at a local minimum than at neighboring points. Figure 5 illustrates these ideas for a local maximum.

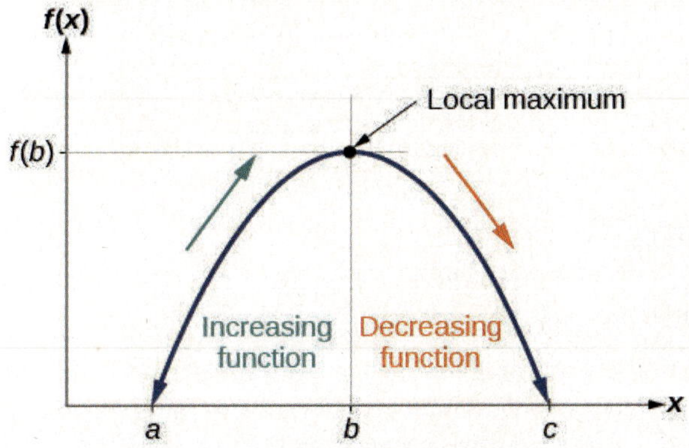

Figure 5 Definition of a local maximum

These observations lead us to a formal definition of local extrema.

Local Minima and Local Maxima

A function f is an **increasing function** on an open interval if $f(b) > f(a)$ for any two input values a and b in the given interval where $b > a$.

A function f is a **decreasing function** on an open interval if $f(b) < f(a)$ for any two input values a and b in the given interval where $b > a$.

A function f has a local maximum at $x = b$ if there exists an interval (a, c) with $a < b < c$ such that, for any x in the interval (a, c), $f(x) \leq f(b)$. Likewise, f has a local minimum at $x = b$ if there exists an interval (a, c) with $a < b < c$ such that, for any x in the interval (a, c), $f(x) \geq f(b)$.

EXAMPLE 7

Finding Increasing and Decreasing Intervals on a Graph

Given the function $p(t)$ in Figure 6, identify the intervals on which the function appears to be increasing.

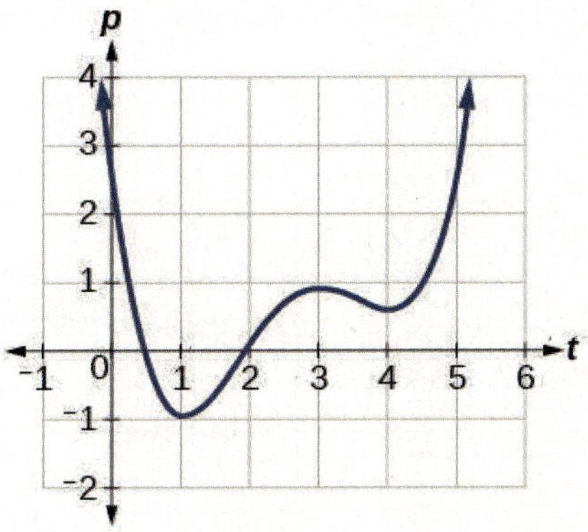

Figure 6

Solution

We see that the function is not constant on any interval. The function is increasing where it slants upward as we move to the right and decreasing where it slants downward as we move to the right. The function appears to be increasing from $t = 1$ to $t = 3$ and from $t = 4$ on.

In interval notation, we would say the function appears to be increasing on the interval (1,3) and the interval $(4, \infty)$.

Analysis

Notice in this example that we used open intervals (intervals that do not include the endpoints), because the function is neither increasing nor decreasing at $t = 1$, $t = 3$, and $t = 4$. These points are the local extrema (two minima and a maximum).

EXAMPLE 8

Finding Local Extrema from a Graph

Graph the function $f(x) = \frac{2}{x} + \frac{x}{3}$. Then use the graph to estimate the local extrema of the function and to determine the intervals on which the function is increasing.

Solution

Using technology, we find that the graph of the function looks like that in Figure 7. It appears there is a low point, or local minimum, between $x = 2$ and $x = 3$, and a mirror-image high point, or local maximum, somewhere between $x = -3$ and $x = -2$.

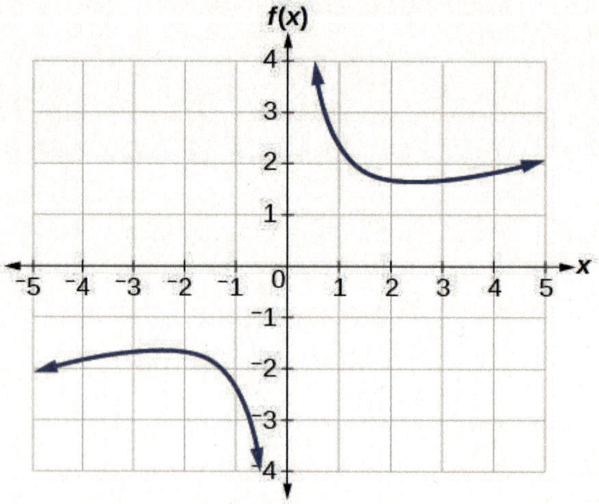

Figure 7

Analysis

Most graphing calculators and graphing utilities can estimate the location of maxima and minima. Figure 8 provides screen images from two different technologies, showing the estimate for the local maximum and minimum.

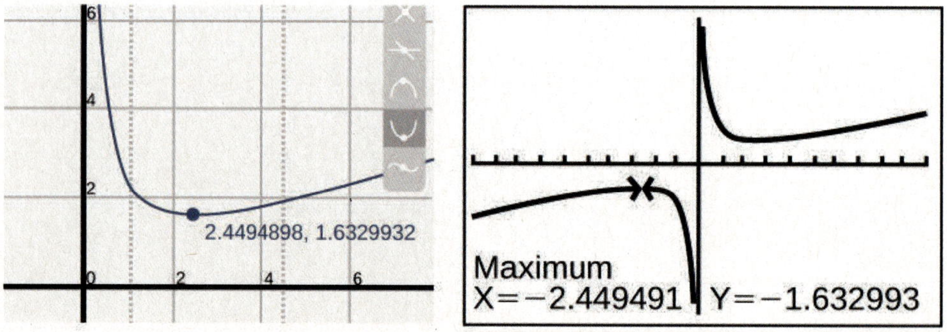

(a) (b)

Figure 8

Based on these estimates, the function is increasing on the interval $(-\infty, -2.449)$ and $(2.449, \infty)$. Notice that, while we expect the extrema to be symmetric, the two different technologies agree only up to four decimals due to the differing approximation algorithms used by each. (The exact location of the extrema is at $\pm\sqrt{6}$, but determining this requires calculus.)

TRY IT #4 Graph the function $f(x) = x^3 - 6x^2 - 15x + 20$ to estimate the local extrema of the function. Use these to determine the intervals on which the function is increasing and decreasing.

EXAMPLE 9

Finding Local Maxima and Minima from a Graph

For the function f whose graph is shown in Figure 9, find all local maxima and minima.

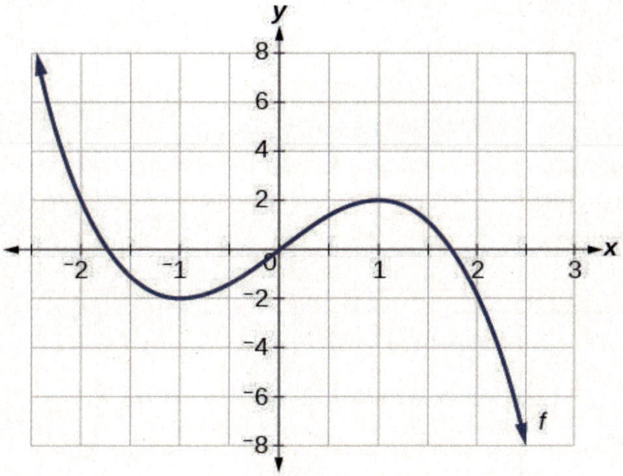

Figure 9

Solution

Observe the graph of f. The graph attains a local maximum at $x = 1$ because it is the highest point in an open interval around $x = 1$. The local maximum is the y -coordinate at $x = 1$, which is 2.

The graph attains a local minimum at $x = -1$ because it is the lowest point in an open interval around $x = -1$. The local minimum is the *y*-coordinate at $x = -1$, which is -2.

Analyzing the Toolkit Functions for Increasing or Decreasing Intervals

We will now return to our toolkit functions and discuss their graphical behavior in Figure 10, Figure 11, and Figure 12.

Function	Increasing/Decreasing	Example
Constant Function $f(x) = c$	Neither increasing nor decreasing	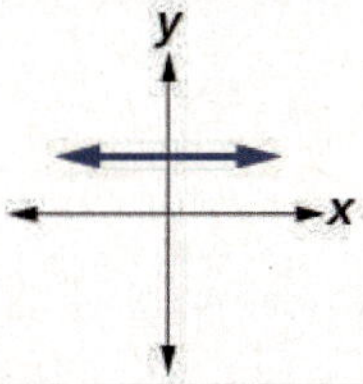
Identity Function $f(x) = x$	Increasing	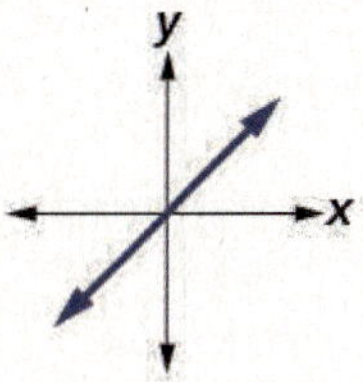
Quadratic Function $f(x) = x^2$	Increasing on $(0, \infty)$ Decreasing on $(-\infty, 0)$ Minimum at $x = 0$	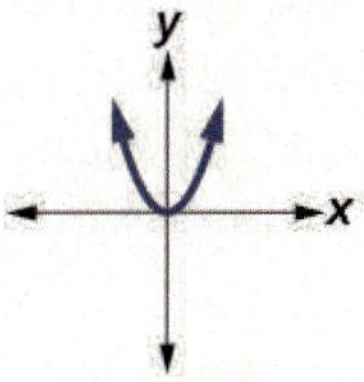

Figure 10

Function	Increasing/Decreasing	Example
Cubic Function $f(x) = x^3$	Increasing	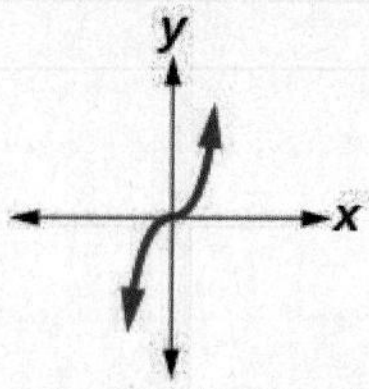
Reciprocal $f(x) = \frac{1}{x}$	Decreasing $(-\infty, 0)\cup(0, \infty)$	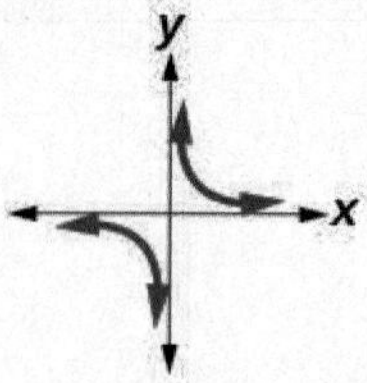
Reciprocal Squared $f(x) = \frac{1}{x^2}$	Increasing on $(-\infty, 0)$ Decreasing on $(0, \infty)$	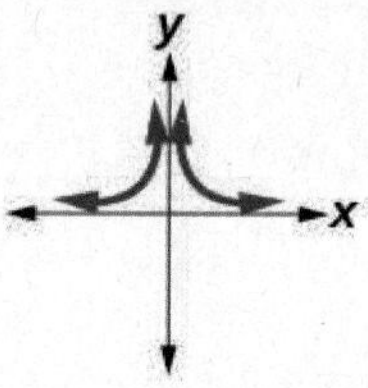

Figure 11

Function	Increasing/Decreasing	Example
Cube Root $f(x) = \sqrt[3]{x}$	Increasing	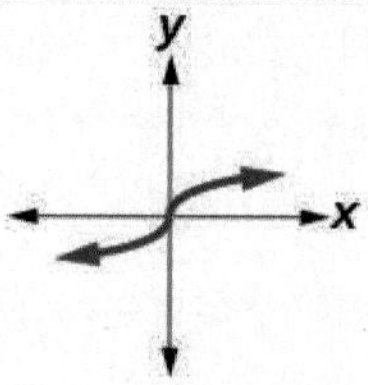
Square Root $f(x) = \sqrt{x}$	Increasing on $(0, \infty)$	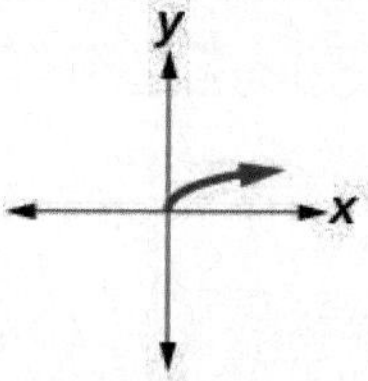
Absolute Value $f(x) = \lvert x \rvert$	Increasing on $(0, \infty)$ Decreasing on $(-\infty, 0)$	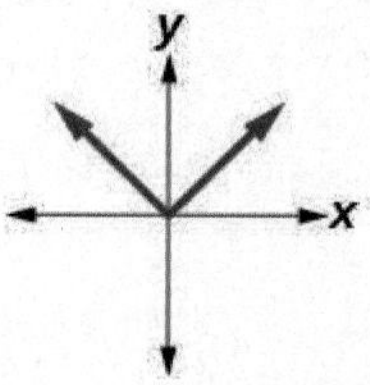

Figure 12

Use A Graph to Locate the Absolute Maximum and Absolute Minimum

There is a difference between locating the highest and lowest points on a graph in a region around an open interval (locally) and locating the highest and lowest points on the graph for the entire domain. The y- coordinates (output) at the highest and lowest points are called the **absolute maximum** and **absolute minimum**, respectively.

To locate absolute maxima and minima from a graph, we need to observe the graph to determine where the graph attains it highest and lowest points on the domain of the function. See Figure 13.

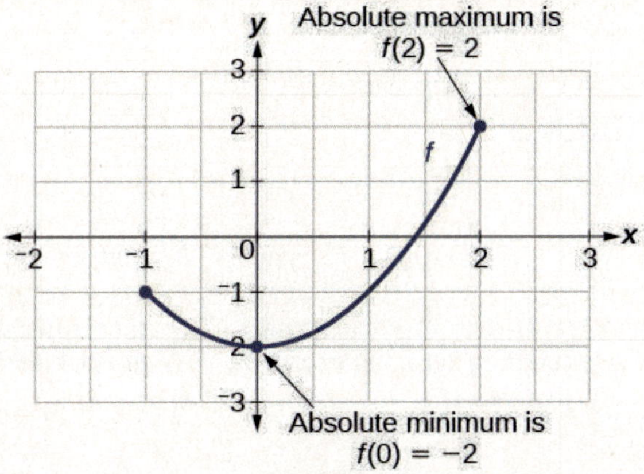

Figure 13

Not every function has an absolute maximum or minimum value. The toolkit function $f(x) = x^3$ is one such function.

Absolute Maxima and Minima

The **absolute maximum** of f at $x = c$ is $f(c)$ where $f(c) \geq f(x)$ for all x in the domain of f.

The **absolute minimum** of f at $x = d$ is $f(d)$ where $f(d) \leq f(x)$ for all x in the domain of f.

EXAMPLE 10

Finding Absolute Maxima and Minima from a Graph

For the function f shown in Figure 14, find all absolute maxima and minima.

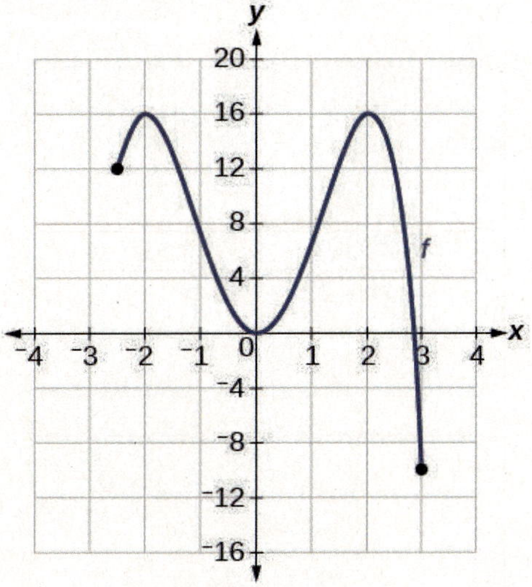

Figure 14

Solution

Observe the graph of f. The graph attains an absolute maximum in two locations, $x = -2$ and $x = 2$, because at these locations, the graph attains its highest point on the domain of the function. The absolute maximum is the *y*-coordinate at $x = -2$ and $x = 2$, which is 16.

The graph attains an absolute minimum at $x = 3$, because it is the lowest point on the domain of the function's graph. The absolute minimum is the *y*-coordinate at $x = 3$, which is -10.

MEDIA

Access this online resource for additional instruction and practice with rates of change.

Average Rate of Change (http://openstax.org/l/aroc)

3.3 SECTION EXERCISES

Verbal

1. Can the average rate of change of a function be constant?

2. If a function f is increasing on (a, b) and decreasing on (b, c), then what can be said about the local extremum of f on (a, c)?

3. How are the absolute maximum and minimum similar to and different from the local extrema?

4. How does the graph of the absolute value function compare to the graph of the quadratic function, $y = x^2$, in terms of increasing and decreasing intervals?

Algebraic

For the following exercises, find the average rate of change of each function on the interval specified for real numbers b or h in simplest form.

5. $f(x) = 4x^2 - 7$ on $[1, \ b]$

6. $g(x) = 2x^2 - 9$ on $[4, \ b]$

7. $p(x) = 3x + 4$ on $[2, \ 2 + h]$

8. $k(x) = 4x - 2$ on $[3, \ 3 + h]$

9. $f(x) = 2x^2 + 1$ on $[x, x + h]$

10. $g(x) = 3x^2 - 2$ on $[x, x + h]$

11. $a(t) = \frac{1}{t+4}$ on $[9, 9 + h]$

12. $b(x) = \frac{1}{x+3}$ on $[1, 1 + h]$

13. $j(x) = 3x^3$ on $[1, 1 + h]$

14. $r(t) = 4t^3$ on $[2, 2 + h]$

15. $\frac{f(x+h)-f(x)}{h}$ given $f(x) = 2x^2 - 3x$ on $[x, x + h]$

Graphical

For the following exercises, consider the graph of f shown in Figure 15.

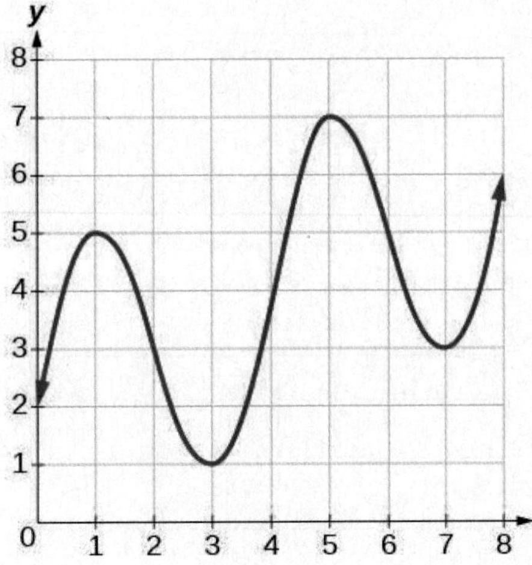

Figure 15

16. Estimate the average rate of change from $x = 1$ to $x = 4$.

17. Estimate the average rate of change from $x = 2$ to $x = 5$.

For the following exercises, use the graph of each function to estimate the intervals on which the function is increasing or decreasing.

18.

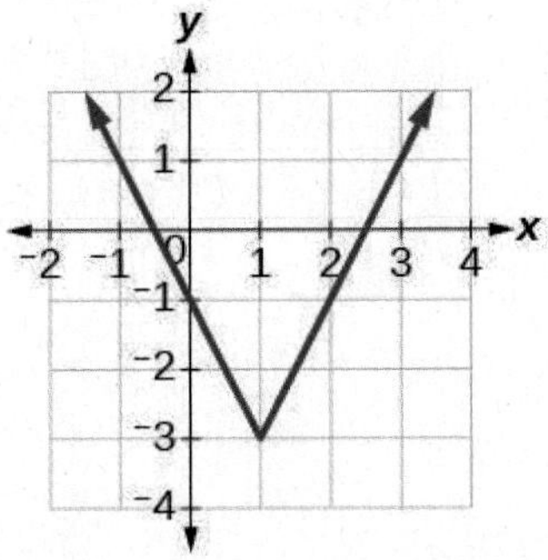

19.

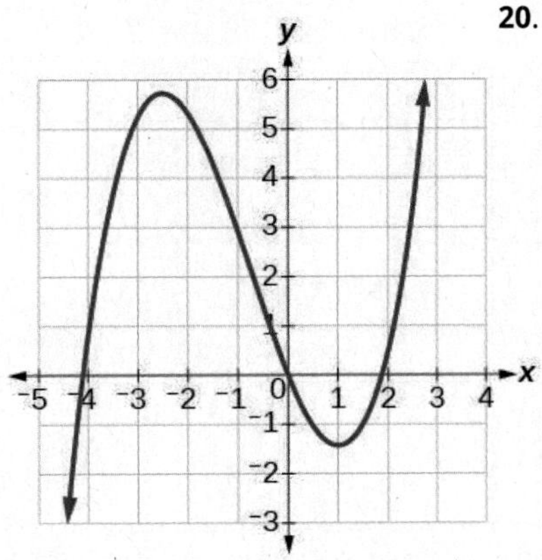

20.

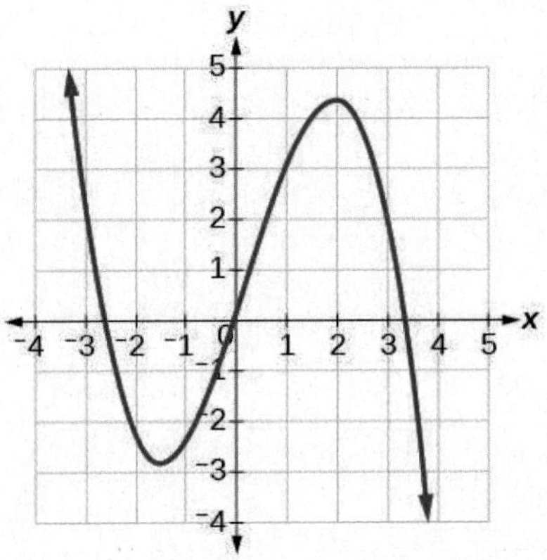

21.

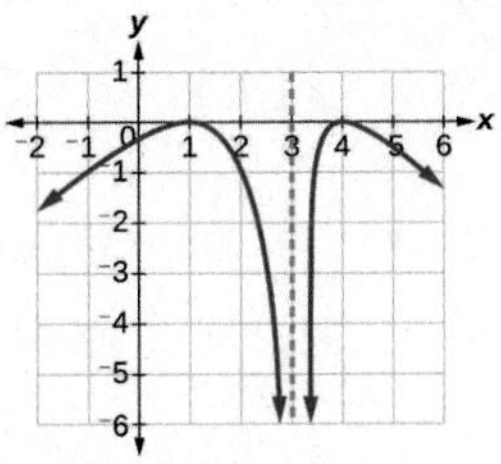

For the following exercises, consider the graph shown in Figure 16.

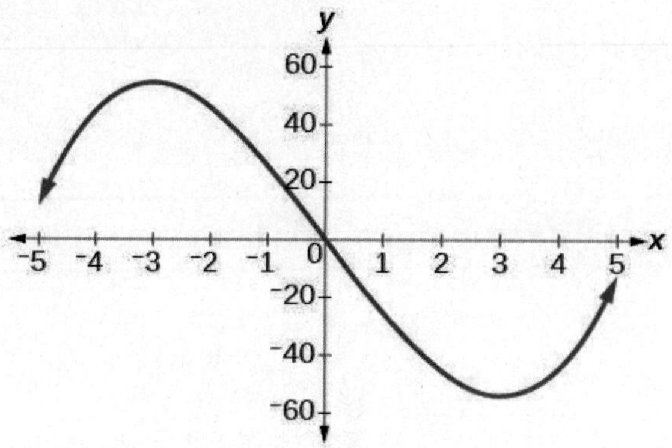

Figure 16

22. Estimate the intervals where the function is increasing or decreasing.

23. Estimate the point(s) at which the graph of f has a local maximum or a local minimum.

For the following exercises, consider the graph in Figure 17.

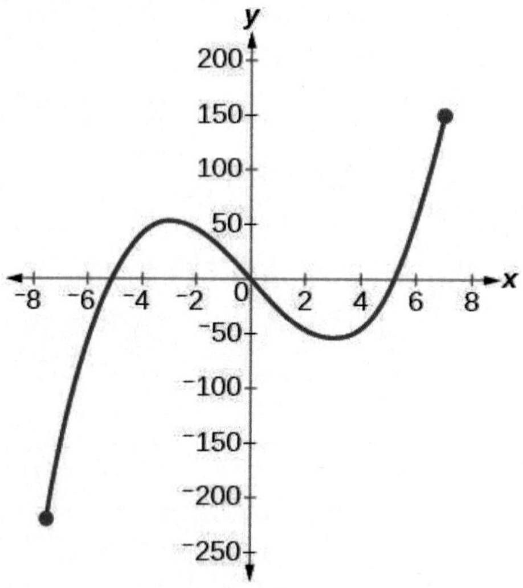

Figure 17

24. If the complete graph of the function is shown, estimate the intervals where the function is increasing or decreasing.

25. If the complete graph of the function is shown, estimate the absolute maximum and absolute minimum.

Numeric

26. Table 3 gives the annual sales (in millions of dollars) of a product from 1998 to 2006. What was the average rate of change of annual sales (a) between 2001 and 2002, and (b) between 2001 and 2004?

Year	Sales (millions of dollars)
1998	201
1999	219
2000	233
2001	243
2002	249
2003	251
2004	249
2005	243
2006	233

Table 3

27. Table 4 gives the population of a town (in thousands) from 2000 to 2008. What was the average rate of change of population (a) between 2002 and 2004, and (b) between 2002 and 2006?

Year	Population (thousands)
2000	87
2001	84
2002	83
2003	80
2004	77
2005	76
2006	78
2007	81
2008	85

Table 4

For the following exercises, find the average rate of change of each function on the interval specified.

28. $f(x) = x^2$ on $[1,\ 5]$

29. $h(x) = 5 - 2x^2$ on $[-2, 4]$

30. $q(x) = x^3$ on $[-4, 2]$

31. $g(x) = 3x^3 - 1$ on $[-3, 3]$

32. $y = \frac{1}{x}$ on $[1,\ 3]$

33. $p(t) = \frac{\left(t^2-4\right)(t+1)}{t^2+3}$ on $[-3, 1]$

34. $k(t) = 6t^2 + \frac{4}{t^3}$ on $[-1, 3]$

Technology

For the following exercises, use a graphing utility to estimate the local extrema of each function and to estimate the intervals on which the function is increasing and decreasing.

35. $f(x) = x^4 - 4x^3 + 5$

36. $h(x) = x^5 + 5x^4 + 10x^3 + 10x^2 - 1$

37. $g(t) = t\sqrt{t+3}$

38. $k(t) = 3t^{\frac{2}{3}} - t$

39. $m(x) = x^4 + 2x^3 - 12x^2 - 10x + 4$

40. $n(x) = x^4 - 8x^3 + 18x^2 - 6x + 2$

Extension

41. The graph of the function f is shown in Figure 18.

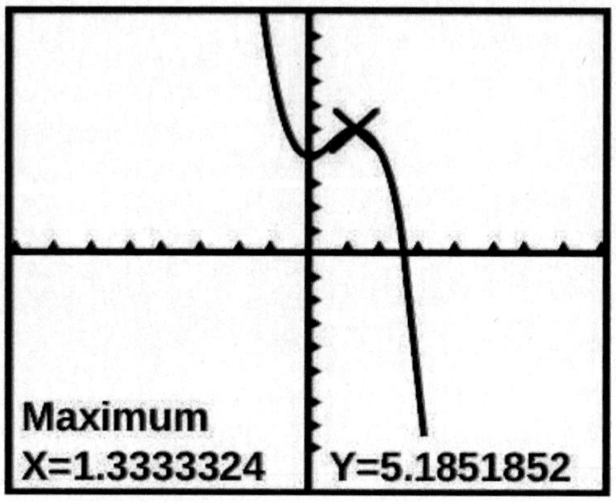

Figure 18

Based on the calculator screen shot, the point $(1.333,\ 5.185)$ is which of the following?

ⓐ a relative (local) maximum of the function

ⓑ the vertex of the function

ⓒ the absolute maximum of the function

ⓓ a zero of the function

42. Let $f(x) = \frac{1}{x}$. Find a number c such that the average rate of change of the function f on the interval $(1, c)$ is $-\frac{1}{4}$.

43. Let $f(x) = \frac{1}{x}$. Find the number b such that the average rate of change of f on the interval $(2, b)$ is $-\frac{1}{10}$.

Real-World Applications

44. At the start of a trip, the odometer on a car read 21,395. At the end of the trip, 13.5 hours later, the odometer read 22,125. Assume the scale on the odometer is in miles. What is the average speed the car traveled during this trip?

45. A driver of a car stopped at a gas station to fill up their gas tank. They looked at their watch, and the time read exactly 3:40 p.m. At this time, they started pumping gas into the tank. At exactly 3:44, the tank was full and the driver noticed that they had pumped 10.7 gallons. What is the average rate of flow of the gasoline into the gas tank?

46. Near the surface of the moon, the distance that an object falls is a function of time. It is given by $d(t) = 2.6667t^2$, where t is in seconds and $d(t)$ is in feet. If an object is dropped from a certain height, find the average velocity of the object from $t = 1$ to $t = 2$.

47. The graph in Figure 19 illustrates the decay of a radioactive substance over t days.

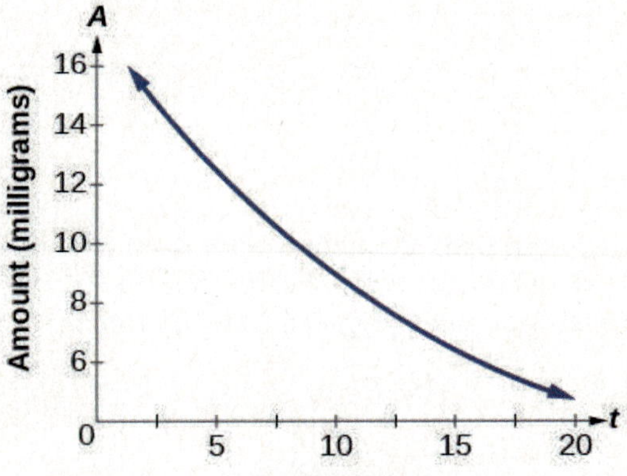

Figure 19

Use the graph to estimate the average decay rate from $t = 5$ to $t = 15$.

3.4 Composition of Functions

Learning Objectives

In this section, you will:

- Combine functions using algebraic operations.
- Create a new function by composition of functions.
- Evaluate composite functions.
- Find the domain of a composite function.
- Decompose a composite function into its component functions.

Suppose we want to calculate how much it costs to heat a house on a particular day of the year. The cost to heat a house will depend on the average daily temperature, and in turn, the average daily temperature depends on the particular day of the year. Notice how we have just defined two relationships: The cost depends on the temperature, and the temperature depends on the day.

Using descriptive variables, we can notate these two functions. The function $C(T)$ gives the cost C of heating a house for a given average daily temperature in T degrees Celsius. The function $T(d)$ gives the average daily temperature on day d of the year. For any given day, $\text{Cost} = C(T(d))$ means that the cost depends on the temperature, which in turns depends on the day of the year. Thus, we can evaluate the cost function at the temperature $T(d)$. For example, we could evaluate $T(5)$ to determine the average daily temperature on the 5th day of the year. Then, we could evaluate the cost function at that temperature. We would write $C(T(5))$.

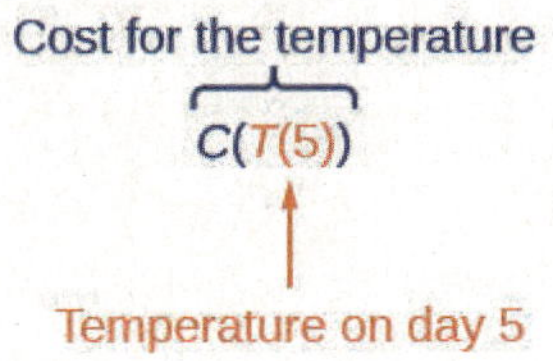

By combining these two relationships into one function, we have performed function composition, which is the focus of this section.

Combining Functions Using Algebraic Operations

Function composition is only one way to combine existing functions. Another way is to carry out the usual algebraic operations on functions, such as addition, subtraction, multiplication and division. We do this by performing the operations with the function outputs, defining the result as the output of our new function.

Suppose we need to add two columns of numbers that represent a husband and wife's separate annual incomes over a period of years, with the result being their total household income. We want to do this for every year, adding only that year's incomes and then collecting all the data in a new column. If $w(y)$ is the wife's income and $h(y)$ is the husband's income in year y, and we want T to represent the total income, then we can define a new function.

$$T(y) = h(y) + w(y)$$

If this holds true for every year, then we can focus on the relation between the functions without reference to a year and write

$$T = h + w$$

Just as for this sum of two functions, we can define difference, product, and ratio functions for any pair of functions that have the same kinds of inputs (not necessarily numbers) and also the same kinds of outputs (which do have to be numbers so that the usual operations of algebra can apply to them, and which also must have the same units or no units when we add and subtract). In this way, we can think of adding, subtracting, multiplying, and dividing functions.

For two functions $f(x)$ and $g(x)$ with real number outputs, we define new functions $f + g,\ f - g,\ fg,$ and $\frac{f}{g}$ by the relations

$$\begin{aligned}(f+g)(x) &= f(x)+g(x)\\ (f-g)(x) &= f(x)-g(x)\\ (fg)(x) &= f(x)g(x)\\ \left(\frac{f}{g}\right)(x) &= \frac{f(x)}{g(x)} \qquad \text{where } g(x) \neq 0\end{aligned}$$

EXAMPLE 1

Performing Algebraic Operations on Functions

Find and simplify the functions $(g - f)(x)$ and $\left(\frac{g}{f}\right)(x)$, given $f(x) = x - 1$ and $g(x) = x^2 - 1$. Are they the same function?

Solution

Begin by writing the general form, and then substitute the given functions.

$$\begin{aligned}(g-f)(x) &= g(x)-f(x)\\ (g-f)(x) &= x^2-1-(x-1)\\ (g-f)(x) &= x^2-x\\ (g-f)(x) &= x(x-1)\end{aligned}$$

$$\begin{aligned}\left(\frac{g}{f}\right)(x) &= \frac{g(x)}{f(x)}\\ \left(\frac{g}{f}\right)(x) &= \frac{x^2-1}{x-1}\\ \left(\frac{g}{f}\right)(x) &= \frac{(x+1)(x-1)}{x-1} \qquad \text{where} x \neq 1\\ \left(\frac{g}{f}\right)(x) &= x+1\end{aligned}$$

No, the functions are not the same.

Note: For $\left(\frac{g}{f}\right)(x)$, the condition $x \neq 1$ is necessary because when $x = 1$, the denominator is equal to 0, which makes the function undefined.

TRY IT #1 Find and simplify the functions $(fg)(x)$ and $(f - g)(x)$.

$$f(x) = x - 1 \quad \text{and} \quad g(x) = x^2 - 1$$

Are they the same function?

Create a Function by Composition of Functions

Performing algebraic operations on functions combines them into a new function, but we can also create functions by composing functions. When we wanted to compute a heating cost from a day of the year, we created a new function that takes a day as input and yields a cost as output. The process of combining functions so that the output of one function becomes the input of another is known as a composition of functions. The resulting function is known as a **composite function**. We represent this combination by the following notation:

$$(f \circ g)(x) = f(g(x))$$

We read the left-hand side as "f composed with g at x," and the right-hand side as "f of g of x." The two sides of the equation have the same mathematical meaning and are equal. The open circle symbol $\circ$ is called the composition operator. We use this operator mainly when we wish to emphasize the relationship between the functions themselves without referring to any particular input value. Composition is a binary operation that takes two functions and forms a new function, much as addition or multiplication takes two numbers and gives a new number. However, it is important not to confuse function composition with multiplication because, as we learned above, in most cases $f(g(x)) \neq f(x)g(x)$.

It is also important to understand the order of operations in evaluating a composite function. We follow the usual convention with parentheses by starting with the innermost parentheses first, and then working to the outside. In the equation above, the function g takes the input x first and yields an output $g(x)$. Then the function f takes $g(x)$ as an input and yields an output $f(g(x))$.

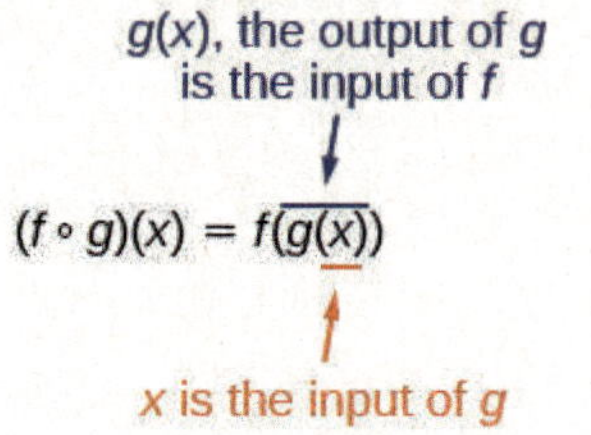

In general, $f \circ g$ and $g \circ f$ are different functions. In other words, in many cases $f(g(x)) \neq g(f(x))$ for all x. We will also see that sometimes two functions can be composed only in one specific order.

For example, if $f(x) = x^2$ and $g(x) = x + 2$, then

$$\begin{aligned} f(g(x)) &= f(x+2) \\ &= (x+2)^2 \\ &= x^2 + 4x + 4 \end{aligned}$$

but

$$\begin{aligned} g(f(x)) &= g\left(x^2\right) \\ &= x^2 + 2 \end{aligned}$$

These expressions are not equal for all values of x, so the two functions are not equal. It is irrelevant that the expressions happen to be equal for the single input value $x = -\frac{1}{2}$.

Note that the range of the inside function (the first function to be evaluated) needs to be within the domain of the outside function. Less formally, the composition has to make sense in terms of inputs and outputs.

Composition of Functions

When the output of one function is used as the input of another, we call the entire operation a composition of functions. For any input x and functions f and g, this action defines a **composite function**, which we write as $f \circ g$ such that

$$(f \circ g)(x) = f(g(x))$$

The domain of the composite function $f \circ g$ is all x such that x is in the domain of g and $g(x)$ is in the domain of f.

It is important to realize that the product of functions fg is not the same as the function composition $f(g(x))$, because, in general, $f(x)g(x) \neq f(g(x))$.

EXAMPLE 2

Determining whether Composition of Functions is Commutative

Using the functions provided, find $f(g(x))$ and $g(f(x))$. Determine whether the composition of the functions is commutative.

$$f(x) = 2x + 1 \qquad g(x) = 3 - x$$

Solution

Let's begin by substituting $g(x)$ into $f(x)$.

$$\begin{aligned} f(g(x)) &= 2(3 - x) + 1 \\ &= 6 - 2x + 1 \\ &= 7 - 2x \end{aligned}$$

Now we can substitute $f(x)$ into $g(x)$.

$$\begin{aligned} g(f(x)) &= 3 - (2x + 1) \\ &= 3 - 2x - 1 \\ &= -2x + 2 \end{aligned}$$

We find that $g(f(x)) \neq f(g(x))$, so the operation of function composition is not commutative.

EXAMPLE 3

Interpreting Composite Functions

The function $c(s)$ gives the number of calories burned completing s sit-ups, and $s(t)$ gives the number of sit-ups a person can complete in t minutes. Interpret $c(s(3))$.

Solution

The inside expression in the composition is $s(3)$. Because the input to the s-function is time, $t = 3$ represents 3 minutes, and $s(3)$ is the number of sit-ups completed in 3 minutes.

Using $s(3)$ as the input to the function $c(s)$ gives us the number of calories burned during the number of sit-ups that can be completed in 3 minutes, or simply the number of calories burned in 3 minutes (by doing sit-ups).

EXAMPLE 4

Investigating the Order of Function Composition

Suppose $f(x)$ gives miles that can be driven in x hours and $g(y)$ gives the gallons of gas used in driving y miles. Which of these expressions is meaningful: $f(g(y))$ or $g(f(x))$?

Solution

The function $y = f(x)$ is a function whose output is the number of miles driven corresponding to the number of hours driven.

$$\text{number of miles} = f\ (\text{number of hours})$$

The function $g(y)$ is a function whose output is the number of gallons used corresponding to the number of miles driven. This means:

$$\text{number of gallons} = g\ (\text{number of miles})$$

The expression $g(y)$ takes miles as the input and a number of gallons as the output. The function $f(x)$ requires a number of hours as the input. Trying to input a number of gallons does not make sense. The expression $f(g(y))$ is meaningless.

The expression $f(x)$ takes hours as input and a number of miles driven as the output. The function $g(y)$ requires a number of miles as the input. Using $f(x)$ (miles driven) as an input value for $g(y)$, where gallons of gas depends on miles driven, does make sense. The expression $g(f(x))$ makes sense, and will yield the number of gallons of gas used, g, driving a certain number of miles, $f(x)$, in x hours.

Q&A Are there any situations where $f(g(y))$ and $g(f(x))$ would both be meaningful or useful expressions?

Yes. For many pure mathematical functions, both compositions make sense, even though they usually produce different new functions. In real-world problems, functions whose inputs and outputs have the same units also may give compositions that are meaningful in either order.

TRY IT #2 The gravitational force on a planet a distance r from the sun is given by the function $G(r)$. The acceleration of a planet subjected to any force F is given by the function $a(F)$. Form a meaningful composition of these two functions, and explain what it means.

Evaluating Composite Functions

Once we compose a new function from two existing functions, we need to be able to evaluate it for any input in its domain. We will do this with specific numerical inputs for functions expressed as tables, graphs, and formulas and with variables as inputs to functions expressed as formulas. In each case, we evaluate the inner function using the starting input and then use the inner function's output as the input for the outer function.

Evaluating Composite Functions Using Tables

When working with functions given as tables, we read input and output values from the table entries and always work from the inside to the outside. We evaluate the inside function first and then use the output of the inside function as the input to the outside function.

EXAMPLE 5

Using a Table to Evaluate a Composite Function

Using Table 1, evaluate $f(g(3))$ and $g(f(3))$.

x	$f(x)$	$g(x)$
1	6	3
2	8	5
3	3	2
4	1	7

Table 1

Solution

To evaluate $f(g(3))$, we start from the inside with the input value 3. We then evaluate the inside expression $g(3)$ using the table that defines the function g : $g(3) = 2$. We can then use that result as the input to the function f, so $g(3)$ is replaced by 2 and we get $f(2)$. Then, using the table that defines the function f, we find that $f(2) = 8$.

$$\begin{aligned} g(3) &= 2 \\ f(g(3)) &= f(2) = 8 \end{aligned}$$

To evaluate $g(f(3))$, we first evaluate the inside expression $f(3)$ using the first table: $f(3) = 3$. Then, using the table for $g,$ we can evaluate

$$g(f(3)) = g(3) = 2$$

Table 2 shows the composite functions $f \circ g$ and $g \circ f$ as tables.

x	$g(x)$	$f(g(x))$	$f(x)$	$g(f(x))$

Table 2

3	2	8	3	2

Table 2

> TRY IT #3 Using Table 1, evaluate $f(g(1))$ and $g(f(4))$.

Evaluating Composite Functions Using Graphs

When we are given individual functions as graphs, the procedure for evaluating composite functions is similar to the process we use for evaluating tables. We read the input and output values, but this time, from the x- and y- axes of the graphs.

HOW TO

Given a composite function and graphs of its individual functions, evaluate it using the information provided by the graphs.

1. Locate the given input to the inner function on the x- axis of its graph.
2. Read off the output of the inner function from the y- axis of its graph.
3. Locate the inner function output on the x- axis of the graph of the outer function.
4. Read the output of the outer function from the y- axis of its graph. This is the output of the composite function.

EXAMPLE 6

Using a Graph to Evaluate a Composite Function

Using Figure 1, evaluate $f(g(1))$.

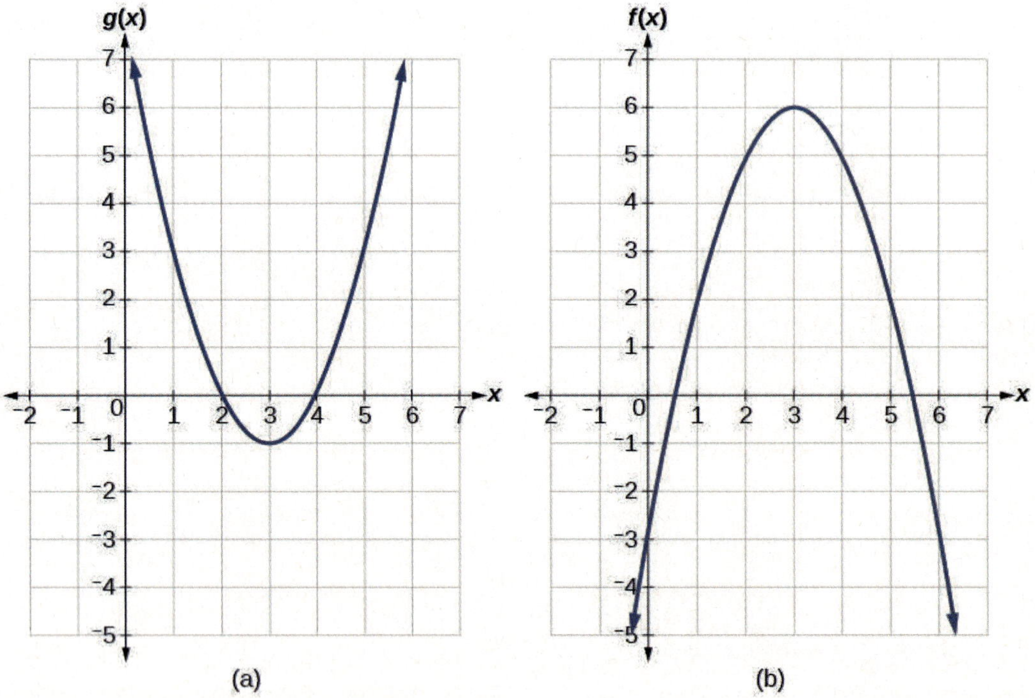

Figure 1

Solution

To evaluate $f(g(1))$, we start with the inside evaluation. See Figure 2.

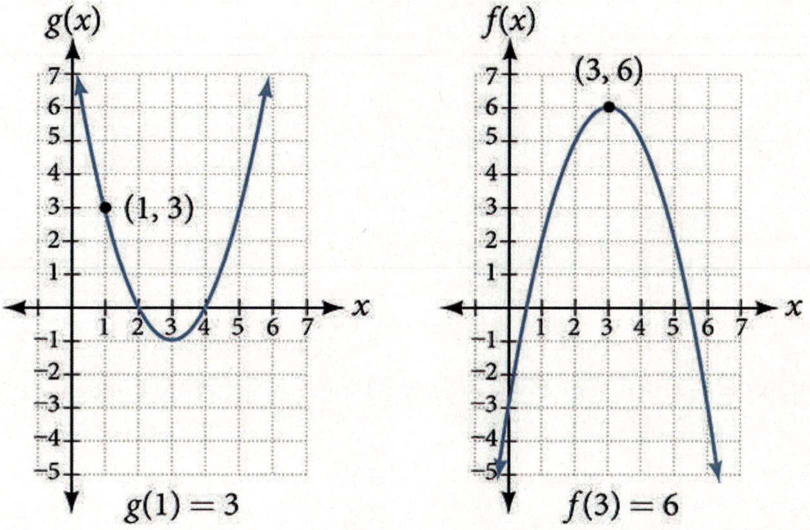

Figure 2

We evaluate $g(1)$ using the graph of $g(x)$, finding the input of 1 on the x- axis and finding the output value of the graph at that input. Here, $g(1) = 3$. We use this value as the input to the function f.

$$f(g(1)) = f(3)$$

We can then evaluate the composite function by looking to the graph of $f(x)$, finding the input of 3 on the x- axis and reading the output value of the graph at this input. Here, $f(3) = 6$, so $f(g(1)) = 6$.

Analysis

Figure 3 shows how we can mark the graphs with arrows to trace the path from the input value to the output value.

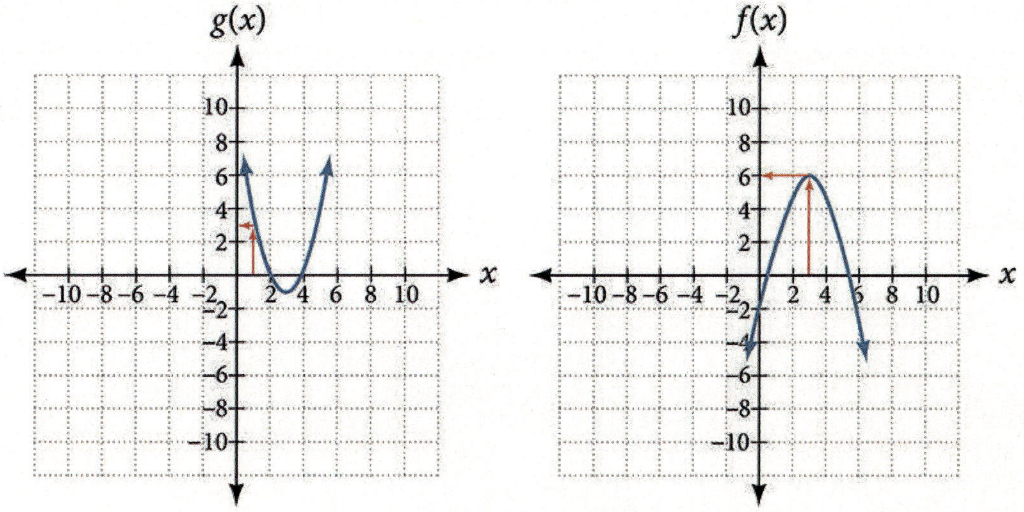

Figure 3

TRY IT #4 Using Figure 1, evaluate $g(f(2))$.

Evaluating Composite Functions Using Formulas

When evaluating a composite function where we have either created or been given formulas, the rule of working from the inside out remains the same. The input value to the outer function will be the output of the inner function, which may be a numerical value, a variable name, or a more complicated expression.

While we can compose the functions for each individual input value, it is sometimes helpful to find a single formula that will calculate the result of a composition $f(g(x))$. To do this, we will extend our idea of function evaluation. Recall that, when we evaluate a function like $f(t) = t^2 - t$, we substitute the value inside the parentheses into the formula wherever we see the input variable.

HOW TO

Given a formula for a composite function, evaluate the function.

1. Evaluate the inside function using the input value or variable provided.
2. Use the resulting output as the input to the outside function.

EXAMPLE 7

Evaluating a Composition of Functions Expressed as Formulas with a Numerical Input

Given $f(t) = t^2 - t$ and $h(x) = 3x + 2$, evaluate $f(h(1))$.

Solution

Because the inside expression is $h(1)$, we start by evaluating $h(x)$ at 1.

$$\begin{aligned} h(1) &= 3(1) + 2 \\ h(1) &= 5 \end{aligned}$$

Then $f(h(1)) = f(5)$, so we evaluate $f(t)$ at an input of 5.

$$\begin{aligned} f(h(1)) &= f(5) \\ f(h(1)) &= 5^2 - 5 \\ f(h(1)) &= 20 \end{aligned}$$

Analysis

It makes no difference what the input variables t and x were called in this problem because we evaluated for specific numerical values.

TRY IT #5 Given $f(t) = t^2 - t$ and $h(x) = 3x + 2$, evaluate

ⓐ $h(f(2))$ ⓑ $h(f(-2))$

Finding the Domain of a Composite Function

As we discussed previously, the domain of a composite function such as $f \circ g$ is dependent on the domain of g and the domain of f. It is important to know when we can apply a composite function and when we cannot, that is, to know the domain of a function such as $f \circ g$. Let us assume we know the domains of the functions f and g separately. If we write the composite function for an input x as $f(g(x))$, we can see right away that x must be a member of the domain of g in order for the expression to be meaningful, because otherwise we cannot complete the inner function evaluation. However, we also see that $g(x)$ must be a member of the domain of f, otherwise the second function evaluation in $f(g(x))$ cannot be completed, and the expression is still undefined. Thus the domain of $f \circ g$ consists of only those inputs in the domain of g that produce outputs from g belonging to the domain of f. Note that the domain of f composed with g is the set of all x such that x is in the domain of g and $g(x)$ is in the domain of f.

Domain of a Composite Function

The domain of a composite function $f(g(x))$ is the set of those inputs x in the domain of g for which $g(x)$ is in the domain of f.

HOW TO

Given a function composition $f(g(x))$, determine its domain.

1. Find the domain of g.
2. Find the domain of f.
3. Find those inputs x in the domain of g for which $g(x)$ is in the domain of f. That is, exclude those inputs x from the domain of g for which $g(x)$ is not in the domain of f. The resulting set is the domain of $f \circ g$.

EXAMPLE 8

Finding the Domain of a Composite Function

Find the domain of

$$(f \circ g)(x) \text{ where } \quad f(x) = \frac{5}{x-1} \text{ and } g(x) = \frac{4}{3x-2}$$

Solution

The domain of $g(x)$ consists of all real numbers except $x = \frac{2}{3}$, since that input value would cause us to divide by 0. Likewise, the domain of f consists of all real numbers except 1. So we need to exclude from the domain of $g(x)$ that value of x for which $g(x) = 1$.

$$\begin{aligned} \frac{4}{3x-2} &= 1 \\ 4 &= 3x-2 \\ 6 &= 3x \\ x &= 2 \end{aligned}$$

So the domain of $f \circ g$ is the set of all real numbers except $\frac{2}{3}$ and 2. This means that

$$x \neq \frac{2}{3} \text{ or } x \neq 2$$

We can write this in interval notation as

$$\left(-\infty, \frac{2}{3}\right) \cup \left(\frac{2}{3}, 2\right) \cup (2, \infty)$$

EXAMPLE 9

Finding the Domain of a Composite Function Involving Radicals

Find the domain of

$$(f \circ g)(x) \text{ where } \quad f(x) = \sqrt{x+2} \text{ and } g(x) = \sqrt{3-x}$$

Solution

Because we cannot take the square root of a negative number, the domain of g is $\left(-\infty, 3\right]$. Now we check the domain of the composite function

$$(f \circ g)(x) = \sqrt{\sqrt{3-x}+2}$$

For $(f \circ g)(x) = \sqrt{\sqrt{3-x}+2}$, $\sqrt{3-x}+2 \geq 0$, since the radicand of a square root must be positive. Since square roots are positive, $\sqrt{3-x} \geq 0$, or, $3-x \geq 0$, which gives a domain of $(-\infty, 3]$.

Analysis

This example shows that knowledge of the range of functions (specifically the inner function) can also be helpful in finding the domain of a composite function. It also shows that the domain of $f \circ g$ can contain values that are not in the domain of f, though they must be in the domain of g.

TRY IT #6 Find the domain of

$$(f \circ g)(x) \text{ where } \quad f(x) = \frac{1}{x-2} \text{ and } g(x) = \sqrt{x+4}$$

Decomposing a Composite Function into its Component Functions

In some cases, it is necessary to decompose a complicated function. In other words, we can write it as a composition of two simpler functions. There may be more than one way to decompose a composite function, so we may choose the decomposition that appears to be most expedient.

EXAMPLE 10

Decomposing a Function

Write $f(x) = \sqrt{5 - x^2}$ as the composition of two functions.

Solution

We are looking for two functions, g and h, so $f(x) = g(h(x))$. To do this, we look for a function inside a function in the formula for $f(x)$. As one possibility, we might notice that the expression $5 - x^2$ is the inside of the square root. We could then decompose the function as

$$h(x) = 5 - x^2 \text{ and } g(x) = \sqrt{x}$$

We can check our answer by recomposing the functions.

$$g(h(x)) = g\left(5 - x^2\right) = \sqrt{5 - x^2}$$

TRY IT #7 Write $f(x) = \frac{4}{3 - \sqrt{4 + x^2}}$ as the composition of two functions.

MEDIA

Access these online resources for additional instruction and practice with composite functions.

Composite Functions (http://openstax.org/l/compfunction)
Composite Function Notation Application (http://openstax.org/l/compfuncnot)
Composite Functions Using Graphs (http://openstax.org/l/compfuncgraph)
Decompose Functions (http://openstax.org/l/decompfunction)
Composite Function Values (http://openstax.org/l/compfuncvalue)

3.4 SECTION EXERCISES

Verbal

1. How does one find the domain of the quotient of two functions, $\frac{f}{g}$?

2. What is the composition of two functions, $f \circ g$?

3. If the order is reversed when composing two functions, can the result ever be the same as the answer in the original order of the composition? If yes, give an example. If no, explain why not.

4. How do you find the domain for the composition of two functions, $f \circ g$?

Algebraic

For the following exercises, determine the domain for each function in interval notation.

5. Given $f(x) = x^2 + 2x$ and $g(x) = 6 - x^2$, find $f + g,\ f - g,\ fg$, and $\frac{f}{g}$.

6. Given $f(x) = -3x^2 + x$ and $g(x) = 5$, find $f + g,\ f - g,\ fg$, and $\frac{f}{g}$.

7. Given $f(x) = 2x^2 + 4x$ and $g(x) = \frac{1}{2x}$, find $f + g,\ f - g,\ fg$, and $\frac{f}{g}$.

8. Given $f(x) = \frac{1}{x-4}$ and $g(x) = \frac{1}{6-x}$, find $f + g,\ f - g,\ fg$, and $\frac{f}{g}$.

9. Given $f(x) = 3x^2$ and $g(x) = \sqrt{x-5}$, find $f + g,\ f - g,\ fg$, and $\frac{f}{g}$.

10. Given $f(x) = \sqrt{x}$ and $g(x) = |x - 3|$, find $\frac{g}{f}$.

11. For the following exercise, find the indicated function given $f(x) = 2x^2 + 1$ and $g(x) = 3x - 5$.

ⓐ $f(g(2))$ ⓑ $f(g(x))$ ⓒ $g(f(x))$
ⓓ $(g \circ g)(x)$ ⓔ $(f \circ f)(-2)$

For the following exercises, use each pair of functions to find $f(g(x))$ and $g(f(x))$. Simplify your answers.

12. $f(x) = x^2 + 1,\ \ g(x) = \sqrt{x+2}$

13. $f(x) = \sqrt{x} + 2,\ \ g(x) = x^2 + 3$

14. $f(x) = |x|,\ \ g(x) = 5x + 1$

15. $f(x) = \sqrt[3]{x},\ \ g(x) = \frac{x+1}{x^3}$

16. $f(x) = \frac{1}{x-6},\ \ g(x) = \frac{7}{x} + 6$

17. $f(x) = \frac{1}{x-4},\ \ g(x) = \frac{2}{x} + 4$

For the following exercises, use each set of functions to find $f(g(h(x)))$. Simplify your answers.

18. $f(x) = x^4 + 6$, $g(x) = x - 6$, and $h(x) = \sqrt{x}$

19. $f(x) = x^2 + 1$, $g(x) = \frac{1}{x}$, and $h(x) = x + 3$

20. Given $f(x) = \frac{1}{x}$ and $g(x) = x - 3$, find the following:

ⓐ $(f \circ g)(x)$
ⓑ the domain of $(f \circ g)(x)$ in interval notation
ⓒ $(g \circ f)(x)$
ⓓ the domain of $(g \circ f)(x)$
ⓔ $\left(\frac{f}{g}\right)(x)$

21. Given $f(x) = \sqrt{2-4x}$ and $g(x) = -\frac{3}{x}$, find the following:

ⓐ $(g \circ f)(x)$
ⓑ the domain of $(g \circ f)(x)$ in interval notation

22. Given the functions $f(x) = \frac{1-x}{x}$ and $g(x) = \frac{1}{1+x^2}$, find the following:

ⓐ $(g \circ f)(x)$ ⓑ $(g \circ f)(2)$

23. Given functions $p(x) = \frac{1}{\sqrt{x}}$ and $m(x) = x^2 - 4$, state the domain of each of the following functions using interval notation:

ⓐ $\frac{p(x)}{m(x)}$ ⓑ $p(m(x))$
ⓒ $m(p(x))$

24. Given functions $q(x) = \frac{1}{\sqrt{x}}$ and $h(x) = x^2 - 9$, state the domain of each of the following functions using interval notation.

ⓐ $\frac{q(x)}{h(x)}$ ⓑ $q(h(x))$ ⓒ $h(q(x))$

25. For $f(x) = \frac{1}{x}$ and $g(x) = \sqrt{x-1}$, write the domain of $(f \circ g)(x)$ in interval notation.

For the following exercises, find functions $f(x)$ and $g(x)$ so the given function can be expressed as $h(x) = f(g(x))$.

26. $h(x) = (x+2)^2$

27. $h(x) = (x-5)^3$

28. $h(x) = \frac{3}{x-5}$

29. $h(x) = \frac{4}{(x+2)^2}$

30. $h(x) = 4 + \sqrt[3]{x}$

31. $h(x) = \sqrt[3]{\frac{1}{2x-3}}$

32. $h(x) = \frac{1}{(3x^2-4)^{-3}}$

33. $h(x) = \sqrt[4]{\frac{3x-2}{x+5}}$

34. $h(x) = \left(\frac{8+x^3}{8-x^3}\right)^4$

35. $h(x) = \sqrt{2x+6}$

36. $h(x) = (5x-1)^3$

37. $h(x) = \sqrt[3]{x-1}$

38. $h(x) = \left|x^2+7\right|$

39. $h(x) = \frac{1}{(x-2)^3}$

40. $h(x) = \left(\frac{1}{2x-3}\right)^2$

41. $h(x) = \sqrt{\frac{2x-1}{3x+4}}$

Graphical

For the following exercises, use the graphs of f, shown in Figure 4, and g, shown in Figure 5, to evaluate the expressions.

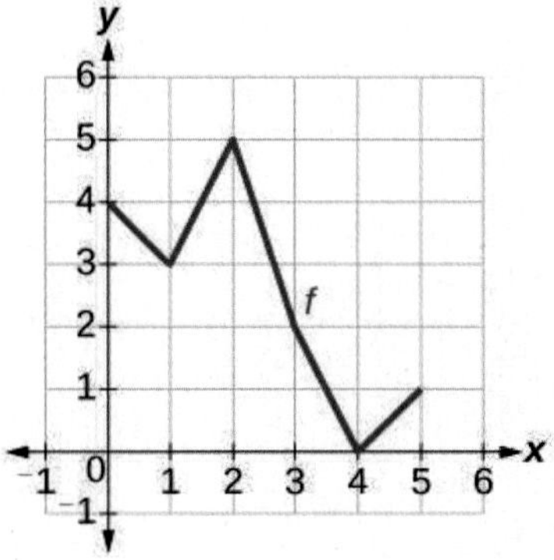

Figure 4

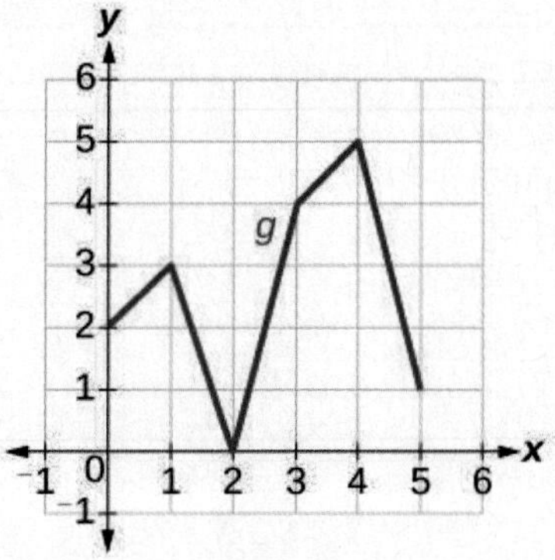

Figure 5

42. $f(g(3))$

43. $f(g(1))$

44. $g(f(1))$

45. $g(f(0))$

46. $f(f(5))$

47. $f(f(4))$

48. $g(g(2))$

49. $g(g(0))$

For the following exercises, use graphs of $f(x)$, shown in Figure 6, $g(x)$, shown in Figure 7, and $h(x)$, shown in Figure 8, to evaluate the expressions.

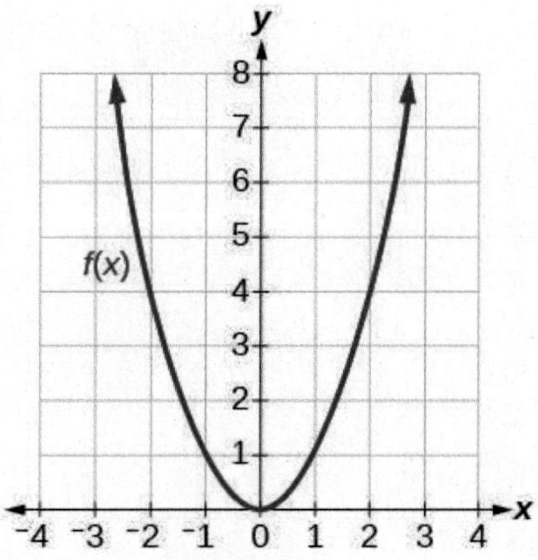

Figure 6

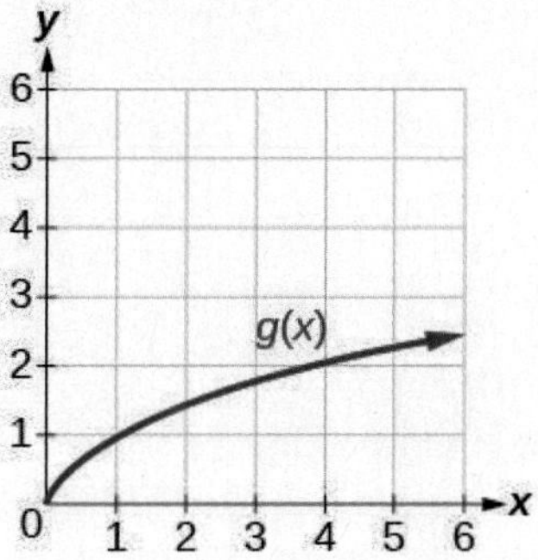

Figure 7

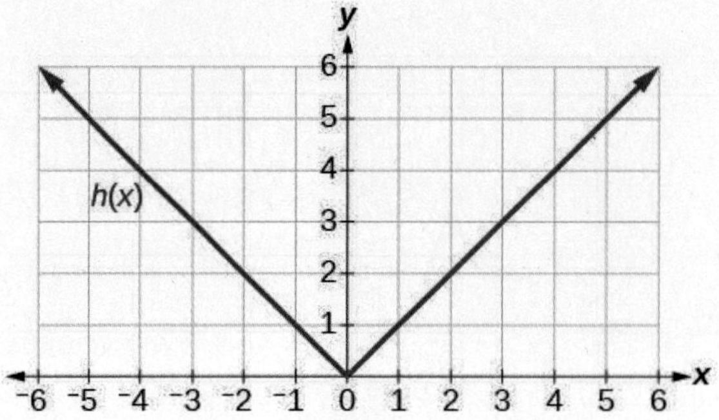

Figure 8

50. $g(f(1))$

51. $g(f(2))$

52. $f(g(4))$

53. $f(g(1))$

54. $f(h(2))$

55. $h(f(2))$

56. $f(g(h(4)))$

57. $f(g(f(-2)))$

Numeric

For the following exercises, use the function values for f and g shown in Table 3 to evaluate each expression.

x	$f(x)$	$g(x)$
0	7	9
1	6	5
2	5	6
3	8	2
4	4	1
5	0	8
6	2	7
7	1	3
8	9	4
9	3	0

Table 3

58. $f(g(8))$

59. $f(g(5))$

60. $g(f(5))$

61. $g(f(3))$

62. $f(f(4))$

63. $f(f(1))$

64. $g(g(2))$

65. $g(g(6))$

For the following exercises, use the function values for f and g shown in Table 4 to evaluate the expressions.

x	$f(x)$	$g(x)$
−3	11	−8
−2	9	−3
−1	7	0
0	5	1
1	3	0
2	1	−3
3	−1	−8

Table 4

66. $(f \circ g)(1)$

67. $(f \circ g)(2)$

68. $(g \circ f)(2)$

69. $(g \circ f)(3)$

70. $(g \circ g)(1)$

71. $(f \circ f)(3)$

For the following exercises, use each pair of functions to find $f\,(g\,(0))$ and $g\,(f(0))$.

72. $f(x) = 4x + 8, \;\; g(x) = 7 - x^2$

73. $f(x) = 5x + 7, \;\; g(x) = 4 - 2x^2$

74. $f(x) = \sqrt{x + 4}, \;\; g(x) = 12 - x^3$

75. $f(x) = \frac{1}{x+2}, \;\; g(x) = 4x + 3$

For the following exercises, use the functions $f(x) = 2x^2 + 1$ and $g(x) = 3x + 5$ to evaluate or find the composite function as indicated.

76. $f\,(g(2))$

77. $f\,(g(x))$

78. $g\,(f(-3))$

79. $(g \circ g)(x)$

Extensions

For the following exercises, use $f(x) = x^3 + 1$ and $g(x) = \sqrt[3]{x - 1}$.

80. Find $(f \circ g)(x)$ and $(g \circ f)(x)$. Compare the two answers.

81. Find $(f \circ g)(2)$ and $(g \circ f)(2)$.

82. What is the domain of $(g \circ f)(x)$?

83. What is the domain of $(f \circ g)(x)$?

84. Let $f(x) = \frac{1}{x}$.

ⓐ Find $(f \circ f)(x)$.

ⓑ Is $(f \circ f)(x)$ for any function f the same result as the answer to part (a) for any function? Explain.

For the following exercises, let $F(x) = (x + 1)^5$, $f(x) = x^5$, *and* $g(x) = x + 1$.

85. True or False: $(g \circ f)(x) = F(x)$.

86. True or False: $(f \circ g)(x) = F(x)$.

For the following exercises, find the composition when $f(x) = x^2 + 2$ *for all* $x \geq 0$ *and* $g(x) = \sqrt{x - 2}$.

87. $(f \circ g)(6)$; $(g \circ f)(6)$

88. $(g \circ f)(a)$; $(f \circ g)(a)$

89. $(f \circ g)(11)$; $(g \circ f)(11)$

Real-World Applications

90. The function $D(p)$ gives the number of items that will be demanded when the price is p. The production cost $C(x)$ is the cost of producing x items. To determine the cost of production when the price is $6, you would do which of the following?

ⓐ Evaluate $D(C(6))$.

ⓑ Evaluate $C(D(6))$.

ⓒ Solve $D(C(x)) = 6$.

ⓓ Solve $C(D(p)) = 6$.

91. The function $A(d)$ gives the pain level on a scale of 0 to 10 experienced by a patient with d milligrams of a pain-reducing drug in her system. The milligrams of the drug in the patient's system after t minutes is modeled by $m(t)$. Which of the following would you do in order to determine when the patient will be at a pain level of 4?

ⓐ Evaluate $A(m(4))$.

ⓑ Evaluate $m(A(4))$.

ⓒ Solve $A(m(t)) = 4$.

ⓓ Solve $m(A(d)) = 4$.

92. A store offers customers a 30% discount on the price x of selected items. Then, the store takes off an additional 15% at the cash register. Write a price function $P(x)$ that computes the final price of the item in terms of the original price x. (Hint: Use function composition to find your answer.)

93. A rain drop hitting a lake makes a circular ripple. If the radius, in inches, grows as a function of time in minutes according to $r(t) = 25\sqrt{t + 2}$, find the area of the ripple as a function of time. Find the area of the ripple at $t = 2$.

94. A forest fire leaves behind an area of grass burned in an expanding circular pattern. If the radius of the circle of burning grass is increasing with time according to the formula $r(t) = 2t + 1$, express the area burned as a function of time, t (minutes).

95. Use the function you found in the previous exercise to find the total area burned after 5 minutes.

96. The radius r, in inches, of a spherical balloon is related to the volume, V, by $r(V) = \sqrt[3]{\frac{3V}{4\pi}}$. Air is pumped into the balloon, so the volume after t seconds is given by $V(t) = 10 + 20t$.

ⓐ Find the composite function $r(V(t))$.

ⓑ Find the *exact* time when the radius reaches 10 inches.

97. The number of bacteria in a refrigerated food product is given by $N(T) = 23T^2 - 56T + 1$, $3 < T < 33$, where T is the temperature of the food. When the food is removed from the refrigerator, the temperature is given by $T(t) = 5t + 1.5$, where t is the time in hours.

ⓐ Find the composite function $N(T(t))$.

ⓑ Find the time (round to two decimal places) when the bacteria count reaches 6752.

3.5 Transformation of Functions

Learning Objectives

In this section, you will:

- Graph functions using vertical and horizontal shifts.
- Graph functions using reflections about the x-axis and the y-axis.
- Determine whether a function is even, odd, or neither from its graph.
- Graph functions using compressions and stretches.
- Combine transformations.

Figure 1 (credit: "Misko"/Flickr)

We all know that a flat mirror enables us to see an accurate image of ourselves and whatever is behind us. When we tilt the mirror, the images we see may shift horizontally or vertically. But what happens when we bend a flexible mirror? Like a carnival funhouse mirror, it presents us with a distorted image of ourselves, stretched or compressed horizontally or vertically. In a similar way, we can distort or transform mathematical functions to better adapt them to describing objects or processes in the real world. In this section, we will take a look at several kinds of transformations.

Graphing Functions Using Vertical and Horizontal Shifts

Often when given a problem, we try to model the scenario using mathematics in the form of words, tables, graphs, and equations. One method we can employ is to adapt the basic graphs of the toolkit functions to build new models for a given scenario. There are systematic ways to alter functions to construct appropriate models for the problems we are trying to solve.

Identifying Vertical Shifts

One simple kind of transformation involves shifting the entire graph of a function up, down, right, or left. The simplest shift is a **vertical shift**, moving the graph up or down, because this transformation involves adding a positive or negative constant to the function. In other words, we add the same constant to the output value of the function regardless of the input. For a function $g(x) = f(x) + k$, the function $f(x)$ is shifted vertically k units. See Figure 2 for an example.

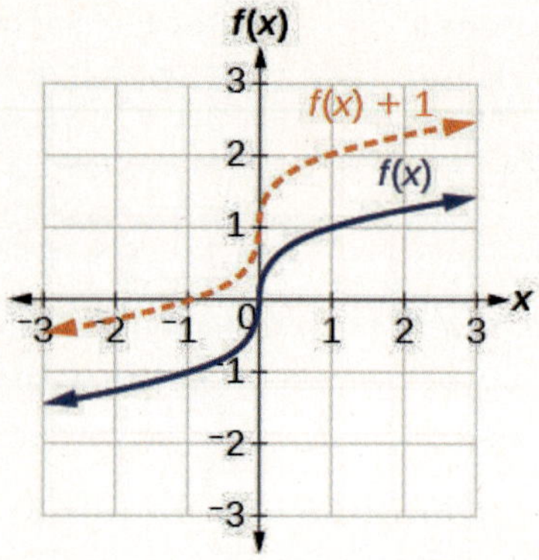

Figure 2 Vertical shift by $k = 1$ of the cube root function $f(x) = \sqrt[3]{x}$.

To help you visualize the concept of a vertical shift, consider that $y = f(x)$. Therefore, $f(x) + k$ is equivalent to $y + k$. Every unit of y is replaced by $y + k$, so the y-value increases or decreases depending on the value of k. The result is a shift upward or downward.

Vertical Shift

Given a function $f(x)$, a new function $g(x) = f(x) + k$, where k is a constant, is a **vertical shift** of the function $f(x)$. All the output values change by k units. If k is positive, the graph will shift up. If k is negative, the graph will shift down.

EXAMPLE 1

Adding a Constant to a Function

To regulate temperature in a green building, airflow vents near the roof open and close throughout the day. Figure 3 shows the area of open vents V (in square feet) throughout the day in hours after midnight, t. During the summer, the facilities manager decides to try to better regulate temperature by increasing the amount of open vents by 20 square feet throughout the day and night. Sketch a graph of this new function.

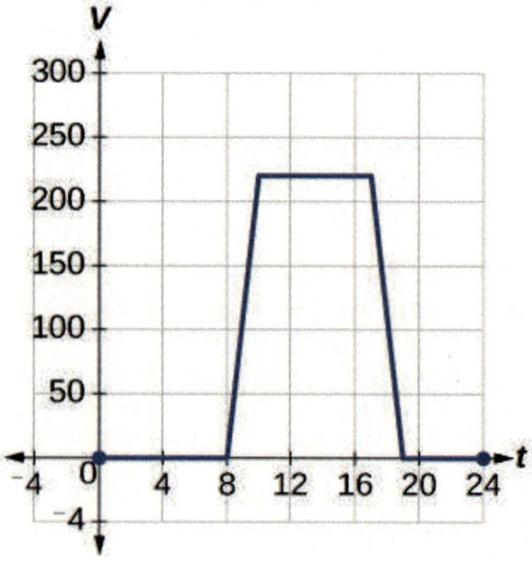

Figure 3

Solution

We can sketch a graph of this new function by adding 20 to each of the output values of the original function. This will have the effect of shifting the graph vertically up, as shown in Figure 4.

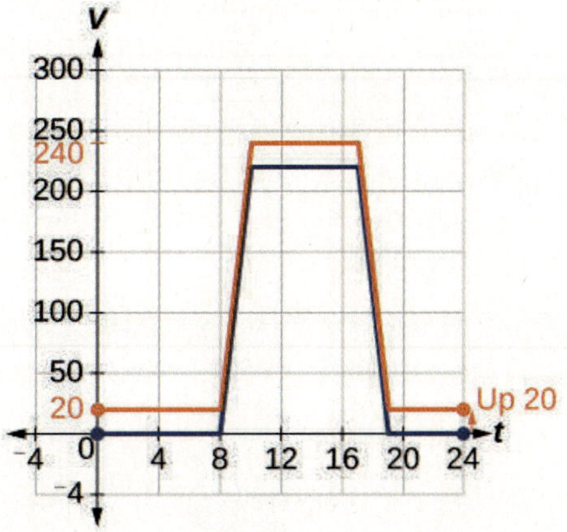

Figure 4

Notice that in Figure 4, for each input value, the output value has increased by 20, so if we call the new function $S(t)$, we could write

$$S(t) = V(t) + 20$$

This notation tells us that, for any value of t, $S(t)$ can be found by evaluating the function V at the same input and then adding 20 to the result. This defines S as a transformation of the function V, in this case a vertical shift up 20 units. Notice that, with a vertical shift, the input values stay the same and only the output values change. See Table 1.

t	0	8	10	17	19	24
$V(t)$	0	0	220	220	0	0
$S(t)$	20	20	240	240	20	20

Table 1

HOW TO

Given a tabular function, create a new row to represent a vertical shift.

1. Identify the output row or column.
2. Determine the magnitude of the shift.
3. Add the shift to the value in each output cell. Add a positive value for up or a negative value for down.

EXAMPLE 2

Shifting a Tabular Function Vertically

A function $f(x)$ is given in Table 2. Create a table for the function $g(x) = f(x) - 3$.

x	2	4	6	8
$f(x)$	1	3	7	11

Table 2

Solution

The formula $g(x) = f(x) - 3$ tells us that we can find the output values of g by subtracting 3 from the output values of f.

For example:

$$\begin{aligned} f(2) &= 1 && \text{Given} \\ g(x) &= f(x) - 3 && \text{Given transformation} \\ g(2) &= f(2) - 3 \\ &= 1 - 3 \\ &= -2 \end{aligned}$$

Subtracting 3 from each $f(x)$ value, we can complete a table of values for $g(x)$ as shown in Table 3.

x	2	4	6	8
$f(x)$	1	3	7	11
$g(x)$	−2	0	4	8

Table 3

Analysis
As with the earlier vertical shift, notice the input values stay the same and only the output values change.

TRY IT #1 The function $h(t) = -4.9t^2 + 30t$ gives the height h of a ball (in meters) thrown upward from the ground after t seconds. Suppose the ball was instead thrown from the top of a 10-m building. Relate this new height function $b(t)$ to $h(t)$, and then find a formula for $b(t)$.

Identifying Horizontal Shifts

We just saw that the vertical shift is a change to the output, or outside, of the function. We will now look at how changes to input, on the inside of the function, change its graph and meaning. A shift to the input results in a movement of the graph of the function left or right in what is known as a **horizontal shift**, shown in Figure 5.

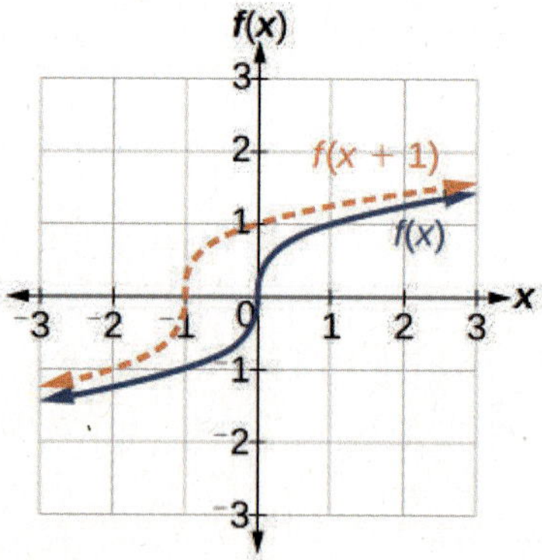

Figure 5 Horizontal shift of the function $f(x) = \sqrt[3]{x}$. Note that $(x + 1)$ means $h = -1$, which shifts the graph to the left, that is, towards *negative* values of x.

For example, if $f(x) = x^2$, then $g(x) = (x - 2)^2$ is a new function. Each input is reduced by 2 prior to squaring the function. The result is that the graph is shifted 2 units to the right, because we would need to increase the prior input by 2 units to yield the same output value as given in f.

Horizontal Shift

Given a function f, a new function $g(x) = f(x - h)$, where h is a constant, is a **horizontal shift** of the function f. If h is positive, the graph will shift right. If h is negative, the graph will shift left.

EXAMPLE 3

Adding a Constant to an Input

Returning to our building airflow example from Figure 3, suppose that in autumn the facilities manager decides that the original venting plan starts too late, and wants to begin the entire venting program 2 hours earlier. Sketch a graph of the new function.

Solution

We can set $V(t)$ to be the original program and $F(t)$ to be the revised program.

$$\begin{aligned} V(t) &= \text{the original venting plan} \\ F(t) &= \text{starting 2 hrs sooner} \end{aligned}$$

In the new graph, at each time, the airflow is the same as the original function V was 2 hours later. For example, in the original function V, the airflow starts to change at 8 a.m., whereas for the function F, the airflow starts to change at 6 a.m. The comparable function values are $V(8) = F(6)$. See Figure 6. Notice also that the vents first opened to $220\ \text{ft}^2$ at 10 a.m. under the original plan, while under the new plan the vents reach $220\ \text{ft}^2$ at 8 a.m., so $V(10) = F(8)$.

In both cases, we see that, because $F(t)$ starts 2 hours sooner, $h = -2$. That means that the same output values are reached when $F(t) = V(t - (-2)) = V(t + 2)$.

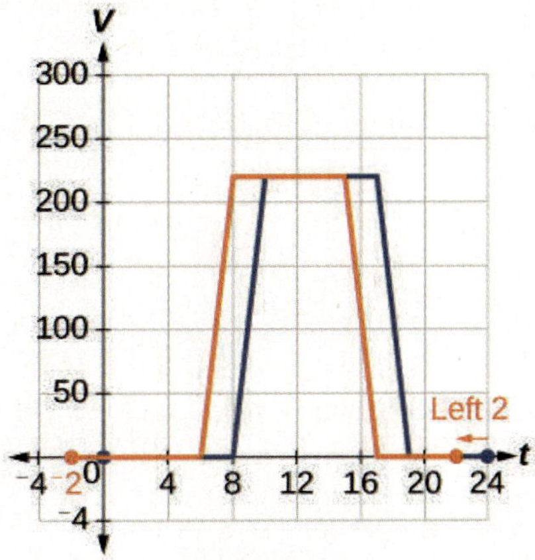

Figure 6

Analysis

Note that $V(t + 2)$ has the effect of shifting the graph to the *left*.

Horizontal changes or "inside changes" affect the domain of a function (the input) instead of the range and often seem counterintuitive. The new function $F(t)$ uses the same outputs as $V(t)$, but matches those outputs to inputs 2 hours earlier than those of $V(t)$. Said another way, we must add 2 hours to the input of V to find the corresponding output for $F: F(t) = V(t + 2)$.

HOW TO

Given a tabular function, create a new row to represent a horizontal shift.

1. Identify the input row or column.
2. Determine the magnitude of the shift.
3. Add the shift to the value in each input cell.

EXAMPLE 4

Shifting a Tabular Function Horizontally

A function $f(x)$ is given in Table 4. Create a table for the function $g(x) = f(x - 3)$.

x	2	4	6	8
$f(x)$	1	3	7	11

Table 4

Solution

The formula $g(x) = f(x - 3)$ tells us that the output values of g are the same as the output value of f when the input value is 3 less than the original value. For example, we know that $f(2) = 1$. To get the same output from the function g, we will need an input value that is 3 *larger*. We input a value that is 3 larger for $g(x)$ because the function takes 3 away before evaluating the function f.

$$\begin{aligned} g(5) &= f(5-3) \\ &= f(2) \\ &= 1 \end{aligned}$$

We continue with the other values to create Table 5.

x	5	7	9	11
$x - 3$	2	4	6	8
$f(x-3)$	1	3	7	11
$g(x)$	1	3	7	11

Table 5

The result is that the function $g(x)$ has been shifted to the right by 3. Notice the output values for $g(x)$ remain the same as the output values for $f(x)$, but the corresponding input values, x, have shifted to the right by 3. Specifically, 2 shifted to 5, 4 shifted to 7, 6 shifted to 9, and 8 shifted to 11.

Analysis

Figure 7 represents both of the functions. We can see the horizontal shift in each point.

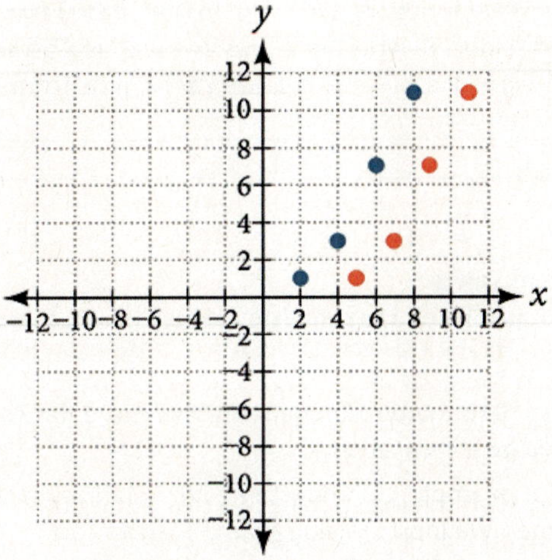

Figure 7

EXAMPLE 5

Identifying a Horizontal Shift of a Toolkit Function

Figure 8 represents a transformation of the toolkit function $f(x) = x^2$. Relate this new function $g(x)$ to $f(x)$, and then find a formula for $g(x)$.

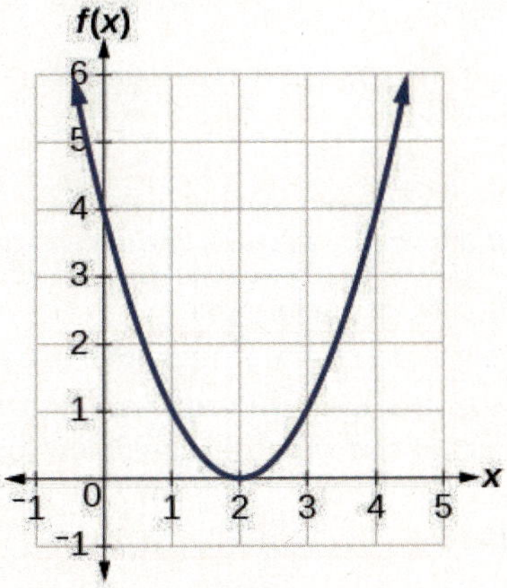

Figure 8

Solution

Notice that the graph is identical in shape to the $f(x) = x^2$ function, but the *x*-values are shifted to the right 2 units. The vertex used to be at (0,0), but now the vertex is at (2,0). The graph is the basic quadratic function shifted 2 units to the right, so

$$g(x) = f(x-2)$$

Notice how we must input the value $x = 2$ to get the output value $y = 0$; the *x*-values must be 2 units larger because of the shift to the right by 2 units. We can then use the definition of the $f(x)$ function to write a formula for $g(x)$ by evaluating $f(x-2)$.

$$\begin{aligned} f(x) &= x^2 \\ g(x) &= f(x-2) \\ g(x) &= f(x-2) = (x-2)^2 \end{aligned}$$

Analysis

To determine whether the shift is $+2$ or -2 , consider a single reference point on the graph. For a quadratic, looking at the vertex point is convenient. In the original function, $f(0) = 0$. In our shifted function, $g(2) = 0$. To obtain the output value of 0 from the function f, we need to decide whether a plus or a minus sign will work to satisfy $g(2) = f(x - 2) = f(0) = 0$. For this to work, we will need to *subtract* 2 units from our input values.

EXAMPLE 6

Interpreting Horizontal versus Vertical Shifts

The function $G(m)$ gives the number of gallons of gas required to drive m miles. Interpret $G(m) + 10$ and $G(m + 10)$.

Solution

$G(m) + 10$ can be interpreted as adding 10 to the output, gallons. This is the gas required to drive m miles, plus another 10 gallons of gas. The graph would indicate a vertical shift.

$G(m + 10)$ can be interpreted as adding 10 to the input, miles. So this is the number of gallons of gas required to drive 10 miles more than m miles. The graph would indicate a horizontal shift.

TRY IT #2 Given the function $f(x) = \sqrt{x}$, graph the original function $f(x)$ and the transformation $g(x) = f(x + 2)$ on the same axes. Is this a horizontal or a vertical shift? Which way is the graph shifted and by how many units?

Combining Vertical and Horizontal Shifts

Now that we have two transformations, we can combine them. Vertical shifts are outside changes that affect the output (y-) values and shift the function up or down. Horizontal shifts are inside changes that affect the input (x-) values and shift the function left or right. Combining the two types of shifts will cause the graph of a function to shift up or down *and* left or right.

HOW TO

Given a function and both a vertical and a horizontal shift, sketch the graph.

1. Identify the vertical and horizontal shifts from the formula.
2. The vertical shift results from a constant added to the output. Move the graph up for a positive constant and down for a negative constant.
3. The horizontal shift results from a constant added to the input. Move the graph left for a positive constant and right for a negative constant.
4. Apply the shifts to the graph in either order.

EXAMPLE 7

Graphing Combined Vertical and Horizontal Shifts

Given $f(x) = |x|$, sketch a graph of $h(x) = f(x + 1) - 3$.

Solution

The function f is our toolkit absolute value function. We know that this graph has a V shape, with the point at the origin. The graph of h has transformed f in two ways: $f(x + 1)$ is a change on the inside of the function, giving a horizontal shift left by 1, and the subtraction by 3 in $f(x + 1) - 3$ is a change to the outside of the function, giving a vertical shift down by 3. The transformation of the graph is illustrated in Figure 9.

Let us follow one point of the graph of $f(x) = |x|$.

- The point $(0, 0)$ is transformed first by shifting left 1 unit: $(0, 0) \to (-1, 0)$
- The point $(-1, 0)$ is transformed next by shifting down 3 units: $(-1, 0) \to (-1, -3)$

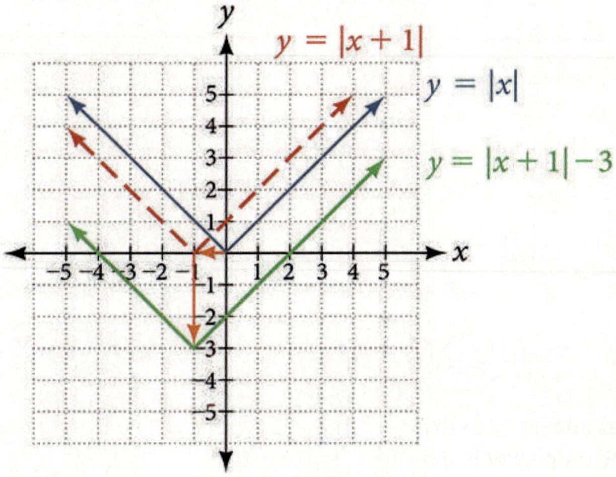

Figure 9

Figure 10 shows the graph of h.

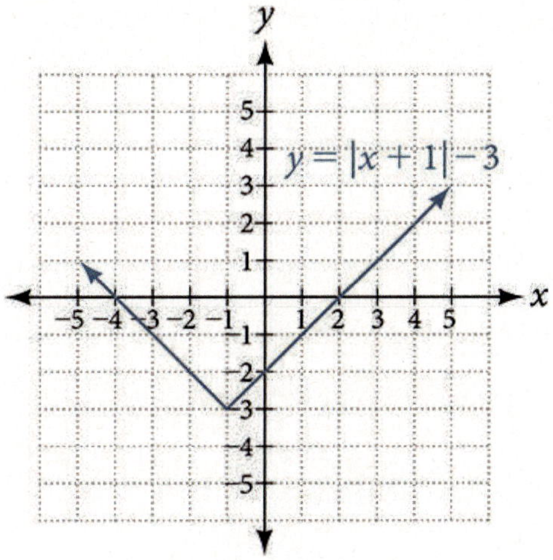

Figure 10

TRY IT #3 Given $f(x) = |x|$, sketch a graph of $h(x) = f(x - 2) + 4$.

EXAMPLE 8

Identifying Combined Vertical and Horizontal Shifts

Write a formula for the graph shown in Figure 11, which is a transformation of the toolkit square root function.

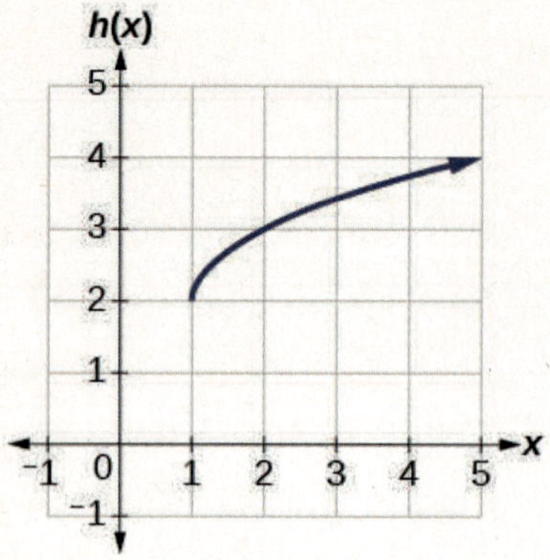

Figure 11

Solution

The graph of the toolkit function starts at the origin, so this graph has been shifted 1 to the right and up 2. In function notation, we could write that as

$$h(x) = f(x - 1) + 2$$

Using the formula for the square root function, we can write

$$h(x) = \sqrt{x - 1} + 2$$

Analysis

Note that this transformation has changed the domain and range of the function. This new graph has domain $[1, \infty)$ and range $[2, \infty)$.

TRY IT #4 Write a formula for a transformation of the toolkit reciprocal function $f(x) = \frac{1}{x}$ that shifts the function's graph one unit to the right and one unit up.

Graphing Functions Using Reflections about the Axes

Another transformation that can be applied to a function is a reflection over the *x*- or *y*-axis. A **vertical reflection** reflects a graph vertically across the *x*-axis, while a **horizontal reflection** reflects a graph horizontally across the *y*-axis. The reflections are shown in Figure 12.

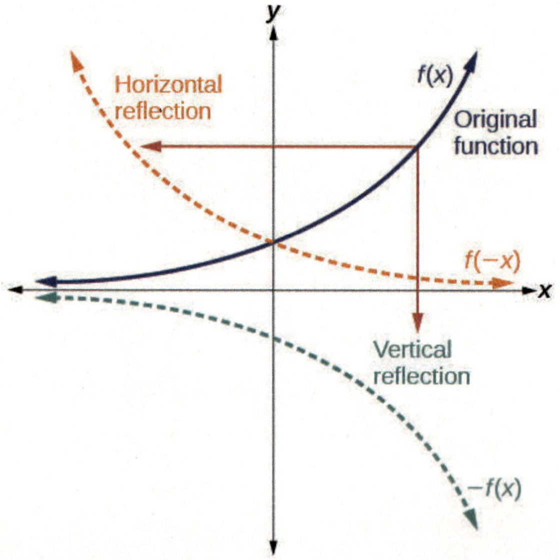

Figure 12 Vertical and horizontal reflections of a function.

Notice that the vertical reflection produces a new graph that is a mirror image of the base or original graph about the

x-axis. The horizontal reflection produces a new graph that is a mirror image of the base or original graph about the y-axis.

Reflections

Given a function $f(x\ )$, a new function $g(x) = -f(x)$ is a **vertical reflection** of the function $f(x)$, sometimes called a reflection about (or over, or through) the x-axis.

Given a function $f(x)$, a new function $g(x) = f(-x)$ is a **horizontal reflection** of the function $f(x)$, sometimes called a reflection about the y-axis.

HOW TO

Given a function, reflect the graph both vertically and horizontally.

1. Multiply all outputs by –1 for a vertical reflection. The new graph is a reflection of the original graph about the x-axis.
2. Multiply all inputs by –1 for a horizontal reflection. The new graph is a reflection of the original graph about the y-axis.

EXAMPLE 9

Reflecting a Graph Horizontally and Vertically
Reflect the graph of $s(t) = \sqrt{t}$ (a) vertically and (b) horizontally.

Solution

(a)

Reflecting the graph vertically means that each output value will be reflected over the horizontal t-axis as shown in Figure 13.

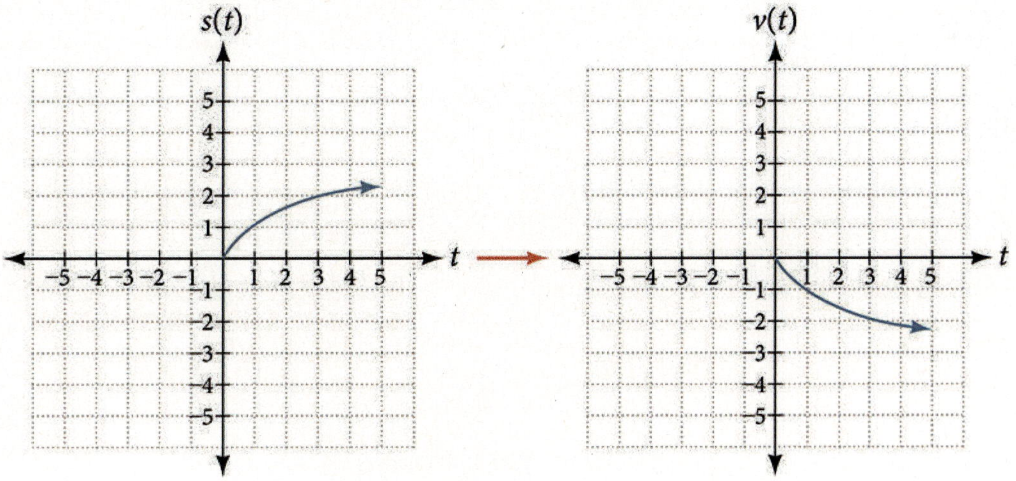

Figure 13 Vertical reflection of the square root function

Because each output value is the opposite of the original output value, we can write

$$V(t) = -s(t) \text{ or } V(t) = -\sqrt{t}$$

Notice that this is an outside change, or vertical shift, that affects the output $s(t)$ values, so the negative sign belongs outside of the function.

ⓑ
Reflecting horizontally means that each input value will be reflected over the vertical axis as shown in Figure 14.

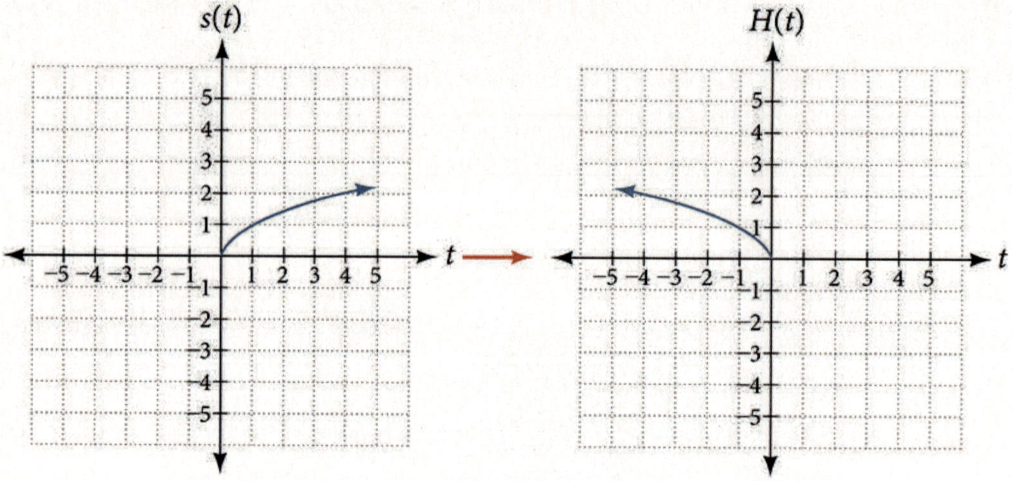

Figure 14 Horizontal reflection of the square root function

Because each input value is the opposite of the original input value, we can write

$$H(t) = s(-t) \text{ or } H(t) = \sqrt{-t}$$

Notice that this is an inside change or horizontal change that affects the input values, so the negative sign is on the inside of the function.

Note that these transformations can affect the domain and range of the functions. While the original square root function has domain $[0, \infty)$ and range $[0, \infty)$, the vertical reflection gives the $V(t)$ function the range $\left(-\infty, \ 0\right]$ and the horizontal reflection gives the $H(t)$ function the domain $\left(-\infty, \ 0\right]$.

TRY IT #5 Reflect the graph of $f(x) = |x - 1|$ (a) vertically and (b) horizontally.

EXAMPLE 10

Reflecting a Tabular Function Horizontally and Vertically
A function $f(x)$ is given as Table 6. Create a table for the functions below.

ⓐ $g(x) = -f(x)$ ⓑ $h(x) = f(-x)$

x	2	4	6	8
$f(x)$	1	3	7	11

Table 6

Solution

ⓐ

For $g(x)$, the negative sign outside the function indicates a vertical reflection, so the x-values stay the same and each output value will be the opposite of the original output value. See Table 7.

x	2	4	6	8
$g(x)$	−1	−3	−7	−11

Table 7

ⓑ

For $h(x)$, the negative sign inside the function indicates a horizontal reflection, so each input value will be the opposite of the original input value and the $h(x)$ values stay the same as the $f(x)$ values. See Table 8.

x	−2	−4	−6	−8
$h(x)$	1	3	7	11

Table 8

TRY IT #6 A function $f(x)$ is given as Table 9. Create a table for the functions below.

ⓐ $g(x) = -f(x)$ ⓑ $h(x) = f(-x)$

x	−2	0	2	4
$f(x)$	5	10	15	20

Table 9

EXAMPLE 11

Applying a Learning Model Equation

A common model for learning has an equation similar to $k(t) = -2^{-t} + 1$, where k is the percentage of mastery that can be achieved after t practice sessions. This is a transformation of the function $f(t) = 2^t$ shown in Figure 15. Sketch a graph of $k(t)$.

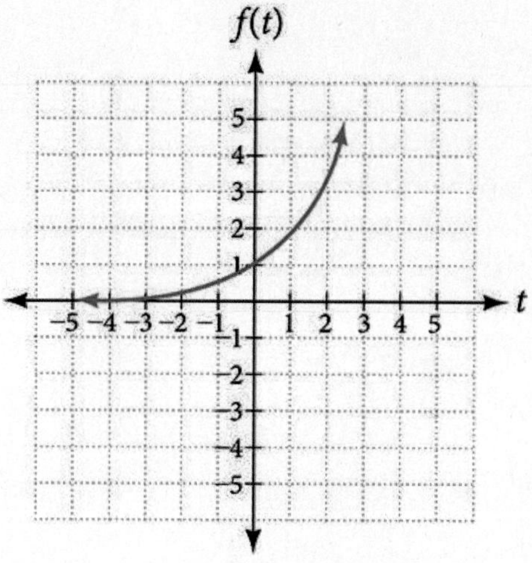

Figure 15

Solution

This equation combines three transformations into one equation.

- A horizontal reflection: $f(-t) = 2^{-t}$
- A vertical reflection: $-f(-t) = -2^{-t}$
- A vertical shift: $-f(-t) + 1 = -2^{-t} + 1$

We can sketch a graph by applying these transformations one at a time to the original function. Let us follow two points through each of the three transformations. We will choose the points (0, 1) and (1, 2).

1. First, we apply a horizontal reflection: (0, 1) (-1, 2).
2. Then, we apply a vertical reflection: (0, -1) (-1, -2)
3. Finally, we apply a vertical shift: (0, 0) (-1, -1)).

This means that the original points, (0,1) and (1,2) become (0,0) and (-1,-1) after we apply the transformations.

In Figure 16, the first graph results from a horizontal reflection. The second results from a vertical reflection. The third results from a vertical shift up 1 unit.

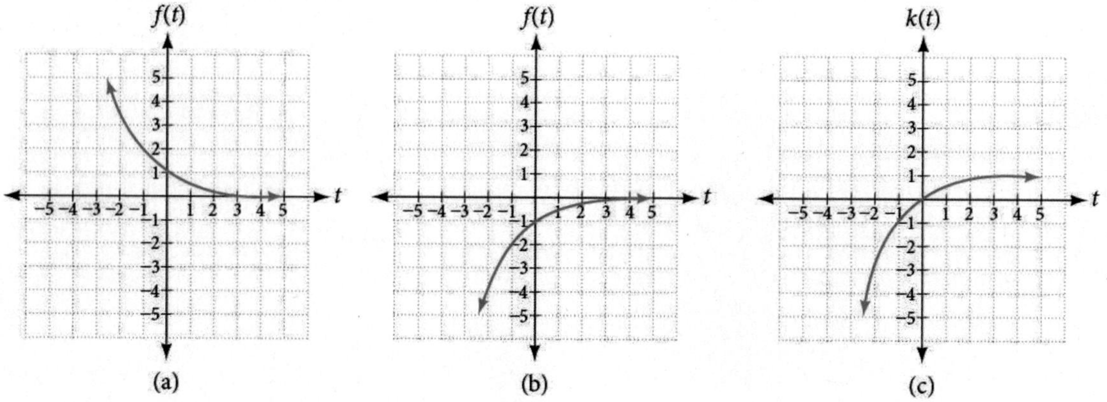

Figure 16

Analysis

As a model for learning, this function would be limited to a domain of $t \geq 0$, with corresponding range $[0, 1)$.

TRY IT #7 Given the toolkit function $f(x) = x^2$, graph $g(x) = -f(x)$ and $h(x) = f(-x)$. Take note of any surprising behavior for these functions.

Determining Even and Odd Functions

Some functions exhibit symmetry so that reflections result in the original graph. For example, horizontally reflecting the toolkit functions $f(x) = x^2$ or $f(x) = |x|$ will result in the original graph. We say that these types of graphs are symmetric about the *y*-axis. A function whose graph is symmetric about the *y*-axis is called an **even function.**

If the graphs of $f(x) = x^3$ or $f(x) = \frac{1}{x}$ were reflected over *both* axes, the result would be the original graph, as shown in Figure 17.

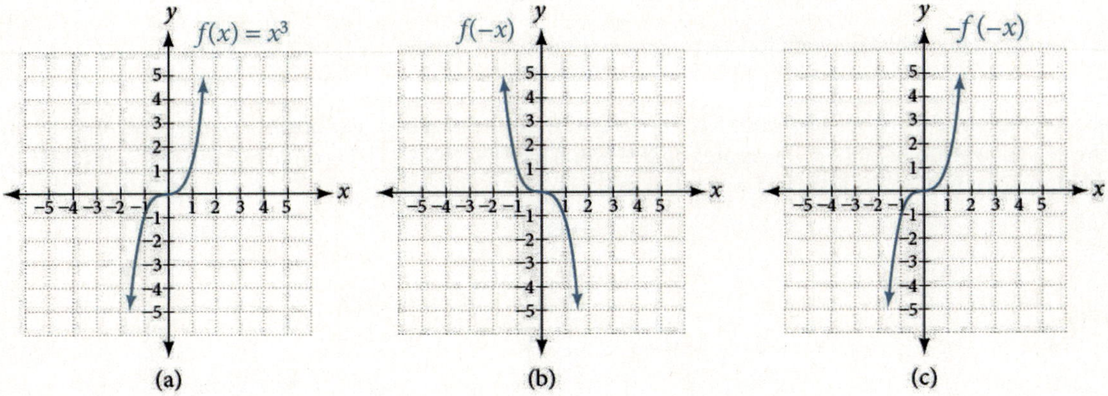

Figure 17 (a) The cubic toolkit function (b) Horizontal reflection of the cubic toolkit function (c) Horizontal and vertical reflections reproduce the original cubic function.

We say that these graphs are symmetric about the origin. A function with a graph that is symmetric about the origin is called an **odd function**.

Note: A function can be neither even nor odd if it does not exhibit either symmetry. For example, $f(x) = 2^x$ is neither even nor odd. Also, the only function that is both even and odd is the constant function $f(x) = 0$.

Even and Odd Functions

A function is called an **even function** if for every input x

$$f(x) = f(-x)$$

The graph of an even function is symmetric about the y- axis.

A function is called an **odd function** if for every input x

$$f(x) = -f(-x)$$

The graph of an odd function is symmetric about the origin.

HOW TO

Given the formula for a function, determine if the function is even, odd, or neither.

1. Determine whether the function satisfies $f(x) = f(-x)$. If it does, it is even.
2. Determine whether the function satisfies $f(x) = -f(-x)$. If it does, it is odd.
3. If the function does not satisfy either rule, it is neither even nor odd.

EXAMPLE 12

Determining whether a Function Is Even, Odd, or Neither

Is the function $f(x) = x^3 + 2x$ even, odd, or neither?

Solution

Without looking at a graph, we can determine whether the function is even or odd by finding formulas for the reflections and determining if they return us to the original function. Let's begin with the rule for even functions.

$$f(-x) = (-x)^3 + 2(-x) = -x^3 - 2x$$

This does not return us to the original function, so this function is not even. We can now test the rule for odd functions.

$$-f(-x) = -\left(-x^3 - 2x\right) = x^3 + 2x$$

Because $-f(-x) = f(x)$, this is an odd function.

Analysis

Consider the graph of f in Figure 18. Notice that the graph is symmetric about the origin. For every point (x, y) on the graph, the corresponding point $(-x, -y)$ is also on the graph. For example, (1, 3) is on the graph of f, and the corresponding point $(-1, -3)$ is also on the graph.

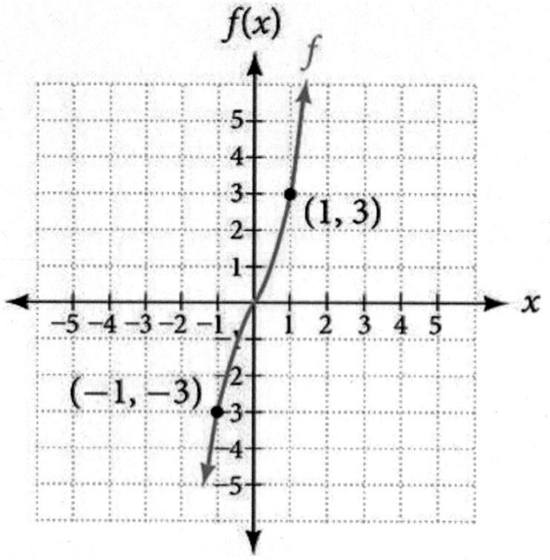

Figure 18

> TRY IT #8 Is the function $f(s) = s^4 + 3s^2 + 7$ even, odd, or neither?

Graphing Functions Using Stretches and Compressions

Adding a constant to the inputs or outputs of a function changed the position of a graph with respect to the axes, but it did not affect the shape of a graph. We now explore the effects of multiplying the inputs or outputs by some quantity.

We can transform the inside (input values) of a function or we can transform the outside (output values) of a function. Each change has a specific effect that can be seen graphically.

Vertical Stretches and Compressions

When we multiply a function by a positive constant, we get a function whose graph is stretched or compressed vertically in relation to the graph of the original function. If the constant is greater than 1, we get a **vertical stretch**; if the constant is between 0 and 1, we get a **vertical compression**. Figure 19 shows a function multiplied by constant factors 2 and 0.5 and the resulting vertical stretch and compression.

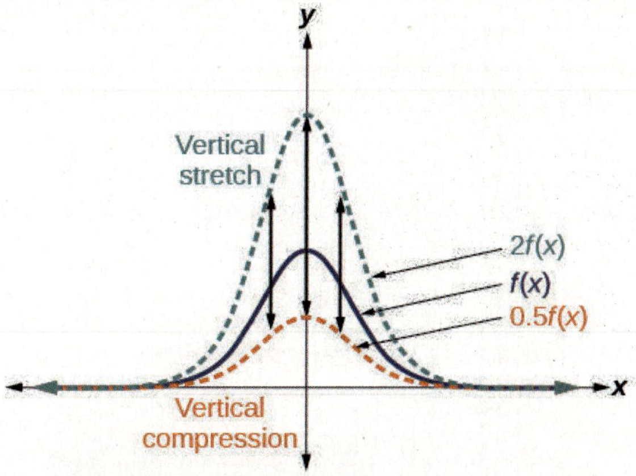

Figure 19 Vertical stretch and compression

Vertical Stretches and Compressions

Given a function $f(x)$, a new function $g(x) = af(x)$, where a is a constant, is a **vertical stretch** or **vertical compression** of the function $f(x)$.

- If $a > 1$, then the graph will be stretched.
- If $0 < a < 1$, then the graph will be compressed.
- If $a < 0$, then there will be combination of a vertical stretch or compression with a vertical reflection.

HOW TO

Given a function, graph its vertical stretch.

1. Identify the value of a.
2. Multiply all range values by a.
3. If $a > 1$, the graph is stretched by a factor of a.

   If $0 < a < 1$, the graph is compressed by a factor of a.

   If $a < 0$, the graph is either stretched or compressed and also reflected about the *x*-axis.

EXAMPLE 13

Graphing a Vertical Stretch

A function $P(t)$ models the population of fruit flies. The graph is shown in Figure 20.

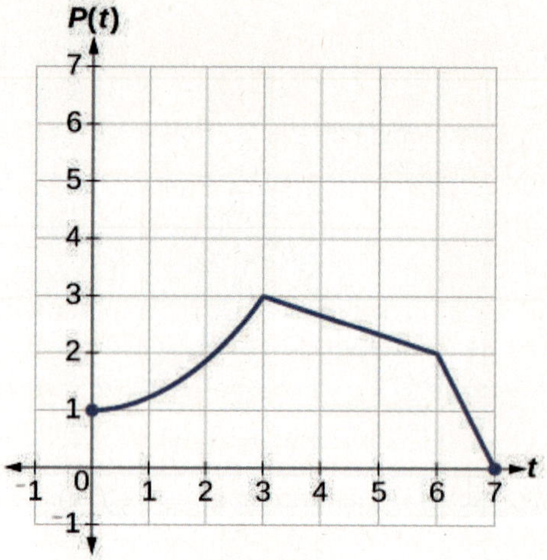

Figure 20

A scientist is comparing this population to another population, Q, whose growth follows the same pattern, but is twice as large. Sketch a graph of this population.

Solution

Because the population is always twice as large, the new population's output values are always twice the original function's output values. Graphically, this is shown in Figure 21.

If we choose four reference points, (0, 1), (3, 3), (6, 2) and (7, 0) we will multiply all of the outputs by 2.

The following shows where the new points for the new graph will be located.

$$\begin{gathered}(0,\ 1) \rightarrow (0,\ 2)\\(3,\ 3) \rightarrow (3,\ 6)\\(6,\ 2) \rightarrow (6,\ 4)\\(7,\ 0) \rightarrow (7,\ 0)\end{gathered}$$

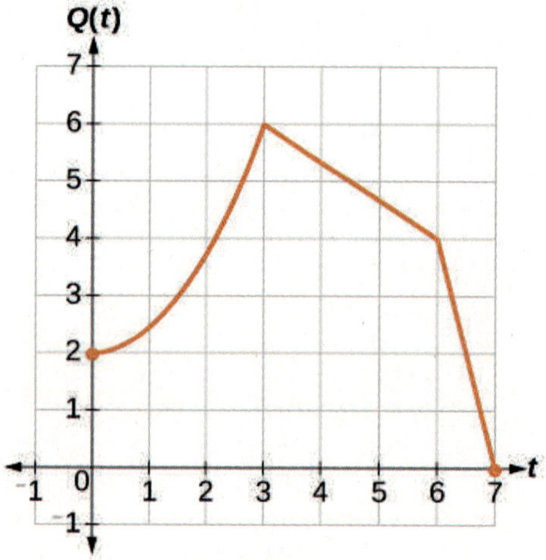

Figure 21

Symbolically, the relationship is written as

$$Q(t) = 2P(t)$$

This means that for any input t, the value of the function Q is twice the value of the function P. Notice that the effect on the graph is a vertical stretching of the graph, where every point doubles its distance from the horizontal axis. The input values, t, stay the same while the output values are twice as large as before.

HOW TO

Given a tabular function and assuming that the transformation is a vertical stretch or compression, create a table for a vertical compression.

1. Determine the value of a.
2. Multiply all of the output values by a.

EXAMPLE 14

Finding a Vertical Compression of a Tabular Function

A function f is given as Table 10. Create a table for the function $g(x) = \frac{1}{2}f(x)$.

x	2	4	6	8
$f(x)$	1	3	7	11

Table 10

Solution

The formula $g(x) = \frac{1}{2}f(x)$ tells us that the output values of g are half of the output values of f with the same inputs. For example, we know that $f(4) = 3$. Then

$$g(4) = \frac{1}{2}f(4) = \frac{1}{2}(3) = \frac{3}{2}$$

We do the same for the other values to produce Table 11.

x	2	4	6	8
$g(x)$	$\frac{1}{2}$	$\frac{3}{2}$	$\frac{7}{2}$	$\frac{11}{2}$

Table 11

Analysis

The result is that the function $g(x)$ has been compressed vertically by $\frac{1}{2}$. Each output value is divided in half, so the graph is half the original height.

TRY IT #9 A function f is given as Table 12. Create a table for the function $g(x) = \frac{3}{4}f(x)$.

x	2	4	6	8
$f(x)$	12	16	20	0

Table 12

EXAMPLE 15

Recognizing a Vertical Stretch

The graph in Figure 22 is a transformation of the toolkit function $f(x) = x^3$. Relate this new function $g(x)$ to $f(x)$, and then find a formula for $g(x)$.

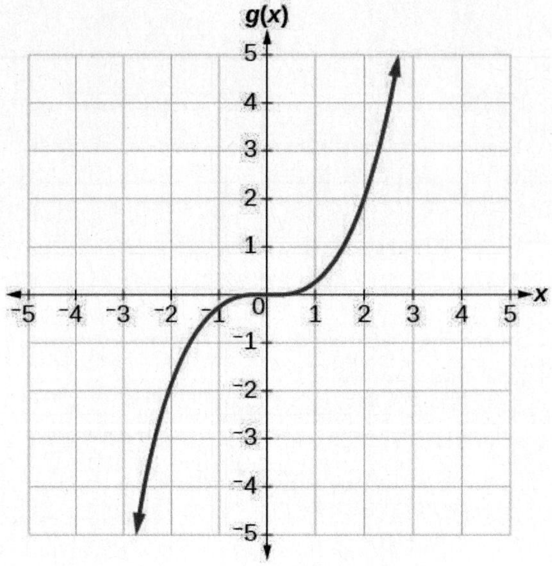

Figure 22

Solution

When trying to determine a vertical stretch or shift, it is helpful to look for a point on the graph that is relatively clear. In this graph, it appears that $g(2) = 2$. With the basic cubic function at the same input, $f(2) = 2^3 = 8$. Based on that, it appears that the outputs of g are $\frac{1}{4}$ the outputs of the function f because $g(2) = \frac{1}{4}f(2)$. From this we can fairly safely conclude that $g(x) = \frac{1}{4}f(x)$.

We can write a formula for g by using the definition of the function f.

$$g(x) = \frac{1}{4}f(x) = \frac{1}{4}x^3$$

TRY IT #10 Write the formula for the function that we get when we stretch the identity toolkit function by a factor of 3, and then shift it down by 2 units.

Horizontal Stretches and Compressions

Now we consider changes to the inside of a function. When we multiply a function's input by a positive constant, we get a function whose graph is stretched or compressed horizontally in relation to the graph of the original function. If the constant is between 0 and 1, we get a **horizontal stretch**; if the constant is greater than 1, we get a **horizontal compression** of the function.

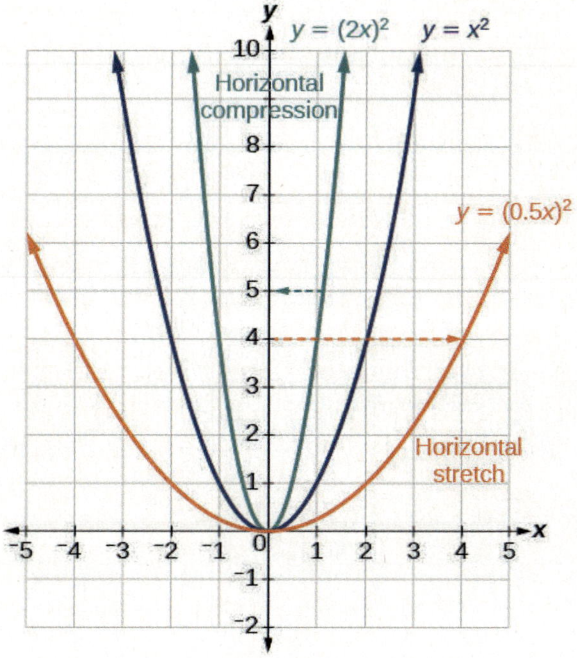

Figure 23

Given a function $y = f(x)$, the form $y = f(bx)$ results in a horizontal stretch or compression. Consider the function $y = x^2$. Observe Figure 23. The graph of $y = (0.5x)^2$ is a horizontal stretch of the graph of the function $y = x^2$ by a factor of 2. The graph of $y = (2x)^2$ is a horizontal compression of the graph of the function $y = x^2$ by a factor of $\frac{1}{2}$.

Horizontal Stretches and Compressions

Given a function $f(x)$, a new function $g(x) = f(bx)$, where b is a constant, is a **horizontal stretch** or **horizontal compression** of the function $f(x)$.

- If $b > 1$, then the graph will be compressed by $\frac{1}{b}$.
- If $0 < b < 1$, then the graph will be stretched by $\frac{1}{b}$.
- If $b < 0$, then there will be combination of a horizontal stretch or compression with a horizontal reflection.

HOW TO

Given a description of a function, sketch a horizontal compression or stretch.

1. Write a formula to represent the function.
2. Set $g(x) = f(bx)$ where $b > 1$ for a compression or $0 < b < 1$ for a stretch.

EXAMPLE 16

Graphing a Horizontal Compression

Suppose a scientist is comparing a population of fruit flies to a population that progresses through its lifespan twice as fast as the original population. In other words, this new population, R, will progress in 1 hour the same amount as the original population does in 2 hours, and in 2 hours, it will progress as much as the original population does in 4 hours. Sketch a graph of this population.

Solution

Symbolically, we could write

$$\begin{aligned} R(1) &= P(2), \\ R(2) &= P(4), \quad \text{and in general,} \\ R(t) &= P(2t). \end{aligned}$$

See Figure 24 for a graphical comparison of the original population and the compressed population.

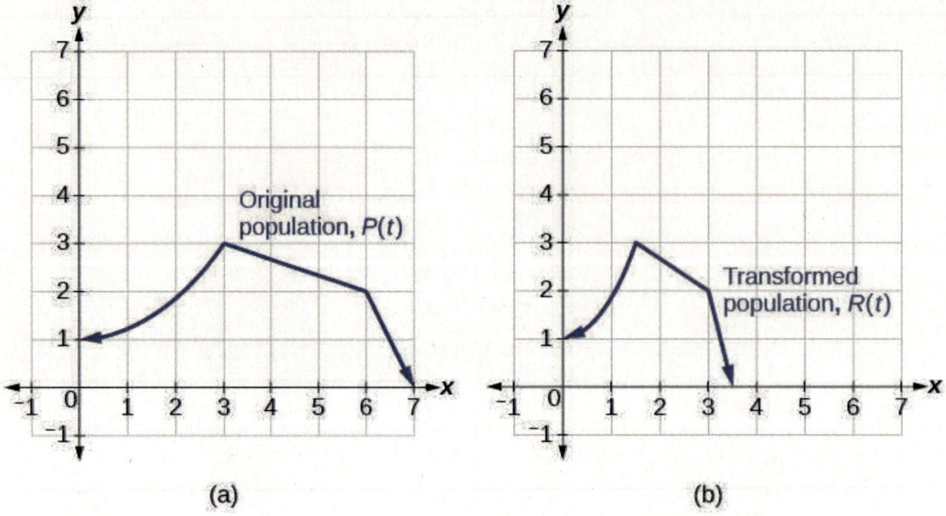

Figure 24 (a) Original population graph (b) Compressed population graph

Analysis

Note that the effect on the graph is a horizontal compression where all input values are half of their original distance from the vertical axis.

EXAMPLE 17

Finding a Horizontal Stretch for a Tabular Function

A function $f(x)$ is given as Table 13. Create a table for the function $g(x) = f\left(\frac{1}{2}x\right)$.

x	2	4	6	8
$f(x)$	1	3	7	11

Table 13

Solution

The formula $g(x) = f\left(\frac{1}{2}x\right)$ tells us that the output values for g are the same as the output values for the function f at an input half the size. Notice that we do not have enough information to determine $g(2)$ because $g(2) = f\left(\frac{1}{2} \cdot 2\right) = f(1)$, and we do not have a value for $f(1)$ in our table. Our input values to g will need to be twice as large to get inputs for f that we can evaluate. For example, we can determine $g(4)$.

$$g(4) = f\left(\frac{1}{2} \cdot 4\right) = f(2) = 1$$

We do the same for the other values to produce Table 14.

x	4	8	12	16

Table 14

$g(x)$	1	3	7	11

Table 14

Figure 25 shows the graphs of both of these sets of points.

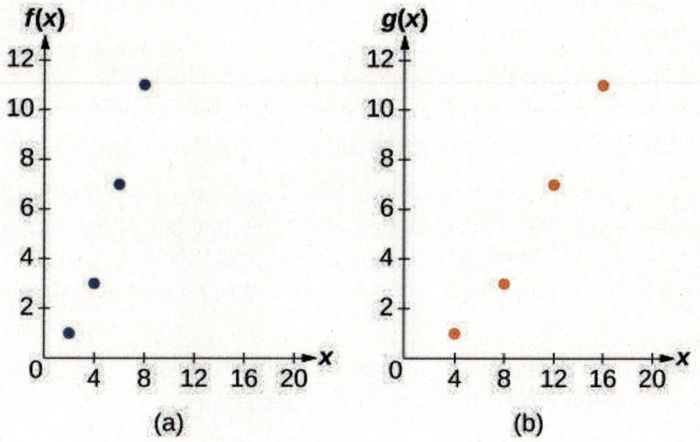

Figure 25

Analysis

Because each input value has been doubled, the result is that the function $g(x)$ has been stretched horizontally by a factor of 2.

EXAMPLE 18

Recognizing a Horizontal Compression on a Graph

Relate the function $g(x)$ to $f(x)$ in Figure 26.

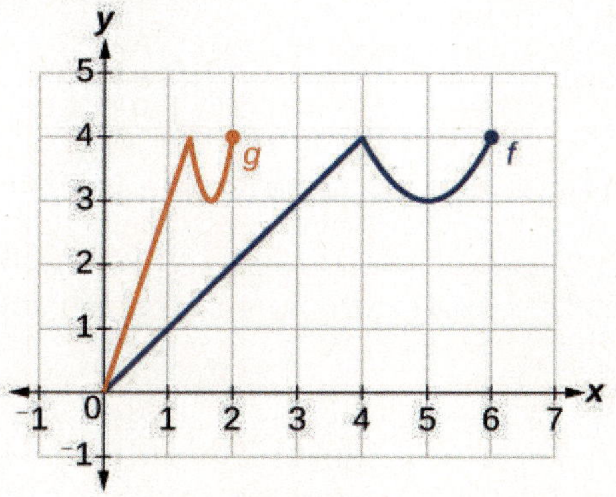

Figure 26

Solution

The graph of $g(x)$ looks like the graph of $f(x)$ horizontally compressed. Because $f(x)$ ends at $(6, 4)$ and $g(x)$ ends at $(2, 4)$, we can see that the x- values have been compressed by $\frac{1}{3}$, because $6\left(\frac{1}{3}\right) = 2$. We might also notice that $g(2) = f(6)$ and $g(1) = f(3)$. Either way, we can describe this relationship as $g(x) = f(3x)$. This is a horizontal compression by $\frac{1}{3}$.

Analysis

Notice that the coefficient needed for a horizontal stretch or compression is the reciprocal of the stretch or compression. So to stretch the graph horizontally by a scale factor of 4, we need a coefficient of $\frac{1}{4}$ in our function: $f\left(\frac{1}{4}x\right)$. This means that the input values must be four times larger to produce the same result, requiring the input to be larger, causing the horizontal stretching.

> TRY IT #11 Write a formula for the toolkit square root function horizontally stretched by a factor of 3.

Performing a Sequence of Transformations

When combining transformations, it is very important to consider the order of the transformations. For example, vertically shifting by 3 and then vertically stretching by 2 does not create the same graph as vertically stretching by 2 and then vertically shifting by 3, because when we shift first, both the original function and the shift get stretched, while only the original function gets stretched when we stretch first.

When we see an expression such as $2f(x) + 3$, which transformation should we start with? The answer here follows nicely from the order of operations. Given the output value of $f(x)$, we first multiply by 2, causing the vertical stretch, and then add 3, causing the vertical shift. In other words, multiplication before addition.

Horizontal transformations are a little trickier to think about. When we write $g(x) = f(2x + 3)$, for example, we have to think about how the inputs to the function g relate to the inputs to the function f. Suppose we know $f(7) = 12$. What input to g would produce that output? In other words, what value of x will allow $g(x) = f(2x + 3) = 12$? We would need $2x + 3 = 7$. To solve for x, we would first subtract 3, resulting in a horizontal shift, and then divide by 2, causing a horizontal compression.

This format ends up being very difficult to work with, because it is usually much easier to horizontally stretch a graph before shifting. We can work around this by factoring inside the function.

$$f(bx + p) = f\left(b\left(x + \frac{p}{b}\right)\right)$$

Let's work through an example.

$$f(x) = (2x + 4)^2$$

We can factor out a 2.

$$f(x) = (2(x + 2))^2$$

Now we can more clearly observe a horizontal shift to the left 2 units and a horizontal compression. Factoring in this way allows us to horizontally stretch first and then shift horizontally.

Combining Transformations

When combining vertical transformations written in the form $af(x) + k$, first vertically stretch by a and then vertically shift by k.

When combining horizontal transformations written in the form $f(bx - h)$, first horizontally shift by $\frac{h}{b}$ and then horizontally stretch by $\frac{1}{b}$.

When combining horizontal transformations written in the form $f(b(x - h))$, first horizontally stretch by $\frac{1}{b}$ and then horizontally shift by h.

Horizontal and vertical transformations are independent. It does not matter whether horizontal or vertical transformations are performed first.

EXAMPLE 19

Finding a Triple Transformation of a Tabular Function

Given Table 15 for the function $f(x)$, create a table of values for the function $g(x) = 2f(3x) + 1$.

x	6	12	18	24
$f(x)$	10	14	15	17

Table 15

Solution

There are three steps to this transformation, and we will work from the inside out. Starting with the horizontal transformations, $f(3x)$ is a horizontal compression by $\frac{1}{3}$, which means we multiply each x- value by $\frac{1}{3}$. See Table 16.

x	2	4	6	8
$f(3x)$	10	14	15	17

Table 16

Looking now to the vertical transformations, we start with the vertical stretch, which will multiply the output values by 2. We apply this to the previous transformation. See Table 17.

x	2	4	6	8
$2f(3x)$	20	28	30	34

Table 17

Finally, we can apply the vertical shift, which will add 1 to all the output values. See Table 18.

x	2	4	6	8
$g(x) = 2f(3x) + 1$	21	29	31	35

Table 18

EXAMPLE 20

Finding a Triple Transformation of a Graph

Use the graph of $f(x)$ in Figure 27 to sketch a graph of $k(x) = f\left(\frac{1}{2}x + 1\right) - 3$.

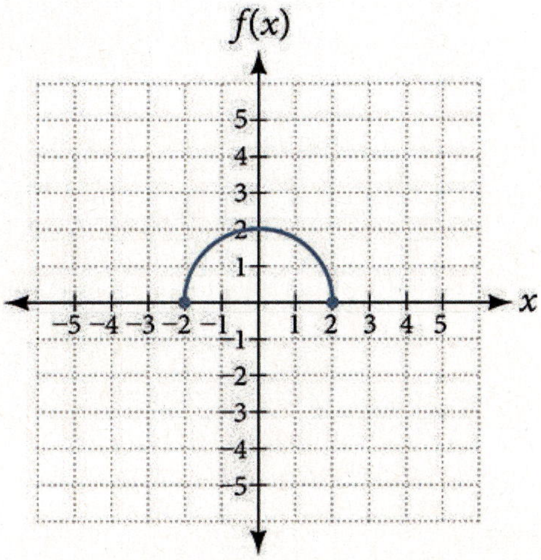

Figure 27

Solution

To simplify, let's start by factoring out the inside of the function.

$$f\left(\frac{1}{2}x + 1\right) - 3 = f\left(\frac{1}{2}(x + 2)\right) - 3$$

By factoring the inside, we can first horizontally stretch by 2, as indicated by the $\frac{1}{2}$ on the inside of the function. Remember that twice the size of 0 is still 0, so the point (0,2) remains at (0,2) while the point (2,0) will stretch to (4,0). See Figure 28.

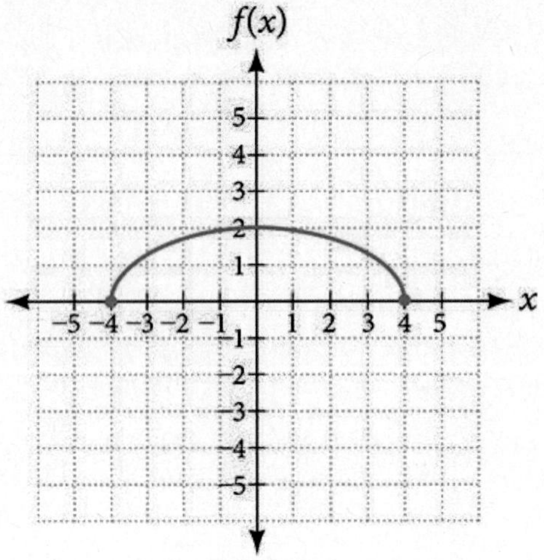

Figure 28

Next, we horizontally shift left by 2 units, as indicated by $x + 2$. See Figure 29.

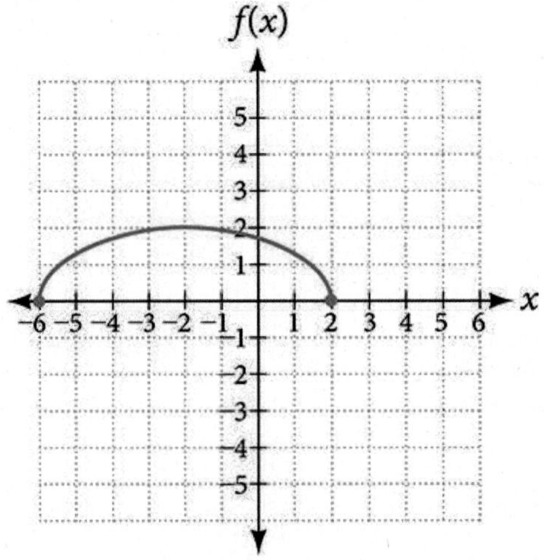

Figure 29

Last, we vertically shift down by 3 to complete our sketch, as indicated by the -3 on the outside of the function. See Figure 30.

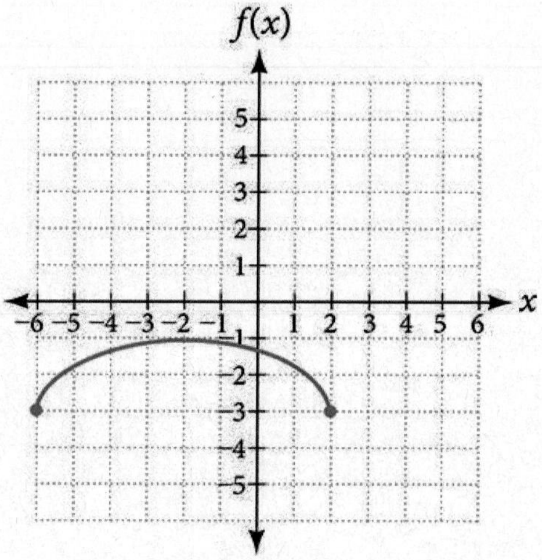

Figure 30

MEDIA

Access this online resource for additional instruction and practice with transformation of functions.

Function Transformations (http://openstax.org/l/functrans)

3.5 SECTION EXERCISES

Verbal

1. When examining the formula of a function that is the result of multiple transformations, how can you tell a horizontal shift from a vertical shift?

2. When examining the formula of a function that is the result of multiple transformations, how can you tell a horizontal stretch from a vertical stretch?

3. When examining the formula of a function that is the result of multiple transformations, how can you tell a horizontal compression from a vertical compression?

4. When examining the formula of a function that is the result of multiple transformations, how can you tell a reflection with respect to the *x*-axis from a reflection with respect to the *y*-axis?

5. How can you determine whether a function is odd or even from the formula of the function?

Algebraic

For the following exercises, write a formula for the function obtained when the graph is shifted as described.

6. $f(x) = \sqrt{x}$ is shifted up 1 unit and to the left 2 units.

7. $f(x) = |x|$ is shifted down 3 units and to the right 1 unit.

8. $f(x) = \frac{1}{x}$ is shifted down 4 units and to the right 3 units.

9. $f(x) = \frac{1}{x^2}$ is shifted up 2 units and to the left 4 units.

For the following exercises, describe how the graph of the function is a transformation of the graph of the original function f.

10. $y = f(x - 49)$

11. $y = f(x + 43)$

12. $y = f(x + 3)$

13. $y = f(x - 4)$

14. $y = f(x) + 5$

15. $y = f(x) + 8$

16. $y = f(x) - 2$

17. $y = f(x) - 7$

18. $y = f(x - 2) + 3$

19. $y = f(x + 4) - 1$

For the following exercises, determine the interval(s) on which the function is increasing and decreasing.

20. $f(x) = 4(x + 1)^2 - 5$

21. $g(x) = 5(x + 3)^2 - 2$

22. $a(x) = \sqrt{-x + 4}$

23. $k(x) = -3\sqrt{x} - 1$

Graphical

For the following exercises, use the graph of $f(x) = 2^x$ shown in Figure 31 to sketch a graph of each transformation of $f(x)$.

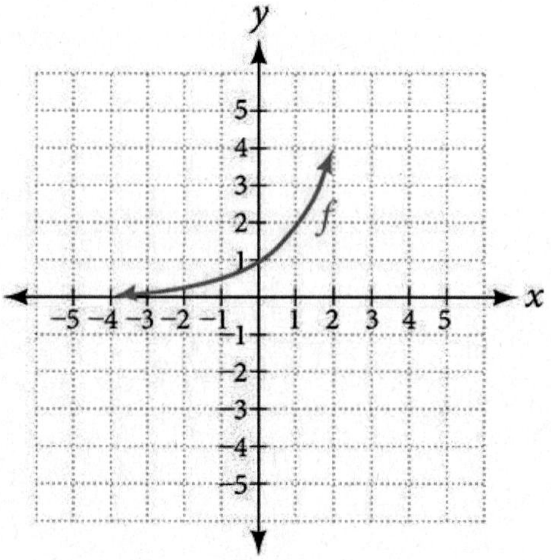

Figure 31

24. $g(x) = 2^x + 1$

25. $h(x) = 2^x - 3$

26. $w(x) = 2^{x-1}$

For the following exercises, sketch a graph of the function as a transformation of the graph of one of the toolkit functions.

27. $f(t) = (t + 1)^2 - 3$

28. $h(x) = |x - 1| + 4$

29. $k(x) = (x - 2)^3 - 1$

30. $m(t) = 3 + \sqrt{t + 2}$

Numeric

31. Tabular representations for the functions f, g, and h are given below. Write $g(x)$ and $h(x)$ as transformations of $f(x)$.

x	−2	−1	0	1	2
$f(x)$	−2	−1	−3	1	2

x	−1	0	1	2	3
$g(x)$	−2	−1	−3	1	2

x	−2	−1	0	1	2
$h(x)$	−1	0	−2	2	3

32. Tabular representations for the functions f, g, and h are given below. Write $g(x)$ and $h(x)$ as transformations of $f(x)$.

x	−2	−1	0	1	2
$f(x)$	−1	−3	4	2	1

x	−3	−2	−1	0	1
$g(x)$	−1	−3	4	2	1

x	−2	−1	0	1	2
$h(x)$	−2	−4	3	1	0

For the following exercises, write an equation for each graphed function by using transformations of the graphs of one of the toolkit functions.

33.

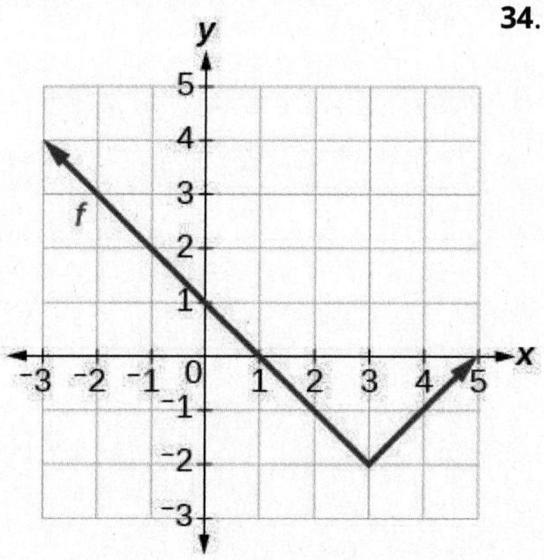

34.

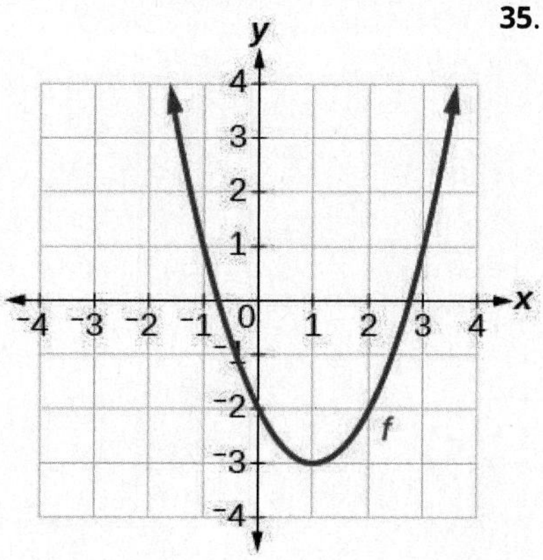

35.

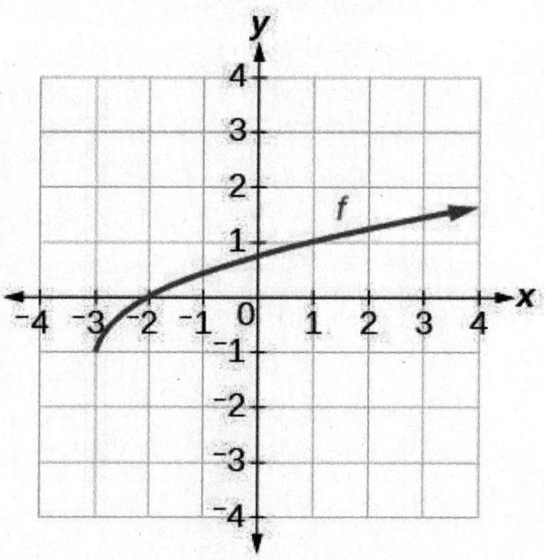

36.

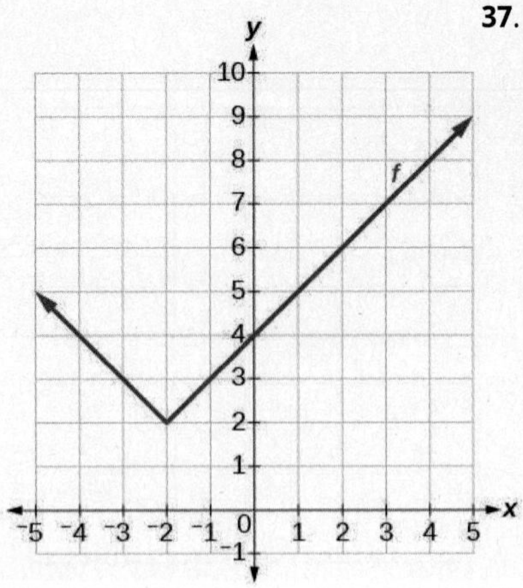

37.

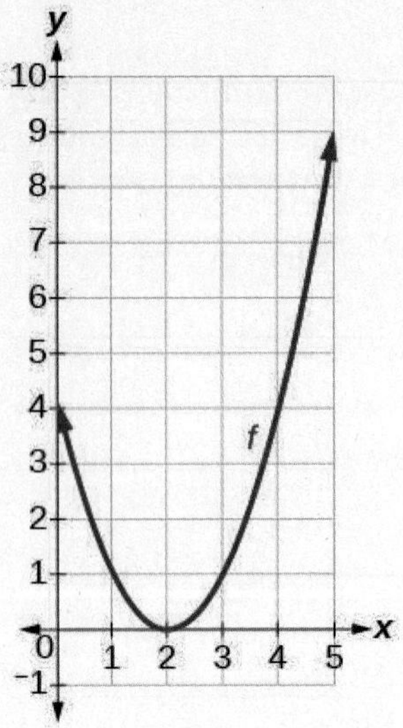

38.

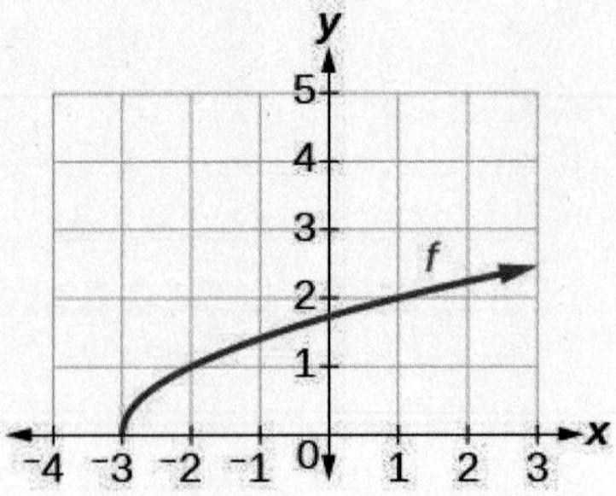

39.

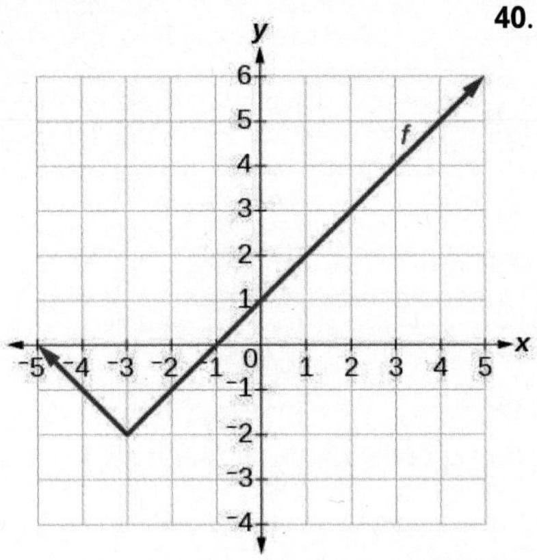

40.

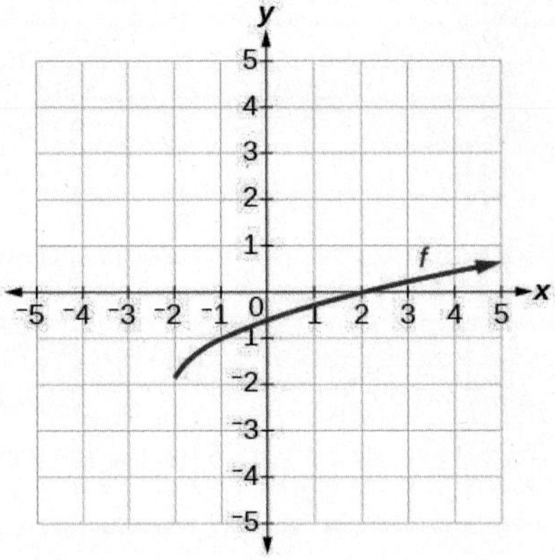

For the following exercises, use the graphs of transformations of the square root function to find a formula for each of the functions.

41.

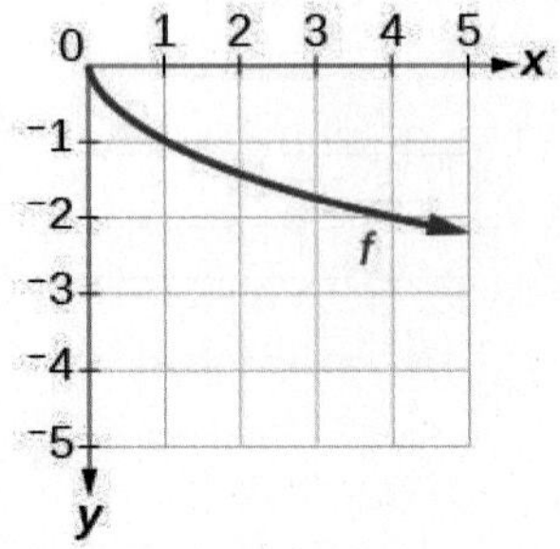

42.

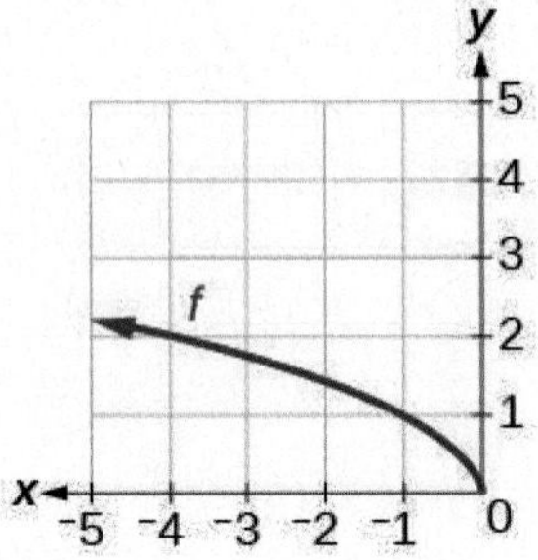

For the following exercises, use the graphs of the transformed toolkit functions to write a formula for each of the resulting functions.

43.

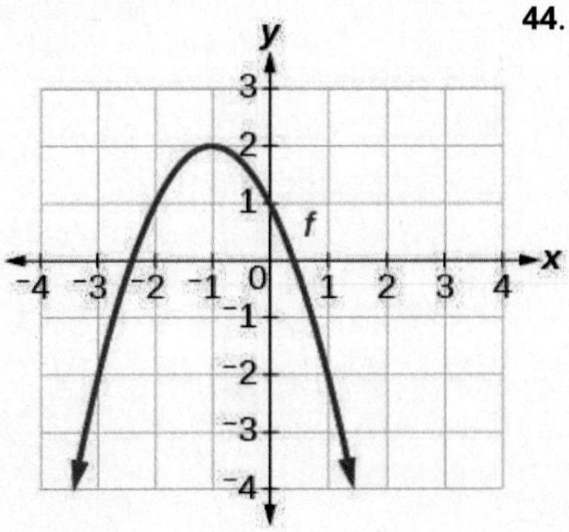

44.

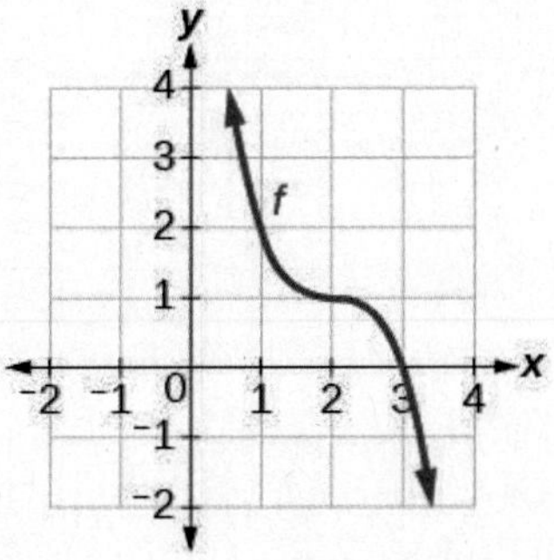

45.

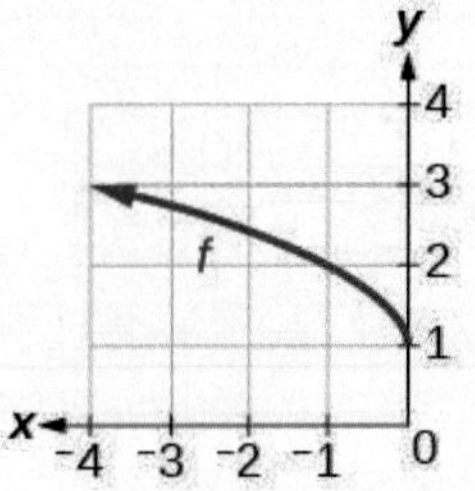

46.

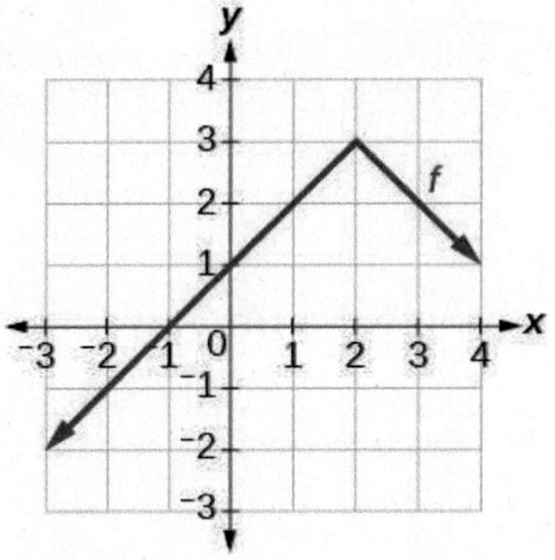

For the following exercises, determine whether the function is odd, even, or neither.

47. $f(x) = 3x^4$

48. $g(x) = \sqrt{x}$

49. $h(x) = \frac{1}{x} + 3x$

50. $f(x) = (x - 2)^2$

51. $g(x) = 2x^4$

52. $h(x) = 2x - x^3$

For the following exercises, describe how the graph of each function is a transformation of the graph of the original function f.

53. $g(x) = -f(x)$

54. $g(x) = f(-x)$

55. $g(x) = 4f(x)$

56. $g(x) = 6f(x)$

57. $g(x) = f(5x)$

58. $g(x) = f(2x)$

59. $g(x) = f\left(\frac{1}{3}x\right)$

60. $g(x) = f\left(\frac{1}{5}x\right)$

61. $g(x) = 3f(-x)$

62. $g(x) = -f(3x)$

For the following exercises, write a formula for the function g that results when the graph of a given toolkit function is transformed as described.

63. The graph of $f(x) = |x|$ is reflected over the y -axis and horizontally compressed by a factor of $\frac{1}{4}$.

64. The graph of $f(x) = \sqrt{x}$ is reflected over the x -axis and horizontally stretched by a factor of 2.

65. The graph of $f(x) = \frac{1}{x^2}$ is vertically compressed by a factor of $\frac{1}{3}$, then shifted to the left 2 units and down 3 units.

66. The graph of $f(x) = \frac{1}{x}$ is vertically stretched by a factor of 8, then shifted to the right 4 units and up 2 units.

67. The graph of $f(x) = x^2$ is vertically compressed by a factor of $\frac{1}{2}$, then shifted to the right 5 units and up 1 unit.

68. The graph of $f(x) = x^2$ is horizontally stretched by a factor of 3, then shifted to the left 4 units and down 3 units.

For the following exercises, describe how the formula is a transformation of a toolkit function. Then sketch a graph of the transformation.

69. $g(x) = 4(x + 1)^2 - 5$

70. $g(x) = 5(x + 3)^2 - 2$

71. $h(x) = -2|x - 4| + 3$

72. $k(x) = -3\sqrt{x} - 1$

73. $m(x) = \frac{1}{2}x^3$

74. $n(x) = \frac{1}{3}|x - 2|$

75. $p(x) = \left(\frac{1}{3}x\right)^3 - 3$

76. $q(x) = \left(\frac{1}{4}x\right)^3 + 1$

77. $a(x) = \sqrt{-x + 4}$

For the following exercises, use the graph in Figure 32 to sketch the given transformations.

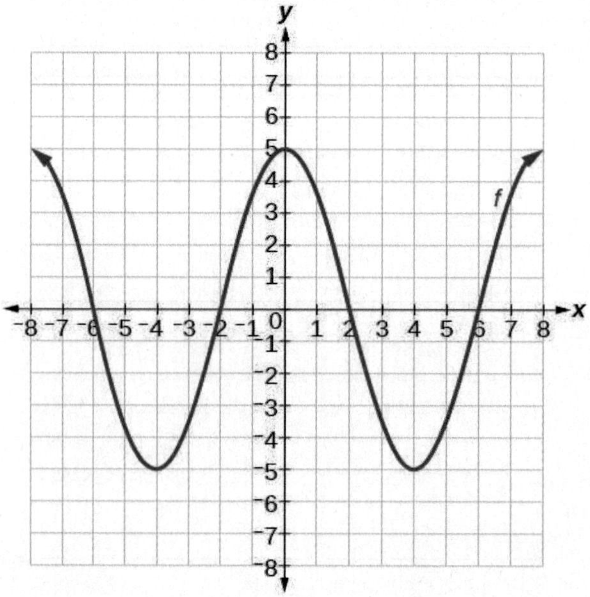

Figure 32

78. $g(x) = f(x) - 2$

79. $g(x) = -f(x)$

80. $g(x) = f(x + 1)$

81. $g(x) = f(x - 2)$

3.6 Absolute Value Functions

Learning Objectives

In this section, you will:

- Graph an absolute value function.
- Solve an absolute value equation.

Figure 1 Distances in deep space can be measured in all directions. As such, it is useful to consider distance in terms of absolute values. (credit: "s58y"/Flickr)

Until the 1920s, the so-called spiral nebulae were believed to be clouds of dust and gas in our own galaxy, some tens of thousands of light years away. Then, astronomer Edwin Hubble proved that these objects are galaxies in their own right, at distances of millions of light years. Today, astronomers can detect galaxies that are billions of light years away. Distances in the universe can be measured in all directions. As such, it is useful to consider distance as an absolute value function. In this section, we will continue our investigation of absolute value functions.

Understanding Absolute Value

Recall that in its basic form $f(x) = |x|$, the absolute value function is one of our toolkit functions. The absolute value function is commonly thought of as providing the distance the number is from zero on a number line. Algebraically, for whatever the input value is, the output is the value without regard to sign. Knowing this, we can use absolute value functions to solve some kinds of real-world problems.

Absolute Value Function

The absolute value function can be defined as a piecewise function

$$f(x) = |x| = \begin{cases} x & \text{if } x \geq 0 \\ -x & \text{if } x < 0 \end{cases}$$

EXAMPLE 1

Using Absolute Value to Determine Resistance

Electrical parts, such as resistors and capacitors, come with specified values of their operating parameters: resistance, capacitance, etc. However, due to imprecision in manufacturing, the actual values of these parameters vary somewhat from piece to piece, even when they are supposed to be the same. The best that manufacturers can do is to try to guarantee that the variations will stay within a specified range, often $\pm 1\%$, $\pm 5\%$, or $\pm 10\%$.

Suppose we have a resistor rated at 680 ohms, $\pm 5\%$. Use the absolute value function to express the range of possible values of the actual resistance.

Solution

We can find that 5% of 680 ohms is 34 ohms. The absolute value of the difference between the actual and nominal resistance should not exceed the stated variability, so, with the resistance R in ohms,

$$|R - 680| \leq 34$$

TRY IT #1 Students who score within 20 points of 80 will pass a test. Write this as a distance from 80 using absolute value notation.

Graphing an Absolute Value Function

The most significant feature of the absolute value graph is the corner point at which the graph changes direction. This point is shown at the origin in Figure 2.

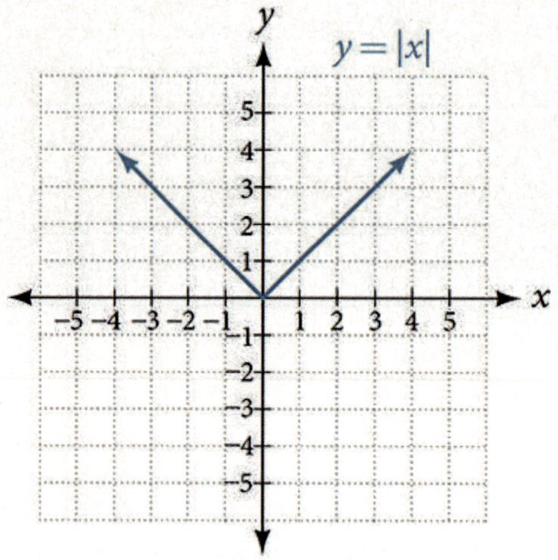

Figure 2

Figure 3 shows the graph of $y = 2|x-3| + 4$. The graph of $y = |x|$ has been shifted right 3 units, vertically stretched by a factor of 2, and shifted up 4 units. This means that the corner point is located at $(3, 4)$ for this transformed function.

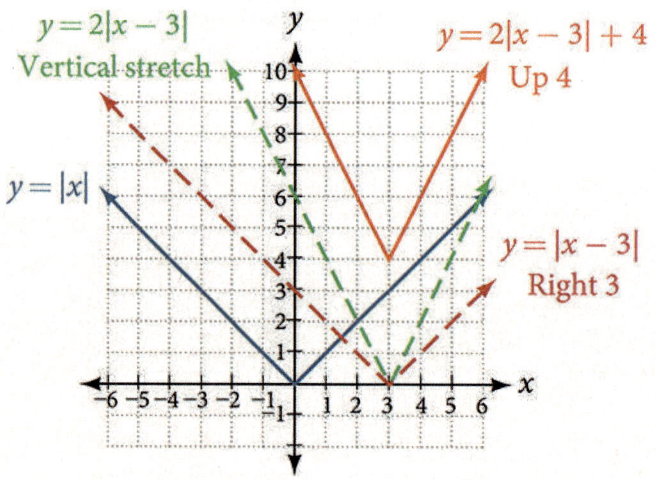

Figure 3

EXAMPLE 2

Writing an Equation for an Absolute Value Function Given a Graph

Write an equation for the function graphed in Figure 4.

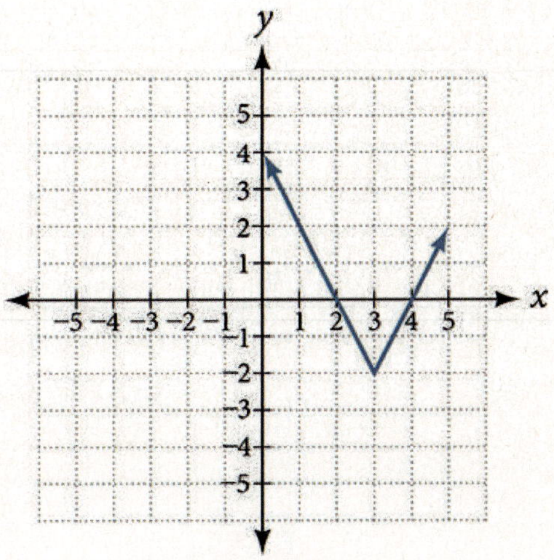

Figure 4

Solution

The basic absolute value function changes direction at the origin, so this graph has been shifted to the right 3 units and down 2 units from the basic toolkit function. See Figure 5.

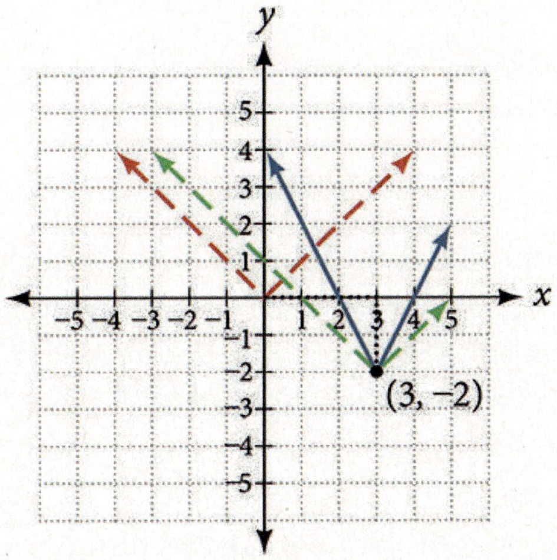

Figure 5

We also notice that the graph appears vertically stretched, because the width of the final graph on a horizontal line is not equal to 2 times the vertical distance from the corner to this line, as it would be for an unstretched absolute value function. Instead, the width is equal to 1 times the vertical distance as shown in Figure 6.

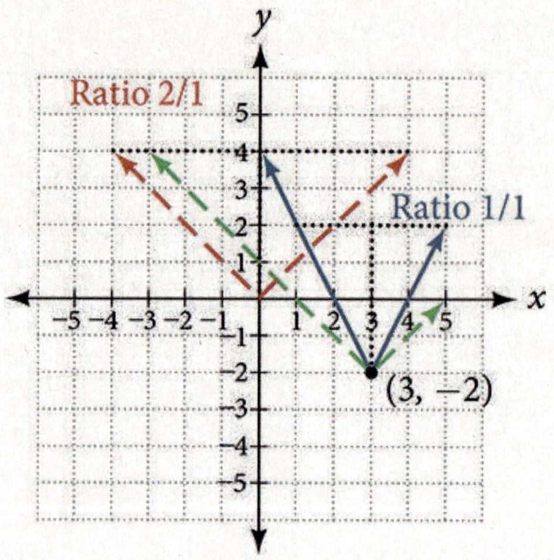

Figure 6

From this information we can write the equation

$$\begin{aligned} f(x) &= 2|x-3|-2, && \text{treating the stretch as } a \text{ vertical stretch,or} \\ f(x) &= |2(x-3)|-2, && \text{treating the stretch as } a \text{ horizontal compression.} \end{aligned}$$

Analysis

Note that these equations are algebraically equivalent—the stretch for an absolute value function can be written interchangeably as a vertical or horizontal stretch or compression. Note also that if the vertical stretch factor is negative, there is also a reflection about the x-axis.

Q&A **If we couldn't observe the stretch of the function from the graphs, could we algebraically determine it?**

Yes. If we are unable to determine the stretch based on the width of the graph, we can solve for the stretch factor by putting in a known pair of values for x and $f(x)$.

$$f(x) = a|x-3|-2$$

Now substituting in the point (1, 2)

$$\begin{aligned} 2 &= a|1-3|-2 \\ 4 &= 2a \\ a &= 2 \end{aligned}$$

TRY IT #2 Write the equation for the absolute value function that is horizontally shifted left 2 units, is vertically flipped, and vertically shifted up 3 units.

Q&A **Do the graphs of absolute value functions always intersect the vertical axis? The horizontal axis?**

Yes, they always intersect the vertical axis. The graph of an absolute value function will intersect the vertical axis when the input is zero.

No, they do not always intersect the horizontal axis. The graph may or may not intersect the horizontal axis, depending on how the graph has been shifted and reflected. It is possible for the absolute value function to intersect the horizontal axis at zero, one, or two points (see [Figure 7](#)).

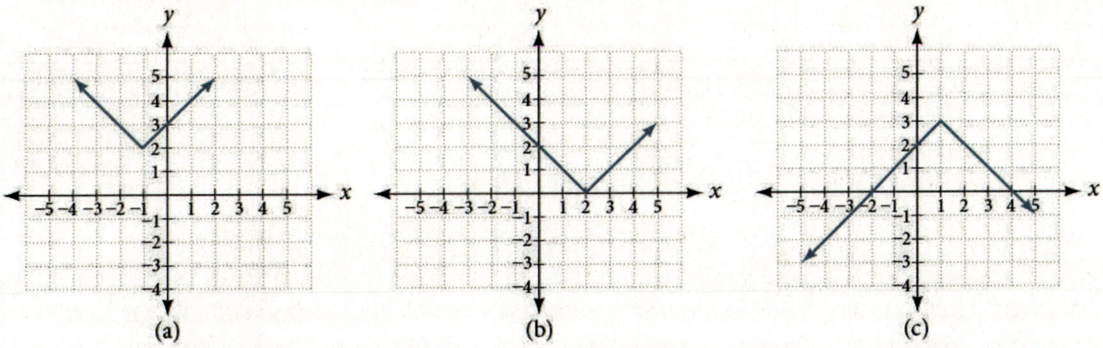

Figure 7 (a) The absolute value function does not intersect the horizontal axis. (b) The absolute value function intersects the horizontal axis at one point. (c) The absolute value function intersects the horizontal axis at two points.

Solving an Absolute Value Equation

In Other Type of Equations, we touched on the concepts of absolute value equations. Now that we understand a little more about their graphs, we can take another look at these types of equations. Now that we can graph an absolute value function, we will learn how to solve an absolute value equation. To solve an equation such as $8 = |2x - 6|$, we notice that the absolute value will be equal to 8 if the quantity inside the absolute value is 8 or -8. This leads to two different equations we can solve independently.

$$\begin{aligned} 2x - 6 &= 8 \\ 2x &= 14 \\ x &= 7 \end{aligned} \qquad \text{or} \qquad \begin{aligned} 2x - 6 &= -8 \\ 2x &= -2 \\ x &= -1 \end{aligned}$$

Knowing how to solve problems involving absolute value functions is useful. For example, we may need to identify numbers or points on a line that are at a specified distance from a given reference point.

An absolute value equation is an equation in which the unknown variable appears in absolute value bars. For example,

$$\begin{aligned} &|x| = 4, \\ &|2x - 1| = 3, \text{ or} \\ &|5x + 2| - 4 = 9 \end{aligned}$$

Solutions to Absolute Value Equations

For real numbers A and B, an equation of the form $|A| = B$, with $B \geq 0$, will have solutions when $A = B$ or $A = -B$. If $B < 0$, the equation $|A| = B$ has no solution.

HOW TO

Given the formula for an absolute value function, find the horizontal intercepts of its graph.

1. Isolate the absolute value term.
2. Use $|A| = B$ to write $A = B$ or $-A = B$, assuming $B > 0$.
3. Solve for x.

EXAMPLE 3

Finding the Zeros of an Absolute Value Function

For the function $f(x) = |4x + 1| - 7$, find the values of x such that $f(x) = 0$.

Solution

$$\begin{aligned} 0 &= |4x+1|-7 && \text{Substitute 0 for } f(x). \\ 7 &= |4x+1| && \text{Isolate the absolute value on one side of the equation.} \end{aligned}$$

$$\begin{array}{rclcrcll} 7 &=& 4x+1 & \text{or} & -7 &=& 4x+1 & \text{Break into two separate equations and solve.} \\ 6 &=& 4x & & -8 &=& 4x & \\ x &=& \frac{6}{4}=1.5 & & x &=& \frac{-8}{4}=-2 & \end{array}$$

The function outputs 0 when $x = \frac{3}{2}$ or $x = -2$. See Figure 8.

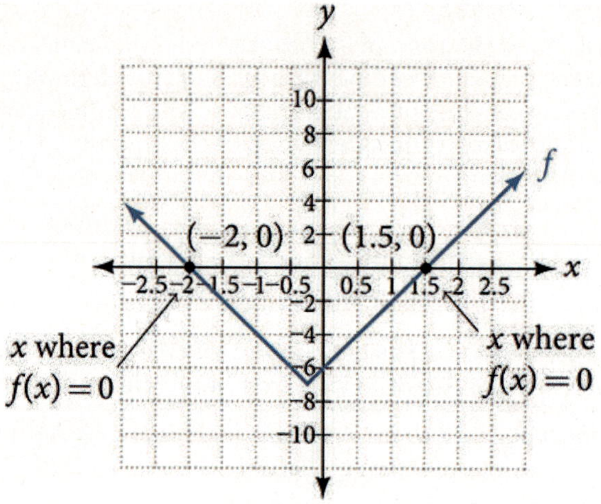

Figure 8

TRY IT #3 For the function $f(x) = |2x - 1| - 3$, find the values of x such that $f(x) = 0$.

Q&A **Should we always expect two answers when solving $|A| = B$?**

No. We may find one, two, or even no answers. For example, there is no solution to $2 + |3x - 5| = 1$.

MEDIA

Access these online resources for additional instruction and practice with absolute value.

Graphing Absolute Value Functions (http://openstax.org/l/graphabsvalue)
Graphing Absolute Value Functions 2 (http://openstax.org/l/graphabsvalue2)

3.6 SECTION EXERCISES

Verbal

1. How do you solve an absolute value equation?

2. How can you tell whether an absolute value function has two *x*-intercepts without graphing the function?

3. When solving an absolute value function, the isolated absolute value term is equal to a negative number. What does that tell you about the graph of the absolute value function?

4. How can you use the graph of an absolute value function to determine the *x*-values for which the function values are negative?

Algebraic

5. Describe all numbers x that are at a distance of 4 from the number 8. Express this set of numbers using absolute value notation.

6. Describe all numbers x that are at a distance of $\frac{1}{2}$ from the number −4. Express this set of numbers using absolute value notation.

7. Describe the situation in which the distance that point x is from 10 is at least 15 units. Express this set of numbers using absolute value notation.

8. Find all function values $f(x)$ such that the distance from $f(x)$ to the value 8 is less than 0.03 units. Express this set of numbers using absolute value notation.

For the following exercises, find the x- and y-intercepts of the graphs of each function.

9. $f(x) = 4\,|x - 3| + 4$

10. $f(x) = -3\,|x - 2| - 1$

11. $f(x) = -2\,|x + 1| + 6$

12. $f(x) = -5|x + 2| + 15$

13. $f(x) = 2|x - 1| - 6$

14. $f(x) = |-2x + 1| - 13$

15. $f(x) = -|x - 9| + 16$

Graphical

For the following exercises, graph the absolute value function. Plot at least five points by hand for each graph.

16. $y = |x - 1|$

17. $y = |x + 1|$

18. $y = |x| + 1$

For the following exercises, graph the given functions by hand.

19. $y = |x| - 2$

20. $y = -|x|$

21. $y = -|x| - 2$

22. $y = -|x - 3| - 2$

23. $f(x) = -|x - 1| - 2$

24. $f(x) = -|x + 3| + 4$

25. $f(x) = 2|x + 3| + 1$

26. $f(x) = 3|x - 2| + 3$

27. $f(x) = |2x - 4| - 3$

28. $f(x) = |3x + 9| + 2$

29. $f(x) = -|x - 1| - 3$

30. $f(x) = -|x + 4| - 3$

31. $f(x) = \frac{1}{2}|x + 4| - 3$

Technology

32. Use a graphing utility to graph $f(x) = 10|x - 2|$ on the viewing window $[0, 4]$. Identify the corresponding range. Show the graph.

33. Use a graphing utility to graph $f(x) = -100|x| + 100$ on the viewing window $[-5, 5]$. Identify the corresponding range. Show the graph.

For the following exercises, graph each function using a graphing utility. Specify the viewing window.

34. $f(x) = -0.1\,|0.1(0.2 - x)| + 0.3$

35. $f(x) = 4 \times 10^9\left|x - \left(5 \times 10^9\right)\right| + 2 \times 10^9$

Extensions

For the following exercises, solve the inequality.

36. If possible, find all values of a such that there are no x- intercepts for $f(x) = 2|x + 1| + a$.

37. If possible, find all values of a such that there are no y-intercepts for $f(x) = 2|x + 1| + a$.

Real-World Applications

38. Cities A and B are on the same east-west line. Assume that city A is located at the origin. If the distance from city A to city B is at least 100 miles and x represents the distance from city B to city A, express this using absolute value notation.

39. The true proportion p of people who give a favorable rating to Congress is 8% with a margin of error of 1.5%. Describe this statement using an absolute value equation.

40. Students who score within 18 points of the number 82 will pass a particular test. Write this statement using absolute value notation and use the variable x for the score.

41. A machinist must produce a bearing that is within 0.01 inches of the correct diameter of 5.0 inches. Using x as the diameter of the bearing, write this statement using absolute value notation.

42. The tolerance for a ball bearing is 0.01. If the true diameter of the bearing is to be 2.0 inches and the measured value of the diameter is x inches, express the tolerance using absolute value notation.

3.7 Inverse Functions

Learning Objectives

In this section, you will:

- Verify inverse functions.
- Determine the domain and range of an inverse function, and restrict the domain of a function to make it one-to-one.
- Find or evaluate the inverse of a function.
- Use the graph of a one-to-one function to graph its inverse function on the same axes.

A reversible heat pump is a climate-control system that is an air conditioner and a heater in a single device. Operated in one direction, it pumps heat out of a house to provide cooling. Operating in reverse, it pumps heat into the building from the outside, even in cool weather, to provide heating. As a heater, a heat pump is several times more efficient than conventional electrical resistance heating.

If some physical machines can run in two directions, we might ask whether some of the function "machines" we have been studying can also run backwards. Figure 1 provides a visual representation of this question. In this section, we will consider the reverse nature of functions.

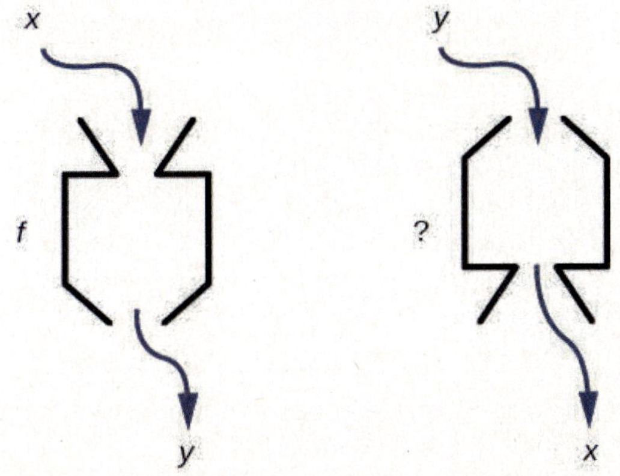

Figure 1 Can a function "machine" operate in reverse?

Verifying That Two Functions Are Inverse Functions

Betty is traveling to Milan for a fashion show and wants to know what the temperature will be. She is not familiar with the Celsius scale. To get an idea of how temperature measurements are related, Betty wants to convert 75 degrees Fahrenheit to degrees Celsius using the formula

$$C = \frac{5}{9}(F - 32)$$

and substitutes 75 for F to calculate

$$\frac{5}{9}(75 - 32) \approx 24°\text{C}$$

Knowing that a comfortable 75 degrees Fahrenheit is about 24 degrees Celsius, Betty gets the week's weather forecast from Figure 2 for Milan, and wants to convert all of the temperatures to degrees Fahrenheit.

Figure 2

At first, Betty considers using the formula she has already found to complete the conversions. After all, she knows her algebra, and can easily solve the equation for F after substituting a value for C. For example, to convert 26 degrees Celsius, she could write

$$\begin{aligned} 26 &= \tfrac{5}{9}(F-32) \\ 26\cdot\tfrac{9}{5} &= F-32 \\ F &= 26\cdot\tfrac{9}{5}+32\approx 79 \end{aligned}$$

After considering this option for a moment, however, she realizes that solving the equation for each of the temperatures will be awfully tedious. She realizes that since evaluation is easier than solving, it would be much more convenient to have a different formula, one that takes the Celsius temperature and outputs the Fahrenheit temperature.

The formula for which Betty is searching corresponds to the idea of an **inverse function**, which is a function for which the input of the original function becomes the output of the inverse function and the output of the original function becomes the input of the inverse function.

Given a function $f(x)$, we represent its inverse as $f^{-1}(x)$, read as "f inverse of x." The raised -1 is part of the notation. It is not an exponent; it does not imply a power of -1 . In other words, $f^{-1}(x)$ does *not* mean $\frac{1}{f(x)}$ because $\frac{1}{f(x)}$ is the reciprocal of f and not the inverse.

The "exponent-like" notation comes from an analogy between function composition and multiplication: just as $a^{-1}a=1$ (1 is the identity element for multiplication) for any nonzero number a, so $f^{-1}\circ f$ equals the identity function, that is,

$$\left(f^{-1}\circ f\right)(x)=f^{-1}\left(f(x)\right)=f^{-1}\left(y\right)=x$$

This holds for all x in the domain of f. Informally, this means that inverse functions "undo" each other. However, just as zero does not have a reciprocal, some functions do not have inverses.

Given a function $f(x)$, we can verify whether some other function $g(x)$ is the inverse of $f(x)$ by checking whether either $g(f(x))=x$ or $f(g(x))=x$ is true. We can test whichever equation is more convenient to work with because they are logically equivalent (that is, if one is true, then so is the other.)

For example, $y=4x$ and $y=\frac{1}{4}x$ are inverse functions.

$$\left(f^{-1}\circ f\right)(x)=f^{-1}\left(4x\right)=\frac{1}{4}(4x)=x$$

and

$$\left(f\circ f^{-1}\right)(x)=f\left(\frac{1}{4}x\right)=4\left(\frac{1}{4}x\right)=x$$

A few coordinate pairs from the graph of the function $y=4x$ are (−2, −8), (0, 0), and (2, 8). A few coordinate pairs from the graph of the function $y=\frac{1}{4}x$ are (−8, −2), (0, 0), and (8, 2). If we interchange the input and output of each coordinate pair of a function, the interchanged coordinate pairs would appear on the graph of the inverse function.

Inverse Function

For any one-to-one function $f(x)=y$, a function $f^{-1}(x)$ is an **inverse function** of f if $f^{-1}(y)=x$. This can also be written as $f^{-1}(f(x))=x$ for all x in the domain of f. It also follows that $f(f^{-1}(x))=x$ for all x in the domain of f^{-1}

if f^{-1} is the inverse of f.

The notation f^{-1} is read "f inverse." Like any other function, we can use any variable name as the input for f^{-1}, so we will often write $f^{-1}(x)$, which we read as "f inverse of x." Keep in mind that

$$f^{-1}(x) \neq \frac{1}{f(x)}$$

and not all functions have inverses.

EXAMPLE 1

Identifying an Inverse Function for a Given Input-Output Pair

If for a particular one-to-one function $f(2) = 4$ and $f(5) = 12$, what are the corresponding input and output values for the inverse function?

Solution

The inverse function reverses the input and output quantities, so if

$$\begin{aligned} f(2) &= 4, \text{ then } f^{-1}(4) = 2; \\ f(5) &= 12, \text{ then } f^{-1}(12) = 5. \end{aligned}$$

Alternatively, if we want to name the inverse function g, then $g(4) = 2$ and $g(12) = 5$.

Analysis

Notice that if we show the coordinate pairs in a table form, the input and output are clearly reversed. See Table 1.

$(x, f(x))$	$(x, g(x))$
$(2, 4)$	$(4, 2)$
$(5, 12)$	$(12, 5)$

Table 1

TRY IT #1 Given that $h^{-1}(6) = 2$, what are the corresponding input and output values of the original function h?

HOW TO

Given two functions $f(x)$ and $g(x)$, test whether the functions are inverses of each other.

1. Determine whether $f(g(x)) = x$ or $g(f(x)) = x$.
2. If either statement is true, then both are true, and $g = f^{-1}$ and $f = g^{-1}$. If either statement is false, then both are false, and $g \neq f^{-1}$ and $f \neq g^{-1}$.

EXAMPLE 2

Testing Inverse Relationships Algebraically

If $f(x) = \frac{1}{x+2}$ and $g(x) = \frac{1}{x} - 2$, is $g = f^{-1}$?

Solution

$$\begin{aligned} g(f(x)) &= \frac{1}{\left(\frac{1}{x+2}\right)} - 2 \\ &= x + 2 - 2 \\ &= x \end{aligned}$$

so

$$g = f^{-1} \text{ and } f = g^{-1}$$

This is enough to answer yes to the question, but we can also verify the other formula.

$$\begin{aligned} f(g(x)) &= \frac{1}{\frac{1}{x}-2+2} \\ &= \frac{1}{\frac{1}{x}} \\ &= x \end{aligned}$$

Analysis

Notice the inverse operations are in reverse order of the operations from the original function.

TRY IT #2 If $f(x) = x^3 - 4$ and $g(x) = \sqrt[3]{x+4}$, is $g = f^{-1}$?

EXAMPLE 3

Determining Inverse Relationships for Power Functions

If $f(x) = x^3$ (the cube function) and $g(x) = \frac{1}{3}x$, is $g = f^{-1}$?

Solution

$$f(g(x)) = \frac{x^3}{27} \neq x$$

No, the functions are not inverses.

Analysis

The correct inverse to the cube is, of course, the cube root $\sqrt[3]{x} = x^{\frac{1}{3}}$, that is, the one-third is an exponent, not a multiplier.

TRY IT #3 If $f(x) = (x-1)^3$ and $g(x) = \sqrt[3]{x} + 1$, is $g = f^{-1}$?

Finding Domain and Range of Inverse Functions

The outputs of the function f are the inputs to f^{-1}, so the range of f is also the domain of f^{-1}. Likewise, because the inputs to f are the outputs of f^{-1}, the domain of f is the range of f^{-1}. We can visualize the situation as in Figure 3.

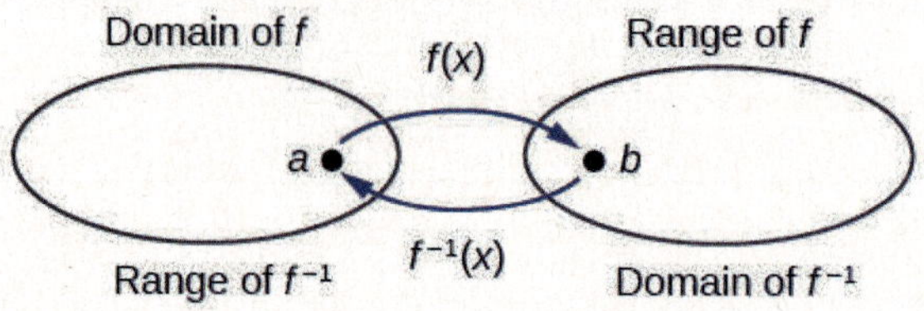

Figure 3 Domain and range of a function and its inverse

When a function has no inverse function, it is possible to create a new function where that new function on a limited domain does have an inverse function. For example, the inverse of $f(x) = \sqrt{x}$ is $f^{-1}(x) = x^2$, because a square "undoes" a square root; but the square is only the inverse of the square root on the domain $\left[0, \infty\right)$, since that is the range of $f(x) = \sqrt{x}$.

We can look at this problem from the other side, starting with the square (toolkit quadratic) function $f(x) = x^2$. If we want to construct an inverse to this function, we run into a problem, because for every given output of the quadratic function, there are two corresponding inputs (except when the input is 0). For example, the output 9 from the quadratic function corresponds to the inputs 3 and –3. But an output from a function is an input to its inverse; if this inverse input corresponds to more than one inverse output (input of the original function), then the "inverse" is not a function at all! To put it differently, the quadratic function is not a one-to-one function; it fails the horizontal line test, so it does not have an inverse function. In order for a function to have an inverse, it must be a one-to-one function.

In many cases, if a function is not one-to-one, we can still restrict the function to a part of its domain on which it is one-to-one. For example, we can make a restricted version of the square function $f(x) = x^2$ with its domain limited to $\left[0, \infty\right)$, which is a one-to-one function (it passes the horizontal line test) and which has an inverse (the square-root function).

If $f(x) = (x-1)^2$ on $\left[1, \infty\right)$, then the inverse function is $f^{-1}(x) = \sqrt{x} + 1$.

- The domain of f = range of f^{-1} = $\left[1, \infty\right)$.
- The domain of f^{-1} = range of f = $\left[0, \infty\right)$.

Q&A **Is it possible for a function to have more than one inverse?**

No. If two supposedly different functions, say, g and h, both meet the definition of being inverses of another function f, then you can prove that $g = h$. We have just seen that some functions only have inverses if we restrict the domain of the original function. In these cases, there may be more than one way to restrict the domain, leading to different inverses. However, on any one domain, the original function still has only one unique inverse.

Domain and Range of Inverse Functions

The range of a function $f(x)$ is the domain of the inverse function $f^{-1}(x)$.

The domain of $f(x)$ is the range of $f^{-1}(x)$.

HOW TO

Given a function, find the domain and range of its inverse.

1. If the function is one-to-one, write the range of the original function as the domain of the inverse, and write the domain of the original function as the range of the inverse.
2. If the domain of the original function needs to be restricted to make it one-to-one, then this restricted domain becomes the range of the inverse function.

EXAMPLE 4

Finding the Inverses of Toolkit Functions

Identify which of the toolkit functions besides the quadratic function are not one-to-one, and find a restricted domain on which each function is one-to-one, if any. The toolkit functions are reviewed in Table 2. We restrict the domain in such a fashion that the function assumes all y-values exactly once.

Constant	Identity	Quadratic	Cubic	Reciprocal
$f(x) = c$	$f(x) = x$	$f(x) = x^2$	$f(x) = x^3$	$f(x) = \frac{1}{x}$
Reciprocal squared	**Cube root**	**Square root**	**Absolute value**	
$f(x) = \frac{1}{x^2}$	$f(x) = \sqrt[3]{x}$	$f(x) = \sqrt{x}$	$f(x) = \|x\|$	

Table 2

Solution

The constant function is not one-to-one, and there is no domain (except a single point) on which it could be one-to-one, so the constant function has no inverse.

The absolute value function can be restricted to the domain $\left[0, \infty\right)$, where it is equal to the identity function.

The reciprocal-squared function can be restricted to the domain $\left(0, \infty\right)$.

Analysis

We can see that these functions (if unrestricted) are not one-to-one by looking at their graphs, shown in Figure 4. They both would fail the horizontal line test. However, if a function is restricted to a certain domain so that it passes the horizontal line test, then in that restricted domain, it can have an inverse.

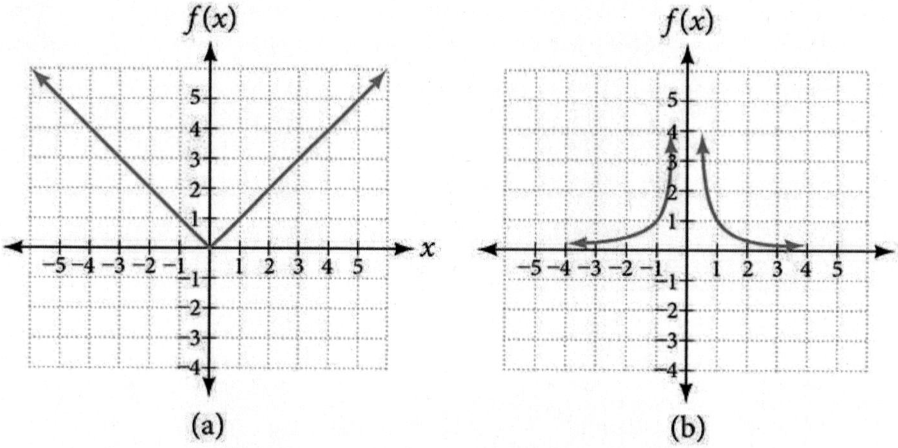

Figure 4 (a) Absolute value (b) Reciprocal square

TRY IT #4 The domain of function f is $(1, \infty)$ and the range of function f is $(-\infty, -2)$. Find the domain and range of the inverse function.

Finding and Evaluating Inverse Functions

Once we have a one-to-one function, we can evaluate its inverse at specific inverse function inputs or construct a complete representation of the inverse function in many cases.

Inverting Tabular Functions

Suppose we want to find the inverse of a function represented in table form. Remember that the domain of a function is the range of the inverse and the range of the function is the domain of the inverse. So we need to interchange the domain and range.

Each row (or column) of inputs becomes the row (or column) of outputs for the inverse function. Similarly, each row (or column) of outputs becomes the row (or column) of inputs for the inverse function.

EXAMPLE 5

Interpreting the Inverse of a Tabular Function

A function $f(t)$ is given in Table 3, showing distance in miles that a car has traveled in t minutes. Find and interpret $f^{-1}(70)$.

t (minutes)	30	50	70	90
$f(t)$ (miles)	20	40	60	70

Table 3

Solution

The inverse function takes an output of f and returns an input for f. So in the expression $f^{-1}(70)$, 70 is an output value of the original function, representing 70 miles. The inverse will return the corresponding input of the original function f, 90 minutes, so $f^{-1}(70) = 90$. The interpretation of this is that, to drive 70 miles, it took 90 minutes.

Alternatively, recall that the definition of the inverse was that if $f(a) = b$, then $f^{-1}(b) = a$. By this definition, if we are given $f^{-1}(70) = a$, then we are looking for a value a so that $f(a) = 70$. In this case, we are looking for a t so that $f(t) = 70$, which is when $t = 90$.

TRY IT #5 Using Table 4, find and interpret ⓐ $f(60)$, and ⓑ $f^{-1}(60)$.

t (minutes)	30	50	60	70	90
$f(t)$ (miles)	20	40	50	60	70

Table 4

Evaluating the Inverse of a Function, Given a Graph of the Original Function

We saw in Functions and Function Notation that the domain of a function can be read by observing the horizontal extent of its graph. We find the domain of the inverse function by observing the *vertical* extent of the graph of the original function, because this corresponds to the horizontal extent of the inverse function. Similarly, we find the range of the inverse function by observing the *horizontal* extent of the graph of the original function, as this is the vertical extent of the inverse function. If we want to evaluate an inverse function, we find its input within its domain, which is all or part of the vertical axis of the original function's graph.

HOW TO

Given the graph of a function, evaluate its inverse at specific points.

1. Find the desired input on the *y*-axis of the given graph.
2. Read the inverse function's output from the *x*-axis of the given graph.

EXAMPLE 6

Evaluating a Function and Its Inverse from a Graph at Specific Points

A function $g(x)$ is given in Figure 5. Find $g(3)$ and $g^{-1}(3)$.

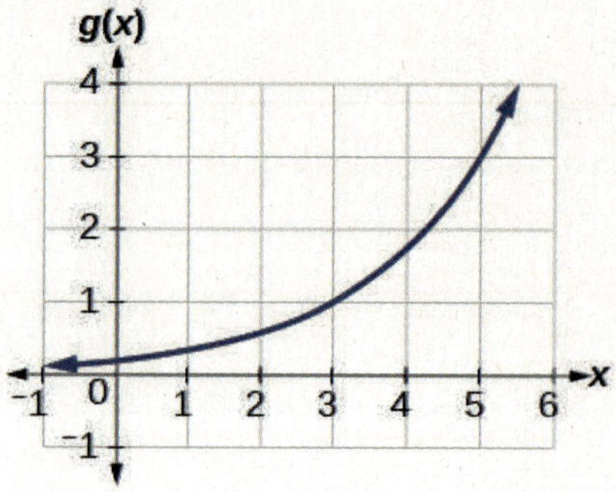

Figure 5

Solution

To evaluate $g(3)$, we find 3 on the *x*-axis and find the corresponding output value on the *y*-axis. The point $(3, 1)$ tells us that $g(3) = 1$.

To evaluate $g^{-1}(3)$, recall that by definition $g^{-1}(3)$ means the value of *x* for which $g(x) = 3$. By looking for the output value 3 on the vertical axis, we find the point $(5, 3)$ on the graph, which means $g(5) = 3$, so by definition, $g^{-1}(3) = 5$. See Figure 6.

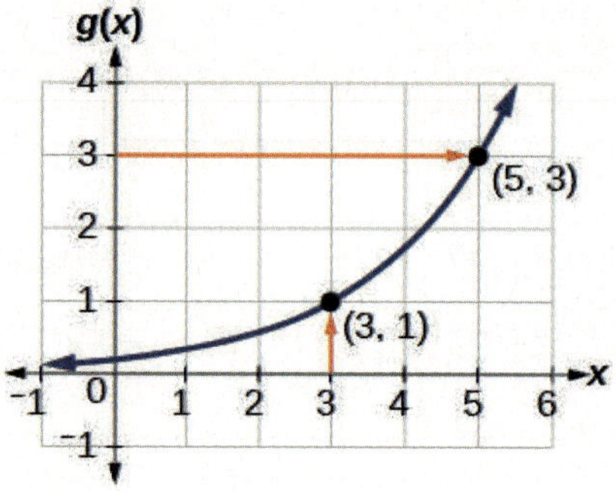

Figure 6

TRY IT #6 Using the graph in Figure 6, ⓐ find $g^{-1}(1)$, and ⓑ estimate $g^{-1}(4)$.

Finding Inverses of Functions Represented by Formulas

Sometimes we will need to know an inverse function for all elements of its domain, not just a few. If the original function is given as a formula—for example, y as a function of x— we can often find the inverse function by solving to obtain x as a function of y.

HOW TO

Given a function represented by a formula, find the inverse.

1. Make sure f is a one-to-one function.
2. Solve for x.
3. Interchange x and y.

EXAMPLE 7

Inverting the Fahrenheit-to-Celsius Function

Find a formula for the inverse function that gives Fahrenheit temperature as a function of Celsius temperature.

$$C = \frac{5}{9}(F - 32)$$

Solution

$$\begin{aligned} C &= \tfrac{5}{9}(F - 32) \\ C \cdot \tfrac{9}{5} &= F - 32 \\ F &= \tfrac{9}{5}C + 32 \end{aligned}$$

By solving in general, we have uncovered the inverse function. If

$$C = h(F) = \frac{5}{9}(F - 32),$$

then

$$F = h^{-1}(C) = \frac{9}{5}C + 32$$

In this case, we introduced a function h to represent the conversion because the input and output variables are descriptive, and writing C^{-1} could get confusing.

TRY IT #7 Solve for x in terms of y given $y = \frac{1}{3}(x - 5)$.

EXAMPLE 8

Solving to Find an Inverse Function

Find the inverse of the function $f(x) = \frac{2}{x-3} + 4$.

Solution

$$\begin{aligned} y &= \tfrac{2}{x-3} + 4 && \text{Set up an equation.} \\ y - 4 &= \tfrac{2}{x-3} && \text{Subtract 4 from both sides.} \\ x - 3 &= \tfrac{2}{y-4} && \text{Multiply both sides by } x - 3 \text{ and divide by } y - 4. \\ x &= \tfrac{2}{y-4} + 3 && \text{Add 3 to both sides.} \end{aligned}$$

So $f^{-1}(y) = \frac{2}{y-4} + 3$ or $f^{-1}(x) = \frac{2}{x-4} + 3$.

Analysis

The domain and range of f exclude the values 3 and 4, respectively. f and f^{-1} are equal at two points but are not the same function, as we can see by creating Table 5.

x	1	2	5	$f^{-1}(y)$
$f(x)$	3	2	5	y

Table 5

EXAMPLE 9

Solving to Find an Inverse with Radicals

Find the inverse of the function $f(x) = 2 + \sqrt{x - 4}$.

Solution

$$
\begin{aligned}
y &= 2+\sqrt{x-4} \\
(y-2)^2 &= x-4 \\
x &= (y-2)^2+4
\end{aligned}
$$

So $f^{-1}(x) = (x-2)^2+4$.

The domain of f is $[4, \infty)$. Notice that the range of f is $[2, \infty)$, so this means that the domain of the inverse function f^{-1} is also $[2, \infty)$.

Analysis

The formula we found for $f^{-1}(x)$ looks like it would be valid for all real x. However, f^{-1} itself must have an inverse (namely, f) so we have to restrict the domain of f^{-1} to $[2, \infty)$ in order to make f^{-1} a one-to-one function. This domain of f^{-1} is exactly the range of f.

TRY IT #8 What is the inverse of the function $f(x) = 2-\sqrt{x}$? State the domains of both the function and the inverse function.

Finding Inverse Functions and Their Graphs

Now that we can find the inverse of a function, we will explore the graphs of functions and their inverses. Let us return to the quadratic function $f(x) = x^2$ restricted to the domain $[0, \infty)$, on which this function is one-to-one, and graph it as in Figure 7.

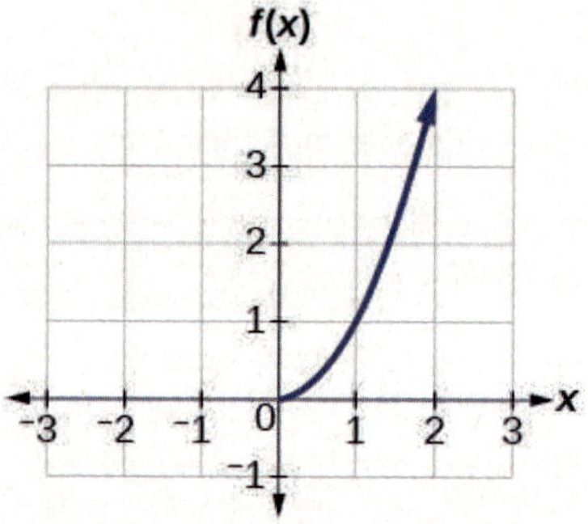

Figure 7 Quadratic function with domain restricted to $[0, \infty)$.

Restricting the domain to $[0, \infty)$ makes the function one-to-one (it will obviously pass the horizontal line test), so it has an inverse on this restricted domain.

We already know that the inverse of the toolkit quadratic function is the square root function, that is, $f^{-1}(x) = \sqrt{x}$. What happens if we graph both f and f^{-1} on the same set of axes, using the x- axis for the input to both f and f^{-1}?

We notice a distinct relationship: The graph of $f^{-1}(x)$ is the graph of $f(x)$ reflected about the diagonal line $y = x$, which we will call the identity line, shown in Figure 8.

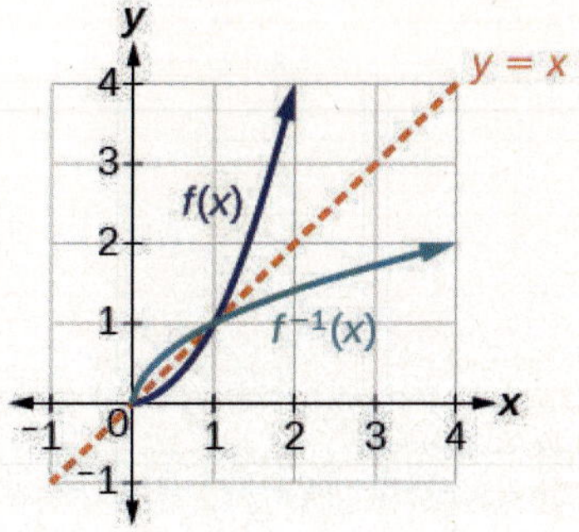

Figure 8 Square and square-root functions on the non-negative domain

This relationship will be observed for all one-to-one functions, because it is a result of the function and its inverse swapping inputs and outputs. This is equivalent to interchanging the roles of the vertical and horizontal axes.

EXAMPLE 10

Finding the Inverse of a Function Using Reflection about the Identity Line

Given the graph of $f(x)$ in Figure 9, sketch a graph of $f^{-1}(x)$.

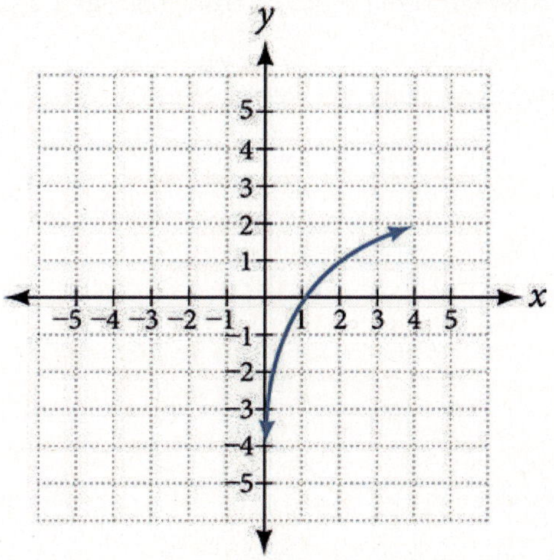

Figure 9

Solution

This is a one-to-one function, so we will be able to sketch an inverse. Note that the graph shown has an apparent domain of $\left(0, \infty\right)$ and range of $\left(-\infty, \infty\right)$, so the inverse will have a domain of $\left(-\infty, \infty\right)$ and range of $\left(0, \infty\right)$.

If we reflect this graph over the line $y = x$, the point $(1, 0)$ reflects to $(0, 1)$ and the point $(4, 2)$ reflects to $(2, 4)$. Sketching the inverse on the same axes as the original graph gives Figure 10.

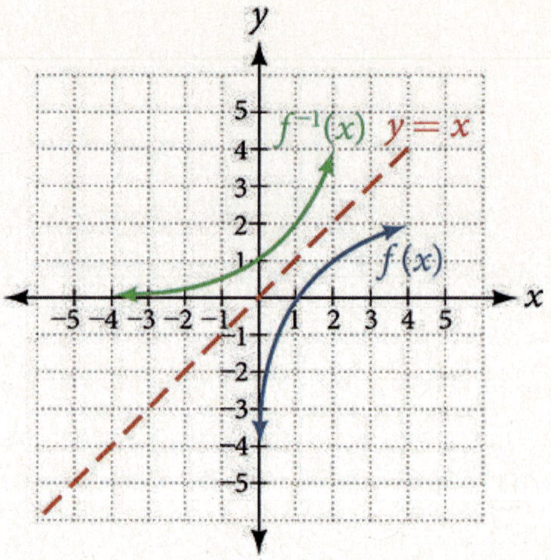

Figure 10 The function and its inverse, showing reflection about the identity line

---

TRY IT #9 Draw graphs of the functions f and f^{-1} from Example 8.

Q&A **Is there any function that is equal to its own inverse?**

Yes. If $f = f^{-1}$, then $f(f(x)) = x$, and we can think of several functions that have this property. The identity function does, and so does the reciprocal function, because

$$\frac{1}{\frac{1}{x}} = x$$

Any function $f(x) = c - x$, where c is a constant, is also equal to its own inverse.

MEDIA

Access these online resources for additional instruction and practice with inverse functions.

Inverse Functions (http://openstax.org/l/inversefunction)
One-to-one Functions (http://openstax.org/l/onetoone)
Inverse Function Values Using Graph (http://openstax.org/l/inversfuncgraph)
Restricting the Domain and Finding the Inverse (http://openstax.org/l/restrictdomain)

3.7 SECTION EXERCISES

Verbal

1. Describe why the horizontal line test is an effective way to determine whether a function is one-to-one?

2. Why do we restrict the domain of the function $f(x) = x^2$ to find the function's inverse?

3. Can a function be its own inverse? Explain.

4. Are one-to-one functions either always increasing or always decreasing? Why or why not?

5. How do you find the inverse of a function algebraically?

Algebraic

6. Show that the function $f(x) = a - x$ is its own inverse for all real numbers a.

For the following exercises, find $f^{-1}(x)$ for each function.

7. $f(x) = x + 3$

8. $f(x) = x + 5$

9. $f(x) = 2 - x$

10. $f(x) = 3 - x$

11. $f(x) = \frac{x}{x+2}$

12. $f(x) = \frac{2x+3}{5x+4}$

For the following exercises, find a domain on which each function f is one-to-one and non-decreasing. Write the domain in interval notation. Then find the inverse of f restricted to that domain.

13. $f(x) = (x + 7)^2$

14. $f(x) = (x - 6)^2$

15. $f(x) = x^2 - 5$

16. Given $f(x) = \frac{x}{2+x}$ and $g(x) = \frac{2x}{1-x}$:

ⓐ Find $f(g(x))$ and $g(f(x))$.

ⓑ What does the answer tell us about the relationship between $f(x)$ and $g(x)$?

For the following exercises, use function composition to verify that $f(x)$ and $g(x)$ are inverse functions.

17. $f(x) = \sqrt[3]{x - 1}$ and $g(x) = x^3 + 1$

18. $f(x) = -3x + 5$ and $g(x) = \frac{x-5}{-3}$

Graphical

For the following exercises, use a graphing utility to determine whether each function is one-to-one.

19. $f(x) = \sqrt{x}$

20. $f(x) = \sqrt[3]{3x + 1}$

21. $f(x) = -5x + 1$

22. $f(x) = x^3 - 27$

For the following exercises, determine whether the graph represents a one-to-one function.

23.

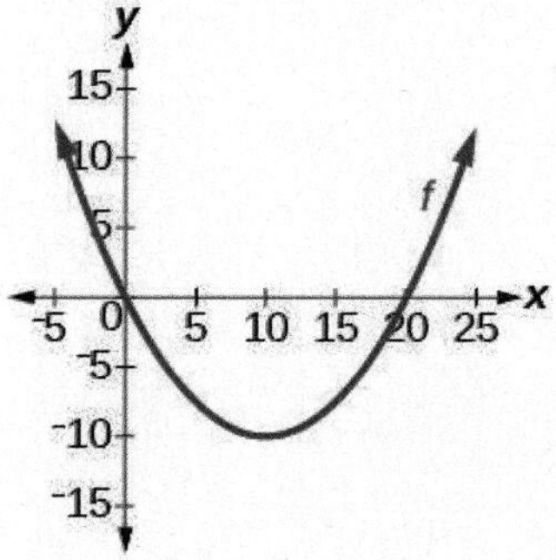

24.

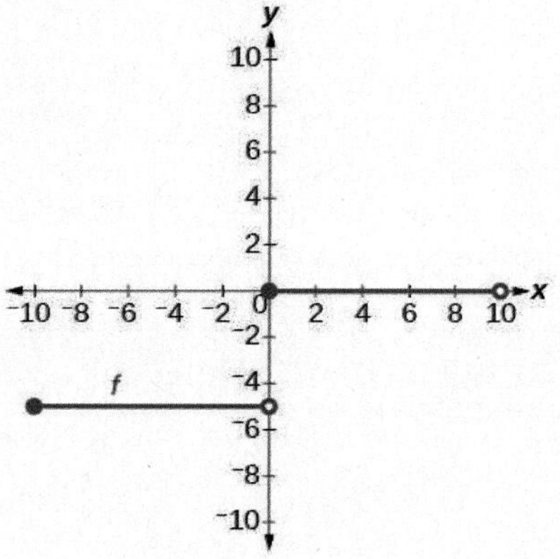

For the following exercises, use the graph of f shown in Figure 11.

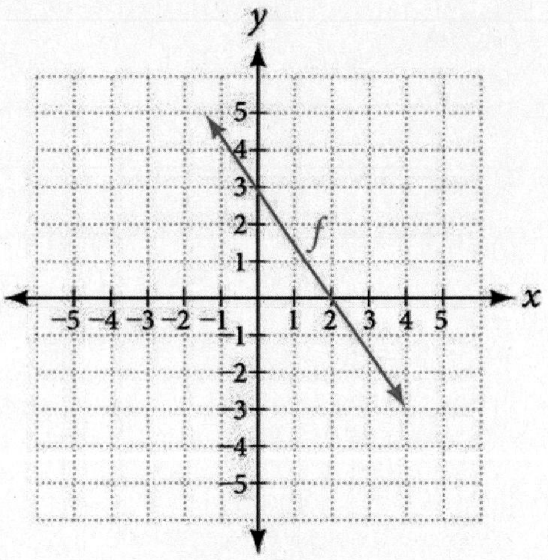

Figure 11

25. Find $f(0)$.

26. Solve $f(x) = 0$.

27. Find $f^{-1}(0)$.

28. Solve $f^{-1}(x) = 0$.

For the following exercises, use the graph of the one-to-one function shown in Figure 12.

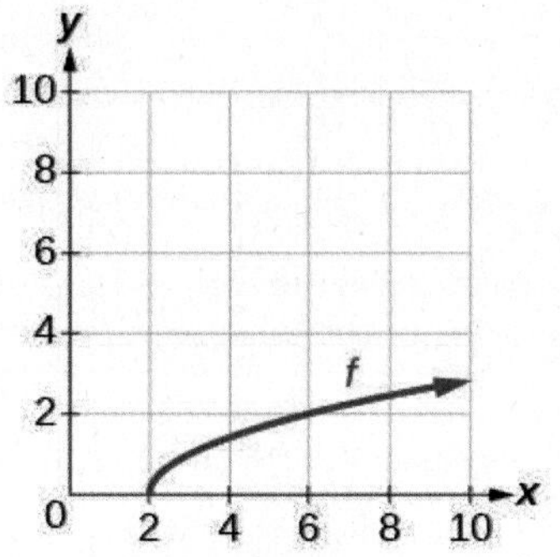

Figure 12

29. Sketch the graph of f^{-1}.

30. Find $f(6)$ and $f^{-1}(2)$.

31. If the complete graph of f is shown, find the domain of f.

32. If the complete graph of f is shown, find the range of f.

Numeric

For the following exercises, evaluate or solve, assuming that the function f is one-to-one.

33. If $f(6) = 7$, find $f^{-1}(7)$.

34. If $f(3) = 2$, find $f^{-1}(2)$.

35. If $f^{-1}(-4) = -8$, find $f(-8)$.

36. If $f^{-1}(-2) = -1$, find $f(-1)$.

For the following exercises, use the values listed in Table 6 to evaluate or solve.

x	0	1	2	3	4	5	6	7	8	9
$f(x)$	8	0	7	4	2	6	5	3	9	1

Table 6

37. Find $f(1)$.

38. Solve $f(x) = 3$.

39. Find $f^{-1}(0)$.

40. Solve $f^{-1}(x) = 7$.

41. Use the tabular representation of f in Table 7 to create a table for $f^{-1}(x)$.

x	3	6	9	13	14
$f(x)$	1	4	7	12	16

Table 7

Technology

For the following exercises, find the inverse function. Then, graph the function and its inverse.

42. $f(x) = \frac{3}{x-2}$

43. $f(x) = x^3 - 1$

44. Find the inverse function of $f(x) = \frac{1}{x-1}$. Use a graphing utility to find its domain and range. Write the domain and range in interval notation.

Real-World Applications

45. To convert from x degrees Celsius to y degrees Fahrenheit, we use the formula $f(x) = \frac{9}{5}x + 32$. Find the inverse function, if it exists, and explain its meaning.

46. The circumference C of a circle is a function of its radius given by $C(r) = 2\pi r$. Express the radius of a circle as a function of its circumference. Call this function $r(C)$. Find $r(36\pi)$ and interpret its meaning.

47. A car travels at a constant speed of 50 miles per hour. The distance the car travels in miles is a function of time, t, in hours given by $d(t) = 50t$. Find the inverse function by expressing the time of travel in terms of the distance traveled. Call this function $t(d)$. Find $t(180)$ and interpret its meaning.

Chapter Review

Key Terms

absolute maximum the greatest value of a function over an interval

absolute minimum the lowest value of a function over an interval

average rate of change the difference in the output values of a function found for two values of the input divided by the difference between the inputs

composite function the new function formed by function composition, when the output of one function is used as the input of another

decreasing function a function is decreasing in some open interval if $f(b) < f(a)$ for any two input values a and b in the given interval where $b > a$

dependent variable an output variable

domain the set of all possible input values for a relation

even function a function whose graph is unchanged by horizontal reflection, $f(x) = f(-x)$, and is symmetric about the y- axis

function a relation in which each input value yields a unique output value

horizontal compression a transformation that compresses a function's graph horizontally, by multiplying the input by a constant $b > 1$

horizontal line test a method of testing whether a function is one-to-one by determining whether any horizontal line intersects the graph more than once

horizontal reflection a transformation that reflects a function's graph across the *y*-axis by multiplying the input by -1

horizontal shift a transformation that shifts a function's graph left or right by adding a positive or negative constant to the input

horizontal stretch a transformation that stretches a function's graph horizontally by multiplying the input by a constant $0 < b < 1$

increasing function a function is increasing in some open interval if $f(b) > f(a)$ for any two input values a and b in the given interval where $b > a$

independent variable an input variable

input each object or value in a domain that relates to another object or value by a relationship known as a function

interval notation a method of describing a set that includes all numbers between a lower limit and an upper limit; the lower and upper values are listed between brackets or parentheses, a square bracket indicating inclusion in the set, and a parenthesis indicating exclusion

inverse function for any one-to-one function $f(x)$, the inverse is a function $f^{-1}(x)$ such that $f^{-1}(f(x)) = x$ for all x in the domain of f; this also implies that $f\left(f^{-1}(x)\right) = x$ for all x in the domain of f^{-1}

local extrema collectively, all of a function's local maxima and minima

local maximum a value of the input where a function changes from increasing to decreasing as the input value increases.

local minimum a value of the input where a function changes from decreasing to increasing as the input value increases.

odd function a function whose graph is unchanged by combined horizontal and vertical reflection, $f(x) = -f(-x)$, and is symmetric about the origin

one-to-one function a function for which each value of the output is associated with a unique input value

output each object or value in the range that is produced when an input value is entered into a function

piecewise function a function in which more than one formula is used to define the output

range the set of output values that result from the input values in a relation

rate of change the change of an output quantity relative to the change of the input quantity

relation a set of ordered pairs

set-builder notation a method of describing a set by a rule that all of its members obey; it takes the form $\{x|\ \text{statement about } x\}$

vertical compression a function transformation that compresses the function's graph vertically by multiplying the output by a constant $0 < a < 1$

vertical line test a method of testing whether a graph represents a function by determining whether a vertical line intersects the graph no more than once

vertical reflection a transformation that reflects a function's graph across the *x*-axis by multiplying the output by -1

vertical shift a transformation that shifts a function's graph up or down by adding a positive or negative constant to the output

vertical stretch a transformation that stretches a function's graph vertically by multiplying the output by a constant $a > 1$

Key Equations

Constant function	$f(x) = c$, where c is a constant
Identity function	$f(x) = x$
Absolute value function	$f(x) = \lvert x \rvert$
Quadratic function	$f(x) = x^2$
Cubic function	$f(x) = x^3$
Reciprocal function	$f(x) = \frac{1}{x}$
Reciprocal squared function	$f(x) = \frac{1}{x^2}$
Square root function	$f(x) = \sqrt{x}$
Cube root function	$f(x) = \sqrt[3]{x}$

Average rate of change	$\frac{\Delta y}{\Delta x} = \frac{f(x_2)-f(x_1)}{x_2-x_1}$

Composite function	$(f \circ g)(x) = f(g(x))$

Vertical shift	$g(x) = f(x) + k$ (up for $k > 0$)
Horizontal shift	$g(x) = f(x - h)$ (right for $h > 0$)
Vertical reflection	$g(x) = -f(x)$
Horizontal reflection	$g(x) = f(-x)$
Vertical stretch	$g(x) = af(x)$ ($a > 0$)
Vertical compression	$g(x) = af(x)$ $(0 < a < 1)$
Horizontal stretch	$g(x) = f(bx)$ $(0 < b < 1)$
Horizontal compression.	$g(x) = f(bx)$ ($b > 1$)

Key Concepts

3.1 Functions and Function Notation

- A relation is a set of ordered pairs. A function is a specific type of relation in which each domain value, or input, leads to exactly one range value, or output. See Example 1 and Example 2.
- Function notation is a shorthand method for relating the input to the output in the form $y = f(x)$. See Example 3 and Example 4.
- In tabular form, a function can be represented by rows or columns that relate to input and output values. See

Example 5.

- To evaluate a function, we determine an output value for a corresponding input value. Algebraic forms of a function can be evaluated by replacing the input variable with a given value. See Example 6 and Example 7.
- To solve for a specific function value, we determine the input values that yield the specific output value. See Example 8.
- An algebraic form of a function can be written from an equation. See Example 9 and Example 10.
- Input and output values of a function can be identified from a table. See Example 11.
- Relating input values to output values on a graph is another way to evaluate a function. See Example 12.
- A function is one-to-one if each output value corresponds to only one input value. See Example 13.
- A graph represents a function if any vertical line drawn on the graph intersects the graph at no more than one point. See Example 14.
- The graph of a one-to-one function passes the horizontal line test. See Example 15.

3.2 Domain and Range

- The domain of a function includes all real input values that would not cause us to attempt an undefined mathematical operation, such as dividing by zero or taking the square root of a negative number.
- The domain of a function can be determined by listing the input values of a set of ordered pairs. See Example 1.
- The domain of a function can also be determined by identifying the input values of a function written as an equation. See Example 2, Example 3, and Example 4.
- Interval values represented on a number line can be described using inequality notation, set-builder notation, and interval notation. See Example 5.
- For many functions, the domain and range can be determined from a graph. See Example 6 and Example 7.
- An understanding of toolkit functions can be used to find the domain and range of related functions. See Example 8, Example 9, and Example 10.
- A piecewise function is described by more than one formula. See Example 11 and Example 12.
- A piecewise function can be graphed using each algebraic formula on its assigned subdomain. See Example 13.

3.3 Rates of Change and Behavior of Graphs

- A rate of change relates a change in an output quantity to a change in an input quantity. The average rate of change is determined using only the beginning and ending data. See Example 1.
- Identifying points that mark the interval on a graph can be used to find the average rate of change. See Example 2.
- Comparing pairs of input and output values in a table can also be used to find the average rate of change. See Example 3.
- An average rate of change can also be computed by determining the function values at the endpoints of an interval described by a formula. See Example 4 and Example 5.
- The average rate of change can sometimes be determined as an expression. See Example 6.
- A function is increasing where its rate of change is positive and decreasing where its rate of change is negative. See Example 7.
- A local maximum is where a function changes from increasing to decreasing and has an output value larger (more positive or less negative) than output values at neighboring input values.
- A local minimum is where the function changes from decreasing to increasing (as the input increases) and has an output value smaller (more negative or less positive) than output values at neighboring input values.
- Minima and maxima are also called extrema.
- We can find local extrema from a graph. See Example 8 and Example 9.
- The highest and lowest points on a graph indicate the maxima and minima. See Example 10.

3.4 Composition of Functions

- We can perform algebraic operations on functions. See Example 1.
- When functions are combined, the output of the first (inner) function becomes the input of the second (outer) function.
- The function produced by combining two functions is a composite function. See Example 2 and Example 3.
- The order of function composition must be considered when interpreting the meaning of composite functions. See Example 4.
- A composite function can be evaluated by evaluating the inner function using the given input value and then evaluating the outer function taking as its input the output of the inner function.
- A composite function can be evaluated from a table. See Example 5.
- A composite function can be evaluated from a graph. See Example 6.
- A composite function can be evaluated from a formula. See Example 7.

- The domain of a composite function consists of those inputs in the domain of the inner function that correspond to outputs of the inner function that are in the domain of the outer function. See Example 8 and Example 9.
- Just as functions can be combined to form a composite function, composite functions can be decomposed into simpler functions.
- Functions can often be decomposed in more than one way. See Example 10.

3.5 Transformation of Functions

- A function can be shifted vertically by adding a constant to the output. See Example 1 and Example 2.
- A function can be shifted horizontally by adding a constant to the input. See Example 3, Example 4, and Example 5.
- Relating the shift to the context of a problem makes it possible to compare and interpret vertical and horizontal shifts. See Example 6.
- Vertical and horizontal shifts are often combined. See Example 7 and Example 8.
- A vertical reflection reflects a graph about the x- axis. A graph can be reflected vertically by multiplying the output by –1.
- A horizontal reflection reflects a graph about the y- axis. A graph can be reflected horizontally by multiplying the input by –1.
- A graph can be reflected both vertically and horizontally. The order in which the reflections are applied does not affect the final graph. See Example 9.
- A function presented in tabular form can also be reflected by multiplying the values in the input and output rows or columns accordingly. See Example 10.
- A function presented as an equation can be reflected by applying transformations one at a time. See Example 11.
- Even functions are symmetric about the y- axis, whereas odd functions are symmetric about the origin.
- Even functions satisfy the condition $f(x) = f(-x)$.
- Odd functions satisfy the condition $f(x) = -f(-x)$.
- A function can be odd, even, or neither. See Example 12.
- A function can be compressed or stretched vertically by multiplying the output by a constant. See Example 13, Example 14, and Example 15.
- A function can be compressed or stretched horizontally by multiplying the input by a constant. See Example 16, Example 17, and Example 18.
- The order in which different transformations are applied does affect the final function. Both vertical and horizontal transformations must be applied in the order given. However, a vertical transformation may be combined with a horizontal transformation in any order. See Example 19 and Example 20.

3.6 Absolute Value Functions

- Applied problems, such as ranges of possible values, can also be solved using the absolute value function. See Example 1.
- The graph of the absolute value function resembles a letter V. It has a corner point at which the graph changes direction. See Example 2.
- In an absolute value equation, an unknown variable is the input of an absolute value function.
- If the absolute value of an expression is set equal to a positive number, expect two solutions for the unknown variable. See Example 3.

3.7 Inverse Functions

- If $g(x)$ is the inverse of $f(x)$, then $g(f(x)) = f(g(x)) = x$. See Example 1, Example 2, and Example 3.
- Only some of the toolkit functions have an inverse. See Example 4.
- For a function to have an inverse, it must be one-to-one (pass the horizontal line test).
- A function that is not one-to-one over its entire domain may be one-to-one on part of its domain.
- For a tabular function, exchange the input and output rows to obtain the inverse. See Example 5.
- The inverse of a function can be determined at specific points on its graph. See Example 6.
- To find the inverse of a formula, solve the equation $y = f(x)$ for x as a function of y. Then exchange the labels x and y. See Example 7, Example 8, and Example 9.
- The graph of an inverse function is the reflection of the graph of the original function across the line $y = x$. See Example 10.

Exercises

Review Exercises

Functions and Function Notation

For the following exercises, determine whether the relation is a function.

1. $\{(a, b), (c, d), (e, d)\}$

2. $\{(5, 2), (6, 1), (6, 2), (4, 8)\}$

3. $y^2 + 4 = x$, for x the independent variable and y the dependent variable

4. Is the graph in Figure 1 a function?

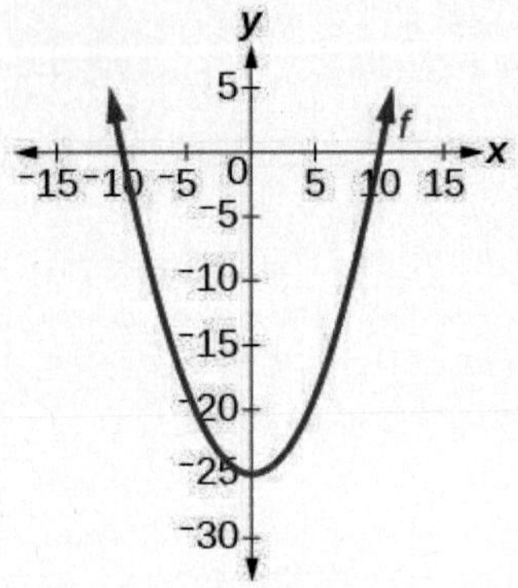

Figure 1

For the following exercises, evaluate $f(-3)$; $f(2)$; $f(-a)$; $-f(a)$; $f(a+h)$.

5. $f(x) = -2x^2 + 3x$

6. $f(x) = 2\,|3x - 1|$

For the following exercises, determine whether the functions are one-to-one.

7. $f(x) = -3x + 5$

8. $f(x) = |x - 3|$

For the following exercises, use the vertical line test to determine if the relation whose graph is provided is a function.

9.

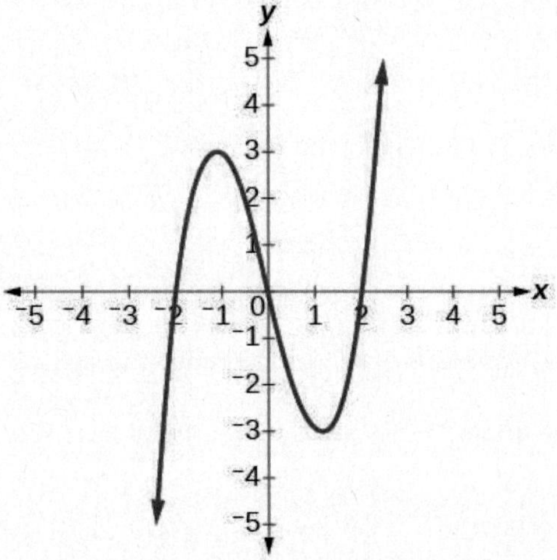

10.

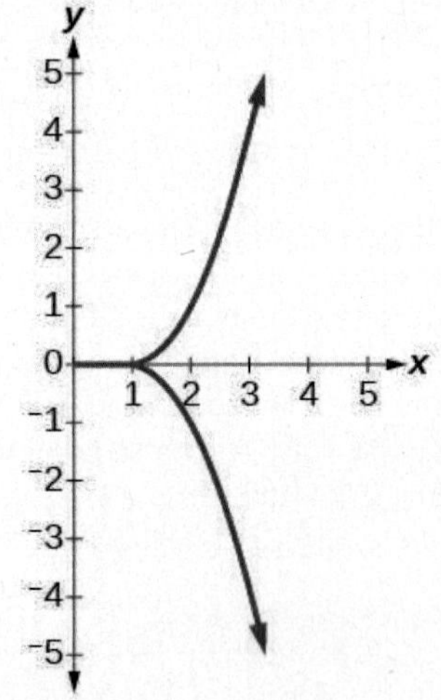

11.

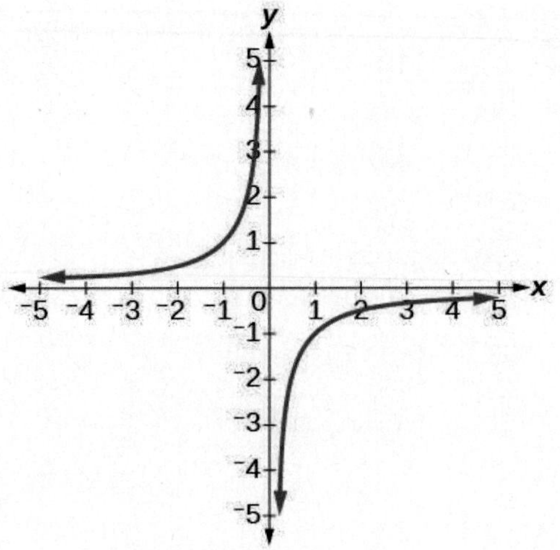

For the following exercises, graph the functions.

12. $f(x) = |x + 1|$

13. $f(x) = x^2 - 2$

For the following exercises, use Figure 2 to approximate the values.

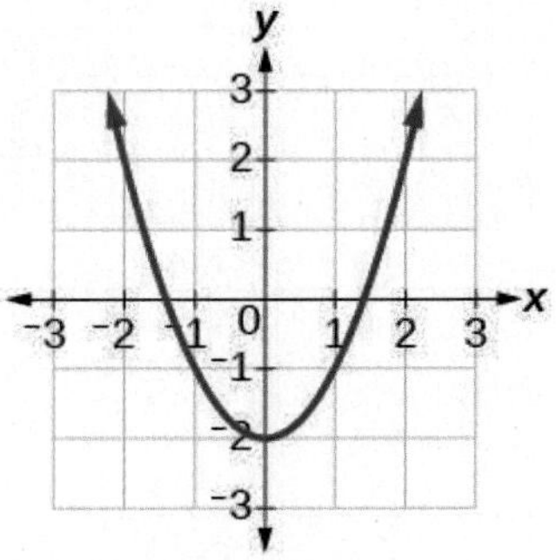

Figure 2

14. $f(2)$

15. $f(-2)$

16. If $f(x) = -2$, then solve for x.

17. If $f(x) = 1$, then solve for x.

For the following exercises, use the function $h(t) = -16t^2 + 80t$ to find the values in simplest form.

18. $\frac{h(2)-h(1)}{2-1}$

19. $\frac{h(a)-h(1)}{a-1}$

Domain and Range

For the following exercises, find the domain of each function, expressing answers using interval notation.

20. $f(x) = \frac{2}{3x+2}$

21. $f(x) = \frac{x-3}{x^2-4x-12}$

22. $f(x) = \frac{\sqrt{x-6}}{\sqrt{x-4}}$

23. Graph this piecewise function:
$$f(x) = \begin{cases} x+1 & x < -2 \\ -2x-3 & x \geq -2 \end{cases}$$

Rates of Change and Behavior of Graphs

For the following exercises, find the average rate of change of the functions from $x = 1$ to $x = 2$.

24. $f(x) = 4x - 3$

25. $f(x) = 10x^2 + x$

26. $f(x) = -\frac{2}{x^2}$

For the following exercises, use the graphs to determine the intervals on which the functions are increasing, decreasing, or constant.

27.

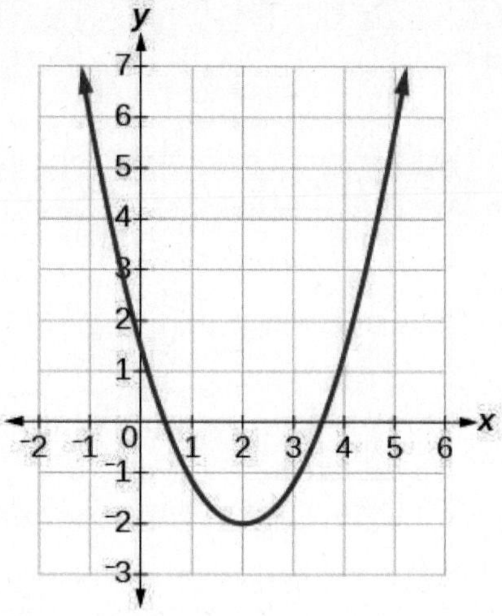

28.

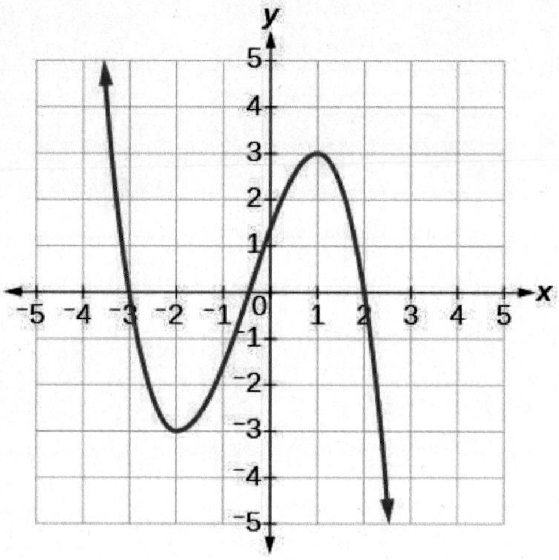

29.

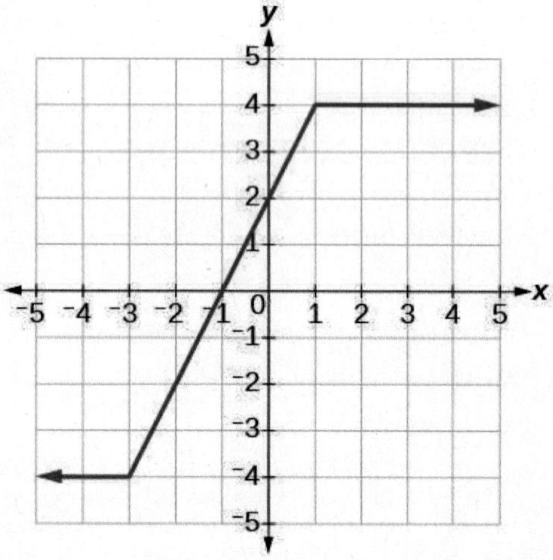

30. Find the local minimum of the function graphed in Exercise 3.27.

31. Find the local extrema for the function graphed in Exercise 3.28.

32. For the graph in Figure 3, the domain of the function is $[-3, 3]$. The range is $[-10, 10]$. Find the absolute minimum of the function on this interval.

33. Find the absolute maximum of the function graphed in Figure 3.

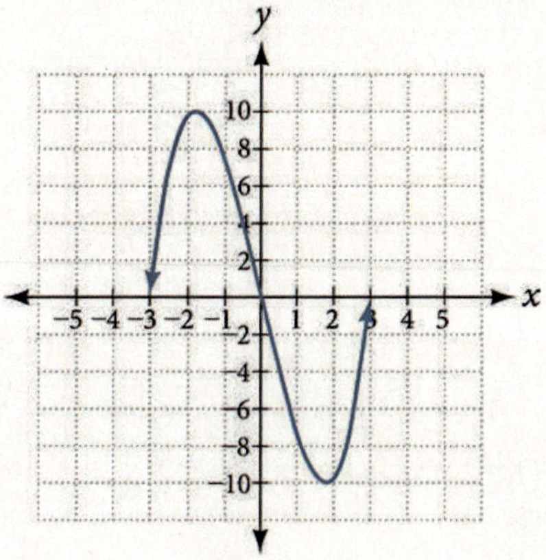

Figure 3

Composition of Functions

For the following exercises, find $(f \circ g)(x)$ and $(g \circ f)(x)$ for each pair of functions.

34. $f(x) = 4 - x, \;\; g(x) = -4x$

35. $f(x) = 3x + 2, \;\; g(x) = 5 - 6x$

36. $f(x) = x^2 + 2x, \;\; g(x) = 5x + 1$

37. $f(x) = \sqrt{x+2}, \;\; g(x) = \frac{1}{x}$

38. $f(x) = \frac{x+3}{2}, \;\; g(x) = \sqrt{1-x}$

For the following exercises, find $(f \circ g)$ and the domain for $(f \circ g)(x)$ for each pair of functions.

39. $f(x) = \frac{x+1}{x+4}, \;\; g(x) = \frac{1}{x}$

40. $f(x) = \frac{1}{x+3}, \;\; g(x) = \frac{1}{x-9}$

41. $f(x) = \frac{1}{x}, \;\; g(x) = \sqrt{x}$

42. $f(x) = \frac{1}{x^2-1}, \;\; g(x) = \sqrt{x+1}$

For the following exercises, express each function H as a composition of two functions f and g where $H(x) = (f \circ g)(x)$.

43. $H(x) = \sqrt{\frac{2x-1}{3x+4}}$

44. $H(x) = \frac{1}{(3x^2-4)^{-3}}$

Transformation of Functions

For the following exercises, sketch a graph of the given function.

45. $f(x) = (x-3)^2$

46. $f(x) = (x+4)^3$

47. $f(x) = \sqrt{x} + 5$

48. $f(x) = -x^3$

49. $f(x) = \sqrt[3]{-x}$

50. $f(x) = 5\sqrt{-x} - 4$

51. $f(x) = 4\left[|x-2| - 6\right]$

52. $f(x) = -(x+2)^2 - 1$

For the following exercises, sketch the graph of the function g if the graph of the function f is shown in Figure 4.

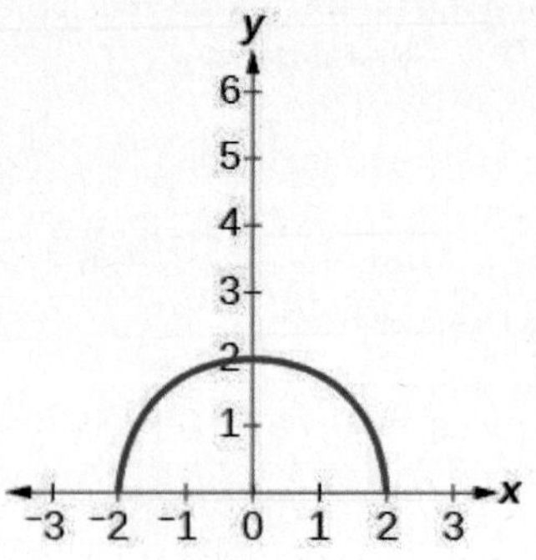

Figure 4

53. $g(x) = f(x - 1)$

54. $g(x) = 3f(x)$

For the following exercises, write the equation for the standard function represented by each of the graphs below.

55.

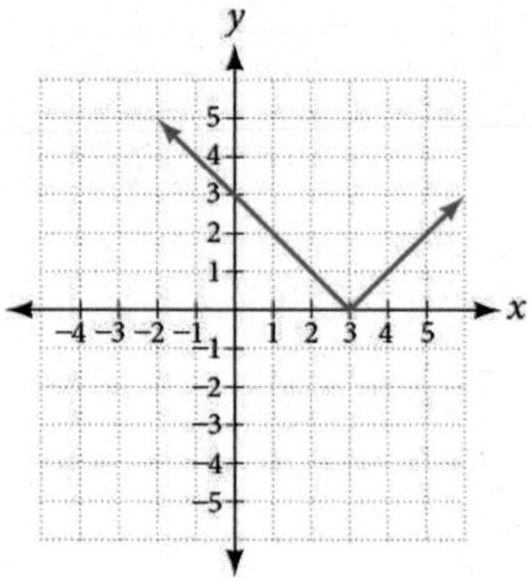

56.

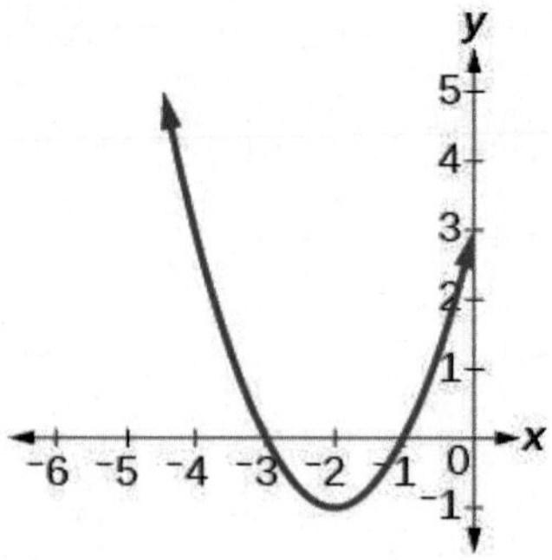

For the following exercises, determine whether each function below is even, odd, or neither.

57. $f(x) = 3x^4$

58. $g(x) = \sqrt{x}$

59. $h(x) = \frac{1}{x} + 3x$

For the following exercises, analyze the graph and determine whether the graphed function is even, odd, or neither.

60.

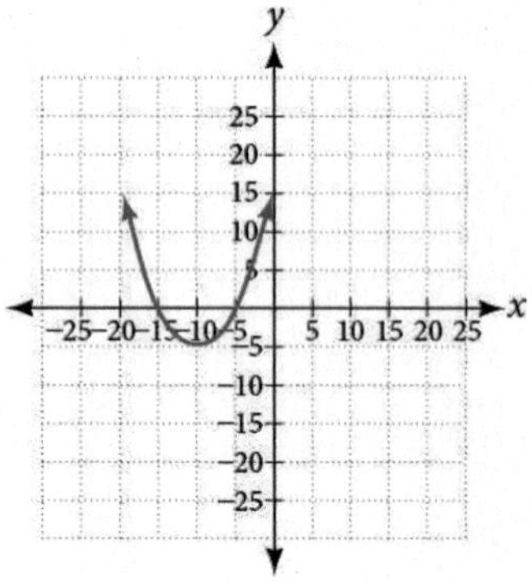

61.

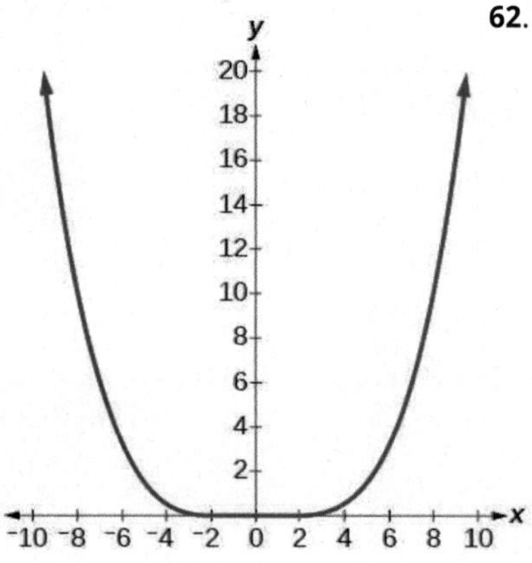

62.

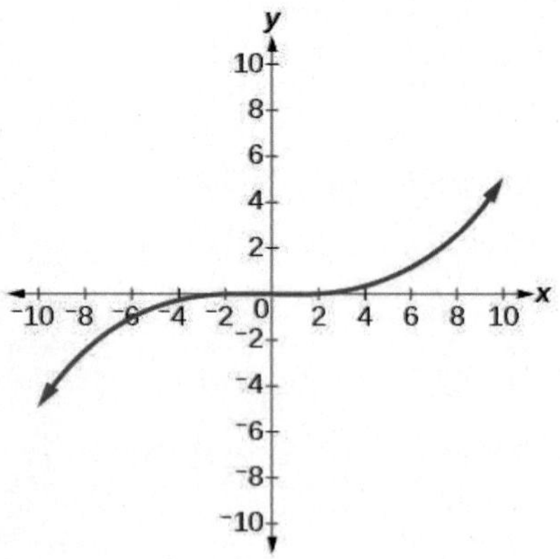

Absolute Value Functions

For the following exercises, write an equation for the transformation of $f(x) = |x|$.

63.

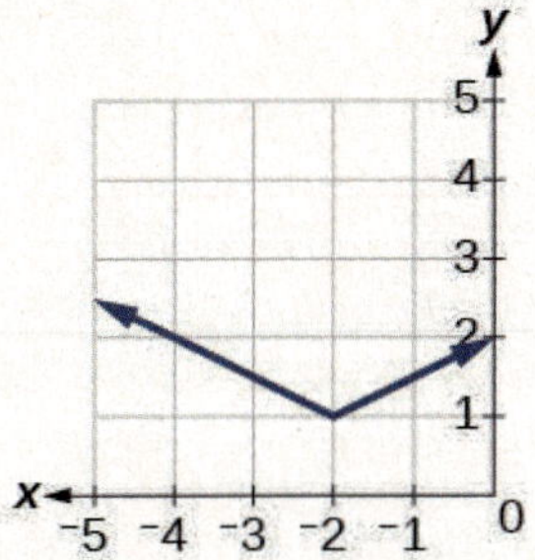

64.

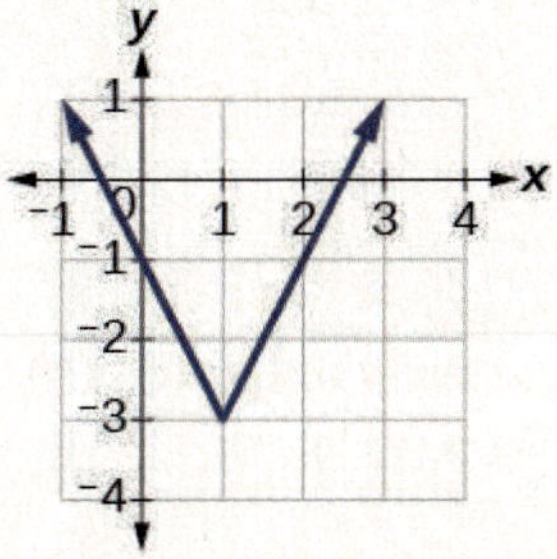

65.

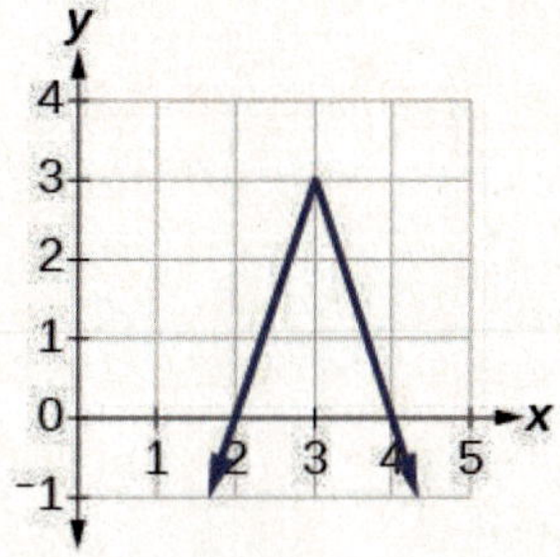

For the following exercises, graph the absolute value function.

66. $f(x) = |x - 5|$

67. $f(x) = -|x - 3|$

68. $f(x) = |2x - 4|$

Inverse Functions

For the following exercises, find $f^{-1}(x)$ for each function.

69. $f(x) = 9 + 10x$

70. $f(x) = \frac{x}{x+2}$

For the following exercise, find a domain on which the function f is one-to-one and non-decreasing. Write the domain in interval notation. Then find the inverse of f restricted to that domain.

71. $f(x) = x^2 + 1$

72. Given $f(x) = x^3 - 5$ and $g(x) = \sqrt[3]{x+5}$:

ⓐ Find $f(g(x))$ and $g(f(x))$.
ⓑ What does the answer tell us about the relationship between $f(x)$ and $g(x)$?

For the following exercises, use a graphing utility to determine whether each function is one-to-one.

73. $f(x) = \frac{1}{x}$

74. $f(x) = -3x^2 + x$

75. If $f(5) = 2$, find $f^{-1}(2)$.

76. If $f(1) = 4$, find $f^{-1}(4)$.

Practice Test

For the following exercises, determine whether each of the following relations is a function.

1. $y = 2x + 8$

2. $\{(2, 1), (3, 2), (-1, 1), (0, -2)\}$

For the following exercises, evaluate the function $f(x) = -3x^2 + 2x$ at the given input.

3. $f(-2)$

4. $f(a)$

5. Show that the function $f(x) = -2(x-1)^2 + 3$ is not one-to-one.

6. Write the domain of the function $f(x) = \sqrt{3-x}$ in interval notation.

7. Given $f(x) = 2x^2 - 5x$, find $f(a+1) - f(1)$ in simplest form.

8. Graph the function $f(x) = \begin{cases} x+1 & \text{if} \quad -2 < x < 3 \\ -x & \text{if} \quad x \geq 3 \end{cases}$

9. Find the average rate of change of the function $f(x) = 3 - 2x^2 + x$ by finding $\frac{f(b)-f(a)}{b-a}$ in simplest form.

For the following exercises, use the functions $f(x) = 3 - 2x^2 + x$ and $g(x) = \sqrt{x}$ to find the composite functions.

10. $(g \circ f)(x)$

11. $(g \circ f)(1)$

12. Express $H(x) = \sqrt[3]{5x^2 - 3x}$ as a composition of two functions, f and g, where $(f \circ g)(x) = H(x)$.

For the following exercises, graph the functions by translating, stretching, and/or compressing a toolkit function.

13. $f(x) = \sqrt{x+6} - 1$

14. $f(x) = \frac{1}{x+2} - 1$

For the following exercises, determine whether the functions are even, odd, or neither.

15. $f(x) = -\frac{5}{x^2} + 9x^6$

16. $f(x) = -\frac{5}{x^3} + 9x^5$

17. $f(x) = \frac{1}{x}$

18. Graph the absolute value function $f(x) = -2|x-1| + 3$.

For the following exercises, find the inverse of the function.

19. $f(x) = 3x - 5$

20. $f(x) = \frac{4}{x+7}$

For the following exercises, use the graph of g shown in Figure 1.

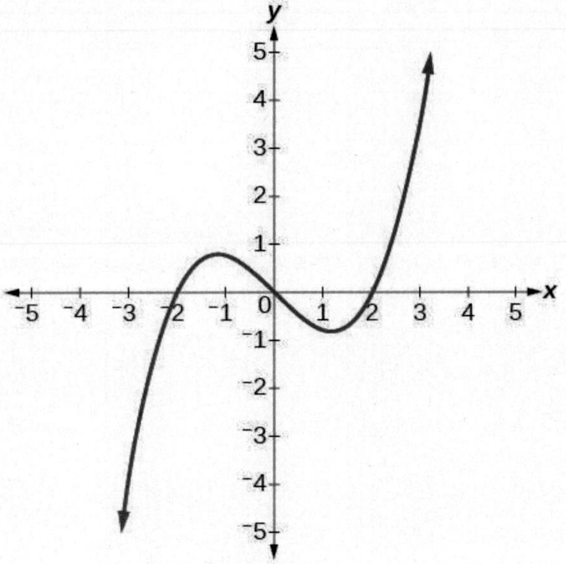

Figure 1

21. On what intervals is the function increasing?

22. On what intervals is the function decreasing?

23. Approximate the local minimum of the function. Express the answer as an ordered pair.

24. Approximate the local maximum of the function. Express the answer as an ordered pair.

For the following exercises, use the graph of the piecewise function shown in Figure 2.

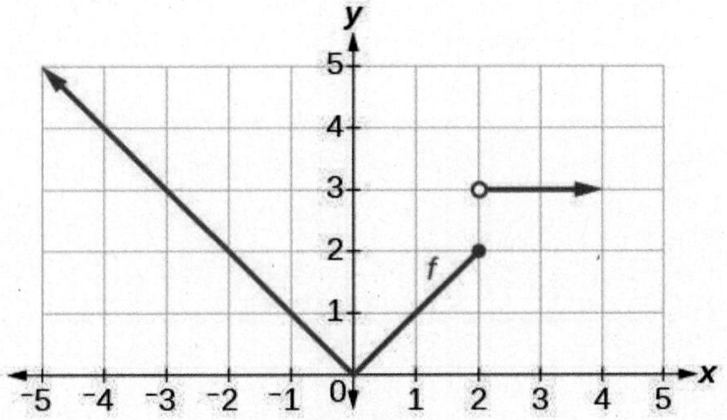

Figure 2

25. Find $f(2)$.

26. Find $f(-2)$.

27. Write an equation for the piecewise function.

For the following exercises, use the values listed in Table 1.

x	0	1	2	3	4	5	6	7	8
$F(x)$	1	3	5	7	9	11	13	15	17

Table 1

28. Find $F(6)$.

29. Solve the equation $F(x) = 5$.

30. Is the graph increasing or decreasing on its domain?

31. Is the function represented by the graph one-to-one?

32. Find $F^{-1}(15)$.

33. Given $f(x) = -2x + 11$, find $f^{-1}(x)$.

4 LINEAR FUNCTIONS

A bamboo forest in China (credit: "JFXie"/Flickr)

Chapter Outline

Introduction to Linear Functions

Imagine placing a plant in the ground one day and finding that it has doubled its height just a few days later. Although it may seem incredible, this can happen with certain types of bamboo species. These members of the grass family are the fastest-growing plants in the world. One species of bamboo has been observed to grow nearly 1.5 inches every hour. [1] In a twenty-four hour period, this bamboo plant grows about 36 inches, or an incredible 3 feet! A constant rate of change, such as the growth cycle of this bamboo plant, is a linear function.

Recall from Functions and Function Notation that a function is a relation that assigns to every element in the domain exactly one element in the range. Linear functions are a specific type of function that can be used to model many real-world applications, such as plant growth over time. In this chapter, we will explore linear functions, their graphs, and how to relate them to data.

4.1 Linear Functions

Learning Objectives

In this section, you will:

- Represent a linear function.
- Determine whether a linear function is increasing, decreasing, or constant.
- Interpret slope as a rate of change.
- Write and interpret an equation for a linear function.
- Graph linear functions.
- Determine whether lines are parallel or perpendicular.
- Write the equation of a line parallel or perpendicular to a given line.

1 http://www.guinnessworldrecords.com/records-3000/fastest-growing-plant/

Figure 1 Shanghai MagLev Train (credit: "kanegen"/Flickr)

Just as with the growth of a bamboo plant, there are many situations that involve constant change over time. Consider, for example, the first commercial maglev train in the world, the Shanghai MagLev Train (Figure 1). It carries passengers comfortably for a 30-kilometer trip from the airport to the subway station in only eight minutes[2].

Suppose a maglev train travels a long distance, and maintains a constant speed of 83 meters per second for a period of time once it is 250 meters from the station. How can we analyze the train's distance from the station as a function of time? In this section, we will investigate a kind of function that is useful for this purpose, and use it to investigate real-world situations such as the train's distance from the station at a given point in time.

Representing Linear Functions

The function describing the train's motion is a linear function, which is defined as a function with a constant rate of change. This is a polynomial of degree 1. There are several ways to represent a linear function, including word form, function notation, tabular form, and graphical form. We will describe the train's motion as a function using each method.

Representing a Linear Function in Word Form

Let's begin by describing the linear function in words. For the train problem we just considered, the following word sentence may be used to describe the function relationship.

- *The train's distance from the station is a function of the time during which the train moves at a constant speed plus its original distance from the station when it began moving at constant speed.*

The speed is the rate of change. Recall that a rate of change is a measure of how quickly the dependent variable changes with respect to the independent variable. The rate of change for this example is constant, which means that it is the same for each input value. As the time (input) increases by 1 second, the corresponding distance (output) increases by 83 meters. The train began moving at this constant speed at a distance of 250 meters from the station.

Representing a Linear Function in Function Notation

Another approach to representing linear functions is by using function notation. One example of function notation is an equation written in the slope-intercept form of a line, where x is the input value, m is the rate of change, and b is the initial value of the dependent variable.

$$\begin{array}{ll} \text{Equation form} & y = mx + b \\ \text{Function notation} & f(x) = mx + b \end{array}$$

In the example of the train, we might use the notation $D(t)$ where the total distance D is a function of the time t. The rate, m, is 83 meters per second. The initial value of the dependent variable b is the original distance from the station, 250 meters. We can write a generalized equation to represent the motion of the train.

$$D(t) = 83t + 250$$

Representing a Linear Function in Tabular Form

A third method of representing a linear function is through the use of a table. The relationship between the distance from the station and the time is represented in Figure 2. From the table, we can see that the distance changes by 83 meters for every 1 second increase in time.

2 http://www.chinahighlights.com/shanghai/transportation/maglev-train.htm

1 second 1 second 1 second

t	0	1	2	3
$D(t)$	250	333	416	499

83 meters 83 meters 83 meters

Figure 2 Tabular representation of the function D showing selected input and output values

Q&A **Can the input in the previous example be any real number?**

No. The input represents time so while nonnegative rational and irrational numbers are possible, negative real numbers are not possible for this example. The input consists of non-negative real numbers.

Representing a Linear Function in Graphical Form

Another way to represent linear functions is visually, using a graph. We can use the function relationship from above, $D(t) = 83t + 250$, to draw a graph as represented in Figure 3. Notice the graph is a line. When we plot a linear function, the graph is always a line.

The rate of change, which is constant, determines the slant, or slope of the line. The point at which the input value is zero is the vertical intercept, or *y*-intercept, of the line. We can see from the graph that the *y*-intercept in the train example we just saw is $(0, 250)$ and represents the distance of the train from the station when it began moving at a constant speed.

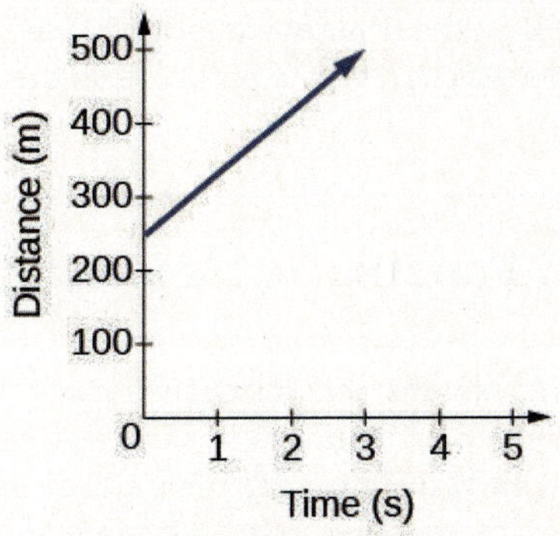

Figure 3 The graph of $D(t) = 83t + 250$. Graphs of linear functions are lines because the rate of change is constant.

Notice that the graph of the train example is restricted, but this is not always the case. Consider the graph of the line $f(x) = 2x + 1$. Ask yourself what numbers can be input to the function. In other words, what is the domain of the function? The domain is comprised of all real numbers because any number may be doubled, and then have one added to the product.

Linear Function

A **linear function** is a function whose graph is a line. Linear functions can be written in the **slope-intercept form** of a line

$$f(x) = mx + b$$

where b is the initial or starting value of the function (when input, $x = 0$), and m is the constant rate of change, or slope of the function. The *y*-intercept is at $(0, b)$.

EXAMPLE 1

Using a Linear Function to Find the Pressure on a Diver

The pressure, P, in pounds per square inch (PSI) on the diver in Figure 4 depends upon her depth below the water surface, d, in feet. This relationship may be modeled by the equation, $P(d) = 0.434d + 14.696$. Restate this function in words.

Figure 4 (credit: Ilse Reijs and Jan-Noud Hutten)

Solution

To restate the function in words, we need to describe each part of the equation. The pressure as a function of depth equals four hundred thirty-four thousandths times depth plus fourteen and six hundred ninety-six thousandths.

Analysis

The initial value, 14.696, is the pressure in PSI on the diver at a depth of 0 feet, which is the surface of the water. The rate of change, or slope, is 0.434 PSI per foot. This tells us that the pressure on the diver increases 0.434 PSI for each foot her depth increases.

Determining Whether a Linear Function Is Increasing, Decreasing, or Constant

The linear functions we used in the two previous examples increased over time, but not every linear function does. A linear function may be increasing, decreasing, or constant. For an increasing function, as with the train example, the output values increase as the input values increase. The graph of an increasing function has a positive slope. A line with a positive slope slants upward from left to right as in Figure 5**(a)**. For a decreasing function, the slope is negative. The output values decrease as the input values increase. A line with a negative slope slants downward from left to right as in Figure 5**(b)**. If the function is constant, the output values are the same for all input values so the slope is zero. A line with a slope of zero is horizontal as in Figure 5**(c)**.

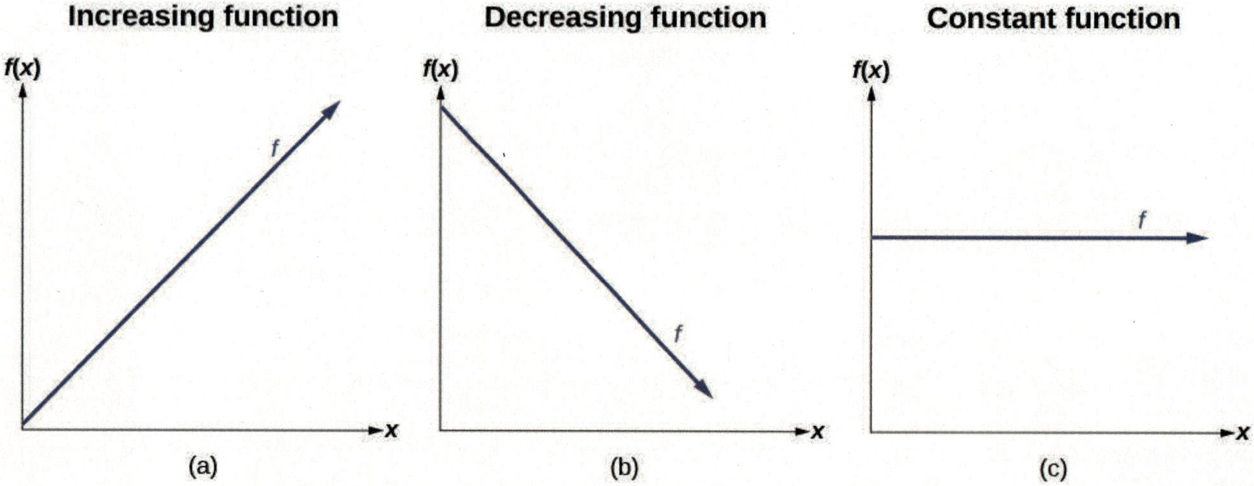

Figure 5

Increasing and Decreasing Functions

The slope determines if the function is an **increasing linear function**, a **decreasing linear function**, or a constant function.

$f(x) = mx + b$ is an increasing function if $m > 0$.
$f(x) = mx + b$ is a decreasing function if $m < 0$.
$f(x) = mx + b$ is a constant function if $m = 0$.

EXAMPLE 2

Deciding Whether a Function Is Increasing, Decreasing, or Constant

Studies from the early 2010s indicated that teens sent about 60 texts a day, while more recent data indicates much higher messaging rates among all users, particularly considering the various apps with which people can communicate.[3]. For each of the following scenarios, find the linear function that describes the relationship between the input value and the output value. Then, determine whether the graph of the function is increasing, decreasing, or constant.

ⓐ The total number of texts a teen sends is considered a function of time in days. The input is the number of days, and output is the total number of texts sent.
ⓑ A person has a limit of 500 texts per month in their data plan. The input is the number of days, and output is the total number of texts remaining for the month.
ⓒ A person has an unlimited number of texts in their data plan for a cost of $50 per month. The input is the number of days, and output is the total cost of texting each month.

Solution

Analyze each function.

ⓐ The function can be represented as $f(x) = 60x$ where x is the number of days. The slope, 60, is positive so the function is increasing. This makes sense because the total number of texts increases with each day.
ⓑ The function can be represented as $f(x) = 500 - 60x$ where x is the number of days. In this case, the slope is negative so the function is decreasing. This makes sense because the number of texts remaining decreases each day and this function represents the number of texts remaining in the data plan after x days.
ⓒ The cost function can be represented as $f(x) = 50$ because the number of days does not affect the total cost. The slope is 0 so the function is constant.

Interpreting Slope as a Rate of Change

In the examples we have seen so far, the slope was provided to us. However, we often need to calculate the slope given input and output values. Recall that given two values for the input, x_1 and x_2, and two corresponding values for the output, y_1 and y_2 —which can be represented by a set of points, (x_1, y_1) and (x_2, y_2) —we can calculate the slope m.

$$m = \frac{\text{change in output (rise)}}{\text{change in input (run)}} = \frac{\Delta y}{\Delta x} = \frac{y_2 - y_1}{x_2 - x_1}$$

Note that in function notation we can obtain two corresponding values for the output y_1 and y_2 for the function f, $y_1 = f(x_1)$ and $y_2 = f(x_2)$, so we could equivalently write

$$m = \frac{f(x_2) - f(x_1)}{x_2 - x_1}$$

Figure 6 indicates how the slope of the line between the points, (x_1, y_1) and (x_2, y_2), is calculated. Recall that the slope measures steepness, or slant. The greater the absolute value of the slope, the steeper the slant is.

3 http://www.cbsnews.com/8301-501465_162-57400228-501465/teens-are-sending-60-texts-a-day-study-says/

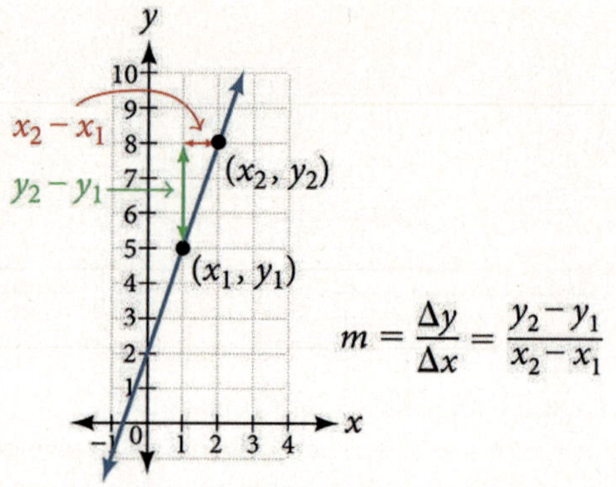

Figure 6 The slope of a function is calculated by the change in y divided by the change in x. It does not matter which coordinate is used as the (x_2, y_2) and which is the (x_1, y_1), as long as each calculation is started with the elements from the same coordinate pair.

Q&A **Are the units for slope always** $\frac{\text{units for the output}}{\text{units for the input}}$ **?**

Yes. Think of the units as the change of output value for each unit of change in input value. An example of slope could be miles per hour or dollars per day. Notice the units appear as a ratio of units for the output per units for the input.

Calculate Slope

The slope, or rate of change, of a function m can be calculated according to the following:

$$m = \frac{\text{change in output (rise)}}{\text{change in input (run)}} = \frac{\Delta y}{\Delta x} = \frac{y_2 - y_1}{x_2 - x_1}$$

where x_1 and x_2 are input values, y_1 and y_2 are output values.

HOW TO

Given two points from a linear function, calculate and interpret the slope.

1. Determine the units for output and input values.
2. Calculate the change of output values and change of input values.
3. Interpret the slope as the change in output values per unit of the input value.

EXAMPLE 3

Finding the Slope of a Linear Function

If $f(x)$ is a linear function, and $(3, -2)$ and $(8, 1)$ are points on the line, find the slope. Is this function increasing or decreasing?

Solution

The coordinate pairs are $(3, -2)$ and $(8, 1)$. To find the rate of change, we divide the change in output by the change in input.

$$m = \frac{\text{change in output}}{\text{change in input}} = \frac{1 - (-2)}{8 - 3} = \frac{3}{5}$$

We could also write the slope as $m = 0.6$. The function is increasing because $m > 0$.

Analysis

As noted earlier, the order in which we write the points does not matter when we compute the slope of the line as long as the first output value, or *y*-coordinate, used corresponds with the first input value, or *x*-coordinate, used. Note that if we had reversed them, we would have obtained the same slope.

$$m = \frac{(-2) - (1)}{3 - 8} = \frac{-3}{-5} = \frac{3}{5}$$

TRY IT #1 If $f(x)$ is a linear function, and $(2, 3)$ and $(0, 4)$ are points on the line, find the slope. Is this function increasing or decreasing?

EXAMPLE 4

Finding the Population Change from a Linear Function

The population of a city increased from 23,400 to 27,800 between 2008 and 2012. Find the change of population per year if we assume the change was constant from 2008 to 2012.

Solution

The rate of change relates the change in population to the change in time. The population increased by $27,800 - 23,400 = 4400$ people over the four-year time interval. To find the rate of change, divide the change in the number of people by the number of years.

$$\frac{4{,}400 \text{ people}}{4 \text{ years}} = 1{,}100 \; \frac{\text{people}}{\text{year}}$$

So the population increased by 1,100 people per year.

Analysis

Because we are told that the population increased, we would expect the slope to be positive. This positive slope we calculated is therefore reasonable.

TRY IT #2 The population of a small town increased from 1,442 to 1,868 between 2009 and 2012. Find the change of population per year if we assume the change was constant from 2009 to 2012.

Writing and Interpreting an Equation for a Linear Function

Recall from Equations and Inequalities that we wrote equations in both the slope-intercept form and the point-slope form. Now we can choose which method to use to write equations for linear functions based on the information we are given. That information may be provided in the form of a graph, a point and a slope, two points, and so on. Look at the graph of the function f in Figure 7.

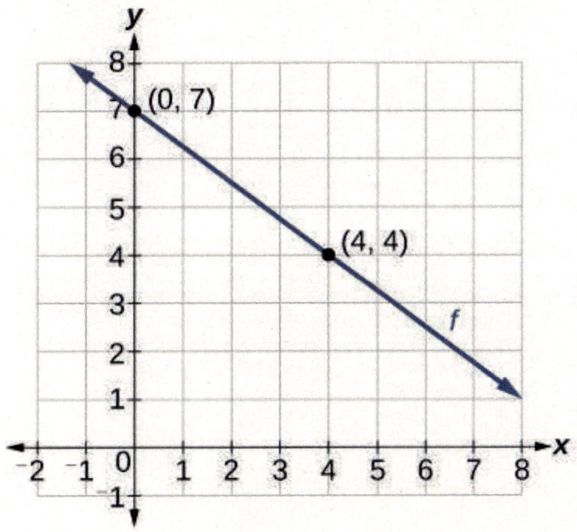

Figure 7

We are not given the slope of the line, but we can choose any two points on the line to find the slope. Let's choose $(0, 7)$ and $(4, 4)$.

$$\begin{aligned} m &= \frac{y_2 - y_1}{x_2 - x_1} \\ &= \frac{4-7}{4-0} \\ &= -\frac{3}{4} \end{aligned}$$

Now we can substitute the slope and the coordinates of one of the points into the point-slope form.

$$\begin{aligned} y - y_1 &= m(x - x_1) \\ y - 4 &= -\frac{3}{4}(x - 4) \end{aligned}$$

If we want to rewrite the equation in the slope-intercept form, we would find

$$\begin{aligned} y - 4 &= -\frac{3}{4}(x - 4) \\ y - 4 &= -\frac{3}{4}x + 3 \\ y &= -\frac{3}{4}x + 7 \end{aligned}$$

If we want to find the slope-intercept form without first writing the point-slope form, we could have recognized that the line crosses the *y*-axis when the output value is 7. Therefore, $b = 7$. We now have the initial value b and the slope m so we can substitute m and b into the slope-intercept form of a line.

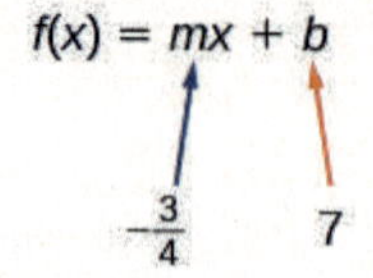

$$f(x) = -\frac{3}{4}x + 7$$

So the function is $f(x) = -\frac{3}{4}x + 7$, and the linear equation would be $y = -\frac{3}{4}x + 7$.

HOW TO

Given the graph of a linear function, write an equation to represent the function.

1. Identify two points on the line.
2. Use the two points to calculate the slope.
3. Determine where the line crosses the *y*-axis to identify the *y*-intercept by visual inspection.
4. Substitute the slope and *y*-intercept into the slope-intercept form of a line equation.

EXAMPLE 5

Writing an Equation for a Linear Function

Write an equation for a linear function given a graph of f shown in Figure 8.

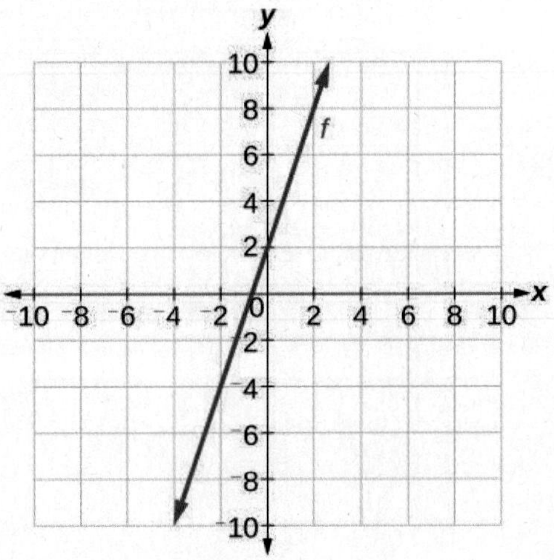

Figure 8

Solution

Identify two points on the line, such as $(0, 2)$ and $(-2, -4)$. Use the points to calculate the slope.

$$\begin{aligned} m &= \frac{y_2 - y_1}{x_2 - x_1} \\ &= \frac{-4-2}{-2-0} \\ &= \frac{-6}{-2} \\ &= 3 \end{aligned}$$

Substitute the slope and the coordinates of one of the points into the point-slope form.

$$\begin{aligned} y - y_1 &= m(x - x_1) \\ y - (-4) &= 3(x - (-2)) \\ y + 4 &= 3(x + 2) \end{aligned}$$

We can use algebra to rewrite the equation in the slope-intercept form.

$$\begin{aligned} y + 4 &= 3(x + 2) \\ y + 4 &= 3x + 6 \\ y &= 3x + 2 \end{aligned}$$

Analysis

This makes sense because we can see from Figure 9 that the line crosses the *y*-axis at the point $(0, \ 2)$, which is the *y*-intercept, so $b = 2$.

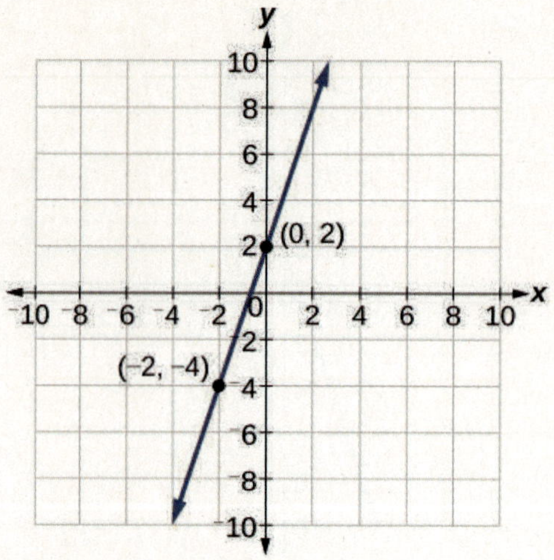

Figure 9

EXAMPLE 6

Writing an Equation for a Linear Cost Function

Suppose Ben starts a company in which he incurs a fixed cost of $1,250 per month for the overhead, which includes his office rent. His production costs are $37.50 per item. Write a linear function C where $C(x)$ is the cost for x items produced in a given month.

Solution

The fixed cost is present every month, $1,250. The costs that can vary include the cost to produce each item, which is $37.50. The variable cost, called the marginal cost, is represented by 37.5. The cost Ben incurs is the sum of these two costs, represented by $C(x) = 1250 + 37.5x$.

Analysis

If Ben produces 100 items in a month, his monthly cost is found by substituting 100 for x.

$$\begin{aligned} C(100) &= 1250 + 37.5(100) \\ &= 5000 \end{aligned}$$

So his monthly cost would be $5,000.

EXAMPLE 7

Writing an Equation for a Linear Function Given Two Points

If f is a linear function, with $f(3) = -2$, and $f(8) = 1$, find an equation for the function in slope-intercept form.

Solution

We can write the given points using coordinates.

$$\begin{aligned} f(3) &= -2 \to (3, -2) \\ f(8) &= 1 \to (8, 1) \end{aligned}$$

We can then use the points to calculate the slope.

$$\begin{aligned} m &= \frac{y_2 - y_1}{x_2 - x_1} \\ &= \frac{1-(-2)}{8-3} \\ &= \frac{3}{5} \end{aligned}$$

Substitute the slope and the coordinates of one of the points into the point-slope form.

$$\begin{aligned} y - y_1 &= m(x - x_1) \\ y - (-2) &= \tfrac{3}{5}(x - 3) \end{aligned}$$

We can use algebra to rewrite the equation in the slope-intercept form.

$$\begin{aligned} y + 2 &= \tfrac{3}{5}(x - 3) \\ y + 2 &= \tfrac{3}{5}x - \tfrac{9}{5} \\ y &= \tfrac{3}{5}x - \tfrac{19}{5} \end{aligned}$$

TRY IT #3 If $f(x)$ is a linear function, with $f(2) = -11$, and $f(4) = -25$, write an equation for the function in slope-intercept form.

Modeling Real-World Problems with Linear Functions

In the real world, problems are not always explicitly stated in terms of a function or represented with a graph. Fortunately, we can analyze the problem by first representing it as a linear function and then interpreting the components of the function. As long as we know, or can figure out, the initial value and the rate of change of a linear function, we can solve many different kinds of real-world problems.

HOW TO

Given a linear function f and the initial value and rate of change, evaluate $f(c)$.

1. Determine the initial value and the rate of change (slope).
2. Substitute the values into $f(x) = mx + b$.
3. Evaluate the function at $x = c$.

EXAMPLE 8

Using a Linear Function to Determine the Number of Songs in a Music Collection

Marcus currently has 200 songs in his music collection. Every month, he adds 15 new songs. Write a formula for the number of songs, N, in his collection as a function of time, t, the number of months. How many songs will he own at the end of one year?

Solution

The initial value for this function is 200 because he currently owns 200 songs, so $N(0) = 200$, which means that $b = 200$.

The number of songs increases by 15 songs per month, so the rate of change is 15 songs per month. Therefore we know that $m = 15$. We can substitute the initial value and the rate of change into the slope-intercept form of a line.

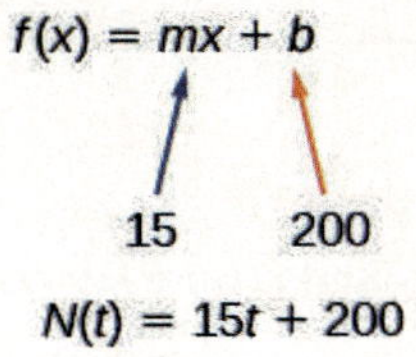

Figure 10

We can write the formula $N(t) = 15t + 200$.

With this formula, we can then predict how many songs Marcus will have at the end of one year (12 months). In other words, we can evaluate the function at $t = 12$.

$$\begin{aligned} N(12) &= 15(12) + 200 \\ &= 180 + 200 \\ &= 380 \end{aligned}$$

Marcus will have 380 songs in 12 months.

Analysis

Notice that N is an increasing linear function. As the input (the number of months) increases, the output (number of songs) increases as well.

EXAMPLE 9

Using a Linear Function to Calculate Salary Based on Commission

Working as an insurance salesperson, Ilya earns a base salary plus a commission on each new policy. Therefore, Ilya's weekly income I, depends on the number of new policies, n, he sells during the week. Last week he sold 3 new policies, and earned \$760 for the week. The week before, he sold 5 new policies and earned \$920. Find an equation for $I(n)$, and interpret the meaning of the components of the equation.

Solution

The given information gives us two input-output pairs: $(3, 760)$ and $(5, 920)$. We start by finding the rate of change.

$$\begin{aligned} m &= \frac{920-760}{5-3} \\ &= \frac{\$160}{2 \text{ policies}} \\ &= \$80 \text{ per policy} \end{aligned}$$

Keeping track of units can help us interpret this quantity. Income increased by \$160 when the number of policies increased by 2, so the rate of change is \$80 per policy. Therefore, Ilya earns a commission of \$80 for each policy sold during the week.

We can then solve for the initial value.

$$\begin{aligned} I(n) &= 80n + b \\ 760 &= 80(3) + b \qquad \text{when } n = 3, I(3) = 760 \\ 760 &- 80(3) = b \\ 520 &= b \end{aligned}$$

The value of b is the starting value for the function and represents Ilya's income when $n = 0$, or when no new policies are sold. We can interpret this as Ilya's base salary for the week, which does not depend upon the number of policies sold.

We can now write the final equation.

$$I(n) = 80n + 520$$

Our final interpretation is that Ilya's base salary is \$520 per week and he earns an additional \$80 commission for each policy sold.

EXAMPLE 10

Using Tabular Form to Write an Equation for a Linear Function

Table 1 relates the number of rats in a population to time, in weeks. Use the table to write a linear equation.

number of weeks, *w*	0	2	4	6
number of rats, *P(w)*	1000	1080	1160	1240

Table 1

Solution

We can see from the table that the initial value for the number of rats is 1000, so $b = 1000$.

Rather than solving for m, we can tell from looking at the table that the population increases by 80 for every 2 weeks that pass. This means that the rate of change is 80 rats per 2 weeks, which can be simplified to 40 rats per week.

$$P(w) = 40w + 1000$$

If we did not notice the rate of change from the table we could still solve for the slope using any two points from the table. For example, using $(2, 1080)$ and $(6, 1240)$

$$\begin{aligned} m &= \frac{1240-1080}{6-2} \\ &= \frac{160}{4} \\ &= 40 \end{aligned}$$

Q&A **Is the initial value always provided in a table of values like Table 1?**

No. Sometimes the initial value is provided in a table of values, but sometimes it is not. If you see an input of 0, then the initial value would be the corresponding output. If the initial value is not provided because there is no value of input on the table equal to 0, find the slope, substitute one coordinate pair and the slope into $f(x) = mx + b$, and solve for b.

TRY IT #4 A new plant food was introduced to a young tree to test its effect on the height of the tree. Table 2 shows the height of the tree, in feet, x months since the measurements began. Write a linear function, $H(x)$, where x is the number of months since the start of the experiment.

x	0	2	4	8	12
***H*(*x*)**	12.5	13.5	14.5	16.5	18.5

Table 2

Graphing Linear Functions

Now that we've seen and interpreted graphs of linear functions, let's take a look at how to create the graphs. There are three basic methods of graphing linear functions. The first is by plotting points and then drawing a line through the points. The second is by using the *y*-intercept and slope. And the third method is by using transformations of the identity function $f(x) = x$.

Graphing a Function by Plotting Points

To find points of a function, we can choose input values, evaluate the function at these input values, and calculate output values. The input values and corresponding output values form coordinate pairs. We then plot the coordinate pairs on a grid. In general, we should evaluate the function at a minimum of two inputs in order to find at least two points on the graph. For example, given the function, $f(x) = 2x$, we might use the input values 1 and 2. Evaluating the function for an input value of 1 yields an output value of 2, which is represented by the point $(1, 2)$. Evaluating the function for an input value of 2 yields an output value of 4, which is represented by the point $(2, 4)$. Choosing three points is often advisable because if all three points do not fall on the same line, we know we made an error.

HOW TO

Given a linear function, graph by plotting points.

1. Choose a minimum of two input values.
2. Evaluate the function at each input value.
3. Use the resulting output values to identify coordinate pairs.
4. Plot the coordinate pairs on a grid.
5. Draw a line through the points.

EXAMPLE 11

Graphing by Plotting Points

Graph $f(x) = -\frac{2}{3}x + 5$ by plotting points.

Solution

Begin by choosing input values. This function includes a fraction with a denominator of 3, so let's choose multiples of 3 as input values. We will choose 0, 3, and 6.

Evaluate the function at each input value, and use the output value to identify coordinate pairs.

$$\begin{array}{ll} x = 0 & f(0) = -\frac{2}{3}(0) + 5 = 5 \Rightarrow (0, 5) \\ x = 3 & f(3) = -\frac{2}{3}(3) + 5 = 3 \Rightarrow (3, 3) \\ x = 6 & f(6) = -\frac{2}{3}(6) + 5 = 1 \Rightarrow (6, 1) \end{array}$$

Plot the coordinate pairs and draw a line through the points. Figure 11 represents the graph of the function $f(x) = -\frac{2}{3}x + 5$.

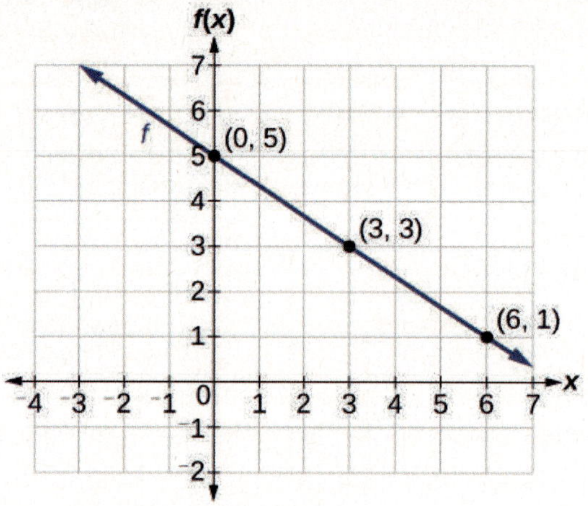

Figure 11 The graph of the linear function $f(x) = -\frac{2}{3}x + 5$.

Analysis

The graph of the function is a line as expected for a linear function. In addition, the graph has a downward slant, which indicates a negative slope. This is also expected from the negative, constant rate of change in the equation for the function.

TRY IT #5 Graph $f(x) = -\frac{3}{4}x + 6$ by plotting points.

Graphing a Function Using y-intercept and Slope

Another way to graph linear functions is by using specific characteristics of the function rather than plotting points. The first characteristic is its *y*-intercept, which is the point at which the input value is zero. To find the *y*-intercept, we can set $x = 0$ in the equation.

The other characteristic of the linear function is its slope.

Let's consider the following function.

$$f(x) = \frac{1}{2}x + 1$$

The slope is $\frac{1}{2}$. Because the slope is positive, we know the graph will slant upward from left to right. The *y*-intercept is the point on the graph when $x = 0$. The graph crosses the *y*-axis at $(0, 1)$. Now we know the slope and the *y*-intercept. We can begin graphing by plotting the point $(0, 1)$. We know that the slope is the change in the *y*-coordinate over the change in the *x*-coordinate. This is commonly referred to as rise over run, $m = \frac{\text{rise}}{\text{run}}$. From our example, we have $m = \frac{1}{2}$, which means that the rise is 1 and the run is 2. So starting from our *y*-intercept $(0, 1)$, we can rise 1 and then run 2, or run 2 and then rise 1. We repeat until we have a few points, and then we draw a line through the points as shown in Figure 12.

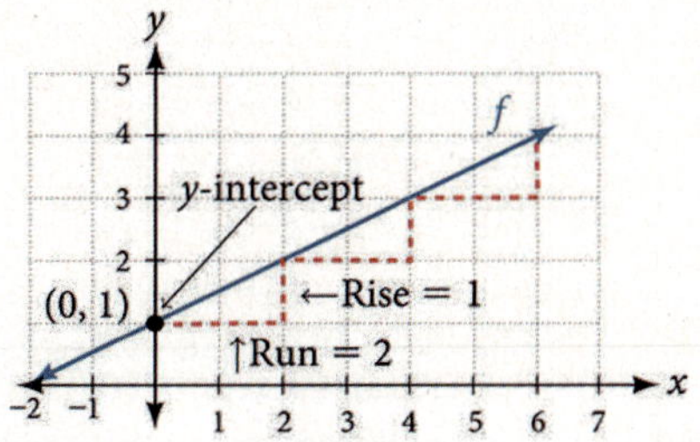

Figure 12

Graphical Interpretation of a Linear Function

In the equation $f(x) = mx + b$

- b is the y-intercept of the graph and indicates the point $(0, b)$ at which the graph crosses the y-axis.
- m is the slope of the line and indicates the vertical displacement (rise) and horizontal displacement (run) between each successive pair of points. Recall the formula for the slope:

$$m = \frac{\text{change in output (rise)}}{\text{change in input (run)}} = \frac{\Delta y}{\Delta x} = \frac{y_2 - y_1}{x_2 - x_1}$$

Q&A **Do all linear functions have *y*-intercepts?**

Yes. All linear functions cross the y-axis and therefore have y-intercepts. (Note: *A vertical line is parallel to the y-axis does not have a y-intercept, but it is not a function.*)

HOW TO

Given the equation for a linear function, graph the function using the *y*-intercept and slope.

1. Evaluate the function at an input value of zero to find the y-intercept.
2. Identify the slope as the rate of change of the input value.
3. Plot the point represented by the y-intercept.
4. Use $\frac{\text{rise}}{\text{run}}$ to determine at least two more points on the line.
5. Sketch the line that passes through the points.

EXAMPLE 12

Graphing by Using the *y*-intercept and Slope
Graph $f(x) = -\frac{2}{3}x + 5$ using the y-intercept and slope.

Solution
Evaluate the function at $x = 0$ to find the y-intercept. The output value when $x = 0$ is 5, so the graph will cross the y-axis at $(0, 5)$.

According to the equation for the function, the slope of the line is $-\frac{2}{3}$. This tells us that for each vertical decrease in the "rise" of -2 units, the "run" increases by 3 units in the horizontal direction. We can now graph the function by first plotting the y-intercept on the graph in Figure 13. From the initial value $(0, 5)$ we move down 2 units and to the right 3 units. We can extend the line to the left and right by repeating, and then drawing a line through the points.

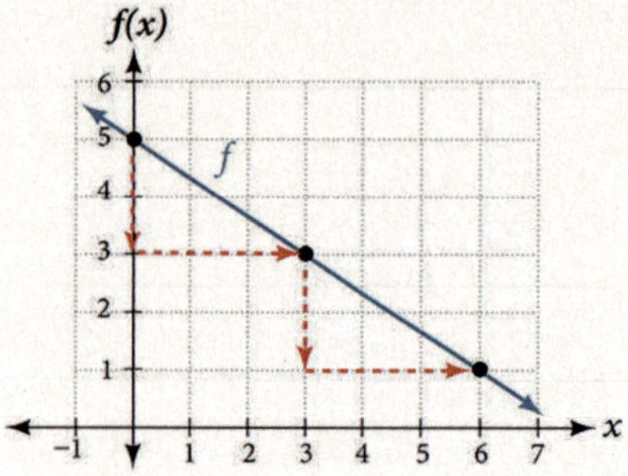

Figure 13 Graph of $f(x) = -2/3x + 5$ and shows how to calculate the rise over run for the slope.

Analysis

The graph slants downward from left to right, which means it has a negative slope as expected.

TRY IT #6 Find a point on the graph we drew in Example 12 that has a negative *x*-value.

Graphing a Function Using Transformations

Another option for graphing is to use a transformation of the identity function $f(x) = x$. A function may be transformed by a shift up, down, left, or right. A function may also be transformed using a reflection, stretch, or compression.

Vertical Stretch or Compression

In the equation $f(x) = mx$, the m is acting as the vertical stretch or compression of the identity function. When m is negative, there is also a vertical reflection of the graph. Notice in Figure 14 that multiplying the equation of $f(x) = x$ by m stretches the graph of f by a factor of m units if $m > 1$ and compresses the graph of f by a factor of m units if $0 < m < 1$. This means the larger the absolute value of m, the steeper the slope.

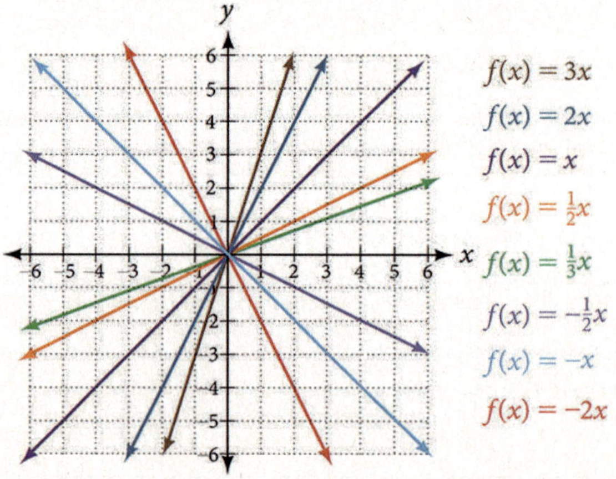

Figure 14 Vertical stretches and compressions and reflections on the function $f(x) = x$

Vertical Shift

In $f(x) = mx + b$, the b acts as the vertical shift, moving the graph up and down without affecting the slope of the line. Notice in Figure 15 that adding a value of b to the equation of $f(x) = x$ shifts the graph of f a total of b units up if b is positive and $|b|$ units down if b is negative.

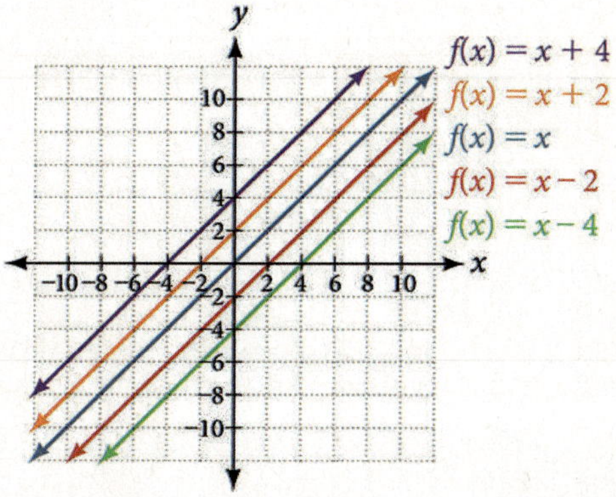

Figure 15 This graph illustrates vertical shifts of the function $f(x) = x$.

Using vertical stretches or compressions along with vertical shifts is another way to look at identifying different types of linear functions. Although this may not be the easiest way to graph this type of function, it is still important to practice each method.

HOW TO

Given the equation of a linear function, use transformations to graph the linear function in the form $f(x) = mx + b$.

1. Graph $f(x) = x$.
2. Vertically stretch or compress the graph by a factor m.
3. Shift the graph up or down b units.

EXAMPLE 13

Graphing by Using Transformations

Graph $f(x) = \frac{1}{2}x - 3$ using transformations.

Solution

The equation for the function shows that $m = \frac{1}{2}$ so the identity function is vertically compressed by $\frac{1}{2}$. The equation for the function also shows that $b = -3$ so the identity function is vertically shifted down 3 units. First, graph the identity function, and show the vertical compression as in Figure 16.

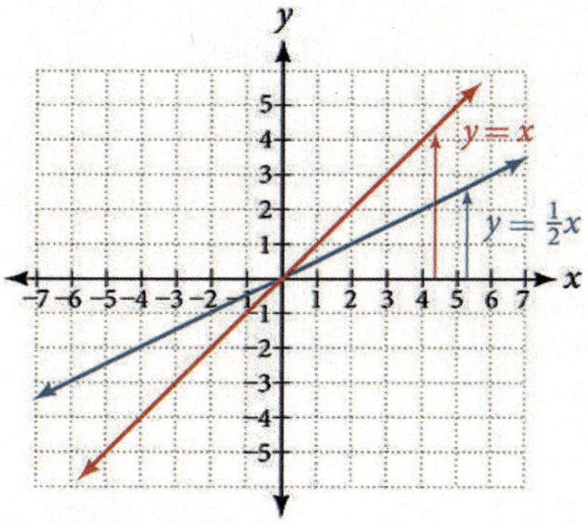

Figure 16 The function, $y = x$, compressed by a factor of $\frac{1}{2}$.

Then show the vertical shift as in Figure 17.

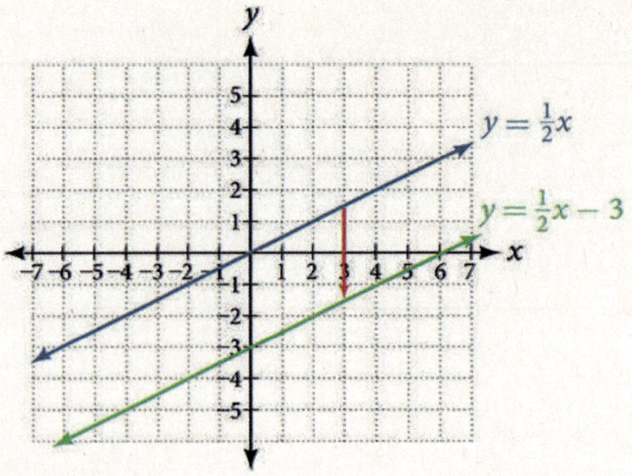

Figure 17 The function $y = \frac{1}{2}x$, shifted down 3 units.

TRY IT #7 Graph $f(x) = 4 + 2x$ using transformations.

Q&A **In Example 15, could we have sketched the graph by reversing the order of the transformations?**

No. The order of the transformations follows the order of operations. When the function is evaluated at a given input, the corresponding output is calculated by following the order of operations. This is why we performed the compression first. For example, following the order: Let the input be 2.

$$\begin{aligned} f(2) &= \tfrac{1}{2}(2) - 3 \\ &= 1 - 3 \\ &= -2 \end{aligned}$$

Writing the Equation for a Function from the Graph of a Line

Earlier, we wrote the equation for a linear function from a graph. Now we can extend what we know about graphing linear functions to analyze graphs a little more closely. Begin by taking a look at Figure 18. We can see right away that the graph crosses the *y*-axis at the point $(0, 4)$ so this is the *y*-intercept.

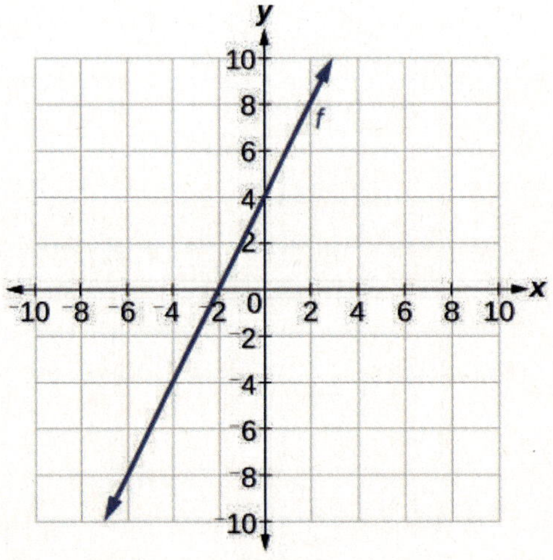

Figure 18

Then we can calculate the slope by finding the rise and run. We can choose any two points, but let's look at the point $(-2, 0)$. To get from this point to the *y*-intercept, we must move up 4 units (rise) and to the right 2 units (run). So the slope must be

$$m = \frac{\text{rise}}{\text{run}} = \frac{4}{2} = 2$$

Substituting the slope and *y*-intercept into the slope-intercept form of a line gives

$$y = 2x + 4$$

HOW TO

Given a graph of linear function, find the equation to describe the function.

1. Identify the *y*-intercept of an equation.
2. Choose two points to determine the slope.
3. Substitute the *y*-intercept and slope into the slope-intercept form of a line.

EXAMPLE 14

Matching Linear Functions to Their Graphs

Match each equation of the linear functions with one of the lines in Figure 19.

ⓐ $f(x) = 2x + 3$ ⓑ $g(x) = 2x - 3$ ⓒ $h(x) = -2x + 3$ ⓓ $j(x) = \frac{1}{2}x + 3$

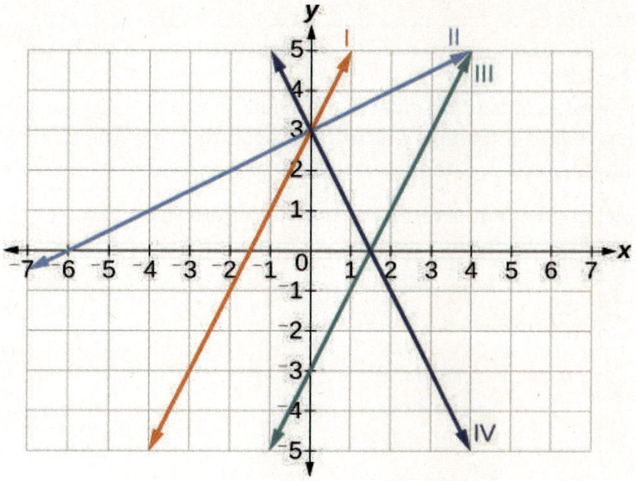

Figure 19

Solution

Analyze the information for each function.

ⓐ This function has a slope of 2 and a *y*-intercept of 3. It must pass through the point (0, 3) and slant upward from left to right. We can use two points to find the slope, or we can compare it with the other functions listed. Function g has the same slope, but a different *y*-intercept. Lines I and III have the same slant because they have the same slope. Line III does not pass through $(0, 3)$ so f must be represented by line I.

ⓑ This function also has a slope of 2, but a *y*-intercept of -3. It must pass through the point $(0, -3)$ and slant upward from left to right. It must be represented by line III.

ⓒ This function has a slope of –2 and a *y*-intercept of 3. This is the only function listed with a negative slope, so it must be represented by line IV because it slants downward from left to right.

ⓓ This function has a slope of $\frac{1}{2}$ and a *y*-intercept of 3. It must pass through the point (0, 3) and slant upward from left to right. Lines I and II pass through $(0, 3)$, but the slope of j is less than the slope of f so the line for j must be flatter. This function is represented by Line II.

Now we can re-label the lines as in Figure 20.

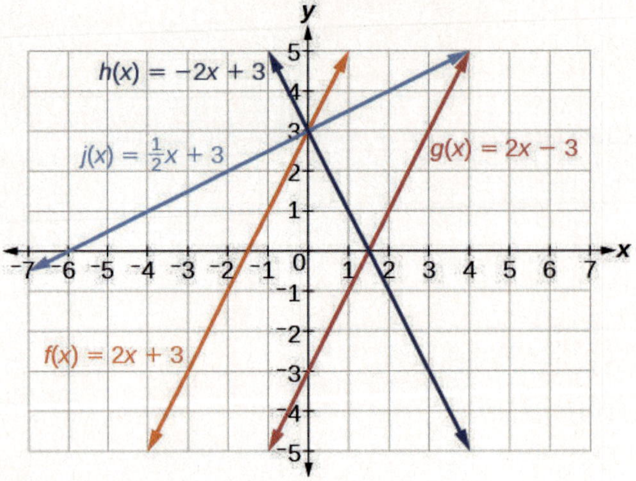

Figure 20

Finding the x-intercept of a Line

So far we have been finding the *y*-intercepts of a function: the point at which the graph of the function crosses the *y*-axis. Recall that a function may also have an *x*-intercept, which is the *x*-coordinate of the point where the graph of the function crosses the *x*-axis. In other words, it is the input value when the output value is zero.

To find the *x*-intercept, set a function $f(x)$ equal to zero and solve for the value of x. For example, consider the function shown.

$$f(x) = 3x - 6$$

Set the function equal to 0 and solve for x.

$$\begin{aligned} 0 &= 3x - 6 \\ 6 &= 3x \\ 2 &= x \\ x &= 2 \end{aligned}$$

The graph of the function crosses the *x*-axis at the point $(2, 0)$.

Q&A **Do all linear functions have *x*-intercepts?**

No. However, linear functions of the form $y = c$, where c is a nonzero real number are the only examples of linear functions with no x-intercept. For example, $y = 5$ is a horizontal line 5 units above the x-axis. This function has no x-intercepts, as shown in ***Figure 21***.

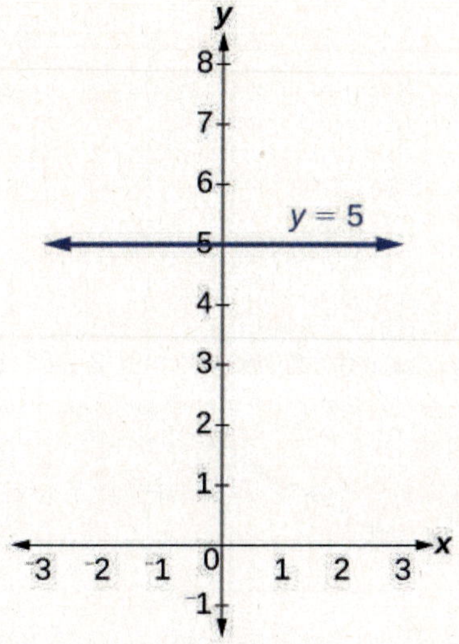

Figure 21

x-intercept

The x-intercept of the function is value of x when $f(x) = 0$. It can be solved by the equation $0 = mx + b$.

EXAMPLE 15

Finding an x-intercept
Find the x-intercept of $f(x) = \frac{1}{2}x - 3$.

Solution
Set the function equal to zero to solve for x.

$$\begin{aligned} 0 &= \tfrac{1}{2}x - 3 \\ 3 &= \tfrac{1}{2}x \\ 6 &= x \\ x &= 6 \end{aligned}$$

The graph crosses the x-axis at the point $(6, 0)$.

Analysis
A graph of the function is shown in Figure 22. We can see that the x-intercept is $(6, 0)$ as we expected.

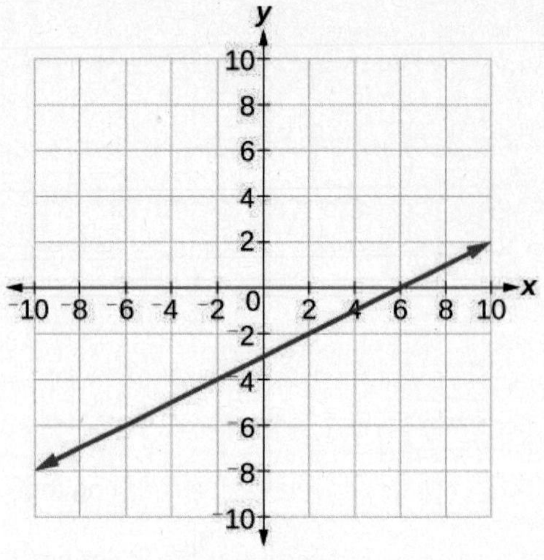

Figure 22

> TRY IT #8 Find the *x*-intercept of $f(x) = \frac{1}{4}x - 4$.

Describing Horizontal and Vertical Lines

There are two special cases of lines on a graph—horizontal and vertical lines. A horizontal line indicates a constant output, or *y*-value. In Figure 23, we see that the output has a value of 2 for every input value. The change in outputs between any two points, therefore, is 0. In the slope formula, the numerator is 0, so the slope is 0. If we use $m = 0$ in the equation $f(x) = mx + b$, the equation simplifies to $f(x) = b$. In other words, the value of the function is a constant. This graph represents the function $f(x) = 2$.

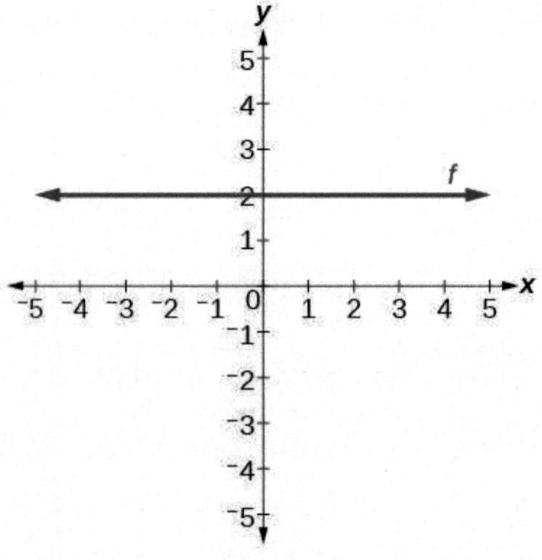

x	−4	−2	0	2	4
y	2	2	2	2	2

Figure 23 A horizontal line representing the function $f(x) = 2$

A vertical line indicates a constant input, or *x*-value. We can see that the input value for every point on the line is 2, but the output value varies. Because this input value is mapped to more than one output value, a vertical line does not represent a function. Notice that between any two points, the change in the input values is zero. In the slope formula, the denominator will be zero, so the slope of a vertical line is undefined.

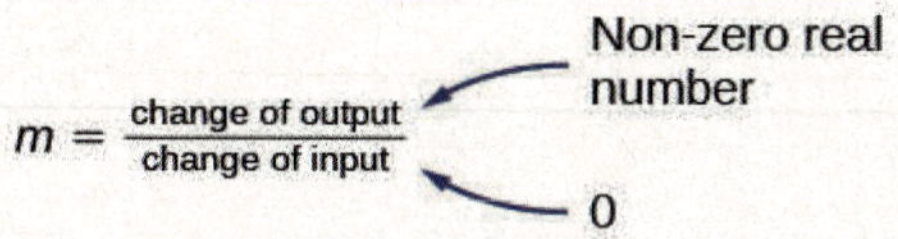

Figure 24 Example of how a line has a vertical slope. 0 in the denominator of the slope.

A vertical line, such as the one in Figure 25, has an *x*-intercept, but no *y*-intercept unless it's the line $x = 0$. This graph represents the line $x = 2$.

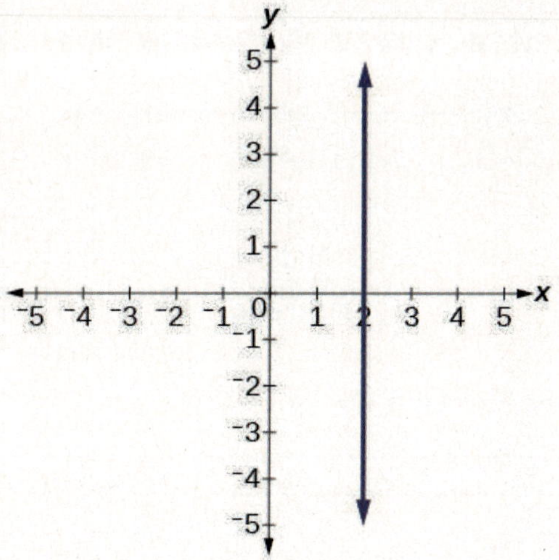

x	2	2	2	2	2
y	−4	−2	0	2	4

Figure 25 The vertical line, $x = 2$, which does not represent a function

Horizontal and Vertical Lines

Lines can be horizontal or vertical.

A **horizontal line** is a line defined by an equation in the form $f(x) = b$.

A **vertical line** is a line defined by an equation in the form $x = a$.

EXAMPLE 16

Writing the Equation of a Horizontal Line

Write the equation of the line graphed in Figure 26.

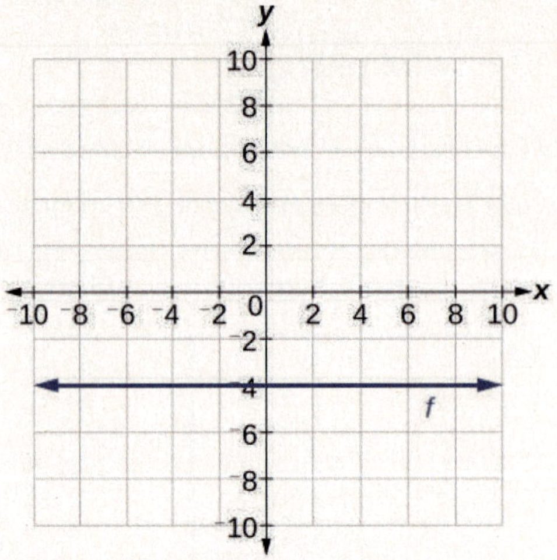

Figure 26

Solution

For any *x*-value, the *y*-value is -4, so the equation is $y = -4$.

EXAMPLE 17

Writing the Equation of a Vertical Line

Write the equation of the line graphed in Figure 27.

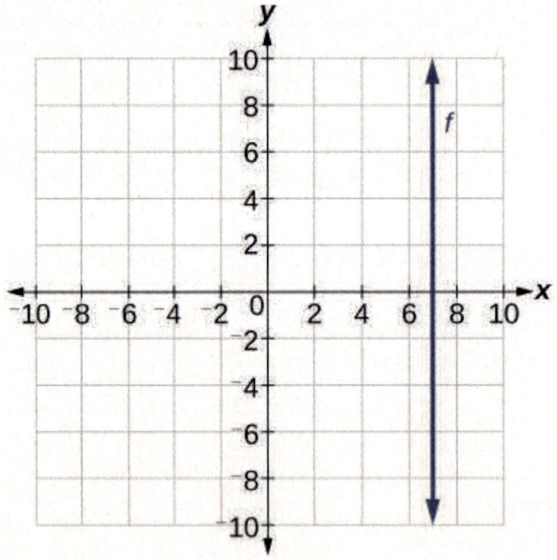

Figure 27

Solution

The constant *x*-value is 7, so the equation is $x = 7$.

Determining Whether Lines are Parallel or Perpendicular

The two lines in Figure 28 are parallel lines: they will never intersect. They have exactly the same steepness, which means their slopes are identical. The only difference between the two lines is the *y*-intercept. If we shifted one line vertically toward the other, they would become coincident.

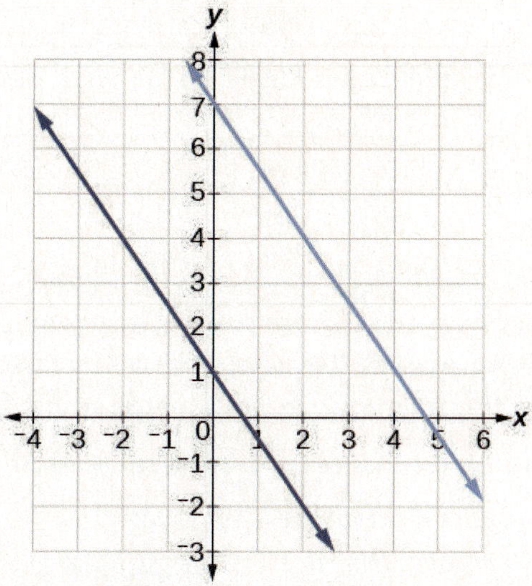

Figure 28 Parallel lines

We can determine from their equations whether two lines are parallel by comparing their slopes. If the slopes are the same and the *y*-intercepts are different, the lines are parallel. If the slopes are different, the lines are not parallel.

$$\left.\begin{array}{l} f(x) = -2x + 6 \\ f(x) = -2x - 4 \end{array}\right\} \text{parallel} \qquad \left.\begin{array}{l} f(x) = 3x + 2 \\ f(x) = 2x + 2 \end{array}\right\} \text{not parallel}$$

Unlike parallel lines, perpendicular lines do intersect. Their intersection forms a right, or 90-degree, angle. The two lines in Figure 29 are perpendicular.

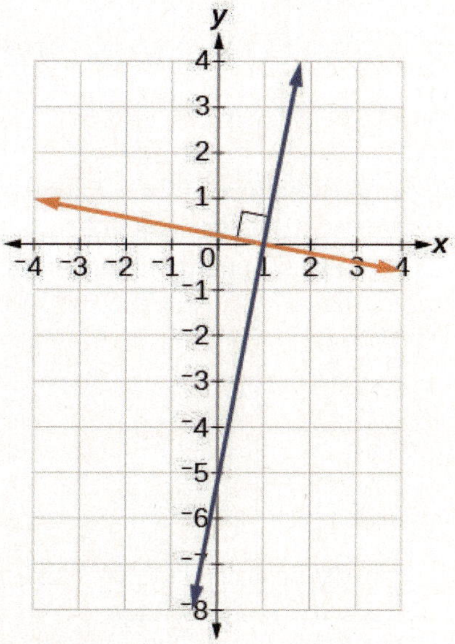

Figure 29 Perpendicular lines

Perpendicular lines do not have the same slope. The slopes of perpendicular lines are different from one another in a specific way. The slope of one line is the negative reciprocal of the slope of the other line. The product of a number and its reciprocal is 1. So, if m_1 and m_2 are negative reciprocals of one another, they can be multiplied together to yield -1.

$$m_1 m_2 = -1$$

To find the reciprocal of a number, divide 1 by the number. So the reciprocal of 8 is $\frac{1}{8}$, and the reciprocal of $\frac{1}{8}$ is 8. To find the negative reciprocal, first find the reciprocal and then change the sign.

As with parallel lines, we can determine whether two lines are perpendicular by comparing their slopes, assuming that the lines are neither horizontal nor vertical. The slope of each line below is the negative reciprocal of the other so the lines are perpendicular.

$$\begin{array}{rcll} f(x) & = & \frac{1}{4}x + 2 & \text{negative reciprocal of } \frac{1}{4} \text{ is } -4 \\ f(x) & = & -4x + 3 & \text{negative reciprocal of } -4 \text{ is } \frac{1}{4} \end{array}$$

The product of the slopes is –1.

$$-4\left(\frac{1}{4}\right) = -1$$

Parallel and Perpendicular Lines

Two lines are **parallel lines** if they do not intersect. The slopes of the lines are the same.

$$f(x) = m_1 x + b_1 \text{ and } g(x) = m_2 x + b_2 \text{ are parallel if and only if } m_1 = m_2$$

If and only if $b_1 = b_2$ and $m_1 = m_2$, we say the lines coincide. Coincident lines are the same line.

Two lines are **perpendicular lines** if they intersect to form a right angle.

$$f(x) = m_1 x + b_1 \text{ and } g(x) = m_2 x + b_2 \text{ are perpendicular if and only if}$$

$$m_1 m_2 = -1, \text{ so } m_2 = -\frac{1}{m_1}$$

EXAMPLE 18

Identifying Parallel and Perpendicular Lines

Given the functions below, identify the functions whose graphs are a pair of parallel lines and a pair of perpendicular lines.

$$\begin{array}{rclcrcl} f(x) & = & 2x + 3 & & h(x) & = & -2x + 2 \\ g(x) & = & \frac{1}{2}x - 4 & & j(x) & = & 2x - 6 \end{array}$$

Solution

Parallel lines have the same slope. Because the functions $f(x) = 2x + 3$ and $j(x) = 2x - 6$ each have a slope of 2, they represent parallel lines. Perpendicular lines have negative reciprocal slopes. Because −2 and $\frac{1}{2}$ are negative reciprocals, the functions $g(x) = \frac{1}{2}x - 4$ and $h(x) = -2x + 2$ represent perpendicular lines.

Analysis

A graph of the lines is shown in Figure 30.

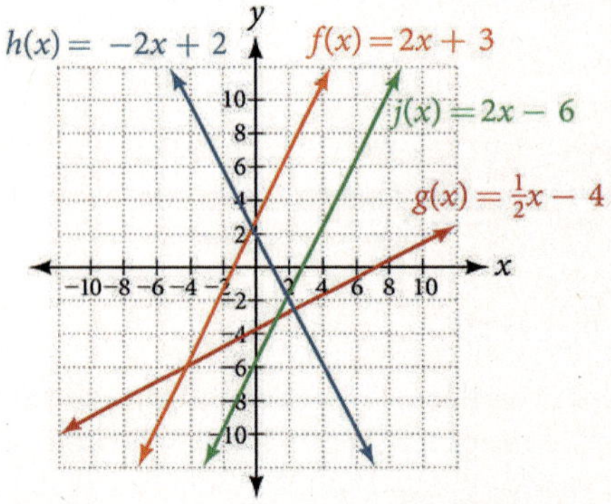

Figure 30

The graph shows that the lines $f(x) = 2x + 3$ and $j(x) = 2x - 6$ are parallel, and the lines $g(x) = \frac{1}{2}x - 4$ and $h(x) = -2x + 2$ are perpendicular.

Writing the Equation of a Line Parallel or Perpendicular to a Given Line

If we know the equation of a line, we can use what we know about slope to write the equation of a line that is either parallel or perpendicular to the given line.

Writing Equations of Parallel Lines

Suppose for example, we are given the equation shown.

$$f(x) = 3x + 1$$

We know that the slope of the line formed by the function is 3. We also know that the y-intercept is $(0, 1)$. Any other line with a slope of 3 will be parallel to $f(x)$. So the lines formed by all of the following functions will be parallel to $f(x)$.

$$\begin{aligned} g(x) &= 3x + 6 \\ h(x) &= 3x + 1 \\ p(x) &= 3x + \tfrac{2}{3} \end{aligned}$$

Suppose then we want to write the equation of a line that is parallel to f and passes through the point $(1, 7)$. This type of problem is often described as a point-slope problem because we have a point and a slope. In our example, we know that the slope is 3. We need to determine which value of b will give the correct line. We can begin with the point-slope form of an equation for a line, and then rewrite it in the slope-intercept form.

$$\begin{aligned} y - y_1 &= m(x - x_1) \\ y - 7 &= 3(x - 1) \\ y - 7 &= 3x - 3 \\ y &= 3x + 4 \end{aligned}$$

So $g(x) = 3x + 4$ is parallel to $f(x) = 3x + 1$ and passes through the point $(1, 7)$.

HOW TO

Given the equation of a function and a point through which its graph passes, write the equation of a line parallel to the given line that passes through the given point.

1. Find the slope of the function.
2. Substitute the given values into either the general point-slope equation or the slope-intercept equation for a line.
3. Simplify.

EXAMPLE 19

Finding a Line Parallel to a Given Line

Find a line parallel to the graph of $f(x) = 3x + 6$ that passes through the point $(3, 0)$.

Solution

The slope of the given line is 3. If we choose the slope-intercept form, we can substitute $m = 3$, $x = 3$, and $f(x) = 0$ into the slope-intercept form to find the y-intercept.

$$\begin{aligned} g(x) &= 3x + b \\ 0 &= 3(3) + b \\ b &= -9 \end{aligned}$$

The line parallel to $f(x)$ that passes through $(3, 0)$ is $g(x) = 3x - 9$.

Analysis

We can confirm that the two lines are parallel by graphing them. Figure 31 shows that the two lines will never intersect.

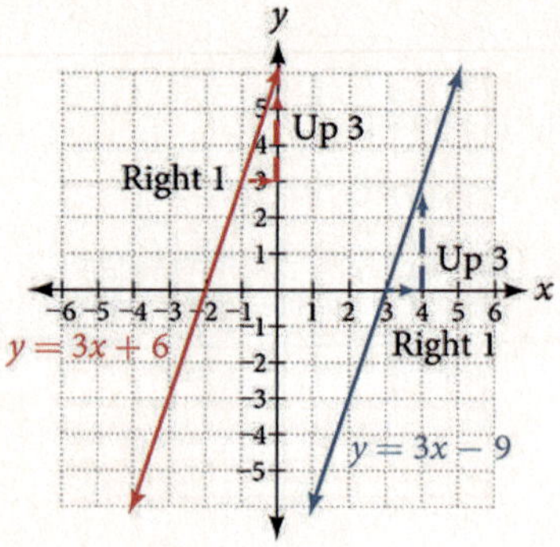

Figure 31

Writing Equations of Perpendicular Lines

We can use a very similar process to write the equation for a line perpendicular to a given line. Instead of using the same slope, however, we use the negative reciprocal of the given slope. Suppose we are given the function shown.

$$f(x) = 2x + 4$$

The slope of the line is 2, and its negative reciprocal is $-\frac{1}{2}$. Any function with a slope of $-\frac{1}{2}$ will be perpendicular to $f(x)$. So the lines formed by all of the following functions will be perpendicular to $f(x)$.

$$\begin{aligned} g(x) &= -\tfrac{1}{2}x + 4 \\ h(x) &= -\tfrac{1}{2}x + 2 \\ p(x) &= -\tfrac{1}{2}x - \tfrac{1}{2} \end{aligned}$$

As before, we can narrow down our choices for a particular perpendicular line if we know that it passes through a given point. Suppose then we want to write the equation of a line that is perpendicular to $f(x)$ and passes through the point $(4, 0)$. We already know that the slope is $-\frac{1}{2}$. Now we can use the point to find the *y*-intercept by substituting the given values into the slope-intercept form of a line and solving for b.

$$\begin{aligned} g(x) &= mx + b \\ 0 &= -\tfrac{1}{2}(4) + b \\ 0 &= -2 + b \\ 2 &= b \\ b &= 2 \end{aligned}$$

The equation for the function with a slope of $-\frac{1}{2}$ and a *y*-intercept of 2 is

$$g(x) = -\frac{1}{2}x + 2$$

So $g(x) = -\frac{1}{2}x + 2$ is perpendicular to $f(x) = 2x + 4$ and passes through the point $(4, 0)$. Be aware that perpendicular lines may not look obviously perpendicular on a graphing calculator unless we use the square zoom feature.

Q&A **A horizontal line has a slope of zero and a vertical line has an undefined slope. These two lines are perpendicular, but the product of their slopes is not –1. Doesn't this fact contradict the definition of perpendicular lines?**

No. For two perpendicular linear functions, the product of their slopes is –1. However, a vertical line is not a function so the definition is not contradicted.

HOW TO

Given the equation of a function and a point through which its graph passes, write the equation of a line

perpendicular to the given line.

1. Find the slope of the function.
2. Determine the negative reciprocal of the slope.
3. Substitute the new slope and the values for x and y from the coordinate pair provided into $g(x) = mx + b$.
4. Solve for b.
5. Write the equation of the line.

EXAMPLE 20

Finding the Equation of a Perpendicular Line

Find the equation of a line perpendicular to $f(x) = 3x + 3$ that passes through the point $(3, 0)$.

Solution

The original line has slope $m = 3$, so the slope of the perpendicular line will be its negative reciprocal, or $-\frac{1}{3}$. Using this slope and the given point, we can find the equation of the line.

$$\begin{aligned} g(x) &= -\tfrac{1}{3}x + b \\ 0 &= -\tfrac{1}{3}(3) + b \\ 1 &= b \\ b &= 1 \end{aligned}$$

The line perpendicular to $f(x)$ that passes through $(3, 0)$ is $g(x) = -\frac{1}{3}x + 1$.

Analysis

A graph of the two lines is shown in Figure 32.

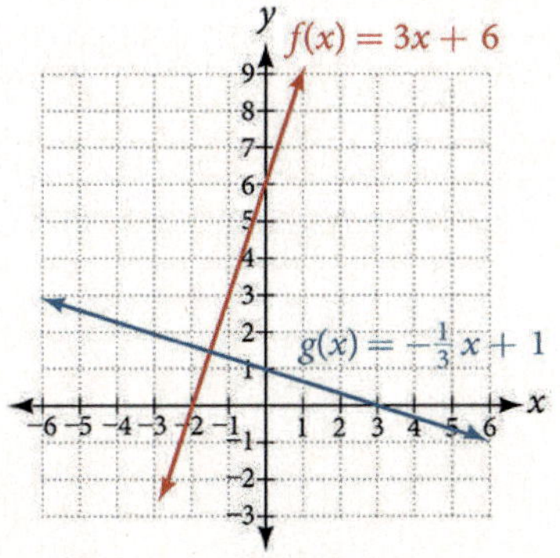

Figure 32

Note that that if we graph perpendicular lines on a graphing calculator using standard zoom, the lines may not appear to be perpendicular. Adjusting the window will make it possible to zoom in further to see the intersection more closely.

TRY IT #9 Given the function $h(x) = 2x - 4$, write an equation for the line passing through $(0, 0)$ that is

ⓐ parallel to $h(x)$ ⓑ perpendicular to $h(x)$

HOW TO

Given two points on a line and a third point, write the equation of the perpendicular line that passes through the point.

1. Determine the slope of the line passing through the points.
2. Find the negative reciprocal of the slope.

3. Use the slope-intercept form or point-slope form to write the equation by substituting the known values.
4. Simplify.

EXAMPLE 21

Finding the Equation of a Line Perpendicular to a Given Line Passing through a Point

A line passes through the points $(-2, 6)$ and $(4, 5)$. Find the equation of a perpendicular line that passes through the point $(4, 5)$.

Solution

From the two points of the given line, we can calculate the slope of that line.

$$\begin{aligned} m_1 &= \frac{5-6}{4-(-2)} \\ &= \frac{-1}{6} \\ &= -\frac{1}{6} \end{aligned}$$

Find the negative reciprocal of the slope.

$$\begin{aligned} m_2 &= \frac{-1}{-\frac{1}{6}} \\ &= -1\left(-\frac{6}{1}\right) \\ &= 6 \end{aligned}$$

We can then solve for the y-intercept of the line passing through the point $(4, 5)$.

$$\begin{aligned} g(x) &= 6x + b \\ 5 &= 6(4) + b \\ 5 &= 24 + b \\ -19 &= b \\ b &= -19 \end{aligned}$$

The equation for the line that is perpendicular to the line passing through the two given points and also passes through point $(4, 5)$ is

$$y = 6x - 19$$

TRY IT #10 A line passes through the points, $(-2, -15)$ and $(2, -3)$. Find the equation of a perpendicular line that passes through the point, $(6, 4)$.

MEDIA

Access this online resource for additional instruction and practice with linear functions.

Linear Functions (http://openstax.org/l/linearfunctions)
Finding Input of Function from the Output and Graph (http://openstax.org/l/findinginput)
Graphing Functions using Tables (http://openstax.org/l/graphwithtable)

4.1 SECTION EXERCISES

Verbal

1. Terry is skiing down a steep hill. Terry's elevation, $E(t)$, in feet after t seconds is given by $E(t) = 3000 - 70t$. Write a complete sentence describing Terry's starting elevation and how it is changing over time.

2. Jessica is walking home from a friend's house. After 2 minutes she is 1.4 miles from home. Twelve minutes after leaving, she is 0.9 miles from home. What is her rate in miles per hour?

3. A boat is 100 miles away from the marina, sailing directly toward it at 10 miles per hour. Write an equation for the distance of the boat from the marina after t hours.

4. If the graphs of two linear functions are perpendicular, describe the relationship between the slopes and the y-intercepts.

5. If a horizontal line has the equation $f(x) = a$ and a vertical line has the equation $x = a$, what is the point of intersection? Explain why what you found is the point of intersection.

Algebraic

For the following exercises, determine whether the equation of the curve can be written as a linear function.

6. $y = \frac{1}{4}x + 6$

7. $y = 3x - 5$

8. $y = 3x^2 - 2$

9. $3x + 5y = 15$

10. $3x^2 + 5y = 15$

11. $3x + 5y^2 = 15$

12. $-2x^2 + 3y^2 = 6$

13. $-\frac{x-3}{5} = 2y$

For the following exercises, determine whether each function is increasing or decreasing.

14. $f(x) = 4x + 3$

15. $g(x) = 5x + 6$

16. $a(x) = 5 - 2x$

17. $b(x) = 8 - 3x$

18. $h(x) = -2x + 4$

19. $k(x) = -4x + 1$

20. $j(x) = \frac{1}{2}x - 3$

21. $p(x) = \frac{1}{4}x - 5$

22. $n(x) = -\frac{1}{3}x - 2$

23. $m(x) = -\frac{3}{8}x + 3$

For the following exercises, find the slope of the line that passes through the two given points.

24. $(2, 4)$ and $(4, 10)$

25. $(1, 5)$ and $(4, 11)$

26. $(-1, 4)$ and $(5, 2)$

27. $(8, -2)$ and $(4, 6)$

28. $(6, 11)$ and $(-4, 3)$

For the following exercises, given each set of information, find a linear equation satisfying the conditions, if possible.

29. $f(-5) = -4$, and $f(5) = 2$

30. $f(-1) = 4$, and $f(5) = 1$

31. Passes through $(2, 4)$ and $(4, 10)$

32. Passes through $(1, 5)$ and $(4, 11)$

33. Passes through $(-1, 4)$ and $(5, 2)$

34. Passes through $(-2, 8)$ and $(4, 6)$

35. *x* intercept at $(-2, 0)$ and *y* intercept at $(0, -3)$

36. *x* intercept at $(-5, 0)$ and *y* intercept at $(0, 4)$

For the following exercises, determine whether the lines given by the equations below are parallel, perpendicular, or neither.

37. $$\begin{aligned} 4x - 7y &= 10 \\ 7x + 4y &= 1 \end{aligned}$$

38. $$\begin{aligned} 3y + x &= 12 \\ -y &= 8x + 1 \end{aligned}$$

39. $$\begin{aligned} 3y + 4x &= 12 \\ -6y &= 8x + 1 \end{aligned}$$

40. $$\begin{aligned} 6x - 9y &= 10 \\ 3x + 2y &= 1 \end{aligned}$$

For the following exercises, find the x- and y-intercepts of each equation.

41. $f(x) = -x + 2$

42. $g(x) = 2x + 4$

43. $h(x) = 3x - 5$

44. $k(x) = -5x + 1$

45. $-2x + 5y = 20$

46. $7x + 2y = 56$

For the following exercises, use the descriptions of each pair of lines given below to find the slopes of Line 1 and Line 2. Is each pair of lines parallel, perpendicular, or neither?

47. Line 1: Passes through $(0, 6)$ and $(3, -24)$

Line 2: Passes through $(-1, 19)$ and $(8, -71)$

48. Line 1: Passes through $(-8, -55)$ and $(10, 89)$

Line 2: Passes through $(9, -44)$ and $(4, -14)$

49. Line 1: Passes through $(2, 3)$ and $(4, -1)$

Line 2: Passes through $(6, 3)$ and $(8, 5)$

50. Line 1: Passes through $(1, 7)$ and $(5, 5)$

Line 2: Passes through $(-1, -3)$ and $(1, 1)$

51. Line 1: Passes through $(2, 5)$ and $(5, -1)$

Line 2: Passes through $(-3, 7)$ and $(3, -5)$

For the following exercises, write an equation for the line described.

52. Write an equation for a line parallel to $f(x) = -5x - 3$ and passing through the point $(2, -12)$.

53. Write an equation for a line parallel to $g(x) = 3x - 1$ and passing through the point $(4, 9)$.

54. Write an equation for a line perpendicular to $h(t) = -2t + 4$ and passing through the point $(-4, -1)$.

55. Write an equation for a line perpendicular to $p(t) = 3t + 4$ and passing through the point $(3, 1)$.

Graphical

For the following exercises, find the slope of the line graphed.

56.

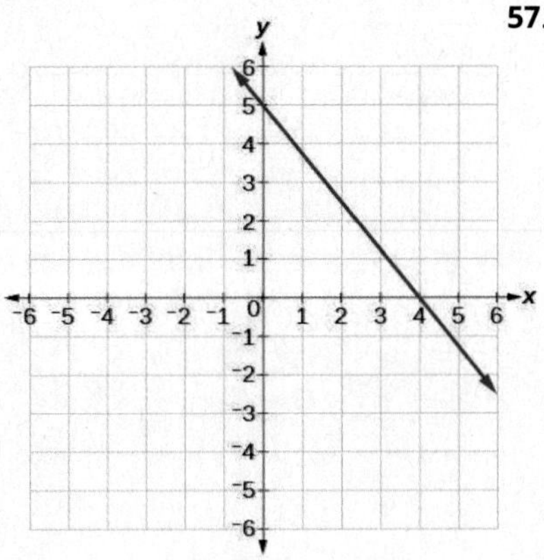

57.

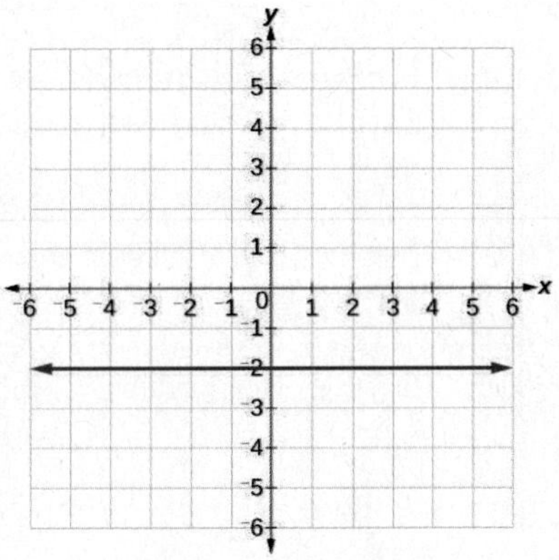

For the following exercises, write an equation for the line graphed.

58.

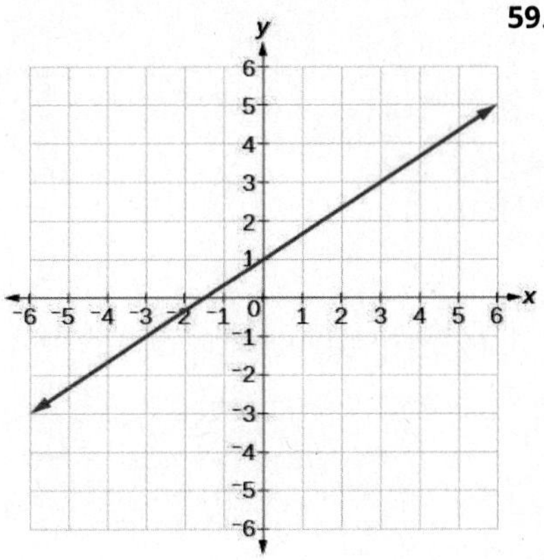

59.

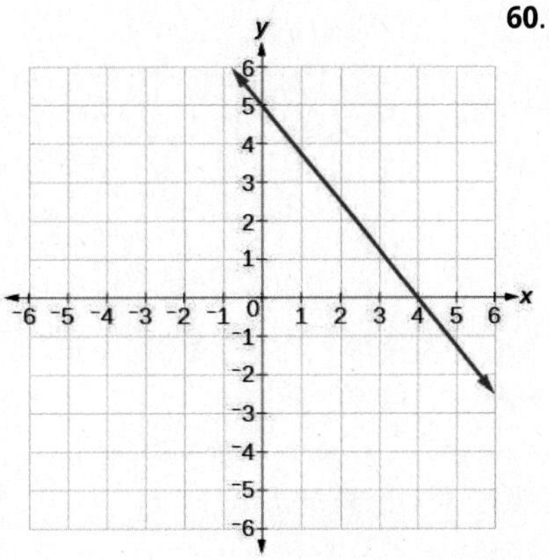

60.

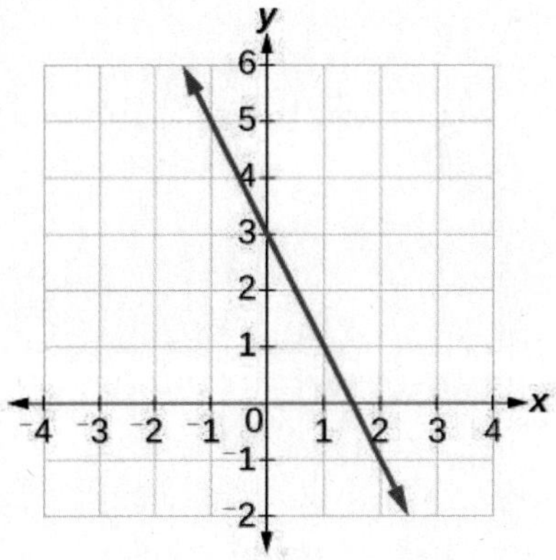

61.

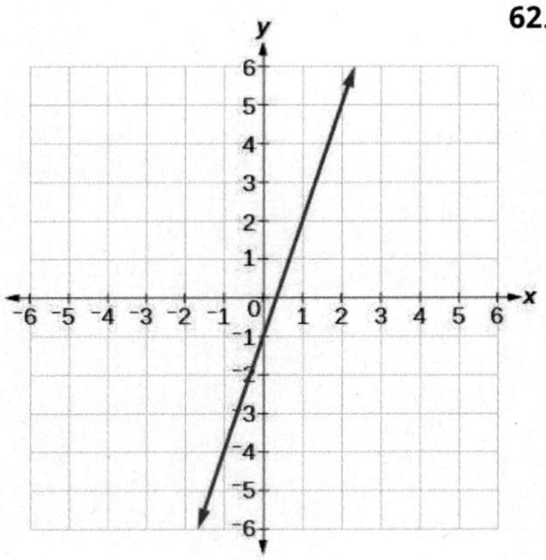

62.

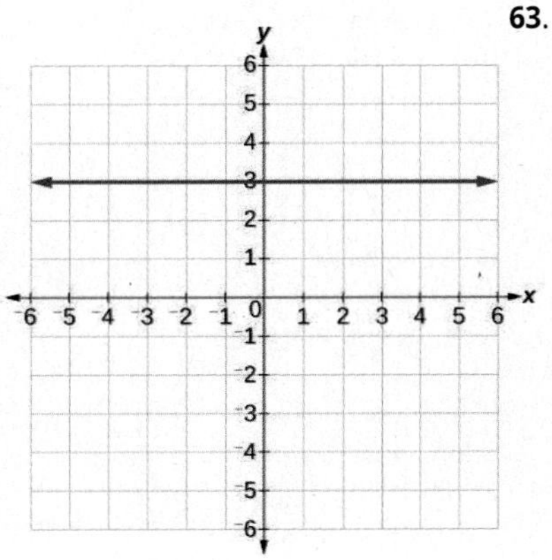

63.

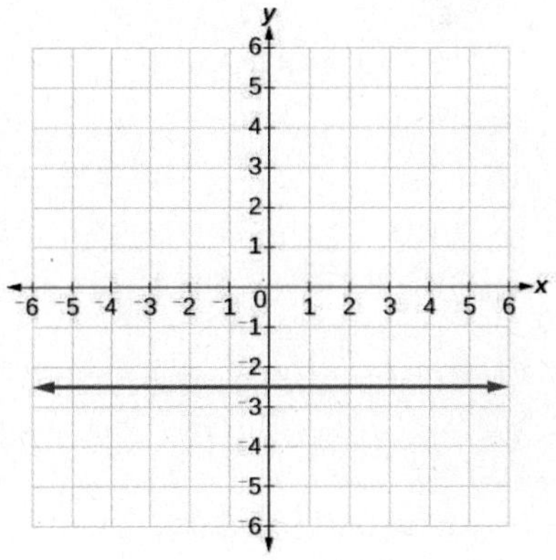

For the following exercises, match the given linear equation with its graph in Figure 33.

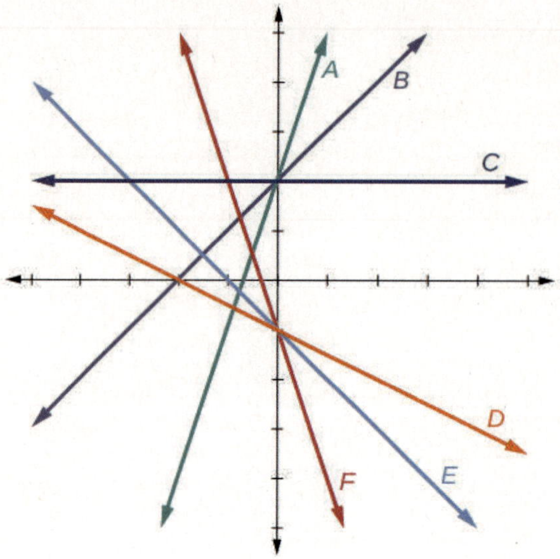

Figure 33

64. $f(x) = -x - 1$

65. $f(x) = -3x - 1$

66. $f(x) = -\frac{1}{2}x - 1$

67. $f(x) = 2$

68. $f(x) = 2 + x$

69. $f(x) = 3x + 2$

For the following exercises, sketch a line with the given features.

70. An x-intercept of $(-4, 0)$ and y-intercept of $(0, -2)$

71. An x-intercept $(-2, 0)$ and y-intercept of $(0, 4)$

72. A y-intercept of $(0, 7)$ and slope $-\frac{3}{2}$

73. A y-intercept of $(0, 3)$ and slope $\frac{2}{5}$

74. Passing through the points $(-6, -2)$ and $(6, -6)$

75. Passing through the points $(-3, -4)$ and $(3, 0)$

For the following exercises, sketch the graph of each equation.

76. $f(x) = -2x - 1$

77. $f(x) = -3x + 2$

78. $f(x) = \frac{1}{3}x + 2$

79. $f(x) = \frac{2}{3}x - 3$

80. $f(t) = 3 + 2t$

81. $p(t) = -2 + 3t$

82. $x = 3$

83. $x = -2$

84. $r(x) = 4$

For the following exercises, write the equation of the line shown in the graph.

85.

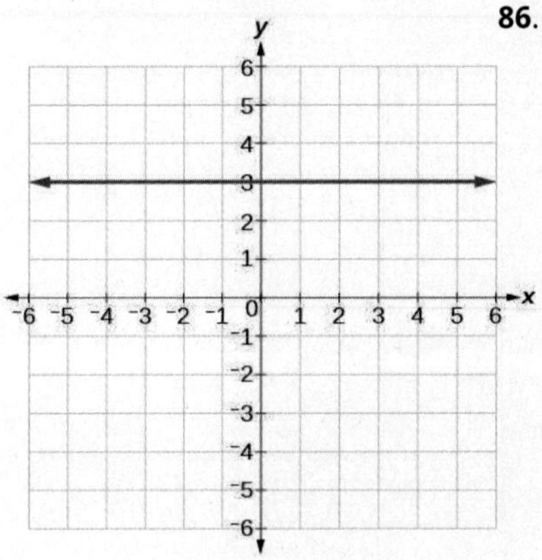

86.

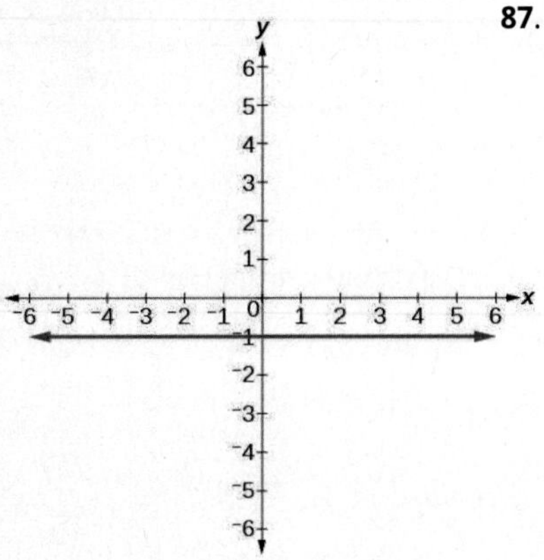

87.

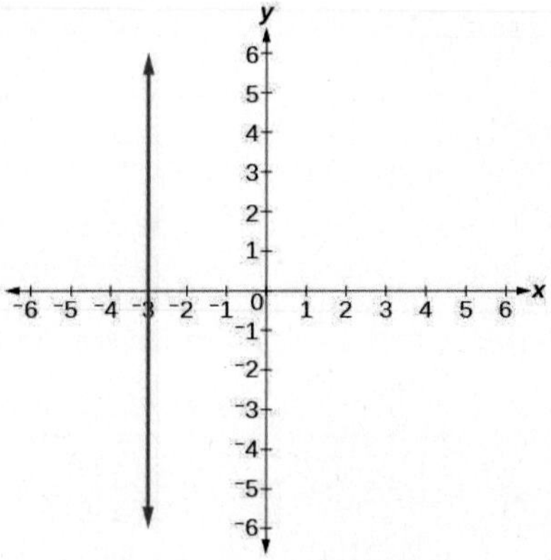

88.

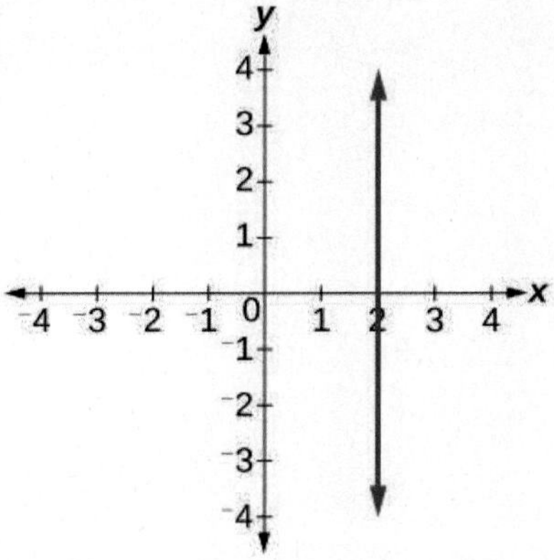

Numeric

For the following exercises, which of the tables could represent a linear function? For each that could be linear, find a linear equation that models the data.

89.

x	0	5	10	15
$g(x)$	5	–10	–25	–40

90.

x	0	5	10	15
$h(x)$	5	30	105	230

91.

x	0	5	10	15
$f(x)$	-5	20	45	70

92.

x	5	10	20	25
$k(x)$	13	28	58	73

93.

x	0	2	4	6
$g(x)$	6	–19	–44	–69

94.

x	2	4	8	10
$h(x)$	13	23	43	53

95.

x	2	4	6	8
$f(x)$	–4	16	36	56

96.

x	0	2	6	8
$k(x)$	6	31	106	231

Technology

For the following exercises, use a calculator or graphing technology to complete the task.

97. If f is a linear function, $f(0.1) = 11.5$, and $f(0.4) = -5.9$, find an equation for the function.

98. Graph the function f on a domain of $[-10, 10] : f(x) = 0.02x - 0.01.$ Enter the function in a graphing utility. For the viewing window, set the minimum value of x to be -10 and the maximum value of x to be 10 .

99. Graph the function f on a domain of $[-10, 10] : fx) = 2,500x + 4,000$

100. Table 3 shows the input, w, and output, k, for a linear function k.

ⓐ Fill in the missing values of the table.
ⓑ Write the linear function k, round to 3 decimal places.

w	-10	5.5	67.5	*b*
k	30	-26	*a*	-44

Table 3

101. Table 4 shows the input, p, and output, q, for a linear function q.

ⓐ Fill in the missing values of the table.
ⓑ Write the linear function k.

p	0.5	0.8	12	*b*
q	400	700	*a*	1,000,000

Table 4

102. Graph the linear function f on a domain of $[-10, 10]$ for the function whose slope is $\frac{1}{8}$ and y-intercept is $\frac{31}{16}$. Label the points for the input values of -10 and 10.

103. Graph the linear function f on a domain of $[-0.1, 0.1]$ for the function whose slope is 75 and y-intercept is -22.5. Label the points for the input values of -0.1 and 0.1.

104. Graph the linear function f where $f(x) = ax + b$ on the same set of axes on a domain of $[-4, 4]$ for the following values of a and b.

ⓐ $a = 2; b = 3$
ⓑ $a = 2; b = 4$
ⓒ $a = 2; b = -4$
ⓓ $a = 2; b = -5$

Extensions

105. Find the value of x if a linear function goes through the following points and has the following slope: $(x, 2), (-4, 6), \quad m = 3$

106. Find the value of y if a linear function goes through the following points and has the following slope: $(10, y), (25, 100), \quad m = -5$

107. Find the equation of the line that passes through the following points:

$(a, \;\; b)$ and $(a, \;\; b + 1)$

108. Find the equation of the line that passes through the following points:

$(2a, b)$ and $(a, b + 1)$

109. Find the equation of the line that passes through the following points:

$(a, 0)$ and (c, d)

110. Find the equation of the line parallel to the line $g(x) = -0.01x+2.01$ through the point $(1, 2)$.

111. Find the equation of the line perpendicular to the line $g(x) = -0.01x+2.01$ through the point $(1, 2)$.

For the following exercises, use the functions $f(x) = -0.1x+200$ and $g(x) = 20x + 0.1$.

112. Find the point of intersection of the lines f and g.

113. Where is $f(x)$ greater than $g(x)$? Where is $g(x)$ greater than $f(x)$?

Real-World Applications

114. At noon, a barista notices that they have $20 in her tip jar. If they maks an average of $0.50 from each customer, how much will the barista have in the tip jar if they serve n more customers during the shift?

115. A gym membership with two personal training sessions costs $125, while gym membership with five personal training sessions costs $260. What is cost per session?

116. A clothing business finds there is a linear relationship between the number of shirts, n, it can sell and the price, p, it can charge per shirt. In particular, historical data shows that 1,000 shirts can be sold at a price of $\$30$, while 3,000 shirts can be sold at a price of $22. Find a linear equation in the form $p(n) = mn + b$ that gives the price p they can charge for n shirts.

117. A phone company charges for service according to the formula: $C(n) = 24 + 0.1n$, where n is the number of minutes talked, and $C(n)$ is the monthly charge, in dollars. Find and interpret the rate of change and initial value.

118. A farmer finds there is a linear relationship between the number of bean stalks, n, she plants and the yield, y, each plant produces. When she plants 30 stalks, each plant yields 30 oz of beans. When she plants 34 stalks, each plant produces 28 oz of beans. Find a linear relationships in the form $y = mn + b$ that gives the yield when n stalks are planted.

119. A city's population in the year 1960 was 287,500. In 1989 the population was 275,900. Compute the rate of growth of the population and make a statement about the population rate of change in people per year.

120. A town's population has been growing linearly. In 2003, the population was 45,000, and the population has been growing by 1,700 people each year. Write an equation, $P(t)$, for the population t years after 2003.

121. Suppose that average annual income (in dollars) for the years 1990 through 1999 is given by the linear function: $I(x) = 1054x + 23{,}286$, where x is the number of years after 1990. Which of the following interprets the slope in the context of the problem?

ⓐ As of 1990, average annual income was $23,286.
ⓑ In the ten-year period from 1990–1999, average annual income increased by a total of $1,054.
ⓒ Each year in the decade of the 1990s, average annual income increased by $1,054.
ⓓ Average annual income rose to a level of $23,286 by the end of 1999.

122. When temperature is 0 degrees Celsius, the Fahrenheit temperature is 32. When the Celsius temperature is 100, the corresponding Fahrenheit temperature is 212. Express the Fahrenheit temperature as a linear function of C, the Celsius temperature, $F(C)$.

ⓐ Find the rate of change of Fahrenheit temperature for each unit change temperature of Celsius.
ⓑ Find and interpret $F(28)$.
ⓒ Find and interpret $F(-40)$.

4.2 Modeling with Linear Functions

Learning Objectives

In this section, you will:

- Build linear models from verbal descriptions.
- Model a set of data with a linear function.

Figure 1 (credit: EEK Photography/Flickr)

Elan is a college student who plans to spend a summer in Seattle. Elan has saved $3,500 for their trip and anticipates spending $400 each week on rent, food, and activities. How can we write a linear model to represent this situation? What would be the *x*-intercept, and what can Elan learn from it? To answer these and related questions, we can create a model using a linear function. Models such as this one can be extremely useful for analyzing relationships and making predictions based on those relationships. In this section, we will explore examples of linear function models.

Building Linear Models from Verbal Descriptions

When building linear models to solve problems involving quantities with a constant rate of change, we typically follow the same problem strategies that we would use for any type of function. Let's briefly review them:

1. Identify changing quantities, and then define descriptive variables to represent those quantities. When appropriate, sketch a picture or define a coordinate system.
2. Carefully read the problem to identify important information. Look for information that provides values for the variables or values for parts of the functional model, such as slope and initial value.
3. Carefully read the problem to determine what we are trying to find, identify, solve, or interpret.
4. Identify a solution pathway from the provided information to what we are trying to find. Often this will involve checking and tracking units, building a table, or even finding a formula for the function being used to model the problem.
5. When needed, write a formula for the function.
6. Solve or evaluate the function using the formula.
7. Reflect on whether your answer is reasonable for the given situation and whether it makes sense mathematically.
8. Clearly convey your result using appropriate units, and answer in full sentences when necessary.

Now let's take a look at the student in Seattle. In Elan's situation, there are two changing quantities: time and money. The amount of money they have remaining while on vacation depends on how long they stay. We can use this information to define our variables, including units.

$$\begin{aligned} &\text{Output: } M, \text{money remaining, in dollars} \\ &\text{Input: } t, \text{time, in weeks} \end{aligned}$$

So, the amount of money remaining depends on the number of weeks: $M(t)$.

We can also identify the initial value and the rate of change.

Initial Value: She saved \$3,500, so \$3,500 is the initial value for M.

Rate of Change: She anticipates spending \$400 each week, so $-$\$400 per week is the rate of change, or slope.

Notice that the unit of dollars per week matches the unit of our output variable divided by our input variable. Also, because the slope is negative, the linear function is decreasing. This should make sense because she is spending money each week.

The rate of change is constant, so we can start with the linear model $M(t) = mt + b$. Then we can substitute the intercept and slope provided.

$$M(t) = mt + b$$

m: -400 b: 3500

$$M(t) = -400t + 3500$$

To find the t-intercept (horizontal axis intercept), we set the output to zero, and solve for the input.

$$\begin{aligned} 0 &= -400t + 3500 \\ t &= \frac{3500}{400} \\ &= 8.75 \end{aligned}$$

The t-intercept (horizontal axis intercept) is 8.75 weeks. Because this represents the input value when the output will be zero, we could say that Elan will have no money left after 8.75 weeks.

When modeling any real-life scenario with functions, there is typically a limited domain over which that model will be valid—almost no trend continues indefinitely. Here the domain refers to the number of weeks. In this case, it doesn't make sense to talk about input values less than zero. A negative input value could refer to a number of weeks before Elan saved \$3,500, but the scenario discussed poses the question once they saved \$3,500 because this is when the trip and subsequent spending starts. It is also likely that this model is not valid after the t-intercept (horizontal axis intercept), unless Elan uses a credit card and goes into debt. The domain represents the set of input values, so the reasonable domain for this function is $0 \le t \le 8.75$.

In this example, we were given a written description of the situation. We followed the steps of modeling a problem to analyze the information. However, the information provided may not always be the same. Sometimes we might be provided with an intercept. Other times we might be provided with an output value. We must be careful to analyze the information we are given, and use it appropriately to build a linear model.

Using a Given Intercept to Build a Model

Some real-world problems provide the vertical axis intercept, which is the constant or initial value. Once the vertical axis intercept is known, the *t*-intercept (horizontal axis intercept) can be calculated. Suppose, for example, that Hannah plans to pay off a no-interest loan from her parents. Her loan balance is $1,000. She plans to pay $250 per month until her balance is $0. The *y*-intercept is the initial amount of her debt, or $1,000. The rate of change, or slope, is -$250 per month. We can then use the slope-intercept form and the given information to develop a linear model.

$$\begin{aligned} f(x) &= mx + b \\ &= -250x + 1000 \end{aligned}$$

Now we can set the function equal to 0, and solve for x to find the *x*-intercept.

$$\begin{aligned} 0 &= -250x + 1000 \\ 1000 &= 250x \\ 4 &= x \\ x &= 4 \end{aligned}$$

The *x*-intercept is the number of months it takes her to reach a balance of $0. The *x*-intercept is 4 months, so it will take Hannah four months to pay off her loan.

Using a Given Input and Output to Build a Model

Many real-world applications are not as direct as the ones we just considered. Instead they require us to identify some aspect of a linear function. We might sometimes instead be asked to evaluate the linear model at a given input or set the equation of the linear model equal to a specified output.

HOW TO

Given a word problem that includes two pairs of input and output values, use the linear function to solve a problem.

1. Identify the input and output values.
2. Convert the data to two coordinate pairs.
3. Find the slope.
4. Write the linear model.
5. Use the model to make a prediction by evaluating the function at a given *x*-value.
6. Use the model to identify an *x*-value that results in a given *y*-value.
7. Answer the question posed.

EXAMPLE 1

Using a Linear Model to Investigate a Town's Population

A town's population has been growing linearly. In 2004, the population was 6,200. By 2009, the population had grown to 8,100. Assume this trend continues.

ⓐ Predict the population in 2013. ⓑ Identify the year in which the population will reach 15,000.

Solution

The two changing quantities are the population size and time. While we could use the actual year value as the input quantity, doing so tends to lead to very cumbersome equations because the *y*-intercept would correspond to the year 0, more than 2000 years ago!

To make computation a little nicer, we will define our input as the number of years since 2004.

$$\text{Input: } t\text{, years since 2004}$$
$$\text{Output: } P(t)\text{, the town's population}$$

To predict the population in 2013 ($t = 9$), we would first need an equation for the population. Likewise, to find when the population would reach 15,000, we would need to solve for the input that would provide an output of 15,000. To write an equation, we need the initial value and the rate of change, or slope.

To determine the rate of change, we will use the change in output per change in input.

$$m = \frac{\text{change in output}}{\text{change in input}}$$

The problem gives us two input-output pairs. Converting them to match our defined variables, the year 2004 would correspond to $t = 0$, giving the point $(0, 6200)$. Notice that through our clever choice of variable definition, we have "given" ourselves the *y*-intercept of the function. The year 2009 would correspond to $t = 5$, giving the point $(5, 8100)$.

The two coordinate pairs are $(0, 6200)$ and $(5, 8100)$. Recall that we encountered examples in which we were provided two points earlier in the chapter. We can use these values to calculate the slope.

$$\begin{aligned} m &= \frac{8100-6200}{5-0} \\ &= \frac{1900}{5} \\ &= 380 \text{ people per year} \end{aligned}$$

We already know the *y*-intercept of the line, so we can immediately write the equation:

$$P(t) = 380t + 6200$$

To predict the population in 2013, we evaluate our function at $t = 9$.

$$\begin{aligned} P(9) &= 380(9) + 6{,}200 \\ &= 9{,}620 \end{aligned}$$

If the trend continues, our model predicts a population of 9,620 in 2013.

To find when the population will reach 15,000, we can set $P(t) = 15000$ and solve for t.

$$\begin{aligned} 15000 &= 380t + 6200 \\ 8800 &= 380t \\ t &\approx 23.158 \end{aligned}$$

Our model predicts the population will reach 15,000 in a little more than 23 years after 2004, or somewhere around the year 2027.

TRY IT #1 A company sells doughnuts. They incur a fixed cost of $25,000 for rent, insurance, and other expenses. It costs $0.25 to produce each doughnut.

ⓐ Write a linear model to represent the cost C of the company as a function of x, the number of doughnuts produced.
ⓑ Find and interpret the *y*-intercept.

TRY IT #2 A city's population has been growing linearly. In 2008, the population was 28,200. By 2012, the population was 36,800. Assume this trend continues.

ⓐ Predict the population in 2014.
ⓑ Identify the year in which the population will reach 54,000.

Using a Diagram to Build a Model

It is useful for many real-world applications to draw a picture to gain a sense of how the variables representing the input and output may be used to answer a question. To draw the picture, first consider what the problem is asking for. Then, determine the input and the output. The diagram should relate the variables. Often, geometrical shapes or figures are drawn. Distances are often traced out. If a right triangle is sketched, the Pythagorean Theorem relates the sides. If a rectangle is sketched, labeling width and height is helpful.

EXAMPLE 2

Using a Diagram to Model Distance Walked

Anna and Emanuel start at the same intersection. Anna walks east at 4 miles per hour while Emanuel walks south at 3 miles per hour. They are communicating with a two-way radio that has a range of 2 miles. How long after they start walking will they fall out of radio contact?

Solution

In essence, we can partially answer this question by saying they will fall out of radio contact when they are 2 miles apart, which leads us to ask a new question:

"How long will it take them to be 2 miles apart"?

In this problem, our changing quantities are time and position, but ultimately we need to know how long will it take for them to be 2 miles apart. We can see that time will be our input variable, so we'll define our input and output variables.

$$\text{Input: } t, \text{time in hours.}$$
$$\text{Output: } A(t), \text{distance in miles, and } E(t), \text{distance in miles}$$

Because it is not obvious how to define our output variable, we'll start by drawing a picture such as Figure 2.

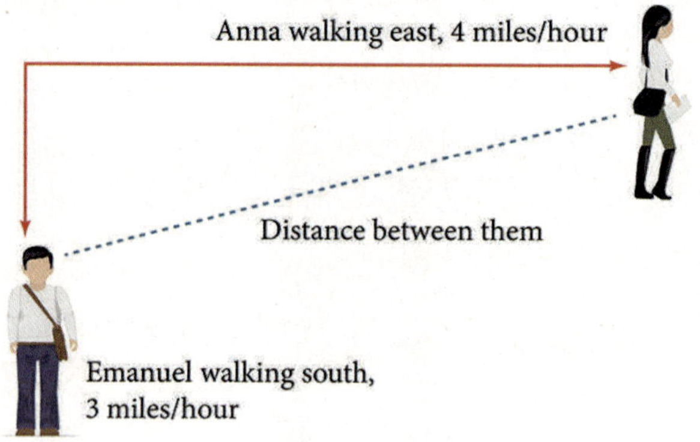

Figure 2

Initial Value: They both start at the same intersection so when $t = 0$, the distance traveled by each person should also be 0. Thus the initial value for each is 0.

Rate of Change: Anna is walking 4 miles per hour and Emanuel is walking 3 miles per hour, which are both rates of change. The slope for A is 4 and the slope for E is 3.

Using those values, we can write formulas for the distance each person has walked.

$$\begin{aligned} A(t) &= 4t \\ E(t) &= 3t \end{aligned}$$

For this problem, the distances from the starting point are important. To notate these, we can define a coordinate system, identifying the "starting point" at the intersection where they both started. Then we can use the variable, A, which we introduced above, to represent Anna's position, and define it to be a measurement from the starting point in the eastward direction. Likewise, can use the variable, E, to represent Emanuel's position, measured from the starting point in the southward direction. Note that in defining the coordinate system, we specified both the starting point of the measurement and the direction of measure.

We can then define a third variable, D, to be the measurement of the distance between Anna and Emanuel. Showing the variables on the diagram is often helpful, as we can see from Figure 3.

Recall that we need to know how long it takes for D, the distance between them, to equal 2 miles. Notice that for any given input t, the outputs $A(t)$, $E(t)$, and $D(t)$ represent distances.

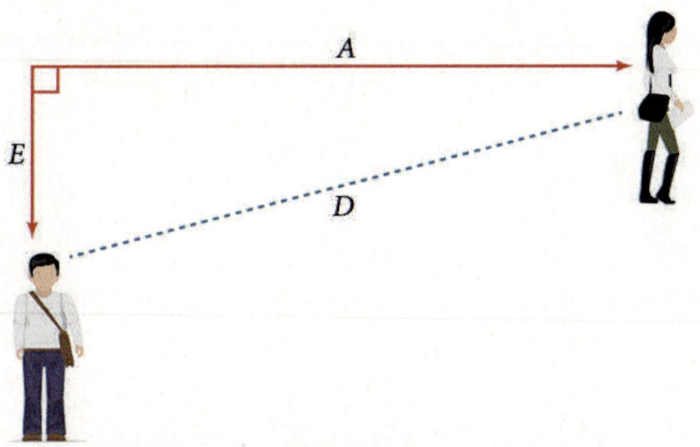

Figure 3

Figure 2 shows us that we can use the Pythagorean Theorem because we have drawn a right angle.

Using the Pythagorean Theorem, we get:

$$\begin{aligned} D(t)^2 &= A(t)^2 + E(t)^2 \\ &= (4t)^2 + (3t)^2 \\ &= 16t^2 + 9t^2 \\ &= 25t^2 \\ D(t) &= \pm\sqrt{25t^2} && \text{Solve for } D(t) \text{ using the square root.} \\ &= \pm 5|t| \end{aligned}$$

In this scenario we are considering only positive values of t, so our distance $D(t)$ will always be positive. We can simplify this answer to $D(t) = 5t$. This means that the distance between Anna and Emanuel is also a linear function. Because D is a linear function, we can now answer the question of when the distance between them will reach 2 miles. We will set the output $D(t) = 2$ and solve for t.

$$\begin{aligned} D(t) &= 2 \\ 5t &= 2 \\ t &= \tfrac{2}{5} = 0.4 \end{aligned}$$

They will fall out of radio contact in 0.4 hour, or 24 minutes.

Q&A **Should I draw diagrams when given information based on a geometric shape?**

Yes. Sketch the figure and label the quantities and unknowns on the sketch.

EXAMPLE 3

Using a Diagram to Model Distance Between Cities

There is a straight road leading from the town of Westborough to Agritown 30 miles east and 10 miles north. Partway down this road, it junctions with a second road, perpendicular to the first, leading to the town of Eastborough. If the town of Eastborough is located 20 miles directly east of the town of Westborough, how far is the road junction from Westborough?

Solution

It might help here to draw a picture of the situation. See Figure 4. It would then be helpful to introduce a coordinate system. While we could place the origin anywhere, placing it at Westborough seems convenient. This puts Agritown at coordinates $(30, 10)$, and Eastborough at $(20, 0)$.

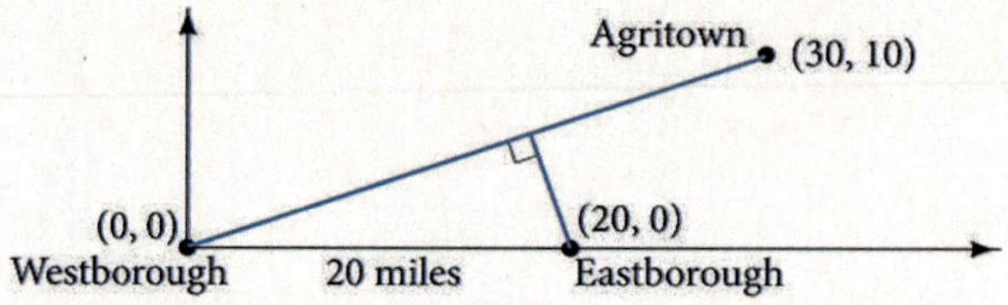

Figure 4

Using this point along with the origin, we can find the slope of the line from Westborough to Agritown.

$$m = \frac{10-0}{30-0} = \frac{1}{3}$$

Now we can write an equation to describe the road from Westborough to Agritown.

$$W(x) = \frac{1}{3}x$$

From this, we can determine the perpendicular road to Eastborough will have slope $m = -3$. Because the town of Eastborough is at the point (20, 0), we can find the equation.

$$\begin{aligned} E(x) &= -3x + b \\ 0 &= -3(20) + b && \text{Substitute } (20,0) \text{ into the equation.} \\ b &= 60 \\ E(x) &= -3x + 60 \end{aligned}$$

We can now find the coordinates of the junction of the roads by finding the intersection of these lines. Setting them equal,

$$\begin{aligned} \tfrac{1}{3}x &= -3x + 60 \\ \tfrac{10}{3}x &= 60 \\ 10x &= 180 \\ x &= 18 && \text{Substitute this back into } W(x). \\ y &= W(18) \\ &= \tfrac{1}{3}(18) \\ &= 6 \end{aligned}$$

The roads intersect at the point (18, 6). Using the distance formula, we can now find the distance from Westborough to the junction.

$$\begin{aligned} \text{distance} &= \sqrt{(x_2 - x_1)^2 + (y_2 - y_1)^2} \\ &= \sqrt{(18-0)^2 + (6-0)^2} \\ &\approx 18.974 \text{ miles} \end{aligned}$$

Analysis

One nice use of linear models is to take advantage of the fact that the graphs of these functions are lines. This means real-world applications discussing maps need linear functions to model the distances between reference points.

TRY IT #3 There is a straight road leading from the town of Timpson to Ashburn 60 miles east and 12 miles north. Partway down the road, it junctions with a second road, perpendicular to the first, leading to the town of Garrison. If the town of Garrison is located 22 miles directly east of the town of Timpson, how far is the road junction from Timpson?

Modeling a Set of Data with Linear Functions

Real-world situations including two or more linear functions may be modeled with a system of linear equations. Remember, when solving a system of linear equations, we are looking for points the two lines have in common. Typically, there are three types of answers possible, as shown in Figure 5.

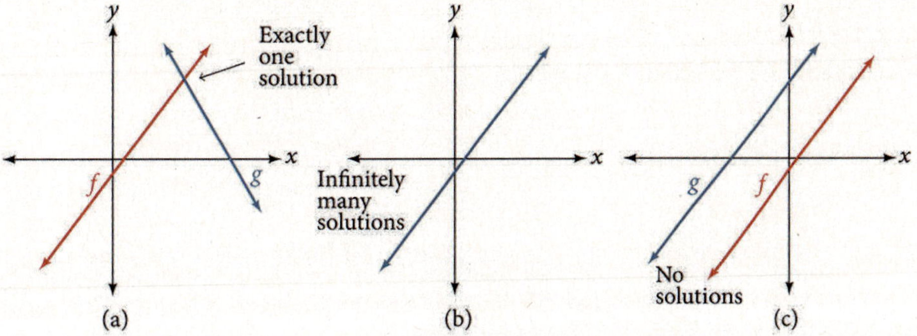

Figure 5

HOW TO

Given a situation that represents a system of linear equations, write the system of equations and identify the solution.

1. Identify the input and output of each linear model.
2. Identify the slope and y-intercept of each linear model.
3. Find the solution by setting the two linear functions equal to another and solving for x, or find the point of intersection on a graph.

EXAMPLE 4

Building a System of Linear Models to Choose a Truck Rental Company

Jamal is choosing between two truck-rental companies. The first, Keep on Trucking, Inc., charges an up-front fee of $20, then 59 cents a mile. The second, Move It Your Way, charges an up-front fee of $16, then 63 cents a mile[4]. When will Keep on Trucking, Inc. be the better choice for Jamal?

Solution

The two important quantities in this problem are the cost and the number of miles driven. Because we have two companies to consider, we will define two functions in Table 1.

Input	d, distance driven in miles
Outputs	$K(d)$: cost, in dollars, for renting from Keep on Trucking $M(d)$ cost, in dollars, for renting from Move It Your Way
Initial Value	Up-front fee: $K(0) = 20$ and $M(0) = 16$
Rate of Change	$K(d) = \$0.59$ /mile and $P(d) = \$0.63$ /mile

Table 1

A linear function is of the form $f(x) = mx + b$. Using the rates of change and initial charges, we can write the equations

$$\begin{aligned} K(d) &= 0.59d + 20 \\ M(d) &= 0.63d + 16 \end{aligned}$$

Using these equations, we can determine when Keep on Trucking, Inc., will be the better choice. Because all we have to make that decision from is the costs, we are looking for when Move It Your Way, will cost less, or when $K(d) < M(d)$. The solution pathway will lead us to find the equations for the two functions, find the intersection, and then see where

4 Rates retrieved Aug 2, 2010 from http://www.budgettruck.com and http://www.uhaul.com/

the $K(d)$ function is smaller.

These graphs are sketched in Figure 6, with $K(d)$ in blue.

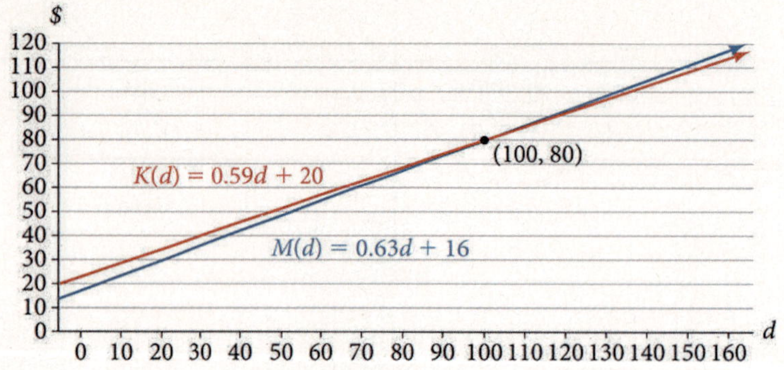

Figure 6

To find the intersection, we set the equations equal and solve:

$$\begin{aligned} K(d) &= M(d) \\ 0.59d + 20 &= 0.63d + 16 \\ 4 &= 0.04d \\ 100 &= d \\ d &= 100 \end{aligned}$$

This tells us that the cost from the two companies will be the same if 100 miles are driven. Either by looking at the graph, or noting that $K(d)$ is growing at a slower rate, we can conclude that Keep on Trucking, Inc. will be the cheaper price when more than 100 miles are driven, that is $d > 100$.

MEDIA

Access this online resource for additional instruction and practice with linear function models.

Interpreting a Linear Function (http://openstax.org/l/interpretlinear)

4.2 SECTION EXERCISES

Verbal

1. Explain how to find the input variable in a word problem that uses a linear function.

2. Explain how to find the output variable in a word problem that uses a linear function.

3. Explain how to interpret the initial value in a word problem that uses a linear function.

4. Explain how to determine the slope in a word problem that uses a linear function.

Algebraic

5. Find the area of a parallelogram bounded by the *y*-axis, the line $x = 3$, the line $f(x) = 1 + 2x$, and the line parallel to $f(x)$ passing through $(2, 7)$.

6. Find the area of a triangle bounded by the *x*-axis, the line $f(x) = 12 - \frac{1}{3}x$, and the line perpendicular to $f(x)$ that passes through the origin.

7. Find the area of a triangle bounded by the *y*-axis, the line $f(x) = 9 - \frac{6}{7}x$, and the line perpendicular to $f(x)$ that passes through the origin.

8. Find the area of a parallelogram bounded by the *x*-axis, the line $g(x) = 2$, the line $f(x) = 3x$, and the line parallel to $f(x)$ passing through $(6, 1)$.

For the following exercises, consider this scenario: A town's population has been decreasing at a constant rate. In 2010 the population was 5,900. By 2012 the population had dropped to 4,700. Assume this trend continues.

9. Predict the population in 2016.

10. Identify the year in which the population will reach 0.

For the following exercises, consider this scenario: A town's population has been increased at a constant rate. In 2010 the population was 46,020. By 2012 the population had increased to 52,070. Assume this trend continues.

11. Predict the population in 2016.

12. Identify the year in which the population will reach 75,000.

For the following exercises, consider this scenario: A town has an initial population of 75,000. It grows at a constant rate of 2,500 per year for 5 years.

13. Find the linear function that models the town's population P as a function of the year, t, where t is the number of years since the model began.

14. Find a reasonable domain and range for the function P.

15. If the function P is graphed, find and interpret the *x*- and *y*-intercepts.

16. If the function P is graphed, find and interpret the slope of the function.

17. When will the population reach 100,000?

18. What is the population in the year 12 years from the onset of the model?

For the following exercises, consider this scenario: The weight of a newborn is 7.5 pounds. The baby gained one-half pound a month for its first year.

19. Find the linear function that models the baby's weight W as a function of the age of the baby, in months, t.

20. Find a reasonable domain and range for the function W.

21. If the function W is graphed, find and interpret the *x*- and *y*-intercepts.

22. If the function *W* is graphed, find and interpret the slope of the function.

23. When did the baby weight 10.4 pounds?

24. What is the output when the input is 6.2?

For the following exercises, consider this scenario: The number of people afflicted with the common cold in the winter months steadily decreased by 205 each year from 2005 until 2010. In 2005, 12,025 people were inflicted.

25. Find the linear function that models the number of people inflicted with the common cold C as a function of the year, t.

26. Find a reasonable domain and range for the function C.

27. If the function C is graphed, find and interpret the x- and y-intercepts.

28. If the function C is graphed, find and interpret the slope of the function.

29. When will the output reach 0?

30. In what year will the number of people be 9,700?

Graphical

For the following exercises, use the graph in Figure 7, which shows the profit, y, in thousands of dollars, of a company in a given year, t, where t represents the number of years since 1980.

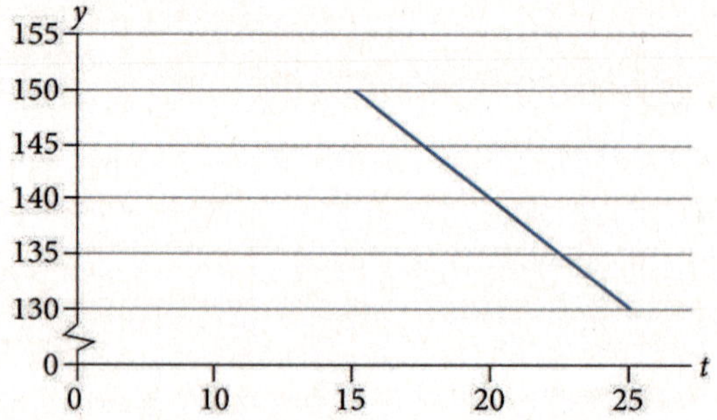

Figure 7

31. Find the linear function y, where y depends on t, the number of years since 1980.

32. Find and interpret the y-intercept.

33. Find and interpret the x-intercept.

34. Find and interpret the slope.

For the following exercises, use the graph in Figure 8, which shows the profit, y, in thousands of dollars, of a company in a given year, t, where t represents the number of years since 1980.

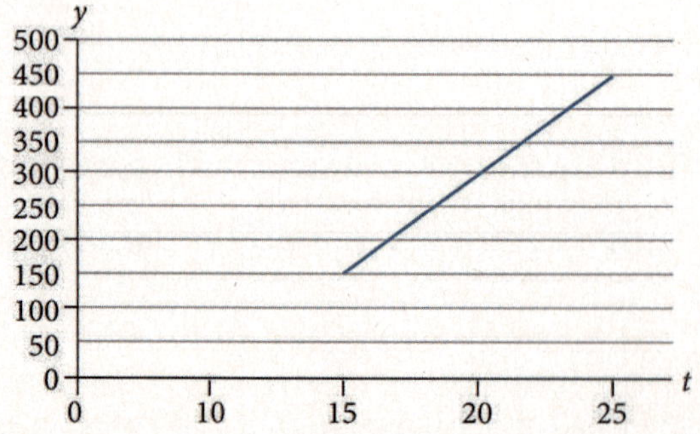

Figure 8

35. Find the linear function y, where y depends on t, the number of years since 1980.

36. Find and interpret the y-intercept.

37. Find and interpret the x-intercept.

38. Find and interpret the slope.

Numeric

For the following exercises, use the median home values in Mississippi and Hawaii (adjusted for inflation) shown in Table 2. Assume that the house values are changing linearly.

Year	Mississippi	Hawaii
1950	$25,200	$74,400
2000	$71,400	$272,700

Table 2

39. In which state have home values increased at a higher rate?

40. If these trends were to continue, what would be the median home value in Mississippi in 2010?

41. If we assume the linear trend existed before 1950 and continues after 2000, the two states' median house values will be (or were) equal in what year? (The answer might be absurd.)

For the following exercises, use the median home values in Indiana and Alabama (adjusted for inflation) shown in Table 3. Assume that the house values are changing linearly.

Year	Indiana	Alabama
1950	$37,700	$27,100
2000	$94,300	$85,100

Table 3

42. In which state have home values increased at a higher rate?

43. If these trends were to continue, what would be the median home value in Indiana in 2010?

44. If we assume the linear trend existed before 1950 and continues after 2000, the two states' median house values will be (or were) equal in what year? (The answer might be absurd.)

Real-World Applications

45. In 2004, a school population was 1001. By 2008 the population had grown to 1697. Assume the population is changing linearly.

ⓐ How much did the population grow between the year 2004 and 2008?
ⓑ How long did it take the population to grow from 1001 students to 1697 students?
ⓒ What is the average population growth per year?
ⓓ What was the population in the year 2000?
ⓔ Find an equation for the population, P, of the school t years after 2000.
ⓕ Using your equation, predict the population of the school in 2011.

46. In 2003, a town's population was 1431. By 2007 the population had grown to 2134. Assume the population is changing linearly.

ⓐ How much did the population grow between the year 2003 and 2007?
ⓑ How long did it take the population to grow from 1431 people to 2134 people?
ⓒ What is the average population growth per year?
ⓓ What was the population in the year 2000?
ⓔ Find an equation for the population, P, of the town t years after 2000.
ⓕ Using your equation, predict the population of the town in 2014.

47. A phone company has a monthly cellular plan where a customer pays a flat monthly fee and then a certain amount of money per minute used for voice or video calling. If a customer uses 410 minutes, the monthly cost will be \$71.50. If the customer uses 720 minutes, the monthly cost will be \$118.

ⓐ Find a linear equation for the monthly cost of the cell plan as a function of x, the number of monthly minutes used.
ⓑ Interpret the slope and y-intercept of the equation.
ⓒ Use your equation to find the total monthly cost if 687 minutes are used.

48. A phone company has a monthly cellular data plan where a customer pays a flat monthly fee of \$10 and then a certain amount of money per megabyte (MB) of data used on the phone. If a customer uses 20 MB, the monthly cost will be \$11.20. If the customer uses 130 MB, the monthly cost will be \$17.80.

ⓐ Find a linear equation for the monthly cost of the data plan as a function of x, the number of MB used.
ⓑ Interpret the slope and y-intercept of the equation.
ⓒ Use your equation to find the total monthly cost if 250 MB are used.

49. In 1991, the moose population in a park was measured to be 4,360. By 1999, the population was measured again to be 5,880. Assume the population continues to change linearly.

ⓐ Find a formula for the moose population, P since 1990.
ⓑ What does your model predict the moose population to be in 2003?

50. In 2003, the owl population in a park was measured to be 340. By 2007, the population was measured again to be 285. The population changes linearly. Let the input be years since 2003.

ⓐ Find a formula for the owl population, P. Let the input be years since 2003.
ⓑ What does your model predict the owl population to be in 2012?

51. The Federal Helium Reserve held about 16 billion cubic feet of helium in 2010 and is being depleted by about 2.1 billion cubic feet each year.

ⓐ Give a linear equation for the remaining federal helium reserves, R, in terms of t, the number of years since 2010.
ⓑ In 2015, what will the helium reserves be?
ⓒ If the rate of depletion doesn't change, in what year will the Federal Helium Reserve be depleted?

52. Suppose the world's oil reserves in 2014 are 1,820 billion barrels. If, on average, the total reserves are decreasing by 25 billion barrels of oil each year:

ⓐ Give a linear equation for the remaining oil reserves, R, in terms of t, the number of years since now.
ⓑ Seven years from now, what will the oil reserves be?
ⓒ If the rate at which the reserves are decreasing is constant, when will the world's oil reserves be depleted?

53. You are choosing between two different prepaid cell phone plans. The first plan charges a rate of 26 cents per minute. The second plan charges a monthly fee of $19.95 *plus* 11 cents per minute. How many minutes would you have to use in a month in order for the second plan to be preferable?

54. You are choosing between two different window washing companies. The first charges $5 per window. The second charges a base fee of $40 plus $3 per window. How many windows would you need to have for the second company to be preferable?

55. When hired at a new job selling jewelry, you are given two pay options:

Option A: Base salary of $17,000 a year with a commission of 12% of your sales

Option B: Base salary of $20,000 a year with a commission of 5% of your sales

How much jewelry would you need to sell for option A to produce a larger income?

56. When hired at a new job selling electronics, you are given two pay options:

Option A: Base salary of $14,000 a year with a commission of 10% of your sales

Option B: Base salary of $19,000 a year with a commission of 4% of your sales

How much electronics would you need to sell for option A to produce a larger income?

57. When hired at a new job selling electronics, you are given two pay options:

Option A: Base salary of $20,000 a year with a commission of 12% of your sales

Option B: Base salary of $26,000 a year with a commission of 3% of your sales

How much electronics would you need to sell for option A to produce a larger income?

58. When hired at a new job selling electronics, you are given two pay options:

Option A: Base salary of $10,000 a year with a commission of 9% of your sales

Option B: Base salary of $20,000 a year with a commission of 4% of your sales

How much electronics would you need to sell for option A to produce a larger income?

4.3 Fitting Linear Models to Data

Learning Objectives

In this section, you will:

- Draw and interpret scatter diagrams.
- Use a graphing utility to find the line of best fit.
- Distinguish between linear and nonlinear relations.
- Fit a regression line to a set of data and use the linear model to make predictions.

A professor is attempting to identify trends among final exam scores. His class has a mixture of students, so he wonders if there is any relationship between age and final exam scores. One way for him to analyze the scores is by creating a diagram that relates the age of each student to the exam score received. In this section, we will examine one such diagram known as a scatter plot.

Drawing and Interpreting Scatter Plots

A scatter plot is a graph of plotted points that may show a relationship between two sets of data. If the relationship is from a linear model, or a model that is nearly linear, the professor can draw conclusions using his knowledge of linear functions. Figure 1 shows a sample scatter plot.

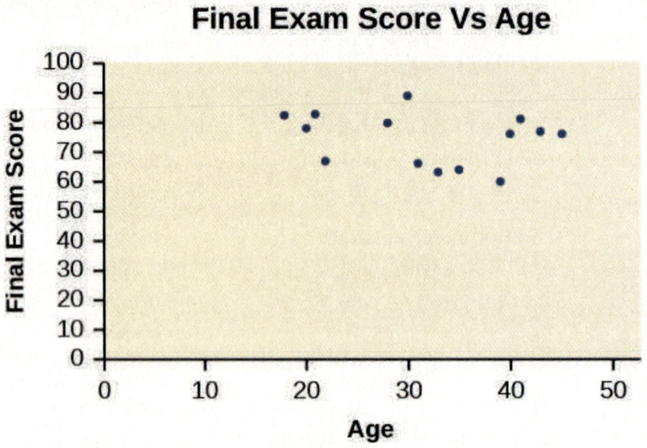

Figure 1 A scatter plot of age and final exam score variables

Notice this scatter plot does *not* indicate a linear relationship. The points do not appear to follow a trend. In other words, there does not appear to be a relationship between the age of the student and the score on the final exam.

EXAMPLE 1

Using a Scatter Plot to Investigate Cricket Chirps

Table 1 shows the number of cricket chirps in 15 seconds, for several different air temperatures, in degrees Fahrenheit[5]. Plot this data, and determine whether the data appears to be linearly related.

Chirps	44	35	20.4	33	31	35	18.5	37	26
Temperature	80.5	70.5	57	66	68	72	52	73.5	53

Table 1 Cricket Chirps vs Air Temperature

Solution

Plotting this data, as depicted in Figure 2 suggests that there may be a trend. We can see from the trend in the data that the number of chirps increases as the temperature increases. The trend appears to be roughly linear, though certainly not perfectly so.

5 Selected data from http://classic.globe.gov/fsl/scientistsblog/2007/10/. Retrieved Aug 3, 2010

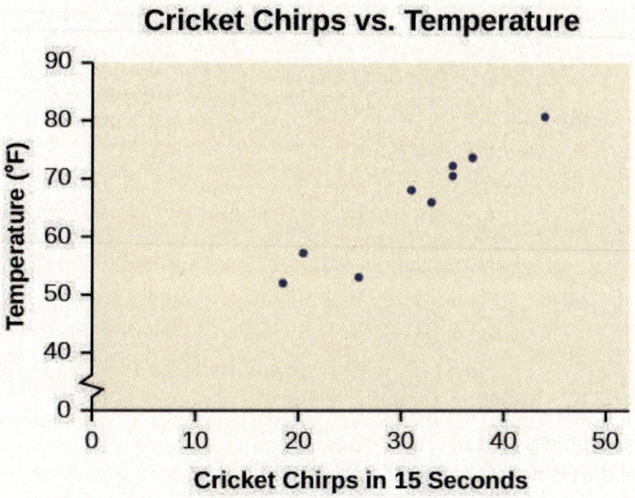

Figure 2

Finding the Line of Best Fit

Once we recognize a need for a linear function to model that data, the natural follow-up question is "what is that linear function?" One way to approximate our linear function is to sketch the line that seems to best fit the data. Then we can extend the line until we can verify the y-intercept. We can approximate the slope of the line by extending it until we can estimate the $\frac{\text{rise}}{\text{run}}$.

EXAMPLE 2

Finding a Line of Best Fit

Find a linear function that fits the data in Table 1 by "eyeballing" a line that seems to fit.

Solution

On a graph, we could try sketching a line. Using the starting and ending points of our hand drawn line, points (0, 30) and (50, 90), this graph has a slope of

$$\begin{aligned} m &= \frac{60}{50} \\ &= 1.2 \end{aligned}$$

and a y-intercept at 30. This gives an equation of

$$T(c) = 1.2c + 30$$

where c is the number of chirps in 15 seconds, and $T(c)$ is the temperature in degrees Fahrenheit. The resulting equation is represented in Figure 3.

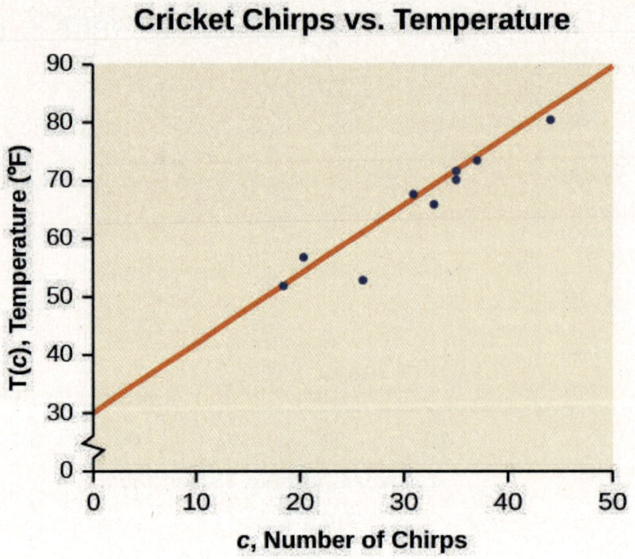

Figure 3

Analysis

This linear equation can then be used to approximate answers to various questions we might ask about the trend.

Recognizing Interpolation or Extrapolation

While the data for most examples does not fall perfectly on the line, the equation is our best guess as to how the relationship will behave outside of the values for which we have data. We use a process known as **interpolation** when we predict a value inside the domain and range of the data. The process of **extrapolation** is used when we predict a value outside the domain and range of the data.

Figure 4 compares the two processes for the cricket-chirp data addressed in Example 2. We can see that interpolation would occur if we used our model to predict temperature when the values for chirps are between 18.5 and 44. Extrapolation would occur if we used our model to predict temperature when the values for chirps are less than 18.5 or greater than 44.

There is a difference between making predictions inside the domain and range of values for which we have data and outside that domain and range. Predicting a value outside of the domain and range has its limitations. When our model no longer applies after a certain point, it is sometimes called **model breakdown**. For example, predicting a cost function for a period of two years may involve examining the data where the input is the time in years and the output is the cost. But if we try to extrapolate a cost when $x = 50$, that is in 50 years, the model would not apply because we could not account for factors fifty years in the future.

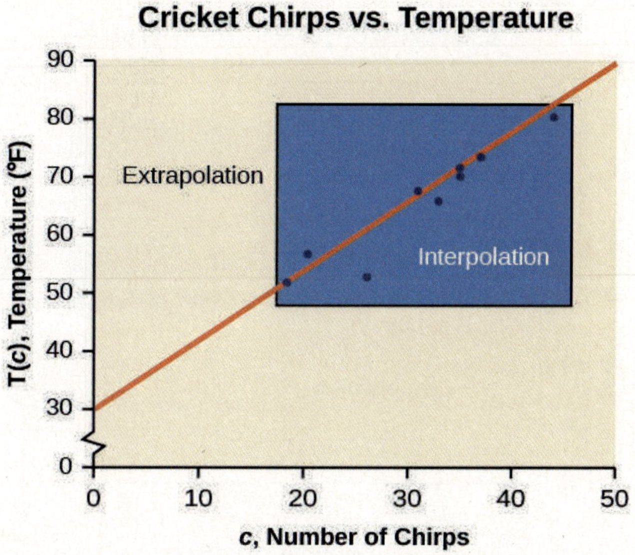

Figure 4 Interpolation occurs within the domain and range of the provided data whereas extrapolation occurs outside.

Interpolation and Extrapolation

Different methods of making predictions are used to analyze data.

The method of **interpolation** involves predicting a value inside the domain and/or range of the data.
The method of **extrapolation** involves predicting a value outside the domain and/or range of the data.
Model breakdown occurs at the point when the model no longer applies.

EXAMPLE 3

Understanding Interpolation and Extrapolation

Use the cricket data from Table 1 to answer the following questions:

ⓐ Would predicting the temperature when crickets are chirping 30 times in 15 seconds be interpolation or extrapolation? Make the prediction, and discuss whether it is reasonable.
ⓑ Would predicting the number of chirps crickets will make at 40 degrees be interpolation or extrapolation? Make the prediction, and discuss whether it is reasonable.

Solution

ⓐ The number of chirps in the data provided varied from 18.5 to 44. A prediction at 30 chirps per 15 seconds is inside the domain of our data, so would be interpolation. Using our model:

$$\begin{aligned} T(30) &= 30 + 1.2(30) \\ &= 66 \text{ degrees} \end{aligned}$$

Based on the data we have, this value seems reasonable.

ⓑ The temperature values varied from 52 to 80.5. Predicting the number of chirps at 40 degrees is extrapolation because 40 is outside the range of our data. Using our model:

$$\begin{aligned} 40 &= 30 + 1.2c \\ 10 &= 1.2c \\ c &\approx 8.33 \end{aligned}$$

We can compare the regions of interpolation and extrapolation using Figure 5.

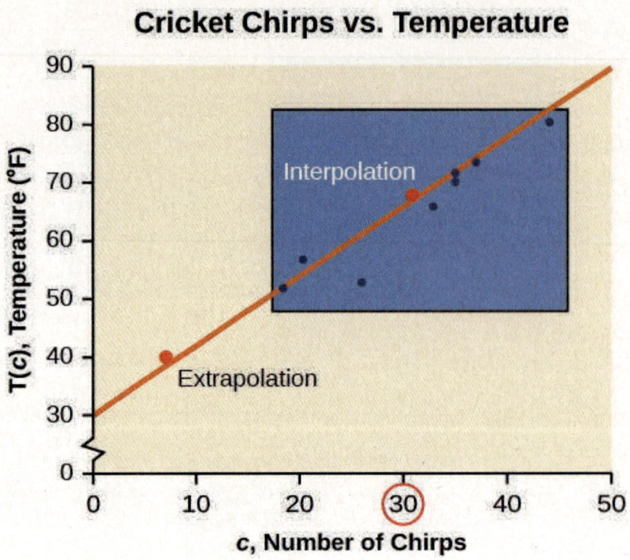

Figure 5

Analysis

Our model predicts the crickets would chirp 8.33 times in 15 seconds. While this might be possible, we have no reason to believe our model is valid outside the domain and range. In fact, generally crickets stop chirping altogether below around 50 degrees.

TRY IT #1 According to the data from Table 1, what temperature can we predict it is if we counted 20 chirps in 15 seconds?

Finding the Line of Best Fit Using a Graphing Utility

While eyeballing a line works reasonably well, there are statistical techniques for fitting a line to data that minimize the differences between the line and data values[6]. One such technique is called **least squares regression** and can be computed by many graphing calculators, spreadsheet software, statistical software, and many web-based calculators[7]. Least squares regression is one means to determine the line that best fits the data, and here we will refer to this method as linear regression.

HOW TO

Given data of input and corresponding outputs from a linear function, find the best fit line using linear regression.

1. Enter the input in List 1 (L1).
2. Enter the output in List 2 (L2).
3. On a graphing utility, select Linear Regression (LinReg).

EXAMPLE 4

Finding a Least Squares Regression Line

Find the least squares regression line using the cricket-chirp data in Table 2.

Solution

1. Enter the input (chirps) in List 1 (L1).

6 Technically, the method minimizes the sum of the squared differences in the vertical direction between the line and the data values.

7 For example, http://www.shodor.org/unchem/math/lls/leastsq.html

2. Enter the output (temperature) in List 2 (L2). See Table 2.

L1	44	35	20.4	33	31	35	18.5	37	26
L2	80.5	70.5	57	66	68	72	52	73.5	53

Table 2

3. On a graphing utility, select Linear Regression (LinReg). Using the cricket chirp data from earlier, with technology we obtain the equation:

$$T(c) = 30.281 + 1.143c$$

Analysis

Notice that this line is quite similar to the equation we "eyeballed" but should fit the data better. Notice also that using this equation would change our prediction for the temperature when hearing 30 chirps in 15 seconds from 66 degrees to:

$$\begin{aligned} T(30) &= 30.281 + 1.143(30) \\ &= 64.571 \\ &\approx 64.6 \text{ degrees} \end{aligned}$$

The graph of the scatter plot with the least squares regression line is shown in Figure 6.

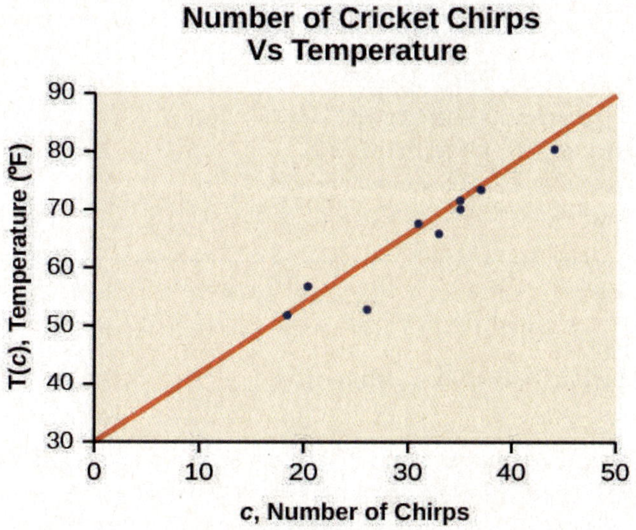

Figure 6

Q&A **Will there ever be a case where two different lines will serve as the best fit for the data?**

No. There is only one best fit line.

Distinguishing Between Linear and Nonlinear Models

As we saw above with the cricket-chirp model, some data exhibit strong linear trends, but other data, like the final exam scores plotted by age, are clearly nonlinear. Most calculators and computer software can also provide us with the correlation coefficient, which is a measure of how closely the line fits the data. Many graphing calculators require the user to turn a "diagnostic on" selection to find the correlation coefficient, which mathematicians label as r The correlation coefficient provides an easy way to get an idea of how close to a line the data falls.

We should compute the correlation coefficient only for data that follows a linear pattern or to determine the degree to which a data set is linear. If the data exhibits a nonlinear pattern, the correlation coefficient for a linear regression is meaningless. To get a sense for the relationship between the value of r and the graph of the data, Figure 7 shows some large data sets with their correlation coefficients. Remember, for all plots, the horizontal axis shows the input and the vertical axis shows the output.

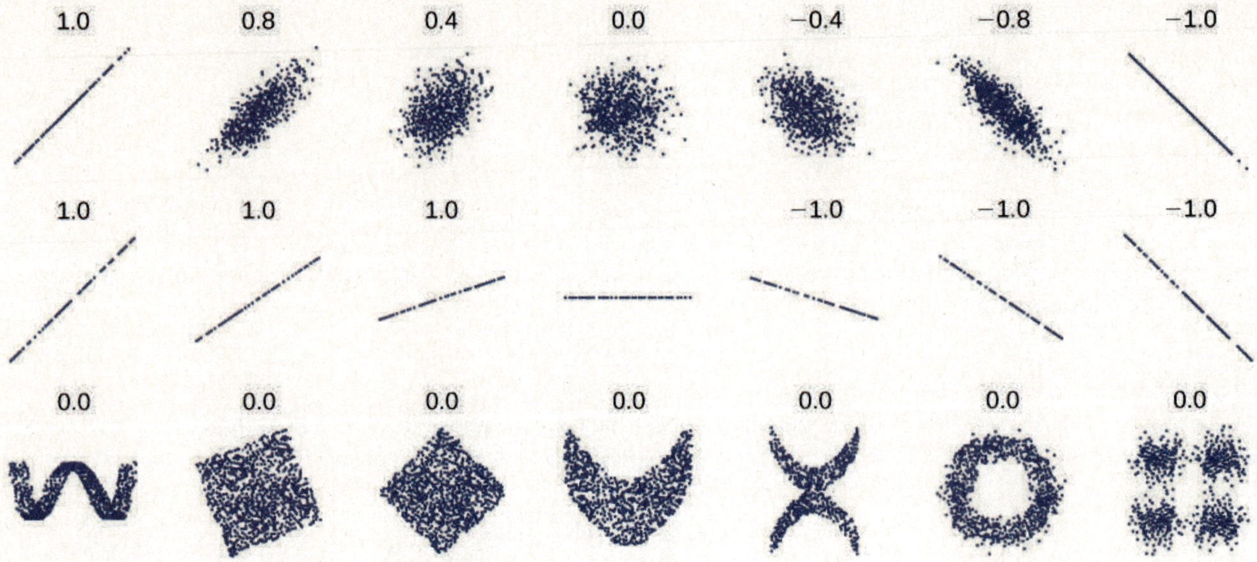

Figure 7 Plotted data and related correlation coefficients. (credit: "DenisBoigelot," Wikimedia Commons)

Correlation Coefficient

The **correlation coefficient** is a value, r, between –1 and 1.

- $r > 0$ suggests a positive (increasing) relationship
- $r < 0$ suggests a negative (decreasing) relationship
- The closer the value is to 0, the more scattered the data.
- The closer the value is to 1 or –1, the less scattered the data is.

EXAMPLE 5

Finding a Correlation Coefficient

Calculate the correlation coefficient for cricket-chirp data in Table 1.

Solution

Because the data appear to follow a linear pattern, we can use technology to calculate r Enter the inputs and corresponding outputs and select the Linear Regression. The calculator will also provide you with the correlation coefficient, $r = 0.9509$. This value is very close to 1, which suggests a strong increasing linear relationship.

Note: For some calculators, the Diagnostics must be turned "on" in order to get the correlation coefficient when linear regression is performed: [2nd]>[0]>[alpha][x–1], then scroll to **DIAGNOSTICSON**.

Fitting a Regression Line to a Set of Data

Once we determine that a set of data is linear using the correlation coefficient, we can use the regression line to make predictions. As we learned above, a regression line is a line that is closest to the data in the scatter plot, which means that only one such line is a best fit for the data.

EXAMPLE 6

Using a Regression Line to Make Predictions

Gasoline consumption in the United States has been steadily increasing. Consumption data from 1994 to 2004 is shown in Table 3.[8] Determine whether the trend is linear, and if so, find a model for the data. Use the model to predict the consumption in 2008.

8 http://www.bts.gov/publications/national_transportation_statistics/2005/html/table_04_10.html

Year	'94	'95	'96	'97	'98	'99	'00	'01	'02	'03	'04
Consumption (billions of gallons)	113	116	118	119	123	125	126	128	131	133	136

Table 3

The scatter plot of the data, including the least squares regression line, is shown in Figure 8.

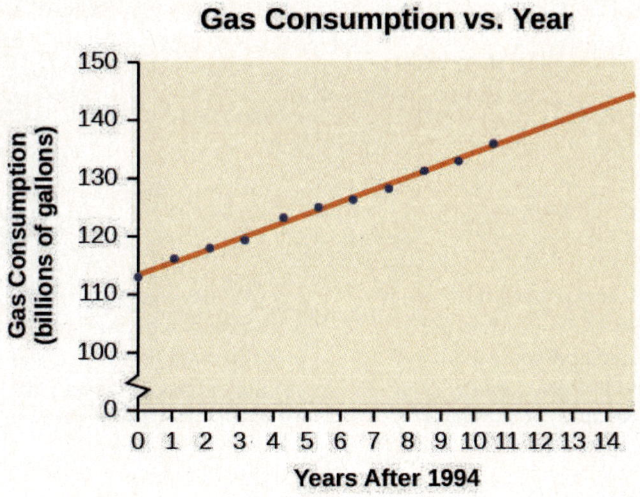

Figure 8

Solution

We can introduce a new input variable, t, representing years since 1994.

The least squares regression equation is:

$$C(t) = 113.318 + 2.209t$$

Using technology, the correlation coefficient was calculated to be 0.9965, suggesting a very strong increasing linear trend.

Using this to predict consumption in 2008 ($t = 14$),

$$\begin{aligned} C(14) &= 113.318 + 2.209(14) \\ &= 144.244 \end{aligned}$$

The model predicts 144.244 billion gallons of gasoline consumption in 2008.

TRY IT #2 Use the model we created using technology in Example 6 to predict the gas consumption in 2011. Is this an interpolation or an extrapolation?

MEDIA

Access these online resources for additional instruction and practice with fitting linear models to data.

Introduction to Regression Analysis (http://openstax.org/l/introregress)
Linear Regression (http://openstax.org/l/linearregress)

4.3 SECTION EXERCISES

Verbal

1. Describe what it means if there is a model breakdown when using a linear model.

2. What is interpolation when using a linear model?

3. What is extrapolation when using a linear model?

4. Explain the difference between a positive and a negative correlation coefficient.

5. Explain how to interpret the absolute value of a correlation coefficient.

Algebraic

6. A regression was run to determine whether there is a relationship between hours of TV watched per day (x) and number of sit-ups a person can do (y). The results of the regression are given below. Use this to predict the number of sit-ups a person who watches 11 hours of TV can do.

$$y = ax + b$$
$$a = -1.341$$
$$b = 32.234$$
$$r = -0.896$$

7. A regression was run to determine whether there is a relationship between the diameter of a tree (x , in inches) and the tree's age (y , in years). The results of the regression are given below. Use this to predict the age of a tree with diameter 10 inches.

$$y = ax + b$$
$$a = 6.301$$
$$b = -1.044$$
$$r = -0.970$$

For the following exercises, draw a scatter plot for the data provided. Does the data appear to be linearly related?

8.

0	2	4	6	8	10
-22	−19	−15	−11	-6	-2

9.

1	2	3	4	5	6
46	50	59	75	100	136

10.

100	250	300	450	600	750
12	12.6	13.1	14	14.5	15.2

11.

1	3	5	7	9	11
1	9	28	65	125	216

12. For the following data, draw a scatter plot. If we wanted to know when the population would reach 15,000, would the answer involve interpolation or extrapolation? Eyeball the line, and estimate the answer.

Year	Population
1990	11,500
1995	12,100
2000	12,700
2005	13,000
2010	13,750

13. For the following data, draw a scatter plot. If we wanted to know when the temperature would reach 28°F, would the answer involve interpolation or extrapolation? Eyeball the line and estimate the answer.

Temperature,°F	16	18	20	25	30
Time, seconds	46	50	54	55	62

Graphical

For the following exercises, match each scatterplot with one of the four specified correlations in Figure 9 and Figure 10.

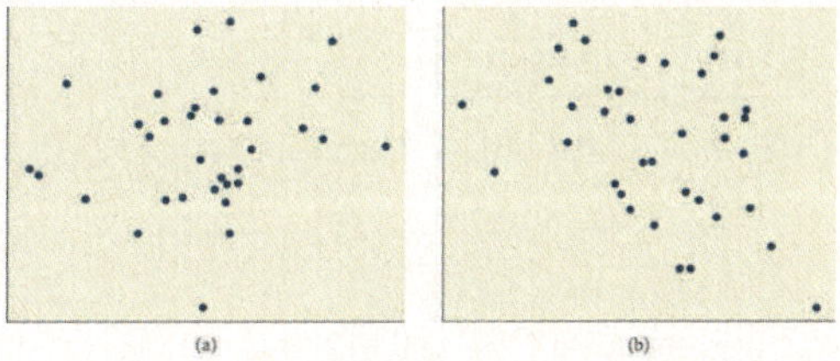

Figure 9

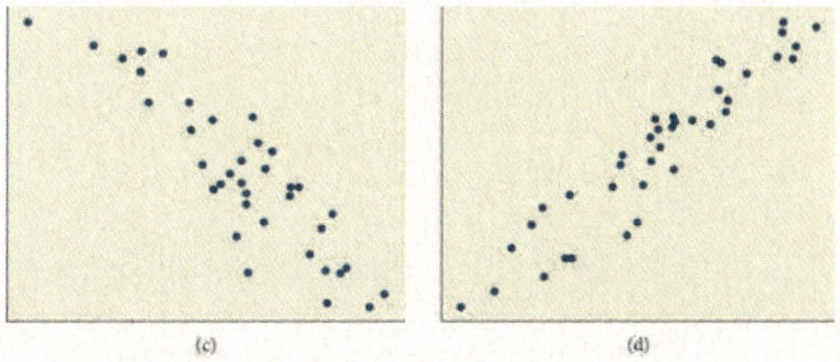

Figure 10

14. $r = 0.95$

15. $r = -0.89$

16. $r = -0.26$

17. $r = -0.39$

For the following exercises, draw a best-fit line for the plotted data.

18.

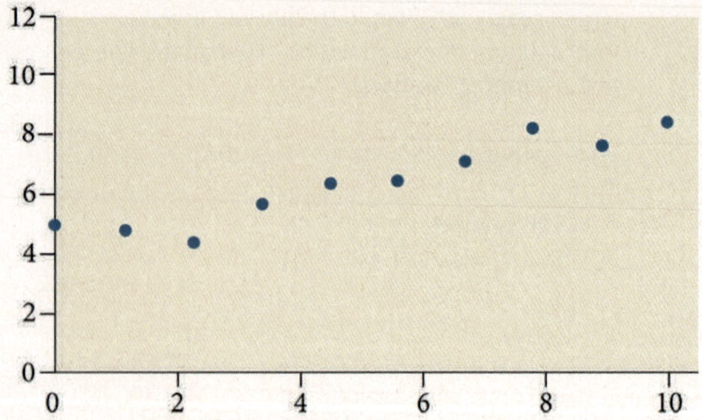

19.

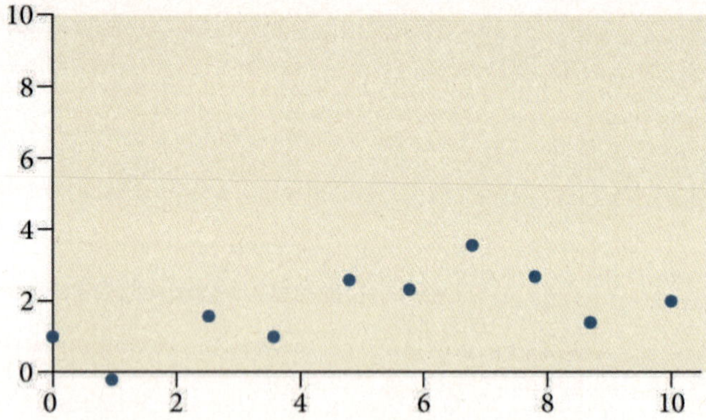

20.

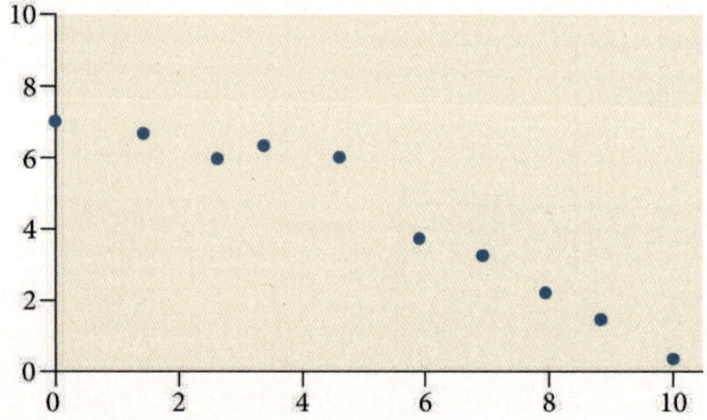

21.

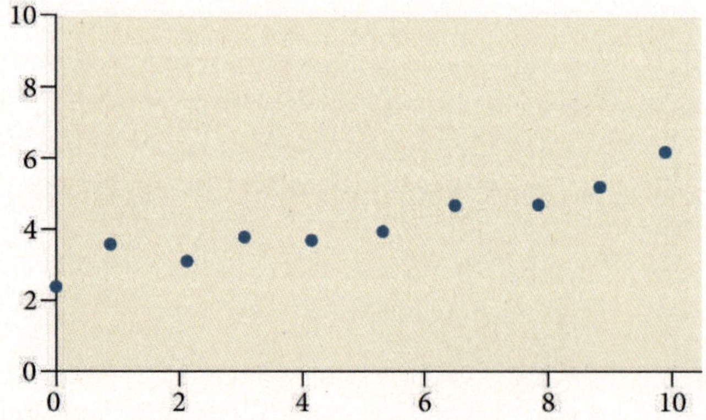

Numeric

22. The U.S. Census tracks the percentage of persons 25 years or older who are college graduates. That data for several years is given in Table 4.[9] Determine whether the trend appears linear. If so, and assuming the trend continues, in what year will the percentage exceed 35%?

Year	Percent Graduates
1990	21.3
1992	21.4
1994	22.2
1996	23.6
1998	24.4
2000	25.6
2002	26.7
2004	27.7
2006	28
2008	29.4

Table 4

23. The U.S. import of wine (in hectoliters) for several years is given in Table 5. Determine whether the trend appears linear. If so, and assuming the trend continues, in what year will imports exceed 12,000 hectoliters?

Year	Imports
1992	2665
1994	2688
1996	3565
1998	4129
2000	4584
2002	5655
2004	6549
2006	7950
2008	8487
2009	9462

Table 5

24. Table 6 shows the year and the number of people unemployed in a particular city for several years. Determine whether the trend appears linear. If so, and assuming the trend continues, in what year will the number of unemployed reach 5?

Year	Number Unemployed
1990	750
1992	670
1994	650
1996	605
1998	550
2000	510
2002	460
2004	420
2006	380
2008	320

Table 6

Technology

For the following exercises, use each set of data to calculate the regression line using a calculator or other technology tool, and determine the correlation coefficient to 3 decimal places of accuracy.

25.

x	8	15	26	31	56
y	23	41	53	72	103

26.

x	5	7	10	12	15
y	4	12	17	22	24

9 Based on data from http://www.census.gov/hhes/socdemo/education/data/cps/historical/index.html. Accessed 5/1/2014.

27.

x	y	x	y
3	21.9	10	18.54
4	22.22	11	15.76
5	22.74	12	13.68
6	22.26	13	14.1
7	20.78	14	14.02
8	17.6	15	11.94
9	16.52	16	12.76

28.

x	y
4	44.8
5	43.1
6	38.8
7	39
8	38
9	32.7
10	30.1
11	29.3
12	27
13	25.8

29.

x	21	25	30	31	40	50
y	17	11	2	–1	–18	-40

30.

x	y
100	2000
80	1798
60	1589
55	1580
40	1390
20	1202

31.

x	900	988	1000	1010	1200	1205
y	70	80	82	84	105	108

Extensions

32. Graph $f(x) = 0.5x + 10$. Pick a set of five ordered pairs using inputs $x = -2, 1, 5, 6, 9$ and use linear regression to verify that the function is a good fit for the data.

33. Graph $f(x) = -2x - 10$. Pick a set of five ordered pairs using inputs $x = -2, 1, 5, 6, 9$ and use linear regression to verify the function.

For the following exercises, consider this scenario: The profit of a company decreased steadily over a ten-year span. The following ordered pairs shows dollars and the number of units sold in hundreds and the profit in thousands of over the ten-year span, (number of units sold, profit) for specific recorded years:

$(46, 1{,}600), (48, 1{,}550), (50, 1{,}505), (52, 1{,}540), (54, 1{,}495)$.

34. Use linear regression to determine a function P where the profit in thousands of dollars depends on the number of units sold in hundreds.

35. Find to the nearest tenth and interpret the *x*-intercept.

36. Find to the nearest tenth and interpret the *y*-intercept.

Real-World Applications

For the following exercises, consider this scenario: The population of a city increased steadily over a ten-year span. The following ordered pairs shows the population and the year over the ten-year span, (population, year) for specific recorded years:

$(2500, 2000), (2650, 2001), (3000, 2003), (3500, 2006), (4200, 2010)$

37. Use linear regression to determine a function y, where the year depends on the population. Round to three decimal places of accuracy.

38. Predict when the population will hit 8,000.

For the following exercises, consider this scenario: The profit of a company increased steadily over a ten-year span. The following ordered pairs show the number of units sold in hundreds and the profit in thousands of over the ten year span, (number of units sold, profit) for specific recorded years:

$(46, 250), (48, 305), (50, 350), (52, 390), (54, 410)$.

39. Use linear regression to determine a function *y*, where the profit in thousands of dollars depends on the number of units sold in hundreds.

40. Predict when the profit will exceed one million dollars.

For the following exercises, consider this scenario: The profit of a company decreased steadily over a ten-year span. The following ordered pairs show dollars and the number of units sold in hundreds and the profit in thousands of over the ten-year span (number of units sold, profit) for specific recorded years:

$(46, 250), (48, 225), (50, 205), (52, 180), (54, 165)$.

41. Use linear regression to determine a function *y*, where the profit in thousands of dollars depends on the number of units sold in hundreds.

42. Predict when the profit will dip below the $25,000 threshold.

Chapter Review

Key Terms

correlation coefficient a value, r, between –1 and 1 that indicates the degree of linear correlation of variables, or how closely a regression line fits a data set.

decreasing linear function a function with a negative slope: If $f(x) = mx + b,$ then $m < 0$

extrapolation predicting a value outside the domain and range of the data

horizontal line a line defined by $f(x) = b,$ where b is a real number. The slope of a horizontal line is 0.

increasing linear function a function with a positive slope: If $f(x) = mx + b,$ then $m > 0.$

interpolation predicting a value inside the domain and range of the data

least squares regression a statistical technique for fitting a line to data in a way that minimizes the differences between the line and data values

linear function a function with a constant rate of change that is a polynomial of degree 1, and whose graph is a straight line

model breakdown when a model no longer applies after a certain point

parallel lines two or more lines with the same slope

perpendicular lines two lines that intersect at right angles and have slopes that are negative reciprocals of each other

point-slope form the equation for a line that represents a linear function of the form $y - y_1 = m(x - x_1)$

slope the ratio of the change in output values to the change in input values; a measure of the steepness of a line

slope-intercept form the equation for a line that represents a linear function in the form $f(x) = mx + b$

vertical line a line defined by $x = a,$ where a is a real number. The slope of a vertical line is undefined.

Key Concepts

4.1 Linear Functions

- Linear functions can be represented in words, function notation, tabular form, and graphical form. See Example 1.
- An increasing linear function results in a graph that slants upward from left to right and has a positive slope. A decreasing linear function results in a graph that slants downward from left to right and has a negative slope. A constant linear function results in a graph that is a horizontal line. See Example 2.
- Slope is a rate of change. The slope of a linear function can be calculated by dividing the difference between *y*-values by the difference in corresponding *x*-values of any two points on the line. See Example 3 and Example 4.
- An equation for a linear function can be written from a graph. See Example 5.
- The equation for a linear function can be written if the slope m and initial value b are known. See Example 6 and Example 7.
- A linear function can be used to solve real-world problems given information in different forms. See Example 8, Example 9, and Example 10.
- Linear functions can be graphed by plotting points or by using the *y*-intercept and slope. See Example 11 and Example 12.
- Graphs of linear functions may be transformed by using shifts up, down, left, or right, as well as through stretches, compressions, and reflections. See Example 13.
- The equation for a linear function can be written by interpreting the graph. See Example 14.
- The *x*-intercept is the point at which the graph of a linear function crosses the *x*-axis. See Example 15.
- Horizontal lines are written in the form, $f(x) = b$. See Example 16.
- Vertical lines are written in the form, $x = b$. See Example 17.
- Parallel lines have the same slope. Perpendicular lines have negative reciprocal slopes, assuming neither is vertical. See Example 18.
- A line parallel to another line, passing through a given point, may be found by substituting the slope value of the line and the *x*- and *y*-values of the given point into the equation, $f(x) = mx + b$, and using the b that results. Similarly, the point-slope form of an equation can also be used. See Example 19.
- A line perpendicular to another line, passing through a given point, may be found in the same manner, with the exception of using the negative reciprocal slope. See Example 20 and Example 21.

4.2 Modeling with Linear Functions

- We can use the same problem strategies that we would use for any type of function.
- When modeling and solving a problem, identify the variables and look for key values, including the slope and *y*-intercept. See Example 1.
- Draw a diagram, where appropriate. See Example 2 and Example 3.
- Check for reasonableness of the answer.

- Linear models may be built by identifying or calculating the slope and using the *y*-intercept.
  - The *x*-intercept may be found by setting $y = 0$, which is setting the expression $mx + b$ equal to 0.
  - The point of intersection of a system of linear equations is the point where the *x*- and *y*-values are the same. See Example 4.
  - A graph of the system may be used to identify the points where one line falls below (or above) the other line.

4.3 Fitting Linear Models to Data

- Scatter plots show the relationship between two sets of data. See Example 1.
- Scatter plots may represent linear or non-linear models.
- The line of best fit may be estimated or calculated, using a calculator or statistical software. See Example 2.
- Interpolation can be used to predict values inside the domain and range of the data, whereas extrapolation can be used to predict values outside the domain and range of the data. See Example 3.
- The correlation coefficient, r, indicates the degree of linear relationship between data. See Example 4.
- A regression line best fits the data. See Example 5.
- The least squares regression line is found by minimizing the squares of the distances of points from a line passing through the data and may be used to make predictions regarding either of the variables. See Example 6.

Exercises

Review Exercises

Linear Functions

1. Determine whether the algebraic equation is linear. $2x + 3y = 7$

2. Determine whether the algebraic equation is linear. $6x^2 - y = 5$

3. Determine whether the function is increasing or decreasing.

   $f(x) = 7x - 2$

4. Determine whether the function is increasing or decreasing.

   $g(x) = -x + 2$

5. Given each set of information, find a linear equation that satisfies the given conditions, if possible.

   Passes through $(7, 5)$ and $(3, 17)$

6. Given each set of information, find a linear equation that satisfies the given conditions, if possible.

   x-intercept at $(6, 0)$ and *y*-intercept at $(0, 10)$

7. Find the slope of the line shown in the graph.

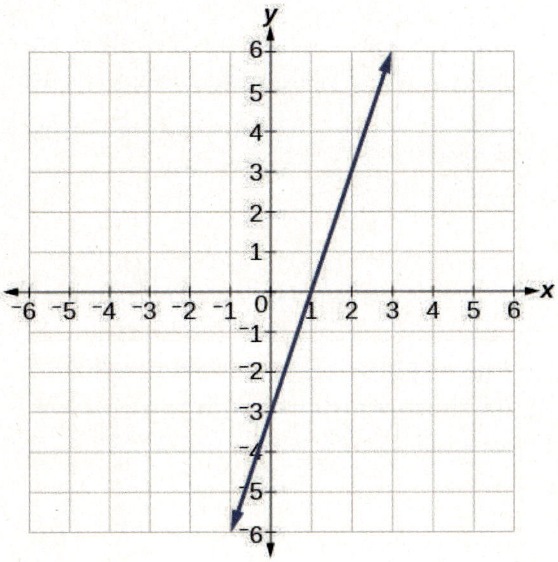

8. Find the slope of the line graphed.

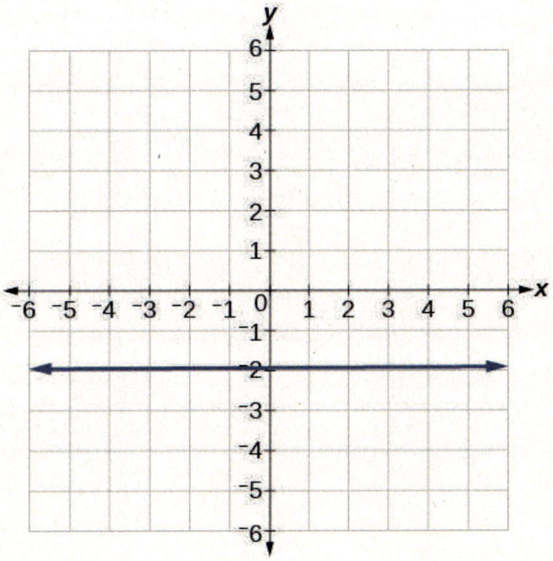

9. Write an equation in slope-intercept form for the line shown.

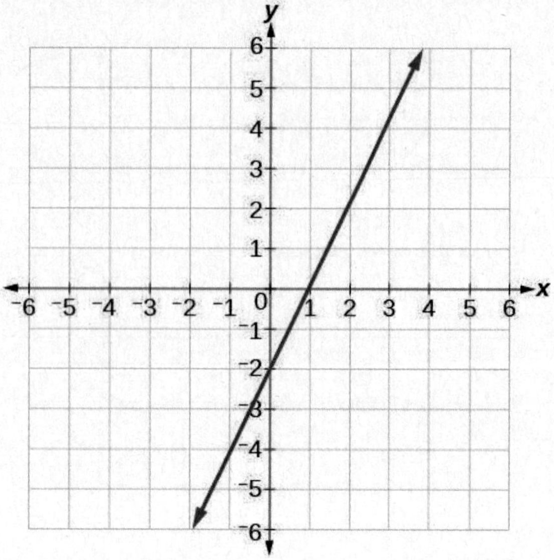

10. Does the following table represent a linear function? If so, find the linear equation that models the data.

x	–4	0	2	10
g(x)	18	–2	–12	–52

11. Does the following table represent a linear function? If so, find the linear equation that models the data.

x	6	8	12	26
g(x)	–8	–12	–18	–46

12. On June 1st, a company has \$4,000,000 profit. If the company then loses 150,000 dollars per day thereafter in the month of June, what is the company's profit n^{th} day after June 1st?

For the following exercises, determine whether the lines given by the equations below are parallel, perpendicular, or neither parallel nor perpendicular:

13. $2x - 6y = 12$
$-x + 3y = 1$

14. $y = \frac{1}{3}x - 2$
$3x + y = -9$

For the following exercises, find the x- and y- intercepts of the given equation

15. $7x + 9y = -63$

16. $f(x) = 2x - 1$

For the following exercises, use the descriptions of the pairs of lines to find the slopes of Line 1 and Line 2. Is each pair of lines parallel, perpendicular, or neither?

17. Line 1: Passes through $(5, 11)$ and $(10, 1)$

Line 2: Passes through $(-1, 3)$ and $(-5, 11)$

18. Line 1: Passes through $(8, -10)$ and $(0, -26)$

Line 2: Passes through $(2, 5)$ and $(4, 4)$

19. Write an equation for a line perpendicular to $f(x) = 5x - 1$ and passing through the point (5, 20).

20. Find the equation of a line with a *y*- intercept of $(0, 2)$ and slope $-\frac{1}{2}$.

21. Sketch a graph of the linear function $f(t) = 2t - 5$.

22. Find the point of intersection for the 2 linear functions: $\begin{aligned} x &= y + 6 \\ 2x - y &= 13 \end{aligned}$.

23. A car rental company offers two plans for renting a car.

Plan A: 25 dollars per day and 10 cents per mile

Plan B: 50 dollars per day with free unlimited mileage

How many miles would you need to drive for plan B to save you money?

Modeling with Linear Functions

24. Find the area of a triangle bounded by the *y* axis, the line $f(x) = 10 - 2x$, and the line perpendicular to f that passes through the origin.

25. A town's population increases at a constant rate. In 2010 the population was 55,000. By 2012 the population had increased to 76,000. If this trend continues, predict the population in 2016.

26. The number of people afflicted with the common cold in the winter months dropped steadily by 50 each year since 2004 until 2010. In 2004, 875 people were inflicted.

Find the linear function that models the number of people afflicted with the common cold *C* as a function of the year, *t*. When will no one be afflicted?

For the following exercises, use the graph in Figure 1 showing the profit, y, in thousands of dollars, of a company in a given year, x, where x represents years since 1980.

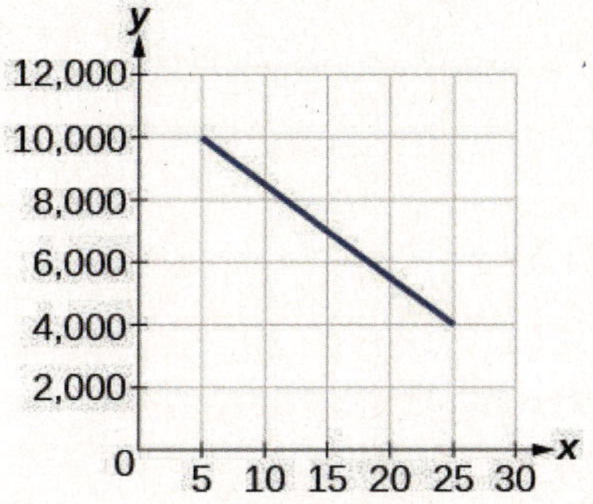

Figure 1

27. Find the linear function *y*, where *y* depends on x, the number of years since 1980.

28. Find and interpret the *y*-intercept.

For the following exercise, consider this scenario: In 2004, a school population was 1,700. By 2012 the population had grown to 2,500.

29. Assume the population is changing linearly.

ⓐ How much did the population grow between the year 2004 and 2012?
ⓑ What is the average population growth per year?
ⓒ Find an equation for the population, P, of the school t years after 2004.

For the following exercises, consider this scenario: In 2000, the moose population in a park was measured to be 6,500. By 2010, the population was measured to be 12,500. Assume the population continues to change linearly.

30. Find a formula for the moose population, P.

31. What does your model predict the moose population to be in 2020?

For the following exercises, consider this scenario: The median home values in subdivisions Pima Central and East Valley (adjusted for inflation) are shown in Table 1. Assume that the house values are changing linearly.

Year	Pima Central	East Valley
1970	32,000	120,250
2010	85,000	150,000

Table 1

32. In which subdivision have home values increased at a higher rate?

33. If these trends were to continue, what would be the median home value in Pima Central in 2015?

Fitting Linear Models to Data

34. Draw a scatter plot for the data in Table 2. Then determine whether the data appears to be linearly related.

0	-105
2	-50
4	1
6	55
8	105
10	160

Table 2

35. Draw a scatter plot for the data in Table 3. If we wanted to know when the population would reach 15,000, would the answer involve interpolation or extrapolation?

Year	Population
1990	5,600
1995	5,950
2000	6,300
2005	6,600
2010	6,900

Table 3

36. Eight students were asked to estimate their score on a 10-point quiz. Their estimated and actual scores are given in Table 4. Plot the points, then sketch a line that fits the data.

Predicted	Actual
6	6
7	7
7	8
8	8
7	9
9	10
10	10
10	9

Table 4

37. Draw a best-fit line for the plotted data.

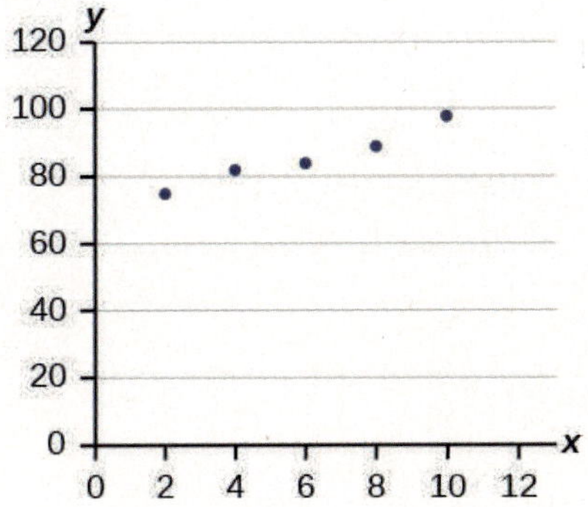

For the following exercises, consider the data in Table 5, which shows the percent of unemployed in a city of people 25 years or older who are college graduates is given below, by year.

Year	2000	2002	2005	2007	2010
Percent Graduates	6.5	7.0	7.4	8.2	9.0

Table 5

38. Determine whether the trend appears to be linear. If so, and assuming the trend continues, find a linear regression model to predict the percent of unemployed in a given year to three decimal places.

39. In what year will the percentage exceed 12%?

40. Based on the set of data given in Table 6, calculate the regression line using a calculator or other technology tool, and determine the correlation coefficient to three decimal places.

x	17	20	23	26	29
y	15	25	31	37	40

Table 6

41. Based on the set of data given in Table 7, calculate the regression line using a calculator or other technology tool, and determine the correlation coefficient to three decimal places.

x	10	12	15	18	20
y	36	34	30	28	22

Table 7

For the following exercises, consider this scenario: The population of a city increased steadily over a ten-year span. The following ordered pairs show the population and the year over the ten-year span (population, year) for specific recorded years:

$(3{,}600, 2000); (4{,}000, 2001); (4{,}700, 2003); (6{,}000, 2006)$

42. Use linear regression to determine a function y, where the year depends on the population, to three decimal places of accuracy.

43. Predict when the population will hit 12,000.

44. What is the correlation coefficient for this model to three decimal places of accuracy?

45. According to the model, what is the population in 2014?

Practice Test

1. Determine whether the following algebraic equation can be written as a linear function. $2x + 3y = 7$

2. Determine whether the following function is increasing or decreasing. $f(x) = -2x + 5$

3. Determine whether the following function is increasing or decreasing. $f(x) = 7x + 9$

4. Find a linear equation that passes through (5, 1) and (3, −9), if possible.

5. Find a linear equation, that has an x intercept at (−4, 0) and a y-intercept at (0, −6), if possible.

6. Find the slope of the line in Figure 1.

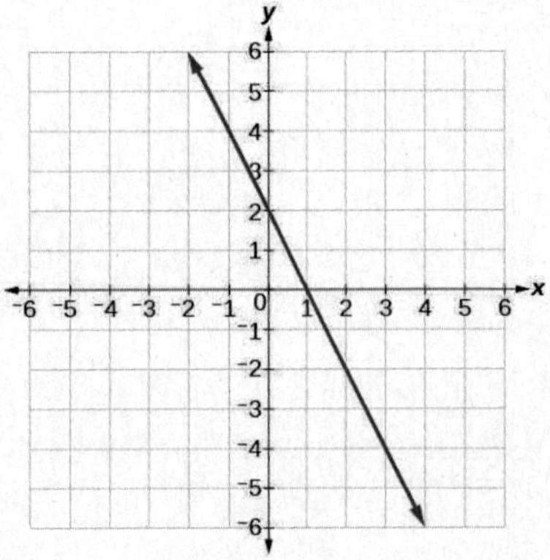

Figure 1

7. Write an equation for line in Figure 2.

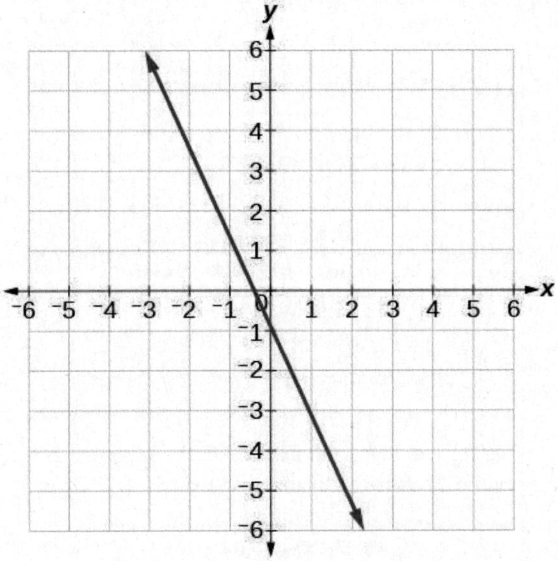

Figure 2

8. Does Table 1 represent a linear function? If so, find a linear equation that models the data.

x	−6	0	2	4
$g(x)$	14	32	38	44

Table 1

9. Does Table 2 represent a linear function? If so, find a linear equation that models the data.

x	1	3	7	11
$g(x)$	4	9	19	12

Table 2

10. At 6 am, an online company has sold 120 items that day. If the company sells an average of 30 items per hour for the remainder of the day, write an expression to represent the number of items that were sold n after 6 am.

For the following exercises, determine whether the lines given by the equations below are parallel, perpendicular, or neither parallel nor perpendicular.

11. $y = \frac{3}{4}x - 9$
$-4x - 3y = 8$

12. $-2x + y = 3$
$3x + \frac{3}{2}y = 5$

13. Find the x- and y-intercepts of the equation $2x + 7y = -14$.

14. Given below are descriptions of two lines. Find the slopes of Line 1 and Line 2. Is the pair of lines parallel, perpendicular, or neither?

Line 1: Passes through $(-2, -6)$ and $(3, 14)$

Line 2: Passes through $(2, 6)$ and $(4, 14)$

15. Write an equation for a line perpendicular to $f(x) = 4x + 3$ and passing through the point $(8, 10)$.

16. Sketch a line with a y-intercept of $(0, 5)$ and slope $-\frac{5}{2}$.

17. Graph of the linear function $f(x) = -x + 6$.

18. For the two linear functions, find the point of intersection:
$x = y + 2$
$2x - 3y = -1$.

19. A car rental company offers two plans for renting a car.

Plan A: $25 per day and $0.10 per mile

Plan B: $40 per day with free unlimited mileage

How many miles would you need to drive for plan B to save you money?

20. Find the area of a triangle bounded by the y axis, the line $f(x) = 12 - 4x$, and the line perpendicular to f that passes through the origin.

21. A town's population increases at a constant rate. In 2010 the population was 65,000. By 2012 the population had increased to 90,000. Assuming this trend continues, predict the population in 2018.

22. The number of people afflicted with the common cold in the winter months dropped steadily by 25 each year since 2002 until 2012. In 2002, 8,040 people were inflicted. Find the linear function that models the number of people afflicted with the common cold C as a function of the year, t. When will less than 6,000 people be afflicted?

For the following exercises, use the graph in Figure 3, showing the profit, y, in thousands of dollars, of a company in a given year, x, where x represents years since 1980.

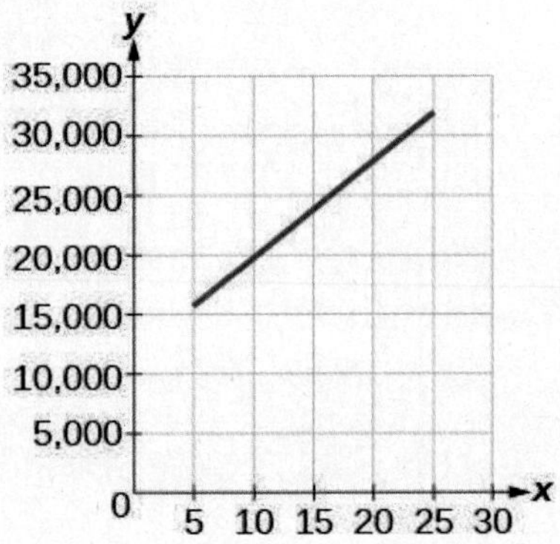

Figure 3

23. Find the linear function y, where y depends on x, the number of years since 1980.

24. Find and interpret the y-intercept.

25. In 2004, a school population was 1250. By 2012 the population had dropped to 875. Assume the population is changing linearly.

ⓐ How much did the population drop between the year 2004 and 2012?
ⓑ What is the average population decline per year?
ⓒ Find an equation for the population, P, of the school t years after 2004.

26. Draw a scatter plot for the data provided in Table 3. Then determine whether the data appears to be linearly related.

0	2	4	6	8	10
–450	–200	10	265	500	755

Table 3

27. Draw a best-fit line for the plotted data.

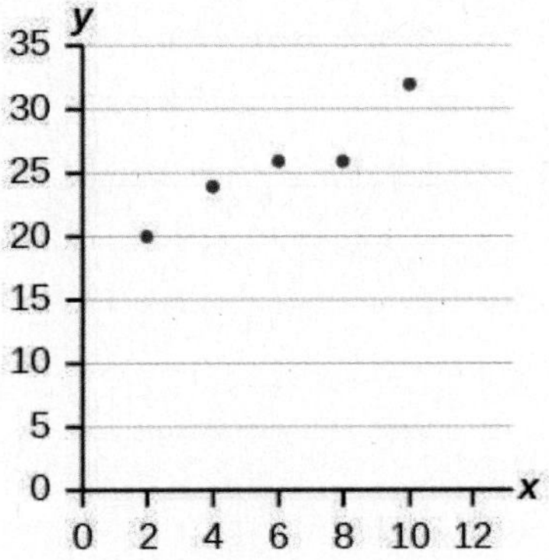

For the following exercises, use Table 4, which shows the percent of unemployed persons 25 years or older who are college graduates in a particular city, by year.

Year	2000	2002	2005	2007	2010
Percent Graduates	8.5	8.0	7.2	6.7	6.4

Table 4

28. Determine whether the trend appears linear. If so, and assuming the trend continues, find a linear regression model to predict the percent of unemployed in a given year to three decimal places.

29. In what year will the percentage drop below 4%?

30. Based on the set of data given in Table 5, calculate the regression line using a calculator or other technology tool, and determine the correlation coefficient. Round to three decimal places of accuracy.

x	16	18	20	24	26
y	106	110	115	120	125

Table 5

For the following exercises, consider this scenario: The population of a city increased steadily over a ten-year span. The following ordered pairs shows the population (in hundreds) and the year over the ten-year span, (population, year) for specific recorded years:

$(4,500, 2000); (4,700, 2001); (5,200, 2003); (5,800, 2006)$

31. Use linear regression to determine a function *y*, where the year depends on the population. Round to three decimal places of accuracy.

32. Predict when the population will hit 20,000.

33. What is the correlation coefficient for this model?

5 POLYNOMIAL AND RATIONAL FUNCTIONS

Whether they think about it in mathematical terms or not, scuba divers must consider the impact of functional relationships in order to remain safe. The gas laws, which are a series of relations and equations that describe the behavior of most gases, play a core role in diving. This diver, near the wreck of a World War II Japanese ocean liner turned troop transport, must remain attentive to gas laws during their dive and as they ascend to the surface. (credit: "Aikoku - Aft Gun": modification of work by montereydiver/flickr)

Chapter Outline

Introduction to Polynomial and Rational Functions

You don't need to dive very deep to feel the effects of pressure. As a person in their neighborhood pool moves eight, ten, twelve feet down, they often feel pain in their ears as a result of water and air pressure differentials. Pressure plays a much greater role at ocean diving depths.

id="scuban">Scuba and free divers are constantly negotiating the effects of pressure in order to experience enjoyable, safe, and productive dives. Gases in a person's respiratory system and diving apparatus interact according to certain physical properties, which upon discovery and evaluation are collectively known as the gas laws. Some are conceptually simple, such as the inverse relationship regarding pressure and volume, and others are more complex. While their formulas seem more straightforward than many you will encounter in this chapter, the gas laws are generally polynomial expressions.

5.1 Quadratic Functions

Learning Objectives

In this section, you will:

- Recognize characteristics of parabolas.
- Understand how the graph of a parabola is related to its quadratic function.
- Determine a quadratic function's minimum or maximum value.
- Solve problems involving a quadratic function's minimum or maximum value.

Figure 1 An array of satellite dishes. (credit: Matthew Colvin de Valle, Flickr)

Curved antennas, such as the ones shown in Figure 1, are commonly used to focus microwaves and radio waves to transmit television and telephone signals, as well as satellite and spacecraft communication. The cross-section of the antenna is in the shape of a parabola, which can be described by a quadratic function.

In this section, we will investigate quadratic functions, which frequently model problems involving area and projectile motion. Working with quadratic functions can be less complex than working with higher degree functions, so they provide a good opportunity for a detailed study of function behavior.

Recognizing Characteristics of Parabolas

The graph of a quadratic function is a U-shaped curve called a parabola. One important feature of the graph is that it has an extreme point, called the **vertex**. If the parabola opens up, the vertex represents the lowest point on the graph, or the minimum value of the quadratic function. If the parabola opens down, the vertex represents the highest point on the graph, or the maximum value. In either case, the vertex is a turning point on the graph. The graph is also symmetric with a vertical line drawn through the vertex, called the **axis of symmetry**. These features are illustrated in Figure 2.

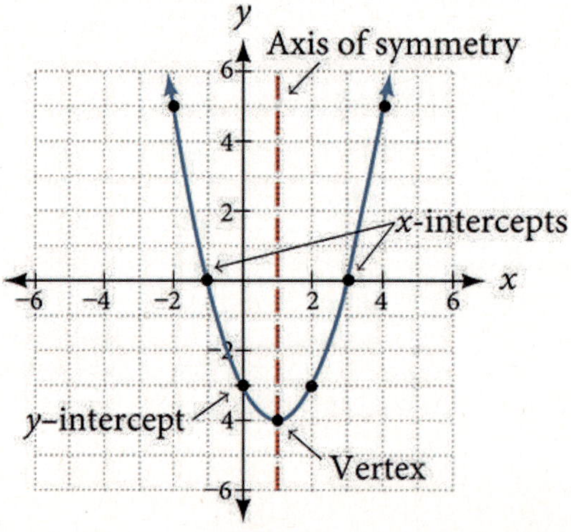

Figure 2

The y-intercept is the point at which the parabola crosses the y-axis. The x-intercepts are the points at which the parabola crosses the x-axis. If they exist, the x-intercepts represent the **zeros,** or **roots**, of the quadratic function, the values of x at which $y = 0$.

EXAMPLE 1

Identifying the Characteristics of a Parabola

Determine the vertex, axis of symmetry, zeros, and y- intercept of the parabola shown in Figure 3.

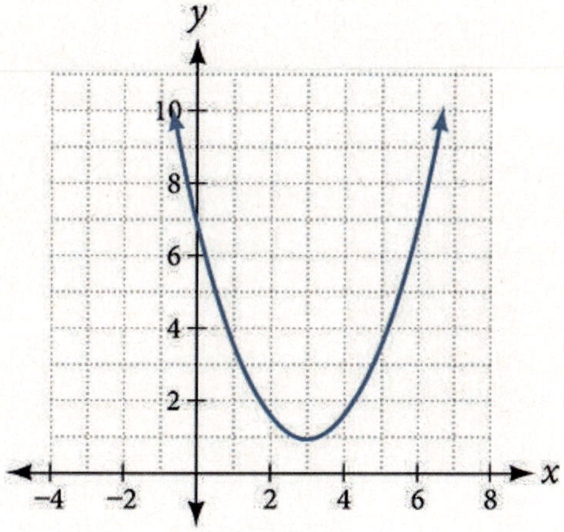

Figure 3

Solution

The vertex is the turning point of the graph. We can see that the vertex is at $(3, 1)$. Because this parabola opens upward, the axis of symmetry is the vertical line that intersects the parabola at the vertex. So the axis of symmetry is $x = 3$. This parabola does not cross the x- axis, so it has no zeros. It crosses the y- axis at $(0, 7)$ so this is the y-intercept.

Understanding How the Graphs of Parabolas are Related to Their Quadratic Functions

The **general form of a quadratic function** presents the function in the form

$$f(x) = ax^2 + bx + c$$

where a, b, and c are real numbers and $a \neq 0$. If $a > 0$, the parabola opens upward. If $a < 0$, the parabola opens downward. We can use the general form of a parabola to find the equation for the axis of symmetry.

The axis of symmetry is defined by $x = -\frac{b}{2a}$. If we use the quadratic formula, $x = \frac{-b \pm \sqrt{b^2 - 4ac}}{2a}$, to solve $ax^2 + bx + c = 0$ for the x- intercepts, or zeros, we find the value of x halfway between them is always $x = -\frac{b}{2a}$, the equation for the axis of symmetry.

Figure 4 represents the graph of the quadratic function written in general form as $y = x^2 + 4x + 3$. In this form, $a = 1, b = 4$, and $c = 3$. Because $a > 0$, the parabola opens upward. The axis of symmetry is $x = -\frac{4}{2(1)} = -2$. This also makes sense because we can see from the graph that the vertical line $x = -2$ divides the graph in half. The vertex always occurs along the axis of symmetry. For a parabola that opens upward, the vertex occurs at the lowest point on the graph, in this instance, $(-2, -1)$. The x- intercepts, those points where the parabola crosses the x- axis, occur at $(-3, 0)$ and $(-1, 0)$.

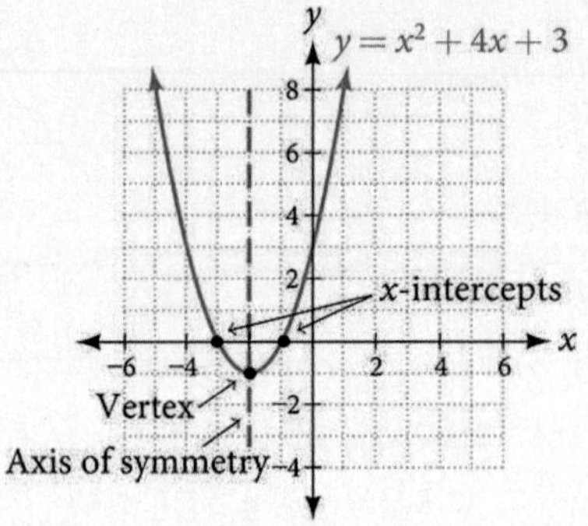

Figure 4

The **standard form of a quadratic function** presents the function in the form

$$f(x) = a(x - h)^2 + k$$

where $(h, \ k)$ is the vertex. Because the vertex appears in the standard form of the quadratic function, this form is also known as the **vertex form of a quadratic function**.

As with the general form, if $a > 0$, the parabola opens upward and the vertex is a minimum. If $a < 0$, the parabola opens downward, and the vertex is a maximum. Figure 5 represents the graph of the quadratic function written in standard form as $y = -3(x + 2)^2 + 4$. Since $x - h = x + 2$ in this example, $h = -2$. In this form, $a = -3$, $h = -2$, and $k = 4$. Because $a < 0$, the parabola opens downward. The vertex is at $(-2, \ 4)$.

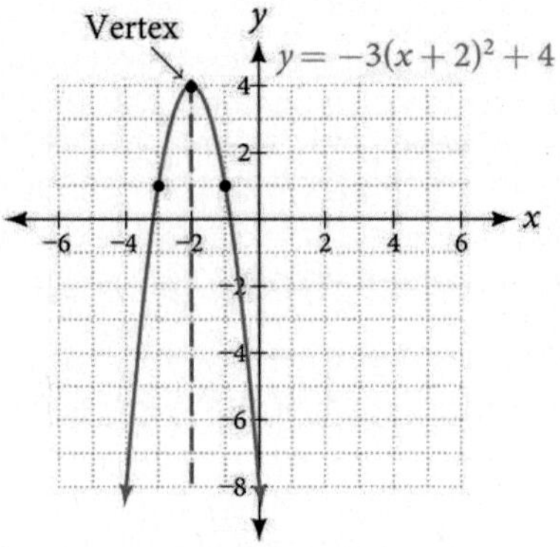

Figure 5

The standard form is useful for determining how the graph is transformed from the graph of $y = x^2$. Figure 6 is the graph of this basic function.

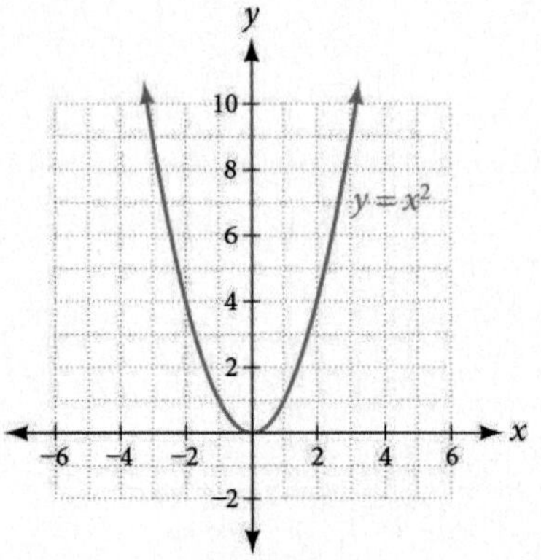

Figure 6

If $k > 0$, the graph shifts upward, whereas if $k < 0$, the graph shifts downward. In Figure 5, $k > 0$, so the graph is shifted 4 units upward. If $h > 0$, the graph shifts toward the right and if $h < 0$, the graph shifts to the left. In Figure 5, $h < 0$, so the graph is shifted 2 units to the left. The magnitude of a indicates the stretch of the graph. If $|a| > 1$, the point associated with a particular x- value shifts farther from the x-axis, so the graph appears to become narrower, and there is a vertical stretch. But if $|a| < 1$, the point associated with a particular x- value shifts closer to the x-axis, so the graph appears to become wider, but in fact there is a vertical compression. In Figure 5, $|a| > 1$, so the graph becomes narrower.

The standard form and the general form are equivalent methods of describing the same function. We can see this by expanding out the general form and setting it equal to the standard form.

$$\begin{aligned} a(x-h)^2 + k &= ax^2 + bx + c \\ ax^2 - 2ahx + (ah^2 + k) &= ax^2 + bx + c \end{aligned}$$

For the linear terms to be equal, the coefficients must be equal.

$$-2ah = b, \text{ so } h = -\frac{b}{2a}$$

This is the axis of symmetry we defined earlier. Setting the constant terms equal:

$$\begin{aligned} ah^2 + k &= c \\ k &= c - ah^2 \\ &= c - a - \left(\tfrac{b}{2a}\right)^2 \\ &= c - \tfrac{b^2}{4a} \end{aligned}$$

In practice, though, it is usually easier to remember that k is the output value of the function when the input is h, so $f(h) = k$.

Forms of Quadratic Functions

A quadratic function is a polynomial function of degree two. The graph of a quadratic function is a parabola.

The **general form of a quadratic function** is $f(x) = ax^2 + bx + c$ where a, b, and c are real numbers and $a \neq 0$.

The **standard form of a quadratic function** is $f(x) = a(x-h)^2 + k$ where $a \neq 0$.

The vertex (h, k) is located at

$$h = -\frac{b}{2a}, \quad k = f(h) = f\left(\frac{-b}{2a}\right)$$

HOW TO

Given a graph of a quadratic function, write the equation of the function in general form.

1. Identify the horizontal shift of the parabola; this value is h. Identify the vertical shift of the parabola; this value is k.
2. Substitute the values of the horizontal and vertical shift for h and k. in the function $f(x) = a(x-h)^2 + k$.
3. Substitute the values of any point, other than the vertex, on the graph of the parabola for x and $f(x)$.
4. Solve for the stretch factor, $|a|$.
5. Expand and simplify to write in general form.

EXAMPLE 2

Writing the Equation of a Quadratic Function from the Graph

Write an equation for the quadratic function g in Figure 7 as a transformation of $f(x) = x^2$, and then expand the formula, and simplify terms to write the equation in general form.

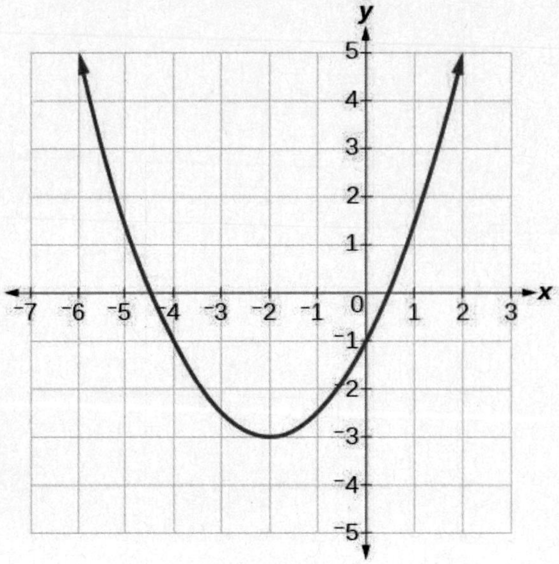

Figure 7

Solution

We can see the graph of g is the graph of $f(x) = x^2$ shifted to the left 2 and down 3, giving a formula in the form $g(x) = a(x - (-2))^2 - 3 = a(x + 2)^2 - 3$.

Substituting the coordinates of a point on the curve, such as $(0, -1)$, we can solve for the stretch factor.

$$\begin{aligned} -1 &= a(0+2)^2 - 3 \\ 2 &= 4a \\ a &= \frac{1}{2} \end{aligned}$$

In standard form, the algebraic model for this graph is $(g)x = \frac{1}{2}(x + 2)^2 - 3$.

To write this in general polynomial form, we can expand the formula and simplify terms.

$$\begin{aligned} g(x) &= \frac{1}{2}(x+2)^2 - 3 \\ &= \frac{1}{2}(x+2)(x+2) - 3 \\ &= \frac{1}{2}(x^2 + 4x + 4) - 3 \\ &= \frac{1}{2}x^2 + 2x + 2 - 3 \\ &= \frac{1}{2}x^2 + 2x - 1 \end{aligned}$$

Notice that the horizontal and vertical shifts of the basic graph of the quadratic function determine the location of the vertex of the parabola; the vertex is unaffected by stretches and compressions.

Analysis

We can check our work using the table feature on a graphing utility. First enter $\mathrm{Y1} = \frac{1}{2}(x + 2)^2 - 3$. Next, select TBLSET, then use $\mathrm{TblStart} = -6$ and $\Delta\mathrm{Tbl} = 2$, and select TABLE. See Table 1.

x	-6	-4	-2	0	2
y	5	–1	–3	–1	5

Table 1

The ordered pairs in the table correspond to points on the graph.

TRY IT #1 A coordinate grid has been superimposed over the quadratic path of a basketball in Figure 8. Find an equation for the path of the ball. Does the shooter make the basket?

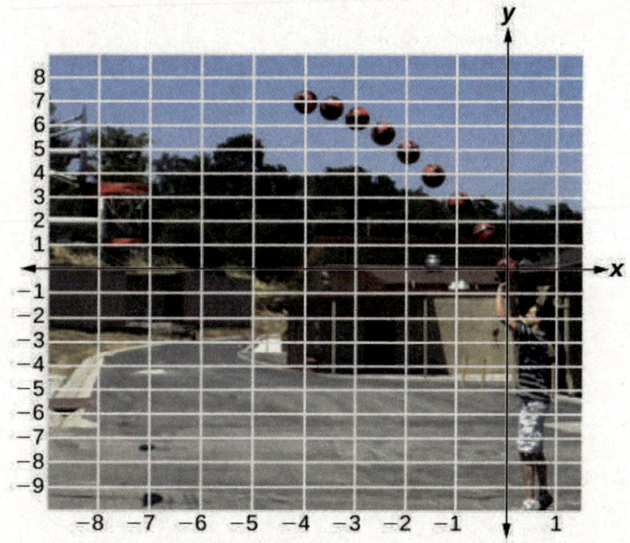

Figure 8 (credit: modification of work by Dan Meyer)

HOW TO

Given a quadratic function in general form, find the vertex of the parabola.

1. Identify a, b, and c.
2. Find h, the x-coordinate of the vertex, by substituting a and b into $h = -\frac{b}{2a}$.
3. Find k, the y-coordinate of the vertex, by evaluating $k = f(h) = f\left(-\frac{b}{2a}\right)$.

EXAMPLE 3

Finding the Vertex of a Quadratic Function

Find the vertex of the quadratic function $f(x) = 2x^2 - 6x + 7$. Rewrite the quadratic in standard form (vertex form).

Solution

The horizontal coordinate of the vertex will be at

$$\begin{aligned} h &= -\frac{b}{2a} \\ &= -\frac{-6}{2(2)} \\ &= \frac{6}{4} \\ &= \frac{3}{2} \end{aligned}$$

The vertical coordinate of the vertex will be at

$$\begin{aligned} k &= f(h) \\ &= f\left(\frac{3}{2}\right) \\ &= 2\left(\frac{3}{2}\right)^2 - 6\left(\frac{3}{2}\right) + 7 \\ &= \frac{5}{2} \end{aligned}$$

Rewriting into standard form, the stretch factor will be the same as the a in the original quadratic. First, find the horizontal coordinate of the vertex. Then find the vertical coordinate of the vertex. Substitute the values into standard form, using the " a " from the general form.

$$\begin{aligned} f(x) &= ax^2 + bx + c \\ f(x) &= 2x^2 - 6x + 7 \end{aligned}$$

The standard form of a quadratic function prior to writing the function then becomes the following:

$$f(x) = 2\left(x - \frac{3}{2}\right)^2 + \frac{5}{2}$$

Analysis

One reason we may want to identify the vertex of the parabola is that this point will inform us where the maximum or minimum value of the output occurs, k, and where it occurs, x.

TRY IT #2 Given the equation $g(x) = 13 + x^2 - 6x$, write the equation in general form and then in standard form.

Finding the Domain and Range of a Quadratic Function

Any number can be the input value of a quadratic function. Therefore, the domain of any quadratic function is all real numbers. Because parabolas have a maximum or a minimum point, the range is restricted. Since the vertex of a parabola will be either a maximum or a minimum, the range will consist of all *y*-values greater than or equal to the *y*-coordinate at the turning point or less than or equal to the *y*-coordinate at the turning point, depending on whether the parabola opens up or down.

Domain and Range of a Quadratic Function

The domain of any quadratic function is all real numbers unless the context of the function presents some restrictions.

The range of a quadratic function written in general form $f(x) = ax^2 + bx + c$ with a positive a value is $f(x) \geq f\left(-\frac{b}{2a}\right)$, or $\left[f\left(-\frac{b}{2a}\right), \infty\right)$; the range of a quadratic function written in general form with a negative a value is $f(x) \leq f\left(-\frac{b}{2a}\right)$, or $\left(-\infty, f\left(-\frac{b}{2a}\right)\right]$.

The range of a quadratic function written in standard form $f(x) = a(x - h)^2 + k$ with a positive a value is $f(x) \geq k$; the range of a quadratic function written in standard form with a negative a value is $f(x) \leq k$.

HOW TO

Given a quadratic function, find the domain and range.

1. Identify the domain of any quadratic function as all real numbers.
2. Determine whether a is positive or negative. If a is positive, the parabola has a minimum. If a is negative, the parabola has a maximum.
3. Determine the maximum or minimum value of the parabola, k.
4. If the parabola has a minimum, the range is given by $f(x) \geq k$, or $\left[k, \infty\right)$. If the parabola has a maximum, the range is given by $f(x) \leq k$, or $\left(-\infty, k\right]$.

EXAMPLE 4

Finding the Domain and Range of a Quadratic Function

Find the domain and range of $f(x) = -5x^2 + 9x - 1$.

Solution

As with any quadratic function, the domain is all real numbers.

Because a is negative, the parabola opens downward and has a maximum value. We need to determine the maximum value. We can begin by finding the x- value of the vertex.

$$\begin{aligned} h &= -\frac{b}{2a} \\ &= -\frac{9}{2(-5)} \\ &= \frac{9}{10} \end{aligned}$$

The maximum value is given by $f(h)$.

$$\begin{aligned} f\left(\frac{9}{10}\right) &= -5\left(\frac{9}{10}\right)^2 + 9\left(\frac{9}{10}\right) - 1 \\ &= \frac{61}{20} \end{aligned}$$

The range is $f(x) \le \frac{61}{20}$, or $\left(-\infty, \frac{61}{20}\right]$.

TRY IT #3 Find the domain and range of $f(x) = 2\left(x - \frac{4}{7}\right)^2 + \frac{8}{11}$.

Determining the Maximum and Minimum Values of Quadratic Functions

The output of the quadratic function at the vertex is the maximum or minimum value of the function, depending on the orientation of the parabola. We can see the maximum and minimum values in Figure 9.

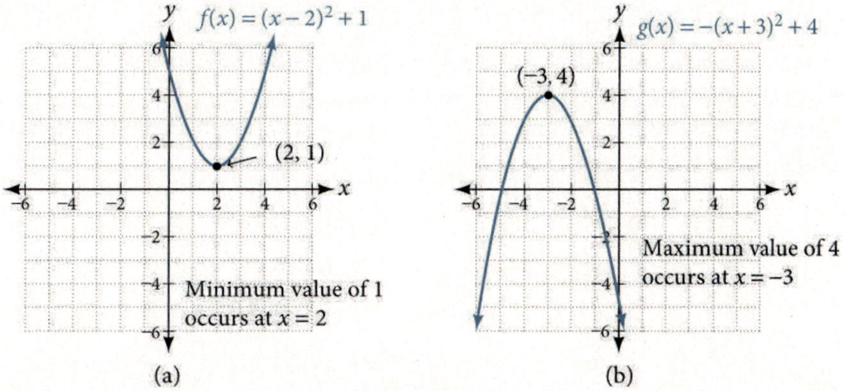

Figure 9

There are many real-world scenarios that involve finding the maximum or minimum value of a quadratic function, such as applications involving area and revenue.

EXAMPLE 5

Finding the Maximum Value of a Quadratic Function

A backyard farmer wants to enclose a rectangular space for a new garden within her fenced backyard. She has purchased 80 feet of wire fencing to enclose three sides, and she will use a section of the backyard fence as the fourth side.

ⓐ Find a formula for the area enclosed by the fence if the sides of fencing perpendicular to the existing fence have length L.

ⓑ What dimensions should she make her garden to maximize the enclosed area?

Solution

Let's use a diagram such as Figure 10 to record the given information. It is also helpful to introduce a temporary variable, W, to represent the width of the garden and the length of the fence section parallel to the backyard fence.

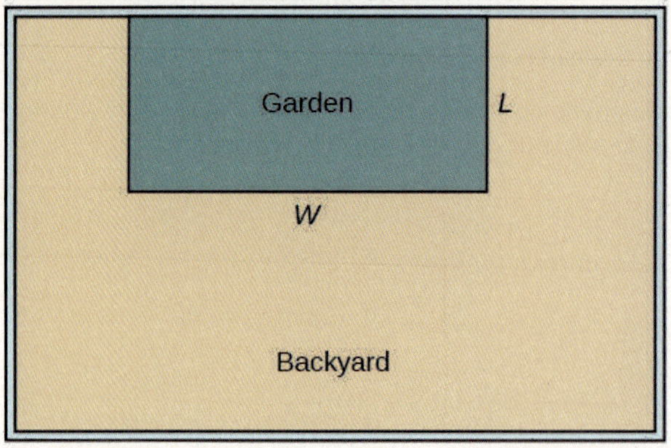

Figure 10

ⓐ We know we have only 80 feet of fence available, and $L + W + L = 80$, or more simply, $2L + W = 80$. This allows us to represent the width, W, in terms of L.

$$W = 80 - 2L$$

Now we are ready to write an equation for the area the fence encloses. We know the area of a rectangle is length multiplied by width, so

$$\begin{aligned} A &= LW = L(80 - 2L) \\ A(L) &= 80L - 2L^2 \end{aligned}$$

This formula represents the area of the fence in terms of the variable length L. The function, written in general form, is

$$A(L) = -2L^2 + 80L.$$

ⓑ The quadratic has a negative leading coefficient, so the graph will open downward, and the vertex will be the maximum value for the area. In finding the vertex, we must be careful because the equation is not written in standard polynomial form with decreasing powers. This is why we rewrote the function in general form above. Since a is the coefficient of the squared term, $a = -2, b = 80$, and $c = 0$.

To find the vertex:

$$\begin{aligned} h &= -\frac{b}{2a} & \qquad k &= A(20) \\ &= -\frac{80}{2(-2)} & \text{and} \qquad &= 80(20) - 2(20)^2 \\ &= 20 & &= 800 \end{aligned}$$

The maximum value of the function is an area of 800 square feet, which occurs when $L = 20$ feet. When the shorter sides are 20 feet, there is 40 feet of fencing left for the longer side. To maximize the area, she should enclose the garden so the two shorter sides have length 20 feet and the longer side parallel to the existing fence has length 40 feet.

Analysis

This problem also could be solved by graphing the quadratic function. We can see where the maximum area occurs on a graph of the quadratic function in Figure 11.

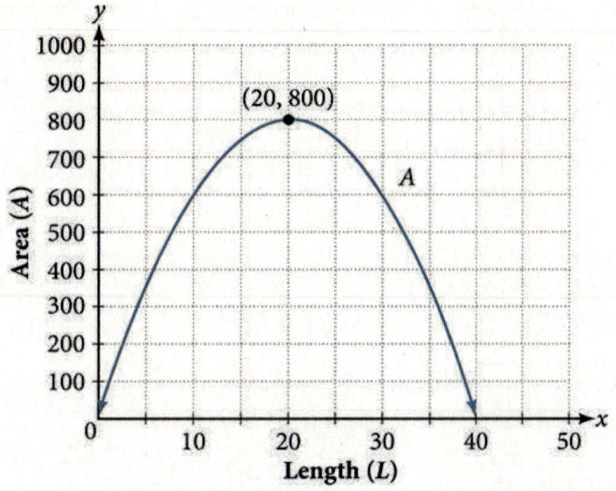

Figure 11

HOW TO

Given an application involving revenue, use a quadratic equation to find the maximum.

1. Write a quadratic equation for a revenue function.
2. Find the vertex of the quadratic equation.
3. Determine the y-value of the vertex.

EXAMPLE 6

Finding Maximum Revenue

The unit price of an item affects its supply and demand. That is, if the unit price goes up, the demand for the item will usually decrease. For example, a local newspaper currently has 84,000 subscribers at a quarterly charge of \$30. Market research has suggested that if the owners raise the price to \$32, they would lose 5,000 subscribers. Assuming that subscriptions are linearly related to the price, what price should the newspaper charge for a quarterly subscription to maximize their revenue?

Solution

Revenue is the amount of money a company brings in. In this case, the revenue can be found by multiplying the price per subscription times the number of subscribers, or quantity. We can introduce variables, p for price per subscription and Q for quantity, giving us the equation $\text{Revenue} = pQ$.

Because the number of subscribers changes with the price, we need to find a relationship between the variables. We know that currently $p = 30$ and $Q = 84{,}000$. We also know that if the price rises to \$32, the newspaper would lose 5,000 subscribers, giving a second pair of values, $p = 32$ and $Q = 79{,}000$. From this we can find a linear equation relating the two quantities. The slope will be

$$\begin{aligned} m &= \frac{79{,}000-84{,}000}{32-30} \\ &= \frac{-5{,}000}{2} \\ &= -2{,}500 \end{aligned}$$

This tells us the paper will lose 2,500 subscribers for each dollar they raise the price. We can then solve for the y-intercept.

$$\begin{aligned} Q &= -2500p + b && \text{Substitute in the point } Q = 84{,}000 \text{ and } p = 30 \\ 84{,}000 &= -2500(30) + b && \text{Solve for } b \\ b &= 159{,}000 \end{aligned}$$

This gives us the linear equation $Q = -2{,}500p + 159{,}000$ relating cost and subscribers. We now return to our revenue equation.

$$\begin{aligned} \text{Revenue} &= pQ \\ \text{Revenue} &= p(-2{,}500p + 159{,}000) \\ \text{Revenue} &= -2{,}500p^2 + 159{,}000p \end{aligned}$$

We now have a quadratic function for revenue as a function of the subscription charge. To find the price that will maximize revenue for the newspaper, we can find the vertex.

$$\begin{aligned} h &= -\frac{159{,}000}{2(-2{,}500)} \\ &= 31.8 \end{aligned}$$

The model tells us that the maximum revenue will occur if the newspaper charges $31.80 for a subscription. To find what the maximum revenue is, we evaluate the revenue function.

$$\begin{aligned} \text{maximum revenue} &= -2{,}500(31.8)^2 + 159{,}000(31.8) \\ &= 2{,}528{,}100 \end{aligned}$$

Analysis

This could also be solved by graphing the quadratic as in Figure 12. We can see the maximum revenue on a graph of the quadratic function.

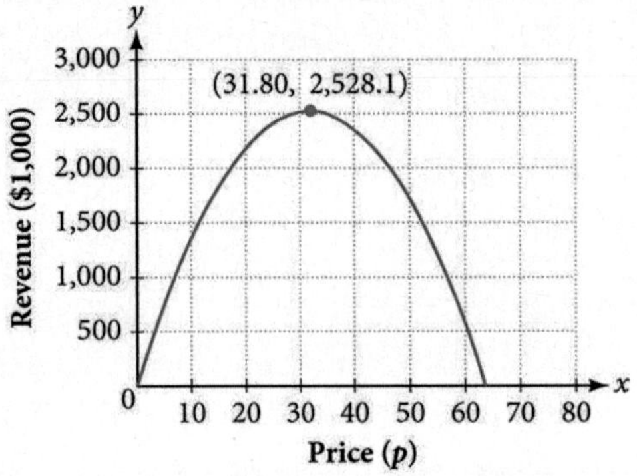

Figure 12

Finding the *x*- and *y*-Intercepts of a Quadratic Function

Much as we did in the application problems above, we also need to find intercepts of quadratic equations for graphing parabolas. Recall that we find the y- intercept of a quadratic by evaluating the function at an input of zero, and we find the x- intercepts at locations where the output is zero. Notice in Figure 13 that the number of x- intercepts can vary depending upon the location of the graph.

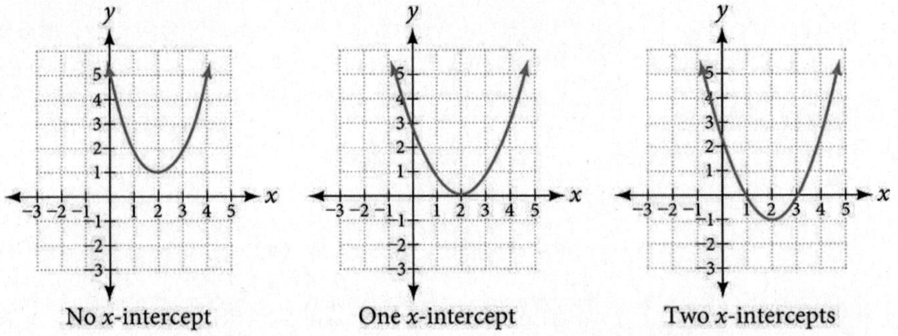

Figure 13 Number of x-intercepts of a parabola

HOW TO

Given a quadratic function $f(x)$, find the y- and x-intercepts.

1. Evaluate $f(0)$ to find the y-intercept.

2. Solve the quadratic equation $f(x) = 0$ to find the *x*-intercepts.

EXAMPLE 7

Finding the *y*- and *x*-Intercepts of a Parabola

Find the *y*- and *x*-intercepts of the quadratic $f(x) = 3x^2 + 5x - 2$.

Solution

We find the *y*-intercept by evaluating $f(0)$.

$$\begin{aligned} f(0) &= 3(0)^2 + 5(0) - 2 \\ &= -2 \end{aligned}$$

So the *y*-intercept is at $(0, -2)$.

For the *x*-intercepts, we find all solutions of $f(x) = 0$.

$$0 = 3x^2 + 5x - 2$$

In this case, the quadratic can be factored easily, providing the simplest method for solution.

$$0 = (3x - 1)(x + 2)$$

So the *x*-intercepts are at $\left(\frac{1}{3}, 0\right)$ and $(-2, 0)$.

Analysis

By graphing the function, we can confirm that the graph crosses the *y*-axis at $(0, -2)$. We can also confirm that the graph crosses the *x*-axis at $\left(\frac{1}{3}, 0\right)$ and $(-2, 0)$. See Figure 14

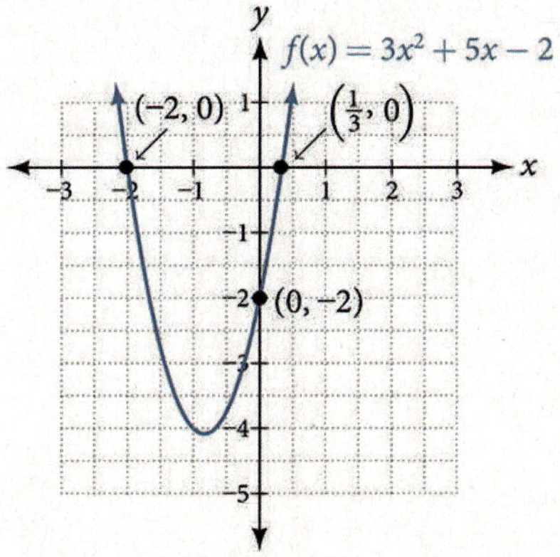

Figure 14

Rewriting Quadratics in Standard Form

In Example 7, the quadratic was easily solved by factoring. However, there are many quadratics that cannot be factored. We can solve these quadratics by first rewriting them in standard form.

HOW TO

Given a quadratic function, find the x- intercepts by rewriting in standard form.

1. Substitute a and b into $h = -\frac{b}{2a}$.
2. Substitute $x = h$ into the general form of the quadratic function to find k.
3. Rewrite the quadratic in standard form using h and k.
4. Solve for when the output of the function will be zero to find the x- intercepts.

EXAMPLE 8

Finding the x-Intercepts of a Parabola

Find the x- intercepts of the quadratic function $f(x) = 2x^2 + 4x - 4$.

Solution

We begin by solving for when the output will be zero.

$$0 = 2x^2 + 4x - 4$$

Because the quadratic is not easily factorable in this case, we solve for the intercepts by first rewriting the quadratic in standard form.

$$f(x) = a(x - h)^2 + k$$

We know that $a = 2$. Then we solve for h and k.

$$\begin{aligned} h &= -\tfrac{b}{2a} & \qquad k &= f(-1) \\ &= -\tfrac{4}{2(2)} & &= 2(-1)^2 + 4(-1) - 4 \\ &= -1 & &= -6 \end{aligned}$$

So now we can rewrite in standard form.

$$f(x) = 2(x + 1)^2 - 6$$

We can now solve for when the output will be zero.

$$\begin{aligned} 0 &= 2(x + 1)^2 - 6 \\ 6 &= 2(x + 1)^2 \\ 3 &= (x + 1)^2 \\ x + 1 &= \pm\sqrt{3} \\ x &= -1 \pm \sqrt{3} \end{aligned}$$

The graph has x-intercepts at $(-1 - \sqrt{3}, 0)$ and $(-1 + \sqrt{3}, 0)$.

We can check our work by graphing the given function on a graphing utility and observing the x- intercepts. See Figure 15.

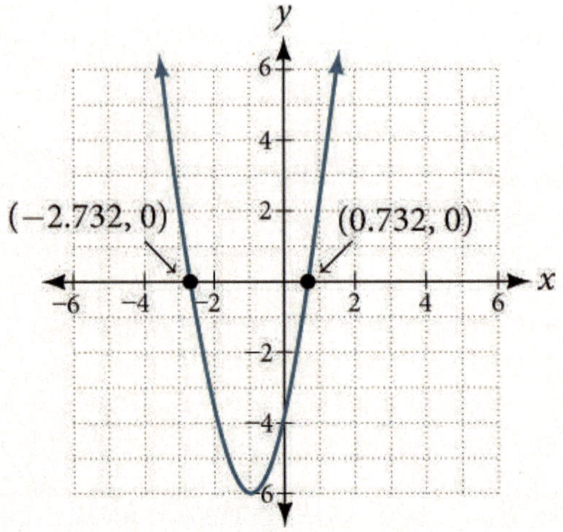

Figure 15

Analysis

We could have achieved the same results using the quadratic formula. Identify $a = 2, b = 4$ and $c = -4$.

$$\begin{aligned} x &= \frac{-b\pm\sqrt{b^2-4ac}}{2a} \\ &= \frac{-4\pm\sqrt{4^2-4(2)(-4)}}{2(2)} \\ &= \frac{-4\pm\sqrt{48}}{4} \\ &= \frac{-4\pm\sqrt{3(16)}}{4} \\ &= -1\pm\sqrt{3} \end{aligned}$$

So the x-intercepts occur at $\left(-1-\sqrt{3},0\right)$ and $\left(-1+\sqrt{3},0\right)$.

> TRY IT #4 In a Try It, we found the standard and general form for the function $g(x) = 13 + x^2 - 6x$. Now find the y- and x-intercepts (if any).

EXAMPLE 9

Applying the Vertex and x-Intercepts of a Parabola

A ball is thrown upward from the top of a 40 foot high building at a speed of 80 feet per second. The ball's height above ground can be modeled by the equation $H(t) = -16t^2 + 80t + 40$.

ⓐ When does the ball reach the maximum height? ⓑ What is the maximum height of the ball?

ⓒ When does the ball hit the ground?

Solution

ⓐ The ball reaches the maximum height at the vertex of the parabola.

$$\begin{aligned} h &= -\frac{80}{2(-16)} \\ &= \frac{80}{32} \\ &= \frac{5}{2} \\ &= 2.5 \end{aligned}$$

The ball reaches a maximum height after 2.5 seconds.

ⓑ To find the maximum height, find the y- coordinate of the vertex of the parabola.

$$\begin{aligned} k &= H\left(-\frac{b}{2a}\right) \\ &= H(2.5) \\ &= -16(2.5)^2 + 80(2.5) + 40 \\ &= 140 \end{aligned}$$

The ball reaches a maximum height of 140 feet.

ⓒ To find when the ball hits the ground, we need to determine when the height is zero, $H(t) = 0$.

We use the quadratic formula.

$$\begin{aligned} t &= \frac{-80 \pm \sqrt{80^2 - 4(-16)(40)}}{2(-16)} \\ &= \frac{-80 \pm \sqrt{8960}}{-32} \end{aligned}$$

Because the square root does not simplify nicely, we can use a calculator to approximate the values of the solutions.

$$t = \frac{-80 - \sqrt{8960}}{-32} \approx 5.458 \quad \text{or} \quad t = \frac{-80 + \sqrt{8960}}{-32} \approx -0.458$$

The second answer is outside the reasonable domain of our model, so we conclude the ball will hit the ground after about 5.458 seconds. See Figure 16.

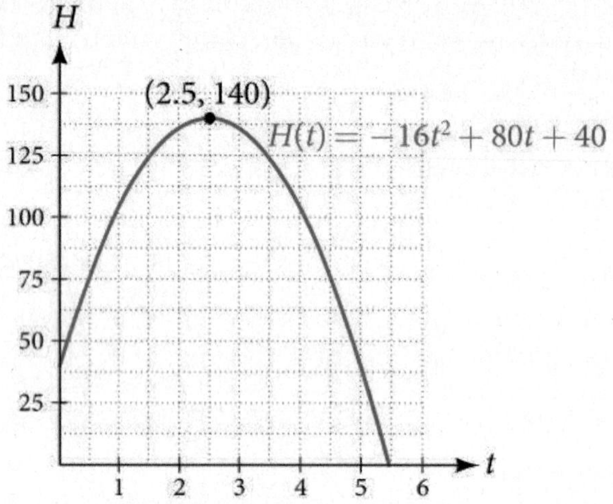

Figure 16

Note that the graph does not represent the physical path of the ball upward and downward. Keep the quantities on each axis in mind while interpreting the graph.

TRY IT #5 A rock is thrown upward from the top of a 112-foot high cliff overlooking the ocean at a speed of 96 feet per second. The rock's height above ocean can be modeled by the equation $H(t) = -16t^2 + 96t + 112$.

ⓐ When does the rock reach the maximum height?

ⓑ What is the maximum height of the rock? ⓒ When does the rock hit the ocean?

MEDIA

Access these online resources for additional instruction and practice with quadratic equations.

Graphing Quadratic Functions in General Form (http://openstax.org/l/graphquadgen)
Graphing Quadratic Functions in Standard Form (http://openstax.org/l/graphquadstan)
Quadratic Function Review (http://openstax.org/l/quadfuncrev)
Characteristics of a Quadratic Function (http://openstax.org/l/characterquad)

5.1 SECTION EXERCISES

Verbal

1. Explain the advantage of writing a quadratic function in standard form.

2. How can the vertex of a parabola be used in solving real-world problems?

3. Explain why the condition of $a \neq 0$ is imposed in the definition of the quadratic function.

4. What is another name for the standard form of a quadratic function?

5. What two algebraic methods can be used to find the horizontal intercepts of a quadratic function?

Algebraic

For the following exercises, rewrite the quadratic functions in standard form and give the vertex.

6. $f(x) = x^2 - 12x + 32$

7. $g(x) = x^2 + 2x - 3$

8. $f(x) = x^2 - x$

9. $f(x) = x^2 + 5x - 2$

10. $h(x) = 2x^2 + 8x - 10$

11. $k(x) = 3x^2 - 6x - 9$

12. $f(x) = 2x^2 - 6x$

13. $f(x) = 3x^2 - 5x - 1$

For the following exercises, determine whether there is a minimum or maximum value to each quadratic function. Find the value and the axis of symmetry.

14. $y(x) = 2x^2 + 10x + 12$

15. $f(x) = 2x^2 - 10x + 4$

16. $f(x) = -x^2 + 4x + 3$

17. $f(x) = 4x^2 + x - 1$

18. $h(t) = -4t^2 + 6t - 1$

19. $f(x) = \frac{1}{2}x^2 + 3x + 1$

20. $f(x) = -\frac{1}{3}x^2 - 2x + 3$

For the following exercises, determine the domain and range of the quadratic function.

21. $f(x) = (x - 3)^2 + 2$

22. $f(x) = -2(x + 3)^2 - 6$

23. $f(x) = x^2 + 6x + 4$

24. $f(x) = 2x^2 - 4x + 2$

25. $k(x) = 3x^2 - 6x - 9$

For the following exercises, use the vertex (h, k) and a point on the graph (x, y) to find the general form of the equation of the quadratic function.

26. $(h, k) = (2, 0), (x, y) = (4, 4)$

27. $(h, k) = (-2, -1), (x, y) = (-4, 3)$

28. $(h, k) = (0, 1), (x, y) = (2, 5)$

29. $(h, k) = (2, 3), (x, y) = (5, 12)$

30. $(h, k) = (-5, 3), (x, y) = (2, 9)$

31. $(h, k) = (3, 2), (x, y) = (10, 1)$

32. $(h, k) = (0, 1), (x, y) = (1, 0)$

33. $(h, k) = (1, 0), (x, y) = (0, 1)$

Graphical

For the following exercises, sketch a graph of the quadratic function and give the vertex, axis of symmetry, and intercepts.

34. $f(x) = x^2 - 2x$

35. $f(x) = x^2 - 6x - 1$

36. $f(x) = x^2 - 5x - 6$

37. $f(x) = x^2 - 7x + 3$

38. $f(x) = -2x^2 + 5x - 8$

39. $f(x) = 4x^2 - 12x - 3$

For the following exercises, write the equation for the graphed quadratic function.

40.

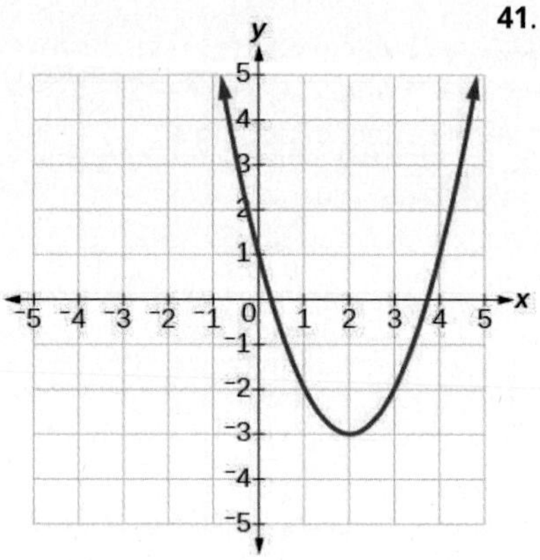

41.

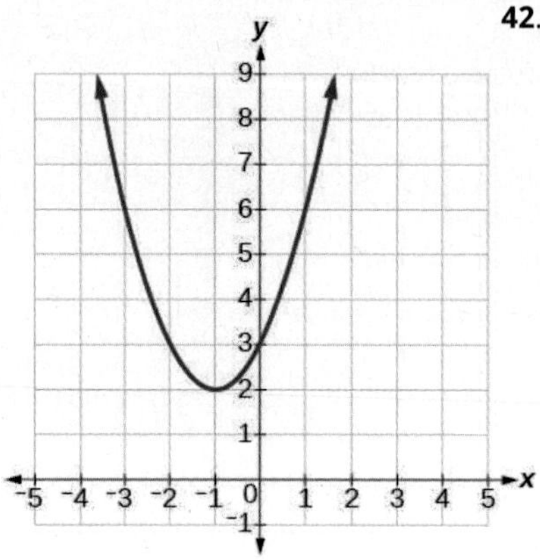

42.

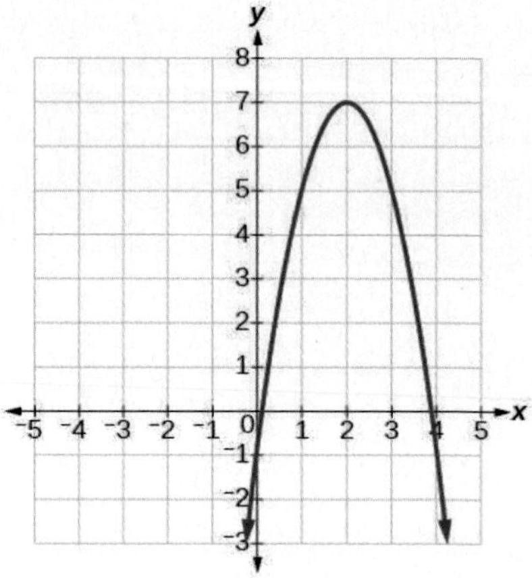

43.

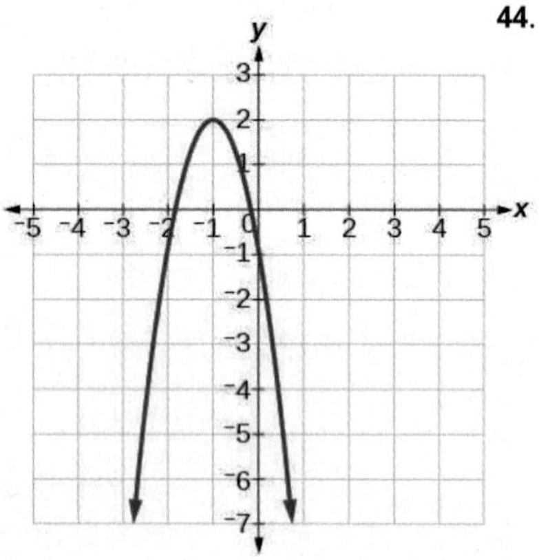

44.

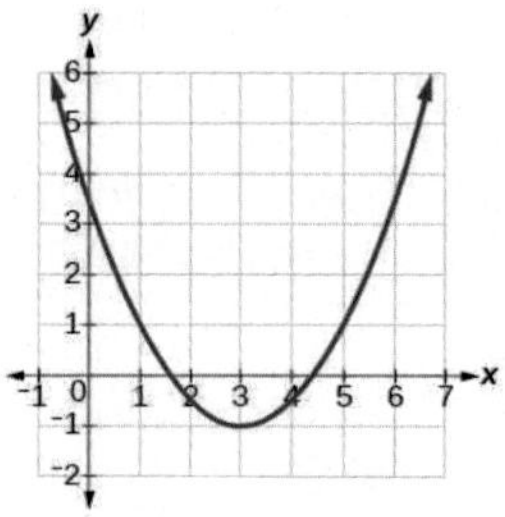

45.

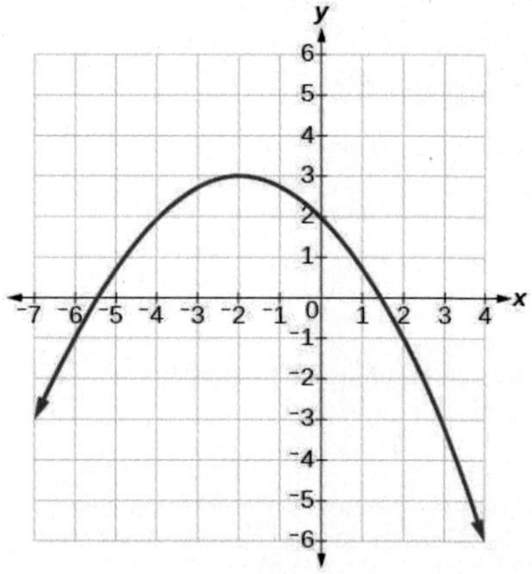

Numeric

For the following exercises, use the table of values that represent points on the graph of a quadratic function. By determining the vertex and axis of symmetry, find the general form of the equation of the quadratic function.

46.

x	-2	-1	0	1	2
y	5	2	1	2	5

47.

x	-2	-1	0	1	2
y	1	0	1	4	9

48.

x	-2	-1	0	1	2
y	-2	1	2	1	-2

49.

x	-2	-1	0	1	2
y	-8	-3	0	1	0

50.

x	-2	-1	0	1	2
y	8	2	0	2	8

Technology

For the following exercises, use a calculator to find the answer.

51. Graph on the same set of axes the functions $f(x) = x^2$, $f(x) = 2x^2$, and $f(x) = \frac{1}{3}x^2$.

What appears to be the effect of changing the coefficient?

52. Graph on the same set of axes $f(x) = x^2$, $f(x) = x^2 + 2$ and $f(x) = x^2$, $f(x) = x^2 + 5$ and $f(x) = x^2 - 3$. What appears to be the effect of adding a constant?

53. Graph on the same set of axes $f(x) = x^2$, $f(x) = (x - 2)^2$, $f(x - 3)^2$, and $f(x) = (x + 4)^2$.

What appears to be the effect of adding or subtracting those numbers?

54. The path of an object projected at a 45 degree angle with initial velocity of 80 feet per second is given by the function $h(x) = \frac{-32}{(80)^2}x^2 + x$ where x is the horizontal distance traveled and $h(x)$ is the height in feet. Use the TRACE feature of your calculator to determine the height of the object when it has traveled 100 feet away horizontally.

55. A suspension bridge can be modeled by the quadratic function $h(x) = .0001x^2$ with $-2000 \leq x \leq 2000$ where $|x|$ is the number of feet from the center and $h(x)$ is height in feet. Use the TRACE feature of your calculator to estimate how far from the center does the bridge have a height of 100 feet.

Extensions

For the following exercises, use the vertex of the graph of the quadratic function and the direction the graph opens to find the domain and range of the function.

56. Vertex $(1, -2)$, opens up.

57. Vertex $(-1, 2)$ opens down.

58. Vertex $(-5, 11)$, opens down.

59. Vertex $(-100, 100)$, opens up.

For the following exercises, write the equation of the quadratic function that contains the given point and has the same shape as the given function.

60. Contains $(1, 1)$ and has shape of $f(x) = 2x^2$. Vertex is on the y- axis.

61. Contains $(-1, 4)$ and has the shape of $f(x) = 2x^2$. Vertex is on the y- axis.

62. Contains $(2, 3)$ and has the shape of $f(x) = 3x^2$. Vertex is on the y- axis.

63. Contains $(1, -3)$ and has the shape of $f(x) = -x^2$. Vertex is on the y- axis.

64. Contains $(4, 3)$ and has the shape of $f(x) = 5x^2$. Vertex is on the y- axis.

65. Contains $(1, -6)$ has the shape of $f(x) = 3x^2$. Vertex has x-coordinate of -1.

Real-World Applications

66. Find the dimensions of the rectangular dog park producing the greatest enclosed area given 200 feet of fencing.

67. Find the dimensions of the rectangular dog park split into 2 pens of the same size producing the greatest possible enclosed area given 300 feet of fencing.

68. Find the dimensions of the rectangular dog park producing the greatest enclosed area split into 3 sections of the same size given 500 feet of fencing.

69. Among all of the pairs of numbers whose sum is 6, find the pair with the largest product. What is the product?

70. Among all of the pairs of numbers whose difference is 12, find the pair with the smallest product. What is the product?

71. Suppose that the price per unit in dollars of a cell phone production is modeled by $p = \$45 - 0.0125x$, where x is in thousands of phones produced, and the revenue represented by thousands of dollars is $R = x \cdot p$. Find the production level that will maximize revenue.

72. A rocket is launched in the air. Its height, in meters above sea level, as a function of time, in seconds, is given by $h(t) = -4.9t^2 + 229t + 234$. Find the maximum height the rocket attains.

73. A ball is thrown in the air from the top of a building. Its height, in meters above ground, as a function of time, in seconds, is given by $h(t) = -4.9t^2 + 24t + 8$. How long does it take to reach maximum height?

74. A soccer stadium holds 62,000 spectators. With a ticket price of \$11, the average attendance has been 26,000. When the price dropped to \$9, the average attendance rose to 31,000. Assuming that attendance is linearly related to ticket price, what ticket price would maximize revenue?

75. A farmer finds that if she plants 75 trees per acre, each tree will yield 20 bushels of fruit. She estimates that for each additional tree planted per acre, the yield of each tree will decrease by 3 bushels. How many trees should she plant per acre to maximize her harvest?

5.2 Power Functions and Polynomial Functions

Learning Objectives

In this section, you will:

- Identify power functions.
- Identify end behavior of power functions.
- Identify polynomial functions.
- Identify the degree and leading coefficient of polynomial functions.

Figure 1 (credit: Jason Bay, Flickr)

Suppose a certain species of bird thrives on a small island. Its population over the last few years is shown in Table 1.

Year	2009	2010	2011	2012	2013
Bird Population	800	897	992	1,083	1,169

Table 1

The population can be estimated using the function $P(t) = -0.3t^3 + 97t + 800$, where $P(t)$ represents the bird population on the island t years after 2009. We can use this model to estimate the maximum bird population and when it will occur. We can also use this model to predict when the bird population will disappear from the island. In this section, we will examine functions that we can use to estimate and predict these types of changes.

Identifying Power Functions

Before we can understand the bird problem, it will be helpful to understand a different type of function. A **power function** is a function with a single term that is the product of a real number, a **coefficient,** and a variable raised to a fixed real number.

As an example, consider functions for area or volume. The function for the area of a circle with radius r is

$$A(r) = \pi r^2$$

and the function for the volume of a sphere with radius r is

$$V(r) = \frac{4}{3}\pi r^3$$

Both of these are examples of power functions because they consist of a coefficient, π or $\frac{4}{3}\pi$, multiplied by a variable r raised to a power.

Power Function

A **power function** is a function that can be represented in the form

$$f(x) = kx^p$$

where k and p are real numbers, and k is known as the **coefficient**.

Q&A **Is $f(x) = 2^x$ a power function?**

No. A power function contains a variable base raised to a fixed power. This function has a constant base raised to a variable power. This is called an exponential function, not a power function.

EXAMPLE 1

Identifying Power Functions

Which of the following functions are power functions?

$$\begin{array}{rcll} f(x) & = & 1 & \text{Constant function} \\ f(x) & = & x & \text{Identify function} \\ f(x) & = & x^2 & \text{Quadratic function} \\ f(x) & = & x^3 & \text{Cubic function} \\ f(x) & = & \frac{1}{x} & \text{Reciprocal function} \\ f(x) & = & \frac{1}{x^2} & \text{Reciprocal squared function} \\ f(x) & = & \sqrt{x} & \text{Square root function} \\ f(x) & = & \sqrt[3]{x} & \text{Cube root function} \end{array}$$

Solution

All of the listed functions are power functions.

The constant and identity functions are power functions because they can be written as $f(x) = x^0$ and $f(x) = x^1$ respectively.

The quadratic and cubic functions are power functions with whole number powers $f(x) = x^2$ and $f(x) = x^3$.

The reciprocal and reciprocal squared functions are power functions with negative whole number powers because they can be written as $f(x) = x^{-1}$ and $f(x) = x^{-2}$.

The square and cube root functions are power functions with fractional powers because they can be written as $f(x) = x^{\frac{1}{2}}$ or $f(x) = x^{\frac{1}{3}}$.

TRY IT #1 Which functions are power functions?

$$\begin{array}{rcl} f(x) & = & 2x \cdot 4x^3 \\ g(x) & = & -x^5 + 5x^3 \\ h(x) & = & \frac{2x^5-1}{3x^2+4} \end{array}$$

Identifying End Behavior of Power Functions

Figure 2 shows the graphs of $f(x) = x^2$, $g(x) = x^4$ and $h(x) = x^6$, which are all power functions with even, whole-number powers. Notice that these graphs have similar shapes, very much like that of the quadratic function in the toolkit. However, as the power increases, the graphs flatten somewhat near the origin and become steeper away from the origin.

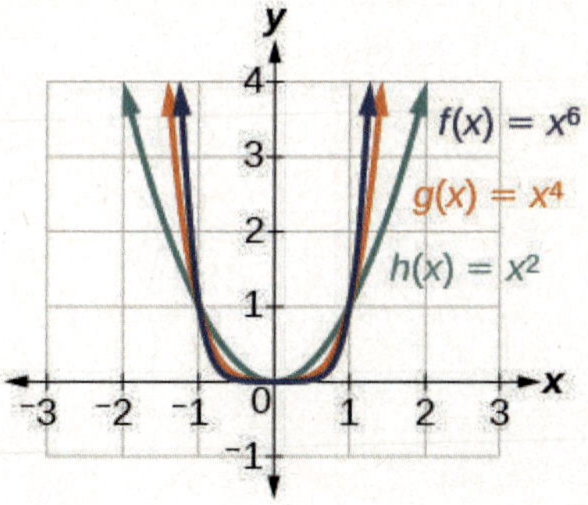

Figure 2 Even-power functions

To describe the behavior as numbers become larger and larger, we use the idea of infinity. We use the symbol ∞ for positive infinity and $-\infty$ for negative infinity. When we say that " x approaches infinity," which can be symbolically written as $x \to \infty$, we are describing a behavior; we are saying that x is increasing without bound.

With the positive even-power function, as the input increases or decreases without bound, the output values become very large, positive numbers. Equivalently, we could describe this behavior by saying that as x approaches positive or negative infinity, the $f(x)$ values increase without bound. In symbolic form, we could write

$$\text{as } x \to \pm\infty, \ f(x) \to \infty$$

Figure 3 shows the graphs of $f(x) = x^3, \ \ g(x) = x^5$, and $h(x) = x^7$, which are all power functions with odd, whole-number powers. Notice that these graphs look similar to the cubic function in the toolkit. Again, as the power increases, the graphs flatten near the origin and become steeper away from the origin.

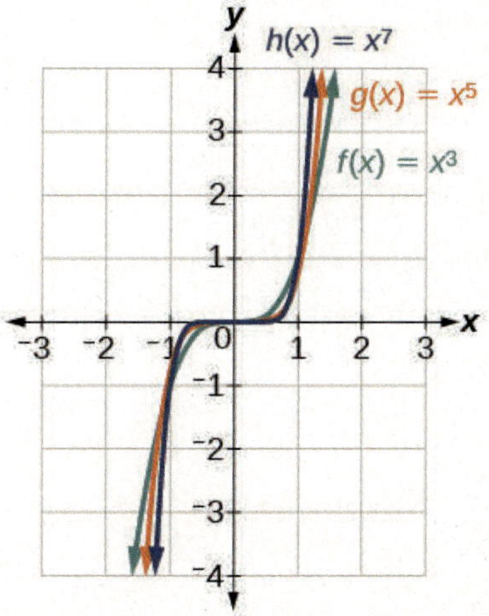

Figure 3 Odd-power functions

These examples illustrate that functions of the form $f(x) = x^n$ reveal symmetry of one kind or another. First, in Figure 2 we see that even functions of the form $f(x) = x^n$, n even, are symmetric about the y- axis. In Figure 3 we see that odd functions of the form $f(x) = x^n$, n odd, are symmetric about the origin.

For these odd power functions, as x approaches negative infinity, $f(x)$ decreases without bound. As x approaches positive infinity, $f(x)$ increases without bound. In symbolic form we write

$$\text{as } x \to -\infty, \ f(x) \to -\infty$$

$$\text{as } x \to \infty, \ f(x) \to \infty$$

The behavior of the graph of a function as the input values get very small ($x \to -\infty$) and get very large ($x \to \infty$) is referred to as the **end behavior** of the function. We can use words or symbols to describe end behavior.

Figure 4 shows the end behavior of power functions in the form $f(x) = kx^n$ where n is a non-negative integer depending on the power and the constant.

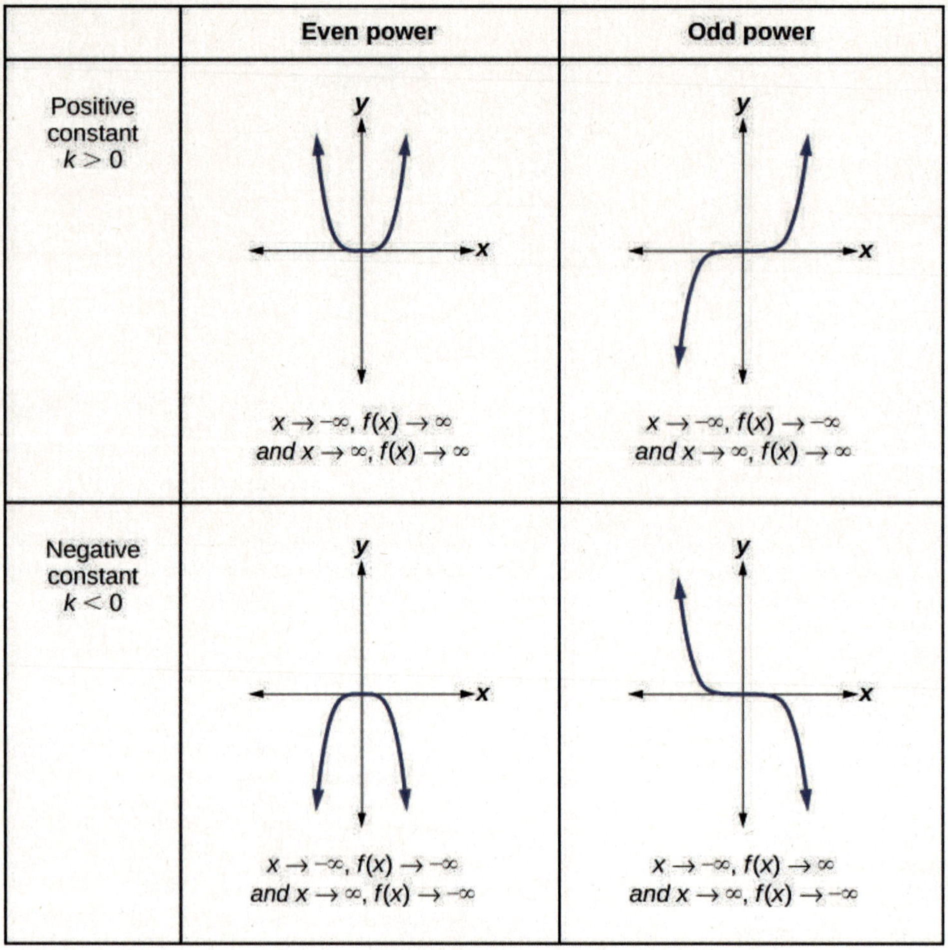

Figure 4

HOW TO

Given a power function $f(x) = kx^n$ where n is a non-negative integer, identify the end behavior.

1. Determine whether the power is even or odd.
2. Determine whether the constant is positive or negative.
3. Use Figure 4 to identify the end behavior.

EXAMPLE 2

Identifying the End Behavior of a Power Function

Describe the end behavior of the graph of $f(x) = x^8$.

Solution

The coefficient is 1 (positive) and the exponent of the power function is 8 (an even number). As x approaches infinity, the output (value of $f(x)$) increases without bound. We write as $x \to \infty$, $f(x) \to \infty$. As x approaches negative infinity, the output increases without bound. In symbolic form, as $x \to -\infty$, $f(x) \to \infty$. We can graphically represent the function as shown in Figure 5.

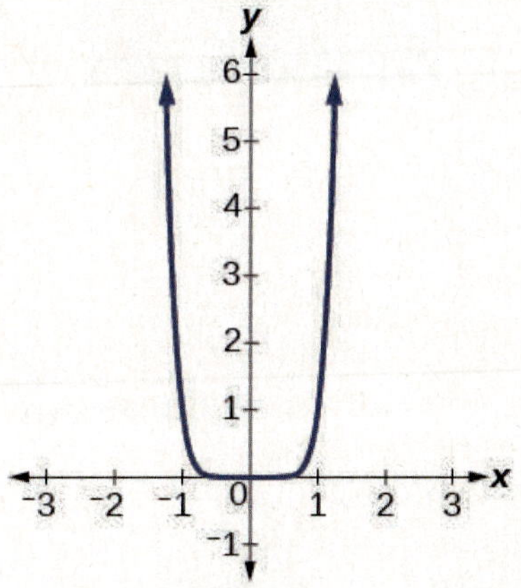

Figure 5

EXAMPLE 3

Identifying the End Behavior of a Power Function.

Describe the end behavior of the graph of $f(x) = -x^9$.

Solution

The exponent of the power function is 9 (an odd number). Because the coefficient is -1 (negative), the graph is the reflection about the x- axis of the graph of $f(x) = x^9$. Figure 6 shows that as x approaches infinity, the output decreases without bound. As x approaches negative infinity, the output increases without bound. In symbolic form, we would write

$$\text{as } x \to -\infty,\ f(x) \to \infty$$

$$\text{as } x \to \infty,\ f(x) \to -\infty$$

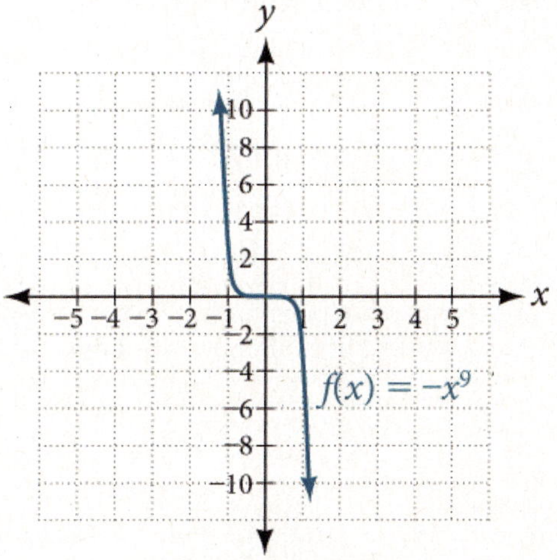

Figure 6

Analysis

We can check our work by using the table feature on a graphing utility.

x	$f(x)$
–10	1,000,000,000

Table 2

x	$f(x)$
-5	1,953,125
0	0
5	−1,953,125
10	−1,000,000,000

Table 2

We can see from Table 2 that, when we substitute very small values for x, the output is very large, and when we substitute very large values for x, the output is very small (meaning that it is a very large negative value).

> **TRY IT** #2 Describe in words and symbols the end behavior of $f(x) = -5x^4$.

Identifying Polynomial Functions

An oil pipeline bursts in the Gulf of Mexico, causing an oil slick in a roughly circular shape. The slick is currently 24 miles in radius, but that radius is increasing by 8 miles each week. We want to write a formula for the area covered by the oil slick by combining two functions. The radius r of the spill depends on the number of weeks w that have passed. This relationship is linear.

$$r(w) = 24 + 8w$$

We can combine this with the formula for the area A of a circle.

$$A(r) = \pi r^2$$

Composing these functions gives a formula for the area in terms of weeks.

$$\begin{aligned} A(w) &= A(r(w)) \\ &= A(24 + 8w) \\ &= \pi(24 + 8w)^2 \end{aligned}$$

Multiplying gives the formula.

$$A(w) = 576\pi + 384\pi w + 64\pi w^2$$

This formula is an example of a **polynomial function**. A polynomial function consists of either zero or the sum of a finite number of non-zero terms, each of which is a product of a number, called the coefficient of the term, and a variable raised to a non-negative integer power.

Polynomial Functions

Let n be a non-negative integer. A **polynomial function** is a function that can be written in the form

$$f(x) = a_n x^n + \ldots + a_2 x^2 + a_1 x + a_0$$

This is called the general form of a polynomial function. Each a_i is a coefficient and can be any real number, but $a_n \neq$. Each expression $a_i x^i$ is a **term of a polynomial function**.

EXAMPLE 4

Identifying Polynomial Functions
Which of the following are polynomial functions?

$$\begin{aligned} f(x) &= 2x^3 \cdot 3x + 4 \\ g(x) &= -x(x^2 - 4) \\ h(x) &= 5\sqrt{x+2} \end{aligned}$$

Solution

The first two functions are examples of polynomial functions because they can be written in the form $f(x) = a_n x^n + ... + a_2 x^2 + a_1 x + a_0$, where the powers are non-negative integers and the coefficients are real numbers.

- $f(x)$ can be written as $f(x) = 6x^4 + 4$.
- $g(x)$ can be written as $g(x) = -x^3 + 4x$.
- $h(x)$ cannot be written in this form and is therefore not a polynomial function.

Identifying the Degree and Leading Coefficient of a Polynomial Function

Because of the form of a polynomial function, we can see an infinite variety in the number of terms and the power of the variable. Although the order of the terms in the polynomial function is not important for performing operations, we typically arrange the terms in descending order of power, or in general form. The **degree** of the polynomial is the highest power of the variable that occurs in the polynomial; it is the power of the first variable if the function is in general form. The **leading term** is the term containing the highest power of the variable, or the term with the highest degree. The **leading coefficient** is the coefficient of the leading term.

Terminology of Polynomial Functions

We often rearrange polynomials so that the powers are descending.

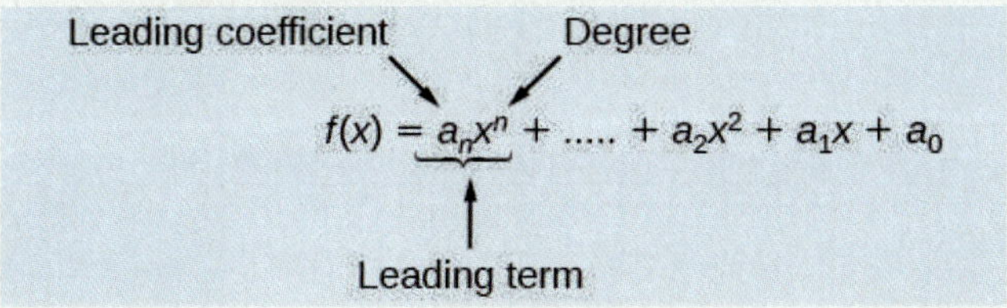

When a polynomial is written in this way, we say that it is in general form.

HOW TO

Given a polynomial function, identify the degree and leading coefficient.

1. Find the highest power of x to determine the degree of the function.
2. Identify the term containing the highest power of x to find the leading term.
3. Identify the coefficient of the leading term.

EXAMPLE 5

Identifying the Degree and Leading Coefficient of a Polynomial Function

Identify the degree, leading term, and leading coefficient of the following polynomial functions.

$$\begin{aligned} f(x) &= 3 + 2x^2 - 4x^3 \\ g(t) &= 5t^5 - 2t^3 + 7t \\ h(p) &= 6p - p^3 - 2 \end{aligned}$$

Solution

For the function $f(x)$, the highest power of x is 3, so the degree is 3. The leading term is the term containing that degree, $-4x^3$. The leading coefficient is the coefficient of that term, -4.

For the function $g(t)$, the highest power of t is 5, so the degree is 5. The leading term is the term containing that degree, $5t^5$. The leading coefficient is the coefficient of that term, 5.

For the function $h(p)$, the highest power of p is 3, so the degree is 3. The leading term is the term containing that degree, $-p^3$. The leading coefficient is the coefficient of that term, -1.

TRY IT #3 Identify the degree, leading term, and leading coefficient of the polynomial $f(x) = 4x^2 - x^6 + 2x - 6$.

Identifying End Behavior of Polynomial Functions

Knowing the degree of a polynomial function is useful in helping us predict its end behavior. To determine its end behavior, look at the leading term of the polynomial function. Because the power of the leading term is the highest, that term will grow significantly faster than the other terms as x gets very large or very small, so its behavior will dominate the graph. For any polynomial, the end behavior of the polynomial will match the end behavior of the power function consisting of the leading term. See Table 3.

Polynomial Function	Leading Term	Graph of Polynomial Function
$f(x) = 5x^4 + 2x^3 - x - 4$	$5x^4$	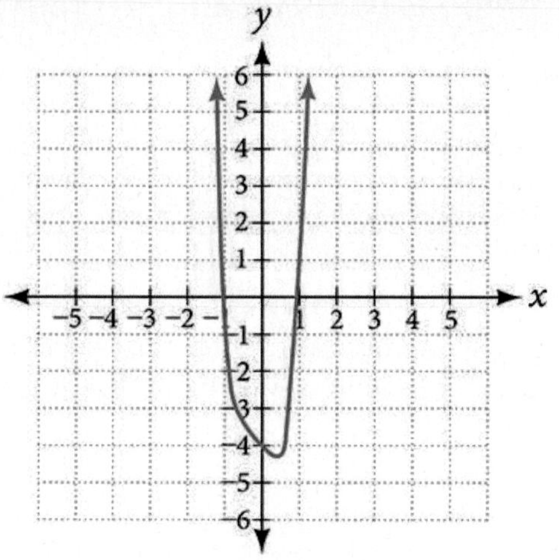
$f(x) = -2x^6 - x^5 + 3x^4 + x^3$	$-2x^6$	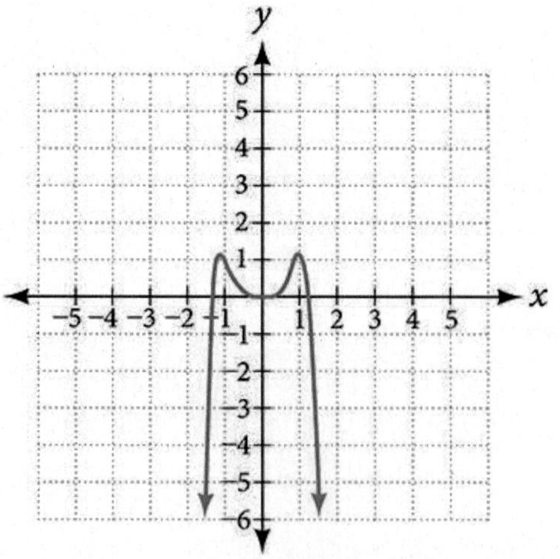

Table 3

Polynomial Function	Leading Term	Graph of Polynomial Function
$f(x) = 3x^5 - 4x^4 + 2x^2 + 1$	$3x^5$	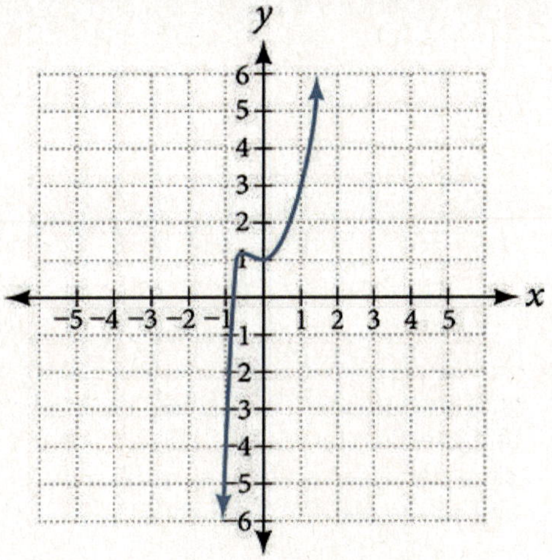
$f(x) = -6x^3 + 7x^2 + 3x + 1$	$-6x^3$	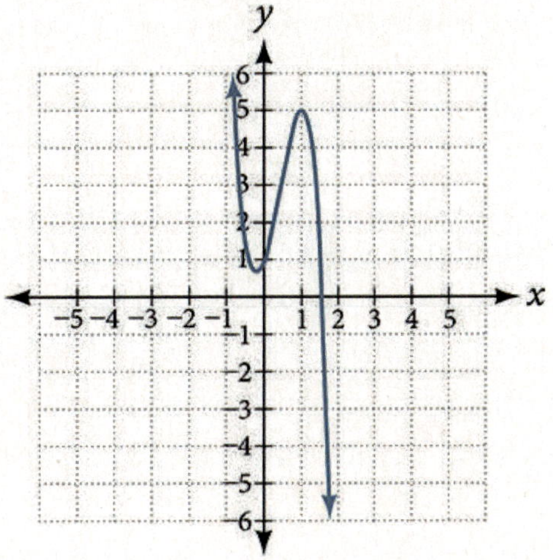

Table 3

EXAMPLE 6

Identifying End Behavior and Degree of a Polynomial Function

Describe the end behavior and determine a possible degree of the polynomial function in Figure 7.

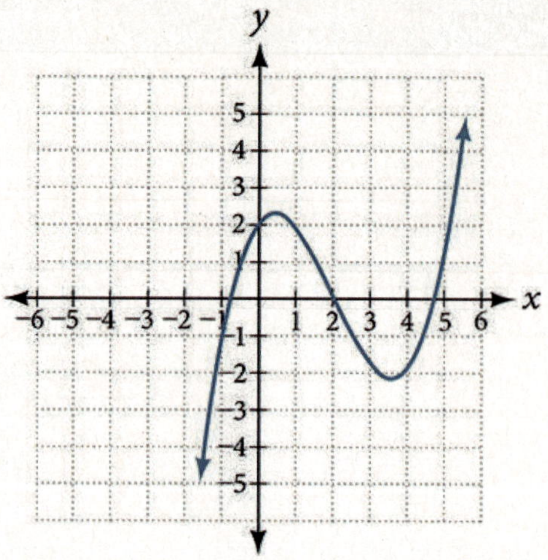

Figure 7

Solution

As the input values x get very large, the output values $f(x)$ increase without bound. As the input values x get very small, the output values $f(x)$ decrease without bound. We can describe the end behavior symbolically by writing

$$\text{as } x \to -\infty,\ f(x) \to -\infty$$

$$\text{as } x \to \infty,\ f(x) \to \infty$$

In words, we could say that as x values approach infinity, the function values approach infinity, and as x values approach negative infinity, the function values approach negative infinity.

We can tell this graph has the shape of an odd degree power function that has not been reflected, so the degree of the polynomial creating this graph must be odd and the leading coefficient must be positive.

TRY IT #4 Describe the end behavior, and determine a possible degree of the polynomial function in Figure 8.

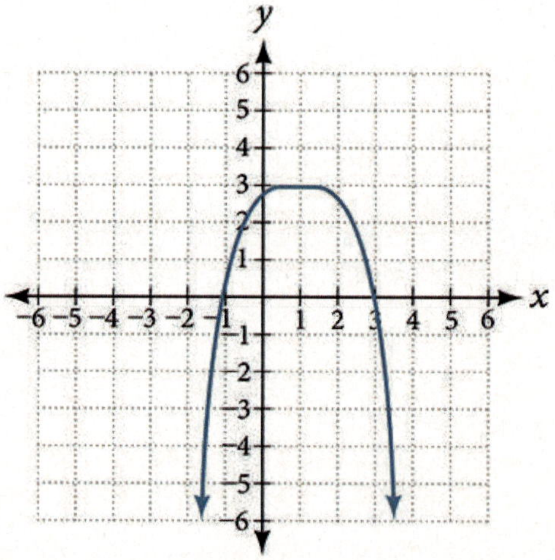

Figure 8

EXAMPLE 7

Identifying End Behavior and Degree of a Polynomial Function

Given the function $f(x) = -3x^2(x-1)(x+4)$, express the function as a polynomial in general form, and determine the leading term, degree, and end behavior of the function.

Solution

Obtain the general form by expanding the given expression for $f(x)$.

$$\begin{aligned} f(x) &= -3x^2(x-1)(x+4) \\ &= -3x^2\left(x^2+3x-4\right) \\ &= -3x^4-9x^3+12x^2 \end{aligned}$$

The general form is $f(x) = -3x^4 - 9x^3 + 12x^2$. The leading term is $-3x^4$; therefore, the degree of the polynomial is 4. The degree is even (4) and the leading coefficient is negative (–3), so the end behavior is

$$\text{as } x \to -\infty,\ f(x) \to -\infty$$

$$\text{as } x \to \infty,\ f(x) \to -\infty$$

TRY IT #5 Given the function $f(x) = 0.2(x-2)(x+1)(x-5)$, express the function as a polynomial in general form and determine the leading term, degree, and end behavior of the function.

Identifying Local Behavior of Polynomial Functions

In addition to the end behavior of polynomial functions, we are also interested in what happens in the "middle" of the function. In particular, we are interested in locations where graph behavior changes. A **turning point** is a point at which the function values change from increasing to decreasing or decreasing to increasing.

We are also interested in the intercepts. As with all functions, the *y*-intercept is the point at which the graph intersects the vertical axis. The point corresponds to the coordinate pair in which the input value is zero. Because a polynomial is a function, only one output value corresponds to each input value so there can be only one *y*-intercept $(0, a_0)$. The *x*-intercepts occur at the input values that correspond to an output value of zero. It is possible to have more than one *x*-intercept. See Figure 9.

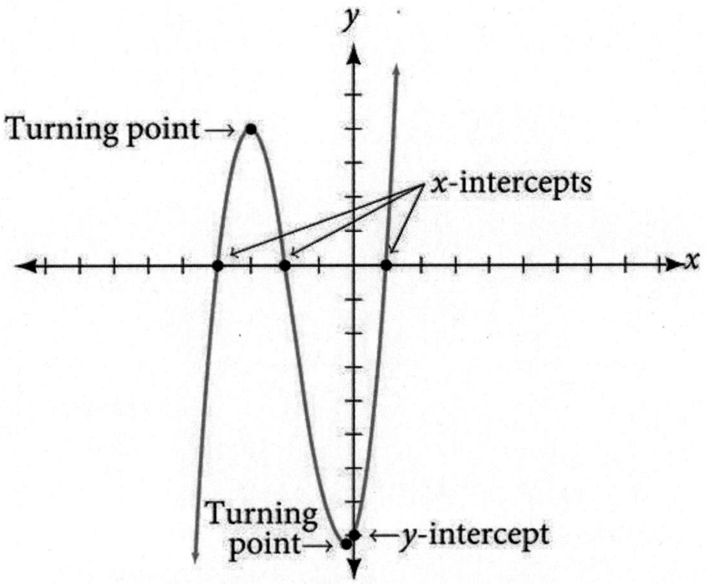

Figure 9

Intercepts and Turning Points of Polynomial Functions

A **turning point** of a graph is a point at which the graph changes direction from increasing to decreasing or decreasing to increasing. The *y*-intercept is the point at which the function has an input value of zero. The *x*-intercepts are the points at which the output value is zero.

HOW TO

Given a polynomial function, determine the intercepts.

1. Determine the *y*-intercept by setting $x = 0$ and finding the corresponding output value.
2. Determine the *x*-intercepts by solving for the input values that yield an output value of zero.

EXAMPLE 8

Determining the Intercepts of a Polynomial Function

Given the polynomial function $f(x) = (x - 2)(x + 1)(x - 4)$, written in factored form for your convenience, determine the *y*- and *x*-intercepts.

Solution

The *y*-intercept occurs when the input is zero so substitute 0 for x.

$$\begin{aligned} f(0) &= f(0) = (0 - 2)(0 + 1)(0 - 4) \\ &= (-2)(1)(-4) \\ &= 8 \end{aligned}$$

The *y*-intercept is (0, 8).

The *x*-intercepts occur when the output is zero.

$$0 = (x - 2)(x + 1)(x - 4)$$

$$\begin{array}{rclcrclcrcl} x - 2 &=& 0 & \text{or} & x + 1 &=& 0 & \text{or} & x - 4 &=& 0 \\ x &=& 2 & \text{or} & x &=& -1 & \text{or} & x &=& 4 \end{array}$$

The *x*-intercepts are $(2, 0), (-1, 0)$, and $(4, 0)$.

We can see these intercepts on the graph of the function shown in Figure 10.

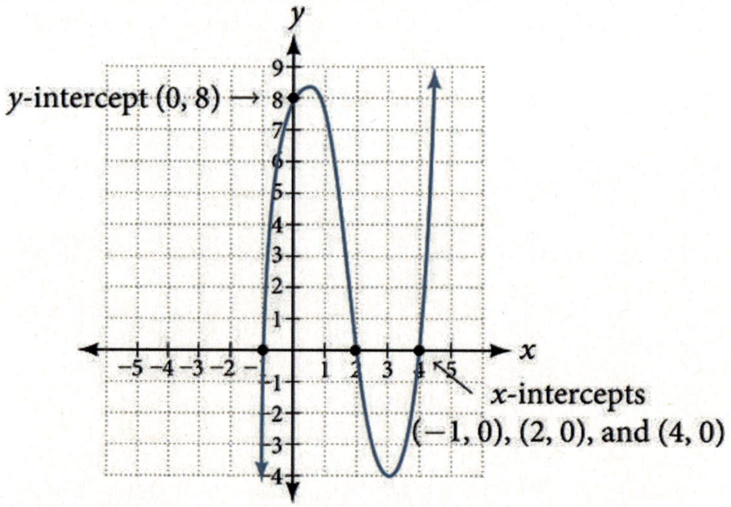

Figure 10

EXAMPLE 9

Determining the Intercepts of a Polynomial Function with Factoring

Given the polynomial function $f(x) = x^4 - 4x^2 - 45$, determine the *y*- and *x*-intercepts.

Solution

The y-intercept occurs when the input is zero.

$$\begin{aligned} f(0) &= (0)^4 - 4(0)^2 - 45 \\ &= -45 \end{aligned}$$

The y-intercept is $(0, -45)$.

The x-intercepts occur when the output is zero. To determine when the output is zero, we will need to factor the polynomial.

$$\begin{aligned} f(x) &= x^4 - 4x^2 - 45 \\ &= \left(x^2 - 9\right)\left(x^2 + 5\right) \\ &= (x-3)(x+3)\left(x^2 + 5\right) \end{aligned}$$

$$0 = (x-3)(x+3)\left(x^2+5\right)$$

$$\begin{array}{ccccc} x - 3 = 0 & \text{or} & x + 3 = 0 & \text{or} & x^2 + 5 = 0 \\ x = 3 & \text{or} & x = -3 & \text{or} & \text{(no real solution)} \end{array}$$

The x-intercepts are $(3, 0)$ and $(-3, 0)$.

We can see these intercepts on the graph of the function shown in Figure 11. We can see that the function is even because $f(x) = f(-x)$.

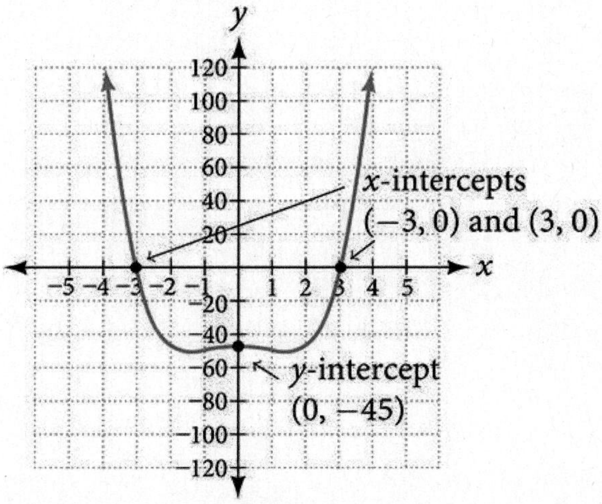

Figure 11

TRY IT #6 Given the polynomial function $f(x) = 2x^3 - 6x^2 - 20x$, determine the y- and x-intercepts.

Comparing Smooth and Continuous Graphs

The degree of a polynomial function helps us to determine the number of x-intercepts and the number of turning points. A polynomial function of nth degree is the product of n factors, so it will have at most n roots or zeros, or x-intercepts. The graph of the polynomial function of degree n must have at most $n-1$ turning points. This means the graph has at most one fewer turning point than the degree of the polynomial or one fewer than the number of factors.

A **continuous function** has no breaks in its graph: the graph can be drawn without lifting the pen from the paper. A **smooth curve** is a graph that has no sharp corners. The turning points of a smooth graph must always occur at rounded curves. The graphs of polynomial functions are both continuous and smooth.

Intercepts and Turning Points of Polynomials

A polynomial of degree n will have, at most, n x-intercepts and $n - 1$ turning points.

EXAMPLE 10

Determining the Number of Intercepts and Turning Points of a Polynomial

Without graphing the function, determine the local behavior of the function by finding the maximum number of *x*-intercepts and turning points for $f(x) = -3x^{10} + 4x^7 - x^4 + 2x^3$.

Solution

The polynomial has a degree of 10, so there are at most 10 *x*-intercepts and at most 9 turning points.

TRY IT #7 Without graphing the function, determine the maximum number of *x*-intercepts and turning points for $f(x) = 108 - 13x^9 - 8x^4 + 14x^{12} + 2x^3$.

EXAMPLE 11

Drawing Conclusions about a Polynomial Function from the Graph

What can we conclude about the polynomial represented by the graph shown in Figure 12 based on its intercepts and turning points?

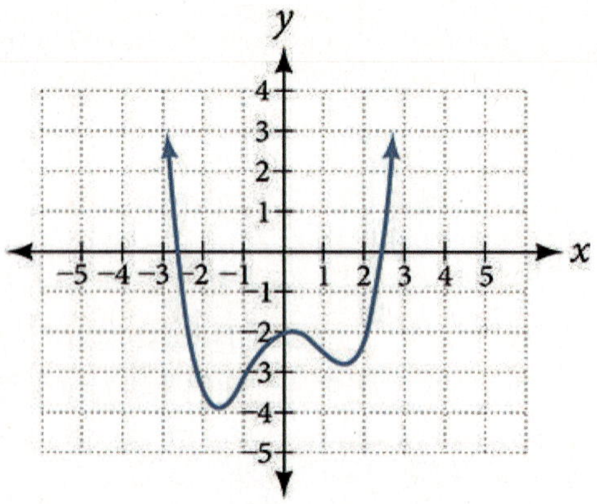

Figure 12

Solution

The end behavior of the graph tells us this is the graph of an even-degree polynomial. See Figure 13.

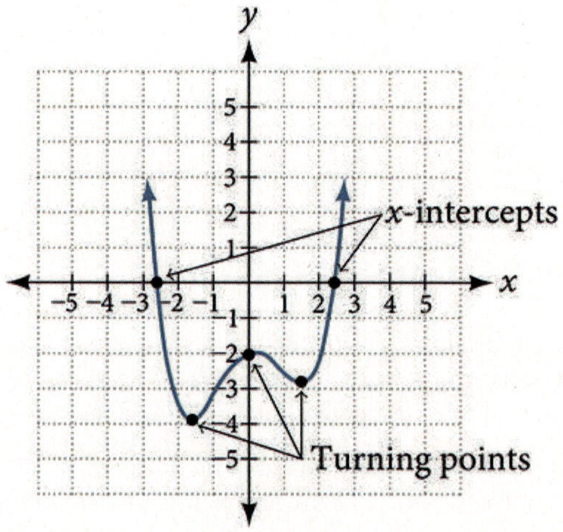

Figure 13

The graph has 2 *x*-intercepts, suggesting a degree of 2 or greater, and 3 turning points, suggesting a degree of 4 or greater. Based on this, it would be reasonable to conclude that the degree is even and at least 4.

TRY IT #8 What can we conclude about the polynomial represented by the graph shown in Figure 14 based on its intercepts and turning points?

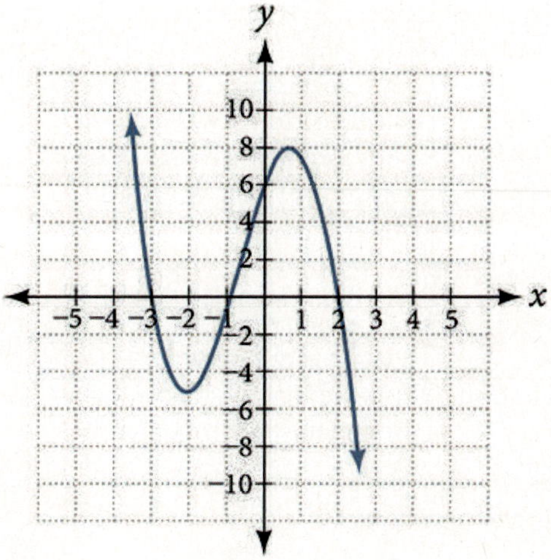

Figure 14

EXAMPLE 12

Drawing Conclusions about a Polynomial Function from the Factors

Given the function $f(x) = -4x\,(x+3)\,(x-4)$, determine the local behavior.

Solution

The *y*-intercept is found by evaluating $f(0)$.

$$\begin{aligned} f(0) &= -4(0)(0+3)(0-4 \\ &= 0 \end{aligned}$$

The *y*-intercept is $(0, 0)$.

The *x*-intercepts are found by determining the zeros of the function.

$$0 = -4x(x+3)(x-4)$$

$$\begin{array}{lllllll} x = 0 & \text{or} & x+3 = 0 & \text{or} & x-4 = 0 \\ x = 0 & \text{or} & x = -3 & \text{or} & x = 4 \end{array}$$

The *x*-intercepts are $(0, 0)$, $(-3, 0)$, and $(4, 0)$.

The degree is 3 so the graph has at most 2 turning points.

TRY IT #9 Given the function $f(x) = 0.2(x-2)(x+1)(x-5)$, determine the local behavior.

MEDIA

Access these online resources for additional instruction and practice with power and polynomial functions.

Find Key Information about a Given Polynomial Function (http://openstax.org/l/keyinfopoly)
End Behavior of a Polynomial Function (http://openstax.org/l/endbehavior)
Turning Points and x- intercepts of Polynomial Functions (http://openstax.org/l/turningpoints)
Least Possible Degree of a Polynomial Function (http://openstax.org/l/leastposdegree)

5.2 SECTION EXERCISES

Verbal

1. Explain the difference between the coefficient of a power function and its degree.

2. If a polynomial function is in factored form, what would be a good first step in order to determine the degree of the function?

3. In general, explain the end behavior of a power function with odd degree if the leading coefficient is positive.

4. What is the relationship between the degree of a polynomial function and the maximum number of turning points in its graph?

5. What can we conclude if, in general, the graph of a polynomial function exhibits the following end behavior? As $x \to -\infty,\ f(x) \to -\infty$ and as $x \to \infty,\ f(x) \to -\infty$.

Algebraic

For the following exercises, identify the function as a power function, a polynomial function, or neither.

6. $f(x) = x^5$

7. $f(x) = \left(x^2\right)^3$

8. $f(x) = x - x^4$

9. $f(x) = \frac{x^2}{x^2-1}$

10. $f(x) = 2x\,(x+2)\,(x-1)^2$

11. $f(x) = 3^{x+1}$

For the following exercises, find the degree and leading coefficient for the given polynomial.

12. $-3x^4$

13. $7 - 2x^2$

14. $-2x^2 - 3x^5 + x - 6$

15. $x\left(4 - x^2\right)(2x+1)$

16. $x^2(2x-3)^2$

For the following exercises, determine the end behavior of the functions.

17. $f\,(x) = x^4$

18. $f\,(x) = x^3$

19. $f\,(x) = -x^4$

20. $f\,(x) = -x^9$

21. $f(x) = -2x^4 - 3x^2 + x - 1$

22. $f(x) = 3x^2 + x - 2$

23. $f(x) = x^2\left(2x^3 - x + 1\right)$

24. $f(x) = (2-x)^7$

For the following exercises, find the intercepts of the functions.

25. $f\,(t) = 2\,(t-1)\,(t+2)\,(t-3)$

26. $g\,(n) = -2\,(3n-1)\,(2n+1)$

27. $f(x) = x^4 - 16$

28. $f(x) = x^3 + 27$

29. $f(x) = x\left(x^2 - 2x - 8\right)$

30. $f(x) = (x+3)\left(4x^2 - 1\right)$

Graphical

For the following exercises, determine the least possible degree of the polynomial function shown.

31.

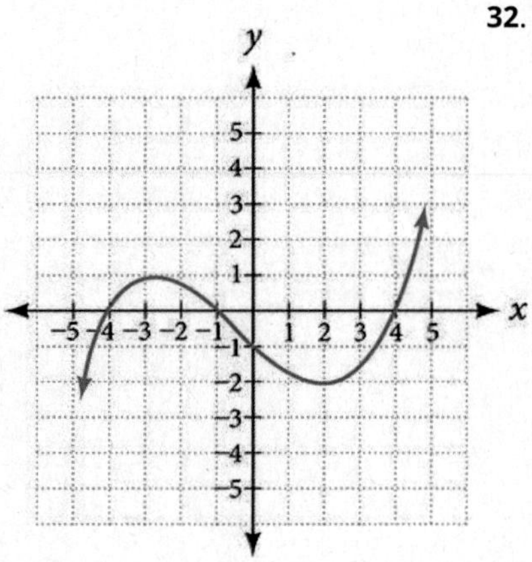

32.

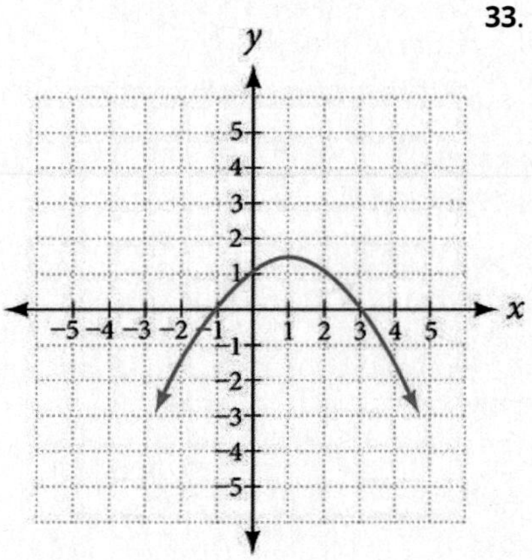

33.

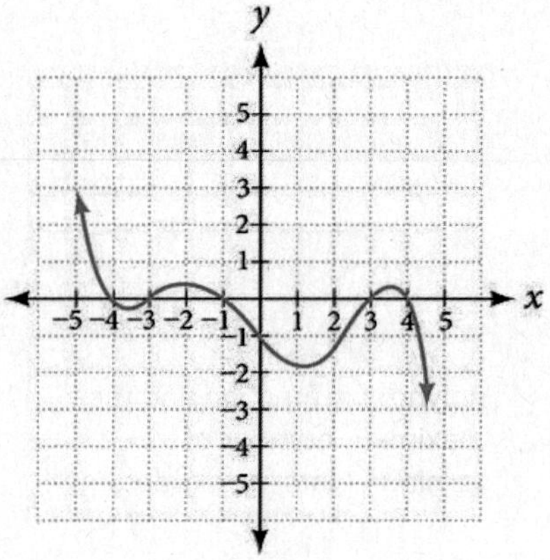

34.

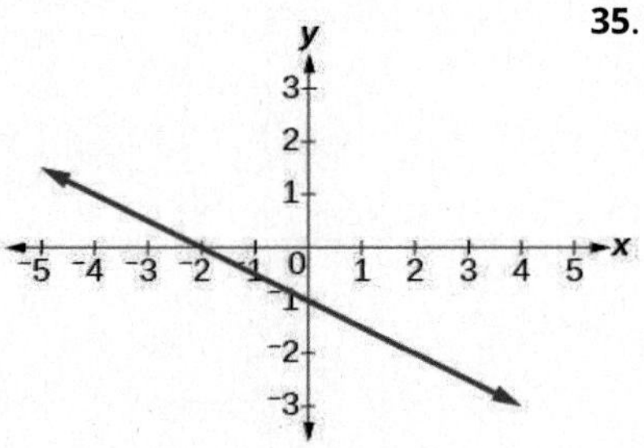

35.

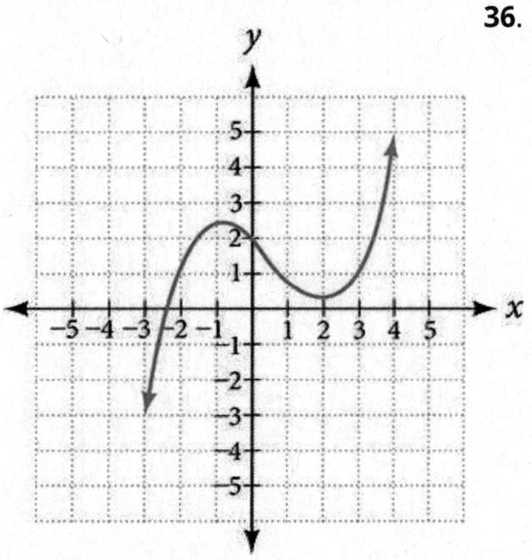

36.

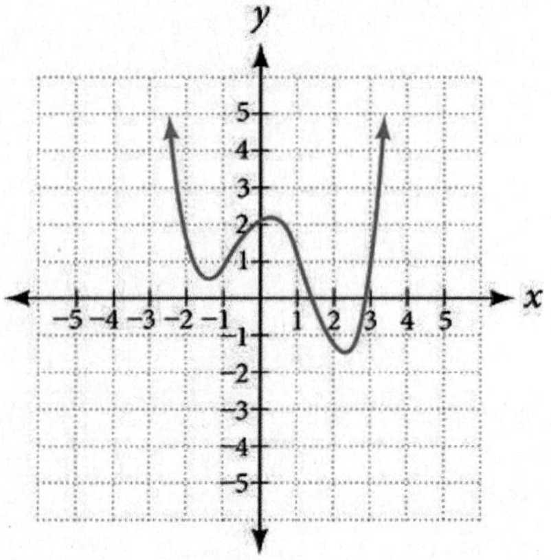

37.

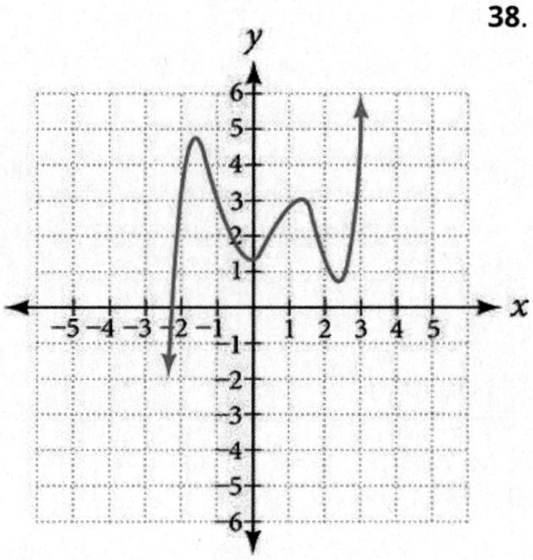

38.

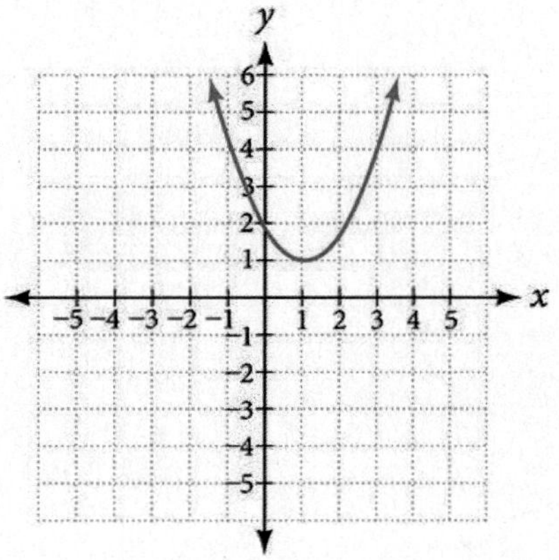

For the following exercises, determine whether the graph of the function provided is a graph of a polynomial function. If so, determine the number of turning points and the least possible degree for the function.

39.

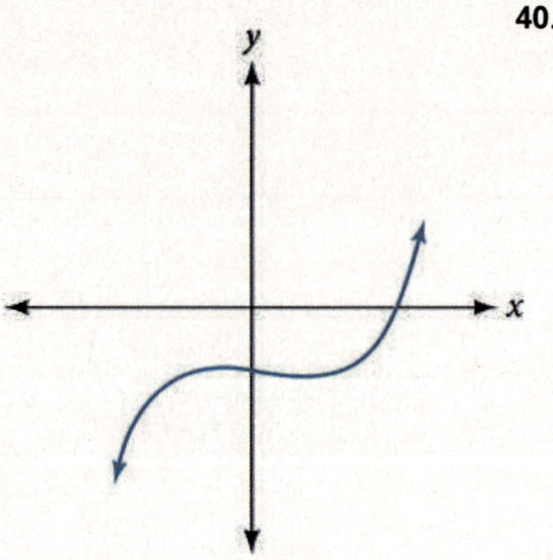

40.

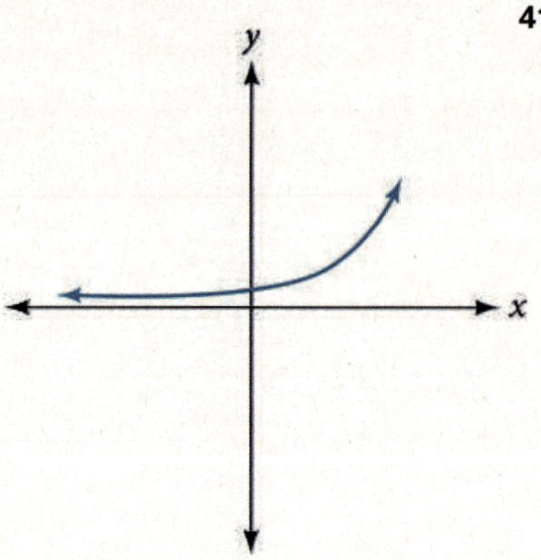

41.

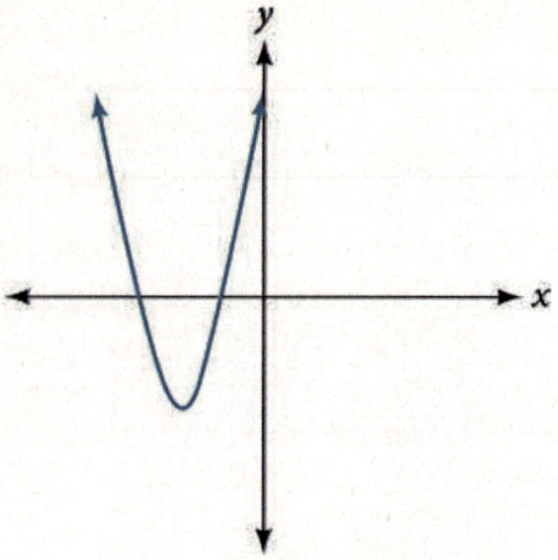

42.

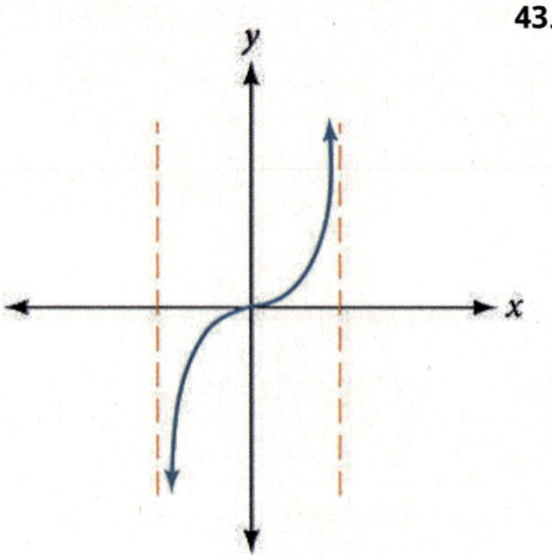

43.

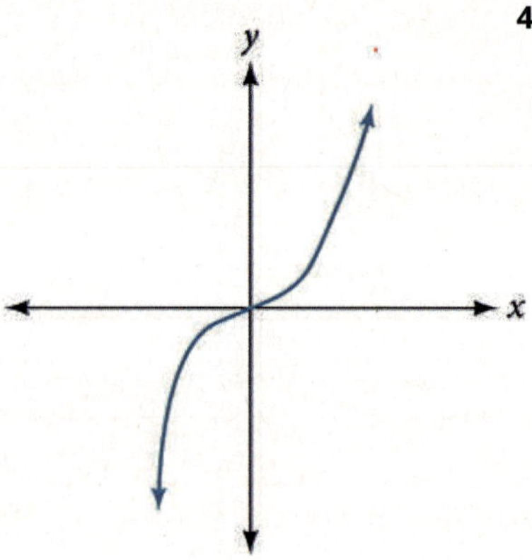

44.

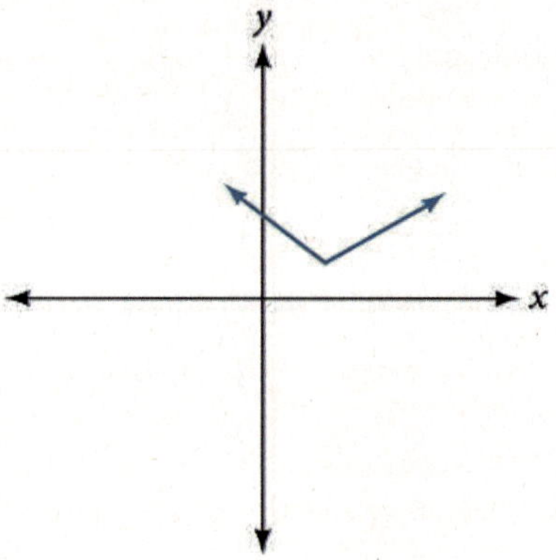

45.

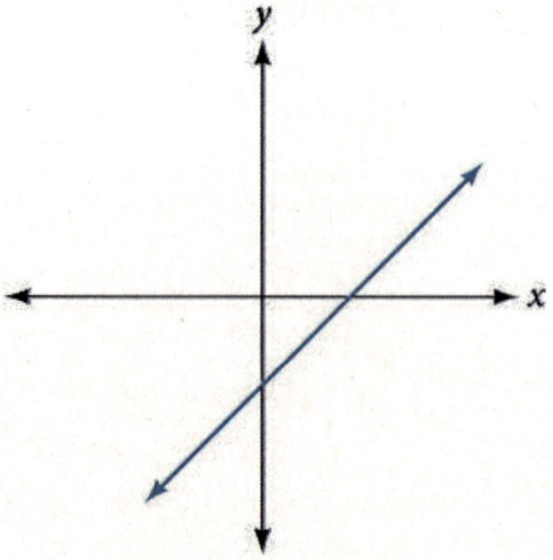

Numeric

For the following exercises, make a table to confirm the end behavior of the function.

46. $f(x) = -x^3$

47. $f(x) = x^4 - 5x^2$

48. $f(x) = x^2(1 - x)^2$

49. $f(x) = (x - 1)(x - 2)(3 - x)$

50. $f(x) = \frac{x^5}{10} - x^4$

Technology

For the following exercises, graph the polynomial functions using a calculator. Based on the graph, determine the intercepts and the end behavior.

51. $f(x) = x^3(x - 2)$

52. $f(x) = x(x - 3)(x + 3)$

53. $f(x) = x(14 - 2x)(10 - 2x)$

54. $f(x) = x(14 - 2x)(10 - 2x)^2$

55. $f(x) = x^3 - 16x$

56. $f(x) = x^3 - 27$

57. $f(x) = x^4 - 81$

58. $f(x) = -x^3 + x^2 + 2x$

59. $f(x) = x^3 - 2x^2 - 15x$

60. $f(x) = x^3 - 0.01x$

Extensions

For the following exercises, use the information about the graph of a polynomial function to determine the function. Assume the leading coefficient is 1 or –1. There may be more than one correct answer.

61. The y- intercept is $(0, -4)$. The x- intercepts are $(-2, 0)$, $(2, 0)$. Degree is 2.

End behavior: as $x \to -\infty$, $f(x) \to \infty$; as $x \to \infty$, $f(x) \to \infty$.

62. The y- intercept is $(0, 9)$. The x- intercepts are $(-3, 0)$, $(3, 0)$. Degree is 2.

End behavior: as $x \to -\infty$, $f(x) \to -\infty$, as $x \to \infty$, $f(x) \to -\infty$.

63. The y- intercept is $(0, 0)$. The x- intercepts are $(0, 0)$, $(2, 0)$. Degree is 3.

End behavior: as $x \to -\infty$, $f(x) \to -\infty$, as $x \to \infty$, $f(x) \to \infty$.

64. The y- intercept is $(0, 1)$. The x- intercept is $(1, 0)$. Degree is 3.

End behavior: as $x \to -\infty$, $f(x) \to \infty$, as $x \to \infty$, $f(x) \to -\infty$.

65. The y- intercept is $(0, 1)$. There is no x- intercept. Degree is 4.

End behavior: as $x \to -\infty$, $f(x) \to \infty$, as $x \to \infty$, $f(x) \to \infty$.

Real-World Applications

For the following exercises, use the written statements to construct a polynomial function that represents the required information.

66. An oil slick is expanding as a circle. The radius of the circle is increasing at the rate of 20 meters per day. Express the area of the circle as a function of d, the number of days elapsed.

67. A cube has an edge of 3 feet. The edge is increasing at the rate of 2 feet per minute. Express the volume of the cube as a function of m, the number of minutes elapsed.

68. A rectangle has a length of 10 inches and a width of 6 inches. If the length is increased by x inches and the width increased by twice that amount, express the area of the rectangle as a function of x.

69. An open box is to be constructed by cutting out square corners of x- inch sides from a piece of cardboard 8 inches by 8 inches and then folding up the sides. Express the volume of the box as a function of x.

70. A rectangle is twice as long as it is wide. Squares of side 2 feet are cut out from each corner. Then the sides are folded up to make an open box. Express the volume of the box as a function of the width (x).

5.3 Graphs of Polynomial Functions

Learning Objectives

In this section, you will:

- Recognize characteristics of graphs of polynomial functions.
- Use factoring to find zeros of polynomial functions.
- Identify zeros and their multiplicities.
- Determine end behavior.
- Understand the relationship between degree and turning points.
- Graph polynomial functions.
- Use the Intermediate Value Theorem.

The revenue in millions of dollars for a fictional cable company from 2006 through 2013 is shown in Table 1.

Year	2006	2007	2008	2009	2010	2011	2012	2013
Revenues	52.4	52.8	51.2	49.5	48.6	48.6	48.7	47.1

Table 1

The revenue can be modeled by the polynomial function

$$R(t) = -0.037t^4 + 1.414t^3 - 19.777t^2 + 118.696t - 205.332$$

where R represents the revenue in millions of dollars and t represents the year, with $t = 6$ corresponding to 2006. Over which intervals is the revenue for the company increasing? Over which intervals is the revenue for the company decreasing? These questions, along with many others, can be answered by examining the graph of the polynomial function. We have already explored the local behavior of quadratics, a special case of polynomials. In this section we will explore the local behavior of polynomials in general.

Recognizing Characteristics of Graphs of Polynomial Functions

Polynomial functions of degree 2 or more have graphs that do not have sharp corners; recall that these types of graphs are called smooth curves. Polynomial functions also display graphs that have no breaks. Curves with no breaks are called continuous. Figure 1 shows a graph that represents a polynomial function and a graph that represents a function that is not a polynomial.

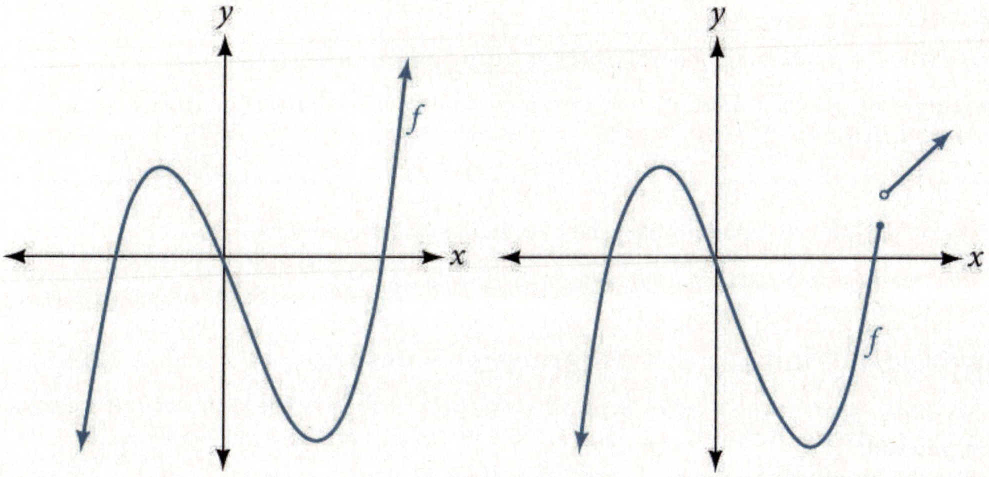

Figure 1

EXAMPLE 1

Recognizing Polynomial Functions

Which of the graphs in Figure 2 represents a polynomial function?

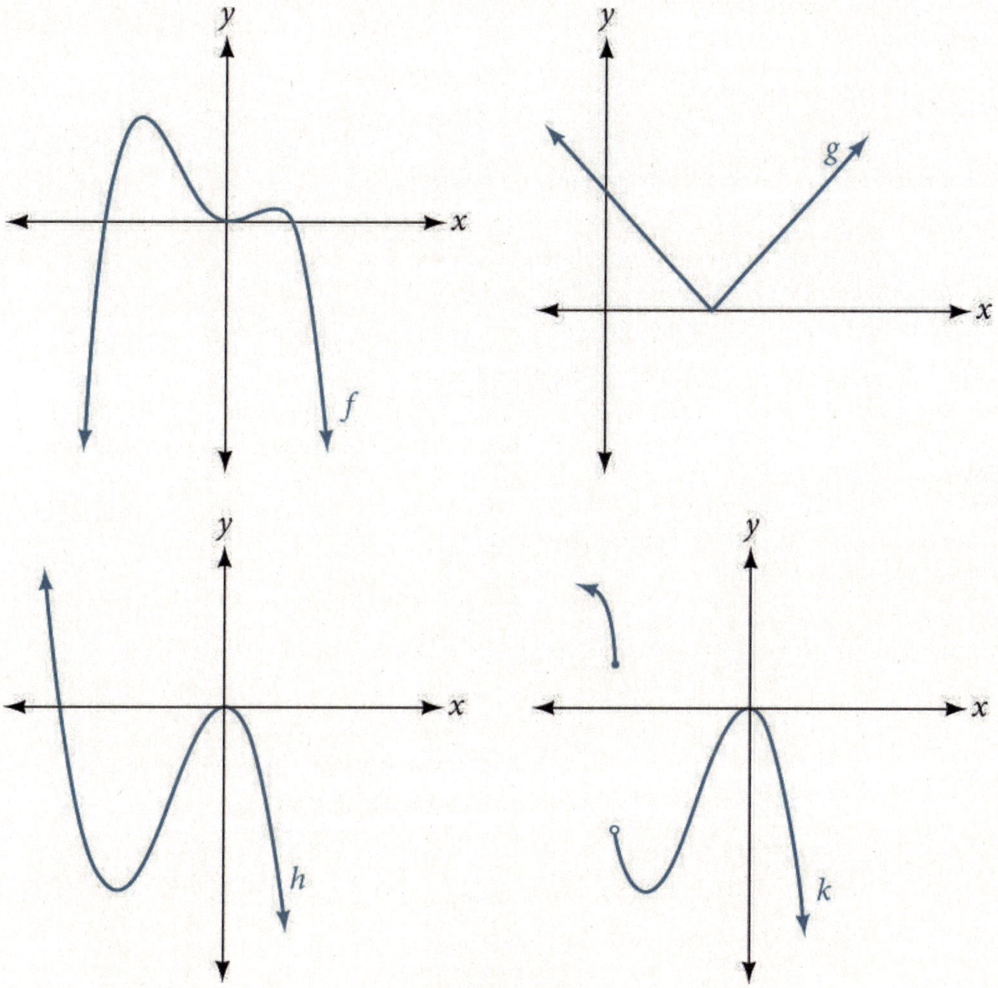

Figure 2

Solution

The graphs of f and h are graphs of polynomial functions. They are smooth and continuous.

The graphs of g and k are graphs of functions that are not polynomials. The graph of function g has a sharp corner. The graph of function k is not continuous.

Q&A **Do all polynomial functions have as their domain all real numbers?**

Yes. Any real number is a valid input for a polynomial function.

Using Factoring to Find Zeros of Polynomial Functions

Recall that if f is a polynomial function, the values of x for which $f(x) = 0$ are called zeros of f. If the equation of the polynomial function can be factored, we can set each factor equal to zero and solve for the zeros.

We can use this method to find x- intercepts because at the x- intercepts we find the input values when the output value is zero. For general polynomials, this can be a challenging prospect. While quadratics can be solved using the relatively simple quadratic formula, the corresponding formulas for cubic and fourth-degree polynomials are not simple enough to remember, and formulas do not exist for general higher-degree polynomials. Consequently, we will limit ourselves to three cases:

1. The polynomial can be factored using known methods: greatest common factor and trinomial factoring.
2. The polynomial is given in factored form.
3. Technology is used to determine the intercepts.

HOW TO

Given a polynomial function f, find the *x*-intercepts by factoring.

1. Set $f(x) = 0$.
2. If the polynomial function is not given in factored form:
   a. Factor out any common monomial factors.
   b. Factor any factorable binomials or trinomials.
3. Set each factor equal to zero and solve to find the x- intercepts.

EXAMPLE 2

Finding the *x*-Intercepts of a Polynomial Function by Factoring

Find the *x*-intercepts of $f(x) = x^6 - 3x^4 + 2x^2$.

Solution

We can attempt to factor this polynomial to find solutions for $f(x) = 0$.

$$\begin{aligned} x^6 - 3x^4 + 2x^2 &= 0 && \text{Factor out the greatest common factor.} \\ x^2\left(x^4 - 3x^2 + 2\right) &= 0 && \text{Factor the trinomial.} \\ x^2\left(x^2 - 1\right)\left(x^2 - 2\right) &= 0 && \text{Set each factor equal to zero.} \end{aligned}$$

$$\begin{array}{rclcrclcrcl} & & & & \left(x^2 - 1\right) & = & 0 & & \left(x^2 - 2\right) & = & 0 \\ x^2 & = & 0 & \text{or} & x^2 & = & 1 & \text{or} & x^2 & = & 2 \\ x & = & 0 & & x & = & \pm 1 & & x & = & \pm\sqrt{2} \end{array}$$

This gives us five *x*-intercepts: $(0,0), (1,0), (-1,0), (\sqrt{2},0)$, and $(-\sqrt{2},0)$. See Figure 3. We can see that this is an even function because it is symmetric about the *y*-axis.

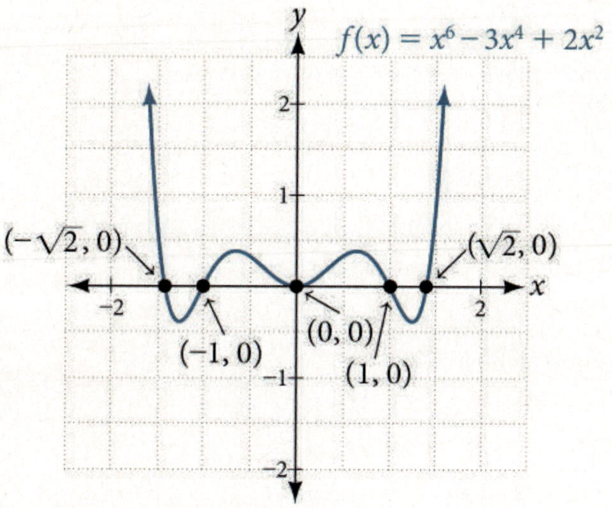

Figure 3

EXAMPLE 3

Finding the x-Intercepts of a Polynomial Function by Factoring

Find the x-intercepts of $f(x) = x^3 - 5x^2 - x + 5$.

Solution

Find solutions for $f(x) = 0$ by factoring.

$$\begin{aligned} x^3 - 5x^2 - x + 5 &= 0 && \text{Factor by grouping.} \\ x^2(x-5) - (x-5) &= 0 && \text{Factor out the common factor.} \\ \left(x^2 - 1\right)(x-5) &= 0 && \text{Factor the difference of squares.} \\ (x+1)(x-1)(x-5) &= 0 && \text{Set each factor equal to zero.} \end{aligned}$$

$$\begin{aligned} x + 1 &= 0 & \text{or} \quad x - 1 &= 0 & \text{or} \quad x - 5 &= 0 \\ x &= -1 & x &= 1 & x &= 5 \end{aligned}$$

There are three x-intercepts: $(-1, 0)$, $(1, 0)$, and $(5, 0)$. See Figure 4.

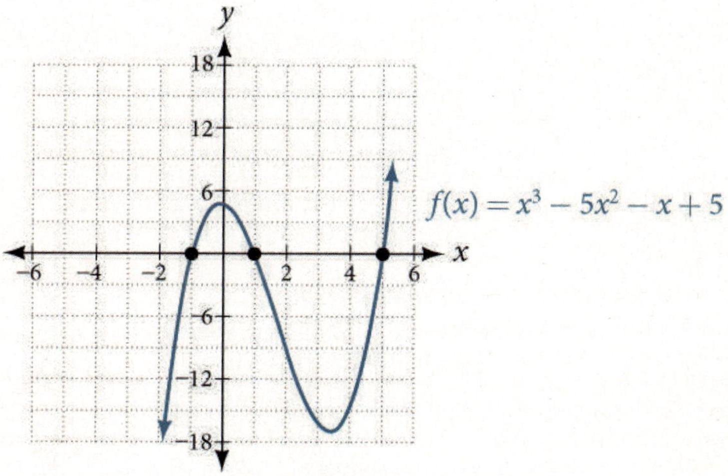

Figure 4

EXAMPLE 4

Finding the y- and x-Intercepts of a Polynomial in Factored Form

Find the y- and x-intercepts of $g(x) = (x-2)^2(2x+3)$.

Solution

The y-intercept can be found by evaluating $g(0)$.

$$\begin{aligned} g(0) &= (0-2)^2(2(0)+3) \\ &= 12 \end{aligned}$$

So the y-intercept is $(0, 12)$.

The x-intercepts can be found by solving $g(x) = 0$.

$$(x-2)^2(2x+3) = 0$$

$$\begin{aligned} (x-2)^2 &= 0 \\ x-2 &= 0 \\ x &= 2 \end{aligned} \qquad \text{or} \qquad \begin{aligned} (2x+3) &= 0 \\ x &= -\tfrac{3}{2} \end{aligned}$$

So the x-intercepts are $(2, 0)$ and $\left(-\frac{3}{2}, 0\right)$.

Analysis

We can always check that our answers are reasonable by using a graphing calculator to graph the polynomial as shown in Figure 5.

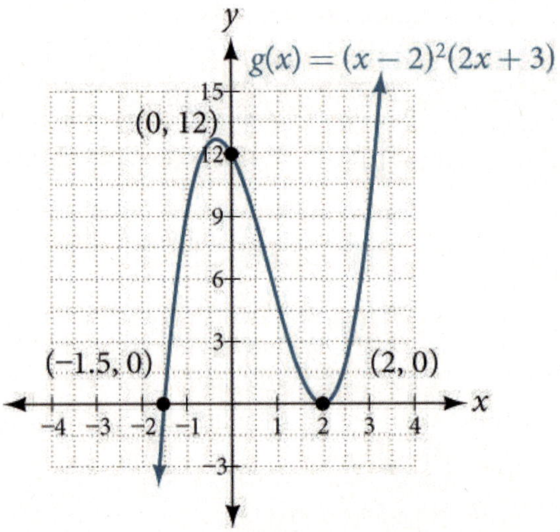

Figure 5

EXAMPLE 5

Finding the x-Intercepts of a Polynomial Function Using a Graph

Find the x-intercepts of $h(x) = x^3 + 4x^2 + x - 6$.

Solution

This polynomial is not in factored form, has no common factors, and does not appear to be factorable using techniques previously discussed. Fortunately, we can use technology to find the intercepts. Keep in mind that some values make graphing difficult by hand. In these cases, we can take advantage of graphing utilities.

Looking at the graph of this function, as shown in Figure 6, it appears that there are x-intercepts at $x = -3, -2,$ and 1.

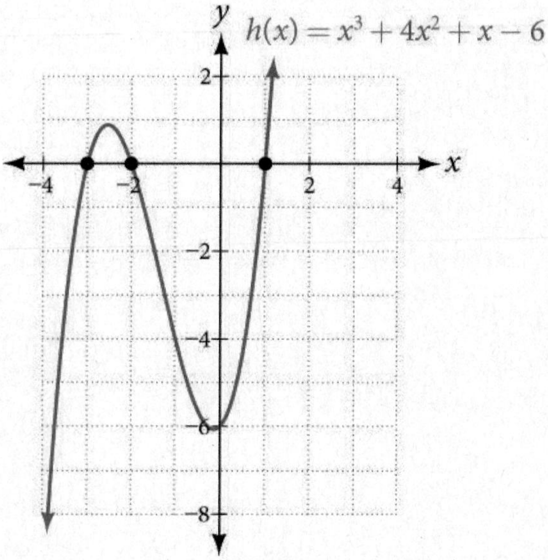

Figure 6

We can check whether these are correct by substituting these values for x and verifying that

$$h(-3) = h(-2) = h(1) = 0$$

Since $h(x) = x^3 + 4x^2 + x - 6$, we have:

$$\begin{aligned} h(-3) &= (-3)^3 + 4(-3)^2 + (-3) - 6 = -27 + 36 - 3 - 6 = 0 \\ h(-2) &= (-2)^3 + 4(-2)^2 + (-2) - 6 = -8 + 16 - 2 - 6 = 0 \\ h(1) &= (1)^3 + 4(1)^2 + (1) - 6 = 1 + 4 + 1 - 6 = 0 \end{aligned}$$

Each x-intercept corresponds to a zero of the polynomial function and each zero yields a factor, so we can now write the polynomial in factored form.

$$\begin{aligned} h(x) &= x^3 + 4x^2 + x - 6 \\ &= (x + 3)(x + 2)(x - 1) \end{aligned}$$

> **TRY IT** #1 Find the y- and x-intercepts of the function $f(x) = x^4 - 19x^2 + 30x$.

Identifying Zeros and Their Multiplicities

Graphs behave differently at various x-intercepts. Sometimes, the graph will cross over the horizontal axis at an intercept. Other times, the graph will touch the horizontal axis and "bounce" off.

Suppose, for example, we graph the function shown.

$$f(x) = (x + 3)(x - 2)^2(x + 1)^3$$

Notice in Figure 7 that the behavior of the function at each of the x-intercepts is different.

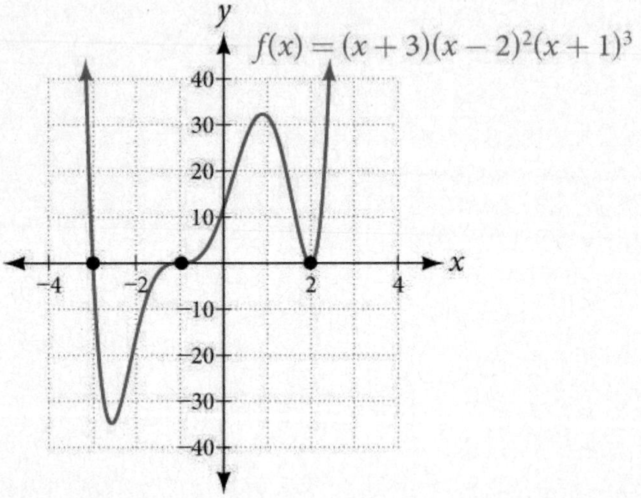

Figure 7 Identifying the behavior of the graph at an x-intercept by examining the multiplicity of the zero.

The x-intercept $x = -3$ is the solution of equation $(x + 3) = 0$. The graph passes directly through the x-intercept at $x = -3$. The factor is linear (has a degree of 1), so the behavior near the intercept is like that of a line—it passes directly through the intercept. We call this a single zero because the zero corresponds to a single factor of the function.

The x-intercept $x = 2$ is the repeated solution of equation $(x - 2)^2 = 0$. The graph touches the axis at the intercept and changes direction. The factor is quadratic (degree 2), so the behavior near the intercept is like that of a quadratic—it bounces off of the horizontal axis at the intercept.

$$(x - 2)^2 = (x - 2)(x - 2)$$

The factor is repeated, that is, the factor $(x - 2)$ appears twice. The number of times a given factor appears in the factored form of the equation of a polynomial is called the **multiplicity**. The zero associated with this factor, $x = 2$, has multiplicity 2 because the factor $(x - 2)$ occurs twice.

The x-intercept $x = -1$ is the repeated solution of factor $(x + 1)^3 = 0$. The graph passes through the axis at the intercept, but flattens out a bit first. This factor is cubic (degree 3), so the behavior near the intercept is like that of a cubic—with the same S-shape near the intercept as the toolkit function $f(x) = x^3$. We call this a triple zero, or a zero with multiplicity 3.

For zeros with even multiplicities, the graphs *touch* or are tangent to the x-axis. For zeros with odd multiplicities, the graphs *cross* or intersect the x-axis. See Figure 8 for examples of graphs of polynomial functions with multiplicity 1, 2, and 3.

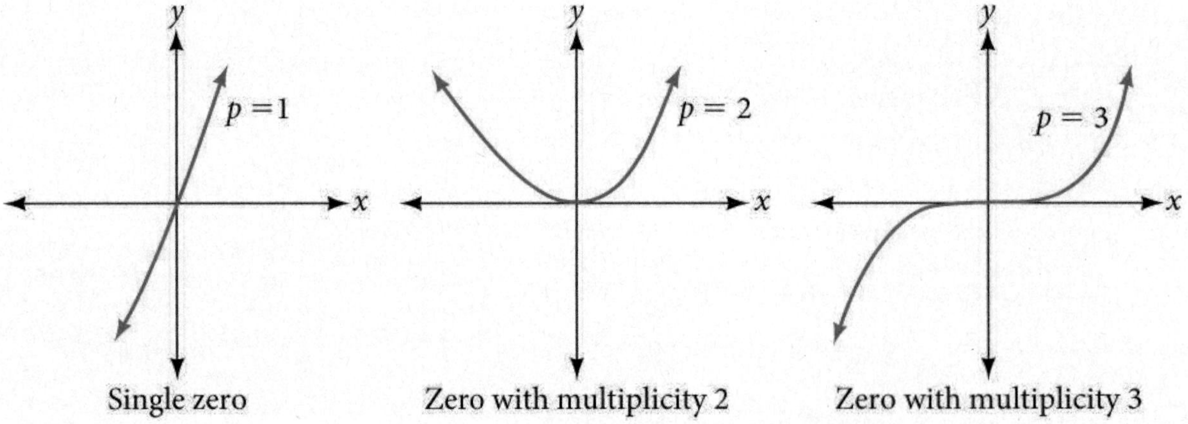

Figure 8

For higher even powers, such as 4, 6, and 8, the graph will still touch and bounce off of the horizontal axis but, for each increasing even power, the graph will appear flatter as it approaches and leaves the x-axis.

For higher odd powers, such as 5, 7, and 9, the graph will still cross through the horizontal axis, but for each increasing odd power, the graph will appear flatter as it approaches and leaves the x-axis.

Graphical Behavior of Polynomials at x-Intercepts

If a polynomial contains a factor of the form $(x-h)^p$, the behavior near the x- intercept h is determined by the power p. We say that $x=h$ is a zero of **multiplicity** p.

The graph of a polynomial function will touch the x-axis at zeros with even multiplicities. The graph will cross the x-axis at zeros with odd multiplicities.

The sum of the multiplicities is the degree of the polynomial function.

HOW TO

Given a graph of a polynomial function of degree n, identify the zeros and their multiplicities.

1. If the graph crosses the x-axis and appears almost linear at the intercept, it is a single zero.
2. If the graph touches the x-axis and bounces off of the axis, it is a zero with even multiplicity.
3. If the graph crosses the x-axis at a zero, it is a zero with odd multiplicity.
4. The sum of the multiplicities is n.

EXAMPLE 6

Identifying Zeros and Their Multiplicities

Use the graph of the function of degree 6 in Figure 9 to identify the zeros of the function and their possible multiplicities.

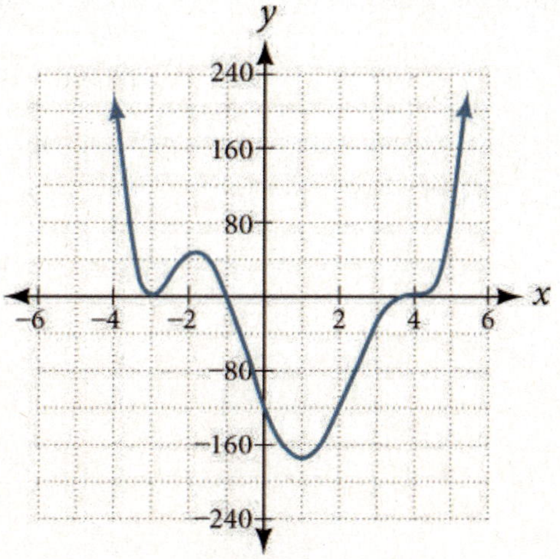

Figure 9

Solution

The polynomial function is of degree 6. The sum of the multiplicities must be 6.

Starting from the left, the first zero occurs at $x=-3$. The graph touches the x-axis, so the multiplicity of the zero must be even. The zero of -3 most likely has multiplicity 2.

The next zero occurs at $x=-1$. The graph looks almost linear at this point. This is a single zero of multiplicity 1.

The last zero occurs at $x=4$. The graph crosses the x-axis, so the multiplicity of the zero must be odd. We know that the multiplicity is likely 3 and that the sum of the multiplicities is 6.

TRY IT #2 Use the graph of the function of degree 9 in Figure 10 to identify the zeros of the function and their multiplicities.

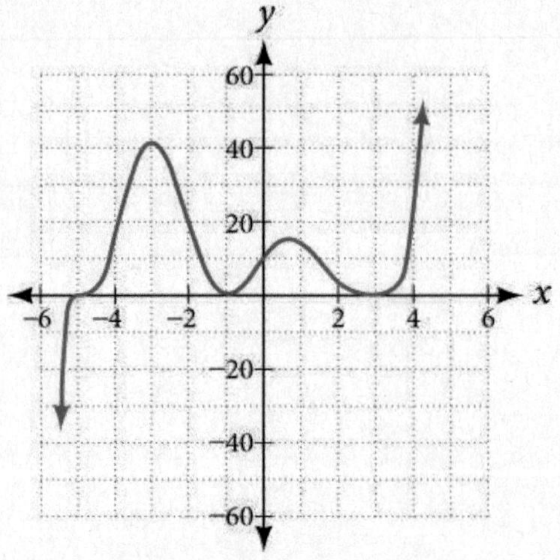

Figure 10

Determining End Behavior

As we have already learned, the behavior of a graph of a polynomial function of the form

$$f(x) = a_n x^n + a_{n-1} x^{n-1} + ... + a_1 x + a_0$$

will either ultimately rise or fall as x increases without bound and will either rise or fall as x decreases without bound. This is because for very large inputs, say 100 or 1,000, the leading term dominates the size of the output. The same is true for very small inputs, say –100 or –1,000.

Recall that we call this behavior the *end behavior* of a function. As we pointed out when discussing quadratic equations, when the leading term of a polynomial function, $a_n x^n$, is an even power function, as x increases or decreases without bound, $f(x)$ increases without bound. When the leading term is an odd power function, as x decreases without bound, $f(x)$ also decreases without bound; as x increases without bound, $f(x)$ also increases without bound. If the leading term is negative, it will change the direction of the end behavior. Figure 11 summarizes all four cases.

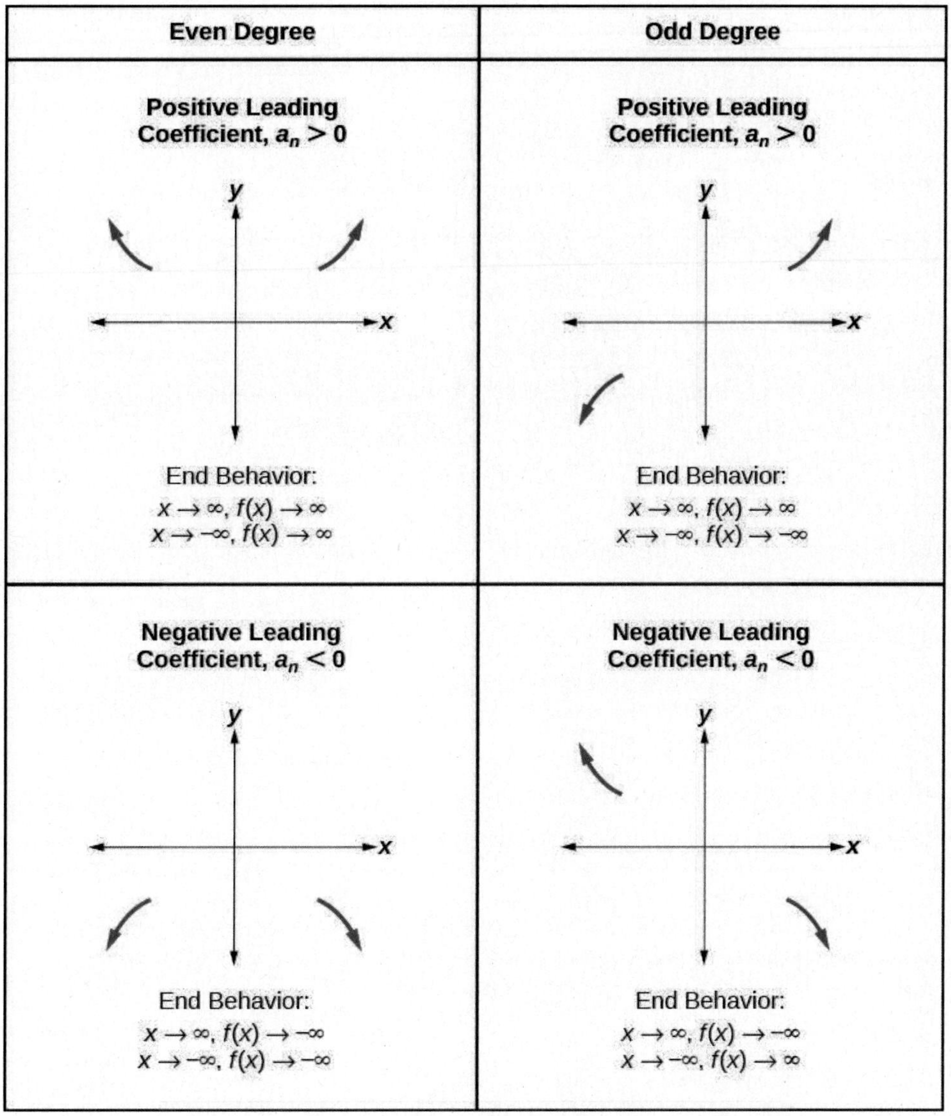

Figure 11

Understanding the Relationship between Degree and Turning Points

In addition to the end behavior, recall that we can analyze a polynomial function's local behavior. It may have a turning point where the graph changes from increasing to decreasing (rising to falling) or decreasing to increasing (falling to rising). Look at the graph of the polynomial function $f(x) = x^4 - x^3 - 4x^2 + 4x$ in Figure 12. The graph has three turning points.

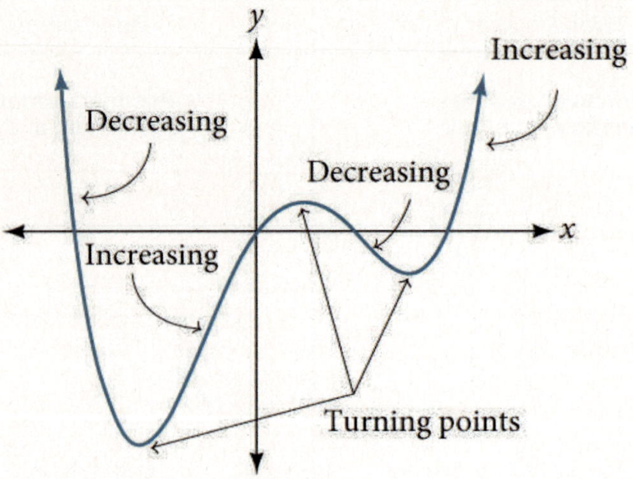

Figure 12

This function f is a 4th degree polynomial function and has 3 turning points. The maximum number of turning points of a polynomial function is always one less than the degree of the function.

Interpreting Turning Points

A turning point is a point of the graph where the graph changes from increasing to decreasing (rising to falling) or decreasing to increasing (falling to rising).

A polynomial of degree n will have at most $n - 1$ turning points.

EXAMPLE 7

Finding the Maximum Number of Turning Points Using the Degree of a Polynomial Function
Find the maximum number of turning points of each polynomial function.

ⓐ $f(x) = -x^3 + 4x^5 - 3x^2 + 1$ ⓑ $f(x) = -(x - 1)^2 \left(1 + 2x^2\right)$

Solution

ⓐ

First, rewrite the polynomial function in descending order: $f(x) = 4x^5 - x^3 - 3x^2 + 1$

Identify the degree of the polynomial function. This polynomial function is of degree 5.

The maximum number of turning points is $5 - 1 = 4$.

ⓑ

First, identify the leading term of the polynomial function if the function were expanded.

$$f(x) = -(x - 1)^2\ (1 + 2x^2)$$
$$a_n = -(x^2)\ (2x^2)\ {-2x^4}$$

Then, identify the degree of the polynomial function. This polynomial function is of degree 4.

The maximum number of turning points is $4 - 1 = 3$.

Graphing Polynomial Functions

We can use what we have learned about multiplicities, end behavior, and turning points to sketch graphs of polynomial functions. Let us put this all together and look at the steps required to graph polynomial functions.

HOW TO

Given a polynomial function, sketch the graph.

1. Find the intercepts.
2. Check for symmetry. If the function is an even function, its graph is symmetrical about the y- axis, that is, $f(-x) = f(x)$. If a function is an odd function, its graph is symmetrical about the origin, that is, $f(-x) = -f(x)$.
3. Use the multiplicities of the zeros to determine the behavior of the polynomial at the x- intercepts.
4. Determine the end behavior by examining the leading term.
5. Use the end behavior and the behavior at the intercepts to sketch a graph.
6. Ensure that the number of turning points does not exceed one less than the degree of the polynomial.
7. Optionally, use technology to check the graph.

EXAMPLE 8

Sketching the Graph of a Polynomial Function

Sketch a graph of $f(x) = -2(x+3)^2(x-5)$.

Solution

This graph has two x-intercepts. At $x = -3$, the factor is squared, indicating a multiplicity of 2. The graph will bounce at this x-intercept. At $x = 5$, the function has a multiplicity of one, indicating the graph will cross through the axis at this intercept.

The y-intercept is found by evaluating $f(0)$.

$$\begin{aligned} f(0) &= -2(0+3)^2(0-5) \\ &= -2 \cdot 9 \cdot (-5) \\ &= 90 \end{aligned}$$

The y-intercept is $(0, 90)$.

Additionally, we can see the leading term, if this polynomial were multiplied out, would be $-2x^3$, so the end behavior is that of a vertically reflected cubic, with the outputs decreasing as the inputs approach infinity, and the outputs increasing as the inputs approach negative infinity. See Figure 13.

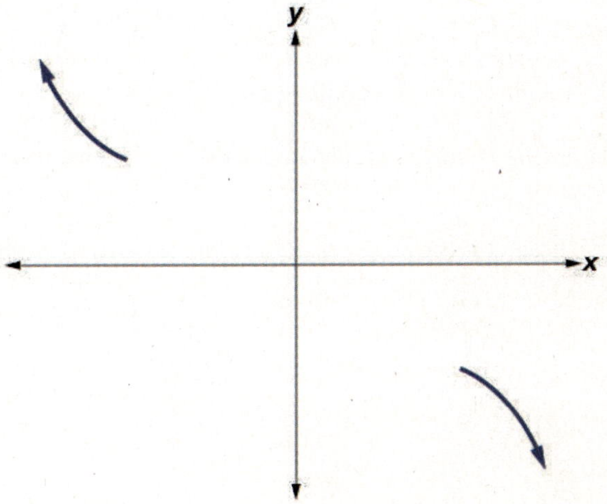

Figure 13

To sketch this, we consider that:

- As $x \to -\infty$ the function $f(x) \to \infty$, so we know the graph starts in the second quadrant and is decreasing toward the x- axis.

- Since $f(-x) = -2(-x+3)^2(-x-5)$ is not equal to $f(x)$, the graph does not display symmetry.
- At $(-3, 0)$, the graph bounces off of the *x*-axis, so the function must start increasing.
  At $(0, 90)$, the graph crosses the *y*-axis at the *y*-intercept. See Figure 14.

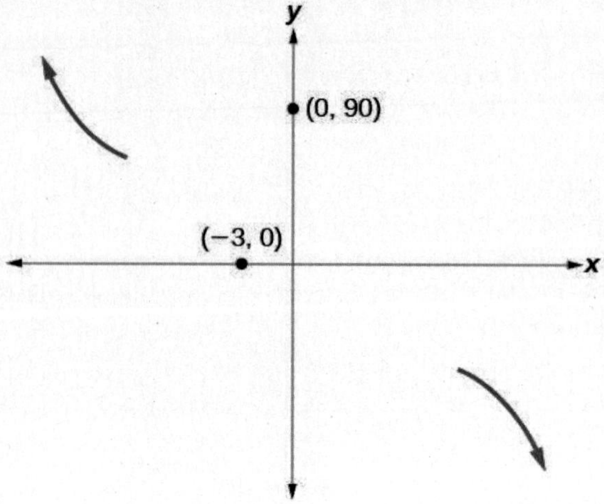

Figure 14

Somewhere after this point, the graph must turn back down or start decreasing toward the horizontal axis because the graph passes through the next intercept at $(5, 0)$. See Figure 15.

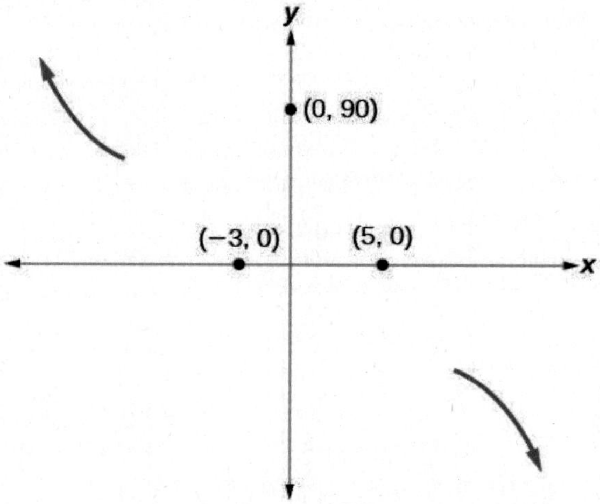

Figure 15

As $x \to \infty$ the function $f(x) \to -\infty$, so we know the graph continues to decrease, and we can stop drawing the graph in the fourth quadrant.

Using technology, we can create the graph for the polynomial function, shown in Figure 16, and verify that the resulting graph looks like our sketch in Figure 15.

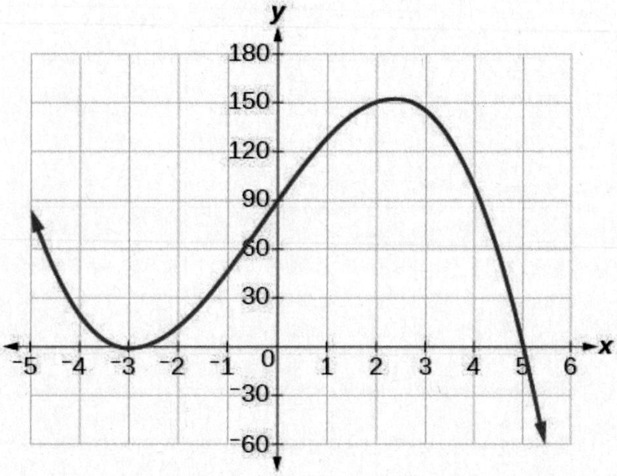

Figure 16 The complete graph of the polynomial function $f(x) = -2(x+3)^2(x-5)$

> **TRY IT** #3 Sketch a graph of $f(x) = \frac{1}{4}x(x-1)^4(x+3)^3$.

Using the Intermediate Value Theorem

In some situations, we may know two points on a graph but not the zeros. If those two points are on opposite sides of the *x*-axis, we can confirm that there is a zero between them. Consider a polynomial function f whose graph is smooth and continuous. The **Intermediate Value Theorem** states that for two numbers a and b in the domain of f, if $a < b$ and $f(a) \neq f(b)$, then the function f takes on every value between $f(a)$ and $f(b)$. (While the theorem is intuitive, the proof is actually quite complicated and requires higher mathematics.) We can apply this theorem to a special case that is useful in graphing polynomial functions. If a point on the graph of a continuous function f at $x = a$ lies above the x- axis and another point at $x = b$ lies below the x- axis, there must exist a third point between $x = a$ and $x = b$ where the graph crosses the x- axis. Call this point $(c,\ f(c))$. This means that we are assured there is a solution c where $f(c) = 0$.

In other words, the Intermediate Value Theorem tells us that when a polynomial function changes from a negative value to a positive value, the function must cross the x- axis. Figure 17 shows that there is a zero between a and b.

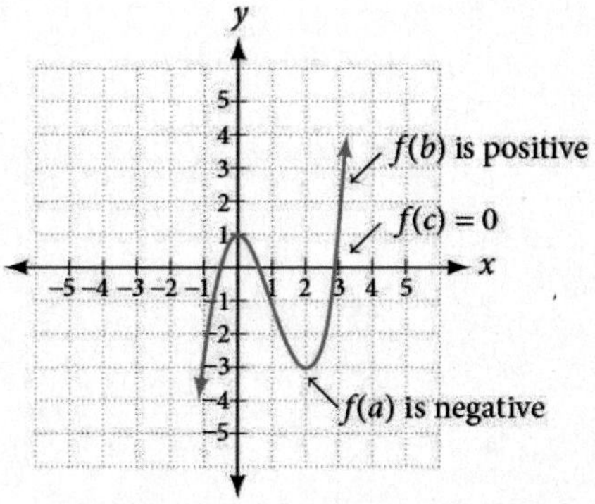

Figure 17 Using the Intermediate Value Theorem to show there exists a zero.

Intermediate Value Theorem

Let f be a polynomial function. The **Intermediate Value Theorem** states that if $f(a)$ and $f(b)$ have opposite signs, then there exists at least one value c between a and b for which $f(c) = 0$.

EXAMPLE 9

Using the Intermediate Value Theorem

Show that the function $f(x) = x^3 - 5x^2 + 3x + 6$ has at least two real zeros between $x = 1$ and $x = 4$.

Solution

As a start, evaluate $f(x)$ at the integer values $x = 1, 2, 3,$ and 4. See Table 2.

x	1	2	3	4
$f(x)$	5	0	–3	2

Table 2

We see that one zero occurs at $x = 2$. Also, since $f(3)$ is negative and $f(4)$ is positive, by the Intermediate Value Theorem, there must be at least one real zero between 3 and 4.

We have shown that there are at least two real zeros between $x = 1$ and $x = 4$.

Analysis

We can also see on the graph of the function in Figure 18 that there are two real zeros between $x = 1$ and $x = 4$.

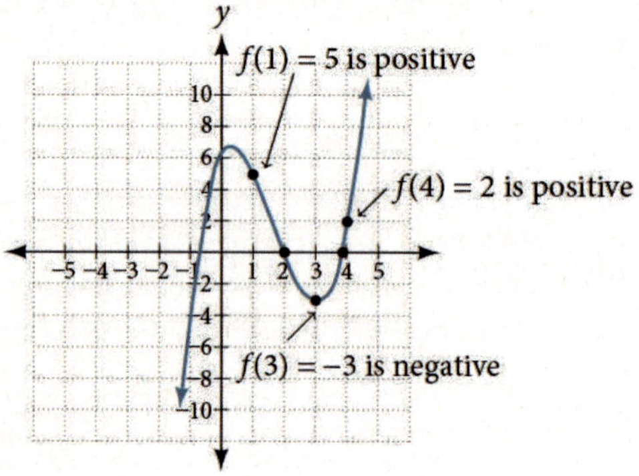

Figure 18

TRY IT #4 Show that the function $f(x) = 7x^5 - 9x^4 - x^2$ has at least one real zero between $x = 1$ and $x = 2$.

Writing Formulas for Polynomial Functions

Now that we know how to find zeros of polynomial functions, we can use them to write formulas based on graphs. Because a polynomial function written in factored form will have an *x*-intercept where each factor is equal to zero, we can form a function that will pass through a set of *x*-intercepts by introducing a corresponding set of factors.

Factored Form of Polynomials

If a polynomial of lowest degree p has horizontal intercepts at $x = x_1, x_2, \ldots, x_n$, then the polynomial can be written in the factored form: $f(x) = a(x - x_1)^{p_1}(x - x_2)^{p_2} \cdots (x - x_n)^{p_n}$ where the powers p_i on each factor can be determined by the behavior of the graph at the corresponding intercept, and the stretch factor a can be determined given a value of the function other than the *x*-intercept.

HOW TO

Given a graph of a polynomial function, write a formula for the function.

1. Identify the x-intercepts of the graph to find the factors of the polynomial.
2. Examine the behavior of the graph at the x-intercepts to determine the multiplicity of each factor.
3. Find the polynomial of least degree containing all the factors found in the previous step.
4. Use any other point on the graph (the y-intercept may be easiest) to determine the stretch factor.

EXAMPLE 10

Writing a Formula for a Polynomial Function from the Graph

Write a formula for the polynomial function shown in Figure 19.

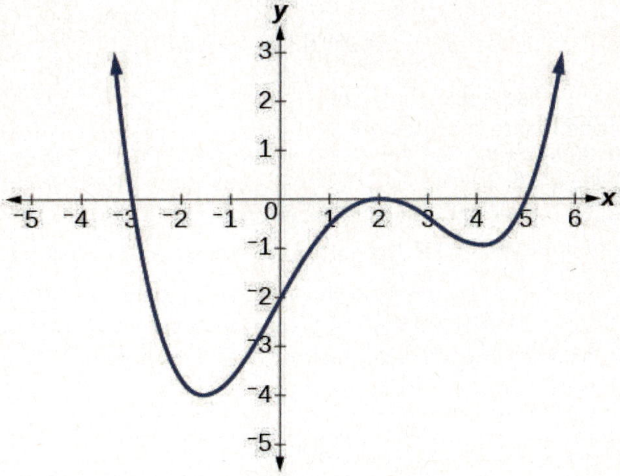

Figure 19

Solution

This graph has three x-intercepts: $x = -3, 2,$ and 5. The y-intercept is located at $(0, 2)$. At $x = -3$ and $x = 5$, the graph passes through the axis linearly, suggesting the corresponding factors of the polynomial will be linear. At $x = 2$, the graph bounces at the intercept, suggesting the corresponding factor of the polynomial will be second degree (quadratic). Together, this gives us

$$f(x) = a(x+3)(x-2)^2(x-5)$$

To determine the stretch factor, we utilize another point on the graph. We will use the y- intercept $(0, -2)$, to solve for a.

$$\begin{aligned} f(0) &= a(0+3)(0-2)^2(0-5) \\ -2 &= a(0+3)(0-2)^2(0-5) \\ -2 &= -60a \\ a &= \frac{1}{30} \end{aligned}$$

The graphed polynomial appears to represent the function $f(x) = \frac{1}{30}(x+3)(x-2)^2(x-5)$.

TRY IT #5 Given the graph shown in Figure 20, write a formula for the function shown.

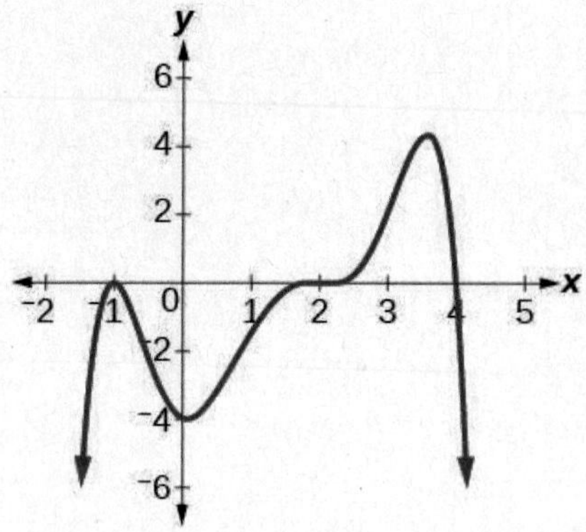

Figure 20

Using Local and Global Extrema

With quadratics, we were able to algebraically find the maximum or minimum value of the function by finding the vertex. For general polynomials, finding these turning points is not possible without more advanced techniques from calculus. Even then, finding where extrema occur can still be algebraically challenging. For now, we will estimate the locations of turning points using technology to generate a graph.

Each turning point represents a local minimum or maximum. Sometimes, a turning point is the highest or lowest point on the entire graph. In these cases, we say that the turning point is a **global maximum** or a **global minimum**. These are also referred to as the absolute maximum and absolute minimum values of the function.

Local and Global Extrema

A local maximum or local minimum at $x = a$ (sometimes called the relative maximum or minimum, respectively) is the output at the highest or lowest point on the graph in an open interval around $x = a$. If a function has a local maximum at a, then $f(a) \geq f(x)$ for all x in an open interval around $x = a$. If a function has a local minimum at a, then $f(a) \leq f(x)$ for all x in an open interval around $x = a$.

A **global maximum** or **global minimum** is the output at the highest or lowest point of the function. If a function has a global maximum at a, then $f(a) \geq f(x)$ for all x. If a function has a global minimum at a, then $f(a) \leq f(x)$ for all x.

We can see the difference between local and global extrema in Figure 21.

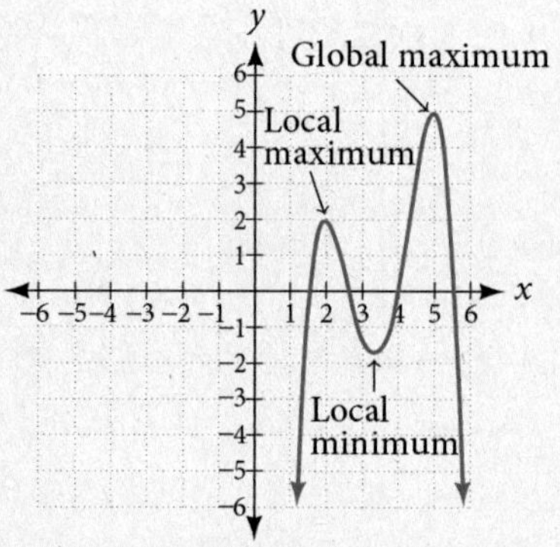

Figure 21

Q&A **Do all polynomial functions have a global minimum or maximum?**

No. Only polynomial functions of even degree have a global minimum or maximum. For example, $f(x) = x$ has neither a global maximum nor a global minimum.

EXAMPLE 11

Using Local Extrema to Solve Applications

An open-top box is to be constructed by cutting out squares from each corner of a 14 cm by 20 cm sheet of plastic and then folding up the sides. Find the size of squares that should be cut out to maximize the volume enclosed by the box.

Solution

We will start this problem by drawing a picture like that in Figure 22, labeling the width of the cut-out squares with a variable, w.

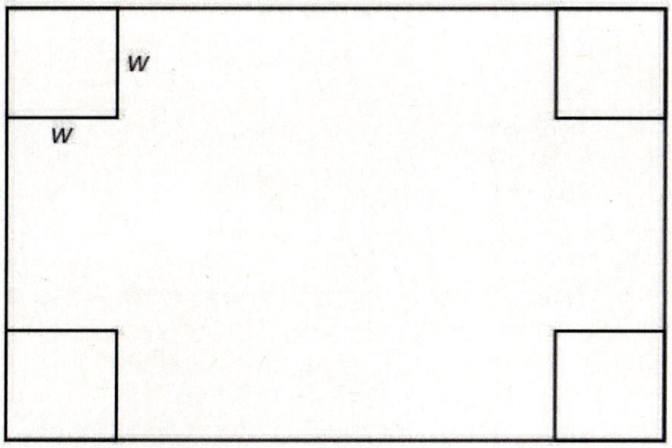

Figure 22

Notice that after a square is cut out from each end, it leaves a $(14 - 2w)$ cm by $(20 - 2w)$ cm rectangle for the base of the box, and the box will be w cm tall. This gives the volume

$$\begin{aligned} V(w) &= (20 - 2w)(14 - 2w)w \\ &= 280w - 68w^2 + 4w^3 \end{aligned}$$

Notice, since the factors are w, $20\text{–}2w$ and $14\text{–}2w$, the three zeros are 10, 7, and 0, respectively. Because a height of 0 cm is not reasonable, we consider the only the zeros 10 and 7. The shortest side is 14 and we are cutting off two squares, so values w may take on are greater than zero or less than 7. This means we will restrict the domain of this function to $0 < w < 7$. Using technology to sketch the graph of $V(w)$ on this reasonable domain, we get a graph like that in Figure 23. We can use this graph to estimate the maximum value for the volume, restricted to values for w that are reasonable for this problem—values from 0 to 7.

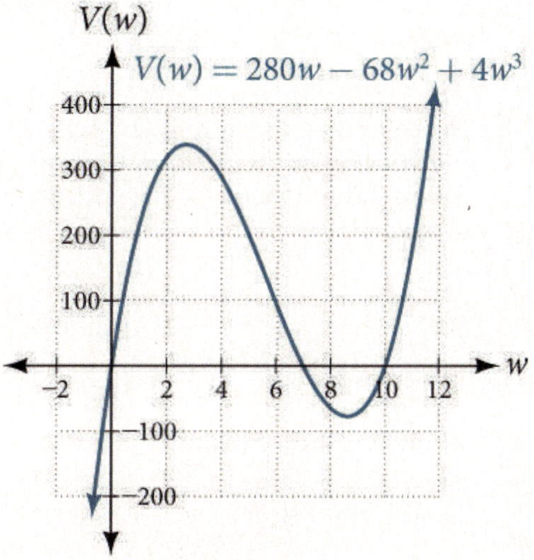

Figure 23

From this graph, we turn our focus to only the portion on the reasonable domain, $[0, \ 7]$. We can estimate the maximum value to be around 340 cubic cm, which occurs when the squares are about 2.75 cm on each side. To improve this estimate, we could use advanced features of our technology, if available, or simply change our window to zoom in on our graph to produce Figure 24.

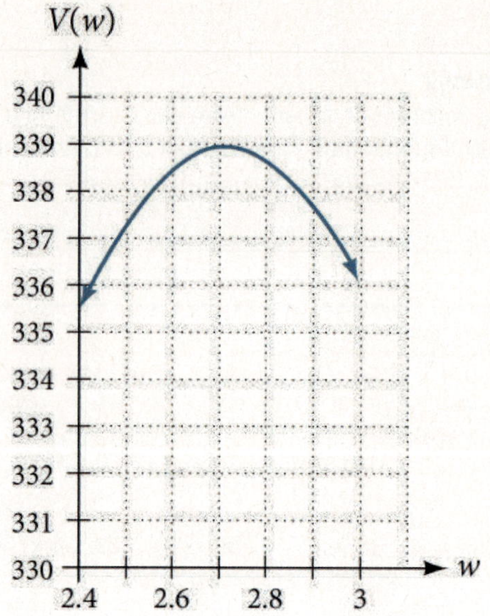

Figure 24

From this zoomed-in view, we can refine our estimate for the maximum volume to about 339 cubic cm, when the squares measure approximately 2.7 cm on each side.

TRY IT #6 Use technology to find the maximum and minimum values on the interval $[-1, 4]$ of the function $f(x) = -0.2(x-2)^3(x+1)^2(x-4)$.

MEDIA

Access the following online resource for additional instruction and practice with graphing polynomial functions.

Intermediate Value Theorem (http://openstax.org/l/ivt)

5.3 SECTION EXERCISES

Verbal

1. What is the difference between an x- intercept and a zero of a polynomial function f?

2. If a polynomial function of degree n has n distinct zeros, what do you know about the graph of the function?

3. Explain how the Intermediate Value Theorem can assist us in finding a zero of a function.

4. Explain how the factored form of the polynomial helps us in graphing it.

5. If the graph of a polynomial just touches the x-axis and then changes direction, what can we conclude about the factored form of the polynomial?

Algebraic

For the following exercises, find the x- or t-intercepts of the polynomial functions.

6. $C(t) = 2(t-4)(t+1)(t-6)$

7. $C(t) = 3(t+2)(t-3)(t+5)$

8. $C(t) = 4t(t-2)^2(t+1)$

9. $C(t) = 2t(t-3)(t+1)^2$

10. $C(t) = 2t^4 - 8t^3 + 6t^2$

11. $C(t) = 4t^4 + 12t^3 - 40t^2$

12. $f(x) = x^4 - x^2$

13. $f(x) = x^3 + x^2 - 20x$

14. $f(x) = x^3 + 6x^2 - 7x$

15. $f(x) = x^3 + x^2 - 4x - 4$

16. $f(x) = x^3 + 2x^2 - 9x - 18$

17. $f(x) = 2x^3 - x^2 - 8x + 4$

18. $f(x) = x^6 - 7x^3 - 8$

19. $f(x) = 2x^4 + 6x^2 - 8$

20. $f(x) = x^3 - 3x^2 - x + 3$

21. $f(x) = x^6 - 2x^4 - 3x^2$

22. $f(x) = x^6 - 3x^4 - 4x^2$

23. $f(x) = x^5 - 5x^3 + 4x$

For the following exercises, use the Intermediate Value Theorem to confirm that the given polynomial has at least one zero within the given interval.

24. $f(x) = x^3 - 9x$, between $x = -4$ and $x = -2$.

25. $f(x) = x^3 - 9x$, between $x = 2$ and $x = 4$.

26. $f(x) = x^5 - 2x$, between $x = 1$ and $x = 2$.

27. $f(x) = -x^4 + 4$, between $x = 1$ and $x = 3$.

28. $f(x) = -2x^3 - x$, between $x = -1$ and $x = 1$.

29. $f(x) = x^3 - 100x + 2$, between $x = 0.01$ and $x = 0.1$

For the following exercises, find the zeros and give the multiplicity of each.

30. $f(x) = (x+2)^3(x-3)^2$

31. $f(x) = x^2(2x+3)^5(x-4)^2$

32. $f(x) = x^3(x-1)^3(x+2)$

33. $f(x) = x^2\left(x^2 + 4x + 4\right)$

34. $f(x) = (2x+1)^3\left(9x^2 - 6x + 1\right)$

35. $f(x) = (3x+2)^5\left(x^2 - 10x + 25\right)$

36. $f(x) = x\left(4x^2 - 12x + 9\right)\left(x^2 + 8x + 16\right)$

37. $f(x) = x^6 - x^5 - 2x^4$

38. $f(x) = 3x^4 + 6x^3 + 3x^2$

39. $f(x) = 4x^5 - 12x^4 + 9x^3$

40. $f(x) = 2x^4\left(x^3 - 4x^2 + 4x\right)$

41. $f(x) = 4x^4\left(9x^4 - 12x^3 + 4x^2\right)$

Graphical

For the following exercises, graph the polynomial functions. Note x- and y- intercepts, multiplicity, and end behavior.

42. $f(x) = (x+3)^2(x-2)$

43. $g(x) = (x+4)(x-1)^2$

44. $h(x) = (x-1)^3(x+3)^2$

45. $k(x) = (x-3)^3(x-2)^2$

46. $m(x) = -2x(x-1)(x+3)$

47. $n(x) = -3x(x+2)(x-4)$

For the following exercises, use the graphs to write the formula for a polynomial function of least degree.

48.

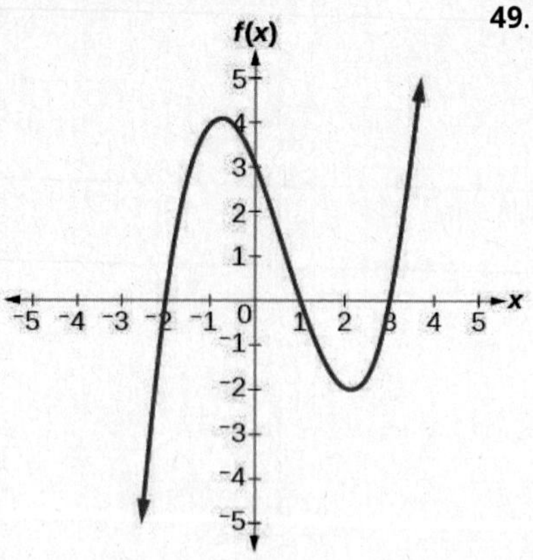

49.

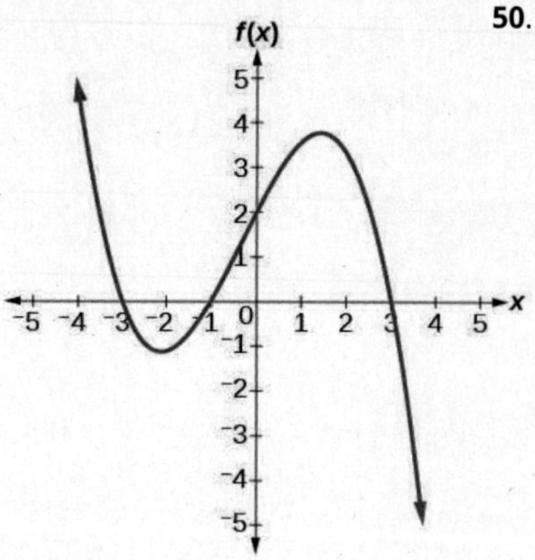

50.

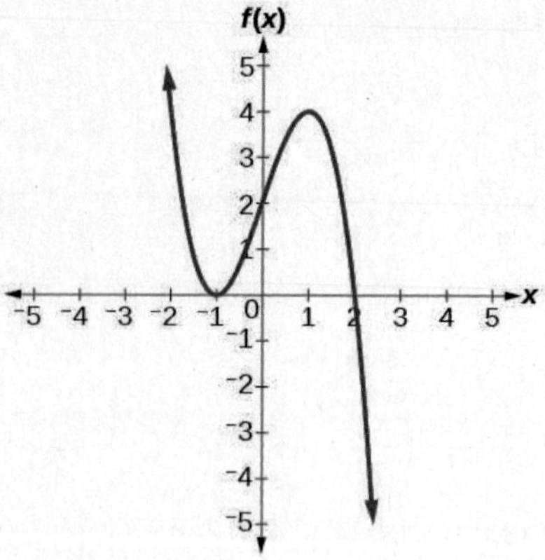

51.

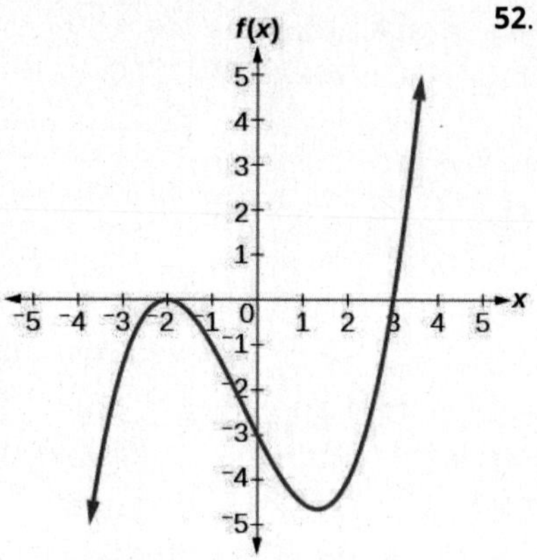

52.

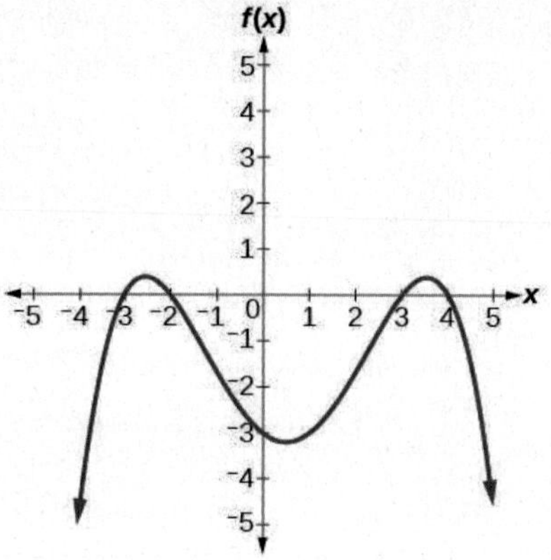

For the following exercises, use the graph to identify zeros and multiplicity.

53.

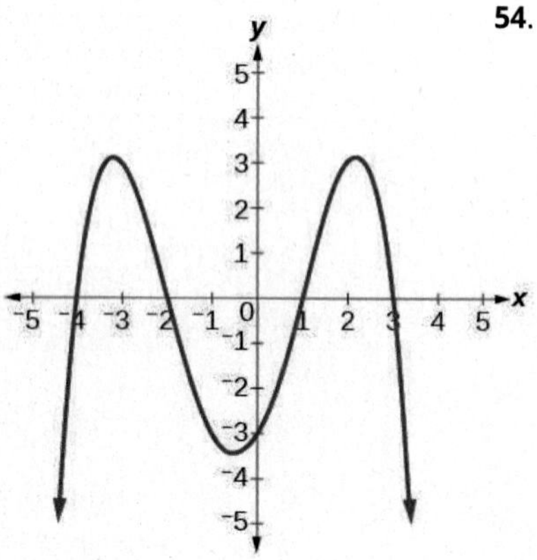

54.

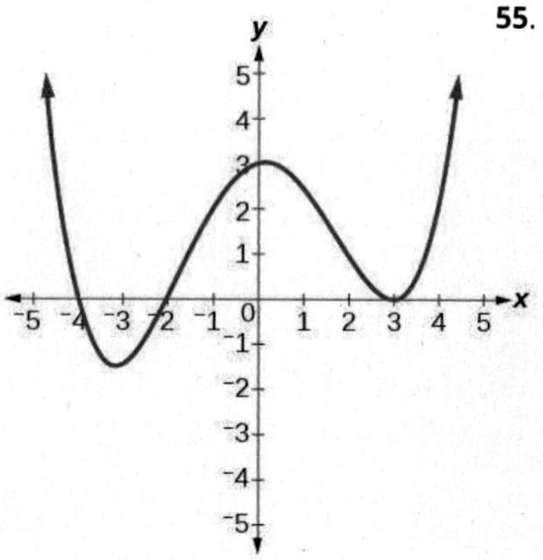

55.

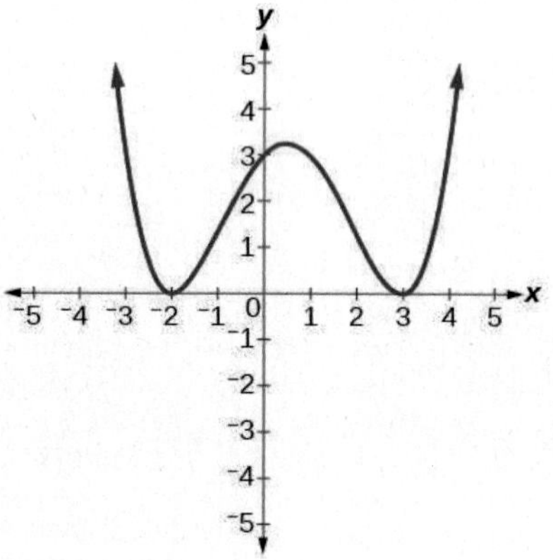

56.

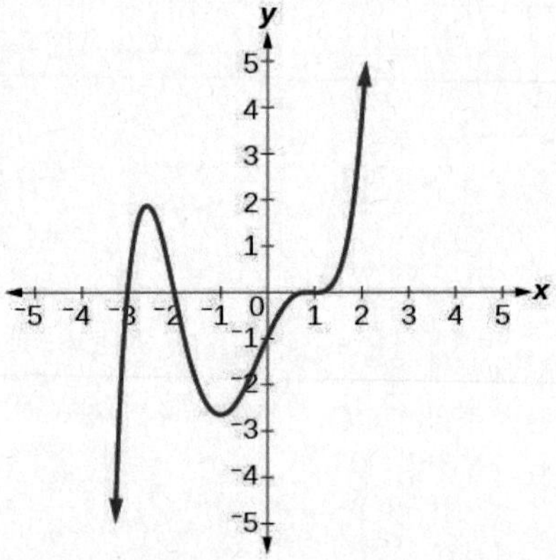

For the following exercises, use the given information about the polynomial graph to write the equation.

57. Degree 3. Zeros at $x = -2$, $x = 1$, and $x = 3$. y-intercept at $(0, -4)$.

58. Degree 3. Zeros at $x = -5$, $x = -2$, and $x = 1$. y-intercept at $(0, 6)$

59. Degree 5. Roots of multiplicity 2 at $x = 3$ and $x = 1$, and a root of multiplicity 1 at $x = -3$. y-intercept at $(0, 9)$

60. Degree 4. Root of multiplicity 2 at $x = 4$, and a roots of multiplicity 1 at $x = 1$ and $x = -2$. y-intercept at $(0, -3)$.

61. Degree 5. Double zero at $x = 1$, and triple zero at $x = 3$. Passes through the point $(2, 15)$.

62. Degree 3. Zeros at $x = 4$, $x = 3$, and $x = 2$. y-intercept at $(0, -24)$.

63. Degree 3. Zeros at $x = -3$, $x = -2$ and $x = 1$. y-intercept at $(0, 12)$.

64. Degree 5. Roots of multiplicity 2 at $x = -3$ and $x = 2$ and a root of multiplicity 1 at $x = -2$.

y-intercept at $(0, 4)$.

65. Degree 4. Roots of multiplicity 2 at $x = \frac{1}{2}$ and roots of multiplicity 1 at $x = 6$ and $x = -2$.

y-intercept at $(0,18)$.

66. Double zero at $x = -3$ and triple zero at $x = 0$. Passes through the point $(1, 32)$.

Technology

For the following exercises, use a calculator to approximate local minima and maxima or the global minimum and maximum.

67. $f(x) = x^3 - x - 1$

68. $f(x) = 2x^3 - 3x - 1$

69. $f(x) = x^4 + x$

70. $f(x) = -x^4 + 3x - 2$

71. $f(x) = x^4 - x^3 + 1$

Extensions

For the following exercises, use the graphs to write a polynomial function of least degree.

72.

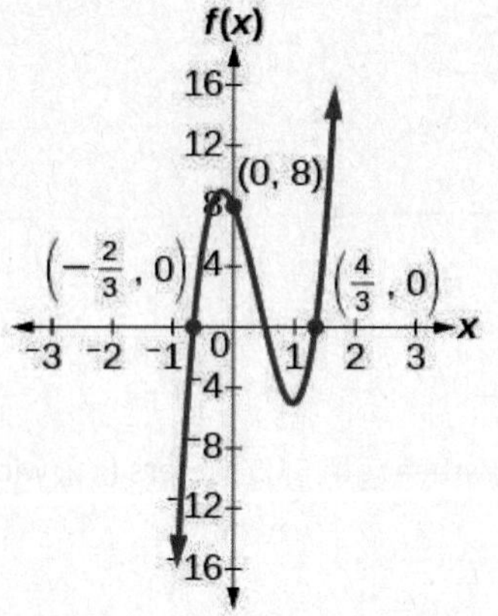

73.

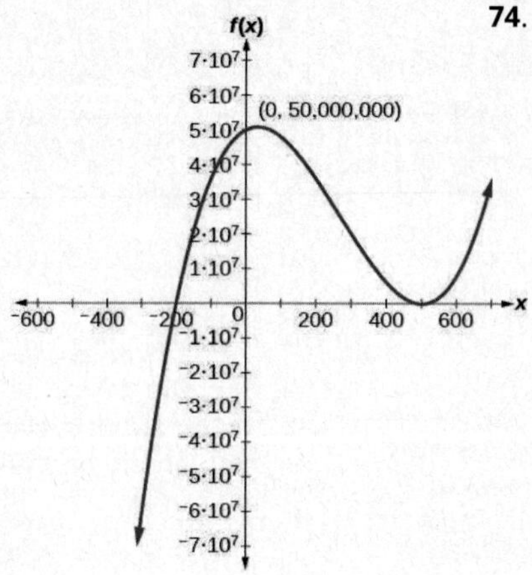

74.

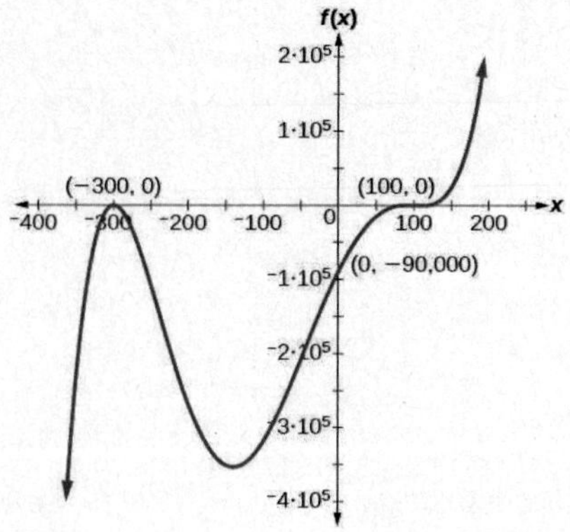

Real-World Applications

For the following exercises, write the polynomial function that models the given situation.

75. A rectangle has a length of 10 units and a width of 8 units. Squares of x by x units are cut out of each corner, and then the sides are folded up to create an open box. Express the volume of the box as a polynomial function in terms of x.

76. Consider the same rectangle of the preceding problem. Squares of $2x$ by $2x$ units are cut out of each corner. Express the volume of the box as a polynomial in terms of x.

77. A square has sides of 12 units. Squares $x + 1$ by $x + 1$ units are cut out of each corner, and then the sides are folded up to create an open box. Express the volume of the box as a function in terms of x.

78. A cylinder has a radius of $x + 2$ units and a height of 3 units greater. Express the volume of the cylinder as a polynomial function.

79. A right circular cone has a radius of $3x + 6$ and a height 3 units less. Express the volume of the cone as a polynomial function. The volume of a cone is $V = \frac{1}{3}\pi r^2 h$ for radius r and height h.

5.4 Dividing Polynomials

Learning Objectives

In this section, you will:

- Use long division to divide polynomials.
- Use synthetic division to divide polynomials.

Figure 1 Lincoln Memorial, Washington, D.C. (credit: Ron Cogswell, Flickr)

The exterior of the Lincoln Memorial in Washington, D.C., is a large rectangular solid with length 61.5 meters (m), width 40 m, and height 30 m.[1] We can easily find the volume using elementary geometry.

$$\begin{aligned} V &= l \cdot w \cdot h \\ &= 61.5 \cdot 40 \cdot 30 \\ &= 73{,}800 \end{aligned}$$

So the volume is 73,800 cubic meters (m^3) . Suppose we knew the volume, length, and width. We could divide to find the height.

$$\begin{aligned} h &= \frac{V}{l \cdot w} \\ &= \frac{73{,}800}{61.5 \cdot 40} \\ &= 30 \end{aligned}$$

As we can confirm from the dimensions above, the height is 30 m. We can use similar methods to find any of the missing dimensions. We can also use the same method if any, or all, of the measurements contain variable expressions. For example, suppose the volume of a rectangular solid is given by the polynomial $3x^4 - 3x^3 - 33x^2 + 54x$. The length of the solid is given by $3x$; the width is given by $x - 2$. To find the height of the solid, we can use polynomial division, which is the focus of this section.

Using Long Division to Divide Polynomials

We are familiar with the long division algorithm for ordinary arithmetic. We begin by dividing into the digits of the dividend that have the greatest place value. We divide, multiply, subtract, include the digit in the next place value position, and repeat. For example, let's divide 178 by 3 using long division.

Long Division

$$\begin{array}{r} 59 \\ 3\overline{)178} \\ \underline{-15\downarrow} \\ 28 \\ \underline{-27} \\ 1 \end{array}$$

Step 1: $5 \times 3 = 15$ and $17 - 15 = 2$
Step 2: Bring down the 8
Step 3: $9 \times 3 = 27$ and $28 - 27 = 1$

Answer: $59\ R\ 1$ or $59\frac{1}{3}$

Another way to look at the solution is as a sum of parts. This should look familiar, since it is the same method used to check division in elementary arithmetic.

$$\begin{aligned} \text{dividend} &= (\text{divisor} \cdot \text{quotient}) + \text{remainder} \\ 178 &= (3 \cdot 59) + 1 \\ &= 177 + 1 \\ &= 178 \end{aligned}$$

We call this the **Division Algorithm** and will discuss it more formally after looking at an example.

Division of polynomials that contain more than one term has similarities to long division of whole numbers. We can write

1 National Park Service. "Lincoln Memorial Building Statistics." http://www.nps.gov/linc/historyculture/lincoln-memorial-building-statistics.htm. Accessed 4/3/2014

a polynomial dividend as the product of the divisor and the quotient added to the remainder. The terms of the polynomial division correspond to the digits (and place values) of the whole number division. This method allows us to divide two polynomials. For example, if we were to divide $2x^3 - 3x^2 + 4x + 5$ by $x + 2$ using the long division algorithm, it would look like this:

$$x + 2\overline{)2x^3 - 3x^2 + 4x + 5}$$

Set up the division problem.

$$\begin{array}{r} 2x^2 \qquad\qquad\quad \\ x + 2\overline{)2x^3 - 3x^2 + 4x + 5} \end{array}$$

$2x^3$ divided by x is $2x^2$.

$$\begin{array}{r} 2x^2 \qquad\qquad\quad \\ x + 2\overline{)2x^3 - 3x^2 + 4x + 5} \\ \underline{-(2x^3 + 4x^2)} \qquad\quad \\ -7x^2 + 4x \quad \end{array}$$

Multiply $x + 2$ by $2x^2$.
Subtract.
Bring down the next term.

$$\begin{array}{r} 2x^2 - 7x \qquad\quad \\ x + 2\overline{)2x^3 - 3x^2 + 4x + 5} \\ \underline{-(2x^3 + 4x^2)} \qquad\quad \\ -7x^2 + 4x \quad \\ \underline{-(-7x^2 - 14x)} \quad \\ 18x + 5 \end{array}$$

$-7x^2$ divided by x is $-7x$.
Multiply $x + 2$ by $-7x$.
Subtract. Bring down the next term.

$$\begin{array}{r} 2x^2 - 7x + 18 \\ x + 2\overline{)2x^3 - 3x^2 + 4x + 5} \\ \underline{-(2x^3 + 4x^2)} \qquad\quad \\ -7x^2 + 4x \quad \\ \underline{-(-7x^2 - 14x)} \quad \\ 18x + 5 \\ \underline{-(18x + 36)} \\ -31 \end{array}$$

$18x$ divided by x is 18.
Multiply $x + 2$ by 18.
Subtract.

We have found

$$\frac{2x^3 - 3x^2 + 4x + 5}{x + 2} = 2x^2 - 7x + 18 - \frac{31}{x + 2}$$

or

$$2x^3 - 3x^2 + 4x + 5 = (x + 2)(2x^2 - 7x + 18) - 31$$

We can identify the dividend, the divisor, the quotient, and the remainder.

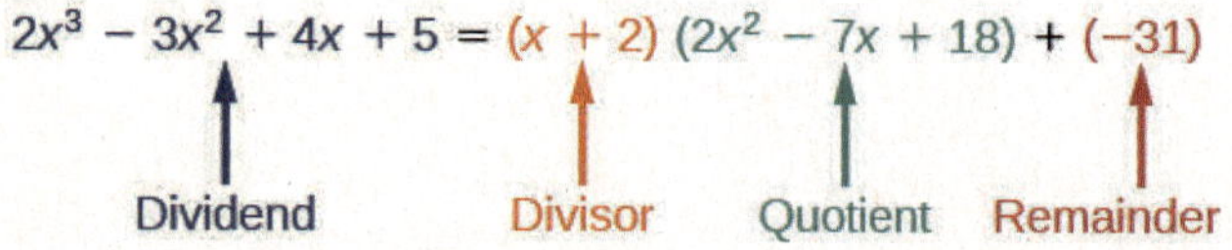

Writing the result in this manner illustrates the Division Algorithm.

The Division Algorithm

The **Division Algorithm** states that, given a polynomial dividend $f(x)$ and a non-zero polynomial divisor $d(x)$ where the degree of $d(x)$ is less than or equal to the degree of $f(x)$, there exist unique polynomials $q(x)$ and $r(x)$ such that

$$f(x) = d(x)q(x) + r(x)$$

$q(x)$ is the quotient and $r(x)$ is the remainder. The remainder is either equal to zero or has degree strictly less than $d(x)$.

If $r(x) = 0$, then $d(x)$ divides evenly into $f(x)$. This means that, in this case, both $d(x)$ and $q(x)$ are factors of $f(x)$.

HOW TO

Given a polynomial and a binomial, use long division to divide the polynomial by the binomial.

1. Set up the division problem.
2. Determine the first term of the quotient by dividing the leading term of the dividend by the leading term of the divisor.
3. Multiply the answer by the divisor and write it below the like terms of the dividend.
4. Subtract the bottom binomial from the top binomial.
5. Bring down the next term of the dividend.
6. Repeat steps 2–5 until reaching the last term of the dividend.
7. If the remainder is non-zero, express as a fraction using the divisor as the denominator.

EXAMPLE 1

Using Long Division to Divide a Second-Degree Polynomial

Divide $5x^2 + 3x - 2$ by $x + 1$.

Solution

$$x+1\overline{)\,5x^2+3x-2} \qquad \text{Set up the division problem.}$$

$$\begin{array}{r} 5x \\ x+1\overline{)\,5x^2+3x-2} \end{array} \qquad 5x^2 \text{ divided by } x \text{ is } 5x.$$

$$\begin{array}{rl} 5x & \\ x+1\overline{)\,5x^2+3x-2} & \\ \underline{-(5x^2+5x)} & \text{Multiply } x+1 \text{ by } 5x. \\ -2x-2 & \text{Subtract. Bring down the next term.} \end{array}$$

$$\begin{array}{rl} 5x-2 & \\ x+1\overline{)\,5x^2+3x-2} & -2x \text{ divided by } x \text{ is } -2. \\ \underline{-(5x^2+5x)} & \\ -2x-2 & \\ \underline{-(-2x-2)} & \text{Multiply } x+1 \text{ by } -2. \\ 0 & \text{Subtract.} \end{array}$$

The quotient is $5x - 2$. The remainder is 0. We write the result as

$$\frac{5x^2 + 3x - 2}{x + 1} = 5x - 2$$

or

$$5x^2 + 3x - 2 = (x + 1)(5x - 2)$$

Analysis

This division problem had a remainder of 0. This tells us that the dividend is divided evenly by the divisor, and that the divisor is a factor of the dividend.

EXAMPLE 2

Using Long Division to Divide a Third-Degree Polynomial

Divide $6x^3 + 11x^2 - 31x + 15$ by $3x - 2$.

Solution

$$\begin{array}{rll}
 & 2x^2 + \ \ 5x - \ 7 & \\
3x - 2 \,\big)\! & \overline{6x^3 + 11x^2 - 31x + 15} & 6x^3 \text{ divided by } 3x \text{ is } 2x^2. \\
 & \underline{-(6x^3 - 4x^2)} & \text{Multiply } 3x - 2 \text{ by } 2x^2. \\
 & 15x^2 - 31x & \text{Subtract. Bring down the next term. } 15x^2 \text{ divided by } 3x \text{ is } 5x. \\
 & \underline{-(15x^2 - 10x)} & \text{Multiply } 3x - 2 \text{ by } 5x. \\
 & -21x + 15 & \text{Subtract. Bring down the next term. } -21x \text{ divided by } 3x \text{ is } -7. \\
 & \underline{-(-21x + 14)} & \text{Multiply } 3x - 2 \text{ by } -7. \\
 & 1 & \text{Subtract. The remainder is 1.}
\end{array}$$

There is a remainder of 1. We can express the result as:

$$\frac{6x^3 + 11x^2 - 31x + 15}{3x - 2} = 2x^2 + 5x - 7 + \frac{1}{3x - 2}$$

Analysis

We can check our work by using the Division Algorithm to rewrite the solution. Then multiply.

$$(3x - 2)(2x^2 + 5x - 7) + 1 = 6x^3 + 11x^2 - 31x + 15$$

Notice, as we write our result,

- the dividend is $6x^3 + 11x^2 - 31x + 15$
- the divisor is $3x - 2$
- the quotient is $2x^2 + 5x - 7$
- the remainder is 1

TRY IT #1 Divide $16x^3 - 12x^2 + 20x - 3$ by $4x + 5$.

Using Synthetic Division to Divide Polynomials

As we've seen, long division of polynomials can involve many steps and be quite cumbersome. **Synthetic division** is a shorthand method of dividing polynomials for the special case of dividing by a linear factor whose leading coefficient is 1.

To illustrate the process, recall the example at the beginning of the section.

Divide $2x^3 - 3x^2 + 4x + 5$ by $x + 2$ using the long division algorithm.

The final form of the process looked like this:

$$\begin{array}{rl}
 & 2x^2 - 7x + 18 \\
x + 2 \,\big)\! & \overline{2x^3 - 3x^2 + 4x + 5} \\
 & \underline{-(2x^3 + 4x^2)} \\
 & -7x^2 + 4x \\
 & \underline{-(-7x^2 - 14x)} \\
 & 18x + 5 \\
 & \underline{-(18x + 36)} \\
 & -31
\end{array}$$

There is a lot of repetition in the table. If we don't write the variables but, instead, line up their coefficients in columns under the division sign and also eliminate the partial products, we already have a simpler version of the entire problem.

$$\begin{array}{r|rrrr}
2 & 2 & -3 & 4 & 5 \\
 & \underline{-2} & \underline{-4} & & \\
 & & -7 & 14 & \\
 & & & \underline{18} & \underline{-36} \\
 & & & & -31
\end{array}$$

Synthetic division carries this simplification even a few more steps. Collapse the table by moving each of the rows up to fill any vacant spots. Also, instead of dividing by 2, as we would in division of whole numbers, then multiplying and subtracting the middle product, we change the sign of the "divisor" to –2, multiply and add. The process starts by bringing down the leading coefficient.

$$\begin{array}{r|rrrr} -2 & 2 & -3 & 4 & 5 \\ & & -4 & 14 & -36 \\ \hline & 2 & -7 & 18 & -31 \end{array}$$

We then multiply it by the "divisor" and add, repeating this process column by column, until there are no entries left. The bottom row represents the coefficients of the quotient; the last entry of the bottom row is the remainder. In this case, the quotient is $2x^2 - 7x + 18$ and the remainder is -31. The process will be made more clear in Example 3.

Synthetic Division

Synthetic division is a shortcut that can be used when the divisor is a binomial in the form $x - k$ where k is a real number. In **synthetic division**, only the coefficients are used in the division process.

HOW TO

Given two polynomials, use synthetic division to divide.

1. Write k for the divisor.
2. Write the coefficients of the dividend.
3. Bring the lead coefficient down.
4. Multiply the lead coefficient by k. Write the product in the next column.
5. Add the terms of the second column.
6. Multiply the result by k. Write the product in the next column.
7. Repeat steps 5 and 6 for the remaining columns.
8. Use the bottom numbers to write the quotient. The number in the last column is the remainder and has degree 0, the next number from the right has degree 1, the next number from the right has degree 2, and so on.

EXAMPLE 3

Using Synthetic Division to Divide a Second-Degree Polynomial

Use synthetic division to divide $5x^2 - 3x - 36$ by $x - 3$.

Solution

Begin by setting up the synthetic division. Write k and the coefficients.

$$\begin{array}{r|rrr} 3 & 5 & -3 & -36 \\ & & & \\ \hline & & & \end{array}$$

Bring down the lead coefficient. Multiply the lead coefficient by k.

$$\begin{array}{r|rrr} 3 & 5 & -3 & -36 \\ & & 15 & \\ \hline & 5 & & \end{array}$$

Continue by adding the numbers in the second column. Multiply the resulting number by k. Write the result in the next column. Then add the numbers in the third column.

$$\begin{array}{r|rrr} 3 & 5 & -3 & -36 \\ & & 15 & 36 \\ \hline & 5 & 12 & 0 \end{array}$$

The result is $5x + 12$. The remainder is 0. So $x - 3$ is a factor of the original polynomial.

Analysis

Just as with long division, we can check our work by multiplying the quotient by the divisor and adding the remainder.

$$(x - 3)(5x + 12) + 0 = 5x^2 - 3x - 36$$

EXAMPLE 4

Using Synthetic Division to Divide a Third-Degree Polynomial

Use synthetic division to divide $4x^3 + 10x^2 - 6x - 20$ by $x + 2$.

Solution

The binomial divisor is $x + 2$ so $k = -2$. Add each column, multiply the result by –2, and repeat until the last column is reached.

$$\begin{array}{r|rrrr} -2 & 4 & 10 & -6 & -20 \\ & & -8 & -4 & 20 \\ \hline & 4 & 2 & -10 & 0 \end{array}$$

The result is $4x^2 + 2x - 10$. The remainder is 0. Thus, $x + 2$ is a factor of $4x^3 + 10x^2 - 6x - 20$.

Analysis

The graph of the polynomial function $f(x) = 4x^3 + 10x^2 - 6x - 20$ in Figure 2 shows a zero at $x = k = -2$. This confirms that $x + 2$ is a factor of $4x^3 + 10x^2 - 6x - 20$.

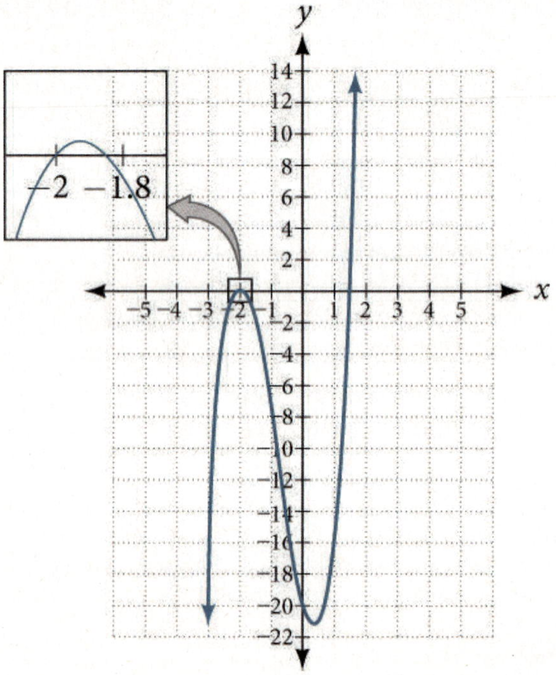

Figure 2

EXAMPLE 5

Using Synthetic Division to Divide a Fourth-Degree Polynomial

Use synthetic division to divide $-9x^4 + 10x^3 + 7x^2 - 6$ by $x - 1$.

Solution

Notice there is no x-term. We will use a zero as the coefficient for that term.

$$\begin{array}{r|rrrrr} 1 & -9 & 10 & 7 & 0 & -6 \\ & & -9 & 1 & 8 & 8 \\ \hline & -9 & 1 & 8 & 8 & 2 \end{array}$$

The result is $-9x^3 + x^2 + 8x + 8 + \frac{2}{x-1}$.

TRY IT #2 Use synthetic division to divide $3x^4 + 18x^3 - 3x + 40$ by $x + 7$.

Using Polynomial Division to Solve Application Problems

Polynomial division can be used to solve a variety of application problems involving expressions for area and volume. We looked at an application at the beginning of this section. Now we will solve that problem in the following example.

EXAMPLE 6

Using Polynomial Division in an Application Problem

The volume of a rectangular solid is given by the polynomial $3x^4 - 3x^3 - 33x^2 + 54x$. The length of the solid is given by $3x$ and the width is given by $x - 2$. Find the height, h, of the solid.

Solution

There are a few ways to approach this problem. We need to divide the expression for the volume of the solid by the expressions for the length and width. Let us create a sketch as in Figure 3.

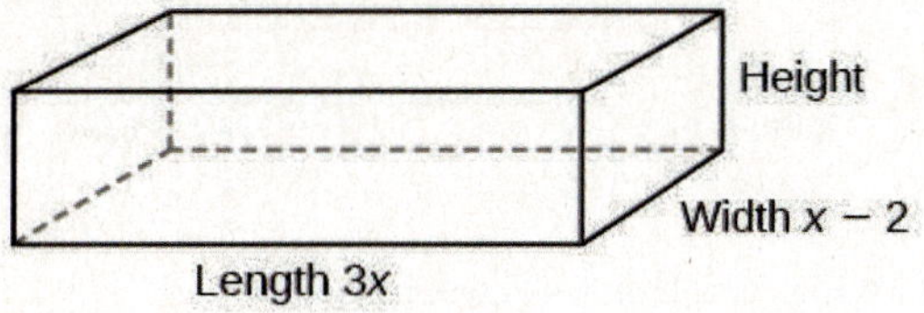

Figure 3

We can now write an equation by substituting the known values into the formula for the volume of a rectangular solid.

$$\begin{aligned} V &= l \cdot w \cdot h \\ 3x^4 - 3x^3 - 33x^2 + 54x &= 3x \cdot (x-2) \cdot h \end{aligned}$$

To solve for h, first divide both sides by $3x$.

$$\begin{aligned} \frac{3x \cdot (x-2) \cdot h}{3x} &= \frac{3x^4 - 3x^3 - 33x^2 + 54x}{3x} \\ (x-2)h &= x^3 - x^2 - 11x + 18 \end{aligned}$$

Now solve for h using synthetic division.

$$h = \frac{x^3 - x^2 - 11x + 18}{x - 2}$$

$$\begin{array}{r|rrrr} 2 & 1 & -1 & -11 & 18 \\ & & 2 & 2 & -18 \\ \hline & 1 & 1 & -9 & 0 \end{array}$$

The quotient is $x^2 + x - 9$ and the remainder is 0. The height of the solid is $x^2 + x - 9$.

TRY IT #3 The area of a rectangle is given by $3x^3 + 14x^2 - 23x + 6$. The width of the rectangle is given by $x + 6$. Find an expression for the length of the rectangle.

MEDIA

Access these online resources for additional instruction and practice with polynomial division.

Dividing a Trinomial by a Binomial Using Long Division (http://openstax.org/l/dividetribild)
Dividing a Polynomial by a Binomial Using Long Division (http://openstax.org/l/dividepolybild)
Ex 2: Dividing a Polynomial by a Binomial Using Synthetic Division (http://openstax.org/l/dividepolybisd2)
Ex 4: Dividing a Polynomial by a Binomial Using Synthetic Division (http://openstax.org/l/dividepolybisd4)

5.4 SECTION EXERCISES

Verbal

1. If division of a polynomial by a binomial results in a remainder of zero, what can be conclude?
2. If a polynomial of degree n is divided by a binomial of degree 1, what is the degree of the quotient?

Algebraic

For the following exercises, use long division to divide. Specify the quotient and the remainder.

3. $(x^2 + 5x - 1) \div (x - 1)$
4. $(2x^2 - 9x - 5) \div (x - 5)$
5. $(3x^2 + 23x + 14) \div (x + 7)$
6. $(4x^2 - 10x + 6) \div (4x + 2)$
7. $(6x^2 - 25x - 25) \div (6x + 5)$
8. $(-x^2 - 1) \div (x + 1)$
9. $(2x^2 - 3x + 2) \div (x + 2)$
10. $(x^3 - 126) \div (x - 5)$
11. $(3x^2 - 5x + 4) \div (3x + 1)$
12. $(x^3 - 3x^2 + 5x - 6) \div (x - 2)$
13. $(2x^3 + 3x^2 - 4x + 15) \div (x + 3)$

For the following exercises, use synthetic division to find the quotient. Ensure the equation is in the form required by synthetic division. (Hint: divide the dividend and divisor by the coefficient of the linear term in the divisor.)

14. $(3x^3 - 2x^2 + x - 4) \div (x + 3)$
15. $(2x^3 - 6x^2 - 7x + 6) \div (x - 4)$
16. $(6x^3 - 10x^2 - 7x - 15) \div (x + 1)$
17. $(4x^3 - 12x^2 - 5x - 1) \div (2x + 1)$
18. $(9x^3 - 9x^2 + 18x + 5) \div (3x - 1)$
19. $(3x^3 - 2x^2 + x - 4) \div (x + 3)$
20. $(-6x^3 + x^2 - 4) \div (2x - 3)$
21. $(2x^3 + 7x^2 - 13x - 3) \div (2x - 3)$
22. $(3x^3 - 5x^2 + 2x + 3) \div (x + 2)$
23. $(4x^3 - 5x^2 + 13) \div (x + 4)$
24. $(x^3 - 3x + 2) \div (x + 2)$
25. $(x^3 - 21x^2 + 147x - 343) \div (x - 7)$
26. $(x^3 - 15x^2 + 75x - 125) \div (x - 5)$
27. $(9x^3 - x + 2) \div (3x - 1)$
28. $(6x^3 - x^2 + 5x + 2) \div (3x + 1)$
29. $(x^4 + x^3 - 3x^2 - 2x + 1) \div (x + 1)$
30. $(x^4 - 3x^2 + 1) \div (x - 1)$
31. $(x^4 + 2x^3 - 3x^2 + 2x + 6) \div (x + 3)$
32. $(x^4 - 10x^3 + 37x^2 - 60x + 36) \div (x - 2)$
33. $(x^4 - 8x^3 + 24x^2 - 32x + 16) \div (x - 2)$
34. $(x^4 + 5x^3 - 3x^2 - 13x + 10) \div (x + 5)$
35. $(x^4 - 12x^3 + 54x^2 - 108x + 81) \div (x - 3)$
36. $(4x^4 - 2x^3 - 4x + 2) \div (2x - 1)$
37. $(4x^4 + 2x^3 - 4x^2 + 2x + 2) \div (2x + 1)$

For the following exercises, use synthetic division to determine whether the first expression is a factor of the second. If it is, indicate the factorization.

38. $x-2,\ \ 4x^3-3x^2-8x+4$

39. $x-2,\ \ 3x^4-6x^3-5x+10$

40. $x+3,\ \ -4x^3+5x^2+8$

41. $x-2,\ \ 4x^4-15x^2-4$

42. $x-\frac{1}{2},\ \ 2x^4-x^3+2x-1$

43. $x+\frac{1}{3},\ \ 3x^4+x^3-3x+1$

Graphical

For the following exercises, use the graph of the third-degree polynomial and one factor to write the factored form of the polynomial suggested by the graph. The leading coefficient is one.

44. Factor is x^2-x+3

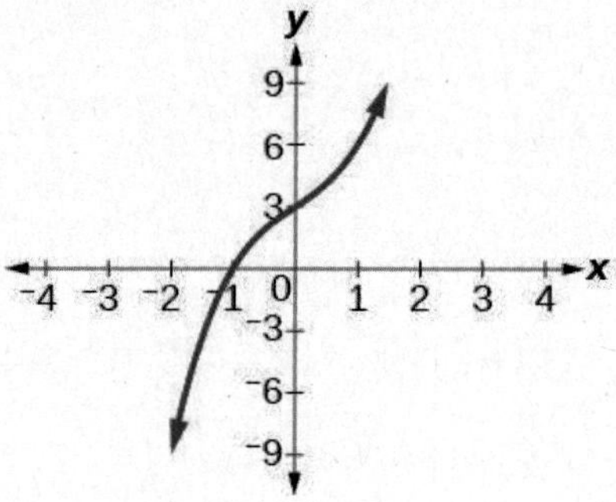

45. Factor is (x^2+2x+4)

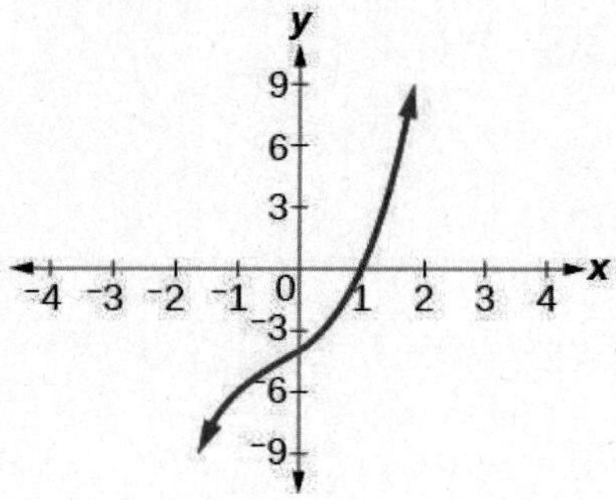

46. Factor is x^2+2x+5

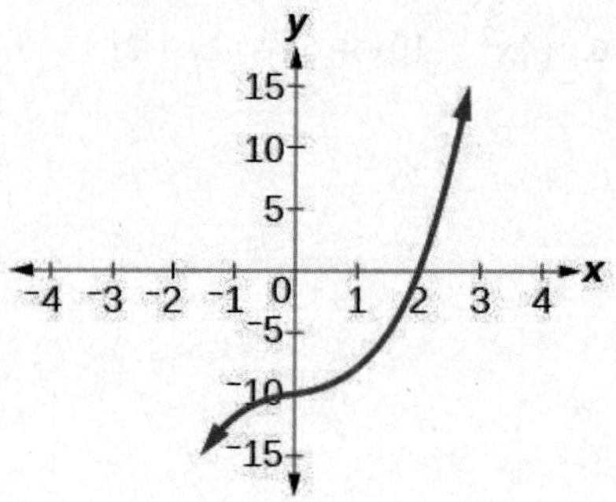

47. Factor is x^2+x+1

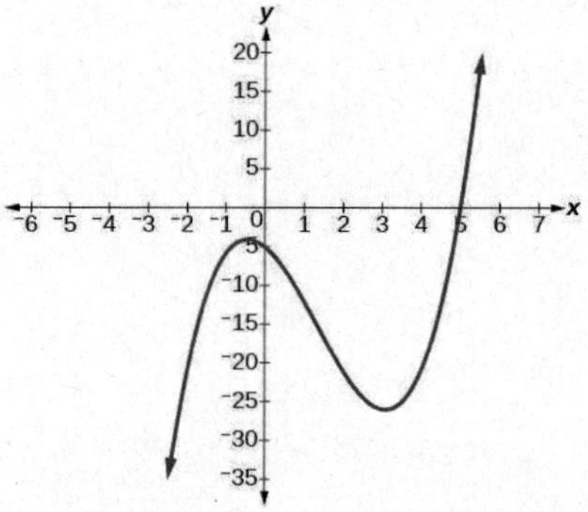

48. Factor is x^2+2x+2

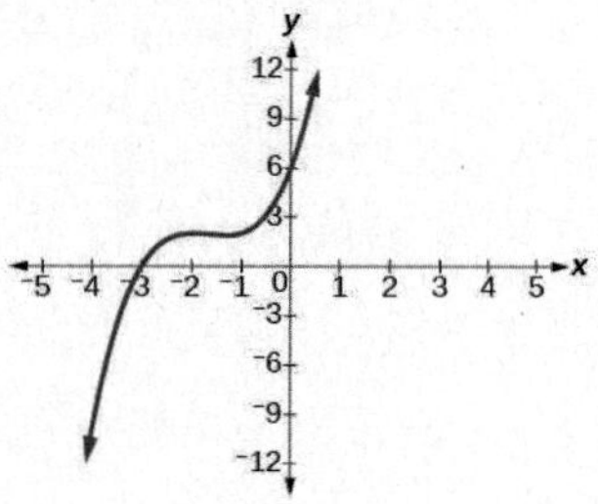

For the following exercises, use synthetic division to find the quotient and remainder.

49. $\frac{4x^3-33}{x-2}$

50. $\frac{2x^3+25}{x+3}$

51. $\frac{3x^3+2x-5}{x-1}$

52. $\frac{-4x^3-x^2-12}{x+4}$

53. $\frac{x^4-22}{x+2}$

Technology

For the following exercises, use a calculator with CAS to answer the questions.

54. Consider $\frac{x^k-1}{x-1}$ with $k=1,\ 2,\ 3$. What do you expect the result to be if $k=4$?

55. Consider $\frac{x^k+1}{x+1}$ for $k=1,\ 3,\ 5$. What do you expect the result to be if $k=7$?

56. Consider $\frac{x^4-k^4}{x-k}$ for $k = 1,\ 2,\ 3$. What do you expect the result to be if $k = 4$?

57. Consider $\frac{x^k}{x+1}$ with $k = 1,\ 2,\ 3$. What do you expect the result to be if $k = 4$?

58. Consider $\frac{x^k}{x-1}$ with $k = 1,\ 2,\ 3$. What do you expect the result to be if $k = 4$?

Extensions

For the following exercises, use synthetic division to determine the quotient involving a complex number.

59. $\frac{x+1}{x-i}$

60. $\frac{x^2+1}{x-i}$

61. $\frac{x+1}{x+i}$

62. $\frac{x^2+1}{x+i}$

63. $\frac{x^3+1}{x-i}$

Real-World Applications

For the following exercises, use the given length and area of a rectangle to express the width algebraically.

64. Length is $x + 5$, area is $2x^2 + 9x - 5$.

65. Length is $2x + 5$, area is $4x^3 + 10x^2 + 6x + 15$

66. Length is $3x-4$, area is $6x^4 - 8x^3 + 9x^2 - 9x - 4$

For the following exercises, use the given volume of a box and its length and width to express the height of the box algebraically.

67. Volume is $12x^3 + 20x^2 - 21x - 36$, length is $2x + 3$, width is $3x - 4$.

68. Volume is $18x^3 - 21x^2 - 40x + 48$, length is $3x-4$, width is $3x-4$.

69. Volume is $10x^3 + 27x^2 + 2x - 24$, length is $5x-4$, width is $2x + 3$.

70. Volume is $10x^3 + 30x^2 - 8x - 24$, length is 2, width is $x + 3$.

For the following exercises, use the given volume and radius of a cylinder to express the height of the cylinder algebraically.

71. Volume is $\pi\left(25x^3 - 65x^2 - 29x - 3\right)$, radius is $5x + 1$.

72. Volume is $\pi\left(4x^3 + 12x^2 - 15x - 50\right)$, radius is $2x + 5$.

73. Volume is $\pi\left(3x^4 + 24x^3 + 46x^2 - 16x - 32\right)$, radius is $x + 4$.

5.5 Zeros of Polynomial Functions

Learning Objectives

In this section, you will:

- Evaluate a polynomial using the Remainder Theorem.
- Use the Factor Theorem to solve a polynomial equation.
- Use the Rational Zero Theorem to find rational zeros.
- Find zeros of a polynomial function.
- Use the Linear Factorization Theorem to find polynomials with given zeros.
- Use Descartes' Rule of Signs.
- Solve real-world applications of polynomial equations

A new bakery offers decorated, multi-tiered cakes for display and cutting at Quinceañera and wedding celebrations, as well as sheet cakes for children's birthday parties and other special occasions to serve most of the guests. The bakery wants the volume of a small sheet cake to be 351 cubic inches. The cake is in the shape of a rectangular solid. They want the length of the cake to be four inches longer than the width of the cake and the height of the cake to be one-third of the width. What should the dimensions of the cake pan be?

This problem can be solved by writing a cubic function and solving a cubic equation for the volume of the cake. In this section, we will discuss a variety of tools for writing polynomial functions and solving polynomial equations.

Evaluating a Polynomial Using the Remainder Theorem

In the last section, we learned how to divide polynomials. We can now use polynomial division to evaluate polynomials using the **Remainder Theorem**. If the polynomial is divided by $x-k$, the remainder may be found quickly by evaluating the polynomial function at k, that is, $f(k)$ Let's walk through the proof of the theorem.

Recall that the Division Algorithm states that, given a polynomial dividend $f(x)$ and a non-zero polynomial divisor $d(x)$, there exist unique polynomials $q(x)$ and $r(x)$ such that

$$f(x) = d(x)q(x) + r(x)$$

and either $r(x) = 0$ or the degree of $r(x)$ is less than the degree of $d(x)$. In practice divisors, $d(x)$ will have degrees less than or equal to the degree of $f(x)$. If the divisor, $d(x)$, is $x - k$, this takes the form

$$f(x) = (x - k)q(x) + r$$

Since the divisor $x - k$ is linear, the remainder will be a constant, r. And, if we evaluate this for $x = k$, we have

$$\begin{aligned} f(k) &= (k-k)q(k) + r \\ &= 0 \cdot q(k) + r \\ &= r \end{aligned}$$

In other words, $f(k)$ is the remainder obtained by dividing $f(x)$ by $x - k$.

The Remainder Theorem

If a polynomial $f(x)$ is divided by $x - k$, then the remainder is the value $f(k)$.

HOW TO

Given a polynomial function f, evaluate $f(x)$ at $x = k$ using the Remainder Theorem.

1. Use synthetic division to divide the polynomial by $x - k$.
2. The remainder is the value $f(k)$.

EXAMPLE 1

Using the Remainder Theorem to Evaluate a Polynomial

Use the Remainder Theorem to evaluate $f(x) = 6x^4 - x^3 - 15x^2 + 2x - 7$ at $x = 2$.

Solution

To find the remainder using the Remainder Theorem, use synthetic division to divide the polynomial by $x-2$.

$$\begin{array}{r|rrrrr} 2 & 6 & -1 & -15 & 2 & -7 \\ & & 12 & 22 & 14 & 32 \\ \hline & 6 & 11 & 7 & 16 & 25 \end{array}$$

The remainder is 25. Therefore, $f(2)=25$.

Analysis

We can check our answer by evaluating $f(2)$.

$$\begin{aligned} f(x) &= 6x^4-x^3-15x^2+2x-7 \\ f(2) &= 6(2)^4-(2)^3-15(2)^2+2(2)-7 \\ &= 25 \end{aligned}$$

TRY IT #1 Use the Remainder Theorem to evaluate $f(x)=2x^5-3x^4-9x^3+8x^2+2$ at $x=-3$.

Using the Factor Theorem to Solve a Polynomial Equation

The **Factor Theorem** is another theorem that helps us analyze polynomial equations. It tells us how the zeros of a polynomial are related to the factors. Recall that the Division Algorithm.

$$f(x)=(x-k)q(x)+r$$

If k is a zero, then the remainder r is $f(k)=0$ and $f(x)=(x-k)q(x)+0$ or $f(x)=(x-k)q(x)$.

Notice, written in this form, $x-k$ is a factor of $f(x)$. We can conclude if k is a zero of $f(x)$, then $x-k$ is a factor of $f(x)$.

Similarly, if $x-k$ is a factor of $f(x)$, then the remainder of the Division Algorithm $f(x)=(x-k)q(x)+r$ is 0. This tells us that k is a zero.

This pair of implications is the Factor Theorem. As we will soon see, a polynomial of degree n in the complex number system will have n zeros. We can use the Factor Theorem to completely factor a polynomial into the product of n factors. Once the polynomial has been completely factored, we can easily determine the zeros of the polynomial.

The Factor Theorem

According to the **Factor Theorem**, k is a zero of $f(x)$ if and only if $(x-k)$ is a factor of $f(x)$.

HOW TO

Given a factor and a third-degree polynomial, use the Factor Theorem to factor the polynomial.

1. Use synthetic division to divide the polynomial by $(x-k)$.
2. Confirm that the remainder is 0.
3. Write the polynomial as the product of $(x-k)$ and the quadratic quotient.
4. If possible, factor the quadratic.
5. Write the polynomial as the product of factors.

EXAMPLE 2

Using the Factor Theorem to Find the Zeros of a Polynomial Expression

Show that $(x+2)$ is a factor of x^3-6x^2-x+30. Find the remaining factors. Use the factors to determine the zeros of the polynomial.

Solution

We can use synthetic division to show that $(x + 2)$ is a factor of the polynomial.

$$\begin{array}{r|rrrr} -2 & 1 & -6 & -1 & 30 \\ & & -2 & 16 & -30 \\ \hline & 1 & -8 & 15 & 0 \end{array}$$

The remainder is zero, so $(x + 2)$ is a factor of the polynomial. We can use the Division Algorithm to write the polynomial as the product of the divisor and the quotient:

$$(x + 2)\left(x^2 - 8x + 15\right)$$

We can factor the quadratic factor to write the polynomial as

$$(x + 2)(x - 3)(x - 5)$$

By the Factor Theorem, the zeros of $x^3 - 6x^2 - x + 30$ are -2, 3, and 5.

TRY IT #2 Use the Factor Theorem to find the zeros of $f(x) = x^3 + 4x^2 - 4x - 16$ given that $(x - 2)$ is a factor of the polynomial.

Using the Rational Zero Theorem to Find Rational Zeros

Another use for the Remainder Theorem is to test whether a rational number is a zero for a given polynomial. But first we need a pool of rational numbers to test. The **Rational Zero Theorem** helps us to narrow down the number of possible rational zeros using the ratio of the factors of the constant term and factors of the leading coefficient of the polynomial

Consider a quadratic function with two zeros, $x = \frac{2}{5}$ and $x = \frac{3}{4}$. By the Factor Theorem, these zeros have factors associated with them. Let us set each factor equal to 0, and then construct the original quadratic function absent its stretching factor.

$x - \frac{2}{5} = 0$ or $x - \frac{3}{4} = 0$	Set each factor equal to 0.
$5x - 2 = 0$ or $4x - 3 = 0$	Multiply both sides of the equation to eliminate fractions.
$f(x) = (5x - 2)(4x - 3)$	Create the quadratic function, multiplying the factors.
$f(x) = 20x^2 - 23x + 6$	Expand the polynomial.
$f(x) = (5 \cdot 4)x^2 - 23x + (2 \cdot 3)$	

Notice that two of the factors of the constant term, 6, are the two numerators from the original rational roots: 2 and 3. Similarly, two of the factors from the leading coefficient, 20, are the two denominators from the original rational roots: 5 and 4.

We can infer that the numerators of the rational roots will always be factors of the constant term and the denominators will be factors of the leading coefficient. This is the essence of the Rational Zero Theorem; it is a means to give us a pool of possible rational zeros.

The Rational Zero Theorem

The **Rational Zero Theorem** states that, if the polynomial $f(x) = a_n x^n + a_{n-1} x^{n-1} + ... + a_1 x + a_0$ has integer coefficients, then every rational zero of $f(x)$ has the form $\frac{p}{q}$ where p is a factor of the constant term a_0 and q is a factor of the leading coefficient a_n.

When the leading coefficient is 1, the possible rational zeros are the factors of the constant term.

HOW TO

Given a polynomial function $f(x)$, use the Rational Zero Theorem to find rational zeros.

1. Determine all factors of the constant term and all factors of the leading coefficient.
2. Determine all possible values of $\frac{p}{q}$, where p is a factor of the constant term and q is a factor of the leading coefficient. Be sure to include both positive and negative candidates.
3. Determine which possible zeros are actual zeros by evaluating each case of $f(\frac{p}{q})$.

EXAMPLE 3

Listing All Possible Rational Zeros

List all possible rational zeros of $f(x) = 2x^4 - 5x^3 + x^2 - 4$.

Solution

The only possible rational zeros of $f(x)$ are the quotients of the factors of the last term, –4, and the factors of the leading coefficient, 2.

The constant term is –4; the factors of –4 are $p = \pm1, \pm2, \pm4$.

The leading coefficient is 2; the factors of 2 are $q = \pm1, \pm2$.

If any of the four real zeros are rational zeros, then they will be of one of the following factors of –4 divided by one of the factors of 2.

$$\frac{p}{q} = \pm\frac{1}{1}, \pm\frac{1}{2} \qquad \frac{p}{q} = \pm\frac{2}{1}, \pm\frac{2}{2} \qquad \frac{p}{q} = \pm\frac{4}{1}, \pm\frac{4}{2}$$

Note that $\frac{2}{2} = 1$ and $\frac{4}{2} = 2$, which have already been listed. So we can shorten our list.

$$\frac{p}{q} = \frac{\text{Factors of the last}}{\text{Factors of the first}} = \pm1, \pm2, \pm4, \pm\frac{1}{2}$$

EXAMPLE 4

Using the Rational Zero Theorem to Find Rational Zeros

Use the Rational Zero Theorem to find the rational zeros of $f(x) = 2x^3 + x^2 - 4x + 1$.

Solution

The Rational Zero Theorem tells us that if $\frac{p}{q}$ is a zero of $f(x)$, then p is a factor of 1 and q is a factor of 2.

$$\begin{aligned} \frac{p}{q} &= \frac{\text{factor of constant term}}{\text{factor of leading coefficient}} \\ &= \frac{\text{factor of 1}}{\text{factor of 2}} \end{aligned}$$

The factors of 1 are ±1 and the factors of 2 are ±1 and ±2. The possible values for $\frac{p}{q}$ are ±1 and $\pm\frac{1}{2}$. These are the possible rational zeros for the function. We can determine which of the possible zeros are actual zeros by substituting these values for x in $f(x)$.

$$\begin{aligned} f(-1) &= 2(-1)^3 + (-1)^2 - 4(-1) + 1 = 4 \\ f(1) &= 2(1)^3 + (1)^2 - 4(1) + 1 = 0 \\ f\left(-\tfrac{1}{2}\right) &= 2\left(-\tfrac{1}{2}\right)^3 + \left(-\tfrac{1}{2}\right)^2 - 4\left(-\tfrac{1}{2}\right) + 1 = 3 \\ f\left(\tfrac{1}{2}\right) &= 2\left(\tfrac{1}{2}\right)^3 + \left(\tfrac{1}{2}\right)^2 - 4\left(\tfrac{1}{2}\right) + 1 = -\tfrac{1}{2} \end{aligned}$$

Of those, $-1, -\frac{1}{2},$ and $\frac{1}{2}$ are not zeros of $f(x)$. 1 is the only rational zero of $f(x)$.

TRY IT #3 Use the Rational Zero Theorem to find the rational zeros of $f(x) = x^3 - 5x^2 + 2x + 1$.

Finding the Zeros of Polynomial Functions

The Rational Zero Theorem helps us to narrow down the list of possible rational zeros for a polynomial function. Once we have done this, we can use synthetic division repeatedly to determine all of the **zeros** of a polynomial function.

HOW TO

Given a polynomial function f, use synthetic division to find its zeros.

1. Use the Rational Zero Theorem to list all possible rational zeros of the function.
2. Use synthetic division to evaluate a given possible zero by synthetically dividing the candidate into the polynomial. If the remainder is 0, the candidate is a zero. If the remainder is not zero, discard the candidate.
3. Repeat step two using the quotient found with synthetic division. If possible, continue until the quotient is a quadratic.
4. Find the zeros of the quadratic function. Two possible methods for solving quadratics are factoring and using the quadratic formula.

EXAMPLE 5

Finding the Zeros of a Polynomial Function with Repeated Real Zeros

Find the zeros of $f(x) = 4x^3 - 3x - 1$.

Solution

The Rational Zero Theorem tells us that if $\frac{p}{q}$ is a zero of $f(x)$, then p is a factor of –1 and q is a factor of 4.

$$\begin{aligned}\frac{p}{q} &= \frac{\text{factor of constant term}}{\text{factor of leading coefficient}}\\ &= \frac{\text{factor of } -1}{\text{factor of } 4}\end{aligned}$$

The factors of -1 are ± 1 and the factors of 4 are $\pm 1, \pm 2,$ and ± 4. The possible values for $\frac{p}{q}$ are $\pm 1,\ \pm\frac{1}{2},$ and $\pm\frac{1}{4}$. These are the possible rational zeros for the function. We will use synthetic division to evaluate each possible zero until we find one that gives a remainder of 0. Let's begin with 1.

$$\begin{array}{r|rrrr} 1 & 4 & 0 & -3 & -1 \\ & & 4 & 4 & 1 \\ \hline & 4 & 4 & 1 & 0 \end{array}$$

Dividing by $(x - 1)$ gives a remainder of 0, so 1 is a zero of the function. The polynomial can be written as

$$(x - 1)(4x^2 + 4x + 1)$$

The quadratic is a perfect square. $f(x)$ can be written as

$$(x - 1)(2x + 1)^2$$

We already know that 1 is a zero. The other zero will have a multiplicity of 2 because the factor is squared. To find the other zero, we can set the factor equal to 0.

$$\begin{aligned}2x + 1 &= 0\\ x &= -\frac{1}{2}\end{aligned}$$

The zeros of the function are 1 and $-\frac{1}{2}$ with multiplicity 2.

Analysis

Look at the graph of the function f in Figure 1. Notice, at $x = -0.5$, the graph bounces off the x-axis, indicating the even multiplicity (2,4,6...) for the zero -0.5. At $x = 1$, the graph crosses the x-axis, indicating the odd multiplicity (1,3,5...) for the zero $x = 1$.

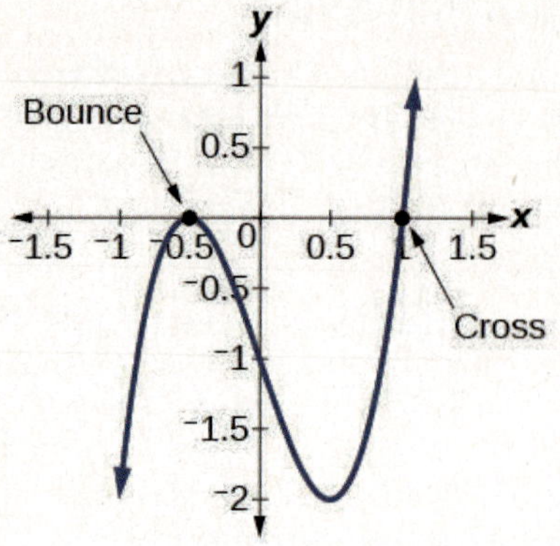

Figure 1

Using the Fundamental Theorem of Algebra

Now that we can find rational zeros for a polynomial function, we will look at a theorem that discusses the number of complex zeros of a polynomial function. The **Fundamental Theorem of Algebra** tells us that every polynomial function has at least one complex zero. This theorem forms the foundation for solving polynomial equations.

Suppose f is a polynomial function of degree four, and $f(x) = 0$. The Fundamental Theorem of Algebra states that there is at least one complex solution, call it c_1. By the Factor Theorem, we can write $f(x)$ as a product of $x - c_1$ and a polynomial quotient. Since $x - c_1$ is linear, the polynomial quotient will be of degree three. Now we apply the Fundamental Theorem of Algebra to the third-degree polynomial quotient. It will have at least one complex zero, call it c_2. So we can write the polynomial quotient as a product of $x - c_2$ and a new polynomial quotient of degree two. Continue to apply the Fundamental Theorem of Algebra until all of the zeros are found. There will be four of them and each one will yield a factor of $f(x)$.

The Fundamental Theorem of Algebra

The **Fundamental Theorem of Algebra** states that, if $f(x)$ is a polynomial of degree $n > 0$, then $f(x)$ has at least one complex zero.

We can use this theorem to argue that, if $f(x)$ is a polynomial of degree $n > 0$, and a is a non-zero real number, then $f(x)$ has exactly n linear factors

$$f(x) = a(x - c_1)(x - c_2)...(x - c_n)$$

where $c_1, c_2, ..., c_n$ are complex numbers. Therefore, $f(x)$ has n roots if we allow for multiplicities.

Q&A **Does every polynomial have at least one imaginary zero?**

No. Real numbers are a subset of complex numbers, but not the other way around. A complex number is not necessarily imaginary. Real numbers are also complex numbers.

EXAMPLE 6

Finding the Zeros of a Polynomial Function with Complex Zeros

Find the zeros of $f(x) = 3x^3 + 9x^2 + x + 3$.

Solution

The Rational Zero Theorem tells us that if $\frac{p}{q}$ is a zero of $f(x)$, then p is a factor of 3 and q is a factor of 3.

$$\begin{aligned} \frac{p}{q} &= \frac{\text{factor of constant term}}{\text{factor of leading coefficient}} \\ &= \frac{\text{factor of 3}}{\text{factor of 3}} \end{aligned}$$

The factors of 3 are ± 1 and ± 3. The possible values for $\frac{p}{q}$, and therefore the possible rational zeros for the function, are $\pm 3, \pm 1,$ and $\pm \frac{1}{3}$. We will use synthetic division to evaluate each possible zero until we find one that gives a remainder of 0. Let's begin with –3.

$$\begin{array}{r|rrrr} -3 & 3 & 9 & 1 & 3 \\ & & -9 & 0 & -3 \\ \hline & 3 & 0 & 1 & 0 \end{array}$$

Dividing by $(x + 3)$ gives a remainder of 0, so –3 is a zero of the function. The polynomial can be written as

$$(x + 3)\left(3x^2 + 1\right)$$

We can then set the quadratic equal to 0 and solve to find the other zeros of the function.

$$\begin{aligned} 3x^2 + 1 &= 0 \\ x^2 &= -\frac{1}{3} \\ x &= \pm\sqrt{-\frac{1}{3}} = \pm\frac{i\sqrt{3}}{3} \end{aligned}$$

The zeros of $f(x)$ are –3 and $\pm\frac{i\sqrt{3}}{3}$.

Analysis

Look at the graph of the function f in Figure 2. Notice that, at $x = -3$, the graph crosses the *x*-axis, indicating an odd multiplicity (1) for the zero $x = -3$. Also note the presence of the two turning points. This means that, since there is a 3rd degree polynomial, we are looking at the maximum number of turning points. So, the end behavior of increasing without bound to the right and decreasing without bound to the left will continue. Thus, all the *x*-intercepts for the function are shown. So either the multiplicity of $x = -3$ is 1 and there are two complex solutions, which is what we found, or the multiplicity at $x = -3$ is three. Either way, our result is correct.

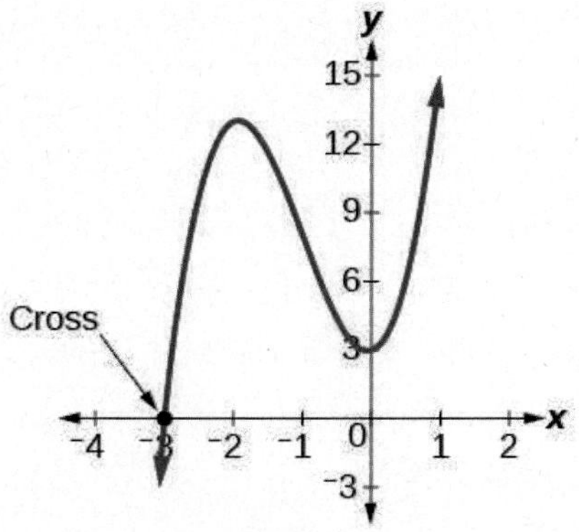

Figure 2

> **TRY IT** #4 Find the zeros of $f(x) = 2x^3 + 5x^2 - 11x + 4$.

Using the Linear Factorization Theorem to Find Polynomials with Given Zeros

A vital implication of the Fundamental Theorem of Algebra, as we stated above, is that a polynomial function of degree n will have n zeros in the set of complex numbers, if we allow for multiplicities. This means that we can factor the polynomial function into n factors. The **Linear Factorization Theorem** tells us that a polynomial function will have the same number of factors as its degree, and that each factor will be in the form $(x - c)$, where c is a complex number.

Let f be a polynomial function with real coefficients, and suppose $a + bi, b \neq 0$, is a zero of $f(x)$. Then, by the Factor Theorem, $x - (a + bi)$ is a factor of $f(x)$. For f to have real coefficients, $x - (a - bi)$ must also be a factor of $f(x)$. This is true because any factor other than $x - (a - bi)$, when multiplied by $x - (a + bi)$, will leave imaginary components in the product. Only multiplication with conjugate pairs will eliminate the imaginary parts and result in real coefficients. In other words, if a polynomial function f with real coefficients has a complex zero $a + bi$, then the complex conjugate $a - bi$ must also be a zero of $f(x)$. This is called the Complex Conjugate Theorem.

Complex Conjugate Theorem

According to the **Linear Factorization Theorem,** a polynomial function will have the same number of factors as its degree, and each factor will be in the form $(x - c)$, where c is a complex number.

If the polynomial function f has real coefficients and a complex zero in the form $a + bi$, then the complex conjugate of the zero, $a - bi$, is also a zero.

HOW TO

Given the zeros of a polynomial function f and a point (c, f(c)) on the graph of f, use the Linear Factorization Theorem to find the polynomial function.

1. Use the zeros to construct the linear factors of the polynomial.
2. Multiply the linear factors to expand the polynomial.
3. Substitute $(c, f(c))$ into the function to determine the leading coefficient.
4. Simplify.

EXAMPLE 7

Using the Linear Factorization Theorem to Find a Polynomial with Given Zeros

Find a fourth degree polynomial with real coefficients that has zeros of -3, 2, i, such that $f(-2) = 100$.

Solution

Because $x = i$ is a zero, by the Complex Conjugate Theorem $x = -i$ is also a zero. The polynomial must have factors of $(x + 3),\ (x - 2),\ (x - i)$, and $(x + i)$. Since we are looking for a degree 4 polynomial, and now have four zeros, we have all four factors. Let's begin by multiplying these factors.

$$\begin{aligned} f(x) &= a(x+3)(x-2)(x-i)(x+i) \\ f(x) &= a\left(x^2+x-6\right)\left(x^2+1\right) \\ f(x) &= a\left(x^4+x^3-5x^2+x-6\right) \end{aligned}$$

We need to find *a* to ensure $f(-2) = 100$. Substitute $x = -2$ and $f(2) = 100$ into $f(x)$.

$$\begin{aligned} 100 &= a((-2)^4+(-2)^3-5(-2)^2+(-2)-6) \\ 100 &= a(-20) \\ -5 &= a \end{aligned}$$

So the polynomial function is

$$f(x) = -5\left(x^4+x^3-5x^2+x-6\right)$$

or

$$f(x) = -5x^4-5x^3+25x^2-5x+30$$

Analysis

We found that both i and $-i$ were zeros, but only one of these zeros needed to be given. If i is a zero of a polynomial with real coefficients, then $-i$ must also be a zero of the polynomial because $-i$ is the complex conjugate of i.

Q&A **If $2 + 3i$ were given as a zero of a polynomial with real coefficients, would $2 - 3i$ also need to be a zero?**

Yes. When any complex number with an imaginary component is given as a zero of a polynomial with real coefficients, the conjugate must also be a zero of the polynomial.

> **TRY IT** #5 Find a third degree polynomial with real coefficients that has zeros of 5 and $-2i$ such that $f(1) = 10$.

Using Descartes' Rule of Signs

There is a straightforward way to determine the possible numbers of positive and negative real zeros for any polynomial function. If the polynomial is written in descending order, **Descartes' Rule of Signs** tells us of a relationship between the number of sign changes in $f(x)$ and the number of positive real zeros. For example, the polynomial function below has one sign change.

$$f(x) = x^4 + x^3 + x^2 + x - 1$$

This tells us that the function must have 1 positive real zero.

There is a similar relationship between the number of sign changes in $f(-x)$ and the number of negative real zeros.

$$f(-x) = (-x)^4 + (-x)^3 + (-x)^2 + (-x) - 1$$
$$f(-x) = +x^4 - x^3 + x^2 - x - 1$$

In this case, $f(-x)$ has 3 sign changes. This tells us that $f(x)$ could have 3 or 1 negative real zeros.

Descartes' Rule of Signs

According to **Descartes' Rule of Signs**, if we let $f(x) = a_n x^n + a_{n-1} x^{n-1} + ... + a_1 x + a_0$ be a polynomial function with real coefficients:

- The number of positive real zeros is either equal to the number of sign changes of $f(x)$ or is less than the number of sign changes by an even integer.
- The number of negative real zeros is either equal to the number of sign changes of $f(-x)$ or is less than the number of sign changes by an even integer.

EXAMPLE 8

Using Descartes' Rule of Signs

Use Descartes' Rule of Signs to determine the possible numbers of positive and negative real zeros for $f(x) = -x^4 - 3x^3 + 6x^2 - 4x - 12$.

Solution

Begin by determining the number of sign changes.

$$f(x) = -x^4 - 3x^3 + 6x^2 - 4x - 12$$

Figure 3

There are two sign changes, so there are either 2 or 0 positive real roots. Next, we examine $f(-x)$ to determine the number of negative real roots.

$$\begin{aligned} f(-x) &= -(-x)^4 - 3(-x)^3 + 6(-x)^2 - 4(-x) - 12 \\ f(-x) &= -x^4 + 3x^3 + 6x^2 + 4x - 12 \end{aligned}$$

$$f(-x) = -x^4 + 3x^3 + 6x^2 + 4x - 12$$

Figure 4

Again, there are two sign changes, so there are either 2 or 0 negative real roots.

There are four possibilities, as we can see in Table 1.

Positive Real Zeros	Negative Real Zeros	Complex Zeros	Total Zeros
2	2	0	4
2	0	2	4
0	2	2	4
0	0	4	4

Table 1

Analysis

We can confirm the numbers of positive and negative real roots by examining a graph of the function. See Figure 5. We can see from the graph that the function has 0 positive real roots and 2 negative real roots.

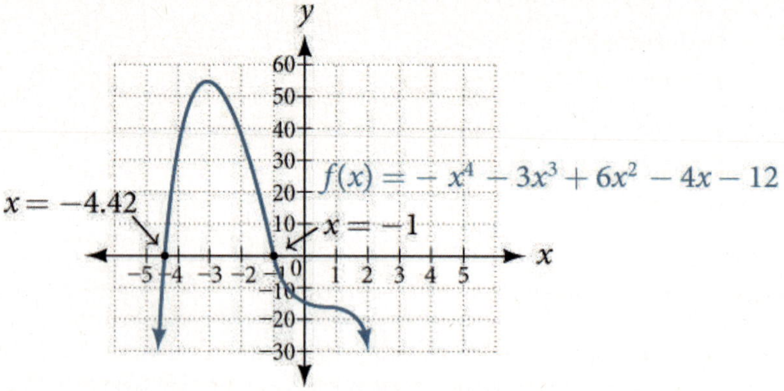

Figure 5

TRY IT #6 Use Descartes' Rule of Signs to determine the maximum possible numbers of positive and negative real zeros for $f(x) = 2x^4 - 10x^3 + 11x^2 - 15x + 12$. Use a graph to verify the numbers of positive and negative real zeros for the function.

Solving Real-World Applications

We have now introduced a variety of tools for solving polynomial equations. Let's use these tools to solve the bakery problem from the beginning of the section.

EXAMPLE 9

Solving Polynomial Equations

A new bakery offers decorated, multi-tiered cakes for display and cutting at Quinceañera and wedding celebrations, as well as sheet cakes for children's birthday parties and other special occasions to serve most of the guests. The bakery wants the volume of a small sheet cake to be 351 cubic inches. The cake is in the shape of a rectangular solid. They want the length of the cake to be four inches longer than the width of the cake and the height of the cake to be one-third of the width. What should the dimensions of the cake pan be?

Solution

Begin by writing an equation for the volume of the cake. The volume of a rectangular solid is given by $V = lwh$. We were given that the length must be four inches longer than the width, so we can express the length of the cake as $l = w + 4$. We were given that the height of the cake is one-third of the width, so we can express the height of the cake as $h = \frac{1}{3}w$. Let's write the volume of the cake in terms of width of the cake.

$$
\begin{aligned}
V &= (w+4)(w)\left(\tfrac{1}{3}w\right) \\
V &= \tfrac{1}{3}w^3 + \tfrac{4}{3}w^2
\end{aligned}
$$

Substitute the given volume into this equation.

$$
\begin{aligned}
351 &= \tfrac{1}{3}w^3 + \tfrac{4}{3}w^2 && \text{Substitute 351 for } V. \\
1053 &= w^3 + 4w^2 && \text{Multiply both sides by 3.} \\
0 &= w^3 + 7w^2 - 1053 && \text{Subtract 1053 from both sides.}
\end{aligned}
$$

Descartes' rule of signs tells us there is one positive solution. The Rational Zero Theorem tells us that the possible rational zeros are $\pm1,\ \pm3,\ \pm9,\ \pm13,\ \pm27,\ \pm39,\ \pm81,\ \pm117,\ \pm351$, and ±1053. We can use synthetic division to test these possible zeros. Only positive numbers make sense as dimensions for a cake, so we need not test any negative values. Let's begin by testing values that make the most sense as dimensions for a small sheet cake. Use synthetic division to check $x = 1$.

$$
\begin{array}{r|rrrr}
1 & 1 & 4 & 0 & -1053 \\
  &   & 1 & 5 & 5 \\ \hline
  & 1 & 5 & 5 & -1048
\end{array}
$$

Since 1 is not a solution, we will check $x = 3$.

$$
\begin{array}{r|rrrr}
3 & 1 & 4 & 0 & -1053 \\
  &   & 3 & 21 & 63 \\ \hline
  & 1 & 7 & 21 & -990
\end{array}
$$

Since 3 is not a solution either, we will test $x = 9$.

$$
\begin{array}{r|rrrr}
9 & 1 & 4 & 0 & -1053 \\
  &   & 9 & 117 & 1053 \\ \hline
  & 1 & 13 & 117 & 0
\end{array}
$$

Synthetic division gives a remainder of 0, so 9 is a solution to the equation. We can use the relationships between the width and the other dimensions to determine the length and height of the sheet cake pan.

$$
l = w + 4 = 9 + 4 = 13 \text{ and } h = \frac{1}{3}w = \frac{1}{3}(9) = 3
$$

The sheet cake pan should have dimensions 13 inches by 9 inches by 3 inches.

TRY IT #7 A shipping container in the shape of a rectangular solid must have a volume of 84 cubic meters. The client tells the manufacturer that, because of the contents, the length of the container must be one meter longer than the width, and the height must be one meter greater than twice the width. What should the dimensions of the container be?

MEDIA

Access these online resources for additional instruction and practice with zeros of polynomial functions.

Real Zeros, Factors, and Graphs of Polynomial Functions (http://openstax.org/l/realzeros)
Complex Factorization Theorem (http://openstax.org/l/factortheorem)
Find the Zeros of a Polynomial Function (http://openstax.org/l/findthezeros)
Find the Zeros of a Polynomial Function 2 (http://openstax.org/l/findthezeros2)
Find the Zeros of a Polynomial Function 3 (http://openstax.org/l/findthezeros3)

5.5 SECTION EXERCISES

Verbal

1. Describe a use for the Remainder Theorem.
2. Explain why the Rational Zero Theorem does not guarantee finding zeros of a polynomial function.
3. What is the difference between rational and real zeros?
4. If Descartes' Rule of Signs reveals a no change of signs or one sign of changes, what specific conclusion can be drawn?
5. If synthetic division reveals a zero, why should we try that value again as a possible solution?

Algebraic

For the following exercises, use the Remainder Theorem to find the remainder.

6. $(x^4 - 9x^2 + 14) \div (x - 2)$
7. $(3x^3 - 2x^2 + x - 4) \div (x + 3)$
8. $(x^4 + 5x^3 - 4x - 17) \div (x + 1)$
9. $(-3x^2 + 6x + 24) \div (x - 4)$
10. $(5x^5 - 4x^4 + 3x^3 - 2x^2 + x - 1) \div (x + 6)$
11. $(x^4 - 1) \div (x - 4)$
12. $(3x^3 + 4x^2 - 8x + 2) \div (x - 3)$
13. $(4x^3 + 5x^2 - 2x + 7) \div (x + 2)$

For the following exercises, use the Factor Theorem to find all real zeros for the given polynomial function and one factor.

14. $f(x) = 2x^3 - 9x^2 + 13x - 6;\ \ x - 1$
15. $f(x) = 2x^3 + x^2 - 5x + 2;\ \ x + 2$
16. $f(x) = 3x^3 + x^2 - 20x + 12;\ \ x + 3$
17. $f(x) = 2x^3 + 3x^2 + x + 6;\ \ x + 2$
18. $f(x) = -5x^3 + 16x^2 - 9;\ \ x - 3$
19. $x^3 + 3x^2 + 4x + 12;\ \ x + 3$
20. $4x^3 - 7x + 3;\ \ x - 1$
21. $2x^3 + 5x^2 - 12x - 30,\quad 2x + 5$

For the following exercises, use the Rational Zero Theorem to find the real solution(s) to each equation.

22. $x^3 - 3x^2 - 10x + 24 = 0$
23. $2x^3 + 7x^2 - 10x - 24 = 0$
24. $x^3 + 2x^2 - 9x - 18 = 0$
25. $x^3 + 5x^2 - 16x - 80 = 0$
26. $x^3 - 3x^2 - 25x + 75 = 0$
27. $2x^3 - 3x^2 - 32x - 15 = 0$
28. $2x^3 + x^2 - 7x - 6 = 0$
29. $2x^3 - 3x^2 - x + 1 = 0$
30. $3x^3 - x^2 - 11x - 6 = 0$
31. $2x^3 - 5x^2 + 9x - 9 = 0$
32. $2x^3 - 3x^2 + 4x + 3 = 0$
33. $x^4 - 2x^3 - 7x^2 + 8x + 12 = 0$

34. $x^4 + 2x^3 - 9x^2 - 2x + 8 = 0$ **35.** $4x^4 + 4x^3 - 25x^2 - x + 6 = 0$ **36.** $2x^4 - 3x^3 - 15x^2 + 32x - 12 = 0$

37. $x^4 + 2x^3 - 4x^2 - 10x - 5 = 0$ **38.** $4x^3 - 3x + 1 = 0$ **39.** $8x^4 + 26x^3 + 39x^2 + 26x + 6$

For the following exercises, find all complex solutions (real and non-real).

40. $x^3 + x^2 + x + 1 = 0$ **41.** $x^3 - 8x^2 + 25x - 26 = 0$ **42.** $x^3 + 13x^2 + 57x + 85 = 0$

43. $3x^3 - 4x^2 + 11x + 10 = 0$ **44.** $x^4 + 2x^3 + 22x^2 + 50x - 75 = 0$ **45.** $2x^3 - 3x^2 + 32x + 17 = 0$

Graphical

For the following exercises, use Descartes' Rule to determine the possible number of positive and negative solutions. Confirm with the given graph.

46. $f(x) = x^3 - 1$ **47.** $f(x) = x^4 - x^2 - 1$ **48.** $f(x) = x^3 - 2x^2 - 5x + 6$

49. $f(x) = x^3 - 2x^2 + x - 1$ **50.** $f(x) = x^4 + 2x^3 - 12x^2 + 14x - 5$ **51.** $f(x) = 2x^3 + 37x^2 + 200x + 300$

52. $f(x) = x^3 - 2x^2 - 16x + 32$ **53.** $f(x) = 2x^4 - 5x^3 - 5x^2 + 5x + 3$

54. $f(x) = 2x^4 - 5x^3 - 14x^2 + 20x + 8$ **55.** $f(x) = 10x^4 - 21x^2 + 11$

Numeric

For the following exercises, list all possible rational zeros for the functions.

56. $f(x) = x^4 + 3x^3 - 4x + 4$ **57.** $f(x) = 2x^3 + 3x^2 - 8x + 5$ **58.** $f(x) = 3x^3 + 5x^2 - 5x + 4$

59. $f(x) = 6x^4 - 10x^2 + 13x + 1$ **60.** $f(x) = 4x^5 - 10x^4 + 8x^3 + x^2 - 8$

Technology

For the following exercises, use your calculator to graph the polynomial function. Based on the graph, find the rational zeros. All real solutions are rational.

61. $f(x) = 6x^3 - 7x^2 + 1$ **62.** $f(x) = 4x^3 - 4x^2 - 13x - 5$ **63.** $f(x) = 8x^3 - 6x^2 - 23x + 6$

64. $f(x) = 12x^4 + 55x^3 + 12x^2 - 117x + 54$ **65.** $f(x) = 16x^4 - 24x^3 + x^2 - 15x + 25$

Extensions

For the following exercises, construct a polynomial function of least degree possible using the given information.

66. Real roots: –1, 1, 3 and $(2, f(2)) = (2, 4)$

67. Real roots: –1, 1 (with multiplicity 2 and 1) and $(2, f(2)) = (2, 4)$

68. Real roots: –2, $\frac{1}{2}$ (with multiplicity 2) and $(-3, f(-3)) = (-3, 5)$

69. Real roots: $-\frac{1}{2}$, 0, $\frac{1}{2}$ and $(-2, f(-2)) = (-2, 6)$

70. Real roots: –4, –1, 1, 4 and $(-2, f(-2)) = (-2, 10)$

Real-World Applications

For the following exercises, find the dimensions of the box described.

71. The length is twice as long as the width. The height is 2 inches greater than the width. The volume is 192 cubic inches.

72. The length, width, and height are consecutive whole numbers. The volume is 120 cubic inches.

73. The length is one inch more than the width, which is one inch more than the height. The volume is 86.625 cubic inches.

74. The length is three times the height and the height is one inch less than the width. The volume is 108 cubic inches.

75. The length is 3 inches more than the width. The width is 2 inches more than the height. The volume is 120 cubic inches.

For the following exercises, find the dimensions of the right circular cylinder described.

76. The radius is 3 inches more than the height. The volume is 16π cubic meters.

77. The height is one less than one half the radius. The volume is 72π cubic meters.

78. The radius and height differ by one meter. The radius is larger and the volume is 48π cubic meters.

79. The radius and height differ by two meters. The height is greater and the volume is 28.125π cubic meters.

80. The radius is $\frac{1}{3}$ meter greater than the height. The volume is $\frac{98}{9}\pi$ cubic meters.

5.6 Rational Functions

Learning Objectives

In this section, you will:

- Use arrow notation.
- Solve applied problems involving rational functions.
- Find the domains of rational functions.
- Identify vertical asymptotes.
- Identify horizontal asymptotes.
- Graph rational functions.

Suppose we know that the cost of making a product is dependent on the number of items, x, produced. This is given by the equation $C(x) = 15{,}000x - 0.1x^2 + 1000$. If we want to know the average cost for producing x items, we would divide the cost function by the number of items, x.

The average cost function, which yields the average cost per item for x items produced, is

$$f(x) = \frac{15{,}000x - 0.1x^2 + 1000}{x}$$

Many other application problems require finding an average value in a similar way, giving us variables in the denominator. Written without a variable in the denominator, this function will contain a negative integer power.

In the last few sections, we have worked with polynomial functions, which are functions with non-negative integers for exponents. In this section, we explore rational functions, which have variables in the denominator.

Using Arrow Notation

We have seen the graphs of the basic reciprocal function and the squared reciprocal function from our study of toolkit functions. Examine these graphs, as shown in Figure 1, and notice some of their features.

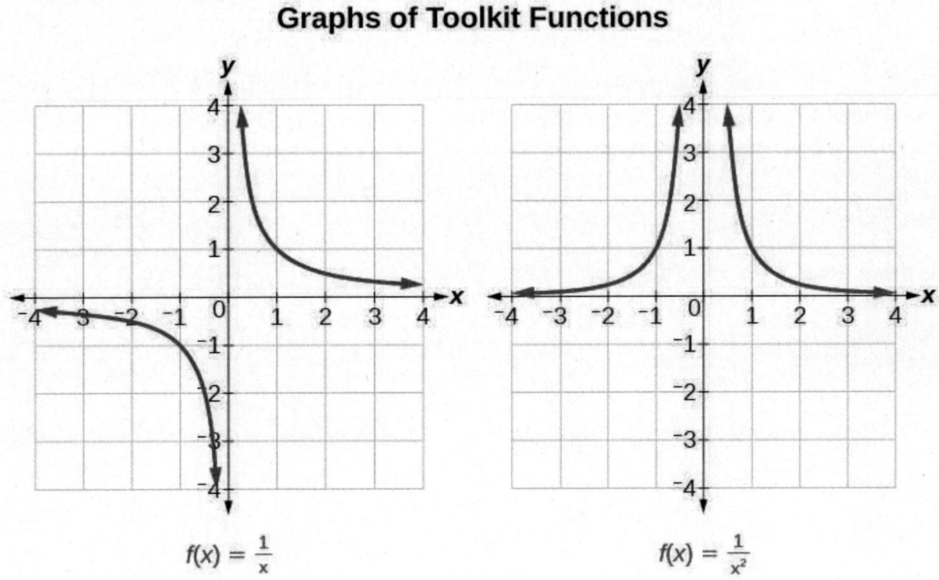

Figure 1

Several things are apparent if we examine the graph of $f(x) = \frac{1}{x}$.

1. On the left branch of the graph, the curve approaches the *x*-axis $(y = 0)$ as $x \to -\infty$.
2. As the graph approaches $x = 0$ from the left, the curve drops, but as we approach zero from the right, the curve rises.
3. Finally, on the right branch of the graph, the curves approaches the *x*-axis $(y = 0)$ as $x \to \infty$.

To summarize, we use **arrow notation** to show that x or $f(x)$ is approaching a particular value. See Table 1.

Symbol	Meaning
$x \to a^-$	x approaches a from the left ($x < a$ but close to a)
$x \to a^+$	x approaches a from the right ($x > a$ but close to a)
$x \to \infty$	x approaches infinity (x increases without bound)
$x \to -\infty$	x approaches negative infinity (x decreases without bound)
$f(x) \to \infty$	the output approaches infinity (the output increases without bound)
$f(x) \to -\infty$	the output approaches negative infinity (the output decreases without bound)
$f(x) \to a$	the output approaches a

Table 1

Local Behavior of $f(x) = \frac{1}{x}$

Let's begin by looking at the reciprocal function, $f(x) = \frac{1}{x}$. We cannot divide by zero, which means the function is undefined at $x = 0$; so zero is not in the domain. As the input values approach zero from the left side (becoming very small, negative values), the function values decrease without bound (in other words, they approach negative infinity). We can see this behavior in Table 2.

x	-0.1	-0.01	-0.001	-0.0001
$f(x) = \frac{1}{x}$	-10	-100	-1000	-10,000

Table 2

We write in arrow notation

$$\text{as } x \to 0^-, f(x) \to -\infty$$

As the input values approach zero from the right side (becoming very small, positive values), the function values increase without bound (approaching infinity). We can see this behavior in Table 3.

x	0.1	0.01	0.001	0.0001
$f(x) = \frac{1}{x}$	10	100	1000	10,000

Table 3

We write in arrow notation

$$\text{As } x \to 0^+,\ f(x) \to \infty.$$

See Figure 2.

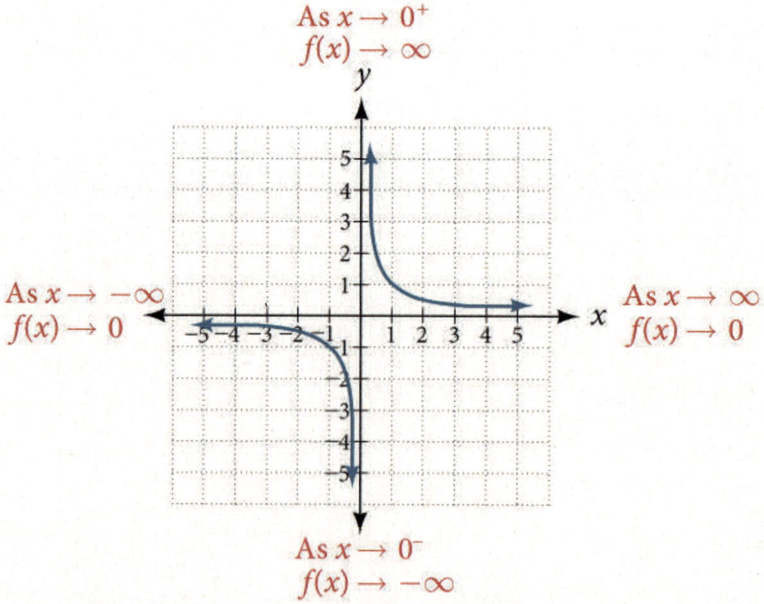

Figure 2

This behavior creates a **vertical asymptote**, which is a vertical line that the graph approaches but never crosses. In this case, the graph is approaching the vertical line $x = 0$ as the input becomes close to zero. See Figure 3.

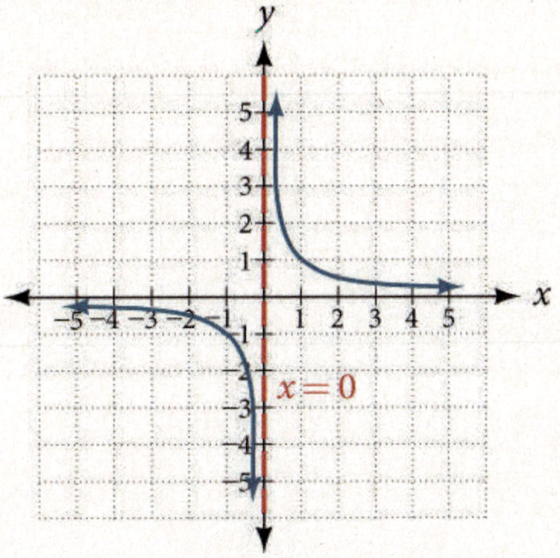

Figure 3

Vertical Asymptote

A **vertical asymptote** of a graph is a vertical line $x = a$ where the graph tends toward positive or negative infinity as the inputs approach a. We write

$$\text{As } x \to a, f(x) \to \pm\infty.$$

End Behavior of $f(x) = \frac{1}{x}$

As the values of x approach infinity, the function values approach 0. As the values of x approach negative infinity, the function values approach 0. See Figure 4. Symbolically, using arrow notation

As $x \to \infty$, $f(x) \to 0$, and as $x \to -\infty$, $f(x) \to 0$.

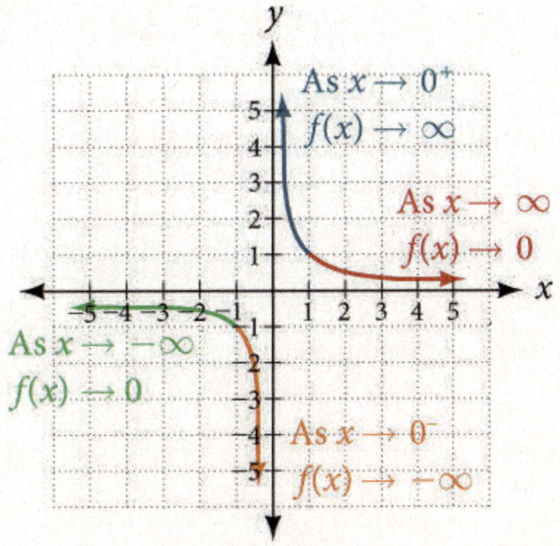

Figure 4

Based on this overall behavior and the graph, we can see that the function approaches 0 but never actually reaches 0; it seems to level off as the inputs become large. This behavior creates a **horizontal asymptote**, a horizontal line that the graph approaches as the input increases or decreases without bound. In this case, the graph is approaching the horizontal line $y = 0$. See Figure 5.

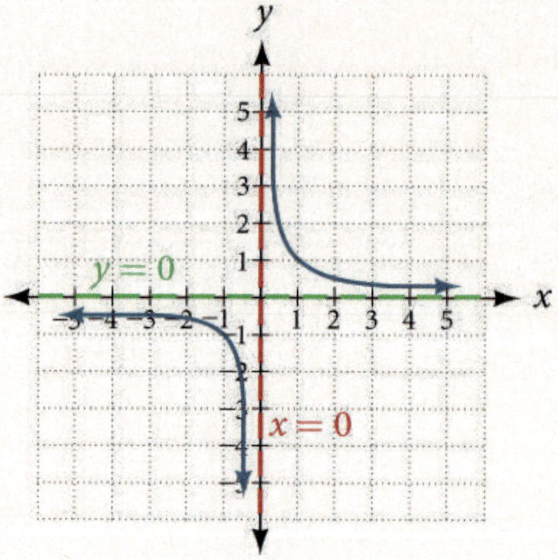

Figure 5

Horizontal Asymptote

A **horizontal asymptote** of a graph is a horizontal line $y = b$ where the graph approaches the line as the inputs increase or decrease without bound. We write

$$\text{As } x \to \infty \text{ or } x \to -\infty, \quad f(x) \to b.$$

EXAMPLE 1

Using Arrow Notation

Use arrow notation to describe the end behavior and local behavior of the function graphed in Figure 6.

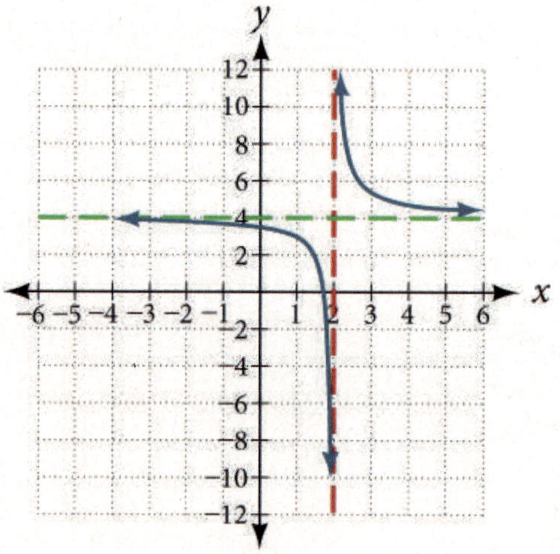

Figure 6

Solution

Notice that the graph is showing a vertical asymptote at $x = 2$, which tells us that the function is undefined at $x = 2$.

$$\text{As } x \to 2^-, f(x) \to -\infty, \quad \text{and as } x \to 2^+, \quad f(x) \to \infty.$$

And as the inputs decrease without bound, the graph appears to be leveling off at output values of 4, indicating a horizontal asymptote at $y = 4$. As the inputs increase without bound, the graph levels off at 4.

$$\text{As } x \to \infty, \quad f(x) \to 4 \text{ and as } x \to -\infty, \quad f(x) \to 4.$$

> **TRY IT** #1 Use arrow notation to describe the end behavior and local behavior for the reciprocal squared function.

EXAMPLE 2

Using Transformations to Graph a Rational Function

Sketch a graph of the reciprocal function shifted two units to the left and up three units. Identify the horizontal and vertical asymptotes of the graph, if any.

Solution

Shifting the graph left 2 and up 3 would result in the function

$$f(x) = \frac{1}{x+2} + 3$$

or equivalently, by giving the terms a common denominator,

$$f(x) = \frac{3x+7}{x+2}$$

The graph of the shifted function is displayed in Figure 7.

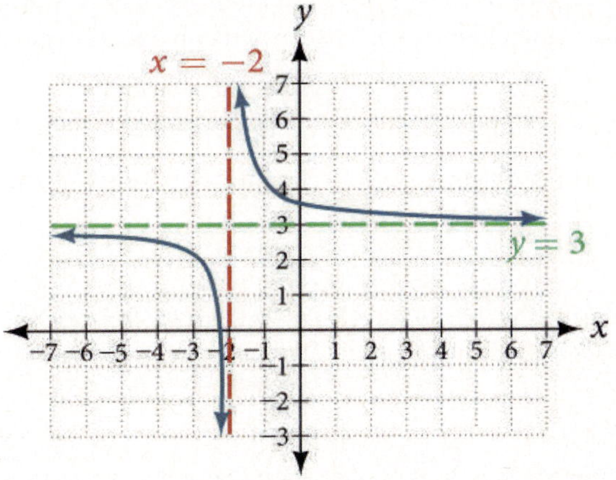

Figure 7

Notice that this function is undefined at $x = -2$, and the graph also is showing a vertical asymptote at $x = -2$.

$$\text{As } x \to -2^-,\ f(x) \to -\infty, \text{ and as } x \to -2^+,\ f(x) \to \infty.$$

As the inputs increase and decrease without bound, the graph appears to be leveling off at output values of 3, indicating a horizontal asymptote at $y = 3$.

$$\text{As } x \to \pm\infty,\ f(x) \to 3.$$

Analysis

Notice that horizontal and vertical asymptotes are shifted left 2 and up 3 along with the function.

> **TRY IT** #2 Sketch the graph, and find the horizontal and vertical asymptotes of the reciprocal squared function that has been shifted right 3 units and down 4 units.

Solving Applied Problems Involving Rational Functions

In Example 2, we shifted a toolkit function in a way that resulted in the function $f(x) = \frac{3x+7}{x+2}$. This is an example of a rational function. A **rational function** is a function that can be written as the quotient of two polynomial functions. Many real-world problems require us to find the ratio of two polynomial functions. Problems involving rates and concentrations often involve rational functions.

Rational Function

A **rational function** is a function that can be written as the quotient of two polynomial functions $P(x)$ and $Q(x)$.

$$f(x) = \frac{P(x)}{Q(x)} = \frac{a_p x^p + a_{p-1} x^{p-1} + \ldots + a_1 x + a_0}{b_q x^q + b_{q-1} x^{q-1} + \ldots + b_1 x + b_0}, Q(x) \neq 0$$

EXAMPLE 3

Solving an Applied Problem Involving a Rational Function

After running out of pre-packaged supplies, a nurse in a refugee camp is preparing an intravenous sugar solution for patients in the camp hospital. A large mixing tank currently contains 100 gallons of distilled water into which 5 pounds of sugar have been mixed. A tap will open pouring 10 gallons per minute of water into the tank at the same time sugar is poured into the tank at a rate of 1 pound per minute. Find the ratio of sugar to water, in pounds per gallon in the tank after 12 minutes. Is that a greater ratio of sugar to water, in pounds per gallon than at the beginning?

Solution

Let t be the number of minutes since the tap opened. Since the water increases at 10 gallons per minute, and the sugar increases at 1 pound per minute, these are constant rates of change. This tells us the amount of water in the tank is changing linearly, as is the amount of sugar in the tank. We can write an equation independently for each:

$$\begin{aligned} \text{water: } W(t) &= 100 + 10t \text{ in gallons} \\ \text{sugar: } S(t) &= 5 + 1t \text{ in pounds} \end{aligned}$$

The ratio of sugar to water, in pounds per gallon, C, will be the ratio of pounds of sugar to gallons of water

$$C(t) = \frac{5 + t}{100 + 10t}$$

The ratio of sugar to water, in pounds per gallon after 12 minutes is given by evaluating $C(t)$ at $t = 12$.

$$\begin{aligned} C(12) &= \frac{5+12}{100+10(12)} \\ &= \frac{17}{220} \end{aligned}$$

This means the ratio of sugar to water, in pounds per gallon is 17 pounds of sugar to 220 gallons of water.

At the beginning, the ratio of sugar to water, in pounds per gallon is

$$\begin{aligned} C(0) &= \frac{5+0}{100+10(0)} \\ &= \frac{1}{20} \end{aligned}$$

Since $\frac{17}{220} \approx 0.08 > \frac{1}{20} = 0.05$, the ratio of sugar to water, in pounds per gallon is greater after 12 minutes than at the beginning.

TRY IT #3 There are 1,200 first-year and 1,500 second-year students at a rally at noon. After 12 p.m., 20 first-year students arrive at the rally every five minutes while 15 second-year students leave the rally. Find the ratio of first-year to second-year students at 1 p.m.

Finding the Domains of Rational Functions

A vertical asymptote represents a value at which a rational function is undefined, so that value is not in the domain of the function. A reciprocal function cannot have values in its domain that cause the denominator to equal zero. In general, to find the domain of a rational function, we need to determine which inputs would cause division by zero.

Domain of a Rational Function

The domain of a rational function includes all real numbers except those that cause the denominator to equal zero.

HOW TO

Given a rational function, find the domain.

1. Set the denominator equal to zero.
2. Solve to find the *x*-values that cause the denominator to equal zero.
3. The domain is all real numbers except those found in Step 2.

EXAMPLE 4

Finding the Domain of a Rational Function

Find the domain of $f(x) = \frac{x+3}{x^2-9}$.

Solution

Begin by setting the denominator equal to zero and solving.

$$\begin{aligned} x^2 - 9 &= 0 \\ x^2 &= 9 \\ x &= \pm 3 \end{aligned}$$

The denominator is equal to zero when $x = \pm 3$. The domain of the function is all real numbers except $x = \pm 3$.

Analysis

A graph of this function, as shown in Figure 8, confirms that the function is not defined when $x = \pm 3$.

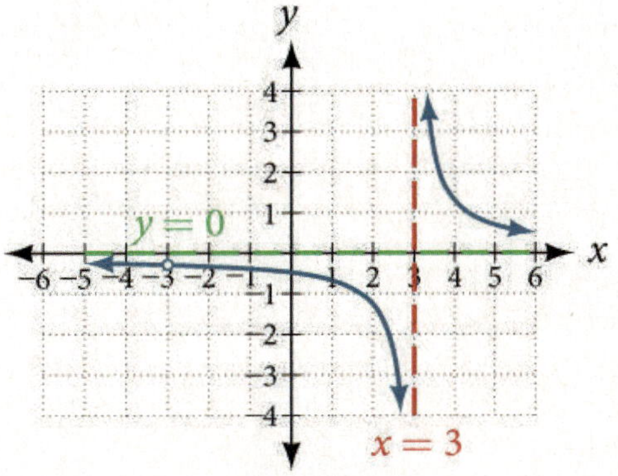

Figure 8

There is a vertical asymptote at $x = 3$ and a hole in the graph at $x = -3$. We will discuss these types of holes in greater detail later in this section.

TRY IT #4 Find the domain of $f(x) = \frac{4x}{5(x-1)(x-5)}$.

Identifying Vertical Asymptotes of Rational Functions

By looking at the graph of a rational function, we can investigate its local behavior and easily see whether there are asymptotes. We may even be able to approximate their location. Even without the graph, however, we can still determine whether a given rational function has any asymptotes, and calculate their location.

Vertical Asymptotes

The vertical asymptotes of a rational function may be found by examining the factors of the denominator that are not common to the factors in the numerator. Vertical asymptotes occur at the zeros of such factors.

HOW TO

Given a rational function, identify any vertical asymptotes of its graph.

1. Factor the numerator and denominator.
2. Note any restrictions in the domain of the function.
3. Reduce the expression by canceling common factors in the numerator and the denominator.
4. Note any values that cause the denominator to be zero in this simplified version. These are where the vertical asymptotes occur.
5. Note any restrictions in the domain where asymptotes do not occur. These are removable discontinuities, or "holes."

EXAMPLE 5

Identifying Vertical Asymptotes

Find the vertical asymptotes of the graph of $k(x) = \frac{5+2x^2}{2-x-x^2}$.

Solution

First, factor the numerator and denominator.

$$\begin{aligned} k(x) &= \frac{5+2x^2}{2-x-x^2} \\ &= \frac{5+2x^2}{(2+x)(1-x)} \end{aligned}$$

To find the vertical asymptotes, we determine where this function will be undefined by setting the denominator equal to zero:

$$\begin{aligned} (2+x)(1-x) &= 0 \\ x &= -2, 1 \end{aligned}$$

Neither $x = -2$ nor $x = 1$ are zeros of the numerator, so the two values indicate two vertical asymptotes. The graph in Figure 9 confirms the location of the two vertical asymptotes.

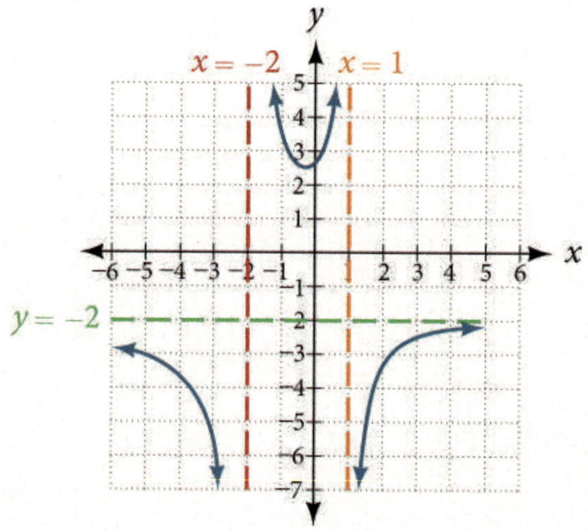

Figure 9

Removable Discontinuities

Occasionally, a graph will contain a hole: a single point where the graph is not defined, indicated by an open circle. We call such a hole a **removable discontinuity**.

For example, the function $f(x) = \frac{x^2-1}{x^2-2x-3}$ may be re-written by factoring the numerator and the denominator.

$$f(x) = \frac{(x+1)(x-1)}{(x+1)(x-3)}$$

Notice that $x + 1$ is a common factor to the numerator and the denominator. The zero of this factor, $x = -1$, is the location of the removable discontinuity. Notice also that $x-3$ is not a factor in both the numerator and denominator. The zero of this factor, $x = 3$, is the vertical asymptote. See Figure 10. [Note that removable discontinuities may not be visible when we use a graphing calculator, depending upon the window selected.]

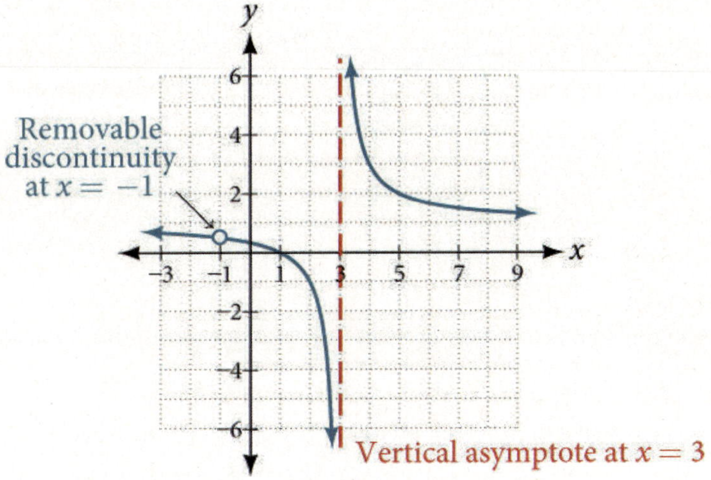

Figure 10

Removable Discontinuities of Rational Functions

A **removable discontinuity** occurs in the graph of a rational function at $x = a$ if a is a zero for a factor in the denominator that is common with a factor in the numerator. We factor the numerator and denominator and check for common factors. If we find any, we set the common factor equal to 0 and solve. This is the location of the removable discontinuity. This is true if the multiplicity of this factor is greater than or equal to that in the denominator. If the multiplicity of this factor is greater in the denominator, then there is still an asymptote at that value.

EXAMPLE 6

Identifying Vertical Asymptotes and Removable Discontinuities for a Graph

Find the vertical asymptotes and removable discontinuities of the graph of $k(x) = \frac{x-2}{x^2-4}$.

Solution

Factor the numerator and the denominator.

$$k(x) = \frac{x-2}{(x-2)(x+2)}$$

Notice that there is a common factor in the numerator and the denominator, $x-2$. The zero for this factor is $x = 2$. This is the location of the removable discontinuity.

Notice that there is a factor in the denominator that is not in the numerator, $x + 2$. The zero for this factor is $x = -2$. The vertical asymptote is $x = -2$. See Figure 11.

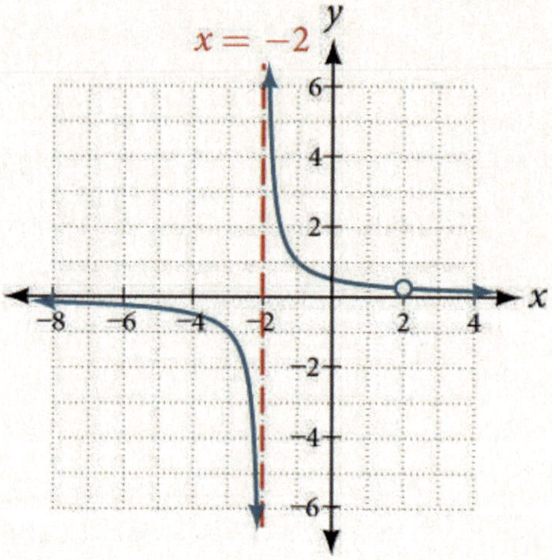

Figure 11

The graph of this function will have the vertical asymptote at $x = -2$, but at $x = 2$ the graph will have a hole.

> **TRY IT** #5 Find the vertical asymptotes and removable discontinuities of the graph of $f(x) = \frac{x^2-25}{x^3-6x^2+5x}$.

Identifying Horizontal Asymptotes of Rational Functions

While vertical asymptotes describe the behavior of a graph as the *output* gets very large or very small, horizontal asymptotes help describe the behavior of a graph as the *input* gets very large or very small. Recall that a polynomial's end behavior will mirror that of the leading term. Likewise, a rational function's end behavior will mirror that of the ratio of the function that is the ratio of the leading terms.

There are three distinct outcomes when checking for horizontal asymptotes:

Case 1: If the degree of the denominator > degree of the numerator, there is a horizontal asymptote at $y = 0$.

$$\text{Example: } f(x) = \frac{4x+2}{x^2+4x-5}$$

In this case, the end behavior is $f(x) \approx \frac{4x}{x^2} = \frac{4}{x}$. This tells us that, as the inputs increase or decrease without bound, this function will behave similarly to the function $g(x) = \frac{4}{x}$, and the outputs will approach zero, resulting in a horizontal asymptote at $y = 0$. See Figure 12. Note that this graph crosses the horizontal asymptote.

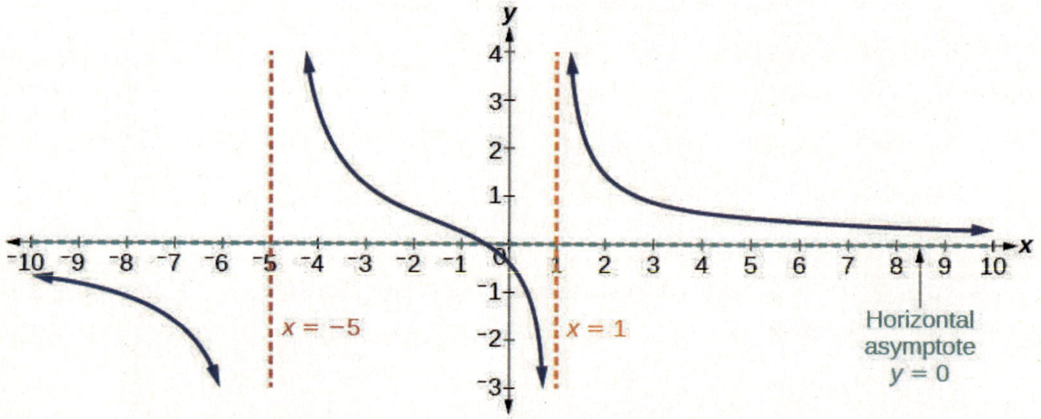

Figure 12 Horizontal asymptote $y = 0$ when $f(x) = \frac{p(x)}{q(x)},\ q(x) \neq 0$ where degree of $p <$ degree of q.

Case 2: If the degree of the denominator < degree of the numerator by one, we get a slant asymptote.

$$\text{Example: } f(x) = \frac{3x^2 - 2x + 1}{x - 1}$$

In this case, the end behavior is $f(x) \approx \frac{3x^2}{x} = 3x$. This tells us that as the inputs increase or decrease without bound, this function will behave similarly to the function $g(x) = 3x$. As the inputs grow large, the outputs will grow and not level off, so this graph has no horizontal asymptote. However, the graph of $g(x) = 3x$ looks like a diagonal line, and since f will behave similarly to g, it will approach a line close to $y = 3x$. This line is a slant asymptote.

To find the equation of the slant asymptote, divide $\frac{3x^2-2x+1}{x-1}$. The quotient is $3x + 1$, and the remainder is 2. The slant asymptote is the graph of the line $g(x) = 3x + 1$. See Figure 13.

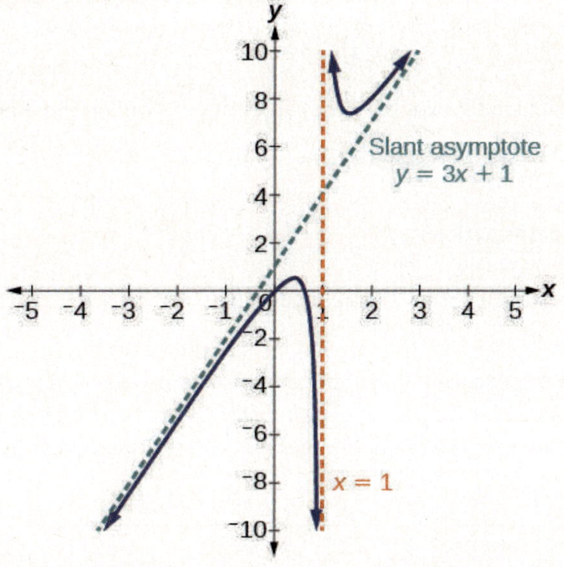

Figure 13 Slant asymptote when $f(x) = \frac{p(x)}{q(x)}$, $q(x) \neq 0$ where degree of $p >$ degree of q by 1.

Case 3: If the degree of the denominator = degree of the numerator, there is a horizontal asymptote at $y = \frac{a_n}{b_n}$, where a_n and b_n are the leading coefficients of $p(x)$ and $q(x)$ for $f(x) = \frac{p(x)}{q(x)}, q(x) \neq 0$.

$$\text{Example: } f(x) = \frac{3x^2 + 2}{x^2 + 4x - 5}$$

In this case, the end behavior is $f(x) \approx \frac{3x^2}{x^2} = 3$. This tells us that as the inputs grow large, this function will behave like the function $g(x) = 3$, which is a horizontal line. As $x \to \pm\infty$, $f(x) \to 3$, resulting in a horizontal asymptote at $y = 3$. See Figure 14. Note that this graph crosses the horizontal asymptote.

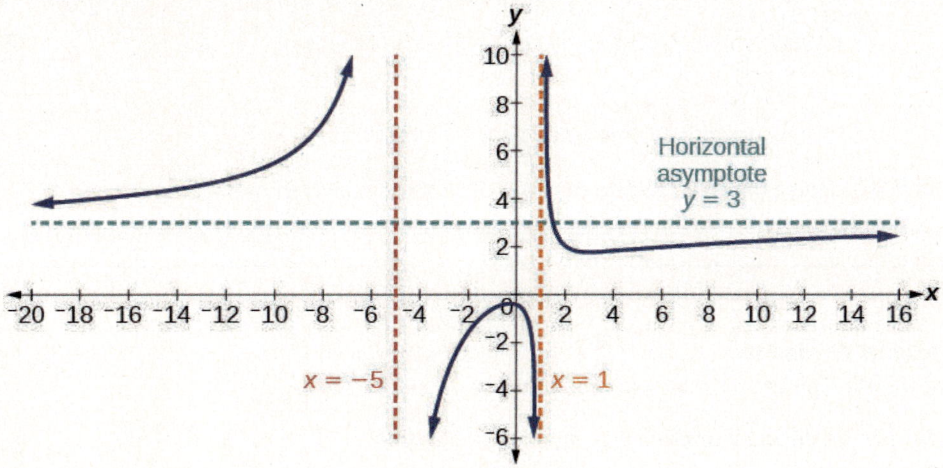

Figure 14 Horizontal asymptote when $f(x) = \frac{p(x)}{q(x)}$, $q(x) \neq 0$ where degree of $p =$ degree of q.

Notice that, while the graph of a rational function will never cross a vertical asymptote, the graph may or may not cross a horizontal or slant asymptote. Also, although the graph of a rational function may have many vertical asymptotes, the graph will have at most one horizontal (or slant) asymptote.

It should be noted that, if the degree of the numerator is larger than the degree of the denominator by more than one, the end behavior of the graph will mimic the behavior of the reduced end behavior fraction. For instance, if we had the function

$$f(x) = \frac{3x^5 - x^2}{x + 3}$$

with end behavior

$$f(x) \approx \frac{3x^5}{x} = 3x^4,$$

the end behavior of the graph would look similar to that of an even polynomial with a positive leading coefficient.

$$x \to \pm\infty, \ f(x) \to \infty$$

Horizontal Asymptotes of Rational Functions

The horizontal asymptote of a rational function can be determined by looking at the degrees of the numerator and denominator.

- Degree of numerator *is less than* degree of denominator: horizontal asymptote at $y = 0$.
- Degree of numerator *is greater than degree of denominator by one*: no horizontal asymptote; slant asymptote.
- Degree of numerator *is equal to* degree of denominator: horizontal asymptote at ratio of leading coefficients.

EXAMPLE 7

Identifying Horizontal and Slant Asymptotes

For the functions listed, identify the horizontal or slant asymptote.

ⓐ $g(x) = \frac{6x^3-10x}{2x^3+5x^2}$ ⓑ $h(x) = \frac{x^2-4x+1}{x+2}$ ⓒ $k(x) = \frac{x^2+4x}{x^3-8}$

Solution

For these solutions, we will use $f(x) = \frac{p(x)}{q(x)}, \ q(x) \neq 0.$

ⓐ $g(x) = \frac{6x^3-10x}{2x^3+5x^2}$: The degree of $p =$ degree of $q = 3,$ so we can find the horizontal asymptote by taking the ratio of the leading terms. There is a horizontal asymptote at $y = \frac{6}{2}$ or $y = 3.$

ⓑ $h(x) = \frac{x^2-4x+1}{x+2}$: The degree of $p = 2$ and degree of $q = 1.$ Since $p > q$ by 1, there is a slant asymptote found at

$$\begin{array}{r|rrr} -2 & 1 & -4 & 1 \\ & & -2 & 12 \\ \hline & 1 & -6 & 13 \end{array}$$

$\frac{x^2-4x+1}{x+2}.$

The quotient is $x-6$ and the remainder is 13. There is a slant asymptote at $y = x-6.$

ⓒ $k(x) = \frac{x^2+4x}{x^3-8}$: The degree of $p = 2$ < degree of $q = 3,$ so there is a horizontal asymptote $y = 0.$

EXAMPLE 8

Identifying Horizontal Asymptotes

In the sugar concentration problem earlier, we created the equation $C(t) = \frac{5+t}{100+10t}.$

Find the horizontal asymptote and interpret it in context of the problem.

Solution

Both the numerator and denominator are linear (degree 1). Because the degrees are equal, there will be a horizontal asymptote at the ratio of the leading coefficients. In the numerator, the leading term is t, with coefficient 1. In the denominator, the leading term is $10t$, with coefficient 10. The horizontal asymptote will be at the ratio of these values:

$$t \to \infty,\ C(t) \to \frac{1}{10}$$

This function will have a horizontal asymptote at $y = \frac{1}{10}$.

This tells us that as the values of t increase, the values of C will approach $\frac{1}{10}$. In context, this means that, as more time goes by, the concentration of sugar in the tank will approach one-tenth of a pound of sugar per gallon of water or $\frac{1}{10}$ pounds per gallon.

EXAMPLE 9

Identifying Horizontal and Vertical Asymptotes

Find the horizontal and vertical asymptotes of the function

$$f(x) = \frac{(x-2)(x+3)}{(x-1)(x+2)(x-5)}$$

Solution

First, note that this function has no common factors, so there are no potential removable discontinuities.

The function will have vertical asymptotes when the denominator is zero, causing the function to be undefined. The denominator will be zero at $x = 1, -2,$ and 5, indicating vertical asymptotes at these values.

The numerator has degree 2, while the denominator has degree 3. Since the degree of the denominator is greater than the degree of the numerator, the denominator will grow faster than the numerator, causing the outputs to tend towards zero as the inputs get large, and so as $x \to \pm\infty,\ f(x) \to 0$. This function will have a horizontal asymptote at $y = 0$. See Figure 15.

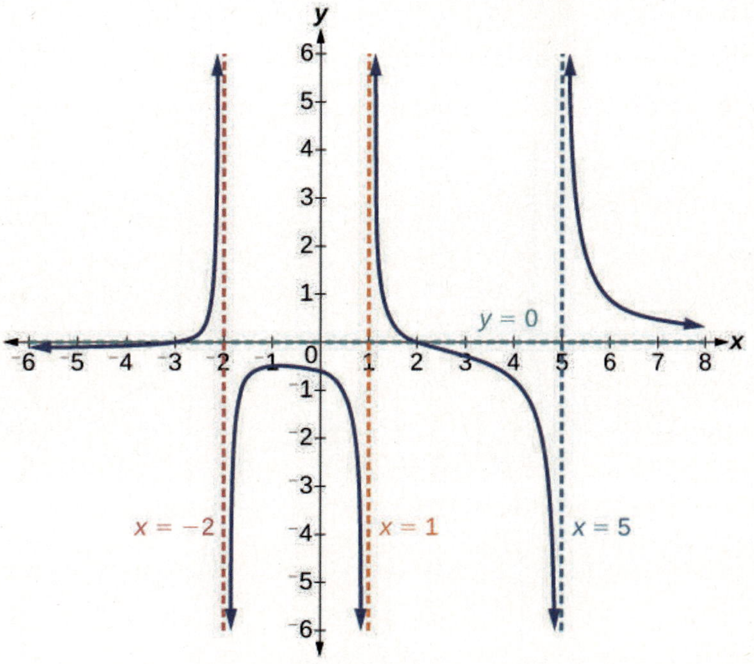

Figure 15

TRY IT #6 Find the vertical and horizontal asymptotes of the function:

$$f(x) = \frac{(2x-1)(2x+1)}{(x-2)(x+3)}$$

Intercepts of Rational Functions

A rational function will have a *y*-intercept at $f(0)$, if the function is defined at zero. A rational function will not have a *y*-intercept if the function is not defined at zero.

Likewise, a rational function will have *x*-intercepts at the inputs that cause the output to be zero. Since a fraction is only equal to zero when the numerator is zero, *x*-intercepts can only occur when the numerator of the rational function is equal to zero.

EXAMPLE 10

Finding the Intercepts of a Rational Function

Find the intercepts of $f(x) = \frac{(x-2)(x+3)}{(x-1)(x+2)(x-5)}$.

Solution

We can find the *y*-intercept by evaluating the function at zero

$$\begin{aligned} f(0) &= \frac{(0-2)(0+3)}{(0-1)(0+2)(0-5)} \\ &= \frac{-6}{10} \\ &= -\frac{3}{5} \\ &= -0.6 \end{aligned}$$

The *x*-intercepts will occur when the function is equal to zero:

$$\begin{aligned} 0 &= \frac{(x-2)(x+3)}{(x-1)(x+2)(x-5)} && \text{This is zero when the numerator is zero.} \\ 0 &= (x-2)(x+3) \\ x &= 2, -3 \end{aligned}$$

The *y*-intercept is $(0, –0.6)$, the *x*-intercepts are $(2, 0)$ and $(–3, 0)$. See Figure 16.

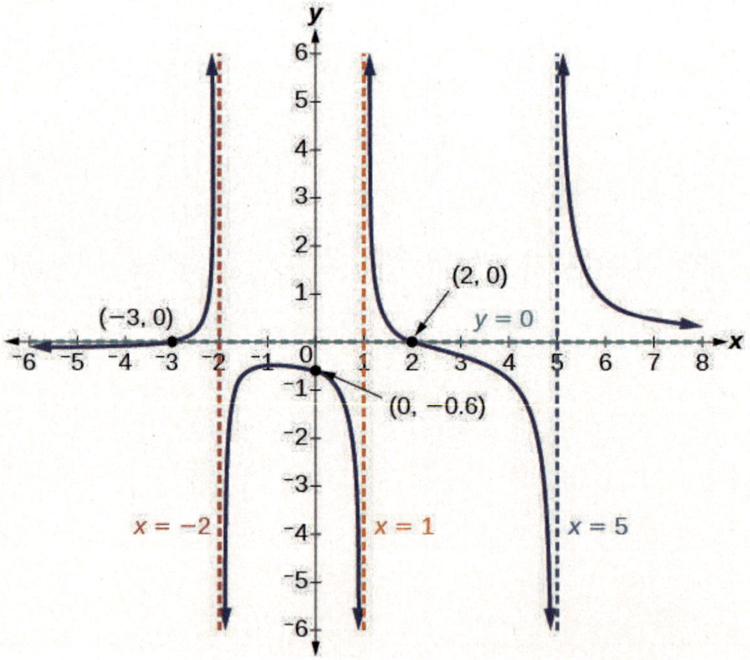

Figure 16

> **TRY IT** #7 Given the reciprocal squared function that is shifted right 3 units and down 4 units, write this as a rational function. Then, find the *x*- and *y*-intercepts and the horizontal and vertical asymptotes.

Graphing Rational Functions

In Example 9, we see that the numerator of a rational function reveals the *x*-intercepts of the graph, whereas the denominator reveals the vertical asymptotes of the graph. As with polynomials, factors of the numerator may have integer powers greater than one. Fortunately, the effect on the shape of the graph at those intercepts is the same as we saw with polynomials.

The vertical asymptotes associated with the factors of the denominator will mirror one of the two toolkit reciprocal functions. When the degree of the factor in the denominator is odd, the distinguishing characteristic is that on one side of the vertical asymptote the graph heads towards positive infinity, and on the other side the graph heads towards negative infinity. See Figure 17.

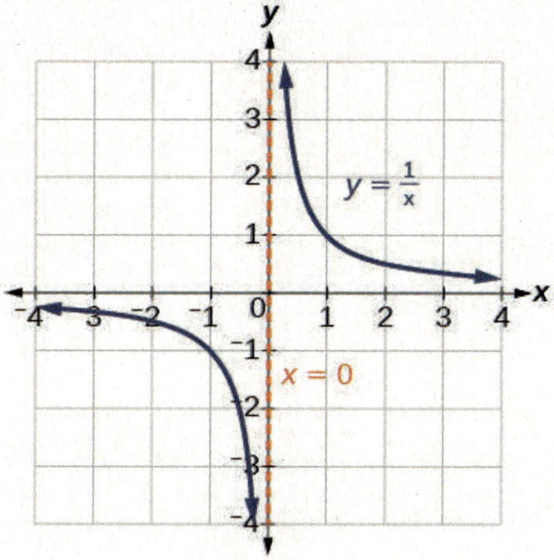

Figure 17

When the degree of the factor in the denominator is even, the distinguishing characteristic is that the graph either heads toward positive infinity on both sides of the vertical asymptote or heads toward negative infinity on both sides. See Figure 18.

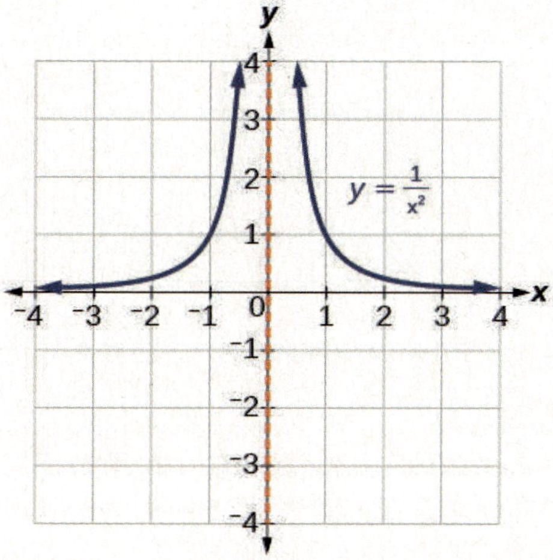

Figure 18

For example, the graph of $f(x) = \frac{(x+1)^2(x-3)}{(x+3)^2(x-2)}$ is shown in Figure 19.

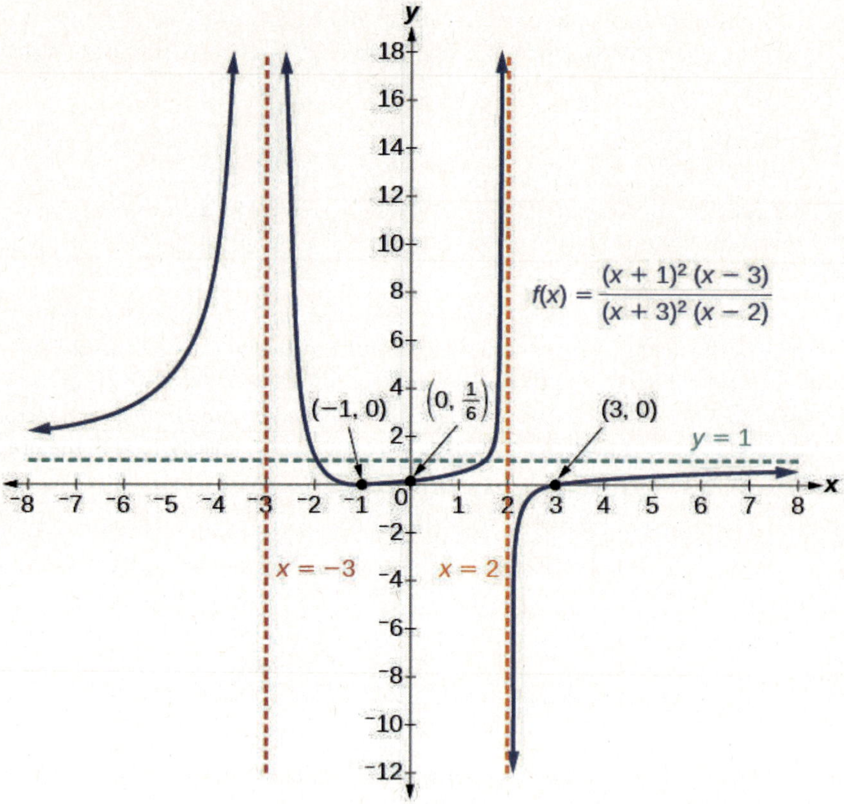

Figure 19

- At the *x*-intercept $x = -1$ corresponding to the $(x + 1)^2$ factor of the numerator, the graph "bounces", consistent with the quadratic nature of the factor.
- At the *x*-intercept $x = 3$ corresponding to the $(x - 3)$ factor of the numerator, the graph passes through the axis as we would expect from a linear factor.
- At the vertical asymptote $x = -3$ corresponding to the $(x + 3)^2$ factor of the denominator, the graph heads towards positive infinity on both sides of the asymptote, consistent with the behavior of the function $f(x) = \frac{1}{x^2}$.
- At the vertical asymptote $x = 2$, corresponding to the $(x - 2)$ factor of the denominator, the graph heads towards positive infinity on the left side of the asymptote and towards negative infinity on the right side.

HOW TO

Given a rational function, sketch a graph.

1. Evaluate the function at 0 to find the *y*-intercept.
2. Factor the numerator and denominator.
3. For factors in the numerator not common to the denominator, determine where each factor of the numerator is zero to find the *x*-intercepts.
4. Find the multiplicities of the *x*-intercepts to determine the behavior of the graph at those points.
5. For factors in the denominator, note the multiplicities of the zeros to determine the local behavior. For those factors not common to the numerator, find the vertical asymptotes by setting those factors equal to zero and then solve.
6. For factors in the denominator common to factors in the numerator, find the removable discontinuities by setting those factors equal to 0 and then solve.
7. Compare the degrees of the numerator and the denominator to determine the horizontal or slant asymptotes.
8. Sketch the graph.

EXAMPLE 11

Graphing a Rational Function

Sketch a graph of $f(x) = \frac{(x+2)(x-3)}{(x+1)^2(x-2)}$.

Solution

We can start by noting that the function is already factored, saving us a step.

Next, we will find the intercepts. Evaluating the function at zero gives the *y*-intercept:

$$\begin{aligned} f(0) &= \frac{(0+2)(0-3)}{(0+1)^2(0-2)} \\ &= 3 \end{aligned}$$

To find the *x*-intercepts, we determine when the numerator of the function is zero. Setting each factor equal to zero, we find *x*-intercepts at $x = -2$ and $x = 3$. At each, the behavior will be linear (multiplicity 1), with the graph passing through the intercept.

We have a *y*-intercept at $(0, 3)$ and *x*-intercepts at $(-2, 0)$ and $(3, 0)$.

To find the vertical asymptotes, we determine when the denominator is equal to zero. This occurs when $x + 1 = 0$ and when $x - 2 = 0$, giving us vertical asymptotes at $x = -1$ and $x = 2$.

There are no common factors in the numerator and denominator. This means there are no removable discontinuities.

Finally, the degree of denominator is larger than the degree of the numerator, telling us this graph has a horizontal asymptote at $y = 0$.

To sketch the graph, we might start by plotting the three intercepts. Since the graph has no *x*-intercepts between the vertical asymptotes, and the *y*-intercept is positive, we know the function must remain positive between the asymptotes, letting us fill in the middle portion of the graph as shown in Figure 20.

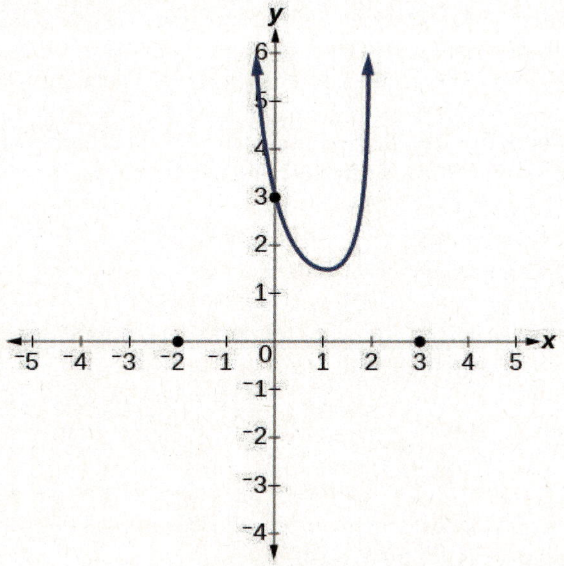

Figure 20

The factor associated with the vertical asymptote at $x = -1$ was squared, so we know the behavior will be the same on both sides of the asymptote. The graph heads toward positive infinity as the inputs approach the asymptote on the right, so the graph will head toward positive infinity on the left as well.

For the vertical asymptote at $x = 2$, the factor was not squared, so the graph will have opposite behavior on either side of the asymptote. See Figure 21. After passing through the *x*-intercepts, the graph will then level off toward an output of zero, as indicated by the horizontal asymptote.

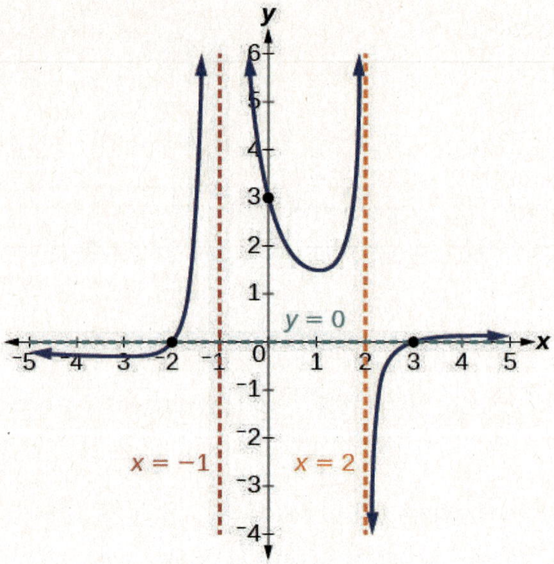

Figure 21

> TRY IT #8 Given the function $f(x) = \frac{(x+2)^2(x-2)}{2(x-1)^2(x-3)}$, use the characteristics of polynomials and rational functions to describe its behavior and sketch the function.

Writing Rational Functions

Now that we have analyzed the equations for rational functions and how they relate to a graph of the function, we can use information given by a graph to write the function. A rational function written in factored form will have an *x*-intercept where each factor of the numerator is equal to zero. (An exception occurs in the case of a removable discontinuity.) As a result, we can form a numerator of a function whose graph will pass through a set of *x*-intercepts by introducing a corresponding set of factors. Likewise, because the function will have a vertical asymptote where each factor of the denominator is equal to zero, we can form a denominator that will produce the vertical asymptotes by introducing a corresponding set of factors.

Writing Rational Functions from Intercepts and Asymptotes

If a rational function has *x*-intercepts at $x = x_1, x_2, ..., x_n$, vertical asymptotes at $x = v_1, v_2, \ldots, v_m$, and no $x_i = \text{any } v_j$, then the function can be written in the form:

$$f(x) = a\frac{(x - x_1)^{p_1}(x - x_2)^{p_2} \cdots (x - x_n)^{p_n}}{(x - v_1)^{q_1}(x - v_2)^{q_2} \cdots (x - v_m)^{q_n}}$$

where the powers p_i or q_i on each factor can be determined by the behavior of the graph at the corresponding intercept or asymptote, and the stretch factor a can be determined given a value of the function other than the *x*-intercept or by the horizontal asymptote if it is nonzero.

HOW TO

Given a graph of a rational function, write the function.

1. Determine the factors of the numerator. Examine the behavior of the graph at the *x*-intercepts to determine the zeroes and their multiplicities. (This is easy to do when finding the "simplest" function with small multiplicities—such as 1 or 3—but may be difficult for larger multiplicities—such as 5 or 7, for example.)
2. Determine the factors of the denominator. Examine the behavior on both sides of each vertical asymptote to

determine the factors and their powers.
3. Use any clear point on the graph to find the stretch factor.

EXAMPLE 12

Writing a Rational Function from Intercepts and Asymptotes

Write an equation for the rational function shown in Figure 22.

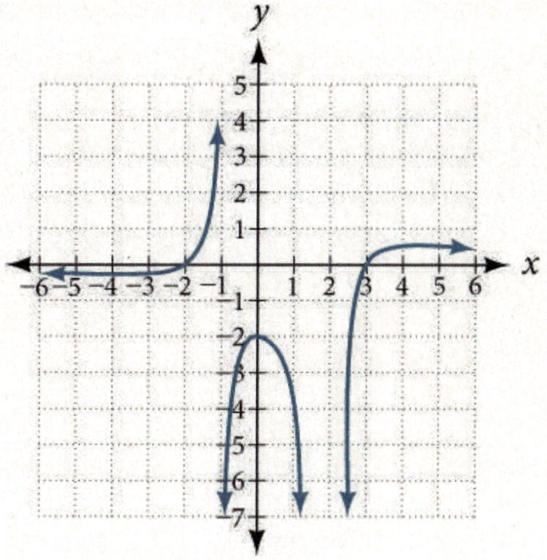

Figure 22

Solution

The graph appears to have x-intercepts at $x = -2$ and $x = 3$. At both, the graph passes through the intercept, suggesting linear factors. The graph has two vertical asymptotes. The one at $x = -1$ seems to exhibit the basic behavior similar to $\frac{1}{x}$, with the graph heading toward positive infinity on one side and heading toward negative infinity on the other. The asymptote at $x = 2$ is exhibiting a behavior similar to $\frac{1}{x^2}$, with the graph heading toward negative infinity on both sides of the asymptote. See Figure 23.

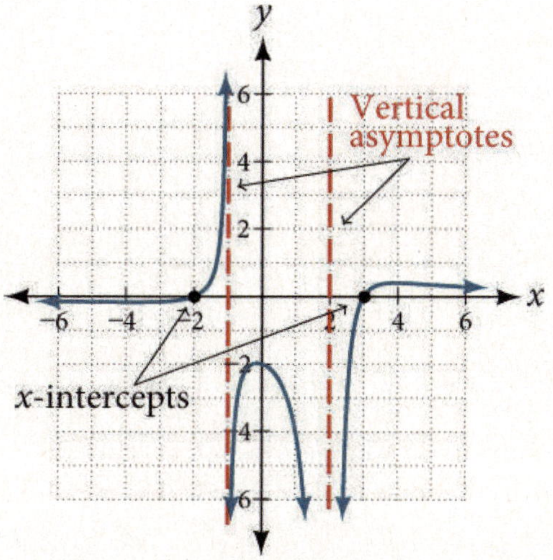

Figure 23

We can use this information to write a function of the form

$$f(x) = a\frac{(x+2)(x-3)}{(x+1)(x-2)^2}$$

To find the stretch factor, we can use another clear point on the graph, such as the y-intercept $(0, -2)$.

$$\begin{aligned} -2 &= a\frac{(0+2)(0-3)}{(0+1)(0-2)^2} \\ -2 &= a\frac{-6}{4} \\ a &= \frac{-8}{-6} = \frac{4}{3} \end{aligned}$$

This gives us a final function of $f(x) = \frac{4(x+2)(x-3)}{3(x+1)(x-2)^2}$.

MEDIA

Access these online resources for additional instruction and practice with rational functions.

Graphing Rational Functions (http://openstax.org/l/graphrational)
Find the Equation of a Rational Function (http://openstax.org/l/equatrational)
Determining Vertical and Horizontal Asymptotes (http://openstax.org/l/asymptote)
Find the Intercepts, Asymptotes, and Hole of a Rational Function (http://openstax.org/l/interasymptote)

5.6 SECTION EXERCISES

Verbal

1. What is the fundamental difference in the algebraic representation of a polynomial function and a rational function?

2. What is the fundamental difference in the graphs of polynomial functions and rational functions?

3. If the graph of a rational function has a removable discontinuity, what must be true of the functional rule?

4. Can a graph of a rational function have no vertical asymptote? If so, how?

5. Can a graph of a rational function have no *x*-intercepts? If so, how?

Algebraic

For the following exercises, find the domain of the rational functions.

6. $f(x) = \frac{x-1}{x+2}$

7. $f(x) = \frac{x+1}{x^2-1}$

8. $f(x) = \frac{x^2+4}{x^2-2x-8}$

9. $f(x) = \frac{x^2+4x-3}{x^4-5x^2+4}$

For the following exercises, find the domain, vertical asymptotes, and horizontal asymptotes of the functions.

10. $f(x) = \frac{4}{x-1}$

11. $f(x) = \frac{2}{5x+2}$

12. $f(x) = \frac{x}{x^2-9}$

13. $f(x) = \frac{x}{x^2+5x-36}$

14. $f(x) = \frac{3+x}{x^3-27}$

15. $f(x) = \frac{3x-4}{x^3-16x}$

16. $f(x) = \frac{x^2-1}{x^3+9x^2+14x}$

17. $f(x) = \frac{x+5}{x^2-25}$

18. $f(x) = \frac{x-4}{x-6}$

19. $f(x) = \frac{4-2x}{3x-1}$

For the following exercises, find the x- and y-intercepts for the functions.

20. $f(x) = \frac{x+5}{x^2+4}$

21. $f(x) = \frac{x}{x^2-x}$

22. $f(x) = \frac{x^2+8x+7}{x^2+11x+30}$

23. $f(x) = \frac{x^2+x+6}{x^2-10x+24}$

24. $f(x) = \frac{94-2x^2}{3x^2-12}$

For the following exercises, describe the local and end behavior of the functions.

25. $f(x) = \frac{x}{2x+1}$

26. $f(x) = \frac{2x}{x-6}$

27. $f(x) = \frac{-2x}{x-6}$

28. $f(x) = \frac{x^2-4x+3}{x^2-4x-5}$

29. $f(x) = \frac{2x^2-32}{6x^2+13x-5}$

For the following exercises, find the slant asymptote of the functions.

30. $f(x) = \frac{24x^2+6x}{2x+1}$

31. $f(x) = \frac{4x^2-10}{2x-4}$

32. $f(x) = \frac{81x^2-18}{3x-2}$

33. $f(x) = \frac{6x^3-5x}{3x^2+4}$

34. $f(x) = \frac{x^2+5x+4}{x-1}$

Graphical

For the following exercises, use the given transformation to graph the function. Note the vertical and horizontal asymptotes.

35. The reciprocal function shifted up two units.

36. The reciprocal function shifted down one unit and left three units.

37. The reciprocal squared function shifted to the right 2 units.

38. The reciprocal squared function shifted down 2 units and right 1 unit.

For the following exercises, find the horizontal intercepts, the vertical intercept, the vertical asymptotes, and the horizontal or slant asymptote of the functions. Use that information to sketch a graph.

39. $p(x) = \frac{2x-3}{x+4}$

40. $q(x) = \frac{x-5}{3x-1}$

41. $s(x) = \frac{4}{(x-2)^2}$

42. $r(x) = \frac{5}{(x+1)^2}$

43. $f(x) = \frac{3x^2-14x-5}{3x^2+8x-16}$

44. $g(x) = \frac{2x^2+7x-15}{3x^2-14x+15}$

45. $a(x) = \frac{x^2+2x-3}{x^2-1}$

46. $b(x) = \frac{x^2-x-6}{x^2-4}$

47. $h(x) = \frac{2x^2+x-1}{x-4}$

48. $k(x) = \frac{2x^2-3x-20}{x-5}$

49. $w(x) = \frac{(x-1)(x+3)(x-5)}{(x+2)^2(x-4)}$

50. $z(x) = \frac{(x+2)^2(x-5)}{(x-3)(x+1)(x+4)}$

For the following exercises, write an equation for a rational function with the given characteristics.

51. Vertical asymptotes at $x = 5$ and $x = -5$, x-intercepts at $(2, 0)$ and $(-1, 0)$, y-intercept at $(0, 4)$

52. Vertical asymptotes at $x = -4$ and $x = -1$, x-intercepts at $(1, 0)$ and $(5, 0)$, y-intercept at $(0, 7)$

53. Vertical asymptotes at $x = -4$ and $x = -5$, x-intercepts at $(4, 0)$ and $(-6, 0)$, Horizontal asymptote at $y = 7$

54. Vertical asymptotes at $x = -3$ and $x = 6$, x-intercepts at $(-2, 0)$ and $(1, 0)$, Horizontal asymptote at $y = -2$

55. Vertical asymptote at $x = -1$, Double zero at $x = 2$, y-intercept at $(0, 2)$

56. Vertical asymptote at $x = 3$, Double zero at $x = 1$, y-intercept at $(0, 4)$

For the following exercises, use the graphs to write an equation for the function.

57.

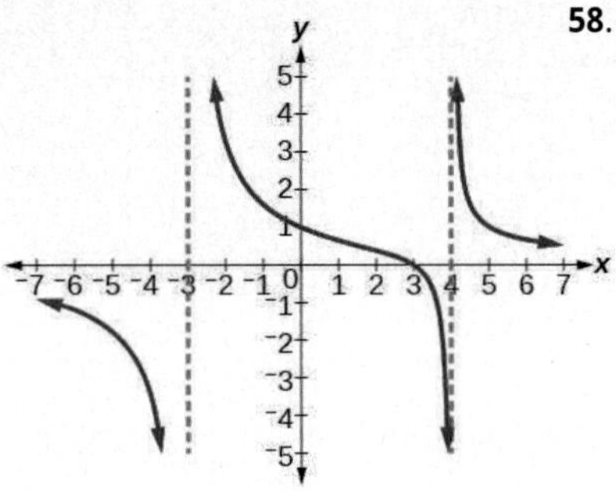

58.

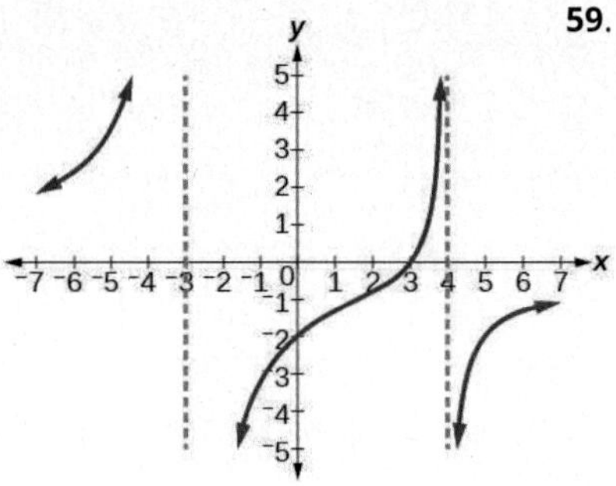

59.

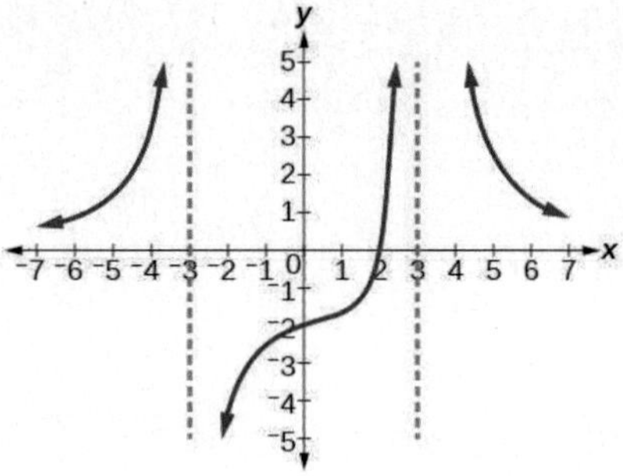

60.

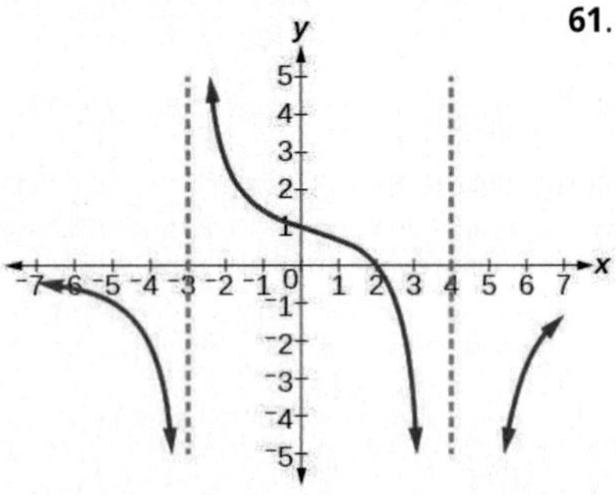

61.

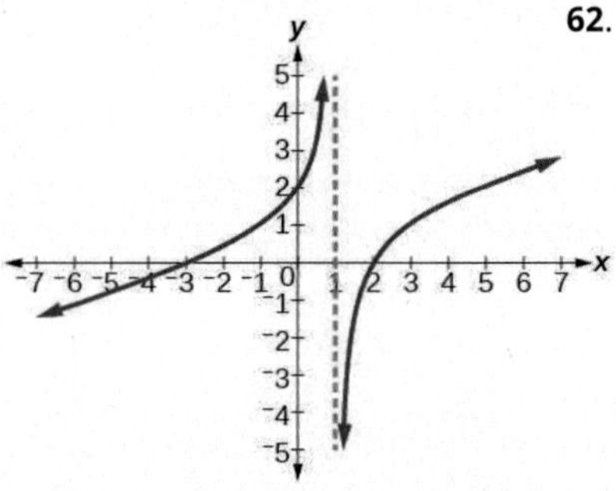

62.

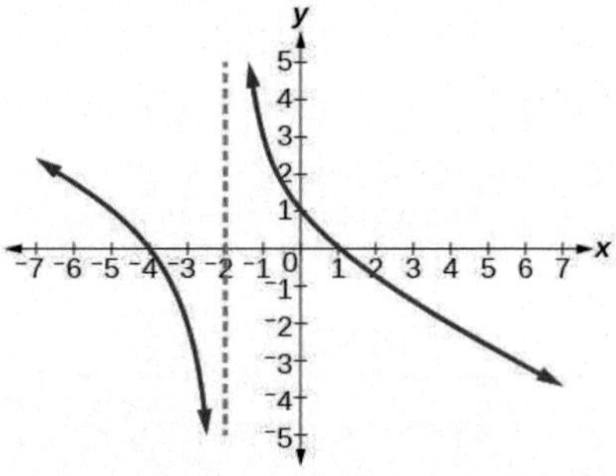

63.

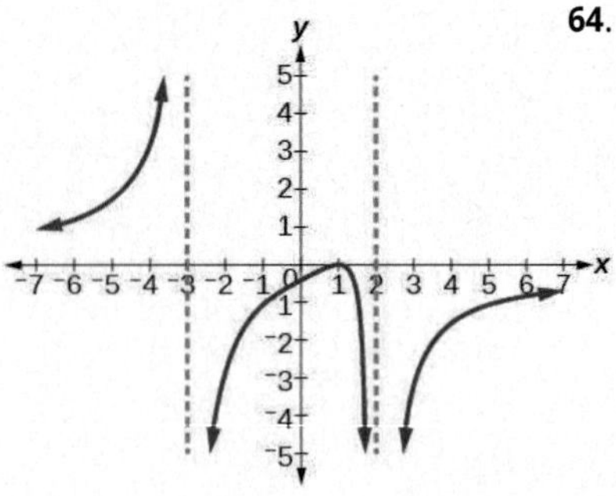

64.

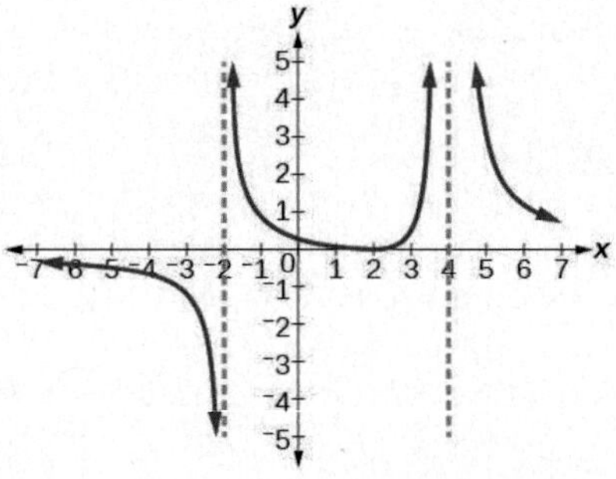

Numeric

For the following exercises, make tables to show the behavior of the function near the vertical asymptote and reflecting the horizontal asymptote

65. $f(x) = \frac{1}{x-2}$

66. $f(x) = \frac{x}{x-3}$

67. $f(x) = \frac{2x}{x+4}$

68. $f(x) = \frac{2x}{(x-3)^2}$

69. $f(x) = \frac{x^2}{x^2+2x+1}$

Technology

For the following exercises, use a calculator to graph $f(x)$. Use the graph to solve $f(x) > 0$.

70. $f(x) = \frac{2}{x+1}$

71. $f(x) = \frac{4}{2x-3}$

72. $f(x) = \frac{2}{(x-1)(x+2)}$

73. $f(x) = \frac{x+2}{(x-1)(x-4)}$

74. $f(x) = \frac{(x+3)^2}{(x-1)^2(x+1)}$

Extensions

For the following exercises, identify the removable discontinuity.

75. $f(x) = \frac{x^2-4}{x-2}$

76. $f(x) = \frac{x^3+1}{x+1}$

77. $f(x) = \frac{x^2+x-6}{x-2}$

78. $f(x) = \frac{2x^2+5x-3}{x+3}$

79. $f(x) = \frac{x^3+x^2}{x+1}$

Real-World Applications

For the following exercises, express a rational function that describes the situation.

80. In the refugee camp hospital, a large mixing tank currently contains 200 gallons of water, into which 10 pounds of sugar have been mixed. A tap will open, pouring 10 gallons of water per minute into the tank at the same time sugar is poured into the tank at a rate of 3 pounds per minute. Find the concentration (pounds per gallon) of sugar in the tank after t minutes.

81. In the refugee camp hospital, a large mixing tank currently contains 300 gallons of water, into which 8 pounds of sugar have been mixed. A tap will open, pouring 20 gallons of water per minute into the tank at the same time sugar is poured into the tank at a rate of 2 pounds per minute. Find the concentration (pounds per gallon) of sugar in the tank after t minutes.

For the following exercises, use the given rational function to answer the question.

82. The concentration C of a drug in a patient's bloodstream t hours after injection is given by $C(t) = \frac{2t}{3+t^2}$. What happens to the concentration of the drug as t increases?

83. The concentration C of a drug in a patient's bloodstream t hours after injection is given by $C(t) = \frac{100t}{2t^2+75}$. Use a calculator to approximate the time when the concentration is highest.

For the following exercises, construct a rational function that will help solve the problem. Then, use a calculator to answer the question.

84. An open box with a square base is to have a volume of 108 cubic inches. Find the dimensions of the box that will have minimum surface area. Let x = length of the side of the base.

85. A rectangular box with a square base is to have a volume of 20 cubic feet. The material for the base costs 30 cents/ square foot. The material for the sides costs 10 cents/square foot. The material for the top costs 20 cents/square foot. Determine the dimensions that will yield minimum cost. Let x = length of the side of the base.

86. A right circular cylinder has volume of 100 cubic inches. Find the radius and height that will yield minimum surface area. Let x = radius.

87. A right circular cylinder with no top has a volume of 50 cubic meters. Find the radius that will yield minimum surface area. Let x = radius.

88. A right circular cylinder is to have a volume of 40 cubic inches. It costs 4 cents/square inch to construct the top and bottom and 1 cent/square inch to construct the rest of the cylinder. Find the radius to yield minimum cost. Let x = radius.

5.7 Inverses and Radical Functions

Learning Objectives

In this section, you will:

- Find the inverse of an invertible polynomial function.
- Restrict the domain to find the inverse of a polynomial function.

Park rangers and other trail managers may construct rock piles, stacks, or other arrangements, usually called cairns, to mark trails or other landmarks. (Rangers and environmental scientists discourage hikers from doing the same, in order to avoid confusion and preserve the habitats of plants and animals.) A cairn in the form of a mound of gravel is in the shape of a cone with the height equal to twice the radius.

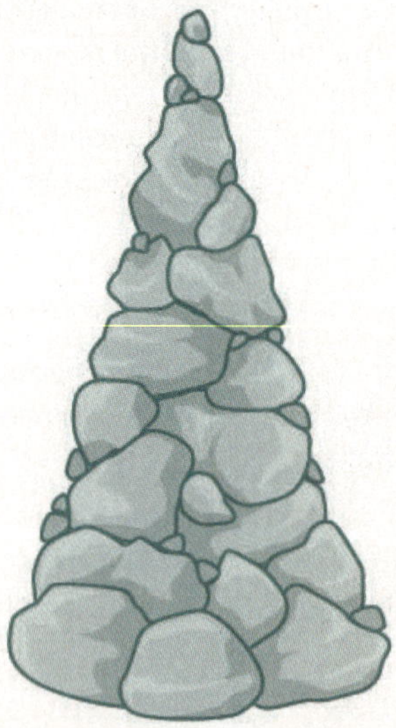

Figure 1

The volume is found using a formula from elementary geometry.

$$\begin{aligned} V &= \tfrac{1}{3}\pi r^2 h \\ &= \tfrac{1}{3}\pi r^2 (2r) \\ &= \tfrac{2}{3}\pi r^3 \end{aligned}$$

We have written the volume V in terms of the radius r. However, in some cases, we may start out with the volume and want to find the radius. For example: A customer purchases 100 cubic feet of gravel to construct a cone shape mound with a height twice the radius. What are the radius and height of the new cone? To answer this question, we use the formula

$$r = \sqrt[3]{\frac{3V}{2\pi}}$$

This function is the inverse of the formula for V in terms of r.

In this section, we will explore the inverses of polynomial and rational functions and in particular the radical functions we encounter in the process.

Finding the Inverse of a Polynomial Function

Two functions f and g are inverse functions if for every coordinate pair in $f, (a, b)$, there exists a corresponding coordinate pair in the inverse function, $g, (b,\ a)$. In other words, the coordinate pairs of the inverse functions have the input and output interchanged. Only one-to-one functions have inverses. Recall that a one-to-one function has a unique output value for each input value and passes the horizontal line test.

For example, suppose the Sustainability Club builds a water runoff collector in the shape of a parabolic trough as shown in Figure 2. We can use the information in the figure to find the surface area of the water in the trough as a function of the depth of the water.

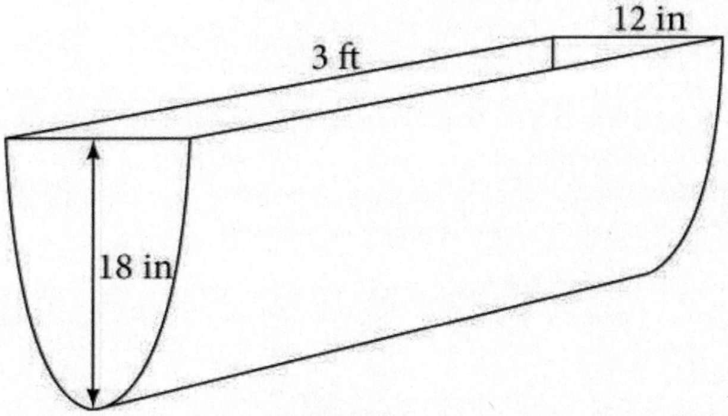

Figure 2

Because it will be helpful to have an equation for the parabolic cross-sectional shape, we will impose a coordinate system at the cross section, with x measured horizontally and y measured vertically, with the origin at the vertex of the parabola. See Figure 3.

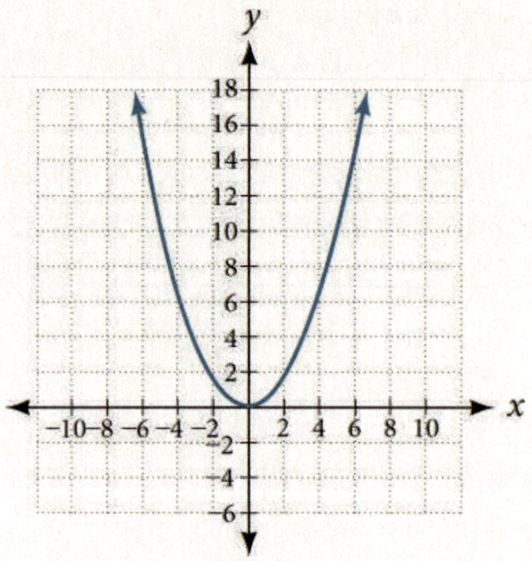

Figure 3

From this we find an equation for the parabolic shape. We placed the origin at the vertex of the parabola, so we know the equation will have form $y(x) = ax^2$. Our equation will need to pass through the point (6, 18), from which we can solve for the stretch factor a.

$$\begin{aligned} 18 &= a6^2 \\ a &= \tfrac{18}{36} \\ &= \tfrac{1}{2} \end{aligned}$$

Our parabolic cross section has the equation

$$y(x) = \frac{1}{2}x^2$$

We are interested in the surface area of the water, so we must determine the width at the top of the water as a function of the water depth. For any depth y, the width will be given by $2x$, so we need to solve the equation above for x and find the inverse function. However, notice that the original function is not one-to-one, and indeed, given any output there are two inputs that produce the same output, one positive and one negative.

To find an inverse, we can restrict our original function to a limited domain on which it *is* one-to-one. In this case, it makes sense to restrict ourselves to positive x values. On this domain, we can find an inverse by solving for the input variable:

$$\begin{aligned} y &= \tfrac{1}{2}x^2 \\ 2y &= x^2 \\ x &= \pm\sqrt{2y} \end{aligned}$$

This is not a function as written. We are limiting ourselves to positive x values, so we eliminate the negative solution, giving us the inverse function we're looking for.

$$y = \frac{x^2}{2}, \quad x > 0$$

Because x is the distance from the center of the parabola to either side, the entire width of the water at the top will be $2x$. The trough is 3 feet (36 inches) long, so the surface area will then be:

$$\begin{aligned} \text{Area} &= l \cdot w \\ &= 36 \cdot 2x \\ &= 72x \\ &= 72\sqrt{2y} \end{aligned}$$

This example illustrates two important points:

1. When finding the inverse of a quadratic, we have to limit ourselves to a domain on which the function is one-to-one.
2. The inverse of a quadratic function is a square root function. Both are toolkit functions and different types of power

functions.

Functions involving roots are often called radical functions. While it is not possible to find an inverse of most polynomial functions, some basic polynomials do have inverses. Such functions are called **invertible functions**, and we use the notation $f^{-1}(x)$.

Warning: $f^{-1}(x)$ is not the same as the reciprocal of the function $f(x)$. This use of "–1" is reserved to denote inverse functions. To denote the reciprocal of a function $f(x)$, we would need to write $(f(x))^{-1} = \frac{1}{f(x)}$.

An important relationship between inverse functions is that they "undo" each other. If f^{-1} is the inverse of a function f, then f is the inverse of the function f^{-1}. In other words, whatever the function f does to x, f^{-1} undoes it—and vice-versa.

$$f^{-1}(f(x)) = x, \quad \text{for all } x \text{ in the domain of } f$$

and

$$f\left(f^{-1}(x)\right) = x, \quad \text{for all } x \text{ in the domain of } f^{-1}$$

Note that the inverse switches the domain and range of the original function.

Verifying Two Functions Are Inverses of One Another

Two functions, f and g, are inverses of one another if for all x in the domain of f and g.

$$g(f(x)) = f(g(x)) = x$$

HOW TO

Given a polynomial function, find the inverse of the function by restricting the domain in such a way that the new function is one-to-one.

1. Replace $f(x)$ with y.
2. Interchange x and y.
3. Solve for y, and rename the function $f^{-1}(x)$.

EXAMPLE 1

Verifying Inverse Functions

Show that $f(x) = \frac{1}{x+1}$ and $f^{-1}(x) = \frac{1}{x} - 1$ are inverses, for $x \neq 0, -1$.

Solution

We must show that $f^{-1}(f(x)) = x$ and $f\left(f^{-1}(x)\right) = x$.

$$\begin{aligned} f^{-1}(f(x)) &= f^{-1}\left(\frac{1}{x+1}\right) \\ &= \frac{1}{\frac{1}{x+1}} - 1 \\ &= (x+1) - 1 \\ &= x \\ f(f^{-1}(x)) &= f\left(\frac{1}{x} - 1\right) \\ &= \frac{1}{\left(\frac{1}{x}-1\right)+1} \\ &= \frac{1}{\frac{1}{x}} \\ &= x \end{aligned}$$

Therefore, $f(x) = \frac{1}{x+1}$ and $f^{-1}(x) = \frac{1}{x} - 1$ are inverses.

TRY IT #1 Show that $f(x) = \frac{x+5}{3}$ and $f^{-1}(x) = 3x - 5$ are inverses.

EXAMPLE 2

Finding the Inverse of a Cubic Function

Find the inverse of the function $f(x) = 5x^3 + 1$.

Solution

This is a transformation of the basic cubic toolkit function, and based on our knowledge of that function, we know it is one-to-one. Solving for the inverse by solving for x.

$$\begin{aligned} y &= 5x^3 + 1 \\ x &= 5y^3 + 1 \\ x - 1 &= 5y^3 \\ \frac{x-1}{5} &= y^3 \\ f^{-1}(x) &= \sqrt[3]{\frac{x-1}{5}} \end{aligned}$$

Analysis

Look at the graph of f and f^{-1}. Notice that one graph is the reflection of the other about the line $y = x$. This is always the case when graphing a function and its inverse function.

Also, since the method involved interchanging x and y, notice corresponding points. If (a, b) is on the graph of f, then (b, a) is on the graph of f^{-1}. Since $(0, 1)$ is on the graph of f, then $(1, 0)$ is on the graph of f^{-1}. Similarly, since $(1, 6)$ is on the graph of f, then $(6, 1)$ is on the graph of f^{-1}. See Figure 4.

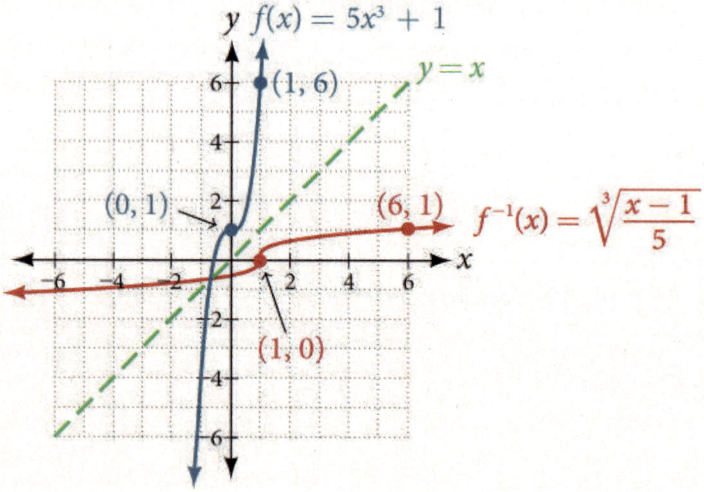

Figure 4

TRY IT #2 Find the inverse function of $f(x) = \sqrt[3]{x+4}$.

Restricting the Domain to Find the Inverse of a Polynomial Function

So far, we have been able to find the inverse functions of cubic functions without having to restrict their domains. However, as we know, not all cubic polynomials are one-to-one. Some functions that are not one-to-one may have their domain restricted so that they are one-to-one, but only over that domain. The function over the restricted domain would then have an inverse function. Since quadratic functions are not one-to-one, we must restrict their domain in order to find their inverses.

Restricting the Domain

If a function is not one-to-one, it cannot have an inverse. If we restrict the domain of the function so that it becomes one-to-one, thus creating a new function, this new function will have an inverse.

HOW TO

Given a polynomial function, restrict the domain of a function that is not one-to-one and then find the inverse.

1. Restrict the domain by determining a domain on which the original function is one-to-one.
2. Replace $f(x)$ with y.
3. Interchange x and y.
4. Solve for y, and rename the function or pair of function $f^{-1}(x)$.
5. Revise the formula for $f^{-1}(x)$ by ensuring that the outputs of the inverse function correspond to the restricted domain of the original function.

EXAMPLE 3

Restricting the Domain to Find the Inverse of a Polynomial Function

Find the inverse function of f:

ⓐ $f(x) = (x-4)^2,\ x \geq 4$ ⓑ $f(x) = (x-4)^2,\ x \leq 4$

Solution

The original function $f(x) = (x-4)^2$ is not one-to-one, but the function is restricted to a domain of $x \geq 4$ or $x \leq 4$ on which it is one-to-one. See Figure 5.

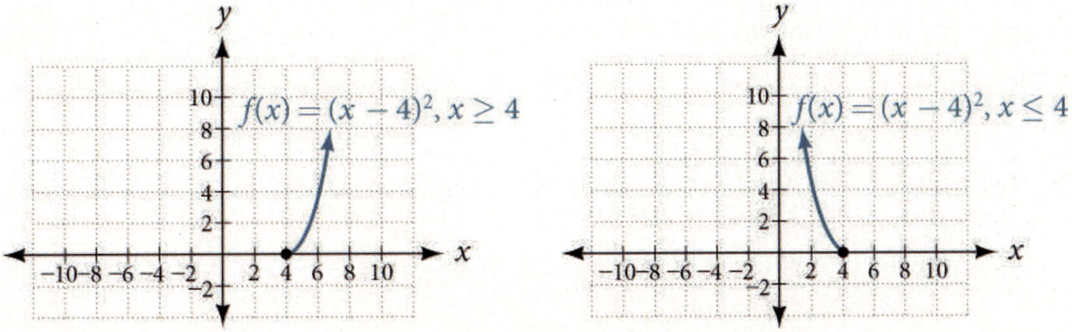

Figure 5

To find the inverse, start by replacing $f(x)$ with the simple variable y.

$$\begin{aligned} y &= (x-4)^2 && \text{Interchange} x \text{and } y. \\ x &= (y-4)^2 && \text{Take the square root.} \\ \pm\sqrt{x} &= y-4 && \text{Add 4 to both sides.} \\ 4\pm\sqrt{x} &= y \end{aligned}$$

This is not a function as written. We need to examine the restrictions on the domain of the original function to determine the inverse. Since we reversed the roles of x and y for the original $f(x)$, we looked at the domain: the values x could assume. When we reversed the roles of x and y, this gave us the values y could assume. For this function, $x \geq 4$, so for the inverse, we should have $y \geq 4$, which is what our inverse function gives.

ⓐ The domain of the original function was restricted to $x \geq 4$, so the outputs of the inverse need to be the same, $f(x) \geq 4$, and we must use the + case:

$$f^{-1}(x) = 4 + \sqrt{x}$$

ⓑ The domain of the original function was restricted to $x \leq 4$, so the outputs of the inverse need to be the same, $f(x) \leq 4$, and we must use the - case:

$$f^{-1}(x) = 4 - \sqrt{x}$$

Analysis

On the graphs in Figure 6, we see the original function graphed on the same set of axes as its inverse function. Notice that together the graphs show symmetry about the line $y = x$. The coordinate pair $(4,\ 0)$ is on the graph of f and the coordinate pair $(0,\ 4)$ is on the graph of f^{-1}. For any coordinate pair, if $(a,\ b)$ is on the graph of f, then $(b,\ a)$ is on the graph of f^{-1}. Finally, observe that the graph of f intersects the graph of f^{-1} on the line $y = x$. Points of intersection for the graphs of f and f^{-1} will always lie on the line $y = x$.

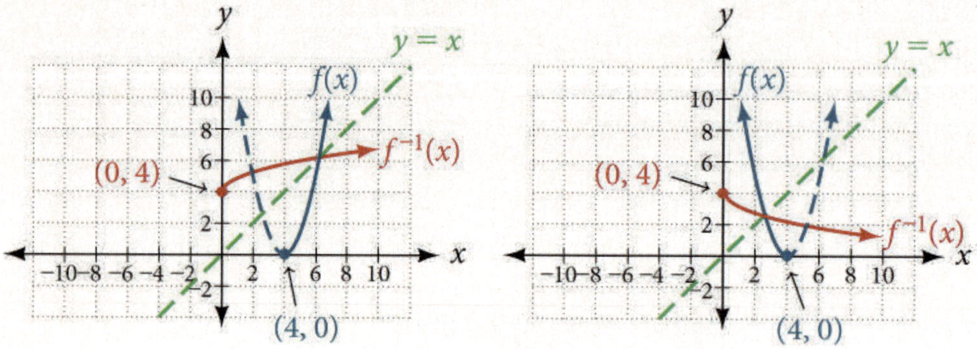

Figure 6

EXAMPLE 4

Finding the Inverse of a Quadratic Function When the Restriction Is Not Specified

Restrict the domain and then find the inverse of

$$f(x) = (x - 2)^2 - 3.$$

Solution

We can see this is a parabola with vertex at $(2, -3)$ that opens upward. Because the graph will be decreasing on one side of the vertex and increasing on the other side, we can restrict this function to a domain on which it will be one-to-one by limiting the domain to $x \geq 2$.

To find the inverse, we will use the vertex form of the quadratic. We start by replacing $f(x)$ with a simple variable, y, then solve for x.

$$\begin{aligned} y &= (x-2)^2 - 3 && \text{Interchange } x \text{ and } y. \\ x &= (y-2)^2 - 3 && \text{Add 3 to both sides.} \\ x+3 &= (y-2)^2 && \text{Take the square root.} \\ \pm\sqrt{x+3} &= y-2 && \text{Add 2 to both sides.} \\ 2 \pm \sqrt{x+3} &= y && \text{Rename the function.} \\ f^{-1}(x) &= 2 \pm \sqrt{x+3} \end{aligned}$$

Now we need to determine which case to use. Because we restricted our original function to a domain of $x \geq 2$, the outputs of the inverse should be the same, telling us to utilize the + case

$$f^{-1}(x) = 2 + \sqrt{x+3}$$

If the quadratic had not been given in vertex form, rewriting it into vertex form would be the first step. This way we may easily observe the coordinates of the vertex to help us restrict the domain.

Analysis

Notice that we arbitrarily decided to restrict the domain on $x \geq 2$. We could just have easily opted to restrict the domain on $x \leq 2$, in which case $f^{-1}(x) = 2 - \sqrt{x+3}$. Observe the original function graphed on the same set of axes as its inverse function in Figure 7. Notice that both graphs show symmetry about the line $y = x$. The coordinate pair $(2, \ -3)$ is on the graph of f and the coordinate pair $(-3, \ 2)$ is on the graph of f^{-1}. Observe from the graph of both functions on the same set of axes that

$$\text{domain of } f = \text{range of } f^{-1} = \left[2, \infty\right)$$

and

$$\text{domain of } f^{-1} = \text{range of } f = \left[-3, \infty\right).$$

Finally, observe that the graph of f intersects the graph of f^{-1} along the line $y = x$.

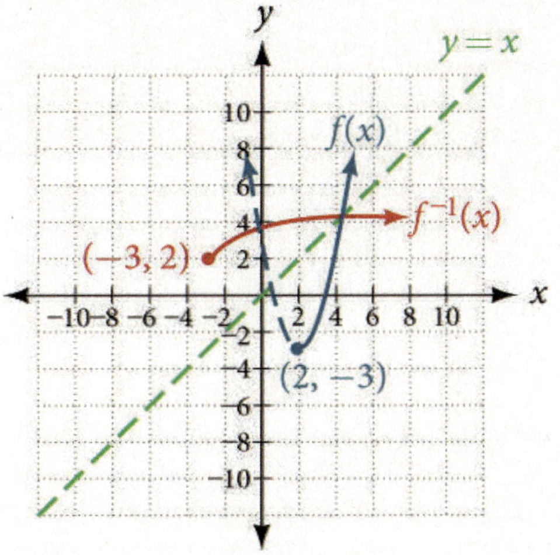

Figure 7

> **TRY IT** #3 Find the inverse of the function $f(x) = x^2 + 1$, on the domain $x \geq 0$.

Solving Applications of Radical Functions

Notice that the functions from previous examples were all polynomials, and their inverses were radical functions. If we want to find the inverse of a radical function, we will need to restrict the domain of the answer because the range of the original function is limited.

HOW TO

Given a radical function, find the inverse.

1. Determine the range of the original function.
2. Replace $f(x)$ with y, then solve for x.
3. If necessary, restrict the domain of the inverse function to the range of the original function.

EXAMPLE 5

Finding the Inverse of a Radical Function

Restrict the domain of the function $f(x) = \sqrt{x-4}$ and then find the inverse.

Solution

Note that the original function has range $f(x) \geq 0$. Replace $f(x)$ with y, then solve for x.

$$\begin{aligned} y &= \sqrt{x-4} && \text{Replace } f(x) \text{ with } y. \\ x &= \sqrt{y-4} && \text{Interchange } x \text{ and } y. \\ x &= \sqrt{y-4} && \text{Square each side.} \\ x^2 &= y-4 && \text{Add 4.} \\ x^2+4 &= y && \text{Rename the function } f^{-1}(x). \\ f^{-1}(x) &= x^2+4 \end{aligned}$$

Recall that the domain of this function must be limited to the range of the original function.

$$f^{-1}(x) = x^2 + 4, x \geq 0$$

Analysis

Notice in Figure 8 that the inverse is a reflection of the original function over the line $y = x$. Because the original function has only positive outputs, the inverse function has only positive inputs.

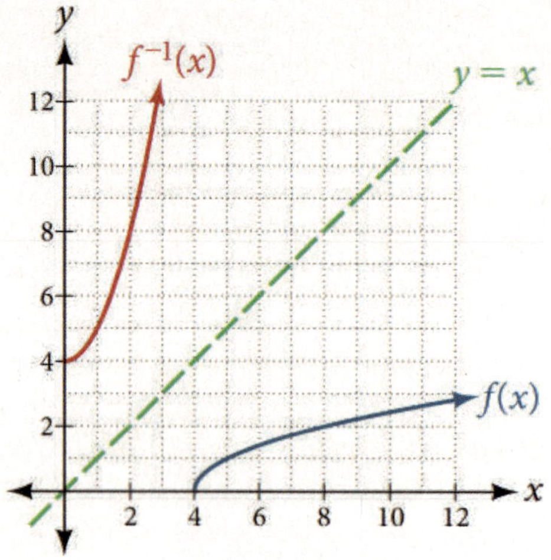

Figure 8

> **TRY IT** #4 Restrict the domain and then find the inverse of the function $f(x) = \sqrt{2x+3}$.

Solving Applications of Radical Functions

Radical functions are common in physical models, as we saw in the section opener. We now have enough tools to be able to solve the problem posed at the start of the section.

EXAMPLE 6

Solving an Application with a Cubic Function

Park rangers construct a mound of gravel in the shape of a cone with the height equal to twice the radius. The volume of the cone in terms of the radius is given by

$$V = \frac{2}{3}\pi r^3$$

Find the inverse of the function $V = \frac{2}{3}\pi r^3$ that determines the volume V of a cone and is a function of the radius r. Then use the inverse function to calculate the radius of such a mound of gravel measuring 100 cubic feet. Use $\pi = 3.14$.

Solution

Start with the given function for V. Notice that the meaningful domain for the function is $r > 0$ since negative radii would not make sense in this context nor would a radius of 0. Also note the range of the function (hence, the domain of the inverse function) is $V > 0$. Solve for r in terms of V, using the method outlined previously. Note that in real-world applications, we do not swap the variables when finding inverses. Instead, we change which variable is considered to be the independent variable.

$$\begin{aligned} V &= \tfrac{2}{3}\pi r^3 \\ r^3 &= \frac{3V}{2\pi} && \text{Solve for } r^3. \\ r &= \sqrt[3]{\frac{3V}{2\pi}} && \text{Solve for } r. \end{aligned}$$

This is the result stated in the section opener. Now evaluate this for $V = 100$ and $\pi = 3.14$.

$$\begin{aligned} r &= \sqrt[3]{\frac{3V}{2\pi}} \\ &= \sqrt[3]{\frac{3 \cdot 100}{2 \cdot 3.14}} \\ &\approx \sqrt[3]{47.7707} \\ &\approx 3.63 \end{aligned}$$

Therefore, the radius is about 3.63 ft.

Determining the Domain of a Radical Function Composed with Other Functions

When radical functions are composed with other functions, determining domain can become more complicated.

EXAMPLE 7

Finding the Domain of a Radical Function Composed with a Rational Function

Find the domain of the function $f(x) = \sqrt{\frac{(x+2)(x-3)}{(x-1)}}$.

Solution

Because a square root is only defined when the quantity under the radical is non-negative, we need to determine where $\frac{(x+2)(x-3)}{(x-1)} \geq 0$. The output of a rational function can change signs (change from positive to negative or vice versa) at *x*-intercepts and at vertical asymptotes. For this equation, the graph could change signs at $x = -2,\ 1,$ and 3.

To determine the intervals on which the rational expression is positive, we could test some values in the expression or sketch a graph. While both approaches work equally well, for this example we will use a graph as shown in Figure 9.

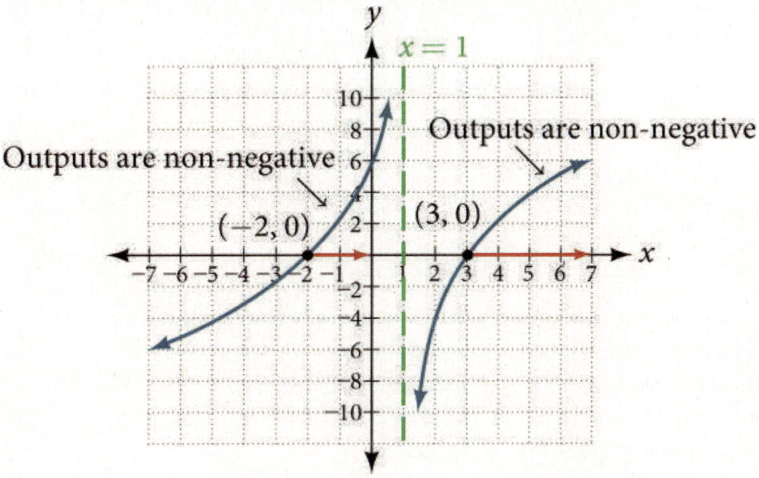

Figure 9

This function has two *x*-intercepts, both of which exhibit linear behavior near the *x*-intercepts. There is one vertical asymptote, corresponding to a linear factor; this behavior is similar to the basic reciprocal toolkit function, and there is no horizontal asymptote because the degree of the numerator is larger than the degree of the denominator. There is a *y*-intercept at $(0, \sqrt{6})$.

From the *y*-intercept and *x*-intercept at $x = -2$, we can sketch the left side of the graph. From the behavior at the asymptote, we can sketch the right side of the graph.

From the graph, we can now tell on which intervals the outputs will be non-negative, so that we can be sure that the original function $f(x)$ will be defined. $f(x)$ has domain $-2 \leq x < 1$ or $x \geq 3$, or in interval notation, $[-2, 1) \cup [3, \infty)$.

Finding Inverses of Rational Functions

As with finding inverses of quadratic functions, it is sometimes desirable to find the inverse of a rational function, particularly of rational functions that are the ratio of linear functions, such as in concentration applications.

EXAMPLE 8

Finding the Inverse of a Rational Function

The function $C = \frac{20+0.4n}{100+n}$ represents the concentration C of an acid solution after n mL of 40% solution has been added to 100 mL of a 20% solution. First, find the inverse of the function; that is, find an expression for n in terms of C. Then use your result to determine how much of the 40% solution should be added so that the final mixture is a 35% solution.

Solution

We first want the inverse of the function in order to determine how many mL we need for a given concentration. We will solve for n in terms of C.

$$\begin{aligned} C &= \frac{20+0.4n}{100+n} \\ C(100+n) &= 20+0.4n \\ 100C+Cn &= 20+0.4n \\ 100C-20 &= 0.4n-Cn \\ 100C-20 &= (0.4n-C)n \\ n &= \frac{100C-20}{0.4-C} \end{aligned}$$

Now evaluate this function at 35%, which is $C = 0.35$.

$$\begin{aligned} n &= \frac{100(0.35)-20}{0.4-0.35} \\ &= \frac{15}{0.05} \\ &= 300 \end{aligned}$$

We can conclude that 300 mL of the 40% solution should be added.

TRY IT #5 Find the inverse of the function $f(x) = \frac{x+3}{x-2}$.

MEDIA

Access these online resources for additional instruction and practice with inverses and radical functions.

Graphing the Basic Square Root Function (http://openstax.org/l/graphsquareroot)
Find the Inverse of a Square Root Function (http://openstax.org/l/inversesquare)
Find the Inverse of a Rational Function (http://openstax.org/l/inverserational)
Find the Inverse of a Rational Function and an Inverse Function Value (http://openstax.org/l/rationalinverse)
Inverse Functions (http://openstax.org/l/inversefunction)

5.7 SECTION EXERCISES

Verbal

1. Explain why we cannot find inverse functions for all polynomial functions.

2. Why must we restrict the domain of a quadratic function when finding its inverse?

3. When finding the inverse of a radical function, what restriction will we need to make?

4. The inverse of a quadratic function will always take what form?

Algebraic

For the following exercises, find the inverse of the function on the given domain.

5. $f(x) = (x - 4)^2, \ [4, \infty)$

6. $f(x) = (x + 2)^2, \ [-2, \infty)$

7. $f(x) = (x + 1)^2 - 3, \ [-1, \infty)$

8. $f(x) = 3x^2 + 5, \ \left(\infty, 0\right]$

9. $f(x) = 12 - x^2, \ [0, \infty)$

10. $f(x) = 9 - x^2, \ [0, \infty)$

11. $f(x) = 2x^2 + 4, \ [0, \infty)$

For the following exercises, find the inverse of the functions.

12. $f(x) = x^3 + 5$

13. $f(x) = 3x^3 + 1$

14. $f(x) = 4 - x^3$

15. $f(x) = 4 - 2x^3$

For the following exercises, find the inverse of the functions.

16. $f(x) = \sqrt{2x + 1}$

17. $f(x) = \sqrt{3 - 4x}$

18. $f(x) = 9 + \sqrt{4x - 4}$

19. $f(x) = \sqrt{6x - 8} + 5$

20. $f(x) = 9 + 2\sqrt[3]{x}$

21. $f(x) = 3 - \sqrt[3]{x}$

22. $f(x) = \frac{2}{x+8}$

23. $f(x) = \frac{3}{x-4}$

24. $f(x) = \frac{x+3}{x+7}$

25. $f(x) = \frac{x-2}{x+7}$

26. $f(x) = \frac{3x+4}{5-4x}$

27. $f(x) = \frac{5x+1}{2-5x}$

28. $f(x) = x^2 + 2x, \ [-1, \infty)$

29. $f(x) = x^2 + 4x + 1, \ [-2, \infty)$

30. $f(x) = x^2 - 6x + 3, \ [3, \infty)$

Graphical

For the following exercises, find the inverse of the function and graph both the function and its inverse.

31. $f(x) = x^2 + 2, \quad x \geq 0$

32. $f(x) = 4 - x^2, \quad x \geq 0$

33. $f(x) = (x + 3)^2, \quad x \geq -3$

34. $f(x) = (x - 4)^2, \quad x \geq 4$

35. $f(x) = x^3 + 3$

36. $f(x) = 1 - x^3$

37. $f(x) = x^2 + 4x, \quad x \geq -2$

38. $f(x) = x^2 - 6x + 1, \quad x \geq 3$

39. $f(x) = \frac{2}{x}$

40. $f(x) = \frac{1}{x^2}, \quad x \geq 0$

For the following exercises, use a graph to help determine the domain of the functions.

41. $f(x) = \sqrt{\frac{(x+1)(x-1)}{x}}$

42. $f(x) = \sqrt{\frac{(x+2)(x-3)}{x-1}}$

43. $f(x) = \sqrt{\frac{x(x+3)}{x-4}}$

44. $f(x) = \sqrt{\frac{x^2-x-20}{x-2}}$

45. $f(x) = \sqrt{\frac{9-x^2}{x+4}}$

Technology

For the following exercises, use a calculator to graph the function. Then, using the graph, give three points on the graph of the inverse with y-coordinates given.

46. $f(x) = x^3 - x - 2, y = 1, 2, 3$

47. $f(x) = x^3 + x - 2, y = 0, 1, 2$

48. $f(x) = x^3 + 3x - 4, y = 0, 1, 2$

49. $f(x) = x^3 + 8x - 4, y = -1, 0, 1$

50. $f(x) = x^4 + 5x + 1, y = -1, 0, 1$

Extensions

For the following exercises, find the inverse of the functions with a, b, c positive real numbers.

51. $f(x) = ax^3 + b$

52. $f(x) = x^2 + bx$

53. $f(x) = \sqrt{ax^2 - b}$

54. $f(x) = \sqrt[3]{ax + b}$

55. $f(x) = \frac{ax+b}{x+c}$

Real-World Applications

For the following exercises, determine the function described and then use it to answer the question.

56. An object dropped from a height of 200 meters has a height, $h(t)$, in meters after t seconds have lapsed, such that $h(t) = 200 - 4.9t^2$. Express t as a function of height, h, and find the time to reach a height of 50 meters.

57. An object dropped from a height of 600 feet has a height, $h(t)$, in feet after t seconds have elapsed, such that $h(t) = 600 - 16t^2$. Express t as a function of height h, and find the time to reach a height of 400 feet.

58. The volume, V, of a sphere in terms of its radius, r, is given by $V(r) = \frac{4}{3}\pi r^3$. Express r as a function of V, and find the radius of a sphere with volume of 200 cubic feet.

59. The surface area, A, of a sphere in terms of its radius, r, is given by $A(r) = 4\pi r^2$. Express r as a function of A, and find the radius of a sphere with a surface area of 1000 square inches.

60. A container holds 100 mL of a solution that is 25 mL acid. If n mL of a solution that is 60% acid is added, the function $C(n) = \frac{25+.6n}{100+n}$ gives the concentration, C, as a function of the number of mL added, n. Express n as a function of C and determine the number of mL that need to be added to have a solution that is 50% acid.

61. The period T, in seconds, of a simple pendulum as a function of its length l, in feet, is given by $T(l) = 2\pi\sqrt{\frac{l}{32.2}}$. Express l as a function of T and determine the length of a pendulum with period of 2 seconds.

62. The volume of a cylinder , V, in terms of radius, r, and height, h, is given by $V = \pi r^2 h$. If a cylinder has a height of 6 meters, express the radius as a function of V and find the radius of a cylinder with volume of 300 cubic meters.

63. The surface area, A, of a cylinder in terms of its radius, r, and height, h, is given by $A = 2\pi r^2 + 2\pi rh$. If the height of the cylinder is 4 feet, express the radius as a function of A and find the radius if the surface area is 200 square feet.

64. The volume of a right circular cone, V, in terms of its radius, r, and its height, h, is given by $V = \frac{1}{3}\pi r^2 h$. Express r in terms of V if the height of the cone is 12 feet and find the radius of a cone with volume of 50 cubic inches.

65. Consider a cone with height of 30 feet. Express the radius, r, in terms of the volume, V, and find the radius of a cone with volume of 1000 cubic feet.

5.8 Modeling Using Variation

Learning Objectives

In this section, you will:

- Solve direct variation problems.
- Solve inverse variation problems.
- Solve problems involving joint variation.

A pre-owned car dealer has just offered their best candidate, Nicole, a position in sales. The position offers 16% commission on her sales. Her earnings depend on the amount of her sales. For instance, if she sells a vehicle for $4,600, she will earn $736. As she considers the offer, she takes into account the typical price of the dealer's cars, the overall market, and how many she can reasonably expect to sell. In this section, we will look at relationships, such as this one, between earnings, sales, and commission rate.

Solving Direct Variation Problems

In the example above, Nicole's earnings can be found by multiplying her sales by her commission. The formula $e = 0.16s$ tells us her earnings, e, come from the product of 0.16, her commission, and the sale price of the vehicle. If we create a table, we observe that as the sales price increases, the earnings increase as well, which should be intuitive. See Table 1.

s , sales price	$e = 0.16s$	Interpretation
$4,600	$e = 0.16(4{,}600) = 736$	A sale of a $4,600 vehicle results in $736 earnings.
$9,200	$e = 0.16(9{,}200) = 1{,}472$	A sale of a $9,200 vehicle results in $1472 earnings.
$18,400	$e = 0.16(18{,}400) = 2{,}944$	A sale of a $18,400 vehicle results in $2944 earnings.

Table 1

Notice that earnings are a multiple of sales. As sales increase, earnings increase in a predictable way. Double the sales of the vehicle from \$4,600 to \$9,200, and we double the earnings from \$736 to \$1,472. As the input increases, the output increases as a multiple of the input. A relationship in which one quantity is a constant multiplied by another quantity is called **direct variation**. Each variable in this type of relationship **varies directly** with the other.

Figure 1 represents the data for Nicole's potential earnings. We say that earnings vary directly with the sales price of the car. The formula $y = kx^n$ is used for direct variation. The value k is a nonzero constant greater than zero and is called the **constant of variation**. In this case, $k = 0.16$ and $n = 1$. We saw functions like this one when we discussed power functions.

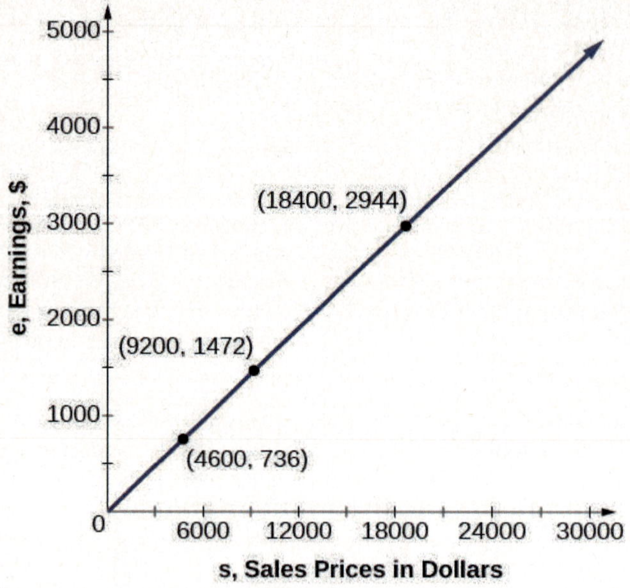

Figure 1

Direct Variation

If x and y are related by an equation of the form

$$y = kx^n$$

then we say that the relationship is **direct variation** and y **varies directly** with, or is proportional to, the nth power of x. In direct variation relationships, there is a nonzero constant ratio $k = \frac{y}{x^n}$, where k is called the **constant of variation**, which help defines the relationship between the variables.

HOW TO

Given a description of a direct variation problem, solve for an unknown.

1. Identify the input, x, and the output, y.
2. Determine the constant of variation. You may need to divide y by the specified power of x to determine the constant of variation.
3. Use the constant of variation to write an equation for the relationship.
4. Substitute known values into the equation to find the unknown.

EXAMPLE 1

Solving a Direct Variation Problem

The quantity y varies directly with the cube of x. If $y = 25$ when $x = 2$, find y when x is 6.

Solution

The general formula for direct variation with a cube is $y = kx^3$. The constant can be found by dividing y by the cube of x.

$$\begin{aligned} k &= \frac{y}{x^3} \\ &= \frac{25}{2^3} \\ &= \frac{25}{8} \end{aligned}$$

Now use the constant to write an equation that represents this relationship.

$$y = \frac{25}{8}x^3$$

Substitute $x = 6$ and solve for y.

$$\begin{aligned} y &= \frac{25}{8}(6)^3 \\ &= 675 \end{aligned}$$

Analysis

The graph of this equation is a simple cubic, as shown in Figure 2.

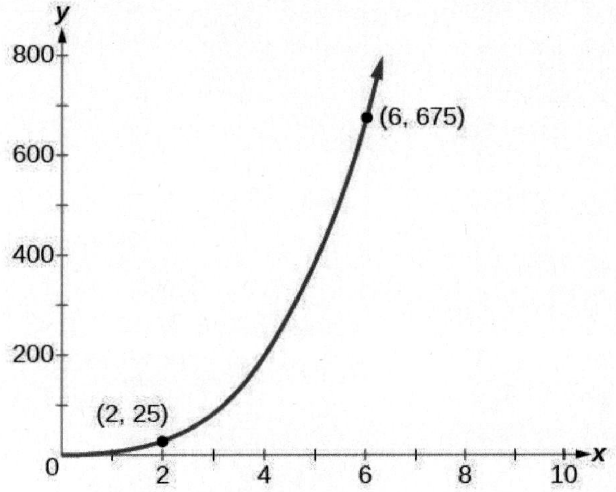

Figure 2

Q&A **Do the graphs of all direct variation equations look like Example 1?**

No. Direct variation equations are power functions—they may be linear, quadratic, cubic, quartic, radical, etc. But all of the graphs pass through $(0,0)$.

TRY IT #1 The quantity y varies directly with the square of x. If $y = 24$ when $x = 3$, find y when x is 4.

Solving Inverse Variation Problems

Water temperature in an ocean varies inversely to the water's depth. The formula $T = \frac{14,000}{d}$ gives us the temperature in degrees Fahrenheit at a depth in feet below Earth's surface. Consider the Atlantic Ocean, which covers 22% of Earth's surface. At a certain location, at the depth of 500 feet, the temperature may be 28°F.

If we create Table 2, we observe that, as the depth increases, the water temperature decreases.

d, depth	$T = \frac{14,000}{d}$	Interpretation
500 ft	$\frac{14,000}{500} = 28$	At a depth of 500 ft, the water temperature is 28° F.

Table 2

d, depth	$T = \frac{14{,}000}{d}$	Interpretation
1000 ft	$\frac{14{,}000}{1000} = 14$	At a depth of 1,000 ft, the water temperature is 14° F.
2000 ft	$\frac{14{,}000}{2000} = 7$	At a depth of 2,000 ft, the water temperature is 7° F.

Table 2

We notice in the relationship between these variables that, as one quantity increases, the other decreases. The two quantities are said to be **inversely proportional** and each term **varies inversely** with the other. Inversely proportional relationships are also called **inverse variations**.

For our example, Figure 3 depicts the inverse variation. We say the water temperature varies inversely with the depth of the water because, as the depth increases, the temperature decreases. The formula $y = \frac{k}{x}$ for inverse variation in this case uses $k = 14{,}000$.

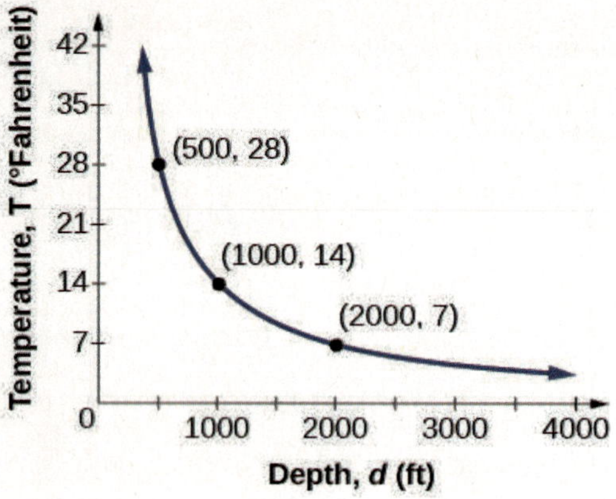

Figure 3

Inverse Variation

If x and y are related by an equation of the form

$$y = \frac{k}{x^n}$$

where k is a nonzero constant, then we say that y **varies inversely** with the nth power of x. In **inversely proportional** relationships, or **inverse variations**, there is a constant multiple $k = x^n y$.

EXAMPLE 2

Writing a Formula for an Inversely Proportional Relationship

A tourist plans to drive 100 miles. Find a formula for the time the trip will take as a function of the speed the tourist drives.

Solution

Recall that multiplying speed by time gives distance. If we let t represent the drive time in hours, and v represent the velocity (speed or rate) at which the tourist drives, then $vt = \text{distance}$. Because the distance is fixed at 100 miles, $vt = 100$ so $t = 100/v$. Because time is a function of velocity, we can write $t(v)$.

$$\begin{aligned} t(v) &= \frac{100}{v} \\ &= 100v^{-1} \end{aligned}$$

We can see that the constant of variation is 100 and, although we can write the relationship using the negative exponent, it is more common to see it written as a fraction. We say that time varies inversely with velocity.

HOW TO

Given a description of an indirect variation problem, solve for an unknown.

1. Identify the input, x, and the output, y.
2. Determine the constant of variation. You may need to multiply y by the specified power of x to determine the constant of variation.
3. Use the constant of variation to write an equation for the relationship.
4. Substitute known values into the equation to find the unknown.

EXAMPLE 3

Solving an Inverse Variation Problem

A quantity y varies inversely with the cube of x. If $y = 25$ when $x = 2$, find y when x is 6.

Solution

The general formula for inverse variation with a cube is $y = \frac{k}{x^3}$. The constant can be found by multiplying y by the cube of x.

$$\begin{aligned} k &= x^3 y \\ &= 2^3 \cdot 25 \\ &= 200 \end{aligned}$$

Now we use the constant to write an equation that represents this relationship.

$$\begin{aligned} y &= \frac{k}{x^3}, \quad k = 200 \\ y &= \frac{200}{x^3} \end{aligned}$$

Substitute $x = 6$ and solve for y.

$$\begin{aligned} y &= \frac{200}{6^3} \\ &= \frac{25}{27} \end{aligned}$$

Analysis

The graph of this equation is a rational function, as shown in Figure 4.

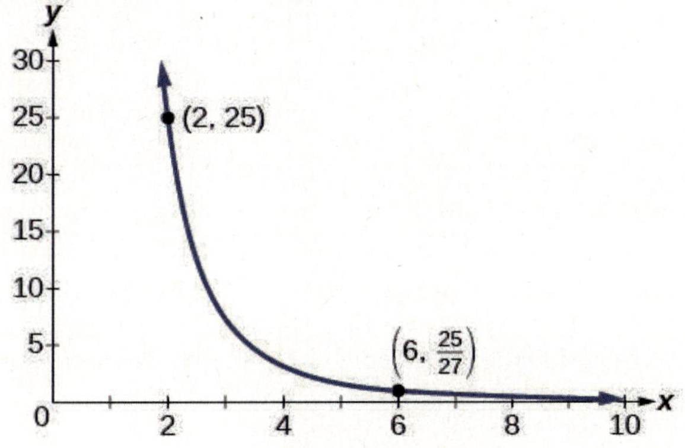

Figure 4

TRY IT #2 A quantity y varies inversely with the square of x. If $y = 8$ when $x = 3$, find y when x is 4.

Solving Problems Involving Joint Variation

Many situations are more complicated than a basic direct variation or inverse variation model. One variable often depends on multiple other variables. When a variable is dependent on the product or quotient of two or more variables, this is called **joint variation**. For example, the cost of busing students for each school trip varies with the number of students attending and the distance from the school. The variable c, cost, varies jointly with the number of students, n, and the distance, d.

Joint Variation

Joint variation occurs when a variable varies directly or inversely with multiple variables.

For instance, if x varies directly with both y and z, we have $x = kyz$. If x varies directly with y and inversely with z, we have $x = \frac{ky}{z}$. Notice that we only use one constant in a joint variation equation.

EXAMPLE 4

Solving Problems Involving Joint Variation

A quantity x varies directly with the square of y and inversely with the cube root of z. If $x = 6$ when $y = 2$ and $z = 8$, find x when $y = 1$ and $z = 27$.

Solution

Begin by writing an equation to show the relationship between the variables.

$$x = \frac{ky^2}{\sqrt[3]{z}}$$

Substitute $x = 6$, $y = 2$, and $z = 8$ to find the value of the constant k.

$$\begin{aligned} 6 &= \frac{k2^2}{\sqrt[3]{8}} \\ 6 &= \frac{4k}{2} \\ 3 &= k \end{aligned}$$

Now we can substitute the value of the constant into the equation for the relationship.

$$x = \frac{3y^2}{\sqrt[3]{z}}$$

To find x when $y = 1$ and $z = 27$, we will substitute values for y and z into our equation.

$$\begin{aligned} x &= \frac{3(1)^2}{\sqrt[3]{27}} \\ &= 1 \end{aligned}$$

TRY IT #3 A quantity x varies directly with the square of y and inversely with z. If $x = 40$ when $y = 4$ and $z = 2$, find x when $y = 10$ and $z = 25$.

MEDIA

Access these online resources for additional instruction and practice with direct and inverse variation.

Direct Variation (http://openstax.org/l/directvariation)
Inverse Variation (http://openstax.org/l/inversevariatio)
Direct and Inverse Variation (http://openstax.org/l/directinverse)

5.8 SECTION EXERCISES

Verbal

1. What is true of the appearance of graphs that reflect a direct variation between two variables?

2. If two variables vary inversely, what will an equation representing their relationship look like?

3. Is there a limit to the number of variables that can vary jointly? Explain.

Algebraic

For the following exercises, write an equation describing the relationship of the given variables.

4. y varies directly as x and when $x = 6, y = 12$.

5. y varies directly as the square of x and when $x = 4, \quad y = 80$.

6. y varies directly as the square root of x and when $x = 36, \quad y = 24$.

7. y varies directly as the cube of x and when $x = 36, \quad y = 24$.

8. y varies directly as the cube root of x and when $x = 27, \quad y = 15$.

9. y varies directly as the fourth power of x and when $x = 1, \quad y = 6$.

10. y varies inversely as x and when $x = 4, \quad y = 2$.

11. y varies inversely as the square of x and when $x = 3, \quad y = 2$.

12. y varies inversely as the cube of x and when $x = 2, \quad y = 5$.

13. y varies inversely as the fourth power of x and when $x = 3, \quad y = 1$.

14. y varies inversely as the square root of x and when $x = 25, \quad y = 3$.

15. y varies inversely as the cube root of x and when $x = 64, \quad y = 5$.

16. y varies jointly with x and z and when $x = 2$ and $z = 3, \quad y = 36$.

17. y varies jointly as x, z, and w and when $x = 1, \quad z = 2, \quad w = 5$, then $y = 100$.

18. y varies jointly as the square of x and the square of z and when $x = 3$ and $z = 4$, then $y = 72$.

19. y varies jointly as x and the square root of z and when $x = 2$ and $z = 25$, then $y = 100$.

20. y varies jointly as the square of x the cube of z and the square root of W. When $x = 1, z = 2$, and $w = 36$, then $y = 48$.

21. y varies jointly as x and z and inversely as w. When $x = 3, \quad z = 5$, and $w = 6$, then $y = 10$.

22. y varies jointly as the square of x and the square root of z and inversely as the cube of w. When $x = 3, z = 4$, and $w = 3$, then $y = 6$.

23. y varies jointly as x and z and inversely as the square root of w and the square of t. When $x = 3, z = 1, w = 25$, and $t = 2$, then $y = 6$.

Numeric

For the following exercises, use the given information to find the unknown value.

24. y varies directly as x. When $x = 3$, then $y = 12$. Find y wneh $x = 20$.

25. y varies directly as the square of x. When $x = 2$, then $y = 16$. Find y when $x = 8$.

26. y varies directly as the cube of x. When $x = 3$, then $y = 5$. Find y when $x = 4$.

27. y varies directly as the square root of x. When $x = 16$, then $y = 4$. Find y when $x = 36$.

28. y varies directly as the cube root of x. When $x = 125$, then $y = 15$. Find y when $x = 1{,}000$.

29. y varies inversely with x. When $x = 3$, then $y = 2$. Find y when $x = 1$.

30. y varies inversely with the square of x. When $x = 4$, then $y = 3$. Find y when $x = 2$.

31. y varies inversely with the cube of x. When $x = 3$, then $y = 1$. Find y when $x = 1$.

32. y varies inversely with the square root of x. When $x = 64$, then $y = 12$. Find y when $x = 36$.

33. y varies inversely with the cube root of x. When $x = 27$, then $y = 5$. Find y when $x = 125$.

34. y varies jointly as $x \text{and} z$. When $x = 4$ and $z = 2$, then $y = 16$. Find y when $x = 3$ and $z = 3$.

35. y varies jointly as $x, z, \text{and} w$. When $x = 2$, $z = 1$, and $w = 12$, then $y = 72$. Find y when $x = 1$, $z = 2$, and $w = 3$.

36. y varies jointly as x and the square of z. When $x = 2$ and $z = 4$, then $y = 144$. Find y when $x = 4$ and $z = 5$.

37. y varies jointly as the square of x and the square root of z. When $x = 2$ and $z = 9$, then $y = 24$. Find y when $x = 3$ and $z = 25$.

38. y varies jointly as x and z and inversely as w. When $x = 5$, $z = 2$, and $w = 20$, then $y = 4$. Find y when $x = 3$ and $z = 8$, and $w = 48$.

39. y varies jointly as the square of x and the cube of z and inversely as the square root of w. When $x = 2$, $z = 2$, and $w = 64$, then $y = 12$. Find y when $x = 1$, $z = 3$, and $w = 4$.

40. y varies jointly as the square of x and of z and inversely as the square root of w and of t. When $x = 2$, $z = 3$, $w = 16$, and $t = 3$, then $y = 1$. Find y when $x = 3$, $z = 2$, $w = 36$, and $t = 5$.

Technology

For the following exercises, use a calculator to graph the equation implied by the given variation.

41. y varies directly with the square of x and when $x = 2, y = 3$.

42. y varies directly as the cube of x and when $x = 2, y = 4$.

43. y varies directly as the square root of x and when $x = 36, y = 2$.

44. y varies inversely with x and when $x = 6, y = 2$.

45. y varies inversely as the square of x and when $x = 1, y = 4$.

Extensions

For the following exercises, use Kepler's Law, which states that the square of the time, T, required for a planet to orbit the Sun varies directly with the cube of the mean distance, a, that the planet is from the Sun.

46. Using Earth's time of 1 year and mean distance of 93 million miles, find the equation relating T and a.

47. Use the result from the previous exercise to determine the time required for Mars to orbit the Sun if its mean distance is 142 million miles.

48. Using Earth's distance of 150 million kilometers, find the equation relating T and a.

49. Use the result from the previous exercise to determine the time required for Venus to orbit the Sun if its mean distance is 108 million kilometers.

50. Using Earth's distance of 1 astronomical unit (A.U.), determine the time for Saturn to orbit the Sun if its mean distance is 9.54 A.U.

Real-World Applications

For the following exercises, use the given information to answer the questions.

51. The distance s that an object falls varies directly with the square of the time, t, of the fall. If an object falls 16 feet in one second, how long for it to fall 144 feet?

52. The velocity v of a falling object varies directly to the time, t, of the fall. If after 2 seconds, the velocity of the object is 64 feet per second, what is the velocity after 5 seconds?

53. The rate of vibration of a string under constant tension varies inversely with the length of the string. If a string is 24 inches long and vibrates 128 times per second, what is the length of a string that vibrates 64 times per second?

54. The volume of a gas held at constant temperature varies indirectly as the pressure of the gas. If the volume of a gas is 1200 cubic centimeters when the pressure is 200 millimeters of mercury, what is the volume when the pressure is 300 millimeters of mercury?

55. The weight of an object above the surface of Earth varies inversely with the square of the distance from the center of Earth. If a body weighs 50 pounds when it is 3960 miles from Earth's center, what would it weigh it were 3970 miles from Earth's center?

56. The intensity of light measured in foot-candles varies inversely with the square of the distance from the light source. Suppose the intensity of a light bulb is 0.08 foot-candles at a distance of 3 meters. Find the intensity level at 8 meters.

57. The current in a circuit varies inversely with its resistance measured in ohms. When the current in a circuit is 40 amperes, the resistance is 10 ohms. Find the current if the resistance is 12 ohms.

58. The force exerted by the wind on a plane surface varies jointly with the square of the velocity of the wind and with the area of the plane surface. If the area of the surface is 40 square feet surface and the wind velocity is 20 miles per hour, the resulting force is 15 pounds. Find the force on a surface of 65 square feet with a velocity of 30 miles per hour.

59. The horsepower (hp) that a shaft can safely transmit varies jointly with its speed (in revolutions per minute (rpm) and the cube of the diameter. If the shaft of a certain material 3 inches in diameter can transmit 45 hp at 100 rpm, what must the diameter be in order to transmit 60 hp at 150 rpm?

60. The kinetic energy K of a moving object varies jointly with its mass m and the square of its velocity v. If an object weighing 40 kilograms with a velocity of 15 meters per second has a kinetic energy of 1000 joules, find the kinetic energy if the velocity is increased to 20 meters per second.

Chapter Review

Key Terms

arrow notation a way to represent symbolically the local and end behavior of a function by using arrows to indicate that an input or output approaches a value

axis of symmetry a vertical line drawn through the vertex of a parabola, that opens up or down, around which the parabola is symmetric; it is defined by $x = -\frac{b}{2a}$.

coefficient a nonzero real number multiplied by a variable raised to an exponent

constant of variation the non-zero value k that helps define the relationship between variables in direct or inverse variation

continuous function a function whose graph can be drawn without lifting the pen from the paper because there are no breaks in the graph

degree the highest power of the variable that occurs in a polynomial

Descartes' Rule of Signs a rule that determines the maximum possible numbers of positive and negative real zeros based on the number of sign changes of $f(x)$ and $f(-x)$

direct variation the relationship between two variables that are a constant multiple of each other; as one quantity increases, so does the other

Division Algorithm given a polynomial dividend $f(x)$ and a non-zero polynomial divisor $d(x)$ where the degree of $d(x)$ is less than or equal to the degree of $f(x)$, there exist unique polynomials $q(x)$ and $r(x)$ such that $f(x) = d(x)q(x) + r(x)$ where $q(x)$ is the quotient and $r(x)$ is the remainder. The remainder is either equal to zero or has degree strictly less than $d(x)$.

end behavior the behavior of the graph of a function as the input decreases without bound and increases without bound

Factor Theorem k is a zero of polynomial function $f(x)$ if and only if $(x - k)$ is a factor of $f(x)$

Fundamental Theorem of Algebra a polynomial function with degree greater than 0 has at least one complex zero

general form of a quadratic function the function that describes a parabola, written in the form $f(x) = ax^2 + bx + c$, where a, b, and c are real numbers and $a \neq 0$.

global maximum highest turning point on a graph; $f(a)$ where $f(a) \geq f(x)$ for all x.

global minimum lowest turning point on a graph; $f(a)$ where $f(a) \leq f(x)$ for all x.

horizontal asymptote a horizontal line $y = b$ where the graph approaches the line as the inputs increase or decrease without bound.

Intermediate Value Theorem for two numbers a and b in the domain of f, if $a < b$ and $f(a) \neq f(b)$, then the function f takes on every value between $f(a)$ and $f(b)$; specifically, when a polynomial function changes from a negative value to a positive value, the function must cross the x- axis

inverse variation the relationship between two variables in which the product of the variables is a constant

inversely proportional a relationship where one quantity is a constant divided by the other quantity; as one quantity increases, the other decreases

invertible function any function that has an inverse function

joint variation a relationship where a variable varies directly or inversely with multiple variables

leading coefficient the coefficient of the leading term

leading term the term containing the highest power of the variable

Linear Factorization Theorem allowing for multiplicities, a polynomial function will have the same number of factors as its degree, and each factor will be in the form $(x - c)$, where c is a complex number

multiplicity the number of times a given factor appears in the factored form of the equation of a polynomial; if a polynomial contains a factor of the form $(x - h)^p$, $x = h$ is a zero of multiplicity p.

polynomial function a function that consists of either zero or the sum of a finite number of non-zero terms, each of which is a product of a number, called the coefficient of the term, and a variable raised to a non-negative integer power.

power function a function that can be represented in the form $f(x) = kx^p$ where k is a constant, the base is a variable, and the exponent, p, is a constant

rational function a function that can be written as the ratio of two polynomials

Rational Zero Theorem the possible rational zeros of a polynomial function have the form $\frac{p}{q}$ where p is a factor of the constant term and q is a factor of the leading coefficient.

Remainder Theorem if a polynomial $f(x)$ is divided by $x - k$, then the remainder is equal to the value $f(k)$

removable discontinuity a single point at which a function is undefined that, if filled in, would make the function continuous; it appears as a hole on the graph of a function

roots in a given function, the values of x at which $y = 0$, also called zeros

smooth curve a graph with no sharp corners
standard form of a quadratic function the function that describes a parabola, written in the form $f(x) = a(x-h)^2 + k$, where $(h, \ k)$ is the vertex
synthetic division a shortcut method that can be used to divide a polynomial by a binomial of the form $x - k$
term of a polynomial function any $a_i x^i$ of a polynomial function in the form $f(x) = a_n x^n + ... + a_2 x^2 + a_1 x + a_0$
turning point the location at which the graph of a function changes direction
varies directly a relationship where one quantity is a constant multiplied by the other quantity
varies inversely a relationship where one quantity is a constant divided by the other quantity
vertex the point at which a parabola changes direction, corresponding to the minimum or maximum value of the quadratic function
vertex form of a quadratic function another name for the standard form of a quadratic function
vertical asymptote a vertical line $x = a$ where the graph tends toward positive or negative infinity as the inputs approach a
zeros in a given function, the values of x at which $y = 0$, also called roots

Key Equations

general form of a quadratic function	$f(x) = ax^2 + bx + c$
standard form of a quadratic function	$f(x) = a(x-h)^2 + k$
general form of a polynomial function	$f(x) = a_n x^n + ... + a_2 x^2 + a_1 x + a_0$
Division Algorithm	$f(x) = d(x)q(x) + r(x)$ where $q(x) \neq 0$
Rational Function	$f(x) = \frac{P(x)}{Q(x)} = \frac{a_p x^p + a_{p-1} x^{p-1} + ... + a_1 x + a_0}{b_q x^q + b_{q-1} x^{q-1} + ... + b_1 x + b_0}, \ Q(x) \neq 0$
Direct variation	$y = kx^n$, k is a nonzero constant.
Inverse variation	$y = \frac{k}{x^n}$, k is a nonzero constant.

Key Concepts

5.1 Quadratic Functions

- A polynomial function of degree two is called a quadratic function.
- The graph of a quadratic function is a parabola. A parabola is a U-shaped curve that can open either up or down.
- The axis of symmetry is the vertical line passing through the vertex. The zeros, or x- intercepts, are the points at which the parabola crosses the x- axis. The y- intercept is the point at which the parabola crosses the y- axis. See Example 1, Example 7, and Example 8.
- Quadratic functions are often written in general form. Standard or vertex form is useful to easily identify the vertex of a parabola. Either form can be written from a graph. See Example 2.
- The vertex can be found from an equation representing a quadratic function. See Example 3.
- The domain of a quadratic function is all real numbers. The range varies with the function. See Example 4.
- A quadratic function's minimum or maximum value is given by the y- value of the vertex.
- The minimum or maximum value of a quadratic function can be used to determine the range of the function and to solve many kinds of real-world problems, including problems involving area and revenue. See Example 5 and Example 6.
- The vertex and the intercepts can be identified and interpreted to solve real-world problems. See Example 9.

5.2 Power Functions and Polynomial Functions

- A power function is a variable base raised to a number power. See Example 1.
- The behavior of a graph as the input decreases beyond bound and increases beyond bound is called the end behavior.
- The end behavior depends on whether the power is even or odd. See Example 2 and Example 3.
- A polynomial function is the sum of terms, each of which consists of a transformed power function with positive whole number power. See Example 4.
- The degree of a polynomial function is the highest power of the variable that occurs in a polynomial. The term containing the highest power of the variable is called the leading term. The coefficient of the leading term is called the leading coefficient. See Example 5.
- The end behavior of a polynomial function is the same as the end behavior of the power function represented by the leading term of the function. See Example 6 and Example 7.
- A polynomial of degree n will have at most n x-intercepts and at most $n - 1$ turning points. See Example 8, Example 9, Example 10, Example 11, and Example 12.

5.3 Graphs of Polynomial Functions

- Polynomial functions of degree 2 or more are smooth, continuous functions. See Example 1.
- To find the zeros of a polynomial function, if it can be factored, factor the function and set each factor equal to zero. See Example 2, Example 3, and Example 4.
- Another way to find the x- intercepts of a polynomial function is to graph the function and identify the points at which the graph crosses the x- axis. See Example 5.
- The multiplicity of a zero determines how the graph behaves at the x- intercepts. See Example 6.
- The graph of a polynomial will cross the horizontal axis at a zero with odd multiplicity.
- The graph of a polynomial will touch the horizontal axis at a zero with even multiplicity.
- The end behavior of a polynomial function depends on the leading term.
- The graph of a polynomial function changes direction at its turning points.
- A polynomial function of degree n has at most $n - 1$ turning points. See Example 7.
- To graph polynomial functions, find the zeros and their multiplicities, determine the end behavior, and ensure that the final graph has at most $n - 1$ turning points. See Example 8 and Example 10.
- Graphing a polynomial function helps to estimate local and global extremas. See Example 11.
- The Intermediate Value Theorem tells us that if $f(a)$ and $f(b)$ have opposite signs, then there exists at least one value c between a and b for which $f(c) = 0$. See Example 9.

5.4 Dividing Polynomials

- Polynomial long division can be used to divide a polynomial by any polynomial with equal or lower degree. See Example 1 and Example 2.
- The Division Algorithm tells us that a polynomial dividend can be written as the product of the divisor and the quotient added to the remainder.
- Synthetic division is a shortcut that can be used to divide a polynomial by a binomial in the form $x - k$. See Example 3, Example 4, and Example 5.
- Polynomial division can be used to solve application problems, including area and volume. See Example 6.

5.5 Zeros of Polynomial Functions

- To find $f(k)$, determine the remainder of the polynomial $f(x)$ when it is divided by $x - k$. This is known as the Remainder Theorem. See Example 1.
- According to the Factor Theorem, k is a zero of $f(x)$ if and only if $(x - k)$ is a factor of $f(x)$. See Example 2.
- According to the Rational Zero Theorem, each rational zero of a polynomial function with integer coefficients will be equal to a factor of the constant term divided by a factor of the leading coefficient. See Example 3 and Example 4.
- When the leading coefficient is 1, the possible rational zeros are the factors of the constant term.
- Synthetic division can be used to find the zeros of a polynomial function. See Example 5.
- According to the Fundamental Theorem, every polynomial function has at least one complex zero. See Example 6.
- Every polynomial function with degree greater than 0 has at least one complex zero.
- Allowing for multiplicities, a polynomial function will have the same number of factors as its degree. Each factor will be in the form $(x - c)$, where c is a complex number. See Example 7.
- The number of positive real zeros of a polynomial function is either the number of sign changes of the function or less than the number of sign changes by an even integer.
- The number of negative real zeros of a polynomial function is either the number of sign changes of $f(-x)$ or less

than the number of sign changes by an even integer. See Example 8.
- Polynomial equations model many real-world scenarios. Solving the equations is easiest done by synthetic division. See Example 9.

5.6 Rational Functions

- We can use arrow notation to describe local behavior and end behavior of the toolkit functions $f(x) = \frac{1}{x}$ and $f(x) = \frac{1}{x^2}$. See Example 1.
- A function that levels off at a horizontal value has a horizontal asymptote. A function can have more than one vertical asymptote. See Example 2.
- Application problems involving rates and concentrations often involve rational functions. See Example 3.
- The domain of a rational function includes all real numbers except those that cause the denominator to equal zero. See Example 4.
- The vertical asymptotes of a rational function will occur where the denominator of the function is equal to zero and the numerator is not zero. See Example 5.
- A removable discontinuity might occur in the graph of a rational function if an input causes both numerator and denominator to be zero. See Example 6.
- A rational function's end behavior will mirror that of the ratio of the leading terms of the numerator and denominator functions. See Example 7, Example 8, Example 9, and Example 10.
- Graph rational functions by finding the intercepts, behavior at the intercepts and asymptotes, and end behavior. See Example 11.
- If a rational function has x-intercepts at $x = x_1, x_2, \ldots, x_n$, vertical asymptotes at $x = v_1, v_2, \ldots, v_m$, and no $x_i = \text{any } v_j$, then the function can be written in the form

$$f(x) = a\frac{(x-x_1)^{p_1}(x-x_2)^{p_2}\cdots(x-x_n)^{p_n}}{(x-v_1)^{q_1}(x-v_2)^{q_2}\cdots(x-v_m)^{q_n}}$$

See Example 12.

5.7 Inverses and Radical Functions

- The inverse of a quadratic function is a square root function.
- If f^{-1} is the inverse of a function f, then f is the inverse of the function f^{-1}. See Example 1.
- While it is not possible to find an inverse of most polynomial functions, some basic polynomials are invertible. See Example 2.
- To find the inverse of certain functions, we must restrict the function to a domain on which it will be one-to-one. See Example 3 and Example 4.
- When finding the inverse of a radical function, we need a restriction on the domain of the answer. See Example 5 and Example 7.
- Inverse and radical and functions can be used to solve application problems. See Example 6 and Example 8.

5.8 Modeling Using Variation

- A relationship where one quantity is a constant multiplied by another quantity is called direct variation. See Example 1.
- Two variables that are directly proportional to one another will have a constant ratio.
- A relationship where one quantity is a constant divided by another quantity is called inverse variation. See Example 2.
- Two variables that are inversely proportional to one another will have a constant multiple. See Example 3.
- In many problems, a variable varies directly or inversely with multiple variables. We call this type of relationship joint variation. See Example 4.

Exercises

Review Exercises

Quadratic Functions

For the following exercises, write the quadratic function in standard form. Then give the vertex and axes intercepts. Finally, graph the function.

1. $f(x) = x^2 - 4x - 5$

2. $f(x) = -2x^2 - 4x$

For the following exercises, find the equation of the quadratic function using the given information.

3. The vertex is $(-2, 3)$ and a point on the graph is $(3, 6)$.

4. The vertex is $(-3, 6.5)$ and a point on the graph is $(2, 6)$.

For the following exercises, complete the task.

5. A rectangular plot of land is to be enclosed by fencing. One side is along a river and so needs no fence. If the total fencing available is 600 meters, find the dimensions of the plot to have maximum area.

6. An object projected from the ground at a 45 degree angle with initial velocity of 120 feet per second has height, h, in terms of horizontal distance traveled, x, given by $h(x) = \frac{-32}{(120)^2}x^2 + x$. Find the maximum height the object attains.

Power Functions and Polynomial Functions

For the following exercises, determine if the function is a polynomial function and, if so, give the degree and leading coefficient.

7. $f(x) = 4x^5 - 3x^3 + 2x - 1$

8. $f(x) = 5^{x+1} - x^2$

9. $f(x) = x^2\left(3 - 6x + x^2\right)$

For the following exercises, determine end behavior of the polynomial function.

10. $f(x) = 2x^4 + 3x^3 - 5x^2 + 7$

11. $f(x) = 4x^3 - 6x^2 + 2$

12. $f(x) = 2x^2(1 + 3x - x^2)$

Graphs of Polynomial Functions

For the following exercises, find all zeros of the polynomial function, noting multiplicities.

13. $f(x) = (x + 3)^2(2x - 1)(x + 1)^3$

14. $f(x) = x^5 + 4x^4 + 4x^3$

15. $f(x) = x^3 - 4x^2 + x - 4$

For the following exercises, based on the given graph, determine the zeros of the function and note multiplicity.

16.

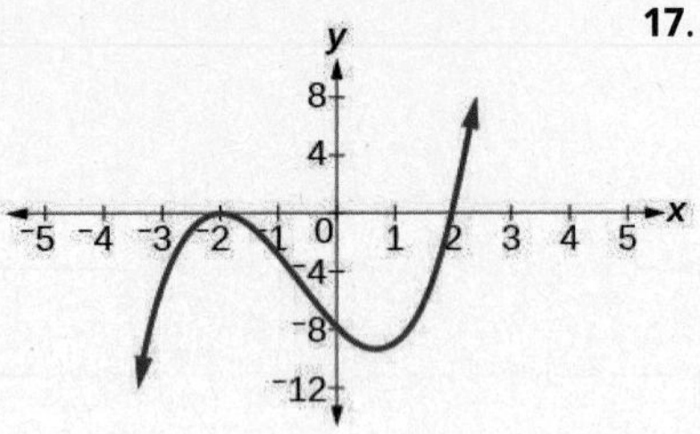

17.

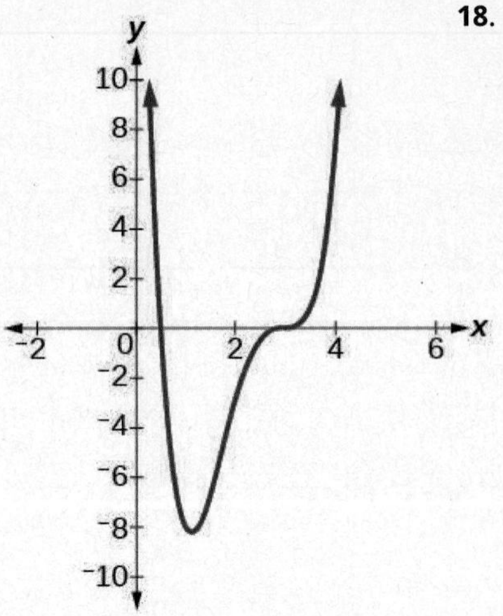

18. Use the Intermediate Value Theorem to show that at least one zero lies between 2 and 3 for the function $f(x) = x^3 - 5x + 1$

Dividing Polynomials

For the following exercises, use long division to find the quotient and remainder.

19. $\frac{x^3-2x^2+4x+4}{x-2}$

20. $\frac{3x^4-4x^2+4x+8}{x+1}$

For the following exercises, use synthetic division to find the quotient. If the divisor is a factor, then write the factored form.

21. $\frac{x^3-2x^2+5x-1}{x+3}$

22. $\frac{x^3+4x+10}{x-3}$

23. $\frac{2x^3+6x^2-11x-12}{x+4}$

24. $\frac{3x^4+3x^3+2x+2}{x+1}$

Zeros of Polynomial Functions

For the following exercises, use the Rational Zero Theorem to help you solve the polynomial equation.

25. $2x^3 - 3x^2 - 18x - 8 = 0$

26. $3x^3 + 11x^2 + 8x - 4 = 0$

27. $2x^4 - 17x^3 + 46x^2 - 43x + 12 = 0$

28. $4x^4 + 8x^3 + 19x^2 + 32x + 12 = 0$

For the following exercises, use Descartes' Rule of Signs to find the possible number of positive and negative solutions.

29. $x^3 - 3x^2 - 2x + 4 = 0$

30. $2x^4 - x^3 + 4x^2 - 5x + 1 = 0$

Rational Functions

For the following exercises, find the intercepts and the vertical and horizontal asymptotes, and then use them to sketch a graph of the function.

31. $f(x) = \frac{x+2}{x-5}$

32. $f(x) = \frac{x^2+1}{x^2-4}$

33. $f(x) = \frac{3x^2-27}{x^2+x-2}$

34. $f(x) = \frac{x+2}{x^2-9}$

For the following exercises, find the slant asymptote.

35. $f(x) = \frac{x^2-1}{x+2}$

36. $f(x) = \frac{2x^3-x^2+4}{x^2+1}$

Inverses and Radical Functions

For the following exercises, find the inverse of the function with the domain given.

37. $f(x) = (x-2)^2,\ \ x \geq 2$

38. $f(x) = (x+4)^2 - 3,\ \ x \geq -4$

39. $f(x) = x^2 + 6x - 2,\ \ x \geq -3$

40. $f(x) = 2x^3 - 3$

41. $f(x) = \sqrt{4x+5} - 3$

42. $f(x) = \frac{x-3}{2x+1}$

Modeling Using Variation

For the following exercises, find the unknown value.

43. y varies directly as the square of x. If when $x = 3,\ \ y = 36$, find y if $x = 4$.

44. y varies inversely as the square root of x If when $x = 25,\ \ y = 2$, find y if $x = 4$.

45. y varies jointly as the cube of x and as z. If when $x = 1$ and $z = 2, y = 6$, find y if $x = 2$ and $z = 3$.

46. y varies jointly as x and the square of z and inversely as the cube of w. If when $x = 3, z = 4$, and $w = 2$, $y = 48$, find y if $x = 4$, $z = 5$, and $w = 3$.

For the following exercises, solve the application problem.

47. The weight of an object above the surface of the earth varies inversely with the distance from the center of the earth. If a person weighs 150 pounds when he is on the surface of the earth (3,960 miles from center), find the weight of the person if he is 20 miles above the surface.

48. The volume V of an ideal gas varies directly with the temperature T and inversely with the pressure P. A cylinder contains oxygen at a temperature of 310 degrees K and a pressure of 18 atmospheres in a volume of 120 liters. Find the pressure if the volume is decreased to 100 liters and the temperature is increased to 320 degrees K.

Practice Test

Give the degree and leading coefficient of the following polynomial function.

1. $f(x) = x^3\left(3 - 6x - 2x^2\right)$

Determine the end behavior of the polynomial function.

2. $f(x) = 8x^3 - 3x^2 + 2x - 4$

3. $f(x) = -2x^2(4 - 3x - 5x^2)$

Write the quadratic function in standard form. Determine the vertex and axes intercepts and graph the function.

4. $f(x) = x^2 + 2x - 8$

Given information about the graph of a quadratic function, find its equation.

5. Vertex $(2, 0)$ and point on graph $(4, 12)$.

Solve the following application problem.

6. A rectangular field is to be enclosed by fencing. In addition to the enclosing fence, another fence is to divide the field into two parts, running parallel to two sides. If 1,200 feet of fencing is available, find the maximum area that can be enclosed.

Find all zeros of the following polynomial functions, noting multiplicities.

7. $f(x) = (x - 3)^3(3x - 1)(x - 1)^2$

8. $f(x) = 2x^6 - 12x^5 + 18x^4$

Based on the graph, determine the zeros of the function and multiplicities.

9.

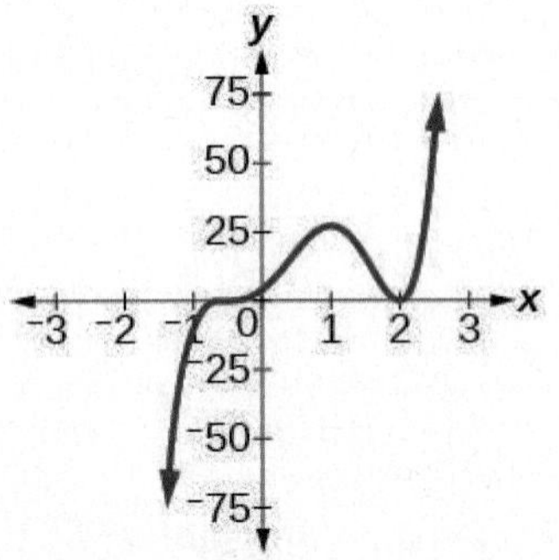

Use long division to find the quotient.

10. $\frac{2x^3+3x-4}{x+2}$

Use synthetic division to find the quotient. If the divisor is a factor, write the factored form.

11. $\frac{x^4+3x^2-4}{x-2}$

12. $\frac{2x^3+5x^2-7x-12}{x+3}$

Use the Rational Zero Theorem to help you find the zeros of the polynomial functions.

13. $f(x) = 2x^3 + 5x^2 - 6x - 9$

14. $f(x) = 4x^4 + 8x^3 + 21x^2 + 17x + 4$

15. $f(x) = 4x^4 + 16x^3 + 13x^2 - 15x - 18$

16. $f(x) = x^5 + 6x^4 + 13x^3 + 14x^2 + 12x + 8$

Given the following information about a polynomial function, find the function.

17. It has a double zero at $x = 3$ and zeros at $x = 1$ and $x = -2$. Its *y*-intercept is $(0, 12)$.

18. It has a zero of multiplicity 3 at $x = \frac{1}{2}$ and another zero at $x = -3$. It contains the point $(1, 8)$.

Use Descartes' Rule of Signs to determine the possible number of positive and negative solutions.

19. $8x^3 - 21x^2 + 6 = 0$

For the following rational functions, find the intercepts and horizontal and vertical asymptotes, and sketch a graph.

20. $f(x) = \frac{x+4}{x^2-2x-3}$

21. $f(x) = \frac{x^2+2x-3}{x^2-4}$

Find the slant asymptote of the rational function.

22. $f(x) = \frac{x^2+3x-3}{x-1}$

Find the inverse of the function.

23. $f(x) = \sqrt{x-2} + 4$

24. $f(x) = 3x^3 - 4$

25. $f(x) = \frac{2x+3}{3x-1}$

Find the unknown value.

26. y varies inversely as the square of x and when $x = 3$, $y = 2$. Find y if $x = 1$.

27. y varies jointly with x and the cube root of z. If when $x = 2$ and $z = 27$, $y = 12$, find y if $x = 5$ and $z = 8$.

Solve the following application problem.

28. The distance a body falls varies directly as the square of the time it falls. If an object falls 64 feet in 2 seconds, how long will it take to fall 256 feet?

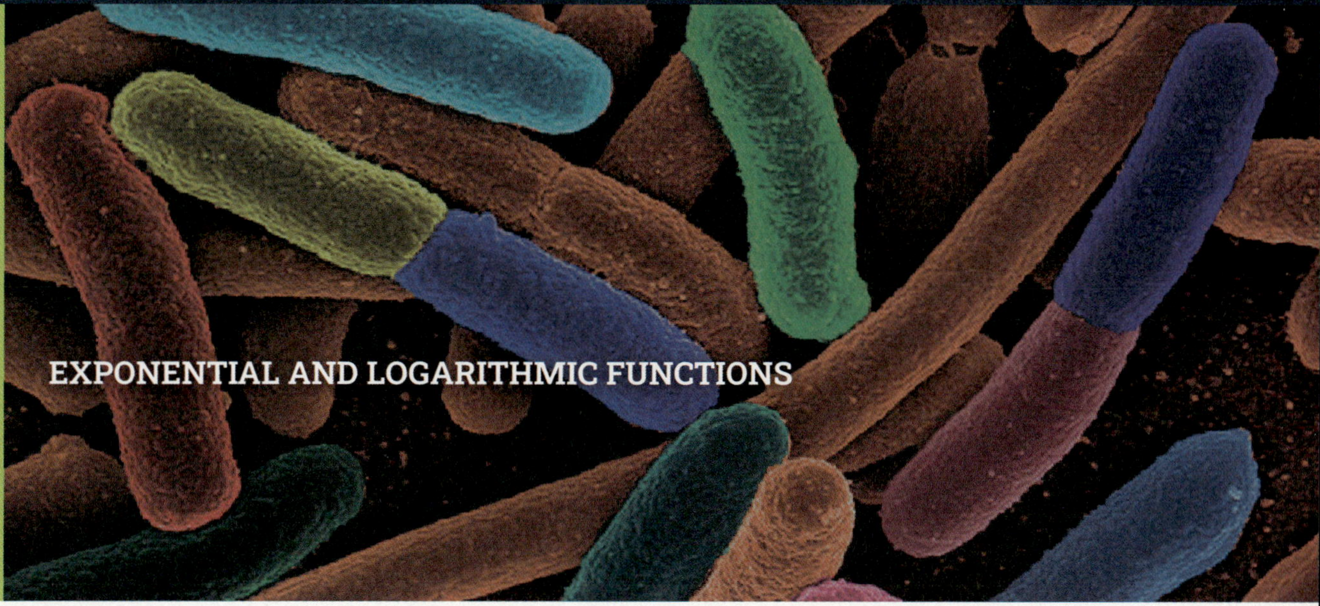

6 EXPONENTIAL AND LOGARITHMIC FUNCTIONS

Electron micrograph of *E.Coli* bacteria (credit: "Mattosaurus," Wikimedia Commons)

Chapter Outline

Introduction to Exponential and Logarithmic Functions

Focus in on a square centimeter of your skin. Look closer. Closer still. If you could look closely enough, you would see hundreds of thousands of microscopic organisms. They are bacteria, and they are not only on your skin, but in your mouth, nose, and even your intestines. In fact, the bacterial cells in your body at any given moment outnumber your own cells. But that is no reason to feel bad about yourself. While some bacteria can cause illness, many are healthy and even essential to the body.

Bacteria commonly reproduce through a process called binary fission, during which one bacterial cell splits into two. When conditions are right, bacteria can reproduce very quickly. Unlike humans and other complex organisms, the time required to form a new generation of bacteria is often a matter of minutes or hours, as opposed to days or years.[1]

For simplicity's sake, suppose we begin with a culture of one bacterial cell that can divide every hour. Table 1 shows the number of bacterial cells at the end of each subsequent hour. We see that the single bacterial cell leads to over one thousand bacterial cells in just ten hours! And if we were to extrapolate the table to twenty-four hours, we would have over 16 million!

Hour	0	1	2	3	4	5	6	7	8	9	10
Bacteria	1	2	4	8	16	32	64	128	256	512	1024

Table 1

1 Todar, PhD, Kenneth. Todar's Online Textbook of Bacteriology. http://textbookofbacteriology.net/growth_3.html.

In this chapter, we will explore exponential functions, which can be used for, among other things, modeling growth patterns such as those found in bacteria. We will also investigate logarithmic functions, which are closely related to exponential functions. Both types of functions have numerous real-world applications when it comes to modeling and interpreting data.

6.1 Exponential Functions

Learning Objectives

In this section, you will:

- Evaluate exponential functions.
- Find the equation of an exponential function.
- Use compound interest formulas.
- Evaluate exponential functions with base e .

India is the second most populous country in the world with a population of about 1.39 billion people in 2021. The population is growing at a rate of about 1.2% each year[2] . If this rate continues, the population of India will exceed China's population by the year 2027. When populations grow rapidly, we often say that the growth is "exponential," meaning that something is growing very rapidly. To a mathematician, however, the term *exponential growth* has a very specific meaning. In this section, we will take a look at *exponential functions*, which model this kind of rapid growth.

Identifying Exponential Functions

When exploring linear growth, we observed a constant rate of change—a constant number by which the output increased for each unit increase in input. For example, in the equation $f(x) = 3x + 4$, the slope tells us the output increases by 3 each time the input increases by 1. The scenario in the India population example is different because we have a *percent* change per unit time (rather than a constant change) in the number of people.

Defining an Exponential Function

A study found that the percent of the population who are vegans in the United States doubled from 2009 to 2011. In 2011, 2.5% of the population was vegan, adhering to a diet that does not include any animal products—no meat, poultry, fish, dairy, or eggs. If this rate continues, vegans will make up 10% of the U.S. population in 2015, 40% in 2019, and 80% in 2021.

What exactly does it mean to *grow exponentially*? What does the word *double* have in common with *percent increase*? People toss these words around errantly. Are these words used correctly? The words certainly appear frequently in the media.

- **Percent change** refers to a *change* based on a *percent* of the original amount.
- **Exponential growth** refers to an *increase* based on a constant multiplicative rate of change over equal increments of time, that is, a *percent* increase of the original amount over time.
- **Exponential decay** refers to a *decrease* based on a constant multiplicative rate of change over equal increments of time, that is, a *percent* decrease of the original amount over time.

For us to gain a clear understanding of exponential growth, let us contrast exponential growth with linear growth. We will construct two functions. The first function is exponential. We will start with an input of 0, and increase each input by 1. We will double the corresponding consecutive outputs. The second function is linear. We will start with an input of 0, and increase each input by 1. We will add 2 to the corresponding consecutive outputs. See Table 1.

x	$f(x) = 2^x$	$g(x) = 2x$
0	1	0
1	2	2
2	4	4
3	8	6

Table 1

2 http://www.worldometers.info/world-population/. Accessed February 24, 2014.

x	$f(x) = 2^x$	$g(x) = 2x$
4	16	8
5	32	10
6	64	12

Table 1

From Table 1 we can infer that for these two functions, exponential growth dwarfs linear growth.

- **Exponential growth** refers to the original value from the range increases by the *same percentage* over equal increments found in the domain.
- **Linear growth** refers to the original value from the range increases by the *same amount* over equal increments found in the domain.

Apparently, the difference between "the same percentage" and "the same amount" is quite significant. For exponential growth, over equal increments, the constant multiplicative rate of change resulted in doubling the output whenever the input increased by one. For linear growth, the constant additive rate of change over equal increments resulted in adding 2 to the output whenever the input was increased by one.

The general form of the exponential function is $f(x) = ab^x$, where a is any nonzero number, b is a positive real number not equal to 1.

- If $b > 1$, the function grows at a rate proportional to its size.
- If $0 < b < 1$, the function decays at a rate proportional to its size.

Let's look at the function $f(x) = 2^x$ from our example. We will create a table (Table 2) to determine the corresponding outputs over an interval in the domain from -3 to 3.

x	-3	-2	-1	0	1	2	3
$f(x) = 2^x$	$2^{-3} = \frac{1}{8}$	$2^{-2} = \frac{1}{4}$	$2^{-1} = \frac{1}{2}$	$2^0 = 1$	$2^1 = 2$	$2^2 = 4$	$2^3 = 8$

Table 2

Let us examine the graph of f by plotting the ordered pairs we observe on the table in Figure 1, and then make a few observations.

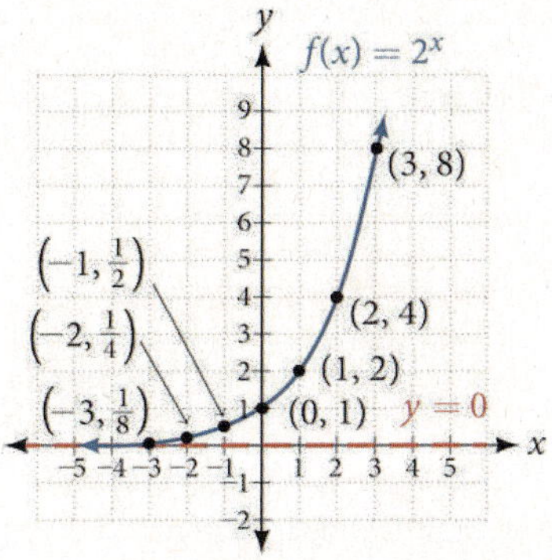

Figure 1

Let's define the behavior of the graph of the exponential function $f(x) = 2^x$ and highlight some its key characteristics.

- the domain is $\left(-\infty, \infty\right)$,
- the range is $\left(0, \infty\right)$,
- as $x \to \infty$, $f(x) \to \infty$,

- as $x \to -\infty$, $f(x) \to 0$,
- $f(x)$ is always increasing,
- the graph of $f(x)$ will never touch the x-axis because base two raised to any exponent never has the result of zero.
- $y = 0$ is the horizontal asymptote.
- the *y*-intercept is 1.

Exponential Function

For any real number x, an exponential function is a function with the form

$$f(x) = ab^x$$

where

- a is a non-zero real number called the initial value and
- b is any positive real number such that $b \neq 1$.
- The domain of f is all real numbers.
- The range of f is all positive real numbers if $a > 0$.
- The range of f is all negative real numbers if $a < 0$.
- The *y*-intercept is $(0, a)$, and the horizontal asymptote is $y = 0$.

EXAMPLE 1

Identifying Exponential Functions

Which of the following equations are *not* exponential functions?

- $f(x) = 4^{3(x-2)}$
- $g(x) = x^3$
- $h(x) = \left(\frac{1}{3}\right)^x$
- $j(x) = (-2)^x$

Solution

By definition, an exponential function has a constant as a base and an independent variable as an exponent. Thus, $g(x) = x^3$ does not represent an exponential function because the base is an independent variable. In fact, $g(x) = x^3$ is a power function.

Recall that the base *b* of an exponential function is always a positive constant, and $b \neq 1$. Thus, $j(x) = (-2)^x$ does not represent an exponential function because the base, -2, is less than 0.

TRY IT #1 Which of the following equations represent exponential functions?

- $f(x) = 2x^2 - 3x + 1$
- $g(x) = 0.875^x$
- $h(x) = 1.75x + 2$
- $j(x) = 1095.6^{-2x}$

Evaluating Exponential Functions

Recall that the base of an exponential function must be a positive real number other than 1. Why do we limit the base *b* to positive values? To ensure that the outputs will be real numbers. Observe what happens if the base is not positive:

- Let $b = -9$ and $x = \frac{1}{2}$. Then $f(x) = f\left(\frac{1}{2}\right) = (-9)^{\frac{1}{2}} = \sqrt{-9}$, which is not a real number.

Why do we limit the base to positive values other than 1? Because base 1 results in the constant function. Observe what happens if the base is 1 :

- Let $b = 1$. Then $f(x) = 1^x = 1$ for any value of x.

To evaluate an exponential function with the form $f(x) = b^x$, we simply substitute x with the given value, and calculate the resulting power. For example:

Let $f(x) = 2^x$. What is $f(3)$?

$$\begin{aligned} f(x) &= 2^x \\ f(3) &= 2^3 && \text{Substitute } x = 3. \\ &= 8 && \text{Evaluate the power.} \end{aligned}$$

To evaluate an exponential function with a form other than the basic form, it is important to follow the order of operations. For example:

Let $f(x) = 30(2)^x$. What is $f(3)$?

$$\begin{aligned} f(x) &= 30(2)^x \\ f(3) &= 30(2)^3 && \text{Substitute } x = 3. \\ &= 30(8) && \text{Simplify the power first.} \\ &= 240 && \text{Multiply.} \end{aligned}$$

Note that if the order of operations were not followed, the result would be incorrect:

$$f(3) = 30(2)^3 \neq 60^3 = 216{,}000$$

EXAMPLE 2

Evaluating Exponential Functions

Let $f(x) = 5(3)^{x+1}$. Evaluate $f(2)$ without using a calculator.

Solution

Follow the order of operations. Be sure to pay attention to the parentheses.

$$\begin{aligned} f(x) &= 5(3)^{x+1} \\ f(2) &= 5(3)^{2+1} && \text{Substitute } x = 2. \\ &= 5(3)^3 && \text{Add the exponents.} \\ &= 5(27) && \text{Simplify the power.} \\ &= 135 && \text{Multiply.} \end{aligned}$$

> **TRY IT** #2 Let $f(x) = 8(1.2)^{x-5}$. Evaluate $f(3)$ using a calculator. Round to four decimal places.

Defining Exponential Growth

Because the output of exponential functions increases very rapidly, the term "exponential growth" is often used in everyday language to describe anything that grows or increases rapidly. However, exponential growth can be defined more precisely in a mathematical sense. If the growth rate is proportional to the amount present, the function models exponential growth.

Exponential Growth

A function that models **exponential growth** grows by a rate proportional to the amount present. For any real number x and any positive real numbers a and b such that $b \neq 1$, an exponential growth function has the form

$$f(x) = ab^x$$

where

- a is the initial or starting value of the function.
- b is the growth factor or growth multiplier per unit x.

In more general terms, we have an *exponential function*, in which a constant base is raised to a variable exponent. To differentiate between linear and exponential functions, let's consider two companies, A and B. Company A has 100 stores and expands by opening 50 new stores a year, so its growth can be represented by the function

$A(x) = 100 + 50x$. Company B has 100 stores and expands by increasing the number of stores by 50% each year, so its growth can be represented by the function $B(x) = 100(1 + 0.5)^x$.

A few years of growth for these companies are illustrated in Table 3.

Year, x	Stores, Company A	Stores, Company B
0	$100 + 50(0) = 100$	$100(1 + 0.5)^0 = 100$
1	$100 + 50(1) = 150$	$100(1 + 0.5)^1 = 150$
2	$100 + 50(2) = 200$	$100(1 + 0.5)^2 = 225$
3	$100 + 50(3) = 250$	$100(1 + 0.5)^3 = 337.5$
x	$A(x) = 100 + 50x$	$B(x) = 100(1 + 0.5)^x$

Table 3

The graphs comparing the number of stores for each company over a five-year period are shown in Figure 2. We can see that, with exponential growth, the number of stores increases much more rapidly than with linear growth.

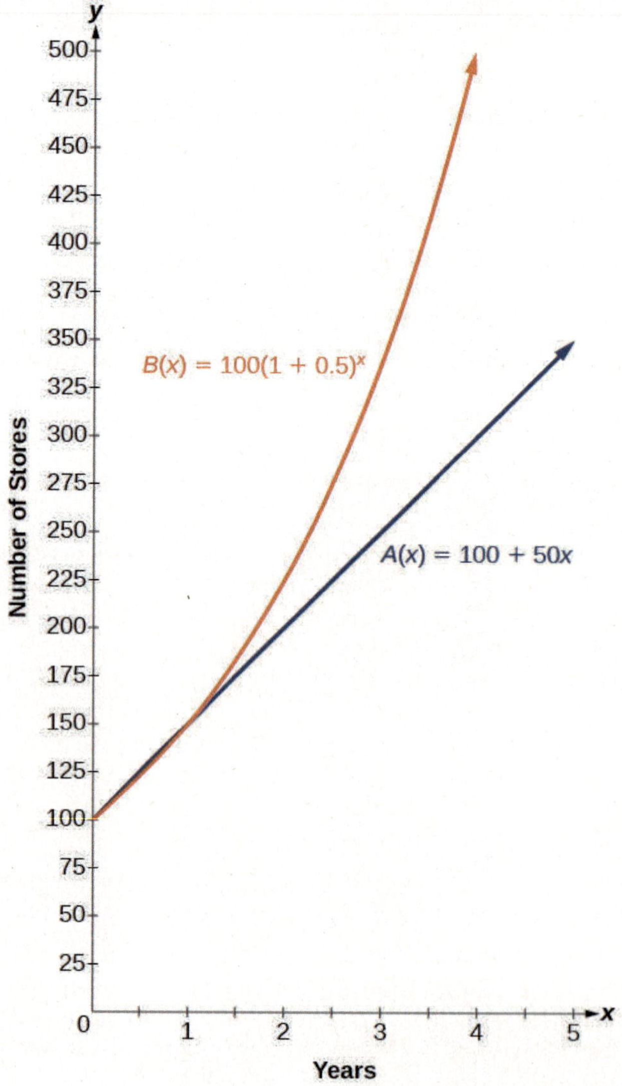

Figure 2 The graph shows the numbers of stores Companies A and B opened over a five-year period.

Notice that the domain for both functions is $[0, \infty)$, and the range for both functions is $[100, \infty)$. After year 1, Company B always has more stores than Company A.

Now we will turn our attention to the function representing the number of stores for Company B, $B(x) = 100(1 + 0.5)^x$. In this exponential function, 100 represents the initial number of stores, 0.50 represents the growth rate, and $1 + 0.5 = 1.5$ represents the growth factor. Generalizing further, we can write this function as $B(x) = 100(1.5)^x$, where 100 is the initial value, 1.5 is called the *base*, and x is called the *exponent*.

EXAMPLE 3

Evaluating a Real-World Exponential Model

At the beginning of this section, we learned that the population of India was about 1.25 billion in the year 2013, with an annual growth rate of about 1.2%. This situation is represented by the growth function $P(t) = 1.25(1.012)^t$, where t is the number of years since 2013. To the nearest thousandth, what will the population of India be in 2031?

Solution

To estimate the population in 2031, we evaluate the models for $t = 18$, because 2031 is 18 years after 2013. Rounding to the nearest thousandth,

$$P(18) = 1.25(1.012)^{18} \approx 1.549$$

There will be about 1.549 billion people in India in the year 2031.

TRY IT #3 The population of China was about 1.39 billion in the year 2013, with an annual growth rate of about 0.6%. This situation is represented by the growth function $P(t) = 1.39(1.006)^t$, where t is the number of years since 2013. To the nearest thousandth, what will the population of China be for the year 2031? How does this compare to the population prediction we made for India in Example 3?

Finding Equations of Exponential Functions

In the previous examples, we were given an exponential function, which we then evaluated for a given input. Sometimes we are given information about an exponential function without knowing the function explicitly. We must use the information to first write the form of the function, then determine the constants a and b, and evaluate the function.

HOW TO

Given two data points, write an exponential model.

1. If one of the data points has the form $(0, a)$, then a is the initial value. Using a, substitute the second point into the equation $f(x) = a(b)^x$, and solve for b.
2. If neither of the data points have the form $(0, a)$, substitute both points into two equations with the form $f(x) = a(b)^x$. Solve the resulting system of two equations in two unknowns to find a and b.
3. Using the a and b found in the steps above, write the exponential function in the form $f(x) = a(b)^x$.

EXAMPLE 4

Writing an Exponential Model When the Initial Value Is Known

In 2006, 80 deer were introduced into a wildlife refuge. By 2012, the population had grown to 180 deer. The population was growing exponentially. Write an exponential function $N(t)$ representing the population (N) of deer over time t.

Solution

We let our independent variable t be the number of years after 2006. Thus, the information given in the problem can be written as input-output pairs: (0, 80) and (6, 180). Notice that by choosing our input variable to be measured as years after 2006, we have given ourselves the initial value for the function, $a = 80$. We can now substitute the second point into the equation $N(t) = 80b^t$ to find b :

$$\begin{aligned} N(t) &= 80b^t & \\ 180 &= 80b^6 & \text{Substitute using point } (6,\ 180). \\ \tfrac{9}{4} &= b^6 & \text{Divide and write in lowest terms.} \\ b &= \left(\tfrac{9}{4}\right)^{\frac{1}{6}} & \text{Isolate } b \text{ using properties of exponents.} \\ b &\approx 1.1447 & \text{Round to 4 decimal places.} \end{aligned}$$

NOTE: *Unless otherwise stated, do not round any intermediate calculations. Then round the final answer to four places for the remainder of this section.*

The exponential model for the population of deer is $N(t) = 80(1.1447)^t$. (Note that this exponential function models short-term growth. As the inputs gets large, the output will get increasingly larger, so much so that the model may not be useful in the long term.)

We can graph our model to observe the population growth of deer in the refuge over time. Notice that the graph in Figure 3 passes through the initial points given in the problem, $(0, 80)$ and $(6, 180)$. We can also see that the domain for the function is $[0, \infty)$, and the range for the function is $[80, \infty)$.

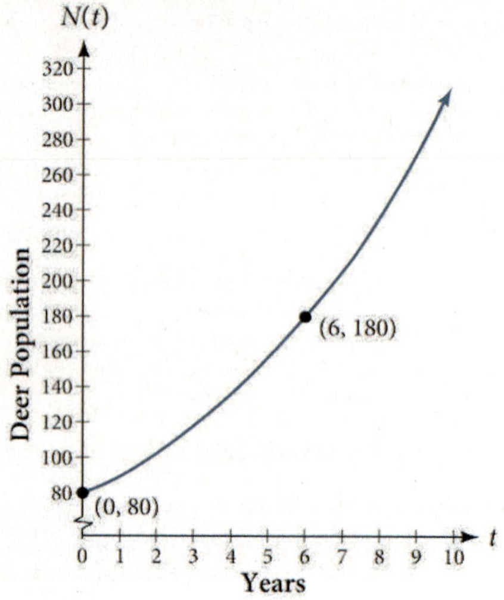

Figure 3 Graph showing the population of deer over time, $N(t) = 80(1.1447)^t$, t years after 2006

TRY IT #4 A wolf population is growing exponentially. In 2011, 129 wolves were counted. By 2013, the population had reached 236 wolves. What two points can be used to derive an exponential equation modeling this situation? Write the equation representing the population N of wolves over time t.

EXAMPLE 5

Writing an Exponential Model When the Initial Value is Not Known

Find an exponential function that passes through the points $(-2, 6)$ and $(2, 1)$.

Solution

Because we don't have the initial value, we substitute both points into an equation of the form $f(x) = ab^x$, and then solve the system for a and b.

- Substituting $(-2, 6)$ gives $6 = ab^{-2}$
- Substituting $(2, 1)$ gives $1 = ab^2$

Use the first equation to solve for a in terms of b:

$$6 = ab^{-2}$$
$$\frac{6}{b^{-2}} = a \quad \text{Divide.}$$
$$a = 6b^2 \quad \text{Use properties of exponents to rewrite the denominator.}$$

Substitute a in the second equation, and solve for b :

$$1 = ab^2$$
$$1 = 6b^2b^2 = 6b^4 \quad \text{Substitute } a.$$
$$b = \left(\frac{1}{6}\right)^{\frac{1}{4}} \quad \text{Use properties of exponents to isolate } b.$$
$$b \approx 0.6389 \quad \text{Round 4 decimal places.}$$

Use the value of b in the first equation to solve for the value of a :

$$a = 6b^2 \approx 6(0.6389)^2 \approx 2.4492$$

Thus, the equation is $f(x) = 2.4492(0.6389)^x$.

We can graph our model to check our work. Notice that the graph in Figure 4 passes through the initial points given in the problem, $(-2, \ 6)$ and $(2, \ 1)$. The graph is an example of an exponential decay function.

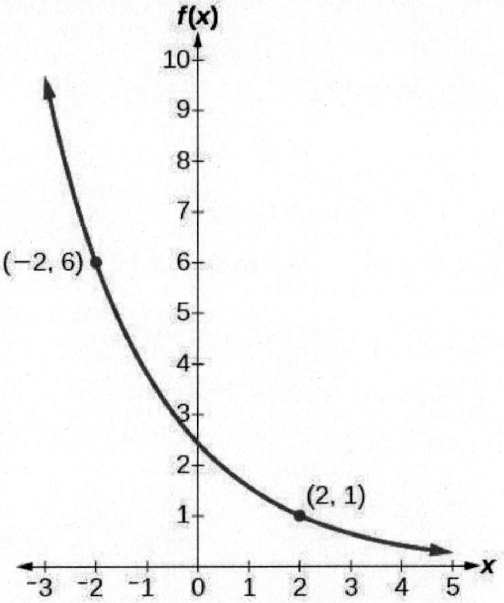

Figure 4 The graph of $f(x) = 2.4492(0.6389)^x$ models exponential decay.

TRY IT #5 Given the two points $(1, 3)$ and $(2, 4.5)$, find the equation of the exponential function that passes through these two points.

Q&A **Do two points always determine a unique exponential function?**

Yes, provided the two points are either both above the x-axis or both below the x-axis and have different x-coordinates. But keep in mind that we also need to know that the graph is, in fact, an exponential function. Not every graph that looks exponential really is exponential. We need to know the graph is based on a model that shows the same percent growth with each unit increase in x, which in many real world cases involves time.

HOW TO

Given the graph of an exponential function, write its equation.

1. First, identify two points on the graph. Choose the *y*-intercept as one of the two points whenever possible. Try to choose points that are as far apart as possible to reduce round-off error.
2. If one of the data points is the *y*-intercept $(0, a)$, then a is the initial value. Using a, substitute the second point into the equation $f(x) = a(b)^x$, and solve for b.
3. If neither of the data points have the form $(0, a)$, substitute both points into two equations with the form $f(x) = a(b)^x$. Solve the resulting system of two equations in two unknowns to find a and b.
4. Write the exponential function, $f(x) = a(b)^x$.

EXAMPLE 6

Writing an Exponential Function Given Its Graph

Find an equation for the exponential function graphed in Figure 5.

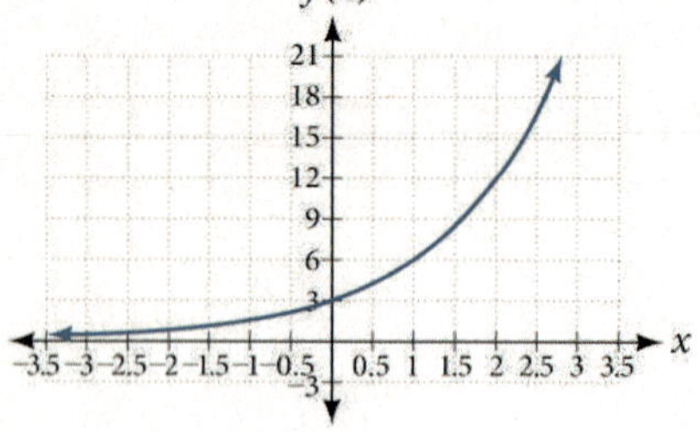

Figure 5

Solution

We can choose the *y*-intercept of the graph, $(0, 3)$, as our first point. This gives us the initial value, $a = 3$. Next, choose a point on the curve some distance away from $(0, 3)$ that has integer coordinates. One such point is $(2, 12)$.

$y = ab^x$	Write the general form of an exponential equation.
$y = 3b^x$	Substitute the initial value 3 for a.
$12 = 3b^2$	Substitute in 12 for y and 2 for x.
$4 = b^2$	Divide by 3.
$b = \pm 2$	Take the square root.

Because we restrict ourselves to positive values of b, we will use $b = 2$. Substitute a and b into the standard form to yield the equation $f(x) = 3(2)^x$.

TRY IT #6 Find an equation for the exponential function graphed in Figure 6.

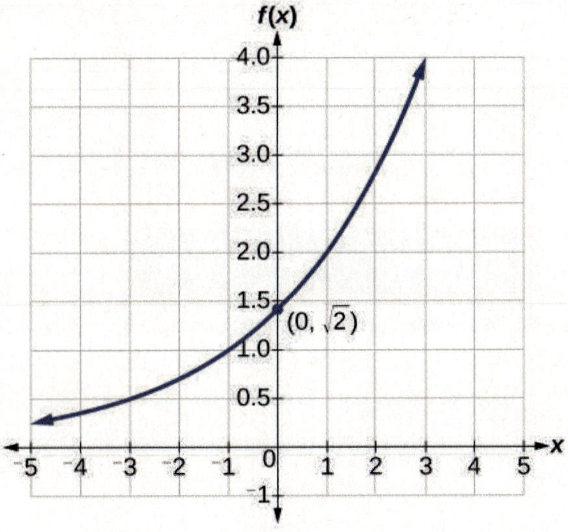

Figure 6

HOW TO

Given two points on the curve of an exponential function, use a graphing calculator to find the equation.

1. Press **[STAT].**
2. Clear any existing entries in columns **L1** or **L2.**
3. In **L1**, enter the *x*-coordinates given.
4. In **L2**, enter the corresponding *y*-coordinates.
5. Press **[STAT]** again. Cursor right to **CALC**, scroll down to **ExpReg (Exponential Regression)**, and press **[ENTER].**
6. The screen displays the values of *a* and *b* in the exponential equation $y = a \cdot b^x$.

EXAMPLE 7

Using a Graphing Calculator to Find an Exponential Function
Use a graphing calculator to find the exponential equation that includes the points $(2, 24.8)$ and $(5, 198.4)$.

Solution
Follow the guidelines above. First press **[STAT]**, **[EDIT]**, **[1: Edit...]**, and clear the lists **L1** and **L2**. Next, in the **L1** column, enter the *x*-coordinates, 2 and 5. Do the same in the **L2** column for the *y*-coordinates, 24.8 and 198.4.

Now press **[STAT]**, **[CALC]**, **[0: ExpReg]** and press **[ENTER]**. The values $a = 6.2$ and $b = 2$ will be displayed. The exponential equation is $y = 6.2 \cdot 2^x$.

TRY IT #7 Use a graphing calculator to find the exponential equation that includes the points (3, 75.98) and (6, 481.07).

Applying the Compound-Interest Formula

Savings instruments in which earnings are continually reinvested, such as mutual funds and retirement accounts, use **compound interest**. The term *compounding* refers to interest earned not only on the original value, but on the accumulated value of the account.

The **annual percentage rate (APR)** of an account, also called the **nominal rate**, is the yearly interest rate earned by an investment account. The term *nominal* is used when the compounding occurs a number of times other than once per year. In fact, when interest is compounded more than once a year, the effective interest rate ends up being *greater* than the nominal rate! This is a powerful tool for investing.

We can calculate the compound interest using the compound interest formula, which is an exponential function of the variables time t, principal P, APR r, and number of compounding periods in a year n :

$$A(t) = P\left(1 + \frac{r}{n}\right)^{nt}$$

For example, observe Table 4, which shows the result of investing $1,000 at 10% for one year. Notice how the value of the account increases as the compounding frequency increases.

Frequency	Value after 1 year
Annually	$1100
Semiannually	$1102.50
Quarterly	$1103.81
Monthly	$1104.71
Daily	$1105.16

Table 4

The Compound Interest Formula

Compound interest can be calculated using the formula

$$A(t) = P\left(1 + \frac{r}{n}\right)^{nt}$$

where

- $A(t)$ is the account value,
- t is measured in years,
- P is the starting amount of the account, often called the principal, or more generally present value,
- r is the annual percentage rate (APR) expressed as a decimal, and
- n is the number of compounding periods in one year.

EXAMPLE 8

Calculating Compound Interest

If we invest $3,000 in an investment account paying 3% interest compounded quarterly, how much will the account be worth in 10 years?

Solution

Because we are starting with $3,000, $P = 3000$. Our interest rate is 3%, so $r = 0.03$. Because we are compounding quarterly, we are compounding 4 times per year, so $n = 4$. We want to know the value of the account in 10 years, so we are looking for $A(10)$, the value when $t = 10$.

$$\begin{aligned} A(t) &= P\left(1 + \tfrac{r}{n}\right)^{nt} && \text{Use the compound interest formula.} \\ A(10) &= 3000\left(1 + \tfrac{0.03}{4}\right)^{4\cdot 10} && \text{Substitute using given values.} \\ &\approx \$4045.05 && \text{Round to two decimal places.} \end{aligned}$$

The account will be worth about $4,045.05 in 10 years.

TRY IT #8 An initial investment of $100,000 at 12% interest is compounded weekly (use 52 weeks in a year). What will the investment be worth in 30 years?

EXAMPLE 9

Using the Compound Interest Formula to Solve for the Principal

A 529 Plan is a college-savings plan that allows relatives to invest money to pay for a child's future college tuition; the account grows tax-free. Lily wants to set up a 529 account for her new granddaughter and wants the account to grow to $40,000 over 18 years. She believes the account will earn 6% compounded semi-annually (twice a year). To the nearest dollar, how much will Lily need to invest in the account now?

Solution

The nominal interest rate is 6%, so $r = 0.06$. Interest is compounded twice a year, so $k = 2$.

We want to find the initial investment, P, needed so that the value of the account will be worth $40,000 in 18 years. Substitute the given values into the compound interest formula, and solve for P.

$$
\begin{array}{lll}
A(t) & = P\left(1+\frac{r}{n}\right)^{nt} & \text{Use the compound interest formula.} \\
40{,}000 & = P\left(1+\frac{0.06}{2}\right)^{2(18)} & \text{Substitute using given values } A,\ r,\ n, \text{ and } t. \\
40{,}000 & = P(1.03)^{36} & \text{Simplify.} \\
\frac{40{,}000}{(1.03)^{36}} & = P & \text{Isolate } P. \\
P & \approx \$13,801 & \text{Divide and round to the nearest dollar.}
\end{array}
$$

Lily will need to invest $13,801 to have $40,000 in 18 years.

> TRY IT #9 Refer to Example 9. To the nearest dollar, how much would Lily need to invest if the account is compounded quarterly?

Evaluating Functions with Base *e*

As we saw earlier, the amount earned on an account increases as the compounding frequency increases. Table 5 shows that the increase from annual to semi-annual compounding is larger than the increase from monthly to daily compounding. This might lead us to ask whether this pattern will continue.

Examine the value of $1 invested at 100% interest for 1 year, compounded at various frequencies, listed in Table 5.

Frequency	$A(n) = \left(1+\frac{1}{n}\right)^n$	Value
Annually	$\left(1+\frac{1}{1}\right)^1$	$2
Semiannually	$\left(1+\frac{1}{2}\right)^2$	$2.25
Quarterly	$\left(1+\frac{1}{4}\right)^4$	$2.441406
Monthly	$\left(1+\frac{1}{12}\right)^{12}$	$2.613035
Daily	$\left(1+\frac{1}{365}\right)^{365}$	$2.714567
Hourly	$\left(1+\frac{1}{8760}\right)^{8760}$	$2.718127
Once per minute	$\left(1+\frac{1}{525600}\right)^{525600}$	$2.718279
Once per second	$\left(1+\frac{1}{31536000}\right)^{31536000}$	$2.718282

Table 5

These values appear to be approaching a limit as n increases without bound. In fact, as n gets larger and larger, the expression $\left(1+\frac{1}{n}\right)^n$ approaches a number used so frequently in mathematics that it has its own name: the letter e. This value is an irrational number, which means that its decimal expansion goes on forever without repeating. Its approximation to six decimal places is shown below.

The Number e

The letter e represents the irrational number

$$\left(1+\frac{1}{n}\right)^n, \text{as } n \text{ increases without bound}$$

The letter e is used as a base for many real-world exponential models. To work with base e, we use the approximation, $e \approx 2.718282$. The constant was named by the Swiss mathematician Leonhard Euler (1707–1783) who first investigated and discovered many of its properties.

EXAMPLE 10

Using a Calculator to Find Powers of e

Calculate $e^{3.14}$. Round to five decimal places.

Solution

On a calculator, press the button labeled $\left[e^x\right]$. The window shows $[e$ ^( ]. Type 3.14 and then close parenthesis, $[)]$. Press [ENTER]. Rounding to 5 decimal places, $e^{3.14} \approx 23.10387$. Caution: Many scientific calculators have an "Exp" button, which is used to enter numbers in scientific notation. It is not used to find powers of e.

> **TRY IT** #10 Use a calculator to find $e^{-0.5}$. Round to five decimal places.

Investigating Continuous Growth

So far we have worked with rational bases for exponential functions. For most real-world phenomena, however, e is used as the base for exponential functions. Exponential models that use e as the base are called *continuous growth or decay models.* We see these models in finance, computer science, and most of the sciences, such as physics, toxicology, and fluid dynamics.

The Continuous Growth/Decay Formula

For all real numbers t, and all positive numbers a and r, continuous growth or decay is represented by the formula

$$A(t) = ae^{rt}$$

where

- a is the initial value,
- r is the continuous growth rate per unit time,
- and t is the elapsed time.

If $r > 0$, then the formula represents continuous growth. If $r < 0$, then the formula represents continuous decay.

For business applications, the continuous growth formula is called the continuous compounding formula and takes the form

$$A(t) = Pe^{rt}$$

where

- P is the principal or the initial invested,
- r is the growth or interest rate per unit time,

- and t is the period or term of the investment.

HOW TO

Given the initial value, rate of growth or decay, and time t, solve a continuous growth or decay function.

1. Use the information in the problem to determine a , the initial value of the function.
2. Use the information in the problem to determine the growth rate r.
   a. If the problem refers to continuous growth, then $r > 0$.
   b. If the problem refers to continuous decay, then $r < 0$.
3. Use the information in the problem to determine the time t.
4. Substitute the given information into the continuous growth formula and solve for $A(t)$.

EXAMPLE 11

Calculating Continuous Growth

A person invested $1,000 in an account earning a nominal 10% per year compounded continuously. How much was in the account at the end of one year?

Solution

Since the account is growing in value, this is a continuous compounding problem with growth rate $r = 0.10$. The initial investment was $1,000, so $P = 1000$. We use the continuous compounding formula to find the value after $t = 1$ year:

$$\begin{aligned} A(t) &= Pe^{rt} && \text{Use the continuous compounding formula.} \\ &= 1000(e)^{0.1} && \text{Substitute known values for } P,\ r, \text{ and } t. \\ &\approx 1105.17 && \text{Use a calculator to approximate.} \end{aligned}$$

The account is worth $1,105.17 after one year.

TRY IT #11 A person invests $100,000 at a nominal 12% interest per year compounded continuously. What will be the value of the investment in 30 years?

EXAMPLE 12

Calculating Continuous Decay

Radon-222 decays at a continuous rate of 17.3% per day. How much will 100 mg of Radon-222 decay to in 3 days?

Solution

Since the substance is decaying, the rate, 17.3% , is negative. So, $r = -0.173$. The initial amount of radon-222 was 100 mg, so $a = 100$. We use the continuous decay formula to find the value after $t = 3$ days:

$$\begin{aligned} A(t) &= ae^{rt} && \text{Use the continuous growth formula.} \\ &= 100e^{-0.173(3)} && \text{Substitute known values for } a,\ r, \text{ and } t. \\ &\approx 59.5115 && \text{Use a calculator to approximate.} \end{aligned}$$

So 59.5115 mg of radon-222 will remain.

TRY IT #12 Using the data in Example 12, how much radon-222 will remain after one year?

MEDIA

Access these online resources for additional instruction and practice with exponential functions.

Exponential Growth Function (http://openstax.org/l/expgrowth)
Compound Interest (http://openstax.org/l/compoundint)

6.1 SECTION EXERCISES

Verbal

1. Explain why the values of an increasing exponential function will eventually overtake the values of an increasing linear function.

2. Given a formula for an exponential function, is it possible to determine whether the function grows or decays exponentially just by looking at the formula? Explain.

3. The Oxford Dictionary defines the word *nominal* as a value that is "stated or expressed but not necessarily corresponding exactly to the real value."[3] Develop a reasonable argument for why the term *nominal rate* is used to describe the annual percentage rate of an investment account that compounds interest.

Algebraic

For the following exercises, identify whether the statement represents an exponential function. Explain.

4. The average annual population increase of a pack of wolves is 25.

5. A population of bacteria decreases by a factor of $\frac{1}{8}$ every 24 hours.

6. The value of a coin collection has increased by 3.25% annually over the last 20 years.

7. For each training session, a personal trainer charges his clients $\$5$ less than the previous training session.

8. The height of a projectile at time t is represented by the function $h(t) = -4.9t^2 + 18t + 40.$

For the following exercises, consider this scenario: For each year t, the population of a forest of trees is represented by the function $A(t) = 115(1.025)^t$. In a neighboring forest, the population of the same type of tree is represented by the function $B(t) = 82(1.029)^t$. (Round answers to the nearest whole number.)

9. Which forest's population is growing at a faster rate?

10. Which forest had a greater number of trees initially? By how many?

11. Assuming the population growth models continue to represent the growth of the forests, which forest will have a greater number of trees after 20 years? By how many?

3 Oxford Dictionary. http://oxforddictionaries.com/us/definition/american_english/nomina.

12. Assuming the population growth models continue to represent the growth of the forests, which forest will have a greater number of trees after 100 years? By how many?

13. Discuss the above results from the previous four exercises. Assuming the population growth models continue to represent the growth of the forests, which forest will have the greater number of trees in the long run? Why? What are some factors that might influence the long-term validity of the exponential growth model?

For the following exercises, determine whether the equation represents exponential growth, exponential decay, or neither. Explain.

14. $y = 300(1 - t)^5$

15. $y = 220(1.06)^x$

16. $y = 16.5(1.025)^{\frac{1}{x}}$

17. $y = 11{,}701(0.97)^t$

For the following exercises, find the formula for an exponential function that passes through the two points given.

18. $(0, 6)$ and $(3, 750)$

19. $(0, 2000)$ and $(2, 20)$

20. $\left(-1, \frac{3}{2}\right)$ and $(3, 24)$

21. $(-2, 6)$ and $(3, 1)$

22. $(3, 1)$ and $(5, 4)$

For the following exercises, determine whether the table could represent a function that is linear, exponential, or neither. If it appears to be exponential, find a function that passes through the points.

23.

x	1	2	3	4
$f(x)$	70	40	10	-20

24.

x	1	2	3	4
$h(x)$	70	49	34.3	24.01

25.

x	1	2	3	4
$m(x)$	80	61	42.9	25.61

26.

x	1	2	3	4
$f(x)$	10	20	40	80

27.

x	1	2	3	4
$g(x)$	-3.25	2	7.25	12.5

For the following exercises, use the compound interest formula, $A(t) = P\left(1 + \frac{r}{n}\right)^{nt}$.

28. After a certain number of years, the value of an investment account is represented by the equation $A = 10,250\left(1 + \frac{0.04}{12}\right)^{120}$. What is the value of the account?

29. What was the initial deposit made to the account in the previous exercise?

30. How many years had the account from the previous exercise been accumulating interest?

31. An account is opened with an initial deposit of $6,500 and earns 3.6% interest compounded semi-annually. What will the account be worth in 20 years?

32. How much more would the account in the previous exercise have been worth if the interest were compounding weekly?

33. Solve the compound interest formula for the principal, P.

34. Use the formula found in the previous exercise to calculate the initial deposit of an account that is worth $\$14,472.74$ after earning 5.5% interest compounded monthly for 5 years. (Round to the nearest dollar.)

35. How much more would the account in the previous two exercises be worth if it were earning interest for 5 more years?

36. Use properties of rational exponents to solve the compound interest formula for the interest rate, r.

37. Use the formula found in the previous exercise to calculate the interest rate for an account that was compounded semi-annually, had an initial deposit of $9,000 and was worth $13,373.53 after 10 years.

38. Use the formula found in the previous exercise to calculate the interest rate for an account that was compounded monthly, had an initial deposit of $5,500, and was worth $38,455 after 30 years.

For the following exercises, determine whether the equation represents continuous growth, continuous decay, or neither. Explain.

39. $y = 3742(e)^{0.75t}$

40. $y = 150(e)^{\frac{3.25}{t}}$

41. $y = 2.25(e)^{-2t}$

42. Suppose an investment account is opened with an initial deposit of $\$12,000$ earning 7.2% interest compounded continuously. How much will the account be worth after 30 years?

43. How much less would the account from Exercise 42 be worth after 30 years if it were compounded monthly instead?

Numeric

For the following exercises, evaluate each function. Round answers to four decimal places, if necessary.

44. $f(x) = 2(5)^x$, for $f(-3)$

45. $f(x) = -4^{2x+3}$, for $f(-1)$

46. $f(x) = e^x$, for $f(3)$

47. $f(x) = -2e^{x-1}$, for $f(-1)$

48. $f(x) = 2.7(4)^{-x+1} + 1.5$, for $f(-2)$

49. $f(x) = 1.2e^{2x} - 0.3$, for $f(3)$

50. $f(x) = -\frac{3}{2}(3)^{-x} + \frac{3}{2}$, for $f(2)$

Technology

For the following exercises, use a graphing calculator to find the equation of an exponential function given the points on the curve.

51. $(0, 3)$ and $(3, 375)$

52. $(3, 222.62)$ and $(10, 77.456)$

53. $(20, 29.495)$ and $(150, 730.89)$

54. $(5, 2.909)$ and $(13, 0.005)$

55. $(11{,}310.035)$ and $(25{,}356365.2)$

Extensions

56. The *annual percentage yield* (APY) of an investment account is a representation of the actual interest rate earned on a compounding account. It is based on a compounding period of one year. Show that the APY of an account that compounds monthly can be found with the formula $\text{APY} = \left(1 + \frac{r}{12}\right)^{12} - 1$.

57. Repeat the previous exercise to find the formula for the APY of an account that compounds daily. Use the results from this and the previous exercise to develop a function $I(n)$ for the APY of any account that compounds n times per year.

58. Recall that an exponential function is any equation written in the form $f(x) = a \cdot b^x$ such that a and b are positive numbers and $b \neq 1$. Any positive number b can be written as $b = e^n$ for some value of n. Use this fact to rewrite the formula for an exponential function that uses the number e as a base.

59. In an exponential decay function, the base of the exponent is a value between 0 and 1. Thus, for some number $b > 1$, the exponential decay function can be written as $f(x) = a \cdot \left(\frac{1}{b}\right)^x$. Use this formula, along with the fact that $b = e^n$, to show that an exponential decay function takes the form $f(x) = a(e)^{-nx}$ for some positive number n.

60. The formula for the amount A in an investment account with a nominal interest rate r at any time t is given by $A(t) = a(e)^{rt}$, where a is the amount of principal initially deposited into an account that compounds continuously. Prove that the percentage of interest earned to principal at any time t can be calculated with the formula $I(t) = e^{rt} - 1$.

Real-World Applications

61. The fox population in a certain region has an annual growth rate of 9% per year. In the year 2012, there were 23,900 fox counted in the area. What is the fox population predicted to be in the year 2020?

62. A scientist begins with 100 milligrams of a radioactive substance that decays exponentially. After 35 hours, 50mg of the substance remains. How many milligrams will remain after 54 hours?

63. In the year 1985, a house was valued at $110,000. By the year 2005, the value had appreciated to $145,000. What was the annual growth rate between 1985 and 2005? Assume that the value continued to grow by the same percentage. What was the value of the house in the year 2010?

64. A car was valued at $38,000 in the year 2007. By 2013, the value had depreciated to $11,000 If the car's value continues to drop by the same percentage, what will it be worth by 2017?

65. Jaylen wants to save $54,000 for a down payment on a home. How much will he need to invest in an account with 8.2% APR, compounding daily, in order to reach his goal in 5 years?

66. Kyoko has $10,000 that she wants to invest. Her bank has several investment accounts to choose from, all compounding daily. Her goal is to have $15,000 by the time she finishes graduate school in 6 years. To the nearest hundredth of a percent, what should her minimum annual interest rate be in order to reach her goal? (*Hint*: solve the compound interest formula for the interest rate.)

67. Alyssa opened a retirement account with 7.25% APR in the year 2000. Her initial deposit was $13,500. How much will the account be worth in 2025 if interest compounds monthly? How much more would she make if interest compounded continuously?

68. An investment account with an annual interest rate of 7% was opened with an initial deposit of $4,000 Compare the values of the account after 9 years when the interest is compounded annually, quarterly, monthly, and continuously.

6.2 Graphs of Exponential Functions

Learning Objectives

- Graph exponential functions.
- Graph exponential functions using transformations.

As we discussed in the previous section, exponential functions are used for many real-world applications such as finance, forensics, computer science, and most of the life sciences. Working with an equation that describes a real-world situation gives us a method for making predictions. Most of the time, however, the equation itself is not enough. We learn a lot about things by seeing their pictorial representations, and that is exactly why graphing exponential equations is a powerful tool. It gives us another layer of insight for predicting future events.

Graphing Exponential Functions

Before we begin graphing, it is helpful to review the behavior of exponential growth. Recall the table of values for a function of the form $f(x) = b^x$ whose base is greater than one. We'll use the function $f(x) = 2^x$. Observe how the output values in Table 1 change as the input increases by 1.

x	-3	-2	-1	0	1	2	3
$f(x) = 2^x$	$\frac{1}{8}$	$\frac{1}{4}$	$\frac{1}{2}$	1	2	4	8

Table 1

Each output value is the product of the previous output and the base, 2. We call the base 2 the *constant ratio*. In fact, for any exponential function with the form $f(x) = ab^x$, b is the constant ratio of the function. This means that as the input increases by 1, the output value will be the product of the base and the previous output, regardless of the value of a.

Notice from the table that

- the output values are positive for all values of x;
- as x increases, the output values increase without bound; and
- as x decreases, the output values grow smaller, approaching zero.

Figure 1 shows the exponential growth function $f(x) = 2^x$.

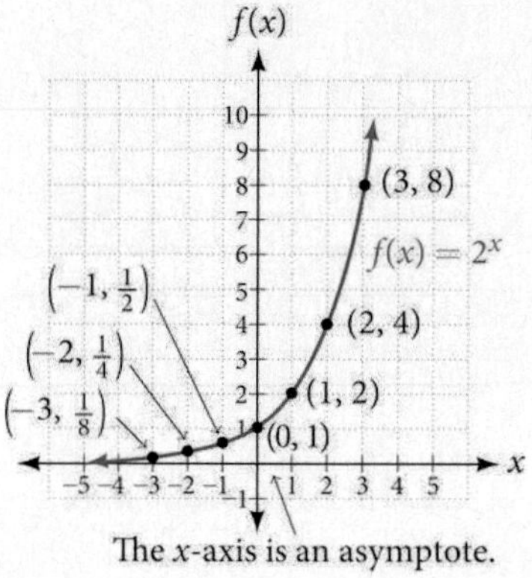

Figure 1 Notice that the graph gets close to the *x*-axis, but never touches it.

The domain of $f(x) = 2^x$ is all real numbers, the range is $\left(0, \infty\right)$, and the horizontal asymptote is $y = 0$.

To get a sense of the behavior of **exponential decay**, we can create a table of values for a function of the form $f(x) = b^x$ whose base is between zero and one. We'll use the function $g(x) = \left(\frac{1}{2}\right)^x$. Observe how the output values in Table 2 change as the input increases by 1.

x	-3	-2	-1	0	1	2	3
$g(x) = \left(\frac{1}{2}\right)^x$	8	4	2	1	$\frac{1}{2}$	$\frac{1}{4}$	$\frac{1}{8}$

Table 2

Again, because the input is increasing by 1, each output value is the product of the previous output and the base, or constant ratio $\frac{1}{2}$.

Notice from the table that

- the output values are positive for all values of x;
- as x increases, the output values grow smaller, approaching zero; and
- as x decreases, the output values grow without bound.

Figure 2 shows the exponential decay function, $g(x) = \left(\frac{1}{2}\right)^x$.

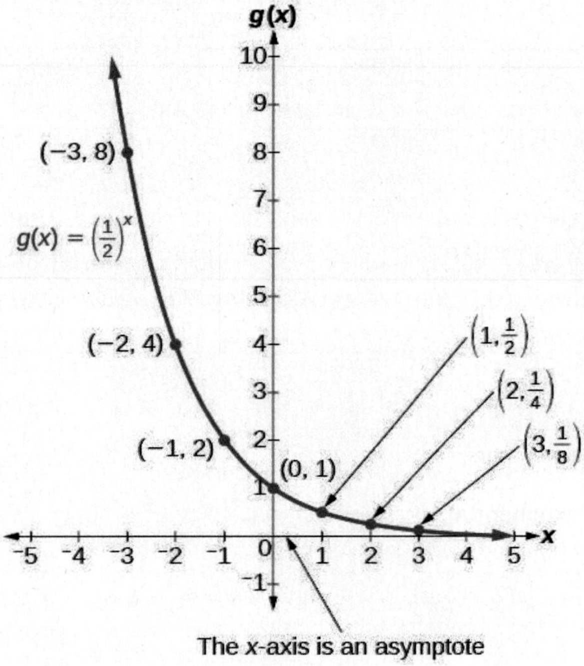

Figure 2

The domain of $g(x) = \left(\frac{1}{2}\right)^x$ is all real numbers, the range is $\left(0, \infty\right)$, and the horizontal asymptote is $y = 0$.

Characteristics of the Graph of the Parent Function $f(x) = b^x$

An exponential function with the form $f(x) = b^x$, $b > 0$, $b \neq 1$, has these characteristics:

- one-to-one function
- horizontal asymptote: $y = 0$
- domain: $(-\infty,\ \infty)$
- range: $(0, \infty)$
- *x*-intercept: none
- *y*-intercept: $(0, 1)$
- increasing if $b > 1$
- decreasing if $b < 1$

Figure 3 compares the graphs of exponential growth and decay functions.

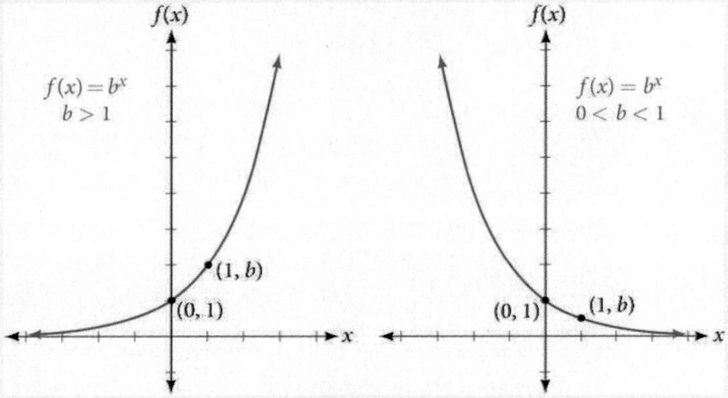

Figure 3

HOW TO

Given an exponential function of the form $f(x) = b^x$, graph the function.

1. Create a table of points.
2. Plot at least 3 point from the table, including the *y*-intercept $(0, 1)$.
3. Draw a smooth curve through the points.
4. State the domain, $\left(-\infty, \infty\right)$, the range, $\left(0, \infty\right)$, and the horizontal asymptote, $y = 0$.

EXAMPLE 1

Sketching the Graph of an Exponential Function of the Form $f(x) = b^x$

Sketch a graph of $f(x) = 0.25^x$. State the domain, range, and asymptote.

Solution

Before graphing, identify the behavior and create a table of points for the graph.

- Since $b = 0.25$ is between zero and one, we know the function is decreasing. The left tail of the graph will increase without bound, and the right tail will approach the asymptote $y = 0$.
- Create a table of points as in Table 3.

x	-3	-2	-1	0	1	2	3
$f(x) = 0.25^x$	64	16	4	1	0.25	0.0625	0.015625

Table 3

- Plot the *y*-intercept, $(0, 1)$, along with two other points. We can use $(-1, 4)$ and $(1, 0.25)$.

Draw a smooth curve connecting the points as in Figure 4.

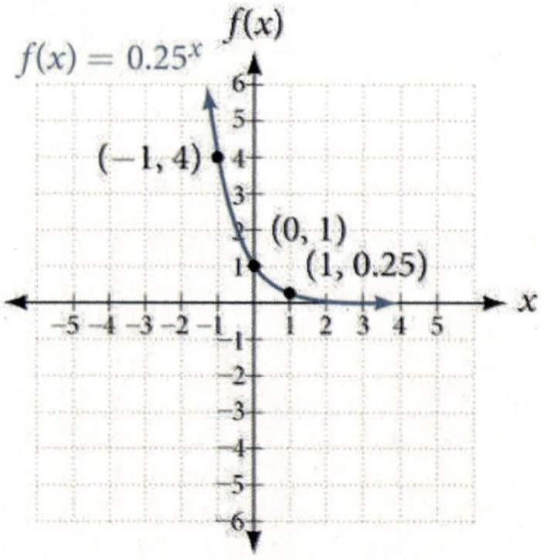

Figure 4

The domain is $\left(-\infty, \infty\right)$; the range is $\left(0, \infty\right)$; the horizontal asymptote is $y = 0$.

TRY IT #1 Sketch the graph of $f(x) = 4^x$. State the domain, range, and asymptote.

Graphing Transformations of Exponential Functions

Transformations of exponential graphs behave similarly to those of other functions. Just as with other parent functions, we can apply the four types of transformations—shifts, reflections, stretches, and compressions—to the parent function $f(x) = b^x$ without loss of shape. For instance, just as the quadratic function maintains its parabolic shape when shifted, reflected, stretched, or compressed, the exponential function also maintains its general shape regardless of the transformations applied.

Graphing a Vertical Shift

The first transformation occurs when we add a constant d to the parent function $f(x) = b^x$, giving us a vertical shift d units in the same direction as the sign. For example, if we begin by graphing a parent function, $f(x) = 2^x$, we can then graph two vertical shifts alongside it, using $d = 3$: the upward shift, $g(x) = 2^x + 3$ and the downward shift, $h(x) = 2^x - 3$. Both vertical shifts are shown in Figure 5.

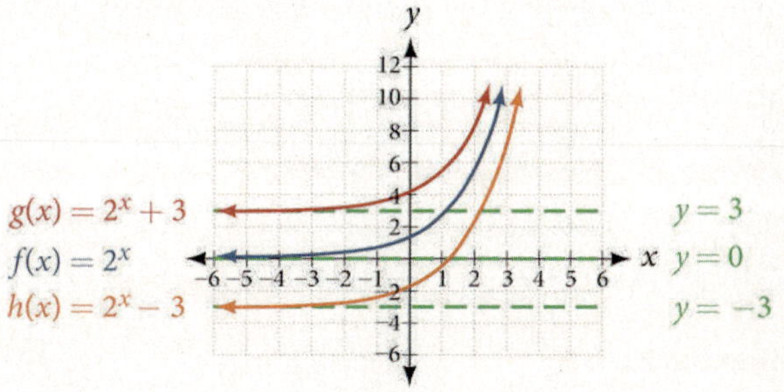

Figure 5

Observe the results of shifting $f(x) = 2^x$ vertically:

- The domain, $\left(-\infty, \infty\right)$ remains unchanged.
- When the function is shifted up 3 units to $g(x) = 2^x + 3$:
  - The *y*-intercept shifts up 3 units to $(0, 4)$.
  - The asymptote shifts up 3 units to $y = 3$.
  - The range becomes $\left(3, \infty\right)$.
- When the function is shifted down 3 units to $h(x) = 2^x - 3$:
  - The *y*-intercept shifts down 3 units to $(0, -2)$.
  - The asymptote also shifts down 3 units to $y = -3$.
  - The range becomes $\left(-3, \infty\right)$.

Graphing a Horizontal Shift

The next transformation occurs when we add a constant c to the input of the parent function $f(x) = b^x$, giving us a horizontal shift c units in the *opposite* direction of the sign. For example, if we begin by graphing the parent function $f(x) = 2^x$, we can then graph two horizontal shifts alongside it, using $c = 3$: the shift left, $g(x) = 2^{x+3}$, and the shift right, $h(x) = 2^{x-3}$. Both horizontal shifts are shown in Figure 6.

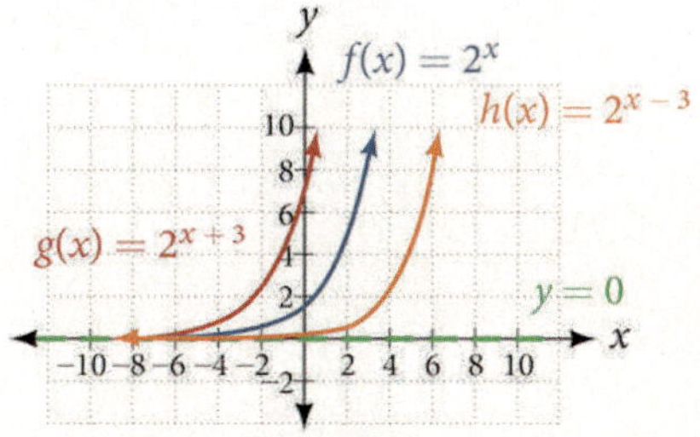

Figure 6

Observe the results of shifting $f(x) = 2^x$ horizontally:

- The domain, $\left(-\infty, \infty\right)$, remains unchanged.
- The asymptote, $y = 0$, remains unchanged.
- The *y*-intercept shifts such that:
  - When the function is shifted left 3 units to $g(x) = 2^{x+3}$, the *y*-intercept becomes $(0, 8)$. This is because $2^{x+3} = (8)\, 2^x$, so the initial value of the function is 8.
  - When the function is shifted right 3 units to $h(x) = 2^{x-3}$, the *y*-intercept becomes $\left(0, \frac{1}{8}\right)$. Again, see that $2^{x-3} = \left(\frac{1}{8}\right) 2^x$, so the initial value of the function is $\frac{1}{8}$.

Shifts of the Parent Function $f(x) = b^x$

For any constants c and d, the function $f(x) = b^{x+c} + d$ shifts the parent function $f(x) = b^x$

- vertically d units, in the *same* direction of the sign of d.
- horizontally c units, in the *opposite* direction of the sign of c.
- The y-intercept becomes $(0, b^c + d)$.
- The horizontal asymptote becomes $y = d$.
- The range becomes $\left(d, \infty\right)$.
- The domain, $\left(-\infty, \infty\right)$, remains unchanged.

HOW TO

Given an exponential function with the form $f(x) = b^{x+c} + d$, graph the translation.

1. Draw the horizontal asymptote $y = d$.
2. Identify the shift as $(-c, d)$. Shift the graph of $f(x) = b^x$ left c units if c is positive, and right c units if c is negative.
3. Shift the graph of $f(x) = b^x$ up d units if d is positive, and down d units if d is negative.
4. State the domain, $\left(-\infty, \infty\right)$, the range, $\left(d, \infty\right)$, and the horizontal asymptote $y = d$.

EXAMPLE 2

Graphing a Shift of an Exponential Function

Graph $f(x) = 2^{x+1} - 3$. State the domain, range, and asymptote.

Solution

We have an exponential equation of the form $f(x) = b^{x+c} + d$, with $b = 2$, $c = 1$, and $d = -3$.

Draw the horizontal asymptote $y = d$, so draw $y = -3$.

Identify the shift as $(-c, d)$, so the shift is $(-1, -3)$.

Shift the graph of $f(x) = b^x$ left 1 units and down 3 units.

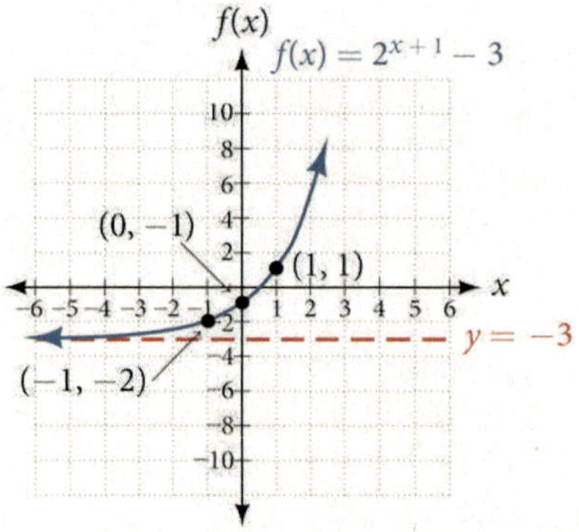

Figure 7

The domain is $\left(-\infty, \infty\right)$; the range is $\left(-3, \infty\right)$; the horizontal asymptote is $y = -3$.

TRY IT #2 Graph $f(x) = 2^{x-1} + 3$. State domain, range, and asymptote.

HOW TO

Given an equation of the form $f(x) = b^{x+c} + d$ for x, use a graphing calculator to approximate the solution.

- Press **[Y=]**. Enter the given exponential equation in the line headed "**Y_1=**".
- Enter the given value for $f(x)$ in the line headed "**Y_2=**".
- Press **[WINDOW]**. Adjust the y-axis so that it includes the value entered for "**Y_2=**".
- Press **[GRAPH]** to observe the graph of the exponential function along with the line for the specified value of $f(x)$.
- To find the value of x, we compute the point of intersection. Press **[2ND]** then **[CALC]**. Select "intersect" and press **[ENTER]** three times. The point of intersection gives the value of x for the indicated value of the function.

EXAMPLE 3

Approximating the Solution of an Exponential Equation

Solve $42 = 1.2(5)^x + 2.8$ graphically. Round to the nearest thousandth.

Solution

Press **[Y=]** and enter $1.2(5)^x + 2.8$ next to **Y_1=**. Then enter 42 next to **Y2=**. For a window, use the values –3 to 3 for x and –5 to 55 for y. Press **[GRAPH]**. The graphs should intersect somewhere near $x = 2$.

For a better approximation, press **[2ND]** then **[CALC]**. Select **[5: intersect]** and press **[ENTER]** three times. The x-coordinate of the point of intersection is displayed as 2.1661943. (Your answer may be different if you use a different window or use a different value for **Guess?**) To the nearest thousandth, $x \approx 2.166$.

TRY IT #3 Solve $4 = 7.85(1.15)^x - 2.27$ graphically. Round to the nearest thousandth.

Graphing a Stretch or Compression

While horizontal and vertical shifts involve adding constants to the input or to the function itself, a stretch or compression occurs when we multiply the parent function $f(x) = b^x$ by a constant $|a| > 0$. For example, if we begin by graphing the parent function $f(x) = 2^x$, we can then graph the stretch, using $a = 3$, to get $g(x) = 3(2)^x$ as shown on the left in Figure 8, and the compression, using $a = \frac{1}{3}$, to get $h(x) = \frac{1}{3}(2)^x$ as shown on the right in Figure 8.

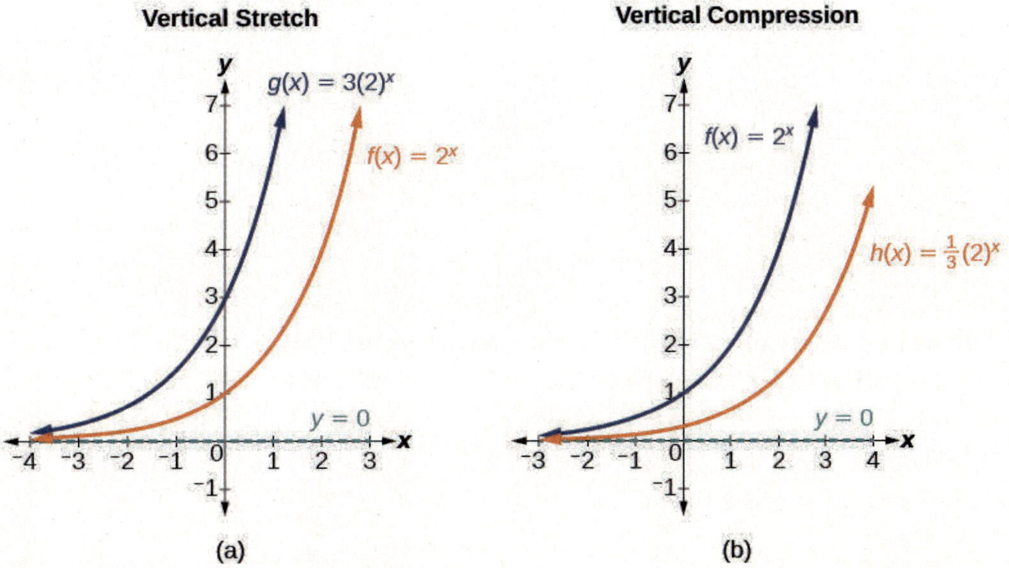

Figure 8 (a) $g(x) = 3(2)^x$ stretches the graph of $f(x) = 2^x$ vertically by a factor of 3. (b) $h(x) = \frac{1}{3}(2)^x$ compresses the graph of $f(x) = 2^x$ vertically by a factor of $\frac{1}{3}$.

Stretches and Compressions of the Parent Function $f(x) = b^x$

For any factor $a > 0$, the function $f(x) = a(b)^x$

- is stretched vertically by a factor of a if $|a| > 1$.
- is compressed vertically by a factor of a if $|a| < 1$.
- has a y-intercept of $(0, a)$.
- has a horizontal asymptote at $y = 0$, a range of $\left(0, \infty\right)$, and a domain of $\left(-\infty, \infty\right)$, which are unchanged from the parent function.

EXAMPLE 4

Graphing the Stretch of an Exponential Function

Sketch a graph of $f(x) = 4\left(\frac{1}{2}\right)^x$. State the domain, range, and asymptote.

Solution

Before graphing, identify the behavior and key points on the graph.

- Since $b = \frac{1}{2}$ is between zero and one, the left tail of the graph will increase without bound as x decreases, and the right tail will approach the x-axis as x increases.
- Since $a = 4$, the graph of $f(x) = \left(\frac{1}{2}\right)^x$ will be stretched by a factor of 4.
- Create a table of points as shown in Table 4.

x	-3	-2	-1	0	1	2	3
$f(x) = 4\left(\frac{1}{2}\right)^x$	32	16	8	4	2	1	0.5

Table 4

- Plot the y-intercept, $(0, 4)$, along with two other points. We can use $(-1, 8)$ and $(1, 2)$.

Draw a smooth curve connecting the points, as shown in Figure 9.

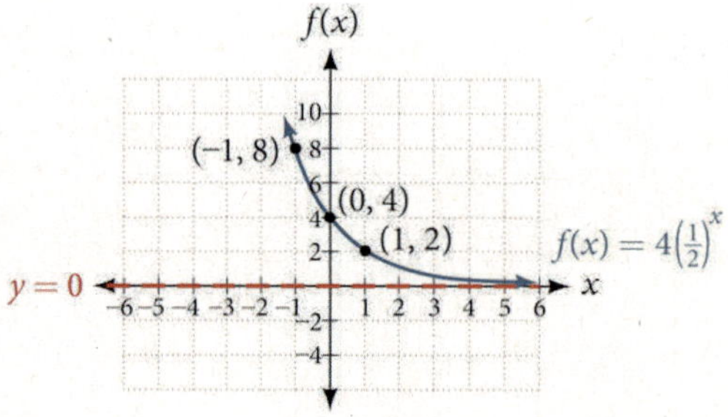

Figure 9

The domain is $\left(-\infty, \infty\right)$; the range is $\left(0, \infty\right)$; the horizontal asymptote is $y = 0$.

TRY IT #4 Sketch the graph of $f(x) = \frac{1}{2}(4)^x$. State the domain, range, and asymptote.

Graphing Reflections

In addition to shifting, compressing, and stretching a graph, we can also reflect it about the x-axis or the y-axis. When we multiply the parent function $f(x) = b^x$ by -1, we get a reflection about the x-axis. When we multiply the input by -1, we get a reflection about the y-axis. For example, if we begin by graphing the parent function $f(x) = 2^x$, we can then graph the two reflections alongside it. The reflection about the x-axis, $g(x) = -2^x$, is shown on the left side of Figure 10, and the reflection about the *y*-axis $h(x) = 2^{-x}$, is shown on the right side of Figure 10.

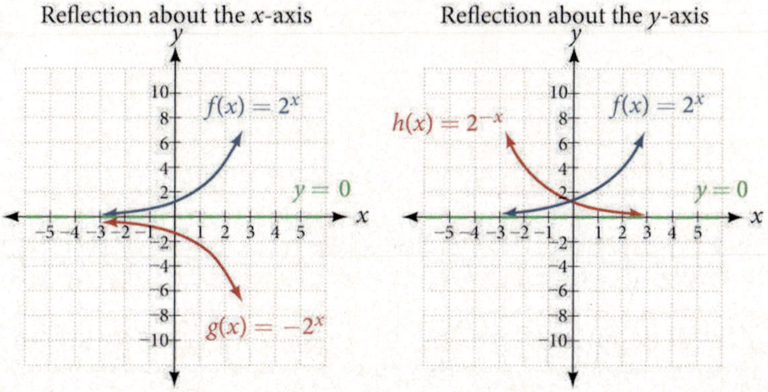

Figure 10 (a) $g(x) = -2^x$ reflects the graph of $f(x) = 2^x$ about the x-axis. (b) $g(x) = 2^{-x}$ reflects the graph of $f(x) = 2^x$ about the *y*-axis.

Reflections of the Parent Function $f(x) = b^x$

The function $f(x) = -b^x$

- reflects the parent function $f(x) = b^x$ about the *x*-axis.
- has a *y*-intercept of $(0, -1)$.
- has a range of $\left(-\infty, 0\right)$.
- has a horizontal asymptote at $y = 0$ and domain of $\left(-\infty, \infty\right)$, which are unchanged from the parent function.

The function $f(x) = b^{-x}$

- reflects the parent function $f(x) = b^x$ about the *y*-axis.
- has a *y*-intercept of $(0, 1)$, a horizontal asymptote at $y = 0$, a range of $\left(0, \infty\right)$, and a domain of $\left(-\infty, \infty\right)$, which are unchanged from the parent function.

EXAMPLE 5

Writing and Graphing the Reflection of an Exponential Function

Find and graph the equation for a function, $g(x)$, that reflects $f(x) = \left(\frac{1}{4}\right)^x$ about the *x*-axis. State its domain, range, and asymptote.

Solution

Since we want to reflect the parent function $f(x) = \left(\frac{1}{4}\right)^x$ about the *x*-axis, we multiply $f(x)$ by -1 to get, $g(x) = -\left(\frac{1}{4}\right)^x$. Next we create a table of points as in Table 5.

x	-3	-2	-1	0	1	2	3
$g(x) = -\left(\frac{1}{4}\right)^x$	-64	-16	-4	-1	-0.25	-0.0625	-0.0156

Table 5

Plot the *y*-intercept, $(0, -1)$, along with two other points. We can use $(-1, -4)$ and $(1, -0.25)$.

Draw a smooth curve connecting the points:

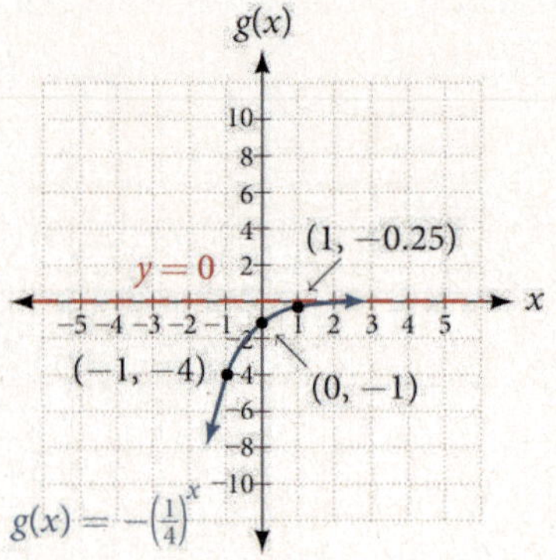

Figure 11

The domain is $\left(-\infty, \infty\right)$; the range is $\left(-\infty, 0\right)$; the horizontal asymptote is $y = 0$.

TRY IT #5 Find and graph the equation for a function, $g(x)$, that reflects $f(x) = 1.25^x$ about the y-axis. State its domain, range, and asymptote.

Summarizing Translations of the Exponential Function

Now that we have worked with each type of translation for the exponential function, we can summarize them in Table 6 to arrive at the general equation for translating exponential functions.

Translations of the Parent Function $f(x) = b^x$	
Translation	**Form**
Shift • Horizontally c units to the left • Vertically d units up	$f(x) = b^{x+c} + d$
Stretch and Compress • Stretch if $\|a\| > 1$ • Compression if $0 < \|a\| < 1$	$f(x) = ab^x$
Reflect about the x-axis	$f(x) = -b^x$
Reflect about the y-axis	$f(x) = b^{-x} = \left(\frac{1}{b}\right)^x$
General equation for all translations	$f(x) = ab^{x+c} + d$

Table 6

Translations of Exponential Functions

A translation of an exponential function has the form

$f(x) = ab^{x+c} + d$

Where the parent function, $y = b^x$, $b > 1$, is

- shifted horizontally c units to the left.
- stretched vertically by a factor of $|a|$ if $|a| > 0$.

- compressed vertically by a factor of $|a|$ if $0 < |a| < 1$.
- shifted vertically d units.
- reflected about the x-axis when $a < 0$.

Note the order of the shifts, transformations, and reflections follow the order of operations.

EXAMPLE 6

Writing a Function from a Description

Write the equation for the function described below. Give the horizontal asymptote, the domain, and the range.

- $f(x) = e^x$ is vertically stretched by a factor of 2 , reflected across the y-axis, and then shifted up 4 units.

Solution

We want to find an equation of the general form $f(x) = ab^{x+c} + d$. We use the description provided to find a, b, c, and d.

- We are given the parent function $f(x) = e^x$, so $b = e$.
- The function is stretched by a factor of 2 , so $a = 2$.
- The function is reflected about the y-axis. We replace x with $-x$ to get: e^{-x}.
- The graph is shifted vertically 4 units, so $d = 4$.

Substituting in the general form we get,

$$\begin{aligned} f(x) &= ab^{x+c} + d \\ &= 2e^{-x+0} + 4 \\ &= 2e^{-x} + 4 \end{aligned}$$

The domain is $\left(-\infty, \infty\right)$; the range is $\left(4, \infty\right)$; the horizontal asymptote is $y = 4$.

TRY IT #6 Write the equation for function described below. Give the horizontal asymptote, the domain, and the range.

- $f(x) = e^x$ is compressed vertically by a factor of $\frac{1}{3}$, reflected across the x-axis and then shifted down 2 units.

MEDIA

Access this online resource for additional instruction and practice with graphing exponential functions.

Graph Exponential Functions (http://openstax.org/l/graphexpfunc)

6.2 SECTION EXERCISES

Verbal

1. What role does the horizontal asymptote of an exponential function play in telling us about the end behavior of the graph?

2. What is the advantage of knowing how to recognize transformations of the graph of a parent function algebraically?

Algebraic

3. The graph of $f(x) = 3^x$ is reflected about the y-axis and stretched vertically by a factor of 4. What is the equation of the new function, $g(x)$? State its y-intercept, domain, and range.

4. The graph of $f(x) = \left(\frac{1}{2}\right)^{-x}$ is reflected about the y-axis and compressed vertically by a factor of $\frac{1}{5}$. What is the equation of the new function, $g(x)$? State its y-intercept, domain, and range.

5. The graph of $f(x) = 10^x$ is reflected about the x-axis and shifted upward 7 units. What is the equation of the new function, $g(x)$? State its y-intercept, domain, and range.

6. The graph of $f(x) = (1.68)^x$ is shifted right 3 units, stretched vertically by a factor of 2, reflected about the x-axis, and then shifted downward 3 units. What is the equation of the new function, $g(x)$? State its y-intercept (to the nearest thousandth), domain, and range.

7. The graph of $f(x)= -\frac{1}{2}\left(\frac{1}{4}\right)^{x-2} + 4$ is shifted downward 4 units, and then shifted left 2 units, stretched vertically by a factor of 4, and reflected about the x-axis. What is the equation of the new function, $g(x)$? State its y-intercept, domain, and range.

Graphical

For the following exercises, graph the function and its reflection about the y-axis on the same axes, and give the y-intercept.

8. $f(x) = 3\left(\frac{1}{2}\right)^x$

9. $g(x) = -2(0.25)^x$

10. $h(x) = 6(1.75)^{-x}$

For the following exercises, graph each set of functions on the same axes.

11. $f(x) = 3\left(\frac{1}{4}\right)^x$, $g(x) = 3(2)^x$, and $h(x) = 3(4)^x$

12. $f(x) = \frac{1}{4}(3)^x$, $g(x) = 2(3)^x$, and $h(x) = 4(3)^x$

For the following exercises, match each function with one of the graphs in Figure 12.

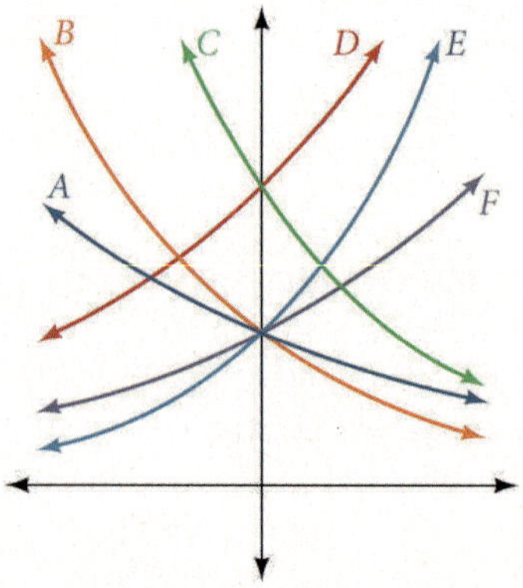

Figure 12

13. $f(x) = 2(0.69)^x$

14. $f(x) = 2(1.28)^x$

15. $f(x) = 2(0.81)^x$

16. $f(x) = 4(1.28)^x$

17. $f(x) = 2(1.59)^x$

18. $f(x) = 4(0.69)^x$

For the following exercises, use the graphs shown in Figure 13. All have the form $f(x) = ab^x$.

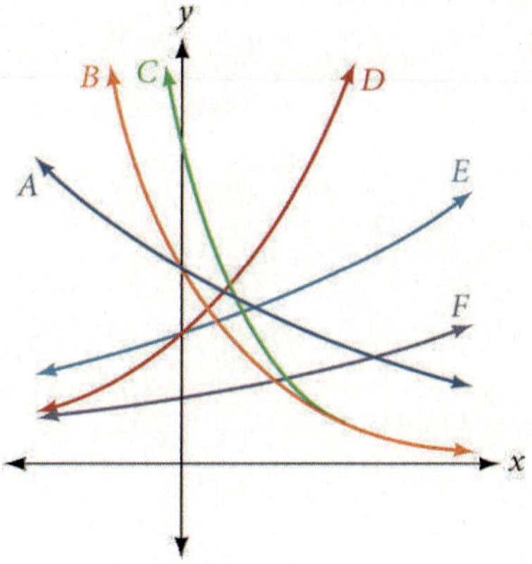

Figure 13

19. Which graph has the largest value for b?

20. Which graph has the smallest value for b?

21. Which graph has the largest value for a?

22. Which graph has the smallest value for a?

For the following exercises, graph the function and its reflection about the x-axis on the same axes.

23. $f(x) = \frac{1}{2}(4)^x$

24. $f(x) = 3(0.75)^x - 1$

25. $f(x) = -4(2)^x + 2$

For the following exercises, graph the transformation of $f(x) = 2^x$. Give the horizontal asymptote, the domain, and the range.

26. $f(x) = 2^{-x}$

27. $h(x) = 2^x + 3$

28. $f(x) = 2^{x-2}$

For the following exercises, describe the end behavior of the graphs of the functions.

29. $f(x) = -5(4)^x - 1$

30. $f(x) = 3\left(\frac{1}{2}\right)^x - 2$

31. $f(x) = 3(4)^{-x} + 2$

For the following exercises, start with the graph of $f(x) = 4^x$. Then write a function that results from the given transformation.

32. Shift $f(x)$ 4 units upward

33. Shift $f(x)$ 3 units downward

34. Shift $f(x)$ 2 units left

35. Shift $f(x)$ 5 units right

36. Reflect $f(x)$ about the x-axis

37. Reflect $f(x)$ about the y-axis

For the following exercises, each graph is a transformation of $y = 2^x$. Write an equation describing the transformation.

38.

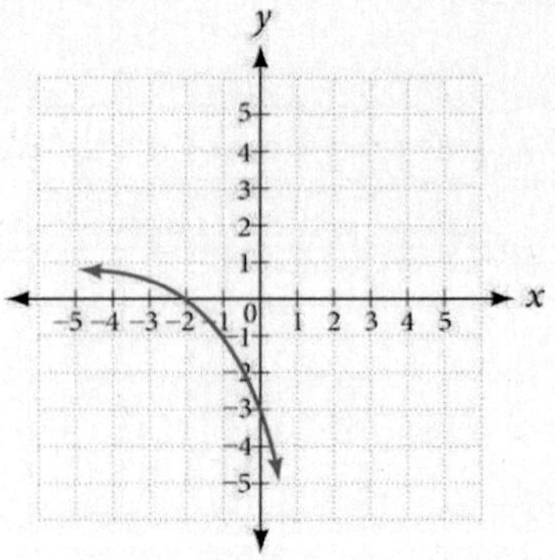

39.

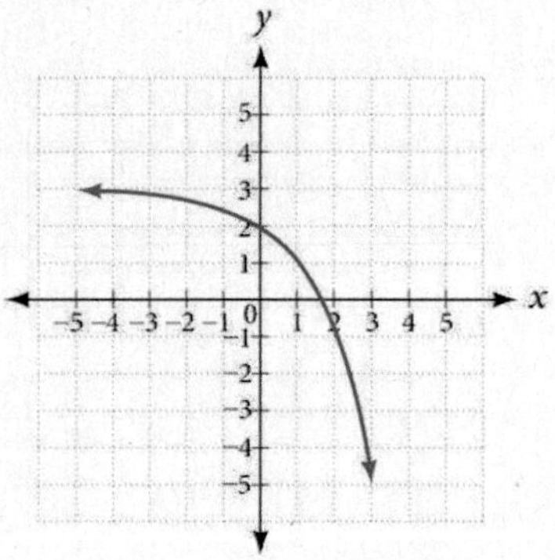

40.

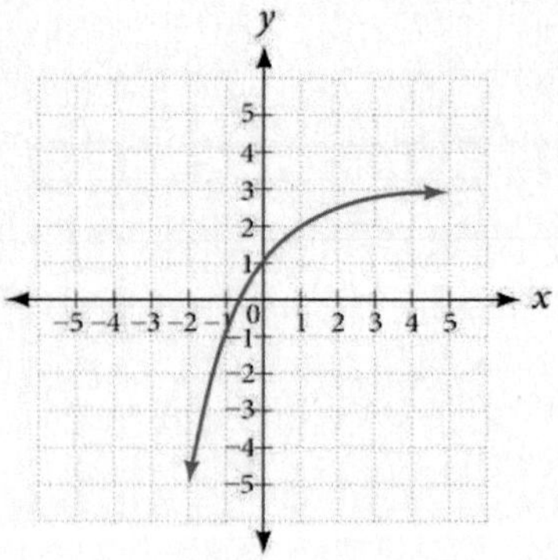

For the following exercises, find an exponential equation for the graph.

41.

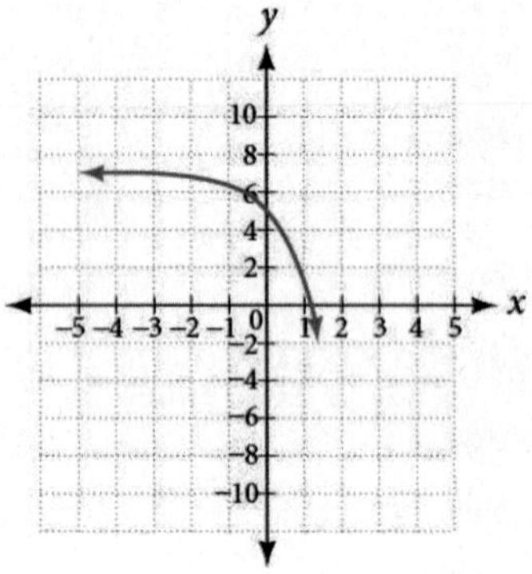

42.

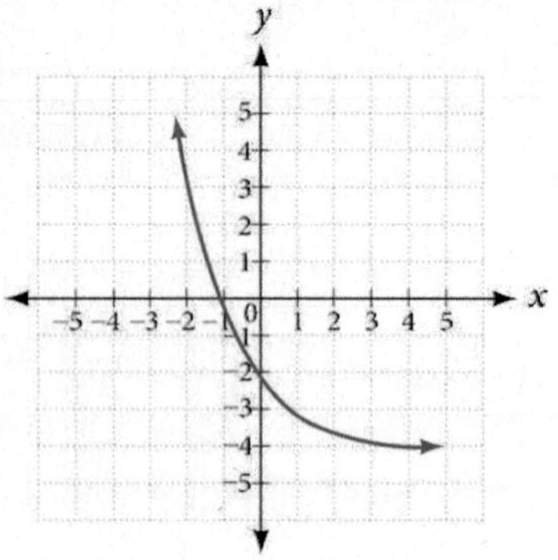

Numeric

For the following exercises, evaluate the exponential functions for the indicated value of x.

43. $g(x) = \frac{1}{3}(7)^{x-2}$ for $g(6)$.

44. $f(x) = 4(2)^{x-1} - 2$ for $f(5)$.

45. $h(x) = -\frac{1}{2}\left(\frac{1}{2}\right)^x + 6$ for $h(-7)$.

Technology

For the following exercises, use a graphing calculator to approximate the solutions of the equation. Round to the nearest thousandth.

46. $-50 = -\left(\frac{1}{2}\right)^{-x}$

47. $116 = \frac{1}{4}\left(\frac{1}{8}\right)^x$

48. $12 = 2(3)^x + 1$

49. $5 = 3\left(\frac{1}{2}\right)^{x-1} - 2$

50. $-30 = -4(2)^{x+2} + 2$

Extensions

51. Explore and discuss the graphs of $F(x) = (b)^x$ and $G(x) = \left(\frac{1}{b}\right)^x$. Then make a conjecture about the relationship between the graphs of the functions b^x and $\left(\frac{1}{b}\right)^x$ for any real number $b > 0$.

52. Prove the conjecture made in the previous exercise.

53. Explore and discuss the graphs of $f(x) = 4^x$, $g(x) = 4^{x-2}$, and $h(x) = \left(\frac{1}{16}\right)4^x$. Then make a conjecture about the relationship between the graphs of the functions b^x and $\left(\frac{1}{b^n}\right)b^x$ for any real number n and real number $b > 0$.

54. Prove the conjecture made in the previous exercise.

6.3 Logarithmic Functions

Learning Objectives

In this section, you will:

- Convert from logarithmic to exponential form.
- Convert from exponential to logarithmic form.
- Evaluate logarithms.
- Use common logarithms.
- Use natural logarithms.

Figure 1 Devastation of March 11, 2011 earthquake in Honshu, Japan. (credit: Daniel Pierce)

In 2010, a major earthquake struck Haiti, destroying or damaging over 285,000 homes[4]. One year later, another, stronger earthquake devastated Honshu, Japan, destroying or damaging over 332,000 buildings,[5] like those shown in Figure 1. Even though both caused substantial damage, the earthquake in 2011 was 100 times stronger than the earthquake in Haiti. How do we know? The magnitudes of earthquakes are measured on a scale known as the Richter Scale. The Haitian earthquake registered a 7.0 on the Richter Scale[6] whereas the Japanese earthquake registered a 9.0.[7]

The Richter Scale is a base-ten logarithmic scale. In other words, an earthquake of magnitude 8 is not twice as great as an earthquake of magnitude 4. It is $10^{8-4} = 10^4 = 10{,}000$ times as great! In this lesson, we will investigate the nature of the Richter Scale and the base-ten function upon which it depends.

Converting from Logarithmic to Exponential Form

In order to analyze the magnitude of earthquakes or compare the magnitudes of two different earthquakes, we need to be able to convert between logarithmic and exponential form. For example, suppose the amount of energy released

4 http://earthquake.usgs.gov/earthquakes/eqinthenews/2010/us2010rja6/#summary. Accessed 3/4/2013.

5 http://earthquake.usgs.gov/earthquakes/eqinthenews/2011/usc0001xgp/#summary. Accessed 3/4/2013.

6 http://earthquake.usgs.gov/earthquakes/eqinthenews/2010/us2010rja6/. Accessed 3/4/2013.

7 http://earthquake.usgs.gov/earthquakes/eqinthenews/2011/usc0001xgp/#details. Accessed 3/4/2013.

from one earthquake were 500 times greater than the amount of energy released from another. We want to calculate the difference in magnitude. The equation that represents this problem is $10^x = 500$, where x represents the difference in magnitudes on the Richter Scale. How would we solve for x?

We have not yet learned a method for solving exponential equations. None of the algebraic tools discussed so far is sufficient to solve $10^x = 500$. We know that $10^2 = 100$ and $10^3 = 1000$, so it is clear that x must be some value between 2 and 3, since $y = 10^x$ is increasing. We can examine a graph, as in Figure 2, to better estimate the solution.

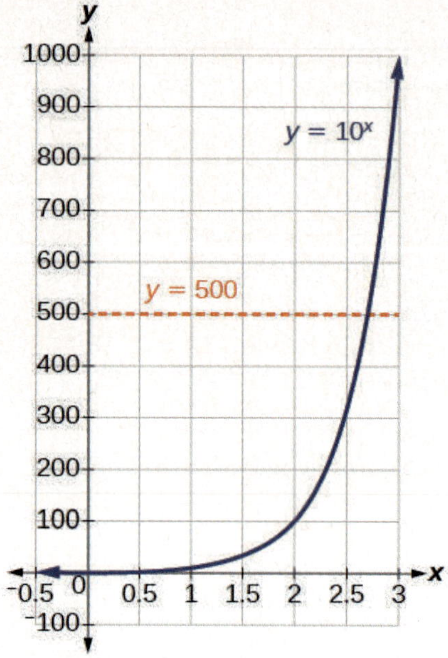

Figure 2

Estimating from a graph, however, is imprecise. To find an algebraic solution, we must introduce a new function. Observe that the graph in Figure 2 passes the horizontal line test. The exponential function $y = b^x$ is one-to-one, so its inverse, $x = b^y$ is also a function. As is the case with all inverse functions, we simply interchange x and y and solve for y to find the inverse function. To represent y as a function of x, we use a logarithmic function of the form $y = \log_b(x)$. The base b **logarithm** of a number is the exponent by which we must raise b to get that number.

We read a logarithmic expression as, "The logarithm with base b of x is equal to y, " or, simplified, "log base b of x is y. " We can also say, " b raised to the power of y is x, " because logs are exponents. For example, the base 2 logarithm of 32 is 5, because 5 is the exponent we must apply to 2 to get 32. Since $2^5 = 32$, we can write $\log_2 32 = 5$. We read this as "log base 2 of 32 is 5."

We can express the relationship between logarithmic form and its corresponding exponential form as follows:

$$\log_b(x) = y \Leftrightarrow b^y = x, b > 0, b \neq 1$$

Note that the base b is always positive.

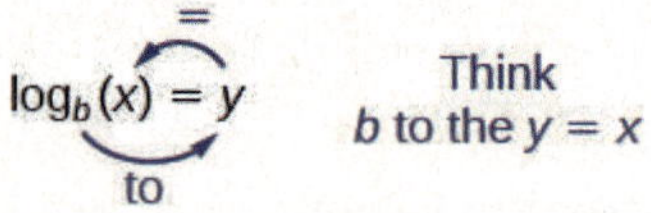

Because logarithm is a function, it is most correctly written as $\log_b(x)$, using parentheses to denote function evaluation, just as we would with $f(x)$. However, when the input is a single variable or number, it is common to see the parentheses dropped and the expression written without parentheses, as $\log_b x$. Note that many calculators require parentheses around the x.

We can illustrate the notation of logarithms as follows:

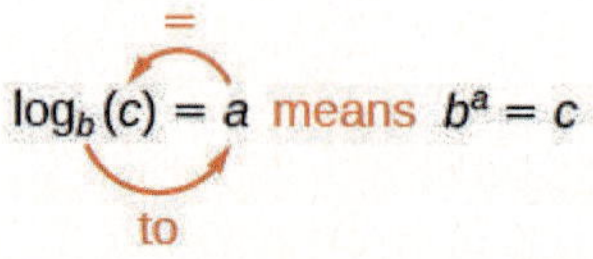

Notice that, comparing the logarithm function and the exponential function, the input and the output are switched. This means $y = \log_b(x)$ and $y = b^x$ are inverse functions.

Definition of the Logarithmic Function

A **logarithm** base b of a positive number x satisfies the following definition.

For $x > 0, b > 0, b \neq 1,$

$$y = \log_b(x) \quad \text{is equivalent to } b^y = x$$

where,

- we read $\log_b(x)$ as, "the logarithm with base b of x " or the "log base b of x."
- the logarithm y is the exponent to which b must be raised to get x.

Also, since the logarithmic and exponential functions switch the x and y values, the domain and range of the exponential function are interchanged for the logarithmic function. Therefore,

- the domain of the logarithm function with base b is $(0, \infty)$.
- the range of the logarithm function with base b is $(-\infty, \infty)$.

Q&A **Can we take the logarithm of a negative number?**

No. Because the base of an exponential function is always positive, no power of that base can ever be negative. We can never take the logarithm of a negative number. Also, we cannot take the logarithm of zero. Calculators may output a log of a negative number when in complex mode, but the log of a negative number is not a real number.

HOW TO

Given an equation in logarithmic form $\log_b(x) = y$, convert it to exponential form.

1. Examine the equation $y = \log_b(x)$ and identify $b, y,$ and$x$.
2. Rewrite $\log_b(x) = y$ as $b^y = x$.

EXAMPLE 1

Converting from Logarithmic Form to Exponential Form

Write the following logarithmic equations in exponential form.

ⓐ $\log_6\left(\sqrt{6}\right) = \frac{1}{2}$ ⓑ $\log_3(9) = 2$

Solution

First, identify the values of $b, y,$ and$x$. Then, write the equation in the form $b^y = x$.

ⓐ $\log_6\left(\sqrt{6}\right) = \frac{1}{2}$

Here, $b = 6, y = \frac{1}{2},$ and $x = \sqrt{6}$. Therefore, the equation $\log_6\left(\sqrt{6}\right) = \frac{1}{2}$ is equivalent to $6^{\frac{1}{2}} = \sqrt{6}$.

ⓑ $\log_3(9) = 2$

Here, $b = 3, y = 2,$ and $x = 9$. Therefore, the equation $\log_3(9) = 2$ is equivalent to $3^2 = 9$.

TRY IT #1 Write the following logarithmic equations in exponential form.

ⓐ $\log_{10}(1{,}000{,}000) = 6$ ⓑ $\log_5(25) = 2$

Converting from Exponential to Logarithmic Form

To convert from exponents to logarithms, we follow the same steps in reverse. We identify the base b, exponent x, and output y. Then we write $x = \log_b(y)$.

EXAMPLE 2

Converting from Exponential Form to Logarithmic Form

Write the following exponential equations in logarithmic form.

a. $2^3 = 8$
b. $5^2 = 25$
c. $10^{-4} = \frac{1}{10,000}$

Solution

First, identify the values of $b, y,$ and x. Then, write the equation in the form $x = \log_b(y)$.

a. $2^3 = 8$
Here, $b = 2$, $x = 3$, and $y = 8$. Therefore, the equation $2^3 = 8$ is equivalent to $\log_2(8) = 3$.

b. $5^2 = 25$
Here, $b = 5$, $x = 2$, and $y = 25$. Therefore, the equation $5^2 = 25$ is equivalent to $\log_5(25) = 2$.

c. $10^{-4} = \frac{1}{10,000}$
Here, $b = 10$, $x = -4$, and $y = \frac{1}{10,000}$. Therefore, the equation $10^{-4} = \frac{1}{10,000}$ is equivalent to $\log_{10}\left(\frac{1}{10,000}\right) = -4$.

TRY IT #2 Write the following exponential equations in logarithmic form.

ⓐ $3^2 = 9$ ⓑ $5^3 = 125$ ⓒ $2^{-1} = \frac{1}{2}$

Evaluating Logarithms

Knowing the squares, cubes, and roots of numbers allows us to evaluate many logarithms mentally. For example, consider $\log_2 8$. We ask, "To what exponent must 2 be raised in order to get 8?" Because we already know $2^3 = 8$, it follows that $\log_2 8 = 3$.

Now consider solving $\log_7 49$ and $\log_3 27$ mentally.

- We ask, "To what exponent must 7 be raised in order to get 49?" We know $7^2 = 49$. Therefore, $\log_7 49 = 2$
- We ask, "To what exponent must 3 be raised in order to get 27?" We know $3^3 = 27$. Therefore, $\log_3 27 = 3$

Even some seemingly more complicated logarithms can be evaluated without a calculator. For example, let's evaluate $\log_{\frac{2}{3}} \frac{4}{9}$ mentally.

- We ask, "To what exponent must $\frac{2}{3}$ be raised in order to get $\frac{4}{9}$? " We know $2^2 = 4$ and $3^2 = 9$, so $\left(\frac{2}{3}\right)^2 = \frac{4}{9}$. Therefore, $\log_{\frac{2}{3}}\left(\frac{4}{9}\right) = 2$.

HOW TO

Given a logarithm of the form $y = \log_b(x)$, evaluate it mentally.

1. Rewrite the argument x as a power of $b : b^y = x$.
2. Use previous knowledge of powers of b identify y by asking, "To what exponent should b be raised in order to get x? "

EXAMPLE 3

Solving Logarithms Mentally
Solve $y = \log_4 (64)$ without using a calculator.

Solution
First we rewrite the logarithm in exponential form: $4^y = 64$. Next, we ask, "To what exponent must 4 be raised in order to get 64?"

We know

$$4^3 = 64$$

Therefore,

$$\log_4(64) = 3$$

TRY IT #3 Solve $y = \log_{121} (11)$ without using a calculator.

EXAMPLE 4

Evaluating the Logarithm of a Reciprocal
Evaluate $y = \log_3 \left(\frac{1}{27}\right)$ without using a calculator.

Solution
First we rewrite the logarithm in exponential form: $3^y = \frac{1}{27}$. Next, we ask, "To what exponent must 3 be raised in order to get $\frac{1}{27}$? "

We know $3^3 = 27$, but what must we do to get the reciprocal, $\frac{1}{27}$? Recall from working with exponents that $b^{-a} = \frac{1}{b^a}$. We use this information to write

$$\begin{aligned} 3^{-3} &= \frac{1}{3^3} \\ &= \frac{1}{27} \end{aligned}$$

Therefore, $\log_3 \left(\frac{1}{27}\right) = -3$.

TRY IT #4 Evaluate $y = \log_2 \left(\frac{1}{32}\right)$ without using a calculator.

Using Common Logarithms

Sometimes we may see a logarithm written without a base. In this case, we assume that the base is 10. In other words, the expression $\log (x)$ means $\log_{10} (x)$. We call a base-10 logarithm a **common logarithm**. Common logarithms are used to measure the Richter Scale mentioned at the beginning of the section. Scales for measuring the brightness of stars and the pH of acids and bases also use common logarithms.

Definition of the Common Logarithm

A **common logarithm** is a logarithm with base 10. We write $\log_{10} (x)$ simply as $\log (x)$. The common logarithm of a positive number x satisfies the following definition.

For $x > 0$,

$$y = \log (x) \quad \text{is equivalent to } 10^y = x$$

We read $\log (x)$ as, "the logarithm with base 10 of x " or "log base 10 of x. "

The logarithm y is the exponent to which 10 must be raised to get x.

HOW TO

Given a common logarithm of the form $y = \log(x)$, evaluate it mentally.

1. Rewrite the argument x as a power of $10 : 10^y = x$.
2. Use previous knowledge of powers of 10 to identify y by asking, "To what exponent must 10 be raised in order to get x? "

EXAMPLE 5

Finding the Value of a Common Logarithm Mentally

Evaluate $y = \log(1000)$ without using a calculator.

Solution

First we rewrite the logarithm in exponential form: $10^y = 1000$. Next, we ask, "To what exponent must 10 be raised in order to get 1000?" We know

$$10^3 = 1000$$

Therefore, $\log(1000) = 3$.

TRY IT #5 Evaluate $y = \log(1{,}000{,}000)$.

HOW TO

Given a common logarithm with the form $y = \log(x)$, evaluate it using a calculator.

1. Press **[LOG]**.
2. Enter the value given for x, followed by **[ )]**.
3. Press **[ENTER]**.

EXAMPLE 6

Finding the Value of a Common Logarithm Using a Calculator

Evaluate $y = \log(321)$ to four decimal places using a calculator.

Solution

- Press **[LOG]**.
- Enter 321, followed by **[ )]**.
- Press **[ENTER]**.

Rounding to four decimal places, $\log(321) \approx 2.5065$.

Analysis

Note that $10^2 = 100$ and that $10^3 = 1000$. Since 321 is between 100 and 1000, we know that $\log(321)$ must be between $\log(100)$ and $\log(1000)$. This gives us the following:

$$\begin{array}{ccccc} 100 & < & 321 & < & 1000 \\ 2 & < & 2.5065 & < & 3 \end{array}$$

TRY IT #6 Evaluate $y = \log(123)$ to four decimal places using a calculator.

EXAMPLE 7

Rewriting and Solving a Real-World Exponential Model

The amount of energy released from one earthquake was 500 times greater than the amount of energy released from another. The equation $10^x = 500$ represents this situation, where x is the difference in magnitudes on the Richter Scale. To the nearest thousandth, what was the difference in magnitudes?

Solution

We begin by rewriting the exponential equation in logarithmic form.

$$\begin{aligned} 10^x &= 500 \\ \log(500) &= x \quad \text{Use the definition of the common log.} \end{aligned}$$

Next we evaluate the logarithm using a calculator:

- Press **[LOG]**.
- Enter 500, followed by **[ )]**.
- Press **[ENTER]**.
- To the nearest thousandth, $\log(500) \approx 2.699$.

The difference in magnitudes was about 2.699.

TRY IT #7 The amount of energy released from one earthquake was $8{,}500$ times greater than the amount of energy released from another. The equation $10^x = 8500$ represents this situation, where x is the difference in magnitudes on the Richter Scale. To the nearest thousandth, what was the difference in magnitudes?

Using Natural Logarithms

The most frequently used base for logarithms is e. Base e logarithms are important in calculus and some scientific applications; they are called **natural logarithms**. The base e logarithm, $\log_e(x)$, has its own notation, $\ln(x)$.

Most values of $\ln(x)$ can be found only using a calculator. The major exception is that, because the logarithm of 1 is always 0 in any base, $\ln 1 = 0$. For other natural logarithms, we can use the $\ln$ key that can be found on most scientific calculators. We can also find the natural logarithm of any power of e using the inverse property of logarithms.

Definition of the Natural Logarithm

A **natural logarithm** is a logarithm with base e. We write $\log_e(x)$ simply as $\ln(x)$. The natural logarithm of a positive number x satisfies the following definition.

For $x > 0$,

$$y = \ln(x) \text{ is equivalent to } e^y = x$$

We read $\ln(x)$ as, "the logarithm with base e of x" or "the natural logarithm of x."

The logarithm y is the exponent to which e must be raised to get x.

Since the functions $y = e^x$ and $y = \ln(x)$ are inverse functions, $\ln(e^x) = x$ for all x and $e^{\ln(x)} = x$ for $x > 0$.

HOW TO

Given a natural logarithm with the form $y = \ln(x)$, evaluate it using a calculator.

1. Press **[LN]**.
2. Enter the value given for x, followed by **[ )]**.
3. Press **[ENTER]**.

EXAMPLE 8

Evaluating a Natural Logarithm Using a Calculator

Evaluate $y = \ln(500)$ to four decimal places using a calculator.

Solution

- Press **[LN]**.
- Enter 500, followed by **[ )]**.
- Press **[ENTER]**.

Rounding to four decimal places, $\ln(500) \approx 6.2146$

TRY IT #8 Evaluate $\ln(-500)$.

MEDIA

Access this online resource for additional instruction and practice with logarithms.

Introduction to Logarithms (http://openstax.org/l/intrologarithms)

6.3 SECTION EXERCISES

Verbal

1. What is a base b logarithm? Discuss the meaning by interpreting each part of the equivalent equations $b^y = x$ and $\log_b x = y$ for $b > 0, b \neq 1$.

2. How is the logarithmic function $f(x) = \log_b x$ related to the exponential function $g(x) = b^x$? What is the result of composing these two functions?

3. How can the logarithmic equation $\log_b x = y$ be solved for x using the properties of exponents?

4. Discuss the meaning of the common logarithm. What is its relationship to a logarithm with base b, and how does the notation differ?

5. Discuss the meaning of the natural logarithm. What is its relationship to a logarithm with base b, and how does the notation differ?

Algebraic

For the following exercises, rewrite each equation in exponential form.

6. $\log_4(q) = m$

7. $\log_a(b) = c$

8. $\log_{16}(y) = x$

9. $\log_x(64) = y$

10. $\log_y(x) = -11$

11. $\log_{15}(a) = b$

12. $\log_y(137) = x$

13. $\log_{13}(142) = a$

14. $\log(v) = t$

15. $\ln(w) = n$

For the following exercises, rewrite each equation in logarithmic form.

16. $4^x = y$

17. $c^d = k$

18. $m^{-7} = n$

19. $19^x = y$

20. $x^{-\frac{10}{13}} = y$

21. $n^4 = 103$

22. $\left(\frac{7}{5}\right)^m = n$

23. $y^x = \frac{39}{100}$

24. $10^a = b$

25. $e^k = h$

For the following exercises, solve for x by converting the logarithmic equation to exponential form.

26. $\log_3(x) = 2$

27. $\log_2(x) = -3$

28. $\log_5(x) = 2$

29. $\log_3(x) = 3$

30. $\log_2(x) = 6$

31. $\log_9(x) = \frac{1}{2}$

32. $\log_{18}(x) = 2$

33. $\log_6(x) = -3$

34. $\log(x) = 3$

35. $\ln(x) = 2$

For the following exercises, use the definition of common and natural logarithms to simplify.

36. $\log(100^8)$

37. $10^{\log(32)}$

38. $2\log(.0001)$

39. $e^{\ln(1.06)}$

40. $\ln\left(e^{-5.03}\right)$

41. $e^{\ln(10.125)} + 4$

Numeric

For the following exercises, evaluate the base b logarithmic expression without using a calculator.

42. $\log_3\left(\frac{1}{27}\right)$

43. $\log_6(\sqrt{6})$

44. $\log_2\left(\frac{1}{8}\right) + 4$

45. $6\log_8(4)$

For the following exercises, evaluate the common logarithmic expression without using a calculator.

46. $\log(10,000)$

47. $\log(0.001)$

48. $\log(1) + 7$

49. $2\log(100^{-3})$

For the following exercises, evaluate the natural logarithmic expression without using a calculator.

50. $\ln(e^{\frac{1}{3}})$

51. $\ln(1)$

52. $\ln(e^{-0.225}) - 3$

53. $25\ln(e^{\frac{2}{5}})$

Technology

For the following exercises, evaluate each expression using a calculator. Round to the nearest thousandth.

54. $\log(0.04)$

55. $\ln(15)$

56. $\ln\left(\frac{4}{5}\right)$

57. $\log(\sqrt{2})$

58. $\ln(\sqrt{2})$

Extensions

59. Is $x = 0$ in the domain of the function $f(x) = \log(x)$? If so, what is the value of the function when $x = 0$? Verify the result.

60. Is $f(x) = 0$ in the range of the function $f(x) = \log(x)$? If so, for what value of x? Verify the result.

61. Is there a number x such that $\ln x = 2$? If so, what is that number? Verify the result.

62. Is the following true: $\frac{\log_3(27)}{\log_4\left(\frac{1}{64}\right)} = -1$? Verify the result.

63. Is the following true: $\frac{\ln\left(e^{1.725}\right)}{\ln(1)} = 1.725$? Verify the result.

Real-World Applications

64. The exposure index EI for a camera is a measurement of the amount of light that hits the image receptor. It is determined by the equation $EI = \log_2\left(\frac{f^2}{t}\right)$, where f is the "f-stop" setting on the camera, and t is the exposure time in seconds. Suppose the f-stop setting is 8 and the desired exposure time is 2 seconds. What will the resulting exposure index be?

65. Refer to the previous exercise. Suppose the light meter on a camera indicates an EI of -2, and the desired exposure time is 16 seconds. What should the f-stop setting be?

66. The intensity levels I of two earthquakes measured on a seismograph can be compared by the formula $\log\frac{I_1}{I_2} = M_1 - M_2$ where M is the magnitude given by the Richter Scale. In August 2009, an earthquake of magnitude 6.1 hit Honshu, Japan. In March 2011, that same region experienced yet another, more devastating earthquake, this time with a magnitude of 9.0.[8] How many times greater was the intensity of the 2011 earthquake? Round to the nearest whole number.

6.4 Graphs of Logarithmic Functions

Learning Objectives

In this section, you will:

- Identify the domain of a logarithmic function.
- Graph logarithmic functions.

8 http://earthquake.usgs.gov/earthquakes/world/historical.php. Accessed 3/4/2014.

In Graphs of Exponential Functions, we saw how creating a graphical representation of an exponential model gives us another layer of insight for predicting future events. How do logarithmic graphs give us insight into situations? Because every logarithmic function is the inverse function of an exponential function, we can think of every output on a logarithmic graph as the input for the corresponding inverse exponential equation. In other words, logarithms give the *cause* for an *effect*.

To illustrate, suppose we invest $\$2500$ in an account that offers an annual interest rate of 5%, compounded continuously. We already know that the balance in our account for any year t can be found with the equation $A = 2500e^{0.05t}$.

But what if we wanted to know the year for any balance? We would need to create a corresponding new function by interchanging the input and the output; thus we would need to create a logarithmic model for this situation. By graphing the model, we can see the output (year) for any input (account balance). For instance, what if we wanted to know how many years it would take for our initial investment to double? Figure 1 shows this point on the logarithmic graph.

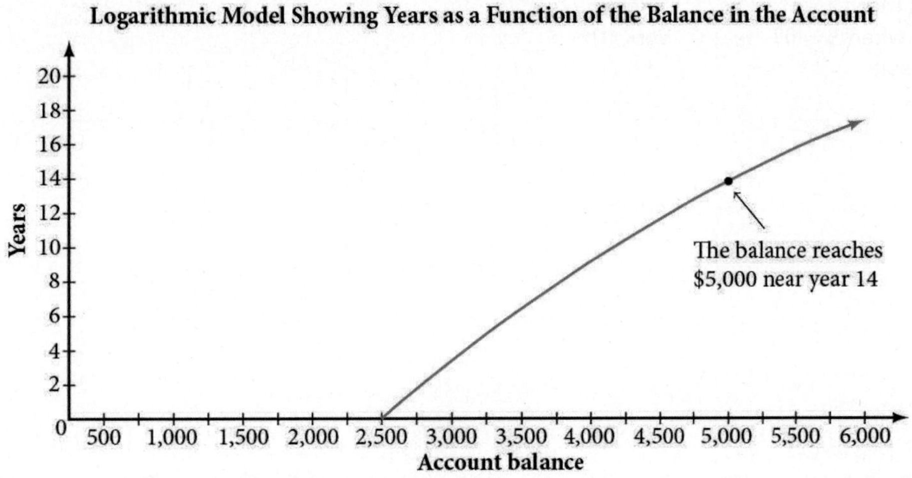

Figure 1

In this section we will discuss the values for which a logarithmic function is defined, and then turn our attention to graphing the family of logarithmic functions.

Finding the Domain of a Logarithmic Function

Before working with graphs, we will take a look at the domain (the set of input values) for which the logarithmic function is defined.

Recall that the exponential function is defined as $y = b^x$ for any real number x and constant $b > 0, b \neq 1$, where

- The domain of y is $\left(-\infty, \infty\right)$.
- The range of y is $\left(0, \infty\right)$.

In the last section we learned that the logarithmic function $y = \log_b(x)$ is the inverse of the exponential function $y = b^x$. So, as inverse functions:

- The domain of $y = \log_b(x)$ is the range of $y = b^x$: $\left(0, \infty\right)$.
- The range of $y = \log_b(x)$ is the domain of $y = b^x$: $\left(-\infty, \infty\right)$.

Transformations of the parent function $y = \log_b(x)$ behave similarly to those of other functions. Just as with other parent functions, we can apply the four types of transformations—shifts, stretches, compressions, and reflections—to the parent function without loss of shape.

In Graphs of Exponential Functions we saw that certain transformations can change the *range* of $y = b^x$. Similarly, applying transformations to the parent function $y = \log_b(x)$ can change the *domain*. When finding the domain of a logarithmic function, therefore, it is important to remember that the domain consists *only of positive real numbers*. That is, the argument of the logarithmic function must be greater than zero.

For example, consider $f(x) = \log_4 (2x - 3)$. This function is defined for any values of x such that the argument, in this case $2x - 3$, is greater than zero. To find the domain, we set up an inequality and solve for x :

$$\begin{aligned} 2x - 3 &> 0 && \text{Show the argument greater than zero.} \\ 2x &> 3 && \text{Add 3.} \\ x &> 1.5 && \text{Divide by 2.} \end{aligned}$$

In interval notation, the domain of $f(x) = \log_4 (2x - 3)$ is $\left(1.5, \infty\right)$.

HOW TO

Given a logarithmic function, identify the domain.

1. Set up an inequality showing the argument greater than zero.
2. Solve for x.
3. Write the domain in interval notation.

EXAMPLE 1

Identifying the Domain of a Logarithmic Shift

What is the domain of $f(x) = \log_2(x + 3)$?

Solution

The logarithmic function is defined only when the input is positive, so this function is defined when $x + 3 > 0$. Solving this inequality,

$$\begin{aligned} x + 3 &> 0 && \text{The input must be positive.} \\ x &> -3 && \text{Subtract 3.} \end{aligned}$$

The domain of $f(x) = \log_2(x + 3)$ is $\left(-3, \infty\right)$.

TRY IT #1 What is the domain of $f(x) = \log_5(x - 2) + 1$?

EXAMPLE 2

Identifying the Domain of a Logarithmic Shift and Reflection

What is the domain of $f(x) = \log(5 - 2x)$?

Solution

The logarithmic function is defined only when the input is positive, so this function is defined when $5-2x > 0$. Solving this inequality,

$$\begin{aligned} 5 - 2x &> 0 && \text{The input must be positive.} \\ -2x &> -5 && \text{Subtract 5.} \\ x &< \tfrac{5}{2} && \text{Divide by } -2 \text{ and switch the inequality.} \end{aligned}$$

The domain of $f(x) = \log(5 - 2x)$ is $\left(-\infty, \frac{5}{2}\right)$.

TRY IT #2 What is the domain of $f(x) = \log(x - 5) + 2$?

Graphing Logarithmic Functions

Now that we have a feel for the set of values for which a logarithmic function is defined, we move on to graphing logarithmic functions. The family of logarithmic functions includes the parent function $y = \log_b (x)$ along with all its transformations: shifts, stretches, compressions, and reflections.

We begin with the parent function $y = \log_b(x)$. Because every logarithmic function of this form is the inverse of an exponential function with the form $y = b^x$, their graphs will be reflections of each other across the line $y = x$. To illustrate this, we can observe the relationship between the input and output values of $y = 2^x$ and its equivalent $x = \log_2(y)$ in Table 1.

x	-3	-2	-1	0	1	2	3
$2^x = y$	$\frac{1}{8}$	$\frac{1}{4}$	$\frac{1}{2}$	1	2	4	8
$\log_2(y) = x$	-3	-2	-1	0	1	2	3

Table 1

Using the inputs and outputs from Table 1, we can build another table to observe the relationship between points on the graphs of the inverse functions $f(x) = 2^x$ and $g(x) = \log_2(x)$. See Table 2.

$f(x) = 2^x$	$\left(-3, \frac{1}{8}\right)$	$\left(-2, \frac{1}{4}\right)$	$\left(-1, \frac{1}{2}\right)$	$(0, 1)$	$(1, 2)$	$(2, 4)$	$(3, 8)$
$g(x) = \log_2(x)$	$\left(\frac{1}{8}, -3\right)$	$\left(\frac{1}{4}, -2\right)$	$\left(\frac{1}{2}, -1\right)$	$(1, 0)$	$(2, 1)$	$(4, 2)$	$(8, 3)$

Table 2

As we'd expect, the *x*- and *y*-coordinates are reversed for the inverse functions. Figure 2 shows the graph of f and g.

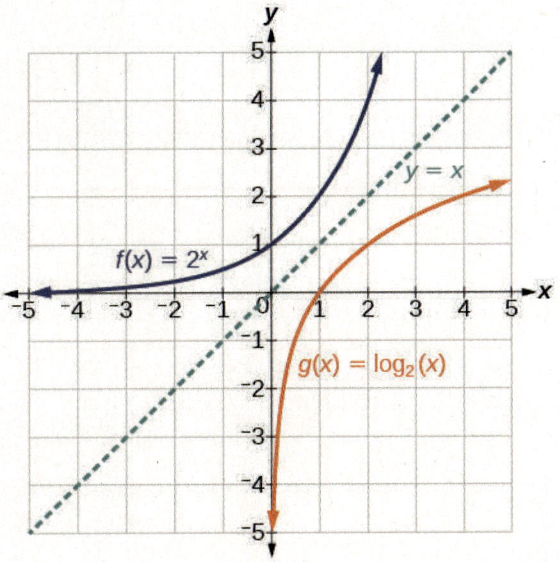

Figure 2 Notice that the graphs of $f(x) = 2^x$ and $g(x) = \log_2(x)$ are reflections about the line $y = x$.

Observe the following from the graph:

- $f(x) = 2^x$ has a *y*-intercept at $(0, 1)$ and $g(x) = \log_2(x)$ has an *x*- intercept at $(1, 0)$.
- The domain of $f(x) = 2^x$, $\left(-\infty, \infty\right)$, is the same as the range of $g(x) = \log_2(x)$.
- The range of $f(x) = 2^x$, $\left(0, \infty\right)$, is the same as the domain of $g(x) = \log_2(x)$.

Characteristics of the Graph of the Parent Function, $f(x) = \log_b(x)$:

For any real number x and constant $b > 0$, $b \neq 1$, we can see the following characteristics in the graph of $f(x) = \log_b(x)$:

- one-to-one function
- vertical asymptote: $x = 0$
- domain: $(0, \infty)$
- range: $\left(-\infty, \infty\right)$
- x-intercept: $(1, 0)$ and key point $(b, 1)$
- y-intercept: none
- increasing if $b > 1$
- decreasing if $0 < b < 1$

See Figure 3.

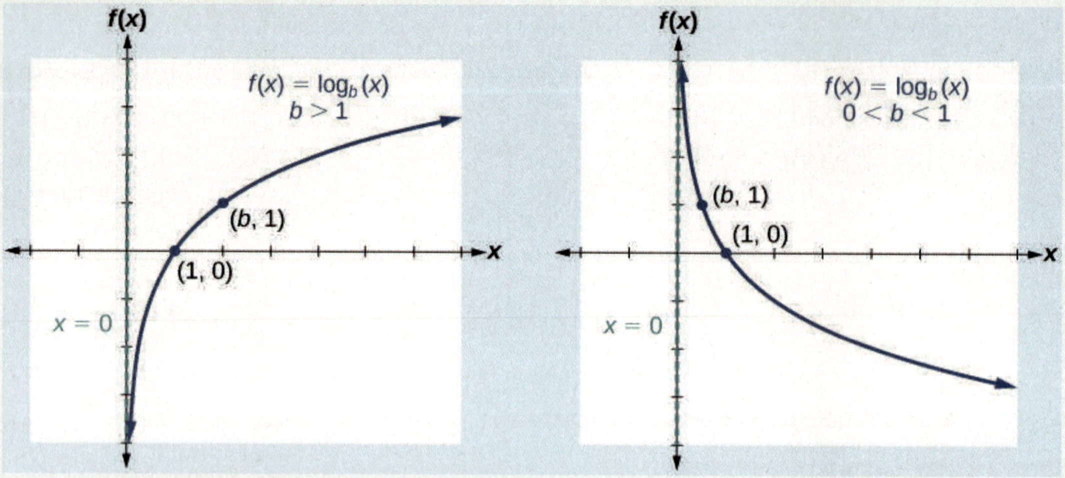

Figure 3

Figure 4 shows how changing the base b in $f(x) = \log_b(x)$ can affect the graphs. Observe that the graphs compress vertically as the value of the base increases. (*Note:* recall that the function $\ln(x)$ has base $e \approx 2.718$.)

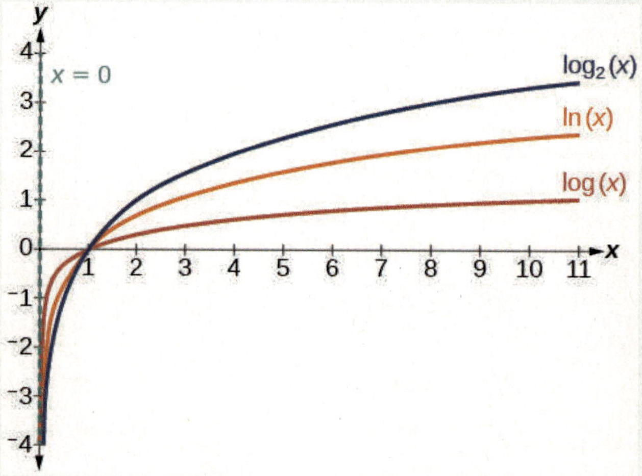

Figure 4 The graphs of three logarithmic functions with different bases, all greater than 1.

HOW TO

Given a logarithmic function with the form $f(x) = \log_b(x)$, graph the function.

1. Draw and label the vertical asymptote, $x = 0$.
2. Plot the x-intercept, $(1, 0)$.

3. Plot the key point $(b, 1)$.
4. Draw a smooth curve through the points.
5. State the domain, $\left(0, \infty\right)$, the range, $\left(-\infty, \infty\right)$, and the vertical asymptote, $x = 0$.

EXAMPLE 3

Graphing a Logarithmic Function with the Form $f(x) = \log_b(x)$.

Graph $f(x) = \log_5(x)$. State the domain, range, and asymptote.

Solution

Before graphing, identify the behavior and key points for the graph.

- Since $b = 5$ is greater than one, we know the function is increasing. The left tail of the graph will approach the vertical asymptote $x = 0$, and the right tail will increase slowly without bound.
- The *x*-intercept is $(1, 0)$.
- The key point $(5, 1)$ is on the graph.
- We draw and label the asymptote, plot and label the points, and draw a smooth curve through the points (see Figure 5).

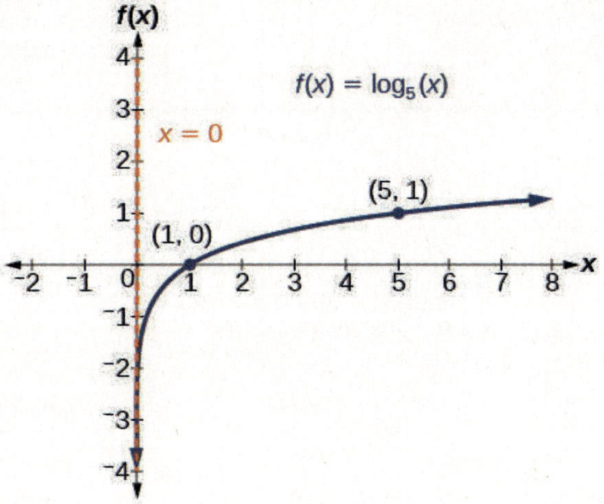

Figure 5

The domain is $\left(0, \infty\right)$, the range is $\left(-\infty, \infty\right)$, and the vertical asymptote is $x = 0$.

TRY IT #3 Graph $f(x) = \log_{\frac{1}{5}}(x)$. State the domain, range, and asymptote.

Graphing Transformations of Logarithmic Functions

As we mentioned in the beginning of the section, transformations of logarithmic graphs behave similarly to those of other parent functions. We can shift, stretch, compress, and reflect the parent function $y = \log_b(x)$ without loss of shape.

Graphing a Horizontal Shift of $f(x) = \log_b(x)$

When a constant c is added to the input of the parent function $f(x) = log_b(x)$, the result is a horizontal shift c units in the *opposite* direction of the sign on c. To visualize horizontal shifts, we can observe the general graph of the parent function $f(x) = \log_b(x)$ and for $c > 0$ alongside the shift left, $g(x) = \log_b(x + c)$, and the shift right, $h(x) = \log_b(x - c)$. See Figure 6.

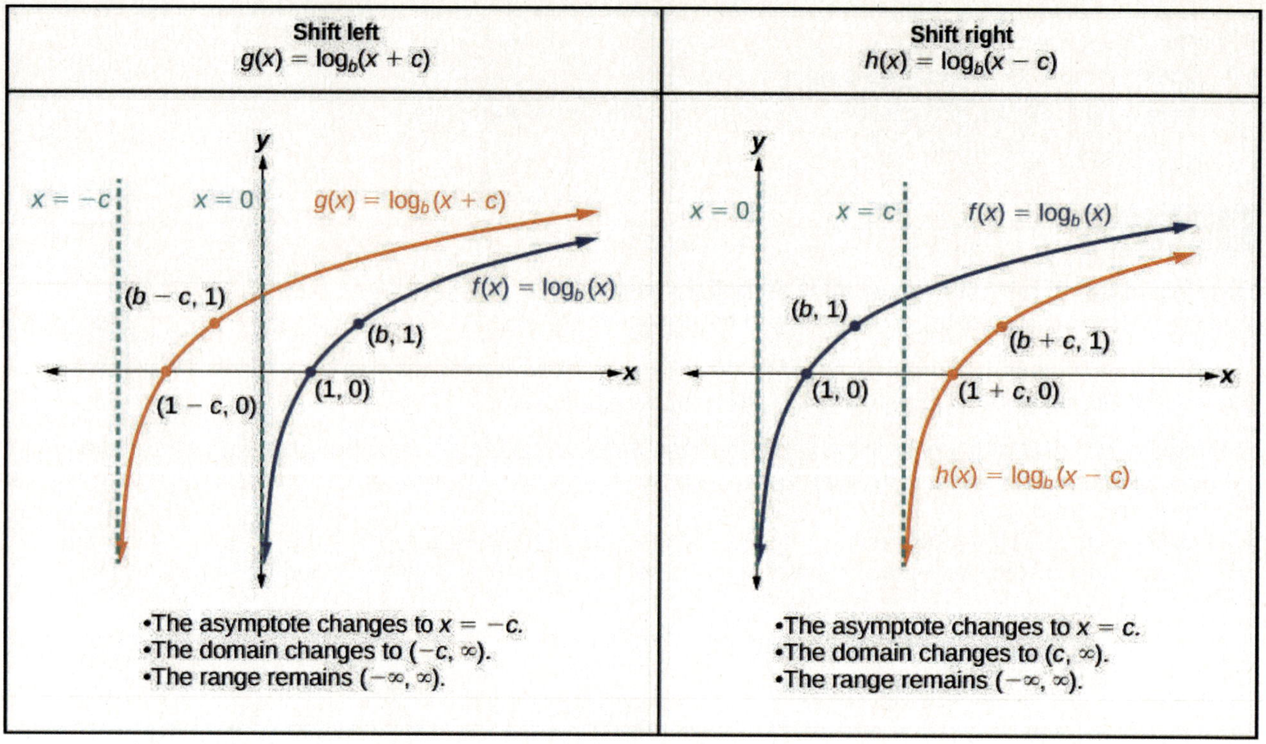

Figure 6

Horizontal Shifts of the Parent Function $f(x) = \log_b(x)$

For any constant c, the function $f(x) = \log_b(x + c)$

- shifts the parent function $y = \log_b(x)$ left c units if $c > 0$.
- shifts the parent function $y = \log_b(x)$ right c units if $c < 0$.
- has the vertical asymptote $x = -c$.
- has domain $\left(-c, \infty\right)$.
- has range $\left(-\infty, \infty\right)$.

HOW TO

Given a logarithmic function with the form $f(x) = \log_b(x + c)$, graph the translation.

1. Identify the horizontal shift:
   a. If $c > 0$, shift the graph of $f(x) = \log_b(x)$ left c units.
   b. If $c < 0$, shift the graph of $f(x) = \log_b(x)$ right c units.
2. Draw the vertical asymptote $x = -c$.
3. Identify three key points from the parent function. Find new coordinates for the shifted functions by subtracting c from the x coordinate.
4. Label the three points.
5. The Domain is $\left(-c, \infty\right)$, the range is $\left(-\infty, \infty\right)$, and the vertical asymptote is $x = -c$.

EXAMPLE 4

Graphing a Horizontal Shift of the Parent Function $y = \log_b(x)$

Sketch the horizontal shift $f(x) = \log_3(x - 2)$ alongside its parent function. Include the key points and asymptotes on the graph. State the domain, range, and asymptote.

Solution

Since the function is $f(x) = \log_3(x - 2)$, we notice $x + (-2) = x - 2$.

Thus $c = -2$, so $c < 0$. This means we will shift the function $f(x) = \log_3(x)$ right 2 units.

The vertical asymptote is $x = -(-2)$ or $x = 2$.

Consider the three key points from the parent function, $\left(\frac{1}{3}, -1\right)$, $(1, 0)$, and $(3, 1)$.

The new coordinates are found by adding 2 to the x coordinates.

Label the points $\left(\frac{7}{3}, -1\right)$, $(3, 0)$, and $(5, 1)$.

The domain is $\left(2, \infty\right)$, the range is $\left(-\infty, \infty\right)$, and the vertical asymptote is $x = 2$.

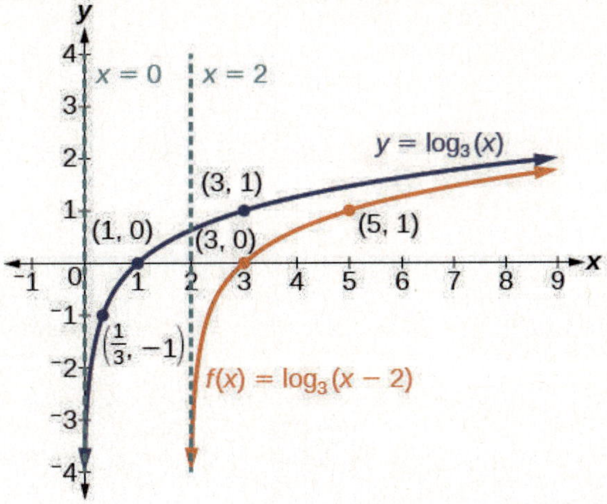

Figure 7

TRY IT #4 Sketch a graph of $f(x) = \log_3(x + 4)$ alongside its parent function. Include the key points and asymptotes on the graph. State the domain, range, and asymptote.

Graphing a Vertical Shift of $y = \log_b(x)$

When a constant d is added to the parent function $f(x) = \log_b(x)$, the result is a vertical shift d units in the direction of the sign on d. To visualize vertical shifts, we can observe the general graph of the parent function $f(x) = \log_b(x)$ alongside the shift up, $g(x) = \log_b(x) + d$ and the shift down, $h(x) = \log_b(x) - d$. See Figure 8.

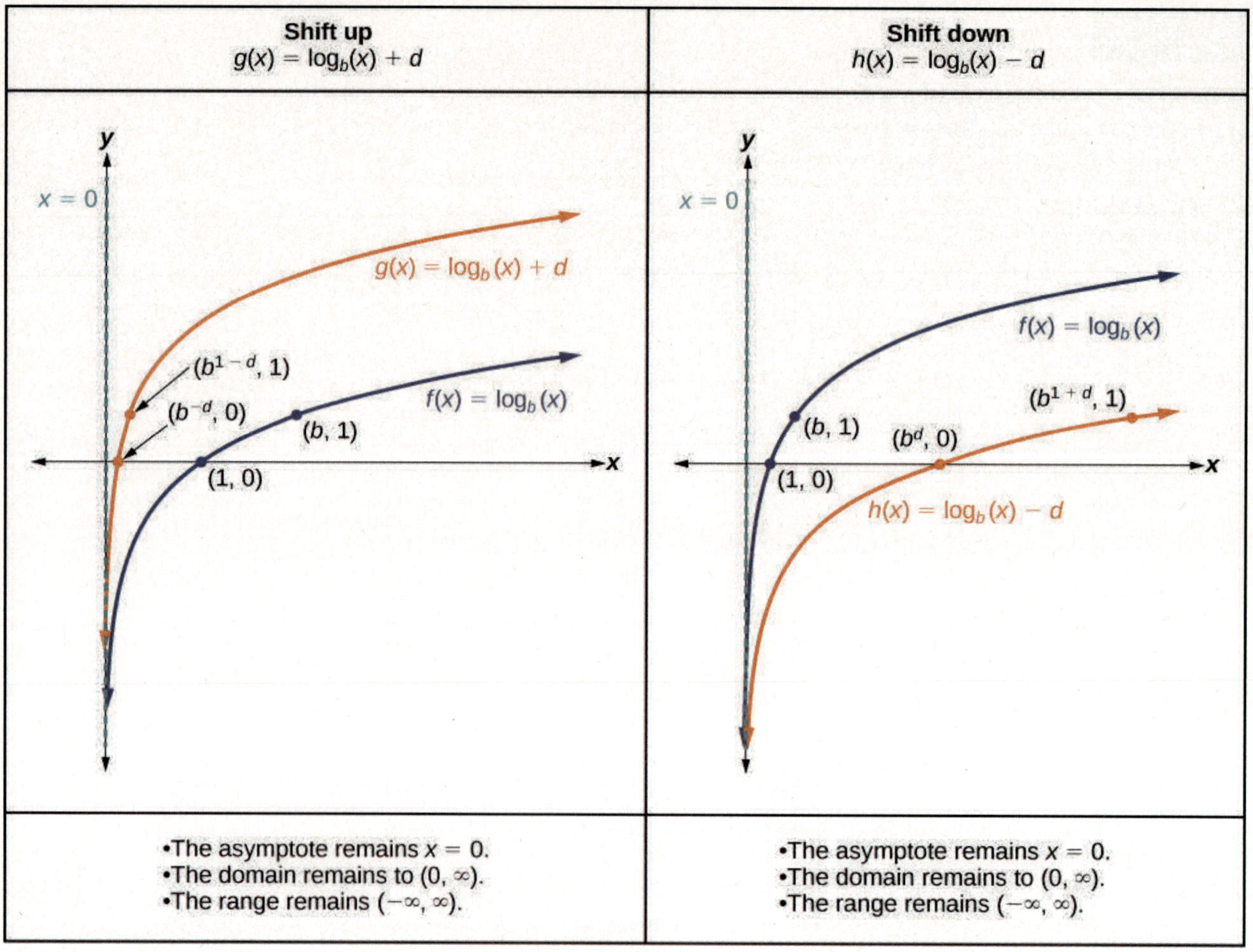

Figure 8

Vertical Shifts of the Parent Function $y = \log_b(x)$

For any constant d, the function $f(x) = \log_b(x) + d$

- shifts the parent function $y = \log_b(x)$ up d units if $d > 0$.
- shifts the parent function $y = \log_b(x)$ down d units if $d < 0$.
- has the vertical asymptote $x = 0$.
- has domain $\left(0, \infty\right)$.
- has range $\left(-\infty, \infty\right)$.

HOW TO

Given a logarithmic function with the form $f(x) = \log_b(x) + d$, graph the translation.

1. Identify the vertical shift:
   - If $d > 0$, shift the graph of $f(x) = \log_b(x)$ up d units.
   - If $d < 0$, shift the graph of $f(x) = \log_b(x)$ down d units.
2. Draw the vertical asymptote $x = 0$.
3. Identify three key points from the parent function. Find new coordinates for the shifted functions by adding d to the y coordinate.

4. Label the three points.
5. The domain is $\left(0,\infty\right)$, the range is $\left(-\infty,\infty\right)$, and the vertical asymptote is $x = 0$.

EXAMPLE 5

Graphing a Vertical Shift of the Parent Function $y = \log_b(x)$

Sketch a graph of $f(x) = \log_3(x) - 2$ alongside its parent function. Include the key points and asymptote on the graph. State the domain, range, and asymptote.

Solution

Since the function is $f(x) = \log_3(x) - 2$, we will notice $d = -2$. Thus $d < 0$.

This means we will shift the function $f(x) = \log_3(x)$ down 2 units.

The vertical asymptote is $x = 0$.

Consider the three key points from the parent function, $\left(\frac{1}{3}, -1\right)$, $(1, 0)$, and $(3, 1)$.

The new coordinates are found by subtracting 2 from the y coordinates.

Label the points $\left(\frac{1}{3}, -3\right)$, $(1, -2)$, and $(3, -1)$.

The domain is $\left(0, \infty\right)$, the range is $\left(-\infty, \infty\right)$, and the vertical asymptote is $x = 0$.

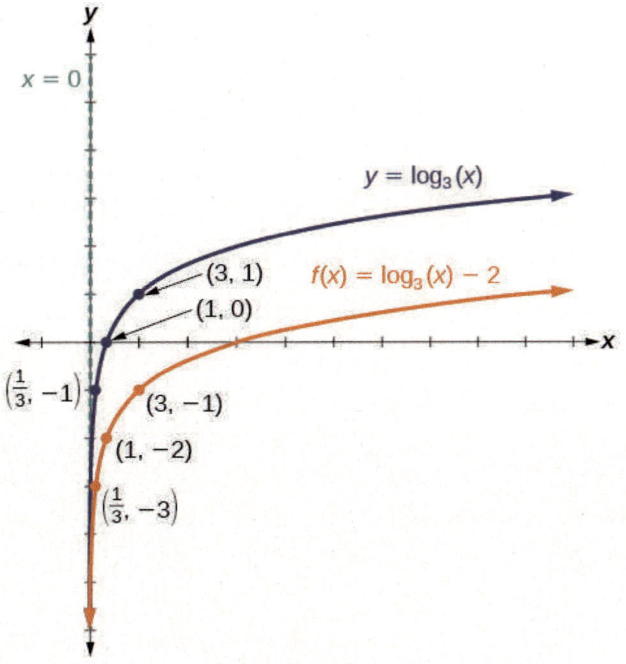

Figure 9

The domain is $\left(0, \infty\right)$, the range is $\left(-\infty, \infty\right)$, and the vertical asymptote is $x = 0$.

> TRY IT #5 Sketch a graph of $f(x) = \log_2(x) + 2$ alongside its parent function. Include the key points and asymptote on the graph. State the domain, range, and asymptote.

Graphing Stretches and Compressions of $y = \log_b(x)$

When the parent function $f(x) = \log_b(x)$ is multiplied by a constant $a > 0$, the result is a vertical stretch or compression of the original graph. To visualize stretches and compressions, we set $a > 1$ and observe the general graph of the parent function $f(x) = \log_b(x)$ alongside the vertical stretch, $g(x) = a\log_b(x)$ and the vertical compression, $h(x) = \frac{1}{a}\log_b(x)$. See Figure 10.

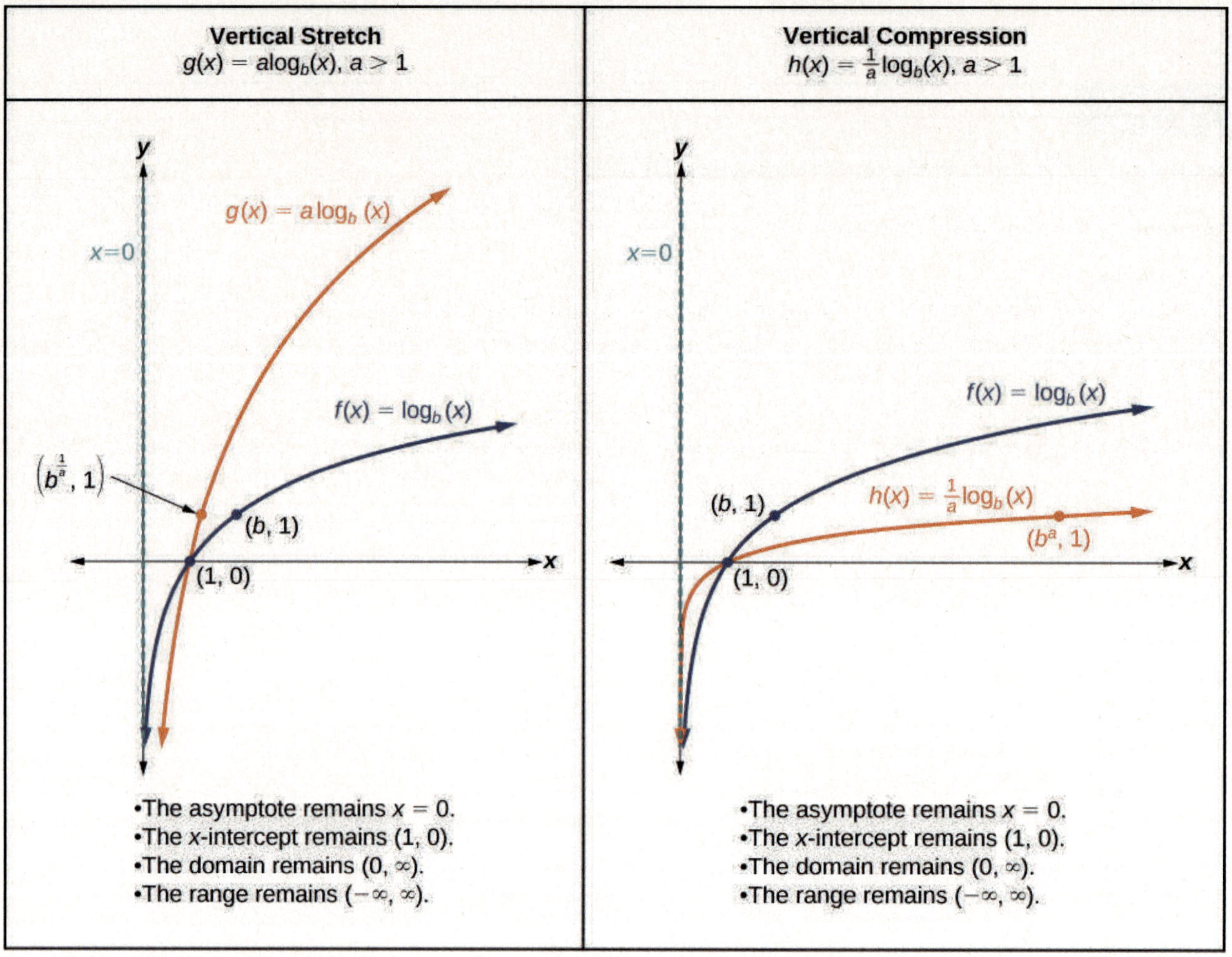

Figure 10

Vertical Stretches and Compressions of the Parent Function $y = \log_b(x)$

For any constant $a > 1$, the function $f(x) = a\log_b(x)$

- stretches the parent function $y = \log_b(x)$ vertically by a factor of a if $a > 1$.
- compresses the parent function $y = \log_b(x)$ vertically by a factor of a if $0 < a < 1$.
- has the vertical asymptote $x = 0$.
- has the x-intercept $(1, 0)$.
- has domain $\left(0, \infty\right)$.
- has range $\left(-\infty, \infty\right)$.

HOW TO

Given a logarithmic function with the form $f(x) = a\log_b(x)$, $a > 0$, graph the translation.

1. Identify the vertical stretch or compressions:
   - If $|a| > 1$, the graph of $f(x) = \log_b(x)$ is stretched by a factor of a units.
   - If $|a| < 1$, the graph of $f(x) = \log_b(x)$ is compressed by a factor of a units.

2. Draw the vertical asymptote $x = 0$.
3. Identify three key points from the parent function. Find new coordinates for the shifted functions by multiplying the y coordinates by a.
4. Label the three points.
5. The domain is $\left(0, \infty\right)$, the range is $\left(-\infty, \infty\right)$, and the vertical asymptote is $x = 0$.

EXAMPLE 6

Graphing a Stretch or Compression of the Parent Function $y = \log_b(x)$

Sketch a graph of $f(x) = 2\log_4(x)$ alongside its parent function. Include the key points and asymptote on the graph. State the domain, range, and asymptote.

Solution

Since the function is $f(x) = 2\log_4(x)$, we will notice $a = 2$.

This means we will stretch the function $f(x) = \log_4(x)$ by a factor of 2.

The vertical asymptote is $x = 0$.

Consider the three key points from the parent function, $\left(\frac{1}{4}, -1\right)$, $(1, 0)$, and $(4, 1)$.

The new coordinates are found by multiplying the y coordinates by 2.

Label the points $\left(\frac{1}{4}, -2\right)$, $(1, 0)$, and $(4, 2)$.

The domain is $\left(0, \infty\right)$, the range is $\left(-\infty, \infty\right)$, and the vertical asymptote is $x = 0$. See Figure 11.

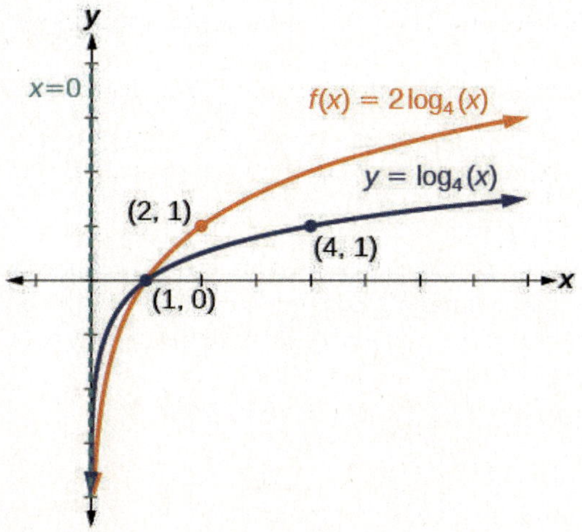

Figure 11

The domain is $\left(0, \infty\right)$, the range is $\left(-\infty, \infty\right)$, and the vertical asymptote is $x = 0$.

TRY IT #6 Sketch a graph of $f(x) = \frac{1}{2}\log_4(x)$ alongside its parent function. Include the key points and asymptote on the graph. State the domain, range, and asymptote.

EXAMPLE 7

Combining a Shift and a Stretch

Sketch a graph of $f(x) = 5\log(x + 2)$. State the domain, range, and asymptote.

Solution

Remember: what happens inside parentheses happens first. First, we move the graph left 2 units, then stretch the function vertically by a factor of 5, as in Figure 12. The vertical asymptote will be shifted to $x = -2$. The x-intercept will be $(-1,0)$. The domain will be $\left(-2, \infty\right)$. Two points will help give the shape of the graph: $(-1, 0)$ and $(8, 5)$. We chose $x = 8$ as the x-coordinate of one point to graph because when $x = 8$, $x + 2 = 10$, the base of the common logarithm.

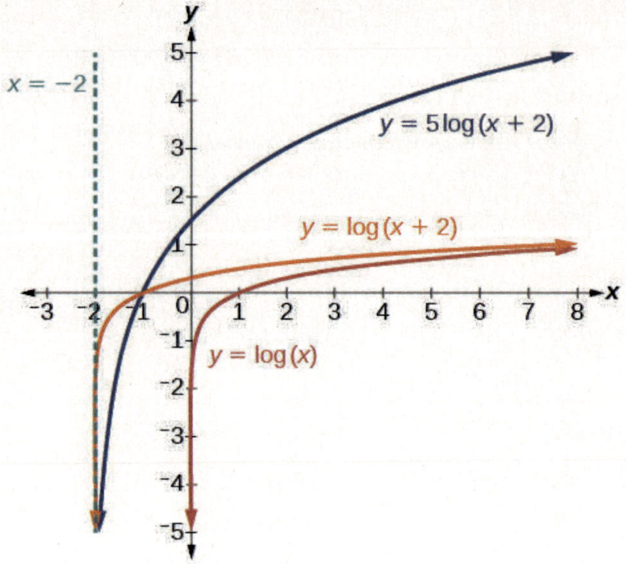

Figure 12

The domain is $\left(-2, \infty\right)$, the range is $\left(-\infty, \infty\right)$, and the vertical asymptote is $x = -2$.

TRY IT #7 Sketch a graph of the function $f(x) = 3\log(x - 2) + 1$. State the domain, range, and asymptote.

Graphing Reflections of $f(x) = \log_b(x)$

When the parent function $f(x) = \log_b(x)$ is multiplied by -1, the result is a reflection about the x-axis. When the *input* is multiplied by -1, the result is a reflection about the y-axis. To visualize reflections, we restrict $b > 1$, and observe the general graph of the parent function $f(x) = \log_b(x)$ alongside the reflection about the x-axis, $g(x) = -\log_b(x)$ and the reflection about the y-axis, $h(x) = \log_b(-x)$.

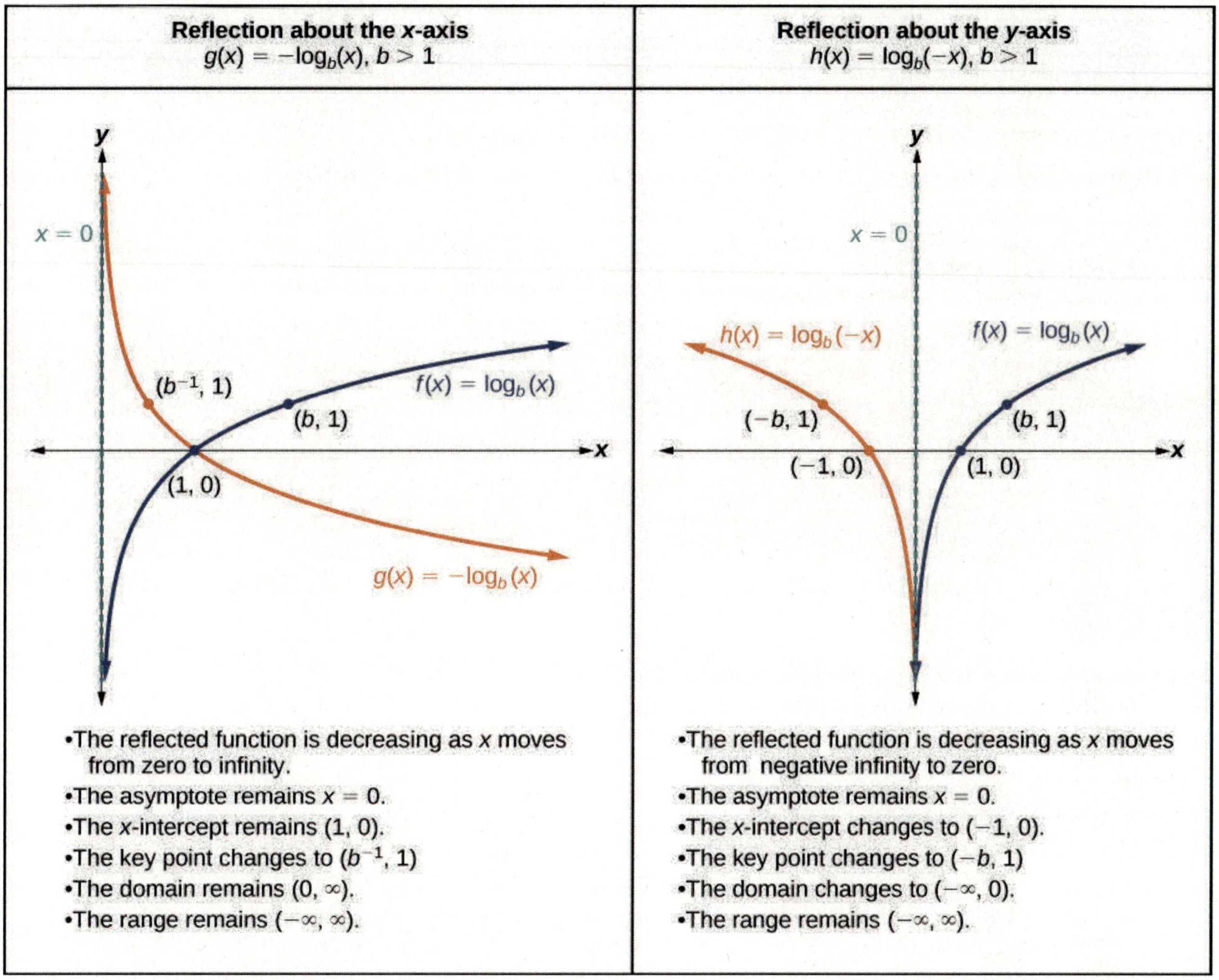

Figure 13

Reflections of the Parent Function $y = \log_b(x)$

The function $f(x) = -\log_b(x)$

- reflects the parent function $y = \log_b(x)$ about the *x*-axis.
- has domain, $(0, \infty)$, range, $(-\infty, \infty)$, and vertical asymptote, $x = 0$, which are unchanged from the parent function.

The function $f(x) = \log_b(-x)$

- reflects the parent function $y = \log_b(x)$ about the *y*-axis.
- has domain $(-\infty, 0)$.
- has range, $(-\infty, \infty)$, and vertical asymptote, $x = 0$, which are unchanged from the parent function.

HOW TO

Given a logarithmic function with the parent function $f(x) = \log_b(x)$, graph a translation.

If $f(x) = -\log_b(x)$	If $f(x) = \log_b(-x)$
1. Draw the vertical asymptote, $x = 0$.	1. Draw the vertical asymptote, $x = 0$.
2. Plot the *x*-intercept, $(1, 0)$.	2. Plot the *x*-intercept, $(1, 0)$.
3. Reflect the graph of the parent function $f(x) = \log_b(x)$ about the *x*-axis.	3. Reflect the graph of the parent function $f(x) = \log_b(x)$ about the *y*-axis.
4. Draw a smooth curve through the points.	4. Draw a smooth curve through the points.
5. State the domain, (0, ∞), the range, (−∞, ∞), and the vertical asymptote $x = 0$.	5. State the domain, (−∞, 0) the range, (−∞, ∞) and the vertical asymptote $x = 0$.

Table 3

EXAMPLE 8

Graphing a Reflection of a Logarithmic Function

Sketch a graph of $f(x) = \log(-x)$ alongside its parent function. Include the key points and asymptote on the graph. State the domain, range, and asymptote.

Solution

Before graphing $f(x) = \log(-x)$, identify the behavior and key points for the graph.

- Since $b = 10$ is greater than one, we know that the parent function is increasing. Since the *input* value is multiplied by -1, f is a reflection of the parent graph about the *y*-axis. Thus, $f(x) = \log(-x)$ will be decreasing as x moves from negative infinity to zero, and the right tail of the graph will approach the vertical asymptote $x = 0$.
- The *x*-intercept is $(-1, 0)$.
- We draw and label the asymptote, plot and label the points, and draw a smooth curve through the points.

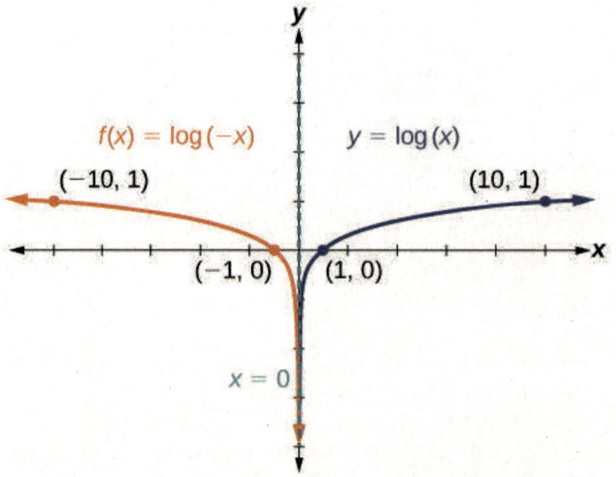

Figure 14

The domain is $(-\infty, 0)$, the range is $(-\infty, \infty)$, and the vertical asymptote is $x = 0$.

TRY IT #8 Graph $f(x) = -\log(-x)$. State the domain, range, and asymptote.

HOW TO

Given a logarithmic equation, use a graphing calculator to approximate solutions.

1. Press **[Y=]**. Enter the given logarithm equation or equations as **Y_1=** and, if needed, **Y_2=**.
2. Press **[GRAPH]** to observe the graphs of the curves and use **[WINDOW]** to find an appropriate view of the graphs, including their point(s) of intersection.
3. To find the value of x, we compute the point of intersection. Press **[2ND]** then **[CALC]**. Select "intersect" and press **[ENTER]** three times. The point of intersection gives the value of x, for the point(s) of intersection.

EXAMPLE 9

Approximating the Solution of a Logarithmic Equation

Solve $4\ln(x) + 1 = -2\ln(x-1)$ graphically. Round to the nearest thousandth.

Solution

Press **[Y=]** and enter $4\ln(x) + 1$ next to **Y_1=**. Then enter $-2\ln(x-1)$ next to **Y_2=**. For a window, use the values 0 to 5 for x and –10 to 10 for y. Press **[GRAPH]**. The graphs should intersect somewhere a little to right of $x = 1$.

For a better approximation, press **[2ND]** then **[CALC]**. Select **[5: intersect]** and press **[ENTER]** three times. The x-coordinate of the point of intersection is displayed as 1.3385297. (Your answer may be different if you use a different window or use a different value for **Guess?**) So, to the nearest thousandth, $x \approx 1.339$.

TRY IT #9 Solve $5\log(x+2) = 4 - \log(x)$ graphically. Round to the nearest thousandth.

Summarizing Translations of the Logarithmic Function

Now that we have worked with each type of translation for the logarithmic function, we can summarize each in Table 4 to arrive at the general equation for translating exponential functions.

Translations of the Parent Function $y = \log_b(x)$	
Translation	**Form**
Shift • Horizontally c units to the left • Vertically d units up	$y = \log_b(x+c) + d$
Stretch and Compress • Stretch if $\lvert a \rvert > 1$ • Compression if $\lvert a \rvert < 1$	$y = a\log_b(x)$
Reflect about the x-axis	$y = -\log_b(x)$
Reflect about the y-axis	$y = \log_b(-x)$
General equation for all translations	$y = a\log_b(x+c) + d$

Table 4

Translations of Logarithmic Functions

All translations of the parent logarithmic function, $y = \log_b(x)$, have the form

$$f(x) = a\log_b(x + c) + d$$

where the parent function, $y = \log_b(x), b > 1$, is

- shifted vertically up d units.
- shifted horizontally to the left c units.
- stretched vertically by a factor of $|a|$ if $|a| > 0$.
- compressed vertically by a factor of $|a|$ if $0 < |a| < 1$.
- reflected about the *x*-axis when $a < 0$.

For $f(x) = \log(-x)$, the graph of the parent function is reflected about the *y*-axis.

EXAMPLE 10

Finding the Vertical Asymptote of a Logarithm Graph

What is the vertical asymptote of $f(x) = -2\log_3(x + 4) + 5$?

Solution

The vertical asymptote is at $x = -4$.

Analysis

The coefficient, the base, and the upward translation do not affect the asymptote. The shift of the curve 4 units to the left shifts the vertical asymptote to $x = -4$.

TRY IT #10 What is the vertical asymptote of $f(x) = 3 + \ln(x - 1)$?

EXAMPLE 11

Finding the Equation from a Graph

Find a possible equation for the common logarithmic function graphed in Figure 15.

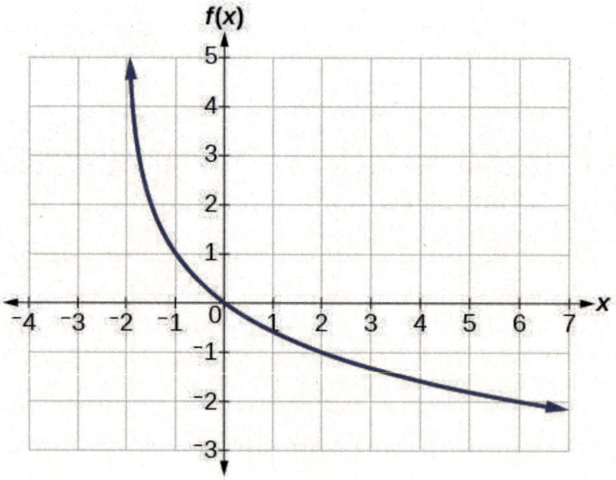

Figure 15

Solution

This graph has a vertical asymptote at $x = -2$ and has been vertically reflected. We do not know yet the vertical shift or the vertical stretch. We know so far that the equation will have form:

$$f(x) = -a\log(x + 2) + k$$

It appears the graph passes through the points $(-1, 1)$ and $(2, -1)$. Substituting $(-1, 1)$,

$$\begin{aligned} 1 &= -a\log(-1+2)+k && \text{Substitute } (-1, 1). \\ 1 &= -a\log(1)+k && \text{Arithmetic.} \\ 1 &= k && \log(1) = 0. \end{aligned}$$

Next, substituting in $(2, -1)$,

$$\begin{aligned} -1 &= -a\log(2+2)+1 && \text{Plug in } (2, -1). \\ -2 &= -a\log(4) && \text{Arithmetic.} \\ a &= \frac{2}{\log(4)} && \text{Solve for } a. \end{aligned}$$

This gives us the equation $f(x) = -\frac{2}{\log(4)}\log(x+2)+1$.

Analysis

We can verify this answer by comparing the function values in Table 5 with the points on the graph in Figure 15.

x	−1	0	1	2	3
$f(x)$	1	0	−0.58496	−1	−1.3219
x	4	5	6	7	8
$f(x)$	−1.5850	−1.8074	−2	−2.1699	−2.3219

Table 5

TRY IT #11 Give the equation of the natural logarithm graphed in Figure 16.

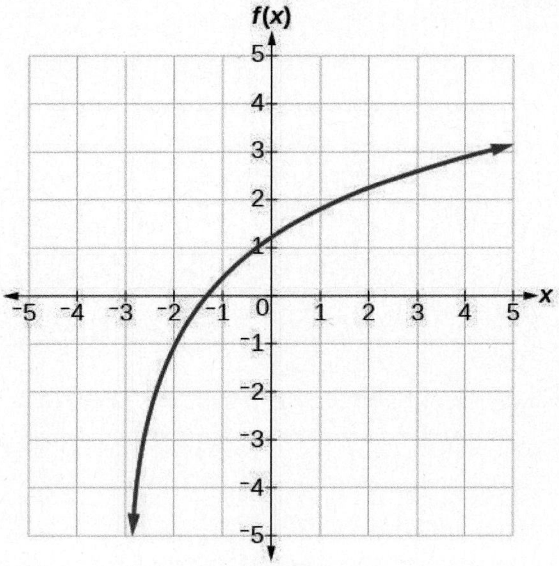

Figure 16

Q&A **Is it possible to tell the domain and range and describe the end behavior of a function just by looking at the graph?**

Yes, if we know the function is a general logarithmic function. For example, look at the graph in Figure 16. The graph approaches $x = -3$ (or thereabouts) more and more closely, so $x = -3$ is, or is very close to, the vertical asymptote. It approaches from the right, so the domain is all points to the right,

$\{x \mid x > -3\}$. *The range, as with all general logarithmic functions, is all real numbers. And we can see the end behavior because the graph goes down as it goes left and up as it goes right. The end behavior is that as* $x \to -3^+, f(x) \to -\infty$ *and as* $x \to \infty, f(x) \to \infty$.

MEDIA

Access these online resources for additional instruction and practice with graphing logarithms.

Graph an Exponential Function and Logarithmic Function (http://openstax.org/l/graphexplog)
Match Graphs with Exponential and Logarithmic Functions (http://openstax.org/l/matchexplog)
Find the Domain of Logarithmic Functions (http://openstax.org/l/domainlog)

6.4 SECTION EXERCISES

Verbal

1. The inverse of every logarithmic function is an exponential function and vice-versa. What does this tell us about the relationship between the coordinates of the points on the graphs of each?

2. What type(s) of translation(s), if any, affect the range of a logarithmic function?

3. What type(s) of translation(s), if any, affect the domain of a logarithmic function?

4. Consider the general logarithmic function $f(x) = \log_b(x)$. Why can't x be zero?

5. Does the graph of a general logarithmic function have a horizontal asymptote? Explain.

Algebraic

For the following exercises, state the domain and range of the function.

6. $f(x) = \log_3(x + 4)$

7. $h(x) = \ln\left(\frac{1}{2} - x\right)$

8. $g(x) = \log_5(2x + 9) - 2$

9. $h(x) = \ln(4x + 17) - 5$

10. $f(x) = \log_2(12 - 3x) - 3$

For the following exercises, state the domain and the vertical asymptote of the function.

11. $f(x) = \log_b(x - 5)$

12. $g(x) = \ln(3 - x)$

13. $f(x) = \log(3x + 1)$

14. $f(x) = 3\log(-x) + 2$

15. $g(x) = -\ln(3x + 9) - 7$

For the following exercises, state the domain, vertical asymptote, and end behavior of the function.

16. $f(x) = \ln(2 - x)$

17. $f(x) = \log\left(x - \frac{3}{7}\right)$

18. $h(x) = -\log(3x - 4) + 3$

19. $g(x) = \ln(2x + 6) - 5$

20. $f(x) = \log_3(15 - 5x) + 6$

For the following exercises, state the domain, range, and x- and y-intercepts, if they exist. If they do not exist, write DNE.

21. $h(x) = \log_4 (x - 1) + 1$

22. $f(x) = \log (5x + 10) + 3$

23. $g(x) = \ln (-x) - 2$

24. $f(x) = \log_2 (x + 2) - 5$

25. $h(x) = 3 \ln (x) - 9$

Graphical

For the following exercises, match each function in Figure 17 with the letter corresponding to its graph.

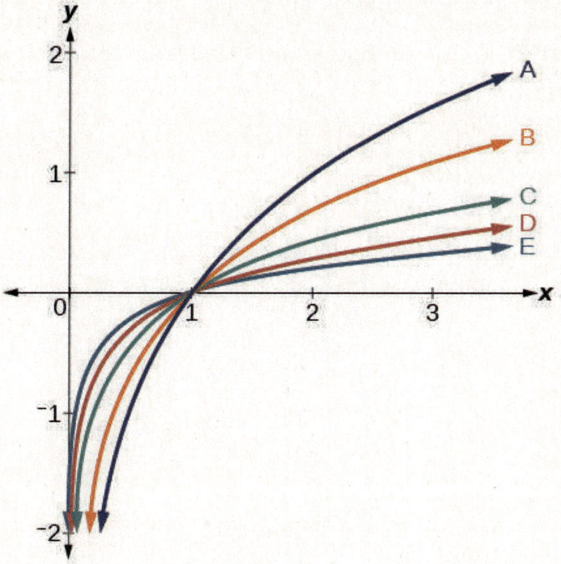

Figure 17

26. $d(x) = \log (x)$

27. $f(x) = \ln(x)$

28. $g(x) = \log_2 (x)$

29. $h(x) = \log_5 (x)$

30. $j(x) = \log_{25} (x)$

For the following exercises, match each function in Figure 18 with the letter corresponding to its graph.

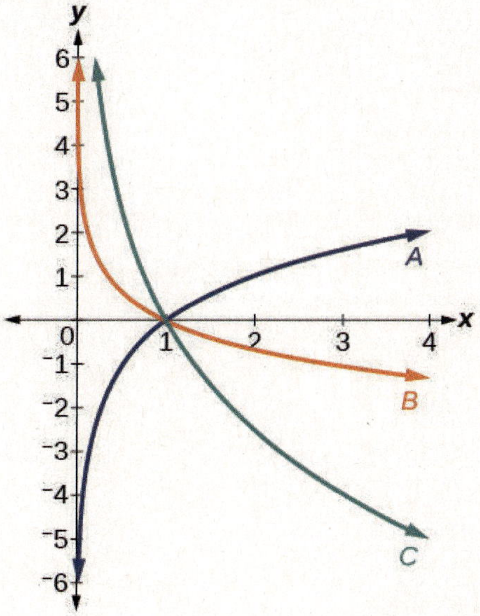

Figure 18

31. $f(x) = \log_{\frac{1}{3}}(x)$

32. $g(x) = \log_2(x)$

33. $h(x) = \log_{\frac{3}{4}}(x)$

For the following exercises, sketch the graphs of each pair of functions on the same axis.

34. $f(x) = \log(x)$ and $g(x) = 10^x$

35. $f(x) = \log(x)$ and $g(x) = \log_{\frac{1}{2}}(x)$

36. $f(x) = \log_4(x)$ and $g(x) = \ln(x)$

37. $f(x) = e^x$ and $g(x) = \ln(x)$

For the following exercises, match each function in Figure 19 with the letter corresponding to its graph.

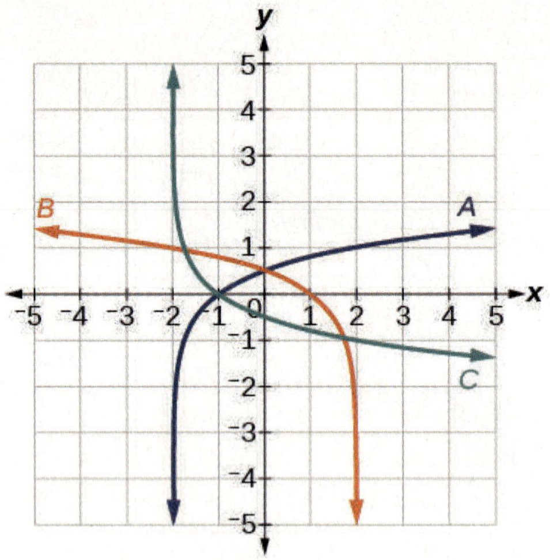

Figure 19

38. $f(x) = \log_4(-x+2)$

39. $g(x) = -\log_4(x+2)$

40. $h(x) = \log_4(x+2)$

For the following exercises, sketch the graph of the indicated function.

41. $f(x) = \log_2(x+2)$

42. $f(x) = 2\log(x)$

43. $f(x) = \ln(-x)$

44. $g(x) = \log(4x+16) + 4$

45. $g(x) = \log(6-3x) + 1$

46. $h(x) = -\frac{1}{2}\ln(x+1) - 3$

For the following exercises, write a logarithmic equation corresponding to the graph shown.

47. Use $y = \log_2(x)$ as the parent function.

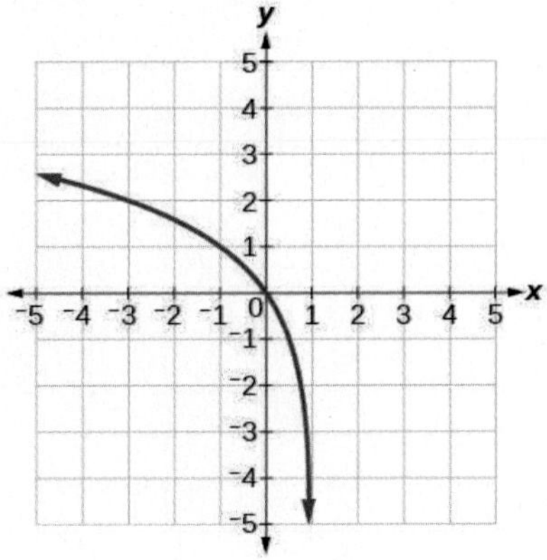

48. Use $f(x) = \log_3(x)$ as the parent function.

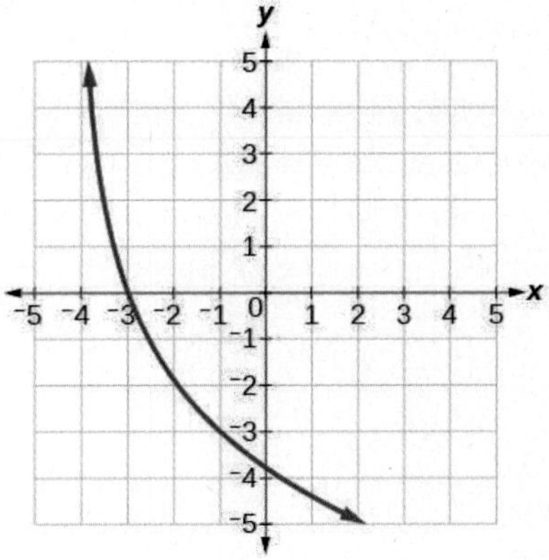

49. Use $f(x) = \log_4(x)$ as the parent function.

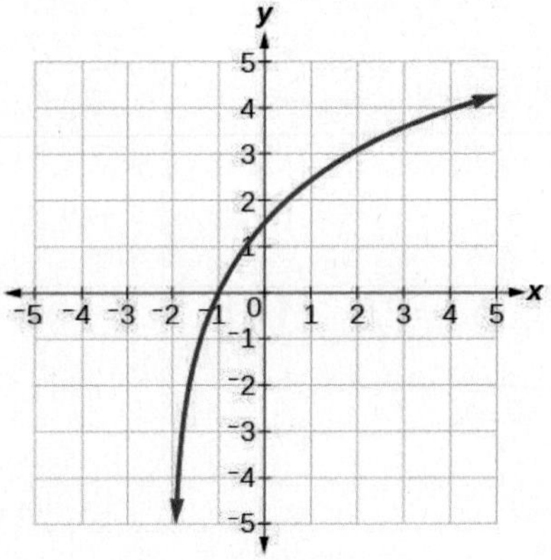

50. Use $f(x) = \log_5(x)$ as the parent function.

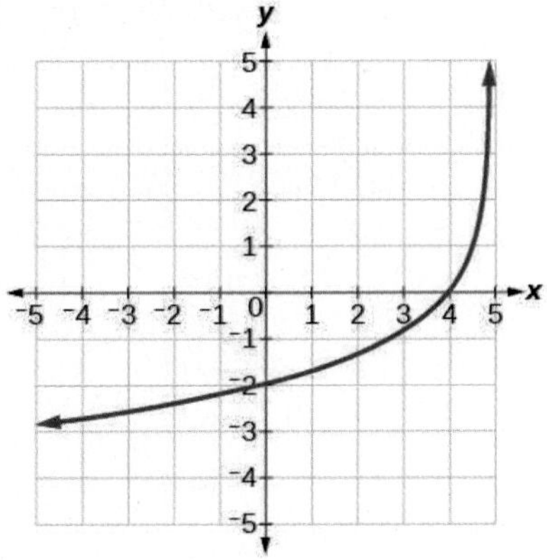

Technology

For the following exercises, use a graphing calculator to find approximate solutions to each equation.

51. $\log(x-1)+2 = \ln(x-1)+2$

52. $\log(2x-3)+2 = -\log(2x-3)+5$

53. $\ln(x-2) = -\ln(x+1)$

54. $2\ln(5x+1) = \frac{1}{2}\ln(-5x)+1$

55. $\frac{1}{3}\log(1-x) = \log(x+1)+\frac{1}{3}$

Extensions

56. Let b be any positive real number such that $b \neq 1$. What must $\log_b 1$ be equal to? Verify the result.

57. Explore and discuss the graphs of $f(x) = \log_{\frac{1}{2}}(x)$ and $g(x) = -\log_2(x)$. Make a conjecture based on the result.

58. Prove the conjecture made in the previous exercise.

59. What is the domain of the function $f(x) = \ln\left(\frac{x+2}{x-4}\right)$? Discuss the result.

60. Use properties of exponents to find the x-intercepts of the function $f(x) = \log\left(x^2 + 4x + 4\right)$ algebraically. Show the steps for solving, and then verify the result by graphing the function.

6.5 Logarithmic Properties

Learning Objectives

In this section, you will:

- Use the product rule for logarithms.
- Use the quotient rule for logarithms.
- Use the power rule for logarithms.
- Expand logarithmic expressions.
- Condense logarithmic expressions.
- Use the change-of-base formula for logarithms.

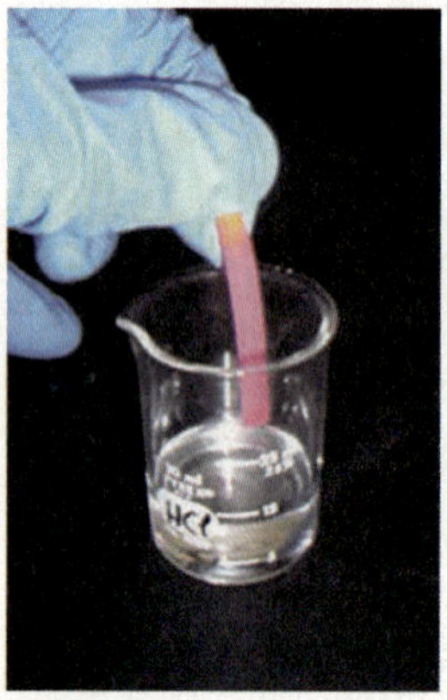

Figure 1 The pH of hydrochloric acid is tested with litmus paper. (credit: David Berardan)

In chemistry, pH is used as a measure of the acidity or alkalinity of a substance. The pH scale runs from 0 to 14. Substances with a pH less than 7 are considered acidic, and substances with a pH greater than 7 are said to be basic. Our bodies, for instance, must maintain a pH close to 7.35 in order for enzymes to work properly. To get a feel for what is acidic and what is basic, consider the following pH levels of some common substances:

- Battery acid: 0.8
- Stomach acid: 2.7
- Orange juice: 3.3
- Pure water: 7 (at 25° C)
- Human blood: 7.35
- Fresh coconut: 7.8
- Sodium hydroxide (lye): 14

To determine whether a solution is acidic or basic, we find its pH, which is a measure of the number of active positive hydrogen ions in the solution. The pH is defined by the following formula, where H^+ is the concentration of hydrogen ion in the solution

$$\begin{aligned} \text{pH} &= -\log([H^+]) \\ &= \log\left(\frac{1}{[H^+]}\right) \end{aligned}$$

The equivalence of $-\log\left(\left[H^+\right]\right)$ and $\log\left(\frac{1}{\left[H^+\right]}\right)$ is one of the logarithm properties we will examine in this section.

Using the Product Rule for Logarithms

Recall that the logarithmic and exponential functions "undo" each other. This means that logarithms have similar properties to exponents. Some important properties of logarithms are given here. First, the following properties are easy to prove.

$$\log_b 1 = 0$$
$$\log_b b = 1$$

For example, $\log_5 1 = 0$ since $5^0 = 1$. And $\log_5 5 = 1$ since $5^1 = 5$.

Next, we have the inverse property.

$$\log_b(b^x) = x$$
$$b^{\log_b x} = x, x > 0$$

For example, to evaluate $\log(100)$, we can rewrite the logarithm as $\log_{10}\left(10^2\right)$, and then apply the inverse property $\log_b(b^x) = x$ to get $\log_{10}\left(10^2\right) = 2$.

To evaluate $e^{\ln(7)}$, we can rewrite the logarithm as $e^{\log_e 7}$, and then apply the inverse property $b^{\log_b x} = x$ to get $e^{\log_e 7} = 7$.

Finally, we have the one-to-one property.

$$\log_b M = \log_b N \quad \text{if and only if} \quad M = N$$

We can use the one-to-one property to solve the equation $\log_3(3x) = \log_3(2x + 5)$ for x. Since the bases are the same, we can apply the one-to-one property by setting the arguments equal and solving for x :

$$\begin{aligned} 3x &= 2x + 5 && \text{Set the arguments equal.} \\ x &= 5 && \text{Subtract } 2x. \end{aligned}$$

But what about the equation $\log_3(3x) + \log_3(2x + 5) = 2$? The one-to-one property does not help us in this instance. Before we can solve an equation like this, we need a method for combining terms on the left side of the equation.

Recall that we use the *product rule of exponents* to combine the product of powers by adding exponents: $x^a x^b = x^{a+b}$. We have a similar property for logarithms, called the **product rule for logarithms**, which says that the logarithm of a product is equal to a sum of logarithms. Because logs are exponents, and we multiply like bases, we can add the exponents. We will use the inverse property to derive the product rule below.

Given any real number x and positive real numbers M, N, and b, where $b \neq 1$, we will show

$$\log_b(MN) = \log_b(M) + \log_b(N).$$

Let $m = \log_b M$ and $n = \log_b N$. In exponential form, these equations are $b^m = M$ and $b^n = N$. It follows that

$$\begin{aligned} \log_b(MN) &= \log_b(b^m b^n) && \text{Substitute for } M \text{ and } N. \\ &= \log_b\left(b^{m+n}\right) && \text{Apply the product rule for exponents.} \\ &= m + n && \text{Apply the inverse property of logs.} \\ &= \log_b(M) + \log_b(N) && \text{Substitute for } m \text{ and } n. \end{aligned}$$

Note that repeated applications of the product rule for logarithms allow us to simplify the logarithm of the product of any number of factors. For example, consider $\log_b(wxyz)$. Using the product rule for logarithms, we can rewrite this logarithm of a product as the sum of logarithms of its factors:

$$\log_b(wxyz) = \log_b w + \log_b x + \log_b y + \log_b z$$

The Product Rule for Logarithms

The **product rule for logarithms** can be used to simplify a logarithm of a product by rewriting it as a sum of individual logarithms.

$$\log_b(MN) = \log_b(M) + \log_b(N) \quad \text{for } b > 0$$

HOW TO

Given the logarithm of a product, use the product rule of logarithms to write an equivalent sum of logarithms.

1. Factor the argument completely, expressing each whole number factor as a product of primes.
2. Write the equivalent expression by summing the logarithms of each factor.

EXAMPLE 1

Using the Product Rule for Logarithms

Expand $\log_3(30x(3x+4))$.

Solution

We begin by factoring the argument completely, expressing 30 as a product of primes.

$$\log_3(30x(3x+4)) = \log_3(2\cdot 3\cdot 5\cdot x\cdot(3x+4))$$

Next we write the equivalent equation by summing the logarithms of each factor.

$$\log_3(30x(3x+4)) = \log_3(2) + \log_3(3) + \log_3(5) + \log_3(x) + \log_3(3x+4)$$

TRY IT #1 Expand $\log_b(8k)$.

Using the Quotient Rule for Logarithms

For quotients, we have a similar rule for logarithms. Recall that we use the *quotient rule of exponents* to combine the quotient of exponents by subtracting: $\frac{x^a}{x^b} = x^{a-b}$. The **quotient rule for logarithms** says that the logarithm of a quotient is equal to a difference of logarithms. Just as with the product rule, we can use the inverse property to derive the quotient rule.

Given any real number x and positive real numbers M, N, and b, where $b \neq 1$, we will show

$$\log_b\left(\frac{M}{N}\right) = \log_b(M) - \log_b(N).$$

Let $m = \log_b M$ and $n = \log_b N$. In exponential form, these equations are $b^m = M$ and $b^n = N$. It follows that

$$\begin{aligned}
\log_b\left(\tfrac{M}{N}\right) &= \log_b\left(\tfrac{b^m}{b^n}\right) && \text{Substitute for } M \text{ and } N.\\
&= \log_b\left(b^{m-n}\right) && \text{Apply the quotient rule for exponents.}\\
&= m - n && \text{Apply the inverse property of logs.}\\
&= \log_b(M) - \log_b(N) && \text{Substitute for } m \text{ and } n.
\end{aligned}$$

For example, to expand $\log\left(\frac{2x^2+6x}{3x+9}\right)$, we must first express the quotient in lowest terms. Factoring and canceling we get,

$$\begin{aligned}
\log\left(\frac{2x^2+6x}{3x+9}\right) &= \log\left(\frac{2x(x+3)}{3(x+3)}\right) && \text{Factor the numerator and denominator.}\\
&= \log\left(\tfrac{2x}{3}\right) && \text{Cancel the common factors.}
\end{aligned}$$

Next we apply the quotient rule by subtracting the logarithm of the denominator from the logarithm of the numerator. Then we apply the product rule.

$$\begin{aligned}
\log\left(\tfrac{2x}{3}\right) &= \log(2x) - \log(3)\\
&= \log(2) + \log(x) - \log(3)
\end{aligned}$$

The Quotient Rule for Logarithms

The **quotient rule for logarithms** can be used to simplify a logarithm or a quotient by rewriting it as the difference of individual logarithms.

$$\log_b\left(\frac{M}{N}\right) = \log_b M - \log_b N$$

HOW TO

Given the logarithm of a quotient, use the quotient rule of logarithms to write an equivalent difference of logarithms.

1. Express the argument in lowest terms by factoring the numerator and denominator and canceling common terms.
2. Write the equivalent expression by subtracting the logarithm of the denominator from the logarithm of the numerator.
3. Check to see that each term is fully expanded. If not, apply the product rule for logarithms to expand completely.

EXAMPLE 2

Using the Quotient Rule for Logarithms

Expand $\log_2\left(\frac{15x(x-1)}{(3x+4)(2-x)}\right)$.

Solution

First we note that the quotient is factored and in lowest terms, so we apply the quotient rule.

$$\log_2\left(\frac{15x(x-1)}{(3x+4)(2-x)}\right) = \log_2(15x(x-1)) - \log_2((3x+4)(2-x))$$

Notice that the resulting terms are logarithms of products. To expand completely, we apply the product rule, noting that the prime factors of the factor 15 are 3 and 5.

$$\begin{aligned}\log_2(15x(x-1)) - \log_2((3x+4)(2-x)) &= [\log_2(3) + \log_2(5) + \log_2(x) + \log_2(x-1)] - [\log_2(3x+4) + \log_2(2-x)] \\ &= \log_2(3) + \log_2(5) + \log_2(x) + \log_2(x-1) - \log_2(3x+4) - \log_2(2-x)\end{aligned}$$

Analysis

There are exceptions to consider in this and later examples. First, because denominators must never be zero, this expression is not defined for $x = -\frac{4}{3}$ and $x = 2$. Also, since the argument of a logarithm must be positive, we note as we observe the expanded logarithm, that $x > 0$, $x > 1$, $x > -\frac{4}{3}$, and $x < 2$. Combining these conditions is beyond the scope of this section, and we will not consider them here or in subsequent exercises.

TRY IT #2 Expand $\log_3\left(\frac{7x^2+21x}{7x(x-1)(x-2)}\right)$.

Using the Power Rule for Logarithms

We've explored the product rule and the quotient rule, but how can we take the logarithm of a power, such as x^2? One method is as follows:

$$\begin{aligned}\log_b\left(x^2\right) &= \log_b(x \cdot x) \\ &= \log_b x + \log_b x \\ &= 2\log_b x\end{aligned}$$

Notice that we used the product rule for logarithms to find a solution for the example above. By doing so, we have derived the **power rule for logarithms**, which says that the log of a power is equal to the exponent times the log of the base. Keep in mind that, although the input to a logarithm may not be written as a power, we may be able to change it to a power. For example,

$$100 = 10^2 \qquad \sqrt{3} = 3^{\frac{1}{2}} \qquad \frac{1}{e} = e^{-1}$$

The Power Rule for Logarithms

The **power rule for logarithms** can be used to simplify the logarithm of a power by rewriting it as the product of the exponent times the logarithm of the base.

$$\log_b\left(M^n\right) = n\log_b M$$

HOW TO

Given the logarithm of a power, use the power rule of logarithms to write an equivalent product of a factor and a logarithm.

1. Express the argument as a power, if needed.
2. Write the equivalent expression by multiplying the exponent times the logarithm of the base.

EXAMPLE 3

Expanding a Logarithm with Powers

Expand $\log_2 x^5$.

Solution

The argument is already written as a power, so we identify the exponent, 5, and the base, x, and rewrite the equivalent expression by multiplying the exponent times the logarithm of the base.

$$\log_2\left(x^5\right) = 5\log_2 x$$

TRY IT #3 Expand $\ln x^2$.

EXAMPLE 4

Rewriting an Expression as a Power before Using the Power Rule

Expand $\log_3(25)$ using the power rule for logs.

Solution

Expressing the argument as a power, we get $\log_3(25) = \log_3\left(5^2\right)$.

Next we identify the exponent, 2, and the base, 5, and rewrite the equivalent expression by multiplying the exponent times the logarithm of the base.

$$\log_3\left(5^2\right) = 2\log_3(5)$$

TRY IT #4 Expand $\ln\left(\frac{1}{x^2}\right)$.

EXAMPLE 5

Using the Power Rule in Reverse

Rewrite $4\ln(x)$ using the power rule for logs to a single logarithm with a leading coefficient of 1.

Solution

Because the logarithm of a power is the product of the exponent times the logarithm of the base, it follows that the product of a number and a logarithm can be written as a power. For the expression $4\ln(x)$, we identify the factor, 4, as the exponent and the argument, x, as the base, and rewrite the product as a logarithm of a power: $4\ln(x) = \ln(x^4)$.

TRY IT #5 Rewrite $2\log_3 4$ using the power rule for logs to a single logarithm with a leading coefficient of 1.

Expanding Logarithmic Expressions

Taken together, the product rule, quotient rule, and power rule are often called "laws of logs." Sometimes we apply more than one rule in order to simplify an expression. For example:

$$\begin{aligned}\log_b\left(\tfrac{6x}{y}\right) &= \log_b(6x) - \log_b y\\ &= \log_b 6 + \log_b x - \log_b y\end{aligned}$$

We can use the power rule to expand logarithmic expressions involving negative and fractional exponents. Here is an alternate proof of the quotient rule for logarithms using the fact that a reciprocal is a negative power:

$$\begin{aligned}\log_b\left(\tfrac{A}{C}\right) &= \log_b\left(AC^{-1}\right)\\ &= \log_b(A) + \log_b\left(C^{-1}\right)\\ &= \log_b A + (-1)\log_b C\\ &= \log_b A - \log_b C\end{aligned}$$

We can also apply the product rule to express a sum or difference of logarithms as the logarithm of a product.

With practice, we can look at a logarithmic expression and expand it mentally, writing the final answer. Remember, however, that we can only do this with products, quotients, powers, and roots—never with addition or subtraction inside the argument of the logarithm.

EXAMPLE 6

Expanding Logarithms Using Product, Quotient, and Power Rules

Rewrite $\ln\left(\frac{x^4 y}{7}\right)$ as a sum or difference of logs.

Solution

First, because we have a quotient of two expressions, we can use the quotient rule:

$$\ln\left(\frac{x^4 y}{7}\right) = \ln\left(x^4 y\right) - \ln(7)$$

Then seeing the product in the first term, we use the product rule:

$$\ln\left(x^4 y\right) - \ln(7) = \ln\left(x^4\right) + \ln(y) - \ln(7)$$

Finally, we use the power rule on the first term:

$$\ln\left(x^4\right) + \ln(y) - \ln(7) = 4\ln(x) + \ln(y) - \ln(7)$$

TRY IT #6 Expand $\log\left(\frac{x^2 y^3}{z^4}\right)$.

EXAMPLE 7

Using the Power Rule for Logarithms to Simplify the Logarithm of a Radical Expression

Expand $\log\left(\sqrt{x}\right)$.

Solution

$$\begin{aligned}\log\left(\sqrt{x}\right) &= \log x^{\left(\frac{1}{2}\right)}\\ &= \frac{1}{2}\log x\end{aligned}$$

TRY IT #7 Expand $\ln\left(\sqrt[3]{x^2}\right)$.

Q&A **Can we expand** $\ln\left(x^2+y^2\right)$?

No. There is no way to expand the logarithm of a sum or difference inside the argument of the logarithm.

EXAMPLE 8

Expanding Complex Logarithmic Expressions

Expand $\log_6\left(\frac{64x^3(4x+1)}{(2x-1)}\right)$.

Solution

We can expand by applying the Product and Quotient Rules.

$$\begin{aligned}\log_6\left(\frac{64x^3(4x+1)}{(2x-1)}\right) &= \log_6 64+\log_6 x^3+\log_6(4x+1)-\log_6(2x-1) && \text{Apply the Quotient Rule.}\\ &= \log_6 2^6+\log_6 x^3+\log_6(4x+1)-\log_6(2x-1) && \text{Simplify by writing } 64 \text{ as } 2^6.\\ &= 6\log_6 2+3\log_6 x+\log_6(4x+1)-\log_6(2x-1) && \text{Apply the Power Rule.}\end{aligned}$$

TRY IT #8 Expand $\ln\left(\frac{\sqrt{(x-1)(2x+1)^2}}{(x^2-9)}\right)$.

Condensing Logarithmic Expressions

We can use the rules of logarithms we just learned to condense sums, differences, and products with the same base as a single logarithm. It is important to remember that the logarithms must have the same base to be combined. We will learn later how to change the base of any logarithm before condensing.

HOW TO

Given a sum, difference, or product of logarithms with the same base, write an equivalent expression as a single logarithm.

1. Apply the power property first. Identify terms that are products of factors and a logarithm, and rewrite each as the logarithm of a power.
2. Next apply the product property. Rewrite sums of logarithms as the logarithm of a product.
3. Apply the quotient property last. Rewrite differences of logarithms as the logarithm of a quotient.

EXAMPLE 9

Using the Product and Quotient Rules to Combine Logarithms

Write $\log_3(5) + \log_3(8) - \log_3(2)$ as a single logarithm.

Solution

Using the product and quotient rules

$$\log_3(5) + \log_3(8) = \log_3(5 \cdot 8) = \log_3(40)$$

This reduces our original expression to

$$\log_3(40) - \log_3(2)$$

Then, using the quotient rule

$$\log_3(40) - \log_3(2) = \log_3\left(\frac{40}{2}\right) = \log_3(20)$$

TRY IT #9 Condense $\log 3 - \log 4 + \log 5 - \log 6$.

EXAMPLE 10

Condensing Complex Logarithmic Expressions

Condense $\log_2\left(x^2\right) + \frac{1}{2}\log_2(x-1) - 3\log_2\left((x+3)^2\right)$.

Solution

We apply the power rule first:

$$\log_2\left(x^2\right) + \frac{1}{2}\log_2(x-1) - 3\log_2\left((x+3)^2\right) = \log_2\left(x^2\right) + \log_2\left(\sqrt{x-1}\right) - \log_2\left((x+3)^6\right)$$

Next we apply the product rule to the sum:

$$\log_2\left(x^2\right) + \log_2\left(\sqrt{x-1}\right) - \log_2\left((x+3)^6\right) = \log_2\left(x^2\sqrt{x-1}\right) - \log_2\left((x+3)^6\right)$$

Finally, we apply the quotient rule to the difference:

$$\log_2\left(x^2\sqrt{x-1}\right) - \log_2\left((x+3)^6\right) = \log_2\frac{x^2\sqrt{x-1}}{(x+3)^6}$$

TRY IT #10 Rewrite $\log(5) + 0.5\log(x) - \log(7x-1) + 3\log(x-1)$ as a single logarithm.

EXAMPLE 11

Rewriting as a Single Logarithm

Rewrite $2\log x - 4\log(x+5) + \frac{1}{x}\log(3x+5)$ as a single logarithm.

Solution

We apply the power rule first:

$$2\log x - 4\log(x+5) + \frac{1}{x}\log(3x+5) = \log\left(x^2\right) - \log(x+5)^4 + \log\left((3x+5)^{x^{-1}}\right)$$

Next we rearrange and apply the product rule to the sum:

$$\log\left(x^2\right) - \log(x+5)^4 + \log\left((3x+5)^{x^{-1}}\right)$$

$$= \log\left(x^2\right) + \log\left((3x+5)^{x^{-1}}\right) - \log(x+5)^4$$

$$= \log\left(x^2(3x+5)^{x^{-1}}\right) - \log(x+5)^4$$

Finally, we apply the quotient rule to the difference:

$$= \log\left(x^2(3x+5)^{x^{-1}}\right) - \log(x+5)^4 = \log\frac{x^2(3x+5)^{x^{-1}}}{(x+5)^4}$$

> **TRY IT** #11 Condense $4(3\log(x) + \log(x+5) - \log(2x+3))$.

EXAMPLE 12

Applying of the Laws of Logs

Recall that, in chemistry, $\text{pH} = -\log[H^+]$. If the concentration of hydrogen ions in a liquid is doubled, what is the effect on pH?

Solution

Suppose C is the original concentration of hydrogen ions, and P is the original pH of the liquid. Then $P = -\log(C)$. If the concentration is doubled, the new concentration is $2C$. Then the pH of the new liquid is

$$\text{pH} = -\log(2C)$$

Using the product rule of logs

$$\text{pH} = -\log(2C) = -(\log(2) + \log(C)) = -\log(2) - \log(C)$$

Since $P = -\log(C)$, the new pH is

$$\text{pH} = P - \log(2) \approx P - 0.301$$

When the concentration of hydrogen ions is doubled, the pH decreases by about 0.301.

> **TRY IT** #12 How does the pH change when the concentration of positive hydrogen ions is decreased by half?

Using the Change-of-Base Formula for Logarithms

Most calculators can evaluate only common and natural logs. In order to evaluate logarithms with a base other than 10 or e, we use the **change-of-base formula** to rewrite the logarithm as the quotient of logarithms of any other base; when using a calculator, we would change them to common or natural logs.

To derive the change-of-base formula, we use the one-to-one property and **power rule for logarithms**.

Given any positive real numbers M, b, and n, where $n \neq 1$ and $b \neq 1$, we show

$$\log_b M = \frac{\log_n M}{\log_n b}$$

Let $y = \log_b M$.By exponentiating both sides with base$b$, we arrive at an exponential form, namely $b^y = M$. It follows that

$$\begin{array}{lll} \log_n(b^y) & = \log_n M & \text{Apply the one-to-one property.} \\ y\log_n b & = \log_n M & \text{Apply the power rule for logarithms.} \\ y & = \frac{\log_n M}{\log_n b} & \text{Isolate } y. \\ \log_b M & = \frac{\log_n M}{\log_n b} & \text{Substitute for } y. \end{array}$$

For example, to evaluate $\log_5 36$ using a calculator, we must first rewrite the expression as a quotient of common or natural logs. We will use the common log.

$$\begin{array}{lll} \log_5 36 & = \frac{\log(36)}{\log(5)} & \text{Apply the change of base formula using base 10.} \\ & \approx 2.2266 & \text{Use a calculator to evaluate to 4 decimal places.} \end{array}$$

The Change-of-Base Formula

The **change-of-base formula** can be used to evaluate a logarithm with any base.

For any positive real numbers $M, b,$ and $n,$ where $n \neq 1$ and $b \neq 1,$

$$\log_b M = \frac{\log_n M}{\log_n b}.$$

It follows that the change-of-base formula can be used to rewrite a logarithm with any base as the quotient of common or natural logs.

$$\log_b M = \frac{\ln M}{\ln b}$$

and

$$\log_b M = \frac{\log M}{\log b}$$

HOW TO

Given a logarithm with the form $\log_b M$, use the change-of-base formula to rewrite it as a quotient of logs with any positive base n, where $n \neq 1$.

1. Determine the new base n, remembering that the common log, $\log(x)$, has base 10, and the natural log, $\ln(x)$, has base e.
2. Rewrite the log as a quotient using the change-of-base formula
   a. The numerator of the quotient will be a logarithm with base n and argument M.
   b. The denominator of the quotient will be a logarithm with base n and argument b.

EXAMPLE 13

Changing Logarithmic Expressions to Expressions Involving Only Natural Logs

Change $\log_5 3$ to a quotient of natural logarithms.

Solution

Because we will be expressing $\log_5 3$ as a quotient of natural logarithms, the new base, $n = e$.

We rewrite the log as a quotient using the change-of-base formula. The numerator of the quotient will be the natural log with argument 3. The denominator of the quotient will be the natural log with argument 5.

$$\begin{aligned} \log_b M &= \frac{\ln M}{\ln b} \\ \log_5 3 &= \frac{\ln 3}{\ln 5} \end{aligned}$$

TRY IT #13 Change $\log_{0.5} 8$ to a quotient of natural logarithms.

Q&A **Can we change common logarithms to natural logarithms?**

Yes. Remember that $\log 9$ *means* $\log_{10} 9$. *So,* $\log 9 = \frac{\ln 9}{\ln 10}$.

EXAMPLE 14

Using the Change-of-Base Formula with a Calculator

Evaluate $\log_2(10)$ using the change-of-base formula with a calculator.

Solution

According to the change-of-base formula, we can rewrite the log base 2 as a logarithm of any other base. Since our calculators can evaluate the natural log, we might choose to use the natural logarithm, which is the log base e.

$$\begin{array}{ll} \log_2 10 = \frac{\ln 10}{\ln 2} & \text{Apply the change of base formula using base } e. \\ \approx 3.3219 & \text{Use a calculator to evaluate to 4 decimal places.} \end{array}$$

TRY IT #14 Evaluate $\log_5(100)$ using the change-of-base formula.

MEDIA

Access these online resources for additional instruction and practice with laws of logarithms.

The Properties of Logarithms (http://openstax.org/l/proplog)
Expand Logarithmic Expressions (http://openstax.org/l/expandlog)
Evaluate a Natural Logarithmic Expression (http://openstax.org/l/evaluatelog)

6.5 SECTION EXERCISES

Verbal

1. How does the power rule for logarithms help when solving logarithms with the form $\log_b\left(\sqrt[n]{x}\right)$?

2. What does the change-of-base formula do? Why is it useful when using a calculator?

Algebraic

For the following exercises, expand each logarithm as much as possible. Rewrite each expression as a sum, difference, or product of logs.

3. $\log_b(7x \cdot 2y)$

4. $\ln(3ab \cdot 5c)$

5. $\log_b\left(\frac{13}{17}\right)$

6. $\log_4\left(\frac{\frac{x}{z}}{w}\right)$

7. $\ln\left(\frac{1}{4^k}\right)$

8. $\log_2(y^x)$

For the following exercises, condense to a single logarithm if possible.

9. $\ln(7) + \ln(x) + \ln(y)$

10. $\log_3(2) + \log_3(a) + \log_3(11) + \log_3(b)$

11. $\log_b(28) - \log_b(7)$

12. $\ln(a) - \ln(d) - \ln(c)$

13. $-\log_b\left(\frac{1}{7}\right)$

14. $\frac{1}{3}\ln(8)$

For the following exercises, use the properties of logarithms to expand each logarithm as much as possible. Rewrite each expression as a sum, difference, or product of logs.

15. $\log\left(\frac{x^{15}y^{13}}{z^{19}}\right)$

16. $\ln\left(\frac{a^{-2}}{b^{-4}c^5}\right)$

17. $\log\left(\sqrt{x^3y^{-4}}\right)$

18. $\ln\left(y\sqrt{\frac{y}{1-y}}\right)$

19. $\log\left(x^2y^3\sqrt[3]{x^2y^5}\right)$

For the following exercises, condense each expression to a single logarithm using the properties of logarithms.

20. $\log\left(2x^4\right)+\log\left(3x^5\right)$

21. $\ln(6x^9)-\ln(3x^2)$

22. $2\log(x)+3\log(x+1)$

23. $\log(x)-\frac{1}{2}\log(y)+3\log(z)$

24. $4\log_7(c)+\frac{\log_7(a)}{3}+\frac{\log_7(b)}{3}$

For the following exercises, rewrite each expression as an equivalent ratio of logs using the indicated base.

25. $\log_7(15)$ to base e

26. $\log_{14}(55.875)$ to base 10

For the following exercises, suppose $\log_5(6)=a$ and $\log_5(11)=b$. Use the change-of-base formula along with properties of logarithms to rewrite each expression in terms of a and b. Show the steps for solving.

27. $\log_{11}(5)$

28. $\log_6(55)$

29. $\log_{11}\left(\frac{6}{11}\right)$

Numeric

For the following exercises, use properties of logarithms to evaluate without using a calculator.

30. $\log_3\left(\frac{1}{9}\right)-3\log_3(3)$

31. $6\log_8(2)+\frac{\log_8(64)}{3\log_8(4)}$

32. $2\log_9(3)-4\log_9(3)+\log_9\left(\frac{1}{729}\right)$

For the following exercises, use the change-of-base formula to evaluate each expression as a quotient of natural logs. Use a calculator to approximate each to five decimal places.

33. $\log_3(22)$

34. $\log_8(65)$

35. $\log_6(5.38)$

36. $\log_4\left(\frac{15}{2}\right)$

37. $\log_{\frac{1}{2}}(4.7)$

Extensions

38. Use the product rule for logarithms to find all x values such that $\log_{12}(2x+6)+\log_{12}(x+2)=2$. Show the steps for solving.

39. Use the quotient rule for logarithms to find all x values such that $\log_6(x+2)-\log_6(x-3)=1$. Show the steps for solving.

40. Can the power property of logarithms be derived from the power property of exponents using the equation $b^x = m$? If not, explain why. If so, show the derivation.

41. Prove that $\log_b(n) = \frac{1}{\log_n(b)}$ for any positive integers $b > 1$ and $n > 1$.

42. Does $\log_{81}(2401) = \log_3(7)$? Verify the claim algebraically.

6.6 Exponential and Logarithmic Equations

Learning Objectives

In this section, you will:

- Use like bases to solve exponential equations.
- Use logarithms to solve exponential equations.
- Use the definition of a logarithm to solve logarithmic equations.
- Use the one-to-one property of logarithms to solve logarithmic equations.
- Solve applied problems involving exponential and logarithmic equations.

Figure 1 Wild rabbits in Australia. The rabbit population grew so quickly in Australia that the event became known as the "rabbit plague." (credit: Richard Taylor, Flickr)

In 1859, an Australian landowner named Thomas Austin released 24 rabbits into the wild for hunting. Because Australia had few predators and ample food, the rabbit population exploded. In fewer than ten years, the rabbit population numbered in the millions.

Uncontrolled population growth, as in the wild rabbits in Australia, can be modeled with exponential functions. Equations resulting from those exponential functions can be solved to analyze and make predictions about exponential growth. In this section, we will learn techniques for solving exponential functions.

Using Like Bases to Solve Exponential Equations

The first technique involves two functions with like bases. Recall that the one-to-one property of exponential functions tells us that, for any real numbers b, S, and T, where $b > 0,\ \ b \neq 1$, $b^S = b^T$ if and only if $S = T$.

In other words, when an exponential equation has the same base on each side, the exponents must be equal. This also applies when the exponents are algebraic expressions. Therefore, we can solve many exponential equations by using the rules of exponents to rewrite each side as a power with the same base. Then, we use the fact that exponential functions are one-to-one to set the exponents equal to one another, and solve for the unknown.

For example, consider the equation $3^{4x-7} = \frac{3^{2x}}{3}$. To solve for x, we use the division property of exponents to rewrite the right side so that both sides have the common base, 3. Then we apply the one-to-one property of exponents by setting the exponents equal to one another and solving for x:

$$\begin{aligned} 3^{4x-7} &= \frac{3^{2x}}{3} && \\ 3^{4x-7} &= \frac{3^{2x}}{3^1} && \text{Rewrite 3 as } 3^1. \\ 3^{4x-7} &= 3^{2x-1} && \text{Use the division property of exponents.} \\ 4x-7 &= 2x-1 && \text{Apply the one-to-one property of exponents.} \\ 2x &= 6 && \text{Subtract } 2x \text{ and add 7 to both sides.} \\ x &= 3 && \text{Divide by 3.} \end{aligned}$$

Using the One-to-One Property of Exponential Functions to Solve Exponential Equations

For any algebraic expressions $S$and T, and any positive real number $b \neq 1$,

$$b^S = b^T \text{ if and only if } S = T$$

HOW TO

Given an exponential equation with the form $b^S = b^T$, where S and T are algebraic expressions with an unknown, solve for the unknown.

1. Use the rules of exponents to simplify, if necessary, so that the resulting equation has the form $b^S = b^T$.
2. Use the one-to-one property to set the exponents equal.
3. Solve the resulting equation, $S = T$, for the unknown.

EXAMPLE 1

Solving an Exponential Equation with a Common Base

Solve $2^{x-1} = 2^{2x-4}$.

Solution

$$\begin{aligned} 2^{x-1} &= 2^{2x-4} && \text{The common base is 2.} \\ x-1 &= 2x-4 && \text{By the one-to-one property the exponents must be equal.} \\ x &= 3 && \text{Solve for } x. \end{aligned}$$

TRY IT #1 Solve $5^{2x} = 5^{3x+2}$.

Rewriting Equations So All Powers Have the Same Base

Sometimes the common base for an exponential equation is not explicitly shown. In these cases, we simply rewrite the terms in the equation as powers with a common base, and solve using the one-to-one property.

For example, consider the equation $256 = 4^{x-5}$. We can rewrite both sides of this equation as a power of 2. Then we apply the rules of exponents, along with the one-to-one property, to solve for x :

$$\begin{aligned} 256 &= 4^{x-5} && \\ 2^8 &= \left(2^2\right)^{x-5} && \text{Rewrite each side as a power with base 2.} \\ 2^8 &= 2^{2x-10} && \text{Use the one-to-one property of exponents.} \\ 8 &= 2x-10 && \text{Apply the one-to-one property of exponents.} \\ 18 &= 2x && \text{Add 10 to both sides.} \\ x &= 9 && \text{Divide by 2.} \end{aligned}$$

HOW TO

Given an exponential equation with unlike bases, use the one-to-one property to solve it.

1. Rewrite each side in the equation as a power with a common base.
2. Use the rules of exponents to simplify, if necessary, so that the resulting equation has the form $b^S = b^T$.
3. Use the one-to-one property to set the exponents equal.
4. Solve the resulting equation, $S = T$, for the unknown.

EXAMPLE 2

Solving Equations by Rewriting Them to Have a Common Base

Solve $8^{x+2} = 16^{x+1}$.

Solution

$8^{x+2} = 16^{x+1}$	
$\left(2^3\right)^{x+2} = \left(2^4\right)^{x+1}$	Write 8 and 16 as powers of 2.
$2^{3x+6} = 2^{4x+4}$	To take a power of a power, multiply exponents.
$3x + 6 = 4x + 4$	Use the one-to-one property to set the exponents equal.
$x = 2$	Solve for x.

TRY IT #2 Solve $5^{2x} = 25^{3x+2}$.

EXAMPLE 3

Solving Equations by Rewriting Roots with Fractional Exponents to Have a Common Base

Solve $2^{5x} = \sqrt{2}$.

Solution

$2^{5x} = 2^{\frac{1}{2}}$	Write the square root of 2 as a power of 2.
$x = \frac{1}{2}$	Use the one-to-one property.
$x = \frac{1}{10}$	Solve for x.

TRY IT #3 Solve $5^x = \sqrt{5}$.

Q&A **Do all exponential equations have a solution? If not, how can we tell if there is a solution during the problem-solving process?**

No. Recall that the range of an exponential function is always positive. While solving the equation, we may obtain an expression that is undefined.

EXAMPLE 4

Solving an Equation with Positive and Negative Powers

Solve $3^{x+1} = -2$.

Solution

This equation has no solution. There is no real value of x that will make the equation a true statement because any power of a positive number is positive.

Analysis

Figure 2 shows that the two graphs do not cross so the left side is never equal to the right side. Thus the equation has no solution.

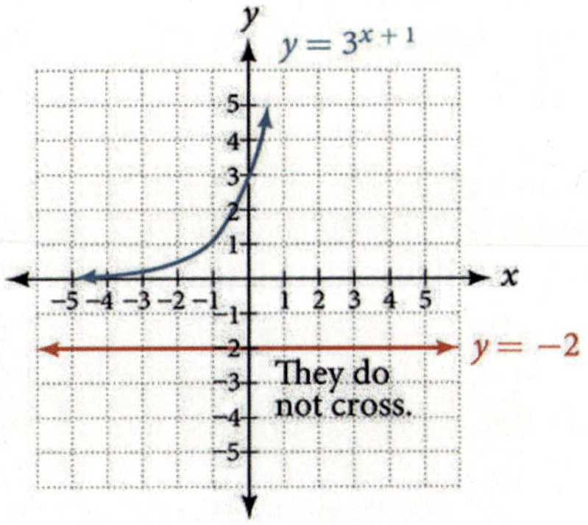

Figure 2

TRY IT #4 Solve $2^x = -100$.

Solving Exponential Equations Using Logarithms

Sometimes the terms of an exponential equation cannot be rewritten with a common base. In these cases, we solve by taking the logarithm of each side. Recall, since $\log(a) = \log(b)$ is equivalent to $a = b$, we may apply logarithms with the same base on both sides of an exponential equation.

HOW TO

Given an exponential equation in which a common base cannot be found, solve for the unknown.

1. Apply the logarithm of both sides of the equation.
   a. If one of the terms in the equation has base 10, use the common logarithm.
   b. If none of the terms in the equation has base 10, use the natural logarithm.
2. Use the rules of logarithms to solve for the unknown.

EXAMPLE 5

Solving an Equation Containing Powers of Different Bases

Solve $5^{x+2} = 4^x$.

Solution

$5^{x+2} = 4^x$	There is no easy way to get the powers to have the same base.
$\ln 5^{x+2} = \ln 4^x$	Take ln of both sides.
$(x+2)\ln 5 = x\ln 4$	Use laws of logs.
$x\ln 5 + 2\ln 5 = x\ln 4$	Use the distributive law.
$x\ln 5 - x\ln 4 = -2\ln 5$	Get terms containing x on one side, terms without x on the other.
$x(\ln 5 - \ln 4) = -2\ln 5$	On the left hand side, factor out an$x$.
$x\ln\left(\frac{5}{4}\right) = \ln\left(\frac{1}{25}\right)$	Use the laws of logs.
$x = \dfrac{\ln\left(\frac{1}{25}\right)}{\ln\left(\frac{5}{4}\right)}$	Divide by the coefficient of x.

TRY IT #5 Solve $2^x = 3^{x+1}$.

Q&A **Is there any way to solve $2^x = 3^x$?**

Yes. The solution is 0.

Equations Containing *e*

One common type of exponential equations are those with base e. This constant occurs again and again in nature, in mathematics, in science, in engineering, and in finance. When we have an equation with a base e on either side, we can use the natural logarithm to solve it.

HOW TO

Given an equation of the form $y = Ae^{kt}$, solve for t.

1. Divide both sides of the equation by A.
2. Apply the natural logarithm of both sides of the equation.
3. Divide both sides of the equation by k.

EXAMPLE 6

Solve an Equation of the Form $y = Ae^{kt}$

Solve $100 = 20e^{2t}$.

Solution

$$\begin{aligned} 100 &= 20e^{2t} && \\ 5 &= e^{2t} && \text{Divide by the coefficient of the power.} \\ \ln 5 &= 2t && \text{Take ln of both sides. Use the fact that } \ln(x) \text{ and } e^x \text{ are inverse functions.} \\ t &= \frac{\ln 5}{2} && \text{Divide by the coefficient of } t. \end{aligned}$$

Analysis

Using laws of logs, we can also write this answer in the form $t = \ln \sqrt{5}$. If we want a decimal approximation of the answer, we use a calculator.

TRY IT #6 Solve $3e^{0.5t} = 11$.

Q&A **Does every equation of the form $y = Ae^{kt}$ have a solution?**

No. There is a solution when $k \neq 0$, *and when y and* A *are either both 0 or neither 0, and they have the same sign. An example of an equation with this form that has no solution is* $2 = -3e^t$.

EXAMPLE 7

Solving an Equation That Can Be Simplified to the Form $y = Ae^{kt}$

Solve $4e^{2x} + 5 = 12$.

Solution

$$4e^{2x} + 5 = 12$$

$$4e^{2x} = 7 \quad \text{Combine like terms.}$$

$$e^{2x} = \frac{7}{4} \quad \text{Divide by the coefficient of the power.}$$

$$2x = \ln\left(\frac{7}{4}\right) \quad \text{Take ln of both sides.}$$

$$x = \frac{1}{2}\ln\left(\frac{7}{4}\right) \quad \text{Solve for } x.$$

TRY IT #7 Solve $3 + e^{2t} = 7e^{2t}$.

Extraneous Solutions

Sometimes the methods used to solve an equation introduce an **extraneous solution**, which is a solution that is correct algebraically but does not satisfy the conditions of the original equation. One such situation arises in solving when the logarithm is taken on both sides of the equation. In such cases, remember that the argument of the logarithm must be positive. If the number we are evaluating in a logarithm function is negative, there is no output.

EXAMPLE 8

Solving Exponential Functions in Quadratic Form

Solve $e^{2x} - e^x = 56$.

Solution

$e^{2x} - e^x = 56$	
$e^{2x} - e^x - 56 = 0$	Get one side of the equation equal to zero.
$(e^x + 7)(e^x - 8) = 0$	Factor by the FOIL method.
$e^x + 7 = 0 \text{ or} e^x - 8 = 0$	If a product is zero, then one factor must be zero.
$e^x = -7 \text{ or } e^x = 8$	Isolate the exponentials.
$e^x = 8$	Reject the equation in which the power equals a negative number.
$x = \ln 8$	Solve the equation in which the power equals a positive number.

Analysis

When we plan to use factoring to solve a problem, we always get zero on one side of the equation, because zero has the unique property that when a product is zero, one or both of the factors must be zero. We reject the equation $e^x = -7$ because a positive number never equals a negative number. The solution $\ln(-7)$ is not a real number, and in the real number system this solution is rejected as an extraneous solution.

TRY IT #8 Solve $e^{2x} = e^x + 2$.

Q&A **Does every logarithmic equation have a solution?**

No. Keep in mind that we can only apply the logarithm to a positive number. Always check for extraneous solutions.

Using the Definition of a Logarithm to Solve Logarithmic Equations

We have already seen that every logarithmic equation $\log_b(x) = y$ is equivalent to the exponential equation $b^y = x$. We can use this fact, along with the rules of logarithms, to solve logarithmic equations where the argument is an algebraic expression.

For example, consider the equation $\log_2(2) + \log_2(3x - 5) = 3$. To solve this equation, we can use rules of logarithms to rewrite the left side in compact form and then apply the definition of logs to solve for x :

$\log_2(2) + \log_2(3x-5) = 3$	
$\log_2(2(3x-5)) = 3$	Apply the product rule of logarithms.
$\log_2(6x-10) = 3$	Distribute.
$2^3 = 6x - 10$	Apply the definition of a logarithm.
$8 = 6x - 10$	Calculate2^3.
$18 = 6x$	Add 10 to both sides.
$x = 3$	Divide by 6.

Using the Definition of a Logarithm to Solve Logarithmic Equations

For any algebraic expression S and real numbers b and c, where $b > 0,\ \ b \neq 1$,

$$\log_b(S) = c \quad \text{if and only if} \quad b^c = S$$

EXAMPLE 9

Using Algebra to Solve a Logarithmic Equation

Solve $2\ln x + 3 = 7$.

Solution

$2\ln x + 3 = 7$	
$2\ln x = 4$	Subtract 3.
$\ln x = 2$	Divide by 2.
$x = e^2$	Rewrite in exponential form.

TRY IT #9 Solve $6 + \ln x = 10$.

EXAMPLE 10

Using Algebra Before and After Using the Definition of the Natural Logarithm

Solve $2\ln(6x) = 7$.

Solution

$2\ln(6x) = 7$	
$\ln(6x) = \frac{7}{2}$	Divide by 2.
$6x = e^{\left(\frac{7}{2}\right)}$	Use the definition of ln.
$x = \frac{1}{6}e^{\left(\frac{7}{2}\right)}$	Divide by 6.

TRY IT #10 Solve $2\ln(x+1) = 10$.

EXAMPLE 11

Using a Graph to Understand the Solution to a Logarithmic Equation

Solve $\ln x = 3$.

Solution

$\ln x = 3$	
$x = e^3$	Use the definition of the natural logarithm.

Figure 3 represents the graph of the equation. On the graph, the x-coordinate of the point at which the two graphs intersect is close to 20. In other words $e^3 \approx 20$. A calculator gives a better approximation: $e^3 \approx 20.0855$.

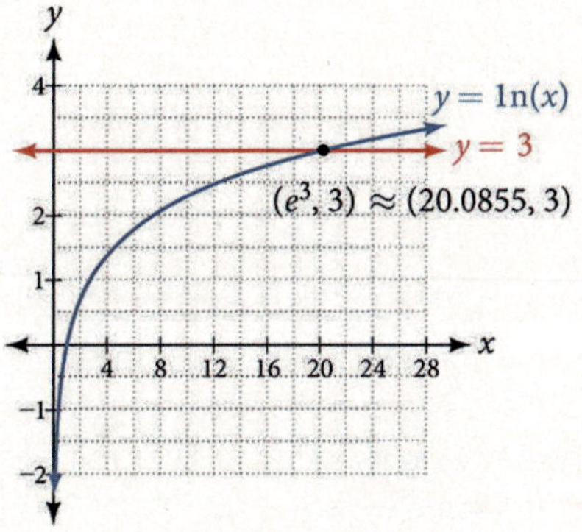

Figure 3 The graphs of $y = \ln x$ and $y = 3$ cross at the point $(e^3, 3)$, which is approximately (20.0855, 3).

TRY IT #11 Use a graphing calculator to estimate the approximate solution to the logarithmic equation $2^x = 1000$ to 2 decimal places.

Using the One-to-One Property of Logarithms to Solve Logarithmic Equations

As with exponential equations, we can use the one-to-one property to solve logarithmic equations. The one-to-one property of logarithmic functions tells us that, for any real numbers $x > 0$, $S > 0$, $T > 0$ and any positive real number b, where $b \neq 1$,

$$\log_b S = \log_b T \text{ if and only if } S = T.$$

For example,

$$\text{If } \log_2(x-1) = \log_2(8), \text{ then } x - 1 = 8.$$

So, if $x - 1 = 8$, then we can solve for x, and we get $x = 9$. To check, we can substitute $x = 9$ into the original equation: $\log_2(9-1) = \log_2(8) = 3$. In other words, when a logarithmic equation has the same base on each side, the arguments must be equal. This also applies when the arguments are algebraic expressions. Therefore, when given an equation with logs of the same base on each side, we can use rules of logarithms to rewrite each side as a single logarithm. Then we use the fact that logarithmic functions are one-to-one to set the arguments equal to one another and solve for the unknown.

For example, consider the equation $\log(3x-2) - \log(2) = \log(x+4)$. To solve this equation, we can use the rules of logarithms to rewrite the left side as a single logarithm, and then apply the one-to-one property to solve for x :

$$\begin{array}{ll} \log(3x-2) - \log(2) = \log(x+4) & \\ \log\left(\frac{3x-2}{2}\right) = \log(x+4) & \text{Apply the quotient rule of logarithms.} \\ \frac{3x-2}{2} = x+4 & \text{Apply the one to one property of a logarithm.} \\ 3x - 2 = 2x + 8 & \text{Multiply both sides of the equation by 2.} \\ x = 10 & \text{Subtract } 2x \text{ and add 2.} \end{array}$$

To check the result, substitute $x = 10$ into $\log(3x-2) - \log(2) = \log(x+4)$.

$$\begin{array}{ll} \log(3(10)-2) - \log(2) = \log((10)+4) & \\ \log(28) - \log(2) = \log(14) & \\ \log\left(\frac{28}{2}\right) = \log(14) & \text{The solution checks.} \end{array}$$

Using the One-to-One Property of Logarithms to Solve Logarithmic Equations

For any algebraic expressions S and T and any positive real number b, where $b \neq 1$,

$$\log_b S = \log_b T \quad \text{if and only if} \quad S = T$$

Note, when solving an equation involving logarithms, always check to see if the answer is correct or if it is an extraneous solution.

HOW TO

Given an equation containing logarithms, solve it using the one-to-one property.

1. Use the rules of logarithms to combine like terms, if necessary, so that the resulting equation has the form $\log_b S = \log_b T$.
2. Use the one-to-one property to set the arguments equal.
3. Solve the resulting equation, $S = T$, for the unknown.

EXAMPLE 12

Solving an Equation Using the One-to-One Property of Logarithms

Solve $\ln(x^2) = \ln(2x + 3)$.

Solution

$\ln(x^2) = \ln(2x+3)$	
$x^2 = 2x + 3$	Use the one-to-one property of the logarithm.
$x^2 - 2x - 3 = 0$	Get zero on one side before factoring.
$(x-3)(x+1) = 0$	Factor using FOIL.
$x - 3 = 0 \text{ or } x + 1 = 0$	If a product is zero, one of the factors must be zero.
$x = 3 \text{ or } x = -1$	Solve for x.

Analysis

There are two solutions: 3 or -1. The solution -1 is negative, but it checks when substituted into the original equation because the argument of the logarithm functions is still positive.

TRY IT #12 Solve $\ln(x^2) = \ln 1$.

Solving Applied Problems Using Exponential and Logarithmic Equations

In previous sections, we learned the properties and rules for both exponential and logarithmic functions. We have seen that any exponential function can be written as a logarithmic function and vice versa. We have used exponents to solve logarithmic equations and logarithms to solve exponential equations. We are now ready to combine our skills to solve equations that model real-world situations, whether the unknown is in an exponent or in the argument of a logarithm.

One such application is in science, in calculating the time it takes for half of the unstable material in a sample of a radioactive substance to decay, called its half-life. Table 1 lists the half-life for several of the more common radioactive substances.

Substance	Use	Half-life
gallium-67	nuclear medicine	80 hours
cobalt-60	manufacturing	5.3 years
technetium-99m	nuclear medicine	6 hours

Table 1

Substance	Use	Half-life
americium-241	construction	432 years
carbon-14	archeological dating	5,715 years
uranium-235	atomic power	703,800,000 years

Table 1

We can see how widely the half-lives for these substances vary. Knowing the half-life of a substance allows us to calculate the amount remaining after a specified time. We can use the formula for radioactive decay:

$$\begin{aligned} A(t) &= A_0 e^{\frac{\ln(0.5)}{T}t} \\ A(t) &= A_0 e^{\ln(0.5)\frac{t}{T}} \\ A(t) &= A_0 (e^{\ln(0.5)})^{\frac{t}{T}} \\ A(t) &= A_0 \left(\tfrac{1}{2}\right)^{\frac{t}{T}} \end{aligned}$$

where

- A_0 is the amount initially present
- T is the half-life of the substance
- t is the time period over which the substance is studied
- $A(t)$ is the amount of the substance present after time t

EXAMPLE 13

Using the Formula for Radioactive Decay to Find the Quantity of a Substance
How long will it take for ten percent of a 1000-gram sample of uranium-235 to decay?

Solution

$$\begin{aligned} y &= 1000e^{\frac{\ln(0.5)}{703,800,000}t} && \\ 900 &= 1000e^{\frac{\ln(0.5)}{703,800,000}t} && \text{After 10\% decays, 900 grams are left.} \\ 0.9 &= e^{\frac{\ln(0.5)}{703,800,000}t} && \text{Divide by 1000.} \\ \ln(0.9) &= \ln\left(e^{\frac{\ln(0.5)}{703,800,000}t}\right) && \text{Take ln of both sides.} \\ \ln(0.9) &= \frac{\ln(0.5)}{703,800,000}t && \ln(e^M) = M \\ t &= 703,800,000 \times \frac{\ln(0.9)}{\ln(0.5)} \text{ years} && \text{Solve for } t. \\ t &\approx 106,979,777 \text{ years} && \end{aligned}$$

Analysis
Ten percent of 1000 grams is 100 grams. If 100 grams decay, the amount of uranium-235 remaining is 900 grams.

TRY IT #13 How long will it take before twenty percent of our 1000-gram sample of uranium-235 has decayed?

MEDIA

Access these online resources for additional instruction and practice with exponential and logarithmic equations.

Solving Logarithmic Equations (http://openstax.org/l/solvelogeq)

Solving Exponential Equations with Logarithms (http://openstax.org/l/solveexplog)

6.6 SECTION EXERCISES

Verbal

1. How can an exponential equation be solved?

2. When does an extraneous solution occur? How can an extraneous solution be recognized?

3. When can the one-to-one property of logarithms be used to solve an equation? When can it not be used?

Algebraic

For the following exercises, use like bases to solve the exponential equation.

4. $4^{-3v-2} = 4^{-v}$

5. $64 \cdot 4^{3x} = 16$

6. $3^{2x+1} \cdot 3^x = 243$

7. $2^{-3n} \cdot \frac{1}{4} = 2^{n+2}$

8. $625 \cdot 5^{3x+3} = 125$

9. $\frac{36^{3b}}{36^{2b}} = 216^{2-b}$

10. $\left(\frac{1}{64}\right)^{3n} \cdot 8 = 2^6$

For the following exercises, use logarithms to solve.

11. $9^{x-10} = 1$

12. $2e^{6x} = 13$

13. $e^{r+10} - 10 = -42$

14. $2 \cdot 10^{9a} = 29$

15. $-8 \cdot 10^{p+7} - 7 = -24$

16. $7e^{3n-5} + 5 = -89$

17. $e^{-3k} + 6 = 44$

18. $-5e^{9x-8} - 8 = -62$

19. $-6e^{9x+8} + 2 = -74$

20. $2^{x+1} = 5^{2x-1}$

21. $e^{2x} - e^x - 132 = 0$

22. $7e^{8x+8} - 5 = -95$

23. $10e^{8x+3} + 2 = 8$

24. $4e^{3x+3} - 7 = 53$

25. $8e^{-5x-2} - 4 = -90$

26. $3^{2x+1} = 7^{x-2}$

27. $e^{2x} - e^x - 6 = 0$

28. $3e^{3-3x} + 6 = -31$

For the following exercises, use the definition of a logarithm to rewrite the equation as an exponential equation.

29. $\log\left(\frac{1}{100}\right) = -2$

30. $\log_{324}(18) = \frac{1}{2}$

For the following exercises, use the definition of a logarithm to solve the equation.

31. $5\log_7 n = 10$

32. $-8\log_9 x = 16$

33. $4 + \log_2(9k) = 2$

34. $2\log(8n+4) + 6 = 10$

35. $10 - 4\ln(9 - 8x) = 6$

For the following exercises, use the one-to-one property of logarithms to solve.

36. $\ln(10 - 3x) = \ln(-4x)$

37. $\log_{13}(5n - 2) = \log_{13}(8 - 5n)$

38. $\log(x + 3) - \log(x) = \log(74)$

39. $\ln(-3x) = \ln\left(x^2 - 6x\right)$

40. $\log_4(6 - m) = \log_4 3m$

41. $\ln(x - 2) - \ln(x) = \ln(54)$

42. $\log_9\left(2n^2 - 14n\right) = \log_9\left(-45 + n^2\right)$

43. $\ln\left(x^2 - 10\right) + \ln(9) = \ln(10)$

For the following exercises, solve each equation for x.

44. $\log(x + 12) = \log(x) + \log(12)$

45. $\ln(x) + \ln(x - 3) = \ln(7x)$

46. $\log_2(7x + 6) = 3$

47. $\ln(7) + \ln\left(2 - 4x^2\right) = \ln(14)$

48. $\log_8(x + 6) - \log_8(x) = \log_8(58)$

49. $\ln(3) - \ln(3 - 3x) = \ln(4)$

50. $\log_3(3x) - \log_3(6) = \log_3(77)$

Graphical

For the following exercises, solve the equation for x, if there is a solution. Then graph both sides of the equation, and observe the point of intersection (if it exists) to verify the solution.

51. $\log_9(x) - 5 = -4$

52. $\log_3(x) + 3 = 2$

53. $\ln(3x) = 2$

54. $\ln(x - 5) = 1$

55. $\log(4) + \log(-5x) = 2$

56. $-7 + \log_3(4 - x) = -6$

57. $\ln(4x - 10) - 6 = -5$

58. $\log(4 - 2x) = \log(-4x)$

59. $\log_{11}\left(-2x^2 - 7x\right) = \log_{11}(x - 2)$

60. $\ln(2x + 9) = \ln(-5x)$

61. $\log_9(3 - x) = \log_9(4x - 8)$

62. $\log\left(x^2 + 13\right) = \log(7x + 3)$

63. $\frac{3}{\log_2(10)} - \log(x - 9) = \log(44)$

64. $\ln(x) - \ln(x + 3) = \ln(6)$

For the following exercises, solve for the indicated value, and graph the situation showing the solution point.

65. An account with an initial deposit of $6,500 earns 7.25% annual interest, compounded continuously. How much will the account be worth after 20 years?

66. The formula for measuring sound intensity in decibels D is defined by the equation $D = 10\log\left(\frac{I}{I_0}\right)$, where I is the intensity of the sound in watts per square meter and $I_0 = 10^{-12}$ is the lowest level of sound that the average person can hear. How many decibels are emitted from a jet plane with a sound intensity of $8.3 \cdot 10^2$ watts per square meter?

67. The population of a small town is modeled by the equation $P = 1650e^{0.5t}$ where t is measured in years. In approximately how many years will the town's population reach 20,000?

Technology

For the following exercises, solve each equation by rewriting the exponential expression using the indicated logarithm. Then use a calculator to approximate the variable to 3 decimal places.

68. $1000(1.03)^t = 5000$ using the common log.

69. $e^{5x} = 17$ using the natural log

70. $3(1.04)^{3t} = 8$ using the common log

71. $3^{4x-5} = 38$ using the common log

72. $50e^{-0.12t} = 10$ using the natural log

For the following exercises, use a calculator to solve the equation. Unless indicated otherwise, round all answers to the nearest ten-thousandth.

73. $7e^{3x-5} + 7.9 = 47$

74. $\ln(3) + \ln(4.4x + 6.8) = 2$

75. $\log(-0.7x - 9) = 1 + 5\log(5)$

76. Atmospheric pressure P in pounds per square inch is represented by the formula $P = 14.7e^{-0.21x}$, where x is the number of miles above sea level. To the nearest foot, how high is the peak of a mountain with an atmospheric pressure of 8.369 pounds per square inch? (*Hint*: there are 5280 feet in a mile)

77. The magnitude *M* of an earthquake is represented by the equation $M = \frac{2}{3}\log\left(\frac{E}{E_0}\right)$ where E is the amount of energy released by the earthquake in joules and $E_0 = 10^{4.4}$ is the assigned minimal measure released by an earthquake. To the nearest hundredth, what would the magnitude be of an earthquake releasing $1.4 \cdot 10^{13}$ joules of energy?

Extensions

78. Use the definition of a logarithm along with the one-to-one property of logarithms to prove that $b^{\log_b x} = x$.

79. Recall the formula for continually compounding interest, $y = Ae^{kt}$. Use the definition of a logarithm along with properties of logarithms to solve the formula for time t such that t is equal to a single logarithm.

80. Recall the compound interest formula $A = a\left(1 + \frac{r}{k}\right)^{kt}$. Use the definition of a logarithm along with properties of logarithms to solve the formula for time t.

81. Newton's Law of Cooling states that the temperature T of an object at any time t can be described by the equation $T = T_s + (T_0 - T_s)e^{-kt}$, where T_s is the temperature of the surrounding environment, T_0 is the initial temperature of the object, and k is the cooling rate. Use the definition of a logarithm along with properties of logarithms to solve the formula for time t such that t is equal to a single logarithm.

6.7 Exponential and Logarithmic Models

Learning Objectives

In this section, you will:

- Model exponential growth and decay.
- Use Newton's Law of Cooling.
- Use logistic-growth models.
- Choose an appropriate model for data.
- Express an exponential model in base e .

Figure 1 A nuclear research reactor inside the Neely Nuclear Research Center on the Georgia Institute of Technology campus (credit: Georgia Tech Research Institute)

We have already explored some basic applications of exponential and logarithmic functions. In this section, we explore some important applications in more depth, including radioactive isotopes and Newton's Law of Cooling.

Modeling Exponential Growth and Decay

In real-world applications, we need to model the behavior of a function. In mathematical modeling, we choose a familiar general function with properties that suggest that it will model the real-world phenomenon we wish to analyze. In the case of rapid growth, we may choose the exponential growth function:

$$y = A_0 e^{kt}$$

where A_0 is equal to the value at time zero, e is Euler's constant, and k is a positive constant that determines the rate (percentage) of growth. We may use the exponential growth function in applications involving **doubling time**, the time it takes for a quantity to double. Such phenomena as wildlife populations, financial investments, biological samples, and natural resources may exhibit growth based on a doubling time. In some applications, however, as we will see when we discuss the logistic equation, the logistic model sometimes fits the data better than the exponential model.

On the other hand, if a quantity is falling rapidly toward zero, without ever reaching zero, then we should probably choose the **exponential decay** model. Again, we have the form $y = A_0 e^{kt}$ where A_0 is the starting value, and e is Euler's constant. Now k is a negative constant that determines the rate of decay. We may use the exponential decay model when we are calculating **half-life**, or the time it takes for a substance to exponentially decay to half of its original quantity. We use half-life in applications involving radioactive isotopes.

In our choice of a function to serve as a mathematical model, we often use data points gathered by careful observation and measurement to construct points on a graph and hope we can recognize the shape of the graph. Exponential growth and decay graphs have a distinctive shape, as we can see in Figure 2 and Figure 3. It is important to remember that, although parts of each of the two graphs seem to lie on the x-axis, they are really a tiny distance above the x-axis.

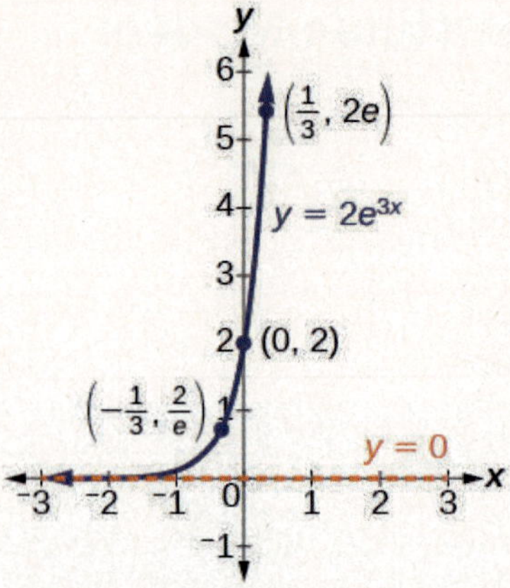

Figure 2 A graph showing exponential growth. The equation is $y = 2e^{3x}$.

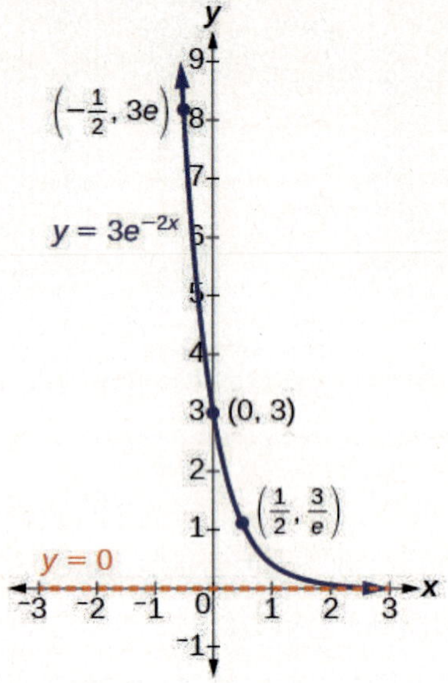

Figure 3 A graph showing exponential decay. The equation is $y = 3e^{-2x}$.

Exponential growth and decay often involve very large or very small numbers. To describe these numbers, we often use orders of magnitude. The **order of magnitude** is the power of ten, when the number is expressed in scientific notation, with one digit to the left of the decimal. For example, the distance to the nearest star, Proxima Centauri, measured in kilometers, is 40,113,497,200,000 kilometers. Expressed in scientific notation, this is $4.01134972 \times 10^{13}$. So, we could describe this number as having order of magnitude 10^{13}.

Characteristics of the Exponential Function, $y = A_0 e^{kt}$

An exponential function with the form $y = A_0 e^{kt}$ has the following characteristics:

- one-to-one function
- horizontal asymptote: $y = 0$
- domain: $(-\infty, \infty)$
- range: $(0, \infty)$
- x intercept: none
- y-intercept: $(0, A_0)$
- increasing if $k > 0$ (see Figure 4)
- decreasing if $k < 0$ (see Figure 4)

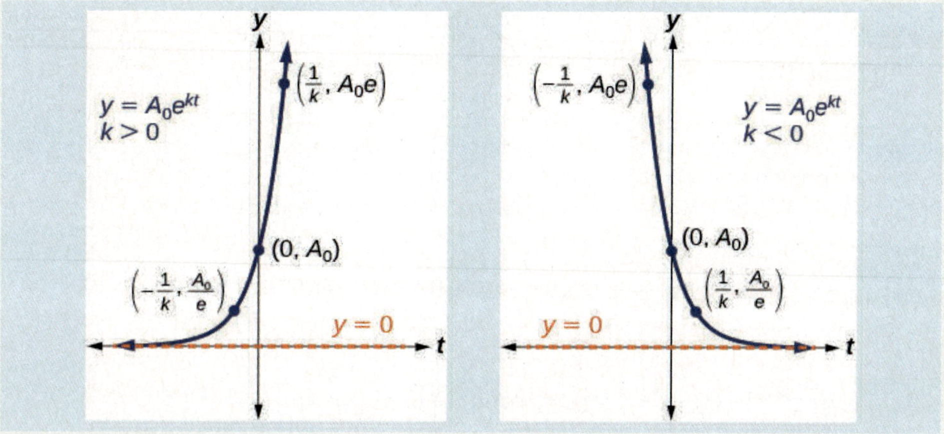

Figure 4 An exponential function models exponential growth when $k > 0$ and exponential decay when $k < 0$.

EXAMPLE 1

Graphing Exponential Growth

A population of bacteria doubles every hour. If the culture started with 10 bacteria, graph the population as a function of time.

Solution

When an amount grows at a fixed percent per unit time, the growth is exponential. To find A_0 we use the fact that A_0 is the amount at time zero, so $A_0 = 10$. To find k, use the fact that after one hour $(t = 1)$ the population doubles from 10 to 20. The formula is derived as follows

$$\begin{aligned} 20 &= 10e^{k \cdot 1} \\ 2 &= e^k && \text{Divide by 10} \\ \ln 2 &= k && \text{Take the natural logarithm} \end{aligned}$$

so $k = \ln(2)$. Thus the equation we want to graph is $y = 10e^{(\ln 2)t} = 10(e^{\ln 2})^t = 10 \cdot 2^t$. The graph is shown in Figure 5.

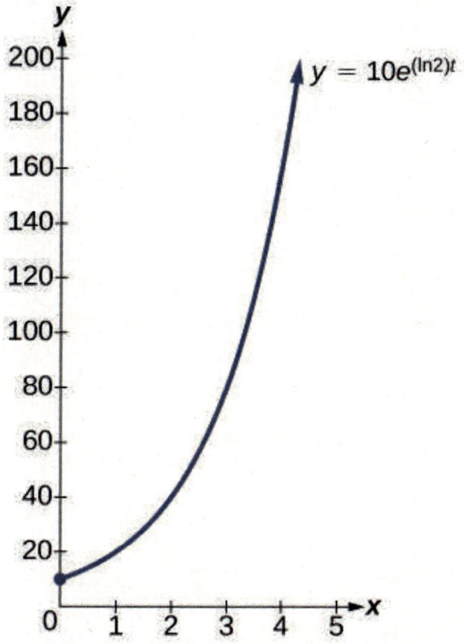

Figure 5 The graph of $y = 10e^{(\ln 2)t}$

Analysis

The population of bacteria after ten hours is 10,240. We could describe this amount is being of the order of magnitude 10^4. The population of bacteria after twenty hours is 10,485,760 which is of the order of magnitude 10^7, so we could say that the population has increased by three orders of magnitude in ten hours.

Half-Life

We now turn to **exponential decay**. One of the common terms associated with exponential decay, as stated above, is **half-life**, the length of time it takes an exponentially decaying quantity to decrease to half its original amount. Every radioactive isotope has a half-life, and the process describing the exponential decay of an isotope is called radioactive decay.

To find the half-life of a function describing exponential decay, solve the following equation:

$$\frac{1}{2}A_0 = A_o e^{kt}$$

We find that the half-life depends only on the constant k and not on the starting quantity A_0.

The formula is derived as follows

$\frac{1}{2}A_0 = A_o e^{kt}$	
$\frac{1}{2} = e^{kt}$	Divide by A_0.
$\ln\left(\frac{1}{2}\right) = kt$	Take the natural log.
$-\ln(2) = kt$	Apply laws of logarithms.
$-\frac{\ln(2)}{k} = t$	Divide by k.

Since t, the time, is positive, k must, as expected, be negative. This gives us the half-life formula

$$t = -\frac{\ln(2)}{k}$$

HOW TO

Given the half-life, find the decay rate.

1. Write $A = A_o e^{kt}$.
2. Replace A by $\frac{1}{2}A_0$ and replace t by the given half-life.
3. Solve to find k. Express k as an exact value (do not round).

Note: *It is also possible to find the decay rate using* $k = -\frac{\ln(2)}{t}$.

EXAMPLE 2

Finding the Function that Describes Radioactive Decay

The half-life of carbon-14 is 5,730 years. Express the amount of carbon-14 remaining as a function of time, t.

Solution

This formula is derived as follows.

$A = A_0 e^{kt}$	The continuous growth formula.
$0.5A_0 = A_0 e^{k \cdot 5730}$	Substitute the half-life for t and $0.5A_0$ for $f(t)$.
$0.5 = e^{5730k}$	Divide by A_0.
$\ln(0.5) = 5730k$	Take the natural log of both sides.
$k = \frac{\ln(0.5)}{5730}$	Divide by the coefficient of k.
$A = A_0 e^{\left(\frac{\ln(0.5)}{5730}\right)t}$	Substitute for r in the continuous growth formula.

The function that describes this continuous decay is $f(t) = A_0 e^{\left(\frac{\ln(0.5)}{5730}\right)t}$. We observe that the coefficient of t,

$\frac{\ln(0.5)}{5730} \approx -1.2097 \times 10^{-4}$ is negative, as expected in the case of exponential decay.

TRY IT #1 The half-life of plutonium-244 is 80,000,000 years. Find a function that gives the amount of plutonium-244 remaining as a function of time, measured in years.

Radiocarbon Dating

The formula for radioactive decay is important in radiocarbon dating, which is used to calculate the approximate date a plant or animal died. Radiocarbon dating was discovered in 1949 by Willard Libby, who won a Nobel Prize for his discovery. It compares the difference between the ratio of two isotopes of carbon in an organic artifact or fossil to the ratio of those two isotopes in the air. It is believed to be accurate to within about 1% error for plants or animals that died within the last 60,000 years.

Carbon-14 is a radioactive isotope of carbon that has a half-life of 5,730 years. It occurs in small quantities in the carbon dioxide in the air we breathe. Most of the carbon on Earth is carbon-12, which has an atomic weight of 12 and is not radioactive. Scientists have determined the ratio of carbon-14 to carbon-12 in the air for the last 60,000 years, using tree rings and other organic samples of known dates—although the ratio has changed slightly over the centuries.

As long as a plant or animal is alive, the ratio of the two isotopes of carbon in its body is close to the ratio in the atmosphere. When it dies, the carbon-14 in its body decays and is not replaced. By comparing the ratio of carbon-14 to carbon-12 in a decaying sample to the known ratio in the atmosphere, the date the plant or animal died can be approximated.

Since the half-life of carbon-14 is 5,730 years, the formula for the amount of carbon-14 remaining after t years is

$$A \approx A_0 e^{\left(\frac{\ln(0.5)}{5730}\right)t}$$

where

- A is the amount of carbon-14 remaining
- A_0 is the amount of carbon-14 when the plant or animal began decaying.

This formula is derived as follows:

$$\begin{array}{ll} A = A_0 e^{kt} & \text{The continuous growth formula.} \\ 0.5A_0 = A_0 e^{k \cdot 5730} & \text{Substitute the half-life for } t \text{ and } 0.5A_0 \text{ for } f(t). \\ 0.5 = e^{5730k} & \text{Divide by } A_0. \\ \ln(0.5) = 5730k & \text{Take the natural log of both sides.} \\ k = \frac{\ln(0.5)}{5730} & \text{Divide by the coefficient of } k. \\ A = A_0 e^{\left(\frac{\ln(0.5)}{5730}\right)t} & \text{Substitute for } k \text{ in the continuous growth formula.} \end{array}$$

To find the age of an object, we solve this equation for t :

$$t = \frac{\ln\left(\frac{A}{A_0}\right)}{-0.000121}$$

Out of necessity, we neglect here the many details that a scientist takes into consideration when doing carbon-14 dating, and we only look at the basic formula. The ratio of carbon-14 to carbon-12 in the atmosphere is approximately 0.0000000001%. Let r be the ratio of carbon-14 to carbon-12 in the organic artifact or fossil to be dated, determined by a method called liquid scintillation. From the equation $A \approx A_0 e^{-0.000121t}$ we know the ratio of the percentage of carbon-14 in the object we are dating to the initial amount of carbon-14 in the object when it was formed is $r = \frac{A}{A_0} \approx e^{-0.000121t}$. We solve this equation for t, to get

$$t = \frac{\ln(r)}{-0.000121}$$

HOW TO

Given the percentage of carbon-14 in an object, determine its age.

1. Express the given percentage of carbon-14 as an equivalent decimal, k.
2. Substitute for k in the equation $t = \frac{\ln(r)}{-0.000121}$ and solve for the age, t.

EXAMPLE 3

Finding the Age of a Bone

A bone fragment is found that contains 20% of its original carbon-14. To the nearest year, how old is the bone?

Solution

We substitute $20\% = 0.20$ for r in the equation and solve for t :

$$\begin{aligned} t &= \frac{\ln(r)}{-0.000121} && \text{Use the general form of the equation.} \\ &= \frac{\ln(0.20)}{-0.000121} && \text{Substitute for } r. \\ &\approx 13301 && \text{Round to the nearest year.} \end{aligned}$$

The bone fragment is about 13,301 years old.

Analysis

The instruments that measure the percentage of carbon-14 are extremely sensitive and, as we mention above, a scientist will need to do much more work than we did in order to be satisfied. Even so, carbon dating is only accurate to about 1%, so this age should be given as 13,301 years $\pm$ 1% or 13,301 years $\pm$ 133 years.

TRY IT #2 Cesium-137 has a half-life of about 30 years. If we begin with 200 mg of cesium-137, will it take more or less than 230 years until only 1 milligram remains?

Calculating Doubling Time

For decaying quantities, we determined how long it took for half of a substance to decay. For growing quantities, we might want to find out how long it takes for a quantity to double. As we mentioned above, the time it takes for a quantity to double is called the **doubling time**.

Given the basic exponential growth equation $A = A_0 e^{kt}$, doubling time can be found by solving for when the original quantity has doubled, that is, by solving $2A_0 = A_0 e^{kt}$.

The formula is derived as follows:

$$\begin{aligned} 2A_0 &= A_0 e^{kt} && \\ 2 &= e^{kt} && \text{Divide by } A_0. \\ \ln 2 &= kt && \text{Take the natural logarithm.} \\ t &= \frac{\ln 2}{k} && \text{Divide by the coefficient of } t. \end{aligned}$$

Thus the doubling time is

$$t = \frac{\ln 2}{k}$$

EXAMPLE 4

Finding a Function That Describes Exponential Growth

According to Moore's Law, the doubling time for the number of transistors that can be put on a computer chip is approximately two years. Give a function that describes this behavior.

Solution

The formula is derived as follows:

$t = \frac{\ln 2}{k}$	The doubling time formula.
$2 = \frac{\ln 2}{k}$	Use a doubling time of two years.
$k = \frac{\ln 2}{2}$	Multiply by k and divide by 2.
$A = A_0 e^{\frac{\ln 2}{2} t}$	Substitute k into the continuous growth formula.

The function is $A_0 e^{\frac{\ln 2}{2} t}$.

TRY IT #3 Recent data suggests that, as of 2013, the rate of growth predicted by Moore's Law no longer holds. Growth has slowed to a doubling time of approximately three years. Find the new function that takes that longer doubling time into account.

Using Newton's Law of Cooling

Exponential decay can also be applied to temperature. When a hot object is left in surrounding air that is at a lower temperature, the object's temperature will decrease exponentially, leveling off as it approaches the surrounding air temperature. On a graph of the temperature function, the leveling off will correspond to a horizontal asymptote at the temperature of the surrounding air. Unless the room temperature is zero, this will correspond to a vertical shift of the generic **exponential decay** function. This translation leads to **Newton's Law of Cooling**, the scientific formula for temperature as a function of time as an object's temperature is equalized with the ambient temperature

$$T(t) = ae^{kt} + T_s$$

This formula is derived as follows:

$T(t) = Ab^{ct} + T_s$	
$T(t) = Ae^{\ln(b^{ct})} + T_s$	Laws of logarithms.
$T(t) = Ae^{ct \ln b} + T_s$	Laws of logarithms.
$T(t) = Ae^{kt} + T_s$	Rename the constant $c \ln b$, calling it k.

Newton's Law of Cooling

The temperature of an object, T, in surrounding air with temperature T_s will behave according to the formula

$$T(t) = Ae^{kt} + T_s$$

where

- t is time
- A is the difference between the initial temperature of the object and the surroundings
- k is a constant, the continuous rate of cooling of the object

HOW TO

Given a set of conditions, apply Newton's Law of Cooling.

1. Set T_s equal to the y-coordinate of the horizontal asymptote (usually the ambient temperature).
2. Substitute the given values into the continuous growth formula $T(t) = Ae^{kt} + T_s$ to find the parameters A and k.
3. Substitute in the desired time to find the temperature or the desired temperature to find the time.

EXAMPLE 5

Using Newton's Law of Cooling

A cheesecake is taken out of the oven with an ideal internal temperature of $165°F$, and is placed into a $35°F$ refrigerator. After 10 minutes, the cheesecake has cooled to $150°F$. If we must wait until the cheesecake has cooled to $70°F$ before we eat it, how long will we have to wait?

Solution

Because the surrounding air temperature in the refrigerator is 35 degrees, the cheesecake's temperature will decay exponentially toward 35, following the equation

$$T(t) = Ae^{kt} + 35$$

We know the initial temperature was 165, so $T(0) = 165$.

$$\begin{aligned} 165 &= Ae^{k0} + 35 && \text{Substitute } (0, 165). \\ A &= 130 && \text{Solve for } A. \end{aligned}$$

We were given another data point, $T(10) = 150$, which we can use to solve for k.

$$\begin{aligned} 150 &= 130e^{k10} + 35 && \text{Substitute } (10, 150). \\ 115 &= 130e^{k10} && \text{Subtract 35.} \\ \frac{115}{130} &= e^{10k} && \text{Divide by 130.} \\ \ln\left(\frac{115}{130}\right) &= 10k && \text{Take the natural log of both sides.} \\ k &= \frac{\ln\left(\frac{115}{130}\right)}{10} \approx -0.0123 && \text{Divide by the coefficient of} k. \end{aligned}$$

This gives us the equation for the cooling of the cheesecake: $T(t) = 130e^{-0.0123t} + 35$.

Now we can solve for the time it will take for the temperature to cool to 70 degrees.

$$\begin{aligned} 70 &= 130e^{-0.0123t} + 35 && \text{Substitute in 70 for} T(t). \\ 35 &= 130e^{-0.0123t} && \text{Subtract 35.} \\ \frac{35}{130} &= e^{-0.0123t} && \text{Divide by 130.} \\ \ln\left(\frac{35}{130}\right) &= -0.0123t && \text{Take the natural log of both sides} \\ t &= \frac{\ln\left(\frac{35}{130}\right)}{-0.0123} \approx 106.68 && \text{Divide by the coefficient of} t. \end{aligned}$$

It will take about 107 minutes, or one hour and 47 minutes, for the cheesecake to cool to $70°F$.

> **TRY IT** #4 A pitcher of water at 40 degrees Fahrenheit is placed into a 70 degree room. One hour later, the temperature has risen to 45 degrees. How long will it take for the temperature to rise to 60 degrees?

Using Logistic Growth Models

Exponential growth cannot continue forever. Exponential models, while they may be useful in the short term, tend to fall apart the longer they continue. Consider an aspiring writer who writes a single line on day one and plans to double the number of lines she writes each day for a month. By the end of the month, she must write over 17 billion lines, or one-half-billion pages. It is impractical, if not impossible, for anyone to write that much in such a short period of time. Eventually, an exponential model must begin to approach some limiting value, and then the growth is forced to slow. For this reason, it is often better to use a model with an upper bound instead of an exponential growth model, though the exponential growth model is still useful over a short term, before approaching the limiting value.

The **logistic growth model** is approximately exponential at first, but it has a reduced rate of growth as the output approaches the model's upper bound, called the **carrying capacity**. For constants a, b, and c, the logistic growth of a population over time t is represented by the model

$$f(t) = \frac{c}{1 + ae^{-bt}}$$

The graph in Figure 6 shows how the growth rate changes over time. The graph increases from left to right, but the growth rate only increases until it reaches its point of maximum growth rate, at which point the rate of increase decreases.

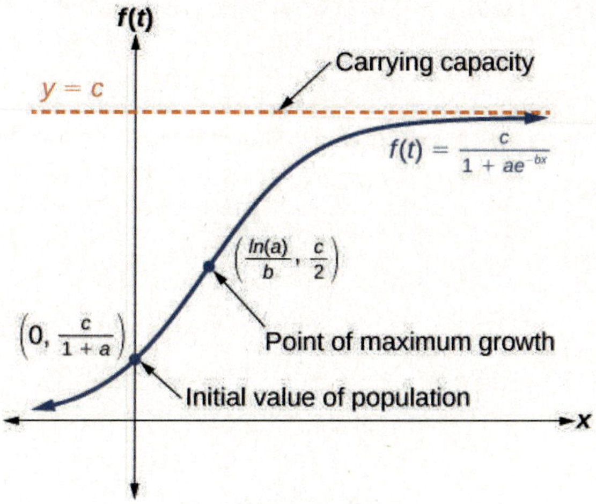

Figure 6

Logistic Growth

The logistic growth model is

$$f(t) = \frac{c}{1 + ae^{-bt}}$$

where

- $\frac{c}{1+a}$ is the initial value
- c is the *carrying capacity*, or *limiting value*
- b is a constant determined by the rate of growth.

EXAMPLE 6

Using the Logistic-Growth Model

An influenza epidemic spreads through a population rapidly, at a rate that depends on two factors: The more people who have the flu, the more rapidly it spreads, and also the more uninfected people there are, the more rapidly it spreads. These two factors make the logistic model a good one to study the spread of communicable diseases. And, clearly, there is a maximum value for the number of people infected: the entire population.

For example, at time $t = 0$ there is one person in a community of 1,000 people who has the flu. So, in that community, at most 1,000 people can have the flu. Researchers find that for this particular strain of the flu, the logistic growth constant is $b = 0.6030$. Estimate the number of people in this community who will have had this flu after ten days. Predict how many people in this community will have had this flu after a long period of time has passed.

Solution

We substitute the given data into the logistic growth model

$$f(t) = \frac{c}{1 + ae^{-bt}}$$

Because at most 1,000 people, the entire population of the community, can get the flu, we know the limiting value is $c = 1000$. To find a, we use the formula that the number of cases at time $t = 0$ is $\frac{c}{1+a} = 1$, from which it follows that $a = 999$. This model predicts that, after ten days, the number of people who have had the flu is $f(t) = \frac{1000}{1+999e^{-0.6030x}} \approx 293.8$. Because the actual number must be a whole number (a person has either had the flu or not) we round to 294. In the long term, the number of people who will contract the flu is the limiting value, $c = 1000$.

Analysis

Remember that, because we are dealing with a virus, we cannot predict with certainty the number of people infected. The model only approximates the number of people infected and will not give us exact or actual values.

The graph in Figure 7 gives a good picture of how this model fits the data.

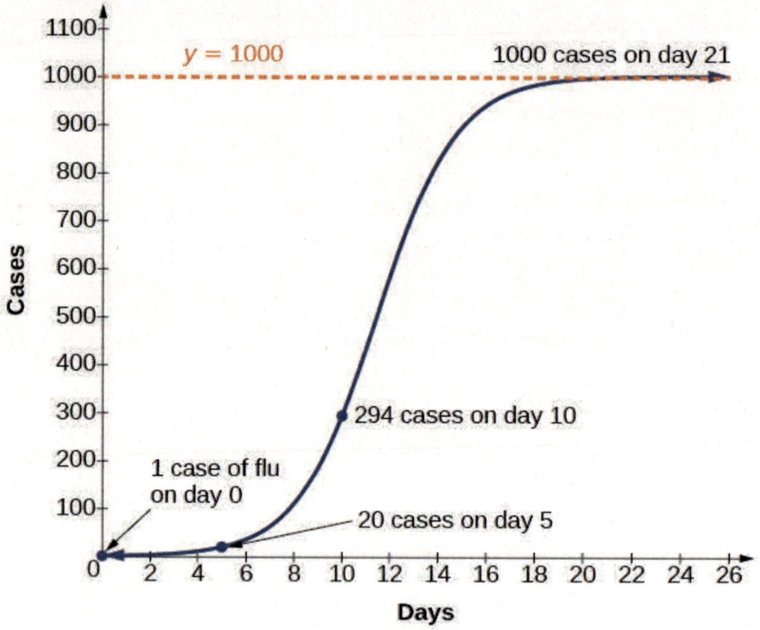

Figure 7 The graph of $f(t) = \frac{1000}{1+999e^{-0.6030x}}$

TRY IT #5 Using the model in Example 6, estimate the number of cases of flu on day 15.

Choosing an Appropriate Model for Data

Now that we have discussed various mathematical models, we need to learn how to choose the appropriate model for the raw data we have. Many factors influence the choice of a mathematical model, among which are experience, scientific laws, and patterns in the data itself. Not all data can be described by elementary functions. Sometimes, a function is chosen that approximates the data over a given interval. For instance, suppose data were gathered on the number of homes bought in the United States from the years 1960 to 2013. After plotting these data in a scatter plot, we notice that the shape of the data from the years 2000 to 2013 follow a logarithmic curve. We could restrict the interval from 2000 to 2010, apply regression analysis using a logarithmic model, and use it to predict the number of home buyers for the year 2015.

Three kinds of functions that are often useful in mathematical models are linear functions, exponential functions, and logarithmic functions. If the data lies on a straight line, or seems to lie approximately along a straight line, a linear model may be best. If the data is non-linear, we often consider an exponential or logarithmic model, though other models, such as quadratic models, may also be considered.

In choosing between an exponential model and a logarithmic model, we look at the way the data curves. This is called the concavity. If we draw a line between two data points, and all (or most) of the data between those two points lies above that line, we say the curve is concave down. We can think of it as a bowl that bends downward and therefore cannot hold water. If all (or most) of the data between those two points lies below the line, we say the curve is concave up. In this case, we can think of a bowl that bends upward and can therefore hold water. An exponential curve, whether rising or falling, whether representing growth or decay, is always concave up away from its horizontal asymptote. A logarithmic curve is always concave away from its vertical asymptote. In the case of positive data, which is the most common case, an exponential curve is always concave up, and a logarithmic curve always concave down.

A logistic curve changes concavity. It starts out concave up and then changes to concave down beyond a certain point, called a point of inflection.

After using the graph to help us choose a type of function to use as a model, we substitute points, and solve to find the parameters. We reduce round-off error by choosing points as far apart as possible.

EXAMPLE 7

Choosing a Mathematical Model

Does a linear, exponential, logarithmic, or logistic model best fit the values listed in Table 1? Find the model, and use a graph to check your choice.

x	1	2	3	4	5	6	7	8	9
y	0	1.386	2.197	2.773	3.219	3.584	3.892	4.159	4.394

Table 1

Solution

First, plot the data on a graph as in Figure 8. For the purpose of graphing, round the data to two decimal places.

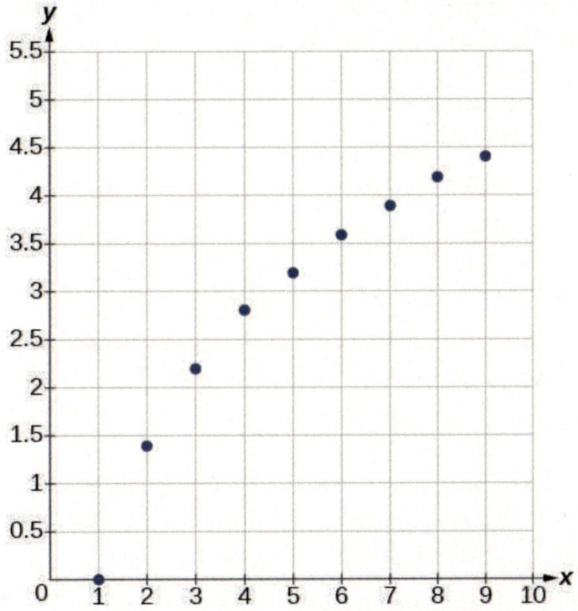

Figure 8

Clearly, the points do not lie on a straight line, so we reject a linear model. If we draw a line between any two of the points, most or all of the points between those two points lie above the line, so the graph is concave down, suggesting a logarithmic model. We can try $y = a\ln(bx)$. Plugging in the first point, $(1,0)$, gives $0 = a\ln b$. We reject the case that $a = 0$ (if it were, all outputs would be 0), so we know $\ln(b) = 0$. Thus $b = 1$ and $y = a\ln(x)$. Next we can use the point $(9,4.394)$ to solve for a :

$$\begin{aligned} y &= a\ln(x) \\ 4.394 &= a\ln(9) \\ a &= \frac{4.394}{\ln(9)} \end{aligned}$$

Because $a = \frac{4.394}{\ln(9)} \approx 2$, an appropriate model for the data is $y = 2\ln(x)$.

To check the accuracy of the model, we graph the function together with the given points as in Figure 9.

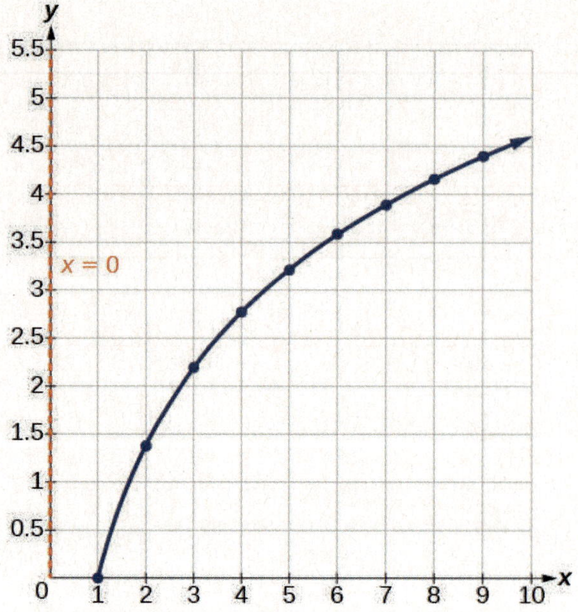

Figure 9 The graph of $y = 2\ln x$.

We can conclude that the model is a good fit to the data.

Compare Figure 9 to the graph of $y = \ln\left(x^2\right)$ shown in Figure 10.

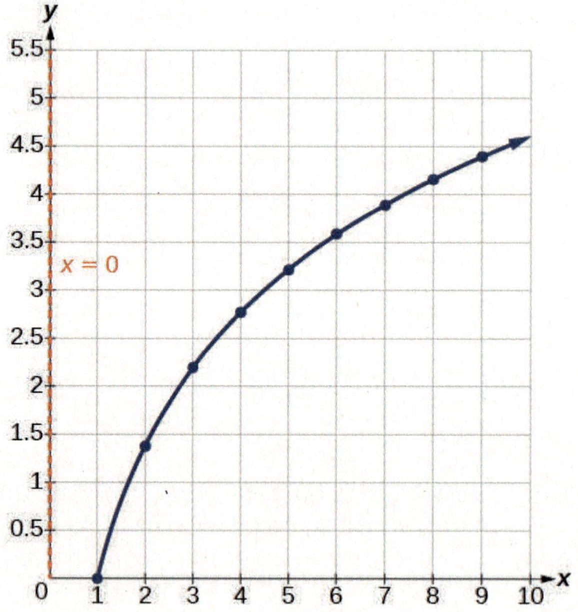

Figure 10 The graph of $y = \ln\left(x^2\right)$

The graphs appear to be identical when $x > 0$. A quick check confirms this conclusion: $y = \ln\left(x^2\right) = 2\ln(x)$ for $x > 0$.

However, if $x < 0$, the graph of $y = \ln\left(x^2\right)$ includes a "extra" branch, as shown in Figure 11. This occurs because, while $y = 2\ln(x)$ cannot have negative values in the domain (as such values would force the argument to be negative), the function $y = \ln\left(x^2\right)$ can have negative domain values.

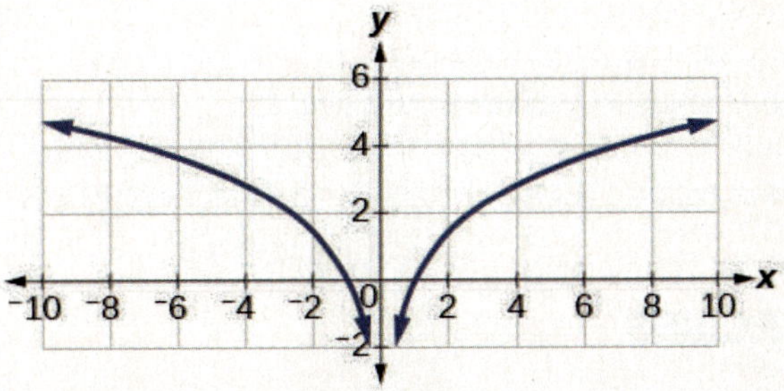

Figure 11

> **TRY IT** #6 Does a linear, exponential, or logarithmic model best fit the data in Table 2? Find the model.

x	1	2	3	4	5	6	7	8	9
y	3.297	5.437	8.963	14.778	24.365	40.172	66.231	109.196	180.034

Table 2

Expressing an Exponential Model in Base e

While powers and logarithms of any base can be used in modeling, the two most common bases are 10 and e. In science and mathematics, the base e is often preferred. We can use laws of exponents and laws of logarithms to change any base to base e.

HOW TO

Given a model with the form $y = ab^x$, change it to the form $y = A_0 e^{kx}$.

1. Rewrite $y = ab^x$ as $y = ae^{\ln(b^x)}$.
2. Use the power rule of logarithms to rewrite y as $y = ae^{x\ln(b)} = ae^{\ln(b)x}$.
3. Note that $a = A_0$ and $k = \ln(b)$ in the equation $y = A_0 e^{kx}$.

EXAMPLE 8

Changing to base e

Change the function $y = 2.5(3.1)^x$ so that this same function is written in the form $y = A_0 e^{kx}$.

Solution

The formula is derived as follows

$$
\begin{aligned}
y &= 2.5(3.1)^x \\
&= 2.5e^{\ln(3.1^x)} && \text{Insert exponential and its inverse.} \\
&= 2.5e^{x\ln 3.1} && \text{Laws of logs.} \\
&= 2.5e^{(\ln 3.1)x} && \text{Commutative law of multiplication}
\end{aligned}
$$

> **TRY IT** #7 Change the function $y = 3(0.5)^x$ to one having e as the base.

MEDIA

Access these online resources for additional instruction and practice with exponential and logarithmic models.

Logarithm Application – pH (http://openstax.org/l/logph)
Exponential Model – Age Using Half-Life (http://openstax.org/l/expmodelhalf)
Newton's Law of Cooling (http://openstax.org/l/newtoncooling)
Exponential Growth Given Doubling Time (http://openstax.org/l/expgrowthdbl)
Exponential Growth – Find Initial Amount Given Doubling Time (http://openstax.org/l/initialdouble)

6.7 SECTION EXERCISES

Verbal

1. With what kind of exponential model would *half-life* be associated? What role does half-life play in these models?

2. What is carbon dating? Why does it work? Give an example in which carbon dating would be useful.

3. With what kind of exponential model would *doubling time* be associated? What role does doubling time play in these models?

4. Define Newton's Law of Cooling. Then name at least three real-world situations where Newton's Law of Cooling would be applied.

5. What is an order of magnitude? Why are orders of magnitude useful? Give an example to explain.

Numeric

6. The temperature of an object in degrees Fahrenheit after t minutes is represented by the equation $T(t) = 68e^{-0.0174t} + 72$. To the nearest degree, what is the temperature of the object after one and a half hours?

For the following exercises, use the logistic growth model $f(x) = \frac{150}{1+8e^{-2x}}$.

7. Find and interpret $f(0)$. Round to the nearest tenth.

8. Find and interpret $f(4)$. Round to the nearest tenth.

9. Find the carrying capacity.

10. Graph the model.

11. Determine whether the data from the table could best be represented as a function that is linear, exponential, or logarithmic. Then write a formula for a model that represents the data.

x	$f(x)$
–2	0.694
–1	0.833
0	1
1	1.2
2	1.44
3	1.728
4	2.074
5	2.488

12. Rewrite $f(x) = 1.68(0.65)^x$ as an exponential equation with base e to five decimal places.

Technology

For the following exercises, enter the data from each table into a graphing calculator and graph the resulting scatter plots. Determine whether the data from the table could represent a function that is linear, exponential, or logarithmic.

13.

x	$f(x)$
1	2
2	4.079
3	5.296
4	6.159
5	6.828
6	7.375
7	7.838
8	8.238
9	8.592
10	8.908

14.

x	$f(x)$
1	2.4
2	2.88
3	3.456
4	4.147
5	4.977
6	5.972
7	7.166
8	8.6
9	10.32
10	12.383

15.

x	$f(x)$
4	9.429
5	9.972
6	10.415
7	10.79
8	11.115
9	11.401
10	11.657
11	11.889
12	12.101
13	12.295

16.

x	$f(x)$
1.25	5.75
2.25	8.75
3.56	12.68
4.2	14.6
5.65	18.95
6.75	22.25
7.25	23.75
8.6	27.8
9.25	29.75
10.5	33.5

For the following exercises, use a graphing calculator and this scenario: the population of a fish farm in t years is modeled by the equation $P(t) = \frac{1000}{1+9e^{-0.6t}}$.

17. Graph the function.

18. What is the initial population of fish?

19. To the nearest tenth, what is the doubling time for the fish population?

20. To the nearest whole number, what will the fish population be after 2 years?

21. To the nearest tenth, how long will it take for the population to reach 900?

22. What is the carrying capacity for the fish population? Justify your answer using the graph of P.

Extensions

23. A substance has a half-life of 2.045 minutes. If the initial amount of the substance was 132.8 grams, how many half-lives will have passed before the substance decays to 8.3 grams? What is the total time of decay?

24. The formula for an increasing population is given by $P(t) = P_0 e^{rt}$ where P_0 is the initial population and $r > 0$. Derive a general formula for the time t it takes for the population to increase by a factor of M.

25. Recall the formula for calculating the magnitude of an earthquake, $M = \frac{2}{3}\log\left(\frac{S}{S_0}\right)$. Show each step for solving this equation algebraically for the seismic moment S.

26. What is the y-intercept of the logistic growth model $y = \frac{c}{1+ae^{-rx}}$? Show the steps for calculation. What does this point tell us about the population?

27. Prove that $b^x = e^{x\ln(b)}$ for positive $b \neq 1$.

Real-World Applications

For the following exercises, use this scenario: A doctor prescribes 125 milligrams of a therapeutic drug that decays by about 30% each hour.

28. To the nearest hour, what is the half-life of the drug?

29. Write an exponential model representing the amount of the drug remaining in the patient's system after t hours. Then use the formula to find the amount of the drug that would remain in the patient's system after 3 hours. Round to the nearest milligram.

30. Using the model found in the previous exercise, find $f(10)$ and interpret the result. Round to the nearest hundredth.

For the following exercises, use this scenario: A tumor is injected with 0.5 grams of Iodine-125, which has a decay rate of 1.15% per day.

31. To the nearest day, how long will it take for half of the Iodine-125 to decay?

32. Write an exponential model representing the amount of Iodine-125 remaining in the tumor after t days. Then use the formula to find the amount of Iodine-125 that would remain in the tumor after 60 days. Round to the nearest tenth of a gram.

33. A scientist begins with 250 grams of a radioactive substance. After 250 minutes, the sample has decayed to 32 grams. Rounding to five decimal places, write an exponential equation representing this situation. To the nearest minute, what is the half-life of this substance?

34. The half-life of Radium-226 is 1590 years. What is the annual decay rate? Express the decimal result to four decimal places and the percentage to two decimal places.

35. The half-life of Erbium-165 is 10.4 hours. What is the hourly decay rate? Express the decimal result to four decimal places and the percentage to two decimal places.

36. A wooden artifact from an archeological dig contains 60 percent of the carbon-14 that is present in living trees. To the nearest year, about how many years old is the artifact? (The half-life of carbon-14 is 5730 years.)

37. A research student is working with a culture of bacteria that doubles in size every twenty minutes. The initial population count was 1350 bacteria. Rounding to five decimal places, write an exponential equation representing this situation. To the nearest whole number, what is the population size after 3 hours?

For the following exercises, use this scenario: A biologist recorded a count of 360 bacteria present in a culture after 5 minutes and 1000 bacteria present after 20 minutes.

38. To the nearest whole number, what was the initial population in the culture?

39. Rounding to six decimal places, write an exponential equation representing this situation. To the nearest minute, how long did it take the population to double?

For the following exercises, use this scenario: A pot of warm soup with an internal temperature of $100°$ Fahrenheit was taken off the stove to cool in a $69°$ F room. After fifteen minutes, the internal temperature of the soup was $95°$ F.

40. Use Newton's Law of Cooling to write a formula that models this situation.

41. To the nearest minute, how long will it take the soup to cool to $80°$ F?

42. To the nearest degree, what will the temperature be after 2 and a half hours?

For the following exercises, use this scenario: A turkey is taken out of the oven with an internal temperature of $165°$F and is allowed to cool in a $75°$F room. After half an hour, the internal temperature of the turkey is $145°$F.

43. Write a formula that models this situation.

44. To the nearest degree, what will the temperature be after 50 minutes?

45. To the nearest minute, how long will it take the turkey to cool to $110°$ F?

For the following exercises, find the value of the number shown on each logarithmic scale. Round all answers to the nearest thousandth.

46.

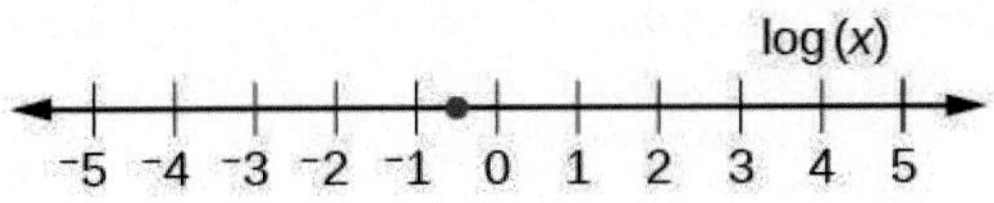

47.

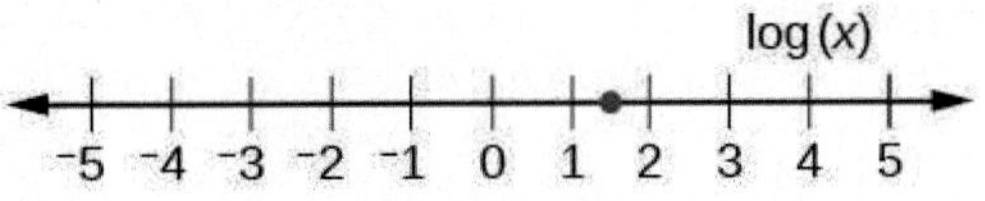

48. Plot each set of approximate values of intensity of sounds on a logarithmic scale: Whisper: $10^{-10}\ \frac{W}{m^2}$, Vacuum: $10^{-4}\ \frac{W}{m^2}$, Jet: $10^2\ \frac{W}{m^2}$

49. Recall the formula for calculating the magnitude of an earthquake, $M = \frac{2}{3}\log\left(\frac{S}{S_0}\right)$. One earthquake has magnitude 3.9 on the MMS scale. If a second earthquake has 750 times as much energy as the first, find the magnitude of the second quake. Round to the nearest hundredth.

For the following exercises, use this scenario: The equation $N(t) = \frac{500}{1+49e^{-0.7t}}$ models the number of people in a town who have heard a rumor after t days.

50. How many people started the rumor?

51. To the nearest whole number, how many people will have heard the rumor after 3 days?

52. As *t* increases without bound, what value does $N(t)$ approach? Interpret your answer.

For the following exercise, choose the correct answer choice.

53. A doctor injects a patient with 13 milligrams of radioactive dye that decays exponentially. After 12 minutes, there are 4.75 milligrams of dye remaining in the patient's system. Which is an appropriate model for this situation?

ⓐ $f(t) = 13(0.0805)^t$ ⓑ $f(t) = 13e^{0.9195t}$

ⓒ $f(t) = 13e^{(-0.0839t)}$

ⓓ $f(t) = \frac{4.75}{1+13e^{-0.83925t}}$

6.8 Fitting Exponential Models to Data

Learning Objectives

In this section, you will:

- Build an exponential model from data.
- Build a logarithmic model from data.
- Build a logistic model from data.

In previous sections of this chapter, we were either given a function explicitly to graph or evaluate, or we were given a set of points that were guaranteed to lie on the curve. Then we used algebra to find the equation that fit the points exactly. In this section, we use a modeling technique called *regression analysis* to find a curve that models data collected from real-world observations. With regression analysis, we don't expect all the points to lie perfectly on the curve. The idea is to find a model that best fits the data. Then we use the model to make predictions about future events.

Do not be confused by the word *model*. In mathematics, we often use the terms *function*, *equation*, and *model* interchangeably, even though they each have their own formal definition. The term *model* is typically used to indicate that the equation or function approximates a real-world situation.

We will concentrate on three types of regression models in this section: exponential, logarithmic, and logistic. Having already worked with each of these functions gives us an advantage. Knowing their formal definitions, the behavior of their graphs, and some of their real-world applications gives us the opportunity to deepen our understanding. As each regression model is presented, key features and definitions of its associated function are included for review. Take a moment to rethink each of these functions, reflect on the work we've done so far, and then explore the ways regression is used to model real-world phenomena.

Building an Exponential Model from Data

As we've learned, there are a multitude of situations that can be modeled by exponential functions, such as investment growth, radioactive decay, atmospheric pressure changes, and temperatures of a cooling object. What do these phenomena have in common? For one thing, all the models either increase or decrease as time moves forward. But that's not the whole story. It's the *way* data increase or decrease that helps us determine whether it is best modeled by an exponential equation. Knowing the behavior of exponential functions in general allows us to recognize when to use exponential regression, so let's review exponential growth and decay.

Recall that exponential functions have the form $y = ab^x$ or $y = A_0e^{kx}$. When performing regression analysis, we use the form most commonly used on graphing utilities, $y = ab^x$. Take a moment to reflect on the characteristics we've already learned about the exponential function $y = ab^x$ (assume $a > 0$) :

- b must be greater than zero and not equal to one.
- The initial value of the model is $y = a$.
  - If $b > 1$, the function models exponential growth. As x increases, the outputs of the model increase slowly at first, but then increase more and more rapidly, without bound.
  - If $0 < b < 1$, the function models exponential decay. As x increases, the outputs for the model decrease rapidly at first and then level off to become asymptotic to the *x*-axis. In other words, the outputs never become equal to or less than zero.

As part of the results, your calculator will display a number known as the *correlation coefficient*, labeled by the variable r, or r^2. (You may have to change the calculator's settings for these to be shown.) The values are an indication of the "goodness of fit" of the regression equation to the data. We more commonly use the value of r^2 instead of r, but the closer either value is to 1, the better the regression equation approximates the data.

Exponential Regression

Exponential regression is used to model situations in which growth begins slowly and then accelerates rapidly without bound, or where decay begins rapidly and then slows down to get closer and closer to zero. We use the command "**ExpReg**" on a graphing utility to fit an exponential function to a set of data points. This returns an equation of the form, $y = ab^x$

Note that:

- b must be non-negative.
- when $b > 1$, we have an exponential growth model.

- when $0 < b < 1$, we have an exponential decay model.

HOW TO

Given a set of data, perform exponential regression using a graphing utility.

1. Use the **STAT** then **EDIT** menu to enter given data.
   a. Clear any existing data from the lists.
   b. List the input values in the L1 column.
   c. List the output values in the L2 column.
2. Graph and observe a scatter plot of the data using the **STATPLOT** feature.
   a. Use **ZOOM** [**9**] to adjust axes to fit the data.
   b. Verify the data follow an exponential pattern.
3. Find the equation that models the data.
   a. Select "**ExpReg**" from the **STAT** then **CALC** menu.
   b. Use the values returned for *a* and *b* to record the model, $y = ab^x$.
4. Graph the model in the same window as the scatterplot to verify it is a good fit for the data.

EXAMPLE 1

Using Exponential Regression to Fit a Model to Data

In 2007, a university study was published investigating the crash risk of alcohol impaired driving. Data from 2,871 crashes were used to measure the association of a person's blood alcohol level (BAC) with the risk of being in an accident. Table 1 shows results from the study [9]. The *relative risk* is a measure of how many times more likely a person is to crash. So, for example, a person with a BAC of 0.09 is 3.54 times as likely to crash as a person who has not been drinking alcohol.

BAC	0	0.01	0.03	0.05	0.07	0.09
Relative Risk of Crashing	1	1.03	1.06	1.38	2.09	3.54
BAC	0.11	0.13	0.15	0.17	0.19	0.21
Relative Risk of Crashing	6.41	12.6	22.1	39.05	65.32	99.78

Table 1

a. Let x represent the BAC level, and let y represent the corresponding relative risk. Use exponential regression to fit a model to these data.
b. After 6 drinks, a person weighing 160 pounds will have a BAC of about 0.16. How many times more likely is a person with this weight to crash if they drive after having a 6-pack of beer? Round to the nearest hundredth.

Solution

a. Using the **STAT** then **EDIT** menu on a graphing utility, list the **BAC** values in L1 and the relative risk values in L2. Then use the **STATPLOT** feature to verify that the scatterplot follows the exponential pattern shown in Figure 1:

9 Source: *Indiana University Center for Studies of Law in Action, 2007*

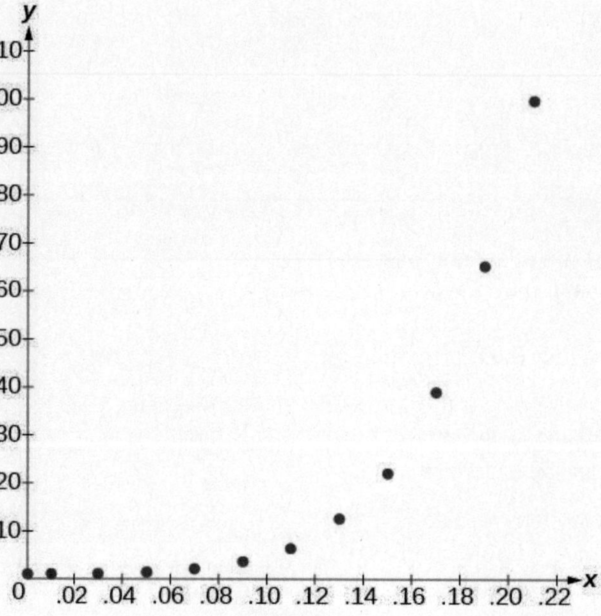

Figure 1

Use the "**ExpReg**" command from the **STAT** then **CALC** menu to obtain the exponential model,

$$y = 0.58304829(2.20720213\text{E}10)^x$$

Converting from scientific notation, we have:

$$y = 0.58304829(22{,}072{,}021{,}300)^x$$

Notice that $r^2 \approx 0.97$ which indicates the model is a good fit to the data. To see this, graph the model in the same window as the scatterplot to verify it is a good fit as shown in Figure 2:

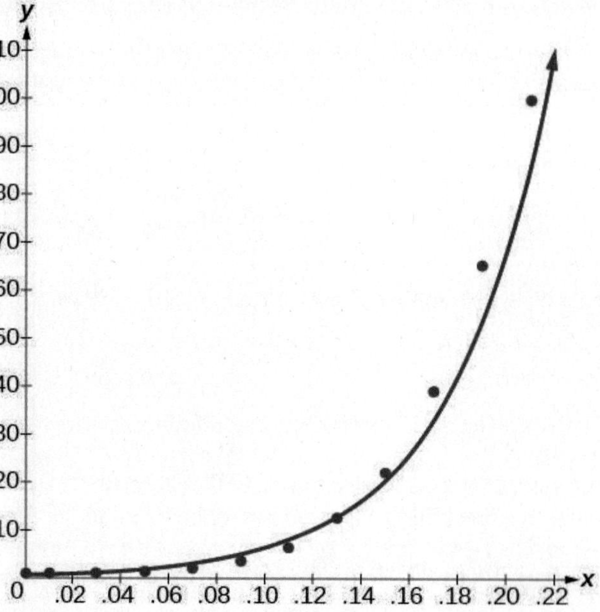

Figure 2

b. Use the model to estimate the risk associated with a BAC of 0.16. Substitute 0.16 for x in the model and solve for y.

$y = 0.58304829(22{,}072{,}021{,}300)^x$	Use the regression model found in part (a).
$= 0.58304829(22{,}072{,}021{,}300)^{0.16}$	Substitute 0.16 for x.
≈ 26.35	Round to the nearest hundredth.

If a 160-pound person drives after having 6 drinks, they are about 26.35 times more likely to crash than if driving while sober.

TRY IT #1 Table 2 shows a recent graduate's credit card balance each month after graduation.

Month	1	2	3	4	5	6	7	8
Debt ($)	620.00	761.88	899.80	1039.93	1270.63	1589.04	1851.31	2154.92

Table 2

ⓐ Use exponential regression to fit a model to these data.

ⓑ If spending continues at this rate, what will the graduate's credit card debt be one year after graduating?

Q&A **Is it reasonable to assume that an exponential regression model will represent a situation indefinitely?**

No. Remember that models are formed by real-world data gathered for regression. It is usually reasonable to make estimates within the interval of original observation (interpolation). However, when a model is used to make predictions, it is important to use reasoning skills to determine whether the model makes sense for inputs far beyond the original observation interval (extrapolation).

Building a Logarithmic Model from Data

Just as with exponential functions, there are many real-world applications for logarithmic functions: intensity of sound, pH levels of solutions, yields of chemical reactions, production of goods, and growth of infants. As with exponential models, data modeled by logarithmic functions are either always increasing or always decreasing as time moves forward. Again, it is the *way* they increase or decrease that helps us determine whether a logarithmic model is best.

Recall that logarithmic functions increase or decrease rapidly at first, but then steadily slow as time moves on. By reflecting on the characteristics we've already learned about this function, we can better analyze real world situations that reflect this type of growth or decay. When performing logarithmic regression analysis, we use the form of the logarithmic function most commonly used on graphing utilities, $y = a + b\ln(x)$. For this function

- All input values, x, must be greater than zero.
- The point $(1, a)$ is on the graph of the model.
- If $b > 0$, the model is increasing. Growth increases rapidly at first and then steadily slows over time.
- If $b < 0$, the model is decreasing. Decay occurs rapidly at first and then steadily slows over time.

Logarithmic Regression

Logarithmic regression is used to model situations where growth or decay accelerates rapidly at first and then slows over time. We use the command "LnReg" on a graphing utility to fit a logarithmic function to a set of data points. This returns an equation of the form,

$$y = a + b\ln(x)$$

Note that

- all input values, x, must be non-negative.
- when $b > 0$, the model is increasing.
- when $b < 0$, the model is decreasing.

HOW TO

Given a set of data, perform logarithmic regression using a graphing utility.

1. Use the **STAT** then **EDIT** menu to enter given data.
   a. Clear any existing data from the lists.
   b. List the input values in the L1 column.
   c. List the output values in the L2 column.
2. Graph and observe a scatter plot of the data using the **STATPLOT** feature.
   a. Use **ZOOM** [9] to adjust axes to fit the data.
   b. Verify the data follow a logarithmic pattern.
3. Find the equation that models the data.
   a. Select **"LnReg"** from the **STAT** then **CALC** menu.
   b. Use the values returned for *a* and *b* to record the model, $y = a + b\ln(x)$.
4. Graph the model in the same window as the scatterplot to verify it is a good fit for the data.

EXAMPLE 2

Using Logarithmic Regression to Fit a Model to Data

Due to advances in medicine and higher standards of living, life expectancy has been increasing in most developed countries since the beginning of the 20th century.

Table 3 shows the average life expectancies, in years, of Americans from 1900–2010[10].

Year	1900	1910	1920	1930	1940	1950
Life Expectancy(Years)	47.3	50.0	54.1	59.7	62.9	68.2
Year	1960	1970	1980	1990	2000	2010
Life Expectancy(Years)	69.7	70.8	73.7	75.4	76.8	78.7

Table 3

ⓐ Let x represent time in decades starting with $x = 1$ for the year 1900, $x = 2$ for the year 1910, and so on. Let y represent the corresponding life expectancy. Use logarithmic regression to fit a model to these data.
ⓑ Use the model to predict the average American life expectancy for the year 2030.

10 Source: *Center for Disease Control and Prevention, 2013*

Solution

ⓐ Using the **STAT** then **EDIT** menu on a graphing utility, list the years using values 1–12 in L1 and the corresponding life expectancy in L2. Then use the **STATPLOT** feature to verify that the scatterplot follows a logarithmic pattern as shown in Figure 3:

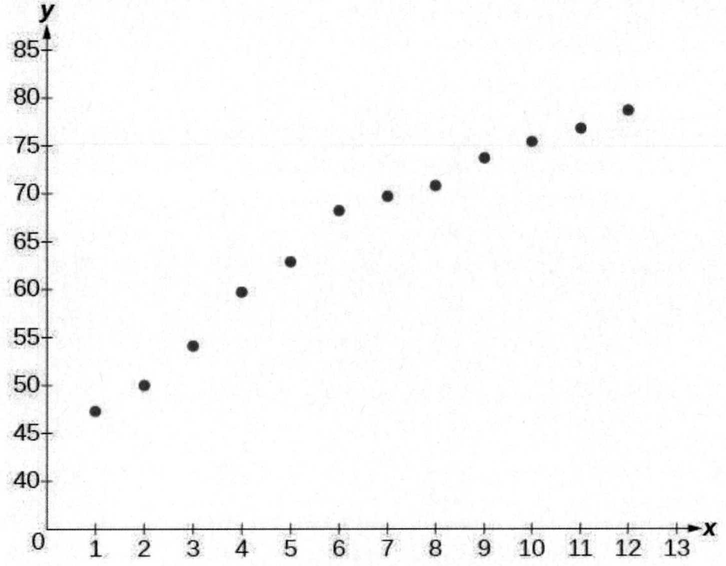

Figure 3

Use the "**LnReg**" command from the **STAT** then **CALC** menu to obtain the logarithmic model,

$$y = 42.52722583 + 13.85752327\ln(x)$$

Next, graph the model in the same window as the scatterplot to verify it is a good fit as shown in Figure 4:

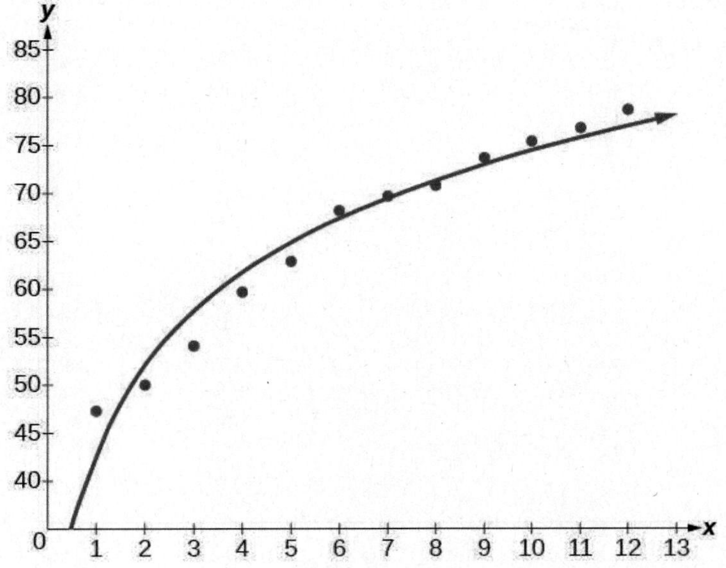

Figure 4

ⓑ To predict the life expectancy of an American in the year 2030, substitute $x = 14$ for the in the model and solve for y :

$$\begin{aligned} y &= 42.52722583 + 13.85752327\ln(x) && \text{Use the regression model found in part (a).} \\ &= 42.52722583 + 13.85752327\ln(14) && \text{Substitute 14 for } x. \\ &\approx 79.1 && \text{Round to the nearest tenth.} \end{aligned}$$

If life expectancy continues to increase at this pace, the average life expectancy of an American will be 79.1 by the year 2030.

> **TRY IT** #2 Sales of a video game released in the year 2000 took off at first, but then steadily slowed as time moved on. Table 4 shows the number of games sold, in thousands, from the years 2000–2010.

Year	2000	2001	2002	2003	2004	2005
Number Sold (thousands)	142	149	154	155	159	161
Year	2006	2007	2008	2009	2010	-
Number Sold (thousands)	163	164	164	166	167	-

Table 4

ⓐ Let x represent time in years starting with $x = 1$ for the year 2000. Let y represent the number of games sold in thousands. Use logarithmic regression to fit a model to these data.
ⓑ If games continue to sell at this rate, how many games will sell in 2015? Round to the nearest thousand.

Building a Logistic Model from Data

Like exponential and logarithmic growth, logistic growth increases over time. One of the most notable differences with logistic growth models is that, at a certain point, growth steadily slows and the function approaches an upper bound, or *limiting value*. Because of this, logistic regression is best for modeling phenomena where there are limits in expansion, such as availability of living space or nutrients.

It is worth pointing out that logistic functions actually model resource-limited exponential growth. There are many examples of this type of growth in real-world situations, including population growth and spread of disease, rumors, and even stains in fabric. When performing logistic regression analysis, we use the form most commonly used on graphing utilities:

$$y = \frac{c}{1 + ae^{-bx}}$$

Recall that:

- $\frac{c}{1+a}$ is the initial value of the model.
- when $b > 0$, the model increases rapidly at first until it reaches its point of maximum growth rate, $\left(\frac{\ln(a)}{b}, \frac{c}{2}\right)$. At that point, growth steadily slows and the function becomes asymptotic to the upper bound $y = c$.
- c is the limiting value, sometimes called the *carrying capacity*, of the model.

Logistic Regression

Logistic regression is used to model situations where growth accelerates rapidly at first and then steadily slows to an upper limit. We use the command "Logistic" on a graphing utility to fit a logistic function to a set of data points. This returns an equation of the form

$$y = \frac{c}{1 + ae^{-bx}}$$

Note that

- The initial value of the model is $\frac{c}{1+a}$.
- Output values for the model grow closer and closer to $y = c$ as time increases.

HOW TO

Given a set of data, perform logistic regression using a graphing utility.

1. Use the **STAT** then **EDIT** menu to enter given data.
   a. Clear any existing data from the lists.
   b. List the input values in the L1 column.
   c. List the output values in the L2 column.
2. Graph and observe a scatter plot of the data using the **STATPLOT** feature.
   a. Use **ZOOM** [9] to adjust axes to fit the data.
   b. Verify the data follow a logistic pattern.
3. Find the equation that models the data.
   a. Select "**Logistic**" from the **STAT** then **CALC** menu.
   b. Use the values returned for a, b, and c to record the model, $y = \frac{c}{1+ae^{-bx}}$.
4. Graph the model in the same window as the scatterplot to verify it is a good fit for the data.

EXAMPLE 3

Using Logistic Regression to Fit a Model to Data

Mobile telephone service has increased rapidly in America since the mid 1990s. Today, almost all residents have cellular service. Table 5 shows the percentage of Americans with cellular service between the years 1995 and 2012 [11].

Year	Americans with Cellular Service (%)	Year	Americans with Cellular Service (%)
1995	12.69	2004	62.852
1996	16.35	2005	68.63
1997	20.29	2006	76.64
1998	25.08	2007	82.47
1999	30.81	2008	85.68
2000	38.75	2009	89.14
2001	45.00	2010	91.86
2002	49.16	2011	95.28
2003	55.15	2012	98.17

Table 5

ⓐ Let x represent time in years starting with $x = 0$ for the year 1995. Let y represent the corresponding percentage of residents with cellular service. Use logistic regression to fit a model to these data.
ⓑ Use the model to calculate the percentage of Americans with cell service in the year 2013. Round to the nearest tenth of a percent.
ⓒ Discuss the value returned for the upper limit, c. What does this tell you about the model? What would the limiting

11 Source: *The World Bank, 2013*

value be if the model were exact?

Solution

ⓐ Using the **STAT** then **EDIT** menu on a graphing utility, list the years using values 0–15 in L1 and the corresponding percentage in L2. Then use the **STATPLOT** feature to verify that the scatterplot follows a logistic pattern as shown in Figure 5:

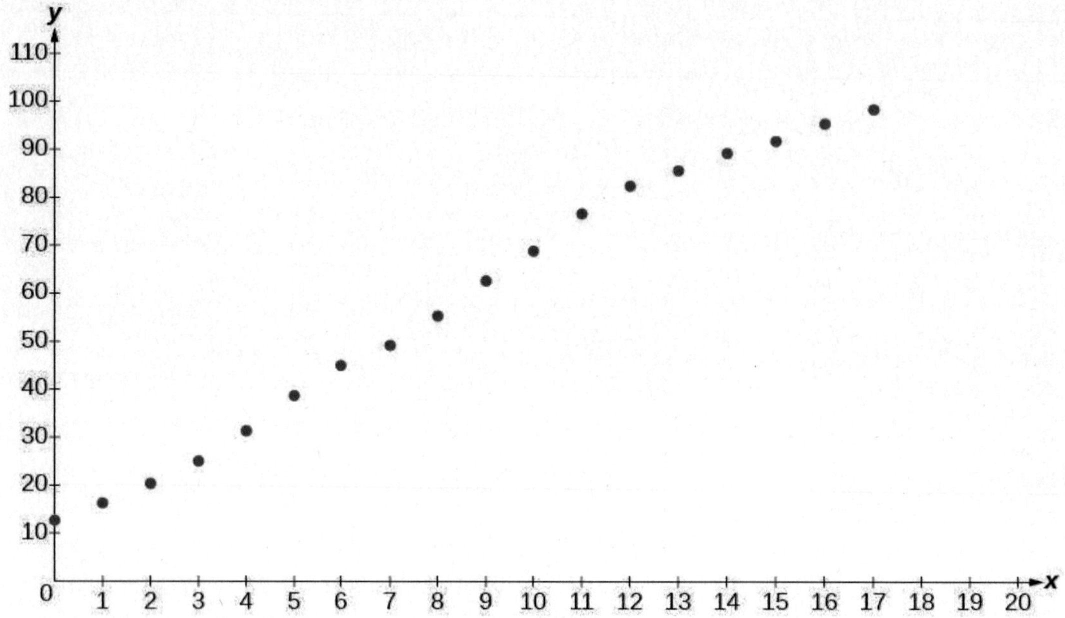

Figure 5

Use the "**Logistic**" command from the **STAT** then **CALC** menu to obtain the logistic model,

$$y = \frac{105.7379526}{1 + 6.88328979e^{-0.2595440013x}}$$

Next, graph the model in the same window as shown in Figure 6 the scatterplot to verify it is a good fit:

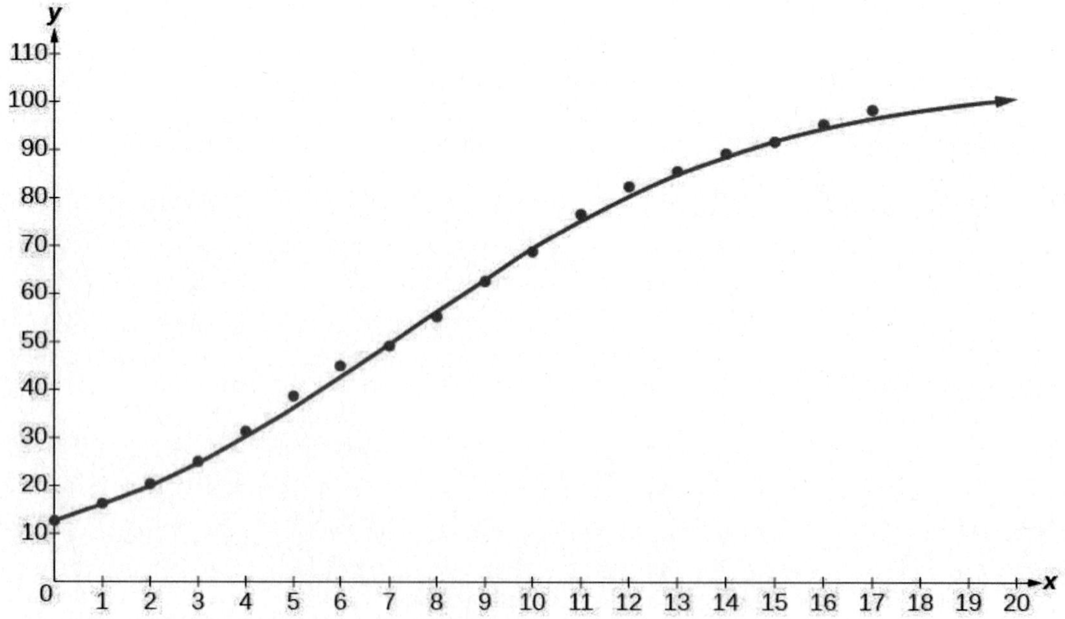

Figure 6

ⓑ

To approximate the percentage of Americans with cellular service in the year 2013, substitute $x = 18$ for the in the model and solve for y :

$$\begin{aligned} y &= \frac{105.7379526}{1+6.88328979e^{-0.2595440013x}} && \text{Use the regression model found in part (a).} \\ &= \frac{105.7379526}{1+6.88328979e^{-0.2595440013(18)}} && \text{Substitute 18 for } x. \\ &\approx 99.3 && \text{Round to the nearest tenth} \end{aligned}$$

According to the model, about 99.3% of Americans had cellular service in 2013.

ⓒ

The model gives a limiting value of about 105. This means that the maximum possible percentage of Americans with cellular service would be 105%, which is impossible. (How could over 100% of a population have cellular service?) If the model were exact, the limiting value would be $c = 100$ and the model's outputs would get very close to, but never actually reach 100%. After all, there will always be someone out there without cellular service!

TRY IT #3 Table 6 shows the population, in thousands, of harbor seals in the Wadden Sea over the years 1997 to 2012.

Year	Seal Population (Thousands)	Year	Seal Population (Thousands)
1997	3.493	2005	19.590
1998	5.282	2006	21.955
1999	6.357	2007	22.862
2000	9.201	2008	23.869
2001	11.224	2009	24.243
2002	12.964	2010	24.344
2003	16.226	2011	24.919
2004	18.137	2012	25.108

Table 6

ⓐ Let x represent time in years starting with $x = 0$ for the year 1997. Let y represent the number of seals in thousands. Use logistic regression to fit a model to these data.

ⓑ Use the model to predict the seal population for the year 2020.

ⓒ To the nearest whole number, what is the limiting value of this model?

MEDIA

Access this online resource for additional instruction and practice with exponential function models.

Exponential Regression on a Calculator (https://openstax.org/l/pregresscalc)

6.8 SECTION EXERCISES

Verbal

1. What situations are best modeled by a logistic equation? Give an example, and state a case for why the example is a good fit.

2. What is a carrying capacity? What kind of model has a carrying capacity built into its formula? Why does this make sense?

3. What is regression analysis? Describe the process of performing regression analysis on a graphing utility.

4. What might a scatterplot of data points look like if it were best described by a logarithmic model?

5. What does the y-intercept on the graph of a logistic equation correspond to for a population modeled by that equation?

Graphical

For the following exercises, match the given function of best fit with the appropriate scatterplot in Figure 7 through Figure 11. Answer using the letter beneath the matching graph.

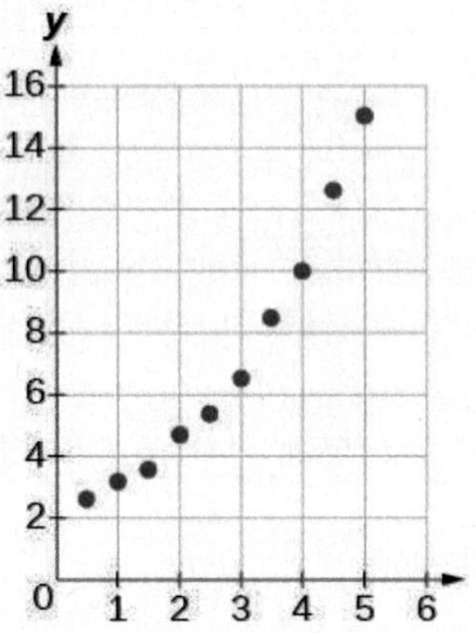

(a)

Figure 7

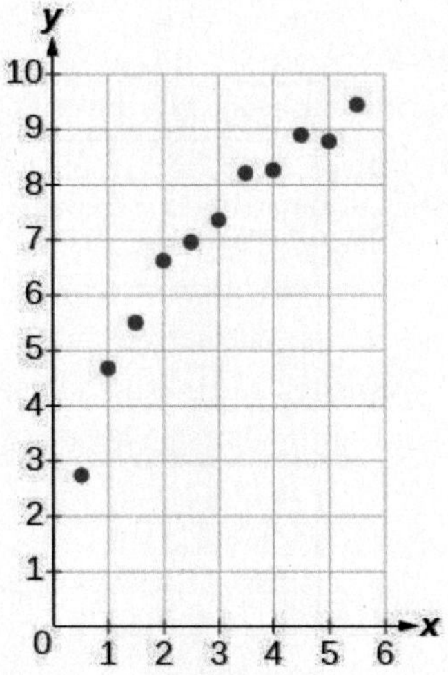

(b)

Figure 8

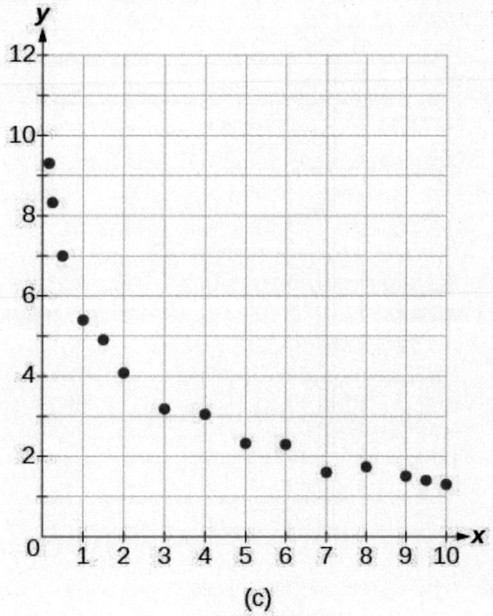

(c)

Figure 9

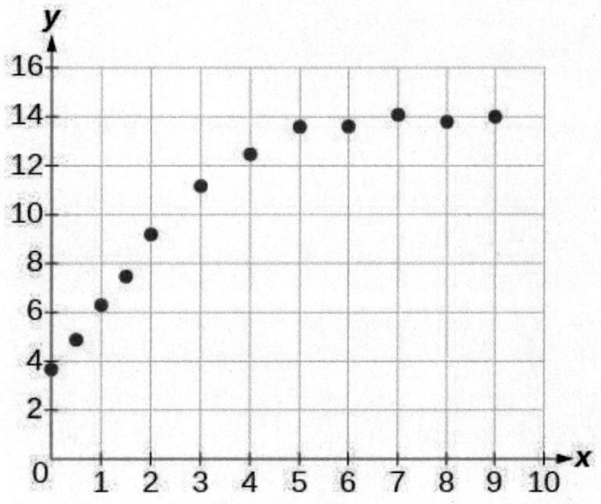

(d)

Figure 10

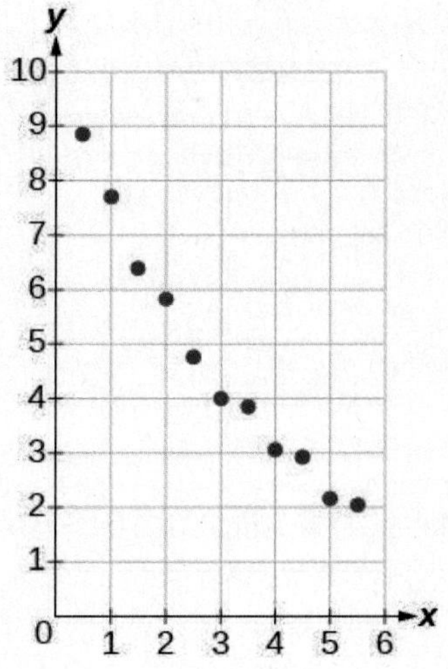

(e)

Figure 11

6. $y = 10.209e^{-0.294x}$

7. $y = 5.598 - 1.912\ln(x)$

8. $y = 2.104(1.479)^x$

9. $y = 4.607 + 2.733\ln(x)$

10. $y = \frac{14.005}{1+2.79e^{-0.812x}}$

Numeric

11. To the nearest whole number, what is the initial value of a population modeled by the logistic equation $P(t) = \frac{175}{1+6.995e^{-0.68t}}$? What is the carrying capacity?

12. Rewrite the exponential model $A(t) = 1550(1.085)^x$ as an equivalent model with base e. Express the exponent to four significant digits.

13. A logarithmic model is given by the equation $h(p) = 67.682 - 5.792\ln(p)$. To the nearest hundredth, for what value of p does $h(p) = 62$?

14. A logistic model is given by the equation $P(t) = \frac{90}{1+5e^{-0.42t}}$. To the nearest hundredth, for what value of t does $P(t) = 45$?

15. What is the y-intercept on the graph of the logistic model given in the previous exercise?

Technology

For the following exercises, use this scenario: The population P of a koi pond over x months is modeled by the function $P(x) = \frac{68}{1+16e^{-0.28x}}$.

16. Graph the population model to show the population over a span of 3 years.

17. What was the initial population of koi?

18. How many koi will the pond have after one and a half years?

19. How many months will it take before there are 20 koi in the pond?

20. Use the intersect feature to approximate the number of months it will take before the population of the pond reaches half its carrying capacity.

For the following exercises, use this scenario: The population P of an endangered species habitat for wolves is modeled by the function $P(x) = \frac{558}{1+54.8e^{-0.462x}}$, where x is given in years.

21. Graph the population model to show the population over a span of 10 years.

22. What was the initial population of wolves transported to the habitat?

23. How many wolves will the habitat have after 3 years?

24. How many years will it take before there are 100 wolves in the habitat?

25. Use the intersect feature to approximate the number of years it will take before the population of the habitat reaches half its carrying capacity.

For the following exercises, refer to Table 7.

x	1	2	3	4	5	6
f(x)	1125	1495	2310	3294	4650	6361

Table 7

26. Use a graphing calculator to create a scatter diagram of the data.

27. Use the regression feature to find an exponential function that best fits the data in the table.

28. Write the exponential function as an exponential equation with base e.

29. Graph the exponential equation on the scatter diagram.

30. Use the intersect feature to find the value of x for which $f(x) = 4000$.

For the following exercises, refer to Table 8.

x	1	2	3	4	5	6
f(x)	555	383	307	210	158	122

Table 8

31. Use a graphing calculator to create a scatter diagram of the data.

32. Use the regression feature to find an exponential function that best fits the data in the table.

33. Write the exponential function as an exponential equation with base e.

34. Graph the exponential equation on the scatter diagram.

35. Use the intersect feature to find the value of x for which $f(x) = 250$.

For the following exercises, refer to Table 9.

x	1	2	3	4	5	6
f(x)	5.1	6.3	7.3	7.7	8.1	8.6

Table 9

36. Use a graphing calculator to create a scatter diagram of the data.

37. Use the LOGarithm option of the REGression feature to find a logarithmic function of the form $y = a + b\ln(x)$ that best fits the data in the table.

38. Use the logarithmic function to find the value of the function when $x = 10$.

39. Graph the logarithmic equation on the scatter diagram.

40. Use the intersect feature to find the value of x for which $f(x) = 7$.

For the following exercises, refer to Table 10.

x	1	2	3	4	5	6	7	8
f(x)	7.5	6	5.2	4.3	3.9	3.4	3.1	2.9

Table 10

41. Use a graphing calculator to create a scatter diagram of the data.

42. Use the **LOG**arithm option of the **REG**ression feature to find a logarithmic function of the form $y = a + b\ln(x)$ that best fits the data in the table.

43. Use the logarithmic function to find the value of the function when $x = 10$.

44. Graph the logarithmic equation on the scatter diagram.

45. Use the intersect feature to find the value of x for which $f(x) = 8$.

For the following exercises, refer to Table 11.

x	1	2	3	4	5	6	7	8	9	10
f(x)	8.7	12.3	15.4	18.5	20.7	22.5	23.3	24	24.6	24.8

Table 11

46. Use a graphing calculator to create a scatter diagram of the data.

47. Use the LOGISTIC regression option to find a logistic growth model of the form $y = \frac{c}{1+ae^{-bx}}$ that best fits the data in the table.

48. Graph the logistic equation on the scatter diagram.

49. To the nearest whole number, what is the predicted carrying capacity of the model?

50. Use the intersect feature to find the value of x for which the model reaches half its carrying capacity.

For the following exercises, refer to Table 12.

x	0	2	4	5	7	8	10	11	15	17
$f(x)$	12	28.6	52.8	70.3	99.9	112.5	125.8	127.9	135.1	135.9

Table 12

51. Use a graphing calculator to create a scatter diagram of the data.

52. Use the LOGISTIC regression option to find a logistic growth model of the form $y = \frac{c}{1+ae^{-bx}}$ that best fits the data in the table.

53. Graph the logistic equation on the scatter diagram.

54. To the nearest whole number, what is the predicted carrying capacity of the model?

55. Use the intersect feature to find the value of x for which the model reaches half its carrying capacity.

Extensions

56. Recall that the general form of a logistic equation for a population is given by $P(t) = \frac{c}{1+ae^{-bt}}$, such that the initial population at time $t = 0$ is $P(0) = P_0$. Show algebraically that $\frac{c-P(t)}{P(t)} = \frac{c-P_0}{P_0}e^{-bt}$.

57. Use a graphing utility to find an exponential regression formula $f(x)$ and a logarithmic regression formula $g(x)$ for the points $(1.5, 1.5)$ and $(8.5, \ 8.5)$. Round all numbers to 6 decimal places. Graph the points and both formulas along with the line $y = x$ on the same axis. Make a conjecture about the relationship of the regression formulas.

58. Verify the conjecture made in the previous exercise. Round all numbers to six decimal places when necessary.

59. Find the inverse function $f^{-1}(x)$ for the logistic function $f(x) = \frac{c}{1+ae^{-bx}}$. Show all steps.

60. Use the result from the previous exercise to graph the logistic model $P(t) = \frac{20}{1+4e^{-0.5t}}$ along with its inverse on the same axis. What are the intercepts and asymptotes of each function?

Chapter Review

Key Terms

annual percentage rate (APR) the yearly interest rate earned by an investment account, also called *nominal rate*

carrying capacity in a logistic model, the limiting value of the output

change-of-base formula a formula for converting a logarithm with any base to a quotient of logarithms with any other base.

common logarithm the exponent to which 10 must be raised to get x; $\log_{10}(x)$ is written simply as $\log(x)$.

compound interest interest earned on the total balance, not just the principal

doubling time the time it takes for a quantity to double

exponential growth a model that grows by a rate proportional to the amount present

extraneous solution a solution introduced while solving an equation that does not satisfy the conditions of the original equation

half-life the length of time it takes for a substance to exponentially decay to half of its original quantity

logarithm the exponent to which b must be raised to get x; written $y = \log_b(x)$

logistic growth model a function of the form $f(x) = \frac{c}{1+ae^{-bx}}$ where $\frac{c}{1+a}$ is the initial value, c is the carrying capacity, or limiting value, and b is a constant determined by the rate of growth

natural logarithm the exponent to which the number e must be raised to get x; $\log_e(x)$ is written as $\ln(x)$.

Newton's Law of Cooling the scientific formula for temperature as a function of time as an object's temperature is equalized with the ambient temperature

nominal rate the yearly interest rate earned by an investment account, also called *annual percentage rate*

order of magnitude the power of ten, when a number is expressed in scientific notation, with one non-zero digit to the left of the decimal

power rule for logarithms a rule of logarithms that states that the log of a power is equal to the product of the exponent and the log of its base

product rule for logarithms a rule of logarithms that states that the log of a product is equal to a sum of logarithms

quotient rule for logarithms a rule of logarithms that states that the log of a quotient is equal to a difference of logarithms

Key Equations

definition of the exponential function	$f(x) = b^x$, where $b > 0$, $b \neq 1$
definition of exponential growth	$f(x) = ab^x$, where $a > 0, b > 0, b \neq 1$
compound interest formula	$A(t) = P\left(1 + \frac{r}{n}\right)^{nt}$, where $A(t)$ is the account value at time t t is the number of years P is the initial investment, often called the principal r is the annual percentage rate (APR), or nominal rate n is the number of compounding periods in one year
continuous growth formula	$A(t) = ae^{rt}$, where t is the number of unit time periods of growth a is the starting amount (in the continuous compounding formula a is replaced with P, the principal) e is the mathematical constant, $e \approx 2.718282$
General Form for the Translation of the Parent Function $f(x) = b^x$	$f(x) = ab^{x+c} + d$

Definition of the logarithmic function	For $x > 0, b > 0, b \neq 1$, $y = \log_b(x)$ if and only if $b^y = x$.
Definition of the common logarithm	For $x > 0$, $y = \log(x)$ if and only if $10^y = x$.
Definition of the natural logarithm	For $x > 0$, $y = \ln(x)$ if and only if $e^y = x$.

General Form for the Translation of the Parent Logarithmic Function $f(x) = \log_b(x)$	$f(x) = a\log_b(x + c) + d$

The Product Rule for Logarithms	$\log_b(MN) = \log_b(M) + \log_b(N)$	
The Quotient Rule for Logarithms	$\log_b\left(\frac{M}{N}\right) = \log_b M - \log_b N$	
The Power Rule for Logarithms	$\log_b(M^n) = n\log_b M$	
The Change-of-Base Formula	$\log_b M = \frac{\log_n M}{\log_n b}$	$n > 0, n \neq 1, b \neq 1$

One-to-one property for exponential functions	For any algebraic expressions S and T and any positive real number b, where $b^S = b^T$ if and only if $S = T$.
Definition of a logarithm	For any algebraic expression S and positive real numbers b and c, where $b \neq 1$, $\log_b(S) = c$ if and only if $b^c = S$.
One-to-one property for logarithmic functions	For any algebraic expressions S and T and any positive real number b, where $b \neq 1$, $\log_b S = \log_b T$ if and only if $S = T$.
Half-life formula	If $A = A_0 e^{kt}$, $k < 0$, the half-life is $t = -\frac{\ln(2)}{k}$.
Carbon-14 dating	$t = \frac{\ln\left(\frac{A}{A_0}\right)}{-0.000121}$. A_0 is the amount of carbon-14 when the plant or animal died. A is the amount of carbon-14 remaining today. t is the age of the fossil in years
Doubling time formula	If $A = A_0 e^{kt}$, $k > 0$, the doubling time is $t = \frac{\ln 2}{k}$
Newton's Law of Cooling	$T(t) = Ae^{kt} + T_s$, where T_s is the ambient temperature, $A = T(0) - T_s$, and k is the continuous rate of cooling.

Key Concepts

6.1 Exponential Functions

- An exponential function is defined as a function with a positive constant other than 1 raised to a variable exponent. See Example 1.

- A function is evaluated by solving at a specific value. See Example 2 and Example 3.
- An exponential model can be found when the growth rate and initial value are known. See Example 4.
- An exponential model can be found when the two data points from the model are known. See Example 5.
- An exponential model can be found using two data points from the graph of the model. See Example 6.
- An exponential model can be found using two data points from the graph and a calculator. See Example 7.
- The value of an account at any time t can be calculated using the compound interest formula when the principal, annual interest rate, and compounding periods are known. See Example 8.
- The initial investment of an account can be found using the compound interest formula when the value of the account, annual interest rate, compounding periods, and life span of the account are known. See Example 9.
- The number e is a mathematical constant often used as the base of real world exponential growth and decay models. Its decimal approximation is $e \approx 2.718282$.
- Scientific and graphing calculators have the key $\left[e^x\right]$ or $\left[\exp(x)\right]$ for calculating powers of e. See Example 10.
- Continuous growth or decay models are exponential models that use e as the base. Continuous growth and decay models can be found when the initial value and growth or decay rate are known. See Example 11 and Example 12.

6.2 Graphs of Exponential Functions

- The graph of the function $f(x) = b^x$ has a y-intercept at $(0,\ 1)$, domain $\left(-\infty,\ \infty\right)$, range $\left(0,\ \infty\right)$, and horizontal asymptote $y = 0$. See Example 1.
- If $b > 1$, the function is increasing. The left tail of the graph will approach the asymptote $y = 0$, and the right tail will increase without bound.
- If $0 < b < 1$, the function is decreasing. The left tail of the graph will increase without bound, and the right tail will approach the asymptote $y = 0$.
- The equation $f(x) = b^x + d$ represents a vertical shift of the parent function $f(x) = b^x$.
- The equation $f(x) = b^{x+c}$ represents a horizontal shift of the parent function $f(x) = b^x$. See Example 2.
- Approximate solutions of the equation $f(x) = b^{x+c} + d$ can be found using a graphing calculator. See Example 3.
- The equation $f(x) = ab^x$, where $a > 0$, represents a vertical stretch if $|a| > 1$ or compression if $0 < |a| < 1$ of the parent function $f(x) = b^x$. See Example 4.
- When the parent function $f(x) = b^x$ is multiplied by -1, the result, $f(x) = -b^x$, is a reflection about the x-axis. When the input is multiplied by -1, the result, $f(x) = b^{-x}$, is a reflection about the y-axis. See Example 5.
- All translations of the exponential function can be summarized by the general equation $f(x) = ab^{x+c} + d$. See Table 3.
- Using the general equation $f(x) = ab^{x+c} + d$, we can write the equation of a function given its description. See Example 6.

6.3 Logarithmic Functions

- The inverse of an exponential function is a logarithmic function, and the inverse of a logarithmic function is an exponential function.
- Logarithmic equations can be written in an equivalent exponential form, using the definition of a logarithm. See Example 1.
- Exponential equations can be written in their equivalent logarithmic form using the definition of a logarithm See Example 2.
- Logarithmic functions with base b can be evaluated mentally using previous knowledge of powers of b. See Example 3 and Example 4.
- Common logarithms can be evaluated mentally using previous knowledge of powers of 10. See Example 5.
- When common logarithms cannot be evaluated mentally, a calculator can be used. See Example 6.
- Real-world exponential problems with base 10 can be rewritten as a common logarithm and then evaluated using a calculator. See Example 7.
- Natural logarithms can be evaluated using a calculator Example 8.

6.4 Graphs of Logarithmic Functions

- To find the domain of a logarithmic function, set up an inequality showing the argument greater than zero, and solve for x. See Example 1 and Example 2
- The graph of the parent function $f(x) = \log_b(x)$ has an x-intercept at $(1, 0)$, domain $\left(0, \infty\right)$, range $\left(-\infty, \infty\right)$, vertical asymptote $x = 0$, and
  - if $b > 1$, the function is increasing.
  - if $0 < b < 1$, the function is decreasing.

See Example 3.
- The equation $f(x) = \log_b(x + c)$ shifts the parent function $y = \log_b(x)$ horizontally
  - left c units if $c > 0$.
  - right c units if $c < 0$.

  See Example 4.
- The equation $f(x) = \log_b(x) + d$ shifts the parent function $y = \log_b(x)$ vertically
  - up d units if $d > 0$.
  - down d units if $d < 0$.

  See Example 5.
- For any constant $a > 0$, the equation $f(x) = a\log_b(x)$
  - stretches the parent function $y = \log_b(x)$ vertically by a factor of a if $|a| > 1$.
  - compresses the parent function $y = \log_b(x)$ vertically by a factor of a if $|a| < 1$.

  See Example 6 and Example 7.
- When the parent function $y = \log_b(x)$ is multiplied by -1, the result is a reflection about the *x*-axis. When the input is multiplied by -1, the result is a reflection about the *y*-axis.
  - The equation $f(x) = -\log_b(x)$ represents a reflection of the parent function about the *x*-axis.
  - The equation $f(x) = \log_b(-x)$ represents a reflection of the parent function about the *y*-axis.

  See Example 8.
  - A graphing calculator may be used to approximate solutions to some logarithmic equations See Example 9.
- All translations of the logarithmic function can be summarized by the general equation $f(x) = a\log_b(x + c) + d$. See Table 4.
- Given an equation with the general form $f(x) = a\log_b(x + c) + d$, we can identify the vertical asymptote $x = -c$ for the transformation. See Example 10.
- Using the general equation $f(x) = a\log_b(x + c) + d$, we can write the equation of a logarithmic function given its graph. See Example 11.

6.5 Logarithmic Properties

- We can use the product rule of logarithms to rewrite the log of a product as a sum of logarithms. See Example 1.
- We can use the quotient rule of logarithms to rewrite the log of a quotient as a difference of logarithms. See Example 2.
- We can use the power rule for logarithms to rewrite the log of a power as the product of the exponent and the log of its base. See Example 3, Example 4, and Example 5.
- We can use the product rule, the quotient rule, and the power rule together to combine or expand a logarithm with a complex input. See Example 6, Example 7, and Example 8.
- The rules of logarithms can also be used to condense sums, differences, and products with the same base as a single logarithm. See Example 9, Example 10, Example 11, and Example 12.
- We can convert a logarithm with any base to a quotient of logarithms with any other base using the change-of-base formula. See Example 13.
- The change-of-base formula is often used to rewrite a logarithm with a base other than 10 and e as the quotient of natural or common logs. That way a calculator can be used to evaluate. See Example 14.

6.6 Exponential and Logarithmic Equations

- We can solve many exponential equations by using the rules of exponents to rewrite each side as a power with the same base. Then we use the fact that exponential functions are one-to-one to set the exponents equal to one another and solve for the unknown.
- When we are given an exponential equation where the bases are explicitly shown as being equal, set the exponents equal to one another and solve for the unknown. See Example 1.
- When we are given an exponential equation where the bases are *not* explicitly shown as being equal, rewrite each side of the equation as powers of the same base, then set the exponents equal to one another and solve for the unknown. See Example 2, Example 3, and Example 4.
- When an exponential equation cannot be rewritten with a common base, solve by taking the logarithm of each side. See Example 5.
- We can solve exponential equations with base e, by applying the natural logarithm of both sides because exponential and logarithmic functions are inverses of each other. See Example 6 and Example 7.
- After solving an exponential equation, check each solution in the original equation to find and eliminate any

extraneous solutions. See Example 8.

- When given an equation of the form $\log_b(S) = c$, where S is an algebraic expression, we can use the definition of a logarithm to rewrite the equation as the equivalent exponential equation $b^c = S$, and solve for the unknown. See Example 9 and Example 10.
- We can also use graphing to solve equations with the form $\log_b(S) = c$. We graph both equations $y = \log_b(S)$ and $y = c$ on the same coordinate plane and identify the solution as the x-value of the intersecting point. See Example 11.
- When given an equation of the form $\log_b S = \log_b T$, where S and T are algebraic expressions, we can use the one-to-one property of logarithms to solve the equation $S = T$ for the unknown. See Example 12.
- Combining the skills learned in this and previous sections, we can solve equations that model real world situations, whether the unknown is in an exponent or in the argument of a logarithm. See Example 13.

6.7 Exponential and Logarithmic Models

- The basic exponential function is $f(x) = ab^x$. If $b > 1$, we have exponential growth; if $0 < b < 1$, we have exponential decay.
- We can also write this formula in terms of continuous growth as $A = A_0e^{kx}$, where A_0 is the starting value. If A_0 is positive, then we have exponential growth when $k > 0$ and exponential decay when $k < 0$. See Example 1.
- In general, we solve problems involving exponential growth or decay in two steps. First, we set up a model and use the model to find the parameters. Then we use the formula with these parameters to predict growth and decay. See Example 2.
- We can find the age, t, of an organic artifact by measuring the amount, k, of carbon-14 remaining in the artifact and using the formula $t = \frac{\ln(k)}{-0.000121}$ to solve for t. See Example 3.
- Given a substance's doubling time or half-time, we can find a function that represents its exponential growth or decay. See Example 4.
- We can use Newton's Law of Cooling to find how long it will take for a cooling object to reach a desired temperature, or to find what temperature an object will be after a given time. See Example 5.
- We can use logistic growth functions to model real-world situations where the rate of growth changes over time, such as population growth, spread of disease, and spread of rumors. See Example 6.
- We can use real-world data gathered over time to observe trends. Knowledge of linear, exponential, logarithmic, and logistic graphs help us to develop models that best fit our data. See Example 7.
- Any exponential function with the form $y = ab^x$ can be rewritten as an equivalent exponential function with the form $y = A_0e^{kx}$ where $k = \ln b$. See Example 8.

6.8 Fitting Exponential Models to Data

- Exponential regression is used to model situations where growth begins slowly and then accelerates rapidly without bound, or where decay begins rapidly and then slows down to get closer and closer to zero.
- We use the command "ExpReg" on a graphing utility to fit function of the form $y = ab^x$ to a set of data points. See Example 1.
- Logarithmic regression is used to model situations where growth or decay accelerates rapidly at first and then slows over time.
- We use the command "LnReg" on a graphing utility to fit a function of the form $y = a + b\ln(x)$ to a set of data points. See Example 2.
- Logistic regression is used to model situations where growth accelerates rapidly at first and then steadily slows as the function approaches an upper limit.
- We use the command "Logistic" on a graphing utility to fit a function of the form $y = \frac{c}{1+ae^{-bx}}$ to a set of data points. See Example 3.

Exercises

Review Exercises

Exponential Functions

1. Determine whether the function $y = 156(0.825)^t$ represents exponential growth, exponential decay, or neither. Explain

2. The population of a herd of deer is represented by the function $A(t) = 205(1.13)^t$, where t is given in years. To the nearest whole number, what will the herd population be after 6 years?

3. Find an exponential equation that passes through the points $(2, 2.25)$ and $(5, 60.75)$.

4. Determine whether Table 1 could represent a function that is linear, exponential, or neither. If it appears to be exponential, find a function that passes through the points.

x	1	2	3	4
f(x)	3	0.9	0.27	0.081

Table 1

5. A retirement account is opened with an initial deposit of \$8,500 and earns 8.12% interest compounded monthly. What will the account be worth in 20 years?

6. Hsu-Mei wants to save \$5,000 for a down payment on a car. To the nearest dollar, how much will she need to invest in an account now with 7.5% APR, compounded daily, in order to reach her goal in 3 years?

7. Does the equation $y = 2.294e^{-0.654t}$ represent continuous growth, continuous decay, or neither? Explain.

8. Suppose an investment account is opened with an initial deposit of $\$10,500$ earning 6.25% interest, compounded continuously. How much will the account be worth after 25 years?

Graphs of Exponential Functions

9. Graph the function $f(x) = 3.5(2)^x$. State the domain and range and give the *y*-intercept.

10. Graph the function $f(x) = 4\left(\frac{1}{8}\right)^x$ and its reflection about the *y*-axis on the same axes, and give the *y*-intercept.

11. The graph of $f(x) = 6.5^x$ is reflected about the y-axis and stretched vertically by a factor of 7. What is the equation of the new function, $g(x)$? State its y-intercept, domain, and range.

12. The graph below shows transformations of the graph of $f(x) = 2^x$. What is the equation for the transformation?

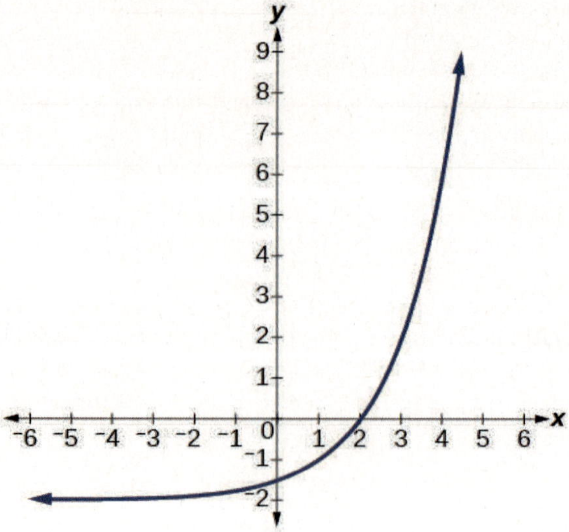

Figure 1

Logarithmic Functions

13. Rewrite $\log_{17}(4913) = x$ as an equivalent exponential equation.

14. Rewrite $\ln(s) = t$ as an equivalent exponential equation.

15. Rewrite $a^{-\frac{2}{5}} = b$ as an equivalent logarithmic equation.

16. Rewrite $e^{-3.5} = h$ as an equivalent logarithmic equation.

17. Solve for x if $\log_{64}(x) = \frac{1}{3}$ by converting the logarithmic equation $\log_{64}(x) = \frac{1}{3}$ to exponential form.

18. Evaluate $\log_5\left(\frac{1}{125}\right)$ without using a calculator.

19. Evaluate $\log(0.000001)$ without using a calculator.

20. Evaluate $\log(4.005)$ using a calculator. Round to the nearest thousandth.

21. Evaluate $\ln\left(e^{-0.8648}\right)$ without using a calculator.

22. Evaluate $\ln\left(\sqrt[3]{18}\right)$ using a calculator. Round to the nearest thousandth.

Graphs of Logarithmic Functions

23. Graph the function $g(x) = \log(7x + 21) - 4$.

24. Graph the function $h(x) = 2\ln(9 - 3x) + 1$.

25. State the domain, vertical asymptote, and end behavior of the function $g(x) = \ln(4x + 20) - 17$.

Logarithmic Properties

26. Rewrite $\ln(7r \cdot 11st)$ in expanded form.

27. Rewrite $\log_8(x) + \log_8(5) + \log_8(y) + \log_8(13)$ in compact form.

28. Rewrite $\log_m\left(\frac{67}{83}\right)$ in expanded form.

29. Rewrite $\ln(z) - \ln(x) - \ln(y)$ in compact form.

30. Rewrite $\ln\left(\frac{1}{x^5}\right)$ as a product.

31. Rewrite $-\log_y\left(\frac{1}{12}\right)$ as a single logarithm.

32. Use properties of logarithms to expand $\log\left(\frac{r^2 s^{11}}{t^{14}}\right)$.

33. Use properties of logarithms to expand $\ln\left(2b\sqrt{\frac{b+1}{b-1}}\right)$.

34. Condense the expression $5\ln(b) + \ln(c) + \frac{\ln(4-a)}{2}$ to a single logarithm.

35. Condense the expression $3\log_7 v + 6\log_7 w - \frac{\log_7 u}{3}$ to a single logarithm.

36. Rewrite $\log_3(12.75)$ to base e.

37. Rewrite $5^{12x-17} = 125$ as a logarithm. Then apply the change of base formula to solve for x using the common log. Round to the nearest thousandth.

Exponential and Logarithmic Equations

38. Solve $216^{3x} \cdot 216^x = 36^{3x+2}$ by rewriting each side with a common base.

39. Solve $\frac{125}{\left(\frac{1}{625}\right)^{-x-3}} = 5^3$ by rewriting each side with a common base.

40. Use logarithms to find the exact solution for $7 \cdot 17^{-9x} - 7 = 49$. If there is no solution, write *no solution.*

41. Use logarithms to find the exact solution for $3e^{6n-2} + 1 = -60$. If there is no solution, write *no solution.*

42. Find the exact solution for $5e^{3x} - 4 = 6$. If there is no solution, write *no solution.*

43. Find the exact solution for $2e^{5x-2} - 9 = -56$. If there is no solution, write *no solution.*

44. Find the exact solution for $5^{2x-3} = 7^{x+1}$. If there is no solution, write *no solution.*

45. Find the exact solution for $e^{2x} - e^x - 110 = 0$. If there is no solution, write *no solution.*

46. Use the definition of a logarithm to solve. $-5\log_7(10n) = 5$.

47. Use the definition of a logarithm to find the exact solution for $9 + 6\ln(a+3) = 33$.

48. Use the one-to-one property of logarithms to find an exact solution for $\log_8(7) + \log_8(-4x) = \log_8(5)$. If there is no solution, write *no solution.*

49. Use the one-to-one property of logarithms to find an exact solution for $\ln(5) + \ln\left(5x^2 - 5\right) = \ln(56)$. If there is no solution, write *no solution.*

50. The formula for measuring sound intensity in decibels D is defined by the equation $D = 10\log\left(\frac{I}{I_0}\right)$, where I is the intensity of the sound in watts per square meter and $I_0 = 10^{-12}$ is the lowest level of sound that the average person can hear. How many decibels are emitted from a large orchestra with a sound intensity of $6.3 \cdot 10^{-3}$ watts per square meter?

51. The population of a city is modeled by the equation $P(t) = 256,114e^{0.25t}$ where t is measured in years. If the city continues to grow at this rate, how many years will it take for the population to reach one million?

52. Find the inverse function f^{-1} for the exponential function $f(x) = 2 \cdot e^{x+1} - 5$.

53. Find the inverse function f^{-1} for the logarithmic function $f(x) = 0.25 \cdot \log_2 \left(x^3 + 1\right)$.

Exponential and Logarithmic Models

For the following exercises, use this scenario: A doctor prescribes 300 milligrams of a therapeutic drug that decays by about 17% each hour.

54. To the nearest minute, what is the half-life of the drug?

55. Write an exponential model representing the amount of the drug remaining in the patient's system after t hours. Then use the formula to find the amount of the drug that would remain in the patient's system after 24 hours. Round to the nearest hundredth of a gram.

For the following exercises, use this scenario: A soup with an internal temperature of $350°$ Fahrenheit was taken off the stove to cool in a $71°\text{F}$ room. After fifteen minutes, the internal temperature of the soup was $175°\text{F}$.

56. Use Newton's Law of Cooling to write a formula that models this situation.

57. How many minutes will it take the soup to cool to $85°\text{F}$?

For the following exercises, use this scenario: The equation $N(t) = \frac{1200}{1+199e^{-0.625t}}$ models the number of people in a school who have heard a rumor after t days.

58. How many people started the rumor?

59. To the nearest tenth, how many days will it be before the rumor spreads to half the carrying capacity?

60. What is the carrying capacity?

For the following exercises, enter the data from each table into a graphing calculator and graph the resulting scatter plots. Determine whether the data from the table would likely represent a function that is linear, exponential, or logarithmic.

61.

x	*f(x)*
1	3.05
2	4.42
3	6.4
4	9.28
5	13.46
6	19.52
7	28.3
8	41.04
9	59.5
10	86.28

62.

x	*f(x)*
0.5	18.05
1	17
3	15.33
5	14.55
7	14.04
10	13.5
12	13.22
13	13.1
15	12.88
17	12.69
20	12.45

63. Find a formula for an exponential equation that goes through the points $(-2, 100)$ and $(0, 4)$. Then express the formula as an equivalent equation with base *e*.

Fitting Exponential Models to Data

64. What is the carrying capacity for a population modeled by the logistic equation $P(t) = \frac{250{,}000}{1+499e^{-0.45t}}$? What is the initial population for the model?

65. The population of a culture of bacteria is modeled by the logistic equation $P(t) = \frac{14{,}250}{1+29e^{-0.62t}}$, where t is in days. To the nearest tenth, how many days will it take the culture to reach 75% of its carrying capacity?

For the following exercises, use a graphing utility to create a scatter diagram of the data given in the table. Observe the shape of the scatter diagram to determine whether the data is best described by an exponential, logarithmic, or logistic model. Then use the appropriate regression feature to find an equation that models the data. When necessary, round values to five decimal places.

66.

x	*f(x)*
1	409.4
2	260.7
3	170.4
4	110.6
5	74
6	44.7
7	32.4
8	19.5
9	12.7
10	8.1

67.

x	*f(x)*
0.15	36.21
0.25	28.88
0.5	24.39
0.75	18.28
1	16.5
1.5	12.99
2	9.91
2.25	8.57
2.75	7.23
3	5.99
3.5	4.81

68.

x	*f(x)*
0	9
2	22.6
4	44.2
5	62.1
7	96.9
8	113.4
10	133.4
11	137.6
15	148.4
17	149.3

Practice Test

1. The population of a pod of bottlenose dolphins is modeled by the function $A(t) = 8(1.17)^t$, where t is given in years. To the nearest whole number, what will the pod population be after 3 years?

2. Find an exponential equation that passes through the points $(0, 4)$ and $(2, 9)$.

3. Drew wants to save $2,500 to go to the next World Cup. To the nearest dollar, how much will he need to invest in an account now with 6.25% APR, compounding daily, in order to reach his goal in 4 years?

4. An investment account was opened with an initial deposit of $9,600 and earns 7.4% interest, compounded continuously. How much will the account be worth after 15 years?

5. Graph the function $f(x) = 5(0.5)^{-x}$ and its reflection across the *y*-axis on the same axes, and give the *y*-intercept.

6. The graph shows transformations of the graph of $f(x) = \left(\frac{1}{2}\right)^x$. What is the equation for the transformation?

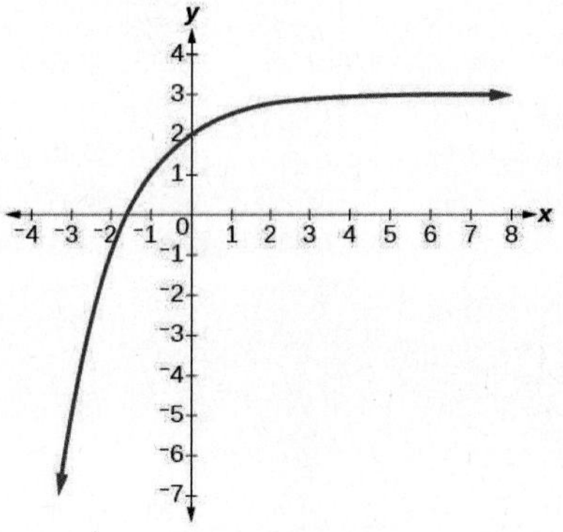

7. Rewrite $\log_{8.5}(614.125) = a$ as an equivalent exponential equation.

8. Rewrite $e^{\frac{1}{2}} = m$ as an equivalent logarithmic equation.

9. Solve for x by converting the logarithmic equation $log_{\frac{1}{7}}(x) = 2$ to exponential form.

10. Evaluate $\log(10{,}000{,}000)$ without using a calculator.

11. Evaluate $\ln(0.716)$ using a calculator. Round to the nearest thousandth.

12. Graph the function $g(x) = \log(12 - 6x) + 3$.

13. State the domain, vertical asymptote, and end behavior of the function $f(x) = \log_5(39 - 13x) + 7$.

14. Rewrite $\log(17a \cdot 2b)$ as a sum.

15. Rewrite $\log_t(96) - \log_t(8)$ in compact form.

16. Rewrite $\log_8\left(a^{\frac{1}{b}}\right)$ as a product.

17. Use properties of logarithm to expand $\ln\left(y^3 z^2 \cdot \sqrt[3]{x-4}\right)$.

18. Condense the expression $4\ln(c) + \ln(d) + \frac{\ln(a)}{3} + \frac{\ln(b+3)}{3}$ to a single logarithm.

19. Rewrite $16^{3x-5} = 1000$ as a logarithm. Then apply the change of base formula to solve for x using the natural log. Round to the nearest thousandth.

20. Solve $\left(\frac{1}{81}\right)^x \cdot \frac{1}{243} = \left(\frac{1}{9}\right)^{-3x-1}$ by rewriting each side with a common base.

21. Use logarithms to find the exact solution for $-9e^{10a-8} - 5 = -41$. If there is no solution, write *no solution*.

22. Find the exact solution for $10e^{4x+2} + 5 = 56$. If there is no solution, write *no solution.*

23. Find the exact solution for $-5e^{-4x-1} - 4 = 64$. If there is no solution, write *no solution.*

24. Find the exact solution for $2^{x-3} = 6^{2x-1}$. If there is no solution, write *no solution.*

25. Find the exact solution for $e^{2x} - e^{x} - 72 = 0$. If there is no solution, write *no solution.*

26. Use the definition of a logarithm to find the exact solution for $4\log(2n) - 7 = -11$

27. Use the one-to-one property of logarithms to find an exact solution for $\log\left(4x^2 - 10\right) + \log(3) = \log(51)$ If there is no solution, write *no solution.*

28. The formula for measuring sound intensity in decibels D is defined by the equation $D = 10\log\left(\frac{I}{I_0}\right)$, where I is the intensity of the sound in watts per square meter and $I_0 = 10^{-12}$ is the lowest level of sound that the average person can hear. How many decibels are emitted from a rock concert with a sound intensity of $4.7 \cdot 10^{-1}$ watts per square meter?

29. A radiation safety officer is working with 112 grams of a radioactive substance. After 17 days, the sample has decayed to 80 grams. Rounding to five significant digits, write an exponential equation representing this situation. To the nearest day, what is the half-life of this substance?

30. Write the formula found in the previous exercise as an equivalent equation with base e. Express the exponent to five significant digits.

31. A bottle of soda with a temperature of 71° Fahrenheit was taken off a shelf and placed in a refrigerator with an internal temperature of 35° F. After ten minutes, the internal temperature of the soda was 63° F. Use Newton's Law of Cooling to write a formula that models this situation. To the nearest degree, what will the temperature of the soda be after one hour?

32. The population of a wildlife habitat is modeled by the equation $P(t) = \frac{360}{1+6.2e^{-0.35t}}$, where t is given in years. How many animals were originally transported to the habitat? How many years will it take before the habitat reaches half its capacity?

33. Enter the data from Table 1 into a graphing calculator and graph the resulting scatter plot. Determine whether the data from the table would likely represent a function that is linear, exponential, or logarithmic.

x	*f(x)*
1	3
2	8.55
3	11.79
4	14.09
5	15.88
6	17.33
7	18.57
8	19.64
9	20.58
10	21.42

Table 1

34. The population of a lake of fish is modeled by the logistic equation $P(t) = \frac{16{,}120}{1+25e^{-0.75t}}$, where t is time in years. To the nearest hundredth, how many years will it take the lake to reach 80% of its carrying capacity?

For the following exercises, use a graphing utility to create a scatter diagram of the data given in the table. Observe the shape of the scatter diagram to determine whether the data is best described by an exponential, logarithmic, or logistic model. Then use the appropriate regression feature to find an equation that models the data. When necessary, round values to five decimal places.

35.

x	*f(x)*
1	20
2	21.6
3	29.2
4	36.4
5	46.6
6	55.7
7	72.6
8	87.1
9	107.2
10	138.1

36.

x	*f(x)*
3	13.98
4	17.84
5	20.01
6	22.7
7	24.1
8	26.15
9	27.37
10	28.38
11	29.97
12	31.07
13	31.43

37.

x	*f(x)*
0	2.2
0.5	2.9
1	3.9
1.5	4.8
2	6.4
3	9.3
4	12.3
5	15
6	16.2
7	17.3
8	17.9

The tide rises and falls at regular, predictable intervals. (credit: Andrea Schaffer, Flickr)

Chapter Outline

Introduction to The Unit Circle: Sine and Cosine Functions

Life is dense with phenomena that repeat in regular intervals. Each day, for example, the tides rise and fall in response to the gravitational pull of the moon. And as a result of the motion of the moon itself, the tides occur with different strengths. Throughout history, many Indigenous peoples have used this regularity to build cultural narratives and direct key activities, such as agriculture, hunting, and fishing. Aboriginal people in the Torres Strait area (the northern tip) of Australia used the tidal peaks to determine the best times to fish. Their elders explain that the stronger spring tides stirred up sediment and obscured fish vision, leaving them more likely to take in lures and resulting in a larger catch.[1]

In mathematics, a function that repeats its values in regular intervals is known as a periodic function. The graphs of such functions show a general shape reflective of a pattern that keeps repeating. This means the graph of the function has the same output at exactly the same place in every cycle. And this translates to all the cycles of the function having exactly the same length. So, if we know all the details of one full cycle of a true periodic function, then we know the state of the function's outputs at all times, future and past. In this chapter, we will investigate various examples of periodic functions.

1 Hamacher, D.W., Tapim, A., Passi, S., and Barsa, J. (2018). Dancing with the stars – astronomy and music in the Torres Strait. In Imagining Other Worlds: Explorations in Astronomy and Culture.

7.1 Angles

Learning Objectives

In this section you will:

- Draw angles in standard position.
- Convert between degrees and radians.
- Find coterminal angles.
- Find the length of a circular arc.
- Use linear and angular speed to describe motion on a circular path.

A golfer swings to hit a ball over a sand trap and onto the green. An airline pilot maneuvers a plane toward a narrow runway. A dress designer creates the latest fashion. What do they all have in common? They all work with angles, and so do all of us at one time or another. Sometimes we need to measure angles exactly with instruments. Other times we estimate them or judge them by eye. Either way, the proper angle can make the difference between success and failure in many undertakings. In this section, we will examine properties of angles.

Drawing Angles in Standard Position

Properly defining an angle first requires that we define a ray. A **ray** is a directed line segment. It consists of one point on a line and all points extending in one direction from that point. The first point is called the endpoint of the ray. We can refer to a specific ray by stating its endpoint and any other point on it. The ray in Figure 1 can be named as ray EF, or in symbol form $\overrightarrow{EF}$.

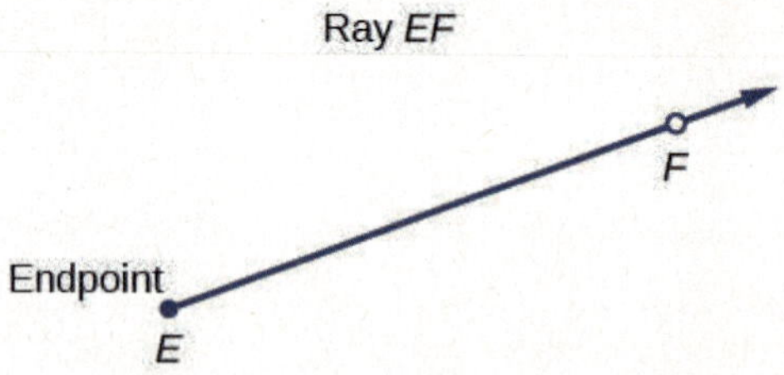

Figure 1

An **angle** is the union of two rays having a common endpoint. The endpoint is called the **vertex** of the angle, and the two rays are the sides of the angle. The angle in Figure 2 is formed from $\overrightarrow{ED}$ and $\overrightarrow{EF}$. Angles can be named using a point on each ray and the vertex, such as angle *DEF*, or in symbol form $\angle DEF$.

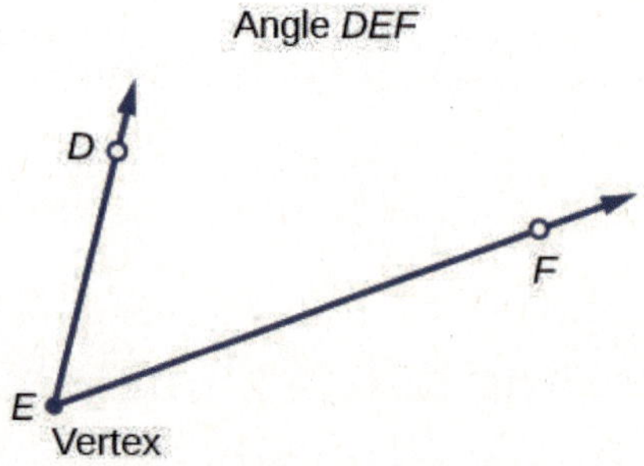

Figure 2

Greek letters are often used as variables for the measure of an angle. Table 1 is a list of Greek letters commonly used to represent angles, and a sample angle is shown in Figure 3.

θ	φ or ϕ	α	β	γ
theta	phi	alpha	beta	gamma

Table 1

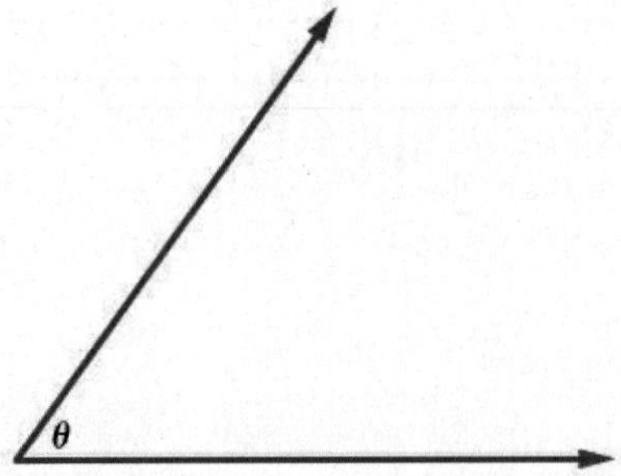

Figure 3 Angle theta, shown as $\angle\theta$

Angle creation is a dynamic process. We start with two rays lying on top of one another. We leave one fixed in place, and rotate the other. The fixed ray is the **initial side,** and the rotated ray is the **terminal side**. In order to identify the different sides, we indicate the rotation with a small arrow close to the vertex as in Figure 4.

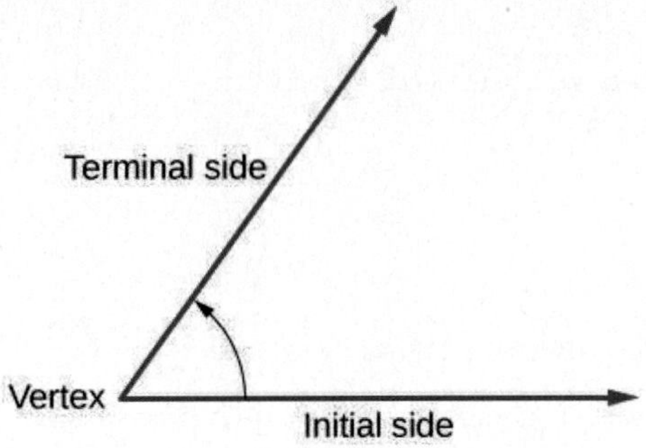

Figure 4

As we discussed at the beginning of the section, there are many applications for angles, but in order to use them correctly, we must be able to measure them. The **measure of an angle** is the amount of rotation from the initial side to the terminal side. Probably the most familiar unit of angle measurement is the degree. One **degree** is $\frac{1}{360}$ of a circular rotation, so a complete circular rotation contains 360 degrees. An angle measured in degrees should always include the unit "degrees" after the number, or include the degree symbol °. For example, $90 \text{ degrees} = 90°$.

To formalize our work, we will begin by drawing angles on an *x-y* coordinate plane. Angles can occur in any position on the coordinate plane, but for the purpose of comparison, the convention is to illustrate them in the same position whenever possible. An angle is in **standard position** if its vertex is located at the origin, and its initial side extends along the positive *x*-axis. See Figure 5.

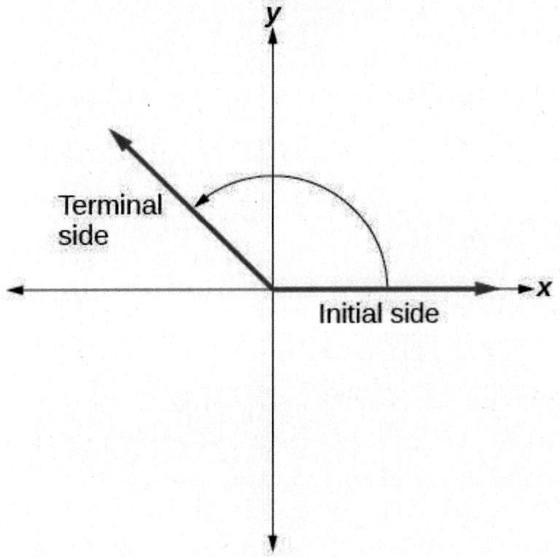

Figure 5

If the angle is measured in a counterclockwise direction from the initial side to the terminal side, the angle is said to be a **positive angle**. If the angle is measured in a clockwise direction, the angle is said to be a **negative angle**.

Drawing an angle in standard position always starts the same way—draw the initial side along the positive *x*-axis. To place the terminal side of the angle, we must calculate the fraction of a full rotation the angle represents. We do that by dividing the angle measure in degrees by $360°$. For example, to draw a $90°$ angle, we calculate that $\frac{90°}{360°} = \frac{1}{4}$. So, the terminal side will be one-fourth of the way around the circle, moving counterclockwise from the positive *x*-axis. To draw a $360°$ angle, we calculate that $\frac{360°}{360°} = 1$. So the terminal side will be 1 complete rotation around the circle, moving counterclockwise from the positive *x*-axis. In this case, the initial side and the terminal side overlap. See Figure 6.

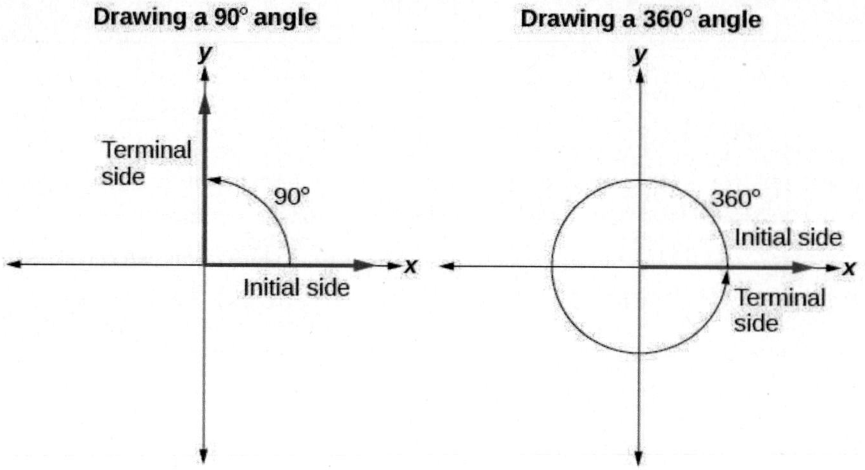

Figure 6

Since we define an angle in standard position by its terminal side, we have a special type of angle whose terminal side lies on an axis, a **quadrantal angle**. This type of angle can have a measure of 0°, 90°, 180°, 270°, or 360°. See Figure 7.

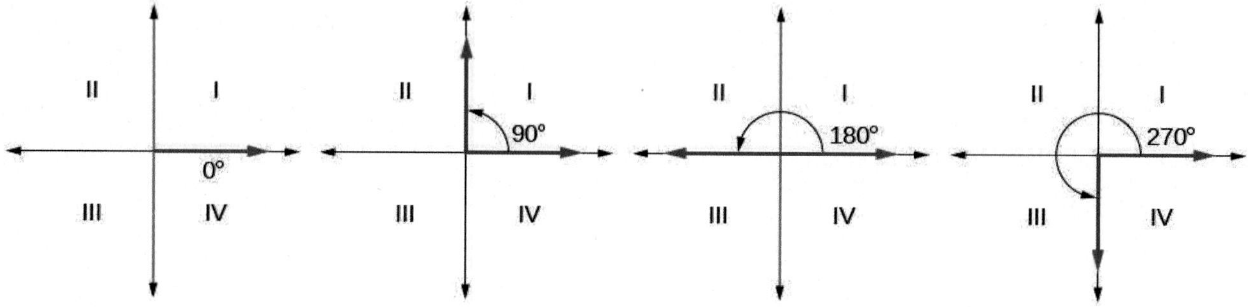

Figure 7 Quadrantal angles have a terminal side that lies along an axis. Examples are shown.

Quadrantal Angles

An angle is a **quadrantal angle** if its terminal side lies on an axis, including 0°, 90°, 180°, 270°, or 360°.

HOW TO

Given an angle measure in degrees, draw the angle in standard position.

1. Express the angle measure as a fraction of $360°$.
2. Reduce the fraction to simplest form.
3. Draw an angle that contains that same fraction of the circle, beginning on the positive *x*-axis and moving counterclockwise for positive angles and clockwise for negative angles.

EXAMPLE 1

Drawing an Angle in Standard Position Measured in Degrees

ⓐ Sketch an angle of $30°$ in standard position. ⓑ Sketch an angle of $-135°$ in standard position.

Solution

ⓐ

Divide the angle measure by $360°$.

$$\frac{30°}{360°} = \frac{1}{12}$$

To rewrite the fraction in a more familiar fraction, we can recognize that

$$\frac{1}{12} = \frac{1}{3}\left(\frac{1}{4}\right)$$

One-twelfth equals one-third of a quarter, so by dividing a quarter rotation into thirds, we can sketch a line at $30°$, as in Figure 8.

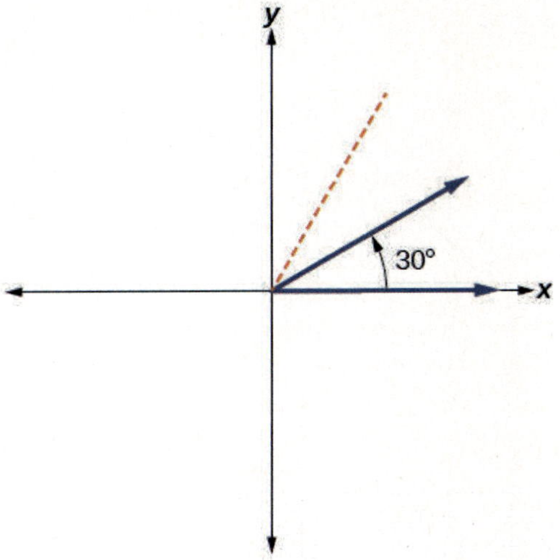

Figure 8

ⓑ
Divide the angle measure by 360°.

$$\frac{-135°}{360°} = -\frac{3}{8}$$

In this case, we can recognize that

$$-\frac{3}{8} = -\frac{3}{2}\left(\frac{1}{4}\right)$$

Three-eighths is one and one-half times a quarter, so we place a line by moving clockwise one full quarter and one-half of another quarter, as in Figure 9.

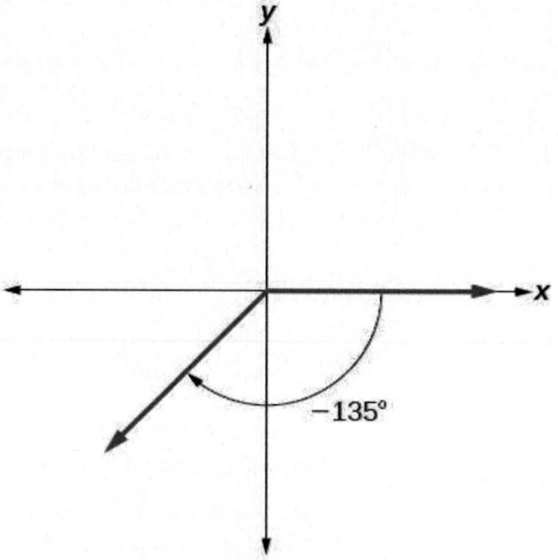

Figure 9

TRY IT #1 Show an angle of 240° on a circle in standard position.

Converting Between Degrees and Radians

Dividing a circle into 360 parts is an arbitrary choice, although it creates the familiar degree measurement. We may choose other ways to divide a circle. To find another unit, think of the process of drawing a circle. Imagine that you stop before the circle is completed. The portion that you drew is referred to as an arc. An arc may be a portion of a full circle, a full circle, or more than a full circle, represented by more than one full rotation. The length of the arc around an entire circle is called the circumference of that circle.

The circumference of a circle is $C = 2\pi r$. If we divide both sides of this equation by r, we create the ratio of the circumference, which is always 2π, to the radius, regardless of the length of the radius. So the circumference of any circle is $2\pi \approx 6.28$ times the length of the radius. That means that if we took a string as long as the radius and used it to measure consecutive lengths around the circumference, there would be room for six full string-lengths and a little more than a quarter of a seventh, as shown in Figure 10.

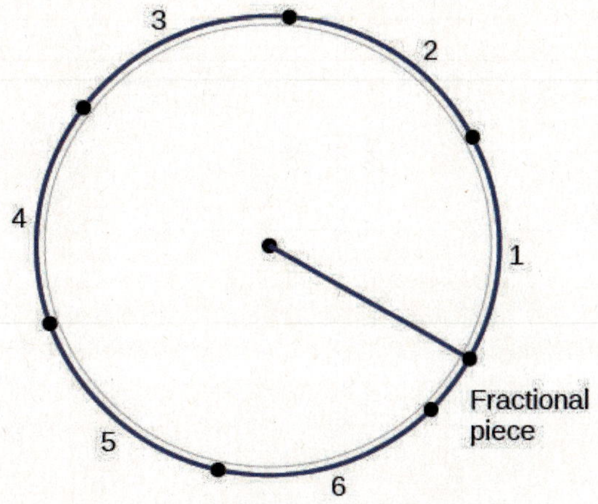

Figure 10

This brings us to our new angle measure. One **radian** is the measure of a central angle of a circle that intercepts an arc equal in length to the radius of that circle. A central angle is an angle formed at the center of a circle by two radii. Because the total circumference equals 2π times the radius, a full circular rotation is 2π radians.

$$\begin{aligned} 2\pi \text{ radians} &= 360° \\ \pi \text{ radians} &= \frac{360°}{2} = 180° \\ 1 \text{ radian} &= \frac{180°}{\pi} \approx 57.3° \end{aligned}$$

See Figure 11. Note that when an angle is described without a specific unit, it refers to radian measure. For example, an angle measure of 3 indicates 3 radians. In fact, radian measure is dimensionless, since it is the quotient of a length (circumference) divided by a length (radius) and the length units cancel.

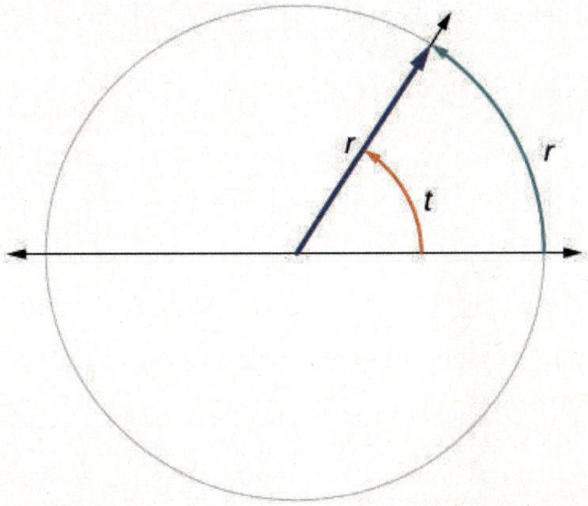

Figure 11 The angle t sweeps out a measure of one radian. Note that the length of the intercepted arc is the same as the length of the radius of the circle.

Relating Arc Lengths to Radius

An **arc length** s is the length of the curve along the arc. Just as the full circumference of a circle always has a constant ratio to the radius, the arc length produced by any given angle also has a constant relation to the radius, regardless of the length of the radius.

This ratio, called the radian measure, is the same regardless of the radius of the circle—it depends only on the angle. This property allows us to define a measure of any angle as the ratio of the arc length s to the radius r. See Figure 12.

$$\begin{aligned} s &= r\theta \\ \theta &= \frac{s}{r} \end{aligned}$$

If $s = r$, then $\theta = \frac{r}{r} = 1 \text{ radian}$.

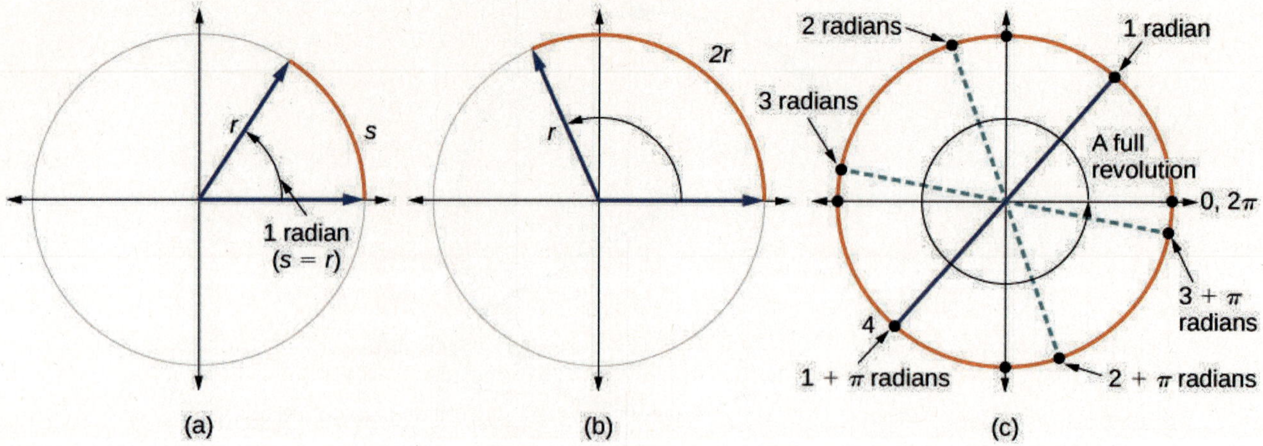

Figure 12 (a) In an angle of 1 radian, the arc length s equals the radius r. (b) An angle of 2 radians has an arc length $s = 2r$. (c) A full revolution is 2π, or about 6.28 radians.

To elaborate on this idea, consider two circles, one with radius 2 and the other with radius 3. Recall the circumference of a circle is $C = 2\pi r$, where r is the radius. The smaller circle then has circumference $2\pi(2) = 4\pi$ and the larger has circumference $2\pi(3) = 6\pi$. Now we draw a $45°$ angle on the two circles, as in Figure 13.

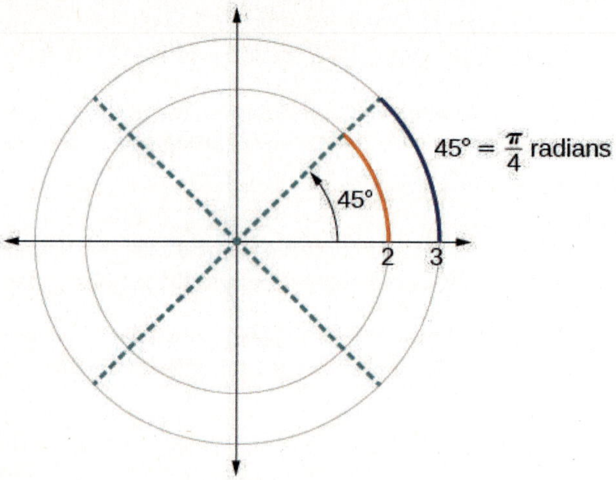

Figure 13 A $45°$ angle contains one-eighth of the circumference of a circle, regardless of the radius.

Notice what happens if we find the ratio of the arc length divided by the radius of the circle.

$$\begin{aligned} \text{Smaller circle: } \frac{\frac{1}{2}\pi}{2} &= \frac{1}{4}\pi \\ \text{Larger circle: } \frac{\frac{3}{4}\pi}{3} &= \frac{1}{4}\pi \end{aligned}$$

Since both ratios are $\frac{1}{4}\pi$, the angle measures of both circles are the same, even though the arc length and radius differ.

Radians

One radian is the measure of the central angle of a circle such that the length of the arc between the initial side and the terminal side is equal to the radius of the circle. A full revolution $(360°)$ equals 2π radians. A half revolution $(180°)$ is equivalent to π radians.

The **radian measure** of an angle is the ratio of the length of the arc subtended by the angle to the radius of the circle. In other words, if s is the length of an arc of a circle, and r is the radius of the circle, then the central angle containing that arc measures $\frac{s}{r}$ radians. In a circle of radius 1, the radian measure corresponds to the length of the arc.

Q&A **A measure of 1 radian looks to be about $60°$. Is that correct?**

Yes. It is approximately $57.3°$. *Because* 2π *radians equals* $360°$, 1 *radian equals* $\frac{360°}{2\pi} \approx 57.3°$.

Using Radians

Because radian measure is the ratio of two lengths, it is a unitless measure. For example, in Figure 12, suppose the radius were 2 inches and the distance along the arc were also 2 inches. When we calculate the radian measure of the angle, the "inches" cancel, and we have a result without units. Therefore, it is not necessary to write the label "radians" after a radian measure, and if we see an angle that is not labeled with "degrees" or the degree symbol, we can assume that it is a radian measure.

Considering the most basic case, the unit circle (a circle with radius 1), we know that 1 rotation equals 360 degrees, $360°$. We can also track one rotation around a circle by finding the circumference, $C = 2\pi r$, and for the unit circle $C = 2\pi$. These two different ways to rotate around a circle give us a way to convert from degrees to radians.

$$\begin{aligned} 1 \text{ rotation} &= 360° &= 2\pi \text{ radians} \\ \tfrac{1}{2} \text{ rotation} &= 180° &= \pi \text{ radians} \\ \tfrac{1}{4} \text{ rotation} &= 90° &= \tfrac{\pi}{2} \text{ radians} \end{aligned}$$

Identifying Special Angles Measured in Radians

In addition to knowing the measurements in degrees and radians of a quarter revolution, a half revolution, and a full revolution, there are other frequently encountered angles in one revolution of a circle with which we should be familiar. It is common to encounter multiples of 30, 45, 60, and 90 degrees. These values are shown in Figure 14. Memorizing these angles will be very useful as we study the properties associated with angles.

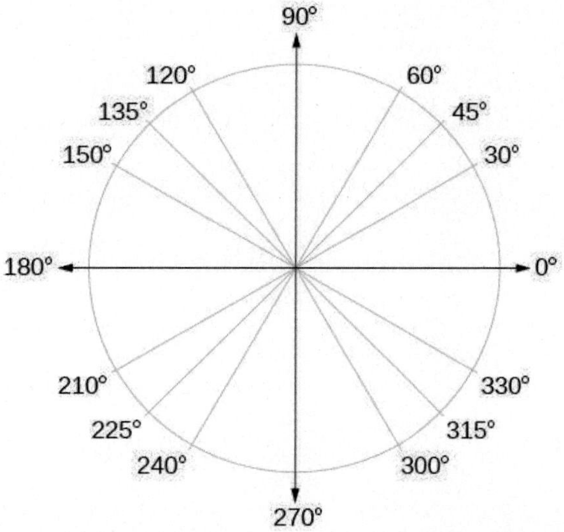

Figure 14 Commonly encountered angles measured in degrees

Now, we can list the corresponding radian values for the common measures of a circle corresponding to those listed in Figure 14, which are shown in Figure 15. Be sure you can verify each of these measures.

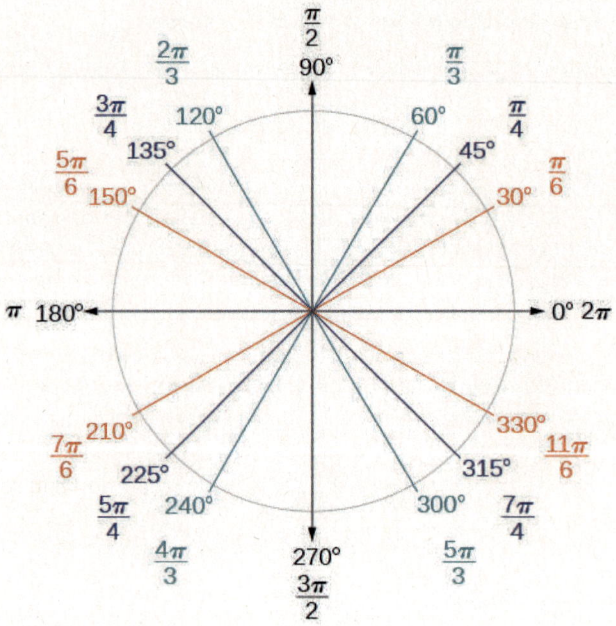

Figure 15 Commonly encountered angles measured in radians

EXAMPLE 2

Finding a Radian Measure

Find the radian measure of one-third of a full rotation.

Solution

For any circle, the arc length along such a rotation would be one-third of the circumference. We know that

$$1 \text{ rotation} = 2\pi r$$

So,

$$\begin{aligned} s &= \tfrac{1}{3}(2\pi r) \\ &= \tfrac{2\pi r}{3} \end{aligned}$$

The radian measure would be the arc length divided by the radius.

$$\begin{aligned} \text{radian measure} &= \frac{\frac{2\pi r}{3}}{r} \\ &= \frac{2\pi r}{3r} \\ &= \frac{2\pi}{3} \end{aligned}$$

TRY IT #2 Find the radian measure of three-fourths of a full rotation.

Converting Between Radians and Degrees

Because degrees and radians both measure angles, we need to be able to convert between them. We can easily do so using a proportion where θ is the measure of the angle in degrees and θ_R is the measure of the angle in radians.

$$\frac{\theta}{180} = \frac{\theta_R}{\pi}$$

This proportion shows that the measure of angle θ in degrees divided by 180 equals the measure of angle θ in radians divided by π. Or, phrased another way, degrees is to 180 as radians is to π.

$$\frac{\text{Degrees}}{180} = \frac{\text{Radians}}{\pi}$$

Converting between Radians and Degrees

To convert between degrees and radians, use the proportion

$$\frac{\theta}{180} = \frac{\theta_R}{\pi}$$

EXAMPLE 3

Converting Radians to Degrees

Convert each radian measure to degrees.

ⓐ $\frac{\pi}{6}$ ⓑ 3

Solution

Because we are given radians and we want degrees, we should set up a proportion and solve it.

ⓐ We use the proportion, substituting the given information.

$$\begin{aligned} \frac{\theta}{180} &= \frac{\theta_R}{\pi} \\ \frac{\theta}{180} &= \frac{\frac{\pi}{6}}{\pi} \\ \theta &= \frac{180}{6} \\ \theta &= 30^\circ \end{aligned}$$

ⓑ We use the proportion, substituting the given information.

$$\begin{aligned} \frac{\theta}{180} &= \frac{\theta_R}{\pi} \\ \frac{\theta}{180} &= \frac{3}{\pi} \\ \theta &= \frac{3(180)}{\pi} \\ \theta &\approx 172^\circ \end{aligned}$$

TRY IT #3 Convert $-\frac{3\pi}{4}$ radians to degrees.

EXAMPLE 4

Converting Degrees to Radians

Convert 15 degrees to radians.

Solution

In this example, we start with degrees and want radians, so we again set up a proportion, but we substitute the given information into a different part of the proportion.

$$\begin{aligned} \frac{\theta}{180} &= \frac{\theta_R}{\pi} \\ \frac{15}{180} &= \frac{\theta_R}{\pi} \\ \frac{15\pi}{180} &= \theta_R \\ \frac{\pi}{12} &= \theta_R \end{aligned}$$

Analysis

Another way to think about this problem is by remembering that $30^\circ = \frac{\pi}{6}$. Because $15^\circ = \frac{1}{2}(30^\circ)$, we can find that $\frac{1}{2}\left(\frac{\pi}{6}\right)$ is $\frac{\pi}{12}$.

TRY IT #4 Convert 126° to radians.

Finding Coterminal Angles

Converting between degrees and radians can make working with angles easier in some applications. For other applications, we may need another type of conversion. Negative angles and angles greater than a full revolution are more awkward to work with than those in the range of $0°$ to $360°$, or 0 to 2π. It would be convenient to replace those out-of-range angles with a corresponding angle within the range of a single revolution.

It is possible for more than one angle to have the same terminal side. Look at Figure 16. The angle of $140°$ is a positive angle, measured counterclockwise. The angle of $-220°$ is a negative angle, measured clockwise. But both angles have the same terminal side. If two angles in standard position have the same terminal side, they are coterminal angles. Every angle greater than $360°$ or less than $0°$ is coterminal with an angle between $0°$ and $360°$, and it is often more convenient to find the coterminal angle within the range of $0°$ to $360°$ than to work with an angle that is outside that range.

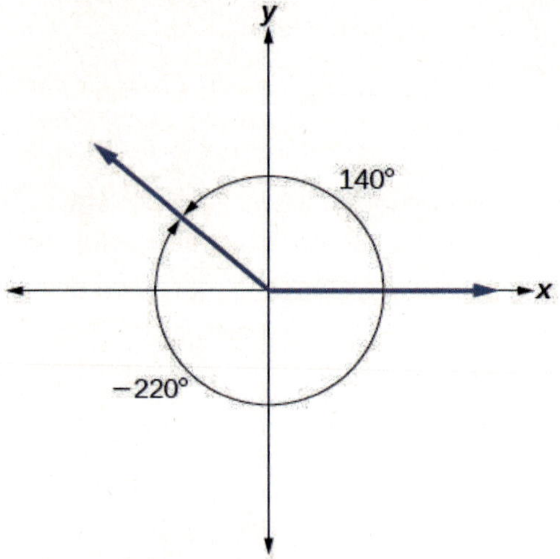

Figure 16 An angle of $140°$ and an angle of $-220°$ are coterminal angles.

Any angle has infinitely many coterminal angles because each time we add $360°$ to that angle—or subtract $360°$ from it—the resulting value has a terminal side in the same location. For example, $100°$ and $460°$ are coterminal for this reason, as is $-260°$.

An angle's reference angle is the measure of the smallest, positive, acute angle t formed by the terminal side of the angle t and the horizontal axis. Thus positive reference angles have terminal sides that lie in the first quadrant and can be used as models for angles in other quadrants. See Figure 17 for examples of reference angles for angles in different quadrants.

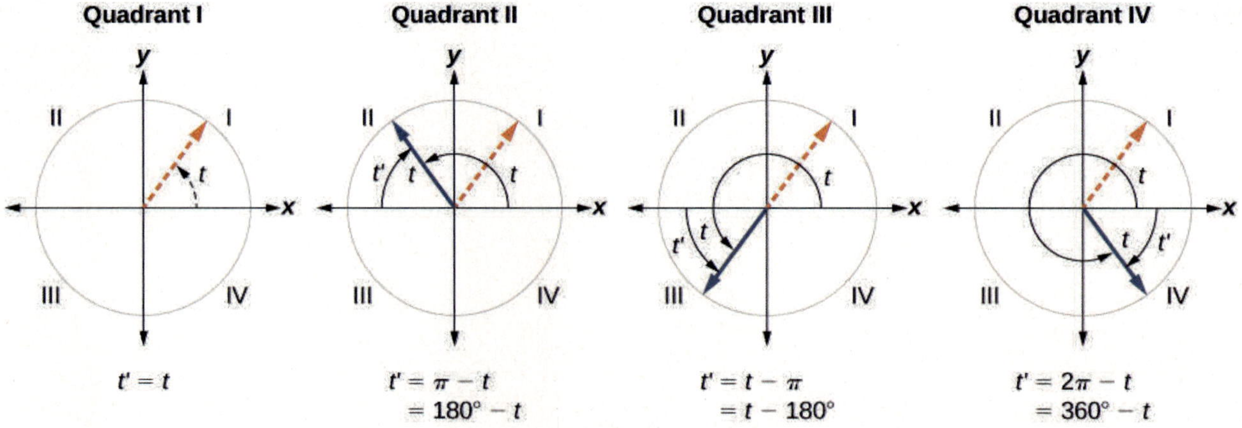

Figure 17

Coterminal and Reference Angles

Coterminal angles are two angles in standard position that have the same terminal side.

An angle's **reference angle** is the size of the smallest acute angle, t', formed by the terminal side of the angle t and the horizontal axis.

HOW TO

Given an angle greater than $360°$, find a coterminal angle between $0°$ and $360°$

1. Subtract $360°$ from the given angle.
2. If the result is still greater than $360°$, subtract $360°$ again till the result is between $0°$ and $360°$.
3. The resulting angle is coterminal with the original angle.

EXAMPLE 5

Finding an Angle Coterminal with an Angle of Measure Greater Than $360°$

Find the least positive angle θ that is coterminal with an angle measuring $800°$, where $0° \leq \theta < 360°$.

Solution

An angle with measure $800°$ is coterminal with an angle with measure $800 - 360 = 440°$, but $440°$ is still greater than $360°$, so we subtract $360°$ again to find another coterminal angle: $440 - 360 = 80°$.

The angle $\theta = 80°$ is coterminal with $800°$. To put it another way, $800°$ equals $80°$ plus two full rotations, as shown in Figure 18.

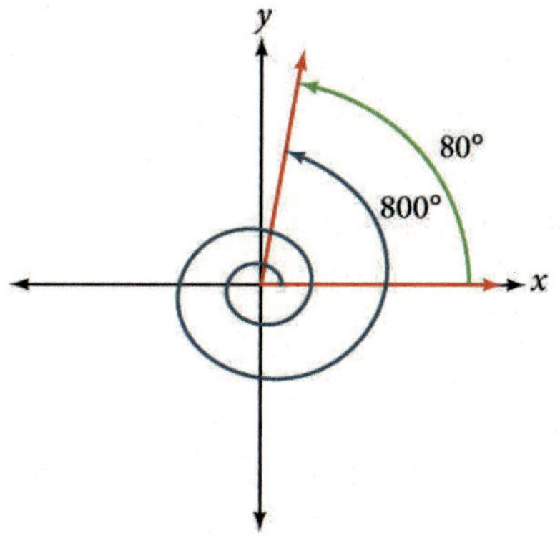

Figure 18

TRY IT #5 Find an angle α that is coterminal with an angle measuring $870°$, where $0° \leq \alpha < 360°$.

HOW TO

Given an angle with measure less than $0°$, find a coterminal angle having a measure between $0°$ and $360°$.

1. Add $360°$ to the given angle.
2. If the result is still less than $0°$, add $360°$ again until the result is between $0°$ and $360°$.
3. The resulting angle is coterminal with the original angle.

EXAMPLE 6

Finding an Angle Coterminal with an Angle Measuring Less Than $0°$

Show the angle with measure $-45°$ on a circle and find a positive coterminal angle α such that $0° \leq \alpha < 360°$.

Solution

Since $45°$ is half of $90°$, we can start at the positive horizontal axis and measure clockwise half of a $90°$ angle.

Because we can find coterminal angles by adding or subtracting a full rotation of $360°$, we can find a positive coterminal angle here by adding $360°$.

$$-45° + 360° = 315°$$

We can then show the angle on a circle, as in Figure 19.

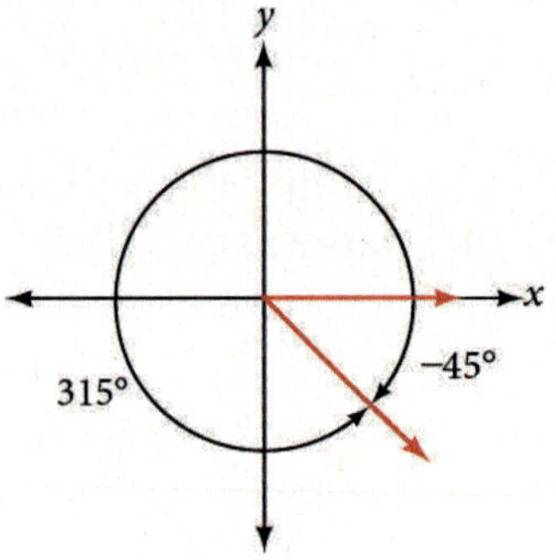

Figure 19

TRY IT #6 Find an angle β that is coterminal with an angle measuring $-300°$ such that $0° \leq \beta < 360°$.

Finding Coterminal Angles Measured in Radians

We can find coterminal angles measured in radians in much the same way as we have found them using degrees. In both cases, we find coterminal angles by adding or subtracting one or more full rotations.

HOW TO

Given an angle greater than 2π, find a coterminal angle between 0 and 2π.

1. Subtract 2π from the given angle.
2. If the result is still greater than 2π, subtract 2π again until the result is between 0 and 2π.
3. The resulting angle is coterminal with the original angle.

EXAMPLE 7

Finding Coterminal Angles Using Radians

Find an angle β that is coterminal with $\frac{19\pi}{4}$, where $0 \leq \beta < 2\pi$.

Solution

When working in degrees, we found coterminal angles by adding or subtracting 360 degrees, a full rotation. Likewise, in radians, we can find coterminal angles by adding or subtracting full rotations of 2π radians:

$$\begin{aligned}\frac{19\pi}{4} - 2\pi &= \frac{19\pi}{4} - \frac{8\pi}{4} \\ &= \frac{11\pi}{4}\end{aligned}$$

The angle $\frac{11\pi}{4}$ is coterminal, but not less than 2π, so we subtract another rotation.

$$\begin{aligned} \frac{11\pi}{4} - 2\pi &= \frac{11\pi}{4} - \frac{8\pi}{4} \\ &= \frac{3\pi}{4} \end{aligned}$$

The angle $\frac{3\pi}{4}$ is coterminal with $\frac{19\pi}{4}$, as shown in Figure 20.

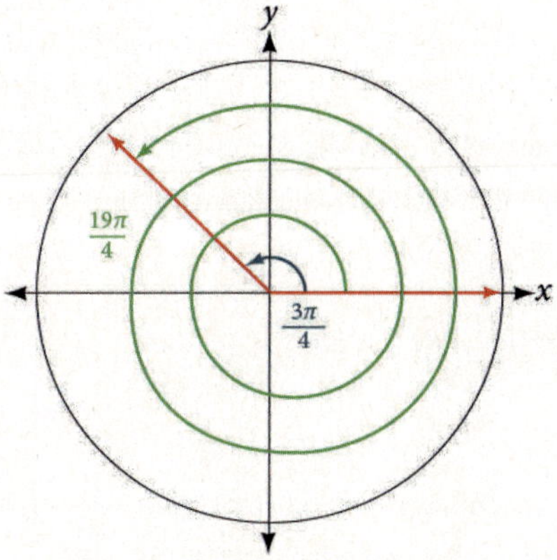

Figure 20

> TRY IT #7 Find an angle of measure θ that is coterminal with an angle of measure $-\frac{17\pi}{6}$ where $0 \leq \theta < 2\pi$.

Determining the Length of an Arc

Recall that the radian measure θ of an angle was defined as the ratio of the arc length s of a circular arc to the radius r of the circle, $\theta = \frac{s}{r}$. From this relationship, we can find arc length along a circle, given an angle.

Arc Length on a Circle

In a circle of radius r, the length of an arc s subtended by an angle with measure θ in radians, shown in Figure 21, is

$$s = r\theta$$

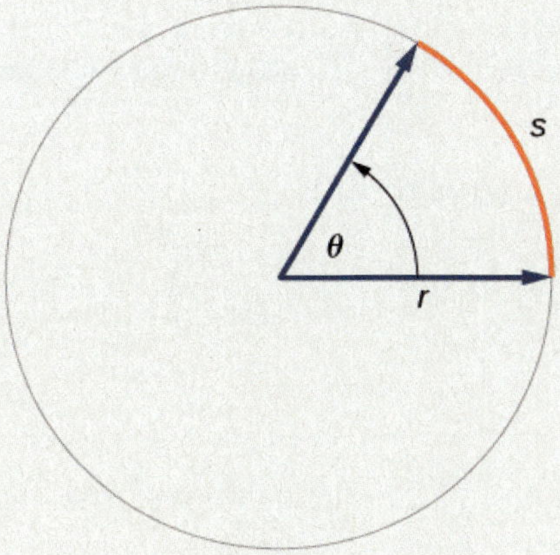

Figure 21

HOW TO

Given a circle of radius r, calculate the length s of the arc subtended by a given angle of measure θ.

1. If necessary, convert θ to radians.

2. Multiply the radius r $\theta: s = r\theta$.

EXAMPLE 8

Finding the Length of an Arc

Assume the orbit of Mercury around the sun is a perfect circle. Mercury is approximately 36 million miles from the sun.

ⓐ In one Earth day, Mercury completes 0.0114 of its total revolution. How many miles does it travel in one day?

ⓑ Use your answer from part (a) to determine the radian measure for Mercury's movement in one Earth day.

Solution

ⓐ Let's begin by finding the circumference of Mercury's orbit.

$$\begin{aligned} C &= 2\pi r \\ &= 2\pi(36 \text{ million miles}) \\ &\approx 226 \text{ million miles} \end{aligned}$$

Since Mercury completes 0.0114 of its total revolution in one Earth day, we can now find the distance traveled.

$$(0.0114)\,226 \text{ million miles} = 2.58 \text{ million miles}$$

ⓑ Now, we convert to radians.

$$\begin{aligned} \text{radian} &= \frac{\text{arclength}}{\text{radius}} \\ &= \frac{2.58 \text{ million miles}}{36 \text{ million miles}} \\ &= 0.0717 \end{aligned}$$

> **TRY IT** #8 Find the arc length along a circle of radius 10 units subtended by an angle of 215°.

Finding the Area of a Sector of a Circle

In addition to arc length, we can also use angles to find the area of a sector of a circle. A sector is a region of a circle bounded by two radii and the intercepted arc, like a slice of pizza or pie. Recall that the area of a circle with radius r can be found using the formula $A = \pi r^2$. If the two radii form an angle of θ, measured in radians, then $\frac{\theta}{2\pi}$ is the ratio of the angle measure to the measure of a full rotation and is also, therefore, the ratio of the area of the sector to the area of the circle. Thus, the **area of a sector** is the fraction $\frac{\theta}{2\pi}$ multiplied by the entire area. (Always remember that this formula only applies if θ is in radians.)

$$\begin{aligned} \text{Area of sector} &= \left(\frac{\theta}{2\pi}\right)\pi r^2 \\ &= \frac{\theta \pi r^2}{2\pi} \\ &= \frac{1}{2}\theta r^2 \end{aligned}$$

Area of a Sector

The **area of a sector** of a circle with radius r subtended by an angle θ, measured in radians, is

$$A = \frac{1}{2}\theta r^2$$

See Figure 22.

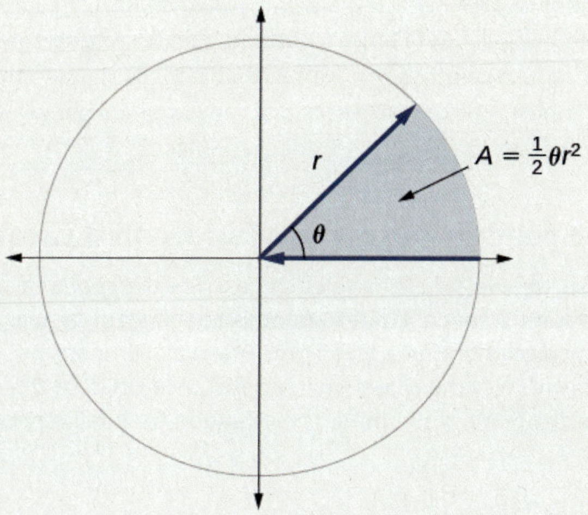

Figure 22 The area of the sector equals half the square of the radius times the central angle measured in radians.

HOW TO

Given a circle of radius r, find the area of a sector defined by a given angle θ.

1. If necessary, convert θ to radians.
2. Multiply half the radian measure of θ by the square of the radius r : $A = \frac{1}{2}\theta r^2$.

EXAMPLE 9

Finding the Area of a Sector

An automatic lawn sprinkler sprays a distance of 20 feet while rotating 30 degrees, as shown in Figure 23. What is the area of the sector of grass the sprinkler waters?

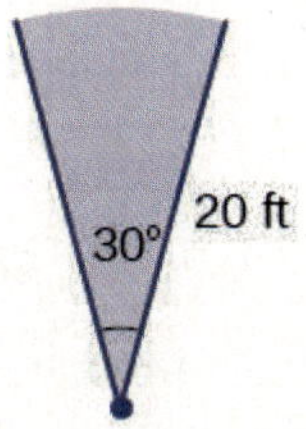

Figure 23 The sprinkler sprays 20 ft within an arc of $30°$.

Solution

First, we need to convert the angle measure into radians. Because 30 degrees is one of our special angles, we already know the equivalent radian measure, but we can also convert:

$$\begin{aligned} 30 \text{ degrees} &= 30 \cdot \frac{\pi}{180} \\ &= \frac{\pi}{6} \text{ radians} \end{aligned}$$

The area of the sector is then

$$\begin{aligned} \text{Area} &= \frac{1}{2}\left(\frac{\pi}{6}\right)(20)^2 \\ &\approx 104.72 \end{aligned}$$

So the area is about 104.72 ft^2.

> **TRY IT** #9 In central pivot irrigation, which creates the field shapes similar to the image at the beginning of Equations and Inequalities, a large irrigation pipe on wheels rotates around a center point. A farmer has a central pivot system with a radius of 400 meters. If water restrictions only allow her to water 150 thousand square meters a day, what angle should she set the system to cover? Write the answer in radian measure to two decimal places.

Use Linear and Angular Speed to Describe Motion on a Circular Path

In addition to finding the area of a sector, we can use angles to describe the speed of a moving object. An object traveling in a circular path has two types of speed. **Linear speed** is speed along a straight path and can be determined by the distance it moves along (its displacement) in a given time interval. For instance, if a wheel with radius 5 inches rotates once a second, a point on the edge of the wheel moves a distance equal to the circumference, or 10π inches, every second. So the linear speed of the point is 10π in./s. The equation for linear speed is as follows where v is linear speed, s is displacement, and t is time.

$$v = \frac{s}{t}$$

Angular speed results from circular motion and can be determined by the angle through which a point rotates in a given time interval. In other words, **angular speed** is angular rotation per unit time. So, for instance, if a gear makes a full rotation every 4 seconds, we can calculate its angular speed as $\frac{360 \text{ degrees}}{4 \text{ seconds}} = 90$ degrees per second. Angular speed can be given in radians per second, rotations per minute, or degrees per hour for example. The equation for angular speed is as follows, where ω (read as omega) is angular speed, θ is the angle traversed, and t is time.

$$\omega = \frac{\theta}{t}$$

Combining the definition of angular speed with the arc length equation, $s = r\theta$, we can find a relationship between angular and linear speeds. The angular speed equation can be solved for θ, giving $\theta = \omega t$. Substituting this into the arc length equation gives:

$$\begin{aligned} s &= r\theta \\ &= r\omega t \end{aligned}$$

Substituting this into the linear speed equation gives:

$$\begin{aligned} v &= \frac{s}{t} \\ &= \frac{r\omega t}{t} \\ &= r\omega \end{aligned}$$

Angular and Linear Speed

As a point moves along a circle of radius r, its **angular speed**, ω, is the angular rotation θ per unit time, t.

$$\omega = \frac{\theta}{t}$$

The **linear speed**, v, of the point can be found as the distance traveled, arc length s, per unit time, t.

$$v = \frac{s}{t}$$

When the angular speed is measured in radians per unit time, linear speed and angular speed are related by the equation

$$v = r\omega$$

This equation states that the angular speed in radians, ω, representing the amount of rotation occurring in a unit of time, can be multiplied by the radius r to calculate the total arc length traveled in a unit of time, which is the definition of linear speed.

HOW TO

Given the amount of angle rotation and the time elapsed, calculate the angular speed.

1. If necessary, convert the angle measure to radians.
2. Divide the angle in radians by the number of time units elapsed: $\omega = \frac{\theta}{t}$.
3. The resulting speed will be in radians per time unit.

Water wheels have been used for thousands of years to transfer the power of flowing water to other devices. The image below depicts the design of the the 3rd century Roman water wheel in Hierapolis, a city in what is now Turkey. Water turned the wheel, which in turn rotated a crank connected to two saws used to cut blocks. These design elements were used in water wheel applications throughout the world, and even provided the underlying principle for the steam engine, invented about 1500 years later.

EXAMPLE 10

Finding Angular Speed

A water wheel, shown in Figure 24, completes 1 rotation every 5 seconds. Find the angular speed in radians per second.

Figure 24

Solution

The wheel completes 1 rotation, or passes through an angle of 2π radians in 5 seconds, so the angular speed would be $\omega = \frac{2\pi}{5} \approx 1.257$ radians per second.

TRY IT #10 A vintage vinyl record is played on a turntable rotating clockwise at a rate of 45 rotations per minute. Find the angular speed in radians per second.

HOW TO

Given the radius of a circle, an angle of rotation, and a length of elapsed time, determine the linear speed.

1. Convert the total rotation to radians if necessary.
2. Divide the total rotation in radians by the elapsed time to find the angular speed: apply $\omega = \frac{\theta}{t}$.

3. Multiply the angular speed by the length of the radius to find the linear speed, expressed in terms of the length unit used for the radius and the time unit used for the elapsed time: apply $v = r\omega$.

EXAMPLE 11

Finding a Linear Speed

A bicycle has wheels 28 inches in diameter. A tachometer determines the wheels are rotating at 180 RPM (revolutions per minute). Find the speed the bicycle is traveling down the road.

Solution

Here, we have an angular speed and need to find the corresponding linear speed, since the linear speed of the outside of the tires is the speed at which the bicycle travels down the road.

We begin by converting from rotations per minute to radians per minute. It can be helpful to utilize the units to make this conversion:

$$180\frac{\cancel{\text{rotations}}}{\text{minute}} \cdot \frac{2\pi\ \text{radians}}{\cancel{\text{rotation}}} = 360\pi\frac{\text{radians}}{\text{minute}}$$

Using the formula from above along with the radius of the wheels, we can find the linear speed:

$$\begin{aligned} v &= (14\ \text{inches})\left(360\pi\frac{\text{radians}}{\text{minute}}\right) \\ &= 5040\pi\frac{\text{inches}}{\text{minute}} \end{aligned}$$

Remember that radians are a unitless measure, so it is not necessary to include them.

Finally, we may wish to convert this linear speed into a more familiar measurement, like miles per hour.

$$5040\pi\frac{\cancel{\text{inches}}}{\cancel{\text{minute}}} \cdot \frac{1\ \cancel{\text{feet}}}{12\ \cancel{\text{inches}}} \cdot \frac{1\ \text{mile}}{5280\ \cancel{\text{feet}}} \cdot \frac{60\ \cancel{\text{minutes}}}{1\ \text{hour}} \approx 14.99\ \text{miles per hour (mph)}$$

TRY IT #11 A satellite is rotating around Earth at 0.25 radian per hour at an altitude of 242 km above Earth. If the radius of Earth is 6378 kilometers, find the linear speed of the satellite in kilometers per hour.

MEDIA

Access these online resources for additional instruction and practice with angles, arc length, and areas of sectors.

Angles in Standard Position (http://openstax.org/l/standardpos)
Angle of Rotation (http://openstax.org/l/angleofrotation)
Coterminal Angles (http://openstax.org/l/coterminal)
Determining Coterminal Angles (http://openstax.org/l/detcoterm)
Positive and Negative Coterminal Angles (http://openstax.org/l/posnegcoterm)
Radian Measure (http://openstax.org/l/radianmeas)
Coterminal Angles in Radians (http://openstax.org/l/cotermrad)
Arc Length and Area of a Sector (http://openstax.org/l/arclength)

7.1 SECTION EXERCISES

Verbal

1. Draw an angle in standard position. Label the vertex, initial side, and terminal side.

2. Explain why there are an infinite number of angles that are coterminal to a certain angle.

3. State what a positive or negative angle signifies, and explain how to draw each.

4. How does radian measure of an angle compare to the degree measure? Include an explanation of 1 radian in your paragraph.

5. Explain the differences between linear speed and angular speed when describing motion along a circular path.

Graphical

For the following exercises, draw an angle in standard position with the given measure.

6. $30°$

7. $300°$

8. $-80°$

9. $135°$

10. $-150°$

11. $\frac{2\pi}{3}$

12. $\frac{7\pi}{4}$

13. $\frac{5\pi}{6}$

14. $\frac{\pi}{2}$

15. $-\frac{\pi}{10}$

16. $415°$

17. $-120°$

18. $-315°$

19. $\frac{22\pi}{3}$

20. $-\frac{\pi}{6}$

21. $-\frac{4\pi}{3}$

For the following exercises, refer to Figure 25. Round to two decimal places.

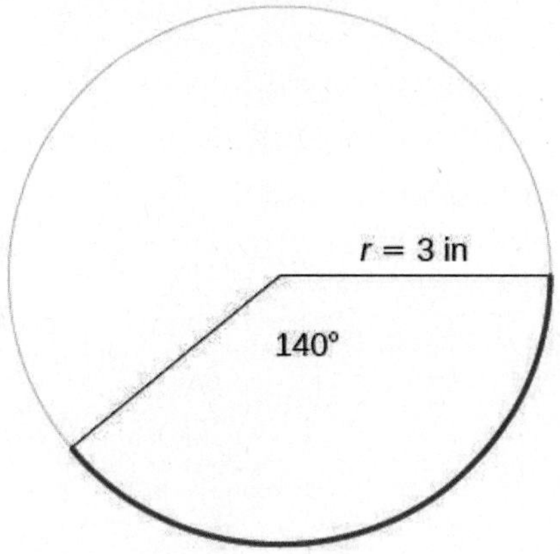

Figure 25

22. Find the arc length.

23. Find the area of the sector.

For the following exercises, refer to Figure 26. Round to two decimal places.

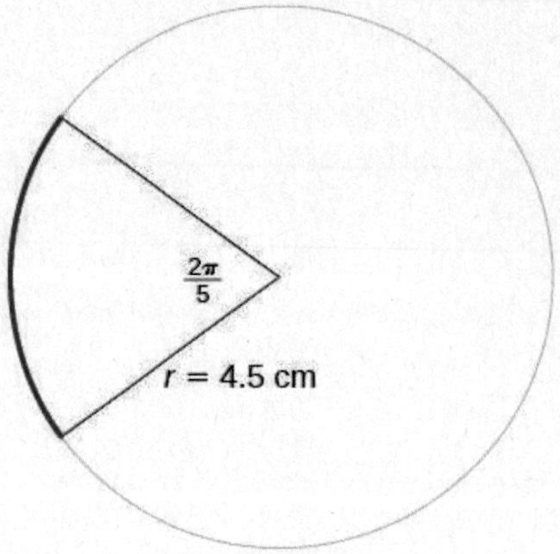

Figure 26

24. Find the arc length.

25. Find the area of the sector.

Algebraic

For the following exercises, convert angles in radians to degrees.

26. $\frac{3\pi}{4}$ radians

27. $\frac{\pi}{9}$ radians

28. $-\frac{5\pi}{4}$ radians

29. $\frac{\pi}{3}$ radians

30. $-\frac{7\pi}{3}$ radians

31. $-\frac{5\pi}{12}$ radians

32. $\frac{11\pi}{6}$ radians

For the following exercises, convert angles in degrees to radians.

33. $90°$

34. $100°$

35. $-540°$

36. $-120°$

37. $180°$

38. $-315°$

39. $150°$

For the following exercises, use the given information to find the length of a circular arc. Round to two decimal places.

40. Find the length of the arc of a circle of radius 12 inches subtended by a central angle of $\frac{\pi}{4}$. radians.

41. Find the length of the arc of a circle of radius 5.02 miles subtended by the central angle of $\frac{\pi}{3}$.

42. Find the length of the arc of a circle of diameter 14 meters subtended by the central angle of $\frac{5\pi}{6}$.

43. Find the length of the arc of a circle of radius 10 centimeters subtended by the central angle of $50°$.

44. Find the length of the arc of a circle of radius 5 inches subtended by the central angle of $220°$.

45. Find the length of the arc of a circle of diameter 12 meters subtended by the central angle is $63°$.

For the following exercises, use the given information to find the area of the sector. Round to four decimal places.

46. A sector of a circle has a central angle of $45°$ and a radius 6 cm.

47. A sector of a circle has a central angle of $30°$ and a radius of 20 cm.

48. A sector of a circle with diameter 10 feet and an angle of $\frac{\pi}{2}$ radians.

49. A sector of a circle with radius of 0.7 inches and an angle of π radians.

For the following exercises, find the angle between $0°$ and $360°$ that is coterminal to the given angle.

50. $-40°$

51. $-110°$

52. $700°$

53. $1400°$

For the following exercises, find the angle between 0 and 2π in radians that is coterminal to the given angle.

54. $-\frac{\pi}{9}$

55. $\frac{10\pi}{3}$

56. $\frac{13\pi}{6}$

57. $\frac{44\pi}{9}$

Real-World Applications

58. A truck with 32-inch diameter wheels is traveling at 60 mi/h. Find the angular speed of the wheels in rad/min. How many revolutions per minute do the wheels make?

59. A bicycle with 24-inch diameter wheels is traveling at 15 mi/h. Find the angular speed of the wheels in rad/min. How many revolutions per minute do the wheels make?

60. A wheel of radius 8 inches is rotating 15°/s. What is the linear speed v, the angular speed in RPM, and the angular speed in rad/s?

61. A wheel of radius 14 inches is rotating 0.5 rad/s. What is the linear speed v, the angular speed in RPM, and the angular speed in deg/s?

62. A computer hard drive disc has diameter of 120 millimeters. When playing audio, the angular speed varies to keep the linear speed constant where the disc is being read. When reading along the outer edge of the disc, the angular speed is about 200 RPM (revolutions per minute). Find the linear speed.

63. When being burned in a writable CD-R drive, the angular speed of a CD is often much faster than when playing audio, but the angular speed still varies to keep the linear speed constant where the disc is being written. When writing along the outer edge of the disc, the angular speed of one drive is about 4800 RPM (revolutions per minute). Find the linear speed if the CD has diameter of 120 millimeters.

64. A person is standing on the equator of Earth (radius 3960 miles). What are their linear and angular speeds?

65. Find the distance along an arc on the surface of Earth that subtends a central angle of 5 minutes $\left(1 \text{ minute} = \frac{1}{60} \text{ degree}\right)$. The radius of Earth is 3960 miles.

66. Find the distance along an arc on the surface of Earth that subtends a central angle of 7 minutes $\left(1 \text{ minute} = \frac{1}{60} \text{ degree}\right)$. The radius of Earth is 3960 miles.

67. Consider a clock with an hour hand and minute hand. What is the measure of the angle the minute hand traces in 20 minutes?

Extensions

68. Two cities have the same longitude. The latitude of city A is 9.00 degrees north and the latitude of city B is 30.00 degree north. Assume the radius of the earth is 3960 miles. Find the distance between the two cities.

69. A city is located at 40 degrees north latitude. Assume the radius of the earth is 3960 miles and the earth rotates once every 24 hours. Find the linear speed of a person who resides in this city.

70. A city is located at 75 degrees north latitude. Assume the radius of the earth is 3960 miles and the earth rotates once every 24 hours. Find the linear speed of a person who resides in this city.

71. Find the linear speed of the moon if the average distance between the earth and moon is 239,000 miles, assuming the orbit of the moon is circular and requires about 28 days. Express answer in miles per hour.

72. A bicycle has wheels 28 inches in diameter. A tachometer determines that the wheels are rotating at 180 RPM (revolutions per minute). Find the speed the bicycle is travelling down the road.

73. A car travels 3 miles. Its tires make 2640 revolutions. What is the radius of a tire in inches?

74. A wheel on a tractor has a 24-inch diameter. How many revolutions does the wheel make if the tractor travels 4 miles?

7.2 Right Triangle Trigonometry

Learning Objectives

In this section you will:

- Use right triangles to evaluate trigonometric functions.
- Find function values for $30°\left(\frac{\pi}{6}\right), 45°\left(\frac{\pi}{4}\right)$, and $60°\left(\frac{\pi}{3}\right)$.
- Use equal cofunctions of complementary angles.
- Use the definitions of trigonometric functions of any angle.
- Use right-triangle trigonometry to solve applied problems.

Mt. Everest, which straddles the border between China and Nepal, is the tallest mountain in the world. Measuring its height is no easy task. In fact, the actual measurement has been a source of controversy for hundreds of years. The measurement process involves the use of triangles and a branch of mathematics known as trigonometry. In this section, we will define a new group of functions known as trigonometric functions, and find out how they can be used to measure heights, such as those of the tallest mountains.

Using Right Triangles to Evaluate Trigonometric Functions

Figure 1 shows a right triangle with a vertical side of length y and a horizontal side has length x. Notice that the triangle is inscribed in a circle of radius 1. Such a circle, with a center at the origin and a radius of 1, is known as a **unit circle**.

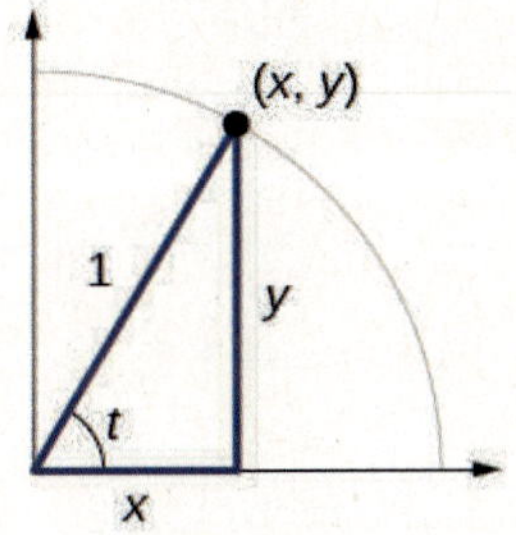

Figure 1

We can define the trigonometric functions in terms an angle t and the lengths of the sides of the triangle. The **adjacent side** is the side closest to the angle, x. (Adjacent means "next to.") The **opposite side** is the side across from the angle, y. The **hypotenuse** is the side of the triangle opposite the right angle, 1. These sides are labeled in Figure 2.

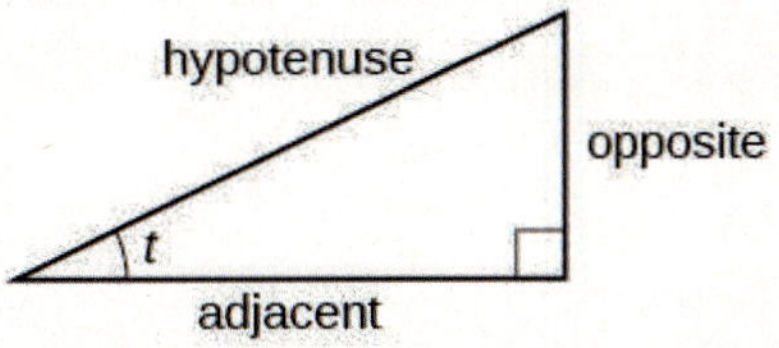

Figure 2 The sides of a right triangle in relation to angle t

Given a right triangle with an acute angle of t, the first three trigonometric functions are listed.

$$\begin{array}{lll} \text{Sine} & \sin t & = \dfrac{\text{opposite}}{\text{hypotenuse}} \\ \text{Cosine} & \cos t & = \dfrac{\text{adjacent}}{\text{hypotenuse}} \\ \text{Tangent} & \tan t & = \dfrac{\text{opposite}}{\text{adjacent}} \end{array}$$

A common mnemonic for remembering these relationships is SohCahToa, formed from the first letters of "**S**ine is **o**pposite over **h**ypotenuse, **C**osine is **a**djacent over **h**ypotenuse, **T**angent is **o**pposite over **a**djacent."

For the triangle shown in Figure 1, we have the following.

$$\begin{aligned} \sin t &= \frac{y}{1} \\ \cos t &= \frac{x}{1} \\ \tan t &= \frac{y}{x} \end{aligned}$$

HOW TO

Given the side lengths of a right triangle and one of the acute angles, find the sine, cosine, and tangent of that angle.

1. Find the sine as the ratio of the opposite side to the hypotenuse.
2. Find the cosine as the ratio of the adjacent side to the hypotenuse.
3. Find the tangent as the ratio of the opposite side to the adjacent side.

EXAMPLE 1

Evaluating a Trigonometric Function of a Right Triangle

Given the triangle shown in Figure 3, find the value of $\cos\ \alpha$.

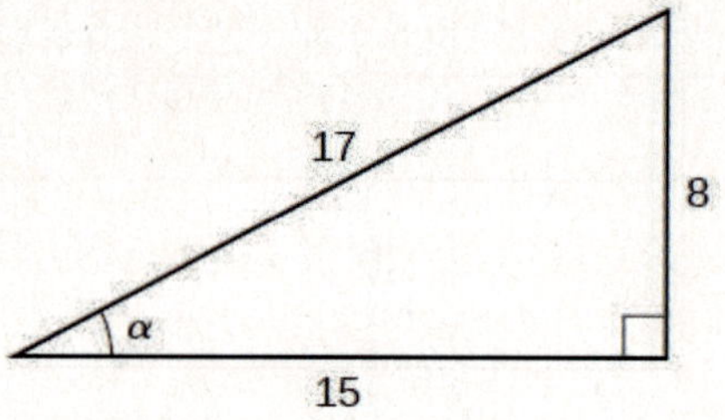

Figure 3

Solution

The side adjacent to the angle is 15, and the hypotenuse of the triangle is 17.

$$\begin{aligned}\cos(\alpha) &= \frac{\text{adjacent}}{\text{hypotenuse}}\\ &= \frac{15}{17}\end{aligned}$$

TRY IT #1 Given the triangle shown in Figure 4, find the value of $\sin\ t$.

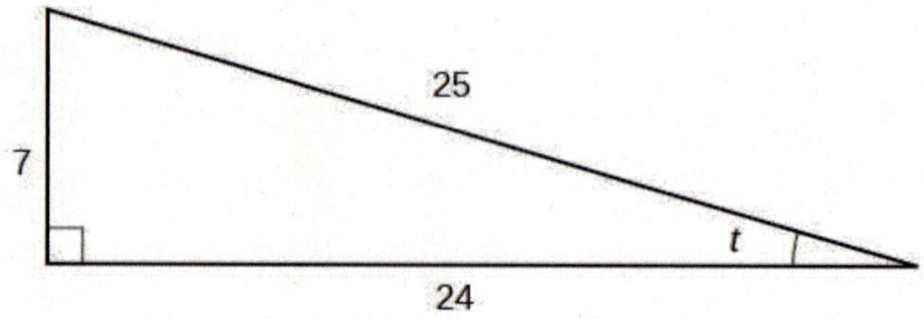

Figure 4

Reciprocal Functions

In addition to sine, cosine, and tangent, there are three more functions. These too are defined in terms of the sides of the triangle.

$$\begin{aligned}&\text{Secant} \quad &\sec t &= \frac{\text{hypotenuse}}{\text{adjacent}}\\ &\text{Cosecant} \quad &\csc t &= \frac{\text{hypotenuse}}{\text{opposite}}\\ &\text{Cotangent} \quad &\cot t &= \frac{\text{adjacent}}{\text{opposite}}\end{aligned}$$

Take another look at these definitions. These functions are the reciprocals of the first three functions.

$$\begin{aligned}\sin t &= \frac{1}{\csc t} \qquad & \csc t &= \frac{1}{\sin t}\\ \cos t &= \frac{1}{\sec t} \qquad & \sec t &= \frac{1}{\cos t}\\ \tan t &= \frac{1}{\cot t} \qquad & \cot t &= \frac{1}{\tan t}\end{aligned}$$

When working with right triangles, keep in mind that the same rules apply regardless of the orientation of the triangle. In fact, we can evaluate the six trigonometric functions of either of the two acute angles in the triangle in Figure 5. The side opposite one acute angle is the side adjacent to the other acute angle, and vice versa.

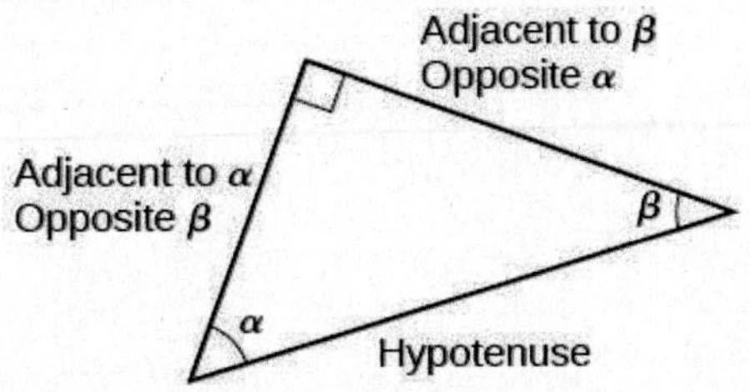

Figure 5 The side adjacent to one angle is opposite the other angle.

Many problems ask for all six trigonometric functions for a given angle in a triangle. A possible strategy to use is to find the sine, cosine, and tangent of the angles first. Then, find the other trigonometric functions easily using the reciprocals.

HOW TO

Given the side lengths of a right triangle, evaluate the six trigonometric functions of one of the acute angles.

1. If needed, draw the right triangle and label the angle provided.
2. Identify the angle, the adjacent side, the side opposite the angle, and the hypotenuse of the right triangle.
3. Find the required function:
   - sine as the ratio of the opposite side to the hypotenuse
   - cosine as the ratio of the adjacent side to the hypotenuse
   - tangent as the ratio of the opposite side to the adjacent side
   - secant as the ratio of the hypotenuse to the adjacent side
   - cosecant as the ratio of the hypotenuse to the opposite side
   - cotangent as the ratio of the adjacent side to the opposite side

EXAMPLE 2

Evaluating Trigonometric Functions of Angles Not in Standard Position

Using the triangle shown in Figure 6, evaluate $\sin\ \alpha, \cos\ \alpha, \tan\ \alpha, \sec\ \alpha, \csc\ \alpha,$ and $\cot\ \alpha.$

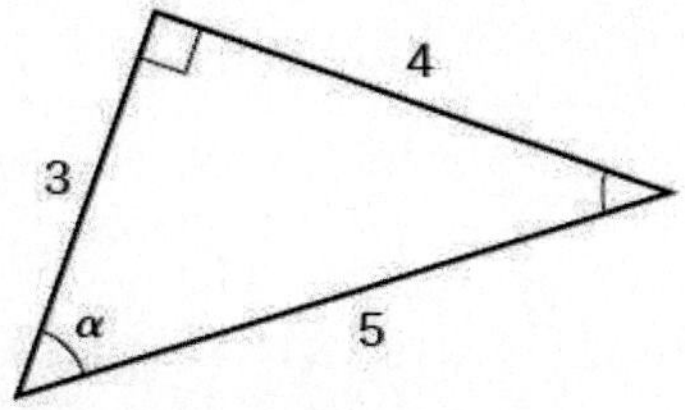

Figure 6

Solution

$$\begin{aligned}
\sin\ \alpha &= \frac{\text{opposite } \alpha}{\text{hypotenuse}} &= \frac{4}{5} \\
\cos\ \alpha &= \frac{\text{adjacent to } \alpha}{\text{hypotenuse}} &= \frac{3}{5} \\
\tan\ \alpha &= \frac{\text{opposite } \alpha}{\text{adjacent to } \alpha} &= \frac{4}{3} \\
\sec\ \alpha &= \frac{\text{hypotenuse}}{\text{adjacent to } \alpha} &= \frac{5}{3} \\
\csc\ \alpha &= \frac{\text{hypotenuse}}{\text{opposite } \alpha} &= \frac{5}{4} \\
\cot\ \alpha &= \frac{\text{adjacent to } \alpha}{\text{opposite } \alpha} &= \frac{3}{4}
\end{aligned}$$

Analysis

Another approach would have been to find sine, cosine, and tangent first. Then find their reciprocals to determine the other functions.

$$\sec \alpha = \frac{1}{\cos \alpha} = \frac{1}{\frac{3}{5}} = \frac{5}{3}$$

$$\csc \alpha = \frac{1}{\sin \alpha} = \frac{1}{\frac{4}{5}} = \frac{5}{4}$$

$$\cot \alpha = \frac{1}{\tan \alpha} = \frac{1}{\frac{4}{3}} = \frac{3}{4}$$

TRY IT #2 Using the triangle shown in Figure 7, evaluate $\sin t, \cos t, \tan t, \sec t, \csc t,$ and $\cot t$.

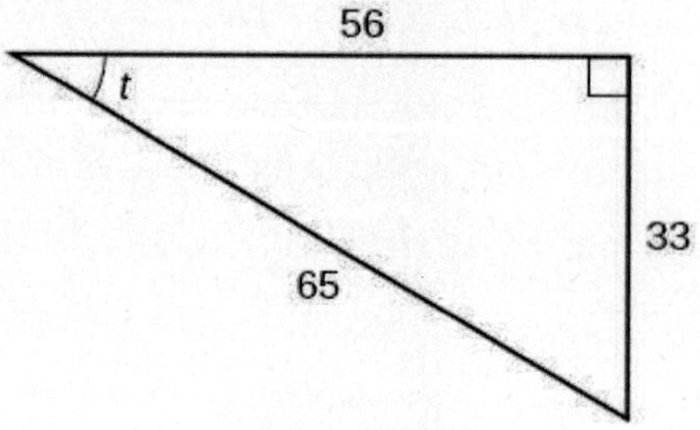

Figure 7

Finding Trigonometric Functions of Special Angles Using Side Lengths

It is helpful to evaluate the trigonometric functions as they relate to the special angles—multiples of $30°, 60°$, and $45°$. Remember, however, that when dealing with right triangles, we are limited to angles between $0°$ and $90°$.

Suppose we have a $30°, 60°, 90°$ triangle, which can also be described as a $\frac{\pi}{6}, \frac{\pi}{3}, \frac{\pi}{2}$ triangle. The sides have lengths in the relation $s, \sqrt{3}s, 2s$. The sides of a $45°, 45°, 90°$ triangle, which can also be described as a $\frac{\pi}{4}, \frac{\pi}{4}, \frac{\pi}{2}$ triangle, have lengths in the relation $s, s, \sqrt{2}s$. These relations are shown in Figure 8.

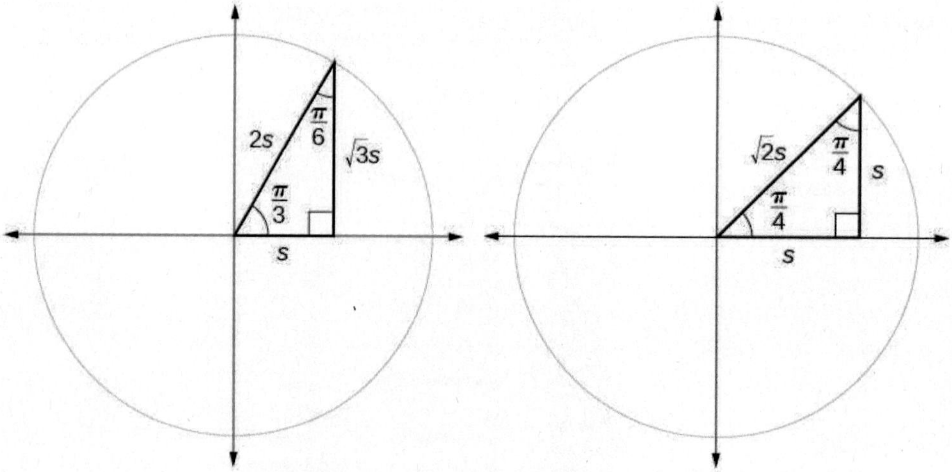

Figure 8 Side lengths of special triangles

We can then use the ratios of the side lengths to evaluate trigonometric functions of special angles.

HOW TO

Given trigonometric functions of a special angle, evaluate using side lengths.

1. Use the side lengths shown in Figure 8 for the special angle you wish to evaluate.
2. Use the ratio of side lengths appropriate to the function you wish to evaluate.

EXAMPLE 3

Evaluating Trigonometric Functions of Special Angles Using Side Lengths

Find the exact value of the trigonometric functions of $\frac{\pi}{3}$, using side lengths.

Solution

$$\begin{aligned}
\sin\left(\frac{\pi}{3}\right) &= \frac{\text{opp}}{\text{hyp}} = \frac{\sqrt{3}s}{2s} = \frac{\sqrt{3}}{2} \\
\cos\left(\frac{\pi}{3}\right) &= \frac{\text{adj}}{\text{hyp}} = \frac{s}{2s} = \frac{1}{2} \\
\tan\left(\frac{\pi}{3}\right) &= \frac{\text{opp}}{\text{adj}} = \frac{\sqrt{3}s}{s} = \sqrt{3} \\
\sec\left(\frac{\pi}{3}\right) &= \frac{\text{hyp}}{\text{adj}} = \frac{2s}{s} = 2 \\
\csc\left(\frac{\pi}{3}\right) &= \frac{\text{hyp}}{\text{opp}} = \frac{2s}{\sqrt{3}s} = \frac{2}{\sqrt{3}} = \frac{2\sqrt{3}}{3} \\
\cot\left(\frac{\pi}{3}\right) &= \frac{\text{adj}}{\text{opp}} = \frac{s}{\sqrt{3}s} = \frac{1}{\sqrt{3}} = \frac{\sqrt{3}}{3}
\end{aligned}$$

TRY IT #3 Find the exact value of the trigonometric functions of $\frac{\pi}{4}$, using side lengths.

Using Equal Cofunction of Complements

If we look more closely at the relationship between the sine and cosine of the special angles, we notice a pattern. In a right triangle with angles of $\frac{\pi}{6}$ and $\frac{\pi}{3}$, we see that the sine of $\frac{\pi}{3}$, namely $\frac{\sqrt{3}}{2}$, is also the cosine of $\frac{\pi}{6}$, while the sine of $\frac{\pi}{6}$, namely $\frac{1}{2}$, is also the cosine of $\frac{\pi}{3}$.

$$\begin{aligned}
\sin\frac{\pi}{3} &= \cos\frac{\pi}{6} = \frac{\sqrt{3}s}{2s} = \frac{\sqrt{3}}{2} \\
\sin\frac{\pi}{6} &= \cos\frac{\pi}{3} = \frac{s}{2s} = \frac{1}{2}
\end{aligned}$$

See Figure 9.

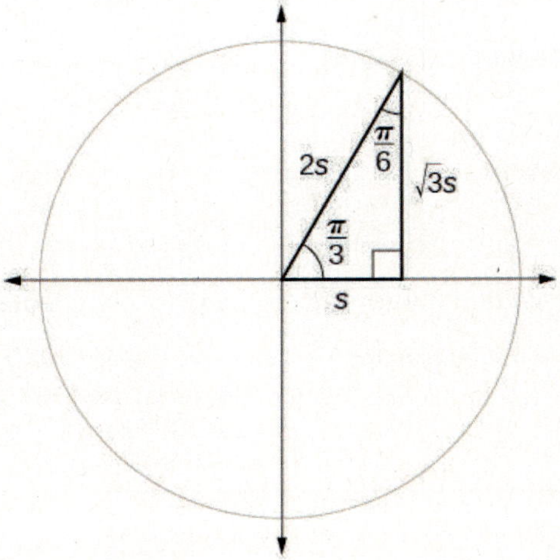

Figure 9 The sine of $\frac{\pi}{3}$ equals the cosine of $\frac{\pi}{6}$ and vice versa.

This result should not be surprising because, as we see from Figure 9, the side opposite the angle of $\frac{\pi}{3}$ is also the side adjacent to $\frac{\pi}{6}$, so $\sin\left(\frac{\pi}{3}\right)$ and $\cos\left(\frac{\pi}{6}\right)$ are exactly the same ratio of the same two sides, $\sqrt{3}s$ and $2s$. Similarly, $\cos\left(\frac{\pi}{3}\right)$ and $\sin\left(\frac{\pi}{6}\right)$ are also the same ratio using the same two sides, s and $2s$.

The interrelationship between the sines and cosines of $\frac{\pi}{6}$ and $\frac{\pi}{3}$ also holds for the two acute angles in any right triangle, since in every case, the ratio of the same two sides would constitute the sine of one angle and the cosine of the other.

Since the three angles of a triangle add to π, and the right angle is $\frac{\pi}{2}$, the remaining two angles must also add up to $\frac{\pi}{2}$. That means that a right triangle can be formed with any two angles that add to $\frac{\pi}{2}$ —in other words, any two complementary angles. So we may state a *cofunction identity*: If any two angles are complementary, the sine of one is the cosine of the other, and vice versa. This identity is illustrated in Figure 10.

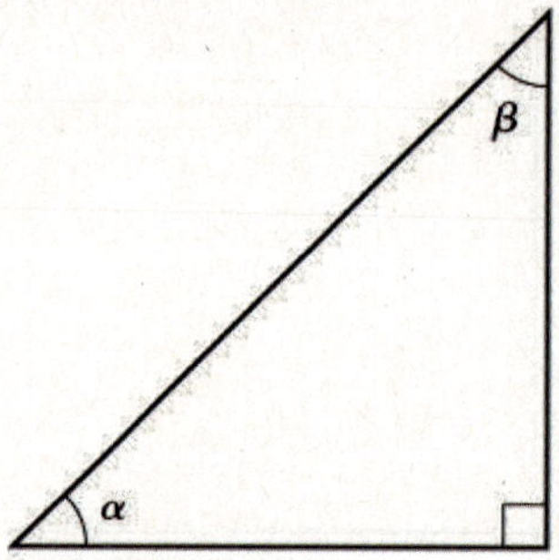

$$\sin\alpha = \cos\beta$$
$$\sin\beta = \cos\alpha$$

Figure 10 Cofunction identity of sine and cosine of complementary angles

Using this identity, we can state without calculating, for instance, that the sine of $\frac{\pi}{12}$ equals the cosine of $\frac{5\pi}{12}$, and that the sine of $\frac{5\pi}{12}$ equals the cosine of $\frac{\pi}{12}$. We can also state that if, for a given angle t, $\cos\ t = \frac{5}{13}$, then $\sin\left(\frac{\pi}{2} - t\right) = \frac{5}{13}$ as well.

Cofunction Identities

The **cofunction identities** in radians are listed in Table 1.

$\cos\ t = \sin\left(\frac{\pi}{2} - t\right)$	$\sin\ t = \cos\left(\frac{\pi}{2} - t\right)$
$\tan\ t = \cot\left(\frac{\pi}{2} - t\right)$	$\cot\ t = \tan\left(\frac{\pi}{2} - t\right)$
$\sec\ t = \csc\left(\frac{\pi}{2} - t\right)$	$\csc\ t = \sec\left(\frac{\pi}{2} - t\right)$

Table 1

HOW TO

Given the sine and cosine of an angle, find the sine or cosine of its complement.

1. To find the sine of the complementary angle, find the cosine of the original angle.
2. To find the cosine of the complementary angle, find the sine of the original angle.

EXAMPLE 4

Using Cofunction Identities

If $\sin\ t = \frac{5}{12}$, find $\cos\left(\frac{\pi}{2} - t\right)$.

Solution

According to the cofunction identities for sine and cosine, we have the following.

$$\sin\ t = \cos\left(\frac{\pi}{2} - t\right)$$

So

$$\cos\left(\frac{\pi}{2} - t\right) = \frac{5}{12}$$

TRY IT #4 If $\csc\left(\frac{\pi}{6}\right) = 2$, find $\sec\left(\frac{\pi}{3}\right)$.

Using Trigonometric Functions

In previous examples, we evaluated the sine and cosine in triangles where we knew all three sides. But the real power of right-triangle trigonometry emerges when we look at triangles in which we know an angle but do not know all the sides.

HOW TO

Given a right triangle, the length of one side, and the measure of one acute angle, find the remaining sides.

1. For each side, select the trigonometric function that has the unknown side as either the numerator or the denominator. The known side will in turn be the denominator or the numerator.
2. Write an equation setting the function value of the known angle equal to the ratio of the corresponding sides.
3. Using the value of the trigonometric function and the known side length, solve for the missing side length.

EXAMPLE 5

Finding Missing Side Lengths Using Trigonometric Ratios

Find the unknown sides of the triangle in Figure 11.

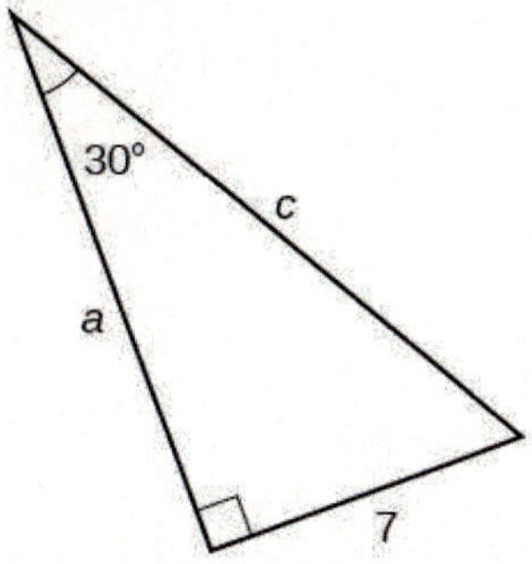

Figure 11

Solution

We know the angle and the opposite side, so we can use the tangent to find the adjacent side.

$$\tan(30°) = \frac{7}{a}$$

We rearrange to solve for a.

$$\begin{aligned} a &= \frac{7}{\tan(30°)} \\ &\approx 12.1 \end{aligned}$$

We can use the sine to find the hypotenuse.

$$\sin(30°) = \frac{7}{c}$$

Again, we rearrange to solve for c.

$$\begin{aligned} c &= \frac{7}{\sin(30°)} \\ &= 14 \end{aligned}$$

> TRY IT #5 A right triangle has one angle of $\frac{\pi}{3}$ and a hypotenuse of 20. Find the unknown sides and angle of the triangle.

Using Right Triangle Trigonometry to Solve Applied Problems

Right-triangle trigonometry has many practical applications. For example, the ability to compute the lengths of sides of a triangle makes it possible to find the height of a tall object without climbing to the top or having to extend a tape measure along its height. We do so by measuring a distance from the base of the object to a point on the ground some distance away, where we can look up to the top of the tall object at an angle. The **angle of elevation** of an object above an observer relative to the observer is the angle between the horizontal and the line from the object to the observer's eye. The right triangle this position creates has sides that represent the unknown height, the measured distance from the base, and the angled line of sight from the ground to the top of the object. Knowing the measured distance to the base of the object and the angle of the line of sight, we can use trigonometric functions to calculate the unknown height.

Similarly, we can form a triangle from the top of a tall object by looking downward. The **angle of depression** of an object below an observer relative to the observer is the angle between the horizontal and the line from the object to the observer's eye. See Figure 12.

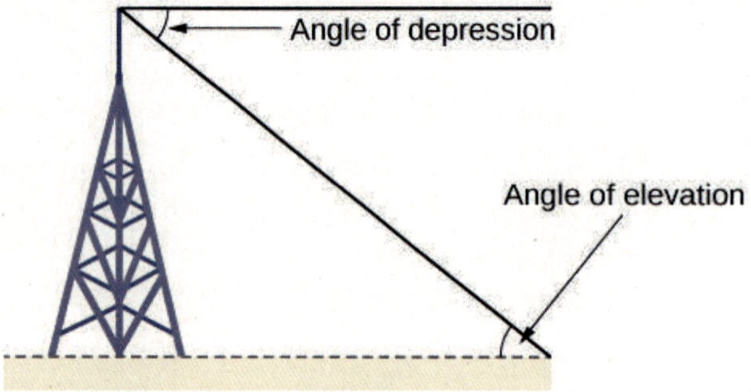

Figure 12

HOW TO

Given a tall object, measure its height indirectly.

1. Make a sketch of the problem situation to keep track of known and unknown information.
2. Lay out a measured distance from the base of the object to a point where the top of the object is clearly visible.
3. At the other end of the measured distance, look up to the top of the object. Measure the angle the line of sight makes with the horizontal.
4. Write an equation relating the unknown height, the measured distance, and the tangent of the angle of the line of sight.
5. Solve the equation for the unknown height.

EXAMPLE 6

Measuring a Distance Indirectly

To find the height of a tree, a person walks to a point 30 feet from the base of the tree. She measures an angle of 57° between a line of sight to the top of the tree and the ground, as shown in Figure 13. Find the height of the tree.

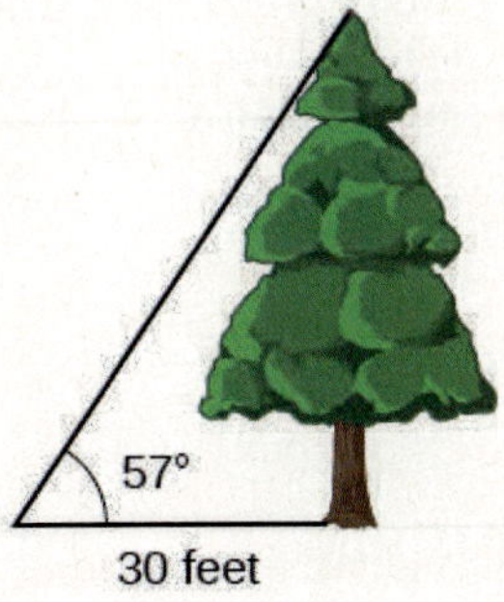

Figure 13

Solution

We know that the angle of elevation is 57° and the adjacent side is 30 ft long. The opposite side is the unknown height.

The trigonometric function relating the side opposite to an angle and the side adjacent to the angle is the tangent. So we will state our information in terms of the tangent of 57°, letting h be the unknown height.

$$\begin{aligned} \tan\theta &= \frac{\text{opposite}}{\text{adjacent}} & \\ \tan(57°) &= \frac{h}{30} & \text{Solve for } h. \\ h &= 30\tan(57°) & \text{Multiply.} \\ h &\approx 46.2 & \text{Use a calculator.} \end{aligned}$$

The tree is approximately 46 feet tall.

TRY IT #6 How long a ladder is needed to reach a windowsill 50 feet above the ground if the ladder rests against the building making an angle of $\frac{5\pi}{12}$ with the ground? Round to the nearest foot.

MEDIA

Access these online resources for additional instruction and practice with right triangle trigonometry.

Finding Trig Functions on Calculator (http://openstax.org/l/findtrigcal)
Finding Trig Functions Using a Right Triangle (http://openstax.org/l/trigrttri)
Relate Trig Functions to Sides of a Right Triangle (http://openstax.org/l/reltrigtri)
Determine Six Trig Functions from a Triangle (http://openstax.org/l/sixtrigfunc)
Determine Length of Right Triangle Side (http://openstax.org/l/rttriside)

7.2 SECTION EXERCISES

Verbal

1. For the given right triangle, label the adjacent side, opposite side, and hypotenuse for the indicated angle.

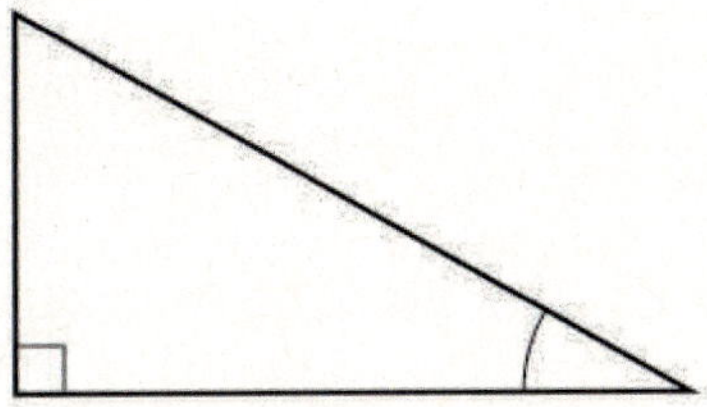

2. When a right triangle with a hypotenuse of 1 is placed in a circle of radius 1, which sides of the triangle correspond to the x- and y-coordinates?

3. The tangent of an angle compares which sides of the right triangle?

4. What is the relationship between the two acute angles in a right triangle?

5. Explain the cofunction identity.

Algebraic

For the following exercises, use cofunctions of complementary angles.

6. $\cos(34°) = \sin(\_\_\_°)$

7. $\cos\left(\frac{\pi}{3}\right) = \sin(\_\_\_)$

8. $\csc(21°) = \sec(\_\_\_°)$

9. $\tan\left(\frac{\pi}{4}\right) = \cot(\_\_\_)$

For the following exercises, find the lengths of the missing sides if side a is opposite angle A, side b is opposite angle B, and side c is the hypotenuse.

10. $\cos B = \frac{4}{5}, a = 10$

11. $\sin B = \frac{1}{2}, a = 20$

12. $\tan A = \frac{5}{12}, b = 6$

13. $\tan A = 100, b = 100$

14. $\sin B = \frac{1}{\sqrt{3}}, a = 2$

15. $a = 5, \measuredangle A = 60°$

16. $c = 12, \measuredangle A = 45°$

Graphical

For the following exercises, use Figure 14 to evaluate each trigonometric function of angle A.

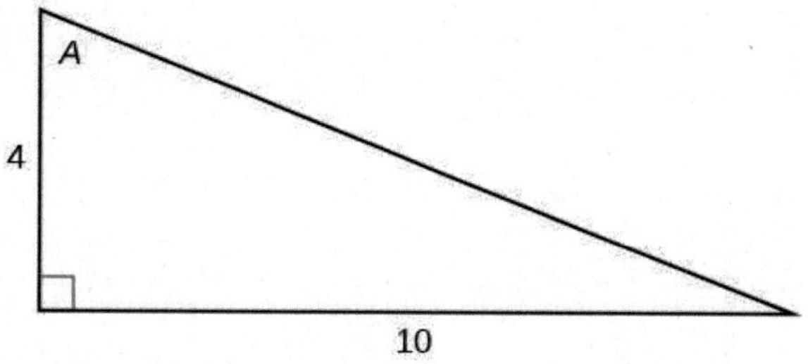

Figure 14

17. $\sin A$

18. $\cos A$

19. $\tan A$

20. $\csc A$

21. $\sec A$

22. $\cot A$

For the following exercises, use Figure 15 to evaluate each trigonometric function of angle A.

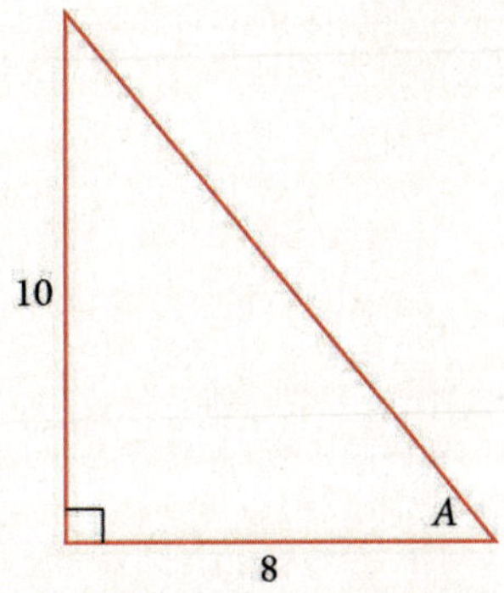

Figure 15

23. $\sin A$ **24.** $\cos A$ **25.** $\tan A$

26. $\csc A$ **27.** $\sec A$ **28.** $\cot A$

For the following exercises, solve for the unknown sides of the given triangle.

29.

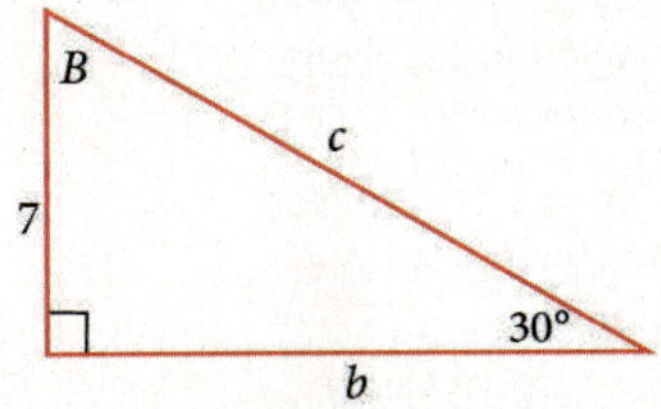

30.

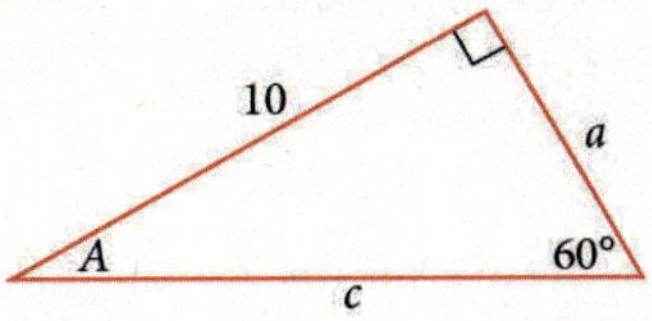

31.

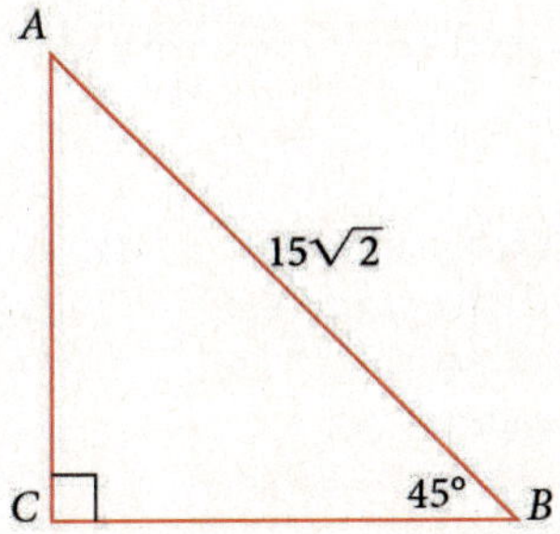

Technology

For the following exercises, use a calculator to find the length of each side to four decimal places.

32.

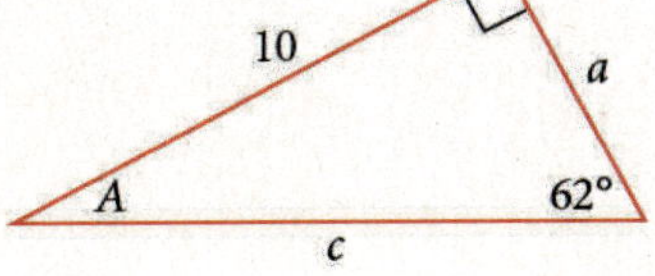

33.

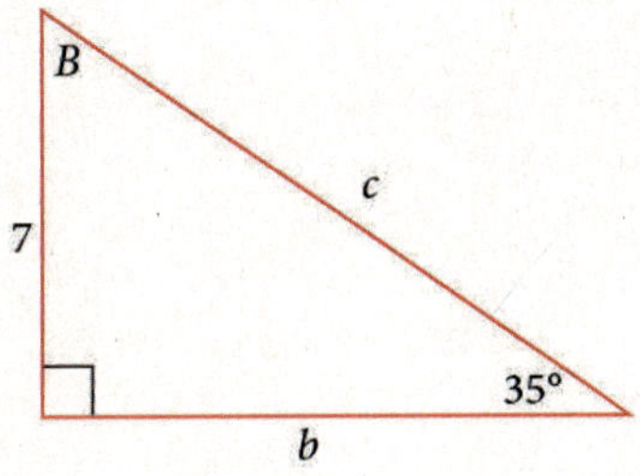

34.

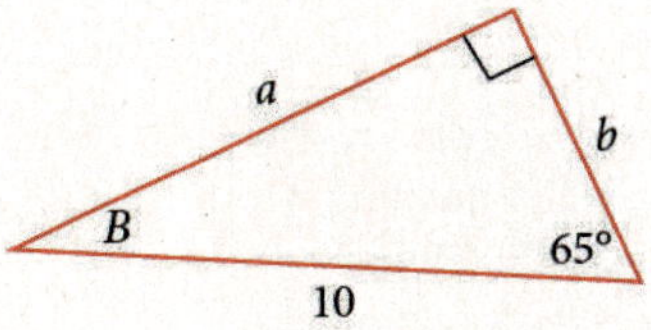

35.

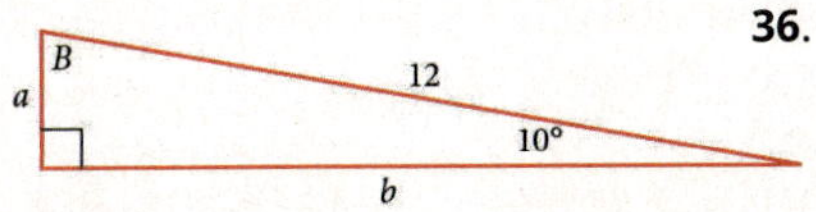

36.

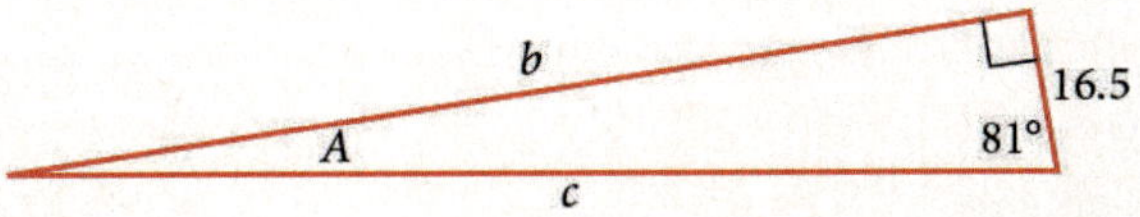

37. $b = 15, \measuredangle B = 15°$ **38.** $c = 200, \measuredangle B = 5°$ **39.** $c = 50, \measuredangle B = 21°$

40. $a = 30, \measuredangle A = 27°$ **41.** $b = 3.5, \measuredangle A = 78°$

Extensions

42. Find x.

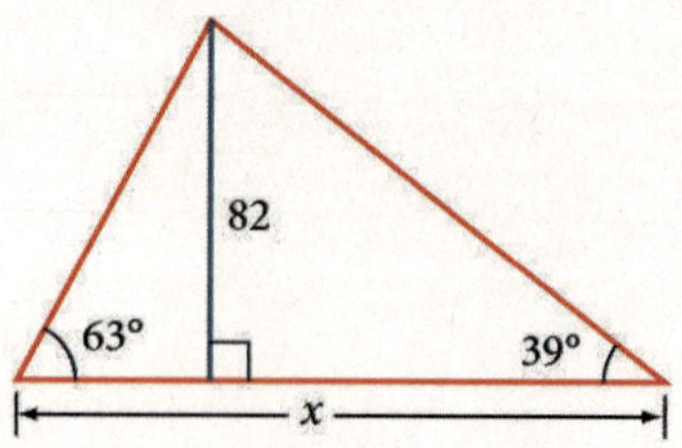

43. Find x.

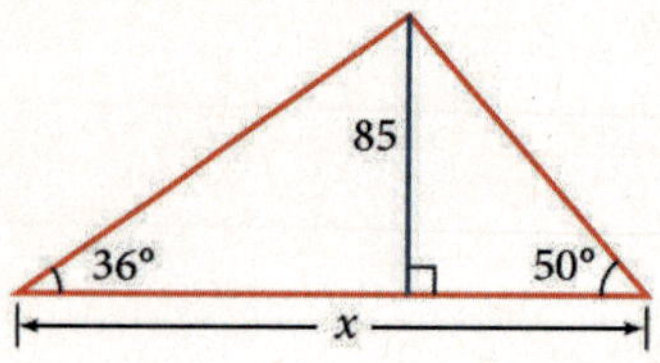

44. Find x.

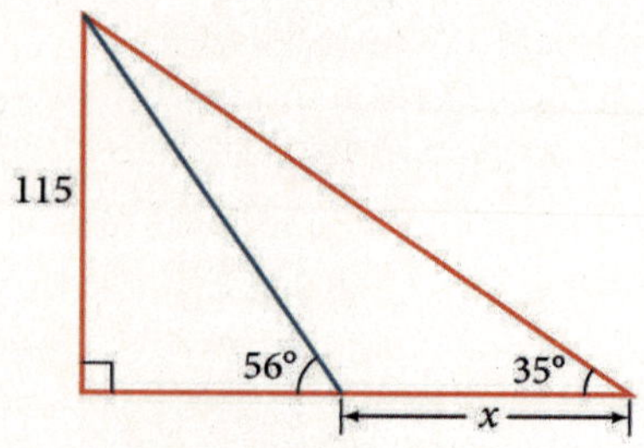

45. Find x.

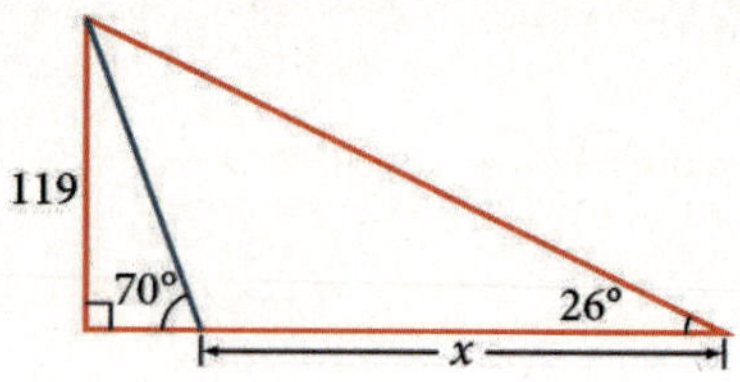

46. A radio tower is located 400 feet from a building. From a window in the building, a person determines that the angle of elevation to the top of the tower is $36°$, and that the angle of depression to the bottom of the tower is $23°$. How tall is the tower?

47. A radio tower is located 325 feet from a building. From a window in the building, a person determines that the angle of elevation to the top of the tower is $43°$, and that the angle of depression to the bottom of the tower is $31°$. How tall is the tower?

48. A 200-foot tall monument is located in the distance. From a window in a building, a person determines that the angle of elevation to the top of the monument is $15°$, and that the angle of depression to the bottom of the monument is $2°$. How far is the person from the monument?

49. A 400-foot tall monument is located in the distance. From a window in a building, a person determines that the angle of elevation to the top of the monument is $18°$, and that the angle of depression to the bottom of the monument is $3°$. How far is the person from the monument?

50. There is an antenna on the top of a building. From a location 300 feet from the base of the building, the angle of elevation to the top of the building is measured to be $40°$. From the same location, the angle of elevation to the top of the antenna is measured to be $43°$. Find the height of the antenna.

51. There is lightning rod on the top of a building. From a location 500 feet from the base of the building, the angle of elevation to the top of the building is measured to be $36°$. From the same location, the angle of elevation to the top of the lightning rod is measured to be $38°$. Find the height of the lightning rod.

Real-World Applications

52. A 33-ft ladder leans against a building so that the angle between the ground and the ladder is $80°$. How high does the ladder reach up the side of the building?

53. A 23-ft ladder leans against a building so that the angle between the ground and the ladder is $80°$. How high does the ladder reach up the side of the building?

54. The angle of elevation to the top of a building in Charlotte is found to be 9 degrees from the ground at a distance of 1 mile from the base of the building. Using this information, find the height of the building.

55. The angle of elevation to the top of a building in Seattle is found to be 2 degrees from the ground at a distance of 2 miles from the base of the building. Using this information, find the height of the building.

56. Assuming that a 370-foot tall giant redwood grows vertically, if I walk a certain distance from the tree and measure the angle of elevation to the top of the tree to be $60°$, how far from the base of the tree am I?

7.3 Unit Circle

Learning Objectives

In this section you will:

- Find function values for the sine and cosine of $30°$ or $\left(\frac{\pi}{6}\right)$, $45°$ or $\left(\frac{\pi}{4}\right)$, and $60°$ or $\left(\frac{\pi}{3}\right)$.
- Identify the domain and range of sine and cosine functions.
- Find reference angles.
- Use reference angles to evaluate trigonometric functions.

Figure 1 The Singapore Flyer was the world's tallest Ferris wheel until being overtaken by the High Roller in Las Vegas and the Ain Dubai in Dubai. (credit: "Vibin JK"/Flickr)

Looking for a thrill? Then consider a ride on the Ain Dubai, the world's tallest Ferris wheel. Located in Dubai, the most populous city and the financial and tourism hub of the United Arab Emirates, the wheel soars to 820 feet, about 1.5 tenths of a mile. Described as an observation wheel, riders enjoy spectacular views of the Burj Khalifa (the world's tallest building) and the Palm Jumeirah (a human-made archipelago home to over 10,000 people and 20 resorts) as they travel from the ground to the peak and down again in a repeating pattern. In this section, we will examine this type of revolving motion around a circle. To do so, we need to define the type of circle first, and then place that circle on a coordinate system. Then we can discuss circular motion in terms of the coordinate pairs.

Finding Trigonometric Functions Using the Unit Circle

We have already defined the trigonometric functions in terms of right triangles. In this section, we will redefine them in terms of the unit circle. Recall that a unit circle is a circle centered at the origin with radius 1, as shown in Figure 2. The angle (in radians) that t intercepts forms an arc of length s. Using the formula $s = rt$, and knowing that $r = 1$, we see that for a unit circle, $s = t$.

The x- and y-axes divide the coordinate plane into four quarters called quadrants. We label these quadrants to mimic the direction a positive angle would sweep. The four quadrants are labeled I, II, III, and IV.

For any angle t, we can label the intersection of the terminal side and the unit circle as by its coordinates, (x, y). The coordinates x and y will be the outputs of the trigonometric functions $f(t) = \cos\ t$ and $f(t) = \sin\ t$, respectively. This means $x = \cos t$ and $y = \sin t$.

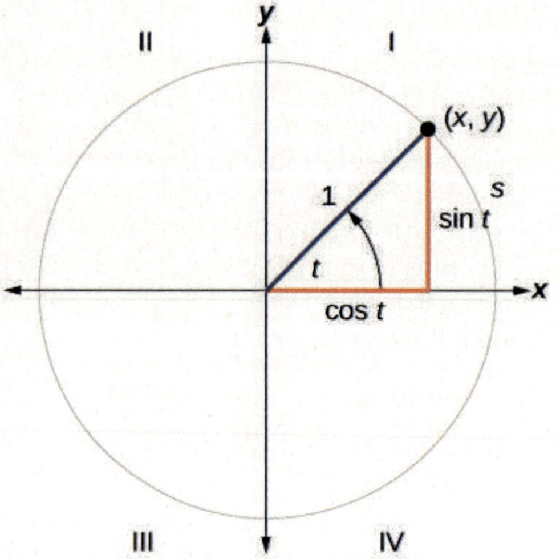

Figure 2 Unit circle where the central angle is t radians

Unit Circle

A **unit circle** has a center at $(0, 0)$ and radius 1. In a unit circle, the length of the intercepted arc is equal to the radian measure of the central angle t.

Let (x, y) be the endpoint on the unit circle of an arc of arc length s. The (x, y) coordinates of this point can be described as functions of the angle.

Defining Sine and Cosine Functions from the Unit Circle

The sine function relates a real number t to the y-coordinate of the point where the corresponding angle intercepts the unit circle. More precisely, the sine of an angle t equals the y-value of the endpoint on the unit circle of an arc of length t. In Figure 2, the sine is equal to y. Like all functions, the **sine function** has an input and an output. Its input is the measure of the angle; its output is the y-coordinate of the corresponding point on the unit circle.

The **cosine function** of an angle t equals the x-value of the endpoint on the unit circle of an arc of length t. In Figure 3, the cosine is equal to x.

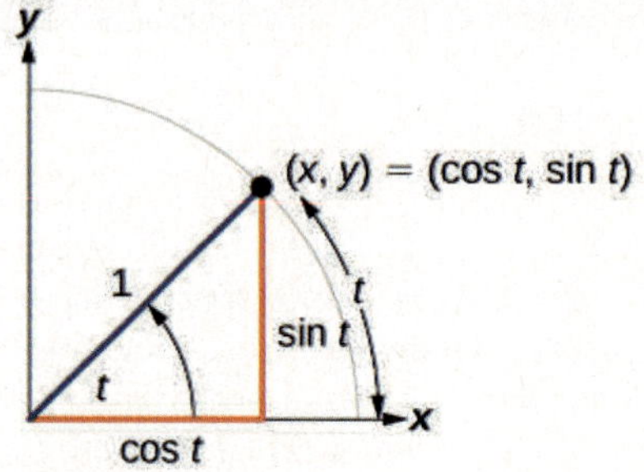

Figure 3

Because it is understood that sine and cosine are functions, we do not always need to write them with parentheses: $\sin\ t$ is the same as $\sin(t)$ and $\cos t$ is the same as $\cos(t)$. Likewise, $\cos^2 t$ is a commonly used shorthand notation for $(\cos(t))^2$.

Be aware that many calculators and computers do not recognize the shorthand notation. When in doubt, use the extra parentheses when entering calculations into a calculator or computer.

Sine and Cosine Functions

If t is a real number and a point (x, y) on the unit circle corresponds to a central angle t, then

$$\cos\ t = x$$
$$\sin\ t = y$$

HOW TO

Given a point $P\,(x, y)$ on the unit circle corresponding to an angle of t, find the sine and cosine.

1. The sine of t is equal to the y-coordinate of point $P : \sin t = y.$
2. The cosine of t is equal to the x-coordinate of point $P : \cos\ t = x.$

EXAMPLE 1

Finding Function Values for Sine and Cosine

Point P is a point on the unit circle corresponding to an angle of t, as shown in Figure 4. Find $\cos(t)$ and $\sin(t)$.

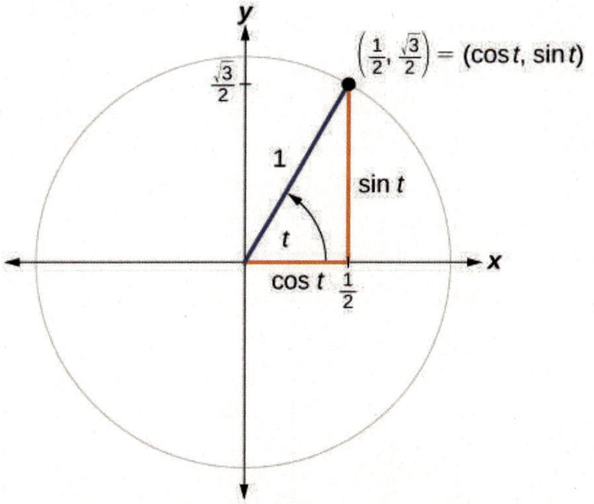

Figure 4

Solution

We know that $\cos\ t$ is the x-coordinate of the corresponding point on the unit circle and $\sin\ t$ is the y-coordinate of the corresponding point on the unit circle. So:

$$\begin{aligned} x &= \cos\ t &= \tfrac{1}{2} \\ y &= \sin\ t &= \tfrac{\sqrt{3}}{2} \end{aligned}$$

TRY IT #1 A certain angle t corresponds to a point on the unit circle at $\left(-\frac{\sqrt{2}}{2}, \frac{\sqrt{2}}{2}\right)$ as shown in Figure 5. Find $\cos\ t$ and $\sin\ t$.

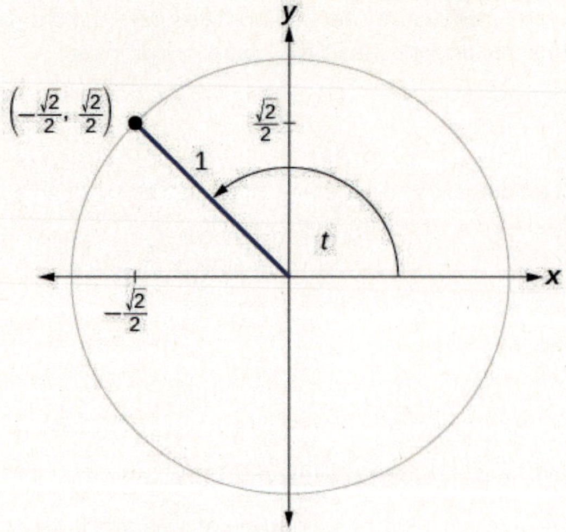

Figure 5

Finding Sines and Cosines of Angles on an Axis

For quadrantral angles, the corresponding point on the unit circle falls on the *x*- or *y*-axis. In that case, we can easily calculate cosine and sine from the values of x and y.

EXAMPLE 2

Calculating Sines and Cosines along an Axis

Find $\cos(90°)$ and $\sin(90°)$.

Solution

Moving $90°$ counterclockwise around the unit circle from the positive *x*-axis brings us to the top of the circle, where the (x, y) coordinates are $(0, 1)$, as shown in Figure 6.

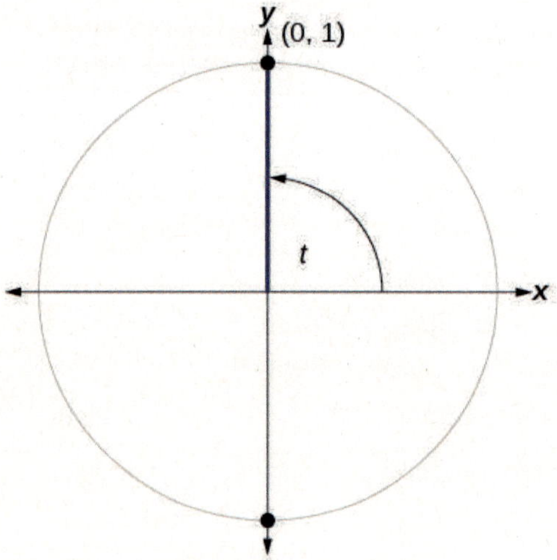

Figure 6

We can then use our definitions of cosine and sine.

$$\begin{aligned} x &= \cos t &= \cos(90°) &= 0 \\ y &= \sin t &= \sin(90°) &= 1 \end{aligned}$$

The cosine of $90°$ is 0; the sine of $90°$ is 1.

TRY IT #2 Find cosine and sine of the angle π.

The Pythagorean Identity

Now that we can define sine and cosine, we will learn how they relate to each other and the unit circle. Recall that the equation for the unit circle is $x^2 + y^2 = 1$. Because $x = \cos\ t$ and $y = \sin\ t$, we can substitute for x and y to get $\cos^2 t + \sin^2 t = 1$. This equation, $\cos^2 t + \sin^2 t = 1$, is known as the Pythagorean Identity. See Figure 7.

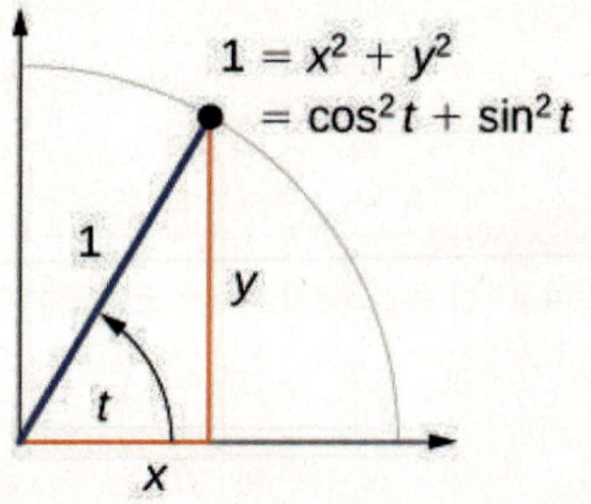

Figure 7

We can use the Pythagorean Identity to find the cosine of an angle if we know the sine, or vice versa. However, because the equation yields two solutions, we need additional knowledge of the angle to choose the solution with the correct sign. If we know the quadrant where the angle is, we can easily choose the correct solution.

Pythagorean Identity

The **Pythagorean Identity** states that, for any real number t,

$$\cos^2 t + \sin^2 t = 1$$

HOW TO

Given the sine of some angle t and its quadrant location, find the cosine of t.

1. Substitute the known value of $\sin\ t$ into the Pythagorean Identity.
2. Solve for $\cos\ t$.
3. Choose the solution with the appropriate sign for the x-values in the quadrant where t is located.

EXAMPLE 3

Finding a Cosine from a Sine or a Sine from a Cosine

If $\sin(t) = \frac{3}{7}$ and t is in the second quadrant, find $\cos(t)$.

Solution

If we drop a vertical line from the point on the unit circle corresponding to t, we create a right triangle, from which we can see that the Pythagorean Identity is simply one case of the Pythagorean Theorem. See Figure 8.

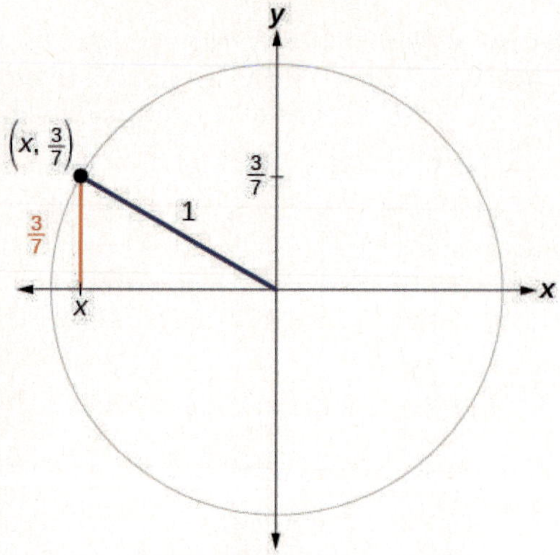

Figure 8

Substituting the known value for sine into the Pythagorean Identity,

$$\begin{aligned} \cos^2(t) + \sin^2(t) &= 1 \\ \cos^2(t) + \frac{9}{49} &= 1 \\ \cos^2(t) &= \frac{40}{49} \\ \cos(t) &= \pm\sqrt{\frac{40}{49}} = \pm\frac{\sqrt{40}}{7} = \pm\frac{2\sqrt{10}}{7} \end{aligned}$$

Because the angle is in the second quadrant, we know the *x*-value is a negative real number, so the cosine is also negative.

$$\cos(t) = -\frac{2\sqrt{10}}{7}$$

TRY IT #3 If $\cos(t) = \frac{24}{25}$ and t is in the fourth quadrant, find $\sin(t)$.

Finding Sines and Cosines of Special Angles

We have already learned some properties of the special angles, such as the conversion from radians to degrees, and we found their sines and cosines using right triangles. We can also calculate sines and cosines of the special angles using the Pythagorean Identity.

Finding Sines and Cosines of $45°$ Angles

First, we will look at angles of $45°$ or $\frac{\pi}{4}$, as shown in Figure 9. A $45°–45°–90°$ triangle is an isosceles triangle, so the *x*- and *y*-coordinates of the corresponding point on the circle are the same. Because the *x*- and *y*-values are the same, the sine and cosine values will also be equal.

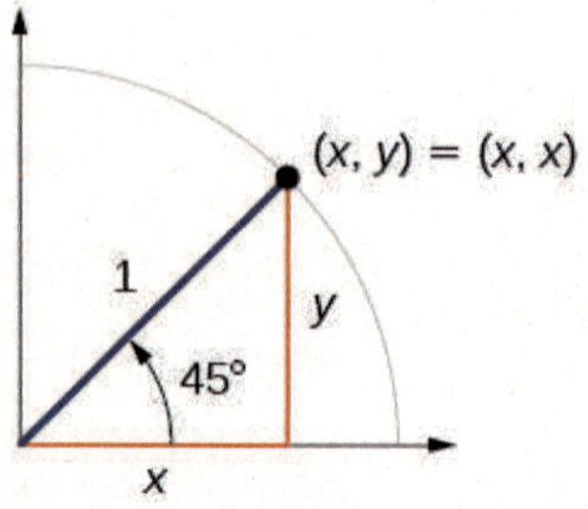

Figure 9

At $t = \frac{\pi}{4}$, which is 45 degrees, the radius of the unit circle bisects the first quadrantal angle. This means the radius lies along the line $y = x$. A unit circle has a radius equal to 1 so the right triangle formed below the line $y = x$ has sides x

and y ($y = x$), and radius = 1. See Figure 10.

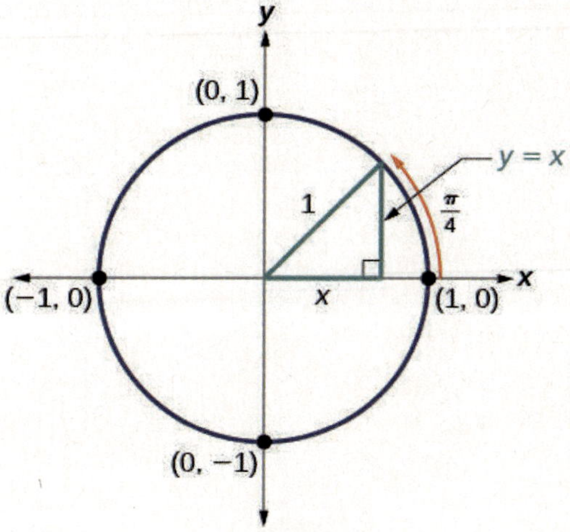

Figure 10

From the Pythagorean Theorem we get

$$x^2 + y^2 = 1$$

We can then substitute $y = x$.

$$x^2 + x^2 = 1$$

Next we combine like terms.

$$2x^2 = 1$$

And solving for x, we get

$$\begin{aligned} x^2 &= \frac{1}{2} \\ x &= \pm\frac{1}{\sqrt{2}} \end{aligned}$$

In quadrant I, $x = \frac{1}{\sqrt{2}}$.

At $t = \frac{\pi}{4}$ or 45 degrees,

$$\begin{aligned} (x, y) &= (x, x) = \left(\frac{1}{\sqrt{2}}, \frac{1}{\sqrt{2}}\right) \\ x &= \frac{1}{\sqrt{2}}, y = \frac{1}{\sqrt{2}} \\ \cos t &= \frac{1}{\sqrt{2}}, \sin t = \frac{1}{\sqrt{2}} \end{aligned}$$

If we then rationalize the denominators, we get

$$\begin{aligned} \cos t &= \frac{1}{\sqrt{2}}\frac{\sqrt{2}}{\sqrt{2}} \\ &= \frac{\sqrt{2}}{2} \\ \sin t &= \frac{1}{\sqrt{2}}\frac{\sqrt{2}}{\sqrt{2}} \\ &= \frac{\sqrt{2}}{2} \end{aligned}$$

Therefore, the (x, y) coordinates of a point on a circle of radius 1 at an angle of $45°$ are $\left(\frac{\sqrt{2}}{2}, \frac{\sqrt{2}}{2}\right)$.

Finding Sines and Cosines of $30°$ and $60°$ Angles

Next, we will find the cosine and sine at an angle of $30°$, or $\frac{\pi}{6}$. First, we will draw a triangle inside a circle with one side at an angle of $30°$, and another at an angle of $-30°$, as shown in Figure 11. If the resulting two right triangles are combined into one large triangle, notice that all three angles of this larger triangle will be $60°$, as shown in Figure 12.

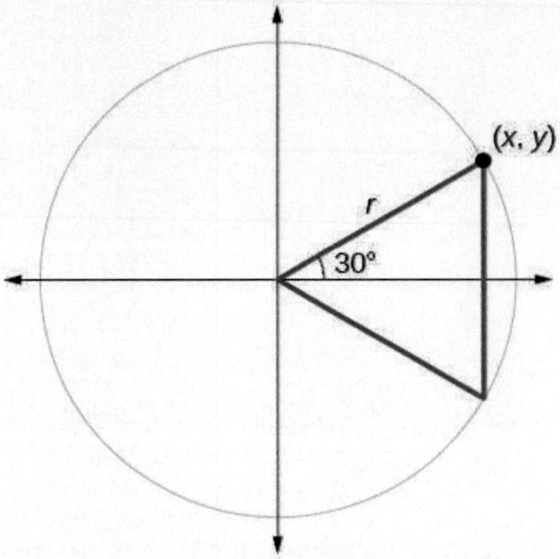

Figure 11

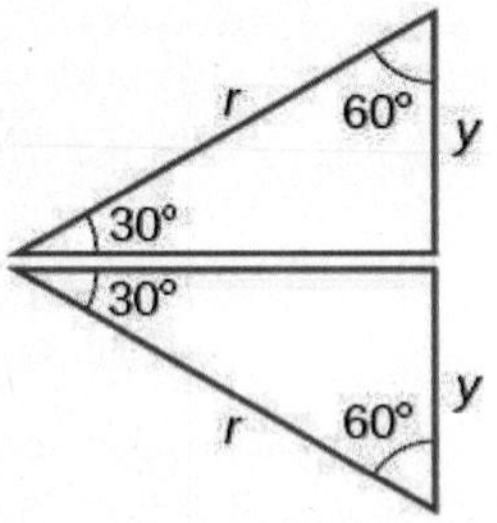

Figure 12

Because all the angles are equal, the sides are also equal. The vertical line has length $2y$, and since the sides are all equal, we can also conclude that $r = 2y$ or $y = \frac{1}{2}r$. Since $\sin\ t = y$,

$$\sin\left(\frac{\pi}{6}\right) = \frac{1}{2}r$$

And since $r = 1$ in our unit circle,

$$\begin{aligned}\sin\left(\frac{\pi}{6}\right) &= \frac{1}{2}(1)\\ &= \frac{1}{2}\end{aligned}$$

Using the Pythagorean Identity, we can find the cosine value.

$$\begin{aligned}\cos^2\left(\frac{\pi}{6}\right) + \sin^2\left(\frac{\pi}{6}\right) &= 1\\ \cos^2\left(\frac{\pi}{6}\right) + \left(\frac{1}{2}\right)^2 &= 1\\ \cos^2\left(\frac{\pi}{6}\right) &= \frac{3}{4} && \text{Use the square root property.}\\ \cos\left(\frac{\pi}{6}\right) &= \frac{\pm\sqrt{3}}{\pm\sqrt{4}} = \frac{\sqrt{3}}{2} && \text{Since } y \text{ is positive, choose the positive root.}\end{aligned}$$

The (x, y) coordinates for the point on a circle of radius 1 at an angle of $30°$ are $\left(\frac{\sqrt{3}}{2}, \frac{1}{2}\right)$. At $t = \frac{\pi}{3}$ $(60°)$, the radius of the unit circle, 1, serves as the hypotenuse of a 30-60-90 degree right triangle, BAD, as shown in Figure 13. Angle A has measure $60°$. At point B, we draw an angle ABC with measure of $60°$. We know the angles in a triangle sum to $180°$, so the measure of angle C is also $60°$. Now we have an equilateral triangle. Because each side of the equilateral triangle ABC is the same length, and we know one side is the radius of the unit circle, all sides must be of length 1.

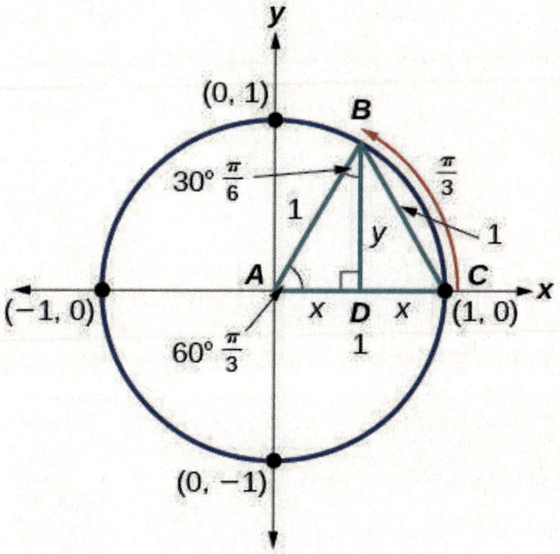

Figure 13

The measure of angle ABD is 30°. Angle ABC is double angle ABD, so its measure is 60°. BD is the perpendicular bisector of AC, so it cuts AC in half. This means that AD is $\frac{1}{2}$ the radius, or $\frac{1}{2}$. Notice that AD is the *x*-coordinate of point B, which is at the intersection of the 60° angle and the unit circle. This gives us a triangle BAD with hypotenuse of 1 and side x of length $\frac{1}{2}$.

From the Pythagorean Theorem, we get

$$x^2 + y^2 = 1$$

Substituting $x = \frac{1}{2}$, we get

$$\left(\frac{1}{2}\right)^2 + y^2 = 1$$

Solving for y, we get

$$\begin{aligned} \frac{1}{4} + y^2 &= 1 \\ y^2 &= 1 - \frac{1}{4} \\ y^2 &= \frac{3}{4} \\ y &= \pm\frac{\sqrt{3}}{2} \end{aligned}$$

Since $t = \frac{\pi}{3}$ has the terminal side in quadrant I where the *y*-coordinate is positive, we choose $y = \frac{\sqrt{3}}{2}$, the positive value.

At $t = \frac{\pi}{3}$ (60°), the (x, y) coordinates for the point on a circle of radius 1 at an angle of $60°$ are $\left(\frac{1}{2}, \frac{\sqrt{3}}{2}\right)$, so we can find the sine and cosine.

$$\begin{aligned} (x, y) &= \left(\frac{1}{2}, \frac{\sqrt{3}}{2}\right) \\ x &= \frac{1}{2}, y = \frac{\sqrt{3}}{2} \\ \cos t &= \frac{1}{2}, \sin t = \frac{\sqrt{3}}{2} \end{aligned}$$

We have now found the cosine and sine values for all of the most commonly encountered angles in the first quadrant of the unit circle. Table 1 summarizes these values.

Angle	0	$\frac{\pi}{6}$, or $30°$	$\frac{\pi}{4}$, or $45°$	$\frac{\pi}{3}$, or $60°$	$\frac{\pi}{2}$, or $90°$

Table 1

Cosine	1	$\frac{\sqrt{3}}{2}$	$\frac{\sqrt{2}}{2}$	$\frac{1}{2}$	0
Sine	0	$\frac{1}{2}$	$\frac{\sqrt{2}}{2}$	$\frac{\sqrt{3}}{2}$	1

Table 1

Figure 14 shows the common angles in the first quadrant of the unit circle.

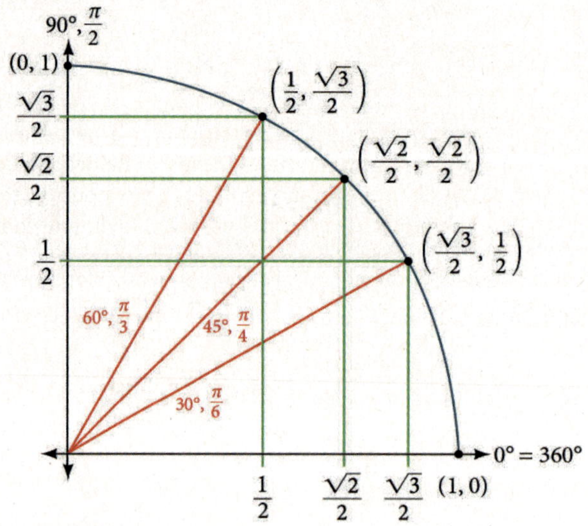

Figure 14

Using a Calculator to Find Sine and Cosine

To find the cosine and sine of angles other than the special angles, we turn to a computer or calculator. **Be aware**: Most calculators can be set into "degree" or "radian" mode, which tells the calculator the units for the input value. When we evaluate $\cos(30)$ on our calculator, it will evaluate it as the cosine of 30 degrees if the calculator is in degree mode, or the cosine of 30 radians if the calculator is in radian mode.

HOW TO

Given an angle in radians, use a graphing calculator to find the cosine.

1. If the calculator has degree mode and radian mode, set it to radian mode.
2. Press the COS key.
3. Enter the radian value of the angle and press the close-parentheses key ")".
4. Press ENTER.

EXAMPLE 4

Using a Graphing Calculator to Find Sine and Cosine

Evaluate $\cos\left(\frac{5\pi}{3}\right)$ using a graphing calculator or computer.

Solution

Enter the following keystrokes:

COS(5 × π ÷ 3) ENTER

$$\cos\left(\frac{5\pi}{3}\right) = 0.5$$

Analysis

We can find the cosine or sine of an angle in degrees directly on a calculator with degree mode. For calculators or software that use only radian mode, we can find the sine of $20°$, for example, by including the conversion factor to radians as part of the input:

$$\text{SIN(} 20 \times \pi \div 180 \text{) ENTER}$$

TRY IT #4 Evaluate $\sin\left(\frac{\pi}{3}\right)$.

Identifying the Domain and Range of Sine and Cosine Functions

Now that we can find the sine and cosine of an angle, we need to discuss their domains and ranges. What are the domains of the sine and cosine functions? That is, what are the smallest and largest numbers that can be inputs of the functions? Because angles smaller than 0 and angles larger than 2π can still be graphed on the unit circle and have real values of x, y, and r, there is no lower or upper limit to the angles that can be inputs to the sine and cosine functions. The input to the sine and cosine functions is the rotation from the positive *x*-axis, and that may be any real number.

What are the ranges of the sine and cosine functions? What are the least and greatest possible values for their output? We can see the answers by examining the unit circle, as shown in Figure 15. The bounds of the *x*-coordinate are $[-1, 1]$. The bounds of the *y*-coordinate are also $[-1, 1]$. Therefore, the range of both the sine and cosine functions is $[-1, 1]$.

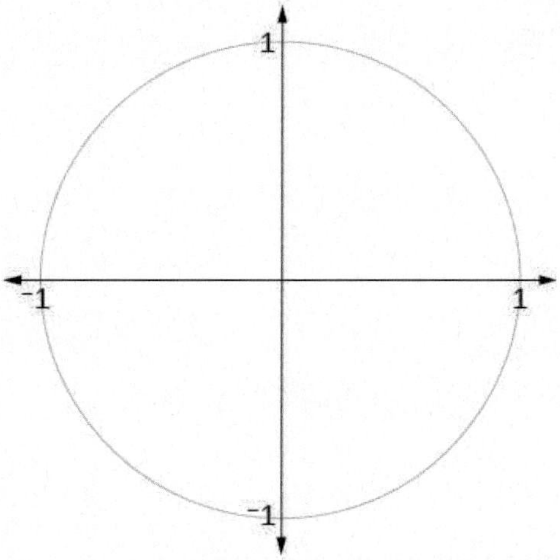

Figure 15

Finding Reference Angles

We have discussed finding the sine and cosine for angles in the first quadrant, but what if our angle is in another quadrant? For any given angle in the first quadrant, there is an angle in the second quadrant with the same sine value. Because the sine value is the *y*-coordinate on the unit circle, the other angle with the same sine will share the same *y*-value, but have the opposite *x*-value. Therefore, its cosine value will be the opposite of the first angle's cosine value.

Likewise, there will be an angle in the fourth quadrant with the same cosine as the original angle. The angle with the same cosine will share the same *x*-value but will have the opposite *y*-value. Therefore, its sine value will be the opposite of the original angle's sine value.

As shown in Figure 16, angle α has the same sine value as angle t; the cosine values are opposites. Angle β has the same cosine value as angle t; the sine values are opposites.

$$\sin(t) = \sin(\alpha) \quad \text{and} \quad \cos(t) = -\cos(\alpha)$$
$$\sin(t) = -\sin(\beta) \quad \text{and} \quad \cos(t) = \cos(\beta)$$

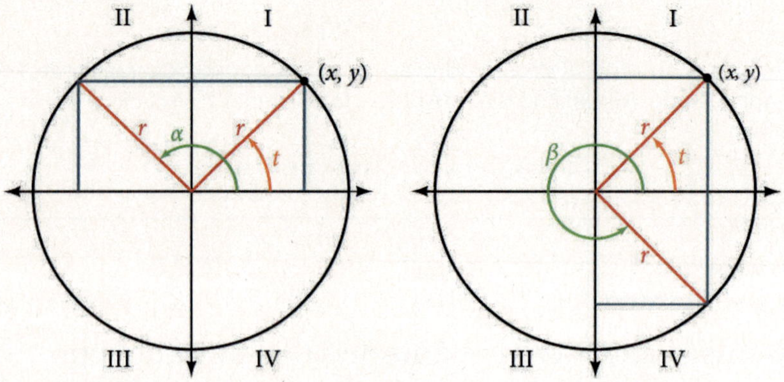

Figure 16

Recall that an angle's reference angle is the acute angle, t, formed by the terminal side of the angle t and the horizontal axis. A reference angle is always an angle between 0 and $90°$, or 0 and $\frac{\pi}{2}$ radians. As we can see from Figure 17, for any angle in quadrants II, III, or IV, there is a reference angle in quadrant I.

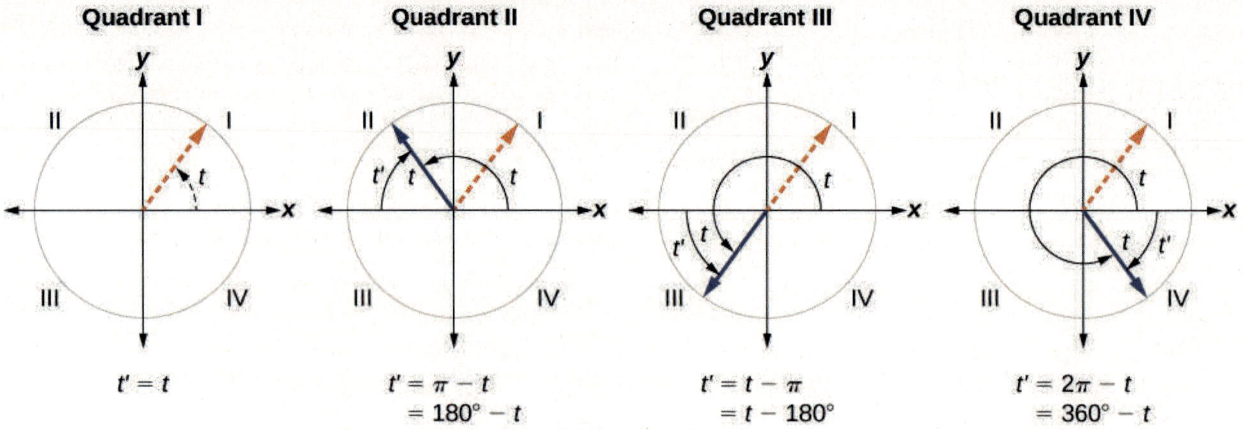

Figure 17

HOW TO

Given an angle between 0 and 2π, find its reference angle.

1. An angle in the first quadrant is its own reference angle.
2. For an angle in the second or third quadrant, the reference angle is $|\pi - t|$ or $|180° - t|$.
3. For an angle in the fourth quadrant, the reference angle is $2\pi - t$ or $360° - t$.
4. If an angle is less than 0 or greater than 2π, add or subtract 2π as many times as needed to find an equivalent angle between 0 and 2π.

EXAMPLE 5

Finding a Reference Angle

Find the reference angle of $225°$ as shown in Figure 18.

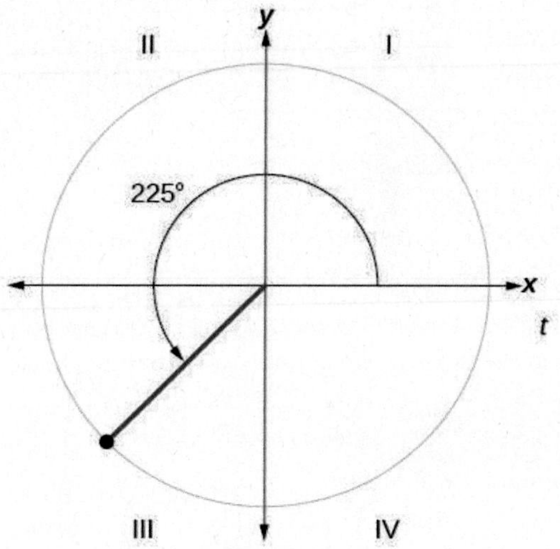

Figure 18

Solution

Because 225° is in the third quadrant, the reference angle is

$$|(180° - 225°)| = |-45°| = 45°$$

TRY IT #5 Find the reference angle of $\frac{5\pi}{3}$.

Using Reference Angles

Now let's take a moment to reconsider the Ferris wheel introduced at the beginning of this section. Suppose a rider snaps a photograph while stopped twenty feet above ground level. The rider then rotates three-quarters of the way around the circle. What is the rider's new elevation? To answer questions such as this one, we need to evaluate the sine or cosine functions at angles that are greater than 90 degrees or at a negative angle. Reference angles make it possible to evaluate trigonometric functions for angles outside the first quadrant. They can also be used to find (x, y) coordinates for those angles. We will use the reference angle of the angle of rotation combined with the quadrant in which the terminal side of the angle lies.

Using Reference Angles to Evaluate Trigonometric Functions

We can find the cosine and sine of any angle in any quadrant if we know the cosine or sine of its reference angle. The absolute values of the cosine and sine of an angle are the same as those of the reference angle. The sign depends on the quadrant of the original angle. The cosine will be positive or negative depending on the sign of the *x*-values in that quadrant. The sine will be positive or negative depending on the sign of the *y*-values in that quadrant.

Using Reference Angles to Find Cosine and Sine

Angles have cosines and sines with the same absolute value as their reference angles. The sign (positive or negative) can be determined from the quadrant of the angle.

HOW TO

Given an angle in standard position, find the reference angle, and the cosine and sine of the original angle.

1. Measure the angle between the terminal side of the given angle and the horizontal axis. That is the reference angle.
2. Determine the values of the cosine and sine of the reference angle.
3. Give the cosine the same sign as the *x*-values in the quadrant of the original angle.

4. Give the sine the same sign as the y-values in the quadrant of the original angle.

EXAMPLE 6

Using Reference Angles to Find Sine and Cosine

ⓐ Using a reference angle, find the exact value of $\cos(150°)$ and $\sin(150°)$.

ⓑ Using the reference angle, find $\cos\frac{5\pi}{4}$ and $\sin\frac{5\pi}{4}$.

Solution

ⓐ $150°$ is located in the second quadrant. The angle it makes with the x-axis is $180° - 150° = 30°$, so the reference angle is $30°$.

This tells us that $150°$ has the same sine and cosine values as $30°$, except for the sign.

$$\cos(30°) = \frac{\sqrt{3}}{2} \quad \text{and} \quad \sin(30°) = \frac{1}{2}$$

Since $150°$ is in the second quadrant, the x-coordinate of the point on the circle is negative, so the cosine value is negative. The y-coordinate is positive, so the sine value is positive.

$$\cos(150°) = -\frac{\sqrt{3}}{2} \quad \text{and} \quad \sin(150°) = \frac{1}{2}$$

ⓑ $\frac{5\pi}{4}$ is in the third quadrant. Its reference angle is $\frac{5\pi}{4} - \pi = \frac{\pi}{4}$. The cosine and sine of $\frac{\pi}{4}$ are both $\frac{\sqrt{2}}{2}$. In the third quadrant, both x and y are negative, so:

$$\cos\frac{5\pi}{4} = -\frac{\sqrt{2}}{2} \quad \text{and} \quad \sin\frac{5\pi}{4} = -\frac{\sqrt{2}}{2}$$

TRY IT #6 ⓐ Use the reference angle of $315°$ to find $\cos(315°)$ and $\sin(315°)$.

ⓑ Use the reference angle of $-\frac{\pi}{6}$ to find $\cos\left(-\frac{\pi}{6}\right)$ and $\sin\left(-\frac{\pi}{6}\right)$.

Using Reference Angles to Find Coordinates

Now that we have learned how to find the cosine and sine values for special angles in the first quadrant, we can use symmetry and reference angles to fill in cosine and sine values for the rest of the special angles on the unit circle. They are shown in Figure 19. Take time to learn the (x, y) coordinates of all of the major angles in the first quadrant.

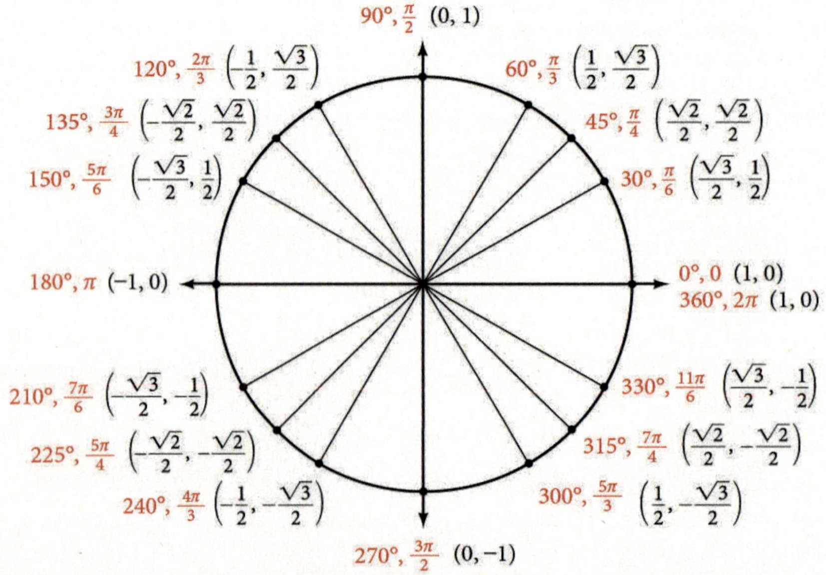

Figure 19 Special angles and coordinates of corresponding points on the unit circle

In addition to learning the values for special angles, we can use reference angles to find (x, y) coordinates of any point on the unit circle, using what we know of reference angles along with the identities

$$x = \cos t$$
$$y = \sin t$$

First we find the reference angle corresponding to the given angle. Then we take the sine and cosine values of the reference angle, and give them the signs corresponding to the *y*- and *x*-values of the quadrant.

HOW TO

Given the angle of a point on a circle and the radius of the circle, find the (x, y) coordinates of the point.

1. Find the reference angle by measuring the smallest angle to the *x*-axis.
2. Find the cosine and sine of the reference angle.
3. Determine the appropriate signs for x and y in the given quadrant.

EXAMPLE 7

Using the Unit Circle to Find Coordinates

Find the coordinates of the point on the unit circle at an angle of $\frac{7\pi}{6}$.

Solution

We know that the angle $\frac{7\pi}{6}$ is in the third quadrant.

First, let's find the reference angle by measuring the angle to the *x*-axis. To find the reference angle of an angle whose terminal side is in quadrant III, we find the difference of the angle and π.

$$\frac{7\pi}{6} - \pi = \frac{\pi}{6}$$

Next, we will find the cosine and sine of the reference angle.

$$\cos\left(\frac{\pi}{6}\right) = \frac{\sqrt{3}}{2} \qquad \sin\left(\frac{\pi}{6}\right) = \frac{1}{2}$$

We must determine the appropriate signs for *x* and *y* in the given quadrant. Because our original angle is in the third quadrant, where both x and y are negative, both cosine and sine are negative.

$$\begin{aligned} \cos\left(\frac{7\pi}{6}\right) &= -\frac{\sqrt{3}}{2} \\ \sin\left(\frac{7\pi}{6}\right) &= -\frac{1}{2} \end{aligned}$$

Now we can calculate the (x, y) coordinates using the identities $x = \cos\ \theta$ and $y = \sin\ \theta$.

The coordinates of the point are $\left(-\frac{\sqrt{3}}{2}, -\frac{1}{2}\right)$ on the unit circle.

TRY IT #7 Find the coordinates of the point on the unit circle at an angle of $\frac{5\pi}{3}$.

MEDIA

Access these online resources for additional instruction and practice with sine and cosine functions.

Trigonometric Functions Using the Unit Circle (http://openstax.org/l/trigunitcir)
Sine and Cosine from the Unit (http://openstax.org/l/sincosuc)
Sine and Cosine from the Unit Circle and Multiples of Pi Divided by Six (http://openstax.org/l/sincosmult)
Sine and Cosine from the Unit Circle and Multiples of Pi Divided by Four (http://openstax.org/l/sincosmult4)
Trigonometric Functions Using Reference Angles (http://openstax.org/l/trigrefang)

7.3 SECTION EXERCISES

Verbal

1. Describe the unit circle.

2. What do the x- and y-coordinates of the points on the unit circle represent?

3. Discuss the difference between a coterminal angle and a reference angle.

4. Explain how the cosine of an angle in the second quadrant differs from the cosine of its reference angle in the unit circle.

5. Explain how the sine of an angle in the second quadrant differs from the sine of its reference angle in the unit circle.

Algebraic

For the following exercises, use the given sign of the sine and cosine functions to find the quadrant in which the terminal point determined by t lies.

6. $\sin(t) < 0$ and $\cos(t) < 0$

7. $\sin(t) > 0$ and $\cos(t) > 0$

8. $\sin(t) > 0$ and $\cos(t) < 0$

9. $\sin(t) > 0$ and $\cos(t) > 0$

For the following exercises, find the exact value of each trigonometric function.

10. $\sin\frac{\pi}{2}$

11. $\sin\frac{\pi}{3}$

12. $\cos\frac{\pi}{2}$

13. $\cos\frac{\pi}{3}$

14. $\sin\frac{\pi}{4}$

15. $\cos\frac{\pi}{4}$

16. $\sin\frac{\pi}{6}$

17. $\sin\pi$

18. $\sin\frac{3\pi}{2}$

19. $\cos\pi$

20. $\cos 0$

21. $\cos\frac{\pi}{6}$

22. $\sin 0$

Numeric

For the following exercises, state the reference angle for the given angle.

23. $240°$

24. $-170°$

25. $100°$

26. $-315°$

27. $135°$

28. $\frac{5\pi}{4}$

29. $\frac{2\pi}{3}$

30. $\frac{5\pi}{6}$

31. $\frac{-11\pi}{3}$

32. $\frac{-7\pi}{4}$

33. $\frac{-\pi}{8}$

For the following exercises, find the reference angle, the quadrant of the terminal side, and the sine and cosine of each angle. If the angle is not one of the angles on the unit circle, use a calculator and round to three decimal places.

34. $225°$

35. $300°$

36. $320°$

37. $135°$

38. $210°$

39. $120°$

40. $250°$

41. $150°$

42. $\frac{5\pi}{4}$

43. $\frac{7\pi}{6}$

44. $\frac{5\pi}{3}$

45. $\frac{3\pi}{4}$

46. $\frac{4\pi}{3}$

47. $\frac{2\pi}{3}$

48. $\frac{5\pi}{6}$

49. $\frac{7\pi}{4}$

For the following exercises, find the requested value.

50. If $\cos(t) = \frac{1}{7}$ and t is in the fourth quadrant, find $\sin(t)$.

51. If $\cos(t) = \frac{2}{9}$ and t is in the first quadrant, find $\sin(t)$.

52. If $\sin(t) = \frac{3}{8}$ and t is in the second quadrant, find $\cos(t)$.

53. If $\sin(t) = -\frac{1}{4}$ and t is in the third quadrant, find $\cos(t)$.

54. Find the coordinates of the point on a circle with radius 15 corresponding to an angle of $220°$.

55. Find the coordinates of the point on a circle with radius 20 corresponding to an angle of $120°$.

56. Find the coordinates of the point on a circle with radius 8 corresponding to an angle of $\frac{7\pi}{4}$.

57. Find the coordinates of the point on a circle with radius 16 corresponding to an angle of $\frac{5\pi}{9}$.

58. State the domain of the sine and cosine functions.

59. State the range of the sine and cosine functions.

Graphical

For the following exercises, use the given point on the unit circle to find the value of the sine and cosine of t.

60.

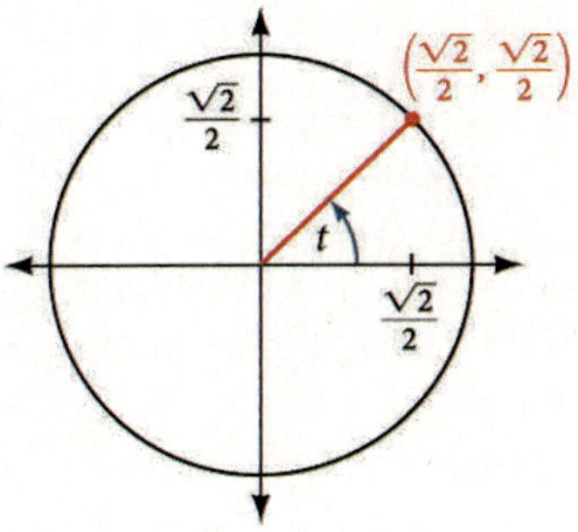

61.

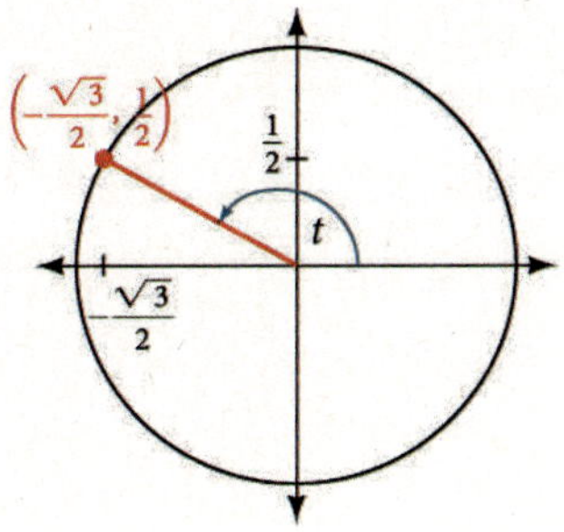

62.

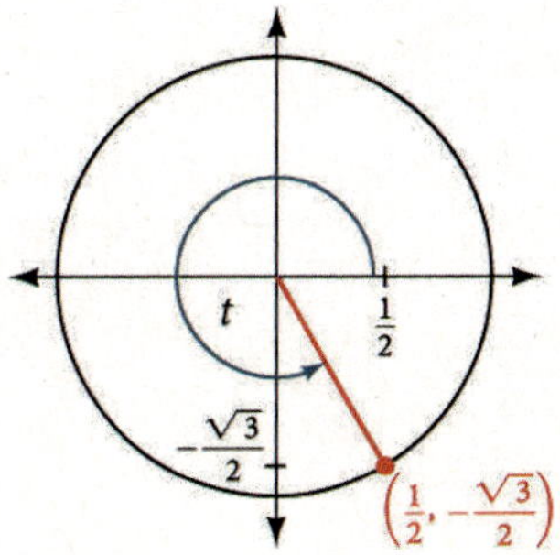

63.

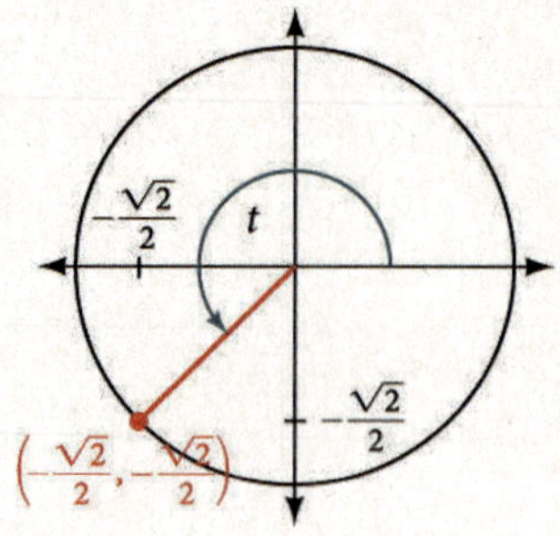

64.

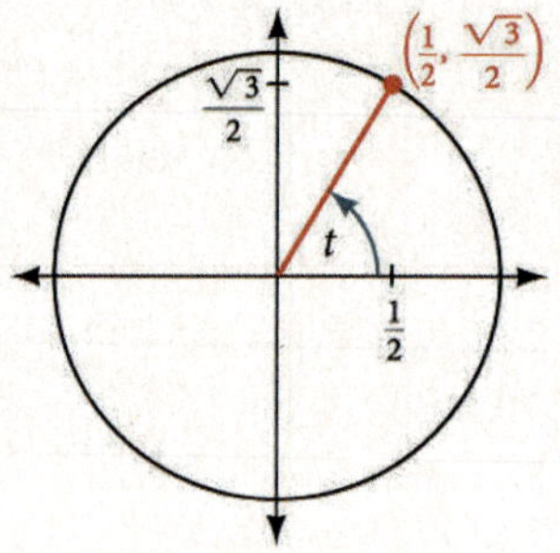

65.

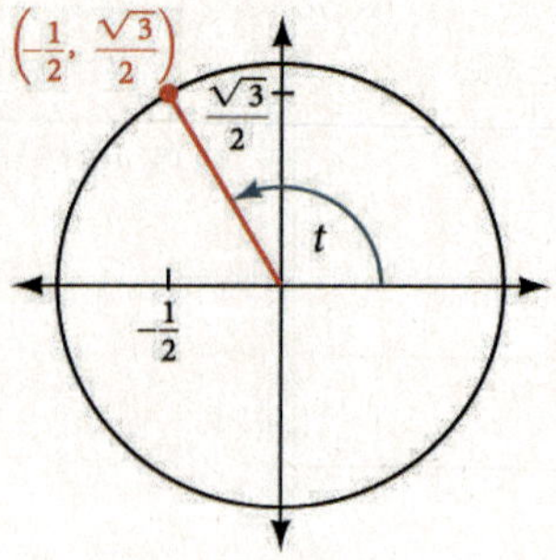

66.

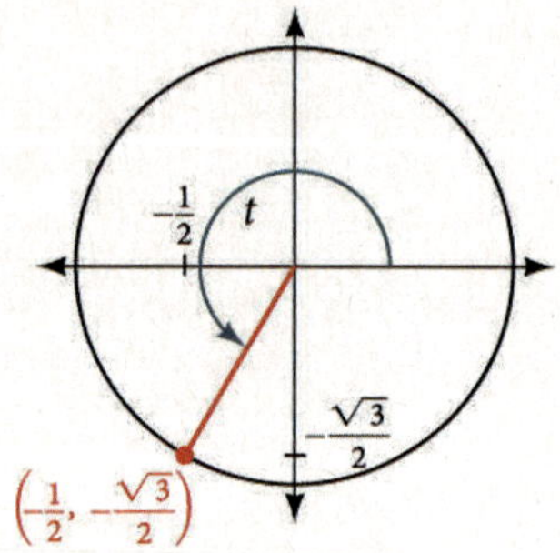

67.

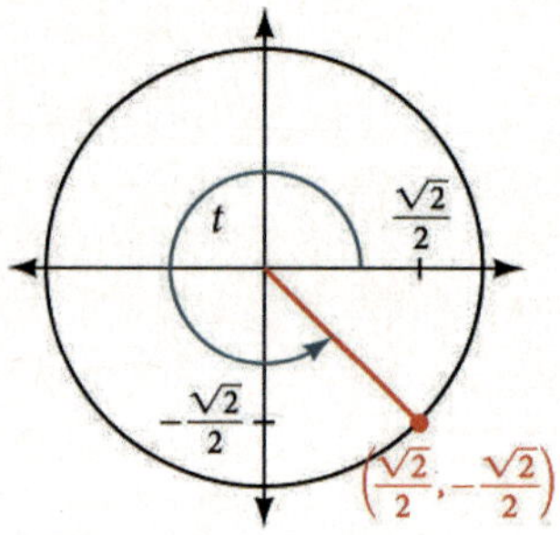

68.

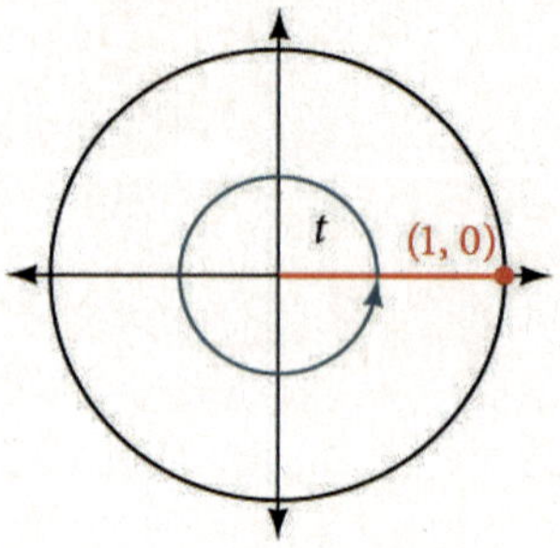

69.

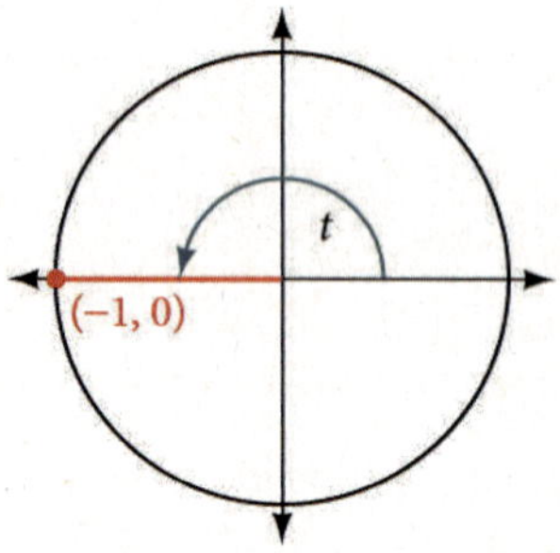

70.

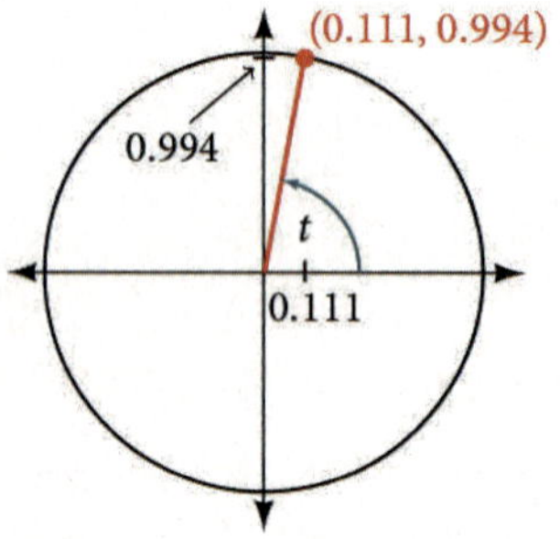

71.

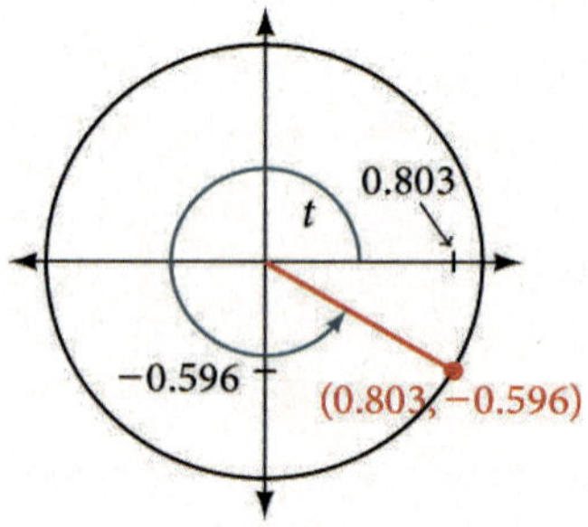

72.

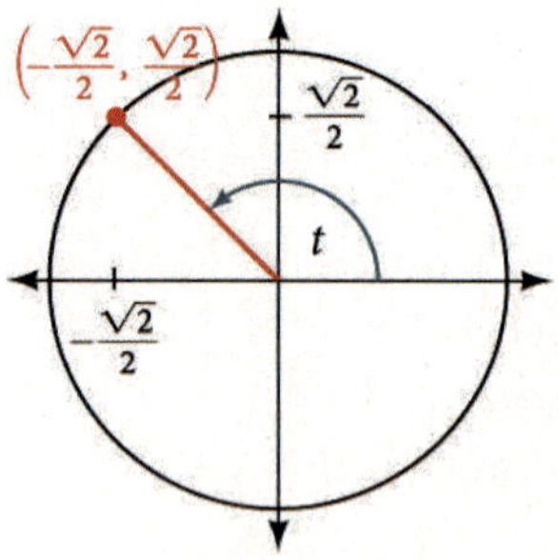

73.

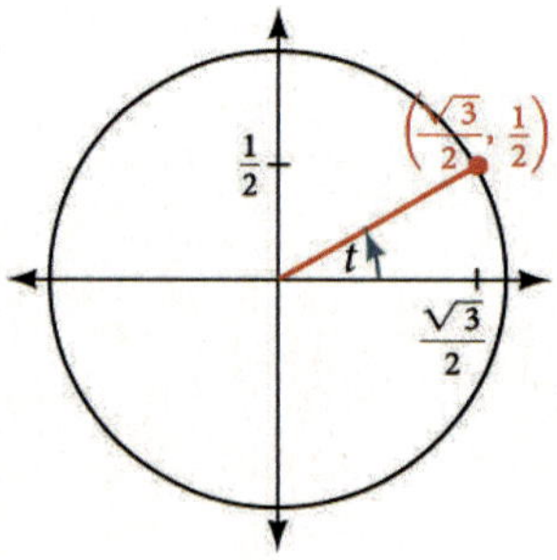

74.

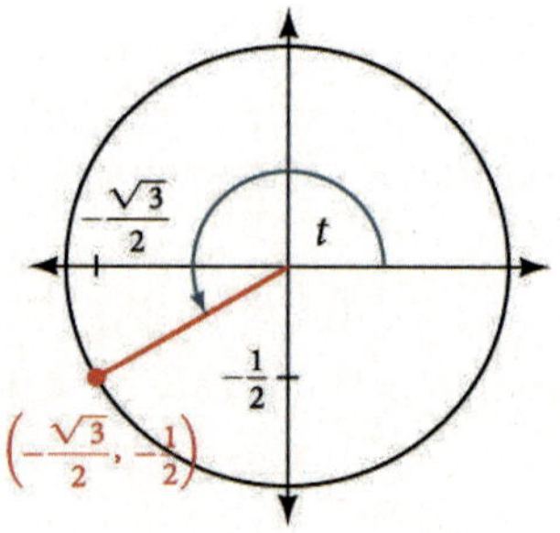

75.

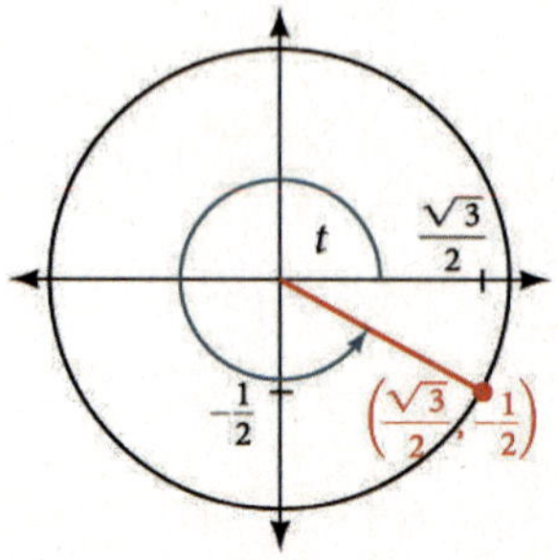

76.

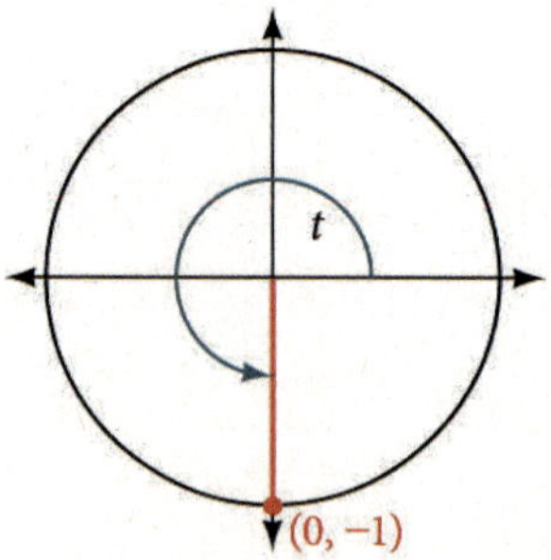

77.

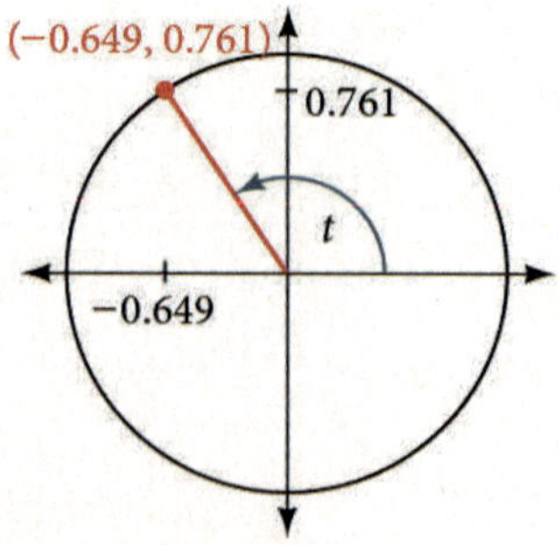

78.

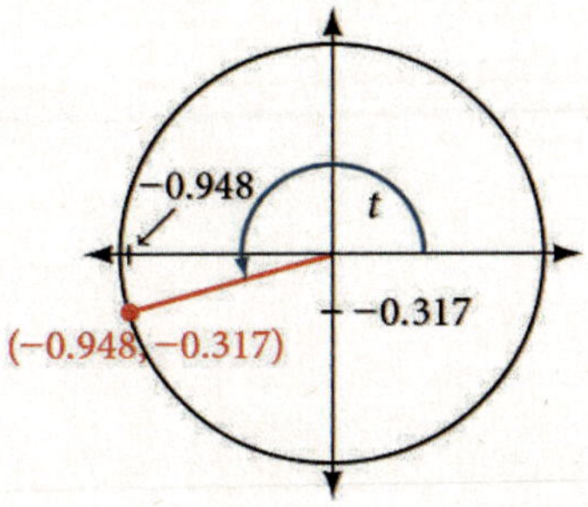

79.

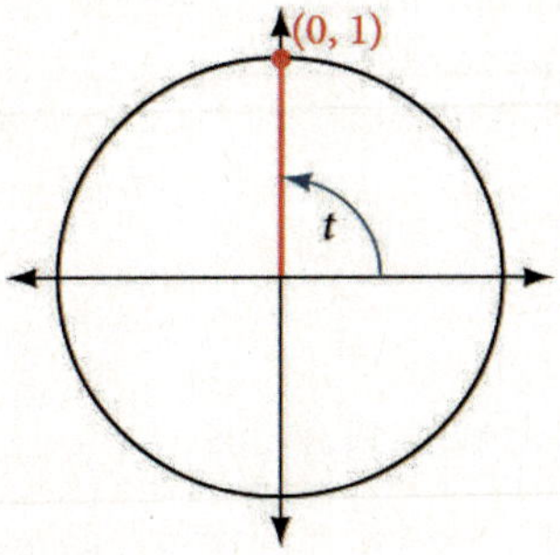

Technology

For the following exercises, use a graphing calculator to evaluate.

80. $\sin \frac{5\pi}{9}$

81. $\cos \frac{5\pi}{9}$

82. $\sin \frac{\pi}{10}$

83. $\cos \frac{\pi}{10}$

84. $\sin \frac{3\pi}{4}$

85. $\cos \frac{3\pi}{4}$

86. $\sin 98°$

87. $\cos 98°$

88. $\cos 310°$

89. $\sin 310°$

Extensions

For the following exercises, evaluate.

90. $\sin\left(\frac{11\pi}{3}\right)\cos\left(\frac{-5\pi}{6}\right)$

91. $\sin\left(\frac{3\pi}{4}\right)\cos\left(\frac{5\pi}{3}\right)$

92. $\sin\left(-\frac{4\pi}{3}\right)\cos\left(\frac{\pi}{2}\right)$

93. $\sin\left(\frac{-9\pi}{4}\right)\cos\left(\frac{-\pi}{6}\right)$

94. $\sin\left(\frac{\pi}{6}\right)\cos\left(\frac{-\pi}{3}\right)$

95. $\sin\left(\frac{7\pi}{4}\right)\cos\left(\frac{-2\pi}{3}\right)$

96. $\cos\left(\frac{5\pi}{6}\right)\cos\left(\frac{2\pi}{3}\right)$

97. $\cos\left(\frac{-\pi}{3}\right)\cos\left(\frac{\pi}{4}\right)$

98. $\sin\left(\frac{-5\pi}{4}\right)\sin\left(\frac{11\pi}{6}\right)$

99. $\sin(\pi)\sin\left(\frac{\pi}{6}\right)$

Real-World Applications

For the following exercises, use this scenario: A child enters a carousel that takes one minute to revolve once around. The child enters at the point $(0, 1)$, that is, on the due north position. Assume the carousel revolves counter clockwise.

100. What are the coordinates of the child after 45 seconds?

101. What are the coordinates of the child after 90 seconds?

102. What are the coordinates of the child after 125 seconds?

103. When will the child have coordinates $(0.707, -0.707)$ if the ride lasts 6 minutes? (There are multiple answers.)

104. When will the child have coordinates $(-0.866, -0.5)$ if the ride lasts 6 minutes?

7.4 The Other Trigonometric Functions

Learning Objectives

In this section you will:

- Find exact values of the trigonometric functions secant, cosecant, tangent, and cotangent of $\frac{\pi}{3}$, $\frac{\pi}{4}$, and $\frac{\pi}{6}$.
- Use reference angles to evaluate the trigonometric functions secant, tangent, and cotangent.
- Use properties of even and odd trigonometric functions.
- Recognize and use fundamental identities.
- Evaluate trigonometric functions with a calculator.

A wheelchair ramp that meets the standards of the Americans with Disabilities Act must make an angle with the ground whose tangent is $\frac{1}{12}$ or less, regardless of its length. A tangent represents a ratio, so this means that for every 1 inch of rise, the ramp must have 12 inches of run. Trigonometric functions allow us to specify the shapes and proportions of objects independent of exact dimensions. We have already defined the sine and cosine functions of an angle. Though sine and cosine are the trigonometric functions most often used, there are four others. Together they make up the set of six trigonometric functions. In this section, we will investigate the remaining functions.

Finding Exact Values of the Trigonometric Functions Secant, Cosecant, Tangent, and Cotangent

We can also define the remaining functions in terms of the unit circle with a point (x, y) corresponding to an angle of t, as shown in Figure 1. As with the sine and cosine, we can use the (x, y) coordinates to find the other functions.

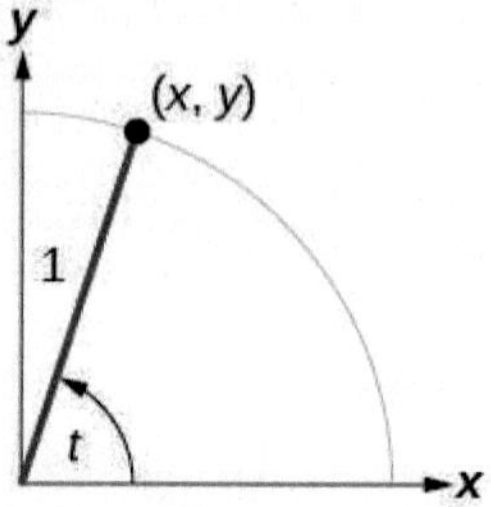

Figure 1

The first function we will define is the tangent. The **tangent** of an angle is the ratio of the *y*-value to the *x*-value of the corresponding point on the unit circle. In Figure 1, the tangent of angle t is equal to $\frac{y}{x}, x \neq 0$. Because the *y*-value is equal to the sine of t, and the *x*-value is equal to the cosine of t, the tangent of angle t can also be defined as $\frac{\sin t}{\cos t}, \cos t \neq 0$. The tangent function is abbreviated as $\tan$. The remaining three functions can all be expressed as reciprocals of functions we have already defined.

- The **secant** function is the reciprocal of the cosine function. In Figure 1, the secant of angle t is equal to $\frac{1}{\cos t} = \frac{1}{x}, x \neq 0$. The secant function is abbreviated as $\sec$.
- The **cotangent** function is the reciprocal of the tangent function. In Figure 1, the cotangent of angle t is equal to $\frac{\cos t}{\sin t} = \frac{x}{y}, y \neq 0$. The cotangent function is abbreviated as $\cot$.
- The **cosecant** function is the reciprocal of the sine function. In Figure 1, the cosecant of angle t is equal to $\frac{1}{\sin t} = \frac{1}{y}, y \neq 0$. The cosecant function is abbreviated as $\csc$.

Tangent, Secant, Cosecant, and Cotangent Functions

If t is a real number and (x, y) is a point where the terminal side of an angle of t radians intercepts the unit circle, then

$$\begin{aligned} \tan t &= \tfrac{y}{x}, x \neq 0 \\ \sec t &= \tfrac{1}{x}, x \neq 0 \\ \csc t &= \tfrac{1}{y}, y \neq 0 \\ \cot t &= \tfrac{x}{y}, y \neq 0 \end{aligned}$$

EXAMPLE 1

Finding Trigonometric Functions from a Point on the Unit Circle

The point $\left(-\frac{\sqrt{3}}{2}, \frac{1}{2}\right)$ is on the unit circle, as shown in Figure 2. Find $\sin\ t, \cos\ t, \tan\ t, \sec\ t, \csc\ t,$ and $\cot\ t$.

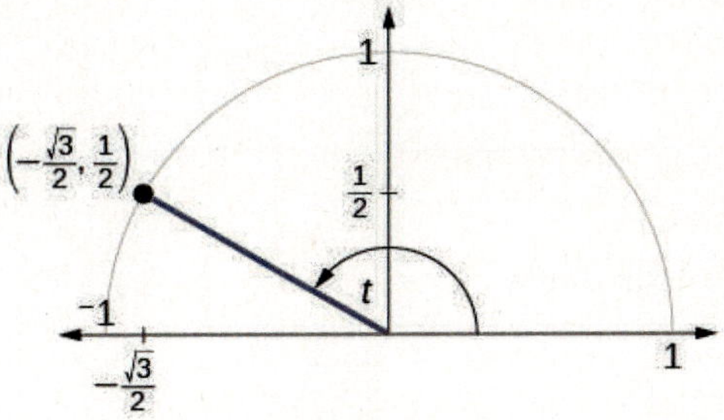

Figure 2

Solution

Because we know the (x, y) coordinates of the point on the unit circle indicated by angle t, we can use those coordinates to find the six functions:

$$\begin{aligned} \sin t &= y &&= \tfrac{1}{2} \\ \cos t &= x &&= -\tfrac{\sqrt{3}}{2} \\ \tan t &= \tfrac{y}{x} &&= \frac{\frac{1}{2}}{-\frac{\sqrt{3}}{2}} = \tfrac{1}{2}\left(-\tfrac{2}{\sqrt{3}}\right) = -\tfrac{1}{\sqrt{3}} = -\tfrac{\sqrt{3}}{3} \\ \sec t &= \tfrac{1}{x} &&= \frac{1}{-\frac{\sqrt{3}}{2}} = -\tfrac{2}{\sqrt{3}} = -\tfrac{2\sqrt{3}}{3} \\ \csc t &= \tfrac{1}{y} &&= \frac{1}{\frac{1}{2}} = 2 \\ \cot t &= \tfrac{x}{y} &&= \frac{-\frac{\sqrt{3}}{2}}{\frac{1}{2}} = -\tfrac{\sqrt{3}}{2}\left(\tfrac{2}{1}\right) = -\sqrt{3} \end{aligned}$$

> **TRY IT** #1 The point $\left(\frac{\sqrt{2}}{2}, -\frac{\sqrt{2}}{2}\right)$ is on the unit circle, as shown in Figure 3. Find $\sin\ t, \cos\ t, \tan\ t, \sec\ t, \csc\ t,$ and $\cot\ t$.

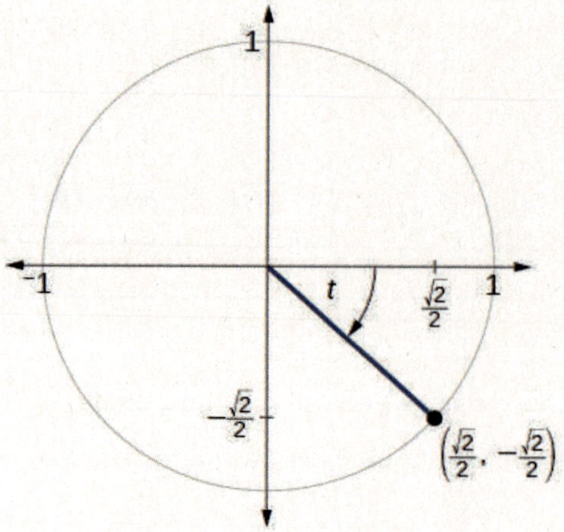

Figure 3

EXAMPLE 2

Finding the Trigonometric Functions of an Angle

Find $\sin t, \cos t, \tan t, \sec t, \csc t,$ and $\cot t$. when $t = \frac{\pi}{6}$.

Solution

We have previously used the properties of equilateral triangles to demonstrate that $\sin \frac{\pi}{6} = \frac{1}{2}$ and $\cos \frac{\pi}{6} = \frac{\sqrt{3}}{2}$. We can use these values and the definitions of tangent, secant, cosecant, and cotangent as functions of sine and cosine to find the remaining function values.

$$\begin{aligned}
\tan \frac{\pi}{6} &= \frac{\sin \frac{\pi}{6}}{\cos \frac{\pi}{6}} \\
&= \frac{\frac{1}{2}}{\frac{\sqrt{3}}{2}} = \frac{1}{\sqrt{3}} = \frac{\sqrt{3}}{3} \\
\sec \frac{\pi}{6} &= \frac{1}{\cos \frac{\pi}{6}} \\
&= \frac{1}{\frac{\sqrt{3}}{2}} = \frac{2}{\sqrt{3}} = \frac{2\sqrt{3}}{3} \\
\csc \frac{\pi}{6} &= \frac{1}{\sin \frac{\pi}{6}} = \frac{1}{\frac{1}{2}} = 2 \\
\cot \frac{\pi}{6} &= \frac{\cos \frac{\pi}{6}}{\sin \frac{\pi}{6}} \\
&= \frac{\frac{\sqrt{3}}{2}}{\frac{1}{2}} = \sqrt{3}
\end{aligned}$$

> **TRY IT** #2 Find $\sin t, \cos t, \tan t, \sec t, \csc t,$ and $\cot t$. when $t = \frac{\pi}{3}$.

Because we know the sine and cosine values for the common first-quadrant angles, we can find the other function values for those angles as well by setting x equal to the cosine and y equal to the sine and then using the definitions of tangent, secant, cosecant, and cotangent. The results are shown in Table 1.

Angle	0	$\frac{\pi}{6}$, or 30°	$\frac{\pi}{4}$, or 45°	$\frac{\pi}{3}$, or 60°	$\frac{\pi}{2}$, or 90°
Cosine	1	$\frac{\sqrt{3}}{2}$	$\frac{\sqrt{2}}{2}$	$\frac{1}{2}$	0
Sine	0	$\frac{1}{2}$	$\frac{\sqrt{2}}{2}$	$\frac{\sqrt{3}}{2}$	1
Tangent	0	$\frac{\sqrt{3}}{3}$	1	$\sqrt{3}$	Undefined
Secant	1	$\frac{2\sqrt{3}}{3}$	$\sqrt{2}$	2	Undefined
Cosecant	Undefined	2	$\sqrt{2}$	$\frac{2\sqrt{3}}{3}$	1
Cotangent	Undefined	$\sqrt{3}$	1	$\frac{\sqrt{3}}{3}$	0

Table 1

Using Reference Angles to Evaluate Tangent, Secant, Cosecant, and Cotangent

We can evaluate trigonometric functions of angles outside the first quadrant using reference angles as we have already done with the sine and cosine functions. The procedure is the same: Find the reference angle formed by the terminal side of the given angle with the horizontal axis. The trigonometric function values for the original angle will be the same as those for the reference angle, except for the positive or negative sign, which is determined by *x*- and *y*-values in the original quadrant. Figure 4 shows which functions are positive in which quadrant.

To help remember which of the six trigonometric functions are positive in each quadrant, we can use the mnemonic phrase "A Smart Trig Class." Each of the four words in the phrase corresponds to one of the four quadrants, starting with quadrant I and rotating counterclockwise. In quadrant I, which is "**A**," **a**ll of the six trigonometric functions are positive. In quadrant II, "**S**mart," only **s**ine and its reciprocal function, cosecant, are positive. In quadrant III, "**T**rig," only **t**angent and its reciprocal function, cotangent, are positive. Finally, in quadrant IV, "**C**lass," only **c**osine and its reciprocal function, secant, are positive.

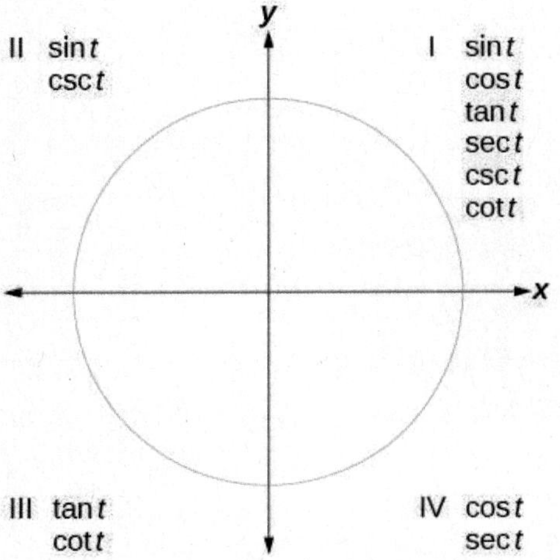

Figure 4 The trigonometric functions are each listed in the quadrants in which they are positive.

HOW TO

Given an angle not in the first quadrant, use reference angles to find all six trigonometric functions.

1. Measure the angle formed by the terminal side of the given angle and the horizontal axis. This is the reference

angle.
2. Evaluate the function at the reference angle.
3. Observe the quadrant where the terminal side of the original angle is located. Based on the quadrant, determine whether the output is positive or negative.

EXAMPLE 3

Using Reference Angles to Find Trigonometric Functions

Use reference angles to find all six trigonometric functions of $-\frac{5\pi}{6}$.

Solution

The angle between this angle's terminal side and the *x*-axis is $\frac{\pi}{6}$, so that is the reference angle. Since $-\frac{5\pi}{6}$ is in the third quadrant, where both x and y are negative, cosine, sine, secant, and cosecant will be negative, while tangent and cotangent will be positive.

$$\begin{aligned}\cos\left(-\frac{5\pi}{6}\right) &= -\frac{\sqrt{3}}{2}, \sin\left(-\frac{5\pi}{6}\right) &= -\frac{1}{2}, \tan\left(-\frac{5\pi}{6}\right) &= \frac{\sqrt{3}}{3},\\ \sec\left(-\frac{5\pi}{6}\right) &= -\frac{2\sqrt{3}}{3}, \csc\left(-\frac{5\pi}{6}\right) &= -2, \cot\left(-\frac{5\pi}{6}\right) &= \sqrt{3}\end{aligned}$$

TRY IT #3 Use reference angles to find all six trigonometric functions of $-\frac{7\pi}{4}$.

Using Even and Odd Trigonometric Functions

To be able to use our six trigonometric functions freely with both positive and negative angle inputs, we should examine how each function treats a negative input. As it turns out, there is an important difference among the functions in this regard.

Consider the function $f(x) = x^2$, shown in Figure 5. The graph of the function is symmetrical about the *y*-axis. All along the curve, any two points with opposite *x*-values have the same function value. This matches the result of calculation: $(4)^2 = (-4)^2, (-5)^2 = (5)^2$, and so on. So $f(x) = x^2$ is an even function, a function such that two inputs that are opposites have the same output. That means $f(-x) = f(x)$.

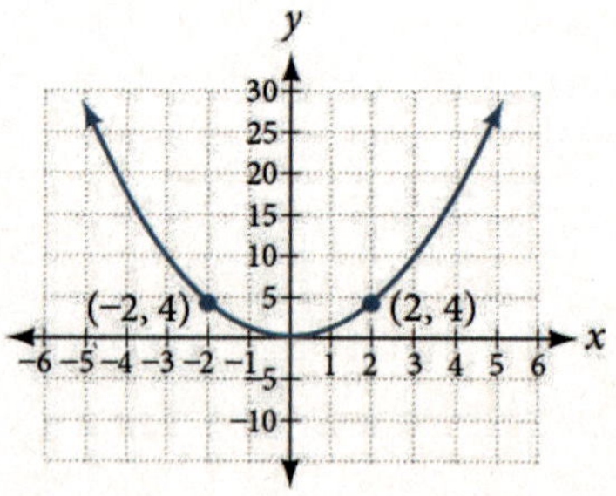

Figure 5 The function $f(x) = x^2$ is an even function.

Now consider the function $f(x) = x^3$, shown in Figure 6. The graph is not symmetrical about the *y*-axis. All along the graph, any two points with opposite *x*-values also have opposite *y*-values. So $f(x) = x^3$ is an odd function, one such that two inputs that are opposites have outputs that are also opposites. That means $f(-x) = -f(x)$.

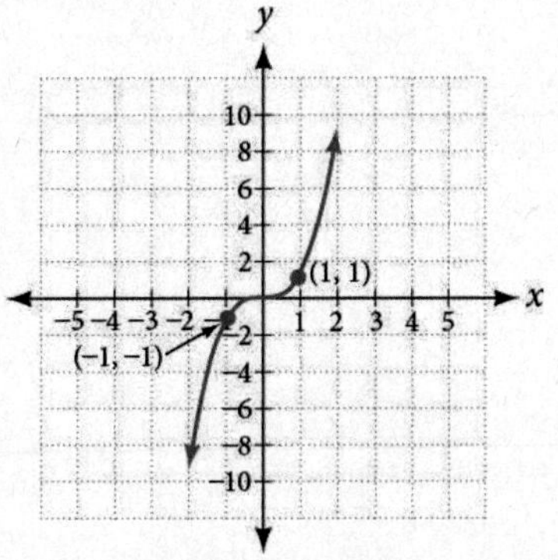

Figure 6 The function $f(x) = x^3$ is an odd function.

We can test whether a trigonometric function is even or odd by drawing a unit circle with a positive and a negative angle, as in Figure 7. The sine of the positive angle is y. The sine of the negative angle is $-y$. The sine function, then, is an odd function. We can test each of the six trigonometric functions in this fashion. The results are shown in Table 2.

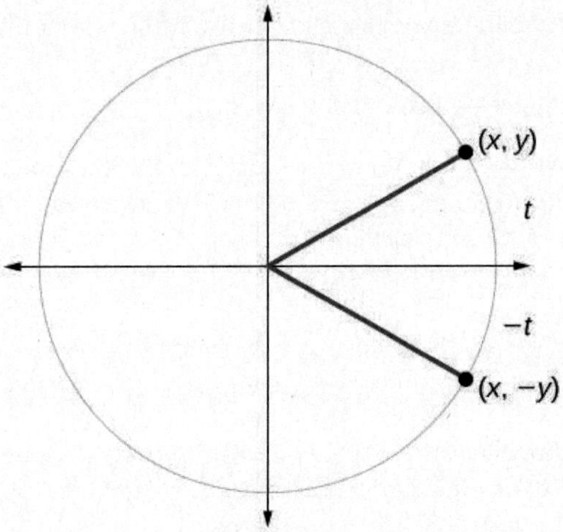

Figure 7

$\sin t = y$ $\sin(-t) = -y$ $\sin t \neq \sin(-t)$	$\cos t = x$ $\cos(-t) = x$ $\cos t = \cos(-t)$	$\tan(t) = \frac{y}{x}$ $\tan(-t) = -\frac{y}{x}$ $\tan t \neq \tan(-t)$
$\sec t = \frac{1}{x}$ $\sec(-t) = \frac{1}{x}$ $\sec t = \sec(-t)$	$\csc t = \frac{1}{y}$ $\csc(-t) = \frac{1}{-y}$ $\csc t \neq \csc(-t)$	$\cot t = \frac{x}{y}$ $\cot(-t) = \frac{x}{-y}$ $\cot t \neq \cot(-t)$

Table 2

Even and Odd Trigonometric Functions

An even function is one in which $f(-x) = f(x)$.

An odd function is one in which $f(-x) = -f(x)$.

Cosine and secant are even:

$$\begin{aligned} \cos(-t) &= \cos t \\ \sec(-t) &= \sec t \end{aligned}$$

Sine, tangent, cosecant, and cotangent are odd:

$$\begin{aligned} \sin(-t) &= -\sin t \\ \tan(-t) &= -\tan t \\ \csc(-t) &= -\csc t \\ \cot(-t) &= -\cot t \end{aligned}$$

EXAMPLE 4

Using Even and Odd Properties of Trigonometric Functions

If the secant of angle t is 2, what is the secant of $-t$?

Solution

Secant is an even function. The secant of an angle is the same as the secant of its opposite. So if the secant of angle t is 2, the secant of $-t$ is also 2.

TRY IT #4 If the cotangent of angle t is $\sqrt{3}$, what is the cotangent of $-t$?

Recognizing and Using Fundamental Identities

We have explored a number of properties of trigonometric functions. Now, we can take the relationships a step further, and derive some fundamental identities. Identities are statements that are true for all values of the input on which they are defined. Usually, identities can be derived from definitions and relationships we already know. For example, the Pythagorean Identity we learned earlier was derived from the Pythagorean Theorem and the definitions of sine and cosine.

Fundamental Identities

We can derive some useful **identities** from the six trigonometric functions. The other four trigonometric functions can be related back to the sine and cosine functions using these basic relationships:

$$\tan\ t = \frac{\sin\ t}{\cos\ t}$$

$$\sec\ t = \frac{1}{\cos\ t}$$

$$\csc\ t = \frac{1}{\sin\ t}$$

$$\cot t = \frac{1}{\tan t} = \frac{\cos t}{\sin t}$$

EXAMPLE 5

Using Identities to Evaluate Trigonometric Functions

ⓐ Given $\sin(45°) = \frac{\sqrt{2}}{2}, \cos(45°) = \frac{\sqrt{2}}{2}$, evaluate $\tan(45°)$.

ⓑ Given $\sin\left(\frac{5\pi}{6}\right) = \frac{1}{2}, \cos\left(\frac{5\pi}{6}\right) = -\frac{\sqrt{3}}{2}$, evaluate $\sec\left(\frac{5\pi}{6}\right)$.

Solution

Because we know the sine and cosine values for these angles, we can use identities to evaluate the other functions.

ⓐ

$$\begin{aligned} \tan(45°) &= \frac{\sin(45°)}{\cos(45°)} \\ &= \frac{\frac{\sqrt{2}}{2}}{\frac{\sqrt{2}}{2}} \\ &= 1 \end{aligned}$$

ⓑ

$$\begin{aligned} \sec\left(\frac{5\pi}{6}\right) &= \frac{1}{\cos\left(\frac{5\pi}{6}\right)} \\ &= \frac{1}{-\frac{\sqrt{3}}{2}} \\ &= \frac{-2\sqrt{3}}{1} \\ &= \frac{-2}{\sqrt{3}} \\ &= -\frac{2\sqrt{3}}{3} \end{aligned}$$

> **TRY IT** #5 Evaluate $\csc\left(\frac{7\pi}{6}\right)$.

EXAMPLE 6

Using Identities to Simplify Trigonometric Expressions

Simplify $\frac{\sec\ t}{\tan\ t}$.

Solution

We can simplify this by rewriting both functions in terms of sine and cosine.

$$\begin{aligned} \frac{\sec t}{\tan t} &= \frac{\frac{1}{\cos t}}{\frac{\sin t}{\cos t}} \\ &= \frac{1}{\cos t} \cdot \frac{\cos t}{\sin t} && \text{Multiply by the reciprocal.} \\ &= \frac{1}{\sin t} = \csc t && \text{Simplify and use the identity.} \end{aligned}$$

By showing that $\frac{\sec\ t}{\tan\ t}$ can be simplified to $\csc\ t$, we have, in fact, established a new identity.

$$\frac{\sec\ t}{\tan\ t} = \csc\ t$$

> **TRY IT** #6 Simplify $(\tan\ t)(\cos\ t)$.

Alternate Forms of the Pythagorean Identity

We can use these fundamental identities to derive alternate forms of the Pythagorean Identity, $\cos^2 t + \sin^2 t = 1$. One form is obtained by dividing both sides by $\cos^2 t$.

$$\begin{aligned} \frac{\cos^2 t}{\cos^2 t} + \frac{\sin^2 t}{\cos^2 t} &= \frac{1}{\cos^2 t} \\ 1 + \tan^2 t &= \sec^2 t \end{aligned}$$

The other form is obtained by dividing both sides by $\sin^2 t$.

$$\begin{aligned} \frac{\cos^2 t}{\sin^2 t} + \frac{\sin^2 t}{\sin^2 t} &= \frac{1}{\sin^2 t} \\ \cot^2 t + 1 &= \csc^2 t \end{aligned}$$

Alternate Forms of the Pythagorean Identity

$$1 + \tan^2 t = \sec^2 t$$

$$\cot^2 t + 1 = \csc^2 t$$

EXAMPLE 7

Using Identities to Relate Trigonometric Functions

If $\cos(t) = \frac{12}{13}$ and t is in quadrant IV, as shown in Figure 8, find the values of the other five trigonometric functions.

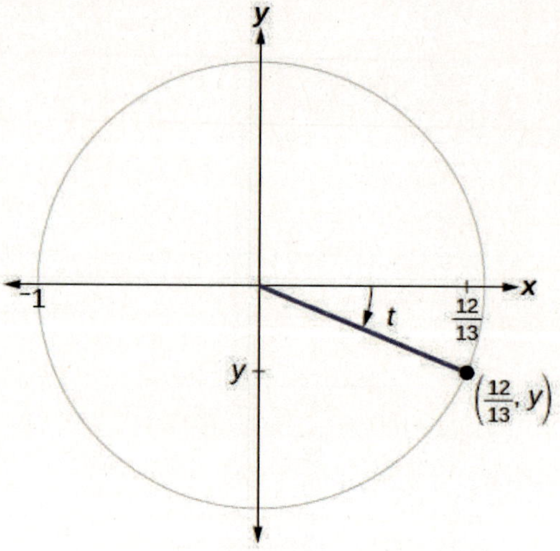

Figure 8

Solution

We can find the sine using the Pythagorean Identity, $\cos^2 t + \sin^2 t = 1$, and the remaining functions by relating them to sine and cosine.

$$\begin{aligned} \left(\tfrac{12}{13}\right)^2 + \sin^2 t &= 1 \\ \sin^2 t &= 1 - \left(\tfrac{12}{13}\right)^2 \\ \sin^2 t &= 1 - \tfrac{144}{169} \\ \sin^2 t &= \tfrac{25}{169} \\ \sin t &= \pm\sqrt{\tfrac{25}{169}} \\ \sin t &= \pm\tfrac{\sqrt{25}}{\sqrt{169}} \\ \sin t &= \pm\tfrac{5}{13} \end{aligned}$$

The sign of the sine depends on the y-values in the quadrant where the angle is located. Since the angle is in quadrant IV, where the y-values are negative, its sine is negative, $-\frac{5}{13}$.

The remaining functions can be calculated using identities relating them to sine and cosine.

$$\begin{aligned} \tan t &= \frac{\sin t}{\cos t} &= \frac{-\frac{5}{13}}{\frac{12}{13}} &= -\frac{5}{12} \\ \sec t &= \frac{1}{\cos t} &= \frac{1}{\frac{12}{13}} &= \frac{13}{12} \\ \csc t &= \frac{1}{\sin t} &= \frac{1}{-\frac{5}{13}} &= \frac{-13}{5} \\ \cot t &= \frac{1}{\tan t} &= \frac{1}{-\frac{5}{12}} &= -\frac{12}{5} \end{aligned}$$

> **TRY IT** #7 If $\sec(t) = -\frac{17}{8}$ and $0 < t < \pi$, find the values of the other five functions.

As we discussed at the beginning of the chapter, a function that repeats its values in regular intervals is known as a periodic function. The trigonometric functions are periodic. For the four trigonometric functions, sine, cosine, cosecant and secant, a revolution of one circle, or 2π, will result in the same outputs for these functions. And for tangent and cotangent, only a half a revolution will result in the same outputs.

Other functions can also be periodic. For example, the lengths of months repeat every four years. If x represents the length time, measured in years, and $f(x)$ represents the number of days in February, then $f(x+4)=f(x)$. This pattern repeats over and over through time. In other words, every four years, February is guaranteed to have the same number of days as it did 4 years earlier. The positive number 4 is the smallest positive number that satisfies this condition and is called the period. A **period** is the shortest interval over which a function completes one full cycle—in this example, the period is 4 and represents the time it takes for us to be certain February has the same number of days.

Period of a Function

The **period** P of a repeating function f is the number representing the interval such that $f(x+P)=f(x)$ for any value of x.

The period of the cosine, sine, secant, and cosecant functions is 2π.

The period of the tangent and cotangent functions is π.

EXAMPLE 8

Finding the Values of Trigonometric Functions

Find the values of the six trigonometric functions of angle t based on Figure 9.

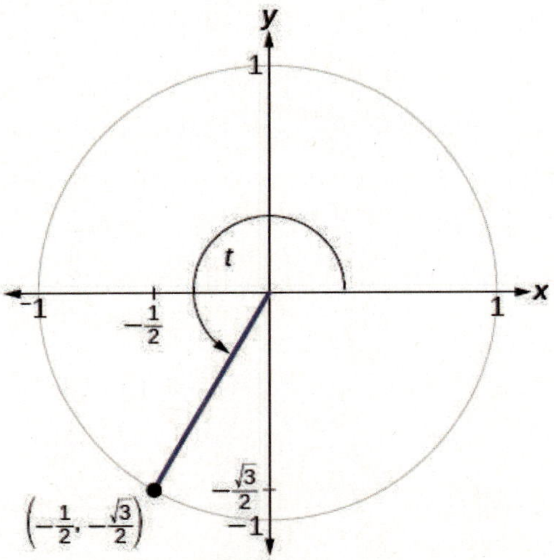

Figure 9

Solution

$$\sin t = y = -\frac{\sqrt{3}}{2}$$

$$\cos t = x = -\frac{1}{2}$$

$$\tan t = \frac{\sin t}{\cos t} = \frac{-\frac{\sqrt{3}}{2}}{-\frac{1}{2}} = \sqrt{3}$$

$$\sec t = \frac{1}{\cos t} = \frac{1}{-\frac{1}{2}} = -2$$

$$\csc t = \frac{1}{\sin t} = \frac{1}{-\frac{\sqrt{3}}{2}} = -\frac{2\sqrt{3}}{3}$$

$$\cot t = \frac{1}{\tan t} = \frac{1}{\sqrt{3}} = \frac{\sqrt{3}}{3}$$

> **TRY IT** #8 Find the values of the six trigonometric functions of angle t based on Figure 10.

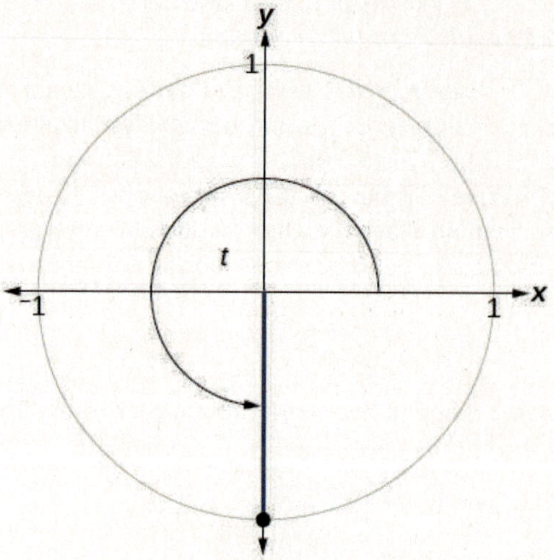

Figure 10

EXAMPLE 9

Finding the Value of Trigonometric Functions

If $\sin(t) = -\frac{\sqrt{3}}{2}$ and $\cos(t) = \frac{1}{2}$, find $\sec(t), \csc(t), \tan(t), \cot(t)$.

Solution

$$\begin{aligned}
\sec t &= \frac{1}{\cos t} &&= \frac{1}{\frac{1}{2}} = 2 \\
\csc t &= \frac{1}{\sin t} &&= \frac{1}{-\frac{\sqrt{3}}{2}} - \frac{2\sqrt{3}}{3} \\
\tan t &= \frac{\sin t}{\cos t} &&= \frac{-\frac{\sqrt{3}}{2}}{\frac{1}{2}} = -\sqrt{3} \\
\cot t &= \frac{1}{\tan t} &&= \frac{1}{-\sqrt{3}} = -\frac{\sqrt{3}}{3}
\end{aligned}$$

> **TRY IT** #9 $\sin(t) = \frac{\sqrt{2}}{2}$ and $\cos(t) = \frac{\sqrt{2}}{2}$, find $\sec(t), \csc(t), \tan(t)$, and $\cot(t)$

Evaluating Trigonometric Functions with a Calculator

We have learned how to evaluate the six trigonometric functions for the common first-quadrant angles and to use them as reference angles for angles in other quadrants. To evaluate trigonometric functions of other angles, we use a scientific or graphing calculator or computer software. If the calculator has a degree mode and a radian mode, confirm the correct mode is chosen before making a calculation.

Evaluating a tangent function with a scientific calculator as opposed to a graphing calculator or computer algebra system is like evaluating a sine or cosine: Enter the value and press the TAN key. For the reciprocal functions, there may not be any dedicated keys that say CSC, SEC, or COT. In that case, the function must be evaluated as the reciprocal of a sine, cosine, or tangent.

If we need to work with degrees and our calculator or software does not have a degree mode, we can enter the degrees multiplied by the conversion factor $\frac{\pi}{180}$ to convert the degrees to radians. To find the secant of $30°$, we could press

$$\text{(for a scientific calculator): } \frac{1}{30 \times \frac{\pi}{180}} \text{ COS}$$

or

$$\text{(for a graphing calculator): } \frac{1}{\cos\left(\frac{30\pi}{180}\right)}$$

HOW TO

Given an angle measure in radians, use a scientific calculator to find the cosecant.

1. If the calculator has degree mode and radian mode, set it to radian mode.
2. Enter: 1/
3. Enter the value of the angle inside parentheses.
4. Press the SIN key.
5. Press the = key.

HOW TO

Given an angle measure in radians, use a graphing utility/calculator to find the cosecant.

- If the graphing utility has degree mode and radian mode, set it to radian mode.
- Enter: 1/
- Press the SIN key.
- Enter the value of the angle inside parentheses.
- Press the ENTER key.

EXAMPLE 10

Evaluating the Cosecant Using Technology

Evaluate the cosecant of $\frac{5\pi}{7}$.

Solution

For a scientific calculator, enter information as follows:

$$\begin{aligned} 1/(5 \times \pi/7) \text{ SIN } &= \\ \csc\left(\tfrac{5\pi}{7}\right) &\approx 1.279 \end{aligned}$$

TRY IT #10 Evaluate the cotangent of $-\frac{\pi}{8}$.

MEDIA

Access these online resources for additional instruction and practice with other trigonometric functions.

Determing Trig Function Values (http://openstax.org/l/trigfuncval)
More Examples of Determining Trig Functions (http://openstax.org/l/moretrigfun)
Pythagorean Identities (http://openstax.org/l/pythagiden)
Trig Functions on a Calculator (http://openstax.org/l/trigcalc)

7.4 SECTION EXERCISES

Verbal

1. On an interval of $[0, 2\pi)$, can the sine and cosine values of a radian measure ever be equal? If so, where?

2. What would you estimate the cosine of π degrees to be? Explain your reasoning.

3. For any angle in quadrant II, if you knew the sine of the angle, how could you determine the cosine of the angle?

4. Describe the secant function.

5. Tangent and cotangent have a period of π. What does this tell us about the output of these functions?

Algebraic

For the following exercises, find the exact value of each expression.

6. $\tan \frac{\pi}{6}$

7. $\sec \frac{\pi}{6}$

8. $\csc \frac{\pi}{6}$

9. $\cot \frac{\pi}{6}$

10. $\tan \frac{\pi}{4}$

11. $\sec \frac{\pi}{4}$

12. $\csc \frac{\pi}{4}$

13. $\cot \frac{\pi}{4}$

14. $\tan \frac{\pi}{3}$

15. $\sec \frac{\pi}{3}$

16. $\csc \frac{\pi}{3}$

17. $\cot \frac{\pi}{3}$

For the following exercises, use reference angles to evaluate the expression.

18. $\tan \frac{5\pi}{6}$

19. $\sec \frac{7\pi}{6}$

20. $\csc \frac{11\pi}{6}$

21. $\cot \frac{13\pi}{6}$

22. $\tan \frac{7\pi}{4}$

23. $\sec \frac{3\pi}{4}$

24. $\csc \frac{5\pi}{4}$

25. $\cot \frac{11\pi}{4}$

26. $\tan \frac{8\pi}{3}$

27. $\sec \frac{4\pi}{3}$

28. $\csc \frac{2\pi}{3}$

29. $\cot \frac{5\pi}{3}$

30. $\tan 225°$

31. $\sec 300°$

32. $\csc 150°$

33. $\cot 240°$

34. $\tan 330°$

35. $\sec 120°$

36. $\csc 210°$

37. $\cot 315°$

38. If $\sin t = \frac{3}{4}$, and t is in quadrant II, find $\cos t, \sec t, \csc t, \tan t,$ and $\cot t$.

39. If $\cos\ t = -\frac{1}{3}$, and t is in quadrant III, find $\sin\ t, \sec\ t, \csc\ t, \tan\ t,$ and $\cot\ t$.

40. If $\tan\ t = \frac{12}{5}$, and $0 \leq t < \frac{\pi}{2}$, find $\sin\ t, \cos\ t, \sec\ t, \csc\ t,$ and $\cot\ t$.

41. If $\sin\ t = \frac{\sqrt{3}}{2}$ and $\cos\ t = \frac{1}{2}$, find $\sec\ t, \csc\ t, \tan\ t,$ and $\cot\ t$.

42. If $\sin\ 40° \approx 0.643$ and $\cos\ 40° \approx 0.766$, find $\sec\ 40°, \csc\ 40°, \tan\ 40°,$ and $\cot\ 40°$.

43. If $\sin t = \frac{\sqrt{2}}{2}$, what is the $\sin(-t)$?

44. If $\cos t = \frac{1}{2}$, what is the $\cos(-t)$?

45. If $\sec t = 3.1$, what is the $\sec(-t)$?

46. If $\csc t = 0.34$, what is the $\csc(-t)$?

47. If $\tan t = -1.4$, what is the $\tan(-t)$?

48. If $\cot t = 9.23$, what is the $\cot(-t)$?

Graphical

For the following exercises, use the angle in the unit circle to find the value of the each of the six trigonometric functions.

49.

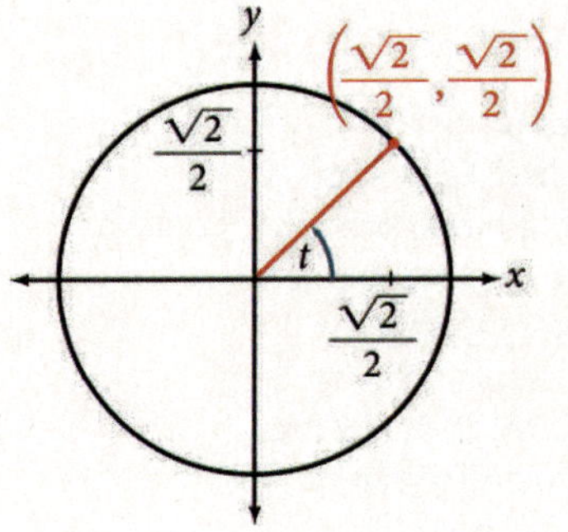

50.

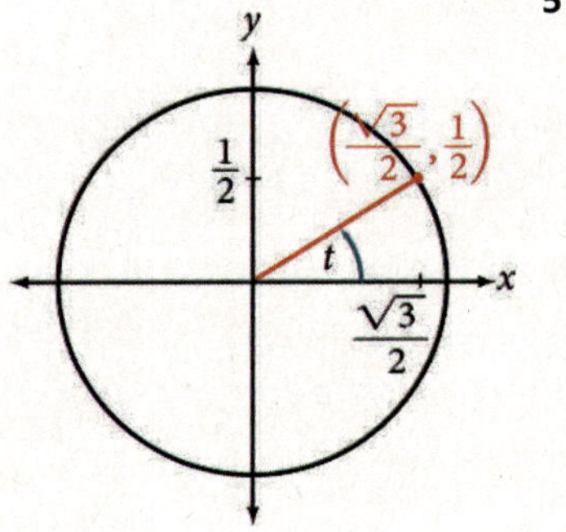

51.

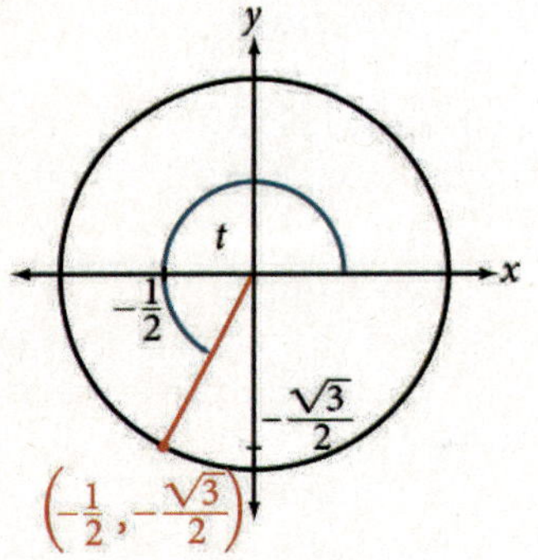

Technology

For the following exercises, use a graphing calculator to evaluate to three decimal places.

52. $\csc\ \frac{5\pi}{9}$

53. $\cot\ \frac{4\pi}{7}$

54. $\sec\ \frac{\pi}{10}$

55. $\tan\ \frac{5\pi}{8}$

56. $\sec\ \frac{3\pi}{4}$

57. $\csc\ \frac{\pi}{4}$

58. $\tan\ 98°$

59. $\cot\ 33°$

60. $\cot\ 140°$

61. $\sec\ 310°$

Extensions

For the following exercises, use identities to evaluate the expression.

62. If $\tan(t) \approx 2.7$, and $\sin(t) \approx 0.94$, find $\cos(t)$.

63. If $\tan(t) \approx 1.3$, and $\cos(t) \approx 0.61$, find $\sin(t)$.

64. If $\csc(t) \approx 3.2$, and $\cos(t) \approx 0.95$, find $\tan(t)$.

65. If $\cot(t) \approx 0.58$, and $\cos(t) \approx 0.5$, find $\csc(t)$.

66. Determine whether the function $f(x) = 2\sin x \cos x$ is even, odd, or neither.

67. Determine whether the function $f(x) = 3\sin^2 x \cos x + \sec x$ is even, odd, or neither.

68. Determine whether the function $f(x) = \sin x - 2\cos^2 x$ is even, odd, or neither.

69. Determine whether the function $f(x) = \csc^2 x + \sec x$ is even, odd, or neither.

For the following exercises, use identities to simplify the expression.

70. $\csc t \tan t$

71. $\frac{\sec t}{\csc t}$

Real-World Applications

72. The amount of sunlight in a certain city can be modeled by the function $h = 15\cos\left(\frac{1}{600}d\right)$, where h represents the hours of sunlight, and d is the day of the year. Use the equation to find how many hours of sunlight there are on February 11, the 42nd day of the year. State the period of the function.

73. The amount of sunlight in a certain city can be modeled by the function $h = 16\cos\left(\frac{1}{500}d\right)$, where h represents the hours of sunlight, and d is the day of the year. Use the equation to find how many hours of sunlight there are on September 24, the 267th day of the year. State the period of the function.

74. The equation $P = 20\sin(2\pi t) + 100$ models the blood pressure, P, where t represents time in seconds. (a) Find the blood pressure after 15 seconds. (b) What are the maximum and minimum blood pressures?

75. The height of a piston, h, in inches, can be modeled by the equation $y = 3\sin x + 1$, where x represents the crank angle. Find the height of the piston when the crank angle is $55°$.

76. The height of a piston, h, in inches, can be modeled by the equation $y = 2\cos x + 5$, where x represents the crank angle. Find the height of the piston when the crank angle is $55°$.

Chapter Review

Key Terms

adjacent side in a right triangle, the side between a given angle and the right angle

angle the union of two rays having a common endpoint

angle of depression the angle between the horizontal and the line from the object to the observer's eye, assuming the object is positioned lower than the observer

angle of elevation the angle between the horizontal and the line from the object to the observer's eye, assuming the object is positioned higher than the observer

angular speed the angle through which a rotating object travels in a unit of time

arc length the length of the curve formed by an arc

area of a sector area of a portion of a circle bordered by two radii and the intercepted arc; the fraction $\frac{\theta}{2\pi}$. multiplied by the area of the entire circle

cosecant the reciprocal of the sine function: on the unit circle, $\csc t = \frac{1}{y}, y \neq 0$

cosine function the *x*-value of the point on a unit circle corresponding to a given angle

cotangent the reciprocal of the tangent function: on the unit circle, $\cot t = \frac{x}{y}, y \neq 0$

coterminal angles description of positive and negative angles in standard position sharing the same terminal side

degree a unit of measure describing the size of an angle as one-360th of a full revolution of a circle

hypotenuse the side of a right triangle opposite the right angle

identities statements that are true for all values of the input on which they are defined

initial side the side of an angle from which rotation begins

linear speed the distance along a straight path a rotating object travels in a unit of time; determined by the arc length

measure of an angle the amount of rotation from the initial side to the terminal side

negative angle description of an angle measured clockwise from the positive *x*-axis

opposite side in a right triangle, the side most distant from a given angle

period the smallest interval P of a repeating function f such that $f(x + P) = f(x)$

positive angle description of an angle measured counterclockwise from the positive *x*-axis

Pythagorean Identity a corollary of the Pythagorean Theorem stating that the square of the cosine of a given angle plus the square of the sine of that angle equals 1

quadrantal angle an angle whose terminal side lies on an axis

radian the measure of a central angle of a circle that intercepts an arc equal in length to the radius of that circle

radian measure the ratio of the arc length formed by an angle divided by the radius of the circle

ray one point on a line and all points extending in one direction from that point; one side of an angle

reference angle the measure of the acute angle formed by the terminal side of the angle and the horizontal axis

secant the reciprocal of the cosine function: on the unit circle, $\sec t = \frac{1}{x}, x \neq 0$

sine function the *y*-value of the point on a unit circle corresponding to a given angle

standard position the position of an angle having the vertex at the origin and the initial side along the positive *x*-axis

tangent the quotient of the sine and cosine: on the unit circle, $\tan t = \frac{y}{x}, x \neq 0$

terminal side the side of an angle at which rotation ends

unit circle a circle with a center at $(0, 0)$ and radius 1

vertex the common endpoint of two rays that form an angle

Key Equations

arc length	$s = r\theta$
area of a sector	$A = \frac{1}{2}\theta r^2$
angular speed	$\omega = \frac{\theta}{t}$
linear speed	$v = \frac{s}{t}$
linear speed related to angular speed	$v = r\omega$

Trigonometric Functions	Sine	$\sin t = \frac{\text{opposite}}{\text{hypotenuse}}$
	Cosine	$\cos t = \frac{\text{adjacent}}{\text{hypotenuse}}$
	Tangent	$\tan t = \frac{\text{opposite}}{\text{adjacent}}$
	Secant	$\sec t = \frac{\text{hypotenuse}}{\text{adjacent}}$
	Cosecant	$\csc t = \frac{\text{hypotenuse}}{\text{opposite}}$
	Cotangent	$\cot t = \frac{\text{adjacent}}{\text{opposite}}$
Reciprocal Trigonometric Functions	$\sin t = \frac{1}{\csc t}$	$\csc t = \frac{1}{\sin t}$
	$\cos t = \frac{1}{\sec t}$	$\sec t = \frac{1}{\cos t}$
	$\tan t = \frac{1}{\cot t}$	$\cot t = \frac{1}{\tan t}$
Cofunction Identities	$\cos t = \sin\left(\frac{\pi}{2} - t\right)$	
	$\sin t = \cos\left(\frac{\pi}{2} - t\right)$	
	$\tan t = \cot\left(\frac{\pi}{2} - t\right)$	
	$\cot t = \tan\left(\frac{\pi}{2} - t\right)$	
	$\sec t = \csc\left(\frac{\pi}{2} - t\right)$	

Cosine	$\cos\ t = x$
Sine	$\sin\ t = y$
Pythagorean Identity	$\cos^2 t + \sin^2 t = 1$

Tangent function	$\tan\ t = \frac{\sin\ t}{\cos\ t}$
Secant function	$\sec\ t = \frac{1}{\cos\ t}$
Cosecant function	$\csc\ t = \frac{1}{\sin\ t}$
Cotangent function	$\cot t = \frac{1}{\tan t} = \frac{\cos t}{\sin t}$

Key Concepts

7.1 Angles

- An angle is formed from the union of two rays, by keeping the initial side fixed and rotating the terminal side. The amount of rotation determines the measure of the angle.
- An angle is in standard position if its vertex is at the origin and its initial side lies along the positive x-axis. A positive angle is measured counterclockwise from the initial side and a negative angle is measured clockwise.
- To draw an angle in standard position, draw the initial side along the positive x-axis and then place the terminal side according to the fraction of a full rotation the angle represents. See Example 1.
- In addition to degrees, the measure of an angle can be described in radians. See Example 2.
- To convert between degrees and radians, use the proportion $\frac{\theta}{180} = \frac{\theta R}{\pi}$. See Example 3 and Example 4.
- Two angles that have the same terminal side are called coterminal angles.
- We can find coterminal angles by adding or subtracting 360° or 2π. See Example 5 and Example 6.

- Coterminal angles can be found using radians just as they are for degrees. See Example 7.
- The length of a circular arc is a fraction of the circumference of the entire circle. See Example 8.
- The area of sector is a fraction of the area of the entire circle. See Example 9.
- An object moving in a circular path has both linear and angular speed.
- The angular speed of an object traveling in a circular path is the measure of the angle through which it turns in a unit of time. See Example 10.
- The linear speed of an object traveling along a circular path is the distance it travels in a unit of time. See Example 11.

7.2 Right Triangle Trigonometry

- We can define trigonometric functions as ratios of the side lengths of a right triangle. See Example 1.
- The same side lengths can be used to evaluate the trigonometric functions of either acute angle in a right triangle. See Example 2.
- We can evaluate the trigonometric functions of special angles, knowing the side lengths of the triangles in which they occur. See Example 3.
- Any two complementary angles could be the two acute angles of a right triangle.
- If two angles are complementary, the cofunction identities state that the sine of one equals the cosine of the other and vice versa. See Example 4.
- We can use trigonometric functions of an angle to find unknown side lengths.
- Select the trigonometric function representing the ratio of the unknown side to the known side. See Example 5.
- Right-triangle trigonometry facilitates the measurement of inaccessible heights and distances.
- The unknown height or distance can be found by creating a right triangle in which the unknown height or distance is one of the sides, and another side and angle are known. See Example 6.

7.3 Unit Circle

- Finding the function values for the sine and cosine begins with drawing a unit circle, which is centered at the origin and has a radius of 1 unit.
- Using the unit circle, the sine of an angle t equals the y-value of the endpoint on the unit circle of an arc of length t whereas the cosine of an angle t equals the x-value of the endpoint. See Example 1.
- The sine and cosine values are most directly determined when the corresponding point on the unit circle falls on an axis. See Example 2.
- When the sine or cosine is known, we can use the Pythagorean Identity to find the other. The Pythagorean Identity is also useful for determining the sines and cosines of special angles. See Example 3.
- Calculators and graphing software are helpful for finding sines and cosines if the proper procedure for entering information is known. See Example 4.
- The domain of the sine and cosine functions is all real numbers.
- The range of both the sine and cosine functions is $[-1, 1]$.
- The sine and cosine of an angle have the same absolute value as the sine and cosine of its reference angle.
- The signs of the sine and cosine are determined from the x- and y-values in the quadrant of the original angle.
- An angle's reference angle is the size angle, t, formed by the terminal side of the angle t and the horizontal axis. See Example 5.
- Reference angles can be used to find the sine and cosine of the original angle. See Example 6.
- Reference angles can also be used to find the coordinates of a point on a circle. See Example 7.

7.4 The Other Trigonometric Functions

- The tangent of an angle is the ratio of the y-value to the x-value of the corresponding point on the unit circle.
- The secant, cotangent, and cosecant are all reciprocals of other functions. The secant is the reciprocal of the cosine function, the cotangent is the reciprocal of the tangent function, and the cosecant is the reciprocal of the sine function.
- The six trigonometric functions can be found from a point on the unit circle. See Example 1.
- Trigonometric functions can also be found from an angle. See Example 2.
- Trigonometric functions of angles outside the first quadrant can be determined using reference angles. See Example 3.
- A function is said to be even if $f(-x) = f(x)$ and odd if $f(-x) = -f(x)$ for all x in the domain of f.
- Cosine and secant are even; sine, tangent, cosecant, and cotangent are odd.
- Even and odd properties can be used to evaluate trigonometric functions. See Example 4.
- The Pythagorean Identity makes it possible to find a cosine from a sine or a sine from a cosine.
- Identities can be used to evaluate trigonometric functions. See Example 5 and Example 6.

- Fundamental identities such as the Pythagorean Identity can be manipulated algebraically to produce new identities. See Example 7.
- The trigonometric functions repeat at regular intervals.
- The period P of a repeating function f is the smallest interval such that $f(x + P) = f(x)$ for any value of x.
- The values of trigonometric functions can be found by mathematical analysis. See Example 8 and Example 9.
- To evaluate trigonometric functions of other angles, we can use a calculator or computer software. See Example 10.

Exercises

Review Exercises

Angles

For the following exercises, convert the angle measures to degrees.

1. $\frac{\pi}{4}$

2. $-\frac{5\pi}{3}$

For the following exercises, convert the angle measures to radians.

3. $-210°$

4. $180°$

5. Find the length of an arc in a circle of radius 7 meters subtended by the central angle of $85°$.

6. Find the area of the sector of a circle with diameter 32 feet and an angle of $\frac{3\pi}{5}$ radians.

For the following exercises, find the angle between $0°$ and $360°$ that is coterminal with the given angle.

7. $420°$

8. $-80°$

For the following exercises, find the angle between 0 and 2π in radians that is coterminal with the given angle.

9. $-\frac{20\pi}{11}$

10. $\frac{14\pi}{5}$

For the following exercises, draw the angle provided in standard position on the Cartesian plane.

11. $-210°$

12. $75°$

13. $\frac{5\pi}{4}$

14. $-\frac{\pi}{3}$

15. Find the linear speed of a point on the equator of the earth if the earth has a radius of 3,960 miles and the earth rotates on its axis every 24 hours. Express answer in miles per hour. Round to the nearest hundredth.

16. A car wheel with a diameter of 18 inches spins at the rate of 10 revolutions per second. What is the car's speed in miles per hour? Round to the nearest hundredth.

Right Triangle Trigonometry

For the following exercises, use side lengths to evaluate.

17. $\cos \frac{\pi}{4}$

18. $\cot \frac{\pi}{3}$

19. $\tan \frac{\pi}{6}$

20. $\cos\left(\frac{\pi}{2}\right) = \sin(\_\_\_°)$

21. $\csc(18°) = \sec(\_\_\_°)$

For the following exercises, use the given information to find the lengths of the other two sides of the right triangle.

22. $\cos B = \frac{3}{5}, a = 6$

23. $\tan A = \frac{5}{9}, b = 6$

For the following exercises, use Figure 1 to evaluate each trigonometric function.

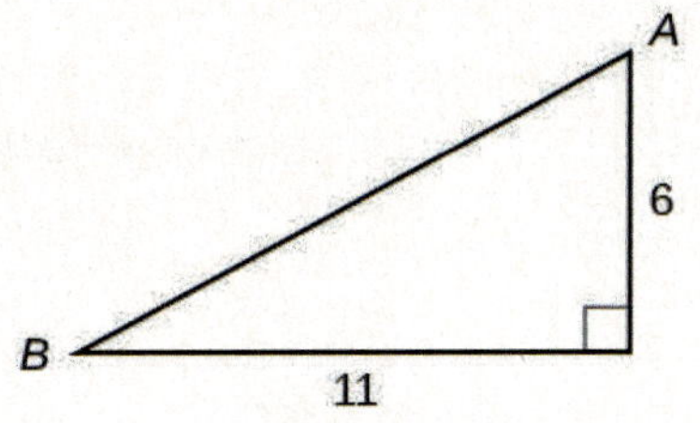

Figure 1

24. $\sin A$

25. $\tan B$

For the following exercises, solve for the unknown sides of the given triangle.

26.

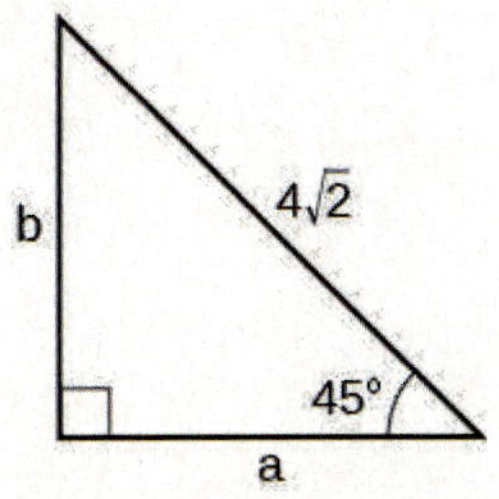

27.

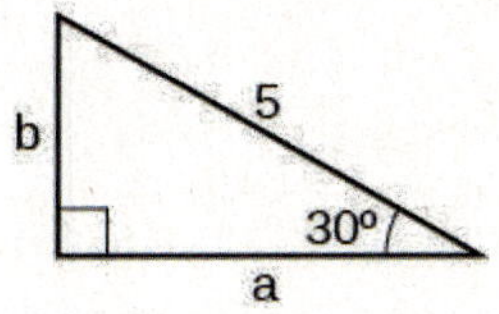

28. A 15-ft ladder leans against a building so that the angle between the ground and the ladder is 70°. How high does the ladder reach up the side of the building? Find the answer to four decimal places.

29. The angle of elevation to the top of a building in Baltimore is found to be 4 degrees from the ground at a distance of 1 mile from the base of the building. Using this information, find the height of the building. Find the answer to four decimal places.

Unit Circle

30. Find the exact value of $\sin \frac{\pi}{3}$.

31. Find the exact value of $\cos \frac{\pi}{4}$.

32. Find the exact value of $\cos \pi$.

33. State the reference angle for $300°$.

34. State the reference angle for $\frac{3\pi}{4}$.

35. Compute cosine of $330°$.

36. Compute sine of $\frac{5\pi}{4}$.

37. State the domain of the sine and cosine functions.

38. State the range of the sine and cosine functions.

The Other Trigonometric Functions

For the following exercises, find the exact value of the given expression.

39. $\cos\ \frac{\pi}{6}$

40. $\tan\ \frac{\pi}{4}$

41. $\csc\ \frac{\pi}{3}$

42. $\sec\ \frac{\pi}{4}$

For the following exercises, use reference angles to evaluate the given expression.

43. $\sec\ \frac{11\pi}{3}$

44. $\sec\ 315°$

45. If $\sec(t) = -2.5$, what is the $\sec(-t)$?

46. If $\tan(t) = -0.6$, what is the $\tan(-t)$?

47. If $\tan(t) = \frac{1}{3}$, find $\tan(t - \pi)$.

48. If $\cos(t) = \frac{\sqrt{2}}{2}$, find $\sin(t + 2\pi)$. There are two possible solutions.

49. Which trigonometric functions are even?

50. Which trigonometric functions are odd?

Practice Test

1. Convert $\frac{5\pi}{6}$ radians to degrees.

2. Convert $-620°$ to radians.

3. Find the length of a circular arc with a radius 12 centimeters subtended by the central angle of $30°$.

4. Find the area of the sector with radius of 8 feet and an angle of $\frac{5\pi}{4}$ radians.

5. Find the angle between $0°$ and $360°$ that is coterminal with $375°$.

6. Find the angle between 0 and 2π in radians that is coterminal with $-\frac{4\pi}{7}$.

7. Draw the angle $315°$ in standard position on the Cartesian plane.

8. Draw the angle $-\frac{\pi}{6}$ in standard position on the Cartesian plane.

9. A carnival has a Ferris wheel with a diameter of 80 feet. The time for the Ferris wheel to make one revolution is 75 seconds. What is the linear speed in feet per second of a point on the Ferris wheel? What is the angular speed in radians per second?

10. Find the missing sides of the triangle $ABC : \sin B = \frac{3}{4}, c = 12$.

11. Find the missing sides of the triangle.

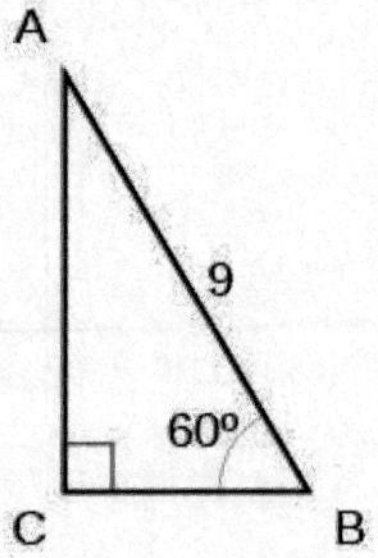

12. The angle of elevation to the top of a building in Chicago is found to be 9 degrees from the ground at a distance of 2000 feet from the base of the building. Using this information, find the height of the building.

13. Find the exact value of $\sin \frac{\pi}{6}$.

14. Compute sine of $240°$.

15. State the domain of the sine and cosine functions.

16. State the range of the sine and cosine functions.

17. Find the exact value of $\cot \frac{\pi}{4}$.

18. Find the exact value of $\tan \frac{\pi}{3}$.

19. Use reference angles to evaluate $\csc \frac{7\pi}{4}$.

20. Use reference angles to evaluate $\tan 210°$.

21. If $\csc t = 0.68$, what is the $\csc(-t)$?

22. If $\cos t = \frac{\sqrt{3}}{2}$, find $\cos(t - 2\pi)$.

23. Find the missing angle: $\cos\left(\frac{\pi}{6}\right) = \sin(\_\_\_)$

8 PERIODIC FUNCTIONS

Dawn colors the sky over the Olare Motorgi Conservancy bordering tha Masai Mara National Reserve in Kenya. (Credit: Modification of "KenyaLive_Day_#02" by Make it Kenya/flickr)

Chapter Outline

Introduction to Periodic Functions

The sun has played a core role in many religions. The ancient Egyptian culture portrayed the sun god, Ra (sometimes written as Re), as undertaking a two-part daily journey, with one portion in the sky (day) and the other through the underworld (night). Surya, the Hindu sun god, traces a similar path through the sky on a chariot pulled by seven horses. While their origins and associated narratives are quite different, both Ra and Surya are primary deities and seen as creators and preservers of life. In many Native American cultures, the sun is core to spiritual and religious practice, but is not always a deity. The Sun Dance, practiced differently by many Native American tribes, was a ceremony that generally paid homage to the sun and, in many cases, tested or expressed the strength of the tribe's people.

As one of the most most prominent natural phenomena and with its close association to giving life, the sun was an obvious subject for reverence. And its regularity, even in ancient times, made it the primary determinant of time. Each day, the sun rises in an easterly direction, approaches some maximum height relative to the celestial equator, and sets in a westerly direction. The celestial equator is an imaginary line that divides the visible universe into two halves in much the same way Earth's equator is an imaginary line that divides the planet into two halves. The exact path the sun appears to follow depends on the exact location on Earth, but each location observes a predictable pattern over time.

The pattern of the sun's motion throughout the course of a year is a periodic function. Creating a visual representation of a periodic function in the form of a graph can help us analyze the properties of the function. In this chapter, we will investigate graphs of sine, cosine, and other trigonometric functions.

8.1 Graphs of the Sine and Cosine Functions

Learning Objectives

In this section, you will:

- Graph variations of $y = \sin(x)$ and $y = \cos(x)$.
- Use phase shifts of sine and cosine curves.

Figure 1 Light can be separated into colors because of its wavelike properties. (credit: "wonderferret"/ Flickr)

White light, such as the light from the sun, is not actually white at all. Instead, it is a composition of all the colors of the rainbow in the form of waves. The individual colors can be seen only when white light passes through an optical prism that separates the waves according to their wavelengths to form a rainbow.

Light waves can be represented graphically by the sine function. In the chapter on Trigonometric Functions (http://openstax.org/books/precalculus-2e/pages/5-introduction-to-trigonometric-functions), we examined trigonometric functions such as the sine function. In this section, we will interpret and create graphs of sine and cosine functions.

Graphing Sine and Cosine Functions

Recall that the sine and cosine functions relate real number values to the *x*- and *y*-coordinates of a point on the unit circle. So what do they look like on a graph on a coordinate plane? Let's start with the sine function. We can create a table of values and use them to sketch a graph. Table 1 lists some of the values for the sine function on a unit circle.

x	0	$\frac{\pi}{6}$	$\frac{\pi}{4}$	$\frac{\pi}{3}$	$\frac{\pi}{2}$	$\frac{2\pi}{3}$	$\frac{3\pi}{4}$	$\frac{5\pi}{6}$	π
$\sin(x)$	0	$\frac{1}{2}$	$\frac{\sqrt{2}}{2}$	$\frac{\sqrt{3}}{2}$	1	$\frac{\sqrt{3}}{2}$	$\frac{\sqrt{2}}{2}$	$\frac{1}{2}$	0

Table 1

Plotting the points from the table and continuing along the *x*-axis gives the shape of the sine function. See Figure 2.

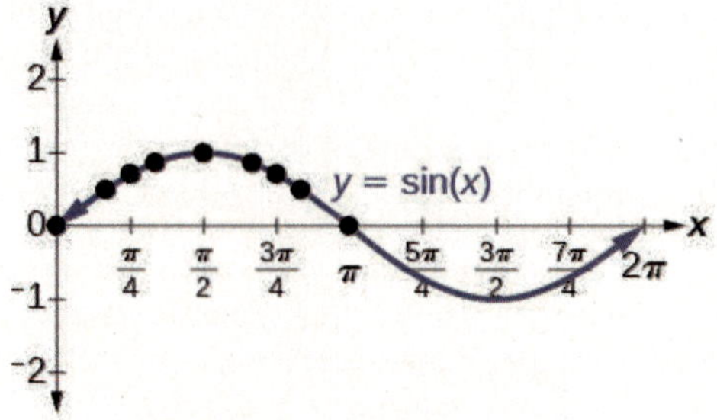

Figure 2 The sine function

Notice how the sine values are positive between 0 and π, which correspond to the values of the sine function in quadrants I and II on the unit circle, and the sine values are negative between π and 2π, which correspond to the values of the sine function in quadrants III and IV on the unit circle. See Figure 3.

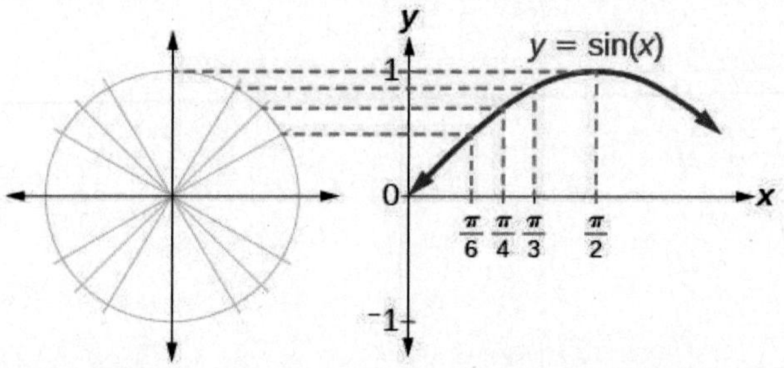

Figure 3 Plotting values of the sine function

Now let's take a similar look at the cosine function. Again, we can create a table of values and use them to sketch a graph. Table 2 lists some of the values for the cosine function on a unit circle.

x	0	$\frac{\pi}{6}$	$\frac{\pi}{4}$	$\frac{\pi}{3}$	$\frac{\pi}{2}$	$\frac{2\pi}{3}$	$\frac{3\pi}{4}$	$\frac{5\pi}{6}$	π
cos (x)	1	$\frac{\sqrt{3}}{2}$	$\frac{\sqrt{2}}{2}$	$\frac{1}{2}$	0	$-\frac{1}{2}$	$-\frac{\sqrt{2}}{2}$	$-\frac{\sqrt{3}}{2}$	-1

Table 2

As with the sine function, we can plots points to create a graph of the cosine function as in Figure 4.

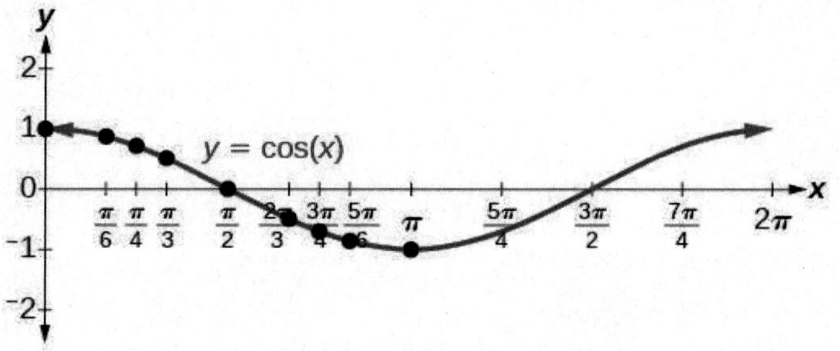

Figure 4 The cosine function

Because we can evaluate the sine and cosine of any real number, both of these functions are defined for all real numbers. By thinking of the sine and cosine values as coordinates of points on a unit circle, it becomes clear that the range of both functions must be the interval $[-1, 1]$.

In both graphs, the shape of the graph repeats after 2π, which means the functions are periodic with a period of 2π. A **periodic function** is a function for which a specific horizontal shift, *P*, results in a function equal to the original function: $f(x + P) = f(x)$ for all values of x in the domain of f. When this occurs, we call the smallest such horizontal shift with $P > 0$ the period of the function. Figure 5 shows several periods of the sine and cosine functions.

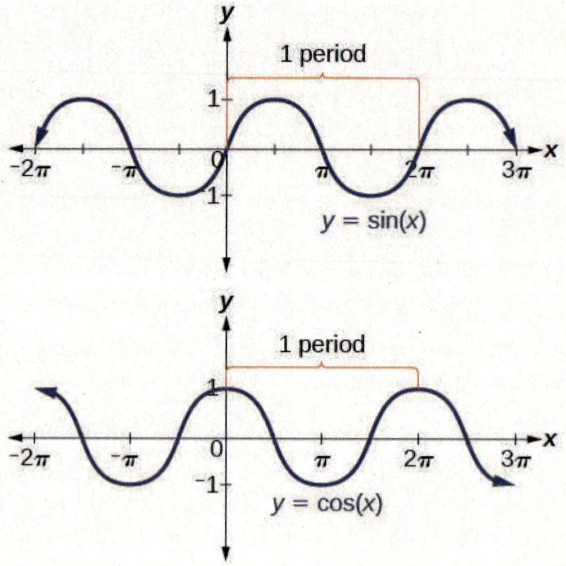

Figure 5

Looking again at the sine and cosine functions on a domain centered at the *y*-axis helps reveal symmetries. As we can see in Figure 6, the sine function is symmetric about the origin. Recall from The Other Trigonometric Functions that we determined from the unit circle that the sine function is an odd function because $\sin(-x) = -\sin\ x$. Now we can clearly see this property from the graph.

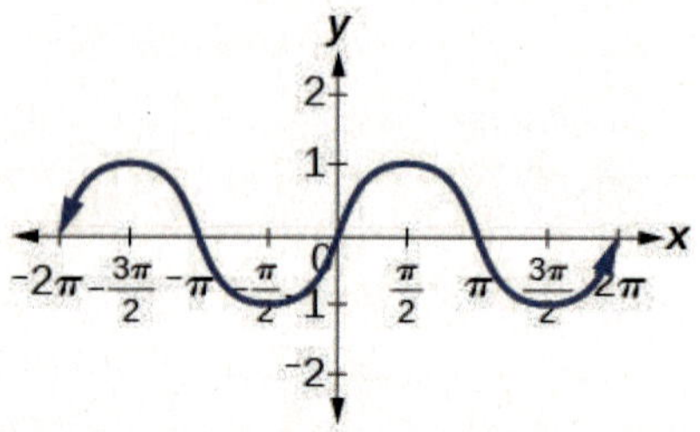

Figure 6 Odd symmetry of the sine function

Figure 7 shows that the cosine function is symmetric about the *y*-axis. Again, we determined that the cosine function is an even function. Now we can see from the graph that $\cos(-x) = \cos\ x$.

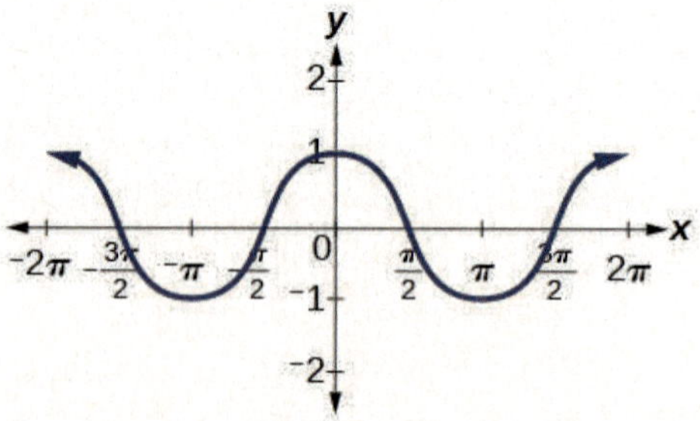

Figure 7 Even symmetry of the cosine function

Characteristics of Sine and Cosine Functions

The sine and cosine functions have several distinct characteristics:

- They are periodic functions with a period of 2π.
- The domain of each function is $\left(-\infty, \infty\right)$ and the range is $[-1, 1]$.
- The graph of $y = \sin\ x$ is symmetric about the origin, because it is an odd function.
- The graph of $y = \cos\ x$ is symmetric about the y-axis, because it is an even function.

Investigating Sinusoidal Functions

As we can see, sine and cosine functions have a regular period and range. If we watch ocean waves or ripples on a pond, we will see that they resemble the sine or cosine functions. However, they are not necessarily identical. Some are taller or longer than others. A function that has the same general shape as a sine or cosine function is known as a **sinusoidal function**. The general forms of sinusoidal functions are

$$y = A\sin(Bx - C) + D$$
and
$$y = A\cos(Bx - C) + D$$

Determining the Period of Sinusoidal Functions

Looking at the forms of sinusoidal functions, we can see that they are transformations of the sine and cosine functions. We can use what we know about transformations to determine the period.

In the general formula, B is related to the period by $P = \frac{2\pi}{|B|}$. If $|B| > 1$, then the period is less than 2π and the function undergoes a horizontal compression, whereas if $|B| < 1$, then the period is greater than 2π and the function undergoes a horizontal stretch. For example, $f(x) = \sin(x)$, $B = 1$, so the period is 2π, which we knew. If $f(x) = \sin(2x)$, then $B = 2$, so the period is π and the graph is compressed. If $f(x) = \sin\left(\frac{x}{2}\right)$, then $B = \frac{1}{2}$, so the period is 4π and the graph is stretched. Notice in Figure 8 how the period is indirectly related to $|B|$.

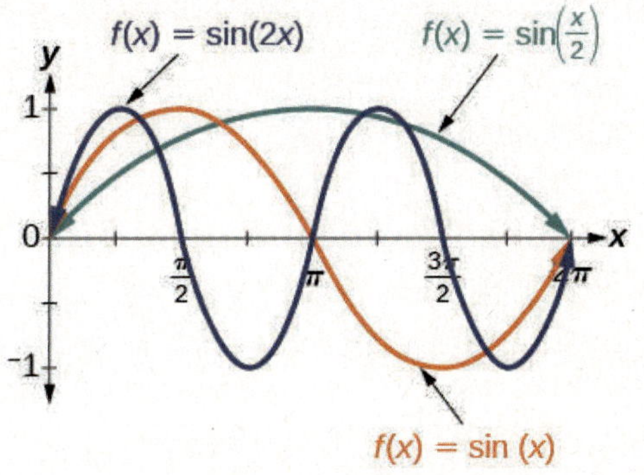

Figure 8

Period of Sinusoidal Functions

If we let $C = 0$ and $D = 0$ in the general form equations of the sine and cosine functions, we obtain the forms

$$y = A\sin(Bx)$$
$$y = A\cos(Bx)$$

The period is $\frac{2\pi}{|B|}$.

EXAMPLE 1

Identifying the Period of a Sine or Cosine Function

Determine the period of the function $f(x) = \sin\left(\frac{\pi}{6}x\right)$.

Solution

Let's begin by comparing the equation to the general form $y = A\sin(Bx)$.

In the given equation, $B = \frac{\pi}{6}$, so the period will be

$$
\begin{aligned}
P &= \frac{2\pi}{|B|} \\
&= \frac{2\pi}{\frac{\pi}{6}} \\
&= 2\pi \cdot \frac{6}{\pi} \\
&= 12
\end{aligned}
$$

> TRY IT #1 Determine the period of the function $g(x) = \cos\left(\frac{x}{3}\right)$.

Determining Amplitude

Returning to the general formula for a sinusoidal function, we have analyzed how the variable B relates to the period. Now let's turn to the variable A so we can analyze how it is related to the **amplitude**, or greatest distance from rest. A represents the vertical stretch factor, and its absolute value $|A|$ is the amplitude. The local maxima will be a distance $|A|$ above the horizontal **midline** of the graph, which is the line $y = D$; because $D = 0$ in this case, the midline is the x-axis. The local minima will be the same distance below the midline. If $|A| > 1$, the function is stretched. For example, the amplitude of $f(x) = 4 \sin x$ is twice the amplitude of $f(x) = 2 \sin x$. If $|A| < 1$, the function is compressed. Figure 9 compares several sine functions with different amplitudes.

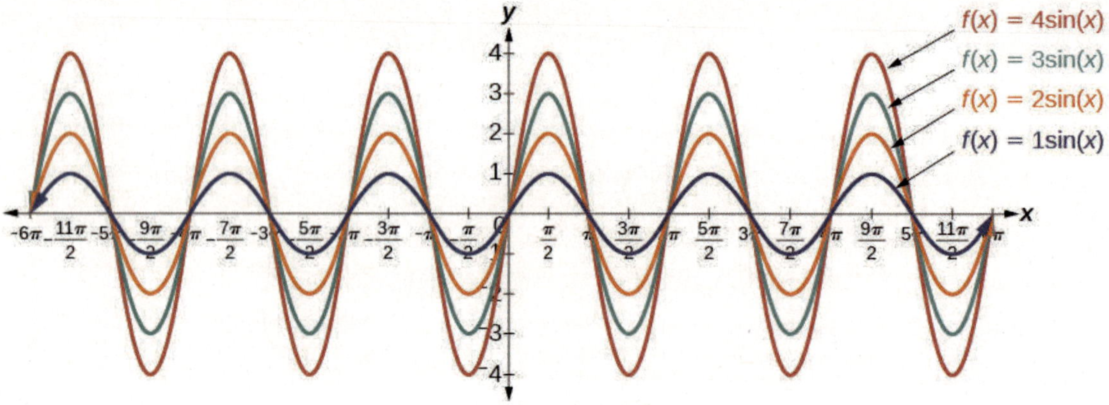

Figure 9

Amplitude of Sinusoidal Functions

If we let $C = 0$ and $D = 0$ in the general form equations of the sine and cosine functions, we obtain the forms

$$y = A\sin(Bx) \text{ and } y = A\cos(Bx)$$

The **amplitude** is $|A|$, which is the vertical height from the **midline** . In addition, notice in the example that

$$|A| = \text{amplitude} = \frac{1}{2}|\text{maximum} - \text{minimum}|$$

EXAMPLE 2

Identifying the Amplitude of a Sine or Cosine Function

What is the amplitude of the sinusoidal function $f(x) = -4\sin(x)$? Is the function stretched or compressed vertically?

Solution

Let's begin by comparing the function to the simplified form $y = A\sin(Bx)$.

In the given function, $A = -4$, so the amplitude is $|A| = |-4| = 4$. The function is stretched.

Analysis

The negative value of A results in a reflection across the *x*-axis of the sine function, as shown in Figure 10.

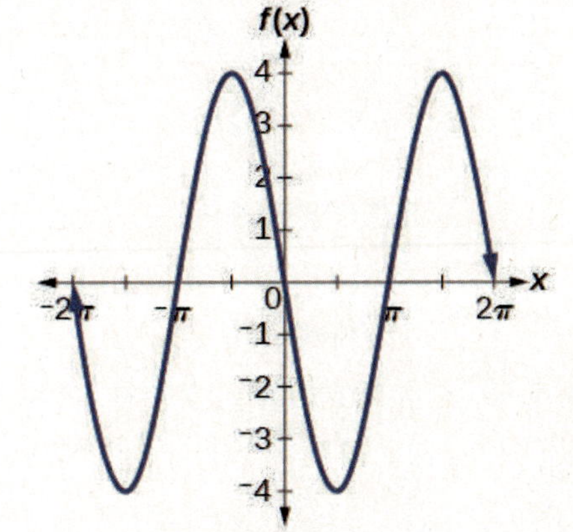

Figure 10

> **TRY IT** #2 What is the amplitude of the sinusoidal function $f(x) = \frac{1}{2}\sin(x)$? Is the function stretched or compressed vertically?

Analyzing Graphs of Variations of $y = \sin x$ and $y = \cos x$

Now that we understand how A and B relate to the general form equation for the sine and cosine functions, we will explore the variables C and D. Recall the general form:

$$y = A\sin(Bx - C) + D \text{ and } y = A\cos(Bx - C) + D$$

$$or$$

$$y = A\sin\left(B\left(x - \frac{C}{B}\right)\right) + D \text{ and } y = A\cos\left(B\left(x - \frac{C}{B}\right)\right) + D$$

The value $\frac{C}{B}$ for a sinusoidal function is called the **phase shift**, or the horizontal displacement of the basic sine or cosine function. If $C > 0$, the graph shifts to the right. If $C < 0$, the graph shifts to the left. The greater the value of $|C|$, the more the graph is shifted. Figure 11 shows that the graph of $f(x) = \sin(x - \pi)$ shifts to the right by π units, which is more than we see in the graph of $f(x) = \sin\left(x - \frac{\pi}{4}\right)$, which shifts to the right by $\frac{\pi}{4}$ units.

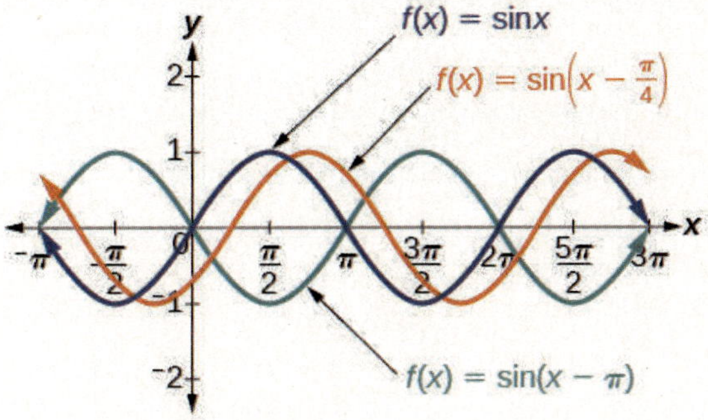

Figure 11

While C relates to the horizontal shift, D indicates the vertical shift from the midline in the general formula for a sinusoidal function. See Figure 12. The function $y = \cos(x) + D$ has its midline at $y = D$.

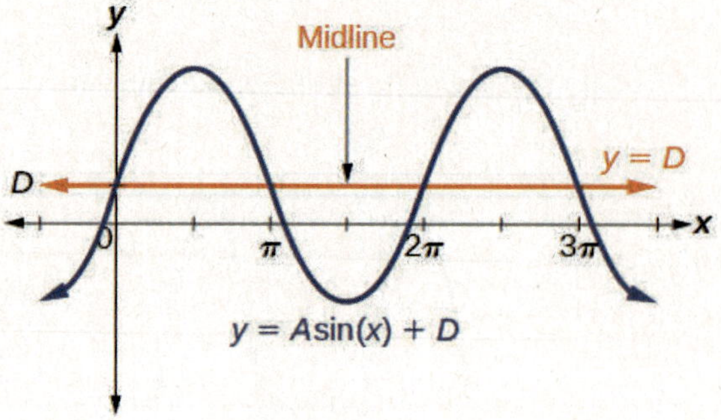

Figure 12

Any value of D other than zero shifts the graph up or down. Figure 13 compares $f(x) = \sin(x)$ with $f(x) = \sin(x) + 2$, which is shifted 2 units up on a graph.

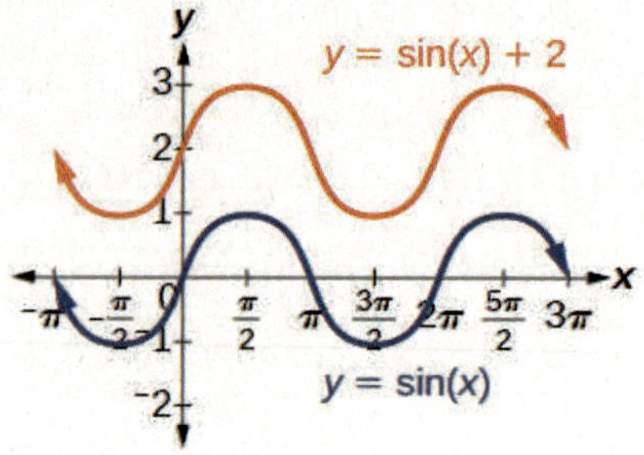

Figure 13

Variations of Sine and Cosine Functions

Given an equation in the form $f(x) = A\sin(Bx - C) + D$ or $f(x) = A\cos(Bx - C) + D$, $\frac{C}{B}$ is the **phase shift** and D is the vertical shift.

EXAMPLE 3

Identifying the Phase Shift of a Function

Determine the direction and magnitude of the phase shift for $f(x) = \sin\left(x + \frac{\pi}{6}\right) - 2$.

Solution

Let's begin by comparing the equation to the general form $y = A\sin(Bx - C) + D$.

In the given equation, notice that $B = 1$ and $C = -\frac{\pi}{6}$. So the phase shift is

$$\begin{aligned} \frac{C}{B} &= -\frac{\frac{\pi}{6}}{1} \\ &= -\frac{\pi}{6} \end{aligned}$$

or $\frac{\pi}{6}$ units to the left.

Analysis

We must pay attention to the sign in the equation for the general form of a sinusoidal function. The equation shows a minus sign before C. Therefore $f(x) = \sin\left(x + \frac{\pi}{6}\right) - 2$ can be rewritten as $f(x) = \sin\left(x - \left(-\frac{\pi}{6}\right)\right) - 2$. If the value of C is negative, the shift is to the left.

TRY IT #3 Determine the direction and magnitude of the phase shift for $f(x) = 3\cos\left(x - \frac{\pi}{2}\right)$.

EXAMPLE 4

Identifying the Vertical Shift of a Function

Determine the direction and magnitude of the vertical shift for $f(x) = \cos(x) - 3$.

Solution

Let's begin by comparing the equation to the general form $y = A\cos(Bx - C) + D$.

In the given equation, $D = -3$ so the shift is 3 units downward.

> TRY IT #4 Determine the direction and magnitude of the vertical shift for $f(x) = 3\sin(x) + 2$.

HOW TO

Given a sinusoidal function in the form $f(x) = A\sin(Bx - C) + D$, identify the midline, amplitude, period, and phase shift.

1. Determine the amplitude as $|A|$.
2. Determine the period as $P = \frac{2\pi}{|B|}$.
3. Determine the phase shift as $\frac{C}{B}$.
4. Determine the midline as $y = D$.

EXAMPLE 5

Identifying the Variations of a Sinusoidal Function from an Equation

Determine the midline, amplitude, period, and phase shift of the function $y = 3\sin(2x) + 1$.

Solution

Let's begin by comparing the equation to the general form $y = A\sin(Bx - C) + D$.

$A = 3$, so the amplitude is $|A| = 3$.

Next, $B = 2$, so the period is $P = \frac{2\pi}{|B|} = \frac{2\pi}{2} = \pi$.

There is no added constant inside the parentheses, so $C = 0$ and the phase shift is $\frac{C}{B} = \frac{0}{2} = 0$.

Finally, $D = 1$, so the midline is $y = 1$.

Analysis

Inspecting the graph, we can determine that the period is π, the midline is $y = 1$, and the amplitude is 3. See Figure 14.

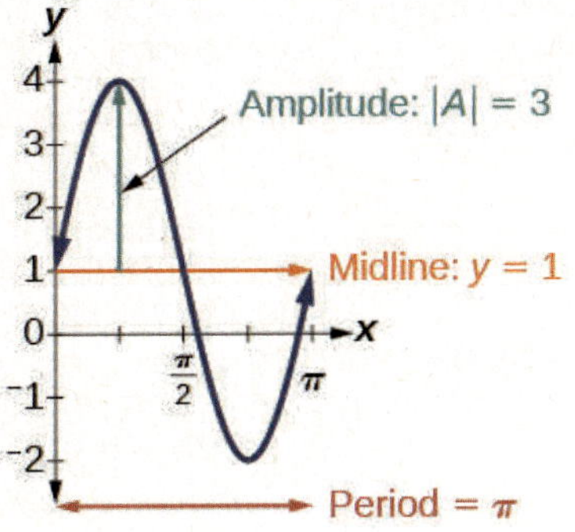

Figure 14

> TRY IT #5 Determine the midline, amplitude, period, and phase shift of the function $y = \frac{1}{2}\cos\left(\frac{x}{3} - \frac{\pi}{3}\right)$.

EXAMPLE 6

Identifying the Equation for a Sinusoidal Function from a Graph

Determine the formula for the cosine function in Figure 15.

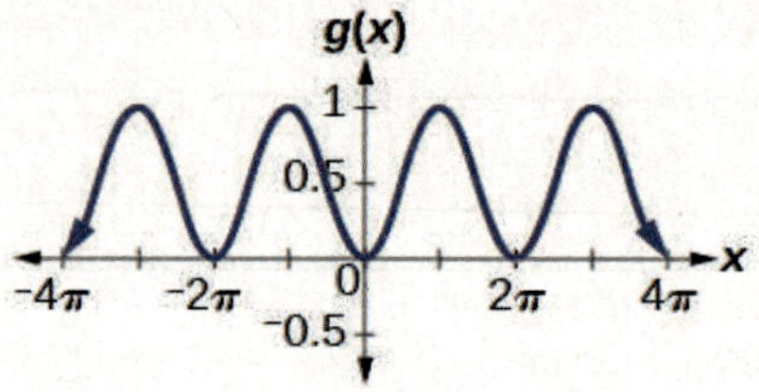

Figure 15

Solution

To determine the equation, we need to identify each value in the general form of a sinusoidal function.

$$y = A\sin(Bx - C) + D$$
$$y = A\cos(Bx - C) + D$$

The graph could represent either a sine or a cosine function that is shifted and/or reflected. When $x = 0$, the graph has an extreme point, $(0, 0)$. Since the cosine function has an extreme point for $x = 0$, let us write our equation in terms of a cosine function.

Let's start with the midline. We can see that the graph rises and falls an equal distance above and below $y = 0.5$. This value, which is the midline, is D in the equation, so $D = 0.5$.

The greatest distance above and below the midline is the amplitude. The maxima are 0.5 units above the midline and the minima are 0.5 units below the midline. So $|A| = 0.5$. Another way we could have determined the amplitude is by recognizing that the difference between the height of local maxima and minima is 1, so $|A| = \frac{1}{2} = 0.5$. Also, the graph is reflected about the *x*-axis so that $A = -0.5$.

The graph is not horizontally stretched or compressed, so $B = 1$; and the graph is not shifted horizontally, so $C = 0$.

Putting this all together,

$$g(x) = -0.5\cos(x) + 0.5$$

TRY IT #6 Determine the formula for the sine function in Figure 16.

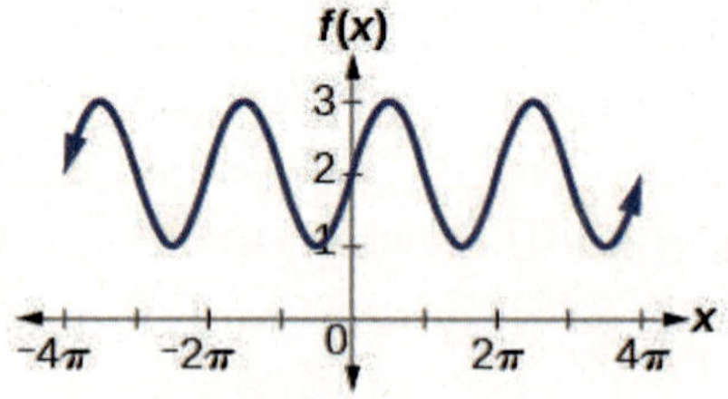

Figure 16

EXAMPLE 7

Identifying the Equation for a Sinusoidal Function from a Graph

Determine the equation for the sinusoidal function in Figure 17.

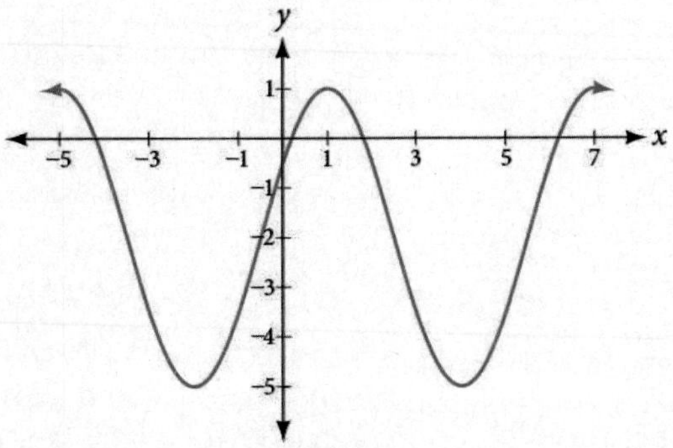

Figure 17

Solution

With the highest value at 1 and the lowest value at -5, the midline will be halfway between at -2. So $D = -2$.

The distance from the midline to the highest or lowest value gives an amplitude of $|A| = 3$.

The period of the graph is 6, which can be measured from the peak at $x = 1$ to the next peak at $x = 7$, or from the distance between the lowest points. Therefore, $P = \frac{2\pi}{|B|} = 6$. Using the positive value for B, we find that

$$B = \frac{2\pi}{P} = \frac{2\pi}{6} = \frac{\pi}{3}$$

So far, our equation is either $y = 3\sin\left(\frac{\pi}{3}x - C\right) - 2$ or $y = 3\cos\left(\frac{\pi}{3}x - C\right) - 2$. For the shape and shift, we have more than one option. We could write this as any one of the following:

- a cosine shifted to the right
- a negative cosine shifted to the left
- a sine shifted to the left
- a negative sine shifted to the right

While any of these would be correct, the cosine shifts are easier to work with than the sine shifts in this case because they involve integer values. So our function becomes

$$y = 3\cos\left(\frac{\pi}{3}x - \frac{\pi}{3}\right) - 2 \text{ or } y = -3\cos\left(\frac{\pi}{3}x + \frac{2\pi}{3}\right) - 2$$

Again, these functions are equivalent, so both yield the same graph.

TRY IT #7 Write a formula for the function graphed in Figure 18.

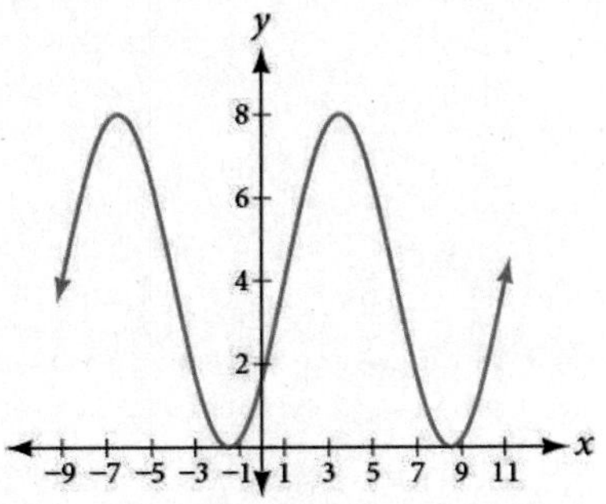

Figure 18

Graphing Variations of $y = \sin x$ and $y = \cos x$

Throughout this section, we have learned about types of variations of sine and cosine functions and used that information to write equations from graphs. Now we can use the same information to create graphs from equations.

Instead of focusing on the general form equations

$$y = A\sin(Bx - C) + D \text{ and } y = A\cos(Bx - C) + D,$$

we will let $C = 0$ and $D = 0$ and work with a simplified form of the equations in the following examples.

HOW TO

Given the function $y = A\sin(Bx)$, sketch its graph.

1. Identify the amplitude, $|A|$.
2. Identify the period, $P = \frac{2\pi}{|B|}$.
3. Start at the origin, with the function increasing to the right if A is positive or decreasing if A is negative.
4. At $x = \frac{\pi}{2|B|}$ there is a local maximum for $A > 0$ or a minimum for $A < 0$, with $y = A$.
5. The curve returns to the x-axis at $x = \frac{\pi}{|B|}$.
6. There is a local minimum for $A > 0$ (maximum for $A < 0$) at $x = \frac{3\pi}{2|B|}$ with $y = -A$.
7. The curve returns again to the x-axis at $x = \frac{2\pi}{|B|}$.

EXAMPLE 8

Graphing a Function and Identifying the Amplitude and Period

Sketch a graph of $f(x) = -2\sin\left(\frac{\pi x}{2}\right)$.

Solution

Let's begin by comparing the equation to the form $y = A\sin(Bx)$.

Step 1. We can see from the equation that $A = -2$, so the amplitude is 2.

$$|A| = 2$$

Step 2. The equation shows that $B = \frac{\pi}{2}$, so the period is

$$\begin{aligned} P &= \frac{2\pi}{\frac{\pi}{2}} \\ &= 2\pi \cdot \frac{2}{\pi} \\ &= 4 \end{aligned}$$

Step 3. Because A is negative, the graph descends as we move to the right of the origin.

Step 4–7. The x-intercepts are at the beginning of one period, $x = 0$, the horizontal midpoints are at $x = 2$ and at the end of one period at $x = 4$.

The quarter points include the minimum at $x = 1$ and the maximum at $x = 3$. A local minimum will occur 2 units below the midline, at $x = 1$, and a local maximum will occur at 2 units above the midline, at $x = 3$. Figure 19 shows the graph of the function.

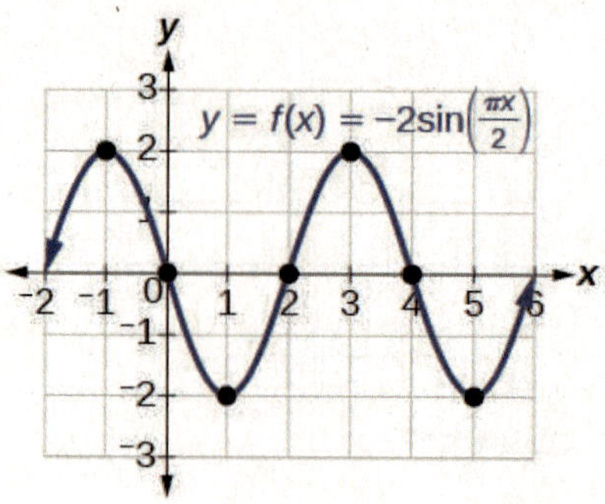

Figure 19

TRY IT #8 Sketch a graph of $g(x) = -0.8\cos(2x)$. Determine the midline, amplitude, period, and phase shift.

HOW TO

Given a sinusoidal function with a phase shift and a vertical shift, sketch its graph.

1. Express the function in the general form $y = A\sin(Bx - C) + D$ or $y = A\cos(Bx - C) + D$.
2. Identify the amplitude, $|A|$.
3. Identify the period, $P = \frac{2\pi}{|B|}$.
4. Identify the phase shift, $\frac{C}{B}$.
5. Draw the graph of $f(x) = A\sin(Bx)$ shifted to the right or left by $\frac{C}{B}$ and up or down by D.

EXAMPLE 9

Graphing a Transformed Sinusoid
Sketch a graph of $f(x) = 3\sin\left(\frac{\pi}{4}x - \frac{\pi}{4}\right)$.

Solution
Step 1. The function is already written in general form: $f(x) = 3\sin\left(\frac{\pi}{4}x - \frac{\pi}{4}\right)$. This graph will have the shape of a sine function, starting at the midline and increasing to the right.
Step 2. $|A| = |3| = 3$. The amplitude is 3.
Step 3. Since $|B| = \left|\frac{\pi}{4}\right| = \frac{\pi}{4}$, we determine the period as follows.

$$P = \frac{2\pi}{|B|} = \frac{2\pi}{\frac{\pi}{4}} = 2\pi \cdot \frac{4}{\pi} = 8$$

The period is 8.

Step 4. Since $C = \frac{\pi}{4}$, the phase shift is

$$\frac{C}{B} = \frac{\frac{\pi}{4}}{\frac{\pi}{4}} = 1.$$

The phase shift is 1 unit.

Step 5. Figure 20 shows the graph of the function.

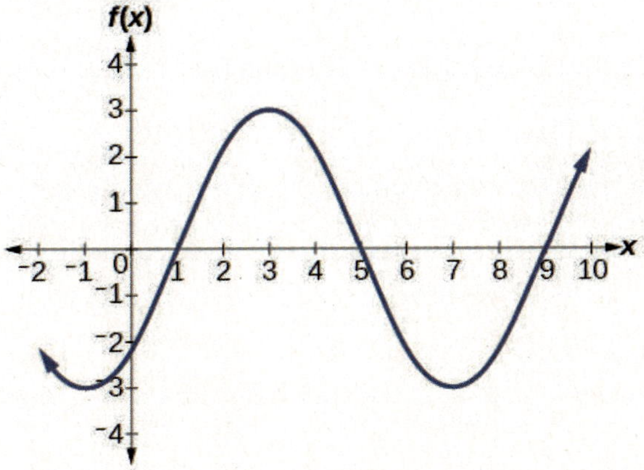

Figure 20 A horizontally compressed, vertically stretched, and horizontally shifted sinusoid

TRY IT #9 Draw a graph of $g(x) = -2\cos\left(\frac{\pi}{3}x + \frac{\pi}{6}\right)$. Determine the midline, amplitude, period, and phase shift.

EXAMPLE 10

Identifying the Properties of a Sinusoidal Function

Given $y = -2\cos\left(\frac{\pi}{2}x + \pi\right) + 3$, determine the amplitude, period, phase shift, and vertical shift. Then graph the function.

Solution

Begin by comparing the equation to the general form and use the steps outlined in Example 9.

$$y = A\cos(Bx - C) + D$$

Step 1. The function is already written in general form.
Step 2. Since $A = -2$, the amplitude is $|A| = 2$.
Step 3. $|B| = \frac{\pi}{2}$, so the period is $P = \frac{2\pi}{|B|} = \frac{2\pi}{\frac{\pi}{2}} = 2\pi \cdot \frac{2}{\pi} = 4$. The period is 4.
Step 4. $C = -\pi$, so we calculate the phase shift as $\frac{C}{B} = \frac{-\pi,}{\frac{\pi}{2}} = -\pi \cdot \frac{2}{\pi} = -2$. The phase shift is -2.
Step 5. $D = 3$, so the midline is $y = 3$, and the vertical shift is up 3.

Since A is negative, the graph of the cosine function has been reflected about the *x*-axis.

Figure 21 shows one cycle of the graph of the function.

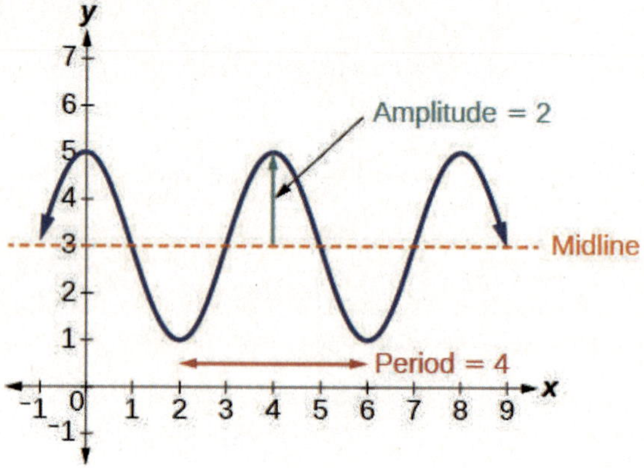

Figure 21

Using Transformations of Sine and Cosine Functions

We can use the transformations of sine and cosine functions in numerous applications. As mentioned at the beginning of the chapter, circular motion can be modeled using either the sine or cosine function.

EXAMPLE 11

Finding the Vertical Component of Circular Motion

A point rotates around a circle of radius 3 centered at the origin. Sketch a graph of the *y*-coordinate of the point as a function of the angle of rotation.

Solution

Recall that, for a point on a circle of radius *r*, the *y*-coordinate of the point is $y = r\ \sin(x)$, so in this case, we get the equation $y(x) = 3\ \sin(x)$. The constant 3 causes a vertical stretch of the *y*-values of the function by a factor of 3, which we can see in the graph in Figure 22.

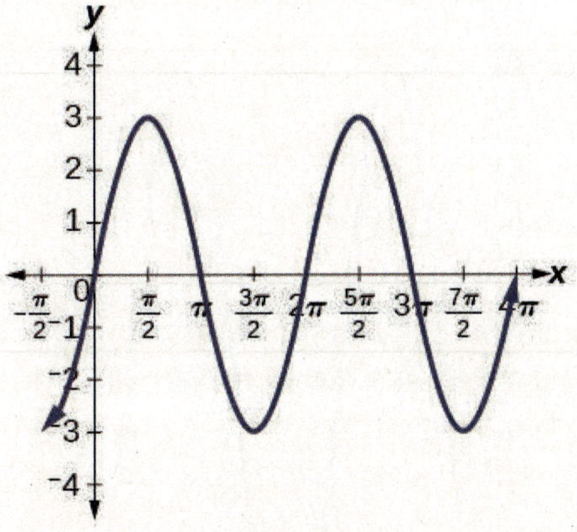

Figure 22

Analysis

Notice that the period of the function is still 2π; as we travel around the circle, we return to the point $(3, 0)$ for $x = 2\pi, 4\pi, 6\pi, \ldots$ Because the outputs of the graph will now oscillate between -3 and 3, the amplitude of the sine wave is 3.

> **TRY IT** #10 What is the amplitude of the function $f(x) = 7\cos(x)$? Sketch a graph of this function.

EXAMPLE 12

Finding the Vertical Component of Circular Motion

A circle with radius 3 ft is mounted with its center 4 ft off the ground. The point closest to the ground is labeled *P*, as shown in Figure 23. Sketch a graph of the height above the ground of the point P as the circle is rotated; then find a function that gives the height in terms of the angle of rotation.

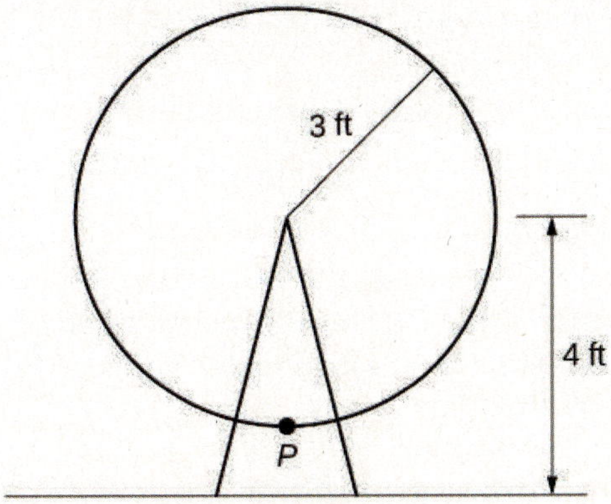

Figure 23

Solution

Sketching the height, we note that it will start 1 ft above the ground, then increase up to 7 ft above the ground, and continue to oscillate 3 ft above and below the center value of 4 ft, as shown in Figure 24.

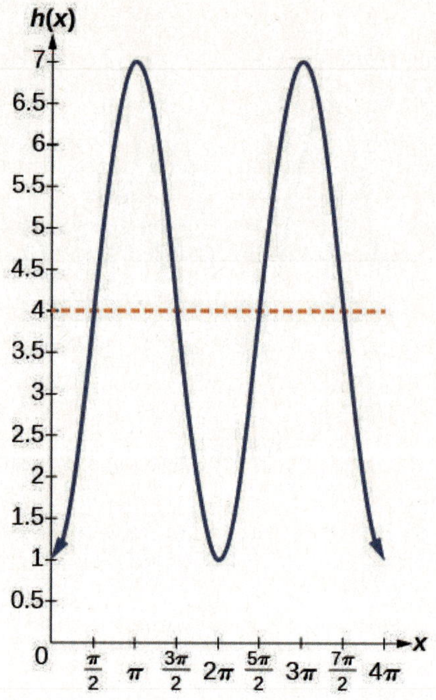

Figure 24

Although we could use a transformation of either the sine or cosine function, we start by looking for characteristics that would make one function easier to use than the other. Let's use a cosine function because it starts at the highest or lowest value, while a sine function starts at the middle value. A standard cosine starts at the highest value, and this graph starts at the lowest value, so we need to incorporate a vertical reflection.

Second, we see that the graph oscillates 3 above and below the center, while a basic cosine has an amplitude of 1, so this graph has been vertically stretched by 3, as in the last example.

Finally, to move the center of the circle up to a height of 4, the graph has been vertically shifted up by 4. Putting these transformations together, we find that

$$y = -3\cos(x) + 4$$

TRY IT #11 A weight is attached to a spring that is then hung from a board, as shown in Figure 25. As the spring oscillates up and down, the position y of the weight relative to the board ranges from –1 in. (at time $x = 0$) to –7 in. (at time $x = \pi$) below the board. Assume the position of y is given as a sinusoidal function of x. Sketch a graph of the function, and then find a cosine function that gives the position y in terms of x.

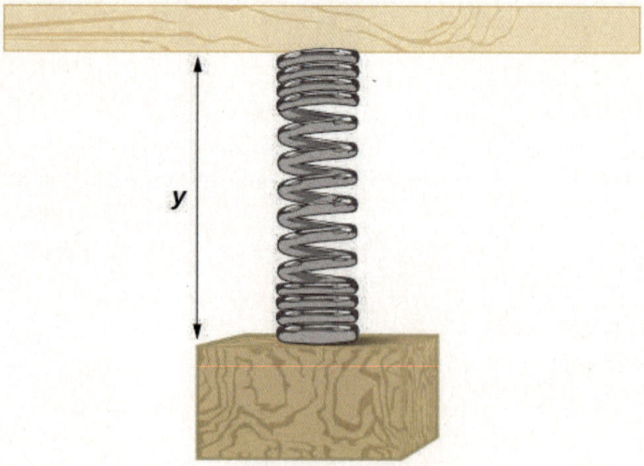

Figure 25

EXAMPLE 13

Determining a Rider's Height on a Ferris Wheel

The London Eye is a huge Ferris wheel with a diameter of 135 meters (443 feet). It completes one rotation every 30 minutes. Riders board from a platform 2 meters above the ground. Express a rider's height above ground as a function of time in minutes.

Solution

With a diameter of 135 m, the wheel has a radius of 67.5 m. The height will oscillate with amplitude 67.5 m above and below the center.

Passengers board 2 m above ground level, so the center of the wheel must be located $67.5 + 2 = 69.5$ m above ground level. The midline of the oscillation will be at 69.5 m.

The wheel takes 30 minutes to complete 1 revolution, so the height will oscillate with a period of 30 minutes.

Lastly, because the rider boards at the lowest point, the height will start at the smallest value and increase, following the shape of a vertically reflected cosine curve.

- Amplitude: 67.5, so $A = 67.5$
- Midline: 69.5, so $D = 69.5$
- Period: 30, so $B = \frac{2\pi}{30} = \frac{\pi}{15}$
- Shape: $-\cos(t)$

An equation for the rider's height would be

$$y = -67.5 \cos\left(\frac{\pi}{15}t\right) + 69.5$$

where t is in minutes and y is measured in meters.

MEDIA

Access these online resources for additional instruction and practice with graphs of sine and cosine functions.

Amplitude and Period of Sine and Cosine (http://openstax.org/l/ampperiod)
Translations of Sine and Cosine (http://openstax.org/l/translasincos)
Graphing Sine and Cosine Transformations (http://openstax.org/l/transformsincos)
Graphing the Sine Function (http://openstax.org/l/graphsinefunc)

8.1 SECTION EXERCISES

Verbal

1. Why are the sine and cosine functions called periodic functions?

2. How does the graph of $y = \sin\ x$ compare with the graph of $y = \cos\ x$? Explain how you could horizontally translate the graph of $y = \sin\ x$ to obtain $y = \cos\ x$.

3. For the equation $A\ \cos(Bx + C) + D$, what constants affect the range of the function and how do they affect the range?

4. How does the range of a translated sine function relate to the equation $y = A\ \sin(Bx + C) + D$?

5. How can the unit circle be used to construct the graph of $f(t) = \sin\ t$?

Graphical

For the following exercises, graph two full periods of each function and state the amplitude, period, and midline. State the maximum and minimum y-values and their corresponding x-values on one period for $x > 0$. Round answers to two decimal places if necessary.

6. $f(x) = 2\sin x$

7. $f(x) = \frac{2}{3}\cos x$

8. $f(x) = -3\sin x$

9. $f(x) = 4\sin x$

10. $f(x) = 2\cos x$

11. $f(x) = \cos(2x)$

12. $f(x) = 2\sin\left(\frac{1}{2}x\right)$

13. $f(x) = 4\cos(\pi x)$

14. $f(x) = 3\cos\left(\frac{6}{5}x\right)$

15. $y = 3\sin(8(x+4)) + 5$

16. $y = 2\sin(3x - 21) + 4$

17. $y = 5\sin(5x + 20) - 2$

For the following exercises, graph one full period of each function, starting at $x = 0$. For each function, state the amplitude, period, and midline. State the maximum and minimum y-values and their corresponding x-values on one period for $x > 0$. State the phase shift and vertical translation, if applicable. Round answers to two decimal places if necessary.

18. $f(t) = 2\sin\left(t - \frac{5\pi}{6}\right)$

19. $f(t) = -\cos\left(t + \frac{\pi}{3}\right) + 1$

20. $f(t) = 4\cos\left(2\left(t + \frac{\pi}{4}\right)\right) - 3$

21. $f(t) = -\sin\left(\frac{1}{2}t + \frac{5\pi}{3}\right)$

22. $f(x) = 4\sin\left(\frac{\pi}{2}(x-3)\right) + 7$

23. Determine the amplitude, midline, period, and an equation involving the sine function for the graph shown in Figure 26.

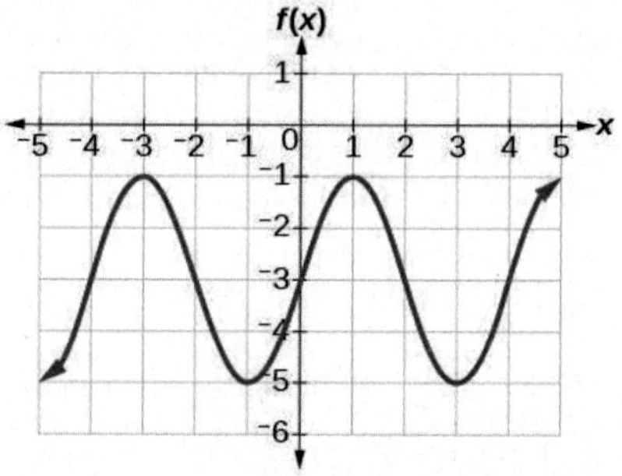

Figure 26

24. Determine the amplitude, period, midline, and an equation involving cosine for the graph shown in Figure 27.

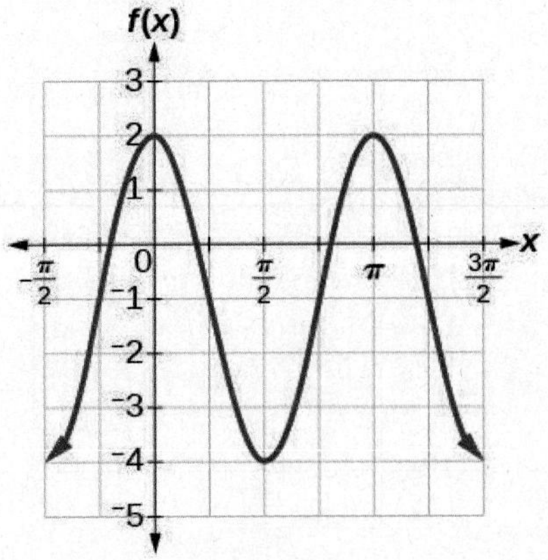

Figure 27

25. Determine the amplitude, period, midline, and an equation involving cosine for the graph shown in Figure 28.

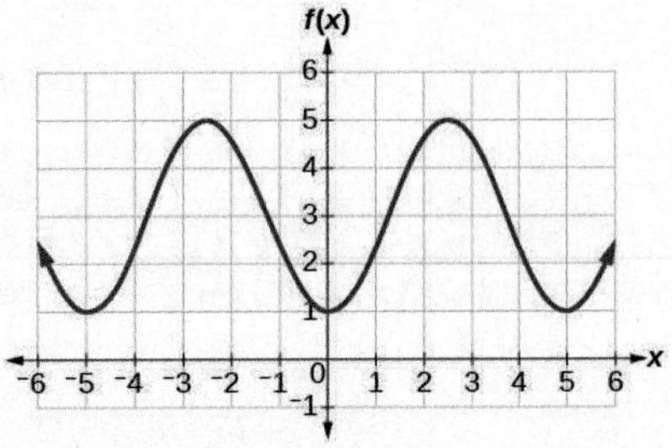

Figure 28

26. Determine the amplitude, period, midline, and an equation involving sine for the graph shown in Figure 29.

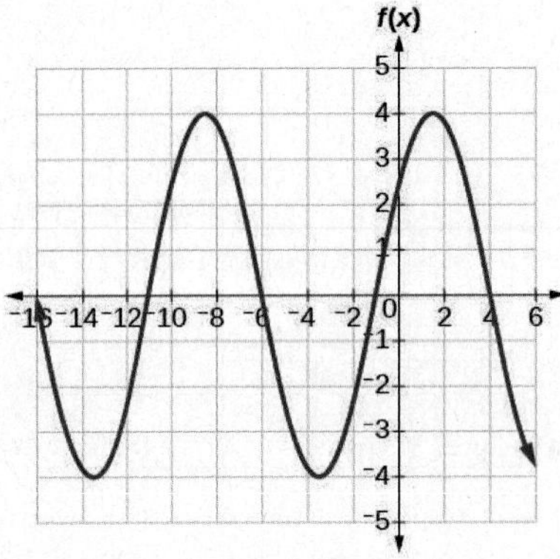

Figure 29

27. Determine the amplitude, period, midline, and an equation involving cosine for the graph shown in Figure 30.

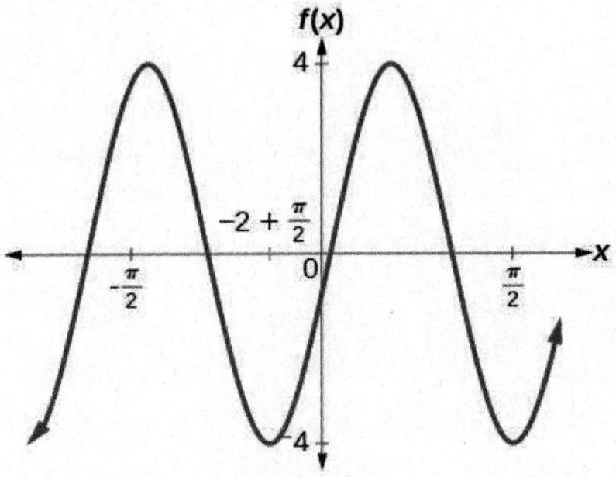

Figure 30

28. Determine the amplitude, period, midline, and an equation involving sine for the graph shown in Figure 31.

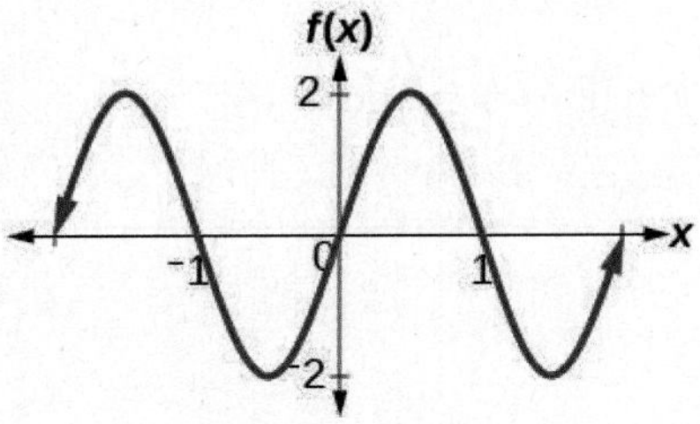

Figure 31

29. Determine the amplitude, period, midline, and an equation involving cosine for the graph shown in Figure 32.

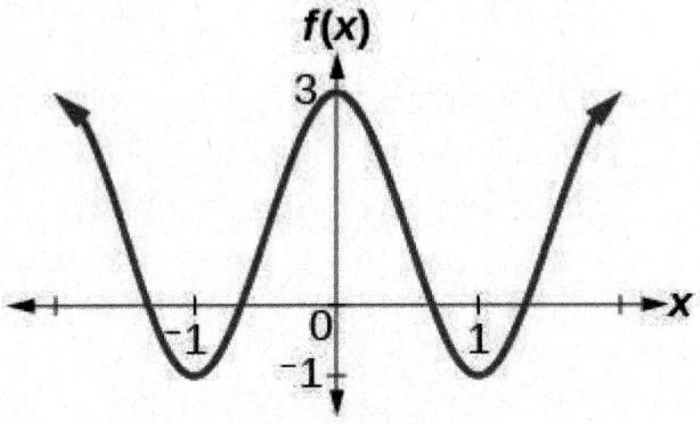

Figure 32

30. Determine the amplitude, period, midline, and an equation involving sine for the graph shown in Figure 33.

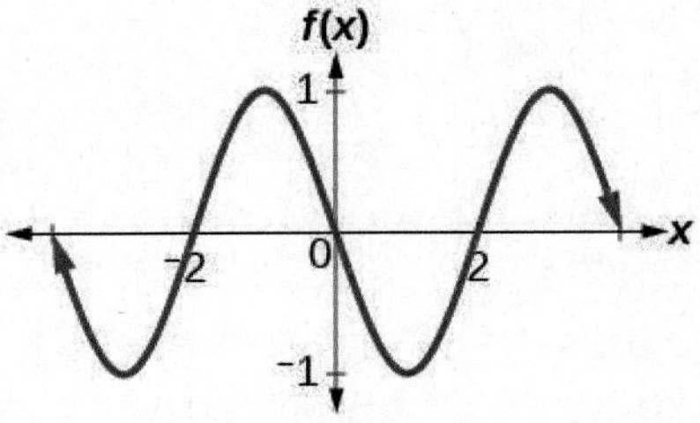

Figure 33

Algebraic

For the following exercises, let $f(x) = \sin\ x$.

31. On $[0, 2\pi\ )$, solve $f\ (x) = 0$.

32. On $[0, 2\pi\ )$, solve $f\ (x) = \frac{1}{2}$.

33. Evaluate $f\ \left(\frac{\pi}{2}\right)$.

34. On $[0, 2\pi)$, $f(x) = \frac{\sqrt{2}}{2}$. Find all values of x.

35. On $[0, 2\pi\ )$, the maximum value(s) of the function occur(s) at what *x*-value(s)?

36. On $[0, 2\pi\ )$, the minimum value(s) of the function occur(s) at what *x*-value(s)?

37. Show that $f(-x) = -f(x)$. This means that $f(x) = \sin\ x$ is an odd function and possesses symmetry with respect to ________________.

For the following exercises, let $f(x) = \cos\ x$.

38. On $[0, 2\pi\ )$, solve the equation $f(x) = \cos\ x = 0$.

39. On $[0, 2\pi\ )$, solve $f(x) = \frac{1}{2}$.

40. On $[0, 2\pi\ )$, find the *x*-intercepts of $f(x) = \cos\ x$.

41. On $[0, 2\pi\ )$, find the *x*-values at which the function has a maximum or minimum value.

42. On $[0, 2\pi\ )$, solve the equation $f(x) = \frac{\sqrt{3}}{2}$.

Technology

43. Graph $h(x) = x + \sin\ x$ on $[0, 2\pi]$. Explain why the graph appears as it does.

44. Graph $h(x) = x + \sin\ x$ on $[-100, 100]$. Did the graph appear as predicted in the previous exercise?

45. Graph $f(x) = x\ \sin\ x$ on $[0, 2\pi]$ and verbalize how the graph varies from the graph of $f(x) = \sin\ x$.

46. Graph $f(x) = x\ \sin\ x$ on the window $[-10, 10]$ and explain what the graph shows.

47. Graph $f(x) = \frac{\sin\ x}{x}$ on the window $[-5\pi, 5\pi]$ and explain what the graph shows.

Real-World Applications

48. A Ferris wheel is 25 meters in diameter and boarded from a platform that is 1 meter above the ground. The six o'clock position on the Ferris wheel is level with the loading platform. The wheel completes 1 full revolution in 10 minutes. The function $h(t)$ gives a person's height in meters above the ground t minutes after the wheel begins to turn.

ⓐ Find the amplitude, midline, and period of $h(t)$.

ⓑ Find a formula for the height function $h(t)$.

ⓒ How high off the ground is a person after 5 minutes?

8.2 Graphs of the Other Trigonometric Functions

Learning Objectives

In this section, you will:

- Analyze the graph of y=tan x.
- Graph variations of y=tan x.
- Analyze the graphs of y=sec x and y=csc x.
- Graph variations of y=sec x and y=csc x.
- Analyze the graph of y=cot x.
- Graph variations of y=cot x.

We know the tangent function can be used to find distances, such as the height of a building, mountain, or flagpole. But what if we want to measure repeated occurrences of distance? Imagine, for example, a fire truck parked next to a warehouse. The rotating light from the truck would travel across the wall of the warehouse in regular intervals. If the input is time, the output would be the distance the beam of light travels. The beam of light would repeat the distance at regular intervals. The tangent function can be used to approximate this distance. Asymptotes would be needed to illustrate the repeated cycles when the beam runs parallel to the wall because, seemingly, the beam of light could appear to extend forever. The graph of the tangent function would clearly illustrate the repeated intervals. In this section, we will explore the graphs of the tangent and other trigonometric functions.

Analyzing the Graph of $y = \tan x$

We will begin with the graph of the tangent function, plotting points as we did for the sine and cosine functions. Recall that

$$\tan x = \frac{\sin x}{\cos x}$$

The period of the tangent function is π because the graph repeats itself on intervals of $k\pi$ where k is a constant. If we graph the tangent function on $-\frac{\pi}{2}$ to $\frac{\pi}{2}$, we can see the behavior of the graph on one complete cycle. If we look at any larger interval, we will see that the characteristics of the graph repeat.

We can determine whether tangent is an odd or even function by using the definition of tangent.

$$\begin{aligned} \tan(-x) &= \frac{\sin(-x)}{\cos(-x)} && \text{Definition of tangent.} \\ &= \frac{-\sin\ x}{\cos\ x} && \text{Sine is an odd function, cosine is even.} \\ &= -\frac{\sin\ x}{\cos\ x} && \text{The quotient of an odd and an even function is odd.} \\ &= -\tan\ x && \text{Definition of tangent.} \end{aligned}$$

Therefore, tangent is an odd function. We can further analyze the graphical behavior of the tangent function by looking at values for some of the special angles, as listed in Table 1.

x	$-\frac{\pi}{2}$	$-\frac{\pi}{3}$	$-\frac{\pi}{4}$	$-\frac{\pi}{6}$	0	$\frac{\pi}{6}$	$\frac{\pi}{4}$	$\frac{\pi}{3}$	$\frac{\pi}{2}$
$\tan(x)$	undefined	$-\sqrt{3}$	-1	$-\frac{\sqrt{3}}{3}$	0	$\frac{\sqrt{3}}{3}$	1	$\sqrt{3}$	undefined

Table 1

These points will help us draw our graph, but we need to determine how the graph behaves where it is undefined. If we look more closely at values when $\frac{\pi}{3} < x < \frac{\pi}{2}$, we can use a table to look for a trend. Because $\frac{\pi}{3} \approx 1.05$ and $\frac{\pi}{2} \approx 1.57$, we will evaluate x at radian measures $1.05 < x < 1.57$ as shown in Table 2.

x	1.3	1.5	1.55	1.56
$\tan\ x$	3.6	14.1	48.1	92.6

Table 2

As x approaches $\frac{\pi}{2}$, the outputs of the function get larger and larger. Because $y = \tan\ x$ is an odd function, we see the corresponding table of negative values in Table 3.

x	-1.3	-1.5	-1.55	-1.56
$\tan\ x$	-3.6	-14.1	-48.1	-92.6

Table 3

We can see that, as x approaches $-\frac{\pi}{2}$, the outputs get smaller and smaller. Remember that there are some values of x for which $\cos\ x = 0$. For example, $\cos\left(\frac{\pi}{2}\right) = 0$ and $\cos\left(\frac{3\pi}{2}\right) = 0$. At these values, the tangent function is undefined, so the graph of $y = \tan\ x$ has discontinuities at $x = \frac{\pi}{2}$ and $\frac{3\pi}{2}$. At these values, the graph of the tangent has vertical asymptotes. Figure 1 represents the graph of $y = \tan\ x$. The tangent is positive from 0 to $\frac{\pi}{2}$ and from π to $\frac{3\pi}{2}$, corresponding to quadrants I and III of the unit circle.

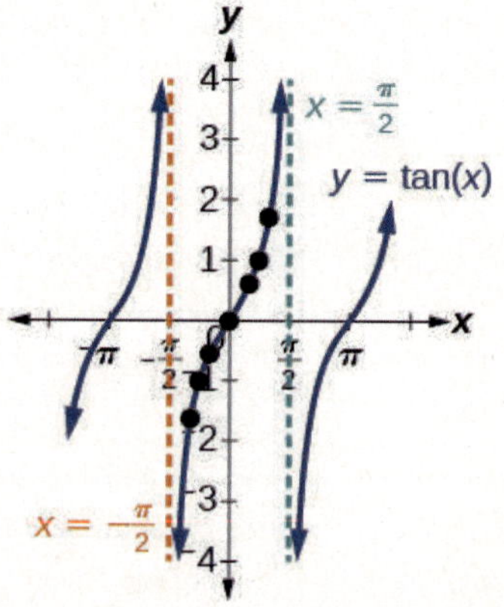

Figure 1 Graph of the tangent function

Graphing Variations of $y = \tan x$

As with the sine and cosine functions, the tangent function can be described by a general equation.

$$y = A\tan(Bx)$$

We can identify horizontal and vertical stretches and compressions using values of A and B. The horizontal stretch can typically be determined from the period of the graph. With tangent graphs, it is often necessary to determine a vertical stretch using a point on the graph.

Because there are no maximum or minimum values of a tangent function, the term *amplitude* cannot be interpreted as it is for the sine and cosine functions. Instead, we will use the phrase *stretching/compressing factor* when referring to the constant A.

Features of the Graph of $y = A\tan(Bx)$

- The stretching factor is $|A|$.
- The period is $P = \frac{\pi}{|B|}$.
- The domain is all real numbers x, where $x \neq \frac{\pi}{2|B|} + \frac{\pi}{|B|}k$ such that k is an integer.
- The range is $(-\infty, \infty)$.
- The asymptotes occur at $x = \frac{\pi}{2|B|} + \frac{\pi}{|B|}k$, where k is an integer.
- $y = A\tan(Bx)$ is an odd function.

Graphing One Period of a Stretched or Compressed Tangent Function

We can use what we know about the properties of the tangent function to quickly sketch a graph of any stretched and/or compressed tangent function of the form $f(x) = A\tan(Bx)$. We focus on a single period of the function including the origin, because the periodic property enables us to extend the graph to the rest of the function's domain if we wish. Our limited domain is then the interval $\left(-\frac{P}{2}, \frac{P}{2}\right)$ and the graph has vertical asymptotes at $\pm\frac{P}{2}$ where $P = \frac{\pi}{B}$. On $\left(-\frac{\pi}{2}, \frac{\pi}{2}\right)$, the graph will come up from the left asymptote at $x = -\frac{\pi}{2}$, cross through the origin, and continue to increase as it approaches the right asymptote at $x = \frac{\pi}{2}$. To make the function approach the asymptotes at the correct rate, we also need to set the vertical scale by actually evaluating the function for at least one point that the graph will pass through. For example, we can use

$$f\left(\frac{P}{4}\right) = A\tan\left(B\frac{P}{4}\right) = A\tan\left(B\frac{\pi}{4B}\right) = A$$

because $\tan\left(\frac{\pi}{4}\right) = 1$.

HOW TO

Given the function $f(x) = A\tan(Bx)$, graph one period.

1. Identify the stretching factor, $|A|$.
2. Identify B and determine the period, $P = \frac{\pi}{|B|}$.
3. Draw vertical asymptotes at $x = -\frac{P}{2}$ and $x = \frac{P}{2}$.
4. For $AB > 0$, the graph approaches the left asymptote at negative output values and the right asymptote at positive output values (reverse for $AB < 0$).
5. Plot reference points at $\left(\frac{P}{4}, A\right)$, $(0, 0)$, and $\left(-\frac{P}{4}, -A\right)$, and draw the graph through these points.

EXAMPLE 1

Sketching a Compressed Tangent

Sketch a graph of one period of the function $y = 0.5\tan\left(\frac{\pi}{2}x\right)$.

Solution

First, we identify A and B.

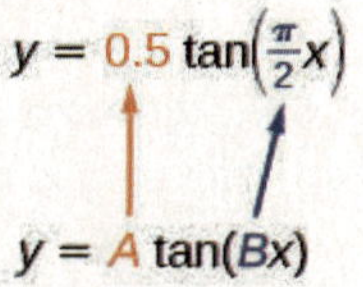

Because $A = 0.5$ and $B = \frac{\pi}{2}$, we can find the stretching/compressing factor and period. The period is $\frac{\pi}{\frac{\pi}{2}} = 2$, so the asymptotes are at $x = \pm 1$. At a quarter period from the origin, we have

$$\begin{aligned} f(0.5) &= 0.5\tan\left(\frac{0.5\pi}{2}\right) \\ &= 0.5\tan\left(\frac{\pi}{4}\right) \\ &= 0.5 \end{aligned}$$

This means the curve must pass through the points $(0.5, 0.5)$, $(0, 0)$, and $(-0.5, -0.5)$. The only inflection point is at the origin. Figure 2 shows the graph of one period of the function.

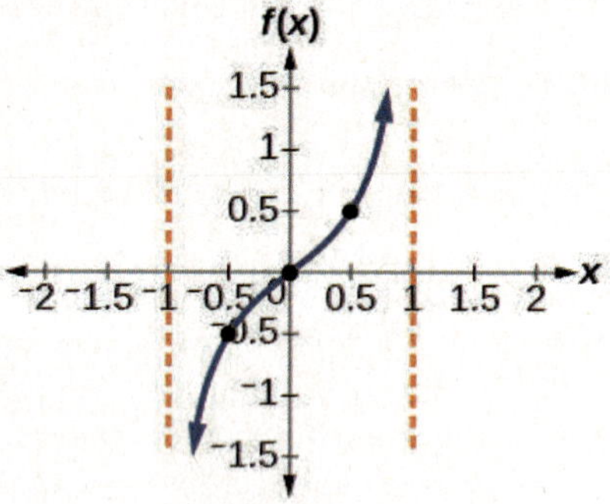

Figure 2

TRY IT #1 Sketch a graph of $f(x) = 3\tan\left(\frac{\pi}{6}x\right)$.

Graphing One Period of a Shifted Tangent Function

Now that we can graph a tangent function that is stretched or compressed, we will add a vertical and/or horizontal (or phase) shift. In this case, we add C and D to the general form of the tangent function.

$$f(x) = A\tan(Bx - C) + D$$

The graph of a transformed tangent function is different from the basic tangent function $\tan x$ in several ways:

Features of the Graph of $y = A\tan(Bx-C)+D$

- The stretching factor is $|A|$.
- The period is $\frac{\pi}{|B|}$.
- The domain is $x \neq \frac{C}{B} + \frac{\pi}{|B|}k$, where k is an integer.
- The range is $(-\infty, \infty)$.
- The vertical asymptotes occur at $x = \frac{C}{B} + \frac{\pi}{2|B|}k$, where k is an odd integer.
- There is no amplitude.
- $y = A\tan(Bx - C) + D$ is an odd function because it is the quotient of odd and even functions (sine and cosine respectively).

HOW TO

Given the function $y = A\tan(Bx - C) + D$, sketch the graph of one period.

1. Express the function given in the form $y = A\tan(Bx - C) + D$.
2. Identify the stretching/compressing factor, $|A|$.
3. Identify B and determine the period, $P = \frac{\pi}{|B|}$.
4. Identify C and determine the phase shift, $\frac{C}{B}$.
5. Draw the graph of $y = A\tan(Bx)$ shifted to the right by $\frac{C}{B}$ and up by D.
6. Sketch the vertical asymptotes, which occur at $x = \frac{C}{B} + \frac{\pi}{2|B|}k$, where k is an odd integer.
7. Plot any three reference points and draw the graph through these points.

EXAMPLE 2

Graphing One Period of a Shifted Tangent Function

Graph one period of the function $y = -2\tan(\pi x + \pi) - 1$.

Solution

Step 1. The function is already written in the form $y = A\tan(Bx - C) + D$.
Step 2. $A = -2$, so the stretching factor is $|A| = 2$.
Step 3. $B = \pi$, so the period is $P = \frac{\pi}{|B|} = \frac{\pi}{\pi} = 1$.
Step 4. $C = -\pi$, so the phase shift is $\frac{C}{B} = \frac{-\pi}{\pi} = -1$.
Step 5-7. The asymptotes are at $x = -\frac{3}{2}$ and $x = -\frac{1}{2}$ and the three recommended reference points are $(-1.25, 1)$, $(-1, -1)$, and $(-0.75, -3)$. The graph is shown in Figure 3.

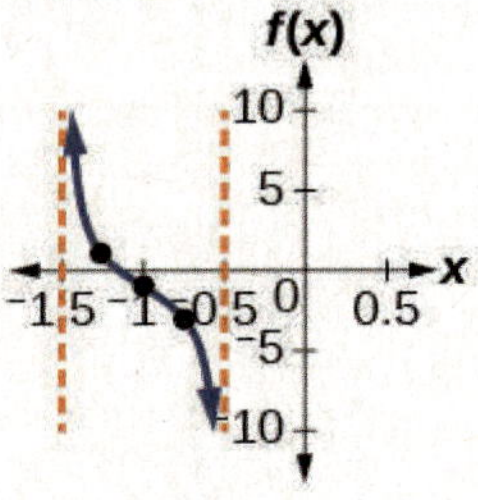

Figure 3

Analysis

Note that this is a decreasing function because $A < 0$.

TRY IT #2 How would the graph in Example 2 look different if we made $A = 2$ instead of -2?

HOW TO

Given the graph of a tangent function, identify horizontal and vertical stretches.

1. Find the period P from the spacing between successive vertical asymptotes or x-intercepts.
2. Write $f(x) = A\tan\left(\frac{\pi}{P}x\right)$.
3. Determine a convenient point $(x, f(x))$ on the given graph and use it to determine A.

EXAMPLE 3

Identifying the Graph of a Stretched Tangent

Find a formula for the function graphed in Figure 4.

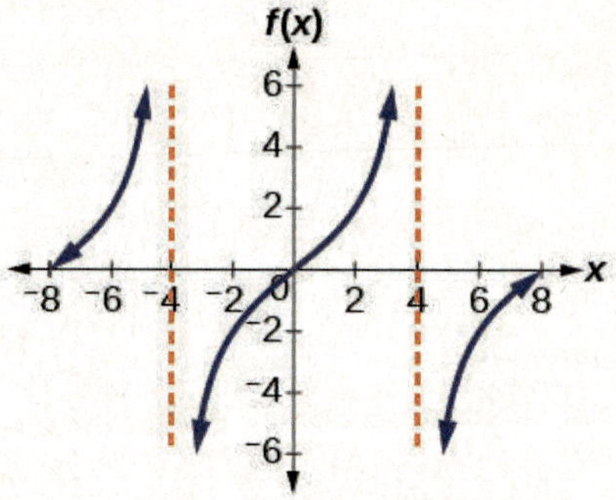

Figure 4 A stretched tangent function

Solution

The graph has the shape of a tangent function.

Step 1. One cycle extends from −4 to 4, so the period is $P = 8$. Since $P = \frac{\pi}{|B|}$, we have $B = \frac{\pi}{P} = \frac{\pi}{8}$.

Step 2. The equation must have the form $f(x) = A \tan\left(\frac{\pi}{8}x\right)$.

Step 3. To find the vertical stretch A, we can use the point $(2, 2)$.

$$2 = A \tan\left(\frac{\pi}{8} \cdot 2\right) = A \tan\left(\frac{\pi}{4}\right)$$

Because $\tan\left(\frac{\pi}{4}\right) = 1$, $A = 2$.

This function would have a formula $f(x) = 2\tan\left(\frac{\pi}{8}x\right)$.

TRY IT #3 Find a formula for the function in Figure 5.

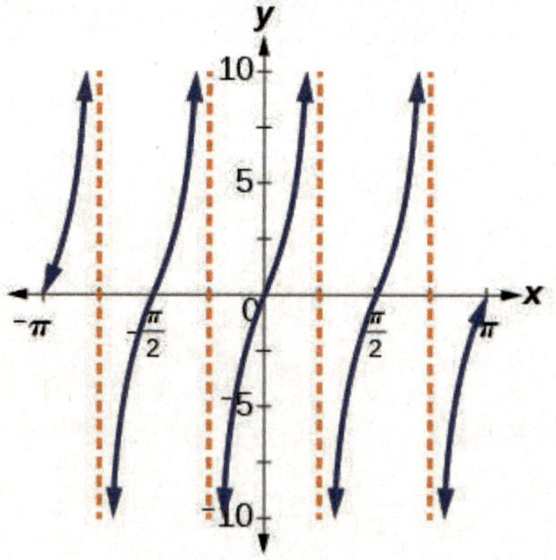

Figure 5

Analyzing the Graphs of $y = \sec x$ and $y = \csc x$

The secant was defined by the reciprocal identity $\sec x = \frac{1}{\cos x}$. Notice that the function is undefined when the cosine is 0, leading to vertical asymptotes at $\frac{\pi}{2}$, $\frac{3\pi}{2}$, etc. Because the cosine is never more than 1 in absolute value, the secant, being the reciprocal, will never be less than 1 in absolute value.

We can graph $y = \sec x$ by observing the graph of the cosine function because these two functions are reciprocals of one another. See Figure 6. The graph of the cosine is shown as a dashed orange wave so we can see the relationship. Where the graph of the cosine function decreases, the graph of the secant function increases. Where the graph of the cosine function increases, the graph of the secant function decreases. When the cosine function is zero, the secant is undefined.

The secant graph has vertical asymptotes at each value of x where the cosine graph crosses the x-axis; we show these in the graph below with dashed vertical lines, but will not show all the asymptotes explicitly on all later graphs involving the secant and cosecant.

Note that, because cosine is an even function, secant is also an even function. That is, $\sec(-x) = \sec x$.

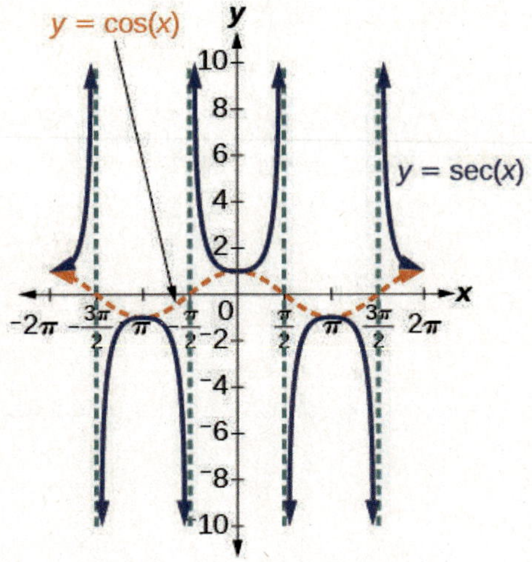

Figure 6 Graph of the secant function, $f(x) = \sec x = \frac{1}{\cos x}$

As we did for the tangent function, we will again refer to the constant $|A|$ as the stretching factor, not the amplitude.

Features of the Graph of $y = A\sec(Bx)$

- The stretching factor is $|A|$.
- The period is $\frac{2\pi}{|B|}$.
- The domain is $x \neq \frac{\pi}{2|B|}k$, where k is an odd integer.
- The range is $(-\infty, -|A|] \cup [|A|, \infty)$.
- The vertical asymptotes occur at $x = \frac{\pi}{2|B|}k$, where k is an odd integer.
- There is no amplitude.
- $y = A\sec(Bx)$ is an even function because cosine is an even function.

Similar to the secant, the cosecant is defined by the reciprocal identity $\csc x = \frac{1}{\sin x}$. Notice that the function is undefined when the sine is 0, leading to a vertical asymptote in the graph at $0, \pi$, etc. Since the sine is never more than 1 in absolute value, the cosecant, being the reciprocal, will never be less than 1 in absolute value.

We can graph $y = \csc x$ by observing the graph of the sine function because these two functions are reciprocals of one another. See Figure 7. The graph of sine is shown as a dashed orange wave so we can see the relationship. Where the graph of the sine function decreases, the graph of the cosecant function increases. Where the graph of the sine function increases, the graph of the cosecant function decreases.

The cosecant graph has vertical asymptotes at each value of x where the sine graph crosses the x-axis; we show these in the graph below with dashed vertical lines.

Note that, since sine is an odd function, the cosecant function is also an odd function. That is, $\csc(-x) = -\csc x$.

The graph of cosecant, which is shown in Figure 7, is similar to the graph of secant.

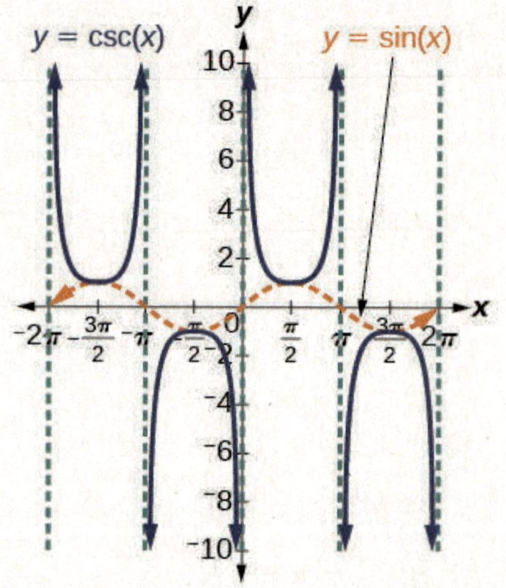

Figure 7 The graph of the cosecant function, $f(x) = \csc x = \frac{1}{\sin x}$

Features of the Graph of $y = A\csc(Bx)$

- The stretching factor is $|A|$.
- The period is $\frac{2\pi}{|B|}$.
- The domain is $x \neq \frac{\pi}{|B|}k$, where k is an integer.
- The range is $\left(-\infty, -|A|\right] \cup \left[|A|, \infty\right)$.
- The asymptotes occur at $x = \frac{\pi}{|B|}k$, where k is an integer.
- $y = A\csc(Bx)$ is an odd function because sine is an odd function.

Graphing Variations of $y = \sec x$ and $y = \csc x$

For shifted, compressed, and/or stretched versions of the secant and cosecant functions, we can follow similar methods to those we used for tangent and cotangent. That is, we locate the vertical asymptotes and also evaluate the functions for a few points (specifically the local extrema). If we want to graph only a single period, we can choose the interval for the period in more than one way. The procedure for secant is very similar, because the cofunction identity means that the secant graph is the same as the cosecant graph shifted half a period to the left. Vertical and phase shifts may be applied to the cosecant function in the same way as for the secant and other functions.The equations become the following.

$$y = A\sec(Bx - C) + D$$

$$y = A\csc(Bx - C) + D$$

Features of the Graph of $y = A\sec(Bx - C) + D$

- The stretching factor is $|A|$.
- The period is $\frac{2\pi}{|B|}$.
- The domain is $x \neq \frac{C}{B} + \frac{\pi}{2|B|}k$, where k is an odd integer.
- The range is $(-\infty, -|A| + D] \cup [|A| + D, \infty)$.
- The vertical asymptotes occur at $x = \frac{C}{B} + \frac{\pi}{2|B|}k$, where k is an odd integer.
- There is no amplitude.
- $y = A\sec(Bx - C) + D$ is an even function because cosine is an even function.

Features of the Graph of $y = A\csc(Bx - C) + D$

- The stretching factor is $|A|$.
- The period is $\frac{2\pi}{|B|}$.
- The domain is $x \neq \frac{C}{B} + \frac{\pi}{|B|}k$, where k is an integer.
- The range is $(-\infty, -|A| + D] \cup [|A| + D, \infty)$.
- The vertical asymptotes occur at $x = \frac{C}{B} + \frac{\pi}{|B|}k$, where k is an integer.
- There is no amplitude.
- $y = A\csc(Bx - C) + D$ is an odd function because sine is an odd function.

HOW TO

Given a function of the form $y = A\sec(Bx)$, graph one period.

1. Express the function given in the form $y = A\sec(Bx)$.
2. Identify the stretching/compressing factor, $|A|$.
3. Identify B and determine the period, $P = \frac{2\pi}{|B|}$.
4. Sketch the graph of $y = A\cos(Bx)$.
5. Use the reciprocal relationship between $y = \cos\ x$ and $y = \sec\ x$ to draw the graph of $y = A\sec(Bx)$.
6. Sketch the asymptotes.
7. Plot any two reference points and draw the graph through these points.

EXAMPLE 4

Graphing a Variation of the Secant Function

Graph one period of $f(x) = 2.5\sec(0.4x)$.

Solution

Step 1. The given function is already written in the general form, $y = A\sec(Bx)$.
Step 2. $A = 2.5$ so the stretching factor is 2.5.
Step 3. $B = 0.4$ so $P = \frac{2\pi}{0.4} = 5\pi$. The period is 5π units.
Step 4. Sketch the graph of the function $g(x) = 2.5\cos(0.4x)$.
Step 5. Use the reciprocal relationship of the cosine and secant functions to draw the cosecant function.
Steps 6–7. Sketch two asymptotes at $x = 1.25\pi$ and $x = 3.75\pi$. We can use two reference points, the local minimum at $(0, 2.5)$ and the local maximum at $(2.5\pi, -2.5)$. Figure 8 shows the graph.

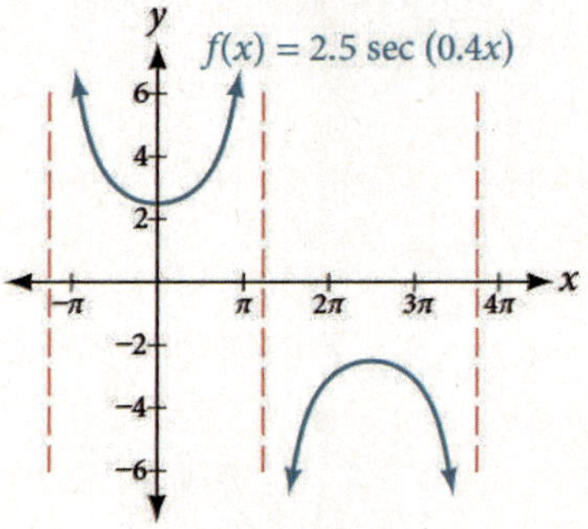

Figure 8

TRY IT #4 Graph one period of $f(x) = -2.5\sec(0.4x)$.

Q&A **Do the vertical shift and stretch/compression affect the secant's range?**

Yes. The range of $f(x) = A\sec(Bx - C) + D$ *is* $\left(-\infty, -|A| + D\right] \cup \left[|A| + D, \infty\right)$.

HOW TO

Given a function of the form $f(x) = A\sec(Bx - C) + D$**, graph one period.**

1. Express the function given in the form $y = A\ \sec(Bx - C) + D$.
2. Identify the stretching/compressing factor, $|A|$.
3. Identify B and determine the period, $\frac{2\pi}{|B|}$.
4. Identify C and determine the phase shift, $\frac{C}{B}$.
5. Draw the graph of $y = A\ \sec(Bx)$, but shift it to the right by $\frac{C}{B}$ and up by D.
6. Sketch the vertical asymptotes, which occur at $x = \frac{C}{B} + \frac{\pi}{2|B|}k$, where k is an odd integer.

EXAMPLE 5

Graphing a Variation of the Secant Function

Graph one period of $y = 4\sec\left(\frac{\pi}{3}x - \frac{\pi}{2}\right) + 1$.

Solution

Step 1. Express the function given in the form $y = 4\sec\left(\frac{\pi}{3}x - \frac{\pi}{2}\right) + 1$.

Step 2. The stretching/compressing factor is $|A| = 4$.

Step 3. The period is

$$\begin{aligned}\frac{2\pi}{|B|} &= \frac{2\pi}{\frac{\pi}{3}} \\ &= \frac{2\pi}{1} \cdot \frac{3}{\pi} \\ &= 6\end{aligned}$$

Step 4. The phase shift is

$$\begin{aligned}\frac{C}{B} &= \frac{\frac{\pi}{2}}{\frac{\pi}{3}} \\ &= \frac{\pi}{2} \cdot \frac{3}{\pi} \\ &= 1.5\end{aligned}$$

Step 5. Draw the graph of $y = A\sec(Bx)$, but shift it to the right by $\frac{C}{B} = 1.5$ and up by $D = 6$.

Step 6. Sketch the vertical asymptotes, which occur at $x = 0, x = 3$, and $x = 6$. There is a local minimum at $(1.5, 5)$ and a local maximum at $(4.5, -3)$. Figure 9 shows the graph.

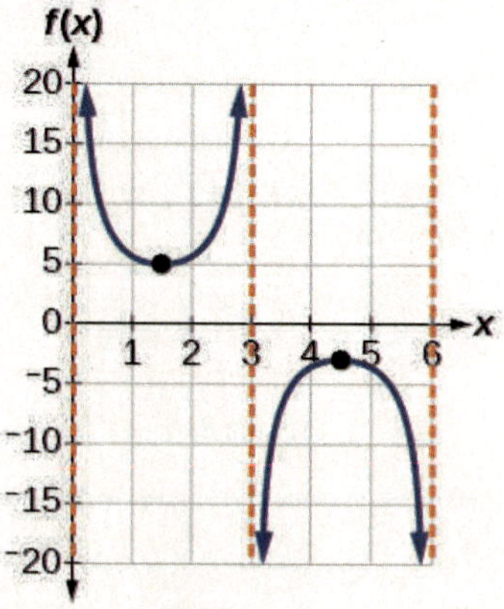

Figure 9

TRY IT #5 Graph one period of $f(x) = -6\sec(4x + 2) - 8$.

Q&A **The domain of $\csc\ x$ was given to be all x such that $x \neq k\pi$ for any integer k. Would the domain of $y = A\csc(Bx - C) + D$ be $x \neq \frac{C+k\pi}{B}$?**

Yes. The excluded points of the domain follow the vertical asymptotes. Their locations show the horizontal shift and compression or expansion implied by the transformation to the original function's input.

HOW TO

Given a function of the form $y = A\csc(Bx)$, graph one period.

1. Express the function given in the form $y = A\csc(Bx)$.
2. $|A|$.
3. Identify B and determine the period, $P = \frac{2\pi}{|B|}$.
4. Draw the graph of $y = A\sin(Bx)$.
5. Use the reciprocal relationship between $y = \sin\ x$ and $y = \csc\ x$ to draw the graph of $y = A\csc(Bx)$.
6. Sketch the asymptotes.
7. Plot any two reference points and draw the graph through these points.

EXAMPLE 6

Graphing a Variation of the Cosecant Function

Graph one period of $f(x) = -3\csc(4x)$.

Solution

Step 1. The given function is already written in the general form, $y = A\csc(Bx)$.
Step 2. $|A| = |-3| = 3$, so the stretching factor is 3.
Step 3. $B = 4$, so $P = \frac{2\pi}{4} = \frac{\pi}{2}$. The period is $\frac{\pi}{2}$ units.
Step 4. Sketch the graph of the function $g(x) = -3\sin(4x)$.
Step 5. Use the reciprocal relationship of the sine and cosecant functions to draw the cosecant function.
Steps 6–7. Sketch three asymptotes at $x = 0$, $x = \frac{\pi}{4}$, and $x = \frac{\pi}{2}$. We can use two reference points, the local maximum at $\left(\frac{\pi}{8}, -3\right)$ and the local minimum at $\left(\frac{3\pi}{8}, 3\right)$. Figure 10 shows the graph.

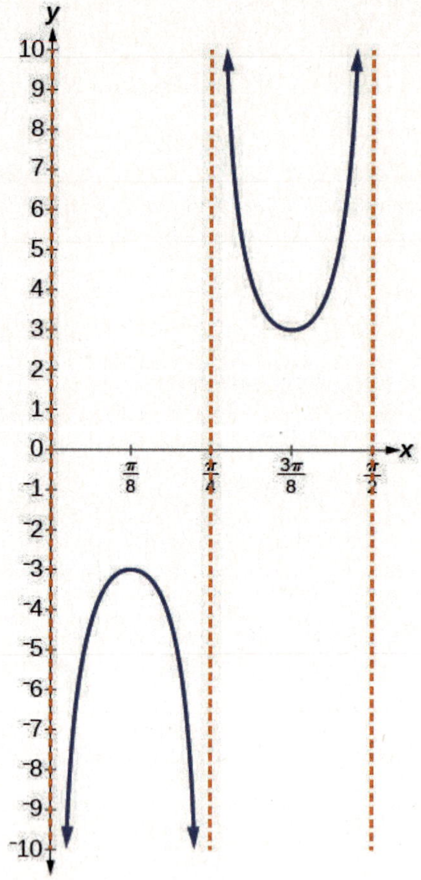

Figure 10

TRY IT #6 Graph one period of $f(x) = 0.5\csc(2x)$.

HOW TO

Given a function of the form $f(x) = A\csc(Bx - C) + D$, graph one period.

1. Express the function given in the form $y = A\csc(Bx - C) + D$.
2. Identify the stretching/compressing factor, $|A|$.
3. Identify B and determine the period, $\frac{2\pi}{|B|}$.
4. Identify C and determine the phase shift, $\frac{C}{B}$.
5. Draw the graph of $y = A\csc(Bx)$ but shift it to the right by $\frac{C}{B}$ and up by D.
6. Sketch the vertical asymptotes, which occur at $x = \frac{C}{B} + \frac{\pi}{|B|}k$, where k is an integer.

EXAMPLE 7

Graphing a Vertically Stretched, Horizontally Compressed, and Vertically Shifted Cosecant

Sketch a graph of $y = 2\csc\left(\frac{\pi}{2}x\right) + 1$. What are the domain and range of this function?

Solution

Step 1. Express the function given in the form $y = 2\csc\left(\frac{\pi}{2}x\right) + 1$.

Step 2. Identify the stretching/compressing factor, $|A| = 2$.

Step 3. The period is $\frac{2\pi}{|B|} = \frac{2\pi}{\frac{\pi}{2}} = \frac{2\pi}{1} \cdot \frac{2}{\pi} = 4.$

Step 4. The phase shift is $\frac{0}{\frac{\pi}{2}} = 0.$

Step 5. Draw the graph of $y = A\csc(Bx)$ but shift it up $D = 1.$

Step 6. Sketch the vertical asymptotes, which occur at $x = 0, x = 2, x = 4.$

The graph for this function is shown in Figure 11.

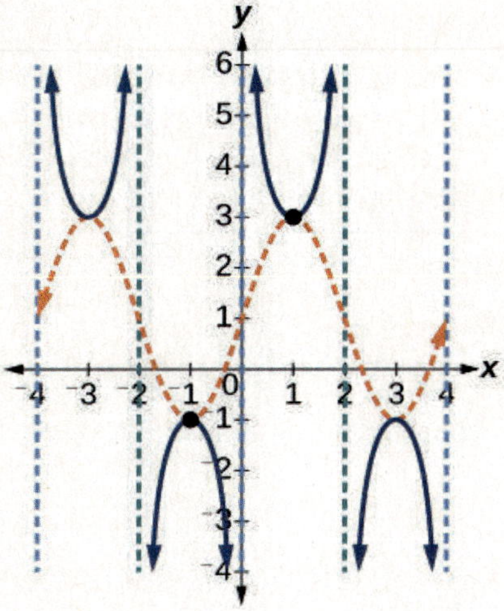

Figure 11 A transformed cosecant function

Analysis

The vertical asymptotes shown on the graph mark off one period of the function, and the local extrema in this interval are shown by dots. Notice how the graph of the transformed cosecant relates to the graph of $f(x) = 2\sin\left(\frac{\pi}{2}x\right) + 1,$ shown as the orange dashed wave.

TRY IT #7 Given the graph of $f(x) = 2\cos\left(\frac{\pi}{2}x\right) + 1$ shown in Figure 12, sketch the graph of $g(x) = 2\sec\left(\frac{\pi}{2}x\right) + 1$ on the same axes.

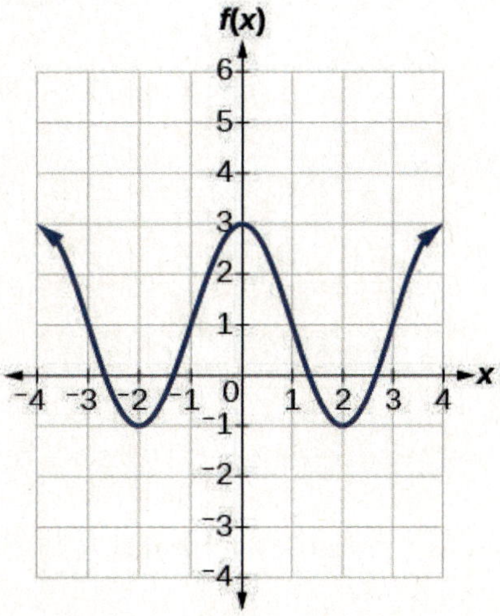

Figure 12

Analyzing the Graph of $y = \cot x$

The last trigonometric function we need to explore is cotangent. The cotangent is defined by the reciprocal identity $\cot\ x = \frac{1}{\tan\ x}$. Notice that the function is undefined when the tangent function is 0, leading to a vertical asymptote in the graph at $0, \pi,$ etc. Since the output of the tangent function is all real numbers, the output of the cotangent function is also all real numbers.

We can graph $y = \cot\ x$ by observing the graph of the tangent function because these two functions are reciprocals of one another. See Figure 13. Where the graph of the tangent function decreases, the graph of the cotangent function increases. Where the graph of the tangent function increases, the graph of the cotangent function decreases.

The cotangent graph has vertical asymptotes at each value of x where $\tan\ x = 0$; we show these in the graph below with dashed lines. Since the cotangent is the reciprocal of the tangent, $\cot\ x$ has vertical asymptotes at all values of x where $\tan\ x = 0$, and $\cot\ x = 0$ at all values of x where $\tan\ x$ has its vertical asymptotes.

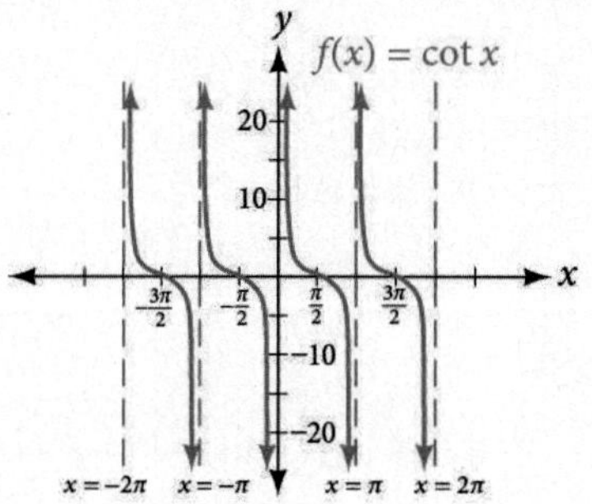

Figure 13 The cotangent function

Features of the Graph of $y = A\cot(Bx)$

- The stretching factor is $|A|$.
- The period is $P = \frac{\pi}{|B|}$.
- The domain is $x \neq \frac{\pi}{|B|}k$, where k is an integer.
- The range is $(-\infty, \infty)$.
- The asymptotes occur at $x = \frac{\pi}{|B|}k$, where k is an integer.
- $y = A\cot(Bx)$ is an odd function.

Graphing Variations of $y = \cot x$

We can transform the graph of the cotangent in much the same way as we did for the tangent. The equation becomes the following.

$$y = A\cot(Bx - C) + D$$

Features of the Graph of $y = A\cot(Bx\text{-}C)+D$

- The stretching factor is $|A|$.
- The period is $\frac{\pi}{|B|}$.
- The domain is $x \neq \frac{C}{B} + \frac{\pi}{|B|}k$, where k is an integer.
- The range is $(-\infty, \infty)$.
- The vertical asymptotes occur at $x = \frac{C}{B} + \frac{\pi}{|B|}k$, where k is an integer.
- There is no amplitude.
- $y = A\cot(Bx)$ is an odd function because it is the quotient of even and odd functions (cosine and sine, respectively)

HOW TO

Given a modified cotangent function of the form $f(x) = A\cot(Bx)$, graph one period.

1. Express the function in the form $f(x) = A\cot(Bx)$.
2. Identify the stretching factor, $|A|$.
3. Identify the period, $P = \frac{\pi}{|B|}$.
4. Draw the graph of $y = A\tan(Bx)$.
5. Plot any two reference points.
6. Use the reciprocal relationship between tangent and cotangent to draw the graph of $y = A\cot(Bx)$.
7. Sketch the asymptotes.

EXAMPLE 8

Graphing Variations of the Cotangent Function

Determine the stretching factor, period, and phase shift of $y = 3\cot(4x)$, and then sketch a graph.

Solution

Step 1. Expressing the function in the form $f(x) = A\cot(Bx)$ gives $f(x) = 3\cot(4x)$.
Step 2. The stretching factor is $|A| = 3$.
Step 3. The period is $P = \frac{\pi}{4}$.
Step 4. Sketch the graph of $y = 3\tan(4x)$.
Step 5. Plot two reference points. Two such points are $\left(\frac{\pi}{16}, 3\right)$ and $\left(\frac{3\pi}{16}, -3\right)$.
Step 6. Use the reciprocal relationship to draw $y = 3\cot(4x)$.
Step 7. Sketch the asymptotes, $x = 0,\ x = \frac{\pi}{4}$.

The blue graph in Figure 14 shows $y = 3\tan(4x)$ and the green graph shows $y = 3\cot(4x)$.

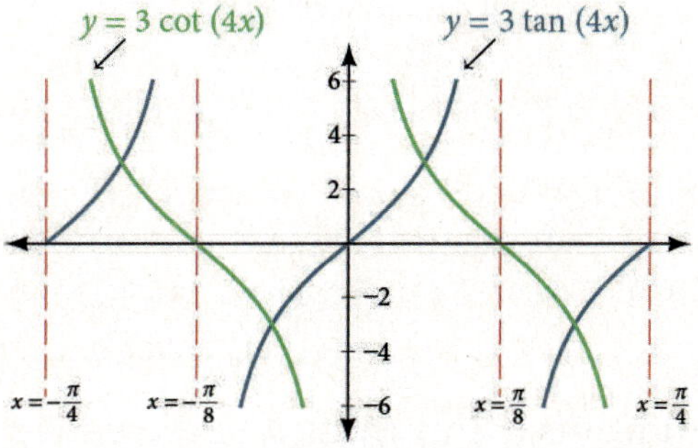

Figure 14

HOW TO

Given a modified cotangent function of the form $f(x) = A\cot(Bx - C) + D$, graph one period.

1. Express the function in the form $f(x) = A\cot(Bx - C) + D$.
2. Identify the stretching factor, $|A|$.
3. Identify the period, $P = \frac{\pi}{|B|}$.
4. Identify the phase shift, $\frac{C}{B}$.
5. Draw the graph of $y = A\tan(Bx)$ shifted to the right by $\frac{C}{B}$ and up by D.
6. Sketch the asymptotes $x = \frac{C}{B} + \frac{\pi}{|B|}k$, where k is an integer.
7. Plot any three reference points and draw the graph through these points.

EXAMPLE 9

Graphing a Modified Cotangent

Sketch a graph of one period of the function $f(x) = 4\cot\left(\frac{\pi}{8}x - \frac{\pi}{2}\right) - 2$.

Solution

Step 1. The function is already written in the general form $f(x) = A\cot(Bx - C) + D$.
Step 2. $A = 4$, so the stretching factor is 4.
Step 3. $B = \frac{\pi}{8}$, so the period is $P = \frac{\pi}{|B|} = \frac{\pi}{\frac{\pi}{8}} = 8$.
Step 4. $C = \frac{\pi}{2}$, so the phase shift is $\frac{C}{B} = \frac{\frac{\pi}{2}}{\frac{\pi}{8}} = 4$.
Step 5. We draw $f(x) = 4\tan\left(\frac{\pi}{8}x - \frac{\pi}{2}\right) - 2$.
Step 6-7. Three points we can use to guide the graph are $(6, 2)$, $(8, -2)$, and $(10, -6)$. We use the reciprocal relationship of tangent and cotangent to draw $f(x) = 4\cot\left(\frac{\pi}{8}x - \frac{\pi}{2}\right) - 2$.
Step 8. The vertical asymptotes are $x = 4$ and $x = 12$.

The graph is shown in Figure 15.

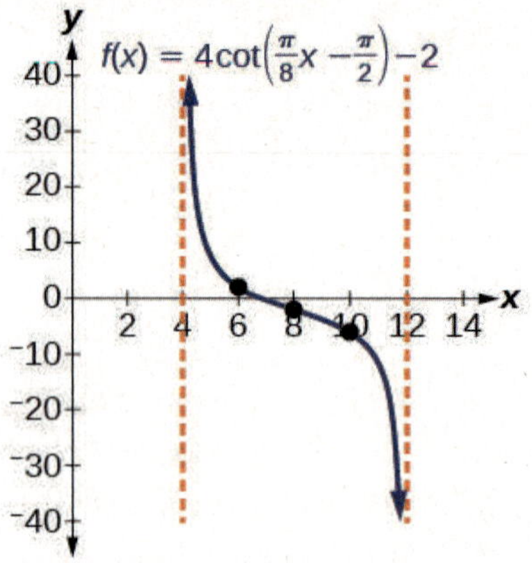

Figure 15 One period of a modified cotangent function

Using the Graphs of Trigonometric Functions to Solve Real-World Problems

Many real-world scenarios represent periodic functions and may be modeled by trigonometric functions. As an example, let's return to the scenario from the section opener. Have you ever observed the beam formed by the rotating light on a fire truck and wondered about the movement of the light beam itself across the wall? The periodic behavior of the distance the light shines as a function of time is obvious, but how do we determine the distance? We can use the tangent function.

EXAMPLE 10

Using Trigonometric Functions to Solve Real-World Scenarios

Suppose the function $y = 5\tan\left(\frac{\pi}{4}t\right)$ marks the distance in the movement of a light beam from the top of a police car across a wall where t is the time in seconds and y is the distance in feet from a point on the wall directly across from the police car.

Ⓐ Find and interpret the stretching factor and period. Ⓑ Graph on the interval $[0, 5]$.
Ⓒ Evaluate $f(1)$ and discuss the function's value at that input.

Solution

ⓐ We know from the general form of $y = A\tan(Bt)$ that $|A|$ is the stretching factor and $\frac{\pi}{B}$ is the period.

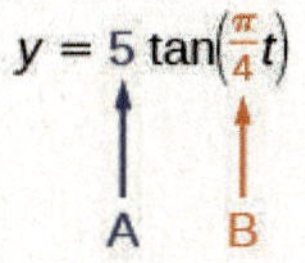

Figure 16

We see that the stretching factor is 5. This means that the beam of light will have moved 5 ft after half the period.

The period is $\frac{\pi}{\frac{\pi}{4}} = \frac{\pi}{1} \cdot \frac{4}{\pi} = 4$. This means that every 4 seconds, the beam of light sweeps the wall. The distance from the spot across from the police car grows larger as the police car approaches.

ⓑ To graph the function, we draw an asymptote at $t = 2$ and use the stretching factor and period. See Figure 17

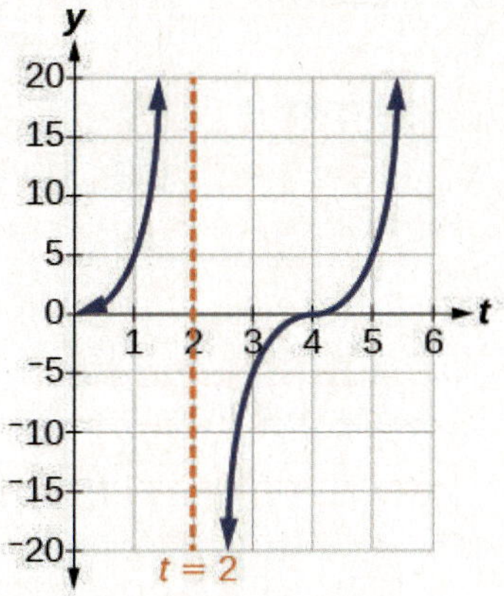

Figure 17

ⓒ period: $f(1) = 5\tan\left(\frac{\pi}{4}(1)\right) = 5(1) = 5$; after 1 second, the beam of has moved 5 ft from the spot across from the police car.

MEDIA

Access these online resources for additional instruction and practice with graphs of other trigonometric functions.

Graphing the Tangent (http://openstax.org/l/graphtangent)
Graphing Cosecant and Secant (http://openstax.org/l/graphcscsec)
Graphing the Cotangent (http://openstax.org/l/graphcot)

8.2 SECTION EXERCISES

Verbal

1. Explain how the graph of the sine function can be used to graph $y = \csc\ x$.

2. How can the graph of $y = \cos\ x$ be used to construct the graph of $y = \sec\ x$?

3. Explain why the period of $\tan\ x$ is equal to π.

4. Why are there no intercepts on the graph of $y = \csc\ x$?

5. How does the period of $y = \csc\ x$ compare with the period of $y = \sin\ x$?

Algebraic

For the following exercises, match each trigonometric function with one of the following graphs.

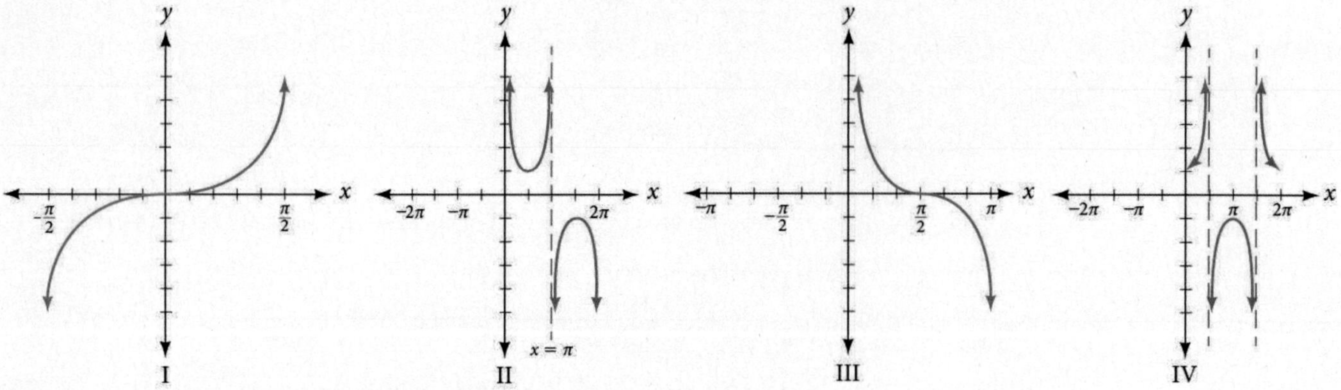

Figure 18

6. $f(x) = \tan\ x$

7. $f(x) = \sec\ x$

8. $f(x) = \csc\ x$

9. $f(x) = \cot\ x$

For the following exercises, find the period and horizontal shift of each of the functions.

10. $f(x) = 2\tan(4x - 32)$

11. $h(x) = 2\sec\left(\frac{\pi}{4}(x + 1)\right)$

12. $m(x) = 6\csc\left(\frac{\pi}{3}x + \pi\right)$

13. If $\tan\ x = -1.5$, find $\tan(-x)$.

14. If $\sec\ x = 2$, find $\sec(-x)$.

15. If $\csc\ x = -5$, find $\csc(-x)$.

16. If $x\sin\ x = 2$, find $(-x)\sin(-x)$.

For the following exercises, rewrite each expression such that the argument x is positive.

17. $\cot(-x)\cos(-x) + \sin(-x)$

18. $\cos(-x) + \tan(-x)\sin(-x)$

Graphical

For the following exercises, sketch two periods of the graph for each of the following functions. Identify the stretching factor, period, and asymptotes.

19. $f(x) = 2\tan(4x - 32)$

20. $h(x) = 2\sec\left(\frac{\pi}{4}(x + 1)\right)$

21. $m(x) = 6\csc\left(\frac{\pi}{3}x + \pi\right)$

22. $j(x) = \tan\left(\frac{\pi}{2}x\right)$

23. $p(x) = \tan\left(x - \frac{\pi}{2}\right)$

24. $f(x) = 4\tan(x)$

25. $f(x) = \tan\left(x + \frac{\pi}{4}\right)$

26. $f(x) = \pi\tan(\pi x - \pi) - \pi$

27. $f(x) = 2\csc(x)$

28. $f(x) = -\frac{1}{4}\csc(x)$

29. $f(x) = 4\sec(3x)$

30. $f(x) = -3\cot(2x)$

31. $f(x) = 7\sec(5x)$

32. $f(x) = \frac{9}{10}\csc(\pi x)$

33. $f(x) = 2\csc\left(x + \frac{\pi}{4}\right) - 1$

34. $f(x) = -\sec\left(x - \frac{\pi}{3}\right) - 2$

35. $f(x) = \frac{7}{5}\csc\left(x - \frac{\pi}{4}\right)$

36. $f(x) = 5\left(\cot\left(x + \frac{\pi}{2}\right) - 3\right)$

For the following exercises, find and graph two periods of the periodic function with the given stretching factor, $|A|$, period, and phase shift.

37. A tangent curve, $A = 1$, period of $\frac{\pi}{3}$; and phase shift $(h,\ k) = \left(\frac{\pi}{4}, 2\right)$

38. A tangent curve, $A = -2$, period of $\frac{\pi}{4}$, and phase shift $(h,\ k) = \left(-\frac{\pi}{4},\ -2\right)$

For the following exercises, find an equation for the graph of each function.

39.

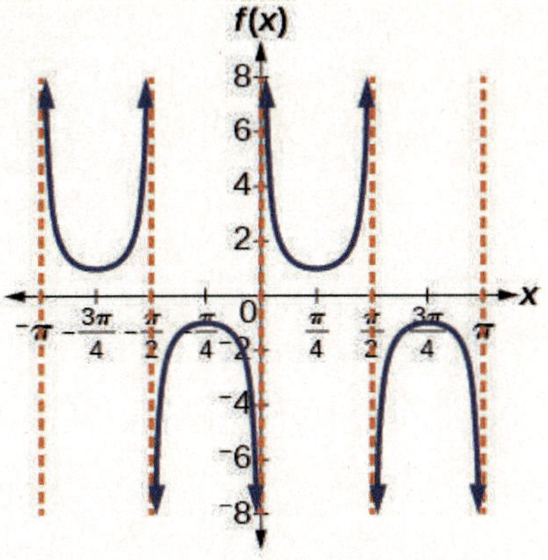

40.

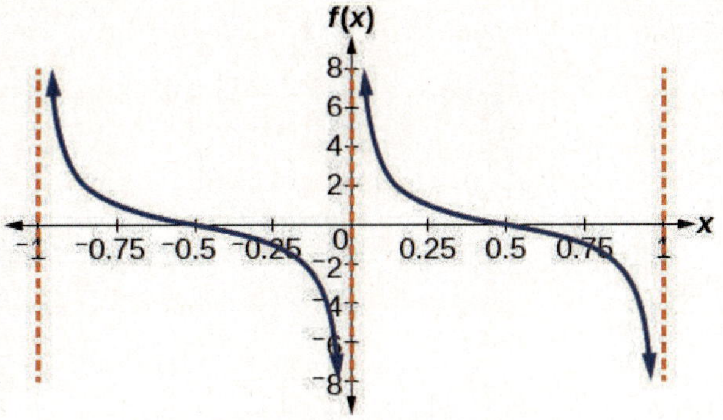

41.

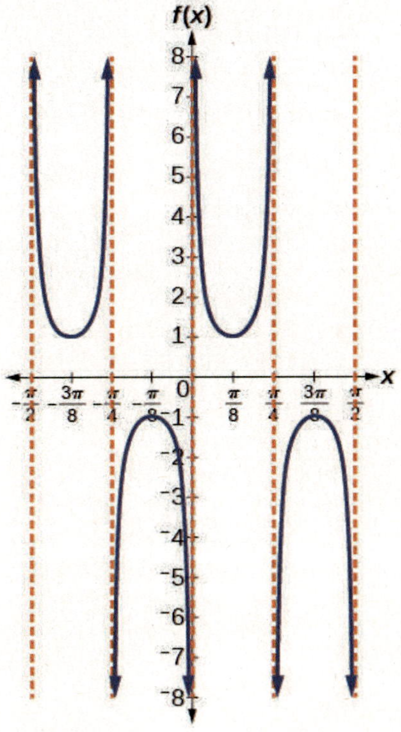

42.

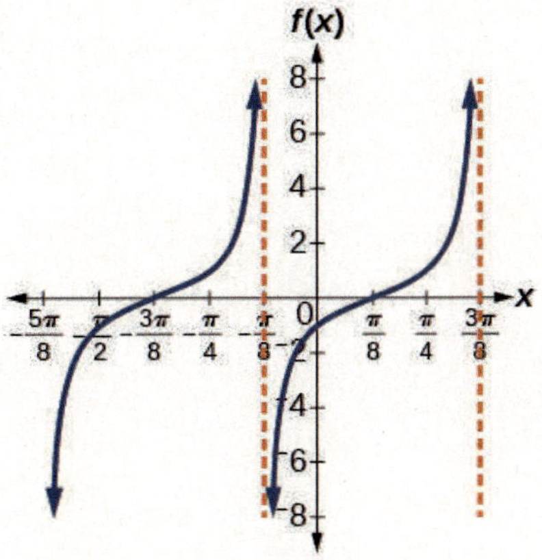

43.

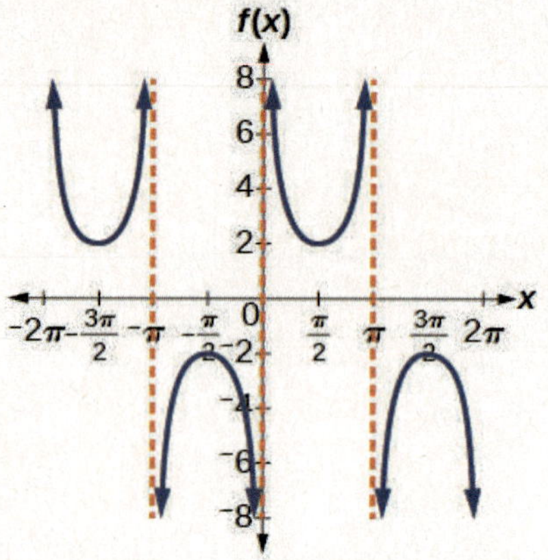

44.

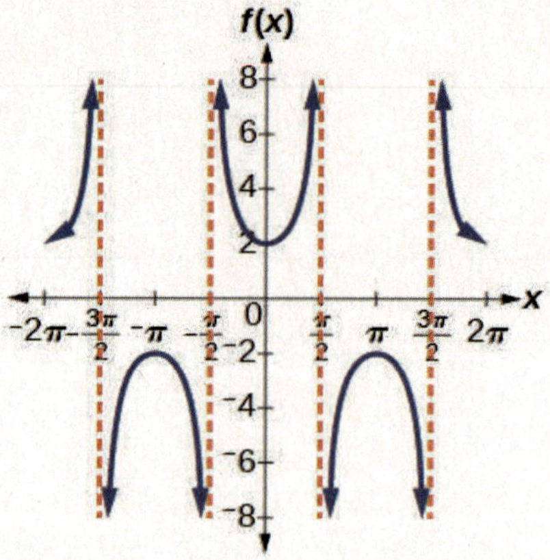

45.

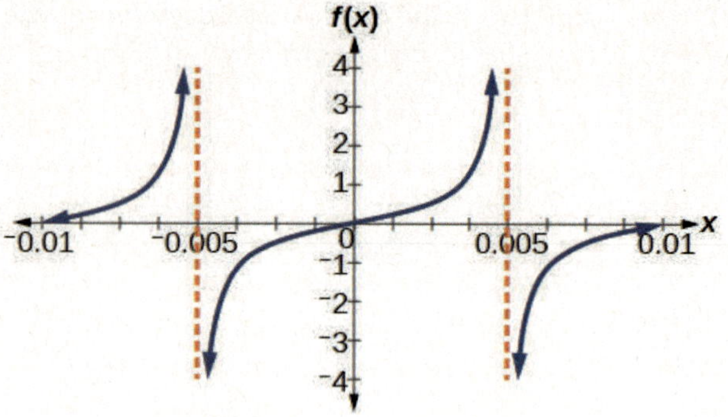

Technology

For the following exercises, use a graphing calculator to graph two periods of the given function. Note: most graphing calculators do not have a cosecant button; therefore, you will need to input $\csc\ x$ *as* $\frac{1}{\sin\ x}$.

46. $f(x) = |\csc(x)|$

47. $f(x) = |\cot(x)|$

48. $f(x) = 2^{\csc(x)}$

49. $f(x) = \frac{\csc(x)}{\sec(x)}$

50. Graph $f(x) = 1 + \sec^2(x) - \tan^2(x)$. What is the function shown in the graph?

51. $f(x) = \sec(0.001x)$

52. $f(x) = \cot(100\pi x)$

53. $f(x) = \sin^2 x + \cos^2 x$

Real-World Applications

54. The function $f(x) = 20\tan\left(\frac{\pi}{10}x\right)$ marks the distance in the movement of a light beam from a police car across a wall for time x, in seconds, and distance $f(x)$, in feet.

ⓐ Graph on the interval $[0,\ 5]$.
ⓑ Find and interpret the stretching factor, period, and asymptote.
ⓒ Evaluate $f(1)$ and $f(2.5)$ and discuss the function's values at those inputs.

55. Standing on the shore of a lake, a fisherman sights a boat far in the distance to his left. Let x, measured in radians, be the angle formed by the line of sight to the ship and a line due north from his position. Assume due north is 0 and x is measured negative to the left and positive to the right. (See Figure 19.) The boat travels from due west to due east and, ignoring the curvature of the Earth, the distance $d(x)$, in kilometers, from the fisherman to the boat is given by the function $d(x) = 1.5\sec(x)$.

ⓐ What is a reasonable domain for $d(x)$?
ⓑ Graph $d(x)$ on this domain.
ⓒ Find and discuss the meaning of any vertical asymptotes on the graph of $d(x)$.
ⓓ Calculate and interpret $d\left(-\frac{\pi}{3}\right)$. Round to the second decimal place.
ⓔ Calculate and interpret $d\left(\frac{\pi}{6}\right)$. Round to the second decimal place.
ⓕ What is the minimum distance between the fisherman and the boat? When does this occur?

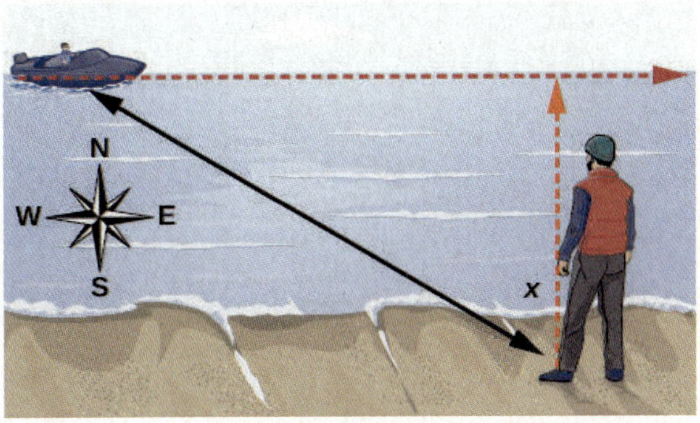

Figure 19

56. A laser rangefinder is locked on a comet approaching Earth. The distance $g(x)$, in kilometers, of the comet after x days, for x in the interval 0 to 30 days, is given by $g(x) = 250{,}000\csc\left(\frac{\pi}{30}x\right)$.

ⓐ Graph $g(x)$ on the interval $[0,\ 30]$.
ⓑ Evaluate $g(5)$ and interpret the information.
ⓒ What is the minimum distance between the comet and Earth? When does this occur? To which constant in the equation does this correspond?
ⓓ Find and discuss the meaning of any vertical asymptotes.

57. A video camera is focused on a rocket on a launching pad 2 miles from the camera. The angle of elevation from the ground to the rocket after x seconds is $\frac{\pi}{120}x$.

ⓐ Write a function expressing the altitude $h(x)$, in miles, of the rocket above the ground after x seconds. Ignore the curvature of the Earth.
ⓑ Graph $h(x)$ on the interval $(0, \ 60)$.
ⓒ Evaluate and interpret the values $h(0)$ and $h(30)$.
ⓓ What happens to the values of $h(x)$ as x approaches 60 seconds? Interpret the meaning of this in terms of the problem.

8.3 Inverse Trigonometric Functions

Learning Objectives

In this section, you will:

- Understand and use the inverse sine, cosine, and tangent functions.
- Find the exact value of expressions involving the inverse sine, cosine, and tangent functions.
- Use a calculator to evaluate inverse trigonometric functions.
- Find exact values of composite functions with inverse trigonometric functions.

For any right triangle, given one other angle and the length of one side, we can figure out what the other angles and sides are. But what if we are given only two sides of a right triangle? We need a procedure that leads us from a ratio of sides to an angle. This is where the notion of an inverse to a trigonometric function comes into play. In this section, we will explore the inverse trigonometric functions.

Understanding and Using the Inverse Sine, Cosine, and Tangent Functions

In order to use inverse trigonometric functions, we need to understand that an inverse trigonometric function "undoes" what the original trigonometric function "does," as is the case with any other function and its inverse. In other words, the domain of the inverse function is the range of the original function, and vice versa, as summarized in Figure 1.

Trig Functions	Inverse Trig Functions
Domain: Measure of an angle	Domain: Ratio
Range: Ratio	Range: Measure of an angle

Figure 1

For example, if $f(x) = \sin\ x$, then we would write $f^{-1}(x) = \sin^{-1}x$. Be aware that $\sin^{-1}x$ does not mean $\frac{1}{\sin x}$. The following examples illustrate the inverse trigonometric functions:

- Since $\sin\left(\frac{\pi}{6}\right) = \frac{1}{2}$, then $\frac{\pi}{6} = \sin^{-1}\left(\frac{1}{2}\right)$.
- Since $\cos(\pi) = -1$, then $\pi = \cos^{-1}(-1)$.
- Since $\tan\left(\frac{\pi}{4}\right) = 1$, then $\frac{\pi}{4} = \tan^{-1}(1)$.

In previous sections, we evaluated the trigonometric functions at various angles, but at times we need to know what angle would yield a specific sine, cosine, or tangent value. For this, we need inverse functions. Recall that, for a one-to-one function, if $f(a) = b$, then an inverse function would satisfy $f^{-1}(b) = a$.

Bear in mind that the sine, cosine, and tangent functions are not one-to-one functions. The graph of each function would fail the horizontal line test. In fact, no periodic function can be one-to-one because each output in its range corresponds to at least one input in every period, and there are an infinite number of periods. As with other functions that are not one-to-one, we will need to restrict the domain of each function to yield a new function that is one-to-one. We choose a domain for each function that includes the number 0. Figure 2 shows the graph of the sine function limited to $\left[-\frac{\pi}{2}, \frac{\pi}{2}\right]$ and the graph of the cosine function limited to $[0, \pi]$.

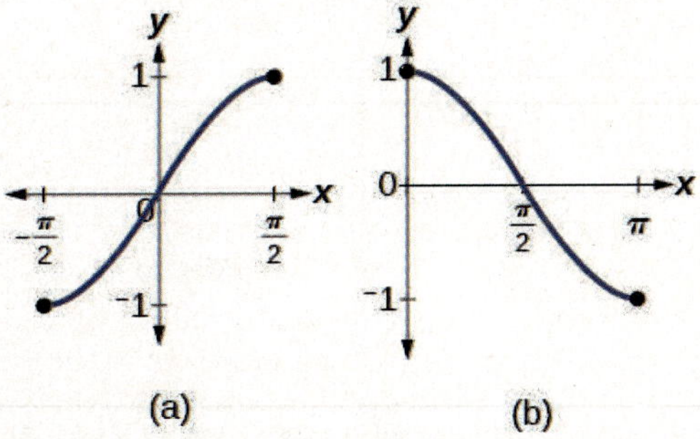

Figure 2 (a) Sine function on a restricted domain of $\left[-\frac{\pi}{2}, \frac{\pi}{2}\right]$; (b) Cosine function on a restricted domain of $[0, \pi]$

Figure 3 shows the graph of the tangent function limited to $\left(-\frac{\pi}{2}, \frac{\pi}{2}\right)$.

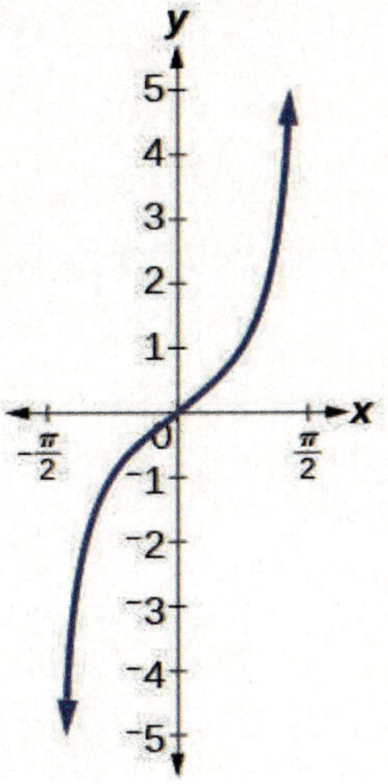

Figure 3 Tangent function on a restricted domain of $\left(-\frac{\pi}{2}, \frac{\pi}{2}\right)$

These conventional choices for the restricted domain are somewhat arbitrary, but they have important, helpful characteristics. Each domain includes the origin and some positive values, and most importantly, each results in a one-to-one function that is invertible. The conventional choice for the restricted domain of the tangent function also has the useful property that it extends from one vertical asymptote to the next instead of being divided into two parts by an asymptote.

On these restricted domains, we can define the inverse trigonometric functions.

- The **inverse sine function** $y = \sin^{-1} x$ means $x = \sin\ y$. The inverse sine function is sometimes called the **arcsine** function, and notated $\arcsin x$.
$$y = \sin^{-1} x \text{ has domain } [-1, 1] \text{ and range } \left[-\frac{\pi}{2}, \frac{\pi}{2}\right]$$
- The **inverse cosine function** $y = \cos^{-1} x$ means $x = \cos\ y$. The inverse cosine function is sometimes called the **arccosine** function, and notated $\arccos\ x$.
$$y = \cos^{-1} x \text{ has domain } [-1, 1] \text{ and range } [0, \pi]$$
- The **inverse tangent function** $y = \tan^{-1} x$ means $x = \tan\ y$. The inverse tangent function is sometimes called the **arctangent** function, and notated $\arctan\ x$.
$$y = \tan^{-1} x \text{ has domain } (-\infty, \infty) \text{ and range } \left(-\frac{\pi}{2}, \frac{\pi}{2}\right)$$

The graphs of the inverse functions are shown in Figure 4, Figure 5, and Figure 6. Notice that the output of each of these inverse functions is a *number,* an angle in radian measure. We see that $\sin^{-1} x$ has domain $[-1, 1]$ and range $\left[-\frac{\pi}{2}, \frac{\pi}{2}\right]$, $\cos^{-1} x$ has domain $[-1,1]$ and range $[0, \pi]$, and $\tan^{-1} x$ has domain of all real numbers and range $\left(-\frac{\pi}{2}, \frac{\pi}{2}\right)$. To find the domain and range of inverse trigonometric functions, switch the domain and range of the original functions. Each graph of the inverse trigonometric function is a reflection of the graph of the original function about the line $y = x$.

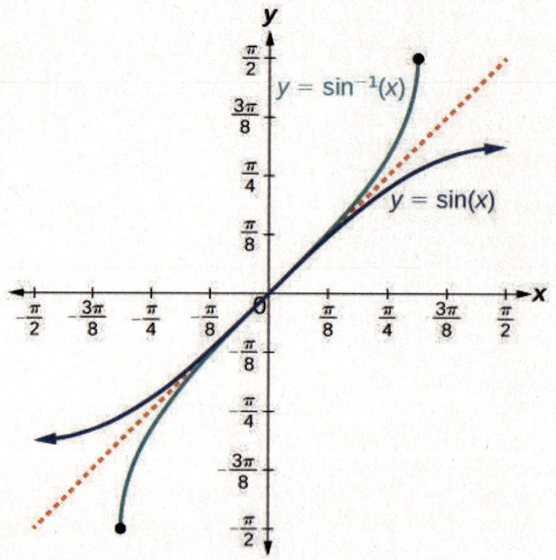

Figure 4 The sine function and inverse sine (or arcsine) function

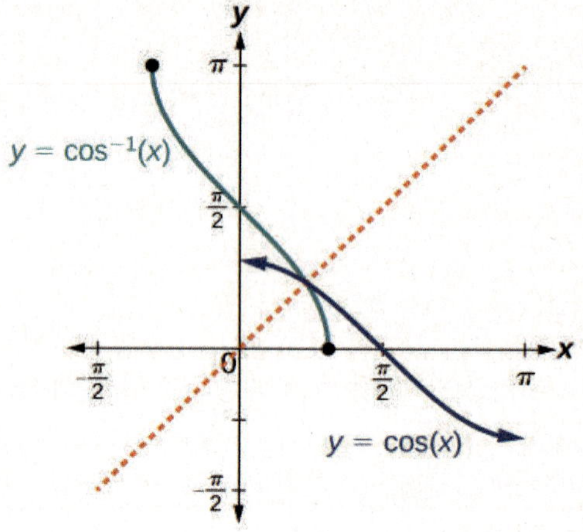

Figure 5 The cosine function and inverse cosine (or arccosine) function

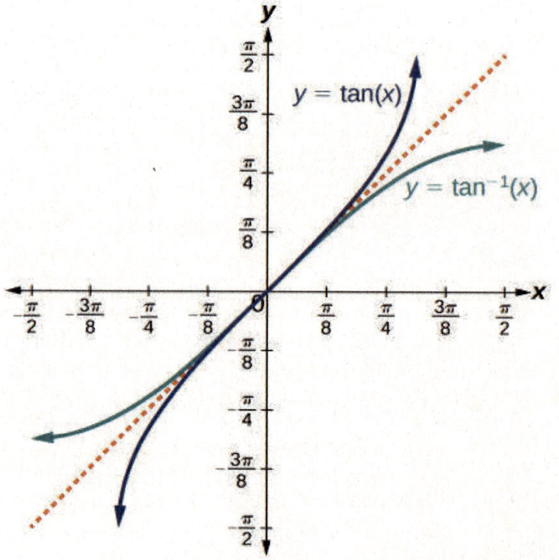

Figure 6 The tangent function and inverse tangent (or arctangent) function

Relations for Inverse Sine, Cosine, and Tangent Functions

For angles in the interval $\left[-\frac{\pi}{2}, \frac{\pi}{2}\right]$, if $\sin\ y = x$, then $\sin^{-1} x = y$.

For angles in the interval $[0, \pi]$, if $\cos\ y = x$, then $\cos^{-1} x = y$.

For angles in the interval $\left(-\frac{\pi}{2}, \frac{\pi}{2}\right)$, if $\tan\ y = x$, then $\tan^{-1} x = y$.

EXAMPLE 1

Writing a Relation for an Inverse Function

Given $\sin\left(\frac{5\pi}{12}\right) \approx 0.96593$, write a relation involving the inverse sine.

Solution

Use the relation for the inverse sine. If $\sin\ y = x$, then $\sin^{-1} x = y$.

In this problem, $x = 0.96593$, and $y = \frac{5\pi}{12}$.

$$\sin^{-1}(0.96593) \approx \frac{5\pi}{12}$$

TRY IT #1 Given $\cos(0.5) \approx 0.8776$, write a relation involving the inverse cosine.

Finding the Exact Value of Expressions Involving the Inverse Sine, Cosine, and Tangent Functions

Now that we can identify inverse functions, we will learn to evaluate them. For most values in their domains, we must evaluate the inverse trigonometric functions by using a calculator, interpolating from a table, or using some other numerical technique. Just as we did with the original trigonometric functions, we can give exact values for the inverse functions when we are using the special angles, specifically $\frac{\pi}{6}$ (30°), $\frac{\pi}{4}$ (45°), and $\frac{\pi}{3}$ (60°), and their reflections into other quadrants.

HOW TO

Given a "special" input value, evaluate an inverse trigonometric function.

1. Find angle x for which the original trigonometric function has an output equal to the given input for the inverse trigonometric function.
2. If x is not in the defined range of the inverse, find another angle y that is in the defined range and has the same sine, cosine, or tangent as x, depending on which corresponds to the given inverse function.

EXAMPLE 2

Evaluating Inverse Trigonometric Functions for Special Input Values

Evaluate each of the following.

ⓐ $\sin^{-1}\left(\frac{1}{2}\right)$ ⓑ $\sin^{-1}\left(-\frac{\sqrt{2}}{2}\right)$ ⓒ $\cos^{-1}\left(-\frac{\sqrt{3}}{2}\right)$ ⓓ $\tan^{-1}(1)$

Solution

ⓐ Evaluating $\sin^{-1}\left(\frac{1}{2}\right)$ is the same as determining the angle that would have a sine value of $\frac{1}{2}$. In other words, what angle x would satisfy $\sin(x) = \frac{1}{2}$? There are multiple values that would satisfy this relationship, such as $\frac{\pi}{6}$ and $\frac{5\pi}{6}$, but we know we need the angle in the interval $\left[-\frac{\pi}{2}, \frac{\pi}{2}\right]$, so the answer will be $\sin^{-1}\left(\frac{1}{2}\right) = \frac{\pi}{6}$. Remember that the inverse is a function, so for each input, we will get exactly one output.

ⓑ To evaluate $\sin^{-1}\left(-\frac{\sqrt{2}}{2}\right)$, we know that $\frac{5\pi}{4}$ and $\frac{7\pi}{4}$ both have a sine value of $-\frac{\sqrt{2}}{2}$, but neither is in the interval $\left[-\frac{\pi}{2}, \frac{\pi}{2}\right]$. For that, we need the negative angle coterminal with $\frac{7\pi}{4}$: $\sin^{-1}(-\frac{\sqrt{2}}{2}) = -\frac{\pi}{4}$.

ⓒ To evaluate $\cos^{-1}\left(-\frac{\sqrt{3}}{2}\right)$, we are looking for an angle in the interval $[0, \pi]$ with a cosine value of $-\frac{\sqrt{3}}{2}$. The angle that satisfies this is $\cos^{-1}\left(-\frac{\sqrt{3}}{2}\right) = \frac{5\pi}{6}$.

ⓓ Evaluating $\tan^{-1}(1)$, we are looking for an angle in the interval $\left(-\frac{\pi}{2}, \frac{\pi}{2}\right)$ with a tangent value of 1. The correct angle is $\tan^{-1}(1) = \frac{\pi}{4}$.

TRY IT #2 Evaluate each of the following.

ⓐ $\sin^{-1}(-1)$ ⓑ $\tan^{-1}(-1)$ ⓒ $\cos^{-1}(-1)$ ⓓ $\cos^{-1}\left(\frac{1}{2}\right)$

Using a Calculator to Evaluate Inverse Trigonometric Functions

To evaluate inverse trigonometric functions that do not involve the special angles discussed previously, we will need to use a calculator or other type of technology. Most scientific calculators and calculator-emulating applications have specific keys or buttons for the inverse sine, cosine, and tangent functions. These may be labeled, for example, **SIN** $^{-1}$, **ARCSIN**, or **ASIN**.

In the previous chapter, we worked with trigonometry on a right triangle to solve for the sides of a triangle given one side and an additional angle. Using the inverse trigonometric functions, we can solve for the angles of a right triangle given two sides, and we can use a calculator to find the values to several decimal places.

In these examples and exercises, the answers will be interpreted as angles and we will use θ as the independent variable. The value displayed on the calculator may be in degrees or radians, so be sure to set the mode appropriate to the application.

EXAMPLE 3

Evaluating the Inverse Sine on a Calculator

Evaluate $\sin^{-1}(0.97)$ using a calculator.

Solution

Because the output of the inverse function is an angle, the calculator will give us a degree value if in degree mode and a radian value if in radian mode. Calculators also use the same domain restrictions on the angles as we are using.

In radian mode, $\sin^{-1}(0.97) \approx 1.3252$. In degree mode, $\sin^{-1}(0.97) \approx 75.93°$. Note that in calculus and beyond we will use radians in almost all cases.

TRY IT #3 Evaluate $\cos^{-1}(-0.4)$ using a calculator.

HOW TO

Given two sides of a right triangle like the one shown in Figure 7, find an angle.

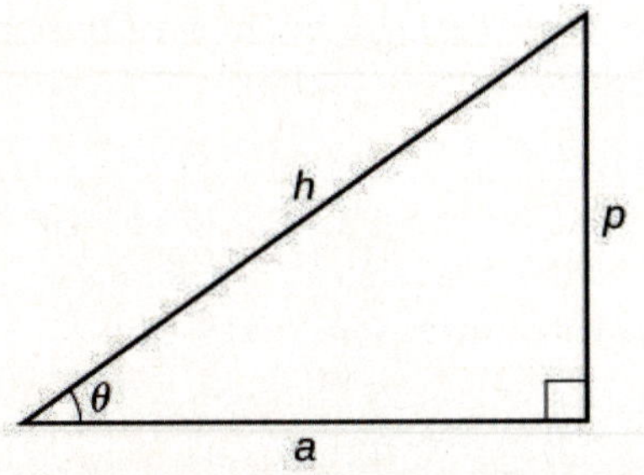

Figure 7

1. If one given side is the hypotenuse of length h and the side of length a adjacent to the desired angle is given, use the equation $\theta = \cos^{-1}\left(\frac{a}{h}\right)$.
2. If one given side is the hypotenuse of length h and the side of length p opposite to the desired angle is given, use the equation $\theta = \sin^{-1}\left(\frac{p}{h}\right)$.
3. If the two legs (the sides adjacent to the right angle) are given, then use the equation $\theta = \tan^{-1}\left(\frac{p}{a}\right)$.

EXAMPLE 4

Applying the Inverse Cosine to a Right Triangle

Solve the triangle in Figure 8 for the angle θ.

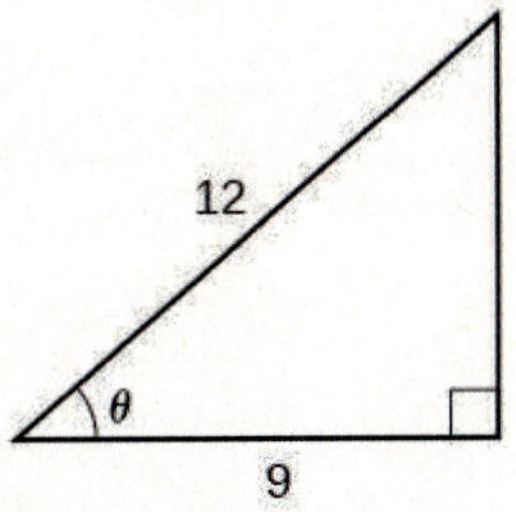

Figure 8

Solution

Because we know the hypotenuse and the side adjacent to the angle, it makes sense for us to use the cosine function.

$\cos\ \theta = \frac{9}{12}$	
$\theta = \cos^{-1}(\frac{9}{12})$	Apply definition of the inverse.
$\theta \approx 0.7227$ or about $41.4096°$	Evaluate.

TRY IT #4 Solve the triangle in Figure 9 for the angle θ.

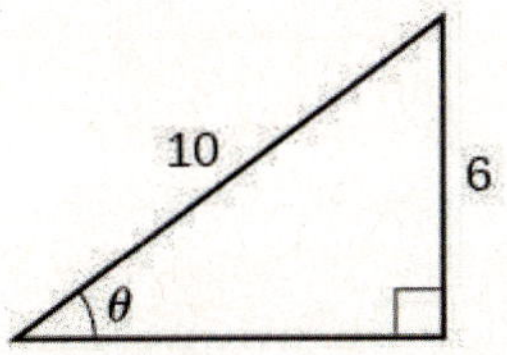

Figure 9

Finding Exact Values of Composite Functions with Inverse Trigonometric Functions

There are times when we need to compose a trigonometric function with an inverse trigonometric function. In these cases, we can usually find exact values for the resulting expressions without resorting to a calculator. Even when the input to the composite function is a variable or an expression, we can often find an expression for the output. To help sort out different cases, let $f(x)$ and $g(x)$ be two different trigonometric functions belonging to the set $\{\sin(x), \cos(x), \tan(x)\}$ and let $f^{-1}(y)$ and $g^{-1}(y)$ be their inverses.

Evaluating Compositions of the Form $f(f^{-1}(y))$ and $f^{-1}(f(x))$

For any trigonometric function, $f\left(f^{-1}(y)\right) = y$ for all y in the proper domain for the given function. This follows from the definition of the inverse and from the fact that the range of f was defined to be identical to the domain of f^{-1}. However, we have to be a little more careful with expressions of the form $f^{-1}(f(x))$.

Compositions of a trigonometric function and its inverse

$$\begin{aligned} \sin(\sin^{-1}x) &= x \text{ for } -1 \leq x \leq 1 \\ \cos(\cos^{-1}x) &= x \text{ for } -1 \leq x \leq 1 \\ \tan(\tan^{-1}x) &= x \text{ for } -\infty < x < \infty \end{aligned}$$

$$\begin{aligned} \sin^{-1}(\sin\ x) &= x \text{ only for } -\tfrac{\pi}{2} \leq x \leq \tfrac{\pi}{2} \\ \cos^{-1}(\cos\ x) &= x \text{ only for } 0 \leq x \leq \pi \\ \tan^{-1}(\tan\ x) &= x \text{ only for } -\tfrac{\pi}{2} < x < \tfrac{\pi}{2} \end{aligned}$$

Q&A **Is it correct that** $\sin^{-1}(\sin\ x) = x$?

No. This equation is correct if x belongs to the restricted domain $\left[-\frac{\pi}{2}, \frac{\pi}{2}\right]$, *but sine is defined for all real input values, and for x outside the restricted interval, the equation is not correct because its inverse always returns a value in* $\left[-\frac{\pi}{2}, \frac{\pi}{2}\right]$. *The situation is similar for cosine and tangent and their inverses. For example,* $\sin^{-1}\left(\sin\left(\frac{3\pi}{4}\right)\right) = \frac{\pi}{4}$.

HOW TO

Given an expression of the form $f^{-1}(f(\theta))$ where $f(\theta) = \sin\ \theta,\ \cos\ \theta,$ or $\tan\ \theta$, evaluate.

1. If θ is in the restricted domain of f, then $f^{-1}(f(\theta)) = \theta$.
2. If not, then find an angle ϕ within the restricted domain of f such that $f(\phi) = f(\theta)$. Then $f^{-1}(f(\theta)) = \phi$.

EXAMPLE 5

Using Inverse Trigonometric Functions

Evaluate the following:

ⓐ $\sin^{-1}\left(\sin\left(\frac{\pi}{3}\right)\right)$ ⓑ $\sin^{-1}\left(\sin\left(\frac{2\pi}{3}\right)\right)$ ⓒ $\cos^{-1}\left(\cos\left(\frac{2\pi}{3}\right)\right)$ ⓓ $\cos^{-1}\left(\cos\left(-\frac{\pi}{3}\right)\right)$

Solution

ⓐ $\frac{\pi}{3}$ is in $\left[-\frac{\pi}{2}, \frac{\pi}{2}\right]$, so $\sin^{-1}\left(\sin\left(\frac{\pi}{3}\right)\right) = \frac{\pi}{3}$.

ⓑ $\frac{2\pi}{3}$ is not in $\left[-\frac{\pi}{2}, \frac{\pi}{2}\right]$, but $\sin\left(\frac{2\pi}{3}\right) = \sin\left(\frac{\pi}{3}\right)$, so $\sin^{-1}\left(\sin\left(\frac{2\pi}{3}\right)\right) = \frac{\pi}{3}$.

ⓒ $\frac{2\pi}{3}$ is in $[0, \pi]$, so $\cos^{-1}\left(\cos\left(\frac{2\pi}{3}\right)\right) = \frac{2\pi}{3}$.

ⓓ $-\frac{\pi}{3}$ is not in $[0, \pi]$, but $\cos\left(-\frac{\pi}{3}\right) = \cos\left(\frac{\pi}{3}\right)$ because cosine is an even function. $\frac{\pi}{3}$ is in $[0, \pi]$, so $\cos^{-1}\left(\cos\left(-\frac{\pi}{3}\right)\right) = \frac{\pi}{3}$.

TRY IT #5 Evaluate $\tan^{-1}\left(\tan\left(\frac{\pi}{8}\right)\right)$ and $\tan^{-1}\left(\tan\left(\frac{11\pi}{9}\right)\right)$.

Evaluating Compositions of the Form $f^{-1}(g(x))$

Now that we can compose a trigonometric function with its inverse, we can explore how to evaluate a composition of a trigonometric function and the inverse of another trigonometric function. We will begin with compositions of the form $f^{-1}(g(x))$. For special values of x, we can exactly evaluate the inner function and then the outer, inverse function. However, we can find a more general approach by considering the relation between the two acute angles of a right triangle where one is θ, making the other $\frac{\pi}{2} - \theta$. Consider the sine and cosine of each angle of the right triangle in Figure 10.

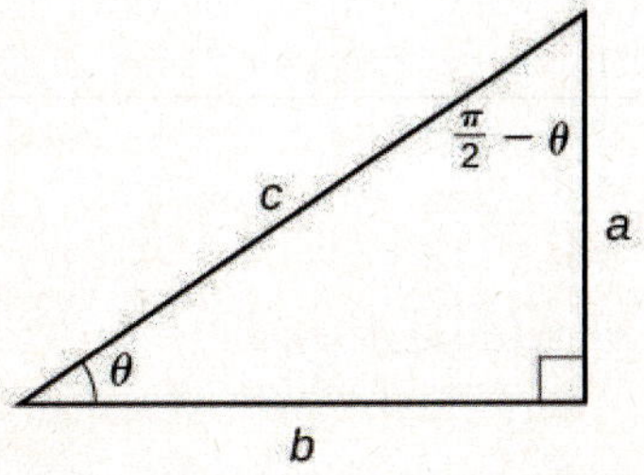

Figure 10 Right triangle illustrating the cofunction relationships

Because $\cos\ \theta = \frac{b}{c} = \sin\left(\frac{\pi}{2} - \theta\right)$, we have $\sin^{-1}(\cos\ \theta) = \frac{\pi}{2} - \theta$ if $0 \le \theta \le \pi$. If θ is not in this domain, then we need to find another angle that has the same cosine as θ and does belong to the restricted domain; we then subtract this angle from $\frac{\pi}{2}$. Similarly, $\sin\ \theta = \frac{a}{c} = \cos\left(\frac{\pi}{2} - \theta\right)$, so $\cos^{-1}(\sin\ \theta) = \frac{\pi}{2} - \theta$ if $-\frac{\pi}{2} \le \theta \le \frac{\pi}{2}$. These are just the function-cofunction relationships presented in another way.

HOW TO

Given functions of the form $\sin^{-1}(\cos\ x)$ and $\cos^{-1}(\sin\ x)$, evaluate them.

1. If x is in $[0, \pi]$, then $\sin^{-1}(\cos\ x) = \frac{\pi}{2} - x$.
2. If x is not in $[0, \pi]$, then find another angle y in $[0, \pi]$ such that $\cos\ y = \cos\ x$.
$$\sin^{-1}(\cos\ x) = \frac{\pi}{2} - y$$
3. If x is in $\left[-\frac{\pi}{2}, \frac{\pi}{2}\right]$, then $\cos^{-1}(\sin\ x) = \frac{\pi}{2} - x$.
4. If x is not in $\left[-\frac{\pi}{2}, \frac{\pi}{2}\right]$, then find another angle y in $\left[-\frac{\pi}{2}, \frac{\pi}{2}\right]$ such that $\sin\ y = \sin\ x$.
$$\cos^{-1}(\sin\ x) = \frac{\pi}{2} - y$$

EXAMPLE 6

Evaluating the Composition of an Inverse Sine with a Cosine

Evaluate $\sin^{-1}\left(\cos\left(\frac{13\pi}{6}\right)\right)$

ⓐ by direct evaluation. ⓑ by the method described previously.

Solution

ⓐ Here, we can directly evaluate the inside of the composition.

$$\begin{aligned}\cos\left(\frac{13\pi}{6}\right) &= \cos\left(\frac{\pi}{6} + 2\pi\right)\\ &= \cos\left(\frac{\pi}{6}\right)\\ &= \frac{\sqrt{3}}{2}\end{aligned}$$

Now, we can evaluate the inverse function as we did earlier.

$$\sin^{-1}\left(\frac{\sqrt{3}}{2}\right) = \frac{\pi}{3}$$

ⓑ We have $x = \frac{13\pi}{6}$, $y = \frac{\pi}{6}$, and

$$\begin{aligned}\sin^{-1}\left(\cos\left(\frac{13\pi}{6}\right)\right) &= \frac{\pi}{2} - \frac{\pi}{6}\\ &= \frac{\pi}{3}\end{aligned}$$

> TRY IT #6 Evaluate $\cos^{-1}\left(\sin\left(-\frac{11\pi}{4}\right)\right)$.

Evaluating Compositions of the Form $f(g^{-1}(x))$

To evaluate compositions of the form $f\left(g^{-1}(x)\right)$, where f and g are any two of the functions sine, cosine, or tangent and x is any input in the domain of g^{-1}, we have exact formulas, such as $\sin\left(\cos^{-1}x\right) = \sqrt{1-x^2}$. When we need to use them, we can derive these formulas by using the trigonometric relations between the angles and sides of a right triangle, together with the use of Pythagoras's relation between the lengths of the sides. We can use the Pythagorean identity, $\sin^2 x + \cos^2 x = 1$, to solve for one when given the other. We can also use the inverse trigonometric functions to find compositions involving algebraic expressions.

EXAMPLE 7

Evaluating the Composition of a Sine with an Inverse Cosine

Find an exact value for $\sin\left(\cos^{-1}\left(\frac{4}{5}\right)\right)$.

Solution

Beginning with the inside, we can say there is some angle such that $\theta = \cos^{-1}\left(\frac{4}{5}\right)$, which means $\cos\ \theta = \frac{4}{5}$, and we are looking for $\sin\ \theta$. We can use the Pythagorean identity to do this.

$$\sin^2\theta + \cos^2\theta = 1 \qquad \text{Use our known value for cosine.}$$
$$\sin^2\theta + \left(\frac{4}{5}\right)^2 = 1 \qquad \text{Solve for sine.}$$
$$\sin^2\theta = 1 - \frac{16}{25}$$
$$\sin\ \theta = \pm\sqrt{\frac{9}{25}} = \pm\frac{3}{5}$$

Since $\theta = \cos^{-1}\left(\frac{4}{5}\right)$ is in quadrant I, $\sin\ \theta$ must be positive, so the solution is $\frac{3}{5}$. See Figure 11.

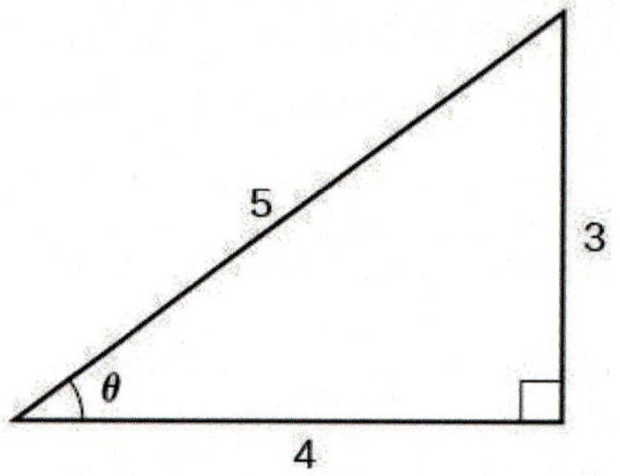

Figure 11 Right triangle illustrating that if $\cos\ \theta = \frac{4}{5}$, then $\sin\ \theta = \frac{3}{5}$

We know that the inverse cosine always gives an angle on the interval $[0, \pi]$, so we know that the sine of that angle must be positive; therefore $\sin\left(\cos^{-1}\left(\frac{4}{5}\right)\right) = \sin\ \theta = \frac{3}{5}$.

> TRY IT #7 Evaluate $\cos\left(\tan^{-1}\left(\frac{5}{12}\right)\right)$.

EXAMPLE 8

Evaluating the Composition of a Sine with an Inverse Tangent

Find an exact value for $\sin\left(\tan^{-1}\left(\frac{7}{4}\right)\right)$.

Solution

While we could use a similar technique as in Example 6, we will demonstrate a different technique here. From the inside, we know there is an angle such that $\tan\ \theta = \frac{7}{4}$. We can envision this as the opposite and adjacent sides on a right triangle, as shown in Figure 12.

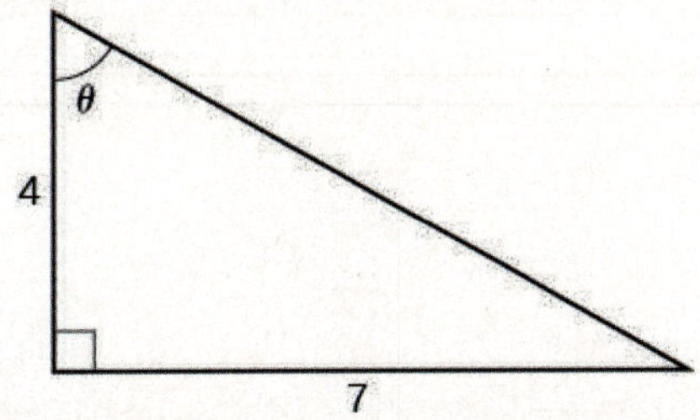

Figure 12 A right triangle with two sides known

Using the Pythagorean Theorem, we can find the hypotenuse of this triangle.

$$4^2 + 7^2 = \text{hypotenuse}^2$$
$$\text{hypotenuse} = \sqrt{65}$$

Now, we can evaluate the sine of the angle as the opposite side divided by the hypotenuse.

$$\sin\ \theta = \frac{7}{\sqrt{65}}$$

This gives us our desired composition.

$$\begin{aligned}\sin\left(\tan^{-1}\left(\tfrac{7}{4}\right)\right) &= \sin\ \theta \\ &= \frac{7}{\sqrt{65}} \\ &= \frac{7\sqrt{65}}{65}\end{aligned}$$

> **TRY IT** #8 Evaluate $\cos\left(\sin^{-1}\left(\frac{7}{9}\right)\right)$.

EXAMPLE 9

Finding the Cosine of the Inverse Sine of an Algebraic Expression

Find a simplified expression for $\cos\left(\sin^{-1}\left(\frac{x}{3}\right)\right)$ for $-3 \le x \le 3$.

Solution

We know there is an angle θ such that $\sin\ \theta = \frac{x}{3}$.

$\sin^2\theta + \cos^2\theta = 1$	Use the Pythagorean Theorem.
$\left(\frac{x}{3}\right)^2 + \cos^2\theta = 1$	Solve for cosine.
$\cos^2\theta = 1 - \frac{x^2}{9}$	
$\cos\theta = \pm\sqrt{\frac{9-x^2}{9}} = \pm\frac{\sqrt{9-x^2}}{3}$	

Because we know that the inverse sine must give an angle on the interval $\left[-\frac{\pi}{2}, \frac{\pi}{2}\right]$, we can deduce that the cosine of that angle must be positive.

$$\cos\left(\sin^{-1}\left(\frac{x}{3}\right)\right) = \frac{\sqrt{9-x^2}}{3}$$

> **TRY IT** #9 Find a simplified expression for $\sin\left(\tan^{-1}(4x)\right)$ for $-\frac{1}{4} \le x \le \frac{1}{4}$.

MEDIA

Access this online resource for additional instruction and practice with inverse trigonometric functions.

Evaluate Expressions Involving Inverse Trigonometric Functions (http://openstax.org/l/evalinverstrig)

8.3 SECTION EXERCISES

Verbal

1. Why do the functions $f(x) = \sin^{-1} x$ and $g(x) = \cos^{-1} x$ have different ranges?

2. Since the functions $y = \cos\ x$ and $y = \cos^{-1} x$ are inverse functions, why is $\cos^{-1}\left(\cos\left(-\frac{\pi}{6}\right)\right)$ not equal to $-\frac{\pi}{6}$?

3. Explain the meaning of $\frac{\pi}{6} = \arcsin(0.5)$.

4. Most calculators do not have a key to evaluate $\sec^{-1}(2)$. Explain how this can be done using the cosine function or the inverse cosine function.

5. Why must the domain of the sine function, $\sin\ x$, be restricted to $\left[-\frac{\pi}{2}, \frac{\pi}{2}\right]$ for the inverse sine function to exist?

6. Discuss why this statement is incorrect: $\arccos(\cos\ x) = x$ for all x.

7. Determine whether the following statement is true or false and explain your answer: $\arccos(-x) = \pi - \arccos\ x$.

Algebraic

For the following exercises, evaluate the expressions.

8. $\sin^{-1}\left(\frac{\sqrt{2}}{2}\right)$

9. $\sin^{-1}\left(-\frac{1}{2}\right)$

10. $\cos^{-1}\left(\frac{1}{2}\right)$

11. $\cos^{-1}\left(-\frac{\sqrt{2}}{2}\right)$

12. $\tan^{-1}(1)$

13. $\tan^{-1}\left(-\sqrt{3}\right)$

14. $\tan^{-1}(-1)$

15. $\tan^{-1}\left(\sqrt{3}\right)$

16. $\tan^{-1}\left(\frac{-1}{\sqrt{3}}\right)$

For the following exercises, use a calculator to evaluate each expression. Express answers to the nearest hundredth.

17. $\cos^{-1}(-0.4)$

18. $\arcsin(0.23)$

19. $\arccos\left(\frac{3}{5}\right)$

20. $\cos^{-1}(0.8)$

21. $\tan^{-1}(6)$

For the following exercises, find the angle θ in the given right triangle. Round answers to the nearest hundredth.

22.

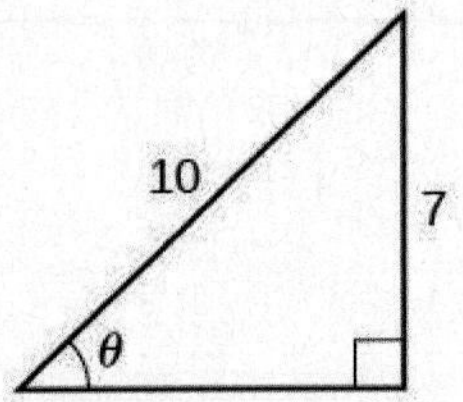

23.

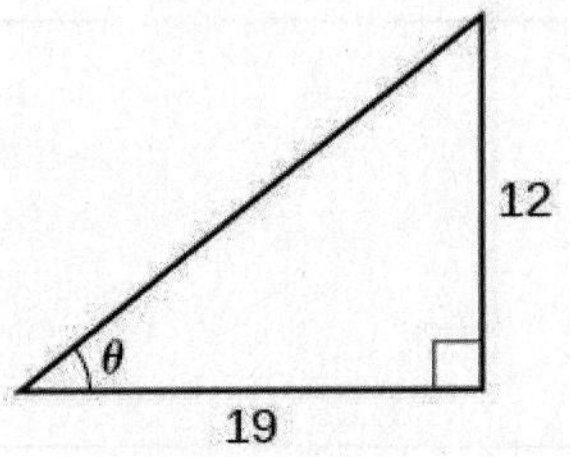

For the following exercises, find the exact value, if possible, without a calculator. If it is not possible, explain why.

24. $\sin^{-1}(\cos(\pi))$

25. $\tan^{-1}(\sin(\pi))$

26. $\cos^{-1}\left(\sin\left(\frac{\pi}{3}\right)\right)$

27. $\tan^{-1}\left(\sin\left(\frac{\pi}{3}\right)\right)$

28. $\sin^{-1}\left(\cos\left(\frac{-\pi}{2}\right)\right)$

29. $\tan^{-1}\left(\sin\left(\frac{4\pi}{3}\right)\right)$

30. $\sin^{-1}\left(\sin\left(\frac{5\pi}{6}\right)\right)$

31. $\tan^{-1}\left(\sin\left(\frac{-5\pi}{2}\right)\right)$

32. $\cos\left(\sin^{-1}\left(\frac{4}{5}\right)\right)$

33. $\sin\left(\cos^{-1}\left(\frac{3}{5}\right)\right)$

34. $\sin\left(\tan^{-1}\left(\frac{4}{3}\right)\right)$

35. $\cos\left(\tan^{-1}\left(\frac{12}{5}\right)\right)$

36. $\cos\left(\sin^{-1}\left(\frac{1}{2}\right)\right)$

For the following exercises, find the exact value of the expression in terms of x with the help of a reference triangle.

37. $\tan\left(\sin^{-1}(x-1)\right)$

38. $\sin\left(\cos^{-1}(1-x)\right)$

39. $\cos\left(\sin^{-1}\left(\frac{1}{x}\right)\right)$

40. $\cos\left(\tan^{-1}(3x-1)\right)$

41. $\tan\left(\sin^{-1}\left(x+\frac{1}{2}\right)\right)$

Extensions

For the following exercises, evaluate the expression without using a calculator. Give the exact value.

42. $$\frac{\sin^{-1}\left(\frac{1}{2}\right)-\cos^{-1}\left(\frac{\sqrt{2}}{2}\right)+\sin^{-1}\left(\frac{\sqrt{3}}{2}\right)-\cos^{-1}(1)}{\cos^{-1}\left(\frac{\sqrt{3}}{2}\right)-\sin^{-1}\left(\frac{\sqrt{2}}{2}\right)+\cos^{-1}\left(\frac{1}{2}\right)-\sin^{-1}(0)}$$

For the following exercises, find the function if $\sin t=\frac{x}{x+1}$.

43. $\cos t$

44. $\sec t$

45. $\cot t$

46. $\cos\left(\sin^{-1}\left(\frac{x}{x+1}\right)\right)$

47. $\tan^{-1}\left(\frac{x}{\sqrt{2x+1}}\right)$

Graphical

48. Graph $y = \sin^{-1} x$ and state the domain and range of the function.

49. Graph $y = \arccos\ x$ and state the domain and range of the function.

50. Graph one cycle of $y = \tan^{-1} x$ and state the domain and range of the function.

51. For what value of x does $\sin\ x = \sin^{-1} x$? Use a graphing calculator to approximate the answer.

52. For what value of x does $\cos\ x = \cos^{-1} x$? Use a graphing calculator to approximate the answer.

Real-World Applications

53. Suppose a 13-foot ladder is leaning against a building, reaching to the bottom of a second-floor window 12 feet above the ground. What angle, in radians, does the ladder make with the building?

54. Suppose you drive 0.6 miles on a road so that the vertical distance changes from 0 to 150 feet. What is the angle of elevation of the road?

55. An isosceles triangle has two congruent sides of length 9 inches. The remaining side has a length of 8 inches. Find the angle that a side of 9 inches makes with the 8-inch side.

56. Without using a calculator, approximate the value of $\arctan(10{,}000)$. Explain why your answer is reasonable.

57. A truss (interior beam structure) for the roof of a house is constructed from two identical right triangles. Each has a base of 12 feet and height of 4 feet. Find the measure of the acute angle adjacent to the 4-foot side.

58. The line $y = \frac{3}{5}x$ passes through the origin in the *x,y*-plane. What is the measure of the angle that the line makes with the positive *x*-axis?

59. The line $y = \frac{-3}{7}x$ passes through the origin in the *x,y*-plane. What is the measure of the angle that the line makes with the negative *x*-axis?

60. What percentage grade should a road have if the angle of elevation of the road is 4 degrees? (The percentage grade is defined as the change in the altitude of the road over a 100-foot horizontal distance. For example a 5% grade means that the road rises 5 feet for every 100 feet of horizontal distance.)

61. A 20-foot ladder leans up against the side of a building so that the foot of the ladder is 10 feet from the base of the building. If specifications call for the ladder's angle of elevation to be between 35 and 45 degrees, does the placement of this ladder satisfy safety specifications?

62. Suppose a 15-foot ladder leans against the side of a house so that the angle of elevation of the ladder is 42 degrees. How far is the foot of the ladder from the side of the house?

Chapter Review

Key Terms

amplitude the vertical height of a function; the constant A appearing in the definition of a sinusoidal function

arccosine another name for the inverse cosine; $\arccos\ x = \cos^{-1} x$

arcsine another name for the inverse sine; $\arcsin\ x = \sin^{-1} x$

arctangent another name for the inverse tangent; $\arctan\ x = \tan^{-1} x$

inverse cosine function the function $\cos^{-1} x$, which is the inverse of the cosine function and the angle that has a cosine equal to a given number

inverse sine function the function $\sin^{-1} x$, which is the inverse of the sine function and the angle that has a sine equal to a given number

inverse tangent function the function $\tan^{-1} x$, which is the inverse of the tangent function and the angle that has a tangent equal to a given number

midline the horizontal line $y = D$, where D appears in the general form of a sinusoidal function

periodic function a function $f(x)$ that satisfies $f(x + P) = f(x)$ for a specific constant P and any value of x

phase shift the horizontal displacement of the basic sine or cosine function; the constant $\frac{C}{B}$

sinusoidal function any function that can be expressed in the form $f(x) = A\sin(Bx - C) + D$ or $f(x) = A\cos(Bx - C) + D$

Key Equations

Sinusoidal functions	$f(x) = A\sin(Bx - C) + D$ $f(x) = A\cos(Bx - C) + D$

Shifted, compressed, and/or stretched tangent function	$y = A\ \tan(Bx - C) + D$
Shifted, compressed, and/or stretched secant function	$y = A\ \sec(Bx - C) + D$
Shifted, compressed, and/or stretched cosecant function	$y = A\ \csc(Bx - C) + D$
Shifted, compressed, and/or stretched cotangent function	$y = A\ \cot(Bx - C) + D$

Key Concepts

8.1 Graphs of the Sine and Cosine Functions

- Periodic functions repeat after a given value. The smallest such value is the period. The basic sine and cosine functions have a period of 2π.
- The function $\sin\ x$ is odd, so its graph is symmetric about the origin. The function $\cos\ x$ is even, so its graph is symmetric about the *y*-axis.
- The graph of a sinusoidal function has the same general shape as a sine or cosine function.
- In the general formula for a sinusoidal function, the period is $P = \frac{2\pi}{|B|}$. See Example 1.
- In the general formula for a sinusoidal function, $|A|$ represents amplitude. If $|A| > 1$, the function is stretched, whereas if $|A| < 1$, the function is compressed. See Example 2.
- The value $\frac{C}{B}$ in the general formula for a sinusoidal function indicates the phase shift. See Example 3.
- The value D in the general formula for a sinusoidal function indicates the vertical shift from the midline. See Example 4.
- Combinations of variations of sinusoidal functions can be detected from an equation. See Example 5.
- The equation for a sinusoidal function can be determined from a graph. See Example 6 and Example 7.
- A function can be graphed by identifying its amplitude and period. See Example 8 and Example 9.
- A function can also be graphed by identifying its amplitude, period, phase shift, and horizontal shift. See Example 10.
- Sinusoidal functions can be used to solve real-world problems. See Example 11, Example 12, and Example 13.

8.2 Graphs of the Other Trigonometric Functions

- The tangent function has period π.
- $f(x) = A\tan(Bx - C) + D$ is a tangent with vertical and/or horizontal stretch/compression and shift. See Example 1, Example 2, and Example 3.
- The secant and cosecant are both periodic functions with a period of 2π. $f(x) = A\sec(Bx - C) + D$ gives a shifted, compressed, and/or stretched secant function graph. See Example 4 and Example 5.
- $f(x) = A\csc(Bx - C) + D$ gives a shifted, compressed, and/or stretched cosecant function graph. See Example 6 and Example 7.
- The cotangent function has period π and vertical asymptotes at $0, \pm\pi, \pm 2\pi, ...$
- The range of cotangent is $\left(-\infty, \infty\right)$, and the function is decreasing at each point in its range.
- The cotangent is zero at $\pm\frac{\pi}{2}, \pm\frac{3\pi}{2}, ...$
- $f(x) = A\cot(Bx - C) + D$ is a cotangent with vertical and/or horizontal stretch/compression and shift. See Example 8 and Example 9.
- Real-world scenarios can be solved using graphs of trigonometric functions. See Example 10.

8.3 Inverse Trigonometric Functions

- An inverse function is one that "undoes" another function. The domain of an inverse function is the range of the original function and the range of an inverse function is the domain of the original function.
- Because the trigonometric functions are not one-to-one on their natural domains, inverse trigonometric functions are defined for restricted domains.
- For any trigonometric function $f(x)$, if $x = f^{-1}(y)$, then $f(x) = y$. However, $f(x) = y$ only implies $x = f^{-1}(y)$ if x is in the restricted domain of f. See Example 1.
- Special angles are the outputs of inverse trigonometric functions for special input values; for example, $\frac{\pi}{4} = \tan^{-1}(1)$ and $\frac{\pi}{6} = \sin^{-1}\left(\frac{1}{2}\right)$. See Example 2.
- A calculator will return an angle within the restricted domain of the original trigonometric function. See Example 3.
- Inverse functions allow us to find an angle when given two sides of a right triangle. See Example 4.
- In function composition, if the inside function is an inverse trigonometric function, then there are exact expressions; for example, $\sin\left(\cos^{-1}(x)\right) = \sqrt{1 - x^2}$. See Example 5.
- If the inside function is a trigonometric function, then the only possible combinations are $\sin^{-1}(\cos x) = \frac{\pi}{2} - x$ if $0 \le x \le \pi$ and $\cos^{-1}(\sin x) = \frac{\pi}{2} - x$ if $-\frac{\pi}{2} \le x \le \frac{\pi}{2}$. See Example 6 and Example 7.
- When evaluating the composition of a trigonometric function with an inverse trigonometric function, draw a reference triangle to assist in determining the ratio of sides that represents the output of the trigonometric function. See Example 8.
- When evaluating the composition of a trigonometric function with an inverse trigonometric function, you may use trig identities to assist in determining the ratio of sides. See Example 9.

Exercises

Review Exercises

Graphs of the Sine and Cosine Functions

For the following exercises, graph the functions for two periods and determine the amplitude or stretching factor, period, midline equation, and asymptotes.

1. $f(x) = -3\cos x + 3$

2. $f(x) = \frac{1}{4}\sin x$

3. $f(x) = 3\cos\left(x + \frac{\pi}{6}\right)$

4. $f(x) = -2\sin\left(x - \frac{2\pi}{3}\right)$

5. $f(x) = 3\sin\left(x - \frac{\pi}{4}\right) - 4$

6. $f(x) = 2\left(\cos\left(x - \frac{4\pi}{3}\right) + 1\right)$

7. $f(x) = 6\sin\left(3x - \frac{\pi}{6}\right) - 1$

8. $f(x) = -100\sin(50x - 20)$

Graphs of the Other Trigonometric Functions

For the following exercises, graph the functions for two periods and determine the amplitude or stretching factor, period, midline equation, and asymptotes.

9. $f(x) = \tan\ x - 4$

10. $f(x) = 2\tan\left(x - \frac{\pi}{6}\right)$

11. $f(x) = -3\tan(4x) - 2$

12. $f(x) = 0.2\cos(0.1x) + 0.3$

For the following exercises, graph two full periods. Identify the period, the phase shift, the amplitude, and asymptotes.

13. $f(x) = \frac{1}{3}\sec\ x$

14. $f(x) = 3\cot\ x$

15. $f(x) = 4\csc(5x)$

16. $f(x) = 8\sec\left(\frac{1}{4}x\right)$

17. $f(x) = \frac{2}{3}\csc\left(\frac{1}{2}x\right)$

18. $f(x) = -\csc(2x + \pi)$

For the following exercises, use this scenario: The population of a city has risen and fallen over a 20-year interval. Its population may be modeled by the following function: $y = 12{,}000 + 8{,}000\sin(0.628x)$, *where the domain is the years since 1980 and the range is the population of the city.*

19. What is the largest and smallest population the city may have?

20. Graph the function on the domain of $[0, 40]$.

21. What are the amplitude, period, and phase shift for the function?

22. Over this domain, when does the population reach 18,000? 13,000?

23. What is the predicted population in 2007? 2010?

For the following exercises, suppose a weight is attached to a spring and bobs up and down, exhibiting symmetry.

24. Suppose the graph of the displacement function is shown in Figure 1, where the values on the x-axis represent the time in seconds and the y-axis represents the displacement in inches. Give the equation that models the vertical displacement of the weight on the spring.

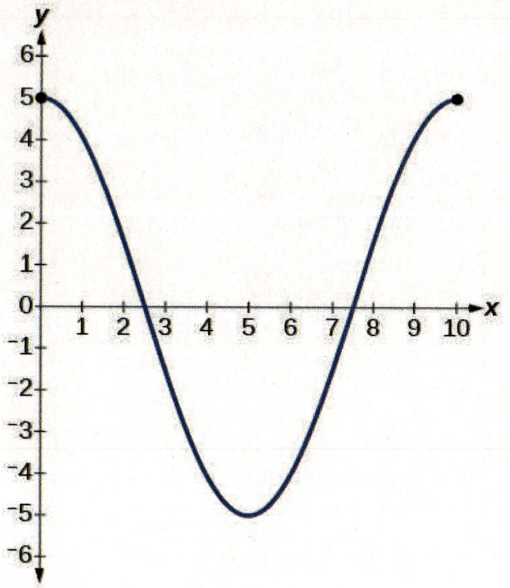

Figure 1

25. At time = 0, what is the displacement of the weight?

26. At what time does the displacement from the equilibrium point equal zero?

27. What is the time required for the weight to return to its initial height of 5 inches? In other words, what is the period for the displacement function?

Inverse Trigonometric Functions

For the following exercises, find the exact value without the aid of a calculator.

28. $\sin^{-1}(1)$

29. $\cos^{-1}\left(\frac{\sqrt{3}}{2}\right)$

30. $\tan^{-1}(-1)$

31. $\cos^{-1}\left(\frac{1}{\sqrt{2}}\right)$

32. $\sin^{-1}\left(\frac{-\sqrt{3}}{2}\right)$

33. $\sin^{-1}\left(\cos\left(\frac{\pi}{6}\right)\right)$

34. $\cos^{-1}\left(\tan\left(\frac{3\pi}{4}\right)\right)$

35. $\sin\left(\sec^{-1}\left(\frac{3}{5}\right)\right)$

36. $\cot\left(\sin^{-1}\left(\frac{3}{5}\right)\right)$

37. $\tan\left(\cos^{-1}\left(\frac{5}{13}\right)\right)$

38. $\sin\left(\cos^{-1}\left(\frac{x}{x+1}\right)\right)$

39. Graph $f(x) = \cos x$ and $f(x) = \sec x$ on the interval $[0, 2\pi)$ and explain any observations.

40. Graph $f(x) = \sin\ x$ and $f(x) = \csc\ x$ and explain any observations.

41. Graph the function $f(x) = \frac{x}{1} - \frac{x^3}{3!} + \frac{x^5}{5!} - \frac{x^7}{7!}$ on the interval $[-1, 1]$ and compare the graph to the graph of $f(x) = \sin\ x$ on the same interval. Describe any observations.

Practice Test

For the following exercises, sketch the graph of each function for two full periods. Determine the amplitude, the period, and the equation for the midline.

1. $f(x) = 0.5 \sin\ x$

2. $f(x) = 5 \cos\ x$

3. $f(x) = 5 \sin\ x$

4. $f(x) = \sin(3x)$

5. $f(x) = -\cos\left(x + \frac{\pi}{3}\right) + 1$

6. $f(x) = 5 \sin\left(3\left(x - \frac{\pi}{6}\right)\right) + 4$

7. $f(x) = 3 \cos\left(\frac{1}{3}x - \frac{5\pi}{6}\right)$

8. $f(x) = \tan(4x)$

9. $f(x) = -2 \tan\left(x - \frac{7\pi}{6}\right) + 2$

10. $f(x) = \pi \cos(3x + \pi)$

11. $f(x) = 5 \csc(3x)$

12. $f(x) = \pi \sec\left(\frac{\pi}{2}x\right)$

13. $f(x) = 2 \csc\left(x + \frac{\pi}{4}\right) - 3$

For the following exercises, determine the amplitude, period, and midline of the graph, and then find a formula for the function.

14. Give in terms of a sine function.

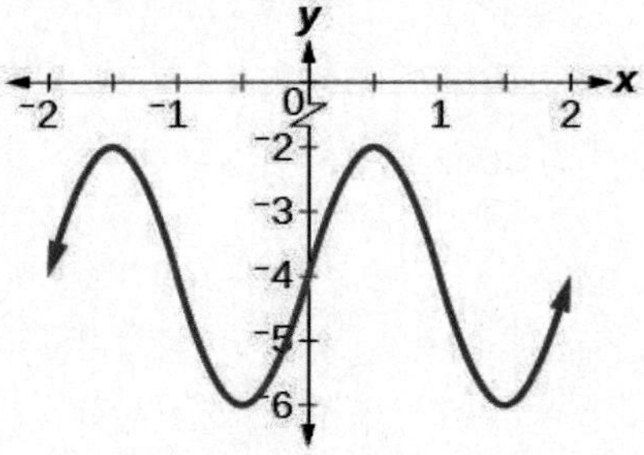

15. Give in terms of a sine function.

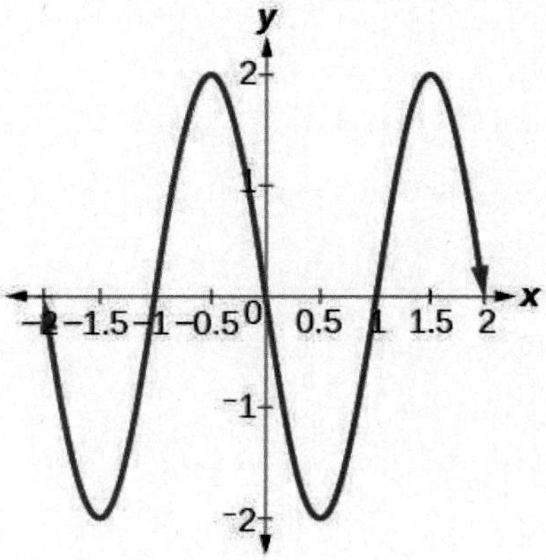

16. Give in terms of a tangent function.

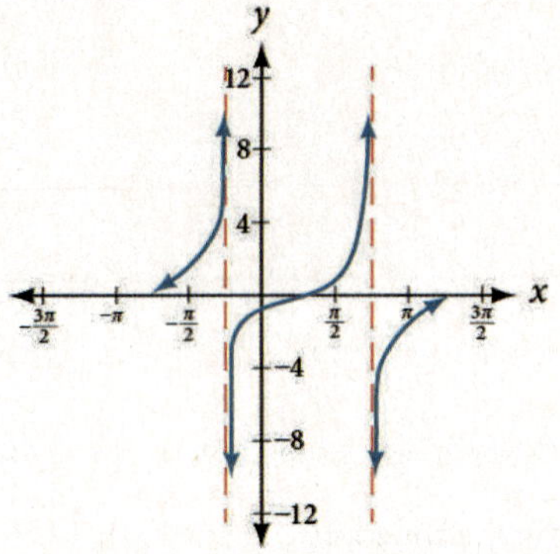

For the following exercises, find the amplitude, period, phase shift, and midline.

17. $y = \sin\left(\frac{\pi}{6}x + \pi\right) - 3$

18. $y = 8\sin\left(\frac{7\pi}{6}x + \frac{7\pi}{2}\right) + 6$

19. The outside temperature over the course of a day can be modeled as a sinusoidal function. Suppose you know the temperature is 68°F at midnight and the high and low temperatures during the day are 80°F and 56°F, respectively. Assuming t is the number of hours since midnight, find a function for the temperature, D, in terms of t.

20. Water is pumped into a storage bin and empties according to a periodic rate. The depth of the water is 3 feet at its lowest at 2:00 a.m. and 71 feet at its highest, which occurs every 5 hours. Write a cosine function that models the depth of the water as a function of time, and then graph the function for one period.

For the following exercises, find the period and horizontal shift of each function.

21. $g(x) = 3\tan(6x + 42)$

22. $n(x) = 4\csc\left(\frac{5\pi}{3}x - \frac{20\pi}{3}\right)$

23. Write the equation for the graph in Figure 1 in terms of the secant function and give the period and phase shift.

24. If $\tan\ x = 3$, find $\tan(-x)$.

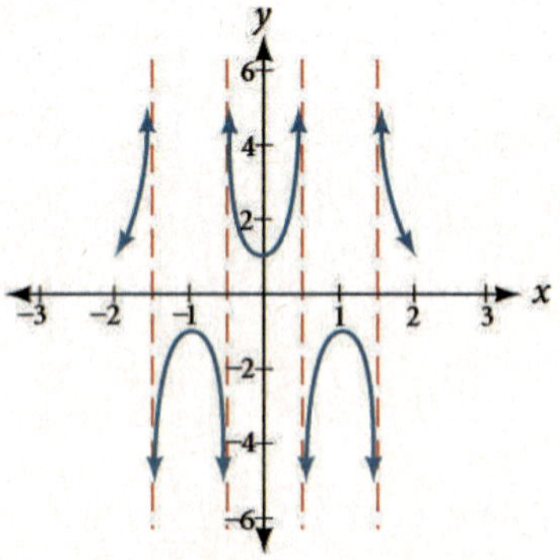

Figure 1

25. If $\sec\ x = 4$, find $\sec(-x)$.

For the following exercises, graph the functions on the specified window and answer the questions.

26. Graph $m(x) = \sin(2x) + \cos(3x)$ on the viewing window $[-10, 10]$ by $[-3, 3]$. Approximate the graph's period.

27. Graph $n(x) = 0.02\sin(50\pi x)$ on the following domains in $x : [0, 1]$ and $[0, 3]$. Suppose this function models sound waves. Why would these views look so different?

28. Graph $f(x) = \frac{\sin x}{x}$ on $[-0.5, 0.5]$ and explain any observations.

For the following exercises, let $f(x) = \frac{3}{5}\cos(6x)$.

29. What is the largest possible value for $f(x)$?

30. What is the smallest possible value for $f(x)$?

31. Where is the function increasing on the interval $[0, 2\pi]$?

For the following exercises, find and graph one period of the periodic function with the given amplitude, period, and phase shift.

32. Sine curve with amplitude 3, period $\frac{\pi}{3}$, and phase shift $(h, k) = \left(\frac{\pi}{4}, 2\right)$

33. Cosine curve with amplitude 2, period $\frac{\pi}{6}$, and phase shift $(h, k) = \left(-\frac{\pi}{4}, 3\right)$

For the following exercises, graph the function. Describe the graph and, wherever applicable, any periodic behavior, amplitude, asymptotes, or undefined points.

34. $f(x) = 5\cos(3x) + 4\sin(2x)$

35. $f(x) = e^{\sin t}$

For the following exercises, find the exact value.

36. $\sin^{-1}\left(\frac{\sqrt{3}}{2}\right)$

37. $\tan^{-1}\left(\sqrt{3}\right)$

38. $\cos^{-1}\left(-\frac{\sqrt{3}}{2}\right)$

39. $\cos^{-1}(\sin(\pi))$

40. $\cos^{-1}\left(\tan\left(\frac{7\pi}{4}\right)\right)$

41. $\cos\left(\sin^{-1}(1 - 2x)\right)$

42. $\cos^{-1}(-0.4)$

43. $\cos\left(\tan^{-1}\left(x^2\right)\right)$

For the following exercises, suppose $\sin t = \frac{x}{x+1}$. Evaluate the following expressions.

44. $\tan t$

45. $\csc t$

46. Given Figure 2, find the measure of angle θ to three decimal places. Answer in radians.

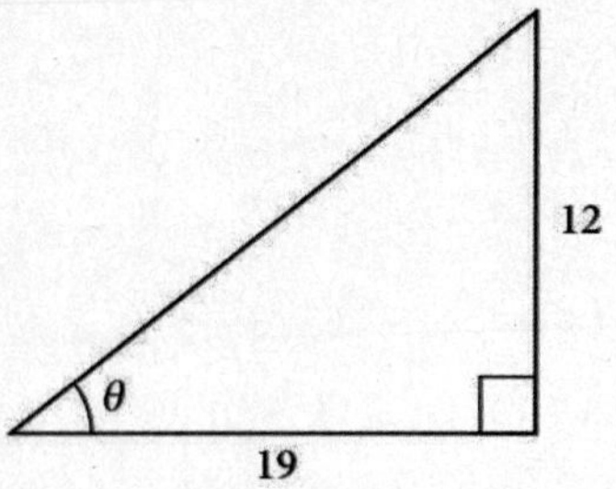

Figure 2

For the following exercises, determine whether the equation is true or false.

47. $\arcsin\left(\sin\left(\frac{5\pi}{6}\right)\right) = \frac{5\pi}{6}$

48. $\arccos\left(\cos\left(\frac{5\pi}{6}\right)\right) = \frac{5\pi}{6}$

49. The grade of a road is 7%. This means that for every horizontal distance of 100 feet on the road, the vertical rise is 7 feet. Find the angle the road makes with the horizontal in radians.

9 TRIGONOMETRIC IDENTITIES AND EQUATIONS

Tennis players can create advantages by changing the angles of their shots. The technology used to decide close calls also relies heavily on mathematics. (credit: modification of "From the 2013 US Open" by Edwin Martinez/flickr)

Chapter Outline

Introduction to Trigonometric Identities and Equations

When we think of tennis as a game of angles, we may imagine players racing up to the net, creating options to deliver powerful cross shots that will leave their opponent stumbling toward the line. This is an exciting and effective method of play, though it brings greater risk.

But while the excitement of the game interplays with all types of geometry, some of the newest innovations make even more use of mathematics. With balls traveling well over 100 miles per hour judges cannot always discern the centimeter or millimeters of difference between a ball that is in or out of bounds. Professional tennis was among the first sports to rely on an advanced tracking system called Hawk-Eye to help make close calls. The system uses several high-resolution cameras that are able to monitor and the ball's movement and its position on the court. Using the images from several cameras at once, the system's computers use trigonometric calculations to triangulate the ball's exact position and, essentially, turn a series of two-dimensional images into a three-dimensional one. Also, since the ball travels faster than the cameras' frame rate, the system also must make predictions to show where a ball is at all times. These technologies generally provide a more accurate game that builds more confidence and fairness. Similar technologies are used for baseball, and automated strike-calling is under discussion.

9.1 Verifying Trigonometric Identities and Using Trigonometric Identities to Simplify Trigonometric Expressions

Learning Objectives

In this section, you will:

- Verify the fundamental trigonometric identities.
- Simplify trigonometric expressions using algebra and the identities.

Figure 1 International passports and travel documents

In espionage movies, we see international spies with multiple passports, each claiming a different identity. However, we know that each of those passports represents the same person. The trigonometric identities act in a similar manner to multiple passports—there are many ways to represent the same trigonometric expression. Just as a spy will choose an Italian passport when traveling to Italy, we choose the identity that applies to the given scenario when solving a trigonometric equation.

In this section, we will begin an examination of the fundamental trigonometric identities, including how we can verify them and how we can use them to simplify trigonometric expressions.

Verifying the Fundamental Trigonometric Identities

Identities enable us to simplify complicated expressions. They are the basic tools of trigonometry used in solving trigonometric equations, just as factoring, finding common denominators, and using special formulas are the basic tools of solving algebraic equations. In fact, we use algebraic techniques constantly to simplify trigonometric expressions. Basic properties and formulas of algebra, such as the difference of squares formula and the perfect squares formula, will simplify the work involved with trigonometric expressions and equations. We already know that all of the trigonometric functions are related because they all are defined in terms of the unit circle. Consequently, any trigonometric identity can be written in many ways.

To verify the trigonometric identities, we usually start with the more complicated side of the equation and essentially rewrite the expression until it has been transformed into the same expression as the other side of the equation. Sometimes we have to factor expressions, expand expressions, find common denominators, or use other algebraic strategies to obtain the desired result. In this first section, we will work with the fundamental identities: the Pythagorean identities, the even-odd identities, the reciprocal identities, and the quotient identities.

We will begin with the **Pythagorean identities** (see Table 1), which are equations involving trigonometric functions based on the properties of a right triangle. We have already seen and used the first of these identifies, but now we will also use additional identities.

Pythagorean Identities		
$\sin^2\theta + \cos^2\theta = 1$	$1 + \cot^2\theta = \csc^2\theta$	$1 + \tan^2\theta = \sec^2\theta$

Table 1

The second and third identities can be obtained by manipulating the first. The identity $1 + \cot^2\theta = \csc^2\theta$ is found by rewriting the left side of the equation in terms of sine and cosine.

Prove: $1 + \cot^2\theta = \csc^2\theta$

$$\begin{aligned} 1+\cot^2\theta &= \left(1+\frac{\cos^2\theta}{\sin^2\theta}\right) && \text{Rewrite the left side.} \\ &= \left(\frac{\sin^2\theta}{\sin^2\theta}\right)+\left(\frac{\cos^2\theta}{\sin^2\theta}\right) && \text{Write both terms with the common denominator.} \\ &= \frac{\sin^2\theta+\cos^2\theta}{\sin^2\theta} \\ &= \frac{1}{\sin^2\theta} \\ &= \csc^2\theta \end{aligned}$$

Similarly, $1+\tan^2\theta=\sec^2\theta$ can be obtained by rewriting the left side of this identity in terms of sine and cosine. This gives

$$\begin{aligned} 1+\tan^2\theta &= 1+\left(\frac{\sin\theta}{\cos\theta}\right)^2 && \text{Rewrite left side.} \\ &= \left(\frac{\cos\theta}{\cos\theta}\right)^2+\left(\frac{\sin\theta}{\cos\theta}\right)^2 && \text{Write both terms with the common denominator.} \\ &= \frac{\cos^2\theta+\sin^2\theta}{\cos^2\theta} \\ &= \frac{1}{\cos^2\theta} \\ &= \sec^2\theta \end{aligned}$$

Recall that we determined which trigonometric functions are odd and which are even. The next set of fundamental identities is the set of **even-odd identities.** The even-odd identities relate the value of a trigonometric function at a given angle to the value of the function at the opposite angle. (See Table 2).

Even-Odd Identities		
$\tan(-\theta)=-\tan\theta$	$\sin(-\theta)=-\sin\theta$	$\cos(-\theta)=\cos\theta$
$\cot(-\theta)=-\cot\theta$	$\csc(-\theta)=-\csc\theta$	$\sec(-\theta)=\sec\theta$

Table 2

Recall that an odd function is one in which $f(-x)=-f(x)$ for all x in the domain of f. The sine function is an odd function because $\sin(-\theta)=-\sin\theta$. The graph of an odd function is symmetric about the origin. For example, consider corresponding inputs of $\frac{\pi}{2}$ and $-\frac{\pi}{2}$. The output of $\sin\left(\frac{\pi}{2}\right)$ is opposite the output of $\sin\left(-\frac{\pi}{2}\right)$. Thus,

$$\begin{aligned} \sin\left(\frac{\pi}{2}\right) &= 1 \\ &\text{and} \\ \sin\left(-\frac{\pi}{2}\right) &= -\sin\left(\frac{\pi}{2}\right) \\ &= -1 \end{aligned}$$

This is shown in Figure 2.

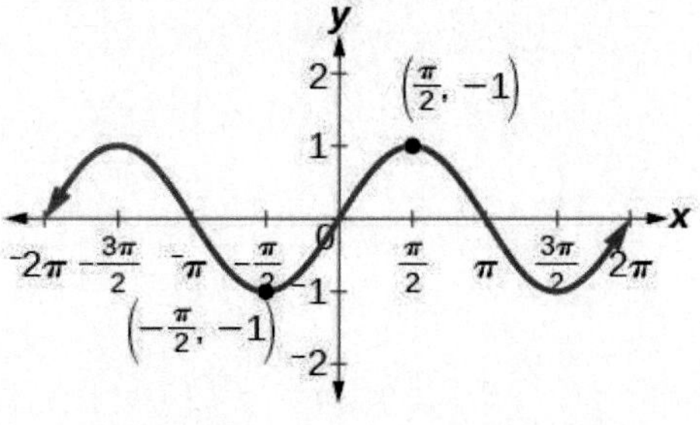

Figure 2 Graph of $y=\sin\theta$

Recall that an even function is one in which

$$f(-x)=f(x) \text{ for all } x \text{ in the domain of } f$$

The graph of an even function is symmetric about the *y*-axis. The cosine function is an even function because $\cos(-\theta)=\cos\theta$. For example, consider corresponding inputs $\frac{\pi}{4}$ and $-\frac{\pi}{4}$. The output of $\cos\left(\frac{\pi}{4}\right)$ is the same as the output of $\cos\left(-\frac{\pi}{4}\right)$. Thus,

$$\begin{aligned}\cos\left(-\frac{\pi}{4}\right) &= \cos\left(\frac{\pi}{4}\right) \\ &\approx 0.707\end{aligned}$$

See Figure 3.

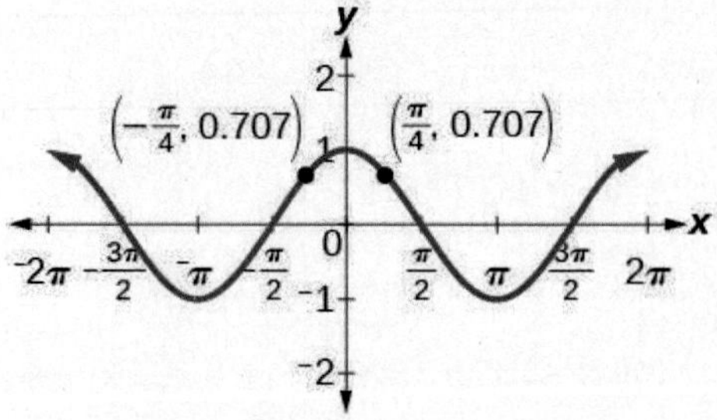

Figure 3 Graph of $y = \cos\ \theta$

For all θ in the domain of the sine and cosine functions, respectively, we can state the following:

- Since $\sin\ (-\ \theta) = -\sin\ \theta$, sine is an odd function.
- Since, $\cos\ (-\ \theta) = \cos\ \theta$, cosine is an even function.

The other even-odd identities follow from the even and odd nature of the sine and cosine functions. For example, consider the tangent identity, $\tan\ (-\ \theta) = -\tan\ \theta$. We can interpret the tangent of a negative angle as $\tan\ (-\ \theta) = \frac{\sin(-\theta)}{\cos(-\theta)} = \frac{-\sin\ \theta}{\cos\ \theta} = -\tan\ \theta$. Tangent is therefore an odd function, which means that $\tan(-\theta) = -\tan(\theta)$ for all θ in the domain of the tangent function.

The cotangent identity, $\cot(-\theta) = -\cot\ \theta$, also follows from the sine and cosine identities. We can interpret the cotangent of a negative angle as $\cot(-\theta) = \frac{\cos(-\theta)}{\sin(-\theta)} = \frac{\cos\ \theta}{-\sin\ \theta} = -\cot\ \theta$. Cotangent is therefore an odd function, which means that $\cot(-\theta) = -\cot(\theta)$ for all θ in the domain of the cotangent function.

The cosecant function is the reciprocal of the sine function, which means that the cosecant of a negative angle will be interpreted as $\csc(-\theta) = \frac{1}{\sin(-\theta)} = \frac{1}{-\sin\ \theta} = -\csc\ \theta$. The cosecant function is therefore odd.

Finally, the secant function is the reciprocal of the cosine function, and the secant of a negative angle is interpreted as $\sec(-\theta) = \frac{1}{\cos(-\theta)} = \frac{1}{\cos\ \theta} = \sec\ \theta$. The secant function is therefore even.

To sum up, only two of the trigonometric functions, cosine and secant, are even. The other four functions are odd, verifying the even-odd identities.

The next set of fundamental identities is the set of **reciprocal identities**, which, as their name implies, relate trigonometric functions that are reciprocals of each other. See Table 3. Recall that we first encountered these identities when defining trigonometric functions from right angles in Right Angle Trigonometry.

Reciprocal Identities	
$\sin\ \theta = \frac{1}{\csc\ \theta}$	$\csc\ \theta = \frac{1}{\sin\ \theta}$
$\cos\ \theta = \frac{1}{\sec\ \theta}$	$\sec\ \theta = \frac{1}{\cos\ \theta}$
$\tan\ \theta = \frac{1}{\cot\ \theta}$	$\cot\ \theta = \frac{1}{\tan\ \theta}$

Table 3

The final set of identities is the set of quotient identities, which define relationships among certain trigonometric functions and can be very helpful in verifying other identities. See Table 4.

Quotient Identities	
$\tan\ \theta = \frac{\sin\ \theta}{\cos\ \theta}$	$\cot\ \theta = \frac{\cos\ \theta}{\sin\ \theta}$

Table 4

The reciprocal and quotient identities are derived from the definitions of the basic trigonometric functions.

Summarizing Trigonometric Identities

The **Pythagorean identities** are based on the properties of a right triangle.

$$\cos^2\theta + \sin^2\theta = 1$$
$$1 + \cot^2\theta = \csc^2\theta$$
$$1 + \tan^2\theta = \sec^2\theta$$

The **even-odd identities** relate the value of a trigonometric function at a given angle to the value of the function at the opposite angle.

$$\tan(-\theta) = -\tan\ \theta$$
$$\cot(-\theta) = -\cot\ \theta$$
$$\sin(-\theta) = -\sin\ \theta$$
$$\csc(-\theta) = -\csc\ \theta$$
$$\cos(-\theta) = \cos\ \theta$$
$$\sec(-\theta) = \sec\ \theta$$

The **reciprocal identities** define reciprocals of the trigonometric functions.

$$\sin\ \theta = \frac{1}{\csc\ \theta}$$
$$\cos\ \theta = \frac{1}{\sec\ \theta}$$
$$\tan\ \theta = \frac{1}{\cot\ \theta}$$
$$\csc\ \theta = \frac{1}{\sin\ \theta}$$
$$\sec\ \theta = \frac{1}{\cos\ \theta}$$
$$\cot\ \theta = \frac{1}{\tan\ \theta}$$

The **quotient identities** define the relationship among the trigonometric functions.

$$\tan\ \theta = \frac{\sin\ \theta}{\cos\ \theta}$$
$$\cot\ \theta = \frac{\cos\ \theta}{\sin\ \theta}$$

EXAMPLE 1

Graphing the Equations of an Identity

Graph both sides of the identity $\cot\ \theta = \frac{1}{\tan\ \theta}$. In other words, on the graphing calculator, graph $y = \cot\ \theta$ and $y = \frac{1}{\tan\ \theta}$.

Solution

See Figure 4.

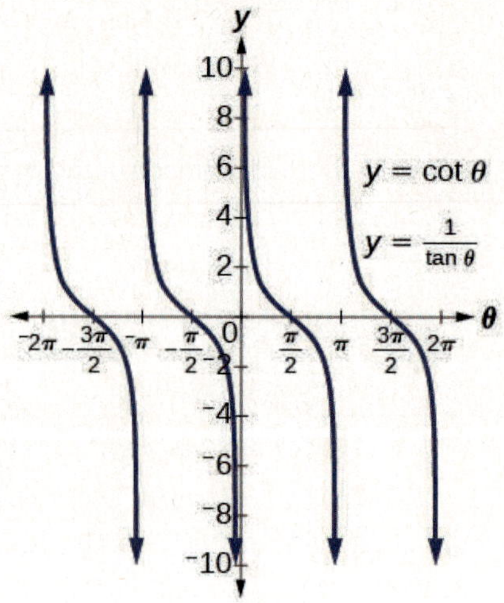

Figure 4

Analysis

We see only one graph because both expressions generate the same image. One is on top of the other. This is a good way to confirm an identity verified with analytical means. If both expressions give the same graph, then they are most likely identities.

HOW TO

Given a trigonometric identity, verify that it is true.

1. Work on one side of the equation. It is usually better to start with the more complex side, as it is easier to simplify than to build.
2. Look for opportunities to factor expressions, square a binomial, or add fractions.
3. Noting which functions are in the final expression, look for opportunities to use the identities and make the proper substitutions.
4. If these steps do not yield the desired result, try converting all terms to sines and cosines.

EXAMPLE 2

Verifying a Trigonometric Identity

Verify $\tan\ \theta \cos\ \theta = \sin\ \theta$.

Solution

We will start on the left side, as it is the more complicated side:

$$\begin{aligned} \tan\ \theta\ \cos\ \theta &= \left(\frac{\sin\ \theta}{\cos\ \theta}\right)\cos\ \theta \\ &= \left(\frac{\sin\ \theta}{\cancel{\cos\ \theta}}\right)\cancel{\cos\ \theta} \\ &= \sin\ \theta \end{aligned}$$

Analysis

This identity was fairly simple to verify, as it only required writing $\tan\ \theta$ in terms of $\sin\ \theta$ and $\cos\ \theta$.

TRY IT #1 Verify the identity $\csc\ \theta\ \cos\ \theta\ \tan\ \theta = 1$.

EXAMPLE 3

Verifying the Equivalency Using the Even-Odd Identities

Verify the following equivalency using the even-odd identities:

$$(1 + \sin\ x)\,[1 + \sin(-x)] = \cos^2 x$$

Solution

Working on the left side of the equation, we have

$$\begin{aligned}(1+\sin\ x)[1+\sin(-x)] &= (1+\sin\ x)(1-\sin\ x) && \text{Since } \sin(-x)\text{=-}\sin\ x\\ &= 1-\sin^2 x && \text{Difference of squares}\\ &= \cos^2 x && \cos^2 x = 1-\sin^2 x\end{aligned}$$

EXAMPLE 4

Verifying a Trigonometric Identity Involving *sec²θ*

Verify the identity $\frac{\sec^2\theta-1}{\sec^2\theta} = \sin^2\theta$

Solution

As the left side is more complicated, let's begin there.

$$\begin{aligned}\frac{\sec^2\theta-1}{\sec^2\theta} &= \frac{\left(\tan^2\theta+1\right)-1}{\sec^2\theta} && \sec^2\theta = \tan^2\theta+1\\ &= \frac{\tan^2\theta}{\sec^2\theta}\\ &= \tan^2\theta\left(\frac{1}{\sec^2\theta}\right)\\ &= \tan^2\theta\left(\cos^2\theta\right) && \cos^2\theta = \frac{1}{\sec^2\theta}\\ &= \left(\frac{\sin^2\theta}{\cos^2\theta}\right)\left(\cos^2\theta\right) && \tan^2\theta = \frac{\sin^2\theta}{\cos^2\theta}\\ &= \left(\frac{\sin^2\theta}{\cancel{\cos^2\theta}}\right)\left(\cancel{\cos^2\theta}\right)\\ &= \sin^2\theta\end{aligned}$$

There is more than one way to verify an identity. Here is another possibility. Again, we can start with the left side.

$$\begin{aligned}\frac{\sec^2\theta-1}{\sec^2\theta} &= \frac{\sec^2\theta}{\sec^2\theta} - \frac{1}{\sec^2\theta}\\ &= 1-\cos^2\theta\\ &= \sin^2\theta\end{aligned}$$

Analysis

In the first method, we used the identity $\sec^2\theta = \tan^2\theta + 1$ and continued to simplify. In the second method, we split the fraction, putting both terms in the numerator over the common denominator. This problem illustrates that there are multiple ways we can verify an identity. Employing some creativity can sometimes simplify a procedure. As long as the substitutions are correct, the answer will be the same.

TRY IT #2 Show that $\frac{\cot\ \theta}{\csc\ \theta} = \cos\ \theta.$

EXAMPLE 5

Creating and Verifying an Identity

Create an identity for the expression $2\ \tan\ \theta\ \sec\ \theta$ by rewriting strictly in terms of sine.

Solution

There are a number of ways to begin, but here we will use the quotient and reciprocal identities to rewrite the expression:

$$\begin{aligned} 2\tan\theta\sec\theta &= 2\left(\frac{\sin\theta}{\cos\theta}\right)\left(\frac{1}{\cos\theta}\right) \\ &= \frac{2\sin\theta}{\cos^2\theta} \\ &= \frac{2\sin\theta}{1-\sin^2\theta} && \text{Substitute } 1-\sin^2\theta \text{ for } \cos^2\theta. \end{aligned}$$

Thus,

$$2\tan\theta\sec\theta = \frac{2\sin\theta}{1-\sin^2\theta}$$

EXAMPLE 6

Verifying an Identity Using Algebra and Even/Odd Identities

Verify the identity:

$$\frac{\sin^2(-\theta)-\cos^2(-\theta)}{\sin(-\theta)-\cos(-\theta)} = \cos\theta - \sin\theta$$

Solution

Let's start with the left side and simplify:

$$\begin{aligned} \frac{\sin^2(\ \theta)-\cos^2(\ \theta)\ -}{\sin(\ \theta)-\cos(\ \theta)-} &= \frac{[\sin(\ \theta)]^2-[\cos(\ \theta)]^2}{\sin(\ \theta)-\cos(\ \theta)-} \\ &= \frac{(-\sin\theta)^2-(\cos\theta)^2}{-\sin\theta-\cos\theta} && \sin(-x) = -\sin x \text{ and } \cos(-x) = \cos x \\ &= \frac{(\sin\theta)^2-(\cos\theta)^2}{-\sin\theta-\cos\theta} && \text{Difference of squares} \\ &= \frac{(\sin\theta-\cos\theta)(\sin\theta+\cos\theta)}{-(\sin\theta+\cos\theta)} \\ &= \frac{(\sin\theta-\cos\theta)(\cancel{\sin\theta+\cos\theta})}{-(\cancel{\sin\theta+\cos\theta})} \\ &= \cos\theta - \sin\theta \end{aligned}$$

> TRY IT #3 Verify the identity $\frac{\sin^2\theta-1}{\tan\theta\sin\theta-\tan\theta} = \frac{\sin\theta+1}{\tan\theta}$.

EXAMPLE 7

Verifying an Identity Involving Cosines and Cotangents

Verify the identity: $(1-\cos^2 x)(1+\cot^2 x) = 1$.

Solution

We will work on the left side of the equation.

$$\begin{aligned} (1-\cos^2 x)(1+\cot^2 x) &= (1-\cos^2 x)\left(1+\frac{\cos^2 x}{\sin^2 x}\right) \\ &= (1-\cos^2 x)\left(\frac{\sin^2 x}{\sin^2 x}+\frac{\cos^2 x}{\sin^2 x}\right) && \text{Find the common denominator.} \\ &= (1-\cos^2 x)\left(\frac{\sin^2 x+\cos^2 x}{\sin^2 x}\right) \\ &= (\sin^2 x)\left(\frac{1}{\sin^2 x}\right) \\ &= 1 \end{aligned}$$

Using Algebra to Simplify Trigonometric Expressions

We have seen that algebra is very important in verifying trigonometric identities, but it is just as critical in simplifying trigonometric expressions before solving. Being familiar with the basic properties and formulas of algebra, such as the difference of squares formula, the perfect square formula, or substitution, will simplify the work involved with trigonometric expressions and equations.

For example, the equation $(\sin\ x + 1)(\sin\ x - 1) = 0$ resembles the equation $(x + 1)(x - 1) = 0$, which uses the factored form of the difference of squares. Using algebra makes finding a solution straightforward and familiar. We can set each factor equal to zero and solve. This is one example of recognizing algebraic patterns in trigonometric expressions or equations.

Another example is the difference of squares formula, $a^2 - b^2 = (a - b)(a + b)$, which is widely used in many areas other than mathematics, such as engineering, architecture, and physics. We can also create our own identities by continually expanding an expression and making the appropriate substitutions. Using algebraic properties and formulas makes many trigonometric equations easier to understand and solve.

EXAMPLE 8

Writing the Trigonometric Expression as an Algebraic Expression

Write the following trigonometric expression as an algebraic expression: $2\cos^2\theta + \cos\ \theta - 1$.

Solution

Notice that the pattern displayed has the same form as a standard quadratic expression, $ax^2 + bx + c$. Letting $\cos\ \theta = x$, we can rewrite the expression as follows:

$$2x^2 + x - 1$$

This expression can be factored as $(2x\text{–} 1)(x + 1)$. If it were set equal to zero and we wanted to solve the equation, we would use the zero factor property and solve each factor for x. At this point, we would replace x with $\cos\ \theta$ and solve for θ.

EXAMPLE 9

Rewriting a Trigonometric Expression Using the Difference of Squares

Rewrite the trigonometric expression using the difference of squares: $4\ \cos^2\theta - 1$.

Solution

Notice that both the coefficient and the trigonometric expression in the first term are squared, and the square of the number 1 is 1. This is the difference of squares.

$$\begin{aligned} 4\ \cos^2\theta - 1 &= (2\ \cos\ \theta)^2 - 1 \\ &= (2\ \cos\ \theta - 1)(2\ \cos\ \theta + 1) \end{aligned}$$

Analysis

If this expression were written in the form of an equation set equal to zero, we could solve each factor using the zero factor property. We could also use substitution like we did in the previous problem and let $\cos\ \theta = x$, rewrite the expression as $4x^2 - 1$, and factor $(2x - 1)(2x + 1)$. Then replace x with $\cos\ \theta$ and solve for the angle.

TRY IT #4 Rewrite the trigonometric expression using the difference of squares: $25 - 9\ \sin^2\ \theta$.

EXAMPLE 10

Simplify by Rewriting and Using Substitution

Simplify the expression by rewriting and using identities:

$$\csc^2\theta - \cot^2\theta$$

Solution

We can start with the Pythagorean identity.

$$1 + \cot^2\theta = \csc^2\theta$$

Now we can simplify by substituting $1 + \cot^2\theta$ for $\csc^2\theta$. We have

$$\begin{aligned} \csc^2\theta - \cot^2\theta &= 1 + \cot^2\theta - \cot^2\theta \\ &= 1 \end{aligned}$$

TRY IT #5 Use algebraic techniques to verify the identity: $\frac{\cos\ \theta}{1+\sin\ \theta} = \frac{1-\sin\ \theta}{\cos\ \theta}$.

(Hint: Multiply the numerator and denominator on the left side by $1 - \sin\ \theta$.)

MEDIA

Access these online resources for additional instruction and practice with the fundamental trigonometric identities.

Fundamental Trigonometric Identities (http://openstax.org/l/funtrigiden)
Verifying Trigonometric Identities (http://openstax.org/l/verifytrigiden)

9.1 SECTION EXERCISES

Verbal

1. We know $g(x) = \cos\ x$ is an even function, and $f(x) = \sin\ x$ and $h(x) = \tan\ x$ are odd functions. What about $G(x) = \cos^2 x$, $F(x) = \sin^2 x$, and $H(x) = \tan^2 x$? Are they even, odd, or neither? Why?

2. Examine the graph of $f(x) = \sec\ x$ on the interval $[-\pi, \pi]$. How can we tell whether the function is even or odd by only observing the graph of $f(x) = \sec\ x$?

3. After examining the reciprocal identity for $\sec\ t$, explain why the function is undefined at certain points.

4. All of the Pythagorean identities are related. Describe how to manipulate the equations to get from $\sin^2 t + \cos^2 t = 1$ to the other forms.

Algebraic

For the following exercises, use the fundamental identities to fully simplify the expression.

5. $\sin\ x\ \cos\ x\ \sec\ x$

6. $\sin(-x)\ \cos(-x)\ \csc(-x)$

7. $\tan\ x\ \sin\ x + \sec\ x\ \cos^2 x$

8. $\csc\ x + \cos\ x\ \cot(-x)$

9. $\frac{\cot\ t+\tan\ t}{\sec(-t)}$

10. $3\ \sin^3\ t\ \csc\ t + \cos^2\ t + 2\ \cos(-t)\cos\ t$

11. $-\tan(-x)\cot(-x)$

12. $\frac{-\sin(x)\cos\ x\ \sec\ x\ \csc\ x\ \tan\ x}{\cot\ x}$

13. $\frac{1+\tan^2\theta}{\csc^2\theta} + \sin^2\theta + \frac{1}{\sec^2\theta}$

14. $\left(\frac{\tan x}{\csc^2 x} + \frac{\tan x}{\sec^2 x}\right)\left(\frac{1+\tan x}{1+\cot x}\right) - \frac{1}{\cos^2 x}$

15. $\frac{1-\cos^2 x}{\tan^2 x} + 2\sin^2 x$

For the following exercises, simplify the first trigonometric expression by writing the simplified form in terms of the second expression.

16. $\frac{\tan x+\cot x}{\csc x}$; $\cos x$

17. $\frac{\sec x+\csc x}{1+\tan x}$; $\sin x$

18. $\frac{\cos x}{1+\sin x} + \tan x$; $\cos x$

19. $\frac{1}{\sin x\cos x} - \cot x$; $\cot x$

20. $\frac{1}{1-\cos x} - \frac{\cos x}{1+\cos x}$; $\csc x$

21. $(\sec x + \csc x)(\sin x + \cos x) - 2 - \cot x$; $\tan x$

22. $\frac{1}{\csc x-\sin x}$; $\sec x$ and $\tan x$

23. $\frac{1-\sin x}{1+\sin x} - \frac{1+\sin x}{1-\sin x}$; $\sec x$ and $\tan x$

24. $\tan x$; $\sec x$

25. $\sec x$; $\cot x$

26. $\sec x$; $\sin x$

27. $\cot x$; $\sin x$

28. $\cot x$; $\csc x$

For the following exercises, verify the identity.

29. $\cos x - \cos^3 x = \cos x \sin^2 x$

30. $\cos x(\tan x - \sec(-x)) = \sin x - 1$

31. $\frac{1+\sin^2 x}{\cos^2 x} = \frac{1}{\cos^2 x} + \frac{\sin^2 x}{\cos^2 x} = 1 + 2\tan^2 x$

32. $(\sin x + \cos x)^2 = 1 + 2\sin x\cos x$

33. $\cos^2 x - \tan^2 x = 2 - \sin^2 x - \sec^2 x$

Extensions

For the following exercises, prove or disprove the identity.

34. $\frac{1}{1+\cos x} - \frac{1}{1-\cos(-x)} = -2\cot x\csc x$

35. $\csc^2 x\left(1 + \sin^2 x\right) = \cot^2 x$

36. $\left(\frac{\sec^2(-x)-\tan^2 x}{\tan x}\right)\left(\frac{2+2\tan x}{2+2\cot x}\right) - 2\sin^2 x = \cos 2x$

37. $\frac{\tan x}{\sec x}\sin(-x) = \cos^2 x$

38. $\frac{\sec(-x)}{\tan x+\cot x} = -\sin(-x)$

39. $\frac{1+\sin x}{\cos x} = \frac{\cos x}{1+\sin(-x)}$

For the following exercises, determine whether the identity is true or false. If false, find an appropriate equivalent expression.

40. $\frac{\cos^2\theta-\sin^2\theta}{1-\tan^2\theta} = \sin^2\theta$

41. $3\sin^2\theta + 4\cos^2\theta = 3 + \cos^2\theta$

42. $\frac{\sec\theta+\tan\theta}{\cot\theta+\cos\theta} = \sec^2\theta$

9.2 Sum and Difference Identities

Learning Objectives

In this section, you will:

- Use sum and difference formulas for cosine.
- Use sum and difference formulas for sine.
- Use sum and difference formulas for tangent.
- Use sum and difference formulas for cofunctions.
- Use sum and difference formulas to verify identities.

Figure 1 Mount McKinley, in Denali National Park, Alaska, rises 20,237 feet (6,168 m) above sea level. It is the highest peak in North America. (credit: Daniel A. Leifheit, Flickr)

How can the height of a mountain be measured? What about the distance from Earth to the sun? Like many seemingly impossible problems, we rely on mathematical formulas to find the answers. The trigonometric identities, commonly used in mathematical proofs, have had real-world applications for centuries, including their use in calculating long distances.

The trigonometric identities we will examine in this section can be traced to a Persian astronomer who lived around 950 AD, but the ancient Greeks discovered these same formulas much earlier and stated them in terms of chords. These are special equations or postulates, true for all values input to the equations, and with innumerable applications.

In this section, we will learn techniques that will enable us to solve problems such as the ones presented above. The formulas that follow will simplify many trigonometric expressions and equations. Keep in mind that, throughout this section, the term *formula* is used synonymously with the word *identity*.

Using the Sum and Difference Formulas for Cosine

Finding the exact value of the sine, cosine, or tangent of an angle is often easier if we can rewrite the given angle in terms of two angles that have known trigonometric values. We can use the special angles, which we can review in the unit circle shown in Figure 2.

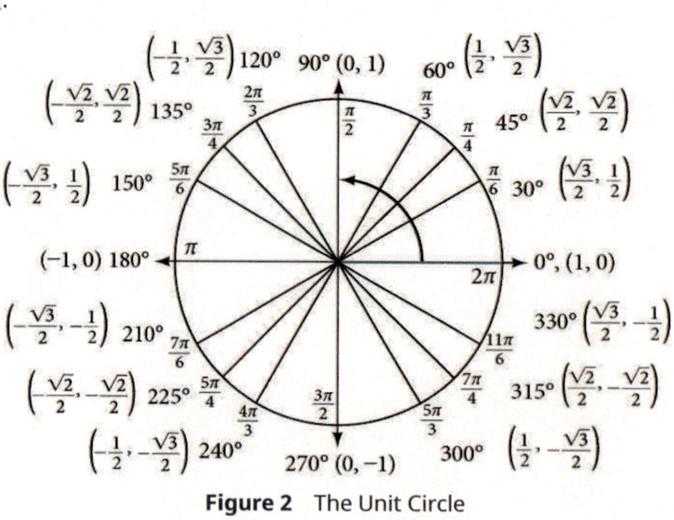

Figure 2 The Unit Circle

We will begin with the sum and difference formulas for cosine, so that we can find the cosine of a given angle if we can break it up into the sum or difference of two of the special angles. See Table 1.

Sum formula for cosine	$\cos(\alpha+\beta)=\cos\ \alpha\ \cos\ \beta-\sin\ \alpha\ \sin\ \beta$
Difference formula for cosine	$\cos(\alpha-\beta)=\cos\ \alpha\ \cos\ \beta+\sin\ \alpha\ \sin\ \beta$

Table 1

First, we will prove the difference formula for cosines. Let's consider two points on the unit circle. See Figure 3. Point P is at an angle α from the positive x-axis with coordinates $(\cos\ \alpha, \sin\ \alpha)$ and point Q is at an angle of β from the positive x-axis with coordinates $(\cos\ \beta, \sin\ \beta)$. Note the measure of angle POQ is $\alpha-\beta$.

Label two more points: A at an angle of $(\alpha-\beta)$ from the positive x-axis with coordinates $(\cos(\alpha-\beta), \sin(\alpha-\beta))$; and point B with coordinates $(1,0)$. Triangle POQ is a rotation of triangle AOB and thus the distance from P to Q is the same as the distance from A to B.

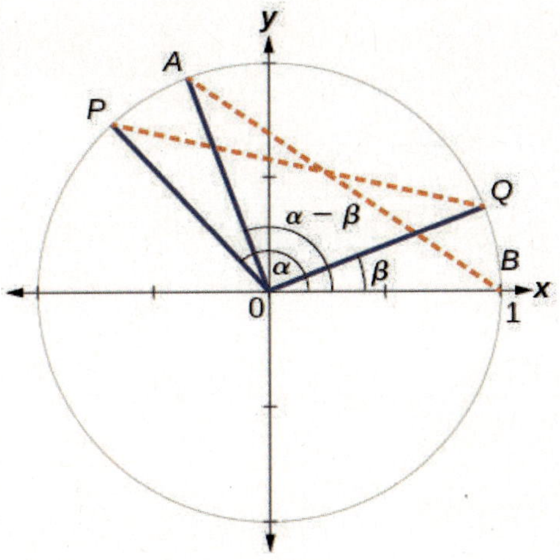

Figure 3

We can find the distance from P to Q using the distance formula.

$$\begin{aligned} d_{PQ} &= \sqrt{(\cos\ \alpha-\cos\ \beta)^2+(\sin\ \alpha-\sin\ \beta)^2} \\ &= \sqrt{\cos^2\alpha-2\ \cos\ \alpha\ \cos\ \beta+\cos^2\beta+\sin^2\alpha-2\ \sin\ \alpha\ \sin\ \beta+\sin^2\beta} \end{aligned}$$

Then we apply the Pythagorean identity and simplify.

$$\begin{aligned} &= \sqrt{\left(\cos^2\alpha+\sin^2\alpha\right)+\left(\cos^2\beta+\sin^2\beta\right)-2\ \cos\ \alpha\ \cos\ \beta-2\ \sin\ \alpha\ \sin\ \beta} \\ &= \sqrt{1+1-2\ \cos\ \alpha\ \cos\ \beta-2\ \sin\ \alpha\ \sin\ \beta} \\ &= \sqrt{2-2\ \cos\ \alpha\ \cos\ \beta-2\ \sin\ \alpha\ \sin\ \beta} \end{aligned}$$

Similarly, using the distance formula we can find the distance from A to B.

$$\begin{aligned} d_{AB} &= \sqrt{(\cos(\alpha-\beta)-1)^2+(\sin(\alpha-\beta)-0)^2} \\ &= \sqrt{\cos^2(\alpha-\beta)-2\ \cos(\alpha-\beta)+1+\sin^2(\alpha-\beta)} \end{aligned}$$

Applying the Pythagorean identity and simplifying we get:

$$\begin{aligned} &= \sqrt{\left(\cos^2(\alpha-\beta)+\sin^2(\alpha-\beta)\right)-2\ \cos(\alpha-\beta)+1} \\ &= \sqrt{1-2\ \cos(\alpha-\beta)+1} \\ &= \sqrt{2-2\ \cos(\alpha-\beta)} \end{aligned}$$

Because the two distances are the same, we set them equal to each other and simplify.

$$\begin{aligned}\sqrt{2-2\cos\alpha\cos\beta-2\sin\alpha\sin\beta} &= \sqrt{2-2\cos(\alpha-\beta)}\\ 2-2\cos\alpha\cos\beta-2\sin\alpha\sin\beta &= 2-2\cos(\alpha-\beta)\end{aligned}$$

Finally we subtract 2 from both sides and divide both sides by -2.

$$\cos\alpha\cos\beta+\sin\alpha\sin\beta=\cos(\alpha-\beta)$$

Thus, we have the difference formula for cosine. We can use similar methods to derive the cosine of the sum of two angles.

Sum and Difference Formulas for Cosine

These formulas can be used to calculate the cosine of sums and differences of angles.

$$\begin{aligned}\cos(\alpha+\beta)&=\cos\alpha\cos\beta-\sin\alpha\sin\beta\\ \cos(\alpha-\beta)&=\cos\alpha\cos\beta+\sin\alpha\sin\beta\end{aligned}$$

HOW TO

Given two angles, find the cosine of the difference between the angles.

1. Write the difference formula for cosine.
2. Substitute the values of the given angles into the formula.
3. Simplify.

EXAMPLE 1

Finding the Exact Value Using the Formula for the Cosine of the Difference of Two Angles

Using the formula for the cosine of the difference of two angles, find the exact value of $\cos\left(\frac{5\pi}{4}-\frac{\pi}{6}\right)$.

Solution

Begin by writing the formula for the cosine of the difference of two angles. Then substitute the given values.

$$\begin{aligned}\cos(\alpha-\beta) &= \cos\alpha\cos\beta+\sin\alpha\sin\beta\\ \cos\left(\frac{5\pi}{4}-\frac{\pi}{6}\right) &= \cos\left(\frac{5\pi}{4}\right)\cos\left(\frac{\pi}{6}\right)+\sin\left(\frac{5\pi}{4}\right)\sin\left(\frac{\pi}{6}\right)\\ &= \left(-\frac{\sqrt{2}}{2}\right)\left(\frac{\sqrt{3}}{2}\right)-\left(\frac{\sqrt{2}}{2}\right)\left(\frac{1}{2}\right)\\ &= -\frac{\sqrt{6}}{4}-\frac{\sqrt{2}}{4}\\ &= \frac{-\sqrt{6}-\sqrt{2}}{4}\end{aligned}$$

Keep in mind that we can always check the answer using a graphing calculator in radian mode.

TRY IT #1 Find the exact value of $\cos\left(\frac{\pi}{3}-\frac{\pi}{4}\right)$.

EXAMPLE 2

Finding the Exact Value Using the Formula for the Sum of Two Angles for Cosine

Find the exact value of $\cos(75°)$.

Solution

As $75°=45°+30°$, we can evaluate $\cos(75°)$ as $\cos(45°+30°)$.

$$\begin{aligned}
\cos(\alpha+\beta) &= \cos\alpha\cos\beta - \sin\alpha\sin\beta \\
\cos(45^\circ+30^\circ) &= \cos(45^\circ)\cos(30^\circ) - \sin(45^\circ)\sin(30^\circ) \\
&= \frac{\sqrt{2}}{2}\left(\frac{\sqrt{3}}{2}\right) - \frac{\sqrt{2}}{2}\left(\frac{1}{2}\right) \\
&= \frac{\sqrt{6}}{4} - \frac{\sqrt{2}}{4} \\
&= \frac{\sqrt{6}-\sqrt{2}}{4}
\end{aligned}$$

Keep in mind that we can always check the answer using a graphing calculator in degree mode.

Analysis

Note that we could have also solved this problem using the fact that $75^\circ = 135^\circ - 60^\circ$.

$$\begin{aligned}
\cos(\alpha-\beta) &= \cos\alpha\cos\beta + \sin\alpha\sin\beta \\
\cos(135^\circ-60^\circ) &= \cos(135^\circ)\cos(60^\circ) + \sin(135^\circ)\sin(60^\circ) \\
&= \left(-\frac{\sqrt{2}}{2}\right)\left(\frac{1}{2}\right) + \left(\frac{\sqrt{2}}{2}\right)\left(\frac{\sqrt{3}}{2}\right) \\
&= \left(-\frac{\sqrt{2}}{4}\right) + \left(\frac{\sqrt{6}}{4}\right) \\
&= \left(\frac{\sqrt{6}-\sqrt{2}}{4}\right)
\end{aligned}$$

TRY IT #2 Find the exact value of $\cos(105^\circ)$.

Using the Sum and Difference Formulas for Sine

The sum and difference formulas for sine can be derived in the same manner as those for cosine, and they resemble the cosine formulas.

Sum and Difference Formulas for Sine

These formulas can be used to calculate the sines of sums and differences of angles.

$$\sin(\alpha+\beta) = \sin\alpha\cos\beta + \cos\alpha\sin\beta$$
$$\sin(\alpha-\beta) = \sin\alpha\cos\beta - \cos\alpha\sin\beta$$

HOW TO

Given two angles, find the sine of the difference between the angles.

1. Write the difference formula for sine.
2. Substitute the given angles into the formula.
3. Simplify.

EXAMPLE 3

Using Sum and Difference Identities to Evaluate the Difference of Angles

Use the sum and difference identities to evaluate the difference of the angles and show that part *a* equals part *b*.

ⓐ $\sin(45^\circ - 30^\circ)$ ⓑ $\sin(135^\circ - 120^\circ)$

Solution

ⓐ Let's begin by writing the formula and substitute the given angles.

$$\begin{aligned} \sin(\alpha - \beta) &= \sin\ \alpha\ \cos\ \beta - \cos\ \alpha\ \sin\ \beta \\ \sin(45° - 30°) &= \sin(45°)\cos(30°) - \cos(45°)\sin(30°) \end{aligned}$$

Next, we need to find the values of the trigonometric expressions.

$$\sin(45°) = \frac{\sqrt{2}}{2},\ \cos(30°) = \frac{\sqrt{3}}{2},\ \cos(45°) = \frac{\sqrt{2}}{2},\ \sin(30°) = \frac{1}{2}$$

Now we can substitute these values into the equation and simplify.

$$\begin{aligned} \sin(45° - 30°) &= \frac{\sqrt{2}}{2}\left(\frac{\sqrt{3}}{2}\right) - \frac{\sqrt{2}}{2}\left(\frac{1}{2}\right) \\ &= \frac{\sqrt{6}-\sqrt{2}}{4} \end{aligned}$$

ⓑ Again, we write the formula and substitute the given angles.

$$\begin{aligned} \sin(\alpha - \beta) &= \sin\ \alpha\ \cos\ \beta - \cos\ \alpha\ \sin\ \beta \\ \sin(135° - 120°) &= \sin(135°)\cos(120°) - \cos(135°)\sin(120°) \end{aligned}$$

Next, we find the values of the trigonometric expressions.

$$\sin(135°) = \frac{\sqrt{2}}{2}, \cos(120°) = -\frac{1}{2}, \cos(135°) = \frac{\sqrt{2}}{2}, \sin(120°) = \frac{\sqrt{3}}{2}$$

Now we can substitute these values into the equation and simplify.

$$\begin{aligned} \sin(135° - 120°) &= \frac{\sqrt{2}}{2}\left(-\frac{1}{2}\right) - \left(-\frac{\sqrt{2}}{2}\right)\left(\frac{\sqrt{3}}{2}\right) \\ &= \frac{-\sqrt{2}+\sqrt{6}}{4} \\ &= \frac{\sqrt{6}-\sqrt{2}}{4} \end{aligned}$$

$$\begin{aligned} \sin(135° - 120°) &= \frac{\sqrt{2}}{2}\left(-\frac{1}{2}\right) - \left(-\frac{\sqrt{2}}{2}\right)\left(\frac{\sqrt{3}}{2}\right) \\ &= \frac{-\sqrt{2}+\sqrt{6}}{4} \\ &= \frac{\sqrt{6}-\sqrt{2}}{4} \end{aligned}$$

EXAMPLE 4

Finding the Exact Value of an Expression Involving an Inverse Trigonometric Function

Find the exact value of $\sin\left(\cos^{-1}\frac{1}{2} + \sin^{-1}\frac{3}{5}\right)$. Then check the answer with a graphing calculator.

Solution

The pattern displayed in this problem is $\sin(\alpha + \beta)$. Let $\alpha = \cos^{-1}\frac{1}{2}$ and $\beta = \sin^{-1}\frac{3}{5}$. Then we can write

$$\begin{aligned} \cos\ \alpha &= \frac{1}{2}, 0 \le \alpha \le \pi \\ \sin\ \beta &= \frac{3}{5}, -\frac{\pi}{2} \le \beta \le \frac{\pi}{2} \end{aligned}$$

We will use the Pythagorean identities to find $\sin\ \alpha$ and $\cos\ \beta$.

$$\begin{aligned}\sin\ \alpha &= \sqrt{1-\cos^2\alpha}\\ &= \sqrt{1-\tfrac{1}{4}}\\ &= \sqrt{\tfrac{3}{4}}\\ &= \tfrac{\sqrt{3}}{2}\\ \cos\ \beta &= \sqrt{1-\sin^2\beta}\\ &= \sqrt{1-\tfrac{9}{25}}\\ &= \sqrt{\tfrac{16}{25}}\\ &= \tfrac{4}{5}\end{aligned}$$

Using the sum formula for sine,

$$\begin{aligned}\sin\left(\cos^{-1}\tfrac{1}{2}+\sin^{-1}\tfrac{3}{5}\right) &= \sin(\alpha+\beta)\\ &= \sin\ \alpha\ \cos\ \beta+\cos\ \alpha\ \sin\ \beta\\ &= \tfrac{\sqrt{3}}{2}\cdot\tfrac{4}{5}+\tfrac{1}{2}\cdot\tfrac{3}{5}\\ &= \tfrac{4\sqrt{3}+3}{10}\end{aligned}$$

Using the Sum and Difference Formulas for Tangent

Finding exact values for the tangent of the sum or difference of two angles is a little more complicated, but again, it is a matter of recognizing the pattern.

Finding the sum of two angles formula for tangent involves taking quotient of the sum formulas for sine and cosine and simplifying. Recall, $\tan\ x=\frac{\sin\ x}{\cos\ x}, \cos\ x\neq 0$.

Let's derive the sum formula for tangent.

$$\begin{aligned}\tan(\alpha+\beta) &= \frac{\sin(\alpha+\beta)}{\cos(\alpha+\beta)}\\ &= \frac{\sin\ \alpha\ \cos\ \beta+\cos\ \alpha\ \sin\ \beta}{\cos\ \alpha\ \cos\ \beta-\sin\ \alpha\ \sin\ \beta}\\ &= \frac{\frac{\sin\ \alpha\ \cos\ \beta+\cos\ \alpha\ \sin\ \beta}{\cos\ \alpha\ \cos\ \beta}}{\frac{\cos\ \alpha\ \cos\ \beta-\sin\ \alpha\ \sin\ \beta}{\cos\ \alpha\ \cos\ \beta}} && \text{Divide the numerator and denominator by } \cos\ \alpha\ \cos\ \beta.\\ &= \frac{\frac{\sin\ \alpha\ \cancel{\cos\ \beta}}{\cos\ \alpha\ \cancel{\cos\ \beta}}+\frac{\cancel{\cos\ \alpha}\ \sin\ \beta}{\cancel{\cos\ \alpha}\ \cos\ \beta}}{\frac{\cancel{\cos\ \alpha}\ \cancel{\cos\ \beta}}{\cancel{\cos\ \alpha}\ \cancel{\cos\ \beta}}-\frac{\sin\ \alpha\ \sin\ \beta}{\cos\ \alpha\ \cos\ \beta}}\\ &= \frac{\frac{\sin\ \alpha}{\cos\ \alpha}+\frac{\sin\ \beta}{\cos\ \beta}}{1-\frac{\sin\ \alpha\ \sin\ \beta}{\cos\ \alpha\ \cos\ \beta}}\\ &= \frac{\tan\ \alpha+\tan\ \beta}{1-\tan\ \alpha\ \tan\ \beta}\end{aligned}$$

We can derive the difference formula for tangent in a similar way.

Sum and Difference Formulas for Tangent

The sum and difference formulas for tangent are:

$$\tan(\alpha+\beta)=\frac{\tan\ \alpha+\tan\ \beta}{1-\tan\ \alpha\ \tan\ \beta}$$

$$\tan(\alpha - \beta) = \frac{\tan\ \alpha - \tan\ \beta}{1 + \tan\ \alpha\ \tan\ \beta}$$

HOW TO

Given two angles, find the tangent of the sum of the angles.

1. Write the sum formula for tangent.
2. Substitute the given angles into the formula.
3. Simplify.

EXAMPLE 5

Finding the Exact Value of an Expression Involving Tangent

Find the exact value of $\tan\left(\frac{\pi}{6} + \frac{\pi}{4}\right)$.

Solution

Let's first write the sum formula for tangent and then substitute the given angles into the formula.

$$\begin{aligned} \tan(\alpha + \beta) &= \frac{\tan\ \alpha + \tan\ \beta}{1 - \tan\ \alpha\ \tan\ \beta} \\ \tan\left(\frac{\pi}{6} + \frac{\pi}{4}\right) &= \frac{\tan\left(\frac{\pi}{6}\right) + \tan\left(\frac{\pi}{4}\right)}{1 - \left(\tan\left(\frac{\pi}{6}\right)\right)\left(\tan\left(\frac{\pi}{4}\right)\right)} \end{aligned}$$

Next, we determine the individual function values within the formula:

$$\tan\left(\frac{\pi}{6}\right) = \frac{1}{\sqrt{3}}, \tan\left(\frac{\pi}{4}\right) = 1$$

So we have

$$\begin{aligned} \tan\left(\frac{\pi}{6} + \frac{\pi}{4}\right) &= \frac{\frac{1}{\sqrt{3}} + 1}{1 - \left(\frac{1}{\sqrt{3}}\right)(1)} \\ &= \frac{\frac{1 + \sqrt{3}}{\sqrt{3}}}{\frac{\sqrt{3} - 1}{\sqrt{3}}} \\ &= \frac{1 + \sqrt{3}}{\sqrt{3}}\left(\frac{\sqrt{3}}{\sqrt{3} - 1}\right) \\ &= \frac{\sqrt{3} + 1}{\sqrt{3} - 1} \end{aligned}$$

TRY IT #3 Find the exact value of $\tan\left(\frac{2\pi}{3} + \frac{\pi}{4}\right)$.

EXAMPLE 6

Finding Multiple Sums and Differences of Angles

Given $\sin\ \alpha = \frac{3}{5}, 0 < \alpha < \frac{\pi}{2}, \cos\ \beta = -\frac{5}{13}, \pi < \beta < \frac{3\pi}{2}$, find

ⓐ $\sin(\alpha + \beta)$ ⓑ $\cos(\alpha + \beta)$ ⓒ $\tan(\alpha + \beta)$ ⓓ $\tan(\alpha - \beta)$

Solution

We can use the sum and difference formulas to identify the sum or difference of angles when the ratio of sine, cosine, or tangent is provided for each of the individual angles. To do so, we construct what is called a reference triangle to help find each component of the sum and difference formulas.

ⓐ To find $\sin(\alpha+\beta)$, we begin with $\sin\alpha=\frac{3}{5}$ and $0<\alpha<\frac{\pi}{2}$. The side opposite α has length 3, the hypotenuse has length 5, and α is in the first quadrant. See Figure 4. Using the Pythagorean Theorem, we can find the length of side a:

$$\begin{aligned} a^2+3^2 &= 5^2 \\ a^2 &= 16 \\ a &= 4 \end{aligned}$$

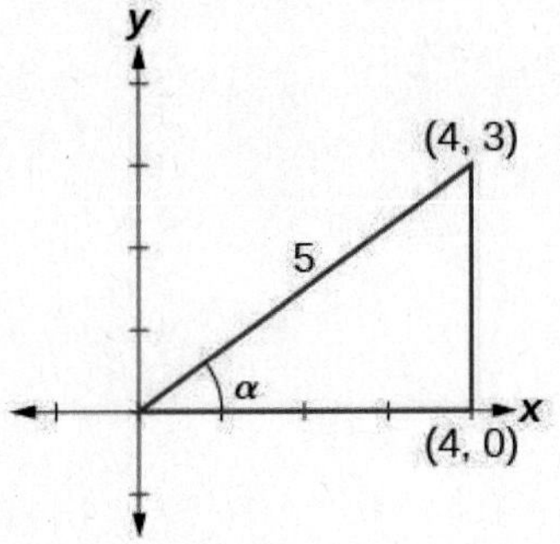

Figure 4

Since $\cos\beta=-\frac{5}{13}$ and $\pi<\beta<\frac{3\pi}{2}$, the side adjacent to β is -5, the hypotenuse is 13, and β is in the third quadrant. See Figure 5. Again, using the Pythagorean Theorem, we have

$$\begin{aligned} (-5)^2+a^2 &= 13^2 \\ 25+a^2 &= 169 \\ a^2 &= 144 \\ a &= \pm 12 \end{aligned}$$

Since β is in the third quadrant, $a=-12$.

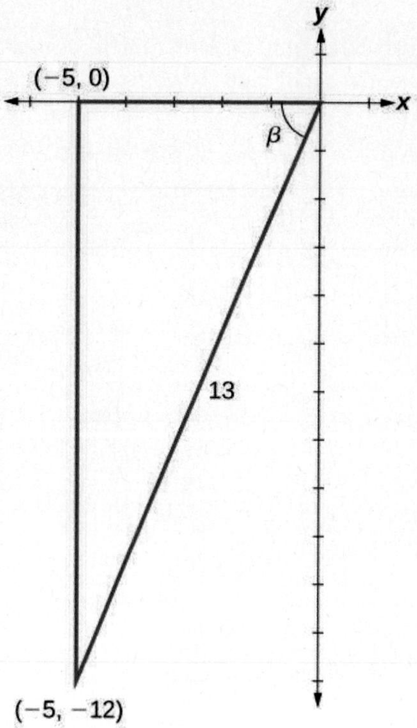

Figure 5

The next step is finding the cosine of α and the sine of β. The cosine of α is the adjacent side over the hypotenuse. We can find it from the triangle in Figure 5: $\cos\ \alpha = \frac{4}{5}$. We can also find the sine of β from the triangle in Figure 5, as opposite side over the hypotenuse: $\sin\ \beta = -\frac{12}{13}$. Now we are ready to evaluate $\sin(\alpha + \beta)$.

$$\begin{aligned} \sin(\alpha+\beta) &= \sin\ \alpha\ \cos\ \beta + \cos\ \alpha\ \sin\ \beta \\ &= \left(\tfrac{3}{5}\right)\left(-\tfrac{5}{13}\right) + \left(\tfrac{4}{5}\right)\left(-\tfrac{12}{13}\right) \\ &= -\tfrac{15}{65} - \tfrac{48}{65} \\ &= -\tfrac{63}{65} \end{aligned}$$

ⓑ We can find $\cos(\alpha + \beta)$ in a similar manner. We substitute the values according to the formula.

$$\begin{aligned} \cos(\alpha+\beta) &= \cos\ \alpha\ \cos\ \beta - \sin\ \alpha\ \sin\ \beta \\ &= \left(\tfrac{4}{5}\right)\left(-\tfrac{5}{13}\right) - \left(\tfrac{3}{5}\right)\left(-\tfrac{12}{13}\right) \\ &= -\tfrac{20}{65} + \tfrac{36}{65} \\ &= \tfrac{16}{65} \end{aligned}$$

ⓒ For $\tan(\alpha + \beta)$, if $\sin\ \alpha = \frac{3}{5}$ and $\cos\ \alpha = \frac{4}{5}$, then

$$\tan\ \alpha = \frac{\frac{3}{5}}{\frac{4}{5}} = \frac{3}{4}$$

If $\sin\ \beta = -\frac{12}{13}$ and $\cos\ \beta = -\frac{5}{13}$, then

$$\tan\ \beta = \frac{\frac{-12}{13}}{\frac{-5}{13}} = \frac{12}{5}$$

Then,

$$\begin{aligned} \tan(\alpha+\beta) &= \frac{\tan\ \alpha+\tan\ \beta}{1-\tan\ \alpha\ \tan\ \beta} \\ &= \frac{\frac{3}{4}+\frac{12}{5}}{1-\frac{3}{4}\left(\frac{12}{5}\right)} \\ &= \frac{\frac{63}{20}}{-\frac{16}{20}} \\ &= -\frac{63}{16} \end{aligned}$$

ⓓ To find $\tan(\alpha-\beta)$, we have the values we need. We can substitute them in and evaluate.

$$\begin{aligned} \tan(\alpha-\beta) &= \frac{\tan\ \alpha-\tan\ \beta}{1+\tan\ \alpha\ \tan\ \beta} \\ &= \frac{\frac{3}{4}-\frac{12}{5}}{1+\frac{3}{4}\left(\frac{12}{5}\right)} \\ &= \frac{-\frac{33}{20}}{\frac{56}{20}} \\ &= -\frac{33}{56} \end{aligned}$$

Analysis

A common mistake when addressing problems such as this one is that we may be tempted to think that α and β are angles in the same triangle, which of course, they are not. Also note that

$$\tan(\alpha+\beta)=\frac{\sin(\alpha+\beta)}{\cos(\alpha+\beta)}$$

Using Sum and Difference Formulas for Cofunctions

Now that we can find the sine, cosine, and tangent functions for the sums and differences of angles, we can use them to do the same for their cofunctions. You may recall from Right Triangle Trigonometry that, if the sum of two positive angles is $\frac{\pi}{2}$, those two angles are complements, and the sum of the two acute angles in a right triangle is $\frac{\pi}{2}$, so they are also complements. In Figure 6, notice that if one of the acute angles is labeled as θ, then the other acute angle must be labeled $\left(\frac{\pi}{2}-\theta\right)$.

Notice also that $\sin\ \theta=\cos\left(\frac{\pi}{2}-\theta\right)$, which is opposite over hypotenuse. Thus, when two angles are complementary, we can say that the sine of θ equals the cofunction of the complement of θ. Similarly, tangent and cotangent are cofunctions, and secant and cosecant are cofunctions.

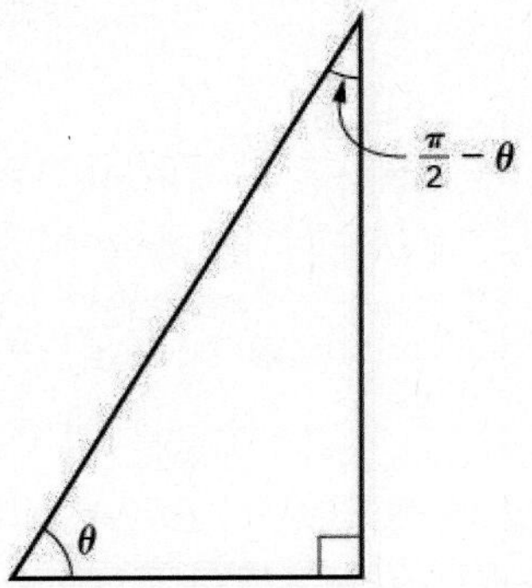

Figure 6

From these relationships, the cofunction identities are formed. Recall that you first encountered these identities in The Unit Circle: Sine and Cosine Functions.

Cofunction Identities

The cofunction identities are summarized in Table 2.

$$\sin\ \theta = \cos\left(\frac{\pi}{2} - \theta\right) \qquad \cos\ \theta = \sin\left(\frac{\pi}{2} - \theta\right)$$

$$\tan\ \theta = \cot\left(\frac{\pi}{2} - \theta\right) \qquad \cot\ \theta = \tan\left(\frac{\pi}{2} - \theta\right)$$

$$\sec\ \theta = \csc\left(\frac{\pi}{2} - \theta\right) \qquad \csc\ \theta = \sec\left(\frac{\pi}{2} - \theta\right)$$

Table 2

Notice that the formulas in the table may also justified algebraically using the sum and difference formulas. For example, using

$$\cos(\alpha - \beta) = \cos\ \alpha \cos\ \beta + \sin\ \alpha \sin\ \beta,$$

we can write

$$\begin{aligned} \cos\left(\frac{\pi}{2} - \theta\right) &= \cos\ \frac{\pi}{2} \cos\ \theta + \sin\ \frac{\pi}{2}\ \sin\ \theta \\ &= (0)\cos\ \theta + (1)\sin\ \theta \\ &= \sin\ \theta \end{aligned}$$

EXAMPLE 7

Finding a Cofunction with the Same Value as the Given Expression

Write $\tan\ \frac{\pi}{9}$ in terms of its cofunction.

Solution

The cofunction of $\tan\ \theta = \cot\left(\frac{\pi}{2} - \theta\right)$. Thus,

$$\begin{aligned} \tan\left(\frac{\pi}{9}\right) &= \cot\left(\frac{\pi}{2} - \frac{\pi}{9}\right) \\ &= \cot\left(\frac{9\pi}{18} - \frac{2\pi}{18}\right) \\ &= \cot\left(\frac{7\pi}{18}\right) \end{aligned}$$

TRY IT #4 Write $\sin\ \frac{\pi}{7}$ in terms of its cofunction.

Using the Sum and Difference Formulas to Verify Identities

Verifying an identity means demonstrating that the equation holds for all values of the variable. It helps to be very familiar with the identities or to have a list of them accessible while working the problems. Reviewing the general rules presented earlier may help simplify the process of verifying an identity.

HOW TO

Given an identity, verify using sum and difference formulas.

1. Begin with the expression on the side of the equal sign that appears most complex. Rewrite that expression until it matches the other side of the equal sign. Occasionally, we might have to alter both sides, but working on only one side is the most efficient.
2. Look for opportunities to use the sum and difference formulas.
3. Rewrite sums or differences of quotients as single quotients.
4. If the process becomes cumbersome, rewrite the expression in terms of sines and cosines.

EXAMPLE 8

Verifying an Identity Involving Sine

Verify the identity $\sin(\alpha+\beta)+\sin(\alpha-\beta)=2\ \sin\ \alpha\ \cos\ \beta.$

Solution

We see that the left side of the equation includes the sines of the sum and the difference of angles.

$$\begin{aligned}\sin(\alpha+\beta) &= \sin\ \alpha\ \cos\ \beta+\cos\ \alpha\ \sin\ \beta\\ \sin(\alpha-\beta) &= \sin\ \alpha\ \cos\ \beta-\cos\ \alpha\ \sin\ \beta\end{aligned}$$

We can rewrite each using the sum and difference formulas.

$$\begin{aligned}\sin(\alpha+\beta)+\sin(\alpha-\beta) &= \sin\ \alpha\ \cos\ \beta+\cos\ \alpha\ \sin\ \beta+\sin\ \alpha\ \cos\ \beta-\cos\ \alpha\ \sin\ \beta\\ &= 2\ \sin\ \alpha\ \cos\ \beta\end{aligned}$$

We see that the identity is verified.

EXAMPLE 9

Verifying an Identity Involving Tangent

Verify the following identity.

$$\frac{\sin(\alpha-\beta)}{\cos\ \alpha\ \cos\ \beta}=\tan\ \alpha-\tan\ \beta$$

Solution

We can begin by rewriting the numerator on the left side of the equation.

$$\begin{aligned}\frac{\sin(\alpha-\beta)}{\cos\ \alpha\ \cos\ \beta} &= \frac{\sin\ \alpha\ \cos\ \beta-\cos\ \alpha\sin\ \beta}{\cos\ \alpha\cos\ \beta}\\ &= \frac{\sin\ \alpha\ \cancel{\cos\ \beta}}{\cos\ \alpha\ \cancel{\cos\ \beta}}-\frac{\cancel{\cos\ \alpha}\ \sin\ \beta}{\cancel{\cos\ \alpha}\ \cos\ \beta} && \text{Rewrite using a common denominator.}\\ &= \frac{\sin\ \alpha}{\cos\ \alpha}-\frac{\sin\ \beta}{\cos\ \beta} && \text{Cancel.}\\ &= \tan\ \alpha-\tan\ \beta && \text{Rewrite in terms of tangent.}\end{aligned}$$

We see that the identity is verified. In many cases, verifying tangent identities can successfully be accomplished by writing the tangent in terms of sine and cosine.

> **TRY IT** #5 Verify the identity: $\tan(\pi-\theta)=-\tan\ \theta.$

EXAMPLE 10

Using Sum and Difference Formulas to Solve an Application Problem

Let L_1 and L_2 denote two non-vertical intersecting lines, and let θ denote the acute angle between L_1 and L_2. See Figure 7. Show that

$$\tan\ \theta=\frac{m_2-m_1}{1+m_1m_2}$$

where m_1 and m_2 are the slopes of L_1 and L_2 respectively. (**Hint:** Use the fact that $\tan\ \theta_1=m_1$ and $\tan\ \theta_2=m_2$.)

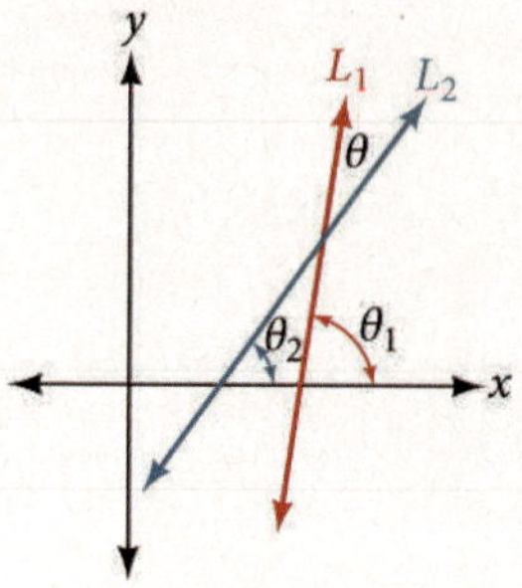

Figure 7

Solution

Using the difference formula for tangent, this problem does not seem as daunting as it might.

$$\begin{aligned}\tan\theta &= \tan(\theta_2-\theta_1)\\ &= \frac{\tan\theta_2-\tan\theta_1}{1+\tan\theta_1\tan\theta_2}\\ &= \frac{m_2-m_1}{1+m_1m_2}\end{aligned}$$

EXAMPLE 11

Investigating a Guy-wire Problem

For a climbing wall, a guy-wire R is attached 47 feet high on a vertical pole. Added support is provided by another guy-wire S attached 40 feet above ground on the same pole. If the wires are attached to the ground 50 feet from the pole, find the angle α between the wires. See Figure 8.

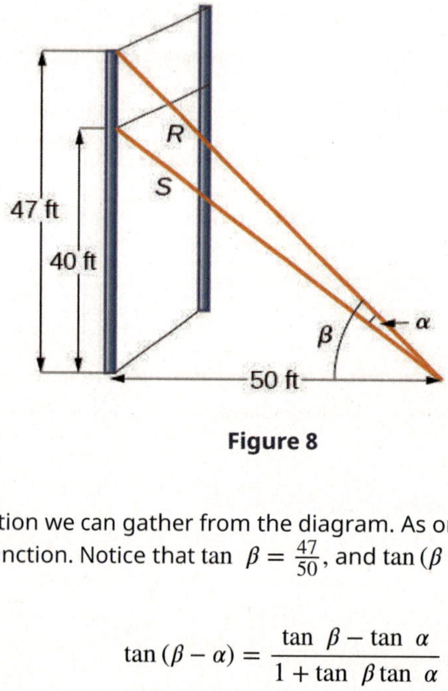

Figure 8

Solution

Let's first summarize the information we can gather from the diagram. As only the sides adjacent to the right angle are known, we can use the tangent function. Notice that $\tan\beta=\frac{47}{50}$, and $\tan(\beta-\alpha)=\frac{40}{50}=\frac{4}{5}$. We can then use difference formula for tangent.

$$\tan(\beta-\alpha)=\frac{\tan\beta-\tan\alpha}{1+\tan\beta\tan\alpha}$$

Now, substituting the values we know into the formula, we have

$$\begin{aligned}\frac{4}{5} &= \frac{\frac{47}{50}-\tan\alpha}{1+\frac{47}{50}\tan\alpha}\\ 4\left(1+\frac{47}{50}\tan\alpha\right) &= 5\left(\frac{47}{50}-\tan\alpha\right)\end{aligned}$$

Use the distributive property, and then simplify the functions.

$$\begin{aligned} 4(1)+4\left(\tfrac{47}{50}\right)\tan\alpha &= 5\left(\tfrac{47}{50}\right)-5\tan\alpha \\ 4+3.76\tan\alpha &= 4.7-5\tan\alpha \\ 5\tan\alpha+3.76\tan\alpha &= 0.7 \\ 8.76\tan\alpha &= 0.7 \\ \tan\alpha &\approx 0.07991 \\ \tan^{-1}(0.07991) &\approx .079741 \end{aligned}$$

Now we can calculate the angle in degrees.

$$\alpha \approx 0.079741\left(\frac{180}{\pi}\right) \approx 4.57°$$

Analysis

Occasionally, when an application appears that includes a right triangle, we may think that solving is a matter of applying the Pythagorean Theorem. That may be partially true, but it depends on what the problem is asking and what information is given.

MEDIA

Access these online resources for additional instruction and practice with sum and difference identities.

Sum and Difference Identities for Cosine (http://openstax.org/l/sumdifcos)
Sum and Difference Identities for Sine (http://openstax.org/l/sumdifsin)
Sum and Difference Identities for Tangent (http://openstax.org/l/sumdiftan)

9.2 SECTION EXERCISES

Verbal

1. Explain the basis for the cofunction identities and when they apply.

2. Is there only one way to evaluate $\cos\left(\frac{5\pi}{4}\right)$? Explain how to set up the solution in two different ways, and then compute to make sure they give the same answer.

3. Explain to someone who has forgotten the even-odd properties of sinusoidal functions how the addition and subtraction formulas can determine this characteristic for $f(x) = \sin(x)$ and $g(x) = \cos(x)$. (Hint: $0 - x = -x$)

Algebraic

For the following exercises, find the exact value.

4. $\cos\left(\frac{7\pi}{12}\right)$

5. $\cos\left(\frac{\pi}{12}\right)$

6. $\sin\left(\frac{5\pi}{12}\right)$

7. $\sin\left(\frac{11\pi}{12}\right)$

8. $\tan\left(-\frac{\pi}{12}\right)$

9. $\tan\left(\frac{19\pi}{12}\right)$

For the following exercises, rewrite in terms of $\sin x$ *and* $\cos x$.

10. $\sin\left(x + \frac{11\pi}{6}\right)$

11. $\sin\left(x - \frac{3\pi}{4}\right)$

12. $\cos\left(x - \frac{5\pi}{6}\right)$

13. $\cos\left(x + \frac{2\pi}{3}\right)$

For the following exercises, simplify the given expression.

14. $\csc\left(\frac{\pi}{2} - t\right)$

15. $\sec\left(\frac{\pi}{2} - \theta\right)$

16. $\cot\left(\frac{\pi}{2} - x\right)$

17. $\tan\left(\frac{\pi}{2} - x\right)$

18. $\sin(2x)\ \cos(5x) - \sin(5x)\ \cos(2x)$

19. $\frac{\tan\left(\frac{3}{2}x\right) - \tan\left(\frac{7}{5}x\right)}{1 + \tan\left(\frac{3}{2}x\right)\tan\left(\frac{7}{5}x\right)}$

For the following exercises, find the requested information.

20. Given that $\sin\ a = \frac{2}{3}$ and $\cos\ b = -\frac{1}{4}$, with a and b both in the interval $\left[\frac{\pi}{2}, \pi\right)$, find $\sin(a + b)$ and $\cos(a - b)$.

21. Given that $\sin\ a = \frac{4}{5}$, and $\cos\ b = \frac{1}{3}$, with a and b both in the interval $\left[0, \frac{\pi}{2}\right)$, find $\sin(a - b)$ and $\cos(a + b)$.

For the following exercises, find the exact value of each expression.

22. $\sin\left(\cos^{-1}(0) - \cos^{-1}\left(\frac{1}{2}\right)\right)$

23. $\cos\left(\cos^{-1}\left(\frac{\sqrt{2}}{2}\right) + \sin^{-1}\left(\frac{\sqrt{3}}{2}\right)\right)$

24. $\tan\left(\sin^{-1}\left(\frac{1}{2}\right) - \cos^{-1}\left(\frac{1}{2}\right)\right)$

Graphical

For the following exercises, simplify the expression, and then graph both expressions as functions to verify the graphs are identical. Confirm your answer using a graphing calculator.

25. $\cos\left(\frac{\pi}{2} - x\right)$

26. $\sin(\pi - x)$

27. $\tan\left(\frac{\pi}{3} + x\right)$

28. $\sin\left(\frac{\pi}{3} + x\right)$

29. $\tan\left(\frac{\pi}{4} - x\right)$

30. $\cos\left(\frac{7\pi}{6} + x\right)$

31. $\sin\left(\frac{\pi}{4} + x\right)$

32. $\cos\left(\frac{5\pi}{4} + x\right)$

For the following exercises, use a graph to determine whether the functions are the same or different. If they are the same, show why. If they are different, replace the second function with one that is identical to the first. (Hint: think $2x = x + x$.)

33. $f(x) = \sin(4x) - \sin(3x)\cos\ x,\ g(x) = \sin\ x\ \cos(3x)$

34. $f(x) = \cos(4x) + \sin\ x\ \sin(3x),\ g(x) = -\cos\ x\ \cos(3x)$

35. $f(x) = \sin(3x)\cos(6x),\ g(x) = -\sin(3x)\cos(6x)$

36. $f(x) = \sin(4x),\ g(x) = \sin(5x)\cos\ x - \cos(5x)\sin\ x$

37. $f(x) = \sin(2x),\ g(x) = 2\ \sin\ x\ \cos\ x$

38. $f(\theta) = \cos(2\theta),\ g(\theta) = \cos^2\theta - \sin^2\theta$

39. $f(\theta) = \tan(2\theta),\ g(\theta) = \frac{\tan\ \theta}{1 + \tan^2\theta}$

40. $f(x) = \sin(3x)\sin\ x, g(x) = \sin^2(2x)\cos^2 x - \cos^2(2x)\sin^2 x$

41. $f(x) = \tan(-x), g(x) = \frac{\tan\ x - \tan(2x)}{1 - \tan\ x\ \tan(2x)}$

Technology

For the following exercises, find the exact value algebraically, and then confirm the answer with a calculator to the fourth decimal point.

42. $\sin(75°)$

43. $\sin(195°)$

44. $\cos(165°)$

45. $\cos(345°)$

46. $\tan(-15°)$

Extensions

For the following exercises, prove the identities provided.

47. $\tan(x + \frac{\pi}{4}) = \frac{\tan\ x + 1}{1 - \tan\ x}$

48. $\frac{\tan(a+b)}{\tan(a-b)} = \frac{\sin\ a\ \cos\ a + \sin\ b\ \cos\ b}{\sin\ a\ \cos\ a - \sin\ b\ \cos\ b}$

49. $\frac{\cos(a+b)}{\cos\ a\ \cos\ b} = 1 - \tan\ a\ \tan\ b$

50. $\cos(x + y)\cos(x - y) = \cos^2 x - \sin^2 y$

51. $\frac{\cos(x+h) - \cos\ x}{h} = \cos\ x \frac{\cos\ h - 1}{h} - \sin\ x \frac{\sin\ h}{h}$

For the following exercises, prove or disprove the statements.

52. $\tan(u + v) = \frac{\tan\ u + \tan\ v}{1 - \tan\ u\ \tan\ v}$

53. $\tan(u - v) = \frac{\tan\ u - \tan\ v}{1 + \tan\ u\ \tan\ v}$

54. $\frac{\tan(x+y)}{1 + \tan\ x\ \tan\ x} = \frac{\tan\ x + \tan\ y}{1 - \tan^2 x\ \tan^2 y}$

55. If α, β, and γ are angles in the same triangle, then prove or disprove $\sin(\alpha + \beta) = \sin\ \gamma$.

56. If α, β, and y are angles in the same triangle, then prove or disprove $\tan\ \alpha + \tan\ \beta + \tan\ \gamma = \tan\ \alpha\ \tan\ \beta\ \tan\ \gamma$

9.3 Double-Angle, Half-Angle, and Reduction Formulas

Learning Objectives

In this section, you will:

- Use double-angle formulas to find exact values.
- Use double-angle formulas to verify identities.
- Use reduction formulas to simplify an expression.
- Use half-angle formulas to find exact values.

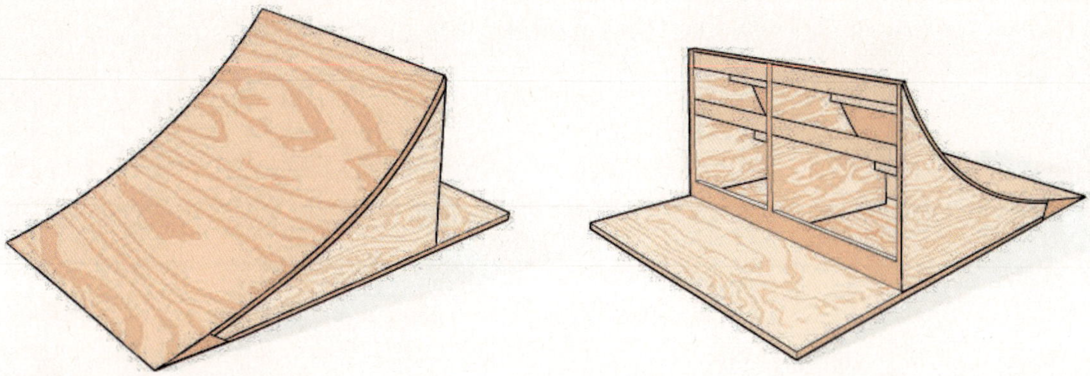

Figure 1 Bicycle and skateboard ramps for advanced riders have a steeper incline than those designed for novices.

Bicycle and skateboard ramps made for competition (see Figure 1) must vary in height depending on the skill level of the competitors. For advanced competitors, the angle formed by the ramp and the ground should be θ such that $\tan\ \theta = \frac{5}{3}$. The angle is divided in half for novices. What is the steepness of the ramp for novices? In this section, we will investigate three additional categories of identities that we can use to answer questions such as this one.

Using Double-Angle Formulas to Find Exact Values

In the previous section, we used addition and subtraction formulas for trigonometric functions. Now, we take another look at those same formulas. The double-angle formulas are a special case of the sum formulas, where $\alpha = \beta$. Deriving the double-angle formula for sine begins with the sum formula,

$$\sin(\alpha + \beta) = \sin\ \alpha\ \cos\ \beta + \cos\ \alpha\ \sin\ \beta$$

If we let $\alpha = \beta = \theta$, then we have

$$\begin{aligned} \sin(\theta + \theta) &= \sin\ \theta\ \cos\ \theta + \cos\ \theta\ \sin\ \theta \\ \sin(2\theta) &= 2\sin\ \theta\ \cos\ \theta \end{aligned}$$

Deriving the double-angle for cosine gives us three options. First, starting from the sum formula, $\cos(\alpha + \beta) = \cos\ \alpha\ \cos\ \beta - \sin\ \alpha\ \sin\ \beta$, and letting $\alpha = \beta = \theta$, we have

$$\begin{aligned} \cos(\theta + \theta) &= \cos\ \theta\ \cos\ \theta - \sin\ \theta\ \sin\ \theta \\ \cos(2\theta) &= \cos^2\theta - \sin^2\theta \end{aligned}$$

Using the Pythagorean properties, we can expand this double-angle formula for cosine and get two more variations. The first variation is:

$$\begin{aligned} \cos(2\theta) &= \cos^2\theta - \sin^2\theta \\ &= \left(1 - \sin^2\theta\right) - \sin^2\theta \\ &= 1 - 2\sin^2\theta \end{aligned}$$

The second variation is:

$$\begin{aligned} \cos(2\theta) &= \cos^2\theta - \sin^2\theta \\ &= \cos^2\theta - \left(1 - \cos^2\theta\right) \\ &= 2\cos^2\theta - 1 \end{aligned}$$

Similarly, to derive the double-angle formula for tangent, replacing $\alpha = \beta = \theta$ in the sum formula gives

$$\begin{aligned} \tan(\alpha + \beta) &= \frac{\tan\ \alpha + \tan\ \beta}{1 - \tan\ \alpha\ \tan\ \beta} \\ \tan(\theta + \theta) &= \frac{\tan\ \theta + \tan\ \theta}{1 - \tan\ \theta\ \tan\ \theta} \\ \tan(2\theta) &= \frac{2\tan\ \theta}{1 - \tan^2\theta} \end{aligned}$$

Double-Angle Formulas

The **double-angle formulas** are summarized as follows:

$$\sin(2\theta) = 2\sin\theta\cos\theta$$

$$\begin{aligned}\cos(2\theta) &= \cos^2\theta - \sin^2\theta \\ &= 1 - 2\sin^2\theta \\ &= 2\cos^2\theta - 1\end{aligned}$$

$$\tan(2\theta) = \frac{2\tan\theta}{1-\tan^2\theta}$$

HOW TO

Given the tangent of an angle and the quadrant in which it is located, use the double-angle formulas to find the exact value.

1. Draw a triangle to reflect the given information.
2. Determine the correct double-angle formula.
3. Substitute values into the formula based on the triangle.
4. Simplify.

EXAMPLE 1

Using a Double-Angle Formula to Find the Exact Value Involving Tangent

Given that $\tan\theta = -\frac{3}{4}$ and θ is in quadrant II, find the following:

ⓐ $\sin(2\theta)$ ⓑ $\cos(2\theta)$ ⓒ $\tan(2\theta)$

Solution

If we draw a triangle to reflect the information given, we can find the values needed to solve the problems on the image. We are given $\tan\theta = -\frac{3}{4}$, such that θ is in quadrant II. The tangent of an angle is equal to the opposite side over the adjacent side, and because θ is in the second quadrant, the adjacent side is on the *x*-axis and is negative. Use the Pythagorean Theorem to find the length of the hypotenuse:

$$\begin{aligned}(-4)^2 + (3)^2 &= c^2 \\ 16 + 9 &= c^2 \\ 25 &= c^2 \\ c &= 5\end{aligned}$$

Now we can draw a triangle similar to the one shown in Figure 2.

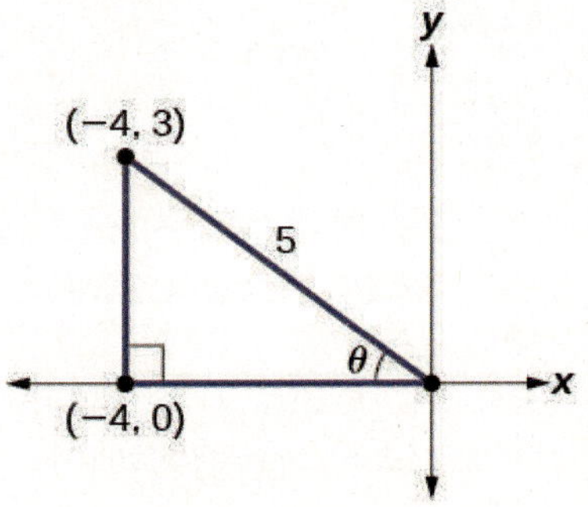

Figure 2

ⓐ Let's begin by writing the double-angle formula for sine.

$$\sin(2\theta) = 2\ \sin\ \theta\ \cos\ \theta$$

We see that we to need to find $\sin\ \theta$ and $\cos\ \theta$. Based on Figure 2, we see that the hypotenuse equals 5, so $\sin\ \theta = \frac{3}{5}$, and $\cos\ \theta = -\frac{4}{5}$. Substitute these values into the equation, and simplify.

Thus,

$$\begin{aligned} \sin(2\theta) &= 2\left(\tfrac{3}{5}\right)\left(-\tfrac{4}{5}\right) \\ &= -\tfrac{24}{25} \end{aligned}$$

ⓑ Write the double-angle formula for cosine.

$$\cos(2\theta) = \cos^2\theta - \sin^2\theta$$

Again, substitute the values of the sine and cosine into the equation, and simplify.

$$\begin{aligned} \cos(2\theta) &= \left(-\tfrac{4}{5}\right)^2 - \left(\tfrac{3}{5}\right)^2 \\ &= \tfrac{16}{25} - \tfrac{9}{25} \\ &= \tfrac{7}{25} \end{aligned}$$

ⓒ Write the double-angle formula for tangent.

$$\tan(2\theta) = \frac{2\ \tan\ \theta}{1 - \tan^2\theta}$$

In this formula, we need the tangent, which we were given as $\tan\ \theta = -\frac{3}{4}$. Substitute this value into the equation, and simplify.

$$\begin{aligned} \tan(2\theta) &= \frac{2\left(-\frac{3}{4}\right)}{1-\left(-\frac{3}{4}\right)^2} \\ &= \frac{-\frac{3}{2}}{1-\frac{9}{16}} \\ &= -\tfrac{3}{2}\left(\tfrac{16}{7}\right) \\ &= -\tfrac{24}{7} \end{aligned}$$

> **TRY IT** #1 Given $\sin\ \alpha = \frac{5}{8}$, with α in quadrant I, find $\cos(2\alpha)$.

EXAMPLE 2

Using the Double-Angle Formula for Cosine without Exact Values

Use the double-angle formula for cosine to write $\cos(6x)$ in terms of $\cos(3x)$.

Solution

$$\begin{aligned} \cos(6x) &= \cos(2(3x)) \\ &= 2\cos^2(3x) - 1 \end{aligned}$$

Analysis

This example illustrates that we can use the double-angle formula without having exact values. It emphasizes that the pattern is what we need to remember and that identities are true for all values in the domain of the trigonometric function.

Using Double-Angle Formulas to Verify Identities

Establishing identities using the double-angle formulas is performed using the same steps we used to derive the sum and difference formulas. Choose the more complicated side of the equation and rewrite it until it matches the other side.

EXAMPLE 3

Using the Double-Angle Formulas to Verify an Identity

Verify the following identity using double-angle formulas:

$$1 + \sin(2\theta) = (\sin\ \theta + \cos\ \theta)^2$$

Solution

We will work on the right side of the equal sign and rewrite the expression until it matches the left side.

$$\begin{aligned}(\sin\ \theta + \cos\ \theta)^2 &= \sin^2\theta + 2\ \sin\ \theta\ \cos\ \theta + \cos^2\theta \\ &= \left(\sin^2\theta + \cos^2\theta\right) + 2\ \sin\ \theta\ \cos\ \theta \\ &= 1 + 2\ \sin\ \theta\ \cos\ \theta \\ &= 1 + \sin(2\theta)\end{aligned}$$

Analysis

This process is not complicated, as long as we recall the perfect square formula from algebra:

$$(a \pm b)^2 = a^2 \pm 2ab + b^2$$

where $a = \sin\ \theta$ and $b = \cos\ \theta$. Part of being successful in mathematics is the ability to recognize patterns. While the terms or symbols may change, the algebra remains consistent.

TRY IT #2 Verify the identity: $\cos^4\theta - \sin^4\theta = \cos(2\theta)$.

EXAMPLE 4

Verifying a Double-Angle Identity for Tangent

Verify the identity:

$$\tan(2\theta) = \frac{2}{\cot\ \theta - \tan\ \theta}$$

Solution

In this case, we will work with the left side of the equation and simplify or rewrite until it equals the right side of the equation.

$$\begin{aligned}\tan(2\theta) &= \frac{2\ \tan\ \theta}{1-\tan^2\theta} && \text{Double-angle formula} \\ &= \frac{2\ \tan\ \theta\left(\frac{1}{\tan\ \theta}\right)}{(1-\tan^2\theta)\left(\frac{1}{\tan\ \theta}\right)} && \text{Multiply by a term that results in desired numerator.} \\ &= \frac{2}{\frac{1}{\tan\ \theta} - \frac{\tan^2\theta}{\tan\ \theta}} \\ &= \frac{2}{\cot\ \theta - \tan\ \theta} && \text{Use reciprocal identity for } \frac{1}{\tan\ \theta}.\end{aligned}$$

Analysis

Here is a case where the more complicated side of the initial equation appeared on the right, but we chose to work the left side. However, if we had chosen the left side to rewrite, we would have been working backwards to arrive at the equivalency. For example, suppose that we wanted to show

$$\frac{2\tan\ \theta}{1 - \tan^2\theta} = \frac{2}{\cot\ \theta - \tan\ \theta}$$

Let's work on the right side.

$$\begin{aligned}\frac{2}{\cot\theta-\tan\theta} &= \frac{2}{\frac{1}{\tan\theta}-\tan\theta}\left(\frac{\tan\theta}{\tan\theta}\right)\\ &= \frac{2\tan\theta}{\frac{1}{\tan\theta}(\tan\theta)-\tan\theta(\tan\theta)}\\ &= \frac{2\tan\theta}{1-\tan^2\theta}\end{aligned}$$

When using the identities to simplify a trigonometric expression or solve a trigonometric equation, there are usually several paths to a desired result. There is no set rule as to what side should be manipulated. However, we should begin with the guidelines set forth earlier.

TRY IT #3 Verify the identity: $\cos(2\theta)\cos\theta = \cos^3\theta - \cos\theta\sin^2\theta.$

Use Reduction Formulas to Simplify an Expression

The double-angle formulas can be used to derive the reduction formulas, which are formulas we can use to reduce the power of a given expression involving even powers of sine or cosine. They allow us to rewrite the even powers of sine or cosine in terms of the first power of cosine. These formulas are especially important in higher-level math courses, calculus in particular. Also called the power-reducing formulas, three identities are included and are easily derived from the double-angle formulas.

We can use two of the three double-angle formulas for cosine to derive the reduction formulas for sine and cosine. Let's begin with $\cos(2\theta) = 1 - 2\sin^2\theta$. Solve for $\sin^2\theta$:

$$\begin{aligned}\cos(2\theta) &= 1-2\sin^2\theta\\ 2\sin^2\theta &= 1-\cos(2\theta)\\ \sin^2\theta &= \frac{1-\cos(2\theta)}{2}\end{aligned}$$

Next, we use the formula $\cos(2\theta) = 2\cos^2\theta - 1$. Solve for $\cos^2\theta$:

$$\begin{aligned}\cos(2\theta) &= 2\cos^2\theta-1\\ 1+\cos(2\theta) &= 2\cos^2\theta\\ \frac{1+\cos(2\theta)}{2} &= \cos^2\theta\end{aligned}$$

The last reduction formula is derived by writing tangent in terms of sine and cosine:

$$\begin{aligned}\tan^2\theta &= \frac{\sin^2\theta}{\cos^2\theta}\\ &= \frac{\frac{1-\cos(2\theta)}{2}}{\frac{1+\cos(2\theta)}{2}} && \text{Substitute the reduction formulas.}\\ &= \left(\frac{1-\cos(2\theta)}{2}\right)\left(\frac{2}{1+\cos(2\theta)}\right)\\ &= \frac{1-\cos(2\theta)}{1+\cos(2\theta)}\end{aligned}$$

Reduction Formulas

The **reduction formulas** are summarized as follows:

$$\sin^2\theta = \frac{1-\cos(2\theta)}{2}$$

$$\cos^2\theta = \frac{1+\cos(2\theta)}{2}$$

$$\tan^2\theta = \frac{1-\cos(2\theta)}{1+\cos(2\theta)}$$

EXAMPLE 5

Writing an Equivalent Expression Not Containing Powers Greater Than 1

Write an equivalent expression for $\cos^4 x$ that does not involve any powers of sine or cosine greater than 1.

Solution

We will apply the reduction formula for cosine twice.

$$\begin{aligned}
\cos^4 x &= (\cos^2 x)^2 \\
&= \left(\frac{1+\cos(2x)}{2}\right)^2 && \text{Substitute reduction formula for } \cos^2 x. \\
&= \frac{1}{4}(1 + 2\cos(2x) + \cos^2(2x)) \\
&= \frac{1}{4} + \frac{1}{2}\cos(2x) + \frac{1}{4}\left(\frac{1+\cos 2(2x)}{2}\right) && \text{Substitute reduction formula for } \cos^2 x. \\
&= \frac{1}{4} + \frac{1}{2}\cos(2x) + \frac{1}{8} + \frac{1}{8}\cos(4x) \\
&= \frac{3}{8} + \frac{1}{2}\cos(2x) + \frac{1}{8}\cos(4x)
\end{aligned}$$

Analysis

The solution is found by using the reduction formula twice, as noted, and the perfect square formula from algebra.

EXAMPLE 6

Using the Power-Reducing Formulas to Prove an Identity

Use the power-reducing formulas to prove

$$\sin^3(2x) = \left[\frac{1}{2}\sin(2x)\right][1 - \cos(4x)]$$

Solution

We will work on simplifying the left side of the equation:

$$\begin{aligned}
\sin^3(2x) &= [\sin(2x)][\sin^2(2x)] \\
&= \sin(2x)\left[\frac{1-\cos(4x)}{2}\right] && \text{Substitute the power-reduction formula.} \\
&= \sin(2x)\left(\frac{1}{2}\right)[1 - \cos(4x)] \\
&= \frac{1}{2}[\sin(2x)][1 - \cos(4x)]
\end{aligned}$$

Analysis

Note that in this example, we substituted

$$\frac{1 - \cos(4x)}{2}$$

for $\sin^2(2x)$. The formula states

$$\sin^2\theta = \frac{1 - \cos(2\theta)}{2}$$

We let $\theta = 2x$, so $2\theta = 4x$.

> **TRY IT** #4 Use the power-reducing formulas to prove that $10\cos^4 x = \frac{15}{4} + 5\cos(2x) + \frac{5}{4}\cos(4x)$.

Using Half-Angle Formulas to Find Exact Values

The next set of identities is the set of **half-angle formulas**, which can be derived from the reduction formulas and we can use when we have an angle that is half the size of a special angle. If we replace θ with $\frac{\alpha}{2}$, the half-angle formula for sine is found by simplifying the equation and solving for $\sin\left(\frac{\alpha}{2}\right)$. Note that the half-angle formulas are preceded by a $\pm$

sign. This does not mean that both the positive and negative expressions are valid. Rather, it depends on the quadrant in which $\frac{\alpha}{2}$ terminates.

The half-angle formula for sine is derived as follows:

$$\begin{aligned} \sin^2\theta &= \frac{1-\cos(2\theta)}{2} \\ \sin^2\left(\frac{\alpha}{2}\right) &= \frac{1-\left(\cos 2\cdot\frac{\alpha}{2}\right)}{2} \\ &= \frac{1-\cos\ \alpha}{2} \\ \sin\left(\frac{\alpha}{2}\right) &= \pm\sqrt{\frac{1-\cos\ \alpha}{2}} \end{aligned}$$

To derive the half-angle formula for cosine, we have

$$\begin{aligned} \cos^2\theta &= \frac{1+\cos(2\theta)}{2} \\ \cos^2\left(\frac{\alpha}{2}\right) &= \frac{1+\cos\left(2\cdot\frac{\alpha}{2}\right)}{2} \\ &= \frac{1+\cos\ \alpha}{2} \\ \cos\left(\frac{\alpha}{2}\right) &= \pm\sqrt{\frac{1+\cos\ \alpha}{2}} \end{aligned}$$

For the tangent identity, we have

$$\begin{aligned} \tan^2\theta &= \frac{1-\cos(2\theta)}{1+\cos(2\theta)} \\ \tan^2\left(\frac{\alpha}{2}\right) &= \frac{1-\cos\left(2\cdot\frac{\alpha}{2}\right)}{1+\cos\left(2\cdot\frac{\alpha}{2}\right)} \\ &= \frac{1-\cos\ \alpha}{1+\cos\ \alpha} \\ \tan\left(\frac{\alpha}{2}\right) &= \pm\sqrt{\frac{1-\cos\ \alpha}{1+\cos\ \alpha}} \end{aligned}$$

Half-Angle Formulas

The **half-angle formulas** are as follows:

$$\sin\left(\frac{\alpha}{2}\right) = \pm\sqrt{\frac{1-\cos\ \alpha}{2}}$$

$$\cos\left(\frac{\alpha}{2}\right) = \pm\sqrt{\frac{1+\cos\ \alpha}{2}}$$

$$\begin{aligned} \tan\left(\frac{\alpha}{2}\right) &= \pm\sqrt{\frac{1-\cos\ \alpha}{1+\cos\ \alpha}} \\ &= \frac{\sin\ \alpha}{1+\cos\ \alpha} \\ &= \frac{1-\cos\ \alpha}{\sin\ \alpha} \end{aligned}$$

EXAMPLE 7

Using a Half-Angle Formula to Find the Exact Value of a Sine Function

Find $\sin(15°)$ using a half-angle formula.

Solution

Since $15° = \frac{30°}{2}$, we use the half-angle formula for sine:

$$\begin{aligned}\sin\frac{30°}{2} &= \sqrt{\frac{1-\cos 30°}{2}} \\ &= \sqrt{\frac{1-\frac{\sqrt{3}}{2}}{2}} \\ &= \sqrt{\frac{\frac{2-\sqrt{3}}{2}}{2}} \\ &= \sqrt{\frac{2-\sqrt{3}}{4}} \\ &= \frac{\sqrt{2-\sqrt{3}}}{2}\end{aligned}$$

Remember that we can check the answer with a graphing calculator.

Analysis

Notice that we used only the positive root because $\sin(15°)$ is positive.

HOW TO

Given the tangent of an angle and the quadrant in which the angle lies, find the exact values of trigonometric functions of half of the angle.

1. Draw a triangle to represent the given information.
2. Determine the correct half-angle formula.
3. Substitute values into the formula based on the triangle.
4. Simplify.

EXAMPLE 8

Finding Exact Values Using Half-Angle Identities

Given that $\tan\ \alpha = \frac{8}{15}$ and α lies in quadrant III, find the exact value of the following:

ⓐ $\sin\left(\frac{\alpha}{2}\right)$ ⓑ $\cos\left(\frac{\alpha}{2}\right)$ ⓒ $\tan\left(\frac{\alpha}{2}\right)$

Solution

Using the given information, we can draw the triangle shown in Figure 3. Using the Pythagorean Theorem, we find the hypotenuse to be 17. Therefore, we can calculate $\sin\ \alpha = -\frac{8}{17}$ and $\cos\ \alpha = -\frac{15}{17}$.

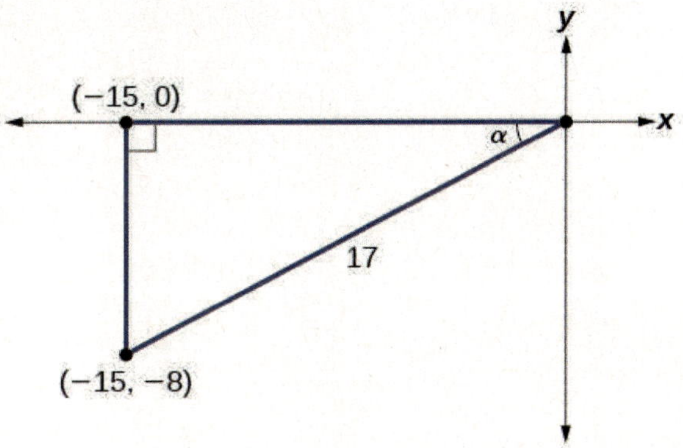

Figure 3

ⓐ Before we start, we must remember that if α is in quadrant III, then $180° < \alpha < 270°$, so $\frac{180°}{2} < \frac{\alpha}{2} < \frac{270°}{2}$. This means that the terminal side of $\frac{\alpha}{2}$ is in quadrant II, since $90° < \frac{\alpha}{2} < 135°$.
To find $\sin\ \frac{\alpha}{2}$, we begin by writing the half-angle formula for sine. Then we substitute the value of the cosine we found from the triangle in Figure 3 and simplify.

$$\begin{aligned}\sin\ \frac{\alpha}{2} &= \pm\sqrt{\frac{1-\cos\ \alpha}{2}}\\ &= \pm\sqrt{\frac{1-\left(-\frac{15}{17}\right)}{2}}\\ &= \pm\sqrt{\frac{\frac{32}{17}}{2}}\\ &= \pm\sqrt{\frac{32}{17}\cdot\frac{1}{2}}\\ &= \pm\sqrt{\frac{16}{17}}\\ &= \pm\frac{4}{\sqrt{17}}\\ &= \frac{4\sqrt{17}}{17}\end{aligned}$$

We choose the positive value of $\sin\ \frac{\alpha}{2}$ because the angle terminates in quadrant II and sine is positive in quadrant II.

ⓑ To find $\cos\ \frac{\alpha}{2}$, we will write the half-angle formula for cosine, substitute the value of the cosine we found from the triangle in Figure 3, and simplify.

$$\begin{aligned}\cos\ \frac{\alpha}{2} &= \pm\sqrt{\frac{1+\cos\ \alpha}{2}}\\ &= \pm\sqrt{\frac{1+\left(-\frac{15}{17}\right)}{2}}\\ &= \pm\sqrt{\frac{\frac{2}{17}}{2}}\\ &= \pm\sqrt{\frac{2}{17}\cdot\frac{1}{2}}\\ &= \pm\sqrt{\frac{1}{17}}\\ &= -\frac{\sqrt{17}}{17}\end{aligned}$$

We choose the negative value of $\cos\ \frac{\alpha}{2}$ because the angle is in quadrant II because cosine is negative in quadrant II.

ⓒ To find $\tan\ \frac{\alpha}{2}$, we write the half-angle formula for tangent. Again, we substitute the value of the cosine we found from the triangle in Figure 3 and simplify.

$$\begin{aligned}\tan\ \frac{\alpha}{2} &= \pm\sqrt{\frac{1-\cos\ \alpha}{1+\cos\ \alpha}}\\ &= \pm\sqrt{\frac{1-(-\frac{15}{17})}{1+(-\frac{15}{17})}}\\ &= \pm\sqrt{\frac{\frac{32}{17}}{\frac{2}{17}}}\\ &= \pm\sqrt{\frac{32}{2}}\\ &= -\sqrt{16}\\ &= -4\end{aligned}$$

We choose the negative value of $\tan\ \frac{\alpha}{2}$ because $\frac{\alpha}{2}$ lies in quadrant II, and tangent is negative in quadrant II.

TRY IT #5 Given that $\sin\ \alpha = -\frac{4}{5}$ and α lies in quadrant IV, find the exact value of $\cos\ \left(\frac{\alpha}{2}\right)$.

EXAMPLE 9

Finding the Measurement of a Half Angle

Now, we will return to the problem posed at the beginning of the section. A bicycle ramp is constructed for high-level competition with an angle of θ formed by the ramp and the ground. Another ramp is to be constructed half as steep for novice competition. If $\tan\ \theta = \frac{5}{3}$ for higher-level competition, what is the measurement of the angle for novice competition?

Solution

Since the angle for novice competition measures half the steepness of the angle for the high level competition, and $\tan\ \theta = \frac{5}{3}$ for high competition, we can find $\cos\ \theta$ from the right triangle and the Pythagorean theorem so that we can use the half-angle identities. See Figure 4.

$$\begin{aligned} 3^2 + 5^2 &= 34 \\ c &= \sqrt{34} \end{aligned}$$

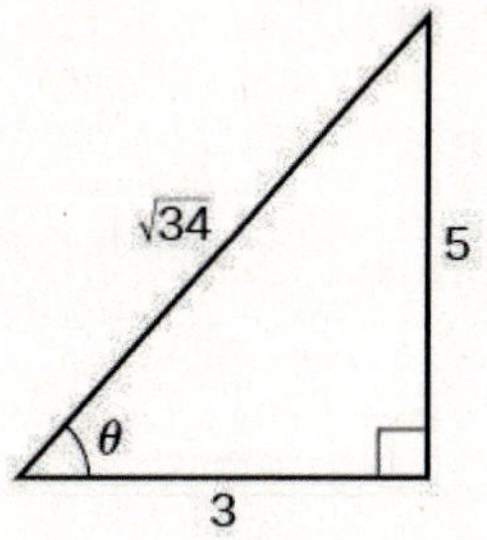

Figure 4

We see that $\cos\ \theta = \frac{3}{\sqrt{34}} = \frac{3\sqrt{34}}{34}$. We can use the half-angle formula for tangent: $\tan\ \frac{\theta}{2} = \sqrt{\frac{1-\cos\ \theta}{1+\cos\ \theta}}$. Since $\tan\ \theta$ is in the first quadrant, so is $\tan\ \frac{\theta}{2}$.

$$\begin{aligned} \tan\ \frac{\theta}{2} &= \sqrt{\frac{1-\frac{3\sqrt{34}}{34}}{1+\frac{3\sqrt{34}}{34}}} \\ &= \sqrt{\frac{\frac{34-3\sqrt{34}}{34}}{\frac{34+3\sqrt{34}}{34}}} \\ &= \sqrt{\frac{34-3\sqrt{34}}{34+3\sqrt{34}}} \\ &\approx 0.57 \end{aligned}$$

We can take the inverse tangent to find the angle: $\tan^{-1}(0.57) \approx 29.7°$. So the angle of the ramp for novice competition is $\approx 29.7°$.

MEDIA

Access these online resources for additional instruction and practice with double-angle, half-angle, and reduction formulas.

Double-Angle Identities (http://openstax.org/l/doubleangiden)
Half-Angle Identities (http://openstax.org/l/halfangleident)

9.3 SECTION EXERCISES

Verbal

1. Explain how to determine the reduction identities from the double-angle identity $\cos(2x) = \cos^2 x - \sin^2 x$.

2. Explain how to determine the double-angle formula for $\tan(2x)$ using the double-angle formulas for $\cos(2x)$ and $\sin(2x)$.

3. We can determine the half-angle formula for $\tan\left(\frac{x}{2}\right) = \frac{\sqrt{1-\cos x}}{\sqrt{1+\cos x}}$ by dividing the formula for $\sin\left(\frac{x}{2}\right)$ by $\cos\left(\frac{x}{2}\right)$. Explain how to determine two formulas for $\tan\left(\frac{x}{2}\right)$ that do not involve any square roots.

4. For the half-angle formula given in the previous exercise for $\tan\left(\frac{x}{2}\right)$, explain why dividing by 0 is not a concern. (Hint: examine the values of $\cos x$ necessary for the denominator to be 0.)

Algebraic

For the following exercises, find the exact values of a) $\sin(2x)$, b) $\cos(2x)$, and c) $\tan(2x)$ without solving for x.

5. If $\sin x = \frac{1}{8}$, and x is in quadrant I.

6. If $\cos x = \frac{2}{3}$, and x is in quadrant I.

7. If $\cos x = -\frac{1}{2}$, and x is in quadrant III.

8. If $\tan x = -8$, and x is in quadrant IV.

For the following exercises, find the values of the six trigonometric functions if the conditions provided hold.

9. $\cos(2\theta) = \frac{3}{5}$ and $90° \le \theta \le 180°$

10. $\cos(2\theta) = \frac{1}{\sqrt{2}}$ and $180° \le \theta \le 270°$

For the following exercises, simplify to one trigonometric expression.

11. $2\sin\left(\frac{\pi}{4}\right) 2\cos\left(\frac{\pi}{4}\right)$

12. $4\sin\left(\frac{\pi}{8}\right)\cos\left(\frac{\pi}{8}\right)$

For the following exercises, find the exact value using half-angle formulas.

13. $\sin\left(\frac{\pi}{8}\right)$

14. $\cos\left(-\frac{11\pi}{12}\right)$

15. $\sin\left(\frac{11\pi}{12}\right)$

16. $\cos\left(\frac{7\pi}{8}\right)$

17. $\tan\left(\frac{5\pi}{12}\right)$

18. $\tan\left(-\frac{3\pi}{12}\right)$

19. $\tan\left(-\frac{3\pi}{8}\right)$

For the following exercises, find the exact values of a) $\sin\left(\frac{x}{2}\right)$, b) $\cos\left(\frac{x}{2}\right)$, and c) $\tan\left(\frac{x}{2}\right)$ without solving for x, when $0° \le x \le 360°$.

20. If $\tan x = -\frac{4}{3}$, and x is in quadrant IV.

21. If $\sin x = -\frac{12}{13}$, and x is in quadrant III.

22. If $\csc x = 7$, and x is in quadrant II.

23. If $\sec\ x = -4$, and x is in quadrant II.

For the following exercises, use Figure 5 to find the requested half and double angles.

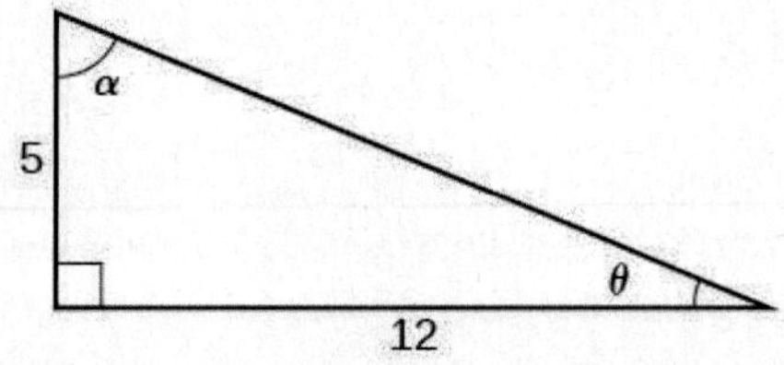

Figure 5

24. Find $\sin(2\theta)$, $\cos(2\theta)$, and $\tan(2\theta)$.

25. Find $\sin(2\alpha)$, $\cos(2\alpha)$, and $\tan(2\alpha)$.

26. Find $\sin\left(\frac{\theta}{2}\right)$, $\cos\left(\frac{\theta}{2}\right)$, and $\tan\left(\frac{\theta}{2}\right)$.

27. Find $\sin\left(\frac{\alpha}{2}\right)$, $\cos\left(\frac{\alpha}{2}\right)$, and $\tan\left(\frac{\alpha}{2}\right)$.

For the following exercises, simplify each expression. Do not evaluate.

28. $\cos^2(28°) - \sin^2(28°)$

29. $2\cos^2(37°) - 1$

30. $1 - 2\sin^2(17°)$

31. $\cos^2(9x) - \sin^2(9x)$

32. $4\ \sin(8x)\ \cos(8x)$

33. $6\ \sin(5x)\ \cos(5x)$

For the following exercises, prove the given identity.

34. $(\sin\ t - \cos\ t)^2 = 1 - \sin(2t)$

35. $\sin(2x) = -2\ \sin(-x)\ \cos(-x)$

36. $\cot\ x - \tan\ x = 2\ \cot(2x)$

37. $\frac{\sin(2\theta)}{1+\cos(2\theta)}\tan^2\theta = \tan^3\ \theta$

For the following exercises, rewrite the expression with an exponent no higher than 1.

38. $\cos^2(5x)$

39. $\cos^2(6x)$

40. $\sin^4(8x)$

41. $\sin^4(3x)$

42. $\cos^2 x\ \sin^4 x$

43. $\cos^4 x\ \sin^2 x$

44. $\tan^2 x\ \sin^2 x$

Technology

For the following exercises, reduce the equations to powers of one, and then check the answer graphically.

45. $\tan^4 x$

46. $\sin^2(2x)$

47. $\sin^2 x\ \cos^2 x$

48. $\tan^2 x\ \sin\ x$

49. $\tan^4 x\ \cos^2 x$

50. $\cos^2 x\ \sin(2x)$

51. $\cos^2(2x)\sin\ x$

52. $\tan^2\left(\frac{x}{2}\right)\ \sin\ x$

For the following exercises, algebraically find an equivalent function, only in terms of $\sin x$ *and/or* $\cos x$, *and then check the answer by graphing both functions.*

53. $\sin(4x)$

54. $\cos(4x)$

Extensions

For the following exercises, prove the identities.

55. $\sin(2x) = \frac{2\tan x}{1+\tan^2 x}$

56. $\cos(2\alpha) = \frac{1-\tan^2\alpha}{1+\tan^2\alpha}$

57. $\tan(2x) = \frac{2\sin x\cos x}{2\cos^2 x-1}$

58. $\left(\sin^2 x - 1\right)^2 = \cos(2x) + \sin^4 x$

59. $\sin(3x) = 3\sin x\cos^2 x - \sin^3 x$

60. $\cos(3x) = \cos^3 x - 3\sin^2 x\cos x$

61. $\frac{1+\cos(2t)}{\sin(2t)-\cos t} = \frac{2\cos t}{2\sin t-1}$

62. $\sin(16x) = 16\sin x\cos x\cos(2x)\cos(4x)\cos(8x)$

63. $\cos(16x) = \left(\cos^2(4x) - \sin^2(4x) - \sin(8x)\right)\left(\cos^2(4x) - \sin^2(4x) + \sin(8x)\right)$

9.4 Sum-to-Product and Product-to-Sum Formulas

Learning Objectives

In this section, you will:

- Express products as sums.
- Express sums as products.

Figure 1 The UCLA marching band (credit: Eric Chan, Flickr).

A band marches down the field creating an amazing sound that bolsters the crowd. That sound travels as a wave that can be interpreted using trigonometric functions. For example, Figure 2 represents a sound wave for the musical note A. In this section, we will investigate trigonometric identities that are the foundation of everyday phenomena such as sound waves.

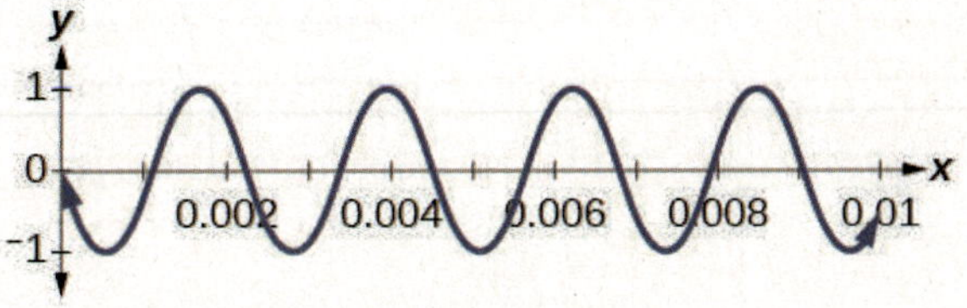

Figure 2

Expressing Products as Sums

We have already learned a number of formulas useful for expanding or simplifying trigonometric expressions, but sometimes we may need to express the product of cosine and sine as a sum. We can use the product-to-sum formulas, which express products of trigonometric functions as sums. Let's investigate the cosine identity first and then the sine identity.

Expressing Products as Sums for Cosine

We can derive the product-to-sum formula from the sum and difference identities for cosine. If we add the two equations, we get:

$$\begin{array}{rcl} \cos\ \alpha\ \cos\ \beta + \sin\ \alpha\ \sin\ \beta & = & \cos(\alpha - \beta) \\ +\ \ \cos\ \alpha\ \cos\ \beta - \sin\ \alpha\ \sin\ \beta & = & \cos(\alpha + \beta) \\ \hline 2\ \cos\ \alpha\ \cos\ \beta & = & \cos(\alpha - \beta) + \cos(\alpha + \beta) \end{array}$$

Then, we divide by 2 to isolate the product of cosines:

$$\cos\ \alpha\ \cos\ \beta = \frac{1}{2}[\cos(\alpha - \beta) + \cos(\alpha + \beta)]$$

HOW TO

Given a product of cosines, express as a sum.

1. Write the formula for the product of cosines.
2. Substitute the given angles into the formula.
3. Simplify.

EXAMPLE 1

Writing the Product as a Sum Using the Product-to-Sum Formula for Cosine

Write the following product of cosines as a sum: $2\ \cos\left(\frac{7x}{2}\right)\ \cos\ \frac{3x}{2}$.

Solution

We begin by writing the formula for the product of cosines:

$$\cos\ \alpha\ \cos\ \beta = \frac{1}{2}[\cos\,(\alpha - \beta) + \cos\,(\alpha + \beta)]$$

We can then substitute the given angles into the formula and simplify.

$$\begin{aligned} 2\ \cos\left(\frac{7x}{2}\right)\cos\left(\frac{3x}{2}\right) &= (2)\left(\frac{1}{2}\right)\left[\cos\left(\frac{7x}{2} - \frac{3x}{2}\right)\right) + \cos\left(\frac{7x}{2} + \frac{3x}{2}\right)\right] \\ &= \left[\cos\left(\frac{4x}{2}\right) + \cos\left(\frac{10x}{2}\right)\right] \\ &= \cos\ 2x + \cos\ 5x \end{aligned}$$

TRY IT #1 Use the product-to-sum formula to write the product as a sum or difference: $\cos(2\theta)\cos(4\theta)$.

Expressing the Product of Sine and Cosine as a Sum

Next, we will derive the product-to-sum formula for sine and cosine from the sum and difference formulas for sine. If we add the sum and difference identities, we get:

$$\begin{array}{rrcl} & \sin(\alpha+\beta) & = & \sin\alpha\cos\beta+\cos\alpha\sin\beta \\ + & \sin(\alpha-\beta) & = & \sin\alpha\cos\beta-\cos\alpha\sin\beta \\ \hline & \sin(\alpha+\beta)+\sin(\alpha-\beta) & = & 2\sin\alpha\cos\beta \end{array}$$

Then, we divide by 2 to isolate the product of cosine and sine:

$$\sin\alpha\cos\beta=\frac{1}{2}[\sin(\alpha+\beta)+\sin(\alpha-\beta)]$$

EXAMPLE 2

Writing the Product as a Sum Containing only Sine or Cosine

Express the following product as a sum containing only sine or cosine and no products: $\sin(4\theta)\cos(2\theta)$.

Solution

Write the formula for the product of sine and cosine. Then substitute the given values into the formula and simplify.

$$\begin{array}{rcl} \sin\alpha\cos\beta & = & \frac{1}{2}[\sin(\alpha+\beta)+\sin(\alpha-\beta)] \\ \sin(4\theta)\cos(2\theta) & = & \frac{1}{2}[\sin(4\theta+2\theta)+\sin(4\theta-2\theta)] \\ & = & \frac{1}{2}[\sin(6\theta)+\sin(2\theta)] \end{array}$$

TRY IT #2 Use the product-to-sum formula to write the product as a sum: $\sin(x+y)\cos(x-y)$.

Expressing Products of Sines in Terms of Cosine

Expressing the product of sines in terms of cosine is also derived from the sum and difference identities for cosine. In this case, we will first subtract the two cosine formulas:

$$\begin{array}{rl} & \cos(\alpha-\beta)=\cos\alpha\cos\beta+\sin\alpha\sin\beta \\ - & \cos(\alpha+\beta)=-(\cos\alpha\cos\beta-\sin\alpha\sin\beta) \\ \hline & \cos(\alpha-\beta)-\cos(\alpha+\beta)=2\sin\alpha\sin\beta \end{array}$$

Then, we divide by 2 to isolate the product of sines:

$$\sin\alpha\sin\beta=\frac{1}{2}[\cos(\alpha-\beta)-\cos(\alpha+\beta)]$$

Similarly we could express the product of cosines in terms of sine or derive other product-to-sum formulas.

The Product-to-Sum Formulas

The **product-to-sum formulas** are as follows:

$$\begin{array}{rcl} \cos\alpha\cos\beta & = & \frac{1}{2}[\cos(\alpha-\beta)+\cos(\alpha+\beta)] \\ \sin\alpha\cos\beta & = & \frac{1}{2}[\sin(\alpha+\beta)+\sin(\alpha-\beta)] \\ \sin\alpha\sin\beta & = & \frac{1}{2}[\cos(\alpha-\beta)-\cos(\alpha+\beta)] \\ \cos\alpha\sin\beta & = & \frac{1}{2}[\sin(\alpha+\beta)-\sin(\alpha-\beta)] \end{array}$$

EXAMPLE 3

Express the Product as a Sum or Difference

Write $\cos(3\theta)\ \cos(5\theta)$ as a sum or difference.

Solution

We have the product of cosines, so we begin by writing the related formula. Then we substitute the given angles and simplify.

$$\begin{aligned}\cos\ \alpha\ \cos\ \beta &= \tfrac{1}{2}[\cos(\alpha-\beta)+\cos(\alpha+\beta)]\\ \cos(3\theta)\cos(5\theta) &= \tfrac{1}{2}[\cos(3\theta-5\theta)+\cos(3\theta+5\theta)]\\ &= \tfrac{1}{2}[\cos(2\theta)+\cos(8\theta)] \qquad \text{Use even-odd identity.}\end{aligned}$$

TRY IT #3 Use the product-to-sum formula to evaluate $\cos\ \frac{11\pi}{12}\ \cos\ \frac{\pi}{12}$.

Expressing Sums as Products

Some problems require the reverse of the process we just used. The sum-to-product formulas allow us to express sums of sine or cosine as products. These formulas can be derived from the product-to-sum identities. For example, with a few substitutions, we can derive the sum-to-product identity for sine. Let $\frac{u+v}{2}=\alpha$ and $\frac{u-v}{2}=\beta$.

Then,

$$\begin{aligned}\alpha+\beta &= \tfrac{u+v}{2}+\tfrac{u-v}{2}\\ &= \tfrac{2u}{2}\\ &= u\end{aligned}$$

$$\begin{aligned}\alpha-\beta &= \tfrac{u+v}{2}-\tfrac{u-v}{2}\\ &= \tfrac{2v}{2}\\ &= v\end{aligned}$$

Thus, replacing α and β in the product-to-sum formula with the substitute expressions, we have

$$\begin{aligned}\sin\ \alpha\ \cos\ \beta &= \tfrac{1}{2}[\sin(\alpha+\beta)+\sin(\alpha-\beta)]\\ \sin\left(\tfrac{u+v}{2}\right)\cos\left(\tfrac{u-v}{2}\right) &= \tfrac{1}{2}[\sin\ u+\sin\ v] \qquad \text{Substitute for}(\alpha+\beta)\text{ and }(\alpha-\beta)\\ 2\ \sin\left(\tfrac{u+v}{2}\right)\cos\left(\tfrac{u-v}{2}\right) &= \sin\ u+\sin\ v\end{aligned}$$

The other sum-to-product identities are derived similarly.

Sum-to-Product Formulas

The **sum-to-product formulas** are as follows:

$$\begin{aligned}\sin\ \alpha+\sin\ \beta &= 2\sin\left(\tfrac{\alpha+\beta}{2}\right)\cos\left(\tfrac{\alpha-\beta}{2}\right)\\ \sin\ \alpha-\sin\ \beta &= 2\sin\left(\tfrac{\alpha-\beta}{2}\right)\cos\left(\tfrac{\alpha+\beta}{2}\right)\\ \cos\ \alpha-\cos\ \beta &= -2\sin\left(\tfrac{\alpha+\beta}{2}\right)\sin\left(\tfrac{\alpha-\beta}{2}\right)\\ \cos\ \alpha+\cos\ \beta &= 2\cos\left(\tfrac{\alpha+\beta}{2}\right)\cos\left(\tfrac{\alpha-\beta}{2}\right)\end{aligned}$$

EXAMPLE 4

Writing the Difference of Sines as a Product

Write the following difference of sines expression as a product: $\sin(4\theta)-\sin(2\theta)$.

Solution

We begin by writing the formula for the difference of sines.

$$\sin\ \alpha - \sin\ \beta = 2\sin\left(\frac{\alpha-\beta}{2}\right)\cos\left(\frac{\alpha+\beta}{2}\right)$$

Substitute the values into the formula, and simplify.

$$\begin{aligned}\sin(4\theta)-\sin(2\theta) &= 2\sin\left(\tfrac{4\theta-2\theta}{2}\right)\ \cos\left(\tfrac{4\theta+2\theta}{2}\right)\\ &= 2\sin\left(\tfrac{2\theta}{2}\right)\ \cos\left(\tfrac{6\theta}{2}\right)\\ &= 2\ \sin\ \theta\ \cos(3\theta)\end{aligned}$$

TRY IT #4 Use the sum-to-product formula to write the sum as a product: $\sin(3\theta)+\sin(\theta)$.

EXAMPLE 5

Evaluating Using the Sum-to-Product Formula

Evaluate $\cos(15°)-\cos(75°)$. Check the answer with a graphing calculator.

Solution

We begin by writing the formula for the difference of cosines.

$$\cos\ \alpha - \cos\ \beta = -2\ \sin\left(\frac{\alpha+\beta}{2}\right)\ \sin\left(\frac{\alpha-\beta}{2}\right)$$

Then we substitute the given angles and simplify.

$$\begin{aligned}\cos(15°)-\cos(75°) &= -2\sin\left(\tfrac{15°+75°}{2}\right)\ \sin\left(\tfrac{15°-75°}{2}\right)\\ &= -2\sin(45°)\ \sin(-30°)\\ &= -2\left(\tfrac{\sqrt{2}}{2}\right)\left(-\tfrac{1}{2}\right)\\ &= \tfrac{\sqrt{2}}{2}\end{aligned}$$

EXAMPLE 6

Proving an Identity

Prove the identity:

$$\frac{\cos(4t)-\cos(2t)}{\sin(4t)+\sin(2t)} = -\tan\ t$$

Solution

We will start with the left side, the more complicated side of the equation, and rewrite the expression until it matches the right side.

$$\begin{aligned}\tfrac{\cos(4t)-\cos(2t)}{\sin(4t)+\sin(2t)} &= \frac{-2\ \sin\left(\tfrac{4t+2t}{2}\right)\ \sin\left(\tfrac{4t-2t}{2}\right)}{2\ \sin\left(\tfrac{4t+2t}{2}\right)\ \cos\left(\tfrac{4t-2t}{2}\right)}\\ &= \frac{-2\ \sin(3t)\sin\ t}{2\ \sin(3t)\cos\ t}\\ &= \frac{-\cancel{2}\ \cancel{\sin(3t)}\ \sin\ t}{\cancel{2}\ \cancel{\sin(3t)}\ \cos\ t}\\ &= -\frac{\sin\ t}{\cos\ t}\\ &= -\tan\ t\end{aligned}$$

Analysis

Recall that verifying trigonometric identities has its own set of rules. The procedures for solving an equation are not the same as the procedures for verifying an identity. When we prove an identity, we pick one side to work on and make substitutions until that side is transformed into the other side.

EXAMPLE 7

Verifying the Identity Using Double-Angle Formulas and Reciprocal Identities

Verify the identity $\csc^2\theta - 2 = \frac{\cos(2\theta)}{\sin^2\theta}$.

Solution

For verifying this equation, we are bringing together several of the identities. We will use the double-angle formula and the reciprocal identities. We will work with the right side of the equation and rewrite it until it matches the left side.

$$\begin{aligned}\frac{\cos(2\theta)}{\sin^2\theta} &= \frac{1-2\sin^2\theta}{\sin^2\theta}\\ &= \frac{1}{\sin^2\theta} - \frac{2\sin^2\theta}{\sin^2\theta}\\ &= \csc^2\theta - 2\end{aligned}$$

TRY IT #5 Verify the identity $\tan\theta\cot\theta - \cos^2\theta = \sin^2\theta$.

MEDIA

Access these online resources for additional instruction and practice with the product-to-sum and sum-to-product identities.

Sum to Product Identities (http://openstax.org/l/sumtoprod)
Sum to Product and Product to Sum Identities (http://openstax.org/l/sumtpptsum)

9.4 SECTION EXERCISES

Verbal

1. Starting with the product to sum formula $\sin\alpha\cos\beta = \frac{1}{2}[\sin(\alpha+\beta)+\sin(\alpha-\beta)]$, explain how to determine the formula for $\cos\alpha\sin\beta$.

2. Provide two different methods of calculating $\cos(195°)\cos(105°)$, one of which uses the product to sum. Which method is easier?

3. Describe a situation where we would convert an equation from a sum to a product and give an example.

4. Describe a situation where we would convert an equation from a product to a sum, and give an example.

Algebraic

For the following exercises, rewrite the product as a sum or difference.

5. $16\sin(16x)\sin(11x)$

6. $20\cos(36t)\cos(6t)$

7. $2\sin(5x)\cos(3x)$

8. $10\cos(5x)\sin(10x)$

9. $\sin(-x)\sin(5x)$

10. $\sin(3x)\cos(5x)$

For the following exercises, rewrite the sum or difference as a product.

11. $\cos(6t)+\cos(4t)$

12. $\sin(3x)+\sin(7x)$

13. $\cos(7x)+\cos(-7x)$

14. $\sin(3x)-\sin(-3x)$

15. $\cos(3x)+\cos(9x)$

16. $\sin h-\sin(3h)$

For the following exercises, evaluate the product for the following using a sum or difference of two functions. Evaluate exactly.

17. $\cos(45°)\cos(15°)$

18. $\cos(45°)\sin(15°)$

19. $\sin(-345°)\sin(-15°)$

20. $\sin(195°)\cos(15°)$

21. $\sin(-45°)\sin(-15°)$

For the following exercises, evaluate the product using a sum or difference of two functions. Leave in terms of sine and cosine.

22. $\cos(23°)\sin(17°)$

23. $2\sin(100°)\sin(20°)$

24. $2\sin(-100°)\sin(-20°)$

25. $\sin(213°)\cos(8°)$

26. $2\cos(56°)\cos(47°)$

For the following exercises, rewrite the sum as a product of two functions. Leave in terms of sine and cosine.

27. $\sin(76°)+\sin(14°)$

28. $\cos(58°)-\cos(12°)$

29. $\sin(101°)-\sin(32°)$

30. $\cos(100°)+\cos(200°)$

31. $\sin(-1°)+\sin(-2°)$

For the following exercises, prove the identity.

32. $\frac{\cos(a+b)}{\cos(a-b)}=\frac{1-\tan a\tan b}{1+\tan a\tan b}$

33. $4\sin(3x)\cos(4x)=2\sin(7x)-2\sin x$

34. $\frac{6\cos(8x)\sin(2x)}{\sin(-6x)}=-3\sin(10x)\csc(6x)+3$

35. $\sin x+\sin(3x)=4\sin x\cos^2 x$

36. $2\left(\cos^3 x-\cos x\sin^2 x\right)=\cos(3x)+\cos x$

37. $2\tan x\cos(3x)=\sec x(\sin(4x)-\sin(2x))$

38. $\cos(a+b)+\cos(a-b)=2\cos a\cos b$

Numeric

For the following exercises, rewrite the sum as a product of two functions or the product as a sum of two functions. Give your answer in terms of sines and cosines. Then evaluate the final answer numerically, rounded to four decimal places.

39. $\cos(58°)+\cos(12°)$

40. $\sin(2°)-\sin(3°)$

41. $\cos(44°)-\cos(22°)$

42. $\cos(176°)\sin(9°)$

43. $\sin(-14°)\sin(85°)$

Technology

For the following exercises, algebraically determine whether each of the given equation is an identity. If it is not an identity, replace the right-hand side with an expression equivalent to the left side. Verify the results by graphing both expressions on a calculator.

44. $2\sin(2x)\sin(3x) = \cos x - \cos(5x)$

45. $\frac{\cos(10\theta)+\cos(6\theta)}{\cos(6\theta)-\cos(10\theta)} = \cot(2\theta)\cot(8\theta)$

46. $\frac{\sin(3x)-\sin(5x)}{\cos(3x)+\cos(5x)} = \tan x$

47. $2\cos(2x)\cos x + \sin(2x)\sin x = 2\sin x$

48. $\frac{\sin(2x)+\sin(4x)}{\sin(2x)-\sin(4x)} = -\tan(3x)\cot x$

For the following exercises, simplify the expression to one term, then graph the original function and your simplified version to verify they are identical.

49. $\frac{\sin(9t)-\sin(3t)}{\cos(9t)+\cos(3t)}$

50. $2\sin(8x)\cos(6x) - \sin(2x)$

51. $\frac{\sin(3x)-\sin x}{\sin x}$

52. $\frac{\cos(5x)+\cos(3x)}{\sin(5x)+\sin(3x)}$

53. $\sin x\cos(15x) - \cos x\sin(15x)$

Extensions

For the following exercises, prove the following sum-to-product formulas.

54. $\sin x - \sin y = 2\sin\left(\frac{x-y}{2}\right)\cos\left(\frac{x+y}{2}\right)$

55. $\cos x + \cos y = 2\cos\left(\frac{x+y}{2}\right)\cos\left(\frac{x-y}{2}\right)$

For the following exercises, prove the identity.

56. $\frac{\sin(6x)+\sin(4x)}{\sin(6x)-\sin(4x)} = \tan(5x)\cot x$

57. $\frac{\cos(3x)+\cos x}{\cos(3x)-\cos x} = -\cot(2x)\cot x$

58. $\frac{\cos(6y)+\cos(8y)}{\sin(6y)-\sin(4y)} = \cot y\cos(7y)\sec(5y)$

59. $\frac{\cos(2y)-\cos(4y)}{\sin(2y)+\sin(4y)} = \tan y$

60. $\frac{\sin(10x)-\sin(2x)}{\cos(10x)+\cos(2x)} = \tan(4x)$

61. $\cos x - \cos(3x) = 4\sin^2 x\cos x$

62. $(\cos(2x) - \cos(4x))^2 + (\sin(4x) + \sin(2x))^2 = 4\sin^2(3x)$

63. $\tan\left(\frac{\pi}{4} - t\right) = \frac{1-\tan t}{1+\tan t}$

9.5 Solving Trigonometric Equations

Learning Objectives

In this section, you will:

- Solve linear trigonometric equations in sine and cosine.
- Solve equations involving a single trigonometric function.
- Solve trigonometric equations using a calculator.
- Solve trigonometric equations that are quadratic in form.
- Solve trigonometric equations using fundamental identities.
- Solve trigonometric equations with multiple angles.
- Solve right triangle problems.

Figure 1 Egyptian pyramids standing near a modern city. (credit: Oisin Mulvihill)

Thales of Miletus (circa 625–547 BC) is known as the founder of geometry. The legend is that he calculated the height of the Great Pyramid of Giza in Egypt using the theory of *similar triangles*, which he developed by measuring the shadow of his staff. He reasoned that when the height of his staff's shadow was exactly equal to the actual height of the staff, then the height of the nearby pyramid's shadow must also be equal to the height of the actual pyramid. Since the structures and their shadows were creating a right triangle with two equal sides, they were similar triangles. By measuring the length of the pyramid's shadow at that moment, he could obtain the height of the pyramid. Based on proportions, this theory has applications in a number of areas, including fractal geometry, engineering, and architecture. Often, the angle of elevation and the angle of depression are found using similar triangles.

In earlier sections of this chapter, we looked at trigonometric identities. Identities are true for all values in the domain of the variable. In this section, we begin our study of trigonometric equations to study real-world scenarios such as the finding the dimensions of the pyramids.

Solving Linear Trigonometric Equations in Sine and Cosine

Trigonometric equations are, as the name implies, equations that involve trigonometric functions. Similar in many ways to solving polynomial equations or rational equations, only specific values of the variable will be solutions, if there are solutions at all. Often we will solve a trigonometric equation over a specified interval. However, just as often, we will be asked to find all possible solutions, and as trigonometric functions are periodic, solutions are repeated within each period. In other words, trigonometric equations may have an infinite number of solutions. Additionally, like rational equations, the domain of the function must be considered before we assume that any solution is valid. The period of both the sine function and the cosine function is 2π. In other words, every 2π units, the *y*-values repeat. If we need to find all possible solutions, then we must add $2\pi k$, where k is an integer, to the initial solution. Recall the rule that gives the format for stating all possible solutions for a function where the period is 2π:

$$\sin\ \theta = \sin(\theta \pm 2k\pi)$$

There are similar rules for indicating all possible solutions for the other trigonometric functions. Solving trigonometric equations requires the same techniques as solving algebraic equations. We read the equation from left to right, horizontally, like a sentence. We look for known patterns, factor, find common denominators, and substitute certain expressions with a variable to make solving a more straightforward process. However, with trigonometric equations, we also have the advantage of using the identities we developed in the previous sections.

EXAMPLE 1

Solving a Linear Trigonometric Equation Involving the Cosine Function

Find all possible exact solutions for the equation $\cos\ \theta = \frac{1}{2}$.

Solution

From the unit circle, we know that

$$\begin{aligned} \cos\ \theta &= \frac{1}{2} \\ \theta &= \frac{\pi}{3}, \frac{5\pi}{3} \end{aligned}$$

These are the solutions in the interval $[0, 2\pi]$. All possible solutions are given by

$$\theta = \frac{\pi}{3} \pm 2k\pi \text{ and } \theta = \frac{5\pi}{3} \pm 2k\pi$$

where k is an integer.

EXAMPLE 2

Solving a Linear Equation Involving the Sine Function

Find all possible exact solutions for the equation $\sin\ t = \frac{1}{2}$.

Solution

Solving for all possible values of t means that solutions include angles beyond the period of 2π. From Figure 2, we can see that the solutions are $t = \frac{\pi}{6}$ and $t = \frac{5\pi}{6}$. But the problem is asking for all possible values that solve the equation. Therefore, the answer is

$$t = \frac{\pi}{6} \pm 2\pi k \text{ and } t = \frac{5\pi}{6} \pm 2\pi k$$

where k is an integer.

HOW TO

Given a trigonometric equation, solve using algebra.

1. Look for a pattern that suggests an algebraic property, such as the difference of squares or a factoring opportunity.
2. Substitute the trigonometric expression with a single variable, such as x or u.
3. Solve the equation the same way an algebraic equation would be solved.
4. Substitute the trigonometric expression back in for the variable in the resulting expressions.
5. Solve for the angle.

EXAMPLE 3

Solve the Linear Trigonometric Equation

Solve the equation exactly: $2\ \cos\ \theta - 3 = -5, 0 \leq \theta < 2\pi$.

Solution

Use algebraic techniques to solve the equation.

$$\begin{aligned} 2\cos\theta - 3 &= -5 \\ 2\cos\theta &= -2 \\ \cos\theta &= -1 \\ \theta &= \pi \end{aligned}$$

TRY IT #1 Solve exactly the following linear equation on the interval $[0, 2\pi)$: $2\sin x + 1 = 0$.

Solving Equations Involving a Single Trigonometric Function

When we are given equations that involve only one of the six trigonometric functions, their solutions involve using algebraic techniques and the unit circle (see Figure 2). We need to make several considerations when the equation involves trigonometric functions other than sine and cosine. Problems involving the reciprocals of the primary trigonometric functions need to be viewed from an algebraic perspective. In other words, we will write the reciprocal function, and solve for the angles using the function. Also, an equation involving the tangent function is slightly different from one containing a sine or cosine function. First, as we know, the period of tangent is π, not 2π. Further, the domain of tangent is all real numbers with the exception of odd integer multiples of $\frac{\pi}{2}$, unless, of course, a problem places its own restrictions on the domain.

EXAMPLE 4

Solving a Problem Involving a Single Trigonometric Function

Solve the problem exactly: $2\sin^2\theta - 1 = 0, 0 \le \theta < 2\pi$.

Solution

As this problem is not easily factored, we will solve using the square root property. First, we use algebra to isolate $\sin\theta$. Then we will find the angles.

$$\begin{aligned} 2\sin^2\theta - 1 &= 0 \\ 2\sin^2\theta &= 1 \\ \sin^2\theta &= \frac{1}{2} \\ \sqrt{\sin^2\theta} &= \pm\sqrt{\frac{1}{2}} \\ \sin\theta &= \pm\frac{1}{\sqrt{2}} = \pm\frac{\sqrt{2}}{2} \\ \theta &= \frac{\pi}{4}, \frac{3\pi}{4}, \frac{5\pi}{4}, \frac{7\pi}{4} \end{aligned}$$

EXAMPLE 5

Solving a Trigonometric Equation Involving Cosecant

Solve the following equation exactly: $\csc\theta = -2, 0 \le \theta < 4\pi$.

Solution

We want all values of θ for which $\csc\theta = -2$ over the interval $0 \le \theta < 4\pi$.

$$\begin{aligned} \csc\theta &= -2 \\ \frac{1}{\sin\theta} &= -2 \\ \sin\theta &= -\frac{1}{2} \\ \theta &= \frac{7\pi}{6}, \frac{11\pi}{6}, \frac{19\pi}{6}, \frac{23\pi}{6} \end{aligned}$$

Analysis

As $\sin\theta = -\frac{1}{2}$, notice that all four solutions are in the third and fourth quadrants.

EXAMPLE 6

Solving an Equation Involving Tangent

Solve the equation exactly: $\tan\left(\theta - \frac{\pi}{2}\right) = 1, 0 \leq \theta < 2\pi$.

Solution

Recall that the tangent function has a period of π. On the interval $[0, \pi)$, and at the angle of $\frac{\pi}{4}$, the tangent has a value of 1. However, the angle we want is $\left(\theta - \frac{\pi}{2}\right)$. Thus, if $\tan\left(\frac{\pi}{4}\right) = 1$, then

$$\begin{aligned} \theta - \frac{\pi}{2} &= \frac{\pi}{4} \\ \theta &= \frac{3\pi}{4} \pm k\pi \end{aligned}$$

Over the interval $[0, 2\pi)$, we have two solutions:

$$\theta = \frac{3\pi}{4} \text{ and } \theta = \frac{3\pi}{4} + \pi = \frac{7\pi}{4}$$

TRY IT #2 Find all solutions for $\tan\ x = \sqrt{3}$.

EXAMPLE 7

Identify all Solutions to the Equation Involving Tangent

Identify all exact solutions to the equation $2(\tan\ x + 3) = 5 + \tan\ x, 0 \leq x < 2\pi$.

Solution

We can solve this equation using only algebra. Isolate the expression $\tan\ x$ on the left side of the equals sign.

$$\begin{aligned} 2(\tan\ x) + 2(3) &= 5 + \tan\ x \\ 2\tan\ x + 6 &= 5 + \tan\ x \\ 2\tan\ x - \tan\ x &= 5 - 6 \\ \tan\ x &= -1 \end{aligned}$$

There are two angles on the unit circle that have a tangent value of -1: $\theta = \frac{3\pi}{4}$ and $\theta = \frac{7\pi}{4}$.

Solve Trigonometric Equations Using a Calculator

Not all functions can be solved exactly using only the unit circle. When we must solve an equation involving an angle other than one of the special angles, we will need to use a calculator. Make sure it is set to the proper mode, either degrees or radians, depending on the criteria of the given problem.

EXAMPLE 8

Using a Calculator to Solve a Trigonometric Equation Involving Sine

Use a calculator to solve the equation $\sin\ \theta = 0.8$, where θ is in radians.

Solution

Make sure mode is set to radians. To find θ, use the inverse sine function. On most calculators, you will need to push the 2ND button and then the SIN button to bring up the $\sin^{-1}$ function. What is shown on the screen is $\sin^{-1}($. The calculator is ready for the input within the parentheses. For this problem, we enter $\sin^{-1}(0.8)$, and press ENTER. Thus, to four decimals places,

$$\sin^{-1}(0.8) \approx 0.9273$$

The solution is

$$\theta \approx 0.9273 \pm 2\pi k$$

The angle measurement in degrees is

$$\begin{aligned} \theta &\approx 53.1^\circ \\ \theta &\approx 180^\circ - 53.1^\circ \\ &\approx 126.9^\circ \end{aligned}$$

Analysis

Note that a calculator will only return an angle in quadrants I or IV for the sine function, since that is the range of the inverse sine. The other angle is obtained by using $\pi - \theta$. Thus, the additional solution is $\approx 2.2143 \pm 2\pi k$

EXAMPLE 9

Using a Calculator to Solve a Trigonometric Equation Involving Secant

Use a calculator to solve the equation $\sec\ \theta = -4$, giving your answer in radians.

Solution

We can begin with some algebra.

$$\begin{aligned} \sec\ \theta &= -4 \\ \frac{1}{\cos\ \theta} &= -4 \\ \cos\ \theta &= -\frac{1}{4} \end{aligned}$$

Check that the MODE is in radians. Now use the inverse cosine function.

$$\begin{aligned} \cos^{-1}\left(-\tfrac{1}{4}\right) &\approx 1.8235 \\ \theta &\approx 1.8235 + 2\pi k \end{aligned}$$

Since $\frac{\pi}{2} \approx 1.57$ and $\pi \approx 3.14$, 1.8235 is between these two numbers, thus $\theta \approx 1.8235$ is in quadrant II. Cosine is also negative in quadrant III. Note that a calculator will only return an angle in quadrants I or II for the cosine function, since that is the range of the inverse cosine. See Figure 2.

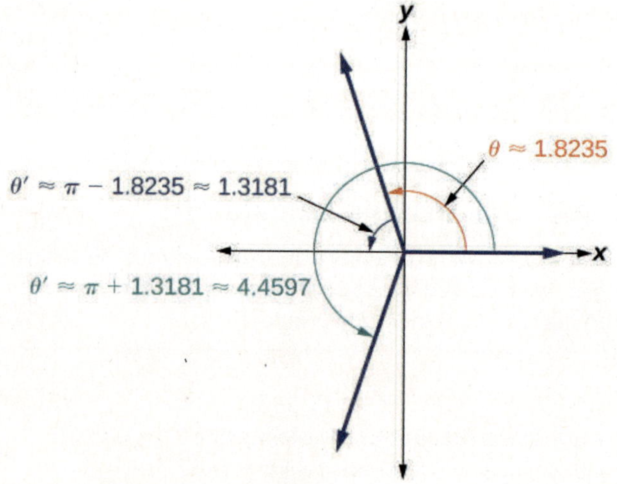

Figure 2

So, we also need to find the measure of the angle in quadrant III. In quadrant II, the reference angle is $\theta\ ' \approx \pi - 1.8235 \approx 1.3181$. The other solution in quadrant III is $\theta\ ' \approx \pi + 1.3181 \approx 4.4597$.

The solutions are $\theta \approx 1.8235 \pm 2\pi k$ and $\theta \approx 4.4597 \pm 2\pi k$.

TRY IT #3 Solve $\cos\ \theta = -0.2$.

Solving Trigonometric Equations in Quadratic Form

Solving a quadratic equation may be more complicated, but once again, we can use algebra as we would for any quadratic equation. Look at the pattern of the equation. Is there more than one trigonometric function in the equation, or is there only one? Which trigonometric function is squared? If there is only one function represented and one of the terms is squared, think about the standard form of a quadratic. Replace the trigonometric function with a variable such as x or u. If substitution makes the equation look like a quadratic equation, then we can use the same methods for solving quadratics to solve the trigonometric equations.

EXAMPLE 10

Solving a Trigonometric Equation in Quadratic Form

Solve the equation exactly: $\cos^2\theta + 3\ \cos\ \theta - 1 = 0, 0 \le \theta < 2\pi$.

Solution

We begin by using substitution and replacing cos θ with x. It is not necessary to use substitution, but it may make the problem easier to solve visually. Let $\cos\ \theta = x$. We have

$$x^2 + 3x - 1 = 0$$

The equation cannot be factored, so we will use the quadratic formula $x = \frac{-b \pm \sqrt{b^2 - 4ac}}{2a}$.

$$\begin{aligned} x &= \frac{-3 \pm \sqrt{(-3)^2 - 4(1)(-1)}}{2} \\ &= \frac{-3 \pm \sqrt{13}}{2} \end{aligned}$$

Replace x with $\cos\ \theta$, and solve.

$$\begin{aligned} \cos\ \theta &= \frac{-3 \pm \sqrt{13}}{2} \\ \theta &= \cos^{-1}\left(\frac{-3 + \sqrt{13}}{2}\right) \end{aligned}$$

Note that only the + sign is used. This is because we get an error when we solve $\theta = \cos^{-1}\left(\frac{-3 - \sqrt{13}}{2}\right)$ on a calculator, since the domain of the inverse cosine function is $[-1, 1]$. However, there is a second solution:

$$\begin{aligned} \theta &= \cos^{-1}\left(\frac{-3 + \sqrt{13}}{2}\right) \\ &\approx 1.26 \end{aligned}$$

This terminal side of the angle lies in quadrant I. Since cosine is also positive in quadrant IV, the second solution is

$$\begin{aligned} \theta &= 2\pi - \cos^{-1}\left(\frac{-3 + \sqrt{13}}{2}\right) \\ &\approx 5.02 \end{aligned}$$

EXAMPLE 11

Solving a Trigonometric Equation in Quadratic Form by Factoring

Solve the equation exactly: $2\ \sin^2\theta - 5\ \sin\ \theta + 3 = 0, 0 \le \theta \le 2\pi$.

Solution

Using grouping, this quadratic can be factored. Either make the real substitution, $\sin\ \theta = u$, or imagine it, as we factor:

$$\begin{aligned} 2\ \sin^2\theta - 5\ \sin\ \theta + 3 &= 0 \\ (2\ \sin\ \theta - 3)(\sin\ \theta - 1) &= 0 \end{aligned}$$

Now set each factor equal to zero.

$$\begin{aligned} 2\sin\theta - 3 &= 0 \\ 2\sin\theta &= 3 \\ \sin\theta &= \frac{3}{2} \end{aligned}$$

$$\begin{aligned} \sin\theta - 1 &= 0 \\ \sin\theta &= 1 \end{aligned}$$

Next solve for θ : $\sin\ \theta \neq \frac{3}{2}$, as the range of the sine function is $[-1, 1]$. However, $\sin\ \theta = 1$, giving the solution $\theta = \frac{\pi}{2}$.

Analysis

Make sure to check all solutions on the given domain as some factors have no solution.

TRY IT #4 Solve $\sin^2\theta = 2\ \cos\ \theta + 2, 0 \leq \theta \leq 2\pi$. [Hint: Make a substitution to express the equation only in terms of cosine.]

EXAMPLE 12

Solving a Trigonometric Equation Using Algebra

Solve exactly:

$$2\ \sin^2\theta + \sin\ \theta = 0; 0 \leq \theta < 2\pi$$

Solution

This problem should appear familiar as it is similar to a quadratic. Let $\sin\ \theta = x$. The equation becomes $2x^2 + x = 0$. We begin by factoring:

$$\begin{aligned} 2x^2 + x &= 0 \\ x(2x + 1) &= 0 \end{aligned}$$

Set each factor equal to zero.

$$\begin{aligned} x &= 0 \\ (2x + 1) &= 0 \\ x &= -\frac{1}{2} \end{aligned}$$

Then, substitute back into the equation the original expression $\sin\ \theta$ for x. Thus,

$$\begin{aligned} \sin\theta &= 0 \\ \\ \theta &= 0, \pi \\ \sin\theta &= -\frac{1}{2} \\ \theta &= \frac{7\pi}{6}, \frac{11\pi}{6} \end{aligned}$$

The solutions within the domain $0 \leq \theta < 2\pi$ are $\theta = 0, \pi, \frac{7\pi}{6}, \frac{11\pi}{6}$.

If we prefer not to substitute, we can solve the equation by following the same pattern of factoring and setting each factor equal to zero.

$$\begin{aligned} 2\sin^2\theta + \sin\theta &= 0 \\ \sin\theta(2\sin\theta + 1) &= 0 \\ \sin\theta &= 0 \\ \theta &= 0, \pi \end{aligned}$$

$$\begin{aligned} 2\sin\theta + 1 &= 0 \\ 2\sin\theta &= -1 \\ \sin\theta &= -\tfrac{1}{2} \\ \theta &= \tfrac{7\pi}{6}, \tfrac{11\pi}{6} \end{aligned}$$

Analysis

We can see the solutions on the graph in Figure 3. On the interval $0 \le \theta < 2\pi$, the graph crosses the *x*-axis four times, at the solutions noted. Notice that trigonometric equations that are in quadratic form can yield up to four solutions instead of the expected two that are found with quadratic equations. In this example, each solution (angle) corresponding to a positive sine value will yield two angles that would result in that value.

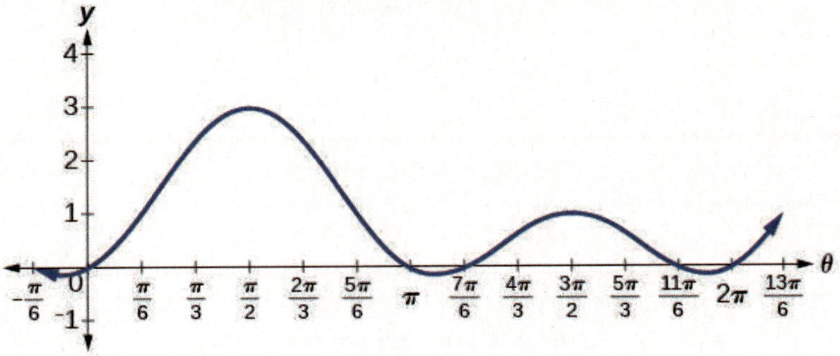

Figure 3

We can verify the solutions on the unit circle in Figure 2 as well.

EXAMPLE 13

Solving a Trigonometric Equation Quadratic in Form

Solve the equation quadratic in form exactly: $2\sin^2\theta - 3\sin\theta + 1 = 0, 0 \le \theta < 2\pi$.

Solution

We can factor using grouping. Solution values of θ can be found on the unit circle.

$$\begin{aligned} (2\sin\theta - 1)(\sin\theta - 1) &= 0 \\ 2\sin\theta - 1 &= 0 \\ \sin\theta &= \tfrac{1}{2} \\ \theta &= \tfrac{\pi}{6}, \tfrac{5\pi}{6} \end{aligned}$$

$$\begin{aligned} \sin\theta &= 1 \\ \theta &= \tfrac{\pi}{2} \end{aligned}$$

TRY IT #5 Solve the quadratic equation $2\cos^2\theta + \cos\theta = 0$.

Solving Trigonometric Equations Using Fundamental Identities

While algebra can be used to solve a number of trigonometric equations, we can also use the fundamental identities because they make solving equations simpler. Remember that the techniques we use for solving are not the same as those for verifying identities. The basic rules of algebra apply here, as opposed to rewriting one side of the identity to match the other side. In the next example, we use two identities to simplify the equation.

EXAMPLE 14

Use Identities to Solve an Equation

Use identities to solve exactly the trigonometric equation over the interval $0 \le x < 2\pi$.

$$\cos\ x\ \cos(2x) + \sin\ x\ \sin(2x) = \frac{\sqrt{3}}{2}$$

Solution

Notice that the left side of the equation is the difference formula for cosine.

$$\begin{aligned} \cos\ x\ \cos(2x) + \sin\ x\ \sin(2x) &= \frac{\sqrt{3}}{2} \\ \cos(x - 2x) &= \frac{\sqrt{3}}{2} && \text{Difference formula for cosine} \\ \cos(-x) &= \frac{\sqrt{3}}{2} && \text{Use the negative angle identity.} \\ \cos\ x &= \frac{\sqrt{3}}{2} \end{aligned}$$

From the unit circle in Figure 2, we see that $\cos\ x = \frac{\sqrt{3}}{2}$ when $x = \frac{\pi}{6}, \frac{11\pi}{6}$.

EXAMPLE 15

Solving the Equation Using a Double-Angle Formula

Solve the equation exactly using a double-angle formula: $\cos(2\ \theta) = \cos\ \theta$.

Solution

We have three choices of expressions to substitute for the double-angle of cosine. As it is simpler to solve for one trigonometric function at a time, we will choose the double-angle identity involving only cosine:

$$\begin{aligned} \cos(2\theta) &= \cos\ \theta \\ 2\cos^2\theta - 1 &= \cos\ \theta \\ 2\ \cos^2\theta - \cos\ \theta - 1 &= 0 \\ (2\ \cos\ \theta + 1)(\cos\ \theta - 1) &= 0 \\ 2\ \cos\ \theta + 1 &= 0 \\ \cos\ \theta &= -\tfrac{1}{2} \\ \\ \cos\ \theta - 1 &= 0 \\ \cos\ \theta &= 1 \end{aligned}$$

So, if $\cos\ \theta = -\frac{1}{2}$, then $\theta = \frac{2\pi}{3} \pm 2\pi k$ and $\theta = \frac{4\pi}{3} \pm 2\pi k$; if $\cos\ \theta = 1$, then $\theta = 0 \pm 2\pi k$.

EXAMPLE 16

Solving an Equation Using an Identity

Solve the equation exactly using an identity: $3\ \cos\ \theta + 3 = 2\ \sin^2\theta, 0 \le \theta < 2\pi$.

Solution

If we rewrite the right side, we can write the equation in terms of cosine:

$$\begin{aligned}
3\cos\theta+3 &= 2\sin^2\theta \\
3\cos\theta+3 &= 2\left(1-\cos^2\theta\right) \\
3\cos\theta+3 &= 2-2\cos^2\theta \\
2\cos^2\theta+3\cos\theta+1 &= 0 \\
(2\cos\theta+1)(\cos\theta+1) &= 0 \\
2\cos\theta+1 &= 0 \\
\cos\theta &= -\tfrac{1}{2} \\
\theta &= \tfrac{2\pi}{3},\tfrac{4\pi}{3} \\
\cos\theta+1 &= 0 \\
\cos\theta &= -1 \\
\theta &= \pi
\end{aligned}$$

Our solutions are $\theta = \frac{2\pi}{3}, \frac{4\pi}{3}, \pi$.

Solving Trigonometric Equations with Multiple Angles

Sometimes it is not possible to solve a trigonometric equation with identities that have a multiple angle, such as $\sin(2x)$ or $\cos(3x)$. When confronted with these equations, recall that $y = \sin(2x)$ is a horizontal compression by a factor of 2 of the function $y = \sin x$. On an interval of 2π, we can graph two periods of $y = \sin(2x)$, as opposed to one cycle of $y = \sin x$. This compression of the graph leads us to believe there may be twice as many x-intercepts or solutions to $\sin(2x) = 0$ compared to $\sin x = 0$. This information will help us solve the equation.

EXAMPLE 17

Solving a Multiple Angle Trigonometric Equation

Solve exactly: $\cos(2x) = \frac{1}{2}$ on $[0, 2\pi)$.

Solution

We can see that this equation is the standard equation with a multiple of an angle. If $\cos(\alpha) = \frac{1}{2}$, we know α is in quadrants I and IV. While $\theta = \cos^{-1}\frac{1}{2}$ will only yield solutions in quadrants I and II, we recognize that the solutions to the equation $\cos\theta = \frac{1}{2}$ will be in quadrants I and IV.

Therefore, the possible angles are $\theta = \frac{\pi}{3}$ and $\theta = \frac{5\pi}{3}$. So, $2x = \frac{\pi}{3}$ or $2x = \frac{5\pi}{3}$, which means that $x = \frac{\pi}{6}$ or $x = \frac{5\pi}{6}$. Does this make sense? Yes, because $\cos\left(2\left(\frac{\pi}{6}\right)\right) = \cos\left(\frac{\pi}{3}\right) = \frac{1}{2}$.

Are there any other possible answers? Let us return to our first step.

In quadrant I, $2x = \frac{\pi}{3}$, so $x = \frac{\pi}{6}$ as noted. Let us revolve around the circle again:

$$\begin{aligned}
2x &= \tfrac{\pi}{3} + 2\pi \\
&= \tfrac{\pi}{3} + \tfrac{6\pi}{3} \\
&= \tfrac{7\pi}{3}
\end{aligned}$$

so $x = \frac{7\pi}{6}$.

One more rotation yields

$$\begin{aligned}
2x &= \tfrac{\pi}{3} + 4\pi \\
&= \tfrac{\pi}{3} + \tfrac{12\pi}{3} \\
&= \tfrac{13\pi}{3}
\end{aligned}$$

$x = \frac{13\pi}{6} > 2\pi$, so this value for x is larger than 2π, so it is not a solution on $[0, 2\pi)$.

In quadrant IV, $2x = \frac{5\pi}{3}$, so $x = \frac{5\pi}{6}$ as noted. Let us revolve around the circle again:

$$\begin{aligned} 2x &= \frac{5\pi}{3} + 2\pi \\ &= \frac{5\pi}{3} + \frac{6\pi}{3} \\ &= \frac{11\pi}{3} \end{aligned}$$

so $x = \frac{11\pi}{6}$.

One more rotation yields

$$\begin{aligned} 2x &= \frac{5\pi}{3} + 4\pi \\ &= \frac{5\pi}{3} + \frac{12\pi}{3} \\ &= \frac{17\pi}{3} \end{aligned}$$

$x = \frac{17\pi}{6} > 2\pi$, so this value for x is larger than 2π, so it is not a solution on $[0, 2\pi)$.

Our solutions are $x = \frac{\pi}{6}, \frac{5\pi}{6}, \frac{7\pi}{6},$ and $\frac{11\pi}{6}$. Note that whenever we solve a problem in the form of $\sin(nx) = c$, we must go around the unit circle n times.

Solving Right Triangle Problems

We can now use all of the methods we have learned to solve problems that involve applying the properties of right triangles and the Pythagorean Theorem. We begin with the familiar Pythagorean Theorem, $a^2 + b^2 = c^2$, and model an equation to fit a situation.

EXAMPLE 18

Using the Pythagorean Theorem to Model an Equation

Use the Pythagorean Theorem, and the properties of right triangles to model an equation that fits the problem.

One of the cables that anchors the center of the London Eye Ferris wheel to the ground must be replaced. The center of the Ferris wheel is 69.5 meters above the ground, and the second anchor on the ground is 23 meters from the base of the Ferris wheel. Approximately how long is the cable, and what is the angle of elevation (from ground up to the center of the Ferris wheel)? See Figure 4.

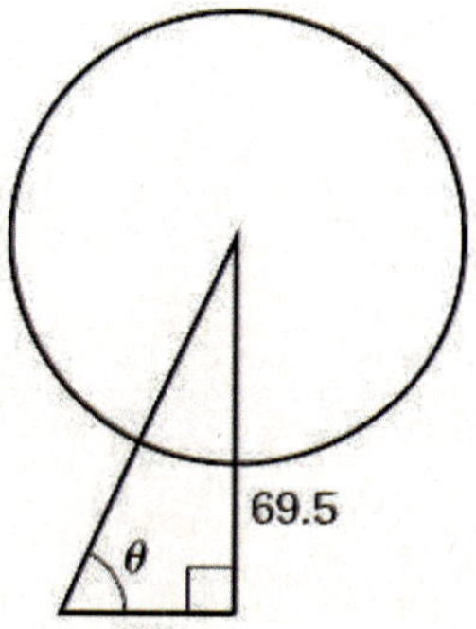

Figure 4

Solution

Using the information given, we can draw a right triangle. We can find the length of the cable with the Pythagorean Theorem.

$$\begin{aligned} a^2 + b^2 &= c^2 \\ (23)^2 + (69.5)^2 &\approx 5359 \\ \sqrt{5359} &\approx 73.2 \text{ m} \end{aligned}$$

The angle of elevation is θ, formed by the second anchor on the ground and the cable reaching to the center of the wheel. We can use the tangent function to find its measure. Round to two decimal places.

$$\tan\,\theta = \frac{69.5}{23}$$

$$\begin{aligned}\tan^{-1}\left(\frac{69.5}{23}\right) &\approx 1.2522\\ &\approx 71.69^\circ\end{aligned}$$

The angle of elevation is approximately $71.7°$, and the length of the cable is 73.2 meters.

EXAMPLE 19

Using the Pythagorean Theorem to Model an Abstract Problem

OSHA safety regulations require that the base of a ladder be placed 1 foot from the wall for every 4 feet of ladder length. Find the angle that a ladder of any length forms with the ground and the height at which the ladder touches the wall.

Solution

For any length of ladder, the base needs to be a distance from the wall equal to one fourth of the ladder's length. Equivalently, if the base of the ladder is "a" feet from the wall, the length of the ladder will be $4a$ feet. See Figure 5.

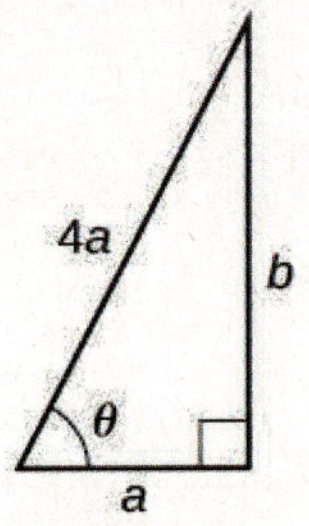

Figure 5

The side adjacent to θ is a and the hypotenuse is $4a$. Thus,

$$\begin{aligned}\cos\,\theta &= \frac{a}{4a} = \frac{1}{4}\\ \cos^{-1}\left(\frac{1}{4}\right) &\approx 75.5^\circ\end{aligned}$$

The elevation of the ladder forms an angle of $75.5°$ with the ground. The height at which the ladder touches the wall can be found using the Pythagorean Theorem:

$$\begin{aligned}a^2 + b^2 &= (4a)^2\\ b^2 &= (4a)^2 - a^2\\ b^2 &= 16a^2 - a^2\\ b^2 &= 15a^2\\ b &= a\sqrt{15}\end{aligned}$$

Thus, the ladder touches the wall at $a\sqrt{15}$ feet from the ground.

MEDIA

Access these online resources for additional instruction and practice with solving trigonometric equations.

Solving Trigonometric Equations I (http://openstax.org/l/solvetrigeqI)
Solving Trigonometric Equations II (http://openstax.org/l/solvetrigeqII)
Solving Trigonometric Equations III (http://openstax.org/l/solvetrigeqIII)
Solving Trigonometric Equations IV (http://openstax.org/l/solvetrigeqIV)
Solving Trigonometric Equations V (http://openstax.org/l/solvetrigeqV)
Solving Trigonometric Equations VI (http://openstax.org/l/solvetrigeqVI)

9.5 SECTION EXERCISES

Verbal

1. Will there always be solutions to trigonometric function equations? If not, describe an equation that would not have a solution. Explain why or why not.

2. When solving a trigonometric equation involving more than one trig function, do we always want to try to rewrite the equation so it is expressed in terms of one trigonometric function? Why or why not?

3. When solving linear trig equations in terms of only sine or cosine, how do we know whether there will be solutions?

Algebraic

For the following exercises, find all solutions exactly on the interval $0 \leq \theta < 2\pi$.

4. $2\sin\theta = -\sqrt{2}$

5. $2\sin\theta = \sqrt{3}$

6. $2\cos\theta = 1$

7. $2\cos\theta = -\sqrt{2}$

8. $\tan\theta = -1$

9. $\tan x = 1$

10. $\cot x + 1 = 0$

11. $4\sin^2 x - 2 = 0$

12. $\csc^2 x - 4 = 0$

For the following exercises, solve exactly on $[0, 2\pi)$.

13. $2\cos\theta = \sqrt{2}$

14. $2\cos\theta = -1$

15. $2\sin\theta = -1$

16. $2\sin\theta = -\sqrt{3}$

17. $2\sin(3\theta) = 1$

18. $2\sin(2\theta) = \sqrt{3}$

19. $2\cos(3\theta) = -\sqrt{2}$

20. $\cos(2\theta) = -\frac{\sqrt{3}}{2}$

21. $2\sin(\pi\theta) = 1$

22. $2\cos\left(\frac{\pi}{5}\theta\right) = \sqrt{3}$

For the following exercises, find all exact solutions on $[0, 2\pi)$.

23. $\sec(x)\sin(x) - 2\sin(x) = 0$

24. $\tan(x) - 2\sin(x)\tan(x) = 0$

25. $2\cos^2 t + \cos(t) = 1$

26. $2\tan^2(t) = 3\sec(t)$

27. $2\sin(x)\cos(x) - \sin(x) + 2\cos(x) - 1 = 0$

28. $\cos^2\theta = \frac{1}{2}$

29. $\sec^2 x = 1$

30. $\tan^2(x) = -1 + 2\tan(-x)$

31. $8\sin^2(x) + 6\sin(x) + 1 = 0$

32. $\tan^5(x) = \tan(x)$

For the following exercises, solve with the methods shown in this section exactly on the interval $[0, 2\pi)$.

33. $\sin(3x)\cos(6x) - \cos(3x)\sin(6x) = -0.9$

34. $\sin(6x)\cos(11x) - \cos(6x)\sin(11x) = -0.1$

35. $\cos(2x)\cos x + \sin(2x)\sin x = 1$

36. $6\sin(2t) + 9\sin t = 0$

37. $9\cos(2\theta) = 9\cos^2\theta - 4$

38. $\sin(2t) = \cos t$

39. $\cos(2t) = \sin t$

40. $\cos(6x) - \cos(3x) = 0$

For the following exercises, solve exactly on the interval $[0, 2\pi)$. Use the quadratic formula if the equations do not factor.

41. $\tan^2 x - \sqrt{3}\tan x = 0$

42. $\sin^2 x + \sin x - 2 = 0$

43. $\sin^2 x - 2\sin x - 4 = 0$

44. $5\cos^2 x + 3\cos x - 1 = 0$

45. $3\cos^2 x - 2\cos x - 2 = 0$

46. $5\sin^2 x + 2\sin x - 1 = 0$

47. $\tan^2 x + 5\tan x - 1 = 0$

48. $\cot^2 x = -\cot x$

49. $-\tan^2 x - \tan x - 2 = 0$

For the following exercises, find exact solutions on the interval $[0, 2\pi)$. Look for opportunities to use trigonometric identities.

50. $\sin^2 x - \cos^2 x - \sin x = 0$

51. $\sin^2 x + \cos^2 x = 0$

52. $\sin(2x) - \sin x = 0$

53. $\cos(2x) - \cos x = 0$

54. $\frac{2\tan x}{2-\sec^2 x} - \sin^2 x = \cos^2 x$

55. $1 - \cos(2x) = 1 + \cos(2x)$

56. $\sec^2 x = 7$

57. $10\sin x\cos x = 6\cos x$

58. $-3\sin t = 15\cos t\sin t$

59. $4\cos^2 x - 4 = 15\cos x$

60. $8\sin^2 x + 6\sin x + 1 = 0$

61. $8\cos^2\theta = 3 - 2\cos\theta$

62. $6\cos^2 x + 7\sin x - 8 = 0$

63. $12\sin^2 t + \cos t - 6 = 0$

64. $\tan x = 3\sin x$

65. $\cos^3 t = \cos t$

Graphical

For the following exercises, algebraically determine all solutions of the trigonometric equation exactly, then verify the results by graphing the equation and finding the zeros.

66. $6\sin^2 x - 5\sin x + 1 = 0$

67. $8\cos^2 x - 2\cos x - 1 = 0$

68. $100\tan^2 x + 20\tan x - 3 = 0$

69. $2\cos^2 x - \cos x + 15 = 0$

70. $20\sin^2 x - 27\sin x + 7 = 0$

71. $2\tan^2 x + 7\tan x + 6 = 0$

72. $130\tan^2 x + 69\tan x - 130 = 0$

Technology

For the following exercises, use a calculator to find all solutions to four decimal places.

73. $\sin x = 0.27$

74. $\sin x = -0.55$

75. $\tan x = -0.34$

76. $\cos\ x = 0.71$

For the following exercises, solve the equations algebraically, and then use a calculator to find the values on the interval $[0, 2\pi)$. *Round to four decimal places.*

77. $\tan^2 x + 3\ \tan\ x - 3 = 0$

78. $6\ \tan^2 x + 13\ \tan\ x = -6$

79. $\tan^2 x - \sec\ x = 1$

80. $\sin^2 x - 2\ \cos^2 x = 0$

81. $2\ \tan^2 x + 9\ \tan\ x - 6 = 0$

82. $4\ \sin^2 x + \sin(2x) \sec\ x - 3 = 0$

Extensions

For the following exercises, find all solutions exactly to the equations on the interval $[0, 2\pi)$.

83. $\csc^2 x - 3\ \csc\ x - 4 = 0$

84. $\sin^2 x - \cos^2 x - 1 = 0$

85. $\sin^2 x\left(1 - \sin^2 x\right) + \cos^2 x\left(1 - \sin^2 x\right) = 0$

86. $3\ \sec^2 x + 2 + \sin^2 x - \tan^2 x + \cos^2 x = 0$

87. $\sin^2 x - 1 + 2\ \cos(2x) - \cos^2 x = 1$

88. $\tan^2 x - 1 - \sec^3 x\ \cos\ x = 0$

89. $\frac{\sin(2x)}{\sec^2 x} = 0$

90. $\frac{\sin(2x)}{2\csc^2 x} = 0$

91. $2\ \cos^2 x - \sin^2 x - \cos\ x - 5 = 0$

92. $\frac{1}{\sec^2 x} + 2 + \sin^2 x + 4\ \cos^2 x = 4$

Real-World Applications

93. An airplane has only enough gas to fly to a city 200 miles northeast of its current location. If the pilot knows that the city is 25 miles north, how many degrees north of east should the airplane fly?

94. If a loading ramp is placed next to a truck, at a height of 4 feet, and the ramp is 15 feet long, what angle does the ramp make with the ground?

95. If a loading ramp is placed next to a truck, at a height of 2 feet, and the ramp is 20 feet long, what angle does the ramp make with the ground?

96. A woman is watching a launched rocket currently 11 miles in altitude. If she is standing 4 miles from the launch pad, at what angle is she looking up from horizontal?

97. An astronaut is in a launched rocket currently 15 miles in altitude. If a man is standing 2 miles from the launch pad, at what angle is the astronaut looking down at him from horizontal? (Hint: this is called the angle of depression.)

98. A woman is standing 8 meters away from a 10-meter tall building. At what angle is she looking to the top of the building?

99. Issa is standing 10 meters away from a 6-meter tall building. Travis is at the top of the building looking down at Issa. At what angle is Travis looking at Issa?

100. A 20-foot tall building has a shadow that is 55 feet long. What is the angle of elevation of the sun?

101. A 90-foot tall building has a shadow that is 2 feet long. What is the angle of elevation of the sun?

102. A spotlight on the ground 3 meters from a 2-meter tall man casts a 6 meter shadow on a wall 6 meters from the man. At what angle is the light?

103. A spotlight on the ground 3 feet from a 5-foot tall woman casts a 15-foot tall shadow on a wall 6 feet from the woman. At what angle is the light?

For the following exercises, find a solution to the following word problem algebraically. Then use a calculator to verify the result. Round the answer to the nearest tenth of a degree.

104. A person does a handstand with their feet touching a wall and their hands 1.5 feet away from the wall. If the person is 6 feet tall, what angle do their feet make with the wall?

105. A person does a handstand with her feet touching a wall and her hands 3 feet away from the wall. If the person is 5 feet tall, what angle do her feet make with the wall?

106. A 23-foot ladder is positioned next to a house. If the ladder slips at 7 feet from the house when there is not enough traction, what angle should the ladder make with the ground to avoid slipping?

Chapter Review

Key Terms

double-angle formulas identities derived from the sum formulas for sine, cosine, and tangent in which the angles are equal

even-odd identities set of equations involving trigonometric functions such that if $f(-x) = -f(x)$, the identity is odd, and if $f(-x) = f(x)$, the identity is even

half-angle formulas identities derived from the reduction formulas and used to determine half-angle values of trigonometric functions

product-to-sum formula a trigonometric identity that allows the writing of a product of trigonometric functions as a sum or difference of trigonometric functions

Pythagorean identities set of equations involving trigonometric functions based on the right triangle properties

quotient identities pair of identities based on the fact that tangent is the ratio of sine and cosine, and cotangent is the ratio of cosine and sine

reciprocal identities set of equations involving the reciprocals of basic trigonometric definitions

reduction formulas identities derived from the double-angle formulas and used to reduce the power of a trigonometric function

sum-to-product formula a trigonometric identity that allows, by using substitution, the writing of a sum of trigonometric functions as a product of trigonometric functions

Key Equations

Pythagorean identities	$\cos^2\theta + \sin^2\theta = 1$ $1 + \cot^2\theta = \csc^2\theta$ $1 + \tan^2\theta = \sec^2\theta$
Even-odd identities	$\tan(-\theta) = -\tan\theta$ $\cot(-\theta) = -\cot\theta$ $\sin(-\theta) = -\sin\theta$ $\csc(-\theta) = -\csc\theta$ $\cos(-\theta) = \cos\theta$ $\sec(-\theta) = \sec\theta$
Reciprocal identities	$\sin\theta = \frac{1}{\csc\theta}$ $\cos\theta = \frac{1}{\sec\theta}$ $\tan\theta = \frac{1}{\cot\theta}$ $\csc\theta = \frac{1}{\sin\theta}$ $\sec\theta = \frac{1}{\cos\theta}$ $\cot\theta = \frac{1}{\tan\theta}$
Quotient identities	$\tan\theta = \frac{\sin\theta}{\cos\theta}$ $\cot\theta = \frac{\cos\theta}{\sin\theta}$

Sum Formula for Cosine	$\cos(\alpha + \beta) = \cos\alpha\cos\beta - \sin\alpha\sin\beta$
Difference Formula for Cosine	$\cos(\alpha - \beta) = \cos\alpha\cos\beta + \sin\alpha\sin\beta$
Sum Formula for Sine	$\sin(\alpha + \beta) = \sin\alpha\cos\beta + \cos\alpha\sin\beta$
Difference Formula for Sine	$\sin(\alpha - \beta) = \sin\alpha\cos\beta - \cos\alpha\sin\beta$

Sum Formula for Tangent	$\tan(\alpha+\beta) = \frac{\tan\ \alpha+\tan\ \beta}{1-\tan\ \alpha\ \tan\ \beta}$
Difference Formula for Tangent	$\tan(\alpha-\beta) = \frac{\tan\ \alpha-\tan\ \beta}{1+\tan\ \alpha\ \tan\ \beta}$
Cofunction identities	$\begin{aligned}\sin\ \theta &= \cos\left(\frac{\pi}{2}-\theta\right)\\ \cos\ \theta &= \sin\left(\frac{\pi}{2}-\theta\right)\\ \tan\ \theta &= \cot\left(\frac{\pi}{2}-\theta\right)\\ \cot\ \theta &= \tan\left(\frac{\pi}{2}-\theta\right)\\ \sec\ \theta &= \csc\left(\frac{\pi}{2}-\theta\right)\\ \csc\ \theta &= \sec\left(\frac{\pi}{2}-\theta\right)\end{aligned}$
Double-angle formulas	$\begin{aligned}\sin(2\theta) &= 2\sin\ \theta\ \cos\ \theta\\ \cos(2\theta) &= \cos^2\theta-\sin^2\theta\\ &= 1-2\sin^2\theta\\ &= 2\cos^2\theta-1\\ \tan(2\theta) &= \frac{2\tan\ \theta}{1-\tan^2\theta}\end{aligned}$
Reduction formulas	$\begin{aligned}\sin^2\theta &= \frac{1-\cos(2\theta)}{2}\\ \cos^2\theta &= \frac{1+\cos(2\theta)}{2}\\ \tan^2\theta &= \frac{1-\cos(2\theta)}{1+\cos(2\theta)}\end{aligned}$
Half-angle formulas	$\begin{aligned}\sin\ \frac{\alpha}{2} &= \pm\sqrt{\frac{1-\cos\ \alpha}{2}}\\ \cos\ \frac{\alpha}{2} &= \pm\sqrt{\frac{1+\cos\ \alpha}{2}}\\ \tan\ \frac{\alpha}{2} &= \pm\sqrt{\frac{1-\cos\ \alpha}{1+\cos\ \alpha}}\\ &= \frac{\sin\ \alpha}{1+\cos\ \alpha}\\ &= \frac{1-\cos\ \alpha}{\sin\ \alpha}\end{aligned}$
Product-to-sum Formulas	$\begin{aligned}\cos\ \alpha\ \cos\ \beta &= \frac{1}{2}[\cos(\alpha-\beta)+\cos(\alpha+\beta)]\\ \sin\ \alpha\ \cos\ \beta &= \frac{1}{2}[\sin(\alpha+\beta)+\sin(\alpha-\beta)]\\ \sin\ \alpha\ \sin\ \beta &= \frac{1}{2}[\cos(\alpha-\beta)-\cos(\alpha+\beta)]\\ \cos\ \alpha\ \sin\ \beta &= \frac{1}{2}[\sin(\alpha+\beta)-\sin(\alpha-\beta)]\end{aligned}$
Sum-to-product Formulas	$\begin{aligned}\sin\ \alpha+\sin\ \beta &= 2\ \sin\left(\frac{\alpha+\beta}{2}\right)\cos\left(\frac{\alpha-\beta}{2}\right)\\ \sin\ \alpha-\sin\ \beta &= 2\ \sin\left(\frac{\alpha-\beta}{2}\right)\cos\left(\frac{\alpha+\beta}{2}\right)\\ \cos\ \alpha-\cos\ \beta &= -2\ \sin\left(\frac{\alpha+\beta}{2}\right)\sin\left(\frac{\alpha-\beta}{2}\right)\\ \cos\ \alpha+\cos\ \beta &= 2\ \cos\left(\frac{\alpha+\beta}{2}\right)\cos\left(\frac{\alpha-\beta}{2}\right)\end{aligned}$

Key Concepts

9.1 Verifying Trigonometric Identities and Using Trigonometric Identities to Simplify Trigonometric Expressions

- There are multiple ways to represent a trigonometric expression. Verifying the identities illustrates how expressions can be rewritten to simplify a problem.
- Graphing both sides of an identity will verify it. See Example 1.
- Simplifying one side of the equation to equal the other side is another method for verifying an identity. See Example 2 and Example 3.
- The approach to verifying an identity depends on the nature of the identity. It is often useful to begin on the more complex side of the equation. See Example 4.
- We can create an identity and then verify it. See Example 5.
- Verifying an identity may involve algebra with the fundamental identities. See Example 6 and Example 7.
- Algebraic techniques can be used to simplify trigonometric expressions. We use algebraic techniques throughout this text, as they consist of the fundamental rules of mathematics. See Example 8, Example 9, and Example 10.

9.2 Sum and Difference Identities

- The sum formula for cosines states that the cosine of the sum of two angles equals the product of the cosines of the angles minus the product of the sines of the angles. The difference formula for cosines states that the cosine of the difference of two angles equals the product of the cosines of the angles plus the product of the sines of the angles.
- The sum and difference formulas can be used to find the exact values of the sine, cosine, or tangent of an angle. See Example 1 and Example 2.
- The sum formula for sines states that the sine of the sum of two angles equals the product of the sine of the first angle and cosine of the second angle plus the product of the cosine of the first angle and the sine of the second angle. The difference formula for sines states that the sine of the difference of two angles equals the product of the sine of the first angle and cosine of the second angle minus the product of the cosine of the first angle and the sine of the second angle. See Example 3.
- The sum and difference formulas for sine and cosine can also be used for inverse trigonometric functions. See Example 4.
- The sum formula for tangent states that the tangent of the sum of two angles equals the sum of the tangents of the angles divided by 1 minus the product of the tangents of the angles. The difference formula for tangent states that the tangent of the difference of two angles equals the difference of the tangents of the angles divided by 1 plus the product of the tangents of the angles. See Example 5.
- The Pythagorean Theorem along with the sum and difference formulas can be used to find multiple sums and differences of angles. See Example 6.
- The cofunction identities apply to complementary angles and pairs of reciprocal functions. See Example 7.
- Sum and difference formulas are useful in verifying identities. See Example 8 and Example 9.
- Application problems are often easier to solve by using sum and difference formulas. See Example 10 and Example 11.

9.3 Double-Angle, Half-Angle, and Reduction Formulas

- Double-angle identities are derived from the sum formulas of the fundamental trigonometric functions: sine, cosine, and tangent. See Example 1, Example 2, Example 3, and Example 4.
- Reduction formulas are especially useful in calculus, as they allow us to reduce the power of the trigonometric term. See Example 5 and Example 6.
- Half-angle formulas allow us to find the value of trigonometric functions involving half-angles, whether the original angle is known or not. See Example 7, Example 8, and Example 9.

9.4 Sum-to-Product and Product-to-Sum Formulas

- From the sum and difference identities, we can derive the product-to-sum formulas and the sum-to-product formulas for sine and cosine.
- We can use the product-to-sum formulas to rewrite products of sines, products of cosines, and products of sine and cosine as sums or differences of sines and cosines. See Example 1, Example 2, and Example 3.
- We can also derive the sum-to-product identities from the product-to-sum identities using substitution.
- We can use the sum-to-product formulas to rewrite sum or difference of sines, cosines, or products sine and cosine as products of sines and cosines. See Example 4.
- Trigonometric expressions are often simpler to evaluate using the formulas. See Example 5.
- The identities can be verified using other formulas or by converting the expressions to sines and cosines. To verify

an identity, we choose the more complicated side of the equals sign and rewrite it until it is transformed into the other side. See Example 6 and Example 7.

9.5 Solving Trigonometric Equations

- When solving linear trigonometric equations, we can use algebraic techniques just as we do solving algebraic equations. Look for patterns, like the difference of squares, quadratic form, or an expression that lends itself well to substitution. See Example 1, Example 2, and Example 3.
- Equations involving a single trigonometric function can be solved or verified using the unit circle. See Example 4, Example 5, and Example 6, and Example 7.
- We can also solve trigonometric equations using a graphing calculator. See Example 8 and Example 9.
- Many equations appear quadratic in form. We can use substitution to make the equation appear simpler, and then use the same techniques we use solving an algebraic quadratic: factoring, the quadratic formula, etc. See Example 10, Example 11, Example 12, and Example 13.
- We can also use the identities to solve trigonometric equation. See Example 14, Example 15, and Example 16.
- We can use substitution to solve a multiple-angle trigonometric equation, which is a compression of a standard trigonometric function. We will need to take the compression into account and verify that we have found all solutions on the given interval. See Example 17.
- Real-world scenarios can be modeled and solved using the Pythagorean Theorem and trigonometric functions. See Example 18.

Exercises

Review Exercises

Solving Trigonometric Equations with Identities

For the following exercises, find all solutions exactly that exist on the interval $[0, 2\pi)$.

1. $\csc^2 t = 3$

2. $\cos^2 x = \frac{1}{4}$

3. $2 \sin \theta = -1$

4. $\tan x \sin x + \sin(-x) = 0$

5. $9 \sin \omega - 2 = 4 \sin^2 \omega$

6. $1 - 2 \tan(\omega) = \tan^2(\omega)$

For the following exercises, use basic identities to simplify the expression.

7. $\sec x \cos x + \cos x - \frac{1}{\sec x}$

8. $\sin^3 x + \cos^2 x \sin x$

For the following exercises, determine if the given identities are equivalent.

9. $\sin^2 x + \sec^2 x - 1 = \frac{(1-\cos^2 x)(1+\cos^2 x)}{\cos^2 x}$

10. $\tan^3 x \csc^2 x \cot^2 x \cos x \sin x = 1$

Sum and Difference Identities

For the following exercises, find the exact value.

11. $\tan\left(\frac{7\pi}{12}\right)$

12. $\cos\left(\frac{25\pi}{12}\right)$

13. $\sin(70°)\cos(25°) - \cos(70°)\sin(25°)$

14. $\cos(83°)\cos(23°) + \sin(83°)\sin(23°)$

For the following exercises, prove the identity.

15. $\cos(4x) - \cos(3x)\cos x = \sin^2 x - 4\cos^2 x \sin^2 x$

16. $\cos(3x) - \cos^3 x = -\cos x \sin^2 x - \sin x \sin(2x)$

For the following exercise, simplify the expression.

17. $\frac{\tan\left(\frac{1}{2}x\right)+\tan\left(\frac{1}{8}x\right)}{1-\tan\left(\frac{1}{8}x\right)\tan\left(\frac{1}{2}x\right)}$

For the following exercises, find the exact value.

18. $\cos\left(\sin^{-1}(0)-\cos^{-1}\left(\frac{1}{2}\right)\right)$

19. $\tan\left(\sin^{-1}(0)+\sin^{-1}\left(\frac{1}{2}\right)\right)$

Double-Angle, Half-Angle, and Reduction Formulas

For the following exercises, find the exact value.

20. Find $\sin(2\theta)$, $\cos(2\theta)$, and $\tan(2\theta)$ given $\cos\ \theta=-\frac{1}{3}$ and θ is in the interval $\left[\frac{\pi}{2},\pi\right]$.

21. Find $\sin(2\theta)$, $\cos(2\theta)$, and $\tan(2\theta)$ given $\sec\ \theta=-\frac{5}{3}$ and θ is in the interval $\left[\frac{\pi}{2},\pi\right]$.

22. $\sin\left(\frac{7\pi}{8}\right)$

23. $\sec\left(\frac{3\pi}{8}\right)$

For the following exercises, use Figure 1 to find the desired quantities.

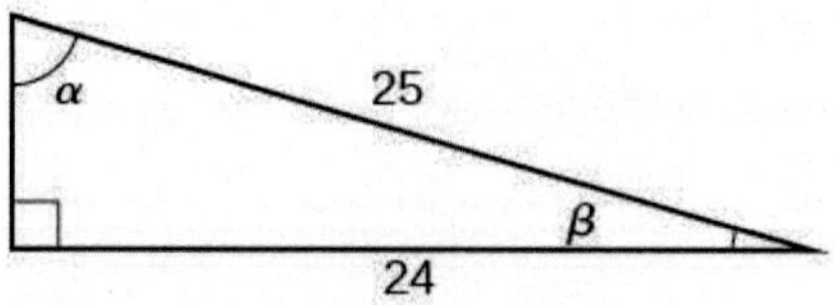

Figure 1

24. $\sin(2\beta)$, $\cos(2\beta)$, $\tan(2\beta)$, $\sin(2\alpha)$, $\cos(2\alpha)$, and $\tan(2\alpha)$

25. $\sin\left(\frac{\beta}{2}\right)$, $\cos\left(\frac{\beta}{2}\right)$, $\tan\left(\frac{\beta}{2}\right)$, $\sin\left(\frac{\alpha}{2}\right)$, $\cos\left(\frac{\alpha}{2}\right)$, and $\tan\left(\frac{\alpha}{2}\right)$

For the following exercises, prove the identity.

26. $\frac{2\cos(2x)}{\sin(2x)}=\cot\ x-\tan\ x$

27. $\cot\ x\ \cos(2x)=-\sin(2x)+\cot\ x$

For the following exercises, rewrite the expression with no powers.

28. $\cos^2 x\sin^4(2x)$

29. $\tan^2 x\ \sin^3 x$

Sum-to-Product and Product-to-Sum Formulas

For the following exercises, evaluate the product for the given expression using a sum or difference of two functions. Write the exact answer.

30. $\cos\left(\frac{\pi}{3}\right)\ \sin\left(\frac{\pi}{4}\right)$

31. $2\ \sin\left(\frac{2\pi}{3}\right)\ \sin\left(\frac{5\pi}{6}\right)$

32. $2\ \cos\left(\frac{\pi}{5}\right)\ \cos\left(\frac{\pi}{3}\right)$

For the following exercises, evaluate the sum by using a product formula. Write the exact answer.

33. $\sin\left(\frac{\pi}{12}\right)-\sin\left(\frac{7\pi}{12}\right)$

34. $\cos\left(\frac{5\pi}{12}\right)+\cos\left(\frac{7\pi}{12}\right)$

For the following exercises, change the functions from a product to a sum or a sum to a product.

35. $\sin(9x)\cos(3x)$

36. $\cos(7x)\cos(12x)$

37. $\sin(11x)+\sin(2x)$

38. $\cos(6x)+\cos(5x)$

Solving Trigonometric Equations

For the following exercises, find all exact solutions on the interval $[0, 2\pi)$.

39. $\tan\ x+1=0$

40. $2\ \sin(2x)+\sqrt{2}=0$

For the following exercises, find all exact solutions on the interval $[0, 2\pi)$.

41. $2\ \sin^2 x-\sin\ x=0$

42. $\cos^2 x-\cos\ x-1=0$

43. $2\ \sin^2 x+5\ \sin\ x+3=0$

44. $\cos\ x-5\ \sin(2x)=0$

45. $\frac{1}{\sec^2 x}+2+\sin^2 x+4\ \cos^2 x=0$

For the following exercises, simplify the equation algebraically as much as possible. Then use a calculator to find the solutions on the interval $[0, 2\pi)$. *Round to four decimal places.*

46. $\sqrt{3}\ \cot^2 x+\cot\ x=1$

47. $\csc^2 x-3\ \csc\ x-4=0$

For the following exercises, graph each side of the equation to find the approximate solutions on the interval $[0, 2\pi)$.

48. $20\ \cos^2 x+21\ \cos\ x+1=0$

49. $\sec^2 x-2\ \sec\ x=15$

Practice Test

For the following exercises, simplify the given expression.

1. $\cos(\text{-}x)\sin\ x\ \cot\ x+\sin^2 x$

2. $\sin(\text{-}x)\cos(\text{-}2x)\text{-}\sin(\text{-}x)\cos(\text{-}2x)$

3. $\csc(\theta)\cot(\theta)\left(\sec^2\theta-1\right)$

4. $\cos^2(\theta)\sin^2(\theta)\left(1+\cot^2(\theta)\right)\left(1+\tan^2(\theta)\right)$

For the following exercises, find the exact value.

5. $\cos\left(\frac{7\pi}{12}\right)$

6. $\tan\left(\frac{3\pi}{8}\right)$

7. $\tan\left(\sin^{-1}\left(\frac{\sqrt{2}}{2}\right)+\tan^{-1}\sqrt{3}\right)$

8. $2\sin\left(\frac{\pi}{4}\right)\sin\left(\frac{\pi}{6}\right)$

9. $\cos\left(\frac{4\pi}{3}+\theta\right)$

10. $\tan\left(-\frac{\pi}{4}+\theta\right)$

For the following exercises, simplify each expression. Do not evaluate.

11. $\cos^2(32°)\tan^2(32°)$

12. $\cot\left(\frac{\theta}{2}\right)$

For the following exercises, find all exact solutions to the equation on $[0, 2\pi)$.

13. $\cos^2 x-\sin^2 x-1=0$

14. $\cos^2 x=\cos\ x$

15. $\cos(2x) + \sin^2 x = 0$

16. $2\sin^2 x - \sin x = 0$

17. Rewrite the expression as a product instead of a sum: $\cos(2x) + \cos(-8x)$.

For the following exercise, rewrite the product as a sum or difference.

18. $8\cos(15x)\sin(3x)$

For the following exercise, rewrite the sum or difference as a product.

19. $2(\sin(8\theta) - \sin(4\theta))$

20. Find all solutions of $\tan(x) - \sqrt{3} = 0.$

21. Find the solutions of $\sec^2 x - 2\sec x = 15$ on the interval $[0, 2\pi)$ algebraically; then graph both sides of the equation to determine the answer.

For the following exercises, find all solutions exactly on the interval $0 \le \theta \le \pi$

22. $2\cos\left(\frac{\theta}{2}\right) = 1$

23. $\sqrt{3}\cot(y) = 1$

24. Find $\sin(2\theta)$, $\cos(2\theta)$, and $\tan(2\theta)$ given $\cot\theta = -\frac{3}{4}$ and θ is on the interval $\left[\frac{\pi}{2}, \pi\right]$.

25. Find $\sin\left(\frac{\theta}{2}\right)$, $\cos\left(\frac{\theta}{2}\right)$, and $\tan\left(\frac{\theta}{2}\right)$ given $\cos\theta = \frac{7}{25}$ and θ is in quadrant IV.

26. Rewrite the expression $\sin^4 x$ with no powers greater than 1.

For the following exercises, prove the identity.

27. $\tan^3 x - \tan x \sec^2 x = \tan(-x)$

28. $\sin(3x) - \cos x \sin(2x) = \cos^2 x \sin x - \sin^3 x$

29. $\frac{\sin(2x)}{\sin x} - \frac{\cos(2x)}{\cos x} = \sec x$

30. Plot the points and find a function of the form $y = A\cos(Bx + C) + D$ that fits the given data.

x	0	1	2	3	4	5
y	−2	2	−2	2	−2	2

31. The displacement $h(t)$ in centimeters of a mass suspended by a spring is modeled by the function $h(t) = \frac{1}{4}\sin(120\pi t)$, where t is measured in seconds. Find the amplitude, period, and frequency of this displacement.

32. A woman is standing 300 feet away from a 2000-foot building. If she looks to the top of the building, at what angle above horizontal is she looking? A worker looks down at her from the 15th floor (1500 feet above her). At what angle is he looking down at her? Round to the nearest tenth of a degree.

33. Two frequencies of sound are played on an instrument governed by the equation $n(t) = 8\ \cos(20\pi t)\cos(1000\pi t)$. What are the period and frequency of the "fast" and "slow" oscillations? What is the amplitude?

34. The average monthly snowfall in a small village in the Himalayas is 6 inches, with the low of 1 inch occurring in July. Construct a function that models this behavior. During what period is there more than 10 inches of snowfall?

35. A spring attached to a ceiling is pulled down 20 cm. After 3 seconds, wherein it completes 6 full periods, the amplitude is only 15 cm. Find the function modeling the position of the spring t seconds after being released. At what time will the spring come to rest? In this case, use 1 cm amplitude as rest.

36. Water levels near a glacier currently average 9 feet, varying seasonally by 2 inches above and below the average and reaching their highest point in January. Due to global warming, the glacier has begun melting faster than normal. Every year, the water levels rise by a steady 3 inches. Find a function modeling the depth of the water t months from now. If the docks are 2 feet above current water levels, at what point will the water first rise above the docks?

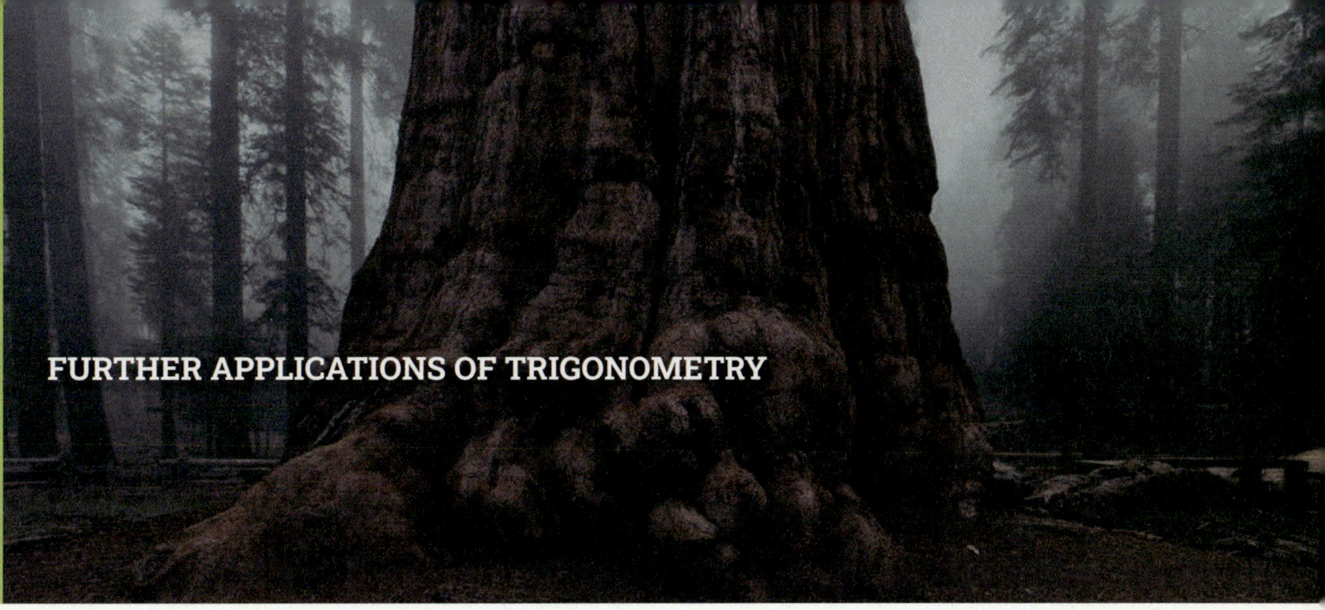

10 FURTHER APPLICATIONS OF TRIGONOMETRY

General Sherman, the world's largest living tree. (credit: Mike Baird, Flickr)

Chapter Outline

Introduction to Further Applications of Trigonometry

The world's largest tree by volume, named General Sherman, stands 274.9 feet tall and resides in Northern California.[1] Just how do scientists know its true height? A common way to measure the height involves determining the angle of elevation, which is formed by the tree and the ground at a point some distance away from the base of the tree. This method is much more practical than climbing the tree and dropping a very long tape measure.

In this chapter, we will explore applications of trigonometry that will enable us to solve many different kinds of problems, including finding the height of a tree. We extend topics we introduced in Trigonometric Functions (http://openstax.org/books/precalculus-2e/pages/5-introduction-to-trigonometric-functions) and investigate applications more deeply and meaningfully.

10.1 Non-right Triangles: Law of Sines

Learning Objectives

In this section, you will:

- Use the Law of Sines to solve oblique triangles.
- Find the area of an oblique triangle using the sine function.
- Solve applied problems using the Law of Sines.

To ensure the safety of over 5,000 U.S. aircraft flying simultaneously during peak times, air traffic controllers monitor and communicate with them after receiving data from the robust radar beacon system. Suppose two radar stations located 20 miles apart each detect an aircraft between them. The angle of elevation measured by the first station is 35 degrees,

1 Source: National Park Service. "The General Sherman Tree." http://www.nps.gov/seki/naturescience/sherman.htm. Accessed April 25, 2014.

whereas the angle of elevation measured by the second station is 15 degrees. How can we determine the altitude of the aircraft? We see in Figure 1 that the triangle formed by the aircraft and the two stations is not a right triangle, so we cannot use what we know about right triangles. In this section, we will find out how to solve problems involving non-right triangles.

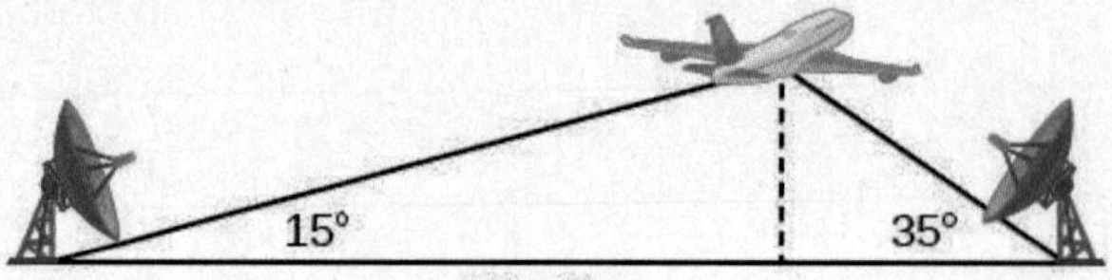

Figure 1

Using the Law of Sines to Solve Oblique Triangles

In any triangle, we can draw an **altitude**, a perpendicular line from one vertex to the opposite side, forming two right triangles. It would be preferable, however, to have methods that we can apply directly to non-right triangles without first having to create right triangles.

Any triangle that is not a right triangle is an **oblique triangle**. Solving an oblique triangle means finding the measurements of all three angles and all three sides. To do so, we need to start with at least three of these values, including at least one of the sides. We will investigate three possible oblique triangle problem situations:

1. **ASA (angle-side-angle)** We know the measurements of two angles and the included side. See Figure 2.

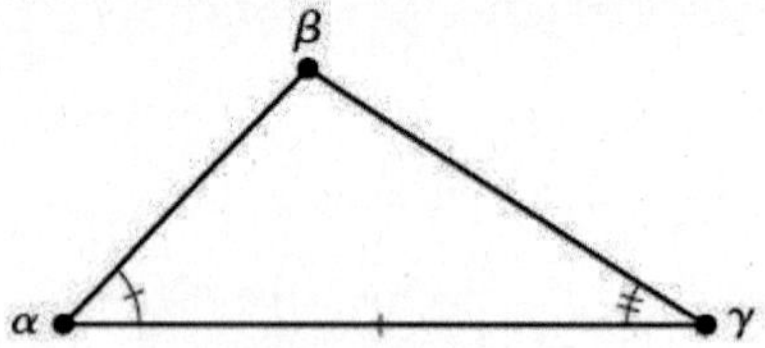

Figure 2

2. **AAS (angle-angle-side)** We know the measurements of two angles and a side that is not between the known angles. See Figure 3.

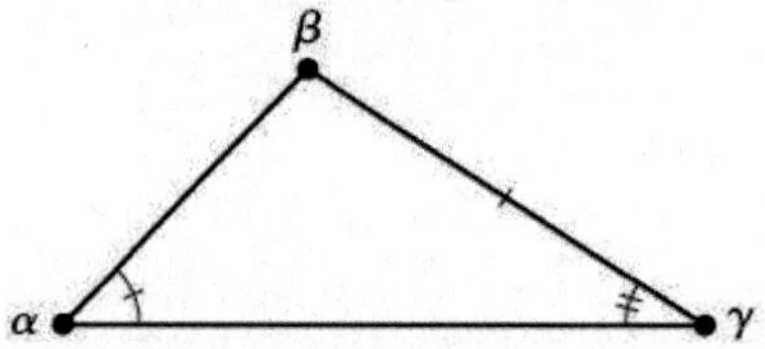

Figure 3

3. **SSA (side-side-angle)** We know the measurements of two sides and an angle that is not between the known sides. See Figure 4.

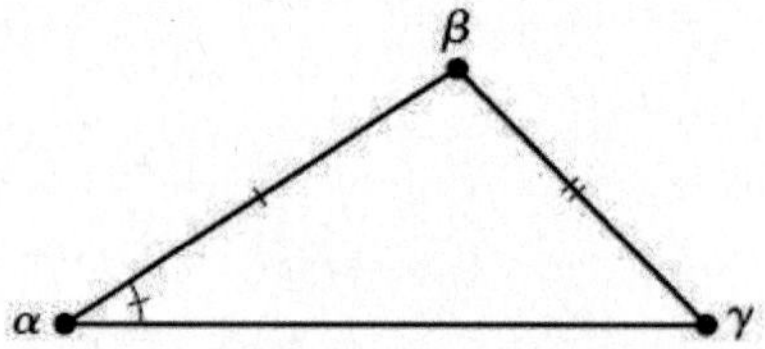

Figure 4

Knowing how to approach each of these situations enables us to solve oblique triangles without having to drop a perpendicular to form two right triangles. Instead, we can use the fact that the ratio of the measurement of one of the angles to the length of its opposite side will be equal to the other two ratios of angle measure to opposite side. Let's see how this statement is derived by considering the triangle shown in Figure 5.

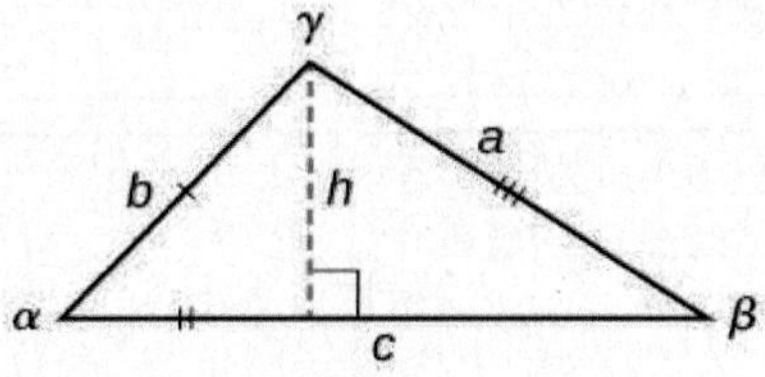

Figure 5

Using the right triangle relationships, we know that $\sin\ \alpha = \frac{h}{b}$ and $\sin\ \beta = \frac{h}{a}$. Solving both equations for h gives two different expressions for h.

$$h = b \sin\ \alpha \text{ and } h = a \sin\ \beta$$

We then set the expressions equal to each other.

$$b \sin\ \alpha = a \sin\ \beta$$

$$\left(\frac{1}{ab}\right)(b \sin\ \alpha) = (a \sin\ \beta)\left(\frac{1}{ab}\right) \qquad \text{Multiply both sides by } \frac{1}{ab}.$$

$$\frac{\sin\ \alpha}{a} = \frac{\sin\ \beta}{b}$$

Similarly, we can compare the other ratios.

$$\frac{\sin\ \alpha}{a} = \frac{\sin\ \gamma}{c} \text{ and } \frac{\sin\ \beta}{b} = \frac{\sin\ \gamma}{c}$$

Collectively, these relationships are called the **Law of Sines**.

$$\frac{\sin\ \alpha}{a} = \frac{\sin\ \beta}{b} = \frac{\sin\ \gamma}{c}$$

Note the standard way of labeling triangles: angle α (alpha) is opposite side a; angle β (beta) is opposite side b; and angle γ (gamma) is opposite side c. See Figure 6.

While calculating angles and sides, be sure to carry the exact values through to the final answer. Generally, final answers are rounded to the nearest tenth, unless otherwise specified.

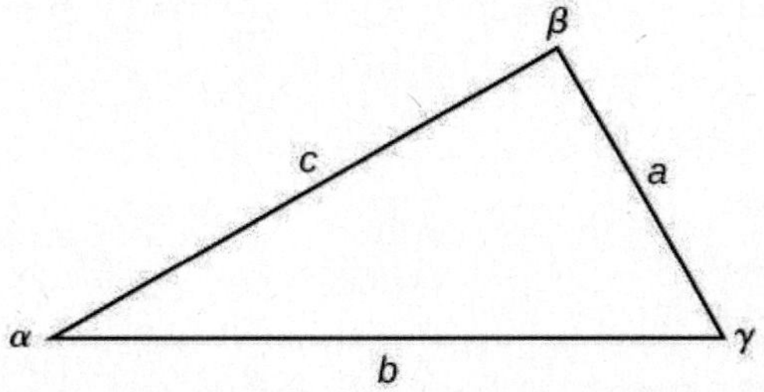

Figure 6

Law of Sines

Given a triangle with angles and opposite sides labeled as in Figure 6, the ratio of the measurement of an angle to the length of its opposite side will be equal to the other two ratios of angle measure to opposite side. All proportions will be equal. The **Law of Sines** is based on proportions and is presented symbolically two ways.

$$\frac{\sin\ \alpha}{a} = \frac{\sin\ \beta}{b} = \frac{\sin\ \gamma}{c}$$

$$\frac{a}{\sin\ \alpha} = \frac{b}{\sin\ \beta} = \frac{c}{\sin\ \gamma}$$

To solve an oblique triangle, use any pair of applicable ratios.

EXAMPLE 1

Solving for Two Unknown Sides and Angle of an AAS Triangle

Solve the triangle shown in Figure 7 to the nearest tenth.

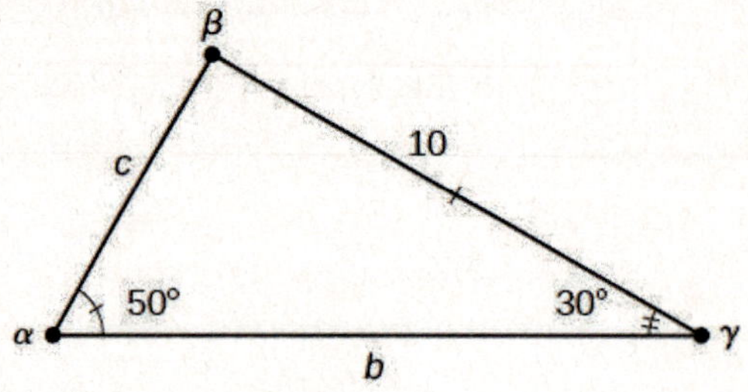

Figure 7

Solution

The three angles must add up to 180 degrees. From this, we can determine that

$$\begin{aligned}\beta &= 180° - 50° - 30° \\ &= 100°\end{aligned}$$

To find an unknown side, we need to know the corresponding angle and a known ratio. We know that angle $\alpha = 50°$ and its corresponding side $a = 10$. We can use the following proportion from the Law of Sines to find the length of c.

$$\frac{\sin(50°)}{10} = \frac{\sin(30°)}{c}$$

$$c\frac{\sin(50°)}{10} = \sin(30°) \qquad \text{Multiply both sides by } c.$$

$$c = \sin(30°)\frac{10}{\sin(50°)} \qquad \text{Multiply by the reciprocal to isolate } c.$$

$$c \approx 6.5$$

Similarly, to solve for b, we set up another proportion.

$$\frac{\sin(50°)}{10} = \frac{\sin(100°)}{b}$$

$$b\sin(50°) = 10\sin(100°) \qquad \text{Multiply both sides by } b.$$

$$b = \frac{10\sin(100°)}{\sin(50°)} \qquad \text{Multiply by the reciprocal to isolate } b.$$

$$b \approx 12.9$$

Therefore, the complete set of angles and sides is

$$\begin{aligned}\alpha &= 50° \quad & a &= 10 \\ \beta &= 100° \quad & b &\approx 12.9 \\ \gamma &= 30° \quad & c &\approx 6.5\end{aligned}$$

TRY IT #1 Solve the triangle shown in Figure 8 to the nearest tenth.

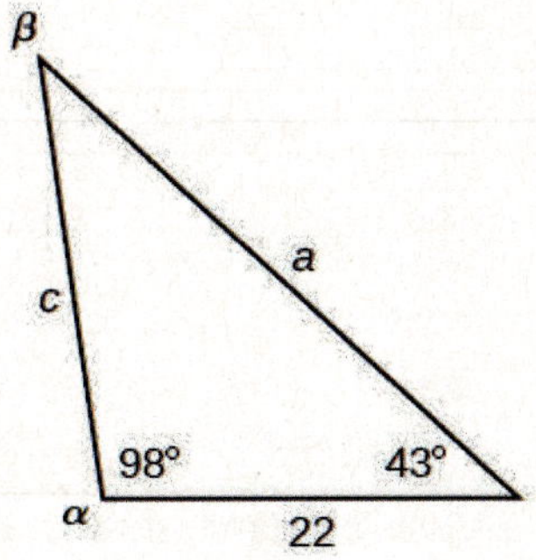

Figure 8

Using The Law of Sines to Solve SSA Triangles

We can use the Law of Sines to solve any oblique triangle, but some solutions may not be straightforward. In some cases, more than one triangle may satisfy the given criteria, which we describe as an **ambiguous case**. Triangles classified as SSA, those in which we know the lengths of two sides and the measurement of the angle opposite one of the given sides, may result in one or two solutions, or even no solution.

Possible Outcomes for SSA Triangles

Oblique triangles in the category SSA may have four different outcomes. Figure 9 illustrates the solutions with the known sides a and b and known angle α.

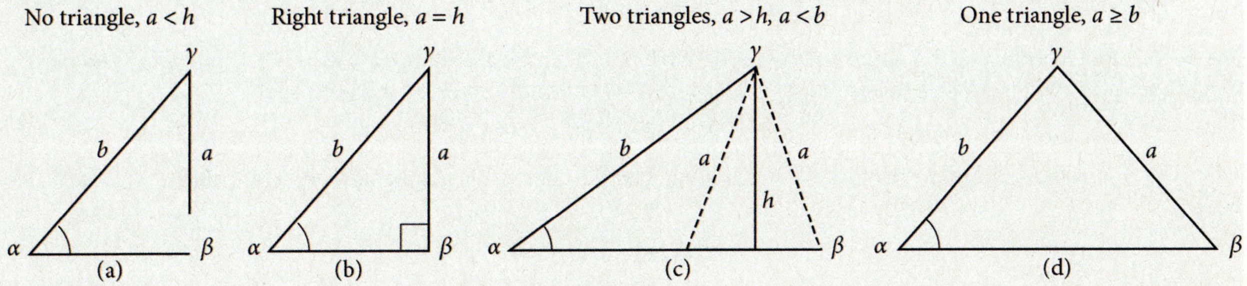

Figure 9

EXAMPLE 2

Solving an Oblique SSA Triangle

Solve the triangle in Figure 10 for the missing side and find the missing angle measures to the nearest tenth.

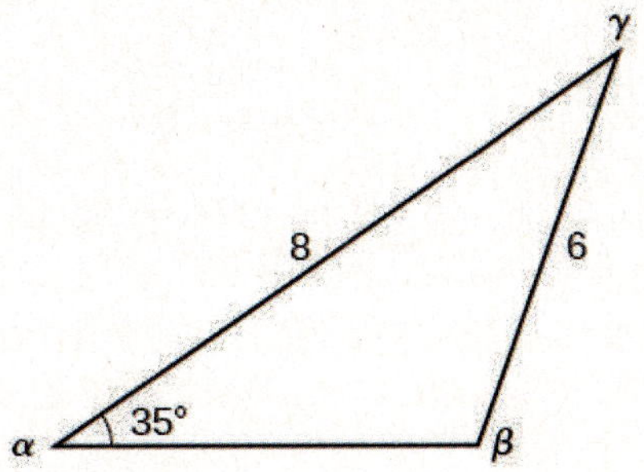

Figure 10

Solution

Use the Law of Sines to find angle β and angle γ, and then side c. Solving for β, we have the proportion

$$\begin{aligned}
\frac{\sin\ \alpha}{a} &= \frac{\sin\ \beta}{b} \\
\frac{\sin(35°)}{6} &= \frac{\sin\ \beta}{8} \\
\frac{8\sin(35°)}{6} &= \sin\ \beta \\
0.7648 &\approx \sin\ \beta \\
\sin^{-1}(0.7648) &\approx 49.9° \\
\beta &\approx 49.9°
\end{aligned}$$

However, in the diagram, angle β appears to be an obtuse angle and may be greater than 90°. How did we get an acute angle, and how do we find the measurement of β? Let's investigate further. Dropping a perpendicular from γ and viewing the triangle from a right angle perspective, we have Figure 11. It appears that there may be a second triangle that will fit the given criteria.

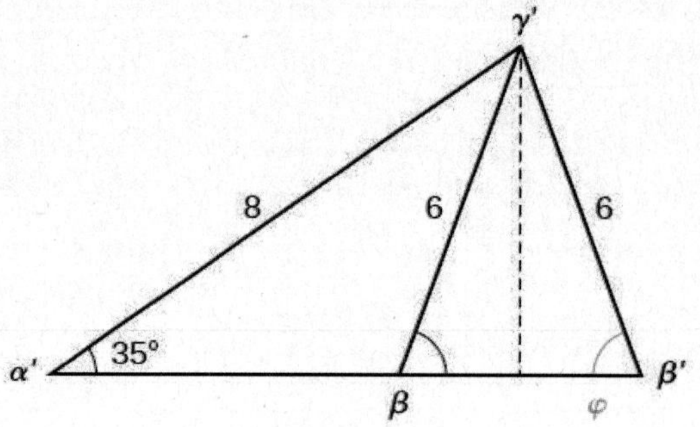

Figure 11

The angle supplementary to β is approximately equal to 49.9°, which means that $\beta = 180° - 49.9° = 130.1°$. (Remember that the sine function is positive in both the first and second quadrants.) Solving for γ, we have

$$\gamma = 180° - 35° - 130.1° \approx 14.9°$$

We can then use these measurements to solve the other triangle. Since γ' is supplementary to the sum of α' and β', we have

$$\gamma' = 180° - 35° - 49.9° \approx 95.1°$$

Now we need to find c and c'.

We have

$$\begin{aligned}
\frac{c}{\sin(14.9°)} &= \frac{6}{\sin(35°)} \\
c &= \frac{6\sin(14.9°)}{\sin(35°)} \approx 2.7
\end{aligned}$$

Finally,

$$\begin{aligned}
\frac{c'}{\sin(95.1°)} &= \frac{6}{\sin(35°)} \\
c' &= \frac{6\sin(95.1°)}{\sin(35°)} \approx 10.4
\end{aligned}$$

To summarize, there are two triangles with an angle of 35°, an adjacent side of 8, and an opposite side of 6, as shown in Figure 12.

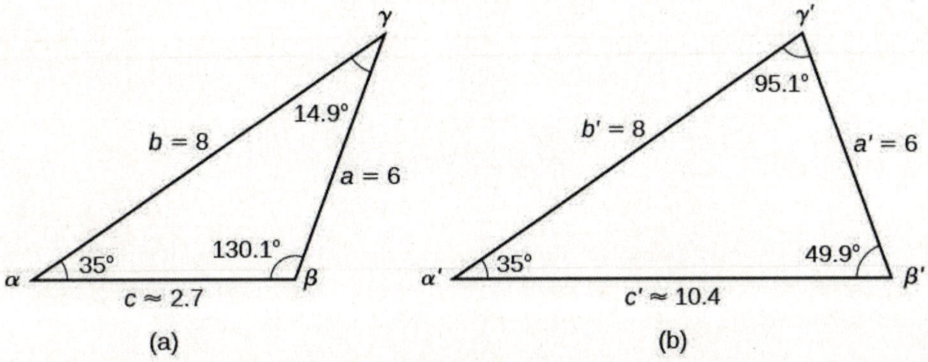

Figure 12

However, we were looking for the values for the triangle with an obtuse angle β. We can see them in the first triangle (a) in Figure 12.

TRY IT #2 Given $\alpha = 80°$, $a = 120$, and $b = 121$, find the missing side and angles. If there is more than one possible solution, show both.

EXAMPLE 3

Solving for the Unknown Sides and Angles of a SSA Triangle

In the triangle shown in Figure 13, solve for the unknown side and angles. Round your answers to the nearest tenth.

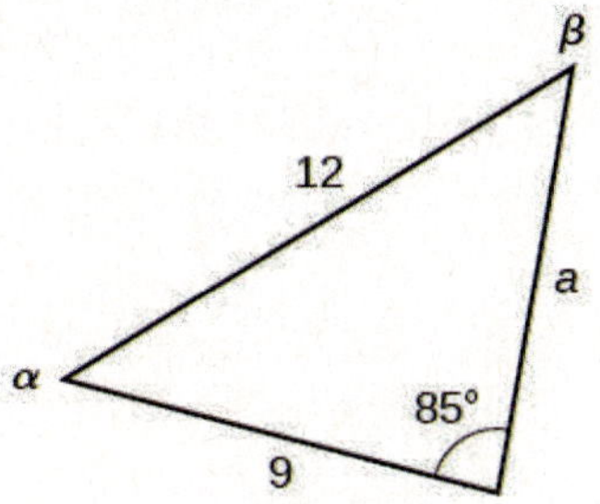

Figure 13

Solution

In choosing the pair of ratios from the Law of Sines to use, look at the information given. In this case, we know the angle $\gamma = 85°$, and its corresponding side $c = 12$, and we know side $b = 9$. We will use this proportion to solve for β.

$$\frac{\sin(85°)}{12} = \frac{\sin\ \beta}{9} \qquad \text{Isolate the unknown.}$$
$$\frac{9\sin(85°)}{12} = \sin\ \beta$$

To find β, apply the inverse sine function. The inverse sine will produce a single result, but keep in mind that there may be two values for β. It is important to verify the result, as there may be two viable solutions, only one solution (the usual case), or no solutions.

$$\beta = \sin^{-1}\left(\frac{9\sin(85°)}{12}\right)$$
$$\beta \approx \sin^{-1}(0.7471)$$
$$\beta \approx 48.3°$$

In this case, if we subtract β from 180°, we find that there may be a second possible solution. Thus, $\beta = 180° - 48.3° \approx 131.7°$. To check the solution, subtract both angles, 131.7° and 85°, from 180°. This gives

$$\alpha = 180° - 85° - 131.7° \approx -36.7°,$$

which is impossible, and so $\beta \approx 48.3°$.

To find the remaining missing values, we calculate $\alpha = 180° - 85° - 48.3° \approx 46.7°$. Now, only side a is needed. Use the Law of Sines to solve for a by one of the proportions.

$$\begin{aligned} \frac{\sin(85°)}{12} &= \frac{\sin(46.7°)}{a} \\ a\frac{\sin(85°)}{12} &= \sin(46.7°) \\ a &= \frac{12\sin(46.7°)}{\sin(85°)} \approx 8.8 \end{aligned}$$

The complete set of solutions for the given triangle is

$$\begin{aligned} &\alpha \approx 46.7° \quad a \approx 8.8 \\ &\beta \approx 48.3° \quad b = 9 \\ &\gamma = 85° \quad c = 12 \end{aligned}$$

> **TRY IT** #3 Given $\alpha = 80°, a = 100, \ b = 10$, find the missing side and angles. If there is more than one possible solution, show both. Round your answers to the nearest tenth.

EXAMPLE 4

Finding the Triangles That Meet the Given Criteria

Find all possible triangles if one side has length 4 opposite an angle of 50°, and a second side has length 10.

Solution

Using the given information, we can solve for the angle opposite the side of length 10. See Figure 14.

$$\begin{aligned} \frac{\sin\ \alpha}{10} &= \frac{\sin(50°)}{4} \\ \sin\ \alpha &= \frac{10\sin(50°)}{4} \\ \sin\ \alpha &\approx 1.915 \end{aligned}$$

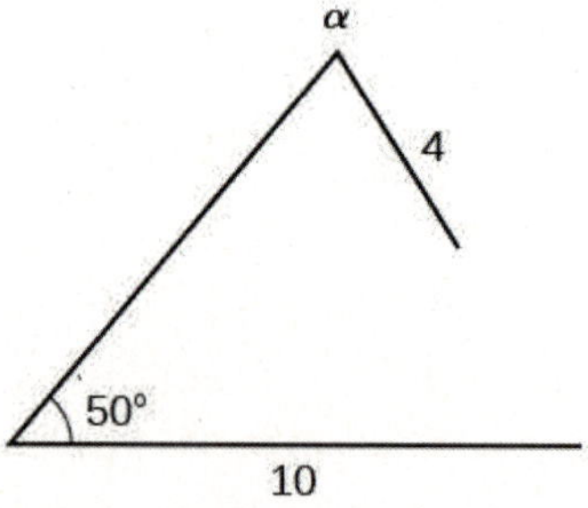

Figure 14

We can stop here without finding the value of α. Because the range of the sine function is $[-1, 1]$, it is impossible for the sine value to be 1.915. In fact, inputting $\sin^{-1}(1.915)$ in a graphing calculator generates an ERROR DOMAIN. Therefore, no triangles can be drawn with the provided dimensions.

> **TRY IT** #4 Determine the number of triangles possible given $a = 31, b = 26, \beta = 48°$.

Finding the Area of an Oblique Triangle Using the Sine Function

Now that we can solve a triangle for missing values, we can use some of those values and the sine function to find the area of an oblique triangle. Recall that the area formula for a triangle is given as $\text{Area} = \frac{1}{2}bh$, where b is base and h is height. For oblique triangles, we must find h before we can use the area formula. Observing the two triangles in Figure

15, one acute and one obtuse, we can drop a perpendicular to represent the height and then apply the trigonometric property $\sin\ \alpha = \frac{\text{opposite}}{\text{hypotenuse}}$ to write an equation for area in oblique triangles. In the acute triangle, we have $\sin\ \alpha = \frac{h}{c}$ or $c \sin\ \alpha = h$. However, in the obtuse triangle, we drop the perpendicular outside the triangle and extend the base b to form a right triangle. The angle used in calculation is α', or $180 - \alpha$.

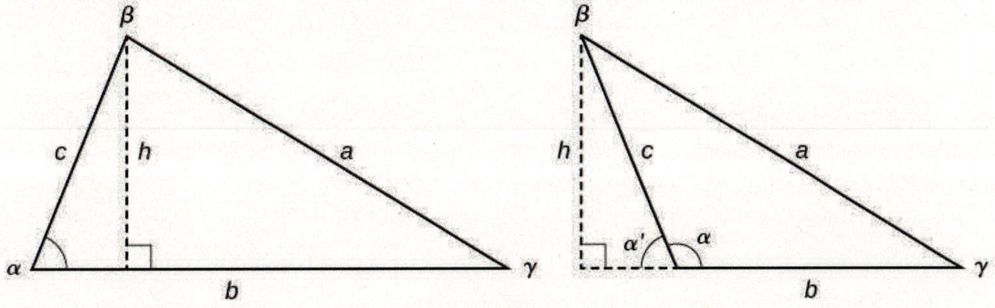

Figure 15

Thus,

$$\text{Area} = \frac{1}{2}(\text{base})(\text{height}) = \frac{1}{2}b(c \sin\ \alpha)$$

Similarly,

$$\text{Area} = \frac{1}{2}a(b \sin\ \gamma) = \frac{1}{2}a(c \sin\ \beta)$$

Area of an Oblique Triangle

The formula for the area of an oblique triangle is given by

$$\begin{aligned}\text{Area} &= \tfrac{1}{2}bc \sin\ \alpha \\ &= \tfrac{1}{2}ac \sin\ \beta \\ &= \tfrac{1}{2}ab \sin\ \gamma\end{aligned}$$

This is equivalent to one-half of the product of two sides and the sine of their included angle.

EXAMPLE 5

Finding the Area of an Oblique Triangle

Find the area of a triangle with sides $a = 90, b = 52$, and angle $\gamma = 102°$. Round the area to the nearest integer.

Solution

Using the formula, we have

$$\begin{aligned}\text{Area} &= \tfrac{1}{2}ab \sin\ \gamma \\ \text{Area} &= \tfrac{1}{2}(90)(52)\sin(102°) \\ \text{Area} &\approx 2289 \text{ square units}\end{aligned}$$

TRY IT #5 Find the area of the triangle given $\beta = 42°$, $a = 7.2$ ft, $c = 3.4$ ft. Round the area to the nearest tenth.

Solving Applied Problems Using the Law of Sines

The more we study trigonometric applications, the more we discover that the applications are countless. Some are flat, diagram-type situations, but many applications in calculus, engineering, and physics involve three dimensions and motion.

EXAMPLE 6

Finding an Altitude

Find the altitude of the aircraft in the problem introduced at the beginning of this section, shown in Figure 16. Round the altitude to the nearest tenth of a mile.

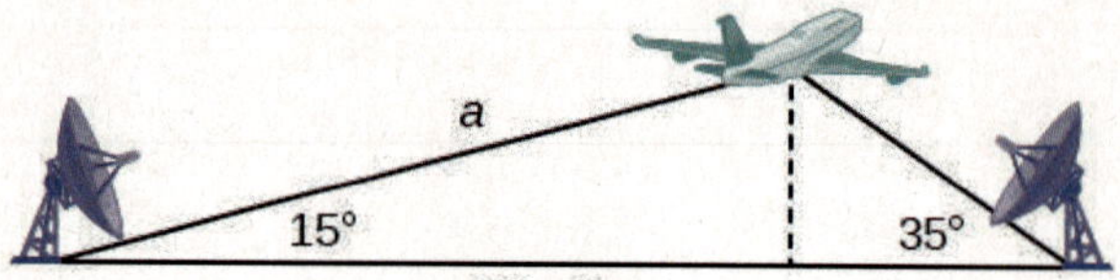

Figure 16

Solution

To find the elevation of the aircraft, we first find the distance from one station to the aircraft, such as the side a, and then use right triangle relationships to find the height of the aircraft, h.

Because the angles in the triangle add up to 180 degrees, the unknown angle must be 180°−15°−35°=130°. This angle is opposite the side of length 20, allowing us to set up a Law of Sines relationship.

$$\begin{aligned}\frac{\sin(130°)}{20} &= \frac{\sin(35°)}{a}\\ a\sin(130°) &= 20\sin(35°)\\ a &= \frac{20\sin(35°)}{\sin(130°)}\\ a &\approx 14.98\end{aligned}$$

The distance from one station to the aircraft is about 14.98 miles.

Now that we know a, we can use right triangle relationships to solve for h.

$$\begin{aligned}\sin(15°) &= \frac{\text{opposite}}{\text{hypotenuse}}\\ \sin(15°) &= \frac{h}{a}\\ \sin(15°) &= \frac{h}{14.98}\\ h &= 14.98\sin(15°)\\ h &\approx 3.88\end{aligned}$$

The aircraft is at an altitude of approximately 3.9 miles.

TRY IT #6 The diagram shown in Figure 17 represents the height of a blimp flying over a football stadium. Find the height of the blimp if the angle of elevation at the southern end zone, point A, is 70°, the angle of elevation from the northern end zone, point B, is 62°, and the distance between the viewing points of the two end zones is 145 yards.

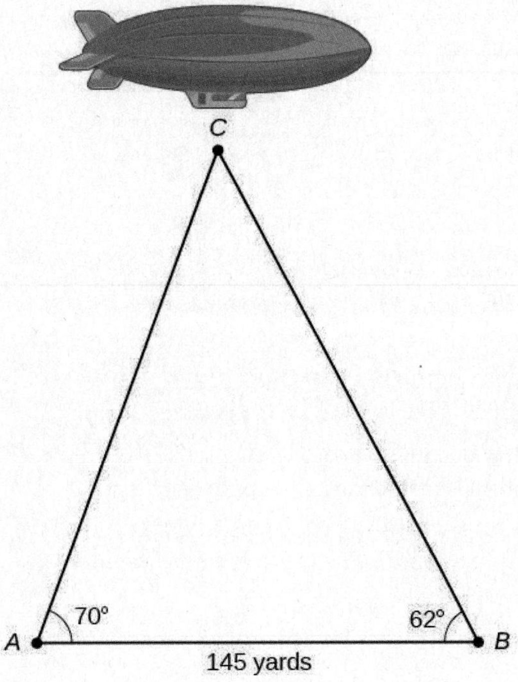

Figure 17

MEDIA

Access these online resources for additional instruction and practice with trigonometric applications.

Law of Sines: The Basics (http://openstax.org/l/sinesbasic)
Law of Sines: The Ambiguous Case (http://openstax.org/l/sinesambiguous)

10.1 SECTION EXERCISES

Verbal

1. Describe the altitude of a triangle.

2. Compare right triangles and oblique triangles.

3. When can you use the Law of Sines to find a missing angle?

4. In the Law of Sines, what is the relationship between the angle in the numerator and the side in the denominator?

5. What type of triangle results in an ambiguous case?

Algebraic

For the following exercises, assume α is opposite side a, β is opposite side b, and γ is opposite side c. Solve each triangle, if possible. Round each answer to the nearest tenth.

6. $\alpha = 43°, \gamma = 69°, a = 20$

7. $\alpha = 35°, \gamma = 73°, c = 20$

8. $\alpha = 60°, \beta = 60°, \gamma = 60°$

9. $a = 4, \alpha = 60°, \beta = 100°$

10. $b = 10, \beta = 95°, \gamma = 30°$

For the following exercises, use the Law of Sines to solve for the missing side for each oblique triangle. Round each answer to the nearest hundredth. Assume that angle A is opposite side a, angle B is opposite side b, and angle C is opposite side c.

11. Find side b when $A = 37°$, $B = 49°, c = 5$.

12. Find side a when $A = 132°, C = 23°, b = 10$.

13. Find side c when $B = 37°, C = 21°, b = 23$.

For the following exercises, assume α is opposite side a, β is opposite side b, and γ is opposite side c. Determine whether there is no triangle, one triangle, or two triangles. Then solve each triangle, if possible. Round each answer to the nearest tenth.

14. $\alpha = 119°, a = 14, b = 26$

15. $\gamma = 113°, b = 10, c = 32$

16. $b = 3.5, c = 5.3, \gamma = 80°$

17. $a = 12, c = 17, \alpha = 35°$

18. $a = 20.5, b = 35.0, \beta = 25°$

19. $a = 7, c = 9, \alpha = 43°$

20. $a = 7, b = 3, \beta = 24°$

21. $b = 13, c = 5, \gamma = 10°$

22. $a = 2.3, c = 1.8, \gamma = 28°$

23. $\beta = 119°, b = 8.2, a = 11.3$

For the following exercises, use the Law of Sines to solve, if possible, the missing side or angle for each triangle or triangles in the ambiguous case. Round each answer to the nearest tenth.

24. Find angle A when $a = 24, b = 5, B = 22°$.

25. Find angle A when $a = 13, b = 6, B = 20°$.

26. Find angle B when $A = 12°, a = 2, b = 9$.

For the following exercises, find the area of the triangle with the given measurements. Round each answer to the nearest tenth.

27. $a = 5, c = 6, \beta = 35°$

28. $b = 11, c = 8, \alpha = 28°$

29. $a = 32, b = 24, \gamma = 75°$

30. $a = 7.2, b = 4.5, \gamma = 43°$

Graphical

For the following exercises, find the length of side x. Round to the nearest tenth.

31.

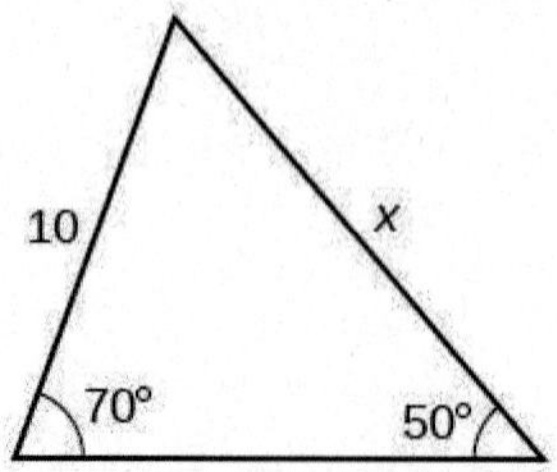

32.

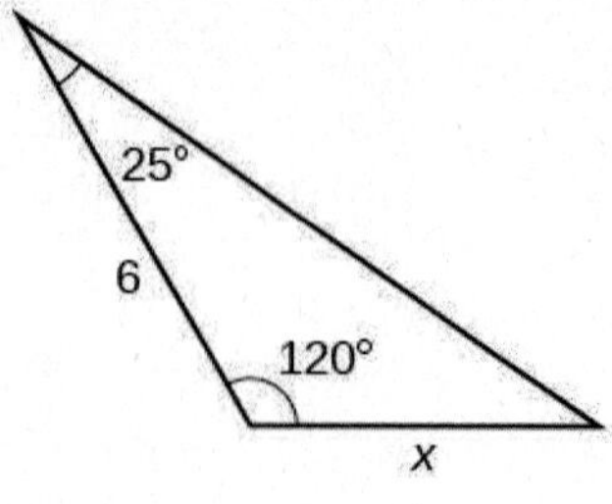

33.

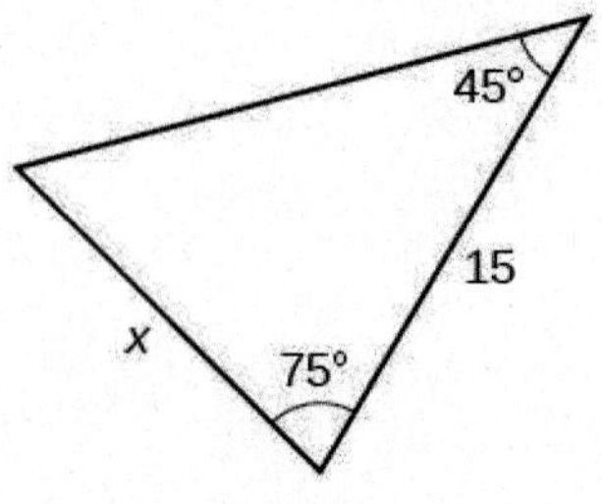

34.

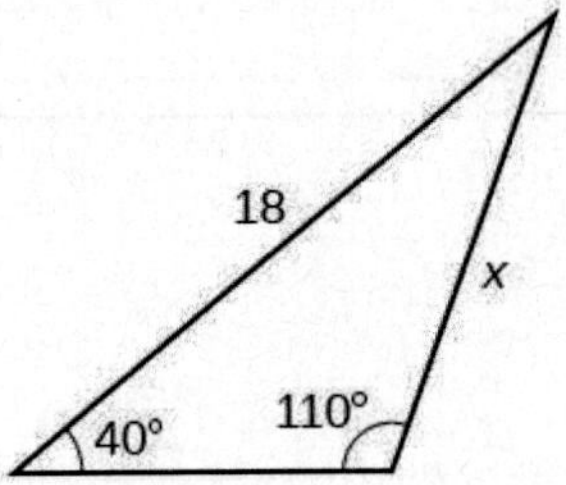

35.

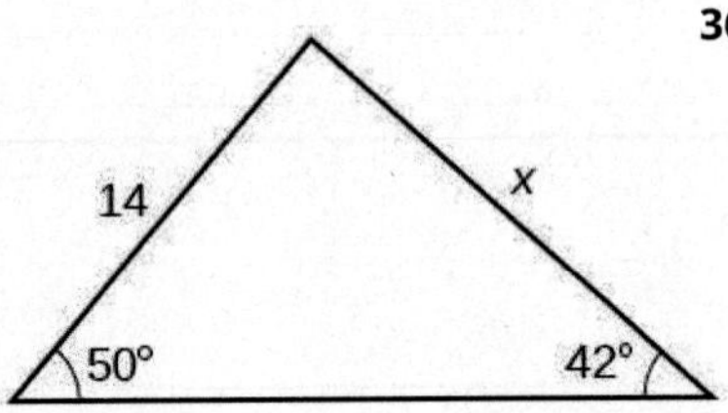

36.

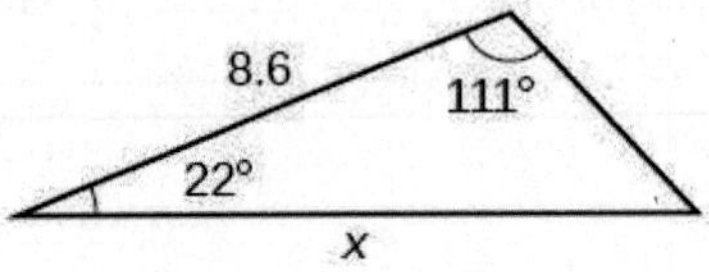

For the following exercises, find the measure of angle x, if possible. Round to the nearest tenth.

37.

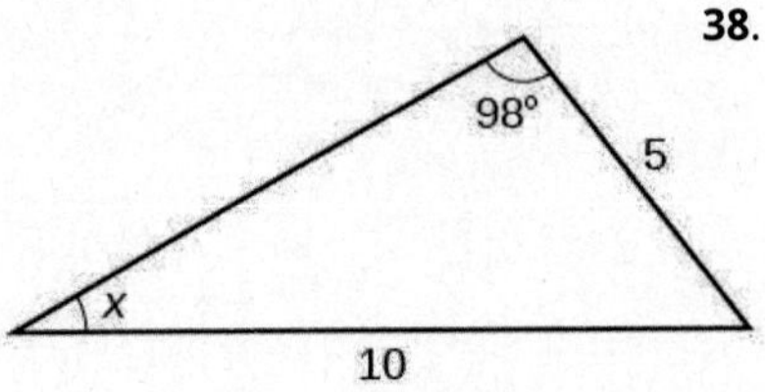

38.

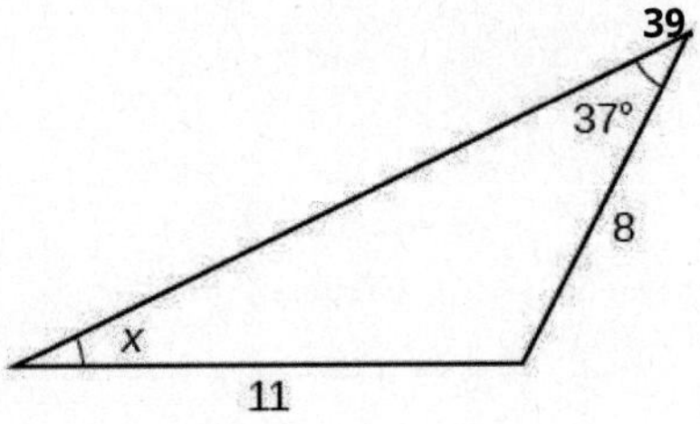

39.

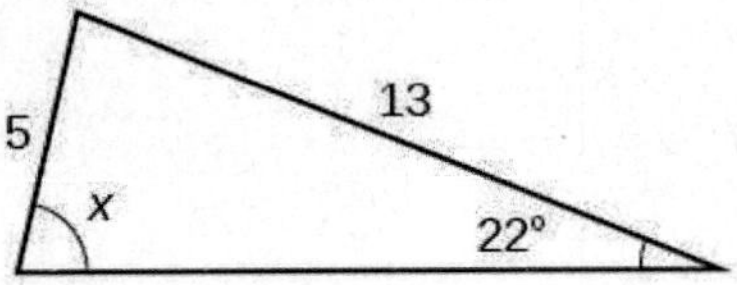

40.

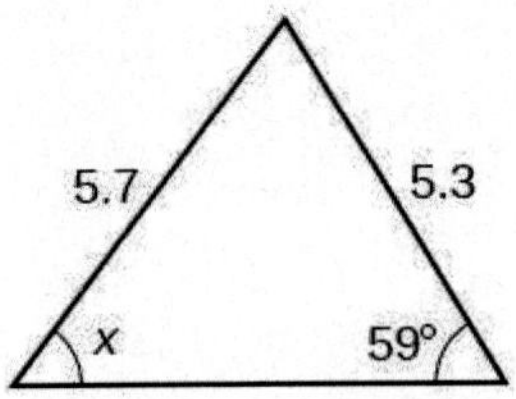

41. Notice that x is an obtuse angle.

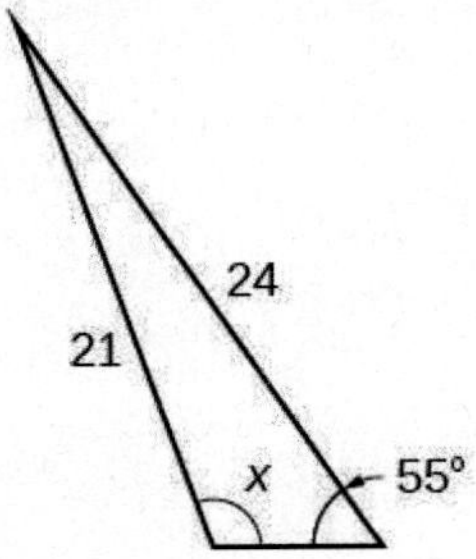

42.

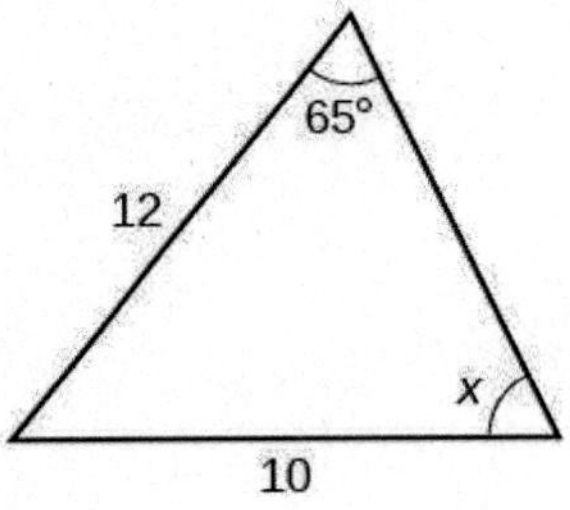

For the following exercise, solve the triangle. Round each answer to the nearest tenth.

43.

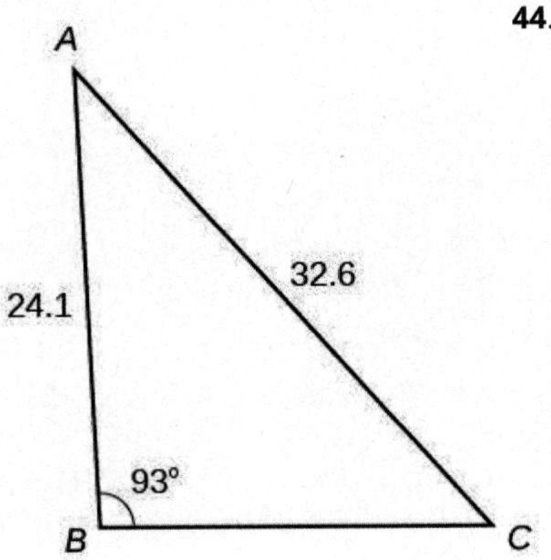

44. For the following exercises, find the area of each triangle. Round each answer to the nearest tenth.

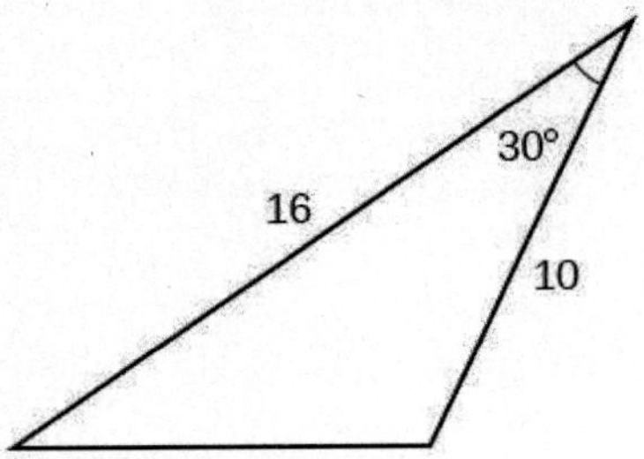

45.

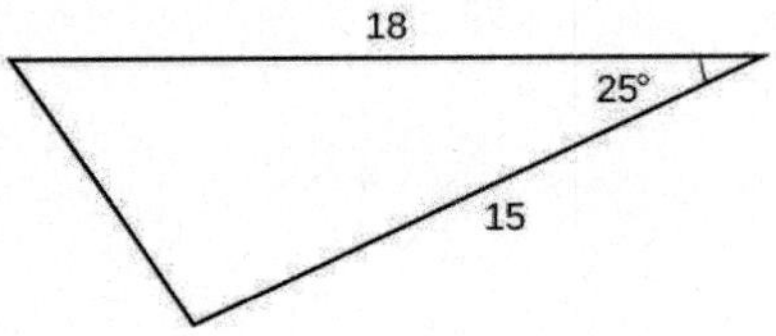

46.

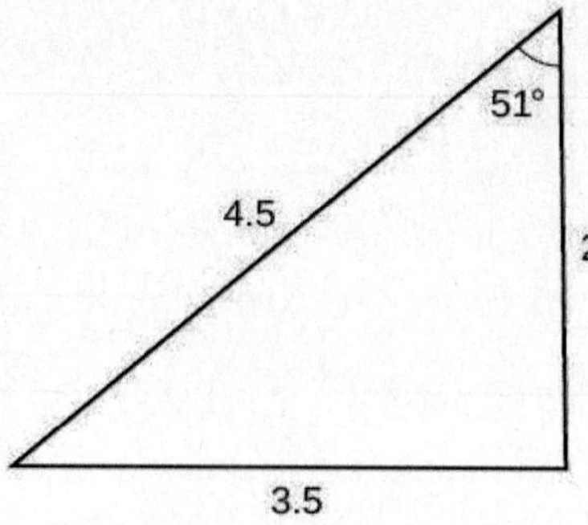

47.

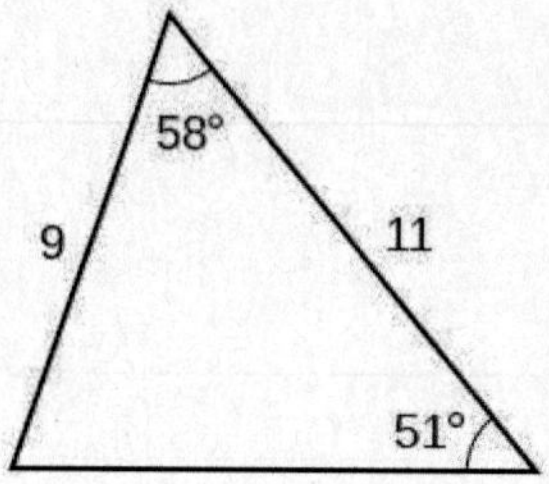

48.

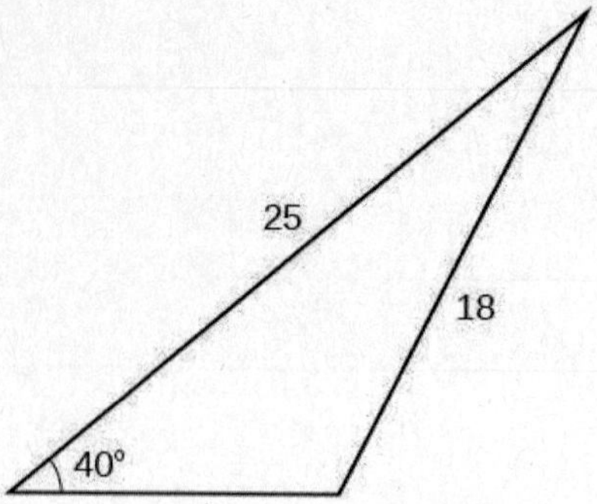

49.

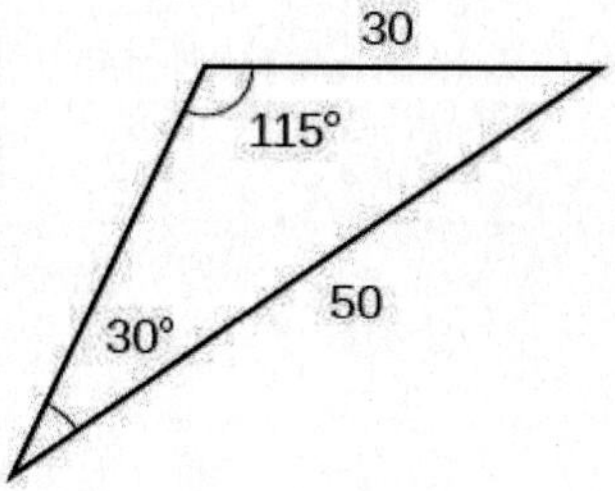

Extensions

50. Find the radius of the circle in Figure 18. Round to the nearest tenth.

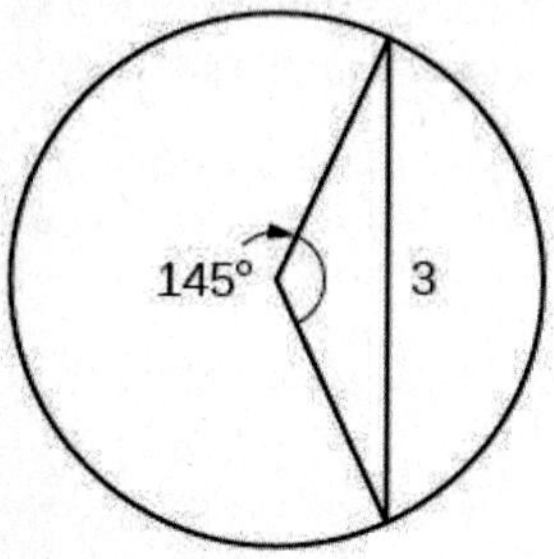

Figure 18

51. Find the diameter of the circle in Figure 19. Round to the nearest tenth.

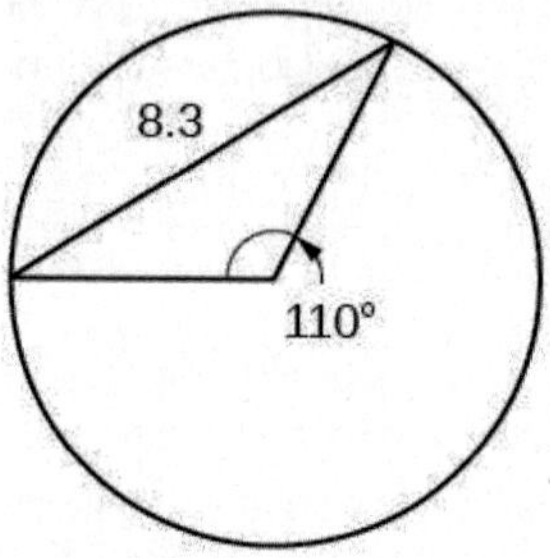

Figure 19

52. Find $m\angle ADC$ in Figure 20. Round to the nearest tenth.

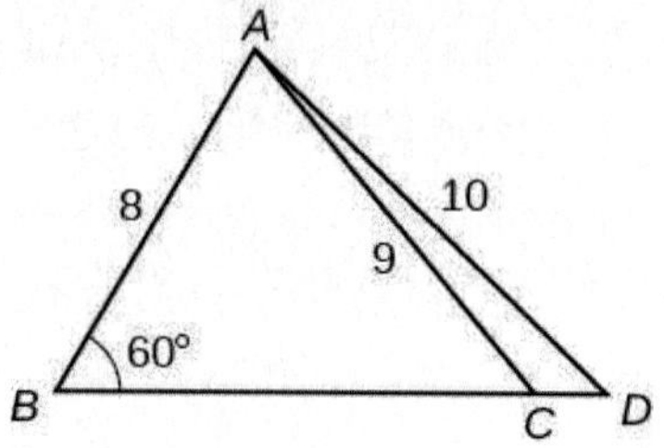

Figure 20

53. Find AD in Figure 21. Round to the nearest tenth.

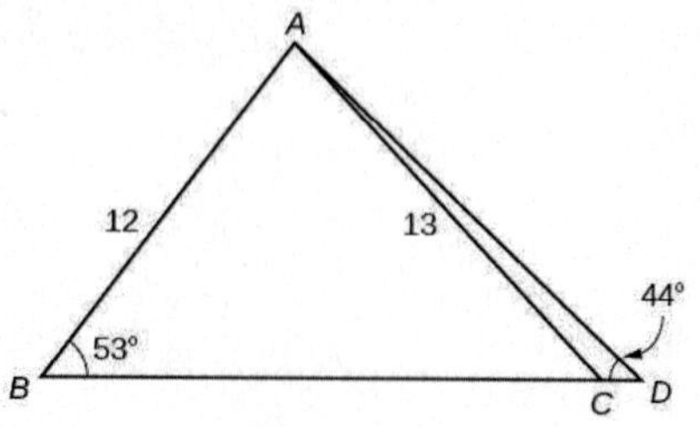

Figure 21

54. Solve both triangles in Figure 22. Round each answer to the nearest tenth.

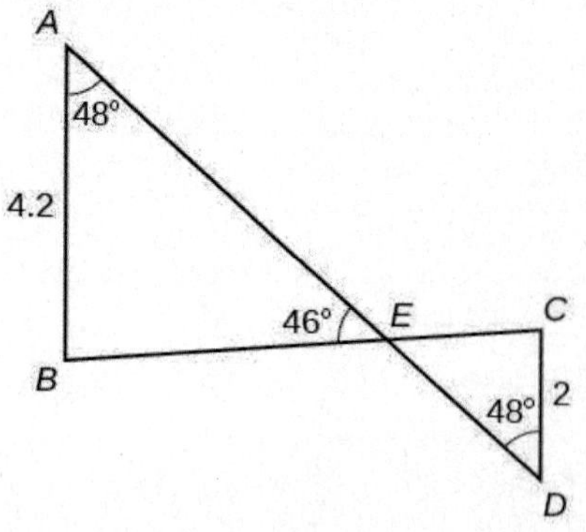

Figure 22

55. Find AB in the parallelogram shown in Figure 23.

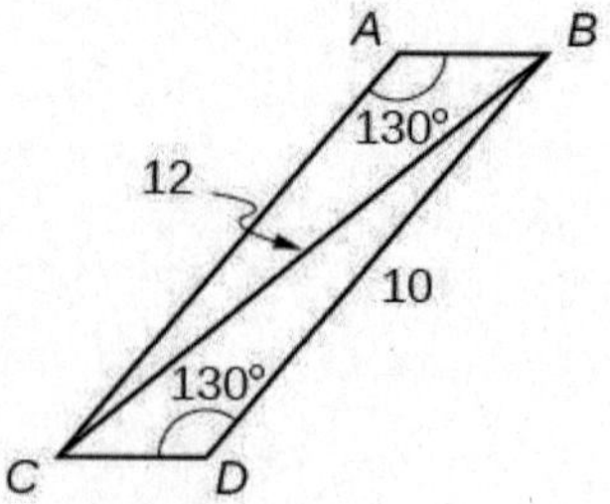

Figure 23

56. Solve the triangle in Figure 24. (Hint: Draw a perpendicular from H to JK). Round each answer to the nearest tenth.

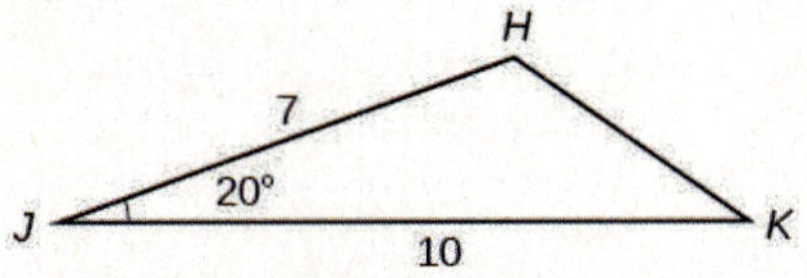

Figure 24

57. Solve the triangle in Figure 25. (Hint: Draw a perpendicular from N to LM). Round each answer to the nearest tenth.

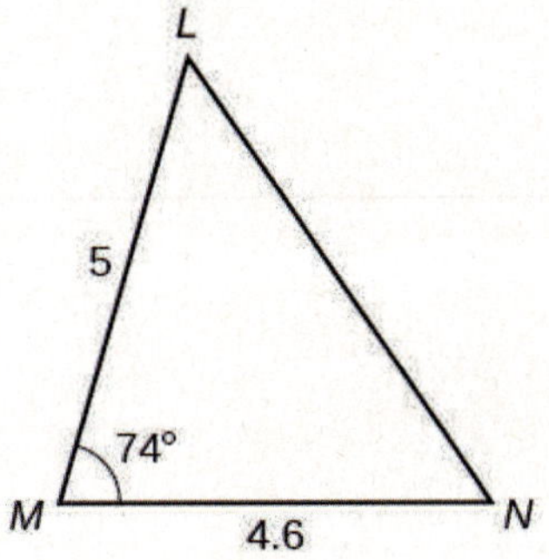

Figure 25

58. In Figure 26, $ABCD$ is not a parallelogram. $\angle m$ is obtuse. Solve both triangles. Round each answer to the nearest tenth.

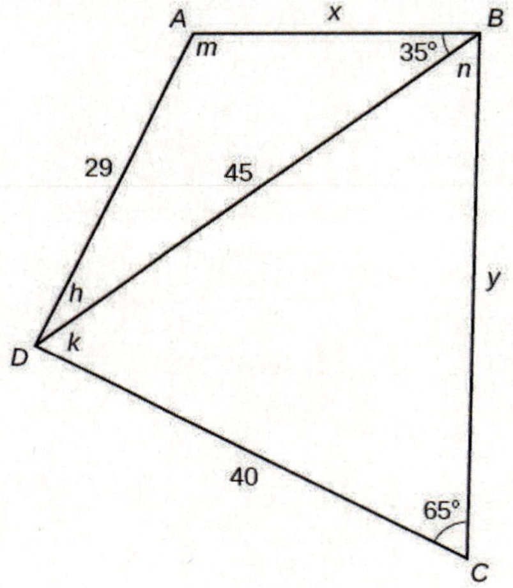

Figure 26

Real-World Applications

59. A pole leans away from the sun at an angle of $7°$ to the vertical, as shown in Figure 27. When the elevation of the sun is $55°$, the pole casts a shadow 42 feet long on the level ground. How long is the pole? Round the answer to the nearest tenth.

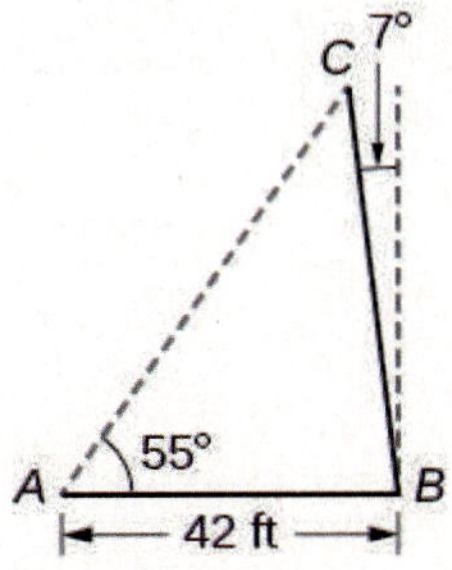

Figure 27

60. To determine how far a boat is from shore, two radar stations 500 feet apart find the angles out to the boat, as shown in Figure 28. Determine the distance of the boat from station A and the distance of the boat from shore. Round your answers to the nearest whole foot.

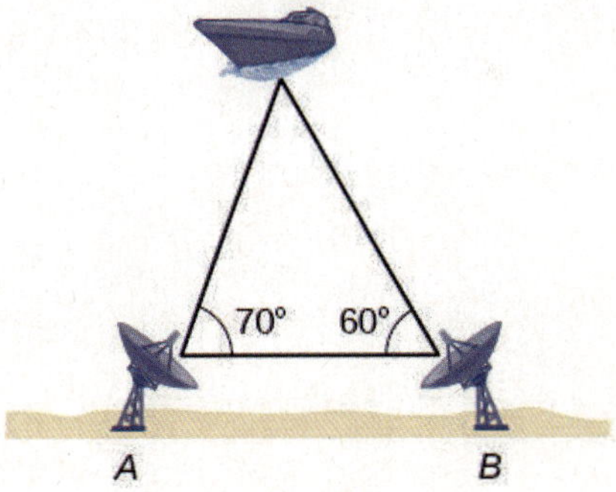

Figure 28

61. Figure 29 shows a satellite orbiting Earth. The satellite passes directly over two tracking stations A and B, which are 69 miles apart. When the satellite is on one side of the two stations, the angles of elevation at A and B are measured to be $83.9°$ and $86.2°$, respectively. How far is the satellite from station A and how high is the satellite above the ground? Round answers to the nearest whole mile.

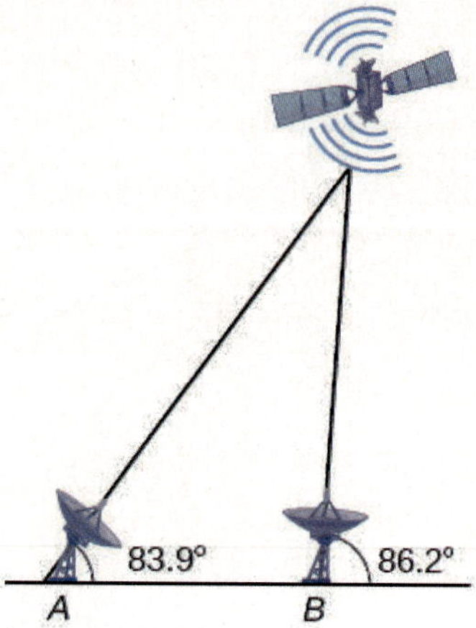

Figure 29

62. A communications tower is located at the top of a steep hill, as shown in Figure 30. The angle of inclination of the hill is $67°$. A guy wire is to be attached to the top of the tower and to the ground, 165 meters downhill from the base of the tower. The angle formed by the guy wire and the hill is $16°$. Find the length of the cable required for the guy wire to the nearest whole meter.

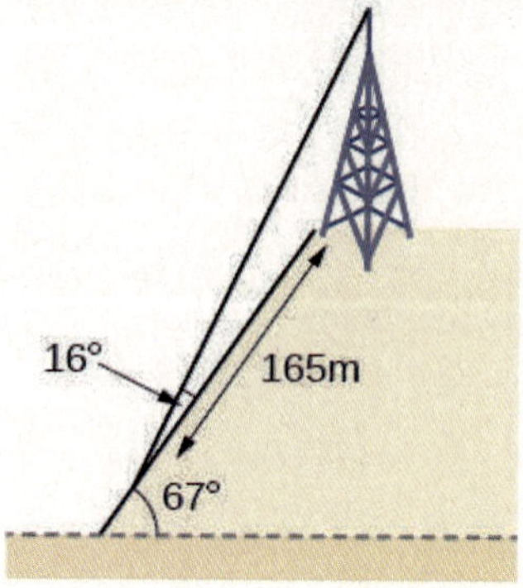

Figure 30

63. The roof of a house is at a $20°$ angle. An 8-foot solar panel is to be mounted on the roof and should be angled $38°$ relative to the horizontal for optimal results. (See Figure 31). How long does the vertical support holding up the back of the panel need to be? Round to the nearest tenth.

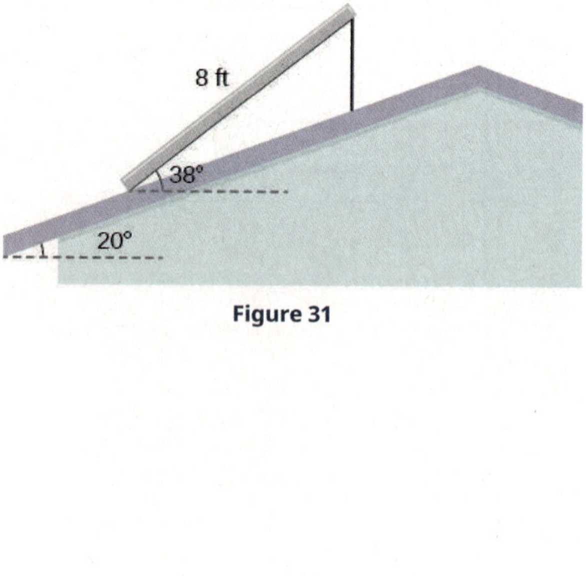

Figure 31

64. Similar to an angle of elevation, an *angle of depression* is the acute angle formed by a horizontal line and an observer's line of sight to an object below the horizontal. A pilot is flying over a straight highway. He determines the angles of depression to two mileposts, 6.6 km apart, to be 37° and 44°, as shown in Figure 32. Find the distance of the plane from point A to the nearest tenth of a kilometer.

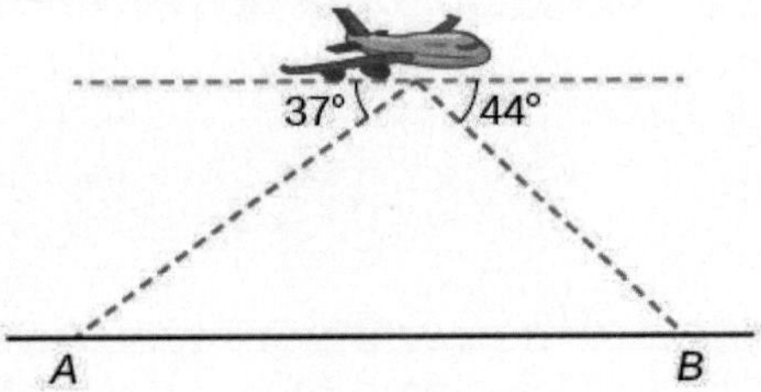

Figure 32

65. A pilot is flying over a straight highway. He determines the angles of depression to two mileposts, 4.3 km apart, to be 32° and 56°, as shown in Figure 33. Find the distance of the plane from point A to the nearest tenth of a kilometer.

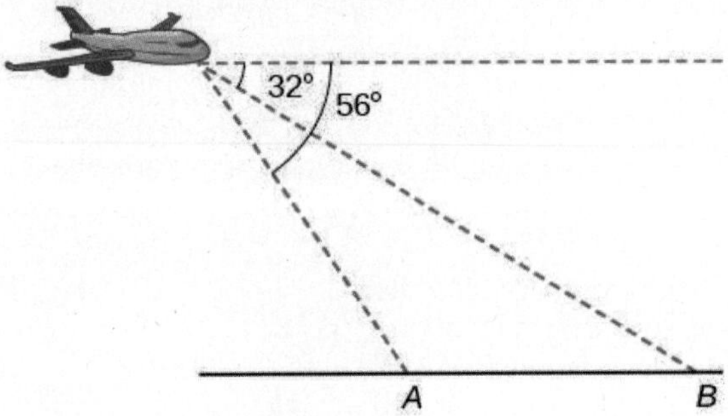

Figure 33

66. In order to estimate the height of a building, two students stand at a certain distance from the building at street level. From this point, they find the angle of elevation from the street to the top of the building to be 39°. They then move 300 feet closer to the building and find the angle of elevation to be 50°. Assuming that the street is level, estimate the height of the building to the nearest foot.

67. In order to estimate the height of a building, two students stand at a certain distance from the building at street level. From this point, they find the angle of elevation from the street to the top of the building to be 35°. They then move 250 feet closer to the building and find the angle of elevation to be 53°. Assuming that the street is level, estimate the height of the building to the nearest foot.

68. Points A and B are on opposite sides of a lake. Point C is 97 meters from A. The measure of angle BAC is determined to be 101°, and the measure of angle ACB is determined to be 53°. What is the distance from A to B, rounded to the nearest whole meter?

69. A man and a woman standing $3\frac{1}{2}$ miles apart spot a hot air balloon at the same time. If the angle of elevation from the man to the balloon is 27°, and the angle of elevation from the woman to the balloon is 41°, find the altitude of the balloon to the nearest foot.

70. Two search teams spot a stranded climber on a mountain. The first search team is 0.5 miles from the second search team, and both teams are at an altitude of 1 mile. The angle of elevation from the first search team to the stranded climber is 15°. The angle of elevation from the second search team to the climber is 22°. What is the altitude of the climber? Round to the nearest tenth of a mile.

71. A street light is mounted on a pole. A 6-foot-tall man is standing on the street a short distance from the pole, casting a shadow. The angle of elevation from the tip of the man's shadow to the top of his head of 28°. A 6-foot-tall woman is standing on the same street on the opposite side of the pole from the man. The angle of elevation from the tip of her shadow to the top of her head is 28°. If the man and woman are 20 feet apart, how far is the street light from the tip of the shadow of each person? Round the distance to the nearest tenth of a foot.

72. Three cities, A, B, and C, are located so that city A is due east of city B. If city C is located 35° west of north from city B and is 100 miles from city A and 70 miles from city B, how far is city A from city B? Round the distance to the nearest tenth of a mile.

73. Two streets meet at an 80° angle. At the corner, a park is being built in the shape of a triangle. Find the area of the park if, along one road, the park measures 180 feet, and along the other road, the park measures 215 feet.

74. Brian's house is on a corner lot. Find the area of the front yard if the edges measure 40 and 56 feet, as shown in Figure 34.

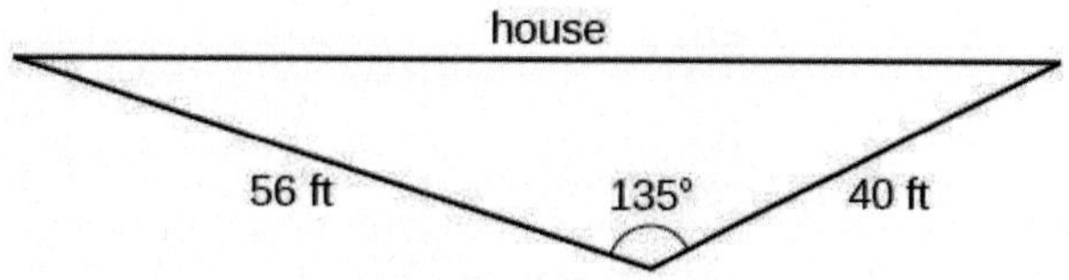

Figure 34

75. The Bermuda triangle is a region of the Atlantic Ocean that connects Bermuda, Florida, and Puerto Rico. Find the area of the Bermuda triangle if the distance from Florida to Bermuda is 1030 miles, the distance from Puerto Rico to Bermuda is 980 miles, and the angle created by the two distances is 62°.

76. A yield sign measures 30 inches on all three sides. What is the area of the sign?

77. Naomi bought a dining table whose top is in the shape of a triangle. Find the area of the table top if two of the sides measure 4 feet and 4.5 feet, and the smaller angles measure 32° and 42°, as shown in Figure 35.

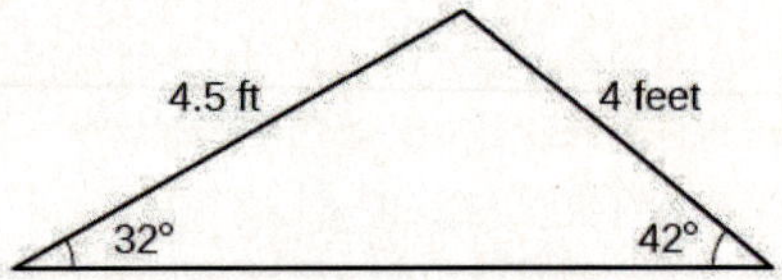

Figure 35

10.2 Non-right Triangles: Law of Cosines

Learning Objectives

In this section, you will:

- Use the Law of Cosines to solve oblique triangles.
- Solve applied problems using the Law of Cosines.
- Use Heron's formula to find the area of a triangle.

Suppose a boat leaves port, travels 10 miles, turns 20 degrees, and travels another 8 miles as shown in Figure 1. How far from port is the boat?

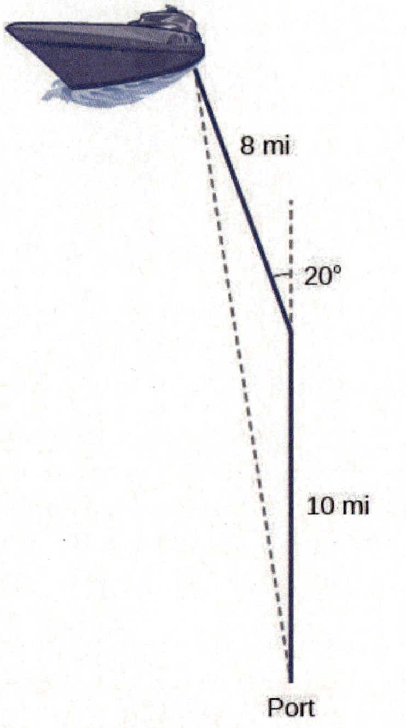

Figure 1

Unfortunately, while the Law of Sines enables us to address many non-right triangle cases, it does not help us with triangles where the known angle is between two known sides, a SAS (side-angle-side) triangle, or when all three sides are known, but no angles are known, a SSS (side-side-side) triangle. In this section, we will investigate another tool for solving oblique triangles described by these last two cases.

Using the Law of Cosines to Solve Oblique Triangles

The tool we need to solve the problem of the boat's distance from the port is the **Law of Cosines**, which defines the relationship among angle measurements and side lengths in oblique triangles. Three formulas make up the Law of Cosines. At first glance, the formulas may appear complicated because they include many variables. However, once the pattern is understood, the Law of Cosines is easier to work with than most formulas at this mathematical level.

Understanding how the Law of Cosines is derived will be helpful in using the formulas. The derivation begins with the **Generalized Pythagorean Theorem**, which is an extension of the Pythagorean Theorem to non-right triangles. Here is how it works: An arbitrary non-right triangle ABC is placed in the coordinate plane with vertex A at the origin, side c drawn along the x-axis, and vertex C located at some point (x, y) in the plane, as illustrated in Figure 2. Generally, triangles exist anywhere in the plane, but for this explanation we will place the triangle as noted.

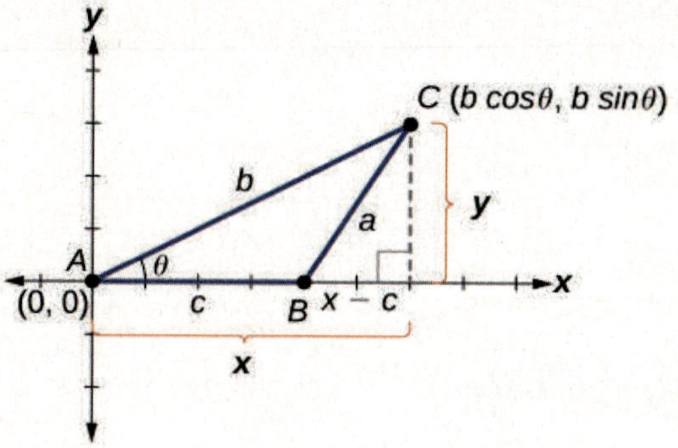

Figure 2

We can drop a perpendicular from C to the x-axis (this is the altitude or height). Recalling the basic trigonometric identities, we know that

$$\cos\,\theta = \frac{x(\text{adjacent})}{b(\text{hypotenuse})} \text{ and } \sin\,\theta = \frac{y(\text{opposite})}{b(\text{hypotenuse})}$$

In terms of θ, $x = b\cos\,\theta$ and $y = b\sin\,\theta$. The (x, y) point located at C has coordinates $(b\cos\,\theta, b\sin\,\theta)$. Using the side $(x - c)$ as one leg of a right triangle and y as the second leg, we can find the length of hypotenuse a using the Pythagorean Theorem. Thus,

$$\begin{aligned}
a^2 &= (x-c)^2 + y^2 & \\
&= (b\cos\,\theta - c)^2 + (b\sin\,\theta)^2 & \text{Substitute } (b\cos\,\theta) \text{ for } x \text{ and } (b\sin\,\theta) \text{ for } y. \\
&= \left(b^2\cos^2\theta - 2bc\cos\,\theta + c^2\right) + b^2\sin^2\theta & \text{Expand the perfect square.} \\
&= b^2\cos^2\theta + b^2\sin^2\theta + c^2 - 2bc\cos\,\theta & \text{Group terms noting that } \cos^2\theta + \sin^2\theta = 1. \\
&= b^2\left(\cos^2\theta + \sin^2\theta\right) + c^2 - 2bc\cos\,\theta & \text{Factor out } b^2. \\
a^2 &= b^2 + c^2 - 2bc\cos\,\theta &
\end{aligned}$$

The formula derived is one of the three equations of the Law of Cosines. The other equations are found in a similar fashion.

Keep in mind that it is always helpful to sketch the triangle when solving for angles or sides. In a real-world scenario, try to draw a diagram of the situation. As more information emerges, the diagram may have to be altered. Make those alterations to the diagram and, in the end, the problem will be easier to solve.

Law of Cosines

The **Law of Cosines** states that the square of any side of a triangle is equal to the sum of the squares of the other two sides minus twice the product of the other two sides and the cosine of the included angle. For triangles labeled as in Figure 3, with angles α, β, and γ, and opposite corresponding sides a, b, and c, respectively, the Law of Cosines is given as three equations.

$$\begin{aligned}
a^2 &= b^2 + c^2 - 2bc \quad \cos\,\alpha \\
b^2 &= a^2 + c^2 - 2ac \quad \cos\,\beta \\
c^2 &= a^2 + b^2 - 2ab \quad \cos\,\gamma
\end{aligned}$$

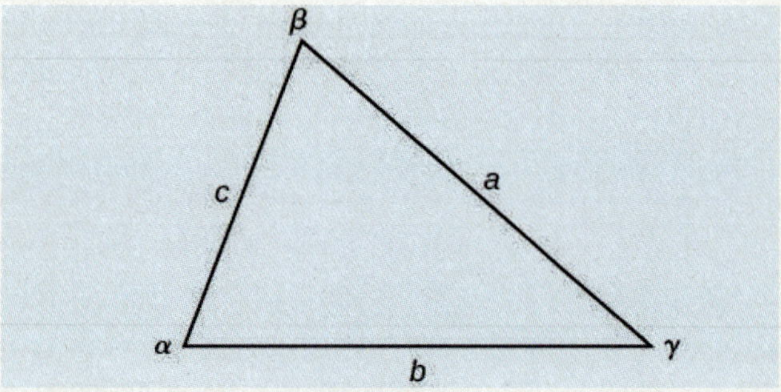

Figure 3

To solve for a missing side measurement, the corresponding opposite angle measure is needed.

When solving for an angle, the corresponding opposite side measure is needed. We can use another version of the Law of Cosines to solve for an angle.

$$\cos\ \alpha = \frac{b^2+c^2-a^2}{2bc}$$
$$\cos\ \beta = \frac{a^2+c^2-b^2}{2ac}$$
$$\cos\ \gamma = \frac{a^2+b^2-c^2}{2ab}$$

HOW TO

Given two sides and the angle between them (SAS), find the measures of the remaining side and angles of a triangle.

1. Sketch the triangle. Identify the measures of the known sides and angles. Use variables to represent the measures of the unknown sides and angles.
2. Apply the Law of Cosines to find the length of the unknown side or angle.
3. Apply the Law of Sines or Cosines to find the measure of a second angle.
4. Compute the measure of the remaining angle.

EXAMPLE 1

Finding the Unknown Side and Angles of a SAS Triangle
Find the unknown side and angles of the triangle in Figure 4.

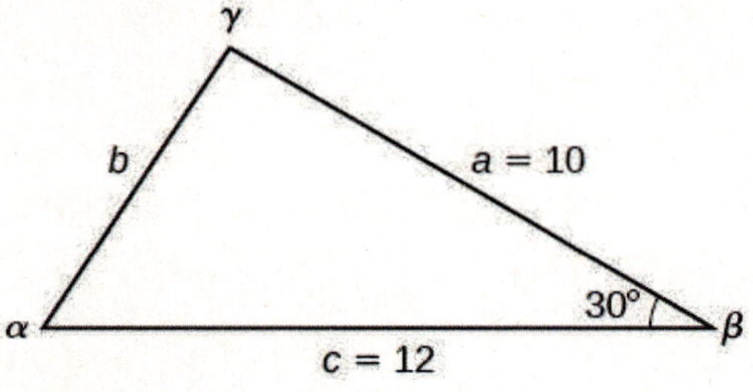

Figure 4

Solution
First, make note of what is given: two sides and the angle between them. This arrangement is classified as SAS and supplies the data needed to apply the Law of Cosines.

Each one of the three laws of cosines begins with the square of an unknown side opposite a known angle. For this example, the first side to solve for is side b, as we know the measurement of the opposite angle β.

$$b^2 = a^2 + c^2 - 2ac\cos\,\beta$$

$$b^2 = 10^2 + 12^2 - 2(10)(12)\cos(30^\circ)$$ Substitute the measurements for the known quantities.

$$b^2 = 100 + 144 - 240\left(\frac{\sqrt{3}}{2}\right)$$ Evaluate the cosine and begin to simplify.

$$b^2 = 244 - 120\sqrt{3}$$

$$b = \sqrt{244 - 120\sqrt{3}}$$ Use the square root property.

$$b \approx 6.013$$

Because we are solving for a length, we use only the positive square root. Now that we know the length b, we can use the Law of Sines to fill in the remaining angles of the triangle. Solving for angle α, we have

$$\frac{\sin\,\alpha}{a} = \frac{\sin\,\beta}{b}$$

$$\frac{\sin\,\alpha}{10} = \frac{\sin(30^\circ)}{6.013}$$

$$\sin\,\alpha = \frac{10\sin(30^\circ)}{6.013}$$ Multiply both sides of the equation by 10.

$$\alpha = \sin^{-1}\left(\frac{10\sin(30^\circ)}{6.013}\right)$$ Find the inverse sine of $\frac{10\sin(30^\circ)}{6.013}$.

$$\alpha \approx 56.3^\circ$$

The other possibility for α would be $\alpha = 180^\circ - 56.3^\circ \approx 123.7^\circ$. In the original diagram, α is adjacent to the longest side, so α is an acute angle and, therefore, 123.7° does not make sense. Notice that if we choose to apply the Law of Cosines, we arrive at a unique answer. We do not have to consider the other possibilities, as cosine is unique for angles between 0° and 180°. Proceeding with $\alpha \approx 56.3^\circ$, we can then find the third angle of the triangle.

$$\gamma = 180^\circ - 30^\circ - 56.3^\circ \approx 93.7^\circ$$

The complete set of angles and sides is

$$\begin{array}{ll} \alpha \approx 56.3^\circ & a = 10 \\ \beta = 30^\circ & b \approx 6.013 \\ \gamma \approx 93.7^\circ & c = 12 \end{array}$$

> **TRY IT** #1 Find the missing side and angles of the given triangle: $\alpha = 30^\circ$, $b = 12$, $c = 24$.

EXAMPLE 2

Solving for an Angle of a SSS Triangle

Find the angle α for the given triangle if side $a = 20$, side $b = 25$, and side $c = 18$.

Solution

For this example, we have no angles. We can solve for any angle using the Law of Cosines. To solve for angle α, we have

$$\begin{aligned}
a^2 &= b^2 + c^2 - 2bc\cos\alpha \\
20^2 &= 25^2 + 18^2 - 2(25)(18)\cos\alpha && \text{Substitute the appropriate measurements.} \\
400 &= 625 + 324 - 900\cos\alpha && \text{Simplify in each step.} \\
400 &= 949 - 900\cos\alpha \\
-549 &= -900\cos\alpha && \text{Isolate } \cos\alpha. \\
\frac{-549}{-900} &= \cos\alpha \\
0.61 &\approx \cos\alpha \\
\cos^{-1}(0.61) &\approx \alpha && \text{Find the inverse cosine.} \\
\alpha &\approx 52.4^\circ
\end{aligned}$$

See Figure 5.

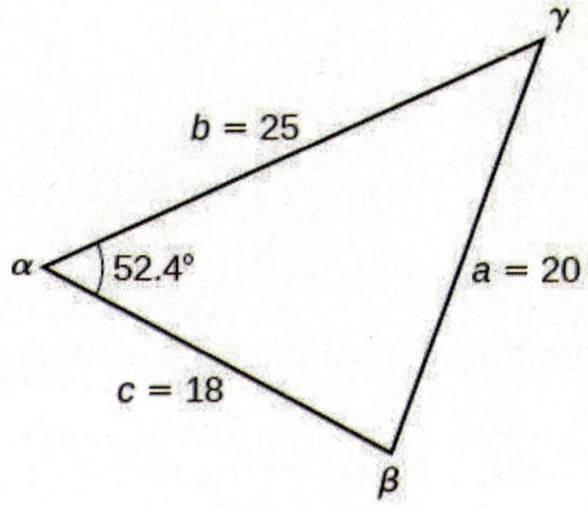

Figure 5

Analysis

Because the inverse cosine can return any angle between 0 and 180 degrees, there will not be any ambiguous cases using this method.

TRY IT #2 Given $a = 5, b = 7,$ and $c = 10$, find the missing angles.

Solving Applied Problems Using the Law of Cosines

Just as the Law of Sines provided the appropriate equations to solve a number of applications, the Law of Cosines is applicable to situations in which the given data fits the cosine models. We may see these in the fields of navigation, surveying, astronomy, and geometry, just to name a few.

EXAMPLE 3

Using the Law of Cosines to Solve a Communication Problem

On many cell phones with GPS, an approximate location can be given before the GPS signal is received. This is accomplished through a process called triangulation, which works by using the distances from two known points. Suppose there are two cell phone towers within range of a cell phone. The two towers are located 6000 feet apart along a straight highway, running east to west, and the cell phone is north of the highway. Based on the signal delay, it can be determined that the signal is 5,050 feet from the first tower and 2,420 feet from the second tower. Determine the position of the cell phone north and east of the first tower, and determine how far it is from the highway.

Solution

For simplicity, we start by drawing a diagram similar to Figure 6 and labeling our given information.

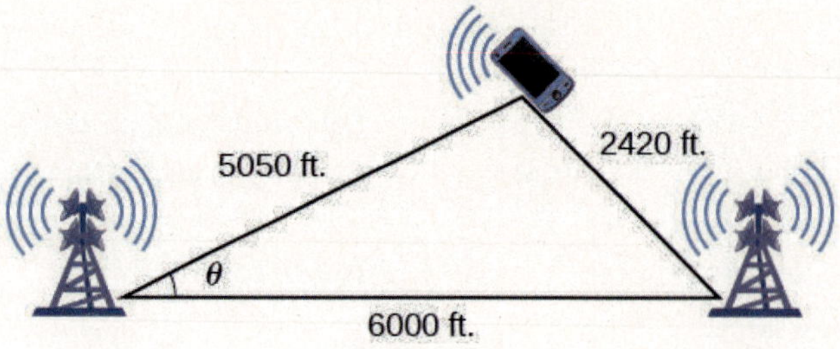

Figure 6

Using the Law of Cosines, we can solve for the angle θ. Remember that the Law of Cosines uses the square of one side to find the cosine of the opposite angle. For this example, let $a = 2{,}420$, $b = 5{,}050$, and $c = 6{,}000$. Thus, θ corresponds to the opposite side $a = 2{,}420$.

$$\begin{aligned} a^2 &= b^2 + c^2 - 2bc \cos\theta \\ (2,420)^2 &= (5,050)^2 + (6,000)^2 - 2(5,050)(6,000) \cos\theta \\ (2,420)^2 - (5,050)^2 - (6,000)^2 &= -2(5,050)(6,000) \cos\theta \\ \frac{(2,420)^2 - (5,050)^2 - (6,000)^2}{-2(5,050)(6,000)} &= \cos\theta \\ \cos\theta &\approx 0.9183 \\ \theta &\approx \cos^{-1}(0.9183) \\ \theta &\approx 23.3^\circ \end{aligned}$$

To answer the questions about the phone's position north and east of the tower, and the distance to the highway, drop a perpendicular from the position of the cell phone, as in Figure 7. This forms two right triangles, although we only need the right triangle that includes the first tower for this problem.

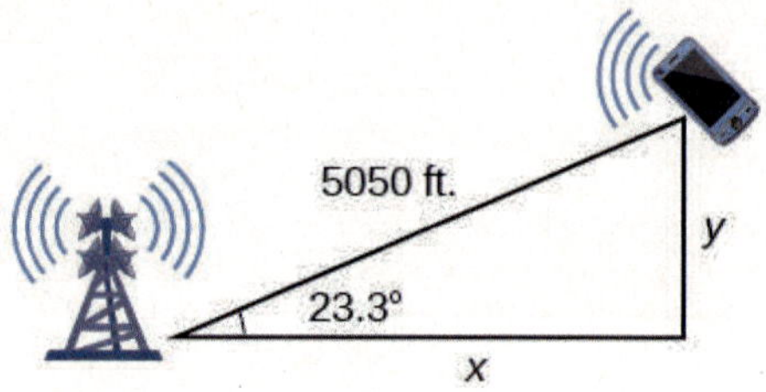

Figure 7

Using the angle $\theta = 23.3^\circ$ and the basic trigonometric identities, we can find the solutions. Thus

$$\begin{aligned} \cos(23.3^\circ) &= \frac{x}{5{,}050} \\ x &= 5{,}050\cos(23.3^\circ) \\ x &\approx 4{,}638.15 \text{ feet} \\ \sin(23.3^\circ) &= \frac{y}{5{,}050} \\ y &= 5{,}050\sin(23.3^\circ) \\ y &\approx 1{,}997.5 \text{ feet} \end{aligned}$$

The cell phone is approximately 4,638 feet east and 1998 feet north of the first tower, and 1998 feet from the highway.

EXAMPLE 4

Calculating Distance Traveled Using a SAS Triangle

Returning to our problem at the beginning of this section, suppose a boat leaves port, travels 10 miles, turns 20 degrees, and travels another 8 miles. How far from port is the boat? The diagram is repeated here in Figure 8.

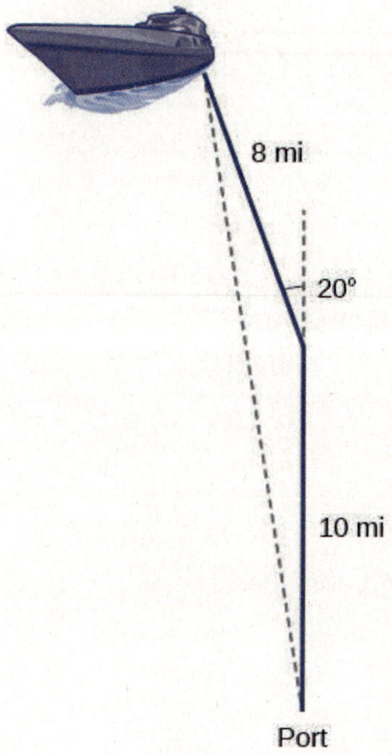

Figure 8

Solution

The boat turned 20 degrees, so the obtuse angle of the non-right triangle is the supplemental angle, $180° - 20° = 160°$. With this, we can utilize the Law of Cosines to find the missing side of the obtuse triangle—the distance of the boat to the port.

$$\begin{aligned} x^2 &= 8^2 + 10^2 - 2(8)(10)\cos(160°) \\ x^2 &= 314.35 \\ x &= \sqrt{314.35} \\ x &\approx 17.7 \text{ miles} \end{aligned}$$

The boat is about 17.7 miles from port.

Using Heron's Formula to Find the Area of a Triangle

We already learned how to find the area of an oblique triangle when we know two sides and an angle. We also know the formula to find the area of a triangle using the base and the height. When we know the three sides, however, we can use Heron's formula instead of finding the height. Heron of Alexandria was a geometer who lived during the first century A.D. He discovered a formula for finding the area of oblique triangles when three sides are known.

Heron's Formula

Heron's formula finds the area of oblique triangles in which sides a, b, and c are known.

$$\text{Area} = \sqrt{s(s-a)(s-b)(s-c)}$$

where $s = \frac{(a+b+c)}{2}$ is one half of the perimeter of the triangle, sometimes called the semi-perimeter.

EXAMPLE 5

Using Heron's Formula to Find the Area of a Given Triangle

Find the area of the triangle in Figure 9 using Heron's formula.

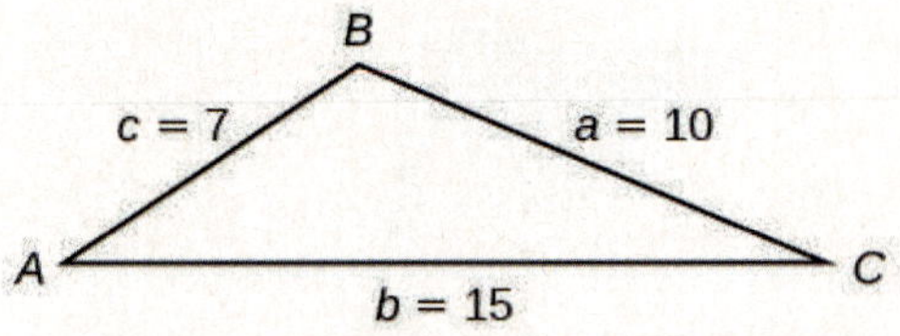

Figure 9

Solution

First, we calculate s.

$$s = \frac{(a+b+c)}{2}$$
$$s = \frac{(10+15+7)}{2} = 16$$

Then we apply the formula.

$$\text{Area} = \sqrt{s(s-a)(s-b)(s-c)}$$
$$\text{Area} = \sqrt{16(16-10)(16-15)(16-7)}$$
$$\text{Area} \approx 29.4$$

The area is approximately 29.4 square units.

TRY IT #3 Use Heron's formula to find the area of a triangle with sides of lengths $a = 29.7$ ft, $b = 42.3$ ft, and $c = 38.4$ ft.

EXAMPLE 6

Applying Heron's Formula to a Real-World Problem

A Chicago city developer wants to construct a building consisting of artist's lofts on a triangular lot bordered by Rush Street, Wabash Avenue, and Pearson Street. The frontage along Rush Street is approximately 62.4 meters, along Wabash Avenue it is approximately 43.5 meters, and along Pearson Street it is approximately 34.1 meters. How many square meters are available to the developer? See Figure 10 for a view of the city property.

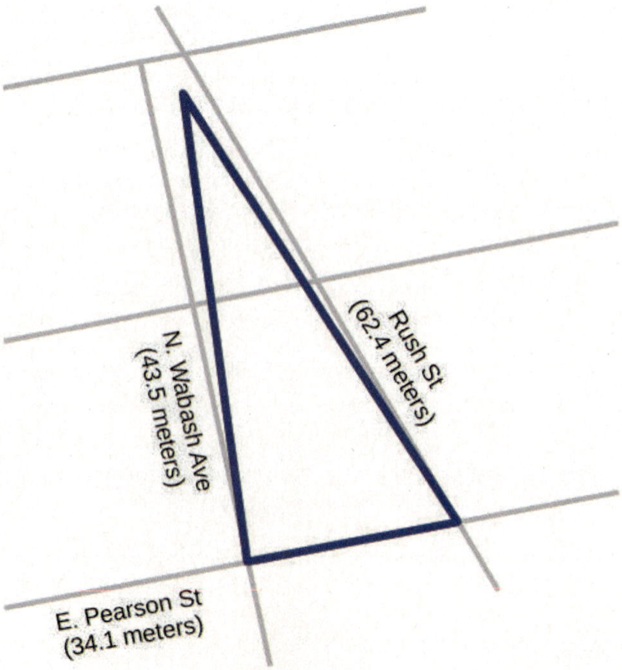

Figure 10

Solution

Find the measurement for s, which is one-half of the perimeter.

$$s = \frac{(62.4+43.5+34.1)}{2}$$
$$s = 70 \text{ m}$$

Apply Heron's formula.

$$\text{Area} = \sqrt{70(70 - 62.4)(70 - 43.5)(70 - 34.1)}$$
$$\text{Area} = \sqrt{506{,}118.2}$$
$$\text{Area} \approx 711.4$$

The developer has about 711.4 square meters.

TRY IT #4 Find the area of a triangle given $a = 4.38 \text{ ft}$, $b = 3.79$ ft, and $c = 5.22$ ft.

MEDIA

Access these online resources for additional instruction and practice with the Law of Cosines.

Law of Cosines (http://openstax.org/l/lawcosines)
Law of Cosines: Applications (http://openstax.org/l/cosineapp)
Law of Cosines: Applications 2 (http://openstax.org/l/cosineapp2)

10.2 SECTION EXERCISES

Verbal

1. If you are looking for a missing side of a triangle, what do you need to know when using the Law of Cosines?

2. If you are looking for a missing angle of a triangle, what do you need to know when using the Law of Cosines?

3. Explain what s represents in Heron's formula.

4. Explain the relationship between the Pythagorean Theorem and the Law of Cosines.

5. When must you use the Law of Cosines instead of the Pythagorean Theorem?

Algebraic

For the following exercises, assume α is opposite side a, β is opposite side b, and γ is opposite side c. If possible, solve each triangle for the unknown side. Round to the nearest tenth.

6. $\gamma = 41.2°, a = 2.49, b = 3.13$

7. $\alpha = 120°, b = 6, c = 7$

8. $\beta = 58.7°, a = 10.6, c = 15.7$

9. $\gamma = 115°, a = 18, b = 23$

10. $\alpha = 119°, a = 26, b = 14$

11. $\gamma = 113°, b = 10, c = 32$

12. $\beta = 67°, a = 49, b = 38$

13. $\alpha = 43.1°, a = 184.2, b = 242.8$

14. $\alpha = 36.6°, a = 186.2, b = 242.2$

15. $\beta = 50°, a = 105, b = 45$

For the following exercises, use the Law of Cosines to solve for the missing angle of the oblique triangle. Round to the nearest tenth.

16. $a = 42, b = 19, c = 30$; find angle A.

17. $a = 14,\ \ b = 13,\ \ c = 20$; find angle C.

18. $a = 16, b = 31, c = 20$; find angle B.

19. $a = 13,\ \ b = 22,\ \ c = 28$; find angle A.

20. $a = 108,\ \ b = 132,\ \ c = 160$; find angle C.

For the following exercises, solve the triangle. Round to the nearest tenth.

21. $A = 35°, b = 8, c = 11$

22. $B = 88°, a = 4.4, c = 5.2$

23. $C = 121°, a = 21, b = 37$

24. $a = 13, b = 11, c = 15$

25. $a = 3.1, b = 3.5, c = 5$

26. $a = 51, b = 25, c = 29$

For the following exercises, use Heron's formula to find the area of the triangle. Round to the nearest hundredth.

27. Find the area of a triangle with sides of length 18 in, 21 in, and 32 in. Round to the nearest tenth.

28. Find the area of a triangle with sides of length 20 cm, 26 cm, and 37 cm. Round to the nearest tenth.

29. $a = \frac{1}{2}\text{ m}, b = \frac{1}{3}\text{ m}, c = \frac{1}{4}\text{ m}$

30. $a = 12.4\text{ ft},\ \ b = 13.7\text{ ft},\ \ c = 20.2\text{ ft}$

31. $a = 1.6\text{ yd},\ \ b = 2.6\text{ yd},\ \ c = 4.1\text{ yd}$

Graphical

For the following exercises, find the length of side x. Round to the nearest tenth.

32.

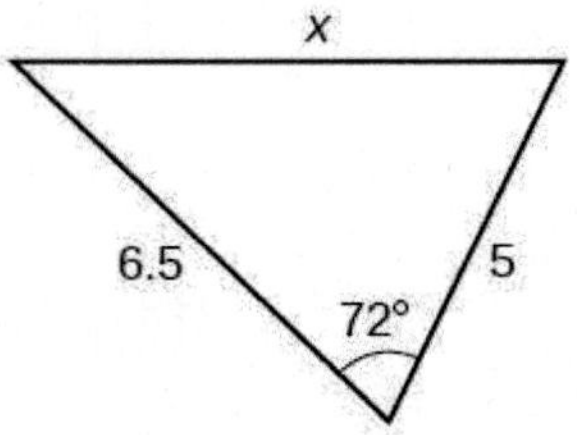

33.

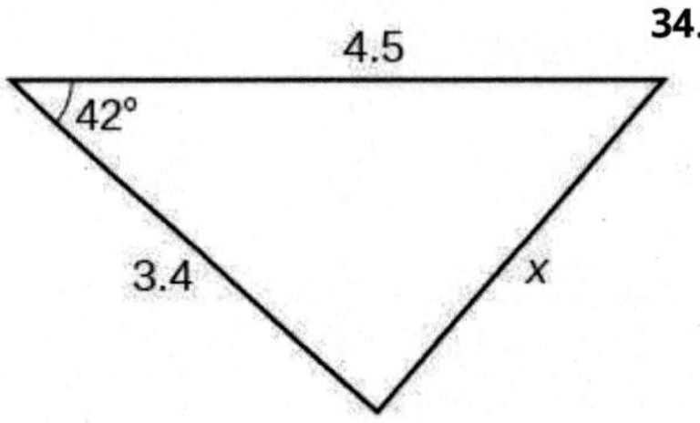

34.

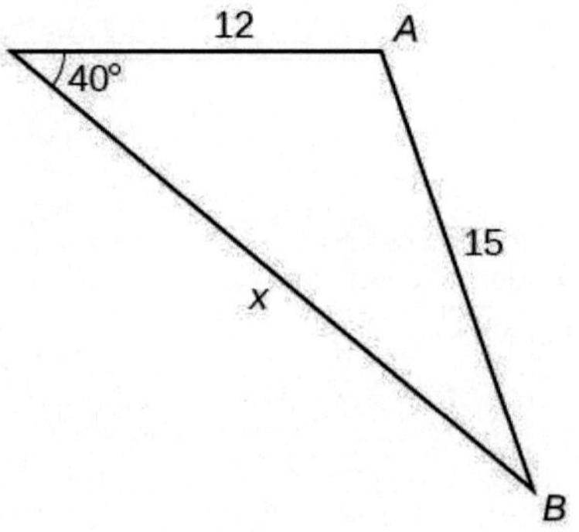

35.

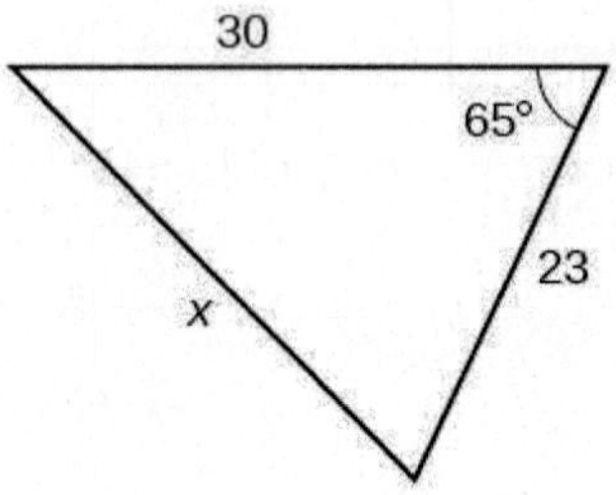

36.

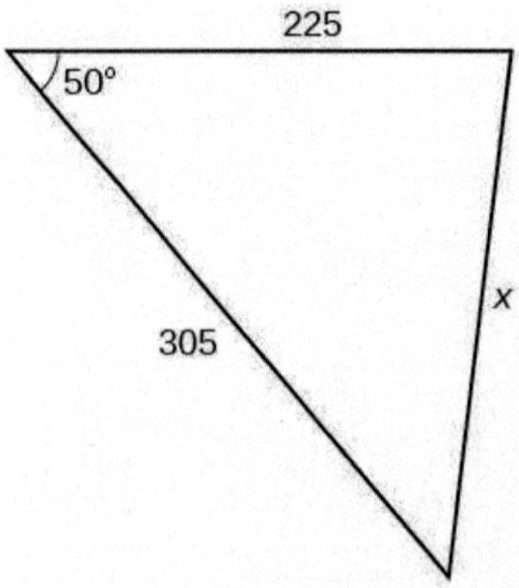

37.

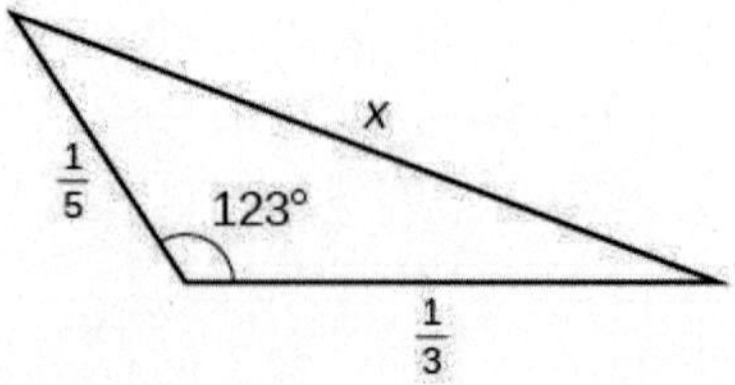

For the following exercises, find the measurement of angle A.

38.

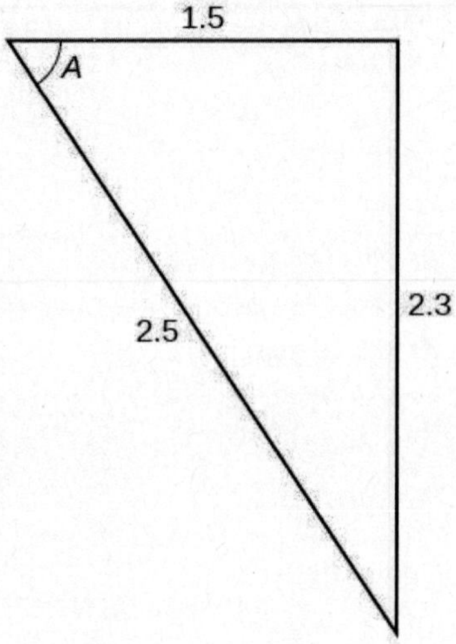

39.

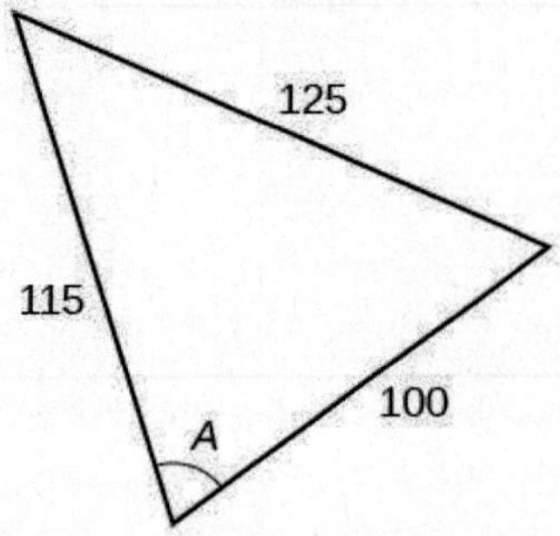

40.

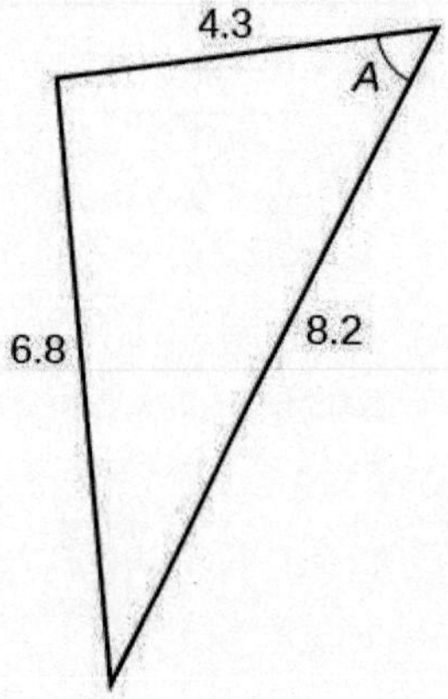

41.

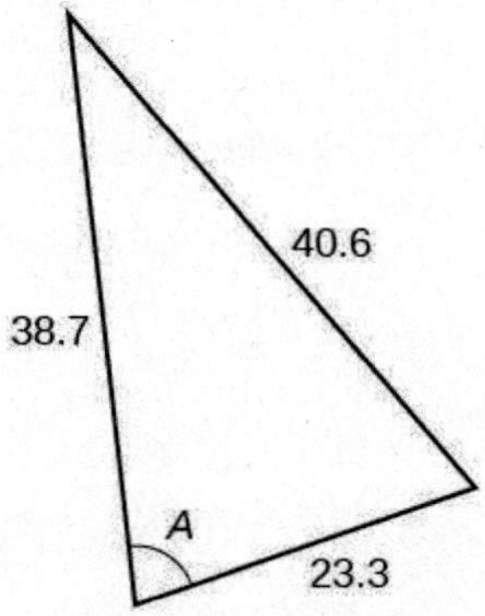

42. Find the measure of each angle in the triangle shown in Figure 11. Round to the nearest tenth.

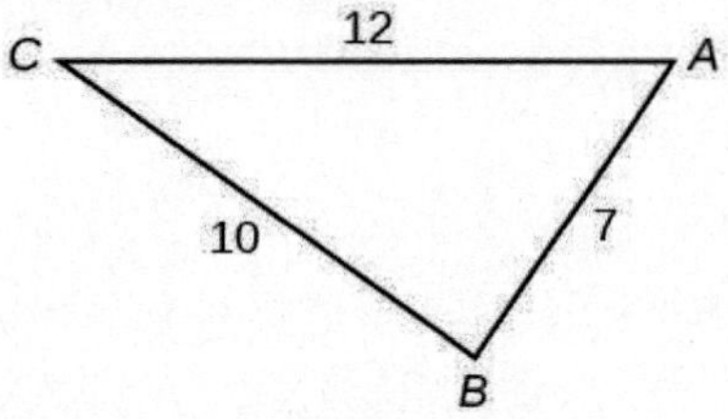

Figure 11

For the following exercises, solve for the unknown side. Round to the nearest tenth.

43.

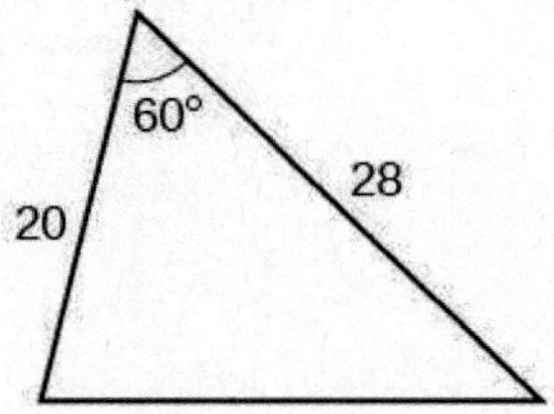

44.

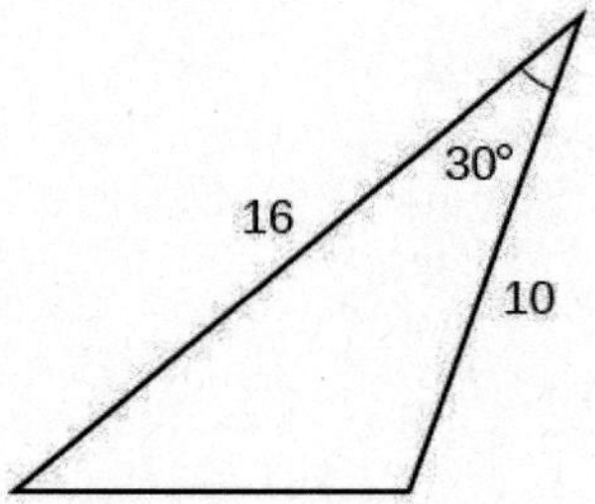

45.

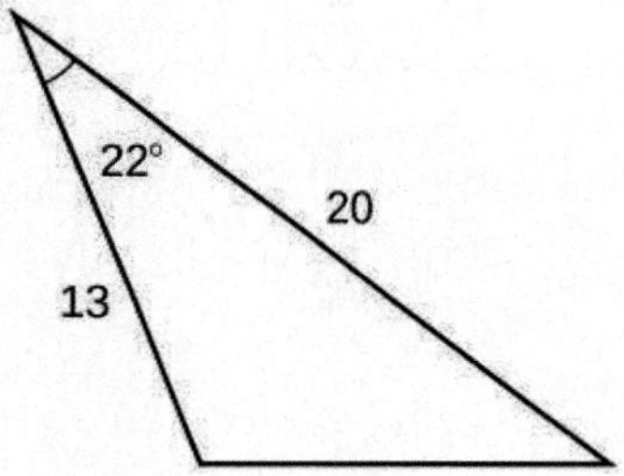

46.

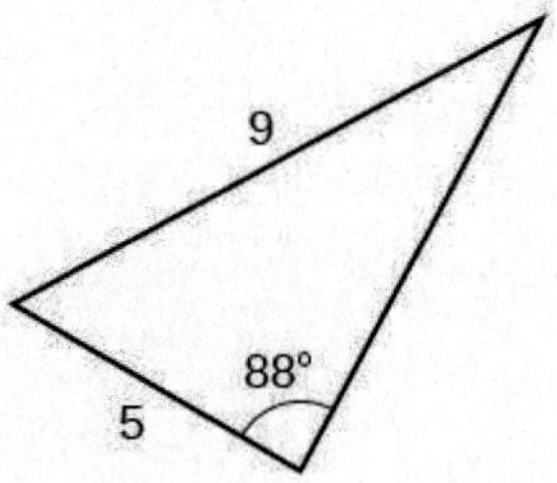

For the following exercises, find the area of the triangle. Round to the nearest hundredth.

47.

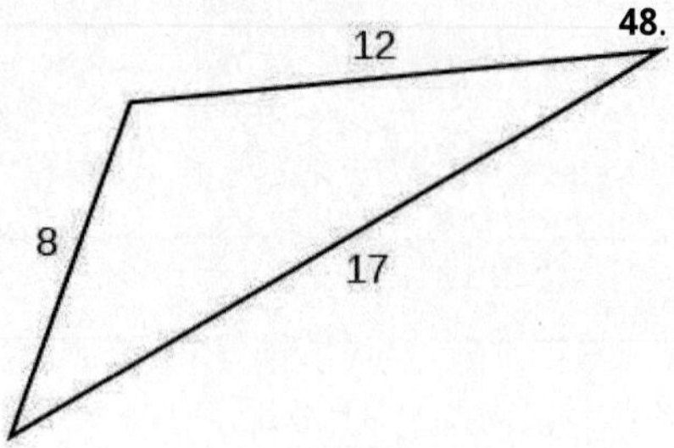

48.

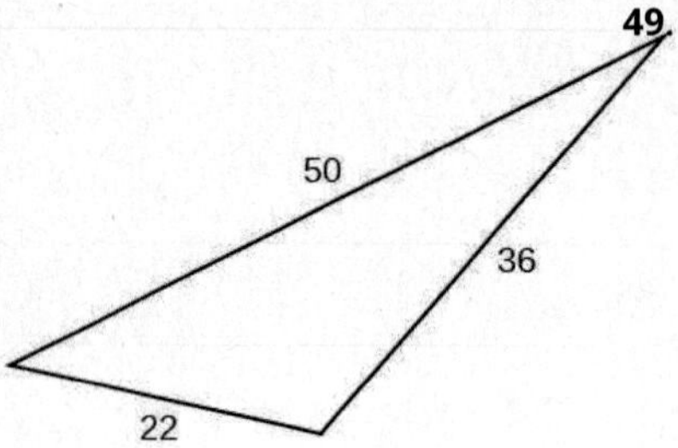

49.

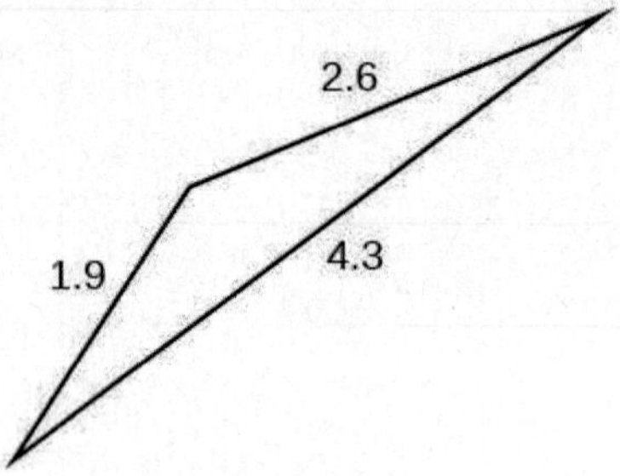

50.

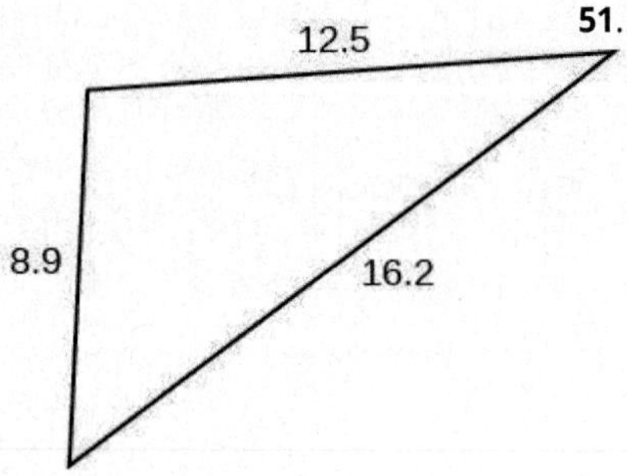

51.

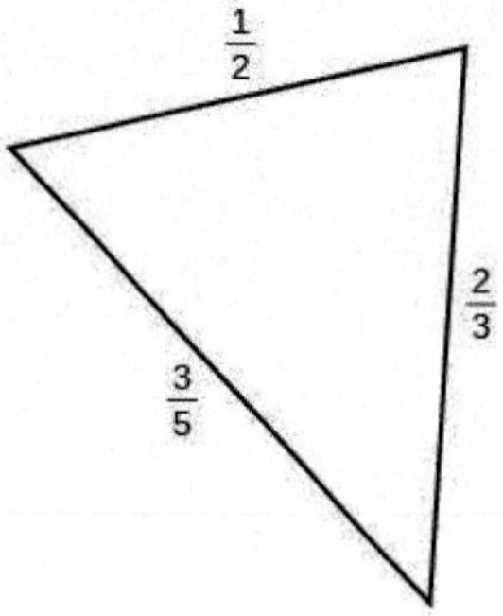

Extensions

52. A parallelogram has sides of length 16 units and 10 units. The shorter diagonal is 12 units. Find the measure of the longer diagonal.

53. The sides of a parallelogram are 11 feet and 17 feet. The longer diagonal is 22 feet. Find the length of the shorter diagonal.

54. The sides of a parallelogram are 28 centimeters and 40 centimeters. The measure of the larger angle is 100°. Find the length of the shorter diagonal.

55. A regular octagon is inscribed in a circle with a radius of 8 inches. (See Figure 12.) Find the perimeter of the octagon.

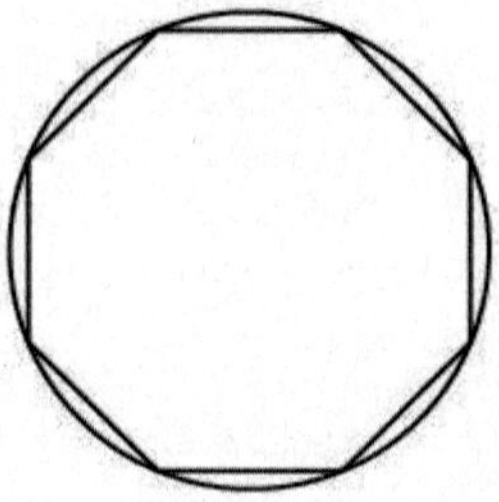

Figure 12

56. A regular pentagon is inscribed in a circle of radius 12 cm. (See Figure 13.) Find the perimeter of the pentagon. Round to the nearest tenth of a centimeter.

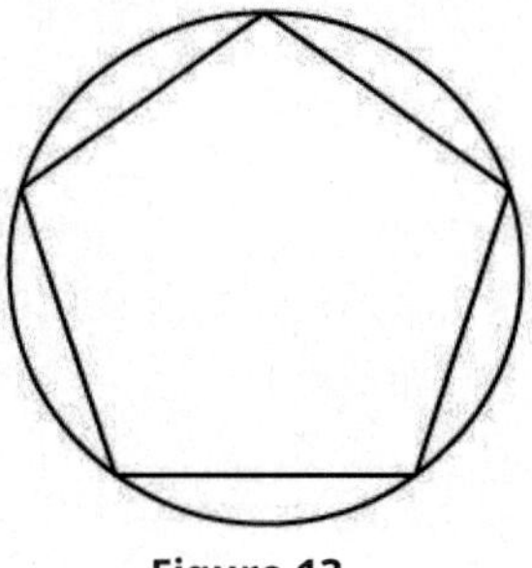

Figure 13

For the following exercises, suppose that $x^2 = 25 + 36 - 60\cos(52)$ *represents the relationship of three sides of a triangle and the cosine of an angle.*

57. Draw the triangle.

58. Find the length of the third side.

For the following exercises, find the area of the triangle.

59.

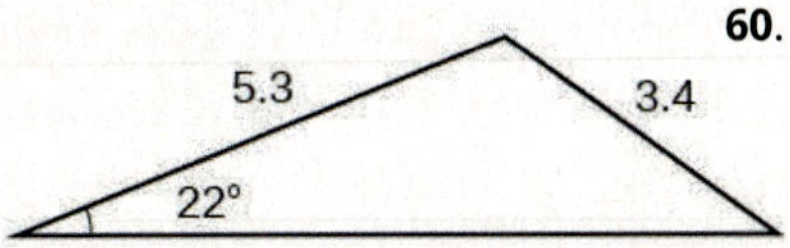

60.

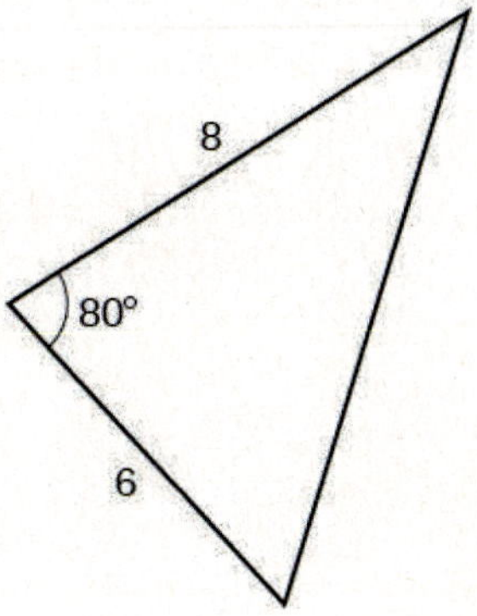

61.

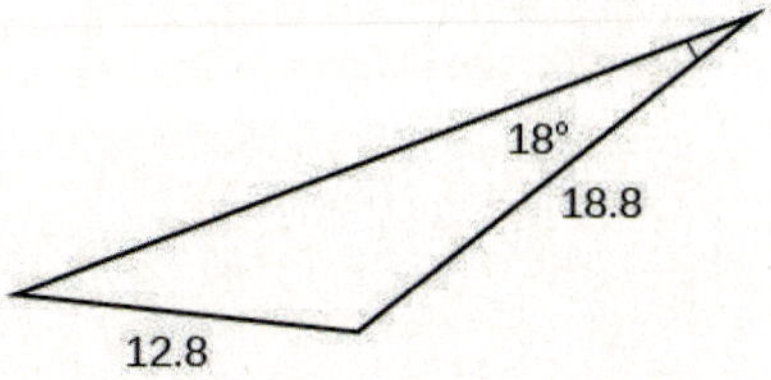

Real-World Applications

62. A surveyor has taken the measurements shown in Figure 14. Find the distance across the lake. Round answers to the nearest tenth.

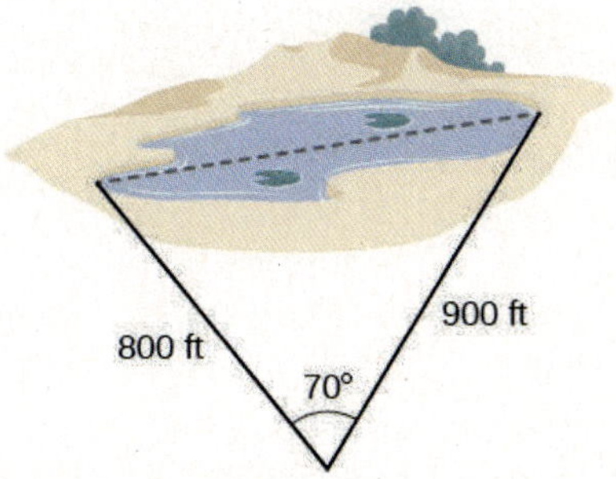

Figure 14

63. A satellite calculates the distances and angle shown in Figure 15 (not to scale). Find the distance between the two cities. Round answers to the nearest tenth.

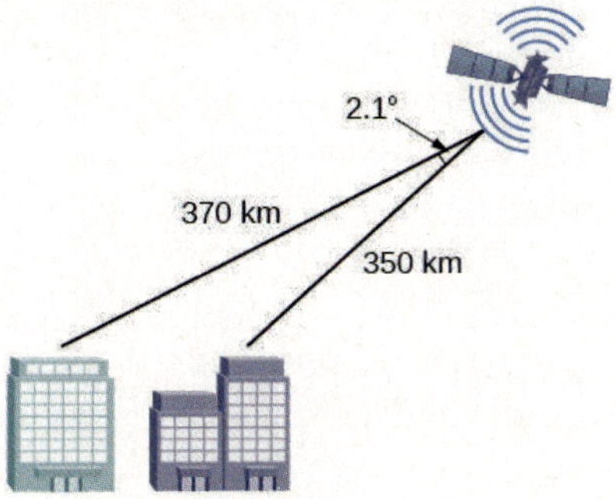

Figure 15

64. An airplane flies 220 miles with a heading of 40°, and then flies 180 miles with a heading of 170°. How far is the plane from its starting point, and at what heading? Round answers to the nearest tenth.

65. A 113-foot tower is located on a hill that is inclined 34° to the horizontal, as shown in Figure 16. A guy-wire is to be attached to the top of the tower and anchored at a point 98 feet uphill from the base of the tower. Find the length of wire needed.

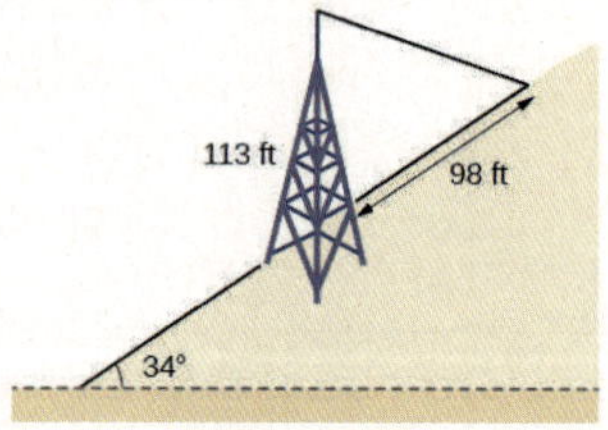

Figure 16

66. Two ships left a port at the same time. One ship traveled at a speed of 18 miles per hour at a heading of 320°. The other ship traveled at a speed of 22 miles per hour at a heading of 194°. Find the distance between the two ships after 10 hours of travel.

67. The graph in Figure 17 represents two boats departing at the same time from the same dock. The first boat is traveling at 18 miles per hour at a heading of 327° and the second boat is traveling at 4 miles per hour at a heading of 60°. Find the distance between the two boats after 2 hours.

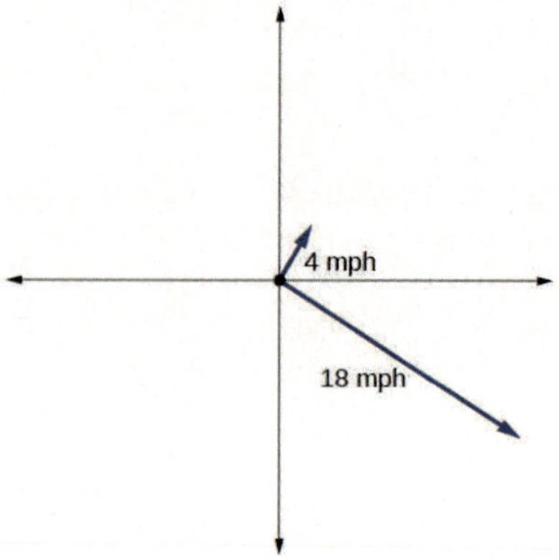

Figure 17

68. A triangular swimming pool measures 40 feet on one side and 65 feet on another side. These sides form an angle that measures 50°. How long is the third side (to the nearest tenth)?

69. A pilot flies in a straight path for 1 hour 30 min. She then makes a course correction, heading 10° to the right of her original course, and flies 2 hours in the new direction. If she maintains a constant speed of 680 miles per hour, how far is she from her starting position?

70. Los Angeles is 1,744 miles from Chicago, Chicago is 714 miles from New York, and New York is 2,451 miles from Los Angeles. Draw a triangle connecting these three cities, and find the angles in the triangle.

71. Philadelphia is 140 miles from Washington, D.C., Washington, D.C. is 442 miles from Boston, and Boston is 315 miles from Philadelphia. Draw a triangle connecting these three cities and find the angles in the triangle.

72. Two planes leave the same airport at the same time. One flies at 20° east of north at 500 miles per hour. The second flies at 30° east of south at 600 miles per hour. How far apart are the planes after 2 hours?

73. Two airplanes take off in different directions. One travels 300 mph due west and the other travels 25° north of west at 420 mph. After 90 minutes, how far apart are they, assuming they are flying at the same altitude?

74. A parallelogram has sides of length 15.4 units and 9.8 units. Its area is 72.9 square units. Find the measure of the longer diagonal.

75. The four sequential sides of a quadrilateral have lengths 4.5 cm, 7.9 cm, 9.4 cm, and 12.9 cm. The angle between the two smallest sides is 117°. What is the area of this quadrilateral?

76. The four sequential sides of a quadrilateral have lengths 5.7 cm, 7.2 cm, 9.4 cm, and 12.8 cm. The angle between the two smallest sides is 106°. What is the area of this quadrilateral?

77. Find the area of a triangular piece of land that measures 30 feet on one side and 42 feet on another; the included angle measures 132°. Round to the nearest whole square foot.

78. Find the area of a triangular piece of land that measures 110 feet on one side and 250 feet on another; the included angle measures 85°. Round to the nearest whole square foot.

10.3 Polar Coordinates

Learning Objectives

In this section, you will:

- Plot points using polar coordinates.
- Convert from polar coordinates to rectangular coordinates.
- Convert from rectangular coordinates to polar coordinates.
- Transform equations between polar and rectangular forms.
- Identify and graph polar equations by converting to rectangular equations.

Over 12 kilometers from port, a sailboat encounters rough weather and is blown off course by a 16-knot wind (see Figure 1). How can the sailor indicate his location to the Coast Guard? In this section, we will investigate a method of representing location that is different from a standard coordinate grid.

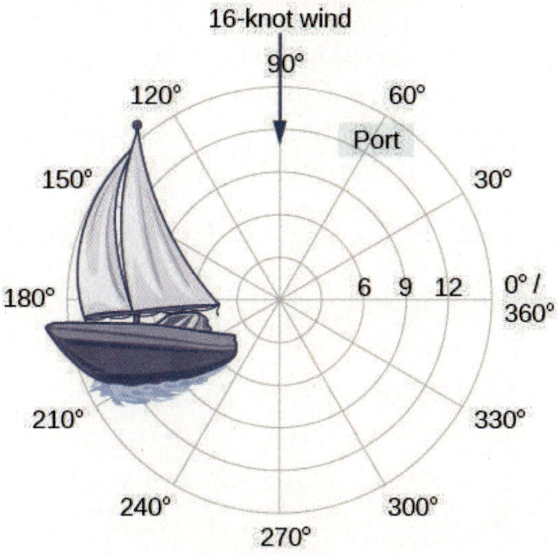

Figure 1

Plotting Points Using Polar Coordinates

When we think about plotting points in the plane, we usually think of rectangular coordinates (x, y) in the Cartesian coordinate plane. However, there are other ways of writing a coordinate pair and other types of grid systems. In this section, we introduce to **polar coordinates**, which are points labeled (r, θ) and plotted on a polar grid. The polar grid is represented as a series of concentric circles radiating out from the **pole**, or the origin of the coordinate plane.

The polar grid is scaled as the unit circle with the positive x-axis now viewed as the **polar axis** and the origin as the pole. The first coordinate r is the radius or length of the directed line segment from the pole. The angle θ, measured in radians, indicates the direction of r. We move counterclockwise from the polar axis by an angle of θ, and measure a directed line segment the length of r in the direction of θ. Even though we measure θ first and then r, the polar point is written with the r-coordinate first. For example, to plot the point $\left(2, \frac{\pi}{4}\right)$, we would move $\frac{\pi}{4}$ units in the counterclockwise direction and then a length of 2 from the pole. This point is plotted on the grid in Figure 2.

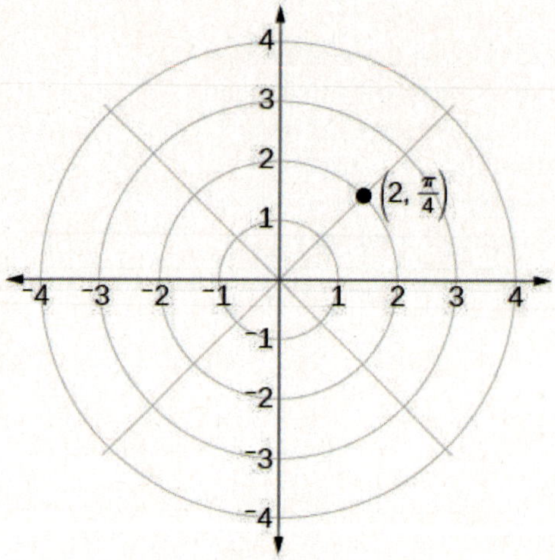

Polar Grid

Figure 2

EXAMPLE 1

Plotting a Point on the Polar Grid

Plot the point $\left(3, \frac{\pi}{2}\right)$ on the polar grid.

Solution

The angle $\frac{\pi}{2}$ is found by sweeping in a counterclockwise direction 90° from the polar axis. The point is located at a length of 3 units from the pole in the $\frac{\pi}{2}$ direction, as shown in Figure 3.

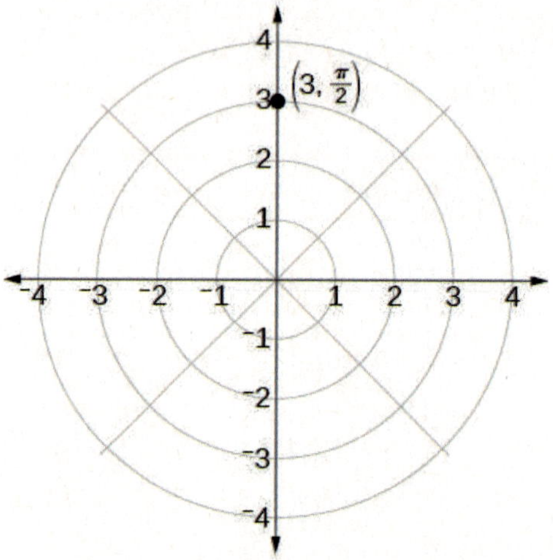

Figure 3

TRY IT #1 Plot the point $\left(2, \frac{\pi}{3}\right)$ in the polar grid.

EXAMPLE 2

Plotting a Point in the Polar Coordinate System with a Negative Component

Plot the point $\left(-2, \frac{\pi}{6}\right)$ on the polar grid.

Solution

We know that $\frac{\pi}{6}$ is located in the first quadrant. However, $r = -2$. We can approach plotting a point with a negative r in two ways:

1. Plot the point $\left(2, \frac{\pi}{6}\right)$ by moving $\frac{\pi}{6}$ in the counterclockwise direction and extending a directed line segment 2 units into the first quadrant. Then retrace the directed line segment back through the pole, and continue 2 units into the third quadrant;
2. Move $\frac{\pi}{6}$ in the counterclockwise direction, and draw the directed line segment from the pole 2 units in the negative direction, into the third quadrant.

See Figure 4(a). Compare this to the graph of the polar coordinate $\left(2, \frac{\pi}{6}\right)$ shown in Figure 4(b).

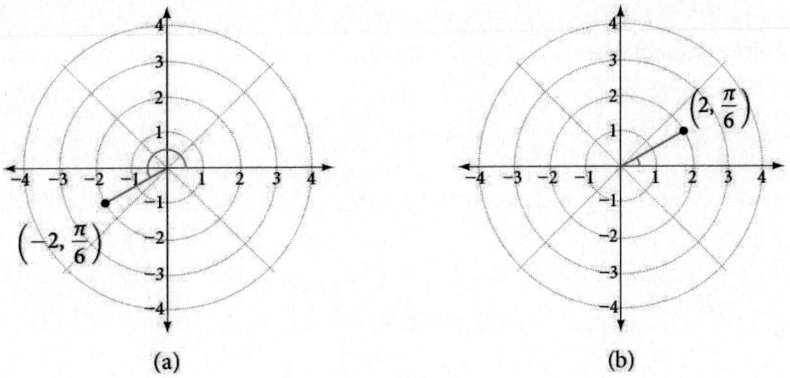

Figure 4

> **TRY IT** #2 Plot the points $\left(3, -\frac{\pi}{6}\right)$ and $\left(2, \frac{9\pi}{4}\right)$ on the same polar grid.

Converting from Polar Coordinates to Rectangular Coordinates

When given a set of polar coordinates, we may need to convert them to rectangular coordinates. To do so, we can recall the relationships that exist among the variables $x, \ y, \ r$, and θ.

$$\cos\ \theta = \frac{x}{r} \rightarrow x = r\cos\ \theta$$
$$\sin\ \theta = \frac{y}{r} \rightarrow y = r\sin\ \theta$$

Dropping a perpendicular from the point in the plane to the *x*-axis forms a right triangle, as illustrated in Figure 5. An easy way to remember the equations above is to think of $\cos\ \theta$ as the adjacent side over the hypotenuse and $\sin\ \theta$ as the opposite side over the hypotenuse.

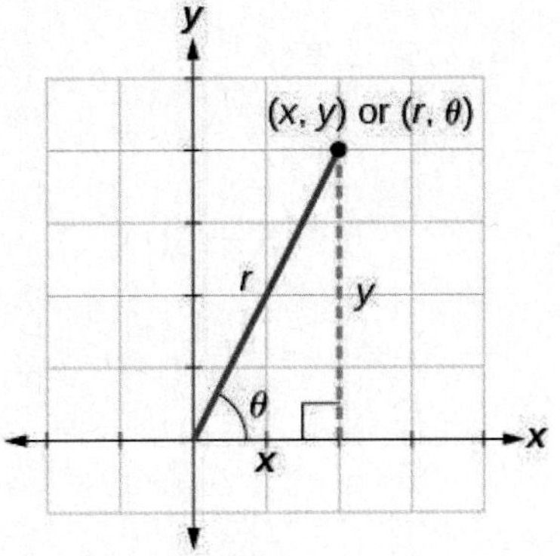

Figure 5

Converting from Polar Coordinates to Rectangular Coordinates

To convert polar coordinates $(r, \ \theta)$ to rectangular coordinates $(x, \ y)$, let

$$\cos\ \theta = \frac{x}{r} \rightarrow x = r\cos\ \theta$$
$$\sin\ \theta = \frac{y}{r} \rightarrow y = r\sin\ \theta$$

HOW TO

Given polar coordinates, convert to rectangular coordinates.

1. Given the polar coordinate (r, θ), write $x = r\cos\ \theta$ and $y = r\sin\ \theta$.
2. Evaluate $\cos\ \theta$ and $\sin\ \theta$.
3. Multiply $\cos\ \theta$ by r to find the x-coordinate of the rectangular form.
4. Multiply $\sin\ \theta$ by r to find the y-coordinate of the rectangular form.

EXAMPLE 3

Writing Polar Coordinates as Rectangular Coordinates

Write the polar coordinates $\left(3, \frac{\pi}{2}\right)$ as rectangular coordinates.

Solution

Use the equivalent relationships.

$$x = r\cos\ \theta$$
$$x = 3\cos\ \tfrac{\pi}{2} = 0$$
$$y = r\sin\ \theta$$
$$y = 3\sin\ \tfrac{\pi}{2} = 3$$

The rectangular coordinates are $(0, 3)$. See Figure 6.

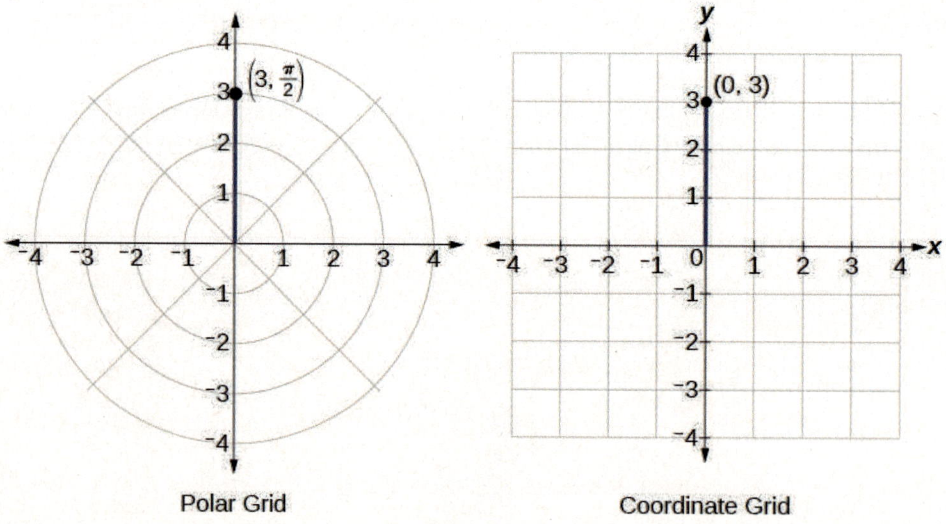

Figure 6

EXAMPLE 4

Writing Polar Coordinates as Rectangular Coordinates

Write the polar coordinates $(-2, 0)$ as rectangular coordinates.

Solution

See Figure 7. Writing the polar coordinates as rectangular, we have

$$x = r\cos\ \theta$$
$$x = -2\cos(0) = -2$$

$$y = r\sin\ \theta$$
$$y = -2\sin(0) = 0$$

The rectangular coordinates are also $(-2, 0)$.

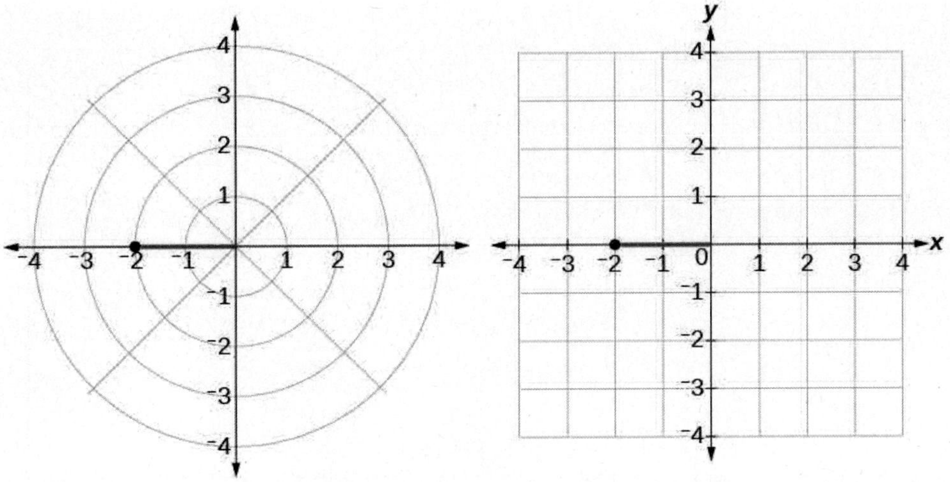

Figure 7

TRY IT #3 Write the polar coordinates $\left(-1, \frac{2\pi}{3}\right)$ as rectangular coordinates.

Converting from Rectangular Coordinates to Polar Coordinates

To convert rectangular coordinates to polar coordinates, we will use two other familiar relationships. With this conversion, however, we need to be aware that a set of rectangular coordinates will yield more than one polar point.

Converting from Rectangular Coordinates to Polar Coordinates

Converting from rectangular coordinates to polar coordinates requires the use of one or more of the relationships illustrated in Figure 8.

$$\cos\ \theta = \frac{x}{r} x = r\cos\ \theta$$
$$\sin\ \theta = \frac{y}{r}\ \text{ or }\ y = r\sin\ \theta$$
$$r^2 = x^2 + y^2$$
$$\tan\ \theta = \frac{y}{x}$$

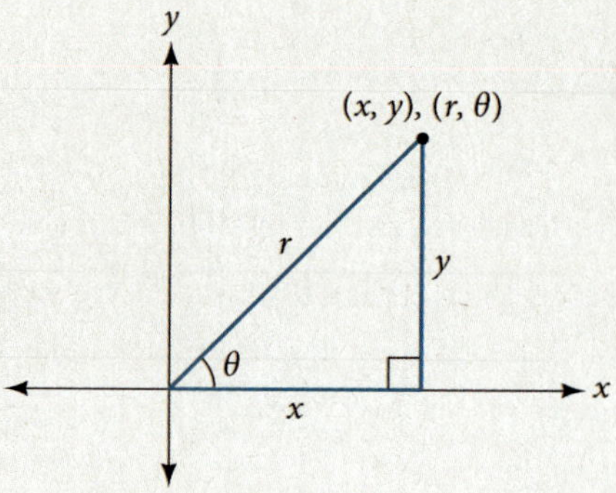

Figure 8

EXAMPLE 5

Writing Rectangular Coordinates as Polar Coordinates
Convert the rectangular coordinates $(3, 3)$ to polar coordinates.

Solution
We see that the original point $(3, 3)$ is in the first quadrant. To find θ, use the formula $\tan\ \theta = \frac{y}{x}$. This gives

$$\begin{aligned}\tan\ \theta &= \tfrac{3}{3}\\ \tan\ \theta &= 1\\ \theta &= \tan^{-1}(1)\\ \theta &= \tfrac{\pi}{4}\end{aligned}$$

To find r, we substitute the values for x and y into the formula $r = \sqrt{x^2 + y^2}$. We know that r must be positive, as $\frac{\pi}{4}$ is in the first quadrant. Thus

$$\begin{aligned}r &= \sqrt{3^2 + 3^2}\\ r &= \sqrt{9 + 9}\\ r &= \sqrt{18} = 3\sqrt{2}\end{aligned}$$

So, $r = 3\sqrt{2}$ and $\theta = \frac{\pi}{4}$, giving us the polar point $\left(3\sqrt{2}, \frac{\pi}{4}\right)$. See Figure 9.

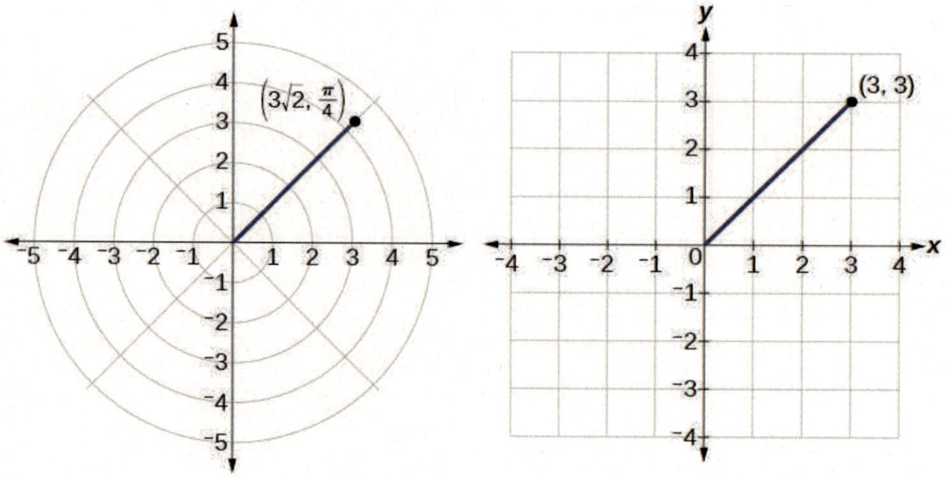

Figure 9

Analysis
There are other sets of polar coordinates that will be the same as our first solution. For example, the points

$\left(-3\sqrt{2}, \frac{5\pi}{4}\right)$ and $\left(3\sqrt{2}, -\frac{7\pi}{4}\right)$ will coincide with the original solution of $\left(3\sqrt{2}, \frac{\pi}{4}\right)$. The point $\left(-3\sqrt{2}, \frac{5\pi}{4}\right)$ indicates a move further counterclockwise by π, which is directly opposite $\frac{\pi}{4}$. The radius is expressed as $-3\sqrt{2}$. However, the angle $\frac{5\pi}{4}$ is located in the third quadrant and, as r is negative, we extend the directed line segment in the opposite direction, into the first quadrant. This is the same point as $\left(3\sqrt{2}, \frac{\pi}{4}\right)$. The point $\left(3\sqrt{2}, -\frac{7\pi}{4}\right)$ is a move further clockwise by $-\frac{7\pi}{4}$, from $\frac{\pi}{4}$. The radius, $3\sqrt{2}$, is the same.

Transforming Equations between Polar and Rectangular Forms

We can now convert coordinates between polar and rectangular form. Converting equations can be more difficult, but it can be beneficial to be able to convert between the two forms. Since there are a number of polar equations that cannot be expressed clearly in Cartesian form, and vice versa, we can use the same procedures we used to convert points between the coordinate systems. We can then use a graphing calculator to graph either the rectangular form or the polar form of the equation.

HOW TO

Given an equation in polar form, graph it using a graphing calculator.

1. Change the **MODE** to **POL**, representing polar form.
2. Press the **Y=** button to bring up a screen allowing the input of six equations: $r_1, r_2, \ldots, r_6$.
3. Enter the polar equation, set equal to r.
4. Press **GRAPH**.

EXAMPLE 6

Writing a Cartesian Equation in Polar Form

Write the Cartesian equation $x^2 + y^2 = 9$ in polar form.

Solution

The goal is to eliminate x and y from the equation and introduce r and θ. Ideally, we would write the equation r as a function of θ. To obtain the polar form, we will use the relationships between (x, y) and (r, θ). Since $x = r\cos\theta$ and $y = r\sin\theta$, we can substitute and solve for r.

$$\begin{aligned}
(r\cos\theta)^2 + (r\sin\theta)^2 &= 9 \\
r^2\cos^2\theta + r^2\sin^2\theta &= 9 \\
r^2(\cos^2\theta + \sin^2\theta) &= 9 \\
r^2(1) &= 9 && \text{Substitute } \cos^2\theta + \sin^2\theta = 1. \\
r &= \pm 3 && \text{Use the square root property.}
\end{aligned}$$

Thus, $x^2 + y^2 = 9$, $r = 3$, and $r = -3$ should generate the same graph. See Figure 10.

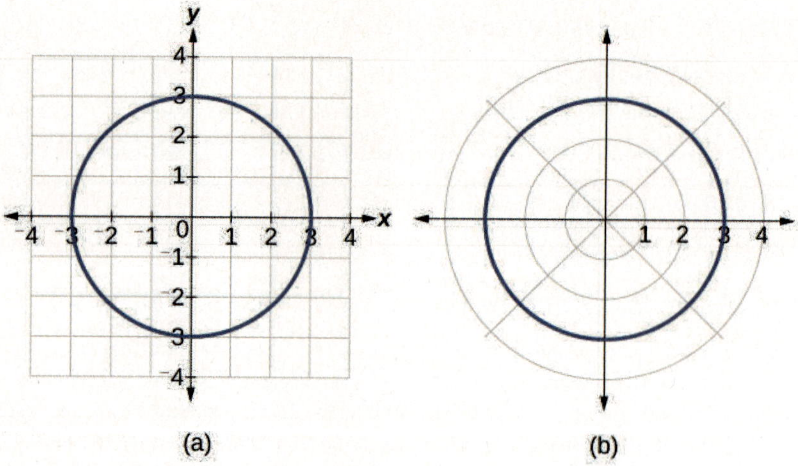

Figure 10 (a) Cartesian form $x^2 + y^2 = 9$ (b) Polar form $r = 3$

To graph a circle in rectangular form, we must first solve for y.

$$\begin{aligned} x^2 + y^2 &= 9 \\ y^2 &= 9 - x^2 \\ y &= \pm\sqrt{9 - x^2} \end{aligned}$$

Note that this is two separate functions, since a circle fails the vertical line test. Therefore, we need to enter the positive and negative square roots into the calculator separately, as two equations in the form $Y_1 = \sqrt{9 - x^2}$ and $Y_2 = -\sqrt{9 - x^2}$. Press **GRAPH.**

EXAMPLE 7

Rewriting a Cartesian Equation as a Polar Equation

Rewrite the Cartesian equation $x^2 + y^2 = 6y$ as a polar equation.

Solution

This equation appears similar to the previous example, but it requires different steps to convert the equation.

We can still follow the same procedures we have already learned and make the following substitutions:

$$\begin{array}{ll} r^2 = 6y & \text{Use } x^2 + y^2 = r^2. \\ r^2 = 6r \sin\ \theta & \text{Substitute } y = r \sin\ \theta. \\ r^2 - 6r \sin\ \theta = 0 & \text{Set equal to 0.} \\ r(r - 6 \sin\ \theta) = 0 & \text{Factor and solve.} \\ r = 0 & \text{We reject } r = 0, \text{ as it only represents one point, } (0, 0). \\ r = 6 \sin\ \theta & \end{array}$$

Therefore, the equations $x^2 + y^2 = 6y$ and $r = 6 \sin\ \theta$ should give us the same graph. See Figure 11.

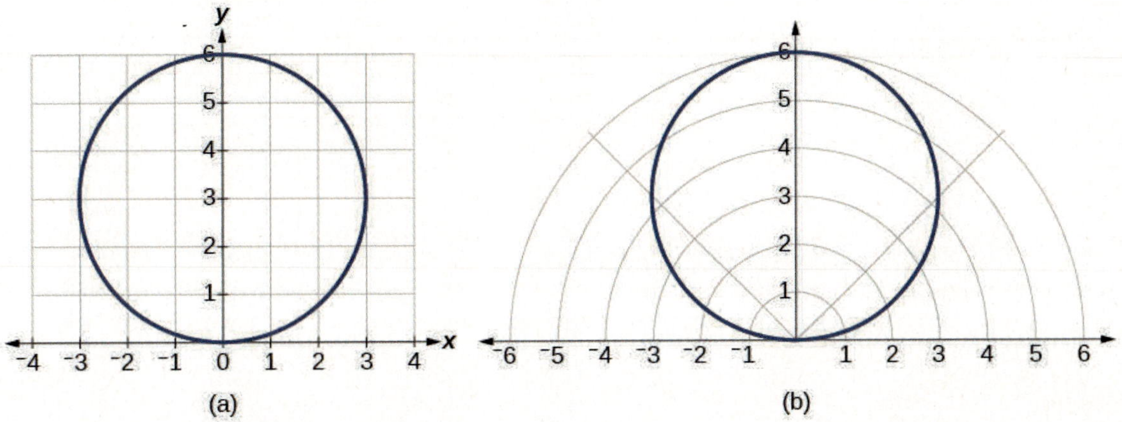

Figure 11 (a) Cartesian form $x^2 + y^2 = 6y$ (b) polar form $r = 6\sin\ \theta$

The Cartesian or rectangular equation is plotted on the rectangular grid, and the polar equation is plotted on the polar grid. Clearly, the graphs are identical.

EXAMPLE 8

Rewriting a Cartesian Equation in Polar Form

Rewrite the Cartesian equation $y = 3x + 2$ as a polar equation.

Solution

We will use the relationships $x = r\cos\ \theta$ and $y = r\sin\ \theta$.

$$\begin{aligned} y &= 3x + 2 \\ r\sin\ \theta &= 3r\cos\ \theta + 2 \\ r\sin\ \theta - 3r\cos\ \theta &= 2 \\ r(\sin\ \theta - 3\cos\ \theta) &= 2 && \text{Isolate } r. \\ r &= \frac{2}{\sin\ \theta - 3\cos\ \theta} && \text{Solve for } r. \end{aligned}$$

> **TRY IT** #4 Rewrite the Cartesian equation $y^2 = 3 - x^2$ in polar form.

Identify and Graph Polar Equations by Converting to Rectangular Equations

We have learned how to convert rectangular coordinates to polar coordinates, and we have seen that the points are indeed the same. We have also transformed polar equations to rectangular equations and vice versa. Now we will demonstrate that their graphs, while drawn on different grids, are identical.

EXAMPLE 9

Graphing a Polar Equation by Converting to a Rectangular Equation

Covert the polar equation $r = 2\sec\ \theta$ to a rectangular equation, and draw its corresponding graph.

Solution

The conversion is

$$\begin{aligned} r &= 2\sec\ \theta \\ r &= \frac{2}{\cos\ \theta} \\ r\cos\ \theta &= 2 \\ x &= 2 \end{aligned}$$

Notice that the equation $r = 2\sec\ \theta$ drawn on the polar grid is clearly the same as the vertical line $x = 2$ drawn on the rectangular grid (see Figure 12). Just as $x = c$ is the standard form for a vertical line in rectangular form, $r = c\sec\ \theta$ is the standard form for a vertical line in polar form.

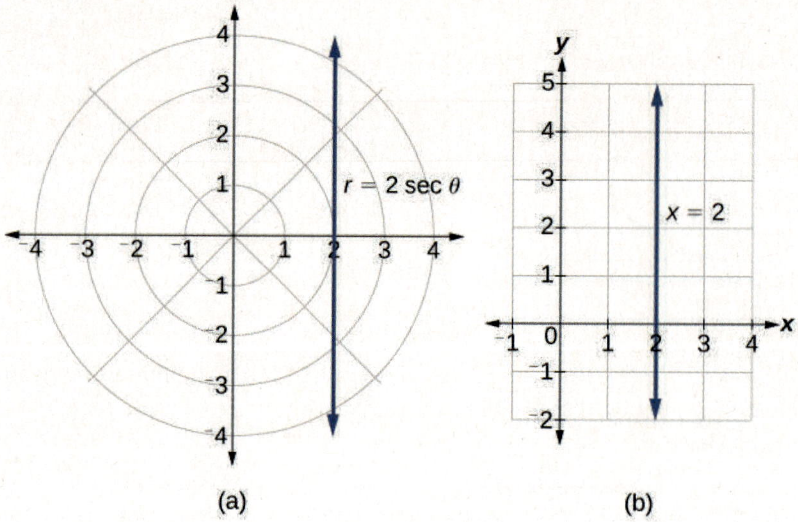

Figure 12 (a) Polar grid (b) Rectangular coordinate system

A similar discussion would demonstrate that the graph of the function $r = 2\csc\ \theta$ will be the horizontal line $y = 2$. In fact, $r = c\csc\ \theta$ is the standard form for a horizontal line in polar form, corresponding to the rectangular form $y = c$.

EXAMPLE 10

Rewriting a Polar Equation in Cartesian Form

Rewrite the polar equation $r = \frac{3}{1-2\cos\ \theta}$ as a Cartesian equation.

Solution

The goal is to eliminate θ and r, and introduce x and y. We clear the fraction, and then use substitution. In order to replace r with x and y, we must use the expression $x^2 + y^2 = r^2$.

$$\begin{aligned} r &= \frac{3}{1-2\ \cos\ \theta} && \\ r\left(1 - 2\left(\tfrac{x}{r}\right)\right) &= 3 && \\ r - 2x &= 3 && \text{Use } \cos\ \theta = \tfrac{x}{r} \textit{ to eliminate } \theta \\ r - 2x &= 3 && \\ r &= 3 + 2x && \text{Isolate } r \\ r^2 &= (3 + 2x)^2 && \text{Square both sides} \\ x^2 + y^2 &= (3 + 2x)^2 && \text{Use } x^2 + y^2 = r^2 \end{aligned}$$

The Cartesian equation is $x^2 + y^2 = (3 + 2x)^2$. However, to graph it, especially using a graphing calculator or computer program, we want to isolate y.

$$\begin{aligned} x^2 + y^2 &= (3 + 2x)^2 \\ y^2 &= (3 + 2x)^2 - x^2 \\ y &= \pm\sqrt{(3 + 2x)^2 - x^2} \end{aligned}$$

When our entire equation has been changed from r and θ to x and y, we can stop, unless asked to solve for y or simplify. See Figure 13.

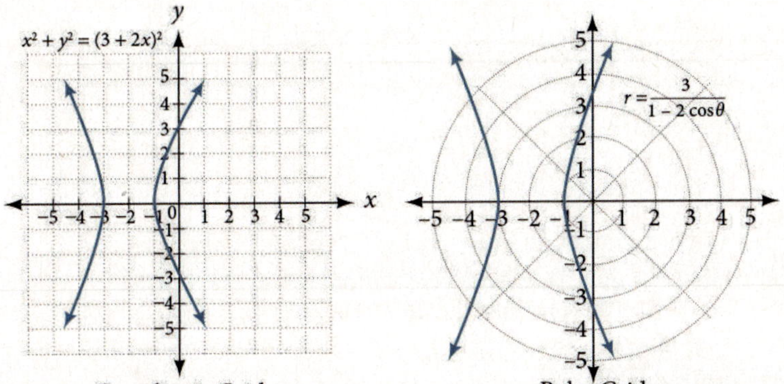

Coordinate Grid

Polar Grid

Figure 13

The "hour-glass" shape of the graph is called a *hyperbola*. Hyperbolas have many interesting geometric features and applications, which we will investigate further in Analytic Geometry.

Analysis

In this example, the right side of the equation can be expanded and the equation simplified further, as shown above. However, the equation cannot be written as a single function in Cartesian form. We may wish to write the rectangular equation in the hyperbola's standard form. To do this, we can start with the initial equation.

$$\begin{aligned}
x^2 + y^2 &= (3 + 2x)^2 \\
x^2 + y^2 - (3 + 2x)^2 &= 0 \\
x^2 + y^2 - \left(9 + 12x + 4x^2\right) &= 0 \\
x^2 + y^2 - 9 - 12x - 4x^2 &= 0 \\
-3x^2 - 12x + y^2 &= 9 && \text{Multiply through by } -1 \\
3x^2 + 12x - y^2 &= -9 \\
3\left(x^2 + 4x+\right) - y^2 &= -9 \\
3\left(x^2 + 4x + 4\right) - y^2 &= -9 + 12 && \textit{Organize} \text{ terms to complete the square for } x \\
3(x + 2)^2 - y^2 &= 3 \\
(x + 2)^2 - \frac{y2}{3} &= 1
\end{aligned}$$

TRY IT #5 Rewrite the polar equation $r = 2\sin\ \theta$ in Cartesian form.

EXAMPLE 11

Rewriting a Polar Equation in Cartesian Form

Rewrite the polar equation $r = \sin(2\theta)$ in Cartesian form.

Solution

$$\begin{aligned}
r &= \sin(2\theta) && \text{Use the double angle identity for sine.} \\
r &= 2\sin\ \theta\cos\ \theta && \text{Use } \cos\ \theta = \tfrac{x}{r} \text{ and } \sin\ \theta = \tfrac{y}{r}. \\
r &= 2\left(\tfrac{x}{r}\right)\left(\tfrac{y}{r}\right) && \text{Simplify.} \\
r &= \frac{2xy}{r^2} && \text{Multiply both sides by } r^2. \\
r^3 &= 2xy \\
\left(\sqrt{x^2 + y^2}\right)^3 &= 2xy && \text{As } x^2 + y^2 = r^2, r = \sqrt{x^2 + y^2}.
\end{aligned}$$

This equation can also be written as

$$\left(x^2 + y^2\right)^{\frac{3}{2}} = 2xy \text{ or } x^2 + y^2 = (2xy)^{\frac{2}{3}}$$

MEDIA

Access these online resources for additional instruction and practice with polar coordinates.

Introduction to Polar Coordinates (http://openstax.org/l/intropolar)
Comparing Polar and Rectangular Coordinates (http://openstax.org/l/polarrect)

10.3 SECTION EXERCISES

Verbal

1. How are polar coordinates different from rectangular coordinates?
2. How are the polar axes different from the *x*- and *y*-axes of the Cartesian plane?
3. Explain how polar coordinates are graphed.
4. How are the points $\left(3, \frac{\pi}{2}\right)$ and $\left(-3, \frac{\pi}{2}\right)$ related?
5. Explain why the points $\left(-3, \frac{\pi}{2}\right)$ and $\left(3, -\frac{\pi}{2}\right)$ are the same.

Algebraic

For the following exercises, convert the given polar coordinates to Cartesian coordinates. Remember to consider the quadrant in which the given point is located when determining θ for the point.

6. $\left(7, \frac{7\pi}{6}\right)$
7. $(5, \pi)$
8. $\left(6, -\frac{\pi}{4}\right)$
9. $\left(-3, \frac{\pi}{6}\right)$
10. $\left(4, \frac{7\pi}{4}\right)$

For the following exercises, convert the given Cartesian coordinates to polar coordinates with $r > 0,\ 0 \le \theta < 2\pi$. Remember to consider the quadrant in which the given point is located.

11. $(4, 2)$
12. $(-4, 6)$
13. $(3, -5)$
14. $(-10, -13)$
15. $(8, 8)$

For the following exercises, convert the given Cartesian equation to a polar equation.

16. $x = 3$
17. $y = 4$
18. $y = 4x^2$
19. $y = 2x^4$
20. $x^2 + y^2 = 4y$
21. $x^2 + y^2 = 3x$
22. $x^2 - y^2 = x$
23. $x^2 - y^2 = 3y$
24. $x^2 + y^2 = 9$
25. $x^2 = 9y$
26. $y^2 = 9x$
27. $9xy = 1$

For the following exercises, convert the given polar equation to a Cartesian equation. Write in the standard form of a conic if possible, and identify the conic section represented.

28. $r = 3\sin\theta$

29. $r = 4\cos\theta$

30. $r = \frac{4}{\sin\theta + 7\cos\theta}$

31. $r = \frac{6}{\cos\theta + 3\sin\theta}$

32. $r = 2\sec\theta$

33. $r = 3\csc\theta$

34. $r = \sqrt{r\cos\theta + 2}$

35. $r^2 = 4\sec\theta\csc\theta$

36. $r = 4$

37. $r^2 = 4$

38. $r = \frac{1}{4\cos\theta - 3\sin\theta}$

39. $r = \frac{3}{\cos\theta - 5\sin\theta}$

Graphical

For the following exercises, find the polar coordinates of the point.

40.

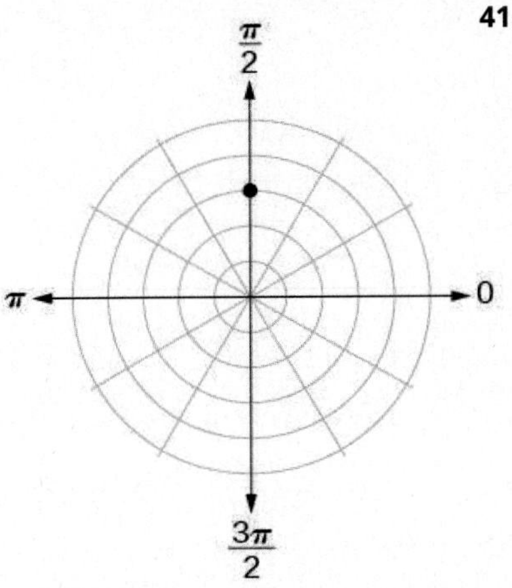

41.

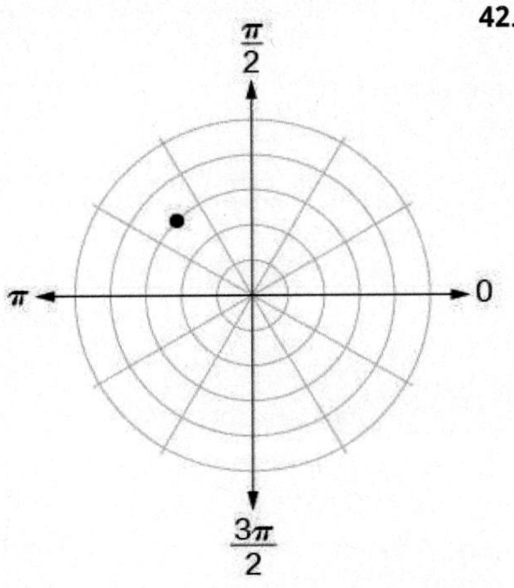

42.

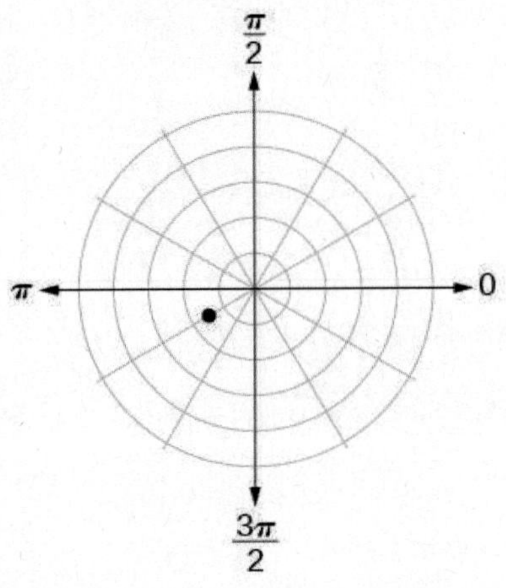

43.

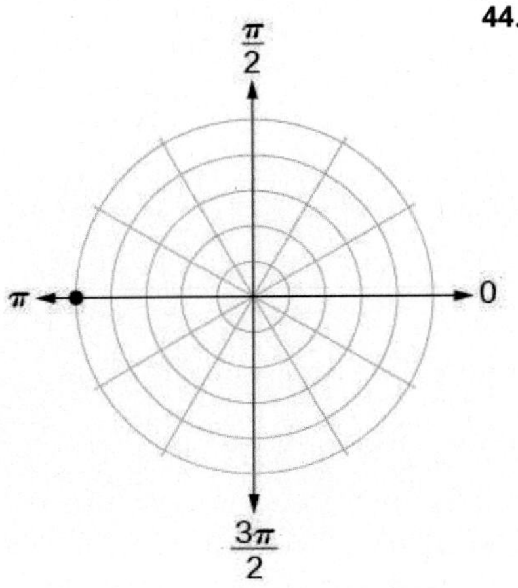

44.

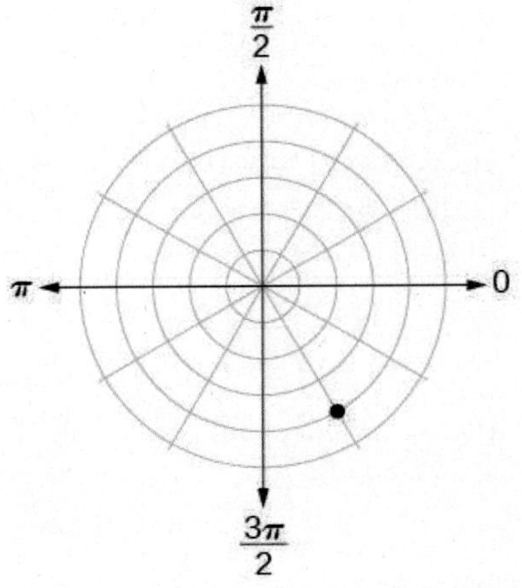

For the following exercises, plot the points.

45. $\left(-2, \frac{\pi}{3}\right)$

46. $\left(-1, -\frac{\pi}{2}\right)$

47. $\left(3.5, \frac{7\pi}{4}\right)$

48. $\left(-4, \frac{\pi}{3}\right)$

49. $\left(5, \frac{\pi}{2}\right)$

50. $\left(4, \frac{-5\pi}{4}\right)$

51. $\left(3, \frac{5\pi}{6}\right)$

52. $\left(-1.5, \frac{7\pi}{6}\right)$

53. $\left(-2, \frac{\pi}{4}\right)$

54. $\left(1, \frac{3\pi}{2}\right)$

For the following exercises, convert the equation from rectangular to polar form and graph on the polar axis.

55. $5x - y = 6$

56. $2x + 7y = -3$

57. $x^2 + (y-1)^2 = 1$

58. $(x+2)^2 + (y+3)^2 = 13$

59. $x = 2$

60. $x^2 + y^2 = 5y$

61. $x^2 + y^2 = 3x$

For the following exercises, convert the equation from polar to rectangular form and graph on the rectangular plane.

62. $r = 6$

63. $r = -4$

64. $\theta = -\frac{2\pi}{3}$

65. $\theta = \frac{\pi}{4}$

66. $r = \sec \theta$

67. $r = -10 \sin \theta$

68. $r = 3 \cos \theta$

Technology

69. Use a graphing calculator to find the rectangular coordinates of $\left(2, -\frac{\pi}{5}\right)$. Round to the nearest thousandth.

70. Use a graphing calculator to find the rectangular coordinates of $\left(-3, \frac{3\pi}{7}\right)$. Round to the nearest thousandth.

71. Use a graphing calculator to find the polar coordinates of $(-7, 8)$ in degrees. Round to the nearest thousandth.

72. Use a graphing calculator to find the polar coordinates of $(3, -4)$ in degrees. Round to the nearest hundredth.

73. Use a graphing calculator to find the polar coordinates of $(-2, 0)$ in radians. Round to the nearest hundredth.

Extensions

74. Describe the graph of $r = a \sec \theta; a > 0$.

75. Describe the graph of $r = a \sec \theta; a < 0$.

76. Describe the graph of $r = a \csc \theta; a > 0$.

77. Describe the graph of $r = a \csc \theta; a < 0$.

78. What polar equations will give an oblique line?

For the following exercise, graph the polar inequality.

79. $r < 4$

80. $0 \leq \theta \leq \frac{\pi}{4}$

81. $\theta = \frac{\pi}{4},\ r \geq 2$

82. $\theta = \frac{\pi}{4},\ r \geq -3$

83. $0 \leq \theta \leq \frac{\pi}{3},\ r < 2$

84. $\frac{-\pi}{6} < \theta \leq \frac{\pi}{3}, -3 < r < 2$

10.4 Polar Coordinates: Graphs

Learning Objectives

In this section you will:

- Test polar equations for symmetry.
- Graph polar equations by plotting points.

The planets move through space in elliptical, periodic orbits about the sun, as shown in Figure 1. They are in constant motion, so fixing an exact position of any planet is valid only for a moment. In other words, we can fix only a planet's *instantaneous* position. This is one application of polar coordinates, represented as (r, θ). We interpret r as the distance from the sun and θ as the planet's angular bearing, or its direction from a fixed point on the sun. In this section, we will focus on the polar system and the graphs that are generated directly from polar coordinates.

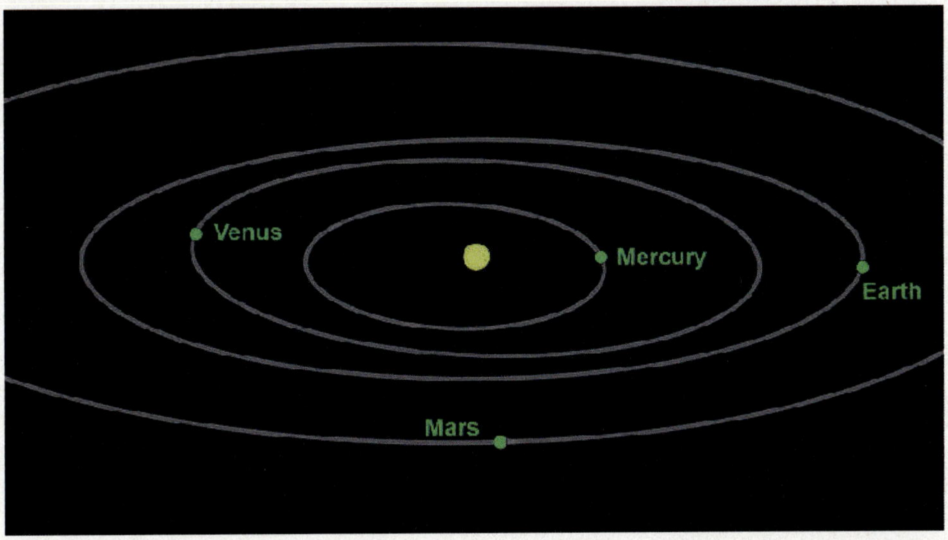

Figure 1 Planets follow elliptical paths as they orbit around the Sun. (credit: modification of work by NASA/JPL-Caltech)

Testing Polar Equations for Symmetry

Just as a rectangular equation such as $y = x^2$ describes the relationship between x and y on a Cartesian grid, a **polar equation** describes a relationship between r and θ on a polar grid. Recall that the coordinate pair (r, θ) indicates that we move counterclockwise from the polar axis (positive x-axis) by an angle of θ, and extend a ray from the pole (origin) r units in the direction of θ. All points that satisfy the polar equation are on the graph.

Symmetry is a property that helps us recognize and plot the graph of any equation. If an equation has a graph that is symmetric with respect to an axis, it means that if we folded the graph in half over that axis, the portion of the graph on one side would coincide with the portion on the other side. By performing three tests, we will see how to apply the properties of symmetry to polar equations. Further, we will use symmetry (in addition to plotting key points, zeros, and maximums of r) to determine the graph of a polar equation.

In the first test, we consider symmetry with respect to the line $\theta = \frac{\pi}{2}$ (y-axis). We replace (r, θ) with $(-r, -\theta)$ to determine if the new equation is equivalent to the original equation. For example, suppose we are given the equation $r = 2\sin\ \theta$;

$$
\begin{aligned}
r &= 2\sin\ \theta \\
-r &= 2\sin(-\theta) && \text{Replace } (r, \theta) \text{ with } (-r, -\theta). \\
-r &= -2\sin\ \theta && \text{Identity: } \sin(-\theta) = -\sin\ \theta. \\
r &= 2\sin\ \theta && \text{Multiply both sides by} -1.
\end{aligned}
$$

This equation exhibits symmetry with respect to the line $\theta = \frac{\pi}{2}$.

In the second test, we consider symmetry with respect to the polar axis (x-axis). We replace (r, θ) with $(r, -\theta)$ or $(-r, \pi - \theta)$ to determine equivalency between the tested equation and the original. For example, suppose we are given the equation $r = 1 - 2\cos\ \theta$.

$$\begin{aligned} r &= 1 - 2\cos\ \theta & \\ r &= 1 - 2\cos(\text{-}\theta) & \text{Replace } (r,\theta) \text{ with } (r,\text{-}\theta). \\ r &= 1 - 2\cos\ \theta & \text{Even/Odd identity} \end{aligned}$$

The graph of this equation exhibits symmetry with respect to the polar axis.

In the third test, we consider symmetry with respect to the pole (origin). We replace (r, θ) with $(-r, \theta)$ to determine if the tested equation is equivalent to the original equation. For example, suppose we are given the equation $r = 2\sin(3\theta\ )$.

$$\begin{aligned} r &= 2\sin(3\theta) \\ -r &= 2\sin(3\theta) \end{aligned}$$

The equation has failed the symmetry test, but that does not mean that it is not symmetric with respect to the pole. Passing one or more of the symmetry tests verifies that symmetry will be exhibited in a graph. However, failing the symmetry tests does not necessarily indicate that a graph will not be symmetric about the line $\theta = \frac{\pi}{2}$, the polar axis, or the pole. In these instances, we can confirm that symmetry exists by plotting reflecting points across the apparent axis of symmetry or the pole. Testing for symmetry is a technique that simplifies the graphing of polar equations, but its application is not perfect.

Symmetry Tests

A **polar equation** describes a curve on the polar grid. The graph of a polar equation can be evaluated for three types of symmetry, as shown in Figure 2.

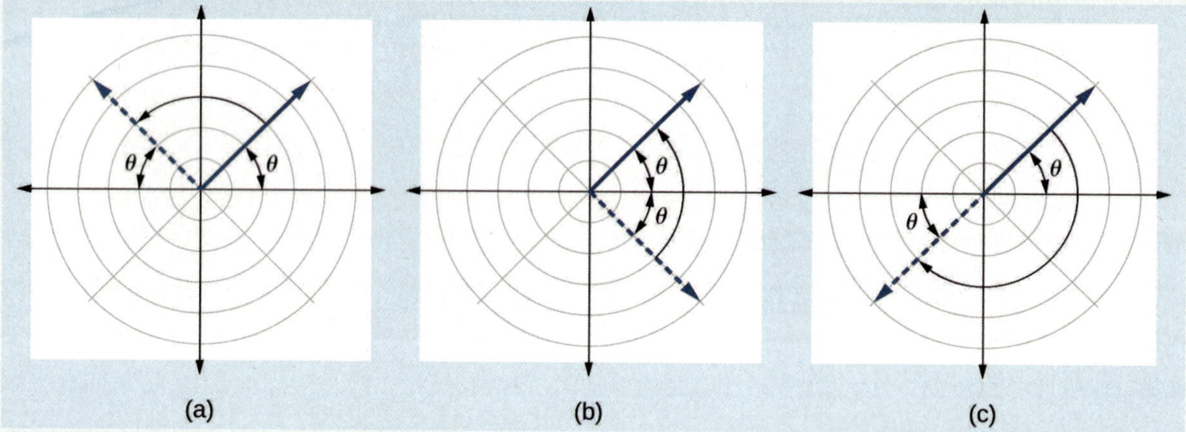

Figure 2 (a) A graph is symmetric with respect to the line $\theta = \frac{\pi}{2}$ (y-axis) if replacing (r, θ) with $(-r, -\theta)$ yields an equivalent equation. (b) A graph is symmetric with respect to the polar axis (x-axis) if replacing (r, θ) with $(r, -\theta)$ or $(-r, \pi-\theta)$ yields an equivalent equation. (c) A graph is symmetric with respect to the pole (origin) if replacing (r, θ) with $(-r, \theta)$ yields an equivalent equation.

HOW TO

Given a polar equation, test for symmetry.

1. Substitute the appropriate combination of components for $(r, \theta) : (-r, -\theta)$ for $\theta = \frac{\pi}{2}$ symmetry; $(r, -\theta)$ for polar axis symmetry; and $(-r, \theta)$ for symmetry with respect to the pole.
2. If the resulting equations are equivalent in one or more of the tests, the graph produces the expected symmetry.

EXAMPLE 1

Testing a Polar Equation for Symmetry

Test the equation $r = 2\sin\ \theta$ for symmetry.

Solution

Test for each of the three types of symmetry.

1) Replacing (r, θ) with $(-r, -\theta)$ yields the same result. Thus, the graph is symmetric with respect to the line $\theta = \frac{\pi}{2}$.	$-r = 2\sin(-\theta)$ $-r = -2\sin\ \theta$ Even-odd identity $r = 2\sin\ \theta$ Multiply by -1 Passed
2) Replacing θ with $-\theta$ does not yield the same equation. Therefore, the graph fails the test and may or may not be symmetric with respect to the polar axis.	$r = 2\sin(-\theta)$ $r = -2\sin\ \theta$ Even-odd identity $r = -2\sin\ \theta \neq 2\sin\ \theta$ Failed
3) Replacing r with $-r$ changes the equation and fails the test. The graph may or may not be symmetric with respect to the pole.	$-r = 2\sin\ \theta$ $r = -2\sin\ \theta \neq 2\sin\ \theta$ Failed

Table 1

Analysis

Using a graphing calculator, we can see that the equation $r = 2\sin\ \theta$ is a circle centered at $(0, 1)$ with radius $r = 1$ and is indeed symmetric to the line $\theta = \frac{\pi}{2}$. We can also see that the graph is not symmetric with the polar axis or the pole. See Figure 3.

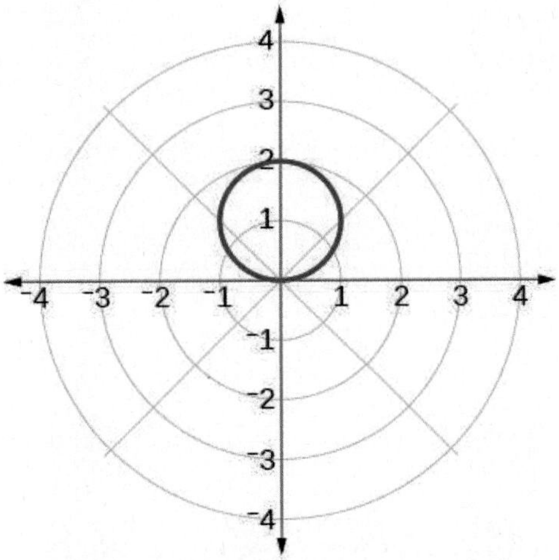

Figure 3

TRY IT #1 Test the equation for symmetry: $r = -2\cos\ \theta$.

Graphing Polar Equations by Plotting Points

To graph in the rectangular coordinate system we construct a table of x and y values. To graph in the polar coordinate system we construct a table of θ and r values. We enter values of θ into a polar equation and calculate r. However, using the properties of symmetry and finding key values of θ and r means fewer calculations will be needed.

Finding Zeros and Maxima

To find the zeros of a polar equation, we solve for the values of θ that result in $r = 0$. Recall that, to find the zeros of polynomial functions, we set the equation equal to zero and then solve for x. We use the same process for polar equations. Set $r = 0$, and solve for θ.

For many of the forms we will encounter, the maximum value of a polar equation is found by substituting those values of θ into the equation that result in the maximum value of the trigonometric functions. Consider $r = 5\cos\ \theta$; the maximum distance between the curve and the pole is 5 units. The maximum value of the cosine function is 1 when $\theta = 0$, so our polar equation is $5\cos\ \theta$, and the value $\theta = 0$ will yield the maximum $|r|$.

Similarly, the maximum value of the sine function is 1 when $\theta = \frac{\pi}{2}$, and if our polar equation is $r = 5\sin\ \theta$, the value $\theta = \frac{\pi}{2}$ will yield the maximum $|r|$. We may find additional information by calculating values of r when $\theta = 0$. These points would be polar axis intercepts, which may be helpful in drawing the graph and identifying the curve of a polar equation.

EXAMPLE 2

Finding Zeros and Maximum Values for a Polar Equation

Using the equation in Example 1, find the zeros and maximum $|r|$ and, if necessary, the polar axis intercepts of $r = 2\sin\ \theta$.

Solution

To find the zeros, set r equal to zero and solve for θ.

$$\begin{aligned}
&2\sin\ \theta = 0\\
&\sin\ \theta = 0\\
&\theta = \sin^{-1}0\\
&\theta = n\pi \qquad \text{where } n \text{ is an integer}
\end{aligned}$$

Substitute any one of the θ values into the equation. We will use 0.

$$\begin{aligned}
r &= 2\sin(0)\\
r &= 0
\end{aligned}$$

The points $(0, 0)$ and $(0, \pm n\pi)$ are the zeros of the equation. They all coincide, so only one point is visible on the graph. This point is also the only polar axis intercept.

To find the maximum value of the equation, look at the maximum value of the trigonometric function $\sin\ \theta$, which occurs when $\theta = \frac{\pi}{2} \pm 2k\pi$ resulting in $\sin\left(\frac{\pi}{2}\right) = 1$. Substitute $\frac{\pi}{2}$ for θ.

$$\begin{aligned}
r &= 2\sin\left(\frac{\pi}{2}\right)\\
r &= 2(1)\\
r &= 2
\end{aligned}$$

Analysis

The point $\left(2, \frac{\pi}{2}\right)$ will be the maximum value on the graph. Let's plot a few more points to verify the graph of a circle. See Table 2 and Figure 4.

θ	$r = 2\sin\ \theta$	r
0	$r = 2\sin(0) = 0$	0
$\frac{\pi}{6}$	$r = 2\sin\left(\frac{\pi}{6}\right) = 1$	1
$\frac{\pi}{3}$	$r = 2\sin\left(\frac{\pi}{3}\right) \approx 1.73$	1.73
$\frac{\pi}{2}$	$r = 2\sin\left(\frac{\pi}{2}\right) = 2$	2
$\frac{2\pi}{3}$	$r = 2\sin\left(\frac{2\pi}{3}\right) \approx 1.73$	1.73

Table 2

θ	$r = 2\sin\ \theta$	r
$\frac{5\pi}{6}$	$r = 2\sin\left(\frac{5\pi}{6}\right) = 1$	1
π	$r = 2\sin(\pi) = 0$	0

Table 2

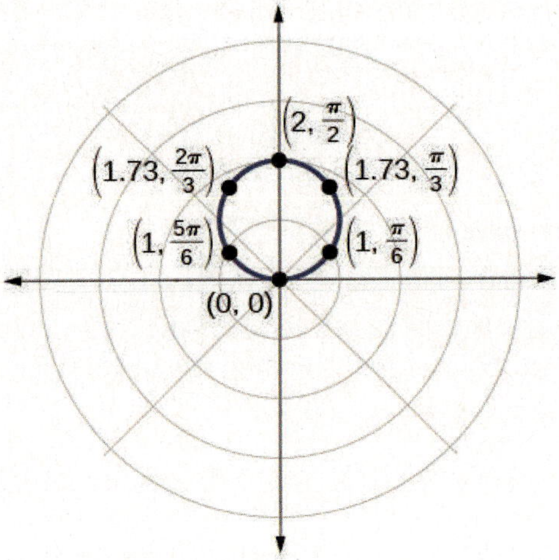

Figure 4

> **TRY IT** #2 Without converting to Cartesian coordinates, test the given equation for symmetry and find the zeros and maximum values of $|r|$: $r = 3\cos\ \theta$.

Investigating Circles

Now we have seen the equation of a circle in the polar coordinate system. In the last two examples, the same equation was used to illustrate the properties of symmetry and demonstrate how to find the zeros, maximum values, and plotted points that produced the graphs. However, the circle is only one of many shapes in the set of polar curves.

There are five classic polar curves: **cardioids, limaçons, lemniscates, rose curves**, and **Archimedes' spirals**. We will briefly touch on the polar formulas for the circle before moving on to the classic curves and their variations.

Formulas for the Equation of a Circle

Some of the formulas that produce the graph of a circle in polar coordinates are given by $r = a\cos\ \theta$ and $r = a\sin\ \theta$, where a is the diameter of the circle or the distance from the pole to the farthest point on the circumference. The radius is $\frac{|a|}{2}$, or one-half the diameter. For $r = a\cos\ \theta$, the center is $\left(\frac{a}{2}, 0\right)$. For $r = a\sin\ \theta$, the center is $\left(\frac{a}{2}, \frac{\pi}{2}\right)$. Figure 5 shows the graphs of these four circles.

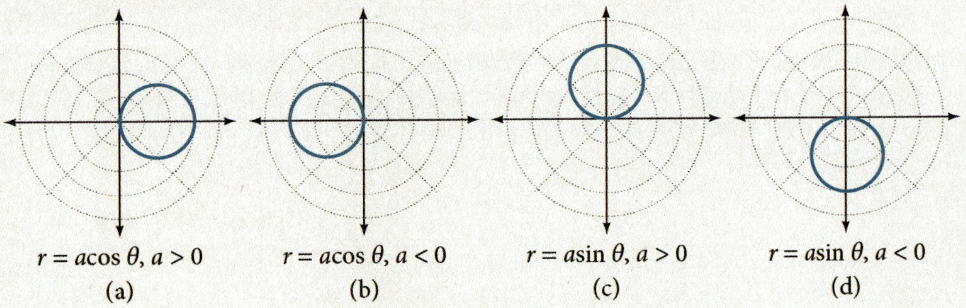

Figure 5

EXAMPLE 3

Sketching the Graph of a Polar Equation for a Circle

Sketch the graph of $r = 4\cos\ \theta$.

Solution

First, testing the equation for symmetry, we find that the graph is symmetric about the polar axis. Next, we find the zeros and maximum $|r|$ for $r = 4\cos\ \theta$. First, set $r = 0$, and solve for θ. Thus, a zero occurs at $\theta = \frac{\pi}{2} \pm k\pi$. A key point to plot is $\left(0, \quad \frac{\pi}{2}\right)$.

To find the maximum value of r, note that the maximum value of the cosine function is 1 when $\theta = 0 \pm 2k\pi$. Substitute $\theta = 0$ into the equation:

$$\begin{aligned} r &= 4\cos\ \theta \\ r &= 4\cos(0) \\ r &= 4(1) = 4 \end{aligned}$$

The maximum value of the equation is 4. A key point to plot is $(4,\ 0)$.

As $r = 4\cos\ \theta$ is symmetric with respect to the polar axis, we only need to calculate r-values for θ over the interval $[0, \pi]$. Points in the upper quadrant can then be reflected to the lower quadrant. Make a table of values similar to Table 3. The graph is shown in Figure 6.

θ	0	$\frac{\pi}{6}$	$\frac{\pi}{4}$	$\frac{\pi}{3}$	$\frac{\pi}{2}$	$\frac{2\pi}{3}$	$\frac{3\pi}{4}$	$\frac{5\pi}{6}$	π
r	4	3.46	2.83	2	0	−2	−2.83	−3.46	−4

Table 3

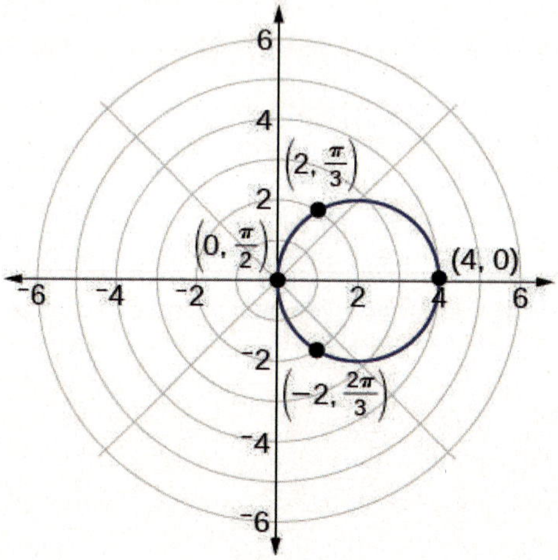

Figure 6

Investigating Cardioids

While translating from polar coordinates to Cartesian coordinates may seem simpler in some instances, graphing the classic curves is actually less complicated in the polar system. The next curve is called a cardioid, as it resembles a heart. This shape is often included with the family of curves called limaçons, but here we will discuss the cardioid on its own.

Formulas for a Cardioid

The formulas that produce the graphs of a **cardioid** are given by $r = a \pm b\cos\ \theta$ and $r = a \pm b\sin\ \theta$ where $a > 0$, $b > 0$, and $\frac{a}{b} = 1$. The cardioid graph passes through the pole, as we can see in Figure 7.

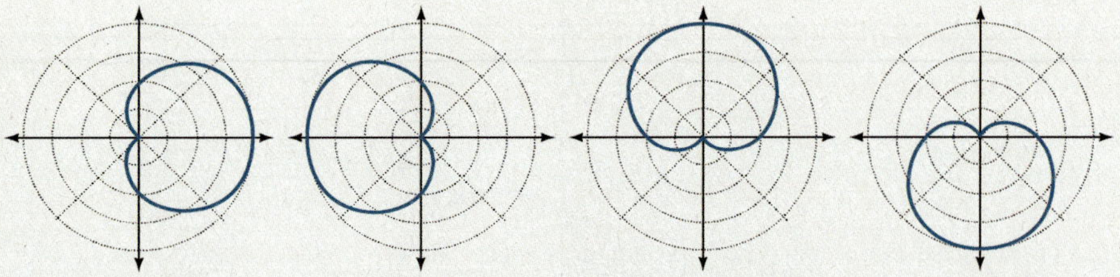

Figure 7

HOW TO

Given the polar equation of a cardioid, sketch its graph.

1. Check equation for the three types of symmetry.
2. Find the zeros. Set $r = 0$.
3. Find the maximum value of the equation according to the maximum value of the trigonometric expression.
4. Make a table of values for r and θ.
5. Plot the points and sketch the graph.

EXAMPLE 4

Sketching the Graph of a Cardioid

Sketch the graph of $r = 2 + 2\cos\ \theta$.

Solution

First, testing the equation for symmetry, we find that the graph of this equation will be symmetric about the polar axis. Next, we find the zeros and maximums. Setting $r = 0$, we have $\theta = \pi + 2k\pi$. The zero of the equation is located at $(0, \pi)$. The graph passes through this point.

The maximum value of $r = 2 + 2\cos\ \theta$ occurs when $\cos\ \theta$ is a maximum, which is when $\cos\ \theta = 1$ or when $\theta = 0$. Substitute $\theta = 0$ into the equation, and solve for r.

$$r = 2 + 2\cos(0)$$
$$r = 2 + 2(1) = 4$$

The point $(4, 0)$ is the maximum value on the graph.

We found that the polar equation is symmetric with respect to the polar axis, but as it extends to all four quadrants, we need to plot values over the interval $[0,\ \pi]$. The upper portion of the graph is then reflected over the polar axis. Next, we make a table of values, as in Table 4, and then we plot the points and draw the graph. See Figure 8.

θ	0	$\frac{\pi}{4}$	$\frac{\pi}{2}$	$\frac{2\pi}{3}$	π
r	4	3.41	2	1	0

Table 4

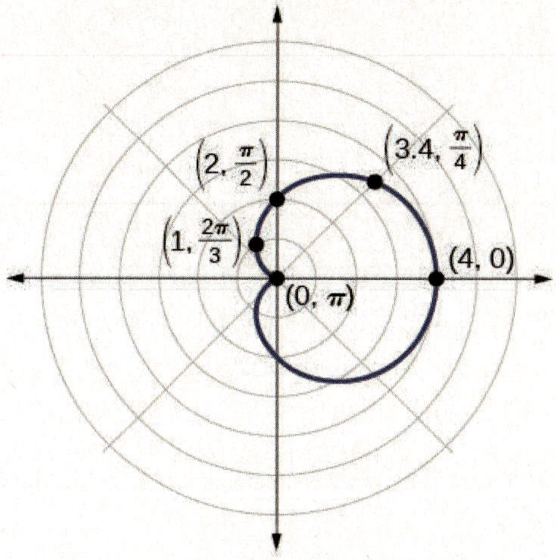

Figure 8

Investigating Limaçons

The word *limaçon* is Old French for "snail," a name that describes the shape of the graph. As mentioned earlier, the cardioid is a member of the limaçon family, and we can see the similarities in the graphs. The other images in this category include the one-loop limaçon and the two-loop (or inner-loop) limaçon. **One-loop limaçons** are sometimes referred to as **dimpled limaçons** when $1 < \frac{a}{b} < 2$ and **convex limaçons** when $\frac{a}{b} \geq 2$.

Formulas for One-Loop Limaçons

The formulas that produce the graph of a dimpled **one-loop limaçon** are given by $r = a \pm b\cos\ \theta$ and $r = a \pm b\sin\ \theta$ where $a > 0,\ b > 0,$ and $1 < \frac{a}{b} < 2$. All four graphs are shown in Figure 9.

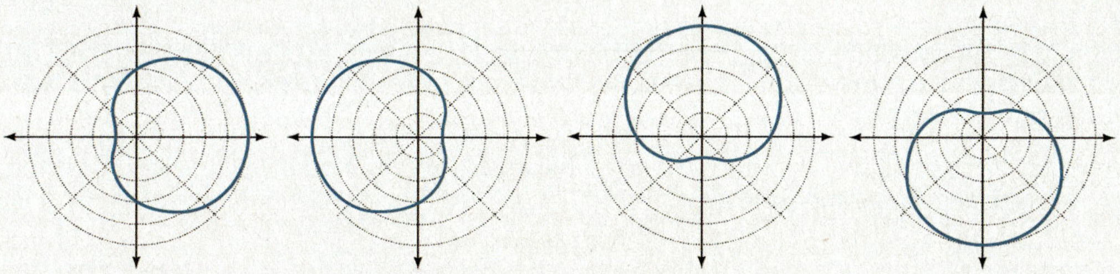

$r = a + b\cos\theta$ (a) | $r = a - b\cos\theta$ (b) | $r = a + b\sin\theta$ (c) | $r = a - b\sin\theta$ (d)

Figure 9 Dimpled limaçons

HOW TO

Given a polar equation for a one-loop limaçon, sketch the graph.

1. Test the equation for symmetry. Remember that failing a symmetry test does not mean that the shape will not exhibit symmetry. Often the symmetry may reveal itself when the points are plotted.
2. Find the zeros.
3. Find the maximum values according to the trigonometric expression.
4. Make a table.
5. Plot the points and sketch the graph.

EXAMPLE 5

Sketching the Graph of a One-Loop Limaçon

Graph the equation $r = 4 - 3\sin\ \theta$.

Solution

First, testing the equation for symmetry, we find that it fails all three symmetry tests, meaning that the graph may or may not exhibit symmetry, so we cannot use the symmetry to help us graph it. However, this equation has a graph that clearly displays symmetry with respect to the line $\theta = \frac{\pi}{2}$, yet it fails all the three symmetry tests. A graphing calculator will immediately illustrate the graph's reflective quality.

Next, we find the zeros and maximum, and plot the reflecting points to verify any symmetry. Setting $r = 0$ results in θ being undefined. What does this mean? How could θ be undefined? The angle θ is undefined for any value of $\sin\ \theta > 1$. Therefore, θ is undefined because there is no value of θ for which $\sin\ \theta > 1$. Consequently, the graph does not pass through the pole. Perhaps the graph does cross the polar axis, but not at the pole. We can investigate other intercepts by calculating r when $\theta = 0$.

$$r(0) = 4 - 3\sin(0)$$
$$r = 4 - 3 \cdot 0 = 4$$

So, there is at least one polar axis intercept at $(4, 0)$.

Next, as the maximum value of the sine function is 1 when $\theta = \frac{\pi}{2}$, we will substitute $\theta = \frac{\pi}{2}$ into the equation and solve for r. Thus, $r = 1$.

Make a table of the coordinates similar to Table 5.

θ	0	$\frac{\pi}{6}$	$\frac{\pi}{3}$	$\frac{\pi}{2}$	$\frac{2\pi}{3}$	$\frac{5\pi}{6}$	π	$\frac{7\pi}{6}$	$\frac{4\pi}{3}$	$\frac{3\pi}{2}$	$\frac{5\pi}{3}$	$\frac{11\pi}{6}$	2π
r	4	2.5	1.4	1	1.4	2.5	4	5.5	6.6	7	6.6	5.5	4

Table 5

The graph is shown in Figure 10.

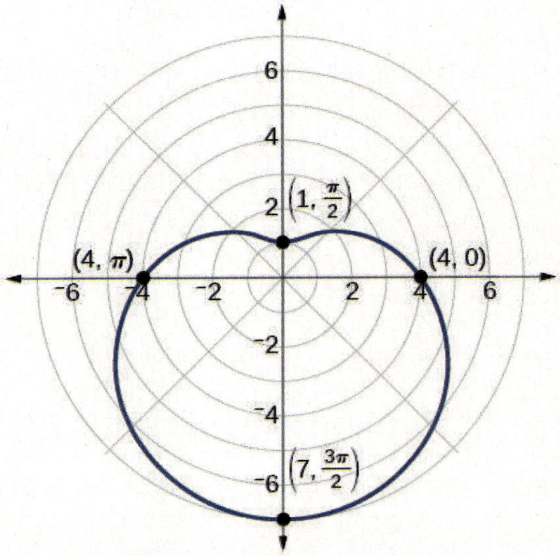

Figure 10 One-loop limaçon

Analysis

This is an example of a curve for which making a table of values is critical to producing an accurate graph. The symmetry tests fail; the zero is undefined. While it may be apparent that an equation involving $\sin\ \theta$ is likely symmetric with respect to the line $\theta = \frac{\pi}{2}$, evaluating more points helps to verify that the graph is correct.

> **TRY IT** #3 Sketch the graph of $r = 3 - 2\cos\ \theta$.

Another type of limaçon, the **inner-loop limaçon**, is named for the loop formed inside the general limaçon shape. It was discovered by the German artist Albrecht Dürer(1471-1528), who revealed a method for drawing the inner-loop limaçon in his 1525 book *Underweysung der Messing*. A century later, the father of mathematician Blaise Pascal, Étienne Pascal(1588-1651), rediscovered it.

Formulas for Inner-Loop Limaçons

The formulas that generate the **inner-loop limaçons** are given by $r = a \pm b\cos\ \theta$ and $r = a \pm b\sin\ \theta$ where $a > 0$, $b > 0$, and $a < b$. The graph of the inner-loop limaçon passes through the pole twice: once for the outer loop, and once for the inner loop. See Figure 11 for the graphs.

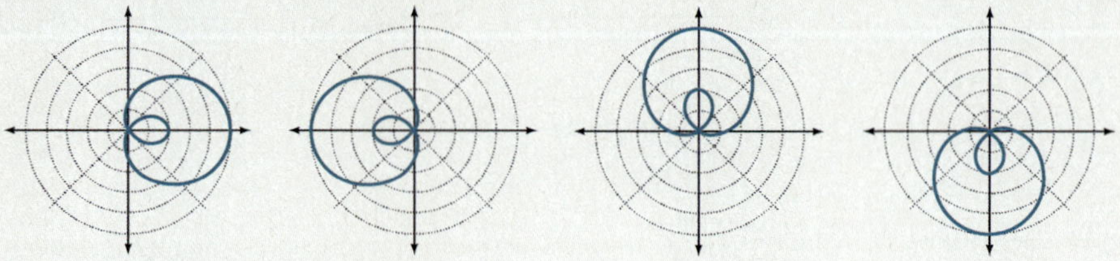

Figure 11

EXAMPLE 6

Sketching the Graph of an Inner-Loop Limaçon

Sketch the graph of $r = 2 + 5\cos\ \theta$.

Solution

Testing for symmetry, we find that the graph of the equation is symmetric about the polar axis. Next, finding the zeros reveals that when $r = 0$, $\theta = 1.98$. The maximum $|r|$ is found when $\cos\ \theta = 1$ or when $\theta = 0$. Thus, the maximum is found at the point (7, 0).

Even though we have found symmetry, the zero, and the maximum, plotting more points will help to define the shape, and then a pattern will emerge.

See Table 6.

θ	0	$\frac{\pi}{6}$	$\frac{\pi}{3}$	$\frac{\pi}{2}$	$\frac{2\pi}{3}$	$\frac{5\pi}{6}$	π	$\frac{7\pi}{6}$	$\frac{4\pi}{3}$	$\frac{3\pi}{2}$	$\frac{5\pi}{3}$	$\frac{11\pi}{6}$	2π
r	7	6.3	4.5	2	−0.5	−2.3	−3	−2.3	−0.5	2	4.5	6.3	7

Table 6

As expected, the values begin to repeat after $\theta = \pi$. The graph is shown in Figure 12.

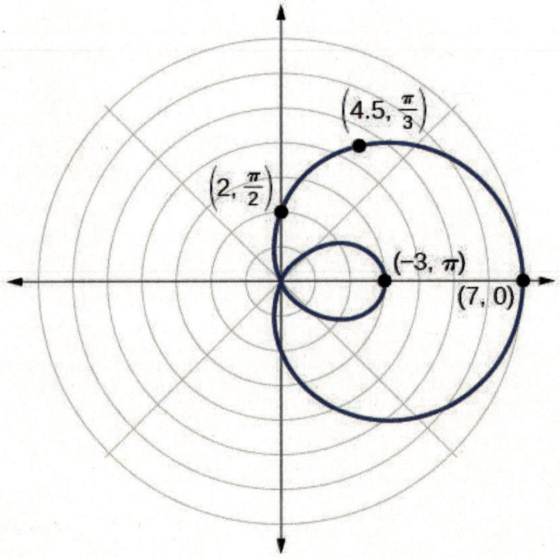

Figure 12 Inner-loop limaçon

Investigating Lemniscates

The lemniscate is a polar curve resembling the infinity symbol ∞ or a figure 8. Centered at the pole, a lemniscate is symmetrical by definition.

Formulas for Lemniscates

The formulas that generate the graph of a **lemniscate** are given by $r^2 = a^2 \cos 2\theta$ and $r^2 = a^2 \sin 2\theta$ where $a \neq 0$. The formula $r^2 = a^2 \sin 2\theta$ is symmetric with respect to the pole. The formula $r^2 = a^2 \cos 2\theta$ is symmetric with respect to the pole, the line $\theta = \frac{\pi}{2}$, and the polar axis. See Figure 13 for the graphs.

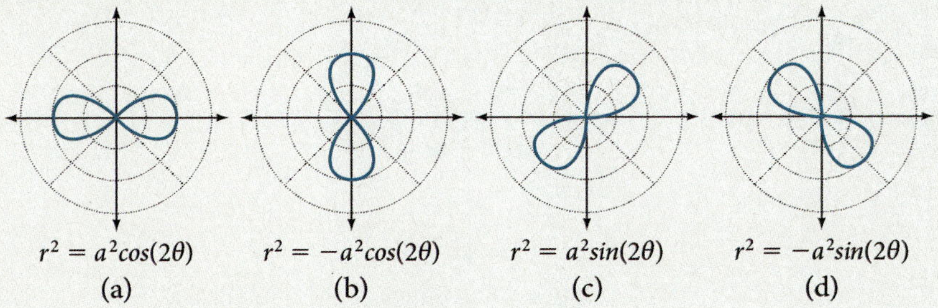

Figure 13

EXAMPLE 7

Sketching the Graph of a Lemniscate

Sketch the graph of $r^2 = 4\cos 2\theta$.

Solution

The equation exhibits symmetry with respect to the line $\theta = \frac{\pi}{2}$, the polar axis, and the pole.

Let's find the zeros. It should be routine by now, but we will approach this equation a little differently by making the substitution $u = 2\theta$.

$$\begin{aligned}
&0 = 4\cos\ 2\theta \\
&0 = 4\cos\ u \\
&0 = \cos\ u \\
&\cos^{-1}0 = \tfrac{\pi}{2} \\
&u = \tfrac{\pi}{2} && \text{Substitute } 2\theta \text{ back in for } u. \\
&2\theta = \tfrac{\pi}{2} \\
&\theta = \tfrac{\pi}{4}
\end{aligned}$$

So, the point $\left(0, \frac{\pi}{4}\right)$ is a zero of the equation.

Now let's find the maximum value. Since the maximum of $\cos\ u = 1$ when $u = 0$, the maximum $\cos\ 2\theta = 1$ when $2\theta = 0$. Thus,

$$\begin{aligned}
r^2 &= 4\cos(0) \\
r^2 &= 4(1) = 4 \\
r &= \pm\sqrt{4}\ \pm 2
\end{aligned}$$

We have a maximum at (2, 0). Since this graph is symmetric with respect to the pole, the line $\theta = \frac{\pi}{2}$, and the polar axis, we only need to plot points in the first quadrant.

Make a table similar to Table 7.

θ	0	$\frac{\pi}{6}$	$\frac{\pi}{4}$
r	± 2	$\pm\sqrt{2}$	0

Table 7

Plot the points on the graph, such as the one shown in Figure 14.

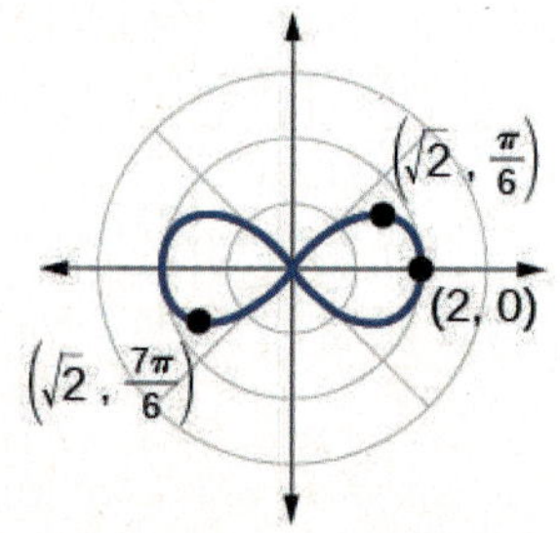

Figure 14 Lemniscate

Analysis

Making a substitution such as $u = 2\theta$ is a common practice in mathematics because it can make calculations simpler. However, we must not forget to replace the substitution term with the original term at the end, and then solve for the unknown.

Some of the points on this graph may not show up using the Trace function on the TI-84 graphing calculator, and the calculator table may show an error for these same points of r. This is because there are no real square roots for these values of θ. In other words, the corresponding r-values of $\sqrt{4\cos(2\theta)}$ are complex numbers because there is a negative number under the radical.

Investigating Rose Curves

The next type of polar equation produces a petal-like shape called a rose curve. Although the graphs look complex, a simple polar equation generates the pattern.

Rose Curves

The formulas that generate the graph of a **rose curve** are given by $r = a\cos\ n\theta$ and $r = a\sin\ n\theta$ where $a \neq 0$. If n is even, the curve has $2n$ petals. If n is odd, the curve has n petals. See Figure 15.

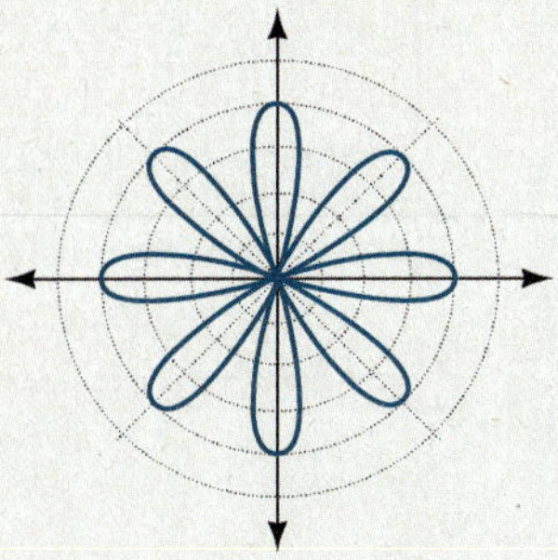

$r = a\cos(n\theta)$, n even

(a)

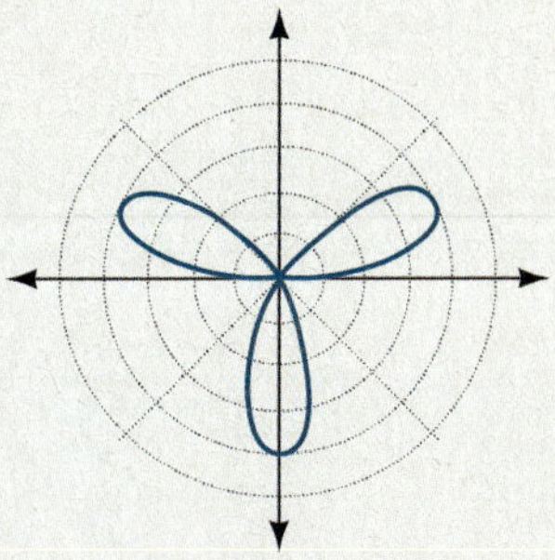

$r = a\sin(n\theta)$, n odd

(b)

Figure 15

EXAMPLE 8

Sketching the Graph of a Rose Curve (*n* Even)

Sketch the graph of $r = 2\cos\ 4\theta$.

Solution

Testing for symmetry, we find again that the symmetry tests do not tell the whole story. The graph is not only symmetric with respect to the polar axis, but also with respect to the line $\theta = \frac{\pi}{2}$ and the pole.

Now we will find the zeros. First make the substitution $u = 4\theta$.

$$\begin{aligned} 0 &= 2\cos\ 4\theta \\ 0 &= \cos\ 4\theta \\ 0 &= \cos\ u \\ \cos^{-1}0 &= u \\ u &= \frac{\pi}{2} \\ 4\theta &= \frac{\pi}{2} \\ \theta &= \frac{\pi}{8} \end{aligned}$$

The zero is $\theta = \frac{\pi}{8}$. The point $\left(0, \frac{\pi}{8}\right)$ is on the curve.

Next, we find the maximum $|r|$. We know that the maximum value of $\cos\ u = 1$ when $\theta = 0$. Thus,

$$\begin{aligned} r &= 2\cos(4 \cdot 0) \\ r &= 2\cos(0) \\ r &= 2(1) = 2 \end{aligned}$$

The point $(2, 0)$ is on the curve.

The graph of the rose curve has unique properties, which are revealed in Table 8.

θ	0	$\frac{\pi}{8}$	$\frac{\pi}{4}$	$\frac{3\pi}{8}$	$\frac{\pi}{2}$	$\frac{5\pi}{8}$	$\frac{3\pi}{4}$
r	2	0	−2	0	2	0	−2

Table 8

As $r = 0$ when $\theta = \frac{\pi}{8}$, it makes sense to divide values in the table by $\frac{\pi}{8}$ units. A definite pattern emerges. Look at the range of r-values: 2, 0, −2, 0, 2, 0, −2, and so on. This represents the development of the curve one petal at a time. Starting at $r = 0$, each petal extends out a distance of $r = 2$, and then turns back to zero $2n$ times for a total of eight petals. See the graph in Figure 16.

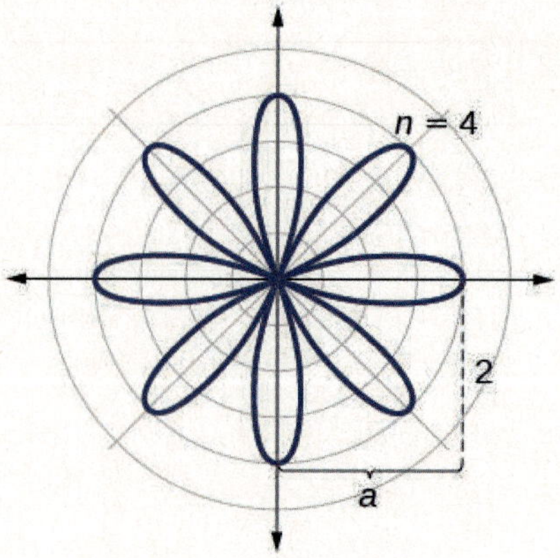

Figure 16 Rose curve, n even

Analysis

When these curves are drawn, it is best to plot the points in order, as in the Table 8. This allows us to see how the graph hits a maximum (the tip of a petal), loops back crossing the pole, hits the opposite maximum, and loops back to the pole. The action is continuous until all the petals are drawn.

TRY IT #4 Sketch the graph of $r = 4\sin(2\theta)$.

EXAMPLE 9

Sketching the Graph of a Rose Curve (n Odd)

Sketch the graph of $r = 2\sin(5\theta)$.

Solution

The graph of the equation shows symmetry with respect to the line $\theta = \frac{\pi}{2}$. Next, find the zeros and maximum. We will want to make the substitution $u = 5\theta$.

$$\begin{gathered} 0 = 2\sin(5\theta) \\ 0 = \sin\ u \\ \sin^{-1}0 = 0 \\ u = 0 \\ 5\theta = 0 \\ \theta = 0 \end{gathered}$$

The maximum value is calculated at the angle where sin θ is a maximum. Therefore,

$$\begin{gathered} r = 2\sin\left(5 \cdot \frac{\pi}{2}\right) \\ r = 2(1) = 2 \end{gathered}$$

Thus, the maximum value of the polar equation is 2. This is the length of each petal. As the curve for n odd yields the same number of petals as n, there will be five petals on the graph. See Figure 17.

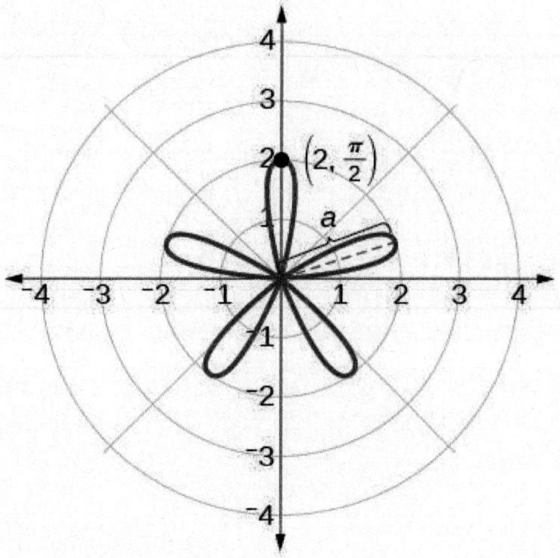

Figure 17 Rose curve, n odd

Create a table of values similar to Table 9.

θ	0	$\frac{\pi}{6}$	$\frac{\pi}{3}$	$\frac{\pi}{2}$	$\frac{2\pi}{3}$	$\frac{5\pi}{6}$	π
r	0	1	−1.73	2	−1.73	1	0

Table 9

TRY IT #5 Sketch the graph of $r = 3\cos(3\theta)$.

Investigating the Archimedes' Spiral

The final polar equation we will discuss is the Archimedes' spiral, named for its discoverer, the Greek mathematician Archimedes (c. 287 BCE-c. 212 BCE), who is credited with numerous discoveries in the fields of geometry and mechanics.

Archimedes' Spiral

The formula that generates the graph of the **Archimedes' spiral** is given by $r = \theta$ for $\theta \geq 0$. As θ increases, r increases at a constant rate in an ever-widening, never-ending, spiraling path. See Figure 18.

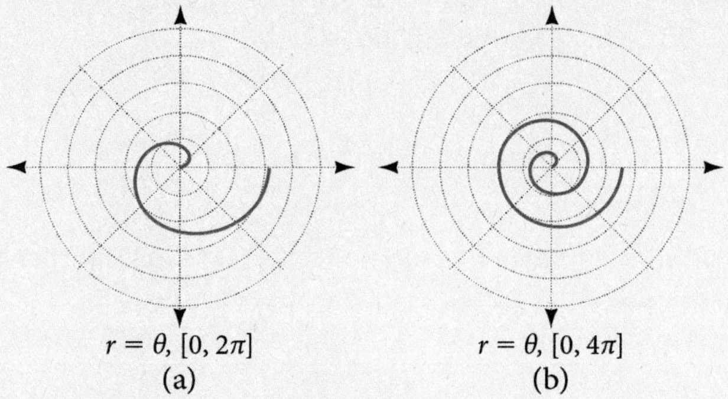

Figure 18

HOW TO

Given an Archimedes' spiral over $[0, 2\pi]$, sketch the graph.

1. Make a table of values for r and θ over the given domain.
2. Plot the points and sketch the graph.

EXAMPLE 10

Sketching the Graph of an Archimedes' Spiral

Sketch the graph of $r = \theta$ over $[0, 2\pi]$.

Solution

As r is equal to θ, the plot of the Archimedes' spiral begins at the pole at the point (0, 0). While the graph hints of symmetry, there is no formal symmetry with regard to passing the symmetry tests. Further, there is no maximum value, unless the domain is restricted.

Create a table such as Table 10.

θ	$\frac{\pi}{4}$	$\frac{\pi}{2}$	π	$\frac{3\pi}{2}$	$\frac{7\pi}{4}$	2π
r	0.785	1.57	3.14	4.71	5.50	6.28

Table 10

Notice that the r-values are just the decimal form of the angle measured in radians. We can see them on a graph in Figure 19.

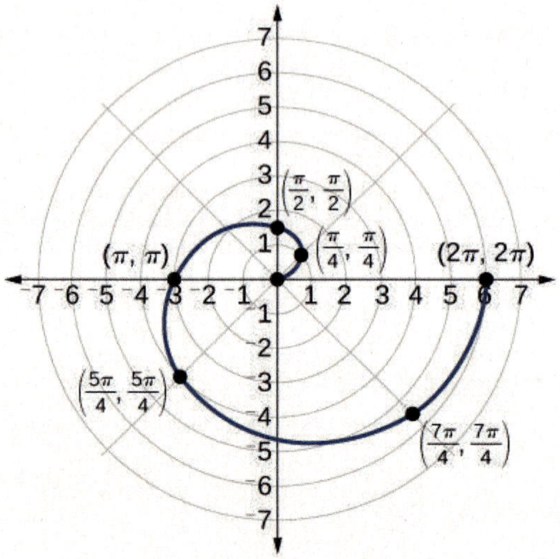

Figure 19 Archimedes' spiral

Analysis

The domain of this polar curve is $[0, 2\pi]$. In general, however, the domain of this function is $\left(-\infty, \infty\right)$. Graphing the equation of the Archimedes' spiral is rather simple, although the image makes it seem like it would be complex.

TRY IT #6 Sketch the graph of $r = -\theta$ over the interval $[0, 4\pi]$.

Summary of Curves

We have explored a number of seemingly complex polar curves in this section. Figure 20 and Figure 21 summarize the graphs and equations for each of these curves.

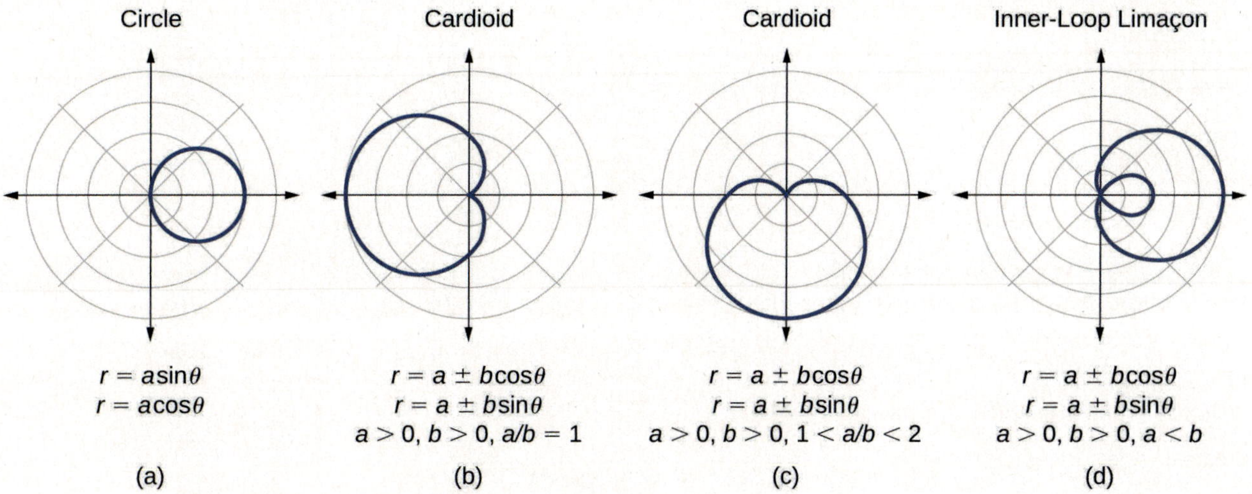

Figure 20

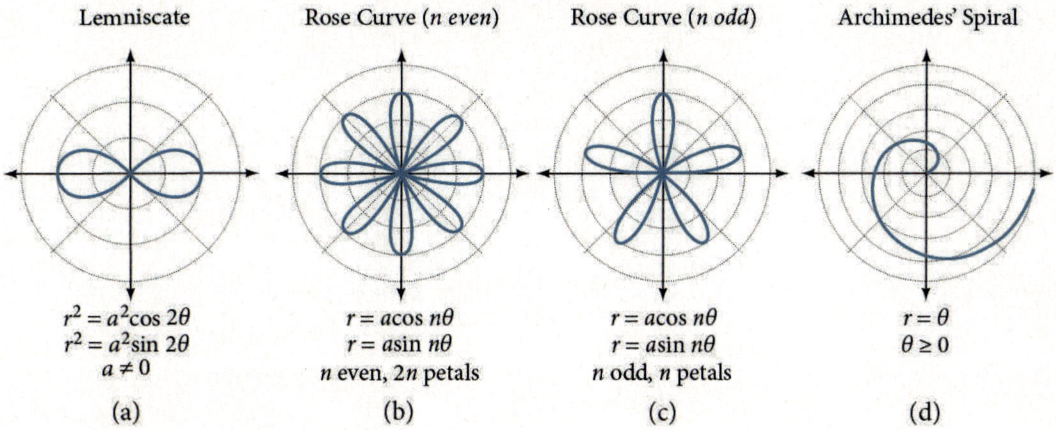

Figure 21

MEDIA

Access these online resources for additional instruction and practice with graphs of polar coordinates.

Graphing Polar Equations Part 1 (http://openstax.org/l/polargraph1)
Graphing Polar Equations Part 2 (http://openstax.org/l/polargraph2)
Animation: The Graphs of Polar Equations (http://openstax.org/l/polaranim)
Graphing Polar Equations on the TI-84 (http://openstax.org/l/polarTI84)

10.4 SECTION EXERCISES

Verbal

1. Describe the three types of symmetry in polar graphs, and compare them to the symmetry of the Cartesian plane.

2. Which of the three types of symmetries for polar graphs correspond to the symmetries with respect to the *x*-axis, *y*-axis, and origin?

3. What are the steps to follow when graphing polar equations?

4. Describe the shapes of the graphs of cardioids, limaçons, and lemniscates.

5. What part of the equation determines the shape of the graph of a polar equation?

Graphical

For the following exercises, test the equation for symmetry.

6. $r = 5\cos\ 3\theta$

7. $r = 3 - 3\cos\ \theta$

8. $r = 3 + 2\sin\ \theta$

9. $r = 3\sin\ 2\theta$

10. $r = 4$

11. $r = 2\theta$

12. $r = 4\cos\ \frac{\theta}{2}$

13. $r = \frac{2}{\theta}$

14. $r = 3\sqrt{1 - \cos^2\theta}$

15. $r = \sqrt{5\sin\ 2\theta}$

For the following exercises, graph the polar equation. Identify the name of the shape.

16. $r = 3\cos\ \theta$

17. $r = 4\sin\ \theta$

18. $r = 2 + 2\cos\ \theta$

19. $r = 2 - 2\cos\ \theta$

20. $r = 5 - 5\sin\ \theta$

21. $r = 3 + 3\sin\ \theta$

22. $r = 3 + 2\sin\ \theta$

23. $r = 7 + 4\sin\ \theta$

24. $r = 4 + 3\cos\ \theta$

25. $r = 5 + 4\cos\ \theta$

26. $r = 10 + 9\cos\ \theta$

27. $r = 1 + 3\sin\ \theta$

28. $r = 2 + 5\sin\ \theta$

29. $r = 5 + 7\sin\ \theta$

30. $r = 2 + 4\cos\ \theta$

31. $r = 5 + 6\cos\ \theta$

32. $r^2 = 36\cos(2\theta)$

33. $r^2 = 10\cos(2\theta)$

34. $r^2 = 4\sin(2\theta)$

35. $r^2 = 10\sin(2\theta)$

36. $r = 3\sin(2\theta)$

37. $r = 3\cos(2\theta)$

38. $r = 5\sin(3\theta)$

39. $r = 4\sin(4\theta)$

40. $r = 4\sin(5\theta)$

41. $r = -\theta$

42. $r = 2\theta$

43. $r = -3\theta$

Technology

For the following exercises, use a graphing calculator to sketch the graph of the polar equation.

44. $r = \frac{1}{\theta}$

45. $r = \frac{1}{\sqrt{\theta}}$

46. $r = 2\sin\ \theta\tan\ \theta$, a cissoid

47. $r = 2\sqrt{1 - \sin^2\theta}$, a hippopede

48. $r = 5 + \cos(4\theta)$

49. $r = 2 - \sin(2\theta)$

50. $r = \theta^2$

51. $r = \theta + 1$

52. $r = \theta\sin\ \theta$

53. $r = \theta\cos\ \theta$

For the following exercises, use a graphing utility to graph each pair of polar equations on a domain of $[0, 4\pi]$ and then explain the differences shown in the graphs.

54. $r = \theta, r = -\theta$

55. $r = \theta, r = \theta + \sin\ \theta$

56. $r = \sin\ \theta + \theta, r = \sin\ \theta - \theta$

57. $r = 2\sin\left(\frac{\theta}{2}\right), r = \theta\sin\left(\frac{\theta}{2}\right)$

58. $r = \sin\left(\cos(3\theta)\right)\ r = \sin(3\theta)$

59. On a graphing utility, graph $r = \sin\left(\frac{16}{5}\theta\right)$ on $[0, 4\pi]$, $[0, 8\pi]$, $[0, 12\pi]$, and $[0, 16\pi]$. Describe the effect of increasing the width of the domain.

60. On a graphing utility, graph and sketch $r = \sin\ \theta + \left(\sin\left(\frac{5}{2}\theta\right)\right)^3$ on $[0, 4\pi]$.

61. On a graphing utility, graph each polar equation. Explain the similarities and differences you observe in the graphs.

$$r_1 = 3\sin(3\theta)$$
$$r_2 = 2\sin(3\theta)$$
$$r_3 = \sin(3\theta)$$

62. On a graphing utility, graph each polar equation. Explain the similarities and differences you observe in the graphs.

$$r_1 = 3 + 3\cos\ \theta$$
$$r_2 = 2 + 2\cos\ \theta$$
$$r_3 = 1 + \cos\ \theta$$

63. On a graphing utility, graph each polar equation. Explain the similarities and differences you observe in the graphs.

$$r_1 = 3\theta$$
$$r_2 = 2\theta$$
$$r_3 = \theta$$

Extensions

For the following exercises, draw each polar equation on the same set of polar axes, and find the points of intersection.

64. $r_1 = 3 + 2\sin\ \theta,\ r_2 = 2$

65. $r_1 = 6 - 4\cos\ \theta,\ r_2 = 4$

66. $r_1 = 1 + \sin\ \theta,\ r_2 = 3\sin\ \theta$

67. $r_1 = 1 + \cos\ \theta,\ r_2 = 3\cos\ \theta$

68. $r_1 = \cos(2\theta),\ r_2 = \sin(2\theta)$

69. $r_1 = \sin^2(2\theta),$ $r_2 = 1 - \cos(4\theta)$

70. $r_1 = \sqrt{3},\ r_2 = 2\sin(\theta)$

71. $r_1{}^2 = \sin\ \theta, r_2{}^2 = \cos\ \theta$

72. $r_1 = 1 + \cos\ \theta,$ $r_2 = 1 - \sin\ \theta$

10.5 Polar Form of Complex Numbers

Learning Objectives

In this section, you will:

- Plot complex numbers in the complex plane.
- Find the absolute value of a complex number.
- Write complex numbers in polar form.
- Convert a complex number from polar to rectangular form.
- Find products of complex numbers in polar form.
- Find quotients of complex numbers in polar form.
- Find powers of complex numbers in polar form.
- Find roots of complex numbers in polar form.

"God made the integers; all else is the work of man." This rather famous quote by nineteenth-century German mathematician Leopold Kronecker sets the stage for this section on the polar form of a complex number. Complex numbers were invented by people and represent over a thousand years of continuous investigation and struggle by mathematicians such as Pythagoras, Descartes, De Moivre, Euler, Gauss, and others. Complex numbers answered questions that for centuries had puzzled the greatest minds in science.

We first encountered complex numbers in Complex Numbers. In this section, we will focus on the mechanics of working with complex numbers: translation of complex numbers from polar form to rectangular form and vice versa, interpretation of complex numbers in the scheme of applications, and application of De Moivre's Theorem.

Plotting Complex Numbers in the Complex Plane

Plotting a complex number $a + bi$ is similar to plotting a real number, except that the horizontal axis represents the real part of the number, a, and the vertical axis represents the imaginary part of the number, bi.

HOW TO

Given a complex number $a + bi$, plot it in the complex plane.

1. Label the horizontal axis as the *real* axis and the vertical axis as the *imaginary axis.*
2. Plot the point in the complex plane by moving a units in the horizontal direction and b units in the vertical direction.

EXAMPLE 1

Plotting a Complex Number in the Complex Plane

Plot the complex number $2 - 3i$ in the complex plane.

Solution

From the origin, move two units in the positive horizontal direction and three units in the negative vertical direction. See Figure 1.

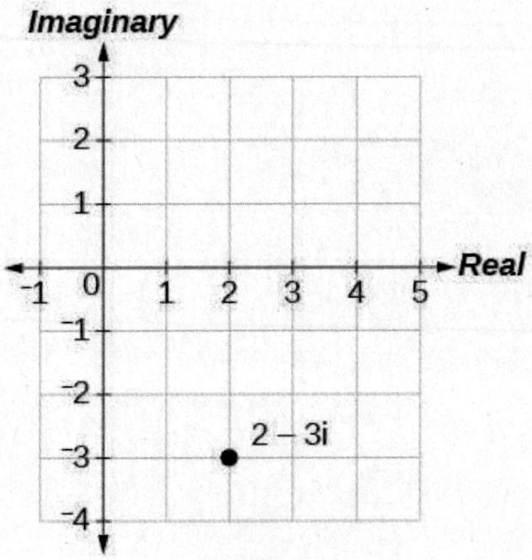

Figure 1

TRY IT #1 Plot the point $1 + 5i$ in the complex plane.

Finding the Absolute Value of a Complex Number

The first step toward working with a complex number in polar form is to find the absolute value. The absolute value of a complex number is the same as its magnitude, or $|z|$. It measures the distance from the origin to a point in the plane. For example, the graph of $z = 2 + 4i$, in Figure 2, shows $|z|$.

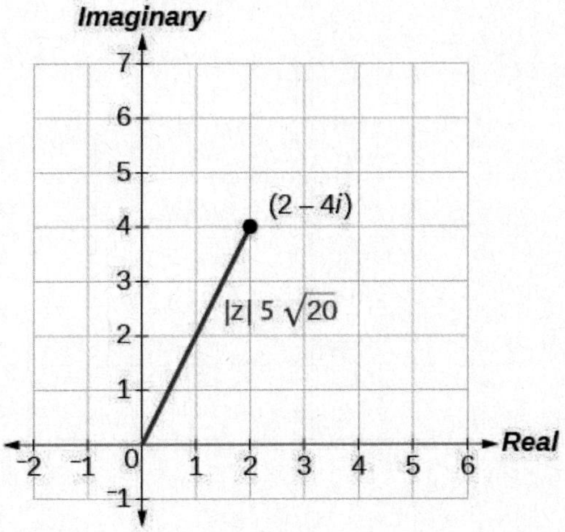

Figure 2

Absolute Value of a Complex Number

Given $z = x + yi$, a complex number, the absolute value of z is defined as

$$|z| = \sqrt{x^2 + y^2}$$

It is the distance from the origin to the point (x, y).

Notice that the absolute value of a real number gives the distance of the number from 0, while the absolute value of a complex number gives the distance of the number from the origin, $(0, \ 0)$.

EXAMPLE 2

Finding the Absolute Value of a Complex Number with a Radical

Find the absolute value of $z = \sqrt{5} - i$.

Solution

Using the formula, we have

$$|z| = \sqrt{x^2 + y^2}$$
$$|z| = \sqrt{\left(\sqrt{5}\right)^2 + (-1)^2}$$
$$|z| = \sqrt{5 + 1}$$
$$|z| = \sqrt{6}$$

See Figure 3.

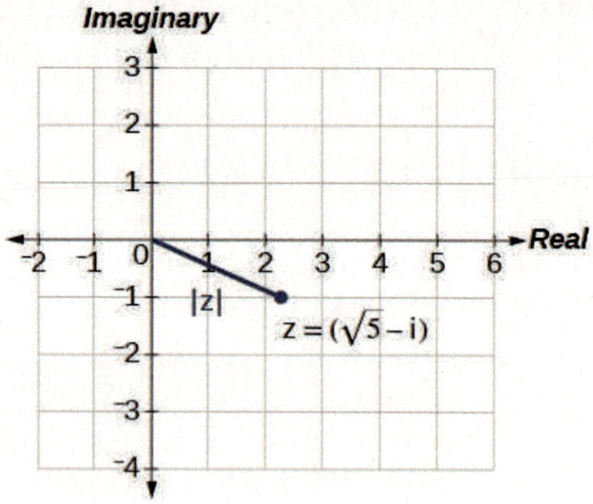

Figure 3

TRY IT #2 Find the absolute value of the complex number $z = 12 - 5i$.

EXAMPLE 3

Finding the Absolute Value of a Complex Number

Given $z = 3 - 4i$, find $|z|$.

Solution

Using the formula, we have

$$|z| = \sqrt{x^2 + y^2}$$
$$|z| = \sqrt{(3)^2 + (-4)^2}$$
$$|z| = \sqrt{9 + 16}$$
$$|z| = \sqrt{25}$$
$$|z| = 5$$

The absolute value z is 5. See Figure 4.

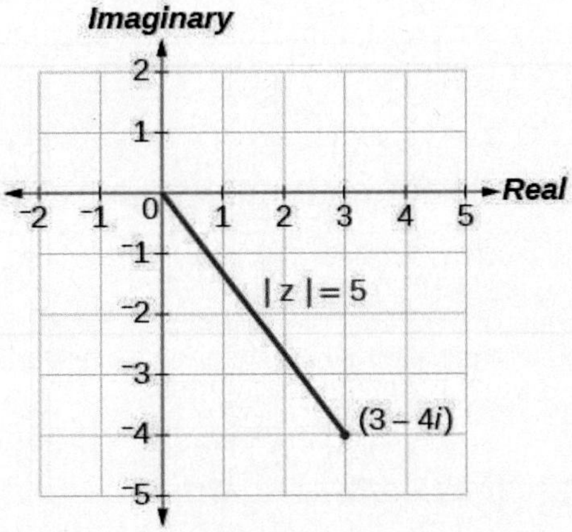

Figure 4

> TRY IT #3 Given $z = 1 - 7i$, find $|z|$.

Writing Complex Numbers in Polar Form

The **polar form of a complex number** expresses a number in terms of an angle θ and its distance from the origin r. Given a complex number in rectangular form expressed as $z = x + yi$, we use the same conversion formulas as we do to write the number in trigonometric form:

$$x = r\cos\ \theta$$
$$y = r\sin\ \theta$$
$$r = \sqrt{x^2 + y^2}$$

We review these relationships in Figure 5.

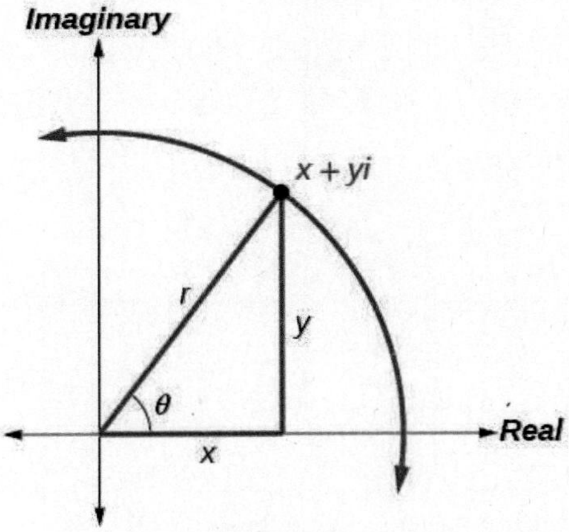

Figure 5

We use the term **modulus** to represent the absolute value of a complex number, or the distance from the origin to the point (x, y). The modulus, then, is the same as r, the radius in polar form. We use θ to indicate the angle of direction (just as with polar coordinates). Substituting, we have

$$z = x + yi$$
$$z = r\cos\ \theta + (r\sin\ \theta)\,i$$
$$z = r(\cos\ \theta + i\sin\ \theta)$$

Polar Form of a Complex Number

Writing a complex number in polar form involves the following conversion formulas:

$$x = r\cos\ \theta$$
$$y = r\sin\ \theta$$
$$r = \sqrt{x^2 + y^2}$$

Making a direct substitution, we have

$$z = x + yi$$
$$z = (r\cos\ \theta) + i\,(r\sin\ \theta)$$
$$z = r\,(\cos\ \theta + i\sin\ \theta)$$

where r is the **modulus** and θ is the **argument**. We often use the abbreviation $r\text{cis}\ \theta$ to represent $r\,(\cos\ \theta + i\sin\ \theta)$.

EXAMPLE 4

Expressing a Complex Number Using Polar Coordinates

Express the complex number $4i$ using polar coordinates.

Solution

On the complex plane, the number $z = 4i$ is the same as $z = 0 + 4i$. Writing it in polar form, we have to calculate r first.

$$r = \sqrt{x^2 + y^2}$$
$$r = \sqrt{0^2 + 4^2}$$
$$r = \sqrt{16}$$
$$r = 4$$

Next, we look at x. If $x = r\cos\ \theta$, and $x = 0$, then $\theta = \frac{\pi}{2}$. In polar coordinates, the complex number $z = 0 + 4i$ can be written as $z = 4\left(\cos\left(\frac{\pi}{2}\right) + i\sin\left(\frac{\pi}{2}\right)\right)$ or $4\text{cis}\left(\frac{\pi}{2}\right)$. See Figure 6.

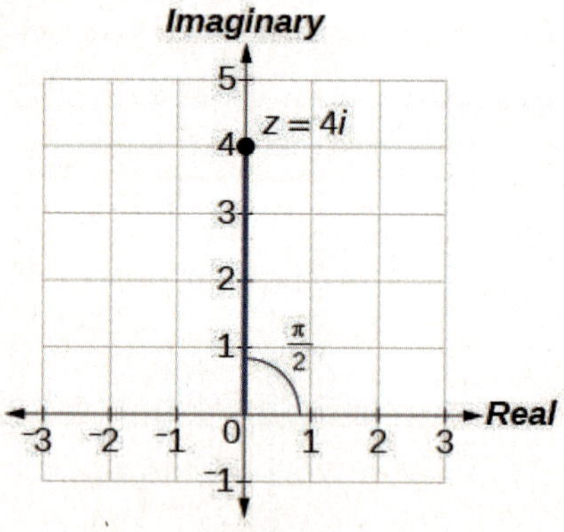

Figure 6

TRY IT #4 Express $z = 3i$ as $r\ \text{cis}\ \theta$ in polar form.

EXAMPLE 5

Finding the Polar Form of a Complex Number

Find the polar form of $-4 + 4i$.

Solution

First, find the value of r.

$$r = \sqrt{x^2 + y^2}$$
$$r = \sqrt{(-4)^2 + (4^2)}$$
$$r = \sqrt{32}$$
$$r = 4\sqrt{2}$$

Find the angle θ using the formula:

$$\cos\ \theta = \frac{x}{r}$$
$$\cos\ \theta = \frac{-4}{4\sqrt{2}}$$
$$\cos\ \theta = -\frac{1}{\sqrt{2}}$$
$$\theta = \cos^{-1}\left(-\frac{1}{\sqrt{2}}\right) = \frac{3\pi}{4}$$

Thus, the solution is $4\sqrt{2}\text{cis}\left(\frac{3\pi}{4}\right)$.

> **TRY IT** #5 Write $z = \sqrt{3} + i$ in polar form.

Converting a Complex Number from Polar to Rectangular Form

Converting a complex number from polar form to rectangular form is a matter of evaluating what is given and using the distributive property. In other words, given $z = r(\cos\ \theta + i\sin\ \theta)$, first evaluate the trigonometric functions $\cos\ \theta$ and $\sin\ \theta$. Then, multiply through by r.

EXAMPLE 6

Converting from Polar to Rectangular Form

Convert the polar form of the given complex number to rectangular form:

$$z = 12\left(\cos\left(\frac{\pi}{6}\right) + i\sin\left(\frac{\pi}{6}\right)\right)$$

Solution

We begin by evaluating the trigonometric expressions.

$$\cos\left(\frac{\pi}{6}\right) = \frac{\sqrt{3}}{2} \text{ and } \sin\left(\frac{\pi}{6}\right) = \frac{1}{2}$$

After substitution, the complex number is

$$z = 12\left(\frac{\sqrt{3}}{2} + \frac{1}{2}i\right)$$

We apply the distributive property:

$$\begin{aligned} z &= 12\left(\frac{\sqrt{3}}{2} + \frac{1}{2}i\right) \\ &= (12)\frac{\sqrt{3}}{2} + (12)\frac{1}{2}i \\ &= 6\sqrt{3} + 6i \end{aligned}$$

The rectangular form of the given point in complex form is $6\sqrt{3} + 6i$.

EXAMPLE 7

Finding the Rectangular Form of a Complex Number

Find the rectangular form of the complex number given $r = 13$ and $\tan\ \theta = \frac{5}{12}$.

Solution

If $\tan\ \theta = \frac{5}{12}$, and $\tan\ \theta = \frac{y}{x}$, we first determine $r = \sqrt{x^2 + y^2} = \sqrt{12^2 + 5^2} = 13$. We then find $\cos\ \theta = \frac{x}{r}$ and $\sin\ \theta = \frac{y}{r}$.

$$\begin{aligned} z &= 13(\cos\ \theta + i \sin\ \theta) \\ &= 13\left(\tfrac{12}{13} + \tfrac{5}{13}i\right) \\ &= 12 + 5i \end{aligned}$$

The rectangular form of the given number in complex form is $12 + 5i$.

TRY IT #6 Convert the complex number to rectangular form:

$$z = 4\left(\cos \frac{11\pi}{6} + i \sin \frac{11\pi}{6}\right)$$

Finding Products of Complex Numbers in Polar Form

Now that we can convert complex numbers to polar form we will learn how to perform operations on complex numbers in polar form. For the rest of this section, we will work with formulas developed by French mathematician Abraham De Moivre (1667-1754). These formulas have made working with products, quotients, powers, and roots of complex numbers much simpler than they appear. The rules are based on multiplying the moduli and adding the arguments.

Products of Complex Numbers in Polar Form

If $z_1 = r_1(\cos\ \theta_1 + i \sin\ \theta_1)$ and $z_2 = r_2(\cos\ \theta_2 + i \sin\ \theta_2)$, then the product of these numbers is given as:

$$\begin{aligned} z_1 z_2 &= r_1 r_2 \left[\cos\left(\theta_1 + \theta_2\right) + i \sin\left(\theta_1 + \theta_2\right)\right] \\ z_1 z_2 &= r_1 r_2 \operatorname{cis}\left(\theta_1 + \theta_2\right) \end{aligned}$$

Notice that the product calls for multiplying the moduli and adding the angles.

EXAMPLE 8

Finding the Product of Two Complex Numbers in Polar Form

Find the product of $z_1 z_2$, given $z_1 = 4(\cos(80°) + i \sin(80°))$ and $z_2 = 2(\cos(145°) + i \sin(145°))$.

Solution

Follow the formula

$$\begin{aligned} z_1 z_2 &= 4 \cdot 2[\cos(80° + 145°) + i \sin(80° + 145°)] \\ z_1 z_2 &= 8[\cos(225°) + i \sin(225°)] \\ z_1 z_2 &= 8\left[\cos\left(\tfrac{5\pi}{4}\right) + i \sin\left(\tfrac{5\pi}{4}\right)\right] \\ z_1 z_2 &= 8\left[-\frac{\sqrt{2}}{2} + i\left(-\frac{\sqrt{2}}{2}\right)\right] \\ z_1 z_2 &= -4\sqrt{2} - 4i\sqrt{2} \end{aligned}$$

Finding Quotients of Complex Numbers in Polar Form

The quotient of two complex numbers in polar form is the quotient of the two moduli and the difference of the two arguments.

Quotients of Complex Numbers in Polar Form

If $z_1 = r_1(\cos\ \theta_1 + i \sin\ \theta_1)$ and $z_2 = r_2(\cos\ \theta_2 + i \sin\ \theta_2)$, then the quotient of these numbers is

$$\frac{z_1}{z_2} = \frac{r_1}{r_2}[\cos(\theta_1 - \theta_2) + i \sin(\theta_1 - \theta_2)], \quad z_2 \neq 0$$
$$\frac{z_1}{z_2} = \frac{r_1}{r_2}\text{cis}(\theta_1 - \theta_2), \quad z_2 \neq 0$$

Notice that the moduli are divided, and the angles are subtracted.

HOW TO

Given two complex numbers in polar form, find the quotient.

1. Divide $\frac{r_1}{r_2}$.
2. Find $\theta_1 - \theta_2$.
3. Substitute the results into the formula: $z = r(\cos\ \theta + i \sin\ \theta)$. Replace r with $\frac{r_1}{r_2}$, and replace θ with $\theta_1 - \theta_2$.
4. Calculate the new trigonometric expressions and multiply through by r.

EXAMPLE 9

Finding the Quotient of Two Complex Numbers

Find the quotient of $z_1 = 2(\cos(213°) + i \sin(213°))$ and $z_2 = 4(\cos(33°) + i \sin(33°))$.

Solution

Using the formula, we have

$$\begin{aligned}
\frac{z_1}{z_2} &= \frac{2}{4}[\cos(213° - 33°) + i \sin(213° - 33°)] \\
\frac{z_1}{z_2} &= \frac{1}{2}[\cos(180°) + i \sin(180°)] \\
\frac{z_1}{z_2} &= \frac{1}{2}[-1 + 0i] \\
\frac{z_1}{z_2} &= -\frac{1}{2} + 0i \\
\frac{z_1}{z_2} &= -\frac{1}{2}
\end{aligned}$$

TRY IT #7 Find the product and the quotient of $z_1 = 2\sqrt{3}(\cos(150°) + i \sin(150°))$ and $z_2 = 2(\cos(30°) + i \sin(30°))$.

Finding Powers of Complex Numbers in Polar Form

Finding powers of complex numbers is greatly simplified using **De Moivre's Theorem**. It states that, for a positive integer n, z^n is found by raising the modulus to the nth power and multiplying the argument by n. It is the standard method used in modern mathematics.

De Moivre's Theorem

If $z = r(\cos\ \theta + i \sin\ \theta)$ is a complex number, then

$$z^n = r^n [\cos(n\theta) + i \sin(n\theta)]$$
$$z^n = r^n \text{cis}(n\theta)$$

where n is a positive integer.

EXAMPLE 10

Evaluating an Expression Using De Moivre's Theorem

Evaluate the expression $(1+i)^5$ using De Moivre's Theorem.

Solution

Since De Moivre's Theorem applies to complex numbers written in polar form, we must first write $(1+i)$ in polar form. Let us find r.

$$\begin{aligned} r &= \sqrt{x^2+y^2} \\ r &= \sqrt{(1)^2+(1)^2} \\ r &= \sqrt{2} \end{aligned}$$

Then we find θ. Using the formula $\tan\ \theta = \frac{y}{x}$ gives

$$\begin{aligned} \tan\ \theta &= \frac{1}{1} \\ \tan\ \theta &= 1 \\ \theta &= \frac{\pi}{4} \end{aligned}$$

Use De Moivre's Theorem to evaluate the expression.

$$\begin{aligned} (a+bi)^n &= r^n[\cos(n\theta)+i\sin(n\theta)] \\ (1+i)^5 &= \left(\sqrt{2}\right)^5\left[\cos\left(5\cdot\frac{\pi}{4}\right)+i\sin\left(5\cdot\frac{\pi}{4}\right)\right] \\ (1+i)^5 &= 4\sqrt{2}\left[\cos\left(\frac{5\pi}{4}\right)+i\sin\left(\frac{5\pi}{4}\right)\right] \\ (1+i)^5 &= 4\sqrt{2}\left[-\frac{\sqrt{2}}{2}+i\left(-\frac{\sqrt{2}}{2}\right)\right] \\ (1+i)^5 &= -4-4i \end{aligned}$$

Finding Roots of Complex Numbers in Polar Form

To find the nth root of a complex number in polar form, we use the nth Root Theorem or De Moivre's Theorem and raise the complex number to a power with a rational exponent. There are several ways to represent a formula for finding nth roots of complex numbers in polar form.

The nth Root Theorem

To find the nth root of a complex number in polar form, use the formula given as

$$z^{\frac{1}{n}} = r^{\frac{1}{n}}\left[\cos\left(\frac{\theta}{n}+\frac{2k\pi}{n}\right)+i\sin\left(\frac{\theta}{n}+\frac{2k\pi}{n}\right)\right]$$

where $k = 0,\ 1,\ 2,\ 3,\ .\ .\ .\ ,\ n-1$. We add $\frac{2k\pi}{n}$ to $\frac{\theta}{n}$ in order to obtain the periodic roots.

EXAMPLE 11

Finding the nth Root of a Complex Number

Evaluate the cube roots of $z = 8\left(\cos\left(\frac{2\pi}{3}\right)+i\sin\left(\frac{2\pi}{3}\right)\right)$.

Solution

We have

$$z^{\frac{1}{3}} = 8^{\frac{1}{3}}\left[\cos\left(\frac{\frac{2\pi}{3}}{3} + \frac{2k\pi}{3}\right) + i\sin\left(\frac{\frac{2\pi}{3}}{3} + \frac{2k\pi}{3}\right)\right]$$

$$z^{\frac{1}{3}} = 2\left[\cos\left(\frac{2\pi}{9} + \frac{2k\pi}{3}\right) + i\sin\left(\frac{2\pi}{9} + \frac{2k\pi}{3}\right)\right]$$

There will be three roots: $k = 0, \quad 1, \quad 2$. When $k = 0$, we have

$$z^{\frac{1}{3}} = 2\left(\cos\left(\frac{2\pi}{9}\right) + i\sin\left(\frac{2\pi}{9}\right)\right)$$

When $k = 1$, we have

$$z^{\frac{1}{3}} = 2\left[\cos\left(\frac{2\pi}{9} + \frac{6\pi}{9}\right) + i\sin\left(\frac{2\pi}{9} + \frac{6\pi}{9}\right)\right] \qquad \text{Add } \frac{2(1)\pi}{3} \text{ to each angle.}$$

$$z^{\frac{1}{3}} = 2\left(\cos\left(\frac{8\pi}{9}\right) + i\sin\left(\frac{8\pi}{9}\right)\right)$$

When $k = 2$, we have

$$z^{\frac{1}{3}} = 2\left[\cos\left(\frac{2\pi}{9} + \frac{12\pi}{9}\right) + i\sin\left(\frac{2\pi}{9} + \frac{12\pi}{9}\right)\right] \qquad \text{Add } \frac{2(2)\pi}{3} \text{ to each angle.}$$

$$z^{\frac{1}{3}} = 2\left(\cos\left(\frac{14\pi}{9}\right) + i\sin\left(\frac{14\pi}{9}\right)\right)$$

Remember to find the common denominator to simplify fractions in situations like this one. For $k = 1$, the angle simplification is

$$\begin{aligned}\frac{\frac{2\pi}{3}}{3} + \frac{2(1)\pi}{3} &= \frac{2\pi}{3}\left(\frac{1}{3}\right) + \frac{2(1)\pi}{3}\left(\frac{3}{3}\right)\\ &= \frac{2\pi}{9} + \frac{6\pi}{9}\\ &= \frac{8\pi}{9}\end{aligned}$$

TRY IT #8 Find the four fourth roots of $16(\cos(120°) + i\sin(120°))$.

MEDIA

Access these online resources for additional instruction and practice with polar forms of complex numbers.

The Product and Quotient of Complex Numbers in Trigonometric Form (http://openstax.org/l/prodquocomplex)
De Moivre's Theorem (http://openstax.org/l/demoivre)

10.5 SECTION EXERCISES

Verbal

1. A complex number is $a + bi$. Explain each part.

2. What does the absolute value of a complex number represent?

3. How is a complex number converted to polar form?

4. How do we find the product of two complex numbers?

5. What is De Moivre's Theorem and what is it used for?

Algebraic

For the following exercises, find the absolute value of the given complex number.

6. $5+\ 3i$

7. $-7+\ i$

8. $-3-3i$

9. $\sqrt{2}-6i$

10. $2i$

11. $2.2-3.1i$

For the following exercises, write the complex number in polar form.

12. $2+2i$

13. $8-4i$

14. $-\frac{1}{2}-\frac{1}{2}\ i$

15. $\sqrt{3}+i$

16. $3i$

For the following exercises, convert the complex number from polar to rectangular form.

17. $z=7\operatorname{cis}\left(\frac{\pi}{6}\right)$

18. $z=2\operatorname{cis}\left(\frac{\pi}{3}\right)$

19. $z=4\operatorname{cis}\left(\frac{7\pi}{6}\right)$

20. $z=7\operatorname{cis}(25°)$

21. $z=3\operatorname{cis}(240°)$

22. $z=\sqrt{2}\operatorname{cis}(100°)$

For the following exercises, find $z_1 z_2$ in polar form.

23. $z_1=2\sqrt{3}\operatorname{cis}(116°);\ \ z_2=2\operatorname{cis}(82°)$

24. $z_1=\sqrt{2}\operatorname{cis}(205°);\ \ z_2=2\sqrt{2}\operatorname{cis}(118°)$

25. $z_1=3\operatorname{cis}(120°);\ \ z_2=\frac{1}{4}\operatorname{cis}(60°)$

26. $z_1=3\operatorname{cis}\left(\frac{\pi}{4}\right);\ \ z_2=5\operatorname{cis}\left(\frac{\pi}{6}\right)$

27. $z_1=\sqrt{5}\operatorname{cis}\left(\frac{5\pi}{8}\right);\ \ z_2=\sqrt{15}\operatorname{cis}\left(\frac{\pi}{12}\right)$

28. $z_1=4\operatorname{cis}\left(\frac{\pi}{2}\right);\ \ z_2=2\operatorname{cis}\left(\frac{\pi}{4}\right)$

For the following exercises, find $\frac{z_1}{z_2}$ in polar form.

29. $z_1=21\operatorname{cis}(135°);\ \ z_2=3\operatorname{cis}(65°)$

30. $z_1=\sqrt{2}\operatorname{cis}(90°);\ \ z_2=2\operatorname{cis}(60°)$

31. $z_1=15\operatorname{cis}(120°);\ \ z_2=3\operatorname{cis}(40°)$

32. $z_1=6\operatorname{cis}\left(\frac{\pi}{3}\right);\ \ z_2=2\operatorname{cis}\left(\frac{\pi}{4}\right)$

33. $z_1=5\sqrt{2}\operatorname{cis}(\pi);\ \ z_2=\sqrt{2}\operatorname{cis}\left(\frac{2\pi}{3}\right)$

34. $z_1=2\operatorname{cis}\left(\frac{3\pi}{5}\right);\ \ z_2=3\operatorname{cis}\left(\frac{\pi}{4}\right)$

For the following exercises, find the powers of each complex number in polar form.

35. Find z^3 when $z=5\operatorname{cis}(45°)$.

36. Find z^4 when $z=2\operatorname{cis}(70°)$.

37. Find z^2 when $z=3\operatorname{cis}(120°)$.

38. Find z^2 when $z=4\operatorname{cis}\left(\frac{\pi}{4}\right)$.

39. Find z^4 when $z=\operatorname{cis}\left(\frac{3\pi}{16}\right)$.

40. Find z^3 when $z=3\operatorname{cis}\left(\frac{5\pi}{3}\right)$.

For the following exercises, evaluate each root.

41. Evaluate the cube root of z when $z = 27\,\text{cis}\,(240°)$.

42. Evaluate the square root of z when $z = 16\,\text{cis}\,(100°)$.

43. Evaluate the cube root of z when $z = 32\,\text{cis}\left(\frac{2\pi}{3}\right)$.

44. Evaluate the square root of z when $z = 32\text{cis}\,(\pi)$.

45. Evaluate the square root of z when $z = 8\,\text{cis}\left(\frac{7\pi}{4}\right)$.

Graphical

For the following exercises, plot the complex number in the complex plane.

46. $2 + 4i$

47. $-3 - 3i$

48. $5 - 4i$

49. $-1 - 5i$

50. $3 + 2i$

51. $2i$

52. -4

53. $6 - 2i$

54. $-2 + i$

55. $1 - 4i$

Technology

For the following exercises, find all answers rounded to the nearest hundredth.

56. Use the rectangular to polar feature on the graphing calculator to change $5 + 5i$ to polar form.

57. Use the rectangular to polar feature on the graphing calculator to change $3 - 2i$ to polar form.

58. Use the rectangular to polar feature on the graphing calculator to change $-3 - 8i$ to polar form.

59. Use the polar to rectangular feature on the graphing calculator to change $4\,\text{cis}\,(120°)$ to rectangular form.

60. Use the polar to rectangular feature on the graphing calculator to change $2\,\text{cis}\,(45°)$ to rectangular form.

61. Use the polar to rectangular feature on the graphing calculator to change $5\,\text{cis}\,(210°)$ to rectangular form.

10.6 Parametric Equations

Learning Objectives

In this section, you will:

- Parameterize a curve.
- Eliminate the parameter.
- Find a rectangular equation for a curve defined parametrically.
- Find parametric equations for curves defined by rectangular equations.

Consider the path a moon follows as it orbits a planet, which simultaneously rotates around the sun, as seen in Figure 1. At any moment, the moon is located at a particular spot relative to the planet. But how do we write and solve the equation for the position of the moon when the distance from the planet, the speed of the moon's orbit around the planet, and the speed of rotation around the sun are all unknowns? We can solve only for one variable at a time.

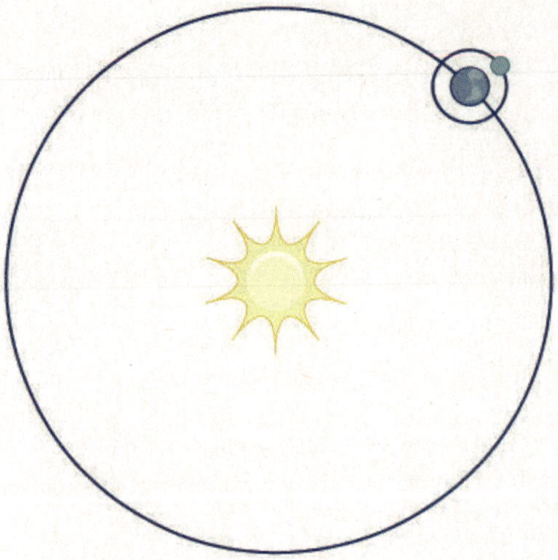

Figure 1

In this section, we will consider sets of equations given by $x(t)$ and $y(t)$ where t is the independent variable of time. We can use these parametric equations in a number of applications when we are looking for not only a particular position but also the direction of the movement. As we trace out successive values of t, the orientation of the curve becomes clear. This is one of the primary advantages of using parametric equations: we are able to trace the movement of an object along a path according to time. We begin this section with a look at the basic components of parametric equations and what it means to parameterize a curve. Then we will learn how to eliminate the parameter, translate the equations of a curve defined parametrically into rectangular equations, and find the parametric equations for curves defined by rectangular equations.

Parameterizing a Curve

When an object moves along a curve—or curvilinear path—in a given direction and in a given amount of time, the position of the object in the plane is given by the x-coordinate and the y-coordinate. However, both x and y vary over time and so are functions of time. For this reason, we add another variable, the **parameter**, upon which both x and y are dependent functions. In the example in the section opener, the parameter is time, t. The x position of the moon at time, t, is represented as the function $x(t)$, and the y position of the moon at time, t, is represented as the function $y(t)$. Together, $x(t)$ and $y(t)$ are called parametric equations, and generate an ordered pair $(x(t),\ y(t))$. Parametric equations primarily describe motion and direction.

When we parameterize a curve, we are translating a single equation in two variables, such as x and y , into an equivalent pair of equations in three variables, x, y, and t. One of the reasons we parameterize a curve is because the parametric equations yield more information: specifically, the direction of the object's motion over time.

When we graph parametric equations, we can observe the individual behaviors of x and of y. There are a number of shapes that cannot be represented in the form $y = f(x)$, meaning that they are not functions. For example, consider the graph of a circle, given as $r^2 = x^2 + y^2$. Solving for y gives $y = \pm\sqrt{r^2 - x^2}$, or two equations: $y_1 = \sqrt{r^2 - x^2}$ and $y_2 = -\sqrt{r^2 - x^2}$. If we graph y_1 and y_2 together, the graph will not pass the vertical line test, as shown in Figure 2. Thus, the equation for the graph of a circle is not a function.

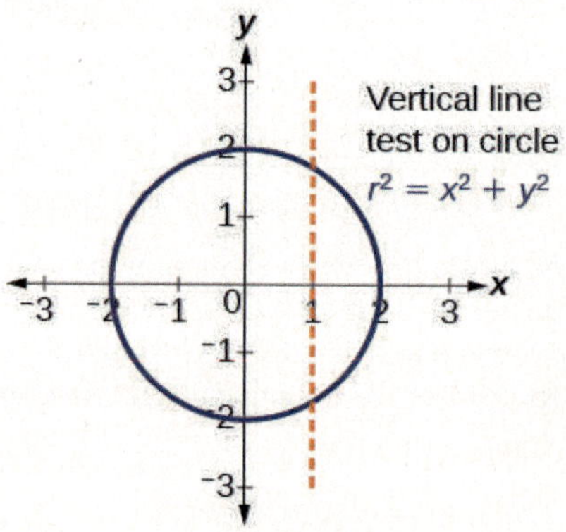

Figure 2

However, if we were to graph each equation on its own, each one would pass the vertical line test and therefore would represent a function. In some instances, the concept of breaking up the equation for a circle into two functions is similar to the concept of creating parametric equations, as we use two functions to produce a non-function. This will become clearer as we move forward.

Parametric Equations

Suppose t is a number on an interval, I. The set of ordered pairs, $(x(t), y(t))$, where $x = f(t)$ and $y = g(t)$, forms a plane curve based on the parameter t. The equations $x = f(t)$ and $y = g(t)$ are the parametric equations.

EXAMPLE 1

Parameterizing a Curve

Parameterize the curve $y = x^2 - 1$ letting $x(t) = t$. Graph both equations.

Solution

If $x(t) = t$, then to find $y(t)$ we replace the variable x with the expression given in $x(t)$. In other words, $y(t) = t^2 - 1$. Make a table of values similar to Table 1, and sketch the graph.

t	$x(t)$	$y(t)$
-4	-4	$y(-4) = (-4)^2 - 1 = 15$
-3	-3	$y(-3) = (-3)^2 - 1 = 8$
-2	-2	$y(-2) = (-2)^2 - 1 = 3$
-1	-1	$y(-1) = (-1)^2 - 1 = 0$
0	0	$y(0) = (0)^2 - 1 = -1$
1	1	$y(1) = (1)^2 - 1 = 0$
2	2	$y(2) = (2)^2 - 1 = 3$
3	3	$y(3) = (3)^2 - 1 = 8$
4	4	$y(4) = (4)^2 - 1 = 15$

Table 1

See the graphs in Figure 3. It may be helpful to use the **TRACE** feature of a graphing calculator to see how the points are generated as t increases.

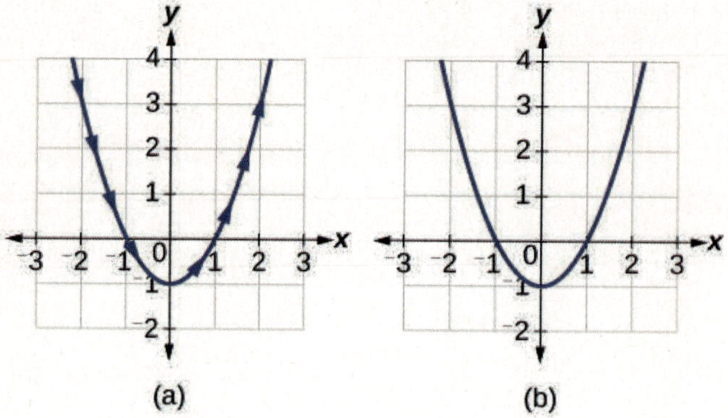

Figure 3 (a) Parametric $y(t) = t^2 - 1$ (b) Rectangular $y = x^2 - 1$

Analysis

The arrows indicate the direction in which the curve is generated. Notice the curve is identical to the curve of $y = x^2 - 1$.

TRY IT #1 Construct a table of values and plot the parametric equations: $x(t) = t - 3$, $y(t) = 2t + 4; \quad -1 \le t \le 2$.

EXAMPLE 2

Finding a Pair of Parametric Equations

Find a pair of parametric equations that models the graph of $y = 1 - x^2$, using the parameter $x(t) = t$. Plot some points and sketch the graph.

Solution

If $x(t) = t$ and we substitute t for x into the y equation, then $y(t) = 1 - t^2$. Our pair of parametric equations is

$$x(t) = t$$
$$y(t) = 1 - t^2$$

To graph the equations, first we construct a table of values like that in Table 2. We can choose values around $t = 0$, from $t = -3$ to $t = 3$. The values in the $x(t)$ column will be the same as those in the t column because $x(t) = t$. Calculate values for the column $y(t)$.

t	$x(t) = t$	$y(t) = 1 - t^2$
-3	-3	$y(-3) = 1 - (-3)^2 = -8$
-2	-2	$y(-2) = 1 - (-2)^2 = -3$
-1	-1	$y(-1) = 1 - (-1)^2 = 0$
0	0	$y(0) = 1 - 0 = 1$
1	1	$y(1) = 1 - (1)^2 = 0$
2	2	$y(2) = 1 - (2)^2 = -3$
3	3	$y(3) = 1 - (3)^2 = -8$

Table 2

The graph of $y = 1 - t^2$ is a parabola facing downward, as shown in Figure 4. We have mapped the curve over the interval $[-3,\ 3]$, shown as a solid line with arrows indicating the orientation of the curve according to t. Orientation refers to the path traced along the curve in terms of increasing values of t. As this parabola is symmetric with respect to the line $x = 0$, the values of x are reflected across the y-axis.

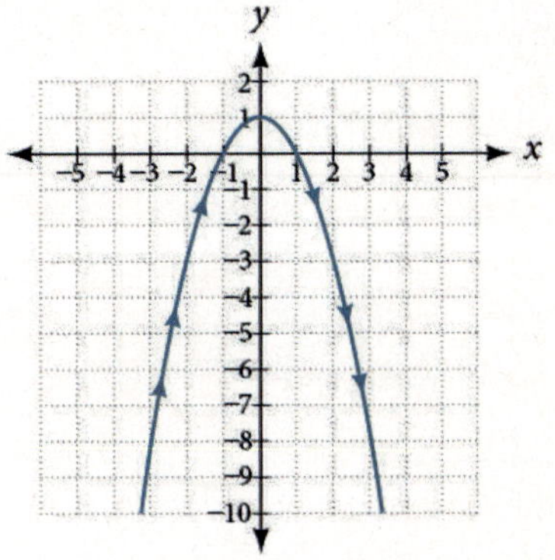

Figure 4

> **TRY IT** #2 Parameterize the curve given by $x = y^3 - 2y$.

EXAMPLE 3

Finding Parametric Equations That Model Given Criteria

An object travels at a steady rate along a straight path $(-5,\ 3)$ to $(3,\ -1)$ in the same plane in four seconds. The coordinates are measured in meters. Find parametric equations for the position of the object.

Solution

The parametric equations are simple linear expressions, but we need to view this problem in a step-by-step fashion. The x-value of the object starts at -5 meters and goes to 3 meters. This means the distance x has changed by 8 meters in 4 seconds, which is a rate of $\frac{8\text{ m}}{4\text{ s}}$, or 2 m/s. We can write the x-coordinate as a linear function with respect to time as $x(t) = 2t - 5$. In the linear function template $y = mx + b$, $2t = mx$ and $-5 = b$.

Similarly, the y-value of the object starts at 3 and goes to -1, which is a change in the distance y of −4 meters in 4 seconds, which is a rate of $\frac{-4\text{ m}}{4\text{ s}}$, or -1m/s. We can also write the y-coordinate as the linear function $y(t) = -t + 3$. Together, these are the parametric equations for the position of the object, where x and y are expressed in meters and t represents time:

$$x(t) = 2t - 5$$
$$y(t) = -t + 3$$

Using these equations, we can build a table of values for t, x, and y (see Table 3). In this example, we limited values of t to non-negative numbers. In general, any value of t can be used.

t	$x(t) = 2t - 5$	$y(t) = -t + 3$
0	$x = 2(0) - 5 = -5$	$y = -(0) + 3 = 3$
1	$x = 2(1) - 5 = -3$	$y = -(1) + 3 = 2$
2	$x = 2(2) - 5 = -1$	$y = -(2) + 3 = 1$
3	$x = 2(3) - 5 = 1$	$y = -(3) + 3 = 0$

Table 3

t	$x(t) = 2t - 5$	$y(t) = -t + 3$
4	$x = 2(4) - 5 = 3$	$y = -(4) + 3 = -1$

Table 3

From this table, we can create three graphs, as shown in Figure 5.

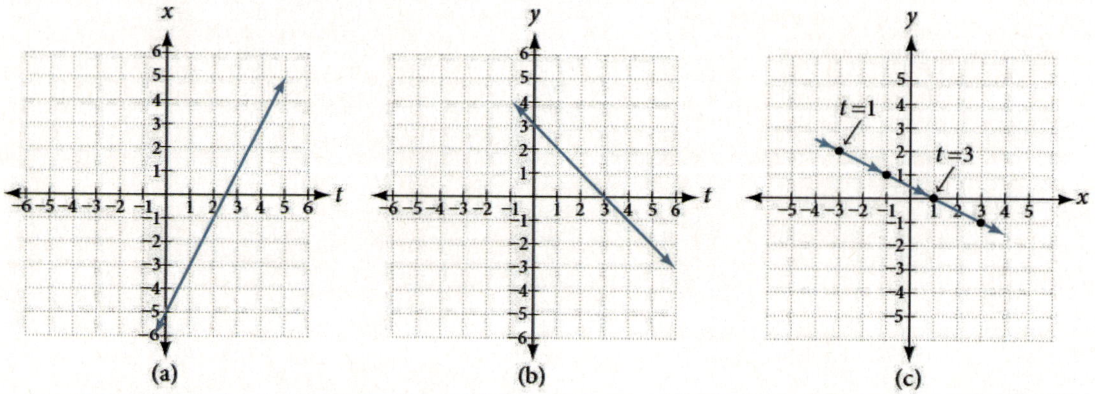

Figure 5 (a) A graph of x vs. t, representing the horizontal position over time. (b) A graph of y vs. t, representing the vertical position over time. (c) A graph of y vs. x, representing the position of the object in the plane at time t.

Analysis

Again, we see that, in Figure 5(c), when the parameter represents time, we can indicate the movement of the object along the path with arrows.

Eliminating the Parameter

In many cases, we may have a pair of parametric equations but find that it is simpler to draw a curve if the equation involves only two variables, such as x and y. Eliminating the parameter is a method that may make graphing some curves easier. However, if we are concerned with the mapping of the equation according to time, then it will be necessary to indicate the orientation of the curve as well. There are various methods for eliminating the parameter t from a set of parametric equations; not every method works for every type of equation. Here we will review the methods for the most common types of equations.

Eliminating the Parameter from Polynomial, Exponential, and Logarithmic Equations

For polynomial, exponential, or logarithmic equations expressed as two parametric equations, we choose the equation that is most easily manipulated and solve for t. We substitute the resulting expression for t into the second equation. This gives one equation in x and y.

EXAMPLE 4

Eliminating the Parameter in Polynomials

Given $x(t) = t^2 + 1$ and $y(t) = 2 + t$, eliminate the parameter, and write the parametric equations as a Cartesian equation.

Solution

We will begin with the equation for y because the linear equation is easier to solve for t.

$$y = 2 + t$$
$$y - 2 = t$$

Next, substitute $y - 2$ for t in $x(t)$.

$$x = t^2 + 1$$
$$x = (y-2)^2 + 1 \qquad \text{Substitute the expression for } t \text{ into } x.$$
$$x = y^2 - 4y + 4 + 1$$
$$x = y^2 - 4y + 5$$
$$x = y^2 - 4y + 5$$

The Cartesian form is $x = y^2 - 4y + 5$.

Analysis

This is an equation for a parabola in which, in rectangular terms, x is dependent on y. From the curve's vertex at $(1, 2)$, the graph sweeps out to the right. See Figure 6. In this section, we consider sets of equations given by the functions $x(t)$ and $y(t)$, where t is the independent variable of time. Notice, both x and y are functions of time; so in general y is not a function of x.

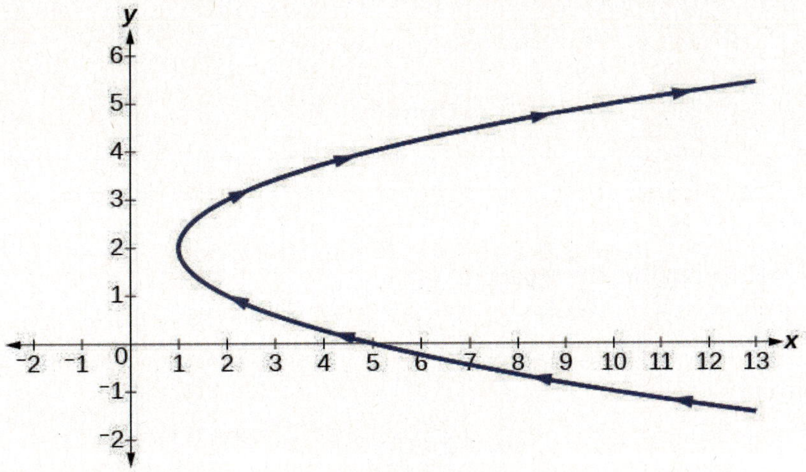

Figure 6

TRY IT #3 Given the equations below, eliminate the parameter and write as a rectangular equation for y as a function of x.

$$x(t) = 2t^2 + 6$$
$$y(t) = 5 - t$$

EXAMPLE 5

Eliminating the Parameter in Exponential Equations

Eliminate the parameter and write as a Cartesian equation: $x(t) = e^{-t}$ and $y(t) = 3e^t$, $t > 0$.

Solution

Isolate e^t.

$$x = e^{-t}$$
$$e^t = \frac{1}{x}$$

Substitute the expression into $y(t)$.

$$y = 3e^t$$
$$y = 3\left(\frac{1}{x}\right)$$
$$y = \frac{3}{x}$$

The Cartesian form is $y = \frac{3}{x}$.

Analysis
The graph of the parametric equation is shown in Figure 7(**a**). The domain is restricted to $t > 0$. The Cartesian equation, $y = \frac{3}{x}$ is shown in Figure 7(**b**) and has only one restriction on the domain, $x \neq 0$.

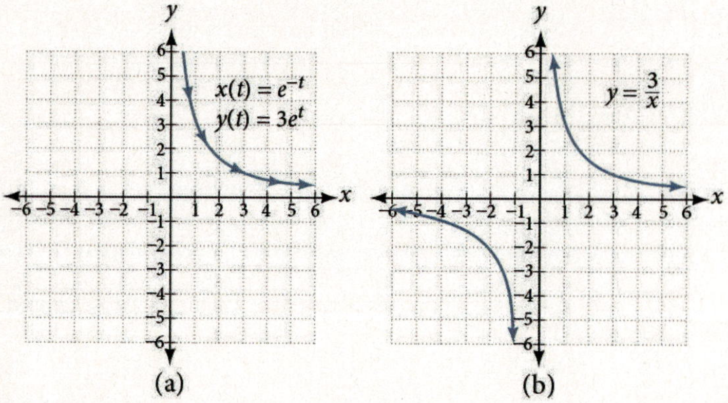

Figure 7

EXAMPLE 6

Eliminating the Parameter in Logarithmic Equations
Eliminate the parameter and write as a Cartesian equation: $x(t) = \sqrt{t} + 2$ and $y(t) = \log(t)$.

Solution
Solve the first equation for t.

$$\begin{aligned} x &= \sqrt{t} + 2 \\ x - 2 &= \sqrt{t} \\ (x-2)^2 &= t \qquad \text{Square both sides.} \end{aligned}$$

Then, substitute the expression for t into the y equation.

$$\begin{aligned} y &= \log(t) \\ y &= \log(x-2)^2 \end{aligned}$$

The Cartesian form is $y = \log(x-2)^2$.

Analysis
To be sure that the parametric equations are equivalent to the Cartesian equation, check the domains. The parametric equations restrict the domain on $x = \sqrt{t} + 2$ to $t > 0$; we restrict the domain on x to $x > 2$. The domain for the parametric equation $y = \log(t)$ is restricted to $t > 0$; we limit the domain on $y = \log(x-2)^2$ to $x > 2$.

TRY IT #4 Eliminate the parameter and write as a rectangular equation.

$$\begin{aligned} x(t) &= t^2 \\ y(t) &= \ln t \quad t > 0 \end{aligned}$$

Eliminating the Parameter from Trigonometric Equations

Eliminating the parameter from trigonometric equations is a straightforward substitution. We can use a few of the familiar trigonometric identities and the Pythagorean Theorem.

First, we use the identities:

$$\begin{aligned} x(t) &= a\cos t \\ y(t) &= b\sin t \end{aligned}$$

Solving for $\cos t$ and $\sin t$, we have

$$\frac{x}{a} = \cos\ t$$
$$\frac{y}{b} = \sin\ t$$

Then, use the Pythagorean Theorem:

$$\cos^2 t + \sin^2 t = 1$$

Substituting gives

$$\cos^2 t + \sin^2 t = \left(\frac{x}{a}\right)^2 + \left(\frac{y}{b}\right)^2 = 1$$

EXAMPLE 7

Eliminating the Parameter from a Pair of Trigonometric Parametric Equations

Eliminate the parameter from the given pair of trigonometric equations where $0 \le t \le 2\pi$ and sketch the graph.

$$x(t) = 4\cos\ t$$
$$y(t) = 3\sin\ t$$

Solution

Solving for $\cos\ t$ and $\sin\ t$, we have

$$x = 4\cos\ t$$
$$\frac{x}{4} = \cos\ t$$
$$y = 3\sin\ t$$
$$\frac{y}{3} = \sin\ t$$

Next, use the Pythagorean identity and make the substitutions.

$$\cos^2 t + \sin^2 t = 1$$
$$\left(\frac{x}{4}\right)^2 + \left(\frac{y}{3}\right)^2 = 1$$
$$\frac{x^2}{16} + \frac{y^2}{9} = 1$$

The graph for the equation is shown in Figure 8.

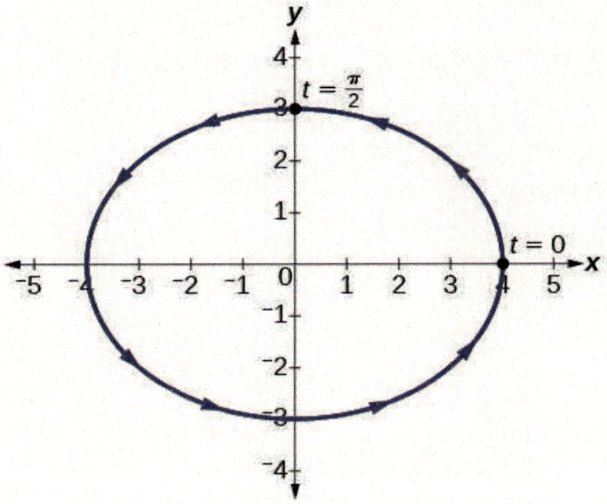

Figure 8

Analysis

Applying the general equations for conic sections (introduced in Analytic Geometry, we can identify $\frac{x^2}{16} + \frac{y^2}{9} = 1$ as an ellipse centered at $(0, 0)$. Notice that when $t = 0$ the coordinates are $(4, 0)$, and when $t = \frac{\pi}{2}$ the coordinates are $(0, 3)$. This shows the orientation of the curve with increasing values of t.

TRY IT #5 Eliminate the parameter from the given pair of parametric equations and write as a Cartesian equation: $x(t) = 2\cos\ t$ and $y(t) = 3\sin\ t$.

Finding Cartesian Equations from Curves Defined Parametrically

When we are given a set of parametric equations and need to find an equivalent Cartesian equation, we are essentially "eliminating the parameter." However, there are various methods we can use to rewrite a set of parametric equations as a Cartesian equation. The simplest method is to set one equation equal to the parameter, such as $x(t) = t$. In this case, $y(t)$ can be any expression. For example, consider the following pair of equations.

$$\begin{aligned} x(t) &= t \\ y(t) &= t^2 - 3 \end{aligned}$$

Rewriting this set of parametric equations is a matter of substituting x for t. Thus, the Cartesian equation is $y = x^2 - 3$.

EXAMPLE 8

Finding a Cartesian Equation Using Alternate Methods

Use two different methods to find the Cartesian equation equivalent to the given set of parametric equations.

$$\begin{aligned} x(t) &= 3t - 2 \\ y(t) &= t + 1 \end{aligned}$$

Solution

Method 1. First, let's solve the x equation for t. Then we can substitute the result into the y equation.

$$\begin{aligned} x &= 3t - 2 \\ x + 2 &= 3t \\ \tfrac{x+2}{3} &= t \end{aligned}$$

Now substitute the expression for t into the y equation.

$$\begin{aligned} y &= t + 1 \\ y &= \left(\tfrac{x+2}{3}\right) + 1 \\ y &= \tfrac{x}{3} + \tfrac{2}{3} + 1 \\ y &= \tfrac{1}{3}x + \tfrac{5}{3} \end{aligned}$$

Method 2. Solve the y equation for t and substitute this expression in the x equation.

$$\begin{aligned} y &= t + 1 \\ y - 1 &= t \end{aligned}$$

Make the substitution and then solve for y.

$$\begin{aligned} x &= 3(y - 1) - 2 \\ x &= 3y - 3 - 2 \\ x &= 3y - 5 \\ x + 5 &= 3y \\ \tfrac{x+5}{3} &= y \\ y &= \tfrac{1}{3}x + \tfrac{5}{3} \end{aligned}$$

TRY IT #6 Write the given parametric equations as a Cartesian equation: $x(t) = t^3$ and $y(t) = t^6$.

Finding Parametric Equations for Curves Defined by Rectangular Equations

Although we have just shown that there is only one way to interpret a set of parametric equations as a rectangular equation, there are multiple ways to interpret a rectangular equation as a set of parametric equations. Any strategy we may use to find the parametric equations is valid if it produces equivalency. In other words, if we choose an expression to represent x, and then substitute it into the y equation, and it produces the same graph over the same domain as the rectangular equation, then the set of parametric equations is valid. If the domain becomes restricted in the set of parametric equations, and the function does not allow the same values for x as the domain of the rectangular equation, then the graphs will be different.

EXAMPLE 9

Finding a Set of Parametric Equations for Curves Defined by Rectangular Equations

Find a set of equivalent parametric equations for $y = (x + 3)^2 + 1$.

Solution

An obvious choice would be to let $x(t) = t$. Then $y(t) = (t + 3)^2 + 1$. But let's try something more interesting. What if we let $x = t + 3$? Then we have

$$\begin{aligned} y &= (x + 3)^2 + 1 \\ y &= ((t + 3) + 3)^2 + 1 \\ y &= (t + 6)^2 + 1 \end{aligned}$$

The set of parametric equations is

$$\begin{aligned} x(t) &= t + 3 \\ y(t) &= (t + 6)^2 + 1 \end{aligned}$$

See Figure 9.

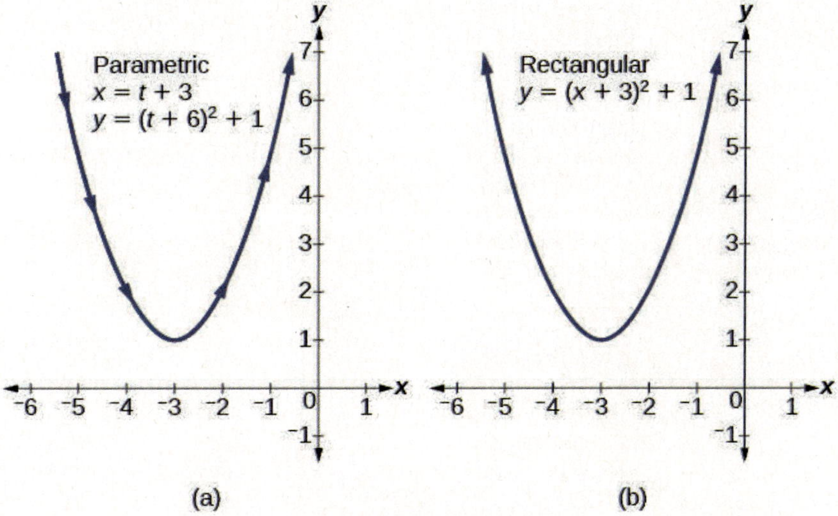

Figure 9

MEDIA

Access these online resources for additional instruction and practice with parametric equations.

Introduction to Parametric Equations (http://openstax.org/l/introparametric)
Converting Parametric Equations to Rectangular Form (http://openstax.org/l/convertpara)

10.6 SECTION EXERCISES

Verbal

1. What is a system of parametric equations?

2. Some examples of a third parameter are time, length, speed, and scale. Explain when time is used as a parameter.

3. Explain how to eliminate a parameter given a set of parametric equations.

4. What is a benefit of writing a system of parametric equations as a Cartesian equation?

5. What is a benefit of using parametric equations?

6. Why are there many sets of parametric equations to represent on Cartesian function?

Algebraic

For the following exercises, eliminate the parameter t to rewrite the parametric equation as a Cartesian equation.

7. $\begin{cases} x(t) = 5 - t \\ y(t) = 8 - 2t \end{cases}$

8. $\begin{cases} x(t) = 6 - 3t \\ y(t) = 10 - t \end{cases}$

9. $\begin{cases} x(t) = 2t + 1 \\ y(t) = 3\sqrt{t} \end{cases}$

10. $\begin{cases} x(t) = 3t - 1 \\ y(t) = 2t^2 \end{cases}$

11. $\begin{cases} x(t) = 2e^t \\ y(t) = 1 - 5t \end{cases}$

12. $\begin{cases} x(t) = e^{-2t} \\ y(t) = 2e^{-t} \end{cases}$

13. $\begin{cases} x(t) = 4\log(t) \\ y(t) = 3 + 2t \end{cases}$

14. $\begin{cases} x(t) = \log(2t) \\ y(t) = \sqrt{t - 1} \end{cases}$

15. $\begin{cases} x(t) = t^3 - t \\ y(t) = 2t \end{cases}$

16. $\begin{cases} x(t) = t - t^4 \\ y(t) = t + 2 \end{cases}$

17. $\begin{cases} x(t) = e^{2t} \\ y(t) = e^{6t} \end{cases}$

18. $\begin{cases} x(t) = t^5 \\ y(t) = t^{10} \end{cases}$

19. $\begin{cases} x(t) = 4\cos t \\ y(t) = 5\sin t \end{cases}$

20. $\begin{cases} x(t) = 3\sin t \\ y(t) = 6\cos t \end{cases}$

21. $\begin{cases} x(t) = 2\cos^2 t \\ y(t) = -\sin t \end{cases}$

22. $\begin{cases} x(t) = \cos t + 4 \\ y(t) = 2\sin^2 t \end{cases}$

23. $\begin{cases} x(t) = t - 1 \\ y(t) = t^2 \end{cases}$

24. $\begin{cases} x(t) = -t \\ y(t) = t^3 + 1 \end{cases}$

25. $\begin{cases} x(t) = 2t - 1 \\ y(t) = t^3 - 2 \end{cases}$

For the following exercises, rewrite the parametric equation as a Cartesian equation by building an x-y table.

26. $\begin{cases} x(t) = 2t - 1 \\ y(t) = t + 4 \end{cases}$

27. $\begin{cases} x(t) = 4 - t \\ y(t) = 3t + 2 \end{cases}$

28. $\begin{cases} x(t) = 2t - 1 \\ y(t) = 5t \end{cases}$

29. $\begin{cases} x(t) = 4t - 1 \\ y(t) = 4t + 2 \end{cases}$

For the following exercises, parameterize (write parametric equations for) each Cartesian equation by setting $x(t) = t$ or by setting $y(t) = t$.

30. $y(x) = 3x^2 + 3$

31. $y(x) = 2\sin x + 1$

32. $x(y) = 3\log(y) + y$

33. $x(y) = \sqrt{y} + 2y$

For the following exercises, parameterize (write parametric equations for) each Cartesian equation by using $x(t) = a\cos t$ and $y(t) = b\sin t$. Identify the curve.

34. $\frac{x^2}{4} + \frac{y^2}{9} = 1$

35. $\frac{x^2}{16} + \frac{y^2}{36} = 1$

36. $x^2 + y^2 = 16$

37. $x^2 + y^2 = 10$

38. Parameterize the line from $(3, 0)$ to $(-2, -5)$ so that the line is at $(3, 0)$ at $t = 0$, and at $(-2, -5)$ at $t = 1$.

39. Parameterize the line from $(-1, 0)$ to $(3, -2)$ so that the line is at $(-1, 0)$ at $t = 0$, and at $(3, -2)$ at $t = 1$.

40. Parameterize the line from $(-1, 5)$ to $(2, 3)$ so that the line is at $(-1, 5)$ at $t = 0$, and at $(2, 3)$ at $t = 1$.

41. Parameterize the line from $(4, 1)$ to $(6, -2)$ so that the line is at $(4, 1)$ at $t = 0$, and at $(6, -2)$ at $t = 1$.

Technology

For the following exercises, use the table feature in the graphing calculator to determine whether the graphs intersect.

42. $\begin{cases} x_1(t) = 3t \\ y_1(t) = 2t - 1 \end{cases}$ and $\begin{cases} x_2(t) = t + 3 \\ y_2(t) = 4t - 4 \end{cases}$

43. $\begin{cases} x_1(t) = t^2 \\ y_1(t) = 2t - 1 \end{cases}$ and $\begin{cases} x_2(t) = -t + 6 \\ y_2(t) = t + 1 \end{cases}$

For the following exercises, use a graphing calculator to complete the table of values for each set of parametric equations.

44. $\begin{cases} x_1(t) = 3t^2 - 3t + 7 \\ y_1(t) = 2t + 3 \end{cases}$

t	x	y
-1		
0		
1		

45. $\begin{cases} x_1(t) = t^2 - 4 \\ y_1(t) = 2t^2 - 1 \end{cases}$

t	x	y
1		
2		
3		

46. $\begin{cases} x_1(t) = t^4 \\ y_1(t) = t^3 + 4 \end{cases}$

t	x	y
-1		
0		
1		
2		

Extensions

47. Find two different sets of parametric equations for $y = (x + 1)^2$.

48. Find two different sets of parametric equations for $y = 3x - 2$.

49. Find two different sets of parametric equations for $y = x^2 - 4x + 4$.

10.7 Parametric Equations: Graphs

Learning Objectives

In this section you will:

- Graph plane curves described by parametric equations by plotting points.
- Graph parametric equations.

While not every fan (or team manager) appreciates it, baseball and many other sports have become dependent on analytics, which involve complex data recording and quantitative evaluation used to understand and predict behavior. The earliest influence of analytics was mostly statistical; more recently, physics and other sciences have come into play. Foremost among these is the focus on launch angle and exit velocity, which when at certain values can almost guarantee a home run. On the other hand, emphasis on launch angle and focusing on home runs rather than overall hitting results in far more outs. Consider the following situation: it is the bottom of the ninth inning, with two outs and two players on base. The home team is losing by two runs. The batter swings and hits the baseball at 140 feet per second and at an angle of approximately $45°$ to the horizontal. How far will the ball travel? Will it clear the fence for a game-winning home run? The outcome may depend partly on other factors (for example, the wind), but mathematicians can model the path of a projectile and predict approximately how far it will travel using parametric equations. In this section, we'll discuss parametric equations and some common applications, such as projectile motion problems.

Figure 1 Parametric equations can model the path of a projectile. (credit: Paul Kreher, Flickr)

Graphing Parametric Equations by Plotting Points

In lieu of a graphing calculator or a computer graphing program, plotting points to represent the graph of an equation is the standard method. As long as we are careful in calculating the values, point-plotting is highly dependable.

HOW TO

Given a pair of parametric equations, sketch a graph by plotting points.

1. Construct a table with three columns: $t, x(t)$, and $y(t)$.
2. Evaluate x and y for values of t over the interval for which the functions are defined.
3. Plot the resulting pairs (x, y).

EXAMPLE 1

Sketching the Graph of a Pair of Parametric Equations by Plotting Points

Sketch the graph of the parametric equations $x(t) = t^2 + 1$, $y(t) = 2 + t$.

Solution

Construct a table of values for t, $x(t)$, and $y(t)$, as in Table 1, and plot the points in a plane.

t	$x(t) = t^2 + 1$	$y(t) = 2 + t$
−5	26	−3
−4	17	−2
−3	10	−1
−2	5	0
−1	2	1
0	1	2
1	2	3
2	5	4
3	10	5
4	17	6
5	26	7

Table 1

The graph is a parabola with vertex at the point $(1, 2)$, opening to the right. See Figure 2.

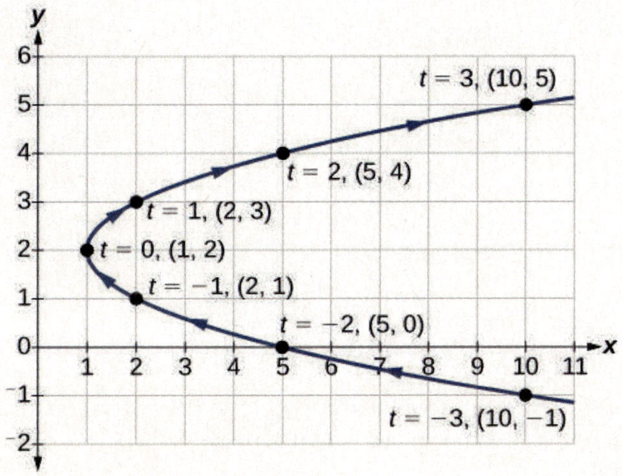

Figure 2

Analysis

As values for t progress in a positive direction from 0 to 5, the plotted points trace out the top half of the parabola. As values of t become negative, they trace out the lower half of the parabola. There are no restrictions on the domain. The arrows indicate direction according to increasing values of t. The graph does not represent a function, as it will fail the vertical line test. The graph is drawn in two parts: the positive values for t, and the negative values for t.

TRY IT #1 Sketch the graph of the parametric equations $x = \sqrt{t}$, $y = 2t + 3$, $0 \le t \le 3$.

EXAMPLE 2

Sketching the Graph of Trigonometric Parametric Equations

Construct a table of values for the given parametric equations and sketch the graph:

$$x = 2\cos\ t$$
$$y = 4\sin\ t$$

Solution

Construct a table like that in Table 2 using angle measure in radians as inputs for t, and evaluating x and y. Using angles with known sine and cosine values for t makes calculations easier.

t	$x = 2\cos\ t$	$y = 4\sin\ t$
0	$x = 2\cos(0) = 2$	$y = 4\sin(0) = 0$
$\frac{\pi}{6}$	$x = 2\cos\left(\frac{\pi}{6}\right) = \sqrt{3}$	$y = 4\sin\left(\frac{\pi}{6}\right) = 2$
$\frac{\pi}{3}$	$x = 2\cos\left(\frac{\pi}{3}\right) = 1$	$y = 4\sin\left(\frac{\pi}{3}\right) = 2\sqrt{3}$
$\frac{\pi}{2}$	$x = 2\cos\left(\frac{\pi}{2}\right) = 0$	$y = 4\sin\left(\frac{\pi}{2}\right) = 4$
$\frac{2\pi}{3}$	$x = 2\cos\left(\frac{2\pi}{3}\right) = -1$	$y = 4\sin\left(\frac{2\pi}{3}\right) = 2\sqrt{3}$
$\frac{5\pi}{6}$	$x = 2\cos\left(\frac{5\pi}{6}\right) = -\sqrt{3}$	$y = 4\sin\left(\frac{5\pi}{6}\right) = 2$
π	$x = 2\cos(\pi) = -2$	$y = 4\sin(\pi) = 0$
$\frac{7\pi}{6}$	$x = 2\cos\left(\frac{7\pi}{6}\right) = -\sqrt{3}$	$y = 4\sin\left(\frac{7\pi}{6}\right) = -2$
$\frac{4\pi}{3}$	$x = 2\cos\left(\frac{4\pi}{3}\right) = -1$	$y = 4\sin\left(\frac{4\pi}{3}\right) = -2\sqrt{3}$
$\frac{3\pi}{2}$	$x = 2\cos\left(\frac{3\pi}{2}\right) = 0$	$y = 4\sin\left(\frac{3\pi}{2}\right) = -4$
$\frac{5\pi}{3}$	$x = 2\cos\left(\frac{5\pi}{3}\right) = 1$	$y = 4\sin\left(\frac{5\pi}{3}\right) = -2\sqrt{3}$
$\frac{11\pi}{6}$	$x = 2\cos\left(\frac{11\pi}{6}\right) = \sqrt{3}$	$y = 4\sin\left(\frac{11\pi}{6}\right) = -2$
2π	$x = 2\cos(2\pi) = 2$	$y = 4\sin(2\pi) = 0$

Table 2

Figure 3 shows the graph.

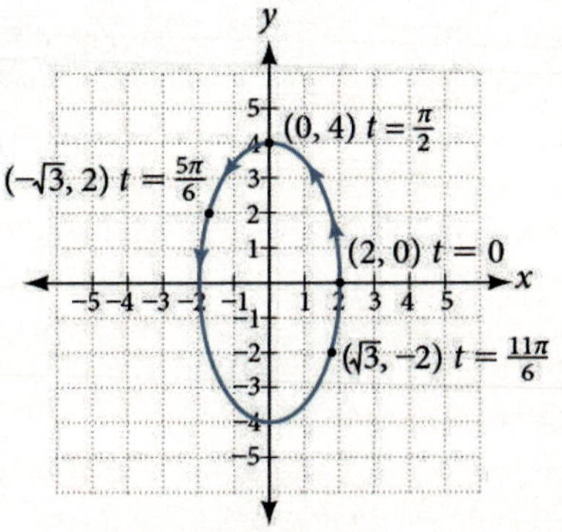

Figure 3

By the symmetry shown in the values of x and y, we see that the parametric equations represent an ellipse. The ellipse is mapped in a counterclockwise direction as shown by the arrows indicating increasing t values.

Analysis

We have seen that parametric equations can be graphed by plotting points. However, a graphing calculator will save some time and reveal nuances in a graph that may be too tedious to discover using only hand calculations.

Make sure to change the mode on the calculator to parametric (PAR). To confirm, the $Y =$ window should show

$$X_{1T} =$$
$$Y_{1T} =$$

instead of $Y_1 =$.

TRY IT #2 Graph the parametric equations: $x = 5\cos\ t$, $y = 3\sin\ t$.

EXAMPLE 3

Graphing Parametric Equations and Rectangular Form Together

Graph the parametric equations $x = 5\cos\ t$ and $y = 2\sin\ t$. First, construct the graph using data points generated from the parametric form. Then graph the rectangular form of the equation. Compare the two graphs.

Solution

Construct a table of values like that in Table 3.

t	$x = 5\cos\ t$	$y = 2\sin\ t$
0	$x = 5\cos(0) = 5$	$y = 2\sin(0) = 0$
1	$x = 5\cos(1) \approx 2.7$	$y = 2\sin(1) \approx 1.7$
2	$x = 5\cos(2) \approx -2.1$	$y = 2\sin(2) \approx 1.8$
3	$x = 5\cos(3) \approx -4.95$	$y = 2\sin(3) \approx 0.28$
4	$x = 5\cos(4) \approx -3.3$	$y = 2\sin(4) \approx -1.5$
5	$x = 5\cos(5) \approx 1.4$	$y = 2\sin(5) \approx -1.9$
−1	$x = 5\cos(-1) \approx 2.7$	$y = 2\sin(-1) \approx -1.7$

Table 3

t	$x = 5\cos t$	$y = 2\sin t$
-2	$x = 5\cos(-2) \approx -2.1$	$y = 2\sin(-2) \approx -1.8$
-3	$x = 5\cos(-3) \approx -4.95$	$y = 2\sin(-3) \approx -0.28$
-4	$x = 5\cos(-4) \approx -3.3$	$y = 2\sin(-4) \approx 1.5$
-5	$x = 5\cos(-5) \approx 1.4$	$y = 2\sin(-5) \approx 1.9$

Table 3

Plot the (x, y) values from the table. See Figure 4.

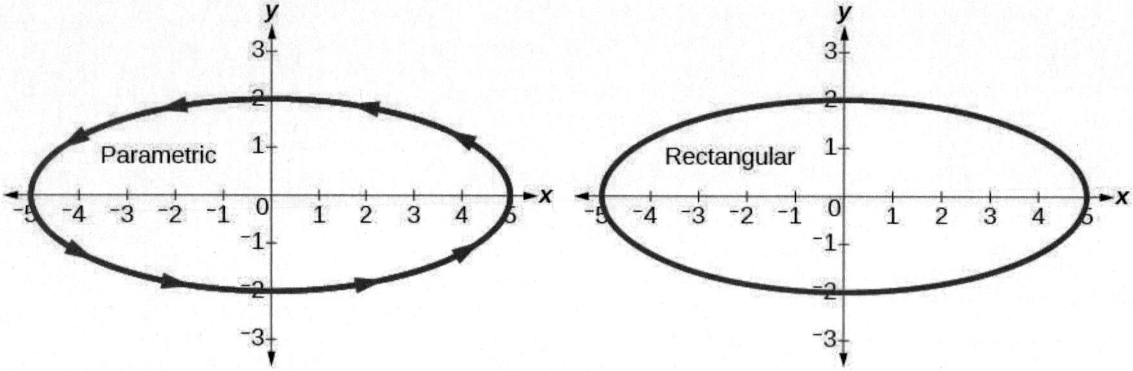

Figure 4

Next, translate the parametric equations to rectangular form. To do this, we solve for t in either $x(t)$ or $y(t)$, and then substitute the expression for t in the other equation. The result will be a function $y(x)$ if solving for t as a function of x, or $x(y)$ if solving for t as a function of y.

$$\begin{aligned} x &= 5\cos t \\ \frac{x}{5} &= \cos t && \text{Solve for } \cos t. \\ y &= 2\sin t && \text{Solve for } \sin t. \\ \frac{y}{2} &= \sin t \end{aligned}$$

Then, use the Pythagorean Theorem.

$$\begin{aligned} \cos^2 t + \sin^2 t &= 1 \\ \left(\frac{x}{5}\right)^2 + \left(\frac{y}{2}\right)^2 &= 1 \\ \frac{x^2}{25} + \frac{y^2}{4} &= 1 \end{aligned}$$

Analysis

In Figure 5, the data from the parametric equations and the rectangular equation are plotted together. The parametric equations are plotted in blue; the graph for the rectangular equation is drawn on top of the parametric in a dashed style colored red. Clearly, both forms produce the same graph.

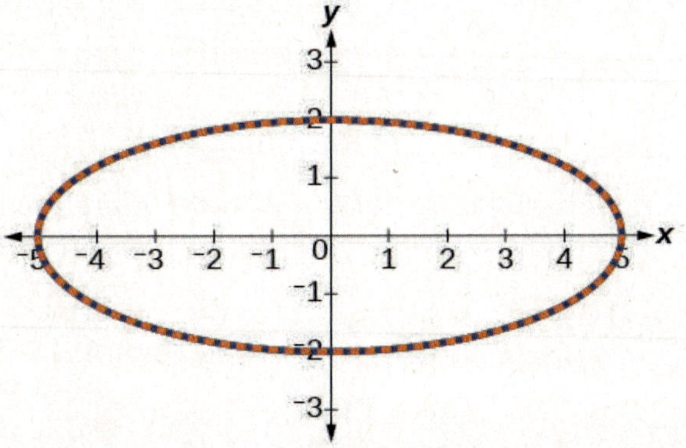

Figure 5

EXAMPLE 4

Graphing Parametric Equations and Rectangular Equations on the Coordinate System

Graph the parametric equations $x = t + 1$ and $y = \sqrt{t}, t \geq 0$, and the rectangular equivalent $y = \sqrt{x - 1}$ on the same coordinate system.

Solution

Construct a table of values for the parametric equations, as we did in the previous example, and graph $y = \sqrt{t}, t \geq 0$ on the same grid, as in Figure 6.

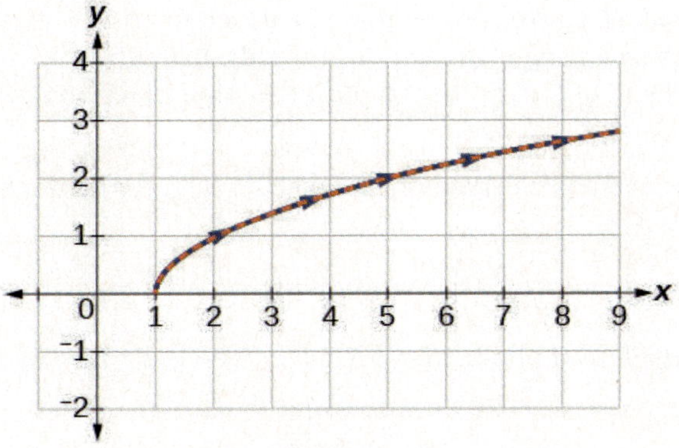

Figure 6

Analysis

With the domain on t restricted, we only plot positive values of t. The parametric data is graphed in blue and the graph of the rectangular equation is dashed in red. Once again, we see that the two forms overlap.

TRY IT #3 Sketch the graph of the parametric equations $x = 2\cos\theta$ and $y = 4\sin\theta$, along with the rectangular equation on the same grid.

Applications of Parametric Equations

Many of the advantages of parametric equations become obvious when applied to solving real-world problems. Although rectangular equations in *x* and *y* give an overall picture of an object's path, they do not reveal the position of an object at a specific time. Parametric equations, however, illustrate how the values of *x* and *y* change depending on *t*, as the location of a moving object at a particular time.

A common application of parametric equations is solving problems involving projectile motion. In this type of motion, an object is propelled forward in an upward direction forming an angle of θ to the horizontal, with an initial speed of v_0, and at a height h above the horizontal.

The path of an object propelled at an inclination of θ to the horizontal, with initial speed v_0, and at a height h above the horizontal, is given by

$$x = (v_0 \cos\theta)t$$
$$y = -\frac{1}{2}gt^2 + (v_0 \sin\theta)t + h$$

where g accounts for the effects of gravity and h is the initial height of the object. Depending on the units involved in the problem, use $g = 32\ \text{ft/s}^2$ or $g = 9.8\ \text{m/s}^2$. The equation for x gives horizontal distance, and the equation for y gives the vertical distance.

HOW TO

Given a projectile motion problem, use parametric equations to solve.

1. The horizontal distance is given by $x = (v_0 \cos\ \theta)\, t$. Substitute the initial speed of the object for v_0.
2. The expression $\cos\ \theta$ indicates the angle at which the object is propelled. Substitute that angle in degrees for $\cos\ \theta$.
3. The vertical distance is given by the formula $y = -\frac{1}{2}gt^2 + (v_0 \sin\ \theta)\, t + h$. The term $-\frac{1}{2}gt^2$ represents the effect of gravity. Depending on units involved, use $g = 32\ \text{ft/s}^2$ or $g = 9.8\ \text{m/s}^2$. Again, substitute the initial speed for v_0, and the height at which the object was propelled for h.
4. Proceed by calculating each term to solve for t.

EXAMPLE 5

Finding the Parametric Equations to Describe the Motion of a Baseball

Solve the problem presented at the beginning of this section. Does the batter hit the game-winning home run? Assume that the ball is hit with an initial velocity of 140 feet per second at an angle of 45° to the horizontal, making contact 3 feet above the ground.

ⓐ Find the parametric equations to model the path of the baseball. ⓑ Where is the ball after 2 seconds?
ⓒ How long is the ball in the air? ⓓ Is it a home run?

Solution

ⓐ

Use the formulas to set up the equations. The horizontal position is found using the parametric equation for x. Thus,

$$x = (v_0 \cos\ \theta)t$$
$$x = (140 \cos(45°))t$$

The vertical position is found using the parametric equation for y. Thus,

$$y = -16t^2 + (v_0 \sin\ \theta)t + h$$
$$y = -16t^2 + (140 \sin(45°))t + 3$$

ⓑ

Substitute 2 into the equations to find the horizontal and vertical positions of the ball.

$$x = (140 \cos(45°))(2)$$
$$x = 198 \text{ feet}$$

$$y = -16(2)^2 + (140 \sin(45°))(2) + 3$$
$$y = 137 \text{ feet}$$

After 2 seconds, the ball is 198 feet away from the batter's box and 137 feet above the ground.

ⓒ

To calculate how long the ball is in the air, we have to find out when it will hit ground, or when $y = 0$. Thus,

$$\begin{aligned} y &= -16t^2 + (140\sin(45°))\,t + 3 \\ y &= 0 && \text{Set } y(t) = 0 \text{ and solve the quadratic.} \\ t &= 6.2173 \end{aligned}$$

When $t = 6.2173$ seconds, the ball has hit the ground. (The quadratic equation can be solved in various ways, but this problem was solved using a computer math program.)

ⓓ

We cannot confirm that the hit was a home run without considering the size of the outfield, which varies from field to field. However, for simplicity's sake, let's assume that the outfield wall is 400 feet from home plate in the deepest part of the park. Let's also assume that the wall is 10 feet high. In order to determine whether the ball clears the wall, we need to calculate how high the ball is when $x = 400$ feet. So we will set $x = 400$, solve for t, and input t into y.

$$\begin{aligned} x &= (140\cos(45°))\,t \\ 400 &= (140\cos(45°))\,t \\ t &= 4.04 \end{aligned} \qquad \begin{aligned} y &= -16(4.04)^2 + (140\sin(45°))\,(4.04) + 3 \\ y &= 141.8 \end{aligned}$$

The ball is 141.8 feet in the air when it soars out of the ballpark. It was indeed a home run. See Figure 7.

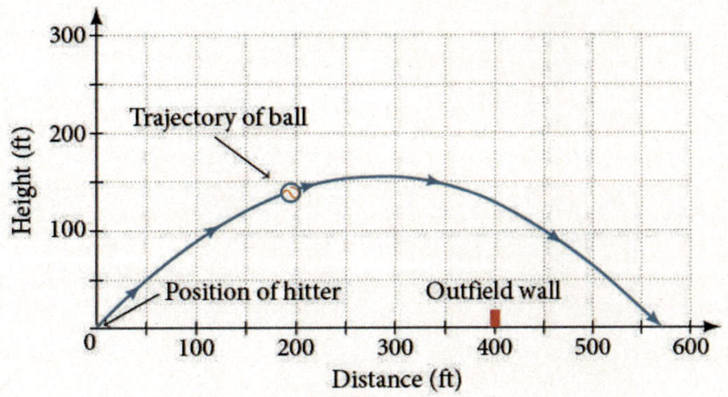

Figure 7

▶ **MEDIA**

Access the following online resource for additional instruction and practice with graphs of parametric equations.

Graphing Parametric Equations on the TI-84 (http://openstax.org/l/graphpara84)

10.7 SECTION EXERCISES

Verbal

1. What are two methods used to graph parametric equations?

2. What is one difference in point-plotting parametric equations compared to Cartesian equations?

3. Why are some graphs drawn with arrows?

4. Name a few common types of graphs of parametric equations.

5. Why are parametric graphs important in understanding projectile motion?

Graphical

For the following exercises, graph each set of parametric equations by making a table of values. Include the orientation on the graph.

6. $\begin{cases} x(t) = t \\ y(t) = t^2 - 1 \end{cases}$

t	-3	-2	-1	0	1	2	3
x							
y							

7. $\begin{cases} x(t) = t - 1 \\ y(t) = t^2 \end{cases}$

t	-3	-2	-1	0	1	2
x						
y						

8. $\begin{cases} x(t) = 2 + t \\ y(t) = 3 - 2t \end{cases}$

t	-2	-1	0	1	2	3
x						
y						

9. $\begin{cases} x(t) = -2 - 2t \\ y(t) = 3 + t \end{cases}$

t	-3	-2	-1	0	1
x					
y					

10. $\begin{cases} x(t) = t^3 \\ y(t) = t + 2 \end{cases}$

t	-2	-1	0	1	2
x					
y					

11. $\begin{cases} x(t) = t^2 \\ y(t) = t + 3 \end{cases}$

t	-2	-1	0	1	2
x					
y					

For the following exercises, sketch the curve and include the orientation.

12. $\begin{cases} x(t) = t \\ y(t) = \sqrt{t} \end{cases}$

13. $\begin{cases} x(t) = -\sqrt{t} \\ y(t) = t \end{cases}$

14. $\begin{cases} x(t) = 5 - |t| \\ y(t) = t + 2 \end{cases}$

15. $\begin{cases} x(t) = -t + 2 \\ y(t) = 5 - |t| \end{cases}$

16. $\begin{cases} x(t) = 4\sin t \\ y(t) = 2\cos t \end{cases}$

17. $\begin{cases} x(t) = 2\sin t \\ y(t) = 4\cos t \end{cases}$

18. $\begin{cases} x(t) = 3\cos^2 t \\ y(t) = -3\sin t \end{cases}$

19. $\begin{cases} x(t) = 3\cos^2 t \\ y(t) = -3\sin^2 t \end{cases}$

20. $\begin{cases} x(t) = \sec t \\ y(t) = \tan t \end{cases}$

21. $\begin{cases} x(t) = \sec t \\ y(t) = \tan^2 t \end{cases}$

22. $\begin{cases} x(t) = \frac{1}{e^{2t}} \\ y(t) = e^{-t} \end{cases}$

For the following exercises, graph the equation and include the orientation. Then, write the Cartesian equation.

23. $\begin{cases} x(t) = t - 1 \\ y(t) = -t^2 \end{cases}$

24. $\begin{cases} x(t) = t^3 \\ y(t) = t + 3 \end{cases}$

25. $\begin{cases} x(t) = 2\cos t \\ y(t) = -\sin t \end{cases}$

26. $\begin{cases} x(t) = 7\cos t \\ y(t) = 7\sin t \end{cases}$

27. $\begin{cases} x(t) = e^{2t} \\ y(t) = -e^{t} \end{cases}$

For the following exercises, graph the equation and include the orientation.

28. $x = t^2,\ y = 3t,\ 0 \le t \le 5$

29. $x = 2t,\ y = t^2,\ -5 \le t \le 5$

30. $x = t,\ y = \sqrt{25 - t^2},\ 0 < t \le 5$

31. $x(t) = -t,\ y(t) = \sqrt{t},\ t \ge 0$

32. $x = -2\cos t,\ y = 6\sin t,\ 0 \le t \le \pi$

33. $x = -\sec t,\ y = \tan t,\ -\frac{\pi}{2} < t < \frac{\pi}{2}$

For the following exercises, use the parametric equations for integers a and b:

$$x(t) = a\cos((a + b)t)$$
$$y(t) = a\cos((a - b)t)$$

34. Graph on the domain $[-\pi, 0]$, where $a = 2$ and $b = 1$, and include the orientation.

35. Graph on the domain $[-\pi, 0]$, where $a = 3$ and $b = 2$, and include the orientation.

36. Graph on the domain $[-\pi, 0]$, where $a = 4$ and $b = 3$, and include the orientation.

37. Graph on the domain $[-\pi, 0]$, where $a = 5$ and $b = 4$, and include the orientation.

38. If a is 1 more than b, describe the effect the values of a and b have on the graph of the parametric equations.

39. Describe the graph if $a = 100$ and $b = 99$.

40. What happens if b is 1 more than a? Describe the graph.

41. If the parametric equations $x(t) = t^2$ and $y(t) = 6 - 3t$ have the graph of a horizontal parabola opening to the right, what would change the direction of the curve?

For the following exercises, describe the graph of the set of parametric equations.

42. $x(t) = -t^2$ and $y(t)$ is linear

43. $y(t) = t^2$ and $x(t)$ is linear

44. $y(t) = -t^2$ and $x(t)$ is linear

45. Write the parametric equations of a circle with center $(0, 0)$, radius 5, and a counterclockwise orientation.

46. Write the parametric equations of an ellipse with center $(0, 0)$, major axis of length 10, minor axis of length 6, and a counterclockwise orientation.

For the following exercises, use a graphing utility to graph on the window $[-3, 3]$ by $[-3, 3]$ on the domain $[0, 2\pi)$ for the following values of a and b, and include the orientation.

$$\begin{cases} x(t) = \sin(at) \\ y(t) = \sin(bt) \end{cases}$$

47. $a = 1, b = 2$

48. $a = 2, b = 1$

49. $a = 3, b = 3$

50. $a = 5, b = 5$

51. $a = 2, b = 5$

52. $a = 5, b = 2$

Technology

For the following exercises, look at the graphs that were created by parametric equations of the form $\begin{cases} x(t) = a\cos(bt) \\ y(t) = c\sin(dt) \end{cases}$. *Use the parametric mode on the graphing calculator to find the values of a, b, c, and d to achieve each graph.*

53.

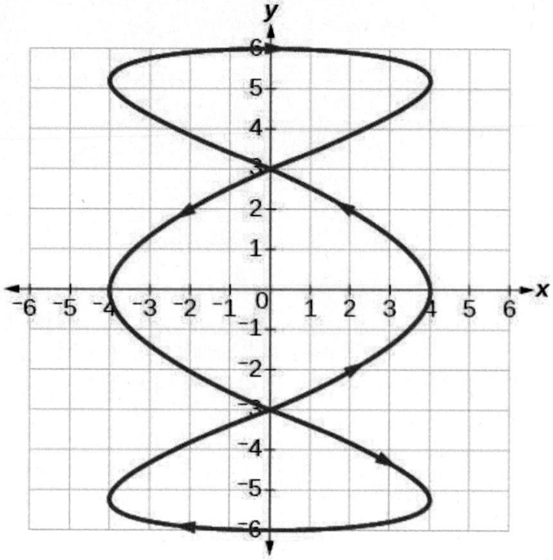

54.

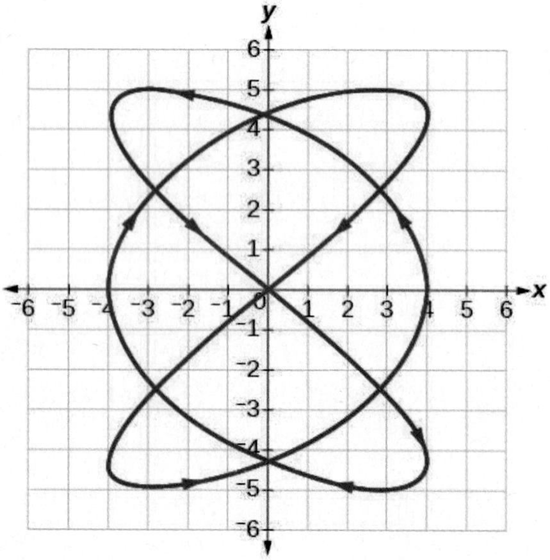

55.

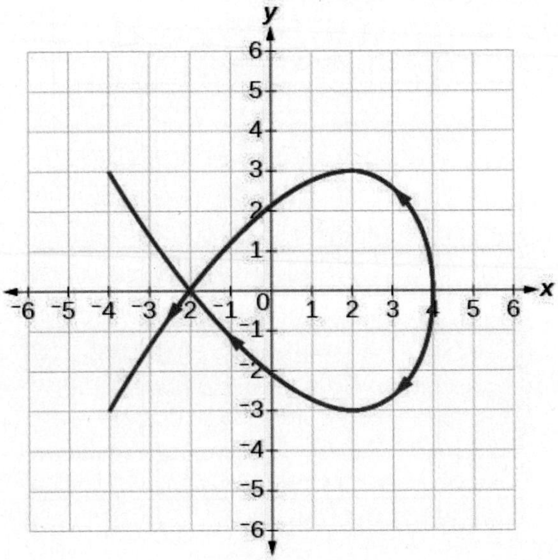

56.

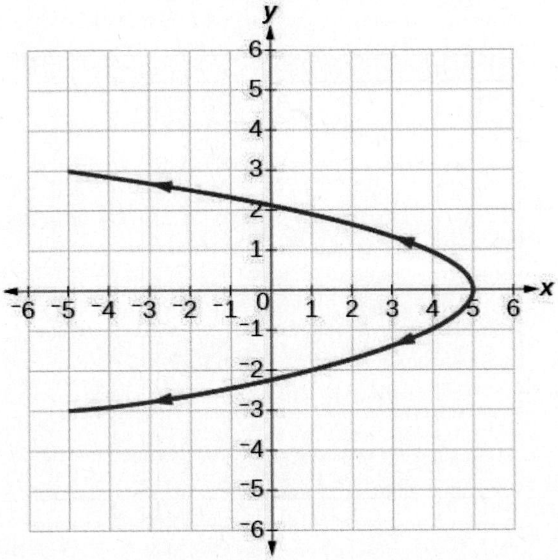

For the following exercises, use a graphing utility to graph the given parametric equations.

a. $\begin{cases} x(t) = \cos t - 1 \\ y(t) = \sin t + t \end{cases}$

b. $\begin{cases} x(t) = \cos t + t \\ y(t) = \sin t - 1 \end{cases}$

c. $\begin{cases} x(t) = t - \sin t \\ y(t) = \cos t - 1 \end{cases}$

57. Graph all three sets of parametric equations on the domain $[0, 2\pi]$.

58. Graph all three sets of parametric equations on the domain $[0, 4\pi]$.

59. Graph all three sets of parametric equations on the domain $[-4\pi, 6\pi]$.

60. The graph of each set of parametric equations appears to "creep" along one of the axes. What controls which axis the graph creeps along?

61. Explain the effect on the graph of the parametric equation when we switched $\sin\ t$ and $\cos\ t$.

62. Explain the effect on the graph of the parametric equation when we changed the domain.

Extensions

63. An object is thrown in the air with vertical velocity of 20 ft/s and horizontal velocity of 15 ft/s. The object's height can be described by the equation $y(t) = -16t^2 + 20t$, while the object moves horizontally with constant velocity 15 ft/s. Write parametric equations for the object's position, and then eliminate time to write height as a function of horizontal position.

64. A skateboarder riding on a level surface at a constant speed of 9 ft/s throws a ball in the air, the height of which can be described by the equation $y(t) = -16t^2 + 10t + 5$. Write parametric equations for the ball's position, and then eliminate time to write height as a function of horizontal position.

For the following exercises, use this scenario: A dart is thrown upward with an initial velocity of 65 ft/s at an angle of elevation of 52°. Consider the position of the dart at any time t. Neglect air resistance.

65. Find parametric equations that model the problem situation.

66. Find all possible values of x that represent the situation.

67. When will the dart hit the ground?

68. Find the maximum height of the dart.

69. At what time will the dart reach maximum height?

For the following exercises, look at the graphs of each of the four parametric equations. Although they look unusual and beautiful, they are so common that they have names, as indicated in each exercise. Use a graphing utility to graph each on the indicated domain.

70. An epicycloid: $\begin{cases} x(t) = 14\cos\ t - \cos(14t) \\ y(t) = 14\sin\ t + \sin(14t) \end{cases}$ on the domain $[0, 2\pi]$.

71. A hypocycloid: $\begin{cases} x(t) = 6\sin\ t + 2\sin(6t) \\ y(t) = 6\cos\ t - 2\cos(6t) \end{cases}$ on the domain $[0, 2\pi]$.

72. A hypotrochoid: $\begin{cases} x(t) = 2\sin\ t + 5\cos(6t) \\ y(t) = 5\cos\ t - 2\sin(6t) \end{cases}$ on the domain $[0, 2\pi]$.

73. A rose: $\begin{cases} x(t) = 5\sin(2t)\sin t \\ y(t) = 5\sin(2t)\cos t \end{cases}$ on the domain $[0, 2\pi]$.

10.8 Vectors

Learning Objectives

In this section you will:

- View vectors geometrically.
- Find magnitude and direction.
- Perform vector addition and scalar multiplication.
- Find the component form of a vector.
- Find the unit vector in the direction of $\mathbf{v}$.
- Perform operations with vectors in terms of $\mathbf{i}$ and $\mathbf{j}$.
- Find the dot product of two vectors.

An airplane is flying at an airspeed of 200 miles per hour headed on a SE bearing of 140°. A north wind (from north to south) is blowing at 16.2 miles per hour, as shown in Figure 1. What are the ground speed and actual bearing of the plane?

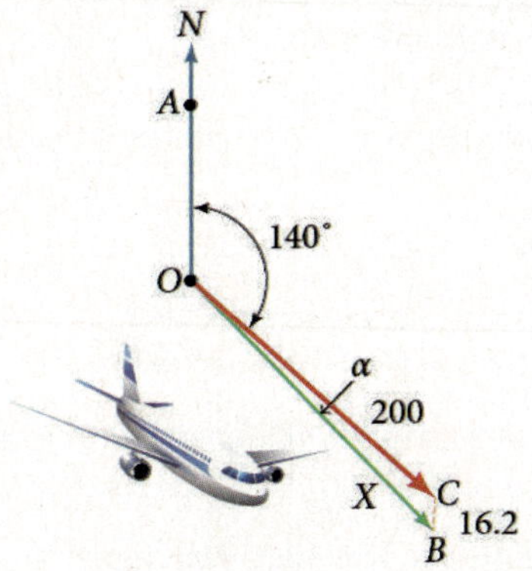

Figure 1

Ground speed refers to the speed of a plane relative to the ground. Airspeed refers to the speed a plane can travel relative to its surrounding air mass. These two quantities are not the same because of the effect of wind. In an earlier section, we used triangles to solve a similar problem involving the movement of boats. Later in this section, we will find the airplane's groundspeed and bearing, while investigating another approach to problems of this type. First, however, let's examine the basics of vectors.

A Geometric View of Vectors

A **vector** is a specific quantity drawn as a line segment with an arrowhead at one end. It has an **initial point**, where it begins, and a **terminal point**, where it ends. A vector is defined by its **magnitude**, or the length of the line, and its direction, indicated by an arrowhead at the terminal point. Thus, a vector is a directed line segment. There are various symbols that distinguish vectors from other quantities:

- Lower case, boldfaced type, with or without an arrow on top such as $\mathbf{v}, \mathbf{u}, w, \vec{\mathbf{v}}, \vec{\mathbf{u}}, \vec{w}$.
- Given initial point P and terminal point Q, a vector can be represented as $\overrightarrow{PQ}$. The arrowhead on top is what indicates that it is not just a line, but a directed line segment.
- Given an initial point of $(0, 0)$ and terminal point $(a, \mathbf{b})$, a vector may be represented as $\langle a, \mathbf{b}\rangle$.

This last symbol $\langle a, \mathbf{b}\rangle$ has special significance. It is called the **standard position**. The position vector has an initial point $(0, 0)$ and a terminal point $(a, \mathbf{b})$. To change any vector into the position vector, we think about the change in the *x*-coordinates and the change in the *y*-coordinates. Thus, if the initial point of a vector $\overrightarrow{CD}$ is $C(x_1, y_1)$ and the terminal point is $D(x_2, y_2)$, then the position vector is found by calculating

$$\begin{aligned}\overrightarrow{A\mathbf{b}} &= \langle x_2 - x_1, y_2 - y_1\rangle \\ &= \langle a, \mathbf{b}\rangle\end{aligned}$$

In Figure 2, we see the original vector $\overrightarrow{CD}$ and the position vector $\overrightarrow{A\mathbf{b}}$.

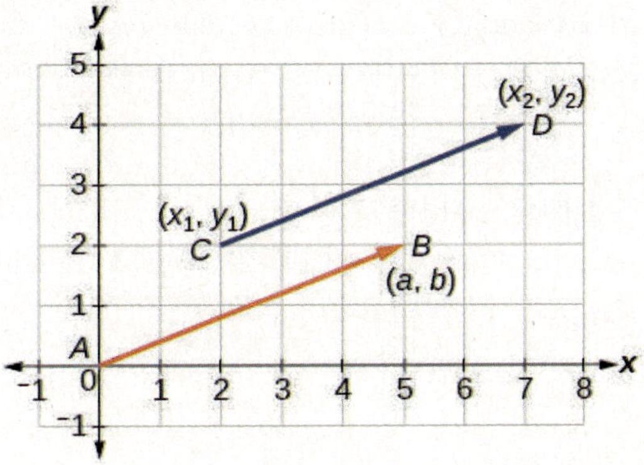

Figure 2

Properties of Vectors

A vector is a directed line segment with an initial point and a terminal point. Vectors are identified by magnitude, or the length of the line, and direction, represented by the arrowhead pointing toward the terminal point. The position vector has an initial point at $(0, 0)$ and is identified by its terminal point $(a, \mathbf{b})$.

EXAMPLE 1

Find the Position Vector

Consider the vector whose initial point is $P(2, 3)$ and terminal point is $Q(6, 4)$. Find the position vector.

Solution

The position vector is found by subtracting one *x*-coordinate from the other *x*-coordinate, and one *y*-coordinate from the other *y*-coordinate. Thus

$$\begin{aligned}\mathbf{v} &= \langle 6-2, 4-3\rangle \\ &= \langle 4, 1\rangle\end{aligned}$$

The position vector begins at $(0, 0)$ and terminates at $(4, 1)$. The graphs of both vectors are shown in Figure 3.

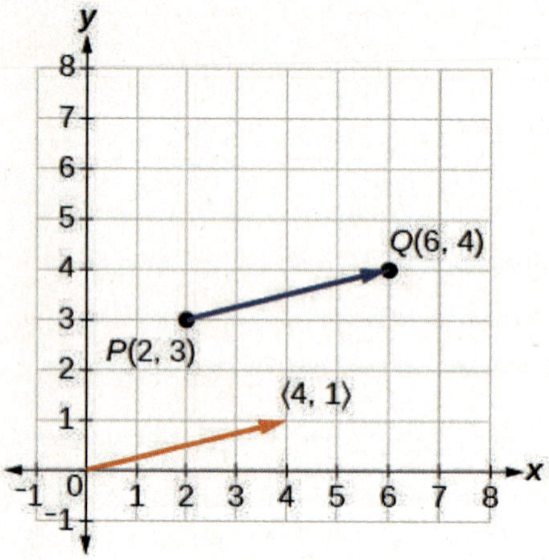

Figure 3

We see that the position vector is $\langle 4, 1\rangle$.

EXAMPLE 2

Drawing a Vector with the Given Criteria and Its Equivalent Position Vector

Find the position vector given that vector $\mathbf{v}$ has an initial point at $(-3, 2)$ and a terminal point at $(4, 5)$, then graph both vectors in the same plane.

Solution

The position vector is found using the following calculation:

$$\begin{aligned}\mathbf{v} &= \langle 4-(-3), 5-2\rangle \\ &= \langle 7, 3\rangle\end{aligned}$$

Thus, the position vector begins at $(0, 0)$ and terminates at $(7, 3)$. See Figure 4.

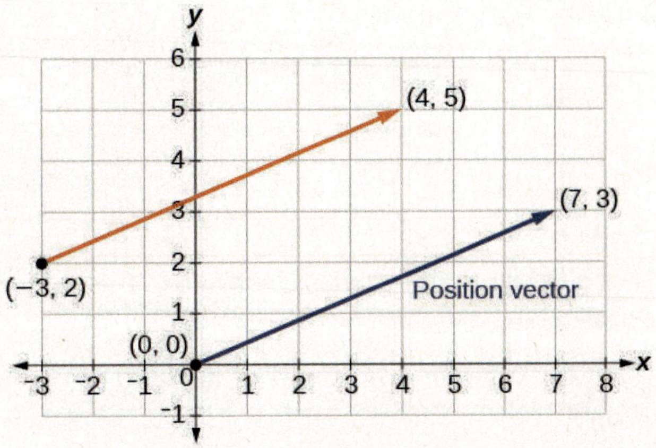

Figure 4

TRY IT #1 Draw a vector $\mathbf{v}$ that connects from the origin to the point $(3, 5)$.

Finding Magnitude and Direction

To work with a vector, we need to be able to find its magnitude and its direction. We find its magnitude using the Pythagorean Theorem or the distance formula, and we find its direction using the inverse tangent function.

Magnitude and Direction of a Vector

Given a position vector $\mathbf{v} = \langle a, \mathbf{b} \rangle$, the magnitude is found by $|\mathbf{v}| = \sqrt{a^2 + \mathbf{b}^2}$. The direction is equal to the angle formed with the *x*-axis, or with the *y*-axis, depending on the application. For a position vector, the direction is found by $\tan\,\theta = \left(\frac{\mathbf{b}}{a}\right) \Rightarrow \theta = \tan^{-1}\left(\frac{\mathbf{b}}{a}\right)$, as illustrated in Figure 5.

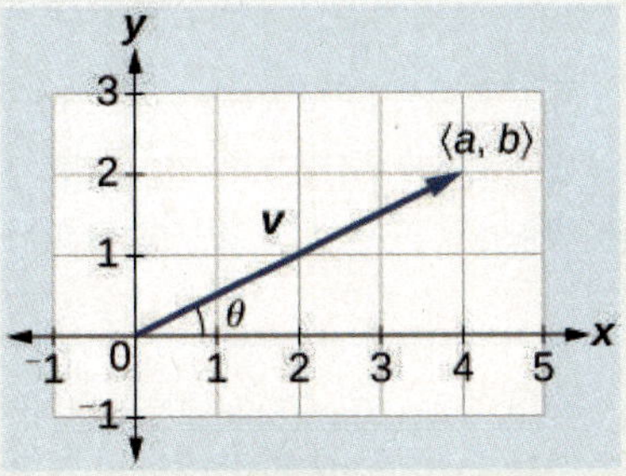

Figure 5

Two vectors ***v*** and ***u*** are considered equal if they have the same magnitude and the same direction. Additionally, if both vectors have the same position vector, they are equal.

EXAMPLE 3

Finding the Magnitude and Direction of a Vector

Find the magnitude and direction of the vector with initial point $P(-8, 1)$ and terminal point $Q(-2, -5)$. Draw the vector.

Solution

First, find the position vector.

$$\begin{aligned}\mathbf{u} &= \langle -2, -(-8), -5-1 \rangle \\ &= \langle 6, -6 \rangle\end{aligned}$$

We use the Pythagorean Theorem to find the magnitude.

$$\begin{aligned}|\mathbf{u}| &= \sqrt{(6)^2 + (-6)^2} \\ &= \sqrt{72} \\ &= 6\sqrt{2}\end{aligned}$$

The direction is given as

$$\begin{aligned}\tan\,\theta &= \tfrac{-6}{6} = -1 \Rightarrow \theta = \tan^{-1}(-1) \\ &= -45^\circ\end{aligned}$$

However, the angle terminates in the fourth quadrant, so we add 360° to obtain a positive angle. Thus, $-45^\circ + 360^\circ = 315^\circ$. See Figure 6.

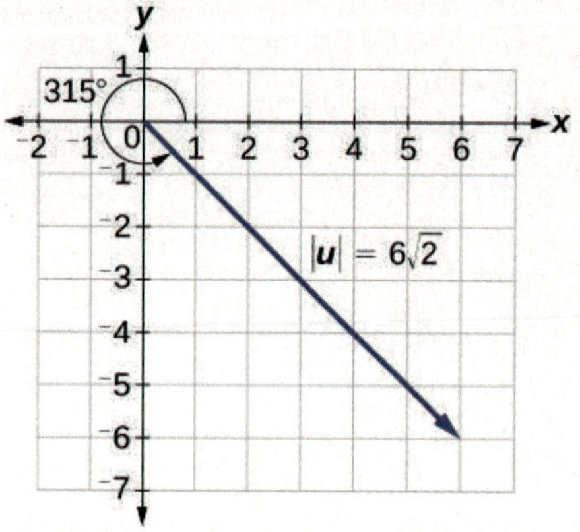

Figure 6

EXAMPLE 4

Showing That Two Vectors Are Equal

Show that vector ***v*** with initial point at $(5, -3)$ and terminal point at $(-1, 2)$ is equal to vector ***u*** with initial point at $(-1, -3)$ and terminal point at $(-7, 2)$. Draw the position vector on the same grid as ***v*** and ***u***. Next, find the magnitude and direction of each vector.

Solution

As shown in Figure 7, draw the vector $\mathbf{v}$ starting at initial $(5, -3)$ and terminal point $(-1, 2)$. Draw the vector $\mathbf{u}$ with initial point $(-1, -3)$ and terminal point $(-7, 2)$. Find the standard position for each.

Next, find and sketch the position vector for ***v*** and ***u***. We have

$$\begin{aligned}\mathbf{v} &= \langle -1 - 5, 2 - (-3)\rangle \\ &= \langle -6, 5\rangle\end{aligned}$$

$$\begin{aligned}\mathbf{u} &= \langle -7 - (-1), 2 - (-3)\rangle \\ &= \langle -6, 5\rangle\end{aligned}$$

Since the position vectors are the same, ***v*** and ***u*** are the same.

An alternative way to check for vector equality is to show that the magnitude and direction are the same for both vectors. To show that the magnitudes are equal, use the Pythagorean Theorem.

$$
\begin{aligned}
|\mathbf{v}| &= \sqrt{(-1-5)^2+(2-(-3))^2} \\
&= \sqrt{(-6)^2+(5)^2} \\
&= \sqrt{36+25} \\
&= \sqrt{61} \\
|\mathbf{u}| &= \sqrt{(-7-(-1))^2+(2-(-3))^2} \\
&= \sqrt{(-6)^2+(5)^2} \\
&= \sqrt{36+25} \\
&= \sqrt{61}
\end{aligned}
$$

As the magnitudes are equal, we now need to verify the direction. Using the tangent function with the position vector gives

$$
\begin{aligned}
\tan\theta &= -\tfrac{5}{6} \Rightarrow \theta = \tan^{-1}\left(-\tfrac{5}{6}\right) \\
&= -39.8^\circ
\end{aligned}
$$

However, we can see that the position vector terminates in the second quadrant, so we add 180°. Thus, the direction is $-39.8^\circ + 180^\circ = 140.2^\circ$.

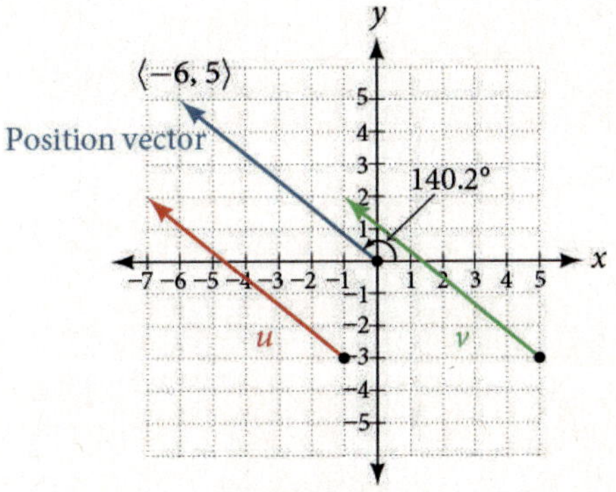

Figure 7

Performing Vector Addition and Scalar Multiplication

Now that we understand the properties of vectors, we can perform operations involving them. While it is convenient to think of the vector $\mathbf{u} = \langle x, y\rangle$ as an arrow or directed line segment from the origin to the point (x, y), vectors can be situated anywhere in the plane. The sum of two vectors ***u*** and ***v***, or **vector addition**, produces a third vector ***u***+ ***v***, the **resultant** vector.

To find ***u*** + ***v***, we first draw the vector ***u***, and from the terminal end of ***u***, we drawn the vector ***v***. In other words, we have the initial point of ***v*** meet the terminal end of ***u***. This position corresponds to the notion that we move along the first vector and then, from its terminal point, we move along the second vector. The sum ***u*** + ***v*** is the resultant vector because it results from addition or subtraction of two vectors. The resultant vector travels directly from the beginning of ***u*** to the end of ***v*** in a straight path, as shown in Figure 8.

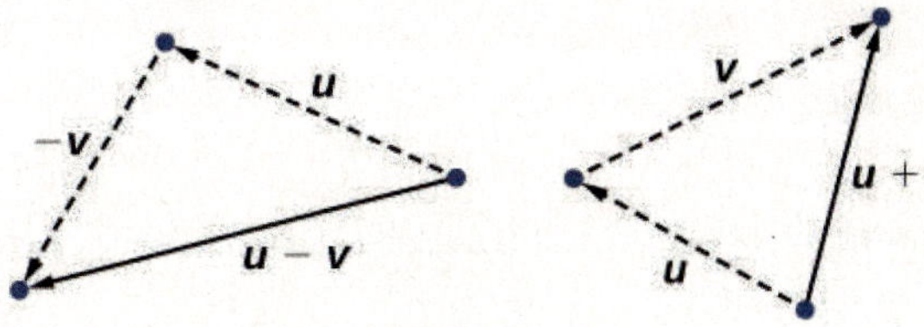

Figure 8

Vector subtraction is similar to vector addition. To find ***u*** − ***v***, view it as ***u*** + (−***v***). Adding −***v*** is reversing direction of ***v*** and adding it to the end of ***u***. The new vector begins at the start of ***u*** and stops at the end point of −***v***. See Figure 9 for a visual that compares vector addition and vector subtraction using parallelograms.

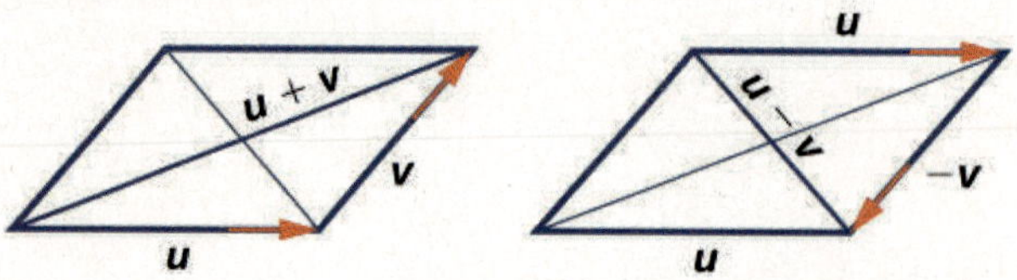

Figure 9

EXAMPLE 5

Adding and Subtracting Vectors

Given $\mathbf{u} = \langle 3, -2\rangle$ and $\mathbf{v} = \langle -1, 4\rangle$, find two new vectors ***u*** + ***v***, and ***u*** - ***v***.

Solution

To find the sum of two vectors, we add the components. Thus,

$$\begin{aligned}\mathbf{u} + \mathbf{v} &= \langle 3, -2\rangle + \langle -1, 4\rangle \\ &= \langle 3 + (-1), -2 + 4\rangle \\ &= \langle 2, 2\rangle\end{aligned}$$

See Figure 10**(a)**.

To find the difference of two vectors, add the negative components of $\mathbf{v}$ to $\mathbf{u}$. Thus,

$$\begin{aligned}\mathbf{u} + (-\mathbf{v}) &= \langle 3, -2\rangle + \langle 1, -4\rangle \\ &= \langle 3 + 1, -2 + (-4)\rangle \\ &= \langle 4, -6\rangle\end{aligned}$$

See Figure 10**(b)**.

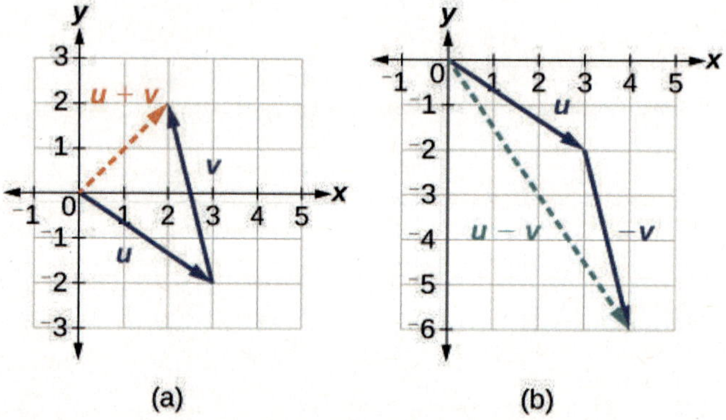

Figure 10 (a) Sum of two vectors (b) Difference of two vectors

Multiplying By a Scalar

While adding and subtracting vectors gives us a new vector with a different magnitude and direction, the process of multiplying a vector by a **scalar**, a constant, changes only the magnitude of the vector or the length of the line. Scalar multiplication has no effect on the direction unless the scalar is negative, in which case the direction of the resulting vector is opposite the direction of the original vector.

Scalar Multiplication

Scalar multiplication involves the product of a vector and a scalar. Each component of the vector is multiplied by the scalar. Thus, to multiply $\mathbf{v} = \langle a, \mathbf{b}\rangle$ by k, we have

$$k\mathbf{v} = \langle ka, k\mathbf{b}\rangle$$

Only the magnitude changes, unless k is negative, and then the vector reverses direction.

EXAMPLE 6

Performing Scalar Multiplication

Given vector $\mathbf{v} = \langle 3, 1 \rangle$, find 3***v***, $\frac{1}{2}\mathbf{v}$, and -***v***.

Solution

See Figure 11 for a geometric interpretation. If $\mathbf{v} = \langle 3, 1 \rangle$, then

$$\begin{aligned} 3\mathbf{v} &= \langle 3 \cdot 3, 3 \cdot 1 \rangle \\ &= \langle 9, 3 \rangle \\ \tfrac{1}{2}\mathbf{v} &= \left\langle \tfrac{1}{2} \cdot 3, \tfrac{1}{2} \cdot 1 \right\rangle \\ &= \left\langle \tfrac{3}{2}, \tfrac{1}{2} \right\rangle \\ -\mathbf{v} &= \langle -3, -1 \rangle \end{aligned}$$

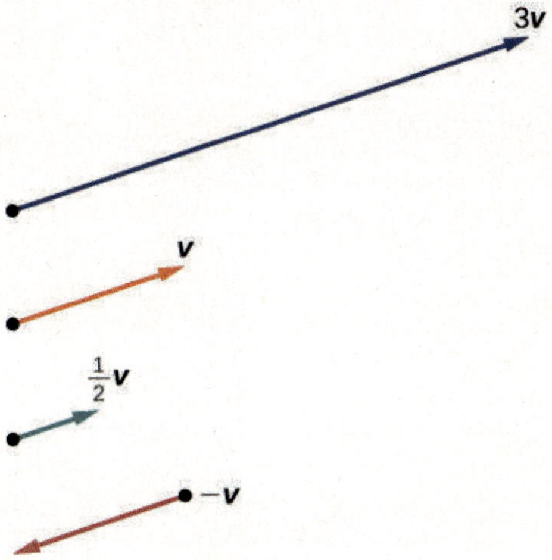

Figure 11

Analysis

Notice that the vector 3***v*** is three times the length of ***v***, $\frac{1}{2}\mathbf{v}$ is half the length of ***v***, and -***v*** is the same length of ***v***, but in the opposite direction.

TRY IT #2 Find the scalar multiple $3\,\mathbf{u}$ given $\mathbf{u} = \langle 5, 4 \rangle$.

EXAMPLE 7

Using Vector Addition and Scalar Multiplication to Find a New Vector

Given $\mathbf{u} = \langle 3, -2 \rangle$ and $\mathbf{v} = \langle -1, 4 \rangle$, find a new vector ***w*** = 3***u*** + 2***v***.

Solution

First, we must multiply each vector by the scalar.

$$\begin{aligned} 3\mathbf{u} &= 3\langle 3, -2 \rangle \\ &= \langle 9, -6 \rangle \\ 2\mathbf{v} &= 2\langle -1, 4 \rangle \\ &= \langle -2, 8 \rangle \end{aligned}$$

Then, add the two together.

$$\begin{aligned} w &= 3\mathbf{u} + 2\mathbf{v} \\ &= \langle 9, -6\rangle + \langle -2, 8\rangle \\ &= \langle 9 - 2, -6 + 8\rangle \\ &= \langle 7, 2\rangle \end{aligned}$$

So, $w = \langle 7, 2\rangle$.

Finding Component Form

In some applications involving vectors, it is helpful for us to be able to break a vector down into its components. Vectors are comprised of two components: the horizontal component is the x direction, and the vertical component is the y direction. For example, we can see in the graph in Figure 12 that the position vector $\langle 2, 3\rangle$ comes from adding the vectors $\mathbf{v}_1$ and $\mathbf{v}_2$. We have $\mathbf{v}_1$ with initial point $(0, 0)$ and terminal point $(2, 0)$.

$$\begin{aligned} \mathbf{v}_1 &= \langle 2 - 0, 0 - 0\rangle \\ &= \langle 2, 0\rangle \end{aligned}$$

We also have $\mathbf{v}_2$ with initial point $(0, 0)$ and terminal point $(0,\ 3)$.

$$\begin{aligned} \mathbf{v}_2 &= \langle 0 - 0, 3 - 0\rangle \\ &= \langle 0, 3\rangle \end{aligned}$$

Therefore, the position vector is

$$\begin{aligned} \mathbf{v} &= \langle 2 + 0, 3 + 0\rangle \\ &= \langle 2, 3\rangle \end{aligned}$$

Using the Pythagorean Theorem, the magnitude of $\mathbf{v}_1$ is 2, and the magnitude of $\mathbf{v}_2$ is 3. To find the magnitude of $\mathbf{v}$, use the formula with the position vector.

$$\begin{aligned} |\mathbf{v}| &= \sqrt{|\mathbf{v}_1|^2 + |\mathbf{v}_2|^2} \\ &= \sqrt{2^2 + 3^2} \\ &= \sqrt{13} \end{aligned}$$

The magnitude of $\mathbf{v}$ is $\sqrt{13}$. To find the direction, we use the tangent function $\tan\ \theta = \frac{y}{x}$.

$$\begin{aligned} \tan\ \theta &= \frac{\mathbf{v}_2}{\mathbf{v}_1} \\ \tan\ \theta &= \frac{3}{2} \\ \theta &= \tan^{-1}\left(\frac{3}{2}\right) = 56.3° \end{aligned}$$

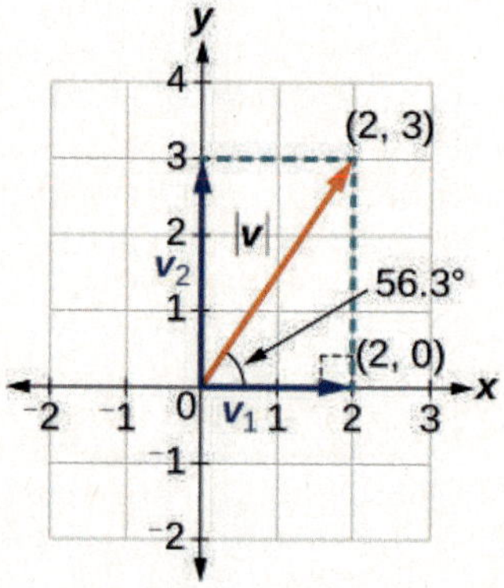

Figure 12

Thus, the magnitude of $\mathbf{v}$ is $\sqrt{13}$ and the direction is $56.3°$ off the horizontal.

EXAMPLE 8

Finding the Components of the Vector

Find the components of the vector $\mathbf{v}$ with initial point $(3, 2)$ and terminal point $(7, 4)$.

Solution

First find the standard position.

$$\begin{aligned}\mathbf{v} &= \langle 7-3, 4-2\rangle \\ &= \langle 4, 2\rangle\end{aligned}$$

See the illustration in Figure 13.

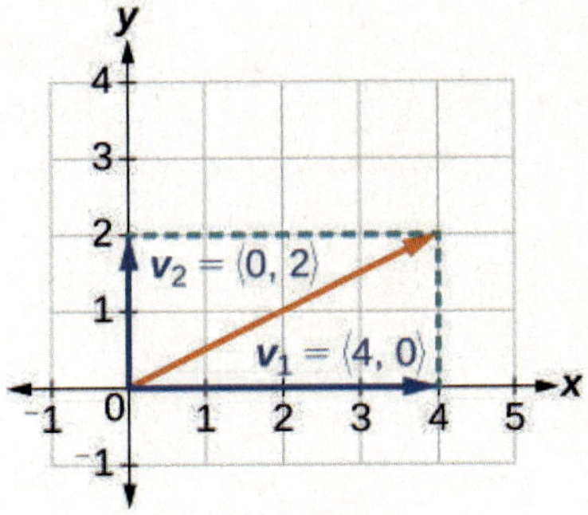

Figure 13

The horizontal component is $\mathbf{v}_1 = \langle 4, 0\rangle$ and the vertical component is $\mathbf{v}_2 = \langle 0, 2\rangle$.

Finding the Unit Vector in the Direction of *v*

In addition to finding a vector's components, it is also useful in solving problems to find a vector in the same direction as the given vector, but of magnitude 1. We call a vector with a magnitude of 1 a **unit vector**. We can then preserve the direction of the original vector while simplifying calculations.

Unit vectors are defined in terms of components. The horizontal unit vector is written as $\mathbf{i} = \langle 1, 0\rangle$ and is directed along the positive horizontal axis. The vertical unit vector is written as $\mathbf{j} = \langle 0, 1\rangle$ and is directed along the positive vertical axis. See Figure 14.

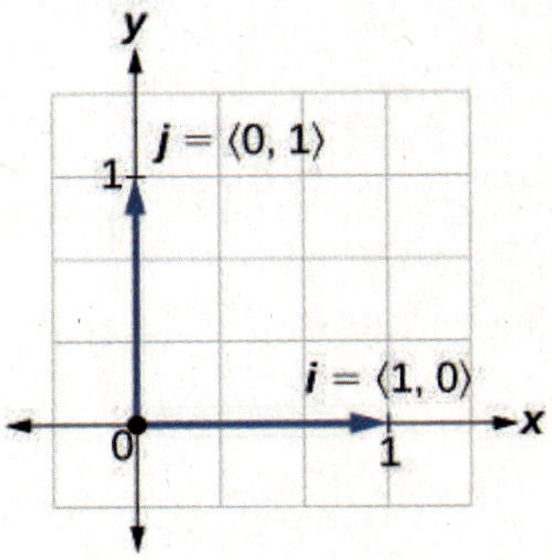

Figure 14

The Unit Vectors

If $\mathbf{v}$ is a nonzero vector, then $\frac{\mathbf{v}}{|\mathbf{v}|}$ is a unit vector in the direction of $\mathbf{v}$. Any vector divided by its magnitude is a unit vector. Notice that magnitude is always a scalar, and dividing by a scalar is the same as multiplying by the reciprocal of the scalar.

EXAMPLE 9

Finding the Unit Vector in the Direction of *v*

Find a unit vector in the same direction as $\mathbf{v} = \langle -5, 12 \rangle$.

Solution

First, we will find the magnitude.

$$\begin{aligned} |\mathbf{v}| &= \sqrt{(-5)^2 + (12)^2} \\ &= \sqrt{25 + 144} \\ &= \sqrt{169} \\ &= 13 \end{aligned}$$

Then we divide each component by $|\mathbf{v}\ |$, which gives a unit vector in the same direction as ***v***:

$$\frac{\mathbf{v}}{|\mathbf{v}|} = -\frac{5}{13}\mathbf{i} + \frac{12}{13}\mathbf{j}$$

or, in component form

$$\frac{\mathbf{v}}{|\mathbf{v}|} = \left\langle -\frac{5}{13}, \frac{12}{13} \right\rangle$$

See Figure 15.

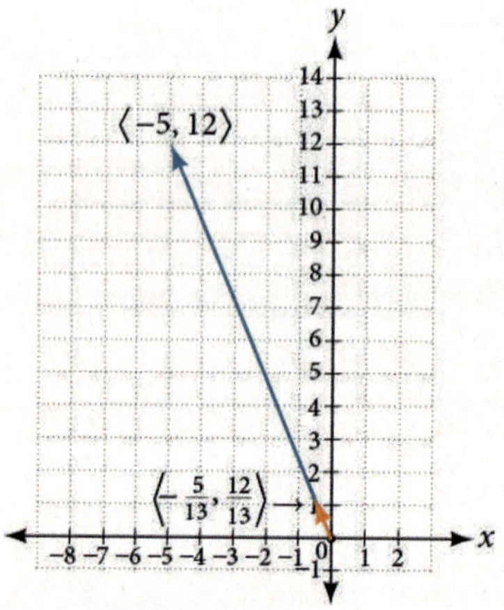

Figure 15

Verify that the magnitude of the unit vector equals 1. The magnitude of $-\frac{5}{13}\mathbf{i} + \frac{12}{13}\mathbf{j}$ is given as

$$\begin{aligned} \sqrt{\left(-\tfrac{5}{13}\right)^2 + \left(\tfrac{12}{13}\right)^2} &= \sqrt{\frac{25}{169} + \frac{144}{169}} \\ &= \sqrt{\frac{169}{169}} = 1 \end{aligned}$$

The vector $\boldsymbol{u} = \frac{5}{13}\ \boldsymbol{i} + \frac{12}{13}\ \boldsymbol{j}$ is the unit vector in the same direction as $\boldsymbol{v} = \langle -5, 12 \rangle$.

Performing Operations with Vectors in Terms of *i* and *j*

So far, we have investigated the basics of vectors: magnitude and direction, vector addition and subtraction, scalar multiplication, the components of vectors, and the representation of vectors geometrically. Now that we are familiar with the general strategies used in working with vectors, we will represent vectors in rectangular coordinates in terms of ***i*** and ***j***.

Vectors in the Rectangular Plane

Given a vector **v** with initial point $P = (x_1, y_1)$ and terminal point $Q = (x_2, y_2)$, **v** is written as

$$\mathbf{v} = (x_2 - x_1)\mathbf{i} + (y_2 - y_1)\mathbf{j}$$

The position vector from $(0, 0)$ to $(a, \mathbf{b})$, where $(x_2 - x_1) = a$ and $(y_2 - y_1) = \mathbf{b}$, is written as $\mathbf{v} = a\mathbf{i} + b\mathbf{j}$. This vector sum is called a linear combination of the vectors ***i*** and ***j***.

The magnitude of $\mathbf{v} = a\mathbf{i} + b\mathbf{j}$ is given as $|\mathbf{v}| = \sqrt{a^2 + \mathbf{b}^2}$. See Figure 16.

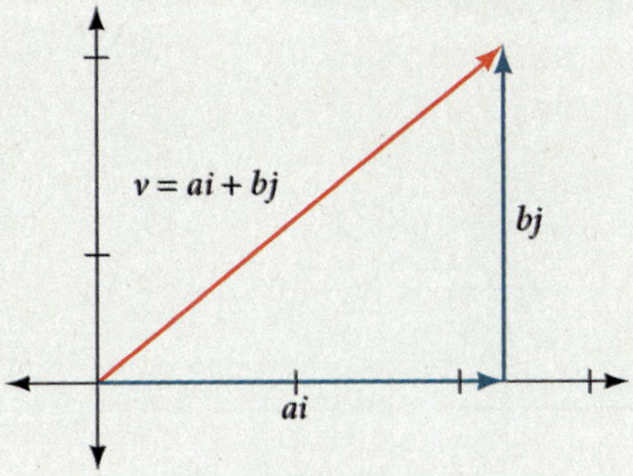

Figure 16

EXAMPLE 10

Writing a Vector in Terms of *i* and *j*

Given a vector **v** with initial point $P = (2, -6)$ and terminal point $Q = (-6, 6)$, write the vector in terms of **i** and **j**.

Solution

Begin by writing the general form of the vector. Then replace the coordinates with the given values.

$$\begin{aligned}\mathbf{v} &= (x_2 - x_1)\mathbf{i} + (y_2 - y_1)\mathbf{j}\\ &= (-6 - 2)\mathbf{i} + (6 - (-6))\mathbf{j}\\ &= -8\mathbf{i} + 12\mathbf{j}\end{aligned}$$

EXAMPLE 11

Writing a Vector in Terms of *i* and *j* Using Initial and Terminal Points

Given initial point $P_1 = (-1, 3)$ and terminal point $P_2 = (2, 7)$, write the vector **v** in terms of **i** and **j**.

Solution

Begin by writing the general form of the vector. Then replace the coordinates with the given values.

$$\begin{aligned}\mathbf{v} &= (x_2 - x_1)\mathbf{i} + (y_2 - y_1)\mathbf{j}\\ \mathbf{v} &= (2 - (-1))\mathbf{i} + (7 - 3)\mathbf{j}\\ &= 3\mathbf{i} + 4\mathbf{j}\end{aligned}$$

TRY IT #3 Write the vector **u** with initial point $P = (-1, 6)$ and terminal point $Q = (7, -5)$ in terms of **i** and **j**.

Performing Operations on Vectors in Terms of *i* and *j*

When vectors are written in terms of **i** and **j**, we can carry out addition, subtraction, and scalar multiplication by performing operations on corresponding components.

Adding and Subtracting Vectors in Rectangular Coordinates

Given $\mathbf{v} = a\mathbf{i} + b\mathbf{j}$ and $\mathbf{u} = c\mathbf{i} + d\mathbf{j}$, then

$$\begin{aligned}\mathbf{v} + \mathbf{u} &= (a + c)\mathbf{i} + (\mathbf{b} + d)\mathbf{j} \\ \mathbf{v} - \mathbf{u} &= (a - c)\mathbf{i} + (\mathbf{b} - d)\mathbf{j}\end{aligned}$$

EXAMPLE 12

Finding the Sum of the Vectors

Find the sum of $\mathbf{v}_1 = 2\mathbf{i} - 3\mathbf{j}$ and $\mathbf{v}_2 = 4\mathbf{i} + 5\mathbf{j}$.

Solution

According to the formula, we have

$$\begin{aligned}\mathbf{v}_1 + \mathbf{v}_2 &= (2 + 4)\mathbf{i} + (-3 + 5)\mathbf{j} \\ &= 6\mathbf{i} + 2\mathbf{j}\end{aligned}$$

Calculating the Component Form of a Vector: Direction

We have seen how to draw vectors according to their initial and terminal points and how to find the position vector. We have also examined notation for vectors drawn specifically in the Cartesian coordinate plane using $\mathbf{i}$ and $\mathbf{j}$. For any of these vectors, we can calculate the magnitude. Now, we want to combine the key points, and look further at the ideas of magnitude and direction.

Calculating direction follows the same straightforward process we used for polar coordinates. We find the direction of the vector by finding the angle to the horizontal. We do this by using the basic trigonometric identities, but with $|\mathbf{v}|$ replacing r.

Vector Components in Terms of Magnitude and Direction

Given a position vector $\mathbf{v} = \langle x, y \rangle$ and a direction angle θ,

$$\begin{aligned}\cos\ \theta &= \frac{x}{|\mathbf{v}|} \quad &\text{and} \quad \sin\ \theta &= \frac{y}{|\mathbf{v}|} \\ x &= |\mathbf{v}| \cos\ \theta & y &= |\mathbf{v}| \sin\ \theta\end{aligned}$$

Thus, $\mathbf{v} = x\mathbf{i} + y\mathbf{j} = |\mathbf{v}| \cos\ \theta\mathbf{i} + |\mathbf{v}| \sin\ \theta\mathbf{j}$, and magnitude is expressed as $|\mathbf{v}| = \sqrt{x^2 + y^2}$.

EXAMPLE 13

Writing a Vector in Terms of Magnitude and Direction

Write a vector with length 7 at an angle of 135° to the positive x-axis in terms of magnitude and direction.

Solution

Using the conversion formulas $x = |\mathbf{v}| \cos\ \theta$ and $y = |\mathbf{v}| \sin\ \theta$, we find that

$$\begin{aligned}x &= 7 \cos(135°) \\ &= -\frac{7\sqrt{2}}{2} \\ y &= 7 \sin(135°) \\ &= \frac{7\sqrt{2}}{2}\end{aligned}$$

This vector can be written as $\mathbf{v} = 7 \cos(135°) + 7 \sin(135°)$ or simplified as

$$\mathbf{v} = -\frac{7\sqrt{2}}{2} + \frac{7\sqrt{2}}{2}$$

TRY IT #4 A vector travels from the origin to the point $(3, 5)$. Write the vector in terms of magnitude and direction.

Finding the Dot Product of Two Vectors

As we discussed earlier in the section, scalar multiplication involves multiplying a vector by a scalar, and the result is a vector. As we have seen, multiplying a vector by a number is called scalar multiplication. If we multiply a vector by a vector, there are two possibilities: the *dot product* and the *cross product*. We will only examine the dot product here; you may encounter the cross product in more advanced mathematics courses.

The dot product of two vectors involves multiplying two vectors together, and the result is a scalar.

Dot Product

The **dot product** of two vectors $\mathbf{v} = \langle a, \mathbf{b} \rangle$ and $\mathbf{u} = \langle c, d \rangle$ is the sum of the product of the horizontal components and the product of the vertical components.

$$\mathbf{v} \cdot \mathbf{u} = ac + \mathbf{b}d$$

To find the angle between the two vectors, use the formula below.

$$\cos\ \theta = \frac{\mathbf{v}}{|\mathbf{v}|} \cdot \frac{\mathbf{u}}{|\mathbf{u}|}$$

EXAMPLE 14

Finding the Dot Product of Two Vectors

Find the dot product of $\mathbf{v} = \langle 5, 12 \rangle$ and $\mathbf{u} = \langle -3, 4 \rangle$.

Solution

Using the formula, we have

$$\begin{aligned} \mathbf{v} \cdot \mathbf{u} &= \langle 5, 12 \rangle \cdot \langle -3, 4 \rangle \\ &= 5 \cdot (-3) + 12 \cdot 4 \\ &= -15 + 48 \\ &= 33 \end{aligned}$$

EXAMPLE 15

Finding the Dot Product of Two Vectors and the Angle between Them

Find the dot product of $\boldsymbol{v}_1 = 5\boldsymbol{i} + 2\boldsymbol{j}$ and $\boldsymbol{v}_2 = 3\boldsymbol{i} + 7\boldsymbol{j}$. Then, find the angle between the two vectors.

Solution

Finding the dot product, we multiply corresponding components.

$$\begin{aligned} \mathbf{v}_1 \cdot \mathbf{v}_2 &= \langle 5, 2 \rangle \cdot \langle 3, 7 \rangle \\ &= 5 \cdot 3 + 2 \cdot 7 \\ &= 15 + 14 \\ &= 29 \end{aligned}$$

To find the angle between them, we use the formula $\cos\ \theta = \frac{\mathbf{v}}{|\mathbf{v}|} \cdot \frac{\mathbf{u}}{|\mathbf{u}|}$.

$$
\begin{aligned}
&\frac{\mathbf{v}}{|\mathbf{v}|}\cdot\frac{\mathbf{u}}{|\mathbf{u}|} = \left\langle \frac{5}{\sqrt{29}} + \frac{2}{\sqrt{29}} \right\rangle \cdot \left\langle \frac{3}{\sqrt{58}} + \frac{7}{\sqrt{58}} \right\rangle \\
&= \frac{5}{\sqrt{29}}\cdot\frac{3}{\sqrt{58}} + \frac{2}{\sqrt{29}}\cdot\frac{7}{\sqrt{58}} \\
&= \frac{15}{\sqrt{1682}} + \frac{14}{\sqrt{1682}} = \frac{29}{\sqrt{1682}} \\
&= 0.707107 \\
&\cos^{-1}(0.707107) = 45^\circ
\end{aligned}
$$

See Figure 17.

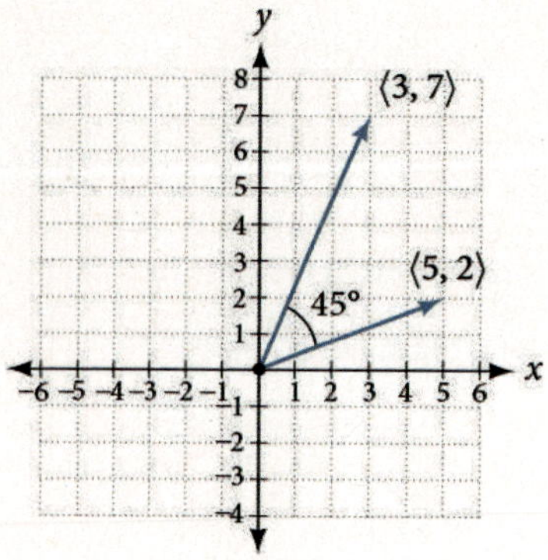

Figure 17

EXAMPLE 16

Finding the Angle between Two Vectors

Find the angle between $\mathbf{u} = \langle -3, 4\rangle$ and $\mathbf{v} = \langle 5, 12\rangle$.

Solution

Using the formula, we have

$$
\begin{aligned}
&\theta = \cos^{-1}\left(\frac{\mathbf{u}}{|\mathbf{u}|}\cdot\frac{\mathbf{v}}{|\mathbf{v}|}\right) \\
&\left(\frac{\mathbf{u}}{|\mathbf{u}|}\cdot\frac{\mathbf{v}}{|\mathbf{v}|}\right) = \frac{-3\mathbf{i}+4\mathbf{j}}{5}\cdot\frac{5\mathbf{i}+12\mathbf{j}}{13} \\
&= \left(-\frac{3}{5}\cdot\frac{5}{13}\right) + \left(\frac{4}{5}\cdot\frac{12}{13}\right) \\
&= -\frac{15}{65} + \frac{48}{65} \\
&= \frac{33}{65} \\
&\theta = \cos^{-1}\left(\frac{33}{65}\right) \\
&= 59.5^\circ
\end{aligned}
$$

See Figure 18.

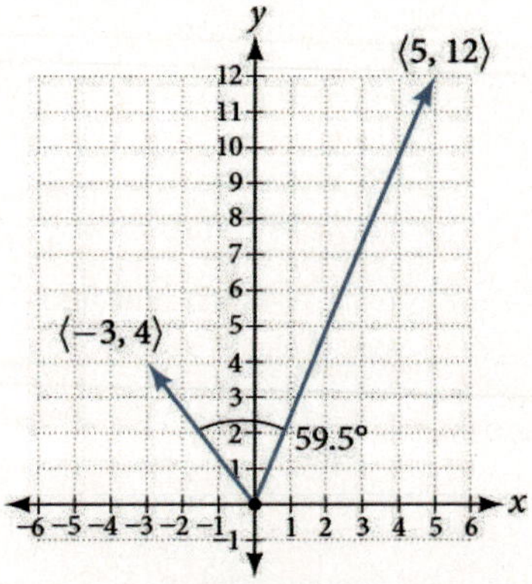

Figure 18

EXAMPLE 17

Finding Ground Speed and Bearing Using Vectors

We now have the tools to solve the problem we introduced in the opening of the section.

An airplane is flying at an airspeed of 200 miles per hour headed on a SE bearing of 140°. A north wind (from north to south) is blowing at 16.2 miles per hour. What are the ground speed and actual bearing of the plane? See Figure 19.

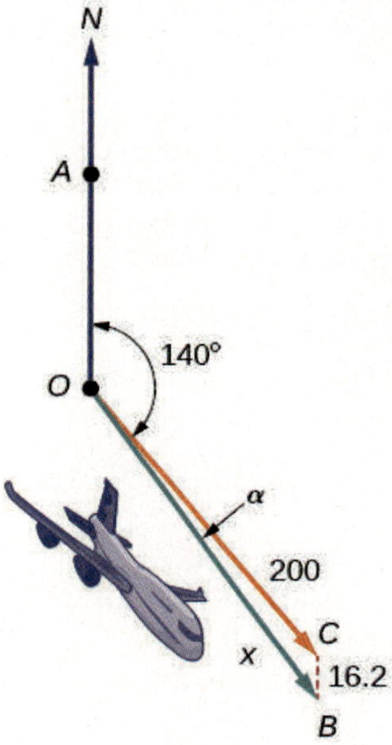

Figure 19

Solution

The ground speed is represented by x in the diagram, and we need to find the angle α in order to calculate the adjusted bearing, which will be $140^\circ + \alpha$.

Notice in Figure 19, that angle $\mathbf{b}CO$ must be equal to angle AOC by the rule of alternating interior angles, so angle $\mathbf{b}CO$ is 140°. We can find x by the Law of Cosines:

$$x^2 = (16.2)^2 + (200)^2 - 2(16.2)(200)\cos(140^\circ)$$
$$x^2 = 45,226.41$$
$$x = \sqrt{45,226.41}$$
$$x = 212.7$$

The ground speed is approximately 213 miles per hour. Now we can calculate the bearing using the Law of Sines.

$$\frac{\sin\ \alpha}{16.2} = \frac{\sin(140°)}{212.7}$$
$$\sin\ \alpha = \frac{16.2\ \sin(140°)}{212.7}$$
$$= 0.04896$$
$$\sin^{-1}(0.04896) = 2.8°$$

Therefore, the plane has a SE bearing of 140°+2.8°=142.8°. The ground speed is 212.7 miles per hour.

MEDIA

Access these online resources for additional instruction and practice with vectors.

Introduction to Vectors (http://openstax.org/l/introvectors)
Vector Operations (http://openstax.org/l/vectoroperation)
The Unit Vector (http://openstax.org/l/unitvector)

10.8 SECTION EXERCISES

Verbal

1. What are the characteristics of the letters that are commonly used to represent vectors?

2. How is a vector more specific than a line segment?

3. What are $\mathbf{i}$ and $\mathbf{j}$, and what do they represent?

4. What is component form?

5. When a unit vector is expressed as $\langle a, \mathbf{b} \rangle$, which letter is the coefficient of the $\mathbf{i}$ and which the $\mathbf{j}$?

Algebraic

6. Given a vector with initial point $(5, 2)$ and terminal point $(-1, -3)$, find an equivalent vector whose initial point is $(0, 0)$. Write the vector in component form $\langle a, \mathbf{b} \rangle$.

7. Given a vector with initial point $(-4, 2)$ and terminal point $(3, -3)$, find an equivalent vector whose initial point is $(0, 0)$. Write the vector in component form $\langle a, \mathbf{b} \rangle$.

8. Given a vector with initial point $(7, -1)$ and terminal point $(-1, -7)$, find an equivalent vector whose initial point is $(0, 0)$. Write the vector in component form $\langle a, \mathbf{b} \rangle$.

For the following exercises, determine whether the two vectors $\mathbf{u}$ *and* $\mathbf{v}$ *are equal, where* $\mathbf{u}$ *has an initial point* P_1 *and a terminal point* P_2 *and* $\mathbf{v}$ *has an initial point* P_3 *and a terminal point* P_4.

9. $P_1 = (5, 1)$, $P_2 = (3, -2)$, $P_3 = (-1, 3)$, and $P_4 = (9, -4)$

10. $P_1 = (2, -3)$, $P_2 = (5, 1)$, $P_3 = (6, -1)$, and $P_4 = (9, 3)$

11. $P_1 = (-1, -1)$, $P_2 = (-4, 5)$, $P_3 = (-10, 6)$, and $P_4 = (-13, 12)$

12. $P_1 = (3, 7)$, $P_2 = (2, 1)$, $P_3 = (1, 2)$, and $P_4 = (-1, -4)$

13. $P_1 = (8, 3)$, $P_2 = (6, 5)$, $P_3 = (11, 8)$, and $P_4 = (9, 10)$

14. Given initial point $P_1 = (-3, 1)$ and terminal point $P_2 = (5, 2)$, write the vector $\mathbf{v}$ in terms of $\mathbf{i}$ and $\mathbf{j}$.

15. Given initial point $P_1 = (6, 0)$ and terminal point $P_2 = (-1, -3)$, write the vector $\mathbf{v}$ in terms of $\mathbf{i}$ and $\mathbf{j}$.

For the following exercises, use the vectors $\boldsymbol{u} = \boldsymbol{i} + 5\boldsymbol{j}$, $\boldsymbol{v} = -2\boldsymbol{i} - 3\boldsymbol{j}$, *and* $\boldsymbol{w} = 4\boldsymbol{i} - \boldsymbol{j}$.

16. Find $\boldsymbol{u} + (\boldsymbol{v} - \boldsymbol{w})$

17. Find $4\boldsymbol{v} + 2\boldsymbol{u}$

For the following exercises, use the given vectors to compute $\mathbf{u} + \mathbf{v}$, $\mathbf{u} - \mathbf{v}$, *and* $2\mathbf{u} - 3\mathbf{v}$.

18. $\mathbf{u} = \langle 2, -3 \rangle, \mathbf{v} = \langle 1, 5 \rangle$

19. $\mathbf{u} = \langle -3, 4 \rangle, \mathbf{v} = \langle -2, 1 \rangle$

20. Let $\boldsymbol{v} = -4\boldsymbol{i} + 3\boldsymbol{j}$. Find a vector that is half the length and points in the same direction as $\mathbf{v}$.

21. Let $\boldsymbol{v} = 5\boldsymbol{i} + 2\boldsymbol{j}$. Find a vector that is twice the length and points in the opposite direction as $\mathbf{v}$.

For the following exercises, find a unit vector in the same direction as the given vector.

22. $\boldsymbol{a} = 3\boldsymbol{i} + 4\boldsymbol{j}$

23. $\boldsymbol{b} = -2\boldsymbol{i} + 5\boldsymbol{j}$

24. $\boldsymbol{c} = 10\boldsymbol{i} - \boldsymbol{j}$

25. $d = -\frac{1}{3}\mathbf{i} + \frac{5}{2}\mathbf{j}$

26. $\boldsymbol{u} = 100\boldsymbol{i} + 200\boldsymbol{j}$

27. $\boldsymbol{u} = -14\boldsymbol{i} + 2\boldsymbol{j}$

For the following exercises, find the magnitude and direction of the vector, $0 \le \theta < 2\pi$.

28. $\langle 0, 4 \rangle$

29. $\langle 6, 5 \rangle$

30. $\langle 2, -5 \rangle$

31. $\langle -4, -6 \rangle$

32. Given $\boldsymbol{u} = 3\boldsymbol{i} - 4\boldsymbol{j}$ and $\boldsymbol{v} = -2\boldsymbol{i} + 3\boldsymbol{j}$, calculate $\mathbf{u} \cdot \mathbf{v}$.

33. Given $\boldsymbol{u} = -\boldsymbol{i} - \boldsymbol{j}$ and $\boldsymbol{v} = \boldsymbol{i} + 5\boldsymbol{j}$, calculate $\mathbf{u} \cdot \mathbf{v}$.

34. Given $\mathbf{u} = \langle -2, 4 \rangle$ and $\mathbf{v} = \langle -3, 1 \rangle$, calculate $\mathbf{u} \cdot \mathbf{v}$.

35. Given $\boldsymbol{u} = \langle -1, 6 \rangle$ and $\boldsymbol{v} = \langle 6, -1 \rangle$, calculate $\mathbf{u} \cdot \mathbf{v}$.

Graphical

For the following exercises, given $\mathbf{v}$, *draw* $\mathbf{v}$, $3\boldsymbol{v}$ *and* $\frac{1}{2}\mathbf{v}$.

36. $\langle 2, -1 \rangle$

37. $\langle -1, 4 \rangle$

38. $\langle -3, -2 \rangle$

For the following exercises, use the vectors shown to sketch **u** + **v**, **u** − **v**, *and* 2**u**.

39.

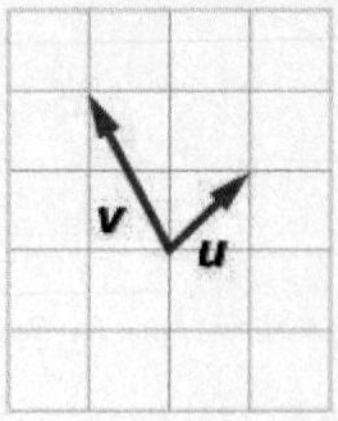

40.

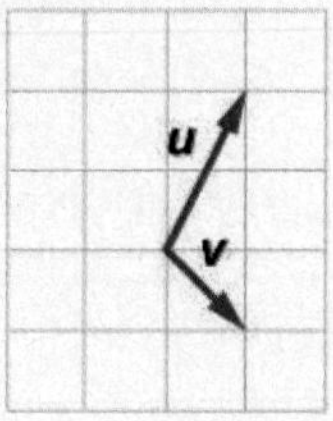

41.

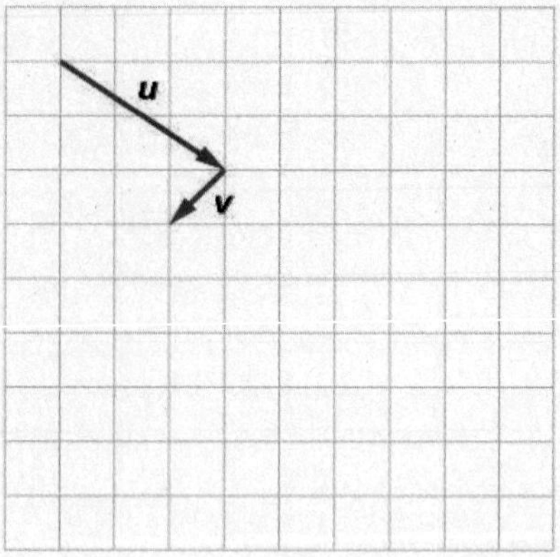

For the following exercises, use the vectors shown to sketch 2**u** + **v**.

42.

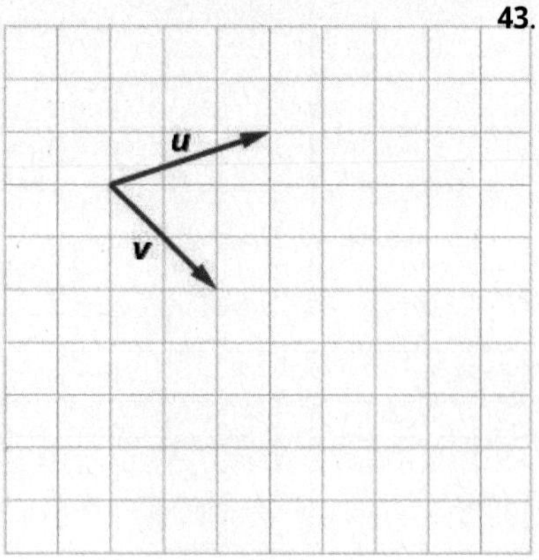

43.

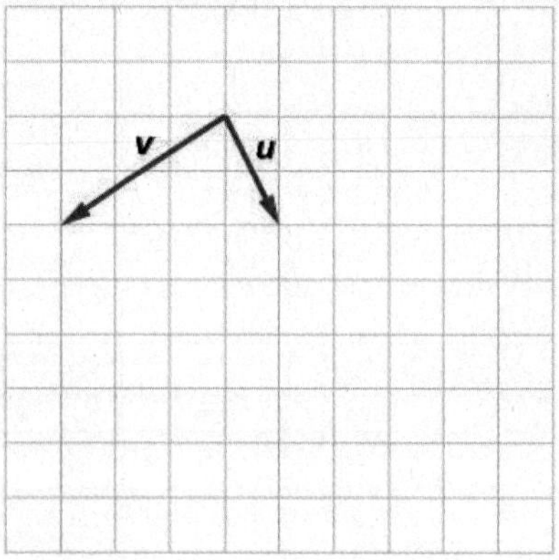

For the following exercises, use the vectors shown to sketch **u** − 3**v**.

44.

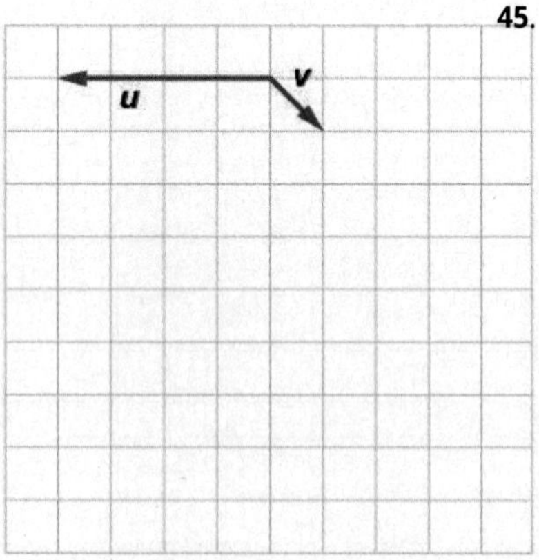

45.

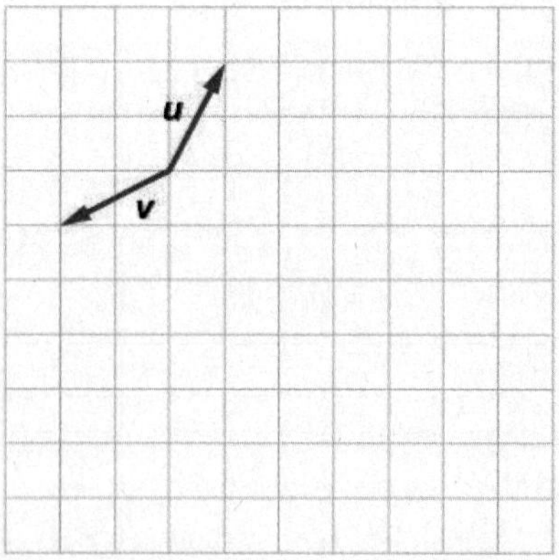

For the following exercises, write the vector shown in component form.

46.

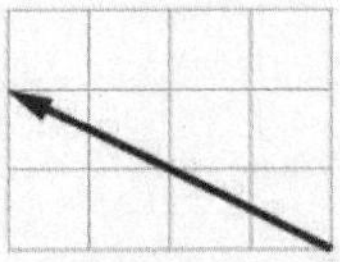

47.

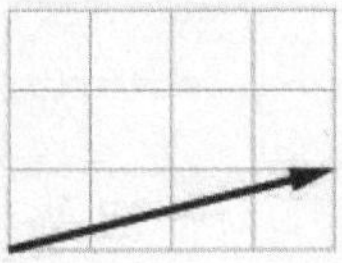

48. Given initial point $P_1 = (2, 1)$ and terminal point $P_2 = (-1, 2)$, write the vector $\mathbf{v}$ in terms of $\mathbf{i}$ and $\mathbf{j}$, then draw the vector on the graph.

49. Given initial point $P_1 = (4, -1)$ and terminal point $P_2 = (-3, 2)$, write the vector $\mathbf{v}$ in terms of $\mathbf{i}$ and $\mathbf{j}$. Draw the points and the vector on the graph.

50. Given initial point $P_1 = (3, 3)$ and terminal point $P_2 = (-3, 3)$, write the vector $\mathbf{v}$ in terms of $\mathbf{i}$ and $\mathbf{j}$. Draw the points and the vector on the graph.

Extensions

For the following exercises, use the given magnitude and direction in standard position, write the vector in component form.

51. $|\mathbf{v}| = 6, \theta = 45°$

52. $|\mathbf{v}| = 8, \theta = 220°$

53. $|\mathbf{v}| = 2, \theta = 300°$

54. $|\mathbf{v}| = 5, \theta = 135°$

55. A 60-pound box is resting on a ramp that is inclined 12°. Rounding to the nearest tenth,

ⓐ Find the magnitude of the normal (perpendicular) component of the force.
ⓑ Find the magnitude of the component of the force that is parallel to the ramp.

56. A 25-pound box is resting on a ramp that is inclined 8°. Rounding to the nearest tenth,

ⓐ Find the magnitude of the normal (perpendicular) component of the force.
ⓑ Find the magnitude of the component of the force that is parallel to the ramp.

57. Find the magnitude of the horizontal and vertical components of a vector with magnitude 8 pounds pointed in a direction of 27° above the horizontal. Round to the nearest hundredth.

58. Find the magnitude of the horizontal and vertical components of the vector with magnitude 4 pounds pointed in a direction of 127° above the horizontal. Round to the nearest hundredth.

59. Find the magnitude of the horizontal and vertical components of a vector with magnitude 5 pounds pointed in a direction of 55° above the horizontal. Round to the nearest hundredth.

60. Find the magnitude of the horizontal and vertical components of the vector with magnitude 1 pound pointed in a direction of 8° above the horizontal. Round to the nearest hundredth.

Real-World Applications

61. A woman leaves home and walks 3 miles west, then 2 miles southwest. How far from home is she, and in what direction must she walk to head directly home?

62. A boat leaves the marina and sails 6 miles north, then 2 miles northeast. How far from the marina is the boat, and in what direction must it sail to head directly back to the marina?

63. A man starts walking from home and walks 4 miles east, 2 miles southeast, 5 miles south, 4 miles southwest, and 2 miles east. How far has he walked? If he walked straight home, how far would he have to walk?

64. A woman starts walking from home and walks 4 miles east, 7 miles southeast, 6 miles south, 5 miles southwest, and 3 miles east. How far has she walked? If she walked straight home, how far would she have to walk?

65. A man starts walking from home and walks 3 miles at 20° north of west, then 5 miles at 10° west of south, then 4 miles at 15° north of east. If he walked straight home, how far would he have to the walk, and in what direction?

66. A woman starts walking from home and walks 6 miles at 40° north of east, then 2 miles at 15° east of south, then 5 miles at 30° south of west. If she walked straight home, how far would she have to walk, and in what direction?

67. An airplane is heading north at an airspeed of 600 km/hr, but there is a wind blowing from the southwest at 80 km/hr. How many degrees off course will the plane end up flying, and what is the plane's speed relative to the ground?

68. An airplane is heading north at an airspeed of 500 km/hr, but there is a wind blowing from the northwest at 50 km/hr. How many degrees off course will the plane end up flying, and what is the plane's speed relative to the ground?

69. An airplane needs to head due north, but there is a wind blowing from the southwest at 60 km/hr. The plane flies with an airspeed of 550 km/hr. To end up flying due north, how many degrees west of north will the pilot need to fly the plane?

70. An airplane needs to head due north, but there is a wind blowing from the northwest at 80 km/hr. The plane flies with an airspeed of 500 km/hr. To end up flying due north, how many degrees west of north will the pilot need to fly the plane?

71. As part of a video game, the point $(5, 7)$ is rotated counterclockwise about the origin through an angle of 35°. Find the new coordinates of this point.

72. As part of a video game, the point $(7, 3)$ is rotated counterclockwise about the origin through an angle of 40°. Find the new coordinates of this point.

73. Two children are throwing a ball back and forth straight across the back seat of a car. The ball is being thrown 10 mph relative to the car, and the car is traveling 25 mph down the road. If one child doesn't catch the ball, and it flies out the window, in what direction does the ball fly (ignoring wind resistance)?

74. Two children are throwing a ball back and forth straight across the back seat of a car. The ball is being thrown 8 mph relative to the car, and the car is traveling 45 mph down the road. If one child doesn't catch the ball, and it flies out the window, in what direction does the ball fly (ignoring wind resistance)?

75. A 50-pound object rests on a ramp that is inclined 19°. Find the magnitude of the components of the force parallel to and perpendicular to (normal) the ramp to the nearest tenth of a pound.

76. Suppose a body has a force of 10 pounds acting on it to the right, 25 pounds acting on it upward, and 5 pounds acting on it 45° from the horizontal. What single force is the resultant force acting on the body?

77. Suppose a body has a force of 10 pounds acting on it to the right, 25 pounds acting on it −135° from the horizontal, and 5 pounds acting on it directed 150° from the horizontal. What single force is the resultant force acting on the body?

78. The condition of equilibrium is when the sum of the forces acting on a body is the zero vector. Suppose a body has a force of 2 pounds acting on it to the right, 5 pounds acting on it upward, and 3 pounds acting on it 45° from the horizontal. What single force is needed to produce a state of equilibrium on the body?

79. Suppose a body has a force of 3 pounds acting on it to the left, 4 pounds acting on it upward, and 2 pounds acting on it 30° from the horizontal. What single force is needed to produce a state of equilibrium on the body? Draw the vector.

Chapter Review

Key Terms

altitude a perpendicular line from one vertex of a triangle to the opposite side, or in the case of an obtuse triangle, to the line containing the opposite side, forming two right triangles

ambiguous case a scenario in which more than one triangle is a valid solution for a given oblique SSA triangle

Archimedes' spiral a polar curve given by $r = \theta$. When multiplied by a constant, the equation appears as $r = a\theta$. As $r = \theta$, the curve continues to widen in a spiral path over the domain.

argument the angle associated with a complex number; the angle between the line from the origin to the point and the positive real axis

cardioid a member of the limaçon family of curves, named for its resemblance to a heart; its equation is given as $r = a \pm b\cos\ \theta$ and $r = a \pm b\sin\ \theta$, where $\frac{a}{b} = 1$

convex limaçon a type of one-loop limaçon represented by $r = a \pm b\cos\ \theta$ and $r = a \pm b\sin\ \theta$ such that $\frac{a}{b} \geq 2$

De Moivre's Theorem formula used to find the nth power or nth roots of a complex number; states that, for a positive integer n, z^n is found by raising the modulus to the nth power and multiplying the angles by n

dimpled limaçon a type of one-loop limaçon represented by $r = a \pm b\cos\ \theta$ and $r = a \pm b\sin\ \theta$ such that $1 < \frac{a}{b} < 2$

dot product given two vectors, the sum of the product of the horizontal components and the product of the vertical components

Generalized Pythagorean Theorem an extension of the Law of Cosines; relates the sides of an oblique triangle and is used for SAS and SSS triangles

initial point the origin of a vector

inner-loop limaçon a polar curve similar to the cardioid, but with an inner loop; passes through the pole twice; represented by $r = a \pm b\cos\ \theta$ and $r = a \pm b\ \sin\ \theta$ where $a < b$

Law of Cosines states that the square of any side of a triangle is equal to the sum of the squares of the other two sides minus twice the product of the other two sides and the cosine of the included angle

Law of Sines states that the ratio of the measurement of one angle of a triangle to the length of its opposite side is equal to the remaining two ratios of angle measure to opposite side; any pair of proportions may be used to solve for a missing angle or side

lemniscate a polar curve resembling a figure 8 and given by the equation $r^2 = a^2\cos\ 2\theta$ and $r^2 = a^2\sin\ 2\theta$, $a \neq 0$

magnitude the length of a vector; may represent a quantity such as speed, and is calculated using the Pythagorean Theorem

modulus the absolute value of a complex number, or the distance from the origin to the point (x, y); also called the amplitude

oblique triangle any triangle that is not a right triangle

one-loop limaçon a polar curve represented by $r = a \pm b\cos\ \theta$ and $r = a \pm b\sin\ \theta$ such that $a > 0, b > 0$, and $\frac{a}{b} > 1$; may be dimpled or convex; does not pass through the pole

parameter a variable, often representing time, upon which x and y are both dependent

polar axis on the polar grid, the equivalent of the positive x-axis on the rectangular grid

polar coordinates on the polar grid, the coordinates of a point labeled (r, θ), where θ indicates the angle of rotation from the polar axis and r represents the radius, or the distance of the point from the pole in the direction of θ

polar equation an equation describing a curve on the polar grid.

polar form of a complex number a complex number expressed in terms of an angle θ and its distance from the origin r; can be found by using conversion formulas $x = r\cos\ \theta,\quad y = r\sin\ \theta$, and $r = \sqrt{x^2 + y^2}$

pole the origin of the polar grid

resultant a vector that results from addition or subtraction of two vectors, or from scalar multiplication

rose curve a polar equation resembling a flower, given by the equations $r = a\cos\ n\theta$ and $r = a\sin\ n\theta$; when n is even there are $2n$ petals, and the curve is highly symmetrical; when n is odd there are n petals.

scalar a quantity associated with magnitude but not direction; a constant

scalar multiplication the product of a constant and each component of a vector

standard position the placement of a vector with the initial point at $(0, 0)$ and the terminal point $(a, \mathbf{b})$, represented by the change in the x-coordinates and the change in the y-coordinates of the original vector

terminal point the end point of a vector, usually represented by an arrow indicating its direction

unit vector a vector that begins at the origin and has magnitude of 1; the horizontal unit vector runs along the x-axis and is defined as $\mathbf{v}_1 = \langle 1, 0 \rangle$ the vertical unit vector runs along the y-axis and is defined as $\mathbf{v}_2 = \langle 0, 1 \rangle$.

vector a quantity associated with both magnitude and direction, represented as a directed line segment with a starting point (initial point) and an end point (terminal point)

vector addition the sum of two vectors, found by adding corresponding components

Key Equations

Law of Sines	$\frac{\sin\alpha}{a} = \frac{\sin\beta}{b} = \frac{\sin\gamma}{c}$ $\frac{a}{\sin\alpha} = \frac{b}{\sin\beta} = \frac{c}{\sin\gamma}$
Area for oblique triangles	$\text{Area} = \frac{1}{2}bc\sin\alpha$ $= \frac{1}{2}ac\sin\beta$ $= \frac{1}{2}ab\sin\gamma$
Law of Cosines	$a^2 = b^2 + c^2 - 2bc\cos\alpha$ $b^2 = a^2 + c^2 - 2ac\cos\beta$ $c^2 = a^2 + b^2 - 2ab\cos\gamma$
Heron's formula	$\text{Area} = \sqrt{s(s-a)(s-b)(s-c)}$ where $s = \frac{(a+b+c)}{2}$
Conversion formulas	$\cos\theta = \frac{x}{r} \rightarrow x = r\cos\theta$ $\sin\theta = \frac{y}{r} \rightarrow y = r\sin\theta$ $r^2 = x^2 + y^2$ $\tan\theta = \frac{y}{x}$

Key Concepts

10.1 Non-right Triangles: Law of Sines

- The Law of Sines can be used to solve oblique triangles, which are non-right triangles.
- According to the Law of Sines, the ratio of the measurement of one of the angles to the length of its opposite side equals the other two ratios of angle measure to opposite side.
- There are three possible cases: ASA, AAS, SSA. Depending on the information given, we can choose the appropriate equation to find the requested solution. See Example 1.
- The ambiguous case arises when an oblique triangle can have different outcomes.
- There are three possible cases that arise from SSA arrangement—a single solution, two possible solutions, and no solution. See Example 2 and Example 3.
- The Law of Sines can be used to solve triangles with given criteria. See Example 4.
- The general area formula for triangles translates to oblique triangles by first finding the appropriate height value. See Example 5.
- There are many trigonometric applications. They can often be solved by first drawing a diagram of the given information and then using the appropriate equation. See Example 6.

10.2 Non-right Triangles: Law of Cosines

- The Law of Cosines defines the relationship among angle measurements and lengths of sides in oblique triangles.
- The Generalized Pythagorean Theorem is the Law of Cosines for two cases of oblique triangles: SAS and SSS. Dropping an imaginary perpendicular splits the oblique triangle into two right triangles or forms one right triangle, which allows sides to be related and measurements to be calculated. See Example 1 and Example 2.
- The Law of Cosines is useful for many types of applied problems. The first step in solving such problems is generally to draw a sketch of the problem presented. If the information given fits one of the three models (the three equations), then apply the Law of Cosines to find a solution. See Example 3 and Example 4.
- Heron's formula allows the calculation of area in oblique triangles. All three sides must be known to apply Heron's formula. See Example 5 and See Example 6.

10.3 Polar Coordinates

- The polar grid is represented as a series of concentric circles radiating out from the pole, or origin.
- To plot a point in the form (r, θ), $\theta > 0$, move in a counterclockwise direction from the polar axis by an angle of θ, and then extend a directed line segment from the pole the length of r in the direction of θ. If θ is negative, move in a clockwise direction, and extend a directed line segment the length of r in the direction of θ. See Example 1.
- If r is negative, extend the directed line segment in the opposite direction of θ. See Example 2.
- To convert from polar coordinates to rectangular coordinates, use the formulas $x = r\cos\theta$ and $y = r\sin\theta$. See Example 3 and Example 4.
- To convert from rectangular coordinates to polar coordinates, use one or more of the formulas: $\cos\theta = \frac{x}{r}$, $\sin\theta = \frac{y}{r}$, $\tan\theta = \frac{y}{x}$, and $r = \sqrt{x^2 + y^2}$. See Example 5.
- Transforming equations between polar and rectangular forms means making the appropriate substitutions based on the available formulas, together with algebraic manipulations. See Example 6, Example 7, and Example 8.
- Using the appropriate substitutions makes it possible to rewrite a polar equation as a rectangular equation, and then graph it in the rectangular plane. See Example 9, Example 10, and Example 11.

10.4 Polar Coordinates: Graphs

- It is easier to graph polar equations if we can test the equations for symmetry with respect to the line $\theta = \frac{\pi}{2}$, the polar axis, or the pole.
- There are three symmetry tests that indicate whether the graph of a polar equation will exhibit symmetry. If an equation fails a symmetry test, the graph may or may not exhibit symmetry. See Example 1.
- Polar equations may be graphed by making a table of values for θ and r.
- The maximum value of a polar equation is found by substituting the value θ that leads to the maximum value of the trigonometric expression.
- The zeros of a polar equation are found by setting $r = 0$ and solving for θ. See Example 2.
- Some formulas that produce the graph of a circle in polar coordinates are given by $r = a\cos\theta$ and $r = a\sin\theta$. See Example 3.
- The formulas that produce the graphs of a cardioid are given by $r = a \pm b\cos\theta$ and $r = a \pm b\sin\theta$, for $a > 0$, $b > 0$, and $\frac{a}{b} = 1$. See Example 4.
- The formulas that produce the graphs of a one-loop limaçon are given by $r = a \pm b\cos\theta$ and $r = a \pm b\sin\theta$ for $1 < \frac{a}{b} < 2$. See Example 5.
- The formulas that produce the graphs of an inner-loop limaçon are given by $r = a \pm b\cos\theta$ and $r = a \pm b\sin\theta$ for $a > 0$, $b > 0$, and $a < b$. See Example 6.
- The formulas that produce the graphs of a lemniscates are given by $r^2 = a^2\cos 2\theta$ and $r^2 = a^2\sin 2\theta$, where $a \neq 0$. See Example 7.
- The formulas that produce the graphs of rose curves are given by $r = a\cos n\theta$ and $r = a\sin n\theta$, where $a \neq 0$; if n is even, there are $2n$ petals, and if n is odd, there are n petals. See Example 8 and Example 9.
- The formula that produces the graph of an Archimedes' spiral is given by $r = \theta$, $\theta \geq 0$. See Example 10.

10.5 Polar Form of Complex Numbers

- Complex numbers in the form $a + bi$ are plotted in the complex plane similar to the way rectangular coordinates are plotted in the rectangular plane. Label the x-axis as the *real* axis and the y-axis as the *imaginary* axis. See Example 1.
- The absolute value of a complex number is the same as its magnitude. It is the distance from the origin to the point: $|z| = \sqrt{a^2 + b^2}$. See Example 2 and Example 3.
- To write complex numbers in polar form, we use the formulas $x = r\cos\theta$, $y = r\sin\theta$, and $r = \sqrt{x^2 + y^2}$. Then, $z = r(\cos\theta + i\sin\theta)$. See Example 4 and Example 5.
- To convert from polar form to rectangular form, first evaluate the trigonometric functions. Then, multiply through by r. See Example 6 and Example 7.
- To find the product of two complex numbers, multiply the two moduli and add the two angles. Evaluate the trigonometric functions, and multiply using the distributive property. See Example 8.
- To find the quotient of two complex numbers in polar form, find the quotient of the two moduli and the difference of the two angles. See Example 9.
- To find the power of a complex number z^n, raise r to the power n, and multiply θ by n. See Example 10.
- Finding the roots of a complex number is the same as raising a complex number to a power, but using a rational exponent. See Example 11.

10.6 Parametric Equations

- Parameterizing a curve involves translating a rectangular equation in two variables, x and y, into two equations in three variables, x, y, and t. Often, more information is obtained from a set of parametric equations. See Example 1, Example 2, and Example 3.
- Sometimes equations are simpler to graph when written in rectangular form. By eliminating t, an equation in x and y is the result.
- To eliminate t, solve one of the equations for t, and substitute the expression into the second equation. See Example 4, Example 5, Example 6, and Example 7.
- Finding the rectangular equation for a curve defined parametrically is basically the same as eliminating the parameter. Solve for t in one of the equations, and substitute the expression into the second equation. See Example 8.
- There are an infinite number of ways to choose a set of parametric equations for a curve defined as a rectangular equation.
- Find an expression for x such that the domain of the set of parametric equations remains the same as the original rectangular equation. See Example 9.

10.7 Parametric Equations: Graphs

- When there is a third variable, a third parameter on which x and y depend, parametric equations can be used.
- To graph parametric equations by plotting points, make a table with three columns labeled t, $x(t)$, and $y(t)$. Choose values for t in increasing order. Plot the last two columns for x and y. See Example 1 and Example 2.
- When graphing a parametric curve by plotting points, note the associated t-values and show arrows on the graph indicating the orientation of the curve. See Example 3 and Example 4.
- Parametric equations allow the direction or the orientation of the curve to be shown on the graph. Equations that are not functions can be graphed and used in many applications involving motion. See Example 5.
- Projectile motion depends on two parametric equations: $x = (v_0 \cos \theta)t$ and $y = -16t^2 + (v_0 \sin \theta)t + h$. Initial velocity is symbolized as v_0. θ represents the initial angle of the object when thrown, and h represents the height at which the object is propelled.

10.8 Vectors

- The position vector has its initial point at the origin. See Example 1.
- If the position vector is the same for two vectors, they are equal. See Example 2.
- Vectors are defined by their magnitude and direction. See Example 3.
- If two vectors have the same magnitude and direction, they are equal. See Example 4.
- Vector addition and subtraction result in a new vector found by adding or subtracting corresponding elements. See Example 5.
- Scalar multiplication is multiplying a vector by a constant. Only the magnitude changes; the direction stays the same. See Example 6 and Example 7.
- Vectors are comprised of two components: the horizontal component along the positive x-axis, and the vertical component along the positive y-axis. See Example 8.
- The unit vector in the same direction of any nonzero vector is found by dividing the vector by its magnitude.
- The magnitude of a vector in the rectangular coordinate system is $|\mathbf{v}| = \sqrt{a^2 + \mathbf{b}^2}$. See Example 9.
- In the rectangular coordinate system, unit vectors may be represented in terms of $\mathbf{i}$ and $\mathbf{j}$ where $\mathbf{i}$ represents the horizontal component and $\mathbf{j}$ represents the vertical component. Then, $\boldsymbol{v} = a\boldsymbol{i} + b\boldsymbol{j}$ is a scalar multiple of $\mathbf{v}$ by real numbers a and $\mathbf{b}$. See Example 10 and Example 11.
- Adding and subtracting vectors in terms of i and j consists of adding or subtracting corresponding coefficients of i and corresponding coefficients of j. See Example 12.
- A vector $v = a\boldsymbol{i} + b\boldsymbol{j}$ is written in terms of magnitude and direction as $\mathbf{v} = |\mathbf{v}| \cos \theta\mathbf{i} + |\mathbf{v}| \sin \theta\mathbf{j}$. See Example 13.
- The dot product of two vectors is the product of the $\mathbf{i}$ terms plus the product of the $\mathbf{j}$ terms. See Example 14.
- We can use the dot product to find the angle between two vectors. Example 15 and Example 16.
- Dot products are useful for many types of physics applications. See Example 17.

Exercises

Review Exercises

Non-right Triangles: Law of Sines

For the following exercises, assume α is opposite side a, β is opposite side $\mathbf{b}$, and γ is opposite side c. Solve each triangle, if possible. Round each answer to the nearest tenth.

1. $\beta = 50°, a = 105, \mathbf{b} = 45$

2. $\alpha = 43.1°, a = 184.2, \mathbf{b} = 242.8$

3. Solve the triangle.

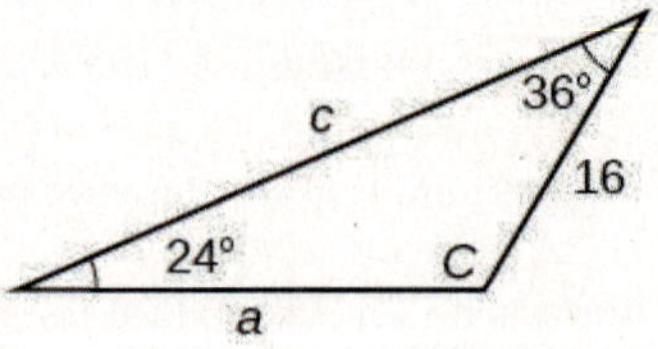

4. Find the area of the triangle.

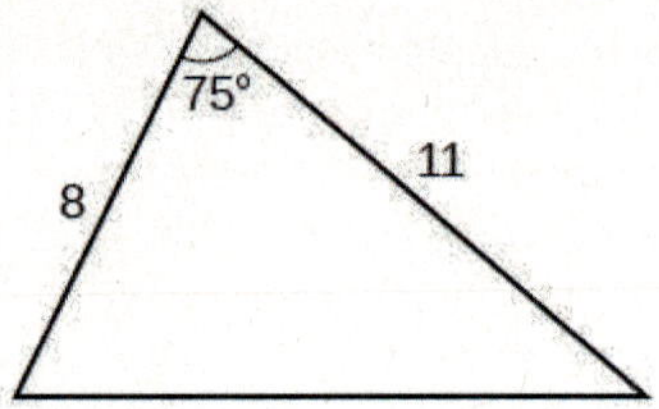

5. A pilot is flying over a straight highway. He determines the angles of depression to two mileposts, 2.1 km apart, to be 25° and 49°, as shown in Figure 1. Find the distance of the plane from point A and the elevation of the plane.

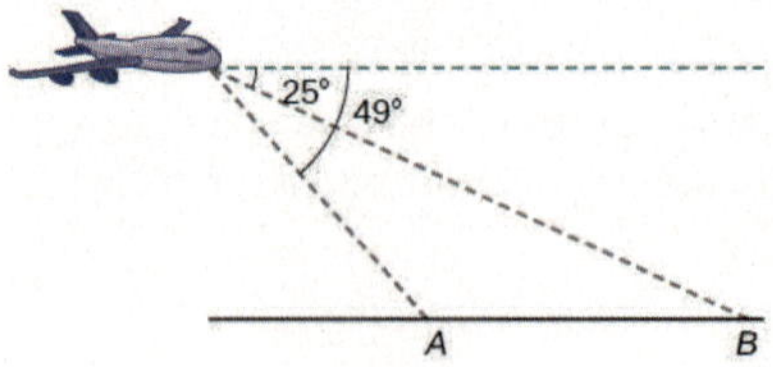

Figure 1

Non-right Triangles: Law of Cosines

6. Solve the triangle, rounding to the nearest tenth, assuming α is opposite side a, β is opposite side $\mathbf{b}$, and γ s opposite side c: $a = 4,\ \mathbf{b} = 6, c = 8$.

7. Solve the triangle in Figure 2, rounding to the nearest tenth.

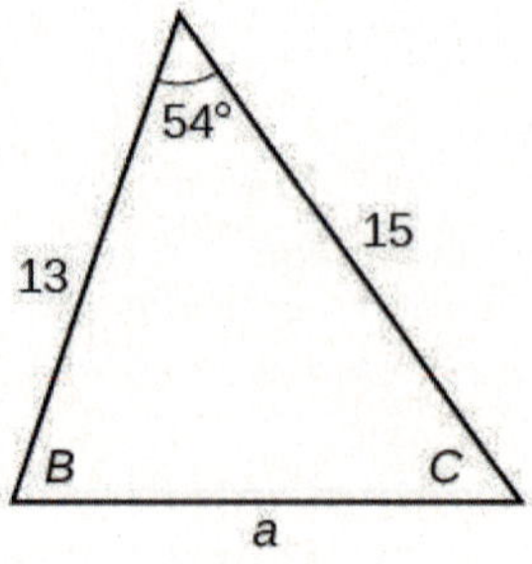

Figure 2

8. Find the area of a triangle with sides of length 8.3, 6.6, and 9.1.

9. To find the distance between two cities, a satellite calculates the distances and angle shown in Figure 3 (not to scale). Find the distance between the cities. Round answers to the nearest tenth.

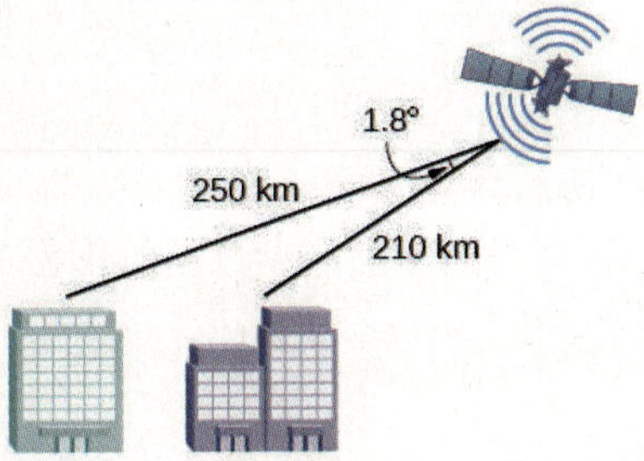

Figure 3

Polar Coordinates

10. Plot the point with polar coordinates $\left(3, \frac{\pi}{6}\right)$.

11. Plot the point with polar coordinates $\left(5, -\frac{2\pi}{3}\right)$

12. Convert $\left(6, -\frac{3\pi}{4}\right)$ to rectangular coordinates.

13. Convert $\left(-2, \frac{3\pi}{2}\right)$ to rectangular coordinates.

14. Convert $(7, -2)$ to polar coordinates.

15. Convert $(-9, -4)$ to polar coordinates.

For the following exercises, convert the given Cartesian equation to a polar equation.

16. $x = -2$

17. $x^2 + y^2 = 64$

18. $x^2 + y^2 = -2y$

For the following exercises, convert the given polar equation to a Cartesian equation.

19. $r = 7\cos\theta$

20. $r = \frac{-2}{4\cos\theta + \sin\theta}$

For the following exercises, convert to rectangular form and graph.

21. $\theta = \frac{3\pi}{4}$

22. $r = 5\sec\theta$

Polar Coordinates: Graphs

For the following exercises, test each equation for symmetry.

23. $r = 4 + 4\sin\theta$

24. $r = 7$

25. Sketch a graph of the polar equation $r = 1 - 5\sin\theta$. Label the axis intercepts.

26. Sketch a graph of the polar equation $r = 5\sin(7\theta)$.

27. Sketch a graph of the polar equation $r = 3 - 3\cos\theta$

Polar Form of Complex Numbers

For the following exercises, find the absolute value of each complex number.

28. $-2+6\mathbf{i}$

29. $4-\ \ 3\mathbf{i}$

Write the complex number in polar form.

30. $5+9\mathbf{i}$

31. $\frac{1}{2}-\frac{\sqrt{3}}{2}\ \mathbf{i}$

For the following exercises, convert the complex number from polar to rectangular form.

32. $z=5\operatorname{cis}\left(\frac{5\pi}{6}\right)$

33. $z=3\operatorname{cis}(40°)$

For the following exercises, find the product $z_1 z_2$ in polar form.

34. $z_1=2\operatorname{cis}(89°)$
$z_2=5\operatorname{cis}(23°)$

35. $z_1=10\operatorname{cis}\left(\frac{\pi}{6}\right)$
$z_2=6\operatorname{cis}\left(\frac{\pi}{3}\right)$

For the following exercises, find the quotient $\frac{z_1}{z_2}$ in polar form.

36. $z_1=12\operatorname{cis}(55°)$
$z_2=3\operatorname{cis}(18°)$

37. $z_1=27\operatorname{cis}\left(\frac{5\pi}{3}\right)$
$z_2=9\operatorname{cis}\left(\frac{\pi}{3}\right)$

For the following exercises, find the powers of each complex number in polar form.

38. Find z^4 when
$z=2\operatorname{cis}(70°)$

39. Find z^2 when
$z=5\operatorname{cis}\left(\frac{3\pi}{4}\right)$

For the following exercises, evaluate each root.

40. Evaluate the cube root of z when $z=64\operatorname{cis}(210°)$.

41. Evaluate the square root of z when $z=25\operatorname{cis}\left(\frac{3\pi}{2}\right)$.

For the following exercises, plot the complex number in the complex plane.

42. $6-2\mathbf{i}$

43. $-1+3\mathbf{i}$

Parametric Equations

For the following exercises, eliminate the parameter t to rewrite the parametric equation as a Cartesian equation.

44. $\begin{cases} x(t) = 3t - 1 \\ y(t) = \sqrt{t} \end{cases}$

45. $\begin{cases} x(t) = -\cos t \\ y(t) = 2\sin^2 t \end{cases}$

46. Parameterize (write a parametric equation for) each Cartesian equation by using $x(t) = a\cos t$ and $y(t) = \mathbf{b}\sin t$ for $\frac{x^2}{25} + \frac{y^2}{16} = 1.$

47. Parameterize the line from $(-2, 3)$ to $(4, 7)$ so that the line is at $(-2, 3)$ at $t = 0$ and $(4, 7)$ at $t = 1$.

Parametric Equations: Graphs

For the following exercises, make a table of values for each set of parametric equations, graph the equations, and include an orientation; then write the Cartesian equation.

48. $\begin{cases} x(t) = 3t^2 \\ y(t) = 2t - 1 \end{cases}$

49. $\begin{cases} x(t) = e^t \\ y(t) = -2e^{5t} \end{cases}$

50. $\begin{cases} x(t) = 3\cos t \\ y(t) = 2\sin t \end{cases}$

51. A ball is launched with an initial velocity of 80 feet per second at an angle of 40° to the horizontal. The ball is released at a height of 4 feet above the ground.

ⓐ Find the parametric equations to model the path of the ball.
ⓑ Where is the ball after 3 seconds?
ⓒ How long is the ball in the air?

Vectors

For the following exercises, determine whether the two vectors, $\mathbf{u}$ and $\mathbf{v}$, are equal, where $\mathbf{u}$ has an initial point P_1 and a terminal point P_2, and $\mathbf{v}$ has an initial point P_3 and a terminal point P_4.

52. $P_1 = (-1, 4), P_2 = (3, 1), P_3 = (5, 5)$ and $P_4 = (9, 2)$

53. $P_1 = (6, 11), P_2 = (-2, 8), P_3 = (0, -1)$ and $P_4 = (-8, 2)$

For the following exercises, use the vectors $\mathbf{u} = 2\mathbf{i} - \mathbf{j}, \mathbf{v} = 4\mathbf{i} - 3\mathbf{j}$, and $w = -2\mathbf{i} + 5\mathbf{j}$ to evaluate the expression.

54. $\boldsymbol{u} - \boldsymbol{v}$

55. $2\boldsymbol{v} - \boldsymbol{u} + \boldsymbol{w}$

For the following exercises, find a unit vector in the same direction as the given vector.

56. $\boldsymbol{a} = 8\boldsymbol{i} - 6\boldsymbol{j}$

57. $\boldsymbol{b} = -3\boldsymbol{i} - \boldsymbol{j}$

For the following exercises, find the magnitude and direction of the vector.

58. $\langle 6, -2 \rangle$

59. $\langle -3, -3 \rangle$

For the following exercises, calculate $\mathbf{u} \cdot \mathbf{v}$.

60. $\boldsymbol{u} = -2\boldsymbol{i} + \boldsymbol{j}$ and $\boldsymbol{v} = 3\boldsymbol{i} + 7\boldsymbol{j}$

61. $\boldsymbol{u} = \boldsymbol{i} + 4\boldsymbol{j}$ and $\boldsymbol{v} = 4\boldsymbol{i} + 3\boldsymbol{j}$

62. Given $\boldsymbol{v} = \langle -3, 4 \rangle$ draw $\boldsymbol{v}$, $2\boldsymbol{v}$, and $\frac{1}{2}\,\boldsymbol{v}$.

63. Given the vectors shown in Figure 4, sketch $\boldsymbol{u} + \boldsymbol{v}$, $\boldsymbol{u} - \boldsymbol{v}$ and $3\boldsymbol{v}$.

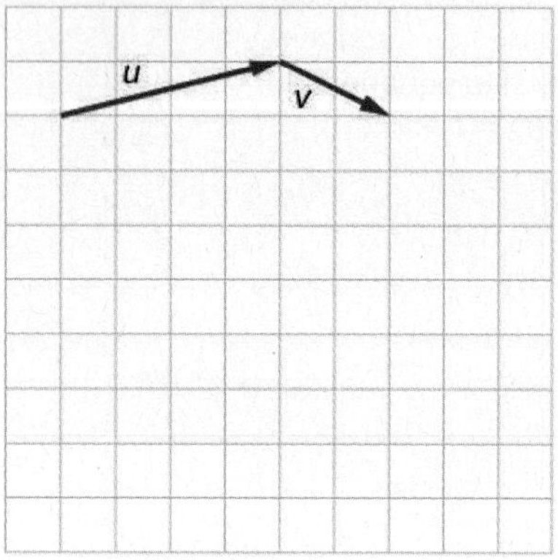

Figure 4

64. Given initial point $P_1 = (3, 2)$ and terminal point $P_2 = (-5, -1)$, write the vector $\mathbf{v}$ in terms of $\mathbf{i}$ and $\mathbf{j}$. Draw the points and the vector on the graph.

Practice Test

1. Assume α is opposite side a, β is opposite side $\mathbf{b}$, and γ is opposite side c. Solve the triangle, if possible, and round each answer to the nearest tenth, given $\beta = 68°, \mathbf{b} = 21, c = 16$.

2. Find the area of the triangle in Figure 1. Round each answer to the nearest tenth.

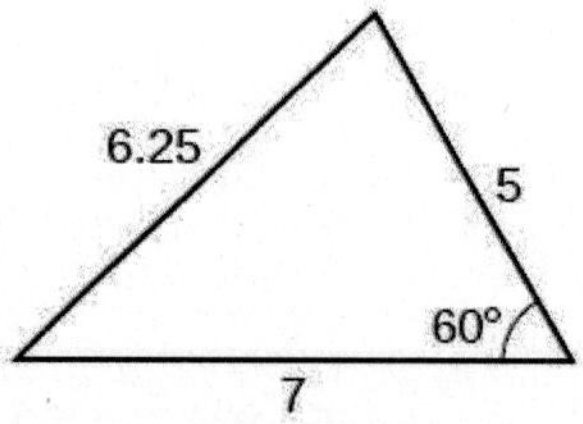

Figure 1

3. A pilot flies in a straight path for 2 hours. He then makes a course correction, heading 15° to the right of his original course, and flies 1 hour in the new direction. If he maintains a constant speed of 575 miles per hour, how far is he from his starting position?

4. Convert $(2, 2)$ to polar coordinates, and then plot the point.

5. Convert $\left(2, \frac{\pi}{3}\right)$ to rectangular coordinates.

6. Convert the polar equation to a Cartesian equation: $x^2 + y^2 = 5\text{y}$.

7. Convert to rectangular form and graph: $r = -3\csc\ \theta$.

8. Test the equation for symmetry: $r = -4\sin(2\theta\ )$.

9. Graph $r = 3 + 3\cos\ \theta$.

10. Graph $r = 3 - 5\sin\ \theta$.

11. Find the absolute value of the complex number $5 - 9\mathbf{i}$.

12. Write the complex number in polar form: $4 + \mathbf{i}$.

13. Convert the complex number from polar to rectangular form: $z = 5\text{cis}\left(\frac{2\pi}{3}\right)$.

Given $z_1 = 8\text{ cis}(36°)$ and $z_2 = 2\text{ cis}(15°)$, evaluate each expression.

14. $z_1 z_2$

15. $\frac{z_1}{z_2}$

16. $(z_2)^3$

17. $\sqrt{z_1}$

18. Plot the complex number $-5 - \mathbf{i}$ in the complex plane.

19. Eliminate the parameter t to rewrite the following parametric equations as a Cartesian equation: $\begin{cases} x(t) = t + 1 \\ y(t) = 2t^2 \end{cases}.$

20. Parameterize (write a parametric equation for) the following Cartesian equation by using $x(t) = a\cos\ t$ and $y(t) = \mathbf{b}\sin\ t$: $\frac{x^2}{36} + \frac{y^2}{100} = 1$.

21. Graph the set of parametric equations and find the Cartesian equation: $\begin{cases} x(t) = -2\sin\ t \\ y(t) = 5\cos\ t \end{cases}.$

22. A ball is launched with an initial velocity of 95 feet per second at an angle of 52° to the horizontal. The ball is released at a height of 3.5 feet above the ground.

ⓐ Find the parametric equations to model the path of the ball.
ⓑ Where is the ball after 2 seconds?
ⓒ How long is the ball in the air?

*For the following exercises, use the vectors **u** = **i** − 3**j** and **v** = 2**i** + 3**j**.*

23. Find 2***u*** − 3***v***.

24. Calculate $\mathbf{u} \cdot \mathbf{v}$.

25. Find a unit vector in the same direction as $\mathbf{v}$.

26. Given vector $\mathbf{v}$ has an initial point $P_1 = (2, 2)$ and terminal point $P_2 = (-1, 0)$, write the vector $\mathbf{v}$ in terms of $\mathbf{i}$ and $\mathbf{j}$. On the graph, draw $\mathbf{v}$, and $-\mathbf{v}$.

11 SYSTEMS OF EQUATIONS AND INEQUALITIES

Enigma machines like this one were used by government and military officials for enciphering and deciphering top-secret communications during World War II. By varying the combinations of the plugboard and the settings of the rotors, encoders could add complex encryption to their messages. Notice that the three rotors each contain 26 pins, one for each letter of the alphabet; later versions had four and five rotors. (credit: modification of "Enigma Machine" by School of Mathematics, University of Manchester/flickr)

Chapter Outline

Introduction to Systems of Equations and Inequalities

At the start of the Second World War, British military and intelligence officers recognized that defeating Nazi Germany would require the Allies to know what the enemy was planning. This task was complicated by the fact that the German military transmitted all of its communications through a presumably uncrackable code created by a machine called Enigma. The Germans had been encoding their messages with this machine since the early 1930s, and were so confident in its security that they used it for everyday military communications as well as highly important strategic messages. Concerned about the increasing military threat, other European nations began working to decipher the Enigma codes. Poland was the first country to make significant advances when it trained and recruited a new group of codebreakers: math students from Poznań University. With the help of intelligence obtained by French spies, Polish mathematicians, led by Marian Rejewski, were able to decipher initial codes and later to understand the wiring of the machines; eventually they create replicas. However, the German military eventually increased the complexity of the machines by adding additional rotors, requiring a new method of decryption.

The machine attached letters on a keyboard to three, four, or five rotors (depending on the version), each with 26 starting positions that could be set prior to encoding; a decryption code (called a cipher key) essentially conveyed these settings to the message recipient, and allowed people to interpret the message using another Enigma machine. Even with the simpler three-rotor scrambler, there were 17,576 different combinations of starting positions (26 x 26 x 26); plus the machine had numerous other methods of introducing variation. Not long after the war started, the British recruited a team of brilliant codebreakers to crack the Enigma code. The codebreakers, led by Alan Turing, used what they knew about the Enigma machine to build a mechanical computer that could crack the code. And that knowledge of what the Germans were planning proved to be a key part of the ultimate Allied victory of Nazi Germany in 1945.

The Enigma is perhaps the most famous cryptographic device ever known. It stands as an example of the pivotal role cryptography has played in society. Now, technology has moved cryptanalysis to the digital world.

Many ciphers are designed using invertible matrices as the method of message transference, as finding the inverse of a matrix is generally part of the process of decoding. In addition to knowing the matrix and its inverse, the receiver must also know the key that, when used with the matrix inverse, will allow the message to be read.

In this chapter, we will investigate matrices and their inverses, and various ways to use matrices to solve systems of equations. First, however, we will study systems of equations on their own: linear and nonlinear, and then partial fractions. We will not be breaking any secret codes here, but we will lay the foundation for future courses.

11.1 Systems of Linear Equations: Two Variables

Learning Objectives

In this section, you will:

- Solve systems of equations by graphing.
- Solve systems of equations by substitution.
- Solve systems of equations by addition.
- Identify inconsistent systems of equations containing two variables.
- Express the solution of a system of dependent equations containing two variables.

Figure 1 (credit: Thomas Sørenes)

A skateboard manufacturer introduces a new line of boards. The manufacturer tracks its costs, which is the amount it spends to produce the boards, and its revenue, which is the amount it earns through sales of its boards. How can the company determine if it is making a profit with its new line? How many skateboards must be produced and sold before a profit is possible? In this section, we will consider linear equations with two variables to answer these and similar questions.

Introduction to Systems of Equations

In order to investigate situations such as that of the skateboard manufacturer, we need to recognize that we are dealing with more than one variable and likely more than one equation. A **system of linear equations** consists of two or more linear equations made up of two or more variables such that all equations in the system are considered simultaneously. To find the unique solution to a system of linear equations, we must find a numerical value for each variable in the system that will satisfy all equations in the system at the same time. Some linear systems may not have a solution and others may have an infinite number of solutions. In order for a linear system to have a unique solution, there must be at least as many equations as there are variables. Even so, this does not guarantee a unique solution.

In this section, we will look at systems of linear equations in two variables, which consist of two equations that contain two different variables. For example, consider the following system of linear equations in two variables.

$$\begin{aligned} 2x + y &= 15 \\ 3x - y &= 5 \end{aligned}$$

The *solution* to a system of linear equations in two variables is any ordered pair that satisfies each equation independently. In this example, the ordered pair (4, 7) is the solution to the system of linear equations. We can verify the solution by substituting the values into each equation to see if the ordered pair satisfies both equations. Shortly we will investigate methods of finding such a solution if it exists.

$$2(4) + (7) = 15 \text{ True}$$
$$3(4) - (7) = 5 \text{ True}$$

In addition to considering the number of equations and variables, we can categorize systems of linear equations by the number of solutions. A **consistent system** of equations has at least one solution. A consistent system is considered to be an **independent system** if it has a single solution, such as the example we just explored. The two lines have different slopes and intersect at one point in the plane. A consistent system is considered to be a **dependent system** if the equations have the same slope and the same *y*-intercepts. In other words, the lines coincide so the equations represent the same line. Every point on the line represents a coordinate pair that satisfies the system. Thus, there are an infinite number of solutions.

Another type of system of linear equations is an **inconsistent system**, which is one in which the equations represent two parallel lines. The lines have the same slope and different *y*-intercepts. There are no points common to both lines; hence, there is no solution to the system.

Types of Linear Systems

There are three types of systems of linear equations in two variables, and three types of solutions.

- An **independent system** has exactly one solution pair (x, y). The point where the two lines intersect is the only solution.
- An **inconsistent system** has no solution. Notice that the two lines are parallel and will never intersect.
- A **dependent system** has infinitely many solutions. The lines are coincident. They are the same line, so every coordinate pair on the line is a solution to both equations.

Figure 2 compares graphical representations of each type of system.

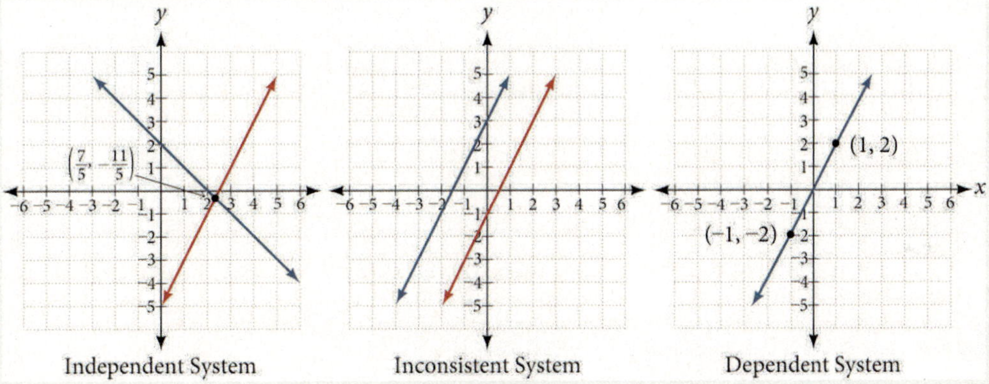

Figure 2

HOW TO

Given a system of linear equations and an ordered pair, determine whether the ordered pair is a solution.

1. Substitute the ordered pair into each equation in the system.
2. Determine whether true statements result from the substitution in both equations; if so, the ordered pair is a solution.

EXAMPLE 1

Determining Whether an Ordered Pair Is a Solution to a System of Equations

Determine whether the ordered pair $(5, 1)$ is a solution to the given system of equations.

$$x + 3y = 8$$
$$2x - 9 = y$$

Solution

Substitute the ordered pair $(5, 1)$ into both equations.

$$(5) + 3(1) = 8$$
$$8 = 8 \qquad \text{True}$$
$$2(5) - 9 = (1)$$
$$1=1 \qquad \text{True}$$

The ordered pair $(5, 1)$ satisfies both equations, so it is the solution to the system.

Analysis

We can see the solution clearly by plotting the graph of each equation. Since the solution is an ordered pair that satisfies both equations, it is a point on both of the lines and thus the point of intersection of the two lines. See Figure 3.

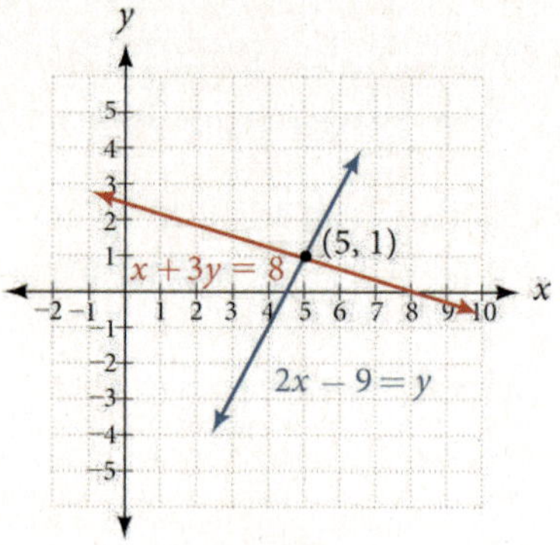

Figure 3

TRY IT #1 Determine whether the ordered pair $(8, 5)$ is a solution to the following system.

$$5x-4y = 20$$
$$2x + 1 = 3y$$

Solving Systems of Equations by Graphing

There are multiple methods of solving systems of linear equations. For a system of linear equations in two variables, we can determine both the type of system and the solution by graphing the system of equations on the same set of axes.

EXAMPLE 2

Solving a System of Equations in Two Variables by Graphing

Solve the following system of equations by graphing. Identify the type of system.

$$2x + y = -8$$
$$x - y = -1$$

Solution

Solve the first equation for y.

$$2x + y = -8$$
$$y = -2x-8$$

Solve the second equation for y.

$$x - y = -1$$
$$y = x + 1$$

Graph both equations on the same set of axes as in Figure 4.

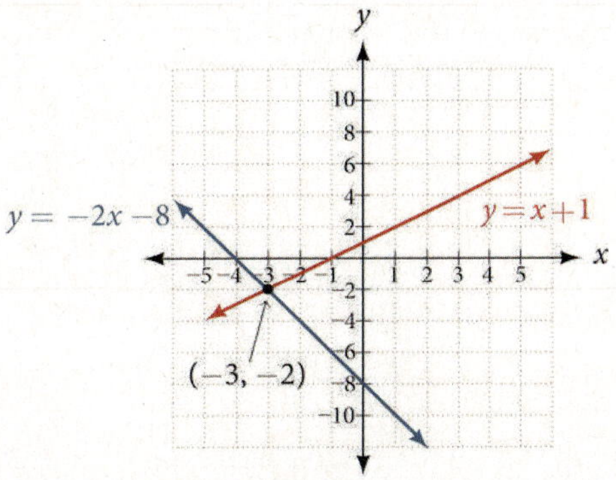

Figure 4

The lines appear to intersect at the point $(-3,-2)$. We can check to make sure that this is the solution to the system by substituting the ordered pair into both equations.

$$\begin{aligned} 2(-3) + (-2) &= -8 \\ -8 &= -8 \qquad \text{True} \\ (-3) - (-2) &= -1 \\ -1 &= -1 \qquad \text{True} \end{aligned}$$

The solution to the system is the ordered pair $(-3,-2)$, so the system is independent.

TRY IT #2 Solve the following system of equations by graphing.

$$\begin{aligned} 2x - 5y &= -25 \\ -4x + 5y &= 35 \end{aligned}$$

Q&A **Can graphing be used if the system is inconsistent or dependent?**

Yes, in both cases we can still graph the system to determine the type of system and solution. If the two lines are parallel, the system has no solution and is inconsistent. If the two lines are identical, the system has infinite solutions and is a dependent system.

Solving Systems of Equations by Substitution

Solving a linear system in two variables by graphing works well when the solution consists of integer values, but if our solution contains decimals or fractions, it is not the most precise method. We will consider two more methods of solving a system of linear equations that are more precise than graphing. One such method is solving a system of equations by the **substitution method**, in which we solve one of the equations for one variable and then substitute the result into the second equation to solve for the second variable. Recall that we can solve for only one variable at a time, which is the reason the substitution method is both valuable and practical.

HOW TO

Given a system of two equations in two variables, solve using the substitution method.

1. Solve one of the two equations for one of the variables in terms of the other.
2. Substitute the expression for this variable into the second equation, then solve for the remaining variable.
3. Substitute that solution into either of the original equations to find the value of the first variable. If possible, write the solution as an ordered pair.
4. Check the solution in both equations.

EXAMPLE 3

Solving a System of Equations in Two Variables by Substitution

Solve the following system of equations by substitution.

$$\begin{aligned} -x + y &= -5 \\ 2x - 5y &= 1 \end{aligned}$$

Solution

First, we will solve the first equation for y.

$$\begin{aligned} -x + y &= -5 \\ y &= x - 5 \end{aligned}$$

Now we can substitute the expression $x-5$ for y in the second equation.

$$\begin{aligned} 2x - 5y &= 1 \\ 2x - 5(x - 5) &= 1 \\ 2x - 5x + 25 &= 1 \\ -3x &= -24 \\ x &= 8 \end{aligned}$$

Now, we substitute $x = 8$ into the first equation and solve for y.

$$\begin{aligned} -(8) + y &= -5 \\ y &= 3 \end{aligned}$$

Our solution is $(8, 3)$.

Check the solution by substituting $(8, 3)$ into both equations.

$$\begin{aligned} -x + y &= -5 \\ -(8) + (3) &= -5 \qquad \text{True} \\ 2x - 5y &= 1 \\ 2(8) - 5(3) &= 1 \qquad \text{True} \end{aligned}$$

TRY IT #3 Solve the following system of equations by substitution.

$$\begin{aligned} x &= y + 3 \\ 4 &= 3x - 2y \end{aligned}$$

Q&A **Can the substitution method be used to solve any linear system in two variables?**

Yes, but the method works best if one of the equations contains a coefficient of 1 or –1 so that we do not have to deal with fractions.

Solving Systems of Equations in Two Variables by the Addition Method

A third method of solving systems of linear equations is the **addition method**. In this method, we add two terms with the same variable, but opposite coefficients, so that the sum is zero. Of course, not all systems are set up with the two terms of one variable having opposite coefficients. Often we must adjust one or both of the equations by multiplication so that one variable will be eliminated by addition.

HOW TO

Given a system of equations, solve using the addition method.

1. Write both equations with *x*- and *y*-variables on the left side of the equal sign and constants on the right.
2. Write one equation above the other, lining up corresponding variables. If one of the variables in the top equation

has the opposite coefficient of the same variable in the bottom equation, add the equations together, eliminating one variable. If not, use multiplication by a nonzero number so that one of the variables in the top equation has the opposite coefficient of the same variable in the bottom equation, then add the equations to eliminate the variable.

3. Solve the resulting equation for the remaining variable.
4. Substitute that value into one of the original equations and solve for the second variable.
5. Check the solution by substituting the values into the other equation.

EXAMPLE 4

Solving a System by the Addition Method

Solve the given system of equations by addition.

$$\begin{aligned} x + 2y &= -1 \\ -x + y &= 3 \end{aligned}$$

Solution

Both equations are already set equal to a constant. Notice that the coefficient of x in the second equation, –1, is the opposite of the coefficient of x in the first equation, 1. We can add the two equations to eliminate x without needing to multiply by a constant.

$$\begin{array}{r} x + 2y = -1 \\ \underline{-x + y = 3} \\ 3y = 2 \end{array}$$

Now that we have eliminated x, we can solve the resulting equation for y.

$$\begin{aligned} 3y &= 2 \\ y &= \tfrac{2}{3} \end{aligned}$$

Then, we substitute this value for y into one of the original equations and solve for x.

$$\begin{aligned} -x + y &= 3 \\ -x + \tfrac{2}{3} &= 3 \\ -x &= 3 - \tfrac{2}{3} \\ -x &= \tfrac{7}{3} \\ x &= -\tfrac{7}{3} \end{aligned}$$

The solution to this system is $\left(-\frac{7}{3}, \frac{2}{3}\right)$.

Check the solution in the first equation.

$$\begin{aligned} x + 2y &= -1 \\ \left(-\tfrac{7}{3}\right) + 2\left(\tfrac{2}{3}\right) &= \\ -\tfrac{7}{3} + \tfrac{4}{3} &= \\ -\tfrac{3}{3} &= \\ -1 &= -1 \qquad \text{True} \end{aligned}$$

Analysis

We gain an important perspective on systems of equations by looking at the graphical representation. See Figure 5 to find that the equations intersect at the solution. We do not need to ask whether there may be a second solution because observing the graph confirms that the system has exactly one solution.

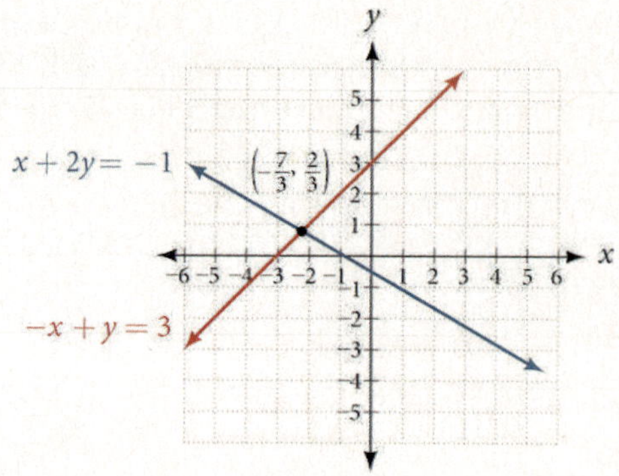

Figure 5

EXAMPLE 5

Using the Addition Method When Multiplication of One Equation Is Required

Solve the given system of equations by the addition method.

$$3x + 5y = -11$$

$$x - 2y = 11$$

Solution

Adding these equations as presented will not eliminate a variable. However, we see that the first equation has $3x$ in it and the second equation has x. So if we multiply the second equation by -3, the x-terms will add to zero.

$$\begin{aligned} &x-2y = 11 \\ &-3(x-2y) = -3(11) && \text{Multiply both sides by } -3. \\ &-3x + 6y = -33 && \text{Use the distributive property.} \end{aligned}$$

Now, let's add them.

$$\begin{aligned} 3x + 5y &= -11 \\ -3x + 6y &= -33 \\ \hline 11y &= -44 \\ y &= -4 \end{aligned}$$

For the last step, we substitute $y = -4$ into one of the original equations and solve for x.

$$\begin{aligned} 3x + 5y &= -11 \\ 3x + 5(-4) &= -11 \\ 3x - 20 &= -11 \\ 3x &= 9 \\ x &= 3 \end{aligned}$$

Our solution is the ordered pair $(3, -4)$. See Figure 6. Check the solution in the original second equation.

$$\begin{aligned} &x - 2y = 11 \\ &(3) - 2(-4) = 3 + 8 \\ &11 = 11 && \text{True} \end{aligned}$$

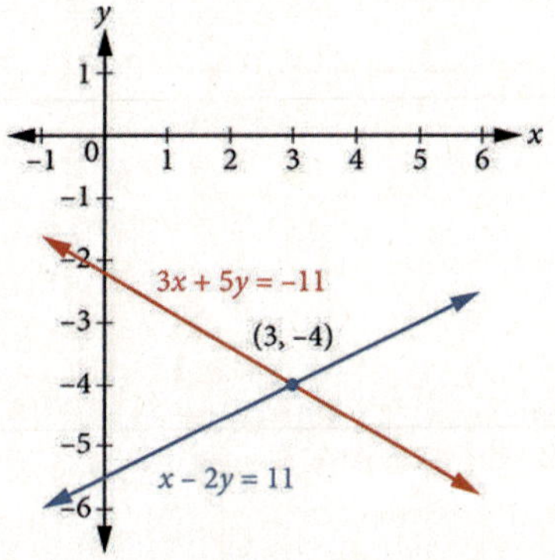

Figure 6

> TRY IT #4 Solve the system of equations by addition.

$$2x - 7y = 2$$
$$3x + y = -20$$

EXAMPLE 6

Using the Addition Method When Multiplication of Both Equations Is Required

Solve the given system of equations in two variables by addition.

$$2x + 3y = -16$$
$$5x - 10y = 30$$

Solution

One equation has $2x$ and the other has $5x$. The least common multiple is $10x$ so we will have to multiply both equations by a constant in order to eliminate one variable. Let's eliminate x by multiplying the first equation by -5 and the second equation by 2.

$$-5(2x + 3y) = -5(-16)$$
$$-10x - 15y = 80$$
$$2(5x - 10y) = 2(30)$$
$$10x - 20y = 60$$

Then, we add the two equations together.

$$-10x - 15y = 80$$
$$\underline{10x - 20y = 60}$$
$$-35y = 140$$
$$y = -4$$

Substitute $y = -4$ into the original first equation.

$$2x + 3(-4) = -16$$
$$2x - 12 = -16$$
$$2x = -4$$
$$x = -2$$

The solution is $(-2, -4)$. Check it in the other equation.

$$\begin{aligned} 5x - 10y &= 30 \\ 5(-2) - 10(-4) &= 30 \\ -10 + 40 &= 30 \\ 30 &= 30 \end{aligned}$$

See Figure 7.

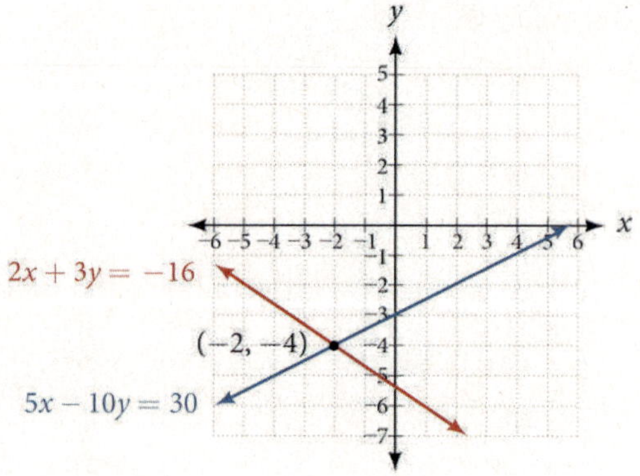

Figure 7

EXAMPLE 7

Using the Addition Method in Systems of Equations Containing Fractions

Solve the given system of equations in two variables by addition.

$$\begin{aligned} \frac{x}{3} + \frac{y}{6} &= 3 \\ \frac{x}{2} - \frac{y}{4} &= 1 \end{aligned}$$

Solution

First clear each equation of fractions by multiplying both sides of the equation by the least common denominator.

$$\begin{aligned} 6\left(\frac{x}{3} + \frac{y}{6}\right) &= 6(3) \\ 2x + y &= 18 \\ 4\left(\frac{x}{2} - \frac{y}{4}\right) &= 4(1) \\ 2x - y &= 4 \end{aligned}$$

Now multiply the second equation by -1 so that we can eliminate the x-variable.

$$\begin{aligned} -1(2x - y) &= -1(4) \\ -2x + y &= -4 \end{aligned}$$

Add the two equations to eliminate the x-variable and solve the resulting equation.

$$\begin{aligned} 2x + y &= 18 \\ \underline{-2x + y} &\underline{= -4} \\ 2y &= 14 \\ y &= 7 \end{aligned}$$

Substitute $y = 7$ into the first equation.

$$\begin{aligned} 2x + (7) &= 18 \\ 2x &= 11 \\ x &= \frac{11}{2} \\ &= 5.5 \end{aligned}$$

The solution is $\left(\frac{11}{2}, 7\right)$. Check it in the other equation.

$$\begin{aligned} \tfrac{x}{2} - \tfrac{y}{4} &= 1 \\ \frac{\frac{11}{2}}{2} - \tfrac{7}{4} &= 1 \\ \tfrac{11}{4} - \tfrac{7}{4} &= 1 \\ \tfrac{4}{4} &= 1 \end{aligned}$$

> **TRY IT** #5 Solve the system of equations by addition.

$$\begin{aligned} 2x + 3y &= 8 \\ 3x + 5y &= 10 \end{aligned}$$

Identifying Inconsistent Systems of Equations Containing Two Variables

Now that we have several methods for solving systems of equations, we can use the methods to identify inconsistent systems. Recall that an inconsistent system consists of parallel lines that have the same slope but different y -intercepts. They will never intersect. When searching for a solution to an inconsistent system, we will come up with a false statement, such as $12 = 0$.

EXAMPLE 8

Solving an Inconsistent System of Equations

Solve the following system of equations.

$$\begin{aligned} x &= 9 - 2y \\ x + 2y &= 13 \end{aligned}$$

Solution

We can approach this problem in two ways. Because one equation is already solved for x, the most obvious step is to use substitution.

$$\begin{aligned} x + 2y &= 13 \\ (9 - 2y) + 2y &= 13 \\ 9 + 0y &= 13 \\ 9 &= 13 \end{aligned}$$

Clearly, this statement is a contradiction because $9 \neq 13$. Therefore, the system has no solution.

The second approach would be to first manipulate the equations so that they are both in slope-intercept form. We manipulate the first equation as follows.

$$\begin{aligned} x &= 9 - 2y \\ 2y &= -x + 9 \\ y &= -\tfrac{1}{2}x + \tfrac{9}{2} \end{aligned}$$

We then convert the second equation expressed to slope-intercept form.

$$\begin{aligned} x + 2y &= 13 \\ 2y &= -x + 13 \\ y &= -\tfrac{1}{2}x + \tfrac{13}{2} \end{aligned}$$

Comparing the equations, we see that they have the same slope but different y-intercepts. Therefore, the lines are parallel and do not intersect.

$$y = -\frac{1}{2}x + \frac{9}{2}$$
$$y = -\frac{1}{2}x + \frac{13}{2}$$

Analysis

Writing the equations in slope-intercept form confirms that the system is inconsistent because all lines will intersect eventually unless they are parallel. Parallel lines will never intersect; thus, the two lines have no points in common. The graphs of the equations in this example are shown in Figure 8.

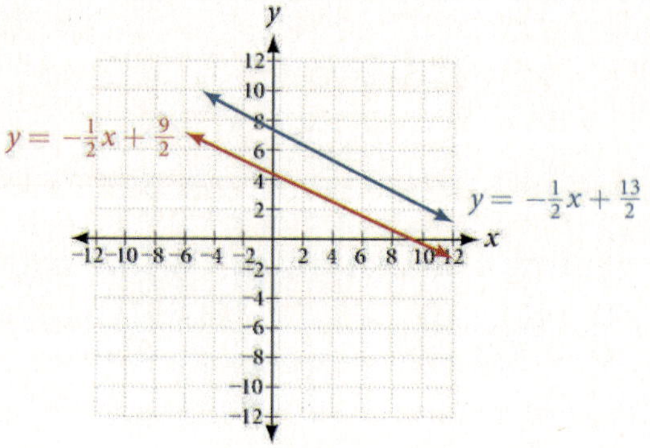

Figure 8

TRY IT #6 Solve the following system of equations in two variables.

$$2y - 2x = 2$$
$$2y - 2x = 6$$

Expressing the Solution of a System of Dependent Equations Containing Two Variables

Recall that a dependent system of equations in two variables is a system in which the two equations represent the same line. Dependent systems have an infinite number of solutions because all of the points on one line are also on the other line. After using substitution or addition, the resulting equation will be an identity, such as $0 = 0$.

EXAMPLE 9

Finding a Solution to a Dependent System of Linear Equations

Find a solution to the system of equations using the addition method.

$$x + 3y = 2$$
$$3x + 9y = 6$$

Solution

With the addition method, we want to eliminate one of the variables by adding the equations. In this case, let's focus on eliminating x. If we multiply both sides of the first equation by -3, then we will be able to eliminate the x -variable.

$$x + 3y = 2$$
$$(-3)(x + 3y) = (-3)(2)$$
$$-3x - 9y = -6$$

Now add the equations.

$$\begin{array}{rl} -3x - 9y & = -6 \\ +\ 3x + 9y & = 6 \\ \hline 0 & = 0 \end{array}$$

We can see that there will be an infinite number of solutions that satisfy both equations.

Analysis

If we rewrote both equations in the slope-intercept form, we might know what the solution would look like before adding. Let's look at what happens when we convert the system to slope-intercept form.

$$\begin{aligned} x + 3y &= 2 \\ 3y &= -x + 2 \\ y &= -\tfrac{1}{3}x + \tfrac{2}{3} \\ 3x + 9y &= 6 \\ 9y &= -3x + 6 \\ y &= -\tfrac{3}{9}x + \tfrac{6}{9} \\ y &= -\tfrac{1}{3}x + \tfrac{2}{3} \end{aligned}$$

See Figure 9. Notice the results are the same. The general solution to the system is $\left(x\ ,\ -\ \tfrac{1}{3}x + \tfrac{2}{3}\right)$.

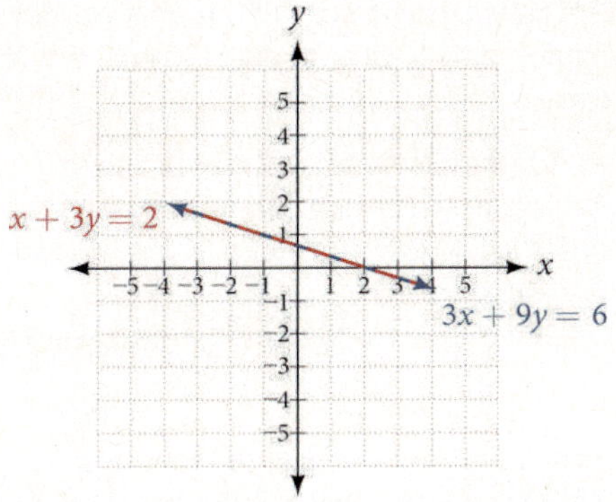

Figure 9

TRY IT #7 Solve the following system of equations in two variables.

$$\begin{aligned} y - 2x &= 5 \\ -3y + 6x &= -15 \end{aligned}$$

Using Systems of Equations to Investigate Profits

Using what we have learned about systems of equations, we can return to the skateboard manufacturing problem at the beginning of the section. The skateboard manufacturer's **revenue function** is the function used to calculate the amount of money that comes into the business. It can be represented by the equation $R = xp$, where $x =$ quantity and $p =$ price. The revenue function is shown in orange in Figure 10.

The **cost function** is the function used to calculate the costs of doing business. It includes fixed costs, such as rent and salaries, and variable costs, such as utilities. The cost function is shown in blue in Figure 10. The x -axis represents quantity in hundreds of units. The y-axis represents either cost or revenue in hundreds of dollars.

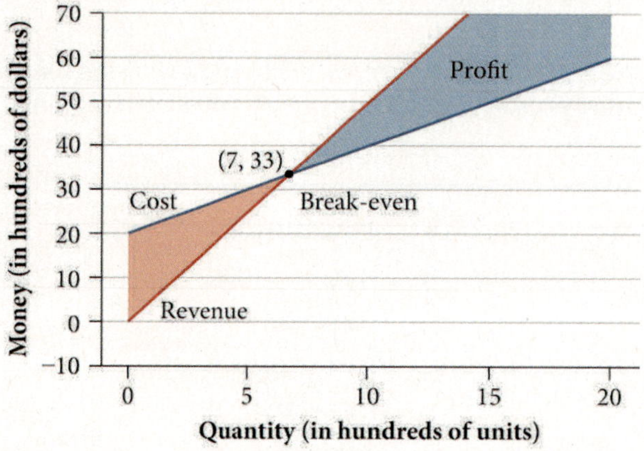

Figure 10

The point at which the two lines intersect is called the **break-even point**. We can see from the graph that if 700 units are produced, the cost is $3,300 and the revenue is also $3,300. In other words, the company breaks even if they produce and sell 700 units. They neither make money nor lose money.

The shaded region to the right of the break-even point represents quantities for which the company makes a profit. The shaded region to the left represents quantities for which the company suffers a loss. The **profit function** is the revenue function minus the cost function, written as $P(x) = R(x) - C(x)$. Clearly, knowing the quantity for which the cost equals the revenue is of great importance to businesses.

EXAMPLE 10

Finding the Break-Even Point and the Profit Function Using Substitution

Given the cost function $C(x) = 0.85x + 35{,}000$ and the revenue function $R(x) = 1.55x$, find the break-even point and the profit function.

Solution

Write the system of equations using y to replace function notation.

$$\begin{aligned} y &= 0.85x + 35{,}000 \\ y &= 1.55x \end{aligned}$$

Substitute the expression $0.85x + 35{,}000$ from the first equation into the second equation and solve for x.

$$\begin{aligned} 0.85x + 35{,}000 &= 1.55x \\ 35{,}000 &= 0.7x \\ 50{,}000 &= x \end{aligned}$$

Then, we substitute $x = 50{,}000$ into either the cost function or the revenue function.

$$1.55\,(50{,}000) = 77{,}500$$

The break-even point is $(50{,}000, 77{,}500)$.

The profit function is found using the formula $P(x) = R(x) - C(x)$.

$$\begin{aligned} P(x) &= 1.55x - (0.85x + 35,000) \\ &= 0.7x - 35,000 \end{aligned}$$

The profit function is $P(x) = 0.7x{-}35{,}000$.

Analysis

The cost to produce 50,000 units is $77,500, and the revenue from the sales of 50,000 units is also $77,500. To make a profit, the business must produce and sell more than 50,000 units. See Figure 11.

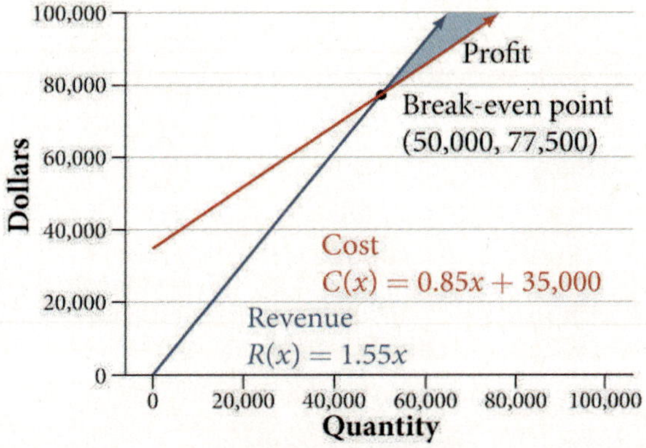

Figure 11

We see from the graph in Figure 12 that the profit function has a negative value until $x = 50{,}000$, when the graph crosses the *x*-axis. Then, the graph emerges into positive *y*-values and continues on this path as the profit function is a straight line. This illustrates that the break-even point for businesses occurs when the profit function is 0. The area to the left of the break-even point represents operating at a loss.

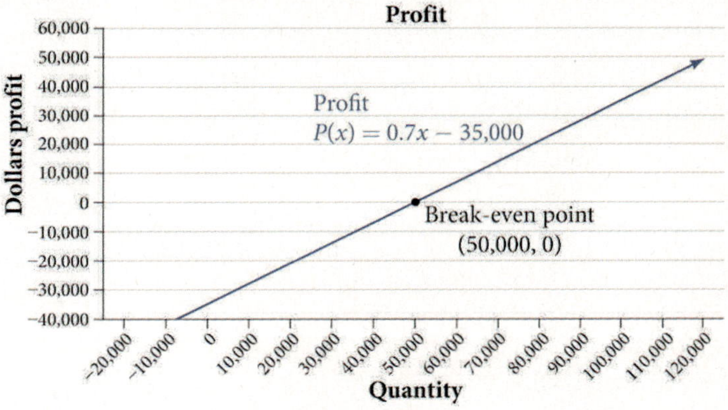

Figure 12

EXAMPLE 11

Writing and Solving a System of Equations in Two Variables

The cost of a ticket to the circus is $\$25.00$ for children and $\$50.00$ for adults. On a certain day, attendance at the circus is $2{,}000$ and the total gate revenue is $\$70{,}000$. How many children and how many adults bought tickets?

Solution

Let *c* = the number of children and *a* = the number of adults in attendance.

The total number of people is $2{,}000$. We can use this to write an equation for the number of people at the circus that day.

$$c + a = 2{,}000$$

The revenue from all children can be found by multiplying $\$25.00$ by the number of children, $25c$. The revenue from all adults can be found by multiplying $\$50.00$ by the number of adults, $50a$. The total revenue is $\$70{,}000$. We can use this to write an equation for the revenue.

$$25c + 50a = 70{,}000$$

We now have a system of linear equations in two variables.

$$\begin{aligned} c + a &= 2{,}000 \\ 25c + 50a &= 70{,}000 \end{aligned}$$

In the first equation, the coefficient of both variables is 1. We can quickly solve the first equation for either c or a. We will solve for a.

$$c + a = 2{,}000$$
$$a = 2{,}000 - c$$

Substitute the expression $2{,}000 - c$ in the second equation for a and solve for c.

$$25c + 50(2{,}000 - c) = 70{,}000$$
$$25c + 100{,}000 - 50c = 70{,}000$$
$$-25c = -30{,}000$$
$$c = 1{,}200$$

Substitute $c = 1{,}200$ into the first equation to solve for a.

$$1{,}200 + a = 2{,}000$$
$$a = 800$$

We find that 1,200 children and 800 adults bought tickets to the circus that day.

TRY IT #8 Meal tickets at the circus cost \$4.00 for children and \$12.00 for adults. If 1,650 meal tickets were bought for a total of \$14,200, how many children and how many adults bought meal tickets?

MEDIA

Access these online resources for additional instruction and practice with systems of linear equations.

Solving Systems of Equations Using Substitution (http://openstax.org/l/syssubst)
Solving Systems of Equations Using Elimination (http://openstax.org/l/syselim)
Applications of Systems of Equations (http://openstax.org/l/sysapp)

11.1 SECTION EXERCISES

Verbal

1. Can a system of linear equations have exactly two solutions? Explain why or why not.

2. If you are performing a break-even analysis for a business and their cost and revenue equations are dependent, explain what this means for the company's profit margins.

3. If you are solving a break-even analysis and get a negative break-even point, explain what this signifies for the company?

4. If you are solving a break-even analysis and there is no break-even point, explain what this means for the company. How should they ensure there is a break-even point?

5. Given a system of equations, explain at least two different methods of solving that system.

Algebraic

For the following exercises, determine whether the given ordered pair is a solution to the system of equations.

6. $\begin{aligned} 5x - y &= 4 \\ x + 6y &= 2 \end{aligned}$ and $(4, 0)$

7. $\begin{aligned} -3x - 5y &= 13 \\ -x + 4y &= 10 \end{aligned}$ and $(-6, 1)$

8. $\begin{aligned} 3x + 7y &= 1 \\ 2x + 4y &= 0 \end{aligned}$ and $(2, 3)$

9. $\begin{aligned} -2x + 5y &= 7 \\ 2x + 9y &= 7 \end{aligned}$ and $(-1, 1)$

10. $\begin{aligned} x + 8y &= 43 \\ 3x - 2y &= -1 \end{aligned}$ and $(3, 5)$

For the following exercises, solve each system by substitution.

11. $\begin{aligned} x + 3y &= 5 \\ 2x + 3y &= 4 \end{aligned}$

12. $\begin{aligned} 3x - 2y &= 18 \\ 5x + 10y &= -10 \end{aligned}$

13. $\begin{aligned} 4x + 2y &= -10 \\ 3x + 9y &= 0 \end{aligned}$

14. $\begin{aligned} 2x + 4y &= -3.8 \\ 9x - 5y &= 1.3 \end{aligned}$

15. $\begin{aligned} -2x + 3y &= 1.2 \\ -3x - 6y &= 1.8 \end{aligned}$

16. $\begin{aligned} x - 0.2y &= 1 \\ -10x + 2y &= 5 \end{aligned}$

17. $\begin{aligned} 3x + 5y &= 9 \\ 30x + 50y &= -90 \end{aligned}$

18. $\begin{aligned} -3x + y &= 2 \\ 12x - 4y &= -8 \end{aligned}$

19. $\begin{aligned} \tfrac{1}{2}x + \tfrac{1}{3}y &= 16 \\ \tfrac{1}{6}x + \tfrac{1}{4}y &= 9 \end{aligned}$

20. $\begin{aligned} -\tfrac{1}{4}x + \tfrac{3}{2}y &= 11 \\ -\tfrac{1}{8}x + \tfrac{1}{3}y &= 3 \end{aligned}$

For the following exercises, solve each system by addition.

21. $\begin{aligned} -2x + 5y &= -42 \\ 7x + 2y &= 30 \end{aligned}$

22. $\begin{aligned} 6x - 5y &= -34 \\ 2x + 6y &= 4 \end{aligned}$

23. $\begin{aligned} 5x - y &= -2.6 \\ -4x - 6y &= 1.4 \end{aligned}$

24. $\begin{aligned} 7x - 2y &= 3 \\ 4x + 5y &= 3.25 \end{aligned}$

25. $\begin{aligned} -x + 2y &= -1 \\ 5x - 10y &= 6 \end{aligned}$

26. $\begin{aligned} 7x + 6y &= 2 \\ -28x - 24y &= -8 \end{aligned}$

27. $\begin{aligned} \tfrac{5}{6}x + \tfrac{1}{4}y &= 0 \\ \tfrac{1}{8}x - \tfrac{1}{2}y &= -\tfrac{43}{120} \end{aligned}$

28. $\begin{aligned} \tfrac{1}{3}x + \tfrac{1}{9}y &= \tfrac{2}{9} \\ -\tfrac{1}{2}x + \tfrac{4}{5}y &= -\tfrac{1}{3} \end{aligned}$

29. $\begin{aligned} -0.2x + 0.4y &= 0.6 \\ x - 2y &= -3 \end{aligned}$

30. $\begin{aligned} -0.1x + 0.2y &= 0.6 \\ 5x - 10y &= 1 \end{aligned}$

For the following exercises, solve each system by any method.

31. $\begin{aligned} 5x + 9y &= 16 \\ x + 2y &= 4 \end{aligned}$

32. $\begin{aligned} 6x - 8y &= -0.6 \\ 3x + 2y &= 0.9 \end{aligned}$

33. $\begin{aligned} 5x - 2y &= 2.25 \\ 7x - 4y &= 3 \end{aligned}$

34. $\begin{aligned} x - \tfrac{5}{12}y &= -\tfrac{55}{12} \\ -6x + \tfrac{5}{2}y &= \tfrac{55}{2} \end{aligned}$

35. $\begin{aligned} 7x - 4y &= \tfrac{7}{6} \\ 2x + 4y &= \tfrac{1}{3} \end{aligned}$

36. $\begin{aligned} 3x + 6y &= 11 \\ 2x + 4y &= 9 \end{aligned}$

37. $\begin{array}{l} \frac{7}{3}x - \frac{1}{6}y = 2 \\ -\frac{21}{6}x + \frac{3}{12}y = -3 \end{array}$

38. $\begin{array}{l} \frac{1}{2}x + \frac{1}{3}y = \frac{1}{3} \\ \frac{3}{2}x + \frac{1}{4}y = -\frac{1}{8} \end{array}$

39. $\begin{array}{l} 2.2x + 1.3y = -0.1 \\ 4.2x + 4.2y = 2.1 \end{array}$

40. $\begin{array}{l} 0.1x + 0.2y = 2 \\ 0.35x - 0.3y = 0 \end{array}$

Graphical

For the following exercises, graph the system of equations and state whether the system is consistent, inconsistent, or dependent and whether the system has one solution, no solution, or infinite solutions.

41. $\begin{array}{l} 3x - y = 0.6 \\ x - 2y = 1.3 \end{array}$

42. $\begin{array}{l} -x + 2y = 4 \\ 2x - 4y = 1 \end{array}$

43. $\begin{array}{l} x + 2y = 7 \\ 2x + 6y = 12 \end{array}$

44. $\begin{array}{l} 3x - 5y = 7 \\ x - 2y = 3 \end{array}$

45. $\begin{array}{l} 3x - 2y = 5 \\ -9x + 6y = -15 \end{array}$

Technology

For the following exercises, use the intersect function on a graphing device to solve each system. Round all answers to the nearest hundredth.

46. $\begin{array}{l} 0.1x + 0.2y = 0.3 \\ -0.3x + 0.5y = 1 \end{array}$

47. $\begin{array}{l} -0.01x + 0.12y = 0.62 \\ 0.15x + 0.20y = 0.52 \end{array}$

48. $\begin{array}{l} 0.5x + 0.3y = 4 \\ 0.25x - 0.9y = 0.46 \end{array}$

49. $\begin{array}{l} 0.15x + 0.27y = 0.39 \\ -0.34x + 0.56y = 1.8 \end{array}$

50. $\begin{array}{l} -0.71x + 0.92y = 0.13 \\ 0.83x + 0.05y = 2.1 \end{array}$

Extensions

For the following exercises, solve each system in terms of $A, B, C, D, E,$ and F where $A – F$ are nonzero numbers. Note that $A \neq B$ and $AE \neq BD$.

51. $\begin{array}{l} x + y = A \\ x - y = B \end{array}$

52. $\begin{array}{l} x + Ay = 1 \\ x + By = 1 \end{array}$

53. $\begin{array}{l} Ax + y = 0 \\ Bx + y = 1 \end{array}$

54. $\begin{array}{l} Ax + By = C \\ x + y = 1 \end{array}$

55. $\begin{array}{l} Ax + By = C \\ Dx + Ey = F \end{array}$

Real-World Applications

For the following exercises, solve for the desired quantity.

56. A stuffed animal business has a total cost of production $C = 12x + 30$ and a revenue function $R = 20x$. Find the break-even point.

57. An Ethiopian restaurant has a cost of production $C(x) = 11x + 120$ and a revenue function $R(x) = 5x$. When does the company start to turn a profit?

58. A cell phone factory has a cost of production $C(x) = 150x + 10,000$ and a revenue function $R(x) = 200x$. What is the break-even point?

59. A musician charges $C(x) = 64x + 20,000$ where x is the total number of attendees at the concert. The venue charges $80 per ticket. After how many people buy tickets does the venue break even, and what is the value of the total tickets sold at that point?

60. A guitar factory has a cost of production $C(x) = 75x + 50,000$. If the company needs to break even after 150 units sold, at what price should they sell each guitar? Round up to the nearest dollar, and write the revenue function.

For the following exercises, use a system of linear equations with two variables and two equations to solve.

61. Find two numbers whose sum is 28 and difference is 13.

62. A number is 9 more than another number. Twice the sum of the two numbers is 10. Find the two numbers.

63. The startup cost for a restaurant is $120,000, and each meal costs $10 for the restaurant to make. If each meal is then sold for $15, after how many meals does the restaurant break even?

64. A moving company charges a flat rate of $150, and an additional $5 for each box. If a taxi service would charge $20 for each box, how many boxes would you need for it to be cheaper to use the moving company, and what would be the total cost?

65. A total of 1,595 first- and second-year college students gathered at a pep rally. The number of first-years exceeded the number of second-years by 15. How many students from each year group were in attendance?

66. 276 students enrolled in an introductory chemistry class. By the end of the semester, 5 times the number of students passed as failed. Find the number of students who passed, and the number of students who failed.

67. There were 130 faculty at a conference. If there were 18 more women than men attending, how many of each gender attended the conference?

68. A jeep and a pickup truck enter a highway running east-west at the same exit heading in opposite directions. The jeep entered the highway 30 minutes before the pickup did, and traveled 7 mph slower than the pickup. After 2 hours from the time the pickup entered the highway, the cars were 306.5 miles apart. Find the speed of each car, assuming they were driven on cruise control and retained the same speed.

69. If a scientist mixed 10% saline solution with 60% saline solution to get 25 gallons of 40% saline solution, how many gallons of 10% and 60% solutions were mixed?

70. An investor earned triple the profits of what they earned last year. If they made $500,000.48 total for both years, how much did the investor earn in profits each year?

71. An investor invested 1.1 million dollars into two land investments. On the first investment, Swan Peak, her return was a 110% increase on the money she invested. On the second investment, Riverside Community, she earned 50% over what she invested. If she earned $1 million in profits, how much did she invest in each of the land deals?

72. If an investor invests a total of $25,000 into two bonds, one that pays 3% simple interest, and the other that pays $2\frac{7}{8}\%$ interest, and the investor earns $737.50 annual interest, how much was invested in each account?

73. If an investor invests $23,000 into two bonds, one that pays 4% in simple interest, and the other paying 2% simple interest, and the investor earns $710.00 annual interest, how much was invested in each account?

74. Blu-rays cost $5.96 more than regular DVDs at All Bets Are Off Electronics. How much would 6 Blu-rays and 2 DVDs cost if 5 Blu-rays and 2 DVDs cost $127.73?

75. A store clerk sold 60 pairs of sneakers. The high-tops sold for $98.99 and the low-tops sold for $129.99. If the receipts for the two types of sales totaled $6,404.40, how many of each type of sneaker were sold?

76. A concert manager counted 350 ticket receipts the day after a concert. The price for a student ticket was $12.50, and the price for an adult ticket was $16.00. The register confirms that $5,075 was taken in. How many student tickets and adult tickets were sold?

77. Admission into an amusement park for 4 children and 2 adults is $116.90. For 6 children and 3 adults, the admission is $175.35. Assuming a different price for children and adults, what is the price of the child's ticket and the price of the adult ticket?

11.2 Systems of Linear Equations: Three Variables

Learning Objectives

In this section, you will:

- Solve systems of three equations in three variables.
- Identify inconsistent systems of equations containing three variables.
- Express the solution of a system of dependent equations containing three variables.

Figure 1 (credit: "Elembis," Wikimedia Commons)

Jordi received an inheritance of $12,000 that he divided into three parts and invested in three ways: in a money-market fund paying 3% annual interest; in municipal bonds paying 4% annual interest; and in mutual funds paying 7% annual interest. Jordi invested $4,000 more in municipal funds than in municipal bonds. He earned $670 in interest the first year. How much did Jordi invest in each type of fund?

Understanding the correct approach to setting up problems such as this one makes finding a solution a matter of following a pattern. We will solve this and similar problems involving three equations and three variables in this section. Doing so uses similar techniques as those used to solve systems of two equations in two variables. However, finding solutions to systems of three equations requires a bit more organization and a touch of visualization.

Solving Systems of Three Equations in Three Variables

In order to solve systems of equations in three variables, known as three-by-three systems, the primary tool we will be using is called Gaussian elimination, named after the prolific German mathematician Karl Friedrich Gauss. While there is no definitive order in which operations are to be performed, there are specific guidelines as to what type of moves can be made. We may number the equations to keep track of the steps we apply. The goal is to eliminate one variable at a time to achieve upper triangular form, the ideal form for a three-by-three system because it allows for straightforward back-substitution to find a solution (x, y, z), which we call an ordered triple. A system in upper triangular form looks like the following:

$$\begin{aligned} Ax + By + Cz &= D \\ Ey + Fz &= G \\ Hz &= K \end{aligned}$$

The third equation can be solved for z, and then we back-substitute to find y and x. To write the system in upper triangular form, we can perform the following operations:

1. Interchange the order of any two equations.
2. Multiply both sides of an equation by a nonzero constant.
3. Add a nonzero multiple of one equation to another equation.

The **solution set** to a three-by-three system is an ordered triple $\{(x, y, z)\}$. Graphically, the ordered triple defines the point that is the intersection of three planes in space. You can visualize such an intersection by imagining any corner in a rectangular room. A corner is defined by three planes: two adjoining walls and the floor (or ceiling). Any point where two walls and the floor meet represents the intersection of three planes.

Number of Possible Solutions

Figure 2 and Figure 3 illustrate possible solution scenarios for three-by-three systems.

- Systems that have a single solution are those which, after elimination, result in a **solution set** consisting of an ordered triple $\{(x, y, z)\}$. Graphically, the ordered triple defines a point that is the intersection of three planes in space.
- Systems that have an infinite number of solutions are those which, after elimination, result in an expression that is always true, such as $0 = 0$. Graphically, an infinite number of solutions represents a line or coincident plane that serves as the intersection of three planes in space.
- Systems that have no solution are those that, after elimination, result in a statement that is a contradiction, such as $3 = 0$. Graphically, a system with no solution is represented by three planes with no point in common.

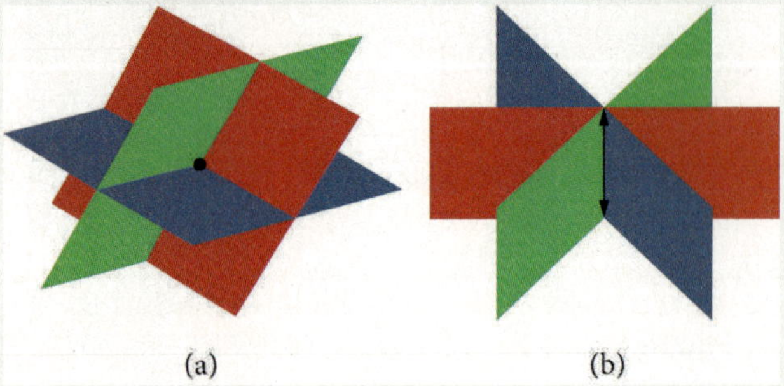

Figure 2 (a)Three planes intersect at a single point, representing a three-by-three system with a single solution. (b) Three planes intersect in a line, representing a three-by-three system with infinite solutions.

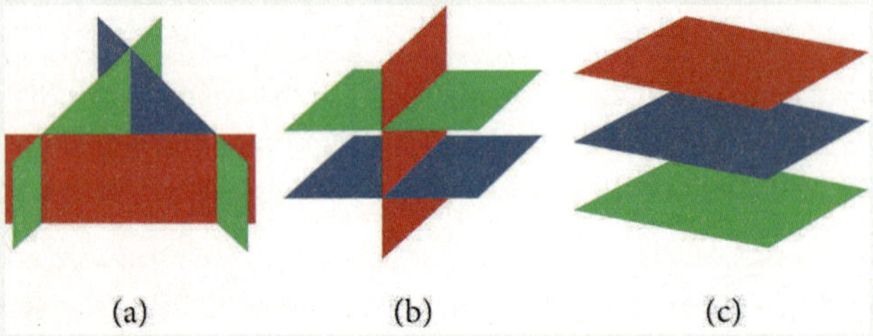

Figure 3 All three figures represent three-by-three systems with no solution. (a) The three planes intersect with each other, but not at a common point. (b) Two of the planes are parallel and intersect with the third plane, but not with each other. (c) All three planes are parallel, so there is no point of intersection.

EXAMPLE 1

Determining Whether an Ordered Triple Is a Solution to a System

Determine whether the ordered triple $(3, -2, 1)$ is a solution to the system.

$$\begin{aligned} x + y + z &= 2 \\ 6x - 4y + 5z &= 31 \\ 5x + 2y + 2z &= 13 \end{aligned}$$

Solution

We will check each equation by substituting in the values of the ordered triple for x, y, and z.

$$\begin{aligned} x + y + z &= 2 \\ (3) + (-2) + (1) &= 2 \\ &\text{True} \end{aligned} \qquad \begin{aligned} 6x - 4y + 5z &= 31 \\ 6(3) - 4(-2) + 5(1) &= 31 \\ 18 + 8 + 5 &= 31 \\ &\text{True} \end{aligned} \qquad \begin{aligned} 5x + 2y + 2z &= 13 \\ 5(3) + 2(-2) + 2(1) &= 13 \\ 15 - 4 + 2 &= 13 \\ &\text{True} \end{aligned}$$

The ordered triple $(3, -2, 1)$ is indeed a solution to the system.

HOW TO

Given a linear system of three equations, solve for three unknowns.

1. Pick any pair of equations and solve for one variable.
2. Pick another pair of equations and solve for the same variable.
3. You have created a system of two equations in two unknowns. Solve the resulting two-by-two system.
4. Back-substitute known variables into any one of the original equations and solve for the missing variable.

EXAMPLE 2

Solving a System of Three Equations in Three Variables by Elimination

Find a solution to the following system:

$$\begin{aligned} x-2y+3z &= 9 \quad (1) \\ -x+3y-z &= -6 \quad (2) \\ 2x-5y+5z &= 17 \quad (3) \end{aligned}$$

Solution

There will always be several choices as to where to begin, but the most obvious first step here is to eliminate x by adding equations (1) and (2).

$$\begin{array}{rl} x-2y+3z=9 & (1) \\ -x+3y-z=-6 & (2) \\ \hline y+2z=3 & (3) \end{array}$$

The second step is multiplying equation (1) by -2 and adding the result to equation (3). These two steps will eliminate the variable x.

$$\begin{array}{rl} -2x+4y-6z=-18 & (1) \text{ multiplied by } -2 \\ 2x-5y+5z=17 & (3) \\ \hline -y-z=-1 & (5) \end{array}$$

In equations (4) and (5), we have created a new two-by-two system. We can solve for z by adding the two equations.

$$\begin{array}{rl} y+2z=3 & (4) \\ -y-z=-1 & (5) \\ \hline z=2 & (6) \end{array}$$

Choosing one equation from each new system, we obtain the upper triangular form:

$$\begin{aligned} x-2y+3z &= 9 \quad (1) \\ y+2z &= 3 \quad (4) \\ z &= 2 \quad (6) \end{aligned}$$

Next, we back-substitute $z=2$ into equation (4) and solve for y.

$$\begin{aligned} y+2(2) &= 3 \\ y+4 &= 3 \\ y &= -1 \end{aligned}$$

Finally, we can back-substitute $z=2$ and $y=-1$ into equation (1). This will yield the solution for x.

$$\begin{aligned} x-2(-1)+3(2) &= 9 \\ x+2+6 &= 9 \\ x &= 1 \end{aligned}$$

The solution is the ordered triple $(1, -1, 2)$. See Figure 4.

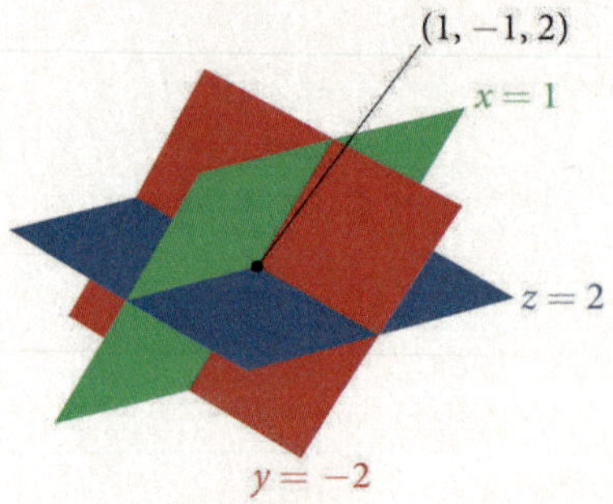

Figure 4

EXAMPLE 3

Solving a Real-World Problem Using a System of Three Equations in Three Variables

In the problem posed at the beginning of the section, Jordi invested his inheritance of $12,000 in three different funds: part in a money-market fund paying 3% interest annually; part in municipal bonds paying 4% annually; and the rest in mutual funds paying 7% annually. Jordi invested $4,000 more in mutual funds than he invested in municipal bonds. The total interest earned in one year was $670. How much did he invest in each type of fund?

Solution

To solve this problem, we use all of the information given and set up three equations. First, we assign a variable to each of the three investment amounts:

$$\begin{aligned} x &= \text{amount invested in money-market fund} \\ y &= \text{amount invested in municipal bonds} \\ z &= \text{amount invested in mutual funds} \end{aligned}$$

The first equation indicates that the sum of the three principal amounts is $12,000.

$$x + y + z = 12{,}000$$

We form the second equation according to the information that Jordi invested $4,000 more in mutual funds than he invested in municipal bonds.

$$z = y + 4{,}000$$

The third equation shows that the total amount of interest earned from each fund equals $670.

$$0.03x + 0.04y + 0.07z = 670$$

Then, we write the three equations as a system.

$$\begin{aligned} x + y + z &= 12{,}000 \\ -y + z &= 4{,}000 \\ 0.03x + 0.04y + 0.07z &= 670 \end{aligned}$$

To make the calculations simpler, we can multiply the third equation by 100. Thus,

$$\begin{aligned} x + y + z &= 12{,}000 \quad (1) \\ -y + z &= 4{,}000 \quad (2) \\ 3x + 4y + 7z &= 67{,}000 \quad (3) \end{aligned}$$

Step 1. Interchange equation (2) and equation (3) so that the two equations with three variables will line up.

$$\begin{aligned} x + y + z &= 12{,}000 \\ 3x + 4y + 7z &= 67{,}000 \\ -y + z &= 4{,}000 \end{aligned}$$

Step 2. Multiply equation (1) by -3 and add to equation (2). Write the result as row 2.

$$\begin{aligned} x + y + z &= 12{,}000 \\ y + 4z &= 31{,}000 \\ -y + z &= 4{,}000 \end{aligned}$$

Step 3. Add equation (2) to equation (3) and write the result as equation (3).

$$\begin{aligned} x + y + z &= 12{,}000 \\ y + 4z &= 31{,}000 \\ 5z &= 35{,}000 \end{aligned}$$

Step 4. Solve for z in equation (3). Back-substitute that value in equation (2) and solve for y. Then, back-substitute the values for z and y into equation (1) and solve for x.

$$\begin{aligned} 5z &= 35{,}000 \\ z &= 7{,}000 \end{aligned}$$

$$\begin{aligned} y + 4(7{,}000) &= 31{,}000 \\ y &= 3{,}000 \end{aligned}$$

$$\begin{aligned} x + 3{,}000 + 7{,}000 &= 12{,}000 \\ x &= 2{,}000 \end{aligned}$$

Jordi invested $2,000 in a money-market fund, $3,000 in municipal bonds, and $7,000 in mutual funds.

TRY IT #1 Solve the system of equations in three variables.

$$\begin{aligned} 2x + y - 2z &= -1 \\ 3x - 3y - z &= 5 \\ x - 2y + 3z &= 6 \end{aligned}$$

Identifying Inconsistent Systems of Equations Containing Three Variables

Just as with systems of equations in two variables, we may come across an inconsistent system of equations in three variables, which means that it does not have a solution that satisfies all three equations. The equations could represent three parallel planes, two parallel planes and one intersecting plane, or three planes that intersect the other two but not at the same location. The process of elimination will result in a false statement, such as $3 = 7$ or some other contradiction.

EXAMPLE 4

Solving an Inconsistent System of Three Equations in Three Variables

Solve the following system.

$$\begin{aligned} x - 3y + z &= 4 \quad (1) \\ -x + 2y - 5z &= 3 \quad (2) \\ 5x - 13y + 13z &= 8 \quad (3) \end{aligned}$$

Solution

Looking at the coefficients of x, we can see that we can eliminate x by adding equation (1) to equation (2).

$$\begin{array}{r} x - 3y + z = 4 \ (1) \\ \underline{-x + 2y - 5z = 3 \ (2)} \\ -y - 4z = 7 \ (4) \end{array}$$

Next, we multiply equation (1) by -5 and add it to equation (3).

$$\begin{array}{ll} -5x + 15y - 5z = -20 & (1) \text{ multiplied by } -5 \\ 5x - 13y + 13z = 8 & (3) \\ \hline 2y + 8z = -12 \quad (5) & \end{array}$$

Then, we multiply equation (4) by 2 and add it to equation (5).

$$\begin{array}{l} -2y - 8z = 14 (4) \text{ multiplied by } 2 \\ 2y + 8z = -12 \ (5) \\ \hline 0 = 2 \end{array}$$

The final equation $0 = 2$ is a contradiction, so we conclude that the system of equations in inconsistent and, therefore, has no solution.

Analysis

In this system, each plane intersects the other two, but not at the same location. Therefore, the system is inconsistent.

TRY IT #2 Solve the system of three equations in three variables.

$$\begin{aligned} x + y + z &= 2 \\ y - 3z &= 1 \\ 2x + y + 5z &= 0 \end{aligned}$$

Expressing the Solution of a System of Dependent Equations Containing Three Variables

We know from working with systems of equations in two variables that a dependent system of equations has an infinite number of solutions. The same is true for dependent systems of equations in three variables. An infinite number of solutions can result from several situations. The three planes could be the same, so that a solution to one equation will be the solution to the other two equations. All three equations could be different but they intersect on a line, which has infinite solutions. Or two of the equations could be the same and intersect the third on a line.

EXAMPLE 5

Finding the Solution to a Dependent System of Equations

Find the solution to the given system of three equations in three variables.

$$\begin{aligned} 2x + y - 3z &= 0 \quad (1) \\ 4x + 2y - 6z &= 0 \quad (2) \\ x - y + z &= 0 \quad (3) \end{aligned}$$

Solution

First, we can multiply equation (1) by -2 and add it to equation (2).

$$\begin{array}{l} -4x - 2y + 6z = 0 \ \text{ equation}(1) \text{ multiplied by } -2 \\ 4x + 2y - 6z = 0 \quad (2) \\ \hline 0 = 0 \end{array}$$

We do not need to proceed any further. The result we get is an identity, $0 = 0$, which tells us that this system has an infinite number of solutions. There are other ways to begin to solve this system, such as multiplying equation (3) by -2, and adding it to equation (1). We then perform the same steps as above and find the same result, $0 = 0$.

When a system is dependent, we can find general expressions for the solutions. Adding equations (1) and (3), we have

$$\begin{array}{r} 2x + y - 3z = 0 \\ x - y + z = 0 \\ \hline 3x - 2z = 0 \end{array}$$

We then solve the resulting equation for z.

$$3x - 2z = 0$$
$$z = \frac{3}{2}x$$

We back-substitute the expression for z into one of the equations and solve for y.

$$2x + y - 3\left(\frac{3}{2}x\right) = 0$$
$$2x + y - \frac{9}{2}x = 0$$
$$y = \frac{9}{2}x - 2x$$
$$y = \frac{5}{2}x$$

So the general solution is $\left(x, \frac{5}{2}x, \frac{3}{2}x\right)$. In this solution, x can be any real number. The values of y and z are dependent on the value selected for x.

Analysis

As shown in Figure 5, two of the planes are the same and they intersect the third plane on a line. The solution set is infinite, as all points along the intersection line will satisfy all three equations.

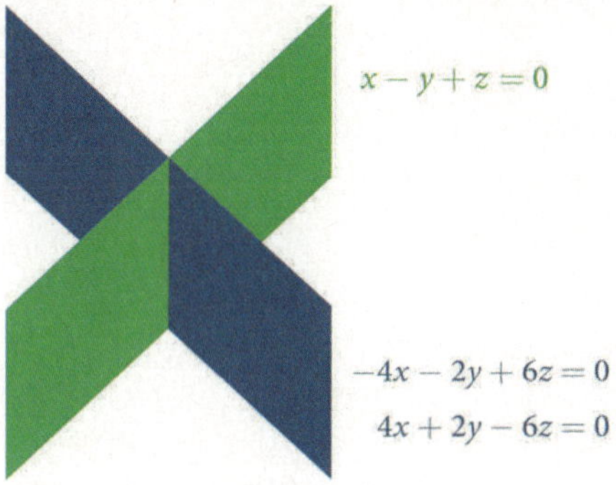

Figure 5

Q&A **Does the generic solution to a dependent system always have to be written in terms of x?**

No, you can write the generic solution in terms of any of the variables, but it is common to write it in terms of x and if needed x and y.

TRY IT #3 Solve the following system.

$$x + y + z = 7$$
$$3x - 2y - z = 4$$
$$x + 6y + 5z = 24$$

MEDIA

Access these online resources for additional instruction and practice with systems of equations in three variables.

Ex 1: System of Three Equations with Three Unknowns Using Elimination (http://openstax.org/l/systhree)
Ex. 2: System of Three Equations with Three Unknowns Using Elimination (http://openstax.org/l/systhelim)

11.2 SECTION EXERCISES

Verbal

1. Can a linear system of three equations have exactly two solutions? Explain why or why not

2. If a given ordered triple solves the system of equations, is that solution unique? If so, explain why. If not, give an example where it is not unique.

3. If a given ordered triple does not solve the system of equations, is there no solution? If so, explain why. If not, give an example.

4. Using the method of addition, is there only one way to solve the system?

5. Can you explain whether there can be only one method to solve a linear system of equations? If yes, give an example of such a system of equations. If not, explain why not.

Algebraic

For the following exercises, determine whether the ordered triple given is the solution to the system of equations.

6. $\begin{array}{r} 2x-6y+6z=-12 \\ x+4y+5z=-1 \\ -x+2y+3z=-1 \end{array}$ and $(0, 1, -1)$

7. $\begin{array}{r} 6x-y+3z=6 \\ 3x+5y+2z=0 \\ x+y=0 \end{array}$ and $(3, -3, -5)$

8. $\begin{array}{r} 6x-7y+z=2 \\ -x-y+3z=4 \\ 2x+y-z=1 \end{array}$ and $(4, 2, -6)$

9. $\begin{array}{r} x-y=0 \\ x-z=5 \\ x-y+z=-1 \end{array}$ and $(4, 4, -1)$

10. $\begin{array}{r} -x-y+2z=3 \\ 5x+8y-3z=4 \\ -x+3y-5z=-5 \end{array}$ and $(4, 1, -7)$

For the following exercises, solve each system by elimination.

11. $\begin{array}{r} 3x-4y+2z=-15 \\ 2x+4y+z=16 \\ 2x+3y+5z=20 \end{array}$

12. $\begin{array}{r} 5x-2y+3z=20 \\ 2x-4y-3z=-9 \\ x+6y-8z=21 \end{array}$

13. $\begin{array}{r} 5x+2y+4z=9 \\ -3x+2y+z=10 \\ 4x-3y+5z=-3 \end{array}$

14. $\begin{array}{r} 4x-3y+5z=31 \\ -x+2y+4z=20 \\ x+5y-2z=-29 \end{array}$

15. $\begin{array}{r} 5x-2y+3z=4 \\ -4x+6y-7z=-1 \\ 3x+2y-z=4 \end{array}$

16. $\begin{array}{r} 4x+6y+9z=0 \\ -5x+2y-6z=3 \\ 7x-4y+3z=-3 \end{array}$

For the following exercises, solve each system by Gaussian elimination.

17. $\begin{array}{r} 2x-y+3z=17 \\ -5x+4y-2z=-46 \\ 2y+5z=-7 \end{array}$

18. $\begin{array}{r} 5x-6y+3z=50 \\ -x+4y=10 \\ 2x-z=10 \end{array}$

19. $\begin{array}{r} 2x+3y-6z=1 \\ -4x-6y+12z=-2 \\ x+2y+5z=10 \end{array}$

20. $$\begin{aligned} 4x + 6y - 2z &= 8 \\ 6x + 9y - 3z &= 12 \\ -2x - 3y + z &= -4 \end{aligned}$$

21. $$\begin{aligned} 2x + 3y - 4z &= 5 \\ -3x + 2y + z &= 11 \\ -x + 5y + 3z &= 4 \end{aligned}$$

22. $$\begin{aligned} 10x + 2y - 14z &= 8 \\ -x - 2y - 4z &= -1 \\ -12x - 6y + 6z &= -12 \end{aligned}$$

23. $$\begin{aligned} x + y + z &= 14 \\ 2y + 3z &= -14 \\ -16y - 24z &= -112 \end{aligned}$$

24. $$\begin{aligned} 5x - 3y + 4z &= -1 \\ -4x + 2y - 3z &= 0 \\ -x + 5y + 7z &= -11 \end{aligned}$$

25. $$\begin{aligned} x + y + z &= 0 \\ 2x - y + 3z &= 0 \\ x - z &= 0 \end{aligned}$$

26. $$\begin{aligned} 3x + 2y - 5z &= 6 \\ 5x - 4y + 3z &= -12 \\ 4x + 5y - 2z &= 15 \end{aligned}$$

27. $$\begin{aligned} x + y + z &= 0 \\ 2x - y + 3z &= 0 \\ x - z &= 1 \end{aligned}$$

28. $$\begin{aligned} 3x - \tfrac{1}{2}y - z &= -\tfrac{1}{2} \\ 4x + z &= 3 \\ -x + \tfrac{3}{2}y &= \tfrac{5}{2} \end{aligned}$$

29. $$\begin{aligned} 6x - 5y + 6z &= 38 \\ \tfrac{1}{5}x - \tfrac{1}{2}y + \tfrac{3}{5}z &= 1 \\ -4x - \tfrac{3}{2}y - z &= -74 \end{aligned}$$

30. $$\begin{aligned} \tfrac{1}{2}x - \tfrac{1}{5}y + \tfrac{2}{5}z &= -\tfrac{13}{10} \\ \tfrac{1}{4}x - \tfrac{2}{5}y - \tfrac{1}{5}z &= -\tfrac{7}{20} \\ -\tfrac{1}{2}x - \tfrac{3}{4}y - \tfrac{1}{2}z &= -\tfrac{5}{4} \end{aligned}$$

31. $$\begin{aligned} -\tfrac{1}{3}x - \tfrac{1}{2}y - \tfrac{1}{4}z &= \tfrac{3}{4} \\ -\tfrac{1}{2}x - \tfrac{1}{4}y - \tfrac{1}{2}z &= 2 \\ -\tfrac{1}{4}x - \tfrac{3}{4}y - \tfrac{1}{2}z &= -\tfrac{1}{2} \end{aligned}$$

32. $$\begin{aligned} \tfrac{1}{2}x - \tfrac{1}{4}y + \tfrac{3}{4}z &= 0 \\ \tfrac{1}{4}x - \tfrac{1}{10}y + \tfrac{2}{5}z &= -2 \\ \tfrac{1}{8}x + \tfrac{1}{5}y - \tfrac{1}{8}z &= 2 \end{aligned}$$

33. $$\begin{aligned} \tfrac{4}{5}x - \tfrac{7}{8}y + \tfrac{1}{2}z &= 1 \\ -\tfrac{4}{5}x - \tfrac{3}{4}y + \tfrac{1}{3}z &= -8 \\ -\tfrac{2}{5}x - \tfrac{7}{8}y + \tfrac{1}{2}z &= -5 \end{aligned}$$

34. $$\begin{aligned} -\tfrac{1}{3}x - \tfrac{1}{8}y + \tfrac{1}{6}z &= -\tfrac{4}{3} \\ -\tfrac{2}{3}x - \tfrac{7}{8}y + \tfrac{1}{3}z &= -\tfrac{23}{3} \\ -\tfrac{1}{3}x - \tfrac{5}{8}y + \tfrac{5}{6}z &= 0 \end{aligned}$$

35. $$\begin{aligned} -\tfrac{1}{4}x - \tfrac{5}{4}y + \tfrac{5}{2}z &= -5 \\ -\tfrac{1}{2}x - \tfrac{5}{3}y + \tfrac{5}{4}z &= \tfrac{55}{12} \\ -\tfrac{1}{3}x - \tfrac{1}{3}y + \tfrac{1}{3}z &= \tfrac{5}{3} \end{aligned}$$

36. $$\begin{aligned} \tfrac{1}{40}x + \tfrac{1}{60}y + \tfrac{1}{80}z &= \tfrac{1}{100} \\ -\tfrac{1}{2}x - \tfrac{1}{3}y - \tfrac{1}{4}z &= -\tfrac{1}{5} \\ \tfrac{3}{8}x + \tfrac{3}{12}y + \tfrac{3}{16}z &= \tfrac{3}{20} \end{aligned}$$

37. $$\begin{aligned} 0.1x - 0.2y + 0.3z &= 2 \\ 0.5x - 0.1y + 0.4z &= 8 \\ 0.7x - 0.2y + 0.3z &= 8 \end{aligned}$$

38. $$\begin{aligned} 0.2x + 0.1y - 0.3z &= 0.2 \\ 0.8x + 0.4y - 1.2z &= 0.1 \\ 1.6x + 0.8y - 2.4z &= 0.2 \end{aligned}$$

39. $$\begin{aligned} 1.1x + 0.7y - 3.1z &= -1.79 \\ 2.1x + 0.5y - 1.6z &= -0.13 \\ 0.5x + 0.4y - 0.5z &= -0.07 \end{aligned}$$

40. $$\begin{aligned} 0.5x - 0.5y + 0.5z &= 10 \\ 0.2x - 0.2y + 0.2z &= 4 \\ 0.1x - 0.1y + 0.1z &= 2 \end{aligned}$$

41. $$\begin{aligned} 0.1x + 0.2y + 0.3z &= 0.37 \\ 0.1x - 0.2y - 0.3z &= -0.27 \\ 0.5x - 0.1y - 0.3z &= -0.03 \end{aligned}$$

42. $$\begin{aligned} 0.5x - 0.5y - 0.3z &= 0.13 \\ 0.4x - 0.1y - 0.3z &= 0.11 \\ 0.2x - 0.8y - 0.9z &= -0.32 \end{aligned}$$

43. $$\begin{aligned} 0.5x + 0.2y - 0.3z &= 1 \\ 0.4x - 0.6y + 0.7z &= 0.8 \\ 0.3x - 0.1y - 0.9z &= 0.6 \end{aligned}$$

44. $$\begin{aligned} 0.3x + 0.3y + 0.5z &= 0.6 \\ 0.4x + 0.4y + 0.4z &= 1.8 \\ 0.4x + 0.2y + 0.1z &= 1.6 \end{aligned}$$

45. $$\begin{aligned} 0.8x + 0.8y + 0.8z &= 2.4 \\ 0.3x - 0.5y + 0.2z &= 0 \\ 0.1x + 0.2y + 0.3z &= 0.6 \end{aligned}$$

Extensions

For the following exercises, solve the system for x, y, and z.

46. $$\begin{aligned} x + y + z &= 3 \\ \frac{x-1}{2} + \frac{y-3}{2} + \frac{z+1}{2} &= 0 \\ \frac{x-2}{3} + \frac{y+4}{3} + \frac{z-3}{3} &= \frac{2}{3} \end{aligned}$$

47. $$\begin{aligned} 5x - 3y - \frac{z+1}{2} &= \frac{1}{2} \\ 6x + \frac{y-9}{2} + 2z &= -3 \\ \frac{x+8}{2} - 4y + z &= 4 \end{aligned}$$

48. $$\begin{aligned} \frac{x+4}{7} - \frac{y-1}{6} + \frac{z+2}{3} &= 1 \\ \frac{x-2}{4} + \frac{y+1}{8} - \frac{z+8}{12} &= 0 \\ \frac{x+6}{3} - \frac{y+2}{3} + \frac{z+4}{2} &= 3 \end{aligned}$$

49. $$\begin{aligned} \frac{x-3}{6} + \frac{y+2}{2} - \frac{z-3}{3} &= 2 \\ \frac{x+2}{4} + \frac{y-5}{2} + \frac{z+4}{2} &= 1 \\ \frac{x+6}{2} - \frac{y-3}{2} + z + 1 &= 9 \end{aligned}$$

50. $$\begin{aligned} \frac{x-1}{3} + \frac{y+3}{4} + \frac{z+2}{6} &= 1 \\ 4x + 3y - 2z &= 11 \\ 0.02x + 0.015y - 0.01z &= 0.065 \end{aligned}$$

Real-World Applications

51. Three even numbers sum up to 108. The smaller is half the larger and the middle number is $\frac{3}{4}$ the larger. What are the three numbers?

52. Three numbers sum up to 147. The smallest number is half the middle number, which is half the largest number. What are the three numbers?

53. At a family reunion, there were only blood relatives, consisting of children, parents, and grandparents, in attendance. There were 400 people total. There were twice as many parents as grandparents, and 50 more children than parents. How many children, parents, and grandparents were in attendance?

54. An animal shelter has a total of 350 animals comprised of cats, dogs, and rabbits. If the number of rabbits is 5 less than one-half the number of cats, and there are 20 more cats than dogs, how many of each animal are at the shelter?

55. Your roommate, Shani, offered to buy groceries for you and your other roommate. The total bill was \$82. She forgot to save the individual receipts but remembered that your groceries were \$0.05 cheaper than half of her groceries, and that your other roommate's groceries were \$2.10 more than your groceries. How much was each of your share of the groceries?

56. Your roommate, John, offered to buy household supplies for you and your other roommate. You live near the border of three states, each of which has a different sales tax. The total amount of money spent was \$100.75. Your supplies were bought with 5% tax, John's with 8% tax, and your third roommate's with 9% sales tax. The total amount of money spent without taxes is \$93.50. If your supplies before tax were \$1 more than half of what your third roommate's supplies were before tax, how much did each of you spend? Give your answer both with and without taxes.

57. Three coworkers work for the same employer. Their jobs are warehouse manager, office manager, and truck driver. The sum of the annual salaries of the warehouse manager and office manager is $82,000. The office manager makes $4,000 more than the truck driver annually. The annual salaries of the warehouse manager and the truck driver total $78,000. What is the annual salary of each of the co-workers?

58. At a carnival, $2,914.25 in receipts were taken at the end of the day. The cost of a child's ticket was $20.50, an adult ticket was $29.75, and a senior citizen ticket was $15.25. There were twice as many senior citizens as adults in attendance, and 20 more children than senior citizens. How many children, adult, and senior citizen tickets were sold?

59. A local band sells out for their concert. They sell all 1,175 tickets for a total purse of $28,112.50. The tickets were priced at $20 for student tickets, $22.50 for children, and $29 for adult tickets. If the band sold twice as many adult as children tickets, how many of each type was sold?

60. In a bag, a child has 325 coins worth $19.50. There were three types of coins: pennies, nickels, and dimes. If the bag contained the same number of nickels as dimes, how many of each type of coin was in the bag?

61. Last year, at Haven's Pond Car Dealership, for a particular model of BMW, Jeep, and Toyota, one could purchase all three cars for a total of $140,000. This year, due to inflation, the same cars would cost $151,830. The cost of the BMW increased by 8%, the Jeep by 5%, and the Toyota by 12%. If the price of last year's Jeep was $7,000 less than the price of last year's BMW, what was the price of each of the three cars last year?

62. When his youngest child moved out, Deandre sold his home and made three investments using gains from the sale. He invested $80,500 into three accounts, one that paid 4% simple interest, one that paid $3\frac{1}{8}\%$ simple interest, and one that paid $2\frac{1}{2}\%$ simple interest. He earned $2,670 interest at the end of one year. If the amount of the money invested in the second account was four times the amount invested in the third account, how much was invested in each account?

63. You inherit one million dollars. You invest it all in three accounts for one year. The first account pays 3% compounded annually, the second account pays 4% compounded annually, and the third account pays 2% compounded annually. After one year, you earn $34,000 in interest. If you invest four times the money into the account that pays 3% compared to 2%, how much did you invest in each account?

64. An entrepreneur sells a portion of their business for one hundred thousand dollars and invests it all in three accounts for one year. The first account pays 4% compounded annually, the second account pays 3% compounded annually, and the third account pays 2% compounded annually. After one year, the entrepreneur earns $3,650 in interest. If they invested five times the money in the account that pays 4% compared to 3%, how much did they invest in each account?

65. The top three countries in oil consumption in a certain year are as follows: the United States, Japan, and China. In millions of barrels per day, the three top countries consumed 39.8% of the world's consumed oil. The United States consumed 0.7% more than four times China's consumption. The United States consumed 5% more than triple Japan's consumption. What percent of the world oil consumption did the United States, Japan, and China consume?[1]

66. The top three countries in oil production in the same year are Saudi Arabia, the United States, and Russia. In millions of barrels per day, the top three countries produced 31.4% of the world's produced oil. Saudi Arabia and the United States combined for 22.1% of the world's production, and Saudi Arabia produced 2% more oil than Russia. What percent of the world oil production did Saudi Arabia, the United States, and Russia produce?[2]

67. The top three sources of oil imports for the United States in the same year were Saudi Arabia, Mexico, and Canada. The three top countries accounted for 47% of oil imports. The United States imported 1.8% more from Saudi Arabia than they did from Mexico, and 1.7% more from Saudi Arabia than they did from Canada. What percent of the United States oil imports were from these three countries?[3]

68. The top three oil producers in the United States in a certain year are the Gulf of Mexico, Texas, and Alaska. The three regions were responsible for 64% of the United States oil production. The Gulf of Mexico and Texas combined for 47% of oil production. Texas produced 3% more than Alaska. What percent of United States oil production came from these regions?[4]

1 "Oil reserves, production and consumption in 2001," accessed April 6, 2014, http://scaruffi.com/politics/oil.html.

2 "Oil reserves, production and consumption in 2001," accessed April 6, 2014, http://scaruffi.com/politics/oil.html.

3 "Oil reserves, production and consumption in 2001," accessed April 6, 2014, http://scaruffi.com/politics/oil.html.

4 "USA: The coming global oil crisis," accessed April 6, 2014, http://www.oilcrisis.com/us/.

69. At one time, in the United States, 398 species of animals were on the endangered species list. The top groups were mammals, birds, and fish, which comprised 55% of the endangered species. Birds accounted for 0.7% more than fish, and fish accounted for 1.5% more than mammals. What percent of the endangered species came from mammals, birds, and fish?

70. Meat consumption in the United States can be broken into three categories: red meat, poultry, and fish. If fish makes up 4% less than one-quarter of poultry consumption, and red meat consumption is 18.2% higher than poultry consumption, what are the percentages of meat consumption?[5]

11.3 Systems of Nonlinear Equations and Inequalities: Two Variables

Learning Objectives

In this section, you will:

- Solve a system of nonlinear equations using substitution.
- Solve a system of nonlinear equations using elimination.
- Graph a nonlinear inequality.
- Graph a system of nonlinear inequalities.

Halley's Comet (Figure 1) orbits the sun about once every 75 years. Its path can be considered to be a very elongated ellipse. Other comets follow similar paths in space. These orbital paths can be studied using systems of equations. These systems, however, are different from the ones we considered in the previous section because the equations are not linear.

Figure 1 Halley's Comet (credit: "NASA Blueshift"/Flickr)

In this section, we will consider the intersection of a parabola and a line, a circle and a line, and a circle and an ellipse. The methods for solving systems of nonlinear equations are similar to those for linear equations.

Solving a System of Nonlinear Equations Using Substitution

A **system of nonlinear equations** is a system of two or more equations in two or more variables containing at least one equation that is not linear. Recall that a linear equation can take the form $Ax + By + C = 0$. Any equation that cannot be written in this form in nonlinear. The substitution method we used for linear systems is the same method we will use for nonlinear systems. We solve one equation for one variable and then substitute the result into the second equation to solve for another variable, and so on. There is, however, a variation in the possible outcomes.

Intersection of a Parabola and a Line

There are three possible types of solutions for a system of nonlinear equations involving a parabola and a line.

5 "The United States Meat Industry at a Glance," accessed April 6, 2014, http://www.meatami.com/ht/d/sp/i/47465/pid/47465.

Possible Types of Solutions for Points of Intersection of a Parabola and a Line

Figure 2 illustrates possible solution sets for a system of equations involving a parabola and a line.

- No solution. The line will never intersect the parabola.
- One solution. The line is tangent to the parabola and intersects the parabola at exactly one point.
- Two solutions. The line crosses on the inside of the parabola and intersects the parabola at two points.

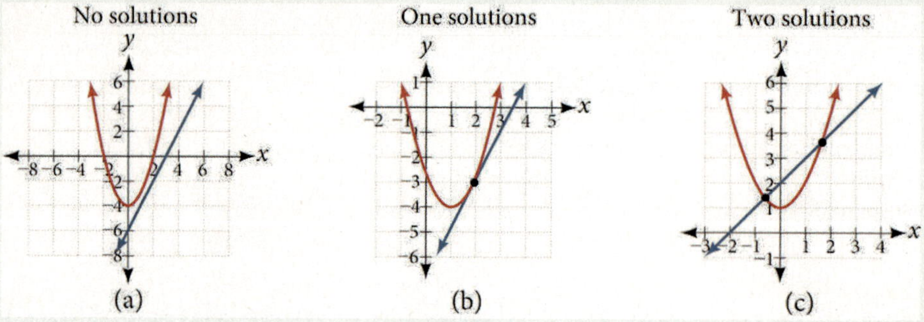

Figure 2

HOW TO

Given a system of equations containing a line and a parabola, find the solution.

1. Solve the linear equation for one of the variables.
2. Substitute the expression obtained in step one into the parabola equation.
3. Solve for the remaining variable.
4. Check your solutions in both equations.

EXAMPLE 1

Solving a System of Nonlinear Equations Representing a Parabola and a Line

Solve the system of equations.

$$\begin{aligned} x - y &= -1 \\ y &= x^2 + 1 \end{aligned}$$

Solution

Solve the first equation for x and then substitute the resulting expression into the second equation.

$$\begin{aligned} x - y &= -1 \\ x &= y - 1 \qquad \text{Solve for } x. \end{aligned}$$

$$\begin{aligned} y &= x^2 + 1 \\ y &= (y-1)^2 + 1 \qquad \text{Substitute expression for } x. \end{aligned}$$

Expand the equation and set it equal to zero.

$$\begin{aligned} y &= (y-1)^2 + 1 \\ &= (y^2 - 2y + 1) + 1 \\ &= y^2 - 2y + 2 \\ 0 &= y^2 - 3y + 2 \\ &= (y-2)(y-1) \end{aligned}$$

Solving for y gives $y = 2$ and $y = 1$. Next, substitute each value for y into the first equation to solve for x. Always substitute the value into the linear equation to check for extraneous solutions.

$$x - y = -1$$
$$x - (2) = -1$$
$$x = 1$$

$$x - (1) = -1$$
$$x = 0$$

The solutions are $(1, 2)$ and $(0, 1)$, which can be verified by substituting these (x, y) values into both of the original equations. See Figure 3.

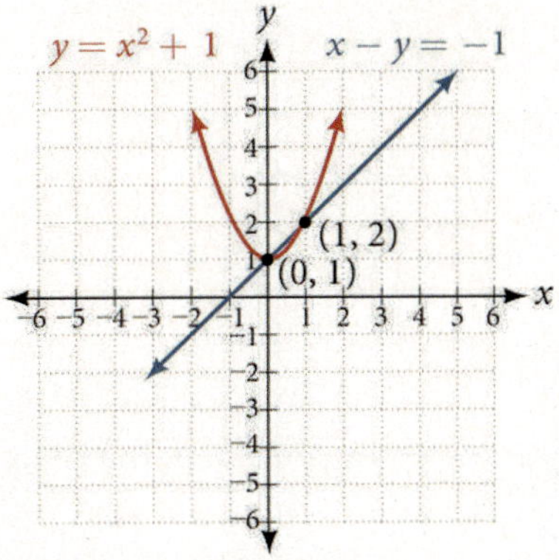

Figure 3

Q&A **Could we have substituted values for y into the second equation to solve for x in Example 1?**

Yes, but because x is squared in the second equation this could give us extraneous solutions for x.

For $y = 1$

$$y = x^2 + 1$$
$$1 = x^2 + 1$$
$$x^2 = 0$$
$$x = \pm\sqrt{0} = 0$$

This gives us the same value as in the solution.

For $y = 2$

$$y = x^2 + 1$$
$$2 = x^2 + 1$$
$$x^2 = 1$$
$$x = \pm\sqrt{1} = \pm 1$$

Notice that -1 is an extraneous solution.

TRY IT #1 Solve the given system of equations by substitution.

$$\begin{aligned} 3x - y &= -2 \\ 2x^2 - y &= 0 \end{aligned}$$

Intersection of a Circle and a Line

Just as with a parabola and a line, there are three possible outcomes when solving a system of equations representing a circle and a line.

Possible Types of Solutions for the Points of Intersection of a Circle and a Line

Figure 4 illustrates possible solution sets for a system of equations involving a circle and a line.

- No solution. The line does not intersect the circle.
- One solution. The line is tangent to the circle and intersects the circle at exactly one point.
- Two solutions. The line crosses the circle and intersects it at two points.

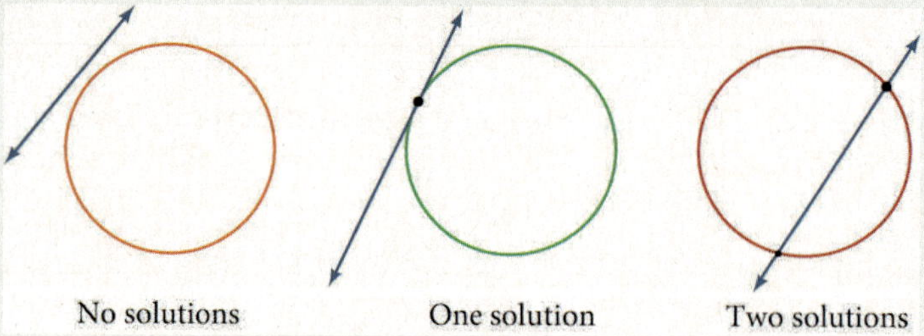

Figure 4

HOW TO

Given a system of equations containing a line and a circle, find the solution.

1. Solve the linear equation for one of the variables.
2. Substitute the expression obtained in step one into the equation for the circle.
3. Solve for the remaining variable.
4. Check your solutions in both equations.

EXAMPLE 2

Finding the Intersection of a Circle and a Line by Substitution

Find the intersection of the given circle and the given line by substitution.

$$x^2 + y^2 = 5$$
$$y = 3x - 5$$

Solution

One of the equations has already been solved for y. We will substitute $y = 3x - 5$ into the equation for the circle.

$$x^2 + (3x - 5)^2 = 5$$
$$x^2 + 9x^2 - 30x + 25 = 5$$
$$10x^2 - 30x + 20 = 0$$

Now, we factor and solve for x.

$$10(x^2 - 3x + 2) = 0$$
$$10(x - 2)(x - 1) = 0$$
$$x = 2$$
$$x = 1$$

Substitute the two x-values into the original linear equation to solve for y.

$$y = 3(2) - 5$$
$$= 1$$

$$y = 3(1) - 5$$
$$= -2$$

The line intersects the circle at $(2, 1)$ and $(1, -2)$, which can be verified by substituting these (x, y) values into both of the original equations. See Figure 5.

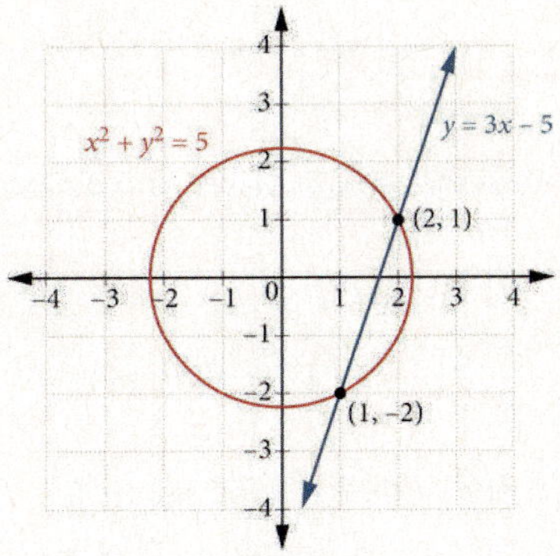

Figure 5

TRY IT #2 Solve the system of nonlinear equations.

$$\begin{aligned} x^2 + y^2 &= 10 \\ x - 3y &= -10 \end{aligned}$$

Solving a System of Nonlinear Equations Using Elimination

We have seen that substitution is often the preferred method when a system of equations includes a linear equation and a nonlinear equation. However, when both equations in the system have like variables of the second degree, solving them using elimination by addition is often easier than substitution. Generally, elimination is a far simpler method when the system involves only two equations in two variables (a two-by-two system), rather than a three-by-three system, as there are fewer steps. As an example, we will investigate the possible types of solutions when solving a system of equations representing a circle and an ellipse.

Possible Types of Solutions for the Points of Intersection of a Circle and an Ellipse

Figure 6 illustrates possible solution sets for a system of equations involving a circle and an ellipse.

- No solution. The circle and ellipse do not intersect. One shape is inside the other or the circle and the ellipse are a distance away from the other.
- One solution. The circle and ellipse are tangent to each other, and intersect at exactly one point.
- Two solutions. The circle and the ellipse intersect at two points.
- Three solutions. The circle and the ellipse intersect at three points.
- Four solutions. The circle and the ellipse intersect at four points.

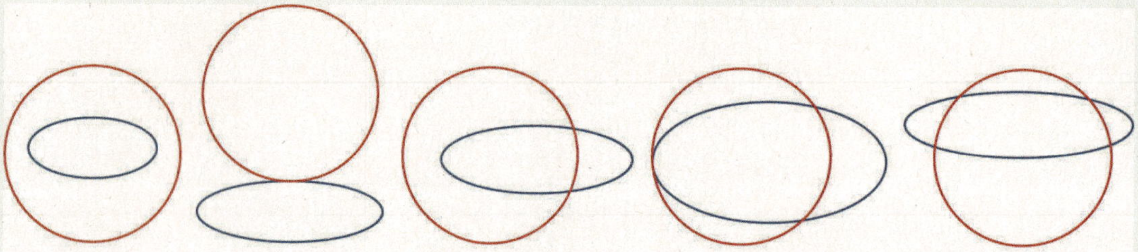

Figure 6

EXAMPLE 3

Solving a System of Nonlinear Equations Representing a Circle and an Ellipse

Solve the system of nonlinear equations.

$$\begin{aligned} x^2 + y^2 &= 26 \quad (1) \\ 3x^2 + 25y^2 &= 100 \quad (2) \end{aligned}$$

Solution

Let's begin by multiplying equation (1) by −3, and adding it to equation (2).

$$\begin{aligned} (-3)(x^2 + y^2) &= (-3)(26) \\ -3x^2 - 3y^2 &= -78 \\ 3x^2 + 25y^2 &= 100 \\ \hline 22y^2 &= 22 \end{aligned}$$

After we add the two equations together, we solve for y.

$$\begin{aligned} y^2 &= 1 \\ y &= \pm\sqrt{1} = \pm 1 \end{aligned}$$

Substitute $y = \pm 1$ into one of the equations and solve for x.

$$\begin{aligned} x^2 + (1)^2 &= 26 \\ x^2 + 1 &= 26 \\ x^2 &= 25 \\ x &= \pm\sqrt{25} = \pm 5 \end{aligned}$$

$$\begin{aligned} x^2 + (-1)^2 &= 26 \\ x^2 + 1 &= 26 \\ x^2 &= 25 = \pm 5 \end{aligned}$$

There are four solutions: $(5, 1)$, $(-5, 1)$, $(5, -1)$, and $(-5, -1)$. See Figure 7.

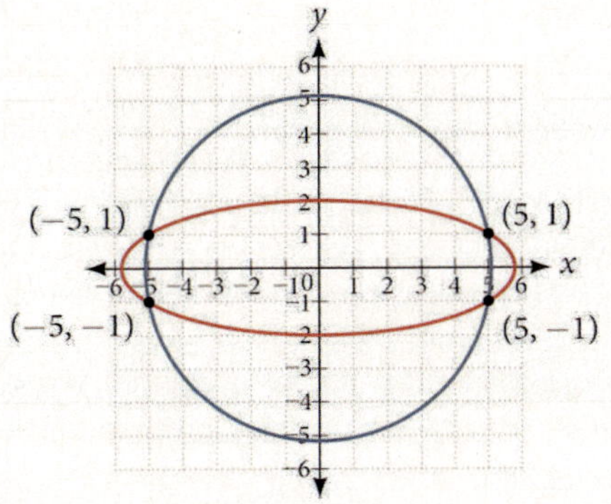

Figure 7

> TRY IT #3 Find the solution set for the given system of nonlinear equations.

$$\begin{aligned} 4x^2 + y^2 &= 13 \\ x^2 + y^2 &= 10 \end{aligned}$$

Graphing a Nonlinear Inequality

All of the equations in the systems that we have encountered so far have involved equalities, but we may also encounter systems that involve inequalities. We have already learned to graph linear inequalities by graphing the corresponding equation, and then shading the region represented by the inequality symbol. Now, we will follow similar steps to graph a nonlinear inequality so that we can learn to solve systems of nonlinear inequalities. A **nonlinear inequality** is an inequality containing a nonlinear expression. Graphing a nonlinear inequality is much like graphing a linear inequality.

Recall that when the inequality is greater than, $y > a$, or less than, $y < a$, the graph is drawn with a dashed line. When the inequality is greater than or equal to, $y \geq a$, or less than or equal to, $y \leq a$, the graph is drawn with a solid line. The graphs will create regions in the plane, and we will test each region for a solution. If one point in the region works, the whole region works. That is the region we shade. See Figure 8.

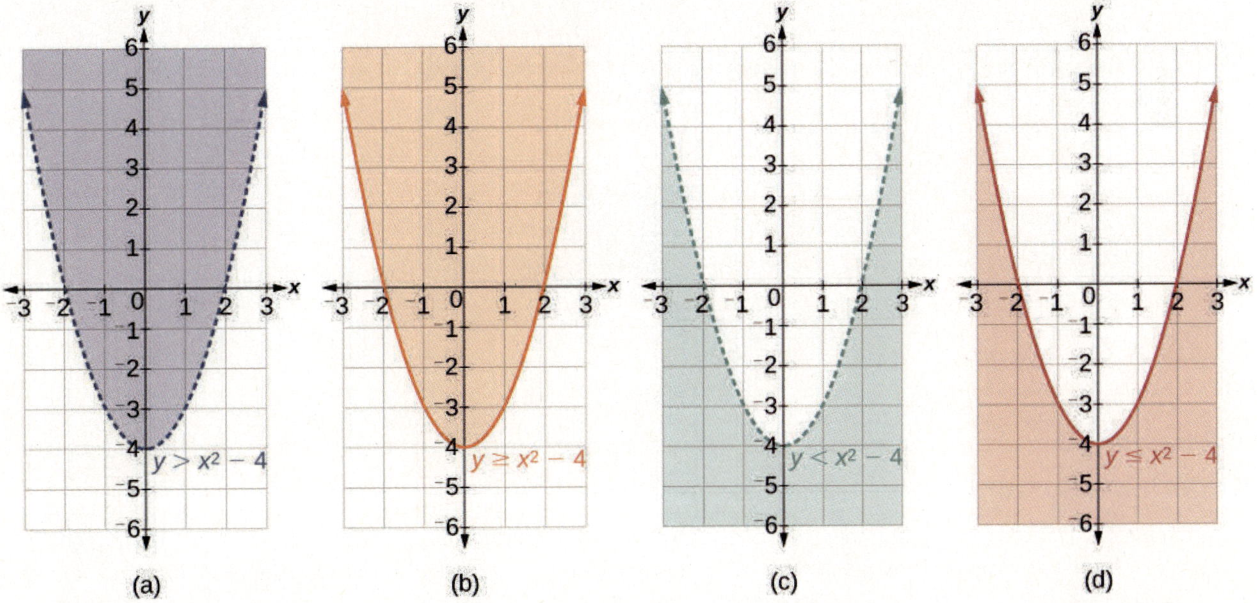

Figure 8 (a) an example of $y > a$; (b) an example of $y \geq a$; (c) an example of $y < a$; (d) an example of $y \leq a$

HOW TO

Given an inequality bounded by a parabola, sketch a graph.

1. Graph the parabola as if it were an equation. This is the boundary for the region that is the solution set.
2. If the boundary is included in the region (the operator is $\leq$ or $\geq$), the parabola is graphed as a solid line.
3. If the boundary is not included in the region (the operator is < or >), the parabola is graphed as a dashed line.
4. Test a point in one of the regions to determine whether it satisfies the inequality statement. If the statement is true, the solution set is the region including the point. If the statement is false, the solution set is the region on the other side of the boundary line.
5. Shade the region representing the solution set.

EXAMPLE 4

Graphing an Inequality for a Parabola
Graph the inequality $y > x^2 + 1$.

Solution

First, graph the corresponding equation $y = x^2 + 1$. Since $y > x^2 + 1$ has a greater than symbol, we draw the graph with a dashed line. Then we choose points to test both inside and outside the parabola. Let's test the points $(0, 2)$ and $(2, 0)$. One point is clearly inside the parabola and the other point is clearly outside.

$$\begin{aligned} y &> x^2 + 1 \\ 2 &> (0)^2 + 1 \\ 2 &> 1 \qquad \text{True} \end{aligned}$$

$$\begin{aligned} 0 &> (2)^2 + 1 \\ 0 &> 5 \qquad \text{False} \end{aligned}$$

The graph is shown in Figure 9. We can see that the solution set consists of all points inside the parabola, but not on the graph itself.

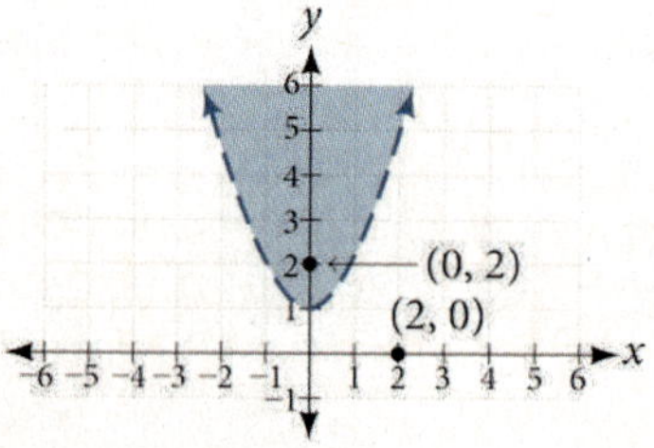

Figure 9

Graphing a System of Nonlinear Inequalities

Now that we have learned to graph nonlinear inequalities, we can learn how to graph systems of nonlinear inequalities. A **system of nonlinear inequalities** is a system of two or more inequalities in two or more variables containing at least one inequality that is not linear. Graphing a system of nonlinear inequalities is similar to graphing a system of linear inequalities. The difference is that our graph may result in more shaded regions that represent a solution than we find in a system of linear inequalities. The solution to a nonlinear system of inequalities is the region of the graph where the shaded regions of the graph of each inequality overlap, or where the regions intersect, called the **feasible region**.

HOW TO

Given a system of nonlinear inequalities, sketch a graph.

1. Find the intersection points by solving the corresponding system of nonlinear equations.
2. Graph the nonlinear equations.
3. Find the shaded regions of each inequality.
4. Identify the feasible region as the intersection of the shaded regions of each inequality or the set of points common to each inequality.

EXAMPLE 5

Graphing a System of Inequalities

Graph the given system of inequalities.

$$x^2 - y \leq 0$$
$$2x^2 + y \leq 12$$

Solution

These two equations are clearly parabolas. We can find the points of intersection by the elimination process: Add both equations and the variable y will be eliminated. Then we solve for x.

$$\begin{aligned} x^2 - y &= 0 \\ 2x^2 + y &= 12 \\ \hline 3x^2 &= 12 \\ x^2 &= 4 \\ x &= \pm 2 \end{aligned}$$

Substitute the x-values into one of the equations and solve for y.

$$\begin{aligned} x^2 - y &= 0 \\ (2)^2 - y &= 0 \\ 4 - y &= 0 \\ y &= 4 \end{aligned}$$

$$\begin{aligned} (-2)^2 - y &= 0 \\ 4 - y &= 0 \\ y &= 4 \end{aligned}$$

The two points of intersection are $(2, 4)$ and $(-2, 4)$. Notice that the equations can be rewritten as follows.

$$\begin{aligned} x^2 - y &\leq 0 \\ x^2 &\leq y \\ y &\geq x^2 \end{aligned}$$

$$\begin{aligned} 2x^2 + y &\leq 12 \\ y &\leq -2x^2 + 12 \end{aligned}$$

Graph each inequality. See Figure 10. The feasible region is the region between the two equations bounded by $2x^2 + y \leq 12$ on the top and $x^2 - y \leq 0$ on the bottom.

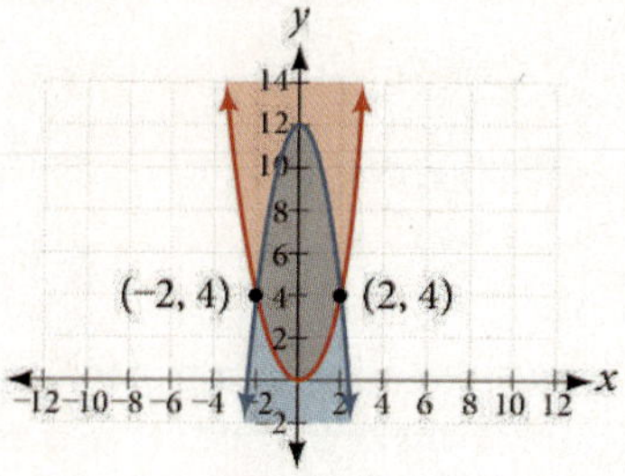

Figure 10

TRY IT #4 Graph the given system of inequalities.

$$\begin{aligned} y &\geq x^2 - 1 \\ x - y &\geq -1 \end{aligned}$$

MEDIA

Access these online resources for additional instruction and practice with nonlinear equations.

Solve a System of Nonlinear Equations Using Substitution (http://openstax.org/l/nonlinsub)
Solve a System of Nonlinear Equations Using Elimination (http://openstax.org/l/nonlinelim)

11.3 SECTION EXERCISES

Verbal

1. Explain whether a system of two nonlinear equations can have exactly two solutions. What about exactly three? If not, explain why not. If so, give an example of such a system, in graph form, and explain why your choice gives two or three answers.

2. When graphing an inequality, explain why we only need to test one point to determine whether an entire region is the solution?

3. When you graph a system of inequalities, will there always be a feasible region? If so, explain why. If not, give an example of a graph of inequalities that does not have a feasible region. Why does it not have a feasible region?

4. If you graph a revenue and cost function, explain how to determine in what regions there is profit.

5. If you perform your break-even analysis and there is more than one solution, explain how you would determine which x-values are profit and which are not.

Algebraic

For the following exercises, solve the system of nonlinear equations using substitution.

6. $\begin{aligned} x + y &= 4 \\ x^2 + y^2 &= 9 \end{aligned}$

7. $\begin{aligned} y &= x - 3 \\ x^2 + y^2 &= 9 \end{aligned}$

8. $\begin{aligned} y &= x \\ x^2 + y^2 &= 9 \end{aligned}$

9. $y = -x$
$x^2 + y^2 = 9$

10. $x = 2$
$x^2 - y^2 = 9$

For the following exercises, solve the system of nonlinear equations using elimination.

11. $4x^2 - 9y^2 = 36$
$4x^2 + 9y^2 = 36$

12. $x^2 + y^2 = 25$
$x^2 - y^2 = 1$

13. $2x^2 + 4y^2 = 4$
$2x^2 - 4y^2 = 25x - 10$

14. $y^2 - x^2 = 9$
$3x^2 + 2y^2 = 8$

15. $x^2 + y^2 + \frac{1}{16} = 2500$
$y = 2x^2$

For the following exercises, use any method to solve the system of nonlinear equations.

16. $-2x^2 + y = -5$
$6x - y = 9$

17. $-x^2 + y = 2$
$-x + y = 2$

18. $x^2 + y^2 = 1$
$y = 20x^2 - 1$

19. $x^2 + y^2 = 1$
$y = -x^2$

20. $2x^3 - x^2 = y$
$y = \frac{1}{2} - x$

21. $9x^2 + 25y^2 = 225$
$(x-6)^2 + y^2 = 1$

22. $x^4 - x^2 = y$
$x^2 + y = 0$

23. $2x^3 - x^2 = y$
$x^2 + y = 0$

For the following exercises, use any method to solve the nonlinear system.

24. $x^2 + y^2 = 9$
$y = 3 - x^2$

25. $x^2 - y^2 = 9$
$x = 3$

26. $x^2 - y^2 = 9$
$y = 3$

27. $x^2 - y^2 = 9$
$x - y = 0$

28. $-x^2 + y = 2$
$-4x + y = -1$

29. $-x^2 + y = 2$
$2y = -x$

30. $x^2 + y^2 = 25$
$x^2 - y^2 = 36$

31. $x^2 + y^2 = 1$
$y^2 = x^2$

32. $16x^2 - 9y^2 + 144 = 0$
$y^2 + x^2 = 16$

33. $3x^2 - y^2 = 12$
$(x-1)^2 + y^2 = 1$

34. $3x^2 - y^2 = 12$
$(x-1)^2 + y^2 = 4$

35. $3x^2 - y^2 = 12$
$x^2 + y^2 = 16$

36. $x^2 - y^2 - 6x - 4y - 11 = 0$
$-x^2 + y^2 = 5$

37. $x^2 + y^2 - 6y = 7$
$x^2 + y = 1$

38. $x^2 + y^2 = 6$
$xy = 1$

Graphical

For the following exercises, graph the inequality.

39. $x^2 + y < 9$

40. $x^2 + y^2 < 4$

For the following exercises, graph the system of inequalities. Label all points of intersection.

41. $x^2 + y < 1$
$y > 2x$

42. $x^2 + y < -5$
$y > 5x + 10$

43. $x^2 + y^2 < 25$
$3x^2 - y^2 > 12$

44. $x^2 - y^2 > -4$
$x^2 + y^2 < 12$

45. $x^2 + 3y^2 > 16$
$3x^2 - y^2 < 1$

Extensions

For the following exercises, graph the inequality.

46. $y \geq e^x$
$y \leq \ln(x) + 5$

47. $y \leq -\log(x)$
$y \leq e^x$

For the following exercises, find the solutions to the nonlinear equations with two variables.

48. $\frac{4}{x^2} + \frac{1}{y^2} = 24$
$\frac{5}{x^2} - \frac{2}{y^2} + 4 = 0$

49. $\frac{6}{x^2} - \frac{1}{y^2} = 8$
$\frac{1}{x^2} - \frac{6}{y^2} = \frac{1}{8}$

50. $x^2 - xy + y^2 - 2 = 0$
$x + 3y = 4$

51. $x^2 - xy - 2y^2 - 6 = 0$
$x^2 + y^2 = 1$

52. $x^2 + 4xy - 2y^2 - 6 = 0$
$x = y + 2$

Technology

For the following exercises, solve the system of inequalities. Use a calculator to graph the system to confirm the answer.

53. $xy < 1$
$y > \sqrt{x}$

54. $x^2 + y < 3$
$y > 2x$

Real-World Applications

For the following exercises, construct a system of nonlinear equations to describe the given behavior, then solve for the requested solutions.

55. Two numbers add up to 300. One number is twice the square of the other number. What are the numbers?

56. The squares of two numbers add to 360. The second number is half the value of the first number squared. What are the numbers?

57. A laptop company has discovered their cost and revenue functions for each day: $C(x) = 3x^2 - 10x + 200$ and $R(x) = -2x^2 + 100x + 50$. If they want to make a profit, what is the range of laptops per day that they should produce? Round to the nearest number which would generate profit.

58. A cell phone company has the following cost and revenue functions: $C(x) = 8x^2 - 600x + 21{,}500$ and $R(x) = -3x^2 + 480x$. What is the range of cell phones they should produce each day so there is profit? Round to the nearest number that generates profit.

11.4 Partial Fractions

Learning Objectives

In this section, you will:

- Decompose $\frac{P(x)}{Q(x)}$, where $Q(x)$ has only nonrepeated linear factors.
- Decompose $\frac{P(x)}{Q(x)}$, where $Q(x)$ has repeated linear factors.
- Decompose $\frac{P(x)}{Q(x)}$, where $Q(x)$ has a nonrepeated irreducible quadratic factor.
- Decompose $\frac{P(x)}{Q(x)}$, where $Q(x)$ has a repeated irreducible quadratic factor.

Earlier in this chapter, we studied systems of two equations in two variables, systems of three equations in three variables, and nonlinear systems. Here we introduce another way that systems of equations can be utilized—the decomposition of rational expressions.

Fractions can be complicated; adding a variable in the denominator makes them even more so. The methods studied in this section will help simplify the concept of a rational expression.

Decomposing $\frac{P(x)}{Q(x)}$ Where *Q(x)* Has Only Nonrepeated Linear Factors

Recall the algebra regarding adding and subtracting rational expressions. These operations depend on finding a common denominator so that we can write the sum or difference as a single, simplified rational expression. In this section, we will look at partial fraction decomposition, which is the undoing of the procedure to add or subtract rational expressions. In other words, it is a return from the single simplified rational expression to the original expressions, called the **partial fraction**.

For example, suppose we add the following fractions:

$$\frac{2}{x-3}+\frac{-1}{x+2}$$

We would first need to find a common denominator, $(x+2)(x-3)$.

Next, we would write each expression with this common denominator and find the sum of the terms.

$$\frac{2}{x-3}\left(\frac{x+2}{x+2}\right)+\frac{-1}{x+2}\left(\frac{x-3}{x-3}\right)=$$
$$\frac{2x+4-x+3}{(x+2)(x-3)}=\frac{x+7}{x^2-x-6}$$

Partial fraction decomposition is the reverse of this procedure. We would start with the solution and rewrite (decompose) it as the sum of two fractions.

$$\underbrace{\frac{x+7}{x^2-x-6}}_{\text{Simplified sum}}=\underbrace{\frac{2}{x-3}+\frac{-1}{x+2}}_{\text{Partial fraction decomposition}}$$

We will investigate rational expressions with linear factors and quadratic factors in the denominator where the degree of the numerator is less than the degree of the denominator. Regardless of the type of expression we are decomposing, the first and most important thing to do is factor the denominator.

When the denominator of the simplified expression contains distinct linear factors, it is likely that each of the original rational expressions, which were added or subtracted, had one of the linear factors as the denominator. In other words, using the example above, the factors of x^2-x-6 are $(x-3)(x+2)$, the denominators of the decomposed rational expression. So we will rewrite the simplified form as the sum of individual fractions and use a variable for each numerator. Then, we will solve for each numerator using one of several methods available for partial fraction decomposition.

Partial Fraction Decomposition of $\frac{P(x)}{Q(x)}$: $Q(x)$ Has Nonrepeated Linear Factors

The partial fraction decomposition of $\frac{P(x)}{Q(x)}$ when $Q(x)$ has nonrepeated linear factors and the degree of $P(x)$ is less than the degree of $Q(x)$ is

$$\frac{P(x)}{Q(x)} = \frac{A_1}{(a_1x + b_1)} + \frac{A_2}{(a_2x + b_2)} + \frac{A_3}{(a_3x + b_3)} + \cdots + \frac{A_n}{(a_nx + b_n)}.$$

HOW TO

Given a rational expression with distinct linear factors in the denominator, decompose it.

1. Use a variable for the original numerators, usually $A, B,$ or C, depending on the number of factors, placing each variable over a single factor. For the purpose of this definition, we use A_n for each numerator
$$\frac{P(x)}{Q(x)} = \frac{A_1}{(a_1x + b_1)} + \frac{A_2}{(a_2x + b_2)} + \cdots + \frac{A_n}{(a_nx + b_n)}$$
2. Multiply both sides of the equation by the common denominator to eliminate fractions.
3. Expand the right side of the equation and collect like terms.
4. Set coefficients of like terms from the left side of the equation equal to those on the right side to create a system of equations to solve for the numerators.

EXAMPLE 1

Decomposing a Rational Function with Distinct Linear Factors

Decompose the given rational expression with distinct linear factors.

$$\frac{3x}{(x+2)(x-1)}$$

Solution

We will separate the denominator factors and give each numerator a symbolic label, like A, B, or C.

$$\frac{3x}{(x+2)(x-1)} = \frac{A}{(x+2)} + \frac{B}{(x-1)}$$

Multiply both sides of the equation by the common denominator to eliminate the fractions:

$$(x+2)(x-1)\left[\frac{3x}{(x+2)(x-1)}\right] = \cancel{(x+2)}(x-1)\left[\frac{A}{\cancel{(x+2)}}\right] + (x+2)\cancel{(x-1)}\left[\frac{B}{\cancel{(x-1)}}\right]$$

The resulting equation is

$$3x = A(x-1) + B(x+2)$$

Expand the right side of the equation and collect like terms.

$$\begin{aligned} 3x &= Ax - A + Bx + 2B \\ 3x &= (A + B)x - A + 2B \end{aligned}$$

Set up a system of equations associating corresponding coefficients.

$$\begin{aligned} 3 &= A + B \\ 0 &= -A + 2B \end{aligned}$$

Add the two equations and solve for B.

$$\begin{aligned} 3 &= A + B \\ 0 &= -A + 2B \\ \hline 3 &= 0 + 3B \\ 1 &= B \end{aligned}$$

Substitute $B = 1$ into one of the original equations in the system.

$$\begin{aligned} 3 &= A + 1 \\ 2 &= A \end{aligned}$$

Thus, the partial fraction decomposition is

$$\frac{3x}{(x+2)(x-1)} = \frac{2}{(x+2)} + \frac{1}{(x-1)}$$

Another method to use to solve for A or B is by considering the equation that resulted from eliminating the fractions and substituting a value for x that will make either the A- or B-term equal 0. If we let $x = 1$, the A- term becomes 0 and we can simply solve for B.

$$\begin{aligned} 3x &= A(x-1) + B(x+2) \\ 3(1) &= A[(1)-1] + B[(1)+2] \\ 3 &= 0 + 3B \\ 1 &= B \end{aligned}$$

Next, either substitute $B = 1$ into the equation and solve for A, or make the B-term 0 by substituting $x = -2$ into the equation.

$$\begin{aligned} 3x &= A(x-1) + B(x+2) \\ 3(-2) &= A[(-2)-1] + B[(-2)+2] \\ -6 &= -3A + 0 \\ \tfrac{-6}{-3} &= A \\ 2 &= A \end{aligned}$$

We obtain the same values for A and B using either method, so the decompositions are the same using either method.

$$\frac{3x}{(x+2)(x-1)} = \frac{2}{(x+2)} + \frac{1}{(x-1)}$$

Although this method is not seen very often in textbooks, we present it here as an alternative that may make some partial fraction decompositions easier. It is known as the Heaviside method, named after Charles Heaviside, a pioneer in the study of electronics.

TRY IT #1 Find the partial fraction decomposition of the following expression.

$$\frac{x}{(x-3)(x-2)}$$

Decomposing $\frac{P(x)}{Q(x)}$ Where $Q(x)$ Has Repeated Linear Factors

Some fractions we may come across are special cases that we can decompose into partial fractions with repeated linear factors. We must remember that we account for repeated factors by writing each factor in increasing powers.

Partial Fraction Decomposition of $\frac{P(x)}{Q(x)}$: $Q(x)$ Has Repeated Linear Factors

The partial fraction decomposition of $\frac{P(x)}{Q(x)}$, when $Q(x)$ has a repeated linear factor occurring n times and the degree of $P(x)$ is less than the degree of $Q(x)$, is

$$\frac{P(x)}{Q(x)} = \frac{A_1}{(ax+b)} + \frac{A_2}{(ax+b)^2} + \frac{A_3}{(ax+b)^3} + \cdots + \frac{A_n}{(ax+b)^n}$$

Write the denominator powers in increasing order.

HOW TO

Given a rational expression with repeated linear factors, decompose it.

1. Use a variable like A, B, or C for the numerators and account for increasing powers of the denominators.
$$\frac{P(x)}{Q(x)} = \frac{A_1}{(ax+b)} + \frac{A_2}{(ax+b)^2} + \ldots + \frac{A_n}{(ax+b)^n}$$
2. Multiply both sides of the equation by the common denominator to eliminate fractions.
3. Expand the right side of the equation and collect like terms.
4. Set coefficients of like terms from the left side of the equation equal to those on the right side to create a system of equations to solve for the numerators.

EXAMPLE 2

Decomposing with Repeated Linear Factors

Decompose the given rational expression with repeated linear factors.

$$\frac{-x^2+2x+4}{x^3-4x^2+4x}$$

Solution

The denominator factors are $x(x-2)^2$. To allow for the repeated factor of $(x-2)$, the decomposition will include three denominators: x, $(x-2)$, and $(x-2)^2$. Thus,

$$\frac{-x^2+2x+4}{x^3-4x^2+4x} = \frac{A}{x} + \frac{B}{(x-2)} + \frac{C}{(x-2)^2}$$

Next, we multiply both sides by the common denominator.

$$x(x-2)^2\left[\frac{-x^2+2x+4}{x(x-2)^2}\right] = \left[\frac{A}{x} + \frac{B}{(x-2)} + \frac{C}{(x-2)^2}\right]x(x-2)^2$$
$$-x^2+2x+4 = A(x-2)^2 + Bx(x-2) + Cx$$

On the right side of the equation, we expand and collect like terms.

$$\begin{aligned} -x^2+2x+4 &= A(x^2-4x+4) + B(x^2-2x) + Cx \\ &= Ax^2 - 4Ax + 4A + Bx^2 - 2Bx + Cx \\ &= (A+B)x^2 + (-4A-2B+C)x + 4A \end{aligned}$$

Next, we compare the coefficients of both sides. This will give the system of equations in three variables:

$$-x^2+2x+4 = (A+B)x^2 + (-4A-2B+C)x + 4A$$
$$\begin{aligned} A+B &= -1 \quad (1) \\ -4A-2B+C &= 2 \quad (2) \\ 4A &= 4 \quad (3) \end{aligned}$$

Solving for A, we have

$$\begin{aligned} 4A &= 4 \\ A &= 1 \end{aligned}$$

Substitute $A = 1$ into equation (1).

$$\begin{aligned} A+B &= -1 \\ (1)+B &= -1 \\ B &= -2 \end{aligned}$$

Then, to solve for C, substitute the values for A and B into equation (2).

$$\begin{aligned} -4A-2B+C &= 2 \\ -4(1)-2(-2)+C &= 2 \\ -4+4+C &= 2 \\ C &= 2 \end{aligned}$$

Thus,

$$\frac{-x^2+2x+4}{x^3-4x^2+4x}=\frac{1}{x}-\frac{2}{(x-2)}+\frac{2}{(x-2)^2}$$

TRY IT #2 Find the partial fraction decomposition of the expression with repeated linear factors.

$$\frac{6x-11}{(x-1)^2}$$

Decomposing $\frac{P(x)}{Q(x)}$, Where *Q(x)* Has a Nonrepeated Irreducible Quadratic Factor

So far, we have performed partial fraction decomposition with expressions that have had linear factors in the denominator, and we applied numerators $A, B,$ or C representing constants. Now we will look at an example where one of the factors in the denominator is a quadratic expression that does not factor. This is referred to as an irreducible quadratic factor. In cases like this, we use a linear numerator such as $Ax+B, Bx+C,$ etc.

Decomposition of $\frac{P(x)}{Q(x)}$: $Q(x)$ Has a Nonrepeated Irreducible Quadratic Factor

The partial fraction decomposition of $\frac{P(x)}{Q(x)}$ such that $Q(x)$ has a nonrepeated irreducible quadratic factor and the degree of $P(x)$ is less than the degree of $Q(x)$ is written as

$$\frac{P(x)}{Q(x)}=\frac{A_1x+B_1}{\left(a_1x^2+b_1x+c_1\right)}+\frac{A_2x+B_2}{\left(a_2x^2+b_2x+c_2\right)}+\cdots+\frac{A_nx+B_n}{\left(a_nx^2+b_nx+c_n\right)}$$

The decomposition may contain more rational expressions if there are linear factors. Each linear factor will have a different constant numerator: $A, B, C,$ and so on.

HOW TO

Given a rational expression where the factors of the denominator are distinct, irreducible quadratic factors, decompose it.

1. Use variables such as $A, B,$ or C for the constant numerators over linear factors, and linear expressions such as $A_1x+B_1, A_2x+B_2,$ etc., for the numerators of each quadratic factor in the denominator.
$$\frac{P(x)}{Q(x)}=\frac{A}{ax+b}+\frac{A_1x+B_1}{\left(a_1x^2+b_1x+c_1\right)}+\frac{A_2x+B_2}{\left(a_2x^2+b_2x+c_2\right)}+\cdots+\frac{A_nx+B_n}{\left(a_nx^2+b_nx+c_n\right)}$$
2. Multiply both sides of the equation by the common denominator to eliminate fractions.
3. Expand the right side of the equation and collect like terms.
4. Set coefficients of like terms from the left side of the equation equal to those on the right side to create a system of equations to solve for the numerators.

EXAMPLE 3

Decomposing $\frac{P(x)}{Q(x)}$ When $Q(x)$ Contains a Nonrepeated Irreducible Quadratic Factor

Find a partial fraction decomposition of the given expression.

$$\frac{8x^2 + 12x-20}{(x+3)\left(x^2+x+2\right)}$$

Solution

We have one linear factor and one irreducible quadratic factor in the denominator, so one numerator will be a constant and the other numerator will be a linear expression. Thus,

$$\frac{8x^2 + 12x-20}{(x+3)\left(x^2+x+2\right)} = \frac{A}{(x+3)} + \frac{Bx+C}{\left(x^2+x+2\right)}$$

We follow the same steps as in previous problems. First, clear the fractions by multiplying both sides of the equation by the common denominator.

$$\begin{aligned}(x+3)(x^2+x+2)\left[\frac{8x^2+12x-20}{(x+3)(x^2+x+2)}\right] &= \left[\frac{A}{(x+3)} + \frac{Bx+C}{(x^2+x+2)}\right](x+3)(x^2+x+2) \\ 8x^2+12x-20 &= A(x^2+x+2)+(Bx+C)(x+3)\end{aligned}$$

Notice we could easily solve for A by choosing a value for x that will make the $Bx+C$ term equal 0. Let $x=-3$ and substitute it into the equation.

$$\begin{aligned}8x^2+12x-20 &= A(x^2+x+2)+(Bx+C)(x+3) \\ 8(-3)^2+12(-3)-20 &= A((-3)^2+(-3)+2)+(B(-3)+C)((-3)+3) \\ 16 &= 8A \\ A &= 2\end{aligned}$$

Now that we know the value of $A,$ substitute it back into the equation. Then expand the right side and collect like terms.

$$\begin{aligned}8x^2+12x-20 &= 2(x^2+x+2)+(Bx+C)(x+3) \\ 8x^2+12x-20 &= 2x^2+2x+4+Bx^2+3B+Cx+3C \\ 8x^2+12x-20 &= (2+B)x^2+(2+3B+C)x+(4+3C)\end{aligned}$$

Setting the coefficients of terms on the right side equal to the coefficients of terms on the left side gives the system of equations.

$$\begin{aligned}2+B &= 8 \quad (1) \\ 2+3B+C &= 12 \quad (2) \\ 4+3C &= -20 \quad (3)\end{aligned}$$

Solve for B using equation (1) and solve for C using equation (3).

$$\begin{aligned}2+B &= 8 \quad (1) \\ B &= 6\end{aligned}$$

$$\begin{aligned}4+3C &= -20 \quad (3) \\ 3C &= -24 \\ C &= -8\end{aligned}$$

Thus, the partial fraction decomposition of the expression is

$$\frac{8x^2 + 12x-20}{(x+3)\left(x^2+x+2\right)} = \frac{2}{(x+3)} + \frac{6x-8}{\left(x^2+x+2\right)}$$

Q&A ***Could we have just set up a system of equations to solve Example 3?***

Yes, we could have solved it by setting up a system of equations without solving for A first. The expansion on the right would be:

$$8x^2 + 12x - 20 = Ax^2 + Ax + 2A + Bx^2 + 3B + Cx + 3C$$
$$8x^2 + 12x - 20 = (A + B)x^2 + (A + 3B + C)x + (2A + 3C)$$

So the system of equations would be:

$$\begin{aligned} A + B &= 8 \\ A + 3B + C &= 12 \\ 2A + 3C &= -20 \end{aligned}$$

TRY IT #3 Find the partial fraction decomposition of the expression with a nonrepeating irreducible quadratic factor.

$$\frac{5x^2 - 6x + 7}{(x-1)\left(x^2 + 1\right)}$$

Decomposing $\frac{P(x)}{Q(x)}$ When $Q(x)$ Has a Repeated Irreducible Quadratic Factor

Now that we can decompose a simplified rational expression with an irreducible quadratic factor, we will learn how to do partial fraction decomposition when the simplified rational expression has repeated irreducible quadratic factors. The decomposition will consist of partial fractions with linear numerators over each irreducible quadratic factor represented in increasing powers.

Decomposition of $\frac{P(x)}{Q(x)}$ When $Q(x)$ Has a Repeated Irreducible Quadratic Factor

The partial fraction decomposition of $\frac{P(x)}{Q(x)}$, when $Q(x)$ has a repeated irreducible quadratic factor and the degree of $P(x)$ is less than the degree of $Q(x)$, is

$$\frac{P(x)}{\left(ax^2 + bx + c\right)^n} = \frac{A_1x + B_1}{\left(ax^2 + bx + c\right)} + \frac{A_2x + B_2}{\left(ax^2 + bx + c\right)^2} + \frac{A_3x + B_3}{\left(ax^2 + bx + c\right)^3} + \cdots + \frac{A_nx + B_n}{\left(ax^2 + bx + c\right)^n}$$

Write the denominators in increasing powers.

HOW TO

Given a rational expression that has a repeated irreducible factor, decompose it.

1. Use variables like $A, B,$ or C for the constant numerators over linear factors, and linear expressions such as $A_1x + B_1, A_2x + B_2,$ etc., for the numerators of each quadratic factor in the denominator written in increasing powers, such as
$$\frac{P(x)}{Q(x)} = \frac{A}{ax + b} + \frac{A_1x + B_1}{(ax^2 + bx + c)} + \frac{A_2x + B_2}{(ax^2 + bx + c)^2} + \cdots + \frac{A_n + B_n}{(ax^2 + bx + c)^n}$$
2. Multiply both sides of the equation by the common denominator to eliminate fractions.
3. Expand the right side of the equation and collect like terms.
4. Set coefficients of like terms from the left side of the equation equal to those on the right side to create a system of equations to solve for the numerators.

EXAMPLE 4

Decomposing a Rational Function with a Repeated Irreducible Quadratic Factor in the Denominator

Decompose the given expression that has a repeated irreducible factor in the denominator.

$$\frac{x^4+x^3+x^2-x+1}{x\left(x^2+1\right)^2}$$

Solution

The factors of the denominator are $x, (x^2+1),$ and $(x^2+1)^2$. Recall that, when a factor in the denominator is a quadratic that includes at least two terms, the numerator must be of the linear form $Ax+B$. So, let's begin the decomposition.

$$\frac{x^4+x^3+x^2-x+1}{x\left(x^2+1\right)^2}=\frac{A}{x}+\frac{Bx+C}{(x^2+1)}+\frac{Dx+E}{(x^2+1)^2}$$

We eliminate the denominators by multiplying each term by $x\left(x^2+1\right)^2$. Thus,

$$x^4+x^3+x^2-x+1=A\left(x^2+1\right)^2+(Bx+C)(x)\left(x^2+1\right)+(Dx+E)(x)$$

Expand the right side.

$$\begin{aligned}x^4+x^3+x^2-x+1&=A(x^4+2x^2+1)+Bx^4+Bx^2+Cx^3+Cx+Dx^2+Ex\\&=Ax^4+2Ax^2+A+Bx^4+Bx^2+Cx^3+Cx+Dx^2+Ex\end{aligned}$$

Now we will collect like terms.

$$x^4+x^3+x^2-x+1=(A+B)x^4+(C)x^3+(2A+B+D)x^2+(C+E)x+A$$

Set up the system of equations matching corresponding coefficients on each side of the equal sign.

$$\begin{aligned}A+B&=1\\C&=1\\2A+B+D&=1\\C+E&=-1\\A&=1\end{aligned}$$

We can use substitution from this point. Substitute $A=1$ into the first equation.

$$\begin{aligned}1+B&=1\\B&=0\end{aligned}$$

Substitute $A=1$ and $B=0$ into the third equation.

$$\begin{aligned}2(1)+0+D&=1\\D&=-1\end{aligned}$$

Substitute $C=1$ into the fourth equation.

$$\begin{aligned}1+E&=-1\\E&=-2\end{aligned}$$

Now we have solved for all of the unknowns on the right side of the equal sign. We have $A=1, B=0, C=1, D=-1,$ and $E=-2$. We can write the decomposition as follows:

$$\frac{x^4+x^3+x^2-x+1}{x\left(x^2+1\right)^2}=\frac{1}{x}+\frac{1}{(x^2+1)}-\frac{x+2}{(x^2+1)^2}$$

TRY IT #4 Find the partial fraction decomposition of the expression with a repeated irreducible quadratic factor.

$$\frac{x^3-4x^2+9x-5}{\left(x^2-2x+3\right)^2}$$

MEDIA

Access these online resources for additional instruction and practice with partial fractions.

Partial Fraction Decomposition (http://openstax.org/l/partdecomp)
Partial Fraction Decomposition With Repeated Linear Factors (http://openstax.org/l/partdecomprlf)
Partial Fraction Decomposition With Linear and Quadratic Factors (http://openstax.org/l/partdecomlqu)

11.4 SECTION EXERCISES

Verbal

1. Can any quotient of polynomials be decomposed into at least two partial fractions? If so, explain why, and if not, give an example of such a fraction

2. Can you explain why a partial fraction decomposition is unique? (Hint: Think about it as a system of equations.)

3. Can you explain how to verify a partial fraction decomposition graphically?

4. You are unsure if you correctly decomposed the partial fraction correctly. Explain how you could double-check your answer.

5. Once you have a system of equations generated by the partial fraction decomposition, can you explain another method to solve it? For example if you had $\frac{7x+13}{3x^2+8x+15} = \frac{A}{x+1} + \frac{B}{3x+5}$, we eventually simplify to $7x + 13 = A(3x + 5) + B(x + 1)$. Explain how you could intelligently choose an x -value that will eliminate either A or B and solve for A and B.

Algebraic

For the following exercises, find the decomposition of the partial fraction for the nonrepeating linear factors.

6. $\frac{5x+16}{x^2+10x+24}$

7. $\frac{3x-79}{x^2-5x-24}$

8. $\frac{-x-24}{x^2-2x-24}$

9. $\frac{10x+47}{x^2+7x+10}$

10. $\frac{x}{6x^2+25x+25}$

11. $\frac{32x-11}{20x^2-13x+2}$

12. $\frac{x+1}{x^2+7x+10}$

13. $\frac{5x}{x^2-9}$

14. $\frac{10x}{x^2-25}$

15. $\frac{6x}{x^2-4}$

16. $\frac{2x-3}{x^2-6x+5}$

17. $\frac{4x-1}{x^2-x-6}$

18. $\frac{4x+3}{x^2+8x+15}$

19. $\frac{3x-1}{x^2-5x+6}$

For the following exercises, find the decomposition of the partial fraction for the repeating linear factors.

20. $\frac{-5x-19}{(x+4)^2}$

21. $\frac{x}{(x-2)^2}$

22. $\frac{7x+14}{(x+3)^2}$

23. $\frac{-24x-27}{(4x+5)^2}$

24. $\frac{-24x-27}{(6x-7)^2}$

25. $\frac{5-x}{(x-7)^2}$

26. $\frac{5x+14}{2x^2+12x+18}$

27. $\frac{5x^2+20x+8}{2x(x+1)^2}$

28. $\frac{4x^2+55x+25}{5x(3x+5)^2}$

29. $\frac{54x^3+127x^2+80x+16}{2x^2(3x+2)^2}$

30. $\frac{x^3-5x^2+12x+144}{x^2\left(x^2+12x+36\right)}$

For the following exercises, find the decomposition of the partial fraction for the irreducible nonrepeating quadratic factor.

31. $\frac{4x^2+6x+11}{(x+2)\left(x^2+x+3\right)}$

32. $\frac{4x^2+9x+23}{(x-1)\left(x^2+6x+11\right)}$

33. $\frac{-2x^2+10x+4}{(x-1)\left(x^2+3x+8\right)}$

34. $\frac{x^2+3x+1}{(x+1)\left(x^2+5x-2\right)}$

35. $\frac{4x^2+17x-1}{(x+3)\left(x^2+6x+1\right)}$

36. $\frac{4x^2}{(x+5)\left(x^2+7x-5\right)}$

37. $\frac{4x^2+5x+3}{x^3-1}$

38. $\frac{-5x^2+18x-4}{x^3+8}$

39. $\frac{3x^2-7x+33}{x^3+27}$

40. $\frac{x^2+2x+40}{x^3-125}$

41. $\frac{4x^2+4x+12}{8x^3-27}$

42. $\frac{-50x^2+5x-3}{125x^3-1}$

43. $\frac{-2x^3-30x^2+36x+216}{x^4+216x}$

For the following exercises, find the decomposition of the partial fraction for the irreducible repeating quadratic factor.

44. $\frac{3x^3+2x^2+14x+15}{\left(x^2+4\right)^2}$

45. $\frac{x^3+6x^2+5x+9}{\left(x^2+1\right)^2}$

46. $\frac{x^3-x^2+x-1}{\left(x^2-3\right)^2}$

47. $\frac{x^2+5x+5}{(x+2)^2}$

48. $\frac{x^3+2x^2+4x}{\left(x^2+2x+9\right)^2}$

49. $\frac{x^2+25}{\left(x^2+3x+25\right)^2}$

50. $\frac{2x^3+11x^2+7x+70}{\left(2x^2+x+14\right)^2}$

51. $\frac{5x+2}{x\left(x^2+4\right)^2}$

52. $\frac{x^4+x^3+8x^2+6x+36}{x\left(x^2+6\right)^2}$

53. $\frac{2x-9}{\left(x^2-x\right)^2}$

54. $\frac{5x^3-2x+1}{\left(x^2+2x\right)^2}$

Extensions

For the following exercises, find the partial fraction expansion.

55. $\frac{x^2+4}{(x+1)^3}$

56. $\frac{x^3-4x^2+5x+4}{(x-2)^3}$

For the following exercises, perform the operation and then find the partial fraction decomposition.

57. $\frac{7}{x+8}+\frac{5}{x-2}-\frac{x-1}{x^2-6x-16}$

58. $\frac{1}{x-4}-\frac{3}{x+6}-\frac{2x+7}{x^2+2x-24}$

59. $\frac{2x}{x^2-16}-\frac{1-2x}{x^2+6x+8}-\frac{x-5}{x^2-4x}$

11.5 Matrices and Matrix Operations

Learning Objectives

In this section, you will:

- Find the sum and difference of two matrices.
- Find scalar multiples of a matrix.
- Find the product of two matrices.

Figure 1 (credit: "SD Dirk," Flickr)

Two club soccer teams, the Wildcats and the Mud Cats, are hoping to obtain new equipment for an upcoming season. Table 1 shows the needs of both teams.

	Wildcats	Mud Cats
Goals	6	10
Balls	30	24
Jerseys	14	20

Table 1

A goal costs \$300; a ball costs \$10; and a jersey costs \$30. How can we find the total cost for the equipment needed for each team? In this section, we discover a method in which the data in the soccer equipment table can be displayed and used for calculating other information. Then, we will be able to calculate the cost of the equipment.

Finding the Sum and Difference of Two Matrices

To solve a problem like the one described for the soccer teams, we can use a matrix, which is a rectangular array of numbers. A row in a matrix is a set of numbers that are aligned horizontally. A column in a matrix is a set of numbers that are aligned vertically. Each number is an entry, sometimes called an element, of the matrix. Matrices (plural) are enclosed in [ ] or ( ), and are usually named with capital letters. For example, three matrices named A, B, and C are shown below.

$$A = \begin{bmatrix} 1 & 2 \\ 3 & 4 \end{bmatrix}, B = \begin{bmatrix} 1 & 2 & 7 \\ 0 & -5 & 6 \\ 7 & 8 & 2 \end{bmatrix}, C = \begin{bmatrix} -1 & 3 \\ 0 & 2 \\ 3 & 1 \end{bmatrix}$$

Describing Matrices

A matrix is often referred to by its size or dimensions: $m \times n$ indicating m rows and n columns. Matrix entries are defined first by row and then by column. For example, to locate the entry in matrix A identified as a_{ij}, we look for the entry in row i, column j. In matrix A, shown below, the entry in row 2, column 3 is a_{23}.

$$A = \begin{bmatrix} a_{11} & a_{12} & a_{13} \\ a_{21} & a_{22} & a_{23} \\ a_{31} & a_{32} & a_{33} \end{bmatrix}$$

A square matrix is a matrix with dimensions $n \times n$, meaning that it has the same number of rows as columns. The 3×3 matrix above is an example of a square matrix.

A row matrix is a matrix consisting of one row with dimensions $1 \times n$.

$$\begin{bmatrix} a_{11} & a_{12} & a_{13} \end{bmatrix}$$

A column matrix is a matrix consisting of one column with dimensions $m \times 1$.

$$\begin{bmatrix} a_{11} \\ a_{21} \\ a_{31} \end{bmatrix}$$

A matrix may be used to represent a system of equations. In these cases, the numbers represent the coefficients of the variables in the system. Matrices often make solving systems of equations easier because they are not encumbered with variables. We will investigate this idea further in the next section, but first we will look at basic matrix operations.

Matrices

A **matrix** is a rectangular array of numbers that is usually named by a capital letter: A, B, C, and so on. Each entry in a matrix is referred to as a_{ij}, such that i represents the row and j represents the column. Matrices are often referred to by their dimensions: $m \times n$ indicating m rows and n columns.

EXAMPLE 1

Finding the Dimensions of the Given Matrix and Locating Entries

Given matrix A :

ⓐ What are the dimensions of matrix A? ⓑ What are the entries at a_{31} and a_{22}?

$$A = \begin{bmatrix} 2 & 1 & 0 \\ 2 & 4 & 7 \\ 3 & 1 & -2 \end{bmatrix}$$

Solution

ⓐ The dimensions are 3×3 because there are three rows and three columns.

ⓑ Entry a_{31} is the number at row 3, column 1, which is 3. The entry a_{22} is the number at row 2, column 2, which is 4. Remember, the row comes first, then the column.

Adding and Subtracting Matrices

We use matrices to list data or to represent systems. Because the entries are numbers, we can perform operations on matrices. We add or subtract matrices by adding or subtracting corresponding entries.

In order to do this, the entries must correspond. Therefore, *addition and subtraction of matrices is only possible when the matrices have the same dimensions.* We can add or subtract a 3×3 matrix and another 3×3 matrix, but we cannot add or subtract a 2×3 matrix and a 3×3 matrix because some entries in one matrix will not have a corresponding entry in the other matrix.

Adding and Subtracting Matrices

Given matrices A and B of like dimensions, addition and subtraction of A and B will produce matrix C or matrix D of the same dimension.

$$A + B = C \text{ such that } a_{ij} + b_{ij} = c_{ij}$$
$$A - B = D \text{ such that } a_{ij} - b_{ij} = d_{ij}$$

Matrix addition is commutative.

$$A + B = B + A$$

It is also associative.

$$(A + B) + C = A + (B + C)$$

EXAMPLE 2

Finding the Sum of Matrices
Find the sum of A and B, given

$$A = \begin{bmatrix} a & b \\ c & d \end{bmatrix} \text{ and } B = \begin{bmatrix} e & f \\ g & h \end{bmatrix}$$

Solution
Add corresponding entries.

$$\begin{aligned} A + B &= \begin{bmatrix} a & b \\ c & d \end{bmatrix} + \begin{bmatrix} e & f \\ g & h \end{bmatrix} \\ &= \begin{bmatrix} a+e & b+f \\ c+g & d+h \end{bmatrix} \end{aligned}$$

EXAMPLE 3

Adding Matrix *A* and Matrix *B*
Find the sum of A and B.

$$A = \begin{bmatrix} 4 & 1 \\ 3 & 2 \end{bmatrix} \text{ and } B = \begin{bmatrix} 5 & 9 \\ 0 & 7 \end{bmatrix}$$

Solution
Add corresponding entries. Add the entry in row 1, column 1, a_{11}, of matrix A to the entry in row 1, column 1, b_{11}, of B. Continue the pattern until all entries have been added.

$$\begin{aligned} A + B &= \begin{bmatrix} 4 & 1 \\ 3 & 2 \end{bmatrix} + \begin{bmatrix} 5 & 9 \\ 0 & 7 \end{bmatrix} \\ &= \begin{bmatrix} 4+5 & 1+9 \\ 3+0 & 2+7 \end{bmatrix} \\ &= \begin{bmatrix} 9 & 10 \\ 3 & 9 \end{bmatrix} \end{aligned}$$

EXAMPLE 4

Finding the Difference of Two Matrices

Find the difference of A and B.

$$A = \begin{bmatrix} -2 & 3 \\ 0 & 1 \end{bmatrix} \text{ and } B = \begin{bmatrix} 8 & 1 \\ 5 & 4 \end{bmatrix}$$

Solution

We subtract the corresponding entries of each matrix.

$$\begin{aligned} A - B &= \begin{bmatrix} -2 & 3 \\ 0 & 1 \end{bmatrix} - \begin{bmatrix} 8 & 1 \\ 5 & 4 \end{bmatrix} \\ &= \begin{bmatrix} -2-8 & 3-1 \\ 0-5 & 1-4 \end{bmatrix} \\ &= \begin{bmatrix} -10 & 2 \\ -5 & -3 \end{bmatrix} \end{aligned}$$

EXAMPLE 5

Finding the Sum and Difference of Two 3 x 3 Matrices

Given A and B :

ⓐ Find the sum. ⓑ Find the difference.

$$A = \begin{bmatrix} 2 & -10 & -2 \\ 14 & 12 & 10 \\ 4 & -2 & 2 \end{bmatrix} \text{ and } B = \begin{bmatrix} 6 & 10 & -2 \\ 0 & -12 & -4 \\ -5 & 2 & -2 \end{bmatrix}$$

Solution

ⓐ Add the corresponding entries.

$$\begin{aligned} A + B &= \begin{bmatrix} 2 & -10 & -2 \\ 14 & 12 & 10 \\ 4 & -2 & 2 \end{bmatrix} + \begin{bmatrix} 6 & 10 & -2 \\ 0 & -12 & -4 \\ -5 & 2 & -2 \end{bmatrix} \\ &= \begin{bmatrix} 2+6 & -10+10 & -2-2 \\ 14+0 & 12-12 & 10-4 \\ 4-5 & -2+2 & 2-2 \end{bmatrix} \\ &= \begin{bmatrix} 8 & 0 & -4 \\ 14 & 0 & 6 \\ -1 & 0 & 0 \end{bmatrix} \end{aligned}$$

ⓑ Subtract the corresponding entries.

$$\begin{aligned} A - B &= \begin{bmatrix} 2 & -10 & -2 \\ 14 & 12 & 10 \\ 4 & -2 & 2 \end{bmatrix} - \begin{bmatrix} 6 & 10 & -2 \\ 0 & -12 & -4 \\ -5 & 2 & -2 \end{bmatrix} \\ &= \begin{bmatrix} 2-6 & -10-10 & -2+2 \\ 14-0 & 12+12 & 10+4 \\ 4+5 & -2-2 & 2+2 \end{bmatrix} \\ &= \begin{bmatrix} -4 & -20 & 0 \\ 14 & 24 & 14 \\ 9 & -4 & 4 \end{bmatrix} \end{aligned}$$

TRY IT #1 Add matrix A and matrix B.

$$A = \begin{bmatrix} 2 & 6 \\ 1 & 0 \\ 1 & -3 \end{bmatrix} \text{ and } B = \begin{bmatrix} 3 & -2 \\ 1 & 5 \\ -4 & 3 \end{bmatrix}$$

Finding Scalar Multiples of a Matrix

Besides adding and subtracting whole matrices, there are many situations in which we need to multiply a matrix by a constant called a scalar. Recall that a scalar is a real number quantity that has magnitude, but not direction. For example, time, temperature, and distance are scalar quantities. The process of scalar multiplication involves multiplying each entry in a matrix by a scalar. A **scalar multiple** is any entry of a matrix that results from scalar multiplication.

Consider a real-world scenario in which a university needs to add to its inventory of computers, computer tables, and chairs in two of the campus labs due to increased enrollment. They estimate that 15% more equipment is needed in both labs. The school's current inventory is displayed in Table 2.

	Lab A	Lab B
Computers	15	27
Computer Tables	16	34
Chairs	16	34

Table 2

Converting the data to a matrix, we have

$$C_{2013} = \begin{bmatrix} 15 & 27 \\ 16 & 34 \\ 16 & 34 \end{bmatrix}$$

To calculate how much computer equipment will be needed, we multiply all entries in matrix C by 0.15.

$$(0.15)C_{2013} = \begin{bmatrix} (0.15)15 & (0.15)27 \\ (0.15)16 & (0.15)34 \\ (0.15)16 & (0.15)34 \end{bmatrix} = \begin{bmatrix} 2.25 & 4.05 \\ 2.4 & 5.1 \\ 2.4 & 5.1 \end{bmatrix}$$

We must round up to the next integer, so the amount of new equipment needed is

$$\begin{bmatrix} 3 & 5 \\ 3 & 6 \\ 3 & 6 \end{bmatrix}$$

Adding the two matrices as shown below, we see the new inventory amounts.

$$\begin{bmatrix} 15 & 27 \\ 16 & 34 \\ 16 & 34 \end{bmatrix} + \begin{bmatrix} 3 & 5 \\ 3 & 6 \\ 3 & 6 \end{bmatrix} = \begin{bmatrix} 18 & 32 \\ 19 & 40 \\ 19 & 40 \end{bmatrix}$$

This means

$$C_{2014} = \begin{bmatrix} 18 & 32 \\ 19 & 40 \\ 19 & 40 \end{bmatrix}$$

Thus, Lab A will have 18 computers, 19 computer tables, and 19 chairs; Lab B will have 32 computers, 40 computer tables, and 40 chairs.

Scalar Multiplication

Scalar multiplication involves finding the product of a constant by each entry in the matrix. Given

$$A = \begin{bmatrix} a_{11} & a_{12} \\ a_{21} & a_{22} \end{bmatrix}$$

the scalar multiple cA is

$$\begin{aligned} cA &= c\begin{bmatrix} a_{11} & a_{12} \\ a_{21} & a_{22} \end{bmatrix} \\ &= \begin{bmatrix} ca_{11} & ca_{12} \\ ca_{21} & ca_{22} \end{bmatrix} \end{aligned}$$

Scalar multiplication is distributive. For the matrices A, B, and C with scalars a and b,

$$a(A + B) = aA + aB$$
$$(a + b)A = aA + bA$$

EXAMPLE 6

Multiplying the Matrix by a Scalar

Multiply matrix A by the scalar 3.

$$A = \begin{bmatrix} 8 & 1 \\ 5 & 4 \end{bmatrix}$$

Solution

Multiply each entry in A by the scalar 3.

$$\begin{aligned} 3A &= 3\begin{bmatrix} 8 & 1 \\ 5 & 4 \end{bmatrix} \\ &= \begin{bmatrix} 3 \cdot 8 & 3 \cdot 1 \\ 3 \cdot 5 & 3 \cdot 4 \end{bmatrix} \\ &= \begin{bmatrix} 24 & 3 \\ 15 & 12 \end{bmatrix} \end{aligned}$$

> **TRY IT** #2 Given matrix B, find $-2B$ where

$$B = \begin{bmatrix} 4 & 1 \\ 3 & 2 \end{bmatrix}$$

EXAMPLE 7

Finding the Sum of Scalar Multiples

Find the sum $3A + 2B$.

$$A = \begin{bmatrix} 1 & -2 & 0 \\ 0 & -1 & 2 \\ 4 & 3 & -6 \end{bmatrix} \text{ and } B = \begin{bmatrix} -1 & 2 & 1 \\ 0 & -3 & 2 \\ 0 & 1 & -4 \end{bmatrix}$$

Solution

First, find $3A$, then $2B$.

$$\begin{aligned} 3A &= \begin{bmatrix} 3 \cdot 1 & 3(-2) & 3 \cdot 0 \\ 3 \cdot 0 & 3(-1) & 3 \cdot 2 \\ 3 \cdot 4 & 3 \cdot 3 & 3(-6) \end{bmatrix} \\ &= \begin{bmatrix} 3 & -6 & 0 \\ 0 & -3 & 6 \\ 12 & 9 & -18 \end{bmatrix} \end{aligned}$$

$$2B = \begin{bmatrix} 2(-1) & 2\cdot 2 & 2\cdot 1 \\ 2\cdot 0 & 2(-3) & 2\cdot 2 \\ 2\cdot 0 & 2\cdot 1 & 2(-4) \end{bmatrix}$$
$$= \begin{bmatrix} -2 & 4 & 2 \\ 0 & -6 & 4 \\ 0 & 2 & -8 \end{bmatrix}$$

Now, add $3A + 2B$.

$$3A + 2B = \begin{bmatrix} 3 & -6 & 0 \\ 0 & -3 & 6 \\ 12 & 9 & -18 \end{bmatrix} + \begin{bmatrix} -2 & 4 & 2 \\ 0 & -6 & 4 \\ 0 & 2 & -8 \end{bmatrix}$$
$$= \begin{bmatrix} 3-2 & -6+4 & 0+2 \\ 0+0 & -3-6 & 6+4 \\ 12+0 & 9+2 & -18-8 \end{bmatrix}$$
$$= \begin{bmatrix} 1 & -2 & 2 \\ 0 & -9 & 10 \\ 12 & 11 & -26 \end{bmatrix}$$

Finding the Product of Two Matrices

In addition to multiplying a matrix by a scalar, we can multiply two matrices. Finding the product of two matrices is only possible when the inner dimensions are the same, meaning that the number of columns of the first matrix is equal to the number of rows of the second matrix. If A is an $m \times r$ matrix and B is an $r \times n$ matrix, then the product matrix AB is an $m \times n$ matrix. For example, the product AB is possible because the number of columns in A is the same as the number of rows in B. If the inner dimensions do not match, the product is not defined.

$$\begin{array}{ccc} A & \cdot & B \\ 2\times 3 & & 3\times 3 \end{array}$$

same

We multiply entries of A with entries of B according to a specific pattern as outlined below. The process of matrix multiplication becomes clearer when working a problem with real numbers.

To obtain the entries in row i of AB, we multiply the entries in row i of A by column j in B and add. For example, given matrices A and B, where the dimensions of A are 2×3 and the dimensions of B are 3×3, the product of AB will be a 2×3 matrix.

$$A = \begin{bmatrix} a_{11} & a_{12} & a_{13} \\ a_{21} & a_{22} & a_{23} \end{bmatrix} \text{ and } B = \begin{bmatrix} b_{11} & b_{12} & b_{13} \\ b_{21} & b_{22} & b_{23} \\ b_{31} & b_{32} & b_{33} \end{bmatrix}$$

Multiply and add as follows to obtain the first entry of the product matrix AB.

1. To obtain the entry in row 1, column 1 of AB, multiply the first row in A by the first column in B, and add.

$$\begin{bmatrix} a_{11} & a_{12} & a_{13} \end{bmatrix} \begin{bmatrix} b_{11} \\ b_{21} \\ b_{31} \end{bmatrix} = a_{11} \cdot b_{11} + a_{12} \cdot b_{21} + a_{13} \cdot b_{31}$$

2. To obtain the entry in row 1, column 2 of AB, multiply the first row of A by the second column in B, and add.

$$\begin{bmatrix} a_{11} & a_{12} & a_{13} \end{bmatrix} \begin{bmatrix} b_{12} \\ b_{22} \\ b_{32} \end{bmatrix} = a_{11} \cdot b_{12} + a_{12} \cdot b_{22} + a_{13} \cdot b_{32}$$

3. To obtain the entry in row 1, column 3 of AB, multiply the first row of A by the third column in B, and add.

$$\begin{bmatrix} a_{11} & a_{12} & a_{13} \end{bmatrix} \begin{bmatrix} b_{13} \\ b_{23} \\ b_{33} \end{bmatrix} = a_{11} \cdot b_{13} + a_{12} \cdot b_{23} + a_{13} \cdot b_{33}$$

We proceed the same way to obtain the second row of AB. In other words, row 2 of A times column 1 of B; row 2 of A times column 2 of B; row 2 of A times column 3 of B. When complete, the product matrix will be

$$AB = \begin{bmatrix} a_{11} \cdot b_{11} + a_{12} \cdot b_{21} + a_{13} \cdot b_{31} & a_{11} \cdot b_{12} + a_{12} \cdot b_{22} + a_{13} \cdot b_{32} & a_{11} \cdot b_{13} + a_{12} \cdot b_{23} + a_{13} \cdot b_{33} \\ a_{21} \cdot b_{11} + a_{22} \cdot b_{21} + a_{23} \cdot b_{31} & a_{21} \cdot b_{12} + a_{22} \cdot b_{22} + a_{23} \cdot b_{32} & a_{21} \cdot b_{13} + a_{22} \cdot b_{23} + a_{23} \cdot b_{33} \end{bmatrix}$$

Properties of Matrix Multiplication

For the matrices A, B, and C the following properties hold.

- Matrix multiplication is associative: $(AB)C = A(BC)$.
- Matrix multiplication is distributive: $C(A + B) = CA + CB,$ $(A + B)C = AC + BC.$

Note that matrix multiplication is not commutative.

EXAMPLE 8

Multiplying Two Matrices

Multiply matrix A and matrix B.

$$A = \begin{bmatrix} 1 & 2 \\ 3 & 4 \end{bmatrix} \text{ and } B = \begin{bmatrix} 5 & 6 \\ 7 & 8 \end{bmatrix}$$

Solution

First, we check the dimensions of the matrices. Matrix A has dimensions 2×2 and matrix B has dimensions 2×2. The inner dimensions are the same so we can perform the multiplication. The product will have the dimensions 2×2.

We perform the operations outlined previously.

$$\begin{aligned} AB &= \begin{bmatrix} 1 & 2 \\ 3 & 4 \end{bmatrix} \cdot \begin{bmatrix} 5 & 6 \\ 7 & 8 \end{bmatrix} \\ &= \begin{bmatrix} 1(5) + 2(7) & 1(6) + 2(8) \\ 3(5) + 4(7) & 3(6) + 4(8) \end{bmatrix} \\ &= \begin{bmatrix} 19 & 22 \\ 43 & 50 \end{bmatrix} \end{aligned}$$

EXAMPLE 9

Multiplying Two Matrices

Given A and B :

ⓐ Find AB. ⓑ Find BA.

$$A = \begin{bmatrix} -1 & 2 & 3 \\ 4 & 0 & 5 \end{bmatrix} \text{ and } B = \begin{bmatrix} 5 & -1 \\ -4 & 0 \\ 2 & 3 \end{bmatrix}$$

Solution

ⓐ As the dimensions of A are 2×3 and the dimensions of B are 3×2, these matrices can be multiplied together because the number of columns in A matches the number of rows in B. The resulting product will be a 2×2 matrix, the number of rows in A by the number of columns in B.

$$\begin{aligned} AB &= \begin{bmatrix} -1 & 2 & 3 \\ 4 & 0 & 5 \end{bmatrix} \begin{bmatrix} 5 & -1 \\ -4 & 0 \\ 2 & 3 \end{bmatrix} \\ &= \begin{bmatrix} -1(5) + 2(-4) + 3(2) & -1(-1) + 2(0) + 3(3) \\ 4(5) + 0(-4) + 5(2) & 4(-1) + 0(0) + 5(3) \end{bmatrix} \\ &= \begin{bmatrix} -7 & 10 \\ 30 & 11 \end{bmatrix} \end{aligned}$$

ⓑ The dimensions of B are 3×2 and the dimensions of A are 2×3. The inner dimensions match so the product is defined and will be a 3×3 matrix.

$$\begin{aligned} BA &= \begin{bmatrix} 5 & -1 \\ -4 & 0 \\ 2 & 3 \end{bmatrix} \begin{bmatrix} -1 & 2 & 3 \\ 4 & 0 & 5 \end{bmatrix} \\ &= \begin{bmatrix} 5(-1) + -1(4) & 5(2) + -1(0) & 5(3) + -1(5) \\ -4(-1) + 0(4) & -4(2) + 0(0) & -4(3) + 0(5) \\ 2(-1) + 3(4) & 2(2) + 3(0) & 2(3) + 3(5) \end{bmatrix} \\ &= \begin{bmatrix} -9 & 10 & 10 \\ 4 & -8 & -12 \\ 10 & 4 & 21 \end{bmatrix} \end{aligned}$$

Analysis

Notice that the products AB and BA are not equal.

$$AB = \begin{bmatrix} -7 & 10 \\ 30 & 11 \end{bmatrix} \neq \begin{bmatrix} -9 & 10 & 10 \\ 4 & -8 & -12 \\ 10 & 4 & 21 \end{bmatrix} = BA$$

This illustrates the fact that matrix multiplication is not commutative.

Q&A **Is it possible for *AB* to be defined but not *BA*?**

Yes, consider a matrix A with dimension 3×4 and matrix B with dimension 4×2. For the product AB the inner dimensions are 4 and the product is defined, but for the product BA the inner dimensions are 2 and 3 so the product is undefined.

EXAMPLE 10

Using Matrices in Real-World Problems

Let's return to the problem presented at the opening of this section. We have Table 3, representing the equipment needs of two soccer teams.

	Wildcats	Mud Cats
Goals	6	10

Table 3

	Wildcats	Mud Cats
Balls	30	24
Jerseys	14	20

Table 3

We are also given the prices of the equipment, as shown in Table 4.

Goal	$300
Ball	$10
Jersey	$30

Table 4

We will convert the data to matrices. Thus, the equipment need matrix is written as

$$E = \begin{bmatrix} 6 & 10 \\ 30 & 24 \\ 14 & 20 \end{bmatrix}$$

The cost matrix is written as

$$C = \begin{bmatrix} 300 & 10 & 30 \end{bmatrix}$$

We perform matrix multiplication to obtain costs for the equipment.

$$\begin{aligned} CE &= \begin{bmatrix} 300 & 10 & 30 \end{bmatrix} \begin{bmatrix} 6 & 10 \\ 30 & 24 \\ 14 & 20 \end{bmatrix} \\ &= \begin{bmatrix} 300(6) + 10(30) + 30(14) & 300(10) + 10(24) + 30(20) \end{bmatrix} \\ &= \begin{bmatrix} 2{,}520 & 3{,}840 \end{bmatrix} \end{aligned}$$

The total cost for equipment for the Wildcats is $2,520, and the total cost for equipment for the Mud Cats is $3,840.

HOW TO

Given a matrix operation, evaluate using a calculator.

1. Save each matrix as a matrix variable $[A], [B], [C], \ldots$
2. Enter the operation into the calculator, calling up each matrix variable as needed.
3. If the operation is defined, the calculator will present the solution matrix; if the operation is undefined, it will display an error message.

EXAMPLE 11

Using a Calculator to Perform Matrix Operations

Find $AB - C$ given

$$A = \begin{bmatrix} -15 & 25 & 32 \\ 41 & -7 & -28 \\ 10 & 34 & -2 \end{bmatrix}, B = \begin{bmatrix} 45 & 21 & -37 \\ -24 & 52 & 19 \\ 6 & -48 & -31 \end{bmatrix}, \text{and } C = \begin{bmatrix} -100 & -89 & -98 \\ 25 & -56 & 74 \\ -67 & 42 & -75 \end{bmatrix}.$$

Solution

On the matrix page of the calculator, we enter matrix A above as the matrix variable $[A]$, matrix B above as the matrix variable $[B]$, and matrix C above as the matrix variable $[C]$.

On the home screen of the calculator, we type in the problem and call up each matrix variable as needed.

$$[A]\,[B] - [C]$$

The calculator gives us the following matrix.

$$\begin{bmatrix} -983 & -462 & 136 \\ 1,820 & 1,897 & -856 \\ -311 & 2,032 & 413 \end{bmatrix}$$

MEDIA

Access these online resources for additional instruction and practice with matrices and matrix operations.

Dimensions of a Matrix (http://openstax.org/l/matrixdimen)
Matrix Addition and Subtraction (http://openstax.org/l/matrixaddsub)
Matrix Operations (http://openstax.org/l/matrixoper)
Matrix Multiplication (http://openstax.org/l/matrixmult)

11.5 SECTION EXERCISES

Verbal

1. Can we add any two matrices together? If so, explain why; if not, explain why not and give an example of two matrices that cannot be added together.

2. Can we multiply any column matrix by any row matrix? Explain why or why not.

3. Can both the products AB and BA be defined? If so, explain how; if not, explain why.

4. Can any two matrices of the same size be multiplied? If so, explain why, and if not, explain why not and give an example of two matrices of the same size that cannot be multiplied together.

5. Does matrix multiplication commute? That is, does $AB = BA$? If so, prove why it does. If not, explain why it does not.

Algebraic

For the following exercises, use the matrices below and perform the matrix addition or subtraction. Indicate if the operation is undefined.

$$A=\begin{bmatrix}1&3\\0&7\end{bmatrix}, B=\begin{bmatrix}2&14\\22&6\end{bmatrix}, C=\begin{bmatrix}1&5\\8&92\\12&6\end{bmatrix}, D=\begin{bmatrix}10&14\\7&2\\5&61\end{bmatrix}, E=\begin{bmatrix}6&12\\14&5\end{bmatrix}, F=\begin{bmatrix}0&9\\78&17\\15&4\end{bmatrix}$$

6. $A+B$

7. $C+D$

8. $A+C$

9. $B-E$

10. $C+F$

11. $D-B$

For the following exercises, use the matrices below to perform scalar multiplication.

$$A=\begin{bmatrix}4&6\\13&12\end{bmatrix}, B=\begin{bmatrix}3&9\\21&12\\0&64\end{bmatrix}, C=\begin{bmatrix}16&3&7&18\\90&5&3&29\end{bmatrix}, D=\begin{bmatrix}18&12&13\\8&14&6\\7&4&21\end{bmatrix}$$

12. $5A$

13. $3B$

14. $-2B$

15. $-4C$

16. $\frac{1}{2}C$

17. $100D$

For the following exercises, use the matrices below to perform matrix multiplication.

$$A=\begin{bmatrix}-1&5\\3&2\end{bmatrix}, B=\begin{bmatrix}3&6&4\\-8&0&12\end{bmatrix}, C=\begin{bmatrix}4&10\\-2&6\\5&9\end{bmatrix}, D=\begin{bmatrix}2&-3&12\\9&3&1\\0&8&-10\end{bmatrix}$$

18. AB

19. BC

20. CA

21. BD

22. DC

23. CB

For the following exercises, use the matrices below to perform the indicated operation if possible. If not possible, explain why the operation cannot be performed.

$$A=\begin{bmatrix}2&-5\\6&7\end{bmatrix}, B=\begin{bmatrix}-9&6\\-4&2\end{bmatrix}, C=\begin{bmatrix}0&9\\7&1\end{bmatrix}, D=\begin{bmatrix}-8&7&-5\\4&3&2\\0&9&2\end{bmatrix}, E=\begin{bmatrix}4&5&3\\7&-6&-5\\1&0&9\end{bmatrix}$$

24. $A+B-C$

25. $4A+5D$

26. $2C+B$

27. $3D+4E$

28. $C-0.5D$

29. $100D-10E$

For the following exercises, use the matrices below to perform the indicated operation if possible. If not possible, explain why the operation cannot be performed. (Hint: $A^2=A\cdot A$)

$$A=\begin{bmatrix}-10&20\\5&25\end{bmatrix}, B=\begin{bmatrix}40&10\\-20&30\end{bmatrix}, C=\begin{bmatrix}-1&0\\0&-1\\1&0\end{bmatrix}$$

30. AB

31. BA

32. CA

33. BC

34. A^2

35. B^2

36. C^2

37. B^2A^2

38. A^2B^2

39. $(AB)^2$

40. $(BA)^2$

For the following exercises, use the matrices below to perform the indicated operation if possible. If not possible, explain why the operation cannot be performed. (Hint: $A^2 = A \cdot A$)

$$A = \begin{bmatrix} 1 & 0 \\ 2 & 3 \end{bmatrix}, B = \begin{bmatrix} -2 & 3 & 4 \\ -1 & 1 & -5 \end{bmatrix}, C = \begin{bmatrix} 0.5 & 0.1 \\ 1 & 0.2 \\ -0.5 & 0.3 \end{bmatrix}, D = \begin{bmatrix} 1 & 0 & -1 \\ -6 & 7 & 5 \\ 4 & 2 & 1 \end{bmatrix}$$

41. AB

42. BA

43. BD

44. DC

45. D^2

46. A^2

47. D^3

48. $(AB)C$

49. $A(BC)$

Technology

For the following exercises, use the matrices below to perform the indicated operation if possible. If not possible, explain why the operation cannot be performed. Use a calculator to verify your solution.

$$A = \begin{bmatrix} -2 & 0 & 9 \\ 1 & 8 & -3 \\ 0.5 & 4 & 5 \end{bmatrix}, B = \begin{bmatrix} 0.5 & 3 & 0 \\ -4 & 1 & 6 \\ 8 & 7 & 2 \end{bmatrix}, C = \begin{bmatrix} 1 & 0 & 1 \\ 0 & 1 & 0 \\ 1 & 0 & 1 \end{bmatrix}$$

50. AB

51. BA

52. CA

53. BC

54. ABC

Extensions

For the following exercises, use the matrix below to perform the indicated operation on the given matrix.

$$B = \begin{bmatrix} 1 & 0 & 0 \\ 0 & 0 & 1 \\ 0 & 1 & 0 \end{bmatrix}$$

55. B^2

56. B^3

57. B^4

58. B^5

59. Using the above questions, find a formula for B^n. Test the formula for B^{201} and B^{202}, using a calculator.

11.6 Solving Systems with Gaussian Elimination

Learning Objectives

In this section, you will:

- Write the augmented matrix of a system of equations.
- Write the system of equations from an augmented matrix.
- Perform row operations on a matrix.
- Solve a system of linear equations using matrices.

Figure 1 German mathematician Carl Friedrich Gauss (1777–1855).

Carl Friedrich Gauss lived during the late 18th century and early 19th century, but he is still considered one of the most prolific mathematicians in history. His contributions to the science of mathematics and physics span fields such as algebra, number theory, analysis, differential geometry, astronomy, and optics, among others. His discoveries regarding matrix theory changed the way mathematicians have worked for the last two centuries.

We first encountered Gaussian elimination in Systems of Linear Equations: Two Variables. In this section, we will revisit this technique for solving systems, this time using matrices.

Writing the Augmented Matrix of a System of Equations

A matrix can serve as a device for representing and solving a system of equations. To express a system in matrix form, we extract the coefficients of the variables and the constants, and these become the entries of the matrix. We use a vertical line to separate the coefficient entries from the constants, essentially replacing the equal signs. When a system is written in this form, we call it an **augmented matrix**.

For example, consider the following 2×2 system of equations.

$$\begin{aligned} 3x + 4y &= 7 \\ 4x - 2y &= 5 \end{aligned}$$

We can write this system as an augmented matrix:

$$\left[\begin{array}{cc|c} 3 & 4 & 7 \\ 4 & -2 & 5 \end{array}\right]$$

We can also write a matrix containing just the coefficients. This is called the **coefficient matrix**.

$$\begin{bmatrix} 3 & 4 \\ 4 & -2 \end{bmatrix}$$

A three-by-three system of equations such as

$$\begin{aligned} 3x - y - z &= 0 \\ x + y &= 5 \\ 2x - 3z &= 2 \end{aligned}$$

has a coefficient matrix

$$\begin{bmatrix} 3 & -1 & -1 \\ 1 & 1 & 0 \\ 2 & 0 & -3 \end{bmatrix}$$

and is represented by the augmented matrix

$$\left[\begin{array}{ccc|c} 3 & -1 & -1 & 0 \\ 1 & 1 & 0 & 5 \\ 2 & 0 & -3 & 2 \end{array}\right]$$

Notice that the matrix is written so that the variables line up in their own columns: *x*-terms go in the first column, *y*-terms in the second column, and *z*-terms in the third column. It is very important that each equation is written in standard form $ax + by + cz = d$ so that the variables line up. When there is a missing variable term in an equation, the coefficient is 0.

HOW TO

Given a system of equations, write an augmented matrix.

1. Write the coefficients of the *x*-terms as the numbers down the first column.
2. Write the coefficients of the *y*-terms as the numbers down the second column.
3. If there are *z*-terms, write the coefficients as the numbers down the third column.
4. Draw a vertical line and write the constants to the right of the line.

EXAMPLE 1

Writing the Augmented Matrix for a System of Equations

Write the augmented matrix for the given system of equations.

$$\begin{aligned} x + 2y - z &= 3 \\ 2x - y + 2z &= 6 \\ x - 3y + 3z &= 4 \end{aligned}$$

Solution

The augmented matrix displays the coefficients of the variables, and an additional column for the constants.

$$\left[\begin{array}{ccc|c} 1 & 2 & -1 & 3 \\ 2 & -1 & 2 & 6 \\ 1 & -3 & 3 & 4 \end{array}\right]$$

TRY IT #1 Write the augmented matrix of the given system of equations.

$$\begin{aligned} 4x - 3y &= 11 \\ 3x + 2y &= 4 \end{aligned}$$

Writing a System of Equations from an Augmented Matrix

We can use augmented matrices to help us solve systems of equations because they simplify operations when the systems are not encumbered by the variables. However, it is important to understand how to move back and forth between formats in order to make finding solutions smoother and more intuitive. Here, we will use the information in an augmented matrix to write the system of equations in standard form.

EXAMPLE 2

Writing a System of Equations from an Augmented Matrix Form

Find the system of equations from the augmented matrix.

$$\left[\begin{array}{ccc|c} 1 & -3 & -5 & -2 \\ 2 & -5 & -4 & 5 \\ -3 & 5 & 4 & 6 \end{array}\right]$$

Solution

When the columns represent the variables x, y, and z,

$$\left[\begin{array}{ccc|c} 1 & -3 & -5 & -2 \\ 2 & -5 & -4 & 5 \\ -3 & 5 & 4 & 6 \end{array}\right] \rightarrow \begin{array}{r} x - 3y - 5z = -2 \\ 2x - 5y - 4z = 5 \\ -3x + 5y + 4z = 6 \end{array}$$

TRY IT #2 Write the system of equations from the augmented matrix.

$$\left[\begin{array}{ccc|c} 1 & -1 & 1 & 5 \\ 2 & -1 & 3 & 1 \\ 0 & 1 & 1 & -9 \end{array}\right]$$

Performing Row Operations on a Matrix

Now that we can write systems of equations in augmented matrix form, we will examine the various **row operations** that can be performed on a matrix, such as addition, multiplication by a constant, and interchanging rows.

Performing row operations on a matrix is the method we use for solving a system of equations. In order to solve the system of equations, we want to convert the matrix to **row-echelon form**, in which there are ones down the **main diagonal** from the upper left corner to the lower right corner, and zeros in every position below the main diagonal as shown.

$$\text{Row-echelon form} \\ \begin{bmatrix} 1 & a & b \\ 0 & 1 & d \\ 0 & 0 & 1 \end{bmatrix}$$

We use row operations corresponding to equation operations to obtain a new matrix that is **row-equivalent** in a simpler form. Here are the guidelines to obtaining row-echelon form.

1. In any nonzero row, the first nonzero number is a 1. It is called a *leading* 1.
2. Any all-zero rows are placed at the bottom on the matrix.
3. Any leading 1 is below and to the right of a previous leading 1.
4. Any column containing a leading 1 has zeros in all other positions in the column.

To solve a system of equations we can perform the following row operations to convert the coefficient matrix to row-echelon form and do back-substitution to find the solution.

1. Interchange rows. (Notation: $R_i \leftrightarrow R_j$)
2. Multiply a row by a constant. (Notation: cR_i)
3. Add the product of a row multiplied by a constant to another row. (Notation: $R_i + cR_j$)

Each of the row operations corresponds to the operations we have already learned to solve systems of equations in three variables. With these operations, there are some key moves that will quickly achieve the goal of writing a matrix in row-echelon form. To obtain a matrix in row-echelon form for finding solutions, we use Gaussian elimination, a method that uses row operations to obtain a 1 as the first entry so that row 1 can be used to convert the remaining rows.

Gaussian Elimination

The **Gaussian elimination** method refers to a strategy used to obtain the row-echelon form of a matrix. The goal is to write matrix A with the number 1 as the entry down the main diagonal and have all zeros below.

$$A = \begin{bmatrix} a_{11} & a_{12} & a_{13} \\ a_{21} & a_{22} & a_{23} \\ a_{31} & a_{32} & a_{33} \end{bmatrix} \xrightarrow{\text{After Gaussian elimination}} A = \begin{bmatrix} 1 & b_{12} & b_{13} \\ 0 & 1 & b_{23} \\ 0 & 0 & 1 \end{bmatrix}$$

The first step of the Gaussian strategy includes obtaining a 1 as the first entry, so that row 1 may be used to alter the rows below.

HOW TO

Given an augmented matrix, perform row operations to achieve row-echelon form.

1. The first equation should have a leading coefficient of 1. Interchange rows or multiply by a constant, if necessary.
2. Use row operations to obtain zeros down the first column below the first entry of 1.
3. Use row operations to obtain a 1 in row 2, column 2.
4. Use row operations to obtain zeros down column 2, below the entry of 1.
5. Use row operations to obtain a 1 in row 3, column 3.
6. Continue this process for all rows until there is a 1 in every entry down the main diagonal and there are only zeros below.
7. If any rows contain all zeros, place them at the bottom.

EXAMPLE 3

Solving a 2×2 System by Gaussian Elimination

Solve the given system by Gaussian elimination.

$$\begin{aligned} 2x + 3y &= 6 \\ x - y &= \tfrac{1}{2} \end{aligned}$$

Solution

First, we write this as an augmented matrix.

$$\left[\begin{array}{cc|c} 2 & 3 & 6 \\ 1 & -1 & \frac{1}{2} \end{array}\right]$$

We want a 1 in row 1, column 1. This can be accomplished by interchanging row 1 and row 2.

$$R_1 \leftrightarrow R_2 \rightarrow \left[\begin{array}{cc|c} 1 & -1 & \frac{1}{2} \\ 2 & 3 & 6 \end{array}\right]$$

We now have a 1 as the first entry in row 1, column 1. Now let's obtain a 0 in row 2, column 1. This can be accomplished by multiplying row 1 by -2, and then adding the result to row 2.

$$-2R_1 + R_2 = R_2 \rightarrow \left[\begin{array}{cc|c} 1 & -1 & \frac{1}{2} \\ 0 & 5 & 5 \end{array}\right]$$

We only have one more step, to multiply row 2 by $\frac{1}{5}$.

$$\frac{1}{5}R_2 = R_2 \rightarrow \left[\begin{array}{cc|c} 1 & -1 & \frac{1}{2} \\ 0 & 1 & 1 \end{array}\right]$$

Use back-substitution. The second row of the matrix represents $y = 1$. Back-substitute $y = 1$ into the first equation.

$$\begin{aligned} x - (1) &= \tfrac{1}{2} \\ x &= \tfrac{3}{2} \end{aligned}$$

The solution is the point $\left(\frac{3}{2}, 1\right)$.

> TRY IT #3 Solve the given system by Gaussian elimination.

$$\begin{aligned}4x + 3y &= 11 \\ x - 3y &= -1\end{aligned}$$

EXAMPLE 4

Using Gaussian Elimination to Solve a System of Equations

Use Gaussian elimination to solve the given 2×2 system of equations.

$$\begin{aligned}2x + y &= 1 \\ 4x + 2y &= 6\end{aligned}$$

Solution

Write the system as an augmented matrix.

$$\left[\begin{array}{cc|c} 2 & 1 & 1 \\ 4 & 2 & 6 \end{array}\right]$$

Obtain a 1 in row 1, column 1. This can be accomplished by multiplying the first row by $\frac{1}{2}$.

$$\frac{1}{2}R_1 = R_1 \rightarrow \left[\begin{array}{cc|c} 1 & \frac{1}{2} & \frac{1}{2} \\ 4 & 2 & 6 \end{array}\right]$$

Next, we want a 0 in row 2, column 1. Multiply row 1 by −4 and add row 1 to row 2.

$$-4R_1 + R_2 = R_2 \rightarrow \left[\begin{array}{cc|c} 1 & \frac{1}{2} & \frac{1}{2} \\ 0 & 0 & 4 \end{array}\right]$$

The second row represents the equation $0 = 4$. Therefore, the system is inconsistent and has no solution.

EXAMPLE 5

Solving a Dependent System

Solve the system of equations.

$$\begin{aligned}3x + 4y &= 12 \\ 6x + 8y &= 24\end{aligned}$$

Solution

Perform row operations on the augmented matrix to try and achieve row-echelon form.

$$A = \left[\begin{array}{cc|c} 3 & 4 & 12 \\ 6 & 8 & 24 \end{array}\right]$$

$$-\frac{1}{2}R_2 + R_1 = R_1 \rightarrow \left[\begin{array}{cc|c} 0 & 0 & 0 \\ 6 & 8 & 24 \end{array}\right]$$

$$R_1 \leftrightarrow R_2 \rightarrow \left[\begin{array}{cc|c} 6 & 8 & 24 \\ 0 & 0 & 0 \end{array}\right]$$

The matrix ends up with all zeros in the last row: $0y = 0$. Thus, there are an infinite number of solutions and the system is classified as dependent. To find the generic solution, return to one of the original equations and solve for y.

$$\begin{aligned} 3x + 4y &= 12 \\ 4y &= 12 - 3x \\ y &= 3 - \tfrac{3}{4}x \end{aligned}$$

So the solution to this system is $\left(x, 3 - \frac{3}{4}x\right)$.

EXAMPLE 6

Performing Row Operations on a 3×3 Augmented Matrix to Obtain Row-Echelon Form

Perform row operations on the given matrix to obtain row-echelon form.

$$\left[\begin{array}{rrr|r} 1 & -3 & 4 & 3 \\ 2 & -5 & 6 & 6 \\ -3 & 3 & 4 & 6 \end{array}\right]$$

Solution

The first row already has a 1 in row 1, column 1. The next step is to multiply row 1 by −2 and add it to row 2. Then replace row 2 with the result.

$$-2R_1 + R_2 = R_2 \rightarrow \left[\begin{array}{rrr|r} 1 & -3 & 4 & 3 \\ 0 & 1 & -2 & 0 \\ -3 & 3 & 4 & 6 \end{array}\right]$$

Next, obtain a zero in row 3, column 1.

$$3R_1 + R_3 = R_3 \rightarrow \left[\begin{array}{rrr|r} 1 & -3 & 4 & 3 \\ 0 & 1 & -2 & 0 \\ 0 & -6 & 16 & 15 \end{array}\right]$$

Next, obtain a zero in row 3, column 2.

$$6R_2 + R_3 = R_3 \rightarrow \left[\begin{array}{rrr|r} 1 & -3 & 4 & 3 \\ 0 & 1 & -2 & 0 \\ 0 & 0 & 4 & 15 \end{array}\right]$$

The last step is to obtain a 1 in row 3, column 3.

$$\frac{1}{4}R_3 = R_3 \rightarrow \left[\begin{array}{rrr|r} 1 & -3 & 4 & 3 \\ 0 & 1 & -2 & 0 \\ 0 & 0 & 1 & \frac{15}{4} \end{array}\right]$$

TRY IT #4 Write the system of equations in row-echelon form.

$$\begin{aligned} x - 2y + 3z &= 9 \\ -x + 3y &= -4 \\ 2x - 5y + 5z &= 17 \end{aligned}$$

Solving a System of Linear Equations Using Matrices

We have seen how to write a system of equations with an augmented matrix, and then how to use row operations and back-substitution to obtain row-echelon form. Now, we will take row-echelon form a step farther to solve a 3 by 3 system of linear equations. The general idea is to eliminate all but one variable using row operations and then back-substitute to solve for the other variables.

EXAMPLE 7

Solving a System of Linear Equations Using Matrices

Solve the system of linear equations using matrices.

$$\begin{aligned} x - y + z &= 8 \\ 2x + 3y - z &= -2 \\ 3x - 2y - 9z &= 9 \end{aligned}$$

Solution

First, we write the augmented matrix.

$$\left[\begin{array}{ccc|c} 1 & -1 & 1 & 8 \\ 2 & 3 & -1 & -2 \\ 3 & -2 & -9 & 9 \end{array}\right]$$

Next, we perform row operations to obtain row-echelon form.

$$-2R_1 + R_2 = R_2 \to \left[\begin{array}{ccc|c} 1 & -1 & 1 & 8 \\ 0 & 5 & -3 & -18 \\ 3 & -2 & -9 & 9 \end{array}\right] \qquad -3R_1 + R_3 = R_3 \to \left[\begin{array}{ccc|c} 1 & -1 & 1 & 8 \\ 0 & 5 & -3 & -18 \\ 0 & 1 & -12 & -15 \end{array}\right]$$

The easiest way to obtain a 1 in row 2 of column 1 is to interchange R_2 and R_3.

$$\text{Interchange } R_2 \text{ and } R_3 \to \left[\begin{array}{cccc} 1 & -1 & 1 & 8 \\ 0 & 1 & -12 & -15 \\ 0 & 5 & -3 & -18 \end{array}\right]$$

Then

$$-5R_2 + R_3 = R_3 \to \left[\begin{array}{ccc|c} 1 & -1 & 1 & 8 \\ 0 & 1 & -12 & -15 \\ 0 & 0 & 57 & 57 \end{array}\right] \qquad -\frac{1}{57}R_3 = R_3 \to \left[\begin{array}{ccc|c} 1 & -1 & 1 & 8 \\ 0 & 1 & -12 & -15 \\ 0 & 0 & 1 & 1 \end{array}\right]$$

The last matrix represents the equivalent system.

$$\begin{aligned} x - y + z &= 8 \\ y - 12z &= -15 \\ z &= 1 \end{aligned}$$

Using back-substitution, we obtain the solution as $(4, -3, 1)$.

EXAMPLE 8

Solving a Dependent System of Linear Equations Using Matrices

Solve the following system of linear equations using matrices.

$$\begin{aligned} -x - 2y + z &= -1 \\ 2x + 3y &= 2 \\ y - 2z &= 0 \end{aligned}$$

Solution

Write the augmented matrix.

$$\left[\begin{array}{ccc|c} -1 & -2 & 1 & -1 \\ 2 & 3 & 0 & 2 \\ 0 & 1 & -2 & 0 \end{array}\right]$$

First, multiply row 1 by -1 to get a 1 in row 1, column 1. Then, perform row operations to obtain row-echelon form.

$$-R_1 \rightarrow \left[\begin{array}{ccc|c} 1 & 2 & -1 & 1 \\ 2 & 3 & 0 & 2 \\ 0 & 1 & -2 & 0 \end{array}\right]$$

$$R_2 \leftrightarrow R_3 \rightarrow \left[\begin{array}{ccc|c} 1 & 2 & -1 & 1 \\ 0 & 1 & -2 & 0 \\ 2 & 3 & 0 & 2 \end{array}\right]$$

$$-2R_1 + R_3 = R_3 \rightarrow \left[\begin{array}{ccc|c} 1 & 2 & -1 & 1 \\ 0 & 1 & -2 & 0 \\ 0 & -1 & 2 & 0 \end{array}\right]$$

$$R_2 + R_3 = R_3 \rightarrow \left[\begin{array}{ccc|c} 1 & 2 & -1 & 2 \\ 0 & 1 & -2 & 1 \\ 0 & 0 & 0 & 0 \end{array}\right]$$

The last matrix represents the following system.

$$\begin{aligned} x + 2y - z &= 1 \\ y - 2z &= 0 \\ 0 &= 0 \end{aligned}$$

We see by the identity $0 = 0$ that this is a dependent system with an infinite number of solutions. We then find the generic solution. By solving the second equation for y and substituting it into the first equation we can solve for z in terms of x.

$$\begin{aligned} x + 2y - z &= 1 \\ y &= 2z \end{aligned}$$

$$\begin{aligned} x + 2(2z) - z &= 1 \\ x + 3z &= 1 \\ z &= \tfrac{1-x}{3} \end{aligned}$$

Now we substitute the expression for z into the second equation to solve for y in terms of x.

$$\begin{aligned} y - 2z &= 0 \\ z &= \tfrac{1-x}{3} \end{aligned}$$

$$\begin{aligned} y - 2\left(\tfrac{1-x}{3}\right) &= 0 \\ y &= \tfrac{2-2x}{3} \end{aligned}$$

The generic solution is $\left(x, \frac{2-2x}{3}, \frac{1-x}{3}\right)$.

TRY IT #5 Solve the system using matrices.

$$\begin{aligned} x + 4y - z &= 4 \\ 2x + 5y + 8z &= 15 \\ x + 3y - 3z &= 1 \end{aligned}$$

Q&A **Can any system of linear equations be solved by Gaussian elimination?**

Yes, a system of linear equations of any size can be solved by Gaussian elimination.

HOW TO

Given a system of equations, solve with matrices using a calculator.

1. Save the augmented matrix as a matrix variable $[A], \ [B], \ [C], \ \ldots$.
2. Use the **ref(** function in the calculator, calling up each matrix variable as needed.

EXAMPLE 9

Solving Systems of Equations with Matrices Using a Calculator

Solve the system of equations.

$$\begin{aligned} 5x + 3y + 9z &= -1 \\ -2x + 3y - z &= -2 \\ -x - 4y + 5z &= 1 \end{aligned}$$

Solution

Write the augmented matrix for the system of equations.

$$\left[\begin{array}{rrr|r} 5 & 3 & 9 & -1 \\ -2 & 3 & -1 & -2 \\ -1 & -4 & 5 & -1 \end{array}\right]$$

On the matrix page of the calculator, enter the augmented matrix above as the matrix variable $[A]$.

$$[A] = \begin{bmatrix} 5 & 3 & 9 & -1 \\ -2 & 3 & -1 & -2 \\ -1 & -4 & 5 & 1 \end{bmatrix}$$

Use the **ref(** function in the calculator, calling up the matrix variable $[A]$.

$$\text{ref}([A])$$

Evaluate.

$$\begin{bmatrix} 1 & \frac{3}{5} & \frac{9}{5} & \frac{1}{5} \\ 0 & 1 & \frac{13}{21} & -\frac{4}{7} \\ 0 & 0 & 1 & -\frac{24}{187} \end{bmatrix} \rightarrow \begin{aligned} x + \tfrac{3}{5}y + \tfrac{9}{5}z &= -\tfrac{1}{5} \\ y + \tfrac{13}{21}z &= -\tfrac{4}{7} \\ z &= -\tfrac{24}{187} \end{aligned}$$

Using back-substitution, the solution is $\left(\frac{61}{187}, -\frac{92}{187}, -\frac{24}{187}\right)$.

EXAMPLE 10

Applying 2 × 2 Matrices to Finance

Carolyn invests a total of \$12,000 in two municipal bonds, one paying 10.5% interest and the other paying 12% interest. The annual interest earned on the two investments last year was \$1,335. How much was invested at each rate?

Solution

We have a system of two equations in two variables. Let $x =$ the amount invested at 10.5% interest, and $y =$ the amount invested at 12% interest.

$$\begin{aligned} x + y &= 12{,}000 \\ 0.105x + 0.12y &= 1{,}335 \end{aligned}$$

As a matrix, we have

$$\left[\begin{array}{cc|c} 1 & 1 & 12{,}000 \\ 0.105 & 0.12 & 1{,}335 \end{array}\right]$$

Multiply row 1 by -0.105 and add the result to row 2.

$$\left[\begin{array}{cc|c} 1 & 1 & 12{,}000 \\ 0 & 0.015 & 75 \end{array}\right]$$

Then,

$$\begin{aligned} 0.015y &= 75 \\ y &= 5{,}000 \end{aligned}$$

So $12{,}000–5{,}000 = 7{,}000$.

Thus, \$5,000 was invested at 12% interest and \$7,000 at 10.5% interest.

EXAMPLE 11

Applying 3 × 3 Matrices to Finance

Ava invests a total of \$10,000 in three accounts, one paying 5% interest, another paying 8% interest, and the third paying 9% interest. The annual interest earned on the three investments last year was \$770. The amount invested at 9% was twice the amount invested at 5%. How much was invested at each rate?

Solution

We have a system of three equations in three variables. Let x be the amount invested at 5% interest, let y be the amount invested at 8% interest, and let z be the amount invested at 9% interest. Thus,

$$\begin{aligned} x + y + z &= 10,000 \\ 0.05x + 0.08y + 0.09z &= 770 \\ 2x - z &= 0 \end{aligned}$$

As a matrix, we have

$$\left[\begin{array}{ccc|c} 1 & 1 & 1 & 10,000 \\ 0.05 & 0.08 & 0.09 & 770 \\ 2 & 0 & -1 & 0 \end{array}\right]$$

Now, we perform Gaussian elimination to achieve row-echelon form.

$$-0.05R_1 + R_2 = R_2 \rightarrow \left[\begin{array}{ccc|c} 1 & 1 & 1 & 10{,}000 \\ 0 & 0.03 & 0.04 & 270 \\ 2 & 0 & -1 & 0 \end{array}\right]$$

$$-2R_1 + R_3 = R_3 \rightarrow \left[\begin{array}{ccc|c} 1 & 1 & 1 & 10{,}000 \\ 0 & 0.03 & 0.04 & 270 \\ 0 & -2 & -3 & -20{,}000 \end{array}\right]$$

$$\frac{1}{0.03}R_2 = R_2 \rightarrow \left[\begin{array}{ccc|c} 0 & 1 & 1 & 10{,}000 \\ 0 & 1 & \frac{4}{3} & 9{,}000 \\ 0 & -2 & -3 & -20{,}000 \end{array}\right]$$

$$2R_2 + R_3 = R_3 \rightarrow \left[\begin{array}{ccc|c} 1 & 1 & 1 & 10{,}000 \\ 0 & 1 & \frac{4}{3} & 9{,}000 \\ 0 & 0 & -\frac{1}{3} & -2{,}000 \end{array}\right]$$

The third row tells us $-\frac{1}{3}z = -2{,}000$; thus $z = 6{,}000$.

The second row tells us $y + \frac{4}{3}z = 9{,}000$. Substituting $z = 6{,}000$, we get

$$\begin{aligned} y + \tfrac{4}{3}(6{,}000) &= 9{,}000 \\ y + 8{,}000 &= 9{,}000 \\ y &= 1{,}000 \end{aligned}$$

The first row tells us $x + y + z = 10,000$. Substituting $y = 1,000$ and $z = 6,000$, we get

$$\begin{aligned} x + 1,000 + 6,000 &= 10{,}000 \\ x &= 3{,}000 \end{aligned}$$

The answer is \$3,000 invested at 5% interest, \$1,000 invested at 8%, and \$6,000 invested at 9% interest.

> **TRY IT** #6 A small shoe company took out a loan of \$1,500,000 to expand their inventory. Part of the money was borrowed at 7%, part was borrowed at 8%, and part was borrowed at 10%. The amount borrowed at 10% was four times the amount borrowed at 7%, and the annual interest on all three loans was \$130,500. Use matrices to find the amount borrowed at each rate.

MEDIA

Access these online resources for additional instruction and practice with solving systems of linear equations using Gaussian elimination.

Solve a System of Two Equations Using an Augmented Matrix (http://openstax.org/l/system2augmat)
Solve a System of Three Equations Using an Augmented Matrix (http://openstax.org/l/system3augmat)
Augmented Matrices on the Calculator (http://openstax.org/l/augmatcalc)

11.6 SECTION EXERCISES

Verbal

1. Can any system of linear equations be written as an augmented matrix? Explain why or why not. Explain how to write that augmented matrix.

2. Can any matrix be written as a system of linear equations? Explain why or why not. Explain how to write that system of equations.

3. Is there only one correct method of using row operations on a matrix? Try to explain two different row operations possible to solve the augmented matrix $\left[\begin{array}{cc|c} 9 & 3 & 0 \\ 1 & -2 & 6 \end{array}\right]$.

4. Can a matrix whose entry is 0 on the diagonal be solved? Explain why or why not. What would you do to remedy the situation?

5. Can a matrix that has 0 entries for an entire row have one solution? Explain why or why not.

Algebraic

For the following exercises, write the augmented matrix for the linear system.

6. $\begin{aligned} 8x - 37y &= 8 \\ 2x + 12y &= 3 \end{aligned}$

7. $\begin{aligned} 16y &= 4 \\ 9x - y &= 2 \end{aligned}$

8. $\begin{aligned} 3x + 2y + 10z &= 3 \\ -6x + 2y + 5z &= 13 \\ 4x + z &= 18 \end{aligned}$

9. $\begin{aligned} x + 5y + 8z &= 19 \\ 12x + 3y &= 4 \\ 3x + 4y + 9z &= -7 \end{aligned}$

10. $\begin{aligned} 6x + 12y + 16z &= 4 \\ 19x - 5y + 3z &= -9 \\ x + 2y &= -8 \end{aligned}$

For the following exercises, write the linear system from the augmented matrix.

11. $\left[\begin{array}{cc|c} -2 & 5 & 5 \\ 6 & -18 & 26 \end{array}\right]$

12. $\left[\begin{array}{cc|c} 3 & 4 & 10 \\ 10 & 17 & 439 \end{array}\right]$

13. $\left[\begin{array}{ccc|c} 3 & 2 & 0 & 3 \\ -1 & -9 & 4 & -1 \\ 8 & 5 & 7 & 8 \end{array}\right]$

14. $\left[\begin{array}{ccc|c} 8 & 29 & 1 & 43 \\ -1 & 7 & 5 & 38 \\ 0 & 0 & 3 & 10 \end{array}\right]$

15. $\left[\begin{array}{ccc|c} 4 & 5 & -2 & 12 \\ 0 & 1 & 58 & 2 \\ 8 & 7 & -3 & -5 \end{array}\right]$

For the following exercises, solve the system by Gaussian elimination.

16. $\left[\begin{array}{cc|c} 1 & 0 & 3 \\ 0 & 0 & 0 \end{array}\right]$

17. $\left[\begin{array}{cc|c} 1 & 0 & 1 \\ 1 & 0 & 2 \end{array}\right]$

18. $\left[\begin{array}{cc|c} 1 & 2 & 3 \\ 4 & 5 & 6 \end{array}\right]$

19. $\left[\begin{array}{cc|c} -1 & 2 & -3 \\ 4 & -5 & 6 \end{array}\right]$

20. $\left[\begin{array}{cc|c} -2 & 0 & 1 \\ 0 & 2 & -1 \end{array}\right]$

21. $\begin{aligned} 2x - 3y &= -9 \\ 5x + 4y &= 58 \end{aligned}$

22. $\begin{aligned} 6x + 2y &= -4 \\ 3x + 4y &= -17 \end{aligned}$

23. $\begin{aligned} 2x + 3y &= 12 \\ 4x + y &= 14 \end{aligned}$

24. $\begin{aligned} -4x - 3y &= -2 \\ 3x - 5y &= -13 \end{aligned}$

25. $\begin{aligned} -5x + 8y &= 3 \\ 10x + 6y &= 5 \end{aligned}$

26. $\begin{aligned} 3x + 4y &= 12 \\ -6x - 8y &= -24 \end{aligned}$

27. $\begin{aligned} -60x + 45y &= 12 \\ 20x - 15y &= -4 \end{aligned}$

28. $\begin{aligned} 11x + 10y &= 43 \\ 15x + 20y &= 65 \end{aligned}$

29. $\begin{aligned} 2x - y &= 2 \\ 3x + 2y &= 17 \end{aligned}$

30. $\begin{aligned} -1.06x - 2.25y &= 5.51 \\ -5.03x - 1.08y &= 5.40 \end{aligned}$

31. $\begin{aligned} \tfrac{3}{4}x - \tfrac{3}{5}y &= 4 \\ \tfrac{1}{4}x + \tfrac{2}{3}y &= 1 \end{aligned}$

32. $\begin{aligned} \tfrac{1}{4}x - \tfrac{2}{3}y &= -1 \\ \tfrac{1}{2}x + \tfrac{1}{3}y &= 3 \end{aligned}$

33. $\left[\begin{array}{ccc|c} 1 & 0 & 0 & 31 \\ 0 & 1 & 1 & 45 \\ 0 & 0 & 1 & 87 \end{array}\right]$

34. $\left[\begin{array}{ccc|c} 1 & 0 & 1 & 50 \\ 1 & 1 & 0 & 20 \\ 0 & 1 & 1 & -90 \end{array}\right]$

35. $\left[\begin{array}{ccc|c} 1 & 2 & 3 & 4 \\ 0 & 5 & 6 & 7 \\ 0 & 0 & 8 & 9 \end{array}\right]$

36. $\left[\begin{array}{ccc|c} -0.1 & 0.3 & -0.1 & 0.2 \\ -0.4 & 0.2 & 0.1 & 0.8 \\ 0.6 & 0.1 & 0.7 & -0.8 \end{array}\right]$

37. $\begin{aligned} -2x + 3y - 2z &= 3 \\ 4x + 2y - z &= 9 \\ 4x - 8y + 2z &= -6 \end{aligned}$

38. $\begin{aligned} x + y - 4z &= -4 \\ 5x - 3y - 2z &= 0 \\ 2x + 6y + 7z &= 30 \end{aligned}$

39. $\begin{aligned} 2x + 3y + 2z &= 1 \\ -4x - 6y - 4z &= -2 \\ 10x + 15y + 10z &= 5 \end{aligned}$

40. $\begin{aligned} x + 2y - z &= 1 \\ -x - 2y + 2z &= -2 \\ 3x + 6y - 3z &= 5 \end{aligned}$

41. $\begin{aligned} x + 2y - z &= 1 \\ -x - 2y + 2z &= -2 \\ 3x + 6y - 3z &= 3 \end{aligned}$

42. $\begin{aligned} x + y &= 2 \\ x + z &= 1 \\ -y - z &= -3 \end{aligned}$

43. $\begin{aligned} x+y+z&=100 \\ x+2z&=125 \\ -y+2z&=25 \end{aligned}$

44. $\begin{aligned} \frac{1}{4}x-\frac{2}{3}z&=-\frac{1}{2} \\ \frac{1}{5}x+\frac{1}{3}y&=\frac{4}{7} \\ \frac{1}{5}y-\frac{1}{3}z&=\frac{2}{9} \end{aligned}$

45. $\begin{aligned} -\frac{1}{2}x+\frac{1}{2}y+\frac{1}{7}z&=-\frac{53}{14} \\ \frac{1}{2}x-\frac{1}{2}y+\frac{1}{4}z&=3 \\ \frac{1}{4}x+\frac{1}{5}y+\frac{1}{3}z&=\frac{23}{15} \end{aligned}$

46. $\begin{aligned} -\frac{1}{2}x-\frac{1}{3}y+\frac{1}{4}z&=-\frac{29}{6} \\ \frac{1}{5}x+\frac{1}{6}y-\frac{1}{7}z&=\frac{431}{210} \\ -\frac{1}{8}x+\frac{1}{9}y+\frac{1}{10}z&=-\frac{49}{45} \end{aligned}$

Extensions

For the following exercises, use Gaussian elimination to solve the system.

47. $\begin{aligned} \frac{x-1}{7}+\frac{y-2}{8}+\frac{z-3}{4}&=0 \\ x+y+z&=6 \\ \frac{x+2}{3}+2y+\frac{z-3}{3}&=5 \end{aligned}$

48. $\begin{aligned} \frac{x-1}{4}-\frac{y+1}{4}+3z&=-1 \\ \frac{x+5}{2}+\frac{y+7}{4}-z&=4 \\ x+y-\frac{z-2}{2}&=1 \end{aligned}$

49. $\begin{aligned} \frac{x-3}{4}-\frac{y-1}{3}+2z&=-1 \\ \frac{x+5}{2}+\frac{y+5}{2}+\frac{z+5}{2}&=8 \\ x+y+z&=1 \end{aligned}$

50. $\begin{aligned} \frac{x-3}{10}+\frac{y+3}{2}-2z&=3 \\ \frac{x+5}{4}-\frac{y-1}{8}+z&=\frac{3}{2} \\ \frac{x-1}{4}+\frac{y+4}{2}+3z&=\frac{3}{2} \end{aligned}$

51. $\begin{aligned} \frac{x-3}{4}-\frac{y-1}{3}+2z&=-1 \\ \frac{x+5}{2}+\frac{y+5}{2}+\frac{z+5}{2}&=7 \\ x+y+z&=1 \end{aligned}$

Real-World Applications

For the following exercises, set up the augmented matrix that describes the situation, and solve for the desired solution.

52. Every day, Angeni's cupcake store sells 5,000 cupcakes in chocolate and vanilla flavors. If the chocolate flavor is 3 times as popular as the vanilla flavor, how many of each cupcake does the store sell per day?

53. At Bakari's competing cupcake store, $4,520 worth of cupcakes are sold daily. The chocolate cupcakes cost $2.25 and the red velvet cupcakes cost $1.75. If the total number of cupcakes sold per day is 2,200, how many of each flavor are sold each day?

54. You invested $10,000 into two accounts: one that has simple 3% interest, the other with 2.5% interest. If your total interest payment after one year was $283.50, how much was in each account after the year passed?

55. You invested $2,300 into account 1, and $2,700 into account 2. If the total amount of interest after one year is $254, and account 2 has 1.5 times the interest rate of account 1, what are the interest rates? Assume simple interest rates.

56. Bikes'R'Us manufactures bikes, which sell for $250. It costs the manufacturer $180 per bike, plus a startup fee of $3,500. After how many bikes sold will the manufacturer break even?

57. A major appliance store has agreed to order vacuums from a startup founded by college engineering students. The store would be able to purchase the vacuums for $86 each, with a delivery fee of $9,200, regardless of how many vacuums are sold. If the store needs to start seeing a profit after 230 units are sold, how much should they charge for the vacuums?

58. The three most popular ice cream flavors are chocolate, strawberry, and vanilla, comprising 83% of the flavors sold at an ice cream shop. If vanilla sells 1% more than twice strawberry, and chocolate sells 11% more than vanilla, how much of the total ice cream consumption are the vanilla, chocolate, and strawberry flavors?

59. At an ice cream shop, three flavors are increasing in demand. Last year, banana, pumpkin, and rocky road ice cream made up 12% of total ice cream sales. This year, the same three ice creams made up 16.9% of ice cream sales. The rocky road sales doubled, the banana sales increased by 50%, and the pumpkin sales increased by 20%. If the rocky road ice cream had one less percent of sales than the banana ice cream, find out the percentage of ice cream sales each individual ice cream made last year.

60. A bag of mixed nuts contains cashews, pistachios, and almonds. There are 1,000 total nuts in the bag, and there are 100 less almonds than pistachios. The cashews weigh 3 g, pistachios weigh 4 g, and almonds weigh 5 g. If the bag weighs 3.7 kg, find out how many of each type of nut is in the bag.

61. A bag of mixed nuts contains cashews, pistachios, and almonds. Originally there were 900 nuts in the bag. 30% of the almonds, 20% of the cashews, and 10% of the pistachios were eaten, and now there are 770 nuts left in the bag. Originally, there were 100 more cashews than almonds. Figure out how many of each type of nut was in the bag to begin with.

11.7 Solving Systems with Inverses

Learning Objectives

In this section, you will:

- Find the inverse of a matrix.
- Solve a system of linear equations using an inverse matrix.

Soriya plans to invest $10,500 into two different bonds to spread out her risk. The first bond has an annual return of 10%, and the second bond has an annual return of 6%. In order to receive an 8.5% return from the two bonds, how much should Soriya invest in each bond? What is the best method to solve this problem?

There are several ways we can solve this problem. As we have seen in previous sections, systems of equations and matrices are useful in solving real-world problems involving finance. After studying this section, we will have the tools to solve the bond problem using the inverse of a matrix.

Finding the Inverse of a Matrix

We know that the multiplicative inverse of a real number a is a^{-1}, and $aa^{-1} = a^{-1}a = \left(\frac{1}{a}\right)a = 1$. For example, $2^{-1} = \frac{1}{2}$ and $\left(\frac{1}{2}\right)2 = 1$. The multiplicative inverse of a matrix is similar in concept, except that the product of matrix A and its inverse A^{-1} equals the identity matrix. The identity matrix is a square matrix containing ones down the main diagonal and zeros everywhere else. We identify identity matrices by I_n where n represents the dimension of the matrix. Observe the following equations.

$$I_2 = \begin{bmatrix} 1 & 0 \\ 0 & 1 \end{bmatrix}$$

$$I_3 = \begin{bmatrix} 1 & 0 & 0 \\ 0 & 1 & 0 \\ 0 & 0 & 1 \end{bmatrix}$$

The identity matrix acts as a 1 in matrix algebra. For example, $AI = IA = A$.

A matrix that has a multiplicative inverse has the properties

$$AA^{-1} = I$$
$$A^{-1}A = I$$

A matrix that has a multiplicative inverse is called an invertible matrix. Only a square matrix may have a multiplicative inverse, as the reversibility, $AA^{-1} = A^{-1}A = I$, is a requirement. Not all square matrices have an inverse, but if A is invertible, then A^{-1} is unique. We will look at two methods for finding the inverse of a 2×2 matrix and a third method that can be used on both 2×2 and 3×3 matrices.

The Identity Matrix and Multiplicative Inverse

The **identity matrix,** I_n, is a square matrix containing ones down the main diagonal and zeros everywhere else.

$$I_2 = \begin{bmatrix} 1 & 0 \\ 0 & 1 \end{bmatrix} \qquad I_3 = \begin{bmatrix} 1 & 0 & 0 \\ 0 & 1 & 0 \\ 0 & 0 & 1 \end{bmatrix}$$

2×2 $\qquad$ 3×3

If A is an $n \times n$ matrix and B is an $n \times n$ matrix such that $AB = BA = I_n$, then $B = A^{-1}$, the **multiplicative inverse of a matrix** A.

EXAMPLE 1

Showing That the Identity Matrix Acts as a 1

Given matrix *A*, show that $AI = IA = A$.

$$A = \begin{bmatrix} 3 & 4 \\ -2 & 5 \end{bmatrix}$$

Solution

Use matrix multiplication to show that the product of A and the identity is equal to the product of the identity and A.

$$AI = \begin{bmatrix} 3 & 4 \\ -2 & 5 \end{bmatrix} \begin{bmatrix} 1 & 0 \\ 0 & 1 \end{bmatrix} = \begin{bmatrix} 3 \cdot 1 + 4 \cdot 0 & 3 \cdot 0 + 4 \cdot 1 \\ -2 \cdot 1 + 5 \cdot 0 & -2 \cdot 0 + 5 \cdot 1 \end{bmatrix} = \begin{bmatrix} 3 & 4 \\ -2 & 5 \end{bmatrix}$$

$$IA = \begin{bmatrix} 1 & 0 \\ 0 & 1 \end{bmatrix} \begin{bmatrix} 3 & 4 \\ -2 & 5 \end{bmatrix} = \begin{bmatrix} 1 \cdot 3 + 0 \cdot (-2) & 1 \cdot 4 + 0 \cdot 5 \\ 0 \cdot 3 + 1 \cdot (-2) & 0 \cdot 4 + 1 \cdot 5 \end{bmatrix} = \begin{bmatrix} 3 & 4 \\ -2 & 5 \end{bmatrix}$$

HOW TO

Given two matrices, show that one is the multiplicative inverse of the other.

1. Given matrix A of order $n \times n$ and matrix B of order $n \times n$ multiply AB.
2. If $AB = I$, then find the product BA. If $BA = I$, then $B = A^{-1}$ and $A = B^{-1}$.

EXAMPLE 2

Showing That Matrix *A* Is the Multiplicative Inverse of Matrix *B*

Show that the given matrices are multiplicative inverses of each other.

$$A = \begin{bmatrix} 1 & 5 \\ -2 & -9 \end{bmatrix}, B = \begin{bmatrix} -9 & -5 \\ 2 & 1 \end{bmatrix}$$

Solution

Multiply AB and BA. If both products equal the identity, then the two matrices are inverses of each other.

$$\begin{aligned} AB &= \begin{bmatrix} 1 & 5 \\ -2 & -9 \end{bmatrix} \cdot \begin{bmatrix} -9 & -5 \\ 2 & 1 \end{bmatrix} \\ &= \begin{bmatrix} 1(-9) + 5(2) & 1(-5) + 5(1) \\ -2(-9) - 9(2) & -2(-5) - 9(1) \end{bmatrix} \\ &= \begin{bmatrix} 1 & 0 \\ 0 & 1 \end{bmatrix} \end{aligned}$$

$$\begin{aligned} BA &= \begin{bmatrix} -9 & -5 \\ 2 & 1 \end{bmatrix} \cdot \begin{bmatrix} 1 & 5 \\ -2 & -9 \end{bmatrix} \\ &= \begin{bmatrix} -9(1) - 5(-2) & -9(5) - 5(-9) \\ 2(1) + 1(-2) & 2(5) + 1(-9) \end{bmatrix} \\ &= \begin{bmatrix} 1 & 0 \\ 0 & 1 \end{bmatrix} \end{aligned}$$

A and B are inverses of each other.

TRY IT #1 Show that the following two matrices are inverses of each other.

$$A = \begin{bmatrix} 1 & 4 \\ -1 & -3 \end{bmatrix}, B = \begin{bmatrix} -3 & -4 \\ 1 & 1 \end{bmatrix}$$

Finding the Multiplicative Inverse Using Matrix Multiplication

We can now determine whether two matrices are inverses, but how would we find the inverse of a given matrix? Since we know that the product of a matrix and its inverse is the identity matrix, we can find the inverse of a matrix by setting up an equation using matrix multiplication.

EXAMPLE 3

Finding the Multiplicative Inverse Using Matrix Multiplication

Use matrix multiplication to find the inverse of the given matrix.

$$A = \begin{bmatrix} 1 & -2 \\ 2 & -3 \end{bmatrix}$$

Solution

For this method, we multiply A by a matrix containing unknown constants and set it equal to the identity.

$$\begin{bmatrix} 1 & -2 \\ 2 & -3 \end{bmatrix} \begin{bmatrix} a & b \\ c & d \end{bmatrix} = \begin{bmatrix} 1 & 0 \\ 0 & 1 \end{bmatrix}$$

Find the product of the two matrices on the left side of the equal sign.

$$\begin{bmatrix} 1 & -2 \\ 2 & -3 \end{bmatrix} \begin{bmatrix} a & b \\ c & d \end{bmatrix} = \begin{bmatrix} 1a-2c & 1b-2d \\ 2a-3c & 2b-3d \end{bmatrix}$$

Next, set up a system of equations with the entry in row 1, column 1 of the new matrix equal to the first entry of the identity, 1. Set the entry in row 2, column 1 of the new matrix equal to the corresponding entry of the identity, which is 0.

$$\begin{aligned} 1a-2c &= 1 \quad R_1 \\ 2a-3c &= 0 \quad R_2 \end{aligned}$$

Using row operations, multiply and add as follows: $(-2)R_1 + R_2 \rightarrow R_2$. Add the equations, and solve for c.

$$\begin{aligned} 1a - 2c &= 1 \\ 0 + 1c &= -2 \\ c &= -2 \end{aligned}$$

Back-substitute to solve for a.

$$\begin{aligned} a-2(-2) &= 1 \\ a + 4 &= 1 \\ a &= -3 \end{aligned}$$

Write another system of equations setting the entry in row 1, column 2 of the new matrix equal to the corresponding entry of the identity, 0. Set the entry in row 2, column 2 equal to the corresponding entry of the identity.

$$\begin{aligned} 1b-2d &= 0 \quad R_1 \\ 2b-3d &= 1 \quad R_2 \end{aligned}$$

Using row operations, multiply and add as follows: $(-2)\,R_1 + R_2 = R_2$. Add the two equations and solve for d.

$$1b-2d = 0$$
$$\frac{0+1d=1}{d=1}$$

Once more, back-substitute and solve for b.

$$\begin{aligned} b-2(1) &= 0 \\ b-2 &= 0 \\ b &= 2 \end{aligned}$$

$$A^{-1} = \begin{bmatrix} -3 & 2 \\ -2 & 1 \end{bmatrix}$$

Finding the Multiplicative Inverse by Augmenting with the Identity

Another way to find the multiplicative inverse is by augmenting with the identity. When matrix A is transformed into I, the augmented matrix I transforms into A^{-1}.

For example, given

$$A = \begin{bmatrix} 2 & 1 \\ 5 & 3 \end{bmatrix}$$

augment A with the identity

$$\left[\begin{array}{cc|cc} 2 & 1 & 1 & 0 \\ 5 & 3 & 0 & 1 \end{array}\right]$$

Perform row operations with the goal of turning A into the identity.

1. Switch row 1 and row 2.

$$\left[\begin{array}{cc|cc} 5 & 3 & 0 & 1 \\ 2 & 1 & 1 & 0 \end{array}\right]$$

2. Multiply row 2 by −2 and add to row 1.

$$\left[\begin{array}{cc|cc} 1 & 1 & -2 & 1 \\ 2 & 1 & 1 & 0 \end{array}\right]$$

3. Multiply row 1 by −2 and add to row 2.

$$\left[\begin{array}{cc|cc} 1 & 1 & -2 & 1 \\ 0 & -1 & 5 & -2 \end{array}\right]$$

4. Add row 2 to row 1.

$$\left[\begin{array}{cc|cc} 1 & 0 & 3 & -1 \\ 0 & -1 & 5 & -2 \end{array}\right]$$

5. Multiply row 2 by -1.

$$\left[\begin{array}{cc|cc} 1 & 0 & 3 & -1 \\ 0 & 1 & -5 & 2 \end{array}\right]$$

The matrix we have found is A^{-1}.

$$A^{-1} = \begin{bmatrix} 3 & -1 \\ -5 & 2 \end{bmatrix}$$

Finding the Multiplicative Inverse of 2×2 Matrices Using a Formula

When we need to find the multiplicative inverse of a 2×2 matrix, we can use a special formula instead of using matrix multiplication or augmenting with the identity.

If A is a 2×2 matrix, such as

$$A = \begin{bmatrix} a & b \\ c & d \end{bmatrix}$$

the multiplicative inverse of A is given by the formula

$$A^{-1} = \frac{1}{ad - bc} \begin{bmatrix} d & -b \\ -c & a \end{bmatrix}$$

where $ad - bc \neq 0$. If $ad - bc = 0$, then A has no inverse.

EXAMPLE 4

Using the Formula to Find the Multiplicative Inverse of Matrix *A*

Use the formula to find the multiplicative inverse of

$$A = \begin{bmatrix} 1 & -2 \\ 2 & -3 \end{bmatrix}$$

Solution

Using the formula, we have

$$A^{-1} = \frac{1}{(1)(-3)-(-2)(2)}\begin{bmatrix}-3 & 2\\ -2 & 1\end{bmatrix}$$
$$= \frac{1}{-3+4}\begin{bmatrix}-3 & 2\\ -2 & 1\end{bmatrix}$$
$$= \begin{bmatrix}-3 & 2\\ -2 & 1\end{bmatrix}$$

Analysis

We can check that our formula works by using one of the other methods to calculate the inverse. Let's augment A with the identity.

$$\left[\begin{array}{cc|cc}1 & -2 & 1 & 0\\ 2 & -3 & 0 & 1\end{array}\right]$$

Perform row operations with the goal of turning A into the identity.

1. Multiply row 1 by −2 and add to row 2.

$$\left[\begin{array}{cc|cc}1 & -2 & 1 & 0\\ 0 & 1 & -2 & 1\end{array}\right]$$

2. Multiply row 1 by 2 and add to row 1.

$$\left[\begin{array}{cc|cc}1 & 0 & -3 & 2\\ 0 & 1 & -2 & 1\end{array}\right]$$

So, we have verified our original solution.

$$A^{-1} = \begin{bmatrix}-3 & 2\\ -2 & 1\end{bmatrix}$$

TRY IT #2 Use the formula to find the inverse of matrix A. Verify your answer by augmenting with the identity matrix.

$$A = \begin{bmatrix}1 & -1\\ 2 & 3\end{bmatrix}$$

EXAMPLE 5

Finding the Inverse of the Matrix, If It Exists

Find the inverse, if it exists, of the given matrix.

$$A = \begin{bmatrix}3 & 6\\ 1 & 2\end{bmatrix}$$

Solution

We will use the method of augmenting with the identity.

$$\left[\begin{array}{cc|cc}3 & 6 & 1 & 0\\ 1 & 2 & 0 & 1\end{array}\right]$$

1. Switch row 1 and row 2.

$$\left[\begin{array}{cc|cc}1 & 3 & 0 & 1\\ 3 & 2 & 1 & 0\end{array}\right]$$

2. Multiply row 1 by −3 and add it to row 2.

$$\left[\begin{array}{cc|cc}1 & 2 & 1 & 0\\ 0 & 0 & -3 & 1\end{array}\right]$$

3. There is nothing further we can do. The zeros in row 2 indicate that this matrix has no inverse.

Finding the Multiplicative Inverse of 3×3 Matrices

Unfortunately, we do not have a formula similar to the one for a 2×2 matrix to find the inverse of a 3×3 matrix. Instead, we will augment the original matrix with the identity matrix and use row operations to obtain the inverse.

Given a 3×3 matrix

$$A = \begin{bmatrix} 2 & 3 & 1 \\ 3 & 3 & 1 \\ 2 & 4 & 1 \end{bmatrix}$$

augment A with the identity matrix

$$A|I = \left[\begin{array}{ccc|ccc} 2 & 3 & 1 & 1 & 0 & 0 \\ 3 & 3 & 1 & 0 & 1 & 0 \\ 2 & 4 & 1 & 0 & 0 & 1 \end{array}\right]$$

To begin, we write the augmented matrix with the identity on the right and A on the left. Performing elementary row operations so that the identity matrix appears on the left, we will obtain the inverse matrix on the right. We will find the inverse of this matrix in the next example.

HOW TO

Given a 3×3 matrix, find the inverse

1. Write the original matrix augmented with the identity matrix on the right.
2. Use elementary row operations so that the identity appears on the left.
3. What is obtained on the right is the inverse of the original matrix.
4. Use matrix multiplication to show that $AA^{-1} = I$ and $A^{-1}A = I$.

EXAMPLE 6

Finding the Inverse of a 3 × 3 Matrix

Given the 3×3 matrix A, find the inverse.

$$A = \begin{bmatrix} 2 & 3 & 1 \\ 3 & 3 & 1 \\ 2 & 4 & 1 \end{bmatrix}$$

Solution

Augment A with the identity matrix, and then begin row operations until the identity matrix replaces A. The matrix on the right will be the inverse of A.

$$\left[\begin{array}{ccc|ccc} 2 & 3 & 1 & 1 & 0 & 0 \\ 3 & 3 & 1 & 0 & 1 & 0 \\ 2 & 4 & 1 & 0 & 0 & 1 \end{array}\right] \xrightarrow{\text{Interchange } R_2 \text{ and } R_1} \left[\begin{array}{ccc|ccc} 3 & 3 & 1 & 0 & 1 & 0 \\ 2 & 3 & 1 & 1 & 0 & 0 \\ 2 & 4 & 1 & 0 & 0 & 1 \end{array}\right]$$

$$-R_2 + R_1 = R_1 \rightarrow \left[\begin{array}{ccc|ccc} 1 & 0 & 0 & -1 & 1 & 0 \\ 2 & 3 & 1 & 1 & 0 & 0 \\ 2 & 4 & 1 & 0 & 0 & 1 \end{array}\right]$$

$$-R_2 + R_3 = R_3 \rightarrow \left[\begin{array}{ccc|ccc} 1 & 0 & 0 & -1 & 1 & 0 \\ 2 & 3 & 1 & 1 & 0 & 0 \\ 0 & 1 & 0 & -1 & 0 & 1 \end{array}\right]$$

$$R_3 \leftrightarrow R_2 \rightarrow \left[\begin{array}{ccc|ccc} 1 & 0 & 0 & -1 & 1 & 0 \\ 0 & 1 & 0 & -1 & 0 & 1 \\ 2 & 3 & 1 & 1 & 0 & 0 \end{array}\right]$$

$$-2R_1 + R_3 = R_3 \rightarrow \left[\begin{array}{ccc|ccc} 1 & 0 & 0 & -1 & 1 & 0 \\ 0 & 1 & 0 & -1 & 0 & 1 \\ 0 & 3 & 1 & 3 & -2 & 0 \end{array}\right]$$

$$-3R_2 + R_3 = R_3 \rightarrow \left[\begin{array}{ccc|ccc} 1 & 0 & 0 & -1 & 1 & 0 \\ 0 & 1 & 0 & -1 & 0 & 1 \\ 0 & 0 & 1 & 6 & -2 & -3 \end{array}\right]$$

Thus,

$$A^{-1} = B = \begin{bmatrix} -1 & 1 & 0 \\ -1 & 0 & 1 \\ 6 & -2 & -3 \end{bmatrix}$$

Analysis

To prove that $B = A^{-1}$, let's multiply the two matrices together to see if the product equals the identity, if $AA^{-1} = I$ and $A^{-1}A = I$.

$$\begin{aligned} AA^{-1} &= \begin{bmatrix} 2 & 3 & 1 \\ 3 & 3 & 1 \\ 2 & 4 & 1 \end{bmatrix} \begin{bmatrix} -1 & 1 & 0 \\ -1 & 0 & 1 \\ 6 & -2 & -3 \end{bmatrix} \\ &= \begin{bmatrix} 2(-1) + 3(-1) + 1(6) & 2(1) + 3(0) + 1(-2) & 2(0) + 3(1) + 1(-3) \\ 3(-1) + 3(-1) + 1(6) & 3(1) + 3(0) + 1(-2) & 3(0) + 3(1) + 1(-3) \\ 2(-1) + 4(-1) + 1(6) & 2(1) + 4(0) + 1(-2) & 2(0) + 4(1) + 1(-3) \end{bmatrix} \\ &= \begin{bmatrix} 1 & 0 & 0 \\ 0 & 1 & 0 \\ 0 & 0 & 1 \end{bmatrix} \end{aligned}$$

$$\begin{aligned} A^{-1}A &= \begin{bmatrix} -1 & 1 & 0 \\ -1 & 0 & 1 \\ 6 & -2 & -3 \end{bmatrix} \begin{bmatrix} 2 & 3 & 1 \\ 3 & 3 & 1 \\ 2 & 4 & 1 \end{bmatrix} \\ &= \begin{bmatrix} -1(2) + 1(3) + 0(2) & -1(3) + 1(3) + 0(4) & -1(1) + 1(1) + 0(1) \\ -1(2) + 0(3) + 1(2) & -1(3) + 0(3) + 1(4) & -1(1) + 0(1) + 1(1) \\ 6(2) + -2(3) + -3(2) & 6(3) + -2(3) + -3(4) & 6(1) + -2(1) + -3(1) \end{bmatrix} \\ &= \begin{bmatrix} 1 & 0 & 0 \\ 0 & 1 & 0 \\ 0 & 0 & 1 \end{bmatrix} \end{aligned}$$

TRY IT #3 Find the inverse of the 3×3 matrix.

$$A = \begin{bmatrix} 2 & -17 & 11 \\ -1 & 11 & -7 \\ 0 & 3 & -2 \end{bmatrix}$$

Solving a System of Linear Equations Using the Inverse of a Matrix

Solving a system of linear equations using the inverse of a matrix requires the definition of two new matrices: X is the matrix representing the variables of the system, and B is the matrix representing the constants. Using matrix multiplication, we may define a system of equations with the same number of equations as variables as

$$AX = B$$

To solve a system of linear equations using an inverse matrix, let A be the coefficient matrix, let X be the variable matrix, and let B be the constant matrix. Thus, we want to solve a system $AX = B$. For example, look at the following system of equations.

$$a_1x + b_1y = c_1$$
$$a_2x + b_2y = c_2$$

From this system, the coefficient matrix is

$$A = \begin{bmatrix} a_1 & b_1 \\ a_2 & b_2 \end{bmatrix}$$

The variable matrix is

$$X = \begin{bmatrix} x \\ y \end{bmatrix}$$

And the constant matrix is

$$B = \begin{bmatrix} c_1 \\ c_2 \end{bmatrix}$$

Then $AX = B$ looks like

$$\begin{bmatrix} a_1 & b_1 \\ a_2 & b_2 \end{bmatrix} \begin{bmatrix} x \\ y \end{bmatrix} = \begin{bmatrix} c_1 \\ c_2 \end{bmatrix}$$

Recall the discussion earlier in this section regarding multiplying a real number by its inverse, $(2^{-1})\ 2 = \left(\frac{1}{2}\right)\ 2 = 1$. To solve a single linear equation $ax = b$ for x, we would simply multiply both sides of the equation by the multiplicative inverse (reciprocal) of a. Thus,

$$\begin{aligned} ax &= b \\ \left(\tfrac{1}{a}\right) ax &= \left(\tfrac{1}{a}\right) b \\ (a^{-1})ax &= (a^{-1})b \\ [(a^{-1})a]x &= (a^{-1})b \\ 1x &= (a^{-1})b \\ x &= (a^{-1})b \end{aligned}$$

The only difference between a solving a linear equation and a system of equations written in matrix form is that finding the inverse of a matrix is more complicated, and matrix multiplication is a longer process. However, the goal is the same—to isolate the variable.

We will investigate this idea in detail, but it is helpful to begin with a 2×2 system and then move on to a 3×3 system.

Solving a System of Equations Using the Inverse of a Matrix

Given a system of equations, write the coefficient matrix A, the variable matrix X, and the constant matrix B. Then

$$AX = B$$

Multiply both sides by the inverse of A to obtain the solution.

$$\begin{aligned} \left(A^{-1}\right) AX &= \left(A^{-1}\right) B \\ \left[\left(A^{-1}\right) A\right] X &= \left(A^{-1}\right) B \\ IX &= \left(A^{-1}\right) B \\ X &= \left(A^{-1}\right) B \end{aligned}$$

Q&A **If the coefficient matrix does not have an inverse, does that mean the system has no solution?**

No, if the coefficient matrix is not invertible, the system could be inconsistent and have no solution, or be dependent and have infinitely many solutions.

EXAMPLE 7

Solving a 2 × 2 System Using the Inverse of a Matrix
Solve the given system of equations using the inverse of a matrix.

$$\begin{aligned} 3x + 8y &= 5 \\ 4x + 11y &= 7 \end{aligned}$$

Solution
Write the system in terms of a coefficient matrix, a variable matrix, and a constant matrix.

$$A = \begin{bmatrix} 3 & 8 \\ 4 & 11 \end{bmatrix}, X = \begin{bmatrix} x \\ y \end{bmatrix}, B = \begin{bmatrix} 5 \\ 7 \end{bmatrix}$$

Then

$$\begin{bmatrix} 3 & 8 \\ 4 & 11 \end{bmatrix} \begin{bmatrix} x \\ y \end{bmatrix} = \begin{bmatrix} 5 \\ 7 \end{bmatrix}$$

First, we need to calculate A^{-1}. Using the formula to calculate the inverse of a 2 by 2 matrix, we have:

$$\begin{aligned} A^{-1} &= \frac{1}{ad-bc} \begin{bmatrix} d & -b \\ -c & a \end{bmatrix} \\ &= \frac{1}{3(11)-8(4)} \begin{bmatrix} 11 & -8 \\ -4 & 3 \end{bmatrix} \\ &= \frac{1}{1} \begin{bmatrix} 11 & -8 \\ -4 & 3 \end{bmatrix} \end{aligned}$$

So,

$$A^{-1} = \begin{bmatrix} 11 & -8 \\ -4 & 3 \end{bmatrix}$$

Now we are ready to solve. Multiply both sides of the equation by A^{-1}.

$$\begin{aligned} &(A^{-1})AX = (A^{-1})B \\ &\begin{bmatrix} 11 & -8 \\ -4 & 3 \end{bmatrix} \begin{bmatrix} 3 & 8 \\ 4 & 11 \end{bmatrix} \begin{bmatrix} x \\ y \end{bmatrix} = \begin{bmatrix} 11 & -8 \\ -4 & 3 \end{bmatrix} \begin{bmatrix} 5 \\ 7 \end{bmatrix} \\ &\begin{bmatrix} 1 & 0 \\ 0 & 1 \end{bmatrix} \begin{bmatrix} x \\ y \end{bmatrix} = \begin{bmatrix} 11(5) + (-8)7 \\ -4(5) + 3(7) \end{bmatrix} \\ &\begin{bmatrix} x \\ y \end{bmatrix} = \begin{bmatrix} -1 \\ 1 \end{bmatrix} \end{aligned}$$

The solution is $(-1, 1)$.

Q&A **Can we solve for X by finding the product BA^{-1}?**

No, recall that matrix multiplication is not commutative, so $A^{-1}B \neq BA^{-1}$. Consider our steps for solving the matrix equation.

$$\begin{aligned} (A^{-1})\,AX &= (A^{-1})\,B \\ [(A^{-1})\,A]\,X &= (A^{-1})\,B \\ IX &= (A^{-1})\,B \\ X &= (A^{-1})\,B \end{aligned}$$

Notice in the first step we multiplied both sides of the equation by A^{-1}, but the A^{-1} was to the left of A on the left side and to the left of B on the right side. Because matrix multiplication is not commutative, order matters.

EXAMPLE 8

Solving a 3 × 3 System Using the Inverse of a Matrix

Solve the following system using the inverse of a matrix.

$$\begin{aligned} 5x+15y+56z &= 35 \\ -4x-11y-41z &= -26 \\ -x-3y-11z &= -7 \end{aligned}$$

Solution

Write the equation $AX = B$.

$$\begin{bmatrix} 5 & 15 & 56 \\ -4 & -11 & -41 \\ -1 & -3 & -11 \end{bmatrix} \begin{bmatrix} x \\ y \\ z \end{bmatrix} = \begin{bmatrix} 35 \\ -26 \\ -7 \end{bmatrix}$$

First, we will find the inverse of A by augmenting with the identity.

$$\left[\begin{array}{ccc|ccc} 5 & 15 & 56 & 1 & 0 & 0 \\ -4 & -11 & -41 & 0 & 1 & 0 \\ -1 & -3 & -11 & 0 & 0 & 1 \end{array}\right]$$

Multiply row 1 by $\frac{1}{5}$.

$$\left[\begin{array}{ccc|ccc} 1 & 3 & \frac{56}{5} & \frac{1}{5} & 0 & 0 \\ -4 & -11 & -41 & 0 & 1 & 0 \\ -1 & -3 & -11 & 0 & 0 & 1 \end{array}\right]$$

Multiply row 1 by 4 and add to row 2.

$$\left[\begin{array}{ccc|ccc} 1 & 3 & \frac{56}{5} & \frac{1}{5} & 0 & 0 \\ 0 & 1 & \frac{19}{5} & \frac{4}{5} & 1 & 0 \\ -1 & -3 & -11 & 0 & 0 & 1 \end{array}\right]$$

Add row 1 to row 3.

$$\left[\begin{array}{ccc|ccc} 1 & 3 & \frac{56}{5} & \frac{1}{5} & 0 & 0 \\ 0 & 1 & \frac{19}{5} & \frac{4}{5} & 1 & 0 \\ 0 & 0 & \frac{1}{5} & \frac{1}{5} & 0 & 1 \end{array}\right]$$

Multiply row 2 by −3 and add to row 1.

$$\left[\begin{array}{ccc|ccc} 1 & 0 & -\frac{1}{5} & -\frac{11}{5} & -3 & 0 \\ 0 & 1 & \frac{19}{5} & \frac{4}{5} & 1 & 0 \\ 0 & 0 & \frac{1}{5} & \frac{1}{5} & 0 & 1 \end{array}\right]$$

Multiply row 3 by 5.

$$\left[\begin{array}{ccc|ccc} 1 & 0 & -\frac{1}{5} & -\frac{11}{5} & -3 & 0 \\ 0 & 1 & \frac{19}{5} & \frac{4}{5} & 1 & 0 \\ 0 & 0 & 1 & 1 & 0 & 5 \end{array}\right]$$

Multiply row 3 by $\frac{1}{5}$ and add to row 1.

$$\left[\begin{array}{ccc|ccc} 1 & 0 & 0 & -2 & -3 & 1 \\ 0 & 1 & \frac{19}{5} & \frac{4}{5} & 1 & 0 \\ 0 & 0 & 1 & 1 & 0 & 5 \end{array}\right]$$

Multiply row 3 by $-\frac{19}{5}$ and add to row 2.

$$\left[\begin{array}{ccc|ccc} 1 & 0 & 0 & -2 & -3 & 1 \\ 0 & 1 & 0 & -3 & 1 & -19 \\ 0 & 0 & 1 & 1 & 0 & 5 \end{array}\right]$$

So,

$$A^{-1} = \begin{bmatrix} -2 & -3 & 1 \\ -3 & 1 & -19 \\ 1 & 0 & 5 \end{bmatrix}$$

Multiply both sides of the equation by A^{-1}. We want $A^{-1}AX = A^{-1}B$:

$$\begin{bmatrix} -2 & -3 & 1 \\ -3 & 1 & -19 \\ 1 & 0 & 5 \end{bmatrix} \begin{bmatrix} 5 & 15 & 56 \\ -4 & -11 & -41 \\ -1 & -3 & -11 \end{bmatrix} \begin{bmatrix} x \\ y \\ z \end{bmatrix} = \begin{bmatrix} -2 & -3 & 1 \\ -3 & 1 & -19 \\ 1 & 0 & 5 \end{bmatrix} \begin{bmatrix} 35 \\ -26 \\ -7 \end{bmatrix}$$

Thus,

$$A^{-1}B = \begin{bmatrix} -70 + 78 - 7 \\ -105 - 26 + 133 \\ 35 + 0 - 35 \end{bmatrix} = \begin{bmatrix} 1 \\ 2 \\ 0 \end{bmatrix}$$

The solution is $(1, 2, 0)$.

TRY IT #4 Solve the system using the inverse of the coefficient matrix.

$$\begin{aligned} 2x - 17y + 11z &= 0 \\ -x + 11y - 7z &= 8 \\ 3y - 2z &= -2 \end{aligned}$$

HOW TO

Given a system of equations, solve with matrix inverses using a calculator.

1. Save the coefficient matrix and the constant matrix as matrix variables $[A]$ and $[B]$.
2. Enter the multiplication into the calculator, calling up each matrix variable as needed.
3. If the coefficient matrix is invertible, the calculator will present the solution matrix; if the coefficient matrix is not invertible, the calculator will present an error message.

EXAMPLE 9

Using a Calculator to Solve a System of Equations with Matrix Inverses

Solve the system of equations with matrix inverses using a calculator

$$\begin{aligned} 2x + 3y + z &= 32 \\ 3x + 3y + z &= -27 \\ 2x + 4y + z &= -2 \end{aligned}$$

Solution

On the matrix page of the calculator, enter the coefficient matrix as the matrix variable $[A]$, and enter the constant matrix as the matrix variable $[B]$.

$$[A] = \begin{bmatrix} 2 & 3 & 1 \\ 3 & 3 & 1 \\ 2 & 4 & 1 \end{bmatrix}, \quad [B] = \begin{bmatrix} 32 \\ -27 \\ -2 \end{bmatrix}$$

On the home screen of the calculator, type in the multiplication to solve for X, calling up each matrix variable as needed.

$$[A]^{-1} \times [B]$$

Evaluate the expression.

$$\begin{bmatrix} -59 \\ -34 \\ 252 \end{bmatrix}$$

MEDIA

Access these online resources for additional instruction and practice with solving systems with inverses.

The Identity Matrix (http://openstax.org/l/identmatrix)
Determining Inverse Matrices (http://openstax.org/l/inversematrix)
Using a Matrix Equation to Solve a System of Equations (http://openstax.org/l/matrixsystem)

11.7 SECTION EXERCISES

Verbal

1. In a previous section, we showed that matrix multiplication is not commutative, that is, $AB \neq BA$ in most cases. Can you explain why matrix multiplication is commutative for matrix inverses, that is, $A^{-1}A = AA^{-1}$?

2. Does every 2×2 matrix have an inverse? Explain why or why not. Explain what condition is necessary for an inverse to exist.

3. Can you explain whether a 2×2 matrix with an entire row of zeros can have an inverse?

4. Can a matrix with an entire column of zeros have an inverse? Explain why or why not.

5. Can a matrix with zeros on the diagonal have an inverse? If so, find an example. If not, prove why not. For simplicity, assume a 2×2 matrix.

Algebraic

In the following exercises, show that matrix A is the inverse of matrix B.

6. $A = \begin{bmatrix} 1 & 0 \\ -1 & 1 \end{bmatrix}, \quad B = \begin{bmatrix} 1 & 0 \\ 1 & 1 \end{bmatrix}$

7. $A = \begin{bmatrix} 1 & 2 \\ 3 & 4 \end{bmatrix}, \quad B = \begin{bmatrix} -2 & 1 \\ \frac{3}{2} & -\frac{1}{2} \end{bmatrix}$

8. $A = \begin{bmatrix} 4 & 5 \\ 7 & 0 \end{bmatrix}, \quad B = \begin{bmatrix} 0 & \frac{1}{7} \\ \frac{1}{5} & -\frac{4}{35} \end{bmatrix}$

9. $A = \begin{bmatrix} -2 & \frac{1}{2} \\ 3 & -1 \end{bmatrix}, \quad B = \begin{bmatrix} -2 & -1 \\ -6 & -4 \end{bmatrix}$

10. $A=\begin{bmatrix}1&0&1\\0&1&-1\\0&1&1\end{bmatrix},\ B=\frac{1}{2}\begin{bmatrix}2&1&-1\\0&1&1\\0&-1&1\end{bmatrix}$

11. $A=\begin{bmatrix}1&2&3\\4&0&2\\1&6&9\end{bmatrix},\ B=\frac{1}{4}\begin{bmatrix}6&0&-2\\17&-3&-5\\-12&2&4\end{bmatrix}$

12. $A=\begin{bmatrix}3&8&2\\1&1&1\\5&6&12\end{bmatrix},\ B=\frac{1}{36}\begin{bmatrix}-6&84&-6\\7&-26&1\\-1&-22&5\end{bmatrix}$

For the following exercises, find the multiplicative inverse of each matrix, if it exists.

13. $\begin{bmatrix}3&-2\\1&9\end{bmatrix}$

14. $\begin{bmatrix}-2&2\\3&1\end{bmatrix}$

15. $\begin{bmatrix}-3&7\\9&2\end{bmatrix}$

16. $\begin{bmatrix}-4&-3\\-5&8\end{bmatrix}$

17. $\begin{bmatrix}1&1\\2&2\end{bmatrix}$

18. $\begin{bmatrix}0&1\\1&0\end{bmatrix}$

19. $\begin{bmatrix}0.5&1.5\\1&-0.5\end{bmatrix}$

20. $\begin{bmatrix}1&0&6\\-2&1&7\\3&0&2\end{bmatrix}$

21. $\begin{bmatrix}0&1&-3\\4&1&0\\1&0&5\end{bmatrix}$

22. $\begin{bmatrix}1&2&-1\\-3&4&1\\-2&-4&-5\end{bmatrix}$

23. $\begin{bmatrix}1&9&-3\\2&5&6\\4&-2&7\end{bmatrix}$

24. $\begin{bmatrix}1&-2&3\\-4&8&-12\\1&4&2\end{bmatrix}$

25. $\begin{bmatrix}\frac{1}{2}&\frac{1}{2}&\frac{1}{2}\\\frac{1}{3}&\frac{1}{4}&\frac{1}{5}\\\frac{1}{6}&\frac{1}{7}&\frac{1}{8}\end{bmatrix}$

26. $\begin{bmatrix}1&2&3\\4&5&6\\7&8&9\end{bmatrix}$

For the following exercises, solve the system using the inverse of a 2×2 matrix.

27. $\begin{aligned}5x-6y&=-61\\4x+3y&=-2\end{aligned}$

28. $\begin{aligned}8x+4y&=-100\\3x-4y&=1\end{aligned}$

29. $\begin{aligned}3x-2y&=6\\-x+5y&=-2\end{aligned}$

30. $\begin{aligned}5x-4y&=-5\\4x+y&=2.3\end{aligned}$

31. $\begin{aligned}-3x-4y&=9\\12x+4y&=-6\end{aligned}$

32. $\begin{aligned}-2x+3y&=\frac{3}{10}\\-x+5y&=\frac{1}{2}\end{aligned}$

33. $\begin{aligned}\frac{8}{5}x-\frac{4}{5}y&=\frac{2}{5}\\-\frac{8}{5}x+\frac{1}{5}y&=\frac{7}{10}\end{aligned}$

34. $\begin{aligned}\frac{1}{2}x+\frac{1}{5}y&=-\frac{1}{4}\\\frac{1}{2}x-\frac{3}{5}y&=-\frac{9}{4}\end{aligned}$

For the following exercises, solve a system using the inverse of a 3×3 matrix.

35. $\begin{aligned}3x-2y+5z&=21\\5x+4y&=37\\x-2y-5z&=5\end{aligned}$

36. $\begin{aligned}4x+4y+4z&=40\\2x-3y+4z&=-12\\-x+3y+4z&=9\end{aligned}$

37. $\begin{aligned}6x-5y-z&=31\\-x+2y+z&=-6\\3x+3y+2z&=13\end{aligned}$

38. $\begin{aligned} 6x-5y+2z &= -4 \\ 2x+5y-z &= 12 \\ 2x+5y+z &= 12 \end{aligned}$

39. $\begin{aligned} 4x-2y+3z &= -12 \\ 2x+2y-9z &= 33 \\ 6y-4z &= 1 \end{aligned}$

40. $\begin{aligned} \tfrac{1}{10}x-\tfrac{1}{5}y+4z &= \tfrac{-41}{2} \\ \tfrac{1}{5}x-20y+\tfrac{2}{5}z &= -101 \\ \tfrac{3}{10}x+4y-\tfrac{3}{10}z &= 23 \end{aligned}$

41. $\begin{aligned} \tfrac{1}{2}x-\tfrac{1}{5}y+\tfrac{1}{5}z &= \tfrac{31}{100} \\ -\tfrac{3}{4}x-\tfrac{1}{4}y+\tfrac{1}{2}z &= \tfrac{7}{40} \\ -\tfrac{4}{5}x-\tfrac{1}{2}y+\tfrac{3}{2}z &= \tfrac{1}{4} \end{aligned}$

42. $\begin{aligned} 0.1x+0.2y+0.3z &= -1.4 \\ 0.1x-0.2y+0.3z &= 0.6 \\ 0.4y+0.9z &= -2 \end{aligned}$

Technology

For the following exercises, use a calculator to solve the system of equations with matrix inverses.

43. $\begin{aligned} 2x-y &= -3 \\ -x+2y &= 2.3 \end{aligned}$

44. $\begin{aligned} -\tfrac{1}{2}x-\tfrac{3}{2}y &= -\tfrac{43}{20} \\ \tfrac{5}{2}x+\tfrac{11}{5}y &= \tfrac{31}{4} \end{aligned}$

45. $\begin{aligned} 12.3x-2y-2.5z &= 2 \\ 36.9x+7y-7.5z &= -7 \\ 8y-5z &= -10 \end{aligned}$

46. $\begin{aligned} 0.5x-3y+6z &= -0.8 \\ 0.7x-2y &= -0.06 \\ 0.5x+4y+5z &= 0 \end{aligned}$

Extensions

For the following exercises, find the inverse of the given matrix.

47. $\begin{bmatrix} 1 & 0 & 1 & 0 \\ 0 & 1 & 0 & 1 \\ 0 & 1 & 1 & 0 \\ 0 & 0 & 1 & 1 \end{bmatrix}$

48. $\begin{bmatrix} -1 & 0 & 2 & 5 \\ 0 & 0 & 0 & 2 \\ 0 & 2 & -1 & 0 \\ 1 & -3 & 0 & 1 \end{bmatrix}$

49. $\begin{bmatrix} 1 & -2 & 3 & 0 \\ 0 & 1 & 0 & 2 \\ 1 & 4 & -2 & 3 \\ -5 & 0 & 1 & 1 \end{bmatrix}$

50. $\begin{bmatrix} 1 & 2 & 0 & 2 & 3 \\ 0 & 2 & 1 & 0 & 0 \\ 0 & 0 & 3 & 0 & 1 \\ 0 & 2 & 0 & 0 & 1 \\ 0 & 0 & 1 & 2 & 0 \end{bmatrix}$

51. $\begin{bmatrix} 1 & 0 & 0 & 0 & 0 & 0 \\ 0 & 1 & 0 & 0 & 0 & 0 \\ 0 & 0 & 1 & 0 & 0 & 0 \\ 0 & 0 & 0 & 1 & 0 & 0 \\ 0 & 0 & 0 & 0 & 1 & 0 \\ 1 & 1 & 1 & 1 & 1 & 1 \end{bmatrix}$

Real-World Applications

For the following exercises, write a system of equations that represents the situation. Then, solve the system using the inverse of a matrix.

52. 2,400 tickets were sold for a basketball game. If the prices for floor 1 and floor 2 were different, and the total amount of money brought in is $64,000, how much was the price of each ticket?

53. In the previous exercise, if you were told there were 400 more tickets sold for floor 2 than floor 1, how much was the price of each ticket?

54. A food drive collected two different types of canned goods, green beans and kidney beans. The total number of collected cans was 350 and the total weight of all donated food was 348 lb, 12 oz. If the green bean cans weigh 2 oz less than the kidney bean cans, how many of each can was donated?

55. Students were asked to bring their favorite fruit to class. 95% of the fruits consisted of banana, apple, and oranges. If oranges were twice as popular as bananas, and apples were 5% less popular than bananas, what are the percentages of each individual fruit?

56. The nursing club held a bake sale to raise money and sold brownies and chocolate chip cookies. They priced the brownies at $1 and the chocolate chip cookies at $0.75. They raised $700 and sold 850 items. How many brownies and how many cookies were sold?

57. A clothing store needs to order new inventory. It has three different types of hats for sale: straw hats, beanies, and cowboy hats. The straw hat is priced at $13.99, the beanie at $7.99, and the cowboy hat at $14.49. If 100 hats were sold this past quarter, $1,119 was taken in by sales, and the amount of beanies sold was 10 more than cowboy hats, how many of each should the clothing store order to replace those already sold?

58. Anna, Percy, and Morgan weigh a combined 370 lb. If Morgan weighs 20 lb more than Percy, and Anna weighs 1.5 times as much as Percy, how much does each person weigh?

59. Three roommates shared a package of 12 ice cream bars, but no one remembers who ate how many. If Micah ate twice as many ice cream bars as Joe, and Albert ate three less than Micah, how many ice cream bars did each roommate eat?

60. A farmer constructed a chicken coop out of chicken wire, wood, and plywood. The chicken wire cost $2 per square foot, the wood $10 per square foot, and the plywood $5 per square foot. The farmer spent a total of $51, and the total amount of materials used was 14 ft^2. He used 3 ft^2 more chicken wire than plywood. How much of each material in did the farmer use?

61. Jay has lemon, orange, and pomegranate trees in his backyard. An orange weighs 8 oz, a lemon 5 oz, and a pomegranate 11 oz. Jay picked 142 pieces of fruit weighing a total of 70 lb, 10 oz. He picked 15.5 times more oranges than pomegranates. How many of each fruit did Jay pick?

11.8 Solving Systems with Cramer's Rule

Learning Objectives

In this section, you will:

- Evaluate 2×2 determinants.
- Use Cramer's Rule to solve a system of equations in two variables.
- Evaluate 3×3 determinants.
- Use Cramer's Rule to solve a system of three equations in three variables.
- Know the properties of determinants.

We have learned how to solve systems of equations in two variables and three variables, and by multiple methods: substitution, addition, Gaussian elimination, using the inverse of a matrix, and graphing. Some of these methods are easier to apply than others and are more appropriate in certain situations. In this section, we will study two more strategies for solving systems of equations.

Evaluating the Determinant of a 2×2 Matrix

A determinant is a real number that can be very useful in mathematics because it has multiple applications, such as calculating area, volume, and other quantities. Here, we will use determinants to reveal whether a matrix is invertible by using the entries of a square matrix to determine whether there is a solution to the system of equations. Perhaps one of the more interesting applications, however, is their use in cryptography. Secure signals or messages are sometimes sent encoded in a matrix. The data can only be decrypted with an invertible matrix and the determinant. For our purposes, we focus on the determinant as an indication of the invertibility of the matrix. Calculating the determinant of a matrix involves following the specific patterns that are outlined in this section.

Find the Determinant of a 2 × 2 Matrix

The **determinant** of a 2×2 matrix, given

$$A = \begin{bmatrix} a & b \\ c & d \end{bmatrix}$$

is defined as

$$\det(A) = \begin{vmatrix} a & b \\ c & d \end{vmatrix} = ad - cb$$

Notice the change in notation. There are several ways to indicate the determinant, including $\det(A)$ and replacing the brackets in a matrix with straight lines, $|A|$.

EXAMPLE 1

Finding the Determinant of a 2 × 2 Matrix

Find the determinant of the given matrix.

$$A = \begin{bmatrix} 5 & 2 \\ -6 & 3 \end{bmatrix}$$

Solution

$$\begin{aligned} \det(A) &= \begin{vmatrix} 5 & 2 \\ -6 & 3 \end{vmatrix} \\ &= 5(3) - (-6)(2) \\ &= 27 \end{aligned}$$

Using Cramer's Rule to Solve a System of Two Equations in Two Variables

We will now introduce a final method for solving systems of equations that uses determinants. Known as Cramer's Rule, this technique dates back to the middle of the 18th century and is named for its innovator, the Swiss mathematician Gabriel Cramer (1704-1752), who introduced it in 1750 in Introduction à l'Analyse des lignes Courbes algébriques. Cramer's Rule is a viable and efficient method for finding solutions to systems with an arbitrary number of unknowns, provided that we have the same number of equations as unknowns.

Cramer's Rule will give us the unique solution to a system of equations, if it exists. However, if the system has no solution or an infinite number of solutions, this will be indicated by a determinant of zero. To find out if the system is inconsistent or dependent, another method, such as elimination, will have to be used.

To understand Cramer's Rule, let's look closely at how we solve systems of linear equations using basic row operations. Consider a system of two equations in two variables.

$$\begin{aligned} a_1x + b_1y &= c_1 \quad (1) \\ a_2x + b_2y &= c_2 \quad (2) \end{aligned}$$

We eliminate one variable using row operations and solve for the other. Say that we wish to solve for x. If equation (2) is multiplied by the opposite of the coefficient of y in equation (1), equation (1) is multiplied by the coefficient of y in equation (2), and we add the two equations, the variable y will be eliminated.

$$\begin{array}{ll} b_2a_1x + b_2b_1y = b_2c_1 & \text{Multiply } R_1 \text{ by } b_2 \\ -b_1a_2x - b_1b_2y = -b_1c_2 & \text{Multiply } R_2 \text{ by } -b_1 \\ \hline b_2a_1x - b_1a_2x = b_2c_1 - b_1c_2 & \end{array}$$

Now, solve for x.

$$\begin{aligned} b_2a_1x - b_1a_2x &= b_2c_1 - b_1c_2 \\ x(b_2a_1 - b_1a_2) &= b_2c_1 - b_1c_2 \\ x &= \frac{b_2c_1 - b_1c_2}{b_2a_1 - b_1a_2} = \frac{\begin{vmatrix} c_1 & b_1 \\ c_2 & b_2 \end{vmatrix}}{\begin{vmatrix} a_1 & b_1 \\ a_2 & b_2 \end{vmatrix}} \end{aligned}$$

Similarly, to solve for y, we will eliminate x.

$$\begin{array}{ll} a_2a_1x + a_2b_1y = a_2c_1 & \text{Multiply } R_1 \text{ by } a_2 \\ -a_1a_2x - a_1b_2y = -a_1c_2 & \text{Multiply } R_2 \text{ by } -a_1 \\ \hline a_2b_1y - a_1b_2y = a_2c_1 - a_1c_2 & \end{array}$$

Solving for y gives

$$a_2 b_1 y - a_1 b_2 y = a_2 c_1 - a_1 c_2$$
$$y(a_2 b_1 - a_1 b_2) = a_2 c_1 - a_1 c_2$$

$$y = \frac{a_2 c_1 - a_1 c_2}{a_2 b_1 - a_1 b_2} = \frac{a_1 c_2 - a_2 c_1}{a_1 b_2 - a_2 b_1} = \frac{\begin{vmatrix} a_1 & c_1 \\ a_2 & c_2 \end{vmatrix}}{\begin{vmatrix} a_1 & b_1 \\ a_2 & b_2 \end{vmatrix}}$$

Notice that the denominator for both x and y is the determinant of the coefficient matrix.

We can use these formulas to solve for x and y, but Cramer's Rule also introduces new notation:

- D : determinant of the coefficient matrix
- D_x : determinant of the numerator in the solution of x
$$x = \frac{D_x}{D}$$
- D_y : determinant of the numerator in the solution of y
$$y = \frac{D_y}{D}$$

The key to Cramer's Rule is replacing the variable column of interest with the constant column and calculating the determinants. We can then express x and y as a quotient of two determinants.

Cramer's Rule for 2×2 Systems

Cramer's Rule is a method that uses determinants to solve systems of equations that have the same number of equations as variables.

Consider a system of two linear equations in two variables.

$$a_1 x + b_1 y = c_1$$
$$a_2 x + b_2 y = c_2$$

The solution using Cramer's Rule is given as

$$x = \frac{D_x}{D} = \frac{\begin{vmatrix} c_1 & b_1 \\ c_2 & b_2 \end{vmatrix}}{\begin{vmatrix} a_1 & b_1 \\ a_2 & b_2 \end{vmatrix}},\quad D \neq 0; \qquad y = \frac{D_y}{D} = \frac{\begin{vmatrix} a_1 & c_1 \\ a_2 & c_2 \end{vmatrix}}{\begin{vmatrix} a_1 & b_1 \\ a_2 & b_2 \end{vmatrix}},\quad D \neq 0.$$

If we are solving for x, the x column is replaced with the constant column. If we are solving for y, the y column is replaced with the constant column.

EXAMPLE 2

Using Cramer's Rule to Solve a 2 × 2 System

Solve the following 2×2 system using Cramer's Rule.

$$12x + 3y = 15$$
$$2x - 3y = 13$$

Solution

Solve for x.

$$x = \frac{D_x}{D} = \frac{\begin{vmatrix} 15 & 3 \\ 13 & -3 \end{vmatrix}}{\begin{vmatrix} 12 & 3 \\ 2 & -3 \end{vmatrix}} = \frac{-45 - 39}{-36 - 6} = \frac{-84}{-42} = 2$$

Solve for y.

$$y = \frac{D_y}{D} = \frac{\begin{vmatrix} 12 & 15 \\ 2 & 13 \end{vmatrix}}{\begin{vmatrix} 12 & 3 \\ 2 & -3 \end{vmatrix}} = \frac{156 - 30}{-36 - 6} = -\frac{126}{42} = -3$$

The solution is $(2, -3)$.

> TRY IT #1 Use Cramer's Rule to solve the 2 × 2 system of equations.

$$\begin{aligned} x + 2y &= -11 \\ -2x + y &= -13 \end{aligned}$$

Evaluating the Determinant of a 3 × 3 Matrix

Finding the determinant of a 2×2 matrix is straightforward, but finding the determinant of a 3×3 matrix is more complicated. One method is to augment the 3×3 matrix with a repetition of the first two columns, giving a 3×5 matrix. Then we calculate the sum of the products of entries *down* each of the three diagonals (upper left to lower right), and subtract the products of entries *up* each of the three diagonals (lower left to upper right). This is more easily understood with a visual and an example.

Find the determinant of the 3×3 matrix.

$$A = \begin{bmatrix} a_1 & b_1 & c_1 \\ a_2 & b_2 & c_2 \\ a_3 & b_3 & c_3 \end{bmatrix}$$

1. Augment A with the first two columns.

$$\det(A) = \left| \begin{matrix} a_1 & b_1 & c_1 \\ a_2 & b_2 & c_2 \\ a_3 & b_3 & c_3 \end{matrix} \middle| \begin{matrix} a_1 & b_1 \\ a_2 & b_2 \\ a_3 & b_3 \end{matrix} \right|$$

2. From upper left to lower right: Multiply the entries down the first diagonal. Add the result to the product of entries down the second diagonal. Add this result to the product of the entries down the third diagonal.
3. From lower left to upper right: Subtract the product of entries up the first diagonal. From this result subtract the product of entries up the second diagonal. From this result, subtract the product of entries up the third diagonal.

det(A) = 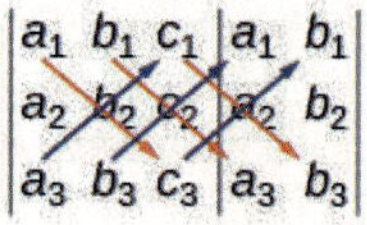

The algebra is as follows:

$$|A| = a_1 b_2 c_3 + b_1 c_2 a_3 + c_1 a_2 b_3 - a_3 b_2 c_1 - b_3 c_2 a_1 - c_3 a_2 b_1$$

EXAMPLE 3

Finding the Determinant of a 3 × 3 Matrix

Find the determinant of the 3 × 3 matrix given

$$A = \begin{bmatrix} 0 & 2 & 1 \\ 3 & -1 & 1 \\ 4 & 0 & 1 \end{bmatrix}$$

Solution

Augment the matrix with the first two columns and then follow the formula. Thus,

$$|A| = \begin{vmatrix} 0 & 2 & 1 \\ 3 & -1 & 1 \\ 4 & 0 & 1 \end{vmatrix}\begin{matrix} 0 & 2 \\ 3 & -1 \\ 4 & 0 \end{matrix}$$
$$= 0(-1)(1) + 2(1)(4) + 1(3)(0) - 4(-1)(1) - 0(1)(0) - 1(3)(2)$$
$$= 0 + 8 + 0 + 4 - 0 - 6$$
$$= 6$$

> **TRY IT** #2 Find the determinant of the 3 × 3 matrix.

$$\det(A) = \begin{vmatrix} 1 & -3 & 7 \\ 1 & 1 & 1 \\ 1 & -2 & 3 \end{vmatrix}$$

Q&A **Can we use the same method to find the determinant of a larger matrix?**

No, this method only works for 2×2 *and* 3×3 *matrices. For larger matrices it is best to use a graphing utility or computer software.*

Using Cramer's Rule to Solve a System of Three Equations in Three Variables

Now that we can find the determinant of a 3 × 3 matrix, we can apply Cramer's Rule to solve a system of three equations in three variables. Cramer's Rule is straightforward, following a pattern consistent with Cramer's Rule for 2 × 2 matrices. As the order of the matrix increases to 3 × 3, however, there are many more calculations required.

When we calculate the determinant to be zero, Cramer's Rule gives no indication as to whether the system has no solution or an infinite number of solutions. To find out, we have to perform elimination on the system.

Consider a 3 × 3 system of equations.

$$a_1x + b_1y + c_1z = d_1$$
$$a_2x + b_2y + c_2z = d_2$$
$$a_3x + b_3y + c_3z = d_3$$

$$x = \frac{D_x}{D}, y = \frac{D_y}{D}, z = \frac{D_z}{D}, D \neq 0$$

where

$$D = \begin{vmatrix} a_1 & b_1 & c_1 \\ a_2 & b_2 & c_2 \\ a_3 & b_3 & c_3 \end{vmatrix}, D_x = \begin{vmatrix} d_1 & b_1 & c_1 \\ d_2 & b_2 & c_2 \\ d_3 & b_3 & c_3 \end{vmatrix}, D_y = \begin{vmatrix} a_1 & d_1 & c_1 \\ a_2 & d_2 & c_2 \\ a_3 & d_3 & c_3 \end{vmatrix}, D_z = \begin{vmatrix} a_1 & b_1 & d_1 \\ a_2 & b_2 & d_2 \\ a_3 & b_3 & d_3 \end{vmatrix}$$

If we are writing the determinant D_x, we replace the x column with the constant column. If we are writing the determinant D_y, we replace the y column with the constant column. If we are writing the determinant D_z, we replace the z column with the constant column. Always check the answer.

EXAMPLE 4

Solving a 3 × 3 System Using Cramer's Rule

Find the solution to the given 3 × 3 system using Cramer's Rule.

$$x + y - z = 6$$
$$3x - 2y + z = -5$$
$$x + 3y - 2z = 14$$

Solution

Use Cramer's Rule.

$$D = \begin{vmatrix} 1 & 1 & -1 \\ 3 & -2 & 1 \\ 1 & 3 & -2 \end{vmatrix}, D_x = \begin{vmatrix} 6 & 1 & -1 \\ -5 & -2 & 1 \\ 14 & 3 & -2 \end{vmatrix}, D_y = \begin{vmatrix} 1 & 6 & -1 \\ 3 & -5 & 1 \\ 1 & 14 & -2 \end{vmatrix}, D_z = \begin{vmatrix} 1 & 1 & 6 \\ 3 & -2 & -5 \\ 1 & 3 & 14 \end{vmatrix}$$

Then,

$$\begin{aligned} x &= \frac{D_x}{D} = \frac{-3}{-3} = 1 \\ y &= \frac{D_y}{D} = \frac{-9}{-3} = 3 \\ z &= \frac{D_z}{D} = \frac{6}{-3} = -2 \end{aligned}$$

The solution is $(1, 3, -2)$.

TRY IT #3 Use Cramer's Rule to solve the 3 × 3 matrix.

$$\begin{aligned} x - 3y + 7z &= 13 \\ x + y + z &= 1 \\ x - 2y + 3z &= 4 \end{aligned}$$

EXAMPLE 5

Using Cramer's Rule to Solve an Inconsistent System

Solve the system of equations using Cramer's Rule.

$$\begin{aligned} 3x - 2y &= 4 \quad (1) \\ 6x - 4y &= 0 \quad (2) \end{aligned}$$

Solution

We begin by finding the determinants D, D_x, and D_y.

$$D = \begin{vmatrix} 3 & -2 \\ 6 & -4 \end{vmatrix} = 3(-4) - 6(-2) = 0$$

We know that a determinant of zero means that either the system has no solution or it has an infinite number of solutions. To see which one, we use the process of elimination. Our goal is to eliminate one of the variables.

1. Multiply equation (1) by -2.
2. Add the result to equation (2).

$$\begin{array}{r} -6x + 4y = -8 \\ 6x - 4y = 0 \\ \hline 0 = -8 \end{array}$$

We obtain the equation $0 = -8$, which is false. Therefore, the system has no solution. Graphing the system reveals two parallel lines. See Figure 1.

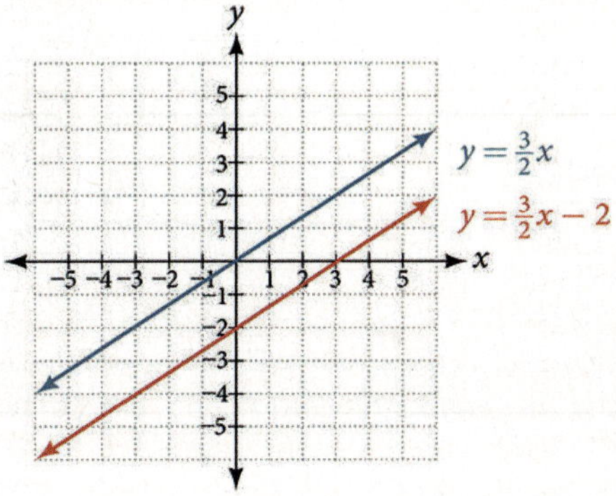

Figure 1

EXAMPLE 6

Use Cramer's Rule to Solve a Dependent System

Solve the system with an infinite number of solutions.

$$\begin{aligned} x - 2y + 3z &= 0 \quad (1) \\ 3x + y - 2z &= 0 \quad (2) \\ 2x - 4y + 6z &= 0 \quad (3) \end{aligned}$$

Solution

Let's find the determinant first. Set up a matrix augmented by the first two columns.

$$\left|\begin{array}{ccc|cc} 1 & -2 & 3 & 1 & -2 \\ 3 & 1 & -2 & 3 & 1 \\ 2 & -4 & 6 & 2 & -4 \end{array}\right|$$

Then,

$$1(1)(6) + (-2)(-2)(2) + 3(3)(-4) - 2(1)(3) - (-4)(-2)(1) - 6(3)(-2) = 0$$

As the determinant equals zero, there is either no solution or an infinite number of solutions. We have to perform elimination to find out.

1. Multiply equation (1) by -2 and add the result to equation (3):

$$\begin{array}{r} -2x + 4y - 6z = 0 \\ 2x - 4y + 6z = 0 \\ \hline 0 = 0 \end{array}$$

2. Obtaining an answer of $0 = 0$, a statement that is always true, means that the system has an infinite number of solutions. Graphing the system, we can see that two of the planes are the same and they both intersect the third plane on a line. See Figure 2.

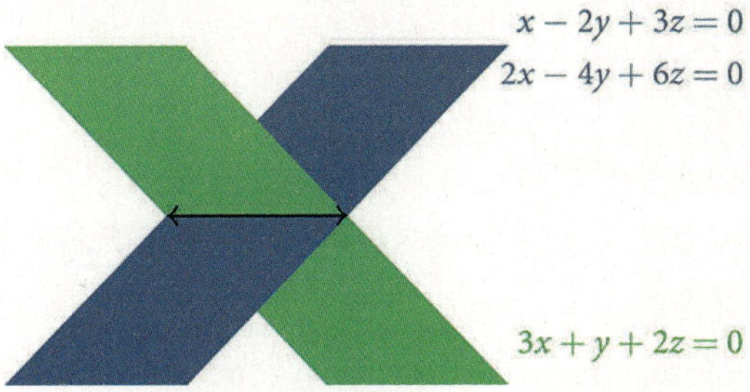

Figure 2

Understanding Properties of Determinants

There are many properties of determinants. Listed here are some properties that may be helpful in calculating the determinant of a matrix.

Properties of Determinants

1. If the matrix is in upper triangular form, the determinant equals the product of entries down the main diagonal.
2. When two rows are interchanged, the determinant changes sign.
3. If either two rows or two columns are identical, the determinant equals zero.
4. If a matrix contains either a row of zeros or a column of zeros, the determinant equals zero.
5. The determinant of an inverse matrix A^{-1} is the reciprocal of the determinant of the matrix A.
6. If any row or column is multiplied by a constant, the determinant is multiplied by the same factor.

EXAMPLE 7

Illustrating Properties of Determinants

Illustrate each of the properties of determinants.

Solution

Property 1 states that if the matrix is in upper triangular form, the determinant is the product of the entries down the main diagonal.

$$A = \begin{bmatrix} 1 & 2 & 3 \\ 0 & 2 & 1 \\ 0 & 0 & -1 \end{bmatrix}$$

Augment A with the first two columns.

$$A = \left[\begin{array}{ccc|cc} 1 & 2 & 3 & 1 & 2 \\ 0 & 2 & 1 & 0 & 2 \\ 0 & 0 & -1 & 0 & 0 \end{array}\right]$$

Then

$$\begin{aligned} \det(A) &= 1(2)(-1) + 2(1)(0) + 3(0)(0) - 0(2)(3) - 0(1)(1) + 1(0)(2) \\ &= -2 \end{aligned}$$

Property 2 states that interchanging rows changes the sign. Given

$$A = \begin{bmatrix} -1 & 5 \\ 4 & -3 \end{bmatrix}, \quad \det(A) = (-1)(-3) - (4)(5) = 3 - 20 = -17$$

$$B = \begin{bmatrix} 4 & -3 \\ -1 & 5 \end{bmatrix}, \quad \det(B) = (4)(5) - (-1)(-3) = 20 - 3 = 17$$

Property 3 states that if two rows or two columns are identical, the determinant equals zero.

$$A = \left[\begin{array}{ccc|cc} 1 & 2 & 2 & 1 & 2 \\ 2 & 2 & 2 & 2 & 2 \\ -1 & 2 & 2 & -1 & 2 \end{array}\right]$$

$$\begin{aligned} \det(A) &= 1(2)(2) + 2(2)(-1) + 2(2)(2) + 1(2)(2) - 2(2)(1) - 2(2)(2) \\ &= 4 - 4 + 8 + 4 - 4 - 8 = 0 \end{aligned}$$

Property 4 states that if a row or column equals zero, the determinant equals zero. Thus,

$$A = \begin{bmatrix} 1 & 2 \\ 0 & 0 \end{bmatrix}, \quad \det(A) = 1(0) - 2(0) = 0$$

Property 5 states that the determinant of an inverse matrix A^{-1} is the reciprocal of the determinant A. Thus,

$$A = \begin{bmatrix} 1 & 2 \\ 3 & 4 \end{bmatrix}, \det(A) = 1(4) - 3(2) = -2$$

$$A^{-1} = \begin{bmatrix} -2 & 1 \\ \frac{3}{2} & -\frac{1}{2} \end{bmatrix}, \det\left(A^{-1}\right) = -2\left(-\tfrac{1}{2}\right) - \left(\tfrac{3}{2}\right)(1) = -\tfrac{1}{2}$$

Property 6 states that if any row or column of a matrix is multiplied by a constant, the determinant is multiplied by the same factor. Thus,

$$A = \begin{bmatrix} 1 & 2 \\ 3 & 4 \end{bmatrix}, \det(A) = 1(4) - 2(3) = -2$$

$$B = \begin{bmatrix} 2(1) & 2(2) \\ 3 & 4 \end{bmatrix}, \det(B) = 2(4) - 3(4) = -4$$

EXAMPLE 8

Using Cramer's Rule and Determinant Properties to Solve a System

Find the solution to the given 3 × 3 system.

$$\begin{aligned} 2x + 4y + 4z &= 2 \quad (1) \\ 3x + 7y + 7z &= -5 \quad (2) \\ x + 2y + 2z &= 4 \quad (3) \end{aligned}$$

Solution

Using Cramer's Rule, we have

$$D = \begin{vmatrix} 2 & 4 & 4 \\ 3 & 7 & 7 \\ 1 & 2 & 2 \end{vmatrix}$$

Notice that the second and third columns are identical. According to Property 3, the determinant will be zero, so there is either no solution or an infinite number of solutions. We have to perform elimination to find out.

1. Multiply equation (3) by –2 and add the result to equation (1).

$$\begin{aligned} -2x - 4y - 4x &= -8 \\ 2x + 4y + 4z &= 2 \\ \hline 0 &= -6 \end{aligned}$$

Obtaining a statement that is a contradiction means that the system has no solution.

MEDIA

Access these online resources for additional instruction and practice with Cramer's Rule.

Solve a System of Two Equations Using Cramer's Rule (http://openstax.org/l/system2cramer)
Solve a Systems of Three Equations using Cramer's Rule (http://openstax.org/l/system3cramer)

11.8 SECTION EXERCISES

Verbal

1. Explain why we can always evaluate the determinant of a square matrix.

2. Examining Cramer's Rule, explain why there is no unique solution to the system when the determinant of your matrix is 0. For simplicity, use a 2×2 matrix.

3. Explain what it means in terms of an inverse for a matrix to have a 0 determinant.

4. The determinant of 2×2 matrix A is 3. If you switch the rows and multiply the first row by 6 and the second row by 2, explain how to find the determinant and provide the answer.

Algebraic

For the following exercises, find the determinant.

5. $\begin{vmatrix} 1 & 2 \\ 3 & 4 \end{vmatrix}$

6. $\begin{vmatrix} -1 & 2 \\ 3 & -4 \end{vmatrix}$

7. $\begin{vmatrix} 2 & -5 \\ -1 & 6 \end{vmatrix}$

8. $\begin{vmatrix} -8 & 4 \\ -1 & 5 \end{vmatrix}$

9. $\begin{vmatrix} 1 & 0 \\ 3 & -4 \end{vmatrix}$

10. $\begin{vmatrix} 10 & 20 \\ 0 & -10 \end{vmatrix}$

11. $\begin{vmatrix} 10 & 0.2 \\ 5 & 0.1 \end{vmatrix}$

12. $\begin{vmatrix} 6 & -3 \\ 8 & 4 \end{vmatrix}$

13. $\begin{vmatrix} -2 & -3 \\ 3.1 & 4,000 \end{vmatrix}$

14. $\begin{vmatrix} -1.1 & 0.6 \\ 7.2 & -0.5 \end{vmatrix}$

15. $\begin{vmatrix} -1 & 0 & 0 \\ 0 & 1 & 0 \\ 0 & 0 & -3 \end{vmatrix}$

16. $\begin{vmatrix} -1 & 4 & 0 \\ 0 & 2 & 3 \\ 0 & 0 & -3 \end{vmatrix}$

17. $\begin{vmatrix} 1 & 0 & 1 \\ 0 & 1 & 0 \\ 1 & 0 & 0 \end{vmatrix}$

18. $\begin{vmatrix} 2 & -3 & 1 \\ 3 & -4 & 1 \\ -5 & 6 & 1 \end{vmatrix}$

19. $\begin{vmatrix} -2 & 1 & 4 \\ -4 & 2 & -8 \\ 2 & -8 & -3 \end{vmatrix}$

20. $\begin{vmatrix} 6 & -1 & 2 \\ -4 & -3 & 5 \\ 1 & 9 & -1 \end{vmatrix}$

21. $\begin{vmatrix} 5 & 1 & -1 \\ 2 & 3 & 1 \\ 3 & -6 & -3 \end{vmatrix}$

22. $\begin{vmatrix} 1.1 & 2 & -1 \\ -4 & 0 & 0 \\ 4.1 & -0.4 & 2.5 \end{vmatrix}$

23. $\begin{vmatrix} 2 & -1.6 & 3.1 \\ 1.1 & 3 & -8 \\ -9.3 & 0 & 2 \end{vmatrix}$

24. $\begin{vmatrix} -\frac{1}{2} & \frac{1}{3} & \frac{1}{4} \\ \frac{1}{5} & -\frac{1}{6} & \frac{1}{7} \\ 0 & 0 & \frac{1}{8} \end{vmatrix}$

For the following exercises, solve the system of linear equations using Cramer's Rule.

25. $\begin{aligned} 2x - 3y &= -1 \\ 4x + 5y &= 9 \end{aligned}$

26. $\begin{aligned} 5x - 4y &= 2 \\ -4x + 7y &= 6 \end{aligned}$

27. $\begin{aligned} 6x - 3y &= 2 \\ -8x + 9y &= -1 \end{aligned}$

28. $\begin{aligned} 2x + 6y &= 12 \\ 5x - 2y &= 13 \end{aligned}$

29. $\begin{aligned} 4x + 3y &= 23 \\ 2x - y &= -1 \end{aligned}$

30. $\begin{aligned} 10x - 6y &= 2 \\ -5x + 8y &= -1 \end{aligned}$

31. $\begin{aligned} 4x - 3y &= -3 \\ 2x + 6y &= -4 \end{aligned}$

32. $\begin{aligned} 4x - 5y &= 7 \\ -3x + 9y &= 0 \end{aligned}$

33. $\begin{aligned} 4x + 10y &= 180 \\ -3x - 5y &= -105 \end{aligned}$

34. $\begin{aligned} 8x - 2y &= -3 \\ -4x + 6y &= 4 \end{aligned}$

For the following exercises, solve the system of linear equations using Cramer's Rule.

35. $\begin{aligned} x + 2y - 4z &= -1 \\ 7x + 3y + 5z &= 26 \\ -2x - 6y + 7z &= -6 \end{aligned}$

36. $\begin{aligned} -5x + 2y - 4z &= -47 \\ 4x - 3y - z &= -94 \\ 3x - 3y + 2z &= 94 \end{aligned}$

37. $\begin{aligned} 4x + 5y - z &= -7 \\ -2x - 9y + 2z &= 8 \\ 5y + 7z &= 21 \end{aligned}$

38. $\begin{aligned} 4x - 3y + 4z &= 10 \\ 5x - 2z &= -2 \\ 3x + 2y - 5z &= -9 \end{aligned}$

39. $\begin{aligned} 4x - 2y + 3z &= 6 \\ -6x + y &= -2 \\ 2x + 7y + 8z &= 24 \end{aligned}$

40. $\begin{aligned} 5x + 2y - z &= 1 \\ -7x - 8y + 3z &= 1.5 \\ 6x - 12y + z &= 7 \end{aligned}$

41. $\begin{aligned} 13x - 17y + 16z &= 73 \\ -11x + 15y + 17z &= 61 \\ 46x + 10y - 30z &= -18 \end{aligned}$

42. $\begin{aligned} -4x - 3y - 8z &= -7 \\ 2x - 9y + 5z &= 0.5 \\ 5x - 6y - 5z &= -2 \end{aligned}$

43. $\begin{aligned} 4x - 6y + 8z &= 10 \\ -2x + 3y - 4z &= -5 \\ x + y + z &= 1 \end{aligned}$

44. $\begin{aligned} 4x - 6y + 8z &= 10 \\ -2x + 3y - 4z &= -5 \\ 12x + 18y - 24z &= -30 \end{aligned}$

Technology

For the following exercises, use the determinant function on a graphing utility.

45. $\begin{vmatrix} 1 & 0 & 8 & 9 \\ 0 & 2 & 1 & 0 \\ 1 & 0 & 3 & 0 \\ 0 & 2 & 4 & 3 \end{vmatrix}$

46. $\begin{vmatrix} 1 & 0 & 2 & 1 \\ 0 & -9 & 1 & 3 \\ 3 & 0 & -2 & -1 \\ 0 & 1 & 1 & -2 \end{vmatrix}$

47. $\begin{vmatrix} \frac{1}{2} & 1 & 7 & 4 \\ 0 & \frac{1}{2} & 100 & 5 \\ 0 & 0 & 2 & 2{,}000 \\ 0 & 0 & 0 & 2 \end{vmatrix}$

48. $\begin{vmatrix} 1 & 0 & 0 & 0 \\ 2 & 3 & 0 & 0 \\ 4 & 5 & 6 & 0 \\ 7 & 8 & 9 & 0 \end{vmatrix}$

Real-World Applications

For the following exercises, create a system of linear equations to describe the behavior. Then, calculate the determinant. Will there be a unique solution? If so, find the unique solution.

49. Two numbers add up to 56. One number is 20 less than the other.

50. Two numbers add up to 104. If you add two times the first number plus two times the second number, your total is 208

51. Three numbers add up to 106. The first number is 3 less than the second number. The third number is 4 more than the first number.

52. Three numbers add to 216. The sum of the first two numbers is 112. The third number is 8 less than the first two numbers combined.

For the following exercises, create a system of linear equations to describe the behavior. Then, solve the system for all solutions using Cramer's Rule.

53. You invest $10,000 into two accounts, which receive 8% interest and 5% interest. At the end of a year, you had $10,710 in your combined accounts. How much was invested in each account?

54. You invest $80,000 into two accounts, $22,000 in one account, and $58,000 in the other account. At the end of one year, assuming simple interest, you have earned $2,470 in interest. The second account receives half a percent less than twice the interest on the first account. What are the interest rates for your accounts?

55. A theater needs to know how many adult tickets and children tickets were sold out of the 1,200 total tickets. If children's tickets are $5.95, adult tickets are $11.15, and the total amount of revenue was $12,756, how many children's tickets and adult tickets were sold?

56. A concert venue sells single tickets for $40 each and couple's tickets for $65. If the total revenue was $18,090 and the 321 tickets were sold, how many single tickets and how many couple's tickets were sold?

57. You decide to paint your kitchen green. You create the color of paint by mixing yellow and blue paints. You cannot remember how many gallons of each color went into your mix, but you know there were 10 gal total. Additionally, you kept your receipt, and know the total amount spent was $29.50. If each gallon of yellow costs $2.59, and each gallon of blue costs $3.19, how many gallons of each color go into your green mix?

58. You sold two types of scarves at a farmers' market and would like to know which one was more popular. The total number of scarves sold was 56, the yellow scarf cost $10, and the purple scarf cost $11. If you had total revenue of $583, how many yellow scarves and how many purple scarves were sold?

59. Your garden produced two types of tomatoes, one green and one red. The red weigh 10 oz, and the green weigh 4 oz. You have 30 tomatoes, and a total weight of 13 lb, 14 oz. How many of each type of tomato do you have?

60. At a market, the three most popular vegetables make up 53% of vegetable sales. Corn has 4% higher sales than broccoli, which has 5% more sales than onions. What percentage does each vegetable have in the market share?

61. At the same market, the three most popular fruits make up 37% of the total fruit sold. Strawberries sell twice as much as oranges, and kiwis sell one more percentage point than oranges. For each fruit, find the percentage of total fruit sold.

62. Three artists performed at a concert venue. The first one charged $15 per ticket, the second artist charged $45 per ticket, and the final one charged $22 per ticket. There were 510 tickets sold, for a total of $12,700. If the first band had 40 more audience members than the second band, how many tickets were sold for each band?

63. A movie theatre sold tickets to three movies. The tickets to the first movie were $5, the tickets to the second movie were $11, and the third movie was $12. 100 tickets were sold to the first movie. The total number of tickets sold was 642, for a total revenue of $6,774. How many tickets for each movie were sold?

For the following exercises, use this scenario: A health-conscious company decides to make a trail mix out of almonds, dried cranberries, and chocolate-covered cashews. The nutritional information for these items is shown in Table 1.

	Fat (g)	Protein (g)	Carbohydrates (g)
Almonds (10)	6	2	3
Cranberries (10)	0.02	0	8
Cashews (10)	7	3.5	5.5

Table 1

64. For the special "low-carb"trail mix, there are 1,000 pieces of mix. The total number of carbohydrates is 425 g, and the total amount of fat is 570.2 g. If there are 200 more pieces of cashews than cranberries, how many of each item is in the trail mix?

65. For the "hiking" mix, there are 1,000 pieces in the mix, containing 390.8 g of fat, and 165 g of protein. If there is the same amount of almonds as cashews, how many of each item is in the trail mix?

66. For the "energy-booster" mix, there are 1,000 pieces in the mix, containing 145 g of protein and 625 g of carbohydrates. If the number of almonds and cashews summed together is equivalent to the amount of cranberries, how many of each item is in the trail mix?

Chapter Review

Key Terms

addition method an algebraic technique used to solve systems of linear equations in which the equations are added in a way that eliminates one variable, allowing the resulting equation to be solved for the remaining variable; substitution is then used to solve for the first variable

augmented matrix a coefficient matrix adjoined with the constant column separated by a vertical line within the matrix brackets

break-even point the point at which a cost function intersects a revenue function; where profit is zero

coefficient matrix a matrix that contains only the coefficients from a system of equations

column a set of numbers aligned vertically in a matrix

consistent system a system for which there is a single solution to all equations in the system and it is an independent system, or if there are an infinite number of solutions and it is a dependent system

cost function the function used to calculate the costs of doing business; it usually has two parts, fixed costs and variable costs

Cramer's Rule a method for solving systems of equations that have the same number of equations as variables using determinants

dependent system a system of linear equations in which the two equations represent the same line; there are an infinite number of solutions to a dependent system

determinant a number calculated using the entries of a square matrix that determines such information as whether there is a solution to a system of equations

entry an element, coefficient, or constant in a matrix

feasible region the solution to a system of nonlinear inequalities that is the region of the graph where the shaded regions of each inequality intersect

Gaussian elimination using elementary row operations to obtain a matrix in row-echelon form

identity matrix a square matrix containing ones down the main diagonal and zeros everywhere else; it acts as a 1 in matrix algebra

inconsistent system a system of linear equations with no common solution because they represent parallel lines, which have no point or line in common

independent system a system of linear equations with exactly one solution pair (x, y)

main diagonal entries from the upper left corner diagonally to the lower right corner of a square matrix

matrix a rectangular array of numbers

multiplicative inverse of a matrix a matrix that, when multiplied by the original, equals the identity matrix

nonlinear inequality an inequality containing a nonlinear expression

partial fraction decomposition the process of returning a simplified rational expression to its original form, a sum or difference of simpler rational expressions

partial fractions the individual fractions that make up the sum or difference of a rational expression before combining them into a simplified rational expression

profit function the profit function is written as $P(x) = R(x) - C(x)$, revenue minus cost

revenue function the function that is used to calculate revenue, simply written as $R = xp$, where $x =$ quantity and $p =$ price

row a set of numbers aligned horizontally in a matrix

row operations adding one row to another row, multiplying a row by a constant, interchanging rows, and so on, with the goal of achieving row-echelon form

row-echelon form after performing row operations, the matrix form that contains ones down the main diagonal and zeros at every space below the diagonal

row-equivalent two matrices A and B are row-equivalent if one can be obtained from the other by performing basic row operations

scalar multiple an entry of a matrix that has been multiplied by a scalar

solution set the set of all ordered pairs or triples that satisfy all equations in a system of equations

substitution method an algebraic technique used to solve systems of linear equations in which one of the two equations is solved for one variable and then substituted into the second equation to solve for the second variable

system of linear equations a set of two or more equations in two or more variables that must be considered simultaneously.

system of nonlinear equations a system of equations containing at least one equation that is of degree larger than one

system of nonlinear inequalities a system of two or more inequalities in two or more variables containing at least one inequality that is not linear

Key Equations

Identity matrix for a 2×2 matrix	$I_2 = \begin{bmatrix} 1 & 0 \\ 0 & 1 \end{bmatrix}$
Identity matrix for a 3×3 matrix	$I_3 = \begin{bmatrix} 1 & 0 & 0 \\ 0 & 1 & 0 \\ 0 & 0 & 1 \end{bmatrix}$
Multiplicative inverse of a 2×2 matrix	$A^{-1} = \frac{1}{ad-bc}\begin{bmatrix} d & -b \\ -c & a \end{bmatrix}$, where $ad - bc \neq 0$

Key Concepts

11.1 Systems of Linear Equations: Two Variables

- A system of linear equations consists of two or more equations made up of two or more variables such that all equations in the system are considered simultaneously.
- The solution to a system of linear equations in two variables is any ordered pair that satisfies each equation independently. See Example 1.
- Systems of equations are classified as independent with one solution, dependent with an infinite number of solutions, or inconsistent with no solution.
- One method of solving a system of linear equations in two variables is by graphing. In this method, we graph the equations on the same set of axes. See Example 2.
- Another method of solving a system of linear equations is by substitution. In this method, we solve for one variable in one equation and substitute the result into the second equation. See Example 3.
- A third method of solving a system of linear equations is by addition, in which we can eliminate a variable by adding opposite coefficients of corresponding variables. See Example 4.
- It is often necessary to multiply one or both equations by a constant to facilitate elimination of a variable when adding the two equations together. See Example 5, Example 6, and Example 7.
- Either method of solving a system of equations results in a false statement for inconsistent systems because they are made up of parallel lines that never intersect. See Example 8.
- The solution to a system of dependent equations will always be true because both equations describe the same line. See Example 9.
- Systems of equations can be used to solve real-world problems that involve more than one variable, such as those relating to revenue, cost, and profit. See Example 10 and Example 11.

11.2 Systems of Linear Equations: Three Variables

- A solution set is an ordered triple $\{(x, y, z)\}$ that represents the intersection of three planes in space. See Example 1.
- A system of three equations in three variables can be solved by using a series of steps that forces a variable to be eliminated. The steps include interchanging the order of equations, multiplying both sides of an equation by a nonzero constant, and adding a nonzero multiple of one equation to another equation. See Example 2.
- Systems of three equations in three variables are useful for solving many different types of real-world problems. See Example 3.
- A system of equations in three variables is inconsistent if no solution exists. After performing elimination operations, the result is a contradiction. See Example 4.
- Systems of equations in three variables that are inconsistent could result from three parallel planes, two parallel planes and one intersecting plane, or three planes that intersect the other two but not at the same location.
- A system of equations in three variables is dependent if it has an infinite number of solutions. After performing elimination operations, the result is an identity. See Example 5.
- Systems of equations in three variables that are dependent could result from three identical planes, three planes intersecting at a line, or two identical planes that intersect the third on a line.

11.3 Systems of Nonlinear Equations and Inequalities: Two Variables

- There are three possible types of solutions to a system of equations representing a line and a parabola: (1) no solution, the line does not intersect the parabola; (2) one solution, the line is tangent to the parabola; and (3) two solutions, the line intersects the parabola in two points. See Example 1.

- There are three possible types of solutions to a system of equations representing a circle and a line: (1) no solution, the line does not intersect the circle; (2) one solution, the line is tangent to the circle; (3) two solutions, the line intersects the circle in two points. See Example 2.
- There are five possible types of solutions to the system of nonlinear equations representing an ellipse and a circle: (1) no solution, the circle and the ellipse do not intersect; (2) one solution, the circle and the ellipse are tangent to each other; (3) two solutions, the circle and the ellipse intersect in two points; (4) three solutions, the circle and ellipse intersect in three places; (5) four solutions, the circle and the ellipse intersect in four points. See Example 3.
- An inequality is graphed in much the same way as an equation, except for > or <, we draw a dashed line and shade the region containing the solution set. See Example 4.
- Inequalities are solved the same way as equalities, but solutions to systems of inequalities must satisfy both inequalities. See Example 5.

11.4 Partial Fractions

- Decompose $\frac{P(x)}{Q(x)}$ by writing the partial fractions as $\frac{A}{a_1 x+b_1} + \frac{B}{a_2 x+b_2}$. Solve by clearing the fractions, expanding the right side, collecting like terms, and setting corresponding coefficients equal to each other, then setting up and solving a system of equations. See Example 1.
- The decomposition of $\frac{P(x)}{Q(x)}$ with repeated linear factors must account for the factors of the denominator in increasing powers. See Example 2.
- The decomposition of $\frac{P(x)}{Q(x)}$ with a nonrepeated irreducible quadratic factor needs a linear numerator over the quadratic factor, as in $\frac{A}{x} + \frac{Bx+C}{(ax^2+bx+c)}$. See Example 3.
- In the decomposition of $\frac{P(x)}{Q(x)}$, where $Q\,(x)$ has a repeated irreducible quadratic factor, when the irreducible quadratic factors are repeated, powers of the denominator factors must be represented in increasing powers as
$$\frac{Ax+B}{(ax^2+bx+c)} + \frac{A_2 x + B_2}{(ax^2+bx+c)^2} + \cdots + \frac{A_n x + B_n}{(ax^2+bx+c)^n}.$$
See Example 4.

11.5 Matrices and Matrix Operations

- A matrix is a rectangular array of numbers. Entries are arranged in rows and columns.
- The dimensions of a matrix refer to the number of rows and the number of columns. A 3×2 matrix has three rows and two columns. See Example 1.
- We add and subtract matrices of equal dimensions by adding and subtracting corresponding entries of each matrix. See Example 2, Example 3, Example 4, and Example 5.
- Scalar multiplication involves multiplying each entry in a matrix by a constant. See Example 6.
- Scalar multiplication is often required before addition or subtraction can occur. See Example 7.
- Multiplying matrices is possible when inner dimensions are the same—the number of columns in the first matrix must match the number of rows in the second.
- The product of two matrices, A and B, is obtained by multiplying each entry in row 1 of A by each entry in column 1 of B; then multiply each entry of row 1 of A by each entry in columns 2 of B, and so on. See Example 8 and Example 9.
- Many real-world problems can often be solved using matrices. See Example 10.
- We can use a calculator to perform matrix operations after saving each matrix as a matrix variable. See Example 11.

11.6 Solving Systems with Gaussian Elimination

- An augmented matrix is one that contains the coefficients and constants of a system of equations. See Example 1.
- A matrix augmented with the constant column can be represented as the original system of equations. See Example 2.
- Row operations include multiplying a row by a constant, adding one row to another row, and interchanging rows.
- We can use Gaussian elimination to solve a system of equations. See Example 3, Example 4, and Example 5.
- Row operations are performed on matrices to obtain row-echelon form. See Example 6.
- To solve a system of equations, write it in augmented matrix form. Perform row operations to obtain row-echelon form. Back-substitute to find the solutions. See Example 7 and Example 8.
- A calculator can be used to solve systems of equations using matrices. See Example 9.
- Many real-world problems can be solved using augmented matrices. See Example 10 and Example 11.

11.7 Solving Systems with Inverses

- An identity matrix has the property $AI = IA = A$. See Example 1.
- An invertible matrix has the property $AA^{-1} = A^{-1}A = I$. See Example 2.
- Use matrix multiplication and the identity to find the inverse of a 2×2 matrix. See Example 3.
- The multiplicative inverse can be found using a formula. See Example 4.
- Another method of finding the inverse is by augmenting with the identity. See Example 5.
- We can augment a 3×3 matrix with the identity on the right and use row operations to turn the original matrix into the identity, and the matrix on the right becomes the inverse. See Example 6.
- Write the system of equations as $AX = B$, and multiply both sides by the inverse of $A : A^{-1}AX = A^{-1}B$. See Example 7 and Example 8.
- We can also use a calculator to solve a system of equations with matrix inverses. See Example 9.

11.8 Solving Systems with Cramer's Rule

- The determinant for $\begin{bmatrix} a & b \\ c & d \end{bmatrix}$ is $ad - bc$. See Example 1.
- Cramer's Rule replaces a variable column with the constant column. Solutions are $x = \frac{D_x}{D}, y = \frac{D_y}{D}$. See Example 2.
- To find the determinant of a 3×3 matrix, augment with the first two columns. Add the three diagonal entries (upper left to lower right) and subtract the three diagonal entries (lower left to upper right). See Example 3.
- To solve a system of three equations in three variables using Cramer's Rule, replace a variable column with the constant column for each desired solution: $x = \frac{D_x}{D}, y = \frac{D_y}{D}, z = \frac{D_z}{D}$. See Example 4.
- Cramer's Rule is also useful for finding the solution of a system of equations with no solution or infinite solutions. See Example 5 and Example 6.
- Certain properties of determinants are useful for solving problems. For example:
  - If the matrix is in upper triangular form, the determinant equals the product of entries down the main diagonal.
  - When two rows are interchanged, the determinant changes sign.
  - If either two rows or two columns are identical, the determinant equals zero.
  - If a matrix contains either a row of zeros or a column of zeros, the determinant equals zero.
  - The determinant of an inverse matrix A^{-1} is the reciprocal of the determinant of the matrix A.
  - If any row or column is multiplied by a constant, the determinant is multiplied by the same factor. See Example 7 and Example 8.

Exercises

Review Exercises

Systems of Linear Equations: Two Variables

For the following exercises, determine whether the ordered pair is a solution to the system of equations.

1. $\begin{array}{l} 3x - y = 4 \\ x + 4y = -3 \end{array}$ and $(-1, 1)$

2. $\begin{array}{l} 6x - 2y = 24 \\ -3x + 3y = 18 \end{array}$ and $(9, 15)$

For the following exercises, use substitution to solve the system of equations.

3. $\begin{array}{l} 10x + 5y = -5 \\ 3x - 2y = -12 \end{array}$

4. $\begin{array}{l} \frac{4}{7}x + \frac{1}{5}y = \frac{43}{70} \\ \frac{5}{6}x - \frac{1}{3}y = -\frac{2}{3} \end{array}$

5. $\begin{array}{l} 5x + 6y = 14 \\ 4x + 8y = 8 \end{array}$

For the following exercises, use addition to solve the system of equations.

6. $\begin{array}{l} 3x + 2y = -7 \\ 2x + 4y = 6 \end{array}$

7. $\begin{array}{l} 3x + 4y = 2 \\ 9x + 12y = 3 \end{array}$

8. $\begin{array}{l} 8x + 4y = 2 \\ 6x - 5y = 0.7 \end{array}$

For the following exercises, write a system of equations to solve each problem. Solve the system of equations.

9. A factory has a cost of production $C(x) = 150x + 15{,}000$ and a revenue function $R(x) = 200x$. What is the break-even point?

10. A performer charges $C(x) = 50x + 10{,}000$, where x is the total number of attendees at a show. The venue charges $75 per ticket. After how many people buy tickets does the venue break even, and what is the value of the total tickets sold at that point?

Systems of Linear Equations: Three Variables

For the following exercises, solve the system of three equations using substitution or addition.

11. $$\begin{aligned} 0.5x - 0.5y &= 10 \\ -0.2y + 0.2x &= 4 \\ 0.1x + 0.1z &= 2 \end{aligned}$$

12. $$\begin{aligned} 5x + 3y - z &= 5 \\ 3x - 2y + 4z &= 13 \\ 4x + 3y + 5z &= 22 \end{aligned}$$

13. $$\begin{aligned} x + y + z &= 1 \\ 2x + 2y + 2z &= 1 \\ 3x + 3y &= 2 \end{aligned}$$

14. $$\begin{aligned} 2x - 3y + z &= -1 \\ x + y + z &= -4 \\ 4x + 2y - 3z &= 33 \end{aligned}$$

15. $$\begin{aligned} 3x + 2y - z &= -10 \\ x - y + 2z &= 7 \\ -x + 3y + z &= -2 \end{aligned}$$

16. $$\begin{aligned} 3x + 4z &= -11 \\ x - 2y &= 5 \\ 4y - z &= -10 \end{aligned}$$

17. $$\begin{aligned} 2x - 3y + z &= 0 \\ 2x + 4y - 3z &= 0 \\ 6x - 2y - z &= 0 \end{aligned}$$

18. $$\begin{aligned} 6x - 4y - 2z &= 2 \\ 3x + 2y - 5z &= 4 \\ 6y - 7z &= 5 \end{aligned}$$

For the following exercises, write a system of equations to solve each problem. Solve the system of equations.

19. Three odd numbers sum up to 61. The smaller is one-third the larger and the middle number is 16 less than the larger. What are the three numbers?

20. A local theatre sells out for their show. They sell all 500 tickets for a total purse of $8,070.00. The tickets were priced at $15 for students, $12 for children, and $18 for adults. If the band sold three times as many adult tickets as children's tickets, how many of each type was sold?

Systems of Nonlinear Equations and Inequalities: Two Variables

For the following exercises, solve the system of nonlinear equations.

21. $$\begin{aligned} y &= x^2 - 7 \\ y &= 5x - 13 \end{aligned}$$

22. $$\begin{aligned} y &= x^2 - 4 \\ y &= 5x + 10 \end{aligned}$$

23. $$\begin{aligned} x^2 + y^2 &= 16 \\ y &= x - 8 \end{aligned}$$

24. $\begin{aligned} x^2 + y^2 &= 25 \\ y &= x^2 + 5 \end{aligned}$

25. $\begin{aligned} x^2 + y^2 &= 4 \\ y - x^2 &= 3 \end{aligned}$

For the following exercises, graph the inequality.

26. $y > x^2 - 1$

27. $\frac{1}{4}x^2 + y^2 < 4$

For the following exercises, graph the system of inequalities.

28. $\begin{aligned} x^2 + y^2 + 2x &< 3 \\ y &> -x^2 - 3 \end{aligned}$

29. $\begin{aligned} x^2 - 2x + y^2 - 4x &< 4 \\ y &< -x + 4 \end{aligned}$

30. $\begin{aligned} x^2 + y^2 &< 1 \\ y^2 &< x \end{aligned}$

Partial Fractions

For the following exercises, decompose into partial fractions.

31. $\frac{-2x+6}{x^2+3x+2}$

32. $\frac{10x+2}{4x^2+4x+1}$

33. $\frac{7x+20}{x^2+10x+25}$

34. $\frac{x-18}{x^2-12x+36}$

35. $\frac{-x^2+36x+70}{x^3-125}$

36. $\frac{-5x^2+6x-2}{x^3+27}$

37. $\frac{x^3-4x^2+3x+11}{(x^2-2)^2}$

38. $\frac{4x^4-2x^3+22x^2-6x+48}{x(x^2+4)^2}$

Matrices and Matrix Operations

For the following exercises, perform the requested operations on the given matrices.

$$A = \begin{bmatrix} 4 & -2 \\ 1 & 3 \end{bmatrix}, B = \begin{bmatrix} 6 & 7 & -3 \\ 11 & -2 & 4 \end{bmatrix}, C = \begin{bmatrix} 6 & 7 \\ 11 & -2 \\ 14 & 0 \end{bmatrix}, D = \begin{bmatrix} 1 & -4 & 9 \\ 10 & 5 & -7 \\ 2 & 8 & 5 \end{bmatrix}, E = \begin{bmatrix} 7 & -14 & 3 \\ 2 & -1 & 3 \\ 0 & 1 & 9 \end{bmatrix}$$

39. $-4A$

40. $10D - 6E$

41. $B + C$

42. AB

43. BA

44. BC

45. CB

46. DE

47. ED

48. EC

49. CE

50. A^3

Solving Systems with Gaussian Elimination

For the following exercises, write the system of linear equations from the augmented matrix. Indicate whether there will be a unique solution.

51. $\left[\begin{array}{ccc|c} 1 & 0 & -3 & 7 \\ 0 & 1 & 2 & -5 \\ 0 & 0 & 0 & 0 \end{array}\right]$

52. $\left[\begin{array}{ccc|c} 1 & 0 & 5 & -9 \\ 0 & 1 & -2 & 4 \\ 0 & 0 & 0 & 3 \end{array}\right]$

For the following exercises, write the augmented matrix from the system of linear equations.

53. $\begin{aligned}-2x + 2y + z &= 7\\ 2x - 8y + 5z &= 0\\ 19x - 10y + 22z &= 3\end{aligned}$

54. $\begin{aligned}4x + 2y - 3z &= 14\\ -12x + 3y + z &= 100\\ 9x - 6y + 2z &= 31\end{aligned}$

55. $\begin{aligned}x + 3z &= 12\\ -x + 4y &= 0\\ y + 2z &= -7\end{aligned}$

For the following exercises, solve the system of linear equations using Gaussian elimination.

56. $\begin{aligned}3x - 4y &= -7\\ -6x + 8y &= 14\end{aligned}$

57. $\begin{aligned}3x - 4y &= 1\\ -6x + 8y &= 6\end{aligned}$

58. $\begin{aligned}-1.1x - 2.3y &= 6.2\\ -5.2x - 4.1y &= 4.3\end{aligned}$

59. $\begin{aligned}2x + 3y + 2z &= 1\\ -4x - 6y - 4z &= -2\\ 10x + 15y + 10z &= 0\end{aligned}$

60. $\begin{aligned}-x + 2y - 4z &= 8\\ 3y + 8z &= -4\\ -7x + y + 2z &= 1\end{aligned}$

Solving Systems with Inverses

For the following exercises, find the inverse of the matrix.

61. $\begin{bmatrix}-0.2 & 1.4\\ 1.2 & -0.4\end{bmatrix}$

62. $\begin{bmatrix}\frac{1}{2} & -\frac{1}{2}\\ -\frac{1}{4} & \frac{3}{4}\end{bmatrix}$

63. $\begin{bmatrix}12 & 9 & -6\\ -1 & 3 & 2\\ -4 & -3 & 2\end{bmatrix}$

64. $\begin{bmatrix}2 & 1 & 3\\ 1 & 2 & 3\\ 3 & 2 & 1\end{bmatrix}$

For the following exercises, find the solutions by computing the inverse of the matrix.

65. $\begin{aligned}0.3x - 0.1y &= -10\\ -0.1x + 0.3y &= 14\end{aligned}$

66. $\begin{aligned}0.4x - 0.2y &= -0.6\\ -0.1x + 0.05y &= 0.3\end{aligned}$

67. $\begin{aligned}4x + 3y - 3z &= -4.3\\ 5x - 4y - z &= -6.1\\ x + z &= -0.7\end{aligned}$

68. $\begin{aligned}-2x - 3y + 2z &= 3\\ -x + 2y + 4z &= -5\\ -2y + 5z &= -3\end{aligned}$

For the following exercises, write a system of equations to solve each problem. Solve the system of equations.

69. Students were asked to bring their favorite fruit to class. 90% of the fruits consisted of banana, apple, and oranges. If oranges were half as popular as bananas and apples were 5% more popular than bananas, what are the percentages of each individual fruit?

70. A school club held a bake sale to raise money and sold brownies and chocolate chip cookies. They priced the brownies at \$2 and the chocolate chip cookies at \$1. They raised \$250 and sold 175 items. How many brownies and how many cookies were sold?

Solving Systems with Cramer's Rule

For the following exercises, find the determinant.

71. $\begin{vmatrix} 100 & 0 \\ 0 & 0 \end{vmatrix}$

72. $\begin{vmatrix} 0.2 & -0.6 \\ 0.7 & -1.1 \end{vmatrix}$

73. $\begin{vmatrix} -1 & 4 & 3 \\ 0 & 2 & 3 \\ 0 & 0 & -3 \end{vmatrix}$

74. $\begin{vmatrix} \sqrt{2} & 0 & 0 \\ 0 & \sqrt{2} & 0 \\ 0 & 0 & \sqrt{2} \end{vmatrix}$

For the following exercises, use Cramer's Rule to solve the linear systems of equations.

75. $\begin{aligned} 4x - 2y &= 23 \\ -5x - 10y &= -35 \end{aligned}$

76. $\begin{aligned} 0.2x - 0.1y &= 0 \\ -0.3x + 0.3y &= 2.5 \end{aligned}$

77. $\begin{aligned} -0.5x + 0.1y &= 0.3 \\ -0.25x + 0.05y &= 0.15 \end{aligned}$

78. $\begin{aligned} x + 6y + 3z &= 4 \\ 2x + y + 2z &= 3 \\ 3x - 2y + z &= 0 \end{aligned}$

79. $\begin{aligned} 4x - 3y + 5z &= -\frac{5}{2} \\ 7x - 9y - 3z &= \frac{3}{2} \\ x - 5y - 5z &= \frac{5}{2} \end{aligned}$

80. $\begin{aligned} \frac{3}{10}x - \frac{1}{5}y - \frac{3}{10}z &= -\frac{1}{50} \\ \frac{1}{10}x - \frac{1}{10}y - \frac{1}{2}z &= -\frac{9}{50} \\ \frac{2}{5}x - \frac{1}{2}y - \frac{3}{5}z &= -\frac{1}{5} \end{aligned}$

Practice Test

Is the following ordered pair a solution to the system of equations?

1. $\begin{aligned} -5x - y &= 12 \\ x + 4y &= 9 \end{aligned}$ with $(-3, 3)$

For the following exercises, solve the systems of linear and nonlinear equations using substitution or elimination. Indicate if no solution exists.

2. $\begin{aligned} \frac{1}{2}x - \frac{1}{3}y &= 4 \\ \frac{3}{2}x - y &= 0 \end{aligned}$

3. $\begin{aligned} -\frac{1}{2}x - 4y &= 4 \\ 2x + 16y &= 2 \end{aligned}$

4. $\begin{aligned} 5x - y &= 1 \\ -10x + 2y &= -2 \end{aligned}$

5. $\begin{aligned} 4x - 6y - 2z &= \frac{1}{10} \\ x - 7y + 5z &= -\frac{1}{4} \\ 3x + 6y - 9z &= \frac{6}{5} \end{aligned}$

6. $\begin{aligned} x + z &= 20 \\ x + y + z &= 20 \\ x + 2y + z &= 10 \end{aligned}$

7. $\begin{aligned} 5x - 4y - 3z &= 0 \\ 2x + y + 2z &= 0 \\ x - 6y - 7z &= 0 \end{aligned}$

8. $\begin{aligned} y &= x^2 + 2x - 3 \\ y &= x - 1 \end{aligned}$

9. $\begin{aligned} y^2 + x^2 &= 25 \\ y^2 - 2x^2 &= 1 \end{aligned}$

For the following exercises, graph the following inequalities.

10. $y < x^2 + 9$

11. $\begin{aligned} x^2 + y^2 &> 4 \\ y &< x^2 + 1 \end{aligned}$

For the following exercises, write the partial fraction decomposition.

12. $\frac{-8x-30}{x^2+10x+25}$

13. $\frac{13x+2}{(3x+1)^2}$

14. $\frac{x^4-x^3+2x-1}{x(x^2+1)^2}$

For the following exercises, perform the given matrix operations.

15. $5\begin{bmatrix} 4 & 9 \\ -2 & 3 \end{bmatrix} + \frac{1}{2}\begin{bmatrix} -6 & 12 \\ 4 & -8 \end{bmatrix}$

16. $\begin{bmatrix} 1 & 4 & -7 \\ -2 & 9 & 5 \\ 12 & 0 & -4 \end{bmatrix}\begin{bmatrix} 3 & -4 \\ 1 & 3 \\ 5 & 10 \end{bmatrix}$

17. $\begin{bmatrix} \frac{1}{2} & \frac{1}{3} \\ \frac{1}{4} & \frac{1}{5} \end{bmatrix}^{-1}$

18. $\det\begin{vmatrix} 0 & 0 \\ 400 & 4{,}000 \end{vmatrix}$

19. $\det\begin{vmatrix} \frac{1}{2} & -\frac{1}{2} & 0 \\ -\frac{1}{2} & 0 & \frac{1}{2} \\ 0 & \frac{1}{2} & 0 \end{vmatrix}$

20. If $\det(A) = -6$, what would be the determinant if you switched rows 1 and 3, multiplied the second row by 12, and took the inverse?

21. Rewrite the system of linear equations as an augmented matrix.

$$\begin{aligned} 14x - 2y + 13z &= 140 \\ -2x + 3y - 6z &= -1 \\ x - 5y + 12z &= 11 \end{aligned}$$

22. Rewrite the augmented matrix as a system of linear equations.

$$\left[\begin{array}{rrr|r} 1 & 0 & 3 & 12 \\ -2 & 4 & 9 & -5 \\ -6 & 1 & 2 & 8 \end{array}\right]$$

For the following exercises, use Gaussian elimination to solve the systems of equations.

23. $\begin{aligned} x - 6y &= 4 \\ 2x - 12y &= 0 \end{aligned}$

24. $\begin{aligned} 2x + y + z &= -3 \\ x - 2y + 3z &= 6 \\ x - y - z &= 6 \end{aligned}$

For the following exercises, use the inverse of a matrix to solve the systems of equations.

25. $\begin{aligned} 4x - 5y &= -50 \\ -x + 2y &= 80 \end{aligned}$

26. $\begin{aligned} \frac{1}{100}x - \frac{3}{100}y + \frac{1}{20}z &= -49 \\ \frac{3}{100}x - \frac{7}{100}y - \frac{1}{100}z &= 13 \\ \frac{9}{100}x - \frac{9}{100}y - \frac{9}{100}z &= 99 \end{aligned}$

For the following exercises, use Cramer's Rule to solve the systems of equations.

27. $\begin{aligned} 200x - 300y &= 2 \\ 400x + 715y &= 4 \end{aligned}$

28. $\begin{aligned} 0.1x + 0.1y - 0.1z &= -1.2 \\ 0.1x - 0.2y + 0.4z &= -1.2 \\ 0.5x - 0.3y + 0.8z &= -5.9 \end{aligned}$

For the following exercises, solve using a system of linear equations.

29. A factory producing cell phones has the following cost and revenue functions: $C(x) = x^2 + 75x + 2{,}688$ and $R(x) = x^2 + 160x$. What is the range of cell phones they should produce each day so there is profit? Round to the nearest number that generates profit.

30. A small fair charges $1.50 for students, $1 for children, and $2 for adults. In one day, three times as many children as adults attended. A total of 800 tickets were sold for a total revenue of $1,050. How many of each type of ticket was sold?

12 ANALYTIC GEOMETRY

The rings of Saturn have produced wonder, as well as misunderstanding, since Galileo first discovered them (he initially thought they were moons). Though they appear to be a series of solid discs even in this 2004 closeup from the Cassini probe, 19th century mathematicians proved that they are made up of billions of small objects clustered together. (credit: modificaion of "Saturn" by NASA/JPL-Caltech/SSI/Kevin M. Gill/flickr)

Chapter Outline

Introduction to Analytic Geometry

The Greek mathematician Menaechmus (c. 380–c. 320 BCE) is generally credited with discovering the shapes formed by the intersection of a plane and a right circular cone. Depending on how he tilted the plane when it intersected the cone, he formed different shapes at the intersection–beautiful shapes with near-perfect symmetry.

It was also said that Aristotle may have had an intuitive understanding of these shapes, as he observed the orbit of the planet to be circular. He presumed that the planets moved in circular orbits around Earth, and for nearly 2000 years this was the commonly held belief.

It was not until the Renaissance movement that Johannes Kepler noticed that the orbits of the planet were not circular in nature. His published law of planetary motion in the 1600s changed our view of the solar system forever. He claimed that the sun was at one end of the orbits, and the planets revolved around the sun in an oval-shaped path.

Other objects in the solar system (and perhaps other systems) follow a similar elliptical path, including the spectacular rings of Saturn. Using this understanding as a basis, 19th century mathematicians like James Clerk Maxwell and Sofya Kovalevskaya showed that despite their appearance through the telescopes of the day (and even in current telescopes), the rings are not solid and continuous, but are rather composed of small particles. Even after the Voyager and Cassini missions have provided close-up and detailed data regarding the ring structures, full understanding of their construction relies heavily on mathematical analysis. Of particular interest are the influences of Saturn's moons and moonlets, and the ways they both disrupt and preserve the ring structure.

In this chapter, we will investigate the two-dimensional figures that are formed when a right circular cone is intersected by a plane. We will begin by studying each of three figures created in this manner. We will develop defining equations for each figure and then learn how to use these equations to solve a variety of problems.

12.1 The Ellipse

Learning Objectives

In this section, you will:

- Write equations of ellipses in standard form.
- Graph ellipses centered at the origin.
- Graph ellipses not centered at the origin.
- Solve applied problems involving ellipses.

Figure 1 The National Statuary Hall in Washington, D.C. (credit: Greg Palmer, Flickr)

Can you imagine standing at one end of a large room and still being able to hear a whisper from a person standing at the other end? The National Statuary Hall in Washington, D.C., shown in Figure 1, is such a room.[1] It is an semi-circular room called a *whispering chamber* because the shape makes it possible for sound to travel along the walls and dome. In this section, we will investigate the shape of this room and its real-world applications, including how far apart two people in Statuary Hall can stand and still hear each other whisper.

Writing Equations of Ellipses in Standard Form

A conic section, or **conic**, is a shape resulting from intersecting a right circular cone with a plane. The angle at which the plane intersects the cone determines the shape, as shown in Figure 2.

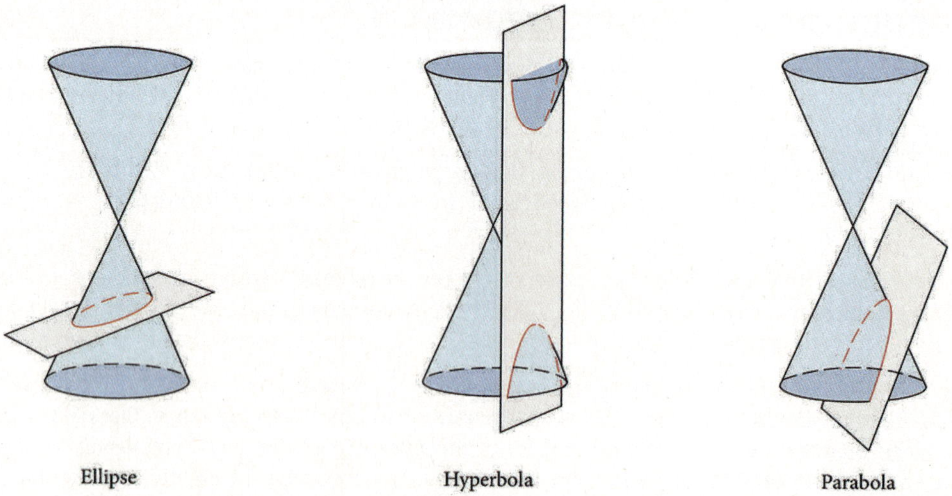

Figure 2

Conic sections can also be described by a set of points in the coordinate plane. Later in this chapter, we will see that the graph of any quadratic equation in two variables is a conic section. The signs of the equations and the coefficients of the variable terms determine the shape. This section focuses on the four variations of the standard form of the equation for the ellipse. An **ellipse** is the set of all points (x, y) in a plane such that the sum of their distances from two fixed points is a constant. Each fixed point is called a **focus** (plural: **foci**).

1 Architect of the Capitol. http://www.aoc.gov. Accessed April 15, 2014.

We can draw an ellipse using a piece of cardboard, two thumbtacks, a pencil, and string. Place the thumbtacks in the cardboard to form the foci of the ellipse. Cut a piece of string longer than the distance between the two thumbtacks (the length of the string represents the constant in the definition). Tack each end of the string to the cardboard, and trace a curve with a pencil held taut against the string. The result is an ellipse. See Figure 3.

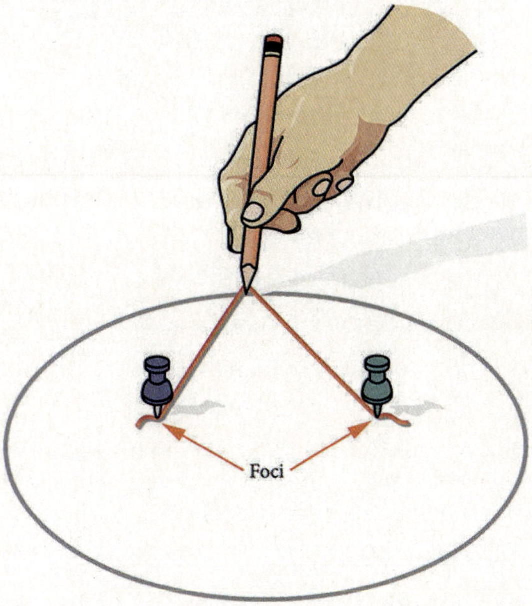

Figure 3

Every ellipse has two axes of symmetry. The longer axis is called the **major axis**, and the shorter axis is called the **minor axis**. Each endpoint of the major axis is the **vertex** of the ellipse (plural: **vertices**), and each endpoint of the minor axis is a **co-vertex** of the ellipse. The **center of an ellipse** is the midpoint of both the major and minor axes. The axes are perpendicular at the center. The foci always lie on the major axis, and the sum of the distances from the foci to any point on the ellipse (the constant sum) is greater than the distance between the foci. See Figure 4.

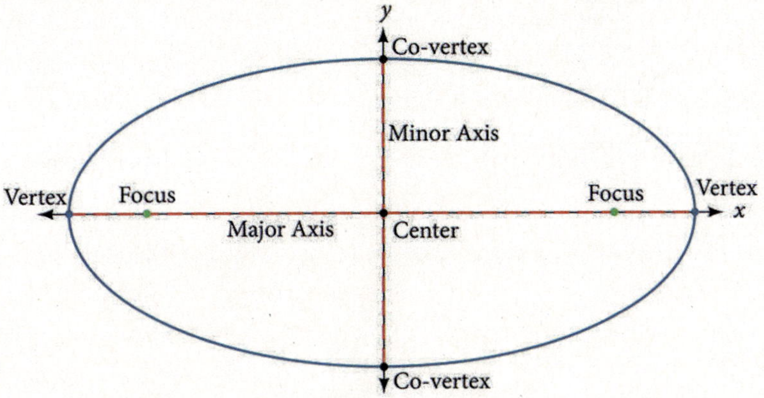

Figure 4

In this section, we restrict ellipses to those that are positioned vertically or horizontally in the coordinate plane. That is, the axes will either lie on or be parallel to the *x*- and *y*-axes. Later in the chapter, we will see ellipses that are rotated in the coordinate plane.

To work with horizontal and vertical ellipses in the coordinate plane, we consider two cases: those that are centered at the origin and those that are centered at a point other than the origin. First we will learn to derive the equations of ellipses, and then we will learn how to write the equations of ellipses in standard form. Later we will use what we learn to draw the graphs.

Deriving the Equation of an Ellipse Centered at the Origin

To derive the equation of an ellipse centered at the origin, we begin with the foci $(-c, 0)$ and $(c, 0)$. The ellipse is the set of all points (x, y) such that the sum of the distances from (x, y) to the foci is constant, as shown in Figure 5.

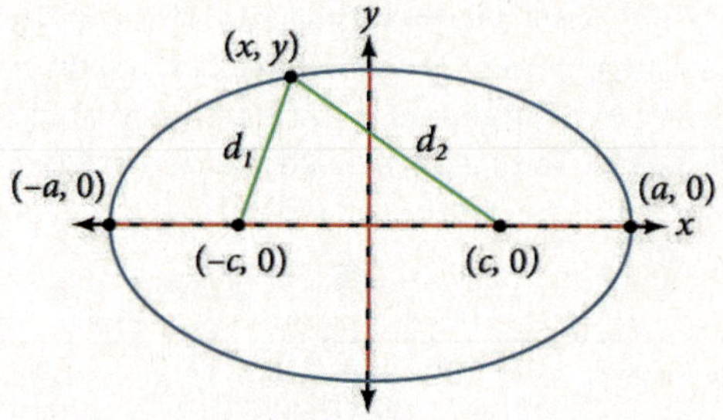

Figure 5

If $(a, 0)$ is a vertex of the ellipse, the distance from $(-c, 0)$ to $(a, 0)$ is $a - (-c) = a + c$. The distance from $(c, 0)$ to $(a, 0)$ is $a - c$. The sum of the distances from the foci to the vertex is

$$(a + c) + (a - c) = 2a$$

If (x, y) is a point on the ellipse, then we can define the following variables:

$$d_1 = \text{the distance from } (-c, 0) \text{ to } (x, y)$$
$$d_2 = \text{the distance from } (c, 0) \text{ to } (x, y)$$

By the definition of an ellipse, $d_1 + d_2$ is constant for any point (x, y) on the ellipse. We know that the sum of these distances is $2a$ for the vertex $(a, 0)$. It follows that $d_1 + d_2 = 2a$ for any point on the ellipse. We will begin the derivation by applying the distance formula. The rest of the derivation is algebraic.

$$d_1 + d_2 = \sqrt{(x - (-c))^2 + (y - 0)^2} + \sqrt{(x - c)^2 + (y - 0)^2} = 2a \quad \text{Distance formula}$$
$$\sqrt{(x + c)^2 + y^2} + \sqrt{(x - c)^2 + y^2} = 2a \quad \text{Simplify expressions.}$$
$$\sqrt{(x + c)^2 + y^2} = 2a - \sqrt{(x - c)^2 + y^2} \quad \text{Move radical to opposite side.}$$
$$(x + c)^2 + y^2 = \left[2a - \sqrt{(x - c)^2 + y^2}\right]^2 \quad \text{Square both sides.}$$
$$x^2 + 2cx + c^2 + y^2 = 4a^2 - 4a\sqrt{(x - c)^2 + y^2} + (x - c)^2 + y^2 \quad \text{Expand the squares.}$$
$$x^2 + 2cx + c^2 + y^2 = 4a^2 - 4a\sqrt{(x - c)^2 + y^2} + x^2 - 2cx + c^2 + y^2 \quad \text{Expand remaining squares.}$$
$$2cx = 4a^2 - 4a\sqrt{(x - c)^2 + y^2} - 2cx \quad \text{Combine like terms.}$$
$$4cx - 4a^2 = -4a\sqrt{(x - c)^2 + y^2} \quad \text{Isolate the radical.}$$
$$cx - a^2 = -a\sqrt{(x - c)^2 + y^2} \quad \text{Divide by 4.}$$
$$\left[cx - a^2\right]^2 = a^2\left[\sqrt{(x - c)^2 + y^2}\right]^2 \quad \text{Square both sides.}$$
$$c^2x^2 - 2a^2cx + a^4 = a^2\left(x^2 - 2cx + c^2 + y^2\right) \quad \text{Expand the squares.}$$
$$c^2x^2 - 2a^2cx + a^4 = a^2x^2 - 2a^2cx + a^2c^2 + a^2y^2 \quad \text{Distribute } a^2.$$
$$a^2x^2 - c^2x^2 + a^2y^2 = a^4 - a^2c^2 \quad \text{Rewrite.}$$
$$x^2\left(a^2 - c^2\right) + a^2y^2 = a^2\left(a^2 - c^2\right) \quad \text{Factor common terms.}$$
$$x^2b^2 + a^2y^2 = a^2b^2 \quad \text{Set } b^2 = a^2 - c^2.$$
$$\frac{x^2b^2}{a^2b^2} + \frac{a^2y^2}{a^2b^2} = \frac{a^2b^2}{a^2b^2} \quad \text{Divide both sides by } a^2b^2.$$
$$\frac{x^2}{a^2} + \frac{y^2}{b^2} = 1 \quad \text{Simplify.}$$

Thus, the standard equation of an ellipse is $\frac{x^2}{a^2} + \frac{y^2}{b^2} = 1$. This equation defines an ellipse centered at the origin. If $a > b$, the ellipse is stretched further in the horizontal direction, and if $b > a$, the ellipse is stretched further in the vertical direction.

Writing Equations of Ellipses Centered at the Origin in Standard Form

Standard forms of equations tell us about key features of graphs. Take a moment to recall some of the standard forms of equations we've worked with in the past: linear, quadratic, cubic, exponential, logarithmic, and so on. By learning to interpret standard forms of equations, we are bridging the relationship between algebraic and geometric representations of mathematical phenomena.

The key features of the ellipse are its center, vertices, co-vertices, foci, and lengths and positions of the major and minor axes. Just as with other equations, we can identify all of these features just by looking at the standard form of the equation. There are four variations of the standard form of the ellipse. These variations are categorized first by the location of the center (the origin or not the origin), and then by the position (horizontal or vertical). Each is presented along with a description of how the parts of the equation relate to the graph. Interpreting these parts allows us to form a mental picture of the ellipse.

Standard Forms of the Equation of an Ellipse with Center (0,0)

The standard form of the equation of an ellipse with center $(0, 0)$ and major axis on the *x-axis* is

$$\frac{x^2}{a^2} + \frac{y^2}{b^2} = 1$$

where

- $a > b$
- the length of the major axis is $2a$
- the coordinates of the vertices are $(\pm a, 0)$
- the length of the minor axis is $2b$
- the coordinates of the co-vertices are $(0, \pm b)$
- the coordinates of the foci are $(\pm c, 0)$, where $c^2 = a^2 - b^2$. See Figure 6 **a**

The standard form of the equation of an ellipse with center $(0, 0)$ and major axis on the *y-axis* is

$$\frac{x^2}{b^2} + \frac{y^2}{a^2} = 1$$

where

- $a > b$
- the length of the major axis is $2a$
- the coordinates of the vertices are $(0, \pm a)$
- the length of the minor axis is $2b$
- the coordinates of the co-vertices are $(\pm b, 0)$
- the coordinates of the foci are $(0, \pm c)$, where $c^2 = a^2 - b^2$. See Figure 6 **b**

Note that the vertices, co-vertices, and foci are related by the equation $c^2 = a^2 - b^2$. When we are given the coordinates of the foci and vertices of an ellipse, we can use this relationship to find the equation of the ellipse in standard form.

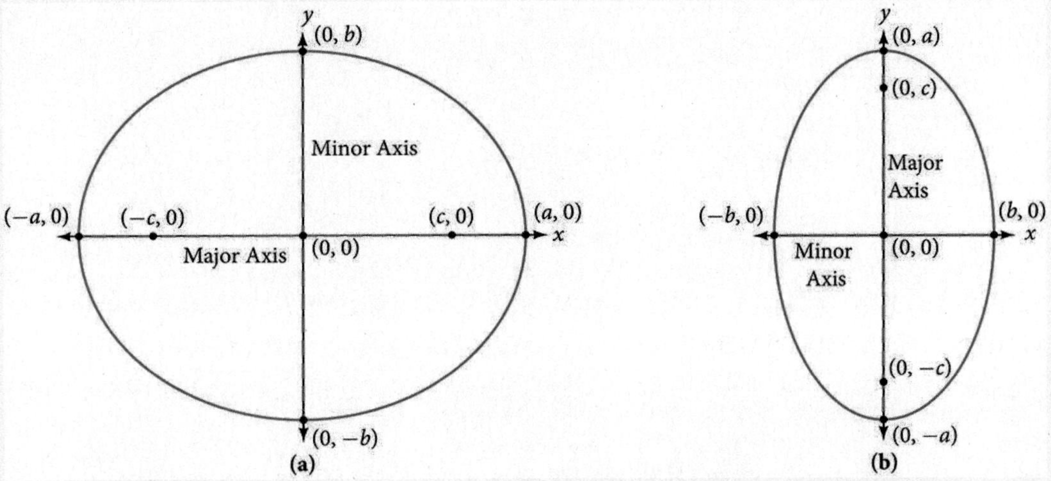

Figure 6 (a) Horizontal ellipse with center $(0, 0)$ (b) Vertical ellipse with center $(0, 0)$

HOW TO

Given the vertices and foci of an ellipse centered at the origin, write its equation in standard form.

1. Determine whether the major axis lies on the *x*- or *y*-axis.
   a. If the given coordinates of the vertices and foci have the form $(\pm a, 0)$ and $(\pm c, 0)$ respectively, then the major axis is the *x*-axis. Use the standard form $\frac{x^2}{a^2} + \frac{y^2}{b^2} = 1$.
   b. If the given coordinates of the vertices and foci have the form $(0, \pm a)$ and $(0, \pm c)$, respectively, then the major axis is the *y*-axis. Use the standard form $\frac{x^2}{b^2} + \frac{y^2}{a^2} = 1$.
2. Use the equation $c^2 = a^2 - b^2$, along with the given coordinates of the vertices and foci, to solve for b^2.
3. Substitute the values for a^2 and b^2 into the standard form of the equation determined in Step 1.

EXAMPLE 1

Writing the Equation of an Ellipse Centered at the Origin in Standard Form
What is the standard form equation of the ellipse that has vertices $(\pm 8, 0)$ and foci $(\pm 5, 0)$?

Solution
The foci are on the *x*-axis, so the major axis is the *x*-axis. Thus, the equation will have the form

$$\frac{x^2}{a^2} + \frac{y^2}{b^2} = 1$$

The vertices are $(\pm 8, 0)$, so $a = 8$ and $a^2 = 64$.

The foci are $(\pm 5, 0)$, so $c = 5$ and $c^2 = 25$.

We know that the vertices and foci are related by the equation $c^2 = a^2 - b^2$. Solving for b^2, we have:

$$\begin{aligned} c^2 &= a^2 - b^2 & \\ 25 &= 64 - b^2 & \text{Substitute for } c^2 \text{ and } a^2. \\ b^2 &= 39 & \text{Solve for } b^2. \end{aligned}$$

Now we need only substitute $a^2 = 64$ and $b^2 = 39$ into the standard form of the equation. The equation of the ellipse is $\frac{x^2}{64} + \frac{y^2}{39} = 1$.

TRY IT #1 What is the standard form equation of the ellipse that has vertices $(0, \pm 4)$ and foci $\left(0, \pm \sqrt{15}\right)$?

Q&A **Can we write the equation of an ellipse centered at the origin given coordinates of just one focus and vertex?**

Yes. Ellipses are symmetrical, so the coordinates of the vertices of an ellipse centered around the origin will always have the form $(\pm a, 0)$ or $(0, \pm a)$. Similarly, the coordinates of the foci will always have the form $(\pm c, 0)$ or $(0, \pm c)$. Knowing this, we can use a and c from the given points, along with the equation $c^2 = a^2 - b^2$, to find b^2.

Writing Equations of Ellipses Not Centered at the Origin

Like the graphs of other equations, the graph of an ellipse can be translated. If an ellipse is translated h units horizontally and k units vertically, the center of the ellipse will be (h, k). This translation results in the standard form of the equation we saw previously, with x replaced by $(x - h)$ and y replaced by $(y - k)$.

Standard Forms of the Equation of an Ellipse with Center (h, k)

The standard form of the equation of an ellipse with center (h, k) and major axis parallel to the x-axis is

$$\frac{(x-h)^2}{a^2}+\frac{(y-k)^2}{b^2}=1$$

where

- $a > b$
- the length of the major axis is $2a$
- the coordinates of the vertices are $(h \pm a, k)$
- the length of the minor axis is $2b$
- the coordinates of the co-vertices are $(h, k \pm b)$
- the coordinates of the foci are $(h \pm c, k)$, where $c^2 = a^2 - b^2$. See Figure 7**a**

The standard form of the equation of an ellipse with center (h, k) and major axis parallel to the y-axis is

$$\frac{(x-h)^2}{b^2}+\frac{(y-k)^2}{a^2}=1$$

where

- $a > b$
- the length of the major axis is $2a$
- the coordinates of the vertices are $(h, k \pm a)$
- the length of the minor axis is $2b$
- the coordinates of the co-vertices are $(h \pm b, k)$
- the coordinates of the foci are $(h, k \pm c)$, where $c^2 = a^2 - b^2$. See Figure 7**b**

Just as with ellipses centered at the origin, ellipses that are centered at a point (h, k) have vertices, co-vertices, and foci that are related by the equation $c^2 = a^2 - b^2$. We can use this relationship along with the midpoint and distance formulas to find the equation of the ellipse in standard form when the vertices and foci are given.

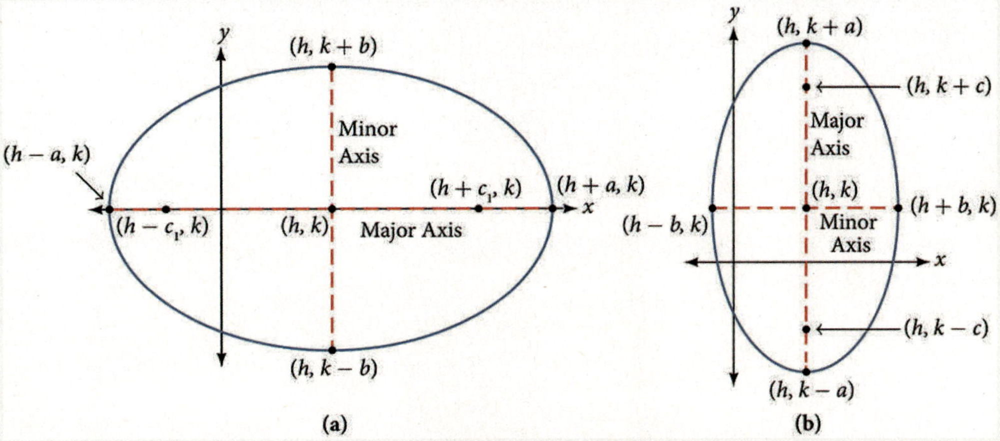

Figure 7 (a) Horizontal ellipse with center (h, k) (b) Vertical ellipse with center (h, k)

HOW TO

Given the vertices and foci of an ellipse not centered at the origin, write its equation in standard form.

1. Determine whether the major axis is parallel to the x- or y-axis.
   a. If the y-coordinates of the given vertices and foci are the same, then the major axis is parallel to the x-axis.

Use the standard form $\frac{(x-h)^2}{a^2} + \frac{(y-k)^2}{b^2} = 1.$

b. If the x-coordinates of the given vertices and foci are the same, then the major axis is parallel to the y-axis. Use the standard form $\frac{(x-h)^2}{b^2} + \frac{(y-k)^2}{a^2} = 1.$

2. Identify the center of the ellipse (h, k) using the midpoint formula and the given coordinates for the vertices.
3. Find a^2 by solving for the length of the major axis, $2a$, which is the distance between the given vertices.
4. Find c^2 using h and k, found in Step 2, along with the given coordinates for the foci.
5. Solve for b^2 using the equation $c^2 = a^2 - b^2$.
6. Substitute the values for h, k, a^2, and b^2 into the standard form of the equation determined in Step 1.

EXAMPLE 2

Writing the Equation of an Ellipse Centered at a Point Other Than the Origin

What is the standard form equation of the ellipse that has vertices $(-2, -8)$ and $(-2, 2)$

and foci $(-2, -7)$ and $(-2, 1)$?

Solution

The x-coordinates of the vertices and foci are the same, so the major axis is parallel to the y-axis. Thus, the equation of the ellipse will have the form

$$\frac{(x-h)^2}{b^2} + \frac{(y-k)^2}{a^2} = 1$$

First, we identify the center, (h, k). The center is halfway between the vertices, $(-2, -8)$ and $(-2, 2)$. Applying the midpoint formula, we have:

$$\begin{aligned}(h, k) &= \left(\frac{-2+(-2)}{2}, \frac{-8+2}{2}\right) \\ &= (-2, -3)\end{aligned}$$

Next, we find a^2. The length of the major axis, $2a$, is bounded by the vertices. We solve for a by finding the distance between the y-coordinates of the vertices.

$$\begin{aligned}2a &= 2 - (-8) \\ 2a &= 10 \\ a &= 5\end{aligned}$$

So $a^2 = 25$.

Now we find c^2. The foci are given by $(h, k \pm c)$. So, $(h, k - c) = (-2, -7)$ and $(h, k + c) = (-2, 1)$. We substitute $k = -3$ using either of these points to solve for c.

$$\begin{aligned}k + c &= 1 \\ -3 + c &= 1 \\ c &= 4\end{aligned}$$

So $c^2 = 16$.

Next, we solve for b^2 using the equation $c^2 = a^2 - b^2$.

$$\begin{aligned}c^2 &= a^2 - b^2 \\ 16 &= 25 - b^2 \\ b^2 &= 9\end{aligned}$$

Finally, we substitute the values found for h, k, a^2, and b^2 into the standard form equation for an ellipse:

$$\frac{(x+2)^2}{9} + \frac{(y+3)^2}{25} = 1$$

> **TRY IT** #2 What is the standard form equation of the ellipse that has vertices $(-3, 3)$ and $(5, 3)$ and foci $\left(1 - 2\sqrt{3}, 3\right)$ and $\left(1 + 2\sqrt{3}, 3\right)$?

Graphing Ellipses Centered at the Origin

Just as we can write the equation for an ellipse given its graph, we can graph an ellipse given its equation. To graph ellipses centered at the origin, we use the standard form $\frac{x^2}{a^2} + \frac{y^2}{b^2} = 1, \;\; a > b$ for horizontal ellipses and $\frac{x^2}{b^2} + \frac{y^2}{a^2} = 1, \;\; a > b$ for vertical ellipses.

HOW TO

Given the standard form of an equation for an ellipse centered at $(0, 0)$, sketch the graph.

1. Use the standard forms of the equations of an ellipse to determine the major axis, vertices, co-vertices, and foci.
   a. If the equation is in the form $\frac{x^2}{a^2} + \frac{y^2}{b^2} = 1$, where $a > b$, then
      - the major axis is the *x*-axis
      - the coordinates of the vertices are $(\pm a, 0)$
      - the coordinates of the co-vertices are $(0, \pm b)$
      - the coordinates of the foci are $(\pm c, 0)$

   b. If the equation is in the form $\frac{x^2}{b^2} + \frac{y^2}{a^2} = 1$, where $a > b$, then
      - the major axis is the *y*-axis
      - the coordinates of the vertices are $(0, \pm a)$
      - the coordinates of the co-vertices are $(\pm b, 0)$
      - the coordinates of the foci are $(0, \pm c)$
2. Solve for c using the equation $c^2 = a^2 - b^2$.
3. Plot the center, vertices, co-vertices, and foci in the coordinate plane, and draw a smooth curve to form the ellipse.

EXAMPLE 3

Graphing an Ellipse Centered at the Origin

Graph the ellipse given by the equation, $\frac{x^2}{9} + \frac{y^2}{25} = 1$. Identify and label the center, vertices, co-vertices, and foci.

Solution

First, we determine the position of the major axis. Because $25 > 9$, the major axis is on the *y*-axis. Therefore, the equation is in the form $\frac{x^2}{b^2} + \frac{y^2}{a^2} = 1$, where $b^2 = 9$ and $a^2 = 25$. It follows that:

- the center of the ellipse is $(0, 0)$
- the coordinates of the vertices are $(0, \pm a) = \left(0, \pm \sqrt{25}\right) = (0, \pm 5)$
- the coordinates of the co-vertices are $(\pm b, 0) = \left(\pm\sqrt{9}, 0\right) = (\pm 3, 0)$
- the coordinates of the foci are $(0, \pm c)$, where $c^2 = a^2 - b^2$ Solving for c, we have:

$$\begin{aligned} c &= \pm\sqrt{a^2 - b^2} \\ &= \pm\sqrt{25 - 9} \\ &= \pm\sqrt{16} \\ &= \pm 4 \end{aligned}$$

Therefore, the coordinates of the foci are $(0, \pm 4)$.

Next, we plot and label the center, vertices, co-vertices, and foci, and draw a smooth curve to form the ellipse. See Figure 8.

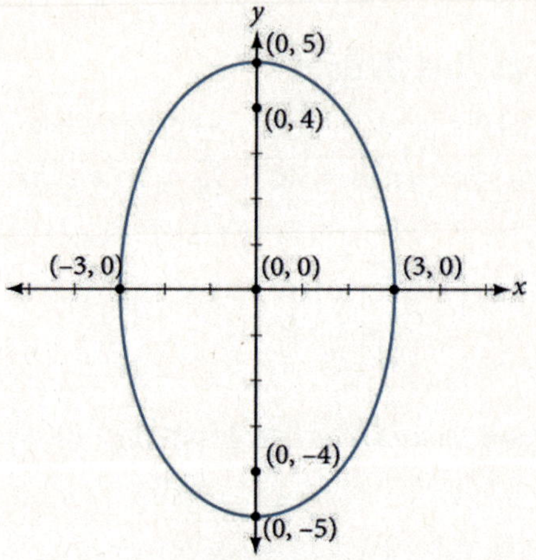

Figure 8

TRY IT #3 Graph the ellipse given by the equation $\frac{x^2}{36} + \frac{y^2}{4} = 1$. Identify and label the center, vertices, co-vertices, and foci.

EXAMPLE 4

Graphing an Ellipse Centered at the Origin from an Equation Not in Standard Form

Graph the ellipse given by the equation $4x^2 + 25y^2 = 100$. Rewrite the equation in standard form. Then identify and label the center, vertices, co-vertices, and foci.

Solution

First, use algebra to rewrite the equation in standard form.

$$\begin{aligned} 4x^2 + 25y^2 &= 100 \\ \frac{4x^2}{100} + \frac{25y^2}{100} &= \frac{100}{100} \\ \frac{x^2}{25} + \frac{y^2}{4} &= 1 \end{aligned}$$

Next, we determine the position of the major axis. Because $25 > 4$, the major axis is on the *x*-axis. Therefore, the equation is in the form $\frac{x^2}{a^2} + \frac{y^2}{b^2} = 1$, where $a^2 = 25$ and $b^2 = 4$. It follows that:

- the center of the ellipse is $(0, 0)$
- the coordinates of the vertices are $(\pm a, 0) = \left(\pm\sqrt{25}, 0\right) = (\pm 5, 0)$
- the coordinates of the co-vertices are $(0, \pm b) = \left(0, \pm\sqrt{4}\right) = (0, \pm 2)$
- the coordinates of the foci are $(\pm c, 0)$, where $c^2 = a^2 - b^2$. Solving for c, we have:

$$\begin{aligned} c &= \pm\sqrt{a^2 - b^2} \\ &= \pm\sqrt{25 - 4} \\ &= \pm\sqrt{21} \end{aligned}$$

Therefore the coordinates of the foci are $\left(\pm\sqrt{21}, 0\right)$.

Next, we plot and label the center, vertices, co-vertices, and foci, and draw a smooth curve to form the ellipse.

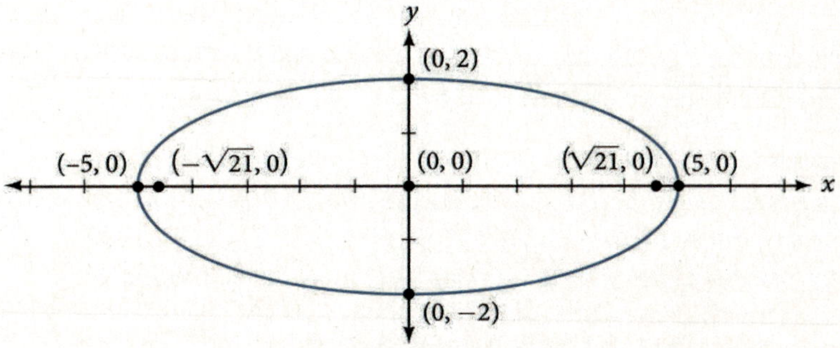

Figure 9

> **TRY IT** #4 Graph the ellipse given by the equation $49x^2 + 16y^2 = 784$. Rewrite the equation in standard form. Then identify and label the center, vertices, co-vertices, and foci.

Graphing Ellipses Not Centered at the Origin

When an ellipse is not centered at the origin, we can still use the standard forms to find the key features of the graph. When the ellipse is centered at some point, (h, k), we use the standard forms $\frac{(x-h)^2}{a^2} + \frac{(y-k)^2}{b^2} = 1,\ \ a > b$ for horizontal ellipses and $\frac{(x-h)^2}{b^2} + \frac{(y-k)^2}{a^2} = 1,\ \ a > b$ for vertical ellipses. From these standard equations, we can easily determine the center, vertices, co-vertices, foci, and positions of the major and minor axes.

HOW TO

Given the standard form of an equation for an ellipse centered at (h, k), sketch the graph.

1. Use the standard forms of the equations of an ellipse to determine the center, position of the major axis, vertices, co-vertices, and foci.
   a. If the equation is in the form $\frac{(x-h)^2}{a^2} + \frac{(y-k)^2}{b^2} = 1$, where $a > b$, then
      - the center is (h, k)
      - the major axis is parallel to the *x*-axis
      - the coordinates of the vertices are $(h \pm a, k)$
      - the coordinates of the co-vertices are $(h, k \pm b)$
      - the coordinates of the foci are $(h \pm c, k)$
   b. If the equation is in the form $\frac{(x-h)^2}{b^2} + \frac{(y-k)^2}{a^2} = 1$, where $a > b$, then
      - the center is (h, k)
      - the major axis is parallel to the *y*-axis
      - the coordinates of the vertices are $(h, k \pm a)$
      - the coordinates of the co-vertices are $(h \pm b, k)$
      - the coordinates of the foci are $(h, k \pm c)$
2. Solve for c using the equation $c^2 = a^2 - b^2$.
3. Plot the center, vertices, co-vertices, and foci in the coordinate plane, and draw a smooth curve to form the ellipse.

EXAMPLE 5

Graphing an Ellipse Centered at (*h*, *k*)

Graph the ellipse given by the equation, $\frac{(x+2)^2}{4} + \frac{(y-5)^2}{9} = 1$. Identify and label the center, vertices, co-vertices, and foci.

Solution

First, we determine the position of the major axis. Because $9 > 4$, the major axis is parallel to the *y*-axis. Therefore, the equation is in the form $\frac{(x-h)^2}{b^2} + \frac{(y-k)^2}{a^2} = 1$, where $b^2 = 4$ and $a^2 = 9$. It follows that:

- the center of the ellipse is $(h, k) = (-2, 5)$
- the coordinates of the vertices are $(h, k \pm a) = (-2, 5 \pm \sqrt{9}) = (-2, 5 \pm 3)$, or $(-2, 2)$ and $(-2, 8)$
- the coordinates of the co-vertices are $(h \pm b, k) = (-2 \pm \sqrt{4}, 5) = (-2 \pm 2, 5)$, or $(-4, 5)$ and $(0, 5)$
- the coordinates of the foci are $(h, k \pm c)$, where $c^2 = a^2 - b^2$. Solving for c, we have:

$$\begin{aligned} c &= \pm\sqrt{a^2 - b^2} \\ &= \pm\sqrt{9-4} \\ &= \pm\sqrt{5} \end{aligned}$$

Therefore, the coordinates of the foci are $\left(-2, 5 - \sqrt{5}\right)$ and $\left(-2, 5 + \sqrt{5}\right)$.

Next, we plot and label the center, vertices, co-vertices, and foci, and draw a smooth curve to form the ellipse.

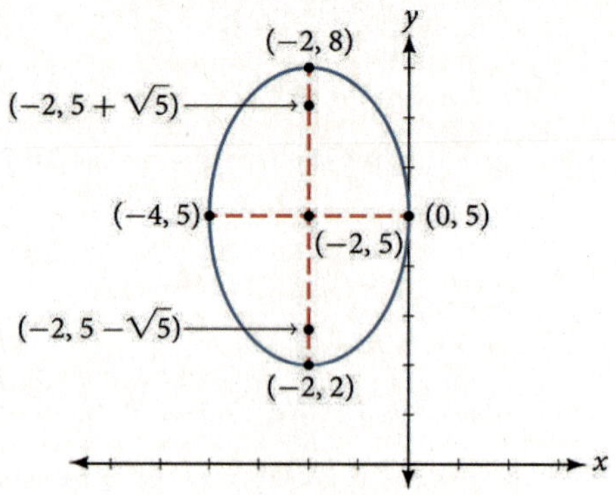

Figure 10

TRY IT #5 Graph the ellipse given by the equation $\frac{(x-4)^2}{36} + \frac{(y-2)^2}{20} = 1$. Identify and label the center, vertices, co-vertices, and foci.

HOW TO

Given the general form of an equation for an ellipse centered at (*h*, *k*), express the equation in standard form.

1. Recognize that an ellipse described by an equation in the form $ax^2 + by^2 + cx + dy + e = 0$ is in general form.
2. Rearrange the equation by grouping terms that contain the same variable. Move the constant term to the opposite side of the equation.
3. Factor out the coefficients of the x^2 and y^2 terms in preparation for completing the square.
4. Complete the square for each variable to rewrite the equation in the form of the sum of multiples of two binomials squared set equal to a constant, $m_1(x-h)^2 + m_2(y-k)^2 = m_3$, where m_1, m_2, and m_3 are constants.
5. Divide both sides of the equation by the constant term to express the equation in standard form.

EXAMPLE 6

Graphing an Ellipse Centered at (*h*, *k*) by First Writing It in Standard Form

Graph the ellipse given by the equation $4x^2 + 9y^2 - 40x + 36y + 100 = 0$. Identify and label the center, vertices, co-vertices, and foci.

Solution

We must begin by rewriting the equation in standard form.

$$4x^2 + 9y^2 - 40x + 36y + 100 = 0$$

Group terms that contain the same variable, and move the constant to the opposite side of the equation.

$$\left(4x^2 - 40x\right) + \left(9y^2 + 36y\right) = -100$$

Factor out the coefficients of the squared terms.

$$4\left(x^2 - 10x\right) + 9\left(y^2 + 4y\right) = -100$$

Complete the square twice. Remember to balance the equation by adding the same constants to each side.

$$4\left(x^2 - 10x + 25\right) + 9\left(y^2 + 4y + 4\right) = -100 + 100 + 36$$

Rewrite as perfect squares.

$$4(x-5)^2 + 9(y+2)^2 = 36$$

Divide both sides by the constant term to place the equation in standard form.

$$\frac{(x-5)^2}{9} + \frac{(y+2)^2}{4} = 1$$

Now that the equation is in standard form, we can determine the position of the major axis. Because $9 > 4$, the major axis is parallel to the x-axis. Therefore, the equation is in the form $\frac{(x-h)^2}{a^2} + \frac{(y-k)^2}{b^2} = 1$, where $a^2 = 9$ and $b^2 = 4$. It follows that:

- the center of the ellipse is $(h, k) = (5, -2)$
- the coordinates of the vertices are $(h \pm a, k) = \left(5 \pm \sqrt{9}, -2\right) = (5 \pm 3, -2)$, or $(2, -2)$ and $(8, -2)$
- the coordinates of the co-vertices are $(h, k \pm b) = \left(5, -2 \pm \sqrt{4}\right) = (5, -2 \pm 2)$, or $(5, -4)$ and $(5, 0)$
- the coordinates of the foci are $(h \pm c, k)$, where $c^2 = a^2 - b^2$. Solving for c, we have:

$$\begin{aligned} c &= \pm\sqrt{a^2 - b^2} \\ &= \pm\sqrt{9 - 4} \\ &= \pm\sqrt{5} \end{aligned}$$

Therefore, the coordinates of the foci are $\left(5 - \sqrt{5}, -2\right)$ and $\left(5+\sqrt{5}, -2\right)$.

Next we plot and label the center, vertices, co-vertices, and foci, and draw a smooth curve to form the ellipse as shown in Figure 11.

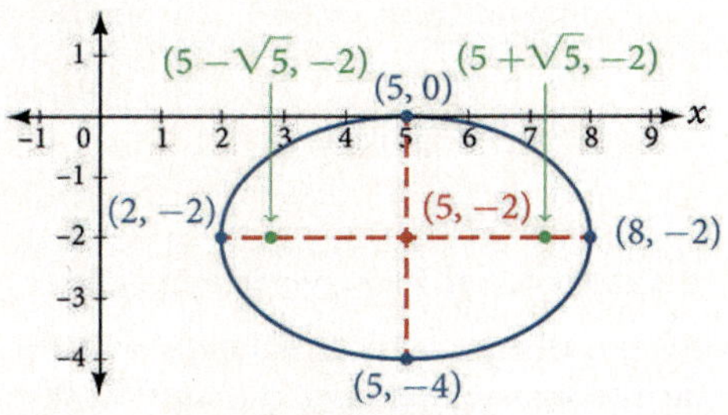

Figure 11

TRY IT #6 Express the equation of the ellipse given in standard form. Identify the center, vertices, co-vertices, and foci of the ellipse.

$$4x^2 + y^2 - 24x + 2y + 21 = 0$$

Solving Applied Problems Involving Ellipses

Many real-world situations can be represented by ellipses, including orbits of planets, satellites, moons and comets, and shapes of boat keels, rudders, and some airplane wings. A medical device called a lithotripter uses elliptical reflectors to break up kidney stones by generating sound waves. Some buildings, called whispering chambers, are designed with elliptical domes so that a person whispering at one focus can easily be heard by someone standing at the other focus. This occurs because of the acoustic properties of an ellipse. When a sound wave originates at one focus of a whispering chamber, the sound wave will be reflected off the elliptical dome and back to the other focus. See Figure 12. In the whisper chamber at the Museum of Science and Industry in Chicago, two people standing at the foci—about 43 feet apart—can hear each other whisper. When these chambers are placed in unexpected places, such as the ones inside Bush International Airport in Houston and Grand Central Terminal in New York City, they can induce surprised reactions among travelers.

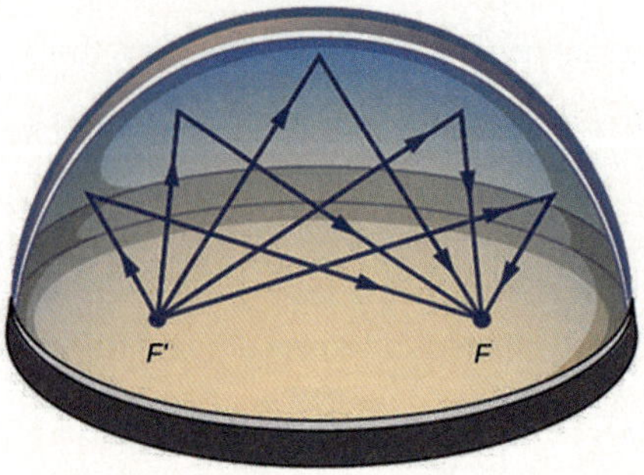

Figure 12 Sound waves are reflected between foci in an elliptical room, called a whispering chamber.

EXAMPLE 7

Locating the Foci of a Whispering Chamber

A large room in an art gallery is a whispering chamber. Its dimensions are 46 feet wide by 96 feet long as shown in Figure 13.

a. What is the standard form of the equation of the ellipse representing the outline of the room? Hint: assume a horizontal ellipse, and let the center of the room be the point $(0, 0)$.
b. If two visitors standing at the foci of this room can hear each other whisper, how far apart are the two visitors? Round to the nearest foot.

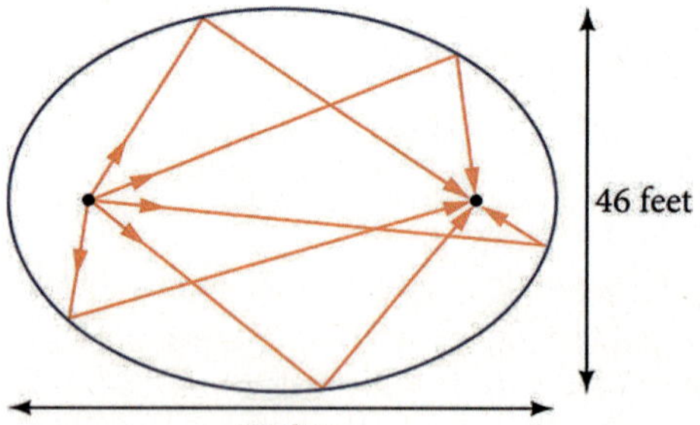

Figure 13

Solution

a. We are assuming a horizontal ellipse with center $(0, 0)$, so we need to find an equation of the form $\frac{x^2}{a^2} + \frac{y^2}{b^2} = 1$, where $a > b$. We know that the length of the major axis, $2a$, is longer than the length of the minor axis, $2b$. So the length of the room, 96, is represented by the major axis, and the width of the room, 46, is represented by the minor axis.
   - Solving for a, we have $2a = 96$, so $a = 48$, and $a^2 = 2304$.
   - Solving for b, we have $2b = 46$, so $b = 23$, and $b^2 = 529$.

   Therefore, the equation of the ellipse is $\frac{x^2}{2304} + \frac{y^2}{529} = 1$.
b. To find the distance between the senators, we must find the distance between the foci, $(\pm c, 0)$, where $c^2 = a^2 - b^2$.

Solving for c, we have:

$c^2 = a^2 - b^2$	
$c^2 = 2304 - 529$	Substitute using the values found in part (a).
$c = \pm\sqrt{2304 - 529}$	Take the square root of both sides.
$c = \pm\sqrt{1775}$	Subtract.
$c \approx \pm 42$	Round to the nearest foot.

The points $(\pm 42, 0)$ represent the foci. Thus, the distance between the senators is $2(42) = 84$ feet.

TRY IT #7 Suppose a whispering chamber is 480 feet long and 320 feet wide.

ⓐ What is the standard form of the equation of the ellipse representing the room? Hint: assume a horizontal ellipse, and let the center of the room be the point $(0, 0)$.
ⓑ If two people are standing at the foci of this room and can hear each other whisper, how far apart are the people? Round to the nearest foot.

MEDIA

Access these online resources for additional instruction and practice with ellipses.

Conic Sections: The Ellipse (http://openstax.org/l/conicellipse)
Graph an Ellipse with Center at the Origin (http://openstax.org/l/grphellorigin)
Graph an Ellipse with Center Not at the Origin (http://openstax.org/l/grphellnot)

12.1 SECTION EXERCISES

Verbal

1. Define an ellipse in terms of its foci.

2. Where must the foci of an ellipse lie?

3. What special case of the ellipse do we have when the major and minor axis are of the same length?

4. For the special case mentioned in the previous question, what would be true about the foci of that ellipse?

5. What can be said about the symmetry of the graph of an ellipse with center at the origin and foci along the y-axis?

Algebraic

For the following exercises, determine whether the given equations represent ellipses. If yes, write in standard form.

6. $2x^2 + y = 4$

7. $4x^2 + 9y^2 = 36$

8. $4x^2 - y^2 = 4$

9. $4x^2 + 9y^2 = 1$

10. $4x^2 - 8x + 9y^2 - 72y + 112 = 0$

For the following exercises, write the equation of an ellipse in standard form, and identify the end points of the major and minor axes as well as the foci.

11. $\frac{x^2}{4} + \frac{y^2}{49} = 1$

12. $\frac{x^2}{100} + \frac{y^2}{64} = 1$

13. $x^2 + 9y^2 = 1$

14. $4x^2 + 16y^2 = 1$

15. $\frac{(x-2)^2}{49} + \frac{(y-4)^2}{25} = 1$

16. $\frac{(x-2)^2}{81} + \frac{(y+1)^2}{16} = 1$

17. $\frac{(x+5)^2}{4} + \frac{(y-7)^2}{9} = 1$

18. $\frac{(x-7)^2}{49} + \frac{(y-7)^2}{49} = 1$

19. $4x^2 - 8x + 9y^2 - 72y + 112 = 0$

20. $9x^2 - 54x + 9y^2 - 54y + 81 = 0$

21. $4x^2 - 24x + 36y^2 - 360y + 864 = 0$

22. $4x^2 + 24x + 16y^2 - 128y + 228 = 0$

23. $4x^2 + 40x + 25y^2 - 100y + 100 = 0$

24. $x^2 + 2x + 100y^2 - 1000y + 2401 = 0$

25. $4x^2 + 24x + 25y^2 + 200y + 336 = 0$

26. $9x^2 + 72x + 16y^2 + 16y + 4 = 0$

For the following exercises, find the foci for the given ellipses.

27. $\frac{(x+3)^2}{25} + \frac{(y+1)^2}{36} = 1$

28. $\frac{(x+1)^2}{100} + \frac{(y-2)^2}{4} = 1$

29. $x^2 + y^2 = 1$

30. $x^2 + 4y^2 + 4x + 8y = 1$

31. $10x^2 + y^2 + 200x = 0$

Graphical

For the following exercises, graph the given ellipses, noting center, vertices, and foci.

32. $\frac{x^2}{25} + \frac{y^2}{36} = 1$

33. $\frac{x^2}{16} + \frac{y^2}{9} = 1$

34. $4x^2 + 9y^2 = 1$

35. $81x^2 + 49y^2 = 1$

36. $\frac{(x-2)^2}{64} + \frac{(y-4)^2}{16} = 1$

37. $\frac{(x+3)^2}{9} + \frac{(y-3)^2}{9} = 1$

38. $\frac{x^2}{2} + \frac{(y+1)^2}{5} = 1$

39. $4x^2 - 8x + 16y^2 - 32y - 44 = 0$

40. $x^2 - 8x + 25y^2 - 100y + 91 = 0$

41. $x^2 + 8x + 4y^2 - 40y + 112 = 0$

42. $64x^2 + 128x + 9y^2 - 72y - 368 = 0$

43. $16x^2 + 64x + 4y^2 - 8y + 4 = 0$

44. $100x^2 + 1000x + y^2 - 10y + 2425 = 0$

45. $4x^2 + 16x + 4y^2 + 16y + 16 = 0$

For the following exercises, use the given information about the graph of each ellipse to determine its equation.

46. Center at the origin, symmetric with respect to the *x*- and *y*-axes, focus at $(4, 0)$, and point on graph $(0, 3)$.

47. Center at the origin, symmetric with respect to the *x*- and *y*-axes, focus at $(0, -2)$, and point on graph $(5, 0)$.

48. Center at the origin, symmetric with respect to the *x*- and *y*-axes, focus at $(3, 0)$, and major axis is twice as long as minor axis.

49. Center $(4, 2)$; vertex $(9, 2)$; one focus: $\left(4 + 2\sqrt{6}, 2\right)$.

50. Center $(3, 5)$; vertex $(3, 11)$; one focus: $\left(3,\ 5+4\sqrt{2}\right)$

51. Center $(-3, 4)$; vertex $(1, 4)$; one focus: $\left(-3 + 2\sqrt{3}, 4\right)$

For the following exercises, given the graph of the ellipse, determine its equation.

52.

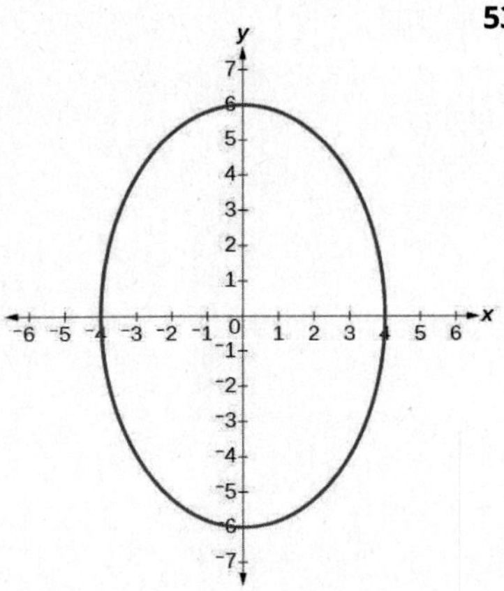

53.

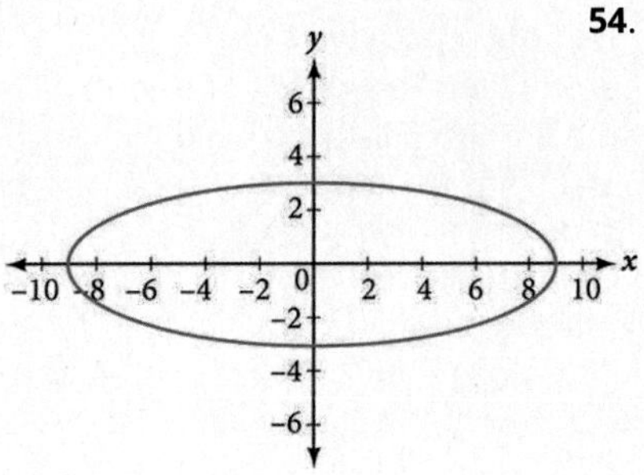

54.

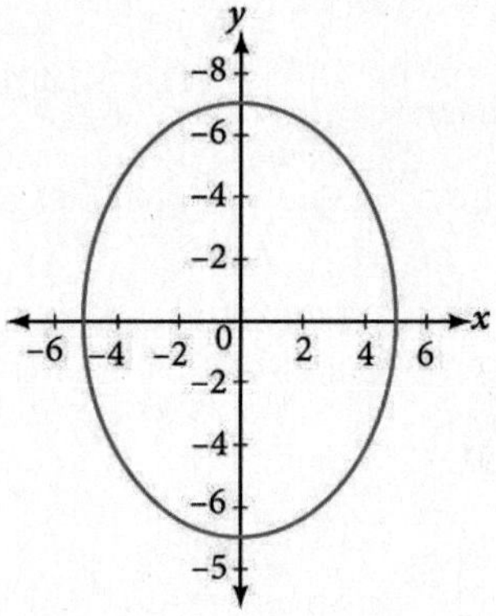

55.

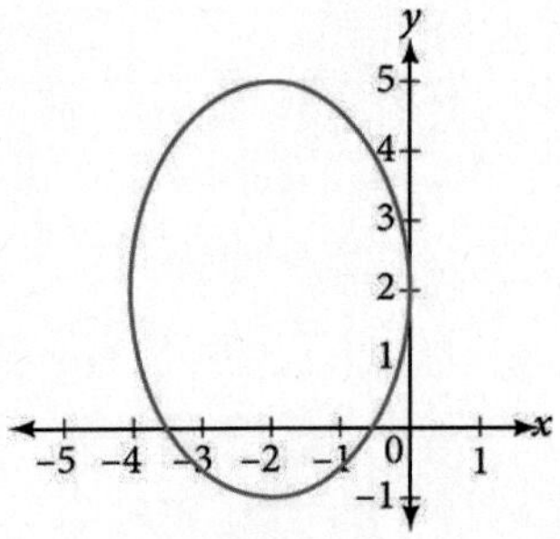

56.

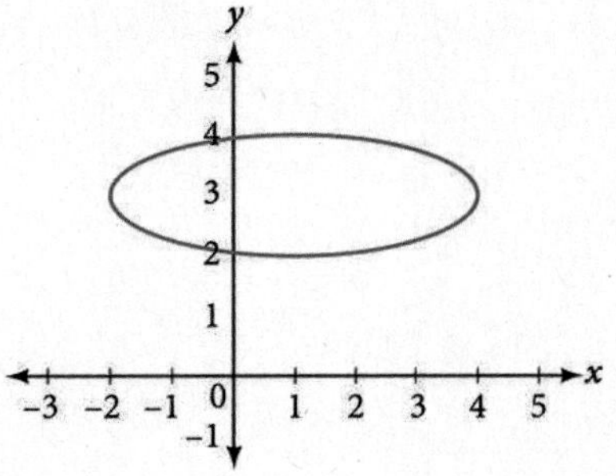

Extensions

For the following exercises, find the area of the ellipse. The area of an ellipse is given by the formula $\text{Area} = a \cdot b \cdot \pi$.

57. $\frac{(x-3)^2}{9} + \frac{(y-3)^2}{16} = 1$

58. $\frac{(x+6)^2}{16} + \frac{(y-6)^2}{36} = 1$

59. $\frac{(x+1)^2}{4} + \frac{(y-2)^2}{5} = 1$

60. $4x^2 - 8x + 9y^2 - 72y + 112 = 0$

61. $9x^2 - 54x + 9y^2 - 54y + 81 = 0$

Real-World Applications

62. Find the equation of the ellipse that will just fit inside a box that is 8 units wide and 4 units high.

63. Find the equation of the ellipse that will just fit inside a box that is four times as wide as it is high. Express in terms of h, the height.

64. An arch has the shape of a semi-ellipse (the top half of an ellipse). The arch has a height of 8 feet and a span of 20 feet. Find an equation for the ellipse, and use that to find the height to the nearest 0.01 foot of the arch at a distance of 4 feet from the center.

65. An arch has the shape of a semi-ellipse. The arch has a height of 12 feet and a span of 40 feet. Find an equation for the ellipse, and use that to find the distance from the center to a point at which the height is 6 feet. Round to the nearest hundredth.

66. A bridge is to be built in the shape of a semi-elliptical arch and is to have a span of 120 feet. The height of the arch at a distance of 40 feet from the center is to be 8 feet. Find the height of the arch at its center.

67. A person in a whispering gallery standing at one focus of the ellipse can whisper and be heard by a person standing at the other focus because all the sound waves that reach the ceiling are reflected to the other person. If a whispering gallery has a length of 120 feet, and the foci are located 30 feet from the center, find the height of the ceiling at the center.

68. A person is standing 8 feet from the nearest wall in a whispering gallery. If that person is at one focus, and the other focus is 80 feet away, what is the length and height at the center of the gallery?

12.2 The Hyperbola

Learning Objectives

In this section, you will:

- Locate a hyperbola's vertices and foci.
- Write equations of hyperbolas in standard form.
- Graph hyperbolas centered at the origin.
- Graph hyperbolas not centered at the origin.
- Solve applied problems involving hyperbolas.

What do paths of comets, supersonic booms, ancient Grecian pillars, and natural draft cooling towers have in common? They can all be modeled by the same type of conic. For instance, when something moves faster than the speed of sound, a shock wave in the form of a cone is created. A portion of a conic is formed when the wave intersects the ground, resulting in a sonic boom. See Figure 1.

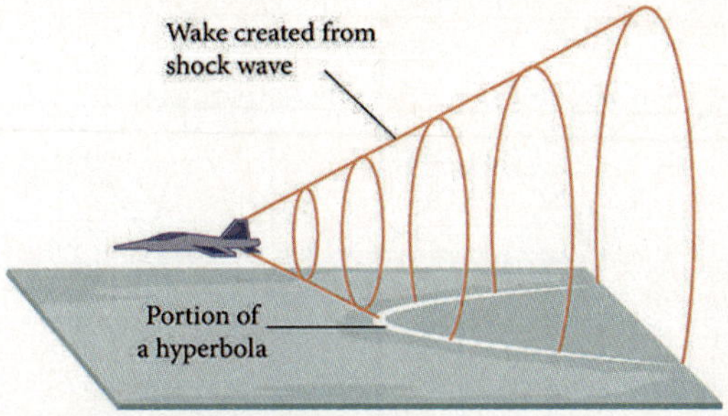

Figure 1 A shock wave intersecting the ground forms a portion of a conic and results in a sonic boom.

Most people are familiar with the sonic boom created by supersonic aircraft, but humans were breaking the sound barrier long before the first supersonic flight. The crack of a whip occurs because the tip is exceeding the speed of sound. The bullets shot from many firearms also break the sound barrier, although the bang of the gun usually supersedes the sound of the sonic boom.

Locating the Vertices and Foci of a Hyperbola

In analytic geometry, a **hyperbola** is a conic section formed by intersecting a right circular cone with a plane at an angle such that both halves of the cone are intersected. This intersection produces two separate unbounded curves that are mirror images of each other. See Figure 2.

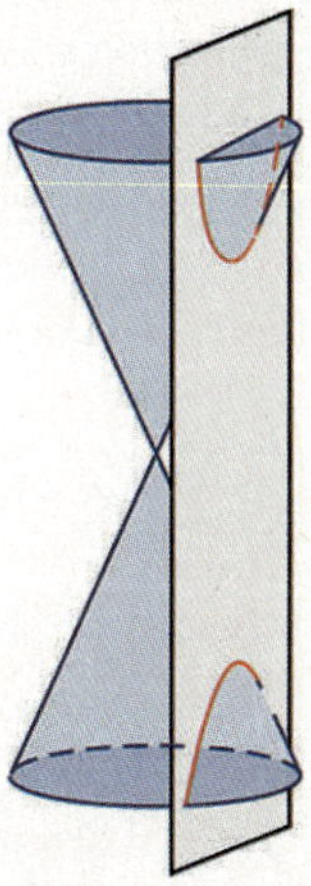

Figure 2 A hyperbola

Like the ellipse, the hyperbola can also be defined as a set of points in the coordinate plane. A hyperbola is the set of all points (x, y) in a plane such that the difference of the distances between (x, y) and the foci is a positive constant.

Notice that the definition of a hyperbola is very similar to that of an ellipse. The distinction is that the hyperbola is defined in terms of the *difference* of two distances, whereas the ellipse is defined in terms of the *sum* of two distances.

As with the ellipse, every hyperbola has two axes of symmetry. The **transverse axis** is a line segment that passes through the center of the hyperbola and has vertices as its endpoints. The foci lie on the line that contains the transverse axis. The **conjugate axis** is perpendicular to the transverse axis and has the co-vertices as its endpoints. The **center of a hyperbola** is the midpoint of both the transverse and conjugate axes, where they intersect. Every hyperbola also has two **asymptotes** that pass through its center. As a hyperbola recedes from the center, its branches approach these asymptotes. The **central rectangle** of the hyperbola is centered at the origin with sides that pass through each vertex and co-vertex; it is a useful tool for graphing the hyperbola and its asymptotes. To sketch the asymptotes of the hyperbola, simply sketch and extend the diagonals of the central rectangle. See Figure 3.

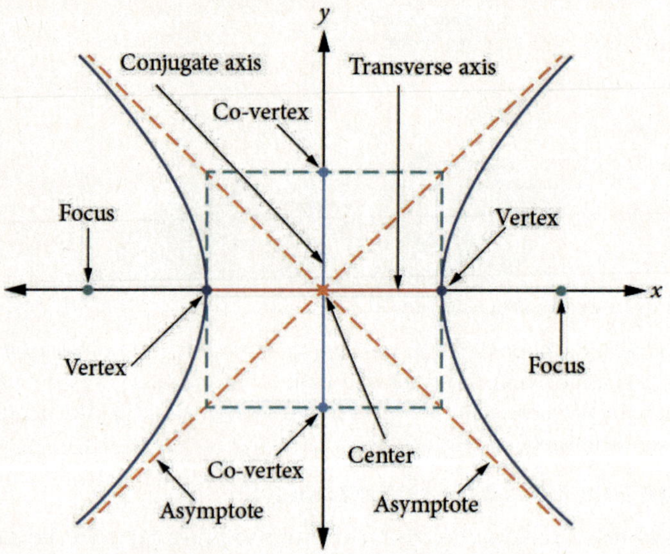

Figure 3 Key features of the hyperbola

In this section, we will limit our discussion to hyperbolas that are positioned vertically or horizontally in the coordinate plane; the axes will either lie on or be parallel to the *x*- and *y*-axes. We will consider two cases: those that are centered at the origin, and those that are centered at a point other than the origin.

Deriving the Equation of a Hyperbola Centered at the Origin

Let $(-c, 0)$ and $(c, 0)$ be the foci of a hyperbola centered at the origin. The hyperbola is the set of all points (x, y) such that the difference of the distances from (x, y) to the foci is constant. See Figure 4.

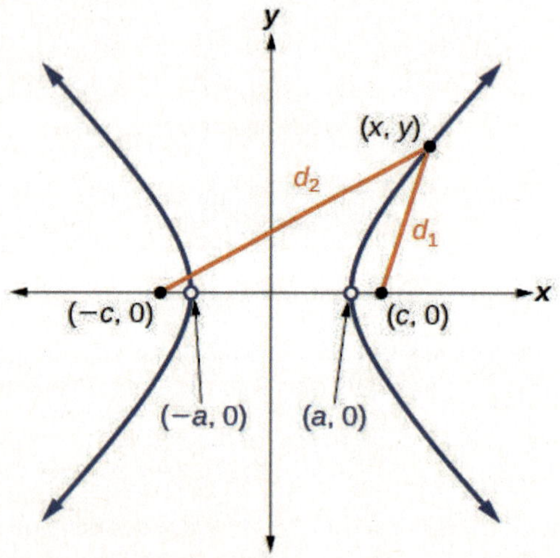

Figure 4

If $(a, 0)$ is a vertex of the hyperbola, the distance from $(-c, 0)$ to $(a, 0)$ is $a - (-c) = a + c$. The distance from $(c, 0)$ to $(a, 0)$ is $c - a$. The difference of the distances from the foci to the vertex is

$$(a + c) - (c - a) = 2a$$

If (x, y) is a point on the hyperbola, we can define the following variables:

$$d_2 = \text{the distance from } (-c, 0) \text{ to } (x, y)$$
$$d_1 = \text{the distance from } (c, 0) \text{ to } (x, y)$$

By definition of a hyperbola, $d_2 - d_1$ is constant for any point (x, y) on the hyperbola. We know that the difference of these distances is $2a$ for the vertex $(a, 0)$. It follows that $d_2 - d_1 = 2a$ for any point on the hyperbola. As with the derivation of the equation of an ellipse, we will begin by applying the distance formula. The rest of the derivation is algebraic. Compare this derivation with the one from the previous section for ellipses.

$d_2 - d_1 = \sqrt{(x-(-c))^2 + (y-0)^2} - \sqrt{(x-c)^2 + (y-0)^2} = 2a$	Distance Formula
$\sqrt{(x+c)^2 + y^2} - \sqrt{(x-c)^2 + y^2} = 2a$	Simplify expressions.
$\sqrt{(x+c)^2 + y^2} = 2a + \sqrt{(x-c)^2 + y^2}$	Move radical to opposite side.
$(x+c)^2 + y^2 = \left(2a + \sqrt{(x-c)^2 + y^2}\right)^2$	Square both sides.
$x^2 + 2cx + c^2 + y^2 = 4a^2 + 4a\sqrt{(x-c)^2 + y^2} + (x-c)^2 + y^2$	Expand the squares.
$x^2 + 2cx + c^2 + y^2 = 4a^2 + 4a\sqrt{(x-c)^2 + y^2} + x^2 - 2cx + c^2 + y^2$	Expand remaining square.
$2cx = 4a^2 + 4a\sqrt{(x-c)^2 + y^2} - 2cx$	Combine like terms.
$4cx - 4a^2 = 4a\sqrt{(x-c)^2 + y^2}$	Isolate the radical.
$cx - a^2 = a\sqrt{(x-c)^2 + y^2}$	Divide by 4.
$\left(cx - a^2\right)^2 = a^2\left(\sqrt{(x-c)^2 + y^2}\right)^2$	Square both sides.
$c^2x^2 - 2a^2cx + a^4 = a^2\left(x^2 - 2cx + c^2 + y^2\right)$	Expand the squares.
$c^2x^2 - 2a^2cx + a^4 = a^2x^2 - 2a^2cx + a^2c^2 + a^2y^2$	Distribute a^2.
$a^4 + c^2x^2 = a^2x^2 + a^2c^2 + a^2y^2$	Combine like terms.
$c^2x^2 - a^2x^2 - a^2y^2 = a^2c^2 - a^4$	Rearrange terms.
$x^2\left(c^2 - a^2\right) - a^2y^2 = a^2\left(c^2 - a^2\right)$	Factor common terms.
$x^2b^2 - a^2y^2 = a^2b^2$	Set $b^2 = c^2 - a^2$.
$\frac{x^2b^2}{a^2b^2} - \frac{a^2y^2}{a^2b^2} = \frac{a^2b^2}{a^2b^2}$	Divide both sides by a^2b^2
$\frac{x^2}{a^2} - \frac{y^2}{b^2} = 1$	

This equation defines a hyperbola centered at the origin with vertices $(\pm a, 0)$ and co-vertices $(0 \pm b)$.

Standard Forms of the Equation of a Hyperbola with Center (0,0)

The standard form of the equation of a hyperbola with center $(0, 0)$ and transverse axis on the *x*-axis is

$$\frac{x^2}{a^2} - \frac{y^2}{b^2} = 1$$

where

- the length of the transverse axis is $2a$
- the coordinates of the vertices are $(\pm a, 0)$
- the length of the conjugate axis is $2b$
- the coordinates of the co-vertices are $(0, \pm b)$
- the distance between the foci is $2c$, where $c^2 = a^2 + b^2$
- the coordinates of the foci are $(\pm c, 0)$
- the equations of the asymptotes are $y = \pm\frac{b}{a}x$

See Figure 5**a**.

The standard form of the equation of a hyperbola with center $(0, 0)$ and transverse axis on the *y*-axis is

$$\frac{y^2}{a^2} - \frac{x^2}{b^2} = 1$$

where

- the length of the transverse axis is $2a$
- the coordinates of the vertices are $(0, \pm a)$
- the length of the conjugate axis is $2b$
- the coordinates of the co-vertices are $(\pm b, 0)$

- the distance between the foci is $2c$, where $c^2 = a^2 + b^2$
- the coordinates of the foci are $(0, \pm c)$
- the equations of the asymptotes are $y = \pm \frac{a}{b}x$

See Figure 5**b**.

Note that the vertices, co-vertices, and foci are related by the equation $c^2 = a^2 + b^2$. When we are given the equation of a hyperbola, we can use this relationship to identify its vertices and foci.

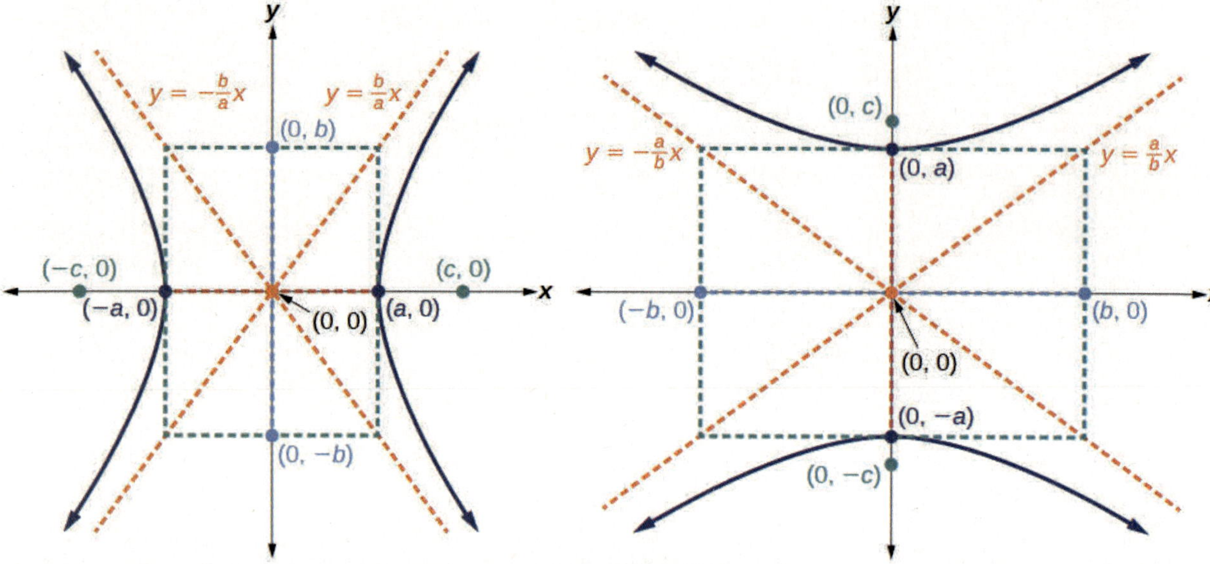

Figure 5 (a) Horizontal hyperbola with center $(0, 0)$ (b) Vertical hyperbola with center $(0, 0)$

HOW TO

Given the equation of a hyperbola in standard form, locate its vertices and foci.

1. Determine whether the transverse axis lies on the *x*- or *y*-axis. Notice that a^2 is always under the variable with the positive coefficient. So, if you set the other variable equal to zero, you can easily find the intercepts. In the case where the hyperbola is centered at the origin, the intercepts coincide with the vertices.
   a. If the equation has the form $\frac{x^2}{a^2} - \frac{y^2}{b^2} = 1$, then the transverse axis lies on the *x*-axis. The vertices are located at $(\pm a, 0)$, and the foci are located at $(\pm c, 0)$.
   b. If the equation has the form $\frac{y^2}{a^2} - \frac{x^2}{b^2} = 1$, then the transverse axis lies on the *y*-axis. The vertices are located at $(0, \pm a)$, and the foci are located at $(0, \pm c)$.
2. Solve for a using the equation $a = \sqrt{a^2}$.
3. Solve for c using the equation $c = \sqrt{a^2 + b^2}$.

EXAMPLE 1

Locating a Hyperbola's Vertices and Foci

Identify the vertices and foci of the hyperbola with equation $\frac{y^2}{49} - \frac{x^2}{32} = 1$.

Solution

The equation has the form $\frac{y^2}{a^2} - \frac{x^2}{b^2} = 1$, so the transverse axis lies on the *y*-axis. The hyperbola is centered at the origin, so the vertices serve as the y-intercepts of the graph. To find the vertices, set $x = 0$, and solve for y.

$$\begin{aligned} 1 &= \frac{y^2}{49} - \frac{x^2}{32} \\ 1 &= \frac{y^2}{49} - \frac{0^2}{32} \\ 1 &= \frac{y^2}{49} \\ y^2 &= 49 \\ y &= \pm\sqrt{49} = \pm 7 \end{aligned}$$

The foci are located at $(0, \pm c)$. Solving for c,

$$c = \sqrt{a^2 + b^2} = \sqrt{49 + 32} = \sqrt{81} = 9$$

Therefore, the vertices are located at $(0, \pm 7)$, and the foci are located at $(0, 9)$.

> **TRY IT** #1 Identify the vertices and foci of the hyperbola with equation $\frac{x^2}{9} - \frac{y^2}{25} = 1$.

Writing Equations of Hyperbolas in Standard Form

Just as with ellipses, writing the equation for a hyperbola in standard form allows us to calculate the key features: its center, vertices, co-vertices, foci, asymptotes, and the lengths and positions of the transverse and conjugate axes. Conversely, an equation for a hyperbola can be found given its key features. We begin by finding standard equations for hyperbolas centered at the origin. Then we will turn our attention to finding standard equations for hyperbolas centered at some point other than the origin.

Hyperbolas Centered at the Origin

Reviewing the standard forms given for hyperbolas centered at $(0, 0)$, we see that the vertices, co-vertices, and foci are related by the equation $c^2 = a^2 + b^2$. Note that this equation can also be rewritten as $b^2 = c^2 - a^2$. This relationship is used to write the equation for a hyperbola when given the coordinates of its foci and vertices.

HOW TO

Given the vertices and foci of a hyperbola centered at $(0, \ 0)$, write its equation in standard form.

1. Determine whether the transverse axis lies on the x- or y-axis.
   a. If the given coordinates of the vertices and foci have the form $(\pm a, 0)$ and $(\pm c, 0)$, respectively, then the transverse axis is the x-axis. Use the standard form $\frac{x^2}{a^2} - \frac{y^2}{b^2} = 1$.
   b. If the given coordinates of the vertices and foci have the form $(0, \pm a)$ and $(0, \pm c)$, respectively, then the transverse axis is the y-axis. Use the standard form $\frac{y^2}{a^2} - \frac{x^2}{b^2} = 1$.
2. Find b^2 using the equation $b^2 = c^2 - a^2$.
3. Substitute the values for a^2 and b^2 into the standard form of the equation determined in Step 1.

EXAMPLE 2

Finding the Equation of a Hyperbola Centered at (0,0) Given its Foci and Vertices

What is the standard form equation of the hyperbola that has vertices $(\pm 6, 0)$ and foci $\left(\pm 2\sqrt{10}, 0\right)$?

Solution

The vertices and foci are on the x-axis. Thus, the equation for the hyperbola will have the form $\frac{x^2}{a^2} - \frac{y^2}{b^2} = 1$.

The vertices are $(\pm 6, 0)$, so $a = 6$ and $a^2 = 36$.

The foci are $\left(\pm 2\sqrt{10}, 0\right)$, so $c = 2\sqrt{10}$ and $c^2 = 40$.

Solving for b^2, we have

$$\begin{aligned} b^2 &= c^2 - a^2 & \\ b^2 &= 40 - 36 & \text{Substitute for } c^2 \text{ and } a^2. \\ b^2 &= 4 & \text{Subtract.} \end{aligned}$$

Finally, we substitute $a^2 = 36$ and $b^2 = 4$ into the standard form of the equation, $\frac{x^2}{a^2} - \frac{y^2}{b^2} = 1$. The equation of the hyperbola is $\frac{x^2}{36} - \frac{y^2}{4} = 1$, as shown in Figure 6.

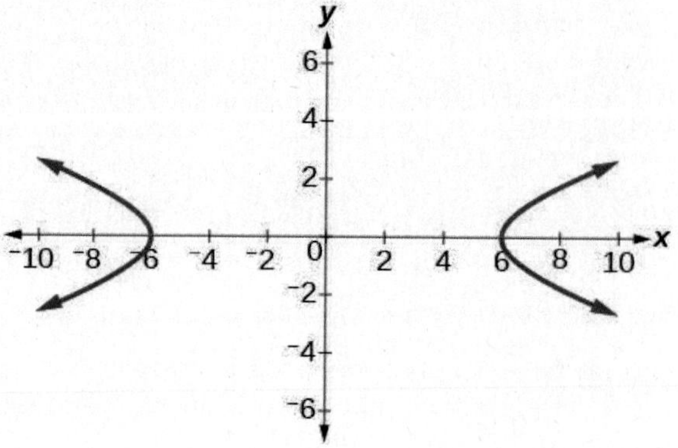

Figure 6

TRY IT #2 What is the standard form equation of the hyperbola that has vertices $(0, \pm 2)$ and foci $\left(0, \pm 2\sqrt{5}\right)$?

Hyperbolas Not Centered at the Origin

Like the graphs for other equations, the graph of a hyperbola can be translated. If a hyperbola is translated h units horizontally and k units vertically, the center of the hyperbola will be (h, k). This translation results in the standard form of the equation we saw previously, with x replaced by $(x - h)$ and y replaced by $(y - k)$.

Standard Forms of the Equation of a Hyperbola with Center (h, k)

The standard form of the equation of a hyperbola with center (h, k) and transverse axis parallel to the x-axis is

$$\frac{(x-h)^2}{a^2} - \frac{(y-k)^2}{b^2} = 1$$

where

- the length of the transverse axis is $2a$
- the coordinates of the vertices are $(h \pm a, k)$
- the length of the conjugate axis is $2b$
- the coordinates of the co-vertices are $(h, k \pm b)$
- the distance between the foci is $2c$, where $c^2 = a^2 + b^2$
- the coordinates of the foci are $(h \pm c, k)$

The asymptotes of the hyperbola coincide with the diagonals of the central rectangle. The length of the rectangle is $2a$ and its width is $2b$. The slopes of the diagonals are $\pm\frac{b}{a}$, and each diagonal passes through the center (h, k). Using the **point-slope formula**, it is simple to show that the equations of the asymptotes are $y = \pm\frac{b}{a}(x - h) + k$. See Figure 7**a**

The standard form of the equation of a hyperbola with center (h, k) and transverse axis parallel to the y-axis is

$$\frac{(y-k)^2}{a^2} - \frac{(x-h)^2}{b^2} = 1$$

where

- the length of the transverse axis is $2a$
- the coordinates of the vertices are $(h, k \pm a)$
- the length of the conjugate axis is $2b$
- the coordinates of the co-vertices are $(h \pm b, k)$
- the distance between the foci is $2c$, where $c^2 = a^2 + b^2$
- the coordinates of the foci are $(h, k \pm c)$

Using the reasoning above, the equations of the asymptotes are $y = \pm\frac{a}{b}(x-h) + k$. See Figure 7**b**.

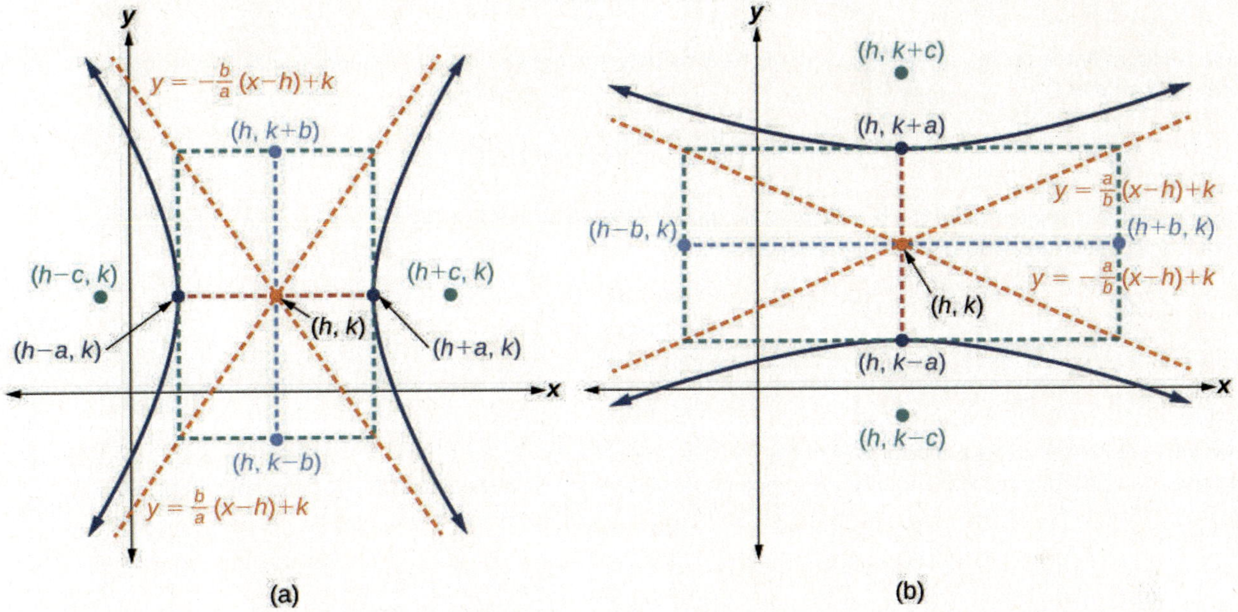

Figure 7 (a) Horizontal hyperbola with center (h, k) (b) Vertical hyperbola with center (h, k)

Like hyperbolas centered at the origin, hyperbolas centered at a point (h, k) have vertices, co-vertices, and foci that are related by the equation $c^2 = a^2 + b^2$. We can use this relationship along with the midpoint and distance formulas to find the standard equation of a hyperbola when the vertices and foci are given.

HOW TO

Given the vertices and foci of a hyperbola centered at (h, k), write its equation in standard form.

1. Determine whether the transverse axis is parallel to the x- or y-axis.
   a. If the y-coordinates of the given vertices and foci are the same, then the transverse axis is parallel to the x-axis. Use the standard form $\frac{(x-h)^2}{a^2} - \frac{(y-k)^2}{b^2} = 1$.
   b. If the x-coordinates of the given vertices and foci are the same, then the transverse axis is parallel to the y-axis. Use the standard form $\frac{(y-k)^2}{a^2} - \frac{(x-h)^2}{b^2} = 1$.
2. Identify the center of the hyperbola, (h, k), using the midpoint formula and the given coordinates for the vertices.
3. Find a^2 by solving for the length of the transverse axis, $2a$, which is the distance between the given vertices.

4. Find c^2 using h and k found in Step 2 along with the given coordinates for the foci.
5. Solve for b^2 using the equation $b^2 = c^2 - a^2$.
6. Substitute the values for h, k, a^2, and b^2 into the standard form of the equation determined in Step 1.

EXAMPLE 3

Finding the Equation of a Hyperbola Centered at (h, k) Given its Foci and Vertices

What is the standard form equation of the hyperbola that has vertices at $(0, -2)$ and $(6, -2)$ and foci at $(-2, -2)$ and $(8, -2)$?

Solution

The *y*-coordinates of the vertices and foci are the same, so the transverse axis is parallel to the *x*-axis. Thus, the equation of the hyperbola will have the form

$$\frac{(x-h)^2}{a^2} - \frac{(y-k)^2}{b^2} = 1$$

First, we identify the center, (h, k). The center is halfway between the vertices $(0, -2)$ and $(6, -2)$. Applying the midpoint formula, we have

$$(h, k) = \left(\frac{0+6}{2}, \frac{-2+(-2)}{2}\right) = (3, -2)$$

Next, we find a^2. The length of the transverse axis, $2a$, is bounded by the vertices. So, we can find a^2 by finding the distance between the *x*-coordinates of the vertices.

$$\begin{aligned} 2a &= |0-6| \\ 2a &= 6 \\ a &= 3 \\ a^2 &= 9 \end{aligned}$$

Now we need to find c^2. The coordinates of the foci are $(h \pm c, k)$. So $(h-c, k) = (-2, -2)$ and $(h+c, k) = (8, -2)$. We can use the *x*-coordinate from either of these points to solve for c. Using the point $(8, -2)$, and substituting $h = 3$,

$$\begin{aligned} h+c &= 8 \\ 3+c &= 8 \\ c &= 5 \\ c^2 &= 25 \end{aligned}$$

Next, solve for b^2 using the equation $b^2 = c^2 - a^2$:

$$\begin{aligned} b^2 &= c^2 - a^2 \\ &= 25 - 9 \\ &= 16 \end{aligned}$$

Finally, substitute the values found for h, k, a^2, and b^2 into the standard form of the equation.

$$\frac{(x-3)^2}{9} - \frac{(y+2)^2}{16} = 1$$

> **TRY IT** #3 What is the standard form equation of the hyperbola that has vertices $(1, -2)$ and $(1, 8)$ and foci $(1, -10)$ and $(1, 16)$?

Graphing Hyperbolas Centered at the Origin

When we have an equation in standard form for a hyperbola centered at the origin, we can interpret its parts to identify the key features of its graph: the center, vertices, co-vertices, asymptotes, foci, and lengths and positions of the transverse and conjugate axes. To graph hyperbolas centered at the origin, we use the standard form $\frac{x^2}{a^2} - \frac{y^2}{b^2} = 1$ for horizontal hyperbolas and the standard form $\frac{y^2}{a^2} - \frac{x^2}{b^2} = 1$ for vertical hyperbolas.

HOW TO

Given a standard form equation for a hyperbola centered at $(0, 0)$, sketch the graph.

1. Determine which of the standard forms applies to the given equation.
2. Use the standard form identified in Step 1 to determine the position of the transverse axis; coordinates for the vertices, co-vertices, and foci; and the equations for the asymptotes.
   a. If the equation is in the form $\frac{x^2}{a^2} - \frac{y^2}{b^2} = 1$, then
      - the transverse axis is on the *x*-axis
      - the coordinates of the vertices are $(\pm a, 0)$
      - the coordinates of the co-vertices are $(0, \pm b)$
      - the coordinates of the foci are $(\pm c, 0)$
      - the equations of the asymptotes are $y = \pm \frac{b}{a} x$

   b. If the equation is in the form $\frac{y^2}{a^2} - \frac{x^2}{b^2} = 1$, then
      - the transverse axis is on the *y*-axis
      - the coordinates of the vertices are $(0, \pm a)$
      - the coordinates of the co-vertices are $(\pm b, 0)$
      - the coordinates of the foci are $(0, \pm c)$
      - the equations of the asymptotes are $y = \pm \frac{a}{b} x$
3. Solve for the coordinates of the foci using the equation $c = \pm\sqrt{a^2 + b^2}$.
4. Plot the vertices, co-vertices, foci, and asymptotes in the coordinate plane, and draw a smooth curve to form the hyperbola.

EXAMPLE 4

Graphing a Hyperbola Centered at (0, 0) Given an Equation in Standard Form

Graph the hyperbola given by the equation $\frac{y^2}{64} - \frac{x^2}{36} = 1$. Identify and label the vertices, co-vertices, foci, and asymptotes.

Solution

The standard form that applies to the given equation is $\frac{y^2}{a^2} - \frac{x^2}{b^2} = 1$. Thus, the transverse axis is on the *y*-axis

The coordinates of the vertices are $(0, \pm a) = \left(0, \pm\sqrt{64}\right) = (0, \pm 8)$

The coordinates of the co-vertices are $(\pm b, 0) = \left(\pm\sqrt{36},\ 0\right) = (\pm 6, 0)$

The coordinates of the foci are $(0, \pm c)$, where $c = \pm\sqrt{a^2 + b^2}$. Solving for c, we have

$$c = \pm\sqrt{a^2 + b^2} = \pm\sqrt{64 + 36} = \pm\sqrt{100} = \pm 10$$

Therefore, the coordinates of the foci are $(0, \pm 10)$

The equations of the asymptotes are $y = \pm\frac{a}{b}x = \pm\frac{8}{6}x = \pm\frac{4}{3}x$

Plot and label the vertices and co-vertices, and then sketch the central rectangle. Sides of the rectangle are parallel to the axes and pass through the vertices and co-vertices. Sketch and extend the diagonals of the central rectangle to show the asymptotes. The central rectangle and asymptotes provide the framework needed to sketch an accurate graph of the hyperbola. Label the foci and asymptotes, and draw a smooth curve to form the hyperbola, as shown in Figure 8.

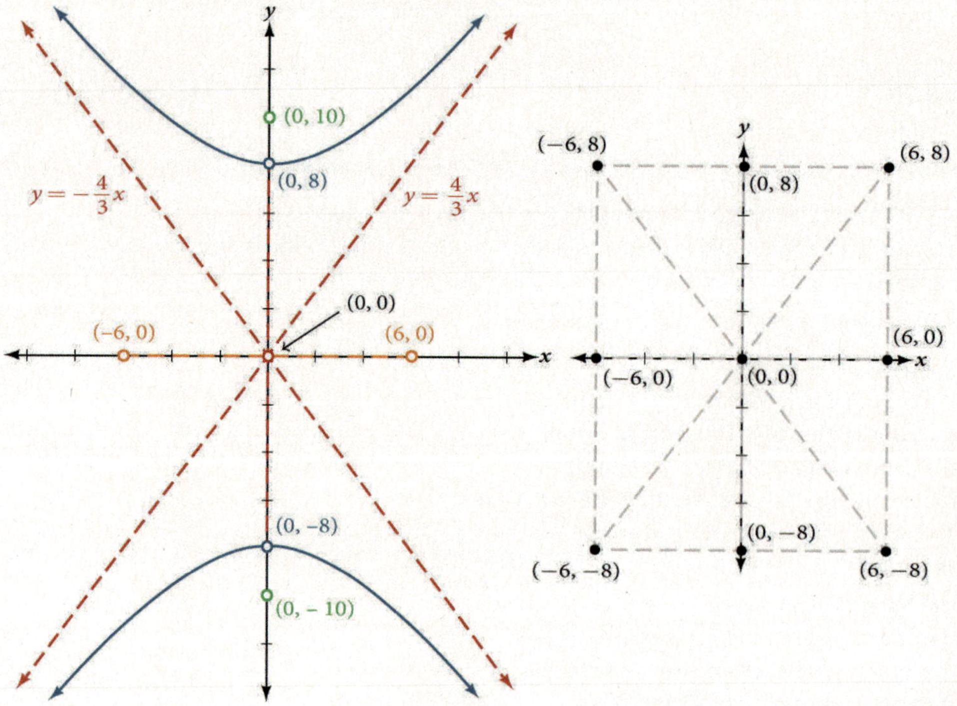

Figure 8

> TRY IT #4 Graph the hyperbola given by the equation $\frac{x^2}{144} - \frac{y^2}{81} = 1$. Identify and label the vertices, co-vertices, foci, and asymptotes.

Graphing Hyperbolas Not Centered at the Origin

Graphing hyperbolas centered at a point (h, k) other than the origin is similar to graphing ellipses centered at a point other than the origin. We use the standard forms $\frac{(x-h)^2}{a^2} - \frac{(y-k)^2}{b^2} = 1$ for horizontal hyperbolas, and $\frac{(y-k)^2}{a^2} - \frac{(x-h)^2}{b^2} = 1$ for vertical hyperbolas. From these standard form equations we can easily calculate and plot key features of the graph: the coordinates of its center, vertices, co-vertices, and foci; the equations of its asymptotes; and the positions of the transverse and conjugate axes.

HOW TO

Given a general form for a hyperbola centered at (h, k), sketch the graph.

1. Convert the general form to that standard form. Determine which of the standard forms applies to the given equation.
2. Use the standard form identified in Step 1 to determine the position of the transverse axis; coordinates for the center, vertices, co-vertices, foci; and equations for the asymptotes.
   a. If the equation is in the form $\frac{(x-h)^2}{a^2} - \frac{(y-k)^2}{b^2} = 1$, then
      - the transverse axis is parallel to the x-axis
      - the center is (h, k)
      - the coordinates of the vertices are $(h \pm a, k)$
      - the coordinates of the co-vertices are $(h, k \pm b)$
      - the coordinates of the foci are $(h \pm c, k)$
      - the equations of the asymptotes are $y = \pm\frac{b}{a}(x - h) + k$

b. If the equation is in the form $\frac{(y-k)^2}{a^2} - \frac{(x-h)^2}{b^2} = 1$, then
   - the transverse axis is parallel to the *y*-axis
   - the center is (h, k)
   - the coordinates of the vertices are $(h, k \pm a)$
   - the coordinates of the co-vertices are $(h \pm b, k)$
   - the coordinates of the foci are $(h, k \pm c)$
   - the equations of the asymptotes are $y = \pm\frac{a}{b}(x - h) + k$

3. Solve for the coordinates of the foci using the equation $c = \pm\sqrt{a^2 + b^2}$.
4. Plot the center, vertices, co-vertices, foci, and asymptotes in the coordinate plane and draw a smooth curve to form the hyperbola.

EXAMPLE 5

Graphing a Hyperbola Centered at (*h*, *k*) Given an Equation in General Form

Graph the hyperbola given by the equation $9x^2 - 4y^2 - 36x - 40y - 388 = 0$. Identify and label the center, vertices, co-vertices, foci, and asymptotes.

Solution

Start by expressing the equation in standard form. Group terms that contain the same variable, and move the constant to the opposite side of the equation.

$$\left(9x^2 - 36x\right) - \left(4y^2 + 40y\right) = 388$$

Factor the leading coefficient of each expression.

$$9\left(x^2 - 4x\right) - 4\left(y^2 + 10y\right) = 388$$

Complete the square twice. Remember to balance the equation by adding the same constants to each side.

$$9\left(x^2 - 4x + 4\right) - 4\left(y^2 + 10y + 25\right) = 388 + 36 - 100$$

Rewrite as perfect squares.

$$9(x-2)^2 - 4(y+5)^2 = 324$$

Divide both sides by the constant term to place the equation in standard form.

$$\frac{(x-2)^2}{36} - \frac{(y+5)^2}{81} = 1$$

The standard form that applies to the given equation is $\frac{(x-h)^2}{a^2} - \frac{(y-k)^2}{b^2} = 1$, where $a^2 = 36$ and $b^2 = 81$, or $a = 6$ and $b = 9$. Thus, the transverse axis is parallel to the *x*-axis. It follows that:

- the center of the ellipse is $(h, k) = (2, -5)$
- the coordinates of the vertices are $(h \pm a, k) = (2 \pm 6, -5)$, or $(-4, -5)$ and $(8, -5)$
- the coordinates of the co-vertices are $(h, k \pm b) = (2, -5 \pm 9)$, or $(2, -14)$ and $(2, 4)$
- the coordinates of the foci are $(h \pm c, k)$, where $c = \pm\sqrt{a^2 + b^2}$. Solving for *c*, we have

$$c = \pm\sqrt{36 + 81} = \pm\sqrt{117} = \pm 3\sqrt{13}$$

Therefore, the coordinates of the foci are $\left(2 - 3\sqrt{13}, -5\right)$ and $\left(2 + 3\sqrt{13}, -5\right)$.

The equations of the asymptotes are $y = \pm\frac{b}{a}(x - h) + k = \pm\frac{3}{2}(x - 2) - 5$.

Next, we plot and label the center, vertices, co-vertices, foci, and asymptotes and draw smooth curves to form the hyperbola, as shown in Figure 9.

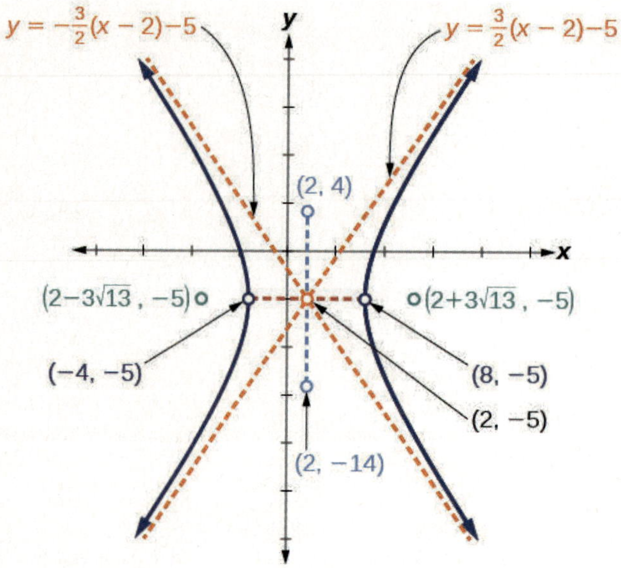

Figure 9

TRY IT #5 Graph the hyperbola given by the standard form of an equation $\frac{(y+4)^2}{100} - \frac{(x-3)^2}{64} = 1$. Identify and label the center, vertices, co-vertices, foci, and asymptotes.

Solving Applied Problems Involving Hyperbolas

As we discussed at the beginning of this section, hyperbolas have real-world applications in many fields, such as astronomy, physics, engineering, and architecture. The design efficiency of hyperbolic cooling towers is particularly interesting. Cooling towers are used to transfer waste heat to the atmosphere and are often touted for their ability to generate power efficiently. Because of their hyperbolic form, these structures are able to withstand extreme winds while requiring less material than any other forms of their size and strength. See Figure 10. For example, a 500-foot tower can be made of a reinforced concrete shell only 6 or 8 inches wide!

Figure 10 Cooling towers at the Drax power station in North Yorkshire, United Kingdom (credit: Les Haines, Flickr)

The first hyperbolic towers were designed in 1914 and were 35 meters high. Today, the tallest cooling towers are in France, standing a remarkable 170 meters tall. In Example 6 we will use the design layout of a cooling tower to find a hyperbolic equation that models its sides.

EXAMPLE 6

Solving Applied Problems Involving Hyperbolas

The design layout of a cooling tower is shown in Figure 11. The tower stands 179.6 meters tall. The diameter of the top is 72 meters. At their closest, the sides of the tower are 60 meters apart.

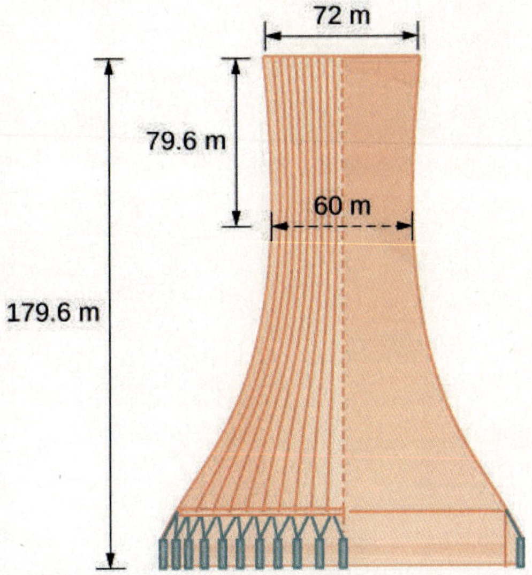

Figure 11 Project design for a natural draft cooling tower

Find the equation of the hyperbola that models the sides of the cooling tower. Assume that the center of the hyperbola—indicated by the intersection of dashed perpendicular lines in the figure—is the origin of the coordinate plane. Round final values to four decimal places.

Solution

We are assuming the center of the tower is at the origin, so we can use the standard form of a horizontal hyperbola centered at the origin: $\frac{x^2}{a^2} - \frac{y^2}{b^2} = 1$, where the branches of the hyperbola form the sides of the cooling tower. We must find the values of a^2 and b^2 to complete the model.

First, we find a^2. Recall that the length of the transverse axis of a hyperbola is $2a$. This length is represented by the distance where the sides are closest, which is given as 60 meters. So, $2a = 60$. Therefore, $a = 30$ and $a^2 = 900$.

To solve for b^2, we need to substitute for x and y in our equation using a known point. To do this, we can use the dimensions of the tower to find some point (x, y) that lies on the hyperbola. We will use the top right corner of the tower to represent that point. Since the *y*-axis bisects the tower, our *x*-value can be represented by the radius of the top, or 36 meters. The *y*-value is represented by the distance from the origin to the top, which is given as 79.6 meters. Therefore,

$$\frac{x^2}{a^2} - \frac{y^2}{b^2} = 1 \quad \text{Standard form of horizontal hyperbola.}$$

$$b^2 = \frac{y^2}{\frac{x^2}{a^2} - 1} \quad \text{Isolate } b^2$$

$$= \frac{(79.6)^2}{\frac{(36)^2}{900} - 1} \quad \text{Substitute for } a^2, x, \text{ and } y$$

$$\approx 14400.3636 \quad \text{Round to four decimal places}$$

The sides of the tower can be modeled by the hyperbolic equation

$$\frac{x^2}{900} - \frac{y^2}{14400.3636} = 1, \text{or } \frac{x^2}{30^2} - \frac{y^2}{120.0015^2} = 1$$

TRY IT #6 A design for a cooling tower project is shown in Figure 12. Find the equation of the hyperbola that models the sides of the cooling tower. Assume that the center of the hyperbola—indicated by the intersection of dashed perpendicular lines in the figure—is the origin of the coordinate plane. Round final values to four decimal places.

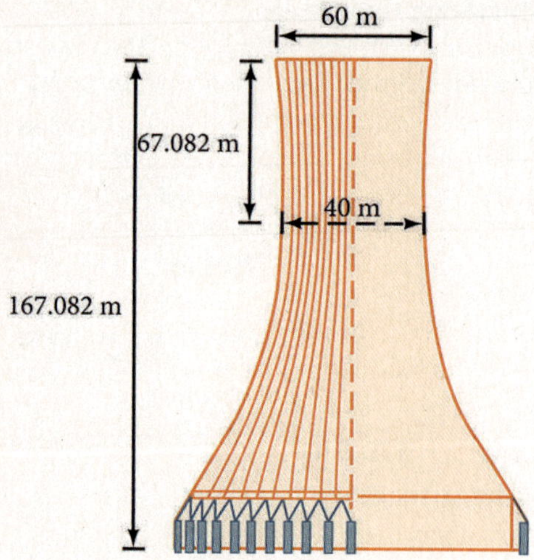

Figure 12

MEDIA

Access these online resources for additional instruction and practice with hyperbolas.

Conic Sections: The Hyperbola Part 1 of 2 (http://openstax.org/l/hyperbola1)
Conic Sections: The Hyperbola Part 2 of 2 (http://openstax.org/l/hyperbola2)
Graph a Hyperbola with Center at Origin (http://openstax.org/l/hyperbolaorigin)
Graph a Hyperbola with Center not at Origin (http://openstax.org/l/hbnotorigin)

12.2 SECTION EXERCISES

Verbal

1. Define a hyperbola in terms of its foci.

2. What can we conclude about a hyperbola if its asymptotes intersect at the origin?

3. What must be true of the foci of a hyperbola?

4. If the transverse axis of a hyperbola is vertical, what do we know about the graph?

5. Where must the center of hyperbola be relative to its foci?

Algebraic

For the following exercises, determine whether the following equations represent hyperbolas. If so, write in standard form.

6. $3y^2 + 2x = 6$

7. $\frac{x^2}{36} - \frac{y^2}{9} = 1$

8. $5y^2 + 4x^2 = 6x$

9. $25x^2 - 16y^2 = 400$

10. $-9x^2 + 18x + y^2 + 4y - 14 = 0$

For the following exercises, write the equation for the hyperbola in standard form if it is not already, and identify the vertices and foci, and write equations of asymptotes.

11. $\frac{x^2}{25} - \frac{y^2}{36} = 1$

12. $\frac{x^2}{100} - \frac{y^2}{9} = 1$

13. $\frac{y^2}{4} - \frac{x^2}{81} = 1$

14. $9y^2 - 4x^2 = 1$

15. $\frac{(x-1)^2}{9} - \frac{(y-2)^2}{16} = 1$

16. $\frac{(y-6)^2}{36} - \frac{(x+1)^2}{16} = 1$

17. $\frac{(x-2)^2}{49} - \frac{(y+7)^2}{49} = 1$

18. $4x^2 - 8x - 9y^2 - 72y + 112 = 0$

19. $-9x^2 - 54x + 9y^2 - 54y + 81 = 0$

20. $4x^2 - 24x - 36y^2 - 360y + 864 = 0$

21. $-4x^2 + 24x + 16y^2 - 128y + 156 = 0$

22. $-4x^2 + 40x + 25y^2 - 100y + 100 = 0$

23. $x^2 + 2x - 100y^2 - 1000y + 2401 = 0$

24. $-9x^2 + 72x + 16y^2 + 16y + 4 = 0$

25. $4x^2 + 24x - 25y^2 + 200y - 464 = 0$

For the following exercises, find the equations of the asymptotes for each hyperbola.

26. $\frac{y^2}{3^2} - \frac{x^2}{3^2} = 1$

27. $\frac{(x-3)^2}{5^2} - \frac{(y+4)^2}{2^2} = 1$

28. $\frac{(y-3)^2}{3^2} - \frac{(x+5)^2}{6^2} = 1$

29. $9x^2 - 18x - 16y^2 + 32y - 151 = 0$

30. $16y^2 + 96y - 4x^2 + 16x + 112 = 0$

Graphical

For the following exercises, sketch a graph of the hyperbola, labeling vertices and foci.

31. $\frac{x^2}{49} - \frac{y^2}{16} = 1$

32. $\frac{x^2}{64} - \frac{y^2}{4} = 1$

33. $\frac{y^2}{9} - \frac{x^2}{25} = 1$

34. $81x^2 - 9y^2 = 1$

35. $\frac{(y+5)^2}{9} - \frac{(x-4)^2}{25} = 1$

36. $\frac{(x-2)^2}{8} - \frac{(y+3)^2}{27} = 1$

37. $\frac{(y-3)^2}{9} - \frac{(x-3)^2}{9} = 1$

38. $-4x^2 - 8x + 16y^2 - 32y - 52 = 0$

39. $x^2 - 8x - 25y^2 - 100y - 109 = 0$

40. $-x^2 + 8x + 4y^2 - 40y + 88 = 0$

41. $64x^2 + 128x - 9y^2 - 72y - 656 = 0$

42. $16x^2 + 64x - 4y^2 - 8y - 4 = 0$

43. $-100x^2 + 1000x + y^2 - 10y - 2575 = 0$

44. $4x^2 + 16x - 4y^2 + 16y + 16 = 0$

For the following exercises, given information about the graph of the hyperbola, find its equation.

45. Vertices at $(3, 0)$ and $(-3, 0)$ and one focus at $(5, 0)$.

46. Vertices at $(0, 6)$ and $(0, -6)$ and one focus at $(0, -8)$.

47. Vertices at $(1, 1)$ and $(11, 1)$ and one focus at $(12, 1)$.

48. Center: $(0, 0)$; vertex: $(0, -13)$; one focus: $\left(0, \sqrt{313}\right)$.

49. Center: $(4, 2)$; vertex: $(9, 2)$; one focus: $\left(4 + \sqrt{26}, 2\right)$.

50. Center: $(3, 5)$; vertex: $(3, 11)$; one focus: $\left(3, 5 + 2\sqrt{10}\right)$.

For the following exercises, given the graph of the hyperbola, find its equation.

51.

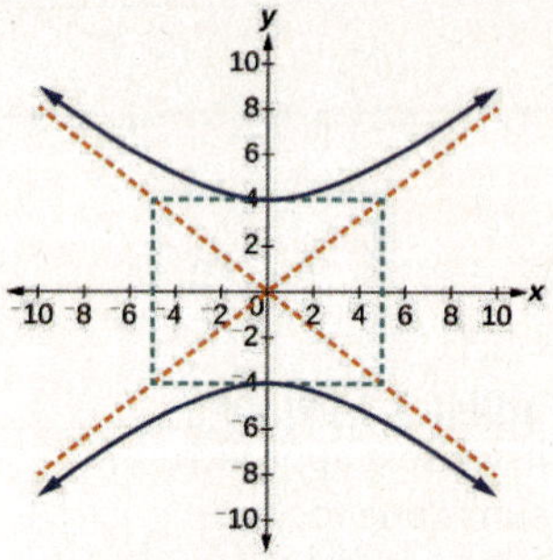

52.

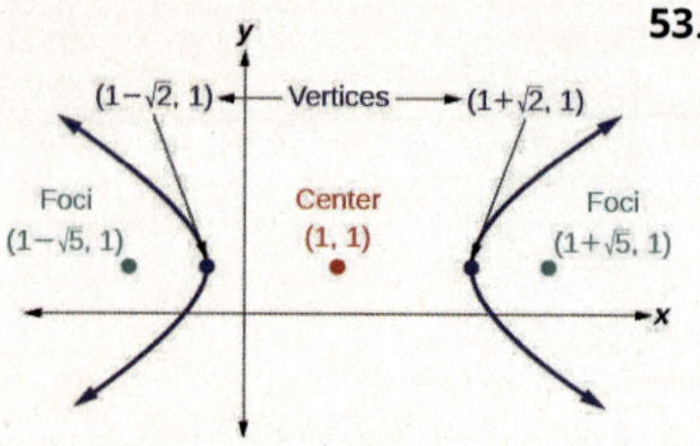

53.

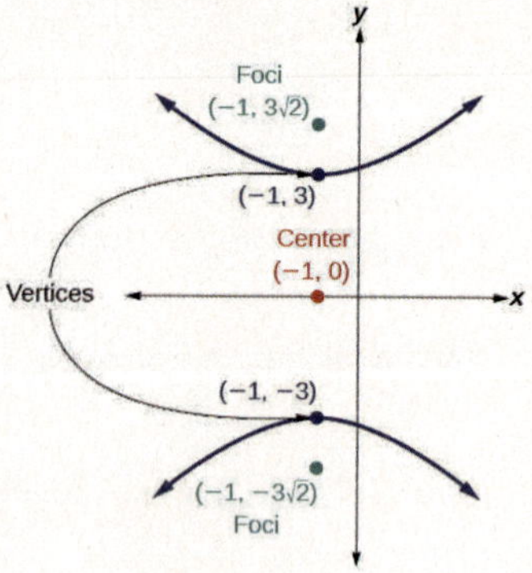

54.

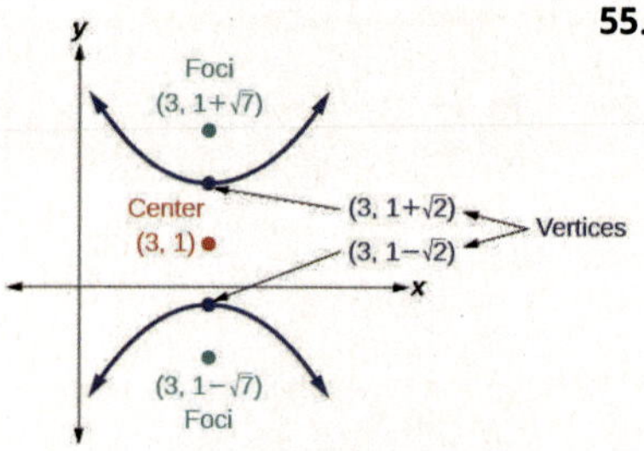

55.

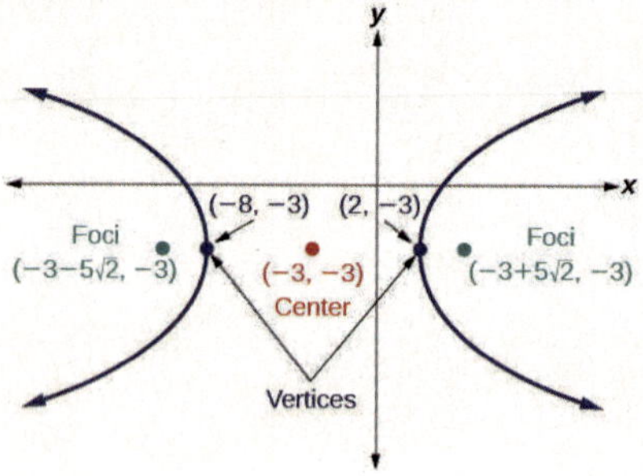

Extensions

For the following exercises, express the equation for the hyperbola as two functions, with y as a function of x. Express as simply as possible. Use a graphing calculator to sketch the graph of the two functions on the same axes.

56. $\frac{x^2}{4} - \frac{y^2}{9} = 1$

57. $\frac{y^2}{9} - \frac{x^2}{1} = 1$

58. $\frac{(x-2)^2}{16} - \frac{(y+3)^2}{25} = 1$

59. $-4x^2 - 16x + y^2 - 2y - 19 = 0$

60. $4x^2 - 24x - y^2 - 4y + 16 = 0$

Real-World Applications

For the following exercises, a hedge is to be constructed in the shape of a hyperbola near a fountain at the center of the yard. Find the equation of the hyperbola and sketch the graph.

61. The hedge will follow the asymptotes $y = x$ and $y = -x$, and its closest distance to the center fountain is 5 yards.

62. The hedge will follow the asymptotes $y = 2x$ and $y = -2x$, and its closest distance to the center fountain is 6 yards.

63. The hedge will follow the asymptotes $y = \frac{1}{2}x$ and $y = -\frac{1}{2}x$, and its closest distance to the center fountain is 10 yards.

64. The hedge will follow the asymptotes $y = \frac{2}{3}x$ and $y = -\frac{2}{3}x$, and its closest distance to the center fountain is 12 yards.

65. The hedge will follow the asymptotes $y = \frac{3}{4}x$ and $y = -\frac{3}{4}x$, and its closest distance to the center fountain is 20 yards.

For the following exercises, assume an object enters our solar system and we want to graph its path on a coordinate system with the sun at the origin and the x-axis as the axis of symmetry for the object's path. Give the equation of the flight path of each object using the given information.

66. The object enters along a path approximated by the line $y = x - 2$ and passes within 1 au (astronomical unit) of the sun at its closest approach, so that the sun is one focus of the hyperbola. It then departs the solar system along a path approximated by the line $y = -x + 2$.

67. The object enters along a path approximated by the line $y = 2x - 2$ and passes within 0.5 au of the sun at its closest approach, so the sun is one focus of the hyperbola. It then departs the solar system along a path approximated by the line $y = -2x + 2$.

68. The object enters along a path approximated by the line $y = 0.5x + 2$ and passes within 1 au of the sun at its closest approach, so the sun is one focus of the hyperbola. It then departs the solar system along a path approximated by the line $y = -0.5x - 2$.

69. The object enters along a path approximated by the line $y = \frac{1}{3}x - 1$ and passes within 1 au of the sun at its closest approach, so the sun is one focus of the hyperbola. It then departs the solar system along a path approximated by the line $y = -\frac{1}{3}x + 1$.

70. The object enters along a path approximated by the line $y = 3x - 9$ and passes within 1 au of the sun at its closest approach, so the sun is one focus of the hyperbola. It then departs the solar system along a path approximated by the line $y = -3x + 9$.

12.3 The Parabola

Learning Objectives

In this section, you will:

- Graph parabolas with vertices at the origin.
- Write equations of parabolas in standard form.
- Graph parabolas with vertices not at the origin.
- Solve applied problems involving parabolas.

Figure 1 Katherine Johnson's pioneering mathematical work in the area of parabolic and other orbital calculations played a significant role in the development of U.S space flight. (credit: NASA)

Katherine Johnson is the pioneering NASA mathematician who was integral to the successful and safe flight and return of many human missions as well as satellites. Prior to the work featured in the movie *Hidden Figures*, she had already made major contributions to the space program. She provided trajectory analysis for the Mercury mission, in which Alan Shepard became the first American to reach space, and she and engineer Ted Sopinski authored a monumental paper regarding placing an object in a precise orbital position and having it return safely to Earth. Many of the orbits she determined were made up of parabolas, and her ability to combine different types of math enabled an unprecedented level of precision. Johnson said, "You tell me when you want it and where you want it to land, and I'll do it backwards and tell you when to take off."

Johnson's work on parabolic orbits and other complex mathematics resulted in successful orbits, Moon landings, and the development of the Space Shuttle program. Applications of parabolas are also critical to other areas of science. Parabolic mirrors (or reflectors) are able to capture energy and focus it to a single point. The advantages of this property are evidenced by the vast list of parabolic objects we use every day: satellite dishes, suspension bridges, telescopes, microphones, spotlights, and car headlights, to name a few. Parabolic reflectors are also used in alternative energy devices, such as solar cookers and water heaters, because they are inexpensive to manufacture and need little maintenance. In this section we will explore the parabola and its uses, including low-cost, energy-efficient solar designs.

Graphing Parabolas with Vertices at the Origin

In The Ellipse, we saw that an ellipse is formed when a plane cuts through a right circular cone. If the plane is parallel to the edge of the cone, an unbounded curve is formed. This curve is a **parabola**. See Figure 2.

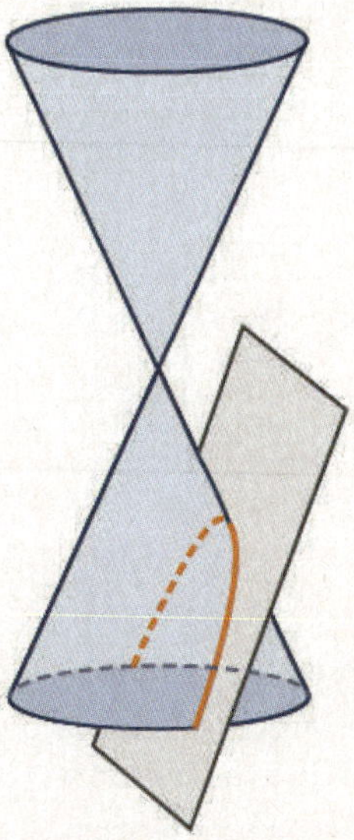

Figure 2 Parabola

Like the ellipse and hyperbola, the parabola can also be defined by a set of points in the coordinate plane. A parabola is the set of all points (x, y) in a plane that are the same distance from a fixed line, called the **directrix**, and a fixed point (the **focus**) not on the directrix.

In Quadratic Functions (http://openstax.org/books/precalculus-2e/pages/3-3-power-functions-and-polynomial-functions), we learned about a parabola's vertex and axis of symmetry. Now we extend the discussion to include other key features of the parabola. See Figure 3. Notice that the axis of symmetry passes through the focus and vertex and is perpendicular to the directrix. The vertex is the midpoint between the directrix and the focus.

The line segment that passes through the focus and is parallel to the directrix is called the **latus rectum**. The endpoints of the latus rectum lie on the curve. By definition, the distance d from the focus to any point P on the parabola is equal to the distance from P to the directrix.

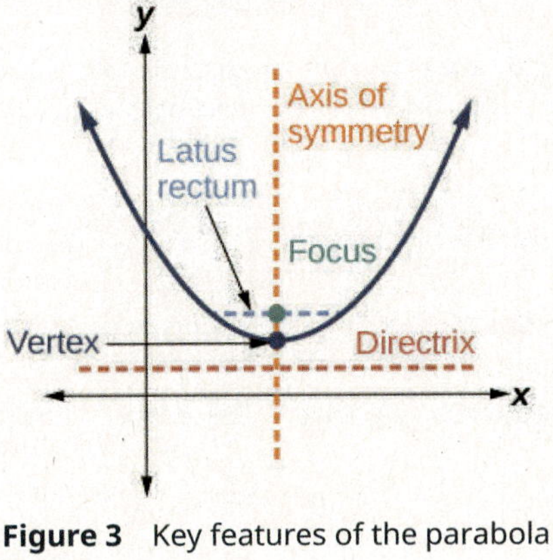

Figure 3 Key features of the parabola

To work with parabolas in the coordinate plane, we consider two cases: those with a vertex at the origin and those with a vertex at a point other than the origin. We begin with the former.

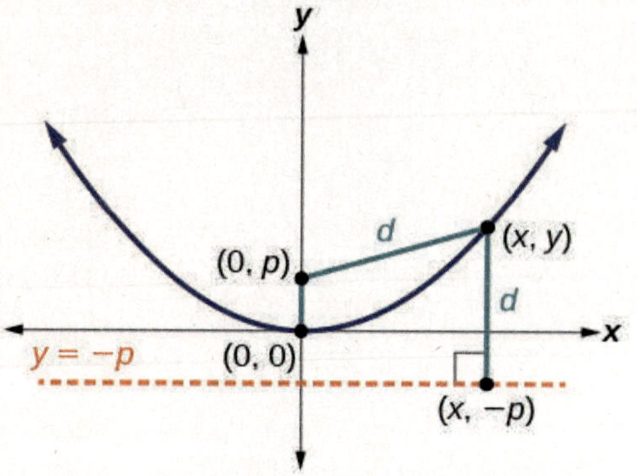

Figure 4

Let (x, y) be a point on the parabola with vertex $(0, 0)$, focus $(0, p)$, and directrix $y = -p$ as shown in Figure 4. The distance d from point (x, y) to point $(x, -p)$ on the directrix is the difference of the *y*-values: $d = y + p$. The distance from the focus $(0, p)$ to the point (x, y) is also equal to d and can be expressed using the distance formula.

$$\begin{aligned} d &= \sqrt{(x-0)^2 + (y-p)^2} \\ &= \sqrt{x^2 + (y-p)^2} \end{aligned}$$

Set the two expressions for d equal to each other and solve for y to derive the equation of the parabola. We do this because the distance from (x, y) to $(0, p)$ equals the distance from (x, y) to $(x, -p)$.

$$\sqrt{x^2 + (y-p)^2} = y + p$$

We then square both sides of the equation, expand the squared terms, and simplify by combining like terms.

$$\begin{aligned} x^2 + (y-p)^2 &= (y+p)^2 \\ x^2 + y^2 - 2py + p^2 &= y^2 + 2py + p^2 \\ x^2 - 2py &= 2py \\ x^2 &= 4py \end{aligned}$$

The equations of parabolas with vertex $(0, 0)$ are $y^2 = 4px$ when the *x*-axis is the axis of symmetry and $x^2 = 4py$ when the *y*-axis is the axis of symmetry. These standard forms are given below, along with their general graphs and key features.

Standard Forms of Parabolas with Vertex (0, 0)

Table 1 and Figure 5 summarize the standard features of parabolas with a vertex at the origin.

Axis of Symmetry	Equation	Focus	Directrix	Endpoints of Latus Rectum
x-axis	$y^2 = 4px$	$(p,\ 0)$	$x = -p$	$(p,\ \pm 2p)$
y-axis	$x^2 = 4py$	$(0,\ p)$	$y = -p$	$(\pm 2p,\ p)$

Table 1

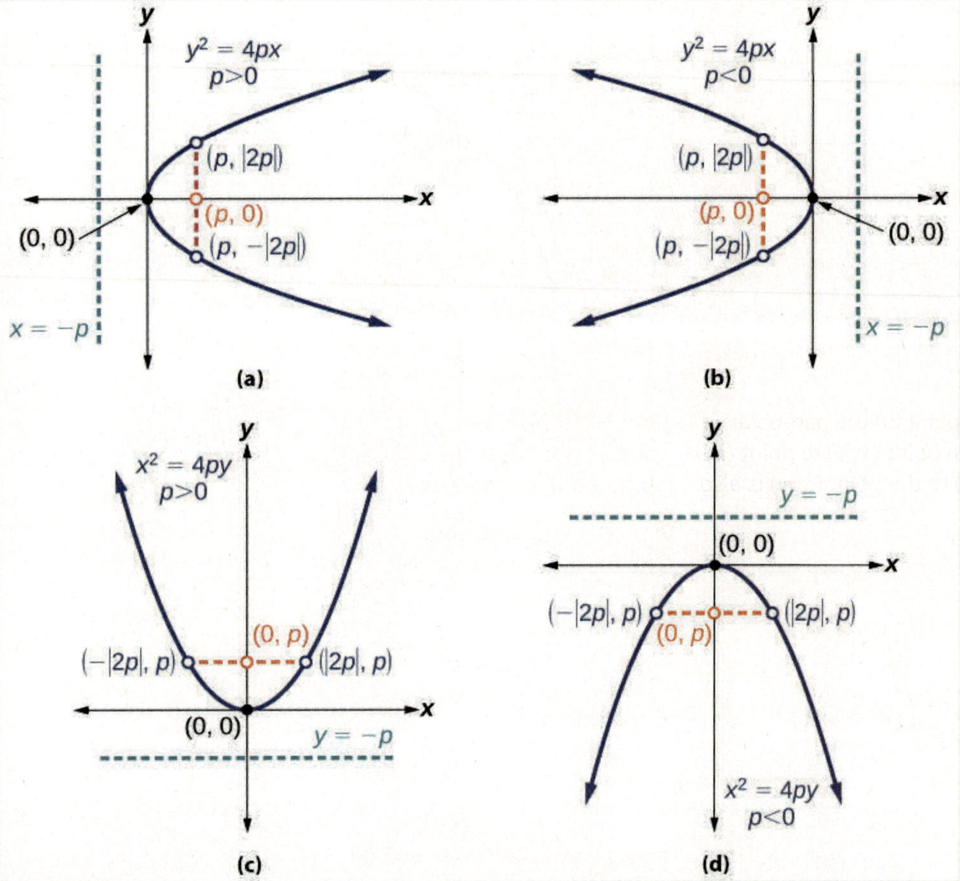

Figure 5 (a) When $p > 0$ and the axis of symmetry is the *x*-axis, the parabola opens right. (b) When $p < 0$ and the axis of symmetry is the *x*-axis, the parabola opens left. (c) When $p > 0$ and the axis of symmetry is the *y*-axis, the parabola opens up. (d) When $p < 0$ and the axis of symmetry is the *y*-axis, the parabola opens down.

The key features of a parabola are its vertex, axis of symmetry, focus, directrix, and latus rectum. See Figure 5. When given a standard equation for a parabola centered at the origin, we can easily identify the key features to graph the parabola.

A line is said to be tangent to a curve if it intersects the curve at exactly one point. If we sketch lines tangent to the parabola at the endpoints of the latus rectum, these lines intersect on the axis of symmetry, as shown in Figure 6.

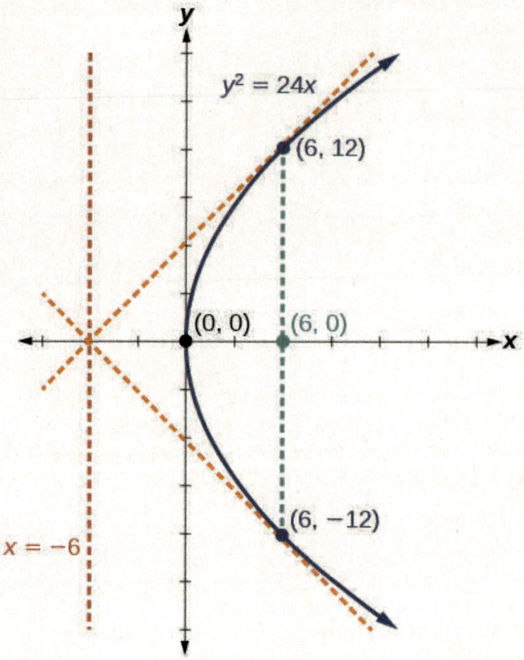

Figure 6

HOW TO

Given a standard form equation for a parabola centered at (0, 0), sketch the graph.

1. Determine which of the standard forms applies to the given equation: $y^2 = 4px$ or $x^2 = 4py$.
2. Use the standard form identified in Step 1 to determine the axis of symmetry, focus, equation of the directrix, and endpoints of the latus rectum.
   a. If the equation is in the form $y^2 = 4px$, then
      - the axis of symmetry is the *x*-axis, $y = 0$
      - set $4p$ equal to the coefficient of *x* in the given equation to solve for p. If $p > 0$, the parabola opens right. If $p < 0$, the parabola opens left.
      - use p to find the coordinates of the focus, $(p, 0)$
      - use p to find the equation of the directrix, $x = -p$
      - use p to find the endpoints of the latus rectum, $(p, \pm 2p)$. Alternately, substitute $x = p$ into the original equation.

   b. If the equation is in the form $x^2 = 4py$, then
      - the axis of symmetry is the *y*-axis, $x = 0$
      - set $4p$ equal to the coefficient of *y* in the given equation to solve for p. If $p > 0$, the parabola opens up. If $p < 0$, the parabola opens down.
      - use p to find the coordinates of the focus, $(0, p)$
      - use p to find equation of the directrix, $y = -p$
      - use p to find the endpoints of the latus rectum, $(\pm 2p, p)$
3. Plot the focus, directrix, and latus rectum, and draw a smooth curve to form the parabola.

EXAMPLE 1

Graphing a Parabola with Vertex (0, 0) and the *x*-axis as the Axis of Symmetry

Graph $y^2 = 24x$. Identify and label the focus, directrix, and endpoints of the latus rectum.

Solution

The standard form that applies to the given equation is $y^2 = 4px$. Thus, the axis of symmetry is the *x*-axis. It follows that:

- $24 = 4p$, so $p = 6$. Since $p > 0$, the parabola opens right
- the coordinates of the focus are $(p, 0) = (6, 0)$
- the equation of the directrix is $x = -p = -6$
- the endpoints of the latus rectum have the same *x*-coordinate at the focus. To find the endpoints, substitute $x = 6$ into the original equation: $(6, \pm 12)$

Next we plot the focus, directrix, and latus rectum, and draw a smooth curve to form the parabola. Figure 7

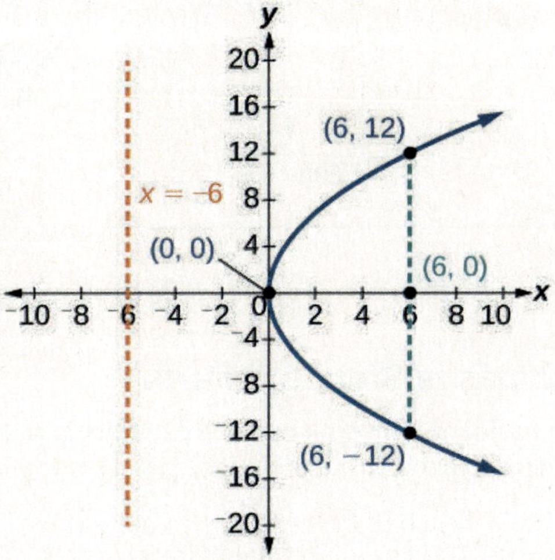

Figure 7

TRY IT #1 Graph $y^2 = -16x$. Identify and label the focus, directrix, and endpoints of the latus rectum.

EXAMPLE 2

Graphing a Parabola with Vertex (0, 0) and the *y*-axis as the Axis of Symmetry

Graph $x^2 = -6y$. Identify and label the focus, directrix, and endpoints of the latus rectum.

Solution

The standard form that applies to the given equation is $x^2 = 4py$. Thus, the axis of symmetry is the *y*-axis. It follows that:

- $-6 = 4p$, so $p = -\frac{3}{2}$. Since $p < 0$, the parabola opens down.
- the coordinates of the focus are $(0, p) = \left(0, -\frac{3}{2}\right)$
- the equation of the directrix is $y = -p = \frac{3}{2}$
- the endpoints of the latus rectum can be found by substituting $y = \frac{3}{2}$ into the original equation, $\left(\pm 3, -\frac{3}{2}\right)$

Next we plot the focus, directrix, and latus rectum, and draw a smooth curve to form the parabola.

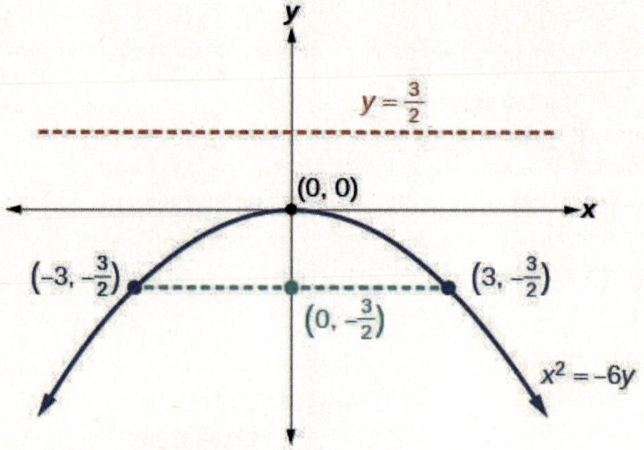

Figure 8

> TRY IT #2 Graph $x^2 = 8y$. Identify and label the focus, directrix, and endpoints of the latus rectum.

Writing Equations of Parabolas in Standard Form

In the previous examples, we used the standard form equation of a parabola to calculate the locations of its key features. We can also use the calculations in reverse to write an equation for a parabola when given its key features.

HOW TO

Given its focus and directrix, write the equation for a parabola in standard form.

1. Determine whether the axis of symmetry is the *x*- or *y*-axis.
   a. If the given coordinates of the focus have the form $(p, 0)$, then the axis of symmetry is the *x*-axis. Use the standard form $y^2 = 4px$.
   b. If the given coordinates of the focus have the form $(0, p)$, then the axis of symmetry is the *y*-axis. Use the standard form $x^2 = 4py$.
2. Multiply $4p$.
3. Substitute the value from Step 2 into the equation determined in Step 1.

EXAMPLE 3

Writing the Equation of a Parabola in Standard Form Given its Focus and Directrix

What is the equation for the parabola with focus $\left(-\frac{1}{2}, 0\right)$ and directrix $x = \frac{1}{2}$?

Solution

The focus has the form $(p, 0)$, so the equation will have the form $y^2 = 4px$.

- Multiplying $4p$, we have $4p = 4\left(-\frac{1}{2}\right) = -2$.
- Substituting for $4p$, we have $y^2 = 4px = -2x$.

Therefore, the equation for the parabola is $y^2 = -2x$.

> TRY IT #3 What is the equation for the parabola with focus $\left(0, \frac{7}{2}\right)$ and directrix $y = -\frac{7}{2}$?

Graphing Parabolas with Vertices Not at the Origin

Like other graphs we've worked with, the graph of a parabola can be translated. If a parabola is translated h units horizontally and k units vertically, the vertex will be (h, k). This translation results in the standard form of the equation we saw previously with x replaced by $(x-h)$ and y replaced by $(y-k)$.

To graph parabolas with a vertex (h, k) other than the origin, we use the standard form $(y-k)^2 = 4p(x-h)$ for parabolas that have an axis of symmetry parallel to the x-axis, and $(x-h)^2 = 4p(y-k)$ for parabolas that have an axis of symmetry parallel to the y-axis. These standard forms are given below, along with their general graphs and key features.

Standard Forms of Parabolas with Vertex (h, k)

Table 2 and Figure 9 summarize the standard features of parabolas with a vertex at a point (h, k).

Axis of Symmetry	Equation	Focus	Directrix	Endpoints of Latus Rectum
$y=k$	$(y-k)^2 = 4p(x-h)$	$(h+p,\ k)$	$x=h-p$	$(h+p,\ k \pm 2p)$
$x=h$	$(x-h)^2 = 4p(y-k)$	$(h,\ k+p)$	$y=k-p$	$(h \pm 2p,\ k+p)$

Table 2

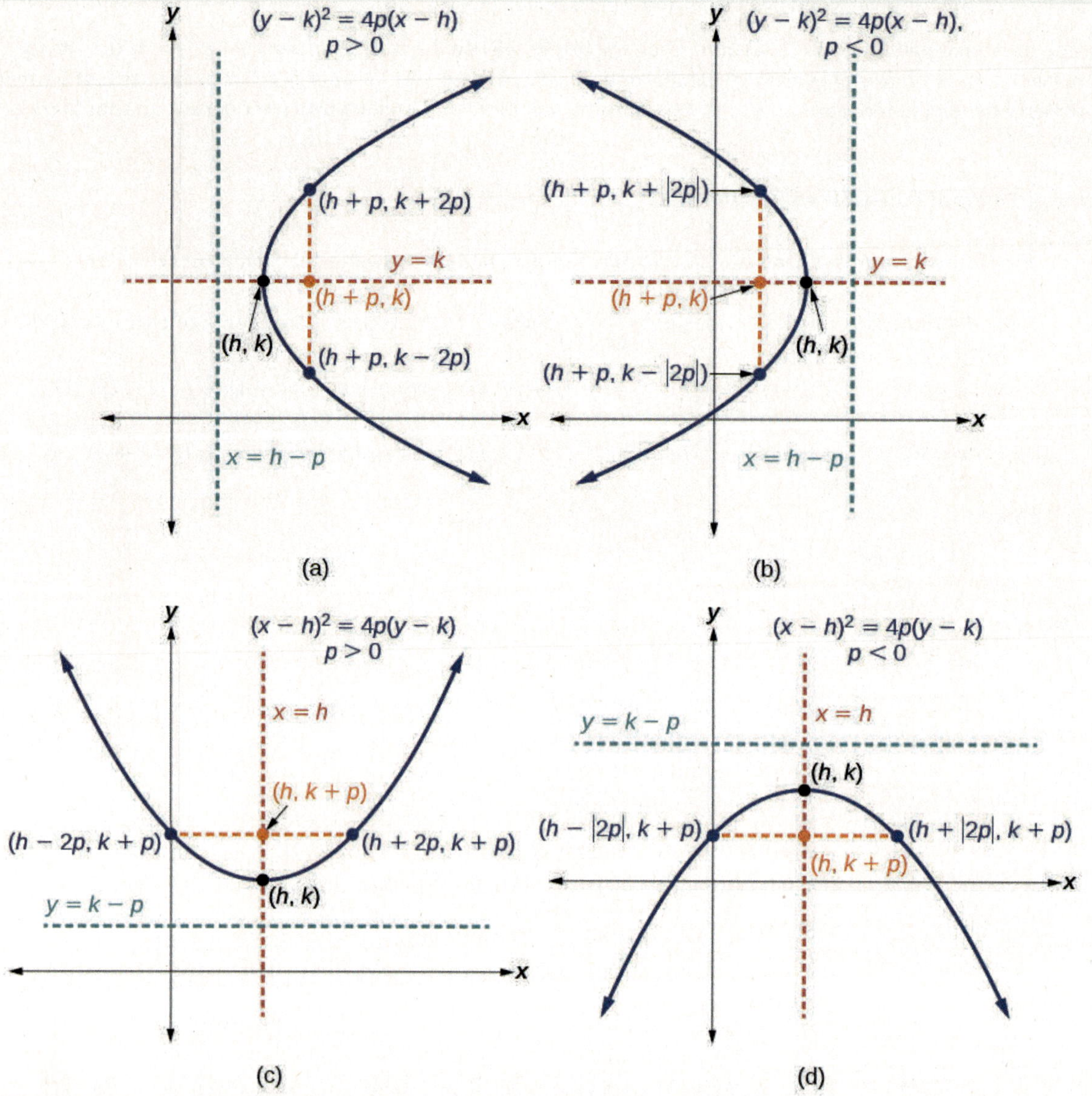

Figure 9 (a) When $p > 0$, the parabola opens right. (b) When $p < 0$, the parabola opens left. (c) When $p > 0$, the parabola opens up. (d) When $p < 0$, the parabola opens down.

HOW TO

Given a standard form equation for a parabola centered at (*h*, *k*), sketch the graph.

1. Determine which of the standard forms applies to the given equation: $(y - k)^2 = 4p(x - h)$ or $(x - h)^2 = 4p(y - k)$.
2. Use the standard form identified in Step 1 to determine the vertex, axis of symmetry, focus, equation of the directrix, and endpoints of the latus rectum.
   a. If the equation is in the form $(y - k)^2 = 4p(x - h)$, then:
      - use the given equation to identify h and k for the vertex, (h, k)
      - use the value of k to determine the axis of symmetry, $y = k$
      - set $4p$ equal to the coefficient of $(x - h)$ in the given equation to solve for p. If $p > 0$, the parabola opens right. If $p < 0$, the parabola opens left.

- use h, k, and p to find the coordinates of the focus, $(h + p, \ k)$
- use h and p to find the equation of the directrix, $x = h - p$
- use h, k, and p to find the endpoints of the latus rectum, $(h + p, k \pm 2p)$

b. If the equation is in the form $(x - h)^2 = 4p(y - k)$, then:
- use the given equation to identify h and k for the vertex, (h, k)
- use the value of h to determine the axis of symmetry, $x = h$
- set $4p$ equal to the coefficient of $(y - k)$ in the given equation to solve for p. If $p > 0$, the parabola opens up. If $p < 0$, the parabola opens down.
- use h, k, and p to find the coordinates of the focus, $(h, \ k + p)$
- use k and p to find the equation of the directrix, $y = k - p$
- use h, k, and p to find the endpoints of the latus rectum, $(h \pm 2p, \ k + p)$

3. Plot the vertex, axis of symmetry, focus, directrix, and latus rectum, and draw a smooth curve to form the parabola.

EXAMPLE 4

Graphing a Parabola with Vertex (*h*, *k*) and Axis of Symmetry Parallel to the *x*-axis

Graph $(y - 1)^2 = -16(x + 3)$. Identify and label the vertex, axis of symmetry, focus, directrix, and endpoints of the latus rectum.

Solution

The standard form that applies to the given equation is $(y - k)^2 = 4p(x - h)$. Thus, the axis of symmetry is parallel to the *x*-axis. It follows that:

- the vertex is $(h, k) = (-3, 1)$
- the axis of symmetry is $y = k = 1$
- $-16 = 4p$, so $p = -4$. Since $p < 0$, the parabola opens left.
- the coordinates of the focus are $(h + p, k) = (-3 + (-4), 1) = (-7, 1)$
- the equation of the directrix is $x = h - p = -3 - (-4) = 1$
- the endpoints of the latus rectum are $(h + p, k \pm 2p) = (-3 + (-4), 1 \pm 2(-4))$, or $(-7, -7)$ and $(-7, 9)$

Next we plot the vertex, axis of symmetry, focus, directrix, and latus rectum, and draw a smooth curve to form the parabola. See Figure 10.

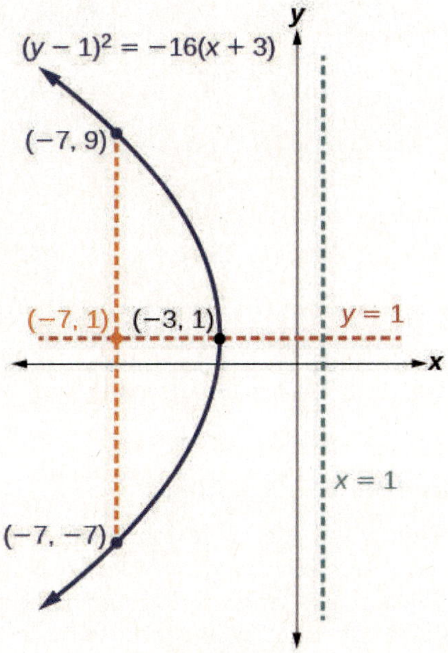

Figure 10

> **TRY IT** #4 Graph $(y+1)^2 = 4(x-8)$. Identify and label the vertex, axis of symmetry, focus, directrix, and endpoints of the latus rectum.

EXAMPLE 5

Graphing a Parabola from an Equation Given in General Form

Graph $x^2 - 8x - 28y - 208 = 0$. Identify and label the vertex, axis of symmetry, focus, directrix, and endpoints of the latus rectum.

Solution

Start by writing the equation of the parabola in standard form. The standard form that applies to the given equation is $(x-h)^2 = 4p(y-k)$. Thus, the axis of symmetry is parallel to the *y*-axis. To express the equation of the parabola in this form, we begin by isolating the terms that contain the variable x in order to complete the square.

$$\begin{aligned} x^2 - 8x - 28y - 208 &= 0 \\ x^2 - 8x &= 28y + 208 \\ x^2 - 8x + 16 &= 28y + 208 + 16 \\ (x-4)^2 &= 28y + 224 \\ (x-4)^2 &= 28(y+8) \\ (x-4)^2 &= 4 \cdot 7 \cdot (y+8) \end{aligned}$$

It follows that:

- the vertex is $(h, k) = (4, -8)$
- the axis of symmetry is $x = h = 4$
- since $p = 7, p > 0$ and so the parabola opens up
- the coordinates of the focus are $(h, k+p) = (4, -8+7) = (4, -1)$
- the equation of the directrix is $y = k - p = -8 - 7 = -15$
- the endpoints of the latus rectum are $(h \pm 2p, k+p) = (4 \pm 2(7), -8+7)$, or $(-10, -1)$ and $(18, -1)$

Next we plot the vertex, axis of symmetry, focus, directrix, and latus rectum, and draw a smooth curve to form the parabola. See Figure 11.

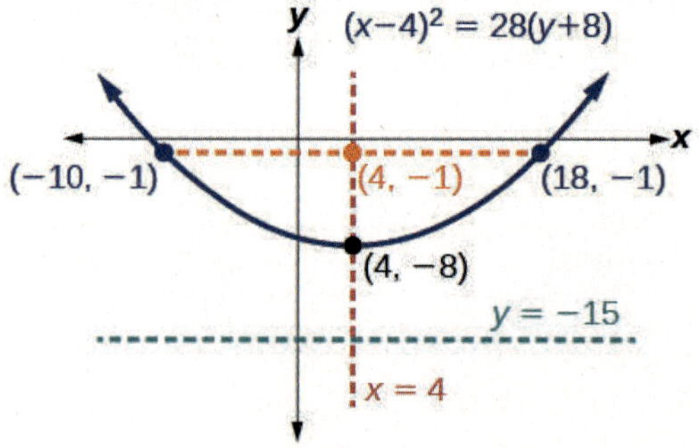

Figure 11

> **TRY IT** #5 Graph $(x+2)^2 = -20(y-3)$. Identify and label the vertex, axis of symmetry, focus, directrix, and endpoints of the latus rectum.

Solving Applied Problems Involving Parabolas

As we mentioned at the beginning of the section, parabolas are used to design many objects we use every day, such as telescopes, suspension bridges, microphones, and radar equipment. Parabolic mirrors, such as the one used to light the Olympic torch, have a very unique reflecting property. When rays of light parallel to the parabola's axis of symmetry are directed toward any surface of the mirror, the light is reflected directly to the focus. See Figure 12. This is why the Olympic torch is ignited when it is held at the focus of the parabolic mirror.

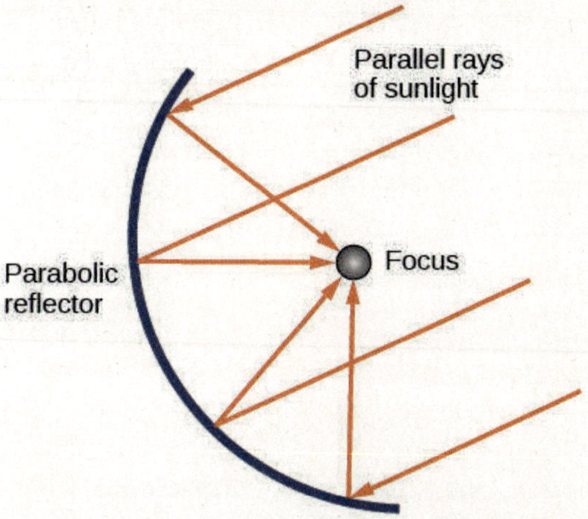

Figure 12 Reflecting property of parabolas

Parabolic mirrors have the ability to focus the sun's energy to a single point, raising the temperature hundreds of degrees in a matter of seconds. Thus, parabolic mirrors are featured in many low-cost, energy efficient solar products, such as solar cookers, solar heaters, and even travel-sized fire starters.

EXAMPLE 6

Solving Applied Problems Involving Parabolas

A cross-section of a design for a travel-sized solar fire starter is shown in Figure 13. The sun's rays reflect off the parabolic mirror toward an object attached to the igniter. Because the igniter is located at the focus of the parabola, the reflected rays cause the object to burn in just seconds.

ⓐ Find the equation of the parabola that models the fire starter. Assume that the vertex of the parabolic mirror is the origin of the coordinate plane.

ⓑ Use the equation found in part ⓐ to find the depth of the fire starter.

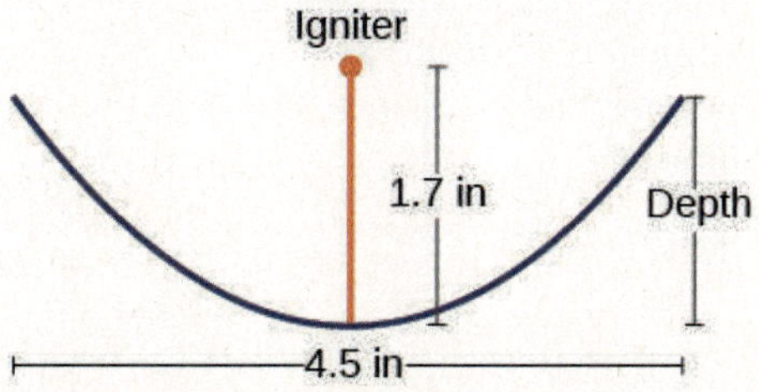

Figure 13 Cross-section of a travel-sized solar fire starter

Solution

ⓐ The vertex of the dish is the origin of the coordinate plane, so the parabola will take the standard form $x^2 = 4py$, where $p > 0$. The igniter, which is the focus, is 1.7 inches above the vertex of the dish. Thus we have $p = 1.7$.

$$x^2 = 4py \quad \text{Standard form of upward-facing parabola with vertex (0,0)}$$
$$x^2 = 4(1.7)y \quad \text{Substitute 1.7 for } p.$$
$$x^2 = 6.8y \quad \text{Multiply.}$$

ⓑ The dish extends $\frac{4.5}{2} = 2.25$ inches on either side of the origin. We can substitute 2.25 for x in the equation from part (a) to find the depth of the dish.

$$x^2 = 6.8y \quad \text{Equation found in part (a).}$$
$$(2.25)^2 = 6.8y \quad \text{Substitute 2.25 for } x.$$
$$y \approx 0.74 \quad \text{Solve for } y.$$

The dish is about 0.74 inches deep.

> **TRY IT** #6 Balcony-sized solar cookers have been designed for families living in India. The top of a dish has a diameter of 1600 mm. The sun's rays reflect off the parabolic mirror toward the "cooker," which is placed 320 mm from the base.
>
> ⓐ Find an equation that models a cross-section of the solar cooker. Assume that the vertex of the parabolic mirror is the origin of the coordinate plane, and that the parabola opens to the right (i.e., has the *x*-axis as its axis of symmetry).
>
> ⓑ Use the equation found in part ⓐ to find the depth of the cooker.

MEDIA

Access these online resources for additional instruction and practice with parabolas.

Conic Sections: The Parabola Part 1 of 2 (http://openstax.org/l/parabola1)
Conic Sections: The Parabola Part 2 of 2 (http://openstax.org/l/parabola2)
Parabola with Vertical Axis (http://openstax.org/l/parabolavertcal)
Parabola with Horizontal Axis (http://openstax.org/l/parabolahoriz)

12.3 SECTION EXERCISES

Verbal

1. Define a parabola in terms of its focus and directrix.

2. If the equation of a parabola is written in standard form and p is positive and the directrix is a vertical line, then what can we conclude about its graph?

3. If the equation of a parabola is written in standard form and p is negative and the directrix is a horizontal line, then what can we conclude about its graph?

4. What is the effect on the graph of a parabola if its equation in standard form has increasing values of p?

5. As the graph of a parabola becomes wider, what will happen to the distance between the focus and directrix?

Algebraic

For the following exercises, determine whether the given equation is a parabola. If so, rewrite the equation in standard form.

6. $y^2 = 4 - x^2$

7. $y = 4x^2$

8. $3x^2 - 6y^2 = 12$

9. $(y - 3)^2 = 8(x - 2)$

10. $y^2 + 12x - 6y - 51 = 0$

For the following exercises, rewrite the given equation in standard form, and then determine the vertex (V), focus (F), and directrix (d) of the parabola.

11. $x = 8y^2$

12. $y = \frac{1}{4}x^2$

13. $y = -4x^2$

14. $x = \frac{1}{8}y^2$

15. $x = 36y^2$

16. $x = \frac{1}{36}y^2$

17. $(x-1)^2 = 4(y-1)$

18. $(y-2)^2 = \frac{4}{5}(x+4)$

19. $(y-4)^2 = 2(x+3)$

20. $(x+1)^2 = 2(y+4)$

21. $(x+4)^2 = 24(y+1)$

22. $(y+4)^2 = 16(x+4)$

23. $y^2 + 12x - 6y + 21 = 0$

24. $x^2 - 4x - 24y + 28 = 0$

25. $5x^2 - 50x - 4y + 113 = 0$

26. $y^2 - 24x + 4y - 68 = 0$

27. $x^2 - 4x + 2y - 6 = 0$

28. $y^2 - 6y + 12x - 3 = 0$

29. $3y^2 - 4x - 6y + 23 = 0$

30. $x^2 + 4x + 8y - 4 = 0$

Graphical

For the following exercises, graph the parabola, labeling the focus and the directrix.

31. $x = \frac{1}{8}y^2$

32. $y = 36x^2$

33. $y = \frac{1}{36}x^2$

34. $y = -9x^2$

35. $(y-2)^2 = -\frac{4}{3}(x+2)$

36. $-5(x+5)^2 = 4(y+5)$

37. $-6(y+5)^2 = 4(x-4)$

38. $y^2 - 6y - 8x + 1 = 0$

39. $x^2 + 8x + 4y + 20 = 0$

40. $3x^2 + 30x - 4y + 95 = 0$

41. $y^2 - 8x + 10y + 9 = 0$

42. $x^2 + 4x + 2y + 2 = 0$

43. $y^2 + 2y - 12x + 61 = 0$

44. $-2x^2 + 8x - 4y - 24 = 0$

For the following exercises, find the equation of the parabola given information about its graph.

45. Vertex is $(0,0)$; directrix is $y = 4$, focus is $(0,-4)$.

46. Vertex is $(0,0)$; directrix is $x = 4$, focus is $(-4,0)$.

47. Vertex is $(2,2)$; directrix is $x = 2 - \sqrt{2}$, focus is $\left(2+\sqrt{2}, 2\right)$.

48. Vertex is $(-2,3)$; directrix is $x = -\frac{7}{2}$, focus is $\left(-\frac{1}{2}, 3\right)$.

49. Vertex is $\left(\sqrt{2}, -\sqrt{3}\right)$; directrix is $x = 2\sqrt{2}$, focus is $\left(0, -\sqrt{3}\right)$.

50. Vertex is $(1,2)$; directrix is $y = \frac{11}{3}$, focus is $\left(1, \frac{1}{3}\right)$.

For the following exercises, determine the equation for the parabola from its graph.

51.

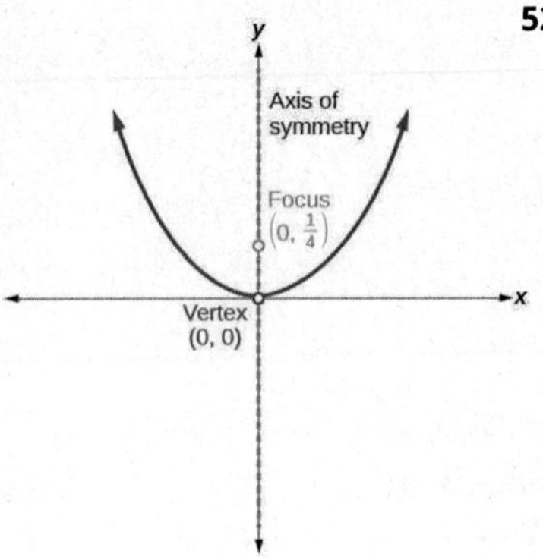

52.

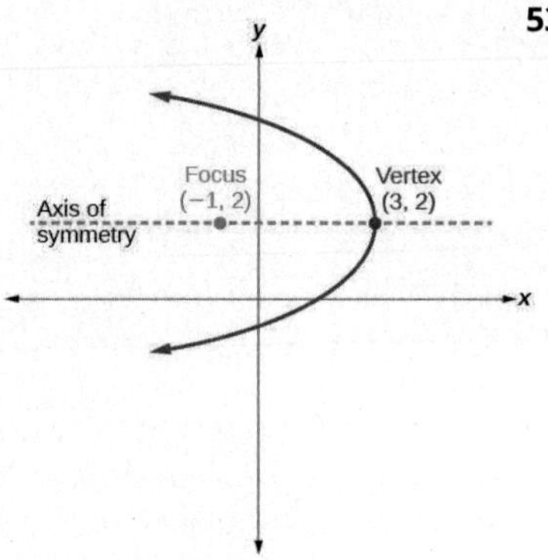

53.

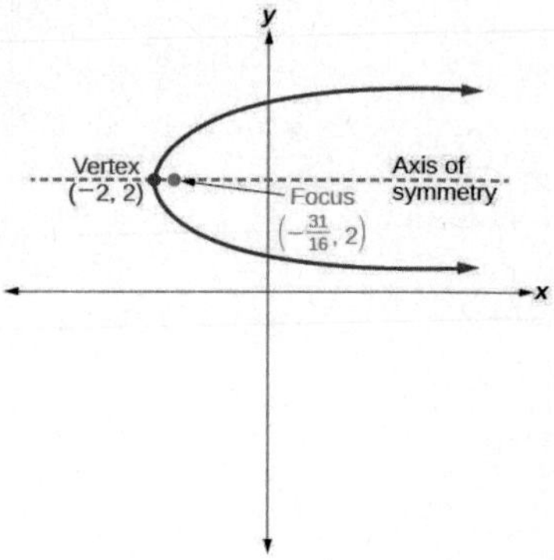

54.

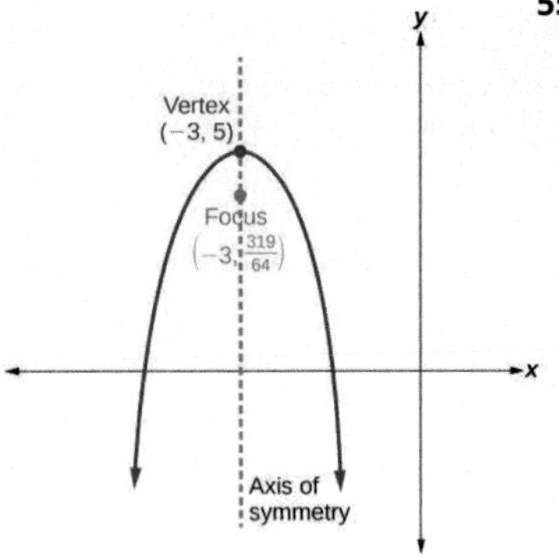

55.

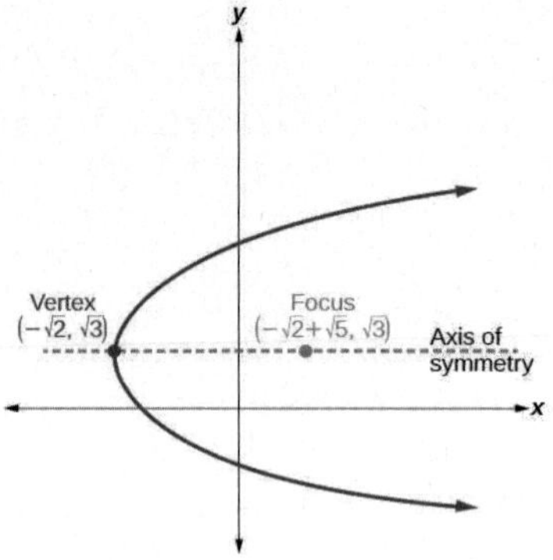

Extensions

For the following exercises, the vertex and endpoints of the latus rectum of a parabola are given. Find the equation.

56. $V(0, 0)$, Endpoints $(2, 1)$, $(-2, 1)$

57. $V(0, 0)$, Endpoints $(-2, 4)$, $(-2, -4)$

58. $V(1, 2)$, Endpoints $(-5, 5)$, $(7, 5)$

59. $V(-3, -1)$, Endpoints $(0, 5)$, $(0, -7)$

60. $V(4, -3)$, Endpoints $\left(5, -\frac{7}{2}\right)$, $\left(3, -\frac{7}{2}\right)$

Real-World Applications

61. The mirror in an automobile headlight has a parabolic cross-section with the light bulb at the focus. On a schematic, the equation of the parabola is given as $x^2 = 4y$. At what coordinates should you place the light bulb?

62. If we want to construct the mirror from the previous exercise such that the focus is located at $(0, 0.25)$, what should the equation of the parabola be?

63. A satellite dish is shaped like a paraboloid of revolution. This means that it can be formed by rotating a parabola around its axis of symmetry. The receiver is to be located at the focus. If the dish is 12 feet across at its opening and 4 feet deep at its center, where should the receiver be placed?

64. Consider the satellite dish from the previous exercise. If the dish is 8 feet across at the opening and 2 feet deep, where should we place the receiver?

65. The reflector in a searchlight is shaped like a paraboloid of revolution. A light source is located 1 foot from the base along the axis of symmetry. If the opening of the searchlight is 3 feet across, find the depth.

66. If the reflector in the searchlight from the previous exercise has the light source located 6 inches from the base along the axis of symmetry and the opening is 4 feet, find the depth.

67. An arch is in the shape of a parabola. It has a span of 100 feet and a maximum height of 20 feet. Find the equation of the parabola, and determine the height of the arch 40 feet from the center.

68. If the arch from the previous exercise has a span of 160 feet and a maximum height of 40 feet, find the equation of the parabola, and determine the distance from the center at which the height is 20 feet.

69. An object is projected so as to follow a parabolic path given by $y = -x^2 + 96x$, where x is the horizontal distance traveled in feet and y is the height. Determine the maximum height the object reaches.

70. For the object from the previous exercise, assume the path followed is given by $y = -0.5x^2 + 80x$. Determine how far along the horizontal the object traveled to reach maximum height.

12.4 Rotation of Axes

Learning Objectives

In this section, you will:

- Identify nondegenerate conic sections given their general form equations.
- Use rotation of axes formulas.
- Write equations of rotated conics in standard form.
- Identify conics without rotating axes.

As we have seen, conic sections are formed when a plane intersects two right circular cones aligned tip to tip and extending infinitely far in opposite directions, which we also call a *cone*. The way in which we slice the cone will determine the type of conic section formed at the intersection. A circle is formed by slicing a cone with a plane perpendicular to the axis of symmetry of the cone. An ellipse is formed by slicing a single cone with a slanted plane not perpendicular to the axis of symmetry. A parabola is formed by slicing the plane through the top or bottom of the double-cone, whereas a hyperbola is formed when the plane slices both the top and bottom of the cone. See Figure 1.

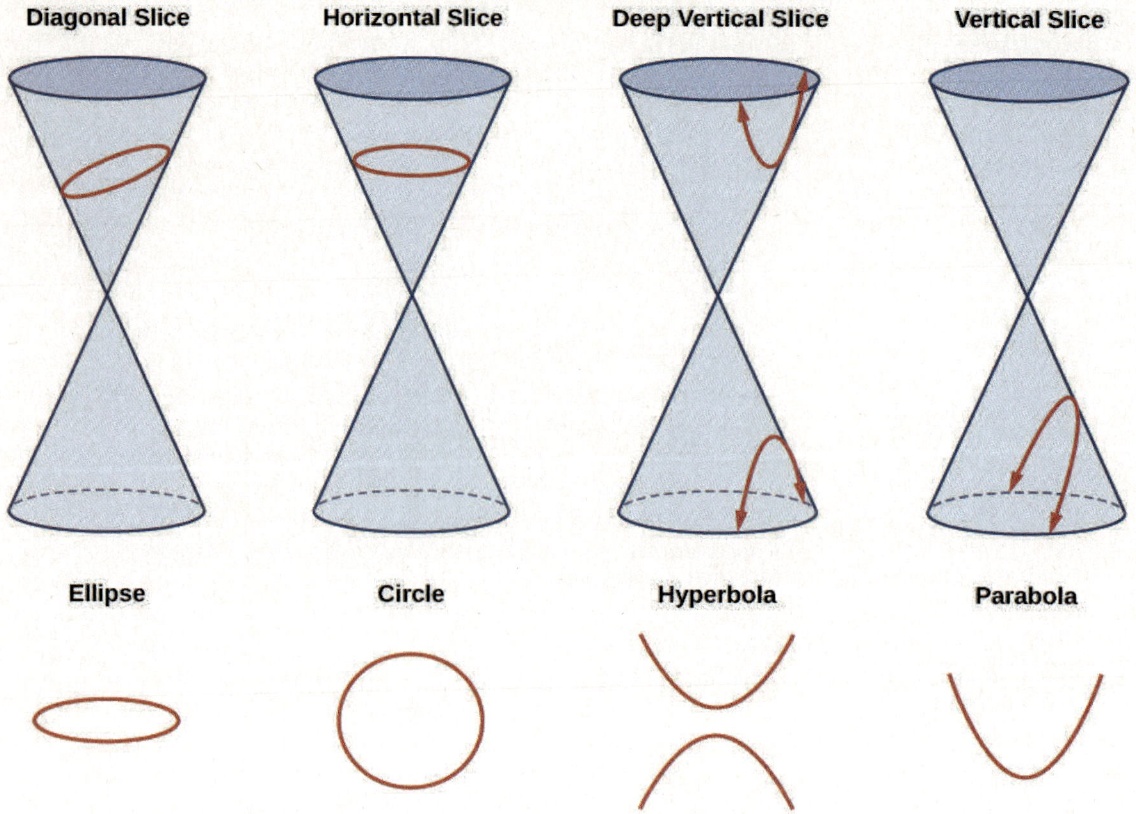

Figure 1 The nondegenerate conic sections

Ellipses, circles, hyperbolas, and parabolas are sometimes called the **nondegenerate conic sections**, in contrast to the **degenerate conic sections**, which are shown in Figure 2. A degenerate conic results when a plane intersects the double cone and passes through the apex. Depending on the angle of the plane, three types of degenerate conic sections are possible: a point, a line, or two intersecting lines.

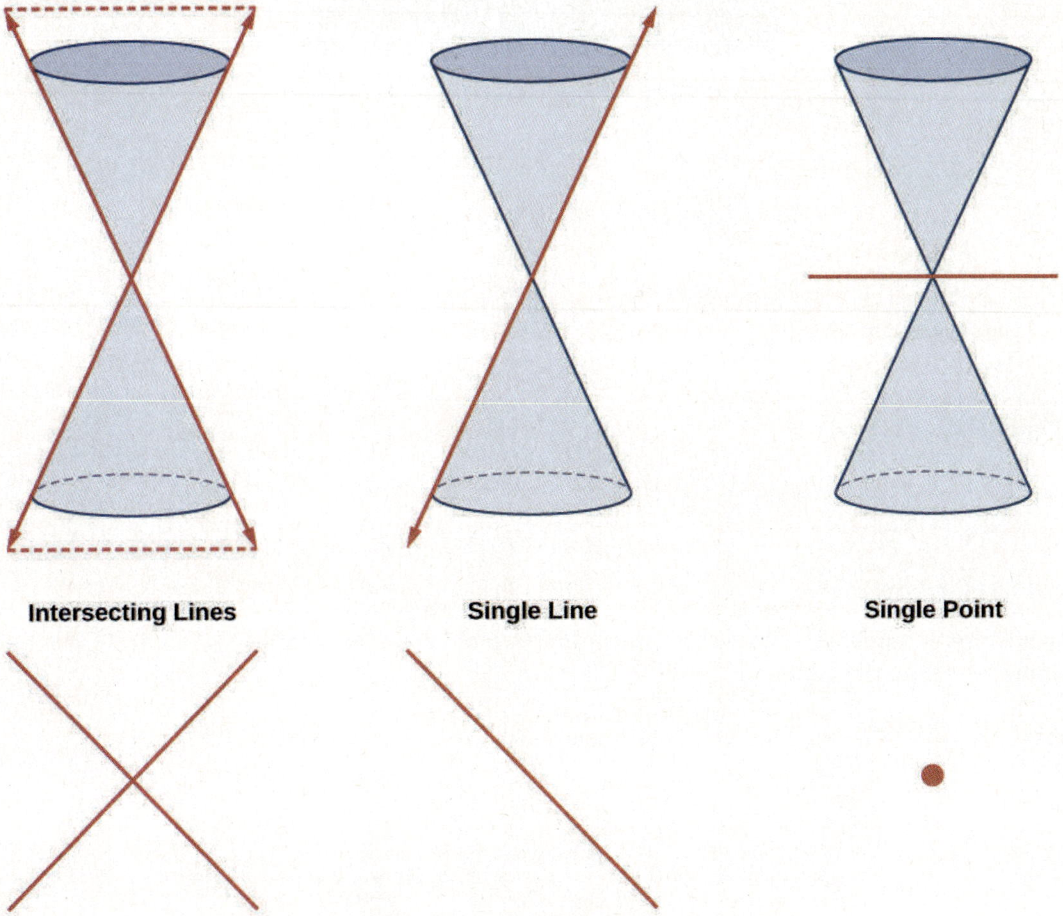

Figure 2 Degenerate conic sections

Identifying Nondegenerate Conics in General Form

In previous sections of this chapter, we have focused on the standard form equations for nondegenerate conic sections. In this section, we will shift our focus to the general form equation, which can be used for any conic. The general form is set equal to zero, and the terms and coefficients are given in a particular order, as shown below.

$$Ax^2 + Bxy + Cy^2 + Dx + Ey + F = 0$$

where A, B, and C are not all zero. We can use the values of the coefficients to identify which type conic is represented by a given equation.

You may notice that the general form equation has an xy term that we have not seen in any of the standard form equations. As we will discuss later, the xy term rotates the conic whenever B is not equal to zero.

Conic Sections	Example
ellipse	$4x^2 + 9y^2 = 1$
circle	$4x^2 + 4y^2 = 1$
hyperbola	$4x^2 - 9y^2 = 1$
parabola	$4x^2 = 9y$ or $4y^2 = 9x$
one line	$4x + 9y = 1$

Table 1

Conic Sections	Example
intersecting lines	$(x-4)(y+4)=0$
parallel lines	$(x-4)(x-9)=0$
a point	$4x^2+4y^2=0$
no graph	$4x^2+4y^2=-1$

Table 1

General Form of Conic Sections

A **conic section** has the general form

$$Ax^2+Bxy+Cy^2+Dx+Ey+F=0$$

where A, B, and C are not all zero.

Table 2 summarizes the different conic sections where $B=0$, and A and C are nonzero real numbers. This indicates that the conic has not been rotated.

ellipse	$Ax^2+Cy^2+Dx+Ey+F=0,\ \ A\neq C$ and $AC>0$
circle	$Ax^2+Cy^2+Dx+Ey+F=0,\ \ A=C$
hyperbola	$Ax^2-Cy^2+Dx+Ey+F=0$ or $-Ax^2+Cy^2+Dx+Ey+F=0$, where A and C are positive
parabola	$Ax^2+Dx+Ey+F=0$ or $Cy^2+Dx+Ey+F=0$

Table 2

HOW TO

Given the equation of a conic, identify the type of conic.

1. Rewrite the equation in the general form, $Ax^2+Bxy+Cy^2+Dx+Ey+F=0$.
2. Identify the values of A and C from the general form.
   a. If A and C are nonzero, have the same sign, and are not equal to each other, then the graph may be an ellipse.
   b. If A and C are equal and nonzero and have the same sign, then the graph may be a circle.
   c. If A and C are nonzero and have opposite signs, then the graph may be a hyperbola.
   d. If either A or C is zero, then the graph may be a parabola.

If *B* = 0, the conic section will have a vertical and/or horizontal axes. If *B* does not equal 0, as shown below, the conic section is rotated. Notice the phrase "may be" in the definitions. That is because the equation may not represent a conic section at all, depending on the values of *A*, *B*, *C*, *D*, *E*, and *F*. For example, the degenerate case of a circle or an ellipse is a point:
$Ax^2+By^2=0$, when *A* and *B* have the same sign.
The degenerate case of a hyperbola is two intersecting straight lines: $Ax^2+By^2=0$, when A and B have opposite signs.
On the other hand, the equation, $Ax^2+By^2+1=0$, when A and B are positive does not represent a graph at

all, since there are no real ordered pairs which satisfy it.

EXAMPLE 1

Identifying a Conic from Its General Form

Identify the graph of each of the following nondegenerate conic sections.

ⓐ $4x^2 - 9y^2 + 36x + 36y - 125 = 0$ ⓑ $9y^2 + 16x + 36y - 10 = 0$ ⓒ $3x^2 + 3y^2 - 2x - 6y - 4 = 0$

ⓓ $-25x^2 - 4y^2 + 100x + 16y + 20 = 0$

Solution

ⓐ Rewriting the general form, we have

$$\begin{array}{c} Ax^2 + Bxy + Cy^2 + Dx + Ey + F = 0 \\ 4x^2 + 0xy + (-9)y^2 + 36x + 36y + (-125) = 0 \end{array}$$

$A = 4$ and $C = -9$, so we observe that A and C have opposite signs. The graph of this equation is a hyperbola.

ⓑ Rewriting the general form, we have

$$\begin{array}{c} Ax^2 + Bxy + Cy^2 + Dx + Ey + F = 0 \\ 0x^2 + 0xy + 9y^2 + 16x + 36y + (-10) = 0 \end{array}$$

$A = 0$ and $C = 9$. We can determine that the equation is a parabola, since A is zero.

ⓒ Rewriting the general form, we have

$$\begin{array}{c} Ax^2 + Bxy + Cy^2 + Dx + Ey + F = 0 \\ 3x^2 + 0xy + 3y^2 + (-2)x + (-6)y + (-4) = 0 \end{array}$$

$A = 3$ and $C = 3$. Because $A = C$, the graph of this equation is a circle.

ⓓ Rewriting the general form, we have

$$\begin{array}{c} Ax^2 + Bxy + Cy^2 + Dx + Ey + F = 0 \\ (-25)x^2 + 0xy + (-4)y^2 + 100x + 16y + 20 = 0 \end{array}$$

$A = -25$ and $C = -4$. Because $AC > 0$ and $A \neq C$, the graph of this equation is an ellipse.

> **TRY IT** #1 Identify the graph of each of the following nondegenerate conic sections.
>
> ⓐ $16y^2 - x^2 + x - 4y - 9 = 0$ ⓑ $16x^2 + 4y^2 + 16x + 49y - 81 = 0$

Finding a New Representation of the Given Equation after Rotating through a Given Angle

Until now, we have looked at equations of conic sections without an xy term, which aligns the graphs with the *x*- and *y*-axes. When we add an xy term, we are rotating the conic about the origin. If the *x*- and *y*-axes are rotated through an angle, say θ, then every point on the plane may be thought of as having two representations: (x, y) on the Cartesian plane with the original *x*-axis and *y*-axis, and (x', y') on the new plane defined by the new, rotated axes, called the x'-axis and y'-axis. See Figure 3.

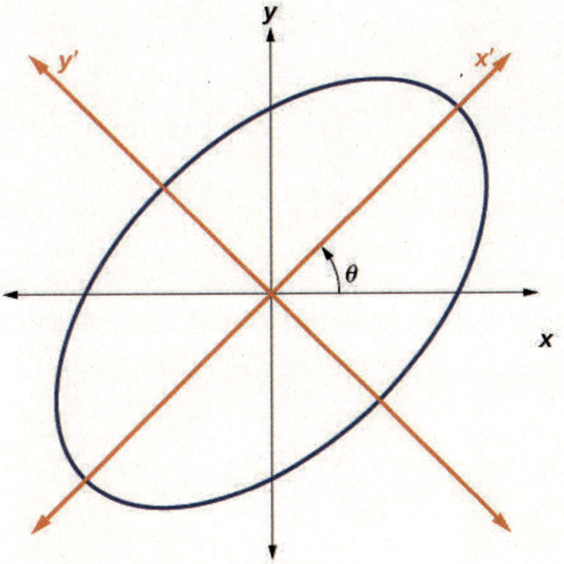

Figure 3 The graph of the rotated ellipse $x^2 + y^2 - xy - 15 = 0$

We will find the relationships between x and y on the Cartesian plane with x' and y' on the new rotated plane. See Figure 4.

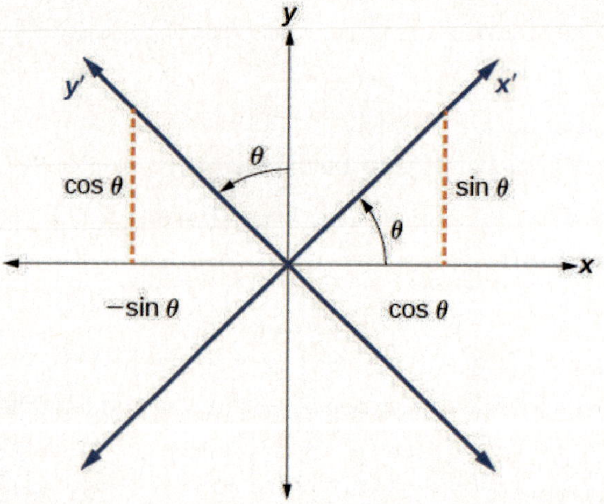

Figure 4 The Cartesian plane with *x*- and *y*-axes and the resulting *x'*− and *y'*−axes formed by a rotation by an angle θ.

The original coordinate *x*- and *y*-axes have unit vectors i and j. The rotated coordinate axes have unit vectors i' and j'. The angle θ is known as the **angle of rotation**. See Figure 5. We may write the new unit vectors in terms of the original ones.

$$i' = \cos\theta i + \sin\theta j$$
$$j' = -\sin\theta i + \cos\theta j$$

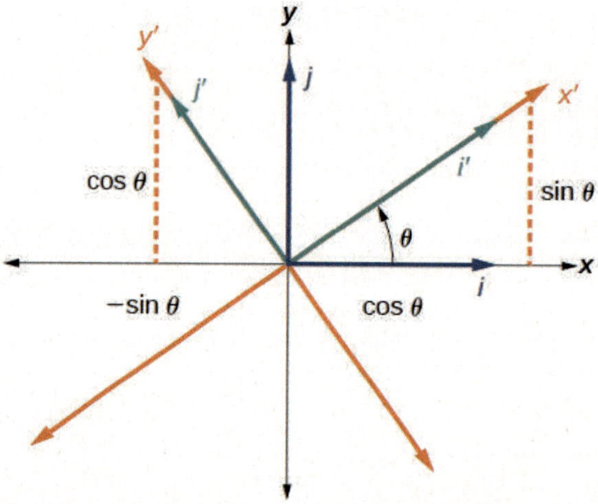

Figure 5 Relationship between the old and new coordinate planes.

Consider a vector u in the new coordinate plane. It may be represented in terms of its coordinate axes.

$u = x'i' + y'j'$	
$u = x'(i\cos\theta + j\sin\theta) + y'(-i\sin\theta + j\cos\theta)$	Substitute.
$u = ix'\cos\theta + jx'\sin\theta - iy'\sin\theta + jy'\cos\theta$	Distribute.
$u = ix'\cos\theta - iy'\sin\theta + jx'\sin\theta + jy'\cos\theta$	Apply commutative property.
$u = (x'\cos\theta - y'\sin\theta)i + (x'\sin\theta + y'\cos\theta)j$	Factor by grouping.

Because $u = x'i' + y'j'$, we have representations of x and y in terms of the new coordinate system.

$$x = x'\cos\theta - y'\sin\theta$$
and
$$y = x'\sin\theta + y'\cos\theta$$

Equations of Rotation

If a point (x, y) on the Cartesian plane is represented on a new coordinate plane where the axes of rotation are formed by rotating an angle θ from the positive x-axis, then the coordinates of the point with respect to the new axes are $\left(x', y'\right)$. We can use the following equations of rotation to define the relationship between (x, y) and $\left(x', y'\right)$:

$$x = x' \cos\ \theta - y' \sin\ \theta$$

and

$$y = x' \sin\ \theta + y' \cos\ \theta$$

HOW TO

Given the equation of a conic, find a new representation after rotating through an angle.

1. Find x and y where $x = x' \cos\ \theta - y' \sin\ \theta$ and $y = x' \sin\ \theta + y' \cos\ \theta$.
2. Substitute the expression for x and y into in the given equation, then simplify.
3. Write the equations with x' and y' in standard form.

EXAMPLE 2

Finding a New Representation of an Equation after Rotating through a Given Angle

Find a new representation of the equation $2x^2 - xy + 2y^2 - 30 = 0$ after rotating through an angle of $\theta = 45°$.

Solution

Find x and y, where $x = x' \cos\ \theta - y' \sin\ \theta$ and $y = x' \sin\ \theta + y' \cos\ \theta$.

Because $\theta = 45°$,

$$\begin{aligned} x &= x' \cos(45°) - y' \sin(45°) \\ x &= x'\left(\frac{1}{\sqrt{2}}\right) - y'\left(\frac{1}{\sqrt{2}}\right) \\ x &= \frac{x' - y'}{\sqrt{2}} \end{aligned}$$

and

$$\begin{aligned} y &= x' \sin(45°) + y' \cos(45°) \\ y &= x'\left(\frac{1}{\sqrt{2}}\right) + y'\left(\frac{1}{\sqrt{2}}\right) \\ y &= \frac{x' + y'}{\sqrt{2}} \end{aligned}$$

Substitute $x = x' \cos\theta - y' \sin\theta$ and $y = x' \sin\ \theta + y' \cos\ \theta$ into $2x^2 - xy + 2y^2 - 30 = 0$.

$$2\left(\frac{x' - y'}{\sqrt{2}}\right)^2 - \left(\frac{x' - y'}{\sqrt{2}}\right)\left(\frac{x' + y'}{\sqrt{2}}\right) + 2\left(\frac{x' + y'}{\sqrt{2}}\right)^2 - 30 = 0$$

Simplify.

$\cancel{2}\frac{(x'-y')(x'-y')}{\cancel{2}} - \frac{(x'-y')(x'+y')}{2} + \cancel{2}\frac{(x'+y')(x'+y')}{\cancel{2}} - 30 = 0$	FOIL method
$x'^2 \cancel{-2x'y'} + y'^2 - \frac{(x'^2-y'^2)}{2} + x'^2 \cancel{+2x'y'} + y'^2 - 30 = 0$	Combine like terms.
$2x'^2 + 2y'^2 - \frac{(x'^2-y'^2)}{2} = 30$	Combine like terms.
$2\left(2x'^2 + 2y'^2 - \frac{(x'^2-y'^2)}{2}\right) = 2(30)$	Multiply both sides by 2.
$4x'^2 + 4y'^2 - (x'^2 - y'^2) = 60$	Simplify.
$4x'^2 + 4y'^2 - x'^2 + y'^2 = 60$	Distribute.
$\frac{3x'^2}{60} + \frac{5y'^2}{60} = \frac{60}{60}$	Set equal to 1.

Write the equations with x' and y' in the standard form.

$$\frac{x'^2}{20} + \frac{y'^2}{12} = 1$$

This equation is an ellipse. Figure 6 shows the graph.

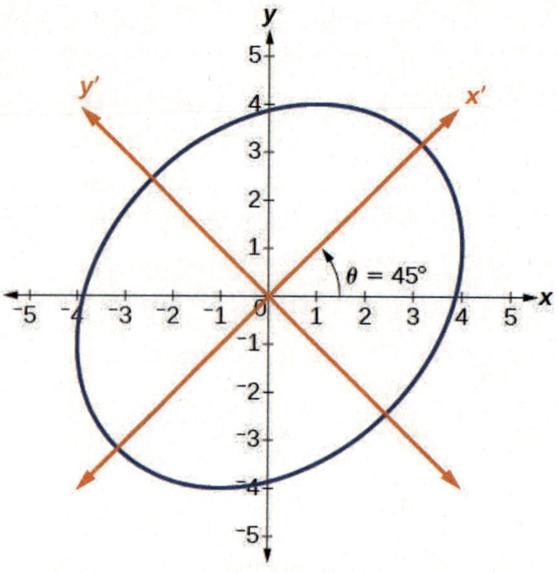

Figure 6

Writing Equations of Rotated Conics in Standard Form

Now that we can find the standard form of a conic when we are given an angle of rotation, we will learn how to transform the equation of a conic given in the form $Ax^2 + Bxy + Cy^2 + Dx + Ey + F = 0$ into standard form by rotating the axes. To do so, we will rewrite the general form as an equation in the x' and y' coordinate system without the $x'y'$ term, by rotating the axes by a measure of θ that satisfies

$$\cot(2\theta) = \frac{A - C}{B}$$

We have learned already that any conic may be represented by the second degree equation

$$Ax^2 + Bxy + Cy^2 + Dx + Ey + F = 0$$

where A, B, and C are not all zero. However, if $B \neq 0$, then we have an xy term that prevents us from rewriting the equation in standard form. To eliminate it, we can rotate the axes by an acute angle θ where $\cot(2\theta) = \frac{A-C}{B}$.

- If $\cot(2\theta) > 0$, then 2θ is in the first quadrant, and θ is between $(0°, 45°)$.
- If $\cot(2\theta) < 0$, then 2θ is in the second quadrant, and θ is between $(45°, 90°)$.
- If $A = C$, then $\theta = 45°$.

HOW TO

Given an equation for a conic in the $x'y'$ system, rewrite the equation without the $x'y'$ term in terms of x' and y', where the x' and y' axes are rotations of the standard axes by θ degrees.

1. Find $\cot(2\theta)$.
2. Find $\sin\ \theta$ and $\cos\ \theta$.
3. Substitute $\sin\ \theta$ and $\cos\ \theta$ into $x = x'\cos\ \theta - y'\sin\ \theta$ and $y = x'\sin\ \theta + y'\cos\ \theta$.
4. Substitute the expression for x and y into in the given equation, and then simplify.
5. Write the equations with x' and y' in the standard form with respect to the rotated axes.

EXAMPLE 3

Rewriting an Equation with respect to the *x'* and *y'* axes without the *x'y'* Term

Rewrite the equation $8x^2 - 12xy + 17y^2 = 20$ in the $x'y'$ system without an $x'y'$ term.

Solution

First, we find $\cot(2\theta)$. See Figure 7.

$$8x^2 - 12xy + 17y^2 = 20 \Rightarrow A = 8,\quad B = -12 \text{ and } C = 17$$
$$\cot(2\theta) = \tfrac{A-C}{B} = \tfrac{8-17}{-12}$$
$$\cot(2\theta) = \tfrac{-9}{-12} = \tfrac{3}{4}$$

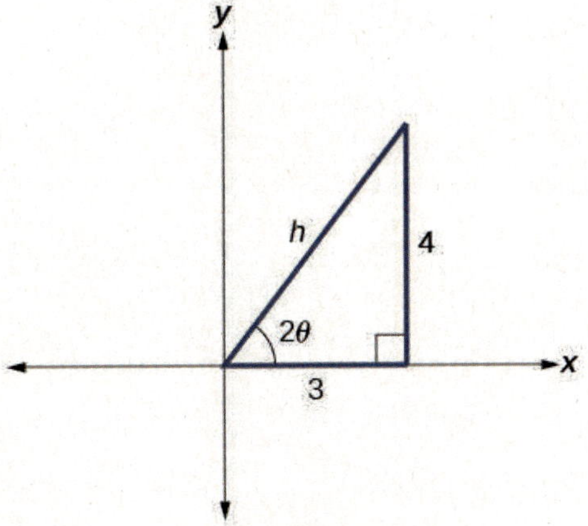

Figure 7

$$\cot(2\theta) = \frac{3}{4} = \frac{\text{adjacent}}{\text{opposite}}$$

So the hypotenuse is

$$3^2 + 4^2 = h^2$$
$$9 + 16 = h^2$$
$$25 = h^2$$
$$h = 5$$

Next, we find $\sin\ \theta$ and $\cos\ \theta$.

$$\sin\ \theta = \sqrt{\frac{1-\cos(2\theta)}{2}} = \sqrt{\frac{1-\frac{3}{5}}{2}} = \sqrt{\frac{\frac{5}{5}-\frac{3}{5}}{2}} = \sqrt{\frac{5-3}{5}\cdot\frac{1}{2}} = \sqrt{\frac{2}{10}} = \sqrt{\frac{1}{5}}$$
$$\sin\ \theta = \frac{1}{\sqrt{5}}$$
$$\cos\ \theta = \sqrt{\frac{1+\cos(2\theta)}{2}} = \sqrt{\frac{1+\frac{3}{5}}{2}} = \sqrt{\frac{\frac{5}{5}+\frac{3}{5}}{2}} = \sqrt{\frac{5+3}{5}\cdot\frac{1}{2}} = \sqrt{\frac{8}{10}} = \sqrt{\frac{4}{5}}$$
$$\cos\ \theta = \frac{2}{\sqrt{5}}$$

Substitute the values of $\sin\ \theta$ and $\cos\ \theta$ into $x = x'\cos\ \theta - y'\sin\ \theta$ and $y = x'\sin\ \theta + y'\cos\ \theta$.

$$x = x'\cos\ \theta - y'\sin\ \theta$$
$$x = x'\left(\frac{2}{\sqrt{5}}\right) - y'\left(\frac{1}{\sqrt{5}}\right)$$
$$x = \frac{2x'-y'}{\sqrt{5}}$$

and

$$y = x'\sin\ \theta + y'\cos\ \theta$$
$$y = x'\left(\frac{1}{\sqrt{5}}\right) + y'\left(\frac{2}{\sqrt{5}}\right)$$
$$y = \frac{x'+2y'}{\sqrt{5}}$$

Substitute the expressions for x and y into in the given equation, and then simplify.

$$8\left(\frac{2x'-y'}{\sqrt{5}}\right)^2 - 12\left(\frac{2x'-y'}{\sqrt{5}}\right)\left(\frac{x'+2y'}{\sqrt{5}}\right) + 17\left(\frac{x'+2y'}{\sqrt{5}}\right)^2 = 20$$
$$8\left(\frac{(2x'-y')(2x'-y')}{5}\right) - 12\left(\frac{(2x'-y')(x'+2y')}{5}\right) + 17\left(\frac{(x'+2y')(x'+2y')}{5}\right) = 20$$
$$8\left(4x'^2 - 4x'y' + y'^2\right) - 12\left(2x'^2 + 3x'y' - 2y'^2\right) + 17\left(x'^2 + 4x'y' + 4y'^2\right) = 100$$
$$32x'^2 - 32x'y' + 8y'^2 - 24x'^2 - 36x'y' + 24y'^2 + 17x'^2 + 68x'y' + 68y'^2 = 100$$
$$25x'^2 + 100y'^2 = 100$$
$$\frac{25}{100}x'^2 + \frac{100}{100}y'^2 = \frac{100}{100}$$

Write the equations with x' and y' in the standard form with respect to the new coordinate system.

$$\frac{x'^2}{4} + \frac{y'^2}{1} = 1$$

Figure 8 shows the graph of the ellipse.

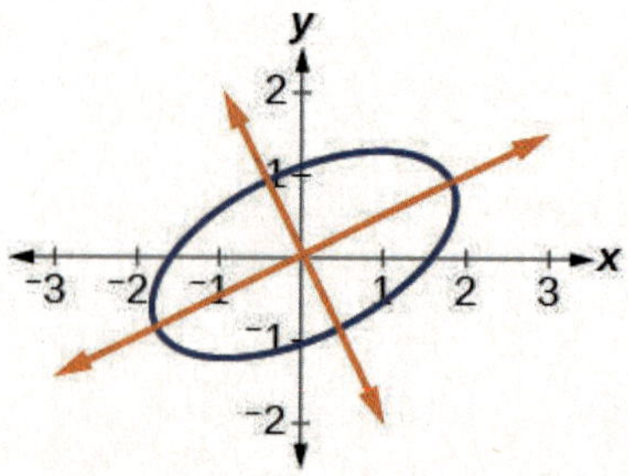

Figure 8

TRY IT #2 Rewrite the $13x^2 - 6\sqrt{3}xy + 7y^2 = 16$ in the $x'y'$ system without the $x'y'$ term.

EXAMPLE 4

Graphing an Equation That Has No x′y′ Terms

Graph the following equation relative to the $x'y'$ system:

$$x^2 + 12xy - 4y^2 = 30$$

Solution

First, we find $\cot(2\theta)$.

$$x^2 + 12xy - 4y^2 = 20 \Rightarrow A = 1,\quad B = 12, \text{ and } C = -4$$

$$\cot(2\theta) = \frac{A-C}{B}$$

$$\cot(2\theta) = \frac{1-(-4)}{12}$$

$$\cot(2\theta) = \frac{5}{12}$$

Because $\cot(2\theta) = \frac{5}{12}$, we can draw a reference triangle as in Figure 9.

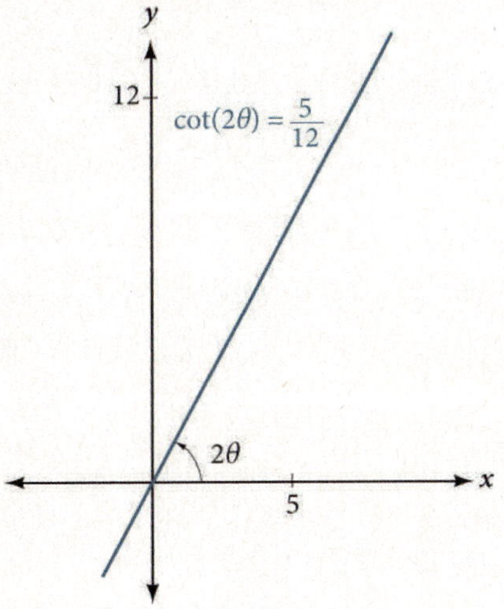

Figure 9

$$\cot(2\theta) = \frac{5}{12} = \frac{\text{adjacent}}{\text{opposite}}$$

Thus, the hypotenuse is

$$5^2 + 12^2 = h^2$$

$$25 + 144 = h^2$$

$$169 = h^2$$

$$h = 13$$

Next, we find $\sin\ \theta$ and $\cos\ \theta$. We will use half-angle identities.

$$\sin\ \theta = \sqrt{\frac{1-\cos(2\theta)}{2}} = \sqrt{\frac{1-\frac{5}{13}}{2}} = \sqrt{\frac{\frac{13}{13}-\frac{5}{13}}{2}} = \sqrt{\frac{8}{13}\cdot\frac{1}{2}} = \frac{2}{\sqrt{13}}$$

$$\cos\ \theta = \sqrt{\frac{1+\cos(2\theta)}{2}} = \sqrt{\frac{1+\frac{5}{13}}{2}} = \sqrt{\frac{\frac{13}{13}+\frac{5}{13}}{2}} = \sqrt{\frac{18}{13}\cdot\frac{1}{2}} = \frac{3}{\sqrt{13}}$$

Now we find x and y.

$$x = x' \cos\ \theta - y' \sin\ \theta$$
$$x = x'\left(\frac{3}{\sqrt{13}}\right) - y'\left(\frac{2}{\sqrt{13}}\right)$$
$$x = \frac{3x' - 2y'}{\sqrt{13}}$$

and

$$y = x' \sin\ \theta + y' \cos\ \theta$$
$$y = x'\left(\frac{2}{\sqrt{13}}\right) + y'\left(\frac{3}{\sqrt{13}}\right)$$
$$y = \frac{2x' + 3y'}{\sqrt{13}}$$

Now we substitute $x = \frac{3x' - 2y'}{\sqrt{13}}$ and $y = \frac{2x' + 3y'}{\sqrt{13}}$ into $x^2 + 12xy - 4y^2 = 30$.

$$\left(\frac{3x' - 2y'}{\sqrt{13}}\right)^2 + 12\left(\frac{3x' - 2y'}{\sqrt{13}}\right)\left(\frac{2x' + 3y'}{\sqrt{13}}\right) - 4\left(\frac{2x' + 3y'}{\sqrt{13}}\right)^2 = 30$$

$$\left(\tfrac{1}{13}\right)\left[(3x' - 2y')^2 + 12(3x' - 2y')(2x' + 3y') - 4(2x' + 3y')^2\right] = 30 \qquad \text{Factor.}$$

$$\left(\tfrac{1}{13}\right)\left[9x'^2 - 12x'y' + 4y'^2 + 12\left(6x'^2 + 5x'y' - 6y'^2\right) - 4\left(4x'^2 + 12x'y' + 9y'^2\right)\right] = 30 \qquad \text{Multiply.}$$

$$\left(\tfrac{1}{13}\right)\left[9x'^2 - 12x'y' + 4y'^2 + 72x'^2 + 60x'y' - 72y'^2 - 16x'^2 - 48x'y' - 36y'^2\right] = 30 \qquad \text{Distribute.}$$

$$\left(\tfrac{1}{13}\right)\left[65x'^2 - 104y'^2\right] = 30 \qquad \text{Combine like terms.}$$

$$65x'^2 - 104y'^2 = 390 \qquad \text{Multiply.}$$

$$\frac{x'^2}{6} - \frac{4y'^2}{15} = 1 \qquad \text{Divide by 390.}$$

Figure 10 shows the graph of the hyperbola $\frac{x'^2}{6} - \frac{4y'^2}{15} = 1$.

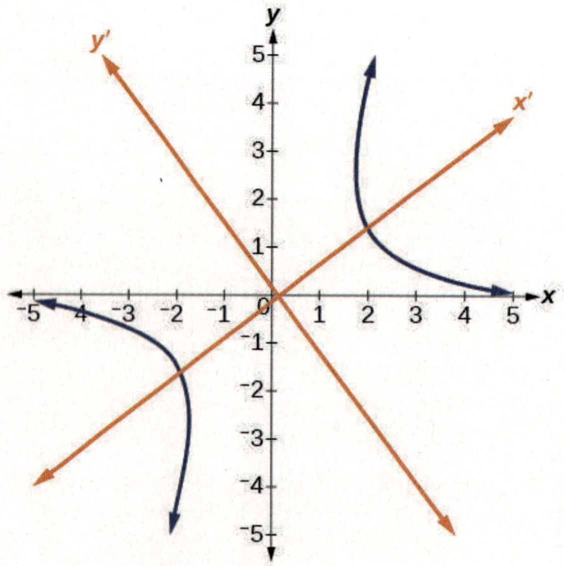

Figure 10

Identifying Conics without Rotating Axes

Now we have come full circle. How do we identify the type of conic described by an equation? What happens when the axes are rotated? Recall, the general form of a conic is

$$Ax^2 + Bxy + Cy^2 + Dx + Ey + F = 0$$

If we apply the rotation formulas to this equation we get the form

$$A'x'^2 + B'x'y' + C'y'^2 + D'x' + E'y' + F' = 0$$

It may be shown that $B^2 - 4AC = B'^2 - 4A'C'$. The expression does not vary after rotation, so we call the expression **invariant.** The discriminant, $B^2 - 4AC$, is invariant and remains unchanged after rotation. Because the discriminant remains unchanged, observing the discriminant enables us to identify the conic section.

Using the Discriminant to Identify a Conic

If the equation $Ax^2 + Bxy + Cy^2 + Dx + Ey + F = 0$ is transformed by rotating axes into the equation $A'x'^2 + B'x'y' + C'y'^2 + D'x' + E'y' + F' = 0$, then $B^2 - 4AC = B'^2 - 4A'C'$.

The equation $Ax^2 + Bxy + Cy^2 + Dx + Ey + F = 0$ is an ellipse, a parabola, or a hyperbola, or a degenerate case of one of these.

If the discriminant, $B^2 - 4AC$, is

- < 0, the conic section is an ellipse
- $= 0$, the conic section is a parabola
- > 0, the conic section is a hyperbola

EXAMPLE 5

Identifying the Conic without Rotating Axes

Identify the conic for each of the following without rotating axes.

ⓐ $5x^2 + 2\sqrt{3}xy + 2y^2 - 5 = 0$ ⓑ $5x^2 + 2\sqrt{3}xy + 12y^2 - 5 = 0$

Solution

ⓐ Let's begin by determining A, B, and C.

$$\underbrace{5}_{A}x^2 + \underbrace{2\sqrt{3}}_{B}xy + \underbrace{2}_{C}y^2 - 5 = 0$$

Now, we find the discriminant.

$$\begin{aligned} B^2 - 4AC &= \left(2\sqrt{3}\right)^2 - 4(5)(2) \\ &= 4(3) - 40 \\ &= 12 - 40 \\ &= -28 < 0 \end{aligned}$$

Therefore, $5x^2 + 2\sqrt{3}xy + 2y^2 - 5 = 0$ represents an ellipse.

ⓑ Again, let's begin by determining A, B, and C.

$$\underbrace{5}_{A}x^2 + \underbrace{2\sqrt{3}}_{B}xy + \underbrace{12}_{C}y^2 - 5 = 0$$

Now, we find the discriminant.

$$\begin{aligned} B^2 - 4AC &= \left(2\sqrt{3}\right)^2 - 4(5)(12) \\ &= 4(3) - 240 \\ &= 12 - 240 \\ &= -228 < 0 \end{aligned}$$

Therefore, $5x^2 + 2\sqrt{3}xy + 12y^2 - 5 = 0$ represents an ellipse.

TRY IT #3 Identify the conic for each of the following without rotating axes.

ⓐ $x^2 - 9xy + 3y^2 - 12 = 0$ ⓑ $10x^2 - 9xy + 4y^2 - 4 = 0$

MEDIA

Access this online resource for additional instruction and practice with conic sections and rotation of axes.

Introduction to Conic Sections (http://openstax.org/l/introconic)

12.4 SECTION EXERCISES

Verbal

1. What effect does the xy term have on the graph of a conic section?

2. If the equation of a conic section is written in the form $Ax^2 + By^2 + Cx + Dy + E = 0$ and $AB = 0$, what can we conclude?

3. If the equation of a conic section is written in the form $Ax^2 + Bxy + Cy^2 + Dx + Ey + F = 0$, and $B^2 - 4AC > 0$, what can we conclude?

4. Given the equation $ax^2 + 4x + 3y^2 - 12 = 0$, what can we conclude if $a > 0$?

5. For the equation $Ax^2 + Bxy + Cy^2 + Dx + Ey + F = 0$, the value of θ that satisfies $\cot(2\theta) = \frac{A-C}{B}$ gives us what information?

Algebraic

For the following exercises, determine which conic section is represented based on the given equation.

6. $9x^2 + 4y^2 + 72x + 36y - 500 = 0$

7. $x^2 - 10x + 4y - 10 = 0$

8. $2x^2 - 2y^2 + 4x - 6y - 2 = 0$

9. $4x^2 - y^2 + 8x - 1 = 0$

10. $4y^2 - 5x + 9y + 1 = 0$

11. $2x^2 + 3y^2 - 8x - 12y + 2 = 0$

12. $4x^2 + 9xy + 4y^2 - 36y - 125 = 0$

13. $3x^2 + 6xy + 3y^2 - 36y - 125 = 0$

14. $-3x^2 + 3\sqrt{3}xy - 4y^2 + 9 = 0$

15. $2x^2 + 4\sqrt{3}xy + 6y^2 - 6x - 3 = 0$

16. $-x^2 + 4\sqrt{2}xy + 2y^2 - 2y + 1 = 0$

17. $8x^2 + 4\sqrt{2}xy + 4y^2 - 10x + 1 = 0$

For the following exercises, find a new representation of the given equation after rotating through the given angle.

18. $3x^2 + xy + 3y^2 - 5 = 0, \theta = 45°$

19. $4x^2 - xy + 4y^2 - 2 = 0, \theta = 45°$

20. $2x^2 + 8xy - 1 = 0, \theta = 30°$

21. $-2x^2 + 8xy + 1 = 0, \theta = 45°$

22. $4x^2 + \sqrt{2}xy + 4y^2 + y + 2 = 0, \theta = 45°$

For the following exercises, determine the angle θ that will eliminate the xy term and write the corresponding equation without the xy term.

23. $x^2 + 3\sqrt{3}xy + 4y^2 + y - 2 = 0$

24. $4x^2 + 2\sqrt{3}xy + 6y^2 + y - 2 = 0$

25. $9x^2 - 3\sqrt{3}xy + 6y^2 + 4y - 3 = 0$

26. $-3x^2 - \sqrt{3}xy - 2y^2 - x = 0$

27. $16x^2 + 24xy + 9y^2 + 6x - 6y + 2 = 0$

28. $x^2 + 4xy + 4y^2 + 3x - 2 = 0$

29. $x^2 + 4xy + y^2 - 2x + 1 = 0$

30. $4x^2 - 2\sqrt{3}xy + 6y^2 - 1 = 0$

Graphical

For the following exercises, rotate through the given angle based on the given equation. Give the new equation and graph the original and rotated equation.

31. $y = -x^2, \theta = -45°$

32. $x = y^2, \theta = 45°$

33. $\frac{x^2}{4} + \frac{y^2}{1} = 1, \theta = 45°$

34. $\frac{y^2}{16} + \frac{x^2}{9} = 1, \theta = 45°$

35. $y^2 - x^2 = 1, \theta = 45°$

36. $y = \frac{x^2}{2}, \theta = 30°$

37. $x = (y - 1)^2, \theta = 30°$

38. $\frac{x^2}{9} + \frac{y^2}{4} = 1, \theta = 30°$

For the following exercises, graph the equation relative to the $x'y'$ system in which the equation has no $x'y'$ term.

39. $xy = 9$

40. $x^2 + 10xy + y^2 - 6 = 0$

41. $x^2 - 10xy + y^2 - 24 = 0$

42. $4x^2 - 3\sqrt{3}xy + y^2 - 22 = 0$

43. $6x^2 + 2\sqrt{3}xy + 4y^2 - 21 = 0$

44. $11x^2 + 10\sqrt{3}xy + y^2 - 64 = 0$

45. $21x^2 + 2\sqrt{3}xy + 19y^2 - 18 = 0$

46. $16x^2 + 24xy + 9y^2 - 130x + 90y = 0$

47. $16x^2 + 24xy + 9y^2 - 60x + 80y = 0$

48. $13x^2 - 6\sqrt{3}xy + 7y^2 - 16 = 0$

49. $4x^2 - 4xy + y^2 - 8\sqrt{5}x - 16\sqrt{5}y = 0$

For the following exercises, determine the angle of rotation in order to eliminate the xy term. Then graph the new set of axes.

50. $6x^2 - 5\sqrt{3}xy + y^2 + 10x - 12y = 0$

51. $6x^2 - 5xy + 6y^2 + 20x - y = 0$

52. $6x^2 - 8\sqrt{3}xy + 14y^2 + 10x - 3y = 0$

53. $4x^2 + 6\sqrt{3}xy + 10y^2 + 20x - 40y = 0$

54. $8x^2 + 3xy + 4y^2 + 2x - 4 = 0$

55. $16x^2 + 24xy + 9y^2 + 20x - 44y = 0$

For the following exercises, determine the value of k based on the given equation.

56. Given $4x^2 + kxy + 16y^2 + 8x + 24y - 48 = 0$, find k for the graph to be a parabola.

57. Given $2x^2 + kxy + 12y^2 + 10x - 16y + 28 = 0$, find k for the graph to be an ellipse.

58. Given $3x^2 + kxy + 4y^2 - 6x + 20y + 128 = 0$, find k for the graph to be a hyperbola.

59. Given $kx^2 + 8xy + 8y^2 - 12x + 16y + 18 = 0$, find k for the graph to be a parabola.

60. Given $6x^2 + 12xy + ky^2 + 16x + 10y + 4 = 0$, find k for the graph to be an ellipse.

12.5 Conic Sections in Polar Coordinates

Learning Objectives

In this section, you will:

- Identify a conic in polar form.
- Graph the polar equations of conics.
- Define conics in terms of a focus and a directrix.

Figure 1 Planets orbiting the sun follow elliptical paths. (credit: NASA Blueshift, Flickr)

Most of us are familiar with orbital motion, such as the motion of a planet around the sun or an electron around an atomic nucleus. Within the planetary system, orbits of planets, asteroids, and comets around a larger celestial body are often elliptical. Comets, however, may take on a parabolic or hyperbolic orbit instead. And, in reality, the characteristics of the planets' orbits may vary over time. Each orbit is tied to the location of the celestial body being orbited and the distance and direction of the planet or other object from that body. As a result, we tend to use polar coordinates to represent these orbits.

In an elliptical orbit, the periapsis is the point at which the two objects are closest, and the apoapsis is the point at which they are farthest apart. Generally, the velocity of the orbiting body tends to increase as it approaches the periapsis and decrease as it approaches the apoapsis. Some objects reach an escape velocity, which results in an infinite orbit. These bodies exhibit either a parabolic or a hyperbolic orbit about a body; the orbiting body breaks free of the celestial body's gravitational pull and fires off into space. Each of these orbits can be modeled by a conic section in the polar coordinate system.

Identifying a Conic in Polar Form

Any conic may be determined by three characteristics: a single focus, a fixed line called the directrix, and the ratio of the distances of each to a point on the graph. Consider the parabola $x = 2 + y^2$ shown in Figure 2.

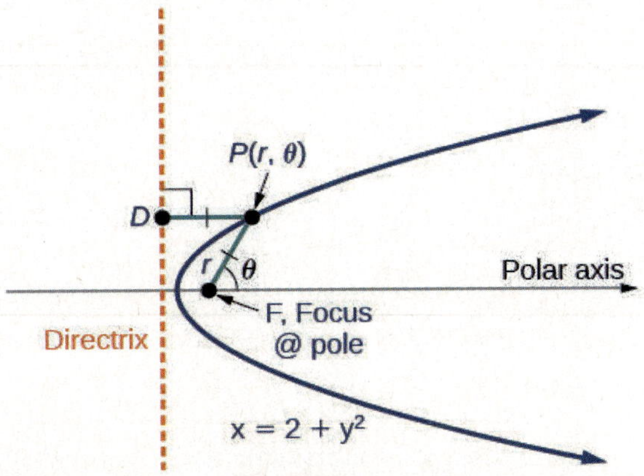

Figure 2

In The Parabola, we learned how a parabola is defined by the focus (a fixed point) and the directrix (a fixed line). In this section, we will learn how to define any conic in the polar coordinate system in terms of a fixed point, the focus $P(r, \theta)$ at the pole, and a line, the directrix, which is perpendicular to the polar axis.

If F is a fixed point, the focus, and D is a fixed line, the directrix, then we can let e be a fixed positive number, called the **eccentricity**, which we can define as the ratio of the distances from a point on the graph to the focus and the point on the graph to the directrix. Then the set of all points P such that $e = \frac{PF}{PD}$ is a conic. In other words, we can define a conic as the set of all points P with the property that the ratio of the distance from P to F to the distance from P to D is equal to the constant e.

For a conic with eccentricity e,

- if $0 \leq e < 1$, the conic is an ellipse
- if $e = 1$, the conic is a parabola
- if $e > 1$, the conic is an hyperbola

With this definition, we may now define a conic in terms of the directrix, $x = \pm p$, the eccentricity e, and the angle θ. Thus, each conic may be written as a **polar equation**, an equation written in terms of r and θ.

The Polar Equation for a Conic

For a conic with a focus at the origin, if the directrix is $x = \pm p$, where p is a positive real number, and the **eccentricity** is a positive real number e, the conic has a **polar equation**

$$r = \frac{ep}{1 \pm e \cos \theta}$$

For a conic with a focus at the origin, if the directrix is $y = \pm p$, where p is a positive real number, and the eccentricity is a positive real number e, the conic has a polar equation

$$r = \frac{ep}{1 \pm e \sin \theta}$$

HOW TO

Given the polar equation for a conic, identify the type of conic, the directrix, and the eccentricity.

1. Multiply the numerator and denominator by the reciprocal of the constant in the denominator to rewrite the equation in standard form.
2. Identify the eccentricity e as the coefficient of the trigonometric function in the denominator.
3. Compare e with 1 to determine the shape of the conic.
4. Determine the directrix as $x = p$ if cosine is in the denominator and $y = p$ if sine is in the denominator. Set ep

equal to the numerator in standard form to solve for x or y.

EXAMPLE 1

Identifying a Conic Given the Polar Form

For each of the following equations, identify the conic with focus at the origin, the directrix, and the eccentricity.

a. $r = \frac{6}{3+2 \sin \theta}$
b. $r = \frac{12}{4+5 \cos \theta}$
c. $r = \frac{7}{2-2 \sin \theta}$

Solution

For each of the three conics, we will rewrite the equation in standard form. Standard form has a 1 as the constant in the denominator. Therefore, in all three parts, the first step will be to multiply the numerator and denominator by the reciprocal of the constant of the original equation, $\frac{1}{c}$, where c is that constant.

a. Multiply the numerator and denominator by $\frac{1}{3}$.

$$r = \frac{6}{3+2\sin \theta} \cdot \frac{\left(\frac{1}{3}\right)}{\left(\frac{1}{3}\right)} = \frac{6\left(\frac{1}{3}\right)}{3\left(\frac{1}{3}\right)+2\left(\frac{1}{3}\right)\sin \theta} = \frac{2}{1+\frac{2}{3} \sin \theta}$$

Because $\sin \theta$ is in the denominator, the directrix is $y = p$. Comparing to standard form, note that $e = \frac{2}{3}$. Therefore, from the numerator,

$$\begin{aligned} 2 &= ep \\ 2 &= \frac{2}{3}p \\ \left(\frac{3}{2}\right)2 &= \left(\frac{3}{2}\right)\frac{2}{3}p \\ 3 &= p \end{aligned}$$

Since $e < 1$, the conic is an ellipse. The eccentricity is $e = \frac{2}{3}$ and the directrix is $y = 3$.

b. Multiply the numerator and denominator by $\frac{1}{4}$.

$$\begin{aligned} r &= \frac{12}{4+5 \cos \theta} \cdot \frac{\left(\frac{1}{4}\right)}{\left(\frac{1}{4}\right)} \\ r &= \frac{12\left(\frac{1}{4}\right)}{4\left(\frac{1}{4}\right)+5\left(\frac{1}{4}\right) \cos \theta} \\ r &= \frac{3}{1+\frac{5}{4} \cos \theta} \end{aligned}$$

Because $\cos \theta$ is in the denominator, the directrix is $x = p$. Comparing to standard form, $e = \frac{5}{4}$. Therefore, from the numerator,

$$\begin{aligned} 3 &= ep \\ 3 &= \frac{5}{4}p \\ \left(\frac{4}{5}\right)3 &= \left(\frac{4}{5}\right)\frac{5}{4}p \\ \frac{12}{5} &= p \end{aligned}$$

Since $e > 1$, the conic is a hyperbola. The eccentricity is $e = \frac{5}{4}$ and the directrix is $x = \frac{12}{5} = 2.4$.

c. Multiply the numerator and denominator by $\frac{1}{2}$.

$$r = \frac{7}{2-2\sin\theta} \cdot \frac{\left(\frac{1}{2}\right)}{\left(\frac{1}{2}\right)}$$

$$r = \frac{7\left(\frac{1}{2}\right)}{2\left(\frac{1}{2}\right)-2\left(\frac{1}{2}\right)\sin\theta}$$

$$r = \frac{\frac{7}{2}}{1-\sin\theta}$$

Because sine is in the denominator, the directrix is $y = -p$. Comparing to standard form, $e = 1$. Therefore, from the numerator,

$$\frac{7}{2} = ep$$
$$\frac{7}{2} = (1)\,p$$
$$\frac{7}{2} = p$$

Because $e = 1$, the conic is a parabola. The eccentricity is $e = 1$ and the directrix is $y = -\frac{7}{2} = -3.5$.

> **TRY IT** #1 Identify the conic with focus at the origin, the directrix, and the eccentricity for $r = \frac{2}{3-\cos\theta}$.

Graphing the Polar Equations of Conics

When graphing in Cartesian coordinates, each conic section has a unique equation. This is not the case when graphing in polar coordinates. We must use the eccentricity of a conic section to determine which type of curve to graph, and then determine its specific characteristics. The first step is to rewrite the conic in standard form as we have done in the previous example. In other words, we need to rewrite the equation so that the denominator begins with 1. This enables us to determine e and, therefore, the shape of the curve. The next step is to substitute values for θ and solve for r to plot a few key points. Setting θ equal to $0, \frac{\pi}{2}, \pi,$ and $\frac{3\pi}{2}$ provides the vertices so we can create a rough sketch of the graph.

EXAMPLE 2

Graphing a Parabola in Polar Form

Graph $r = \frac{5}{3+3\cos\theta}$.

Solution

First, we rewrite the conic in standard form by multiplying the numerator and denominator by the reciprocal of 3, which is $\frac{1}{3}$.

$$r = \frac{5}{3+3\cos\theta} = \frac{5\left(\frac{1}{3}\right)}{3\left(\frac{1}{3}\right)+3\left(\frac{1}{3}\right)\cos\theta}$$

$$r = \frac{\frac{5}{3}}{1+\cos\theta}$$

Because $e = 1$, we will graph a parabola with a focus at the origin. The function has a $\cos\theta$, and there is an addition sign in the denominator, so the directrix is $x = p$.

$$\frac{5}{3} = ep$$
$$\frac{5}{3} = (1)p$$
$$\frac{5}{3} = p$$

The directrix is $x = \frac{5}{3}$.

Plotting a few key points as in Table 1 will enable us to see the vertices. See Figure 3.

	A	B	C	D
θ	0	$\frac{\pi}{2}$	π	$\frac{3\pi}{2}$
$r = \frac{5}{3+3\cos\theta}$	$\frac{5}{6} \approx 0.83$	$\frac{5}{3} \approx 1.67$	undefined	$\frac{5}{3} \approx 1.67$

Table 1

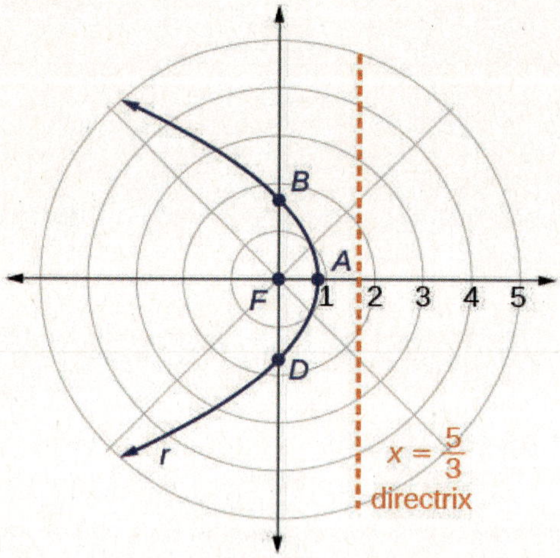

Figure 3

Analysis
We can check our result with a graphing utility. See Figure 4.

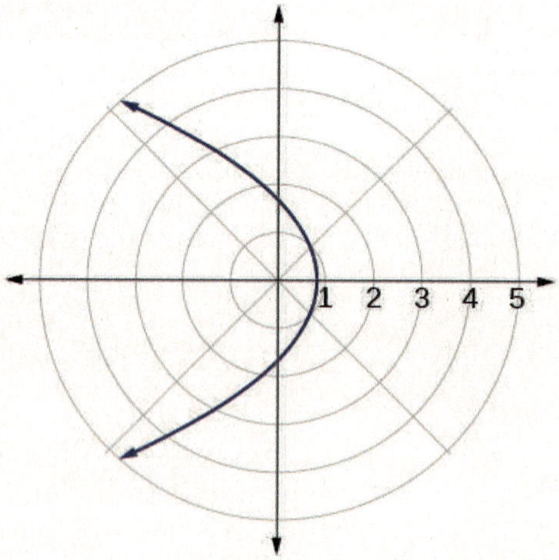

Figure 4

EXAMPLE 3

Graphing a Hyperbola in Polar Form
Graph $r = \frac{8}{2-3\sin\theta}$.

Solution

First, we rewrite the conic in standard form by multiplying the numerator and denominator by the reciprocal of 2, which is $\frac{1}{2}$.

$$r = \frac{8}{2-3\sin\ \theta} = \frac{8\left(\frac{1}{2}\right)}{2\left(\frac{1}{2}\right)-3\left(\frac{1}{2}\right)\sin\ \theta}$$

$$r = \frac{4}{1-\frac{3}{2}\ \sin\ \theta}$$

Because $e = \frac{3}{2}, e > 1$, so we will graph a hyperbola with a focus at the origin. The function has a $\sin\ \theta$ term and there is a subtraction sign in the denominator, so the directrix is $y = -p$.

$$4 = ep$$
$$4 = \left(\frac{3}{2}\right)p$$
$$4\left(\frac{2}{3}\right) = p$$
$$\frac{8}{3} = p$$

The directrix is $y = -\frac{8}{3}$.

Plotting a few key points as in Table 2 will enable us to see the vertices. See Figure 5.

	A	B	C	D
θ	0	$\frac{\pi}{2}$	π	$\frac{3\pi}{2}$
$r = \frac{8}{2-3\sin\ \theta}$	4	-8	4	$\frac{8}{5} = 1.6$

Table 2

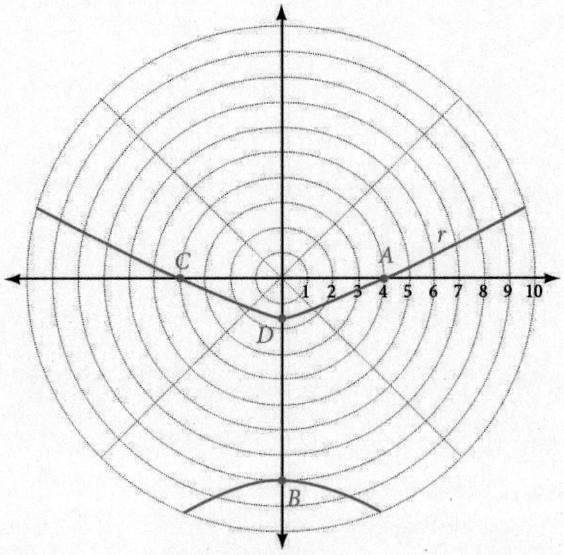

Figure 5

EXAMPLE 4

Graphing an Ellipse in Polar Form

Graph $r = \frac{10}{5-4 \cos \theta}$.

Solution

First, we rewrite the conic in standard form by multiplying the numerator and denominator by the reciprocal of 5, which is $\frac{1}{5}$.

$$r = \frac{10}{5-4\cos \theta} = \frac{10\left(\frac{1}{5}\right)}{5\left(\frac{1}{5}\right)-4\left(\frac{1}{5}\right)\cos \theta}$$

$$r = \frac{2}{1-\frac{4}{5} \cos \theta}$$

Because $e = \frac{4}{5}, e < 1$, so we will graph an ellipse with a focus at the origin. The function has a $\cos \theta$, and there is a subtraction sign in the denominator, so the directrix is $x = -p$.

$$2 = ep$$
$$2 = \left(\frac{4}{5}\right)p$$
$$2\left(\frac{5}{4}\right) = p$$
$$\frac{5}{2} = p$$

The directrix is $x = -\frac{5}{2}$.

Plotting a few key points as in Table 3 will enable us to see the vertices. See Figure 6.

	A	B	C	D
θ	0	$\frac{\pi}{2}$	π	$\frac{3\pi}{2}$
$r = \frac{10}{5-4 \cos \theta}$	10	2	$\frac{10}{9} \approx 1.1$	2

Table 3

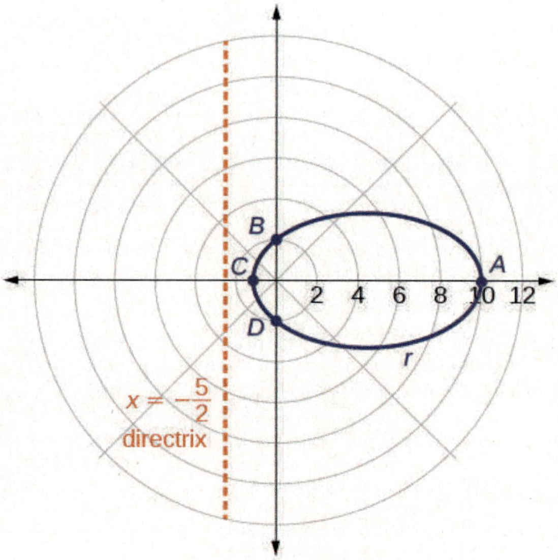

Figure 6

Analysis

We can check our result using a graphing utility. See Figure 7.

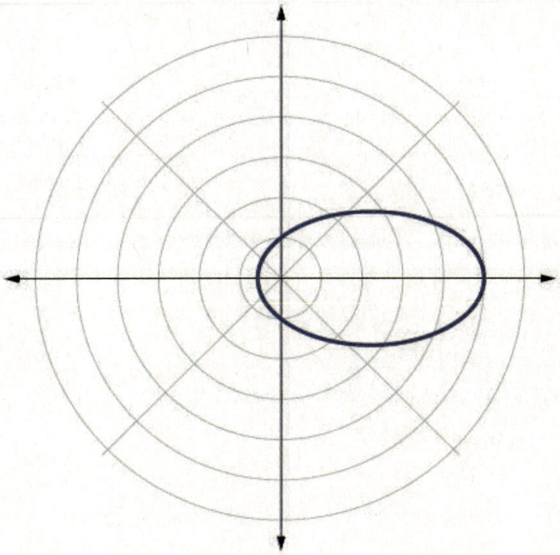

Figure 7 $r = \frac{10}{5-4\ \cos\ \theta}$ graphed on a viewing window of $[-3, 12, 1]$ by $[-4, 4, 1], \theta\ \min = \ 0$ and $\theta\ \max = \ 2\pi$.

TRY IT #2 Graph $r = \frac{2}{4-\cos\ \theta}$.

Defining Conics in Terms of a Focus and a Directrix

So far we have been using polar equations of conics to describe and graph the curve. Now we will work in reverse; we will use information about the origin, eccentricity, and directrix to determine the polar equation.

HOW TO

Given the focus, eccentricity, and directrix of a conic, determine the polar equation.

1. Determine whether the directrix is horizontal or vertical. If the directrix is given in terms of y, we use the general polar form in terms of sine. If the directrix is given in terms of x, we use the general polar form in terms of cosine.
2. Determine the sign in the denominator. If $p < 0$, use subtraction. If $p > 0$, use addition.
3. Write the coefficient of the trigonometric function as the given eccentricity.
4. Write the absolute value of p in the numerator, and simplify the equation.

EXAMPLE 5

Finding the Polar Form of a Vertical Conic Given a Focus at the Origin and the Eccentricity and Directrix

Find the polar form of the conic given a focus at the origin, $e = 3$ and directrix $y = -2$.

Solution

The directrix is $y = -p$, so we know the trigonometric function in the denominator is sine.

Because $y = -2, -2 < 0$, so we know there is a subtraction sign in the denominator. We use the standard form of

$$r = \frac{ep}{1 - e\ \sin\ \theta}$$

and $e = 3$ and $|-2| = 2 = p$.

Therefore,

$$r = \frac{(3)(2)}{1-3 \sin \theta}$$
$$r = \frac{6}{1-3 \sin \theta}$$

EXAMPLE 6

Finding the Polar Form of a Horizontal Conic Given a Focus at the Origin and the Eccentricity and Directrix

Find the polar form of a conic given a focus at the origin, $e = \frac{3}{5}$, and directrix $x = 4$.

Solution

Because the directrix is $x = p$, we know the function in the denominator is cosine. Because $x = 4, 4 > 0$, so we know there is an addition sign in the denominator. We use the standard form of

$$r = \frac{ep}{1 + e \cos \theta}$$

and $e = \frac{3}{5}$ and $|4| = 4 = p$.

Therefore,

$$r = \frac{\left(\frac{3}{5}\right)(4)}{1+\frac{3}{5} \cos \theta}$$
$$r = \frac{\frac{12}{5}}{1+\frac{3}{5} \cos \theta}$$
$$r = \frac{\frac{12}{5}}{1\left(\frac{5}{5}\right)+\frac{3}{5} \cos \theta}$$
$$r = \frac{\frac{12}{5}}{\frac{5}{5}+\frac{3}{5} \cos \theta}$$
$$r = \frac{12}{5} \cdot \frac{5}{5+3 \cos \theta}$$
$$r = \frac{12}{5+3 \cos \theta}$$

TRY IT #3 Find the polar form of the conic given a focus at the origin, $e = 1$, and directrix $x = -1$.

EXAMPLE 7

Converting a Conic in Polar Form to Rectangular Form

Convert the conic $r = \frac{1}{5-5 \sin \theta}$ to rectangular form.

Solution

We will rearrange the formula to use the identities $r = \sqrt{x^2 + y^2}, x = r \cos \theta,$ and $y = r \sin \theta.$

$$r = \frac{1}{5-5 \sin \theta}$$

$r \cdot (5 - 5 \sin \theta) = \frac{1}{5-5 \sin \theta} \cdot (5 - 5 \sin \theta)$	Eliminate the fraction.
$5r - 5r \sin \theta = 1$	Distribute.
$5r = 1 + 5r \sin \theta$	Isolate $5r$.
$25r^2 = (1 + 5r \sin \theta)^2$	Square both sides.
$25(x^2 + y^2) = (1 + 5y)^2$	Substitute $r = \sqrt{x^2 + y^2}$ and $y = r \sin \theta$.
$25x^2 + 25y^2 = 1 + 10y + 25y^2$	Distribute and use FOIL.
$25x^2 - 10y = 1$	Rearrange terms and set equal to 1.

> **TRY IT** #4 Convert the conic $r = \frac{2}{1+2 \cos \theta}$ to rectangular form.

MEDIA

Access these online resources for additional instruction and practice with conics in polar coordinates.

Polar Equations of Conic Sections (http://openstax.org/l/determineconic)
Graphing Polar Equations of Conics - 1 (http://openstax.org/l/graphconic1)
Graphing Polar Equations of Conics - 2 (http://openstax.org/l/graphconic2)

12.5 SECTION EXERCISES

Verbal

1. Explain how eccentricity determines which conic section is given.

2. If a conic section is written as a polar equation, what must be true of the denominator?

3. If a conic section is written as a polar equation, and the denominator involves $\sin \theta$, what conclusion can be drawn about the directrix?

4. If the directrix of a conic section is perpendicular to the polar axis, what do we know about the equation of the graph?

5. What do we know about the focus/foci of a conic section if it is written as a polar equation?

Algebraic

For the following exercises, identify the conic with a focus at the origin, and then give the directrix and eccentricity.

6. $r = \frac{6}{1-2 \cos \theta}$

7. $r = \frac{3}{4-4 \sin \theta}$

8. $r = \frac{8}{4-3 \cos \theta}$

9. $r = \frac{5}{1+2 \sin \theta}$

10. $r = \frac{16}{4+3 \cos \theta}$

11. $r = \frac{3}{10+10 \cos \theta}$

12. $r = \frac{2}{1-\cos \theta}$

13. $r = \frac{4}{7+2 \cos \theta}$

14. $r(1 - \cos \theta) = 3$

15. $r(3 + 5 \sin \theta) = 11$

16. $r(4 - 5 \sin \theta) = 1$

17. $r(7 + 8 \cos \theta) = 7$

For the following exercises, convert the polar equation of a conic section to a rectangular equation.

18. $r = \frac{4}{1+3\sin\theta}$

19. $r = \frac{2}{5-3\sin\theta}$

20. $r = \frac{8}{3-2\cos\theta}$

21. $r = \frac{3}{2+5\cos\theta}$

22. $r = \frac{4}{2+2\sin\theta}$

23. $r = \frac{3}{8-8\cos\theta}$

24. $r = \frac{2}{6+7\cos\theta}$

25. $r = \frac{5}{5-11\sin\theta}$

26. $r(5+2\cos\theta) = 6$

27. $r(2-\cos\theta) = 1$

28. $r(2.5-2.5\sin\theta) = 5$

29. $r = \frac{6\sec\theta}{-2+3\sec\theta}$

30. $r = \frac{6\csc\theta}{3+2\csc\theta}$

For the following exercises, graph the given conic section. If it is a parabola, label the vertex, focus, and directrix. If it is an ellipse, label the vertices and foci. If it is a hyperbola, label the vertices and foci.

31. $r = \frac{5}{2+\cos\theta}$

32. $r = \frac{2}{3+3\sin\theta}$

33. $r = \frac{10}{5-4\sin\theta}$

34. $r = \frac{3}{1+2\cos\theta}$

35. $r = \frac{8}{4-5\cos\theta}$

36. $r = \frac{3}{4-4\cos\theta}$

37. $r = \frac{2}{1-\sin\theta}$

38. $r = \frac{6}{3+2\sin\theta}$

39. $r(1+\cos\theta) = 5$

40. $r(3-4\sin\theta) = 9$

41. $r(3-2\sin\theta) = 6$

42. $r(6-4\cos\theta) = 5$

For the following exercises, find the polar equation of the conic with focus at the origin and the given eccentricity and directrix.

43. Directrix: $x = 4;\ e = \frac{1}{5}$

44. Directrix: $x = -4;\ e = 5$

45. Directrix: $y = 2;\ e = 2$

46. Directrix: $y = -2;\ e = \frac{1}{2}$

47. Directrix: $x = 1;\ e = 1$

48. Directrix: $x = -1;\ e = 1$

49. Directrix: $x = -\frac{1}{4};\ e = \frac{7}{2}$

50. Directrix: $y = \frac{2}{5};\ e = \frac{7}{2}$

51. Directrix: $y = 4;\ e = \frac{3}{2}$

52. Directrix: $x = -2;\ e = \frac{8}{3}$

53. Directrix: $x = -5;\ e = \frac{3}{4}$

54. Directrix: $y = 2;\ e = 2.5$

55. Directrix: $x = -3;\ e = \frac{1}{3}$

Extensions

Recall from Rotation of Axes that equations of conics with an xy term have rotated graphs. For the following exercises, express each equation in polar form with r as a function of θ.

56. $xy = 2$

57. $x^2 + xy + y^2 = 4$

58. $2x^2 + 4xy + 2y^2 = 9$

59. $16x^2 + 24xy + 9y^2 = 4$

60. $2xy + y = 1$

Chapter Review

Key Terms

angle of rotation an acute angle formed by a set of axes rotated from the Cartesian plane where, if $\cot(2\theta) > 0$, then θ is between $(0°, 45°)$; if $\cot(2\theta) < 0$, then θ is between $(45°, 90°)$; and if $\cot(2\theta) = 0$, then $\theta = 45°$

center of a hyperbola the midpoint of both the transverse and conjugate axes of a hyperbola

center of an ellipse the midpoint of both the major and minor axes

conic section any shape resulting from the intersection of a right circular cone with a plane

conjugate axis the axis of a hyperbola that is perpendicular to the transverse axis and has the co-vertices as its endpoints

degenerate conic sections any of the possible shapes formed when a plane intersects a double cone through the apex. Types of degenerate conic sections include a point, a line, and intersecting lines.

directrix a line perpendicular to the axis of symmetry of a parabola; a line such that the ratio of the distance between the points on the conic and the focus to the distance to the directrix is constant

eccentricity the ratio of the distances from a point P on the graph to the focus F and to the directrix D represented by $e = \frac{PF}{PD}$, where e is a positive real number

ellipse the set of all points (x, y) in a plane such that the sum of their distances from two fixed points is a constant

foci plural of focus

focus (of a parabola) a fixed point in the interior of a parabola that lies on the axis of symmetry

focus (of an ellipse) one of the two fixed points on the major axis of an ellipse such that the sum of the distances from these points to any point (x, y) on the ellipse is a constant

hyperbola the set of all points (x, y) in a plane such that the difference of the distances between (x, y) and the foci is a positive constant

latus rectum the line segment that passes through the focus of a parabola parallel to the directrix, with endpoints on the parabola

major axis the longer of the two axes of an ellipse

minor axis the shorter of the two axes of an ellipse

nondegenerate conic section a shape formed by the intersection of a plane with a double right cone such that the plane does not pass through the apex; nondegenerate conics include circles, ellipses, hyperbolas, and parabolas

parabola the set of all points (x, y) in a plane that are the same distance from a fixed line, called the directrix, and a fixed point (the focus) not on the directrix

polar equation an equation of a curve in polar coordinates r and θ

transverse axis the axis of a hyperbola that includes the foci and has the vertices as its endpoints

Key Equations

Horizontal ellipse, center at origin	$\frac{x^2}{a^2} + \frac{y^2}{b^2} = 1, \quad a > b$
Vertical ellipse, center at origin	$\frac{x^2}{b^2} + \frac{y^2}{a^2} = 1, \quad a > b$
Horizontal ellipse, center (h, k)	$\frac{(x-h)^2}{a^2} + \frac{(y-k)^2}{b^2} = 1, \quad a > b$
Vertical ellipse, center (h, k)	$\frac{(x-h)^2}{b^2} + \frac{(y-k)^2}{a^2} = 1, \quad a > b$

Hyperbola, center at origin, transverse axis on *x*-axis	$\frac{x^2}{a^2} - \frac{y^2}{b^2} = 1$
Hyperbola, center at origin, transverse axis on *y*-axis	$\frac{y^2}{a^2} - \frac{x^2}{b^2} = 1$
Hyperbola, center at (h, k), transverse axis parallel to *x*-axis	$\frac{(x-h)^2}{a^2} - \frac{(y-k)^2}{b^2} = 1$

Hyperbola, center at (h, k), transverse axis parallel to y-axis	$\frac{(y-k)^2}{a^2} - \frac{(x-h)^2}{b^2} = 1$
Parabola, vertex at origin, axis of symmetry on x-axis	$y^2 = 4px$
Parabola, vertex at origin, axis of symmetry on y-axis	$x^2 = 4py$
Parabola, vertex at (h, k), axis of symmetry on x-axis	$(y-k)^2 = 4p(x-h)$
Parabola, vertex at (h, k), axis of symmetry on y-axis	$(x-h)^2 = 4p(y-k)$
General Form equation of a conic section	$Ax^2 + Bxy + Cy^2 + Dx + Ey + F = 0$
Rotation of a conic section	$x = x' \cos \theta - y' \sin \theta$ $y = x' \sin \theta + y' \cos \theta$
Angle of rotation	θ, where $\cot(2\theta) = \frac{A-C}{B}$

Key Concepts

12.1 The Ellipse

- An ellipse is the set of all points (x, y) in a plane such that the sum of their distances from two fixed points is a constant. Each fixed point is called a focus (plural: foci).
- When given the coordinates of the foci and vertices of an ellipse, we can write the equation of the ellipse in standard form. See Example 1 and Example 2.
- When given an equation for an ellipse centered at the origin in standard form, we can identify its vertices, co-vertices, foci, and the lengths and positions of the major and minor axes in order to graph the ellipse. See Example 3 and Example 4.
- When given the equation for an ellipse centered at some point other than the origin, we can identify its key features and graph the ellipse. See Example 5 and Example 6.
- Real-world situations can be modeled using the standard equations of ellipses and then evaluated to find key features, such as lengths of axes and distance between foci. See Example 7.

12.2 The Hyperbola

- A hyperbola is the set of all points (x, y) in a plane such that the difference of the distances between (x, y) and the foci is a positive constant.
- The standard form of a hyperbola can be used to locate its vertices and foci. See Example 1.
- When given the coordinates of the foci and vertices of a hyperbola, we can write the equation of the hyperbola in standard form. See Example 2 and Example 3.
- When given an equation for a hyperbola, we can identify its vertices, co-vertices, foci, asymptotes, and lengths and positions of the transverse and conjugate axes in order to graph the hyperbola. See Example 4 and Example 5.
- Real-world situations can be modeled using the standard equations of hyperbolas. For instance, given the dimensions of a natural draft cooling tower, we can find a hyperbolic equation that models its sides. See Example 6.

12.3 The Parabola

- A parabola is the set of all points (x, y) in a plane that are the same distance from a fixed line, called the directrix, and a fixed point (the focus) not on the directrix.
- The standard form of a parabola with vertex $(0, 0)$ and the x-axis as its axis of symmetry can be used to graph the parabola. If $p > 0$, the parabola opens right. If $p < 0$, the parabola opens left. See Example 1.
- The standard form of a parabola with vertex $(0, 0)$ and the y-axis as its axis of symmetry can be used to graph the parabola. If $p > 0$, the parabola opens up. If $p < 0$, the parabola opens down. See Example 2.
- When given the focus and directrix of a parabola, we can write its equation in standard form. See Example 3.
- The standard form of a parabola with vertex (h, k) and axis of symmetry parallel to the x-axis can be used to graph

the parabola. If $p > 0$, the parabola opens right. If $p < 0$, the parabola opens left. See Example 4.

- The standard form of a parabola with vertex (h, k) and axis of symmetry parallel to the *y*-axis can be used to graph the parabola. If $p > 0$, the parabola opens up. If $p < 0$, the parabola opens down. See Example 5.
- Real-world situations can be modeled using the standard equations of parabolas. For instance, given the diameter and focus of a cross-section of a parabolic reflector, we can find an equation that models its sides. See Example 6.

12.4 Rotation of Axes

- Four basic shapes can result from the intersection of a plane with a pair of right circular cones connected tail to tail. They include an ellipse, a circle, a hyperbola, and a parabola.
- A nondegenerate conic section has the general form $Ax^2 + Bxy + Cy^2 + Dx + Ey + F = 0$ where A, B and C are not all zero. The values of A, B, and C determine the type of conic. See Example 1.
- Equations of conic sections with an xy term have been rotated about the origin. See Example 2.
- The general form can be transformed into an equation in the x' and y' coordinate system without the $x'y'$ term. See Example 3 and Example 4.
- An expression is described as invariant if it remains unchanged after rotating. Because the discriminant is invariant, observing it enables us to identify the conic section. See Example 5.

12.5 Conic Sections in Polar Coordinates

- Any conic may be determined by a single focus, the corresponding eccentricity, and the directrix. We can also define a conic in terms of a fixed point, the focus $P(r, \theta)$ at the pole, and a line, the directrix, which is perpendicular to the polar axis.
- A conic is the set of all points $e = \frac{PF}{PD}$, where eccentricity e is a positive real number. Each conic may be written in terms of its polar equation. See Example 1.
- The polar equations of conics can be graphed. See Example 2, Example 3, and Example 4.
- Conics can be defined in terms of a focus, a directrix, and eccentricity. See Example 5 and Example 6.
- We can use the identities $r = \sqrt{x^2 + y^2}$, $x = r \cos \theta$, and $y = r \sin \theta$ to convert the equation for a conic from polar to rectangular form. See Example 7.

Exercises

Review Exercises

The Ellipse

For the following exercises, write the equation of the ellipse in standard form. Then identify the center, vertices, and foci.

1. $\frac{x^2}{25} + \frac{y^2}{64} = 1$

2. $\frac{(x-2)^2}{100} + \frac{(y+3)^2}{36} = 1$

3. $9x^2 + y^2 + 54x - 4y + 76 = 0$

4. $9x^2 + 36y^2 - 36x + 72y + 36 = 0$

For the following exercises, graph the ellipse, noting center, vertices, and foci.

5. $\frac{x^2}{36} + \frac{y^2}{9} = 1$

6. $\frac{(x-4)^2}{25} + \frac{(y+3)^2}{49} = 1$

7. $4x^2 + y^2 + 16x + 4y - 44 = 0$

8. $2x^2 + 3y^2 - 20x + 12y + 38 = 0$

For the following exercises, use the given information to find the equation for the ellipse.

9. Center at $(0, 0)$, focus at $(3, 0)$, vertex at $(-5, 0)$

10. Center at $(2, -2)$, vertex at $(7, -2)$, focus at $(4, -2)$

11. A whispering gallery is to be constructed such that the foci are located 35 feet from the center. If the length of the gallery is to be 100 feet, what should the height of the ceiling be?

The Hyperbola

For the following exercises, write the equation of the hyperbola in standard form. Then give the center, vertices, and foci.

12. $\frac{x^2}{81} - \frac{y^2}{9} = 1$

13. $\frac{(y+1)^2}{16} - \frac{(x-4)^2}{36} = 1$

14. $9y^2 - 4x^2 + 54y - 16x + 29 = 0$

15. $3x^2 - y^2 - 12x - 6y - 9 = 0$

For the following exercises, graph the hyperbola, labeling vertices and foci.

16. $\frac{x^2}{9} - \frac{y^2}{16} = 1$

17. $\frac{(y-1)^2}{49} - \frac{(x+1)^2}{4} = 1$

18. $x^2 - 4y^2 + 6x + 32y - 91 = 0$

19. $2y^2 - x^2 - 12y - 6 = 0$

For the following exercises, find the equation of the hyperbola.

20. Center at $(0, 0)$, vertex at $(0, 4)$, focus at $(0, -6)$

21. Foci at $(3, 7)$ and $(7, 7)$, vertex at $(6, 7)$

The Parabola

For the following exercises, write the equation of the parabola in standard form. Then give the vertex, focus, and directrix.

22. $y^2 = 12x$

23. $(x + 2)^2 = \frac{1}{2}(y - 1)$

24. $y^2 - 6y - 6x - 3 = 0$

25. $x^2 + 10x - y + 23 = 0$

For the following exercises, graph the parabola, labeling vertex, focus, and directrix.

26. $x^2 + 4y = 0$

27. $(y - 1)^2 = \frac{1}{2}(x + 3)$

28. $x^2 - 8x - 10y + 46 = 0$

29. $2y^2 + 12y + 6x + 15 = 0$

For the following exercises, write the equation of the parabola using the given information.

30. Focus at $(-4, 0)$; directrix is $x = 4$

31. Focus at $\left(2, \frac{9}{8}\right)$; directrix is $y = \frac{7}{8}$

32. A cable TV receiving dish is the shape of a paraboloid of revolution. Find the location of the receiver, which is placed at the focus, if the dish is 5 feet across at its opening and 1.5 feet deep.

Rotation of Axes

For the following exercises, determine which of the conic sections is represented.

33. $16x^2 + 24xy + 9y^2 + 24x - 60y - 60 = 0$

34. $4x^2 + 14xy + 5y^2 + 18x - 6y + 30 = 0$

35. $4x^2 + xy + 2y^2 + 8x - 26y + 9 = 0$

For the following exercises, determine the angle θ that will eliminate the xy term, and write the corresponding equation without the xy term.

36. $x^2 + 4xy - 2y^2 - 6 = 0$

37. $x^2 - xy + y^2 - 6 = 0$

For the following exercises, graph the equation relative to the $x'y'$ system in which the equation has no $x'y'$ term.

38. $9x^2 - 24xy + 16y^2 - 80x - 60y + 100 = 0$

39. $x^2 - xy + y^2 - 2 = 0$

40. $6x^2 + 24xy - y^2 - 12x + 26y + 11 = 0$

Conic Sections in Polar Coordinates

For the following exercises, given the polar equation of the conic with focus at the origin, identify the eccentricity and directrix.

41. $r = \frac{10}{1-5 \cos \theta}$

42. $r = \frac{6}{3+2 \cos \theta}$

43. $r = \frac{1}{4+3 \sin \theta}$

44. $r = \frac{3}{5-5 \sin \theta}$

For the following exercises, graph the conic given in polar form. If it is a parabola, label the vertex, focus, and directrix. If it is an ellipse or a hyperbola, label the vertices and foci.

45. $r = \frac{3}{1-\sin \theta}$

46. $r = \frac{8}{4+3 \sin \theta}$

47. $r = \frac{10}{4+5 \cos \theta}$

48. $r = \frac{9}{3-6 \cos \theta}$

For the following exercises, given information about the graph of a conic with focus at the origin, find the equation in polar form.

49. Directrix is $x = 3$ and eccentricity $e = 1$

50. Directrix is $y = -2$ and eccentricity $e = 4$

Practice Test

For the following exercises, write the equation in standard form and state the center, vertices, and foci.

1. $\frac{x^2}{9} + \frac{y^2}{4} = 1$

2. $9y^2 + 16x^2 - 36y + 32x - 92 = 0$

For the following exercises, sketch the graph, identifying the center, vertices, and foci.

3. $\frac{(x-3)^2}{64} + \frac{(y-2)^2}{36} = 1$

4. $2x^2 + y^2 + 8x - 6y - 7 = 0$

5. Write the standard form equation of an ellipse with a center at $(1, 2)$, vertex at $(7, 2)$, and focus at $(4, 2)$.

6. A whispering gallery is to be constructed with a length of 150 feet. If the foci are to be located 20 feet away from the wall, how high should the ceiling be?

For the following exercises, write the equation of the hyperbola in standard form, and give the center, vertices, foci, and asymptotes.

7. $\frac{x^2}{49} - \frac{y^2}{81} = 1$

8. $16y^2 - 9x^2 + 128y + 112 = 0$

For the following exercises, graph the hyperbola, noting its center, vertices, and foci. State the equations of the asymptotes.

9. $\frac{(x-3)^2}{25} - \frac{(y+3)^2}{1} = 1$

10. $y^2 - x^2 + 4y - 4x - 18 = 0$

11. Write the standard form equation of a hyperbola with foci at $(1, 0)$ and $(1, 6)$, and a vertex at $(1, 2)$.

For the following exercises, write the equation of the parabola in standard form, and give the vertex, focus, and equation of the directrix.

12. $y^2 + 10x = 0$

13. $3x^2 - 12x - y + 11 = 0$

For the following exercises, graph the parabola, labeling the vertex, focus, and directrix.

14. $(x - 1)^2 = -4(y + 3)$

15. $y^2 + 8x - 8y + 40 = 0$

16. Write the equation of a parabola with a focus at $(2, 3)$ and directrix $y = -1$.

17. A searchlight is shaped like a paraboloid of revolution. If the light source is located 1.5 feet from the base along the axis of symmetry, and the depth of the searchlight is 3 feet, what should the width of the opening be?

For the following exercises, determine which conic section is represented by the given equation, and then determine the angle θ that will eliminate the xy term.

18. $3x^2 - 2xy + 3y^2 = 4$

19. $x^2 + 4xy + 4y^2 + 6x - 8y = 0$

For the following exercises, rewrite in the $x'y'$ system without the $x'y'$ term, and graph the rotated graph.

20. $11x^2 + 10\sqrt{3}xy + y^2 = 4$

21. $16x^2 + 24xy + 9y^2 - 125x = 0$

For the following exercises, identify the conic with focus at the origin, and then give the directrix and eccentricity.

22. $r = \frac{3}{2 - \sin\ \theta}$

23. $r = \frac{5}{4 + 6\ \cos\ \theta}$

For the following exercises, graph the given conic section. If it is a parabola, label vertex, focus, and directrix. If it is an ellipse or a hyperbola, label vertices and foci.

24. $r = \frac{12}{4 - 8\ \sin\ \theta}$

25. $r = \frac{2}{4 + 4\ \sin\ \theta}$

26. Find a polar equation of the conic with focus at the origin, eccentricity of $e = 2$, and directrix: $x = 3$.

13 SEQUENCES, PROBABILITY, AND COUNTING THEORY

(credit: Robert Couse-Baker, Flickr.)

Chapter Outline

Introduction to Sequences, Probability and Counting Theory

A lottery winner has some big decisions to make regarding what to do with the winnings. Buy a new home? A luxury convertible? A cruise around the world?

The likelihood of winning the lottery is slim, but we all love to fantasize about what we could buy with the winnings. One of the first things a lottery winner has to decide is whether to take the winnings in the form of a lump sum or as a series of regular payments, called an annuity, over an extended period of time.

This decision is often based on many factors, such as tax implications, interest rates, and investment strategies. There are also personal reasons to consider when making the choice, and one can make many arguments for either decision. However, most lottery winners opt for the lump sum.

In this chapter, we will explore the mathematics behind situations such as these. We will take an in-depth look at annuities. We will also look at the branch of mathematics that would allow us to calculate the number of ways to choose lottery numbers and the probability of winning.

13.1 Sequences and Their Notations

Learning Objectives

In this section, you will:

- Write the terms of a sequence defined by an explicit formula.
- Write the terms of a sequence defined by a recursive formula.
- Use factorial notation.

A video game company launches an exciting new advertising campaign. They predict the number of online visits to their website, or hits, will double each day. The model they are using shows 2 hits the first day, 4 hits the second day, 8 hits the third day, and so on. See Table 1.

Day	1	2	3	4	5	...
Hits	2	4	8	16	32	...

Table 1

If their model continues, how many hits will there be at the end of the month? To answer this question, we'll first need to know how to determine a list of numbers written in a specific order. In this section, we will explore these kinds of ordered lists.

Writing the Terms of a Sequence Defined by an Explicit Formula

One way to describe an ordered list of numbers is as a **sequence.** A sequence is a function whose domain is a subset of the counting numbers. The sequence established by the number of hits on the website is

$$\{2, 4, 8, 16, 32, \ldots\}.$$

The ellipsis (...) indicates that the sequence continues indefinitely. Each number in the sequence is called a **term**. The first five terms of this sequence are 2, 4, 8, 16, and 32.

Listing all of the terms for a sequence can be cumbersome. For example, finding the number of hits on the website at the end of the month would require listing out as many as 31 terms. A more efficient way to determine a specific term is by writing a formula to define the sequence.

One type of formula is an explicit formula, which defines the terms of a sequence using their position in the sequence. Explicit formulas are helpful if we want to find a specific term of a sequence without finding all of the previous terms. We can use the formula to find the nth term of the sequence, where n is any positive number. In our example, each number in the sequence is double the previous number, so we can use powers of 2 to write a formula for the nth term.

$$\{2,\ 4,\ 8,\ 16,\ 32,\ \ldots,\ ?,\ \ldots\}$$
$$\downarrow\ \downarrow\ \downarrow\ \downarrow\ \downarrow\qquad\downarrow$$
$$\{2^1,\ 2^2,\ 2^3,\ 2^4,\ 2^5,\ \ldots,\ 2^n\ \ldots\}$$

The first term of the sequence is $2^1 = 2$, the second term is $2^2 = 4$, the third term is $2^3 = 8$, and so on. The nth term of the sequence can be found by raising 2 to the nth power. An explicit formula for a sequence is named by a lower case letter a, b, c... with the subscript n. The explicit formula for this sequence is

$$a_n = 2^n.$$

Now that we have a formula for the nth term of the sequence, we can answer the question posed at the beginning of this section. We were asked to find the number of hits at the end of the month, which we will take to be 31 days. To find the number of hits on the last day of the month, we need to find the 31st term of the sequence. We will substitute 31 for n in the formula.

$$\begin{aligned} a_{31} &= 2^{31} \\ &= 2{,}147{,}483{,}648 \end{aligned}$$

If the doubling trend continues, the company will get 2,147,483,648 hits on the last day of the month. That is over 2.1 billion hits! The huge number is probably a little unrealistic because it does not take consumer interest and competition into account. It does, however, give the company a starting point from which to consider business decisions.

Another way to represent the sequence is by using a table. The first five terms of the sequence and the nth term of the sequence are shown in Table 2.

n	1	2	3	4	5	n
nth term of the sequence, a_n	2	4	8	16	32	2^n

Table 2

Graphing provides a visual representation of the sequence as a set of distinct points. We can see from the graph in Figure 1 that the number of hits is rising at an exponential rate. This particular sequence forms an exponential function.

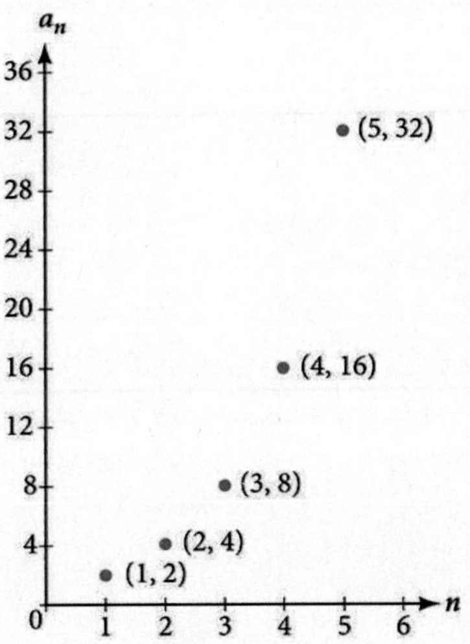

Figure 1

Lastly, we can write this particular sequence as

$$\{2, 4, 8, 16, 32, \ldots, 2^n, \ldots\}.$$

A sequence that continues indefinitely is called an **infinite sequence**. The domain of an infinite sequence is the set of counting numbers. If we consider only the first 10 terms of the sequence, we could write

$$\{2, 4, 8, 16, 32, \ldots, 2^n, \ldots, 1024\}.$$

This sequence is called a **finite sequence** because it does not continue indefinitely.

Sequence

A **sequence** is a function whose domain is the set of positive integers. A **finite sequence** is a sequence whose domain consists of only the first n positive integers. The numbers in a sequence are called **terms**. The variable a with a number subscript is used to represent the terms in a sequence and to indicate the position of the term in the sequence.

$$a_1, a_2, a_3, \ldots, a_n, \ldots$$

We call a_1 the first term of the sequence, a_2 the second term of the sequence, a_3 the third term of the sequence, and so on. The term a_n is called the **nth term of the sequence**, or the general term of the sequence. An **explicit formula** defines the nth term of a sequence using the position of the term. A sequence that continues indefinitely is an **infinite sequence**.

Q&A **Does a sequence always have to begin with a_1?**

No. In certain problems, it may be useful to define the initial term as a_0 instead of a_1. In these problems, the domain of the function includes 0.

HOW TO

Given an explicit formula, write the first n terms of a sequence.

1. Substitute each value of n into the formula. Begin with $n = 1$ to find the first term, a_1.
2. To find the second term, a_2, use $n = 2$.
3. Continue in the same manner until you have identified all n terms.

EXAMPLE 1

Writing the Terms of a Sequence Defined by an Explicit Formula

Write the first five terms of the sequence defined by the explicit formula $a_n = -3n + 8$.

Solution

Substitute $n = 1$ into the formula. Repeat with values 2 through 5 for n.

$$\begin{array}{ll} n = 1 & a_1 = -3(1) + 8 = 5 \\ n = 2 & a_2 = -3(2) + 8 = 2 \\ n = 3 & a_3 = -3(3) + 8 = -1 \\ n = 4 & a_4 = -3(4) + 8 = -4 \\ n = 5 & a_5 = -3(5) + 8 = -7 \end{array}$$

The first five terms are $\{5, \ 2, \ -1, \ -4, \ -7\}$.

Analysis

The sequence values can be listed in a table. A table, such as Table 3, is a convenient way to input the function into a graphing utility.

n	1	2	3	4	5
a_n	5	2	-1	-4	-7

Table 3

A graph can be made from this table of values. From the graph in Figure 2, we can see that this sequence represents a linear function, but notice the graph is not continuous because the domain is over the positive integers only.

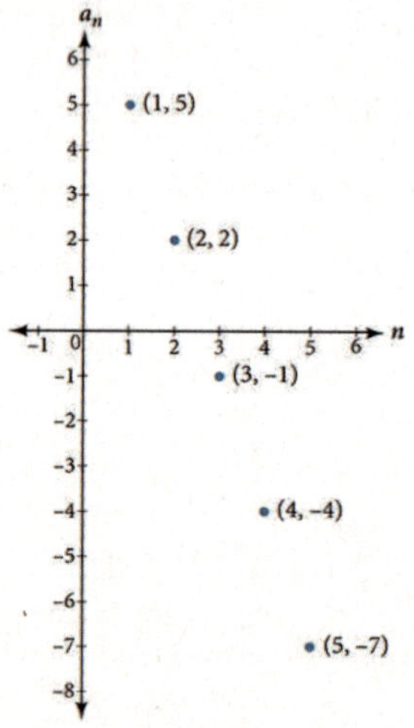

Figure 2

TRY IT #1 Write the first five terms of the sequence defined by the explicit formula $t_n = 5n - 4$.

Investigating Alternating Sequences

Sometimes sequences have terms that are alternate. In fact, the terms may actually alternate in sign. The steps to finding terms of the sequence are the same as if the signs did not alternate. However, the resulting terms will not show increase or decrease as n increases. Let's take a look at the following sequence.

$$\{2, -4, 6, -8\}$$

Notice the first term is greater than the second term, the second term is less than the third term, and the third term is greater than the fourth term. This trend continues forever. Do not rearrange the terms in numerical order to interpret the sequence.

HOW TO

Given an explicit formula with alternating terms, write the first n terms of a sequence.

1. Substitute each value of n into the formula. Begin with $n = 1$ to find the first term, a_1. The sign of the term is given by the $(-1)^n$ in the explicit formula.
2. To find the second term, a_2, use $n = 2$.
3. Continue in the same manner until you have identified all n terms.

EXAMPLE 2

Writing the Terms of an Alternating Sequence Defined by an Explicit Formula

Write the first five terms of the sequence.

$$a_n = \frac{(-1)^n n^2}{n+1}$$

Solution

Substitute $n = 1$, $n = 2$, and so on in the formula.

$$\begin{array}{ll} n = 1 & a_1 = \frac{(-1)^1 1^2}{1+1} = -\frac{1}{2} \\ n = 2 & a_2 = \frac{(-1)^2 2^2}{2+1} = \frac{4}{3} \\ n = 3 & a_3 = \frac{(-1)^3 3^2}{3+1} = -\frac{9}{4} \\ n = 4 & a_4 = \frac{(-1)^4 4^2}{4+1} = \frac{16}{5} \\ n = 5 & a_5 = \frac{(-1)^5 5^2}{5+1} = -\frac{25}{6} \end{array}$$

The first five terms are $\left\{-\frac{1}{2}, \frac{4}{3}, -\frac{9}{4}, \frac{16}{5}, -\frac{25}{6}\right\}$.

Analysis

The graph of this function, shown in Figure 3, looks different from the ones we have seen previously in this section because the terms of the sequence alternate between positive and negative values.

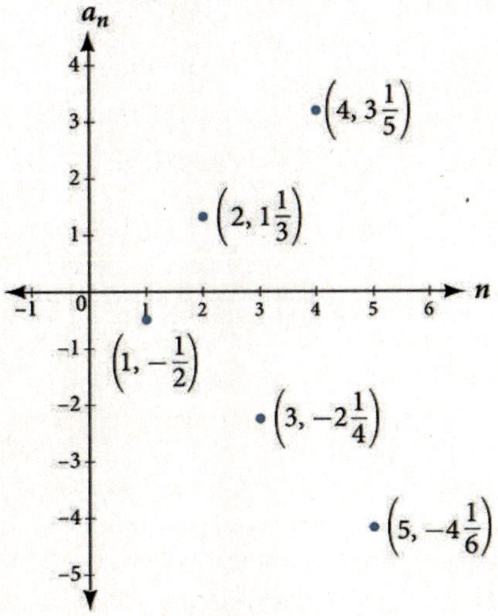

Figure 3

Q&A **In Example 2, does the (–1) to the power of n account for the oscillations of signs?**

Yes, the power might be $n, n+1, n-1$, and so on, but any odd powers will result in a negative term, and any even power will result in a positive term.

TRY IT #2 Write the first five terms of the sequence.

$$a_n = \frac{4n}{(-2)^n}$$

Investigating Piecewise Explicit Formulas

We've learned that sequences are functions whose domain is over the positive integers. This is true for other types of functions, including some piecewise functions. Recall that a piecewise function is a function defined by multiple subsections. A different formula might represent each individual subsection.

HOW TO

Given an explicit formula for a piecewise function, write the first n terms of a sequence

1. Identify the formula to which $n = 1$ applies.
2. To find the first term, a_1, use $n = 1$ in the appropriate formula.
3. Identify the formula to which $n = 2$ applies.
4. To find the second term, a_2, use $n = 2$ in the appropriate formula.
5. Continue in the same manner until you have identified all n terms.

EXAMPLE 3

Writing the Terms of a Sequence Defined by a Piecewise Explicit Formula

Write the first six terms of the sequence.

$$a_n = \begin{cases} n^2 & \text{if } n \text{ is not divisible by 3} \\ \frac{n}{3} & \text{if } n \text{ is divisible by 3} \end{cases}$$

Solution

Substitute $n = 1, n = 2$, and so on in the appropriate formula. Use n^2 when n is not a multiple of 3. Use $\frac{n}{3}$ when n is a multiple of 3.

$$\begin{aligned}
a_1 &= 1^2 = 1 && \text{1 is not a multiple of 3. Use } n^2. \\
a_2 &= 2^2 = 4 && \text{2 is not a multiple of 3. Use } n^2. \\
a_3 &= \tfrac{3}{3} = 1 && \text{3 is a multiple of 3. Use } \tfrac{n}{3}. \\
a_4 &= 4^2 = 16 && \text{4 is not a multiple of 3. Use } n^2. \\
a_5 &= 5^2 = 25 && \text{5 is not a multiple of 3. Use } n^2. \\
a_6 &= \tfrac{6}{3} = 2 && \text{6 is a multiple of 3. Use } \tfrac{n}{3}.
\end{aligned}$$

The first six terms are $\{1,\ 4,\ 1,\ 16,\ 25,\ 2\}$.

Analysis

Every third point on the graph shown in Figure 4 stands out from the two nearby points. This occurs because the sequence was defined by a piecewise function.

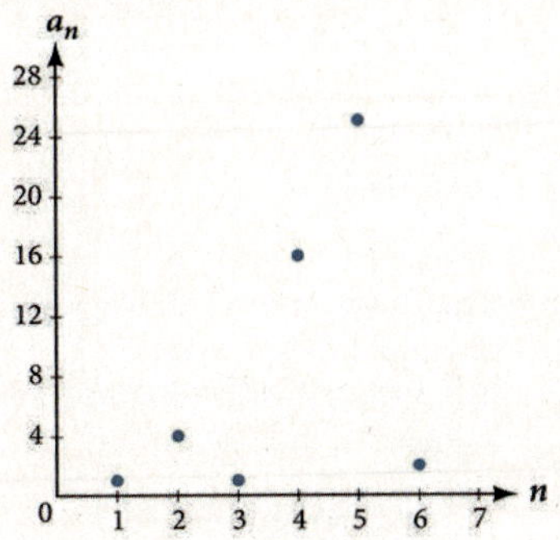

Figure 4

TRY IT #3 Write the first six terms of the sequence.

$$a_n = \begin{cases} 2n^3 & \text{if } n \text{ is odd} \\ \frac{5n}{2} & \text{if } n \text{ is even} \end{cases}$$

Finding an Explicit Formula

Thus far, we have been given the explicit formula and asked to find a number of terms of the sequence. Sometimes, the explicit formula for the nth term of a sequence is not given. Instead, we are given several terms from the sequence. When this happens, we can work in reverse to find an explicit formula from the first few terms of a sequence. The key to finding an explicit formula is to look for a pattern in the terms. Keep in mind that the pattern may involve alternating terms, formulas for numerators, formulas for denominators, exponents, or bases.

HOW TO

Given the first few terms of a sequence, find an explicit formula for the sequence.

1. Look for a pattern among the terms.
2. If the terms are fractions, look for a separate pattern among the numerators and denominators.
3. Look for a pattern among the signs of the terms.
4. Write a formula for a_n in terms of n. Test your formula for $n = 1$, $n = 2$, and $n = 3$.

EXAMPLE 4

Writing an Explicit Formula for the nth Term of a Sequence

Write an explicit formula for the nth term of each sequence.

ⓐ $\left\{-\frac{2}{11}, \frac{3}{13}, -\frac{4}{15}, \frac{5}{17}, -\frac{6}{19}, \ldots\right\}$ ⓑ $\left\{-\frac{2}{25}, -\frac{2}{125}, -\frac{2}{625}, -\frac{2}{3,125}, -\frac{2}{15,625}, \ldots\right\}$ ⓒ $\{e^4, e^5, e^6, e^7, e^8, \ldots\}$

Solution

Look for the pattern in each sequence.

ⓐ The terms alternate between positive and negative. We can use $(-1)^n$ to make the terms alternate. The numerator can be represented by $n + 1$. The denominator can be represented by $2n + 9$.

$$a_n = \frac{(-1)^n(n+1)}{2n+9}$$

ⓑ

The terms are all negative.

$\left\{-\frac{2}{25}, -\frac{2}{125}, -\frac{2}{625}, -\frac{2}{3{,}125}, -\frac{2}{15{,}125}, \ldots\right\}$ The numerator is 2.

$\left\{-\frac{2}{5^2}, -\frac{2}{5^3}, -\frac{2}{5^4}, -\frac{2}{5^6}, -\frac{2}{5^7}, \ldots, -\frac{2}{5^n}\right\}$ The denominators are increasing powers of 5.

So we know that the fraction is negative, the numerator is 2, and the denominator can be represented by 5^{n+1}.

$$a_n = -\frac{2}{5^{n+1}}$$

ⓒ

The terms are powers of e. For $n = 1$, the first term is e^4 so the exponent must be $n + 3$.

$$a_n = e^{n+3}$$

> **TRY IT** #4 Write an explicit formula for the nth term of the sequence.

$$\{9, \ -81, \ 729, \ -6{,}561, \ 59{,}049, \ \ldots\}$$

> **TRY IT** #5 Write an explicit formula for the nth term of the sequence.

$$\left\{-\frac{3}{4}, -\frac{9}{8}, -\frac{27}{12}, -\frac{81}{16}, -\frac{243}{20}, \ldots\right\}$$

> **TRY IT** #6 Write an explicit formula for the nth term of the sequence.

$$\left\{\frac{1}{e^2}, \ \frac{1}{e}, \ 1, \ e, \ e^2, \ldots\right\}$$

Writing the Terms of a Sequence Defined by a Recursive Formula

Sequences occur naturally in the growth patterns of nautilus shells, pinecones, tree branches, and many other natural structures. We may see the sequence in the leaf or branch arrangement, the number of petals of a flower, or the pattern of the chambers in a nautilus shell. Their growth follows the Fibonacci sequence, a famous sequence in which each term can be found by adding the preceding two terms. The numbers in the sequence are 1, 1, 2, 3, 5, 8, 13, 21, 34,.... Other examples from the natural world that exhibit the Fibonacci sequence are the Calla Lily, which has just one petal, the Black-Eyed Susan with 13 petals, and different varieties of daisies that may have 21 or 34 petals.

Each term of the Fibonacci sequence depends on the terms that come before it. The Fibonacci sequence cannot easily be written using an explicit formula. Instead, we describe the sequence using a **recursive formula**, a formula that defines the terms of a sequence using previous terms.

A recursive formula always has two parts: the value of an initial term (or terms), and an equation defining a_n in terms of preceding terms. For example, suppose we know the following:

$$\begin{aligned} a_1 &= 3 \\ a_n &= 2a_{n-1} - 1 \text{ for } n \geq 2 \end{aligned}$$

We can find the subsequent terms of the sequence using the first term.

$$\begin{aligned} a_1 &= 3 \\ a_2 &= 2a_1 - 1 = 2(3) - 1 = 5 \\ a_3 &= 2a_2 - 1 = 2(5) - 1 = 9 \\ a_4 &= 2a_3 - 1 = 2(9) - 1 = 17 \end{aligned}$$

So the first four terms of the sequence are $\{3, \ 5, \ 9, \ 17\}$.

The recursive formula for the Fibonacci sequence states the first two terms and defines each successive term as the sum of the preceding two terms.

$$\begin{aligned} a_1 &= 1 \\ a_2 &= 1 \\ a_n &= a_{n-1} + a_{n-2} \text{ for } n \geq 3 \end{aligned}$$

To find the tenth term of the sequence, for example, we would need to add the eighth and ninth terms. We were told previously that the eighth and ninth terms are 21 and 34, so

$$a_{10} = a_9 + a_8 = 34 + 21 = 55$$

Recursive Formula

A **recursive formula** is a formula that defines each term of a sequence using preceding term(s). Recursive formulas must always state the initial term, or terms, of the sequence.

Q&A **Must the first two terms always be given in a recursive formula?**

No. The Fibonacci sequence defines each term using the two preceding terms, but many recursive formulas define each term using only one preceding term. These sequences need only the first term to be defined.

HOW TO

Given a recursive formula with only the first term provided, write the first n terms of a sequence.

1. Identify the initial term, a_1, which is given as part of the formula. This is the first term.
2. To find the second term, a_2, substitute the initial term into the formula for a_{n-1}. Solve.
3. To find the third term, a_3, substitute the second term into the formula. Solve.
4. Repeat until you have solved for the nth term.

EXAMPLE 5

Writing the Terms of a Sequence Defined by a Recursive Formula

Write the first five terms of the sequence defined by the recursive formula.

$$\begin{aligned} a_1 &= 9 \\ a_n &= 3a_{n-1} - 20, \text{ for } n \geq 2 \end{aligned}$$

Solution

The first term is given in the formula. For each subsequent term, we replace a_{n-1} with the value of the preceding term.

$$\begin{array}{ll} n = 1 & a_1 = 9 \\ n = 2 & a_2 = 3a_1 - 20 = 3(9) - 20 = 27 - 20 = 7 \\ n = 3 & a_3 = 3a_2 - 20 = 3(7) - 20 = 21 - 20 = 1 \\ n = 4 & a_4 = 3a_3 - 20 = 3(1) - 20 = 3 - 20 = -17 \\ n = 5 & a_5 = 3a_4 - 20 = 3(-17) - 20 = -51 - 20 = -71 \end{array}$$

The first five terms are $\{9,\ 7,\ 1,\ -17,\ -71\}$. See Figure 5.

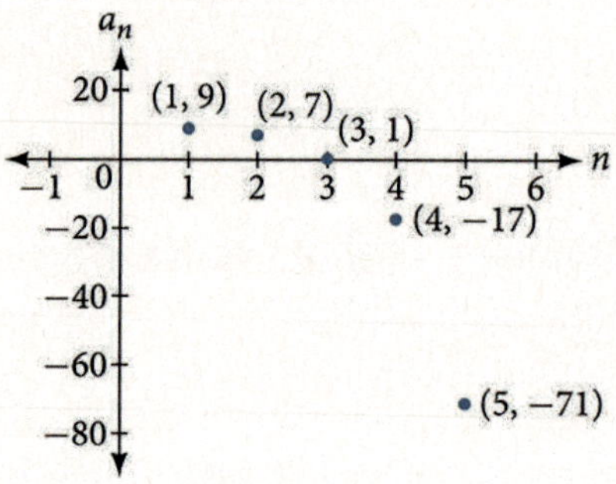

Figure 5

TRY IT #7 Write the first five terms of the sequence defined by the recursive formula.

$$\begin{aligned} a_1 &= 2 \\ a_n &= 2a_{n-1} + 1, \text{ for } n \geq 2 \end{aligned}$$

HOW TO

Given a recursive formula with two initial terms, write the first n terms of a sequence.

1. Identify the initial term, a_1, which is given as part of the formula.
2. Identify the second term, a_2, which is given as part of the formula.
3. To find the third term, substitute the initial term and the second term into the formula. Evaluate.
4. Repeat until you have evaluated the nth term.

EXAMPLE 6

Writing the Terms of a Sequence Defined by a Recursive Formula

Write the first six terms of the sequence defined by the recursive formula.

$$\begin{aligned} a_1 &= 1 \\ a_2 &= 2 \\ a_n &= 3a_{n-1} + 4a_{n-2}, \text{ for } n \geq 3 \end{aligned}$$

Solution

The first two terms are given. For each subsequent term, we replace a_{n-1} and a_{n-2} with the values of the two preceding terms.

$$\begin{array}{ll} n = 3 & a_3 = 3a_2 + 4a_1 = 3(2) + 4(1) = 10 \\ n = 4 & a_4 = 3a_3 + 4a_2 = 3(10) + 4(2) = 38 \\ n = 5 & a_5 = 3a_4 + 4a_3 = 3(38) + 4(10) = 154 \\ n = 6 & a_6 = 3a_5 + 4a_4 = 3(154) + 4(38) = 614 \end{array}$$

The first six terms are $\{1,2,10,38,154,614\}$. See Figure 6.

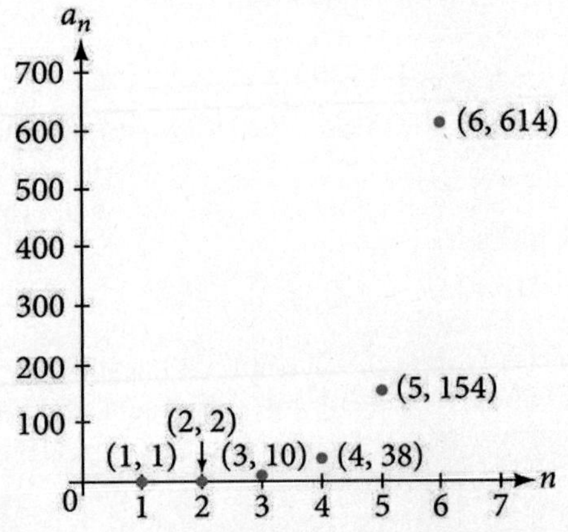

Figure 6

TRY IT #8 Write the first 8 terms of the sequence defined by the recursive formula.

$$\begin{aligned} a_1 &= 0 \\ a_2 &= 1 \\ a_3 &= 1 \\ a_n &= \frac{a_{n-1}}{a_{n-2}} + a_{n-3}, \text{ for } n \geq 4 \end{aligned}$$

Using Factorial Notation

The formulas for some sequences include products of consecutive positive integers. ***n* factorial**, written as $n!$, is the product of the positive integers from 1 to n. For example,

$$\begin{aligned} 4! &= 4 \cdot 3 \cdot 2 \cdot 1 = 24 \\ 5! &= 5 \cdot 4 \cdot 3 \cdot 2 \cdot 1 = 120 \end{aligned}$$

An example of formula containing a factorial is $a_n = (n+1)!$. The sixth term of the sequence can be found by substituting 6 for n.

$$a_6 = (6+1)! = 7! = 7 \cdot 6 \cdot 5 \cdot 4 \cdot 3 \cdot 2 \cdot 1 = 5040$$

The factorial of any whole number n is $n(n-1)!$ We can therefore also think of $5!$ as $5 \cdot 4!$.

n Factorial

***n* factorial** is a mathematical operation that can be defined using a recursive formula. The factorial of n, denoted $n!$, is defined for a positive integer n as:

$$\begin{aligned} 0! &= 1 \\ 1! &= 1 \\ n! &= n(n-1)(n-2)\cdots(2)(1), \text{ for } n \geq 2 \end{aligned}$$

The special case $0!$ is defined as $0! = 1$.

Q&A **Can factorials always be found using a calculator?**

No. Factorials get large very quickly—faster than even exponential functions! When the output gets too large for the calculator, it will not be able to calculate the factorial.

EXAMPLE 7

Writing the Terms of a Sequence Using Factorials

Write the first five terms of the sequence defined by the explicit formula $a_n = \frac{5n}{(n+2)!}$.

Solution

Substitute $n = 1, n = 2$, and so on in the formula.

$$\begin{array}{ll} n=1 & a_1 = \frac{5(1)}{(1+2)!} = \frac{5}{3!} = \frac{5}{3 \cdot 2 \cdot 1} = \frac{5}{6} \\ n=2 & a_2 = \frac{5(2)}{(2+2)!} = \frac{10}{4!} = \frac{10}{4 \cdot 3 \cdot 2 \cdot 1} = \frac{5}{12} \\ n=3 & a_3 = \frac{5(3)}{(3+2)!} = \frac{15}{5!} = \frac{15}{5 \cdot 4 \cdot 3 \cdot 2 \cdot 1} = \frac{1}{8} \\ n=4 & a_4 = \frac{5(4)}{(4+2)!} = \frac{20}{6!} = \frac{20}{6 \cdot 5 \cdot 4 \cdot 3 \cdot 2 \cdot 1} = \frac{1}{36} \\ n=5 & a_5 = \frac{5(5)}{(5+2)!} = \frac{25}{7!} = \frac{25}{7 \cdot 6 \cdot 5 \cdot 4 \cdot 3 \cdot 2 \cdot 1} = \frac{5}{1{,}008} \end{array}$$

The first five terms are $\left\{\frac{5}{6}, \frac{5}{12}, \frac{1}{8}, \frac{1}{36}, \frac{5}{1{,}008}\right\}$.

Analysis

Figure 7 shows the graph of the sequence. Notice that, since factorials grow very quickly, the presence of the factorial term in the denominator results in the denominator becoming much larger than the numerator as n increases. This means the quotient gets smaller and, as the plot of the terms shows, the terms are decreasing and nearing zero.

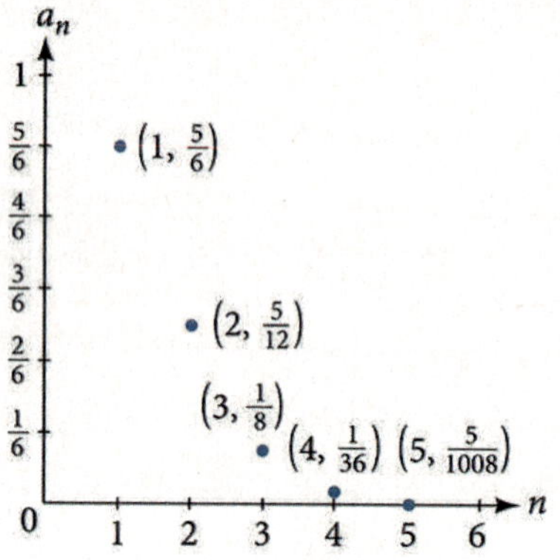

Figure 7

TRY IT #9 Write the first five terms of the sequence defined by the explicit formula $a_n = \frac{(n+1)!}{2n}$.

MEDIA

Access this online resource for additional instruction and practice with sequences.

Finding Terms in a Sequence (http://openstax.org/l/findingterms)

13.1 SECTION EXERCISES

Verbal

1. Discuss the meaning of a sequence. If a finite sequence is defined by a formula, what is its domain? What about an infinite sequence?

2. Describe three ways that a sequence can be defined.

3. Is the ordered set of even numbers an infinite sequence? What about the ordered set of odd numbers? Explain why or why not.

4. What happens to the terms a_n of a sequence when there is a negative factor in the formula that is raised to a power that includes n? What is the term used to describe this phenomenon?

5. What is a factorial, and how is it denoted? Use an example to illustrate how factorial notation can be beneficial.

Algebraic

For the following exercises, write the first four terms of the sequence.

6. $a_n = 2^n - 2$

7. $a_n = -\frac{16}{n+1}$

8. $a_n = -(-5)^{n-1}$

9. $a_n = \frac{2^n}{n^3}$

10. $a_n = \frac{2n+1}{n^3}$

11. $a_n = 1.25 \cdot (-4)^{n-1}$

12. $a_n = -4 \cdot (-6)^{n-1}$

13. $a_n = \frac{n^2}{2n+1}$

14. $a_n = (-10)^n + 1$

15. $a_n = -\left(\frac{4 \cdot (-5)^{n-1}}{5}\right)$

For the following exercises, write the first eight terms of the piecewise sequence.

16. $a_n = \begin{cases} (-2)^n - 2 & \text{if } n \text{ is even} \\ (3)^{n-1} & \text{if } n \text{ is odd} \end{cases}$

17. $a_n = \begin{cases} \frac{n^2}{2n+1} & \text{if } n \le 5 \\ n^2 - 5 & \text{if } n > 5 \end{cases}$

18. $a_n = \begin{cases} (2n+1)^2 & \text{if } n \text{ is divisible by } 4 \\ \frac{2}{n} & \text{if } n \text{ is not divisible by } 4 \end{cases}$

19. $a_n = \begin{cases} -0.6 \cdot 5^{n-1} & \text{if } n \text{ is prime or } 1 \\ 2.5 \cdot (-2)^{n-1} & \text{if } n \text{ is composite} \end{cases}$

20. $a_n = \begin{cases} 4(n^2 - 2) & \text{if } n \le 3 \text{ or } n > 6 \\ \frac{n^2-2}{4} & \text{if } 3 < n \le 6 \end{cases}$

For the following exercises, write an explicit formula for each sequence.

21. $4, 7, 12, 19, 28, \ldots$

22. $-4, 2, -10, 14, -34, \ldots$

23. $1, 1, \frac{4}{3}, 2, \frac{16}{5}, \ldots$

24. $0, \frac{1-e^1}{1+e^2}, \frac{1-e^2}{1+e^3}, \frac{1-e^3}{1+e^4}, \frac{1-e^4}{1+e^5}, \ldots$

25. $1, -\frac{1}{2}, \frac{1}{4}, -\frac{1}{8}, \frac{1}{16}, \ldots$

For the following exercises, write the first five terms of the sequence.

26. $a_1 = 9, \ a_n = a_{n-1} + n$

27. $a_1 = 3, \ a_n = (-3)\,a_{n-1}$

28. $a_1 = -4, \ a_n = \frac{a_{n-1}+2n}{a_{n-1}-1}$

29. $a_1 = -1, \ a_n = \frac{(-3)^{n-1}}{a_{n-1}-2}$

30. $a_1 = -30, \ a_n = \left(2 + a_{n-1}\right)\left(\frac{1}{2}\right)^n$

For the following exercises, write the first eight terms of the sequence.

31. $a_1 = \frac{1}{24}, \ a_2 = 1, \ a_n = \left(2a_{n-2}\right)\left(3a_{n-1}\right)$

32. $a_1 = -1, \ a_2 = 5, \ a_n = a_{n-2}\left(3 - a_{n-1}\right)$

33. $a_1 = 2, \ a_2 = 10, \ a_n = \frac{2(a_{n-1}+2)}{a_{n-2}}$

For the following exercises, write a recursive formula for each sequence.

34. $-2.5, -5, -10, -20, -40, \ldots$

35. $-8, -6, -3, 1, 6, \ldots$

36. $2, \ 4, \ 12, \ 48, \ 240, \ \ldots$

37. $35, \ 38, \ 41, \ 44, \ 47, \ \ldots$

38. $15, 3, \frac{3}{5}, \frac{3}{25}, \frac{3}{125}, \ldots$

For the following exercises, evaluate the factorial.

39. $6!$

40. $\left(\frac{12}{6}\right)!$

41. $\frac{12!}{6!}$

42. $\frac{100!}{99!}$

For the following exercises, write the first four terms of the sequence.

43. $a_n = \frac{n!}{n^2}$

44. $a_n = \frac{3 \cdot n!}{4 \cdot n!}$

45. $a_n = \frac{n!}{n^2-n-1}$

46. $a_n = \frac{100 \cdot n}{n(n-1)!}$

Graphical

For the following exercises, graph the first five terms of the indicated sequence

47. $a_n = \frac{(-1)^n}{n} + n$

48. $a_n = \begin{cases} \frac{4+n}{2n} & \text{if } n \text{ is even} \\ 3+n & \text{if } n \text{ is odd} \end{cases}$

49. $a_1 = 2, \ a_n = \left(-a_{n-1} + 1\right)^2$

50. $a_1 = 1, \ a_n = a_{n-1} + 8$

51. $a_n = \frac{(n+1)!}{(n-1)!}$

For the following exercises, write an explicit formula for the sequence using the first five points shown on the graph.

52.

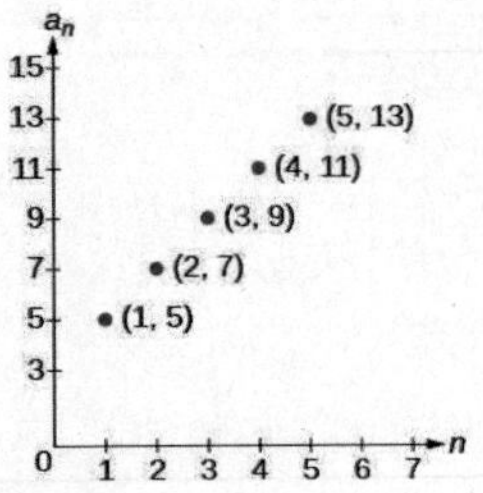

53.

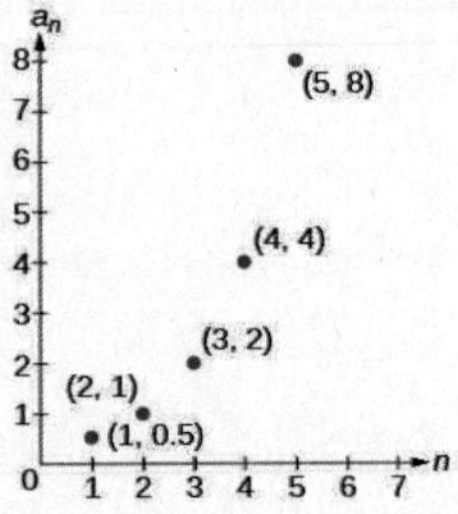

54.

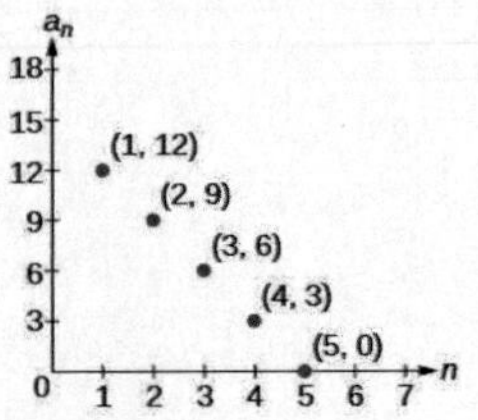

For the following exercises, write a recursive formula for the sequence using the first five points shown on the graph.

55.

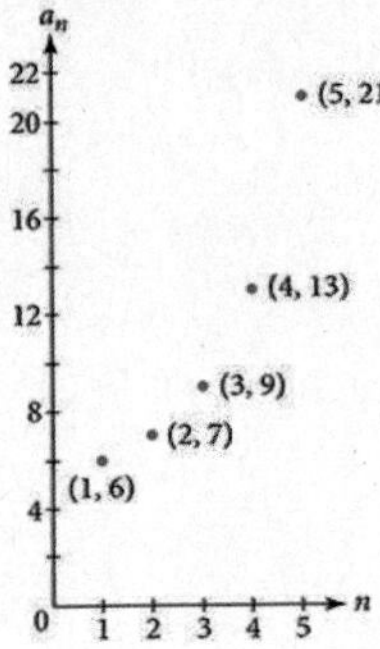

56.

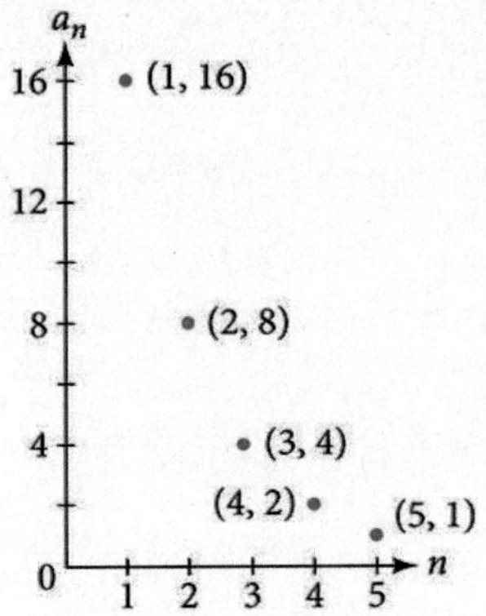

Technology

Follow these steps to evaluate a sequence defined recursively using a graphing calculator:

- *On the home screen, key in the value for the initial term a_1 and press **[ENTER]**.*
- *Enter the recursive formula by keying in all numerical values given in the formula, along with the key strokes **[2ND] ANS** for the previous term a_{n-1}. Press **[ENTER]**.*
- *Continue pressing **[ENTER]** to calculate the values for each successive term.*

For the following exercises, use the steps above to find the indicated term or terms for the sequence.

57. Find the first five terms of the sequence $a_1 = \frac{87}{111}$, $a_n = \frac{4}{3}a_{n-1} + \frac{12}{37}$. Use the >**Frac** feature to give fractional results.

58. Find the 15th term of the sequence $a_1 = 625$, $a_n = 0.8a_{n-1} + 18$.

59. Find the first five terms of the sequence $a_1 = 2$, $a_n = 2^{[(a_{n-1})-1]} + 1$.

60. Find the first ten terms of the sequence $a_1 = 8$, $a_n = \frac{(a_{n-1}+1)!}{a_{n-1}!}$.

61. Find the tenth term of the sequence $a_1 = 2$, $a_n = na_{n-1}$

Follow these steps to evaluate a finite sequence defined by an explicit formula. Using a TI-84, do the following.

- *In the home screen, press **[2ND] LIST**.*
- *Scroll over to **OPS** and choose **"seq("** from the dropdown list. Press **[ENTER]**.*
- *In the line headed **"Expr:"** type in the explicit formula, using the $[X,T,\theta,n]$ button for n*

- *In the line headed **"Variable:"** type in the variable used on the previous step.*
- *In the line headed **"start:"** key in the value of n that begins the sequence.*
- *In the line headed **"end:"** key in the value of n that ends the sequence.*
- *Press **[ENTER]** 3 times to return to the home screen. You will see the sequence syntax on the screen. Press **[ENTER]** to see the list of terms for the finite sequence defined. Use the right arrow key to scroll through the list of terms.*

Using a TI-83, do the following.

- *In the home screen, press **[2ND] LIST**.*
- *Scroll over to **OPS** and choose **"seq("** from the dropdown list. Press **[ENTER]**.*
- *Enter the items in the order **"Expr"**, **"Variable"**, **"start"**, **"end"** separated by commas. See the instructions above for the description of each item.*
- *Press **[ENTER]** to see the list of terms for the finite sequence defined. Use the right arrow key to scroll through the list of terms.*

For the following exercises, use the steps above to find the indicated terms for the sequence. Round to the nearest thousandth when necessary.

62. List the first five terms of the sequence $a_n = -\frac{28}{9}n + \frac{5}{3}$.

63. List the first six terms of the sequence $a_n = \frac{n^3 - 3.5n^2 + 4.1n - 1.5}{2.4n}$.

64. List the first five terms of the sequence $a_n = \frac{15n \cdot (-2)^{n-1}}{47}$

65. List the first four terms of the sequence $a_n = 5.7^n + 0.275\,(n-1)!$

66. List the first six terms of the sequence $a_n = \frac{n!}{n}$.

Extensions

67. Consider the sequence defined by $a_n = -6 - 8n$. Is $a_n = -421$ a term in the sequence? Verify the result.

68. What term in the sequence $a_n = \frac{n^2 + 4n + 4}{2(n+2)}$ has the value 41? Verify the result.

69. Find a recursive formula for the sequence 1, 0, −1, −1, 0, 1, 1, 0, −1, −1, 0, 1, 1, ... (*Hint*: find a pattern for a_n based on the first two terms.)

70. Calculate the first eight terms of the sequences $a_n = \frac{(n+2)!}{(n-1)!}$ and $b_n = n^3 + 3n^2 + 2n$, and then make a conjecture about the relationship between these two sequences.

71. Prove the conjecture made in the preceding exercise.

13.2 Arithmetic Sequences

Learning Objectives

In this section, you will:

- Find the common difference for an arithmetic sequence.
- Write terms of an arithmetic sequence.
- Use a recursive formula for an arithmetic sequence.
- Use an explicit formula for an arithmetic sequence.

Companies often make large purchases, such as computers and vehicles, for business use. The book-value of these supplies decreases each year for tax purposes. This decrease in value is called depreciation. One method of calculating depreciation is straight-line depreciation, in which the value of the asset decreases by the same amount each year.

As an example, consider a woman who starts a small contracting business. She purchases a new truck for $25,000. After five years, she estimates that she will be able to sell the truck for $8,000. The loss in value of the truck will therefore be $17,000, which is $3,400 per year for five years. The truck will be worth $21,600 after the first year; $18,200 after two years; $14,800 after three years; $11,400 after four years; and $8,000 at the end of five years. In this section, we will consider specific kinds of sequences that will allow us to calculate depreciation, such as the truck's value.

Finding Common Differences

The values of the truck in the example are said to form an **arithmetic sequence** because they change by a constant amount each year. Each term increases or decreases by the same constant value called the **common difference** of the sequence. For this sequence, the common difference is –3,400.

$$-3{,}400 \quad -3{,}400 \quad -3{,}400 \quad -3{,}400 \quad -3{,}400$$

$$\{25000, \quad 21600, \quad 18200, \quad 14800, \quad 11400, \quad 8000\}$$

The sequence below is another example of an arithmetic sequence. In this case, the constant difference is 3. You can choose any term of the sequence, and add 3 to find the subsequent term.

$$+3 \quad +3 \quad +3 \quad +3$$

$$\{3, \quad 6, \quad 9, \quad 12, \quad 15, \ldots\}$$

Arithmetic Sequence

An **arithmetic sequence** is a sequence that has the property that the difference between any two consecutive terms is a constant. This constant is called the **common difference**. If a_1 is the first term of an arithmetic sequence and d is the common difference, the sequence will be:

$$\{a_n\} = \{a_1, a_1 + d, a_1 + 2d, a_1 + 3d, \ldots\}$$

EXAMPLE 1

Finding Common Differences

Is each sequence arithmetic? If so, find the common difference.

ⓐ $\{1, 2, 4, 8, 16, \ldots\}$ ⓑ $\{-3, 1, 5, 9, 13, \ldots\}$

Solution

Subtract each term from the subsequent term to determine whether a common difference exists.

ⓐ The sequence is not arithmetic because there is no common difference.

2 - 1 = 1 4 - 2 = 2 8 - 4 = 4 16 - 8 = 8

ⓑ The sequence is arithmetic because there is a common difference. The common difference is 4.

1 - (-3) = 4 5 - 1 = 4 9 - 5 = 4 13 - 9 = 4

Analysis

The graph of each of these sequences is shown in Figure 1. We can see from the graphs that, although both sequences show growth, a is not linear whereas b is linear. Arithmetic sequences have a constant rate of change so their graphs will always be points on a line.

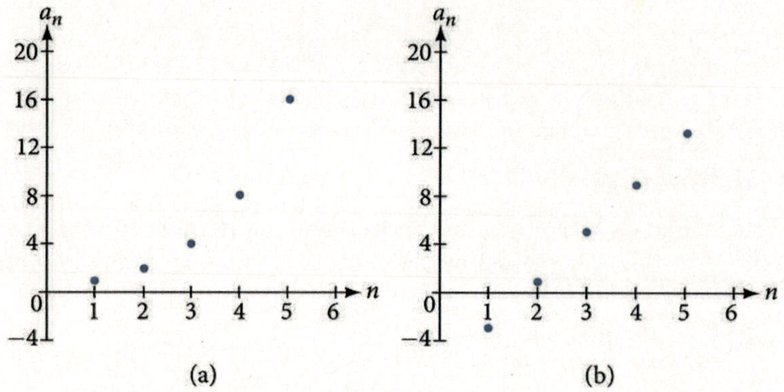

Figure 1

Q&A **If we are told that a sequence is arithmetic, do we have to subtract every term from the following term to find the common difference?**

No. If we know that the sequence is arithmetic, we can choose any one term in the sequence, and subtract it from the subsequent term to find the common difference.

TRY IT #1 Is the given sequence arithmetic? If so, find the common difference.

$$\{18, \quad 16, \quad 14, \quad 12, \quad 10, \ldots\}$$

TRY IT #2 Is the given sequence arithmetic? If so, find the common difference.

$$\{1, \quad 3, \quad 6, \quad 10, \quad 15, \ldots\}$$

Writing Terms of Arithmetic Sequences

Now that we can recognize an arithmetic sequence, we will find the terms if we are given the first term and the common difference. The terms can be found by beginning with the first term and adding the common difference repeatedly. In addition, any term can also be found by plugging in the values of n and d into formula below.

$$a_n = a_1 + (n - 1)d$$

HOW TO

Given the first term and the common difference of an arithmetic sequence, find the first several terms.

1. Add the common difference to the first term to find the second term.
2. Add the common difference to the second term to find the third term.
3. Continue until all of the desired terms are identified.
4. Write the terms separated by commas within brackets.

EXAMPLE 2

Writing Terms of Arithmetic Sequences

Write the first five terms of the arithmetic sequence with $a_1 = 17$ and $d = -3$.

Solution

Adding -3 is the same as subtracting 3. Beginning with the first term, subtract 3 from each term to find the next term.

The first five terms are $\{17, \quad 14, \quad 11, \quad 8, \quad 5\}$

Analysis

As expected, the graph of the sequence consists of points on a line as shown in Figure 2.

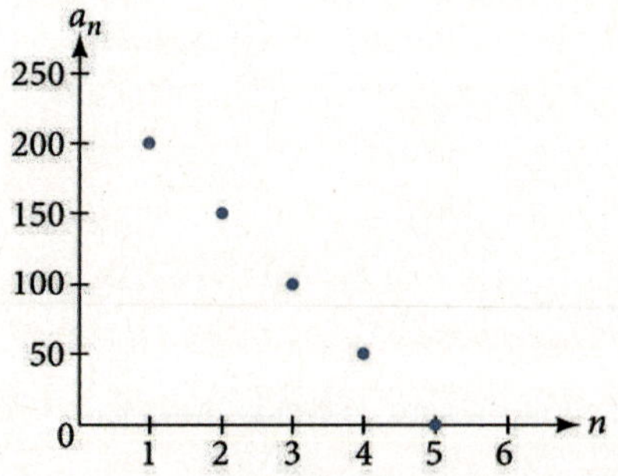

Figure 2

TRY IT #3 List the first five terms of the arithmetic sequence with $a_1 = 1$ and $d = 5$.

HOW TO

Given any first term and any other term in an arithmetic sequence, find a given term.

1. Substitute the values given for a_1, a_n, n into the formula $a_n = a_1 + (n-1)d$ to solve for d.
2. Find a given term by substituting the appropriate values for a_1, n, and d into the formula $a_n = a_1 + (n-1)d$.

EXAMPLE 3

Writing Terms of Arithmetic Sequences

Given $a_1 = 8$ and $a_4 = 14$, find a_5.

Solution

The sequence can be written in terms of the initial term 8 and the common difference d.

$$\{8, 8+d, 8+2d, 8+3d\}$$

We know the fourth term equals 14; we know the fourth term has the form $a_1 + 3d = 8 + 3d$.

We can find the common difference d.

$$a_n = a_1 + (n-1)d$$
$$a_4 = a_1 + 3d$$
$$a_4 = 8 + 3d \quad \text{Write the fourth term of the sequence in terms of } a_1 \text{ and } d.$$
$$14 = 8 + 3d \quad \text{Substitute 14 for } a_4.$$
$$d = 2 \quad \text{Solve for the common difference.}$$

Find the fifth term by adding the common difference to the fourth term.

$$a_5 = a_4 + 2 = 16$$

Analysis

Notice that the common difference is added to the first term once to find the second term, twice to find the third term, three times to find the fourth term, and so on. The tenth term could be found by adding the common difference to the first term nine times or by using the equation $a_n = a_1 + (n-1)d$.

TRY IT #4 Given $a_3 = 7$ and $a_5 = 17$, find a_2.

Using Recursive Formulas for Arithmetic Sequences

Some arithmetic sequences are defined in terms of the previous term using a recursive formula. The formula provides an algebraic rule for determining the terms of the sequence. A recursive formula allows us to find any term of an arithmetic sequence using a function of the preceding term. Each term is the sum of the previous term and the common difference. For example, if the common difference is 5, then each term is the previous term plus 5. As with any recursive formula, the first term must be given.

$$a_n = a_{n-1} + d \qquad n \geq 2$$

Recursive Formula for an Arithmetic Sequence

The recursive formula for an arithmetic sequence with common difference d is:

$$a_n = a_{n-1} + d \qquad n \geq 2$$

HOW TO

Given an arithmetic sequence, write its recursive formula.

1. Subtract any term from the subsequent term to find the common difference.
2. State the initial term and substitute the common difference into the recursive formula for arithmetic sequences.

EXAMPLE 4

Writing a Recursive Formula for an Arithmetic Sequence

Write a recursive formula for the arithmetic sequence.

$$\{-18,\ -7, 4, 15, 26, \ldots\}$$

Solution

The first term is given as -18 . The common difference can be found by subtracting the first term from the second term.

$$d = -7 - (-18) = 11$$

Substitute the initial term and the common difference into the recursive formula for arithmetic sequences.

$$\begin{aligned} a_1 &= -18 \\ a_n &= a_{n-1} + 11, \quad \text{for } n \geq 2 \end{aligned}$$

Analysis

We see that the common difference is the slope of the line formed when we graph the terms of the sequence, as shown in Figure 3. The growth pattern of the sequence shows the constant difference of 11 units.

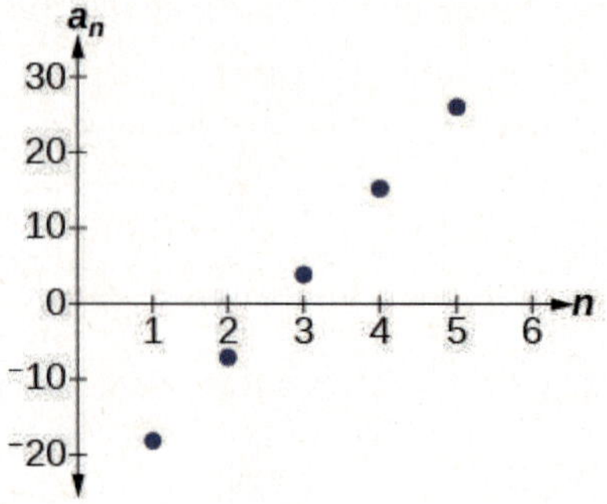

Figure 3

Q&A **Do we have to subtract the first term from the second term to find the common difference?**

No. We can subtract any term in the sequence from the subsequent term. It is, however, most common to subtract the first term from the second term because it is often the easiest method of finding the common difference.

TRY IT #5 Write a recursive formula for the arithmetic sequence.

$$\{25, 37, 49, 61, \ldots\}$$

Using Explicit Formulas for Arithmetic Sequences

We can think of an arithmetic sequence as a function on the domain of the natural numbers; it is a linear function because it has a constant rate of change. The common difference is the constant rate of change, or the slope of the function. We can construct the linear function if we know the slope and the vertical intercept.

$$a_n = a_1 + d(n - 1)$$

To find the *y*-intercept of the function, we can subtract the common difference from the first term of the sequence. Consider the following sequence.

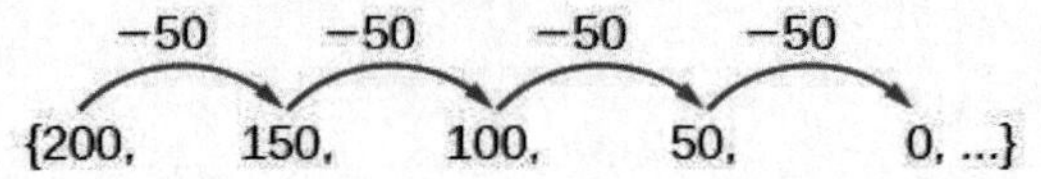

The common difference is -50 , so the sequence represents a linear function with a slope of -50 . To find the y -intercept, we subtract -50 from 200 : $200 - (-50) = 200 + 50 = 250$. You can also find the y -intercept by graphing the function and determining where a line that connects the points would intersect the vertical axis. The graph is shown in Figure 4.

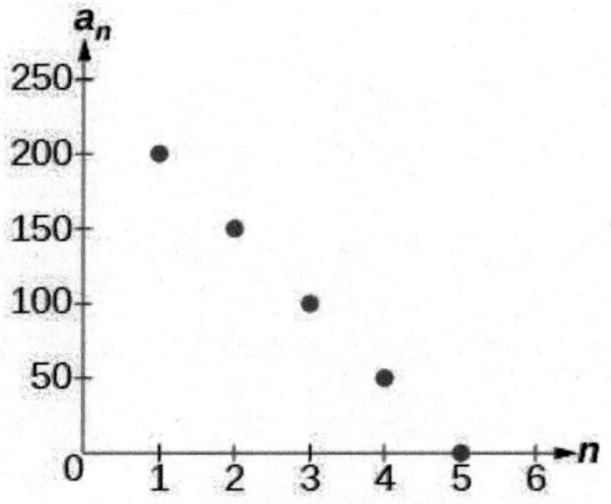

Figure 4

Recall the slope-intercept form of a line is $y = mx + b$. When dealing with sequences, we use a_n in place of y and n in place of x. If we know the slope and vertical intercept of the function, we can substitute them for m and b in the slope-intercept form of a line. Substituting -50 for the slope and 250 for the vertical intercept, we get the following equation:

$$a_n = -50n + 250$$

We do not need to find the vertical intercept to write an explicit formula for an arithmetic sequence. Another explicit formula for this sequence is $a_n = 200 - 50(n - 1)$, which simplifies to $a_n = -50n + 250$.

Explicit Formula for an Arithmetic Sequence

An explicit formula for the nth term of an arithmetic sequence is given by

$$a_n = a_1 + d(n - 1)$$

HOW TO

Given the first several terms for an arithmetic sequence, write an explicit formula.

1. Find the common difference, $a_2 - a_1$.

2. Substitute the common difference and the first term into $a_n = a_1 + d(n-1)$.

EXAMPLE 5

Writing the *n*th Term Explicit Formula for an Arithmetic Sequence

Write an explicit formula for the arithmetic sequence.

$$\{2, 12, 22, 32, 42, \ldots\}$$

Solution

The common difference can be found by subtracting the first term from the second term.

$$\begin{aligned} d &= a_2 - a_1 \\ &= 12 - 2 \\ &= 10 \end{aligned}$$

The common difference is 10. Substitute the common difference and the first term of the sequence into the formula and simplify.

$$\begin{aligned} a_n &= 2 + 10(n-1) \\ a_n &= 10n - 8 \end{aligned}$$

Analysis

The graph of this sequence, represented in Figure 5, shows a slope of 10 and a vertical intercept of -8 .

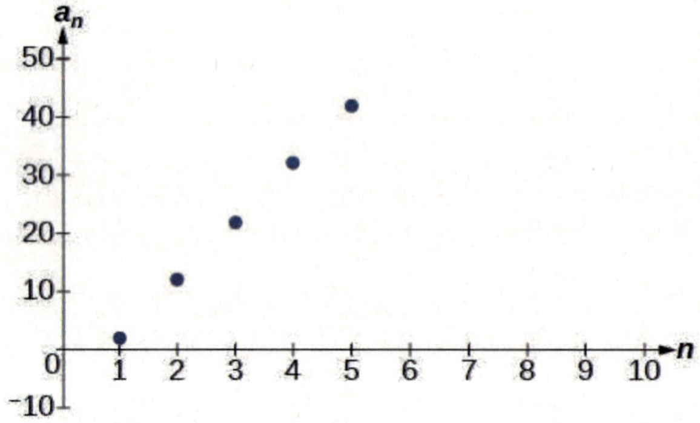

Figure 5

TRY IT #6 Write an explicit formula for the following arithmetic sequence.

$$\{50, 47, 44, 41, \ldots\}$$

Finding the Number of Terms in a Finite Arithmetic Sequence

Explicit formulas can be used to determine the number of terms in a finite arithmetic sequence. We need to find the common difference, and then determine how many times the common difference must be added to the first term to obtain the final term of the sequence.

HOW TO

Given the first three terms and the last term of a finite arithmetic sequence, find the total number of terms.

1. Find the common difference d.
2. Substitute the common difference and the first term into $a_n = a_1 + d(n-1)$.
3. Substitute the last term for a_n and solve for n.

EXAMPLE 6

Finding the Number of Terms in a Finite Arithmetic Sequence

Find the number of terms in the finite arithmetic sequence.

$$\{8, 1, -6, ..., -41\}$$

Solution

The common difference can be found by subtracting the first term from the second term.

$$1 - 8 = -7$$

The common difference is -7 . Substitute the common difference and the initial term of the sequence into the nth term formula and simplify.

$$\begin{aligned} a_n &= a_1 + d(n-1) \\ a_n &= 8 + (-7)(n-1) \\ a_n &= 15 - 7n \end{aligned}$$

Substitute -41 for a_n and solve for n

$$\begin{aligned} -41 &= 15 - 7n \\ 8 &= n \end{aligned}$$

There are eight terms in the sequence.

> TRY IT #7 Find the number of terms in the finite arithmetic sequence.

$$\{6, 11, 16, ..., 56\}$$

Solving Application Problems with Arithmetic Sequences

In many application problems, it often makes sense to use an initial term of a_0 instead of a_1. In these problems, we alter the explicit formula slightly to account for the difference in initial terms. We use the following formula:

$$a_n = a_0 + dn$$

EXAMPLE 7

Solving Application Problems with Arithmetic Sequences

A five-year old child receives an allowance of $1 each week. His parents promise him an annual increase of $2 per week.

ⓐ Write a formula for the child's weekly allowance in a given year.

ⓑ What will the child's allowance be when he is 16 years old?

Solution

ⓐ

The situation can be modeled by an arithmetic sequence with an initial term of 1 and a common difference of 2.

Let A be the amount of the allowance and n be the number of years after age 5. Using the altered explicit formula for an arithmetic sequence we get:

$$A_n = 1 + 2n$$

ⓑ

We can find the number of years since age 5 by subtracting.

$$16 - 5 = 11$$

We are looking for the child's allowance after 11 years. Substitute 11 into the formula to find the child's allowance at age 16.

$$A_{11} = 1 + 2(11) = 23$$

The child's allowance at age 16 will be $23 per week.

TRY IT #8 A woman decides to go for a 10-minute run every day this week and plans to increase the time of her daily run by 4 minutes each week. Write a formula for the time of her run after n weeks. How long will her daily run be 8 weeks from today?

MEDIA

Access this online resource for additional instruction and practice with arithmetic sequences.

Arithmetic Sequences (http://openstax.org/l/arithmeticseq)

13.2 SECTION EXERCISES

Verbal

1. What is an arithmetic sequence?

2. How is the common difference of an arithmetic sequence found?

3. How do we determine whether a sequence is arithmetic?

4. What are the main differences between using a recursive formula and using an explicit formula to describe an arithmetic sequence?

5. Describe how linear functions and arithmetic sequences are similar. How are they different?

Algebraic

For the following exercises, find the common difference for the arithmetic sequence provided.

6. $\{5, 11, 17, 23, 29, ...\}$

7. $\left\{0, \frac{1}{2}, 1, \frac{3}{2}, 2, ...\right\}$

For the following exercises, determine whether the sequence is arithmetic. If so find the common difference.

8. $\{11.4, 9.3, 7.2, 5.1, 3, ...\}$

9. $\{4, 16, 64, 256, 1024, ...\}$

For the following exercises, write the first five terms of the arithmetic sequence given the first term and common difference.

10. $a_1 = -25\,, d = -9$

11. $a_1 = 0\,, d = \frac{2}{3}$

For the following exercises, write the first five terms of the arithmetic series given two terms.

12. $a_1 = 17, \ \ a_7 = -31$

13. $a_{13} = -60, \ \ a_{33} = -160$

For the following exercises, find the specified term for the arithmetic sequence given the first term and common difference.

14. First term is 3, common difference is 4, find the 5th term.

15. First term is 4, common difference is 5, find the 4th term.

16. First term is 5, common difference is 6, find the 8th term.

17. First term is 6, common difference is 7, find the 6th term.

18. First term is 7, common difference is 8, find the 7th term.

For the following exercises, find the first term given two terms from an arithmetic sequence.

19. Find the first term or a_1 of an arithmetic sequence if $a_6 = 12$ and $a_{14} = 28$.

20. Find the first term or a_1 of an arithmetic sequence if $a_7 = 21$ and $a_{15} = 42$.

21. Find the first term or a_1 of an arithmetic sequence if $a_8 = 40$ and $a_{23} = 115$.

22. Find the first term or a_1 of an arithmetic sequence if $a_9 = 54$ and $a_{17} = 102$.

23. Find the first term or a_1 of an arithmetic sequence if $a_{11} = 11$ and $a_{21} = 16$.

For the following exercises, find the specified term given two terms from an arithmetic sequence.

24. $a_1 = 33$ and $a_7 = -15$. Find a_4.

25. $a_3 = -17.1$ and $a_{10} = -15.7$. Find a_{21}.

For the following exercises, use the recursive formula to write the first five terms of the arithmetic sequence.

26. $a_1 = 39;\ a_n = a_{n-1} - 3$

27. $a_1 = -19;\ a_n = a_{n-1} - 1.4$

For the following exercises, write a recursive formula for each arithmetic sequence.

28. $a = \{40, 60, 80, ...\}$

29. $a = \{17, 26, 35, ...\}$

30. $a = \{-1, 2, 5, ...\}$

31. $a = \{12, 17, 22, ...\}$

32. $a = \{-15, -7, 1, ...\}$

33. $a = \{8.9, 10.3, 11.7, ...\}$

34. $a = \{-0.52, -1.02, -1.52, ...\}$

35. $a = \left\{\frac{1}{5}, \frac{9}{20}, \frac{7}{10}, ...\right\}$

36. $a = \left\{-\frac{1}{2}, -\frac{5}{4}, -2, ...\right\}$

37. $a = \left\{\frac{1}{6}, -\frac{11}{12}, -2, ...\right\}$

For the following exercises, write a recursive formula for the given arithmetic sequence, and then find the specified term.

38. $a = \{7, 4, 1, ...\}$; Find the 17th term.

39. $a = \{4, 11, 18, ...\}$; Find the 14th term.

40. $a = \{2, 6, 10, ...\}$; Find the 12th term.

For the following exercises, use the explicit formula to write the first five terms of the arithmetic sequence.

41. $a_n = 24 - 4n$

42. $a_n = \frac{1}{2}n - \frac{1}{2}$

For the following exercises, write an explicit formula for each arithmetic sequence.

43. $a = \{3, 5, 7, ...\}$

44. $a = \{32, 24, 16, ...\}$

45. $a = \{-5, 95, 195, ...\}$

46. $a = \{-17, -217, -417, ...\}$

47. $a = \{1.8, 3.6, 5.4, ...\}$

48. $a = \{-18.1, -16.2, -14.3, ...\}$

49. $a = \{15.8, 18.5, 21.2, ...\}$

50. $a = \left\{\frac{1}{3}, -\frac{4}{3}, -3, ...\right\}$

51. $a = \left\{0, \frac{1}{3}, \frac{2}{3}, ...\right\}$

52. $a = \left\{-5, -\frac{10}{3}, -\frac{5}{3}, \ldots\right\}$

For the following exercises, find the number of terms in the given finite arithmetic sequence.

53. $a = \{3, -4, -11, \ldots, -60\}$

54. $a = \{1.2, 1.4, 1.6, \ldots, 3.8\}$

55. $a = \left\{\frac{1}{2}, 2, \frac{7}{2}, \ldots, 8\right\}$

Graphical

For the following exercises, determine whether the graph shown represents an arithmetic sequence.

56.

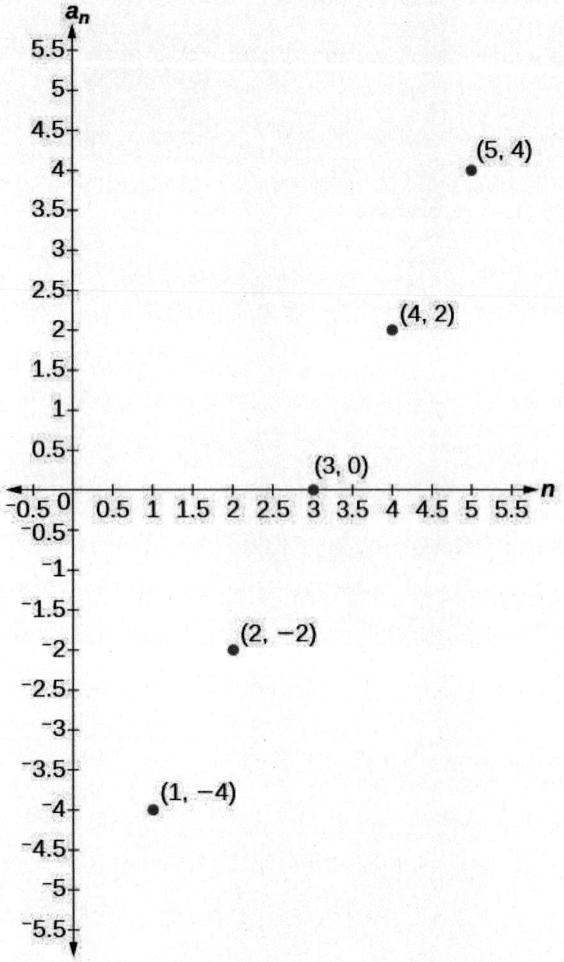

57.

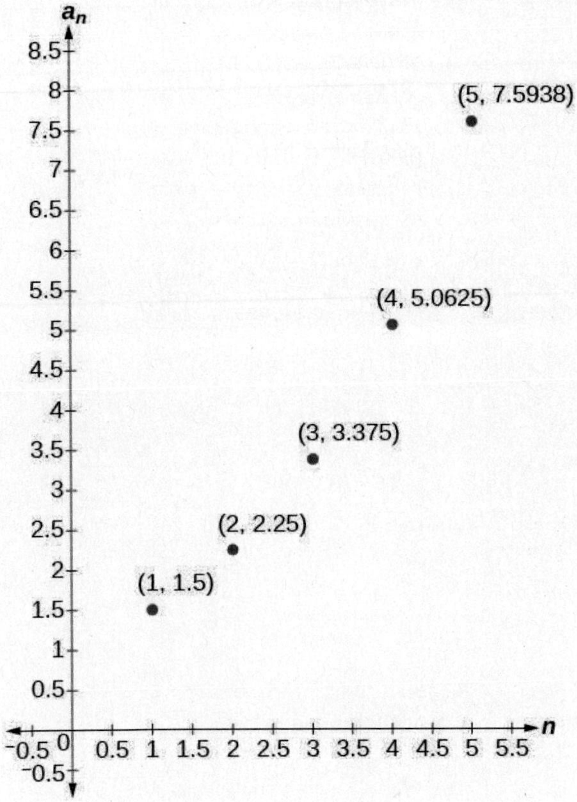

For the following exercises, use the information provided to graph the first 5 terms of the arithmetic sequence.

58. $a_1 = 0, d = 4$

59. $a_1 = 9; a_n = a_{n-1} - 10$

60. $a_n = -12 + 5n$

Technology

For the following exercises, follow the steps to work with the arithmetic sequence $a_n = 3n - 2$ using a graphing calculator:

- *Press **[MODE]***
  - *Select SEQ in the fourth line*
  - *Select DOT in the fifth line*
  - *Press **[ENTER]***
- *Press **[Y=]***
  - *nMin is the first counting number for the sequence. Set $n\text{Min} = 1$*
  - *$u(n)$ is the pattern for the sequence. Set $u(n) = 3n - 2$*
  - *$u(n\text{Min})$ is the first number in the sequence. Set $u(n\text{Min}) = 1$*
- *Press **[2ND]** then **[WINDOW]** to go to **TBLSET***
  - *Set $\text{TblStart} = 1$*
  - *Set $\Delta\text{Tbl} = 1$*
  - *Set Indpnt: Auto and Depend: Auto*
- *Press **[2ND]** then **[GRAPH]** to go to the **TABLE***

61. What are the first seven terms shown in the column with the heading $u(n)$?

62. Use the scroll-down arrow to scroll to $n = 50$. What value is given for $u(n)$?

63. Press **[WINDOW]**. Set $n\text{Min} = 1$, $n\text{Max} = 5$, $x\text{Min} = 0$, $x\text{Max} = 6$, $y\text{Min} = -1$, and $y\text{Max} = 14$. Then press **[GRAPH]**. Graph the sequence as it appears on the graphing calculator.

For the following exercises, follow the steps given above to work with the arithmetic sequence $a_n = \frac{1}{2}n + 5$ using a graphing calculator.

64. What are the first seven terms shown in the column with the heading $u(n)$ in the TABLE feature?

65. Graph the sequence as it appears on the graphing calculator. Be sure to adjust the WINDOW settings as needed.

Extensions

66. Give two examples of arithmetic sequences whose 4th terms are 9.

67. Give two examples of arithmetic sequences whose 10th terms are 206.

68. Find the 5th term of the arithmetic sequence $\{9b, 5b, b, \ldots\}$.

69. Find the 11th term of the arithmetic sequence $\{3a - 2b, a + 2b, -a + 6b \ldots\}$.

70. At which term does the sequence $\{5.4, 14.5, 23.6, \ldots\}$ exceed 151?

71. At which term does the sequence $\left\{\frac{17}{3}, \frac{31}{6}, \frac{14}{3}, \ldots\right\}$ begin to have negative values?

72. For which terms does the finite arithmetic sequence $\left\{\frac{5}{2}, \frac{19}{8}, \frac{9}{4}, \ldots, \frac{1}{8}\right\}$ have integer values?

73. Write an arithmetic sequence using a recursive formula. Show the first 4 terms, and then find the 31st term.

74. Write an arithmetic sequence using an explicit formula. Show the first 4 terms, and then find the 28th term.

13.3 Geometric Sequences

Learning Objectives

In this section, you will:

- Find the common ratio for a geometric sequence.
- List the terms of a geometric sequence.
- Use a recursive formula for a geometric sequence.
- Use an explicit formula for a geometric sequence.

Many jobs offer an annual cost-of-living increase to keep salaries consistent with inflation. Suppose, for example, a recent college graduate finds a position as a sales manager earning an annual salary of \$26,000. He is promised a 2% cost of living increase each year. His annual salary in any given year can be found by multiplying his salary from the previous year by 102%. His salary will be \$26,520 after one year; \$27,050.40 after two years; \$27,591.41 after three years; and so on. When a salary increases by a constant rate each year, the salary grows by a constant factor. In this section, we will review sequences that grow in this way.

Finding Common Ratios

The yearly salary values described form a **geometric sequence** because they change by a constant factor each year. Each term of a geometric sequence increases or decreases by a constant factor called the **common ratio**. The sequence below is an example of a geometric sequence because each term increases by a constant factor of 6. Multiplying any term of the sequence by the common ratio 6 generates the subsequent term.

$$\{1, \overset{\times 6}{\curvearrowright} 6, \overset{\times 6}{\curvearrowright} 36, \overset{\times 6}{\curvearrowright} 216, \overset{\times 6}{\curvearrowright} 1{,}296, \ldots\}$$

Definition of a Geometric Sequence

A **geometric sequence** is one in which any term divided by the previous term is a constant. This constant is called the **common ratio** of the sequence. The common ratio can be found by dividing any term in the sequence by the previous term. If a_1 is the initial term of a geometric sequence and r is the common ratio, the sequence will be

$$\{a_1, \ a_1 r, \ a_1 r^2, \ a_1 r^3, \ \ldots\}.$$

HOW TO

Given a set of numbers, determine if they represent a geometric sequence.

1. Divide each term by the previous term.
2. Compare the quotients. If they are the same, a common ratio exists and the sequence is geometric.

EXAMPLE 1

Finding Common Ratios

Is the sequence geometric? If so, find the common ratio.

ⓐ $1, 2, 4, 8, 16, \ldots$ ⓑ $48, 12, 4, 2, \ldots$

Solution

Divide each term by the previous term to determine whether a common ratio exists.

ⓐ $\frac{2}{1} = 2$ $\frac{4}{2} = 2$ $\frac{8}{4} = 2$ $\frac{16}{8} = 2$

The sequence is geometric because there is a common ratio. The common ratio is 2.

ⓑ $\frac{12}{48} = \frac{1}{4}$ $\frac{4}{12} = \frac{1}{3}$ $\frac{2}{4} = \frac{1}{2}$

The sequence is not geometric because there is not a common ratio.

Analysis

The graph of each sequence is shown in Figure 1. It seems from the graphs that both (a) and (b) appear have the form of the graph of an exponential function in this viewing window. However, we know that (a) is geometric and so this interpretation holds, but (b) is not.

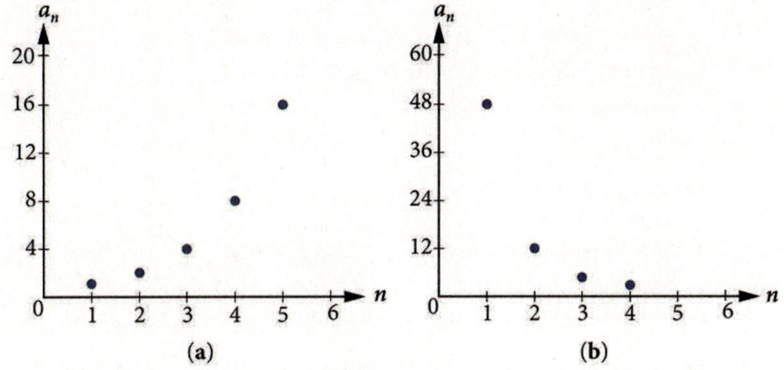

Figure 1

Q&A **If you are told that a sequence is geometric, do you have to divide every term by the previous term to find the common ratio?**

No. If you know that the sequence is geometric, you can choose any one term in the sequence and divide it by the previous term to find the common ratio.

TRY IT #1 Is the sequence geometric? If so, find the common ratio.

$$5, 10, 15, 20, \ldots$$

TRY IT #2 Is the sequence geometric? If so, find the common ratio.

$$100, 20, 4, \frac{4}{5}, \ldots$$

Writing Terms of Geometric Sequences

Now that we can identify a geometric sequence, we will learn how to find the terms of a geometric sequence if we are given the first term and the common ratio. The terms of a geometric sequence can be found by beginning with the first term and multiplying by the common ratio repeatedly. For instance, if the first term of a geometric sequence is $a_1 = -2$ and the common ratio is $r = 4$, we can find subsequent terms by multiplying $-2 \cdot 4$ to get -8 then multiplying the result $-8 \cdot 4$ to get -32 and so on.

$$\begin{aligned} a_1 &= -2 \\ a_2 &= (-2 \cdot 4) = -8 \\ a_3 &= (-8 \cdot 4) = -32 \\ a_4 &= (-32 \cdot 4) = -128 \end{aligned}$$

The first four terms are $\{-2, -8, -32, -128\}$.

HOW TO

Given the first term and the common factor, find the first four terms of a geometric sequence.

1. Multiply the initial term, a_1, by the common ratio to find the next term, a_2.
2. Repeat the process, using $a_n = a_2$ to find a_3 and then a_3 to find $a_{4,}$ until all four terms have been identified.
3. Write the terms separated by commons within brackets.

EXAMPLE 2

Writing the Terms of a Geometric Sequence

List the first four terms of the geometric sequence with $a_1 = 5$ and $r = -2$.

Solution

Multiply a_1 by -2 to find a_2. Repeat the process, using a_2 to find a_3, and so on.

$$\begin{aligned} a_1 &= 5 \\ a_2 &= -2a_1 = -10 \\ a_3 &= -2a_2 = 20 \\ a_4 &= -2a_3 = -40 \end{aligned}$$

The first four terms are $\{5, -10, 20, -40\}$.

TRY IT #3 List the first five terms of the geometric sequence with $a_1 = 18$ and $r = \frac{1}{3}$.

Using Recursive Formulas for Geometric Sequences

A recursive formula allows us to find any term of a geometric sequence by using the previous term. Each term is the product of the common ratio and the previous term. For example, suppose the common ratio is 9. Then each term is nine times the previous term. As with any recursive formula, the initial term must be given.

Recursive Formula for a Geometric Sequence

The recursive formula for a geometric sequence with common ratio r and first term a_1 is

$$a_n = ra_{n-1}, n \geq 2$$

HOW TO

Given the first several terms of a geometric sequence, write its recursive formula.

1. State the initial term.
2. Find the common ratio by dividing any term by the preceding term.
3. Substitute the common ratio into the recursive formula for a geometric sequence.

EXAMPLE 3

Using Recursive Formulas for Geometric Sequences

Write a recursive formula for the following geometric sequence.

$$\{6, 9, 13.5, 20.25, ...\}$$

Solution

The first term is given as 6. The common ratio can be found by dividing the second term by the first term.

$$r = \frac{9}{6} = 1.5$$

Substitute the common ratio into the recursive formula for geometric sequences and define a_1.

$$\begin{aligned} a_n &= ra_{n-1} \\ a_n &= 1.5a_{n-1} \quad \text{for } n \geq 2 \\ a_1 &= 6 \end{aligned}$$

Analysis

The sequence of data points follows an exponential pattern. The common ratio is also the base of an exponential function as shown in Figure 2

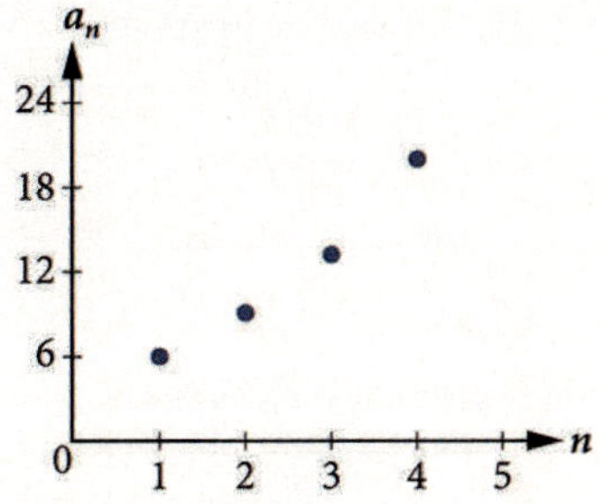

Figure 2

Q&A **Do we have to divide the second term by the first term to find the common ratio?**

No. We can divide any term in the sequence by the previous term. It is, however, most common to divide the second term by the first term because it is often the easiest method of finding the common ratio.

TRY IT #4 Write a recursive formula for the following geometric sequence.

$$\left\{2, \frac{4}{3}, \frac{8}{9}, \frac{16}{27}, \ldots\right\}$$

Using Explicit Formulas for Geometric Sequences

Because a geometric sequence is an exponential function whose domain is the set of positive integers, and the common ratio is the base of the function, we can write explicit formulas that allow us to find particular terms.

$$a_n = a_1 r^{n-1}$$

Let's take a look at the sequence $\{18, 36, 72, 144, 288, \ldots\}$. This is a geometric sequence with a common ratio of 2 and an exponential function with a base of 2. An explicit formula for this sequence is

$$a_n = 18 \cdot 2^{n-1}$$

The graph of the sequence is shown in Figure 3.

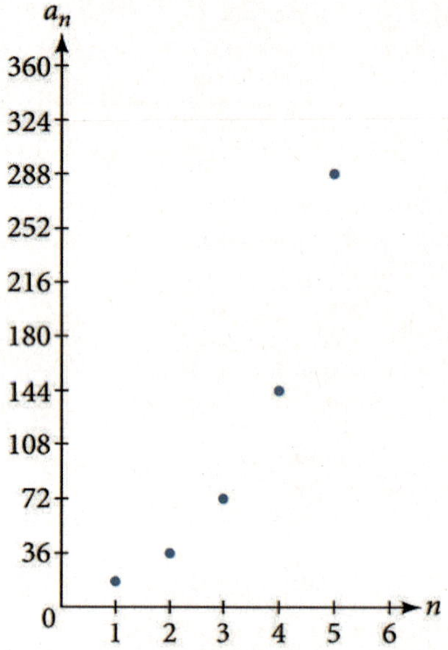

Figure 3

Explicit Formula for a Geometric Sequence

The n th term of a geometric sequence is given by the explicit formula:

$$a_n = a_1 r^{n-1}$$

EXAMPLE 4

Writing Terms of Geometric Sequences Using the Explicit Formula

Given a geometric sequence with $a_1 = 3$ and $a_4 = 24$, find a_2.

Solution

The sequence can be written in terms of the initial term and the common ratio r.

$$3, 3r, 3r^2, 3r^3, \ldots$$

Find the common ratio using the given fourth term.

$$\begin{aligned} a_n &= a_1 r^{n-1} && \\ a_4 &= 3r^3 && \text{Write the fourth term of sequence in terms of } \alpha_1 \text{ and } r \\ 24 &= 3r^3 && \text{Substitute 24 for } a_4 \\ 8 &= r^3 && \text{Divide} \\ r &= 2 && \text{Solve for the common ratio} \end{aligned}$$

Find the second term by multiplying the first term by the common ratio.

$$\begin{aligned} a_2 &= 2a_1 \\ &= 2(3) \\ &= 6 \end{aligned}$$

Analysis

The common ratio is multiplied by the first term once to find the second term, twice to find the third term, three times to find the fourth term, and so on. The tenth term could be found by multiplying the first term by the common ratio nine times or by multiplying by the common ratio raised to the ninth power.

> **TRY IT** #5 Given a geometric sequence with $a_2 = 4$ and $a_3 = 32$, find a_6.

EXAMPLE 5

Writing an Explicit Formula for the n th Term of a Geometric Sequence

Write an explicit formula for the nth term of the following geometric sequence.

$$\{2, 10, 50, 250, \ldots\}$$

Solution

The first term is 2. The common ratio can be found by dividing the second term by the first term.

$$\frac{10}{2} = 5$$

The common ratio is 5. Substitute the common ratio and the first term of the sequence into the formula.

$$\begin{aligned} a_n &= a_1 r^{(n-1)} \\ a_n &= 2 \cdot 5^{n-1} \end{aligned}$$

The graph of this sequence in Figure 4 shows an exponential pattern.

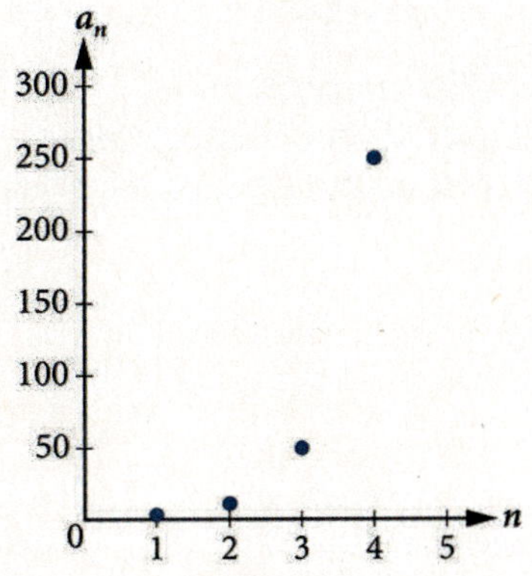

Figure 4

> **TRY IT** #6 Write an explicit formula for the following geometric sequence.

$$\{-1, 3, -9, 27, \ldots\}$$

Solving Application Problems with Geometric Sequences

In real-world scenarios involving geometric sequences, we may need to use an initial term of a_0 instead of a_1. In these problems, we can alter the explicit formula slightly by using the following formula:

$$a_n = a_0 r^n$$

EXAMPLE 6

Solving Application Problems with Geometric Sequences

In 2013, the number of students in a small school is 284. It is estimated that the student population will increase by 4% each year.

ⓐ Write a formula for the student population. ⓑ Estimate the student population in 2020.

Solution

ⓐ

The situation can be modeled by a geometric sequence with an initial term of 284. The student population will be 104% of the prior year, so the common ratio is 1.04.

Let P be the student population and n be the number of years after 2013. Using the explicit formula for a geometric sequence we get

$$P_n = 284 \cdot 1.04^n$$

ⓑ

We can find the number of years since 2013 by subtracting.

$$2020 - 2013 = 7$$

We are looking for the population after 7 years. We can substitute 7 for n to estimate the population in 2020.

$$P_7 = 284 \cdot 1.04^7 \approx 374$$

The student population will be about 374 in 2020.

TRY IT #7 A business starts a new website. Initially the number of hits is 293 due to the curiosity factor. The business estimates the number of hits will increase by 2.6% per week.

ⓐ Write a formula for the number of hits. ⓑ Estimate the number of hits in 5 weeks.

MEDIA

Access these online resources for additional instruction and practice with geometric sequences.

Geometric Sequences (http://openstax.org/l/geometricseq)
Determine the Type of Sequence (http://openstax.org/l/sequencetype)
Find the Formula for a Sequence (http://openstax.org/l/sequenceformula)

13.3 SECTION EXERCISES

Verbal

1. What is a geometric sequence?

2. How is the common ratio of a geometric sequence found?

3. What is the procedure for determining whether a sequence is geometric?

4. What is the difference between an arithmetic sequence and a geometric sequence?

5. Describe how exponential functions and geometric sequences are similar. How are they different?

Algebraic

For the following exercises, find the common ratio for the geometric sequence.

6. $1, 3, 9, 27, 81, ...$

7. $-0.125, 0.25, -0.5, 1, -2, ...$

8. $-2, -\frac{1}{2}, -\frac{1}{8}, -\frac{1}{32}, -\frac{1}{128}, ...$

For the following exercises, determine whether the sequence is geometric. If so, find the common ratio.

9. $-6, -12, -24, -48, -96, ...$

10. $5, 5.2, 5.4, 5.6, 5.8, ...$

11. $-1, \frac{1}{2}, -\frac{1}{4}, \frac{1}{8}, -\frac{1}{16}, ...$

12. $6, 8, 11, 15, 20, ...$

13. $0.8, 4, 20, 100, 500, ...$

For the following exercises, write the first five terms of the geometric sequence, given the first term and common ratio.

14. $a_1 = 8, \quad r = 0.3$

15. $a_1 = 5, \quad r = \frac{1}{5}$

For the following exercises, write the first five terms of the geometric sequence, given any two terms.

16. $a_7 = 64, \quad a_{10} = 512$

17. $a_6 = 25, \quad a_8 = 6.25$

For the following exercises, find the specified term for the geometric sequence, given the first term and common ratio.

18. The first term is 2, and the common ratio is 3. Find the 5th term.

19. The first term is 16 and the common ratio is $-\frac{1}{3}$. Find the 4th term.

For the following exercises, find the specified term for the geometric sequence, given the first four terms.

20. $a_n = \{-1, 2, -4, 8, ...\}$. Find a_{12}.

21. $a_n = \left\{-2, \frac{2}{3}, -\frac{2}{9}, \frac{2}{27}, ...\right\}$. Find a_7.

For the following exercises, write the first five terms of the geometric sequence.

22. $a_1 = -486, \quad a_n = -\frac{1}{3}\, a_{n-1}$

23. $a_1 = 7, \quad a_n = 0.2a_{n-1}$

For the following exercises, write a recursive formula for each geometric sequence.

24. $a_n = \{-1, 5, -25, 125, ...\}$

25. $a_n = \{-32, -16, -8, -4, ...\}$

26. $a_n = \{14, 56, 224, 896, ...\}$

27. $a_n = \{10, -3, 0.9, -0.27, ...\}$

28. $a_n = \{0.61, 1.83, 5.49, 16.47, ...\}$

29. $a_n = \left\{\frac{3}{5}, \frac{1}{10}, \frac{1}{60}, \frac{1}{360}, ...\right\}$

30. $a_n = \left\{-2, \frac{4}{3}, -\frac{8}{9}, \frac{16}{27}, ...\right\}$

31. $a_n = \left\{\frac{1}{512}, -\frac{1}{128}, \frac{1}{32}, -\frac{1}{8}, ...\right\}$

For the following exercises, write the first five terms of the geometric sequence.

32. $a_n = -4 \cdot 5^{n-1}$

33. $a_n = 12 \cdot \left(-\frac{1}{2}\right)^{n-1}$

For the following exercises, write an explicit formula for each geometric sequence.

34. $a_n = \{-2, -4, -8, -16, ...\}$

35. $a_n = \{1, 3, 9, 27, ...\}$

36. $a_n = \{-4, -12, -36, -108, ...\}$

37. $a_n = \{0.8, -4, 20, -100, ...\}$

38. $a_n = \{-1.25, -5, -20, -80, ...\}$

39. $a_n = \left\{-1, -\frac{4}{5}, -\frac{16}{25}, -\frac{64}{125}, ...\right\}$

40. $a_n = \left\{2, \frac{1}{3}, \frac{1}{18}, \frac{1}{108}, ...\right\}$

41. $a_n = \left\{3, -1, \frac{1}{3}, -\frac{1}{9}, ...\right\}$

For the following exercises, find the specified term for the geometric sequence given.

42. Let $a_1 = 4$, $a_n = -3a_{n-1}$. Find a_8.

43. Let $a_n = -\left(-\frac{1}{3}\right)^{n-1}$. Find a_{12}.

For the following exercises, find the number of terms in the given finite geometric sequence.

44. $a_n = \{-1, 3, -9, ..., 2187\}$

45. $a_n = \left\{2, 1, \frac{1}{2}, ..., \frac{1}{1024}\right\}$

Graphical

For the following exercises, determine whether the graph shown represents a geometric sequence.

46.

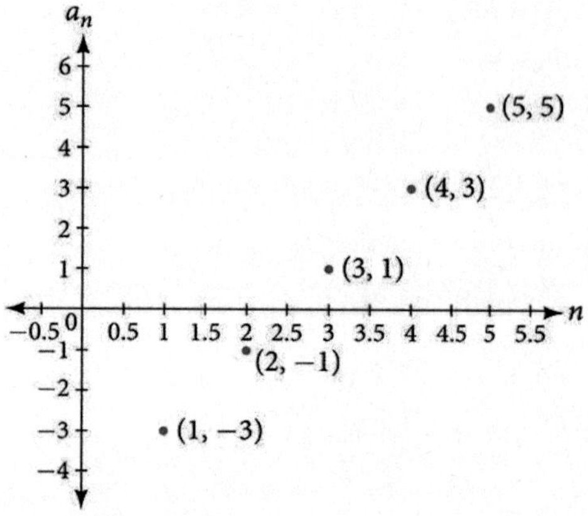

47.

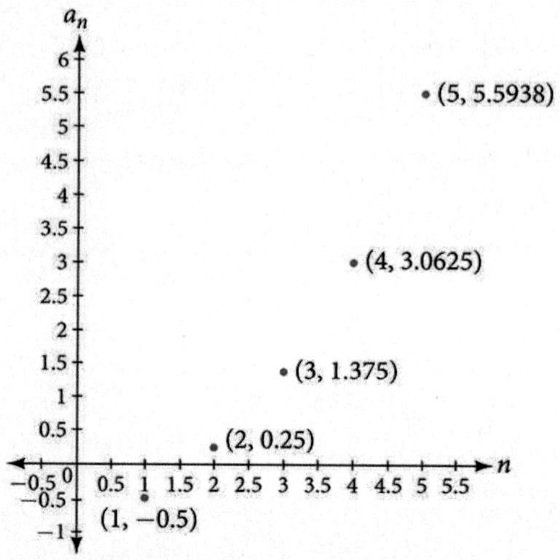

For the following exercises, use the information provided to graph the first five terms of the geometric sequence.

48. $a_1 = 1, \quad r = \frac{1}{2}$

49. $a_1 = 3, \quad a_n = 2a_{n-1}$

50. $a_n = 27 \cdot 0.3^{n-1}$

Extensions

51. Use recursive formulas to give two examples of geometric sequences whose 3rd terms are 200.

52. Use explicit formulas to give two examples of geometric sequences whose 7th terms are 1024.

53. Find the 5th term of the geometric sequence $\{b, 4b, 16b, ...\}$.

54. Find the 7th term of the geometric sequence $\{64a(-b), 32a(-3b), 16a(-9b), ...\}$.

55. At which term does the sequence $\{10, 12, 14.4, 17.28, \ ...\}$ exceed 100?

56. At which term does the sequence $\left\{\frac{1}{2187}, \frac{1}{729}, \frac{1}{243}, \frac{1}{81} \ ...\right\}$ begin to have integer values?

57. For which term does the geometric sequence $a_n = -36\left(\frac{2}{3}\right)^{n-1}$ first have a non-integer value?

58. Use the recursive formula to write a geometric sequence whose common ratio is an integer. Show the first four terms, and then find the 10th term.

59. Use the explicit formula to write a geometric sequence whose common ratio is a decimal number between 0 and 1. Show the first 4 terms, and then find the 8th term.

60. Is it possible for a sequence to be both arithmetic and geometric? If so, give an example.

13.4 Series and Their Notations

Learning Objectives

In this section, you will:

- Use summation notation.
- Use the formula for the sum of the first n terms of an arithmetic series.
- Use the formula for the sum of the first n terms of a geometric series.
- Use the formula for the sum of an infinite geometric series.
- Solve annuity problems.

A parent decides to start a college fund for their daughter. They plan to invest $50 in the fund each month. The fund pays 6% annual interest, compounded monthly. How much money will they have saved when their daughter is ready to start college in 6 years? In this section, we will learn how to answer this question. To do so, we need to consider the amount of money invested and the amount of interest earned.

Using Summation Notation

To find the total amount of money in the college fund and the sum of the amounts deposited, we need to add the amounts deposited each month and the amounts earned monthly. The sum of the terms of a sequence is called a **series**. Consider, for example, the following series.

$$3 + 7 + 11 + 15 + 19 + ...$$

The ***n*th partial sum** of a series is the sum of a finite number of consecutive terms beginning with the first term. The notation S_n represents the partial sum.

$$\begin{aligned} S_1 &= 3 \\ S_2 &= 3 + 7 = 10 \\ S_3 &= 3 + 7 + 11 = 21 \\ S_4 &= 3 + 7 + 11 + 15 = 36 \end{aligned}$$

Summation notation is used to represent series. Summation notation is often known as sigma notation because it uses the Greek capital letter sigma, Σ, to represent the sum. Summation notation includes an explicit formula and specifies the first and last terms in the series. An explicit formula for each term of the series is given to the right of the sigma. A variable called the **index of summation** is written below the sigma. The index of summation is set equal to the **lower limit of summation**, which is the number used to generate the first term in the series. The number above the sigma, called the **upper limit of summation**, is the number used to generate the last term in a series.

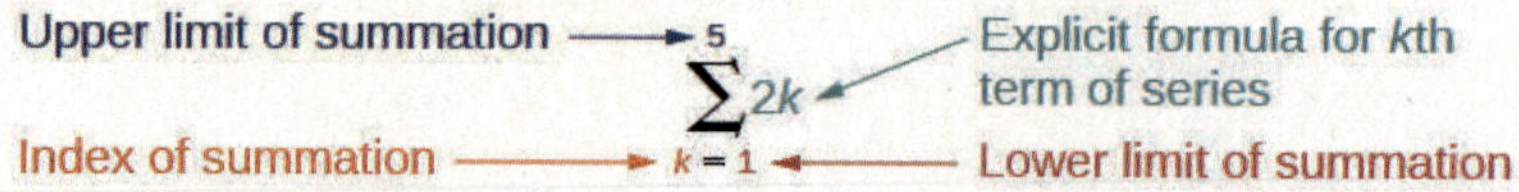

If we interpret the given notation, we see that it asks us to find the sum of the terms in the series $a_k = 2k$ for $k = 1$ through $k = 5$. We can begin by substituting the terms for k and listing out the terms of this series.

$$\begin{aligned} a_1 &= 2(1) = 2 \\ a_2 &= 2(2) = 4 \\ a_3 &= 2(3) = 6 \\ a_4 &= 2(4) = 8 \\ a_5 &= 2(5) = 10 \end{aligned}$$

We can find the sum of the series by adding the terms:

$$\sum_{k=1}^{5} 2k = 2 + 4 + 6 + 8 + 10 = 30$$

Summation Notation

The sum of the first n terms of a **series** can be expressed in **summation notation** as follows:

$$\sum_{k=1}^{n} a_k$$

This notation tells us to find the sum of a_k from $k = 1$ to $k = n$.

k is called the **index of summation**, 1 is the **lower limit of summation**, and n is the **upper limit of summation**.

Q&A **Does the lower limit of summation have to be 1?**

No. The lower limit of summation can be any number, but 1 is frequently used. We will look at examples with lower limits of summation other than 1.

HOW TO

Given summation notation for a series, evaluate the value.

1. Identify the lower limit of summation.
2. Identify the upper limit of summation.
3. Substitute each value of k from the lower limit to the upper limit into the formula.
4. Add to find the sum.

EXAMPLE 1

Using Summation Notation

Evaluate $\sum_{k=3}^{7} k^2$.

Solution

According to the notation, the lower limit of summation is 3 and the upper limit is 7. So we need to find the sum of k^2 from $k = 3$ to $k = 7$. We find the terms of the series by substituting $k = 3,4,5,6,$ and 7 into the function k^2. We add the terms to find the sum.

$$\begin{aligned}\sum_{k=3}^{7} k^2 &= 3^2 + 4^2 + 5^2 + 6^2 + 7^2 \\ &= 9 + 16 + 25 + 36 + 49 \\ &= 135\end{aligned}$$

> **TRY IT** #1 Evaluate $\sum_{k=2}^{5}(3k-1)$.

Using the Formula for Arithmetic Series

Just as we studied special types of sequences, we will look at special types of series. Recall that an arithmetic sequence is a sequence in which the difference between any two consecutive terms is the common difference, d. The sum of the terms of an arithmetic sequence is called an **arithmetic series**. We can write the sum of the first n terms of an arithmetic series as:

$$S_n = a_1 + (a_1 + d) + (a_1 + 2d) + ... + (a_n - d) + a_n.$$

We can also reverse the order of the terms and write the sum as

$$S_n = a_n + (a_n - d) + (a_n - 2d) + ... + (a_1 + d) + a_1.$$

If we add these two expressions for the sum of the first n terms of an arithmetic series, we can derive a formula for the sum of the first n terms of any arithmetic series.

$$\begin{array}{rl} S_n = & a_1 + (a_1 + d) + (a_1 + 2d) + ... + (a_n - d) + a_n \\ + \quad S_n = & a_n + (a_n - d) + (a_n - 2d) + ... + (a_1 + d) + a_1 \\ \hline 2S_n = & (a_1 + a_n) + (a_1 + a_n) + ... + (a_1 + a_n) \end{array}$$

Because there are n terms in the series, we can simplify this sum to

$$2S_n = n(a_1 + a_n).$$

We divide by 2 to find the formula for the sum of the first n terms of an arithmetic series.

$$S_n = \frac{n(a_1 + a_n)}{2}$$

Formula for the Sum of the First n Terms of an Arithmetic Series

An **arithmetic series** is the sum of the terms of an arithmetic sequence. The formula for the sum of the first n terms of an arithmetic sequence is

$$S_n = \frac{n(a_1 + a_n)}{2}$$

HOW TO

Given terms of an arithmetic series, find the sum of the first n terms.

1. Identify a_1 and a_n.
2. Determine n.
3. Substitute values for a_1, a_n, and n into the formula $S_n = \frac{n(a_1 + a_n)}{2}$.
4. Simplify to find S_n.

EXAMPLE 2

Finding the First *n* Terms of an Arithmetic Series

Find the sum of each arithmetic series.

ⓐ $5 + 8 + 11 + 14 + 17 + 20 + 23 + 26 + 29 + 32$ ⓑ $20 + 15 + 10 + \ldots + -50$ ⓒ $\sum_{k=1}^{12} 3k - 8$

Solution

ⓐ

We are given $a_1 = 5$ and $a_n = 32$.

Count the number of terms in the sequence to find $n = 10$.

Substitute values for a_1, a_n, and n into the formula and simplify.

$$S_n = \frac{n(a_1 + a_n)}{2}$$
$$S_{10} = \frac{10(5+32)}{2} = 185$$

ⓑ

We are given $a_1 = 20$ and $a_n = -50$.

Use the formula for the general term of an arithmetic sequence to find n.

$$\begin{aligned} a_n &= a_1 + (n-1)d \\ -50 &= 20 + (n-1)(-5) \\ -70 &= (n-1)(-5) \\ 14 &= n - 1 \\ 15 &= n \end{aligned}$$

Substitute values for a_1, a_n, n into the formula and simplify.

$$S_n = \frac{n(a_1 + a_n)}{2}$$
$$S_{15} = \frac{15(20-50)}{2} = -225$$

ⓒ

To find a_1, substitute $k = 1$ into the given explicit formula.

$$\begin{aligned} a_k &= 3k - 8 \\ a_1 &= 3(1) - 8 = -5 \end{aligned}$$

We are given that $n = 12$. To find a_{12}, substitute $k = 12$ into the given explicit formula.

$$\begin{aligned} a_k &= 3k - 8 \\ a_{12} &= 3(12) - 8 = 28 \end{aligned}$$

Substitute values for a_1, a_n, and n into the formula and simplify.

$$S_n = \frac{n(a_1 + a_n)}{2}$$
$$S_{12} = \frac{12(-5+28)}{2} = 138$$

Use the formula to find the sum of each arithmetic series.

TRY IT #2 $1.4 + 1.6 + 1.8 + 2.0 + 2.2 + 2.4 + 2.6 + 2.8 + 3.0 + 3.2 + 3.4$

TRY IT #3 $13 + 21 + 29 + \ldots + 69$

> **TRY IT** #4 $\sum_{k=1}^{10} 5 - 6k$

EXAMPLE 3

Solving Application Problems with Arithmetic Series

On the Sunday after a minor surgery, a woman is able to walk a half-mile. Each Sunday, she walks an additional quarter-mile. After 8 weeks, what will be the total number of miles she has walked?

Solution

This problem can be modeled by an arithmetic series with $a_1 = \frac{1}{2}$ and $d = \frac{1}{4}$. We are looking for the total number of miles walked after 8 weeks, so we know that $n = 8$, and we are looking for S_8. To find a_8, we can use the explicit formula for an arithmetic sequence.

$$a_n = a_1 + d(n-1)$$
$$a_8 = \frac{1}{2} + \frac{1}{4}(8-1) = \frac{9}{4}$$

We can now use the formula for arithmetic series.

$$S_n = \frac{n(a_1 + a_n)}{2}$$
$$S_8 = \frac{8(\frac{1}{2} + \frac{9}{4})}{2} = 11$$

She will have walked a total of 11 miles.

> **TRY IT** #5 A man earns \$100 in the first week of June. Each week, he earns \$12.50 more than the previous week. After 12 weeks, how much has he earned?

Using the Formula for Geometric Series

Just as the sum of the terms of an arithmetic sequence is called an arithmetic series, the sum of the terms in a geometric sequence is called a **geometric series**. Recall that a geometric sequence is a sequence in which the ratio of any two consecutive terms is the common ratio, r. We can write the sum of the first n terms of a geometric series as

$$S_n = a_1 + ra_1 + r^2 a_1 + ... + r^{n-1} a_1.$$

Just as with arithmetic series, we can do some algebraic manipulation to derive a formula for the sum of the first n terms of a geometric series. We will begin by multiplying both sides of the equation by r.

$$rS_n = ra_1 + r^2 a_1 + r^3 a_1 + ... + r^n a_1$$

Next, we subtract this equation from the original equation.

$$\begin{array}{r} S_n = a_1 + ra_1 + r^2 a_1 + ... + r^{n-1} a_1 \\ -rS_n = -(ra_1 + r^2 a_1 + r^3 a_1 + ... + r^n a_1) \\ \hline (1-r)S_n = a_1 - r^n a_1 \end{array}$$

Notice that when we subtract, all but the first term of the top equation and the last term of the bottom equation cancel out. To obtain a formula for S_n, divide both sides by $(1-r)$.

$$S_n = \frac{a_1(1-r^n)}{1-r} \quad r \neq 1$$

Formula for the Sum of the First n Terms of a Geometric Series

A **geometric series** is the sum of the terms in a geometric sequence. The formula for the sum of the first n terms of a geometric sequence is represented as

$$S_n = \frac{a_1(1-r^n)}{1-r} \quad r \neq 1$$

HOW TO

Given a geometric series, find the sum of the first *n* terms.

1. Identify $a_1, \ r,$ and $n.$
2. Substitute values for $a_1, \ r,$ and n into the formula $S_n = \frac{a_1(1-r^n)}{1-r}.$
3. Simplify to find $S_n.$

EXAMPLE 4

Finding the First *n* Terms of a Geometric Series

Use the formula to find the indicated partial sum of each geometric series.

ⓐ S_{11} for the series $8 + \text{-}4 + 2 + \ldots$ ⓑ $\sum_{k=1}^{6} 3 \cdot 2^k$

Solution

ⓐ

$a_1 = 8$, and we are given that $n = 11.$

We can find r by dividing the second term of the series by the first.

$$r = \frac{-4}{8} = -\frac{1}{2}$$

Substitute values for $a_1, \ r,$ and n into the formula and simplify.

$$S_n = \frac{a_1(1-r^n)}{1-r}$$

$$S_{11} = \frac{8\left(1-\left(-\frac{1}{2}\right)^{11}\right)}{1-\left(-\frac{1}{2}\right)} \approx 5.336$$

ⓑ

Find a_1 by substituting $k = 1$ into the given explicit formula.

$$a_1 = 3 \cdot 2^1 = 6$$

We can see from the given explicit formula that $r = 2$. The upper limit of summation is 6, so $n = 6$.

Substitute values for $a_1, \ r,$ and n into the formula, and simplify.

$$S_n = \frac{a_1(1-r^n)}{1-r}$$

$$S_6 = \frac{6(1-2^6)}{1-2} = 378$$

Use the formula to find the indicated partial sum of each geometric series.

TRY IT #6 S_{20} for the series $1{,}000 + 500 + 250 + \ldots$

TRY IT #7 $\sum_{k=1}^{8} 3^k$

EXAMPLE 5

Solving an Application Problem with a Geometric Series

At a new job, an employee's starting salary is $26,750. He receives a 1.6% annual raise. Find his total earnings at the end of 5 years.

Solution

The problem can be represented by a geometric series with $a_1 = 26,750$; $n = 5$; and $r = 1.016$. Substitute values for a_1, r, and n into the formula and simplify to find the total amount earned at the end of 5 years.

$$S_n = \frac{a_1(1-r^n)}{1-r}$$

$$S_5 = \frac{26{,}750(1-1.016^5)}{1-1.016} \approx 138{,}099.03$$

He will have earned a total of $138,099.03 by the end of 5 years.

> **TRY IT** #8 At a new job, an employee's starting salary is $32,100. She receives a 2% annual raise. How much will she have earned by the end of 8 years?

Using the Formula for the Sum of an Infinite Geometric Series

Thus far, we have looked only at finite series. Sometimes, however, we are interested in the sum of the terms of an infinite sequence rather than the sum of only the first n terms. An **infinite series** is the sum of the terms of an infinite sequence. An example of an infinite series is $2 + 4 + 6 + 8 + ...$

This series can also be written in summation notation as $\sum_{k=1}^{\infty} 2k$, where the upper limit of summation is infinity. Because the terms are not tending to zero, the sum of the series increases without bound as we add more terms. Therefore, the sum of this infinite series is not defined. When the sum is not a real number, we say the series **diverges**.

Determining Whether the Sum of an Infinite Geometric Series is Defined

If the terms of an infinite geometric sequence approach 0, the sum of an infinite geometric series can be defined. The terms in this series approach 0:

$$1 + 0.2 + 0.04 + 0.008 + 0.0016 + ...$$

The common ratio $r = 0.2$. As n gets very large, the values of r^n get very small and approach 0. Each successive term affects the sum less than the preceding term. As each succeeding term gets closer to 0, the sum of the terms approaches a finite value. The terms of any infinite geometric series with $-1 < r < 1$ approach 0; the sum of a geometric series is defined when $-1 < r < 1$.

Determining Whether the Sum of an Infinite Geometric Series is Defined

The sum of an infinite series is defined if the series is geometric and $-1 < r < 1$.

HOW TO

Given the first several terms of an infinite series, determine if the sum of the series exists.

1. Find the ratio of the second term to the first term.
2. Find the ratio of the third term to the second term.
3. Continue this process to ensure the ratio of a term to the preceding term is constant throughout. If so, the series is geometric.
4. If a common ratio, r, was found in step 3, check to see if $-1 < r < 1$. If so, the sum is defined. If not, the sum is not defined.

EXAMPLE 6

Determining Whether the Sum of an Infinite Series is Defined
Determine whether the sum of each infinite series is defined.

ⓐ $12 + 8 + 4 + \ldots$ ⓑ $\frac{3}{4} + \frac{1}{2} + \frac{1}{3} + \ldots$ ⓒ $\sum_{k=1}^{\infty} 27 \cdot \left(\frac{1}{3}\right)^k$ ⓓ $\sum_{k=1}^{\infty} 5k$

Solution

ⓐ The ratio of the second term to the first is $\frac{2}{3}$, which is not the same as the ratio of the third term to the second, $\frac{1}{2}$. The series is not geometric.

ⓑ The ratio of the second term to the first is the same as the ratio of the third term to the second. The series is geometric with a common ratio of $\frac{2}{3}$. The sum of the infinite series is defined.

ⓒ The given formula is exponential with a base of $\frac{1}{3}$; the series is geometric with a common ratio of $\frac{1}{3}$. The sum of the infinite series is defined.

ⓓ The given formula is not exponential; the series is not geometric because the terms are increasing, and so cannot yield a finite sum.

Determine whether the sum of the infinite series is defined.

TRY IT #9 $\frac{1}{3} + \frac{1}{2} + \frac{3}{4} + \frac{9}{8} + \ldots$

TRY IT #10 $24 + (-12) + 6 + (-3) + \ldots$

TRY IT #11 $\sum_{k=1}^{\infty} 15 \cdot (-0.3)^k$

Finding Sums of Infinite Series

When the sum of an infinite geometric series exists, we can calculate the sum. The formula for the sum of an infinite series is related to the formula for the sum of the first n terms of a geometric series.

$$S_n = \frac{a_1(1 - r^n)}{1 - r}$$

We will examine an infinite series with $r = \frac{1}{2}$. What happens to r^n as n increases?

$$\left(\frac{1}{2}\right)^2 = \frac{1}{4}$$
$$\left(\frac{1}{2}\right)^3 = \frac{1}{8}$$
$$\left(\frac{1}{2}\right)^4 = \frac{1}{16}$$

The value of r^n decreases rapidly. What happens for greater values of n?

$$\left(\frac{1}{2}\right)^{10} = \frac{1}{1{,}024}$$
$$\left(\frac{1}{2}\right)^{20} = \frac{1}{1{,}048{,}576}$$
$$\left(\frac{1}{2}\right)^{30} = \frac{1}{1{,}073{,}741{,}824}$$

As n gets very large, r^n gets very small. We say that, as n increases without bound, r^n approaches 0. As r^n approaches 0, $1 - r^n$ approaches 1. When this happens, the numerator approaches a_1. This give us a formula for the sum of an infinite geometric series.

Formula for the Sum of an Infinite Geometric Series

The formula for the sum of an infinite geometric series with $-1 < r < 1$ is

$$S = \frac{a_1}{1-r}$$

HOW TO

Given an infinite geometric series, find its sum.

1. Identify a_1 and r.
2. Confirm that $-1 < r < 1$.
3. Substitute values for a_1 and r into the formula, $S = \frac{a_1}{1-r}$.
4. Simplify to find S.

EXAMPLE 7

Finding the Sum of an Infinite Geometric Series

Find the sum, if it exists, for the following:

ⓐ $10 + 9 + 8 + 7 + \ldots$ ⓑ $248.6 + 99.44 + 39.776 + \ \ldots$ ⓒ $\sum_{k=1}^{\infty} 4{,}374 \cdot (-\frac{1}{3})^{k-1}$

ⓓ $\sum_{k=1}^{\infty} \frac{1}{9} \cdot (\frac{4}{3})^{k}$

Solution

ⓐ There is not a constant ratio; the series is not geometric.

ⓑ

There is a constant ratio; the series is geometric. $a_1 = 248.6$ and $r = \frac{99.44}{248.6} = 0.4$, so the sum exists. Substitute $a_1 = 248.6$ and $r = 0.4$ into the formula and simplify to find the sum:

$$S = \frac{a_1}{1-r}$$
$$S = \frac{248.6}{1-0.4} = 414.\overline{3}$$

ⓒ

The formula is exponential, so the series is geometric with $r =-\frac{1}{3}$. Find a_1 by substituting $k = 1$ into the given explicit formula:

$$a_1 = 4{,}374 \cdot (-\frac{1}{3})^{1-1} = 4{,}374$$

Substitute $a_1 = 4{,}374$ and $r = -\frac{1}{3}$ into the formula, and simplify to find the sum:

$$S = \frac{a_1}{1-r}$$
$$S = \frac{4{,}374}{1-(-\frac{1}{3})} = 3{,}280.5$$

ⓓ The formula is exponential, so the series is geometric, but $r > 1$. The sum does not exist.

EXAMPLE 8

Finding an Equivalent Fraction for a Repeating Decimal

Find an equivalent fraction for the repeating decimal $0.\overline{3}$

Solution

We notice the repeating decimal $0.\overline{3} = 0.333...$ so we can rewrite the repeating decimal as a sum of terms.

$$0.\overline{3} = 0.3 + 0.03 + 0.003 + ...$$

Looking for a pattern, we rewrite the sum, noticing that we see the first term multiplied to 0.1 in the second term, and the second term multiplied to 0.1 in the third term.

$$0.\overline{3} = 0.3 + (0.1)\underbrace{(0.3)}_{\text{First Term}} + (0.1)\underbrace{(0.1)(0.3)}_{\text{Second Term}}$$

Notice the pattern; we multiply each consecutive term by a common ratio of 0.1 starting with the first term of 0.3. So, substituting into our formula for an infinite geometric sum, we have

$$S_n = \frac{a_1}{1-r} = \frac{0.3}{1-0.1} = \frac{0.3}{0.9} = \frac{1}{3}.$$

Find the sum, if it exists.

TRY IT #12 $2 + \frac{2}{3} + \frac{2}{9} + ...$

TRY IT #13 $\sum_{k=1}^{\infty} 0.76k + 1$

TRY IT #14 $\sum_{k=1}^{\infty} \left(-\frac{3}{8}\right)^k$

Solving Annuity Problems

At the beginning of the section, we looked at a problem in which a parent invested a set amount of money each month into a college fund for six years. An **annuity** is an investment in which the purchaser makes a sequence of periodic, equal payments. To find the amount of an annuity, we need to find the sum of all the payments and the interest earned. In the example, the parent invests $50 each month. This is the value of the initial deposit. The account paid 6% annual interest, compounded monthly. To find the interest rate per payment period, we need to divide the 6% annual percentage interest (APR) rate by 12. So the monthly interest rate is 0.5%. We can multiply the amount in the account each month by 100.5% to find the value of the account after interest has been added.

We can find the value of the annuity right after the last deposit by using a geometric series with $a_1 = 50$ and $r = 100.5\% = 1.005$. After the first deposit, the value of the annuity will be $50. Let us see if we can determine the amount in the college fund and the interest earned.

We can find the value of the annuity after n deposits using the formula for the sum of the first n terms of a geometric series. In 6 years, there are 72 months, so $n = 72$. We can substitute $a_1 = 50,\ r = 1.005,$ and $n = 72$ into the formula, and simplify to find the value of the annuity after 6 years.

$$S_{72} = \frac{50(1 - 1.005^{72})}{1 - 1.005} \approx 4{,}320.44$$

After the last deposit, the parent will have a total of $4,320.44 in the account. Notice, the parent made 72 payments of $50 each for a total of $72(50) = \$3{,}600$. This means that because of the annuity, the parent earned $720.44 interest in their college fund.

HOW TO

Given an initial deposit and an interest rate, find the value of an annuity.

1. Determine a_1, the value of the initial deposit.
2. Determine n, the number of deposits.
3. Determine r.
   a. Divide the annual interest rate by the number of times per year that interest is compounded.
   b. Add 1 to this amount to find r.
4. Substitute values for a_1, r, and n into the formula for the sum of the first n terms of a geometric series, $S_n = \frac{a_1(1-r^n)}{1-r}$.
5. Simplify to find S_n, the value of the annuity after n deposits.

EXAMPLE 9

Solving an Annuity Problem

A deposit of $100 is placed into a college fund at the beginning of every month for 10 years. The fund earns 9% annual interest, compounded monthly, and paid at the end of the month. How much is in the account right after the last deposit?

Solution

The value of the initial deposit is $100, so $a_1 = 100$. A total of 120 monthly deposits are made in the 10 years, so $n = 120$. To find r, divide the annual interest rate by 12 to find the monthly interest rate and add 1 to represent the new monthly deposit.

$$r = 1 + \frac{0.09}{12} = 1.0075$$

Substitute $a_1 = 100,\ r = 1.0075,$ and $n = 120$ into the formula for the sum of the first n terms of a geometric series, and simplify to find the value of the annuity.

$$S_{120} = \frac{100(1 - 1.0075^{120})}{1 - 1.0075} \approx 19{,}351.43$$

So the account has $19,351.43 after the last deposit is made.

TRY IT #15 At the beginning of each month, $200 is deposited into a retirement fund. The fund earns 6% annual interest, compounded monthly, and paid into the account at the end of the month. How much is in the account if deposits are made for 10 years?

MEDIA

Access these online resources for additional instruction and practice with series.

Arithmetic Series (http://openstax.org/l/arithmeticser)
Geometric Series (http://openstax.org/l/geometricser)
Summation Notation (http://openstax.org/l/sumnotation)

13.4 SECTION EXERCISES

Verbal

1. What is an nth partial sum?

2. What is the difference between an arithmetic sequence and an arithmetic series?

3. What is a geometric series?

4. How is finding the sum of an infinite geometric series different from finding the nth partial sum?

5. What is an annuity?

Algebraic

For the following exercises, express each description of a sum using summation notation.

6. The sum of terms $m^2 + 3m$ from $m = 1$ to $m = 5$

7. The sum from of $n = 0$ to $n = 4$ of $5n$

8. The sum of $6k - 5$ from $k = -2$ to $k = 1$

9. The sum that results from adding the number 4 five times

For the following exercises, express each arithmetic sum using summation notation.

10. $5 + 10 + 15 + 20 + 25 + 30 + 35 + 40 + 45 + 50$

11. $10 + 18 + 26 + \ldots + 162$

12. $\frac{1}{2} + 1 + \frac{3}{2} + 2 + \ldots + 4$

For the following exercises, use the formula for the sum of the first n terms of each arithmetic sequence.

13. $\frac{3}{2} + 2 + \frac{5}{2} + 3 + \frac{7}{2}$

14. $19 + 25 + 31 + \ldots + 73$

15. $3.2 + 3.4 + 3.6 + \ldots + 5.6$

For the following exercises, express each geometric sum using summation notation.

16. $1 + 3 + 9 + 27 + 81 + 243 + 729 + 2187$

17. $8 + 4 + 2 + \ldots + 0.125$

18. $-\frac{1}{6} + \frac{1}{12} - \frac{1}{24} + \ldots + \frac{1}{768}$

For the following exercises, use the formula for the sum of the first n terms of each geometric sequence, and then state the indicated sum.

19. $9 + 3 + 1 + \frac{1}{3} + \frac{1}{9}$

20. $\sum_{n=1}^{9} 5 \cdot 2^{n-1}$

21. $\sum_{a=1}^{11} 64 \cdot 0.2^{a-1}$

For the following exercises, determine whether the infinite series has a sum. If so, write the formula for the sum. If not, state the reason.

22. $12 + 18 + 24 + 30 + \ldots$

23. $2 + 1.6 + 1.28 + 1.024 + \ldots$

24. $\sum_{m=1}^{\infty} 4^{m-1}$

25. $\sum_{\infty}^{k=1} -\left(-\frac{1}{2}\right)^{k-1}$

Graphical

For the following exercises, use the following scenario. Javier makes monthly deposits into a savings account. He opened the account with an initial deposit of $50. Each month thereafter he increased the previous deposit amount by $20.

26. Graph the arithmetic sequence showing one year of Javier's deposits.

27. Graph the arithmetic series showing the monthly sums of one year of Javier's deposits.

For the following exercises, use the geometric series $\sum_{k=1}^{\infty} \left(\frac{1}{2}\right)^k$.

28. Graph the first 7 partial sums of the series.

29. What number does S_n seem to be approaching in the graph? Find the sum to explain why this makes sense.

Numeric

For the following exercises, find the indicated sum.

30. $\sum_{a=1}^{14} a$

31. $\sum_{n=1}^{6} n(n-2)$

32. $\sum_{k=1}^{17} k^2$

33. $\sum_{k=1}^{7} 2^k$

For the following exercises, use the formula for the sum of the first n terms of an arithmetic series to find the sum.

34. $-1.7 + -0.4 + 0.9 + 2.2 + 3.5 + 4.8$

35. $6 + \frac{15}{2} + 9 + \frac{21}{2} + 12 + \frac{27}{2} + 15$

36. $-1 + 3 + 7 + ... + 31$

37. $\sum_{k=1}^{11} \left(\frac{k}{2} - \frac{1}{2}\right)$

For the following exercises, use the formula for the sum of the first n terms of a geometric series to find the partial sum.

38. S_6 for the series $-2 - 10 - 50 - 250...$

39. S_7 for the series $0.4 - 2 + 10 - 50...$

40. $\sum_{k=1}^{9} 2^{k-1}$

41. $\sum_{n=1}^{10} -2 \cdot \left(\frac{1}{2}\right)^{n-1}$

For the following exercises, find the sum of the infinite geometric series.

42. $4 + 2 + 1 + \frac{1}{2}...$

43. $-1 - \frac{1}{4} - \frac{1}{16} - \frac{1}{64}...$

44. $\sum_{\infty}^{k=1} 3 \cdot \left(\frac{1}{4}\right)^{k-1}$

45. $\sum_{n=1}^{\infty} 4.6 \cdot 0.5^{n-1}$

For the following exercises, determine the value of the annuity for the indicated monthly deposit amount, the number of deposits, and the interest rate.

46. Deposit amount: $\$50$; total deposits: 60; interest rate: 5%, compounded monthly

47. Deposit amount: $\$150$; total deposits: 24; interest rate: 3%, compounded monthly

48. Deposit amount: $\$450$; total deposits: 60; interest rate: 4.5%, compounded quarterly

49. Deposit amount: $\$100$; total deposits: 120; interest rate: 10%, compounded semi-annually

Extensions

50. The sum of terms $50 - k^2$ from $k = x$ through 7 is 115. What is *x*?

51. Write an explicit formula for a_k such that $\sum_{k=0}^{6} a_k = 189$. Assume this is an arithmetic series.

52. Find the smallest value of *n* such that $\sum_{k=1}^{n} (3k-5) > 100.$

53. How many terms must be added before the series $-1 - 3 - 5 - 7....$ has a sum less than -75?

54. Write $0.\overline{65}$ as an infinite geometric series using summation notation. Then use the formula for finding the sum of an infinite geometric series to convert $0.\overline{65}$ to a fraction.

55. The sum of an infinite geometric series is five times the value of the first term. What is the common ratio of the series?

56. To get the best loan rates available, the Coleman family want to save enough money to place 20% down on a $160,000 home. They plan to make monthly deposits of $125 in an investment account that offers 8.5% annual interest compounded semi-annually. Will the Colemans have enough for a 20% down payment after five years of saving? How much money will they have saved?

57. Karl has two years to save $\$10,000$ to buy a used car when he graduates. To the nearest dollar, what would his monthly deposits need to be if he invests in an account offering a 4.2% annual interest rate that compounds monthly?

Real-World Applications

58. Keisha devised a week-long study plan to prepare for finals. On the first day, she plans to study for 1 hour, and each successive day she will increase her study time by 30 minutes. How many hours will Keisha have studied after one week?

59. A boulder rolled down a mountain, traveling 6 feet in the first second. Each successive second, its distance increased by 8 feet. How far did the boulder travel after 10 seconds?

60. A scientist places 50 cells in a petri dish. Every hour, the population increases by 1.5%. What will the cell count be after 1 day?

61. A pendulum travels a distance of 3 feet on its first swing. On each successive swing, it travels $\frac{3}{4}$ the distance of the previous swing. What is the total distance traveled by the pendulum when it stops swinging?

62. Rachael deposits $1,500 into a retirement fund each year. The fund earns 8.2% annual interest, compounded monthly. If she opened her account when she was 19 years old, how much will she have by the time she is 55? How much of that amount will be interest earned?

13.5 Counting Principles

Learning Objectives

In this section, you will:

- Solve counting problems using the Addition Principle.
- Solve counting problems using the Multiplication Principle.
- Solve counting problems using permutations involving n distinct objects.
- Solve counting problems using combinations.
- Find the number of subsets of a given set.
- Solve counting problems using permutations involving n non-distinct objects.

A new company sells customizable cases for tablets and smartphones. Each case comes in a variety of colors and can be personalized for an additional fee with images or a monogram. A customer can choose not to personalize or could choose to have one, two, or three images or a monogram. The customer can choose the order of the images and the letters in the monogram. The company is working with an agency to develop a marketing campaign with a focus on the huge number of options they offer. Counting the possibilities is challenging!

We encounter a wide variety of counting problems every day. There is a branch of mathematics devoted to the study of counting problems such as this one. Other applications of counting include secure passwords, horse racing outcomes, and college scheduling choices. We will examine this type of mathematics in this section.

Using the Addition Principle

The company that sells customizable cases offers cases for tablets and smartphones. There are 3 supported tablet models and 5 supported smartphone models. The **Addition Principle** tells us that we can add the number of tablet options to the number of smartphone options to find the total number of options. By the Addition Principle, there are 8 total options, as we can see in Figure 1.

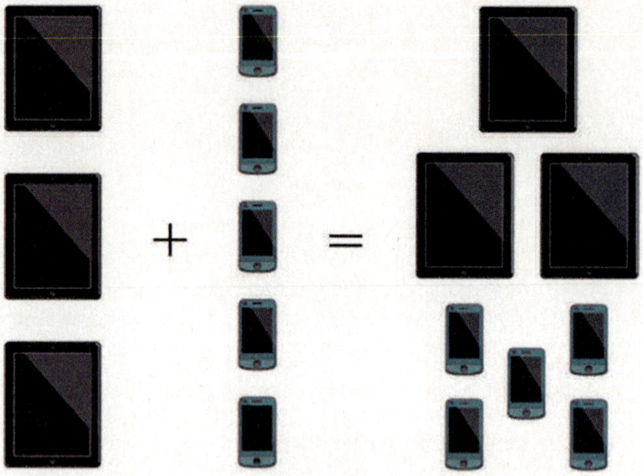

Figure 1

The Addition Principle

According to the **Addition Principle**, if one event can occur in m ways and a second event with no common outcomes can occur in n ways, then the first *or* second event can occur in $m + n$ ways.

EXAMPLE 1

Using the Addition Principle

There are 2 vegetarian entrée options and 5 meat entrée options on a dinner menu. What is the total number of entrée options?

Solution

We can add the number of vegetarian options to the number of meat options to find the total number of entrée options.

Vegetarian + Vegetarian + Meat + Meat + Meat + Meat + Meat

↓ ↓ ↓ ↓ ↓ ↓ ↓

Option 1 + Option 2 + Option 3 + Option 4 + Option 5 + Option 6 + Option 7

There are 7 total options.

TRY IT #1 A student is shopping for a new computer. He is deciding among 3 desktop computers and 4 laptop computers. What is the total number of computer options?

Using the Multiplication Principle

The **Multiplication Principle** applies when we are making more than one selection. Suppose we are choosing an appetizer, an entrée, and a dessert. If there are 2 appetizer options, 3 entrée options, and 2 dessert options on a fixed-price dinner menu, there are a total of 12 possible choices of one each as shown in the tree diagram in Figure 2.

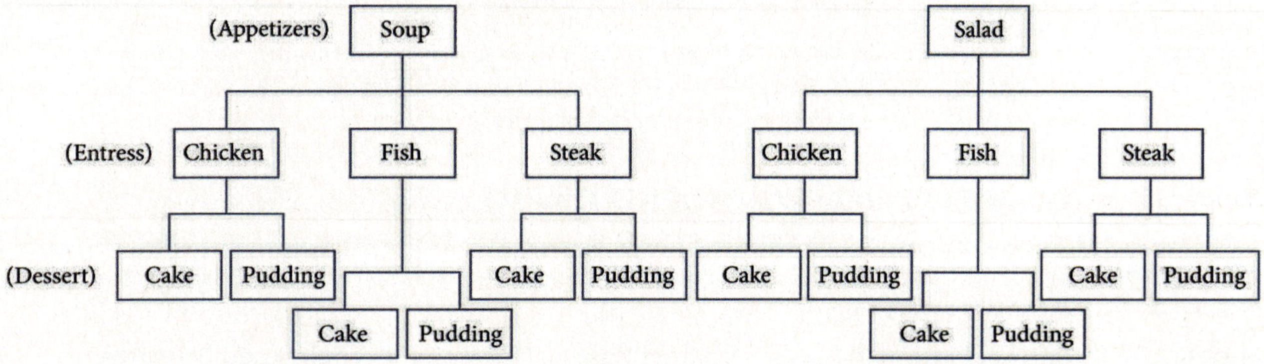

Figure 2

The possible choices are:

1. soup, chicken, cake
2. soup, chicken, pudding
3. soup, fish, cake
4. soup, fish, pudding
5. soup, steak, cake
6. soup, steak, pudding
7. salad, chicken, cake
8. salad, chicken, pudding
9. salad, fish, cake
10. salad, fish, pudding
11. salad, steak, cake
12. salad, steak, pudding

We can also find the total number of possible dinners by multiplying.

We could also conclude that there are 12 possible dinner choices simply by applying the Multiplication Principle.

$$\begin{array}{ccccccc} \#\text{ of appetizer options} & \times & \#\text{ of entree options} & \times & \#\text{ of dessert options} & & \\ 2 & \times & 3 & \times & 2 & = 12 \end{array}$$

The Multiplication Principle

According to the **Multiplication Principle**, if one event can occur in m ways and a second event can occur in n ways after the first event has occurred, then the two events can occur in $m \times n$ ways. This is also known as the **Fundamental Counting Principle**.

EXAMPLE 2

Using the Multiplication Principle

Diane packed 2 skirts, 4 blouses, and a sweater for her business trip. She will need to choose a skirt and a blouse for each outfit and decide whether to wear the sweater. Use the Multiplication Principle to find the total number of possible outfits.

Solution

To find the total number of outfits, find the product of the number of skirt options, the number of blouse options, and the number of sweater options.

$$\begin{array}{cccccc} \#\text{ of skirt options} & \times & \#\text{ of blouse options} & \times & \#\text{ of sweater options} & \\ 2 & \times & 4 & \times & 2 & = 16 \end{array}$$

There are 16 possible outfits.

TRY IT #2 A restaurant offers a breakfast special that includes a breakfast sandwich, a side dish, and a beverage. There are 3 types of breakfast sandwiches, 4 side dish options, and 5 beverage choices. Find the total number of possible breakfast specials.

Finding the Number of Permutations of *n* Distinct Objects

The Multiplication Principle can be used to solve a variety of problem types. One type of problem involves placing objects in order. We arrange letters into words and digits into numbers, line up for photographs, decorate rooms, and more. An ordering of objects is called a **permutation**.

Finding the Number of Permutations of *n* Distinct Objects Using the Multiplication Principle

To solve permutation problems, it is often helpful to draw line segments for each option. That enables us to determine the number of each option so we can multiply. For instance, suppose we have four paintings, and we want to find the number of ways we can hang three of the paintings in order on the wall. We can draw three lines to represent the three places on the wall.

$$\underline{\qquad} \times \underline{\qquad} \times \underline{\qquad}$$

There are four options for the first place, so we write a 4 on the first line.

$$\underline{\quad 4 \quad} \times \underline{\qquad} \times \underline{\qquad}$$

After the first place has been filled, there are three options for the second place so we write a 3 on the second line.

$$\underline{\quad 4 \quad} \times \underline{\quad 3 \quad} \times \underline{\qquad}$$

After the second place has been filled, there are two options for the third place so we write a 2 on the third line. Finally, we find the product.

$$\underline{\quad 4 \quad} \times \underline{\quad 3 \quad} \times \underline{\quad 2 \quad} = 24$$

There are 24 possible permutations of the paintings.

HOW TO

Given *n* distinct options, determine how many permutations there are.

1. Determine how many options there are for the first situation.
2. Determine how many options are left for the second situation.
3. Continue until all of the spots are filled.
4. Multiply the numbers together.

EXAMPLE 3

Finding the Number of Permutations Using the Multiplication Principle

At a swimming competition, nine swimmers compete in a race.

ⓐ How many ways can they place first, second, and third?

ⓑ How many ways can they place first, second, and third if a swimmer named Ariel wins first place? (Assume there is only one contestant named Ariel.)

ⓒ How many ways can all nine swimmers line up for a photo?

Solution

ⓐ Draw lines for each place.

options for 1st place × options for 2nd place × options for 3rd place

There are 9 options for first place. Once someone has won first place, there are 8 remaining options for second place. Once first and second place have been won, there are 7 remaining options for third place.

$$\underline{9} \times \underline{8} \times \underline{7} = 504$$

Multiply to find that there are 504 ways for the swimmers to place.

ⓑ Draw lines for describing each place.

options for 1st place × options for 2nd place × options for 3rd place

We know Ariel must win first place, so there is only 1 option for first place. There are 8 remaining options for second place, and then 7 remaining options for third place.

$$\underline{1} \times \underline{8} \times \underline{7} = 56$$

Multiply to find that there are 56 ways for the swimmers to place if Ariel wins first.

ⓒ

Draw lines for describing each place in the photo.

$$\_\_ \times \_\_ \times \_\_ \times \_\_ \times \_\_ \times \_\_ \times \_\_ \times \_\_ \times \_\_$$

There are 9 choices for the first spot, then 8 for the second, 7 for the third, 6 for the fourth, and so on until only 1 person remains for the last spot.

$$\underline{9} \times \underline{8} \times \underline{7} \times \underline{6} \times \underline{5} \times \underline{4} \times \underline{3} \times \underline{2} \times \underline{1} = 362{,}880$$

There are 362,880 possible permutations for the swimmers to line up.

Analysis

Note that in part c, we found there were 9! ways for 9 people to line up. The number of permutations of n distinct objects can always be found by $n!$.

A family of five is having portraits taken. Use the Multiplication Principle to find the following.

TRY IT #3 How many ways can the family line up for the portrait?

TRY IT #4 How many ways can the photographer line up 3 family members?

TRY IT #5 How many ways can the family line up for the portrait if the parents are required to stand on each end?

Finding the Number of Permutations of *n* Distinct Objects Using a Formula

For some permutation problems, it is inconvenient to use the Multiplication Principle because there are so many numbers to multiply. Fortunately, we can solve these problems using a formula. Before we learn the formula, let's look at two common notations for permutations. If we have a set of n objects and we want to choose r objects from the set in order, we write $P(n, r)$. Another way to write this is nP_r, a notation commonly seen on computers and calculators. To calculate $P(n, r)$, we begin by finding $n!$, the number of ways to line up all n objects. We then divide by $(n - r)!$ to cancel out the $(n - r)$ items that we do not wish to line up.

Let's see how this works with a simple example. Imagine a club of six people. They need to elect a president, a vice president, and a treasurer. Six people can be elected president, any one of the five remaining people can be elected vice president, and any of the remaining four people could be elected treasurer. The number of ways this may be done is $6 \times 5 \times 4 = 120$. Using factorials, we get the same result.

$$\frac{6!}{3!} = \frac{6 \cdot 5 \cdot 4 \cdot 3!}{3!} = 6 \cdot 5 \cdot 4 = 120$$

There are 120 ways to select 3 officers in order from a club with 6 members. We refer to this as a permutation of 6 taken 3 at a time. The general formula is as follows.

$$P(n, r) = \frac{n!}{(n-r)!}$$

Note that the formula stills works if we are choosing all n objects and placing them in order. In that case we would be dividing by $(n-n)!$ or $0!$, which we said earlier is equal to 1. So the number of permutations of n objects taken n at a time is $\frac{n!}{1}$ or just $n!$.

Formula for Permutations of n Distinct Objects

Given n distinct objects, the number of ways to select r objects from the set in order is

$$P(n, r) = \frac{n!}{(n-r)!}$$

HOW TO

Given a word problem, evaluate the possible permutations.

1. Identify n from the given information.
2. Identify r from the given information.
3. Replace n and r in the formula with the given values.
4. Evaluate.

EXAMPLE 4

Finding the Number of Permutations Using the Formula

A professor is creating an exam of 9 questions from a test bank of 12 questions. How many ways can she select and arrange the questions?

Solution

Substitute $n = 12$ and $r = 9$ into the permutation formula and simplify.

$$P(n, r) = \frac{n!}{(n-r)!}$$
$$P(12, 9) = \frac{12!}{(12-9)!} = \frac{12!}{3!} = 79{,}833{,}600$$

There are 79,833,600 possible permutations of exam questions!

Analysis

We can also use a calculator to find permutations. For this problem, we would enter 12, press the $_nP_r$ function, enter 9, and then press the equal sign. The $_nP_r$ function may be located under the MATH menu with probability commands.

Q&A **Could we have solved Example 4 using the Multiplication Principle?**

Yes. We could have multiplied $12 \cdot 11 \cdot 10 \cdot 9 \cdot 8 \cdot 7 \cdot 6 \cdot 5 \cdot 4$ *to find the same answer.*

A play has a cast of 7 actors preparing to make their curtain call. Use the permutation formula to find the following.

TRY IT #6 How many ways can the 7 actors line up?

TRY IT #7 How many ways can 5 of the 7 actors be chosen to line up?

Find the Number of Combinations Using the Formula

So far, we have looked at problems asking us to put objects in order. There are many problems in which we want to select a few objects from a group of objects, but we do not care about the order. When we are selecting objects and the order does not matter, we are dealing with **combinations**. A selection of r objects from a set of n objects where the order does not matter can be written as $C(n, r)$. Just as with permutations, $C(n, r)$ can also be written as ${}_nC_r$. In this case, the general formula is as follows.

$$C(n, r) = \frac{n!}{r!(n-r)!}$$

An earlier problem considered choosing 3 of 4 possible paintings to hang on a wall. We found that there were 24 ways to select 3 of the 4 paintings in order. But what if we did not care about the order? We would expect a smaller number because selecting paintings 1, 2, 3 would be the same as selecting paintings 2, 3, 1. To find the number of ways to select 3 of the 4 paintings, disregarding the order of the paintings, divide the number of permutations by the number of ways to order 3 paintings. There are $3! = 3 \cdot 2 \cdot 1 = 6$ ways to order 3 paintings. There are $\frac{24}{6}$, or 4 ways to select 3 of the 4 paintings. This number makes sense because every time we are selecting 3 paintings, we are *not* selecting 1 painting. There are 4 paintings we could choose *not* to select, so there are 4 ways to select 3 of the 4 paintings.

Formula for Combinations of n Distinct Objects

Given n distinct objects, the number of ways to select r objects from the set is

$$C(n, r) = \frac{n!}{r!(n-r)!}$$

HOW TO

Given a number of options, determine the possible number of combinations.

1. Identify n from the given information.
2. Identify r from the given information.
3. Replace n and r in the formula with the given values.
4. Evaluate.

EXAMPLE 5

Finding the Number of Combinations Using the Formula

A fast food restaurant offers five side dish options. Your meal comes with two side dishes.

ⓐ How many ways can you select your side dishes? ⓑ How many ways can you select 3 side dishes?

Solution

ⓐ We want to choose 2 side dishes from 5 options.

$$C(5, 2) = \frac{5!}{2!(5-2)!} = 10$$

ⓑ We want to choose 3 side dishes from 5 options.

$$C(5, 3) = \frac{5!}{3!(5-3)!} = 10$$

Analysis

We can also use a graphing calculator to find combinations. Enter 5, then press ${}_nC_r$, enter 3, and then press the equal sign. The ${}_nC_r$, function may be located under the MATH menu with probability commands.

Q&A **Is it a coincidence that parts (a) and (b) in Example 5 have the same answers?**

No. When we choose r objects from n objects, we are ***not*** *choosing* $(n-r)$ *objects. Therefore,* $C(n, r) = C(n, n-r)$.

> **TRY IT** #8 An ice cream shop offers 10 flavors of ice cream. How many ways are there to choose 3 flavors for a banana split?

Finding the Number of Subsets of a Set

We have looked only at combination problems in which we chose exactly r objects. In some problems, we want to consider choosing every possible number of objects. Consider, for example, a pizza restaurant that offers 5 toppings. Any number of toppings can be ordered. How many different pizzas are possible?

To answer this question, we need to consider pizzas with any number of toppings. There is $C(5,0) = 1$ way to order a pizza with no toppings. There are $C(5,1) = 5$ ways to order a pizza with exactly one topping. If we continue this process, we get

$$C(5,0) + C(5,1) + C(5,2) + C(5,3) + C(5,4) + C(5,5) = 32$$

There are 32 possible pizzas. This result is equal to 2^5.

We are presented with a sequence of choices. For each of the n objects we have two choices: include it in the subset or not. So for the whole subset we have made n choices, each with two options. So there are a total of $2 \cdot 2 \cdot 2 \cdot \ldots \cdot 2$ possible resulting subsets, all the way from the empty subset, which we obtain when we say "no" each time, to the original set itself, which we obtain when we say "yes" each time.

Formula for the Number of Subsets of a Set

A set containing n distinct objects has 2^n subsets.

EXAMPLE 6

Finding the Number of Subsets of a Set

A restaurant offers butter, cheese, chives, and sour cream as toppings for a baked potato. How many different ways are there to order a potato?

Solution

We are looking for the number of subsets of a set with 4 objects. Substitute $n = 4$ into the formula.

$$\begin{aligned} 2^n &= 2^4 \\ &= 16 \end{aligned}$$

There are 16 possible ways to order a potato.

> **TRY IT** #9 A sundae bar at a wedding has 6 toppings to choose from. Any number of toppings can be chosen. How many different sundaes are possible?

Finding the Number of Permutations of n Non-Distinct Objects

We have studied permutations where all of the objects involved were distinct. What happens if some of the objects are indistinguishable? For example, suppose there is a sheet of 12 stickers. If all of the stickers were distinct, there would be $12!$ ways to order the stickers. However, 4 of the stickers are identical stars, and 3 are identical moons. Because all of the objects are not distinct, many of the $12!$ permutations we counted are duplicates. The general formula for this situation is as follows.

$$\frac{n!}{r_1!r_2!\ldots r_k!}$$

In this example, we need to divide by the number of ways to order the 4 stars and the ways to order the 3 moons to find the number of unique permutations of the stickers. There are $4!$ ways to order the stars and $3!$ ways to order the moon.

$$\frac{12!}{4!3!} = 3{,}326{,}400$$

There are 3,326,400 ways to order the sheet of stickers.

Formula for Finding the Number of Permutations of n Non-Distinct Objects

If there are n elements in a set and r_1 are alike, r_2 are alike, r_3 are alike, and so on through r_k, the number of permutations can be found by

$$\frac{n!}{r_1!r_2!\ldots r_k!}$$

EXAMPLE 7

Finding the Number of Permutations of n Non-Distinct Objects

Find the number of rearrangements of the letters in the word DISTINCT.

Solution

There are 8 letters. Both I and T are repeated 2 times. Substitute $n = 8,\ r_1 = 2,$ and $r_2 = 2$ into the formula.

$$\frac{8!}{2!2!} = 10{,}080$$

There are 10,080 arrangements.

TRY IT #10 Find the number of rearrangements of the letters in the word CARRIER.

MEDIA

Access these online resources for additional instruction and practice with combinations and permutations.

Combinations (http://openstax.org/l/combinations)
Permutations (http://openstax.org/l/permutations)

13.5 SECTION EXERCISES

Verbal

For the following exercises, assume that there are n ways an event A can happen, m ways an event B can happen, and that A and B are non-overlapping.

1. Use the Addition Principle of counting to explain how many ways event A or B can occur.

2. Use the Multiplication Principle of counting to explain how many ways event A and B can occur.

Answer the following questions.

3. When given two separate events, how do we know whether to apply the Addition Principle or the Multiplication Principle when calculating possible outcomes? What conjunctions may help to determine which operations to use?

4. Describe how the permutation of n objects differs from the permutation of choosing r objects from a set of n objects. Include how each is calculated.

5. What is the term for the arrangement that selects r objects from a set of n objects when the order of the r objects is not important? What is the formula for calculating the number of possible outcomes for this type of arrangement?

Numeric

For the following exercises, determine whether to use the Addition Principle or the Multiplication Principle. Then perform the calculations.

6. Let the set $A = \{-5, -3, -1, 2, 3, 4, 5, 6\}$. How many ways are there to choose a negative or an even number from A?

7. Let the set $B = \{-23, -16, -7, -2, 20, 36, 48, 72\}$. How many ways are there to choose a positive or an odd number from A?

8. How many ways are there to pick a red ace or a club from a standard card playing deck?

9. How many ways are there to pick a paint color from 5 shades of green, 4 shades of blue, or 7 shades of yellow?

10. How many outcomes are possible from tossing a pair of coins?

11. How many outcomes are possible from tossing a coin and rolling a 6-sided die?

12. How many two-letter strings—the first letter from A and the second letter from B— can be formed from the sets $A = \{b, c, d\}$ and $B = \{a, e, i, o, u\}$?

13. How many ways are there to construct a string of 3 digits if numbers can be repeated?

14. How many ways are there to construct a string of 3 digits if numbers cannot be repeated?

For the following exercises, compute the value of the expression.

15. $P(5, 2)$

16. $P(8, 4)$

17. $P(3, 3)$

18. $P(9, 6)$

19. $P(11, 5)$

20. $C(8, 5)$

21. $C(12, 4)$

22. $C(26, 3)$

23. $C(7, 6)$

24. $C(10, 3)$

For the following exercises, find the number of subsets in each given set.

25. $\{1, 2, 3, 4, 5, 6, 7, 8, 9, 10\}$

26. $\{a, b, c, \ldots, z\}$

27. A set containing 5 distinct numbers, 4 distinct letters, and 3 distinct symbols

28. The set of even numbers from 2 to 28

29. The set of two-digit numbers between 1 and 100 containing the digit 0

For the following exercises, find the distinct number of arrangements.

30. The letters in the word "juggernaut"

31. The letters in the word "academia"

32. The letters in the word "academia" that begin and end in "a"

33. The symbols in the string #,#,#,@,@,$,$,$,%,%,%,%

34. The symbols in the string #,#,#,@,@,$,$,$,%,%,%,% that begin and end with "%"

Extensions

35. The set, S consists of 900,000,000 whole numbers, each being the same number of digits long. How many digits long is a number from S? (*Hint:* use the fact that a whole number cannot start with the digit 0.)

36. The number of 5-element subsets from a set containing n elements is equal to the number of 6-element subsets from the same set. What is the value of n? (*Hint:* the order in which the elements for the subsets are chosen is not important.)

37. Can $C(n, r)$ ever equal $P(n, r)$? Explain.

38. Suppose a set A has 2,048 subsets. How many distinct objects are contained in A?

39. How many arrangements can be made from the letters of the word "mountains" if all the vowels must form a string?

Real-World Applications

40. A family consisting of 2 parents and 3 children is to pose for a picture with 2 family members in the front and 3 in the back.

ⓐ How many arrangements are possible with no restrictions?
ⓑ How many arrangements are possible if the parents must sit in the front?
ⓒ How many arrangements are possible if the parents must be next to each other?

41. A cell phone company offers 6 different voice packages and 8 different data packages. Of those, 3 packages include both voice and data. How many ways are there to choose either voice or data, but not both?

42. In horse racing, a "trifecta" occurs when a bettor wins by selecting the first three finishers in the exact order (1st place, 2nd place, and 3rd place). How many different trifectas are possible if there are 14 horses in a race?

43. A wholesale T-shirt company offers sizes small, medium, large, and extra-large in organic or non-organic cotton and colors white, black, gray, blue, and red. How many different T-shirts are there to choose from?

44. Hector wants to place billboard advertisements throughout the county for his new business. How many ways can Hector choose 15 neighborhoods to advertise in if there are 30 neighborhoods in the county?

45. An art store has 4 brands of paint pens in 12 different colors and 3 types of ink. How many paint pens are there to choose from?

46. How many ways can a committee of 3 freshmen and 4 juniors be formed from a group of 8 freshmen and 11 juniors?

47. How many ways can a baseball coach arrange the order of 9 batters if there are 15 players on the team?

48. A conductor needs 5 cellists and 5 violinists to play at a diplomatic event. To do this, he ranks the orchestra's 10 cellists and 16 violinists in order of musical proficiency. What is the ratio of the total cellist rankings possible to the total violinist rankings possible?

49. A motorcycle shop has 10 choppers, 6 bobbers, and 5 café racers—different types of vintage motorcycles. How many ways can the shop choose 3 choppers, 5 bobbers, and 2 café racers for a weekend showcase?

50. A skateboard shop stocks 10 types of board decks, 3 types of trucks, and 4 types of wheels. How many different skateboards can be constructed?

51. Just-For-Kicks Sneaker Company offers an online customizing service. How many ways are there to design a custom pair of Just-For-Kicks sneakers if a customer can choose from a basic shoe up to 11 customizable options?

52. A car wash offers the following optional services to the basic wash: clear coat wax, triple foam polish, undercarriage wash, rust inhibitor, wheel brightener, air freshener, and interior shampoo. How many washes are possible if any number of options can be added to the basic wash?

53. Suni bought 20 plants to arrange along the border of her garden. How many distinct arrangements can she make if the plants are comprised of 6 tulips, 6 roses, and 8 daisies?

54. How many unique ways can a string of Christmas lights be arranged from 9 red, 10 green, 6 white, and 12 gold color bulbs?

13.6 Binomial Theorem

Learning Objectives

In this section, you will:

- Apply the Binomial Theorem.

A polynomial with two terms is called a binomial. We have already learned to multiply binomials and to raise binomials to powers, but raising a binomial to a high power can be tedious and time-consuming. In this section, we will discuss a shortcut that will allow us to find $(x + y)^n$ without multiplying the binomial by itself n times.

Identifying Binomial Coefficients

In Counting Principles, we studied combinations. In the shortcut to finding $(x+y)^n$, we will need to use combinations to find the coefficients that will appear in the expansion of the binomial. In this case, we use the notation $\binom{n}{r}$ instead of $C(n,r)$, but it can be calculated in the same way. So

$$\binom{n}{r} = C(n,r) = \frac{n!}{r!(n-r)!}$$

The combination $\binom{n}{r}$ is called a **binomial coefficient**. An example of a binomial coefficient is $\binom{5}{2} = C(5,2) = 10.$

Binomial Coefficients

If n and r are integers greater than or equal to 0 with $n \geq r$, then the **binomial coefficient** is

$$\binom{n}{r} = C(n,r) = \frac{n!}{r!(n-r)!}$$

Q&A **Is a binomial coefficient always a whole number?**

Yes. Just as the number of combinations must always be a whole number, a binomial coefficient will always be a whole number.

EXAMPLE 1

Finding Binomial Coefficients

Find each binomial coefficient.

ⓐ $\binom{5}{3}$ ⓑ $\binom{9}{2}$ ⓒ $\binom{9}{7}$

Solution

Use the formula to calculate each binomial coefficient. You can also use the nC_r function on your calculator.

$$\binom{n}{r} = C(n,r) = \frac{n!}{r!(n-r)!}$$

ⓐ $\binom{5}{3} = \frac{5!}{3!(5-3)!} = \frac{5 \cdot 4 \cdot 3!}{3!2!} = 10$ ⓑ $\binom{9}{2} = \frac{9!}{2!(9-2)!} = \frac{9 \cdot 8 \cdot 7!}{2!7!} = 36$ ⓒ $\binom{9}{7} = \frac{9!}{7!(9-7)!} = \frac{9 \cdot 8 \cdot 7!}{7!2!} = 36$

Analysis

Notice that we obtained the same result for parts (b) and (c). If you look closely at the solution for these two parts, you will see that you end up with the same two factorials in the denominator, but the order is reversed, just as with combinations.

$$\binom{n}{r} = \binom{n}{n-r}$$

TRY IT #1 Find each binomial coefficient.

ⓐ $\binom{7}{3}$ ⓑ $\binom{11}{4}$

Using the Binomial Theorem

When we expand $(x+y)^n$ by multiplying, the result is called a **binomial expansion**, and it includes binomial coefficients. If we wanted to expand $(x+y)^{52}$, we might multiply $(x+y)$ by itself fifty-two times. This could take hours! If we examine some simple binomial expansions, we can find patterns that will lead us to a shortcut for finding more complicated binomial expansions.

$$\begin{aligned}(x+y)^2 &= x^2+2xy+y^2\\(x+y)^3 &= x^3+3x^2y+3xy^2+y^3\\(x+y)^4 &= x^4+4x^3y+6x^2y^2+4xy^3+y^4\end{aligned}$$

First, let's examine the exponents. With each successive term, the exponent for x decreases and the exponent for y increases. The sum of the two exponents is n for each term.

Next, let's examine the coefficients. Notice that the coefficients increase and then decrease in a symmetrical pattern. The coefficients follow a pattern:

$$\binom{n}{0}, \binom{n}{1}, \binom{n}{2}, \ldots, \binom{n}{n}.$$

These patterns lead us to the **Binomial Theorem**, which can be used to expand any binomial.

$$\begin{aligned}(x+y)^n &= \sum_{k=0}^{n}\binom{n}{k}x^{n-k}y^k\\ &= x^n+\binom{n}{1}x^{n-1}y+\binom{n}{2}x^{n-2}y^2+\ldots+\binom{n}{n-1}xy^{n-1}+y^n\end{aligned}$$

Another way to see the coefficients is to examine the expansion of a binomial in general form, $x+y$, to successive powers 1, 2, 3, and 4.

$$\begin{aligned}(x+y)^1 &= x+y\\(x+y)^2 &= x^2+2xy+y^2\\(x+y)^3 &= x^3+3x^2y+3xy^2+y^3\\(x+y)^4 &= x^4+4x^3y+6x^2y^2+4xy^3+y^4\end{aligned}$$

Can you guess the next expansion for the binomial $(x+y)^5$?

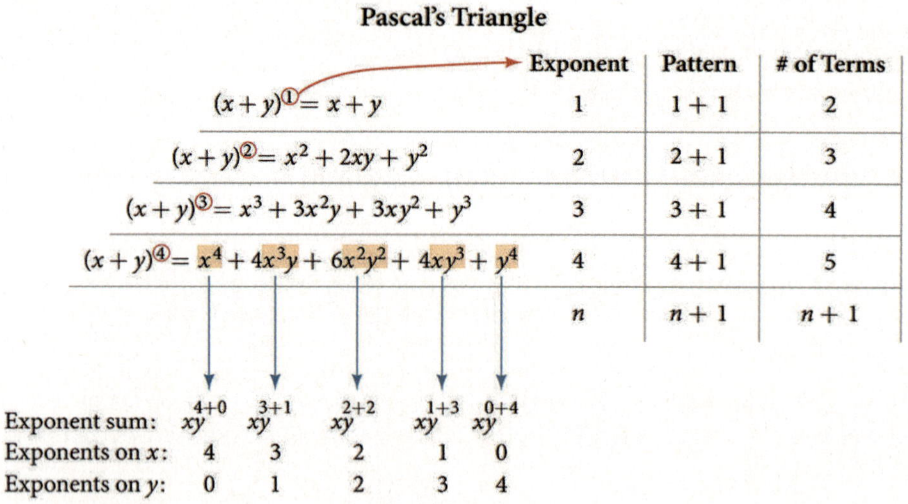

Figure 1

See Figure 1, which illustrates the following:

- There are $n+1$ terms in the expansion of $(x+y)^n$.
- The degree (or sum of the exponents) for each term is n.
- The powers on x begin with n and decrease to 0.
- The powers on y begin with 0 and increase to n.
- The coefficients are symmetric.

To determine the expansion on $(x+y)^5$, we see $n=5$, thus, there will be 5+1 = 6 terms. Each term has a combined degree of 5. In descending order for powers of x, the pattern is as follows:

- Introduce x^5, and then for each successive term reduce the exponent on x by 1 until $x^0=1$ is reached.
- Introduce $y^0=1$, and then increase the exponent on y by 1 until y^5 is reached.

$$x^5,\quad x^4y,\quad x^3y^2,\quad x^2y^3,\quad xy^4,\quad y^5$$

The next expansion would be

$$(x+y)^5 = x^5 + 5x^4y + 10x^3y^2 + 10x^2y^3 + 5xy^4 + y^5.$$

But where do those coefficients come from? The binomial coefficients are symmetric. We can see these coefficients in an array known as Pascal's Triangle, shown in Figure 2. Pascal didn't invent the triangle. The underlying principles had been developed and written about for over 1500 years, first by the Indian mathematician (and poet) Pingala in the second century BCE. Others throughout Asia and Europe worked with the concepts throughout, and the triangle was first published in its graphical form by Omar Khayyam, an Iranian mathematician and astronomer, for whom the triangle is named in Iran. French mathematician Blaise Pascal repopularized it when he republished it and used it to solve a number of probability problems.

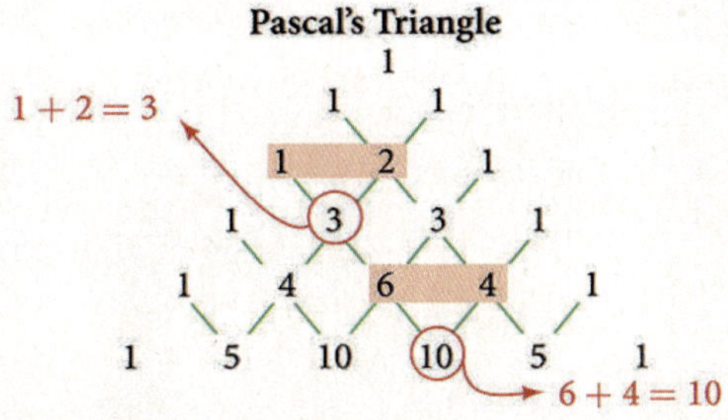

Figure 2

To generate Pascal's Triangle, we start by writing a 1. In the row below, row 2, we write two 1's. In the 3rd row, flank the ends of the rows with 1's, and add $1+1$ to find the middle number, 2. In the nth row, flank the ends of the row with 1's. Each element in the triangle is the sum of the two elements immediately above it.

To see the connection between Pascal's Triangle and binomial coefficients, let us revisit the expansion of the binomials in general form.

$$1 \longrightarrow (x+y)^0 = 1$$
$$1 \quad 1 \longrightarrow (x+y)^1 = x + y$$
$$1 \quad 2 \quad 1 \longrightarrow (x+y)^2 = x^2 + 2xy + y^2$$
$$1 \quad 3 \quad 3 \quad 1 \longrightarrow (x+y)^3 = x^3 + 3x^2y + 3xy^2 + y^3$$
$$1 \quad 4 \quad 6 \quad 4 \quad 1 \longrightarrow (x+y)^4 = x^4 + 4x^3y + 6x^2y^2 + 4xy^3 + y^4$$
$$1 \quad 5 \quad 10 \quad 10 \quad 5 \quad 1 \longrightarrow (x+y)^5 = x^5 + 5x^4y + 10x^3y^2 + 10x^2y^3 + 5xy^4 + y^5$$

The Binomial Theorem

The **Binomial Theorem** is a formula that can be used to expand any binomial.

$$\begin{aligned}(x+y)^n &= \sum_{k=0}^{n} \binom{n}{k} x^{n-k} y^k \\ &= x^n + \binom{n}{1} x^{n-1} y + \binom{n}{2} x^{n-2} y^2 + \ldots + \binom{n}{n-1} x y^{n-1} + y^n\end{aligned}$$

HOW TO

Given a binomial, write it in expanded form.

1. Determine the value of n according to the exponent.
2. Evaluate the $k = 0$ through $k = n$ using the Binomial Theorem formula.
3. Simplify.

EXAMPLE 2

Expanding a Binomial

Write in expanded form.

ⓐ $(x+y)^5$ ⓑ $(3x-y)^4$

Solution

ⓐ Substitute $n=5$ into the formula. Evaluate the $k=0$ through $k=5$ terms. Simplify.

$$\begin{aligned}(x+y)^5 &= \binom{5}{0}x^5y^0 + \binom{5}{1}x^4y^1 + \binom{5}{2}x^3y^2 + \binom{5}{3}x^2y^3 + \binom{5}{4}x^1y^4 + \binom{5}{5}x^0y^5 \\ (x+y)^5 &= x^5 + 5x^4y + 10x^3y^2 + 10x^2y^3 + 5xy^4 + y^5\end{aligned}$$

ⓑ Substitute $n=4$ into the formula. Evaluate the $k=0$ through $k=4$ terms. Notice that $3x$ is in the place that was occupied by x and that $-y$ is in the place that was occupied by y. So we substitute them. Simplify.

$$\begin{aligned}(3x-y)^4 &= \binom{4}{0}(3x)^4(-y)^0 + \binom{4}{1}(3x)^3(-y)^1 + \binom{4}{2}(3x)^2(-y)^2 + \binom{4}{3}(3x)^1(-y)^3 + \binom{4}{4}(3x)^0(-y)^4 \\ (3x-y)^4 &= 81x^4 - 108x^3y + 54x^2y^2 - 12xy^3 + y^4\end{aligned}$$

Analysis

Notice the alternating signs in part b. This happens because $(-y)$ raised to odd powers is negative, but $(-y)$ raised to even powers is positive. This will occur whenever the binomial contains a subtraction sign.

TRY IT #2 Write in expanded form.

ⓐ $(x-y)^5$ ⓑ $(2x+5y)^3$

Using the Binomial Theorem to Find a Single Term

Expanding a binomial with a high exponent such as $(x+2y)^{16}$ can be a lengthy process.

Sometimes we are interested only in a certain term of a binomial expansion. We do not need to fully expand a binomial to find a single specific term.

Note the pattern of coefficients in the expansion of $(x+y)^5$.

$$(x+y)^5 = x^5 + \binom{5}{1}x^4y + \binom{5}{2}x^3y^2 + \binom{5}{3}x^2y^3 + \binom{5}{4}xy^4 + y^5$$

The second term is $\binom{5}{1}x^4y$. The third term is $\binom{5}{2}x^3y^2$. We can generalize this result.

$$\binom{n}{r}x^{n-r}y^r$$

The (r+1)th Term of a Binomial Expansion

The $(r+1)$th term of the binomial expansion of $(x+y)^n$ is:

$$\binom{n}{r}x^{n-r}y^r$$

HOW TO

Given a binomial, write a specific term without fully expanding.

1. Determine the value of n according to the exponent.

2. Determine $(r + 1)$.
3. Determine r.
4. Replace r in the formula for the $(r + 1)$th term of the binomial expansion.

EXAMPLE 3

Writing a Given Term of a Binomial Expansion

Find the tenth term of $(x + 2y)^{16}$ without fully expanding the binomial.

Solution

Because we are looking for the tenth term, $r + 1 = 10$, we will use $r = 9$ in our calculations.

$$\binom{n}{r} x^{n-r} y^r$$

$$\binom{16}{9} x^{16-9}(2y)^9 = 5{,}857{,}280x^7 y^9$$

TRY IT #3 Find the sixth term of $(3x - y)^9$ without fully expanding the binomial.

MEDIA

Access these online resources for additional instruction and practice with binomial expansion.

The Binomial Theorem (http://openstax.org/l/binomialtheorem)
Binomial Theorem Example (http://openstax.org/l/btexample)

13.6 SECTION EXERCISES

Verbal

1. What is a binomial coefficient, and how it is calculated?

2. What role do binomial coefficients play in a binomial expansion? Are they restricted to any type of number?

3. What is the Binomial Theorem and what is its use?

4. When is it an advantage to use the Binomial Theorem? Explain.

Algebraic

For the following exercises, evaluate the binomial coefficient.

5. $\binom{6}{2}$

6. $\binom{5}{3}$

7. $\binom{7}{4}$

8. $\binom{9}{7}$

9. $\binom{10}{9}$

10. $\binom{25}{11}$

11. $\binom{17}{6}$

12. $\binom{200}{199}$

For the following exercises, use the Binomial Theorem to expand each binomial.

13. $(4a - b)^3$

14. $(5a + 2)^3$

15. $(3a + 2b)^3$

16. $(2x + 3y)^4$

17. $(4x + 2y)^5$

18. $(3x - 2y)^4$

19. $(4x - 3y)^5$

20. $\left(\frac{1}{x} + 3y\right)^5$

21. $(x^{-1} + 2y^{-1})^4$

22. $(\sqrt{x} - \sqrt{y})^5$

For the following exercises, use the Binomial Theorem to write the first three terms of each binomial.

23. $(a + b)^{17}$

24. $(x - 1)^{18}$

25. $(a - 2b)^{15}$

26. $(x - 2y)^8$

27. $(3a + b)^{20}$

28. $(2a + 4b)^7$

29. $(x^3 - \sqrt{y})^8$

For the following exercises, find the indicated term of each binomial without fully expanding the binomial.

30. The fourth term of $(2x - 3y)^4$

31. The fourth term of $(3x - 2y)^5$

32. The third term of $(6x - 3y)^7$

33. The eighth term of $(7 + 5y)^{14}$

34. The seventh term of $(a + b)^{11}$

35. The fifth term of $(x - y)^7$

36. The tenth term of $(x - 1)^{12}$

37. The ninth term of $(a - 3b^2)^{11}$

38. The fourth term of $\left(x^3 - \frac{1}{2}\right)^{10}$

39. The eighth term of $\left(\frac{y}{2} + \frac{2}{x}\right)^9$

Graphical

For the following exercises, use the Binomial Theorem to expand the binomial $f(x) = (x + 3)^4$. Then find and graph each indicated sum on one set of axes.

40. Find and graph $f_1(x)$, such that $f_1(x)$ is the first term of the expansion.

41. Find and graph $f_2(x)$, such that $f_2(x)$ is the sum of the first two terms of the expansion.

42. Find and graph $f_3(x)$, such that $f_3(x)$ is the sum of the first three terms of the expansion.

43. Find and graph $f_4(x)$, such that $f_4(x)$ is the sum of the first four terms of the expansion.

44. Find and graph $f_5(x)$, such that $f_5(x)$ is the sum of the first five terms of the expansion.

Extensions

45. In the expansion of $(5x+3y)^n$, each term has the form $\binom{n}{k} a^{n-k} b^k$, where k successively takes on the value $0, 1, 2, \ldots, n$. If $\binom{n}{k} = \binom{7}{2}$, what is the corresponding term?

46. In the expansion of $(a+b)^n$, the coefficient of $a^{n-k} b^k$ is the same as the coefficient of which other term?

47. Consider the expansion of $(x+b)^{40}$. What is the exponent of b in the kth term?

48. Find $\binom{n}{k-1} + \binom{n}{k}$ and write the answer as a binomial coefficient in the form $\binom{n}{k}$. Prove it.
Hint: Use the fact that, for any integer p, such that $p \geq 1$, $p! = p(p-1)!$.

49. Which expression cannot be expanded using the Binomial Theorem? Explain.
   - $(x^2 - 2x + 1)$
   - $(\sqrt{a} + 4\sqrt{a} - 5)^8$
   - $(x^3 + 2y^2 - z)^5$
   - $(3x^2 - \sqrt{2y^3})^{12}$

13.7 Probability

Learning Objectives

In this section, you will:

- Construct probability models.
- Compute probabilities of equally likely outcomes.
- Compute probabilities of the union of two events.
- Use the complement rule to find probabilities.
- Compute probability using counting theory.

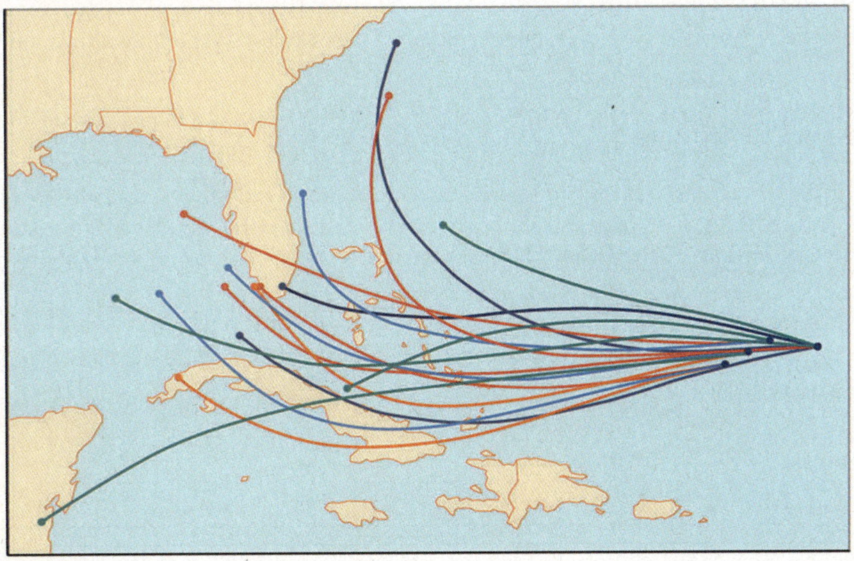

Figure 1 An example of a "spaghetti model," which can be used to predict possible paths of a tropical storm.[1]

Residents of the Southeastern United States are all too familiar with charts, known as spaghetti models, such as the one in Figure 1. They combine a collection of weather data to predict the most likely path of a hurricane. Each colored line represents one possible path. The group of squiggly lines can begin to resemble strands of spaghetti, hence the name. In this section, we will investigate methods for making these types of predictions.

Constructing Probability Models

Suppose we roll a six-sided number cube. Rolling a number cube is an example of an **experiment**, or an activity with an observable result. The numbers on the cube are possible results, or **outcomes**, of this experiment. The set of all possible outcomes of an experiment is called the **sample space** of the experiment. The sample space for this experiment is $\{1, 2, 3, 4, 5, 6\}$. An **event** is any subset of a sample space.

The likelihood of an event is known as **probability**. The probability of an event p is a number that always satisfies $0 \le p \le 1$, where 0 indicates an impossible event and 1 indicates a certain event. A **probability model** is a mathematical description of an experiment listing all possible outcomes and their associated probabilities. For instance, if there is a 1% chance of winning a raffle and a 99% chance of losing the raffle, a probability model would look much like Table 1.

Outcome	Probability
Winning the raffle	1%
Losing the raffle	99%

Table 1

The sum of the probabilities listed in a probability model must equal 1, or 100%.

HOW TO

Given a probability event where each event is equally likely, construct a probability model.

1. Identify every outcome.
2. Determine the total number of possible outcomes.
3. Compare each outcome to the total number of possible outcomes.

EXAMPLE 1

Constructing a Probability Model

Construct a probability model for rolling a single, fair die, with the event being the number shown on the die.

Solution

Begin by making a list of all possible outcomes for the experiment. The possible outcomes are the numbers that can be rolled: 1, 2, 3, 4, 5, and 6. There are six possible outcomes that make up the sample space.

Assign probabilities to each outcome in the sample space by determining a ratio of the outcome to the number of possible outcomes. There is one of each of the six numbers on the cube, and there is no reason to think that any particular face is more likely to show up than any other one, so the probability of rolling any number is $\frac{1}{6}$.

Outcome	Roll of 1	Roll of 2	Roll of 3	Roll of 4	Roll of 5	Roll of 6
Probability	$\frac{1}{6}$	$\frac{1}{6}$	$\frac{1}{6}$	$\frac{1}{6}$	$\frac{1}{6}$	$\frac{1}{6}$

Table 2

1 The figure is for illustrative purposes only and does not model any particular storm.

Q&A **Do probabilities always have to be expressed as fractions?**

No. Probabilities can be expressed as fractions, decimals, or percents. Probability must always be a number between 0 and 1, inclusive of 0 and 1.

TRY IT #1 Construct a probability model for tossing a fair coin.

Computing Probabilities of Equally Likely Outcomes

Let S be a sample space for an experiment. When investigating probability, an event is any subset of S. When the outcomes of an experiment are all equally likely, we can find the probability of an event by dividing the number of outcomes in the event by the total number of outcomes in S. Suppose a number cube is rolled, and we are interested in finding the probability of the event "rolling a number less than or equal to 4." There are 4 possible outcomes in the event and 6 possible outcomes in S, so the probability of the event is $\frac{4}{6} = \frac{2}{3}$.

Computing the Probability of an Event with Equally Likely Outcomes

The probability of an event E in an experiment with sample space S with equally likely outcomes is given by

$$P(E) = \frac{\text{number of elements in } E}{\text{number of elements in } S} = \frac{n(E)}{n(S)}$$

E is a subset of S, so it is always true that $0 \le P(E) \le 1$.

EXAMPLE 2

Computing the Probability of an Event with Equally Likely Outcomes

A six-sided number cube is rolled. Find the probability of rolling an odd number.

Solution

The event "rolling an odd number" contains three outcomes. There are 6 equally likely outcomes in the sample space. Divide to find the probability of the event.

$$P(E) = \frac{3}{6} = \frac{1}{2}$$

TRY IT #2 A number cube is rolled. Find the probability of rolling a number greater than 2.

Computing the Probability of the Union of Two Events

We are often interested in finding the probability that one of multiple events occurs. Suppose we are playing a card game, and we will win if the next card drawn is either a heart or a king. We would be interested in finding the probability of the next card being a heart or a king. The **union of two events** E and F, written $E \cup F$, is the event that occurs if either or both events occur.

$$P(E \cup F) = P(E) + P(F) - P(E \cap F)$$

Suppose the spinner in Figure 2 is spun. We want to find the probability of spinning orange or spinning a b.

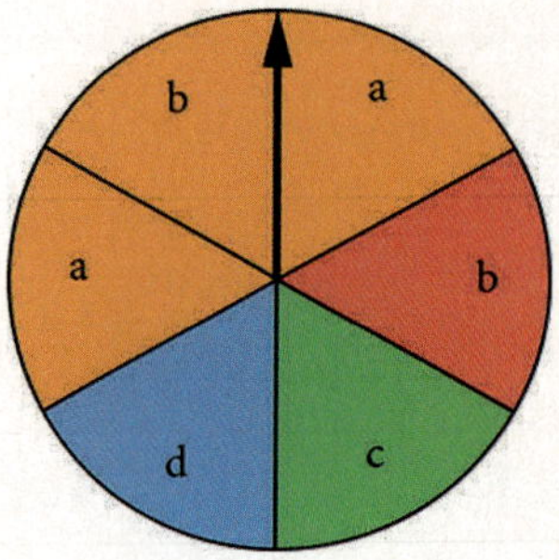

Figure 2

There are a total of 6 sections, and 3 of them are orange. So the probability of spinning orange is $\frac{3}{6} = \frac{1}{2}$. There are a total of 6 sections, and 2 of them have a b. So the probability of spinning a b is $\frac{2}{6} = \frac{1}{3}$. If we added these two probabilities, we would be counting the sector that is both orange and a b twice. To find the probability of spinning an orange or a b, we need to subtract the probability that the sector is both orange and has a b.

$$\frac{1}{2} + \frac{1}{3} - \frac{1}{6} = \frac{2}{3}$$

The probability of spinning orange or a b is $\frac{2}{3}$.

Probability of the Union of Two Events

The probability of the union of two events E and F (written $E \cup F$) equals the sum of the probability of E and the probability of F minus the probability of E and F occurring together (which is called the intersection of E and F and is written as $E \cap F$).

$$P(E \cup F) = P(E) + P(F) - P(E \cap F)$$

EXAMPLE 3

Computing the Probability of the Union of Two Events

A card is drawn from a standard deck. Find the probability of drawing a heart or a 7.

Solution

A standard deck contains an equal number of hearts, diamonds, clubs, and spades. So the probability of drawing a heart is $\frac{1}{4}$. There are four 7s in a standard deck, and there are a total of 52 cards. So the probability of drawing a 7 is $\frac{1}{13}$.

The only card in the deck that is both a heart and a 7 is the 7 of hearts, so the probability of drawing both a heart and a 7 is $\frac{1}{52}$. Substitute $P(H) = \frac{1}{4}$, $P(7) = \frac{1}{13}$, and $P(H \cap 7) = \frac{1}{52}$ into the formula.

$$\begin{aligned} P(E \cup F) &= P(E) + P(F) - P(E \cap F) \\ &= \frac{1}{4} + \frac{1}{13} - \frac{1}{52} \\ &= \frac{4}{13} \end{aligned}$$

The probability of drawing a heart or a 7 is $\frac{4}{13}$.

TRY IT #3 A card is drawn from a standard deck. Find the probability of drawing a red card or an ace.

Computing the Probability of Mutually Exclusive Events

Suppose the spinner in Figure 2 is spun again, but this time we are interested in the probability of spinning an orange or a d. There are no sectors that are both orange and contain a d, so these two events have no outcomes in common. Events are said to be **mutually exclusive events** when they have no outcomes in common. Because there is no overlap, there is nothing to subtract, so the general formula is

$$P(E \cup F) = P(E) + P(F)$$

Notice that with mutually exclusive events, the intersection of E and F is the empty set. The probability of spinning an orange is $\frac{3}{6} = \frac{1}{2}$ and the probability of spinning a d is $\frac{1}{6}$. We can find the probability of spinning an orange or a d simply by adding the two probabilities.

$$\begin{aligned} P(E \cup F) &= P(E) + P(F) \\ &= \tfrac{1}{2} + \tfrac{1}{6} \\ &= \tfrac{2}{3} \end{aligned}$$

The probability of spinning an orange or a d is $\frac{2}{3}$.

Probability of the Union of Mutually Exclusive Events

The probability of the union of two *mutually exclusive* events E and F is given by

$$P(E \cup F) = P(E) + P(F)$$

HOW TO

Given a set of events, compute the probability of the union of mutually exclusive events.

1. Determine the total number of outcomes for the first event.
2. Find the probability of the first event.
3. Determine the total number of outcomes for the second event.
4. Find the probability of the second event.
5. Add the probabilities.

EXAMPLE 4

Computing the Probability of the Union of Mutually Exclusive Events

A card is drawn from a standard deck. Find the probability of drawing a heart or a spade.

Solution

The events "drawing a heart" and "drawing a spade" are mutually exclusive because they cannot occur at the same time. The probability of drawing a heart is $\frac{1}{4}$, and the probability of drawing a spade is also $\frac{1}{4}$, so the probability of drawing a heart or a spade is

$$\frac{1}{4} + \frac{1}{4} = \frac{1}{2}$$

TRY IT #4 A card is drawn from a standard deck. Find the probability of drawing an ace or a king.

Using the Complement Rule to Compute Probabilities

We have discussed how to calculate the probability that an event will happen. Sometimes, we are interested in finding the probability that an event will *not* happen. The **complement of an event** E, denoted E', is the set of outcomes in the sample space that are not in E. For example, suppose we are interested in the probability that a horse will lose a race. If event W is the horse winning the race, then the complement of event W is the horse losing the race.

To find the probability that the horse loses the race, we need to use the fact that the sum of all probabilities in a probability model must be 1.

$$P(E') = 1 - P(E)$$

The probability of the horse winning added to the probability of the horse losing must be equal to 1. Therefore, if the probability of the horse winning the race is $\frac{1}{9}$, the probability of the horse losing the race is simply

$$1 - \frac{1}{9} = \frac{8}{9}$$

The Complement Rule

The probability that the **complement of an event** will occur is given by

$$P(E') = 1 - P(E)$$

EXAMPLE 5

Using the Complement Rule to Calculate Probabilities

Two six-sided number cubes are rolled.

ⓐ Find the probability that the sum of the numbers rolled is less than or equal to 3.

ⓑ Find the probability that the sum of the numbers rolled is greater than 3.

Solution

The first step is to identify the sample space, which consists of all the possible outcomes. There are two number cubes, and each number cube has six possible outcomes. Using the Multiplication Principle, we find that there are 6×6, or 36 total possible outcomes. So, for example, 1-1 represents a 1 rolled on each number cube.

1-1	1-2	1-3	1-4	1-5	1-6
2-1	2-2	2-3	2-4	2-5	2-6
3-1	3-2	3-3	3-4	3-5	3-6
4-1	4-2	4-3	4-4	4-5	4-6
5-1	5-2	5-3	5-4	5-5	5-6
6-1	6-2	6-3	6-4	6-5	6-6

Table 3

ⓐ We need to count the number of ways to roll a sum of 3 or less. These would include the following outcomes: 1-1, 1-2, and 2-1. So there are only three ways to roll a sum of 3 or less. The probability is

$$\frac{3}{36} = \frac{1}{12}$$

ⓑ Rather than listing all the possibilities, we can use the Complement Rule. Because we have already found the probability of the complement of this event, we can simply subtract that probability from 1 to find the probability that the sum of the numbers rolled is greater than 3.

$$\begin{aligned} P(E') &= 1 - P(E) \\ &= 1 - \tfrac{1}{12} \\ &= \tfrac{11}{12} \end{aligned}$$

> **TRY IT** #5 Two number cubes are rolled. Use the Complement Rule to find the probability that the sum is less than 10.

Computing Probability Using Counting Theory

Many interesting probability problems involve counting principles, permutations, and combinations. In these problems, we will use permutations and combinations to find the number of elements in events and sample spaces. These problems can be complicated, but they can be made easier by breaking them down into smaller counting problems.

Assume, for example, that a store has 8 cellular phones and that 3 of those are defective. We might want to find the probability that a couple purchasing 2 phones receives 2 phones that are not defective. To solve this problem, we need to calculate all of the ways to select 2 phones that are not defective as well as all of the ways to select 2 phones. There are 5 phones that are not defective, so there are $C(5,2)$ ways to select 2 phones that are not defective. There are 8 phones, so there are $C(8,2)$ ways to select 2 phones. The probability of selecting 2 phones that are not defective is:

$$\begin{aligned}\frac{\text{ways to select 2 phones that are not defective}}{\text{ways to select 2 phones}} &= \frac{C(5,2)}{C(8,2)} \\ &= \frac{10}{28} \\ &= \frac{5}{14}\end{aligned}$$

EXAMPLE 6

Computing Probability Using Counting Theory

A child randomly selects 5 toys from a bin containing 3 bunnies, 5 dogs, and 6 bears.

ⓐ Find the probability that only bears are chosen. ⓑ Find the probability that 2 bears and 3 dogs are chosen.

ⓒ Find the probability that at least 2 dogs are chosen.

Solution

ⓐ We need to count the number of ways to choose only bears and the total number of possible ways to select 5 toys. There are 6 bears, so there are $C(6,5)$ ways to choose 5 bears. There are 14 toys, so there are $C(14,5)$ ways to choose any 5 toys.

$$\frac{C(6,5)}{C(14,5)} = \frac{6}{2{,}002} = \frac{3}{1{,}001}$$

ⓑ We need to count the number of ways to choose 2 bears and 3 dogs and the total number of possible ways to select 5 toys. There are 6 bears, so there are $C(6,2)$ ways to choose 2 bears. There are 5 dogs, so there are $C(5,3)$ ways to choose 3 dogs. Since we are choosing both bears and dogs at the same time, we will use the Multiplication Principle. There are $C(6,2) \cdot C(5,3)$ ways to choose 2 bears and 3 dogs. We can use this result to find the probability.

$$\frac{C(6,2)C(5,3)}{C(14,5)} = \frac{15 \cdot 10}{2{,}002} = \frac{75}{1{,}001}$$

ⓒ It is often easiest to solve "at least" problems using the Complement Rule. We will begin by finding the probability that fewer than 2 dogs are chosen. If less than 2 dogs are chosen, then either no dogs could be chosen, or 1 dog could be chosen.

When no dogs are chosen, all 5 toys come from the 9 toys that are not dogs. There are $C(9,5)$ ways to choose toys from the 9 toys that are not dogs. Since there are 14 toys, there are $C(14,5)$ ways to choose the 5 toys from all of the toys.

$$\frac{C(9,5)}{C(14,5)} = \frac{63}{1{,}001}$$

If there is 1 dog chosen, then 4 toys must come from the 9 toys that are not dogs, and 1 must come from the 5 dogs. Since we are choosing both dogs and other toys at the same time, we will use the Multiplication Principle. There are $C(5,1) \cdot C(9,4)$ ways to choose 1 dog and 1 other toy.

$$\frac{C(5,1)C(9,4)}{C(14,5)} = \frac{5 \cdot 126}{2{,}002} = \frac{315}{1{,}001}$$

Because these events would not occur together and are therefore mutually exclusive, we add the probabilities to find the probability that fewer than 2 dogs are chosen.

$$\frac{63}{1{,}001} + \frac{315}{1{,}001} = \frac{378}{1{,}001}$$

We then subtract that probability from 1 to find the probability that at least 2 dogs are chosen.

$$1 - \frac{378}{1{,}001} = \frac{623}{1{,}001}$$

TRY IT #6 A child randomly selects 3 gumballs from a container holding 4 purple gumballs, 8 yellow gumballs, and 2 green gumballs.

ⓐ Find the probability that all 3 gumballs selected are purple.
ⓑ Find the probability that no yellow gumballs are selected.
ⓒ Find the probability that at least 1 yellow gumball is selected.

MEDIA

Access these online resources for additional instruction and practice with probability.

Introduction to Probability (http://openstax.org/l/introprob)
Determining Probability (http://openstax.org/l/determineprob)

13.7 SECTION EXERCISES

Verbal

1. What term is used to express the likelihood of an event occurring? Are there restrictions on its values? If so, what are they? If not, explain.

2. What is a sample space?

3. What is an experiment?

4. What is the difference between events and outcomes? Give an example of both using the sample space of tossing a coin 50 times.

5. The *union of two sets* is defined as a set of elements that are present in at least one of the sets. How is this similar to the definition used for the *union of two events* from a probability model? How is it different?

Numeric

For the following exercises, use the spinner shown in Figure 3 to find the probabilities indicated.

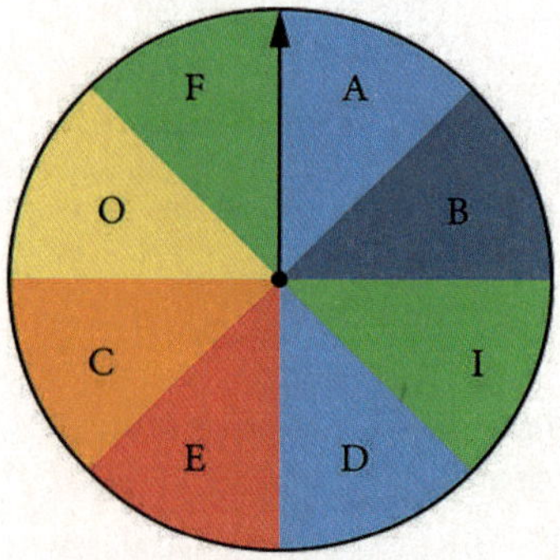

Figure 3

6. Landing on red

7. Landing on a vowel

8. Not landing on blue

9. Landing on purple or a vowel

10. Landing on blue or a vowel

11. Landing on green or blue

12. Landing on yellow or a consonant

13. Not landing on yellow or a consonant

For the following exercises, two coins are tossed.

14. What is the sample space?

15. Find the probability of tossing two heads.

16. Find the probability of tossing exactly one tail.

17. Find the probability of tossing at least one tail.

For the following exercises, four coins are tossed.

18. What is the sample space?

19. Find the probability of tossing exactly two heads.

20. Find the probability of tossing exactly three heads.

21. Find the probability of tossing four heads or four tails.

22. Find the probability of tossing all tails.

23. Find the probability of tossing not all tails.

24. Find the probability of tossing exactly two heads or at least two tails.

25. Find the probability of tossing either two heads or three heads.

For the following exercises, one card is drawn from a standard deck of 52 cards. Find the probability of drawing the following:

26. A club

27. A two

28. Six or seven

29. Red six

30. An ace or a diamond

31. A non-ace

32. A heart or a non-jack

For the following exercises, two dice are rolled, and the results are summed.

33. Construct a table showing the sample space of outcomes and sums.

34. Find the probability of rolling a sum of 3.

35. Find the probability of rolling at least one four or a sum of 8.

36. Find the probability of rolling an odd sum less than 9.

37. Find the probability of rolling a sum greater than or equal to 15.

38. Find the probability of rolling a sum less than 15.

39. Find the probability of rolling a sum less than 6 or greater than 9.

40. Find the probability of rolling a sum between 6 and 9, inclusive.

41. Find the probability of rolling a sum of 5 or 6.

42. Find the probability of rolling any sum other than 5 or 6.

For the following exercises, a coin is tossed, and a card is pulled from a standard deck. Find the probability of the following:

43. A head on the coin or a club

44. A tail on the coin or red ace

45. A head on the coin or a face card

46. No aces

For the following exercises, use this scenario: a bag of M&Ms contains 12 blue, 6 brown, 10 orange, 8 yellow, 8 red, and 4 green M&Ms. Reaching into the bag, a person grabs 5 M&Ms.

47. What is the probability of getting all blue M&Ms?

48. What is the probability of getting 4 blue M&Ms?

49. What is the probability of getting 3 blue M&Ms?

50. What is the probability of getting no brown M&Ms?

Extensions

Use the following scenario for the exercises that follow: In the game of Keno, a player starts by selecting 20 numbers from the numbers 1 to 80. After the player makes his selections, 20 winning numbers are randomly selected from numbers 1 to 80. A win occurs if the player has correctly selected 3, 4, or 5 of the 20 winning numbers. (Round all answers to the nearest hundredth of a percent.)

51. What is the percent chance that a player selects exactly 3 winning numbers?

52. What is the percent chance that a player selects exactly 4 winning numbers?

53. What is the percent chance that a player selects all 5 winning numbers?

54. What is the percent chance of winning?

55. How much less is a player's chance of selecting 3 winning numbers than the chance of selecting either 4 or 5 winning numbers?

Real-World Applications

Use this data for the exercises that follow: In 2013, there were roughly 317 million citizens in the United States, and about 40 million were elderly (aged 65 and over).[2]

56. If you meet a U.S. citizen, what is the percent chance that the person is elderly? (Round to the nearest tenth of a percent.)

57. If you meet five U.S. citizens, what is the percent chance that exactly one is elderly? (Round to the nearest tenth of a percent.)

58. If you meet five U.S. citizens, what is the percent chance that three are elderly? (Round to the nearest tenth of a percent.)

2 United States Census Bureau. http://www.census.gov

59. If you meet five U.S. citizens, what is the percent chance that four are elderly? (Round to the nearest thousandth of a percent.)

60. It is predicted that by 2030, one in five U.S. citizens will be elderly. How much greater will the chances of meeting an elderly person be at that time? What policy changes do you foresee if these statistics hold true?

Chapter Review

Key Terms

Addition Principle if one event can occur in m ways and a second event with no common outcomes can occur in n ways, then the first or second event can occur in $m + n$ ways

annuity an investment in which the purchaser makes a sequence of periodic, equal payments

arithmetic sequence a sequence in which the difference between any two consecutive terms is a constant

arithmetic series the sum of the terms in an arithmetic sequence

binomial coefficient the number of ways to choose r objects from n objects where order does not matter; equivalent to $C(n, r)$, denoted $\binom{n}{r}$

binomial expansion the result of expanding $(x + y)^n$ by multiplying

Binomial Theorem a formula that can be used to expand any binomial

combination a selection of objects in which order does not matter

common difference the difference between any two consecutive terms in an arithmetic sequence

common ratio the ratio between any two consecutive terms in a geometric sequence

complement of an event the set of outcomes in the sample space that are not in the event E

diverge a series is said to diverge if the sum is not a real number

event any subset of a sample space

experiment an activity with an observable result

explicit formula a formula that defines each term of a sequence in terms of its position in the sequence

finite sequence a function whose domain consists of a finite subset of the positive integers $\{1, 2, \ldots n\}$ for some positive integer n

Fundamental Counting Principle if one event can occur in m ways and a second event can occur in n ways after the first event has occurred, then the two events can occur in $m \times n$ ways; also known as the Multiplication Principle

geometric sequence a sequence in which the ratio of a term to a previous term is a constant

geometric series the sum of the terms in a geometric sequence

index of summation in summation notation, the variable used in the explicit formula for the terms of a series and written below the sigma with the lower limit of summation

infinite sequence a function whose domain is the set of positive integers

infinite series the sum of the terms in an infinite sequence

lower limit of summation the number used in the explicit formula to find the first term in a series

Multiplication Principle if one event can occur in m ways and a second event can occur in n ways after the first event has occurred, then the two events can occur in $m \times n$ ways; also known as the Fundamental Counting Principle

mutually exclusive events events that have no outcomes in common

n factorial the product of all the positive integers from 1 to n

nth partial sum the sum of the first n terms of a sequence

nth term of a sequence a formula for the general term of a sequence

outcomes the possible results of an experiment

permutation a selection of objects in which order matters

probability a number from 0 to 1 indicating the likelihood of an event

probability model a mathematical description of an experiment listing all possible outcomes and their associated probabilities

recursive formula a formula that defines each term of a sequence using previous term(s)

sample space the set of all possible outcomes of an experiment

sequence a function whose domain is a subset of the positive integers

series the sum of the terms in a sequence

summation notation a notation for series using the Greek letter sigma; it includes an explicit formula and specifies the first and last terms in the series

term a number in a sequence

union of two events the event that occurs if either or both events occur

upper limit of summation the number used in the explicit formula to find the last term in a series

Key Equations

Formula for a factorial	$0! = 1$ $1! = 1$ $n! = n(n-1)(n-2)\cdots(2)(1)$, for $n \geq 2$
recursive formula for nth term of an arithmetic sequence	$a_n = a_{n-1} + d, n \geq 2$
explicit formula for nth term of an arithmetic sequence	$a_n = a_1 + d(n-1)$
recursive formula for *nth* term of a geometric sequence	$a_n = ra_{n-1}, n \geq 2$
explicit formula for *nth* term of a geometric sequence	$a_n = a_1 r^{n-1}$
sum of the first n terms of an arithmetic series	$S_n = \frac{n(a_1+a_n)}{2}$
sum of the first n terms of a geometric series	$S_n = \frac{a_1(1-r^n)}{1-r}, r \neq 1$
sum of an infinite geometric series with $-1 < r < 1$	$S_n = \frac{a_1}{1-r}, r \neq 1$
number of permutations of n distinct objects taken r at a time	$P(n,r) = \frac{n!}{(n-r)!}$
number of combinations of n distinct objects taken r at a time	$C(n,r) = \frac{n!}{r!(n-r)!}$
number of permutations of n non-distinct objects	$\frac{n!}{r_1!r_2!\ldots r_k!}$
Binomial Theorem	$(x+y)^n = \sum_{k-0}^{n} \binom{n}{k} x^{n-k} y^k$
$(r+1)th$ term of a binomial expansion	$\binom{n}{r} x^{n-r} y^r$
probability of an event with equally likely outcomes	$P(E) = \frac{n(E)}{n(S)}$
probability of the union of two events	$P(E \cup F) = P(E) + P(F) - P(E \cap F)$
probability of the union of mutually exclusive events	$P(E \cup F) = P(E) + P(F)$
probability of the complement of an event	$P(E') = 1 - P(E)$

Key Concepts

13.1 Sequences and Their Notations

- A sequence is a list of numbers, called terms, written in a specific order.
- Explicit formulas define each term of a sequence using the position of the term. See Example 1, Example 2, and Example 3.
- An explicit formula for the nth term of a sequence can be written by analyzing the pattern of several terms. See Example 4.
- Recursive formulas define each term of a sequence using previous terms.
- Recursive formulas must state the initial term, or terms, of a sequence.
- A set of terms can be written by using a recursive formula. See Example 5 and Example 6.
- A factorial is a mathematical operation that can be defined recursively.
- The factorial of n is the product of all integers from 1 to n See Example 7.

13.2 Arithmetic Sequences

- An arithmetic sequence is a sequence where the difference between any two consecutive terms is a constant.
- The constant between two consecutive terms is called the common difference.
- The common difference is the number added to any one term of an arithmetic sequence that generates the subsequent term. See Example 1.
- The terms of an arithmetic sequence can be found by beginning with the initial term and adding the common difference repeatedly. See Example 2 and Example 3.
- A recursive formula for an arithmetic sequence with common difference d is given by $a_n = a_{n-1} + d, n \geq 2$. See Example 4.
- As with any recursive formula, the initial term of the sequence must be given.
- An explicit formula for an arithmetic sequence with common difference d is given by $a_n = a_1 + d(n-1)$. See Example 5.
- An explicit formula can be used to find the number of terms in a sequence. See Example 6.
- In application problems, we sometimes alter the explicit formula slightly to $a_n = a_0 + dn$. See Example 7.

13.3 Geometric Sequences

- A geometric sequence is a sequence in which the ratio between any two consecutive terms is a constant.
- The constant ratio between two consecutive terms is called the common ratio.
- The common ratio can be found by dividing any term in the sequence by the previous term. See Example 1.
- The terms of a geometric sequence can be found by beginning with the first term and multiplying by the common ratio repeatedly. See Example 2 and Example 4.
- A recursive formula for a geometric sequence with common ratio r is given by $a_n = ra_{n-1}$ for $n \geq 2$.
- As with any recursive formula, the initial term of the sequence must be given. See Example 3.
- An explicit formula for a geometric sequence with common ratio r is given by $a_n = a_1 r^{n-1}$. See Example 5.
- In application problems, we sometimes alter the explicit formula slightly to $a_n = a_0 r^n$. See Example 6.

13.4 Series and Their Notations

- The sum of the terms in a sequence is called a series.
- A common notation for series is called summation notation, which uses the Greek letter sigma to represent the sum. See Example 1.
- The sum of the terms in an arithmetic sequence is called an arithmetic series.
- The sum of the first n terms of an arithmetic series can be found using a formula. See Example 2 and Example 3.
- The sum of the terms in a geometric sequence is called a geometric series.
- The sum of the first n terms of a geometric series can be found using a formula. See Example 4 and Example 5.
- The sum of an infinite series exists if the series is geometric with $-1 < r < 1$.
- If the sum of an infinite series exists, it can be found using a formula. See Example 6, Example 7, and Example 8.
- An annuity is an account into which the investor makes a series of regularly scheduled payments. The value of an annuity can be found using geometric series. See Example 9.

13.5 Counting Principles

- If one event can occur in m ways and a second event with no common outcomes can occur in n ways, then the first or second event can occur in $m + n$ ways. See Example 1.
- If one event can occur in m ways and a second event can occur in n ways after the first event has occurred, then the two events can occur in $m \times n$ ways. See Example 2.

- A permutation is an ordering of n objects.
- If we have a set of n objects and we want to choose r objects from the set in order, we write $P(n, r)$.
- Permutation problems can be solved using the Multiplication Principle or the formula for $P(n, r)$. See Example 3 and Example 4.
- A selection of objects where the order does not matter is a combination.
- Given n distinct objects, the number of ways to select r objects from the set is $C(n, r)$ and can be found using a formula. See Example 5.
- A set containing n distinct objects has 2^n subsets. See Example 6.
- For counting problems involving non-distinct objects, we need to divide to avoid counting duplicate permutations. See Example 7.

13.6 Binomial Theorem

- $\binom{n}{r}$ is called a binomial coefficient and is equal to $C(n, r)$. See Example 1.
- The Binomial Theorem allows us to expand binomials without multiplying. See Example 2.
- We can find a given term of a binomial expansion without fully expanding the binomial. See Example 3.

13.7 Probability

- Probability is always a number between 0 and 1, where 0 means an event is impossible and 1 means an event is certain.
- The probabilities in a probability model must sum to 1. See Example 1.
- When the outcomes of an experiment are all equally likely, we can find the probability of an event by dividing the number of outcomes in the event by the total number of outcomes in the sample space for the experiment. See Example 2.
- To find the probability of the union of two events, we add the probabilities of the two events and subtract the probability that both events occur simultaneously. See Example 3.
- To find the probability of the union of two mutually exclusive events, we add the probabilities of each of the events. See Example 4.
- The probability of the complement of an event is the difference between 1 and the probability that the event occurs. See Example 5.
- In some probability problems, we need to use permutations and combinations to find the number of elements in events and sample spaces. See Example 6.

Exercises

Review Exercises

Sequences and Their Notation

1. Write the first four terms of the sequence defined by the recursive formula $a_1 = 2, \quad a_n = a_{n-1} + n.$

2. Evaluate $\frac{6!}{(5-3)!3!}$.

3. Write the first four terms of the sequence defined by the explicit formula $a_n = 10^n + 3.$

4. Write the first four terms of the sequence defined by the explicit formula $a_n = \frac{n!}{n(n+1)}$.

Arithmetic Sequences

5. Is the sequence $\frac{4}{7}, \frac{47}{21}, \frac{82}{21}, \frac{39}{7}, \ldots$ arithmetic? If so, find the common difference.

6. Is the sequence $2, 4, 8, 16, \ \ldots$ arithmetic? If so, find the common difference.

7. An arithmetic sequence has the first term $a_1 = 18$ and common difference $d = -8$. What are the first five terms?

8. An arithmetic sequence has terms $a_3 = 11.7$ and $a_8 = -14.6$. What is the first term?

9. Write a recursive formula for the arithmetic sequence $-20, -10, 0, 10, \ldots$

10. Write a recursive formula for the arithmetic sequence $0, \ -\frac{1}{2}, \ -1, \ -\frac{3}{2}, \ldots,$ and then find the 31st term.

11. Write an explicit formula for the arithmetic sequence $\frac{7}{8}, \frac{29}{24}, \frac{37}{24}, \frac{15}{8}, \ldots$

12. How many terms are in the finite arithmetic sequence $12, 20, 28, \ldots, 172$?

Geometric Sequences

13. Find the common ratio for the geometric sequence $2.5, \ 5, \ 10, \ 20, \ldots$

14. Is the sequence 4, 16, 28, 40 ... geometric? If so find the common ratio. If not, explain why.

15. A geometric sequence has terms $a_7 = 16{,}384$ and $a_9 = 262{,}144$. What are the first five terms?

16. A geometric sequence has the first term $a_1 = -3$ and common ratio $r = \frac{1}{2}$. What is the 8th term?

17. What are the first five terms of the geometric sequence $a_1 = 3, \ a_n = 4 \cdot a_{n-1}$?

18. Write a recursive formula for the geometric sequence $1, \ \frac{1}{3}, \ \frac{1}{9}, \ \frac{1}{27}, \ldots$

19. Write an explicit formula for the geometric sequence $-\frac{1}{5}, \ -\frac{1}{15}, \ -\frac{1}{45}, \ -\frac{1}{135}, \ldots$

20. How many terms are in the finite geometric sequence $-5, \ -\frac{5}{3}, \ -\frac{5}{9}, \ldots, \ -\frac{5}{59{,}049}$?

Series and Their Notation

21. Use summation notation to write the sum of terms $\frac{1}{2}m + 5$ from $m = 0$ to $m = 5$.

22. Use summation notation to write the sum that results from adding the number 13 twenty times.

23. Use the formula for the sum of the first n terms of an arithmetic series to find the sum of the first eleven terms of the arithmetic series 2.5, 4, 5.5, ...

24. A ladder has 15 tapered rungs, the lengths of which increase by a common difference. The first rung is 5 inches long, and the last rung is 20 inches long. What is the sum of the lengths of the rungs?

25. Use the formula for the sum of the first *n* terms of a geometric series to find S_9 for the series $12,\ 6,\ 3,\ \frac{3}{2}, \ldots$

26. The fees for the first three years of a hunting club membership are given in Table 1. If fees continue to rise at the same rate, how much will the total cost be for the first ten years of membership?

Year	Membership Fees
1	\$1500
2	\$1950
3	\$2535

Table 1

27. Find the sum of the infinite geometric series $\sum_{k=1}^{\infty} 45 \cdot \left(-\frac{1}{3}\right)^{k-1}$.

28. A ball has a bounce-back ratio of $\frac{3}{5}$ the height of the previous bounce. Write a series representing the total distance traveled by the ball, assuming it was initially dropped from a height of 5 feet. What is the total distance? (*Hint*: the total distance the ball travels on each bounce is the sum of the heights of the rise and the fall.)

29. Alejandro deposits \$80 of his monthly earnings into an annuity that earns 6.25% annual interest, compounded monthly. How much money will he have saved after 5 years?

30. The twins Hoa and Binh both opened retirement accounts on their 21st birthday. Hoa deposits \$4,800.00 each year, earning 5.5% annual interest, compounded monthly. Binh deposits \$3,600.00 each year, earning 8.5% annual interest, compounded monthly. Which twin will earn the most interest by the time they are 55 years old? How much more?

Counting Principles

31. How many ways are there to choose a number from the set $\{-10, -6, 4, 10, 12, 18, 24, 32\}$ that is divisible by either 4 or 6?

32. In a group of 20 musicians, 12 play piano, 7 play trumpet, and 2 play both piano and trumpet. How many musicians play either piano or trumpet?

33. How many ways are there to construct a 4-digit code if numbers can be repeated?

34. A palette of water color paints has 3 shades of green, 3 shades of blue, 2 shades of red, 2 shades of yellow, and 1 shade of black. How many ways are there to choose one shade of each color?

35. Calculate $P(18, 4)$.

36. In a group of 5 first-year, 10 second-year, 3 third-year, and 2 fourth-year students, how many ways can a president, vice president, and treasurer be elected?

37. Calculate $C(15, 6)$.

38. A coffee shop has 7 Guatemalan roasts, 4 Cuban roasts, and 10 Costa Rican roasts. How many ways can the shop choose 2 Guatemalan, 2 Cuban, and 3 Costa Rican roasts for a coffee tasting event?

39. How many subsets does the set $\{1, 3, 5, \ldots, 99\}$ have?

40. A day spa charges a basic day rate that includes use of a sauna, pool, and showers. For an extra charge, guests can choose from the following additional services: massage, body scrub, manicure, pedicure, facial, and straight-razor shave. How many ways are there to order additional services at the day spa?

41. How many distinct ways can the word DEADWOOD be arranged?

42. How many distinct rearrangements of the letters of the word DEADWOOD are there if the arrangement must begin and end with the letter D?

Binomial Theorem

43. Evaluate the binomial coefficient $\binom{23}{8}$.

44. Use the Binomial Theorem to expand $\left(3x + \frac{1}{2}y\right)^6$.

45. Use the Binomial Theorem to write the first three terms of $(2a + b)^{17}$.

46. Find the fourth term of $(3a^2 - 2b)^{11}$ without fully expanding the binomial.

Probability

For the following exercises, assume two die are rolled.

47. Construct a table showing the sample space.

48. What is the probability that a roll includes a 2?

49. What is the probability of rolling a pair?

50. What is the probability that a roll includes a 2 or results in a pair?

51. What is the probability that a roll doesn't include a 2 or result in a pair?

52. What is the probability of rolling a 5 or a 6?

53. What is the probability that a roll includes neither a 5 nor a 6?

For the following exercises, use the following data: An elementary school survey found that 350 of the 500 students preferred soda to milk. Suppose 8 children from the school are attending a birthday party. (Show calculations and round to the nearest tenth of a percent.)

54. What is the percent chance that all the children attending the party prefer soda?

55. What is the percent chance that at least one of the children attending the party prefers milk?

56. What is the percent chance that exactly 3 of the children attending the party prefer soda?

57. What is the percent chance that exactly 3 of the children attending the party prefer milk?

Practice Test

1. Write the first four terms of the sequence defined by the recursive formula $a =- 14, a_n = \frac{2+a_{n-1}}{2}$.

2. Write the first four terms of the sequence defined by the explicit formula $a_n = \frac{n^2-n-1}{n!}$.

3. Is the sequence $0.3,\ 1.2,\ 2.1,\ 3, \ldots$ arithmetic? If so find the common difference.

4. An arithmetic sequence has the first term $a_1 = -4$ and common difference $d =- \frac{4}{3}$. What is the 6th term?

5. Write a recursive formula for the arithmetic sequence $-2,\ -\frac{7}{2},\ -5,\ -\frac{13}{2}, \ldots$ and then find the 22nd term.

6. Write an explicit formula for the arithmetic sequence $15.6,\ 15,\ 14.4,\ 13.8, \ldots$ and then find the 32nd term.

7. Is the sequence $-2, -1, -\frac{1}{2}, -\frac{1}{4}, \ldots$ geometric? If so find the common ratio. If not, explain why.

8. What is the 11th term of the geometric sequence $-1.5, -3, -6, -12, \ldots$?

9. Write a recursive formula for the geometric sequence $1,\ -\frac{1}{2},\ \frac{1}{4},\ -\frac{1}{8}, \ldots$

10. Write an explicit formula for the geometric sequence $4,\ -\frac{4}{3},\ \frac{4}{9},\ -\frac{4}{27},\ldots$

11. Use summation notation to write the sum of terms $3k^2 - \frac{5}{6}k$ from $k = -3$ to $k = 15$.

12. A community baseball stadium has 10 seats in the first row, 13 seats in the second row, 16 seats in the third row, and so on. There are 56 rows in all. What is the seating capacity of the stadium?

13. Use the formula for the sum of the first n terms of a geometric series to find $\sum_{k=1}^{7} -0.2 \cdot (-5)^{k-1}$.

14. Find the sum of the infinite geometric series $\sum_{k=1}^{\infty} \frac{1}{3} \cdot \left(-\frac{1}{5}\right)^{k-1}$.

15. Ramla deposits \$3,600 into a retirement fund each year. The fund earns 7.5% annual interest, compounded monthly. If she opened her account when she was 20 years old, how much will she have by the time she's 55? How much of that amount was interest earned?

16. In a competition of 50 professional ballroom dancers, 22 compete in the fox-trot competition, 18 compete in the tango competition, and 6 compete in both the fox-trot and tango competitions. How many dancers compete in the fox-trot or tango competitions?

17. A buyer of a new sedan can custom order the car by choosing from 5 different exterior colors, 3 different interior colors, 2 sound systems, 3 motor designs, and either manual or automatic transmission. How many choices does the buyer have?

18. To allocate annual bonuses, a manager must choose his top four employees and rank them first to fourth. In how many ways can he create the "Top-Four" list out of the 32 employees?

19. A music group needs to choose 3 songs to play at the annual Battle of the Bands. How many ways can they choose their set if have 15 songs to pick from?

20. A self-serve frozen yogurt shop has 8 candy toppings and 4 fruit toppings to choose from. How many ways are there to top a frozen yogurt?

21. How many distinct ways can the word EVANESCENCE be arranged if the anagram must end with the letter E?

22. Use the Binomial Theorem to expand $\left(\frac{3}{2}x - \frac{1}{2}y\right)^5$.

23. Find the seventh term of $\left(x^2 - \frac{1}{2}\right)^{13}$ without fully expanding the binomial.

For the following exercises, use the spinner in Figure 1.

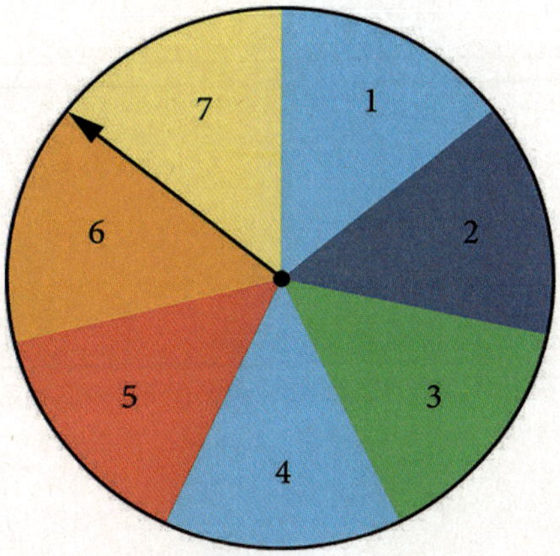

Figure 1

24. Construct a probability model showing each possible outcome and its associated probability. (Use the first letter for colors.)

25. What is the probability of landing on an odd number?

26. What is the probability of landing on blue?

27. What is the probability of landing on blue or an odd number?

28. What is the probability of landing on anything other than blue or an odd number?

29. A bowl of candy holds 16 peppermint, 14 butterscotch, and 10 strawberry flavored candies. Suppose a person grabs a handful of 7 candies. What is the percent chance that exactly 3 are butterscotch? (Show calculations and round to the nearest tenth of a percent.)

A PROOFS, IDENTITIES, AND TOOLKIT FUNCTIONS

Important Proofs and Derivations

Product Rule

$\log_a xy = \log_a x + \log_a y$

Proof:

Let $m = \log_a x$ and $n = \log_a y$.

Write in exponent form.

$x = a^m$ and $y = a^n$.

Multiply.

$xy = a^m a^n = a^{m+n}$

$$\begin{aligned} a^{m+n} &= xy \\ \log_a(xy) &= m+n \\ &= \log_a x + \log_b y \end{aligned}$$

Change of Base Rule

$\log_a b = \frac{\log_c b}{\log_c a}$

$\log_a b = \frac{1}{\log_b a}$

where x and y are positive, and $a > 0, a \neq 1$.

Proof:

Let $x = \log_a b$.

Write in exponent form.

$a^x = b$

Take the $\log_c$ of both sides.

$$\begin{aligned} \log_c a^x &= \log_c b \\ x\log_c a &= \log_c b \\ x &= \frac{\log_c b}{\log_c a} \\ \log_a b &= \frac{\log_c b}{\log_c a} \end{aligned}$$

When $c = b$,

$\log_a b = \frac{\log_b b}{\log_b a} = \frac{1}{\log_b a}$

Heron's Formula

$A = \sqrt{s(s-a)(s-b)(s-c)}$

where $s = \frac{a+b+c}{2}$

Proof:

Let a, b, and c be the sides of a triangle, and h be the height.

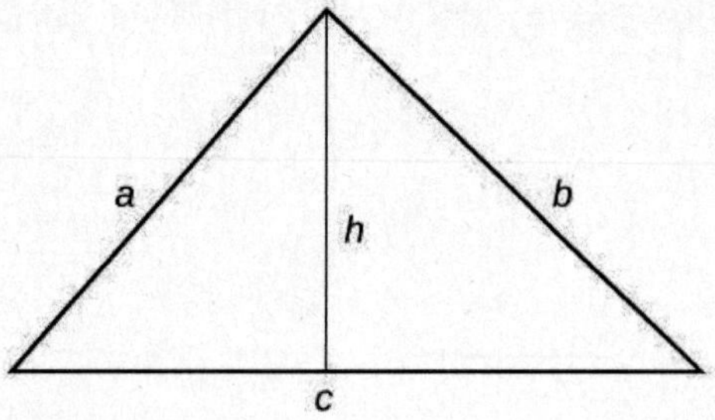

So $s = \frac{a+b+c}{2}$.

We can further name the parts of the base in each triangle established by the height such that $p + q = c$.

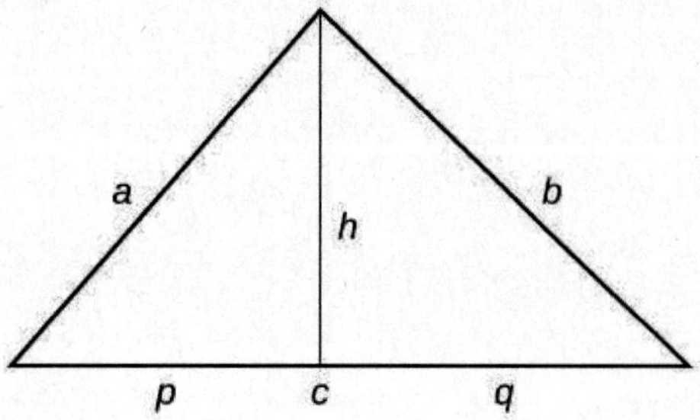

Using the Pythagorean Theorem, $h^2 + p^2 = a^2$ and $h^2 + q^2 = b^2$.

Since $q = c - p$, then $q^2 = (c - p)^2$. Expanding, we find that $q^2 = c^2 - 2cp + p^2$.

We can then add h^2 to each side of the equation to get $h^2 + q^2 = h^2 + c^2 - 2cp + p^2$.

Substitute this result into the equation $h^2 + q^2 = b^2$ yields $b^2 = h^2 + c^2 - 2cp + p^2$.

Then replacing $h^2 + p^2$ with a^2 gives $b^2 = a^2 - 2cp + c^2$.

Solve for p to get

$$p = \frac{a^2+b^2-c^2}{2c}$$

Since $h^2 = a^2 - p^2$, we get an expression in terms of a, b, and c.

$$\begin{aligned}
h^2 &= a^2 - p^2 \\
&= (a + p)(a - p) \\
&= \left[a + \frac{\left(a^2+c^2-b^2\right)}{2c}\right]\left[a - \frac{\left(a^2+c^2-b^2\right)}{2c}\right] \\
&= \frac{\left(2ac+a^2+c^2-b^2\right)\left(2ac-a^2-c^2+b^2\right)}{4c^2} \\
&= \frac{\left((a+c)^2-b^2\right)\left(b^2-(a-c)^2\right)}{4c^2} \\
&= \frac{(a+b+c)(a+c-b)(b+a-c)(b-a+c)}{4c^2} \\
&= \frac{(a+b+c)(-a+b+c)(a-b+c)(a+b-c)}{4c^2} \\
&= \frac{2s\cdot(2s-a)\cdot(2s-b)(2s-c)}{4c^2}
\end{aligned}$$

Therefore,

$$\begin{aligned}
h^2 &= \frac{4s(s-a)(s-b)(s-c)}{c^2} \\
h &= \frac{2\sqrt{s(s-a)(s-b)(s-c)}}{c}
\end{aligned}$$

And since $A = \frac{1}{2}ch$, then

$$\begin{aligned} A &= \frac{1}{2}c\frac{2\sqrt{s(s-a)(s-b)(s-c)}}{c} \\ &= \sqrt{s(s-a)(s-b)(s-c)} \end{aligned}$$

Properties of the Dot Product

$u \cdot v = v \cdot u$

Proof:

$$\begin{aligned} u \cdot v &= \langle u_1, u_2, ...u_n \rangle \cdot \langle v_1, v_2, ...v_n \rangle \\ &= u_1 v_1 + u_2 v_2 + ... + u_n v_n \\ &= v_1 u_1 + v_2 u_2 + ... + v_n v_n \\ &= \langle v_1, v_2, ...v_n \rangle \cdot \langle u_1, u_2, ...u_n \rangle \\ &= v \cdot u \end{aligned}$$

$u \cdot (v + w) = u \cdot v + u \cdot w$

Proof:

$$\begin{aligned} u \cdot (v + w) &= \langle u_1, u_2, ...u_n \rangle \cdot (\langle v_1, v_2, ...v_n \rangle + \langle w_1, w_2, ...w_n \rangle) \\ &= \langle u_1, u_2, ...u_n \rangle \cdot \langle v_1 + w_1, v_2 + w_2, ...v_n + w_n \rangle \\ &= \langle u_1(v_1 + w_1), u_2(v_2 + w_2), ...u_n(v_n + w_n) \rangle \\ &= \langle u_1 v_1 + u_1 w_1, u_2 v_2 + u_2 w_2, ...u_n v_n + u_n w_n \rangle \\ &= \langle u_1 v_1, u_2 v_2, ..., u_n v_n \rangle + \langle u_1 w_1, u_2 w_2, ..., u_n w_n \rangle \\ &= \langle u_1, u_2, ...u_n \rangle \cdot \langle v_1, v_2, ...v_n \rangle + \langle u_1, u_2, ...u_n \rangle \cdot \langle w_1, w_2, ...w_n \rangle \\ &= u \cdot v + u \cdot w \end{aligned}$$

$u \cdot u = |u|^2$

Proof:

$$\begin{aligned} u \cdot u &= \langle u_1, u_2, ...u_n \rangle \cdot \langle u_1, u_2, ...u_n \rangle \\ &= u_1 u_1 + u_2 u_2 + ... + u_n u_n \\ &= {u_1}^2 + {u_2}^2 + ... + {u_n}^2 \\ &= |\langle u_1, u_2, ...u_n \rangle|^2 \\ &= v \cdot u \end{aligned}$$

Standard Form of the Ellipse centered at the Origin

$$1 = \frac{x^2}{a^2} + \frac{y^2}{b^2}$$

Derivation

An ellipse consists of all the points for which the sum of distances from two foci is constant:

$$\sqrt{(x-(-c))^2 + (y-0)^2} + \sqrt{(x-c)^2 + (y-0)^2} = \text{constant}$$

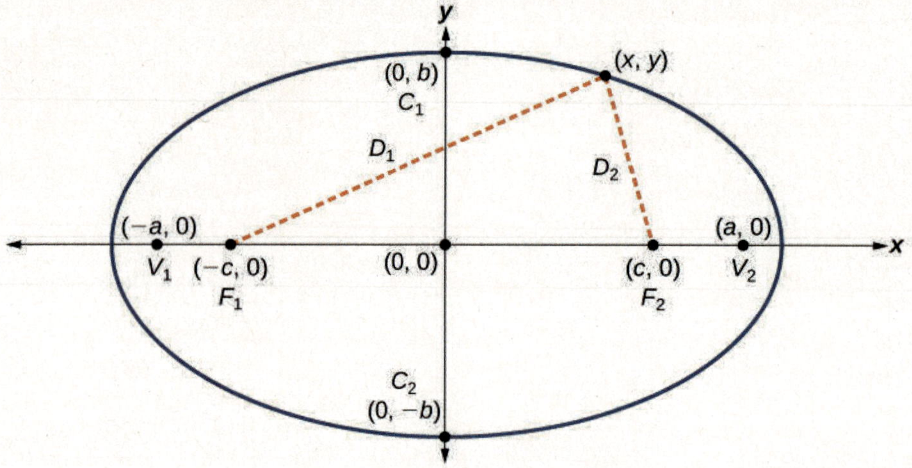

Consider a vertex.

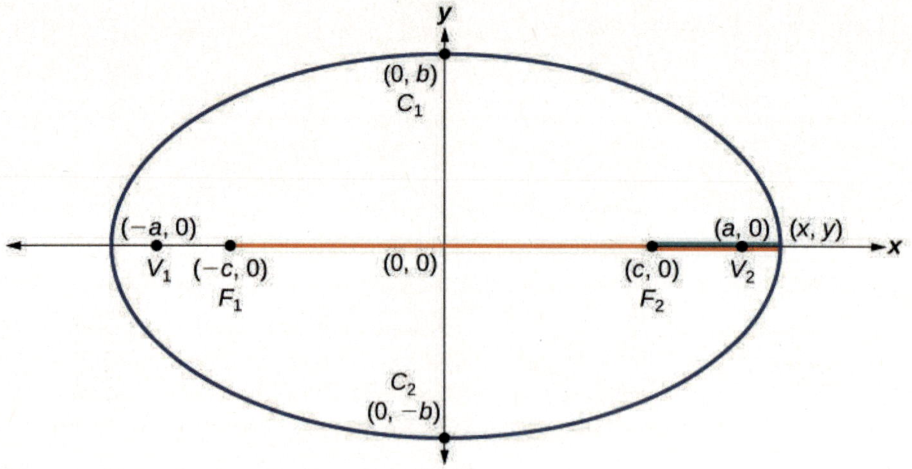

Then, $\sqrt{(x-(-c))^2+(y-0)^2}+\sqrt{(x-c)^2+(y-0)^2}=2a$

Consider a covertex.

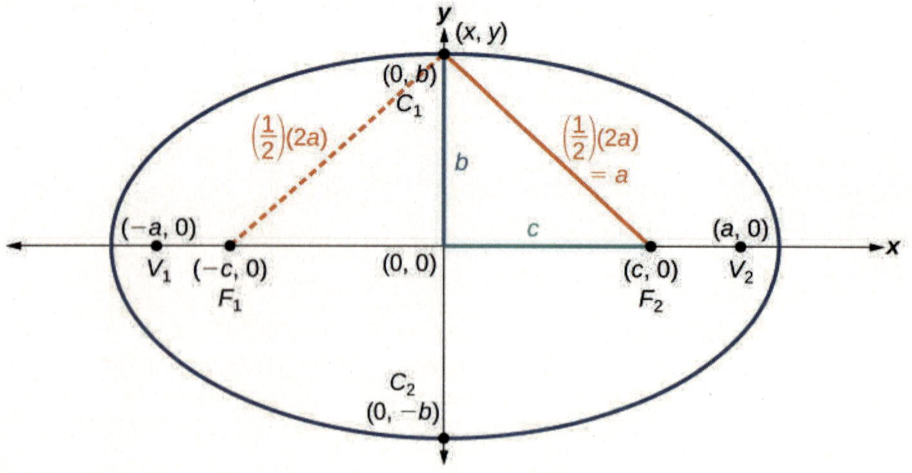

Then $b^2+c^2=a^2$.

$$\begin{aligned}
\sqrt{(x-(-c))^2+(y-0)^2}+\sqrt{(x-c)^2+(y-0)^2} &= 2a \\
\sqrt{(x+c)^2+y^2} &= 2a-\sqrt{(x-c)^2+y^2} \\
(x+c)^2+y^2 &= \left(2a-\sqrt{(x-c)^2+y^2}\right)^2 \\
x^2+2cx+c^2+y^2 &= 4a^2-4a\sqrt{(x-c)^2+y^2}+(x-c)^2+y^2 \\
x^2+2cx+c^2+y^2 &= 4a^2-4a\sqrt{(x-c)^2+y^2}+x^2-2cx+y^2 \\
2cx &= 4a^2-4a\sqrt{(x-c)^2+y^2}-2cx \\
4cx-4a^2 &= 4a\sqrt{(x-c)^2+y^2} \\
-\frac{1}{4a}\left(4cx-4a^2\right) &= \sqrt{(x-c)^2+y^2} \\
a-\frac{c}{a}x &= \sqrt{(x-c)^2+y^2} \\
a^2-2xc+\frac{c^2}{a^2}x^2 &= (x-c)^2+y^2 \\
a^2-2xc+\frac{c^2}{a^2}x^2 &= x^2-2xc+c^2+y^2 \\
a^2+\frac{c^2}{a^2}x^2 &= x^2+c^2+y^2 \\
a^2+\frac{c^2}{a^2}x^2 &= x^2+c^2+y^2 \\
a^2-c^2 &= x^2-\frac{c^2}{a^2}x^2+y^2 \\
a^2-c^2 &= x^2\left(1-\frac{c^2}{a^2}\right)+y^2
\end{aligned}$$

Let $1=\frac{a^2}{a^2}$.

$$\begin{aligned}
a^2-c^2 &= x^2\left(\frac{a^2-c^2}{a^2}\right)+y^2 \\
1 &= \frac{x^2}{a^2}+\frac{y^2}{a^2-c^2}
\end{aligned}$$

Because $b^2+c^2=a^2$, then $b^2=a^2-c^2$.

$$\begin{aligned}
1 &= \frac{x^2}{a^2}+\frac{y^2}{a^2-c^2} \\
1 &= \frac{x^2}{a^2}+\frac{y^2}{b^2}
\end{aligned}$$

Standard Form of the Hyperbola

$1=\frac{x^2}{a^2}-\frac{y^2}{b^2}$

Derivation

A hyperbola is the set of all points in a plane such that the absolute value of the difference of the distances between two fixed points is constant.

Diagram 1

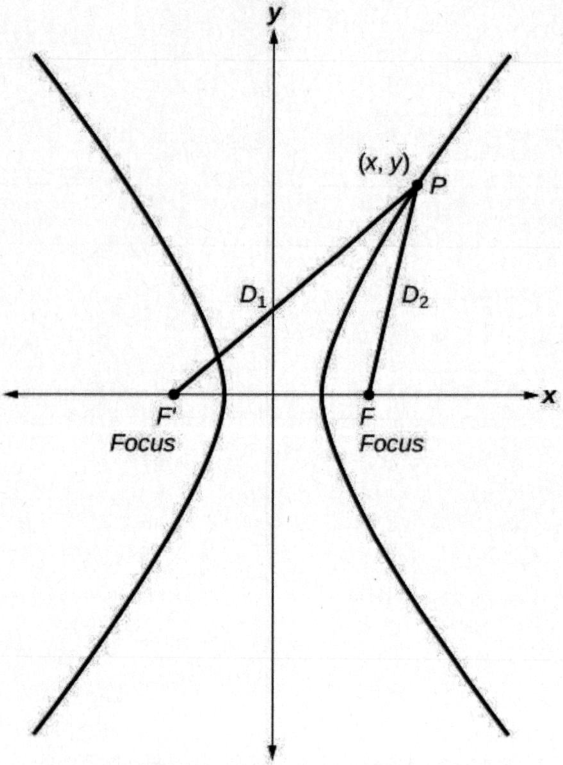

Diagram 2

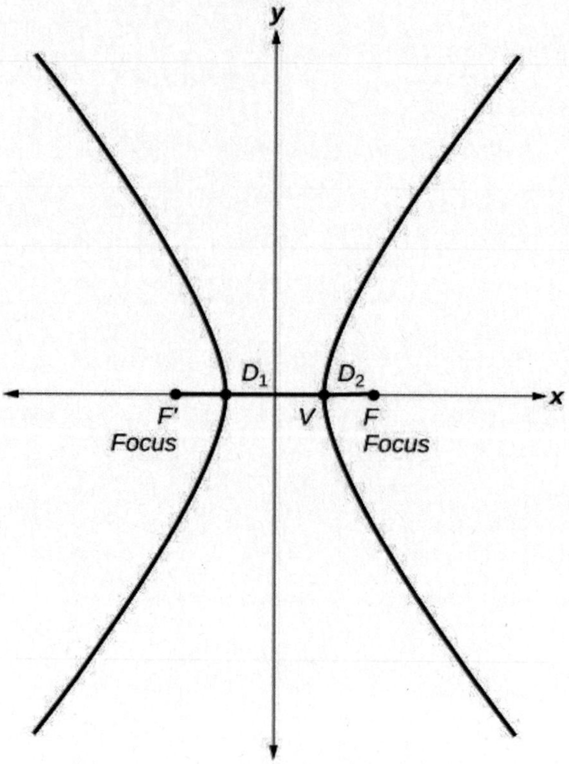

Diagram 1: The difference of the distances from Point *P* to the foci is constant:

$$\sqrt{(x-(-c))^2+(y-0)^2}-\sqrt{(x-c)^2+(y-0)^2}=\text{constant}$$

Diagram 2: When the point is a vertex, the difference is $2a$.

$$\sqrt{(x-(-c))^2+(y-0)^2}-\sqrt{(x-c)^2+(y-0)^2}=2a$$

$$\begin{aligned}
\sqrt{(x-(-c))^2+(y-0)^2}-\sqrt{(x-c)^2+(y-0)^2} &= 2a\\
\sqrt{(x+c)^2+y^2}-\sqrt{(x-c)^2+y^2} &= 2a\\
\sqrt{(x+c)^2+y^2} &= 2a+\sqrt{(x-c)^2+y^2}\\
(x+c)^2+y^2 &= \left(2a+\sqrt{(x-c)^2+y^2}\right)\\
x^2+2cx+c^2+y^2 &= 4a^2+4a\sqrt{(x-c)^2+y^2}\\
x^2+2cx+c^2+y^2 &= 4a^2+4a\sqrt{(x-c)^2+y^2}+x^2-2cx+y^2\\
2cx &= 4a^2+4a\sqrt{(x-c)^2+y^2}-2cx\\
4cx-4a^2 &= 4a\sqrt{(x-c)^2+y^2}\\
cx-a^2 &= a\sqrt{(x-c)^2+y^2}\\
\left(cx-a^2\right)^2 &= a^2\left((x-c)^2+y^2\right)\\
c^2x^2-2a^2c^2x^2+a^4 &= a^2x^2-2a^2c^2x^2+a^2c^2+a^2y^2\\
c^2x^2+a^4 &= a^2x^2+a^2c^2+a^2y^2\\
a^4-a^2c^2 &= a^2x^2-c^2x^2+a^2y^2\\
a^2\left(a^2-c^2\right) &= \left(a^2-c^2\right)x^2+a^2y^2\\
a^2\left(a^2-c^2\right) &= \left(c^2-a^2\right)x^2-a^2y^2
\end{aligned}$$

Define b as a positive number such that $b^2=c^2-a^2$.

$$\begin{aligned} a^2 b^2 &= b^2 x^2 - a^2 y^2 \\ \frac{a^2 b^2}{a^2 b^2} &= \frac{b^2 x^2}{a^2 b^2} - \frac{a^2 y^2}{a^2 b^2} \\ 1 &= \frac{x^2}{a^2} - \frac{y^2}{b^2} \end{aligned}$$

Trigonometric Identities

Pythagorean Identities	$\cos^2\theta + \sin^2\theta = 1$ $1 + \tan^2\theta = \sec^2\theta$ $1 + \cot^2\theta = \csc^2\theta$
Even-Odd Identities	$\cos(-\theta) = \cos\theta$ $\sec(-\theta) = \sec\theta$ $\sin(-\theta) = -\sin\theta$ $\tan(-\theta) = -\tan\theta$ $\csc(-\theta) = -\csc\theta$ $\cot(-\theta) = -\cot\theta$
Cofunction Identities	$\cos\theta = \sin\left(\frac{\pi}{2} - \theta\right)$ $\sin\theta = \cos\left(\frac{\pi}{2} - \theta\right)$ $\tan\theta = \cot\left(\frac{\pi}{2} - \theta\right)$ $\cot\theta = \tan\left(\frac{\pi}{2} - \theta\right)$ $\sec\theta = \csc\left(\frac{\pi}{2} - \theta\right)$ $\csc\theta = \sec\left(\frac{\pi}{2} - \theta\right)$
Fundamental Identities	$\tan\theta = \frac{\sin\theta}{\cos\theta}$ $\sec\theta = \frac{1}{\cos\theta}$ $\csc\theta = \frac{1}{\sin\theta}$ $\cot\theta = \frac{1}{\tan\theta} = \frac{\cos\theta}{\sin\theta}$
Sum and Difference Identities	$\cos(\alpha + \beta) = \cos\alpha\cos\beta - \sin\alpha\sin\beta$ $\cos(\alpha - \beta) = \cos\alpha\cos\beta + \sin\alpha\sin\beta$ $\sin(\alpha + \beta) = \sin\alpha\cos\beta + \cos\alpha\sin\beta$ $\sin(\alpha - \beta) = \sin\alpha\cos\beta - \cos\alpha\sin\beta$ $\tan(\alpha + \beta) = \frac{\tan\alpha + \tan\beta}{1 - \tan\alpha\tan\beta}$ $\tan(\alpha - \beta) = \frac{\tan\alpha - \tan\beta}{1 + \tan\alpha\tan\beta}$
Double-Angle Formulas	$\sin(2\theta) = 2\sin\theta\cos\theta$ $\cos(2\theta) = \cos^2\theta - \sin^2\theta$ $\cos(2\theta) = 1 - 2\sin^2\theta$ $\cos(2\theta) = 2\cos^2\theta - 1$ $\tan(2\theta) = \frac{2\tan\theta}{1 - \tan^2\theta}$

Table A1

Half-Angle Formulas	$\sin\ \frac{\alpha}{2} = \pm\sqrt{\frac{1-\cos\ \alpha}{2}}$ $\cos\ \frac{\alpha}{2} = \pm\sqrt{\frac{1+\cos\ \alpha}{2}}$ $\tan\ \frac{\alpha}{2} = \pm\sqrt{\frac{1-\cos\ \alpha}{1+\cos\ \alpha}}$ $\tan\ \frac{\alpha}{2} = \frac{\sin\ \alpha}{1+\cos\ \alpha}$ $\tan\ \frac{\alpha}{2} = \frac{1-\cos\ \alpha}{\sin\ \alpha}$
Reduction Formulas	$\sin^2\theta = \frac{1-\cos(2\theta)}{2}$ $\cos^2\theta = \frac{1+\cos(2\theta)}{2}$ $\tan^2\theta = \frac{1-\cos(2\theta)}{1+\cos(2\theta)}$
Product-to-Sum Formulas	$\cos\ \alpha\ \cos\ \beta = \frac{1}{2}[\cos(\alpha-\beta)+\cos(\alpha+\beta)]$ $\sin\ \alpha\ \cos\ \beta = \frac{1}{2}[\sin(\alpha+\beta)+\sin(\alpha-\beta)]$ $\sin\ \alpha\ \sin\ \beta = \frac{1}{2}[\cos(\alpha-\beta)-\cos(\alpha+\beta)]$ $\cos\ \alpha\ \sin\ \beta = \frac{1}{2}[\sin(\alpha+\beta)-\sin(\alpha-\beta)]$
Sum-to-Product Formulas	$\sin\ \alpha+\sin\ \beta = 2\ \sin\left(\frac{\alpha+\beta}{2}\right)\ \cos\left(\frac{\alpha-\beta}{2}\right)$ $\sin\ \alpha-\sin\ \beta = 2\ \sin\left(\frac{\alpha-\beta}{2}\right)\ \cos\left(\frac{\alpha+\beta}{2}\right)$ $\cos\ \alpha-\cos\ \beta = -2\ \sin\left(\frac{\alpha+\beta}{2}\right)\ \sin\left(\frac{\alpha-\beta}{2}\right)$ $\cos\ \alpha+\cos\ \beta = 2\ \cos\left(\frac{\alpha+\beta}{2}\right)\ \cos\left(\frac{\alpha-\beta}{2}\right)$
Law of Sines	$\frac{\sin\ \alpha}{a} = \frac{\sin\ \beta}{b} = \frac{\sin\ \gamma}{c}$ $\frac{a}{\sin\ \alpha} = \frac{b}{\sin\ \beta} = \frac{c}{\sin\ \gamma}$
Law of Cosines	$a^2 = b^2 + c^2 - 2bc\ \cos\ \alpha$ $b^2 = a^2 + c^2 - 2ac\ \cos\ \beta$ $c^2 = a^2 + b^2 - 2ab\ \cos\ \gamma$

Table A1

ToolKit Functions

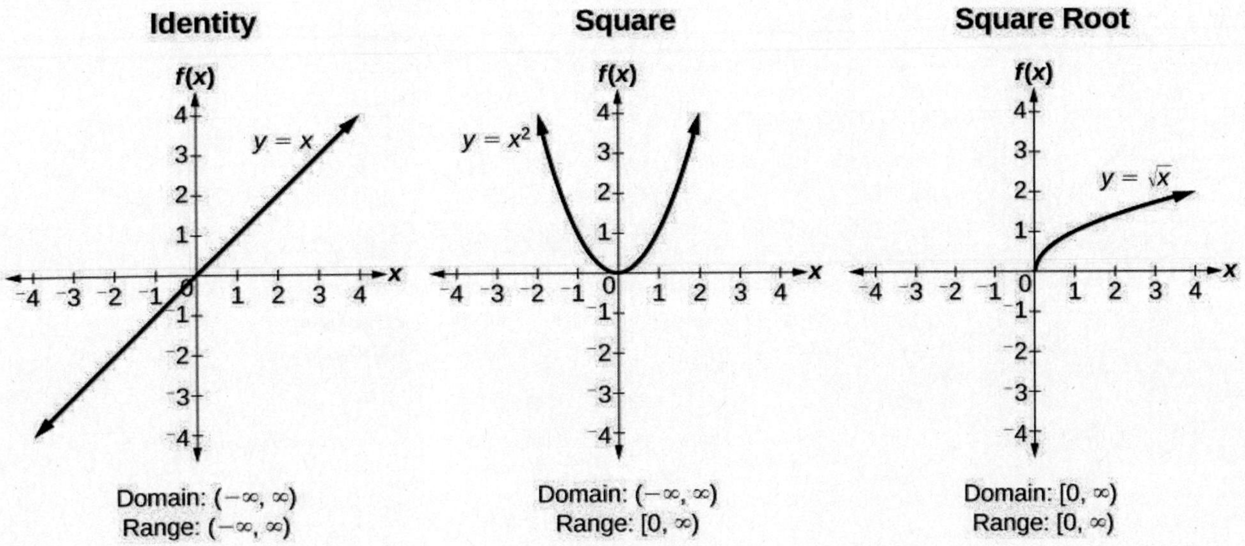

Figure A1

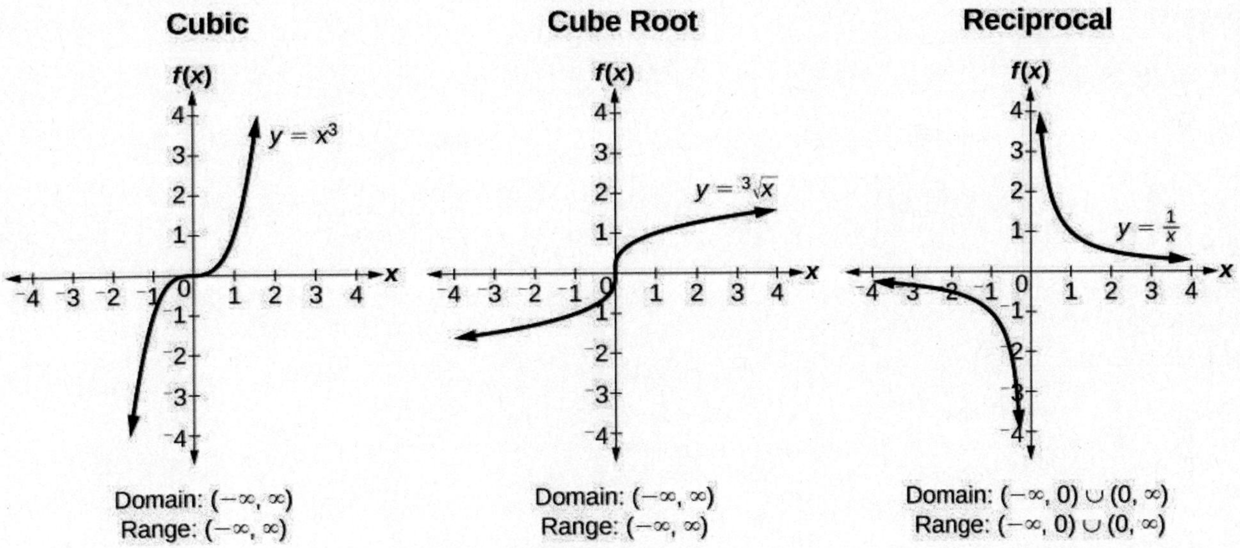

Figure A2

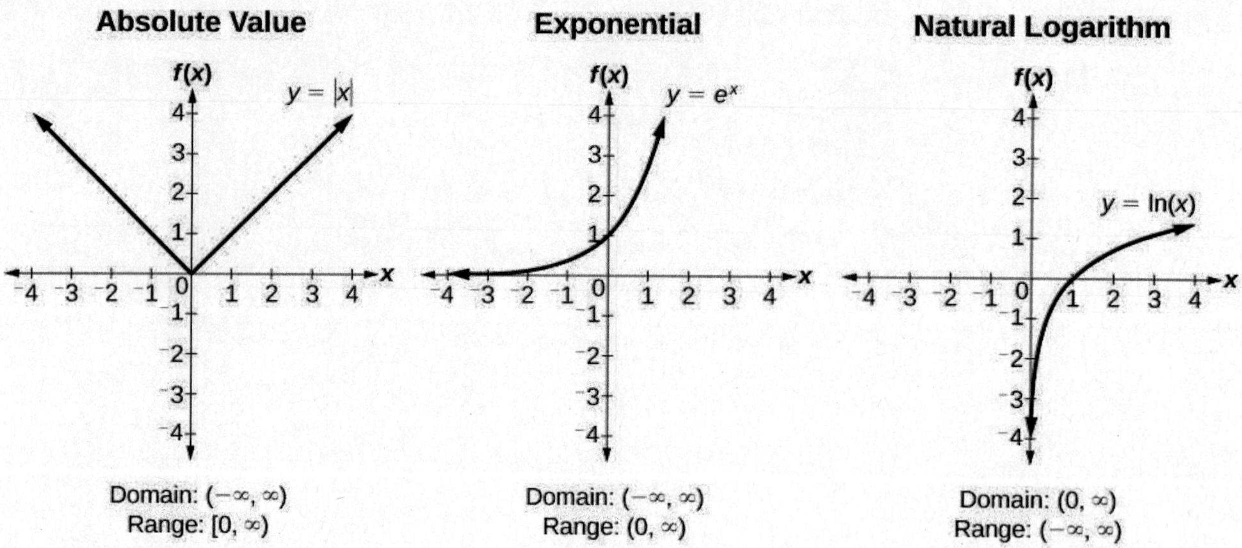

Figure A3

Trigonometric Functions

Unit Circle

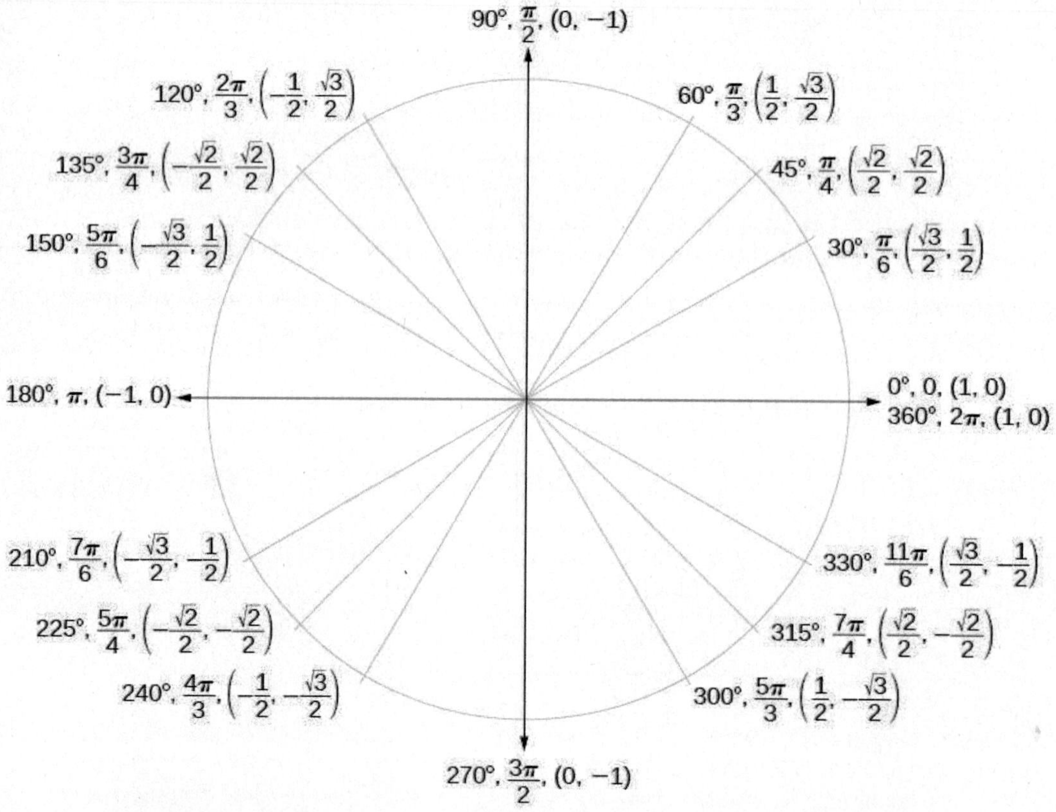

Figure A4

Angle	0	$\frac{\pi}{6}$, or 30°	$\frac{\pi}{4}$, or 45°	$\frac{\pi}{3}$, or 60°	$\frac{\pi}{2}$, or 90°
Cosine	1	$\frac{\sqrt{3}}{2}$	$\frac{\sqrt{2}}{2}$	$\frac{1}{2}$	0

Table A2

Angle	0	$\frac{\pi}{6}$, or 30°	$\frac{\pi}{4}$, or 45°	$\frac{\pi}{3}$, or 60°	$\frac{\pi}{2}$, or 90°
Sine	0	$\frac{1}{2}$	$\frac{\sqrt{2}}{2}$	$\frac{\sqrt{3}}{2}$	1
Tangent	0	$\frac{\sqrt{3}}{3}$	1	$\sqrt{3}$	Undefined
Secant	1	$\frac{2\sqrt{3}}{3}$	$\sqrt{2}$	2	Undefined
Cosecant	Undefined	2	$\sqrt{2}$	$\frac{2\sqrt{3}}{3}$	1
Cotangent	Undefined	$\sqrt{3}$	1	$\frac{\sqrt{3}}{3}$	0

Table A2

Answer Key

Chapter 1

Try It

1.1 Real Numbers: Algebra Essentials

1. ⓐ $\frac{11}{1}$ ⓑ $\frac{3}{1}$ ⓒ $-\frac{4}{1}$

2. ⓐ 4 (or 4.0), terminating; ⓑ $0.\overline{615384}$, repeating; ⓒ –0.85, terminating

3. ⓐ rational and repeating; ⓑ rational and terminating; ⓒ irrational; ⓓ rational and terminating; ⓔ irrational

4. ⓐ positive, irrational; right ⓑ negative, rational; left ⓒ positive, rational; right ⓓ negative, irrational; left ⓔ positive, rational; right

5.

	N	*W*	*I*	*Q*	*Q'*
a. $-\frac{35}{7}$			X	X	
b. 0		X	X	X	
c. $\sqrt{169}$	X	X	X	X	
d. $\sqrt{24}$					X
e. 4.763763763...				X	

6. ⓐ 10 ⓑ 2 ⓒ 4.5 ⓓ 25 ⓔ 26

7. ⓐ 11, commutative property of multiplication, associative property of multiplication, inverse property of multiplication, identity property of multiplication; ⓑ 33, distributive property; ⓒ 26, distributive property; ⓓ $\frac{4}{9}$, commutative property of addition, associative property of addition, inverse property of addition, identity property of addition; ⓔ 0, distributive property, inverse property of addition, identity property of addition

8.

	Constants	Variables
a. $2\pi r(r+h)$	$2, \pi$	r, h
b. 2(L + W)	2	L, W
c. $4y^3+y$	4	y

9. ⓐ 5; ⓑ 11; ⓒ 9; ⓓ 26

10. ⓐ 4; ⓑ 11; ⓒ $\frac{121}{3}\pi$; ⓓ 1728; ⓔ 3

11. 1,152 cm^2

12. ⓐ $-2y-2z$ or $-2(y+z)$; ⓑ $\frac{2}{t}-1$; ⓒ $3pq-4p+q$; ⓓ $7r-2s+6$

13. $A=P(1+rt)$

1.2 Exponents and Scientific Notation

1. ⓐ k^{15} ⓑ $\left(\frac{2}{y}\right)^5$ ⓒ t^{14}
2. ⓐ s^7 ⓑ $(-3)^5$ ⓒ $\left(ef^2\right)^2$
3. ⓐ $(3y)^{24}$ ⓑ t^{35} ⓒ $(-g)^{16}$
4. ⓐ 1 ⓑ $\frac{1}{2}$ ⓒ 1 ⓓ 1
5. ⓐ $\frac{1}{(-3t)^6}$ ⓑ $\frac{1}{f^3}$ ⓒ $\frac{2}{5k^3}$
6. ⓐ $t^{-5} = \frac{1}{t^5}$ ⓑ $\frac{1}{25}$
7. ⓐ $g^{10}h^{15}$ ⓑ $125t^3$ ⓒ $-27y^{15}$ ⓓ $\frac{1}{a^{18}b^{21}}$ ⓔ $\frac{r^{12}}{s^8}$
8. ⓐ $\frac{b^{15}}{c^3}$ ⓑ $\frac{625}{u^{32}}$ ⓒ $\frac{-1}{w^{105}}$ ⓓ $\frac{q^{24}}{p^{32}}$ ⓔ $\frac{1}{c^{20}d^{12}}$
9. ⓐ $\frac{v^6}{8u^3}$ ⓑ $\frac{1}{x^3}$ ⓒ $\frac{e^4}{f^4}$ ⓓ $\frac{27r}{s}$ ⓔ 1 ⓕ $\frac{16h^{10}}{49}$
10. ⓐ $\$1.52 \times 10^5$ ⓑ 7.158×10^9 ⓒ $\$8.55 \times 10^{13}$ ⓓ 3.34×10^{-9} ⓔ 7.15×10^{-8}
11. ⓐ $703,000$ ⓑ $-816,000,000,000$ ⓒ $-0.000\ 000\ 000\ 000\ 39$ ⓓ 0.000008
12. ⓐ -8.475×10^6 ⓑ 8×10^{-8} ⓒ 2.976×10^{13} ⓓ -4.3×10^6 ⓔ $\approx 1.24 \times 10^{15}$
13. Number of cells: 3×10^{13}; length of a cell: 8×10^{-6} m; total length: 2.4×10^8 m or $240,000,000$ m.

1.3 Radicals and Rational Exponents

1. ⓐ 15 ⓑ 3 ⓒ 4 ⓓ 17
2. $5\,|x|\,|y|\,\sqrt{2yz}$. Notice the absolute value signs around *x* and *y*? That's because their value must be positive!
3. $10\,|x|$
4. $\frac{x\sqrt{2}}{3y^2}$. We do not need the absolute value signs for y^2 because that term will always be nonnegative.
5. $b^4\sqrt{3ab}$
6. $13\sqrt{5}$
7. 0
8. $6\sqrt{6}$
9. $14-7\sqrt{3}$
10. ⓐ -6 ⓑ 6 ⓒ $88\sqrt[3]{9}$
11. $\left(\sqrt{9}\right)^5 = 3^5 = 243$
12. $x(5y)^{\frac{9}{2}}$
13. $28x^{\frac{23}{15}}$

1.4 Polynomials

1. The degree is 6, the leading term is $-x^6$, and the leading coefficient is -1.
2. $2x^3 + 7x^2 - 4x - 3$
3. $-11x^3 - x^2 + 7x - 9$

4. $3x^4 - 10x^3 - 8x^2 + 21x + 14$

5. $3x^2 + 16x - 35$

6. $16x^2 - 8x + 1$

7. $4x^2 - 49$

8. $6x^2 + 21xy - 29x - 7y + 9$

1.5 Factoring Polynomials

1. $(b^2 - a)(x + 6)$

2. $(x-6)(x-1)$

3. ⓐ $(2x + 3)(x + 3)$
ⓑ $(3x-1)(2x + 1)$

4. $(7x-1)^2$

5. $(9y + 10)(9y - 10)$

6. $(6a + b)(36a^2 - 6ab + b^2)$

7. $(10x - 1)\left(100x^2 + 10x + 1\right)$

8. $(5a-1)^{-\frac{1}{4}}(17a-2)$

1.6 Rational Expressions

1. $\frac{1}{x+6}$

2. $\frac{(x+5)(x+6)}{(x+2)(x+4)}$

3. 1

4. $\frac{2(x-7)}{(x+5)(x-3)}$

5. $\frac{x^2-y^2}{xy^2}$

1.1 Section Exercises

1. irrational number. The square root of two does not terminate, and it does not repeat a pattern. It cannot be written as a quotient of two integers, so it is irrational.

3. The Associative Properties state that the sum or product of multiple numbers can be grouped differently without affecting the result. This is because the same operation is performed (either addition or subtraction), so the terms can be re-ordered.

5. -6

7. -2

9. -9

11. 9

13. -2

15. 4

17. 0

19. 9

21. 25

23. -6

25. 17

27. 4

29. 14

31. -66

33. -12

35. -44

37. -2

39. $-14y - 11$

41. $-4b + 1$

43. $43z - 3$

45. $9y + 45$

47. $-6b + 6$

49. $\frac{16x}{3}$

51. $9x$

53. $\frac{1}{2}(40-10)+5$

55. irrational number

57. $g+400-2(600)=1200$

59. inverse property of addition

61. 68.4

63. true

65. irrational

67. rational

1.2 Section Exercises

1. No, the two expressions are not the same. An exponent tells how many times you multiply the base. So 2^3 is the same as $2 \times 2 \times 2$, which is 8. 3^2 is the same as 3×3, which is 9.

3. It is a method of writing very small and very large numbers.

5. 81

7. 243

9. $\frac{1}{16}$

11. $\frac{1}{11}$

13. 1

15. 4^9

17. 12^{40}

19. $\frac{1}{7^9}$

21. 3.14×10^{-5}

23. 16,000,000,000

25. a^4

27. b^6c^8

29. ab^2d^3

31. m^4

33. $\frac{q^5}{p^6}$

35. $\frac{y^{21}}{x^{14}}$

37. 25

39. $72a^2$

41. $\frac{c^3}{b^9}$

43. $\frac{y}{81z^6}$

45. 0.00135 m

47. 1.0995×10^{12}

49. 0.00000000003397 in.

51. 12,230,590,464 m^{66}

53. $\frac{a^{14}}{1296}$

55. $\frac{n}{a^9c}$

57. $\frac{1}{a^6b^6c^6}$

59. 0.000000000000000000000000000000000662606957

1.3 Section Exercises

1. When there is no index, it is assumed to be 2 or the square root. The expression would only be equal to the radicand if the index were 1.

3. The principal square root is the nonnegative root of the number.

5. 16

7. 10

9. 14

11. $7\sqrt{2}$

13. $\frac{9\sqrt{5}}{5}$

15. 25

17. $\sqrt{2}$

19. $2\sqrt{6}$

21. $5\sqrt{6}$

23. $6\sqrt{35}$

25. $\frac{2}{15}$

27. $\frac{6\sqrt{10}}{19}$

29. $-\frac{1+\sqrt{17}}{2}$

31. $7\sqrt[3]{2}$

33. $15\sqrt{5}$

35. $20x^2$

37. $7\sqrt{p}$

39. $17m^2\sqrt{m}$

41. $2b\sqrt{a}$

43. $\frac{15x}{7}$

45. $5y^4\sqrt{2}$

47. $\frac{4\sqrt{7d}}{7d}$

49. $\frac{2\sqrt{2}+2\sqrt{6x}}{1-3x}$

51. $-w\sqrt{2w}$

53. $\frac{3\sqrt{x}-\sqrt{3x}}{2}$

55. $5n^5\sqrt{5}$

57. $\frac{9\sqrt{m}}{19m}$

59. $\frac{2}{3d}$

61. $\frac{3\sqrt[4]{2x^2}}{2}$

63. $6z\sqrt[3]{2}$

65. 500 feet

67. $\frac{-5\sqrt{2}-6}{7}$

69. $\frac{\sqrt{mnc}}{a^9cmn}$

71. $\frac{2\sqrt{2}x+\sqrt{2}}{4}$

73. $\frac{\sqrt{3}}{3}$

1.4 Section Exercises

1. The statement is true. In standard form, the polynomial with the highest value exponent is placed first and is the leading term. The degree of a polynomial is the value of the highest exponent, which in standard form is also the exponent of the leading term.

3. Use the distributive property, multiply, combine like terms, and simplify.

5. 2

7. 8

9. 2

11. $4x^2 + 3x + 19$

13. $3w^2 + 30w + 21$

15. $11b^4 - 9b^3 + 12b^2 - 7b + 8$

17. $24x^2 - 4x - 8$

19. $24b^4 - 48b^2 + 24$

21. $99v^2 - 202v + 99$

23. $8n^3 - 4n^2 + 72n - 36$

25. $9y^2 - 42y + 49$

27. $16p^2 + 72p + 81$

29. $9y^2 - 36y + 36$

31. $16c^2 - 1$

33. $225n^2 - 36$

35. $-16m^2 + 16$

37. $121q^2 - 100$

39. $16t^4 + 4t^3 - 32t^2 - t + 7$

41. $y^3 - 6y^2 - y + 18$

43. $3p^3 - p^2 - 12p + 10$

45. $a^2 - b^2$

47. $16t^2 - 40tu + 25u^2$

49. $4t^2 + x^2 + 4t - 5tx - x$

51. $24r^2 + 22rd - 7d^2$

53. $32x^2 - 4x - 3\text{ m}^2$

55. $32t^3 - 100t^2 + 40t + 38$

57. $a^4 + 4a^3c - 16ac^3 - 16c^4$

1.5 Section Exercises

1. The terms of a polynomial do not have to have a common factor for the entire polynomial to be factorable. For example, $4x^2$ and $-9y^2$ don't have a common factor, but the whole polynomial is still factorable: $4x^2 - 9y^2 = (2x + 3y)(2x - 3y)$.

3. Divide the x term into the sum of two terms, factor each portion of the expression separately, and then factor out the GCF of the entire expression.

5. $7m$

7. $10m^3$

9. y

11. $(2a - 3)(a + 6)$

13. $(3n - 11)(2n + 1)$

15. $(p + 1)(2p - 7)$

17. $(5h + 3)(2h - 3)$

19. $(9d - 1)(d - 8)$

21. $(12t + 13)(t - 1)$

23. $(4x + 10)(4x - 10)$

25. $(11p + 13)(11p - 13)$

27. $(19d + 9)(19d - 9)$

29. $(12b + 5c)(12b - 5c)$

31. $(7n + 12)^2$

33. $(15y + 4)^2$

35. $(5p - 12)^2$

37. $(x + 6)\left(x^2 - 6x + 36\right)$

39. $(5a + 7)\left(25a^2 - 35a + 49\right)$

41. $(4x - 5)\left(16x^2 + 20x + 25\right)$

43. $(5r + 12s)\left(25r^2 - 60rs + 144s^2\right)$

45. $(2c + 3)^{-\frac{1}{4}}(-7c - 15)$

47. $(x + 2)^{-\frac{2}{5}}(19x + 10)$

49. $(2z - 9)^{-\frac{3}{2}}(27z - 99)$

51. $(14x - 3)(7x + 9)$

53. $(3x + 5)(3x - 5)$

55. $(2x+5)^2(2x-5)^2$

57. $\left(4z^2+49a^2\right)(2z+7a)(2z-7a)$

59. $\frac{1}{(4x+9)(4x-9)(2x+3)}$

1.6 Section Exercises

1. You can factor the numerator and denominator to see if any of the terms can cancel one another out.

3. True. Multiplication and division do not require finding the LCD because the denominators can be combined through those operations, whereas addition and subtraction require like terms.

5. $\frac{y+5}{y+6}$

7. $3b+3$

9. $\frac{x+4}{2x+2}$

11. $\frac{a+3}{a-3}$

13. $\frac{3n-8}{7n-3}$

15. $\frac{c-6}{c+6}$

17. 1

19. $\frac{d^2-25}{25d^2-1}$

21. $\frac{t+5}{t+3}$

23. $\frac{6x-5}{6x+5}$

25. $\frac{p+6}{4p+3}$

27. $\frac{2d+9}{d+11}$

29. $\frac{12b+5}{3b-1}$

31. $\frac{4y-1}{y+4}$

33. $\frac{10x+4y}{xy}$

35. $\frac{9a-7}{a^2-2a-3}$

37. $\frac{2y^2-y+9}{y^2-y-2}$

39. $\frac{5z^2+z+5}{z^2-z-2}$

41. $\frac{x+2xy+y}{x+xy+y+1}$

43. $\frac{2b+7a}{ab^2}$

45. $\frac{18+ab}{4b}$

47. $a-b$

49. $\frac{3c^2+3c-2}{2c^2+5c+2}$

51. $\frac{15x+7}{x-1}$

53. $\frac{x+9}{x-9}$

55. $\frac{1}{y+2}$

57. 4

Review Exercises

1. −5

3. 53

5. $y=24$

7. $32m$

9. whole

11. irrational

13. 16

15. $3a^6$

17. $\frac{x^3}{32y^3}$

19. a

21. 1.634×10^7

23. 14

25. $5\sqrt{3}$

27. $\frac{4\sqrt{2}}{5}$

29. $\frac{7\sqrt{2}}{50}$

31. $10\sqrt{3}$

33. -3

35. $3x^3 + 4x^2 + 6$

37. $5x^2 - x + 3$

39. $k^2 - 3k - 18$

41. $x^3 + x^2 + x + 1$

43. $3a^2 + 5ab - 2b^2$

45. $9p$

47. $4a^2$

49. $(4a - 3)(2a + 9)$

51. $(x + 5)^2$

53. $(2h - 3k)^2$

55. $(p + 6)\left(p^2 - 6p + 36\right)$

57. $(4q - 3p)\left(16q^2 + 12pq + 9p^2\right)$

59. $(p + 3)^{\frac{1}{3}}(-5p - 24)$

61. $\frac{x+3}{x-4}$

63. $\frac{1}{2}$

65. $\frac{m+2}{m-3}$

67. $\frac{6x+10y}{xy}$

69. $\frac{1}{6}$

Practice Test

1. rational

3. $x = -2$

5. 3,141,500

7. 16

9. 9

11. $2x$

13. 21

15. $\frac{3\sqrt{x}}{4}$

17. $21\sqrt{6}$

19. $13q^3 - 4q^2 - 5q$

21. $n^3 - 6n^2 + 12n - 8$

23. $(4x + 9)(4x - 9)$

25. $(3c - 11)\left(9c^2 + 33c + 121\right)$

27. $\frac{4z-3}{2z-1}$

29. $\frac{3a+2b}{3b}$

Chapter 2

Try It

2.1 The Rectangular Coordinate Systems and Graphs

1.

x	$y = \frac{1}{2}x + 2$	(x, y)
-2	$y = \frac{1}{2}(-2) + 2 = 1$	$(-2, 1)$
-1	$y = \frac{1}{2}(-1) + 2 = \frac{3}{2}$	$\left(-1, \frac{3}{2}\right)$
0	$y = \frac{1}{2}(0) + 2 = 2$	$(0, 2)$
1	$y = \frac{1}{2}(1) + 2 = \frac{5}{2}$	$\left(1, \frac{5}{2}\right)$
2	$y = \frac{1}{2}(2) + 2 = 3$	$(2, 3)$

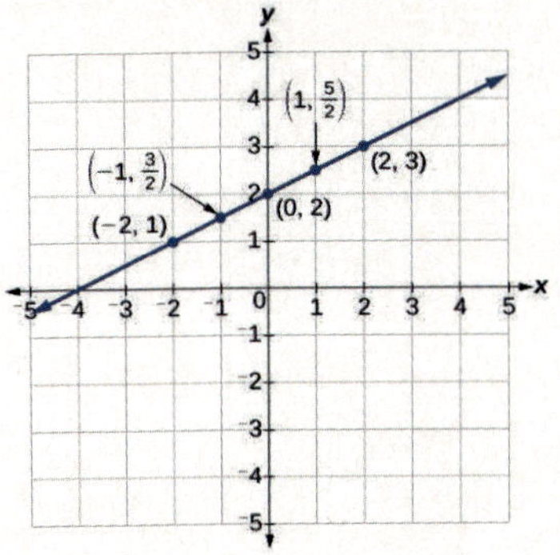

2. x-intercept is $(4, 0)$; y-intercept is $(0, 3)$.

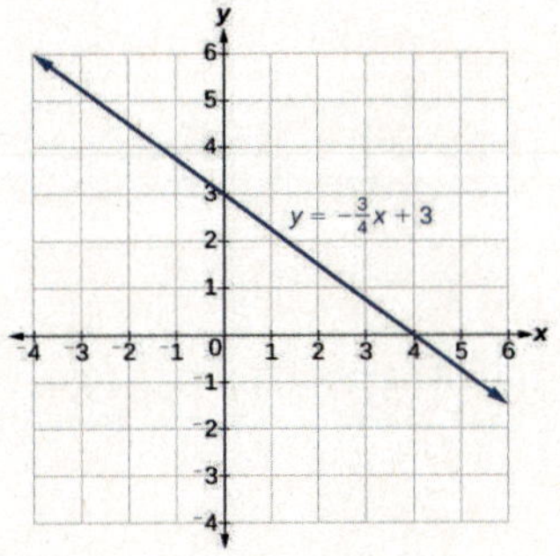

3. $\sqrt{125} = 5\sqrt{5}$

4. $\left(-5, \frac{5}{2}\right)$

2.2 Linear Equations in One Variable

1. $x = -5$

2. $x = -3$

3. $x = \frac{10}{3}$

4. $x = 1$

5. $x = -\frac{7}{17}$. Excluded values are $x = -\frac{1}{2}$ and $x = -\frac{1}{3}$.

6. $x = \frac{1}{3}$

7. $m = -\frac{2}{3}$

8. $y = 4x-3$

9. $x + 3y = 2$

10. Horizontal line: $y = 2$

11. Parallel lines: equations are written in slope-intercept form.

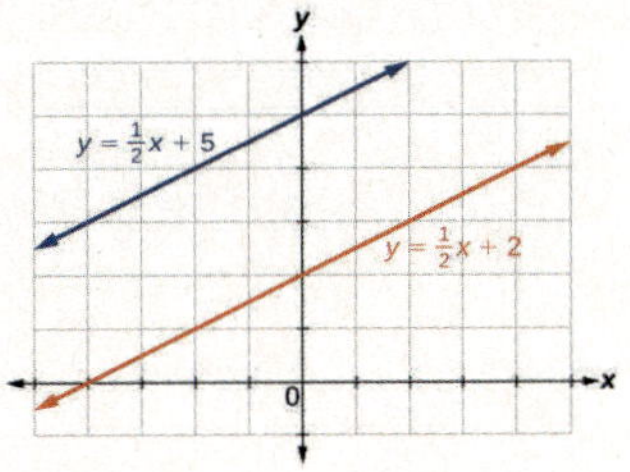

12. $y = 5x + 3$

2.3 Models and Applications

1. 11 and 25
2. $C = 2.5x + 3,650$
3. 45 mi/h
4. $L = 37$ cm, $W = 18$ cm
5. 250 ft^2

2.4 Complex Numbers

1. $\sqrt{-24} = 0 + 2i\sqrt{6}$
2. 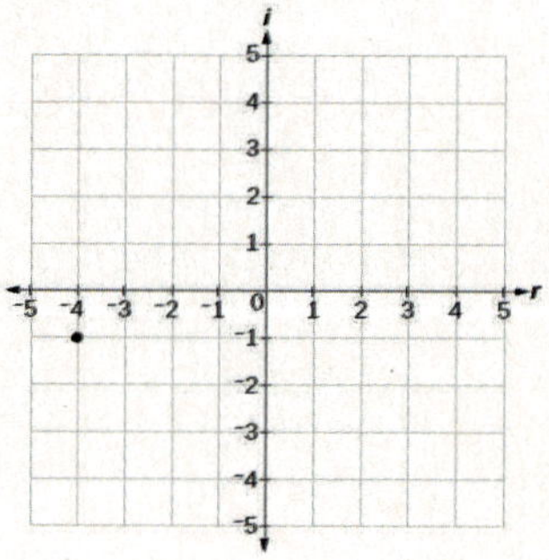

3. $(3-4i) - (2 + 5i) = 1-9i$
4. $\frac{5}{2} - i$
5. $18 + i$
6. $-3-4i$
7. -1

2.5 Quadratic Equations

1. $(x - 6)(x + 1) = 0; x = 6, x = -1$
2. $(x-7)(x + 3) = 0, x = 7,$ $x = -3.$
3. $(x + 5)(x-5) = 0, x = -5,$ $x = 5.$
4. $(3x + 2)(4x + 1) = 0,$ $x = -\frac{2}{3}, x = -\frac{1}{4}$
5. $x = 0, x = -10, x = -1$
6. $x = 4 \pm \sqrt{5}$
7. $x = 3 \pm \sqrt{22}$
8. $x = -\frac{2}{3}, x = \frac{1}{3}$
9. 5 units

2.6 Other Types of Equations

1. $\frac{1}{4}$
2. 25
3. $\{-1\}$
4. $0, \frac{1}{2}, -\frac{1}{2}$
5. 1; extraneous solution $-\frac{2}{9}$
6. -2; extraneous solution -1
7. $-1, \frac{3}{2}$
8. $-3, 3, -i, i$
9. $2, 12$
10. $-1, 0$ is not a solution.

2.7 Linear Inequalities and Absolute Value Inequalities

1. $[-3, 5]$

2. $\left(-\infty, -2\right) \cup \left[3, \infty\right)$

3. $x < 1$

4. $x \geq -5$

5. $\left(2, \infty\right)$

6. $\left[-\frac{3}{14}, \infty\right)$

7. $6 < x \leq 9$ or $(6, 9]$

8. $\left(-\frac{1}{8}, \frac{1}{2}\right)$

9. $|x-2| \leq 3$

10. $k \leq 1$ or $k \geq 7$; in interval notation, this would be $(-\infty, 1] \cup [7, \infty)$.

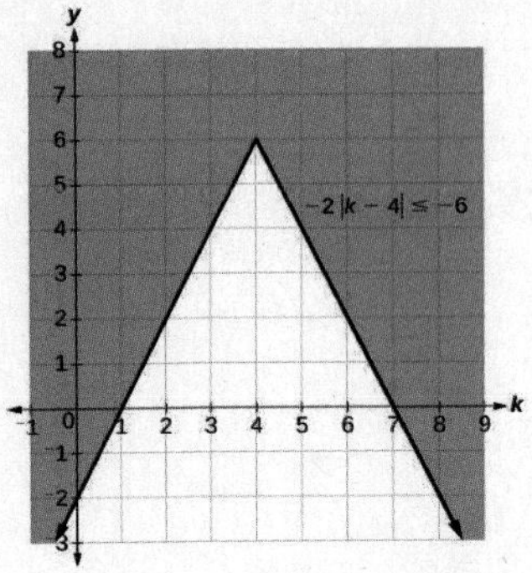

2.1 Section Exercises

1. Answers may vary. Yes. It is possible for a point to be on the *x*-axis or on the *y*-axis and therefore is considered to NOT be in one of the quadrants.

3. The *y*-intercept is the point where the graph crosses the *y*-axis.

5. The *x*-intercept is $(2, 0)$ and the *y*-intercept is $(0, 6)$.

7. The *x*-intercept is $(2, 0)$ and the *y*-intercept is $(0, -3)$.

9. The *x*-intercept is $(3, 0)$ and the *y*-intercept is $\left(0, \frac{9}{8}\right)$.

11. $y = 4 - 2x$

13. $y = \frac{5-2x}{3}$

15. $y = 2x - \frac{4}{5}$

17. $d = \sqrt{74}$

19. $d = \sqrt{36} = 6$

21. $d \approx 62.97$

23. $\left(3, \frac{-3}{2}\right)$

25. $(2, -1)$

27. $(0, 0)$

29. $y = 0$

31.

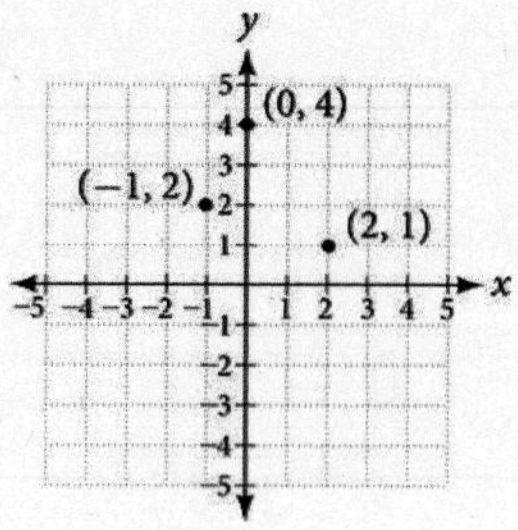

not collinear

33. A: $(-3, 2)$, B: $(1, 3)$, C: $(4, 0)$

35.

x	y
-3	1
0	2
3	3
6	4

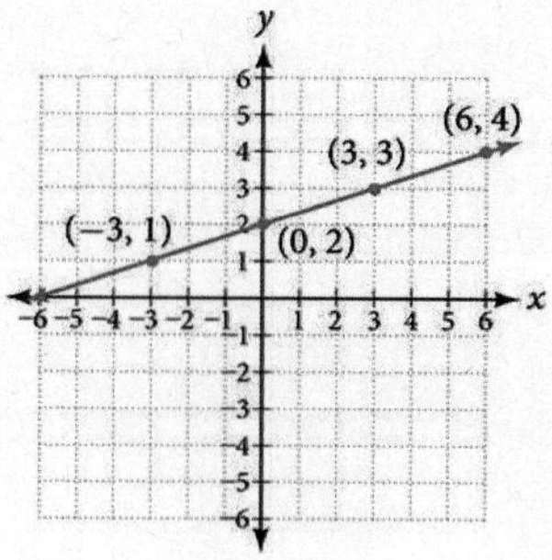

37.

x	y
-3	0
0	1.5
3	3

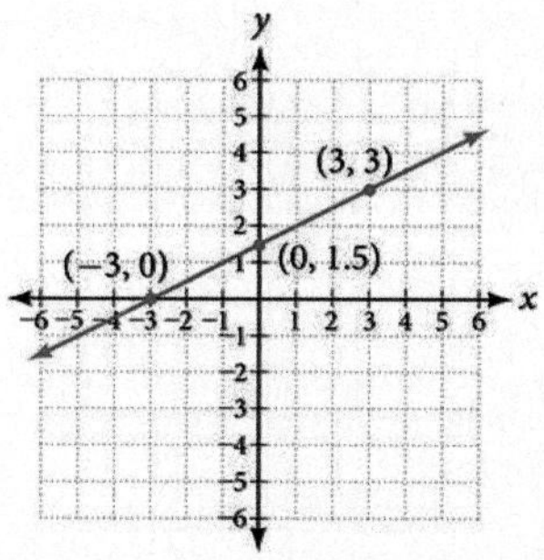

39.

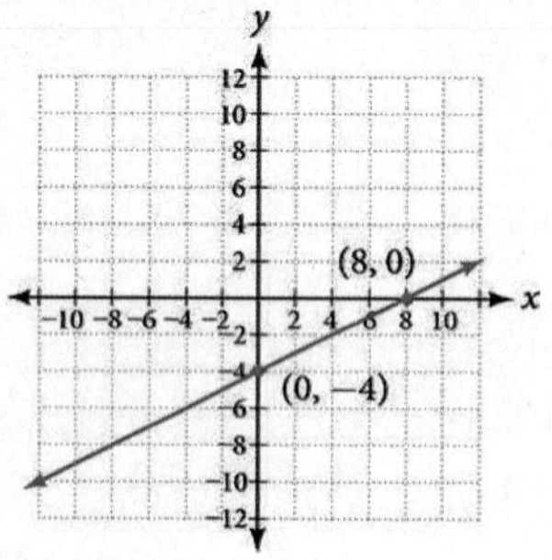

41.

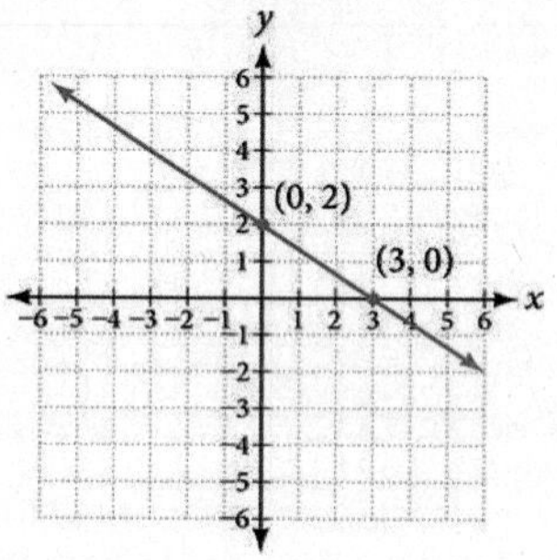

43. $d = 8.246$

45. $d = 5$

47. $(-3, 4)$

49. $x = 0$ $y = -2$

51. $x = 0.75y = 0$

53. $x = -1.667y = 0$

55. $15 - 11.2 = 3.8$mi shorter

57. 6.042

59. Midpoint of each diagonal is the same point $(2, -2)$. Note this is a characteristic of rectangles, but not other quadrilaterals.

61. 37mi

63. 54 ft

2.2 Section Exercises

1. It means they have the same slope.

3. The exponent of the x variable is 1. It is called a first-degree equation.

5. If we insert either value into the equation, they make an expression in the equation undefined (zero in the denominator).

7. $x = 2$

9. $x = \frac{2}{7}$

11. $x = 6$

13. $x = 3$

15. $x = -14$

17. $x \neq -4; x = -3$

19. $x \neq 1$; when we solve this we get $x = 1$, which is excluded, therefore NO solution

21. $x \neq 0; x = -\frac{5}{2}$

23. $y = -\frac{4}{5}x + \frac{14}{5}$

25. $y = -\frac{3}{4}x + 2$

27. $y = \frac{1}{2}x + \frac{5}{2}$

29. $y = -3x - 5$

31. $y = 7$

33. $y = -4$

35. $8x + 5y = 7$

37.

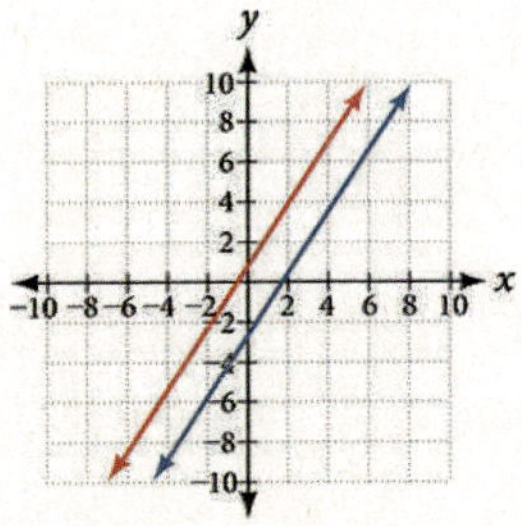

Parallel

39.

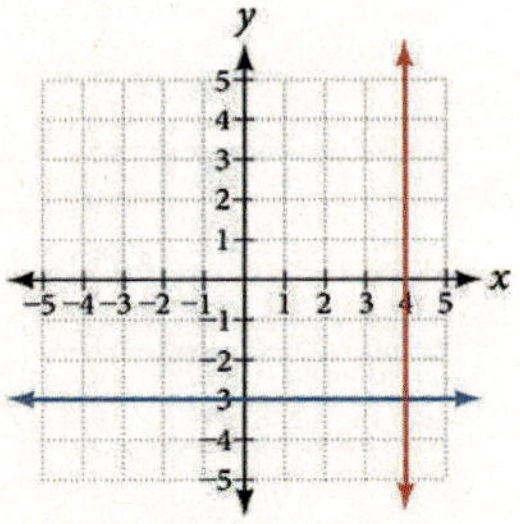

Perpendicular

41. $m = -\frac{9}{7}$

43. $m = \frac{3}{2}$

45. $m_1 = -\frac{1}{3}, \quad m_2 = 3; \quad \text{Perpendicular.}$

47. $y = 0.245x - 45.662.$ Answers may vary. $y_{\min} = -50, \quad y_{\max} = -40$

49. $y = -2.333x + 6.667.$ Answers may vary. $y_{\min} = -10, \quad y_{\max} = 10$

51. $y = -\frac{A}{B}x + \frac{C}{B}$

53. The slope for $(-1, 1)$ to $(0, 4)$ is 3.
The slope for $(-1, 1)$ to $(2, 0)$ is $\frac{-1}{3}$.
The slope for $(2, 0)$ to $(3, 3)$ is 3.
The slope for $(0, 4)$ to $(3, 3)$ is $\frac{-1}{3}$.
Yes they are perpendicular.

55. 30 ft

57. $57.50

59. 220 mi

2.3 Section Exercises

1. Answers may vary. Possible answers: We should define in words what our variable is representing. We should declare the variable. A heading.

3. $2,000 - x$

5. $v + 10$

7. Ann: 23; Beth: 46

9. $20 + 0.05m$

11. 300 min

13. $90 + 40P$

15. 6 devices

17. $50,000 - x$

19. 4 h

21. She traveled for 2 h at 20 mi/h, or 40 miles.

23. \$5,000 at 8% and \$15,000 at 12%

25. $B = 100 + .05x$

27. Plan A

29. $R = 9$

31. $r = \frac{4}{5}$ or 0.8

33. $W = \frac{P-2L}{2} = \frac{58-2(15)}{2} = 14$

35. $f = \frac{pq}{p+q} = \frac{8(13)}{8+13} = \frac{104}{21}$

37. $m = \frac{-5}{4}$

39. $h = \frac{2A}{b_1+b_2}$

41. length = 360 ft; width = 160 ft

43. 405 mi

45. $A = 88 \text{ in.}^2$

47. 28.7

49. $h = \frac{V}{\pi r^2}$

51. $r = \sqrt{\frac{V}{\pi h}}$

53. $C = 12\pi$

2.4 Section Exercises

1. Add the real parts together and the imaginary parts together.

3. Possible answer: i times i equals -1, which is not imaginary.

5. $-8 + 2i$

7. $14 + 7i$

9. $-\frac{23}{29} + \frac{15}{29}i$

11.

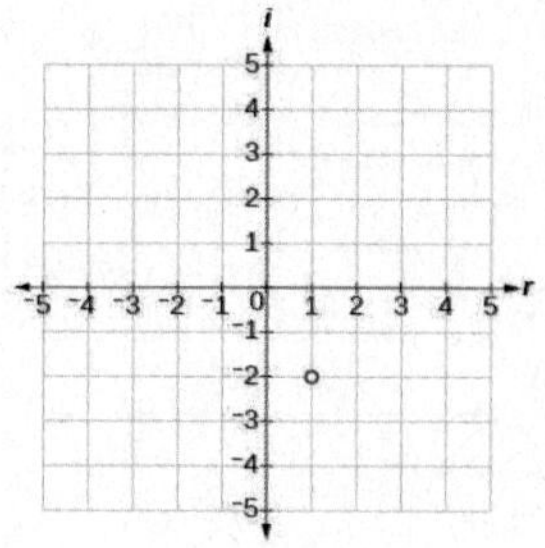

13.

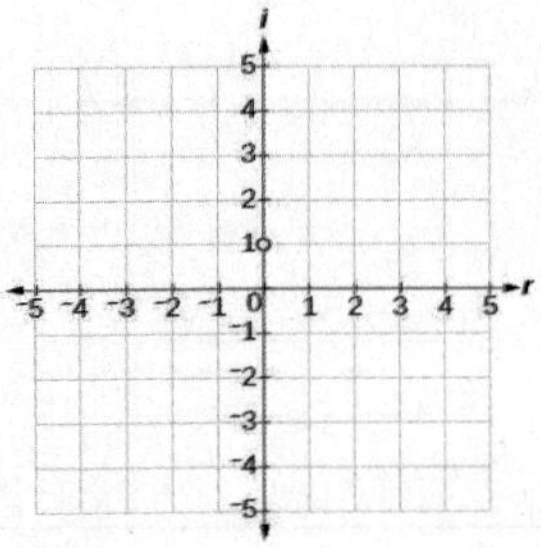

15. $8 - i$

17. $-11 + 4i$

19. $2{-}5i$

21. $6 + 15i$

23. $-16 + 32i$

25. $-4{-}7i$

27. 25

29. $2 - \frac{2}{3}i$

31. $4 - 6i$

33. $\frac{2}{5} + \frac{11}{5}i$

35. $15i$

37. $1 + i\sqrt{3}$

39. 1

41. -1

43. 128i

45. $\left(\frac{\sqrt{3}}{2} + \frac{1}{2}i\right)^6 = -1$

47. $3i$

49. 0

51. $5{-}5i$

53. $-2i$

55. $\frac{9}{2} - \frac{9}{2}i$

2.5 Section Exercises

1. It is a second-degree equation (the highest variable exponent is 2).

3. We want to take advantage of the zero property of multiplication in the fact that if $a \cdot b = 0$ then it must follow that each factor separately offers a solution to the product being zero: $a = 0 \;\; or \;\; \mathrm{b} = 0.$

5. One, when no linear term is present (no *x* term), such as $x^2 = 16$. Two, when the equation is already in the form $(ax + b)^2 = d$.

7. $x = 6, x = 3$

9. $x = \frac{-5}{2}, x = \frac{-1}{3}$

11. $x = 5, x = -5$

13. $x = \frac{-3}{2}, x = \frac{3}{2}$

15. $x = -2, 3$

17. $x = 0, x = \frac{-3}{7}$

19. $x = -6, x = 6$

21. $x = 6, x = -4$

23. $x = 1, x = -2$

25. $x = -2, x = 11$

27. $x = 3 \pm \sqrt{22}$

29. $z = \frac{2}{3}, z = -\frac{1}{2}$

31. $x = \frac{3 \pm \sqrt{17}}{4}$

33. Not real

35. One rational

37. Two real; rational

39. $x = \frac{-1 \pm \sqrt{17}}{2}$

41. $x = \frac{5 \pm \sqrt{13}}{6}$

43. $x = \frac{-1 \pm \sqrt{17}}{8}$

45. $x \approx 0.131$ and $x \approx 2.535$

47. $x \approx -6.7$ and $x \approx 1.7$

49.

$$\begin{aligned} ax^2 + bx + c &= 0 \\ x^2 + \tfrac{b}{a}x &= \tfrac{-c}{a} \\ x^2 + \tfrac{b}{a}x + \tfrac{b^2}{4a^2} &= \tfrac{-c}{a} + \tfrac{b}{4a^2} \\ \left(x + \tfrac{b}{2a}\right)^2 &= \tfrac{b^2 - 4ac}{4a^2} \\ x + \tfrac{b}{2a} &= \pm\sqrt{\tfrac{b^2 - 4ac}{4a^2}} \\ x &= \tfrac{-b \pm \sqrt{b^2 - 4ac}}{2a} \end{aligned}$$

51. $x(x + 10) = 119$; 7 ft. and 17 ft.

53. maximum at $x = 70$

55. The quadratic equation would be $(100x - 0.5x^2) - (60x + 300) = 300.$ The two values of x are 20 and 60.

57. 3 feet

2.6 Section Exercises

1. This is not a solution to the radical equation, it is a value obtained from squaring both sides and thus changing the signs of an equation which has caused it not to be a solution in the original equation.

3. They are probably trying to enter negative 9, but taking the square root of -9 is not a real number. The negative sign is in front of this, so your friend should be taking the square root of 9, cubing it, and then putting the negative sign in front, resulting in -27.

5. A rational exponent is a fraction: the denominator of the fraction is the root or index number and the numerator is the power to which it is raised.

7. $x = 81$

9. $x = 17$

11. $x = 8, \quad x = 27$

13. $x = -2, 1, -1$

15. $y = 0, \ \frac{3}{2}, \ \frac{-3}{2}$

17. $m = 1, -1$

19. $x = \frac{2}{5}, \pm 3i$

21. $x = 32$

23. $t = \frac{44}{3}$

25. $x = 3$

27. $x = -2$

29. $x = 4, \frac{-4}{3}$

31. $x = \frac{-5}{4}, \frac{7}{4}$

33. $x = 3, -2$

35. $x = -5$

37. $x = 1, -1, 3, -3$

39. $x = 2, -2$

41. $x = 1, 5$

43. $x \geq 0$

45. $x = 4, 6, -6, -8$

47. 10 in.

49. 90 kg

2.7 Section Exercises

1. When we divide both sides by a negative it changes the sign of both sides so the sense of the inequality sign changes.

3. $\left(-\infty, \infty\right)$

5. We start by finding the *x*-intercept, or where the function = 0. Once we have that point, which is $(3, 0)$, we graph to the right the straight line graph $y = x-3$, and then when we draw it to the left we plot positive *y* values, taking the absolute value of them.

7. $\left(-\infty, \frac{3}{4}\right]$

9. $\left[-\frac{13}{2}, \infty\right)$

11. $\left(-\infty, 3\right)$

13. $\left(-\infty, -\frac{37}{3}\right]$

15. All real numbers $\left(-\infty, \infty\right)$

17. $\left(-\infty, -\frac{10}{3}\right) \cup \left(4, \infty\right)$

19. $\left(-\infty, -4\right] \cup \left[8, +\infty\right)$

21. No solution

23. $(-5, 11)$

25. $[6, 12]$

27. $[-10, 12]$

29.
$$\begin{array}{ll} x > -6 \text{ and } x > -2 & \text{Take the intersection of two sets.} \\ x > -2, \quad (-2, +\infty) & \end{array}$$

31.
$$\begin{array}{ll} x < -3 \quad \text{or} \quad x \geq 1 & \text{Take the union of the two sets.} \\ (-\infty, -3) \cup \quad [1, \infty) & \end{array}$$

33. $\left(-\infty, -1\right) \cup \left(3, \infty\right)$

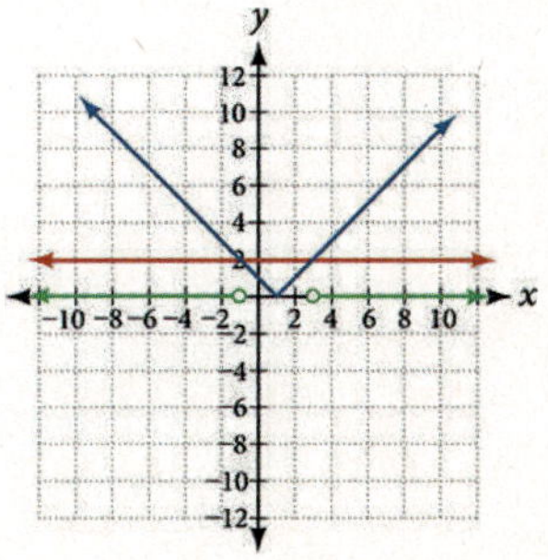

35. $[-11, -3]$

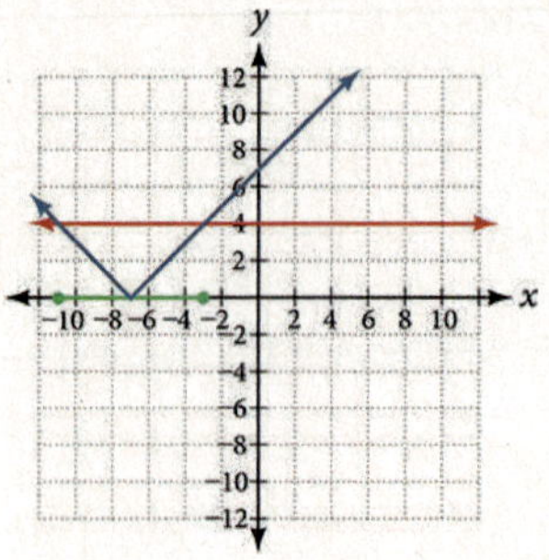

37. It is never less than zero. No solution.

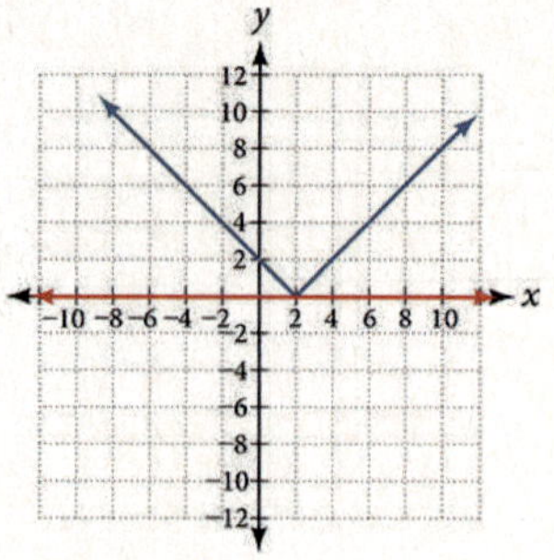

39. Where the blue line is above the orange line; point of intersection is $x = -3$.

$\left(-\infty, -3\right)$

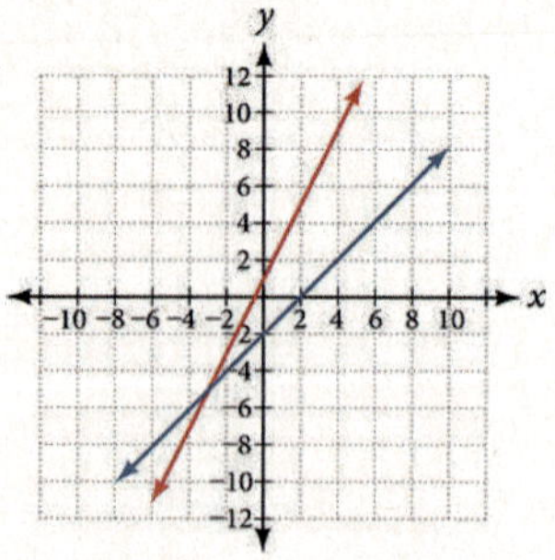

41. Where the blue line is above the orange line; always. All real numbers.

$(-\infty, -\infty)$

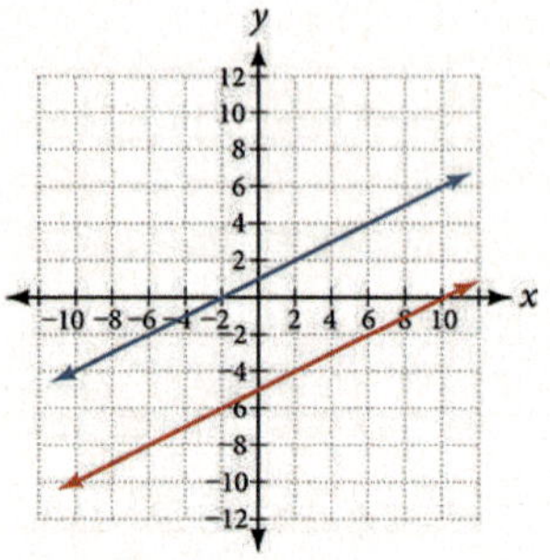

43. $(-1, 3)$

45. $\left(-\infty, 4\right)$

47. $\{x | x < 6\}$

49. $\{x | -3 \le x < 5\}$

51. $(-2, 1]$

53. $\left(-\infty, 4\right]$

55. Where the blue is below the orange; always. All real numbers. $(-\infty, +\infty)$.

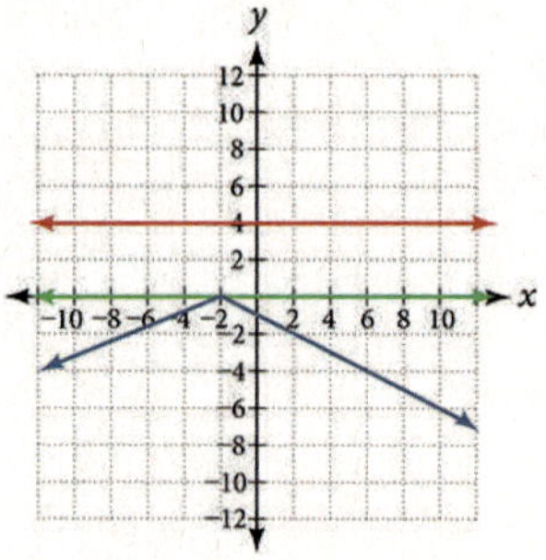

57. Where the blue is below the orange; $(1, 7)$.

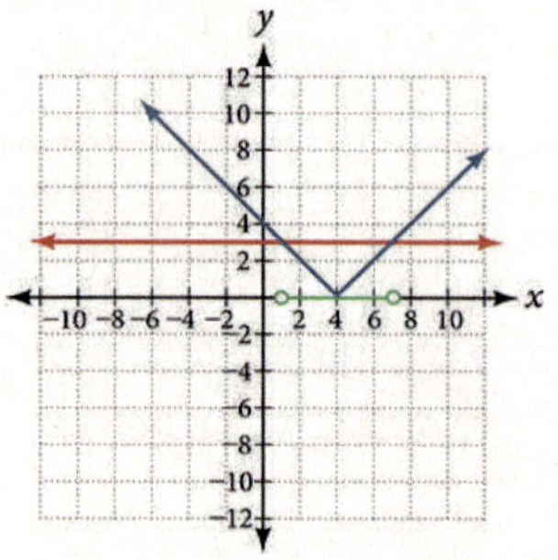

59. $x = 2, \frac{-4}{5}$

61. $(-7, 5]$

63. $80 \leq T \leq 120$
$1,600 \leq 20T \leq 2,400$
$[1,600, 2,400]$

Review Exercises

1. *x*-intercept: $(3, 0)$; *y*-intercept: $(0, -4)$

3. $y = \frac{5}{3}x + 4$

5. $\sqrt{72} = 6\sqrt{2}$

7. 620.097

9. midpoint is $\left(2, \frac{23}{2}\right)$

11.

x	*y*
0	−2
3	2
6	6

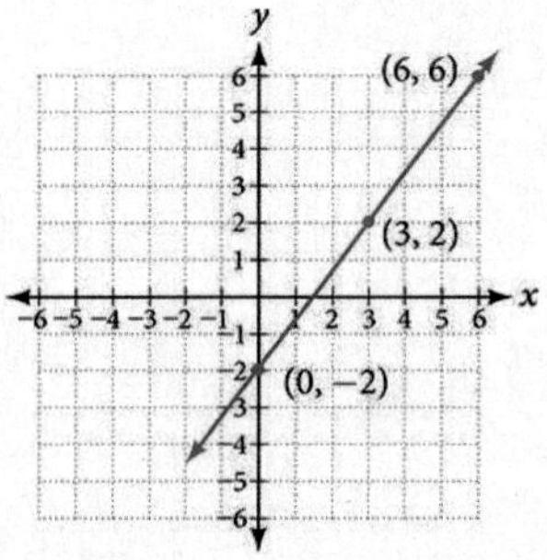

13. $x = 4$

15. $x = \frac{12}{7}$

17. No solution

19. $y = \frac{1}{6}x + \frac{4}{3}$

21. $y = \frac{2}{3}x + 6$

23. females 17, males 56

25. 84 mi

27. $x = -\frac{3}{4} \pm \frac{i\sqrt{47}}{4}$

29. horizontal component -2; vertical component -1

31. $7 + 11i$

33. $16i$

35. $-16 - 30i$

37. $-4 - i\sqrt{10}$

39. $x = 7 - 3i$

41. $x = -1, -5$

43. $x = 0, \frac{9}{7}$

45. $x = 10, -2$

47. $x = \frac{-1 \pm \sqrt{5}}{4}$

49. $x = \frac{2}{5}, \frac{-1}{3}$

51. $x = 5 \pm 2\sqrt{7}$

53. $x = 0, 256$

55. $x = 0, \pm\sqrt{2}$

57. $x = -2$

59. $x = \frac{11}{2}, \frac{-17}{2}$

61. $(-\infty, 4)$

63. $\left[\frac{-10}{3}, 2\right]$

65. No solution

67. $\left(-\frac{4}{3}, \frac{1}{5}\right)$

69. Where the blue is below the orange line; point of intersection is $x = 3.5$.

$(3.5, \infty)$

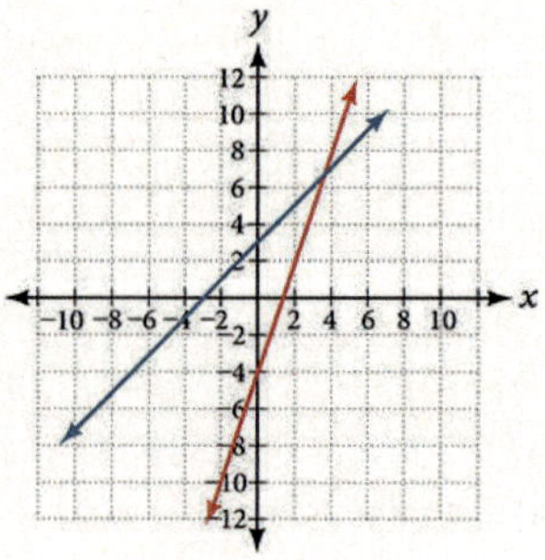

Practice Test

1. $y = \frac{3}{2}x + 2$

x	y
0	2
2	5
4	8

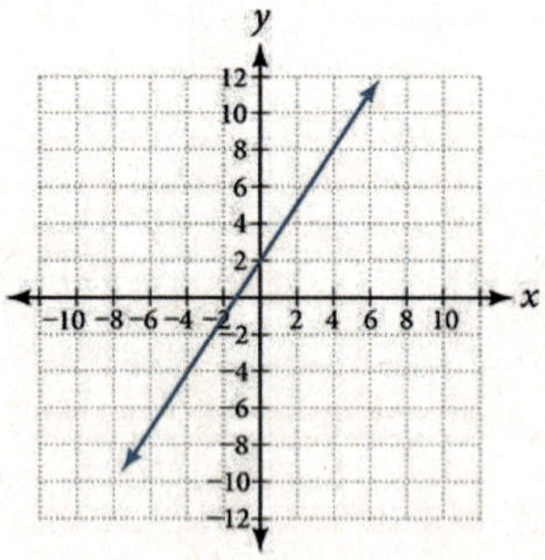

3. $(0, -3)$ $(4, 0)$

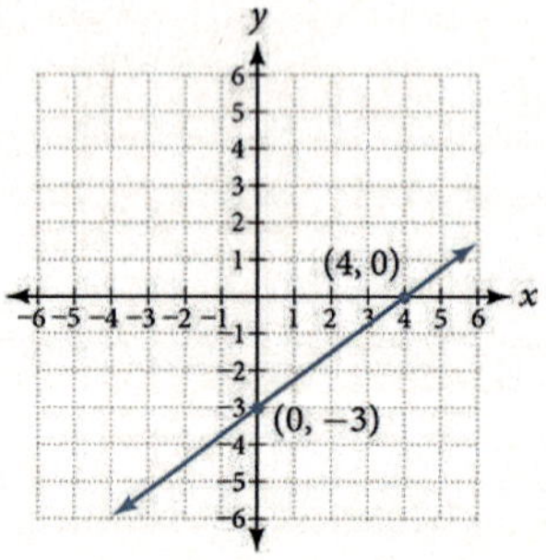

5. $(-\infty, 9]$

7. $x = -15$

9. $x \neq -4, 2;\ x = \frac{-5}{2}, 1$

11. $x = \frac{3 \pm \sqrt{3}}{2}$

13. $(-4, 1)$

15. $y = \frac{-5}{9}x - \frac{2}{9}$

17. $y = \frac{5}{2}x - 4$

19. $14i$

21. $\frac{5}{13} - \frac{14}{13}i$

23. $x = 2, \frac{-4}{3}$

25. $x = \frac{1}{2} \pm \frac{\sqrt{2}}{2}$

27. 4

29. $x = \frac{1}{2}, 2, -2$

Chapter 3

Try It

3.1 Functions and Function Notation

1. ⓐ yes
ⓑ yes (Note: If two players had been tied for, say, 4th place, then the name would not have been a function of rank.)

2. $w = f(d)$

3. yes

4. $g(5) = 1$

5. $m = 8$

6. $y = f(x) = \frac{\sqrt[3]{x}}{2}$

7. $g(1) = 8$

8. $x = 0$ or $x = 2$

9. ⓐ yes, because each bank account has a single balance at any given time;
ⓑ no, because several bank account numbers may have the same balance;
ⓒ no, because the same output may correspond to more than one input.

10. ⓐ Yes, letter grade is a function of percent grade;
ⓑ No, it is not one-to-one. There are 100 different percent numbers we could get but only about five possible letter grades, so there cannot be only one percent number that corresponds to each letter grade.

11. yes

12. No, because it does not pass the horizontal line test.

3.2 Domain and Range

1. $\{-5,\ 0,\ 5,\ 10,\ 15\}$

2. $\left(-\infty, \infty\right)$

3. $\left(-\infty, \frac{1}{2}\right) \cup \left(\frac{1}{2}, \infty\right)$

4. $\left[-\frac{5}{2}, \infty\right)$

5. ⓐ values that are less than or equal to –2, or values that are greater than or equal to –1 and less than 3
ⓑ
$\{x|x \le -2 \text{ or } -1 \le x < 3\}$
ⓒ $(-\infty, -2] \cup [-1, 3)$

6. domain =[1950,2002] range = [47,000,000,89,000,000]

7. domain: $(-\infty, 2]$; range: $(-\infty, 0]$

8.

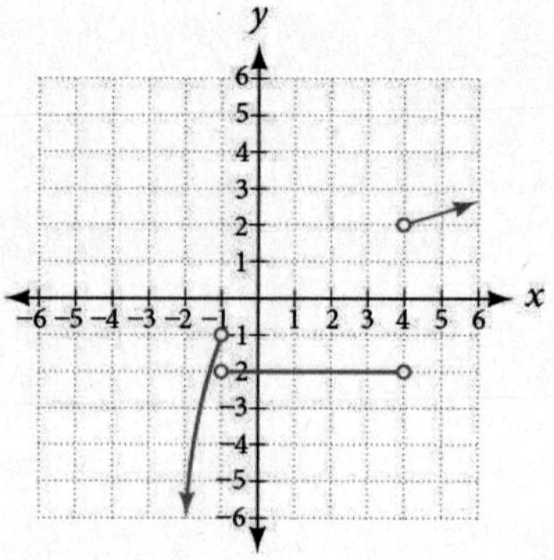

3.3 Rates of Change and Behavior of Graphs

1. $\frac{\$2.84-\$2.31}{5\text{ years}} = \frac{\$0.53}{5\text{ years}} = \$0.106$ per year.

2. $\frac{1}{2}$

3. $a + 7$

4. The local maximum appears to occur at $(-1, 28)$, and the local minimum occurs at $(5, -80)$. The function is increasing on $(-\infty, -1) \cup (5, \infty)$ and decreasing on $(-1, 5)$.

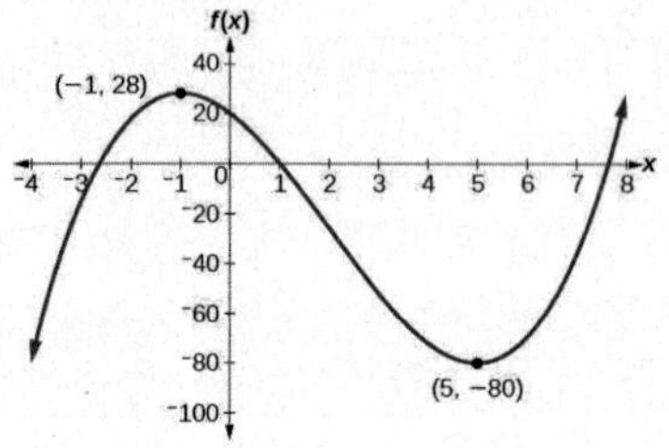

3.4 Composition of Functions

1. $(fg)(x) = f(x)g(x) = (x-1)(x^2-1) = x^3 - x^2 - x + 1$
$(f-g)(x) = f(x) - g(x) = (x-1) - (x^2-1) = x - x^2$

No, the functions are not the same.

2. A gravitational force is still a force, so $a(G(r))$ makes sense as the acceleration of a planet at a distance r from the Sun (due to gravity), but $G(a(F))$ does not make sense.

3. $f(g(1)) = f(3) = 3$ and $g(f(4)) = g(1) = 3$

4. $g(f(2)) = g(5) = 3$

5. ⓐ 8 ⓑ 20

6. $[-4, 0) \cup \left(0, \infty\right)$

7. Possible answer:

$g(x) = \sqrt{4 + x^2}$
$h(x) = \frac{4}{3-x}$
$f = h \circ g$

3.5 Transformation of Functions

1. $b(t) = h(t) + 10 = -4.9t^2 + 30t + 10$

2. The graphs of $f(x)$ and $g(x)$ are shown below. The transformation is a horizontal shift. The function is shifted to the left by 2 units.

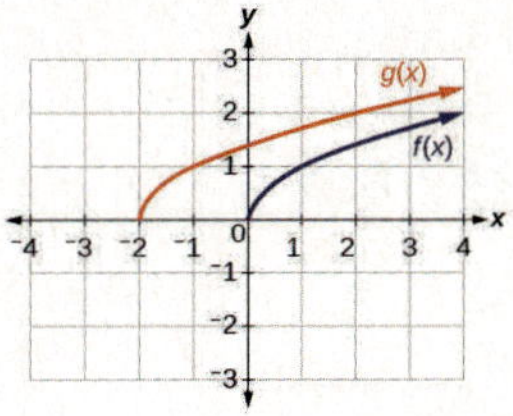

3.

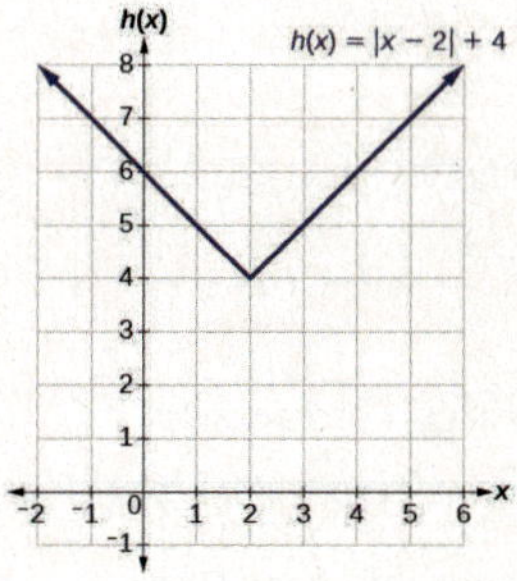

4. $g(x) = \frac{1}{x-1} + 1$

5. ⓐ

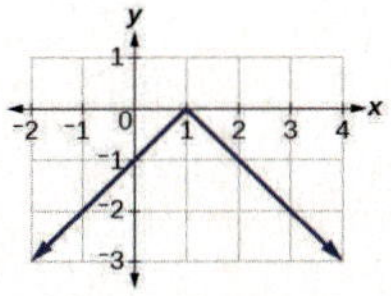

ⓑ

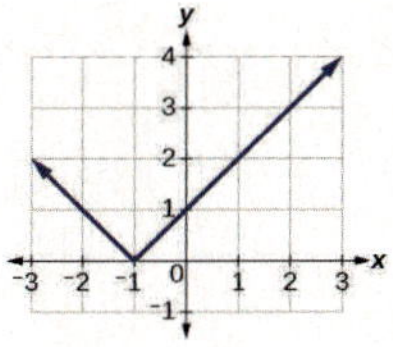

6. ⓐ

$g(x) = -f(x)$

x	-2	0	2	4
$g(x)$	-5	-10	-15	-20

ⓑ

$h(x) = f(-x)$

x	-2	0	2	4
$h(x)$	15	10	5	unknown

7.

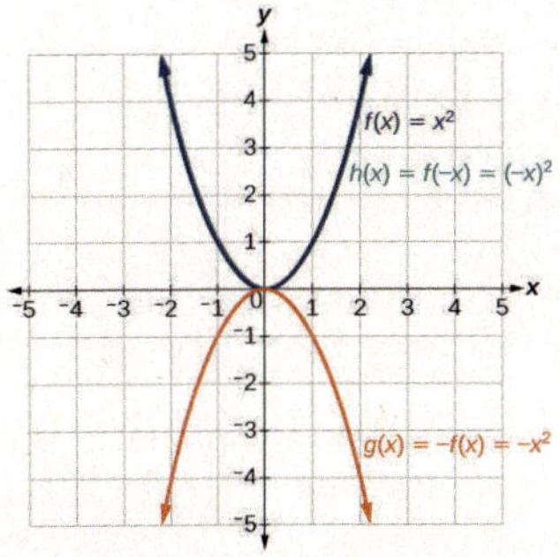

Notice: $g(x) = f(-x)$ looks the same as $f(x)$.

8. even

9.

x	2	4	6	8
$g(x)$	9	12	15	0

10. $g(x) = 3x - 2$

11. $g(x) = f\left(\frac{1}{3}x\right)$ so using the square root function we get $g(x) = \sqrt{\frac{1}{3}x}$

3.6 Absolute Value Functions

1. using the variable p for passing, $|p - 80| \leq 20$

2. $f(x) = -|x + 2| + 3$

3. $x = -1$ or $x = 2$

3.7 Inverse Functions

1. $h(2) = 6$

2. Yes

3. Yes

4. The domain of function f^{-1} is $(-\infty, -2)$ and the range of function f^{-1} is $(1, \infty)$.

5. ⓐ $f(60) = 50$. In 60 minutes, 50 miles are traveled.
ⓑ $f^{-1}(60) = 70$. To travel 60 miles, it will take 70 minutes.

6. ⓐ 3 ⓑ 5.6

7. $x = 3y + 5$

8. $f^{-1}(x) = (2 - x)^2$; domain of f : $\left[0, \infty\right)$; domain of f^{-1} : $\left(-\infty, 2\right]$

9.

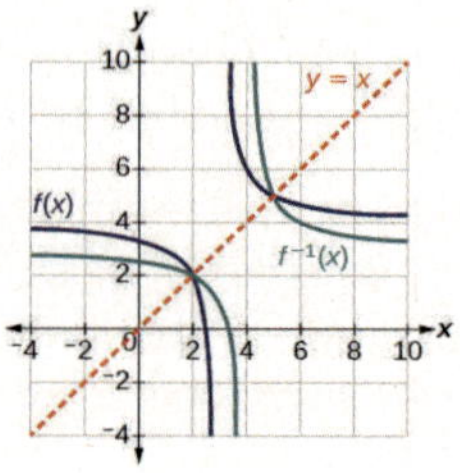

3.1 Section Exercises

1. A relation is a set of ordered pairs. A function is a special kind of relation in which no two ordered pairs have the same first coordinate.

3. When a vertical line intersects the graph of a relation more than once, that indicates that for that input there is more than one output. At any particular input value, there can be only one output if the relation is to be a function.

5. When a horizontal line intersects the graph of a function more than once, that indicates that for that output there is more than one input. A function is one-to-one if each output corresponds to only one input.

7. function

9. function

11. function

13. function

15. function

17. function

19. function

21. function

23. function

25. not a function

27. $f(-3) = -11;$
$f(2) = -1;$
$f(-a) = -2a - 5;$
$-f(a) = -2a + 5;$
$f(a + h) = 2a + 2h - 5$

29. $f(-3) = \sqrt{5} + 5;$
$f(2) = 5;$
$f(-a) = \sqrt{2 + a} + 5;$
$-f(a) = -\sqrt{2 - a} - 5;$
$f(a + h) = \sqrt{2 - a - h} + 5$

31. $f(-3) = 2; f(2) = 1 - 3 = -2;$
$f(-a) = |-a - 1| - |-a + 1|;$
$-f(a) = -|a - 1| + |a + 1|;$
$f(a + h) = |a + h - 1| - |a + h + 1|$

33. $\frac{g(x)-g(a)}{x-a} = x + a + 2, \quad x \neq a$

35. a. $f(-2) = 14$; b. $x = 3$

37. a. $f(5) = 10$; b. $x = -1$ or $x = 4$

39. ⓐ $f(t) = 6 - \frac{2}{3}t;$
ⓑ $f(-3) = 8;$ ⓒ $t = 6$

41. not a function

43. function

45. function

47. function

49. function

51. function

53. ⓐ $f(0) = 1;$
ⓑ $f(x) = -3, x = -2$ or $x = 2$

55. not a function so it is also not a one-to-one function

57. one-to- one function

59. function, but not one-to-one

61. function

63. function

65. not a function

67. $f(x) = 1, x = 2$

69. $f(-2) = 14; \quad f(-1) = 11; \quad f(0) = 8; \quad f(1) = 5; \quad f(2) = 2$

71. $f(-2) = 4; \quad f(-1) = 4.414; \quad f(0) = 4.732; \quad f(1) = 5; \quad f(2) = 5.236$

73. $f(-2) = \frac{1}{9}; \quad f(-1) = \frac{1}{3}; \quad f(0) = 1; \quad f(1) = 3; \quad f(2) = 9$

75. 20

77. $[0, \ 100]$

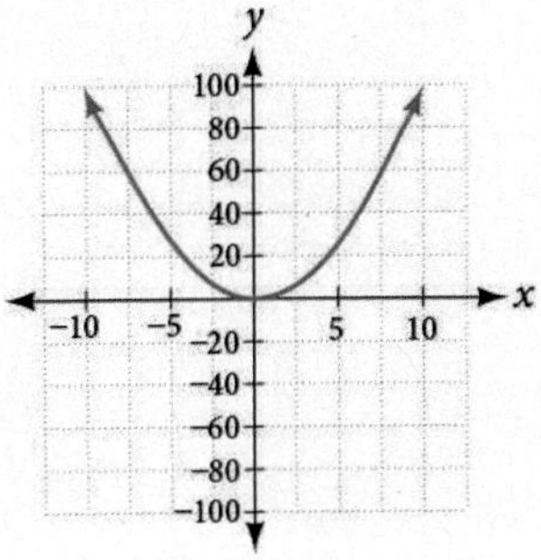

79. $[-0.001, \ 0.001]$

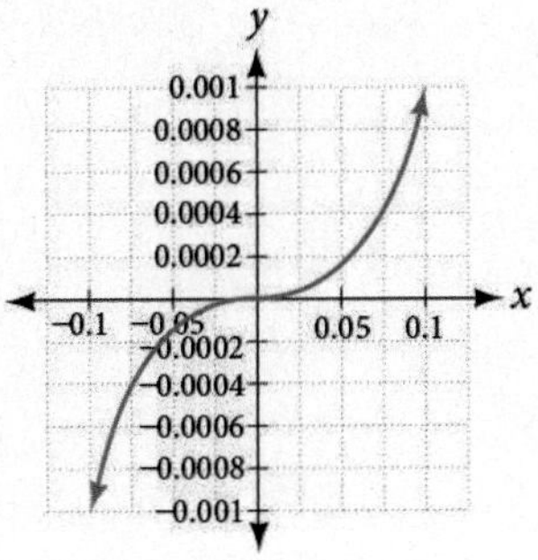

81. $[-1,000,000, \ 1,000,000]$

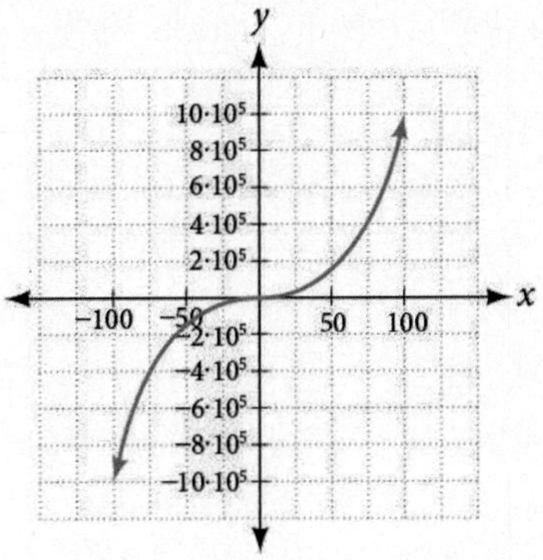

83. $[0,\ 10]$

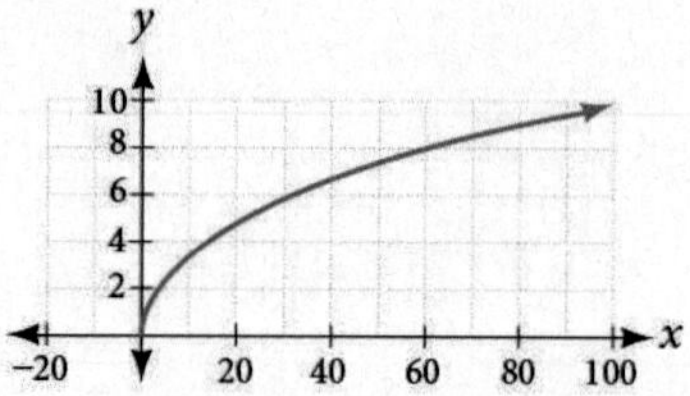

85. $[-0.1, 0.1]$

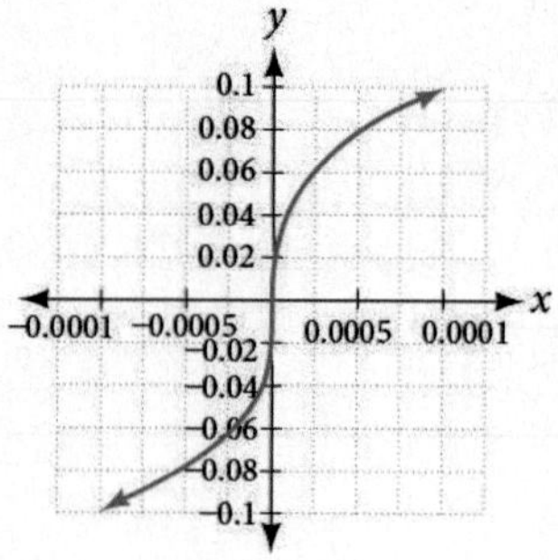

87. $[-100,\ 100]$

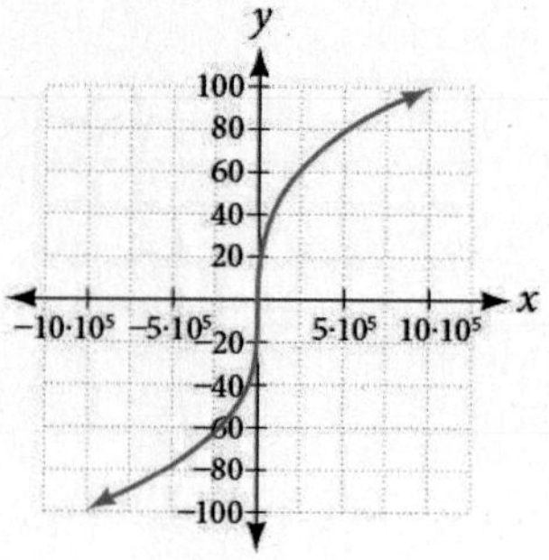

89. ⓐ $g(5000) = 50$; ⓑ The number of cubic yards of dirt required for a garden of 100 square feet is 1.

91. ⓐ The height of a rocket above ground after 1 second is 200 ft. ⓑ The height of a rocket above ground after 2 seconds is 350 ft.

3.2 Section Exercises

1. The domain of a function depends upon what values of the independent variable make the function undefined or imaginary.

3. There is no restriction on x for $f(x) = \sqrt[3]{x}$ because you can take the cube root of any real number. So the domain is all real numbers, $(-\infty, \infty)$. When dealing with the set of real numbers, you cannot take the square root of negative numbers. So x -values are restricted for $f(x) = \sqrt{x}$ to nonnegative numbers and the domain is $[0, \infty)$.

5. Graph each formula of the piecewise function over its corresponding domain. Use the same scale for the x -axis and y -axis for each graph. Indicate inclusive endpoints with a solid circle and exclusive endpoints with an open circle. Use an arrow to indicate $-\infty$ or ∞. Combine the graphs to find the graph of the piecewise function.

7. $(-\infty, \infty)$

9. $(-\infty, 3]$

11. $(-\infty, \infty)$

13. $(-\infty, \infty)$

15. $(-\infty, -\frac{1}{2}) \cup (-\frac{1}{2}, \infty)$

17. $(-\infty, -11) \cup (-11, 2) \cup (2, \infty)$

19. $(-\infty, -3) \cup (-3, 5) \cup (5, \infty)$

21. $(-\infty, 5)$

23. $[6, \infty)$

25. $\left(-\infty, -9\right) \cup (-9, 9) \cup \left(9, \infty\right)$

27. domain: $(2, 8]$, range $[6, 8)$

29. domain: $[-4,\ 4]$, range: $[0,\ 2]$

31. domain: $[-5,\ 3)$, range: $[0, 2]$

33. domain: $(-\infty, 1]$, range: $[0, \infty)$

35. domain: $\left[-6, -\frac{1}{6}\right] \cup \left[\frac{1}{6}, 6\right]$; range: $\left[-6, -\frac{1}{6}\right] \cup \left[\frac{1}{6}, 6\right]$

37. domain: $[-3, \infty)$; range: $[0, \infty)$

39. domain: $(-\infty, \infty)$

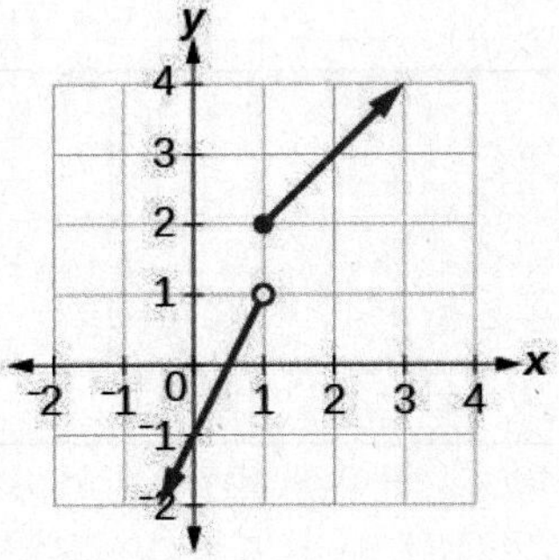

41. domain: $(-\infty, \infty)$

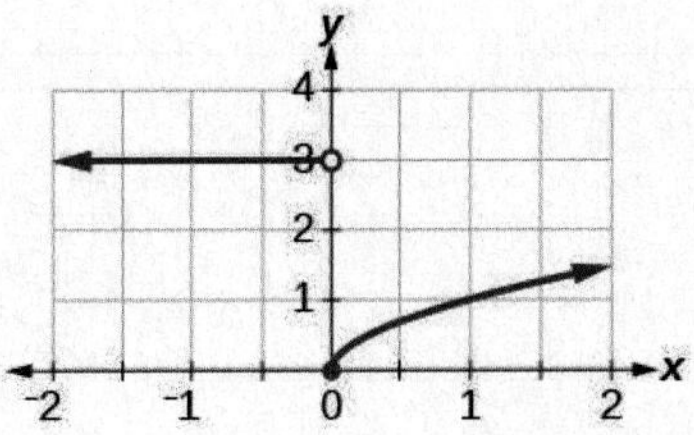

43. domain: $(-\infty, \infty)$

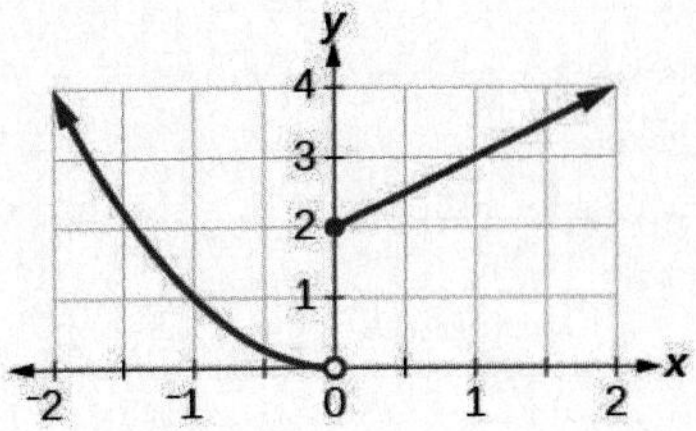

45. domain: $(-\infty, \infty)$

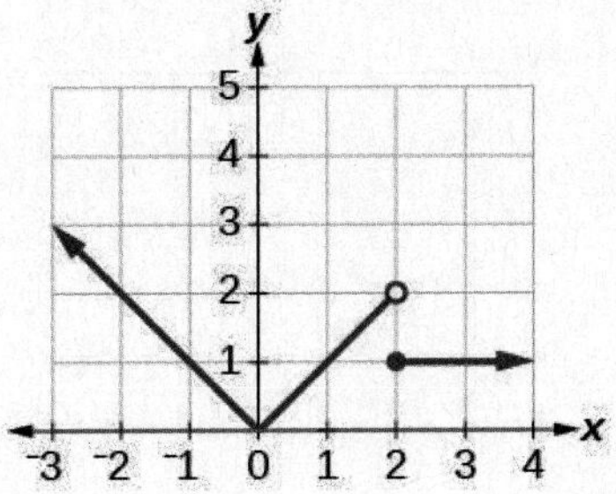

47. $f(-3) = 1;\ \ f(-2) = 0;\ \ f(-1) = 0;\ \ f(0) = 0$

49. $f(-1) = -4;\ \ f(0) = 6;\ \ f(2) = 20;\ \ f(4) = 34$

51. $f(-1) = -5;\ \ f(0) = 3;\ \ f(2) = 3;\ \ f(4) = 16$

53. domain: $(-\infty, 1) \cup (1, \infty)$

55.

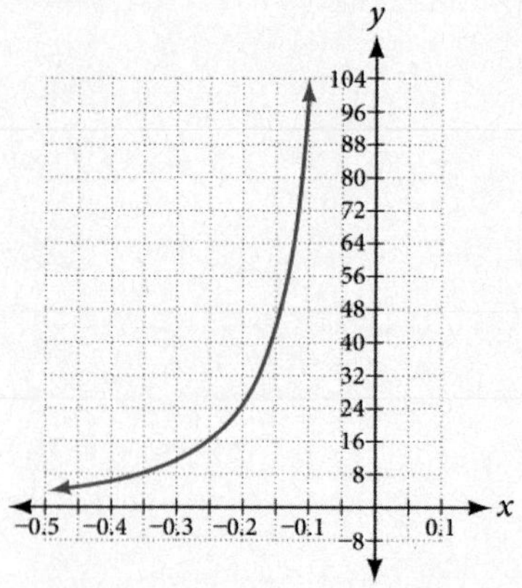

window: $[-0.5, -0.1]$; range: $[4,\ 100]$

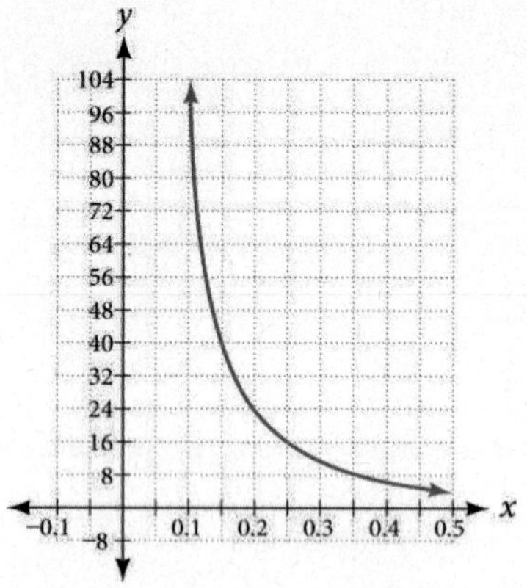

window: $[0.1,\ 0.5]$; range: $[4,\ 100]$

57. $[0,\ 8]$

59. Many answers. One function is $f(x) = \frac{1}{\sqrt{x-2}}$.

61. ⓐ The fixed cost is \$500. ⓑ The cost of making 25 items is \$750. ⓒ The domain is [0, 100] and the range is [500, 1500].

3.3 Section Exercises

1. Yes, the average rate of change of all linear functions is constant.

3. The absolute maximum and minimum relate to the entire graph, whereas the local extrema relate only to a specific region around an open interval.

5. $4(b+1)$

7. 3

9. $4x + 2h$

11. $\frac{-1}{13(13+h)}$

13. $3h^2 + 9h + 9$

15. $4x + 2h - 3$

17. $\frac{4}{3}$

19. increasing on $\left(-\infty, -2.5\right) \cup \left(1, \infty\right)$, decreasing on $(-2.5,\ 1)$

21. increasing on $\left(-\infty, 1\right) \cup (3, 4)$, decreasing on $(1, 3) \cup \left(4, \infty\right)$

23. local maximum: $(-3,\ 60)$, local minimum: $(3,\ -60)$

25. absolute maximum at approximately $(7,\ 150)$, absolute minimum at approximately $(-7.5,\ -220)$

27. ⓐ -3000 ⓑ -1250

29. -4

31. 27

33. -0.167

35. Local minimum at $(3, -22)$, decreasing on $(-\infty,\ 3)$, increasing on $(3,\ \infty)$

37. Local minimum at $(-2, -2)$, decreasing on $(-3, -2)$, increasing on $(-2,\ \infty)$

39. Local maximum at $(-0.39,\ 6)$, local minima at $(-3.15, -47)$ and $(2.04, -32)$, decreasing on $(-\infty, -3.15)$ and $(-0.39,\ 2.04)$, increasing on $(-3.15,\ -0.39)$ and $(2.04,\ \infty)$

41. A

43. $b = 5$

45. 2.7 gallons per minute

47. approximately -0.6 milligrams per day

3.4 Section Exercises

1. Find the numbers that make the function in the denominator g equal to zero, and check for any other domain restrictions on f and g, such as an even-indexed root or zeros in the denominator.

3. Yes. Sample answer: Let $f(x) = x + 1$ and $g(x) = x - 1$. Then $f(g(x)) = f(x - 1) = (x - 1) + 1 = x$ and $g(f(x)) = g(x + 1) = (x + 1) - 1 = x$. So $f \circ g = g \circ f$.

5. $(f+g)(x) = 2x+6$, domain: $(-\infty, \infty)$

$(f-g)(x) = 2x^2+2x-6$, domain: $(-\infty, \infty)$

$(fg)(x) = -x^4 - 2x^3 + 6x^2 + 12x$, domain: $(-\infty, \infty)$

$\left(\frac{f}{g}\right)(x) = \frac{x^2+2x}{6-x^2}$, domain: $(-\infty, -\sqrt{6}) \cup (-\sqrt{6}, \sqrt{6}) \cup (\sqrt{6}, \infty)$

7. $(f+g)(x) = \frac{4x^3+8x^2+1}{2x}$, domain: $(-\infty, 0) \cup (0, \infty)$

$(f-g)(x) = \frac{4x^3+8x^2-1}{2x}$, domain: $(-\infty, 0) \cup (0, \infty)$

$(fg)(x) = x+2$, domain: $(-\infty, 0) \cup (0, \infty)$

$\left(\frac{f}{g}\right)(x) = 4x^3 + 8x^2$, domain: $(-\infty, 0) \cup (0, \infty)$

9. $(f+g)(x) = 3x^2 + \sqrt{x-5}$, domain: $[5, \infty)$

$(f-g)(x) = 3x^2 - \sqrt{x-5}$, domain: $[5, \infty)$

$(fg)(x) = 3x^2\sqrt{x-5}$, domain: $[5, \infty)$

$\left(\frac{f}{g}\right)(x) = \frac{3x^2}{\sqrt{x-5}}$, domain: $(5, \infty)$

11. ⓐ 3 ⓑ $f(g(x)) = 2(3x-5)^2 + 1$
ⓒ $f(g(x)) = 6x^2 - 2$
ⓓ
$(g \circ g)(x) = 3(3x-5) - 5 = 9x - 20$
ⓔ $(f \circ f)(-2) = 163$

13. $f(g(x)) = \sqrt{x^2+3} + 2$, $g(f(x)) = x + 4\sqrt{x} + 7$

15. $f(g(x)) = \sqrt[3]{\frac{x+1}{x^3}} = \frac{\sqrt[3]{x+1}}{x}$, $g(f(x)) = \frac{\sqrt[3]{x}+1}{x}$

17. $(f \circ g)(x) = \frac{1}{\frac{2}{x}+4-4} = \frac{x}{2}$, $(g \circ f)(x) = 2x - 4$

19. $f(g(h(x))) = \left(\frac{1}{x+3}\right)^2 + 1$

21. ⓐ $(g \circ f)(x) = -\frac{3}{\sqrt{2-4x}}$
ⓑ $\left(-\infty, \frac{1}{2}\right)$

23. ⓐ $(0, 2) \cup (2, \infty)$;
ⓑ $(-\infty, -2) \cup (2, \infty)$;
ⓒ $(0, \infty)$

25. $(1, \infty)$

27. sample: $f(x) = x^3$
$g(x) = x - 5$

29. sample: $f(x) = \frac{4}{x}$
$g(x) = (x+2)^2$

31. sample: $f(x) = \sqrt[3]{x}$
$g(x) = \frac{1}{2x-3}$

33. sample: $f(x) = \sqrt[4]{x}$
$g(x) = \frac{3x-2}{x+5}$

35. sample: $f(x) = \sqrt{x}$
$g(x) = 2x + 6$

37. sample: $f(x) = \sqrt[3]{x}$
$g(x) = (x-1)$

39. sample: $f(x) = x^3$
$g(x) = \frac{1}{x-2}$

41. sample: $f(x) = \sqrt{x}$
$g(x) = \frac{2x-1}{3x+4}$

43. 2

45. 5

47. 4

49. 0

51. 2

53. 1

55. 4

57. 4

59. 9

61. 4

63. 2

65. 3

67. 11

69. 0

71. 7

73. $f(g(0)) = 27,\ g(f(0)) = -94$

75. $f(g(0)) = \frac{1}{5},\ g(f(0)) = 5$

77. $18x^2 + 60x + 51$

79. $g \circ g(x) = 9x + 20$

81. 2

83. $(-\infty, \infty)$

85. False

87. $(f \circ g)(6) = 6$; $(g \circ f)(6) = 6$

89. $(f \circ g)(11) = 11$, $(g \circ f)(11) = 11$

91. c

93. $A(t) = \pi\left(25\sqrt{t+2}\right)^2$ and $A(2) = \pi\left(25\sqrt{4}\right)^2 = 2500\pi$ square inches

95. $A(5) = \pi(2(5)+1)^2 = 121\pi$ square units

97. ⓐ $N(T(t)) = 23(5t+1.5)^2 - 56(5t+1.5) + 1$ ⓑ 3.38 hours

3.5 Section Exercises

1. A horizontal shift results when a constant is added to or subtracted from the input. A vertical shifts results when a constant is added to or subtracted from the output.

3. A horizontal compression results when a constant greater than 1 is multiplied by the input. A vertical compression results when a constant between 0 and 1 is multiplied by the output.

5. For a function f, substitute $(-x)$ for (x) in $f(x)$. Simplify. If the resulting function is the same as the original function, $f(-x) = f(x)$, then the function is even. If the resulting function is the opposite of the original function, $f(-x) = -f(x)$, then the original function is odd. If the function is not the same or the opposite, then the function is neither odd nor even.

7. $g(x) = |x-1| - 3$

9. $g(x) = \frac{1}{(x+4)^2} + 2$

11. The graph of $f(x+43)$ is a horizontal shift to the left 43 units of the graph of f.

13. The graph of $f(x-4)$ is a horizontal shift to the right 4 units of the graph of f.

15. The graph of $f(x)+8$ is a vertical shift up 8 units of the graph of f.

17. The graph of $f(x)-7$ is a vertical shift down 7 units of the graph of f.

19. The graph of $f(x+4)-1$ is a horizontal shift to the left 4 units and a vertical shift down 1 unit of the graph of f.

21. decreasing on $(-\infty, -3)$ and increasing on $(-3, \infty)$

23. decreasing on $(0,\ \infty)$

25.

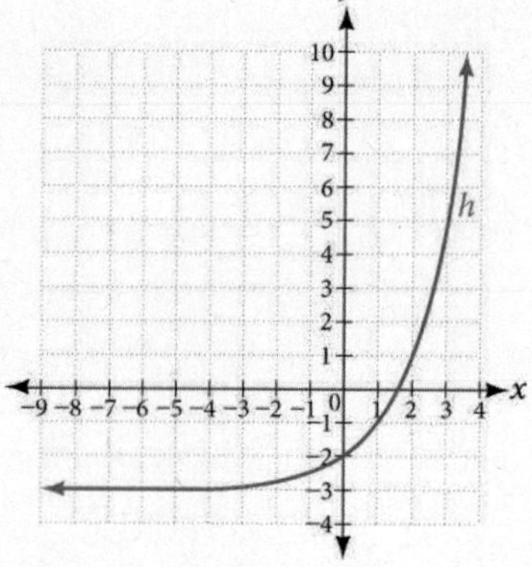

27.

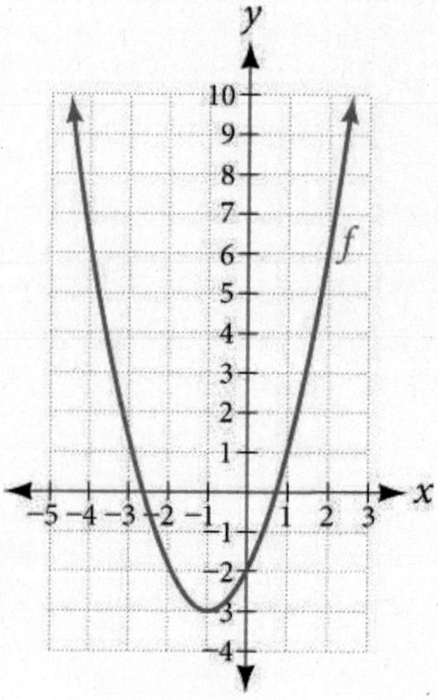

29.

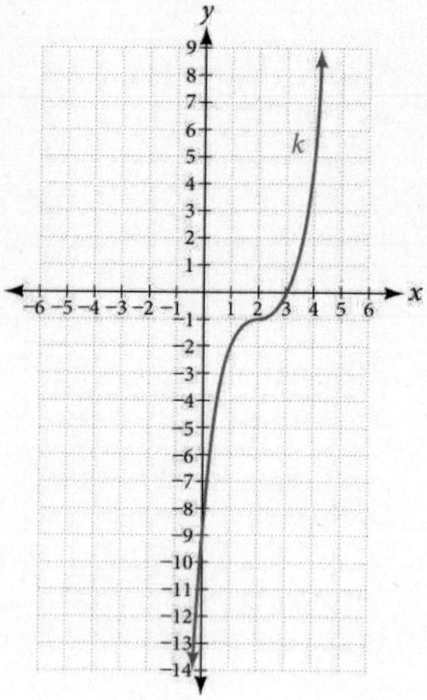

31. $g(x) = f(x-1),\ \ h(x) = f(x) + 1$

33. $f(x) = |x-3| - 2$

35. $f(x) = \sqrt{x+3} - 1$

37. $f(x) = (x-2)^2$

39. $f(x) = |x+3| - 2$

41. $f(x) = -\sqrt{x}$

43. $f(x) = -(x+1)^2 + 2$

45. $f(x) = \sqrt{-x} + 1$

47. even

49. odd

51. even

53. The graph of g is a vertical reflection (across the x -axis) of the graph of f.

55. The graph of g is a vertical stretch by a factor of 4 of the graph of f.

57. The graph of g is a horizontal compression by a factor of $\frac{1}{5}$ of the graph of f.

59. The graph of g is a horizontal stretch by a factor of 3 of the graph of f.

61. The graph of g is a horizontal reflection across the y -axis and a vertical stretch by a factor of 3 of the graph of f.

63. $g(x) = |-4x|$

65. $g(x) = \frac{1}{3(x+2)^2} - 3$

67. $g(x) = \frac{1}{2}(x - 5)^2 + 1$

69. The graph of the function $f(x) = x^2$ is shifted to the left 1 unit, stretched vertically by a factor of 4, and shifted down 5 units.

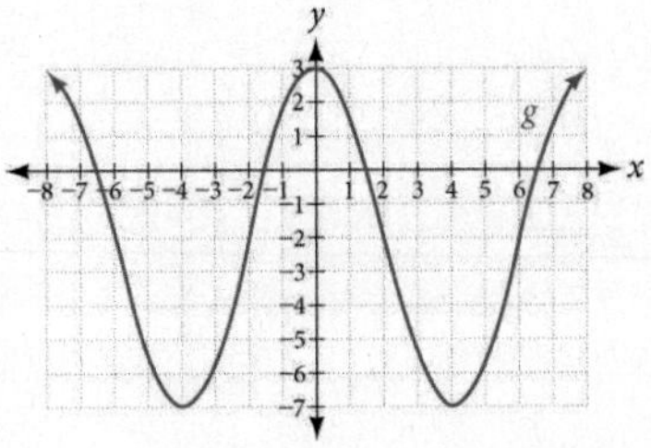

71. The graph of $f(x) = |x|$ is stretched vertically by a factor of 2, shifted horizontally 4 units to the right, reflected across the horizontal axis, and then shifted vertically 3 units up.

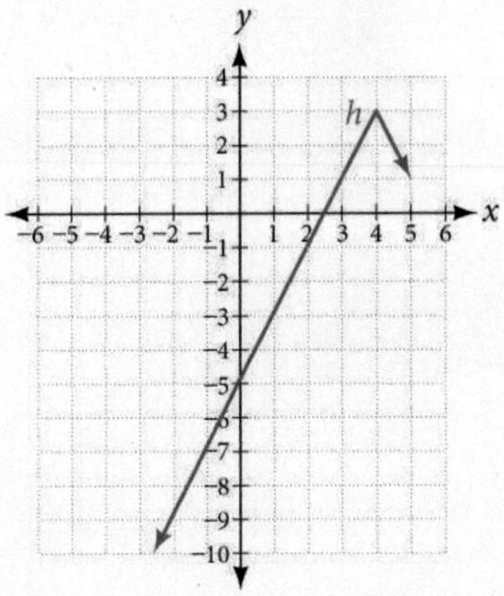

73. The graph of the function $f(x) = x^3$ is compressed vertically by a factor of $\frac{1}{2}$.

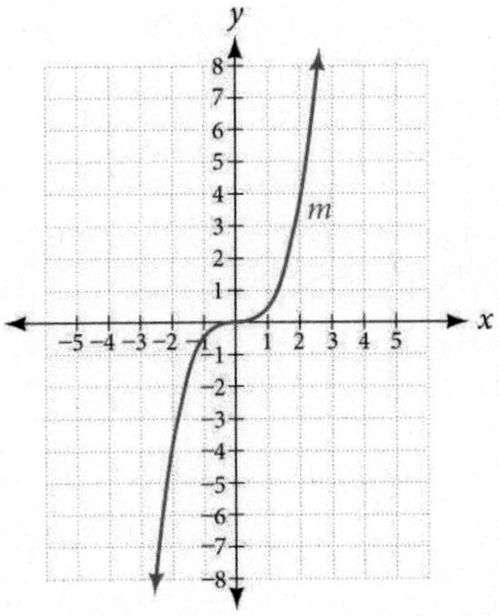

75. The graph of the function is stretched horizontally by a factor of 3 and then shifted vertically downward by 3 units.

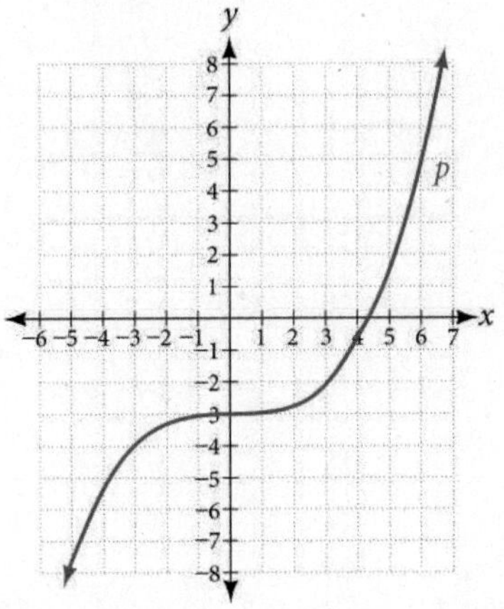

77. The graph of $f(x) = \sqrt{x}$ is shifted right 4 units and then reflected across the vertical line $x = 4$.

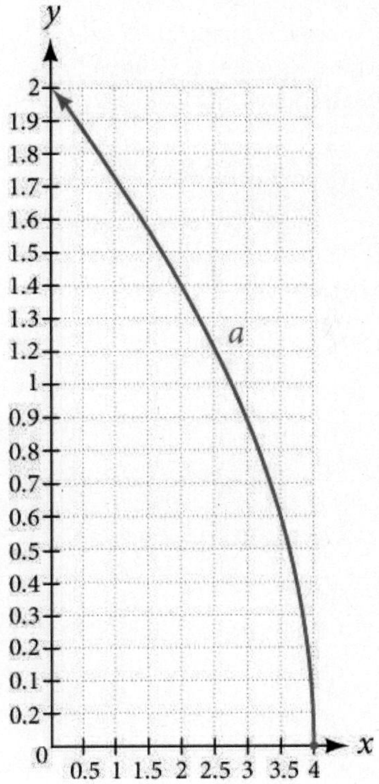

79.

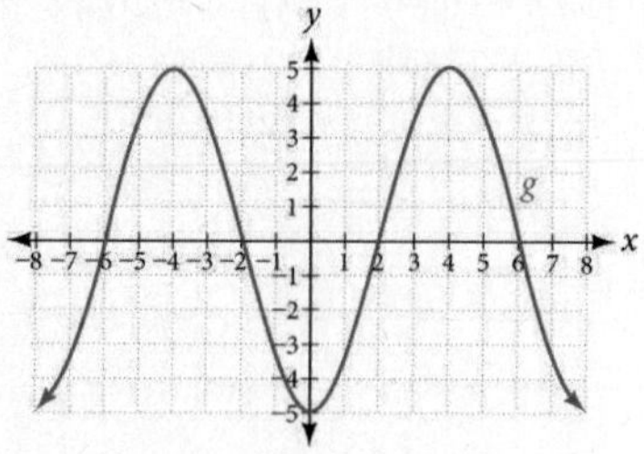

81.

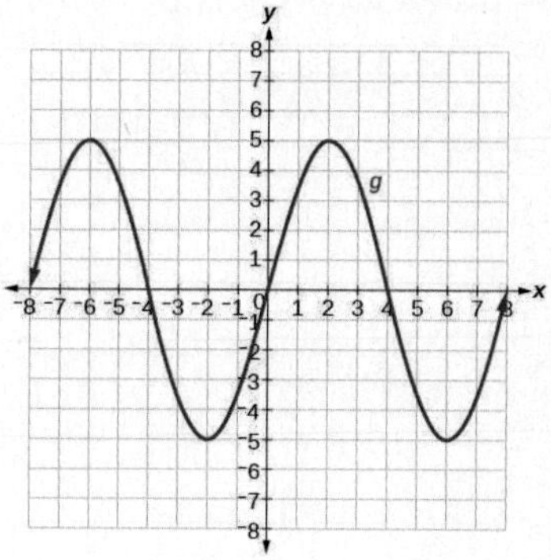

3.6 Section Exercises

1. Isolate the absolute value term so that the equation is of the form $|A| = B$. Form one equation by setting the expression inside the absolute value symbol, A, equal to the expression on the other side of the equation, B. Form a second equation by setting A equal to the opposite of the expression on the other side of the equation, $-B$. Solve each equation for the variable.

3. The graph of the absolute value function does not cross the x -axis, so the graph is either completely above or completely below the x -axis.

5. The distance from x to 8 can be represented using the absolute value statement: | x − 8 | = 4.

7. | x − 10 | ≥ 15

9. There are no x-intercepts.

11. (−4, 0) and (2, 0)

13. $(0, -4), (4, 0), (-2, 0)$

15. $(0, 7), (25, 0), (-7, 0)$

17.

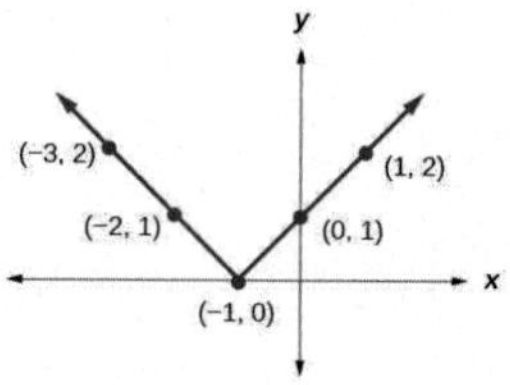

19.

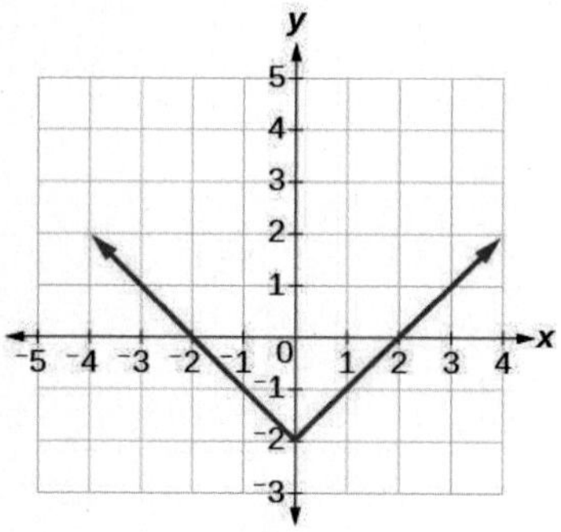

21.

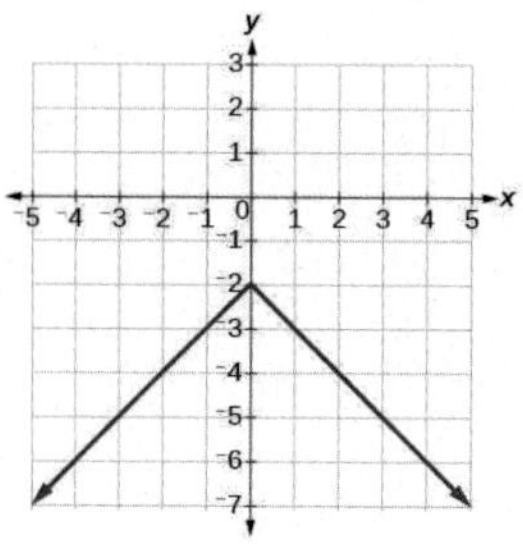

23.

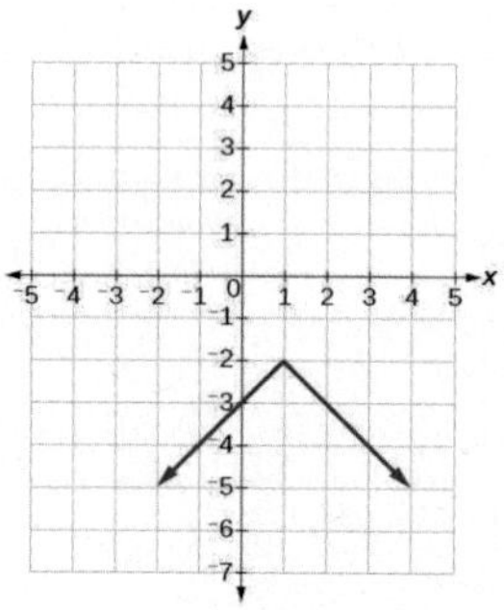

25.

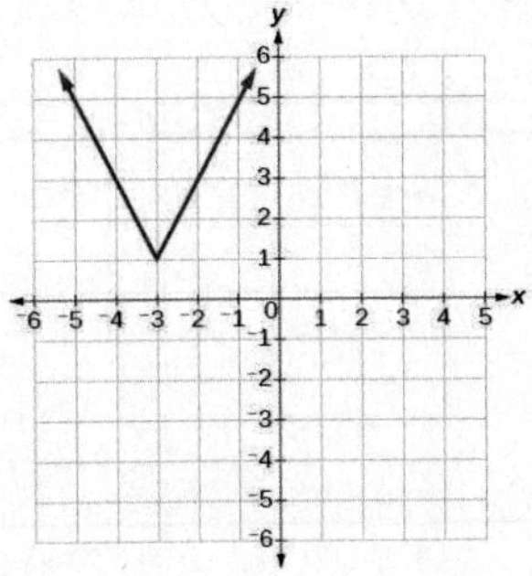

27.

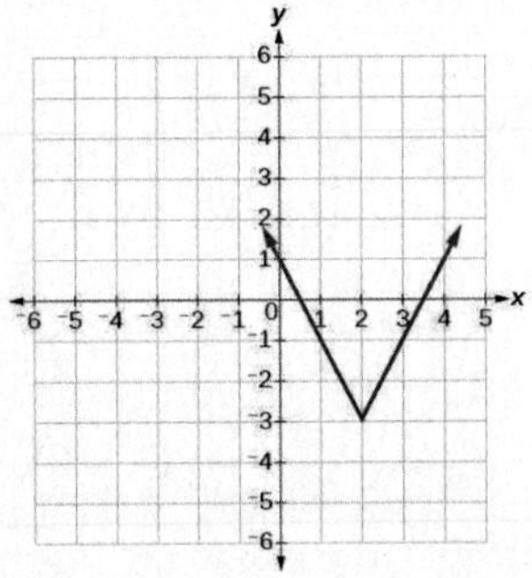

29.

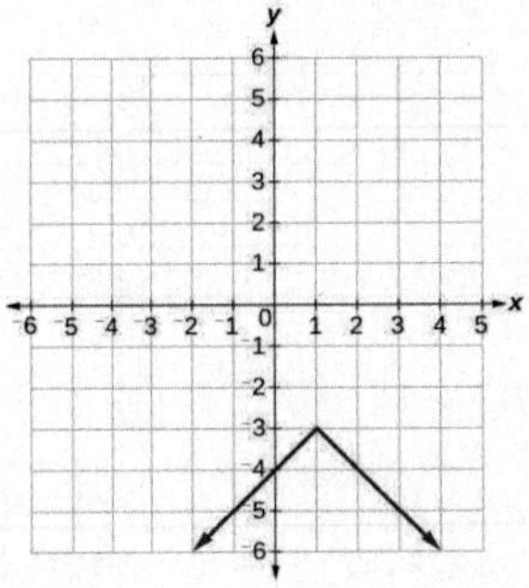

31.

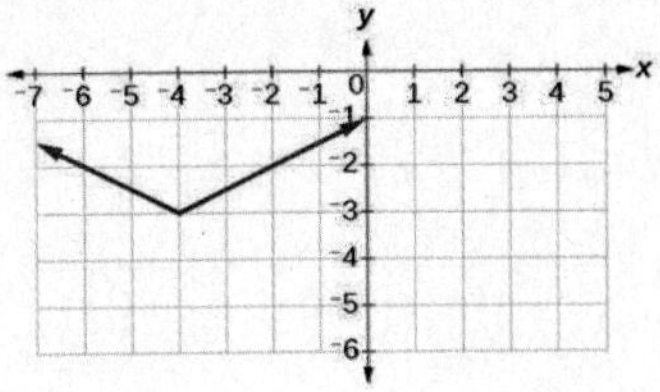

33. range: $[-400, 100]$

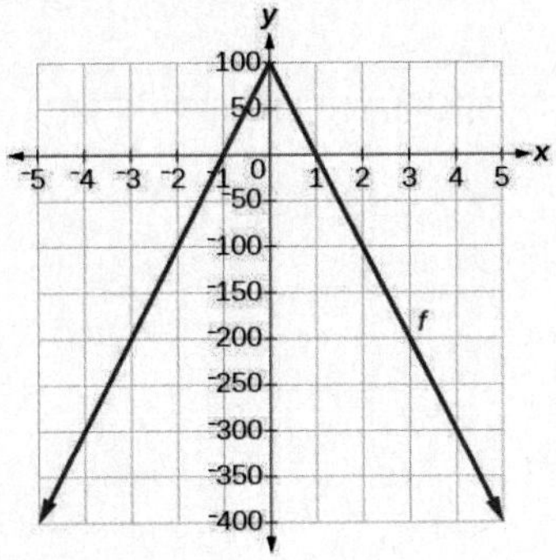

35.

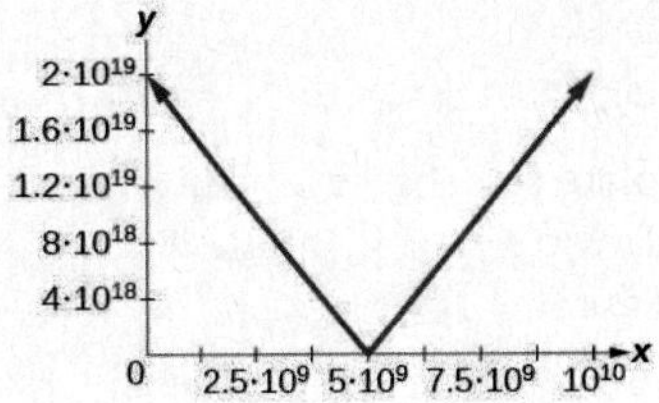

37. There is no solution for a that will keep the function from having a y-intercept. The absolute value function always crosses the y-intercept when $x = 0$.

39. $|p - 0.08| \leq 0.015$

41. $|x - 5.0| \leq 0.01$

3.7 Section Exercises

1. Each output of a function must have exactly one output for the function to be one-to-one. If any horizontal line crosses the graph of a function more than once, that means that y-values repeat and the function is not one-to-one. If no horizontal line crosses the graph of the function more than once, then no y-values repeat and the function is one-to-one.

3. Yes. For example, $f(x) = \frac{1}{x}$ is its own inverse.

5. Given a function $y = f(x)$, solve for x in terms of y. Interchange the x and y. Solve the new equation for y. The expression for y is the inverse, $y = f^{-1}(x)$.

7. $f^{-1}(x) = x - 3$

9. $f^{-1}(x) = 2 - x$

11. $f^{-1}(x) = \frac{-2x}{x-1}$

13. domain of $f(x):\ [-7, \infty);\ f^{-1}(x) = \sqrt{x} - 7$

15. domain of $f(x):\ [0, \infty);\ f^{-1}(x) = \sqrt{x+5}$

16. ⓐ $f(g(x)) = x$ and $g(f(x)) = x$. ⓑ This tells us that f and g are inverse functions

17. $f(g(x)) = x,\ \ g(f(x)) = x$

19. one-to-one

21. one-to-one

23. not one-to-one

25. 3

27. 2

29.

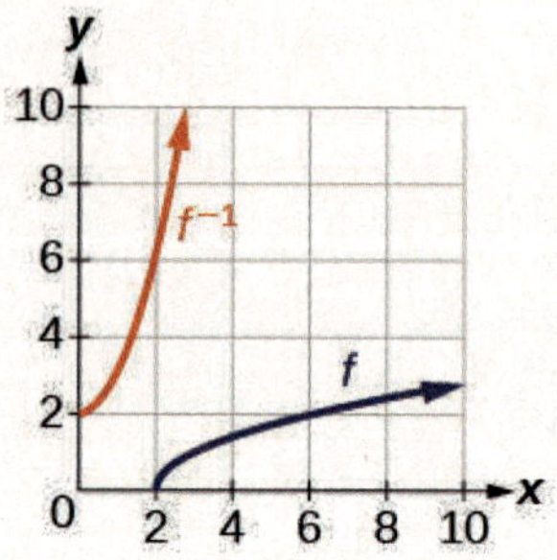

31. $[2, 10]$

33. 6

35. −4

37. 0

39. 1

41.

x	1	4	7	12	16
$f^{-1}(x)$	3	6	9	13	14

43. $f^{-1}(x) = (1 + x)^{1/3}$

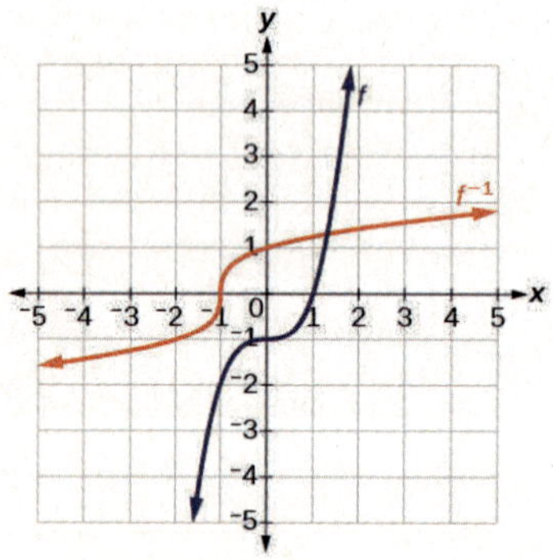

45. $f^{-1}(x) = \frac{5}{9}(x - 32)$. Given the Fahrenheit temperature, x, this formula allows you to calculate the Celsius temperature.

47. $t(d) = \frac{d}{50}$, $t(180) = \frac{180}{50}$. The time for the car to travel 180 miles is 3.6 hours.

Review Exercises

1. function

3. not a function

5. $f(-3) = -27$; $f(2) = -2$; $f(-a) = -2a^2 - 3a$; $-f(a) = 2a^2 - 3a$; $f(a + h) = -2a^2 + 3a - 4ah + 3h - 2h^2$

7. one-to-one

9. function

11. function

13.

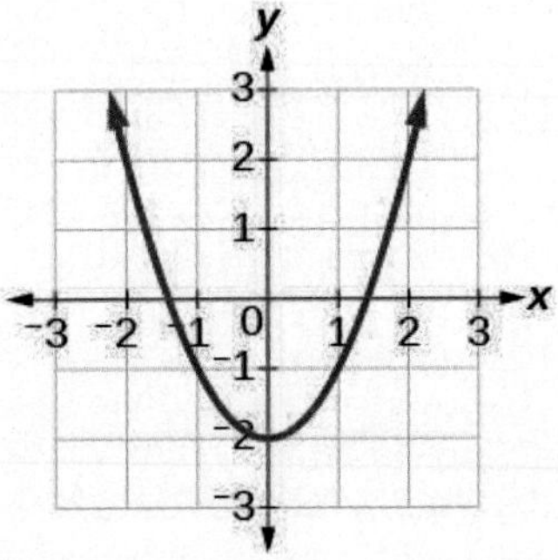

15. 2

17. $x = -1.8$ or or $x = 1.8$

19. $\frac{-64+80a-16a^2}{-1+a} = -16a + 64$

21. $\left(-\infty, -2\right) \cup (-2, 6) \cup \left(6, \infty\right)$

23.

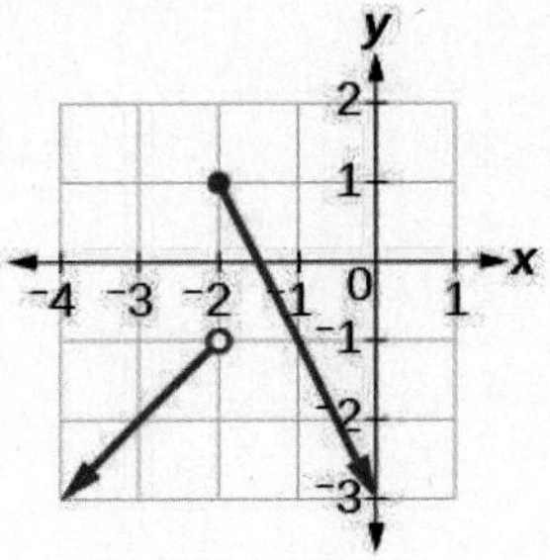

25. 31

27. increasing $\left(2, \infty\right)$; decreasing $(-\infty, 2)$

29. increasing $(-3, 1)$; constant $(-\infty, -3) \cup \left(1, \infty\right)$

31. local minimum $(-2, -3)$; local maximum $(1, 3)$

33. $(-1.8, 10)$

35. $(f \circ g)(x) = 17 - 18x;\ (g \circ f)(x) = -7 - 18x$

37. $(f \circ g)(x) = \sqrt{\frac{1}{x} + 2};\ (g \circ f)(x) = \frac{1}{\sqrt{x+2}}$

39. $(f \circ g)(x) = \frac{1+x}{1+4x},\ x \neq 0,\ x \neq -\frac{1}{4}$

41. $(f \circ g)(x) = \frac{1}{\sqrt{x}},\ x > 0$

43. sample: $g(x) = \frac{2x-1}{3x+4};\ f(x) = \sqrt{x}$

45.

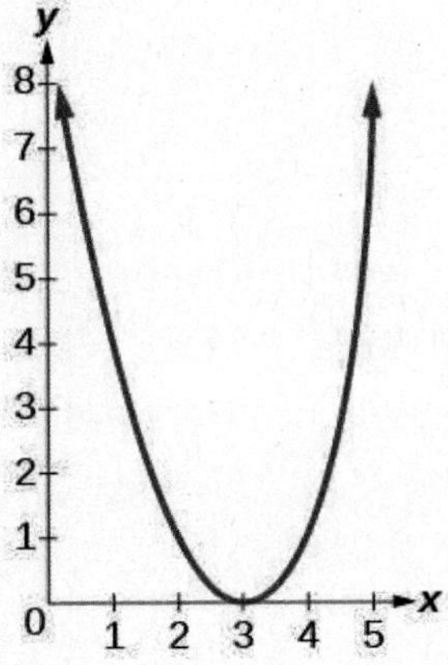

47.

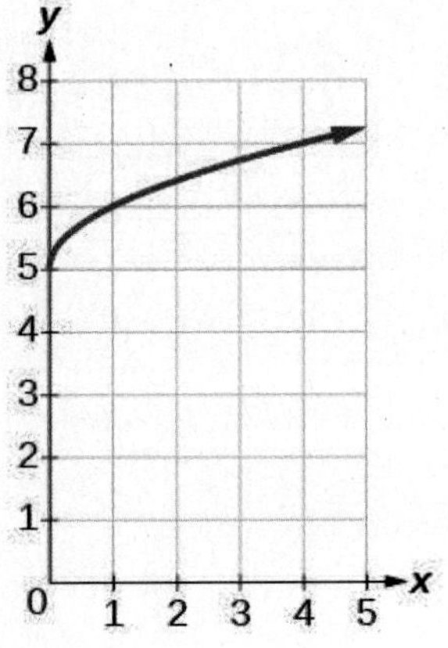

49.

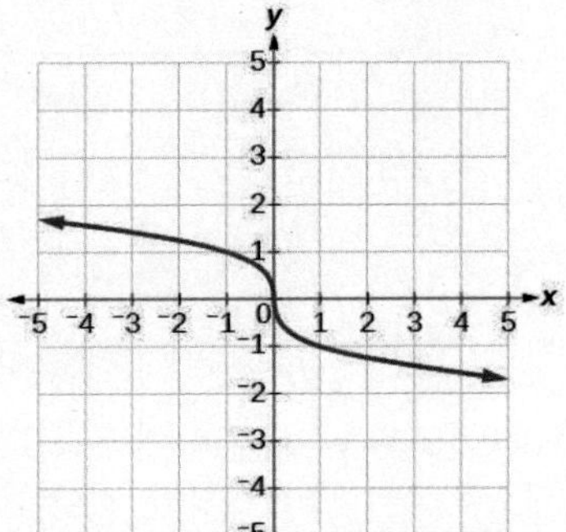

51.

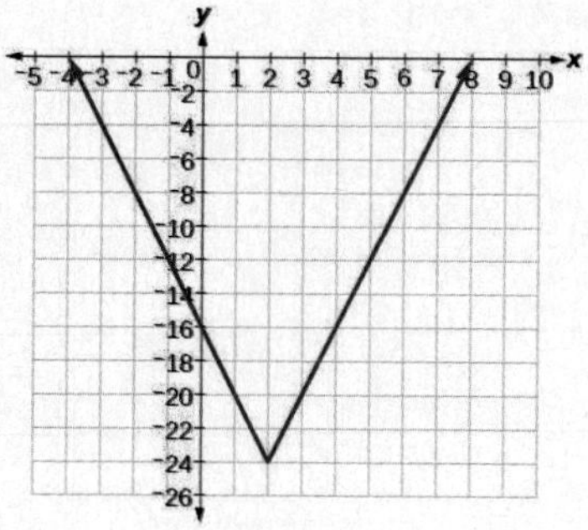

53.

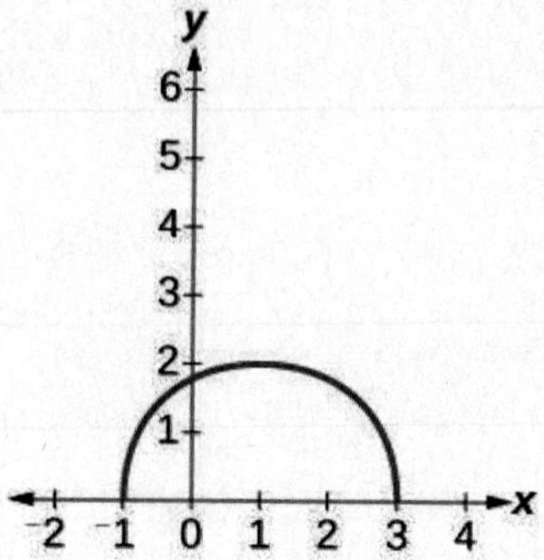

55. $f(x) = |x - 3|$

57. even

59. odd

61. even

63. $f(x) = \frac{1}{2}|x + 2| + 1$

65. $f(x) = -3\,|x - 3| + 3$

67.

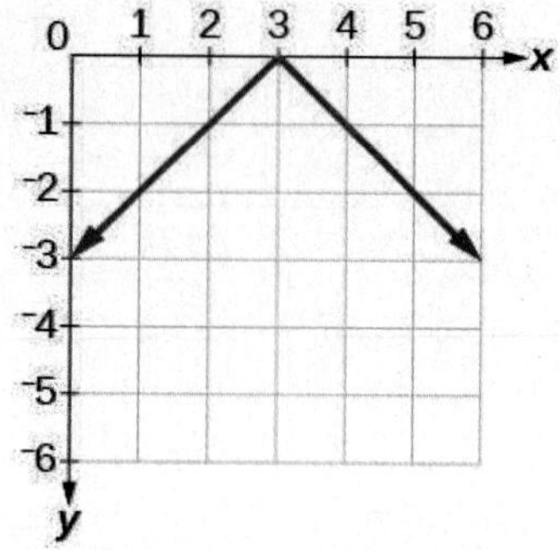

69. $f^{-1}(x) = \frac{x-9}{10}$

71. $f^{-1}(x) = \sqrt{x - 1}$

73. The function is one-to-one.

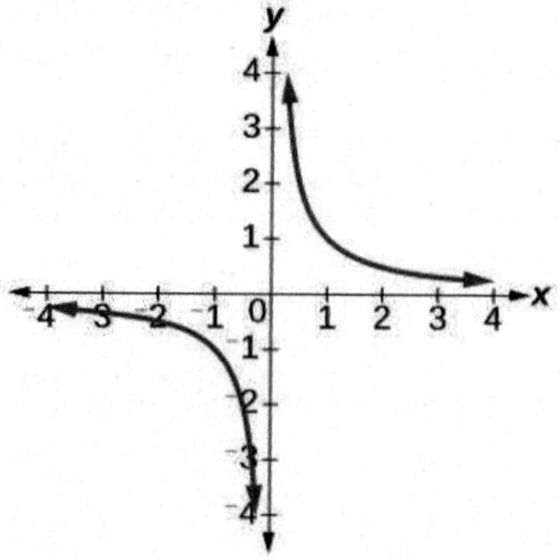

75. 5

Practice Test

1. The relation is a function.

3. −16

5. The graph is a parabola and the graph fails the horizontal line test.

7. $2a^2 - a$

9. $-2(a + b) + 1$

11. $\sqrt{2}$

13.

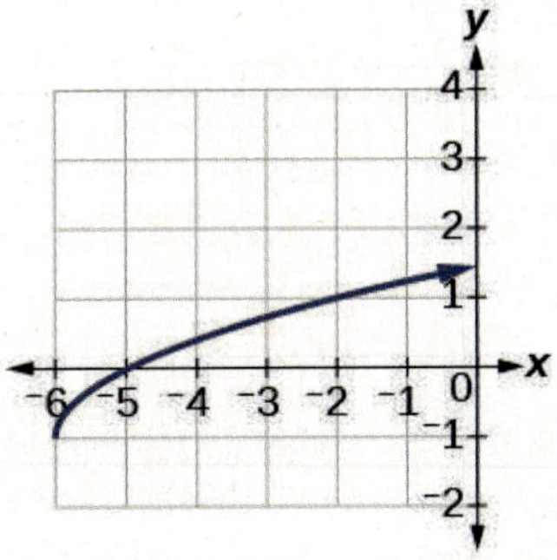

15. even

17. odd

19. $f^{-1}(x) = \frac{x+5}{3}$

21. $(-\infty, -1.1)$ and $(1.1, \infty)$

23. $(1.1, -0.9)$

25. $f(2) = 2$

27. $f(x) = \begin{cases} |x| & \text{if } x \leq 2 \\ 3 & \text{if } x > 2 \end{cases}$

29. $x = 2$

31. yes

33. $f^{-1}(x) = -\frac{x-11}{2}$

Chapter 4

Try It

4.1 Linear Functions

1. $m = \frac{4-3}{0-2} = \frac{1}{-2} = -\frac{1}{2}$; decreasing because $m < 0$.

2. $$m = \frac{1,868 - 1,442}{2,012 - 2,009} = \frac{426}{3} = 142 \text{ people per year}$$

3. $y = -7x + 3$

4. $H(x) = 0.5x + 12.5$

5.

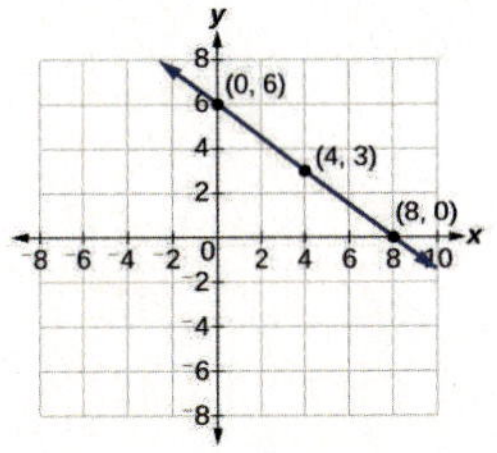

6. Possible answers include $(-3, 7)$, $(-6, 9)$, or $(-9, 11)$.

7.

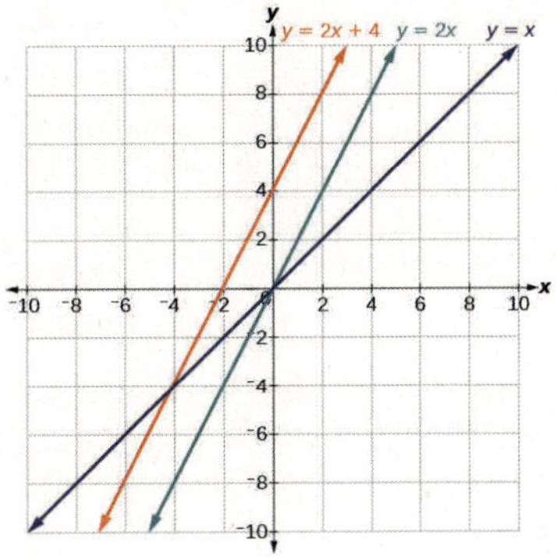

8. $(16, \ 0)$

9. ⓐ $f(x) = 2x$; ⓑ $g(x) = -\frac{1}{2}x$

10. $y = -\frac{1}{3}x + 6$

4.2 Modeling with Linear Functions

1. ⓐ $C(x) = 0.25x + 25,000$ ⓑ The y-intercept is $(0, 25,000)$. If the company does not produce a single doughnut, they still incur a cost of $25,000.

2. ⓐ 41,100 ⓑ 2020

3. 21.57 miles

4.3 Fitting Linear Models to Data

1. $54°F$

2. 150.871 billion gallons; extrapolation

4.1 Section Exercises

1. Terry starts at an elevation of 3000 feet and descends 70 feet per second.

3. $d(t) = 100 - 10t$

5. The point of intersection is $(a,\ a)$. This is because for the horizontal line, all of the y coordinates are a and for the vertical line, all of the x coordinates are a. The point of intersection is on both lines and therefore will have these two characteristics.

7. Yes

9. Yes

11. No

13. Yes

15. Increasing

17. Decreasing

19. Decreasing

21. Increasing

23. Decreasing

25. 2

27. -2

29. $y = \frac{3}{5}x - 1$

31. $y = 3x - 2$

33. $y = -\frac{1}{3}x + \frac{11}{3}$

35. $y = -1.5x - 3$

37. perpendicular

39. parallel

41. $f(0) = -(0) + 2$
$f(0) = 2$
$y - \text{int} : (0, 2)$
$0 = -x + 2$
$x - \text{int} : (2, 0)$

43.
$$h(0) = 3(0) - 5$$
$$h(0) = -5$$
$$y - \text{int} : (0, -5)$$
$$0 = 3x - 5$$
$$x - \text{int} : \left(\tfrac{5}{3}, 0\right)$$

45.
$$-2x + 5y = 20$$
$$-2(0) + 5y = 20$$
$$5y = 20$$
$$y = 4$$
$$y - \text{int} : (0, 4)$$
$$-2x + 5(0) = 20$$
$$x = -10$$
$$x - \text{int} : (-10, 0)$$

47. Line 1: $m = -10$ Line 2: $m = -10$ Parallel

49. Line 1: $m = -2$ Line 2: $m = 1$ Neither

51. Line 1 : $m = -2$ Line 2 : $m = -2$ Parallel

53. $y = 3x - 3$

55. $y = -\frac{1}{3}t + 2$

57. 0

59. $y = -\frac{5}{4}x + 5$

61. $y = 3x - 1$

63. $y = -2.5$

65. F

67. C

69. A

71.

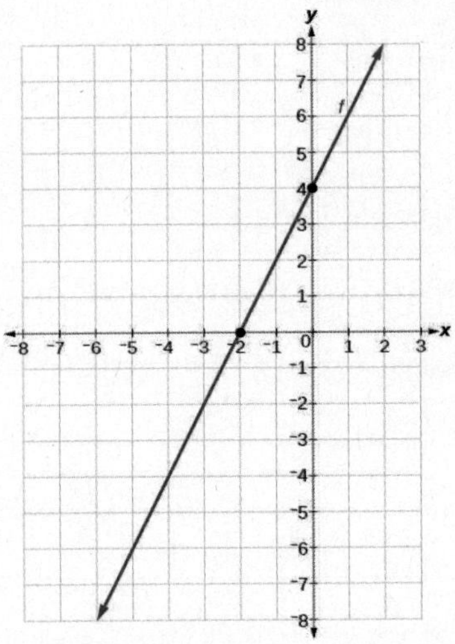

73.

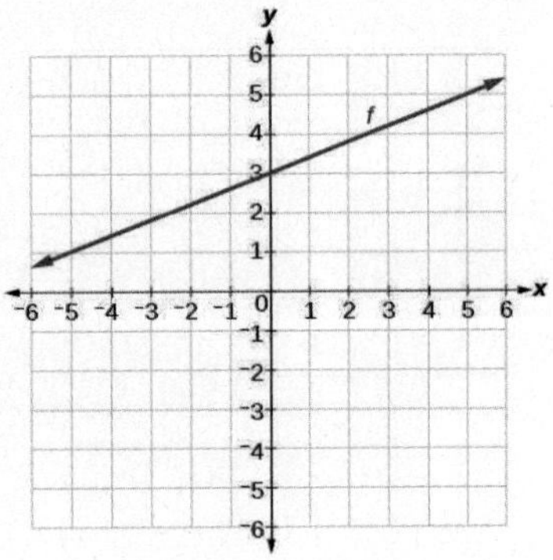

75.

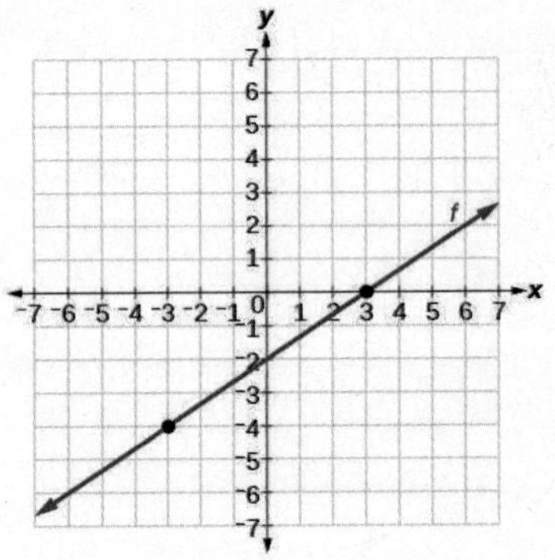

77.

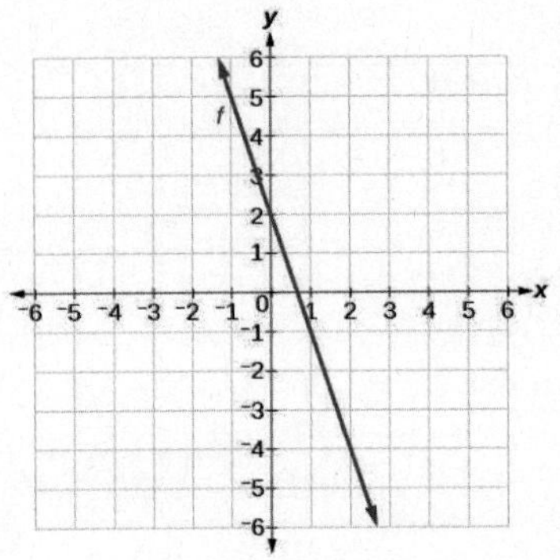

79.

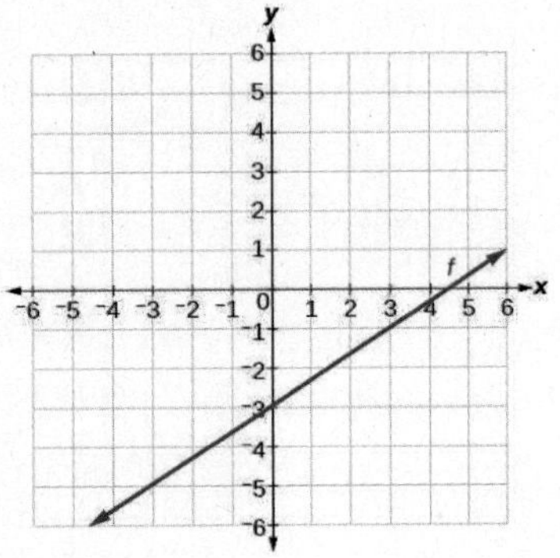

81.

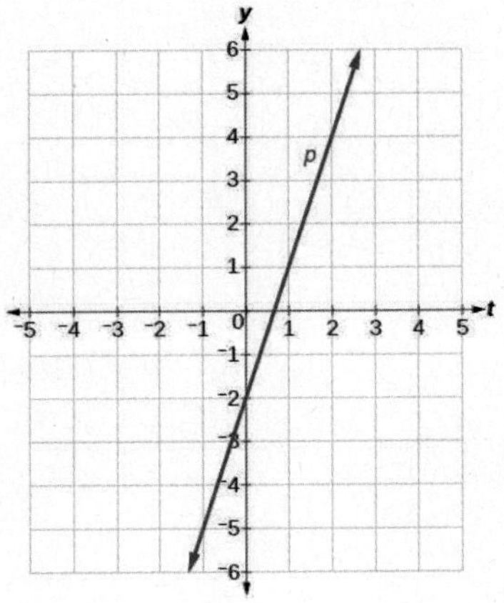

83.

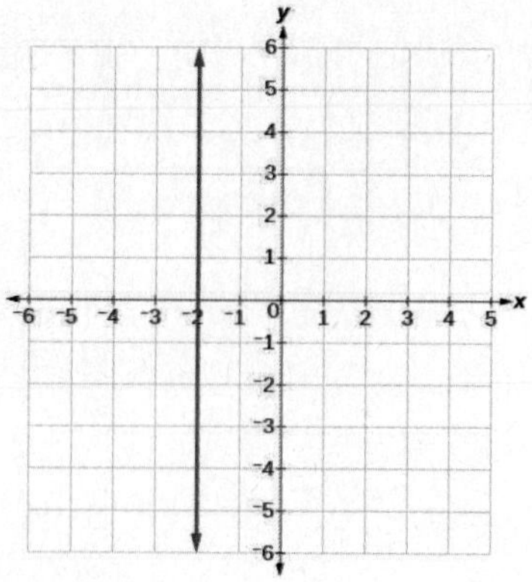

85. $y = 3$

87. $x = -3$

89. Linear, $g(x) = -3x + 5$

91. Linear, $f(x) = 5x - 5$

93. Linear, $g(x) = -\frac{25}{2}x + 6$

95. Linear, $f(x) = 10x - 24$

97. $f(x) = -58x + 17.3$

99.

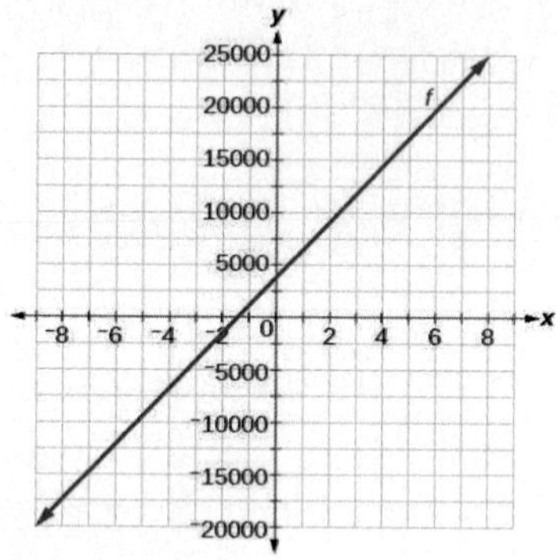

101. ⓐ
$a = 11{,}900, b = 1000.1$
ⓑ $q(p) = 1000p - 100$

103.

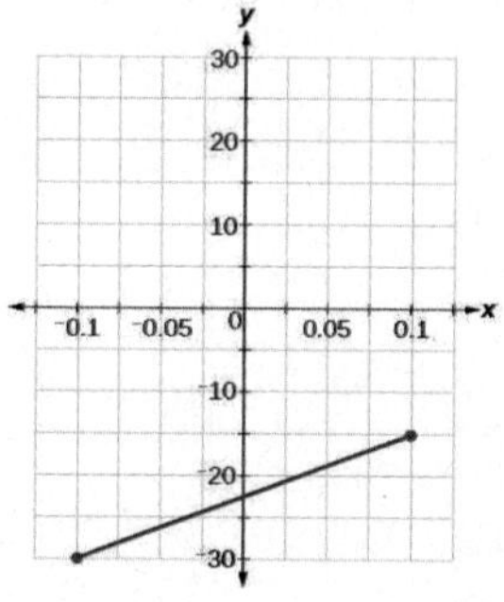

105. $y = -\frac{16}{3}$

107. $x = a$

109. $y = \frac{d}{c-a}x - \frac{ad}{c-a}$

111. $y = 100x - 98$

113. $x < \frac{1999}{201}, x > \frac{1999}{201}$

115. $45 per training session.

117. The rate of change is 0.1. For every additional minute talked, the monthly charge increases by $0.1 or 10 cents. The initial value is 24. When there are no minutes talked, initially the charge is $24.

119. The slope is –400. this means for every year between 1960 and 1989, the population dropped by 400 per year in the city.

121. C

4.2 Section Exercises

1. Determine the independent variable. This is the variable upon which the output depends.

3. To determine the initial value, find the output when the input is equal to zero.

5. 6 square units

7. 20.01 square units

9. 2,300

11. 64,170

13. $P(t) = 75,000 + 2500t$

15. (–30, 0) Thirty years before the start of this model, the town had no citizens. (0, 75,000) Initially, the town had a population of 75,000.

17. Ten years after the model began

19. $W(t) = 0.5t + 7.5$

21. $(-15, 0)$: The *x*-intercept is not a plausible set of data for this model because it means the baby weighed 0 pounds 15 months prior to birth. $(0, 7.5)$: The baby weighed 7.5 pounds at birth.

23. At age 5.8 months

25. $C(t) = 12,025 - 205t$

27. $(58.7, 0)$: In roughly 59 years, the number of people inflicted with the common cold would be 0. $(0, 12, 025)$ Initially there were 12,025 people afflicted by the common cold.

29. 2063

31. $y = -2t + 180$

33. In 2070, the company's profit will be zero.

35. $y = 30t - 300$

37. (10, 0) In the year 1990, the company's profits were zero

39. Hawaii

41. During the year 1933

43. $105,620

45. ⓐ 696 people ⓑ 4 years ⓒ 174 people per year ⓓ 305 people ⓔ P(t) = 305 + 174t ⓕ 2,219 people

47. ⓐ C(x) = 0.15x + 10 ⓑ The flat monthly fee is $10 and there is a $0.15 fee for each additional minute used ⓒ $113.05

49. P(t) = 190t + 4,360

51. ⓐ $R(t) = -2.1t + 16$
ⓑ 5.5 billion cubic feet
ⓒ During the year 2017

53. More than 133 minutes

55. More than $42,857.14 worth of jewelry

57. More than $66,666.67 in sales

4.3 Section Exercises

1. When our model no longer applies, after some value in the domain, the model itself doesn't hold.

3. We predict a value outside the domain and range of the data.

5. The closer the number is to 1, the less scattered the data, the closer the number is to 0, the more scattered the data.

7. 61.966 years

9.

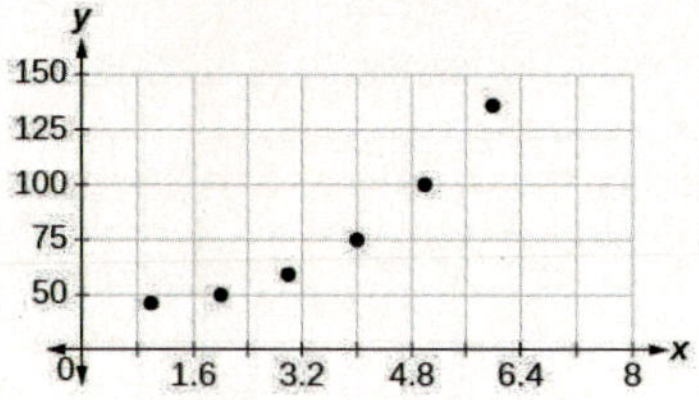

No.

11.

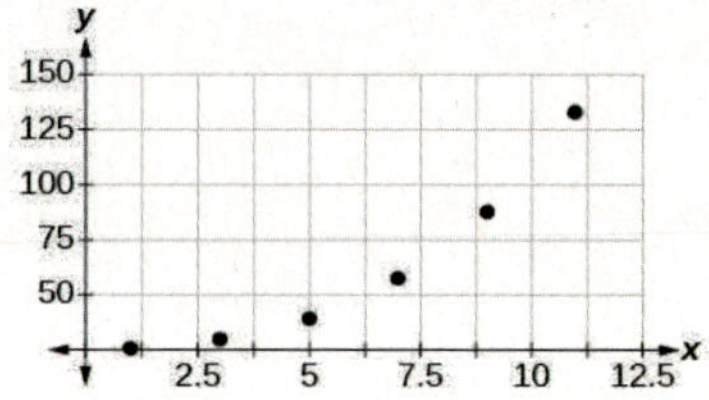

No.

13.

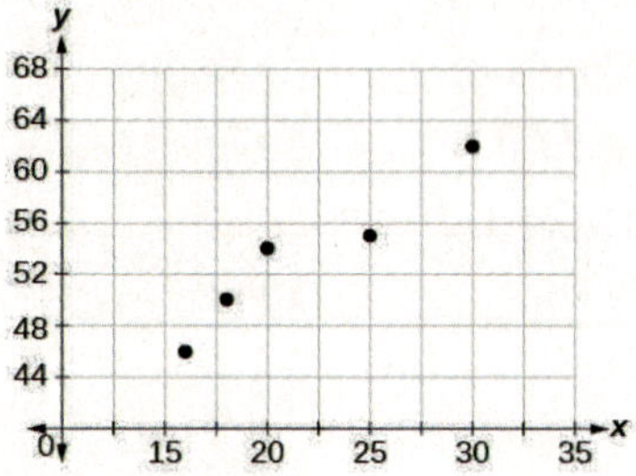

Interpolation. About $60°F$.

15. This value of r indicates a strong negative correlation or slope, so C

17. This value of r indicates a weak negative correlation, so B

19.

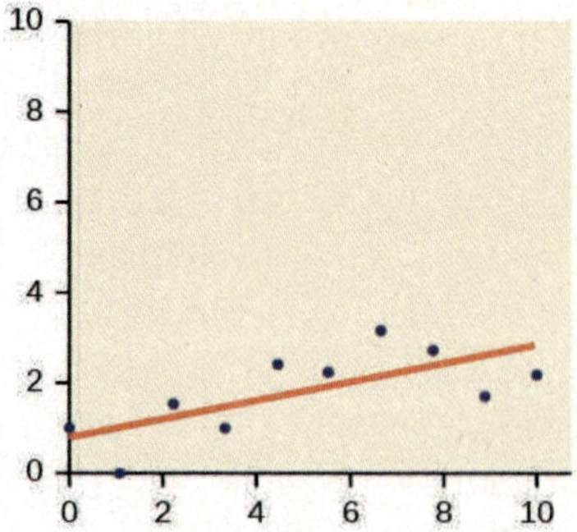

21.

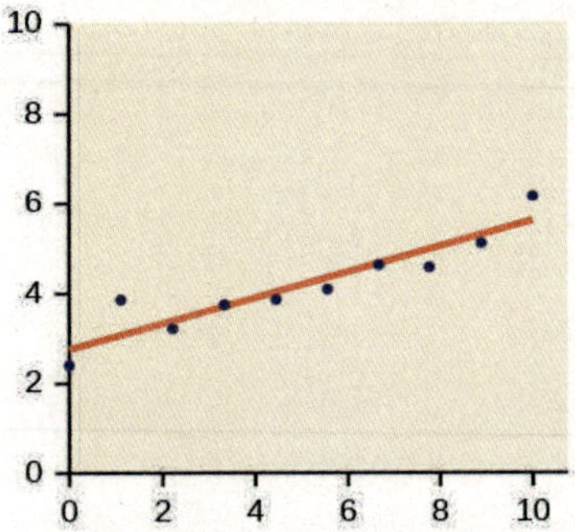

23. Yes, trend appears linear because $r = 0.985$ and will exceed 12,000 near midyear, 2016, 24.6 years since 1992.

25. $y = 1.640x + 13.800$, $r = 0.987$

27. $y = -0.962x + 26.86, \quad r = -0.965$

29. $y = -1.981x + 60.197$; $r = -0.998$

31. $y = 0.121x - 38.841, r = 0.998$

33. $(-2, -6), (1, -12), (5, -20), (6, -22), (9, -28)$; Yes, the function is a good fit.

35. $(189.8, 0)$ If 18,980 units are sold, the company will have a profit of zero dollars.

37. $y = 0.00587x + 1985.41$

39. $y = 20.25x - 671.5$

41. $y = -10.75x + 742.50$

Review Exercises

1. Yes

3. Increasing

5. $y = -3x + 26$

7. 3

9. $y = 2x - 2$

11. Not linear.

13. parallel

15. $(-9, 0); (0, -7)$

17. Line 1: $m = -2$; Line 2: $m = -2$; Parallel

19. $y = -0.2x + 21$

21.

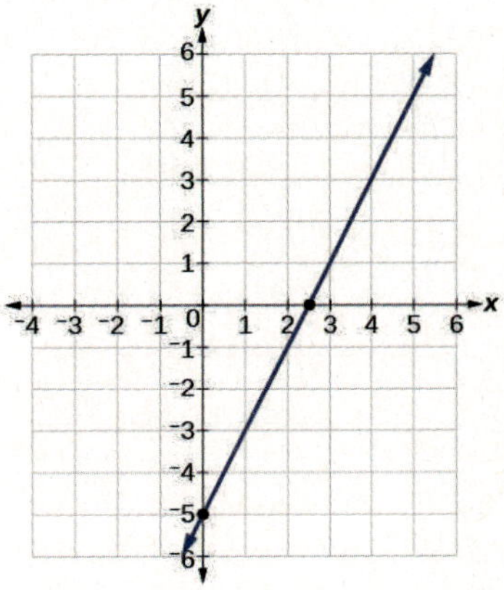

23. More than 250

25. 118,000

27. $y = -300x + 11,500$

29. ⓐ 800
ⓑ 100 students per year
ⓒ $P(t) = 100t + 1700$

31. 18,500

33. $91,625

35. Extrapolation

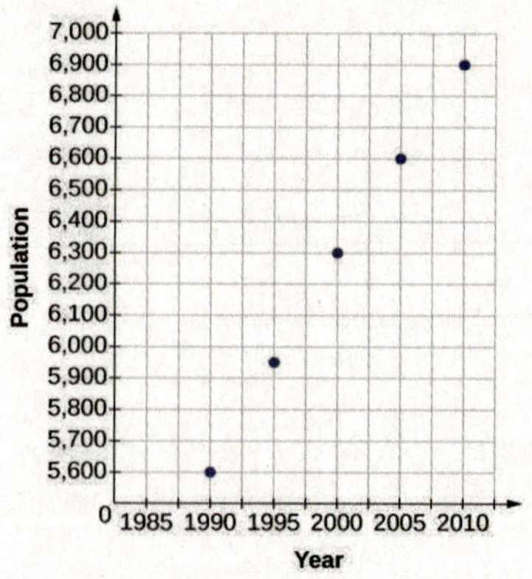

37.

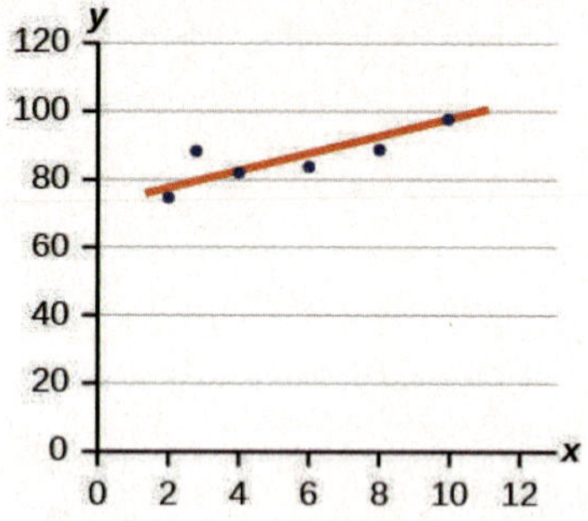

39. 2023

41. $y = -1.294x + 49.412;\ r = -0.974$

43. 2027

45. 7,660

Practice Test

1. Yes

3. Increasing

5. $y = -1.5x - 6$

7. $y = -2x - 1$

9. No

11. Perpendicular

13. $(-7, 0); (0, -2)$

15. $y = -0.25x + 12$

17.

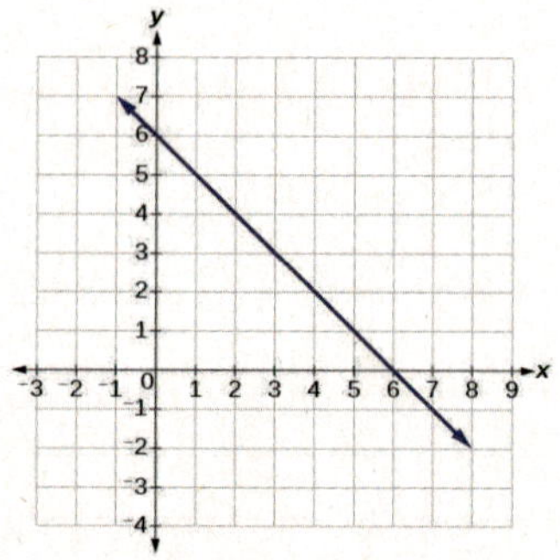

Slope = −1 and y-intercept = 6

19. 150

21. 165,000

23. $y = 875x + 10,625$

25. ⓐ 375
ⓑ dropped an average of 46.875, or about 47 people per year
ⓒ y = −46.875t + 1250

27.

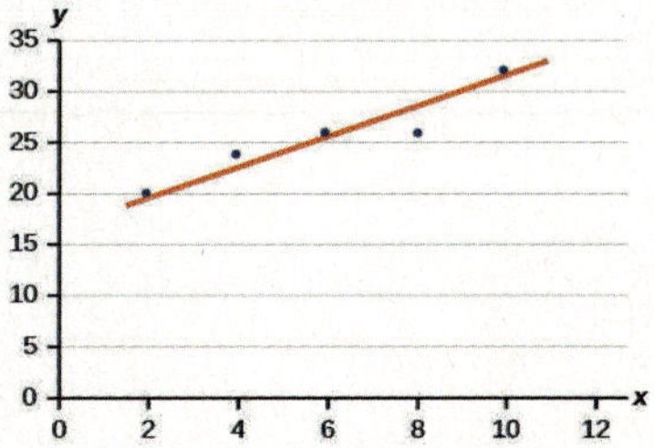

29. In early 2018

31. y = 0.00455x + 1979.5

33. r = 0.999

Chapter 5

Try It

5.1 Quadratic Functions

1. The path passes through the origin and has vertex at $(-4,\ 7)$, so $h(x) = -\frac{7}{16}(x+4)^2 + 7$. To make the shot, $h(-7.5)$ would need to be about 4 but $h(-7.5) \approx 1.64$; he doesn't make it.

2. $g(x) = x^2 - 6x + 13$ in general form; $g(x) = (x-3)^2 + 4$ in standard form

3. The domain is all real numbers. The range is $f(x) \geq \frac{8}{11}$, or $\left[\frac{8}{11}, \infty\right)$.

4. *y*-intercept at (0, 13), No *x*-intercepts

5. ⓐ 3 seconds ⓑ 256 feet ⓒ 7 seconds

5.2 Power Functions and Polynomial Functions

1. $f(x)$ is a power function because it can be written as $f(x) = 8x^5$. The other functions are not power functions.

2. As x approaches positive or negative infinity, $f(x)$ decreases without bound: as

$$x \to \pm\infty,\ f(x) \to -\infty$$

because of the negative coefficient.

3. The degree is 6. The leading term is $-x^6$. The leading coefficient is -1.

4. As

$$x \to \infty,\ f(x) \to -\infty;\ as\ x \to -\infty,\ f(x) \to -\infty.$$

It has the shape of an even degree power function with a negative coefficient.

5. The leading term is $0.2x^3$, so it is a degree 3 polynomial. As x approaches positive infinity, $f(x)$ increases without bound; as x approaches negative infinity, $f(x)$ decreases without bound.

6. y-intercept $(0, 0)$; x-intercepts $(0, 0), (-2, 0)$, and $(5, 0)$

7. There are at most 12 x-intercepts and at most 11 turning points.

8. The end behavior indicates an odd-degree polynomial function; there are 3 x-intercepts and 2 turning points, so the degree is odd and at least 3. Because of the end behavior, we know that the lead coefficient must be negative.

9. The x- intercepts are $(2, 0), (-1, 0)$, and $(5, 0)$, the y-intercept is $(0, 2)$, and the graph has at most 2 turning points.

5.3 Graphs of Polynomial Functions

1. y-intercept $(0, 0)$; x-intercepts $(0, 0), (-5, 0), (2, 0)$, and $(3, 0)$

2. The graph has a zero of –5 with multiplicity 3, a zero of -1 with multiplicity 2, and a zero of 3 with multiplicity 4.

3. 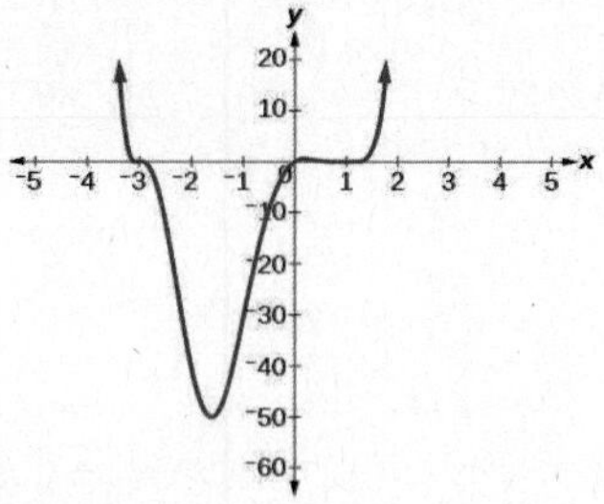

4. Because f is a polynomial function and since $f(1)$ is negative and $f(2)$ is positive, there is at least one real zero between $x = 1$ and $x = 2$.

5. $f(x) = -\frac{1}{8}(x-2)^3(x+1)^2(x-4)$

6. The minimum occurs at approximately the point $(0, -6.5)$, and the maximum occurs at approximately the point $(3.5, 7)$.

5.4 Dividing Polynomials

1. $4x^2 - 8x + 15 - \frac{78}{4x+5}$

2. $3x^3 - 3x^2 + 21x - 150 + \frac{1,090}{x+7}$

3. $3x^2 - 4x + 1$

5.5 Zeros of Polynomial Functions

1. $f(-3) = -412$

2. The zeros are 2, –2, and –4.

3. There are no rational zeros.

4. The zeros are $-4, \frac{1}{2}$, and 1.

5. $f(x) = -\frac{1}{2}x^3 + \frac{5}{2}x^2 - 2x + 10$

6. There must be 4, 2, or 0 positive real roots and 0 negative real roots. The graph shows that there are 2 positive real zeros and 0 negative real zeros.

7. 3 meters by 4 meters by 7 meters

5.6 Rational Functions

1. End behavior: as $x \to \pm\infty,\ f(x) \to 0$; Local behavior: as $x \to 0,\ f(x) \to \infty$ (there are no *x*- or *y*-intercepts)

2.

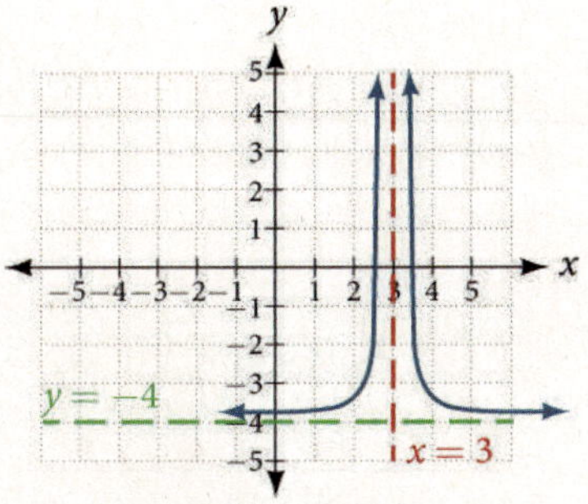

The function and the asymptotes are shifted 3 units right and 4 units down. As $x \to 3, f(x) \to \infty$, and as $x \to \pm\infty, f(x) \to -4$.

The function is $f(x) = \frac{1}{(x-3)^2} - 4$.

3. $\frac{12}{11}$

4. The domain is all real numbers except $x = 1$ and $x = 5$.

5. Removable discontinuity at $x = 5$. Vertical asymptotes: $x = 0,\ \ x = 1$.

6. Vertical asymptotes at $x = 2$ and $x = -3$; horizontal asymptote at $y = 4$.

7. For the transformed reciprocal squared function, we find the rational form.

$$f(x) = \frac{1}{(x-3)^2} - 4 = \frac{1-4(x-3)^2}{(x-3)^2} = \frac{1-4(x^2-6x+9)}{(x-3)(x-3)} = \frac{-4x^2+24x-35}{x^2-6x+9}$$

Because the numerator is the same degree as the denominator we know that as $x \to \pm\infty,\ f(x) \to -4$; so $y = -4$ is the horizontal asymptote. Next, we set the denominator equal to zero, and find that the vertical asymptote is $x = 3$, because as $x \to 3, f(x) \to \infty$. We then set the numerator equal to 0 and find the *x*-intercepts are at $(2.5, 0)$ and $(3.5, 0)$. Finally, we evaluate the function at 0 and find the *y*-intercept to be at $\left(0, \frac{-35}{9}\right)$.

8. Horizontal asymptote at $y = \frac{1}{2}$. Vertical asymptotes at $x = 1$ and $x = 3$. *y*-intercept at $\left(0, \frac{4}{3}.\right)$

x-intercepts at $(2, 0)$ and $(-2, 0)$. $(-2, 0)$ is a zero with multiplicity 2, and the graph bounces off the *x*-axis at this point. $(2, 0)$ is a single zero and the graph crosses the axis at this point.

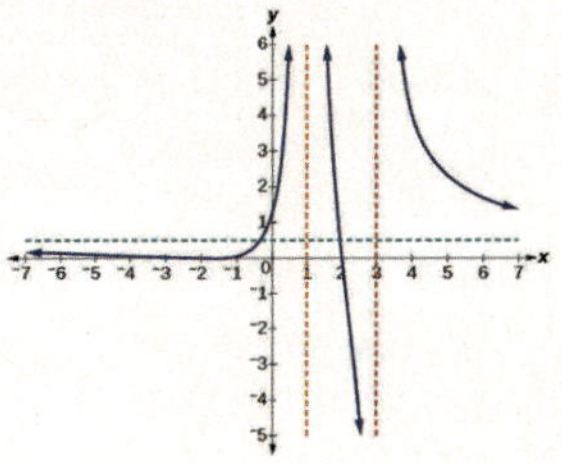

5.7 Inverses and Radical Functions

1. $f^{-1}(f(x)) = f^{-1}\left(\frac{x+5}{3}\right) = 3\left(\frac{x+5}{3}\right) - 5 = (x-5)+5 = x$ and $f\left(f^{-1}(x)\right) = f(3x-5) = \frac{(3x-5)+5}{3} = \frac{3x}{3} = x$

2. $f^{-1}(x) = x^3 - 4$

3. $f^{-1}(x) = \sqrt{x-1}$

4. $f^{-1}(x) = \frac{x^2-3}{2}, x \geq 0$

5. $f^{-1}(x) = \frac{2x+3}{x-1}$

5.8 Modeling Using Variation

1. $\frac{128}{3}$

2. $\frac{9}{2}$

3. $x = 20$

5.1 Section Exercises

1. When written in that form, the vertex can be easily identified.

3. If $a = 0$ then the function becomes a linear function.

5. If possible, we can use factoring. Otherwise, we can use the quadratic formula.

7. $g(x) = (x+1)^2 - 4$, Vertex $(-1, -4)$

9. $f(x) = \left(x + \frac{5}{2}\right)^2 - \frac{33}{4}$, Vertex $\left(-\frac{5}{2}, -\frac{33}{4}\right)$

11. $f(x) = 3(x-1)^2 - 12$, Vertex $(1, -12)$

13. $f(x) = 3\left(x - \frac{5}{6}\right)^2 - \frac{37}{12}$, Vertex $\left(\frac{5}{6}, -\frac{37}{12}\right)$

15. Minimum is $-\frac{17}{2}$ and occurs at $\frac{5}{2}$. Axis of symmetry is $x = \frac{5}{2}$.

17. Minimum is $-\frac{17}{16}$ and occurs at $-\frac{1}{8}$. Axis of symmetry is $x = -\frac{1}{8}$.

19. Minimum is $-\frac{7}{2}$ and occurs at -3. Axis of symmetry is $x = -3$.

21. Domain is $\left(-\infty, \infty\right)$. Range is $[2, \infty)$.

23. Domain is $\left(-\infty, \infty\right)$. Range is $[-5, \infty)$.

25. Domain is $\left(-\infty, \infty\right)$. Range is $[-12, \infty)$.

27. $f(x) = x^2 + 4x + 3$

29. $f(x) = x^2 - 4x + 7$

31. $f(x) = -\frac{1}{49}x^2 + \frac{6}{49}x + \frac{89}{49}$

33. $f(x) = x^2 - 2x + 1$

35. Vertex: (3, −10), axis of symmetry: x = 3, intercepts: $\left(3 + \sqrt{10}, 0\right)$ and $\left(3 - \sqrt{10}, 0\right)$

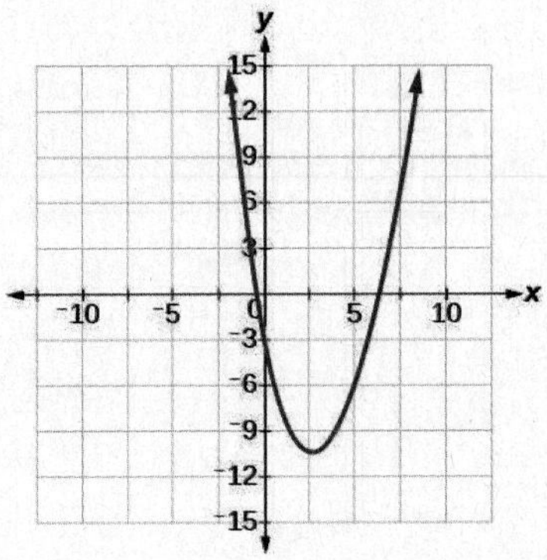

37. Vertex: $\left(\frac{3}{2}, -12\right)$, axis of symmetry: $x = \frac{3}{2}$, intercept: $\left(\frac{3+2\sqrt{3}}{2}, 0\right)$ and $\left(\frac{3-2\sqrt{3}}{2}, 0\right)$

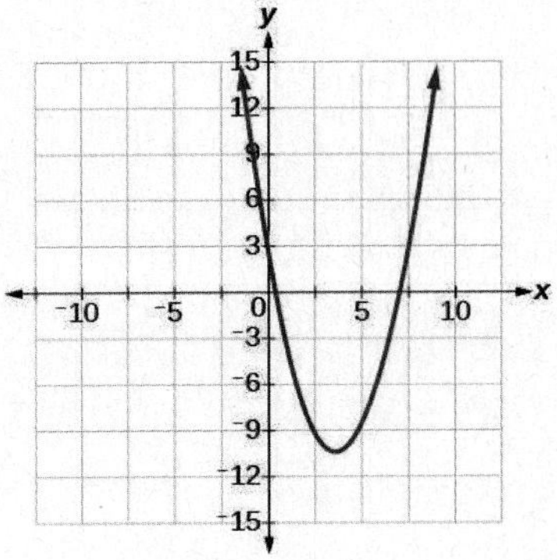

39.

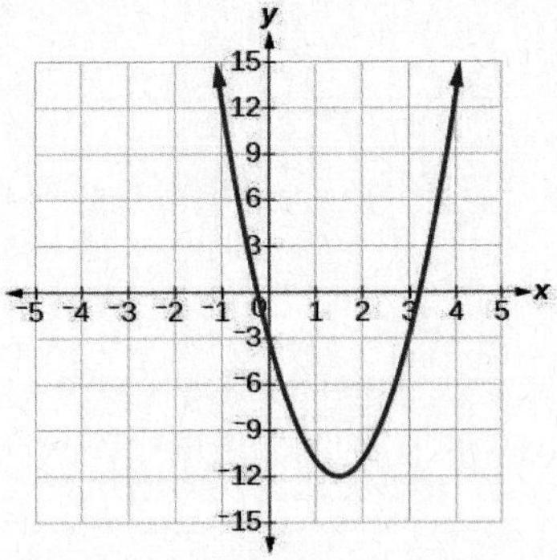

41. $f(x) = x^2 + 2x + 3$

43. $f(x) = -3x^2 - 6x - 1$

45. $f(x) = -\frac{1}{4}x^2 - x + 2$

47. $f(x) = x^2 + 2x + 1$

49. $f(x) = -\frac{x}{2} + 2x$

50. $f(x) = 2x^2$

53. The graph is shifted to the right or left (a horizontal shift).

55. The suspension bridge has 1,000 feet distance from the center.

57. Domain is (-∞, ∞). Range is $\left(-\infty, 2\right]$.

59. Domain: $\left(-\infty, \infty\right)$; range: $\left[100, \infty\right)$

61. $f(x) = 2x^2 + 2$

63. $f(x) = -x^2 - 2$

65. $f(x) = 3x^2 + 6x - 15$

67. 75 feet by 50 feet

69. 3 and 3; product is 9

71. The revenue reaches the maximum value when 1800 thousand phones are produced.

73. 2.449 seconds

75. 41 trees per acre

5.2 Section Exercises

1. The coefficient of the power function is the real number that is multiplied by the variable raised to a power. The degree is the highest power appearing in the function.

3. As x decreases without bound, so does $f(x)$. As x increases without bound, so does $f(x)$.

5. The polynomial function is of even degree and leading coefficient is negative.

7. Power function

9. Neither

11. Neither

13. Degree = 2, Coefficient = -2

15. Degree =4, Coefficient = -2

17. As $x \to \infty$,
$$f(x) \to \infty, \text{ as } x \to -\infty, \ f(x) \to \infty$$

19. As $x \to -\infty$,
$$f(x) \to -\infty, \text{ as } x \to \infty, \ f(x) \to -\infty$$

21. As $x \to -\infty$,
$$f(x) \to -\infty, \text{ as } x \to \infty, \ f(x) \to -\infty$$

23. As $x \to \infty$,
$$f(x) \to \infty, \text{ as } x \to -\infty, \ f(x) \to -\infty$$

25. y-intercept is $(0, 12)$, t-intercepts are $(1, 0)$; $(-2, 0)$; and $(3, 0)$.

27. y-intercept is $(0, -16)$. x-intercepts are $(2, 0)$ and $(-2, 0)$.

29. y-intercept is $(0, 0)$. x-intercepts are $(0, 0)$, $(4, 0)$, and $(-2, 0)$.

31. 3

33. 5

35. 3

37. 5

39. Yes. Number of turning points is 2. Least possible degree is 3.

41. Yes. Number of turning points is 1. Least possible degree is 2.

43. Yes. Number of turning points is 0. Least possible degree is 1.

45. Yes. Number of turning points is 0. Least possible degree is 1.

47.

x	$f(x)$
10	9,500
100	99,950,000
–10	9,500
–100	99,950,000

As $x \to -\infty$,
$$f(x) \to \infty, \text{ as } x \to \infty, \ f(x) \to \infty$$

49.

x	$f(x)$
10	-504
100	-941,094
–10	1,716
–100	1,061,106

As $x \to -\infty$,
$$f(x) \to \infty, \text{ as } x \to \infty, \ f(x) \to -\infty$$

51.

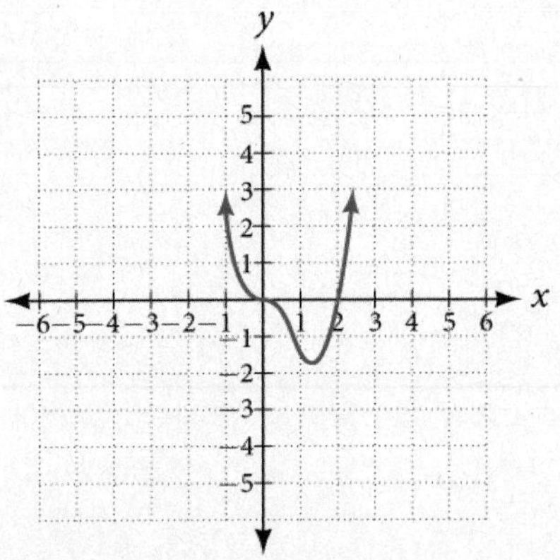

The y- intercept is $(0,\ 0)$. The x- intercepts are $(0,\ 0)$, $(2,\ 0)$. As $x \to -\infty$,

$f(x) \to \infty$, as $x \to \infty$, $f(x) \to \infty$

53.

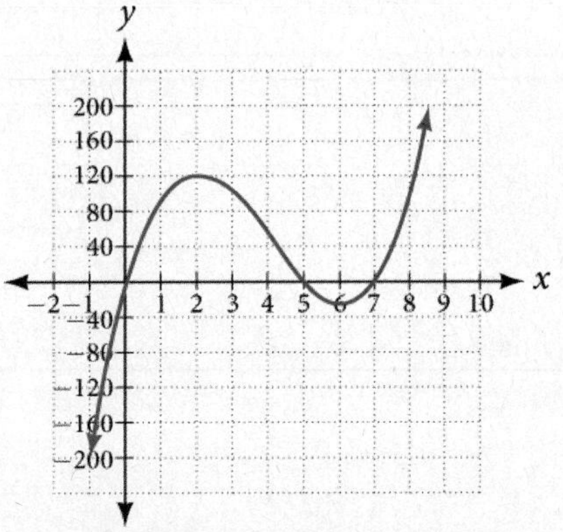

The y- intercept is $(0, 0)$. The x- intercepts are $(0,\ 0)$, $(5,\ 0)$, $(7,\ 0)$. As $x \to -\infty$,

$f(x) \to -\infty$, as $x \to \infty$, $f(x) \to \infty$

55.

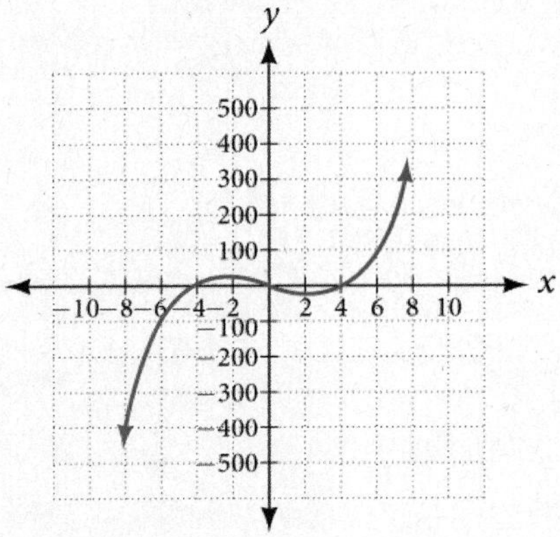

The y- intercept is $(0,\ 0)$. The x- intercept is $(-4,\ 0)$, $(0,\ 0)$, $(4,\ 0)$.

$As\ x \to -\infty$,

$f(x) \to -\infty$, as $x \to \infty$, $f(x) \to \infty$

57.

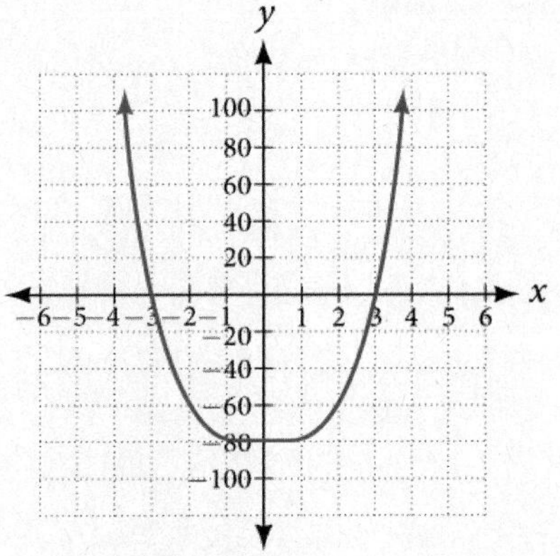

The y- intercept is $(0,\ -81)$. The x- intercept are $(3,\ 0)$, $(-3,\ 0)$. As $x \to -\infty$,

$f(x) \to \infty$, as $x \to \infty$, $f(x) \to \infty$

59.

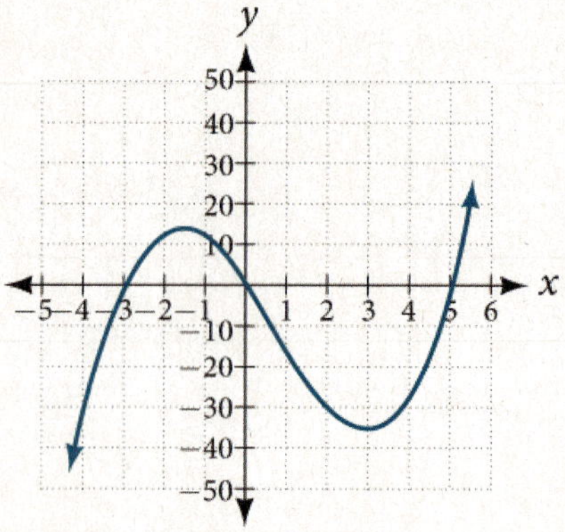

The y- intercept is $(0,\ 0)$. The x- intercepts are $(-3,\ 0),\ \ (0,\ 0),\ \ (5,\ 0)$. As $x \to -\infty,$

$f(x) \to -\infty,\ \text{as}\ x \to \infty,\ f(x) \to \infty$

61. $f(x) = x^2 - 4$

63. $f(x) = x^3 - 4x^2 + 4x$

65. $f(x) = x^4 + 1$

67. $V(m) = 8m^3 + 36m^2 + 54m + 27$

69. $V(x) = 4x^3 - 32x^2 + 64x$

5.3 Section Exercises

1. The x- intercept is where the graph of the function crosses the x- axis, and the zero of the function is the input value for which $f(x) = 0$.

3. If we evaluate the function at a and at b and the sign of the function value changes, then we know a zero exists between a and b.

5. There will be a factor raised to an even power.

7. $(-2, 0), (3, 0), (-5, 0)$

9. $(3, 0), (-1, 0), (0, 0)$

11. $(0, 0),\ \ (-5, 0),\ \ (2, 0)$

13. $(0, 0),\ \ (-5, 0),\ \ (4, 0)$

15. $(2, 0),\ \ (-2, 0),\ \ (-1, 0)$

17. $(-2, 0),\ \ (2, 0),\ \ \left(\frac{1}{2}, 0\right)$

19. $(1, 0),\ \ (-1, 0)$

21. $(0, 0),\ \ (\sqrt{3}, 0),\ \ (-\sqrt{3}, 0)$

23. $(0, 0),\ \ (1, 0),\ (-1, 0),\ \ (2, 0),\ \ (-2, 0)$

25. $f(2) =-10$ and $f(4) = 28$. Sign change confirms.

27. $f(1) = 3$ and $f(3) =-77$. Sign change confirms.

29. $f(0.01) = 1.000001$ and $f(0.1) =-7.999$. Sign change confirms.

31. 0 with multiplicity 2, $-\frac{3}{2}$ with multiplicity 5, 4 with multiplicity 2

33. 0 with multiplicity 2, -2 with multiplicity 2

35. $-\frac{2}{3}$ with multiplicity 5, 5 with multiplicity 2

37. 0 with multiplicity 4, 2 with multiplicity 1, −1 with multiplicity 1

39. $\frac{3}{2}$ with multiplicity 2, 0 with multiplicity 3

41. 0 with multiplicity 6, $\frac{2}{3}$ with multiplicity 2

43. x-intercepts, $(1, 0)$ with multiplicity 2, $(-4, 0)$ with multiplicity 1, y- intercept $(0, 4)$. As $x \to -\infty$, $f(x) \to -\infty$, as $x \to \infty$, $f(x) \to \infty$.

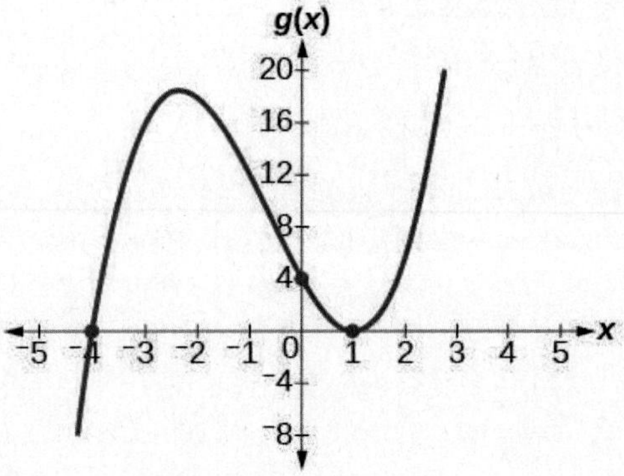

45. x-intercepts $(3, 0)$ with multiplicity 3, $(2, 0)$ with multiplicity 2, y- intercept $(0, -108)$. As $x \to -\infty$, $f(x) \to -\infty$, as $x \to \infty$, $f(x) \to \infty$.

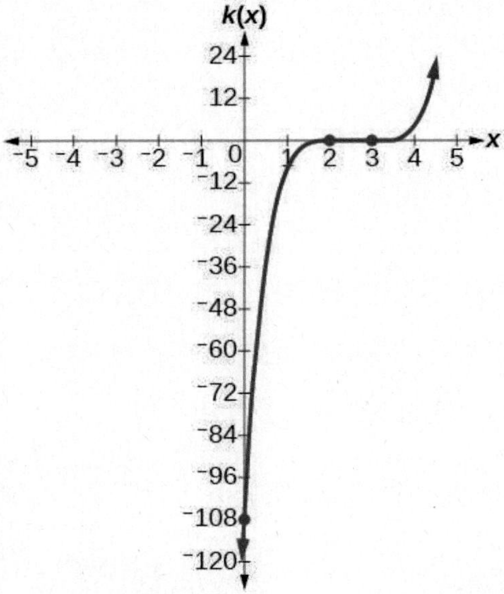

47. x-intercepts $(0, 0)$, $(-2, 0)$, $(4, 0)$ with multiplicity 1, y- intercept $(0, 0)$. As $x \to -\infty$, $f(x) \to \infty$, as $x \to \infty$, $f(x) \to -\infty$.

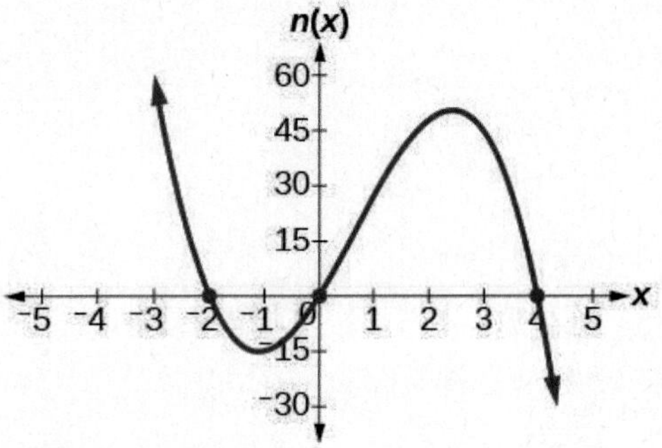

49. $f(x) = -\frac{2}{9}(x-3)(x+1)(x+3)$

51. $f(x) = \frac{1}{4}(x+2)^2(x-3)$

53. -4, -2, 1, 3 with multiplicity 1

55. -2, 3 each with multiplicity 2

57. $f(x) = -\frac{2}{3}(x+2)(x-1)(x-3)$

59. $f(x) = \frac{1}{3}(x-3)^2(x-1)^2(x+3)$

61. $f(x) = -15(x-1)^2(x-3)^3$

63. $f(x) = -2(x+3)(x+2)(x-1)$

65. $f(x) = -\frac{3}{2}(2x-1)^2(x-6)(x+2)$

67. local max $(-.58, -.62)$, local min $(.58, -1.38)$

69. global min $(-.63, -.47)$

71. global min $(.75, .89)$

73. $f(x) = (x-500)^2(x+200)$

75. $f(x) = 4x^3 - 36x^2 + 80x$

77. $f(x) = 4x^3 - 36x^2 + 60x + 100$

79. $f(x) = 9\pi(x^3 + 5x^2 + 8x + 4)$

5.4 Section Exercises

1. The binomial is a factor of the polynomial.

3. $x + 6 + \frac{5}{x-1}$, quotient: $x + 6$, remainder: 5

5. $3x + 2$, quotient: $3x + 2$, remainder: 0

7. $x - 5$, quotient: $x - 5$, remainder: 0

9. $2x - 7 + \frac{16}{x+2}$, quotient: $2x - 7$, remainder: 16

11. $x - 2 + \frac{6}{3x+1}$, quotient: $x - 2$, remainder: 6

13. $2x^2 - 3x + 5$, quotient: $2x^2 - 3x + 5$, remainder: 0

15. $2x^2 + 2x + 1 + \frac{10}{x-4}$

17. $2x^2 - 7x + 1 - \frac{2}{2x+1}$

19. $3x^2 - 11x + 34 - \frac{106}{x+3}$

21. $x^2 + 5x + 1$

23. $4x^2 - 21x + 84 - \frac{323}{x+4}$

25. $x^2 - 14x + 49$

27. $3x^2 + x + \frac{2}{3x-1}$

29. $x^3 - 3x + 1$

31. $x^3 - x^2 + 2$

33. $x^3 - 6x^2 + 12x - 8$

35. $x^3 - 9x^2 + 27x - 27$

37. $2x^3 - 2x + 2$

39. Yes $(x - 2)(3x^3 - 5)$

41. Yes $(x - 2)(4x^3 + 8x^2 + x + 2)$

43. No

45. $(x - 1)(x^2 + 2x + 4)$

47. $(x - 5)(x^2 + x + 1)$

49. Quotient: $4x^2 + 8x + 16$, remainder: -1

51. Quotient: $3x^2 + 3x + 5$, remainder: 0

53. Quotient: $x^3 - 2x^2 + 4x - 8$, remainder: -6

55. $x^6 - x^5 + x^4 - x^3 + x^2 - x + 1$

57. $x^3 - x^2 + x - 1 + \frac{1}{x+1}$

59. $1 + \frac{1+i}{x-i}$

61. $1 + \frac{1-i}{x+i}$

63. $x^2 - ix - 1 + \frac{1-i}{x-i}$

65. $2x^2 + 3$

67. $2x + 3$

69. $x + 2$

71. $x - 3$

73. $3x^2 - 2$

5.5 Section Exercises

1. The theorem can be used to evaluate a polynomial.

3. Rational zeros can be expressed as fractions whereas real zeros include irrational numbers.

5. Polynomial functions can have repeated zeros, so the fact that number is a zero doesn't preclude it being a zero again.

7. -106

9. 0

11. 255

13. -1

15. $-2,\ 1,\ \frac{1}{2}$

17. -2

19. -3

21. $-\frac{5}{2},\ \sqrt{6},\ -\sqrt{6}$

23. $2,\ -4,\ -\frac{3}{2}$

25. $4,\ -4,\ -5$

27. $5,\ -3,\ -\frac{1}{2}$

29. $\frac{1}{2},\ \frac{1+\sqrt{5}}{2},\ \frac{1-\sqrt{5}}{2}$

31. $\frac{3}{2}$

33. $2,\ 3,\ -1,\ -2$

35. $\frac{1}{2},\ -\frac{1}{2},\ 2,\ -3$

37. $-1,\ -1,\ \sqrt{5},\ -\sqrt{5}$

39. $-\frac{3}{4},\ -\frac{1}{2}$

41. $2,\ 3+2i,\ 3-2i$

43. $-\frac{2}{3},\ 1+2i,\ 1-2i$

45. $-\frac{1}{2},\ 1+4i,\ 1-4i$

47. 1 positive, 1 negative

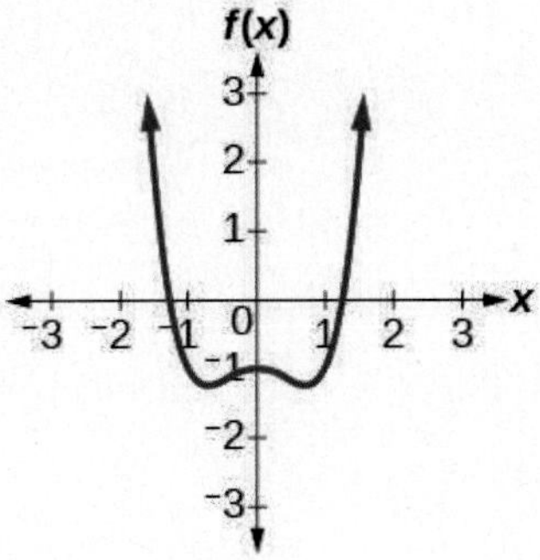

49. 3 or 1 positive, 0 negative

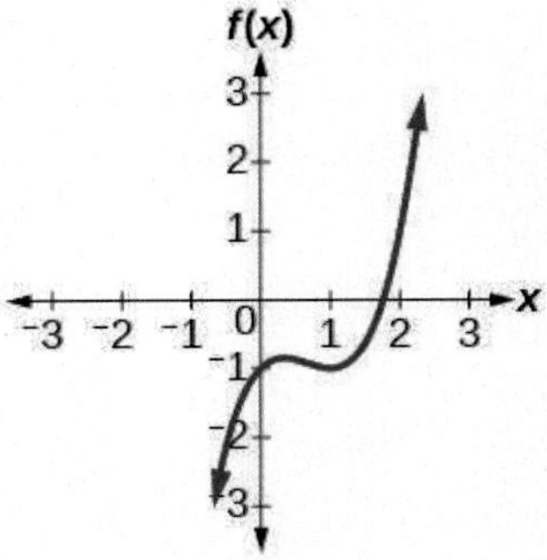

51. 0 positive, 3 or 1 negative

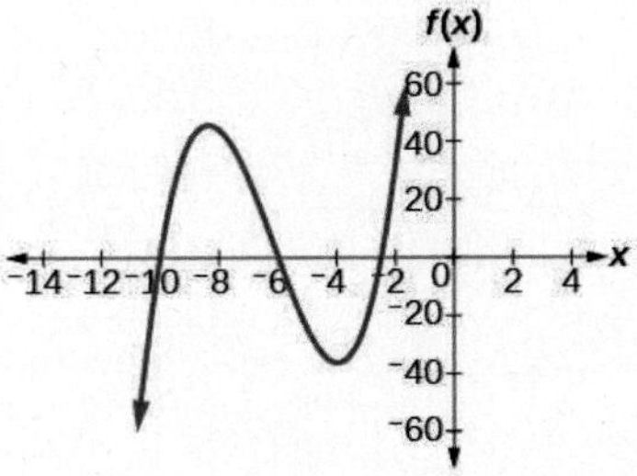

53. 2 or 0 positive, 2 or 0 negative

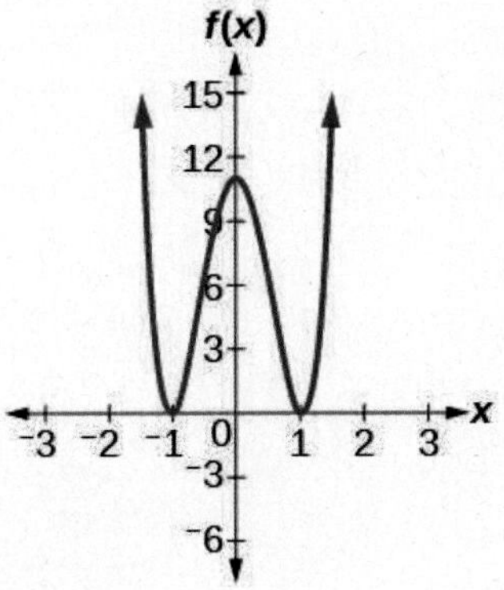

55. 2 or 0 positive, 2 or 0 negative

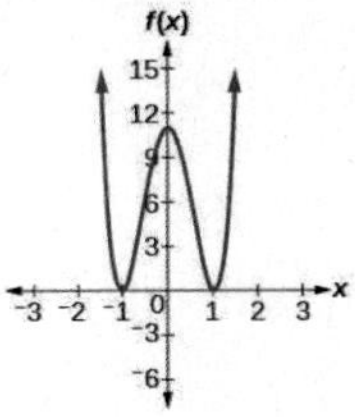

57. $\pm 5,\ \pm 1,\ \pm\frac{5}{2},\ \pm\frac{1}{2}$

59. $\pm 1,\ \pm\frac{1}{2},\ \pm\frac{1}{3},\ \pm\frac{1}{6}$

61. $1,\ \frac{1}{2},\ -\frac{1}{3}$

63. $2,\ \frac{1}{4},\ -\frac{3}{2}$

65. $\frac{5}{4}$

67. $f(x) = \frac{4}{9}\left(x^3 + x^2 - x - 1\right)$

69. $f(x) = -\frac{1}{5}\left(4x^3 - x\right)$

71. 8 by 4 by 6 inches

73. 5.5 by 4.5 by 3.5 inches

75. 8 by 5 by 3 inches

77. Radius = 6 meters, Height = 2 meters

79. Radius = 2.5 meters, Height = 4.5 meters

5.6 Section Exercises

1. The rational function will be represented by a quotient of polynomial functions.

3. The numerator and denominator must have a common factor.

5. Yes. The numerator of the formula of the functions would have only complex roots and/or factors common to both the numerator and denominator.

7. All reals $x \neq -1,\ 1$

9. All reals $x \neq -1,\ -2,\ 1,\ 2$

11. V.A. at $x = -\frac{2}{5}$; H.A. at $y = 0$; Domain is all reals $x \neq -\frac{2}{5}$

13. V.A. at $x = 4,\ -9$; H.A. at $y = 0$; Domain is all reals $x \neq 4,\ -9$

15. V.A. at $x = 0,\ 4,\ -4$; H.A. at $y = 0$; Domain is all reals $x \neq 0, 4,\ -4$

17. V.A. at $x = 5$; H.A. at $y = 0$; Domain is all reals $x \neq 5, -5$

19. V.A. at $x = \frac{1}{3}$; H.A. at $y = -\frac{2}{3}$; Domain is all reals $x \neq \frac{1}{3}$.

21. none

23. x-intercepts none, y-intercept $\left(0, \frac{1}{4}\right)$

25. Local behavior:
$x \to -\frac{1}{2}^{+}, f(x) \to -\infty, x \to -\frac{1}{2}^{-}, f(x) \to \infty$

End behavior: $x \to \pm\infty, f(x) \to \frac{1}{2}$

27. Local behavior:
$x \to 6^{+}, f(x) \to -\infty, x \to 6^{-}, f(x) \to \infty,$

End behavior: $x \to \pm\infty, f(x) \to -2$

29. Local behavior: $x \to -\frac{1}{3}^{+}, f(x) \to \infty, x \to -\frac{1}{3}^{-}, f(x) \to -\infty, x \to \frac{5}{2}^{-}, f(x) \to \infty, x \to \frac{5}{2}^{+}, f(x) \to -\infty$

End behavior: $x \to \pm\infty,\ f(x) \to \frac{1}{3}$

31. $y = 2x + 4$

33. $y = 2x$

35. $V.A.\ \ x = 0, H.A.\ \ y = 2$

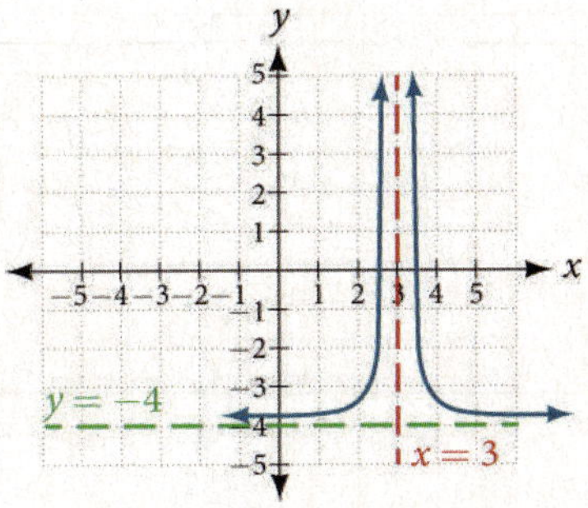

37. $V.A.\ \ x = 2,\ \ H.A.\ \ y = 0$

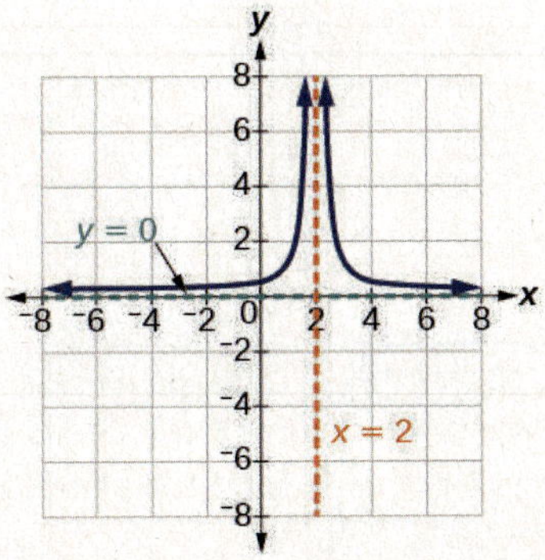

39. $V.A.\ \ x = -4,\ \ H.A.\ \ y = 2; \left(\frac{3}{2}, 0\right); \left(0, -\frac{3}{4}\right)$

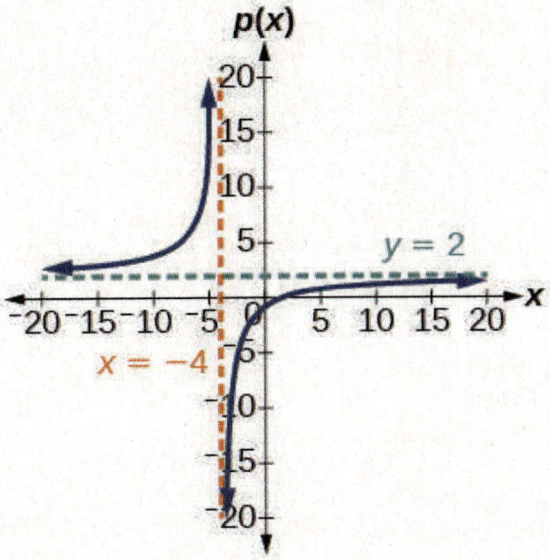

41. $V.A.\ \ x = 2,\ \ H.A.\ \ y = 0,\ \ (0, 1)$

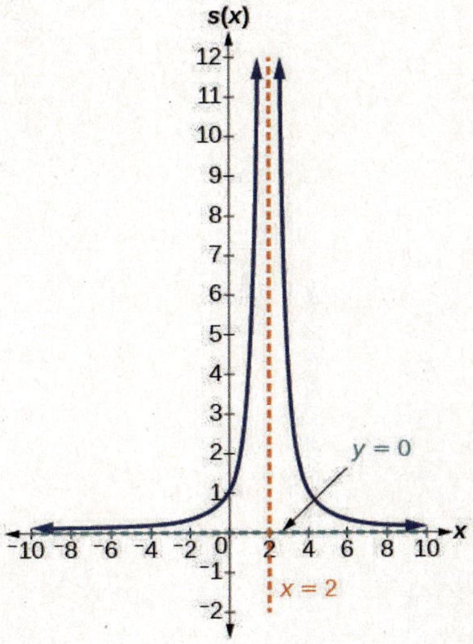

43. $V.A.\ \ x = -4,\ \ x = \frac{4}{3},\ \ H.A.\ \ y = 1; (5, 0); \left(-\frac{1}{3}, 0\right); \left(0, \frac{5}{16}\right)$

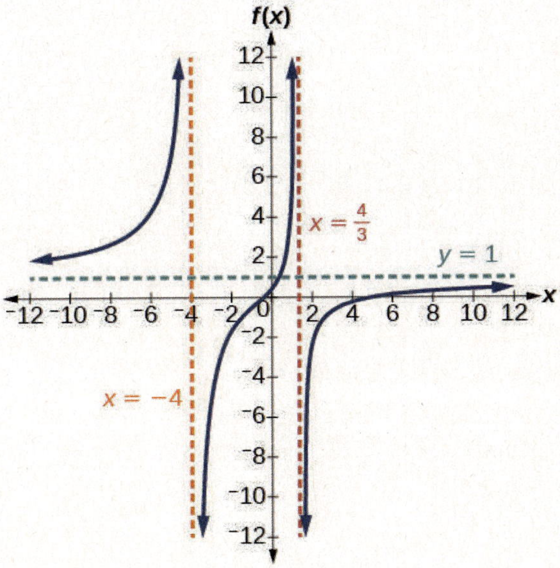

45. $V.A.\ \ x = -1,\ \ H.A.\ \ y = 1; (-3,0); (0,3)$

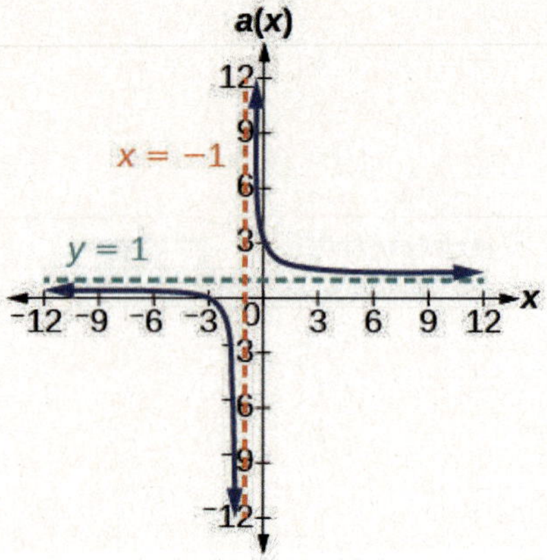

47. $V.A.\ \ x = 4,\ \ S.A.\ \ y = 2x + 9; (-1,0); \left(\frac{1}{2},0\right); \left(0,\frac{1}{4}\right)$

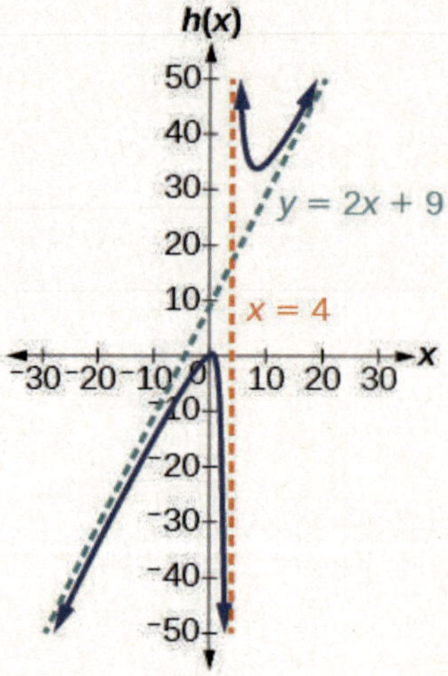

49. $V.A.\ \ x = -2,\ \ x = 4,\ \ H.A.\ \ y = 1, (1,0); (5,0); (-3,0); \left(0, -\frac{15}{16}\right)$

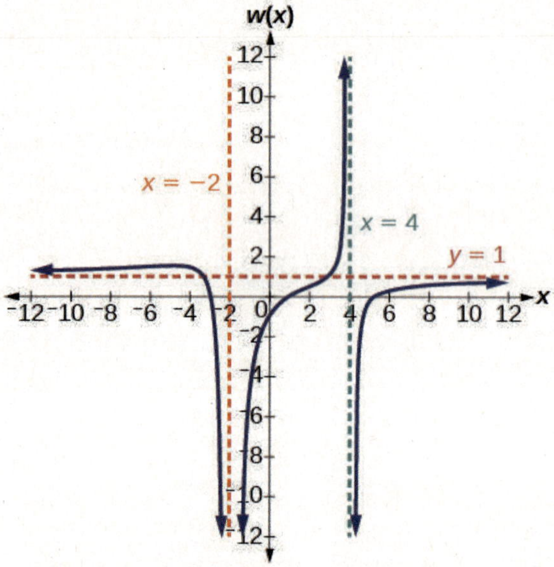

51. $y = 50\frac{x^2-x-2}{x^2-25}$

53. $y = 7\frac{x^2+2x-24}{x^2+9x+20}$

55. $y = \frac{1}{2}\frac{x^2-4x+4}{x+1}$

57. $y = 4\frac{x-3}{x^2-x-12}$

59. $y = \frac{27(x-2)}{(x+3)(x-3)^2}$

61. $y = \frac{1}{3}\frac{x^2+x-6}{x-1}$

63. $y = -6\frac{(x-1)^2}{(x+3)(x-2)^2}$

65.

x	2.01	2.001	2.0001	1.99	1.999
y	100	1,000	10,000	−100	−1,000

x	10	100	1,000	10,000	100,000
y	.125	.0102	.001	.0001	.00001

Vertical asymptote $x = 2$, Horizontal asymptote $y = 0$

67.

x	−4.1	−4.01	−4.001	−3.99	−3.999
y	82	802	8,002	−798	−7998

x	10	100	1,000	10,000	100,000
y	1.4286	1.9331	1.992	1.9992	1.999992

Vertical asymptote $x = -4$, Horizontal asymptote $y = 2$

69.

x	−.9	−.99	−.999	−1.1	−1.01
y	81	9,801	998,001	121	10,201

x	10	100	1,000	10,000	100,000
y	.82645	.9803	.998	.9998	

Vertical asymptote $x = -1$, Horizontal asymptote $y = 1$

71. $\left(\frac{3}{2}, \infty\right)$

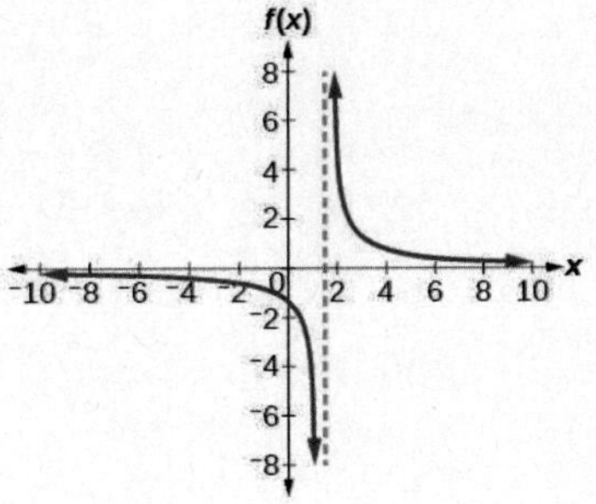

73. $(-2, 1) \cup (4, \infty)$

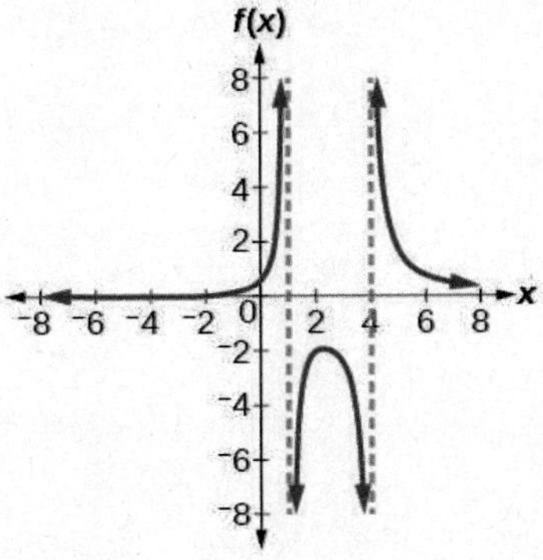

75. $(2, 4)$

77. $(2, 5)$

79. $(-1, 1)$

81. $C(t) = \frac{8+2t}{300+20t}$

83. After about 6.12 hours.

85. $A(x) = 50x^2 + \frac{800}{x}$. 2 by 2 by 5 feet.

87. $A(x) = \pi x^2 + \frac{100}{x}$. Radius = 2.52 meters.

5.7 Section Exercises

1. It can be too difficult or impossible to solve for x in terms of y.

3. We will need a restriction on the domain of the answer.

5. $f^{-1}(x) = \sqrt{x} + 4$

7. $f^{-1}(x) = \sqrt{x+3} - 1$

9. $f^{-1}(x) = \sqrt{12 - x}$

11. $f^{-1}(x) = \pm\sqrt{\frac{x-4}{2}}$

13. $f^{-1}(x) = \sqrt{\frac{x-1}{3}3}$

15. $f^{-1}(x) = \sqrt{\frac{4-x}{2}3}$

17. $f^{-1}(x) = \frac{3-x^2}{4}, \quad \left[0, \infty\right)$

19. $f^{-1}(x) = \frac{(x-5)^2+8}{6}$

21. $f^{-1}(x) = (3 - x)^2$

23. $f^{-1}(x) = \frac{4x+3}{x}$

25. $f^{-1}(x) = \frac{7x-3}{1-x}$

27. $f^{-1}(x) = \frac{2x-1}{5x+5}$

29. $f^{-1}(x) = \sqrt{x+3} - 2$

31.

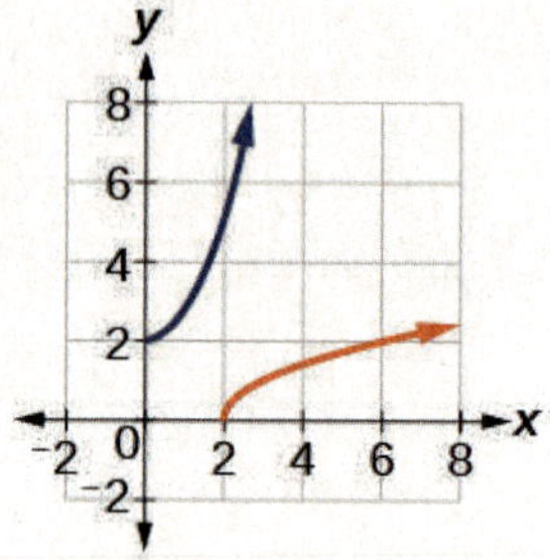

$f^{-1}(x) = \sqrt{x-2}$

33.

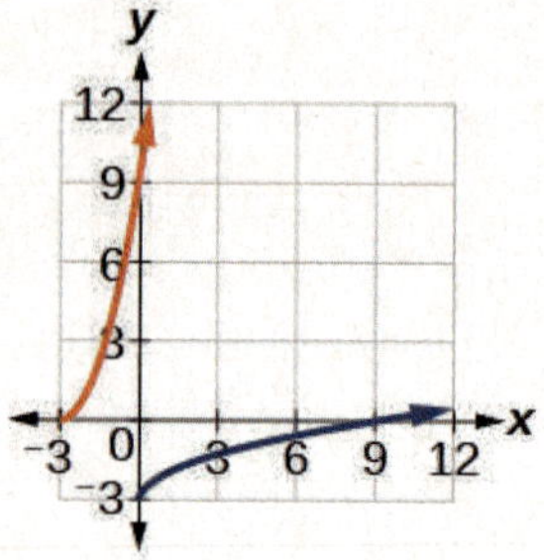

$f^{-1}(x) = \sqrt{x-3}$

35.

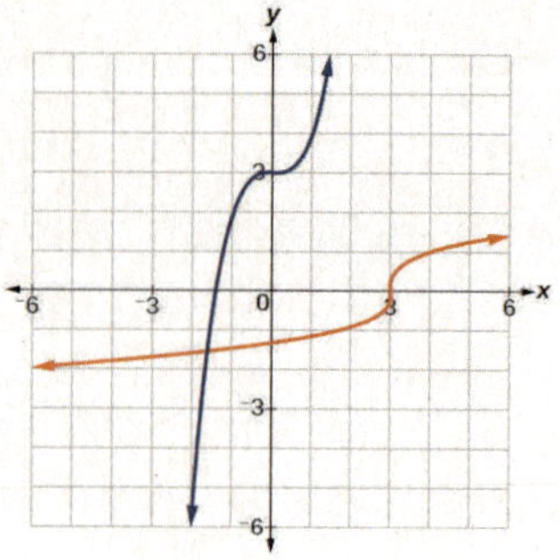

$f^{-1}(x) = \sqrt[3]{x-3}$

37.

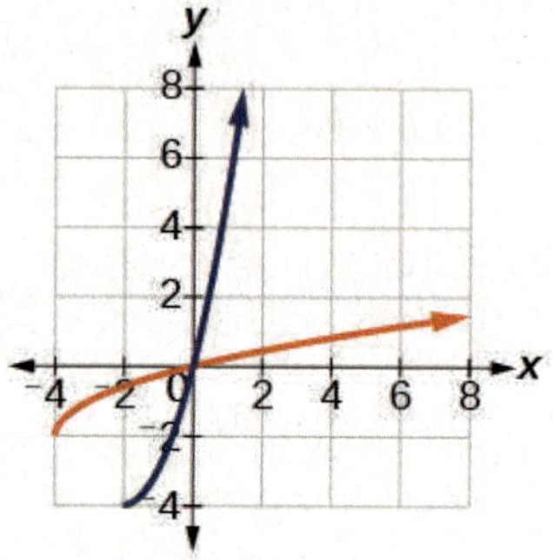

$f^{-1}(x) = \sqrt{x+4} - 2$

39.

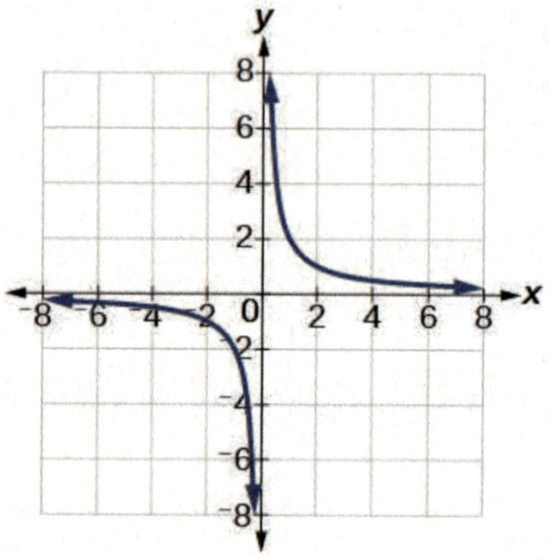

41.

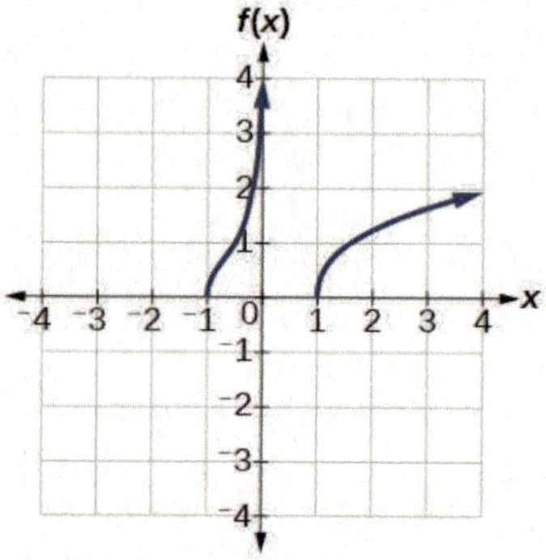

$[-1, 0) \cup [1, \infty)$

43.

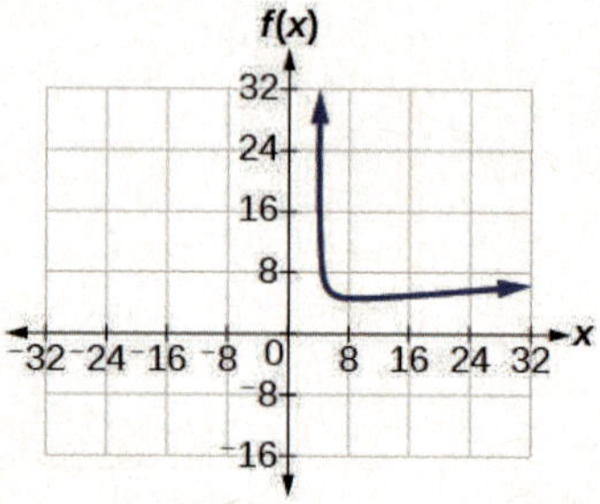

$[-3, 0] \cup (4, \infty)$

45.

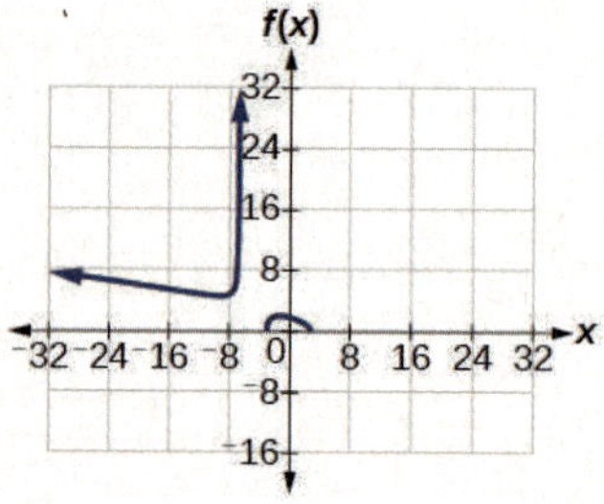

$[-\infty, -4] \cdot [-3, 3]$

47.

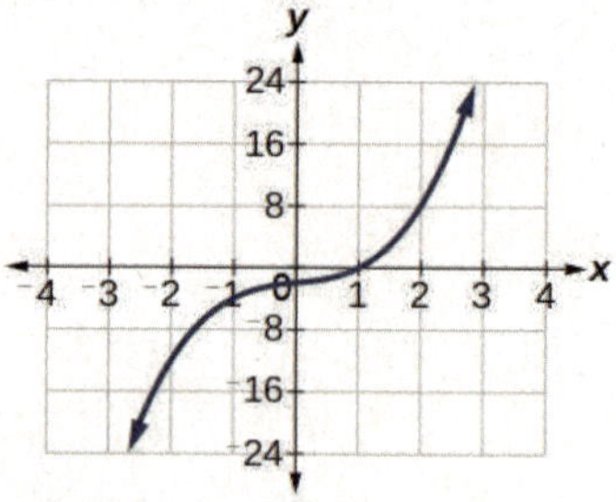

$(-2, 0), (0, 1), (8, 2)$

49.

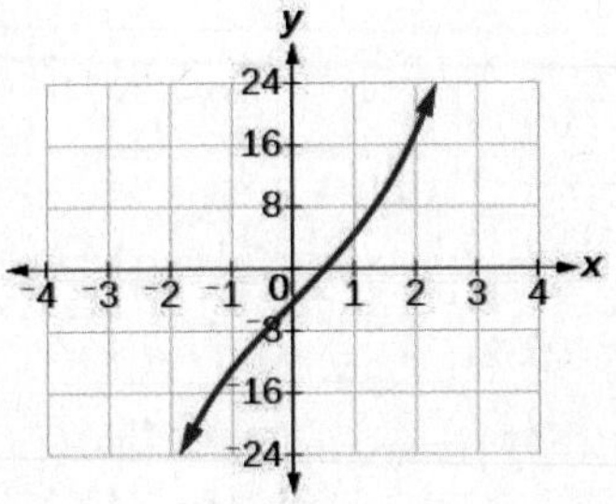

$(-13, -1), (-4, 0), (5, 1)$

51. $f^{-1}(x) = \sqrt[3]{\frac{x-b}{a}}$

53. $f^{-1}(x) = \frac{\sqrt{x^2-b}}{a}$

55. $f^{-1}(x) = \frac{cx-b}{a-x}$

57. $t(h) = \sqrt{\frac{600-h}{16}}$, 3.54 seconds

59. $r(A) = \sqrt{\frac{A}{4\pi}}$, $\approx$ 8.92 in.

61. $l(T) = 32.2\left(\frac{T}{2\pi}\right)$, $\approx$ 3.26 ft

63. $r(A) = \sqrt{\frac{A+2\pi}{8\pi}}$, -2, 3.99 ft

65. $r(V) = \sqrt{\frac{V}{10\pi}}$, ≈ 5.64 ft

5.8 Section Exercises

1. The graph will have the appearance of a power function.

3. No. Multiple variables may jointly vary.

5. $y = 5x^2$

7. $y = \frac{1}{1944}x^3$

9. $y = 6x^4$

11. $y = \frac{18}{x^2}$

13. $y = \frac{81}{x^4}$

15. $y = \frac{20}{\sqrt[3]{x}}$

17. $y = 10xzw$

19. $y = 10x\sqrt{z}$

21. $y = 4\frac{xz}{w}$

23. $y = 40\frac{xz}{\sqrt{w}t^2}$

25. $y = 256$

27. $y = 6$

29. $y = 6$

31. $y = 27$

33. $y = 3$

35. $y = 18$

37. $y = 90$

39. $y = \frac{81}{2}$

41. $y = \frac{3}{4}x^2$

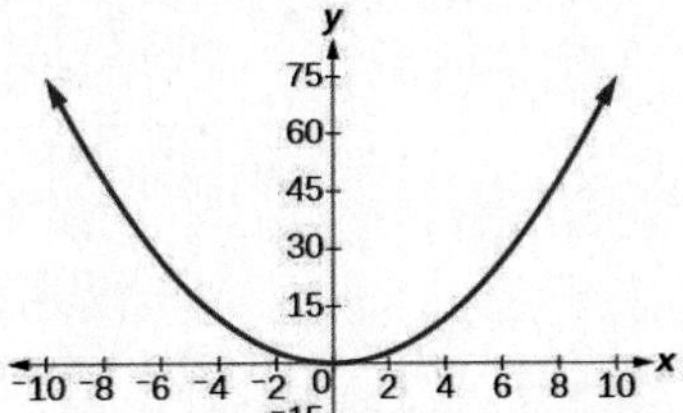

43. $y = \frac{1}{3}\sqrt{x}$

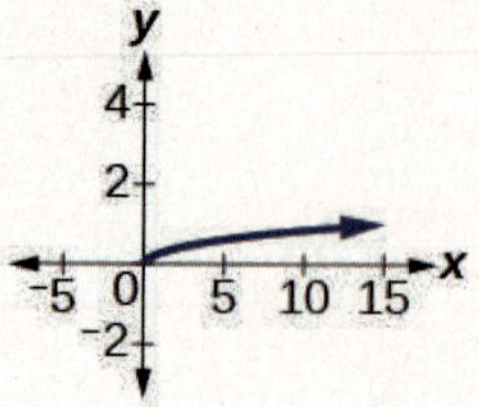

45. $y = \frac{4}{x^2}$

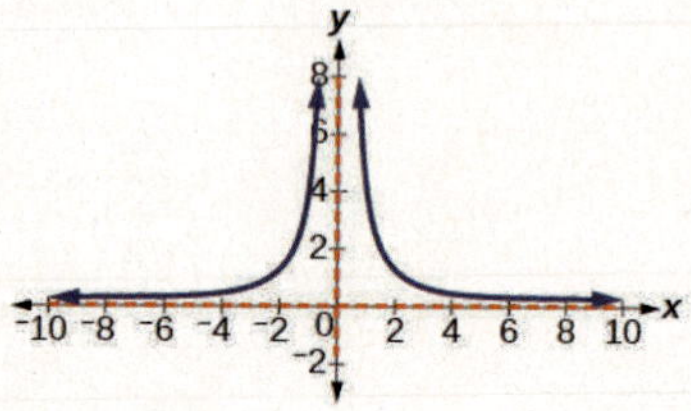

47. 1.89 years

49. 0.61 years

51. 3 seconds

53. 48 inches

55. 49.75 pounds

57. 33.33 amperes

59. 2.88 inches

Review Exercises

1. $f(x) = (x-2)^2 - 9$ vertex $(2,-9)$, intercepts $(5,0)$; $(-1,0)$; $(0,-5)$

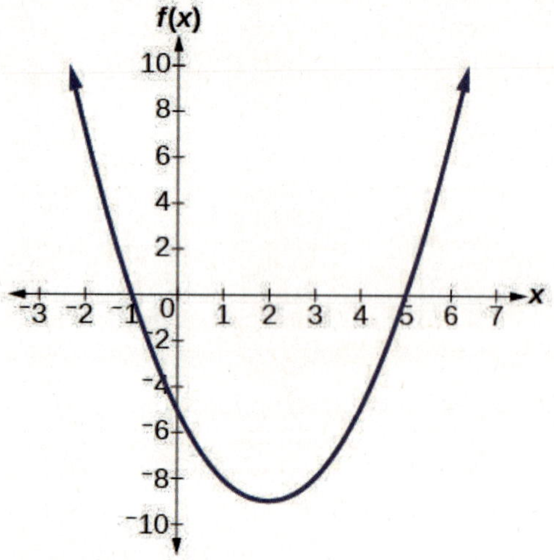

3. $f(x) = \frac{3}{25}(x+2)^2 + 3$

5. 300 meters by 150 meters, the longer side parallel to river.

7. Yes, degree = 5, leading coefficient = 4

9. Yes, degree = 4, leading coefficient = 1

11. As $x \to -\infty,\ f(x) \to -\infty,$ as $x \to \infty,\ f(x) \to \infty$

13. -3 with multiplicity 2, $-\frac{1}{2}$ with multiplicity 1, -1 with multiplicity 3

15. 4 with multiplicity 1

17. $\frac{1}{2}$ with multiplicity 1, 3 with multiplicity 3

19. $x^2 + 4$ with remainder 12

21. $x^2 - 5x + 20 - \frac{61}{x+3}$

23. $2x^2 - 2x - 3$, so factored form is $(x+4)(2x^2 - 2x - 3)$

25. $\left\{-2,\ 4,\ -\frac{1}{2}\right\}$

27. $\{1, 3, 4, \frac{1}{2}\}$

29. 0 or 2 positive, 1 negative

31. Intercepts $(-2, 0)$and $\left(0, -\frac{2}{5}\right)$, Asymptotes $x = 5$ and $y = 1$.

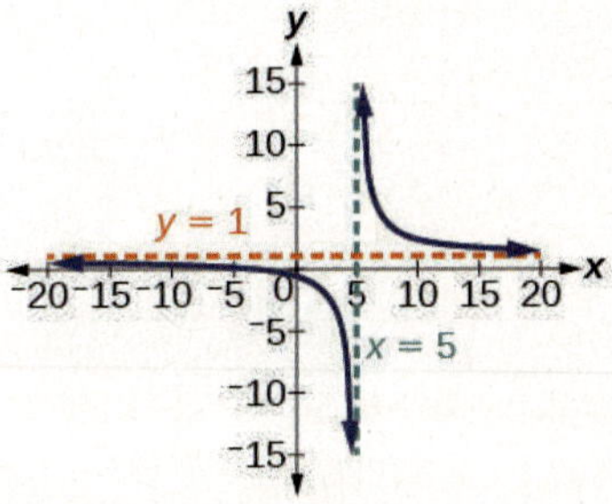

33. Intercepts (3, 0), (-3, 0), and $\left(0, \frac{27}{2}\right)$, Asymptotes $x = 1,\ x = -2,\ y = 3$.

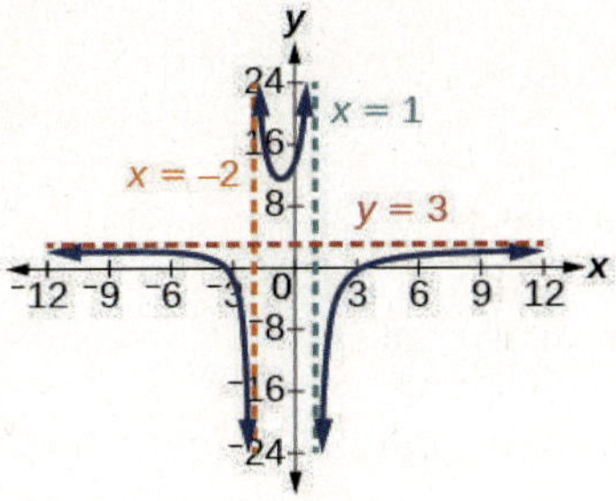

35. $y = \ x - 2$

37. $f^{-1}(x) = \sqrt{x} + 2$

39. $f^{-1}(x) = \sqrt{x + 11} - 3$

41. $f^{-1}(x) = \frac{(x+3)^2 - 5}{4},\ x \geq -3$

43. $y = 64$

45. $y \ = \ 72$

47. 148.5 pounds

Practice Test

1. Degree: 5, leading coefficient: −2

3. $\text{As} x \to -\infty, f(x) \to \infty, \text{As} x \to \infty, f(x) \to \infty$

5. $f(x) = 3(x - 2)^2$

7. 3 with multiplicity 3, $\frac{1}{3}$ with multiplicity 1, 1 with multiplicity 2

9. $-\frac{1}{2}$ with multiplicity 3, 2 with multiplicity 2

11. $x^3 + 2x^2 + 7x + 14 + \frac{26}{x-2}$

13. $\left\{-3, -1, \frac{3}{2}\right\}$

15. 1, −2, and − $\frac{3}{2}$ (multiplicity 2)

17. $f(x) = -\frac{2}{3}(x-3)^2(x-1)(x+2)$

19. 2 or 0 positive, 1 negative

21. $(-3,0)(1,0)\left(0,\frac{3}{4}\right)$

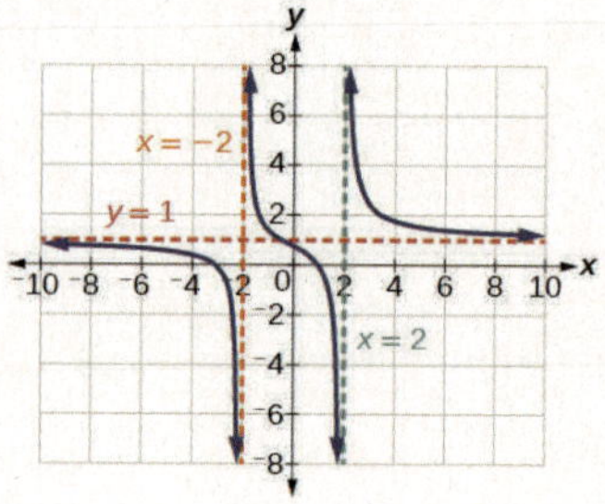

23. $f^{-1}(x) = (x-4)^2 + 2, x \geq 4$

25. $f^{-1}(x) = \frac{x+3}{3x-2}$

27. $y = 20$

Chapter 6

Try It

6.1 Exponential Functions

1. $g(x) = 0.875^x$ and $j(x) = 1095.6^{-2x}$ represent exponential functions.

2. 5.5556

3. About 1.548 billion people; by the year 2031, India's population will exceed China's by about 0.001 billion, or 1 million people.

4. $(0, 129)$ and $(2, 236)$; $N(t) = 129(1.3526)^t$

5. $f(x) = 2(1.5)^x$

6. $f(x) = \sqrt{2}\left(\sqrt{2}\right)^x$. Answers may vary due to round-off error. The answer should be very close to $1.4142(1.4142)^x$.

7. $y \approx 12 \cdot 1.85^x$

8. about $3,644,675.88

9. $13,693

10. $e^{-0.5} \approx 0.60653$

11. $3,659,823.44

12. 3.77E-26 (This is calculator notation for the number written as 3.77×10^{-26} in scientific notation. While the output of an exponential function is never zero, this number is so close to zero that for all practical purposes we can accept zero as the answer.)

6.2 Graphs of Exponential Functions

1. The domain is $\left(-\infty, \infty\right)$; the range is $\left(0, \infty\right)$; the horizontal asymptote is $y = 0$.

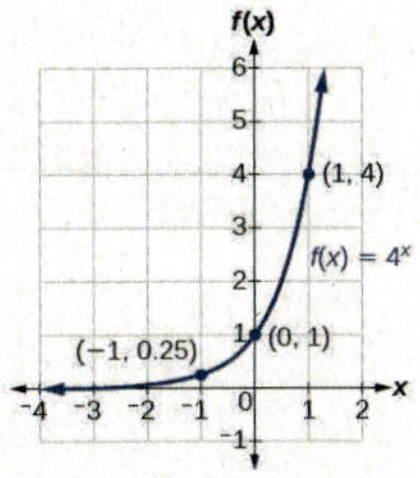

2. The domain is $\left(-\infty, \infty\right)$; the range is $\left(3, \infty\right)$; the horizontal asymptote is $y = 3$.

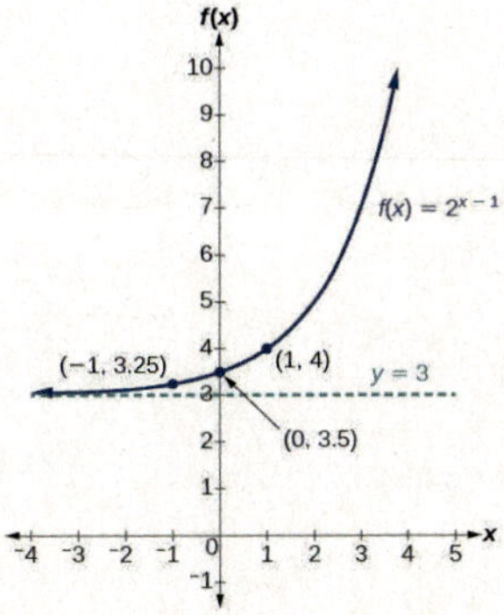

3. $x \approx -1.608$

4. The domain is $\left(-\infty, \infty\right)$; the range is $\left(0, \infty\right)$; the horizontal asymptote is $y = 0$.

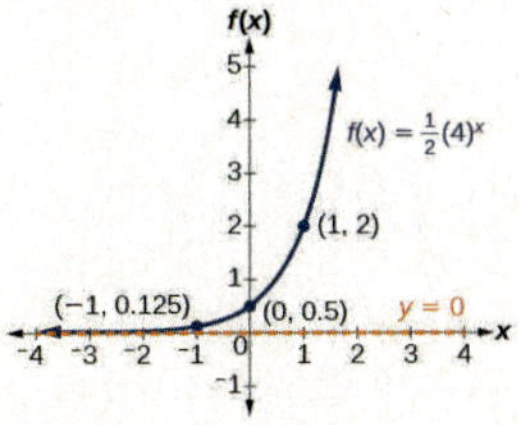

5. The domain is $\left(-\infty, \infty\right)$; the range is $\left(0, \infty\right)$; the horizontal asymptote is $y = 0$.

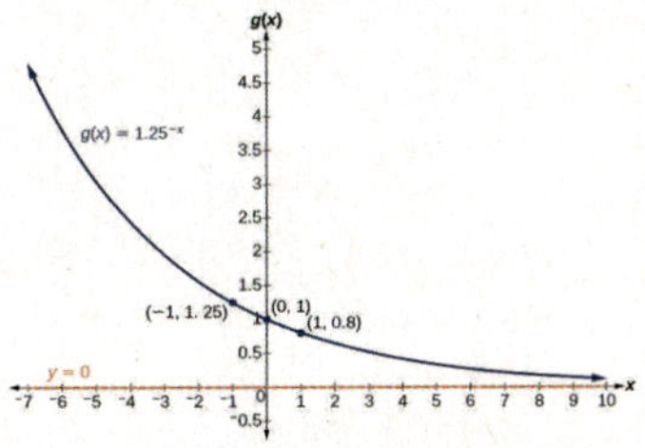

6. $f(x) = -\frac{1}{3}e^x - 2$; the domain is $\left(-\infty, \infty\right)$; the range is $\left(-\infty, -2\right)$; the horizontal asymptote is $y = -2$.

6.3 Logarithmic Functions

1. ⓐ $\log_{10}(1{,}000{,}000) = 6$ is equivalent to $10^6 = 1{,}000{,}000$
ⓑ $\log_5(25) = 2$ is equivalent to $5^2 = 25$

2. ⓐ $3^2 = 9$ is equivalent to $\log_3(9) = 2$
ⓑ $5^3 = 125$ is equivalent to $\log_5(125) = 3$
ⓒ $2^{-1} = \frac{1}{2}$ is equivalent to $\log_2\left(\frac{1}{2}\right) = -1$

3. $\log_{121}(11) = \frac{1}{2}$ (recalling that $\sqrt{121} = (121)^{\frac{1}{2}} = 11$)

4. $\log_2\left(\frac{1}{32}\right) = -5$

5. $\log(1{,}000{,}000) = 6$

6. $\log(123) \approx 2.0899$

7. The difference in magnitudes was about 3.929.

8. It is not possible to take the logarithm of a negative number in the set of real numbers.

6.4 Graphs of Logarithmic Functions

1. $(2, \infty)$

2. $(5, \infty)$

3.

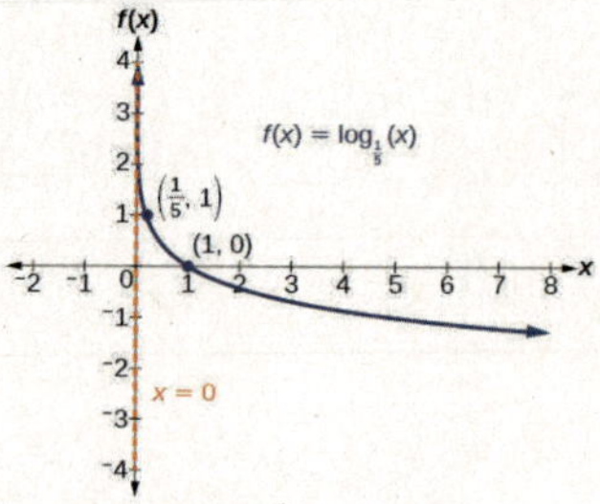

The domain is $(0, \infty)$, the range is $(-\infty, \infty)$, and the vertical asymptote is $x = 0$.

4.

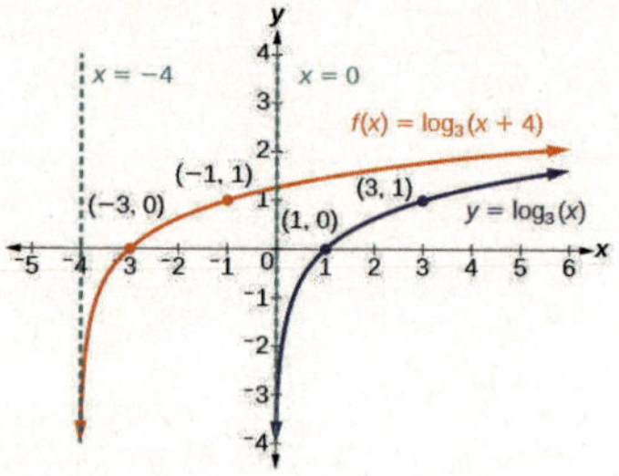

The domain is $(-4, \infty)$, the range $(-\infty, \infty)$, and the asymptote $x = -4$.

5.

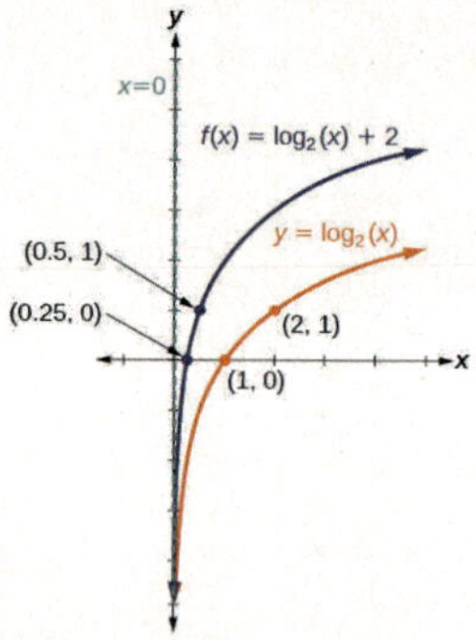

The domain is $(0, \infty)$, the range is $(-\infty, \infty)$, and the vertical asymptote is $x = 0$.

6.

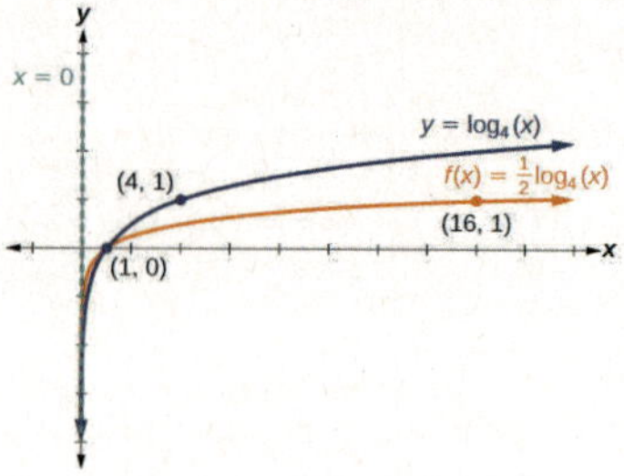

The domain is $(0, \infty)$, the range is $(-\infty, \infty)$, and the vertical asymptote is $x = 0$.

7.

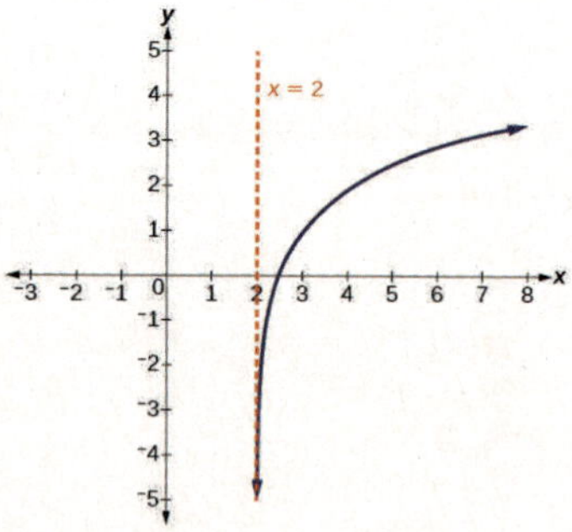

The domain is $(2, \infty)$, the range is $(-\infty, \infty)$, and the vertical asymptote is $x = 2$.

8.

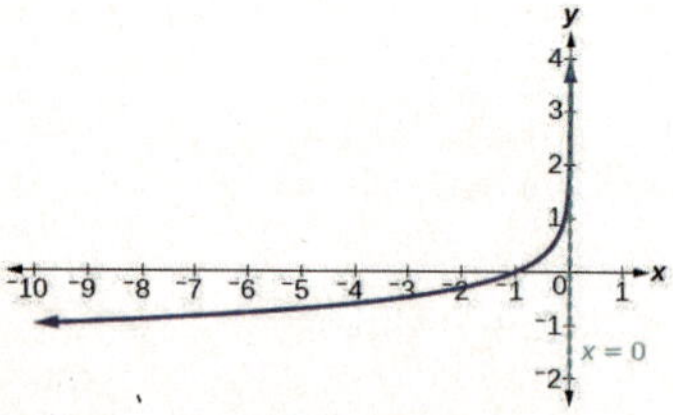

The domain is $(-\infty, 0)$, the range is $(-\infty, \infty)$, and the vertical asymptote is $x = 0$.

9. $x \approx 3.049$

10. $x = 1$

11. $f(x) = 2\ln(x + 3) - 1$

6.5 Logarithmic Properties

1. $\log_b 2 + \log_b 2 + \log_b 2 + \log_b k = 3\log_b 2 + \log_b k$
2. $\log_3(x+3) - \log_3(x-1) - \log_3(x-2)$
3. $2\ln x$
4. $-2\ln(x)$
5. $\log_3 16$
6. $2\log x + 3\log y - 4\log z$
7. $\frac{2}{3}\ln x$
8. $\frac{1}{2}\ln(x-1) + \ln(2x+1) - \ln(x+3) - \ln(x-3)$
9. $\log\left(\frac{3\cdot 5}{4\cdot 6}\right)$; can also be written $\log\left(\frac{5}{8}\right)$ by reducing the fraction to lowest terms.
10. $\log\left(\frac{5(x-1)^3\sqrt{x}}{(7x-1)}\right)$
11. $\log\frac{x^{12}(x+5)^4}{(2x+3)^4}$; this answer could also be written $\log\left(\frac{x^3(x+5)}{(2x+3)}\right)^4$.
12. The pH increases by about 0.301.
13. $\frac{\ln 8}{\ln 0.5}$
14. $\frac{\ln 100}{\ln 5} \approx \frac{4.6051}{1.6094} = 2.861$

6.6 Exponential and Logarithmic Equations

1. $x = -2$
2. $x = -1$
3. $x = \frac{1}{2}$
4. The equation has no solution.
5. $x = \frac{\ln 3}{\ln\left(2/3\right)}$
6. $t = 2\ln\left(\frac{11}{3}\right)$ or $\ln\left(\frac{11}{3}\right)^2$
7. $t = \ln\left(\frac{1}{\sqrt{2}}\right) = -\frac{1}{2}\ln(2)$
8. $x = \ln 2$
9. $x = e^4$
10. $x = e^5 - 1$
11. $x \approx 9.97$
12. $x = 1$ or $x = -1$
13. $t = 703{,}800{,}000 \times \frac{\ln(0.8)}{\ln(0.5)}$ years $\approx$ 226,572,993 years.

6.7 Exponential and Logarithmic Models

1. $f(t) = A_0 e^{-0.0000000087t}$
2. less than 230 years, 229.3157 to be exact
3. $f(t) = A_0 e^{\frac{\ln 2}{3}t}$
4. 6.026 hours
5. 895 cases on day 15
6. Exponential. $y = 2e^{0.5x}$.
7. $y = 3e^{(\ln 0.5)x}$

6.8 Fitting Exponential Models to Data

1. ⓐ The exponential regression model that fits these data is $y = 522.88585984(1.19645256)^x$. ⓑ If spending continues at this rate, the graduate's credit card debt will be \$4,499.38 after one year.

2. ⓐ The logarithmic regression model that fits these data is $y = 141.91242949 + 10.45366573\ln(x)$ ⓑ If sales continue at this rate, about 171,000 games will be sold in the year 2015.

3. ⓐ The logistic regression model that fits these data is $y = \frac{25.65665979}{1+6.113686306e^{-0.3852149008x}}$. ⓑ If the population continues to grow at this rate, there will be about 25,634 seals in 2020. ⓒ To the nearest whole number, the carrying capacity is 25,657.

6.1 Section Exercises

1. Linear functions have a constant rate of change. Exponential functions increase based on a percent of the original.

3. When interest is compounded, the percentage of interest earned to principal ends up being greater than the annual percentage rate for the investment account. Thus, the annual percentage rate does not necessarily correspond to the real interest earned, which is the very definition of *nominal.*

5. exponential; the population decreases by a proportional rate. .

7. not exponential; the charge decreases by a constant amount each visit, so the statement represents a linear function. .

9. The forest represented by the function $B(t) = 82(1.029)^t$.

11. After $t = 20$ years, forest A will have 43 more trees than forest B.

13. Answers will vary. Sample response: For a number of years, the population of forest A will increasingly exceed forest B, but because forest B actually grows at a faster rate, the population will eventually become larger than forest A and will remain that way as long as the population growth models hold. Some factors that might influence the long-term validity of the exponential growth model are drought, an epidemic that culls the population, and other environmental and biological factors.

15. exponential growth; The growth factor, 1.06, is greater than 1.

17. exponential decay; The decay factor, 0.97, is between 0 and 1.

19. $f(x) = 2000(0.1)^x$

21. $f(x) = \left(\frac{1}{6}\right)^{-\frac{3}{5}}\left(\frac{1}{6}\right)^{\frac{x}{5}} \approx 2.93(0.699)^x$

23. Linear

25. Neither

27. Linear

29. $\$10,250$

31. $\$13,268.58$

33. $P = A(t) \cdot \left(1 + \frac{r}{n}\right)^{-nt}$

35. $\$4,572.56$

37. 4%

39. continuous growth; the growth rate is greater than 0.

41. continuous decay; the growth rate is less than 0.

43. $\$669.42$

45. $f(-1) = -4$

47. $f(-1) \approx -0.2707$

49. $f(3) \approx 483.8146$

51. $y = 3 \cdot 5^x$

53. $y \approx 18 \cdot 1.025^x$

55. $y \approx 0.2 \cdot 1.95^x$

57. $\text{APY} = \frac{A(t)-a}{a} = \frac{a\left(1+\frac{r}{365}\right)^{365(1)}-a}{a} = \frac{a\left[\left(1+\frac{r}{365}\right)^{365}-1\right]}{a} = \left(1+\frac{r}{365}\right)^{365} - 1$; $I(n) = \left(1+\frac{r}{n}\right)^n - 1$

59. Let f be the exponential decay function $f(x) = a \cdot \left(\frac{1}{b}\right)^x$ such that $b > 1$. Then for some number $n > 0$, $f(x) = a \cdot \left(\frac{1}{b}\right)^x = a\left(b^{-1}\right)^x = a\left(\left(e^n\right)^{-1}\right)^x = a\left(e^{-n}\right)^x = a(e)^{-nx}$.

61. $47,622$ fox

63. 1.39%; $\$155,368.09$

65. $\$35,838.76$

67. $\$82,247.78$; $\$449.75$

6.2 Section Exercises

1. An asymptote is a line that the graph of a function approaches, as x either increases or decreases without bound. The horizontal asymptote of an exponential function tells us the limit of the function's values as the independent variable gets either extremely large or extremely small.

3. $g(x) = 4(3)^{-x}$; y-intercept: $(0, 4)$; Domain: all real numbers; Range: all real numbers greater than 0.

5. $g(x) = -10^x + 7$; y-intercept: $(0, 6)$; Domain: all real numbers; Range: all real numbers less than 7.

7. $g(x) = 2\left(\frac{1}{4}\right)^x$; y-intercept: $(0,\ 2)$; Domain: all real numbers; Range: all real numbers greater than 0.

9. 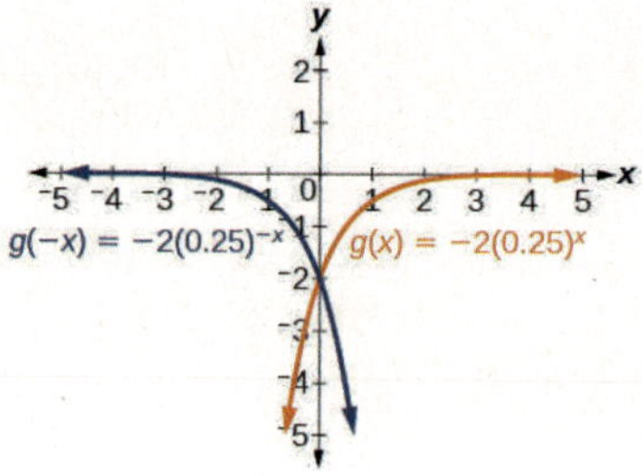

y-intercept: $(0, -2)$

11. 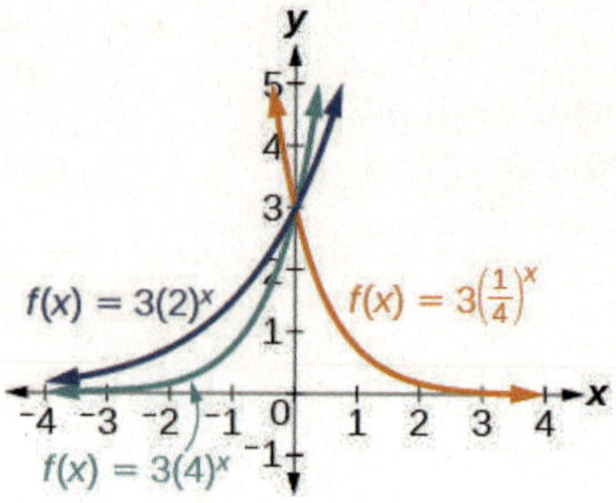

13. B

15. A

17. E

19. D

21. C

23. 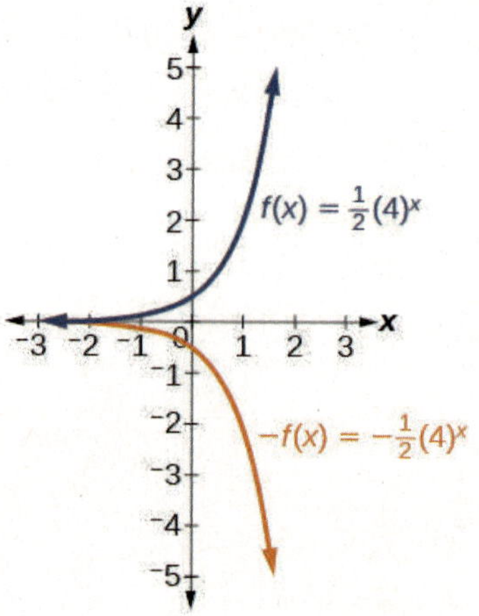

25.

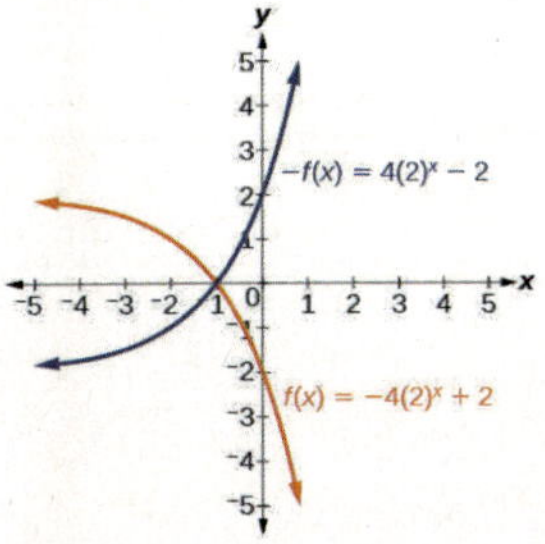

27.

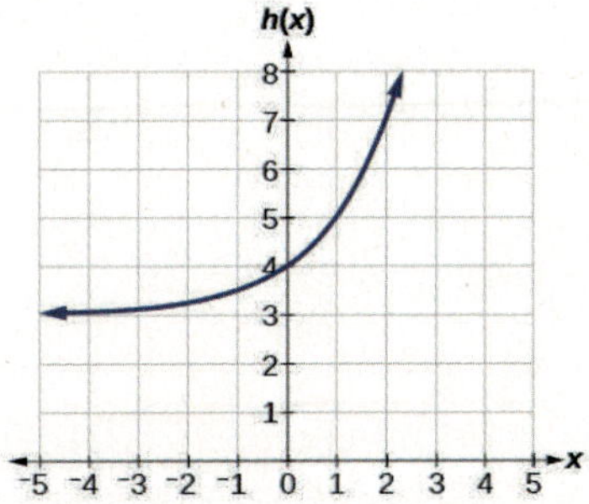

Horizontal asymptote: $h(x) = 3$; Domain: all real numbers; Range: all real numbers strictly greater than 3.

29. As $x \to \infty$, $f(x) \to -\infty$;
As $x \to -\infty$, $f(x) \to -1$

31. As $x \to \infty$, $f(x) \to 2$;
As $x \to -\infty$, $f(x) \to \infty$

33. $f(x) = 4^x - 3$

35. $f(x) = 4^{x-5}$

37. $f(x) = 4^{-x}$

39. $y = -2^x + 3$

41. $y = -2(3)^x + 7$

43. $g(6) = 800 + \frac{1}{3} \approx 800.3333$

45. $h(-7) = -58$

47. $x \approx -2.953$

49. $x \approx -0.222$

51. The graph of $G(x) = \left(\frac{1}{b}\right)^x$ is the refelction about the y-axis of the graph of $F(x) = b^x$; For any real number $b > 0$ and function $f(x) = b^x$, the graph of $\left(\frac{1}{b}\right)^x$ is the the reflection about the y-axis, $F(-x)$.

53. The graphs of $g(x)$ and $h(x)$ are the same and are a horizontal shift to the right of the graph of $f(x)$; For any real number n, real number $b > 0$, and function $f(x) = b^x$, the graph of $\left(\frac{1}{b^n}\right) b^x$ is the horizontal shift $f(x - n)$.

6.3 Section Exercises

1. A logarithm is an exponent. Specifically, it is the exponent to which a base b is raised to produce a given value. In the expressions given, the base b has the same value. The exponent, y, in the expression b^y can also be written as the logarithm, $\log_b x$, and the value of x is the result of raising b to the power of y.

3. Since the equation of a logarithm is equivalent to an exponential equation, the logarithm can be converted to the exponential equation $b^y = x$, and then properties of exponents can be applied to solve for x.

5. The natural logarithm is a special case of the logarithm with base b in that the natural log always has base e. Rather than notating the natural logarithm as $\log_e(x)$, the notation used is $\ln(x)$.

7. $a^c = b$

9. $x^y = 64$

11. $15^b = a$

13. $13^a = 142$

15. $e^n = w$

17. $\log_c(k) = d$

19. $\log_{19} y = x$

21. $\log_n (103) = 4$

23. $\log_y \left(\frac{39}{100}\right) = x$

25. $\ln(h) = k$

27. $x = 2^{-3} = \frac{1}{8}$

29. $x = 3^3 = 27$

31. $x = 9^{\frac{1}{2}} = 3$

33. $x = 6^{-3} = \frac{1}{216}$

35. $x = e^2$

37. 32

39. 1.06

41. 14.125

43. $\frac{1}{2}$

45. 4

47. −3

49. −12

51. 0

53. 10

55. 2.708

57. 0.151

59. No, the function has no defined value for $x = 0$. To verify, suppose $x = 0$ is in the domain of the function $f(x) = \log(x)$. Then there is some number n such that $n = \log(0)$. Rewriting as an exponential equation gives: $10^n = 0$, which is impossible since no such real number n exists. Therefore, $x = 0$ is *not* the domain of the function $f(x) = \log(x)$.

61. Yes. Suppose there exists a real number x such that $\ln x = 2$. Rewriting as an exponential equation gives $x = e^2$, which is a real number. To verify, let $x = e^2$. Then, by definition, $\ln(x) = \ln\left(e^2\right) = 2$.

63. No; $\ln(1) = 0$, so $\frac{\ln\left(e^{1.725}\right)}{\ln(1)}$ is undefined.

65. 2

6.4 Section Exercises

1. Since the functions are inverses, their graphs are mirror images about the line $y = x$. So for every point (a, b) on the graph of a logarithmic function, there is a corresponding point (b, a) on the graph of its inverse exponential function.

3. Shifting the function right or left and reflecting the function about the y-axis will affect its domain.

5. No. A horizontal asymptote would suggest a limit on the range, and the range of any logarithmic function in general form is all real numbers.

7. Domain: $\left(-\infty, \frac{1}{2}\right)$; Range: $\left(-\infty, \infty\right)$

9. Domain: $\left(-\frac{17}{4}, \infty\right)$; Range: $\left(-\infty, \infty\right)$

11. Domain: $\left(5, \infty\right)$; Vertical asymptote: $x = 5$

13. Domain: $\left(-\frac{1}{3}, \infty\right)$; Vertical asymptote: $x = -\frac{1}{3}$

15. Domain: $\left(-3, \infty\right)$; Vertical asymptote: $x = -3$

17. Domain: $\left(\frac{3}{7}, \infty\right)$; Vertical asymptote: $x = \frac{3}{7}$; End behavior: as $x \to \left(\frac{3}{7}\right)^+, f(x) \to -\infty$ and as $x \to \infty, f(x) \to \infty$

19. Domain: $\left(-3, \infty\right)$; Vertical asymptote: $x = -3$; End behavior: as $x \to -3^+$, $f(x) \to -\infty$ and as $x \to \infty$, $f(x) \to \infty$

21. Domain: $\left(1, \infty\right)$; Range: $\left(-\infty, \infty\right)$; Vertical asymptote: $x = 1$; x-intercept: $\left(\frac{5}{4}, 0\right)$; y-intercept: DNE

23. Domain: $\left(-\infty, 0\right)$; Range: $\left(-\infty, \infty\right)$; Vertical asymptote: $x = 0$; x-intercept: $\left(-e^2, 0\right)$; y-intercept: DNE

25. Domain: $\left(0, \infty\right)$; Range: $\left(-\infty, \infty\right)$; Vertical asymptote: $x = 0$; x-intercept: $\left(e^3, 0\right)$; y-intercept: DNE

27. B

29. C

31. B

33. C

35.

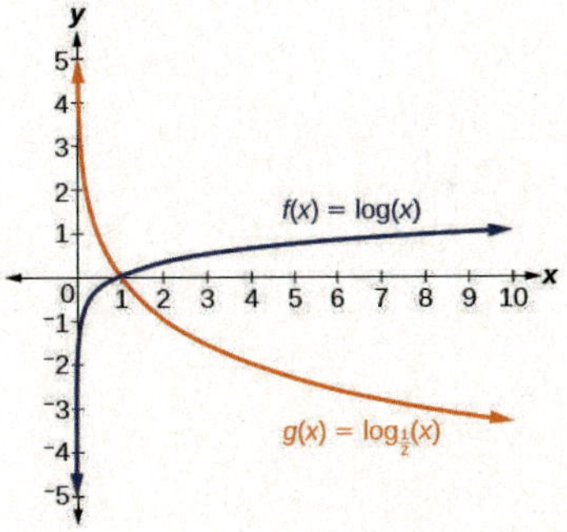

37.

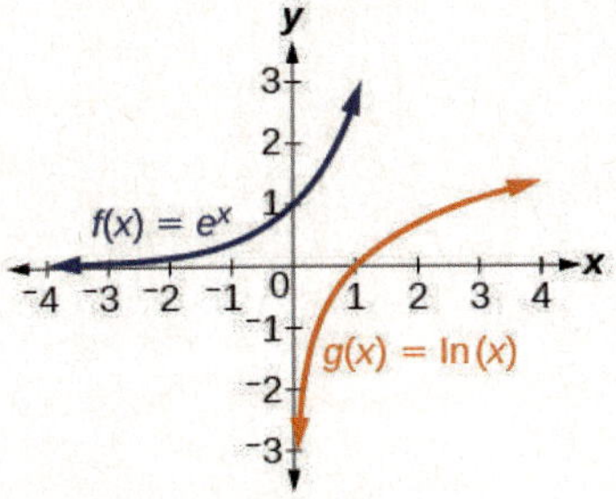

39. C

41.

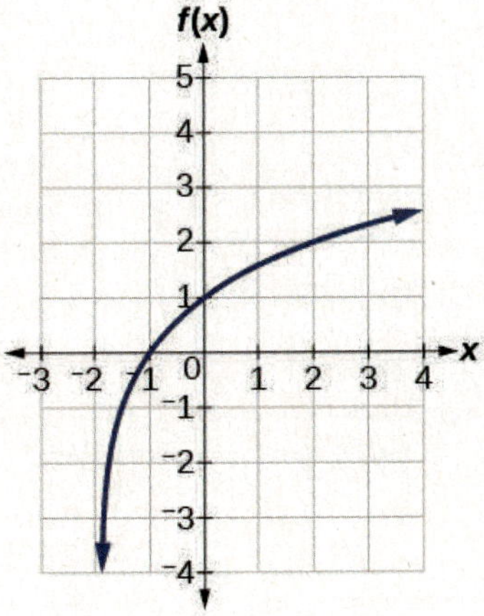

43.

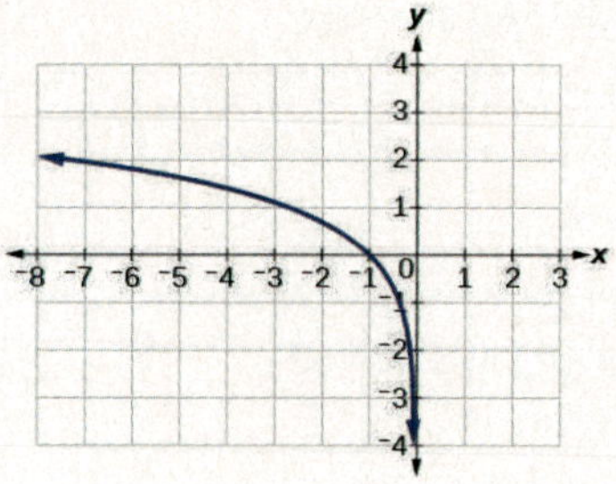

45.

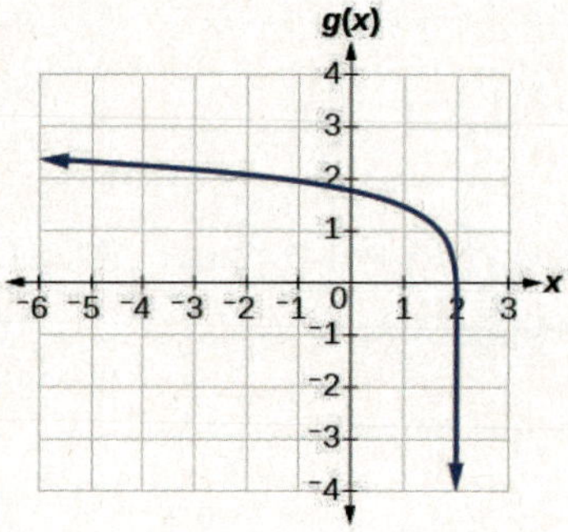

47. $f(x) = \log_2(-(x-1))$

49. $f(x) = 3\log_4(x+2)$

51. $x = 2$

53. $x \approx 2.303$

55. $x \approx -0.472$

57. The graphs of $f(x) = \log_{\frac{1}{2}}(x)$ and $g(x) = -\log_2(x)$ appear to be the same; Conjecture: for any positive base $b \neq 1$, $\log_b(x) = -\log_{\frac{1}{b}}(x)$.

59. Recall that the argument of a logarithmic function must be positive, so we determine where $\frac{x+2}{x-4} > 0$. From the graph of the function $f(x) = \frac{x+2}{x-4}$, note that the graph lies above the x-axis on the interval $\left(-\infty, -2\right)$ and again to the right of the vertical asymptote, that is $\left(4, \infty\right)$. Therefore, the domain is $\left(-\infty, -2\right) \cup \left(4, \infty\right)$.

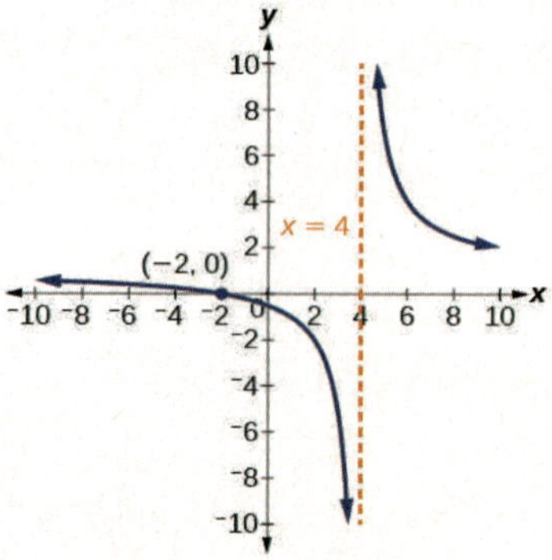

6.5 Section Exercises

1. Any root expression can be rewritten as an expression with a rational exponent so that the power rule can be applied, making the logarithm easier to calculate. Thus, $\log_b\left(x^{\frac{1}{n}}\right) = \frac{1}{n}\log_b(x)$.

3. $\log_b(2) + \log_b(7) + \log_b(x) + \log_b(y)$

5. $\log_b(13) - \log_b(17)$

7. $-k\ln(4)$

9. $\ln(7xy)$

11. $\log_b(4)$

13. $\log_b(7)$

15. $15\log(x) + 13\log(y) - 19\log(z)$

17. $\frac{3}{2}\log(x) - 2\log(y)$

19. $\frac{8}{3}\log(x) + \frac{14}{3}\log(y)$

21. $\ln(2x^7)$

23. $\log\left(\frac{xz^3}{\sqrt{y}}\right)$

25. $\log_7(15) = \frac{\ln(15)}{\ln(7)}$

27. $\log_{11}(5) = \frac{\log_5(5)}{\log_5(11)} = \frac{1}{b}$

29. $\log_{11}\left(\frac{6}{11}\right) = \frac{\log_5\left(\frac{6}{11}\right)}{\log_5(11)} = \frac{\log_5(6)-\log_5(11)}{\log_5(11)} = \frac{a-b}{b} = \frac{a}{b} - 1$

31. 3

33. 2.81359

35. 0.93913

37. −2.23266

39. $x = 4$; By the quotient rule:
$\log_6(x+2) - \log_6(x-3) = \log_6\left(\frac{x+2}{x-3}\right) = 1.$

Rewriting as an exponential equation and solving for x :

$$\begin{aligned} 6^1 &= \frac{x+2}{x-3} \\ 0 &= \frac{x+2}{x-3} - 6 \\ 0 &= \frac{x+2}{x-3} - \frac{6(x-3)}{(x-3)} \\ 0 &= \frac{x+2-6x+18}{x-3} \\ 0 &= \frac{x-4}{x-3} \\ x &= 4 \end{aligned}$$

Checking, we find that
$\log_6(4+2) - \log_6(4-3) = \log_6(6) - \log_6(1)$
is defined, so $x = 4$.

41. Let b and n be positive integers greater than 1. Then, by the change-of-base formula,
$\log_b(n) = \frac{\log_n(n)}{\log_n(b)} = \frac{1}{\log_n(b)}.$

6.6 Section Exercises

1. Determine first if the equation can be rewritten so that each side uses the same base. If so, the exponents can be set equal to each other. If the equation cannot be rewritten so that each side uses the same base, then apply the logarithm to each side and use properties of logarithms to solve.

3. The one-to-one property can be used if both sides of the equation can be rewritten as a single logarithm with the same base. If so, the arguments can be set equal to each other, and the resulting equation can be solved algebraically. The one-to-one property cannot be used when each side of the equation cannot be rewritten as a single logarithm with the same base.

5. $x = -\frac{1}{3}$

7. $n = -1$

9. $b = \frac{6}{5}$

11. $x = 10$

13. No solution

15. $p = \log\left(\frac{17}{8}\right) - 7$

17. $k = -\frac{\ln(38)}{3}$

19. $x = \frac{\ln\left(\frac{38}{3}\right) - 8}{9}$

21. $x = \ln 12$

23. $x = \frac{\ln\left(\frac{3}{5}\right) - 3}{8}$

25. no solution

27. $x = \ln(3)$

29. $10^{-2} = \frac{1}{100}$

31. $n = 49$

33. $k = \frac{1}{36}$

35. $x = \frac{9-e}{8}$

37. $n = 1$

39. No solution

41. No solution

43. $x = \pm\frac{10}{3}$

45. $x = 10$

47. $x = 0$

49. $x = \frac{3}{4}$

51. $x = 9$

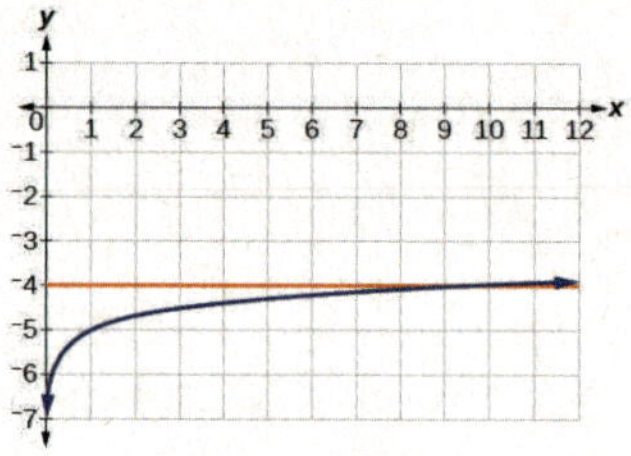

53. $x = \frac{e^2}{3} \approx 2.5$

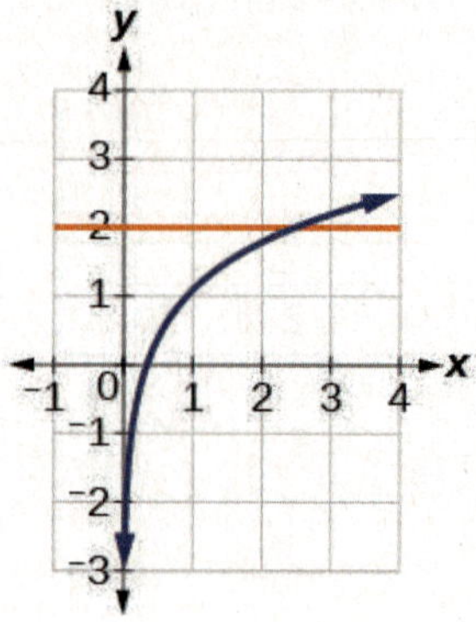

55. $x = -5$

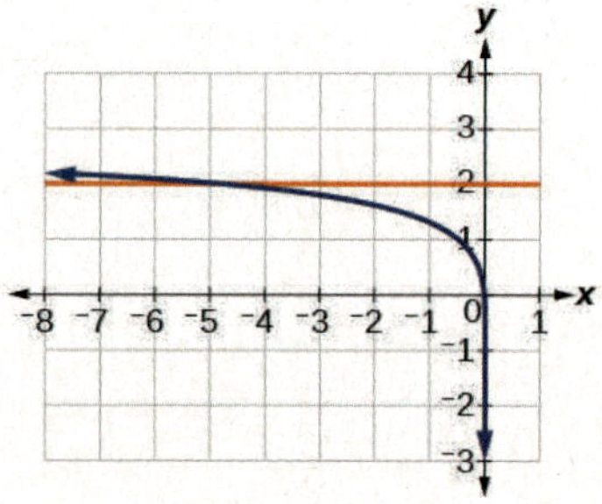

57. $x = \frac{e+10}{4} \approx 3.2$

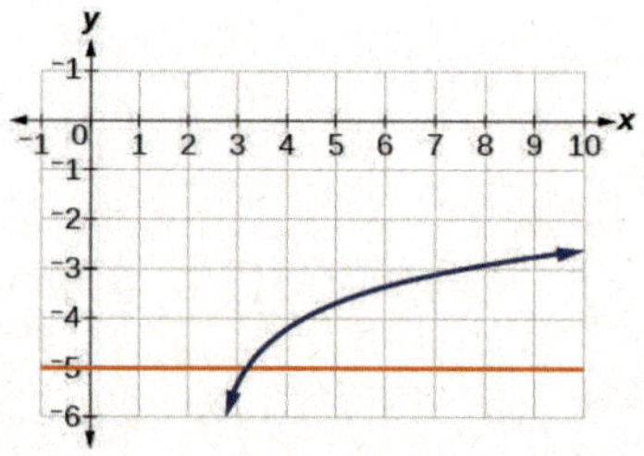

59. No solution

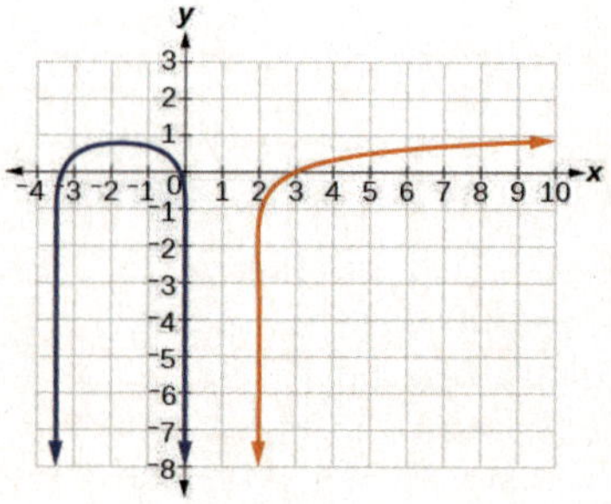

61. $x = \frac{11}{5} \approx 2.2$

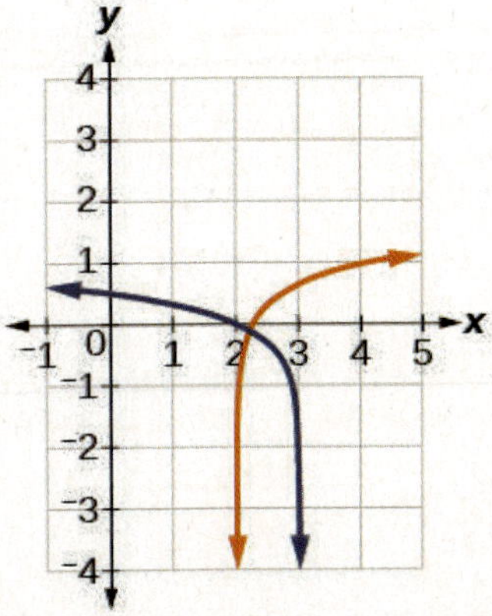

63. $x = \frac{101}{11} \approx 9.2$

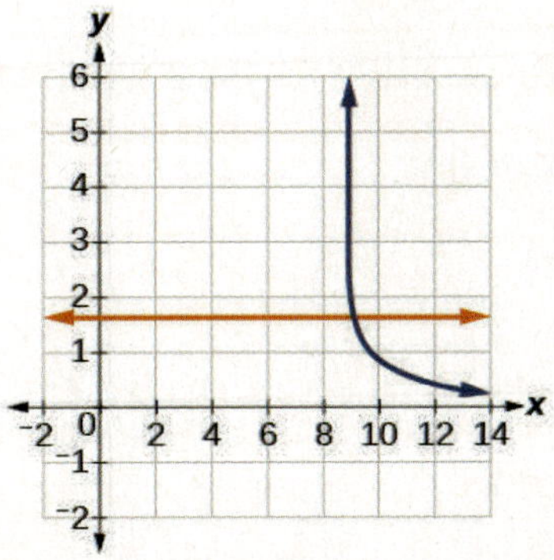

65. about $\$27,710.24$

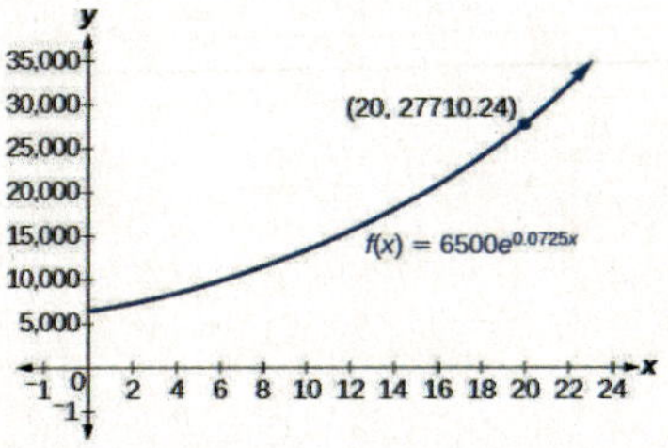

67. about 5 years

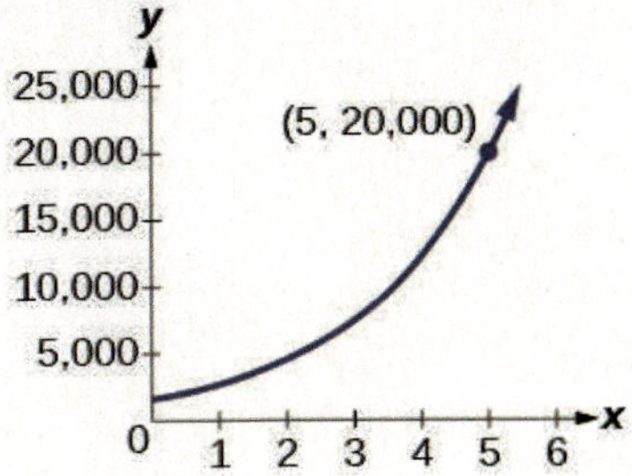

69. $\frac{\ln(17)}{5} \approx 0.567$

71. $x = \frac{\log(38)+5\log(3)}{4\log(3)} \approx 2.078$

73. $x \approx 2.2401$

75. $x \approx -44655.7143$

77. about 5.83

79. $t = \ln\left(\left(\frac{y}{A}\right)^{\frac{1}{k}}\right)$

81. $t = \ln\left(\left(\frac{T-T_s}{T_0-T_s}\right)^{-\frac{1}{k}}\right)$

6.7 Section Exercises

1. Half-life is a measure of decay and is thus associated with exponential decay models. The half-life of a substance or quantity is the amount of time it takes for half of the initial amount of that substance or quantity to decay.

3. Doubling time is a measure of growth and is thus associated with exponential growth models. The doubling time of a substance or quantity is the amount of time it takes for the initial amount of that substance or quantity to double in size.

5. An order of magnitude is the nearest power of ten by which a quantity exponentially grows. It is also an approximate position on a logarithmic scale; Sample response: Orders of magnitude are useful when making comparisons between numbers that differ by a great amount. For example, the mass of Saturn is 95 times greater than the mass of Earth. This is the same as saying that the mass of Saturn is about 10^2 times, or *2 orders of magnitude* greater, than the mass of Earth.

7. $f(0) \approx 16.7$; The amount initially present is about 16.7 units.

9. 150

11. exponential; $f(x) = 1.2^x$

13. logarithmic

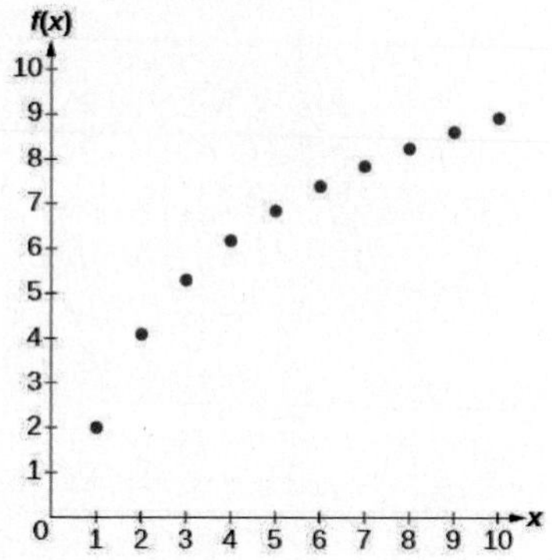

15. logarithmic

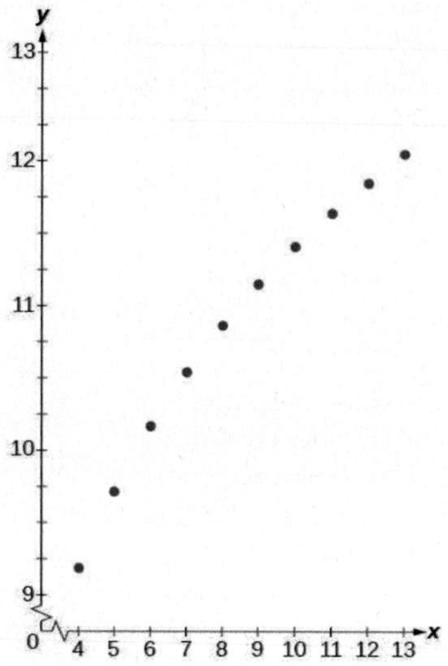

17.

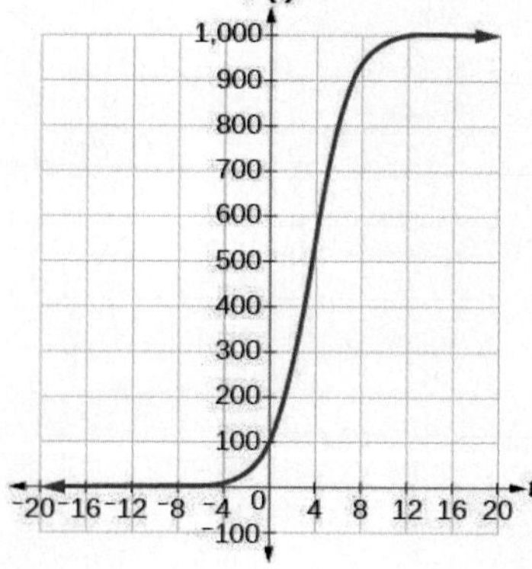

19. about 1.4 years

21. about 7.3 years

23. 4 half-lives; 8.18 minutes

25.

$$M = \frac{2}{3}\log\left(\frac{S}{S_0}\right)$$
$$\log\left(\frac{S}{S_0}\right) = \frac{3}{2}M$$
$$\frac{S}{S_0} = 10^{\frac{3M}{2}}$$
$$S = S_0 10^{\frac{3M}{2}}$$

27. Let $y = b^x$ for some non-negative real number b such that $b \neq 1$. Then,

$$\ln(y) = \ln(b^x)$$
$$\ln(y) = x\ln(b)$$
$$e^{\ln(y)} = e^{x\ln(b)}$$
$$y = e^{x\ln(b)}$$

29. $A = 125e^{(-0.3567t)}$; $A \approx 43$ mg

31. about 60 days

33. $A(t) = 250e^{(-0.00822t)}$; half-life: about 84 minutes

35. $r \approx -0.0667$, So the hourly decay rate is about 6.67%

37. $f(t) = 1350e^{(0.03466t)}$; after 3 hours: $P(180) \approx 691,200$

39. $f(t) = 256e^{(0.068110t)}$; doubling time: about 10 minutes

41. about 88 minutes

43. $T(t) = 90e^{(-0.008377t)} + 75$, where t is in minutes.

45. about 113 minutes

47. $\log(x) = 1.5; \quad x \approx 31.623$

49. MMS magnitude: 5.82

51. $N(3) \approx 71$

53. C

6.8 Section Exercises

1. Logistic models are best used for situations that have limited values. For example, populations cannot grow indefinitely since resources such as food, water, and space are limited, so a logistic model best describes populations.

3. Regression analysis is the process of finding an equation that best fits a given set of data points. To perform a regression analysis on a graphing utility, first list the given points using the STAT then EDIT menu. Next graph the scatter plot using the STAT PLOT feature. The shape of the data points on the scatter graph can help determine which regression feature to use. Once this is determined, select the appropriate regression analysis command from the STAT then CALC menu.

5. The y-intercept on the graph of a logistic equation corresponds to the initial population for the population model.

7. C

9. B

11. $P(0) = 22$; 175

13. $p \approx 2.67$

15. y-intercept: $(0, 15)$

17. 4 koi

19. about 6.8 months.

21.

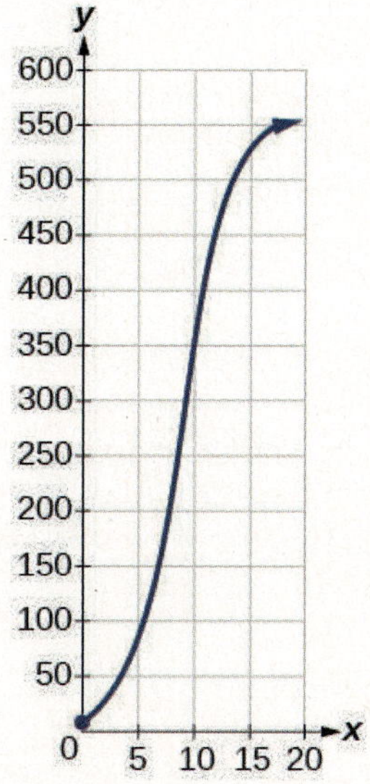

23. About 38 wolves

25. About 8.7 years

27. $f(x) = 776.682(1.426)^x$

29.

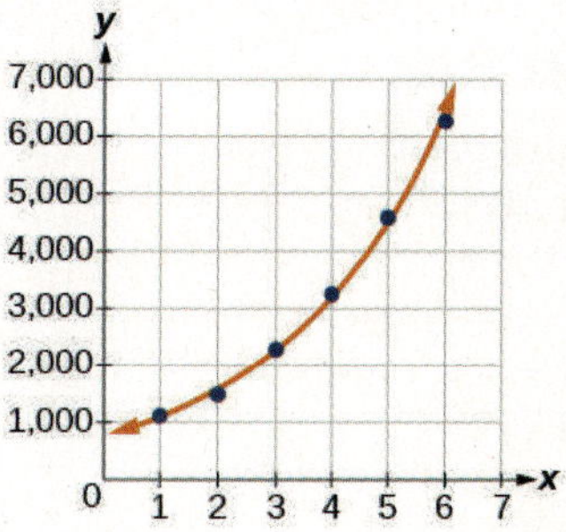

31.

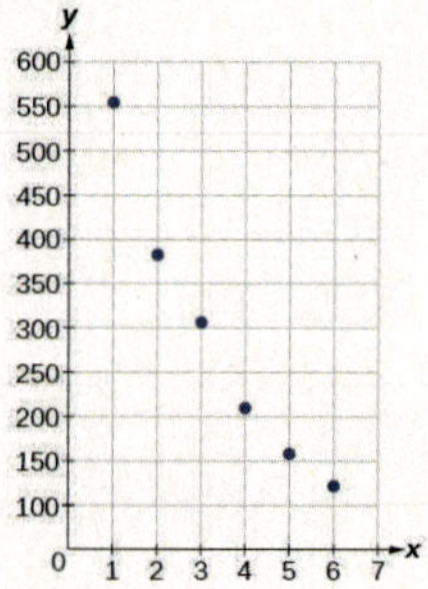

33. $f(x) = 731.92e^{-0.3038x}$

35. When $f(x) = 250, x \approx 3.6$

37. $y = 5.063 + 1.934\log(x)$

39.

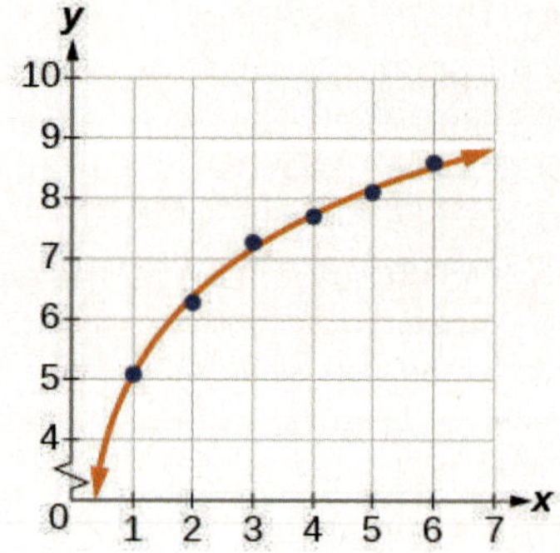

41.

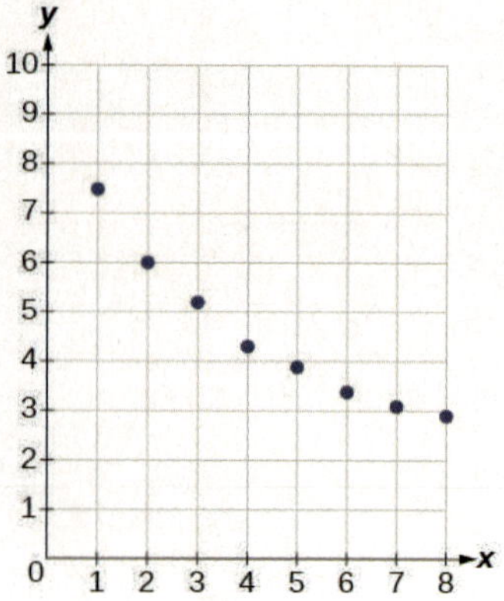

43. When $f(10) \approx 2.3$

45. When $f(x) = 8, x \approx 0.82$

47. $f(x) = \frac{25.081}{1+3.182e^{-0.545x}}$

49. About 25

51.

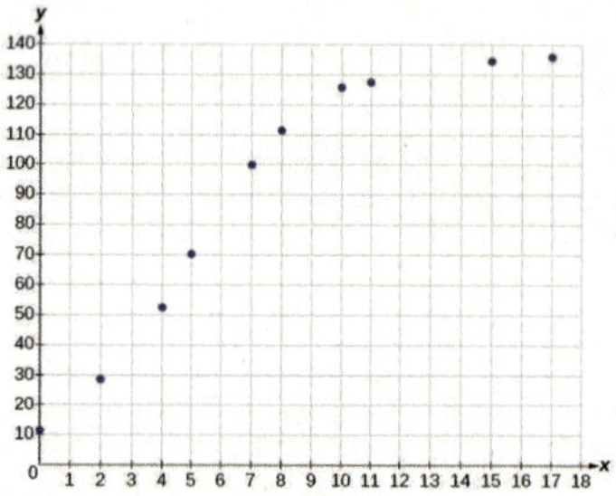

53.

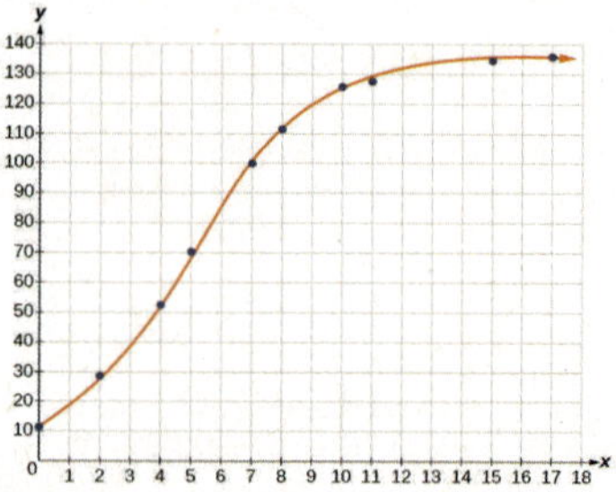

55. When $f(x) = 68, x \approx 4.9$

57. $f(x) = 1.034341(1.281204)^x$; $g(x) = 4.035510$; the regression curves are symmetrical about $y = x$, so it appears that they are inverse functions.

59. $f^{-1}(x) = \frac{\ln(a)-\ln(\frac{c}{x}-1)}{b}$

Review Exercises

1. exponential decay; The growth factor, 0.825, is between 0 and 1.

3. $y = 0.25(3)^x$

5. $\$42,888.18$

7. continuous decay; the growth rate is negative.

9. domain: all real numbers; range: all real numbers strictly greater than zero; y-intercept: (0, 3.5);

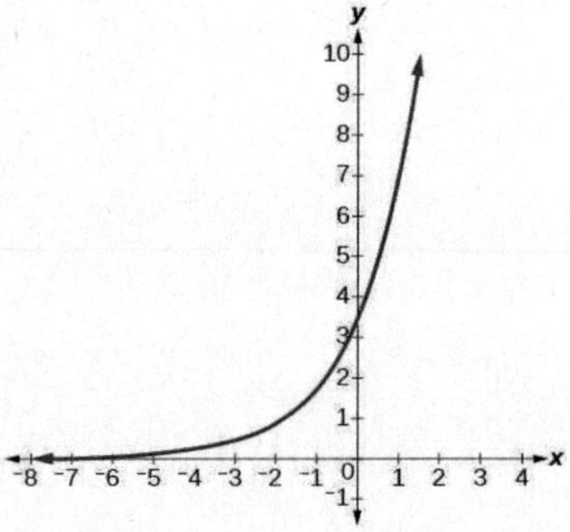

11. $g(x) = 7(6.5)^{-x}$; y-intercept: $(0,\ 7)$; Domain: all real numbers; Range: all real numbers greater than 0.

13. $17^x = 4913$

15. $\log_a b = -\frac{2}{5}$

17. $x = 64^{\frac{1}{3}} = 4$

19. $\log(0.000001) = -6$

21. $\ln\left(e^{-0.8648}\right) = -0.8648$

23.

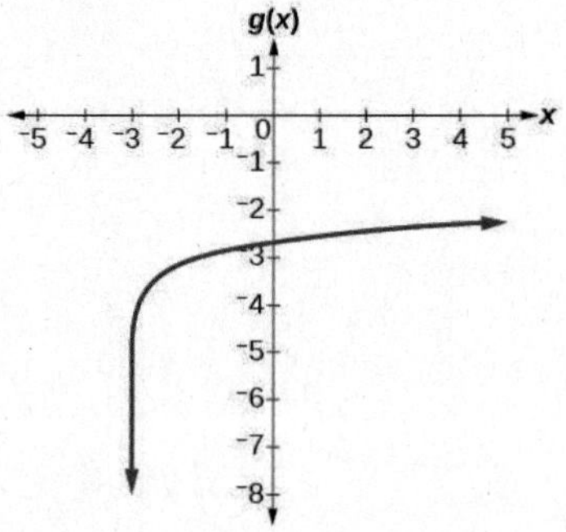

25. Domain: $x > -5$; Vertical asymptote: $x = -5$; End behavior: as $x \to -5^+, f(x) \to -\infty$ and as $x \to \infty, f(x) \to \infty$.

27. $\log_8(65xy)$

29. $\ln\left(\frac{z}{xy}\right)$

31. $\log_y(12)$

33. $\ln(2) + \ln(b) + \frac{\ln(b+1)-\ln(b-1)}{2}$

35. $\log_7\left(\frac{v^3 w^6}{\sqrt[3]{u}}\right)$

37. $x = \frac{\frac{\log(125)}{\log(5)}+17}{12} = \frac{5}{3}$

39. $x = -3$

41. no solution

43. no solution

45. $x = \ln(11)$

47. $a = e^4 - 3$

49. $x = \pm\frac{9}{5}$

51. about 5.45 years

53. $f^{-1}(x) = \sqrt[3]{2^{4x} - 1}$

55. $f(t) = 300(0.83)^t$; $f(24) \approx 3.43 \quad g$

57. about 45 minutes

59. about 8.5 days

61. exponential

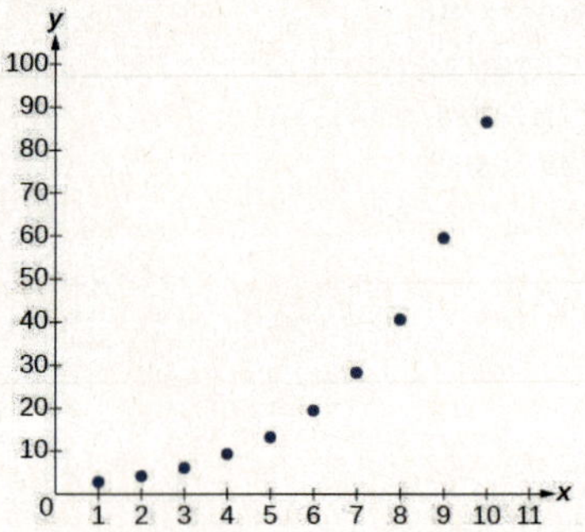

63. $y = 4(0.2)^x$;
$y = 4e^{-1.609438x}$

65. about 7.2 days

67. logarithmic;
$y = 16.68718 - 9.71860\ln(x)$

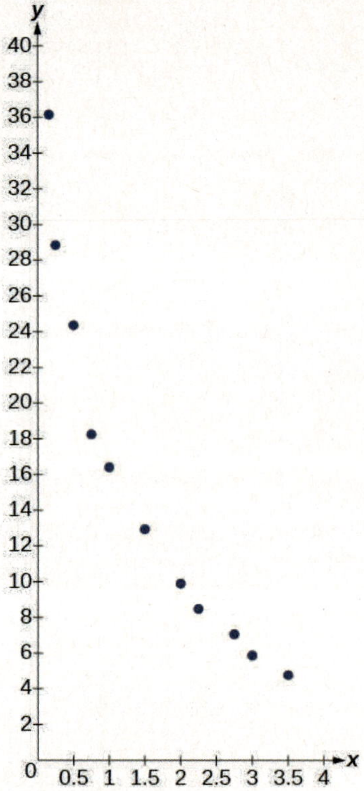

Practice Test

1. About 13 dolphins.

3. $1,947

5. y-intercept: $(0,\ 5)$

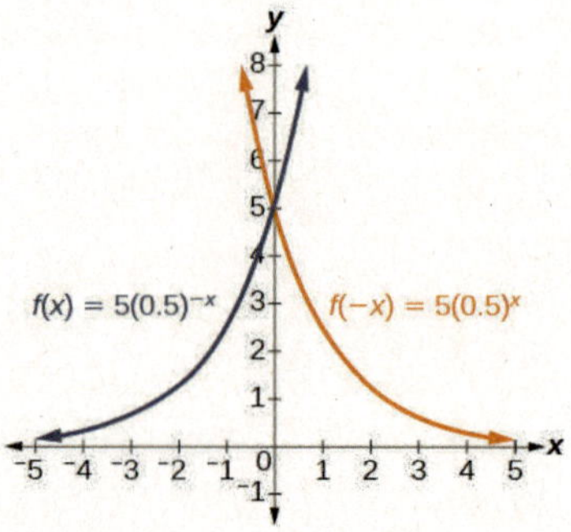

7. $8.5^a = 614.125$

9. $x = \left(\frac{1}{7}\right)^2 = \frac{1}{49}$

11. $\ln(0.716) \approx -0.334$

13. Domain: $x < 3$; Vertical asymptote: $x = 3$; End behavior:

$x \to 3^-, f(x) \to -\infty$ and

$x \to -\infty, f(x) \to \infty$

15. $\log_t(12)$

17. $3\ln(y) + 2\ln(z) + \frac{\ln(x-4)}{3}$

19. $x = \frac{\frac{\ln(1000)}{\ln(16)}+5}{3} \approx 2.497$

21. $a = \frac{\ln(4)+8}{10}$

23. no solution

25. $x = \ln(9)$

27. $x = \pm\frac{3\sqrt{3}}{2}$

29. $f(t) = 112e^{-.019792t}$; half-life: about 35 days

31. $T(t) = 36e^{-0.025131t} + 35; T(60) \approx 43^{\circ}\text{F}$

33. logarithmic

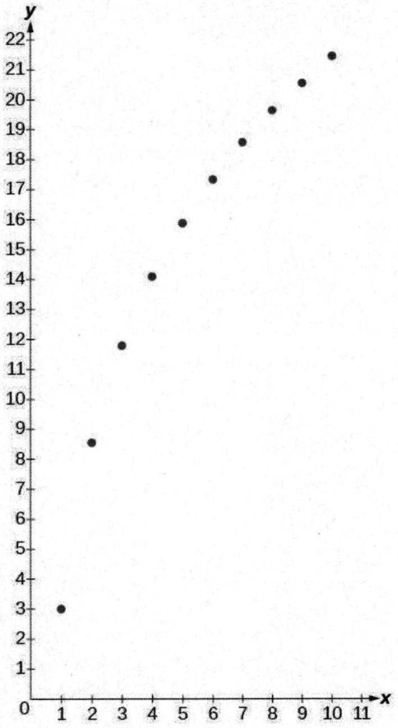

35. exponential;
$y = 15.10062(1.24621)^x$

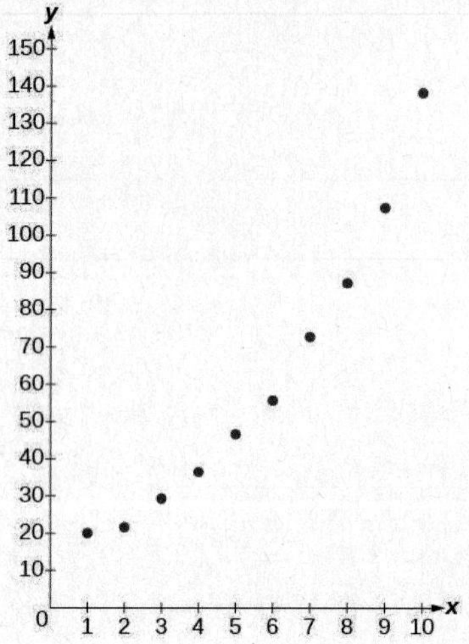

37. logistic;
$y = \frac{18.41659}{1+7.54644e^{-0.68375x}}$

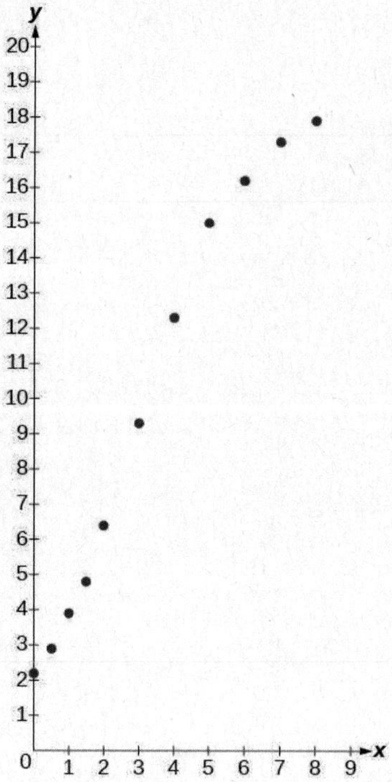

Chapter 7

Try It

7.1 Angles

1.

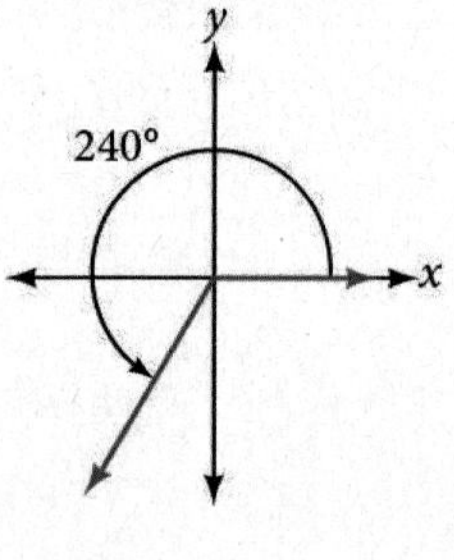

2. $\frac{3\pi}{2}$

3. $-135°$

4. $\frac{7\pi}{10}$

5. $\alpha = 150°$

6. $\beta = 60°$

7. $\frac{7\pi}{6}$

8. $\frac{215\pi}{18} = 37.525$ units

9. 1.88

10. $\frac{-3\pi}{2}$ rad/s

11. 1655 kilometers per hour

7.2 Right Triangle Trigonometry

1. $\frac{7}{25}$

2. $\sin t = \frac{33}{65}, \quad \cos t = \frac{56}{65}, \quad \tan t = \frac{33}{56},$
$\sec t = \frac{65}{56}, \quad \csc t = \frac{65}{33}, \quad \cot t = \frac{56}{33}$

3. $\sin\left(\frac{\pi}{4}\right) = \frac{\sqrt{2}}{2}, \cos\left(\frac{\pi}{4}\right) = \frac{\sqrt{2}}{2}, \tan\left(\frac{\pi}{4}\right) = 1,$
$\sec\left(\frac{\pi}{4}\right) = \sqrt{2}, \csc\left(\frac{\pi}{4}\right) = \sqrt{2}, \cot\left(\frac{\pi}{4}\right) = 1$

4. 2

5. adjacent $= 10$; opposite $= 10\sqrt{3}$; missing angle is $\frac{\pi}{6}$

6. About 52 ft

7.3 Unit Circle

1. $\cos(t) = -\frac{\sqrt{2}}{2}, \sin(t) = \frac{\sqrt{2}}{2}$

2. $\cos(\pi) = -1, \sin(\pi) = 0$

3. $\sin(t) = -\frac{7}{25}$

4. approximately 0.866025403

5. $\frac{\pi}{3}$

6. ⓐ
$\cos(315°) = \frac{\sqrt{2}}{2}, \quad \sin(315°) = \frac{-\sqrt{2}}{2}$
ⓑ
$\cos\left(-\frac{\pi}{6}\right) = \frac{\sqrt{3}}{2}, \sin\left(-\frac{\pi}{6}\right) = -\frac{1}{2}$

7. $\left(\frac{1}{2}, -\frac{\sqrt{3}}{2}\right)$

7.4 The Other Trigonometric Functions

1. $\sin t = -\frac{\sqrt{2}}{2} \cos t = \frac{\sqrt{2}}{2}, \tan t = -1, sect = \sqrt{2}, \csc t = -\sqrt{2}, \cot t = -1$

2. $\sin\frac{\pi}{3} = \frac{\sqrt{3}}{2}, \cos\frac{\pi}{3} = \frac{1}{2}, \tan\frac{\pi}{3} = \sqrt{3}, sec\frac{\pi}{3} = 2, csc\frac{\pi}{3} = \frac{2\sqrt{3}}{3}, cot\frac{\pi}{3} = \frac{\sqrt{3}}{3}$

3. $\sin\left(\frac{-7\pi}{4}\right) = \frac{\sqrt{2}}{2}, \cos\left(\frac{-7\pi}{4}\right) = \frac{\sqrt{2}}{2}, \tan\left(\frac{-7\pi}{4}\right) = 1,$
$\sec\left(\frac{-7\pi}{4}\right) = \sqrt{2}, \csc\left(\frac{-7\pi}{4}\right) = \sqrt{2}, \cot\left(\frac{-7\pi}{4}\right) = 1$

4. $-\sqrt{3}$

5. -2

6. $\sin t$

7. $\cos t = -\frac{8}{17}, \quad \sin t = \frac{15}{17}, \quad \tan t = -\frac{15}{8}$
$\csc t = \frac{17}{15}, \quad \cot t = -\frac{8}{15}$

8. $\sin t = -1, \cos t = 0, \tan t =$ Undefined
$\sec t =$ Undefined, $\csc t = -1, \cot t = 0$

9. $\sec t = \sqrt{2}, \csc t = \sqrt{2}, \tan t = 1, \cot t = 1$

10. ≈ -2.414

7.1 Section Exercises

1.

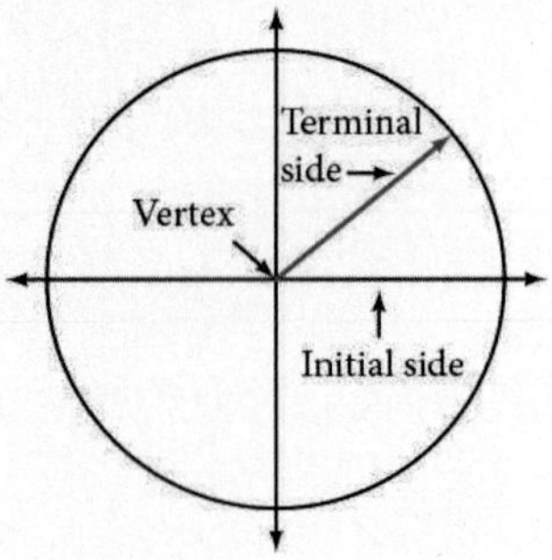

3. Whether the angle is positive or negative determines the direction. A positive angle is drawn in the counterclockwise direction, and a negative angle is drawn in the clockwise direction.

5. Linear speed is a measurement found by calculating distance of an arc compared to time. Angular speed is a measurement found by calculating the angle of an arc compared to time.

7.

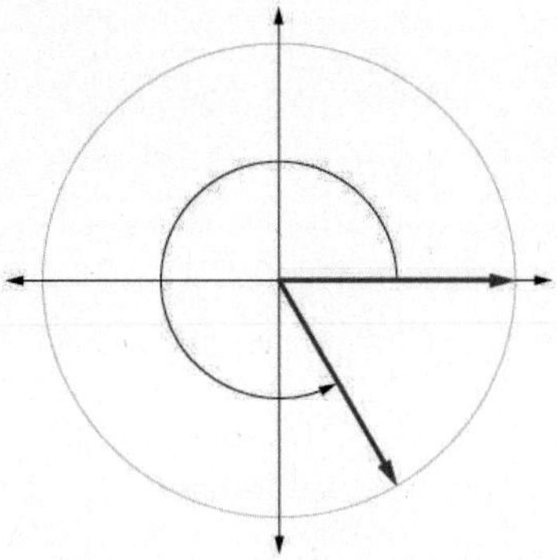

9.

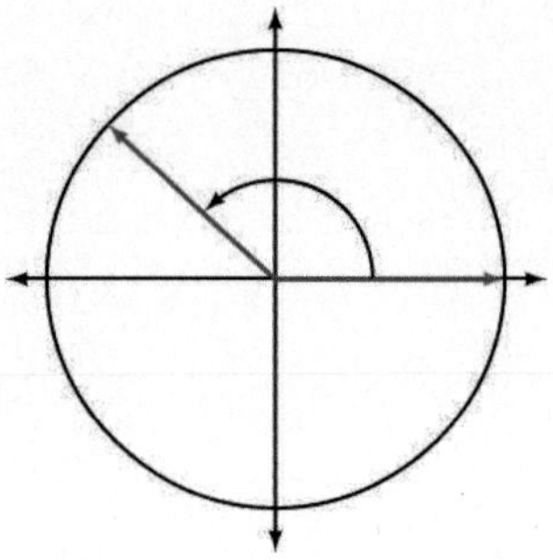

11.

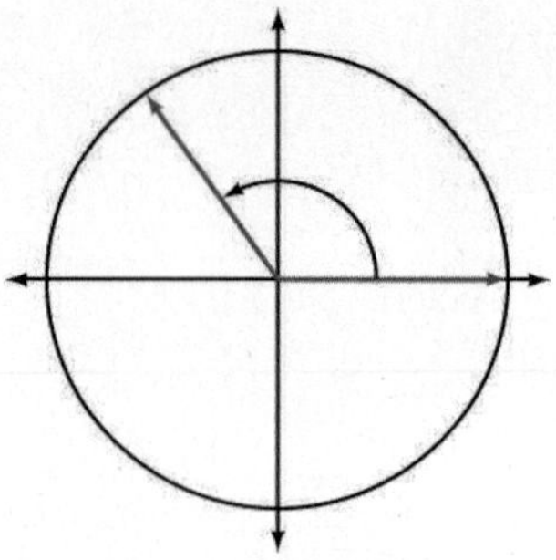

13.

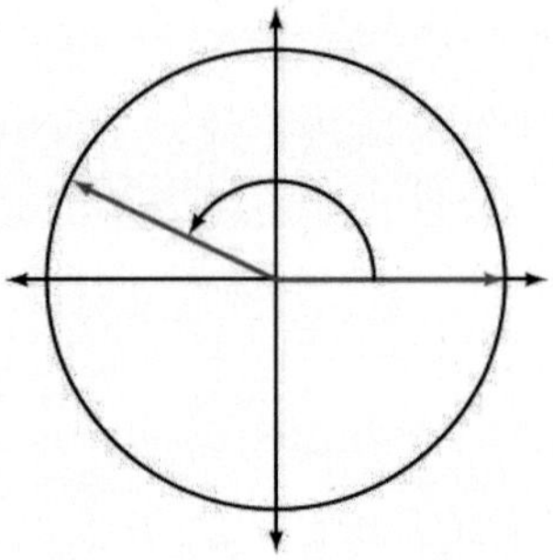

15.

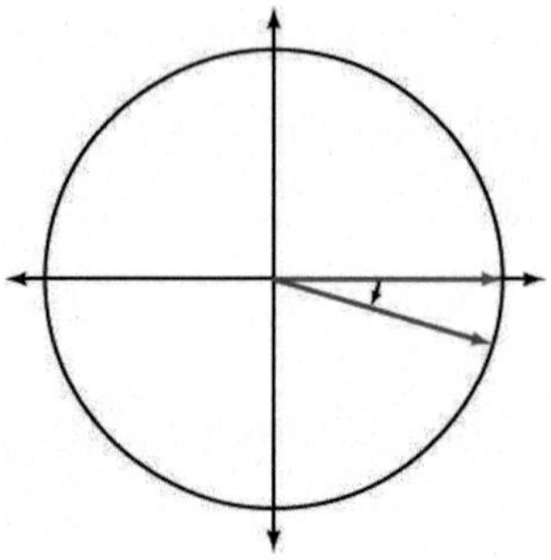

17. $240°$

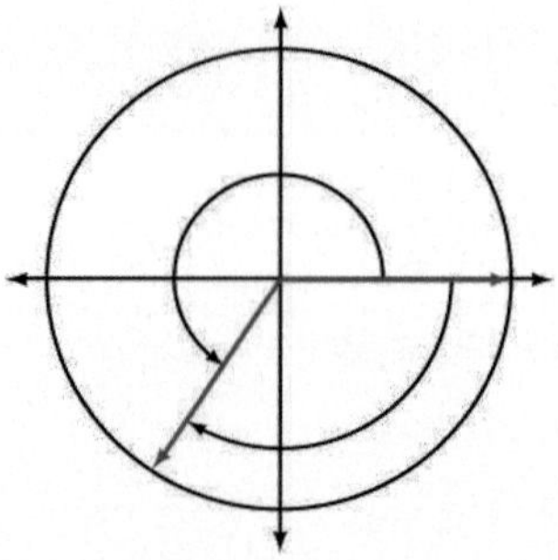

19. $\frac{4\pi}{3}$

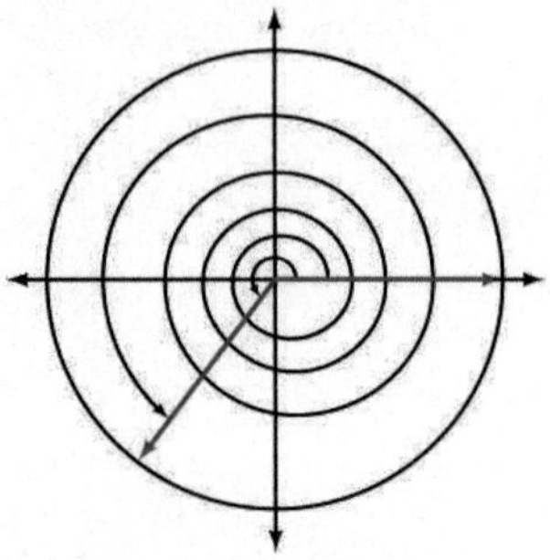

21. $\frac{2\pi}{3}$

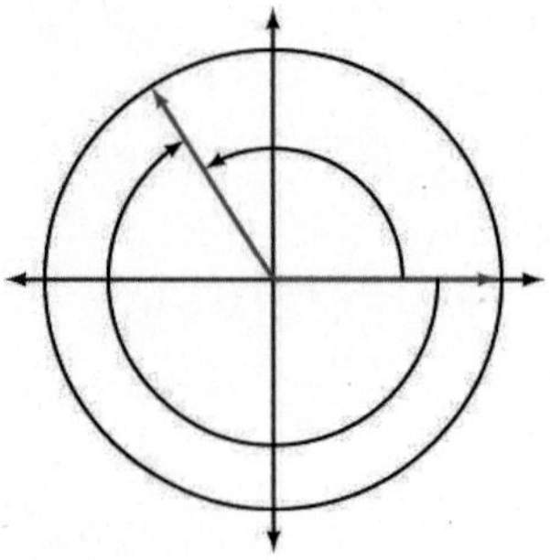

23. $\frac{7\pi}{2} \approx 11.00 \text{ in}^2$

25. $\frac{81\pi}{20} \approx 12.72\ \text{cm}^2$

27. $20°$

29. $60°$

31. $-75°$

33. $\frac{\pi}{2}$ radians

35. -3π radians

37. π radians

39. $\frac{5\pi}{6}$ radians

41. $\frac{5.02\pi}{3} \approx 5.26$ miles

43. $\frac{25\pi}{9} \approx 8.73$ centimeters

45. $\frac{21\pi}{10} \approx 6.60$ meters

47. 104.7198 cm^2

49. 0.7697 in^2

51. $250°$

53. $320°$

55. $\frac{4\pi}{3}$

57. $\frac{8\pi}{9}$

59. 1320 rad/min 210.085 RPM

61. 7 in./s, 4.77 RPM , 28.65 deg/s

63. $1,809,557.37\ \text{mm/min} = 30.16\ \text{m/s}$

65. 5.76 miles

67. $120°$

69. 794 miles per hour

71. 2,234 miles per hour

73. 11.5 inches

7.2 Section Exercises

1.

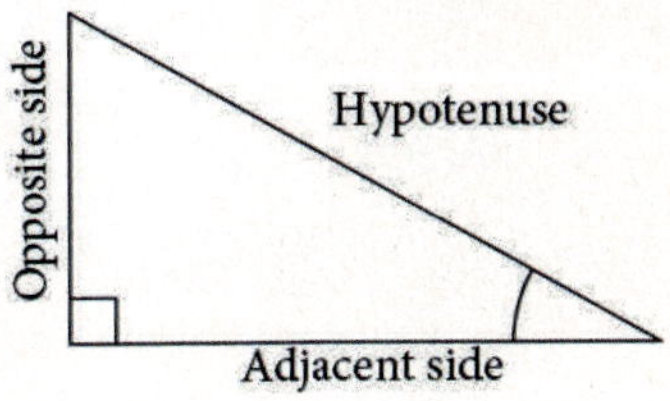

3. The tangent of an angle is the ratio of the opposite side to the adjacent side.

5. For example, the sine of an angle is equal to the cosine of its complement; the cosine of an angle is equal to the sine of its complement.

7. $\frac{\pi}{6}$

9. $\frac{\pi}{4}$

11. $b = \frac{20\sqrt{3}}{3}, c = \frac{40\sqrt{3}}{3}$

13. $a = 10{,}000, c = 10{,}00.5$

15. $b = \frac{5\sqrt{3}}{3}, c = \frac{10\sqrt{3}}{3}$

17. $\frac{5\sqrt{29}}{29}$

19. $\frac{5}{2}$

21. $\frac{\sqrt{29}}{2}$

23. $\frac{5\sqrt{41}}{41}$

25. $\frac{5}{4}$

27. $\frac{\sqrt{41}}{4}$

29. $c = 14, b = 7\sqrt{3}$

31. $a = 15, b = 15$

33. $b = 9.9970, c = 12.2041$

35. $a = 2.0838, b = 11.8177$

37. $a = 55.9808, c = 57.9555$

39. $a = 46.6790, b = 17.9184$

41. $a = 16.4662, c = 16.8341$

43. 188.3159

45. 200.6737

47. 498.3471 ft

49. 1060.09 ft

51. 27.372 ft

53. 22.6506 ft

55. 368.7633 ft

7.3 Section Exercises

1. The unit circle is a circle of radius 1 centered at the origin.

3. Coterminal angles are angles that share the same terminal side. A reference angle is the size of the smallest acute angle, *t*, formed by the terminal side of the angle *t* and the horizontal axis.

5. The sine values are equal.

7. I

9. IV

11. $\frac{\sqrt{3}}{2}$

13. $\frac{1}{2}$

15. $\frac{\sqrt{2}}{2}$

17. 0

19. -1

21. $\frac{\sqrt{3}}{2}$

23. $60°$

25. $80°$

27. $45°$

29. $\frac{\pi}{3}$

31. $\frac{\pi}{3}$

33. $\frac{\pi}{8}$

35. $60°$, Quadrant IV, $\sin(300°) = -\frac{\sqrt{3}}{2}, \cos(300°) = \frac{1}{2}$

37. $45°$, Quadrant II, $\sin(135°) = \frac{\sqrt{2}}{2}, \cos(135°) = -\frac{\sqrt{2}}{2}$

39. $60°$, Quadrant II, $\sin(120°) = \frac{\sqrt{3}}{2}, \cos(120°) = -\frac{1}{2}$

41. $30°$, Quadrant II, $\sin(150°) = \frac{1}{2}, \cos(150°) = -\frac{\sqrt{3}}{2}$

43. $\frac{\pi}{6}$, Quadrant III, $\sin\left(\frac{7\pi}{6}\right) = -\frac{1}{2}, \cos\left(\frac{7\pi}{6}\right) = -\frac{\sqrt{3}}{2}$

45. $\frac{\pi}{4}$, Quadrant II, $\sin\left(\frac{3\pi}{4}\right) = \frac{\sqrt{2}}{2}, \cos\left(\frac{4\pi}{3}\right) = -\frac{\sqrt{2}}{2}$

47. $\frac{\pi}{3}$, Quadrant II, $\sin\left(\frac{2\pi}{3}\right) = \frac{\sqrt{3}}{2}, \cos\left(\frac{2\pi}{3}\right) = -\frac{1}{2}$

49. $\frac{\pi}{4}$, Quadrant IV, $\sin\left(\frac{7\pi}{4}\right) = -\frac{\sqrt{2}}{2}, \cos\left(\frac{7\pi}{4}\right) = \frac{\sqrt{2}}{2}$

51. $\frac{\sqrt{77}}{9}$

53. $-\frac{\sqrt{15}}{4}$

55. $\left(-10, 10\sqrt{3}\right)$

57. $(-2.778, 15.757)$

59. $[-1, 1]$

61. $\sin t = \frac{1}{2}, \cos t = -\frac{\sqrt{3}}{2}$

63. $\sin t = -\frac{\sqrt{2}}{2}, \cos t = -\frac{\sqrt{2}}{2}$

65. $\sin t = \frac{\sqrt{3}}{2}, \cos t = -\frac{1}{2}$

67. $\sin t = -\frac{\sqrt{2}}{2}, \cos t = \frac{\sqrt{2}}{2}$

69. $\sin t = 0,\ \cos t = -1$

71. $\sin t = -0.596,\ \cos t = 0.803$

73. $\sin t = \frac{1}{2}, \cos t = \frac{\sqrt{3}}{2}$

75. $\sin t = -\frac{1}{2}, \cos t = \frac{\sqrt{3}}{2}$

77. $\sin t = 0.761, \cos t = -0.649$

79. $\sin t = 1, \cos t = 0$

81. -0.1736

83. 0.9511

85. -0.7071

87. -0.1392

89. -0.7660

91. $\frac{\sqrt{2}}{4}$

93. $-\frac{\sqrt{6}}{4}$

95. $\frac{\sqrt{2}}{4}$

97. $\frac{\sqrt{2}}{4}$

99. 0

101. $(0, -1)$

103. 37.5 seconds, 97.5 seconds, 157.5 seconds, 217.5 seconds, 277.5 seconds, 337.5 seconds

7.4 Section Exercises

1. Yes, when the reference angle is $\frac{\pi}{4}$ and the terminal side of the angle is in quadrants I and III. Thus, a $x = \frac{\pi}{4}, \frac{5\pi}{4}$, the sine and cosine values are equal.

3. Substitute the sine of the angle in for y in the Pythagorean Theorem $x^2 + y^2 = 1$. Solve for x and take the negative solution.

5. The outputs of tangent and cotangent will repeat every π units.

7. $\frac{2\sqrt{3}}{3}$

9. $\sqrt{3}$

11. $\sqrt{2}$

13. 1

15. 2

17. $\frac{\sqrt{3}}{3}$

19. $-\frac{2\sqrt{3}}{3}$

21. $\sqrt{3}$

23. $-\sqrt{2}$

25. -1

27. -2

29. $-\frac{\sqrt{3}}{3}$

31. 2

33. $\frac{\sqrt{3}}{3}$

35. -2

37. -1

39. $\sin\ t = -\frac{2\sqrt{2}}{3}$, $\sec\ t = -3$, $\csc\ t = -\frac{3\sqrt{2}}{4}$, $\tan\ t = 2\sqrt{2}$, $\cot\ t = \frac{\sqrt{2}}{4}$

41. $\sec t = 2$, $\csc t = \frac{2\sqrt{3}}{3}$, $\tan t = \sqrt{3}$, $\cot t = \frac{\sqrt{3}}{3}$

43. $-\frac{\sqrt{2}}{2}$

45. 3.1

47. 1.4

49. $\sin t = \frac{\sqrt{2}}{2}, \cos t = \frac{\sqrt{2}}{2},$ $\tan t = 1, \cot t = 1,$ $\sec t = \sqrt{2}, \csc t = \sqrt{2}$

51. $\sin t = -\frac{\sqrt{3}}{2},$ $\cos t = -\frac{1}{2} \tan t = \sqrt{3},$ $\cot t = \frac{\sqrt{3}}{3}, \sec t = -2,$ $\csc t = -\frac{2\sqrt{3}}{3}$

53. -0.228

55. -2.414

57. 1.414

59. 1.540

61. 1.556

63. $\sin(t) \approx 0.79$

65. $\csc t \approx 1.16$

67. even

69. even

71. $\frac{\sin\ t}{\cos\ t} = \tan\ t$

73. 13.77 hours, period: 1000π

75. 3.46 inches

Review Exercises

1. $45°$

3. $-\frac{7\pi}{6}$

5. 10.385 meters

7. $60°$

9. $\frac{2\pi}{11}$

11.

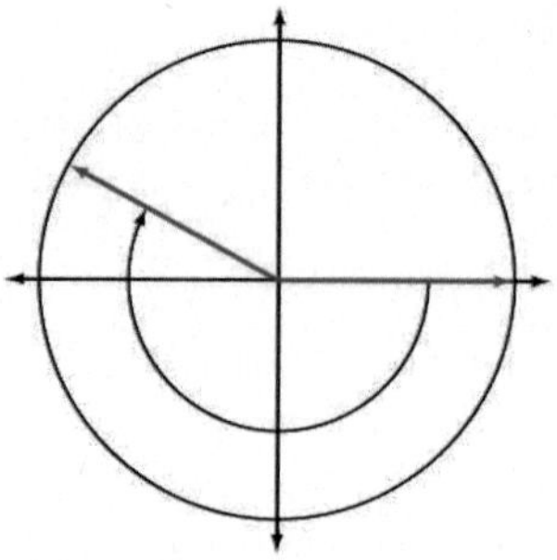

13.

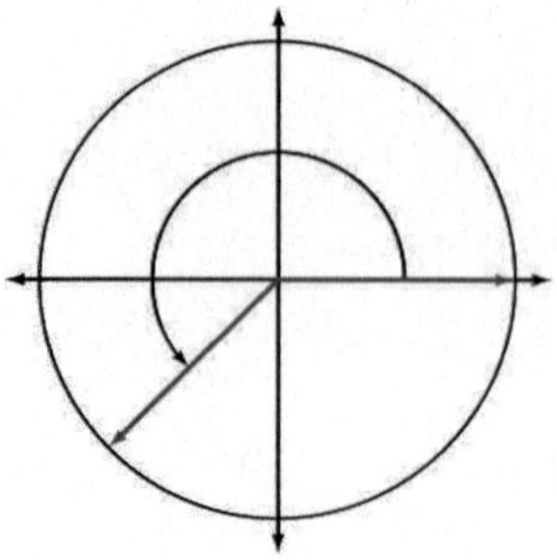

15. 1036.73 miles per hour

17. $\frac{\sqrt{2}}{2}$

19. $\frac{\sqrt{3}}{3}$

21. $72°$

23. $a = \frac{10}{3}, c = \frac{2\sqrt{106}}{3}$

25. $\frac{6}{11}$

27. $a = \frac{5\sqrt{3}}{2}, b = \frac{5}{2}$

29. 369.2136 ft

31. $\frac{\sqrt{2}}{2}$

33. $60°$

35. $\frac{\sqrt{3}}{2}$

37. all real numbers

39. $\frac{\sqrt{3}}{2}$

41. $\frac{2\sqrt{3}}{3}$

43. 2

45. –2.5

47. $\frac{1}{3}$

49. cosine, secant

Practice Test

1. $150°$

3. 6.283 centimeters

5. $15°$

7.

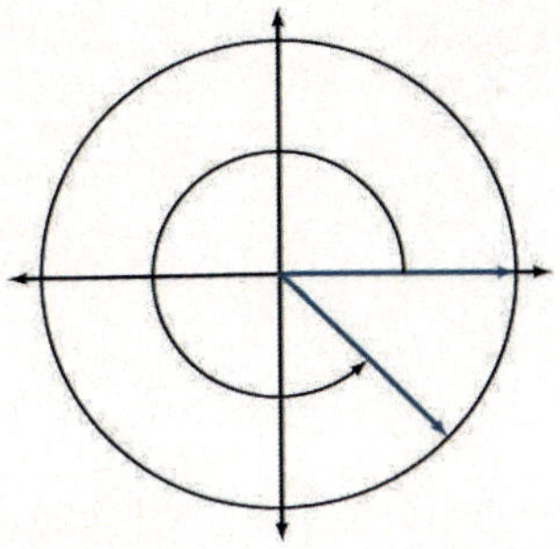

9. 3.351 feet per second, $\frac{2\pi}{75}$ radians per second

11. $a = \frac{9}{2}, b = \frac{9\sqrt{3}}{2}$

13. $\frac{1}{2}$

15. real numbers

17. 1

19. $-\sqrt{2}$

21. –0.68

23. $\frac{\pi}{3}$

Chapter 8

Try It

8.1 Graphs of the Sine and Cosine Functions

1. 6π

2. $\frac{1}{2}$ compressed

3. $\frac{\pi}{2}$; right

4. 2 units up

5. midline: $y = 0$; amplitude: $|A| = \frac{1}{2}$; period: $P = \frac{2\pi}{|B|} = 6\pi$; phase shift: $\frac{C}{B} = \pi$

6. $f(x) = \sin(x) + 2$

7. two possibilities:
$y = 4\sin\left(\frac{\pi}{5}x - \frac{\pi}{5}\right) + 4$ or
$y = -4\sin\left(\frac{\pi}{5}x + \frac{4\pi}{5}\right) + 4$

8.

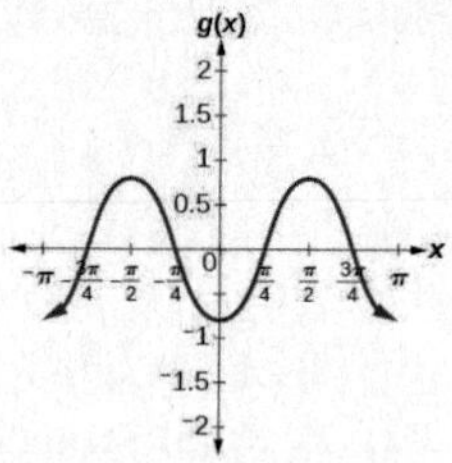

midline: $y = 0$; amplitude: $|A| = 0.8$; period: $P = \frac{2\pi}{|B|} = \pi$; phase shift: $\frac{C}{B} = 0$ or none

9.

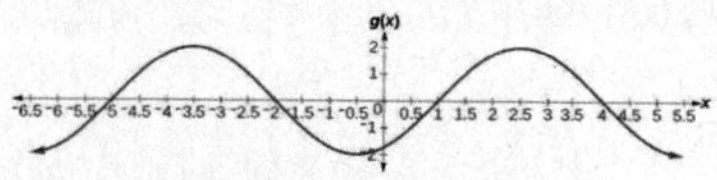

midline: $y = 0$; amplitude: $|A| = 2$; period: $P = \frac{2\pi}{|B|} = 6$; phase shift: $\frac{C}{B} = -\frac{1}{2}$

10. 7

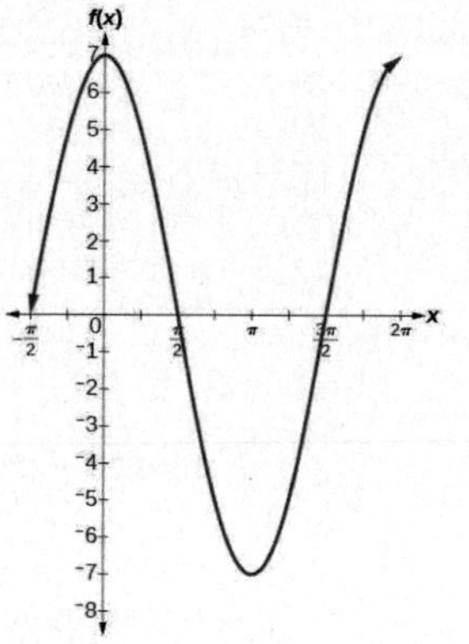

11. $y = 3\cos(x) - 4$

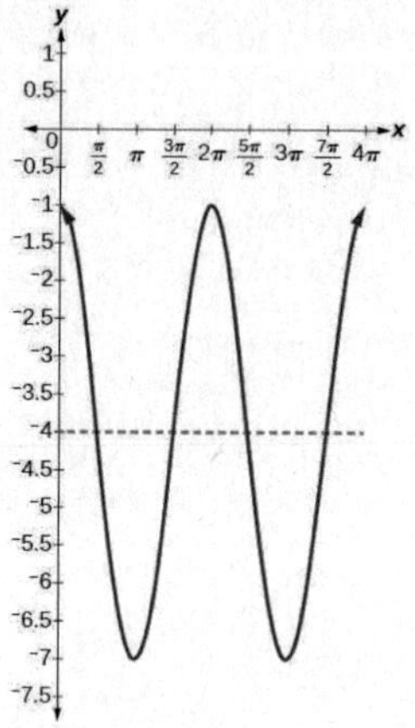

8.2 Graphs of the Other Trigonometric Functions

1.

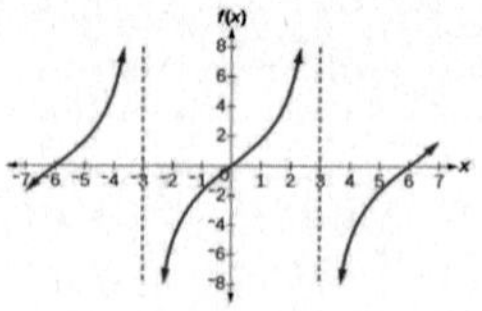

2. It would be reflected across the line $y = -1$, becoming an increasing function.

3. $g(x) = 4\tan(2x)$

4. This is a vertical reflection of the preceding graph because A is negative.

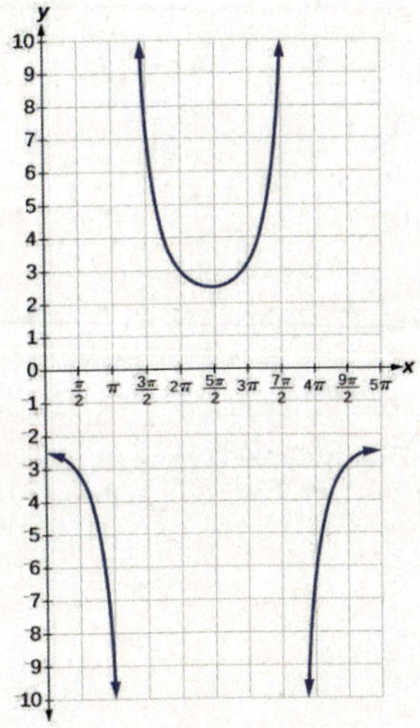

5.

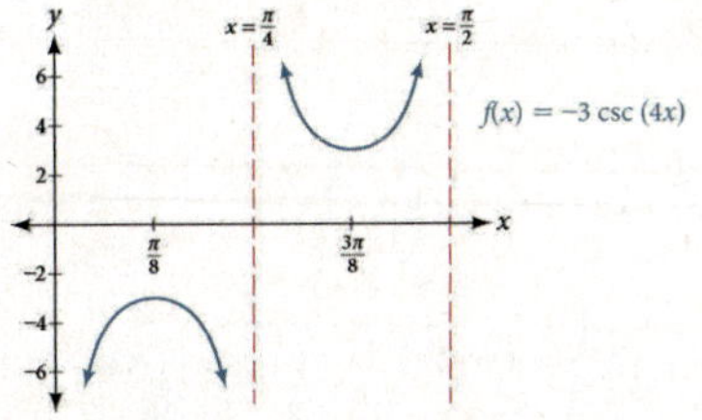

6.

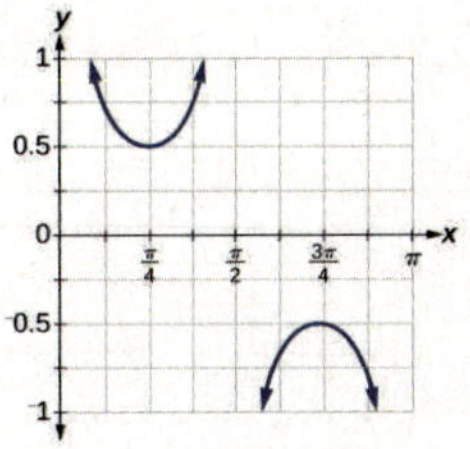

7.

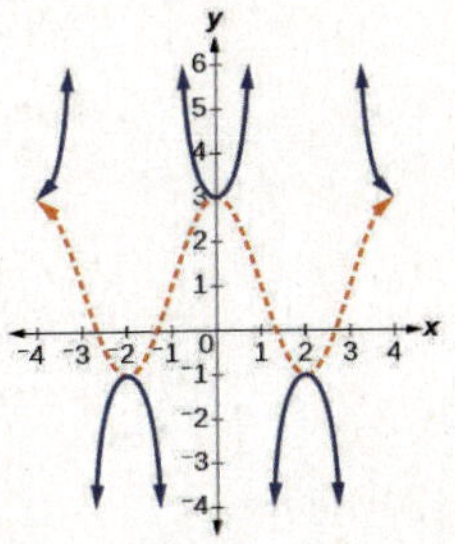

8.3 Inverse Trigonometric Functions

1. $\arccos(0.8776) \approx 0.5$

2. ⓐ $-\frac{\pi}{2}$; ⓑ $-\frac{\pi}{4}$; ⓒ π; ⓓ $\frac{\pi}{3}$

3. 1.9823 or 113.578°

4. $\sin^{-1}(0.6) = 36.87° = 0.6435$ radians

5. $\frac{\pi}{8}$; $\frac{2\pi}{9}$

6. $\frac{3\pi}{4}$

7. $\frac{12}{13}$

8. $\frac{4\sqrt{2}}{9}$

9. $\frac{4x}{\sqrt{16x^2+1}}$

8.1 Section Exercises

1. The sine and cosine functions have the property that $f(x+P) = f(x)$ for a certain P. This means that the function values repeat for every P units on the x-axis.

3. The absolute value of the constant A (amplitude) increases the total range and the constant D (vertical shift) shifts the graph vertically.

5. At the point where the terminal side of t intersects the unit circle, you can determine that the sin t equals the y-coordinate of the point.

7.

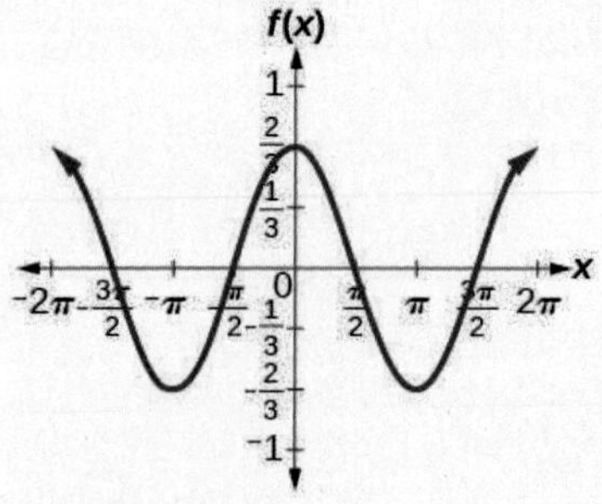

amplitude: $\frac{2}{3}$; period: 2π; midline: $y = 0$; maximum: $y = \frac{2}{3}$ occurs at $x = 0$; minimum: $y = -\frac{2}{3}$ occurs at $x = \pi$; for one period, the graph starts at 0 and ends at 2π

9.

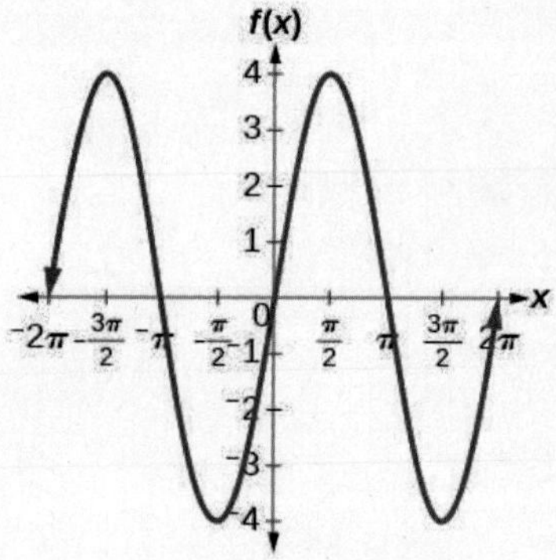

amplitude: 4; period: 2π; midline: $y = 0$; maximum $y = 4$ occurs at $x = \frac{\pi}{2}$; minimum: $y = -4$ occurs at $x = \frac{3\pi}{2}$; one full period occurs from $x = 0$ to $x = 2\pi$

11.

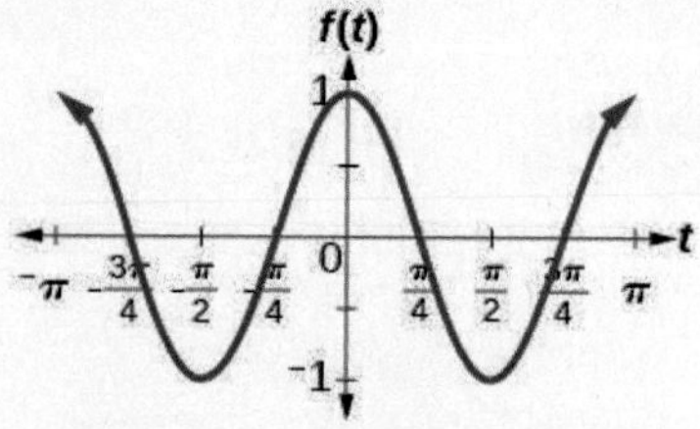

amplitude: 1; period: π; midline: $y = 0$; maximum: $y = 1$ occurs at $x = \pi$; minimum: $y = -1$ occurs at $x = \frac{\pi}{2}$; one full period is graphed from $x = 0$ to $x = \pi$

13.

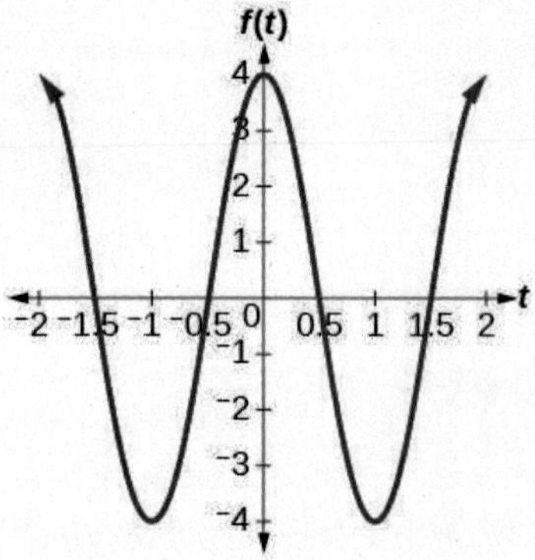

amplitude: 4; period: 2; midline: $y = 0$; maximum: $y = 4$ occurs at $x = 0$; minimum: $y = -4$ occurs at $x = 1$

15.

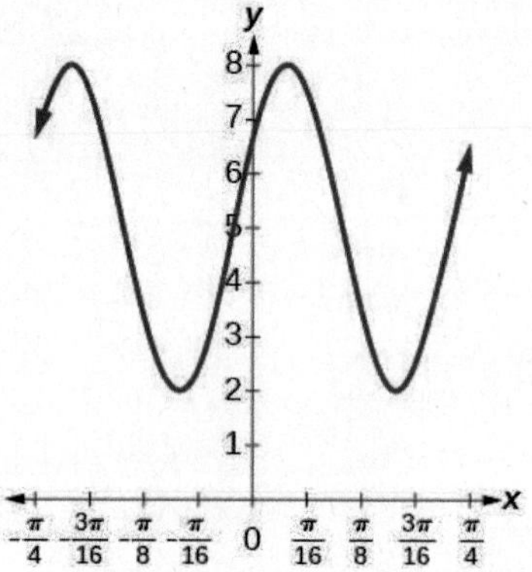

amplitude: 3; period: $\frac{\pi}{4}$; midline: $y = 5$; maximum: $y = 8$ occurs at $x = 0.12$; minimum: $y = 2$ occurs at $x = 0.516$; horizontal shift: -4; vertical translation 5; one period occurs from $x = 0$ to $x = \frac{\pi}{4}$

17.

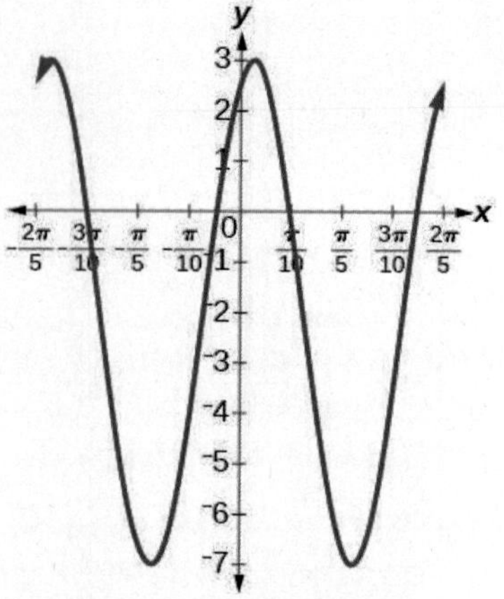

amplitude: 5; period: $\frac{2\pi}{5}$; midline: $y = -2$; maximum: $y = 3$ occurs at $x = 0.08$; minimum: $y = -7$ occurs at $x = 0.71$; phase shift: -4; vertical translation: -2; one full period can be graphed on $x = 0$ to $x = \frac{2\pi}{5}$

19.

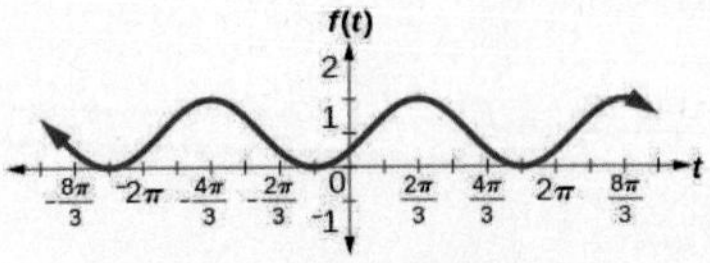

amplitude: 1 ; period: 2π; midline: $y = 1$; maximum: $y = 2$ occurs at $x = 2.09$; maximum: $y = 2$ occurs at $t = 2.09$; minimum: $y = 0$ occurs at $t = 5.24$; phase shift: $-\frac{\pi}{3}$; vertical translation: 1; one full period is from $t = 0$ to $t = 2\pi$

21.

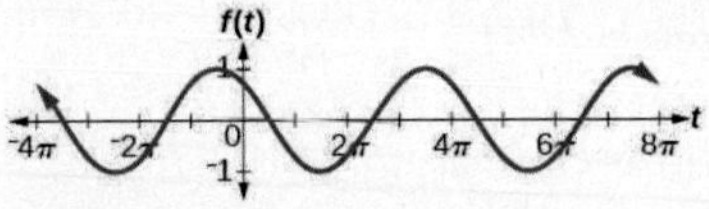

amplitude: 1; period: 4π; midline: $y = 0$; maximum: $y = 1$ occurs at $t = 11.52$; minimum: $y = -1$ occurs at $t = 5.24$; phase shift: $-\frac{10\pi}{3}$; vertical shift: 0

23. amplitude: 2; midline: $y = -3$; period: 4; equation: $f(x) = 2\sin\left(\frac{\pi}{2}x\right) - 3$

25. amplitude: 2; period: 5; midline: $y = 3$; equation: $f(x) = -2\cos\left(\frac{2\pi}{5}x\right) + 3$

27. amplitude: 4; period: 2; midline: $y = 0$; equation: $f(x) = -4\cos\left(\pi\left(x - \frac{\pi}{2}\right)\right)$

29. amplitude: 2; period: 2; midline $y = 1$; equation: $f(x) = 2\cos(\pi x) + 1$

31. $0, \pi$

33. $\sin\left(\frac{\pi}{2}\right) = 1$

35. $\frac{\pi}{2}$

37. $f(x) = \sin x$ is symmetric

39. $\frac{\pi}{3}, \frac{5\pi}{3}$

41. Maximum: 1 at $x = 0$; minimum: -1 at $x = \pi$

43. A linear function is added to a periodic sine function. The graph does not have an amplitude because as the linear function increases without bound the combined function $h(x) = x + \sin x$ will increase without bound as well. The graph is bounded between the graphs of $y = x + 1$ and $y = x - 1$ because sine oscillates between −1 and 1.

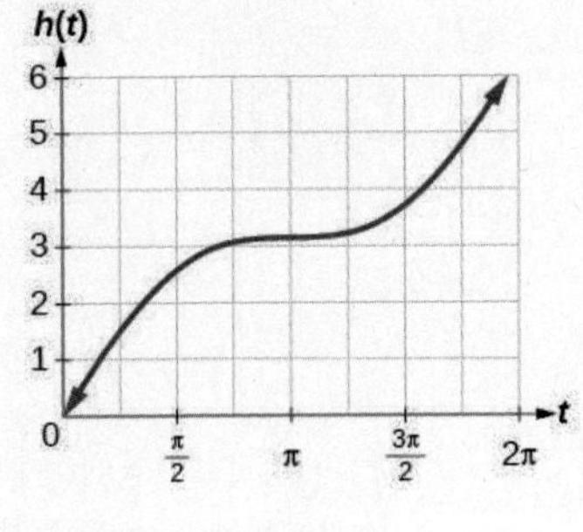

45. There is no amplitude because the function is not bounded.

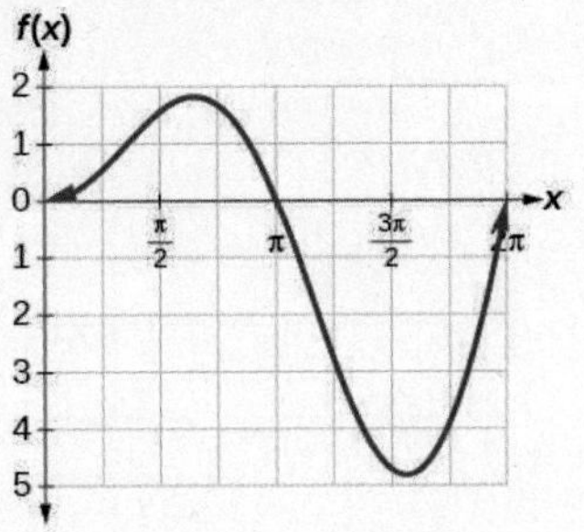

47. The graph is symmetric with respect to the y-axis and there is no amplitude because the function's bounds decrease as $|x|$ grows. There appears to be a horizontal asymptote at $y = 0$.

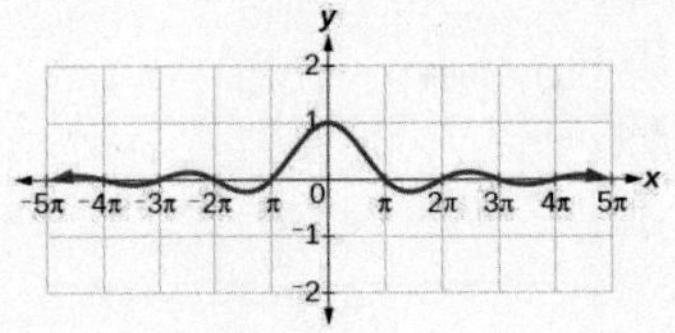

8.2 Section Exercises

1. Since $y = \csc\ x$ is the reciprocal function of $y = \sin\ x$, you can plot the reciprocal of the coordinates on the graph of $y = \sin\ x$ to obtain the y-coordinates of $y = \csc\ x$. The x-intercepts of the graph $y = \sin\ x$ are the vertical asymptotes for the graph of $y = \csc\ x$.

3. Answers will vary. Using the unit circle, one can show that $\tan(x + \pi) = \tan\ x$.

5. The period is the same: 2π.

7. IV

9. III

11. period: 8; horizontal shift: 1 unit to left

13. 1.5

15. 5

17. $-\cot x \cos x - \sin x$

19.

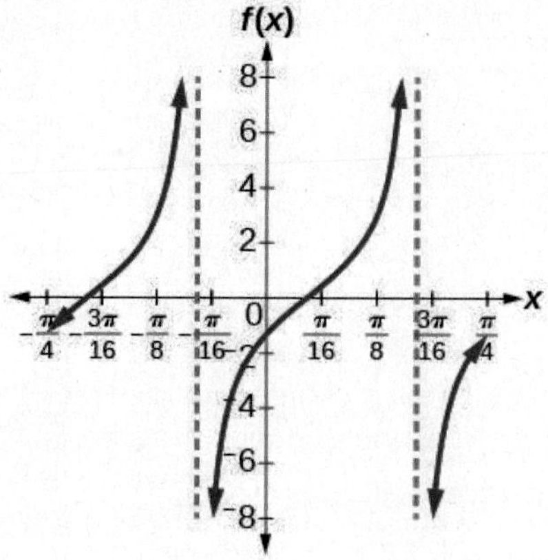

stretching factor: 2; period: $\frac{\pi}{4}$;
asymptotes:
$x = \frac{1}{4}\left(\frac{\pi}{2} + \pi k\right) + 8,$ where k is an integer

21.

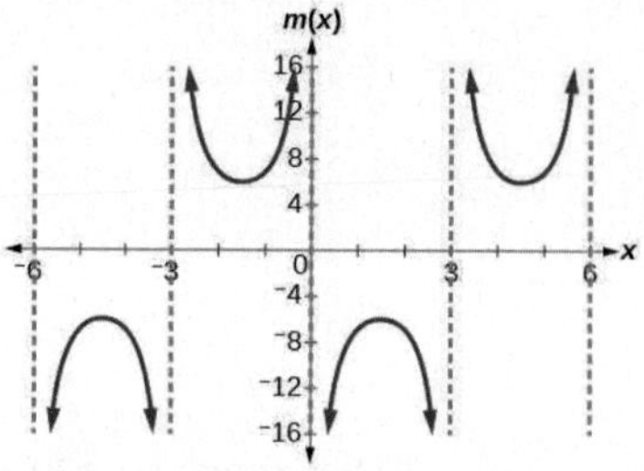

stretching factor: 6; period: 6;
asymptotes:
$x = 3k,$ where k is an integer

23.

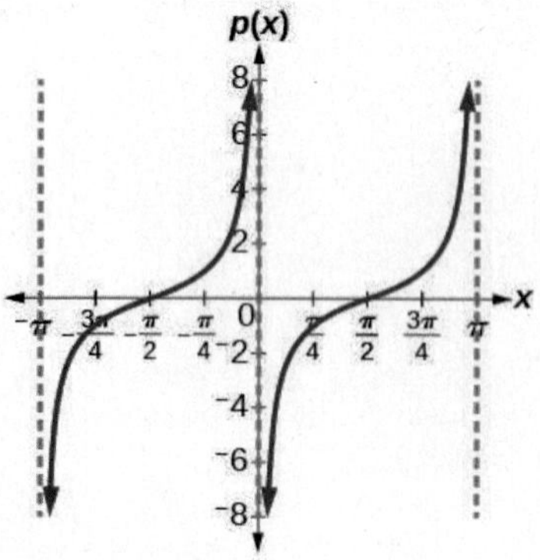

stretching factor: 1; period: π;
asymptotes:
$x = \pi k,$ where k is an integer

25.

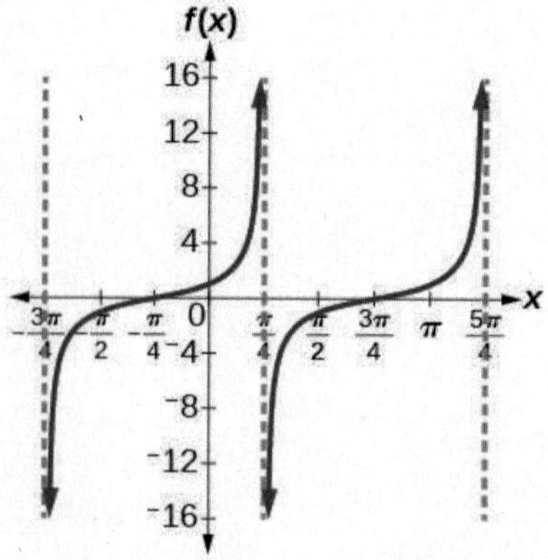

Stretching factor: 1; period: π;
asymptotes:
$x = \frac{\pi}{4} + \pi k,$ where k is an integer

27.

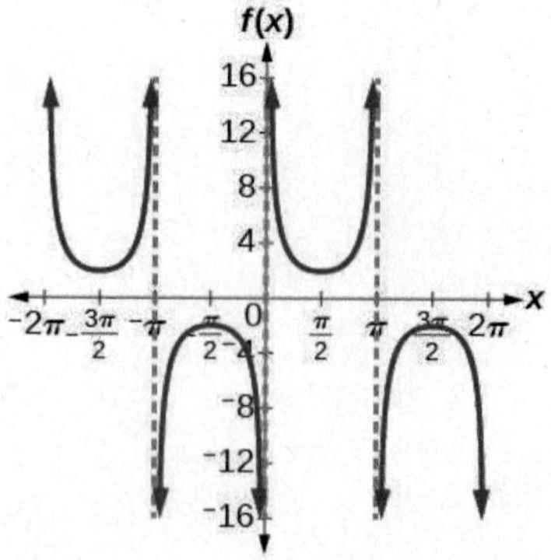

stretching factor: 2; period: 2π;
asymptotes:
$x = \pi k,$ where k is an integer

29.

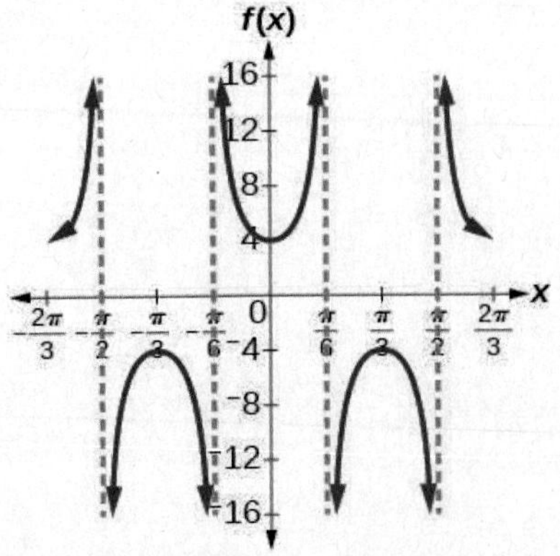

stretching factor: 4; period: $\frac{2\pi}{3}$;
asymptotes:
$x = \frac{\pi}{6}k$, where k is an odd integer

31.

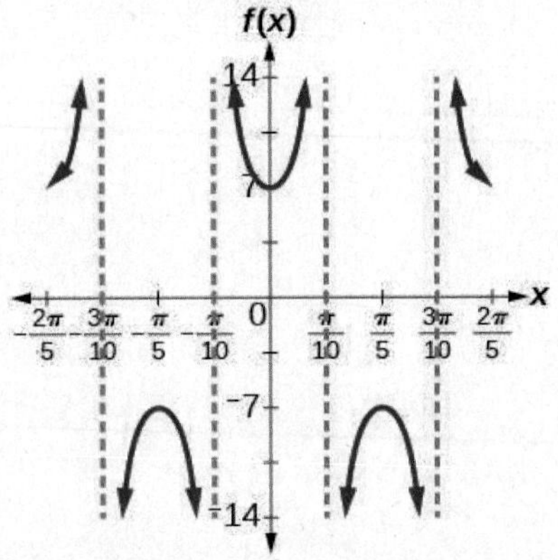

stretching factor: 7; period: $\frac{2\pi}{5}$;
asymptotes:
$x = \frac{\pi}{10}k$, where k is an odd integer

33.

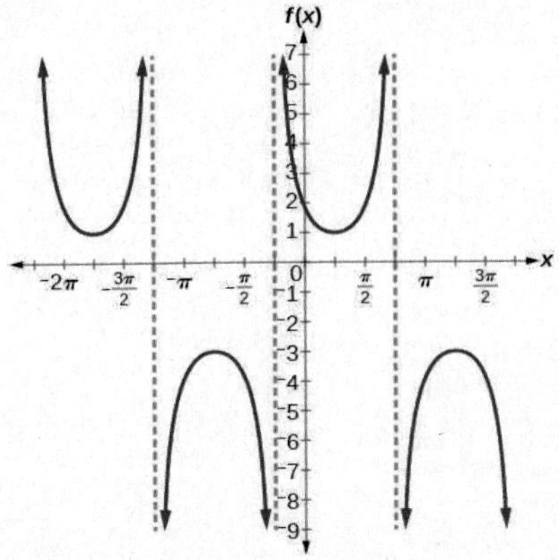

stretching factor: 2; period: 2π;
asymptotes:
$x = -\frac{\pi}{4} + \pi k$, where k is an integer

35.

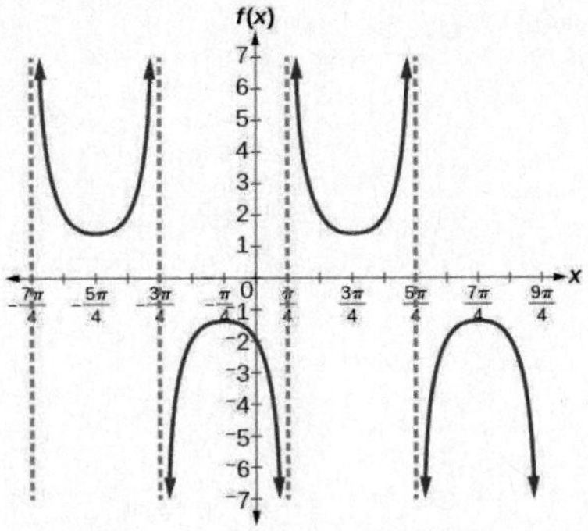

stretching factor: $\frac{7}{5}$; period: 2π;
asymptotes:
$x = \frac{\pi}{4} + \pi k$, where k is an integer

37. $y = \tan\left(3\left(x - \frac{\pi}{4}\right)\right) + 2$

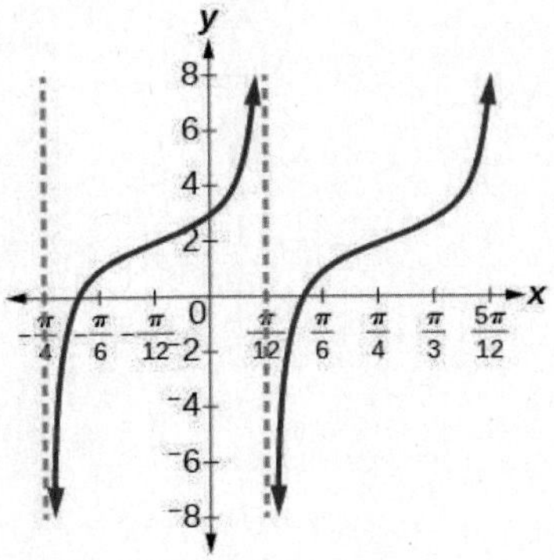

39. $f(x) = \csc(2x)$

41. $f(x) = \csc(4x)$

43. $f(x) = 2\csc x$

45. $f(x) = \frac{1}{2}\tan(100\pi x)$

47.

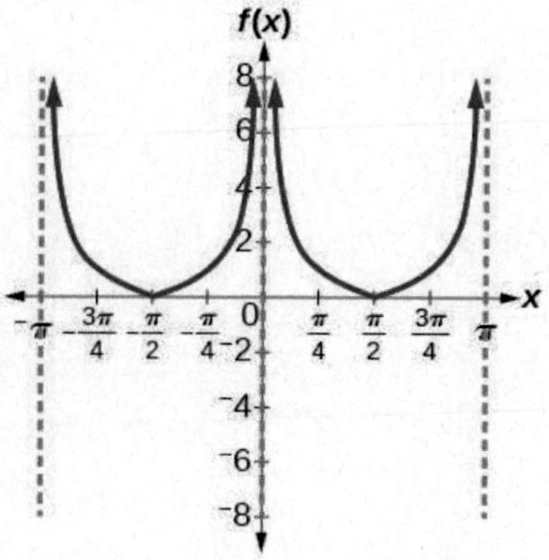

49.

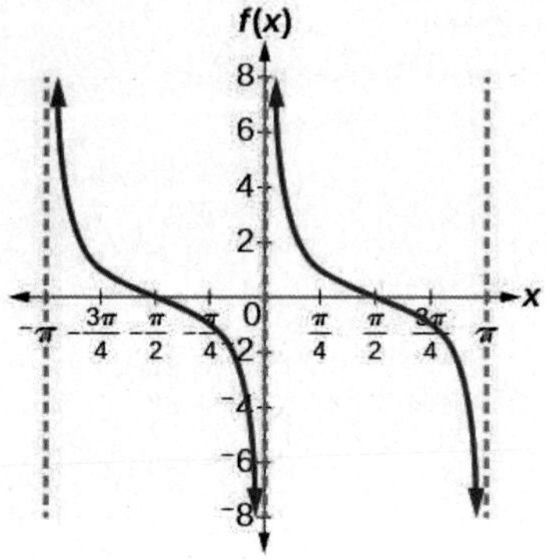

51.

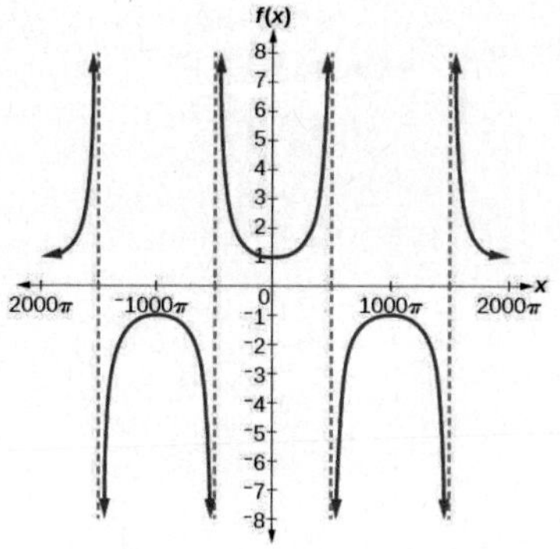

53.

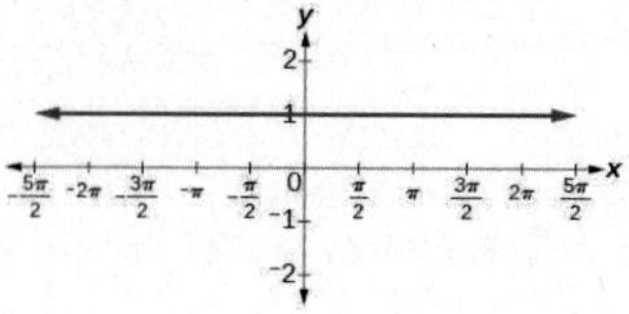

55. ⓐ $\left(-\frac{\pi}{2}, \frac{\pi}{2}\right)$;
ⓑ

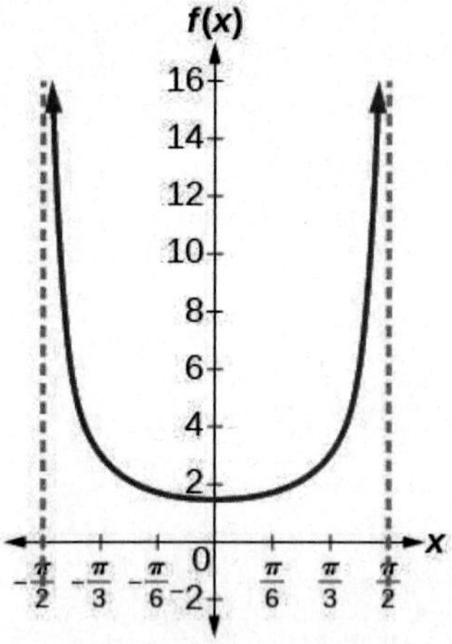

ⓒ $x = -\frac{\pi}{2}$ and $x = \frac{\pi}{2}$; the distance grows without bound as $|x|$ approaches $\frac{\pi}{2}$ —i.e., at right angles to the line representing due north, the boat would be so far away, the fisherman could not see it;
ⓓ 3; when $x = -\frac{\pi}{3}$, the boat is 3 km away;
ⓔ 1.73; when $x = \frac{\pi}{6}$, the boat is about 1.73 km away;
ⓕ 1.5 km; when $x = 0$

57. ⓐ $h(x) = 2\tan\left(\frac{\pi}{120}x\right)$;
ⓑ

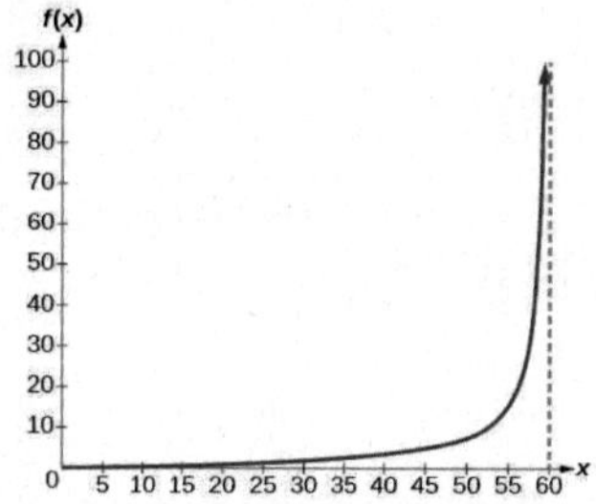

ⓒ $h(0) = 0$: after 0 seconds, the rocket is 0 mi above the ground; $h(30) = 2$: after 30 seconds, the rockets is 2 mi high;
ⓓ As x approaches 60 seconds, the values of $h(x)$ grow increasingly large. The distance to the rocket is growing so large that the camera can no longer track it.

8.3 Section Exercises

1. The function $y = \sin x$ is one-to-one on $\left[-\frac{\pi}{2}, \frac{\pi}{2}\right]$; thus, this interval is the range of the inverse function of $y = \sin x$, $f(x) = \sin^{-1} x$. The function $y = \cos x$ is one-to-one on $[0, \pi]$; thus, this interval is the range of the inverse function of $y = \cos x, f(x) = \cos^{-1} x$.

3. $\frac{\pi}{6}$ is the radian measure of an angle between $-\frac{\pi}{2}$ and $\frac{\pi}{2}$ whose sine is 0.5.

5. In order for any function to have an inverse, the function must be one-to-one and must pass the horizontal line test. The regular sine function is not one-to-one unless its domain is restricted in some way. Mathematicians have agreed to restrict the sine function to the interval $\left[-\frac{\pi}{2}, \frac{\pi}{2}\right]$ so that it is one-to-one and possesses an inverse.

7. True . The angle, θ_1 that equals $\arccos(-x)$, $x > 0$, will be a second quadrant angle with reference angle, θ_2, where θ_2 equals $\arccos x$, $x > 0$. Since θ_2 is the reference angle for θ_1, $\theta_2 = \pi - \theta_1$ and $\arccos(-x)$ = $\pi - \arccos x$-

9. $-\frac{\pi}{6}$

11. $\frac{3\pi}{4}$

13. $-\frac{\pi}{3}$

15. $\frac{\pi}{3}$

17. 1.98

19. 0.93

21. 1.41

23. 0.56 radians

25. 0

27. 0.71

29. -0.71

31. $-\frac{\pi}{4}$

33. 0.8

35. $\frac{5}{13}$

37. $\frac{x-1}{\sqrt{-x^2+2x}}$

39. $\frac{\sqrt{x^2-1}}{x}$

41. $\frac{x+0.5}{\sqrt{-x^2-x+\frac{3}{4}}}$

43. $\frac{\sqrt{2x+1}}{x+1}$

45. $\frac{\sqrt{2x+1}}{x}$

47. t

49.

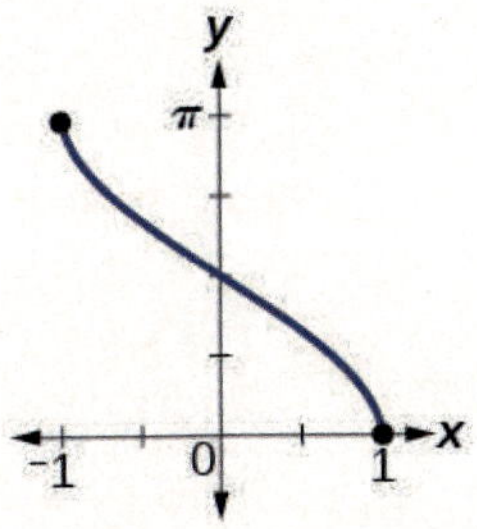

domain $[-1, 1]$; range $[0, \pi]$

51. approximately $x = 0.00$

53. 0.395 radians

55. 1.11 radians

57. 1.25 radians

59. 0.405 radians

61. No. The angle the ladder makes with the horizontal is 60 degrees.

Review Exercises

1. amplitude: 3; period: 2π; midline: $y = 3$; no asymptotes

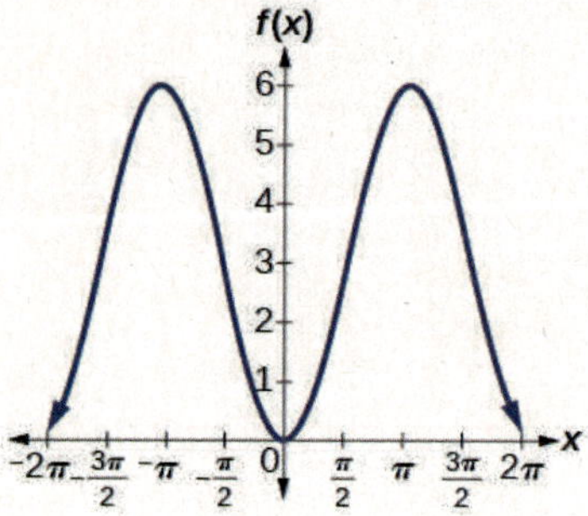

3. amplitude: 3; period: 2π; midline: $y = 0$; no asymptotes

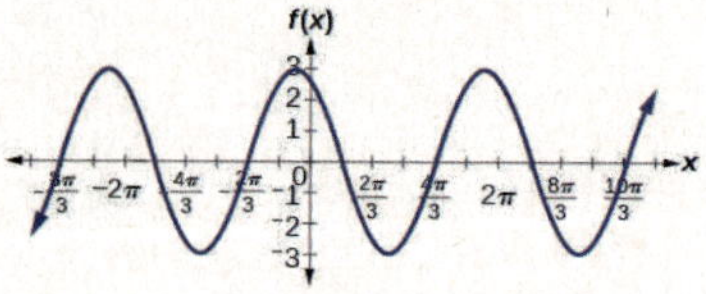

5. amplitude: 3; period: 2π; midline: $y = -4$; no asymptotes

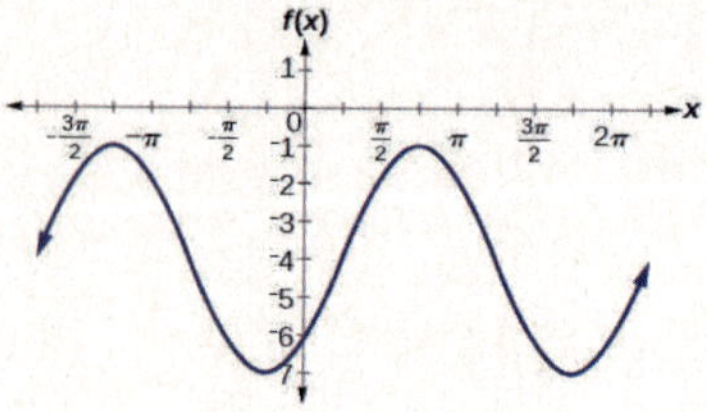

7. amplitude: 6; period: $\frac{2\pi}{3}$; midline: $y = -1$; no asymptotes

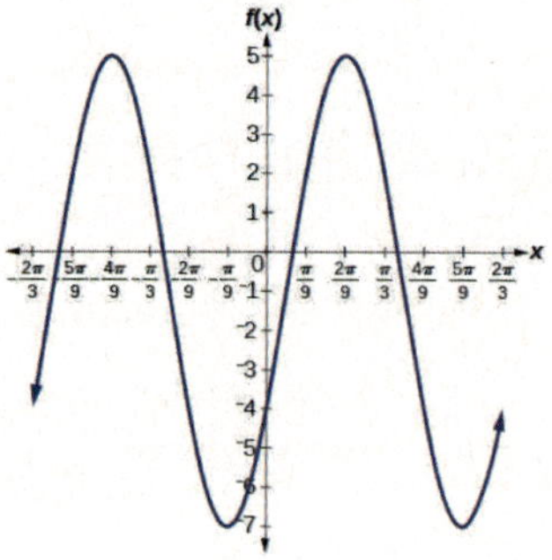

9. stretching factor: none; period: π; midline: $y = -4$; asymptotes: $x = \frac{\pi}{2} + \pi k$, where k is an integer

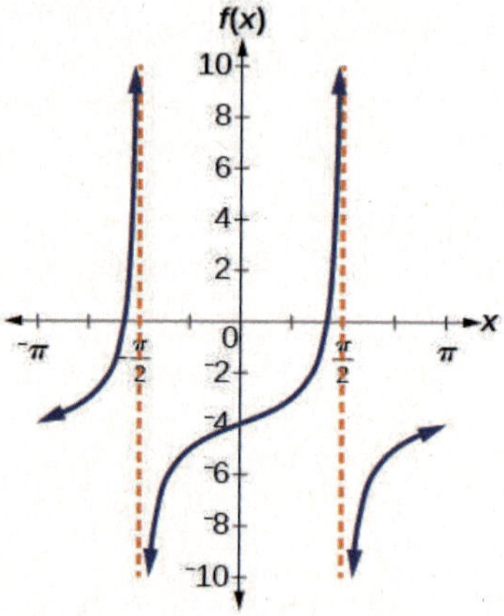

11. stretching factor: 3; period: $\frac{\pi}{4}$; midline: $y = -2$; asymptotes: $x = \frac{\pi}{8} + \frac{\pi}{4}k$, where k is an integer

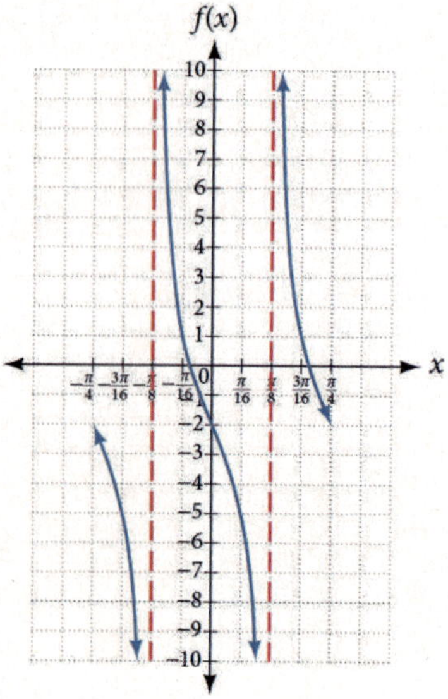

13. amplitude: none; period: 2π; no phase shift; asymptotes: $x = \frac{\pi}{2}k$, where k is an odd integer

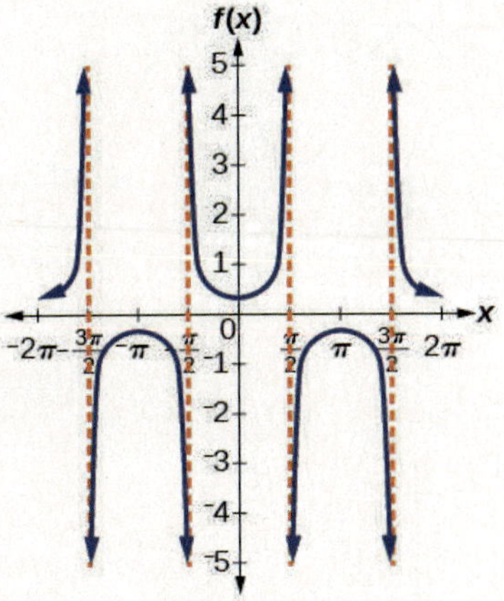

15. amplitude: none; period: $\frac{2\pi}{5}$; no phase shift; asymptotes: $x = \frac{\pi}{5}k$, where k is an integer

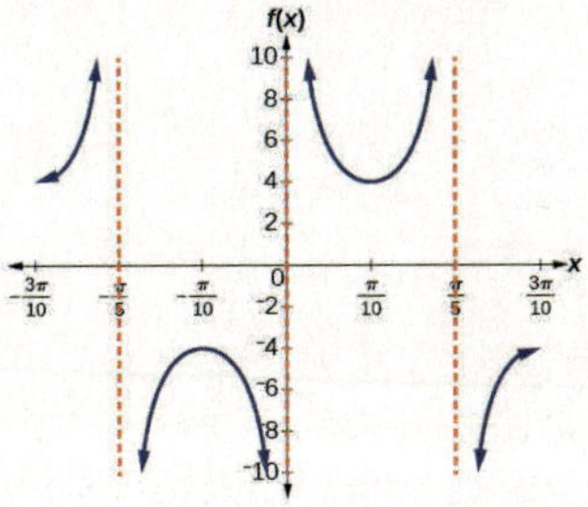

17. amplitude: none; period: 4π; no phase shift; asymptotes: $x = 2\pi k$, where k is an integer

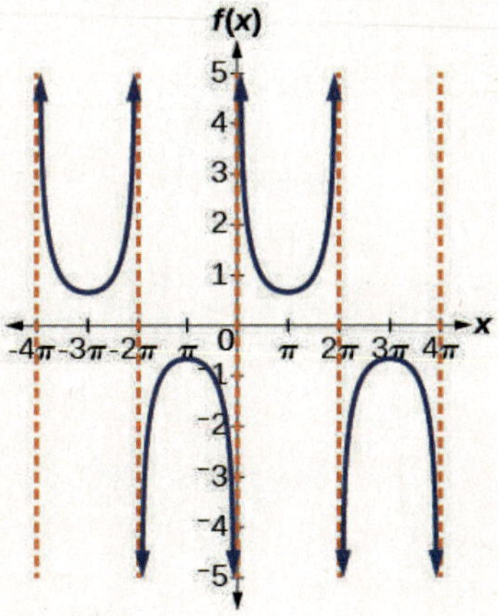

19. largest: 20,000; smallest: 4,000

21. amplitude: 8,000; period: 10; phase shift: 0

23. In 2007, the predicted population is 4,413. In 2010, the population will be 11,924.

25. 5 in.

27. 10 seconds

29. $\frac{\pi}{6}$

31. $\frac{\pi}{4}$

33. $\frac{\pi}{3}$

35. No solution

37. $\frac{12}{5}$

39. The graphs are not symmetrical with respect to the line $y = x$. They are symmetrical with respect to the y-axis.

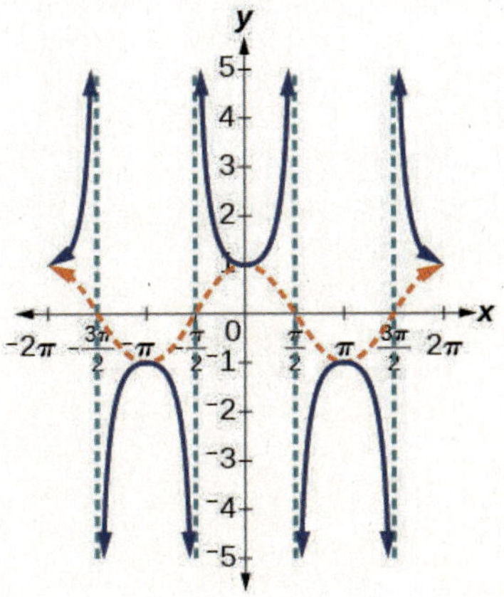

41. The graphs appear to be identical.

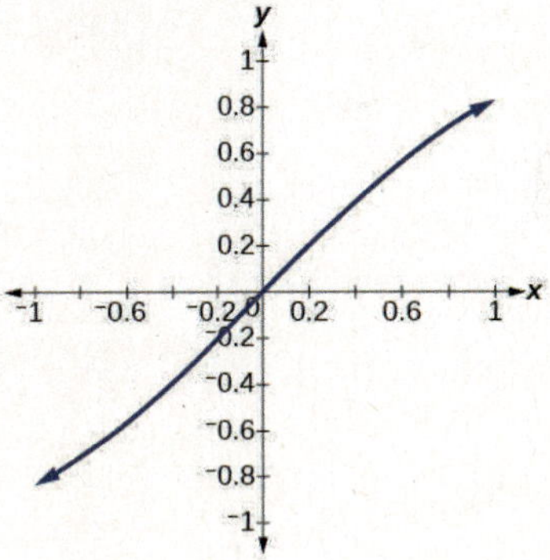

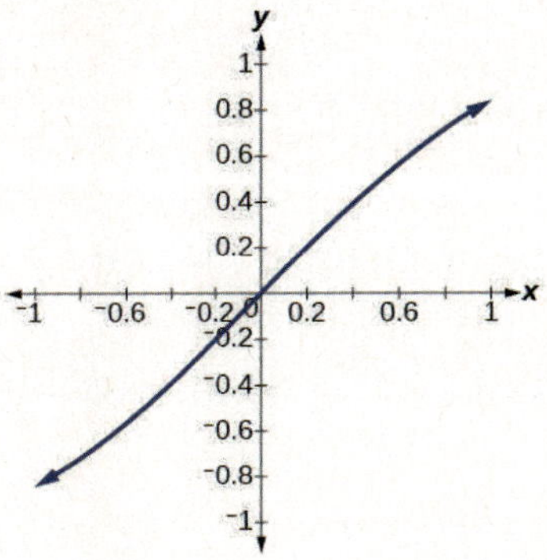

Practice Test

1. amplitude: 0.5; period: 2π; midline $y = 0$

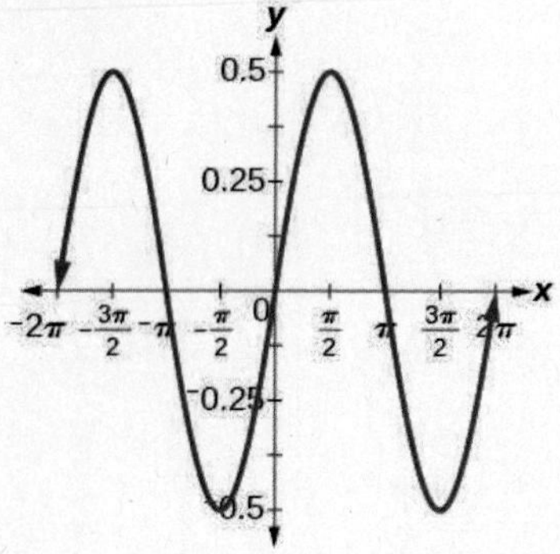

3. amplitude: 5; period: 2π; midline: $y = 0$

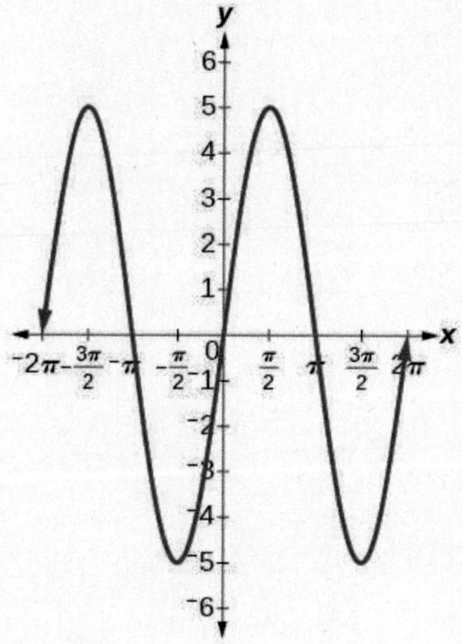

5. amplitude: 1; period: 2π; midline: $y = 1$

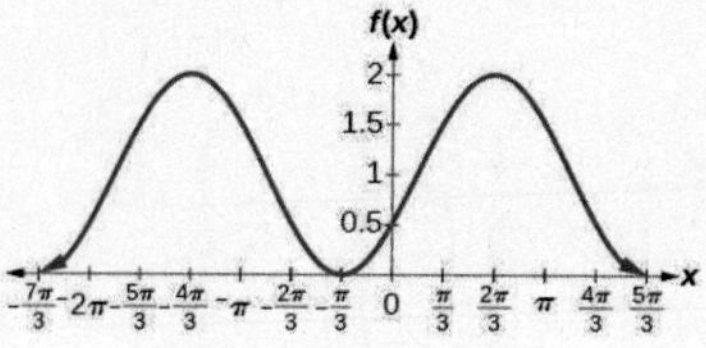

7. amplitude: 3; period: 6π; midline: $y = 0$

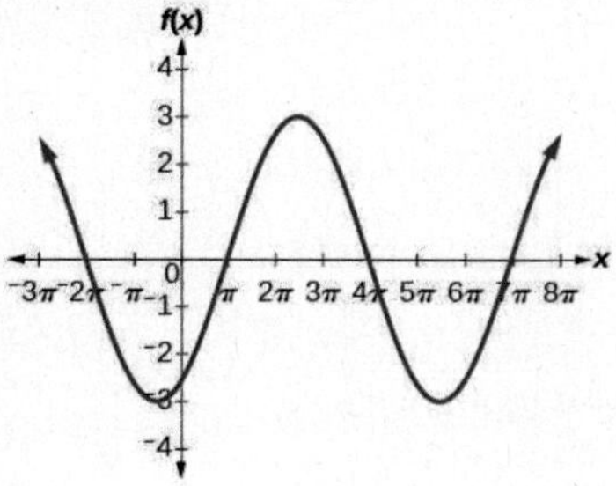

9. amplitude: none; period: π; midline: $y = 0$, asymptotes: $x = \frac{2\pi}{3} + \pi k$, where k is an integer

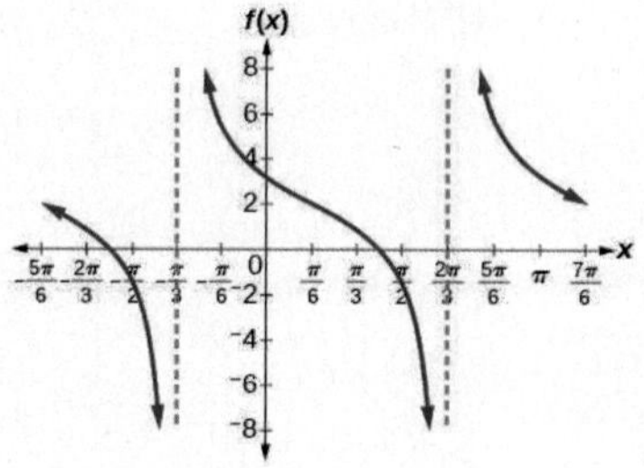

11. amplitude: none; period: $\frac{2\pi}{3}$; midline: $y = 0$, asymptotes: $x = \frac{\pi}{3}k$, where k is an integer

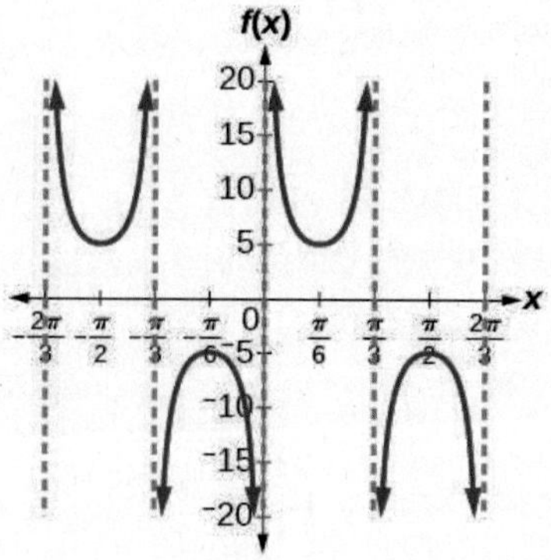

13. amplitude: none; period: 2π; midline: $y = -3$

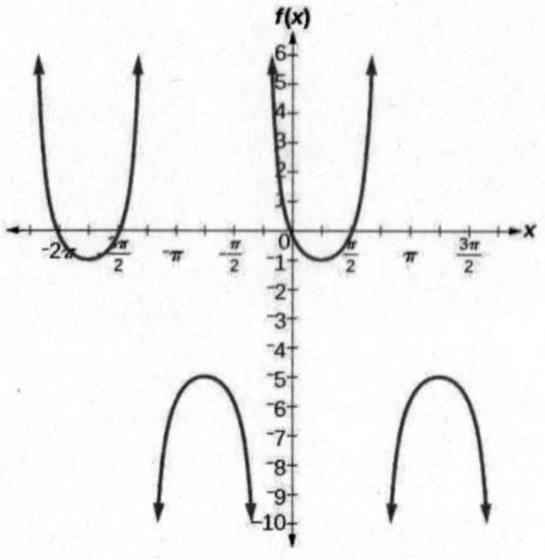

15. amplitude: 2; period: 2; midline: $y = 0$; $f(x) = 2\sin(\pi(x-1))$

17. amplitude: 1; period: 12; phase shift: -6; midline $y = -3$

19. $D(t) = 68 - 12\sin\left(\frac{\pi}{12}x\right)$

21. period: $\frac{\pi}{6}$; horizontal shift: -7

23. $f(x) = \sec(\pi x)$; period: 2; phase shift: 0

25. 4

27. The views are different because the period of the wave is $\frac{1}{25}$. Over a bigger domain, there will be more cycles of the graph.

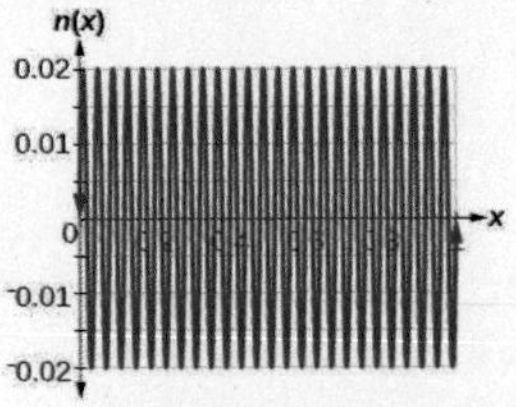

29. $\frac{3}{5}$

31. On the approximate intervals $(0.5, 1), (1.6, 2.1), (2.6, 3.1), (3.7, 4.2), (4.7, 5.2), (5.6, 6.28)$

33. $f(x) = 2\cos\left(12\left(x + \frac{\pi}{4}\right)\right) + 3$

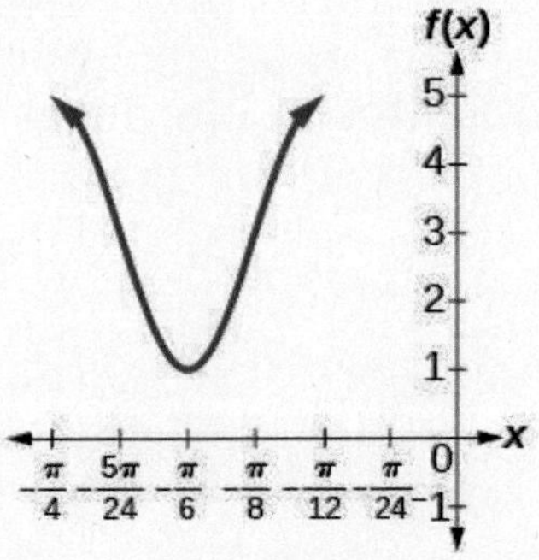

35. This graph is periodic with a period of 2π.

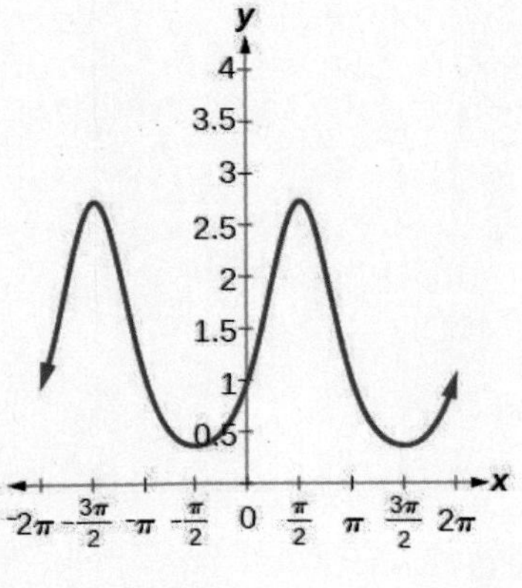

37. $\frac{\pi}{3}$

39. $\frac{\pi}{2}$

41. $\sqrt{1 - (1 - 2x)^2}$

43. $\frac{1}{\sqrt{1+x^4}}$

45. $\frac{x+1}{x}$

47. False

49. approximately 0.07 radians

Chapter 9

Try It

9.1 Verifying Trigonometric Identities and Using Trigonometric Identities to Simplify Trigonometric Expressions

1.
$$\begin{aligned}\csc\theta\cos\theta\tan\theta &= \left(\frac{1}{\sin\theta}\right)\cos\theta\left(\frac{\sin\theta}{\cos\theta}\right)\\ &= \frac{\cos\theta}{\sin\theta}\left(\frac{\sin\theta}{\cos\theta}\right)\\ &= \frac{\sin\theta\cos\theta}{\sin\theta\cos\theta}\\ &= 1\end{aligned}$$

2.
$$\begin{aligned}\frac{\cot\theta}{\csc\theta} &= \frac{\frac{\cos\theta}{\sin\theta}}{\frac{1}{\sin\theta}}\\ &= \frac{\cos\theta}{\sin\theta}\cdot\frac{\sin\theta}{1}\\ &= \cos\theta\end{aligned}$$

3.
$$\begin{aligned}\frac{\sin^2\theta-1}{\tan\theta\sin\theta-\tan\theta} &= \frac{(\sin\theta+1)(\sin\theta-1)}{\tan\theta(\sin\theta-1)}\\ &= \frac{\sin\theta+1}{\tan\theta}\end{aligned}$$

4. This is a difference of squares formula: $25-9\sin^2\theta=(5-3\sin\theta)(5+3\sin\theta)$.

5.
$$\begin{aligned}\frac{\cos\theta}{1+\sin\theta}\left(\frac{1-\sin\theta}{1-\sin\theta}\right) &= \frac{\cos\theta(1-\sin\theta)}{1-\sin^2\theta}\\ &= \frac{\cos\theta(1-\sin\theta)}{\cos^2\theta}\\ &= \frac{1-\sin\theta}{\cos\theta}\end{aligned}$$

9.2 Sum and Difference Identities

1. $\frac{\sqrt{2}+\sqrt{6}}{4}$

2. $\frac{\sqrt{2}-\sqrt{6}}{4}$

3. $\frac{1-\sqrt{3}}{1+\sqrt{3}}$

4. $\cos\left(\frac{5\pi}{14}\right)$

5.
$$\begin{aligned}\tan(\pi-\theta) &= \frac{\tan(\pi)-\tan\theta}{1+\tan(\pi)\tan\theta}\\ &= \frac{0-\tan\theta}{1+0\cdot\tan\theta}\\ &= -\tan\theta\end{aligned}$$

9.3 Double-Angle, Half-Angle, and Reduction Formulas

1. $\cos(2\alpha)=\frac{7}{32}$

2. $\cos^4\theta-\sin^4\theta=\left(\cos^2\theta+\sin^2\theta\right)\left(\cos^2\theta-\sin^2\theta\right)=\cos(2\theta)$

3. $\cos(2\theta)\cos\theta=\left(\cos^2\theta-\sin^2\theta\right)\cos\theta=\cos^3\theta-\cos\theta\sin^2\theta$

4.
$$\begin{aligned}10\cos^4x &= 10\left(\cos^2x\right)^2\\ &= 10\left[\frac{1+\cos(2x)}{2}\right]^2 && \text{Substitute reduction formula for } \cos^2x.\\ &= \frac{10}{4}\left[1+2\cos(2x)+\cos^2(2x)\right]\\ &= \frac{10}{4}+\frac{10}{2}\cos(2x)+\frac{10}{4}\left(\frac{1+\cos 2(2x)}{2}\right) && \text{Substitute reduction formula for } \cos^2x.\\ &= \frac{10}{4}+\frac{10}{2}\cos(2x)+\frac{10}{8}+\frac{10}{8}\cos(4x)\\ &= \frac{30}{8}+5\cos(2x)+\frac{10}{8}\cos(4x)\\ &= \frac{15}{4}+5\cos(2x)+\frac{5}{4}\cos(4x)\end{aligned}$$

5. $-\frac{2}{\sqrt{5}}$

9.4 Sum-to-Product and Product-to-Sum Formulas

1. $\frac{1}{2}(\cos 6\theta + \cos 2\theta)$

2. $\frac{1}{2}(\sin 2x + \sin 2y)$

3. $\frac{-2-\sqrt{3}}{4}$

4. $2\sin(2\theta)\cos(\theta)$

5.
$$\begin{aligned}\tan\ \theta\ \cot\ \theta - \cos^2\theta &= \left(\frac{\sin\ \theta}{\cos\ \theta}\right)\left(\frac{\cos\ \theta}{\sin\ \theta}\right) - \cos^2\theta \\ &= 1 - \cos^2\theta \\ &= \sin^2\theta\end{aligned}$$

9.5 Solving Trigonometric Equations

1. $x = \frac{7\pi}{6}, \frac{11\pi}{6}$

2. $\frac{\pi}{3} \pm \pi k$

3. $\theta \approx 1.7722 \pm 2\pi k$ and $\theta \approx 4.5110 \pm 2\pi k$

4. $\cos\ \theta = -1, \theta = \pi$

5. $\frac{\pi}{2}, \frac{2\pi}{3}, \frac{4\pi}{3}, \frac{3\pi}{2}$

9.1 Section Exercises

1. All three functions, F,G, and H, are even.

   This is because $F(-x) = \sin(-x)\sin(-x) = (-\sin\ x)(-\sin\ x) = \sin^2 x = F(x), G(-x) = \cos(-x)\cos(-x) = \cos\ x\cos\ x = \cos^2 x = G(x)$ and $H(-x) = \tan(-x)\tan(-x) = (-\tan\ x)(-\tan\ x) = \tan^2 x = H(x)$.

3. When $\cos\ t = 0$, then $\sec\ t = \frac{1}{0}$, which is undefined.

5. $\sin\ x$

7. $\sec\ x$

9. $\csc\ t$

11. -1

13. $\sec^2 x$

15. $\sin^2 x + 1$

17. $\frac{1}{\sin\ x}$

19. $\frac{1}{\cot\ x}$

21. $\tan\ x$

23. $-4\sec\ x\tan\ x$

25. $\pm\sqrt{\frac{1}{\cot^2 x} + 1}$

27. $\frac{\pm\sqrt{1-\sin^2 x}}{\sin\ x}$

29. Answers will vary. Sample proof:

$$\begin{aligned}\cos\ x - \cos^3 x &= \cos\ x\left(1 - \cos^2 x\right) \\ &= \cos\ x\sin^2 x\end{aligned}$$

31. Answers will vary. Sample proof:

$$\frac{1+\sin^2 x}{\cos^2 x} = \frac{1}{\cos^2 x} + \frac{\sin^2 x}{\cos^2 x} = \sec^2 x + \tan^2 x = \tan^2 x + 1 + \tan^2 x = 1 + 2\tan^2 x$$

33. Answers will vary. Sample proof:

$$\cos^2 x - \tan^2 x = 1 - \sin^2 x - \left(\sec^2 x - 1\right) = 1 - \sin^2 x - \sec^2 x + 1 = 2 - \sin^2 x - \sec^2 x$$

35. False

37. False

39. Proved with negative and Pythagorean identities

41. True
$$3\sin^2\theta + 4\cos^2\theta = 3\sin^2\theta + 3\cos^2\theta + \cos^2\theta = 3\left(\sin^2\theta + \cos^2\theta\right) + \cos^2\theta = 3 + \cos^2\theta$$

9.2 Section Exercises

1. The cofunction identities apply to complementary angles. Viewing the two acute angles of a right triangle, if one of those angles measures x, the second angle measures $\frac{\pi}{2} - x$. Then $\sin x = \cos\left(\frac{\pi}{2} - x\right)$. The same holds for the other cofunction identities. The key is that the angles are complementary.

3. $\sin(-x) = -\sin x$, so $\sin x$ is odd.
$\cos(-x) = \cos(0 - x) = \cos x$, so $\cos x$ is even.

5. $\frac{\sqrt{2}+\sqrt{6}}{4}$

7. $\frac{\sqrt{6}-\sqrt{2}}{4}$

9. $-2 - \sqrt{3}$

11. $-\frac{\sqrt{2}}{2}\sin x - \frac{\sqrt{2}}{2}\cos x$

13. $-\frac{1}{2}\cos x - \frac{\sqrt{3}}{2}\sin x$

15. $\csc\theta$

17. $\cot x$

19. $\tan\left(\frac{x}{10}\right)$

21.
$$\begin{aligned}\sin(a-b) &= \left(\tfrac{4}{5}\right)\left(\tfrac{1}{3}\right) - \left(\tfrac{3}{5}\right)\left(\tfrac{2\sqrt{2}}{3}\right) &= \tfrac{4-6\sqrt{2}}{15}\\ \cos(a+b) &= \left(\tfrac{3}{5}\right)\left(\tfrac{1}{3}\right) - \left(\tfrac{4}{5}\right)\left(\tfrac{2\sqrt{2}}{3}\right) &= \tfrac{3-8\sqrt{2}}{15}\end{aligned}$$

23. $\frac{\sqrt{2}-\sqrt{6}}{4}$

25. $\sin x$

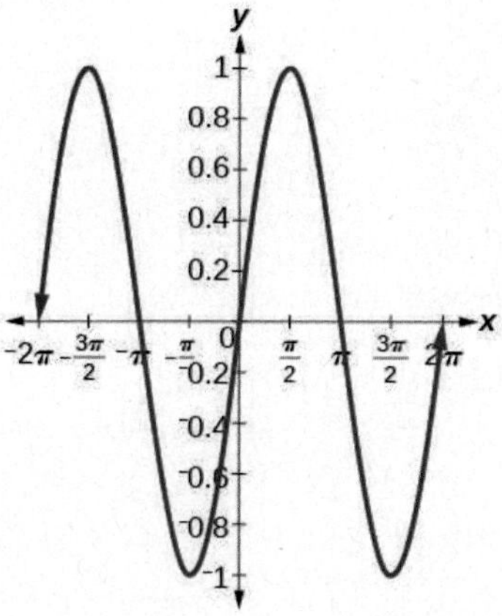

27. $\cot\left(\frac{\pi}{6} - x\right)$

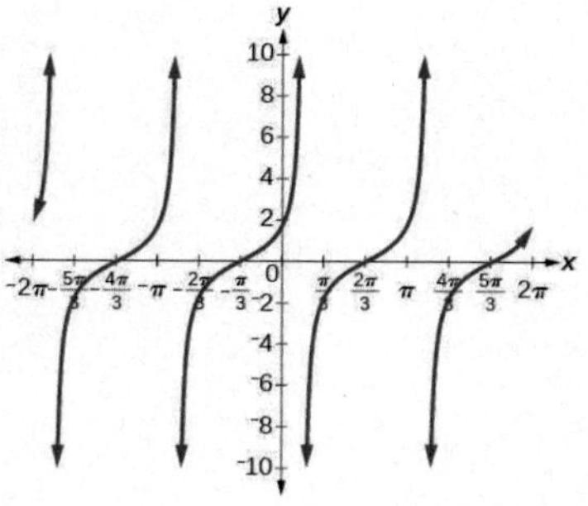

29. $\cot\left(\frac{\pi}{4}+x\right)$

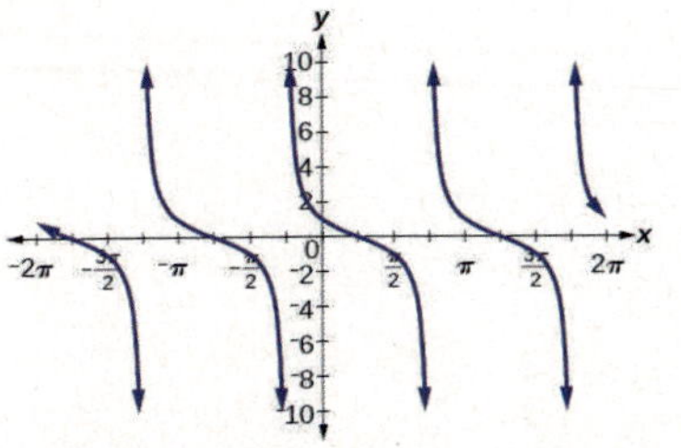

31. $\frac{\sin x}{\sqrt{2}}+\frac{\cos x}{\sqrt{2}}$

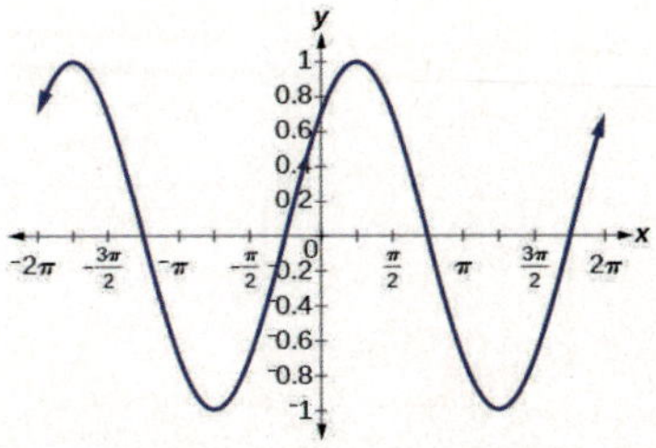

33. They are the same.

35. They are the different, try $g(x)=\sin(9x)-\cos(3x)\sin(6x)$.

37. They are the same.

39. They are the different, try $g(\theta)=\frac{2\tan\theta}{1-\tan^2\theta}$.

41. They are different, try $g(x)=\frac{\tan x-\tan(2x)}{1+\tan x\tan(2x)}$.

43. $-\frac{\sqrt{3}-1}{2\sqrt{2}}$, or -0.2588

45. $\frac{1+\sqrt{3}}{2\sqrt{2}}$, or 0.9659

47.
$$\begin{aligned}\tan\left(x+\frac{\pi}{4}\right) &= \\ \frac{\tan x+\tan\left(\frac{\pi}{4}\right)}{1-\tan x\tan\left(\frac{\pi}{4}\right)} &= \\ \frac{\tan x+1}{1-\tan x(1)} &= \frac{\tan x+1}{1-\tan x}\end{aligned}$$

49.
$$\begin{aligned}\frac{\cos(a+b)}{\cos a\cos b} &= \\ \frac{\cos a\cos b}{\cos a\cos b}-\frac{\sin a\sin b}{\cos a\cos b} &= 1-\tan a\tan b\end{aligned}$$

51.
$$\begin{aligned}\frac{\cos(x+h)-\cos x}{h} &= \\ \frac{\cos x\cosh-\sin x\sinh-\cos x}{h} &= \\ \frac{\cos x(\cosh-1)-\sin x\sinh}{h} &= \cos x\frac{\cos h-1}{h}-\sin x\frac{\sin h}{h}\end{aligned}$$

53. True

55. True. Note that

$$\sin(\alpha+\beta)=\sin(\pi-\gamma)$$

and expand the right hand side.

9.3 Section Exercises

1. Use the Pythagorean identities and isolate the squared term.

3. $\frac{1-\cos x}{\sin x}, \frac{\sin x}{1+\cos x}$, multiplying the top and bottom by $\sqrt{1-\cos x}$ and $\sqrt{1+\cos x}$, respectively.

5. a) $\frac{3\sqrt{7}}{32}$ b) $\frac{31}{32}$ c) $\frac{3\sqrt{7}}{31}$

7. a) $\frac{\sqrt{3}}{2}$ b) $-\frac{1}{2}$ c) $-\sqrt{3}$

9. $\cos\theta=-\frac{2\sqrt{5}}{5}, \sin\theta=\frac{\sqrt{5}}{5}, \tan\theta=-\frac{1}{2}, \csc\theta=\sqrt{5}, \sec\theta=-\frac{\sqrt{5}}{2}, \cot\theta=-2$

11. $2\sin\left(\frac{\pi}{2}\right)$

13. $\frac{\sqrt{2-\sqrt{2}}}{2}$

15. $\frac{\sqrt{2-\sqrt{3}}}{2}$

17. $2+\sqrt{3}$

19. $-1-\sqrt{2}$

21. a) $\frac{3\sqrt{13}}{13}$ b) $-\frac{2\sqrt{13}}{13}$ c) $-\frac{3}{2}$

23. a) $\frac{\sqrt{10}}{4}$ b) $\frac{\sqrt{6}}{4}$ c) $\frac{\sqrt{15}}{3}$

25. $\frac{120}{169}, -\frac{119}{169}, -\frac{120}{119}$

27. $\frac{2\sqrt{13}}{13}, \frac{3\sqrt{13}}{13}, \frac{2}{3}$

29. $\cos(74°)$

31. $\cos(18x)$

33. $3\sin(10x)$

35. $-2\sin(-x)\cos(-x) = -2(-\sin(x)\cos(x)) = \sin(2x)$

37.
$$\begin{aligned}\frac{\sin(2\theta)}{1+\cos(2\theta)}\tan^2\theta &= \frac{2\sin(\theta)\cos(\theta)}{1+\cos^2\theta-\sin^2\theta}\tan^2\theta = \\ \frac{2\sin(\theta)\cos(\theta)}{2\cos^2\theta}\tan^2\theta &= \frac{\sin(\theta)}{\cos\ \theta}\tan^2\theta = \\ \cot(\theta)\tan^2\theta &= \tan^3\ \theta\end{aligned}$$

39. $\frac{1+\cos(12x)}{2}$

41. $\frac{3+\cos(12x)-4\cos(6x)}{8}$

43. $\frac{2+\cos(2x)-2\cos(4x)-\cos(6x)}{32}$

45. $\frac{3+\cos(4x)-4\cos(2x)}{3+\cos(4x)+4\cos(2x)}$

47. $\frac{1-\cos(4x)}{8}$

49. $\frac{3+\cos(4x)-4\cos(2x)}{4(\cos(2x)+1)}$

51. $\frac{(1+\cos(4x))\sin\ x}{2}$

53. $4\sin\ x\cos\ x\left(\cos^2 x-\sin^2 x\right)$

55.
$$\frac{2\tan\ x}{1+\tan^2 x} = \frac{\frac{2\sin\ x}{\cos\ x}}{1+\frac{\sin^2 x}{\cos^2 x}} = \frac{\frac{2\sin\ x}{\cos\ x}}{\frac{\cos^2 x+\sin^2 x}{\cos^2 x}} =$$
$$\frac{2\sin\ x}{\cos\ x}\cdot\frac{\cos^2 x}{1} = 2\sin\ x\cos\ x = \sin(2x)$$

57. $\frac{2\sin\ x\cos\ x}{2\cos^2 x-1} = \frac{\sin(2x)}{\cos(2x)} = \tan(2x)$

59.
$$\begin{aligned}\sin(x+2x) &= \sin\ x\cos(2x)+\sin(2x)\cos\ x \\ &= \sin\ x\left(\cos^2 x-\sin^2 x\right)+2\sin\ x\cos\ x\cos\ x \\ &= \sin\ x\cos^2 x-\sin^3 x+2\sin\ x\cos^2 x \\ &= 3\sin\ x\cos^2 x-\sin^3 x\end{aligned}$$

61.
$$\begin{aligned}\frac{1+\cos(2t)}{\sin(2t)-\cos t} &= \frac{1+2\cos^2 t-1}{2\sin t\cos t-\cos t} \\ &= \frac{2\cos^2 t}{\cos t(2\sin t-1)} \\ &= \frac{2\cos t}{2\sin t-1}\end{aligned}$$

63.
$$\begin{aligned}\left(\cos^2(4x)-\sin^2(4x)-\sin(8x))(\cos^2(4x)-\sin^2(4x)+\sin(8x)\right) &= \\ &= (\cos(8x)-\sin(8x))(\cos(8x)+\sin(8x)) \\ &= \cos^2(8x)-\sin^2(8x) \\ &= \cos(16x)\end{aligned}$$

9.4 Section Exercises

1. Substitute α into cosine and β into sine and evaluate.

3. Answers will vary. There are some equations that involve a sum of two trig expressions where when converted to a product are easier to solve. For example: $\frac{\sin(3x)+\sin\ x}{\cos\ x} = 1.$ When converting the numerator to a product the equation becomes: $\frac{2\ \sin(2x)\cos\ x}{\cos\ x} = 1$

5. $8(\cos(5x)-\cos(27x))$

7. $\sin(2x) + \sin(8x)$

9. $\frac{1}{2}(\cos(6x) - \cos(4x))$

11. $2\cos(5t)\cos t$

13. $2\cos(7x)$

15. $2\cos(6x)\cos(3x)$

17. $\frac{1}{4}\left(1+\sqrt{3}\right)$

19. $\frac{1}{4}\left(\sqrt{3}-2\right)$

21. $\frac{1}{4}\left(\sqrt{3}-1\right)$

23. $\cos(80°) - \cos(120°)$

25. $\frac{1}{2}(\sin(221°) + \sin(205°))$

27. $\sqrt{2}\cos(31°)$

29. $2\cos(66.5°)\sin(34.5°)$

31. $2\sin(-1.5°)\cos(0.5°)$

33. $2\sin(7x) - 2\sin x = 2\sin(4x+3x) - 2\sin(4x-3x) =$
$2(\sin(4x)\cos(3x) + \sin(3x)\cos(4x)) - 2(\sin(4x)\cos(3x) - \sin(3x)\cos(4x)) =$
$2\sin(4x)\cos(3x) + 2\sin(3x)\cos(4x)) - 2\sin(4x)\cos(3x) + 2\sin(3x)\cos(4x)) =$
$4\sin(3x)\cos(4x)$

35. $$\begin{aligned}\sin x + \sin(3x) &= 2\sin\left(\tfrac{4x}{2}\right)\cos\left(\tfrac{-2x}{2}\right) = \\ 2\sin(2x)\cos x &= 2(2\sin x\cos x)\cos x = \\ &4\sin x\cos^2 x\end{aligned}$$

37. $2\tan x\cos(3x) = \frac{2\sin x\cos(3x)}{\cos x} = \frac{2(.5(\sin(4x)-\sin(2x)))}{\cos x} =$
$\frac{1}{\cos x}(\sin(4x) - \sin(2x)) = \sec x(\sin(4x) - \sin(2x))$

39. $2\cos(35°)\cos(23°), 1.5081$

41. $-2\sin(33°)\sin(11°), -0.2078$

43. $\frac{1}{2}(\cos(99°) - \cos(71°)), -0.2410$

45. It is an identity.

47. It is not an identity, but $2\cos^3 x$ is.

49. $\tan(3t)$

51. $2\cos(2x)$

53. $-\sin(14x)$

55. Start with $\cos x + \cos y$. Make a substitution and let $x = \alpha + \beta$ and let $y = \alpha - \beta$, so $\cos x + \cos y$ becomes

$$\cos(\alpha+\beta) + \cos(\alpha-\beta) = \cos\alpha\cos\beta - \sin\alpha\sin\beta + \cos\alpha\cos\beta + \sin\alpha\sin\beta = 2\cos\alpha\cos\beta$$

Since $x = \alpha + \beta$ and $y = \alpha - \beta$, we can solve for α and β in terms of x and y and substitute in for $2\cos\alpha\cos\beta$ and get $2\cos\left(\frac{x+y}{2}\right)\cos\left(\frac{x-y}{2}\right)$.

57. $\frac{\cos(3x)+\cos x}{\cos(3x)-\cos x} = \frac{2\cos(2x)\cos x}{-2\sin(2x)\sin x} = -\cot(2x)\cot x$

59. $$\begin{aligned}\frac{\cos(2y)-\cos(4y)}{\sin(2y)+\sin(4y)} &= \frac{-2\sin(3y)\sin(-y)}{2\sin(3y)\cos y} = \\ \frac{2\sin(3y)\sin(y)}{2\sin(3y)\cos y} &= \tan y\end{aligned}$$

61. $\cos x - \cos(3x) = -2\sin(2x)\sin(-x) =$
$2(2\sin x\cos x)\sin x = 4\sin^2 x\cos x$

63. $\tan\left(\frac{\pi}{4} - t\right) = \frac{\tan\left(\frac{\pi}{4}\right)-\tan t}{1+\tan\left(\frac{\pi}{4}\right)\tan(t)} = \frac{1-\tan t}{1+\tan t}$

9.5 Section Exercises

1. There will not always be solutions to trigonometric function equations. For a basic example, $\cos(x) = -5$.

3. If the sine or cosine function has a coefficient of one, isolate the term on one side of the equals sign. If the number it is set equal to has an absolute value less than or equal to one, the equation has solutions, otherwise it does not. If the sine or cosine does not have a coefficient equal to one, still isolate the term but then divide both sides of the equation by the leading coefficient. Then, if the number it is set equal to has an absolute value greater than one, the equation has no solution.

5. $\frac{\pi}{3}, \frac{2\pi}{3}$

7. $\frac{3\pi}{4}, \frac{5\pi}{4}$

9. $\frac{\pi}{4}, \frac{5\pi}{4}$

11. $\frac{\pi}{4}, \frac{3\pi}{4}, \frac{5\pi}{4}, \frac{7\pi}{4}$

13. $\frac{\pi}{4}, \frac{7\pi}{4}$

15. $\frac{7\pi}{6}, \frac{11\pi}{6}$

17. $\frac{\pi}{18}, \frac{5\pi}{18}, \frac{13\pi}{18}, \frac{17\pi}{18}, \frac{25\pi}{18}, \frac{29\pi}{18}$

19. $\frac{3\pi}{12}, \frac{5\pi}{12}, \frac{11\pi}{12}, \frac{13\pi}{12}, \frac{19\pi}{12}, \frac{21\pi}{12}$

21. $\frac{1}{6}, \frac{5}{6}, \frac{13}{6}, \frac{17}{6}, \frac{25}{6}, \frac{29}{6}, \frac{37}{6}$

23. $0, \frac{\pi}{3}, \pi, \frac{5\pi}{3}$

25. $\frac{\pi}{3}, \pi, \frac{5\pi}{3}$

27. $\frac{\pi}{3}, \frac{3\pi}{2}, \frac{5\pi}{3}$

29. $0, \pi$

31. $\pi - \sin^{-1}\left(-\frac{1}{4}\right), \frac{7\pi}{6}, \frac{11\pi}{6}, 2\pi + \sin^{-1}\left(-\frac{1}{4}\right)$

33. $\frac{1}{3}\left(\sin^{-1}\left(\frac{9}{10}\right)\right)$,
$\frac{\pi}{3} - \frac{1}{3}\left(\sin^{-1}\left(\frac{9}{10}\right)\right)$,
$\frac{2\pi}{3} + \frac{1}{3}\left(\sin^{-1}\left(\frac{9}{10}\right)\right)$,
$\pi - \frac{1}{3}\left(\sin^{-1}\left(\frac{9}{10}\right)\right)$,
$\frac{4\pi}{3} + \frac{1}{3}\left(\sin^{-1}\left(\frac{9}{10}\right)\right)$,
$\frac{5\pi}{3} - \frac{1}{3}\left(\sin^{-1}\left(\frac{9}{10}\right)\right)$

35. 0

37. $\frac{\pi}{6}, \frac{5\pi}{6}, \frac{7\pi}{6}, \frac{11\pi}{6}$

39. $\frac{3\pi}{2}, \frac{\pi}{6}, \frac{5\pi}{6}$

41. $0, \frac{\pi}{3}, \pi, \frac{4\pi}{3}$

43. There are no solutions.

45. $\cos^{-1}\left(\frac{1}{3}\left(1 - \sqrt{7}\right)\right)$,
$2\pi - \cos^{-1}\left(\frac{1}{3}\left(1 - \sqrt{7}\right)\right)$

47. $\tan^{-1}\left(\frac{1}{2}\left(\sqrt{29} - 5\right)\right)$,
$\pi + \tan^{-1}\left(\frac{1}{2}\left(-\sqrt{29} - 5\right)\right)$,
$\pi + \tan^{-1}\left(\frac{1}{2}\left(\sqrt{29} - 5\right)\right)$,
$2\pi + \tan^{-1}\left(\frac{1}{2}\left(-\sqrt{29} - 5\right)\right)$

49. There are no solutions.

51. There are no solutions.

53. $0, \frac{2\pi}{3}, \frac{4\pi}{3}$

55. $\frac{\pi}{4}, \frac{3\pi}{4}, \frac{5\pi}{4}, \frac{7\pi}{4}$

57. $\sin^{-1}\left(\frac{3}{5}\right), \frac{\pi}{2}, \pi - \sin^{-1}\left(\frac{3}{5}\right), \frac{3\pi}{2}$

59. $\cos^{-1}\left(-\frac{1}{4}\right), 2\pi - \cos^{-1}\left(-\frac{1}{4}\right)$

61. $\frac{\pi}{3}, \cos^{-1}\left(-\frac{3}{4}\right),$
$2\pi - \cos^{-1}\left(-\frac{3}{4}\right), \frac{5\pi}{3}$

63. $\cos^{-1}\left(\frac{3}{4}\right), \cos^{-1}\left(-\frac{2}{3}\right),$
$2\pi - \cos^{-1}\left(-\frac{2}{3}\right),$
$2\pi - \cos^{-1}\left(\frac{3}{4}\right)$

65. $0, \frac{\pi}{2}, \pi, \frac{3\pi}{2}$

67. $\frac{\pi}{3}, \cos^{-1}\left(-\frac{1}{4}\right),$
$2\pi - \cos^{-1}\left(-\frac{1}{4}\right), \frac{5\pi}{3}$

69. There are no solutions.

71. $\pi + \tan^{-1}(-2),$
$\pi + \tan^{-1}\left(-\frac{3}{2}\right),$
$2\pi + \tan^{-1}(-2),$
$2\pi + \tan^{-1}\left(-\frac{3}{2}\right)$

73. $2\pi k + 0.2734, 2\pi k + 2.8682$

75. $\pi k - 0.3277$

77. $0.6694, 1.8287, 3.8110, 4.9703$

79. $1.0472, 3.1416, 5.2360$

81. $0.5326, 1.7648, 3.6742, 4.9064$

83. $\sin^{-1}\left(\frac{1}{4}\right), \pi - \sin^{-1}\left(\frac{1}{4}\right), \frac{3\pi}{2}$

85. $\frac{\pi}{2}, \frac{3\pi}{2}$

87. There are no solutions.

89. $0, \frac{\pi}{2}, \pi, \frac{3\pi}{2}$

91. There are no solutions.

93. $7.2°$

95. $5.7°$

97. $82.4°$

99. $31.0°$

101. $88.7°$

103. $59.0°$

105. $36.9°$

Review Exercises

1. $\sin^{-1}\left(\frac{\sqrt{3}}{3}\right),$
$\pi - \sin^{-1}\left(\frac{\sqrt{3}}{3}\right),$
$\pi + \sin^{-1}\left(\frac{\sqrt{3}}{3}\right),$
$2\pi - \sin^{-1}\left(\frac{\sqrt{3}}{3}\right)$

3. $\frac{7\pi}{6}, \frac{11\pi}{6}$

5. $\sin^{-1}\left(\frac{1}{4}\right), \pi - \sin^{-1}\left(\frac{1}{4}\right)$

7. 1

9. Yes

11. $-2 - \sqrt{3}$

13. $\frac{\sqrt{2}}{2}$

15.

$$\begin{aligned}\cos(4x) - \cos(3x)\cos x &= \cos(2x + 2x) - \cos(x + 2x)\cos x \\ &= \cos(2x)\cos(2x) - \sin(2x)\sin(2x) - \cos x \cos(2x)\cos x + \sin x \sin(2x)\cos x \\ &= \left(\cos^2 x - \sin^2 x\right)^2 - 4\cos^2 x \sin^2 x - \cos^2 x\left(\cos^2 x - \sin^2 x\right) + \sin x(2)\sin x \cos x \cos x \\ &= \left(\cos^2 x - \sin^2 x\right)^2 - 4\cos^2 x \sin^2 x - \cos^2 x\left(\cos^2 x - \sin^2 x\right) + 2\sin^2 x \cos^2 x \\ &= \cos^4 x - 2\cos^2 x \sin^2 x + \sin^4 x - 4\cos^2 x \sin^2 x - \cos^4 x + \cos^2 x \sin^2 x + 2\sin^2 x \cos^2 x \\ &= \sin^4 x - 4\cos^2 x \sin^2 x + \cos^2 x \sin^2 x \\ &= \sin^2 x\left(\sin^2 x + \cos^2 x\right) - 4\cos^2 x \sin^2 x \\ &= \sin^2 x - 4\cos^2 x \sin^2 x\end{aligned}$$

17. $\tan\left(\frac{5}{8}x\right)$

19. $\frac{\sqrt{3}}{3}$

21. $-\frac{24}{25}, -\frac{7}{25}, \frac{24}{7}$

23. $\sqrt{2\left(2 + \sqrt{2}\right)}$

25. $\frac{\sqrt{2}}{10}, \frac{7\sqrt{2}}{10}, \frac{1}{7}, \frac{3}{5}, \frac{4}{5}, \frac{3}{4}$

27.

$$\begin{aligned}\cot x\cos(2x) &= \cot x\left(1 - 2\sin^2 x\right) \\ &= \cot x - \frac{\cos x}{\sin x}(2)\sin^2 x \\ &= -2\sin x\cos x + \cot x \\ &= -\sin(2x) + \cot x\end{aligned}$$

29. $\frac{10\sin x - 5\sin(3x) + \sin(5x)}{8(\cos(2x) + 1)}$

31. $\frac{\sqrt{3}}{2}$

33. $-\frac{\sqrt{2}}{2}$

35. $\frac{1}{2}(\sin(6x) + \sin(12x))$

37. $2\sin\left(\frac{13}{2}x\right)\cos\left(\frac{9}{2}x\right)$

39. $\frac{3\pi}{4}, \frac{7\pi}{4}$

41. $0, \frac{\pi}{6}, \frac{5\pi}{6}, \pi$

43. $\frac{3\pi}{2}$

45. No solution

47. $0.2527, 2.8889, 4.7124$

49. $1.3694, 1.9106, 4.3726, 4.9137$

Practice Test

1. 1

3. $\sec(\theta)$

5. $\frac{\sqrt{2} - \sqrt{6}}{4}$

7. $-2 - \sqrt{3}$

9. $-\frac{1}{2}\cos\theta + \frac{\sqrt{3}}{2}\sin\theta$

11. $\frac{1 - \cos(64°)}{2}$

13. $0, \pi$

15. $\frac{\pi}{2}, \frac{3\pi}{2}$

17. $2\cos(3x)\cos(5x)$

19. $4\sin(2\theta)\cos(6\theta)$

21. $x = \cos^{-1}\left(\frac{1}{5}\right)$

23. $\frac{\pi}{3}$

25. $\frac{3}{5}, -\frac{4}{5}, -\frac{3}{4}$

27.

$$\begin{aligned}\tan^3 x - \tan x\sec^2 x &= \tan x\left(\tan^2 x - \sec^2 x\right) \\ &= \tan x\left(\tan^2 x - \left(1 + \tan^2 x\right)\right) \\ &= \tan x\left(\tan^2 x - 1 - \tan^2 x\right) \\ &= -\tan x = \tan(-x) = \tan(-x)\end{aligned}$$

29. $$\begin{aligned}\frac{\sin(2x)}{\sin x}-\frac{\cos(2x)}{\cos x} &= \frac{2\sin x\cos x}{\sin x}-\frac{2\cos^2 x-1}{\cos x}\\ &= 2\cos x-2\cos x+\frac{1}{\cos x}\\ &= \frac{1}{\cos x}=\sec x=\sec x\end{aligned}$$

31. Amplitude: $\frac{1}{4}$, period: $\frac{1}{60}$, frequency: 60 Hz

33. Amplitude: 8, fast period: $\frac{1}{500}$, fast frequency: 500 Hz, slow period: $\frac{1}{10}$, slow frequency: 10 Hz

35. $D(t)=20(0.9086)^t\cos(4\pi t)$
, 31 second

Chapter 10

Try It

10.1 Non-right Triangles: Law of Sines

1. $$\begin{aligned}\alpha &= 98^\circ & a &= 34.6\\ \beta &= 39^\circ & b &= 22\\ \gamma &= 43^\circ & c &= 23.8\end{aligned}$$

2. Solution 1

$$\begin{aligned}\alpha &= 80^\circ & a &= 120\\ \beta &\approx 83.2^\circ & b &= 121\\ \gamma &\approx 16.8^\circ & c &\approx 35.2\end{aligned}$$

Solution 2

$$\begin{aligned}\alpha' &= 80^\circ & a' &= 120\\ \beta' &\approx 96.8^\circ & b' &= 121\\ \gamma' &\approx 3.2^\circ & c' &\approx 6.8\end{aligned}$$

3. $\beta\approx 5.7^\circ, \gamma\approx 94.3^\circ, c\approx 101.3$

4. two

5. about 8.2 square feet

6. 161.9 yd.

10.2 Non-right Triangles: Law of Cosines

1. $a\approx 14.9, \beta\approx 23.8^\circ, \gamma\approx 126.2^\circ$.

2. $\alpha\approx 27.7^\circ, \beta\approx 40.5^\circ, \gamma\approx 111.8^\circ$

3. Area = 552 square feet

4. about 8.15 square feet

10.3 Polar Coordinates

1.

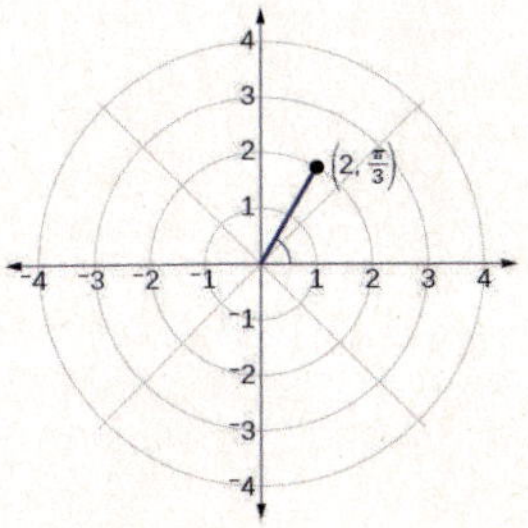

2.

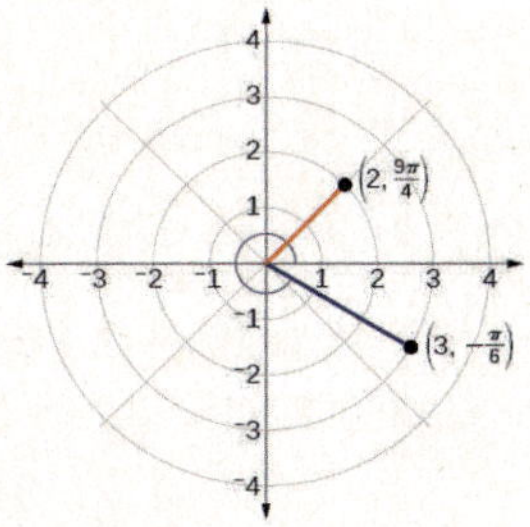

3. $(x, y)=\left(\frac{1}{2}, -\frac{\sqrt{3}}{2}\right)$

4. $r = \sqrt{3}$

5. $x^2 + y^2 = 2y$ or, in the standard form for a circle, $x^2 + (y - 1)^2 = 1$

10.4 Polar Coordinates: Graphs

1. The equation fails the symmetry test with respect to the line $\theta = \frac{\pi}{2}$ and with respect to the pole. It passes the polar axis symmetry test.

2. Tests will reveal symmetry about the polar axis. The zero is $\left(0, \frac{\pi}{2}\right)$, and the maximum value is $(3, 0)$.

3.

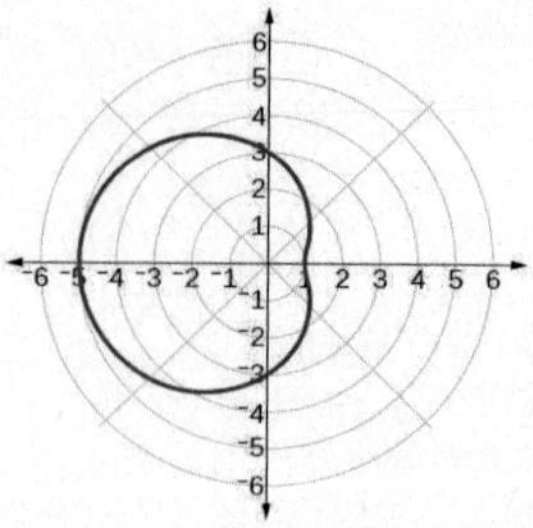

4. The graph is a rose curve, n even

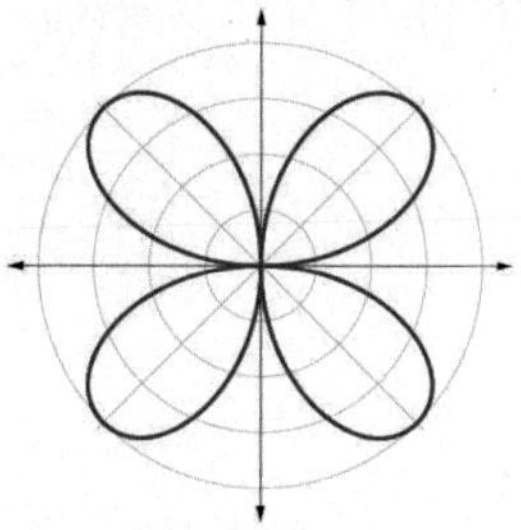

5.

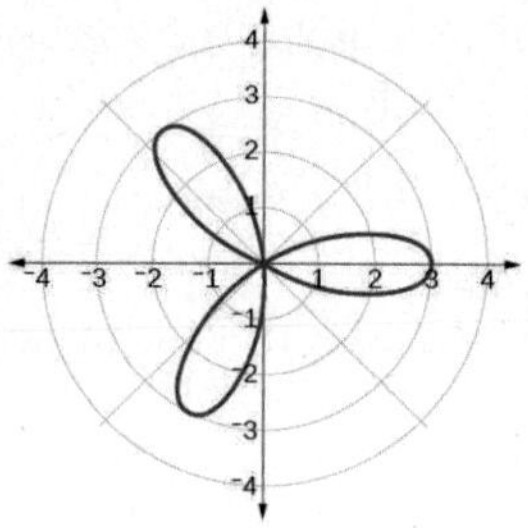

Rose curve, n odd

6.

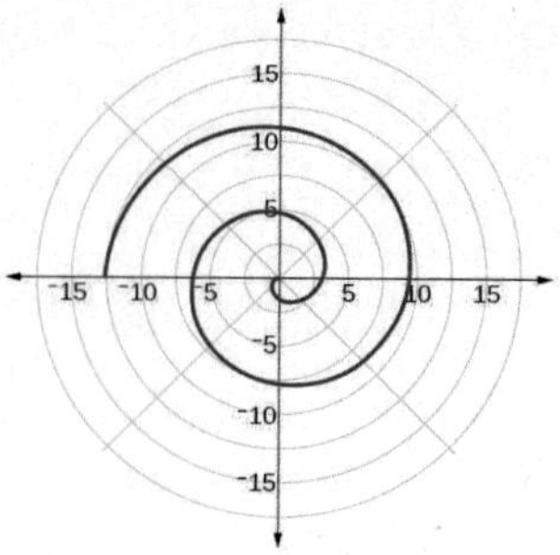

10.5 Polar Form of Complex Numbers

1.

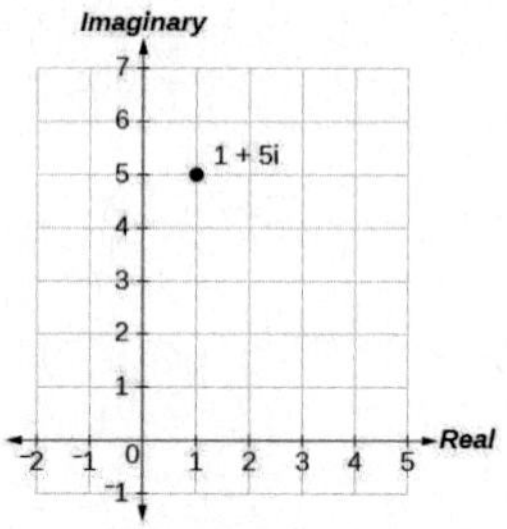

2. 13

3. $|z| = \sqrt{50} = 5\sqrt{2}$

4. $z = 3\left(\cos\left(\frac{\pi}{2}\right) + i\sin\left(\frac{\pi}{2}\right)\right)$

5. $z = 2\left(\cos\left(\frac{\pi}{6}\right) + i\sin\left(\frac{\pi}{6}\right)\right)$

6. $z = 2\sqrt{3} - 2i$

7. $z_1 z_2 = -4\sqrt{3}; \frac{z_1}{z_2} = -\frac{\sqrt{3}}{2} + \frac{3}{2}i$

8. $$\begin{aligned} z_0 &= 2(\cos(30°) + i\sin(30°)) \\ z_1 &= 2(\cos(120°) + i\sin(120°)) \\ z_2 &= 2(\cos(210°) + i\sin(210°)) \\ z_3 &= 2(\cos(300°) + i\sin(300°)) \end{aligned}$$

10.6 Parametric Equations

1.

t	$x(t)$	$y(t)$
−1	−4	2
0	−3	4
1	−2	6
2	−1	8

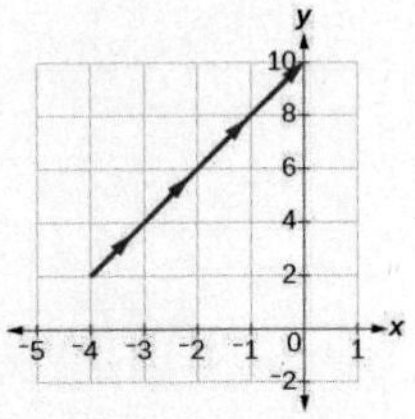

2. $x(t) = t^3 - 2t$
$y(t) = t$

3. $y = 5 - \sqrt{\frac{1}{2}x - 3}$

4. $y = \ln\sqrt{x}$

5. $\frac{x^2}{4} + \frac{y^2}{9} = 1$

6. $y = x^2$

10.7 Parametric Equations: Graphs

1.

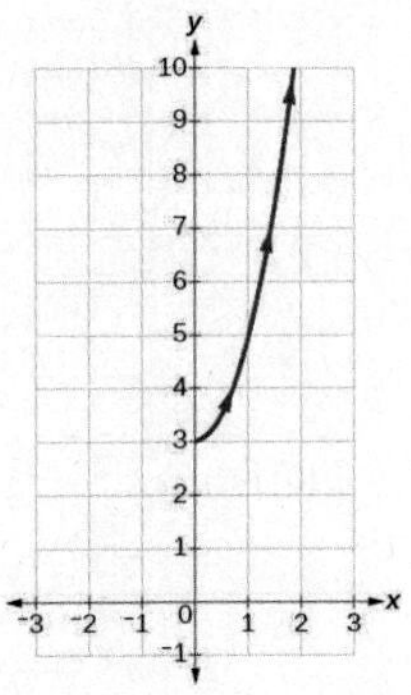

2.

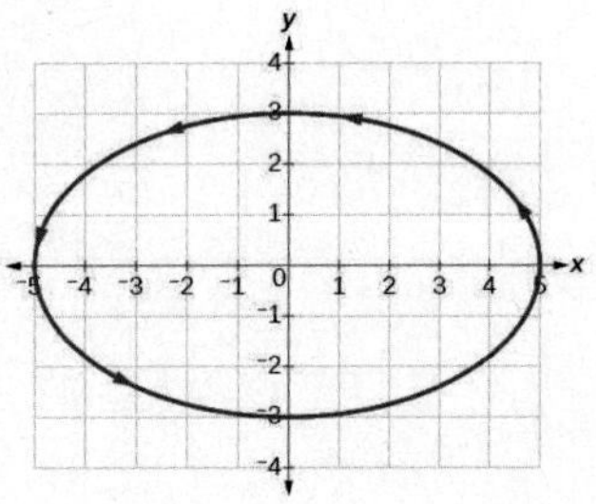

3. The graph of the parametric equations is in red and the graph of the rectangular equation is drawn in blue dots on top of the parametric equations.

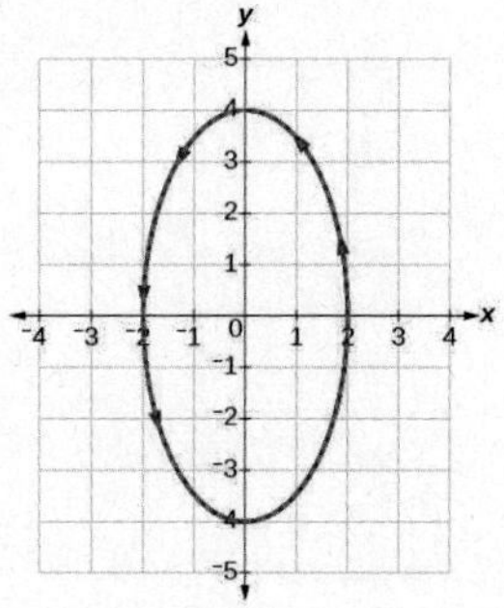

10.8 Vectors

1.

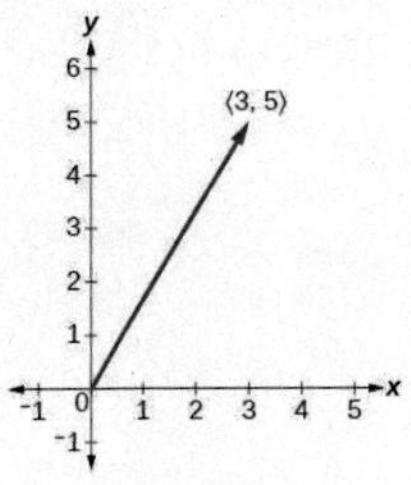

2. $3\mathbf{u} = \langle 15, 12\rangle$

3. $\mathbf{u} = 8\mathbf{i} - 11\mathbf{j}$

4. $\mathbf{v} = \sqrt{34}\cos(59°)\mathbf{i} + \sqrt{34}\sin(59°)\mathbf{j}$

Magnitude = $\sqrt{34}$

$\theta = \tan^{-1}\left(\frac{5}{3}\right) = 59.04°$

10.1 Section Exercises

1. The altitude extends from any vertex to the opposite side or to the line containing the opposite side at a 90° angle.

3. When the known values are the side opposite the missing angle and another side and its opposite angle.

5. A triangle with two given sides and a non-included angle.

7. $\beta = 72°, a \approx 12.0, b \approx 19.9$

9. $\gamma = 20°, b \approx 4.5, c \approx 1.6$

11. $b \approx 3.78$

13. $c \approx 13.70$

15. one triangle, $\alpha \approx 50.3°, \beta \approx 16.7°, a \approx 26.7$

17. two triangles, $\gamma \approx 54.3°, \beta \approx 90.7°, b \approx 20.9$ or $\gamma' \approx 125.7°, \beta' \approx 19.3°, b' \approx 6.9$

19. two triangles, $\beta \approx 75.7°,\ \gamma \approx 61.3°, b \approx 9.9$ or $\beta' \approx 18.3°, \gamma' \approx 118.7°, b' \approx 3.2$

21. two triangles, $\alpha \approx 143.2°, \beta \approx 26.8°, a \approx 17.3$ or $\alpha' \approx 16.8°, \beta' \approx 153.2°, a' \approx 8.3$

23. no triangle possible

25. $A \approx 47.8°$ or $A' \approx 132.2°$

27. 8.6

29. 370.9

31. 12.3

33. 12.2

35. 16.0

37. 29.7°

39. $x = 76.9°$or $x = 103.1°$

41. 110.6°

43. $A \approx 39.4,\ C \approx 47.6,\ BC \approx 20.7$

45. 57.1

47. 42.0

49. 430.2

51. 10.1

53. $AD \approx\ 13.8$

55. $AB \approx 2.8$

57. $L \approx 49.7,\ N \approx 56.3,\ LN \approx 5.8$

59. 51.4 feet

61. The distance from the satellite to station A is approximately 1716 miles. The satellite is approximately 1706 miles above the ground.

63. 2.6 ft

65. 5.6 km

67. 371 ft

69. 5936 ft

71. 24.1 ft

73. 19,056 ft^2

75. 445,624 square miles

77. 8.65 ft^2

10.2 Section Exercises

1. two sides and the angle opposite the missing side.

3. s is the semi-perimeter, which is half the perimeter of the triangle.

5. The Law of Cosines must be used for any oblique (non-right) triangle.

7. 11.3

9. 34.7

11. 26.7

13. 257.4

15. not possible

17. 95.5°

19. 26.9°

21. $B \approx 45.9°, C \approx 99.1°, a \approx 6.4$

23. $A \approx 20.6°, B \approx 38.4°, c \approx 51.1$

25. $A \approx 37.8°, B \approx 43.8, C \approx 98.4°$

27. 177.56 in^2

29. 0.04 m^2

31. 0.91 yd^2

33. 3.0

35. 29.1

37. 0.5

39. 70.7°

41. 77.4°

43. 25.0

45. 9.3

47. 43.52

49. 1.41

51. 0.14

53. 18.3

55. 48.98

57.

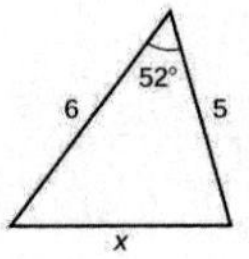

59. 7.62

61. 85.1

63. 24.0 km

65. 99.9 ft

67. 37.3 miles

69. 2371 miles

71.

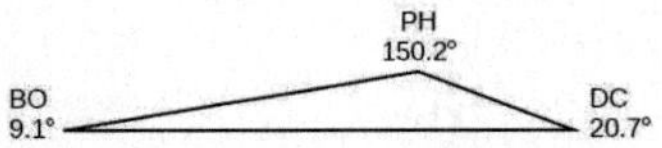

73. 292.4 miles

75. 65.4 cm^2

77. 468 ft^2

10.3 Section Exercises

1. For polar coordinates, the point in the plane depends on the angle from the positive x-axis and distance from the origin, while in Cartesian coordinates, the point represents the horizontal and vertical distances from the origin. For each point in the coordinate plane, there is one representation, but for each point in the polar plane, there are infinite representations.

3. Determine θ for the point, then move r units from the pole to plot the point. If r is negative, move r units from the pole in the opposite direction but along the same angle. The point is a distance of r away from the origin at an angle of θ from the polar axis.

5. The point $\left(-3, \frac{\pi}{2}\right)$ has a positive angle but a negative radius and is plotted by moving to an angle of $\frac{\pi}{2}$ and then moving 3 units in the negative direction. This places the point 3 units down the negative y-axis. The point $\left(3, -\frac{\pi}{2}\right)$ has a negative angle and a positive radius and is plotted by first moving to an angle of $-\frac{\pi}{2}$ and then moving 3 units down, which is the positive direction for a negative angle. The point is also 3 units down the negative y-axis.

7. $(-5, 0)$

9. $\left(-\frac{3\sqrt{3}}{2}, -\frac{3}{2}\right)$

11. $\left(2\sqrt{5},\ 0.464\right)$

13. $\left(\sqrt{34}, 5.253\right)$

15. $\left(8\sqrt{2}, \frac{\pi}{4}\right)$

17. $r = 4\csc\theta$

19. $r = \sqrt[3]{\frac{sin\theta}{2cos^4\theta}}$

21. $r = 3\cos\theta$

23. $r = \frac{3\sin\theta}{\cos(2\theta)}$

25. $r = \frac{9\sin\theta}{\cos^2\theta}$

27. $r = \sqrt{\frac{1}{9\cos\theta\sin\theta}}$

29. $x^2 + y^2 = 4x$ or $\frac{(x-2)^2}{4} + \frac{y^2}{4} = 1$; circle

31. $3y + x = 6$; line

33. $y = 3$; line

35. $xy = 4$; hyperbola

37. $x^2 + y^2 = 4$; circle

39. $x - 5y = 3$; line

41. $\left(3, \frac{3\pi}{4}\right)$

43. $(5, \pi)$

45.

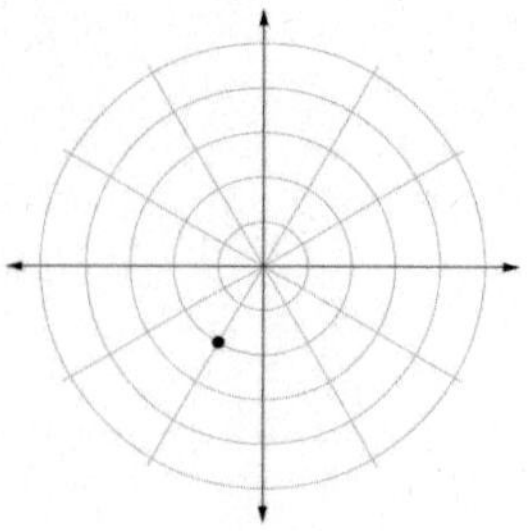

47.

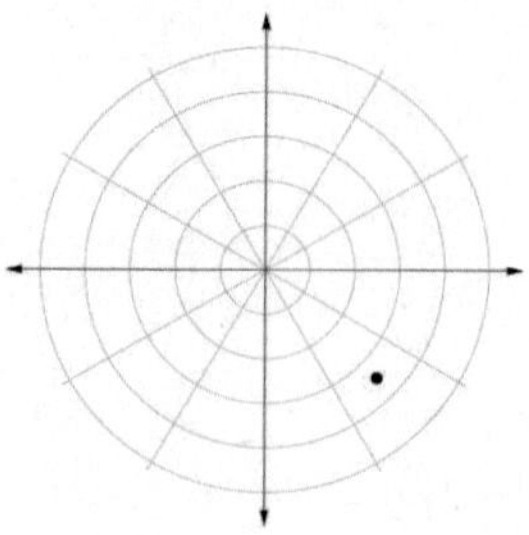

49.

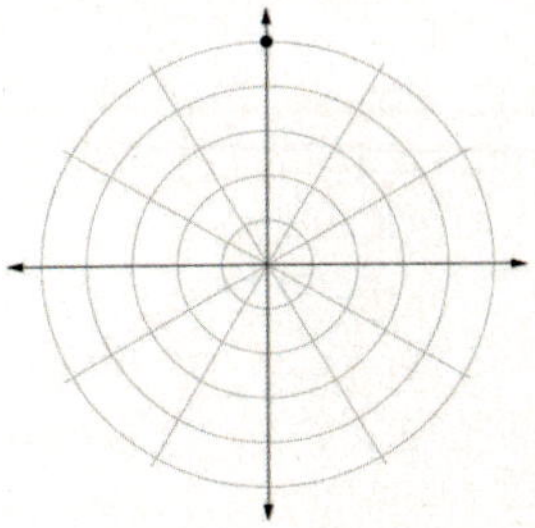

51.

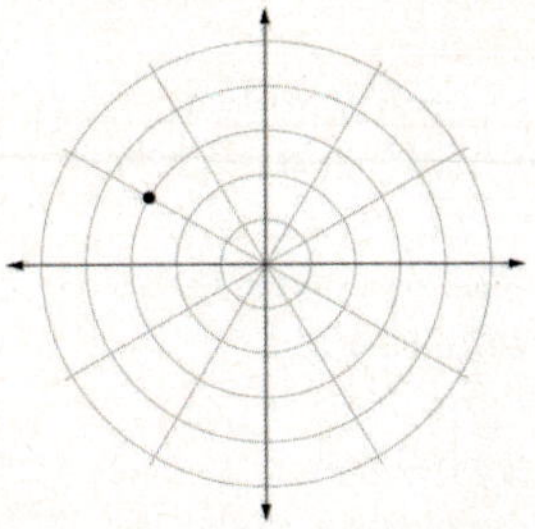

53.

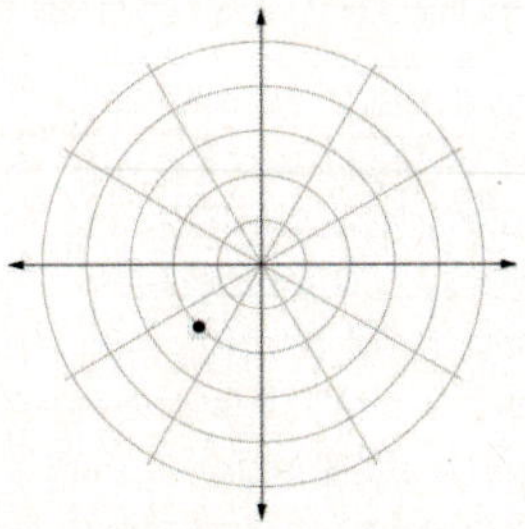

55. $r = \frac{6}{5\cos\theta - \sin\theta}$

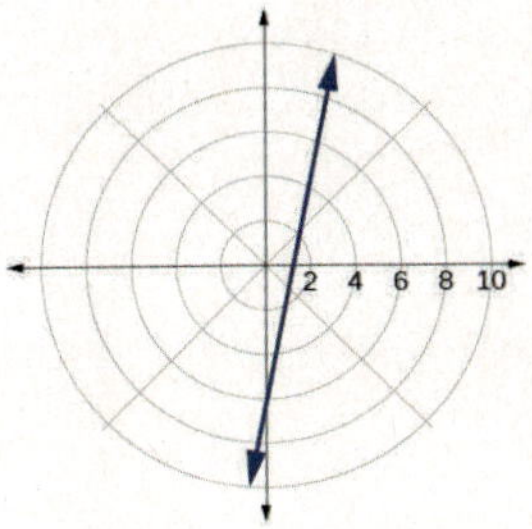

57. $r = 2\sin\theta$

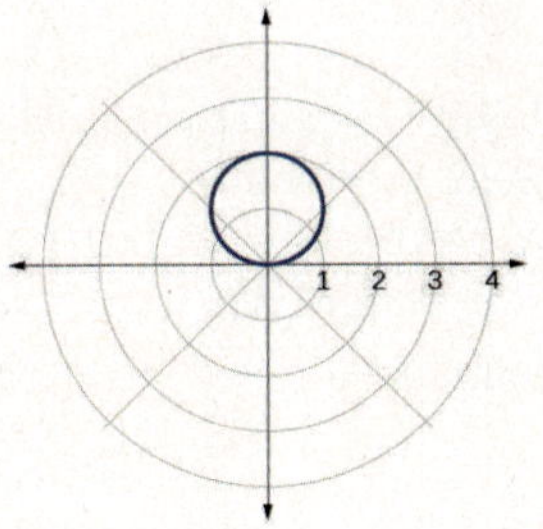

59. $r = \frac{2}{\cos\theta}$

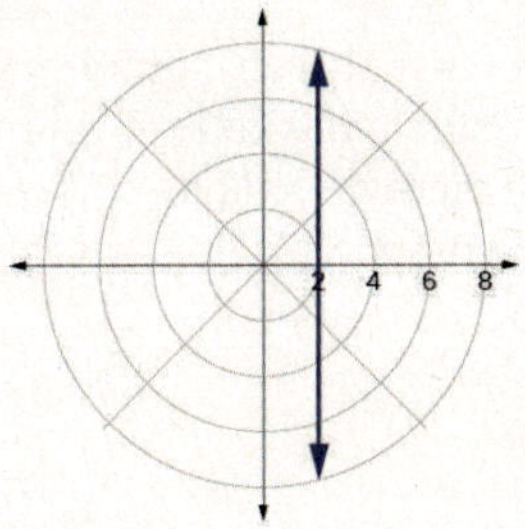

61. $r = 3\cos\theta$

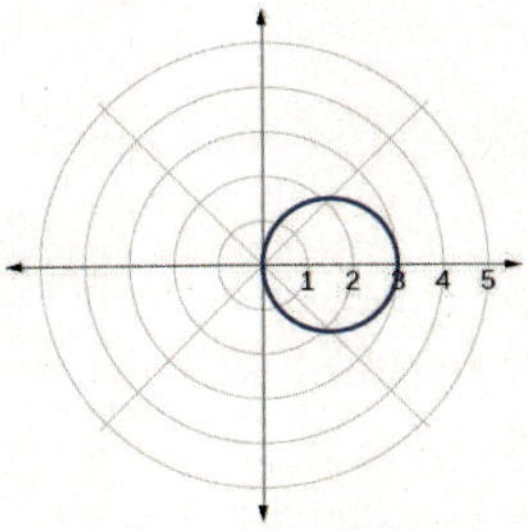

63. $x^2 + y^2 = 16$

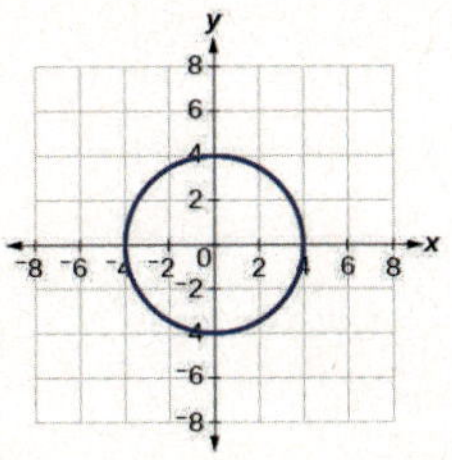

65. $y = x$

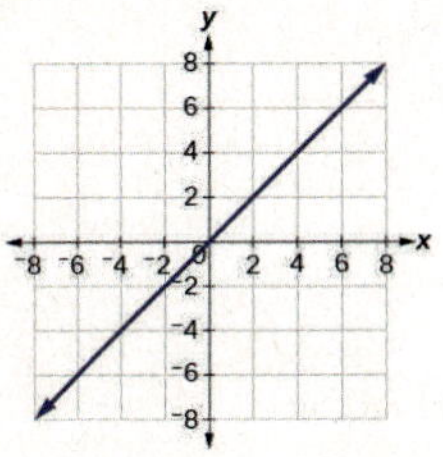

67. $x^2 + (y+5)^2 = 25$

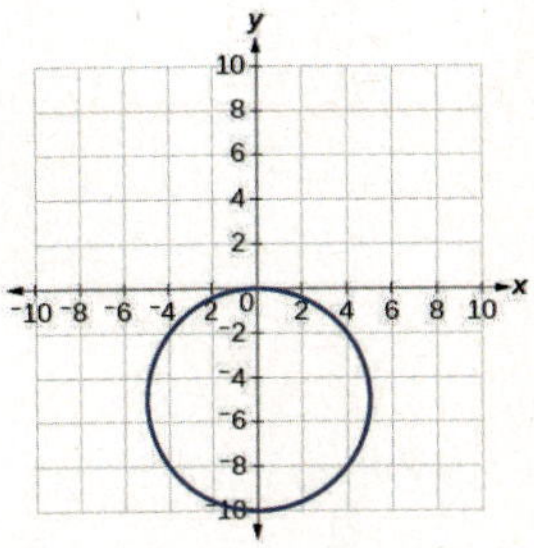

69. $(1.618, -1.176)$

71. $(10.630, 131.186°)$

73. $(2, 3.14)$ *or* $(2, \pi)$

75. A vertical line with a units left of the y-axis.

77. A horizontal line with a units below the x-axis.

79.

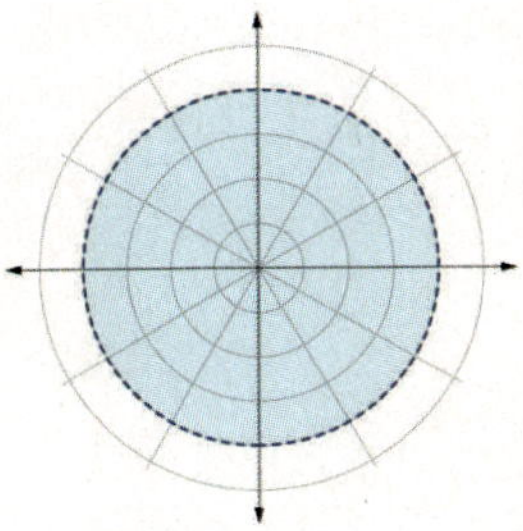

81.

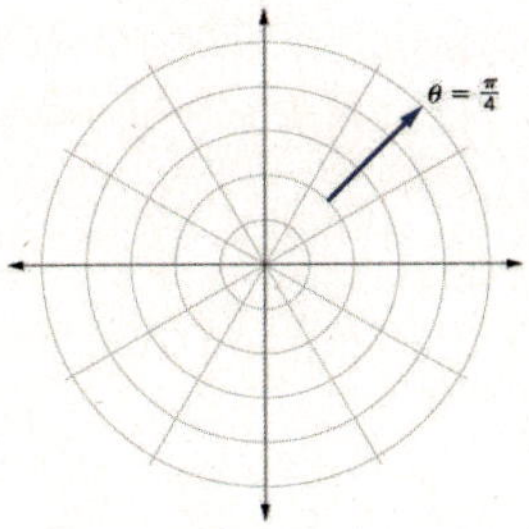

83.

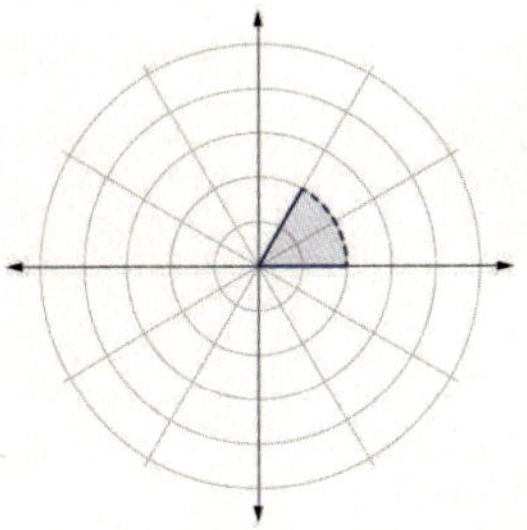

10.4 Section Exercises

1. Symmetry with respect to the polar axis is similar to symmetry about the x-axis, symmetry with respect to the pole is similar to symmetry about the origin, and symmetric with respect to the line $\theta = \frac{\pi}{2}$ is similar to symmetry about the y-axis.

3. Test for symmetry; find zeros, intercepts, and maxima; make a table of values. Decide the general type of graph, cardioid, limaçon, lemniscate, etc., then plot points at $\theta = 0,\ \frac{\pi}{2}, \pi$ and $\frac{3\pi}{2}$, and sketch the graph.

5. The shape of the polar graph is determined by whether or not it includes a sine, a cosine, and constants in the equation.

7. symmetric with respect to the polar axis

9. symmetric with respect to the polar axis, symmetric with respect to the line $\theta = \frac{\pi}{2}$, symmetric with respect to the pole

11. no symmetry

13. no symmetry

15. symmetric with respect to the pole

17. circle

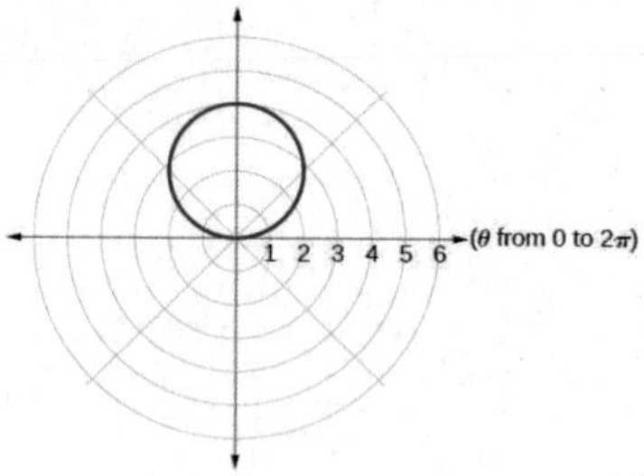

19. cardioid

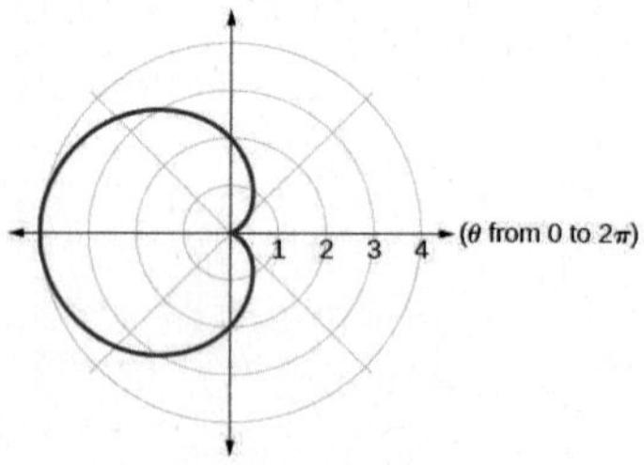

21. cardioid

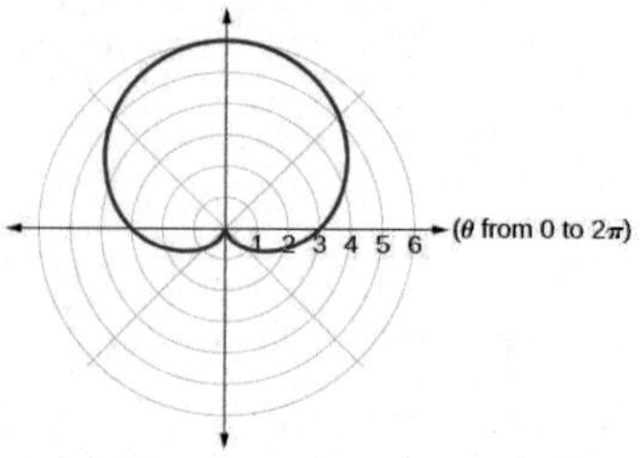

23. one-loop/dimpled limaçon

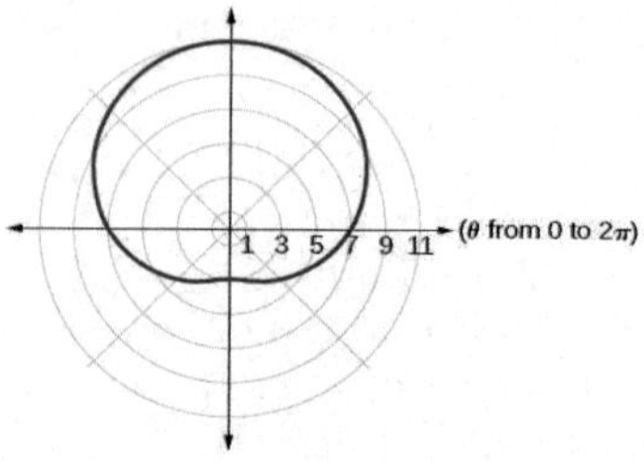

25. one-loop/dimpled limaçon

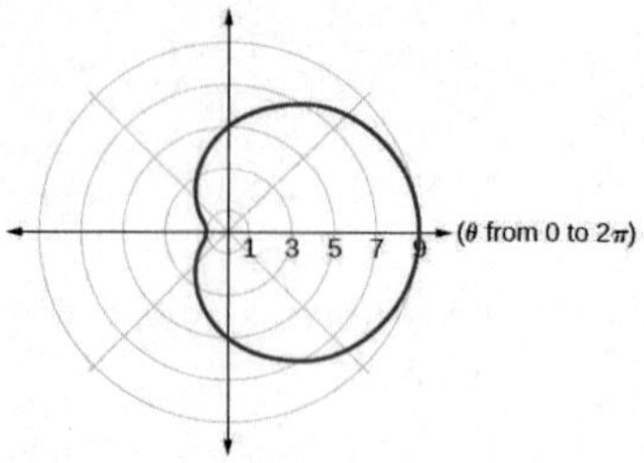

27. inner loop/two-loop limaçon

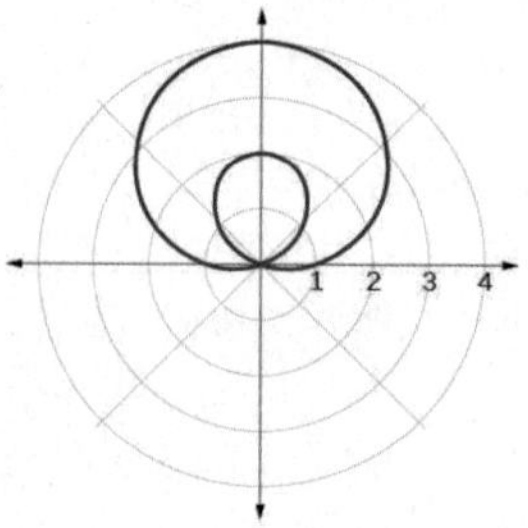

29. inner loop/two-loop limaçon

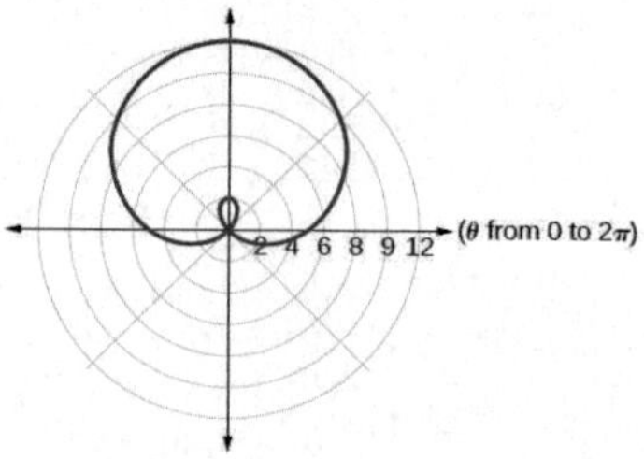

31. inner loop/two-loop limaçon

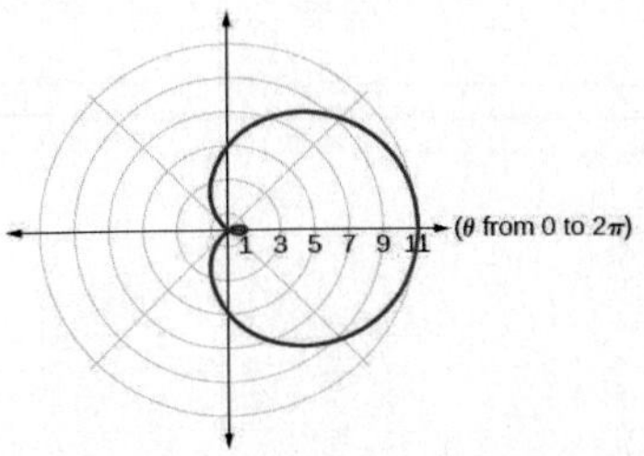

33. lemniscate

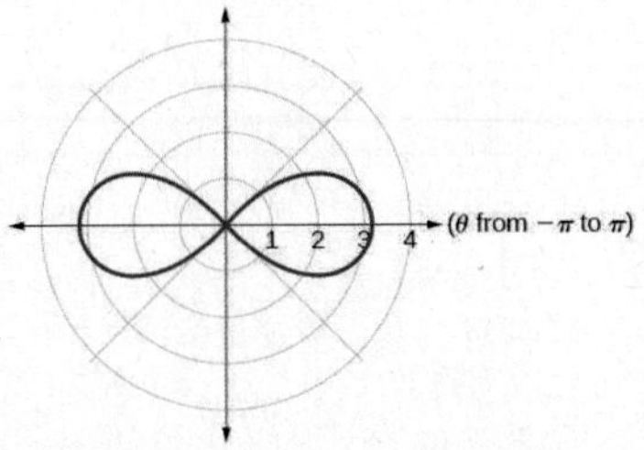

35. lemniscate

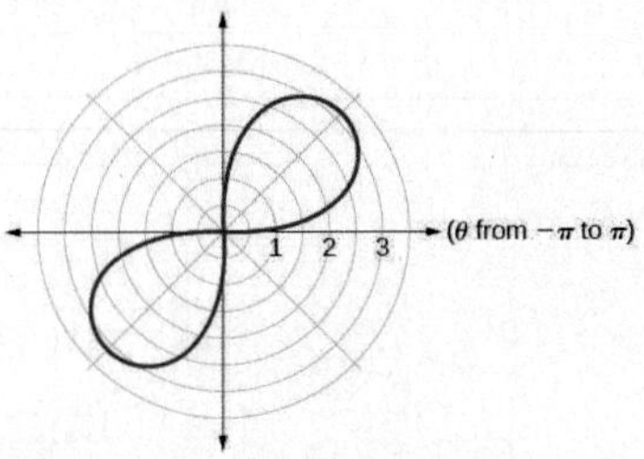

37. rose curve

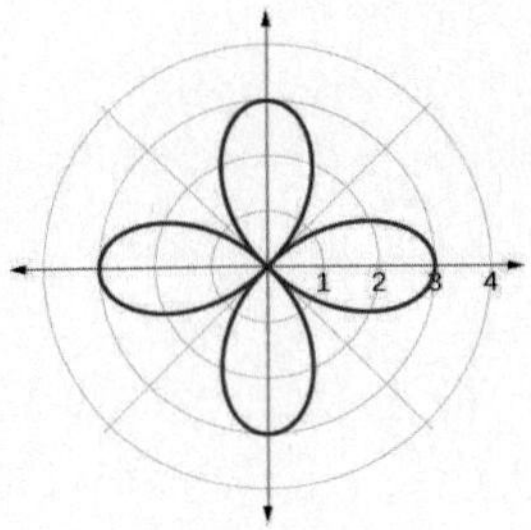

39. rose curve

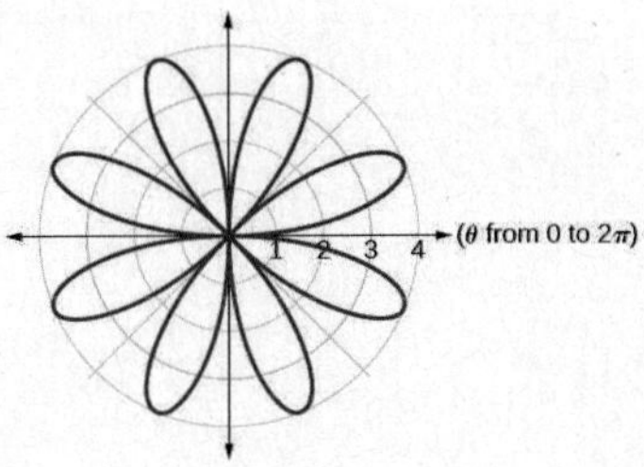

41. Archimedes' spiral

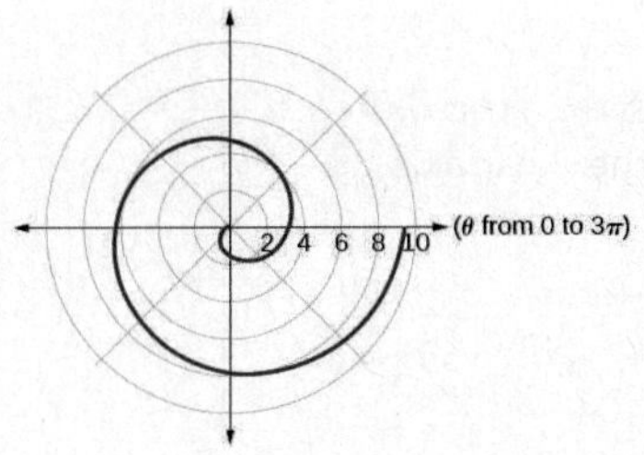

43. Archimedes' spiral

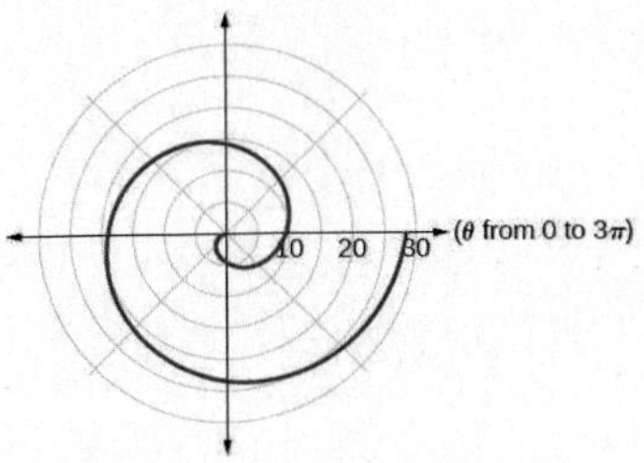

45.

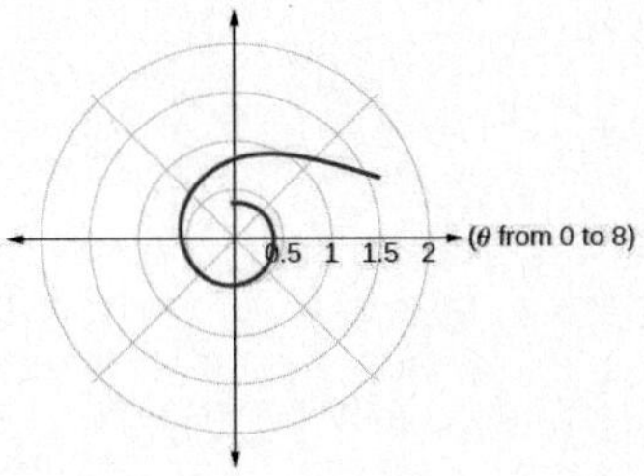

47.

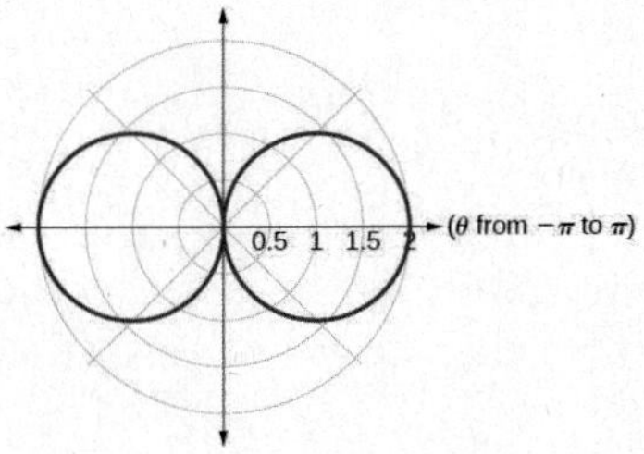

49.

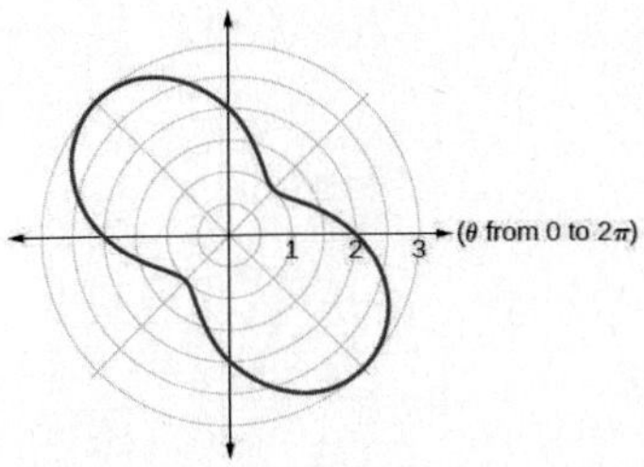

51.

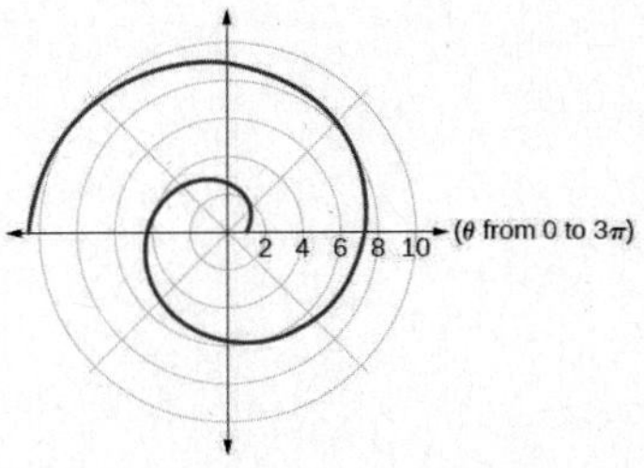

53.

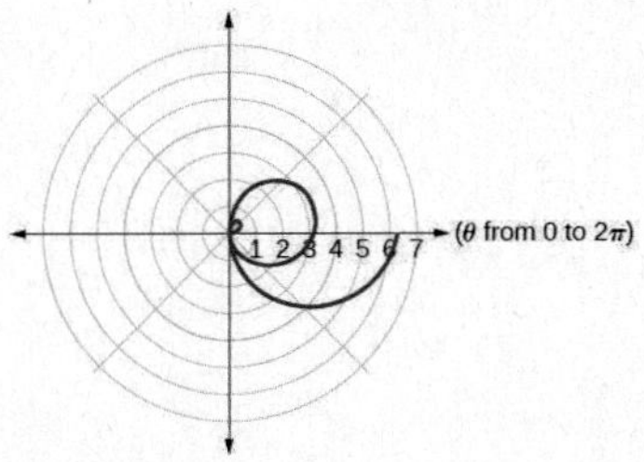

55. They are both spirals, but not quite the same.

57. Both graphs are curves with 2 loops. The equation with a coefficient of θ has two loops on the left, the equation with a coefficient of 2 has two loops side by side. Graph these from 0 to 4π to get a better picture.

59. When the width of the domain is increased, more petals of the flower are visible.

61. The graphs are three-petal, rose curves. The larger the coefficient, the greater the curve's distance from the pole.

63. The graphs are spirals. The smaller the coefficient, the tighter the spiral.

65. $\left(4, \frac{\pi}{3}\right), \left(4, \frac{5\pi}{3}\right)$

67. $\left(\frac{3}{2}, \frac{\pi}{3}\right), \left(\frac{3}{2}, \frac{5\pi}{3}\right)$

69. $\left(0, \frac{\pi}{2}\right), (0, \pi), \left(0, \frac{3\pi}{2}\right), (0, 2\pi)$

71. $\left(\frac{\sqrt[4]{8}}{2}, \frac{\pi}{4}\right), \left(\frac{\sqrt[4]{8}}{2}, \frac{5\pi}{4}\right)$ and at $\theta = \frac{3\pi}{4}, \frac{7\pi}{4}$ since r is squared

10.5 Section Exercises

1. a is the real part, b is the imaginary part, and $i = \sqrt{-1}$

3. Polar form converts the real and imaginary part of the complex number in polar form using $x = r\cos\theta$ and $y = r\sin\theta$.

5. $z^n = r^n(\cos(n\theta) + i\sin(n\theta))$ It is used to simplify polar form when a number has been raised to a power.

7. $5\sqrt{2}$

9. $\sqrt{38}$

11. $\sqrt{14.45}$

13. $4\sqrt{5}\operatorname{cis}(333.4°)$

15. $2\operatorname{cis}\left(\frac{\pi}{6}\right)$

17. $\frac{7\sqrt{3}}{2} + i\frac{7}{2}$

19. $-2\sqrt{3} - 2i$

21. $-1.5 - i\frac{3\sqrt{3}}{2}$

23. $4\sqrt{3}\operatorname{cis}(198°)$

25. $\frac{3}{4}\operatorname{cis}(180°)$

27. $5\sqrt{3}\operatorname{cis}\left(\frac{17\pi}{24}\right)$

29. $7\operatorname{cis}(70°)$

31. $5\operatorname{cis}(80°)$

33. $5\operatorname{cis}\left(\frac{\pi}{3}\right)$

35. $125\operatorname{cis}(135°)$

37. $9\operatorname{cis}(240°)$

39. $\operatorname{cis}\left(\frac{3\pi}{4}\right)$

41. $3\operatorname{cis}(80°), 3\operatorname{cis}(200°), 3\operatorname{cis}(320°)$

43. $2\sqrt[3]{4}\operatorname{cis}\left(\frac{2\pi}{9}\right), 2\sqrt[3]{4}\operatorname{cis}\left(\frac{8\pi}{9}\right), 2\sqrt[3]{4}\operatorname{cis}\left(\frac{14\pi}{9}\right)$

45. $2\sqrt{2}\operatorname{cis}\left(\frac{7\pi}{8}\right), 2\sqrt{2}\operatorname{cis}\left(\frac{15\pi}{8}\right)$

47.

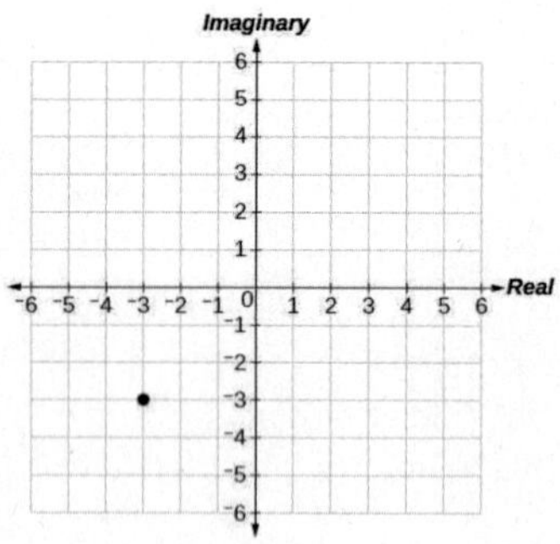

49.

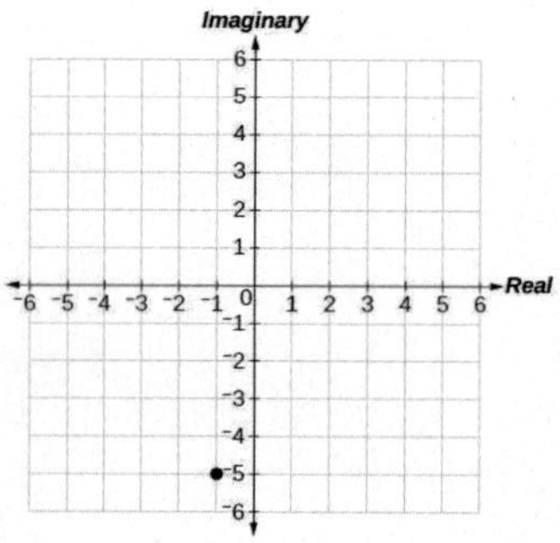

51.

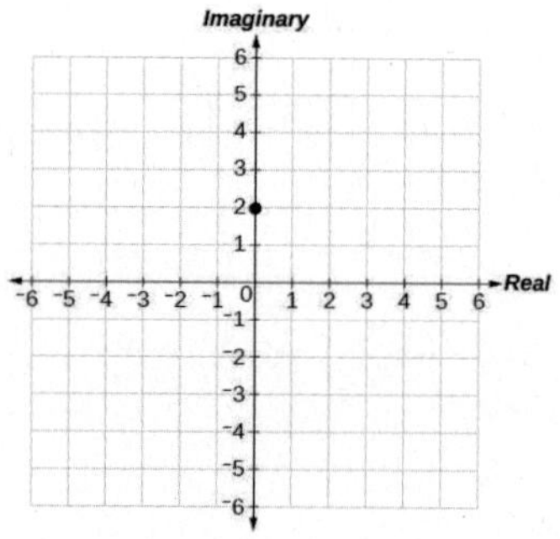

53.

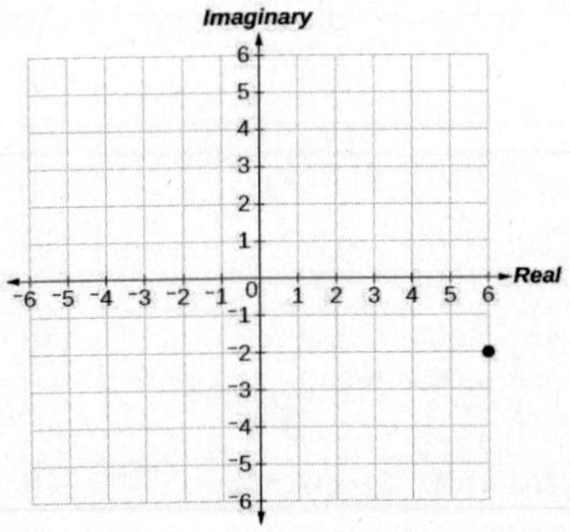

55.

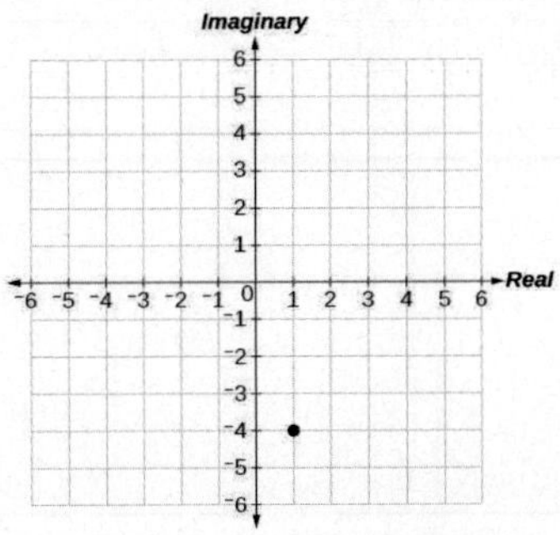

57. $3.61e^{-0.59i}$

59. $-2 + 3.46i$

61. $-4.33 - 2.50i$

10.6 Section Exercises

1. A pair of functions that is dependent on an external factor. The two functions are written in terms of the same parameter. For example, $x = f(t)$ and $y = f(t)$.

3. Choose one equation to solve for t, substitute into the other equation and simplify.

5. Some equations cannot be written as functions, like a circle. However, when written as two parametric equations, separately the equations are functions.

7. $y = -2 + 2x$

9. $y = 3\sqrt{\frac{x-1}{2}}$

11. $x = 2e^{\frac{1-y}{5}}$ or $y = 1 - 5ln\left(\frac{x}{2}\right)$

13. $x = 4\log\left(\frac{y-3}{2}\right)$

15. $x = \left(\frac{y}{2}\right)^3 - \frac{y}{2}$

17. $y = x^3$

19. $\left(\frac{x}{4}\right)^2 + \left(\frac{y}{5}\right)^2 = 1$

21. $y^2 = 1 - \frac{1}{2}x$

23. $y = x^2 + 2x + 1$

25. $y = \left(\frac{x+1}{2}\right)^3 - 2$

27. $y = -3x + 14$

29. $y = x + 3$

31. $\begin{cases} x(t) = t \\ y(t) = 2\sin t + 1 \end{cases}$

33. $\begin{cases} x(t) = \sqrt{t} + 2t \\ y(t) = t \end{cases}$

35. $\begin{cases} x(t) = 4\cos t \\ y(t) = 6\sin t \end{cases}$; Ellipse

37. $\begin{cases} x(t) = \sqrt{10}\cos t \\ y(t) = \sqrt{10}\sin t \end{cases}$; Circle

39. $\begin{cases} x(t) = -1 + 4t \\ y(t) = -2t \end{cases}$

41. $\begin{cases} x(t) = 4 + 2t \\ y(t) = 1 - 3t \end{cases}$

43. yes, at $t = 2$

45.

t	x	y
1	-3	1
2	0	7
3	5	17

47. answers may vary: $\begin{cases} x(t) = t - 1 \\ y(t) = t^2 \end{cases}$ and $\begin{cases} x(t) = t + 1 \\ y(t) = (t+2)^2 \end{cases}$

49. answers may vary: ,

$$\begin{cases} x(t) = t \\ y(t) = t^2 - 4t + 4 \end{cases} \text{ and } \begin{cases} x(t) = t + 2 \\ y(t) = t^2 \end{cases}$$

10.7 Section Exercises

1. plotting points with the orientation arrow and a graphing calculator

3. The arrows show the orientation, the direction of motion according to increasing values of t.

5. The parametric equations show the different vertical and horizontal motions over time.

7.

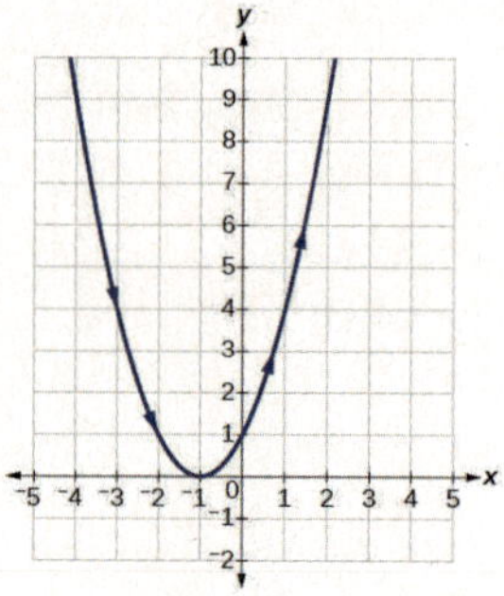

9.

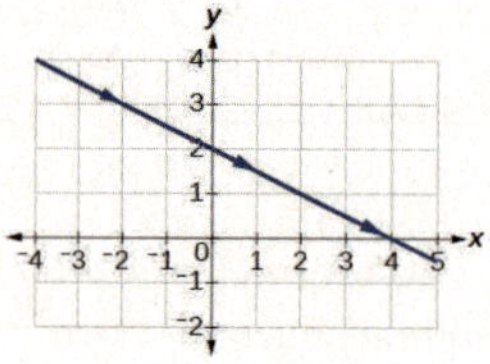

11.

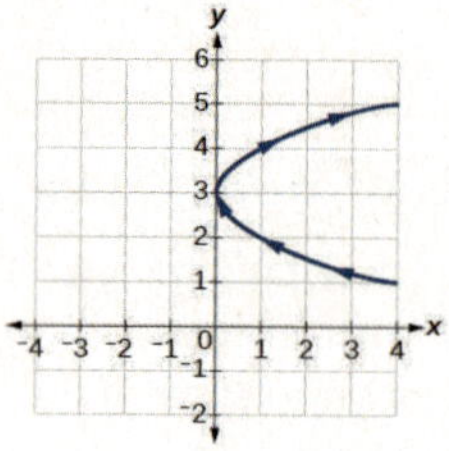

13.

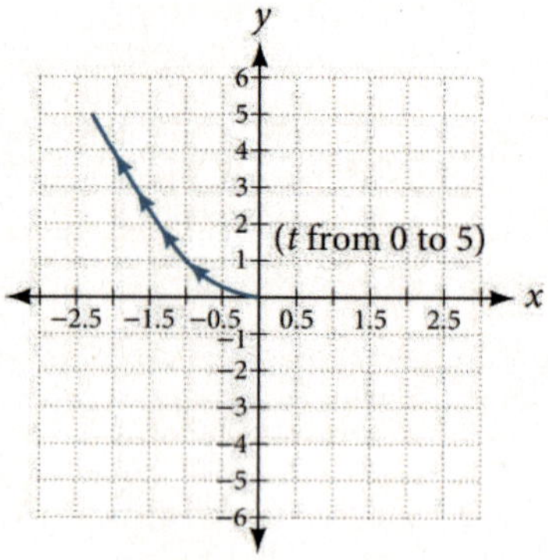

15.

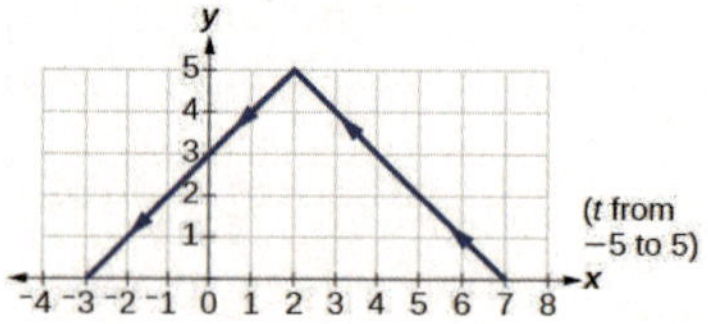

17.

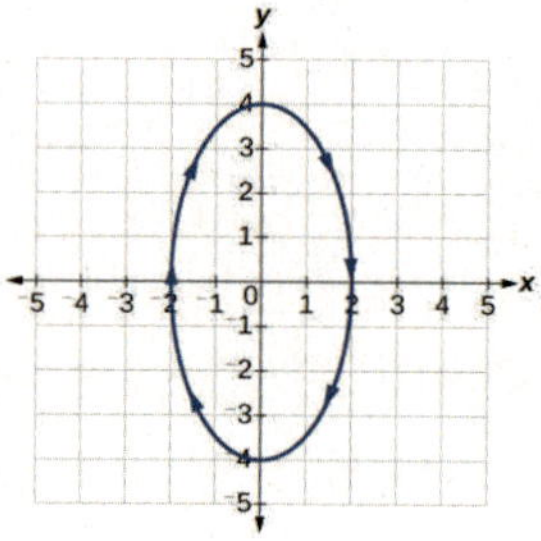

19.

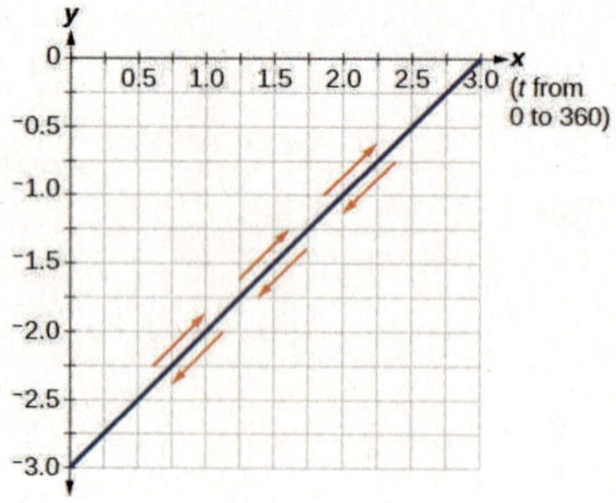

21.

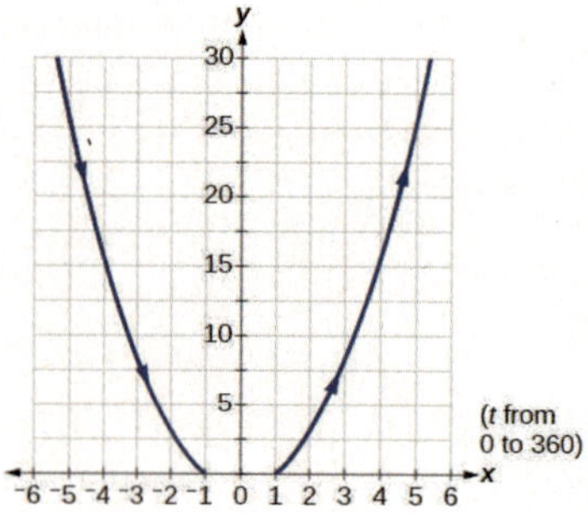

23.

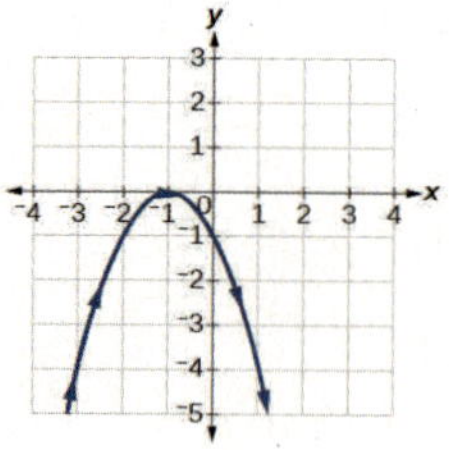

25.

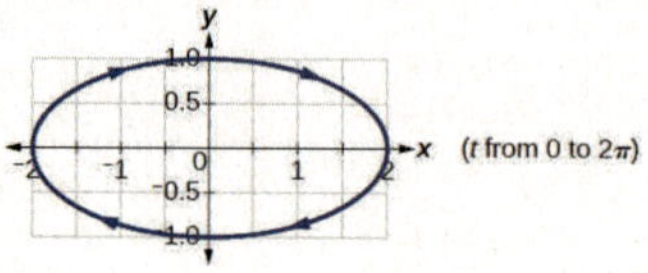

27.

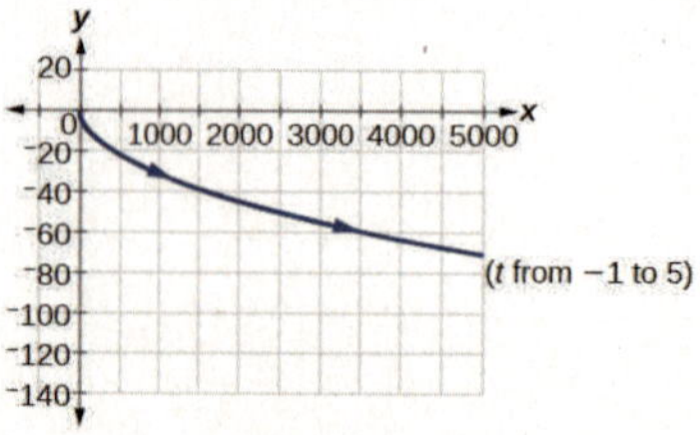

29.

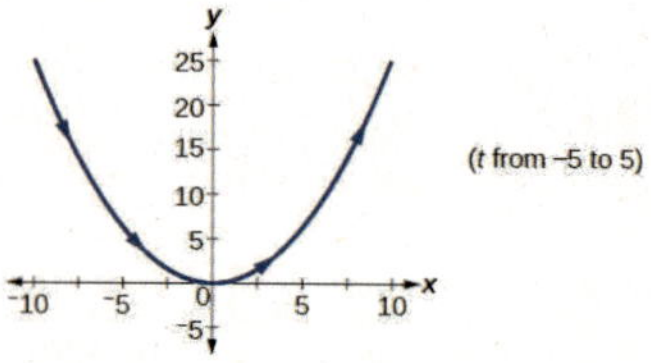

31.

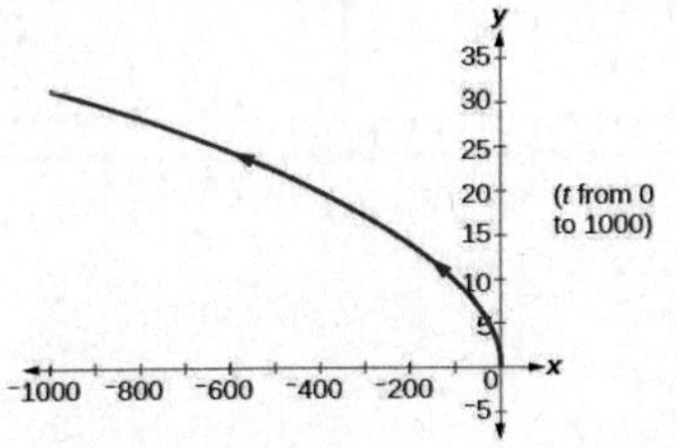

33.

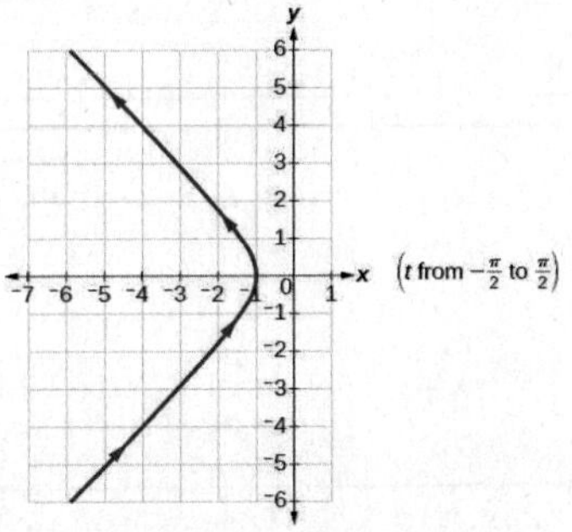

35.

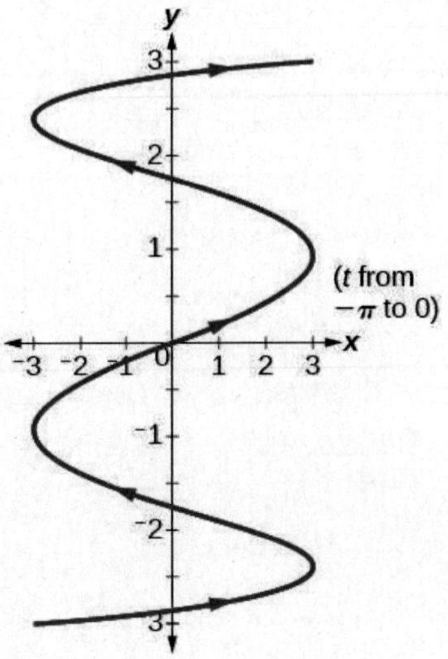

37.

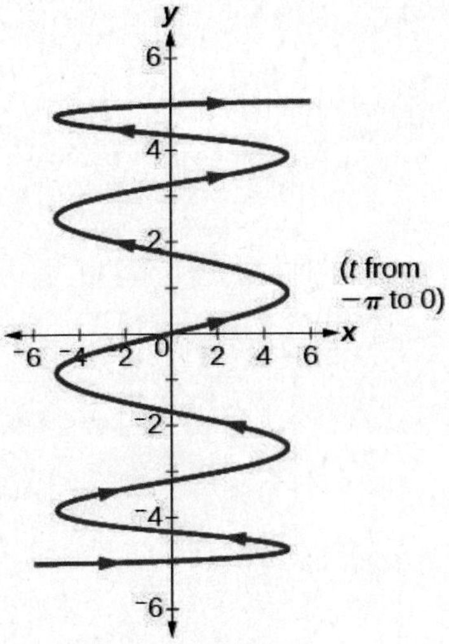

39. There will be 100 back-and-forth motions.

41. Take the opposite of the $x(t)$ equation.

43. The parabola opens up.

45. $\begin{cases} x(t) = 5\cos t \\ y(t) = 5\sin t \end{cases}$

47.

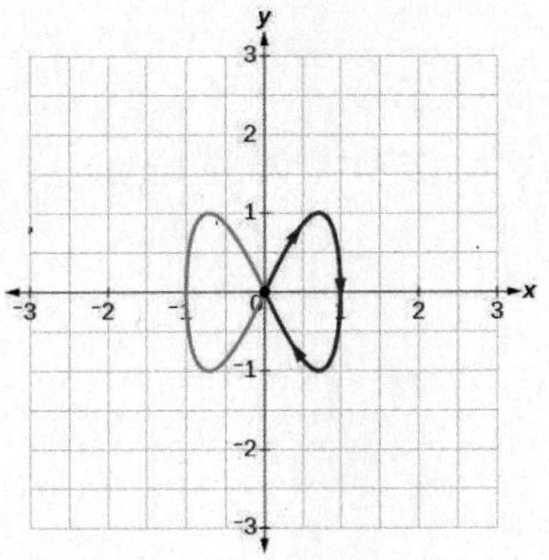

49.

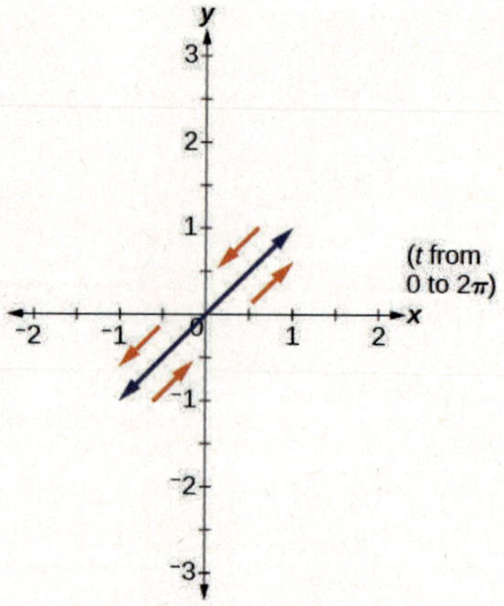

51.

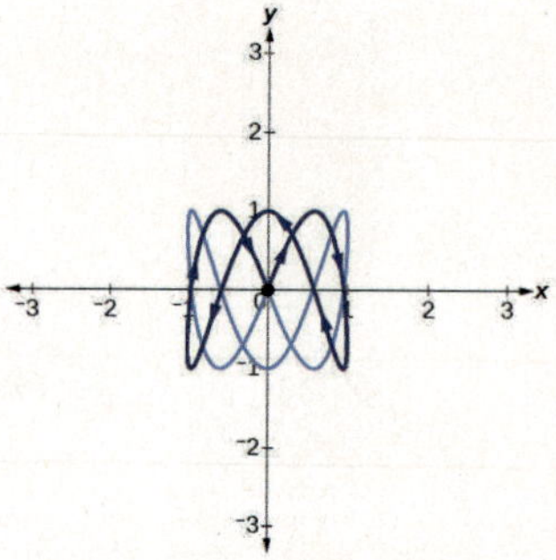

53. $a = 4, b = 3, c = 6, d = 1$

55. $a = 4, b = 2, c = 3, d = 3$

57.

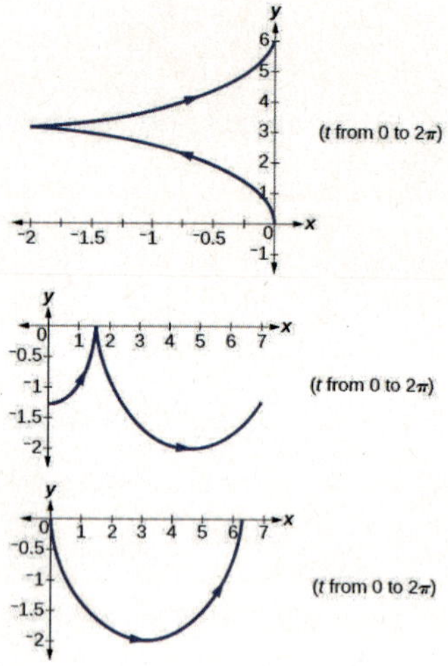

59.

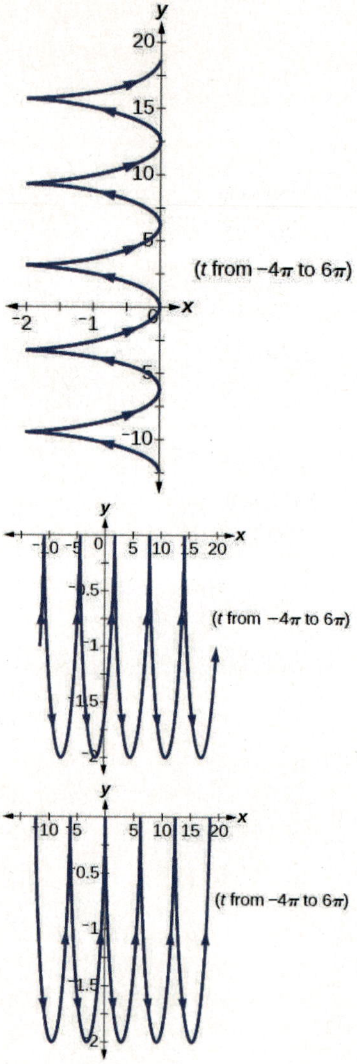

61. The y-intercept changes.

63. $y(x) = -16\left(\frac{x}{15}\right)^2 + 20\left(\frac{x}{15}\right)$

65. $\begin{cases} x(t) = 64t\cos(52°) \\ y(t) = -16t^2 + 64t\sin(52°) \end{cases}$

67. approximately 3.2 seconds

69. 1.6 seconds

71.

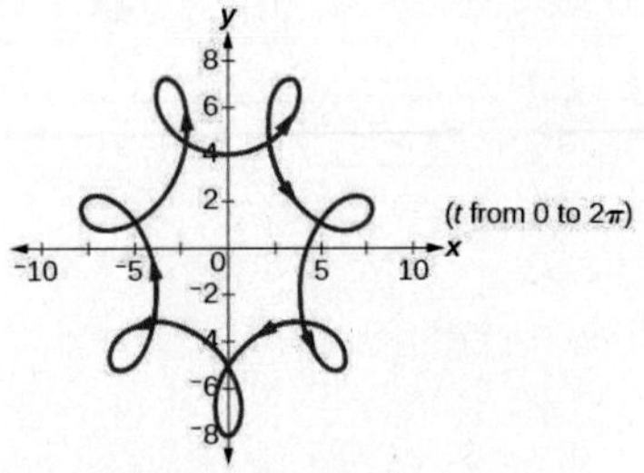

73.

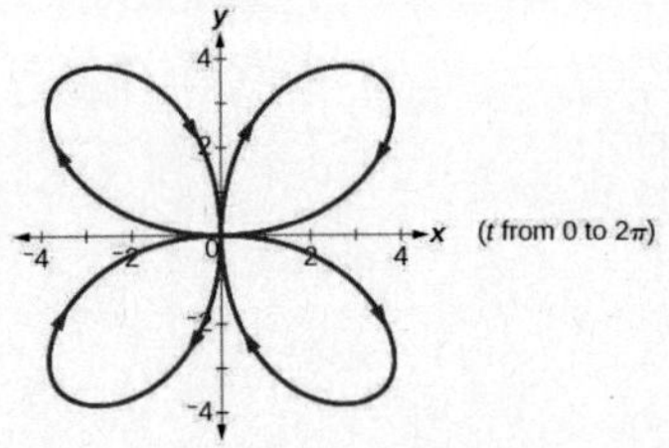

10.8 Section Exercises

1. lowercase, bold letter, usually **u**, **v**, *w*

3. They are unit vectors. They are used to represent the horizontal and vertical components of a vector. They each have a magnitude of 1.

5. The first number always represents the coefficient of the **i**, and the second represents the **j**.

7. $\langle 7, -5 \rangle$

9. not equal

11. equal

13. equal

15. $7\mathbf{i} - 3\mathbf{j}$

17. $-6\mathbf{i} - 2\mathbf{j}$

19. $\mathbf{u} + \mathbf{v} = \langle -5, 5 \rangle, \mathbf{u} - \mathbf{v} = \langle -1, 3 \rangle, 2\mathbf{u} - 3\mathbf{v} = \langle 0, 5 \rangle$

21. $-10\mathbf{i} - 4\mathbf{j}$

23. $-\frac{2\sqrt{29}}{29}\mathbf{i} + \frac{5\sqrt{29}}{29}\mathbf{j}$

25. $-\frac{2\sqrt{229}}{229}\mathbf{i} + \frac{15\sqrt{229}}{229}\mathbf{j}$

27. $-\frac{7\sqrt{2}}{10}\mathbf{i} + \frac{\sqrt{2}}{10}\mathbf{j}$

29. $|\mathbf{v}| = 7.810, \theta = 39.806°$

31. $|\mathbf{v}| = 7.211, \theta = 236.310°$

33. -6

35. -12

37.

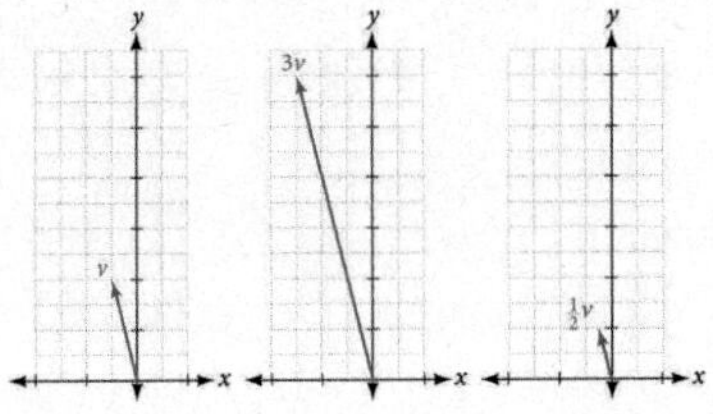

39.

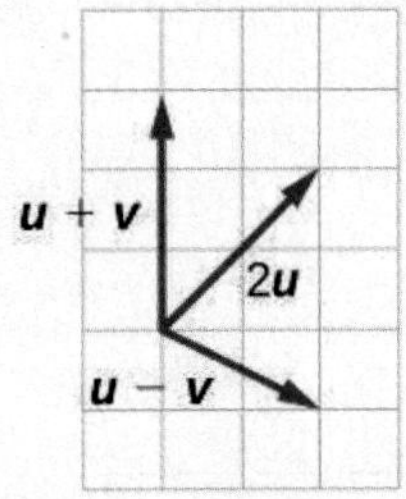

41.

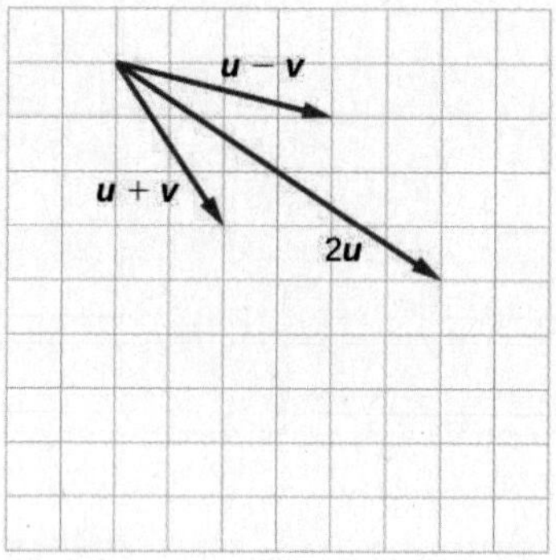

43.

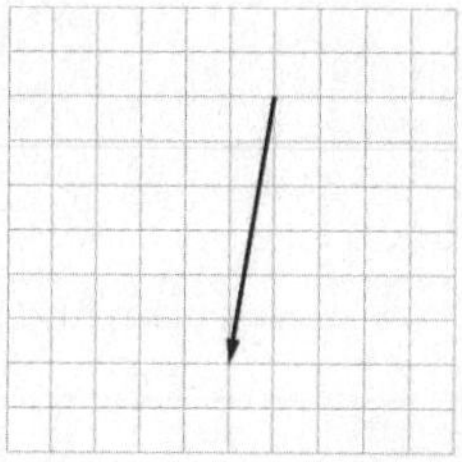

45.

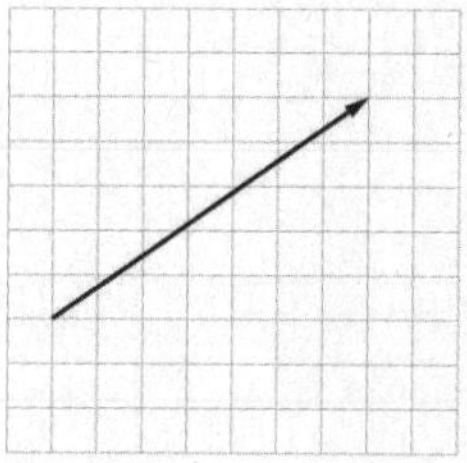

47. $\langle 4, 1 \rangle$

49. $\mathbf{v} = -7\mathbf{i} + 3\mathbf{j}$

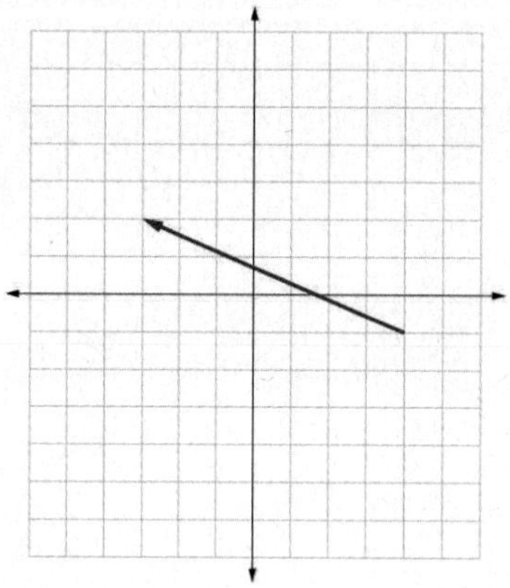

51. $3\sqrt{2}\mathbf{i} + 3\sqrt{2}\mathbf{j}$

53. $\mathbf{i} - \sqrt{3}\mathbf{j}$

55. ⓐ 58.7 ⓑ 12.5

57. $x = 7.13$ pounds, $y = 3.63$ pounds

59. $x = 2.87$ pounds, $y = 4.10$ pounds

61. 4.635 miles, 17.764° N of E

63. 17 miles. 10.318 miles

65. Distance: 2.868. Direction: 86.474° North of West, or 3.526° West of North

67. 4.924°. 659 km/hr

69. 4.424°

71. $(0.081, 8.602)$

73. 21.801°, relative to the car's forward direction

75. parallel: 16.28, perpendicular: 47.28 pounds

77. 19.35 pounds, 231.54° from the horizontal

79. 5.1583 pounds, 75.8° from the horizontal

Review Exercises

1. Not possible

3. $C = 120°, a = 23.1, c = 34.1$

5. distance of the plane from point A : 2.2 km, elevation of the plane: 1.6 km

7. $\mathbf{b} = 71.0°, C = 55.0°, a = 12.8$

9. 40.6 km

11.

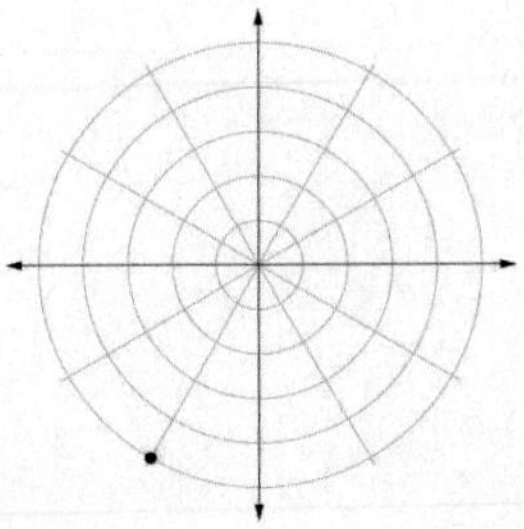

13. $(0, 2)$

15. $(9.8489, 203.96°)$

17. $r = 8$

19. $x^2 + y^2 = 7x$

21. $y = -x$

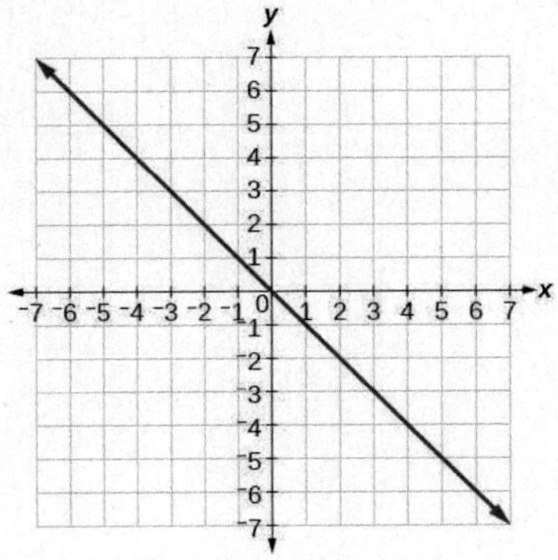

23. symmetric with respect to the line $\theta = \frac{\pi}{2}$

25.

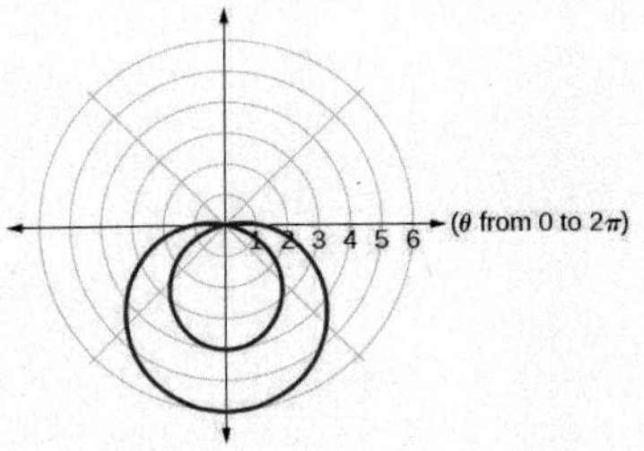

27.

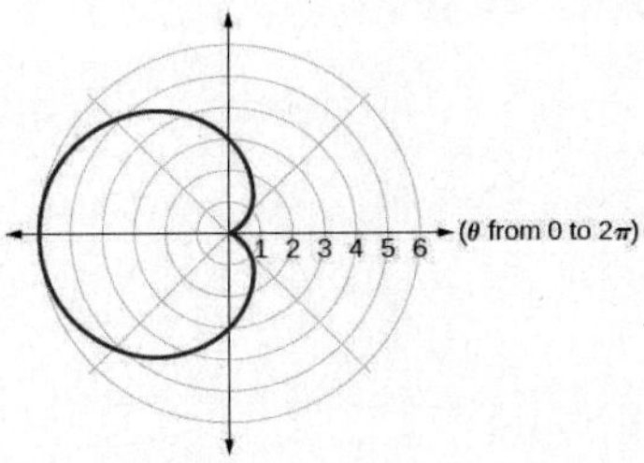

29. 5

31. $\text{cis}\left(-\frac{\pi}{3}\right)$

33. $2.3 + 1.9\mathbf{i}$

35. $60\,\text{cis}\left(\frac{\pi}{2}\right)$

37. $3\,\text{cis}\left(\frac{4\pi}{3}\right)$

39. $25\,\text{cis}\left(\frac{3\pi}{2}\right)$

41. $5\,\text{cis}\left(\frac{3\pi}{4}\right), 5\,\text{cis}\left(\frac{7\pi}{4}\right)$

43.

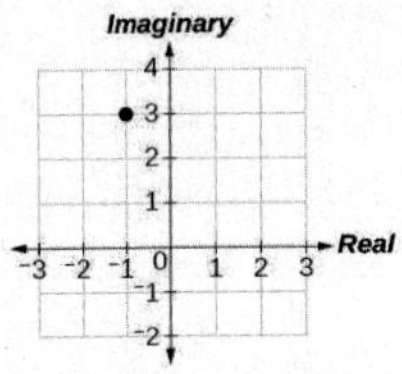

45. $x^2 + \frac{1}{2}y = 1$

47. $\begin{cases} x(t) = -2 + 6t \\ y(t) = 3 + 4t \end{cases}$

49. $y = -2x^5$

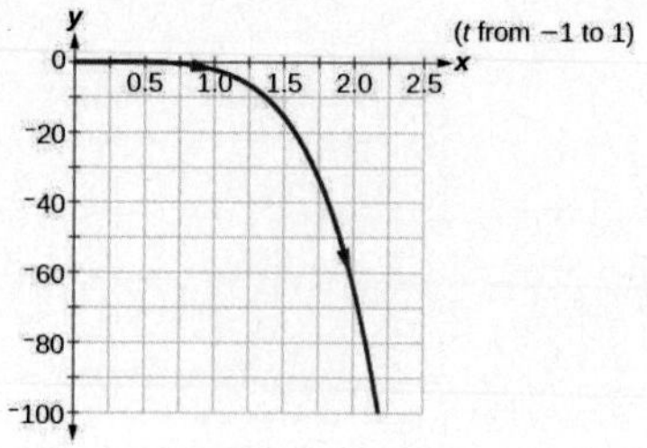

51. a. $\begin{cases} x(t) = (80\cos(40°))\,t \\ y(t) = -16t^2 + (80\sin(40°))\,t + 4 \end{cases}$

b. The ball is 14 feet high and 184 feet from where it was launched.

c. 3.3 seconds

53. not equal

55. $4i$

57. $-\frac{3\sqrt{10}}{10}\,\boldsymbol{i} - \frac{\sqrt{10}}{10}\,\boldsymbol{j}$

59. Magnitude: $3\sqrt{2}$,
Direction: 225°

61. 16

63.

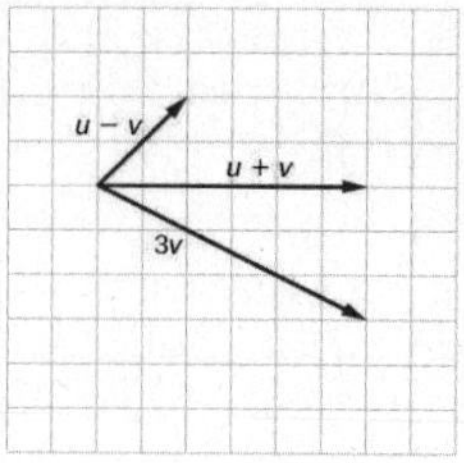

Practice Test

1. $\alpha = 67.1°, \gamma = 44.9°, a = 20.9$

3. 1712 miles

5. $\left(1, \sqrt{3}\right)$

7. $y = -3$

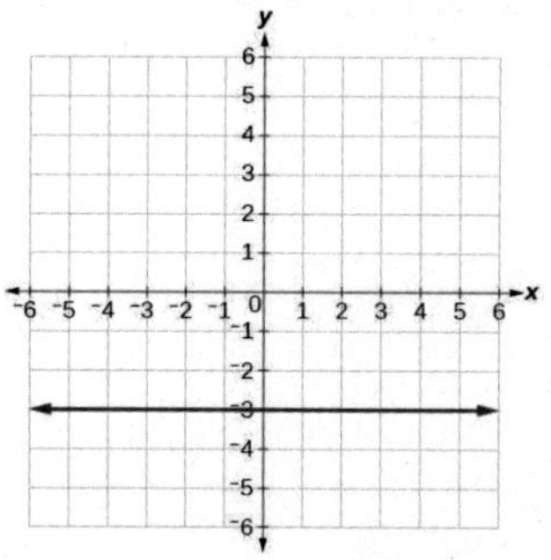

9.

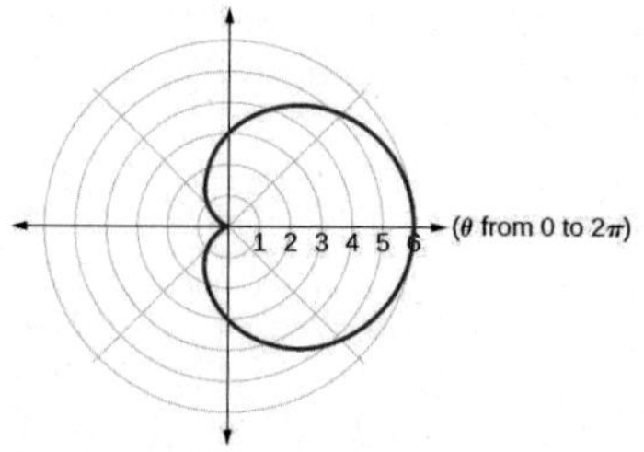

11. $\sqrt{106}$

13. $\frac{-5}{2} + \mathrm{i}\frac{5\sqrt{3}}{2}$

15. $4\,\text{cis}\,(21°)$

17. $2\sqrt{2}\,\text{cis}\,(18°), 2\sqrt{2}\,\text{cis}\,(198°)$

19. $y = 2(x - 1)^2$

21.

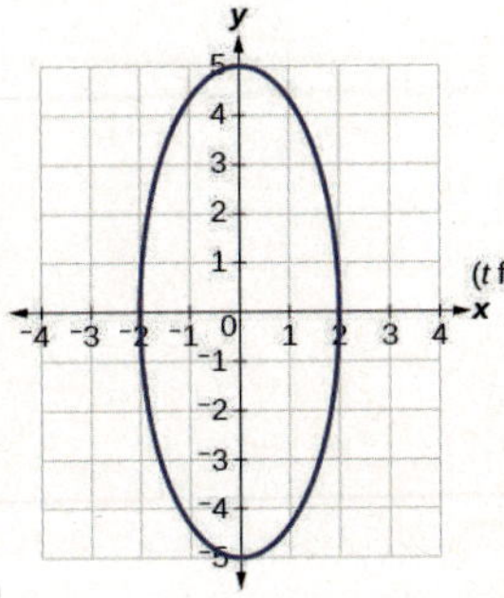

23. $-4i - 15j$

25. $\frac{2\sqrt{13}}{13}\mathbf{i} + \frac{3\sqrt{13}}{13}\mathbf{j}$

Chapter 11

Try It

11.1 Systems of Linear Equations: Two Variables

1. Not a solution.

2. The solution to the system is the ordered pair $(-5, 3)$.

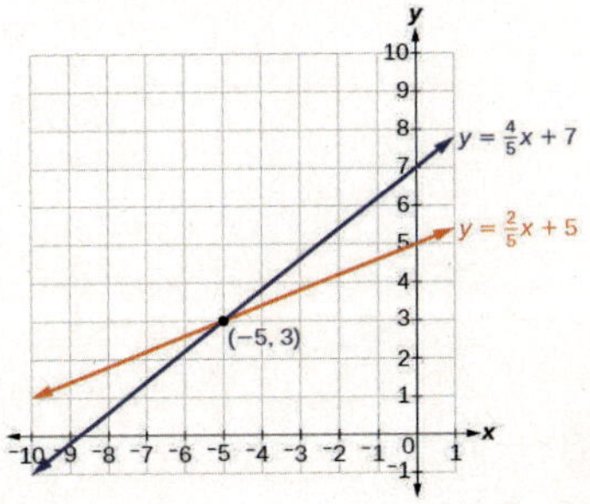

3. $(-2, -5)$

4. $(-6, -2)$

5. $(10, -4)$

6. No solution. It is an inconsistent system.

7. The system is dependent so there are infinite solutions of the form $(x, 2x + 5)$.

8. 700 children, 950 adults

11.2 Systems of Linear Equations: Three Variables

1. $(1, -1, 1)$

2. No solution.

3. Infinite number of solutions of the form $(x, 4x{-}11, -5x + 18)$.

11.3 Systems of Nonlinear Equations and Inequalities: Two Variables

1. $\left(-\frac{1}{2}, \frac{1}{2}\right)$ and $(2, 8)$

2. $(-1, 3)$

3. $\{(1, 3), (1, -3), (-1, 3), (-1, -3)\}$

4. 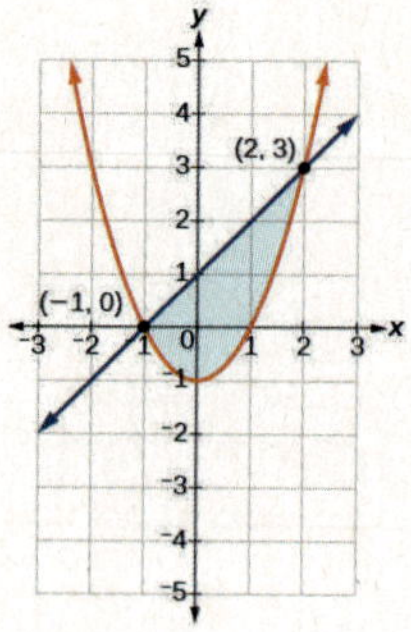

11.4 Partial Fractions

1. $\frac{3}{x-3}-\frac{2}{x-2}$

2. $\frac{6}{x-1}-\frac{5}{(x-1)^2}$

3. $\frac{3}{x-1}+\frac{2x-4}{x^2+1}$

4. $\frac{x-2}{x^2-2x+3}+\frac{2x+1}{\left(x^2-2x+3\right)^2}$

11.5 Matrices and Matrix Operations

1. $A+B=\begin{bmatrix}2 & 6\\ 1 & 0\\ 1 & -3\end{bmatrix}+\begin{bmatrix}3 & -2\\ 1 & 5\\ -4 & 3\end{bmatrix}$
$=\begin{bmatrix}2+3 & 6+(-2)\\ 1+1 & 0+5\\ 1+(-4) & -3+3\end{bmatrix}=\begin{bmatrix}5 & 4\\ 2 & 5\\ -3 & 0\end{bmatrix}$

2. $-2B=\begin{bmatrix}-8 & -2\\ -6 & -4\end{bmatrix}$

11.6 Solving Systems with Gaussian Elimination

1. $\left[\begin{array}{cc|c}4 & -3 & 11\\ 3 & 2 & 4\end{array}\right]$

2. $\begin{array}{r}x-y+z=5\\ 2x-y+3z=1\\ y+z=-9\end{array}$

3. $(2,\ 1)$

4. $\left[\begin{array}{ccc|c}1 & -\frac{5}{2} & \frac{5}{2} & \frac{17}{2}\\ 0 & 1 & 5 & 9\\ 0 & 0 & 1 & 2\end{array}\right]$

5. $(1,\ 1,\ 1)$

6. $150,000 at 7%, $750,000 at 8%, $600,000 at 10%

11.7 Solving Systems with Inverses

1. $AB=\begin{bmatrix}1 & 4\\ -1 & -3\end{bmatrix}\begin{bmatrix}-3 & -4\\ 1 & 1\end{bmatrix}=\begin{bmatrix}1(-3)+4(1) & 1(-4)+4(1)\\ -1(-3)+-3(1) & -1(-4)+-3(1)\end{bmatrix}=\begin{bmatrix}1 & 0\\ 0 & 1\end{bmatrix}$
$BA=\begin{bmatrix}-3 & -4\\ 1 & 1\end{bmatrix}\begin{bmatrix}1 & 4\\ -1 & -3\end{bmatrix}=\begin{bmatrix}-3(1)+-4(-1) & -3(4)+-4(-3)\\ 1(1)+1(-1) & 1(4)+1(-3)\end{bmatrix}=\begin{bmatrix}1 & 0\\ 0 & 1\end{bmatrix}$

2. $A^{-1} = \begin{bmatrix} \frac{3}{5} & \frac{1}{5} \\ -\frac{2}{5} & \frac{1}{5} \end{bmatrix}$

3. $A^{-1} = \begin{bmatrix} 1 & 1 & 2 \\ 2 & 4 & -3 \\ 3 & 6 & -5 \end{bmatrix}$

4. $X = \begin{bmatrix} 4 \\ 38 \\ 58 \end{bmatrix}$

11.8 Solving Systems with Cramer's Rule

1. $(3, -7)$

2. -10

3. $\left(-2, \frac{3}{5}, \frac{12}{5}\right)$

11.1 Section Exercises

1. No, you can either have zero, one, or infinitely many. Examine graphs.

3. This means there is no realistic break-even point. By the time the company produces one unit they are already making profit.

5. You can solve by substitution (isolating x or y), graphically, or by addition.

7. Yes

9. Yes

11. $(-1, 2)$

13. $(-3, 1)$

15. $\left(-\frac{3}{5}, 0\right)$

17. No solutions exist.

19. $\left(\frac{72}{5}, \frac{132}{5}\right)$

21. $(6, -6)$

23. $\left(-\frac{1}{2}, \frac{1}{10}\right)$

25. No solutions exist.

27. $\left(-\frac{1}{5}, \frac{2}{3}\right)$

29. $\left(x, \frac{x+3}{2}\right)$

31. $(-4, 4)$

33. $\left(\frac{1}{2}, \frac{1}{8}\right)$

35. $\left(\frac{1}{6}, 0\right)$

37. $(x, 2(7x-6))$

39. $\left(-\frac{5}{6}, \frac{4}{3}\right)$

41. Consistent with one solution

43. Consistent with one solution

45. Dependent with infinitely many solutions

47. $(-3.08, 4.91)$

49. $(-1.52, 2.29)$

51. $\left(\frac{A+B}{2}, \frac{A-B}{2}\right)$

53. $\left(\frac{-1}{A-B}, \frac{A}{A-B}\right)$

55. $\left(\frac{CE-BF}{BD-AE}, \frac{AF-CD}{BD-AE}\right)$

57. They never turn a profit.

59. $(1, 250, 100, 000)$

61. The numbers are 7.5 and 20.5.

63. 24,000

65. 790 second-year students, 805 first-year students

67. 56 men, 74 women

69. 10 gallons of 10% solution, 15 gallons of 60% solution

71. Swan Peak: $750,000, Riverside: $350,000

73. $12,500 in the first account, $10,500 in the second account.

75. High-tops: 45, Low-tops: 15

77. Infinitely many solutions. We need more information.

11.2 Section Exercises

1. No, there can be only one, zero, or infinitely many solutions.

3. Not necessarily. There could be zero, one, or infinitely many solutions. For example, $(0, 0, 0)$ is not a solution to the system below, but that does not mean that it has no solution.

$$\begin{aligned} 2x + 3y - 6z &= 1 \\ -4x - 6y + 12z &= -2 \\ x + 2y + 5z &= 10 \end{aligned}$$

5. Every system of equations can be solved graphically, by substitution, and by addition. However, systems of three equations become very complex to solve graphically so other methods are usually preferable.

7. No

9. Yes

11. $(-1, 4, 2)$

13. $\left(-\frac{85}{107}, \frac{312}{107}, \frac{191}{107}\right)$

15. $\left(1, \frac{1}{2}, 0\right)$

17. $(4, -6, 1)$

19. $\left(x, \frac{1}{27}(65-16x), \frac{x+28}{27}\right)$

21. $\left(-\frac{45}{13}, \frac{17}{13}, -2\right)$

23. No solutions exist

25. $(0, 0, 0)$

27. $\left(\frac{4}{7}, -\frac{1}{7}, -\frac{3}{7}\right)$

29. $(7, 20, 16)$

31. $(-6, 2, 1)$

33. $(5, 12, 15)$

35. $(-5, -5, -5)$

37. $(10, 10, 10)$

39. $\left(\frac{1}{2}, \frac{1}{5}, \frac{4}{5}\right)$

41. $\left(\frac{1}{2}, \frac{2}{5}, \frac{4}{5}\right)$

43. $(2, 0, 0)$

45. $(1, 1, 1)$

47. $\left(\frac{128}{557}, \frac{23}{557}, \frac{28}{557}\right)$

49. $(6, -1, 0)$

51. 24, 36, 48

53. 70 grandparents, 140 parents, 190 children

55. Your share was \$19.95, Shani's share was \$40, and your other roommate's share was \$22.05.

57. There are infinitely many solutions; we need more information

59. 500 students, 225 children, and 450 adults

61. The BMW was \$49,636, the Jeep was \$42,636, and the Toyota was \$47,727.

63. \$400,000 in the account that pays 3% interest, \$500,000 in the account that pays 4% interest, and \$100,000 in the account that pays 2% interest.

65. The United States consumed 26.3%, Japan 7.1%, and China 6.4% of the world's oil.

67. Saudi Arabia imported 16.8%, Canada imported 15.1%, and Mexico 15.0%

69. Birds were 19.3%, fish were 18.6%, and mammals were 17.1% of endangered species

11.3 Section Exercises

1. A nonlinear system could be representative of two circles that overlap and intersect in two locations, hence two solutions. A nonlinear system could be representative of a parabola and a circle, where the vertex of the parabola meets the circle and the branches also intersect the circle, hence three solutions.

3. No. There does not need to be a feasible region. Consider a system that is bounded by two parallel lines. One inequality represents the region above the upper line; the other represents the region below the lower line. In this case, no points in the plane are located in both regions; hence there is no feasible region.

5. Choose any number between each solution and plug into $C(x)$ and $R(x)$. If $C(x) < R(x)$, then there is profit.

7. $(0,-3),(3,0)$

9. $\left(-\frac{3\sqrt{2}}{2},\frac{3\sqrt{2}}{2}\right),\left(\frac{3\sqrt{2}}{2},-\frac{3\sqrt{2}}{2}\right)$

11. $(-3,0),(3,0)$

13. $\left(\frac{1}{4},-\frac{\sqrt{62}}{8}\right),\left(\frac{1}{4},\frac{\sqrt{62}}{8}\right)$

15. $\left(-\frac{\sqrt{398}}{4},\frac{199}{4}\right),\left(\frac{\sqrt{398}}{4},\frac{199}{4}\right)$

17. $(0,2),(1,3)$

19. $\left(-\sqrt{\frac{1}{2}\left(\sqrt{5}-1\right)},\frac{1}{2}\left(1-\sqrt{5}\right)\right),\left(\sqrt{\frac{1}{2}\left(\sqrt{5}-1\right)},\frac{1}{2}\left(1-\sqrt{5}\right)\right)$

21. $(5,0)$

23. $(0,0)$

25. $(3,0)$

27. No Solutions Exist

29. No Solutions Exist

31. $\left(-\frac{\sqrt{2}}{2},-\frac{\sqrt{2}}{2}\right),\left(-\frac{\sqrt{2}}{2},\frac{\sqrt{2}}{2}\right),\left(\frac{\sqrt{2}}{2},-\frac{\sqrt{2}}{2}\right),\left(\frac{\sqrt{2}}{2},\frac{\sqrt{2}}{2}\right)$

33. $(2,0)$

35. $\left(-\sqrt{7},-3\right),\left(-\sqrt{7},3\right),\left(\sqrt{7},-3\right),\left(\sqrt{7},3\right)$

37. $\left(-\sqrt{\frac{1}{2}\left(\sqrt{73}-5\right)},\frac{1}{2}\left(7-\sqrt{73}\right)\right),\left(\sqrt{\frac{1}{2}\left(\sqrt{73}-5\right)},\frac{1}{2}\left(7-\sqrt{73}\right)\right)$

39.

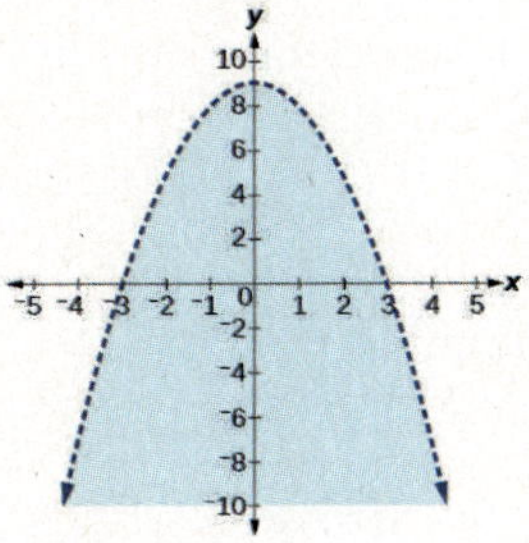

41.

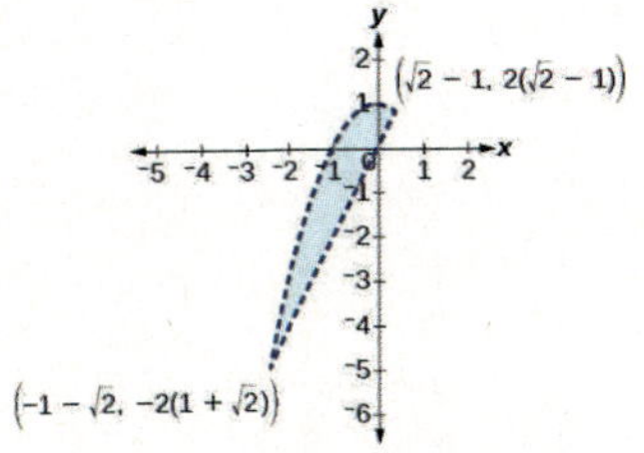

43.

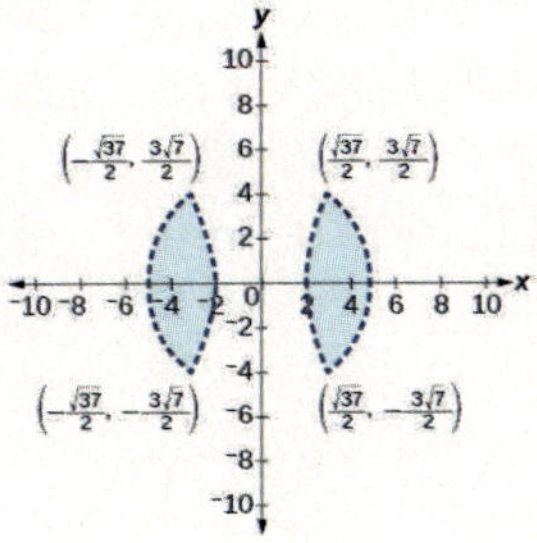

45.

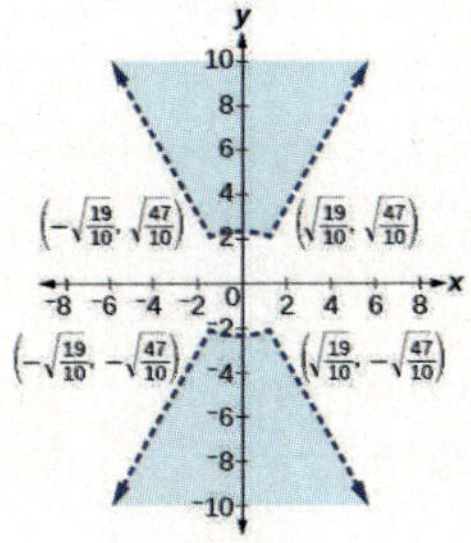

47.

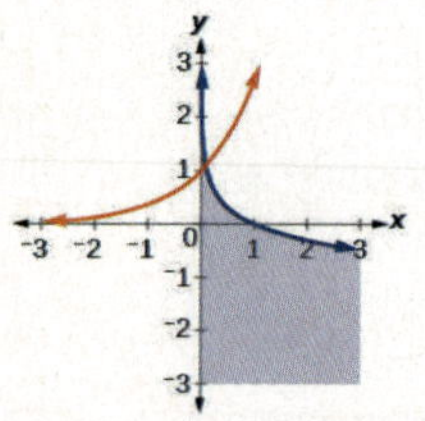

49. $\left(-2\sqrt{\frac{70}{383}},-2\sqrt{\frac{35}{29}}\right),\left(-2\sqrt{\frac{70}{383}},2\sqrt{\frac{35}{29}}\right),\left(2\sqrt{\frac{70}{383}},-2\sqrt{\frac{35}{29}}\right),\left(2\sqrt{\frac{70}{383}},2\sqrt{\frac{35}{29}}\right)$

51. No Solution Exists

53. $x = 0, y > 0$ and $0 < x < 1, \sqrt{x} < y < \frac{1}{x}$

55. 12, 288

57. 2–20 computers

11.4 Section Exercises

1. No, a quotient of polynomials can only be decomposed if the denominator can be factored. For example, $\frac{1}{x^2+1}$ cannot be decomposed because the denominator cannot be factored.

3. Graph both sides and ensure they are equal.

5. If we choose $x = -1$, then the *B*-term disappears, letting us immediately know that $A = 3$. We could alternatively plug in $x = -\frac{5}{3}$, giving us a *B*-value of -2.

7. $\frac{8}{x+3} - \frac{5}{x-8}$

9. $\frac{1}{x+5} + \frac{9}{x+2}$

11. $\frac{3}{5x-2} + \frac{4}{4x-1}$

13. $\frac{5}{2(x+3)} + \frac{5}{2(x-3)}$

15. $\frac{3}{x+2} + \frac{3}{x-2}$

17. $\frac{9}{5(x+2)} + \frac{11}{5(x-3)}$

19. $\frac{8}{x-3} - \frac{5}{x-2}$

21. $\frac{1}{x-2} + \frac{2}{(x-2)^2}$

23. $-\frac{6}{4x+5} + \frac{3}{(4x+5)^2}$

25. $-\frac{1}{x-7} - \frac{2}{(x-7)^2}$

27. $\frac{4}{x} - \frac{3}{2(x+1)} + \frac{7}{2(x+1)^2}$

29. $\frac{4}{x} + \frac{2}{x^2} - \frac{3}{3x+2} + \frac{7}{2(3x+2)^2}$

31. $\frac{x+1}{x^2+x+3} + \frac{3}{x+2}$

33. $\frac{4-3x}{x^2+3x+8} + \frac{1}{x-1}$

35. $\frac{2x-1}{x^2+6x+1} + \frac{2}{x+3}$

37. $\frac{1}{x^2+x+1} + \frac{4}{x-1}$

39. $\frac{2}{x^2-3x+9} + \frac{3}{x+3}$

41. $-\frac{1}{4x^2+6x+9} + \frac{1}{2x-3}$

43. $\frac{1}{x} + \frac{1}{x+6} - \frac{4x}{x^2-6x+36}$

45. $\frac{x+6}{x^2+1} + \frac{4x+3}{\left(x^2+1\right)^2}$

47. $\frac{x+1}{x+2} + \frac{2x+3}{(x+2)^2}$

49. $\frac{1}{x^2+3x+25} - \frac{3x}{\left(x^2+3x+25\right)^2}$

51. $\frac{1}{8x} - \frac{x}{8\left(x^2+4\right)} + \frac{10-x}{2\left(x^2+4\right)^2}$

53. $-\frac{16}{x} - \frac{9}{x^2} + \frac{16}{x-1} - \frac{7}{(x-1)^2}$

55. $\frac{1}{x+1} - \frac{2}{(x+1)^2} + \frac{5}{(x+1)^3}$

57. $\frac{5}{x-2} - \frac{3}{10(x+2)} + \frac{7}{x+8} - \frac{7}{10(x-8)}$

59. $-\frac{5}{4x} - \frac{5}{2(x+2)} + \frac{11}{2(x+4)} + \frac{5}{4(x+4)}$

11.5 Section Exercises

1. No, they must have the same dimensions. An example would include two matrices of different dimensions. One cannot add the following two matrices because the first is a 2×2 matrix and the second is a 2×3 matrix. $\begin{bmatrix} 1 & 2 \\ 3 & 4 \end{bmatrix} + \begin{bmatrix} 6 & 5 & 4 \\ 3 & 2 & 1 \end{bmatrix}$ has no sum.

3. Yes, if the dimensions of A are $m \times n$ and the dimensions of B are $n \times m$, both products will be defined.

5. Not necessarily. To find AB, we multiply the first row of A by the first column of B to get the first entry of AB. To find BA, we multiply the first row of B by the first column of A to get the first entry of BA. Thus, if those are unequal, then the matrix multiplication does not commute.

7. $\begin{bmatrix} 11 & 19 \\ 15 & 94 \\ 17 & 67 \end{bmatrix}$

9. $\begin{bmatrix} -4 & 2 \\ 8 & 1 \end{bmatrix}$

11. Undidentified; dimensions do not match

13. $\begin{bmatrix} 9 & 27 \\ 63 & 36 \\ 0 & 192 \end{bmatrix}$

15. $\begin{bmatrix} -64 & -12 & -28 & -72 \\ -360 & -20 & -12 & -116 \end{bmatrix}$

17. $\begin{bmatrix} 1,800 & 1,200 & 1,300 \\ 800 & 1,400 & 600 \\ 700 & 400 & 2,100 \end{bmatrix}$

19. $\begin{bmatrix} 20 & 102 \\ 28 & 28 \end{bmatrix}$

21. $\begin{bmatrix} 60 & 41 & 2 \\ -16 & 120 & -216 \end{bmatrix}$

23. $\begin{bmatrix} -68 & 24 & 136 \\ -54 & -12 & 64 \\ -57 & 30 & 128 \end{bmatrix}$

25. Undefined; dimensions do not match.

27. $\begin{bmatrix} -8 & 41 & -3 \\ 40 & -15 & -14 \\ 4 & 27 & 42 \end{bmatrix}$

29. $\begin{bmatrix} -840 & 650 & -530 \\ 330 & 360 & 250 \\ -10 & 900 & 110 \end{bmatrix}$

31. $\begin{bmatrix} -350 & 1,050 \\ 350 & 350 \end{bmatrix}$

33. Undefined; inner dimensions do not match.

35. $\begin{bmatrix} 1,400 & 700 \\ -1,400 & 700 \end{bmatrix}$

37. $\begin{bmatrix} 332,500 & 927,500 \\ -227,500 & 87,500 \end{bmatrix}$

39. $\begin{bmatrix} 490,000 & 0 \\ 0 & 490,000 \end{bmatrix}$

41. $\begin{bmatrix} -2 & 3 & 4 \\ -7 & 9 & -7 \end{bmatrix}$

43. $\begin{bmatrix} -4 & 29 & 21 \\ -27 & -3 & 1 \end{bmatrix}$

45. $\begin{bmatrix} -3 & -2 & -2 \\ -28 & 59 & 46 \\ -4 & 16 & 7 \end{bmatrix}$

47. $\begin{bmatrix} 1 & -18 & -9 \\ -198 & 505 & 369 \\ -72 & 126 & 91 \end{bmatrix}$

49. $\begin{bmatrix} 0 & 1.6 \\ 9 & -1 \end{bmatrix}$

51. $\begin{bmatrix} 2 & 24 & -4.5 \\ 12 & 32 & -9 \\ -8 & 64 & 61 \end{bmatrix}$

53. $\begin{bmatrix} 0.5 & 3 & 0.5 \\ 2 & 1 & 2 \\ 10 & 7 & 10 \end{bmatrix}$

55. $\begin{bmatrix} 1 & 0 & 0 \\ 0 & 1 & 0 \\ 0 & 0 & 1 \end{bmatrix}$

57. $\begin{bmatrix} 1 & 0 & 0 \\ 0 & 1 & 0 \\ 0 & 0 & 1 \end{bmatrix}$

59. $B^n = \begin{cases} \begin{bmatrix} 1 & 0 & 0 \\ 0 & 1 & 0 \\ 0 & 0 & 1 \end{bmatrix}, & n \text{ even}, \\ \begin{bmatrix} 1 & 0 & 0 \\ 0 & 0 & 1 \\ 0 & 1 & 0 \end{bmatrix}, & n \text{ odd}. \end{cases}$

11.6 Section Exercises

1. Yes. For each row, the coefficients of the variables are written across the corresponding row, and a vertical bar is placed; then the constants are placed to the right of the vertical bar.

3. No, there are numerous correct methods of using row operations on a matrix. Two possible ways are the following: (1) Interchange rows 1 and 2. Then $R_2 = R_2 - 9R_1$. (2) $R_2 = R_1 - 9R_2$. Then divide row 1 by 9.

5. No. A matrix with 0 entries for an entire row would have either zero or infinitely many solutions.

7. $\left[\begin{array}{cc|c} 0 & 16 & 4 \\ 9 & -1 & 2 \end{array}\right]$

9. $\left[\begin{array}{ccc|c} 1 & 5 & 8 & 16 \\ 12 & 3 & 0 & 4 \\ 3 & 4 & 9 & -7 \end{array}\right]$

11. $\begin{array}{c} -2x + 5y = 5 \\ 6x - 18y = 26 \end{array}$

13. $\begin{array}{l} 3x + 2y = 3 \\ -x - 9y + 4z = -1 \\ 8x + 5y + 7z = 8 \end{array}$

15. $\begin{array}{r} 4x + 5y - 2z = 12 \\ y + 58z = 2 \\ 8x + 7y - 3z = -5 \end{array}$

17. No solutions

19. $(-1, -2)$

21. $(6, 7)$

23. $(3, 2)$

25. $\left(\frac{1}{5}, \frac{1}{2}\right)$

27. $\left(x, \frac{4}{15}(5x + 1)\right)$

29. $(3, 4)$

31. $\left(\frac{196}{39}, -\frac{5}{13}\right)$

33. $(31, -42, 87)$

35. $\left(\frac{21}{40}, \frac{1}{20}, \frac{9}{8}\right)$

37. $\left(\frac{18}{13}, \frac{15}{13}, -\frac{15}{13}\right)$

39. $\left(x, y, \frac{1}{2}(1 - 2x - 3y)\right)$

41. $\left(x, -\frac{x}{2}, -1\right)$

43. $(125, -25, 0)$

45. $(8, 1, -2)$

47. $(1, 2, 3)$

49. $\left(x, \frac{31}{28} - \frac{3x}{4}, \frac{1}{28}(-7x - 3)\right)$

51. No solutions exist.

53. 860 red velvet, 1,340 chocolate

55. 4% for account 1, 6% for account 2

57. $126

59. Banana was 3%, pumpkin was 7%, and rocky road was 2%

61. 100 almonds, 200 cashews, 600 pistachios

11.7 Section Exercises

1. If A^{-1} is the inverse of A, then $AA^{-1} = I$, the identity matrix. Since A is also the inverse of A^{-1}, $A^{-1}A = I$. You can also check by proving this for a 2×2 matrix.

3. No, because ad and bc are both 0, so $ad - bc = 0$, which requires us to divide by 0 in the formula.

5. Yes. Consider the matrix $\begin{bmatrix} 0 & 1 \\ 1 & 0 \end{bmatrix}$. The inverse is found with the following calculation:
$A^{-1} = \frac{1}{0(0)-1(1)} \begin{bmatrix} 0 & -1 \\ -1 & 0 \end{bmatrix} = \begin{bmatrix} 0 & 1 \\ 1 & 0 \end{bmatrix}$.

7. $AB = BA = \begin{bmatrix} 1 & 0 \\ 0 & 1 \end{bmatrix} = I$

9. $AB = BA = \begin{bmatrix} 1 & 0 \\ 0 & 1 \end{bmatrix} = I$

11. $AB = BA = \begin{bmatrix} 1 & 0 & 0 \\ 0 & 1 & 0 \\ 0 & 0 & 1 \end{bmatrix} = I$

13. $\frac{1}{29} \begin{bmatrix} 9 & 2 \\ -1 & 3 \end{bmatrix}$

15. $\frac{1}{69} \begin{bmatrix} -2 & 7 \\ 9 & 3 \end{bmatrix}$

17. There is no inverse

19. $\frac{4}{7} \begin{bmatrix} 0.5 & 1.5 \\ 1 & -0.5 \end{bmatrix}$

21. $\frac{1}{17} \begin{bmatrix} -5 & 5 & -3 \\ 20 & -3 & 12 \\ 1 & -1 & 4 \end{bmatrix}$

23. $\frac{1}{209} \begin{bmatrix} 47 & -57 & 69 \\ 10 & 19 & -12 \\ -24 & 38 & -13 \end{bmatrix}$

25. $\begin{bmatrix} 18 & 60 & -168 \\ -56 & -140 & 448 \\ 40 & 80 & -280 \end{bmatrix}$

27. $(-5, 6)$

29. $(2, 0)$

31. $\left(\frac{1}{3}, -\frac{5}{2}\right)$

33. $\left(-\frac{2}{3}, -\frac{11}{6}\right)$

35. $\left(7, \frac{1}{2}, \frac{1}{5}\right)$

37. $(5, 0, -1)$

39. $\frac{1}{34}(-35, -97, -154)$

41. $\frac{1}{690}(65, -1136, -229)$

43. $\left(-\frac{37}{30}, \frac{8}{15}\right)$

45. $\left(\frac{10}{123}, -1, \frac{2}{5}\right)$

47. $\frac{1}{2} \begin{bmatrix} 2 & 1 & -1 & -1 \\ 0 & 1 & 1 & -1 \\ 0 & -1 & 1 & 1 \\ 0 & 1 & -1 & 1 \end{bmatrix}$

49. $\frac{1}{39} \begin{bmatrix} 3 & 2 & 1 & -7 \\ 18 & -53 & 32 & 10 \\ 24 & -36 & 21 & 9 \\ -9 & 46 & -16 & -5 \end{bmatrix}$

51. $\begin{bmatrix} 1 & 0 & 0 & 0 & 0 & 0 \\ 0 & 1 & 0 & 0 & 0 & 0 \\ 0 & 0 & 1 & 0 & 0 & 0 \\ 0 & 0 & 0 & 1 & 0 & 0 \\ 0 & 0 & 0 & 0 & 1 & 0 \\ -1 & -1 & -1 & -1 & -1 & 1 \end{bmatrix}$

53. Infinite solutions.

55. 50% oranges, 25% bananas, 20% apples

57. 10 straw hats, 50 beanies, 40 cowboy hats

59. Micah ate 6, Joe ate 3, and Albert ate 3.

61. 124 oranges, 10 lemons, 8 pomegranates

11.8 Section Exercises

1. A determinant is the sum and products of the entries in the matrix, so you can always evaluate that product—even if it does end up being 0.

3. The inverse does not exist.

5. -2

7. 7

9. -4

11. 0

13. $-7,990.7$

15. 3

17. -1

19. 224

21. 15

23. -17.03

25. $(1,1)$

27. $\left(\frac{1}{2},\frac{1}{3}\right)$

29. $(2,5)$

31. $\left(-1,-\frac{1}{3}\right)$

33. $(15,12)$

35. $(1,3,2)$

37. $(-1,0,3)$

39. $\left(\frac{1}{2},1,2\right)$

41. $(2,1,4)$

43. Infinite solutions

45. 24

47. 1

49. Yes; 18, 38

51. Yes; 33, 36, 37

53. $7,000 in first account, $3,000 in second account.

55. 120 children, 1,080 adult

57. 4 gal yellow, 6 gal blue

59. 13 green tomatoes, 17 red tomatoes

61. Strawberries 18%, oranges 9%, kiwi 10%

63. 100 for movie 1, 230 for movie 2, 312 for movie 3

65. 300 almonds, 400 cranberries, 300 cashews

Review Exercises

1. No

3. $(-2,3)$

5. $(4,-1)$

7. No solutions exist.

9. $(300,60,000)$

11. Infinite solutions

13. No solutions exist.

15. $(-1,-2,3)$

17. $\left(x,\frac{8x}{5},\frac{14x}{5}\right)$

19. 11, 17, 33

21. $(2,-3),(3,2)$

23. No solution

25. No solution

27.

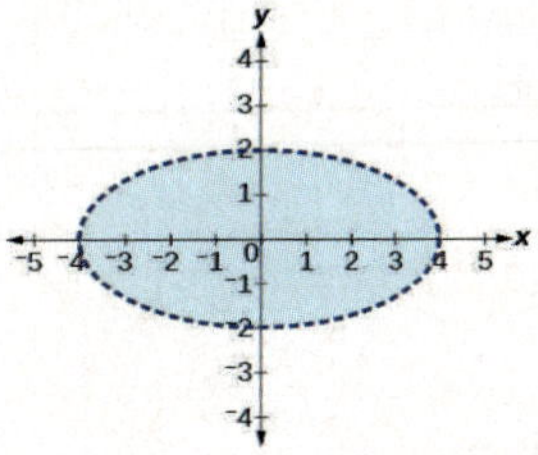

29.

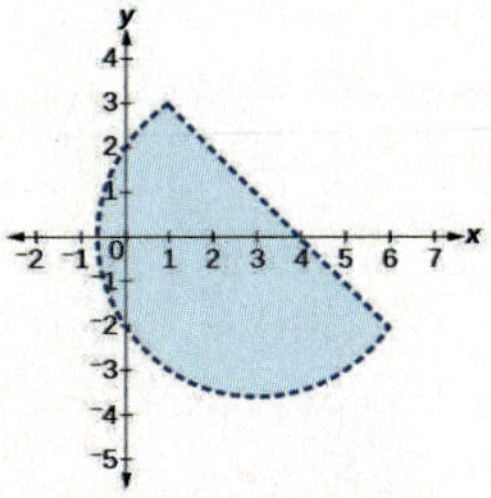

31. $\frac{2}{x+2}, \frac{-4}{x+1}$

33. $\frac{7}{x+5}, \frac{-15}{(x+5)^2}$

35. $\frac{3}{x-5}, \frac{-4x+1}{x^2+5x+25}$

37. $\frac{x-4}{(x^2-2)}, \frac{5x+3}{(x^2-2)^2}$

39. $\begin{bmatrix} -16 & 8 \\ -4 & -12 \end{bmatrix}$

41. undefined; dimensions do not match

43. undefined; inner dimensions do not match

45. $\begin{bmatrix} 113 & 28 & 10 \\ 44 & 81 & -41 \\ 84 & 98 & -42 \end{bmatrix}$

47. $\begin{bmatrix} -127 & -74 & 176 \\ -2 & 11 & 40 \\ 28 & 77 & 38 \end{bmatrix}$

49. undefined; inner dimensions do not match

51. $\begin{aligned} x - 3z &= 7 \\ y + 2z &= -5 \end{aligned}$ with infinite solutions

53. $\left[\begin{array}{ccc|c} -2 & 2 & 1 & 7 \\ 2 & -8 & 5 & 0 \\ 19 & -10 & 22 & 3 \end{array}\right]$

55. $\left[\begin{array}{ccc|c} 1 & 0 & 3 & 12 \\ -1 & 4 & 0 & 0 \\ 0 & 1 & 2 & -7 \end{array}\right]$

57. No solutions exist.

59. No solutions exist.

61. $\frac{1}{8}\begin{bmatrix} 2 & 7 \\ 6 & 1 \end{bmatrix}$

63. No inverse exists.

65. $(-20, 40)$

67. $(-1, 0.2, 0.3)$

69. 17% oranges, 34% bananas, 39% apples

71. 0

73. 6

75. $\left(6, \frac{1}{2}\right)$

77. $(x, 5x + 3)$

79. $\left(0, 0, -\frac{1}{2}\right)$

Practice Test

1. Yes

3. No solutions exist.

5. $\frac{1}{20}(10, 5, 4)$

7. $\left(x, \frac{16x}{5} - \frac{13x}{5}\right)$

9. $(-2\sqrt{2}, -\sqrt{17}), \left(-2\sqrt{2}, \sqrt{17}\right), \left(2\sqrt{2}, -\sqrt{17}\right), \left(2\sqrt{2}, \sqrt{17}\right)$

11.

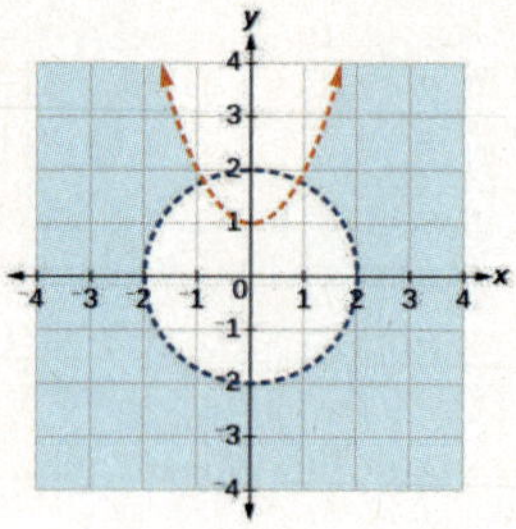

13. $\frac{5}{3x+1} - \frac{2x+3}{(3x+1)^2}$

15. $\begin{bmatrix} 17 & 51 \\ -8 & 11 \end{bmatrix}$

17. $\begin{bmatrix} 12 & -20 \\ -15 & 30 \end{bmatrix}$

19. $-\frac{1}{8}$

21. $\left[\begin{array}{ccc|c} 14 & -2 & 13 & 140 \\ -2 & 3 & -6 & -1 \\ 1 & -5 & 12 & 11 \end{array}\right]$

23. No solutions exist.

25. $(100, 90)$

27. $\left(\frac{1}{100}, 0\right)$

29. 32 or more cell phones per day

Chapter 12

Try It

12.1 The Ellipse

1. $x^2 + \frac{y^2}{16} = 1$

2. $\frac{(x-1)^2}{16} + \frac{(y-3)^2}{4} = 1$

3. center: $(0, 0)$; vertices: $(\pm 6, 0)$; co-vertices: $(0, \pm 2)$; foci: $\left(\pm 4\sqrt{2}, 0\right)$

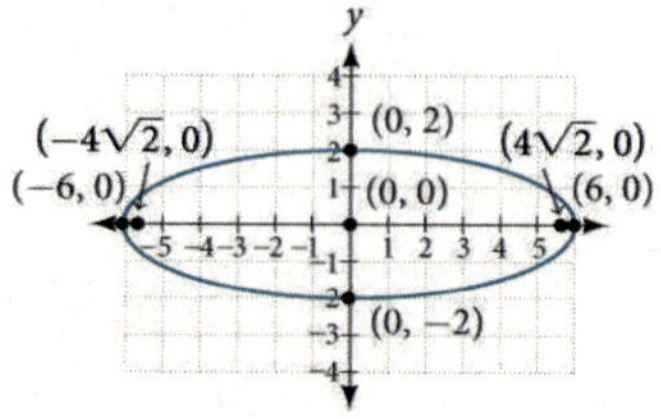

4. Standard form: $\frac{x^2}{16} + \frac{y^2}{49} = 1$; center: $(0, 0)$; vertices: $(0, \pm 7)$; co-vertices: $(\pm 4, 0)$; foci: $\left(0, \pm\sqrt{33}\right)$

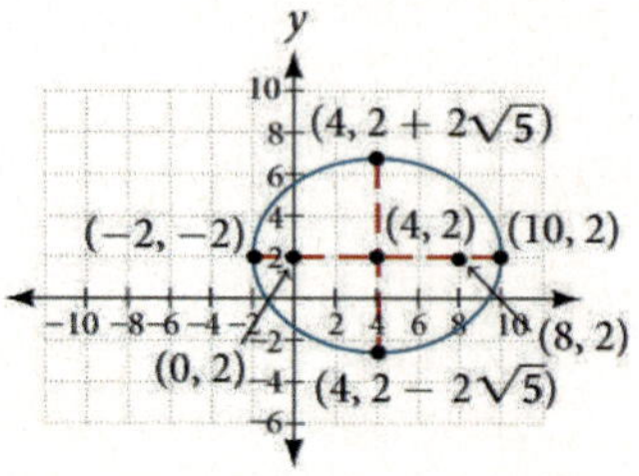

5. Center: $(4, 2)$; vertices: $(-2, 2)$ and $(10, 2)$; co-vertices: $\left(4, 2 - 2\sqrt{5}\right)$ and $\left(4, 2 + 2\sqrt{5}\right)$; foci: $(0, 2)$ and $(8, 2)$

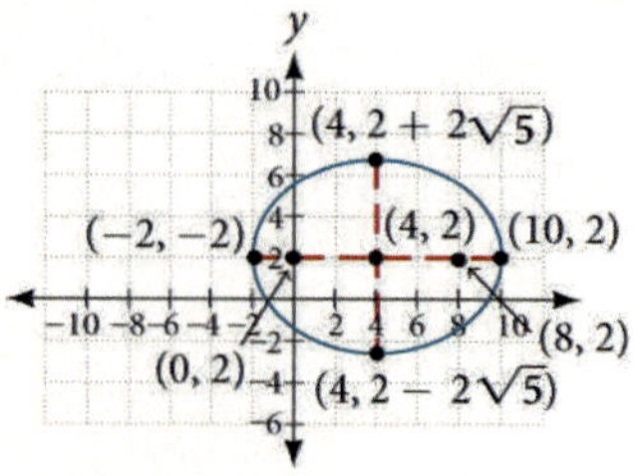

6. $\frac{(x-3)^2}{4} + \frac{(y+1)^2}{16} = 1$; center: $(3, -1)$; vertices: $(3, -5)$ and $(3, 3)$; co-vertices: $(1, -1)$ and $(5, -1)$; foci: $\left(3, -1 - 2\sqrt{3}\right)$ and $\left(3, -1+2\sqrt{3}\right)$

7. ⓐ $\frac{x^2}{57{,}600} + \frac{y^2}{25{,}600} = 1$
ⓑ The people are standing 358 feet apart.

12.2 The Hyperbola

1. Vertices: $(\pm 3, 0)$; Foci: $\left(\pm\sqrt{34}, 0\right)$

2. $\frac{y^2}{4} - \frac{x^2}{16} = 1$

3. $\frac{(y-3)^2}{25} + \frac{(x-1)^2}{144} = 1$

4. vertices: $(\pm 12, 0)$; co-vertices: $(0, \pm 9)$; foci: $(\pm 15, 0)$; asymptotes: $y = \pm\frac{3}{4}x$;

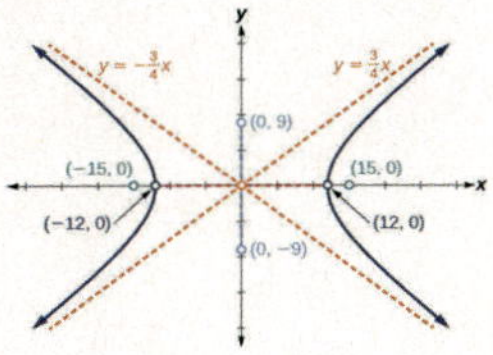

5. center: $(3, -4)$; vertices: $(3, -14)$ and $(3, 6)$; co-vertices: $(-5, -4)$; and $(11, -4)$; foci: $\left(3, -4 - 2\sqrt{41}\right)$ and $\left(3, -4 + 2\sqrt{41}\right)$; asymptotes: $y = \pm\frac{5}{4}(x - 3) - 4$

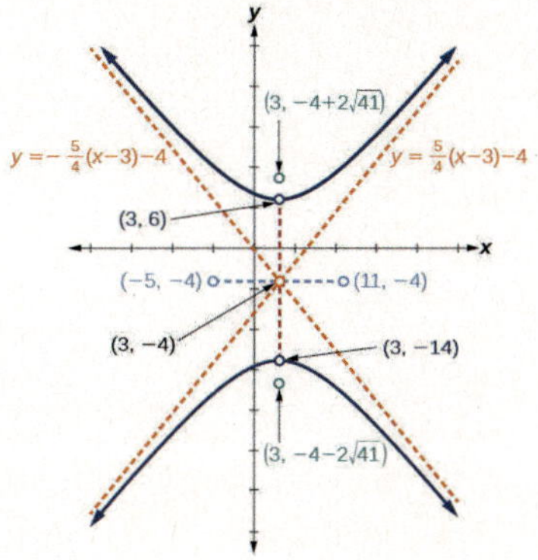

6. The sides of the tower can be modeled by the hyperbolic equation.
$\frac{x^2}{400} - \frac{y^2}{3600} = 1 \text{or} \frac{x^2}{20^2} - \frac{y^2}{60^2} = 1.$

12.3 The Parabola

1. Focus: $(-4, 0)$; Directrix: $x = 4$; Endpoints of the latus rectum: $(-4, \pm 8)$

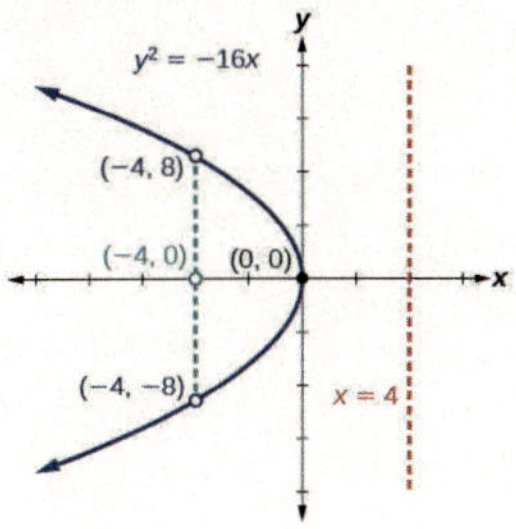

2. Focus: $(0, 2)$; Directrix: $y = -2$; Endpoints of the latus rectum: $(\pm 4, 2)$.

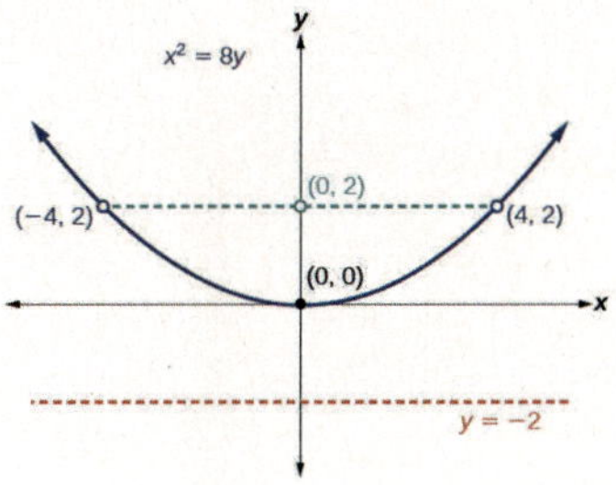

3. $x^2 = 14y$.

4. Vertex: $(8, -1)$; Axis of symmetry: $y = -1$; Focus: $(9, -1)$; Directrix: $x = 7$; Endpoints of the latus rectum: $(9, -3)$ and $(9, 1)$.

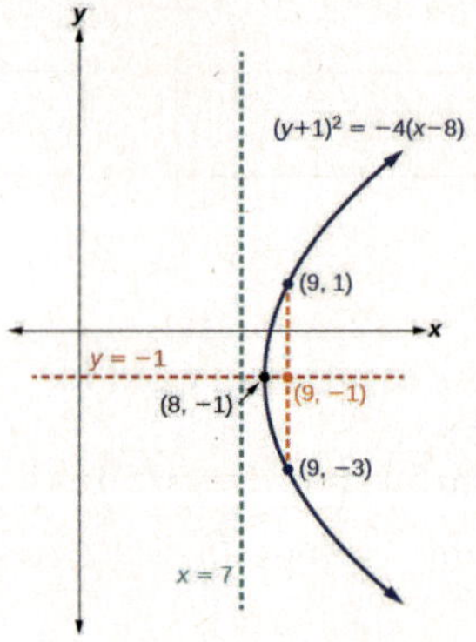

5. Vertex: $(-2, 3)$; Axis of symmetry: $x = -2$; Focus: $(-2, -2)$; Directrix: $y = 8$; Endpoints of the latus rectum: $(-12, -2)$ and $(8, -2)$.

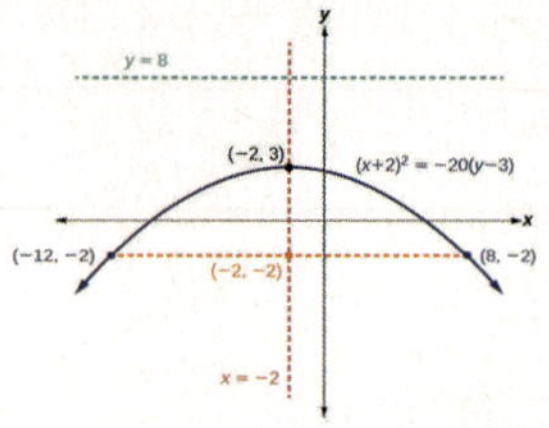

6. ⓐ $y^2 = 1280x$ ⓑ The depth of the cooker is 500 mm

12.4 Rotation of Axes

1. ⓐ hyperbola ⓑ ellipse

2. $\frac{x'^2}{4} + \frac{y'^2}{1} = 1$

3. ⓐ hyperbola ⓑ ellipse

12.5 Conic Sections in Polar Coordinates

1. ellipse; $e = \frac{1}{3}$; $x = -2$

2.

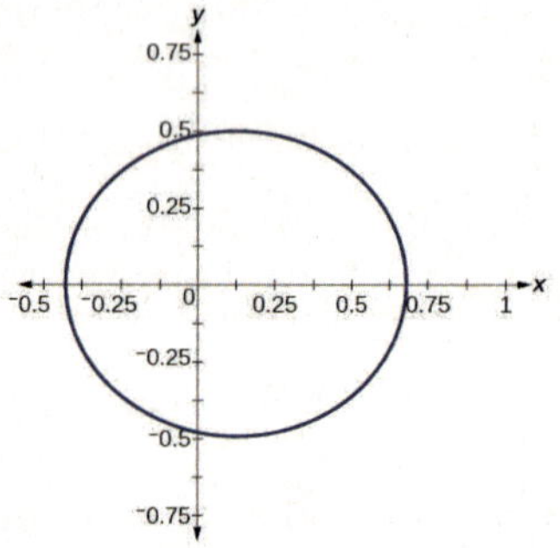

3. $r = \frac{1}{1-\cos\theta}$

4. $4 - 8x + 3x^2 - y^2 = 0$

12.1 Section Exercises

1. An ellipse is the set of all points in the plane the sum of whose distances from two fixed points, called the foci, is a constant.

3. This special case would be a circle.

5. It is symmetric about the x-axis, y-axis, and the origin.

7. yes; $\frac{x^2}{3^2}+\frac{y^2}{2^2}=1$

9. yes; $\frac{x^2}{\left(\frac{1}{2}\right)^2}+\frac{y^2}{\left(\frac{1}{3}\right)^2}=1$

11. $\frac{x^2}{2^2}+\frac{y^2}{7^2}=1$; Endpoints of major axis $(0,7)$ and $(0,-7)$. Endpoints of minor axis $(2,0)$ and $(-2,0)$. Foci at $\left(0,3\sqrt{5}\right),\left(0,-3\sqrt{5}\right)$.

13. $\frac{x^2}{(1)^2}+\frac{y^2}{\left(\frac{1}{3}\right)^2}=1$; Endpoints of major axis $(1,0)$ and $(-1,0)$. Endpoints of minor axis $\left(0,\frac{1}{3}\right),\left(0,-\frac{1}{3}\right)$. Foci at $\left(\frac{2\sqrt{2}}{3},0\right),\left(-\frac{2\sqrt{2}}{3},0\right)$.

15. $\frac{(x-2)^2}{7^2}+\frac{(y-4)^2}{5^2}=1$; Endpoints of major axis $(9,4),(-5,4)$. Endpoints of minor axis $(2,9),(2,-1)$. Foci at $\left(2+2\sqrt{6},4\right),\left(2-2\sqrt{6},4\right)$.

17. $\frac{(x+5)^2}{2^2}+\frac{(y-7)^2}{3^2}=1$; Endpoints of major axis $(-5,10),(-5,4)$. Endpoints of minor axis $(-3,7),(-7,7)$. Foci at $\left(-5,7+\sqrt{5}\right),\left(-5,7-\sqrt{5}\right)$.

19. $\frac{(x-1)^2}{3^2}+\frac{(y-4)^2}{2^2}=1$; Endpoints of major axis $(4,4),(-2,4)$. Endpoints of minor axis $(1,6),(1,2)$. Foci at $\left(1+\sqrt{5},4\right),\left(1-\sqrt{5},4\right)$.

21. $\frac{(x-3)^2}{\left(3\sqrt{2}\right)^2}+\frac{(y-5)^2}{\left(\sqrt{2}\right)^2}=1$; Endpoints of major axis $\left(3+3\sqrt{2},5\right),\left(3-3\sqrt{2},5\right)$. Endpoints of minor axis $\left(3,5+\sqrt{2}\right),\left(3,5-\sqrt{2}\right)$. Foci at $(7,5),(-1,5)$.

23. $\frac{(x+5)^2}{(5)^2}+\frac{(y-2)^2}{(2)^2}=1$; Endpoints of major axis $(0,2),(-10,2)$. Endpoints of minor axis $(-5,4),(-5,0)$. Foci at $\left(-5+\sqrt{21},2\right),\left(-5-\sqrt{21},2\right)$.

25. $\frac{(x+3)^2}{(5)^2}+\frac{(y+4)^2}{(2)^2}=1$; Endpoints of major axis $(2,-4),(-8,-4)$. Endpoints of minor axis $(-3,-2),(-3,-6)$. Foci at $\left(-3+\sqrt{21},-4\right),\left(-3-\sqrt{21},-4\right)$.

27. Foci $\left(-3,-1+\sqrt{11}\right),\left(-3,-1-\sqrt{11}\right)$

29. Focus $(0,0)$

31. Foci $(-10,30),(-10,-30)$

33. Center $(0,0)$, Vertices $(4,0),(-4,0),(0,3),(0,-3)$, Foci $\left(\sqrt{7},0\right),\left(-\sqrt{7},0\right)$

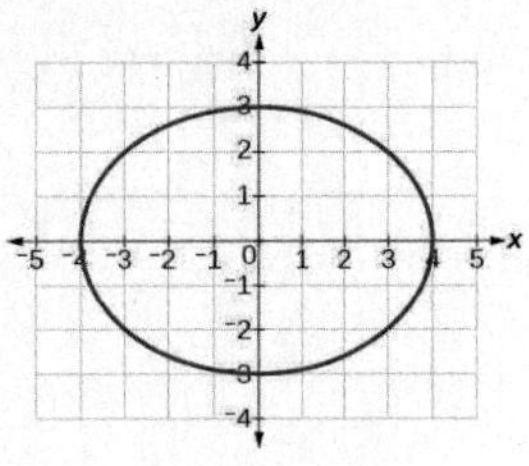

35. Center $(0,0)$, Vertices $\left(\frac{1}{9},0\right),\left(-\frac{1}{9},0\right),\left(0,\frac{1}{7}\right),\left(0,-\frac{1}{7}\right)$, Foci $\left(0,\frac{4\sqrt{2}}{63}\right),\left(0,-\frac{4\sqrt{2}}{63}\right)$

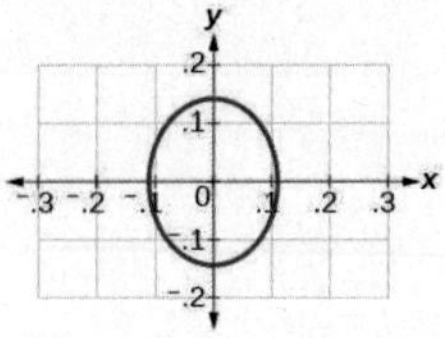

37. Center $(-3,3)$, Vertices $(0,3),(-6,3),(-3,0),(-3,6)$, Focus $(-3,3)$

Note that this ellipse is a circle. The circle has only one focus, which coincides with the center.

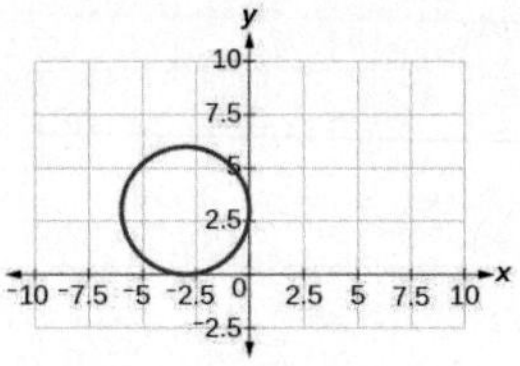

39. Center $(1,1)$, Vertices $(5,1),(-3,1),(1,3),(1,-1)$, Foci $\left(1,1+2\sqrt{3}\right),\left(1,1-2\sqrt{3}\right)$

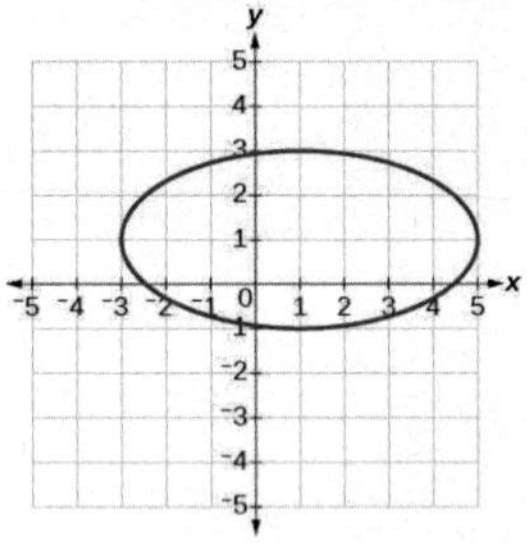

41. Center $(-4,5)$, Vertices $(-2,5),(-6,4),(-4,6),(-4,4)$, Foci $\left(-4+\sqrt{3},5\right),\left(-4-\sqrt{3},5\right)$

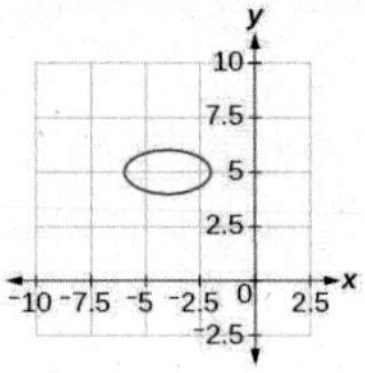

43. Center $(-2,1)$, Vertices $(0,1),(-4,1),(-2,5),(-2,-3)$, Foci $\left(-2,1+2\sqrt{3}\right),\left(-2,1-2\sqrt{3}\right)$

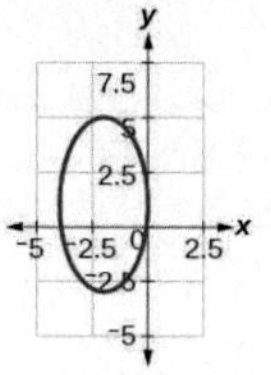

45. Center $(-2,-2)$, Vertices $(0,-2),(-4,-2),(-2,0),(-2,-4)$, Focus $(-2,-2)$

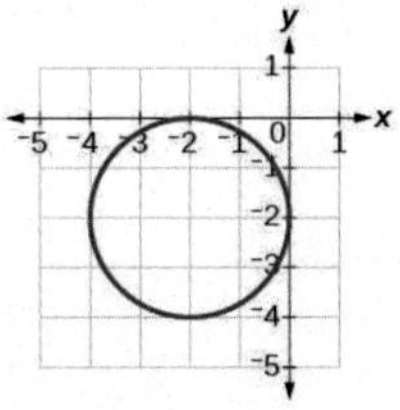

47. $\frac{x^2}{25}+\frac{y^2}{29}=1$

49. $\frac{(x-4)^2}{25}+\frac{(y-2)^2}{1}=1$

51. $\frac{(x+3)^2}{16}+\frac{(y-4)^2}{4}=1$

53. $\frac{x^2}{81}+\frac{y^2}{9}=1$

55. $\frac{(x+2)^2}{4}+\frac{(y-2)^2}{9}=1$

57. $\text{Area}=12\pi$ square units

59. $\text{Area}=2\sqrt{5}\pi$ square units.

61. $\text{Area}=9\pi$ square units.

63. $\frac{x^2}{4h^2}+\frac{y^2}{\frac{1}{4}h^2}=1$

65. $\frac{x^2}{400}+\frac{y^2}{144}=1$. Distance = 17.32 feet

67. Approximately 51.96 feet

12.2 Section Exercises

1. A hyperbola is the set of points in a plane the difference of whose distances from two fixed points (foci) is a positive constant.

3. The foci must lie on the transverse axis and be in the interior of the hyperbola.

5. The center must be the midpoint of the line segment joining the foci.

7. yes $\frac{x^2}{6^2} - \frac{y^2}{3^2} = 1$

9. yes $\frac{x^2}{4^2} - \frac{y^2}{5^2} = 1$

11. $\frac{x^2}{5^2} - \frac{y^2}{6^2} = 1$; vertices: $(5, 0), (-5, 0)$; foci: $\left(\sqrt{61}, 0\right), \left(-\sqrt{61}, 0\right)$; asymptotes: $y = \frac{6}{5}x, y = -\frac{6}{5}x$

13. $\frac{y^2}{2^2} - \frac{x^2}{9^2} = 1$; vertices: $(0, 2), (0, -2)$; foci: $\left(0, \sqrt{85}\right), \left(0, -\sqrt{85}\right)$; asymptotes: $y = \frac{2}{9}x, y = -\frac{2}{9}x$

15. $\frac{(x-1)^2}{3^2} - \frac{(y-2)^2}{4^2} = 1$; vertices: $(4, 2), (-2, 2)$; foci: $(6, 2), (-4, 2)$; asymptotes: $y = \frac{4}{3}(x - 1) + 2, y = -\frac{4}{3}(x - 1) + 2$

17. $\frac{(x-2)^2}{7^2} - \frac{(y+7)^2}{7^2} = 1$; vertices: $(9, -7), (-5, -7)$; foci: $\left(2 + 7\sqrt{2}, -7\right), \left(2 - 7\sqrt{2}, -7\right)$; asymptotes: $y = x - 9, y = -x - 5$

19. $\frac{(x+3)^2}{3^2} - \frac{(y-3)^2}{3^2} = 1$; vertices: $(0, 3), (-6, 3)$; foci: $\left(-3 + 3\sqrt{2}, 1\right), \left(-3 - 3\sqrt{2}, 1\right)$; asymptotes: $y = x + 6, y = -x$

21. $\frac{(y-4)^2}{2^2} - \frac{(x-3)^2}{4^2} = 1$; vertices: $(3, 6), (3, 2)$; foci: $\left(3, 4 + 2\sqrt{5}\right), \left(3, 4 - 2\sqrt{5}\right)$; asymptotes: $y = \frac{1}{2}(x - 3) + 4, y = -\frac{1}{2}(x - 3) + 4$

23. $\frac{(y+5)^2}{7^2} - \frac{(x+1)^2}{70^2} = 1$; vertices: $(-1, 2), (-1, -12)$; foci: $\left(-1, -5 + 7\sqrt{101}\right), \left(-1, -5 - 7\sqrt{101}\right)$; asymptotes: $y = \frac{1}{10}(x + 1) - 5, y = -\frac{1}{10}(x + 1) - 5$

25. $\frac{(x+3)^2}{5^2} - \frac{(y-4)^2}{2^2} = 1$; vertices: $(2, 4), (-8, 4)$; foci: $\left(-3 + \sqrt{29}, 4\right), \left(-3 - \sqrt{29}, 4\right)$; asymptotes: $y = \frac{2}{5}(x + 3) + 4, y = -\frac{2}{5}(x + 3) + 4$

27. $y = \frac{2}{5}(x - 3) - 4, y = -\frac{2}{5}(x - 3) - 4$

29. $y=\frac{3}{4}(x-1)+1,\ y=-\frac{3}{4}(x-1)+1$

31.

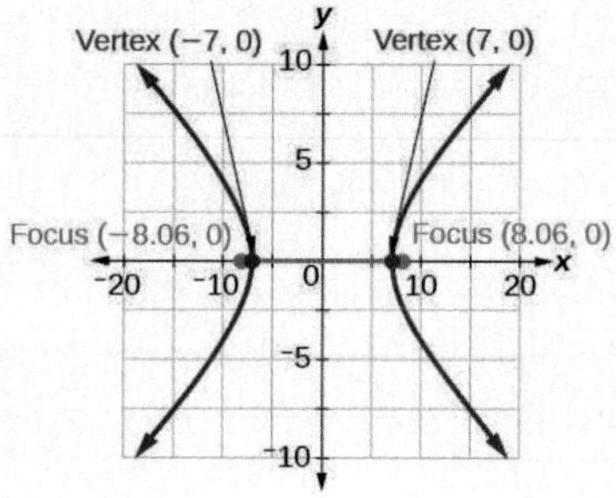

33.

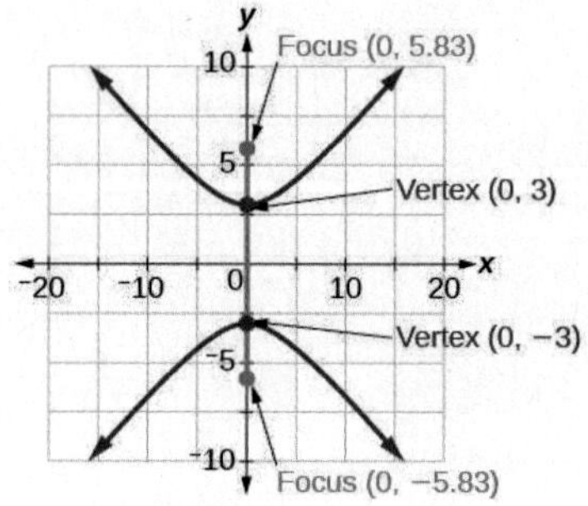

35.

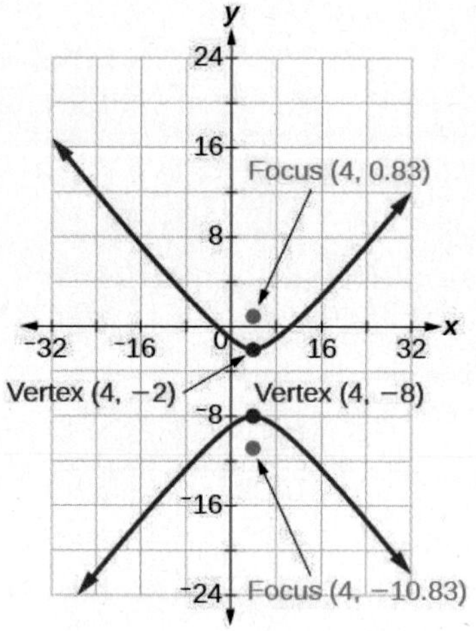

37.

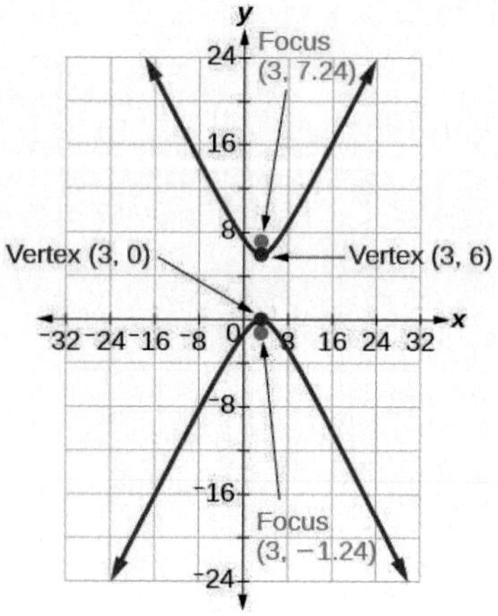

39.

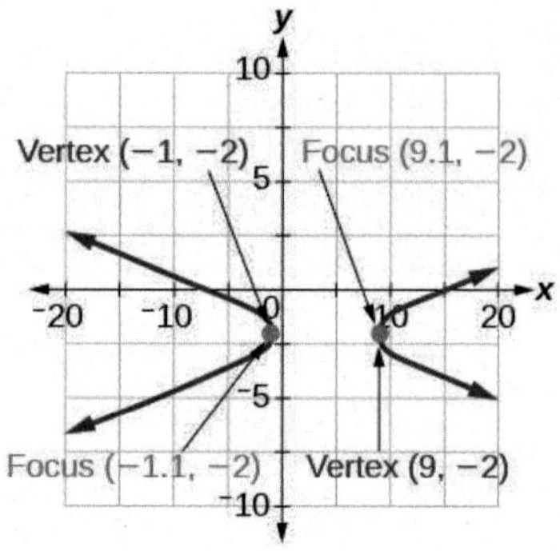

41.

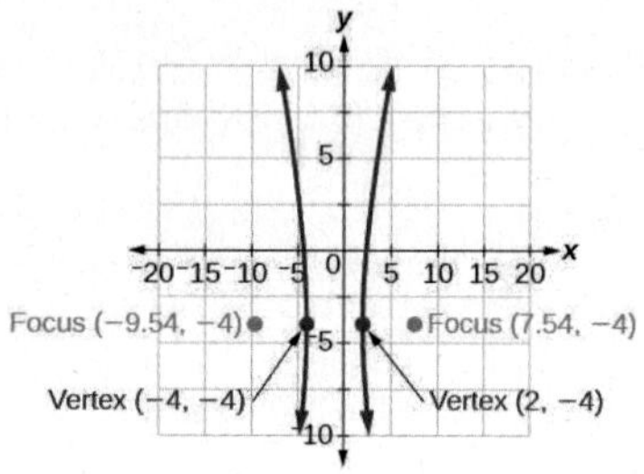

43.

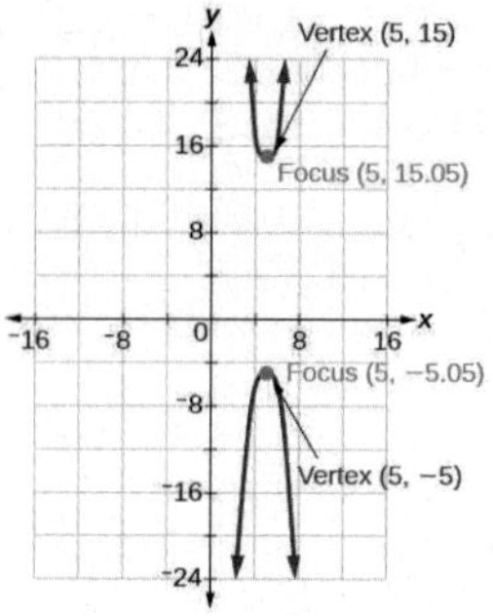

45. $\frac{x^2}{9}-\frac{y^2}{16}=1$

47. $\frac{(x-6)^2}{25}-\frac{(y-1)^2}{11}=1$

49. $\frac{(x-4)^2}{25}-\frac{(y-2)^2}{1}=1$

51. $\frac{y^2}{16}-\frac{x^2}{25}=1$

53. $\frac{y^2}{9}-\frac{(x+1)^2}{9}=1$

55. $\frac{(x+3)^2}{25}-\frac{(y+3)^2}{25}=1$

57. $y(x) = 3\sqrt{x^2+1}, y(x) = -3\sqrt{x^2+1}$

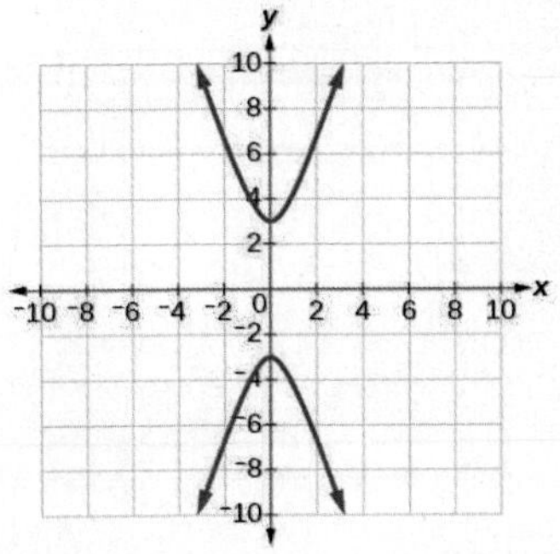

59. $y(x) = 1 + 2\sqrt{x^2+4x+5}, y(x) = 1 - 2\sqrt{x^2+4x+5}$

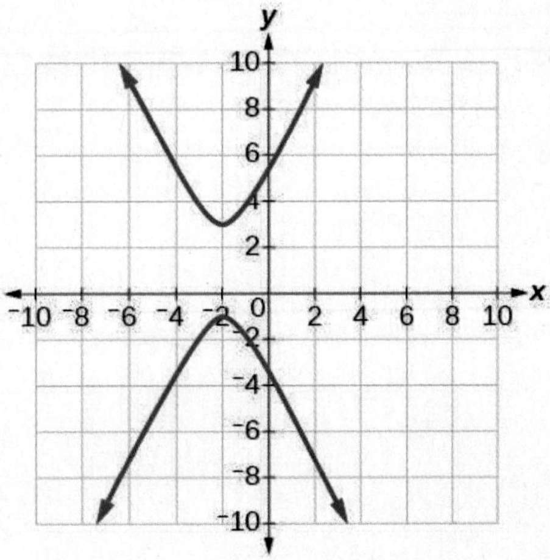

61. $\frac{x^2}{25} - \frac{y^2}{25} = 1$

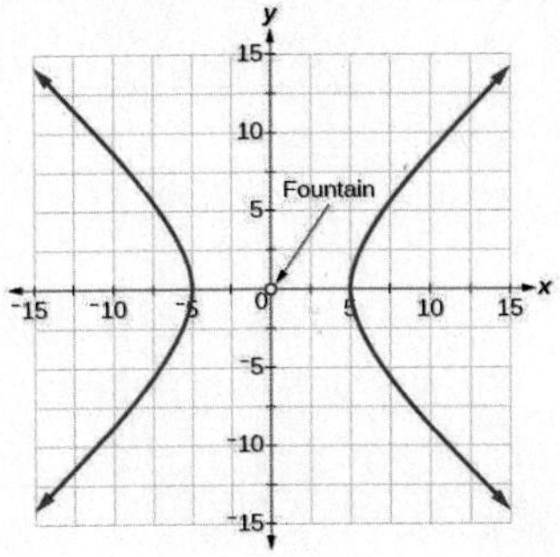

63. $\frac{x^2}{100} - \frac{y^2}{25} = 1$

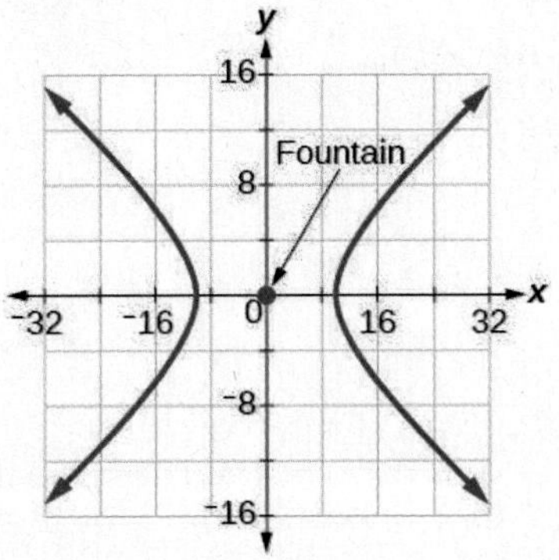

65. $\frac{x^2}{400} - \frac{y^2}{225} = 1$

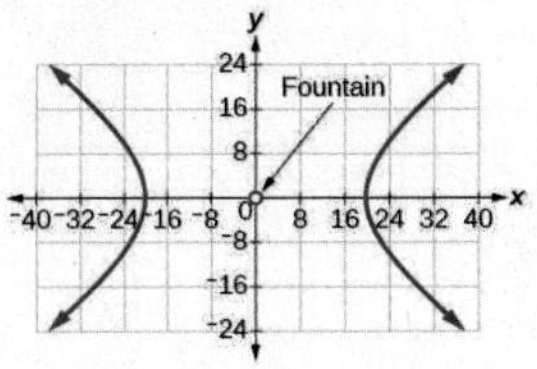

67. $4(x-1)^2 - y2^2 = 16$

69. $\frac{(x-h)^2}{a^2} - \frac{(y-k)^2}{b^2} = \left(x-3\right)^2 - 9y^2 = 4$

12.3 Section Exercises

1. A parabola is the set of points in the plane that lie equidistant from a fixed point, the focus, and a fixed line, the directrix.

3. The graph will open down.

5. The distance between the focus and directrix will increase.

7. yes $x^2 = 4\left(\frac{1}{16}\right)y$

9. yes $(y-3)^2 = 4(2)(x-2)$

11. $y^2 = \frac{1}{8}x, V: (0,0); F: \left(\frac{1}{32}, 0\right); d: x = -\frac{1}{32}$

13. $x^2 = -\frac{1}{4}y, V: (0,0); F: \left(0, -\frac{1}{16}\right); d: y = \frac{1}{16}$

15. $y^2 = \frac{1}{36}x, V: (0,0); F: \left(\frac{1}{144}, 0\right); d: x = -\frac{1}{144}$

17. $(x-1)^2 = 4(y-1), V: (1,1); F: (1,2); d: y = 0$

19. $(y-4)^2 = 2(x+3), V: (-3,4); F: \left(-\frac{5}{2}, 4\right); d: x = -\frac{7}{2}$

21. $(x+4)^2 = 24(y+1), V: (-4,-1); F: (-4,5); d: y = -7$

23. $(y-3)^2 = -12(x+1), V: (-1,3); F: (-4,3); d: x = 2$

25. $(x-5)^2 = \frac{4}{5}(y+3), V: (5,-3); F: \left(5, -\frac{14}{5}\right); d: y = -\frac{16}{5}$

27. $(x-2)^2 = -2(y-5), V: (2,5); F: \left(2, \frac{9}{2}\right); d: y = \frac{11}{2}$

29. $(y-1)^2 = \frac{4}{3}(x-5), V: (5,1); F: \left(\frac{16}{3}, 1\right); d: x = \frac{14}{3}$

31.

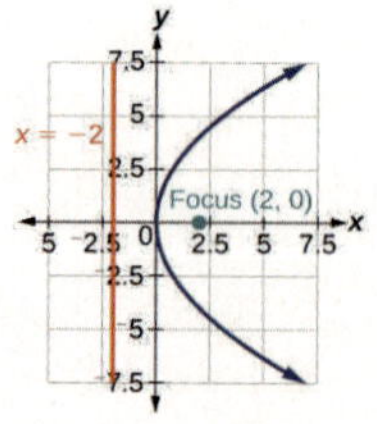

33.

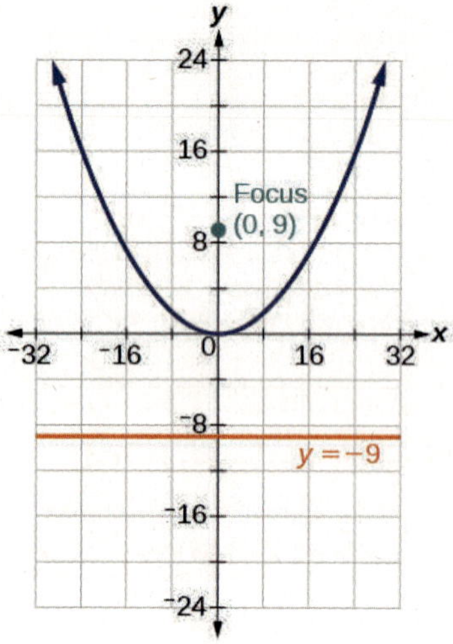

35.

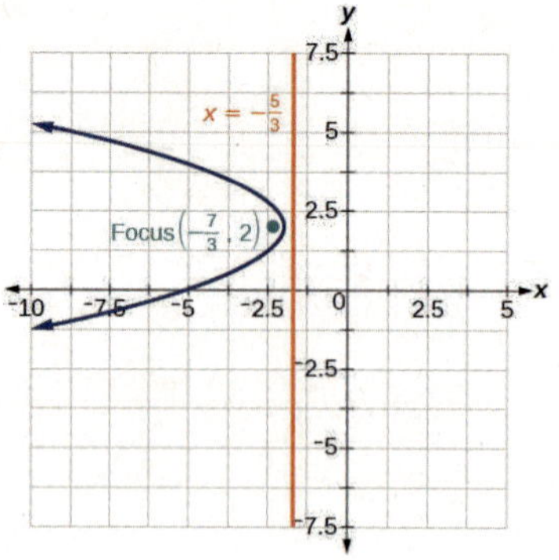

37.

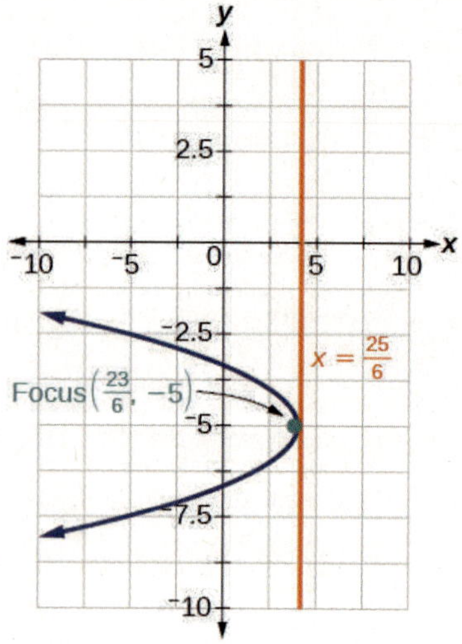

39.

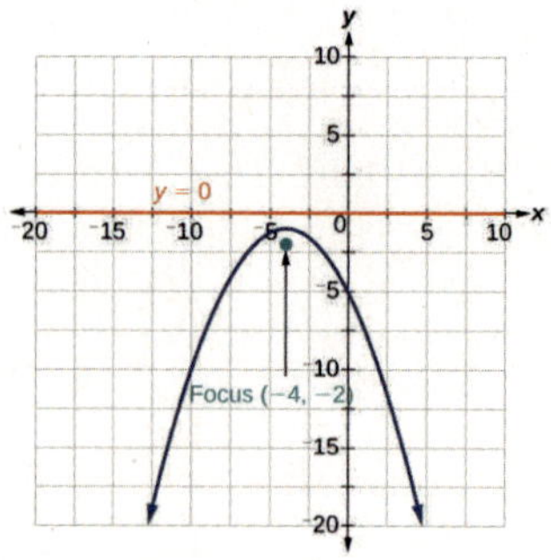

41.

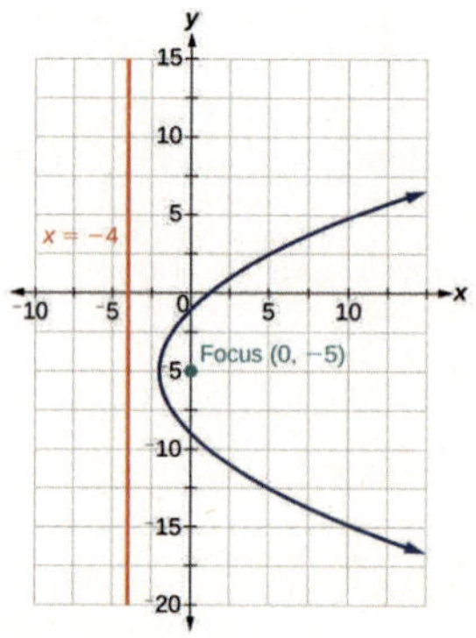

43.

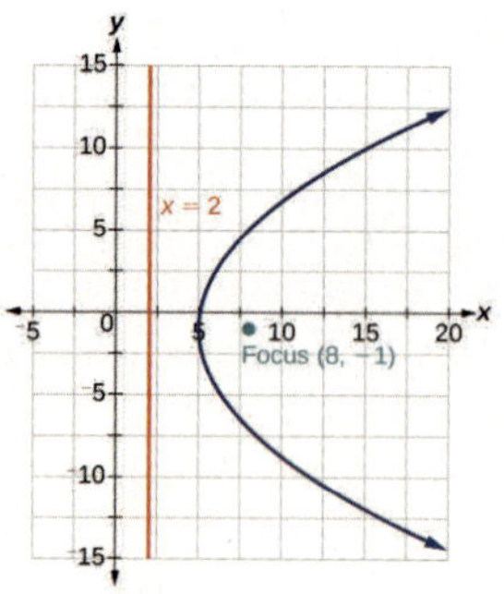

45. $x^2 = -16y$

47. $(y-2)^2 = 4\sqrt{2}(x-2)$

49. $\left(y+\sqrt{3}\right)^2 = -4\sqrt{2}\left(x-\sqrt{2}\right)$

51. $x^2 = y$

53. $(y-2)^2 = \frac{1}{4}(x+2)$

55. $\left(y-\sqrt{3}\right)^2 = 4\sqrt{5}\left(x+\sqrt{2}\right)$

57. $y^2 = -8x$

59. $(y+1)^2 = 12(x+3)$

61. $(0, 1)$

63. At the point 2.25 feet above the vertex.

65. 0.5625 feet

67. $x^2 = -125(y-20)$, height is 7.2 feet

69. 2304 feet

12.4 Section Exercises

1. The xy term causes a rotation of the graph to occur.

3. The conic section is a hyperbola.

5. It gives the angle of rotation of the axes in order to eliminate the xy term.

7. $AB = 0$, parabola

9. $AB = -4 < 0$, hyperbola

11. $AB = 6 > 0$, ellipse

13. $B^2 - 4AC = 0$, parabola

15. $B^2 - 4AC = 0$, parabola

17. $B^2 - 4AC = -96 < 0$, ellipse

19. $7x'^2 + 9y'^2 - 4 = 0$

21. $3x'^2 + 2x'y' - 5y'^2 + 1 = 0$

23. $\theta = 60°, 11x'^2 - y'^2 + \sqrt{3}x' + y' - 4 = 0$

25. $\theta = 150°, 21x'^2 + 9y'^2 + 4x' - 4\sqrt{3}y' - 6 = 0$

27. $\theta \approx 36.9°, 125x'^2 + 6x' - 42y' + 10 = 0$

29. $\theta = 45°, 3x'^2 - y'^2 - \sqrt{2}x' + \sqrt{2}y' + 1 = 0$

31. $\frac{\sqrt{2}}{2}(x' + y') = \frac{1}{2}(x' - y')^2$

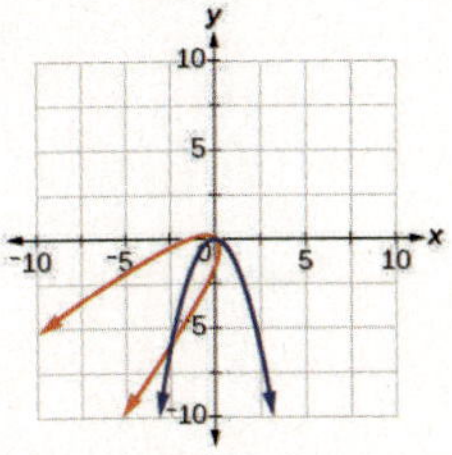

33. $\frac{(x'-y')^2}{8} + \frac{(x'+y')^2}{2} = 1$

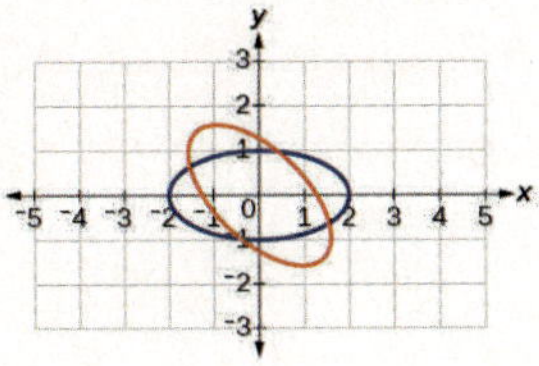

35. $\frac{(x'+y')^2}{2} - \frac{(x'-y')^2}{2} = 1$

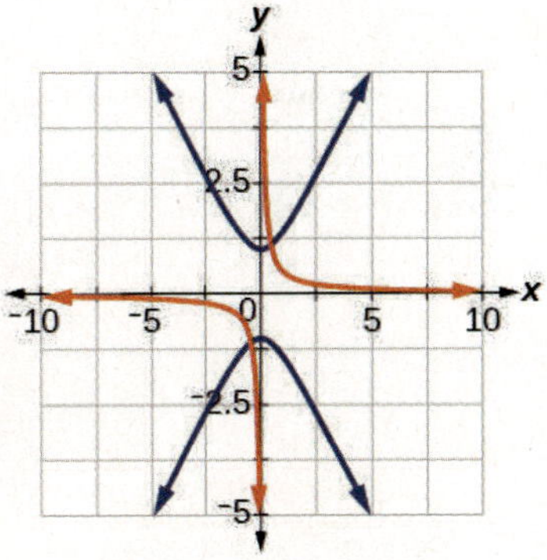

37. $\frac{\sqrt{3}}{2}x' - \frac{1}{2}y' = \left(\frac{1}{2}x' + \frac{\sqrt{3}}{2}y' - 1\right)^2$

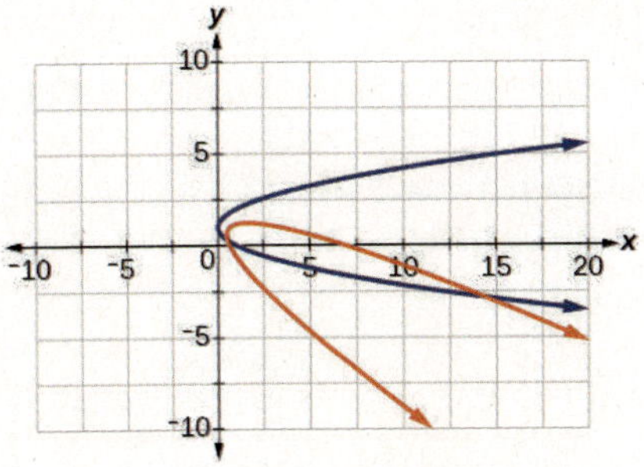

39.

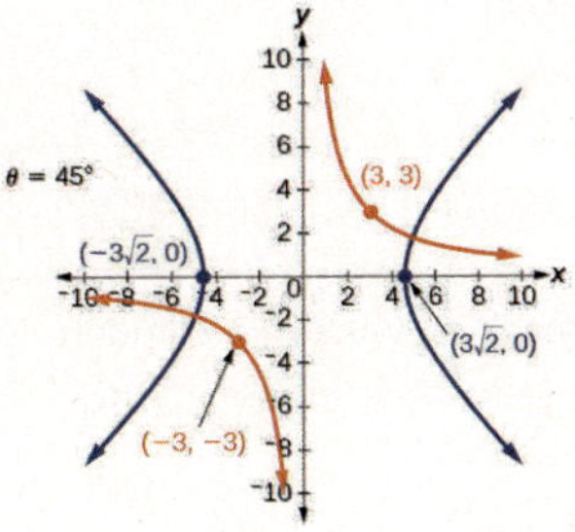

41.

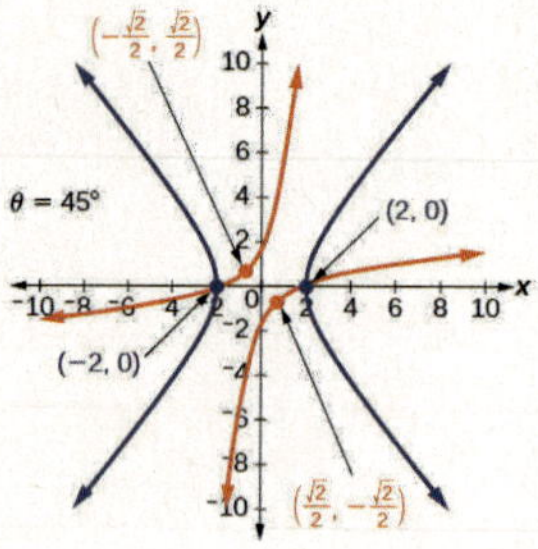

43.

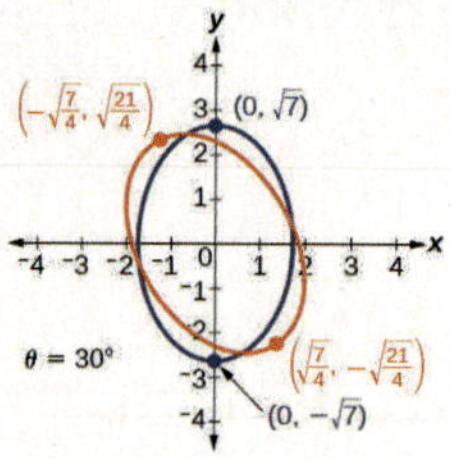

45.

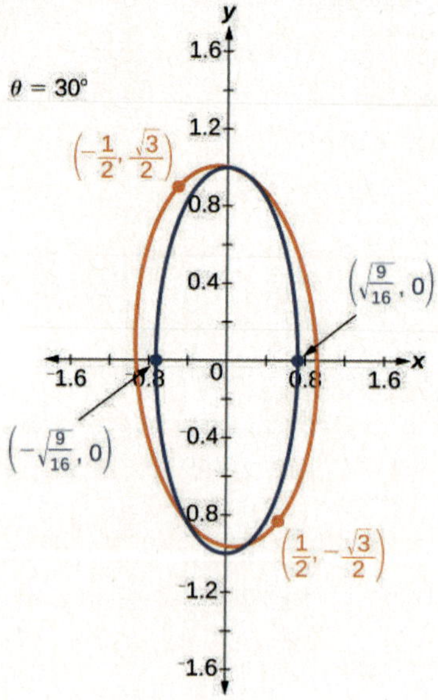

47.

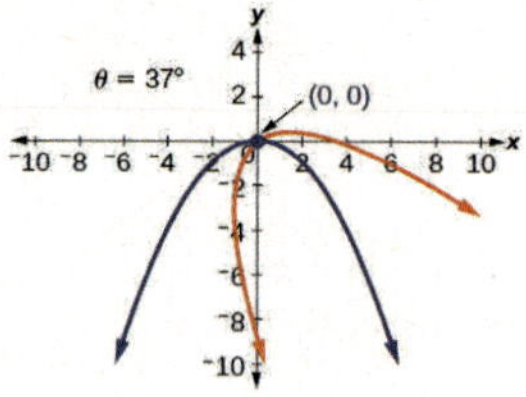

49.

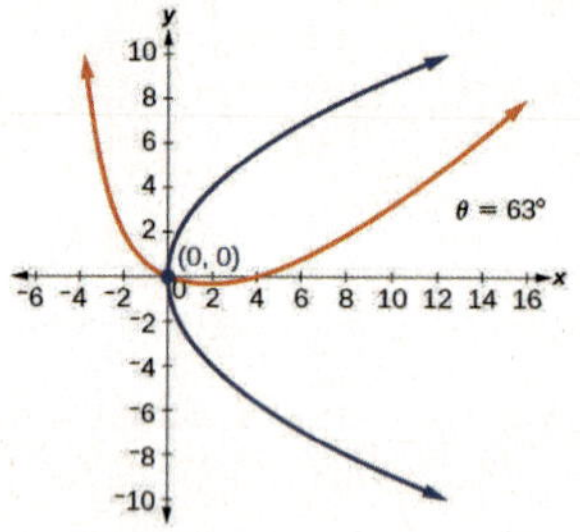

51. $\theta = 45°$

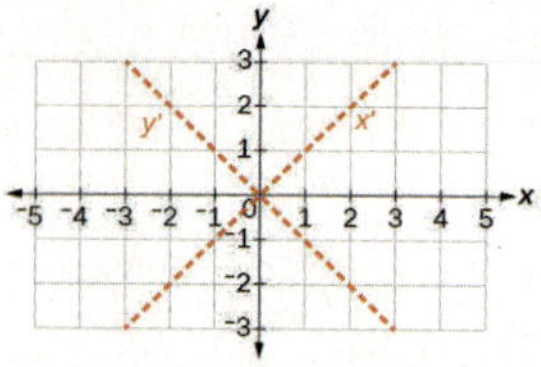

53. $\theta = 60°$

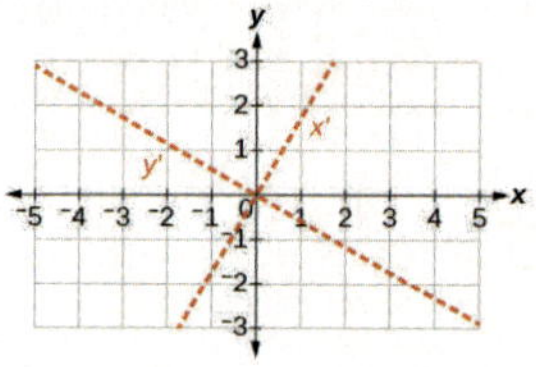

55. $\theta \approx 36.9°$

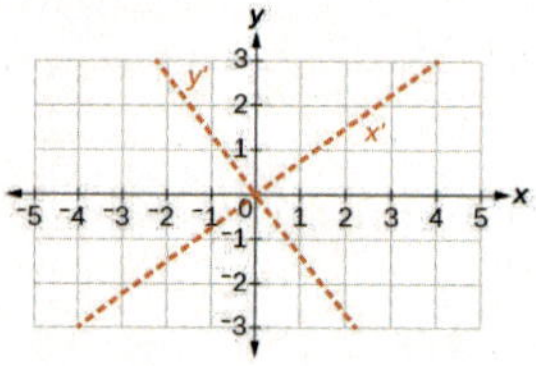

57. $-4\sqrt{6} < k < 4\sqrt{6}$

59. $k = 2$

12.5 Section Exercises

1. If eccentricity is less than 1, it is an ellipse. If eccentricity is equal to 1, it is a parabola. If eccentricity is greater than 1, it is a hyperbola.

3. The directrix will be parallel to the polar axis.

5. One of the foci will be located at the origin.

7. Parabola with $e = 1$ and directrix $\frac{3}{4}$ units below the pole.

9. Hyperbola with $e = 2$ and directrix $\frac{5}{2}$ units above the pole.

11. Parabola with $e = 1$ and directrix $\frac{3}{10}$ units to the right of the pole.

13. Ellipse with $e = \frac{2}{7}$ and directrix 2 units to the right of the pole.

15. Hyperbola with $e = \frac{5}{3}$ and directrix $\frac{11}{5}$ units above the pole.

17. Hyperbola with $e = \frac{8}{7}$ and directrix $\frac{7}{8}$ units to the right of the pole.

19. $25x^2 + 16y^2 - 12y - 4 = 0$

21. $21x^2 - 4y^2 - 30x + 9 = 0$

23. $64y^2 = 48x + 9$

25. $96y^2 - 25x^2 + 110y + 25 = 0$

27. $3x^2 + 4y^2 - 2x - 1 = 0$

29. $5x^2 + 9y^2 - 24x - 36 = 0$

31.

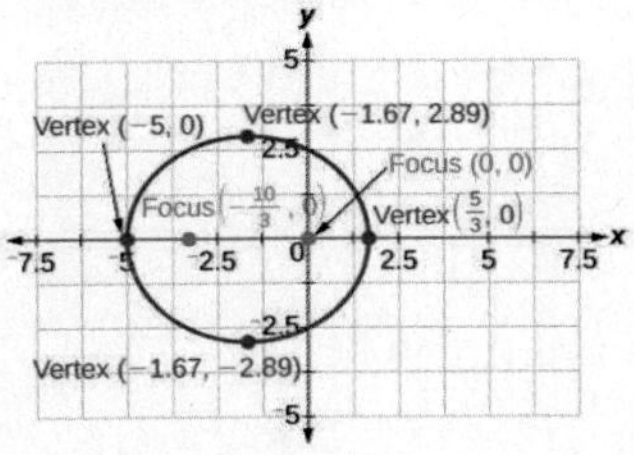

33.

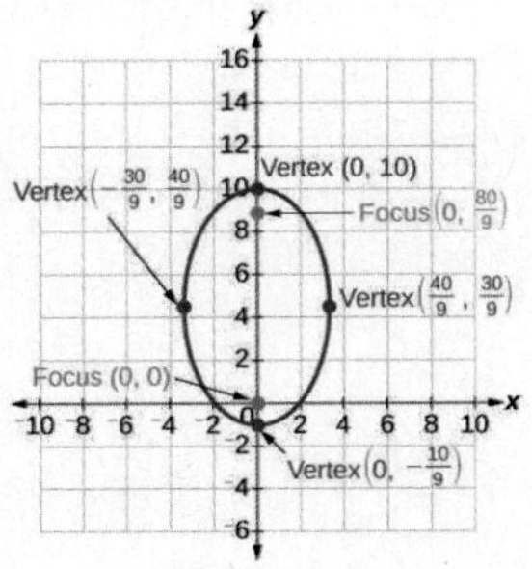

35.

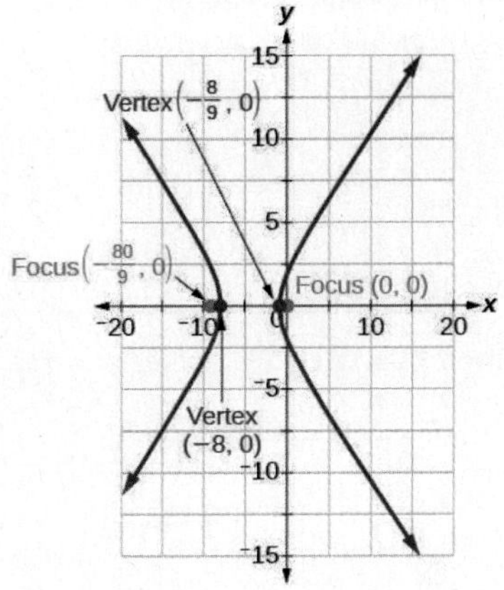

37.

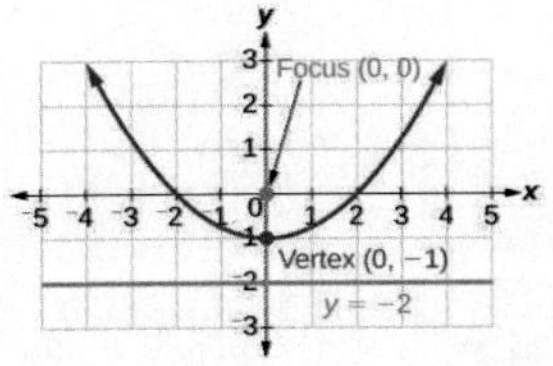

39.

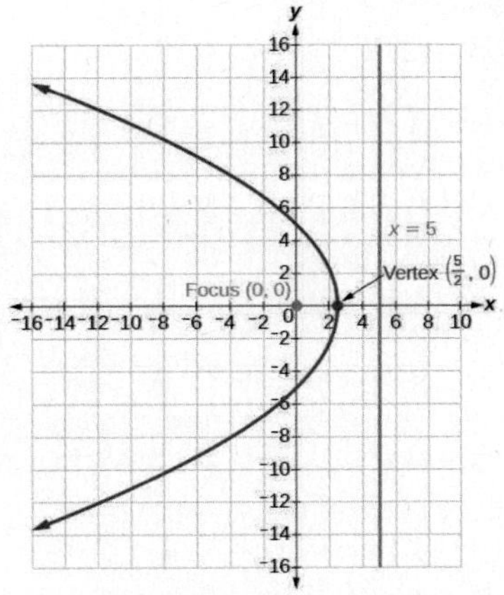

41.

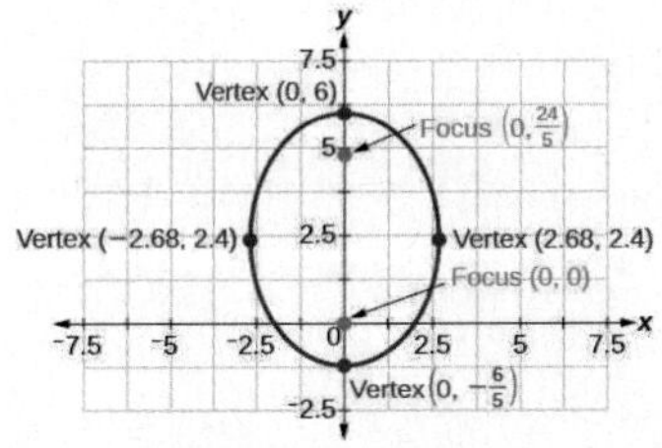

43. $r = \frac{4}{5 + \cos\theta}$

45. $r = \frac{4}{1 + 2\sin\theta}$

47. $r = \frac{1}{1 + \cos\theta}$

49. $r = \frac{7}{8 - 28\cos\theta}$

51. $r = \frac{12}{2 + 3\sin\theta}$

53. $r = \frac{15}{4 - 3\cos\theta}$

55. $r = \frac{3}{3 - 3\cos\theta}$

57. $r = \pm\frac{2}{\sqrt{1 + \sin\theta\cos\theta}}$

59. $r = \pm\frac{2}{4\cos\theta + 3\sin\theta}$

Review Exercises

1. $\frac{x^2}{5^2} + \frac{y^2}{8^2} = 1$; center: $(0, 0)$; vertices: $(5, 0), (-5, 0), (0, 8), (0, -8)$; foci: $\left(0, \sqrt{39}\right), \left(0, -\sqrt{39}\right)$

3. $\frac{(x+3)^2}{1^2} + \frac{(y-2)^2}{3^2} = 1 \quad (-3, 2); \quad (-2, 2), (-4, 2), (-3, 5), (-3, -1); \quad \left(-3, 2 + 2\sqrt{2}\right), \left(-3, 2 - 2\sqrt{2}\right)$

5. center: $(0, 0)$; vertices: $(6, 0), (-6, 0), (0, 3), (0, -3)$; foci: $\left(3\sqrt{3}, 0\right), \left(-3\sqrt{3}, 0\right)$

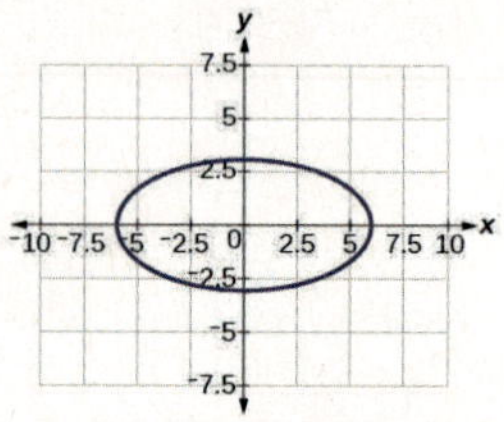

7. center: $(-2, -2)$; vertices: $(2, -2), (-6, -2), (-2, 6), (-2, -10)$; foci: $\left(-2, -2 + 4\sqrt{3},\right), \left(-2, -2 - 4\sqrt{3}\right)$

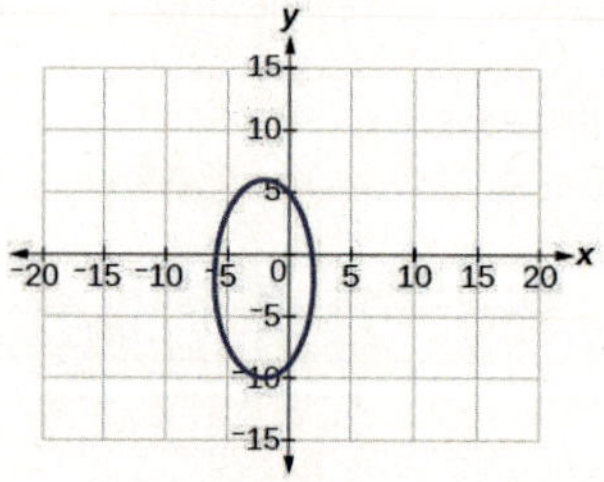

9. $\frac{x^2}{25} + \frac{y^2}{16} = 1$

11. Approximately 35.71 feet

13. $\frac{(y+1)^2}{4^2} - \frac{(x-4)^2}{6^2} = 1$; center: $(4, -1)$; vertices: $(4, 3), (4, -5)$; foci: $\left(4, -1 + 2\sqrt{13}\right), \left(4, -1 - 2\sqrt{13}\right)$

15. $\frac{(x-2)^2}{2^2} - \frac{(y+3)^2}{\left(2\sqrt{3}\right)^2} = 1$; center: $(2, -3)$; vertices: $(4, -3), (0, -3)$; foci: $(6, -3), (-2, -3)$

17.

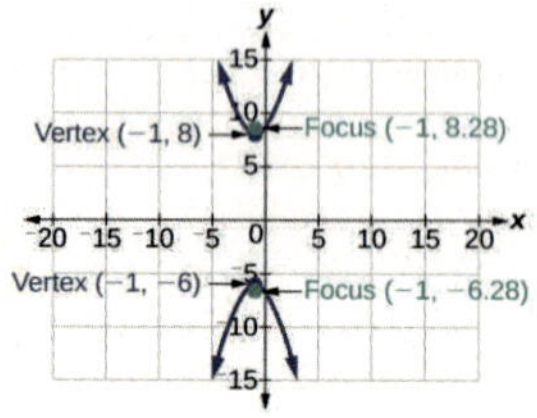

19.

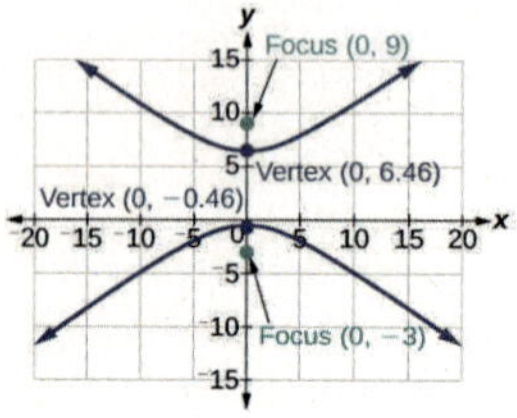

21. $\frac{(x-5)^2}{1} - \frac{(y-7)^2}{3} = 1$

23. $(x + 2)^2 = \frac{1}{2}(y - 1)$; vertex: $(-2, 1)$; focus: $\left(-2, \frac{9}{8}\right)$; directrix: $y = \frac{7}{8}$

25. $(x + 5)^2 = (y + 2)$; vertex: $(-5, -2)$; focus: $\left(-5, -\frac{7}{4}\right)$; directrix: $y = -\frac{9}{4}$

27.

29.

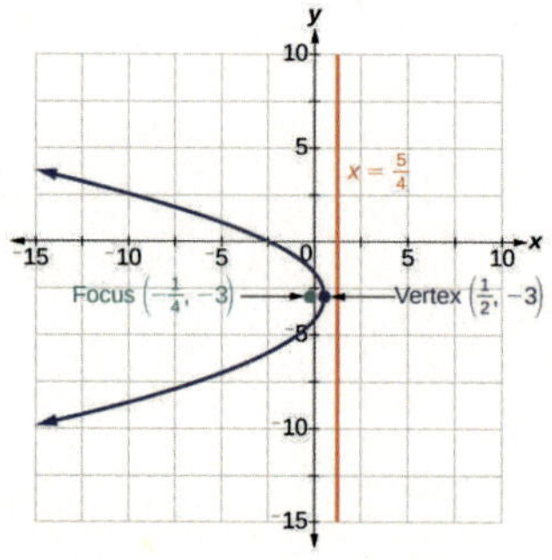

31. $(x - 2)^2 = \left(\frac{1}{2}\right)(y - 1)$

33. $B^2 - 4AC = 0$, parabola

35. $B^2 - 4AC = -31 < 0$, ellipse

37. $\theta = 45°, x'^2 + 3y'^2 - 12 = 0$

39. $\theta = 45°$

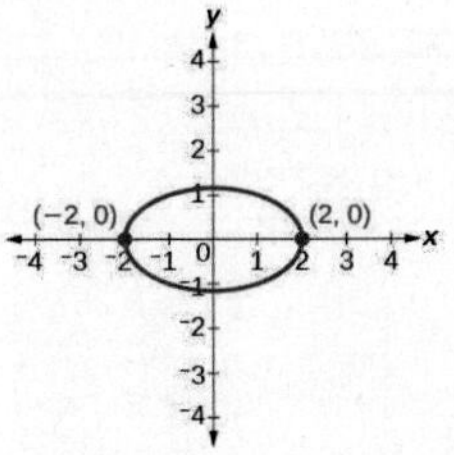

41. Hyperbola with $e = 5$ and directrix 2 units to the left of the pole.

43. Ellipse with $e = \frac{3}{4}$ and directrix $\frac{1}{3}$ unit above the pole.

45.

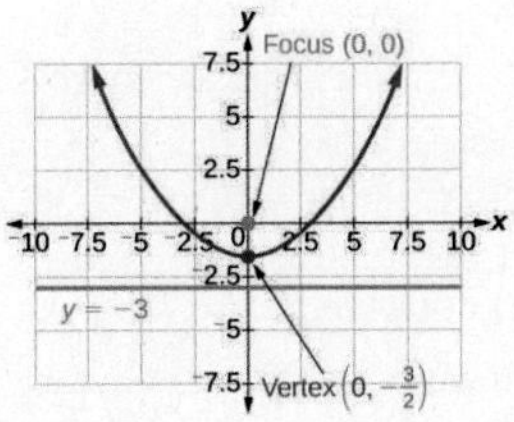

47.

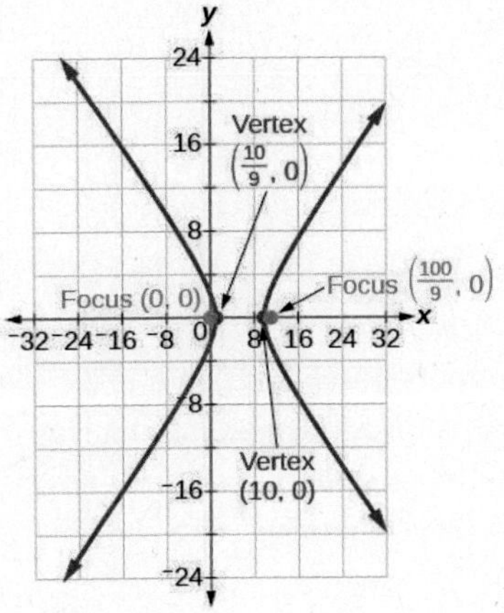

49. $r = \frac{3}{1+\cos\ \theta}$

Practice Test

1. $\frac{x^2}{3^2} + \frac{y^2}{2^2} = 1$; center: $(0, 0)$; vertices: $(3, 0), (-3, 0), (0, 2), (0, -2)$; foci: $\left(\sqrt{5}, 0\right), \left(-\sqrt{5}, 0\right)$

3. center: $(3, 2)$; vertices: $(11, 2), (-5, 2), (3, 8), (3, -4)$; foci: $\left(3 + 2\sqrt{7}, 2\right), \left(3 - 2\sqrt{7}, 2\right)$

5. $\frac{(x-1)^2}{36} + \frac{(y-2)^2}{27} = 1$

7. $\frac{x^2}{7^2} - \frac{y^2}{9^2} = 1$; center: $(0, 0)$; vertices $(7, 0), (-7, 0)$; foci: $\left(\sqrt{130}, 0\right), \left(-\sqrt{130}, 0\right)$; asymptotes: $y = \pm\frac{9}{7}x$

9. center: $(3, -3)$; vertices: $(8, -3), (-2, -3)$; foci: $\left(3 + \sqrt{26}, -3\right), \left(3 - \sqrt{26}, -3\right)$; asymptotes: $y = \pm\frac{1}{5}(x - 3) - 3$

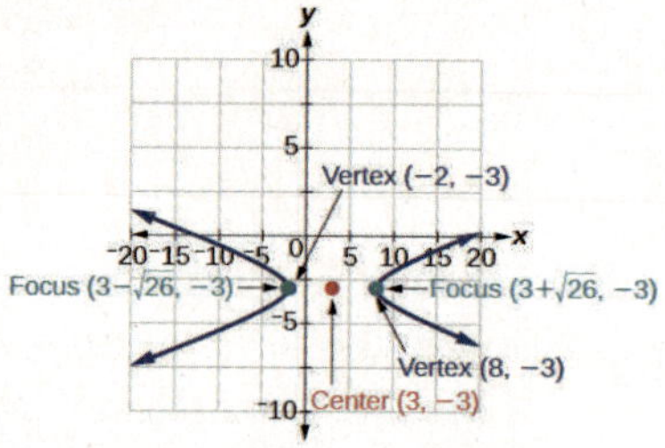

11. $\frac{(y-3)^2}{1} - \frac{(x-1)^2}{8} = 1$

13. $(x - 2)^2 = \frac{1}{3}(y + 1)$; vertex: $(2, -1)$; focus: $\left(2, -\frac{11}{12}\right)$; directrix: $y = -\frac{13}{12}$

15.

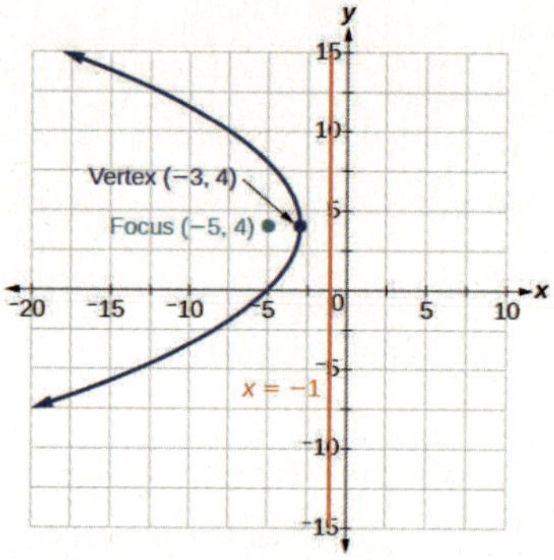

17. Approximately 8.49 feet

19. parabola; $\theta \approx 63.4°$

21. $x'^2 - 4x' + 3y' = 0$

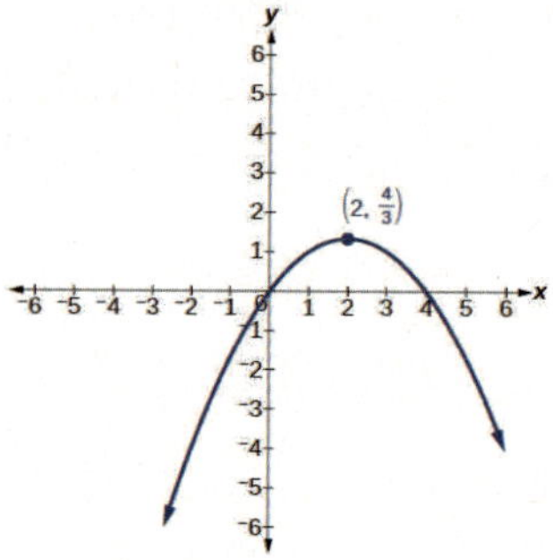

23. Hyperbola with $e = \frac{3}{2}$, and directrix $\frac{5}{6}$ units to the right of the pole.

25.

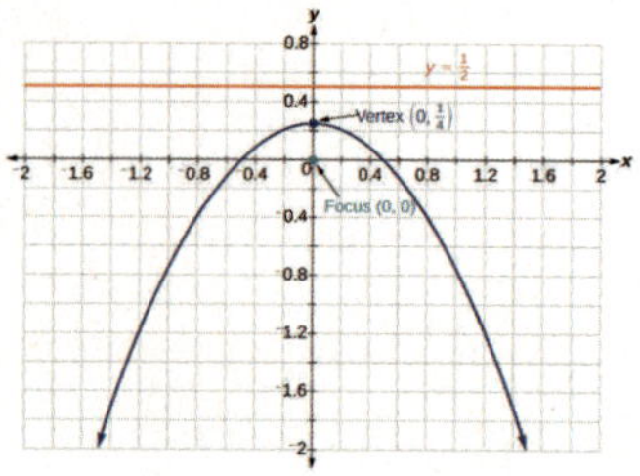

Chapter 13

Try It

13.1 Sequences and Their Notations

1. The first five terms are $\{1, 6,\ 11,\ 16,\ 21\}$.
2. The first five terms are $\left\{-2,\ 2,\ -\frac{3}{2},\ 1,\ -\frac{5}{8}\right\}$.
3. The first six terms are $\{2,\ 5,\ 54,\ 10,\ 250,\ 15\}$.
4. $a_n = (-1)^{n+1} 9^n$
5. $a_n = -\frac{3^n}{4n}$
6. $a_n = e^{n-3}$
7. $\{2,\ 5,\ 11,\ 23,\ 47\}$
8. $\left\{0,\ 1,\ 1,\ 1,\ 2,\ 3,\ \frac{5}{2},\ \frac{17}{6}\right\}$.
9. The first five terms are $\left\{1,\ \frac{3}{2},\ 4,\ 15,\ 72\right\}$.

13.2 Arithmetic Sequences

1. The sequence is arithmetic. The common difference is -2.
2. The sequence is not arithmetic because $3 - 1 \neq 6 - 3$.
3. $\{1,\ 6,\ 11,\ 16,\ 21\}$
4. $a_2 = 2$
5. $a_1 = 25$
   $a_n = a_{n-1} + 12, \quad \text{for } n \geq 2$
6. $a_n = 53 - 3n$
7. There are 11 terms in the sequence.
8. The formula is $T_n = 10 + 4n$, and it will take her 42 minutes.

13.3 Geometric Sequences

1. The sequence is not geometric because $\frac{10}{5} \neq \frac{15}{10}$.
2. The sequence is geometric. The common ratio is $\frac{1}{5}$.
3. $\left\{18, 6, 2, \frac{2}{3}, \frac{2}{9}\right\}$
4. $a_1 = 2$
   $a_n = \frac{2}{3} a_{n-1} \quad \text{for } n \geq 2$
5. $a_6 = 16{,}384$
6. $a_n = -(-3)^{n-1}$
7. ⓐ $P_n = 293 \cdot 1.026a^n$
   ⓑ The number of hits will be about 333.

13.4 Series and Their Notations

1. 38
2. 26.4
3. 328
4. −280
5. \$2,025
6. $\approx 2{,}000.00$
7. 9,840
8. \$275,513.31
9. The sum is not defined.

10. The sum of the infinite series is defined.

11. The sum of the infinite series is defined.

12. 3

13. The series is not geometric.

14. $-\frac{3}{11}$

15. $32,775.87

13.5 Counting Principles

1. 7

2. There are 60 possible breakfast specials.

3. 120

4. 60

5. 12

6. $P(7,7) = 5,040$

7. $P(7,5) = 2,520$

8. $C(10,3) = 120$

9. 64 sundaes

10. 840

13.6 Binomial Theorem

1. ⓐ 35 ⓑ 330

2. ⓐ $x^5 - 5x^4y + 10x^3y^2 - 10x^2y^3 + 5xy^4 - y^5$ ⓑ $8x^3 + 60x^2y + 150xy^2 + 125y^3$

3. $-10,206x^4y^5$

13.7 Probability

1.

Outcome	Probability
Heads	$\frac{1}{2}$
Tails	$\frac{1}{2}$

2. $\frac{2}{3}$

3. $\frac{7}{13}$

4. $\frac{2}{13}$

5. $\frac{5}{6}$

6. a. $\frac{1}{91}$; b. $\frac{5}{91}$; c. $\frac{86}{91}$

13.1 Section Exercises

1. A sequence is an ordered list of numbers that can be either finite or infinite in number. When a finite sequence is defined by a formula, its domain is a subset of the non-negative integers. When an infinite sequence is defined by a formula, its domain is all positive or all non-negative integers.

3. Yes, both sets go on indefinitely, so they are both infinite sequences.

5. A factorial is the product of a positive integer and all the positive integers below it. An exclamation point is used to indicate the operation. Answers may vary. An example of the benefit of using factorial notation is when indicating the product It is much easier to write than it is to write out $13 \cdot 12 \cdot 11 \cdot 10 \cdot 9 \cdot 8 \cdot 7 \cdot 6 \cdot 5 \cdot 4 \cdot 3 \cdot 2 \cdot 1.$

7. First four terms: $-8, \ -\frac{16}{3}, \ -4, \ -\frac{16}{5}$

9. First four terms: $2, \ \frac{1}{2}, \ \frac{8}{27}, \ \frac{1}{4}$.

11. First four terms: $1.25, \ -5, \ 20, \ -80$.

13. First four terms: $\frac{1}{3}, \ \frac{4}{5}, \ \frac{9}{7}, \ \frac{16}{9}$.

15. First four terms: $-\frac{4}{5}, \ 4, \ -20, \ 100$

17. $\frac{1}{3}, \ \frac{4}{5}, \ \frac{9}{7}, \ \frac{16}{9}, \ \frac{25}{11}, \ 31, \ 44, \ 59$

19. $-0.6, -3, -15, -20, -375, -80, -9375, -320$

21. $a_n = n^2 + 3$

23. $a_n = \frac{2^n}{2n}$ or $\frac{2^{n-1}}{n}$

25. $a_n = \left(-\frac{1}{2}\right)^{n-1}$

27. First five terms: $3, \ -9, \ 27, \ -81, \ 243$

29. First five terms: $-1, \ 1, \ -9, \ \frac{27}{11}, \ \frac{891}{5}$

31. $\frac{1}{24}, \ 1, \frac{1}{4}, \ \frac{3}{2}, \ \frac{9}{4}, \ \frac{81}{4}, \ \frac{2187}{8}, \ \frac{531{,}441}{16}$

33. $2, \ 10, \ 12, \ \frac{14}{5}, \ \frac{4}{5}, \ 2, \ 10, \ 12$

35. $a_1 = -8, a_n = a_{n-1} + n$

37. $a_1 = 35, a_n = a_{n-1} + 3$

39. 720

41. 665, 280

43. First four terms: $1, \frac{1}{2}, \frac{2}{3}, \frac{3}{2}$

45. First four terms: $-1, 2, \frac{6}{5}, \frac{24}{11}$

47.

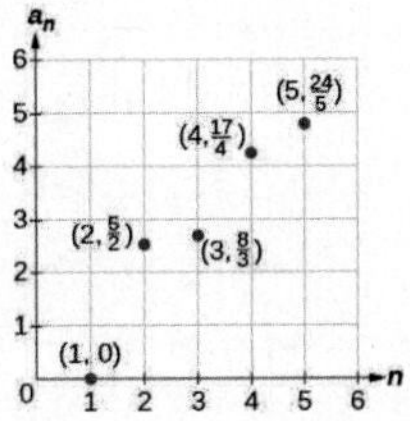

49.

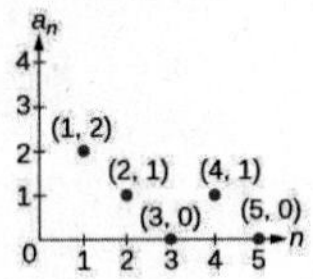

51.

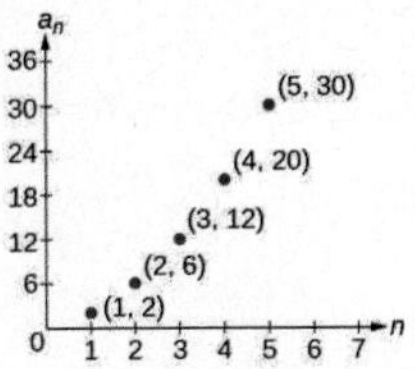

53. $a_n = 2^{n-2}$

55. $a_1 = 6, \ a_n = 2a_{n-1} - 5$

57. First five terms: $\frac{29}{37}, \frac{152}{111}, \frac{716}{333}, \frac{3188}{999}, \frac{13724}{2997}$

59. First five terms: 2, 3, 5, 17, 65537

61. $a_{10} = 7{,}257{,}600$

63. First six terms: 0.042, 0.146, 0.875, 2.385, 4.708

65. First four terms: 5.975, 2.765, 185.743, 1057.25, 6023.521

67. If $a_n = -421$ is a term in the sequence, then solving the equation $-421 = -6 - 8n$ for n will yield a non-negative integer. However, if $-421 = -6 - 8n$, then $n = 51.875$ so $a_n = -421$ is not a term in the sequence.

69. $a_1 = 1, a_2 = 0, a_n = a_{n-1} - a_{n-2}$

71. $\frac{(n+2)!}{(n-1)!} = \frac{(n+2)\cdot(n+1)\cdot(n)\cdot(n-1)\cdot\ldots\cdot3\cdot2\cdot1}{(n-1)\cdot\ldots\cdot3\cdot2\cdot1} = n(n+1)(n+2) = n^3 + 3n^2 + 2n$

13.2 Section Exercises

1. A sequence where each successive term of the sequence increases (or decreases) by a constant value.

3. We find whether the difference between all consecutive terms is the same. This is the same as saying that the sequence has a common difference.

5. Both arithmetic sequences and linear functions have a constant rate of change. They are different because their domains are not the same; linear functions are defined for all real numbers, and arithmetic sequences are defined for natural numbers or a subset of the natural numbers.

7. The common difference is $\frac{1}{2}$

9. The sequence is not arithmetic because $16 - 4 \neq 64 - 16$.

11. $0, \frac{2}{3}, \frac{4}{3}, 2, \frac{8}{3}$

13. $0, -5, -10, -15, -20$

15. $a_4 = 19$

17. $a_6 = 41$

19. $a_1 = 2$

21. $a_1 = 5$

23. $a_1 = 6$

25. $a_{21} = -13.5$

27. $-19, -20.4, -21.8, -23.2, -24.6$

29. $a_1 = 17;\ a_n = a_{n-1} + 9 \quad n \geq 2$

31. $a_1 = 12;\ a_n = a_{n-1} + 5 \quad n \geq 2$

33. $a_1 = 8.9;\ a_n = a_{n-1} + 1.4 \quad n \geq 2$

35. $a_1 = \frac{1}{5};\ a_n = a_{n-1} + \frac{1}{4} \quad n \geq 2$

37. ${}_1 = \frac{1}{6};\ a_n = a_{n-1} - \frac{13}{12} \quad n \geq 2$

39. $a_1 = 4;\ a_n = a_{n-1} + 7;\ a_{14} = 95$

41. First five terms: $20, 16, 12, 8, 4$.

43. $a_n = 1 + 2n$

45. $a_n = -105 + 100n$

47. $a_n = 1.8n$

49. $a_n = 13.1 + 2.7n$

51. $a_n = \frac{1}{3}n - \frac{1}{3}$

53. There are 10 terms in the sequence.

55. There are 6 terms in the sequence.

57. The graph does not represent an arithmetic sequence.

59.

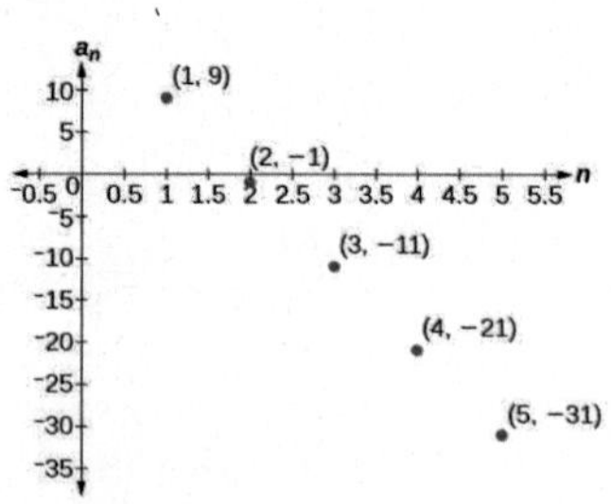

61. $1, 4, 7, 10, 13, 16, 19$

63.

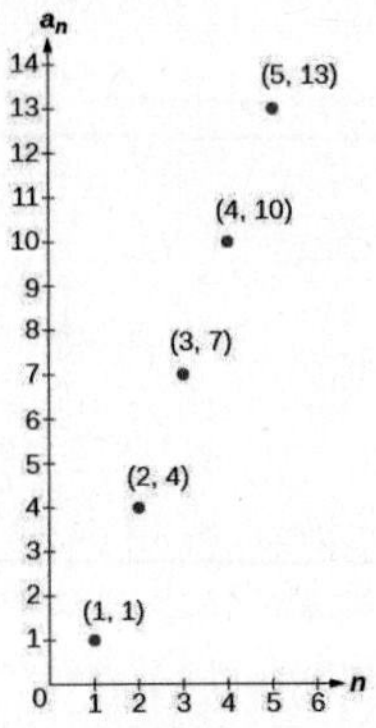

65.

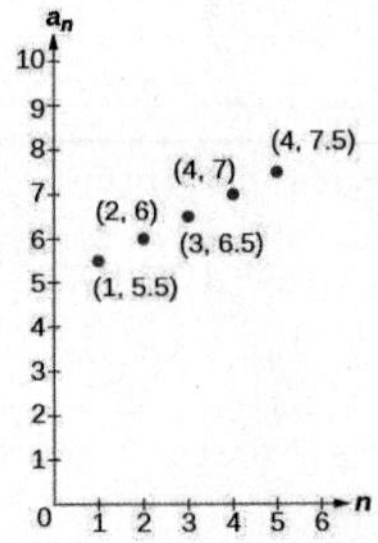

67. Answers will vary. Examples: $a_n = 20.6n$ and $a_n = 2 + 20.4\text{n}$.

69. $a_{11} = -17a + 38b$

71. The sequence begins to have negative values at the 13th term, $a_{13} = -\frac{1}{3}$

73. Answers will vary. Check to see that the sequence is arithmetic. Example: Recursive formula: $a_1 = 3, a_n = a_{n-1} - 3$. First 4 terms: $3, 0, -3, -6 \quad a_{31} = -87$

13.3 Section Exercises

1. A sequence in which the ratio between any two consecutive terms is constant.

3. Divide each term in a sequence by the preceding term. If the resulting quotients are equal, then the sequence is geometric.

5. Both geometric sequences and exponential functions have a constant ratio. However, their domains are not the same. Exponential functions are defined for all real numbers, and geometric sequences are defined only for positive integers. Another difference is that the base of a geometric sequence (the common ratio) can be negative, but the base of an exponential function must be positive.

7. The common ratio is -2

9. The sequence is geometric. The common ratio is 2.

11. The sequence is geometric. The common ratio is $-\frac{1}{2}$.

13. The sequence is geometric. The common ratio is 5.

15. $5, 1, \frac{1}{5}, \frac{1}{25}, \frac{1}{125}$

17. $800, 400, 200, 100, 50$

19. $a_4 = -\frac{16}{27}$

21. $a_7 = -\frac{2}{729}$

23. $7, 1.4, 0.28, 0.056, 0.0112$

25. $a_1 = -32, \quad a_n = \frac{1}{2}a_{n-1}$

27. $a_1 = 10, \quad a_n = -0.3a_{n-1}$

29. $a_1 = \frac{3}{5}, \quad a_n = \frac{1}{6}a_{n-1}$

31. $a_1 = \frac{1}{512}, a_n = -4a_{n-1}$

33. $12, -6, 3, -\frac{3}{2}, \frac{3}{4}$

35. $a_n = 3^{n-1}$

37. $a_n = 0.8 \cdot (-5)^{n-1}$

39. $a_n = -\left(\frac{4}{5}\right)^{n-1}$

41. $a_n = 3 \cdot \left(-\frac{1}{3}\right)^{n-1}$

43. $a_{12} = \frac{1}{177,147}$

45. There are 12 terms in the sequence.

47. The graph does not represent a geometric sequence.

49.

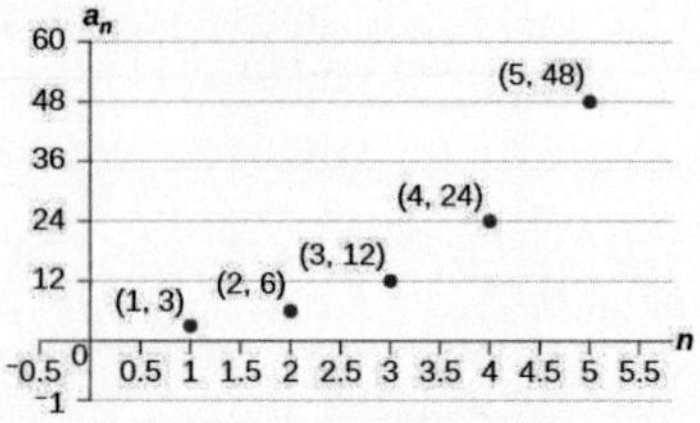

51. Answers will vary. Examples: $a_1 = 800, \quad a_n = 0.5a_{n-1}$ and $a_1 = 12.5, \quad a_n = 4a_{n-1}$

53. $a_5 = 256b$

55. The sequence exceeds 100 at the 14th term, $a_{14} \approx 107$.

57. $a_4 = -\frac{32}{3}$ is the first non-integer value

59. Answers will vary. Example: Explicit formula with a decimal common ratio: $a_n = 400 \cdot 0.5^{n-1}$; First 4 terms: $400, 200, 100, 50; \quad a_8 = 3.125$

13.4 Section Exercises

1. An nth partial sum is the sum of the first n terms of a sequence.

3. A geometric series is the sum of the terms in a geometric sequence.

5. An annuity is a series of regular equal payments that earn a constant compounded interest.

7. $\sum_{n=0}^{4} 5n$

9. $\sum_{k=1}^{5} 4$

11. $\sum_{k=1}^{20} 8k + 2$

13. $S_5 = \frac{5\left(\frac{3}{2}+\frac{7}{2}\right)}{2}$

15. $S_{13} = \frac{13(3.2+5.6)}{2}$

17. $\sum_{k=1}^{7} 8 \cdot 0.5^{k-1}$

19. $S_5 = \frac{9\left(1-\left(\frac{1}{3}\right)^5\right)}{1-\frac{1}{3}} = \frac{121}{9} \approx 13.44$

21. $S_{11} = \frac{64\left(1-0.2^{11}\right)}{1-0.2} = \frac{781,249,984}{9,765,625} \approx 80$

23. The series is defined. $S = \frac{2}{1-0.8}$

25. The series is defined. $S = \frac{-1}{1-\left(-\frac{1}{2}\right)}$

27.

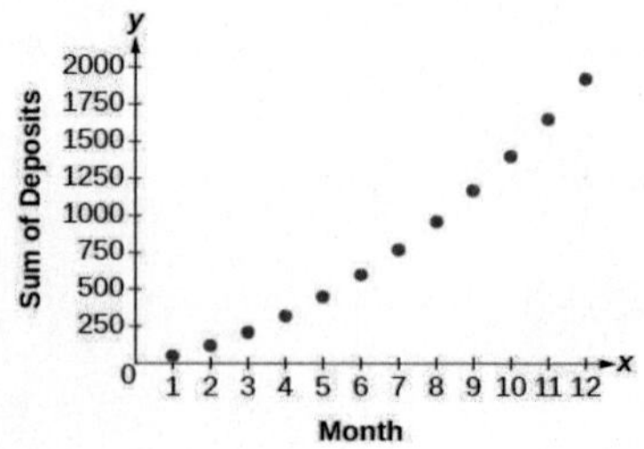

29. Sample answer: The graph of S_n seems to be approaching 1. This makes sense because $\sum_{k=1}^{\infty}\left(\frac{1}{2}\right)^k$ is a defined infinite geometric series with $S = \frac{\frac{1}{2}}{1-\left(\frac{1}{2}\right)} = 1.$

31. 49

33. 254

35. $S_7 = \frac{147}{2}$

37. $S_{11} = \frac{55}{2}$

39. $S_7 = 5208.4$

41. $S_{10} = -\frac{1023}{256}$

43. $S = -\frac{4}{3}$

45. $S = 9.2$

47. $3,705.42

49. $695,823.97

51. $a_k = 30 - k$

53. 9 terms

55. $r = \frac{4}{5}$

57. $400 per month

59. 420 feet

61. 12 feet

13.5 Section Exercises

1. There are $m + n$ ways for either event A or event B to occur.

3. The addition principle is applied when determining the total possible of outcomes of either event occurring. The multiplication principle is applied when determining the total possible outcomes of both events occurring. The word "or" usually implies an addition problem. The word "and" usually implies a multiplication problem.

5. A combination; $C(n, r) = \frac{n!}{(n-r)!r!}$

7. $4 + 2 = 6$

9. $5 + 4 + 7 = 16$

11. $2 \times 6 = 12$

13. $10^3 = 1000$

15. $P(5, 2) = 20$

17. $P(3, 3) = 6$

19. $P(11, 5) = 55,440$

21. $C(12, 4) = 495$

23. $C(7, 6) = 7$

25. $2^{10} = 1024$

27. $2^{12} = 4096$

29. $2^9 = 512$

31. $\frac{8!}{3!} = 6720$

33. $\frac{12!}{3!2!3!4!}$

35. 9

37. Yes, for the trivial cases $r = 0$ and $r = 1$. If $r = 0$, then $C(n, r) = P(n, r) = 1.$ If $r = 1$, then $r = 1$, $C(n, r) = P(n, r) = n.$

39. $\frac{6!}{2!} \times 4! = 8640$

41. $6 - 3 + 8 - 3 = 8$

43. $4 \times 2 \times 5 = 40$

45. $4 \times 12 \times 3 = 144$

47. $P(15, 9) = 1,816,214,400$

49. $C(10, 3) \times C(6, 5) \times C(5, 2) = 7,200$

51. $2^{11} = 2048$

53. $\frac{20!}{6!6!8!} = 116,396,280$

13.6 Section Exercises

1. A binomial coefficient is an alternative way of denoting the combination $C(n, r)$. It is defined as $\binom{n}{r} = C(n, r) = \frac{n!}{r!(n-r)!}$.

3. The Binomial Theorem is defined as $(x + y)^n = \sum_{k=0}^{n} \binom{n}{k} x^{n-k} y^k$ and can be used to expand any binomial.

5. 15

7. 35

9. 10

11. 12,376

13. $64a^3 - 48a^2b + 12ab^2 - b^3$

15. $27a^3 + 54a^2b + 36ab^2 + 8b^3$

17. $1024x^5 + 2560x^4y + 2560x^3y^2 + 1280x^2y^3 + 320xy^4 + 32y^5$

19. $1024x^5 - 3840x^4y + 5760x^3y^2 - 4320x^2y^3 + 1620xy^4 - 243y^5$

21. $\frac{1}{x^4} + \frac{8}{x^3y} + \frac{24}{x^2y^2} + \frac{32}{xy^3} + \frac{16}{y^4}$

23. $a^{17} + 17a^{16}b + 136a^{15}b^2$

25. $a^{15} - 30a^{14}b + 420a^{13}b^2$

27. $3,486,784,401a^{20} + 23,245,229,340a^{19}b + 73,609,892,910a^{18}b^2$

29. $x^{24} - 8x^{21}\sqrt{y} + 28x^{18}y$

31. $-720x^2y^3$

33. $220,812,466,875,000y^7$

35. $35x^3y^4$

37. $1,082,565a^3b^{16}$

39. $\frac{1152y^2}{x^7}$

41. $f_2(x) = x^4 + 12x^3$

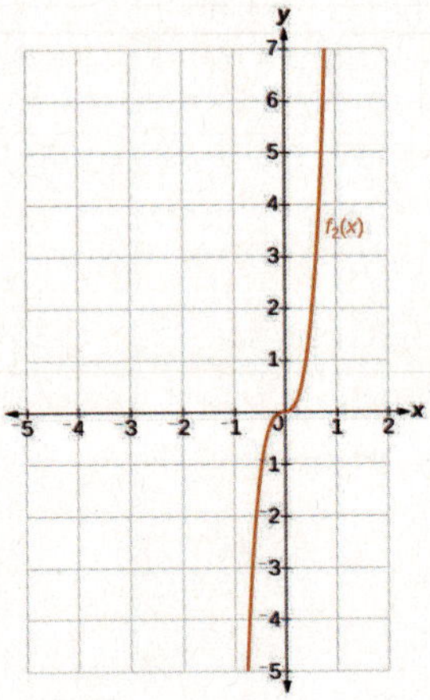

43. $f_4(x) = x^4 + 12x^3 + 54x^2 + 108x$

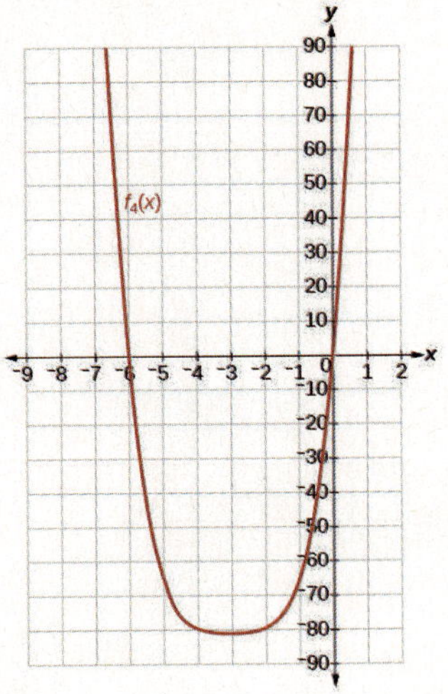

45. $590,625x^5y^2$

47. $k - 1$

49. The expression $(x^3 + 2y^2 - z)^5$ cannot be expanded using the Binomial Theorem because it cannot be rewritten as a binomial.

13.7 Section Exercises

1. probability; The probability of an event is restricted to values between 0 and 1, inclusive of 0 and 1.

3. An experiment is an activity with an observable result.

5. The probability of the *union of two events* occurring is a number that describes the likelihood that at least one of the events from a probability model occurs. In both a union of sets A and B and a union of events A and B, the union includes either A or B or both. The difference is that a union of sets results in another set, while the union of events is a probability, so it is always a numerical value between 0 and 1.

7. $\frac{1}{2}$.

9. $\frac{5}{8}$.

11. $\frac{1}{2}$.

13. $\frac{3}{8}$.

15. $\frac{1}{4}$.

17. $\frac{3}{4}$.

19. $\frac{3}{8}$.

21. $\frac{1}{8}$.

23. $\frac{15}{16}$.

25. $\frac{5}{8}$.

27. $\frac{1}{13}$.

29. $\frac{1}{26}$.

31. $\frac{12}{13}$.

33.

	1	2	3	4	5	6
1	(1,1) 2	(1,2) 3	(1,3) 4	(1,4) 5	(1,5) 6	(1,6) 7
2	(2,1) 3	(2,2) 4	(2,3) 5	(2,4) 6	(2,5) 7	(2,6) 8
3	(3,1) 4	(3,2) 5	(3,3) 6	(3,4) 7	(3,5) 8	(3,6) 9
4	(4,1) 5	(4,2) 6	(4,3) 7	(4,4) 8	(4,5) 9	(4,6) 10
5	(5,1) 6	(5,2) 7	(5,3) 8	(5,4) 9	(5,5) 10	(5,6) 11
6	(6,1) 7	(6,2) 8	(6,3) 9	(6,4) 10	(6,5) 11	(6,6) 12

35. $\frac{5}{12}$.

37. 0.

39. $\frac{4}{9}$.

41. $\frac{1}{4}$.

43. $\frac{5}{8}$

45. $\frac{8}{13}$

47. $\frac{C(12,5)}{C(48,5)} = \frac{1}{2162}$

49. $\frac{C(12,3)C(36,2)}{C(48,5)} = \frac{175}{2162}$

51. $\frac{C(20,3)C(60,17)}{C(80,20)} \approx 12.49\%$

53. $\frac{C(20,5)C(60,15)}{C(80,20)} \approx 23.33\%$

55. $20.50 + 23.33 - 12.49 = 31.34\%$

57. $\frac{C(40000000,1)C(277000000,4)}{C(317000000,5)} = 36.78\%$

59. $\frac{C(40000000,4)C(277000000,1)}{C(317000000,5)} = 0.11\%$

Review Exercises

1. $2, 4, 7, 11$

3. $13, 103, 1003, 10003$

5. The sequence is arithmetic. The common difference is $d = \frac{5}{3}$.

7. $18, 10, 2, -6, -14$

9. $a_1 = -20,\ \ a_n = a_{n-1} + 10$

11. $a_n = \frac{1}{3}n + \frac{13}{24}$

13. $r = 2$

15. 4, 16, 64, 256, 1024

17. $3,\ \ 12,\ \ 48,\ \ 192,\ \ 768$

19. $a_n = -\frac{1}{5} \cdot \left(\frac{1}{3}\right)^{n-1}$

21. $\sum_{m=0}^{5} \left(\frac{1}{2}m + 5\right).$

23. $S_{11} = 110$

25. $S_9 \approx 23.95$

27. $S = \frac{135}{4}$

29. $5,617.61

31. 6

33. $10^4 = 10{,}000$

35. $P(18, 4) = 73{,}440$

37. $C\,(15, 6) = 5005$

39. $2^{50} = 1.13 \times 10^{15}$

41. $\frac{8!}{3!2!} = 3360$

43. $490{,}314$

45. $131{,}072a^{17} + 1{,}114{,}112a^{16}b + 4{,}456{,}448a^{15}b^2$

47.

	1	2	3	4	5	6
1	1,1	1,2	1,3	1,4	1,5	1,6
2	2,1	2,2	2,3	2,4	2,5	2,6
3	3,1	3,2	3,3	3,4	3,5	3,6
4	4,1	4,2	4,3	4,4	4,5	4,6
5	5,1	5,2	5,3	5,4	5,5	5,6
6	6,1	6,2	6,3	6,4	6,5	6,6

49. $\frac{1}{6}$

51. $\frac{5}{9}$

53. $\frac{4}{9}$

55. $1 - \frac{C(350,8)}{C(500,8)} \approx 94.4\%$

57. $\frac{C(150,3)C(350,5)}{C(500,8)} \approx 25.6\%$

Practice Test

1. $-14, -6, -2, 0$

3. The sequence is arithmetic. The common difference is $d = 0.9$.

5. $a_1 = -2,\ a_n = a_{n-1} - \frac{3}{2};\ a_{22} = -\frac{67}{2}$

7. The sequence is geometric. The common ratio is $r = \frac{1}{2}$.

9. $a_1 = 1,\ a_n = -\frac{1}{2} \cdot a_{n-1}$

11. $\sum_{k=-3}^{15} \left(3k^2 - \frac{5}{6}k\right)$

13. $S_7 = -2604.2$

15. Total in account: $\$140,355.75$; Interest earned: $\$14,355.75$

17. $5 \times 3 \times 2 \times 3 \times 2 = 180$

19. $C(15,3) = 455$

21. $\frac{10!}{2!3!2!} = 151{,}200$

23. $\frac{429x^{14}}{16}$

25. $\frac{4}{7}$

27. $\frac{5}{7}$

29. $\frac{C(14,3)C(26,4)}{C(40,7)} \approx 29.2\%$

Index